The 18th edition of AMERICAN MEN & WOMEN OF SCIENCE was
prepared by the R.R. Bowker Database Publishing Group.

Stephen L. Torpie, Managing Editor
Judy Redel, Managing Editor, Research
Richard D. Lanam, Senior Editor
Tanya Hurst, Research Manager
Karen Hallard, Beth Tanis, Associate Editors

Peter Simon, Vice President, Database Publishing Group
Dean Hollister, Director, Database Planning
Edgar Adcock, Jr., Editorial Director, Directories

American Men & Women of Science

1992-93 • 18th Edition

A Biographical Directory of Today's Leaders in Physical, Biological and Related Sciences.

Volume 8 • Discipline Index

R. R. BOWKER
New Providence, New Jersey

124247

Published by R.R. Bowker, a division of Reed Publishing, (USA) Inc.
Copyright© 1992 by Reed Publishing (USA) Inc. All rights reserved. Except as permitted under the Copyright Act of 1976, no part of *American Men and Women Of Science* may be reproduced or transmitted in any form or by any means stored in any information storage and retrieval system, without prior written permission of R.R. Bowker, 121 Chanlon Road, New Providence, New Jersey, 07974.

International Standard Book Number
            Set: 0-8352-3074-0
      Volume I: 0-8352-3075-9
     Volume II: 0-8352-3076-7
    Volume III: 0-8352-3077-5
     Volume IV: 0-8352-3078-3
      Volume V: 0-8352-3079-1
     Volume VI: 0-8352-3080-5
    Volume VII: 0-8352-3081-3
   Volume VIII: 0-8352-3082-1

International Standard Serial Number: 0192-8570
Library of Congress Catalog Card Number: 6-7326
Printed and bound in the United States of America.

8 Volume Set

ISBN 0-8352-3074-0

9 780835 230742

Contents

BIOLOGICAL SCIENCES

CHEMISTRY

COMPUTER SCIENCES

ENGINEERING

ENVIRONMENTAL, EARTH & MARINE SCIENCES

MATHEMATICS

MEDICAL & HEALTH SCIENCES

PHYSICS & ASTRONOMY

OTHER PROFESSIONAL FIELDS

Advisory Committee

Dr. Robert F. Barnes
    Executive Vice President
American Society of Agronomy

Dr. John Kistler Crum
    Executive Director
American Chemical Society

Dr. Charles Henderson Dickens
    Section Head, Survey & Analysis Section
Division of Science Resource Studies
National Science Foundation

Mr. Alan Edward Fechter
    Executive Director
Office of Scientific & Engineering Personnel
National Academy of Science

Dr. Oscar Nicolas Garcia
    Prof Electrical Engineering
Electrical Engineering & Computer Science Department
George Washington University

Dr. Charles George Groat
    Executive Director
American Geological Institute

Dr. Richard E. Hallgren
    Executive Director
American Meteorological Society

Dr. Michael J. Jackson
    Executive Director
Federation of American Societies for Experimental
Biology

Dr. William Howard Jaco
    Executive Director
American Mathematical Society

Dr. Shirley Mahaley Malcom
    Head, Directorate for Education and Human
    Resources Programs
American Association for the Advancement of Science

Mr. Daniel Melnick
    Sr Advisor Research Methodologies
Sciences Resources Studies Division
National Science Foundation

Ms. Beverly Fearn Porter
    Division Manager
Education & Employment Statistics Division
American Institute of Physics

Dr. Terrence R. Russell
    Manager
Office of Professional Services
American Chemical Society

Dr. Irwin Walter Sandberg
    Holder, Cockrell Family Regent Chair
Department of Electrical & Computer Engineering
University of Texas

Dr. William Eldon Splinter
    Interim Vice Chancellor for Research,
    Dean, Graduate Studies
University of Nebraska

Ms. Betty M. Vetter
    Executive Director, Science Manpower Commission
Commission on Professionals in Science &
Technology

Dr. Dael Lee Wolfe
    Professor Emeritus
Graduate School of Public Affairs
University of Washington

Preface

American Men and Women Of Science remains without peer as a chronicle of North American scientific endeavor and achievement. The present work is the eighteenth edition since it was first compiled as *American Men of Science* by J. Mckeen Cattell in 1906. In its eighty-six year history *American Men & Women of Science* has profiled the careers of over 300,000 scientists and engineers. Since the first edition, the number of American scientists and the fields they pursue have grown immensely. This edition alone lists full biographies for 122,817 engineers and scientists, 7021 of which are listed for the first time. Although the book has grown, our stated purpose is the same as when Dr. Cattell first undertook the task of producing a biographical directory of active American scientists. It was his intention to record educational, personal and career data which would make "a contribution to the organization of science in America" and "make men [and women] of science acquainted with one another and with one another's work." It is our hope that this edition will fulfill these goals.

The biographies of engineers and scientists constitute seven of the eight volumes and provide birthdates, birthplaces, field of specialty, education, honorary degrees, professional and concurrent experience, awards, memberships, research information and adresses for each entrant when applicable. The eighth volume, the discipline index, organizes biographees by field of activity. This index, adapted from the National Science Foundation's Taxonomy of Degree and Employment Specialties, classifies entrants by 171 subject specialties listed in the table of contents of Volume 8. For the first time, the index classifies scientists and engineers by state within each subject specialty, allowing the user to more easily locate a scientist in a given area. Also new to this edition is the inclusion of statistical information, and recipients of theNobel Prizes, the Craaford Prize, the Charles Stark

Draper Prize, and the National Medals of Science and Technology received since the last edition.

While the scientific fields covered by *American Men and Women Of Science* are comprehensive, no attempt has been made to include all American scientists. Entrants are meant to be limited to those who have made significant contributions in their field. The names of new entrants were submitted for consideration at the editors' request by current entrants and by leaders of academic, government and private research programs and associations. Those included met the following criteria:

1. Distinguished achievement, by reason of experience, training or accomplishment, including contributions to the literature, coupled with continuing activity in scientific work;

    or

2. Research activity of high quality in science as evidenced by publication in reputable scientific journals; or for those whose work cannot be published due to governmental or industrial security, research activity of high quality in science as evidenced by the judgement of the individual's peers;

    or

3. Attainment of a position of substantial responsibility requiring scientific training and experience.

This edition profiles living scientists in the physical and biological fields, as well as public health scientists, engineers, mathematicians, statisticians, and computer scientists. The information is collected by means of direct communication whenever possible. All entrants receive forms for corroboration and updating. New entrants receive questionaires and verification proofs before publication. The information submitted by entrants is included as completely as possible within the boundaries of editorial and space restrictions. If an entrant does not return the form and his or her current location can be verified in secondary sources, the full entry is repeated. References to the previous edition are given for those who do not return forms and cannot be located, but who are presumed to be still active in science or engineering. Entrants known to be deceased are noted as such and a reference to the previous edition is given. Scientists and engineers who are not citizens of the United States or Canada are included if a significant portion of their work was performed in North America.

The information in AMWS is also available on CD-ROM as part of *SciTech Reference Plus*. In addition to the convenience of searching scientists and engineers, *SciTech Reference Plus* also includes *The Directory of American Research & Technology*, *Corporate Technology Directory*, sci-tech and medical books and serials from *Books in Print* and *Bowker International Series*. *American Men and Women Of Science* is available for online searching through the subscription services of DIALOG Information Services, Inc. (3460 Hillview Ave, Palo Alto, CA 94304) and ORBIT Search Service (800 Westpark Dr, McLean, VA 22102). Both CD-ROM and the on-line subscription services allow all elements of an entry, including field of interest, experience, and location, to be accessed by key word. Tapes and mailing lists are also available through Cahners Direct Mail (John Panza, List Manager, Bowker Files 245 W 17th St, New York, NY, 10011, Tel: 800-537-7930).

A project as large as publishing *American Men and Women Of Science* involves the efforts of a great many people. The editors take this opportunity to thank the eighteenth edition advisory committee for their guidance, encouragement and support. Appreciation is also expressed to the many scientific societies who provided their membership lists for the purpose of locating former entrants whose addresses had changed, and to the tens of thousands of scientists across the country who took time to provide us with biographical information. We also wish to thank Bruce Glaunert, Bonnie Walton, Val Lowman, Debbie Wilson, Mervaine Ricks and all those whose care and devotion to accurate research and editing assured successful production of this edition.

Comments, suggestions and nominations for the nineteenth edition are encouraged and should be directed to The Editors, *American Men and Women Of Science*, R.R. Bowker, 121 Chanlon Road, New Providence, New Jersey, 07974.

Edgar H. Adcock, Jr.
Editorial Director

Statistics

Statistical distribution of entrants in *American Men & Women of Science* is illustrated on the following five pages. The regional scheme for geographical analysis is diagrammed in the map below. A table enumerating the geographic distribution can be found on page xvi, following the charts. The statistics are compiled by tallying all occurrences of a major index subject. Each scientist may choose to be indexed under as many as four categories; thus, the total number of subject references is greater than the number of entrants in *AMWS*.

All Disciplines

	Number	Percent
Northeast	58,325	34.99
Southeast	39,769	23.86
North Central	19,846	11.91
South Central	12,156	7.29
Mountain	11,029	6.62
Pacific	25,550	15.33
TOTAL	**166,675**	**100.00**

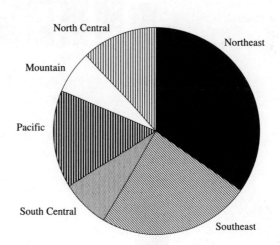

Age Distribution of American Men & Women of Science

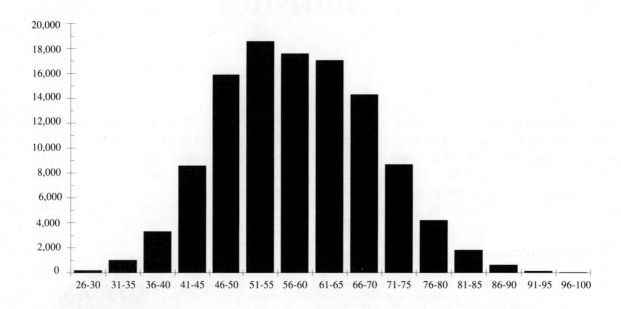

Number of Scientists in Each Discipline of Study

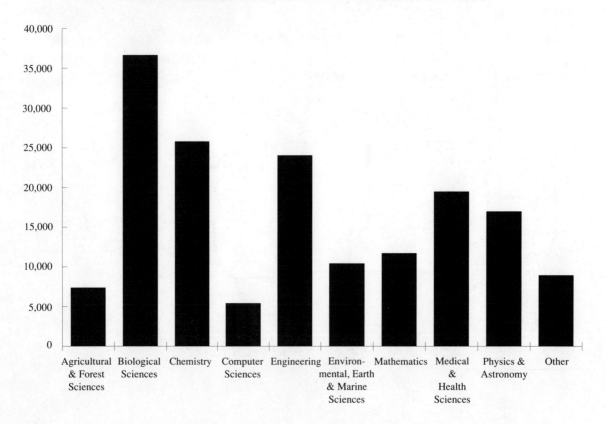

Agricultural & Forest Sciences

	Number	Percent
Northeast	1,574	21.39
Southeast	1,991	27.05
North Central	1,170	15.90
South Central	609	8.27
Mountain	719	9.77
Pacific	1,297	17.62
TOTAL	**7,360**	**100.00**

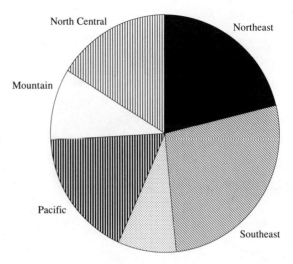

Biological Sciences

	Number	Percent
Northeast	12,162	33.23
Southeast	9,054	24.74
North Central	5,095	13.92
South Central	2,806	7.67
Mountain	2,038	5.57
Pacific	5,449	14.89
TOTAL	**36,604**	**100.00**

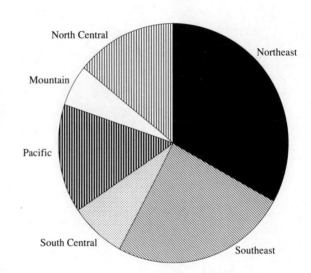

Chemistry

	Number	Percent
Northeast	10,343	40.15
Southeast	6,124	23.77
North Central	3,022	11.73
South Central	1,738	6.75
Mountain	1,300	5.05
Pacific	3,233	12.55
TOTAL	**25,760**	**100.00**

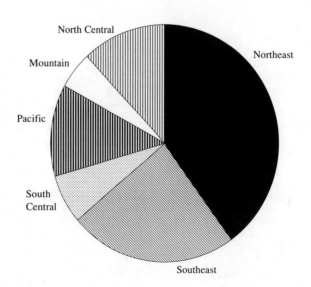

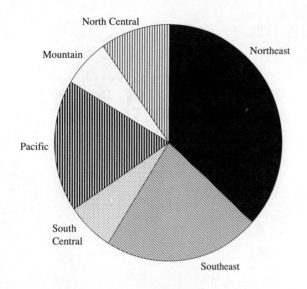

Computer Sciences

	Number	Percent
Northeast	1,987	36.76
Southeast	1,200	22.20
North Central	511	9.45
South Central	360	6.66
Mountain	372	6.88
Pacific	976	18.05
TOTAL	**5,406**	**100.00**

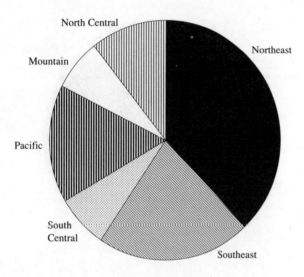

Engineering

	Number	Percent
Northeast	9,122	38.01
Southeast	5,202	21.68
North Central	2,510	10.46
South Central	1,710	7.13
Mountain	1,646	6.86
Pacific	3,807	15.86
TOTAL	**23,997**	**100.00**

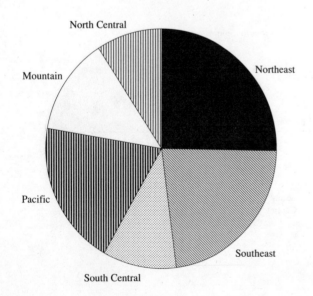

Environmental, Earth & Marine Sciences

	Number	Percent
Northeast	2,657	25.48
Southeast	2,361	22.64
North Central	953	9.14
South Central	1,075	10.31
Mountain	1,359	13.03
Pacific	2,022	19.39
TOTAL	**10,427**	**100.00**

Mathematics

	Number	Percent
Northeast	4,211	35.92
Southeast	2,609	22.26
North Central	1,511	12.89
South Central	884	7.54
Mountain	718	6.13
Pacific	1,789	15.26
TOTAL	**11,722**	**100.00**

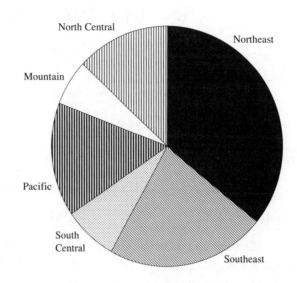

Medical & Health Sciences

	Number	Percent
Northeast	7,115	36.53
Southeast	5,004	25.69
North Central	2,577	13.23
South Central	1,516	7.78
Mountain	755	3.88
Pacific	2,509	12.88
TOTAL	**19,476**	**100.00**

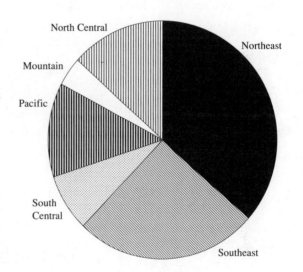

Physics & Astronomy

	Number	Percent
Northeast	5,961	35.12
Southeast	3,670	21.62
North Central	1,579	9.30
South Central	918	5.41
Mountain	1,607	9.47
Pacific	3,238	19.08
TOTAL	**16,973**	**100.00**

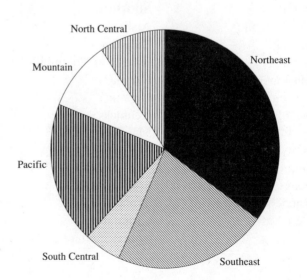

Geographic Distribution of Scientists by Discipline

	Northeast	Southeast	North Central	South Central	Mountain	Pacific	TOTAL
Agricultural & Forest Sciences	1,574	1,991	1,170	609	719	1,297	**7,360**
Biological Sciences	12,162	9,054	5,095	2,806	2,038	5,449	**36,604**
Chemistry	10,343	6,124	3,022	1,738	1,300	3,233	**25,760**
Computer Sciences	1,987	1,200	511	360	372	976	**5,406**
Engineering	9,122	5,202	2,510	1,710	1,646	3,807	**23,997**
Environmental, Earth & Marine Sciences	2,657	2,361	953	1,075	1,359	2,022	**10,427**
Mathematics	4,211	2,609	1,511	884	718	1,789	**11,722**
Medical & Health Sciences	7,115	5,004	2,577	1,516	755	2,509	**19,476**
Physics & Astronomy	5,961	3,670	1,579	918	1,607	3,238	**16,973**
Other Professional Fields	3,193	2,554	918	540	515	1,230	**8,950**
TOTAL	**58,325**	**39,769**	**19,846**	**12,156**	**11,029**	**25,550**	**166,675**

Geographic Definitions

Northeast
Connecticut
Indiana
Maine
Massachusetts
Michigan
New Hampshire
New Jersey
New York
Ohio
Pennsylvania
Rhode Island
Vermont

Southeast
Alabama
Delaware
District of Columbia
Florida
Georgia
Kentucky
Maryland
Mississippi
North Carolina
South Carolina
Tennessee
Virginia
West Virginia

North Central
Illinois
Iowa
Kansas
Minnesota
Missouri
Nebraska
North Dakota
South Dakota
Wisconsin

South Central
Arkansas
Louisiana
Texas
Oklahoma

Mountain
Arizona
Colorado
Idaho
Montana
Nevada
New Mexico
Utah
Wyoming

Pacific
Alaska
California
Hawaii
Oregon
Washington

Abbreviations

AAAS—American Association for the Advancement of Science
abnorm—abnormal
abstr—abstract
acad—academic, academy
acct—Account, accountant, accounting
acoust—acoustic(s), acoustical
ACTH—adrenocorticotrophic hormone
actg—acting
activ—activities, activity
addn—addition(s), additional
Add—Address
adj—adjunct, adjutant
adjust—adjustment
Adm—Admiral
admin—administration, administrative
adminr—administrator(s)
admis—admission(s)
adv—adviser(s), advisory
advan—advance(d), advancement
advert—advertisement, advertising
AEC—Atomic Energy Commission
aerodyn—aerodynamic
aeronaut—aeronautic(s), aeronautical
aerophys—aerophsical, aerophysics
aesthet—aesthetic
AFB—Air Force Base
affil—affiliate(s), affiliation
agr—agricultural, agriculture
agron—agronomic, agronomical, agronomy
agrost—agrostologic, agrostological, agrostology
agt—agent
AID—Agency for International Development
Ala—Alabama
allergol—allergological, allergology
alt—alternate
Alta—Alberta
Am—America, American
AMA—American Medical Association
anal—analysis, analytic, analytical
analog—analogue
anat—anatomic, anatomical, anatomy
anesthesiol—anesthesiology
angiol—angiology
Ann—Annal(s)
ann—annual
anthrop—anthropological, anthropology
anthropom—anthropometric, anthropometrical, anthropometry
antiq—antiquary, antiquities, antiquity
antiqn—antiquarian

apicult—apicultural, apiculture
APO—Army Post Office
app—appoint, appointed
appl—applied
appln—application
approx—approximate(ly)
Apr—April
apt—apartment(s)
aquacult—aquaculture
arbit—arbitration
arch—archives
archaeol—archaeological, archaeology
archit—architectural, architecture
Arg—Argentina, Argentine
Ariz—Arizona
Ark—Arkansas
artil—artillery
asn—association
assoc(s)—associate(s), associated
asst(s)—assistant(s), assistantship(s)
assyriol—Assyriology
astrodyn—astrodynamics
astron—astronomical, astronomy
astronaut—astonautical, astronautics
astronr—astronomer
astrophys—astrophysical, astrophysics
attend—attendant, attending
atty—attorney
audiol—audiology
Aug—August
auth—author
AV—audiovisual
Ave—Avenue
avicult—avicultural, aviculture

b—born
bact—bacterial, bacteriologic, bacteriological, bacteriology
BC—British Colombia
bd—board
behav—behavior(al)
Belg—Belgian, Belgium
Bibl—biblical
bibliog—bibliographic, bibliographical, bibliography
bibliogr-bibliographer
biochem-biochemical, biochemistry
biog—biographical, biography
biol—biological, biology
biomed—biomedical, biomedicine
biomet—biometric(s), biometrical, biometry
biophys—biophysical, biophysics

bk(s)—book(s)
bldg-building
Blvd—Boulevard
Bor—Borough
bot—botanical, botany
br—branch(es)
Brig—Brigadier
Brit—Britain, British
Bro(s)—Brother(s)
byrol—byrology
bull—Bulletin
bur—bureau
bus—business
BWI—British West Indies

c—children
Calif—California
Can—Canada, Canadian
cand—candidate
Capt—Captain
cardiol-cardiology
cardiovasc—cardiovascular
cartog—cartographic, cartographical, cartography
cartogr-cartographer
Cath—Catholic
CEngr—Corp of Engineers
cent—central
Cent Am—Central American
cert—certificate(s), certification, certified
chap—chapter
chem—chemical(s), chemistry
chemother—chemotherapy
chg—change
chmn—chairman
citricult—citriculture
class—classical
climat—climatological, climatology
clin(s)—clinic(s), clinical
cmndg—commanding
Co—County
co—Companies, Company
co-auth—coauthor
co-dir—co-director
co-ed—co-editor
co-educ—coeducation, coeducational
col(s)—college(s), collegiate, colonel
collab—collaboration, collaborative
collabr—collaborator
Colo—Colorado
com—commerce, commercial
Comdr—Commander

ABBREVIATIONS

commun—communicable, communication(s)
comn(s)—commission(s), commissioned
comndg—commanding
comnr—commissioner
comp—comparitive
compos—composition
comput—computation, computer(s), computing
comt(s)—committee(s)
conchol—conchology
conf—conference
cong—congress, congressional
Conn—Connecticut
conserv—conservation, conservatory
consol—consolidated, consolidation
const—constitution, constitutional
construct—construction, constructive
consult(s)—consult, consultant(s), consultantship(s), consultation, consulting
contemp—contemporary
contrib—contribute, contributing, contribution(s)
contribr—contributor
conv—convention
coop—cooperating, cooperation, cooperative
coord—coordinate(d), coordinating, coordination
coordr—coordinator
corp—corporate, corporation(s)
corresp—correspondence, correspondent, corresponding
coun—council, counsel, counseling
counr—councilor, counselor
criminol—criminological, criminology
cryog—cryogenic(s)
crystallog—crystallographic, crystallographical, crystallography
crystallogr—crystallographer
Ct—Court
Ctr—Center
cult—cultural, culture
cur—curator
curric—curriculum
cybernet—cybernetic(s)
cytol—cytological, cytology
Czech—Czechoslovakia

DC—District of Columbia
Dec—December
Del—Delaware
deleg—delegate, delegation
delinq—delinquency, delinquent
dem—democrat(s), democratic
demog—demographic, demography
demogr—demographer
demonstr—demonstrator
dendrol—dendrologic, dendrological, dendrology
dent—dental, dentistry
dep—deputy
dept—department
dermat—dermatologic, dermatological, dermatology
develop—developed, developing, development, developmental
diag—diagnosis, diagnostic
dialectol-dialectological, dialectology
dict—dictionaries, dictionary
Dig—Digest

dipl—diploma, diplomate
dir(s)-director(s), directories, directory
dis—disease(s), disorders
Diss Abst—Dissertation Abstracts
dist—district
distrib—distributed, distribution, distributive
distribr—distributor(s)
div—division, divisional, divorced
DNA-deoxyribonucleic acid
doc—document(s), documentary, documentation
Dom—Dominion
Dr—Drive

E—east
ecol—ecological, ecology
econ(s)—economic(s), economical, economy
economet—econometric(s)
ECT—electroconvulsive or electroshock therapy
ed—edition(s), editor(s), editorial
ed bd—editorial board
educ—education, educational
educr—educator(s)
EEG—electroencephalogram, electroencephalographic, electroencephalography
Egyptol—Egyptology
EKG—electrocardiogram
elec—elecvtric, electrical, electricity
electrochem-electrochemical, electrochemistry
electroph—electrophysical, electrophysics
elem—elementary
embryol—embryologic, embryological, embryology
emer—emeriti, emeritus
employ—employment
encour—encouragement
encycl—encyclopedia
endocrinol—endocrinologic, endocrinology
eng—engineering
Eng—England, English
engr(s)—engineer(s)
enol—enology
Ens—Ensign
entom—entomological, entomology
environ-environment(s), environmental
enzym—enzymology
epidemiol—epideiologic, epidemiological, epidemiology
equip—equipment
ERDA—Energy Research & Development Administration
ESEA—Elementary & Secondary Education Act
espec—especially
estab—established, establishment(s)
ethnog—ethnographic, ethnographical, ethnography
ethnogr—ethnographer
ethnol—ethnologic, ethnological, ethnology
Europ—European
eval—evaluation
Evangel—Evangelical
eve—evening
exam—examination(s), examining
examr—examiner
except—exceptional

exec(s)—executive(s)
exeg-exegeses, exegesis, exegetic, exegetical
exhib(s)—exhibition(s), exhibit(s)
exp—experiment, experimental
exped(s)—expedition(s)
explor—exploration(s), exploratory
expos—exposition
exten—extension

fac—faculty
facil—facilities, facility
Feb—February
fed—federal
fedn—federation
fel(s)—fellow(s), fellowship(s)
fermentol—fermentology
fertil—fertility, fertilization
Fla—Florida
floricult—floricultural, floriculture
found—foundation
FPO—Fleet Post Office
Fr—French
Ft—Fort

Ga—Georgia
gastroenterol—gastroenterological, gastroenterology
gen—general
geneal—genealogical, genealogy
geod—geodesy, geodetic
geog—geographic, geographical, geography
geogr—geographer
geol—geologic, geological, geology
geom—geometric, geometrical, geometry
geomorphol—geomorphologic, geomorphology
geophys—geophysical, geophysics
Ger—German, Germanic, Germany
geriat—geriatric
geront—gerontological, gerontology
GES—Gesellschaft
glaciol—glaciology
gov—governing, governor(s)
govt—government, governmental
grad—graduate(d)
Gt Brit—Great Britain
guid—guidance
gym—gymnasium
gynec—gynecologic, gynecological, gynecology

handbk(s)—handbook(s)
helminth—helminthology
hemat—hematologic, hematological, hematology
herpet—herpetologic, herpetological, herpetology
HEW—Department of Health, Education & Welfare
Hisp—Hispanic, Hispania
hist—historic, historical, history
histol—histological, histology
HM—Her Majesty
hochsch—hochschule
homeop—homeopathic, homeopathy
hon(s)—honor(s), honorable, honorary
hort—horticultural, horticulture
hosp(s)—hospital(s), hospitalization
hq—headquarters

HumRRO—Human Resources Research Office
husb—husbandry
Hwy—Highway
hydraul—hydraulic(s)
hydrodyn—hydrodynamic(s)
hydrol—hydrologic, hydrological, hydrologics
hyg—hygiene, hygienic(s)
hypn—hypnosis

ichthyol—ichthyological, ichthyology
Ill—Illinois
illum—illuminating, illumination
illus—illustrate, illustrated, illustration
illusr—illustrator
immunol—immunologic, immunological, immunology
Imp—Imperial
improv—improvement
Inc—Incorporated
in-chg—in charge
incl—include(s), including
Ind—Indiana
indust(s)—industrial, industries, industry
Inf—infantry
info—information
inorg—inorganic
ins—insurance
inst(s)—institute(s), institution(s)
instnl—institutional(ized)
instr(s)—instruct, instruction, instructor(s)
instrnl—instructional
int—international
intel—intellligence
introd—introduction
invert—invertebrate
invest(s)—investigation(s)
investr—investigator
irrig-irrigation
Ital—Italian

J—Journal
Jan—January
Jct-Junction
jour—journal, journalism
jr—junior
jurisp—jurisprudence
juv—juvenile

Kans—Kansas
Ky—Kentucky

La—Louisiana
lab(s)—laboratories, laboratory
lang—language(s)
laryngol—larygological, laryngology
lect—lecture(s)
lectr—lecturer(s)
legis—legislation, legislative, legislature
lett—letter(s)
lib—liberal
libr—libraries, library
librn—librarian
lic—license(d)
limnol—limnological, limnology
ling—linguistic(s), linguistical
lit—literary, literature
lithol—lithologic, lithological, lithology

Lt—Lieutenant
Ltd—Limited

m—married
mach—machine(s), machinery
mag—magazine(s)
maj—major
malacol—alacology
mammal—mammalogy
Man—Manitoba
Mar—March
Mariol—Mariology
Mass—Massachusetts
mat—material(s)
mat med—materia medica
math—mathematic(s), mathematical
Md—Maryland
mech—mechanic(s), mechanical
med—medical, medicinal, medicine
Mediter—Mediterranean
Mem—Memorial
mem—member(s), membership(s)
ment—mental(ly)
metab—metabolic, metabolism
metall—metallurgic, metallurgical, metallurgy
metallog—metallographic, metallography
metallogr—metallographer
metaphys—metaphysical, metaphysics
meteorol—meteorological, meteorology
metrol—metrological, metrology
metrop—metropolitan
Mex—Mexican, Mexico
mfg—manufacturing
mfr—manufacturer
mgr—manager
mgt—management
Mich—Michigan
microbiol—microbiological, microbiology
micros—microscopic, microscopical, microscopy
mid—middle
mil—military
mineral—mineralogical, mineralogy
Minn—Minnesota
Miss—Mississippi
mkt—market, marketing
Mo—Missouri
mod—modern
monogr—monograph
Mont—Montana
morphol—morphological, morphology
Mt—Mount
mult—multiple
munic—municipal, municipalities
mus—museum(s)
musicol—musicological, musicology
mycol—mycologic, mycology

N—north
NASA—National Aeronautics & Space Administration
nat—national, naturalized
NATO—North Atlantic Treaty Organization
navig—navigation(al)
NB—New Brunswick
NC—North Carolina
NDak—North Dakota
NDEA—National Defense Education Act
Nebr—Nebraska

nematol—nematological, nematology
nerv—nervous
Neth—Netherlands
neurol—neurological, neurology
neuropath—neuropathological, neuropathology
neuropsychiat—neuropsychiatric, neuropsychiatry
neurosurg—neurosurgical, neurosurgery
Nev—Nevada
New Eng—New England
New York—New York City
Nfld—Newfoundland
NH—New Hampshire
NIH—National Institute of Health
NIMH—National Institute of Mental Health
NJ—New Jersey
NMex—New Mexico
No—Number
nonres—nonresident
norm—normal
Norweg—Norwegian
Nov—November
NS—Nova Scotia
NSF—National Science Foundation
NSW—New South Wales
numis—numismatic(s)
nutrit—nutrition, nutritional
NY—New York State
NZ—New Zealand

observ—observatories, observatory
obstet—obstetric(s), obstetrical
occas—occasional(ly)
occup—occupation, occupational
oceanog—oceanographic, oceanographical, oceanography
oceanogr—oceanographer
Oct—October
odontol—odontology
OEEC—Organization for European Economic Cooperation
off—office, official
Okla—Oklahoma
olericult—olericulture
oncol—oncologic, oncology
Ont—Ontario
oper(s)—operation(s), operational, operative
ophthal—ophthalmologic, ophthalmological, ophthalmology
optom—optometric, optometrical, optometry
ord—ordnance
Ore—Oregon
org—organic
orgn—organization(s), organizational
orient—oriental
ornith—ornithological, ornithology
orthod—orthodontia, orthodontic(s)
orthop—orthopedic(s)
osteop—osteopathic, osteopathy
otol—otological, otology
otolaryngol—otolaryngological, otolaryngology
otorhinol—otorhinologic, otorhinology

Pa—Pennsylvania
Pac—Pacific
paleobot—paleobotanical, paleontology
paleont—paleontology

ABBREVIATIONS

Pan-Am—Pan-American
parisitol—parasitology
partic—participant, participating
path—pathologic, pathological, pathology
pedag—pedagogic(s), pedagogical, pedagogy
pediat—pediatric(s)
PEI—Prince Edward Islands
penol—penological, penology
periodont—periodontal, periodontic(s)
petrog—petrographic, petrographical, petrography
petrogr—petrographer
petrol—petroleum, petrologic, petrological, petrology
pharm—pharmacy
pharmaceut—pharmaceutic(s), pharmaceutical(s)
pharmacog—pharmacognosy
pharamacol—pharmacologic, pharmacological, pharmacology
phenomenol—phenomenologic(al), phenomenology
philol—philological, philology
philos—philosophic, philosophical, philosophy
photog—photographic, photography
photogeog—photogeographic, photogeography
photogr—photographer(s)
photogram—photogrammetric, photogrammetry
photom—photometric, photometrical, photometry
phycol—phycology
phys—physical
physiog—physiographic, physiographical, physiography
physiol—physiological, phsysiology
Pkwy—Parkway
Pl—Place
polit—political, politics
polytech—polytechnic(s)
pomol—pomological, pomology
pontif—pontifical
pop—population
Port—Portugal, Portuguese
Pos:—Position
postgrad—postgraduate
PQ—Province of Quebec
PR—Puerto Rico
pract—practice
practr—practitioner
prehist—prehistoric, prehistory
prep—preparation, preparative, preparatory
pres—president
Presby—Presbyterian
preserv—preservation
prev—prevention, preventive
prin—principal
prob(s)—problem(s)
proc—proceedings
proctol—proctologic, proctological, proctology
prod—product(s), production, productive
prof—professional, professor, professorial
Prof Exp—Professional Experience
prog(s)—program(s), programmed, programming
proj—project(s), projection(al), projective

prom—promotion
protozool—protozoology
Prov—Province, Provincial
psychiat-psychiatric, psychiatry
psychoanal—psychoanalysis, psychoanalytic, psychoanalytical
psychol—psychological, psychology
psychomet—psychometric(s)
psychopath-psychopathologic, psychopathology
psychophys—psychophysical, psychophysics
psychophysiol—psychophysiological, psychophysiology
psychosom—psychosomtic(s)
psychother—psychoterapeutic(s), psychotherapy
Pt—Point
pub—public
publ—publication(s), publish(ed), publisher, publishing
pvt—private

Qm—Quartermaster
Qm Gen—Quartermaster General
qual—qualitative, quality
quant—quantitative
quart—quarterly
Que—Quebec

radiol—radiological, radiology
RAF—Royal Air Force
RAFVR—Royal Air Force Volunteer Reserve
RAMC—Royal Army Medical Corps
RAMCR—Royal Army Medical Corps Reserve
RAOC—Royal Army Ornance Corps
RASC—Royal Army Service Corps

RASCR—Royal Army Service Corps Reserve
RCAF—Royal Canadian Air Force
RCAFR—Royal Canadian Air Force Reserve
RCAFVR—Royal Canadian Air Force Volunteer Reserve
RCAMC—Royal Canadian Army Medical Corps
RCAMCR—Royal Canadian Army Medical Corps Reserve
RCASC—Royal Canadian Army Service Corps
RCASCR—Royal Canadian Army Service Corps Reserve
RCEME—Royal Canadian Electrical & Mechanical Engineers
RCN—Royal Canadian Navy
RCNR—Royal Canadian Naval Reserve
RCNVR—Royal Canadian Naval Volunteer Reserve
Rd—Road
RD—Rural Delivery
rec—record(s), recording
redevelop—redevelopment
ref—reference(s)
refrig—refrigeration
regist—register(ed), registration
registr—registrar
regt—regiment(al)
rehab—rehabilitation
rel(s)—relation(s), relative
relig—religion, religious

REME—Royal Electrical & Mechanical Engineers
rep—represent, representative
Repub—Republic
req—requirements
res—research, reserve
rev—review, revised, revision
RFD—Rural Free Delivery
rhet-rhetoric, rhetorical
RI—Rhode Island
Rm—Room
RM—Royal Marines
RN—Royal Navy
RNA—ribonucleic acid
RNR—Royal Naval Reserve
RNVR—Royal Naval Volunteer Reserve
roentgenol—roentgenologic, roentgenological, roentgenology
RR—Railroad, Rural Route
Rte—Route
Russ—Russian
rwy—railway

S—south
SAfrica—South Africa
SAm—South America, South American
sanit—sanitary, sanitation
Sask—Saskatchewan
SC—South Carolina
Scand—Scandinavia(n)
sch(s)—school(s)
scholar—scholarship
sci—science(s), scientific
SDak—South Dakota
SEATO—Southeast Asia Treaty Organization
sec—secondary
sect—section
secy—secretary
seismog—seismograph, seismographic, seismography
seismogr—seismographer
seismol—seismological, seismology
sem—seminar, seminary
Sen—Senator, Senatorial
Sept—September
ser—serial, series
serol—serologic, serological, serology
serv—service(s), serving
silvicult—silvicultural, silviculture
soc(s)—societies, society
soc sci—social science
sociol—sociologic, sociological, sociology
Span—Spanish
spec—special
specif—specification(s)
spectrog—spectrograph, spectrographic, spectrography
spectrogr—spectrographer
spectrophotom—spectrophotometer, spectrophotometric, spectrophotometry
spectros—spectroscopic, spectroscopy
speleol—speleological, speleology
Sq—Square
sr—senior
St—Saint, Street(s)
sta(s)—station(s)
stand—standard(s), standardization
statist—statistical, statistics
Ste—Sainte

steril—sterility
stomatol—stomatology
stratig—stratigraphic, stratigraphy
stratigr—stratigrapher
struct—structural, structure(s)
stud—student(ship)
subcomt—subcommittee
subj—subject
subsid—subsidiary
substa—substation
super—superior
suppl—supplement(s), supplemental,
    supplementary
supt—superintendent
supv—supervising, supervision
supvr—supervisor
supvry—supervisory
surg—surgery, surgical
surv—survey, surveying
survr—surveyor
Swed—Swedish
Switz—Switzerland
symp—symposia, symposium(s)
syphil—syphilology
syst(s)—system(s), systematic(s), systematical

taxon—taxonomic, taxonomy
tech—technical, technique(s)
technol—technologic(al), technology
tel—telegraph(y), telephone
temp—temporary
Tenn—Tennessee
Terr—Terrace
Tex—Texas
textbk(s)—textbook(s)
text ed—text edition
theol—theological, theology
theoret—theoretic(al)
ther—therapy
therapeut—therapeutic(s)
thermodyn—thermodynamic(s)
topog—topographic, topographical,
    topography
topogr—topographer
toxicol—toxicologic, toxicological,
    toxicology
trans—transactions
transl—translated, translation(s)
translr—translator(s)
transp—transport, transportation

treas—treasurer, treasury
treat—treatment
trop—tropical
tuberc—tuberculosis
TV—television
Twp—Township

UAR—United Arab Republic
UK—United Kingdom
UN—United Nations
undergrad—undergraduate
unemploy—unemployment
UNESCO—United Nations Educational
    Scientific & Cultural Organization
UNICEF—United Nations International
    Childrens Fund
univ(s)—universities, university
UNRRA—United Nations Relief &
    Rehabilitation Administration
UNRWA—United Nations Relief & Works
    Agency
urol—urologic, urological, urology
US—United States
USAAF—US Army Air Force
USAAFR—US Army Air Force Reserve
USAF—US Air Force
USAFR—US Air Force Reserve
USAID—US Agency for International
    Development
USAR—US Army Reserve
USCG—US Coast Guard
USCGR—US Coast Guard Reserve
USDA—US Department of Agriculture
USMC—US Marine Corps
USMCR—US Marine Corps Reserve
USN—US Navy
USNAF—US Naval Air Force
USNAFR—US Naval Air Force Reserve
USNR—US Naval Reserve
USPHS—US Public Health Service
USPHSR—US Public Health Service Reserve
USSR—Union of Soviet Socialist Republics

Va—Virginia
var—various
veg—vegetable(s), vegetation
vent—ventilating, ventilation
vert—vertebrate
Vet—Veteran(s)

vet—veterinarian, veterinary
VI—Virgin Islands
vinicult—viniculture
virol—virological, virology
vis—visiting
voc—vocational
vocab—vocabulary
vol(s)—voluntary, volunteer(s), volume(s)
vpres—vice president
vs—versus
Vt—Vermont

W—west
Wash—Washington
WHO—World Health Organization
WI—West Indies
wid—widow, widowed, widower
Wis—Wisconsin
WVa—West Virginia
Wyo—Wyoming

Yearbk(s)—Yearbook(s)
YMCA—Young Men's Christian Association
YMHA—Young Men's Hebrew Association
Yr(s)—Year(s)
YT—Yukon Territory
YWCA—Young Women's Christian
    Association
YWHA—Young Women's Hebrew
    Association

zool—zoological, zoology

DISCIPLINE INDEX

AGRICULTURAL & FOREST SCIENCES

Agricultural Business & Management

ARIZONA
Gordon, Richard Seymour

ARKANSAS
Daniels, L B

CALIFORNIA
Kushner, Arthur S
Thomas, Paul Clarence

COLORADO
Gholson, Larry Estie

DELAWARE
LaRossa, Robert Alan

DISTRICT OF COLUMBIA
Jennings, Vivan M
Tallent, William Hugh

FLORIDA
Nichols, Robert Loring
Peart, Robert McDermand

GEORGIA
Hilton, James Lee

IDAHO
Douglas, Dexter Richard

ILLINOIS
Prescott, Jon Michael

INDIANA
Hoefer, Raymond H

IOWA
Colvin, Thomas Stuart
Harbaugh, Daniel David

KANSAS
Kastner, Curtis Lynn

LOUISIANA
Hensley, Sess D

MARYLAND
Gadsby, Dwight Maxon
Howard, Joseph H

MICHIGAN
Castenson, Roger R

MINNESOTA
Boehlje, Michael Dean

MISSOURI
Carlson, Wayne C
Houghton, John M
Yates-Parker, Nancy L

OHIO
Dembowski, Peter Vincent
Sabourin, Thomas Donald

SOUTH CAROLINA
Fischer, James Roland

TEXAS
Benedict, John Howard, Jr
Davis, Bob

UTAH
Butcher, John Edward

PUERTO RICO
Rodriguez, Jorge Luis

Agricultural Economics

ALABAMA
Bonsi, Conrad K
Drake, Albert Estern
Sutherland, William Neil

ARIZONA
Johnson, Randall Arthur
Lord, William B

ARKANSAS
Havener, Robert D
Headley, Joseph Charles
Parsch, Lucas Dean

CALIFORNIA
Abenes, Fiorello Bigornia
Johnston, Warren E
Lin, Robert I-San
Niles, James Alfred
Sage, Orrin Grant, Jr
Siebert, Jerome Bernard
Squires, Dale Edward
Wallender, Wesley William

CONNECTICUT
Sieckhaus, John Francis

DELAWARE
Elterich, G Joachim
LaRossa, Robert Alan

DISTRICT OF COLUMBIA
Beer, Charles
Harrington, David Holman
Jennings, Vivan M
Mellor, John Williams

FLORIDA
Barnard, Donald Roy
Meltzer, Martin Isaac
Norval, Richard Andrew
Thomason, David Morton
Ward, Ronald Wayne

GEORGIA
Bhagia, Gobind Shewakram
Chiang, Tze I
Freeman, Jere Evans
Purcell, Joseph Carroll

HAWAII
Ching, Chauncey T K

IDAHO
Michalson, Edgar Lloyd

ILLINOIS
Seitz, Wesley Donald

INDIANA
Thompson, Robert Lee

IOWA
Colvin, Thomas Stuart
Kliebenstein, James Bernard

KANSAS
Hess, Carroll V

LOUISIANA
Decossas, Kenneth Miles

MARYLAND
Gadsby, Dwight Maxon
Just, Richard
Weiss, Michael David

MICHIGAN
Gillingham, James Clark
Manderscheid, Lester Vincent
Simard, Albert Joseph

MINNESOTA
Boehlje, Michael Dean
Munson, Robert Dean

Ruttan, Vernon W

MISSISSIPPI
Hurt, Verner C

MISSOURI
Brandt, Jon Alan
Hardin, Clifford Morris

NEBRASKA
Miller, William Lloyd

NEW JERSEY
Burns, David Jerome

NEW YORK
Herdt, Robert William
Jabbur, Ramzi Jibrail
Thompson, John C, Jr
Todaro, Michael P

NORTH CAROLINA
Benrud, Charles Harris
Johnson, Thomas

NORTH DAKOTA
Anderson, Donald E

OHIO
Adams, Dale W
Baldwin, Eldon Dean
Chern, Wen Shyong
Erven, Bernard Lee
Forster, D Lynn
Hushak, Leroy J
Larson, Donald W
Lee, Warren Ford
McCormick, Francis B
Meyer, Richard Lee
Rask, Norman
Shaudys, Edgar T
Vertrees, Robert Layman
Walker, Francis Edwin

OKLAHOMA
Plaxico, James Samuel

OREGON
Buccola, Steven Thomas
Miller, Stanley Frank
Nelson, A Gene
Schmisseur, Wilson Edward
Workman, William Glenn

PENNSYLVANIA
Alter, Theodore Roberts
Epp, Donald James
Kelly, B(ernard) Wayne

SOUTH DAKOTA
Dobbs, Thomas Lawrence

TENNESSEE
Jumper, Sidney Roberts
Papas, Andreas Michael
Rawlins, Nolan Omri
Williamson, Handy, Jr

TEXAS
Davis, Bob
Eddleman, Bobby R
Eubank, Randall Lester
Lacewell, Ronald Dale
Nixon, Donald Merwin
Whitson, Robert Edd

UTAH
Downing, Kenton Benson
Grimshaw, Paul R
Wallentine, Max V
Workman, John Paul

VIRGINIA
Cummings, Ralph Waldo, Jr
Halvorson, Lloyd Chester
Pontius, Steven Kent
Sprague, Lucian Matthew

Woods, William Fred

WEST VIRGINIA
Barr, Alfred L

WISCONSIN
Buse, Reuben Charles
Luby, Patrick Joseph
Schmidt, John Richard

WYOMING
Kearl, Willis Gordon

PUERTO RICO
Gonzalez, Gladys
Lewis, Allen Rogers

ALBERTA
Sonntag, Bernard H

ONTARIO
Menzie, Elmer Lyle

QUEBEC
Coffin, Harold Garth

OTHER COUNTRIES
Bywater, Anthony Colin
Malik, Mazhar Ali Khan

Agriculture, General

ALABAMA
Guthrie, Richard Lafayette
Haaland, Ronald L
Johnson, Clarence Eugene
Johnson, Loyd
Mack, Timothy Patrick

ALASKA
Cochran, Verlan Leyerl

ARIZONA
Day, Arden Dexter
Erickson, Eric Herman, Jr
Gitlin, Harris Martlin
Kidwell, Margaret Gale
Terry, Lucy Irene

ARKANSAS
Andrews, Luther David
Bartlett, Frank David
Brown, A Hayden, Jr
Brown, Connell Jean
Clower, Dan Fredric
Horan, Francis E
Motes, Dennis Roy
Oosterhuis, Derrick M
Pitts, Donald James

CALIFORNIA
Abenes, Fiorello Bigornia
Ayars, James Earl
Benes, Norman Stanley
Bertoldi, Gilbert LeRoy
Calpouzos, Lucas
Christensen, Allen Clare
Donnan, William W
Estilai, Ali
Garber, Richard Hammerle
Henrick, Clive Arthur
Hess, Frederick Dan
Humaydan, Hasib Shaheen
Karinen, Arthur Eli
Lovatt, Carol Jean
Moore, Andrew Brooks
Myers, George Scott, Jr
Prend, Joseph
Raju, Namboori Bhaskara
Rammer, Irwyn Alden
Risch, Stephen John
Rousek, Edwin J
Sage, Orrin Grant, Jr
Sanders, John Stephen
Silk, Margaret Wendy Kuhn
Thomason, Ivan J

Agriculture, General (cont)

Thompson, Chester Ray
Treloar, Alan Edward
Whitlock, Gaylord Purcell
Willemsen, Roger Wayne

COLORADO
Ball, Wilbur Perry
Dickenson, Donald Dwight
Gholson, Larry Estie
Helmerick, Robert Howard
Knutson, Kenneth Wayne
Lamm, Warren Dennis
Luebs, Ralph Edward
Thomas, William Robb

CONNECTICUT
Ahrens, John Frederick
John, Hugo Herman

DELAWARE
Green, Jerome
Smith, Constance Meta

DISTRICT OF COLUMBIA
Beer, Charles
Brown, Lester R
Harris, Clare I
Jennings, Vivan M
Khan, Mohamed Shaheed
Mellor, John Williams
Nickle, David Allan
Parochetti, James V
Plowman, Ronald Dean
Roskoski, Joann Pearl
Strommen, Norton Duane
Wilson, William Mark Dunlop

FLORIDA
Denmark, Harold Anderson
Kretschmer, Albert Emil, Jr
McCloud, Darell Edison
Osborne, Lance Smith
Price, Donald Ray
Simonet, Donald Edward
Smart, Grover Cleveland, Jr
Snyder, Fred Calvin
Stoffella, Peter Joseph
Woodruff, Robert Eugene

GEORGIA
Amos, Henry Estill
Hilton, James Lee
Kanemasu, Edward Tsukasa
McMurray, Birch Lee
Minton, Norman A
Todd, James Wyatt

HAWAII
Bartosik, Alexander Michael
Warner, John Northrup

IDAHO
Bondurant, James A(llison)
Callihan, Robert Harold
McCaffrey, Joseph Peter
Michalson, Edgar Lloyd
Osgood, Charles Edgar

ILLINOIS
Bentley, Orville George
Bristol, Benton Keith
Changnon, Stanley A, Jr
Clark, Jimmy Howard
Easter, Robert Arnold
Janghorbani, Morteza
Kirby, Hilliard Walker
Knake, Ellery Louis
Olsson, Nils Ove
Parsons, Carl Michael
Princen, Lambertus Henricus
Spahr, Sidney Louis
Troyer, Alvah Forrest
Wolff, Robert L
Yoerger, Roger R

INDIANA
Bauman, Thomas Trost
Freeman, Verne Crawford
Gallun, Robert Louis
Gehring, Perry James
Ivaturi, Rao Venkata Krishna
Jantz, O K
Mitchell, Cary Arthur

IOWA
Black, Charles Allen
Harbaugh, Daniel David

KANSAS
Call, Edward Prior
Larson, Vernon C
Whitney, Wendell Keith

KENTUCKY
Householder, William Allen
Little, Charles Oran
Norfleet, Morris L
Sigafus, Roy Edward
Vogel, Willis Gene

LOUISIANA
Board, James Ellery

MAINE
Barton, Barbara Ann

MARYLAND
Benbrook, Charles M
Christy, Alfred Lawrence
Cleland, Charles Frederick
Crosby, Edwin Andrew
Duke, James A
Fried, Maurice
Gadsby, Dwight Maxon
Hopkins, Homer Thawley
Kearney, Philip C
Krizek, Donald Thomas
Matthews, Benjamin F
Menn, Julius Joel
Minnifield, Nita Michele
Nelson, Elton Glen
Ross, Philip
Soto, Gerardo H
Teramura, Alan Hiroshi
Tjio, Joe Hin

MASSACHUSETTS
Botticelli, Charles Robert
Coppinger, Raymond Parke
Deubert, Karl Heinz
Devlin, Robert Martin
Furth, David George
Kepper, Robert Edgar
Redington, Charles Bahr
Rohde, Richard Allen

MICHIGAN
Bird, George W
Henneman, Harold Albert
Hull, Jerome, Jr
Isleib, Donald Richard
Maley, Wayne A
Miller, James Ray
Thomas, John William
Uebersax, Mark Alan

MINNESOTA
Marx, George Donald
Oelke, Ervin Albert
Preiss, Frederick John
Pryor, Gordon Roy
Rehm, George W

MISSISSIPPI
Hardee, Dicky Dan
Hodges, Harry Franklin
Knight, William Eric
Mutchler, Calvin Kendal
Robinson, A(ugust) R(obert)
Tomlinson, James Everett

MISSOURI
Miles, Randall Jay
Pfander, William Harvey
Schumacher, Richard William
Wochok, Zachary Stephen

MONTANA
Morrill, Wendell Lee

NEBRASKA
Dillon, Roy Dean
Knox, Ellis Gilbert
Moore, Kenneth J
Raun, Earle Spangler
Von Bargen, Kenneth Louis

NEW JERSEY
Bahr, James Theodore
Cheng, Kang
Dyer, Judith Gretchen
Eaglesham, Allan Robert James
Joiner, Robert Russell
Markle, George Michael
Marusich, Wilbur Lewis
Stevens, Merwin Allen
Witherell, Peter Charles

NEW YORK
Amundson, Robert Gale
Bayer, George Herbert
Bellvé, Anthony Rex
Butler, Karl Douglas, Sr
Huntington, David Hans
Jacobson, Jay Stanley
Nittler, LeRoy Walter
Raffensperger, Edgar M
Robinson, Terence Lee
Smalley, Ralph Ray
Thomas, Everett Dake
Topp, William Carl
Verma, Ram S

NORTH CAROLINA
Anderson, Thomas Ernest
Blake, Thomas Lewis
Carlson, William Theodore
Ganapathy, Seetha N
Holm, Robert E
Lalor, William Francis
Marco, Gino Joseph
Ross, Richard Henry, Jr

NORTH DAKOTA
Joachim, Frank G

OHIO
Bowers, John Dalton
Cooper, Tommye
Gauntt, William Amor
Hock, Arthur George
Jalil, Mazhar
McCracken, John David
Niemczyk, Harry D
Reddy, Padala Vykuntha
Ritchie, Austin E
Wilson, George Rodger
Wonderling, Thomas Franklin

OKLAHOMA
Klatt, Arthur Raymond
Melouk, Hassan A
Rice, Charles Edward

OREGON
Arnold, Roy Gary
Howard, William Weaver
Miller, Stanley Frank
Shock, Clinton C
Smiley, Richard Wayne
Weiser, Conrad John

PENNSYLVANIA
Aller, Harold Ernest
Bergman, Ernest L
Cerbulis, Janis
Heller, Paul R
Hull, Larry Allen
Montgomery, Ronald Eugene
Russo, Joseph Martin
Wicker, Robert Kirk

SOUTH CAROLINA
Gossett, Billy Joe
Johnson, Albert Wayne
Kittrell, Benjamin Upchurch
Manley, Donald Gene
Nolan, Clifford N
Schoenike, Roland Ernest

TENNESSEE
Armistead, Willis William
Caron, Richard Edward
Chiang, Thomas M
Graveel, John Gerard
Hathcock, Bobby Ray
Helweg, Otto Jennings
Young, Lawrence Dale

TEXAS
Agan, Raymond John
Arkin, Gerald Franklin
Bowling, Clarence C
Brewer, Beverly Sparks
Brokaw, Bryan Edward
Butler, Ogbourne Duke, Jr
Calub, Alfonso deGuzman
Eddleman, Bobby R
Frederiksen, Richard Allan
Greene, Donald Miller
Hoffman, Robert A
Kirk, Ivan Wayne
McDonald, Lynn Dale
Patel, Mayur
Richardson, Arthur Jerold
Shotwell, Thomas Knight
Sij, John William
Stewart, Bobby Alton
Thompson, Granville Berry
Wilson, Richard Hansel

UTAH
Albrechtsen, Rulon S
Chase, Richard Lyle
Clark, C Elmer
Haws, Byron Austin
Thorup, Richard M
Wallentine, Max V

VERMONT
Dritschilo, William

VIRGINIA
Drake, Charles Roy
Frahm, Richard R
Hinckley, Alden Dexter
Pontius, Steven Kent
Van Krey, Harry P

WASHINGTON
Butt, Billy Arthur
Peabody, Dwight Van Dorn, Jr
Smith, Samuel H
Toba, H(achiro) Harold

WISCONSIN
Barnes, Robert F
Delorit, Richard John
Grunewald, Ralph
Holm, LeRoy George
Jorgensen, Neal A
Koval, Charles Francis
Stang, Elden James
Steinhart, Carol Elder
Willis, Harold Lester

PUERTO RICO
Rodriguez, Jorge Luis

ALBERTA
Hadziyev, Dimitri
Harper, Alexander Maitland
Krahn, Thomas Richard
Moyer, James Robert
Young, Bruce Arthur

BRITISH COLUMBIA
Borden, John Harvey
Mackauer, Manfred
Neilsen, Gerald Henry

MANITOBA
Grant, Cynthia Ann
Palaniswamy, Pachagounder

NEWFOUNDLAND
Lim, Kiok-Puan

NOVA SCOTIA
Embree, Charles Gordon
Warman, Philip Robert

ONTARIO
Anderson, Terry Ross
Arnison, Paul Grenville
Asculai, Samuel Simon
Biswas, Asit Kumar
Catling, Paul Miles
Chong, Calvin
Coote, Denis Richard
Emmons, Douglas Byron
Farnworth, Edward Robert
Hoffman, Douglas Weir
Hofstra, Gerald
Hutchinson, Thomas C
Kevan, Peter Graham
Lister, Earl Edward
Svoboda, Josef
Switzer, Clayton Macfie
Turk, Fateh (Frank) M

QUEBEC
Belloncik, Serge
Bertrand, Forest
Coffin, Harold Garth
Dufour, Jacques John
Lapp, Wayne Stanley
Matte, J Jacques
Willemot, Claude

SASKATCHEWAN
Smith, Allan Edward

OTHER COUNTRIES
Edelman, Marvin
Gregory, Peter
Kaltsikes, Pantouses John
Muniappan, Rangaswamy Naicker
Sakamoto, Clarence M

Agronomy

ALABAMA
Donnelly, Edward Daniel
Ensminger, Leonard Elroy
Gill, William Robert
Haaland, Ronald L
Mortvedt, John Jacob
Peterson, Curtis Morris
Rajanna, Bettaiya
Rogers, Howard Topping
Russel, Darrell Arden
Simmons, Charles Ferdinand
Sutherland, William Neil
Thomas, Winfred
Trouse, Albert Charles
Truelove, Bryan
Ward, Coleman Younger

ALASKA
Mitchell, William Warren
Taylor, Roscoe L
Wooding, Frank James

ARIZONA
Ahlgren, Henry Lawrence
Allen, Stephen Gregory
Brickbauer, Elwood Arthur
Briggs, Robert Eugene
Davis, Charles Homer
Day, Arden Dexter
Dennis, Robert E
Feaster, Carl Vance
Fink, Dwayne Harold
Fisher, Warner Douglass
Guinn, Gene
Hawk, Virgil Brown
Jackson, Ernest Baker
Jordan, Gilbert LeRoy
Kimball, Bruce Arnold
Kneebone, William Robert
Loper, Gerald Milton
McAlister, Dean Ferdinand
Meggitt, William Fredric
Metcalfe, Darrel Seymour
Minch, Edwin Wilton
Morton, Howard LeRoy
Richardson, Grant Lee
Rohweder, Dwayne A

Rubis, David Daniel
Schonhorst, Melvin Herman
Smith, Dale
Upchurch, Robert Phillip
Webster, Orrin John
Wierenga, Peter J

ARKANSAS
Bartlett, Frank David
Bourland, Freddie Marshall
Collins, Frederick Clinton
Collister, Earl Harold
Frans, Robert Earl
Hinkle, Dale Albert
King, John William
Lavy, Terry Lee
Motes, Dennis Roy
Musick, Gerald Joe
Oosterhuis, Derrick M
Porter, Owen Archuel
Smith, Roy Jefferson, Jr
Spooner, Arthur Elmon
Stutte, Charles A
Talbert, Ronald Edward
Timmermann, Dan, Jr
Waddle, Bradford Avon
West, Charles Patrick

CALIFORNIA
Anderson, Lars William James
Andres, Lloyd A
Ashton, Floyd Milton
Barker, LeRoy N
Beard, Benjamin H
Bodman, Geoffrey Baldwin
Bradford, Willis Warren
Brecht, Patrick Ernest
Breidenbach, Rowland William
Brooks, William Hamilton
Carter, Lark Poland
Caulder, Jerry Dale
Chu, Chang-Chi
Crawford, Robert Field
Davis, Larry Alan
Domingo, Wayne Elwin
Doner, Harvey Ervin
Embleton, Tom William
Epstein, Emanuel
Estilai, Ali
Ferguson, David B
Foster, Ken Wood
Gerik, James Stephen
Grantz, David Arthur
Hagan, William Leonard
Hall, Anthony Elmitt
Harwood, Richard Roland
Hesse, Walter Herman
Hile, Mahlon Malcolm Schallig
Hills, F Jackson
Hooks, James A
Isom, William Howard
Johnson, Carl William
Lambert, Royce Leone
Laude, Horton Meyer
Lewellen, Robert Thomas
Loomis, Robert Simpson
Lorenz, Oscar Anthony
Mikkelsen, Duane Soren
Murphy, Alfred Henry
Norris, Robert Francis
Omid, Ahmad
Oster, James Donald
Peterson, Maurice Lewellen
Phillips, Donald Arthur
Plant, Richard E
Rawal, Kanti M
Ririe, David
Ritenour, Gary Lee
Robinson, Frank Ernest
Sanders, John Stephen
Schaller, Charles William
Schieferstein, Robert Harold
Schmitz, George William
Shih, Ching-Yuan G
Sims, William Lynn
Smith, Paul Gordon
Teuber, Larry Ross
Thomas, Richard Sanborn
Thorup, James Tat
Tilton, Varien Russell
Van Elswyk, Marinus, Jr
Williams, William Arnold
Worker, George F, Jr
Yuen, Wing
Zary, Keith Wilfred

COLORADO
Akeson, Walter Roy
Crumpacker, David Wilson
Cuany, Robin Louis
Danielson, Robert Eldon
Dickenson, Donald Dwight
Dittberner, Phillip Lynn
Fly, Claude Lee
Halvorson, Ardell David
Haus, Thilo Enoch
Hecker, Richard Jacob
Helmerick, Robert Howard
Keim, Wayne Franklin
Klute, Arnold
Ladd, Sheldon Lane
Larsen, Arnold Lewis
Litzenberger, Samuel Cameron
Luebs, Ralph Edward

McGinnies, William Joseph
Mickelson, Rome H
Norstadt, Fred A
Oldemeyer, Robert King
Quick, James S
Sabey, Burns Roy
Schweizer, Edward E
Shaw, Robert Blaine
Shawcroft, Roy Wayne
Siemer, Eugene Glen
Soltanpour, Parviz Neil
Sullivan, Edward Francis
Tsuchiya, Takumi
Widner, Jimmy Newton
Willis, Wayne O
Wood, Donald Roy
Youngman, Vern E
Zimdahl, Robert Lawrence

CONNECTICUT
Washko, Walter William

DELAWARE
Boyer, John Strickland
Fieldhouse, Donald John
Green, Jerome
Hill, Gideon D
Jones, Edward Raymond
Mitchell, William H
Wittenbach, Vernon Arie
Wolf, Dale E

DISTRICT OF COLUMBIA
Allen, James Ralston
Eisa, Hamdy Mahmoud
Islam, Nurul
Mayes, McKinley
Quebedeaux, Bruno, Jr
Rumburg, Charles Buddy
Schmidt, Berlie Louis
Thomas, Walter Ivan
Wiggans, Samuel Claude

FLORIDA
Albrecht, Stephan LaRowe
Aldrich, Samuel Roy
Anderson, Stanley Robert
Barnett, Ronald David
Bennette, Jerry Mac
Boote, Kenneth Jay
Bowes, George Ernest
Boyd, Frederick Tilghman
Braids, Olin Capron
Brecke, Barry John
Calvert, David Victor
Cantliffe, Daniel James
Chandler, Robert Flint, Jr
Collins, Mary Elizabeth
Dudeck, Albert Eugene
Dunavin, Leonard Sypret, Jr
Everett, Paul Harrison
Forbes, Richard Brainard
Gammon, Nathan, Jr
Gardner, Franklin Pierce
Gilreath, James Preston
Gorbet, Daniel Wayne
Green, Victor Eugene, Jr
Gull, Dwain D
Guzman, Victor Lionel
Hammond, Luther Carlisle
Haramaki, Chiko
Henderson, Lawrence J
Hinson, Kuell
Holder, David Gordon
Horner, Earl Stewart
Johnson, Walter Lee
Kidder, Gerald
Knauft, David A
Kretschmer, Albert Emil, Jr
Lutrick, Monroe Cornelius
McCloud, Darell Edison
McKently, Alexandra H
Matthews, David Livingston
Mislevy, Paul
Monson, Warren Glenn
Nichols, Robert Loring
Norden, Allan James
Orsenigo, Joseph Reuter
Owens, Clarence Burgess
Peacock, Hugh Anthony
Pfahler, Paul Leighton
Prine, Gordon Madison
Quesenberry, Kenneth Hays
Rich, Jimmy Ray
Roberts, Donald Ray
Rodgers, Earl Gilbert
Ruelke, Otto Charles
Scudder, Walter Tredwell
Snyder, George Heft
Stanley, Robert Lee, Jr
Stoffella, Peter Joseph
Teare, Iwan Dale
Thompson, Buford Dale
West, Sherlie Hill
Whitty, Elmo Benjamin
Wolf, Benjamin
Woofter, Harvey Darrell
Workman, Ralph Burns

GEORGIA
Anderson, Oscar Emmett
Ashley, Doyle Allen
Bailey, George William
Boswell, Fred Carlen

Brown, Acton Richard
Brown, Ronald Harold
Burgoa, Benali
Burton, Glenn Willard
Cummins, David Gray
Dibb, David Walter
Douglas, Charles Francis
Dowler, Clyde Cecil
Duncan, Ronny Rush
Eastin, Emory Ford
Forbes, Ian
Frere, Maurice Herbert
Gaines, Tinsley Powell
Gascho, Gary John
Hammons, Ray Otto
Hardcastle, Willis Santford
Hoogenboom, Gerrit
Hoveland, Carl Soren
Jellum, Milton Delbert
Johnson, Jerry Wayne
Kays, Stanley J
Miller, John David
Mills, Harry Arvin
Mixon, Aubrey Clifton
Morey, Darrell Dorr
Murray, Calvin Clyde
Nyczepir, Andrew Peter
Tan, Kim H
Todd, James Wyatt
Weaver, James B, Jr
Whitehead, Marvin Delbert
Widstrom, Neil Wayne
Wilkinson, Robert Eugene
Wilkinson, Stanley R
Wofford, Irvin Mirle
Wood, Bruce Wade
Worley, Ray Edward
Younts, Sanford Eugene

HAWAII
Bartholomew, Duane P
Brewbaker, James Lynn
Cushing, Robert Leavitt
De la Pena, Ramon Serrano
Fox, Robert Lee
Hagihara, Harold Haruo
Hepton, Anthony
Nishimoto, Roy Katsuto
Osgood, Robert Vernon
Rotar, Peter P
Silva, James Anthony
Thompson, John R
Warner, John Northrup
Whitney, Arthur Sheldon

IDAHO
Callihan, Robert Harold
Carter, John Newton
Dwelle, Robert Bruce
Haderlie, Lloyd Conn
Hansen, Leon A
Kleinkopf, Gale Eugene
Lee, Gary Albert
Lehrsch, Gary Allen
Murray, Glen A
Myers, James Robert
O'Keeffe, Lawrence Eugene
Osgood, Charles Edgar
Pearson, Lorentz Clarence
Schweitzer, Leland Ray
Sojka, Robert E
Thill, Donald Cecil
Wright, James Louis

ILLINOIS
Alexander, Denton Eugene
Bernard, Richard Lawson
Blair, Louis Curtis
Boast, Charles Warren
Brown, Lindsay Dietrich
Buettner, Mark Roland
Burger, Ambrose William
Dudley, John Wesley
Elkins, Donald Marcum
Fink, Rodney James
Flood, Brian Robert
Fuess, Frederick William, III
Garwood, Douglas Leon
Graffis, Don Warren
Hadley, Henry Hultman
Harper, James Eugene
Hauptmann, Randal Mark
Heermann, Ruben Martin
Heichel, Gary Harold
Hoeft, Robert Gene
Holt, Donald Alexander
Howell, Robert Wayne
Huck, Morris Glen
Jackobs, Joseph Alden
Jansen, Ivan John
Johnson, Richard Ray
Jugenheimer, Robert William
Kapusta, George
King, S(anford) MacCallum
Kirby, Hilliard Walker
Knake, Ellery Louis
Kurtz, Lester Touby
Leffler, Harry Rex
McGlamery, Marshal Dean
Miller, Darrell Alvin
Myers, Oval, Jr
Nicholaides, John J, III
Nickell, Cecil D
Olsen, Farrel John

Portis, Archie Ray, Jr
Portz, Herbert Lester
Reetz, Harold Frank, Jr
Robison, Norman Glenn
Roskamp, Gordon Keith
Slife, Fred Warren
Spencer, Jack T
Sprague, George Frederick
Steele, Leon
Stickler, Fred Charles
Stoller, Edward W
Thorne, Marlowe Driggs
Troyer, Alvah Forrest
Van Riper, Gordon Everett
Wilcox, Wesley Crain

INDIANA
Alvey, David Dale
Bauman, Thomas Trost
Baumgardner, Marion F
Bucholtz, Dennis Lee
Case, Vernon Wesley
Castleberry, Ron M
Christmas, Ellsworth P
Crane, Paul Levi
Daniel, William Hugh
Glover, David Val
Hilst, Arvin Rudolph
Hoefer, Raymond H
Johannsen, Christian Jakob
Kidd, Frank Alan
Kohnke, Helmut
Lechtenberg, Victor L
Mannering, Jerry Vincent
Mengel, David Bruce
Nelson, Werner Lind
Nielsen, Niels Christian
Nyquist, Wyman Ellsworth
Ohm, Herbert Willis
Palmer, Robert Gerald
Parka, Stanley John
Perkins, A Thomas
Peterson, John Booth
Probst, Albert Henry
Reiss, William Dean
Robinson, Glenn Hugh
Ross, Merrill Arthur, Jr
Schreiber, Marvin Mandel
Shockey, William Lee
Slavik, Nelson Sigman
Steinhardt, Gary Carl
Swearingin, Marvin Laverne
Tsai, Chia-Yin
Vorst, James J
Warren, George Frederick
Williams, James Lovon, Jr
Wright, William Leland
Zimmerman, Lester J

IOWA
Addink, Sylvan
Atkins, Richard Elton
Burris, Joseph Stephen
Buxton, Dwayne Revere
Carlson, Irving Theodore
Carlson, Richard Eugene
Cavalieri, Anthony Joseph, II
Colvin, Thomas Stuart
Cruse, Richard M
Dalton, Lonnie Gene
Duvick, Donald Nelson
Fawcett, Richard Steven
Frederick, Lloyd Randall
Frey, Nicholas Martin
George, John Ronald
Green, Detroy Edward
Hall, Charles Virdus
Hatfield, Jerry Lee
Horton, Robert, Jr
Hutchcroft, Charles Dennett
Kalton, Robert Rankin
Karlen, Douglas Lawrence
Martinson, Charlie Anton
Newlin, Owen Jay
Pearce, Robert Brent
Pesek, John Thomas, Jr
Roath, William Wesley
Shibles, Richard Marwood
Skrdla, Willis Howard
Staniforth, David William
Stritzel, Joseph Andrew
Svec, Leroy Vernon
Thompson, Harvey E
Tomes, Dwight Travis
Voss, Regis D
Webb, John Raymond
Whigham, David Keith
Woolley, Donald Grant
Wych, Robert Dale

KANSAS
Casady, Alfred Jackson
Erickson, John Robert
Feltner, Kurt C
Follett, Roy Hunter
Ham, George Eldon
Harris, Wallace Wayne
Heyne, Elmer George
Hooker, Mark L
Jacobs, Hyde Spencer
Kirkham, M B
Murphy, Larry S
Nilson, Erick Bogseth
Norwood, Charles Arthur

Agronomy (cont)

Phillips, William Maurice
Posler, Gerry Lynn
Skidmore, Edward Lyman
Sorensen, Edgar Lavell
Stahlman, Phillip Wayne
Sunderman, Herbert D
Thien, Stephen John
Vanderlip, Richard L

KENTUCKY
Bitzer, Morris Jay
Collins, Michael
Egli, Dennis B
Gentry, Claude Edwin
Hiatt, Andrew Jackson
Kasperbauer, Michael J
Lacefield, Garry Dale
Massey, Herbert Fane, Jr
Mikulcik, John D
Niffenegger, Daniel Arvid
Poneleit, Charles Gustav
Sims, John Leonidas
Smiley, Jones Hazelwood
Taylor, Norman Linn
Taylor, Timothy H
Wells, Kenneth Lincoln
Yungbluth, Thomas Alan

LOUISIANA
Ayo, Donald Joseph
Board, James Ellery
Boquet, Donald James
Borsari, Bruno
Caffey, Horace Rouse
Davis, Johnny Henry
Dunigan, Edward P
Faw, Wade Farris
Foret, James A
Harlan, Jack Rodney
Henderson, Merlin Theodore
Hoff, Bert John
Jones, Jack Earl
Martin, Freddie Anthony
Miller, Russell Lee
Owings, Addison Davis
Rogers, Robert Larry
Roussel, John S
Sedberry, Joseph E, Jr
Tipton, Kenneth Warren

MAINE
Langille, Alan Ralph

MARYLAND
Aycock, Marvin Kenneth, Jr
Bandel, Vernon Allan
Beste, Charles Edward
Bruns, Herbert Arnold
Campbell, Travis Austin
Christy, Alfred Lawrence
Coffman, Charles Benjamin
Decker, Alvin Morris, Jr
Devine, Thomas Edward
Dickerson, Chester T, Jr
Elgin, James H, Jr
Fertig, Stanford Newton
Glenn, Barbara Peterson
Graumann, Hugo Oswalt
Hanson, Angus Alexander
Hornick, Sharon B
Klingman, Dayton L
Lamp, William Owen
Leffel, Robert Cecil
McKee, Claude Gibbons
Marten, Gordon C
Moseman, John Gustav
Mulchi, Charles Lee
Nelson, Elton Glen
Porter, Wayne Melvin
Ronningen, Thomas Spooner
Sammons, David James
Schnappinger, Melvin Gerhardt, Jr
Shaw, Warren Cleaton
Smith, David Harrison, Jr
Snow, Philip Anthony
Soto, Gerardo H
Steffens, George Louis
Walker, Alan Kent
Weber, Deane Fay
Weiss, Martin George
Woodstock, Lowell Willard
Young, Alvin L

MICHIGAN
Foth, Henry Donald
Inselberg, Edgar
Johnston, Taylor Jimmie
Moline, Waldemar John
Payne, Kenyon Thomas
Penner, Donald
Robertson, G Philip
Rossman, Elmer Chris
Schillinger, John Andrew, Jr
Southwick, Lawrence
Tesar, Milo B
Thomas, John William
Vitosh, Maurice Lee
Watson, Andrew John
Wolt, Jeffrey Duaine

MINNESOTA
Andow, David A
Bauer, Marvin E
Behrens, Richard
Burnside, Orvin C
Cardwell, Vernon Bruce
Cheng, Hwei-Hsien
Comstock, Verne Edward
Davis, David Warren
Elling, Laddie Joe
Gengenbach, Burle Gene
Goodding, John Alan
Hess, Delbert Coy
Hueg, William Frederick, Jr
Jones, Robert James
Lambert, Jean William
Lebsock, Kenneth L
Lofgren, James R
Lueschen, William Everett
Miller, Gerald R
Mock, James Joseph
Munson, Robert Dean
Otto, Harley John
Pryor, Gordon Roy
Rehm, George W
Rines, Howard Wayne
Robinson, Robert George
Stucker, Robert Evan
Sullivan, Timothy Paul
Thompson, Roy Lloyd
Warnes, Dennis Daniel
Wedin, Walter F

MISSISSIPPI
Andrews, Cecil Hunter
Bagley, Clyde Pattison
Baker, Ralph Stanley
Berry, Charles Dennis
Bowman, Donald Houts
Bunch, Harry Dean
Cole, Avean Wayne
Coleman, Otto Harvey
Creech, Roy G
Davis, Richard Richardson
Hodges, Harry Franklin
Kincade, Robert Tyrus
Knight, William Eric
Kurtz, Mark Edward
Lancaster, James D
Lee, Charles Richard
McWhorter, Chester Gray
Manning, Cleo Willard
Maxwell, James Donald
Peterson, Harold LeRoy
Ranney, Carleton David
Scott, Gene E
Spiers, James Monroe
Triplett, Glover Brown, Jr
Vadhwa, Om Parkash
Watson, Clarence Ellis, Jr
Watson, Vance H
Wise, Louis Neal

MISSOURI
Aldrich, Richard John
Anand, Satish Chandra
Bader, Kenneth L
Beckett, Jack Brown
Blevins, Dale Glenn
Carlson, Wayne C
Cavanah, Lloyd (Earl)
Chowdhury, Ikbalur Rashid
Davidson, Steve Edwin
Donald, William Waldie
Fletchall, Oscar Hale
Franson, Raymond Lee
Graham, James Carl
Gustafson, John Perry
Helsel, Zane Roger
Houghton, John M
Justus, Norman Edward
Kerr, Harold Delbert
Miles, Randall Jay
Mitchell, Roger L
Nelson, Curtis Jerome
Paul, Kamalendu Bikash
Peters, Elroy John
Poehlman, John Milton
Radke, Rodney Owen
Richards, Graydon Edward
Sappenfield, William Paul
Schumacher, Richard William
Sleper, David Allen
Tai, William
Wang, Maw Shiu
Wilson, Clyde Livingston
Woodruff, Clarence Merrill
Zech, Arthur Conrad
Zuber, Marcus Stanley

MONTANA
Aase, Jan Kristian
Blake, Tom
Brown, Jarvis Howard
Ditterline, Raymond Lee
Eslick, Robert Freeman
Jackson, Grant D
McCoy, Thomas Joseph
McGuire, Charles Francis
McNeal, Francis H
Martin, John Munson
Miller, Dwane Gene

NEBRASKA
Baltensperger, David Dwight
Blad, Blaine L
Compton, William A
Doran, John Walsh
Eastin, Jerry Dean
Eastin, John A
Flowerday, Albert Dale
Frolik, Elvin Frank
Furrer, John D
Gorz, Herman Jacob
Hanway, Donald Grant
Higley, Leon George
Johnson, Virgil Allen
Knox, Ellis Gilbert
Lewis, David Thomas
Louda, Svata Mary
McGill, David Park
Maranville, Jerry Wesley
Massengale, M A
Miller, Willie
Moore, Kenneth J
Moser, Lowell E
Nelson, Darrell Wayne
Nelson, Lenis Alton
Nichols, James T
Power, James Francis
Raun, Earle Spangler
Roeth, Frederick Warren
Ross, William Max
Schmidt, John Wesley
Smeltzer, Dale Gardner
Specht, James Eugene
Swartzendruber, Dale
Williams, James Henry, Jr

NEVADA
Evans, Raymond Arthur
Gilbert, Dewayne Everett
Jensen, Edwin Harry
Leedy, Clark D

NEW JERSEY
Daie, Jaleh
Gruenhagen, Richard Dale
Ilnicki, Richard Demetry
Justin, James Robert
Marrese, Richard John
Meade, John Arthur
Sprague, Milton Alan
Wilbur, Robert Daniel

NEW MEXICO
Anderson, Dean Mauritz
Banks, Philip Alan
Barnes, Carl Eldon
Chen, David J
Daugherty, LeRoy Arthur
Davis, Dick D
Finkner, Morris Dale
Finkner, Ralph Eugene
Fowler, James Lowell
Gould, Walter Leonard
Malm, Norman R
Melton, Billy Alexander, Jr
Phillips, Gregory Conrad

NEW YORK
Archimovich, Alexander S
Bergstrom, Gary Carlton
Boehle, John, Jr
Drosdoff, Matthew
Fick, Gary Warren
Grunes, David Leon
Kelly, William Cary
Kennedy, Wilbert Keith
Leopold, Aldo Carl
Linscott, Dean L
Lucey, Robert Francis
Minotti, Peter Lee
Murphy, Royse Peak
Obendorf, Ralph Louis
Pardee, William Durley
Plate, Henry
Pratt, Arthur John
Reid, William Shaw
Reisch, Bruce Irving
Robinson, Terence Lee
Sandsted, Roger France
Sirois, David Leon
Sorrells, Mark Earl
Sweet, Robert Dean
Szabo, Steve Stanley
Thomas, Everett Dake
Topoleski, Leonard Daniel
Vittum, Morrill Thayer
Wright, Madison Johnston

NORTH CAROLINA
Blake, Carl Thomas
Carter, Thomas Edward, Jr
Chamblee, Douglas Scales
Clapp, John Garland, Jr
Coble, Harold Dean
Collins, Henry A
Cook, Maurice Gayle
Corbin, Frederick Thomas
Cowett, Everett R
Ellis, John Fletcher
Fike, William Thomas, Jr
Flint, Elizabeth Parker
Friedrich, James Wayne
Gilbert, William Best
Gross, Harry Douglass

NORTH DAKOTA
Hay, Russell Earl, Jr
Hebert, Teddy T
Jones, Guy Langston
Lewis, William Mason
McCollum, Robert Edmund
McLaughlin, Foil William
Miner, Gordon Stanley
Oblinger, Diana Gelene
O'Neal, Thomas Denny
Patterson, Robert Preston
Rogerson, Asa Benjamin
Sanchez, Pedro Antonio
Thompson, Donald Loraine
Timothy, David Harry
Webb, Burleigh C
Wernsman, Earl Allen
Wilson, Lorenzo George
Worsham, Arch Douglas
Wynne, Johnny Calvin
York, Alan Clarence
Zublena, Joseph Peter

NORTH DAKOTA
Bauer, Armand
Carter, Jack Franklin
Deckard, Edward Lee
Doney, Devon Lyle
Foster, Albert Earl
Frank, Albert Bernard
Hammond, James Jacob
Hofmann, Lenat
Joachim, Frank G
Lucken, Karl Allen
Lund, Hartvig Roald
Meyer, Dwain Wilber
Nalewaja, John Dennis
Seiler, Gerald Joseph
Smith, Glenn Sanborn
Weiss, Michael John
Whited, Dean Allen

OHIO
Bendixen, Leo E
Calhoun, Frank Gilbert
Eckert, Donald James
Findley, William Ray, Jr
Gauntt, William Amor
Haghiri, Faz
Henderlong, Paul Robert
Herr, Donald Edward
Hock, Arthur George
Hoy, Casey William
Hurto, Kirk Allen
Jackobs, John Joseph
Lafever, Howard N
Lal, Rattan
Liu, Ting-Ting Y
Long, John A
McCracken, John David
Martin, David P
Miller, Frederick Powell
Niemczyk, Harry D
Parsons, John Lawrence
Schmidt, Walter Harold
Schwer, Joseph Francis
Smeck, Neil Edward
Stoner, Clinton Dale
Streeter, John Gemmil
Stroube, Edward W
Teater, Robert Woodson
Waldron, Acie Chandler
Watson, Maurice E
Wiebold, William John
Zimmerman, Tommy Lynn

OKLAHOMA
Ahring, Robert M
Bates, Richard Pierce
Croy, Lavoy I
Edwards, Lewis Hiram
Greer, Howard A L
Huffine, Wayne Winfield
Klatt, Arthur Raymond
McMurphy, Wilfred E
Murray, Jay Clarence
Nofziger, David Lynn
Powell, Jerrel B
Reeves, Homer Eugene
Santelmann, Paul William
Stritzke, Jimmy Franklin
Tucker, Billy Bob
Verhalen, Laval
Weibel, Dale Eldon

OREGON
Appleby, Arnold Pierce
Brun, William Alexander
Calhoun, Wheeler, Jr
Chilcote, David Owen
Crabtree, Garvin (Dudley)
Dade, Philip Eugene
Fendall, Roger K
Foote, Wilson Hoover
Frakes, Rodney Vance
Hampton, Richard Owen
Hannaway, David Bryon
Jackson, Thomas Lloyd
Johnson, Malcolm Julius
Kronstad, Warren Ervind
Lee, William Orvid
Lund, Steve
Miller, Stanley Frank
Moss, Dale Nelson
Neiland, Bonita J

Rykbost, Kenneth Albert
Simantel, Gerald M
Simonson, Gerald Herman
Smiley, Richard Wayne
Vomocil, James Arthur
Young, J Lowell
Yungen, John A

PENNSYLVANIA
Baker, Dale E
Baylor, John E
Berg, Clyde C
Bergman, Ernest L
Bosshart, Robert Perry
Burns, Allan Fielding
Cleveland, Richard Warren
Cole, Richard H
Duich, Joseph M
Fales, Steven Lewis
Fritton, Daniel Dale
Hall, Jon K
Harrington, Joseph Donald
Hartwig, Nathan Leroy
Hinish, Wilmer Wayne
Johnson, Melvin Walter, Jr
Jung, Gerald Alvin
Kendall, William Anderson
Knievel, Daniel Paul
Krueger, Charles Robert
McKee, Guy William
Marriott, Lawrence Frederick
Risius, Marvin Leroy
Shenk, John Stoner
Shipp, Raymond Francis
Sprague, Howard Bennett
Waddington, Donald Van Pelt
Watschke, Thomas Lee
Wheeler, Donald Alsop

RHODE ISLAND
Duff, Dale Thomas
Hull, Richard James
Skogley, Conrad Richard
Wakefield, Robert Chester

SOUTH CAROLINA
Alexander, Paul Marion
Alston, Jimmy Albert
Chapman, Stephen R
Craddock, Garnet Roy
Deal, Elwyn Ernest
Franklin, Ralph E
Gossett, Billy Joe
Harvey, Lawrence Harmon
Jutras, Michel Wilfrid
Kittrell, Benjamin Upchurch
McClain, Eugene Fredrick
Nolan, Clifford N
Quisenberry, Virgil L
Shipe, Emerson Russell
Singh, Raghbir
Stringer, William Clayton
Wallace, Susan Ulmer

SOUTH DAKOTA
Einhellig, Frank Arnold
Kantack, Benjamin H
Moore, Raymond A
Reeves, Dale Leslie
Walgenbach, David D
Wells, Darrell Gibson

TENNESSEE
Bell, Frank F
Bond, Andrew
Brown, James Melton
Callahan, Lloyd Milton
Eskew, David Lewis
Ewing, John Arthur
Flanagan, Theodore Ross
Foutch, Harley Wayne
Fribourg, Henry August
Graveel, John Gerard
Hayes, Robert M
Josephson, Leonard Melvin
Krueger, William Arthur
Large, Richard L
Luxmoore, Robert John
Naylor, Gerald Wayne
Pulido, Miguel L
Ramey, Harmon Hobson, Jr
Reich, Vernon Henry
Reynolds, John Horace
Skold, Laurence Nelson
Springer, Maxwell Elsworth
West, Dennis R

TEXAS
Arnold, James Darrell
Beard, James B
Bennett, William Frederick
Bovey, Rodney William
Brigham, Raymond Dale
Calub, Alfonso deGuzman
Chandler, James Michael
Collier, Jesse Wilton
Cook, Charles Garland
Craigmiles, Julian Pryor
Duble, Richard Lee
Evans, Raeford G
Gantz, Ralph Lee
Gavande, Sampat A
Hagstrom, Gerow Richard
Hauser, Victor La Vern

Havelka, Ulysses D
Ho, Clara Lin
Holt, Ethan Cleddy
Hopper, Norman Wayne
Hossner, Lloyd Richard
Jones, Ordie Reginal
Kramer, Nicholas William
Lewis, Robert Donald
Lingle, Sarah Elizabeth
Maas, Stephen Joseph
McBee, George Gilbert
McDonald, Lynn Dale
McIlrath, William Oliver
Matches, Arthur Gerald
Niles, George Alva
Osler, Robert Donald
Palmer, Rupert Dewitt
Potts, Richard Carmechial
Reid, Donald J
Riewe, Marvin Edmund
Rouquette, Francis Marion, Jr
Runge, Edward C A
Scifres, Charles Joel
Sij, John William
Smith, Gerald Ray
Smith, Olin Dail
Staten, Raymond Dale
Steiner, Jean Louise
Stewart, Bobby Alton
Swakon, Doreen H D
Taylor, Richard Melvin
Trogdon, William Oren
Van Bavel, Cornelius H M
Vietor, Donald Melvin
Voigt, Paul Warren
Wiegand, Craig Loren
Wiese, Allen F
Wilding, Lawrence Paul
Winter, Steven Ray
Young, Arthur Wesley

UTAH
Albrechtsen, Rulon S
Anderson, Jay LaMar
Asay, Kay Harris
Belcher, Bascom Anthony
Chase, Richard Lyle
Dewey, Wade G
Hamson, Alvin Russell
Horrocks, Rodney Dwain
Jeffery, Larry S
Rasmussen, V Philip, Jr
Robison, Laren R
Rumbaugh, Melvin Dale
Thorup, Richard M

VERMONT
Clark, Neri Anthony
Murphy, William Michael
Wood, Glen Meredith

VIRGINIA
Albrecht, Herbert Richard
Allen, Vivien Gore
Bingham, Samuel Wayne
Buckardt, Henry Lloyd
Buss, Glenn Richard
Carson, Eugene Watson, Jr
Coartney, James S
Conn, Richard Leslie
Dorschner, Kenneth Peter
Griffith, William Kirk
Harrison, Robert Louis
Hatzios, Kriton Kleanthis
Hutcheson, Thomas Barksdale, Jr
Kroontje, Wybe
Lewis, Cornelius Crawford
O'Neill, Eileen Jane
Parrish, David Joe
Sagaral, Erasmo G
Schmidt, Richard Edward
Smith, Townsend Jackson
Taylor, Lincoln Homer
Thiel, Thomas J
Wolf, Dale Duane

WASHINGTON
Allan, Robert Emerson
Bertramson, Bertram Rodney
Boyd, Charles Curtis
Chevalier, Peggy
Comes, Richard Durward
Dawson, Jean Howard
Ensign, Ronald D
Evans, David W
George, Donald Wayne
Hiller, Larry Keith
Johnson, Corwin McGillivray
Kleinhofs, Andris
Knowles, Paulden Ford
Lauer, David Allan
Leng, Earl Reece
Maguire, James Dale
Morrison, Kenneth Jess
Morrow, Larry Alan
Ogg, Alex Grant, Jr
Papendick, Robert I
Peabody, Dwight Van Dorn, Jr
Raese, John Thomas
Rasmussen, Lowell W
Schrader, Lawrence Edwin
Schwendiman, John Leo
Swan, Dean George
Warner, Robert Lewis

Zimmermann, Charles Edward

WEST VIRGINIA
Baker, Barton Scofield
Balasko, John Allan
Singh, Rabindar Nath
Welker, William V, Jr

WISCONSIN
Albrecht, Kenneth Adrian
Andrew, Robert Harry
Barnes, Robert F
Bula, Raymond J
Casler, Michael Darwin
Delorit, Richard John
Doersch, Ronald Ernest
Doll, Jerry Dennis
Drolsom, Paul Newell
Duke, Stanley Houston
Forsberg, Robert Arnold
Goodman, Robert Merwin
Greub, Louis John
Gritton, Earl Thomas
Hagedorn, Donald James
Harpstead, Milo I
Harvey, Robert Gordon, Jr
Higgs, Roger L
Ihrke, Charles Albert
Jackson, Marion Leroy
Miller, Robert W
Oplinger, Edward Scott
Paulson, William H
Pawlisch, Paul E
Peterson, David Maurice
Strohm, Jerry Lee
Tracy, William Francis

WYOMING
Alley, Harold Pugmire
Hart, Richard Harold
Humburg, Neil Edward
Koch, David William
Kolp, Bernard J
Lang, Robert Lee
Miller, Stephen Douglas
Schuman, Gerald E

PUERTO RICO
Rodriguez, Jorge Luis
Salas-Quintana, Salvador

ALBERTA
Andrew, William Treleaven
Coen, Gerald Marvin
Hanna, Michael Ross
Harper, Alexander Maitland
McAndrew, David Wayne
McElgunn, James Douglas
Muendel, Hans-Henning
Vanden Born, William Henry
Walton, Peter Dawson
Wilson, Donald Benjamin

BRITISH COLUMBIA
Ashford, Ross
Holl, Frederick Brian
Stout, Darryl Glen

MANITOBA
Giesbrecht, John
Grant, Cynthia Ann
Hobbs, James Arthur
McGinnis, Robert Cameron
Stobbe, Elmer Henry

NEWFOUNDLAND
McKenzie, David Bruce

NOVA SCOTIA
Bubar, John Stephen
Faught, John Brian
Warman, Philip Robert

ONTARIO
Alex, Jack Franklin
Buzzell, Richard Irving
Campbell, Kenneth Wilford
Chang, Fa Yan
Dhesi, Nazar Singh
Fedak, George
Ho, Keh Ming
Hope, Hugh Johnson
Hume, David John
Kannenberg, Lyndon William
O'Toole, James J
Saidak, Walter John
Sampson, Dexter Reid
Singh, Surinder Shah
Stoskopf, N C
Switzer, Clayton Macfie
Tan, Chin Sheng
Thomas, Ronald Leslie
Warren, Francis Shirley
Zilkey, Bryan Frederick

PRINCE EDWARD ISLAND
Choo, Thin-Meiw
Christie, Bertram Rodney
Kunelius, Heikki Tapani
MacLeod, John Alexander
MacLeod, LLoyd Beck
White, Ronald Paul, Sr

QUEBEC
Chung, Young Sup
Furlan, Valentin
Gasser, Heinz
Gervais, Paul
Klinck, Harold Rutherford
Matte, J Jacques
St Pierre, Jean Claude
Watson, Alan Kemball

SASKATCHEWAN
Austenson, Herman Milton
Cohen, Roger D H
Coupland, Robert Thomas
Harris, Peter
Hay, James Robert
Rossnagel, Brian Gordon
Scoles, Graham John
Slinkard, Alfred Eugene
Tanino, Karen Kikumi
Waddington, John
Zilke, Samuel

OTHER COUNTRIES
Britten, Edward James
Brockman, Frank Elliot
Chomchalow, Narong
Couto, Walter
Hittle, Carl Nelson
Izuno, Takumi
Muchovej, James John
Simkins, Charles Abraham
Turner, Fred, Jr

Animal Husbandry

ALABAMA
Harris, Ralph Rogers
Kuhlers, Daryl Lynn
McGuire, John Albert
Moss, Buelon Rexford
Patterson, Troy B
Sewell, Raymond F
Sharma, Udhishtra Deva
Vernon, Eugene Haworth
Wilkening, Marvin C

ALASKA
Brundage, Arthur Lain
Husby, Fredric Martin

ARIZONA
Brown, William Hedrick
Hale, William Harris
Huber, John Talmage
Johnson, Randall Arthur
Murdock, Fenoi R
Rice, Richard W
Schuh, James Donald
Swingle, Roy Spencer
Taysom, Elvin David
Theurer, Clark Brent

ARKANSAS
Andrews, Luther David
Brown, A Hayden, Jr
Brunson, Clayton (Cody)
Gyles, Nicholas Roy
Harris, Grover Cleveland, Jr
Kellogg, David Wayne
McGilliard, A Dare
Noland, Paul Robert
Stephenson, Edward Luther
Tave, Douglas

CALIFORNIA
Abenes, Fiorello Bigornia
Adams, Thomas Edwards
Anderson, Gary Bruce
Anderson, Russell K
Bauer, Robert Steven
Bradford, G Eric
Burrill, Melinda Jane
Carroll, Floyd Dale
Christensen, Allen Clare
Colvin, Harry Walter, Jr
Dollar, Alexander M
Durrant, Barbara Susan
Gall, Graham A E
Garrett, William Norbert
Halloran, Hobart Rooker
Hedgecock, Dennis
Hixson, Floyd Marcus
Jacobs, John Allen
Kennington, Mack Humpherys
Kutches, Alexander Joseph
Laben, Robert Cochrane
Lofgreen, Glen Pehr
Myers, George Scott, Jr
Prokop, Michael Joseph
Rahlmann, Donald Frederick
Siegrist, Jacob C
Sinha, Yagya Nand
Torell, Donald Theodore
Touchberry, Robert Walton
Vohra, Pran Nath
Waterhouse, Howard N
Weir, William Carl

COLORADO
Johnson, Donald Eugene
Lamm, Warren Dennis
Swanson, Vern Bernard

Animal Husbandry (cont)

Voss, James Leo
Ward, Gerald Madison

CONNECTICUT
Brown, Lynn Ranney
Cowan, William Allen
Eaton, Hamilton Dean
Gaunya, William Stephen
Kinsman, Donald Markham
Schake, Lowell Martin
Stake, Paul Erik

DELAWARE
Haenlein, George Friedrich Wilhelm
Hesseltine, Wilbur R
Link, Bernard Alvin

DISTRICT OF COLUMBIA
Ampy, Franklin
Cooper, George Everett
Dua, Prem Nath
Gray, Richard C
Hardison, Wesley Aurel
Plowman, Ronald Dean
Wilson, William Mark Dunlop

FLORIDA
Ammerman, Clarence Bailey
Baker, Frank Sloan, Jr
Carpenter, James Woodford
Chapman, Herbert L, Jr
Combs, George Ernest
Conrad, Joseph H
Cox, Dennis Henry
Davis, George Kelso
Evans, Lee E
Gaunt, Stanley Newkirk
Hembry, Foster Glen
Koger, Marvin
Loggins, Phillip Edwards
Loosli, John Kasper
McDowell, Lee Russell
Sanford, Malcolm Thomas
Shirley, Ray Louis
Sosa, Omelio, Jr
Thomason, David Morton
Wakeman, Donald Lee

GEORGIA
Benyshek, Larry L
Cullison, Arthur Edison
Daniel, O'Dell G
Flatt, William Perry
Green, Willard Wynn
Hill, Gary Martin
Jensen, Leo Stanley
Mabry, John William
Neathery, Milton White
Newton, George Larry
Powell, George Wythe
Seerley, Robert Wayne
Stuedemann, John Alfred

HAWAII
Weems, Charles William
Wyban, James A

IDAHO
Bull, Richard C

ILLINOIS
Arthur, Robert David
Carr, Tommy Russell
Clark, Jimmy Howard
Corbin, James Edward
Easter, Robert Arnold
Garrigus, Upson Stanley
Gomes, Wayne Reginald
Grossman, Michael
Harrison, Paul C
Hausler, Carl Louis
Heuberger, Glen (Louis)
Hollis, Gilbert Ray
Hutjens, Michael Francis
Jensen, Aldon Homan
Klay, Robert Frank
Lewis, John Morgan
Lodge, James Robert
McKeith, Floyd Kenneth
Madsen, Fred Christian
Monson, William Joye
Neagle, Lyle H
Pothoven, Marvin Arlo
Ricketts, Gary Eugene
Romack, Frank Eldon
Shanks, Roger D
Spahr, Sidney Louis
Thomas, David Lee
Vetter, Richard L
Vint, Larry Francis
Wagner, William Charles

INDIANA
Albright, Jack Lawrence
Baumgardt, Billy Ray
Boston, Andrew Chester
Brown, Herbert
Elkin, Robert Glenn
Forsyth, Dale Marvin
Gard, Don Irvin
Jones, Hobart Wayne
Krider, Jake Luther

Morter, Raymond Lione
Orr, Donald Eugene, Jr
Perry, Tilden Wayne
Raun, Arthur Phillip
Robb, Thomas Wilbern
Rogler, John Charles
Stewart, Terry Sanford
Stob, Martin
Thomas, Elvin Elbert
Wolfrom, Glen Wallace

IOWA
Brackelsberg, Paul O
Ewan, Richard Colin
Ewing, Solon Alexander
Freeman, Albert Eugene
Harbaugh, Daniel David
Harville, David Arthur
Hodson, Harold H, Jr
Hoffman, Mark Peter
Holden, Palmer Joseph
Jacobson, Norman Leonard
Jurgens, Marshall Herman
Kenealy, Michael Douglas
Mente, Glen Allen
Morrical, Daniel Gene
Parrish, Frederick Charles, Jr
Rohlf, Marvin Euguene
Sell, Jerry Lee
Stevermer, Emmett J
Warner, Donald R
Willham, Richard Lewis
Wunder, William W
Young, Jerry Wesley
Zimmerman, Dean R

KANSAS
Brethour, John Raymond
Craig, James Verne
Epp, Melvin David
Harbers, Leniel H
Kropf, Donald Harris
Schalles, Robert R
Smith, Edgar Fitzhugh
Stevenson, Jeffrey Smith
Woods, Walter Ralph

KENTUCKY
Brown, Leonard D
Buck, Charles Frank
Hays, Virgil Wilford
Knapp, Frederick Whiton
Lane, Gary (Thomas)
Ramsay, John Martin
Stephens, Noel, Jr
Thrift, Frederick Aaron
Tucker, Ray Edwin
Varney, William York
Whiteker, McElwyn D

LOUISIANA
Cason, James Lee
Falcon, Carroll James
Franke, Donald Edward
Gassie, Edward William
Robertson, George Leven
Rusoff, Louis Leon

MAINE
Barton, Barbara Ann
Dickey, Howard Chester
Les, Edwin Paul
Musgrave, Stanley Dean

MARYLAND
Alderson, Norris Eugene
Batson, David Banks
Berry, Bradford William
Engster, Henry Martin
Glenn, Barbara Peterson
Graber, George
Moody, Edward Grant
Norman, Howard Duane
Powell, Rex Lynn
Pursel, Vernon George
Putnam, Paul A
Rubin, Max
Smith, F Harrell
Stricklin, William Ray
Terrill, Clair Elman
Vandersall, John Henry
Wildt, David Edwin

MASSACHUSETTS
Borton, Anthony
Fenner, Heinrich
Lyford, Sidney John, Jr
McColl, James Renfrew

MICHIGAN
Aulerich, Richard J
Cook, Robert Merold
Deans, Robert Jack
Emery, Roy Saltsman
Henneman, Harold Albert
Hoefer, Jacob A
Kratzer, D Dal
Lawson, David Michael
McGilliard, Lon Dee
Magee, William Thomas
Mao, Ivan Ling
Miller, Curtis C
Miller, Elwyn Ritter
Nelson, Ronald Harvey

Ritchie, Harlan
Rust, Steven Ronald
Stinson, Al Worth
Subramanian, Marappa G
Tucker, Herbert Allen

MINNESOTA
Boylan, William J
Christians, Charles J
Conlin, Bernard Joseph
Goodrich, Richard Douglas
Hansen, Leslie Bennett
Hanson, Lester Eugene
Jordan, Robert Manseau
Meiske, Jay C
Miller, Kenneth Philip
Oelberg, Thomas Jonathon
Pettigrew, James Eugene, Jr
Rempel, William Ewert
Rust, Joseph William
Wilcox, Clifford LaVar
Young, Charles Wesley

MISSISSIPPI
Bagley, Clyde Pattison
Baker, Bryan, Jr
Fuquay, John Wade
Lindley, Charles Edward
Thomas, Charles Hill
Tomlinson, James Everett

MISSOURI
Anderson, Ralph Robert
Day, Billy Neil
Harmon, Bud Gene
Hill, Jack Filson
Kohlmeier, Ronald Harold
Lasley, John Foster
Merilan, Charles Preston
Meyer, William Ellis
Padgitt, Dennis Darrell
Sewell, Homer B
Smith, Keith James
Smith, Nathan Elbert
Spining, Arthur Milton, III
Stephenson, Alfred Benjamin
Thrasher, George W
Williamson, James Lawrence

MONTANA
Burris, Martin Joe
Dynes, J Robert
Keyes, Everett A
Kress, Donnie Duane
Newman, Clarence Walter
Smith, Ervin Paul
Thomas, Oscar Otto

NEBRASKA
Aberle, Elton D
Adams, Charles Henry
Ahlschwede, William T
Baltensperger, David Dwight
Bennett, Gary Lee
Clanton, Donald Cather
Cundiff, Larry Verl
Dearborn, Delwyn D
Doane, Ted H
Ford, Johny Joe
Fritschen, Robert David
Gleaves, Earl William
Gregory, Keith Edward
Guyer, Paul Quentin
Hunsley, Roger Eugene
Koch, Robert Milton
Larson, Larry Lee
Lewis, Austin James
McGaugh, John Wesley
Nielsen, Merlyn Keith
Omtvedt, Irvin T
Peo, Ernest Ramy, Jr
Van Vleck, Lloyd Dale
Ward, John K
Weichenthal, Burton Arthur
Young, Lawrence Dale

NEVADA
Cunha, Tony Joseph

NEW HAMPSHIRE
Keener, Harry Allan
Skoglund, Winthrop Charles

NEW JERSEY
Cox, James Lee
Eggert, Robert Glenn
Evans, Joseph Liston
Grover, Gary James
Horton, Glyn Michael John
Ingle, Donald Lee
Shor, Aaron Louis

NEW MEXICO
Andrew, James F
Francis, David Wesson
Holland, Lewis
Nelson, Arnold Bernard

NEW YORK
Ainslie, Harry Robert
Auclair, Walter
Bauman, Dale E
Blake, Robert Wesley
Chase, Larry Eugene

Elliot, John Murray
Foote, Robert Hutchinson
Glinsky, Martin Jay
Hintz, Harold Franklin
Hogue, Douglas Emerson
Merrill, William George
Oltenacu, Elizabeth Allison Branford
Pollak, Emil John
Porter, Gilbert Harris
Poston, Hugh Arthur
Randy, Harry Anthony
Reimers, Thomas John
Rowoth, Olin Arthur
Schano, Edward Arthur
Slack, Nelson Hosking
Stockbridge, Robert R
Stouffer, James Ray
Thomas, Everett Dake
Turk, Kenneth Leroy
VanSoest, Peter John
Warner, Richard G
Wellington, George Harvey

NORTH CAROLINA
Allen, Neil Keith
Burns, Joseph Charles
Clawson, Albert J
Colvard, Dean Wallace
Cornwell, John Calhoun
Glazener, Edward Walker
Gonder, Eric Charles
Hyatt, George, Jr
Johnson, George Andrew
Johnson, William Lawrence
Jones, James Robert
Knott, Fred Nelson
Leatherwood, James M
Litton, George Washington
McDaniel, Benjamin Thomas
McDowell, Robert E, Jr
Olson, Leonard Carl
Porterfield, Ira Deward
Poulton, Bruce R
Simons, Elwyn LaVerne
Vandemark, Noland Leroy
Whitlow, Lon Weidner
Young, Robert John

NORTH DAKOTA
Dinusson, William Erling
Edgerly, Charles George Morgan
Fisher, George Robert
Harrold, Robert Lee
Johnson, LaDon Jerome
Karn, James Frederick

OHIO
Anderson, Marlin Dean
Cline, Jack Henry
Coleman, Marilyn A
Danke, Richard John
Davis, Michael E
Dickinson, Frank N
Gauntt, William Amor
Harvey, Walter Robert
Hibbs, John William
Hitchcock, John Paul
Johnson, George Robert
Kottman, Roy Milton
Ludwick, Thomas Murrell
Mahan, Donald C
Mattingly, Steele F
Murray, Finnie Ardrey, Jr
Newland, Herman William
Palmquist, Donald Leonard
Plimpton, Rodney F, Jr
Prince, Terry Jamison
Schmidt, Glen Henry
Tyznik, William John
Wilson, George Rodger
Wilson, Richard Ferrin

OKLAHOMA
Browning, Charles Benton
Buchanan, David Shane
Lu, Christopher D
Luce, William Glenn
MARTIN, JERRY, Jr
Owens, Fredric Newell
Tillman, Allen Douglas
Totusek, Robert
Turman, Elbert Jerome
Walters, Lowell Eugene
Wells, Milton Ernest
Whatley, James Arnold
Whiteman, Joe V

OREGON
Arscott, George Henry
Cheeke, Peter Robert
England, David Charles
Landers, John Herbert, Jr
Miller, Stanley Frank
Skovlin, Jon Matthew
Weswig, Paul Henry

PENNSYLVANIA
Adams, Richard Sanford
Chalupa, William Victor
Cowan, Robert Lee
Dreibelbis, John Adam
Elkins, William L
Hagen, Daniel Russell
Hedde, Richard Duane

Mashaly, Magdi Mohamed
Morrow, David Austin
Muller, Lawrence Dean
Roush, William Burdette
Sherritt, Grant Wilson
Varga, Gabriella Anne
Wangsness, Paul Jerome
Wilson, Lowell L
Ziegler, John Henry, Jr

RHODE ISLAND
Nippo, Murn Marcus

SOUTH CAROLINA
Birrenkott, Glen Peter, Jr
Dickey, Joseph Freeman
Edwards, Robert Lee
Lee, Daniel Dixon, Jr
Skelley, George Calvin, Jr
Thompson, Carl Eugene

SOUTH DAKOTA
Briggs, Hilton Marshall
Britzman, Darwin Gene
Bush, Leon F
Embry, Lawrence Bryan
Emerick, Royce Jasper
Musson, Alfred Lyman
Romans, John Richard
Schingoethe, David John
Wahlstrom, Richard Carl

TENNESSEE
Alexander, Robert Allen
Bletner, James Karl
Chamberlain, Charles Calvin
Griffin, Sumner Albert
Huggins, James Anthony
McFee, Alfred Frank
Miller, James Kincheloe
Scott, Mack Tommie
Shrode, Robert Ray

TEXAS
Bassett, James Wilbur
Butler, Ogbourne Duke, Jr
Byers, Floyd Michael
Calhoun, Millard Clayton
Cartwright, Thomas Campbell
Cassard, Daniel W
Cole, Noel Andy
Coleman, John Dee
Crenshaw, David Brooks
Dahl, Billie Eugene
Davis, Gordon Wayne
DuBose, Leo Edwin
Godbee, Richard Greene, II
Greathouse, Terrence Ray
Hanlon, Roger Thomas
Harms, Paul G
Horton, Otis Howard
Hudson, Frank Alden
Hutcheson, David Paul
Krueger, Willie Frederick
Lane, Alfred Glen
Lewis, Roscoe Warfield
Lippke, Hagen
Mason, Tim Robert
Menzies, Carl Stephen
Preston, Rodney LeRoy
Riewe, Marvin Edmund
Self, Hazzle Layfette
Shelton, James Maurice
Sherrod, Lloyd B
Sudweeks, Earl Max
Thompson, Granville Berry

UTAH
Arave, Clive W
Bennett, James Austin
Gardner, Robert Wayne
Johnston, Norman Paul
Lamb, Robert Cardon
Matthews, Doyle Jensen
Park, Robert Lynn
Richards, Clyde Rich
Slade, Larry Malcom
Stoddard, George Edward
Thomas, Don Wylie
Wallentine, Max V

VERMONT
Balch, Donald James
Dowe, Thomas Whitfield
Foss, Donald C
Smith, Albert Matthews

VIRGINIA
Acker, Duane Calvin
Benson, Robert Haynes
Bovard, Kenly Paul
Carr, Scott Bligh
Eller, Arthur L, Jr
Etgen, William M
Fontenot, Joseph Paul
Frahm, Richard R
Kelly, Robert Frank
Kornegay, Ervin Thaddeus
Mitchell, George Ernest, Jr
Pearson, Ronald Earl
Raun, Ned S
Webb, Kenneth Emerson, Jr
White, John Marvin

WASHINGTON
Froseth, John Allen
Hillers, Joe Karl
Kincaid, Ronald Lee
Kromann, Rodney P
Sasser, Lyle Blaine

WEST VIRGINIA
Cochrane, Robert Lowe
Dunbar, Robert Standish, Jr
Horvath, Donald James
Inskeep, Emmett Keith
Kidder, Harold Edward
McClung, Marvin Richard
Thomas, Roy Orlando
Welch, James Alexander
Zinn, Dale Wendel

WISCONSIN
Bitgood, J(ohn) James
Chapman, Arthur Barclay
Dierschke, Donald Joe
Dollahon, James Clifford
Grummer, Robert Henry
Heidenreich, Charles John
Howard, W Terry
Leibbrandt, Vernon Dean
Miller, Paul Dean
Niedermeier, Robert Paul
Rutledge, Jackie Joe
Walton, Robert Eugene
Wittwer, Leland S

WYOMING
Botkin, Merwin P
Fisser, Herbert George
Hinds, Frank Crossman
Kercher, Conrad J
Nelms, George E

PUERTO RICO
Oliver-Padilla, Fernando Luis

ALBERTA
Church, Robert Bertram
Coulter, Glenn Hartman
Lawson, John Edwin
Mears, Gerald John
Newman, John Alexander
Yeh, Francis Cho-hao

BRITISH COLUMBIA
Blair, Robert
Bowden, David Merle
Hill, Arthur Thomas
Owen, Bruce Douglas
Peterson, Raymond Glen
Stringam, Elwood Williams
Waldern, Donald E

MANITOBA
Castell, Adrian George
Elliot, James I
Grandhi, Raja Ratnam
Ingalls, Jesse Ray
Seale, Marvin Ernest

NEW BRUNSWICK
Bush, Roy Sidney
Nicholson, J W G

NOVA SCOTIA
Cock, Lorne M
MacRae, Herbert F

ONTARIO
Batra, Tilak Raj
Buchanan-Smith, Jock Gordon
Burnside, Edward Blair
Burton, John Heslop
Chambers, James Robert
Evans, Essi H
Gowe, Robb Shelton
Heaney, David Paul
McAllister, Alan Jackson
Mallory, Frank Fenson
Morrison, William D
Nagai, Jiro
Rennie, James Clarence
Stone, John Bruce
Straus, Jozef

QUEBEC
Besner, Michel
Donefer, Eugene
Fahmy, Mohamed Hamed
Farmer, Chantal
Holtmann, Wilfried
Matte, J Jacques
Robert, Suzanne
Roy, Gabriel L

SASKATCHEWAN
Babiuk, Lorne A
Cohen, Roger D H
Howell, William Edwin
Knipfel, Jerry Earl

OTHER COUNTRIES
De Alba Martinez, Jorge
Makita, Takashi
Szebenyi, Emil

Animal Science & Nutrition

ALABAMA
Brewer, Robert Nelson
Cannon, Robert Young
Clark, Alfred James
Frobish, Lowell T
Haaland, Ronald L
Harris, Ralph Rogers
Hawkins, George Elliott, Jr
Moran, Edwin T, Jr
Moss, Buelon Rexford
Sauberlich, Howerde Edwin
Sewell, Raymond F
Voitle, Robert Allen
Wilkening, Marvin C

ALASKA
Husby, Fredric Martin
Tomlin, Don C

ARIZONA
Anderson, Jay Oscar
Brown, William Hedrick
Foreman, Charles Frederick
Hale, William Harris
Huber, John Talmage
Johnson, Randall Arthur
Mulder, John Bastian
Murdock, Fenoi R
Peng, Yeh-Shan
Rice, Richard W
Roberts, William Kenneth
Schmutz, Ervin Marcell
Schuh, James Donald
Sleeth, Rhule Bailey
Stull, John Warren
Swingle, Roy Spencer
Taylor, Richard G
Theurer, Clark Brent

ARKANSAS
Brown, Connell Jean
Daniels, L B
Harris, Grover Cleveland, Jr
Havener, Robert D
Keene, James H
Kellogg, David Wayne
Lee, Kwang
McGilliard, A Dare
Nelson, Talmadge Seab
Noland, Paul Robert
Stallcup, Odie Talmadge
Stephenson, Edward Luther
West, Charles Patrick
Yates, Jerome Douglas

CALIFORNIA
Abenes, Fiorello Bigornia
Anderson, Gary Bruce
Anderson, Russell K
Baldwin, Ransom Leland
Briggs, George McSpadden
Burri, Betty Jane
Cherms, Frank Llewellyn, Jr
Christensen, Allen Clare
Colvin, Harry Walter, Jr
Connor, John Michael
Dollar, Alexander M
Ernst, Ralph Ambrose
Garrett, William Norbert
Gordon, Malcolm Stephen
Grau, Charles Richard
Grill, Herman, Jr
Halloran, Hobart Rooker
Heitman, Hubert, Jr
Jacob, Mary
Johnson, Herman Leonall
Kashyap, Tapeshwar S
Kutches, Alexander Joseph
Law, George Robert John
Leong, Kam Choy
Lofgreen, Glen Pehr
Lucas, David Owen
Moberg, Gary Philip
Myers, George Scott, Jr
Prokop, Michael Joseph
Rahlmann, Donald Frederick
Scott, Karen Christine
Sharp, Gerald Duane
Starkey, Eugene Edward
Talhouk, Rabih Shakib
Trei, John Earl
Waterhouse, Howard N
Weir, William Carl
West, John Wyatt
Wilson, Barry William

COLORADO
Avens, John Stewart
Cramer, David Alan
Johnson, Donald Eugene
Kienholz, Eldon W
Lamm, Warren Dennis
Martin, Jack E
Matsushima, John K
Miller, Byron F
Moreng, Robert Edward
Nett, T M
Nockels, Cheryl Ferris
Reinbold, George W
Rouse, George Elverton
Seidel, George Elias, Jr
Smith, Gary Chester

CONNECTICUT
Brown, Lynn Ranney
Eaton, Hamilton Dean
Kinsman, Donald Markham
Pierro, Louis John
Riesen, John William
Rishell, William Arthur
Stake, Paul Erik

DELAWARE
Cartwright, Aubrey Lee, Jr
Haenlein, George Friedrich Wilhelm
Rasmussen, Arlette Irene
Reitnour, Clarence Melvin
Salsbury, Robert Lawrence
Thoroughgood, Carolyn A

DISTRICT OF COLUMBIA
Hardison, Wesley Aurel
Lunchick, Curt
Morck, Timothy Anton
Wilson, George Donald
Zook, Bernard Charles

FLORIDA
Ammerman, Clarence Bailey
Baker, Frank Sloan, Jr
Bauernfeind, Jacob (Jack) C(hristopher)
Bertrand, Joseph E
Black, Alex
Boehme, Werner Richard
Chapman, Herbert L, Jr
Combs, George Ernest
Conrad, Joseph H
Courtney, Charles Hill
Cox, Dennis Henry
Damron, Bobby Leon
Davis, George Kelso
Douglas, Carroll Reece
Fry, Jack L
Gaunt, Stanley Newkirk
Gentry, Robert Francis
Hale, Kirk Kermit, Jr
Hammond, Andrew Charles
Harms, Robert Henry
Harris, Barney, Jr
Head, H Herbert
Hembry, Foster Glen
Hentges, James Franklin, Jr
Janky, Douglas Michael
Loggins, Phillip Edwards
Loosli, John Kasper
McDowell, Lee Russell
Meltzer, Martin Isaac
Miles, Richard David
Moore, John Edward
Natzke, Roger Paul
Norval, Richard Andrew
Ott, Edgar Alton
Pate, Findlay Moye
Prebluda, Harry Jacob
Seals, Rupert Grant
Shirley, Ray Louis
Smith, Kenneth Leroy
Thomason, David Morton
Van Horn, Harold H, Jr
Wallace, Harold Dean
Warnick, Alvin Cropper
Wilcox, Charles Julian
Wilkowske, Howard Hugo
Wing, James Marvin
Zikakis, John Philip

GEORGIA
Amos, Henry Estill
Barb, C Richard
Britton, Walter Martin
Cherry, Jerry Arthur
Cullison, Arthur Edison
Flatt, William Perry
Fuller, Henry Lester
Hausman, Gary J
Hill, Gary Martin
Jensen, Leo Stanley
Kraeling, Robert Russell
Mabry, John William
McCullough, Marshall Edward
McGinnis, Charles Henry, Jr
Miller, William Jack
Moinuddin, Jessie Fischer
Neathery, Milton White
Neville, Walter Edward, Jr
Newton, George Larry
Parrish, John W, Jr
Powell, George Wythe
Reid, Willard Malcolm
Shutze, John V
Stuedemann, John Alfred
Sutton, William Wallace
Washburn, Kenneth W
Wyatt, Roger Dale

HAWAII
Atkinson, Shannon K C
Koshi, James H
Olbrich, Steven Emil
Ross, Ernest

IDAHO
Bull, Richard C
Lehrer, William Peter, Jr
Montoure, John Ernest
Petersen, Charlie Frederick
Ross, Richard Henry

Young, Robert John

NORTH DAKOTA
Davison, Kenneth Lewis
Dinusson, William Erling
Harrold, Robert Lee
Karn, James Frederick
Park, Chung Sun

OHIO
Allaire, Francis Raymond
Anderson, Marlin Dean
Cline, Jack Henry
Danke, Richard John
Dehority, Burk Allyn
Fechheimer, Nathan S
Gauntt, William Amor
Harper, Willie James
Henningfield, Mary Frances
Hibbs, John William
Hines, Harold C
Hitchcock, John Paul
Latshaw, J David
Linkenheimer, Wayne Henry
Mahan, Donald C
Mikolajcik, Emil Michael
Naber, Edward Carl
Palmquist, Donald Leonard
Prince, Terry Jamison
Smith, Kenneth Larry
Staubus, John Reginald
Stephens, James Fred
Tyznik, William John

OKLAHOMA
Alexander, Joseph Walker
Berry, Joe Gene
Browning, Charles Benton
Bush, Linville John
Johnson, Ronald Roy
Lu, Christopher D
Luce, William Glenn
MARTIN, JERRY, Jr
Nelson, Eldon Carl
Noble, Robert Lee
Owens, Fredric Newell
Thayer, Rollin Harold
Totusek, Robert
Walters, Lowell Eugene

OREGON
Adams, Frank William
Anglemier, Allen Francis
Arscott, George Henry
Bernier, Paul Emile
Church, David Calvin
Ducsay, Charles Andrew
Harper, James Arthur
Holleman, Kendrick Alfred
Oldfield, James Edmund
Schmisseur, Wilson Edward
Weswig, Paul Henry

PENNSYLVANIA
Adams, Richard Sanford
Baumrucker, Craig Richard
Benedict, Robert Curtis
Buss, Edward George
Chalupa, William Victor
Cowan, Robert Lee
Crosby, Lon Owen
Farrell, Harold Maron, Jr
Flipse, Robert Joseph
Hagen, Daniel Russell
Hargrove, George Lynn
Harner, James Philip
Heald, Charles William
Hedde, Richard Duane
Hershberger, Truman Verne
Kao, Race Li-Chan
Keene, Owen David
Kesler, Earl Marshall
Krueger, Charles Robert
Lillie, Robert Jones
McKinstry, Donald Michael
Parish, Roger Cook
Piesco, Nicholas Peter
Ramberg, Charles F, Jr
Roush, William Burdette
Schneider, Henry Peter
Shellenberger, Paul Robert
Varga, Gabriella Anne
Vasilatos-Younken, Regina
Wangsness, Paul Jerome
Wideman, Robert Frederick, Jr

RHODE ISLAND
Cosgrove, Clifford James
Donovan, Gerald Alton
Durfee, Wayne King
Meade, Thomas Leroy
Nippo, Murn Marcus

SOUTH CAROLINA
Barnett, Bobby Dale
Birrenkott, Glen Peter, Jr
Edwards, Robert Lee
Hughes, Buddy Lee
Jones, Jack Edenfield
Lee, Daniel Dixon, Jr
Wise, Milton Bee

SOUTH DAKOTA
Britzman, Darwin Gene

Bush, Leon F
Carlson, Charles Wendell
Costello, William James
Embry, Lawrence Bryan
Emerick, Royce Jasper
Johnson, Marvin Melrose
Males, James Robert
Parsons, John G
Schingoethe, David John
Slyter, Arthur Lowell

TENNESSEE
Barth, Karl M
Bell, Marvin Carl
Chamberlain, Charles Calvin
Chen, Thomas Tien
Demott, Bobby Joe
Herting, David Clair
Miller, James Kincheloe
Montgomery, Monty J
Papas, Andreas Michael
Richardson, Don Orland
Sachan, Dileep Singh
White, Alan Wayne

TEXAS
Abrams, Steven Allen
Albin, Robert Custer
Brokaw, Bryan Edward
Brown, Murray Allison
Butler, Ogbourne Duke, Jr
Byers, Floyd Michael
Calhoun, Millard Clayton
Cassard, Daniel W
Cole, Noel Andy
Coleman, John Dee
Coppock, Carl Edward
Cross, Hiram Russell
Curl, Sam E
DuBose, Leo Edwin
Ellis, William C
Fineg, Jerry
Godbee, Richard Greene, II
Greathouse, Terrence Ray
Helm, Raymond E
Hutcheson, David Paul
Jacob, Horace S
Johnson, Larry
Lane, Alfred Glen
Lewis, Roscoe Warfield
Lippke, Hagen
Mason, Tim Robert
Mellor, David Bridgwood
Mersmann, Harry John
Pond, Wilson Gideon
Preston, Rodney LeRoy
Ramsey, Clovis Boyd
Richardson, Richard Harvey
Rosborough, John Paul
Rowland, Lenton O, Jr
Ryan, Cecil Benjamin
Scholl, Philip Jon
Sherrod, Lloyd B
Standlee, William Jasper
Stewart, Gordon Arnold
Sudweeks, Earl Max
Summers, William Allen, Jr
Swakon, Doreen H D
Thompson, Granville Berry
Trevino, Gilberto Stephenson
Varner, Larry Weldon
Whiteside, Charles Hugh
Will, Paul Arthur

UTAH
Anderson, Melvin Joseph
Butcher, John Edward
Dobson, Donald C
Draper, Carroll Isaac
Ellis, LeGrande Clark
Ernstrom, Carl Anthon
Fonnesbeck, Paul Vance
Gardner, Robert Wayne
Hendricks, Deloy G
Lamb, Robert Cardon
Malechek, John Charles
Park, Robert Lynn
Richardson, Gary Haight
Slade, Larry Malcom
Stoddard, George Edward
Vallentine, John Franklin
Wallentine, Max V
Warnick, Robert Eldredge

VERMONT
Carew, Lyndon Belmont, Jr
Simmons, Kenneth Rogers
Smith, Albert Matthews
Welch, James Graham

VIRGINIA
Acker, Duane Calvin
Allen, Vivien Gore
Boyd, Earl Neal
Bragg, Denver Dayton
Carr, Scott Bligh
Denbow, Donald Michael
Eller, Arthur L, Jr
Etgen, William M
Fontenot, Joseph Paul
Green, George G
Gwazdauskas, Francis Charles
Herbein, Joseph Henry, Jr
Hohenboken, William Daniel

Holmes, Clayton Ernest
Jones, Gerald Murray
Lincicome, David Richard
McGilliard, Michael Lon
Mitchell, George Ernest, Jr
Pearson, Ronald Earl
Polan, Carl E
Potter, Lawrence Merle
Raun, Ned S
Sheppard, Alan Jonathan
Swiger, Louis Andre
Vinson, William Ellis
Webb, Kenneth Emerson, Jr
Wisman, Everett Lee

WASHINGTON
Andrews, Daniel Keller
Bearse, Gordon Everett
Beck, Thomas W
Becker, Walter Alvin
Brekke, Clark Joseph
Carlson, James Roy
Froseth, John Allen
Hardy, Ronald W
Heinemann, Wilton Walter
Kincaid, Ronald Lee
Kromann, Rodney P
Leung, Frederick C
Luedecke, Lloyd O
Reeves, Jerry John
Sasser, Lyle Blaine
Thomas, John M
Towner, Richard Henry
Van Hoosier, Gerald L, Jr

WEST VIRGINIA
Anderson, Gerald Clifton
Becker, Stanley Leonard
Belding, Ralph Cedric
Hansard, Samuel L, II
Horvath, Donald James
Peterson, Ronald A
Thomas, Roy Orlando

WISCONSIN
Albrecht, Kenneth Adrian
Arrington, Louis Carroll
Bird, Herbert Roderick
Bradley, Robert Lester, Jr
Broderick, Glen Allen
Cook, Mark Eric
Ehle, Fred Robert
Greaser, Marion Lewis
Greger, Janet L
Hoffman, William F
Howard, W Terry
Jorgensen, Neal A
Larsen, Howard James
Leibbrandt, Vernon Dean
Maurer, Arthur James
Mertens, David Roy
Satter, Larry Dean
Schaefer, Daniel M
Schultz, Loris Henry
Sendelbach, Anton G
Sunde, Milton Lester
Winder, William Charles
Wittwer, Leland S

WYOMING
Hinds, Frank Crossman
Kercher, Conrad J

PUERTO RICO
Oliver-Padilla, Fernando Luis

ALBERTA
Bowland, John Patterson
Cheng, Kuo-Joan
Clandinin, Donald Robert
Mathison, Gary W(ayne)
Newman, John Alexander
Robblee, Alexander (Robinson)
Young, Bruce Arthur

BRITISH COLUMBIA
Bailey, Charles Basil Mansfield
Beames, R M
Blair, Robert
Bowden, David Merle
Cheng, Kimberly Ming-Tak
Hill, Arthur Thomas
Hunt, John R
Majak, Walter
March, Beryl Elizabeth
Owen, Bruce Douglas
Quinton, Dee Arlington
Waldern, Donald E

MANITOBA
Ayles, George Burton
Castell, Adrian George
Elliot, James I
Grandhi, Raja Ratnam
Kondra, Peter Alexander
Strain, John Henry

NEW BRUNSWICK
Bush, Roy Sidney
Friars, Gerald W
McQueen, Ralph Edward
Nicholson, J W G

NOVA SCOTIA
Cock, Lorne M

ONTARIO
Ainsworth, Louis
Burton, John Heslop
Emmons, Douglas Byron
Evans, Essi H
Fortin, Andre Francis
Gavora, Jan Samuel
Gowe, Robb Shelton
Hamilton, Robert Milton Gregory
Heaney, David Paul
Irvine, Donald McLean
Ivan, Michael
Lin, Ching Y
Morrison, Spencer Horton
Morrison, William D
Osmond, Daniel Harcourt
Pang, Cho Yat
Phillips, William Ernest John
Pigden, Wallace James
Trenholm, Harold Locksley
Usborne, William Ronald
Wolynetz, Mark Stanley

PRINCE EDWARD ISLAND
Winter, Karl A

QUEBEC
Barrette, Daniel Claude
Belzile, Rene
Besner, Michel
Buckland, Roger Basil
De la Noue, Joel Jean-Louis
Donefer, Eugene
Farmer, Chantal
Gariepy, Claude
Matte, J Jacques

SASKATCHEWAN
Beacom, Stanley Ernest
Bell, John Milton
Cohen, Roger D H
Crawford, Roy Douglas
Jones, Graham Alfred
Knipfel, Jerry Earl
Korsrud, Gary Olaf
Patience, John Francis
Salmon, Raymond Edward

OTHER COUNTRIES
Bywater, Anthony Colin
Fairweather-Tait, Susan Jane
Fitzhugh, Henry Allen
Jahn, Ernesto
Mullenax, Charles Howard
Nakagawa, Shizutoshi
Santidrian, Santiago
Schmeling, Sheila Kay
Shore, Laurence Stuart
Szebenyi, Emil

Fish & Wildlife Sciences

ALABAMA
Bayne, David Roberge
Best, Troy Lee
Boschung, Herbert Theodore
Bradley, James T
Brady, Yolanda J
Causey, Miles Keith
Duncan, Bryan Lee
Grizzle, John Manuel
Grover, John Harris
Holler, Nicholas Robert
Huntley, Jimmy Charles
Lawrence, John Medlock
Lovell, Richard Thomas
Lovshin, Leonard Louis, Jr
Mirarchi, Ralph Edward
Moss, Donovan Dean
Paxton, Ralph
Phelps, Ronald P
Plumb, John Alfred
Rogers, Wilmer Alexander
Rosene, Walter, Jr
Rouse, David B
Shell, Eddie Wayne
Smitherman, Renford Oneal

ALASKA
Ahlgren, Molly O
Amend, Donald Ford
Babcock, Malin Marie
Baglin, Raymond Eugene, Jr
Behrend, Donald Fraser
Clark, John Harlan
Dahlberg, Michael Lee
Dean, Frederick Chamberlain
Fagen, Robert
Fay, Francis Hollis
Gard, Richard
Gharrett, Anthony John
Hanley, Thomas Andrew
Hauser, William Joseph
Heifetz, Jonathan
Helle, John Harold (Jack)
Kessel, Brina
Klein, David Robert
LaPerriere, Jacqueline Doyle
Leon, Kenneth Allen
Mathisen, Ole Alfred

Fish & Wildlife Sciences (cont)

Meehan, William Robert
Morrison, John Albert
Olsen, James Calvin
Paul, Augustus John, III
Pella, Jerome Jacob
Reynolds, James Blair
Rockwell, Julius, Jr
Rogers, John Gilbert, Jr
Samson, Fred Burton
Schoen, John Warren
Smoker, William Alexander
Smoker, William Williams
Snyder, George Richard
Straty, Richard Robert
Thomas, Gary Lee

ARIZONA
Beck, John R
Gallizioli, Steve
Gerking, Shelby Delos
Johnson, Raymond Roy
Krausman, Paul Richard
Maughan, Owen Eugene
Minckley, Wendell Lee
Patton, David Roger
Post, George
Rinne, John Norman
Schnell, Jay Heist
Shaw, William Wesley
Smith, Andrew Thomas
Smith, Norman Sherrill
Sowls, Lyle Kenneth
Van Riper, Charles, III

ARKANSAS
Collins, Richard Arlen
Dupree, Harry K
Grizzell, Roy Ames, Jr
Jenkins, Robert M
Kilambi, Raj Varad
Tave, Douglas

CALIFORNIA
Abramson, Norman Jay
Allen, George Herbert
Anderson, Daniel William
Bakun, Andrew
Barrett, Izadore
Baxter, John Lewis
Bayliff, William Henry
Botzler, Richard George
Bray, Richard Newton
Brewer, Gary David
Brownell, Robert Leo, Jr
Cech, Joseph Jerome, Jr
Chadwick, Harold King
Chamberlain, Dilworth Woolley
Chen, Lo-Chai
Chesemore, David Lee
Davis, Gary Everett
Dieterich, Robert Arthur
Ebert, Earl Ernest
Eldridge, Maxwell Bruce
Embree, James Willard, Jr
Erman, Don Coutre
Eschmeyer, William Neil
Farris, David Allen
Follett, Wilbur Irving
Fridley, Robert B
Gall, Graham A E
Gerrodette, Timothy
Gilmer, David Seeley
Gordon, Malcolm Stephen
Harris, Stanley Warren
Hassler, Thomas J
Hedgecock, Dennis
Hendricks, Lawrence Joseph
Hewston, John G
Horn, Michael Hastings
Howard, Walter Egner
Hyde, Kenneth Martin
Iwamoto, Tomio
Jacobsen, Nadine Klecha
Jansen, Henricus Cornelis
Kask, John Laurence
Kerstetter, Theodore Harvey
Klawe, Witold L
Koo, Ted Swei-Yen
Kope, Robert Glenn
Krejsa, Richard Joseph
Kutilek, Michael Joseph
Laurs, Robert Michael
Lenarz, William Henry
McCullough, Dale Richard
Messersmith, James David
Mitchell, John Alexander
Moberg, Gary Philip
Morris, Robert Wharton
Mossman, Archie Stanton
Mulroy, Thomas Wilkinson
Neal, Richard Allan
Ohlendorf, Harry Max
Orcutt, Harold George
Raveling, Dennis Graff
Ridenhour, Richard Lewis
Roelofs, Terry Dean
Ryder, Oliver A
Sakagawa, Gary Toshio
Shelton, William Lee
Smith, Paul Edward
Springer, Paul Frederick
Swift, Camm Churchill

Taylor, Leighton Robert, Jr
Vanicek, C David
Vilkitis, James Richard
Walter, Hartmut
Weintraub, Joel D
Wilcox, Bruce Alexander
Williams, Daniel Frank
Woodruff, David Scott

COLORADO
Asherin, Duane Arthur
Bailey, James Allen
Bradshaw, William Newman
Braun, Clait E
Carlson, Clarence Albert, Jr
Crouch, Glenn LeRoy
Curnow, Richard Dennis
Dittberner, Phillip Lynn
Flickinger, Stephen Albert
Glover, Fred Arthur
Green, Jeffrey Scott
Gregory, Richard Wallace
Hagen, Harold Kolstoe
Hein, Dale Arthur
Keith, James Oliver
Knopf, Fritz L
Lehner, Philip Nelson
Martin, Stephen George
Oldemeyer, John Lee
Parker, H Dennison
Reidinger, Russell Frederick, Jr
Ryder, Ronald Arch
Schmidt, John Lancaster
Smith, Dwight Raymond
Stalnaker, Clair B
Sublette, James Edward
Thompson, Daniel Quale
Thorpe, Bert Duane

CONNECTICUT
Barske, Philip
Katz, Max
Lund, William Albert, Jr
Maguder, Theodore Leo, Jr
Nielsen, Svend Woge
Renfro, William Charles

DELAWARE
Roth, Roland Ray
Schuler, Victor Joseph

DISTRICT OF COLUMBIA
Blockstein, David Edward
Buffington, John Douglas
Domning, Daryl Paul
Field, Ronald James
Knapp, Leslie W
Lovejoy, Thomas E
Nickum, John Gerald
Policansky, David J
Post, Boyd Wallace
Rogers, John Patrick
Strauss, Michael S
Sweeney, James Michael
Tyler, James Chase
Weitzman, Stanley Howard
Wise, John P

FLORIDA
Black, Kenneth Eldon
Bortone, Stephen Anthony
Branch, Lyn Clarke
Brown, Bradford E
Byrd, Isaac Burlin
Christman, Steven Philip
Davis, John Armstrong
Feddern, Henry A
Forrester, Donald Jason
Gilbert, Carter Rowell
Grimes, Churchill Bragaw
Higman, James B
Iversen, Edwin Severin
Jones, Albert Cleveland
Kale, Herbert William, II
Kitchens, Wiley M
Labisky, Ronald Frank
Loesch, Harold Carl
McCann, James Alwyn
Marion, Wayne Richard
Meryman, Charles Dale
Pagan-Font, Francisco Alfredo
Posner, Gerald Seymour
Powers, Joseph Edward
Prince, Eric D
Rathjen, Warren Francis
Robins, Charles Richard
Roessler, Martin A
Wisby, Warren Jensen
Witham, P(hilip) Ross
Woods, Charles Arthur

GEORGIA
Andrews, Charles Lawrence
Briggs, John Carmon
Callaham, Mac A
Johnson, Albert Sydney, III
Marchinton, Robert Larry
Miller, George C
Payne, Jerry Allen
Schultz, Donald Paul
Scott, Donald Charles
Stober, Quentin Jerome
Winger, Parley Vernon

HAWAII
Atkinson, Shannon K C
Bardach, John E
Boehlert, George Walter
Fast, Arlo Wade
Greenfield, David Wayne
Shomura, Richard Sunao
Skillman, Robert Allen
Stone, Charles Porter
Winget, Robert Newell

IDAHO
Ables, Ernest D
Congleton, James Lee
Griffith, John Spencer
Hostetter, Donald Lee
Hungerford, Kenneth Eugene
Keller, Barry Lee
Laundre, John William
MacFarland, Craig George
Nass, Roger Donald
Peek, James Merrell
Platts, William Sidney

ILLINOIS
Anderson, William Leno
Bjorklund, Richard Guy
Buck, David Homer
Burger, George Vanderkarr
Burr, Brooks Milo
Durham, Leonard
Feldhamer, George Alan
Heltne, Paul Gregory
Jahn, Lawrence A
Klimstra, Willard David
Kozicky, Edward Louis
Larimore, Richard Weldon
Le Febvre, Eugene Allen
Meldrim, John Waldo
Nixon, Charles Melville
Page, Lawrence Merle
Paulson, Glenn
Prabhudesai, Mukund M
Roseberry, John L
Thomerson, Jamie E
Waring, George Houstoun, IV

INDIANA
Goetz, Frederick William, Jr
Kirkpatrick, Charles Milton
Kirkpatrick, Ralph Donald
Montague, Fredrick Howard, Jr
Schaible, Robert Hilton
Spacie, Anne
Weeks, Harmon Patrick, Jr

IOWA
Best, Louis Brown
Carlander, Kenneth Dixon
Clark, William Richard
Dinsmore, James Jay
Dowell, Virgil Eugene
Klaas, Erwin Eugene
Menzel, Bruce Willard
Moorman, Robert Bruce
Ramsey, John Scott
Summerfelt, Robert C
Vohs, Paul Anthony, Jr
Wilson, Nixon Albert

KANSAS
Choate, Jerry Ronald
Eddy, Thomas A
Halazon, George Christ
Hopkins, Theodore Louis
Kaufman, Donald Wayne
Klaassen, Harold Eugene
Thompson, Max Clyde

KENTUCKY
Branson, Branley Allan
Bryant, William Stanley
Pearson, William Dean
Thompson, Marvin Pete
Timmons, Thomas Joseph
Williams, John C

LOUISIANA
Chabreck, Robert Henry
Cordes, Carroll Lloyd
Culley, Dudley Dean, Jr
Davis, Billy J
Fingerman, Sue Whitsell
Gunning, Gerald Eugene
Hamilton, Robert Bruce
Hastings, Robert Wayne
Suttkus, Royal Dallas
Wakeman, John Marshall
Wright, Vernon Lee

MAINE
Coulter, Malcolm Wilford
Crawford, Hewlette Spencer, Jr
De Witt, Hugh Hamilton
Hanks, Robert William
Mendall, Howard Lewis
Owen, Ray Bucklin, Jr
Shumway, Sandra Elisabeth
Smith, Roderick MacDowell
Stanley, Jon G
Warner, Kendall

MARYLAND
Adams, Lowell William

HAWAII

Albers, Peter Heinz
Barbehenn, Kyle Ray
Brandt, Stephen Bernard
Chou, Nelson Shih-Toon
Clark, Donald Ray, Jr
Flyger, Vagn Flemming
Fuller, Mark Roy
Gipson, Philip
Griswold, Bernard Lee
Heath, Robert Gardner
Hetrick, Frank M
Hill, Elwood Fayette
Hodgdon, Harry Edward
Hoffman, David J
Houde, Edward Donald
Leedy, Daniel Loney
Nichols, James Dale
Pattee, Oliver Henry
Robbins, Chandler Seymour
Rothschild, Brian James
Russek-Cohen, Estelle
Sparling, Donald Wesley, Jr
Tillman, Michael Francis
Wiemeyer, Stanley Norton
Williams, Walter Ford

MASSACHUSETTS
Booke, Henry Edward
Boreman, John George, Jr
Castro, Gonzalo
Clark, Stephen Howard
Finn, John Thomas
Godfrey, Paul Jeffrey
Greeley, Frederick
Heyerdahl, Eugene Gerhardt
Larson, Joseph Stanley
Moir, Ronald Brown, Jr
Richards, F Paul
Ropes, John Warren
Ross, Michael Ralph

MICHIGAN
Behmer, David J
Belyea, Glenn Young
Bennett, Carl Leroy
Borgeson, David P
Brown, Edward Herriot, Jr
Burton, Thomas Maxie
Cowan, Archibald B
D'Itri, Frank M
Fetterolf, Carlos de la Mesa, Jr
Fink, William Lee
Foster, Neal Robert
Hart, John Henderson
Hartman, Wilbur Lee
Hatch, Richard Wallace
Haufler, Jonathan B
Kevern, Niles Russell
Krull, John Norman
Kutkuhn, Joseph Henry
Lagler, Karl Frank
Langenau, Edward E, Jr
Latta, William Carl
Lenon, Herbert Lee
Mac, Michael John
Manny, Bruce Andrew
Miller, Robert Rush
Passino, Dora R May
Patriarche, Mercer Harding
Peterson, Rolf Olin
Petrides, George Athan
Prince, Harold Hoopes
Robinson, William Laughlin
Sloan, Norman F
Stauffer, Thomas Miel
Sylvester, Joseph Robert
Tack, Peter Isaac
Taylor, Mitchell Kerry
Taylor, William Waller
Williams, Wells Eldon
Winterstein, Scott Richard

MINNESOTA
Adelman, Ira Robert
Arthur, John W
Broderius, Steven James
Buech, Richard Reed
Collins, Hollie L
Gullion, Gordon W
Hokanson, Kenneth Eric Fabian
Jannett, Frederick Joseph, Jr
Jordan, Peter Albion
Kuehn, Jerome H
Lange, Robert Echlin, Jr
McCann, Lester J
McConville, David Raymond
Mech, Lucyan David
Meyer, Fred Paul
Mount, Donald I
Rogers, Lynn Leroy
Spigarelli, Steven Alan
Swanson, Gustav Adolph
Tanner, Ward Dean, Jr
Waters, Thomas Frank
Williams, Steven Frank

MISSISSIPPI
Arner, Dale H
Brown, Robert Dale
Brown, Robert Dale
Dawson, Charles Eric
Regier, Lloyd Wesley
Robinson, Edwin Hollis
Ross, Stephen Thomas

Wakeley, James Stuart

MISSOURI
Baskett, Thomas Sebree
Drobney, Ronald DeLoss
Elder, William Hanna
Faaborg, John Raynor
Finger, Terry Richard
Fredrickson, Leigh H
Fritzell, Erik Kenneth
Funk, John Leon
Hanson, Willis Dale
Hunn, Joseph Bruce
Kuhns, John Farrell
Mosher, Donna Patricia
Pflieger, William Leo
Schoettger, Richard A

MONTANA
Ball, Irvin Joseph
Foresman, Kerry Ryan
Fritts, Steven Hugh
Gould, William Robert, III
Harris, John Tom
Lyon, Leonard Jack
Mackie, Richard John
Metzgar, Lee Hollis
Mitchell, Lawrence Gustave
O'Gara, Bartholomew Willis
Rounds, Burton Ward
Schumacher, Robert E
Taber, Richard Douglas
Tennant, Donald L

NEBRASKA
Bliese, John C W
Case, Ronald Mark
Hergenrader, Gary Lee

NEVADA
Davis, Phillip Burton
Paulson, Larry Jerome

NEW HAMPSHIRE
Barry, William James
Mautz, William Ward
Sower, Stacia Ann
Wilson, Donald Alfred

NEW JERSEY
Applegate, James Edward
Landau, Matthew Paul
Pacheco, Anthony Louis

NEW MEXICO
Davis, Charles A
Hales, Donald Caleb
Holechek, Jerry Lee
Howard, Volney Ward, Jr
Huey, William S
Johnson, James Edward
Lewis, James Chester
Schemnitz, Sanford David

NEW YORK
Brett, Betty Lou Hilton
Brocke, Rainer H
Brothers, Edward Bruce
Cahn, Phyllis Hofstein
Christian, John Jermyn
Dooley, James Keith
Dovel, William Lawrence
Elrod, Joseph Harrison
Greenhall, Arthur Merwin
Ketola, H George
Kiviat, Erik
Lawrence, William Mason
McHugh, John Laurence
McNeil, Richard Jerome
Madison, Dale Martin
Maestrone, Gianpaolo
Muessig, Paul Henry
Muller-Schwarze, Dietland
Neeper, Ralph Arnold
Oglesby, Ray Thurmond
Osterberg, Donald Mouretz
Perlmutter, Alfred
Porter, William Frank
Reuter, Gerald Louis
Ringler, Neil Harrison
Rockwell, Robert Franklin
Severinghaus, Charles William
Tillman, Robert Erwin
VanDruff, Larry Wayne
Walker, Roland
Wilkins, Bruce Tabor
Windsor, Donald Arthur

NORTH CAROLINA
Bolen, Eric George
Brown, Richard Dean
Cope, Oliver Brewern
Crowder, Larry Bryant
Dieter, Michael Phillip
Di Giulio, Richard Thomas
Doerr, Phillip David
Fairchild, Homer Eaton
Hayne, Don William
Huntsman, Gene Raymond
Lewis, Robert Minturn
Lewis, William Madison
Lindquist, David Gregory
Link, Garnett William, Jr
Manooch, Charles Samuel, III

Merriner, John Vennor
Nicholson, William Robert
Peters, David Stewart
Pritchard, G Ian
Rulifson, Roger Allen
Schwartz, Frank Joseph
Steinhagen, William Herrick
Sullivan, Arthur Lyon
Sykes, James Enoch
Towell, William Earnest
West, Jerry Lee

NORTH DAKOTA
Crawford, Richard Dwight
Krapu, Gary Lee

OHIO
Barbour, Clyde D
Berra, Tim Martin
Bookhout, Theodore Arnold
Cole, Charles Franklyn
Crites, John Lee
Culver, David Alan
Eastman, Joseph
Good, Ernest Eugene
Harder, John Dwight
Holeski, Paul Michael
Kinney, Edward Coyle, Jr
Lewis, Michael Anthony
McLean, Edward Bruce
Peterle, Tony J
White, Andrew Michael

OKLAHOMA
Alexander, Joseph Walker
Chapman, Brian Richard
Hill, Loren Gilbert
Kilpatrick, Earl Buddy
Matthews, William John
Namminga, Harold Eugene
Shaw, James Harlan
Summers, Gregory Lawson
Toetz, Dale W

OREGON
Bond, Carl Eldon
Campbell, Charles J
Chalk, David Eugene
Crawford, John Arthur
Dealy, John Edward
De Witt, John William, Jr
Doudoroff, Peter
Eicher, George J
Erickson, Ray Charles
Fryer, John Louis
Garton, Ronald Ray
Hall, James Dane
Hamilton, James Arthur Roy
Harville, John Patrick
Hedtke, James Lee
Hendricks, Jerry Dean
Horton, Howard Franklin
Isaacson, Dennis Lee
Krueger, William Clement
Linn, DeVon Wayne
Liss, William John
Markle, Douglas Frank
Marriage, Lowell Dean
Maser, Chris
Meslow, E Charles
Olla, Bori Liborio
Olson, Robert Eldon
Perry, Lorin Edward
Quast, Jay Charles
Rasmussen, Lois E Little
Richards, Richard Earl
Schneider, Phillip William, Jr
Schoning, Robert Whitney
Schreck, Carl Bernhard
Skovlin, Jon Matthew
Thomas, Jack Ward
Tubb, Richard Arnold
Wagner, Harry Henry
Wick, William Quentin

PENNSYLVANIA
Arnold, Dean Edward
Clark, Richard James
Eckroat, Larry Raymond
Feuer, Robert Charles
Marquis, David Alan
Mathur, Dilip
Snyder, Donald Benjamin
Yahner, Richard Howard

RHODE ISLAND
Durfee, Wayne King
Golet, Francis Charles
Gould, Walter Phillip
Husband, Thomas Paul
Laurence, Geoffrey Cameron
Lazell, James Draper
Mayer, Garry Franklin
Saila, Saul Bernhard

SOUTH CAROLINA
Burrell, Victor Gregory, Jr
Ely, Berten E, III
Harlow, Richard Fessenden
Hays, Sidney Brooks
Johnson, Robert Karl
Kennamer, James Earl
Muska, Carl Frank
Smith, Theodore Isaac Jogues

Sweeney, John Robert
Wood, Gene Wayne

SOUTH DAKOTA
Berry, Charles Richard, Jr
Hamilton, Steven J
Keenlyne, Kent Douglas
Scalet, Charles George
Schmulbach, James C
Uresk, Daniel William

TENNESSEE
Bulow, Frank Joseph
Cada, Glenn Francis
Chance, Charles Jackson
Dearden, Boyd L
Dimmick, Ralph W
Eddlemon, Gerald Kirk
Etnier, David Allen
Kroodsma, Roger Lee
Loar, James M
McCoy, Ralph Hines
Pelton, Michael Ramsay
Sliger, Wilburn Andrew
Webb, J(ohn) Warren
Wilson, James Larry

TEXAS
Anderson, Richard Orr
Blankenship, Lytle Houston
Caillouet, Charles W, Jr
Chaney, Allan Harold
Darnell, Rezneat Milton
Dibble, John Thomas
Dickson, James Gary
Drawe, D Lynn
Fulbright, Timothy Edward
Halls, Lowell Keith
Hellier, Thomas Robert, Jr
Hubbs, Clark
Lewis, Donald Howard
Lieb, Carl Sears
Linton, Thomas LaRue
McEachran, John D
Martin, Floyd Douglas
Mathews, Nancy Ellen
Neill, William Harold
Parker, Nick Charles
Schmidly, David James
Scudday, James Franklin
Stransky, John Janos
Strawn, Robert Kirk
Telfair, Raymond Clark, II
Thill, Ronald E
Varner, Larry Weldon
Weller, Milton Webster
Whiteside, Bobby Gene
Zagata, Michael DeForest

UTAH
Bissonette, John Alfred
Bulkley, Ross Vivian
Chapman, Joseph Alan
Conover, Michael Robert
Dueser, Raymond D
Flinders, Jerran T
Helm, William Thomas
Kadlec, John A
Knowlton, Frederick Frank
Mangum, Fredrick Anthony
Porter, Richard Dee
Rajagopal, P K
Sigler, John William
Sigler, William Franklin
Urness, Philip Joel

VIRGINIA
Batts, Billy Stuart
Berryman, Jack Holmes
Burreson, Eugene M
Chittenden, Mark Eustace, Jr
Cordes, Donald Ormond
Cross, Gerald Howard
Dane, Charles Warren
Davis, William Spencer
Fox, William Walter, Jr
Giles, Robert H, Jr
Haven, Dexter Stearns
Hosner, John Frank
Howe, Marshall Atherton
Jahn, Laurence R
Kirkpatrick, Roy Lee
Linzey, Donald Wayne
Loesch, Joseph G
McGinnes, Burd Sheldon
Neves, Richard Joseph
Reed, James Robert, Jr
Scanlon, Patrick Francis
Scattergood, Leslie Wayne
Short, Henry Laughton
Sprague, Lucian Matthew
Streeter, Robert Glen
Talbot, Lee Merriam
Vaughan, Michael Ray
Walker, Charles R

WASHINGTON
Allen, Julia Natalia
Anderson, James Jay
Aron, William I
Banko, Winston Edgar
Barker, Morris Wayne
Bax, Nicholas John
Bell, Milo C

Bevan, Donald Edward
Burgner, Robert Louis
Catts, Elmer Paul
Clark, William Greer
Dauble, Dennis Deene
Dickhoff, Walton William
Dixon, Kenneth Randall
Emlen, John Merritt
Evermann, James Frederick
Fickeisen, Duane H
Fredin, Reynold A
Fukuhara, Francis M
Gilbert, Frederick Franklin
Ginn, Thomas Clifford
Gunderson, Donald Raymond
Hanson, Wayne Carlyle
Hardy, Ronald W
Hayes, Murray Lawrence
Heinle, Donald Roger
Henry, Kenneth Albin
Jefferts, Keith Bartlett
Karr, James Richard
Laevastu, Taivo
Landolt, Marsha LaMerle
Larkins, Herbert Anthony
Low, Loh-Lee
Maltzeff, Eugene M
Manuwal, David Allen
Marshall, William Hampton
Mearns, Alan John
Nakatani, Roy E
Novotny, Anthony James
Pauley, Gilbert Buckhannan
Raymond, Howard Lawrence
Rogers, Donald Eugene
Roos, John Francis
Royce, William Francis
Salo, Ernest Olavi
Scheffer, Victor B
Seymour, Allyn H
Shaklee, James Brooker
Silliman, Ralph Parks
Stauffer, Gary Dean
Thompson, Richard Baxter
Thorne, Richard Eugene
Utter, Fred Madison
Wedemeyer, Gary Alvin
Whitney, Richard Ralph
Woelke, Charles Edward

WEST VIRGINIA
Bullock, Graham Lambert
Michael, Edwin Daryl
Samuel, David Evan
Shalaway, Scott D
Smith, Robert Leo

WISCONSIN
Anderson, Raymond Kenneth
Baylis, Jeffrey Rowe
Copes, Frederick Albert
Dahlgren, Robert Bernard
Friend, Milton
Hamerstrom, Frederick Nathan
Hickey, Joseph James
Hine, Ruth Louise
Hunt, Robert L
Keith, Lloyd Burrows
McCabe, Robert Albert
Magnuson, John Joseph
Mahadeva, Madhu Narayan
Rongstad, Orrin James
Ruff, Robert LaVerne
Rusch, Donald Harold
Schmitz, William Robert
Stearns, Forest
Temple, Stanley A
Trainer, Daniel Olney

WYOMING
Baxter, George T
Hayden-Wing, Larry Dean
Stanton, Nancy Lea
Wade, Dale A

PUERTO RICO
Wilson, Marcia Hammerquist

ALBERTA
Speller, Stanley Wayne
Stirling, Ian G

BRITISH COLUMBIA
Albright, Lawrence John
Brett, John Roland
Bunnell, Frederick Lindsley
Cheng, Kimberly Ming-Tak
Donaldson, Edward Mossop
Hourston, Alan Stewart
Jamieson, Glen Stewart
Kennedy, William Alexander
Lane, Edwin David
Larkin, Peter Anthony
Lauzier, Raymond B
Lindsey, Casimir Charles
Margolis, Leo
Marliave, Jeffrey Burton
Newman, Murray Arthur
Quinton, Dee Arlington
Routledge, Richard Donovan
Tyler, Albert Vincent
Wilimovsky, Norman Joseph
Zwickel, Fred Charles

Fish & Wildlife Sciences (cont)

MANITOBA
Bodaly, Richard Andrew
Franzin, William Gilbert
Johnson, Lionel
Lawler, George Herbert
Patalas, Kazimierz
Preston, William Burton
Pruitt, William O, Jr
Ward, Fredrick James

NEW BRUNSWICK
Anderson, John Murray
Day, Lewis Rodman
Erskine, Anthony J
Frantsi, Christopher
Haya, Katsuji
Kohler, Carl
Paim, Uno

NEWFOUNDLAND
Khan, Rasul Azim
Mercer, Malcolm Clarence
Ni, I-Hsun
Wroblewski, Joseph S

NOVA SCOTIA
Bowen, William Donald
Dodds, Donald Gilbert
Doyle, Roger Whitney
Freeman, Harry Cleveland
Harrington, Fred Haddox
Hodder, Vincent MacKay
Kimmins, Warwick Charles
Longhurst, Alan R
Odense, Paul Holger
Rojo, Alfonso
Uthe, John Frederick

ONTARIO
Buckner, Charles Henry
Casselman, John Malcolm
Coad, Brian William
Colby, Peter J
Crossman, Edwin John
Emery, Alan Roy
Fraser, James Millan
Fry, Frederick Ernest Joseph
Gibson, George
Harvey, Harold H
Ihssen, Peter Edowald
Kessler, Dan
Lavigne, David M
Law, Cecil E
McCauley, Robert William
MacCrimmon, Hugh Ross
Momot, Walter Thomas
Power, Geoffrey
Powles, Percival Mount
Regier, Henry Abraham
Reynolds, John Keith
Ryder, Richard Armitage
Savan, Milton
Smith, Donald Alan
Stasko, Aivars B
Tait, James Simpson
Tibbles, John James
Weatherley, Alan Harold

QUEBEC
Besner, Michel
Grant, James William Angus
Johnston, C Edward
Lapp, Wayne Stanley
Leduc, Gerard
Leggett, William C
Magnin, Etienne Nicolas
Seguin, Louis-Roch
Shaw, Robert Fletcher
Titman, Rodger Donaldson

SASKATCHEWAN
Flood, Peter Frederick
Forsyth, Douglas John
Irvine, Donald Grant
Mitchell, George Joseph
Wobeser, Gary Arthur

OTHER COUNTRIES
Allen, Robin Leslie
Craig, John Frank
Crompton, David William Thomasson
Davis, William Jackson
Holsworth, William Norton
Koslow, Julian Anthony
May, Robert Carlyle
Schmeling, Sheila Kay
Williams, Francis

Food Science & Technology

ALABAMA
Ammerman, Gale Richard
Dalvi, Ramesh R
Huffman, Dale L

ARIZONA
Bessey, Paul Mack
Kline, Ralph Willard
McNamara, Donald J
Peng, Yeh-Shan

ARKANSAS
Crandall, Philip Glen
Horan, Francis E
Snyder, Harry E

CALIFORNIA
Amerine, Maynard Andrew
Bernhard, Richard Allan
Bolin, Harold R
Boulton, Roger Brett
Chen, Ming Sheng
Chen, Tung-Shan
Coughlin, James Robert
Dunkley, Walter Lewis
Fleming, Sharon Elayne
Frankel, Edwin N
Gerber, Bernard Robert
Goettlich Riemann, Wilhelmina Maria
    Anna
Gump, Barry Hemphill
Hadley, Jeffery A
Josephson, Ronald Victor
Kean, Chester Eugene
Lee, Chi-Hang
Lipton, Werner Jacob
Lu, Nancy Chao
Maier, V(incent) P(aul)
Meads, Philip F
Michener, H(arold) David
Moore, Andrew Brooks
Nielsen, Milo Alfred
O'Donnell, Joseph Allen, III
Pedersen, Leo Damborg
Prater, Arthur Nickolaus
Rabinowitz, Israel Nathan
Singleton, Vernon LeRoy
Somogyi, Laszlo P
Son, Chung Hyun
Stone, Herbert
Swisher, Horton Edward
Toma, Ramses Barsoum
Yuen, Wing

COLORADO
Avens, John Stewart
Deane, Darrell Dwight
Hess, Dexter Winfield
Pyler, Richard Ernst
Thomas, William Robb

CONNECTICUT
Bibeau, Thomas Clifford
Bluhm, Leslie
Hall, Kenneth Noble
Kinsman, Donald Markham

DELAWARE
Hoover, Dallas Gene
Islam, Mir Nazrul
Kinney, Anthony John
Wright, Leon Wendell

DISTRICT OF COLUMBIA
Baker, Charles Wesley
Bernard, Dane Thomas
Bigelow, Sanford Walker
Flora, Lewis Franklin
McBride, Gordon Williams
Miller, David Jacob
Tallent, William Hugh
Wilson, George Donald

FLORIDA
Attaway, John Allen
Bauernfeind, Jacob (Jack) C(hristopher)
Green, Nancy R
Hardenburg, Robert Earle
Jackel, Simon Samuel
Matthews, Richard Finis
Nagy, Steven
Sathe, Shridhar Krishna
Tamplin, Mark Lewis
Weingartner, Karl Ernst

GEORGIA
Brackett, Robert E
Chiang, Tze I
Dull, Gerald G
Highland, Henry Arthur
Jen, Joseph Jwu-Shan
Leffingwell, John C
Loewenstein, Morrison
Pusey, P Lawrence
Robertson, James Aldred
Schuler, George Albert
Wilson, David Merl

HAWAII
Nip, Wai Kit
Williams, David Douglas F
Wood, Betty J

IDAHO
Hibbs, Robert A

ILLINOIS
Abbott, Thomas Paul
Bookwalter, George Norman
Chen, Shepley S
Dunn, Dorothy Fay
Emken, Edward Allen
Empen, Joseph A
Flowers, Russell Sherwood, Jr
Hahn, Richard Ray

Kuhajek, Eugene James
McKeith, Floyd Kenneth
Martin, Scott Elmore
Mateles, Richard I
Mounts, Timothy Lee
Muck, George A
Richmond, Patricia Ann
Sheldon, Victor Lawrence
Singh, Laxman
Soucie, William George
Standing, Charles Nicholas
Thompson, Arthur Robert
Trelease, Richard Davis
Wargel, Robert Joseph
Yackel, Walter Carl

INDIANA
Anderson, David Bennett
BeMiller, James Noble
Butler, Larry G
Etzel, James Edward
Heinicke, Herbert Raymond
Hoff, Johan Eduard
Ivaturi, Rao Venkata Krishna
Katz, Frances R
Nelson, Philip Edwin
Nirschl, Joseph Peter

IOWA
Johnson, Lawrence Alan
Murphy, Patricia A
Topel, David Glen

KANSAS
Bates, Lynn Shannon
Grunewald, Katharine Klevesahl
Kastner, Curtis Lynn
Ponte, Joseph G, Jr
Scheid, Harold E

KENTUCKY
Kemp, James Dillon
Knapp, Frederick Whiton
Maisch, Weldon Frederick
Massik, Michael
Moody, William Glenn
Packett, Leonard Vasco

LOUISIANA
Broeg, Charles Burton
Constantin, Roysell Joseph
Culley, Dudley Dean, Jr
Decossas, Kenneth Miles
Hackney, Cameron Ray
Hegsted, Maren
Johnsen, Peter Berghsey
Kadan, Ranjit Singh
Kuan, Shia Shiong
Shih, Frederick F

MARYLAND
Allison, Richard Gall
Berry, Bradford William
Butrum, Ritva Rauanheimo
Crosby, Edwin Andrew
Falci, Kenneth Joseph
Gilmore, Thomas Meyer
Lee, Yuen San
Metcalfe, Dean Darrel
Nelson, John Howard
Rechcigl, Miloslav, Jr
Salwin, Harold
Webb, Byron H
Westhoff, Dennis Charles

MASSACHUSETTS
Brandler, Philip
Cardello, Armand Vincent
Gorfien, Harold
Learson, Robert Joseph
Salant, Abner Samuel
Taub, Irwin A(llen)

MICHIGAN
Gray, Ian
Hollingsworth, Cornelia Ann
Markakis, Pericles
Muentener, Donald Arthur
Rippen, Alvin Leonard
Saldanha, Leila Genevieve
Smith, Denise Myrtle

MINNESOTA
Allen, C Eugene
Atwell, William Alan
Chandan, Ramesh Chandra
Fritsch, Carl Walter
Hamre, Melvin L
Lonergan, Dennis Arthur
Oelberg, Thomas Jonathon
Pour-El, Akiva
Seeley, Robert
Sullivan, Betty J
Thenen, Shirley Warnock

MISSISSIPPI
Rogers, Robert Wayne
Wang, Ping-Lieh Thomas

MISSOURI
Gnaedinger, Richard H
Gordon, Dennis T
Hedrick, Harold Burdette
Lambeth, Victor Neal

Marshall, Robert T
Meyer, William Ellis
Palamand, Suryanarayana Rao
Prewitt, Larry R
Scallet, Barrett Lerner
Sidoti, Daniel Robert

MONTANA
Richards, Geoffrey Norman

NEBRASKA
Bullerman, Lloyd Bernard
Taylor, Steve L
Wallen, Stanley Eugene

NEW JERSEY
Adler, Irwin L
Busk, Grant Curtis, Jr
Chiang, Bin-Yea
Giddings, George Gosselin
Ho, Chi-Tang
Hsu, Kenneth Hsuehchia
Hu, Ching-Yeh
Klemann, Lawrence Paul
Kleyn, Dick Henry
Smith, Robert Ewing
Tibbetts, Merrick Sawyer
Trivedi, Nayan B

NEW MEXICO
Del Valle, Francisco Rafael

NEW YORK
Aries, Robert Sancier
Bourne, Malcolm Cornelius
Durst, Richard Allen
Efthymiou, Constantine John
Gravani, Robert Bernard
Grimm, Charles Henry
Kahan, Sidney
Liu, Fu-Wen (Frank)
Moline, Sheldon Walter
Plane, Robert Allen
Shaevel, Morton Leonard
Sherbon, John Walter
Shuler, Michael Louis
Siebert, Karl Joseph
Todaro, Michael P
Tzeng, Chu

NORTH CAROLINA
Fleming, Henry Pridgen
Guld, Harvey Joseph
Humphries, Ervin G(rigg)
Johnson, George Andrew
Lineback, David R
Perfetti, Patricia F
Reed, Gerald
Webb, Neil Broyles

NORTH DAKOTA
Brown, Guendoline

OHIO
Baur, Fredric John, Jr
Brooks, James Reed
Darragh, Richard T
Flautt, Thomas Joseph, Jr
Herum, Floyd L(yle)
Irmiter, Theodore Ferer
Latshaw, J David
Lee, Ken
Lindamood, John Benford
Litchfield, John Hyland
Martin, James Harold
Peng, Andrew Chung Yen
Sabharwal, Kulbir
Wilcox, Joseph Clifford
Williams, Robert Mack

OKLAHOMA
Gilliland, Stanley Eugene
Lu, Christopher D
Olson, Harold Cecil
Seideman, Walter E
Thompson, David Russell

OREGON
Clark, James Orie, II
Holmes, Zoe Ann
Wyatt, Carolyn Jane

PENNSYLVANIA
Benedict, Robert Curtis
Birdsall, Marion Ivens
Buchanan, Robert Lester
Fox, Jay B, Jr
Gadek, Frank Joseph
Mast, Morris G
Roseman, Arnold S(aul)
Rothbart, Herbert Lawrence
Somkuti, George A
Thompson, Donald B
Wagner, J Robert
Wolff, Ivan A

RHODE ISLAND
Cosgrove, Clifford James
Lee, Tung-Ching

SOUTH CAROLINA
Godfrey, W Lynn
Morr, Charles Vernon
Richards, Gary Paul

Surak, John Godfrey

SOUTH DAKOTA
Schingoethe, David John

TENNESSEE
Davidson, Philip Michael
Penfield, Marjorie Porter

TEXAS
Brittin, Dorothy Helen Clark
Butler, Ogbourne Duke, Jr
Chen, Anthony Hing
Engler, Cady Roy
Gardner, Frederick Albert
Greene, Donald Miller
Phillips, Guy Frank
Ramsey, Clovis Boyd
Smith, Malcolm Crawford, Jr
Will, Paul Arthur
Yeh, Lee-Chuan Caroline

UTAH
Hendricks, Deloy G
Johnson, John Hal
Lee, Steve S

VERMONT
Atherton, Henry Vernon
Bartel, Lavon L

VIRGINIA
Byrd, Daniel Madison, III
Fellers, David Anthony
LaBudde, Robert Arthur
Marcy, Joseph Edwin
Pierson, Merle Dean
Rippen, Thomas Edward
Stern, Joseph Aaron
Stewart, Kent Kallam

WASHINGTON
Ewart, Hugh Wallace, Jr
Padilla, Andrew, Jr
Pigott, George M
Powers, Joseph Robert
Rasco, Barbara A
Stansby, Maurice Earl
Widmaier, Robert George

WISCONSIN
Greger, Janet L
Kostenbader, Kenneth David, Jr
Kueper, Theodore Vincent
Lindsay, Robert Clarence
Matthews, Mary Eileen
Ogden, Robert Verl
Price, Walter Van Vranken
Scheusner, Dale Lee
Sebree, Bruce Randall

ALBERTA
Koopmans, Henry Sjoerd
Schroder, David John
Stiles, Michael Edgecombe
Whitehouse, Ronald Leslie S

BRITISH COLUMBIA
Liao, Ping-huang
Li-Chan, Eunice
Reed, William J

MANITOBA
Iverson, Stuart Leroy
Jayas, Digvir Singh
Kovacs, Miklos I P
Szekely, Joseph George

NEWFOUNDLAND
Shahidi, Fereidoon

ONTARIO
Chambers, James Robert
Chu, Chun-Lung (George)
Collins, Frank William, Jr
Diosady, Levente Laszlo
Fortin, Andre Francis
Gibson, Rosalind Susan
Lips, Hilaire John
Pigden, Wallace James
Ratnayake, Walisundera Mudiyanselage Nimal
Rubin, Leon Julius
Sarwar, Ghulam
Thompson, Lilian Umale
Usborne, William Ronald
Yeung, David Lawrence

QUEBEC
Bélanger, Jacqueline M R
Britten, Michel
Doyon, Gilles Joseph
Gariepy, Claude
Lamarche, François
Picard, Gaston Arthur
Sood, Vijay Kumar

SASKATCHEWAN
Ingham, Steven Charles
Ingledew, William Michael
Knipfel, Jerry Earl
Sosulski, Frank Walter

OTHER COUNTRIES
Filadelfi-Keszi, Mary Ann Stephanie
Mela, David Jason
Santidrian, Santiago

Forestry

ALABAMA
Beals, Harold Oliver
Bengtson, George Wesley
Boyer, William Davis
Davis, Terry Chaffin
DeBrunner, Louis Earl
Gilmore, Alvan Ray
Goggans, James F
Gulden, Michael Stanley
Hodgkins, Earl Joseph
Holt, William Robert
Hool, James N
Huntley, Jimmy Charles
Johnson, Evert William
Larsen, Harry Stites
Livingston, Knox W
Lyle, Everett Samuel, Jr
Tang, Ruen Chiu
Thompson, Emmett Frank

ALASKA
Hanley, Thomas Andrew
Juday, Glenn Patrick
Larson, Frederic Roger
Rogers, James Joseph
Slaughter, Charles Wesley
Smoker, William Alexander
Werner, Richard Allen

ARIZONA
Ahlgren, Clifford Elmer
Avery, Charles Carrington
Baker, Malchus Brooks, Jr
Boggess, William Randolph
Davis, Edwin Alden
Gilbertson, Robert Lee
Klemmedson, James Otto
Knorr, Philip Noel
Kurmes, Ernest A
Landrum, Leslie Roger
Lord, William B
McPherson, Guy Randall
Minor, Charles Oscar
Ostrom, Carl Eric
Patton, David Roger
Robinson, Dan D
Robinson, William James
Ronco, Frank, Jr
Wagle, Robert Fay
Zwolinski, Malcolm John

ARKANSAS
Baker, James Bert
Bower, David Roy Eugene
Chacko, Rosy J
Grizzell, Roy Ames, Jr
Ku, Timothy Tao
Shade, Elwood B
Strub, Mike Robert
Yeiser, Jimmie Lynn

CALIFORNIA
Barbour, Michael G
Beall, Frank Carroll
Bensch, Klaus George
Benseler, Rolf Wilhelm
Bohm, Howard A
Brooks, William Hamilton
Callaham, Robert Zina
Cobb, Fields White, Jr
Cockrell, Robert Alexander
Colwell, Robert Neil
Cooper, Charles F
Critchfield, William Burke
Davis, Lawrence S
Eriksen, Clyde Hedman
Fairfax, Sally Kirk
Freeland, Forrest Dean, Jr
Fridley, Robert B
Goltz, Robert W
Helms, John Andrew
Houpis, James Louis Joseph
Jansen, Henricus Cornelis
Kelley, Allen Frederick, Jr
Kilgore, Bruce Moody
Kliejunas, John Thomas
Koonce, Andrea Lavender
Kope, Robert Glenn
Kozlowski, Theodore Thomas
Ledig, F Thomas
Levy, Barnet M
Libby, William John, (Jr)
McDonald, Philip Michael
McKillop, William L M
Murphy, Alfred Henry
Paine, Lee Alfred
Partain, Gerald Lavern
Pitelka, Louis Frank
Powers, Robert Field
Roberts, Stephen Winston
Schniewind, Arno Peter
Storey, Theodore George
Teeguarden, Dennis Earl
Thomas, Walter Dill, Jr
Thornburgh, Dale A
Ulevitch, Richard Joel

Van Wagtendonk, Jan Willem
Vaux, Henry James
Voeks, Robert Allen
Volk, Thomas Lewis
Weatherspoon, Charles Phillip
Welch, Robin Ivor
Wilcox, W Wayne
Williams, Carroll Burns, Jr
Wilson, Carl C
Young, Donald Raymond
Ziemer, Robert Ruhl
Zinke, Paul Joseph

COLORADO
Barney, Charles Wesley
Dana, Robert Watson
Davidson, Ross Wallace, Jr
Fechner, Gilbert Henry
Fox, Douglas Gary
Garstka, Walter U(rban)
Hawksworth, Frank Goode
Hoffer, Roger M(ilton)
Kaufmann, Merrill R
Kovner, Jacob L
Martinelli, Mario, Jr
Mogren, Edwin Walfred
Shaw, Charles Gardner, III
Shepperd, Wayne Delbert
Shuler, Craig Edward
Striffler, William D
Tabler, Ronald Dwight
Tsuchiya, Takumi
Wangaard, Frederick Field

CONNECTICUT
Berlyn, Graeme Pierce
Damman, Antoni Willem Hermanus
Furnival, George Mason
Gordon, John C
Gould, Ernest Morton, Jr
Houston, David Royce
John, Hugo Herman
Mergen, Francois
Smith, David Martyn
Stephens, George Robert
Voigt, Garth Kenneth
Wargo, Philip Matthew
Winer, Herbert Isaac

DELAWARE
Kundt, John Fred

DISTRICT OF COLUMBIA
Beuter, John H
Brandt, Robert William
Burns, Russell MacBain
Callahan, James Thomas
Challinor, David
Lyon, Robert Lyndon
Myers, Wayne Lawrence
Post, Boyd Wallace
Roskoski, Joann Pearl
Ross, Eldon Wayne
Smith, Richard S, Jr

FLORIDA
Blakeslee, George M
Bruns, Paul Eric
Fatzinger, Carl Warren
Fisher, Jack Bernard
Goddard, Ray Everett
Harper, Verne Lester
Hetrick, Lawrence Andrew
Hsu, Tsong-Han
Huffman, Jacob Brainard
Janos, David Paul
Kaufman, Clemens Marcus
Long, Alan Jack
Mace, Arnett C, Jr
Marion, Wayne Richard
Reid, Charles Phillip Patrick
Smith, Wayne H
Stroh, Robert Carl
Sullivan, Edward T
Swinford, Kenneth Roberts
White, Timothy Lee

GEORGIA
Bongarten, Bruce C
Broerman, F S
Campbell, William Andrew
Chen, Dillion Tong-ting
Clark, James Samuel
Dinus, Ronald John
Hargreaves, Leon Abraham, Jr
Ike, Albert Francis
Johnson, Albert Sydney, III
Kemp, Arne K
Kraus, John Franklyn
Nord, John C
Pepper, William Donald
Phelps, William Robert
Pienaar, Leon Visser
Skeen, James Norman
Sluder, Earl Ray
Sommer, Harry Edward
Steinbeck, Klaus
Tillman, Larry Jaubert
Ware, Kenneth Dale
Woessner, Ronald Arthur
Wood, Bruce Wade

HAWAII
Baker, Harold Lawrence

Brewbaker, James Lynn
Hamilton, Lawrence Stanley
Merriam, Robert Arnold
Skolmen, Roger Godfrey

IDAHO
Ehrenreich, John Helmuth
Fins, Lauren
Hendee, John Clare
Loewenstein, Howard
MacFarland, Craig George
Megahan, Walter Franklin
Ulliman, Joseph James

ILLINOIS
Ashby, William Clark
Aubertin, Gerald Martin
Brown, Sandra
Fierer, Joshua A
Guiher, John Kenneth
Holland, Israel Irving
Kirby, Hilliard Walker
Lorenz, Ralph William
Myers, Charles Christopher
Rink, George
Van Sambeek, Jerome William
Walters, Charles Sebastian
Williams, David James

INDIANA
Beers, Thomas Wesley
Beineke, Walter Frank
Bramble, William Clark
Byrnes, William Richard
Cassens, Daniel Lee
Chaney, William R
Holt, Harvey Allen
Hunt, Michael O'Leary
Kidd, Frank Alan
Knudson, Douglas Marvin
Merritt, Clair
Moser, John William, Jr
Parker, George Ralph
Riffle, Jerry William
Senft, John Franklin
Suddarth, Stanley Kendrick

IOWA
Countryman, David Wayne
Hall, Richard Brian
Hart, Elwood Roy
Hopkins, Frederick Sherman, Jr
Mack, Michael J
McNabb, Harold Sanderson, Jr
Manwiller, Floyd George
Prestemon, Dean R
Thomson, George Willis

KANSAS
Howe, Virgil K
Van Haverbeke, David F

KENTUCKY
Beatty, Oren Alexander
Coltharp, George B
Olson, James Robert
Rothwell, Frederick Mirvan

LOUISIANA
Adams, John Clyde
Allen, Arthur (Silsby)
Baldwin, Virgil Clark, Jr
Barnett, James P
Bryant, Robert L
Burns, Paul Yoder
Carpenter, Stanley Barton
Carter, Mason Carlton
Choong, Elvin T
Dell, Tommy Ray
Hough, Walter Andrew
Jewell, Frederick Forbes, Sr
Linnartz, Norwin Eugene
Lorio, Peter Leonce, Jr
Oliver, Abe D, Jr
Siegel, William Carl
Teate, James Lamar

MAINE
Ashley, Marshall Douglas
Blum, Barton Morrill
Brown, Gregory Neil
Campana, Richard John
Frank, Robert McKinley, Jr
Griffin, Ralph Hawkins
Knight, Fred Barrows
Nutting, Albert Deane
Young, Harold Edle

MARYLAND
Adams, Lowell William
Brush, Grace Somers
Edmonds, Robert L
Genys, John B
Knox, Robert Gaylord
Martin, John A
Row, Clark

MASSACHUSETTS
Abbott, Herschel George
Arganbright, Donald G
Art, Henry Warren
Ashton, Peter Shaw
Bolotin, Moshe
Bond, Robert Sumner

Forestry (cont)

Foster, Charles Henry Wheelwright
Larson, Joseph Stanley
MacConnell, William Preston
MacDougall, Edward Bruce
McKenzie, Malcolm Arthur
Mader, Donald Lewis
Page, Alan Cameron
Wilson, Brayton F

MICHIGAN
Barnes, Burton Verne
Belcher, Robert Orange
Bourdo, Eric A, Jr
Carow, John
Chappelle, Daniel Eugene
Crowther, C Richard
Fogel, Robert Dale
Frayer, Warren Edward
Freeman, Fred W
Hanover, James W
Hart, John Henderson
Hesterberg, Gene Arthur
James, Lee Morton
Karnosky, David Frank
Kenaga, Duane Leroy
Koelling, Melvin R
Lutz, Harold John
Marty, Robert Joseph
Meteer, James William
Miller, Roswell Kenfield
Preston, Stephen Boylan
Ruby, John L
Simard, Albert Joseph
Simmons, Gary Adair
Sliker, Alan
Sloan, Norman F
Sun, Bernard Ching-Huey
White, Donald Perry
Witter, John Allen

MINNESOTA
Alm, Alvin Arthur
Bakuzis, Egolfs Voldemars
Blanchette, Robert Anthony
Boelter, Don Howard
Brown, Bruce Antone
Buech, Richard Reed
Ek, Alan Ryan
Erickson, Robert W
French, David W
Gertjejansen, Roland O
Gregersen, Hans Miller
Grigal, David F
Hallgren, Alvin Roland
Hann, Robert A
Heinselman, Miron L
Hendricks, Lewis T
Irving, Frank Dunham
Jokela, Jalmer John
Kaufert, Frank Henry
Leary, Rolfe Albert
Miller, William Eldon
Rudolf, Paul Otto
Skok, Richard Arnold
Sucoff, Edward Ira
Zasada, Zigmond Anthony

MISSISSIPPI
Amburgey, Terry L
Elam, William Warren
Foil, Robert Rodney
Hodges, John Deavours
Lyon, Duane Edgar
McCracken, Francis Irvin
Matthes, Ralph Kenneth, Jr
Moak, James Emanuel
Solomon, James Doyle
Switzer, George Lester
Thompson, Warren Slater

MISSOURI
Brown, Merton F
Duncan, Donald Pendleton
Everson, Alan Ray
Kurtz, William Boyce
Pallardy, Stephen Gerard
Sander, Ivan Lee
Schlesinger, Richard Cary
Settergren, Carl David
Smith, Richard Chandler

MONTANA
Arno, Stephen Francis
Bay, Roger Rudolph
Blake, George Marston
Brown, James Kerr
Carlson, Clinton E
Lotan, James
Rounds, Burton Ward
Schmidt, Wyman Carl
Shearer, Raymond Charles
Stark, Nellie May
Steele, Robert Wilbur

NEBRASKA
Arnold, R Keith
Hergenrader, Gary Lee

NEVADA
Fox, Carl Alan
Meeuwig, Richard O'Bannon
Taylor, George Evans, Jr

NEW HAMPSHIRE
Baldwin, Henry Ives
Barrett, James Passmore
Federer, C Anthony
Funk, David Truman
Garrett, Peter Wayne
Safford, Lawrence Oliver
Wilson, Donald Alfred
Winch, Fred Everett, Jr

NEW JERSEY
Green, Edwin James
Hubbell, Stephen Philip
Koepp, Stephen John

NEW MEXICO
Aldon, Earl F
Gosz, James Roman
Grover, Herbert David
Heiner, Terry Charles
Hughes, Jay Melvin
Reifsnyder, William Edward

NEW YORK
Amidon, Thomas Edward
Amundson, Robert Gale
Blackmon, Bobby Glenn
Cohen, Martin William
Cote, Wilfred Arthur, Jr
Cunia, Tiberius
Eschner, Arthur Richard
Gladstone, William Turnbull
Greller, Andrew M
Herrington, Lee Pierce
Horn, Allen Frederick, Jr
Kohut, Robert John
Kraemer, J Hugo
Larson, Charles Conrad
Laurence, John A
Lee, Charles Northam
Leopold, Donald Joseph
Lowe, Josiah L(incoln)
McDonnell, Mark Jeffery
McNeil, Richard Jerome
Maynard, Charles Alvin
Meyer, Robert Walter
Palmer, James F
Resch, Helmuth
Shone, Robert Tilden
Silverborg, Savel Benhard
Stiteler, William Merle, III
Tonelli, John P, Jr
Weinstein, David Alan
Whaley, Ross Samuel
White, Edwin Henry

NORTH CAROLINA
Barden, Lawrence Samuel
Barefoot, Aldos Cortez, Jr
Booth, Donald Campbell
Brotak, Edward Allen
Bruck, Robert Ian
Bryant, Ralph Clement
Cooper, Arthur Wells
Cowling, Ellis Brevier
Davey, Charles Bingham
Dickerson, Willard Addison
Douglass, James Edward
Ellwood, Eric Louis
Eslyn, Wallace Eugene
Hafley, William LeRoy
Harper, James Douglas
Hart, Clarence Arthur
Haskill, John Stephen
Hastings, Felton Leo
Hellmers, Henry
Hepting, George Henry
Kellison, Robert Clay
Knoerr, Kenneth Richard
Kress, Lance Whitaker
Levi, Michael Phillip
Lewis, Gordon Depew
Loomis, Robert Morgan
Mott, Ralph Lionel
Namkoong, Gene
Peet, Robert Krug
Perry, Thomas Oliver
Richardson, Curtis John
Saylor, LeRoy C
Stambaugh, William James
Steensen, Donald H J
Stillwell, Harold Daniel
Sullivan, Arthur Lyon
Swank, Wayne T
Swift, Lloyd Wesley, Jr
Tombaugh, Larry William
Towell, William Earnest
Zobel, Bruce John

NORTH DAKOTA
Lieberman, Diana Dale

OHIO
Boerner, Ralph E J
Brown, James Harold
Cobbe, Thomas James
Cullis, Christopher Ashley
Garland, Hereford
Gatherum, Gordon Elwood
Gorchov, David Louis
Jensen, Keith Frank
Kriebel, Howard Burtt
Larson, Merlyn Milfred
Runkle, James Reade

Snyder, Gary Wayne
Sydnor, Thomas Davis
Vimmerstedt, John P
Whitmore, Frank William

OKLAHOMA
Eikenbary, Raymond Darrell
Langwig, John Edward
Lewis, David Kent
Sturgeon, Edward Earl
Walker, Nathaniel
Wittwer, Robert Frederick

OREGON
Adams, Thomas C
Bell, John Frederick
Briegleb, Philip Anthes
Brown, George Wallace
Dealy, John Edward
Demaree, Thomas L
Ethington, Robert Loren
Ferrell, William Kreiter
Hann, David William
Hansen, Everett Mathew
Heninger, Ronald Lee
Kelsey, Rick Guy
Krahmer, Robert Lee
Landsberg, Johanna D (Joan)
McKimmy, Milford D
Maser, Chris
Mason, Richard Randolph
Merriam, Lawrence Campbell, Jr
Minore, Don
Molina, Randolph John
Myrold, David Douglas
Nelson, Earl Edward
Newton, Michael
Norris, Logan Allen
Owston, Peyton Wood
Paine, David Philip
Ripple, William John
Roth, Lewis Franklin
Rottink, Bruce Allan
Ryan, Roger Baker
Scheffer, Theodore Comstock
Schmidt, Clifford LeRoy
Skovlin, Jon Matthew
Stein, William Ivo
Stoltenberg, Carl H
Stout, Benjamin Boreman
Sutherland, Charles F
Tappeiner, John Cummings, II
Thielges, Bart A
Trappe, James Martin
Walstad, John Daniel
Wheeler, Richard Hunting
Winjum, Jack Keith
Zaerr, Joe Benjamin

PENNSYLVANIA
Alexander, Stuart David
Blankenhorn, Paul Richard
Bowersox, Todd Wilson
Burns, Denver P
Cameron, Edward Alan
Gerhold, Henry Dietrich
Gray, John Lewis
Hutnik, Russell James
Jayne, Benjamin A
Klein, Edward Lawrence
Labosky, Peter, Jr
McDermott, Robert Emmet
Marquis, David Alan
Melton, Rex Eugene
Merrill, William
Shipman, Robert Dean
Sopper, William Edward
Steiner, Kim Carlyle
Sullivan, Alfred Dewitt
Towers, Barry

RHODE ISLAND
Brown, James Henry, Jr
Gould, Walter Phillip

SOUTH CAROLINA
Adams, Daniel Otis
Allen, Robert Max
Cool, Bingham Mercur
Crosby, Emory Spear
Dunn, Benjamin Allen
Haines, Harry Caum
Hook, Donal D
McGregor, William Henry Davis
Nelson, Eric Alan
Schoenike, Roland Ernest
Shain, William Arthur
Taras, Michael Andrew
Van Lear, David Hyde

SOUTH DAKOTA
Boldt, Charles Eugene
Collins, Paul Everett
Draeger, William Charles

TENNESSEE
Barnett, Paul Edward
Cheston, Charles Edward
Core, Harold Addison
Dale, Virginia House
Kuo, Chung-Ming
Luxmoore, Robert John
Mignery, Arnold Louis
Miller, Neil Austin

Patric, James Holton
Ramke, Thomas Franklin
Schneider, Gary
Sharp, John Buckner
Taft, Kingsley Arter, Jr
Thor, Eyvind
Weaver, George Thomas
Wells, Garland Ray
Woods, Frank Wilson

TEXAS
Adair, Kent Thomas
Baker, Robert Donald
Bilan, M Victor
Chang, Mingteh
Dickson, James Gary
Ervin, Hollis Edward
Lowry, Gerald Lafayette
McGrath, William Thomas
Merrifield, Robert G
Shreve, Loy William
Spurr, Stephen Hopkins
Stransky, John Janos
Thill, Ronald E
Van Buijtenen, Johannes Petrus
Venketeswaran, S
Walker, Laurence Colton
Watterston, Kenneth Gordon
Zagata, Michael DeForest

UTAH
Bollinger, William Hugh
Croft, Alfred Russell
DeByle, Norbert V
Downing, Kenton Benson
Fisher, Richard Forrest
Hammond, Mary Elizabeth Hale
Hart, George Emerson, Jr
Krebill, Richard G
Lanner, Ronald Martin
Lassen, Laurence E

VERMONT
Echelberger, Herbert Eugene
Forcier, Lawrence Kenneth
Freeman, Jeffrey VanDuyne
Gregory, Robert Aaron
Tyree, Melvin Thomas
Westing, Arthur H
Whitmore, Roy Alvin, Jr
Wilkinson, Ronald Craig

VIRGINIA
Barber, John Clark
Burkhart, Harold Eugene
Glasser, Wolfgang Gerhard
Hall, Otis F
Hosner, John Frank
Ifju, Geza
Kreh, Richard Edward
Krugman, Stanley Liebert
Little, Elbert Luther, Jr
McElwee, Robert L
McGinnes, Burd Sheldon
Morison, Ian George
Nelson, Thomas Charles
Smith, David William
Sullivan, John Dennis
Youngs, Robert Leland

WASHINGTON
Adams, Darius Mainard
Agee, James Kent
Bare, Barry Bruce
Bethel, James Samuel
Bethlahmy, Nedavia
Boyd, Charles Curtis
Brown, John Henry
Chapman, Roger Charles
Curtis, Robert Orin
Daniels, Jess Donald
DeBell, Dean Shaffer
Dingle, Richard William
Driver, Charles Henry
Fahnestock, George Reeder
Farnum, Peter
Franklin, Jerry Forest
Gaines, Edward M(cCulloch)
Gara, Robert I
Gessel, Stanley Paul
Hatheway, William Howell
Heebner, Charles Frederick
Heilman, Paul E
Johnson, Norman Elden
Kimmey, James William
Kutscha, Norman Paul
McCarthy, Joseph L(ePage)
Maloney, Thomas M
Megraw, Robert Arthur
Noskowiak, Arthur Fredrick
Pierce, William R
Ritchie, Gary Alan
Rustagi, Krishna Prasad
Satterlund, Donald Robert
Schreuder, Gerard Fritz
Scott, David Robert Main
Sharpe, Grant William
Shirley, Frank Connard
Steinbrenner, Eugene Clarence
Stettler, Reinhard Friederich
Stonecypher, Roy W
Tanaka, Yasuomi
Waggener, Thomas Runyan

WEST VIRGINIA
Biggs, Alan Richard
Carvell, Kenneth Llewellyn
Cech, Franklin Charles
Crist, John Benjamin
Gillespie, William Harry
Koch, Christian Burdick
Peters, Penn A
Smith, Walton Ramsay
Stephenson, Steven Lee
White, David Evans
Wiant, Harry Vernon, Jr
Zinn, Gary William

WISCONSIN
Buongiorno, Joseph
Cunningham, Gordon Rowe
Dobbins, Thomas Edward
Einspahr, Dean William
Engelhard, Robert J
Fleischer, Herbert Oswald
Guries, Raymond Paul
Highley, Terry L
Hillier, Richard David
Johnson, Morris Alfred
Kubler, Hans
Larsen, Michael John
Larson, Philip Rodney
Lee, Chen Hui
Nelson, Neil Douglas
Patton, Robert Franklin
Quirk, John Thomas
Rietveld, Willis James
Splitter, Gary Allen
Stearns, Forest
Vick, Charles Booker

WYOMING
DePuit, Edward J

PUERTO RICO
Wadsworth, Frank H

ALBERTA
Bella, Imre E
Dancik, Bruce Paul
Pluth, Donald John
Singh, Teja
Swanson, Robert Harold
Wong, Horne Richard
Yeh, Francis Cho-hao

BRITISH COLUMBIA
Annas, Richard Morris
Banister, Eric Wilton
Borden, John Harvey
Chase, William Henry
Cottell, Philip L
Dobbs, Robert Curry
Drew, T John
Ekramoddoullah, Abul Kalam M
Etheridge, David Elliott
Gardner, Joseph Arthur Frederick
Golding, Douglas Lawrence
Haddock, Philip George
Hatton, John Victor
Holl, Frederick Brian
Hunt, Richard Stanley
Illingworth, Keith
Kennedy, Robert William
Kimmins, James Peter (Hamish)
Kozak, Antal
Lester, Donald Thomas
McLean, John Alexander
Manville, John Fieve
Meagher, Michael Desmond
Mitchell, Kenneth John
Owens, John N
Pollard, Douglas Frederick William
Prasad, Raghubir (Raj)
Promnitz, Lawrence Charles
Reed, William J
Routledge, Richard Donovan
Shamoun, Simon Francis
Smith, Richard Barrie
Thirgood, Jack Vincent
Trussell, Paul Chandos
Weetman, Gordon Frederick
Worrall, John Gatland

NEW BRUNSWICK
Anderson, John Murray
Baskerville, Gordon Lawson
Bonga, Jan Max
Eidt, Douglas Conrad
Eveleigh, Eldon Spencer
Fowler, Donald Paige
Ker, John William
MacLean, David Andrew
Magasi, Laszlo P
Powell, Graham Reginald
Van Groenewoud, Herman

NEWFOUNDLAND
Bajzak, Denes

NOVA SCOTIA
Cooper, John (Hanwell)

ONTARIO
Armson, Kenneth Avery
Basham, Jack T
Buckner, Charles Henry
Burger, Dionys

Carmean, Willard Handy
Davidson, Alexander Grant
Duncan, John Robert
Farmer, Robert E, Jr
Farrar, John Laird
Foster, Neil William
Ginns, James Herbert
Jeglum, John Karl
Jorgensen, Erik
Kayll, Albert James
Kondo, Edward Shinichi
Love, David Vaughan
Morrison, Ian Kenneth
Newnham, Robert Montague
Nordin, Vidar John
Redmond, Douglas Rollen
Rice, Peter (Franklin)
Setliff, Edson Carmack
Shoemaker, Robert Alan
Van Wagner, Charles Edward
Whitney, Roy Davidson
Yeatman, Christopher William

QUEBEC
Bigras, Francine Jeanne
Bourchier, Robert James
Brassard, Andre
Corriveau, Armand Gerard
Furlan, Valentin
Girouard, Ronald Maurice
Grandtner, Miroslav Marian
Hardy, Yvan J
Koran, Zoltan
Lachance, Denis
Laflamme, Gaston
Lafond, Andre
Lavallee, Andre
O'Neil, Louis C
Richard, Claude
Winget, Carl Henry

OTHER COUNTRIES
Budowski, Gerardo
Hocking, Drake
Justines, Gustavo
Nienstaedt, Hans

Horticulture

ALABAMA
Amling, Harry James
Ammerman, Gale Richard
Biswas, Prosanto K
Chambliss, Oyette Lavaughn
Cochis, Thomas
Gilliam, Charles Homer
Greenleaf, Walter Helmuth
Marvel, Mason E
Norton, Joseph Daniel
Orr, Henry Porter
Perkins, Donald Young
Reynolds, Charles William
Sanderson, Kenneth Chapman
Sapra, Val T
Sayers, Earl Roger
Ward, Coleman Younger
Whatley, Booker Tillman

ARIZONA
Alcorn, Stanley Marcus
Bessey, Paul Mack
Crosswhite, Carol D
Crosswhite, Frank Samuel
Fazio, Steve
Foster, Robert Edward, II
Gartner, John Bernard
Hitz, Chester W
Johnson, Herbert Windal
Kneebone, William Robert
Muramoto, Hiroshi
Oekber, Norman Fred
Paris, Clark Davis
Rodney, David Ross
Schonhorst, Melvin Herman
Sharples, George Carroll
Thompson, Anson Ellis
Tucker, Thomas Curtis
Voigt, Robert Lee
Wright, Daniel Craig

ARKANSAS
Bradley, George Alexander
Einert, Alfred Erwin
Kattan, Ahmed A
Moore, James Norman
Morris, Justin Roy
Motes, Dennis Roy
Pitts, Donald James
Reed, Hazell
Rom, Roy Curt
York, John Owen

CALIFORNIA
Alley, Curtis J
Amerine, Maynard Andrew
Angell, Frederick Franklyn
Baldy, Marian Wendorf
Baldy, Richard Wallace
Barker, LeRoy N
Bergh, Berthold Orphie
Beyer, Edgar Herman
Bitters, Willard Paul
Bowerman, Earl Harry

Brewer, Robert Franklin
Bringhurst, Royce S
Catlin, Peter Bostwick
Claypool, Lawrence Leonard
Cook, James Arthur
Crane, Julian Coburn
Davis, Elmo Warren
Dawson, Kerry J
Dempsey, Wesley Hugh
Einset, John William
Ellstrom, Norman Carl
Embleton, Tom William
Epstein, Emanuel
Foster, William J
Fridley, Robert B
Furuta, Tokuji
Gibeault, Victor Andrew
Griggs, William Holland
Hagan, William Leonard
Hanson, George Peter
Harding, James A
Harrington, James Foster
Harris, Richard Wilson
Hartmann, Hudson Thomas
Hess, Charles
Hooks, James A
Houck, Laurie Gerald
Humaydan, Hasib Shaheen
Johannessen, George Andrew
Jones, Winston William
Kader, Adel Abdel
Kester, Dale Emmert
Koch, Gary Marlin
Kofranek, Anton Miles
Kohl, Harry Charles, Jr
Lewellen, Robert Thomas
Lilleland, Omund
Lipton, Werner Jacob
Little, Thomas Morton
Lorenz, Oscar Anthony
Lundeen, Glen Alfred
MacDonald, James Douglas
McNall, Lester R
Madison, John Herbert, Jr
Madore, Monica Agnes
Martin, George C
Matkin, Oris Arthur
Morris, Leonard Leslie
Munns, Donald Neville
Muth, Gilbert Jerome
Nichols, Courtland Geoffrey
Ornduff, Robert
Pollack, Bernard Leonard
Pollard, James Edward
Pratt, Harlan Kelley
Prend, Joseph
Puri, Yesh Paul
Rains, Donald W
Ramming, David Wilbur
Rappaport, Lawrence
Rawal, Kanti M
Reuther, Walter
Rick, Charles Madeira, Jr
Robinson, Frank Ernest
Romney, Evan M
Rubatzky, Vincent E
Ryder, Edward Jonas
Ryugo, Kay
Sacher, Robert Francis
Saltveit, Mikal Endre, Jr
Sanders, John Stephen
Simpson, John Ernest
Soost, Robert Kenneth
Stoner, Martin Franklin
Taylor, Oliver Clifton
Thomas, Paul Clarence
Thomas, Walter Dill, Jr
Thompson, David J
Thomson, John Ansel Armstrong
Tisserat, Brent Howard
Uriu, Kiyoto
Wallace, Arthur
Watson, Donald Pickett
Webb, Albert Dinsmoor
White, Thomas Gailand
Wilkins, Harold
Wyatt, Colen Charles
Yamaguchi, Masatoshi
Zary, Keith Wilfred
Zink, Frank W

COLORADO
Alford, Donald Kay
Basham, Charles W
Brink, Kenneth Maurice
DeMooy, Cornelis Jacobus
Ells, James E
Feucht, James Roger
Hanan, Joe John
Havis, John Ralph
Holm, David Garth
Klett, James Elmer
Knutson, Kenneth Wayne
Moore, Frank Devitt, III
Nabors, Murray Wayne
Stushnoff, Cecil
Townsend, Charley E
Yun, Young Mok

CONNECTICUT
Ahrens, John Frederick
Ashley, Richard Allan
Carpenter, Edwin David
Jaynes, Richard Andrus

Kennard, William Crawford
Koontz, Harold Vivien
Koths, Jay Sanford
Poincelot, Raymond Paul, Jr
Stephens, George Robert
Waxman, Sidney

DELAWARE
Dunham, Charles W
Fieldhouse, Donald John
Green, Jerome
Long, James Delbert
Riggleman, James Dale
Smith, Constance Meta

DISTRICT OF COLUMBIA
Burton, David Lee
Dudley, Theodore R
Eisa, Hamdy Mahmoud
Marlowe, George Albert, Jr
Tompkins, Daniel Reuben
Wiggans, Samuel Claude

FLORIDA
Albrigo, Leo Gene
Bassett, Mark Julian
Brolmann, John Bernardus
Bryan, Herbert Harris
Bullard, Ervin Trowbridge
Burch, Derek George
Burdine, Howard William
Campbell, Carl Walter
Cantliffe, Daniel James
Carpenter, William John
Chandler, Robert Flint, Jr
Childers, Norman Franklin
Childs, William Henry
Conover, Charles Albert
Crall, James Monroe
Davis, Ralph Lanier
Dean, Charles Edgar
Denmark, Harold Anderson
Drinkwater, William Otho
Dusky, Joan Agatha
Elmstrom, Gary William
Emino, Everett Raymond
Fisher, Jack Bernard
Fitzpatrick, George
Gammon, Nathan, Jr
Gerber, John Francis
Gilreath, James Preston
Gorbet, Daniel Wayne
Gray, Dennis John
Guzman, Victor Lionel
Hall, Chesley Barker
Haramaki, Chiko
Harbaugh, Brent Kalen
Hardenburg, Robert Earle
Hatton, Thurman Timbrook, Jr
Hearn, Charles Jackson
Huber, Donald John
Joiner, Jasper Newton
Kender, Walter John
Koo, Robert Chung Jen
Locascio, Salvadore J
Lyrene, Paul Magnus
McConnell, Dennis Brooks
McElwee, Edgar Warren
Marousky, Francis John
Martsolf, J David
Matthews, David Livingston
Montelaro, James
Mortensen, John Alan
Mustard, Margaret Jean
Orth, Paul Gerhardt
Ozaki, Henry Yoshio
Parsons, Lawrence Reed
Phillips, Richard Lee
Pieringer, Arthur Paul
Poole, Richard Turk, Jr
Popenoe, John
Quesenberry, Kenneth Hays
Reitz, Herman J
Rhodes, Ashby Marshall
Robitaille, Henry Arthur
Scudder, Walter Tredwell
Sheehan, Thomas John
Sherman, Wayne Bush
Showalter, Robert Kenneth
Soule, James
Stall, William Martin
Stern, William Louis
Stoffella, Peter Joseph
Tai, Peter Yao-Po
Thompson, Buford Dale
Ting, Sik Vung
Tucker, David Patrick Hislop
Verkade, Stephen Dunning
Wardowski, Wilfred Francis, II
Waters, Willie Estel
Wheaton, Thomas Adair
Wilfret, Gary Joe
Williams, Tom Vare
Wiltbank, William Joseph
Wolf, Benjamin
Woltz, Shreve Simpson

GEORGIA
Austin, Max E
Brown, Acton Richard
Daniell, Jeff Walter
Dekazos, Elias Demetrios
Donoho, Clive Wellington, Jr
Dull, Gerald G

Horticulture (cont)

OKLAHOMA
Bates, Richard Pierce
Campbell, Raymond Earl
Merkle, Owen George
Payne, Richard N
Taliaferro, Charles M
Wann, Elbert Van
Whitcomb, Carl Erwin

OREGON
Apple, Spencer Butler, Jr
Baggett, James Ronald
Bullock, Richard Melvin
Cameron, H Ronald
Cameron, James Wagner
Compton, Oliver Cecil
Coyier, Duane L
Crabtree, Garvin (Dudley)
Evans, Harold J
Frazier, William Allen
Fuchigami, Leslie H
Garren, Ralph, Jr
Hampton, Richard Owen
Jarrell, Wesley Michael
Lagerstedt, Harry Bert
Lawrence, Francis Joseph
Lombard, Porter Bronson
Mack, Harry John
Martin, Lloyd W
Mielke, Eugene Albert
Mok, Machteld Cornelia
Roberts, Alfred Nathan
Rohde, Charles Raymond
Rykbost, Kenneth Albert
Smith, Orrin Ernest
Stevenson, Elmer Clark
Thompson, Maxine Marie
Ticknor, Robert Lewis
Volland, Leonard Allan
Weiser, Conrad John
Westwood, Melvin (Neil)
Yungen, John A

PENNSYLVANIA
Armstrong, Robert John
Beach, Neil William
Beattie, James Monroe
Beelman, Robert B
Bergman, Ernest L
Blumenfield, David
Cole, Richard H
Daum, Donald Richard
Evensen, Kathleen Brown
Feuer, Robert Charles
Haeseler, Carl W
Heller, Paul R
Hill, Richard Ray, Jr
Hull, Larry Allen
Johnson, Littleton Wales
Marshall, Harold Gene
Russo, Joseph Martin
Shannon, Jack Corum
Shenk, John Stoner
Smith, Cyril Beverley
Stinson, Richard Floyd
Tukey, Loren Davenport
Warner, Harlow Lester
Watschke, Thomas Lee
White, John W
Witham, Francis H

RHODE ISLAND
Gough, Robert Edward
Hull, Richard James
Larmie, Walter Esmond
McGuire, John J
Shaw, Richard John

SOUTH CAROLINA
Alexander, Paul Marion
Alston, Jimmy Albert
Courtney, William Henry, III
Dukes, Philip Duskin
Fery, Richard Lee
Halfacre, Robert Gordon
Lineberger, Robert Daniel
Nolan, Clifford N
Ogle, Wayne LeRoy
Pope, Daniel Townsend
Senn, Taze Leonard
Sims, Ernest Theodore, Jr
Skelton, Bobby Joe
Skelton, Thomas Eugene

SOUTH DAKOTA
Cholick, Fred Andrew
Kahler, Alex L
Lunden, Allyn Oscar
Nelson, Donald Carl
Peterson, Ronald M
Prashar, Paul D
Reeves, Dale Leslie

TENNESSEE
Collins, Jimmie Lee
Coorts, Gerald Duane
Downes, John D
Halterlein, Anthony J
Hathcock, Bobby Ray
Josephson, Leonard Melvin
Sams, Carl Earnest
Smith, Rufus Albert, Jr
Southards, Carroll J
Swingle, Homer Dale

TEXAS
Bailey, Leo L
Barham, Warren Sandusky
Bennett, William Frederick
Bockholt, Anton John
Brewer, Beverly Sparks
Calub, Alfonso deGuzman
Davies, Frederick T, Jr
Dibble, John Thomas
Fucik, John Edward
Hensz, Richard Albert
Holt, Ethan Cleddy
Hunter, Richard Edmund
Lipe, John Arthur
McDaniel, Milton Edward
McWilliams, Edward Lacaze
Maunder, A Bruce
Miller, Julian Creighton, Jr
Nightingale, Arthur Esten
Niles, George Alva
Parsons, Jerry Montgomery
Paterson, Donald Robert
Reed, David William
Reinert, James A
Shreve, Loy William
Smith, Dudley Templeton
Storey, James Benton
Tereshkovich, George
Thompson, Tommy Earl
Willingham, Francis Fries, Jr

UTAH
Anderson, Jay LaMar
Hamson, Alvin Russell
Jeffery, Larry S
Larsen, Robert Paul
Locy, Robert Donald
Rasmussen, Harry Paul
Reimschussel, Ernest F
Seeley, Schuyler Drannan
Thomas, James H
Thorup, Richard M
Vest, Hyrum Grant, Jr
Walker, David Rudger
Walser, Ronald Herman

VERMONT
Bailey, Catherine Hayes
Klein, Deana Tarson
Pellett, Norman Eugene

VIRGINIA
Ambler, John Edward
Barden, John Allan
Borchers, Edward Alan
Brubaker, Kenton Kaylor
Cocking, W Dean
Drake, Charles Roy
Fretz, Thomas Alvin
Horsburgh, Robert Laurie
Hortick, Harvey J
Milbocker, Daniel Clement
Rollins, Howard A, Jr
Smeal, Paul Lester
Starling, Thomas Madison

WASHINGTON
Ackley, William Benton
Aichele, Murit Dean
Bienz, Darrel Rudolph
Butler, Jackie Dean
Clark, James Richard
Clore, Walter Joseph
Dean, Bill Bryan
Doughty, Charles Carter
Eighme, Lloyd Elwyn
Ewart, Hugh Wallace, Jr
Heilman, Paul E
Hiller, Larry Keith
Iritani, W M
Johnson, Charles Robert
Larsen, Fenton E
Martin, Mark Wayne
Norton, Robert Alan
Patterson, Max E
Peabody, Dwight Van Dorn, Jr
Proebsting, Edward Louis, Jr
Ryan, George Frisbie
Schekel, Kurt Anthony
Tukey, Harold Bradford, Jr
Williams, Max W
Withner, Carl Leslie, Jr
Wott, John Arthur

WEST VIRGINIA
Brown, Mark Wendell
Horton, Billy D
Welker, William V, Jr

WISCONSIN
Beck, Gail Edwin
Bingham, Edwin Theodore
Binning, Larry Keith
Bula, Raymond J
Curwen, David
Dana, Malcolm Niven
Duewer, Elizabeth Ann
Duewer, Raymond George
Gilbert, Franklin Andrew, Sr
Hagedorn, Donald James
Harrison, Helen Connolly
Hasselkus, Edward R
Holm, LeRoy George
Hopen, Herbert

Hougas, Robert Wayne
McCown, Brent Howard
Peloquin, Stanley J
Peterson, Clinton E
Smith, Richard R
Stang, Elden James
Tibbitts, Theodore William
Tracy, William Francis

WYOMING
Bohnenblust, Kenneth E
Kolp, Bernard J

PUERTO RICO
Goyal, Megh R
Salas-Quintana, Salvador

ALBERTA
Andrews, John Edwin
Krahn, Thomas Richard
Lynch, Dermot Roborg
McAndrew, David Wayne
Toop, Edgar Wesley

BRITISH COLUMBIA
Ballantyne, David John
Dabbs, Donald Henry
Daubeny, Hugh Alexander
Denby, Lyall Gordon
Eaton, George Walter
Fisher, Francis John Fulton
Looney, Norman E
Neilsen, Gerald Henry
Norton, Colin Russell
Putt, Eric Douglas
Quamme, Harvey Allen
Struble, Dean L

MANITOBA
Helgason, Sigurdur Bjorn
LaCroix, Lucien Joseph
Larter, Edward Nathan
McKenzie, Ronald Ian Hector
Staniforth, Richard John
Stefansson, Baldur Rosmund
Townley-Smith, Thomas Frederick
Wiebe, John

NEW BRUNSWICK
Coleman, Warren Kent
Young, Donald Alcoe

NOVA SCOTIA
Embree, Charles Gordon
Webster, David Henry

ONTARIO
Andersen, Emil Thorvald
Anstey, Thomas Herbert
Bishop, Charles Johnson
Childers, Walter Robert
Chu, Chun-Lung (George)
Elfving, Donald Carl
Fuleki, Tibor
Harney, Patricia Marie
Ingratta, Frank Jerry
Kerr, Ernest Andrew
Layne, Richard C
Liptay, Albert
Loughton, Arthur
Marks, Charles Frank
Miles, Neil Wayne
Miller, Sherwood Robert
Nonnecke, Ib Libner
Olson, Arthur Olaf
O'Sullivan, John
Proctor, John Thomas Arthur
Riekels, Jerald Wayne
Svejda, Felicitas Julia
Tan, Chin Sheng
Webb, David Thomas

PRINCE EDWARD ISLAND
Cutcliffe, Jack Alexander
Gupta, Umesh C

QUEBEC
Bigras, Francine Jeanne
Doyon, Gilles Joseph
Gasser, Heinz
Gauthier, Fernand Marcel
Girouard, Ronald Maurice
Klinck, Harold Rutherford
Lawson, Norman C

SASKATCHEWAN
Downey, Richard Keith
Goplen, Bernard Peter
Harvey, Bryan Laurence
Lawrence, Thomas
Maginnes, Edward Alexander
Tanino, Karen Kikumi

OTHER COUNTRIES
Carballo-Quiros, Alfredo
Curtis, Byrd Collins
Filadelfi-Keszi, Mary Ann Stephanie
Hogan, Le Moyne
Holle, Miguel
Khush, Gurdev S
Muniappan, Rangaswamy Naicker
Oertli, Johann Jakob

Phytopathology

ALABAMA
Backman, Paul Anthony
Chambliss, Oyette Lavaughn
Clark, Edward Maurice
Gudauskas, Robert Thomas
Haaland, Ronald L
Jacobsen, Barry James
Weete, John Donald

ARIZONA
Alcorn, Stanley Marcus
Yohem, Karin Hummell

ARKANSAS
Jones, John Paul
Kim, Kyung Soo
Mason, Curtis Leonel
Reed, Hazell
Riggs, Robert D
Scott, Howard Allen

CALIFORNIA
Adaskaveg, James Elliott
Calavan, Edmond Clair
Calpouzos, Lucas
Campbell, Robert Noe
Cobb, Fields White, Jr
Eckard, Kathleen
Erwin, Donald C
Gerik, James Stephen
Gilmore, James Eugene
Goff, Lynda June
Hackney, Robert Ward
Hagan, William Leonard
Hildebrand, Donald Clair
Houck, Laurie Gerald
Jorgenson, Edsel Carpenter
Kliejunas, John Thomas
Lewellen, Robert Thomas
Lucas, David Owen
MacDonald, James Douglas
Mankau, Reinhold
Noffsinger, Ella Mae
Ogawa, Joseph Minoru
Pitts, Robert Gary
Sanders, John Stephen
Schaad, Norman W
Schlegel, David Edward
Starr, Mortimer Paul
Thomason, Ivan J
Tsao, Peter Hsing-tsuen
VanBruggen, Ariena H C
Weinhold, Albert Raymond
Wilcox, W Wayne
Wilhelm, Stephen
Zary, Keith Wilfred

COLORADO
Dickenson, Donald Dwight
Hawksworth, Frank Goode
Helmerick, Robert Howard
Hill, Joseph Paul
Knutson, Kenneth Wayne
Ruppel, Earl George
Yun, Young Mok

CONNECTICUT
Anagnostakis, Sandra Lee
Aylor, Donald Earl
Huang, Liang Hsiung
LaMondia, James A
Smith, Victoria Lynn
Walton, Gerald Steven
Wargo, Philip Matthew

DELAWARE
Carnahan, James Elliot
Carroll, Robert Buck
Delp, Charles Joseph
Fieldhouse, Donald John
Howard, Richard James
Smith, Constance Meta

DISTRICT OF COLUMBIA
Gilbert, Richard Gene
Jackson, Curtis Rukes
Khan, Mohamed Shaheed
Krigsvold, Dale Thomas
Smith, Richard S, Jr

FLORIDA
Agrios, George Nicholas
Blakeslee, George M
Chase, Ann Renee
Cline, Kenneth Charles
Davis, Michael Jay
Engelhard, Arthur William
Giblin-Davis, Robin Michael
Gottwald, Timothy R
Gray, Dennis John
Howard, Charles Marion
Jones, John Paul
Miller, Thomas
Osborne, Lance Smith
Peacock, Hugh Anthony
Purcifull, Dan Elwood
Rich, Jimmy Ray
Ringel, Samuel Morris
Roberts, Daniel Altman
Smart, Grover Cleveland, Jr
Tsai, James Hsi-Cho

LOUISIANA
Owings, Addison Davis

MARYLAND
Campbell, Travis Austin
Galletta, Gene John
Maas, John Lewis
Ng, Timothy J

MASSACHUSETTS
Wilkes, Hilbert Garrison, Jr

MISSISSIPPI
Berry, Charles Dennis
Creech, Roy G

MISSOURI
Guilfoyle, Thomas J
Sappenfield, William Paul

MONTANA
Blake, Tom
Ditterline, Raymond Lee
McCoy, Thomas Joseph

NEBRASKA
Gardner, Charles Olda
Specht, James Eugene

NEVADA
Thyr, Billy Dale

NEW HAMPSHIRE
Kiang, Yun-Tzu

NEW YORK
Cummins, James Nelson
Earle, Elizabeth Deutsch
Last, Robert L
Valentine, Fredrick Arthur

NORTH CAROLINA
Oblinger, Diana Gelene

OKLAHOMA
Banks, Donald Jack
Edwards, Lewis Hiram
Verhalen, Laval

PENNSYLVANIA
Ayers, John E
Berg, Clyde C

SOUTH CAROLINA
Courtney, William Henry, III
Fery, Richard Lee
Jones, Alfred
Thomas, Claude Earle

TENNESSEE
Allen, Freddie Lewis
Conger, Bob Vernon

TEXAS
Bird, Luther Smith
Cook, Charles Garland
Nguyen, Henry Thien

VIRGINIA
Buss, Glenn Richard
Reed, Joseph
Sagaral, Erasmo G

WASHINGTON
Konzak, Calvin Francis
Peterson, Clarence James, Jr

WISCONSIN
Cramer, Jane Harris
Guries, Raymond Paul
Nelson, Neil Douglas
Simon, Philipp William

PUERTO RICO
Rodriguez, Jorge Luis
Salas-Quintana, Salvador

ALBERTA
Woods, Donald Leslie

MANITOBA
Aung, Taing
Kerber, Erich Rudolph

NOVA SCOTIA
Aalders, Lewis Eldon
Chen, Lawrence Chien-Ming

ONTARIO
Burrows, Vernon Douglas
Buzzell, Richard Irving
Campbell, Kenneth Wilford
Kasha, Kenneth John
Reinbergs, Ernests
Warwick, Suzanne Irene
Yeatman, Christopher William

PRINCE EDWARD ISLAND
Choo, Thin-Meiw

SASKATCHEWAN
Baker, Robert John
Knott, Douglas Ronald

Range Science & Management

ALASKA
Hanley, Thomas Andrew

ARIZONA
Davis, Edwin Alden
Johnsen, Thomas Norman, Jr
Jordan, Gilbert LeRoy
Klemmedson, James Otto
McPherson, Guy Randall
Morton, Howard LeRoy
Schmutz, Ervin Marcell
Severson, Keith Edward
Smith, Edwin Lamar, Jr
Springfield, Harry Wayne

CALIFORNIA
Connor, John Michael
Dawson, Kerry J
Green, Lisle Royal
Heady, Harold Franklin
Jansen, Henricus Cornelis
Love, Robert Merton
Murphy, Alfred Henry
Walters, Roland Dick
Williams, William Arnold

COLORADO
Asherin, Duane Arthur
Bement, Robert Earl
Branson, Farrel Allen
Costello, David Francis
Cuany, Robin Louis
Dittberner, Phillip Lynn
Driscoll, Richard Stark
Hansen, Richard M
Hyder, Donald N
Lauenroth, William Karl
McGinnies, William Joseph
Oldemeyer, John Lee
Redente, Edward Francis
Rittenhouse, Larry Ronald
Shaw, Robert Blaine
Smith, Dwight Raymond
Woodmansee, Robert George

FLORIDA
Marion, Wayne Richard

IDAHO
Callihan, Robert Harold
Ehrenreich, John Helmuth
Pearson, Lorentz Clarence
Sharp, Lee Ajax
Wight, Jerald Ross

IOWA
Clark, William Richard
Morrical, Daniel Gene

KANSAS
Hooker, Mark L

KENTUCKY
Baker, John P
Vogel, Willis Gene

LOUISIANA
Grelen, Harold Eugene
Linnartz, Norwin Eugene
Pearson, Henry Alexander

MISSISSIPPI
Bagley, Clyde Pattison

MISSOURI
Lewis, James Kelley
Mosher, Donna Patricia
Shiflet, Thomas Neal

MONTANA
Brown, James Kerr
Heitschmidt, Rodney Keith
MacNeil, Michael
Morris, Melvin Solomon
Rounds, Burton Ward
Taylor, John Edgar

NEBRASKA
Baltensperger, David Dwight
Nichols, James T
Power, James Francis
Waller, Steven Scobee

NEVADA
Davis, Phillip Burton
Eckert, Richard Edgar, Jr
Evans, Raymond Arthur
Gifford, Gerald F

NEW JERSEY
Quinn, James Amos
Stevens, Merwin Allen

NEW MEXICO
Aldon, Earl F
Allred, Kelly Wayne
Anderson, Dean Mauritz
Donart, Gary B
Dwyer, Don D
Fowler, James Lowell
Gibbens, Robert Parker
Herbel, Carlton Homer

Holechek, Jerry Lee
Pieper, Rex Delane
Risser, Paul Gillan

NORTH DAKOTA
Barker, William T
Frank, Albert Bernard
Hofmann, Lenat
Karn, James Frederick
Nyren, Paul Eric
Ries, Ronald Edward

OKLAHOMA
Crockett, Jerry J
Sims, Phillip Leon

OREGON
Bedell, Thomas Erwin
Buckhouse, John Chapple
Chalk, David Eugene
Crawford, John Arthur
Dealy, John Edward
Kelsey, Rick Guy
Krueger, William Clement
Maser, Chris
Skovlin, Jon Matthew
Spears, Brian Merle

SOUTH DAKOTA
Bjugstad, Ardell Jerome
Uresk, Daniel William

TEXAS
Arnold, James Darrell
Boutton, Thomas William
Bovey, Rodney William
Briske, David D
Britton, Carlton M
Dahl, Billie Eugene
Drawe, D Lynn
Fulbright, Timothy Edward
Grumbles, Jim Bob
Halls, Lowell Keith
Harris, Venoia M
Hauser, Victor La Vern
Kothmann, Merwyn Mortimer
Landers, Roger Q, Jr
McCully, Wayne Gunter
Schuster, Joseph L
Scifres, Charles Joel
Sosebee, Ronald Eugene
Swakon, Doreen H D
Thill, Ronald E
Varner, Larry Weldon
White, Larry Dale
Whitson, Robert Edd
Wright, Henry Albert

UTAH
Blauer, Aaron Clyde
Box, Thadis Wayne
Butcher, John Edward
Croft, Alfred Russell
Dobrowolski, James Phillip
Johnson, Douglas Allan
McKell, Cyrus Milo
Malechek, John Charles
Mueggler, Walter Frank
Urness, Philip Joel
Vallentine, John Franklin
Wood, Benjamin W
Workman, John Paul
Yorks, Terence Preston

VIRGINIA
Allen, Vivien Gore
Cook, Charles Wayne

WASHINGTON
Goebel, Carl Jerome
Roche, Ben F, Jr
Schwendiman, John Leo

WISCONSIN
Barnes, Robert F
Jorgensen, Neal A

WYOMING
DePuit, Edward J
Fisser, Herbert George
Hart, Richard Harold
Hayden-Wing, Larry Dean
Kearl, Willis Gordon
Laycock, William Anthony
Powell, Jeff
Schuman, Gerald E
Stanton, Nancy Lea
Sturges, David L
Tigner, James Robert

PUERTO RICO
Wilson, Marcia Hammerquist

ALBERTA
Bailey, Arthur W

BRITISH COLUMBIA
Quinton, Dee Arlington

SASKATCHEWAN
Cohen, Roger D H
Coupland, Robert Thomas
Waddington, John

OTHER COUNTRIES
Hocking, Drake

Soils & Soil Science

ALABAMA
Adams, Fred
Allen, Seward Ellery
Chien, Sen Hsiung
Cope, John Thomas, Jr
Diamond, Ray Byford
Engelstad, Orvis P
Evans, Clyde Edsel
Griffin, Robert Alfred
Guthrie, Richard Lafayette
Hajek, Benjamin F
Hill, Walter Andrew
Hiltbold, Arthur Edward, Jr
Hood, Joseph
Lyle, Everett Samuel, Jr
Molz, Fred John, III
Mortvedt, John Jacob
Rogers, Howard Topping
Simmons, Charles Ferdinand
Struchtemeyer, Roland August
Sutherland, William Neil
Ward, Coleman Younger

ALASKA
Cochran, Verlan Leyerl
Drew, James
Laughlin, Winston Means
Sparrow, Elena Bautista

ARIZONA
Bohn, Hinrich Lorenz
Buras, Nathan
Campbell, Ralph Edmund
Dutt, Gordon Richard
Gardner, Bryant Rogers
Grier, Charles Crocker
Heald, Walter Roland
Kimball, Bruce Arnold
Klemmedson, James Otto
Klopatek, Jeffrey Matthew
Martin, William Paxman
Morris, Gene Ray
Nakayama, Francis Shigeru
Rauschkolb, Roy Simpson
Ray, Howard Eugene
Robinson, Daniel Owen
Shaw, Ellsworth
Stroehlein, Jack Lee
Sultan, Hassan Ahmed
Tucker, Thomas Curtis
Warrick, Arthur W
Whitt, Darnell Moses
Wierenga, Peter J
Wilson, Lorne Graham

ARKANSAS
Allen, Arthur Lee
Baker, James Bert
Bartlett, Frank David
Brown, Donald A
Lavy, Terry Lee
Legg, Joseph Ogden
Pitts, Donald James
Porter, Owen Archuel
Scott, Hubert Donovan
Tennille, Aubrey W
Thompson, Lyell
Wells, Bobby R
West, Charles Patrick
Wolf, Duane Carl

CALIFORNIA
Baker, Warren J
Bodman, Geoffrey Baldwin
Borchardt, Glenn (Arnold)
Broadbent, Francis Everett
Brown, Arthur Lloyd
Brownell, James Richard
Cannell, Glen H
Cate, Robert Bancroft
Cliath, Mark Marshall
Cook, James Arthur
Crawford, Robert Field
Dalton, Francis Norbert
Doner, Harvey Ervin
Duniway, John Mason
Embleton, Tom William
Farmer, Walter Joseph
Focht, Dennis Douglass
Gerik, James Stephen
Gersper, Paul Logan
Goldberg, Sabine Ruth
Grimes, Donald Wilburn
Harmen, Raymond A
Harte, John
Hauxwell, Donald Lawrence
James, Ronald Valdemar
Jones, Milton Bennion
Jury, William Austin
Lambert, Royce Leone
Letey, John, Jr
Liebhardt, William C
Lund, Lanny Jack
Lunt, Owen Raynal
Maas, Eugene Vernon
McAuliffe, Clayton Doyle
Madison, John Herbert, Jr
Martin, James Paxman

Soils & Soil Science (cont)

Tung, Fred Fu
Vitosh, Maurice Lee
White, Donald Perry
Whiteside, Eugene Perry
Wolt, Jeffrey Duaine

MINNESOTA
Adams, Russell S, Jr
Arneman, Harold Frederick
Blake, George Rowland
Boelter, Don Howard
Brown, Bruce Antone
Carlson, Robert Marvin
Cheng, Hwei-Hsien
Clapp, C(harles) Edward
Clapp, Thomas Wright
Dowdy, Robert H
Farnham, Rouse Smith
Foster, George Rainey
Grava, Janis (John)
Grigal, David F
Gupta, Satish Chander
Holt, Robert F
Larson, William Earl
Latterell, Joseph J
Linden, Dennis Robert
Malzer, Gary Lee
Munson, Robert Dean
Olness, Alan
Overdahl, Curtis J
Pryor, Gordon Roy
Randall, Gyles Wade
Rehm, George W
Rust, Richard Henry
Tilman, G David

MISSISSIPPI
Balam, Baxish Singh
Bardsley, Charles Edward
George, Kalankamary Pily
Grissinger, Earl H
Meyer, Lawrence Donald
Mutchler, Calvin Kendal
Nash, Victor E
Nelson, Lyle Engnar
Perry, Edward Belk
Peterson, Harold LeRoy
Pettry, David Emory
Robinson, A(ugust) R(obert)
Shockley, W(oodland) G(ray)
Whisler, Frank Duane

MISSOURI
Blanchar, Robert W
Brown, James Richard
Chowdhury, Ikbalur Rashid
Clare, Stewart
Cox, Gene Spracher
Deming, John Miley
Folks, Homer Clifton
Hasan, Syed Eqbal
Henderson, Gray Stirling
Lambeth, Victor Neal
Marshall, Charles Edmund
Miles, Randall Jay
Ponder, Felix, Jr
Radke, Rodney Owen
Richards, Graydon Edward
Schmidt, Norbert Otto
Shrader, William D
Wilson, Clyde Livingston
Woodruff, Clarence Merrill

MONTANA
Aase, Jan Kristian
Asleson, Johan Arnold
Bauder, James Warren
Brown, Paul Lawson
Ferguson, Albert Hayden
Jackson, Grant D
Nielsen, Gerald Alan
Olsen, Ralph A
Skogley, Earl O

NEBRASKA
Blad, Blaine L
Chesnin, Leon
Doran, John Walsh
Garey, Carroll Laverne
Gast, Robert Gale
Holzhey, Charles Steven
Knox, Ellis Gilbert
Leviticus, Louis I
Lewis, David Thomas
Long, Daryl Clyde
Lynn, Warren Clark
Miller, Willie
Nelson, Darrell Wayne
Nettleton, Wiley Dennis
Power, James Francis
Ross, Sam Jones, Jr
Sander, Donald Henry
Skopp, Joseph Michael
Sorensen, Robert Carl
Swartzendruber, Dale
Volk, Bob G
Wiese, Richard Anton

NEVADA
Bower, Charles Arthur
Cameron, Roy (Eugene)
Gilbert, Dewayne Everett
Leedy, Clark D
Miller, Watkins Wilford

Peterson, Frederick Forney
Thornburn, Thomas H(ampton)

NEW HAMPSHIRE
Federer, C Anthony
Harter, Robert Duane
McDowell, William H
McKim, Harlan L
Safford, Lawrence Oliver

NEW JERSEY
Alderfer, Russell Brunner
Douglas, Lowell Arthur
Eck, Paul
Esrig, Melvin I
Jumikis, Alfreds Richards
Monahan, Edward James
Padhi, Sally Bulpitt
Pramer, David
Tedrow, John Charles Fremont
Toth, Stephen John
Weissmann, Gerd Friedrich Horst

NEW MEXICO
Daugherty, LeRoy Arthur
Essington, Edward Herbert
Fowler, Eric Beaumont
Gile, Leland Henry
Grover, Herbert David
Johnson, Gordon Verle
Liddell, Craig Mason
Lindemann, William Conrad
Melton, Billy Alexander, Jr
Morin, George Cardinal Albert
O'Connor, George Albert
Polzer, Wilfred L
Triandafilidis, George Emmanuel
Tromble, John M

NEW YORK
Alexander, Martin
App, Alva A
Bouldin, David Ritchey
Cline, Marlin George
Drosdoff, Matthew
Francis, Arokiasamy Joseph
Gluck, Ronald Monroe
Grunes, David Leon
Harrison, William Paul
Lathwell, Douglas J
Linkins, Arthur Edward
Miller, Robert Demorest
Minotti, Peter Lee
Misiaszek, Edward T
Reid, William Shaw
Scott, Thomas Walter
Shannon, Stanton
Smalley, Ralph Ray
Stoll, Robert D
Stotzky, Guenther
Wagenet, Robert Jeffrey
Weinstein, David Alan
Welch, Ross Maynard
White, Edwin Henry
Zwarun, Andrew Alexander

NORTH CAROLINA
Baird, Jack Vernon
Bohannon, Robert Arthur
Buol, Stanley Walter
Cassel, D Keith
Cook, Maurice Gayle
Cox, Frederick Russell
Cummings, George August
Daniels, Raymond Bryant
Davey, Charles Bingham
Gilliam, James Wendell
Hassan, Awatif E
Jackson, William Addison
Kamprath, Eugene John
King, Larry Dean
Knight, Clifford Burnham
McCants, Charles Bernard
McCollum, Robert Edmund
McCracken, Ralph Joseph
Miner, Gordon Stanley
Nelson, Paul Victor
Overcash, Michael Ray
Reed, William Edward
Sanchez, Pedro Antonio
Shafer, Steven Ray
Shelton, James Edward
Spinks, Daniel Owen
Van Eck, Willem Adolph
Wahls, Harvey E(dward)
Weber, Jerome Bernard
Weed, Sterling Barg
Wollum, Arthur George, II
Woodhouse, William Walton, Jr
Zublena, Joseph Peter

NORTH DAKOTA
Bauer, Armand
Doll, Eugene Carter
Enz, John Walter
Hofmann, Lenat
Merrill, Stephen Day
Vasey, Edfred H
Zubriski, Joseph Cazimer

OHIO
Basile, Robert Manlius
Calhoun, Frank Gilbert
Darrow, Robert A

Eckert, Donald James
Fausey, Norman Ray
Haghiri, Faz
Hall, Geroge Frederick
Himes, Frank Lawrence
Hock, Arthur George
Hutchinson, Frederick Edward
Lal, Rattan
Logan, Terry James
McLean, Eugene Otis
Miller, Frederick Powell
Munn, David Alan
Ryan, James Anthony
Stoner, Clinton Dale
Sutton, Paul
Taylor, George Stanley
Vimmerstedt, John P
Watson, Maurice E
Zimmerman, Tommy Lynn

OKLAHOMA
Ahring, Robert M
Berg, William Albert
Enfield, Carl George
Gray, Fenton
Johnson, Gordon V
Jones, Randall Jefferies
Laguros, Joakim George
Law, James Pierce, Jr
Lynd, Julian Quentin
Menzel, Ronald George
Morrill, Lawrence George
Nofziger, David Lynn
Puls, Robert W
Reed, Lester W
Smith, Samuel Joseph
Stone, John Floyd

OREGON
Allmaras, Raymond Richard
Cheney, Horace Bellatti
Cochran, Patrick Holmes
Dyrness, Christen Theodore
Elliott, Lloyd Floren
Huddleston, James Herbert
Jackson, Thomas Lloyd
Jarrell, Wesley Michael
Kling, Gerald Fairchild
Landsberg, Johanna D (Joan)
Lu, Kuo Chin
Moore, Duane Grey
Myrold, David Douglas
Parsons, Roger Bruce
Ramig, Robert E
Retallack, Gregory John
Rickman, Ronald Wayne
Rykbost, Kenneth Albert
Simonson, Gerald Herman
Volk, Veril Van
Vomocil, James Arthur
Young, J Lowell
Youngberg, Chester Theodore
Yungen, John A

PENNSYLVANIA
Baker, Dale E
Banerjee, Sushanta Kumar
Bollag, Jean-Marc
Bosshart, Robert Perry
Burns, Allan Fielding
Cheng, Cheng-Yin
Cunningham, Robert Lester
Fox, Richard Henry
Fritton, Daniel Dale
Hall, Jon K
Heddleson, Milford Raynord
Hunter, Albert Sinclair
Johnson, Leon Joseph
Koerner, Robert M
Komarneni, Sridhar
Levin, Michael H(oward)
Loughry, Frank Glade
Marriott, Lawrence Frederick
Petersen, Gary Walter
Pionke, Harry Bernhard
Senft, Joseph Philip
Waddington, Donald Van Pelt
White, John W

RHODE ISLAND
Miller, Robert Harold
Wright, William Ray

SOUTH CAROLINA
Corey, John Charles
Franklin, Ralph E
Hawkins, Richard Horace
Jones, Ulysses Simpson, Jr
Kittrell, Benjamin Upchurch
Lane, Carl Leaton
Ligon, James T(eddie)
Nolan, Clifford N
Sandhu, Shingara Singh
Smith, Bill Ross
Van Lear, David Hyde

SOUTH DAKOTA
Carson, Paul LLewellyn
Fine, Lawrence Oliver
Horton, Maurice Lee
Kohl, Robert A
Moldenhauer, William Calvin
Wiersma, Daniel

TENNESSEE
Brode, William Edward
Buntley, George Jule
Chiang, Thomas M
Foss, John E
Francis, Chester Wayne
Freitag, Dean R(ichard)
Graveel, John Gerard
Kohland, William Francis
Large, Richard L
Lee, Suk Young
Lessman, Gary M
Lewis, Russell J
Lietzke, David Albert
Luxmoore, Robert John
Madhavan, Kunchithapatham
Naddy, Badie Ihrahim
Seatz, Lloyd Frank
Smalley, Glendon William
Springer, Maxwell Elsworth
Tamura, Tsuneo
Wang, Richard Hsu-Shien

TEXAS
Adams, John Edgar
Allen, Bonnie L
Anderson, Duwayne Marlo
Anderson, Warren Boyd
Armstrong, James Clyde
Bloodworth, M(orris) E(lkins)
Burnett, Earl
Dibble, John Thomas
Dixon, Joe Boris
Dregne, Harold Ernest
Eck, Harold Victor
Elward-Berry, Julianne
Gavande, Sampat A
Gerard, Cleveland Joseph
Hauser, Victor La Vern
Heyman, Louis
Hipp, Billy Wayne
Ho, Clara Lin
Hoover, William L
Hossner, Lloyd Richard
Jones, Ordie Reginal
Kovar, John Alvis
Kunze, George William
Loeppert, Richard Henry, Jr
Louden, L Richard
Lowry, Gerald Lafayette
Melton, James Ray
Milford, Murray Hudson
Miyamoto, Seiichi
Neher, David Daniel
Newland, Leo Winburne
Nichols, Joe Dean
Nossaman, Norman L
Onken, Arthur Blake
Ott, Billy Joe
Rao, Shankaranarayana Ramohallinanjunda
Rubink, William Louis
Runge, Edward C A
Runkles, Jack Ralph
Stanford, Geoffrey
Steiner, Jean Louise
Stewart, Bobby Alton
Tackett, Jesse Lee
Taylor, Howard Melvin
Thompson, Louis Jean
Trogdon, William Oren
Unger, Paul Walter
Walker, Laurence Colton
Watterston, Kenneth Gordon
Weaver, Richard Wayne
Wendt, Charles William
Wiegand, Craig Loren
Wilding, Lawrence Paul
Young, Arthur Wesley
Zuberer, David Alan

UTAH
Belnap, Jayne
Croft, Alfred Russell
Dobrowolski, James Phillip
Fisher, Richard Forrest
Fletcher, Joel Eugene
Hanks, Ronald John
Hargreaves, George H(enry)
Harper, Kimball T
Hill, Archie Clyde
James, David Winston
Jeffery, Larry S
Jurinak, Jerome Joseph
Keller, Jack
Miller, Raymond Woodruff
Muir, Melvin K
Nelson, Sheldon Douglas
Olsen, Edwin Carl, III
Openshaw, Martin David
Peterson, Howard Boyd
Rasmussen, V Philip, Jr
Sidle, Roy Carl
Skujins, John Janis
Smith, R L
Southard, Alvin Reid
Terry, Richard Ellis
Thorup, Richard M
Wood, Timothy E
Youd, Thomas Leslie

VERMONT
Bartlett, Richmond J
Magdoff, Frederick Robin

Soils & Soil Science (cont)

Murphy, William Michael

VIRGINIA
Allen, Vivien Gore
Carstea, Dumitru
Dawson, Murray Drayton
Duke, Everette Loranza
Finkl, Charles William, II
Grove, Thurman Lee
Hagedorn, Charles
Hakala, William Walter
Harlan, Phillip Walker
Herd, Darrell Gilbert
Krebs, Robert Dixon
Kreh, Richard Edward
Larew, H(iram) Gordon
McClung, Andrew Colin
Malcolm, John Lowrie
Martens, David Charles
Neal, John Lloyd, Jr
O'Neill, Eileen Jane
Pasko, Thomas Joseph, Jr
Reneau, Raymond B, Jr
Rickert, David A
Thiel, Thomas J
Zelazny, Lucian Walter

WASHINGTON
Bezdicek, David Fred
Boyd, Charles Curtis
Campbell, Gaylon Sanford
Cole, Dale Warren
Cykler, John Freuler
Engibous, James Charles
Gardner, Walter Hale
Gessel, Stanley Paul
Gilkeson, Raymond Allen
Hausenbuiller, Robert Lee
Hedges, John Ivan
Heilman, Paul E
Jensen, Creighton Randall
Kittrick, James Allen
Klock, Glen Orval
Koehler, Fred Eugene
Kuo, Shiou
Lauer, David Allan
Phillips, Steven J
Prunty, Lyle Delmar
Rai, Dhanpat
Reid, Preston Harding
Relyea, John Franklin
Rieger, Samuel
Routson, Ronald C
Schwendiman, John Leo
Steinbrenner, Eugene Clarence
Wildung, Raymond Earl
Wooldridge, David Dilley

WEST VIRGINIA
Keefer, Robert Faris
Sencindiver, John Coe
Singh, Rabindar Nath

WISCONSIN
Barnes, Robert F
Beatty, Marvin Theodore
Bubenzer, Gary Dean
Bundy, Larry Gene
Cain, John Manford
Corey, Richard Boardman
Edil, Tuncer Berat
Eidt, Robert C
Gerloff, Gerald Carl
Harpstead, Milo I
Hensler, Ronald Fred
Hole, Francis Doan
Jackson, Marion Leroy
Larsen, Michael John
McIntosh, Thomas Henry
Murdock, John Thomas
Naik, Tarun Ratilal
Norman, John Matthew
Paulson, William H
Peterson, Arthur Edwin
Schneider, Allan Frank
Stelly, Matthias
Tanner, Champ Bean
Walsh, Leo Marcellus
Willis, Harold Lester

WYOMING
Heil, Robert Dean
Hough, Hugh Walter
Neibling, William Howard
Schuman, Gerald E
Smith, James Lee
Stanton, Nancy Lea
Williams, Stephen Earl

PUERTO RICO
Goyal, Megh R
Lugo-Lopez, Miguel Angel
Ritchey, Kenneth Dale

ALBERTA
Barendregt, Rene William
Chang, Chi
Clementz, David Michael
Coen, Gerald Marvin
Cook, Fred D
Dormaar, Johan Frederik
Dudas, Marvin Joseph

Foscolos, Anthony E
McAndrew, David Wayne
McGill, William Bruce
McMillan, Neil John
Nielsen, Kenneth Fred
Pawluk, Steve
Pluth, Donald John
Rennie, Robert John
Rice, Wendell Alfred
Robertson, James Alexander
Sommerfeldt, Theron G
Webster, Gordon Ritchie

BRITISH COLUMBIA
Davis, Roderick Leigh
Finn, William Daniel Liam
Kowalenko, Charles Grant
Lavkulich, Leslie Michael
Lowe, Lawrence E
Neilsen, Gerald Henry
Russell, Glenn C
Slaymaker, Herbert Olav
Smith, Richard Barrie
Stevenson, David Stuart

MANITOBA
Bailey, Loraine Dolar
Grant, Cynthia Ann
Hedlin, Robert Arthur
Hobbs, James Arthur
Rudgers, Lawrence Alton
Shaykewich, Carl Francis
Smith, Robert Edward
Soper, Robert Joseph

NEW BRUNSWICK
Chow, Thien Lien
Harries, Hinrich
Krause, Helmut
Van Groenewoud, Herman

NEWFOUNDLAND
Hampson, Michael Chisnall
McKenzie, David Bruce

NOVA SCOTIA
Blatt, Carl Roger
Corke, Charles Thomas
Warman, Philip Robert
Wright, James R

ONTARIO
Armson, Kenneth Avery
Bates, Thomas Edward
Beauchamp, Eric G
Behan-Pelletier, Valerie Mary
Bowman, Bruce T
Bunting, Brian Talbot
Clark, John S
Cooper, George S
De Kimpe, Christian Robert
Dickinson, William Trevor
Elrick, David Emerson
Foster, Neil William
Gillham, Robert Winston
Gould, William Douglas
Halstead, Ronald Lawrence
Hoffman, Douglas Weir
Ivarson, Karl C
Jeglum, John Karl
King, Roger Hatton
Lemon, Edgar Rothwell
Mack, Alexander Ross
McKeague, Justin Alexander
McKenney, Donald Joseph
Mathur, Sukhdev Prashad
Matthews, Burton Clare
Miller, Murray Henry
Morrison, Ian Kenneth
Protz, Richard
Quigley, Robert Murvin
Richards, Norval Richard
Schnitzer, Morris
Singh, Surinder Shah
Stebelsky, Ihor
Tan, Chin Sheng
Topp, G Clarke
Tu, Chin Ming
Wall, Gregory John
Williams, Peter J
Winterhalder, Keith

PRINCE EDWARD ISLAND
Gupta, Umesh C
MacLeod, John Alexander
White, Ronald Paul, Sr

QUEBEC
Allen, Sandra Lee
Bonn, Ferdinand J
Bordeleau, Lucien Mario
Bourget, Sylvio-J
Broughton, Robert Stephen
Cescas, Michel Pierre
Laverdiere, Marc Richard
MacKenzie, Angus Finley
Widden, Paul Rodney
Yong, R(aymond)

SASKATCHEWAN
Acton, Donald Findlay
Anderson, Darwin Wayne
Beaton, James Duncan
Chinn, Stanley H F

Hanley, Thomas ODonnell
Huang, P M
Janke, Wilfred Edwin
Nelson, Louise Mary
Nuttall, Wesley Ford
Pretty, Kenneth McAlpine
Rennie, Donald Andrews
St Arnaud, Roland Joseph
Stewart, John Wray Black

OTHER COUNTRIES
Araujo, Jose Emilio Goncalves
Brockman, Frank Elliot
Brydon, James Emerson
Couto, Walter
Dirksen, Christiaan
Forsythe, Warren M
Muchovej, James John
Mugwira, Luke Makore
Muniappan, Rangaswamy Naicker
Oertli, Johann Jakob
Simkins, Charles Abraham
Spencer, John Francis Theodore
Turner, Fred, Jr
Ugolini, Fiorenzo Cesare

Other Agricultural & Forest Sciences

ALABAMA
Campbell, Robert Terry
Clark, Edward Maurice
Giordano, Paul M
Guthrie, Richard Lafayette
Hill, Walter Andrew

ALASKA
Juday, Glenn Patrick

ARIZONA
Baron, William Robert
Kauffeld, Norbert M
Knorr, Philip Noel
Waller, Gordon David

ARKANSAS
Zeide, Boris

CALIFORNIA
Adaskaveg, James Elliott
Akesson, Norman B(erndt)
Belisle, Barbara Wolfanger
Bodman, Geoffrey Baldwin
Fisher, Theodore William
Gary, Norman Erwin
Laidlaw, Harry Hyde, Jr
Orman, Charles
Shelly, John Richard
Taber, Stephen, III
Trevelyan, Benjamin John

COLORADO
Driscoll, Richard Stark
Hansen, Richard M
Shepperd, Wayne Delbert
Tengerdy, Robert Paul
Woodmansee, Robert George

CONNECTICUT
Anagnostakis, Sandra Lee
Connor, Lawrence John
John, Hugo Herman
Newton, David C
Stephens, George Robert

DELAWARE
Caron, Dewey Maurice
Cupery, Willis Eli
Kundt, John Fred

DISTRICT OF COLUMBIA
Cuatrecasas, José
Lunchick, Curt
Post, Boyd Wallace
Rosen, Howard Neal

FLORIDA
Chandler, Robert Flint, Jr
Dressler, Robert Louis
Gerber, John Francis
Littell, Ramon Clarence
Lucansky, Terry Wayne
Miller, James Woodell

GEORGIA
Broerman, F S
Chiang, Tze I

HAWAII
Schmitt, Donald Peter

IDAHO
Callihan, Robert Harold
Michalson, Edgar Lloyd
Thill, Donald Cecil

ILLINOIS
Meyer, Martin Marinus, Jr
Perino, Janice Vinyard
Sinclair, James Burton
Singer, Rolf
Van Sambeek, Jerome William
Wesely, Marvin Larry

INDIANA
Bing, Richard F
Gerwick, Ben Clifford, III
Kidd, Frank Alan
Paschke, John Donald

IOWA
Isely, Duane
Iwig, Mark Michael
Rohlf, Marvin Euguene
Shaw, Robert Harold

KANSAS
Bark, Laurence Dean

KENTUCKY
Olson, James Robert
Vogel, Willis Gene

LOUISIANA
Baker, John Bee
Harbo, John Russell

MARYLAND
Bauer, Peter
Clark, Donald Ray, Jr
Coffman, Charles Benjamin
Foudin, Arnold S
Grau, Fred V
Rango, Albert
Uhart, Michael Scott

MASSACHUSETTS
Kornfield, Jack I
Shawcross, William Edgerton

MICHIGAN
Bourdo, Eric A, Jr
Fulbright, Dennis Wayne

MINNESOTA
Allen, C Eugene
Dahlberg, Duane Arlen
Pour-El, Akiva

MISSISSIPPI
Collins, Johnnie B
Collison, Clarence H
Lloyd, Edwin Phillips

MISSOURI
Bullock, J Bruce
Rash, Jay Justen

MONTANA
Behan, Mark Joseph
Caprio, Joseph Michael
Jackson, Grant D
Latham, Don Jay

NEBRASKA
Knox, Ellis Gilbert

NEW HAMPSHIRE
Baldwin, Henry Ives

NEW JERSEY
Berenbaum, Morris Benjamin

NEW MEXICO
Phillips, Gregory Conrad

NEW YORK
Meyer, Robert Walter
Morse, Roger Alfred
Pack, Albert Boyd
Reisch, Bruce Irving
Schoenly, Kenneth George
Stark, John Howard
Tonelli, John P, Jr
Wolf, Walter Alan

NORTH CAROLINA
Axtell, Richard Charles
Cubberley, Adrian H
Flint, Elizabeth Parker
Thomas, Richard Joseph

OHIO
Hein, Richard William
McCracken, John David
Merritt, Robert Edward
Nelson, Eric V
Vertrees, Robert Layman
Watson, Stanley Arthur

OREGON
Chalk, David Eugene
Denison, William Clark
Ho, Iwan
Landsberg, Johanna D (Joan)
Mason, Richard Randolph
Shock, Clinton C

PENNSYLVANIA
Berthold, Robert, Jr
Daum, Donald Richard
Heller, Paul R
Labosky, Peter, Jr
Press, Linda Seghers
Rogerson, Thomas Dean
Russo, Joseph Martin
Soboczenski, Edward John
Staetz, Charles Alan

RHODE ISLAND
Durfee, Wayne King

SOUTH CAROLINA
Hon, David Nyok-Sai
Quisenberry, Virgil L
Turner, John Lindsey

TENNESSEE
Baldocchi, Dennis D
Blatteis, Clark Martin

TEXAS
Gates, Charles Edgar
Hillery, Herbert Vincent
Humphries, James Edward, Jr
McCully, Wayne Gunter
Smith, Dudley Templeton

VERMONT
Laing, Frederick M

VIRGINIA
Hamrick, Joseph Thomas
Munson, Arvid W
Nichols, James Robbs
Pienkowski, Robert Louis

WASHINGTON
Campbell, Gaylon Sanford
Chiang, Karl Kiu-Kao
Clement, Stephen LeRoy
Lim, Young Woon (Peter)
McCarthy, Joseph L(ePage)
Noel, Jan Christina
Peabody, Dwight Van Dorn, Jr
Tichy, Robert J
Tukey, Harold Bradford, Jr
Zuiches, James J

WEST VIRGINIA
Smith, Walton Ramsay

WISCONSIN
Cramer, Jane Harris
Hole, Francis Doan
Holm, LeRoy George
Lillesand, Thomas Martin
Makela, Lloyd Edward
Norman, John Matthew
Springer, Edward L(ester)

WYOMING
Humburg, Neil Edward

ALBERTA
Wong, Horne Richard

BRITISH COLUMBIA
Gardner, Joseph Arthur Frederick
Paszner, Laszlo

MANITOBA
Palaniswamy, Pachagounder
Pepper, Evan Harold
Singh, Harwant

NEW BRUNSWICK
Coleman, Warren Kent

ONTARIO
Baier, Wolfgang
Bright, Donald Edward
Gillespie, Terry James
Redhead, Scott Alan
Reeds, Lloyd George
Robertson, George Wilber
Smith, Maurice Vernon
Summers, John David
Townsend, Gordon Frederick
Yan, Maxwell Menuhin

QUEBEC
Furlan, Valentin
Stewart, Robin Kenny

SASKATCHEWAN
Kartha, Kutty Krishnan
Ripley, Earle Allison

OTHER COUNTRIES
Hocking, Drake
Holsworth, William Norton
Sleeter, Thomas David

BIOLOGICAL SCIENCES

Anatomy

ALABAMA
Aboukarsh, Naama A Dk
Abrahamson, Dale Raymond
Al-lami, Fadhil
Bhatnagar, Yogendra Mohan
Brown, Jerry William
Chibuzo, Gregory Anenonu
Chronister, Robert Blair
Gardner, William Albert, Jr
Gray, Bruce William
Hamel, Earl Gregory, Jr
Hand, George Samuel, Jr
Hoffman, Henry Harland
Holloway, Clarke L

Hovde, Christian Arneson
Jenkins, Ronald Lee
Kayes, Stephen Geoffrey
Kincaid, Steven Alan
Krista, Laverne Mathew
Marchase, Richard Banfield
Rodning, Charles Bernard
Shackleford, John Murphy
Speed, Edwin Maurice
West, Seymour S
Wilborn, Walter Harrison
Williams, Raymond Crawford
Wyss, James Michael

ALASKA
Ebbesson, Sven O E
Fay, Francis Hollis
Guthrie, Russell Dale

ARIZONA
Angevine, Jay Bernard, Jr
Bernays, Elizabeth Anna
Bertke, Eldridge Melvin
Chiasson, Robert Breton
Hendrix, Mary J C
Krutzsch, Philip Henry
LeBouton, Albert V
McCauley, William John
McCuskey, Robert Scott
Markle, Ronald A
Nicolls, Ken E
Spofford, Walter Richardson, II
Tarby, Theodore John

ARKANSAS
Bentley, Cleo L
Burns, Edward Robert
Cave, Mac Donald
Chang, Louis Wai-Wah
McMillan, Harlan L
Marvin, Horace Newell
Morgans, Leland Foster
Pauly, John Edward
Powell, Ervin William
Scheving, Lawrence Einar
Sherman, Jerome Kalman
Soloff, Bernard Leroy
Tank, Patrick Wayne
Uyeda, Carl Kaoru

CALIFORNIA
Ahmad, Nazir
Akin, Gwynn Collins
Alberch, Pere
Alden, Roland Herrick
Alvarino de Leira, Angeles
Arcadi, John Albert
Armstrong, Rosa Mae
Asling, Clarence Willet
Baker, Mary Ann
Benes, Elinor Simson
Bernard, George W
Bernick, Sol
Beuchat, Carol Ann
Bevelander, Gerrit
Bok, P Dean
Bowers, Roger Raymond
Boyne, Philip John
Breisch, Eric Alan
Burnside, Mary Beth
Callison, George
Campbell, John Howland
Case, Norman Mondell
Chamberlain, Jack G
Chase, Robert A
Chow, Kao Liang
Clemente, Carmine Domenic
Collins, Robert C
Cullen, Michael Joseph
Dalgleish, Arthur E
Dearden, Lyle Conway
Diamond, Ivan
Dixon, Andrew Derart
Dodge, Alice Hribal
Dougherty, Harry L
Drewes, Robert Clifton
Eiserling, Frederick A
Enders, Allen Coffin
Erickson, Kent L
Erpino, Michael James
Fallon, James Harry
Faulkin, Leslie J, Jr
Fierstine, Harry Lee
Fisher, Robin Scott
Fletcher, William H
Garoutte, Bill Charles
Gifford, Ernest Milton
Giolli, Roland A
Glass, Laurel Ellen
Globus, Albert
Goldstein, Abraham M B
Gray, Constance Helen
Gray, Donald James
Greenleaf, Robert Dale
Hayashida, Tetsuo
Hayden, Jess, Jr
Hess, Robert William
Hooker, William Mead
Hungerford, Gerald Fred
Hunt, Guy Marion, Jr
Hunter, Robert L
Jenkins, Floyd Albert
Jones, Edward George
Kahn, Raymond Henry

Karten, Harvey J
Keller, Raymond E
King, Barry Frederick
Kitchell, Ralph Lloyd
Klouda, Mary Ann Aberle
Ko, Chien-Ping
Kruger, Lawrence
Laham, Quentin Nadime
LaVail, Jennifer Hart
LaVail, Matthew Maurice
Lianides, Sylvia Panagos
Long, John Arthur
Lu, John Kuew-Hsiung
Lutt, Carl J
McKinley, Michael P
McMillan, Paul Junior
Magoun, Horace Winchell
Marchand, E Roger
Marshall, John Foster
Maxwell, David Samuel
Menees, James H
Mensah, Patricia Lucas
Monie, Ian Whitelaw
Monroe, Barbara Samson Granger
Mortenson, Theadore Hampton
Murad, Turhon Allen
Murphy, Henry D
Myrick, Albert Charles, Jr
Nelson, Gayle Herbert
Newman, Bertha L
Nishioka, Richard Seiji
Northcutt, Richard Glenn
Orr, Beatrice Yewer
Outzen, Henry Clair, Jr
Palade, George E
Paule, Wendelin Joseph
Plopper, Charles George
Ralston, Henry James, III
Reaven, Eve P
Ribak, Charles Eric
Richard, Christopher Alan
Roberts, Walter Herbert B
Rost, Thomas Lowell
Ruibal, Rodolfo
Sartoris, David John
Saunders, John Bertrand De Cusance Morant
Sawyer, Charles Henry
Scheibel, Arnold Bernard
Schmucker, Douglas Lees
Schooley, Caroline Naus
Schultz, Robert Lowell
Schweisthal, Michael Robert
Sechrist, John William
Shook, Brenda Lee
Slavin, Bernard Geoffrey
Snow, Mikel Henry
Srebnik, Herbert Harry
Stilwell, Donald Lonson
Strautz, Robert Lee
Swett, John Emery
Taylor, Anna Newman
Templeton, McCormick
Towers, Bernard
Turgeon, Judith Lee
Turner, Robert Stuart
Tyler, Walter Steele
Vaughn, James E, Jr
Vijayan, Vijaya Kumari
Wakefield, Caroline Leone
Wangler, Roger Dean
Waters, James Frederick
Wilson, Doris Burda
Wimer, Cynthia Crosby
Winer, Jeffery Allan
Wood, Richard Lyman
Young, Richard Wain
Younoszai, Rafi
Zakhary, Rizkalla
Zamenhof, Stephen
Zimmerman, Emery Gilroy

COLORADO
Billenstien, Dorothy Corinne
Charney, Michael
Davis, Robert Wilson
Dubin, Mark William
Fifkova, Eva
Finger, Thomas Emanuel
Frandson, Rowen Dale
Hanken, James
Jafek, Bruce William
Lebel, Jack Lucien
Liechty, Richard Dale
Meyer, Hermann
Moury, John David
Palmer, Michael Rule
Plakke, Ronald Keith
Rash, John Edward
Roper, Stephen David
Rudnick, Michael Dennis
Schulter-Ellis, Frances Pierce
Solomon, Gordon Charles
Whitlock, David Graham
Willson, John Tucker

CONNECTICUT
Ariyan, Stephan
Barrnett, Russell Joffree
Barry, Michael Anhalt
Blackburn, Daniel Glenn
Chatt, Allen Barrett
Constantine-Paton, Martha
Cooperstein, Sherwin Jerome

Crelin, Edmund Slocum
Forbes, Thomas Rogers
Galton, Peter Malcolm
Grasso, Joseph Anthony
Grimm-Jorgensen, Yvonne
Hand, Arthur Ralph
Matheson, Dale Whitney
Morest, Donald Kent
Mugnaini, Enrico
Roos, Henry
Schwartz, Ilsa Roslow
Watkins, Dudley T
Yaeger, James Amos

DELAWARE
Ruben, Regina Lansing
Skeen, Leslie Carlisle
Song, Jiakun

DISTRICT OF COLUMBIA
Albert, Ernest Narinder
Allan, Frank Duane
Baldwin, Kate M
Ball, William David
Bernor, Raymond Louis
Bernstein, Jerald Jack
Bulger, Ruth Ellen
Campbell, Carlos Boyd Godfrey
Cobb, William Montague
Crisp, Thomas Mitchell, Jr
Domning, Daryl Paul
Emry, Robert John
Ericksen, Mary Frances
Forman, David S
Grand, Theodore I
Hakim, Raziel Samuel
Hayek, Lee-Ann Collins
Hayes, Raymond L, Jr
Herman, Barbara Helen
Hussain, Syed Taseer
Jester, James Vincent
Johnson, Thomas Nick
Kapur, Shakti Prakash
Kelly, Douglas Elliott
Koering, Marilyn Jean
Kromer, Lawrence Frederick
Murphy, James John
Norman, Wesley P
Rapisardi, Salvatore C
Slaby, Frank J
Snell, Richard Saxon
Telford, Ira Rockwood
Warner, Louise
Whitmore, Frank Clifford, Jr
Young, John Karl

FLORIDA
Allen, Ted Tipton
Bennett, Gudrun Staub
Berman, Irwin
Bressler, Steven L
Brown, Hugh Keith
Bunge, Richard Paul
Cameron, Don Frank
Chen, L T
Clendenin, Martha Anne
Cutts, James Harry
DeRousseau, C(arol) Jean
Dwornik, Julian Jonathan
Ericson, Grover Charles
Feldherr, Carl M
Gfeller, Eduard
Goldberg, Stephen
Gordon, Kenneth Richard
Gorniak, Gerard Charles
Hinkley, Robert Edwin, Jr
Hope, George Marion
Kallenbach, Ernst Adolf Theodor
Krishan, Awtar
Larkin, Lynn Haydock
Leonard, Christiana Morison
Maue-Dickson, Wilma
Nolan, Michael Francis
Pearl, Gary Steven
Phelps, Christopher Prine
Reynolds, John Elliott, III
Rhodin, Johannes A G
Romrell, Lynn John
Ross, Michael H
Samuelson, Don Arthur
Saporta, Samuel
Schnitzlein, Harold Norman
Shireman, Rachel Baker
Suzuki, Howard Kazuro
Taylor, George Thomas
Weber, James Edward
Woods, Charles Arthur
Woolfenden, Glen Everett
Wyneken, Jeanette
Yakaitis-Surbis, Albina Ann

GEORGIA
Adkison, Claudia R
Bennett, Sara Neville
Benoit, Peter Wells
Binnicker, Pamela Caroline
Black, Asa C, Jr
Black, John B
Bockman, Dale Edward
Colborn, Gene Louis
Diboll, Alfred
Doetsch, Gernot Siegmar
Edwards, Betty F
English, Arthur William

Anatomy (cont)

Golarz-De Bourne, Maria Nelly
Goodale, Fairfield
Gray, Stephen Wood
Gulati, Adarsh Kumar
Hawkins, Isaac Kinney
Holland, Robert Campbell
Jacobs, Virgil Leon
Kirby, Margaret Loewy
Lo, Woo-Kuen
McKenzie, John Ward
Manocha, Sohan Lall
Mortenson, Leonard Earl
Mulroy, Michael Joseph
Paulsen, Douglas F
Puchtler, Holde
Scott, John Watts, Jr
Sutin, Jerome
Tigges, Johannes
Welter, Dave Allen
Westerfield, Clifford
Wolf, Steven L

HAWAII
Berger, Andrew John
Diamond, Milton
Hammer, Ronald Page, Jr
Kleinfeld, Ruth Grafman
Nelson, Marita Lee

IDAHO
DeSantis, Mark Edward
Eroschenko, Victor Paul
Fuller, Eugene George
Laundre, John William
Stephens, Trent Dee

ILLINOIS
Allin, Edgar Francis
Aydelotte, Margaret Beesley
Bachop, William Earl
Bakkum, Barclay W
Baron, David Alan
Bieler, Rüdiger
Blaha, Gordon C
Bodley, Herbert Daniel, II
Buschmann, MaryBeth Tank
Buschmann, Robert J
Caspary, Donald M
Chiakulas, John James
Cho, Yongock
Combs, Clarence Murphy
Connors, Natalie Ann
Corruccini, Robert Spencer
Costa, Raymond Lincoln, Jr
Cralley, John Clement
Daniels, Edward William
De Bruyn, Peter Paul Henry
Dinsmore, Charles Earle
DuBrul, E Lloyd
Durica, Thomas Edward
Eisenberg, Brenda Russell
Engel, John Jay
Engel, Milton Baer
Farbman, Albert Irving
Foote, Florence Martindale
Forman, G Lawrence
Gaik, Geraldine Catherine
Geinisman, Yuri
Gibbs, Daniel
Gowgiel, Joseph Michael
Greaves, Walter Stalker
Greenough, William Tallant
Griffiths, Thomas Alan
Grimm, Arthur F
Hasegawa, Junji
Hast, Malcolm Howard
Heltne, Paul Gregory
Holmes, Edward Bruce
Holmes, Kenneth Robert
Kasprow, Barbara Ann
Kernis, Marten Murray
Khodadad, Jena Khadem
Kiely, Michael Lawrence
Krieg, Wendell Jordan
Lavelle, Arthur
LaVelle, Faith Wilson
Leeson, Charles Roland
Leven, Robert Maynard
McCandless, David Wayne
Maibenco, Helen Craig
Mixter, Russell Lowell
Monsen, Harry
Moticka, Edward James
Nakajima, Yasuko
Naples, Virginia L
O'Morchoe, Charles C C
O'Morchoe, Patricia Jean
Orr, Mary Faith
Paparo, Anthony A
Phelps, Creighton Halstead
Rabuck, David Glenn
Rafferty, Nancy S
Reed, Charles Allen
Rezak, Michael
Romack, Frank Eldon
Rosenberger, Alfred L
Safanie, Alvin H
Saper, Clifford B
Scapino, Robert Peter
Seale, Raymond Ulric
Siegel, Jonathan Howard
Simon, Mark Robert

Simpson, Sidney Burgess, Jr
Singer, Ronald
Straus, Helen Lorna Puttkammer
Sturtevant, Ruthann Patterson
Sweeny, Lauren J
Thaemert, Jona Carl
Thomas, Carolyn Eyster
Thurow, Gordon Ray
Towns, Clarence, Jr
Tuttle, Russell Howard
Ulinski, Philip Steven
Velardo, Joseph Thomas
Waltenbaugh, Carl
Walter, Robert John
Williams, Thomas Alan
Zaki, Abd El-Moneim Emam
Zalisko, Edward John
Zimmerman, Roger Paul

INDIANA
Anderson, William John
Babbs, Charles Frederick
Bayer, Shirley Ann
Blevins, Charles Edward
Das, Gopal Dwarka
Davis, Grayson Steven
Dial, Norman Arnold
Hafner, Gary Stuart
Hinsman, Edward James
Hoversland, Roger Carl
Hullinger, Ronald Loral
Jersild, Ralph Alvin, Jr
Katzberg, Allan Alfred
Kennedy, Duncan Tilly
McKibben, John Scott
Mescher, Anthony Louis
Mizell, Sherwin
Morré, Dorothy Marie
Murphy, Robert Carl
Murray, Raymond Gorbold
Peterson, Richard George
Pietsch, Paul Andrew
Schmedtje, John Frederick
Schroeder, Dolores Margaret
Sever, David Michael
Shellhamer, Robert Howard
Stromberg, Melvin Willard
Stump, John Edward
Van Sickle, David C
Witzmann, Frank A

IOWA
Adams, Donald Robert
Bergman, Ronald Arly
Bhalla, Ramesh C
Carithers, Jeanine Rutherford
Christensen, George Curtis
Ciochon, Russell Lynn
Coulter, Joe Dan
Dawson, David Lynn
Dellmann, H Dieter
Ghoshal, Nani Gopal
Heidger, Paul McClay, Jr
Jacobs, Richard M
Kessel, Richard Glen
Kirkland, Willis L
Kollros, Jerry John
McInroy, Elmer Eastwood
Maynard, Jerry Allen
Meetz, Gerald David
Mennega, Aaldert
Schelper, Robert Lawrence
Searls, James Collier
Shaw, Gaylord Edward
Stromer, Marvin Henry
Thompson, Sue Ann
Tranel, Daniel T
Williams, Terence Heaton

KANSAS
Beary, Dexter F
Besharse, Joseph Culp
Chapman, Albert Lee
Dunn, Jon D
Enders, George Crandell
Foltz, Floyd Mathew
Gattone, Vincent H, II
Hung, Kuen-Shan
Keller, Leland Edward
Klein, Robert Melvin
Klemm, Robert David
Lambson, Roger O
Lavia, Lynn Alan
Mohn, Melvin P
Redick, Mark Lankford
Vacca, Linda Lee
Walker, Richard Francis

KENTUCKY
Bhatnagar, Kunwar Prasad
Campbell, Ferrell Rulon
Cotter, William Bryan, Jr
Demski, Leo Stanley
Farrar, William Wesley
Fontaine, Julia Clare
Fowler, Ira
Fuller, Peter McAfee
Gillilan, Lois Adell
Gregg, Robert Vincent
Hamon, J Hill
Herbener, George Henry
Longley, James Baird
Matulionis, Daniel H
Moody, William Glenn

Nettleton, G(ary) Stephen
Nikitovitch-Winer, Miroslava B
Rink, Richard Donald
Smith, Stephen D
Swartz, Frank Joseph
Swigart, Richard Hanawalt
Traurig, Harold H

LOUISIANA
Allen, Emory Raworth
Beal, John Anthony
Brizzee, Kenneth Raymond
Carpenter, Stanley Barton
Clawson, Robert Charles
Constantinides, Paris
Davenport, William Daniel, Jr
Dyer, Robert Frank
Garcia, Meredith Mason
Gasser, Raymond Frank
Green, Jeffrey David
Harper, Jon William
Hibbs, Richard Guythal
Homberger, Dominique Gabrielle
Jewell, Frederick Forbes, Sr
Kasten, Frederick H
Layman, Don Lee
Low, Frank Norman
Matthews, Murray Albert
Nickerson, Stephen Clark
Reed, Adrian Faragher
Robinson, Roy Garland, Jr
Ruby, John Robert
Sarphie, Theodore G
Seltzer, Benjamin
Silverman, Harold
Specian, Robert David
Titkemeyer, Charles William
Turner, Hugh Michael
Vaupel, Martin Robert
Walker, Leon Bryan, Jr
Weber, Joseph T
Zimny, Marilyn Lucile

MAINE
Bell, Allen L
Cole, Wilbur Vose
Hinds, James Wadsworth
Minkoff, Eli Cooperman

MARYLAND
Anderson, Larry Douglas
Barry, Ronald Everett, Jr
Bell, Mary
Benevento, Louis Anthony
Brightman, Milton Wilfred
Broadwell, Richard Dow
Church, Lloyd Eugene
Cotton, William Robert
Cowan, W Maxwell
De Monastério, Francisco M
Donati, Edward Joseph
Eglitis, Martin Alexandris
Ennist, David L
Ferrans, Victor Joaquin
Freed, Michael Abraham
Fuson, Roger Baker
Gartner, Leslie Paul
Gobel, Stephen
Green, Martin David
Greenhouse, Gerald Alan
Greulich, Richard Curtice
Grewe, John Mitchell
Gross, James Harrison
Harding, Fann
Harkins, Rosemary Knighton
Hein, Rosemary Ruth
Hiatt, James Lee
Hotton, Nicholas, III
Hutchins, Grover MacGregor
Kalt, Marvin Robert
Kibbey, Maura Christine
Leach, Berton Joe
Liebelt, Annabel Glockler
MacLean, Paul Donald
Margolis, Ronald Neil
Meszler, Richard M
Moreira, Jorge Eduardo
Nelson, Ralph Francis
Oberdorfer, Michael Douglas
Petrali, John Patrick
Platt, William Rady
Rioch, David McKenzie
Robison, Wilbur Gerald, Jr
Rose, Kenneth David
Seibel, Werner
Shear, Charles Robert
Sheridan, Michael N
Strum, Judy May
Yellin, Herbert
Zirkin, Barry Ronald

MASSACHUSETTS
Ahlberg, Henry David
Albertini, David Fred
Appel, Michael Clayton
Atema, Jelle
Balogh, Karoly
Baratz, Robert Sears
Begg, David A
Belt, Warner Duane
Bond, George Walter
Bridges, Robert Stafford
Campbell, William (Aloysius)
Chlapowski, Francis Joseph

Clark, Sam Lillard, Jr
Coombs, Margery Chalifoux
Dacheux, Ramon F, II
Damassa, David Allen
Dittmer, John Edward
Dorey, Cheryl Kathleen
Douglas, William J
El-Bermani, Al-Walid I
Eldred, William D
Feldman, Martin Leonard
Frommer, Jack
George, Stephen Anthony
Gibbons, Michael Francis, Jr
Giffin, Emily Buchholtz
Gipson, Ilene Kay
Gonnella, Patricia Anne
Graybiel, Ann M
Greep, Roy Orval
Gustafson, Alvar Walter
Haidak, Gerald Lewis
Hay, Elizabeth Dexter
Henry, Joseph L
Hoar, Richard Morgan
Hoffmann, Joan Carol
Ito, Susumu
Jacobson, Stanley
Kronman, Joseph Henry
Lambertsen, Richard H
Maran, Janice Wengerd
Marieb, Elaine Nicpon
Marks, Sandy Cole, Jr
Miller, James Albert, Jr
Murnane, Thomas William
Murphy, Richard Arthur
Murthy, A S Krishna
Nandy, Kalidas
Nauta, Walle J H
Nixon, Charles William
Payne, Bertram R
Peters, Alan
Pfister, Richard Charles
Pino, Richard M
Raviola d'Elia, Giuseppina E(nrica)
Remmel, Ronald Sylvester
Rollason, Herbert Duncan
Sidman, Richard Leon
Smith, Dennis Matthew
Sorokin, Sergei Pitirimovitch
Stein, Otto Ludwig
Struthers, Robert Claflin
Susi, Frank Robert
Vanderburg, Charles R
Vaughan, Deborah Whittaker
Welch, Gary William
West, Christopher Drane
White, Edward Lewis

MICHIGAN
Albright, Raymond Gerard
Alcala, Jose Ramon
Al Saadi, A Amir
Al-Saadi, Abdul A
Avery, James Knuckey
Bernstein, Maurice Harry
Bivins, Brack Allen
Boving, Bent Giede
Brown, Esther Marie
Brown, Roger E
Burkel, William E
Buss, Jack Theodore
Carlson, Bruce Martin
Carlson, David Sten
Castelli, Walter Andrew
Christensen, A(lbert) Kent
Connelly, Thomas George
Coye, Robert Dudley
Coyle, Peter
Dapson, Richard W
Diaz, Fernando G
Eichler, Victor B
Falls, William McKenzie
Fischer, Theodore Vernon
Fisher, Don Lowell
Fisher, Leslie John
Floyd, Alton David
Friar, Robert Edsel
Fritts-Williams, Mary Louise Monica
Froiland, Thomas Gordon
Garg, Bhagwan D
Getchell, Thomas Vincent
Glover, Roy Andrew
Hatton, Glenn Irwin
Henry, Raymond Leo
Houts, Larry Lee
Huelke, Donald Fred
Hurst, Edith Marie Maclennan
Jampel, Robert Steven
Jenkins, Thomas William
Johnson, John Irwin, Jr
Kim, Sun-Kee
Lasker, Gabriel (Ward)
Lew, Gloria Maria
Lillie, John Howard
MacCallum, Donald Kenneth
McNamara, James Alyn, Jr
Meyer, David Bernard
Mizeres, Nicholas James
Moosman, Darvan Albert
Nag, Asish Chandra
Newman, Sarah Winans
Person, Steven John
Plagge, James Clarence
Pourcho, Roberta Grace
Pysh, Joseph John

Radin, Eric Leon
Rafols, Jose Antonio
Rieck, Norman Wilbur
Rupp, Ralph Russell
Seefeldt, Vern Dennis
Sippel, Theodore Otto
Strachan, Donald Stewart
Tosney, Kathryn W
Townsend, Samuel Franklin
Trachtenberg, Michael Carl
Tweedle, Charles David
Walker, Bruce Edward
Webb, R Clinton
Wood, Pauline J
Woodburne, Russell Thomas

MINNESOTA
Alsum, Donald James
Bauer, Gustav Eric
Beitz, Alvin James
Carmichael, Stephen Webb
Czarnecki, Caroline Mary Anne
Dapkus, David Conrad
Dixit, Padmakar Kashinath
Elde, Robert Philip
Erickson, James Eldred
Erlandsen, Stanley L
Fletcher, Thomas Francis
Forbes, Donna Jean
Goble, Frans Cleon
Hancock, Peter Adrian
Heggestad, Carl B
Hegre, Orion Donald
Low, Walter Cheney
McCann, Lester J
Meyer, Rita A
Saccoman, Frank (Michael)
Severson, Arlen Raynold
Seybold, Virginia Susan (Dick)
Sicard, Raymond Edward
Smithberg, Morris
Sorenson, Robert Lowell
Sundberg, Ruth Dorothy
Theisen, Charles Thomas
Thompson, Edward William
Todt, William Lynn
Welter, Alphonse Nicholas
Yoss, Robert Eugene

MISSISSIPPI
Ajemian, Martin
Ashburn, Allen David
Ball, Carroll Raybourne
Haines, Duane Edwin
Lynch, James Carlyle
Martin, Billy Joe
Roy, William Arthur
Smith, Byron Colman
Spann, Charles Henry
Walker, James Frederick
Williams, William Lane

MISSOURI
Beetham, Karen Lorraine
Beringer, Theodore Michael
Brown, Herbert Ensign
Butterworth, Bernard Bert
Cooper, Margaret Hardesty
Crummy, Pressley Lee
Decker, John D
Elftman, Alice G
Elftman, Herbert (Oliver)
Gavan, James Anderson
Gibbs, Finley P
Goodge, William Russell
Harvey, Joseph Eldon
Highstein, Stephen Morris
Jacobs, Allen Wayne
Julyan, Frederick John
Kort, Margaret Alexander
Krause, William John
Lake, Lorraine Frances
Loewy, Arthur D(eCosta)
Lowrance, Edward Walton
McClure, Robert Charles
McClure, Theodore Dean
McLaughlin, Carol Lynn
Massopust, Leo Carl, Jr
Menton, David Norman
Moffatt, David John
Momberg, Harold Leslie
Nyquist-Battie, Cynthia
Paull, Willis K, Jr
Peterson, Roy Reed
Price, Joseph Levering
Quay, Wilbur Brooks
Rana, Mohammed Waheeduz-Zaman
Rasmussen, David Tab
Ring, John Robert
Sayegh, Fayez S
Schreiweis, Donald Otto
Smith, Richard Jay
Spiro, Thomas
Taylor, John Joseph
Tolbert, Daniel Lee
Trotter, Mildred
Tumosa, Nina Jean
Woolsey, Thomas Allen
Yeager, Vernon LeRoy
Young, Paul Andrew

MONTANA
Fawcett, Don Wayne
Phillips, Dwight Edward

NEBRASKA
Baumel, Julian Joseph
Binhammer, Robert T
Buell, Katherine Mayhew
Crouse, David Austin
Dalley, Arthur Frederick, II
Dossel, William Edward
Earle, Alvin Mathews
Fawcett, James Davidson
Fougeron, Myron George
Gardner, Paul Jay
Harn, Stanton Douglas
Hill, Marvin Francis
Holyoke, Edward Augustus
Leuschen, M Patricia
Littledike, Ernest Travis
Macaluso, Sister Mary Christelle
Metcalf, William Kenneth
Sharp, John Graham
Skultety, Francis Miles
Sullivan, James Michael
Todd, Gordon Livingston
Warr, William Bruce

NEVADA
Marlow, Ronald William
Schneider, Lawrence Kruse
Stratton, Clifford James

NEW HAMPSHIRE
Chambers, Wilbert Franklin
Meader, Ralph Gibson
Musiek, Frank Edward
Sokol, Hilda Weyl
Stahl, Barbara Jaffe

NEW JERSEY
Agnish, Narsingh Dev
Alger, Elizabeth A
Auletta, Carol Spence
Bagnell, Carol A
Berendsen, Peter Barney
Boccabella, Anthony Vincent
DeFouw, David O
DeProspo, Nicholas Dominick
Eastwood, Abraham Bagot
Feder, Harvey Herman
Feldman, Susan C
Ford, Daniel Morgan
Gagna, Claude Eugene
Gilani, Shamshad H
Gona, Amos G
Gona, Ophelia Delaine
Halpern, Myron Herbert
Hart, Nathan Hoult
Hess, Arthur
Hollinshead, May B
Hsu, Linda
Kleinschuster, Stephen J, III
Kmetz, John Michael
Kozam, George
Krauthamer, George Michael
Laemle, Lois K
Leung, Christopher Chung-Kit
McAuliffe, William Geoffrey
Macdonald, Gordon J
Malamed, Sasha
Mele, Frank Michael
Rhoten, William Blocher
Saiff, Edward Ira
Seiden, David
Tesoriero, John Vincent
Wallace, Edith Winchell
Weis, Peddrick
Wilson, Frank Joseph
Yu, Mang Chung

NEW MEXICO
Bourne, Earl Whitfield
Gray, Edwin R
Kelley, Robert Otis
Martin, William Clarence
Napolitano, Leonard Michael
Saland, Linda C
Trotter, John Allen

NEW YORK
Aldridge, William Gordon
Alexander, A Allan
Ambron, Richard Thomas
Ames, Ira Harold
April, Ernest W
Aschner, Michael
Baden, Ernest
Baker, Robert George
Becker, Norwin Howard
Beckert, William Henry
Bedford, John Michael
Benzo, Camillo Anthony
Bergmann, Louis Lawrence
Bielat, Kenneth L
Black, Virginia H
Bothner, Richard Charles
Boucher, Louis Jack
Brandt, Philip Williams
Brody, Harold
Brophy, Mary O'Reilly
Brownscheidle, Carol Mary
Cabot, John Boit, II
Carr, Malcolm Wallace
Carriere, Rita Margaret
Centola, Grace Marie
Cohan, Christopher Scott
Cohn, Deirdre Arline

Coleman, Paul David
Colman, David Russell
Cummings, John Francis
DeLahunta, Alexander
Demeter, Steven
Dewey, Maynard Merle
De Zeeuw, Carl Henri
Dibennardo, Robert
Di Stefano, Henry Saverio
Doolittle, Richard L
Dornfest, Burton S
Doty, Stephen Bruce
Drakontides, Anna Barbara
Durr, David P
Eckert, Barry S
Edds, Kenneth Tiffany
Edmonds, Richard H
Ely, Charles A
Emmel, Victor Meyer
Engbretson, Gustav Alan
Evans, Howard Edward
Evans, Lance Saylor
Feagans, William Marion
Feinman, Max L
Felten, David L
Fetcho, Joseph Robert
Firriolo, Domenic
Fleagle, John G
Flood, Dorothy Garnett
Fruhman, George Joshua
Garcia, Alfredo Mariano
Gershon, Michael David
Gil, Joan
Glomski, Chester Anthony
Goodman, Donald Charles
Goodwin, Robert Earl
Gray, James Clarke
Greenberger, Lee M
Gresik, Edward William
Grob, Howard Shea
Grosso, Leonard
Habel, Robert Earl
Hansen, John Theodore
Hayes, Everett Russell
Henrikson, Ray Charles
Hiemae, Karen Marion
Hillman, Dean Elof
Hirschman, Albert
Horn, Eugene Harold
Horst, G Roy
Imai, Hideshige
Jakway, Jacqueline Sinks
Jimenez-Marin, Daniel
Jones, Oliver Perry
Kallen, Frank Clements
Kaye, Gordon I
Kaye, Nancy Weber
Kiely, Lawrence J
Kinzey, Warren Glenford
Krause, David Wilfred
Lamberg, Stanley Lawrence
Lazarow, Paul B
Lemanski, Larry Frederick
Levitan, Max
Lewis, Carmie Perrotta
Loy, Rebekah
Lund, Richard
Lutton, John D
McCune, Amy Reed
Maderson, Paul F A
Martin, Kathryn Helen
Mazurkiewicz, Joseph Edward
Mendel, Frank C
Miller, Richard Avery
Miller, Sue Ann
Minor, Ronald R
Mitchell, Ormond Glenn
Monheit, Alan G
Moss-Salentijn, Letty
Motzkin, Shirley M
Noback, Charles Robert
Notter, Mary Frances
Osinchak, Joseph
Padawer, Jacques
Pasik, Pedro
Pasik, Tauba
Penney, David P
Pentney, Roberta Pierson
Pesetsky, Irwin
Pfeiffer, Carroll Athey
Pham, Tuan Duc
Pough, Frederick Harvey
Prutkin, Lawrence
Rasweiler, John Jacob, IV
Rhodes, Rondell H
Ringler, Neil Harrison
Riss, Walter
Robertson, Douglas Reed
Roel, Lawrence Edmund
Ruggiero, David A
Sack, Wolfgang Otto
Sansone, Frances Marie
Satir, Peter
Scharrer, Berta Vogel
Schecter, Arnold Joel
Schmidt, John Thomas
Schuel, Herbert
Severin, Charles Matthew
Sherman, Burton Stuart
Shriver, Joyce Elizabeth
Silverman, Ann Judith
Singh, Inder Jit
Sladek, John Richard, Jr
Smeriglio, Alfred John

Smith, Richard Andrew
Spence, Alexander Perkins
Springer, Alan David
Stempak, Jerome G
Stern, Jack Tuteur, Jr
Strominger, Norman Lewis
Susman, Randall Lee
Swan, Roy Craig, Jr
Szabo, Piroska Ludwig
Terzakis, John A
Tichauer, Erwin Rudolph
Tieman, Suzannah Bliss
Udin, Susan Boymel
Wahlert, John Howard
Ware, Carolyn Bogardus
Warfel, John Hiatt
Waterhouse, Joseph Stallard
Webber, Richard Harry
Wellmann, Klaus Friedrich
Wenk, Eugene J
West, William T
Whiting, Anne Margaret
Yazulla, Stephen
Zanetti, Nina Clare

NORTH CAROLINA
Ankel-Simons, Friderun Annursel
Becker, Roland Frederick
Bell, Mary Allison
Berberian, Paul Anthony
Black, Betty Lynne
Bo, Walter John
Brody, Arnold R
Carter, Charleata A
Cornwell, John Calhoun
Daniel, Hal J
Duke, Kenneth Lindsay
Eddy, Edward Mitchell
Everett, John Wendell
Garris, David Roy
Harrison, Frederick Williams
Henson, O'Dell Williams, Jr
Johnson, Franklin M
Johnston, Malcolm Campbell
Klintworth, Gordon K
Koch, William Edward
Lauder, Jean Miles
Lawrence, Irvin E, Jr
Lay, Douglas M
Lemasters, John J
McCreight, Charles Edward
MacRae, Edith Krugelis
Miller, Inglis J, Jr
Montgomery, Royce Lee
O'Steen, Wendall Keith
Peach, Roy
Putnam, Jeremiah (Jerry) L
Reedy, Michael K
Robertson, George Gordon
Robertson, James David
Sadler, Thomas William
Singleton, Mary Clyde
Smallwood, James Edgar
Smith, Bradley Richard
Sulik, Kathleen Kay
White, Raymond Petrie, Jr
White, Richard Alan
Wolk, Robert George

NORTH DAKOTA
Ollerich, Dwayne A
Ries, Ronald Edward

OHIO
Ackerman, Gustave Adolph, Jr
Albernaz, Jose Geraldo
Alley, Keith Edward
Barnes, Karen Louise
Batten, Bruce Edgar
Bloch, Edward Henry
Cardell, Robert Ridley, Jr
Chantell, Charles J
Ching, Melvin Chung Hing
Clark, David Lee
Crafts, Roger Conant
Cruce, William L R
Crutcher, Keith A
DiDio, Liberato John Alphonse
Diesem, Charles D
Dimlich, Ruth Van Weenen
Drake, Richard Lee
Dutta, Hiran M
Eastman, Joseph
Egar, Margaret Wells
Eglitis, Irma
Ely, Daniel Lee
Enlow, Donald Hugh
Fucci, Donald James
Gaughran, George Richard Lawrence
Gilloteaux, Jacques Jean-Marie A
Glenn, Loyd Lee
Goldstein, David Louis
Hall, James Lawrence
Hamlett, William Cornelius
Hayes, Thomas G
Hering, Thomas M
Hikida, Robert Seiichi
Hilliard, Stephen Dale
Hines, Margaret H
Hopfer, Ulrich
Hostetler, Jeptha Ray
Houston, Willie Walter, Jr
Humbertson, Albert O, Jr
Israel, Harry, III

Anatomy (cont)

Johnson, Thomas Raymond
Keller, Jeffrey Thomas
King, James S
King, John Edward
Latimer, Bruce Millikin
Liebelt, Robert Arthur
Lohse, Carleton Leslie
Martin, George Franklin, Jr
Meineke, Howard Albert
Moore, Fenton Daniel
Morse, Dennis Ervin
Neiman, Gary Scott
Nokes, Richard Francis
Oyen, Ordean James
Pansky, Ben
Patton, Nancy Jane
Peterson, Ellengene Hodges
Rogers, Richard C
St Pierre, Ronald Leslie
Saksena, Vishnu P
Saul, Frank Philip
Shipley, Michael Thomas
Showers, Mary Jane C
Singer, Marcus
Stuesse, Sherry Lynn
Sucheston, Martha Elaine
Taslitz, Norman
Tornheim, Patricia Anne
Venzke, Walter George
Watanabe, Michiko
Williams, Benjamin Hayden
Wismar, Beth Louise

OKLAHOMA
Allison, John Everett
Begovac, Paul C
Breazile, James E
Chung, Kyung Won
Coalson, Robert Ellis
Daron, Garman Harlow
Dugan, Kimiko Hatta
Faulkner, Kenneth Keith
Felts, William Joseph Lawrence
Grubb, Randall Barth
Howes, Robert Ingersoll, Jr
Kerley, Michael A
Knisely, William Hagerman
Lhotka, John Francis
Martin, Loren Gene
Ownby, Charlotte Ledbetter
Papka, Raymond Edward
Russell, Scott D
Snow, Clyde Collins
Staley, Theodore Earnest Leon
Whitmore, Mary (Elizabeth) Rowe
Wickham, M Gary
Wood, Joe George

OREGON
Bartley, Murray Hill, Jr
Bell, Curtis Calvin
Carlisle, Kay Susan
Critchlow, Burtis Vaughn
Dow, Robert Stone
Edmonds, Elaine S
Fahrenbach, Wolf Henrich
Gunberg, David Leo
Hillman, Stanley Severin
Marshall, Frederick James
Niles, Nelson Robinson
Quinton-Cox, Robert
Smith, Catherine Agnes
Weaver, Morris Eugene
Wilson, Marlene Moore
Zimmerman, Earl Abram
Zingeser, Maurice Roy

PENNSYLVANIA
Aker, Franklin David
Albertine, Kurt H
Allen, Theresa O
Amenta, Peter Sebastian
Aston-Jones, Gary Stephen
Beasley, Andrew Bowie
Bennett, Henry Stanley
Bennett, Marvin Herbert
Black, Mark Morris
Boyd, Robert B
Cameron, William Edward
Carpenter, Malcolm Breckenridge
Caso, Louis Victor
Cauna, Nikolajs
Colony-Cokely, Pamela
Conway, John Richard
Coulter, Herbert David, Jr
Crabill, Edward Vaughn
Crouse, Gail
De Groat, William C
DeGroat, William Chesney, Jr
De Pace, Dennis Michael
Dropp, John Jerome
Fish, Frank Eliot
Flexner, Louis Barkhouse
Gay, Carol Virginia Lovejoy
George-Weinstein, Mindy
Greene, Charlotte Helen
Greene, Robert Morris
Grove, Alvin Russell, Jr
Haroian, Alan James
Hartman, Mary Ellen
Hausberger, Franz X
Hilfer, Saul Robert

Johnson, Elmer Marshall
Johnson, Robert Joseph
Kanczak, Norbert M
Karlsson, Ulf Lennart
Kennedy, Michael Craig
Kesner, Michael H
Kochhar, Devendra M
Kriebel, Richard Marvin
Kvist, Tage Nielsen
Ladman, Aaron J(ulius)
Langdon, Herbert Lincoln
Lewis, Michael Edward
McCallister, Lawrence P
McLaughlin, Patricia J
Malewitz, Thomas Donald
Masters, Edwin M
Meyer, R Peter
Miselis, Richard Robert
Monson, Frederick Carlton
Morrison, Adrian Russel
Moskowitz, Norman
Mundell, Robert David
Munger, Bryce Leon
Nachmias, Vivianne T
Nemeth, Andrew Martin
Niewenhuis, Robert James
Niu, Mann Chiang
Ontell, Marcia
Paavola, Laurie Gail
Pelleg, Amir
Pepe, Frank Albert
Phillips, Steven Jones
Piesco, Nicholas Peter
Pratt, Neal Edwin
Pritchard, Hayden N
Raikow, Robert Jay
Ramasastry, Sai Sudarshan
Rapp, Robert
Rosenthal, Theodore Bernard
Ross, Leonard Lester
Rothman, Richard Harrison
Sachs, Howard George
Sanger, Jean M
Schuit, Kenneth Edward
Sedar, Albert William
Shea, John Raymond Michael, Jr
Short, John Albert
Siegel, Michael Ian
Smith, Allen Anderson
Sodicoff, Marvin
Stere, Athleen Jacobs
Sterling, Peter
Surmacz, Cynthia Ann
Swanson, Ernest Allen, Jr
Tobin, Thomas Vincent
Troyer, John Robert
Truex, Raymond Carl
Vergona, Kathleen Anne Dobrosielski
Volz, John Edward
Warner, Francis James
Weiss, Leon
Weston, John Colby
Wideman, Robert Frederick, Jr
Wolfe, Allan Frederick
Zaccaria, Robert Anthony
Zagon, Ian Stuart

RHODE ISLAND
Calabresi, Paul
Dolyak, Frank
Ebner, Ford Francis
Erikson, George Emil
Goss, Richard Johnson
Gough, Robert Edward
Janis, Christine Marie
Lehmkuhle, Stephen W
Ripley, Robert Clarence
Strauss, Elliott William
Yevich, Paul Peter

SOUTH CAROLINA
Augustine, James Robert
Debacker, Hilda Spodheim
Dougherty, William J
Fox, Richard Shirley
Hays, Ruth Lanier
Lockard, Isabel
Metcalf, Isaac Stevens Haistead
Odor, Dorothy Louise
Poteat, William Louis
Simson, Jo Anne V
Weymouth, Richard J
Worthington, Ward Curtis, Jr

SOUTH DAKOTA
Haertel, John David
Martin, James Edward
Moore, Josephine Carroll
Naughten, John Charles
Neufeld, Daniel Arthur
Parke, Wesley Wilkin
Rinker, George Clark
Settles, Harry Emerson

TENNESSEE
Anderson, Ted L
Atnip, Robert Lee
Aulsebrook, Lucille Hagan
Bernstorf, Earl Cranston
Bruesch, Simon Rulin
Burt, Alvin Miller, III
Corliss, Clark Edward
Donaldson, Donald Jay
Elberger, Andrea June

Evans, James Spurgeon
Fedinec, Alexander
Gotcher, Jack Everett
Harwood, Thomas Riegel
Hasty, David Long
Hossler, Fred E
Huggins, James Anthony
King, Lloyd Elijah, Jr
Lawler, James E
Murrell, Leonard Richard
Nanney, Lillian Bradley
Peppler, Richard Douglas
Powers, J Bradley
Pulliam, James Augustus
Richardson, Elisha Roscoe
Rieke, Garl Kalman
Roberts, Lee Knight
Saxon, James Glenn
Schultz, Terry Wayne
Skalko, Richard G(allant)
Wilcox, Harry Hammond
Wilson, Jack Lowery
Young, Joseph Marvin

TEXAS
Adrian, Erle Keys, Jr
Ashby, Jon Kenneth
Ashworth, Robert David
Azizi, Sayed Ausim
Banks, William Joseph, Jr
Blanton, Patricia Louise
Bratton, Gerald Roy
Burghardt, Robert Casey
Burton, Alexis Lucien
Callas, Gerald
Cameron, Ivan Lee
Cannon, Marvin Samuel
Carpenter, Robert James, Jr
Cavazos, Lauro Fred
Coggeshall, Richard E
Cominsky, Nell Catherine
Croley, Thomas Edgar
Dafny, Nachum
Dennison, David Kee
Dill, Russell Eugene
Duncan, Donald
Dung, H C
Fife, William Paul
Finerty, John Charles
Frederick, Jeanne M
Geoghegan, William David
George, Fredrick William
German, Dwight Charles
Gibson, Kathleen Rita
Gonyea, William Joseph
Gonzalez-Lima, Francisco
Grimes, L Nichols
Guthrie, Mary Duncum
Hall, Charles Eric
Halpern, Salmon Reclus
Harrison, Frank
Herbert, Damon Charles
Hild, Walter J
Hill, Ronald Stewart
Houston, Marshall Lee
Kendall, Michael Welt
Kennedy, Joseph Patrick
Leppi, Theodore John
Litke, Larry Lavoe
Lucas, Edgar Arthur
Matthews, James Lester
Moore, Richard Dana
Neaves, William Barlow
Norman, Reid Lynn
Payer, Andrew Francis
Pierce, Jack Robert
Pinero, Gerald Joseph
Rozier, Carolyn K
Russell, Glenn Vinton
Samorajski, Thaddeus
Sampson, Herschel Wayne
Samson, Willis Kendrick
Sauerland, Eberhardt Karl
Schunder, Mary Cothran
Schwartz, Colin John
Seliger, William George
Sis, Raymond Francis
Skjonsby, Harold Samuel
Steele, David Gentry
Taylor, Alan Neil
Tebo, Heyl Gremmer
Throckmorton, Gaylord Scott
Trulson, Michael E
VanderWiel, Carole Jean
Vaughan, Mary Kathleen
Warner, Marlene Ryan
Wilczynski, Walter
Williams, Darryl Marlowe
Williams, Fred Eugene
Williams, Vick Franklin
Willis, William Darrell, Jr
Winborn, William Burt
Wordinger, Robert James
Yee, John Alan
Yollick, Bernard Lawrence
Young, Margaret Claire
Zwaan, Johan Thomas

UTAH
Chapman, Arthur Owen
Creel, Donnell Joseph
Jee, Webster Shew Shun
Miller, Scott Cannon
Nabors, Charles J, Jr

Rodin, Martha Kinscher
Schoenwolf, Gary Charles
Shultz, Leila McReynolds
Stensaas, Larry J
Stensaas, Suzanne Sperling
Stevens, Walter
Van De Graaff, Kent Marshall

VERMONT
Ariano, Marjorie Ann
Krupp, Patricia Powers
Stultz, Walter Alva
Wells, Joseph
Young, William Johnson, II

VIRGINIA
AcKerman, Larry Joseph
Astruc, Juan
Atkins, David Lynn
Banker, Gary A
Brangan, Pamela J
Breil, Sandra J
Brownson, Robert Henry
Brunjes, Peter Crawford
Butler, Ann Benedict
Carson, Keith Alan
Corwin, Jeffrey Todd
Deck, James David
DeSesso, John Michael
Dessouky, Dessouky Ahmad
Evans, Francis Gaynor
Fenner-Crisp, Penelope Ann
Flickinger, Charles John
Grimm, James K
Haar, Jack Luther
Harris, Thomas Mason
Heath, Everett
Isaacson, Robert John
Jeffrey, Jackson Eugene
Jollie, William Pucette
Jordan, Robert Lawrence
Laurie, Gordon William
Leichnetz, George Robert
Muir, William A
Osborne, Paul James
Owers, Noel Oscar
Potts, Malcolm
Pullen, Edwin Wesley
Quattropani, Steven L
Russi, Simon
Sandow, Bruce Arnold
Savitzky, Alan Howard
Scott, David Evans
Seibel, Hugo Rudolf
Sirica, Alphonse Eugene
Snodgrass, Michael Jens
Steeves, Harrison Ross, III
Suter, Daniel B
Talbert, George Brayton

WASHINGTON
Baskin, Denis George
Broderson, Stevan Hardy
Churchill, Lynn
Davies, Jack
DeVito, June Logan
Frederickson, Richard Gordon
Graney, Daniel O
Halbert, Sheridan A
Hampton, James C
Harris, Roger Mason
Hendrickson, Anita Elizabeth
Herring, Susan Weller
Holbrook, Karen Ann
Kashiwa, Herbert Koro
Landau, Barbara Ruth
Luchtel, Daniel Lee
Luft, John Herman
Lundy, John Kent
Odland, George Fisher
Pietsch, Theodore Wells
Ratzlaff, Marc Henry
Reh, Thomas Andrew
Rieke, William Oliver
Rosse, Cornelius
Sundsten, John Wallin
Tamarin, Arnold
Worthman, Robert Paul
Zamora, Cesario Siasoco

WEST VIRGINIA
Beresford, William Anthony
Bowdler, Anthony John
Brown, Paul B
Burns, John Thomas
Chang, William Wei-Lien
Culberson, James Lee
Fix, James D
Friedman, Morton Henry
Hiloowala, Rumy Ardeshir
Martin, William David
Overman, Dennis Orton
Pinkstaff, Carlin Adam
Reilly, Frank Daniel
Reyer, Randall William
Walker, Elizabeth Reed
Williams, Leah Ann

WISCONSIN
Anderson, Frank David
Anderson, John Walberg
Austin, Bert Peter
Benjamin, Hiram Bernard
Bersu, Edward Thorwald

Bolender, David Leslie
Curtis, Robin Livingstone
Fallon, John Francis
Flanigan, Norbert James
Gardner, Weston Deuain
Greaser, Marion Lewis
Jensen, Richard Harvey
Jones, Helena Speiser
Kaplan, Stanley
Kemnitz, Joseph William
Lalley, Peter Michael
Larson, Sanford J
Long, Charles Alan
Long, Sally Yates
Lough, John William, Jr
Markwald, Roger R
Mihailoff, Gregory A
Oaks, John Adams
Oertel, Donata (Mrs Bill M Sugden)
Pettersen, James Clark
Riley, Danny Arthur
Rouse, Thomas C
Royce, George James
Schultz, Edward
Sether, Lowell Albert
Slautterback, David Buell
Snook, Theodore
Snyder, Virginia
Sobkowicz, Hanna Maria
Terasawa, EI
Weil, Michael Ray
Werner, Joan Kathleen

WYOMING
Lillegraven, Jason Arthur

PUERTO RICO
Garcia-Castro, Ivette
Kicliter, Ernest Earl, Jr
Luckett, Winter Patrick
Norvell, John Edmondson, III
Virkki, Niilo

ALBERTA
Cookson, Francis Bernard
Currie, Philip John
Famiglietti, Edward Virgil, Jr
Holland, Graham Rex
Leeson, Thomas Sydney
Levene, Cyril
Osler, Margaret J
Russell, Anthony Patrick
Sperber, Geoffrey Hilliard
Spira, Arthur William
Stell, William Kenyon
Wyse, John Patrick Henry

BRITISH COLUMBIA
Bressler, Bernard Harvey
Friedman, Constance Livingstone
Friz, Carl T
Hollenberg, Martin James
Jasch, Laura Gwendolyn
Salt, Walter Raymond
Schwarz, Dietrich Walter Friedrich
Semba, Kazue
Slonecker, Charles Edward
Todd, Mary Elizabeth
Webber, William A

MANITOBA
Bertalanffy, Felix D
Braekevelt, Charlie Roger
Gibson, Maurice Henry Lindsay
Klass, Alan Arnold
Nathaniel, Edward J H
Persaud, Trivedi Vidhya Nandan
Weisman, Harvey

NOVA SCOTIA
Armour, John Andrew
Chapman, David MacLean
Cooper, John (Hanwell)
Dickson, Douglas Howard
Hansell, Margaret Mary
Hopkins, David Alan
Leslie, Ronald Allan
Rutherford, John Garvey
Saunders, Richard L de C H
Vethamany, Victor Gladstone
Walker, Joan Marion

ONTARIO
Anderson, James Edward
Barr, Murray Llewellyn
Basmajian, John V
Buchanan, George Dale
Buck, Robert Crawforth
Butler, Richard Gordon
Carr, David Harvey
Ciriello, John
Craigie, Edward Horne
De Bold, Adolfo J
Fallding, Margaret Hurlstone Hardy
Galil, Khadry Ahmed
Gammal, Elias Bichara
Geissinger, Hans Dieter
Govind, Choonilal Keshav
Habowsky, Joseph Edmund Johannes
Ham, Arthur Worth
Hinke, Joseph Anthony Michael
Irons, Margaret Jean
Jande, Sohan Singh
Kiernan, John Alan

McMillan, Donald Burley
Melcher, Antony Henry
Metuzals, Janis
Montemurro, Donald Gilbert
Moore, Keith Leon
Romero-Sierra, Cesar Aurelio
Saunders, Shelley Rae
Singh, Roderick Pataudi
Sistek, Vladimir
Ten Cate, Arnold Richard
Winterbottom, Richard

PRINCE EDWARD ISLAND
Singh, Amreek

QUEBEC
Bendayan, Moise
Bisaillon, Andre
Bois, Pierre
Bolduc, Reginald J
Briere, Normand
Clermont, Yves Wilfred
Colonnier, Marc
Courville, Jacques
Daoust, Roger
Dykes, Robert William
Gagnon, Real
Garon, Olivier
Isler, Henri Gustave
Jones, Barbara Ellen
Kopriwa, Beatrix Markus
Leblond, Charles Philippe
Nadler, Norman Jacob
Olivier, Andre
Osmond, Dennis Gordon
Parent, Andre
Paris, David Leonard
Pereira, Gerard P
Pierard, Jean Arthur
Poirier, Louis
Ramon-Moliner, Enrique
Richer, Claude-Lise
Simard, Therese Gabrielle
Stephens, Heather R
Warshawsky, Hershey

SASKATCHEWAN
Brandell, Bruce Reeves
Fedoroff, Sergey
Flood, Peter Frederick
Munkacsi, Istvan
Newstead, James Duncan MacInnes

OTHER COUNTRIES
Armato, Ubaldo
Ben-Ze'ev, Avri
Bondareff, William
Brunser, Oscar
Donato, Rosario Francesco
Fahimi, Hossein Dariush
Ganchrow, Donald
Gertz, Samuel David
Guillery, Rainer Walter
Hashimoto, Paulo Hitonari
Helander, Herbert Dick Ferdinand
Hughes, Abbie Angharad
Ishikawa, Hiroshi
Makita, Takashi
Moscovici, Mauricio
Nakane, Paul K
Oxnard, Charles Ernest
Saito, Takuma
Stallard, Richard E
Szebenyi, Emil
Wei, Stephen Hon Yin
Wenger, Byron Sylvester
Yamada, Eichi

Bacteriology

ALABAMA
Brady, Yolanda J
Hiramoto, Raymond Natsuo
Lauerman, Lloyd Herman, Jr
Montgomery, John R
Mora, Emilio Chavez
Wood, David Oliver

ARIZONA
Archer, Stanley J
Fuller, Wallace Hamilton
Gerba, Charles Peter
Ludovici, Peter Paul
Munch, Theodore
Reeves, Henry Courtland

ARKANSAS
Bates, Joseph H
England, James Donald
Johnson, George Thomas

CALIFORNIA
Ames, Giovanna Ferro-Luzzi
Andreoli, Anthony Joseph
Bayne, Henry Godwin
Bibel, Debra Jan
Biberstein, Ernst Ludwig
Bramhall, John Shepherd
Brown, Robert Lee
Brunke, Karen J
Buggs, Charles Wesley
Calderone, Julius G
Caputi, Roger William

Carlucci, Angelo Francis
Catlin, B Wesley
Chen, Benjamin P P
Chou, Tsong-Wen
Clark, Alvin John
Cline, Thomas Warren
Cooper, Robert Chauncey
Davenport, Calvin Armstrong
Delk, Ann Stevens
Eveland, Warren C
Fulco, Armand J
Glembotski, Christopher Charles
Gold, William
Goodman, Richard E
Gunsalus, Robert Philip
Halasz, Nicholas Alexis
Hartmann, Floyd Wellington
Heck, Joseph Gerard
Hemmingsen, Barbara Bruff
Hildebrand, Donald Clair
Hochstein, Lawrence I
Hoeprich, Paul Daniel
Hoke, Carolyn Kay
Ito, Keith A
Janda, John Michael
Knudson, Gregory Blair
Kustu, Sydney Govons
McClelland, Michael
Marshall, Rosemarie
Martinez, Rafael Juan
Moore, Harold Beveridge
Moyed, Harris S
Nassos, Patricia Saima
Page, Larry J
Phillips, Donald Arthur
Pickett, Morris John
Pressman, Ralph
Raffel, Sidney
Raj, Harkisan D
Reizer, Jonathan
Rittenberg, Sydney Charles
Rokeach, Luis Alberto
Romig, William Robert
Rosenberg, Steven Loren
Rosenthal, Sol Roy
Ruby, Edward George
Schiller, Neal Leander
Simons, Robert W
Smith, Kendric Charles
Speck, Reinhard Staniford
Stevens, Ronald Henry
Taylor, Barry L
Trivett, Terrence Lynn
Uyeda, Charles Tsuneo
VanBruggen, Ariena H C
Vickrey, Herta Miller
Wall, Thomas Randolph
Wedberg, Stanley Edward
Weimberg, Ralph
Weiser, Russel Shively
Woodhour, Allen F
Wu, William Gay
York, Charles James
Zerez, Charles Raymond
Zlotnik, Albert
ZoBell, Claude E

COLORADO
Boyd, William Lee
Deane, Darrell Dwight
Harold, Ruth L
Harrison, Monty DeVerl
Morse, Helvise Glessner
Nagy, Julius G
Olson, John Melvin
Patterson, David
Reinbold, George W
Smith, Ralph E
Thomas, William Robb
Ulrich, John August
Vasil, Michael Lawrence

CONNECTICUT
Abeling, Edwin John
Buck, John David
Celesk, Roger A
Dingman, Douglas Wayne
Gorrell, Thomas Earl
Huang, Liang Hsiung
Low, Kenneth Brooks, Jr
Lynch, John Edward
O'Neill, James F
Rupp, W Dean
Tourtellotte, Mark Eton
Wyckoff, Delaphine Grace Rosa

DELAWARE
Eleuterio, Marianne Kingsbury

DISTRICT OF COLUMBIA
Binn, Leonard Norman
Cutchins, Ernest Charles
DeCicco, Benedict Thomas
Fanning, George Richard
Hill, Walter Ernest
Hugh, Rudolph
Kennedy, Eugene Richard
Pittman, Margaret
Rothman, Sara Weinstein
Stokes, Gerald V
Venkatesam, Malabi M

FLORIDA
Albrecht, Stephan LaRowe

Bacus, James Nevill
Betz, John Vianney
Carlson, Harve J
Clark, William Burton, IV
Donovick, Richard
Duggan, Dennis E
Ellias, Loretta Christine
Gabridge, Michael Gregory
Goihman-Yahr, Mauricio
Gottwald, Timothy R
Martinez, Octavio Vincent
Paradise, Lois Jean
Pates, Anne Louise
Shanmugam, Keelnatham
Thirunavukkarasu
Slack, John Madison
Smith, Paul Howard
Stuy, Johan Harrie
Wilkowske, Howard Hugo

GEORGIA
Barbaree, James Martin
Brackett, Robert E
Bryan, Frank Leon
Caruthers, John Quincy
Cherry, William Bailey
Clemens, William Bryson
Dailey, Harry A
Dowell, Vulus Raymond, Jr
Doyle, Michael Patrick
Goldsmith, Edward
Good, Robert Campbell
Hall, Charles Thomas
Hamdy, Mostafa Kamal
Hancock, Kenneth Farrell
Harrell, William Knox
Jones, Gilda Lynn
Kubica, George P
Morse, Stephen Allen
Pine, Leo
Preblud, Stephen Robert
Reinhardt, Donald Joseph
Sawyer, Richard Trevor
Smith, Peter Byrd
Taylor, Gerald C
Tzianabos, Theodore
Van Eseltine, William Parker
Wiebe, William John
Williams, Joy Elizabeth P
Woodley, Charles Lamar
Ziegler, Harry Kirk

HAWAII
Berger, Leslie Ralph
Patterson, Harry Robert
Phillips, John Howell, Jr

IDAHO
Beck, Sidney M
Crawford, Donald Lee
Crawford, Ronald L
Farrell, Larry Don
Gilmour, Campbell Morrison
Hibbs, Robert A
Hostetter, Donald Lee
Jones, Lewis William
Teresa, George Washington
Winston, Vern

ILLINOIS
Ehrlich, Richard
Goldin, Milton
Henderson, Thomas Otis
Holzer, Timothy J
Klinger, Lawrence Edward
Klubek, Brian Paul
Laffler, Thomas G
McKinley, Vicky L
Mathews, Herbert Lester
Murphy, Richard Allan
Myers, Ronald Berl
Passman, Frederick Jay
Perry, Dennis
Perry, Eugene Arthur
Sames, Richard William
Shapiro, James Alan
Shen, Linus Liang-nene
Simonson, Lloyd Grant
Tripathy, Deoki Nandan
Turek, Fred William
Weiler, William Alexander
Woese, Carl R

INDIANA
Boyer, Ernest Wendell
Claybaugh, Glenn Alan
Daily, Fay Kenoyer
Driesens, Robert James
Elliott, Robert A
Enders, George Leonhard, Jr
Hopper, Samuel Hersey
Keith, Paula Myers
Kinzel, Jerry J
Koch, Arthur Louis
Kory, Mitchell
Levinthal, Mark
Millar, Wayne Norval
Morter, Raymond Lione
Shockey, William Lee
Van Frank, Richard Mark
Wagner, Morris
Welker, George W
Wilson, Raphael

Bacteriology (cont)

IOWA
Hartman, Paul Arthur
Hug, Daniel Hartz
Lockhart, William Raymond
Mickelson, Milo Norval
Quinn, Loyd Yost
Rebstock, Theodore Lynn
Richardson, Robert Louis
Stanton, Thaddeus Brian
Thiermann, Alejandro Bories
Thurston, John Robert
Wessman, Garner Elmer

KANSAS
Bailie, Wayne E
Cho, Cheng T
Dykstra, Mark Allan
Haas, Herbert Frank
Iandolo, John Joseph
Johnson, Donovan Earl
Miller, Glendon Richard
Minocha, Harish C
Warren, Halleck Burkett, Jr

KENTUCKY
Edwards, Ogden Frazelle
Greenberg, Richard Aaron
Lillich, Thomas Tyler
Stuart, James Glen

LOUISIANA
Bornside, George Harry
Boyd, Frank McCalla
Domingue, Gerald James
Johnson, Mary Knettles
Titkemeyer, Charles William
Willis, William Hillman

MAINE
Buck, Charles Elon
Gershman, Melvin
Highlands, Matthew Edward
Jerkofsky, Maryann
Pratt, Darrell Bradford
Tjepkema, John Dirk

MARYLAND
Abramson, I Jerome
Adams, James Miller
Burchard, Robert P
Chakrabarti, Siba Gopal
Chanock, Robert Merritt
Doetsch, Raymond Nicholas
Elespuru, Rosalie K
Ellinghausen, Herman Charles, Jr
Falkenstein, Kathy Fay
Foulds, John Douglas
Galloway, Sheila Margaret
Garges, Susan
Gherna, Robert Larry
Goldenberg, Martin Irwin
Gordon, Lance Kenneth
Gordon, Ruth Evelyn
Hanks, John Harold
Henry, Timothy James
Hill, James Carroll
Hochstein, Herbert Donald
Johnson-Winegar, Anna
Jones, Morris Thompson
Joseph, Sammy William
Krieg, Richard Edward, Jr
McMacken, Roger
Mattick, Joseph Francis
Mohler, Irvin C, Jr
Molenda, John R
Noyes, Howard Ellis
Rosner, Judah Leon
Seto, Belinda P L
Stamper, Hugh Blair
Wagner, David Darley
Williams, Noreen
Wittenberger, Charles Louis

MASSACHUSETTS
Amos, Harold
Canale-Parola, Ercole
Craven, Donald Edward
Crisley, Francis Daniel
Deitz, William Harris
Gibbons, Ronald J
Gilroy, James Joseph
Girard, Kenneth Francis
Kashket, Eva Ruth
Litsky, Warren
Luria, Salvador Edward
MacMahon, Harold Edward
Madore, Bernadette
Pfister, Richard Charles
Reyes, Victor E
Reynolds, John Theodore
Riley, Monica
Sarkar, Nilima
Sbarra, Anthony J
Stern, Ivan J
Thomas, Gail B
Watson, Stanley W

MICHIGAN
Alegnani, William Charles
Breznak, John Allen
Carrick, Lee
Colingsworth, Donald Rudolph

Daniels, Peter John Lovell
Eisenstein, Barry I
Freter, Rolf Gustav
Jackson, Matthew Paul
Jay, James Monroe
Jones, Lily Ann
Juni, Elliot
Loesche, Walter J
McCullough, Willard George
Martin, Joseph Patrick, Jr
Mattman, Lida Holmes
Muentener, Donald Arthur
Mulks, Martha Huard
Pecoraro, Vincent L
Portwood, Lucile Mitchell
Rossmoore, Harold W
Sebek, Oldrich Karel
Shelef, Leora Aya
Wheeler, Albert Harold
Yancey, Robert John, Jr

MINNESOTA
Bauer, Henry
Beggs, William H
Berube, Robert
Haas, Larry Alfred
Johnson, Russell Clarence
Lonergan, Dennis Arthur
McKay, Larry Lee
Mizuno, William George
Needham, Gerald Morton
Porter, Frederic Edwin
Surdy, Ted E
Tsuchiya, Henry Mitsumasa
Wagenaar, Raphael Omer

MISSISSIPPI
Arceneaux, Joseph Lincoln
Brown, Lewis Raymond
Wolverton, Billy Charles

MISSOURI
Barnekow, Russell George, Jr
Eisenstark, Abraham
Elvin-Lewis, Memory P F
Gossling, Jennifer
Granoff, Dan Martin
Hufham, James Birk
Lozeron, Homer A
McCune, Emmett L
Miles, Donald Orval
Misfeldt, Michael Lee
Moulton, Robert Henry
Murray, Patrick Robert
Sargentini, Neil Joseph
Schmidt, Donald Arthur
Wells, Frank Edward

MONTANA
Lackman, David Buell
Lozano, Edgardo A
Myers, Lyle Leslie
Stoenner, Herbert George

NEBRASKA
Rodriguez-Sierra, Jorge F
Thompson, Thomas Leo
Underdahl, Norman Russell
Vidaver, Anne Marie Kopecky

NEW HAMPSHIRE
Adler, Frank Leo
Adler, Louise Tale
Jacobs, Nicholas Joseph
Kerwin, Richard Martin
Rodgers, Frank Gerald

NEW JERSEY
Beskid, George
Bruno, Charles Frank
Coriell, Lewis L
Frances, Saul
Gaffar, Abdul
Green, Erika Ana
Iannarone, Michael
Imaeda, Tamotsu
Johnson, Layne Mark
Kestenbaum, Richard Charles
Koepp, Leila H
Koft, Bernard Waldemar
Kronenwett, Frederick Rudolph
Miller, A(nna) Kathrine
Morneweck, Samuel
Mukai, Cromwell Daisaku
Peterson, Arthur Carl
Regna, Peter P
Rossow, Peter William
Seneca, Harry
Shearer, Marcia Cathrine (Epple)
Shirk, Richard Jay
Solotorovsky, Morris
Tate, Robert Lee, III
Tierno, Philip M, Jr
Umbreit, Wayne William
Weinstein, David

NEW MEXICO
Hageman, James Howard
Roubicek, Rudolf V
Scaletti, Joseph Victor

NEW YORK
Alexander-Jackson, Eleanor Gertrude
Andersen, Kenneth J

Anderson, Kenneth Ellsworth
Beach, David H
Bergstrom, Gary Carlton
Bopp, Lawrence Howard
Bradner, William Turnbull
Bukhari, Ahmad Iqbal
Cardenas, Raúl R, Jr
Chiulli, Angelo Joseph
Corpe, William Albert
Cowell, James Leo
Cravitz, Leo
Delwiche, Eugene Albert
Dennis, Emery Westervelt
Dickey, Robert Shaft
Efthymiou, Constantine John
Ellner, Paul Daniel
Enea, Vincenzo
Fabricant, Catherine G
Fendler, Janos Hugo
Gilardi, Gerald Leland
Goodhue, Charles Thomas
Gordon, Morris Aaron
Hipp, Sally Sloan
Kite, Joseph Hiram, Jr
Lambert, Reginald Max
Lazaroff, Norman
McCarty, Maclyn
MacDonald, Russell Earl
Manjula, Belur N
Marquis, Robert E
Miller, Terry Lynn
Mohn, James Frederic
Murphy, Timothy F
Pierre, Leon L
Poindexter, Jeanne Stove
Rhee, G-Yull
Roll, David E
Scher, William
Shapiro, Caren Knight
Shayegani, Mehdi
Shively, Carl E
Silverman, Morris
Splittstoesser, Clara Quinnell
Splittstoesser, Don Frederick
Steinberg, Bernard Albert
Steinman, Howard Mark
Stevens, Roy Harris
Trowbridge, Richard Stuart
Wolin, Meyer Jerome
Young, William Donald, Jr
Zengel, Janice Marie

NORTH CAROLINA
Curtis, Susan Julia
Elkan, Gerald Hugh
Evans, James Brainerd
Fulghum, Robert Schmidt
Gray, Walter C(larke)
Hopfer, Roy L
Hulcher, Frank H
Hutt, Randy
Jordan, Thomas L
King, Kendall Willard
Knuckles, Joseph Lewis
Kull, Frederick Charles, Sr
Lattuada, Charles P
Matthysse, Ann Gale
Moody, Max Dale
Schwab, John Harris
Shepard, Maurice Charles
Silverman, Myron Simeon
Sizemore, Ronald Kelly
Sparling, Philip Frederick
Straughn, William Ringgold, Jr
Tarver, Fred Russell, Jr
Tove, Shirley Ruth
Whitmire, Carrie Ella
Wyrick, Priscilla Blakeney

NORTH DAKOTA
Berryhill, David Lee
Fillipi, Gordon Michael
McMahon, Kenneth James

OHIO
Baker, Dwight Dee
Bigley, Nancy Jane
Coplin, David Louis
Frea, James Irving
Frey, James R
Hunter, John Earl
Ichida, Allan A
Kagen, Herbert Paul
Kolodziej, Bruno J
Krampitz, Lester Orville
Lawrence, James Vantine
Long, Sterling K(rueger)
Modrzakowski, Malcolm Charles
Rhodes, Judith Carol
Sobota, Anthony E
Somerson, Norman L
Tabita, F Robert
Tuovinen, Olli Heikki
Walker, Richard V

OKLAHOMA
Anglin, J Hill, Jr
Booth, James Samuel
Durham, Norman Nevill
Essenberg, Richard Charles
Flournoy, Dayl Jean
Foutch, Gary Lynn
Hitzman, Donald Oliver
McCallum, Roderick Eugene

Olson, Harold Cecil
Sokatch, John Robert

OREGON
Anderson, Arthur W
Arp, Daniel James
Barry, Arthur Leland
Charlton, David Berry
Dooley, Douglas Charles
Elliott, Lloyd Floren
Ream, Lloyd Walter, Jr
Sandine, William Ewald
Wyatt, Carolyn Jane

PENNSYLVANIA
Actor, Paul
Anderson, Theodore Gustave
Austrian, Robert
Buchanan, Robert Lester
Burdash, Nicholas Michael
Carls, Ralph A
Carpenter, Charles Patten
Castric, Peter Allen
Cinquina, Carmela Louise
Crowell, Richard Lane
Debroy, Chitrita
Dickerson, William H
Duncan, Charles Lee
Fett, William Frederick
Fletcher, Ronald D
Garfinkle, Barry David
Grainger, Thomas Hutcheson, Jr
Gregory, Francis Joseph
Herbert, Michael
Kruse, Conrad Edward
Lee, John Cheung Han
Lehman, Ernest Dale
Levine, Elliot Myron
Lindemeyer, Rochelle G
Lindstrom, Eugene Shipman
Long, Walter Kyle, Jr
Mandel, John Herbert
Miller, Arthur James
Milligan, Wilbert Harvey, III
Millman, Irving
Morton, Harry E
Osborne, William Wesley
Panos, Charles
Phillips, Allen Thurman
Poupard, James Arthur
Rest, Richard Franklin
Shockman, Gerald David
Stone, Joseph Louis
Uscavage, Joseph Peter
Witlin, Bernard
Yurchenco, John Alfonso
Zimmerman, Leonard Norman

RHODE ISLAND
Jones, Kenneth Wayne
Krasner, Robert Irving
Landy, Arthur
Laux, David Charles
Traxler, Richard Warwick
Wood, Norris Philip

SOUTH CAROLINA
Bryson, Thomas Allan
Graber, Charles David
Pannell, Lolita

SOUTH DAKOTA
Ferguson, Michael William
Howard, Ronald M
Kirkbride, Clyde Arnold
Prescott, Lansing M
Sword, Christopher Patrick
West, Thomas Patrick

TENNESSEE
Adler, Howard Irving
Bibb, William Robert
Blaser, Martin Jack
Bozeman, Samuel Richmond
Chiang, Thomas M
Draughon, Frances Ann
Freeman, Bob A
Gaby, William Lawrence
Gartner, T Kent
Goulding, Charles Edwin, Jr
Holtman, Darlington Frank
Johnson, Rother Rodenious
Rightsel, Wilton Adair
Savage, Dwayne Cecil
Shockley, Thomas E
Stevens, Stanley Edward, Jr
Woychik, Richard P

TEXAS
Bennett, Edward Owen
Blouse, Louis E, Jr
Boley, Robert B
Calkins, Harmon Eldred
Collisson, Ellen Whited
Cook, Paul Fabyan
Cooper, Ronda Fern
Crawford, Gladys P
Dempsey, Walter B
Friend, Patric Lee
Goldschmidt, Millicent
Hejtmancik, Kelly Erwin
Ippen-Ihler, Karin Ann
Joys, Terence Michael
Kester, Andrew Stephen

Kiel, Johnathan Lloyd
LaBree, Theodore Robert
Lankford, Charles Ely
LeBlanc, Donald Joseph
Lyles, Sanders Truman
Matney, Thomas Stull
Packchanian, Ardzroony (Arthur)
Phillips, Guy Frank
Pike, Robert Merrett
Robertstad, Gordon Wesley
Rolfe, Rial Dewitt
Schmidt, Jerome P
Schneider, Dennis Ray
Sistrunk, William Allen
Smith, William Russell
Spellman, Craig William
Sweet, Charles Edward
Temeyer, Kevin Bruce
Walter, Ronald Bruce
Widner, William Richard
Wilson, Joe Bransford
Wood, Robert Charles

UTAH
Anderson, Anne Joyce
Carter, Paul Bearnson
Georgopoulos, Constantine Panos
Hayes, Sheldon P
Lighton, John R B
Nicholes, Paul Scott
Roth, John R
Spendlove, John Clifton
Torres, Anthony R
Wright, Donald N

VERMONT
Schaeffer, Warren Ira

VIRGINIA
Claus, George William
Edlich, Richard French
Hsu, Hsiu-Sheng
Moore, Lillian Virginia Holdeman
Nagarkatti, Prakash S
Newman, Jack Huff
O'Kane, Daniel Joseph
Raizen, Carol Eileen
Sarber, Raymond William
Tankersley, Robert Walker, Jr
Welshimer, Herbert Jefferson
White, Alan George Castle
Wilkins, Tracy Dale
Wise, Robert Irby
Yousten, Allan A

WASHINGTON
Barnes, Glover William
Drake, Charles Hadley
Grootes-Reuvecamp, Grada Alijda
Hall, Elizabeth Rose
Hazelbauer, Gerald Lee
Hurd, Robert Charles
Johnstone, Donald Lee
Luedecke, Lloyd O
Naylor, Harry Brooks
Randall, Linda Lea
Staley, James Trotter
Wildung, Raymond Earl

WEST VIRGINIA
Bissonnette, Gary Kent
Bullock, Graham Lambert
Chisler, John Adam
Gain, Ronald Ellsworth
Gerencser, Mary Ann (Aiken)
Gerencser, Vincent Frederic
Woolridge, Robert Leonard

WISCONSIN
Collins, Mary Lynne Perille
Czuprynski, Charles Joseph
Deibel, Robert Howard
Foster, Edwin Michael
Kushnaryov, Vladimir Michael
McIntosh, Elaine Nelson
Myers, Charles R
Ogden, Robert Verl
Qureshi, Nilofer
Rice, Marion McBurney
Rouf, Mohammed Abdur
Schaefer, Daniel M
Snudden, Birdell Harry
Taylor, Jerry Lynn
Whitehead, Howard Allan
Zabransky, Ronald Joseph

WYOMING
Maki, Leroy Robert

ALBERTA
Cheng, Kuo-Joan
Eggert, Frank Michael
Holmes, R H Lavergne
Jellard, Charles H
Langford, Edgar Verden
Madiyalakan, Ragupathy
Rennie, Robert John
Schroder, David John
Stirrat, James Hill
Whitehouse, Ronald Leslie S

BRITISH COLUMBIA
Bismanis, Jekabs Edwards
Campbell, Jack James Ramsay

De Boer, Solke Harmen
Syeklocha, Delfa
Trussell, Paul Chandos

MANITOBA
Lukow, Odean Michelin
Wilt, John Charles
Wiseman, Gordon Marcy

NEW BRUNSWICK
Robinson, John

NOVA SCOTIA
Simpson, Frederick James
Stewart, James Edward

ONTARIO
Beaulieu, J A
Blais, Burton W
Cipera, John Dominik
Clarke, Anthony John
Fitz-James, Philip Chester
Fraser, Ann Davina Elizabeth
Garcia, Manuel Mariano
Goepfert, John McDonnell
Hauschild, Andreas H W
Jones, John Bryan
Jordan, David Carlyle
Kushner, Donn Jean
McKenzie, Allister Roy
Murray, Robert George Everitt
Murray, William Douglas
Rao, Salem S
Roslycky, Eugene Bohdan
Sandham, Herbert James
Schneider, Henry
Speckmann, Gunter Wilhelm-Otto
Stokes, Pamela Mary
Trick, Charles Gordon
Tryphonas, Helen
Willoughby, Donald S
Wilson, Robert James

QUEBEC
Bordeleau, Lucien Mario
Fournier, Michel
Frappier, Armand
Portelance, Vincent Damien
Yaphe, Wilfred

SASKATCHEWAN
Nelson, Louise Mary

OTHER COUNTRIES
Brody, Edward Norman
Dutta-Roy, Asim Kanti
Ha, Tai-You
Hosokawa, Keiichi
Modabber, Farrokh Z
Roy, Raman K
Wu, Albert M

Behavior-Ethology

ALABAMA
Appel, Arthur Gary
Dudley, Susan D
Holler, Nicholas Robert
Moyer, Kenneth Evan
Pegram, George Vernon, Jr
Sparks, David Lee
Taub, Edward
Wyss, James Michael

ALASKA
Ahlgren, Molly O
Fagen, Robert
Helle, John Harold (Jack)

ARIZONA
Balda, Russell Paul
Bernays, Elizabeth Anna
Cheal, MaryLou
Hildebrand, John Grant, III
Klein, Sherwin Jared
Mulder, John Bastian
Rutowski, Ronald Lee
Schnell, Jay Heist
Slobodchikoff, Constantine Nicholas
Smith, Andrew Thomas
Smith, Robert Lloyd
Topoff, Howard Ronald
Wheeler, Lawrence

ARKANSAS
Amlaner, Charles Joseph, Jr
Dykman, Roscoe A
Holson, Ralph Robert
Newton, Joseph Emory O'Neal
Paule, Merle Gale
Reddy, RamaKrishna Pashuvula
Smith, Kimberly Gray

CALIFORNIA
Anderson, John Richard
Arnold, Arthur Palmer
Avila, Vernon Lee
Barlow, George Webber
Beach, Frank Ambrose
Benzer, Seymour
Bradbury, Jack W
Brown, Judith Adele
Bryson, George Gardner

Carterette, Edward Calvin Hayes
Chadwick, Nanette Elizabeth
Chambers, Kathleen Camille
Chapman, Loring Frederick
Chappell, Mark Allen
Cody, Martin Leonard
Cogswell, Howard Lyman
Collias, Nicholas Elias
Collier, Gerald
Courchesne, Eric
Cox, Cathleen Ruth
Cummings, William Charles
Daunton, Nancy Gottlieb
Davis, Edward E
Davis, William Jackson
Dingle, Richard Douglas Hugh
Efron, Robert
Endler, John Arthur
Enright, James Thomas
Farish, Donald James
Ferguson, James Mecham
Fisler, George Frederick
Flanigan, William Francis, Jr
Frey, Dennis Frederick
Galin, David
Gallistel, Charles Ransom
Gambs, Roger Duane
Gary, Norman Erwin
Haas, Richard
Hand, Judith Latta
Hardt, James Victor
Hart, Benjamin Leslie
Howard, Walter Egner
Hrdy, Sarah Blaffer
Jennrich, Ellen Coutlee
Jevning, Ron
Johnson, Chris Alan
Kavanau, Julian Lee
Kimsey, Lynn Siri
Kistner, David Harold
Knapp, Theodore Martin
Krekorian, Charles O'Neil
Kryter, Karl David
Levinson, Arthur David
Lindburg, Donald Gilson
Loeblich, Karen Elizabeth
Lovich, Jeffrey Edward
Lukens, Herbert Richard, Jr
McGaugh, James L
MacMillen, Richard Edward
Martin, James Tillison
Moriarty, Daniel Delmar, Jr
Morin, James Gunnar
Mossman, Archie Stanton
Naitoh, Paul Yoshimasa
Nelson, Keith
Ono, Joyce Kazuyo
Page, Robert Eugene, Jr
Powell, Jerry Alan
Price, Mary Vaughan
Randall, Janet Ann
Reiss, Diana
Riesen, Austin Herbert
Ritzmann, Ronald Fred
Rotenberry, John Thomas
Sassenrath, Ethelda Norberg
Schell, Anne McCall
Schusterman, Ronald Jay
Schwab, Ernest Roe
Shook, Brenda Lee
Simmel, Edward Clemens
Soltysik, Szczesny Stefan
Squire, Larry Ryan
Stamps, Judy Ann
Stein, Larry
Storms, Lowell H
Tenaza, Richard Reuben
Thompson, Paul O
Thompson, Richard Frederick
Vail, Patrick Virgil
Vehrencamp, Sandra Lee
Waser, Nickolas Merritt
Weinberger, Norman Malcolm
Weinrich, James D
Wenzel, Bernice Martha
Wenzel, Zita Marta
Williston, John Stoddard
Wimer, Cynthia Crosby
Wood, David Lee
Zaidel, Eran
Zuk, Marlene

COLORADO
Audesirk, Gerald Joseph
Audesirk, Teresa Eck
Bekoff, Anne C
Bekoff, Marc
Bernstein, Stephen
Bond, Richard Randolph
Breed, Michael Dallam
DeFries, John Clarence
Drummond, Boyce Alexander, III
Eaton, Robert Charles
Fall, Michael William
Green, Jeffrey Scott
Krear, Harry Robert
Kulkosky, Paul Joseph
Laudenslager, Mark LeRoy
Lehner, Philip Nelson
Puck, Mary Hill
Purcell, Kenneth
Sharpless, Seth Kinman
Tomback, Diana Francine
Troxell, Wade Oakes

Watkins, Linda Rothblum
Wiens, John Anthony
Wilson, James Russell

CONNECTICUT
Barry, Michael Anhalt
Block, Bartley C
Craig, Catherine Lee
Davis, Michael
Goldman, Bruce Dale
Henry, Charles Stuart
Lynch, Carol Becker
Maier, Chris Thomas
Maxson, Stephen C
Miller, Neal Elgar
Moehlman, Patricia des Roses
Newton, David C
Novick, Alvin
Rebuffe-Scrive, Marielle Francoise
Redmond, Donald Eugene, Jr
Rettenmeyer, Carl William
Roberts, Mervin Francis
Russock, Howard Israel
Sachs, Benjamin David
Snyder, Daniel Raphael
Stevens, Joseph Charles
Tobias, Jerry Vernon
Weinstein, Curt David
Wells, Kentwood David
Wilson, William August

DELAWARE
Graham, Frances Keesler
Tam, Sang William
Wood, Thomas Kenneth

DISTRICT OF COLUMBIA
Barrows, Edward Myron
Blockstein, David Edward
Fox, Michael Wilson
Gould, Edwin
Jacobs, William Wood, Jr
Jacobsen, Frederick Marius
Kleiman, Devra Gail
Malcom, Shirley Mahaley
Nickle, David Allan
Ralls, Katherine Smith
Raslear, Thomas G
Reichman, Omer James
Rubinoff, Roberta Wolff
Wolfe, Thomas Lee
Yaniv, Simone Liliane
Young, John Karl

FLORIDA
Adams, Ralph M
Allen, Ted Tipton
Berg, William Keith
Branch, Lyn Clarke
Brockmann, Helen Jane
Brown, John Lott
Calkins, Carrol Otto
Cohen, Sanford I
Dawson, William Woodson
De Lorge, John Oldham
Dewsbury, Donald Allen
Eisdorfer, Carl
Eisenberg, John Frederick
Elfner, Lloyd F
Evoy, William (Harrington)
Gleeson, Richard Alan
Goldstein, Mark Kane
Gruber, Samuel Harvey
Gude, Richard Hunter
Herrnkind, William Frank
Hope, George Marion
Kaufmann, John Henry
Kenshalo, Daniel Ralph
Knowlton, Nancy
Levey, Douglas J
Lillywhite, Harvey B
Lounibos, Leon Philip
Mackenzie, Richard Stanley
Myrberg, Arthur August, Jr
Phillips, Michael Ian
Porter, Sanford Dee
Raizada, Mohan K
Roubik, David Ward
Rowland, Neil Edward
Schneiderman, Neil
Spielberger, Charles Donald
Teas, Donald Claude
Tingle, Frederic Carley
Vander Meer, Robert Kenneth
Vierck, Charles John, Jr
Walker, Thomas Jefferson
Williams, Theodore P
Winters, Ray Wyatt
Woolfenden, Glen Everett
Wyneken, Jeanette

GEORGIA
Bernstein, Irwin Samuel
Byrd, Larry Donald
Dragoin, William Bailey
Dusenbery, David Brock
Hartlage, Lawrence Clifton
Jackson, William James
McDaniel, William Franklin
Marchinton, Robert Larry
Matthews, Robert Wendell
Mulligan, Benjamin Edward
Nadler, Ronald D
Pulliam, H Ronald

Phillips, Robert Rhodes
Pickering, Thomas G
Porter, William Frank
Rabe, Ausma
Reith, Maarten E A
Ringler, Neil Harrison
Ritter, Walter Paul
Roelofs, Wendell L
Rosenblum, Leonard Allen
Sackeim, Harold A
Salvi, Richard J
Sank, Diane
Schneider, Kathryn Claire (Johnson)
Schupf, Nicole
Sechzer, Jeri Altneu
Seeley, Thomas Dyer
Sherman, Paul Willard
Slaughter, John Sim
Smith, Charles James
Sobczak, Thomas Victor
Stewart, Margaret McBride
Sullivan, Daniel Joseph
Thomas, Garth Johnson
Verrillo, Ronald Thomas
Walcott, Charles
Wasserman, Marvin
Wineburg, Elliot N
Wolf, Larry Louis
Zitrin, Charlotte Marker

NORTH CAROLINA
Biggs, Walter Clark, Jr
Brown, Richard Dean
Case, Verna Miller
Crowder, Larry Bryant
Dixon, Norman Rex
Eason, Robert Gaston
Edens, Frank Wesley
Eichelman, Burr S, Jr
Erickson, Robert Porter
Gottlieb, Gilbert
Hall, Warren G
Hall, William Charles
Howard, James Lawrence
Kamykowski, Daniel
King, Richard Austin
Lewis, Mark Henry
Logan, Cheryl Ann
Moseley, Lynn Johnson
Mueller, Helmut Charles
Myers, Robert Durant
Oppenheim, Ronald William
Powell, Roger Allen
Rulifson, Roger Allen
Soderquist, David Richard
Surwit, Richard Samuel
Vandenbergh, John Garry
Weigl, Peter Douglas
Weiss, Jay M
Wiley, Richard Haven, Jr
Wolk, Robert George

NORTH DAKOTA
Fivizzani, Albert John, Jr
Gladue, Brian Anthony
Penland, James Granville

OHIO
Beal, Kathleen Grabaskas
Brose, David Stephen
Burtt, Edward Howland, Jr
Case, Denis Stephen
Corson, Samuel Abraham
DeNelsky, Garland
Downhower, Jerry F
Ely, Daniel Lee
Keck, Max Johann
Knight, Walter Rea
Lacey, Beatrice Cates
McLean, Edward Bruce
Mudry, Karen Michele
Mulick, James Anton
Orcutt, Frederic Scott, Jr
Patterson, Michael Milton
Pettijohn, Terry Frank
Radabaugh, Dennis Charles
Rovner, Jerome Sylvan
Rushforth, Norman B
Shipley, Michael Thomas
Stoffer, Richard Lawrence
Svendsen, Gerald Eugene
Swift, Michael Crane
Taylor, Douglas Hiram
Uetz, George William
Valentine, Barry Dean
Vessey, Stephen H
Vorhees, Charles V

OKLAHOMA
Baird, Troy Alan
Chapman, Brian Richard
Fox, Stanley Forrest
Leonard, Janet Louise
Lynn, Robert Thomas
Mares, Michael Allen
Mock, Douglas Wayne
Schaefer, Carl Francis
Vestal, Bedford Mather
Wells, Harrington
Young, Sharon Clairene

OREGON
Betts, Burr Joseph
Colvin, Dallas Verne

Cross, Stephen P
Eaton, Gordon Gray
Fay, Warren Henry
Fernald, Russell Dawson
Hixon, Mark A
Mason, Robert Thomas
Olla, Bori Liborio
Phillips, David
Phoenix, Charles Henry
Richards, Richard Earl
Ryker, Lee Chester

PENNSYLVANIA
Allen, Theresa O
Angstadt, Robert B
Balling, Jan Walter
Beauchamp, Gary Keith
Brenner, Frederic J
Bridger, Wagner H
Carey, Michael Dean
Carey, William Bacon
Clark, Richard James
Cole, James Edward
Cutler, Winnifred Berg
Cutt, Roger Alan
DeHaven-Hudkins, Diane Louise
Dinges, David Francis
Epstein, Alan Neil
Goldstein, E Bruce
Graves, Hannon B
Green, Barry George
Hart, David Dickinson
Hoffman, Daniel Lewis
Hurvich, Leo Maurice
Keiper, Ronald R
Kendall, Philip C
Klinger, Thomas Scott
Kurland, Jeffrey Arnold
Lewis, Michael Edward
Ode, Philip E
Pion, Lawrence V
Poplawsky, Alex James
Prezant, Robert Steven
Ray, William J
Schultz, Jack C
Settle, Richard Gregg
Sidie, James Michael
Singer, Alan G
Smith, W John
Steele, Craig William
Stellar, Eliot
Stewart, Charles Newby
Thompson, Roger Kevin Russell
Ward, Ingeborg L
Wert, Jonathan Maxwell, Jr
Williams, Timothy C
Williamson, Craig Edward
Wysocki, Charles Joseph
Yahner, Richard Howard

RHODE ISLAND
Janis, Christine Marie
Lehmkuhle, Stephen W
Lipsitt, Lewis Paeff
Pratt, David Mariotti
Riggs, Lorrin Andrews
Sanberg, Paul Ronald
Waage, Jonathan King

SOUTH CAROLINA
Bildstein, Keith Louis
Coleman, James Roland
Forsythe, Dennis Martin
Gauthreaux, Sidney Anthony
Kaiser, Charles Frederick
Lacher, Thomas Edward, Jr
Sears, Harold Frederick

SOUTH DAKOTA
Schlenker, Evelyn Heymann

TENNESSEE
Altemeier, William Arthur, III
Ambrose, Harrison William, III
Bealer, Steven Lee
Burghardt, Gordon Martin
Elberger, Andrea June
Greenberg, Neil
Kaas, Jon Howard
Lubar, Joel F
Powers, J Bradley

TEXAS
Beaver, Bonnie Veryle
Belk, Gene Denton
Brodie, Edmund Darrell, Jr
Buskirk, Ruth Elizabeth
Coelho, Anthony Mendes, Jr
Collins, Anita Marguerite
Crawford, Morris Lee Jackson
Crews, David Pafford
Dalterio, Susan Linda
Easley, Stephen Phillip
Formanowicz, Daniel Robert, Jr
Foster, John Robert
Gonzalez-Lima, Francisco
Greenberg, Les Paul
Hamilton, Charles R
Hanlon, Roger Thomas
Ingold, Donald Alfred
Johnson, Patricia Ann J
McCarley, Wardlow Howard
Malin, David Herbert
Marks, Gerald A

Mathews, Nancy Ellen
Moushegian, George
Richerson, Jim Vernon
Rogers, Walter Russell
Rubink, William Louis
Rylander, Michael Kent
Schreiber, Robert Alan
Sherry, Clifford Joseph
Sperling, Harry George
Stein, Jerry Michael
Suddick, Richard Phillips
Vinson, S Bradleigh
Whitsett, Johnson Mallory, II
Wilczynski, Walter
Willig, Michael Robert
Wright, Anthony Aune

UTAH
Arave, Clive W
Balph, David Finley
Balph, Martha Hatch
Beck, Edward C
Cheney, Carl D
Conover, Michael Robert
Davidson, Diane West
Fleming, Donovan Ernest
Hsiao, Ting Hun
Kesner, Raymond Pierre
Messina, Frank James
Schenkenberg, Thomas

VERMONT
Heinrich, Bernd
Schall, Joseph Julian
Simmons, Kenneth Rogers

VIRGINIA
Balster, Robert L(ouis)
Block, Gene David
Brown, Luther Park
Brunjes, Peter Crawford
Davis, Joel L
Fashing, Norman James
Fine, Michael Lawrence
Fink, Linda Susan
Gold, Paul Ernest
Grimm, James K
Gullotta, Frank Paul
Harrell, Ruth Flinn
Hornbuckle, Phyllis Ann
Jenssen, Thomas Alan
King, H(enry) E(ugene)
Lenhardt, Martin Louis
McCarty, Richard Charles
Marcellini, Dale Leroy
Pienkowski, Robert Louis
Pribram, Karl Harry
Rosenblum, William I

WASHINGTON
Anderson, James Jay
Bowden, Douglas Mchose
Cook, Paul Pakes, Jr
Dudley, Donald Larry
Galusha, Joseph G, Jr
Gentry, Roger Lee
Gilbert, Frederick Franklin
Gray, Robert H
Kenagy, George James
Kenney, Nancy Jane
Loughlin, Thomas Richard
Lovely, Richard Herbert
Martin, Dennis John
Monan, Gerald E
Paulson, Dennis Roy
Pearson, Walter Howard
Pietsch, Theodore Wells
Shepherd, Linda Jean
Simpson, John Barclay
Smith, Orville Auverne

WEST VIRGINIA
Constantz, George Doran
Covalt-Dunning, Dorothy
Dunning, Dorothy Covalt
Marshall, Joseph Andrew
Muul, Illar

WISCONSIN
Boese, Gilbert Karyle
Cleeland, Charles Samuel
Curtis, Robin Livingstone
Ficken, Millicent Sigler
Ficken, Robert W
Goldstein, Robert
Gottfried, Bradley M
Goy, Robert William
Hailman, Jack Parker
Hekmat, Hamid Moayed
Kemnitz, Joseph William
Kent, Raymond D
Minock, Michael Edward
Norris, Dale Melvin, Jr
Schenk, Roy Urban
Snowdon, Charles Thomas
Strier, Karen Barbara
Takahashi, Lorey K
Vomachka, Archie Joel
Wilson, Richard Howard
Yasukawa, Ken

WYOMING
Boyce, Mark S
Denniston, Rollin H, II

Lockwood, Jeffrey Alan
Reynolds, Richard Truman
Rose, James David

PUERTO RICO
Lewis, Allen Rogers
Shapiro, Douglas York

ALBERTA
Byers, John Robert
Geist, Valerius
Koopmans, Henry Sjoerd
Murie, Jan O
Sainsbury, Robert Stephen
Seghers, Benoni Hendrik
Steiner, Andre Louis
Szabo, Tibor Imre
Veale, Warren Lorne

BRITISH COLUMBIA
Borden, John Harvey
Cheng, Kimberly Ming-Tak
Craig, Kenneth Denton
Dill, Lawrence Michael
Greenwood, Donald Dean
Healey, Michael Charles
Hill, Arthur Thomas
Phillips, Anthony George
Semba, Kazue
Steeves, John Douglas
Ward, Lawrence McCue

MANITOBA
Evans, Roger Malcolm
Hughes, Kenneth Russell
Kingsley, Michael Charles Stephen
Nance, Dwight Maurice
Palaniswamy, Pachagounder
Zach, Reto

NEW BRUNSWICK
McKenzie, Joseph Addison
Seabrook, William Davidson

NEWFOUNDLAND
Green, John M

NOVA SCOTIA
Brown, Richard George Bolney
Fentress, John Carroll
Harrington, Fred Haddox
Hopkins, David Alan
Van Houten, Ronald G

ONTARIO
Bailey, Edward D
Berrill, Michael
Brooks, Ronald James
Cade, William Henry
Caplan, Paula Joan
Collins, Nicholas Clark
Cornell, James Morris
Dagg, Anne Innis
Darling, Donald Christopher
Donald, Merlin Wilfred
Falls, James Bruce
Gray, David Robert
Keenleyside, Miles Hugh Alston
Kovacs, Kit M
Leung, Lai-Wo Stan
Mallory, Frank Fenson
Mogenson, Gordon James
Montgomerie, Robert Dennis
Morris, Ralph Dennis
Noakes, David Lloyd George
Robertson, Raleigh John
Rollman, Gary Bernard
Stouffer, James L
Vanderwolf, Cornelius Hendrik
Winter, Peter

QUEBEC
Chouinard, Guy
Cloutier, Conrad Francois
Ferron, Jean H
Fitzgerald, Gerard John
Grant, James William Angus
Green, David M(artin)
Hilton, Donald Frederick James
McNeil, Jeremy Nichol
Milner, Brenda (Atkinson)
Rau, Manfred Ernst
Robert, Suzanne
Sanborne, Paul Michael
Titman, Rodger Donaldson

SASKATCHEWAN
Doane, John Frederick
Flood, Peter Frederick
Mitchell, George Joseph
Oliphant, Lynn Wesley

OTHER COUNTRIES
Biederman-Thorson, Marguerite Ann
Ganchrow, Donald
Glickstein, Mitchell
Hanukoglu, Israel
Hisada, Mituhiko
Holldobler, Berthold Karl
Holman, Richard Bruce
Hummel, Hans Eckhardt
Jacobson, Bertil
Moynihan, Martin Humphrey
Nevo, Eviatar

Behavior-Ethology (cont)

Oxnard, Charles Ernest
Walther, Fritz R
Wettstein, Joseph G

Biochemistry

ALABAMA
Abrahamson, Dale Raymond
Alexander, Herman Davis
Aull, John Louis
Ayling, June E
Baker, John Rowland
Baliga, B Surendra
Ball, Laurence Andrew
Baugh, Charles M
Becker, Gerald Leonard
Benos, Dale John
Bhatnagar, Yogendra Mohan
Bhown, Ajit Singh
Birkedal Hansen, Henning
Bradley, James T
Brown, Alfred Ellis
Burns, Moore J
Butler, William Thomas
Carlson, Gerald Lowell
Chang, Chi Hsiung
Chapatwala, Kirit D
Cherry, Joe H
Christian, Samuel Terry
Clarke, Benjamin L
Cook, William Joseph
Daron, Harlow Hoover
Digerness, Stanley B
Elgavish, Ada S
Elliott, Howard Clyde
Fendley, Ted Wyatt
Flodin, N(estor) W(inston)
Freeman, Bruce Alan
Glaze, Robert P
Goodman, Steven Richard
Guidry, Clyde R
Haggard, James Herbert
Hall, Leo McAloon
Hardman, John Kemper
Harvey, Stephen Craig
Higgins, N Patrick
Hill, Donald Lynch
Huggins, Clyde Griffin
Johnson, Brian John
Kaplan, Ronald S
Kochakian, Charles Daniel
Krumdieck, Carlos L
Lebowitz, Jacob
Lehman, Robert Harold
Lindsay, Raymond H
Long, Calvin Lee
Longenecker, Herbert Eugene
Lorincz, Andrew Endre
Lutwak, Leo
McCord, Joe Milton
McKibbin, John Mead
Magargal, Wells Wrisley, II
Marchase, Richard Banfield
Meezan, Elias
Mego, John L
Melius, Paul
Miller, Edward Joseph
Moore, Robert Blaine
Niedermeier, William
Otto, David A
Parish, Edward James
Paxton, Ralph
Pillion, Dennis Joseph
Prejean, Joe David
Pruitt, Kenneth M
Rhodes, Richard Kent
Rivers, Douglas Bernard
Sani, Brahma Porinchu
Sauberlich, Howerde Edwin
Schaffer, Stephen Ward
Schrohenloher, Ralph Edward
Schutzbach, John Stephen
Segrest, Jere Palmer
Short, William Arthur
Siegal, Gene Philip
Smith, Robert C
Smith-Somerville, Harriett Elizabeth
Strada, Samuel Joseph
Strength, Delphin Ralph
Suling, William John
Svacha, Anna Johnson
Taylor, Kenneth Boivin
Thompson, Jerry Nelson
Thompson, Wynelle Doggett
Tomana, Milan
Urry, Dan Wesley
Volker, Joseph Francis
Weete, John Donald
Wells, Robert Dale
Wheeler, Glynn Pearce
Whikehart, David Ralph
Wilkoff, Lee Joseph
Wingo, William Jacob
Winkler, Herbert H
Winters, Alvin L
Wooten, Marie W
Yarbrough, James David
Yeilding, Lerena Wade

ALASKA
Behrisch, Hans Werner
Button, Don K
Duffy, Lawrence Kevin
Gharrett, Anthony John
Nakada, Henry Isao
Stekoll, Michael Steven
Wennerstrom, David E

ARIZONA
Allen, John Rybolt
Barstow, Leon E
Bieber, Allan Leroy
Bowden, George Timothy
Brush, James S
Caldwell, Roger Lee
Carter, Herbert Edmund
Cassidy, Suzanne Bletterman
Chandler, Douglas Edwin
Cress, Anne Elizabeth
Cronin, John Read
Cusanovich, Michael A
Doubek, Dennis Lee
Eckardt, Robert E
Eskelson, Cleamond D
Fowler, Dona Jane
Fuller, Wallace Hamilton
Goll, Darrel Eugene
Gordon, Richard Seymour
Guerriero, Vincent, Jr
Halpert, James Robert
Harkins, Kristi R
Harris, Joseph
Hazel, Jeffrey Ronald
Hendrix, Donald Louis
Hewlett, Martinez Joseph
Holmes, William Farrar
Jensen, Richard Grant
Kay, Marguerite M B
Kazal, Louis Anthony
Koldovsky, Otakar
Krahl, Maurice Edward
Larkins, Brian Allen
Law, John Harold
Lei, David Kai Yui
Lei, David Kai Yui
Liddell, Robert William, Jr
Lipke, William G
Little, John Wesley
Loper, Gerald Milton
Luchsinger, Wayne Wesley
Lukas, Ronald John
McLean, Katharine Weidman
Maddy, Kenneth Hilton
Marchalonis, John Jacob
Mardian, James K W
Mason, Merle
Milch, Lawrence Jacques
Moore, Ana Maria
Moore, Thomas Andrew
Mosher, Richard Arthur
Moss, Lloyd Kent
Peng, Yeh-Shan
Peterson, James Douglas
Potts, Albert Mintz
Price, Ralph Lorin
Reeves, Henry Courtland
Reid, Bobby Leroy
Rupley, John Allen
Shull, James Jay
Siemankowski, Raymond Francis
Spizizen, John
Taylor, Richard G
Thomas, Joseph James
Tischler, Marc Eliot
Trelease, Richard Norman
Varnell, Thomas Raymond
Vermaas, Willem F J
Vestling, Carl Swensson
Ward, Samuel
Weber, Charles Walter
Wells, Michael Arthur
Whiting, Frank M
Yall, Irving
Yamamura, Henry Ichiro
Yohem, Karin Hummell

ARKANSAS
Allaben, William Thomas
Benson, Ann Marie
Beranek, David T
Bhuvaneswaran, Chidambaram
Blackwell, Richard Quentin
Brewster, Marjorie Ann
Chowdhury, Parimal
Cornett, Lawrence Eugene
De Luca, Donald Carl
Elbein, Alan D
Epstein, Joshua
Evans, Frederick Earl
Fink, Louis Maier
Forsythe, Richard Hamilton
Greenman, David Lewis
Harris, Grover Cleveland, Jr
Harrison, Robert Walker, III
Heflich, Robert Henry
Jackson, Carlton Darnell
Jones, Robin Richard
Kadlubar, Fred F
Kraemer, Louise Margaret
Lane, Forrest Eugene
Leakey, Julian Edwin Arundell
Light, Kim Edward
Millett, Francis Spencer

Morris, Manford D
Nelson, Charles A
Paule, Merle Gale
Pynes, Gene Dale
Sanders, Louis Lee
Sheehan, Daniel Michael
Shideler, Robert Weaver
Smith, Carroll Ward
Smith, William Grady
Snyder, Harry E
Stallcup, Odie Talmadge
Steinmeier, Robert C
Stone, Joseph
Vesely, David Lynn
Wadkins, Charles L
Waldroup, Park William
Winter, Charles Gordon
Yeh, Yunchi
York, John Lyndal

CALIFORNIA
Abbott, Mitchel Theodore
Abel, Carlos Alberto
Abrahamson, Lila
Abrash, Henry I
Ackrell, Brian A C
Aftergood, Lilla
Ahuja, Jagan N
Alexander, Nicholas Michael
Alfin-Slater, Roslyn Berniece
Allen, Charles Freeman
Alousi, Adawia A
Altieri, Dario Carlo
Amer, Mohamed Samir
Ames, Bruce Nathan
Ames, Giovanna Ferro-Luzzi
Amy, Nancy Klein
Anand, Rajen S(ingh)
Anderson, Thomas Alexander
Andreoli, Anthony Joseph
Appleman, M Michael
Arai, Ken-Ichi
Arfin, Stuart Michael
Arnon, Daniel I(srael)
Arroyave, Guillermo
Ashmore, Charles Robert
Atkinson, Daniel Edward
Ayengar, Padmasini (Mrs Frederick
    Aladjem)
Azhar, Salman
Babior, Bernard M
Bagdasarian, Andranik
Bailey-Serres, Julia
Baker, Nome
Balch, William E
Baldeschwieler, John Dickson
Ballou, Clinton Edward
Barankiewicz, Jerzy Andrzej
Barker, Horace Albert
Barnes, Paul Richard
Barnet, Harry Nathan
Bartlett, Grant Rogers
Bartley, John C
Bartnicki-Garcia, Salomon
Bassham, James Alan
Bauer, Roger Duane
Baumgarten, Werner Andreas
Baxter, Claude Frederick
Baxter, John Darling
Beckendorf, Steven K
Becker, David Morris
Becker, Joseph F
Bell, Jack Perkins
Beltz, Richard Edward
Benemann, John Rudiger
Benisek, William Frank
Benjamini, Eliezer
Bennett, Edward Leigh
Bennett, Raymond Dudley
Bennett, William Franklin
Benson, Andrew Alm
Benson, James R
Benson, Robert Leland
Berg, Paul
Berk, Arnold J
Berry, Edward Alan
Bertolami, Charles Nicholas
Bessman, Samuel Paul
Bhatnagar, Rajendra Sahai
Bhattacharya, Prabir
Bidlack, Wayne Ross
Bigler, William Norman
Bikle, Daniel David
Bishop, John Michael
Bjorkman, Pamela J
Black, Arthur Leo
Blackburn, Elizabeth Helen
Blankenship, James W
Blankenship, James William
Bokoch, Gary M
Bondar, Richard Jay Laurent
Bondy, Stephen Claude
Bonting, Sjoerd Lieuwe
Bonura, Thomas
Bourne, Henry R
Bowen, Charles E
Bower, Annette
Bowman, Barry J
Boyd, James Brown
Bradbury, E Morton
Bradfield, Robert B
Bradshaw, Ralph Alden
Bramhall, John Shepherd
Brandon, David Lawrence

Breidenbach, Rowland William
Brody, Stuart
Brostoff, Steven Warren
Brown, Jeanette Snyder
Brown, Joan Heller
Brown, William Duane
Brownell, Anna Gale
Brunke, Karen J
Brunton, Laurence
Brutlag, Douglas Lee
Bryant, Peter James
Buchanan, Bob Branch
Burd, John Frederick
Burgess, Teresa Lynn
Burton, Louis
Buttlaire, Daniel Howard
Buzin, Carolyn Hattox
Byard, James Leonard
Byerley, Lauri Olson
Byers, Sanford Oscar
Cabot, Myles Clayton
Calarco, Patricia G
Camien, Merrill Nelson
Campagnoni, Anthony Thomas
Campbell, Alice del Campillo
Campbell, Judith Lynn
Cantor, Charles Robert
Cape, Ronald Elliot
Caporaso, Fredric
Carbon, John Anthony
Carew, Thomas Edward
Carle, Glenn Clifford
Carlisle, Edith M
Carlson, Don Marvin
Carroll, Edward James, Jr
Carson, Virginia Rosalie Gottschall
Carsten, Mary E
Castanera, Esther Goossen
Castles, James Joseph, Jr
Celniker, Susan Elizabeth
Chaffee, Rowand R J
Chamberlin, Michael John
Chan, Bock G
Chan, Sham-Yuen
Chang, Ding
Chang, Eppie Sheng
Chang, Ernest Sun-Mei
Chang, Jaw-Kang
Chao, Li-Pen
Chapman, David J
Charlang, Gisela Wohlrab
Chaykin, Sterling
Chen, Chiadao
Chen, Ming Sheng
Chen, Stephen Shi-Hua
Chen, Tung-Shan
Cheng, Sze-Chuh
Cherkin, Arthur
Chou, Tsong-Wen
Christensen, Halvor Niels
Chuang, Ronald Yan-Li
Civen, Morton
Clarke, Steven Gerard
Clayton, Raymond Brazenor
Clegg, James S
Cline, Thomas Warren
Clough, Wendy Glasgow
Cohen, Natalie Shulman
Cohen, Robert Elliot
Cohlberg, Jeffrey Allan
Cohn, Major Lloyd
Collins, James Francis
Collins, Robert C
Colvin, Harry Walter, Jr
Conn, Eric Edward
Conrad, Harry Edward
Conrad, Herbert M
Cooper, Geoffrey Kenneth
Cooper, James Burgess
Cotman, Carl Wayne
Coty, William Allen
Cozzarelli, Nicholas Robert
Criddle, Richard S
Cross, John William
Cunningham, Dennis Dean
Curnutte, John Tolliver, III
Dahms, Arthur Stephen
Dahmus, Michael E
Danenberg, Peter V
Dang, Peter Hung-Chen
Daniel, Louise Jane
Dasgupta, Asim
David, Gary Samuel
Davies, Huw M
Davis, Frank French
Davis, Rowland Hallowell
Davis, William Ellsmore, Jr
Dekker, Charles Abram
DeLange, Robert J
Delk, Ann Stevens
DeLuca, Marlene
Delwiche, Constant Collin
DeMet, Edward Michael
Dennis, Edward A
Des Marais, David John
DeVenuto, Frank
DeVillis, Jean
Diamond, Ivan
Dietz, George William, Jr
Doolittle, Russell F
Downing, Michael Richard
Dragon, Elizabeth Alice Oosterom
Draper, Roy Douglas
Dugaiczyk, Achilles

Dukes, Peter Paul
Dunaway, Marietta
Dunn, Michael F
Du Pont, Frances Marguerite
Durrum, Emmett Leigh
Durzan, Donald John
Duzgunes, Nejat
Dyer, Denzel Leroy
Eaks, Irving Leslie
Ecker, David John
Eckhart, Walter
Eiduson, Samuel
Eiler, John Joseph
Einset, John William
Einstein, Elizabeth Roboz
Ellis, Stanley
Ellman, George Leon
Emr, Scott David
Endahl, Gerald Leroy
Eng, Lawrence F
Engvall, Eva Susanna
Eppstein, Deborah Anne
Epstein, Charles Joseph
Etzler, Marilynn Edith
Evard, Rene
Fahey, Robert C
Fan, Hsing Yun
Felton, James Steven
Fenimore, David Clarke
Feramisco, James Robert
Fernandez, Alberto Antonio
Fessler, John Hans
Fessler, Liselotte I
Fielding, Christopher J
Fife, Thomas Harley
Filner, Philip
Fineberg, Richard Arnold
Fink, Kathryn Ferguson
Fink, Robert M
Finkle, Bernard Joseph
Fiorindo, Robert Philip
Fischer, Imre A
Fisher, Knute Adrian
Fishman, William Harold
Flanagan, Steven Douglas
Flashner, Michael
Fleming, James Emmitt
Fluharty, Arvan Lawrence
Forrest, Irene Stephanie
Fortes, George (Peter Alexander)
Fox, C Fred
Fox, Joan Elizabeth Bothwell
Francke, Uta
Frankel, Edwin N
Freer, Stephan T
Frey, Terrence G
Friedberg, Errol Clive
Friedkin, Morris Enton
Friedlander, Martin
Fujimoto, George Iwao
Fukuda, Minoru
Fukushima, David Kenzo
Fukuyama, Thomas T
Fulco, Armand J
Gadol, Nancy
Gaertner, Frank Herbert
Garcia, Eugene N
Garfin, David Edward
Garibaldi, John Attilio
Gautsch, James Willard
Geiger, Paul Jerome
Gelfand, David H
Geller, Edward
Gerber, Louis P
Gerhart, John C
Gibson, Thomas Richard
Gigliotti, Helen Jean
Ginsberg, Theodore
Giorgio, Anthony Joseph
Giri, Shri N
Glabe, Charles G
Glasky, Alvin Jerald
Glazer, Alexander Namiot
Glembotski, Christopher Charles
Glitz, Dohn George
Gluck, Louis
Glushko, Victor
Gochman, Nathan
Gold, William
Goldbloom, David Ellis
Gonzalez, Elma
Goodman, Michael Gordon
Goodman, Myron F
Gordon, Adrienne Sue
Gordon, Harold Thomas
Gorelick, Kenneth J
Gralla, Jay Douglas
Gray, Gary M
Green, Melvin Howard
Greenberg, David Morris
Greene, Frank Clemson
Greenleaf, Robert Dale
Griffin, John Henry
Griffith, Michael James
Griffith, Owen Malcolm
Grill, Herman, Jr
Grodsky, Gerold Morton
Gruenwedel, Dieter Wolfgang
Gunning, Barbara E
Gunsalus, Robert Philip
Gupta, Rishab Kumar
Guthrie, Christine
Haard, Norman F
Haberland, Margaret Elizabeth

Haen, Peter John
Hafeman, Dean Gary
Hainski, Martha Barrionuevo
Hajdu, Joseph
Hall, Michael Oakley
Hammerschlag, Richard
Hammes, Gordon G
Hanawalt, Philip Courtland
Hankinson, Oliver
Harary, Isaac
Harding, Boyd W
Harper, Elvin
Harper, Harold Anthony
Harper, Judith Jean
Hatefi, Youssef
Hatfield, G Wesley
Hathaway, Gary Michael
Hawkes, Susan Patricia
Hawkes, Wayne Christian
Hayman, Ernest Paul
Hearn, Walter Russell
Heasley, Victor Lee
Hector, Mina Fisher
Hedrick, Jerry Leo
Heftmann, Erich
Hegenauer, Jack C
Heinrich, Milton Rollin
Heinrichs, W LeRoy
Helinski, Donald Raymond
Helwig, Harold Lavern
Henderson, Gary Borgar
Hendren, Richard Wayne
Henry, Helen L
Henry, Richard Joseph
Herber, Raymond
Herschman, Harvey R
Hershey, John William Baker
Hess, Frederick Dan
Hessinger, David Alwyn
Hill, Robert
Hines, Leonard Russell
Hintz, Marie I
Hjelmeland, Leonard M
Hoagland, Vincent DeForest, Jr
Hoch, Sallie O'Neil
Hochstein, Paul Eugene
Hogan, Christopher James
Hoke, Glenn Dale
Holden, Joseph Thaddeus
Hollander, Leonore
Holley, Robert William
Holmes, David G
Holmquist, Walter Richard
Holten, Darold Duane
Holten, Virginia Zewe
Horowitz, Norman Harold
Hosoda, Junko
Hostetler, Karl Yoder
Howard, Bruce David
Howard, Russell John
Hsiao, Theodore Ching-Teh
Hsieh, Philip Kwok-Young
Hsueh, Aaron Jen Wang
Hubbard, Richard W
Hubbard, William Jack
Huennekens, Frank Matthew, Jr
Huffaker, Ray C
Hugli, Tony Edward
Hullar, Theodore Lee
Hunter, Tony
Hyman, Bradley Clark
Hyman, Richard W
Ibsen, Kenneth Howard
Insel, Paul Anthony
Itano, Harvey Akio
Jacob, Mary
Jacobson, Gail M
Jariwalla, Raxit Jayantilal
Jelinek, Bohdan
Jewett, Sandra Lynne
Johansson, Mats W
Johnson, Paul Hickok
Johnson, Randolph Mellus
Jones, Oliver William
Jones, Theodore Harold Douglas
Jordan, Mary Ann
Josephson, Ronald Victor
Jukes, Thomas Hughes
Kaback, Howard Ronald
Kaiser, Armin Dale
Kallman, Burton Jay
Kalman, Sumner Myron
Kalra, Vijay Kumar
Kammen, Harold Oscar
Kan, Yuet Wai
Kane, John Power
Kaneko, Jiro Jerry
Kang, Kenneth S
Kan-Mitchell, June
Kaplan, Phyllis Deen
Kaplan, Stanley A
Karasek, Marvin A
Katz, Joseph
Kavenoff, Ruth
Kay, Robert Eugene
Keating, Eugene Kneeland
Kedes, Laurence H
Kelley, Darshan Singh
Kelley, Leon A
Kelly, Regis Baker
Kennedy, Mary Bernadette
Kenney, William Clark
Khatra, Balwant Singh
Kientz, Marvin L

Kihara, Hayato
Kim, Sung-Hou
Kingdon, Henry Shannon
Kinoshita, Jin Harold
Kirsch, Jack Frederick
Kirschbaum, Joel Bruce
Kirschner, Marc Wallace
Kishimoto, Yasuo
Klain, George John
Kliewer, Walter Mark
Klouda, Mary Ann Aberle
Kohler, George Oscar
Koobs, Dick Herman
Korenman, Stanley G
Korn, David
Kornberg, Arthur
Kornberg, Roger David
Koshland, Daniel Edward, Jr
Koths, Kirston Edward
Krauss, Ronald
Kun, Ernest
Kunkee, Ralph Edward
Kuo, Harng-Shen
Kurnick, Nathaniel Bertrand
Kurtzman, Ralph Harold, Jr
Kushinsky, Stanley
Kuwahara, Steven Sadao
Kyte, Jack Ernst
Labavitch, John Marcus
Lad, Pramod Madhusudan
Lanchantin, Gerard Francis
Landolfi, Nicholas F
Lanyi, Janos K
Largman, Corey
Lascelles, June
Last, Jerold Alan
Lawrence, Paul J
Lebherz, Herbert G
Lee, Amy Shiu
Leffert, Hyam Lerner
Lehman, Israel Robert
Lehrer, Harris Irving
Lei, Shau-Ping Laura
Lembach, Kenneth James
Lenhoff, Howard Maer
Leong, Kam Choy
Leung, Peter
Leung, Philip Min Bun
Levintow, Leon
Levy, Daniel
Lewis, James Clement
Lewis, Urban James
Lianides, Sylvia Panagos
Lieberman, Jack
Lin, Jiann-Tsyh
Lindquist, Robert Nels
Ling, Nicholas Chi-Kwan
Linn, Stuart Michael
Lipsick, Joseph Steven
Litman, David Jay
Liu, Edwin H
Lolley, Richard Newton
Lönnerdal, Bo L
Lopo, Alina C
Lovatt, Carol Jean
Lovelace, C James
Ludowieg, Julio
Lyon, Irving
McCaman, Marilyn Wales
McClelland, Michael
McClure, William Owen
McElroy, William David
McGaughey, Charles Gilbert
McIntire, Junius Merlin
McIntire, William S
McKee, Ralph Wendell
McLaughlin, Calvin Sturgis
McMillan, Paul Junior
McNamee, Mark G
Maggio, Edward Thomas
Maier, V(incent) P(aul)
Makarem, Anis H
Makker, Sudesh Paul
Malkin, Harold Marshall
Malkin, Richard
Mann, Lewis Theodore, Jr
Manning, Jerry Edsel
Mansour, Tag Eldin
Marangos, Paul Jerome
Marcus, Frank
Markland, Francis Swaby, Jr
Marsh, James Lawrence
Martin, George Steven
Massie, Barry Michael
Matthews, Harry Roy
Matthews, Thomas Robert
Mattson, Fred Hugh
Mayron, Lewis Walter
Mazelis, Mendel
Mead, James Franklyn
Mealey, Edward H
Meares, Claude Francis
Meehan, Thomas (Dennis)
Meikle, Richard William
Melchior, Jacklyn Butler
Melis, Anastasios
Mendelson, Robert Alexander, Jr
Merchant, Sabeeha
Merriam, Esther Virginia
Meyerowitz, Elliot Martin
Miljanich, George Paul
Miller, Harrison
Miller, Jon Philip
Miller, Lloyd George

Miller, William Walter
Mitchell, Herschel Kenworthy
Miyada, Don Shuso
Mizuno, Nobuko S(himotori)
Mochizuki, Diane Yukiko
Moffitt, Robert Allan
Mohan, Chandra
Mohrenweiser, Harvey Walter
Moldave, Kivie
Moore, Gerald L
Morales, Daniel Richard
Morehouse, Margaret Gulick
Mosteller, Raymond Dee
Moyed, Harris S
Mudd, John Brian
Mueller, Peter Klaus
Mule, Salvatore Joseph
Murphy, Alexander James
Murphy, Terence Martin
Nacht, Sergio
Nagel, Glenn M
Nambiar, Krishnan P
Nassos, Patricia Saima
Neidleman, Saul L
Nelson, Gary Joe
Nemere, Ilka M
Neufeld, Elizabeth Fondal
Newbrun, Ernest
Newell, Gordon Wilfred
Newton, William Edward
Ngo, That Tjien
Nicolson, Margery O'Neal
Nielsen, Milo Alfred
Nikaido, Hiroshi
Nimni, Marcel Efraim
Noller, Harry Francis, Jr
Noltmann, Ernst August
Norman, Anthony Westcott
Nussenbaum, Siegfried Fred
Nyhan, William Leo
O'Connor, Daniel Thomas
O'Donnell, Joseph Allen, III
Ohms, Jack Ivan
Oldham, Susan Banks
Oliphant, Edward Eugene
Olsen, Charles Edward
Olson, Alfred C
Omaye, Stanley Teruo
Oppenheimer, Norman Joseph
Ortiz de Montellano, Paul Richard
Osborn, Terry Wayne
Ottke, Robert Crittenden
Ottoboni, M(inna) Alice
Overby, Lacy Rasco
Owades, Joseph Lawrence
Oyama, Jiro
Oyama, Vance I
Packer, Lester
Paech, Christian
Paik, Young-Ki
Palmer, Grant H
Palmer, John Warren
Papahadjopoulos, Demetrios Panayotis
Papkoff, Harold
Parthasarathy, Sampath
Patterson, Paul H
Patton, John Stuart
Paulson, James Carsten
Peckham, William Dierolf
Penhoet, Edward Etienne
Perlgut, Louis E
Perrault, Jacques
Peters, James Milton
Peters, John Henry
Petersen, Gene
Peterson, Charles Marquis
Peterson, Neal Alfred
Pharriss, Bruce Bailey
Phelps, Michael Edward
Philippart, Michel Paul
Phillips, David Richard
Phillips, Donald Arthur
Phleger, Charles Frederick
Pierce, John Grissim
Pierschbacher, Michael Dean
Piez, Karl Anton
Pigiet, Vincent P
Pilgeram, Laurence Oscar
Pirkle, Hubert Chaille
Piszkiewicz, Dennis
Pitesky, Isadore
Platzer, Edward George
Poling, Stephen Michael
Popjak, George Joseph
Powanda, Michael Christopher
Prescott, Benjamin
Prestayko, Archie William
Price, Paul Arms
Prusiner, Stanley Ben
Pryer, Nancy Kathryn
Quail, Peter Hugh
Quintanilha, Alexandre Tiedtke
Quistad, Gary Bennet
Rabinowitz, Israel Nathan
Rabinowitz, Jesse Charles
Rabovsky, Jean
Rae-Venter, Barbara
Raijman, Luisa J
Rall, Stanley Carlton, Jr
Ramachandran, Janakiraman
Rao, Ananda G
Recsei, Paul Andor
Reese, Floyd Ernest
Reese, Robert Trafton

Biochemistry (cont)

Register, Ulma Doyle
Reid, Ted Warren
Reiness, Gary
Reizer, Jonathan
Reizer, Jonathan
Remsen, Joyce F
Rho, Joon H
Rice, Robert Hafling
Rich, Terrell L
Richards, John Hall
Richmond, Jonas Edward
Rinderknecht, Heinrich
Rittenhouse, Harry George
Roberts, Eugene
Roberts, Martin
Roberts, Sidney
Robinson, James McOmber
Rodgers, Richard Michael
Roepke, Raymond Rollin
Rogers, Quinton Ray
Rogoff, Martin Harold
Rokeach, Luis Alberto
Roll, Paul M
Romani, Roger Joseph
Rome, Leonard H
Rosenberg, Abraham
Rosenberg, Steven Loren
Rosenfeld, Ron Gershon
Rosenquist, Grace Link
Rossi, John Joseph
Rothman, Stephen Sutton
Rowley, Gerald L(eroy)
Rowley, Rodney Ray
Roy-Burman, Pradip
Rubenstein, Kenneth E
Rucker, Robert Blain
Rusay, Ronald Joseph
Russell, Percy J
Rutter, William J
Sabry, Zakaria I
Sachs, George
Sadava, David Eric
Sadee, Wolfgang
Saier, Milton H, Jr
Saltman, Paul David
Salzberg, David Aaron
Sammak, Paul J
Sampson, David Ashmore
Samuel, Charles Edward
Sanchez, Albert
Santi, Daniel V
Sayre, Francis Warren
Scala, James
Schachman, Howard Kapnek
Schaffer, Frederick Leland
Schaffer, Sheldon Arthur
Schapiro, Harriette Charlotte
Schekman, Randy W
Schelar, Virginia Mae
Scheve, Larry Gerard
Schiffman, Sandra
Schimke, Robert T
Schleich, Thomas W
Schmid, Peter
Schmid, Rudi
Schneir, Michael Lewis
Schnitzer, Jan Eugeneusz
Schotz, Michael C
Schraer, Rosemary
Schroeder, Duane David
Schwimmer, Sigmund
Scogin, Ron Lynn
Searle, Gilbert Leslie
Sears, Duane William
Seegers, Walter Henry
Seegmiller, Jarvis Edwin
Segall, Paul Edward
Segel, Irwin Harvey
Selassie, Cynthia R
Serat, William Felkner
Sevall, Jack Sanders
Shah, Shantilal Nathubhai
Shannon, Leland Marion
Shapiro, Lucille
Shaw, Kenneth Noel Francis
Shen, Che-Kun James
Sherman, Linda Arlene
Shih, Jean Chen
Shimizu, C Susan
Shively, John Ernest
Shneour, Elie Alexis
Shooter, Eric Manvers
Shore, Virgie Guinn
Shrawder, Elsie June
Shugarman, Peter Melvin
Shum, Archie Chue
Sie, Edward Hsien Choh
Siegman, Fred Stephen
Sigman, David Stephan
Siiteri, Pentti Kasper
Simmons, Norman Stanley
Simoni, Robert Dario
Simons, Robert W
Simonsen, Donald Howard
Singer, Thomas Peter
Sinha, Yagya Nand
Sinibaldi, Ralph Michael
Sinsheimer, Robert Louis
Siperstein, Marvin David
Slater, Grant Gay
Slavkin, Harold Charles
Smith, Andrew Philip

Smith, Cassandra Lynn
Smith, Charles G
Smith, Emil L
Smith, Helene Sheila
Smith, Kendric Charles
Smith, Kevin Malcolm
Smith, Marion Edmonds
Smith, Martyn Thomas
Smith, Roberts Angus
Smith, Steven Sidney
Smith, Stuart
Smuckler, Edward Aaron
Snow, John Thomas
Snowdowne, Kenneth William
Song, Moon K
Spanis, Curt William
Sparling, Mary Lee
Spray, Clive Robert
Springer, Wayne Richard
Spudich, James Anthony
Stallcup, Michael R
Stanczyk, Frank Zygmunt
Stanley, Wendell Meredith, Jr
Stanton, Hubert Coleman
Steinberg, Daniel
Stellwagen, Robert Harwood
Sterling, Rex Elliott
Stern, Robert
Sternberg, Moshe
Stevens, Ronald Henry
Stewart, Charles Jack
Still, Gerald G
Stokstad, Evan Ludvig Robert
Stone, Deborah Bennett
Strehler, Bernard Louis
Strother, Allen
Stroynowski, Iwona T
Stubbs, John Dorton
Stumpf, Paul Karl
Subramani, Suresh
Sullivan, Sean Michael
Sussman, Howard H
Sweetman, Lawrence
Swendseid, Marian Edna
Sy, Jose
Szego, Clara Marian
Taborsky, George
Tallman, John Gary
Tappel, Aloys Louis
Tarver, Harold
Taylor, Robert Thomas
Taylor, Susan Serota
Teresi, Joseph Dominic
Thiele, Elizabeth Henriette
Thomas, Dudley Watson
Thomas, Heriberto Victor
Thomas, Richard Sanborn
Thomas, Robert E
Thompson, Chester Ray
Thomson, John Ansel Armstrong
Thornber, James Philip
Thorner, Jeremy William
Tjian, Robert Tse Nan
Tokes, Zoltan Andras
Tomlinson, Geraldine Ann
Tomlinson, Raymond Valentine
Townsend, R Reid
Tramell, Paul Richard
Traugh, Jolinda Ann
Traut, Robert Rush
Traylor, Patricia Shizuko
Trifunac, Natalia Pisker
Troy, Frederic Arthur
Tsao, Constance S
Tsuboi, Kenneth Kaz
Tsuji, Frederick Ichiro
Turgeon, Judith Lee
Ulevitch, Richard Joel
Valentine, Raymond Carlyle
Vandlen, Richard Lee
Varki, Ajit Pothan
Varmus, Harold Elliot
Varon, Silvio Salomone
Vehar, Gordon Allen
Vice, John Leonard
Vickery, Larry Edward
Victoria, Edward Jess, Jr
Vilker, Vincent Lee
Villarejo, Merna
Villarreal, Luis Perez
Vohra, Pran Nath
Volcani, Benjamin Elazari
Vold, Barbara Schneider
Von HUngen, Kern
Vreeland, Valerie Jane
Vreman, Hendrik Jan
Wade, Michael James
Wadman, W Hugh
Walbot, Virginia Elizabeth
Walker, Howard David
Wallace, Joan M
Wallace, Robert Allan
Walter, Harry
Wang, Ching Chung
Wang, Howard Hao
Warner, Robert Collett
Warner, Thomas Garrie
Wasterlain, Claude Guy
Watanabe, Ronald S
Watson, John Alfred
Wax, Harry
Webb, David Ritchie
Weber, Arthur L
Weber, Bruce Howard

Weber, Heather R(oss) Wilson
Weber, Thomas Byrnes
Wedding, Randolph Townsend
Wegner, Marcus Immanuel
Wehr, Carl Timothy
Weimberg, Ralph
Weinberg, Barbara Lee Huberman
Weisgraber, Karl Heinrich
Weiss, Richard Louis
Wergedal, Jon E
West, Charles Allen
Westall, Frederick Charles
Whitaker, John Robert
White, Thomas James
Whitlock, Gaylord Purcell
Wiktorowicz, John Edward
Wilcox, Gary Lynn
Wilcox, Ronald Bruce
Wilkinson, David Ian
Williams, Mary Ann
Wilson, Allan Charles
Wilson, Jerry Lee
Wilson, Leslie
Wilson, Lowell D
Winkelhake, Jeffrey Lee
Wolf, George
Wolfgram, Frederick John
Wong, Kin-Ping
Wood, Peter Douglas
Wood, William Irwin
Wood, Willis Avery
Wu, Chuen-Shang C
Wu, Chung
Yager, Janice L Winter
Yamaguchi, Masatoshi
Yau-Young, Annie O
Yguerabide, Juan
Yoshida, Akira
Young, Janis Dillaha
Yu, David Tak Yan
Yu, Sharon S M
Yuwiler, Arthur
Zabin, Irving
Zaffaroni, Alejandro
Zahnley, James Curry
Zamenhof, Stephen
Zeichner-David, Margarita
Zerez, Charles Raymond
Ziboh, Vincent Azubike
Ziccardi, Robert John
Zill, Leonard Peter
Zuckekandl, Emile

COLORADO

Abrams, Adolph
Akeson, Walter Roy
Allen, Kenneth G D
Allen, Robert H
Arend, William Phelps
Azari, Parviz
Bailey, David Tiffany
Bamburg, James Robert
Barrett, Dennis
Berens, Randolph Lee
Bowden, Joe Allen
Bowles, Jean Alyce
Brennan, Patrick Joseph
Brooks Springs, Suzanne Beth
Brown, Jerry L
Bublitz, Clark
Caughey, Winslow Spaulding
Chan, Laurence Kwong-fai
Charkey, Lowell William
Charney, Michael
Church, Brooks Davis
Cooper, Dermot M F
Corcoran, John William
Deitrich, Richard Adam
Dever, John E, Jr
Dobersen, Michael J
Downing, Mancourt
Dyckes, Douglas Franz
Eley, James H
Erwin, Virgil Gene
Fahrney, David Emory
Fitzpatrick, Francis Anthony
Frerman, Frank Edward
Froede, Harry Curt
Gamborg, Oluf Lind
Giclas, Patricia C
Glode, Leonard Michael
Grainger, Robert Ball
Gramera, Robert Eugene
Grieve, Robert B
Ham, Richard George
Hamar, Dwayne Walter
Harold, Franklin Marcel
Harrill, Inez Kemble
Harrison, Merle E(dward)
Hesterberg, Thomas William
Hibler, Charles Phillip
Hinman, Norman Dean
Horwitz, Kathryn Bloch
James, Gordon Thomas
Jangaard, Norman Olaf
Jansen, Gustav Richard
Jones, Carol A
Kano-Sueoka, Tamiko
Kerr, Sylvia Jean
Kinsky, Stephen Charles
Kloppel, Thomas Mathew
Langan, Thomas Augustine
Lee, Virginia Ann
Lin, Leu-Fen Hou

Linden, James Carl
Lowndes, Joseph M
McCord, Joe M
McHenry, Charles S
Mackenzie, Cosmo Glenn
Maga, Joseph Andrew
Malkinson, Alvin Maynard
Martin, Jack E
Martin, Susan Scott
Masken, James Frederick
Mathies, James Crosby
Metzger, H Peter
Moore, Frank Archer
Mykles, Donald Lee
Olson, John Melvin
Patterson, David
Pearson, John Richard
Pizer, Lewis Ivan
Prentice, Neville
Quissell, David Olin
Reiss, Oscar Kully
Rhoads, William Denham
Roberts, Walden Kay
Ross, Cleon Walter
Saidel, Leo James
Seeds, Nicholas Warren
Seibert, Michael
Sinensky, Michael
Skelton, Marilyn Mae
Sneider, Thomas W
Solomons, Clive (Charles)
Stewart, John Morrow
Stifel, Fred B
Storey, Richard Drake
Tolbert, Bert Mills
Tomasi, Gordon Ernest
Tu, Anthony T
Virtue, Robert
Weliky, Irving
Wenger, David Arthur
Wilson, Irwin B
Winstead, Jack Alan
Woody, A-Young Moon
Woody, Robert Wayne
Yarus, Michael J

CONNECTICUT

Adelberg, Edward Allen
Amacher, David E
Armitage, Ian MacLeod
Aronson, Peter S
Bausher, Larry Paul
Behrman, Harold R
Bluhm, Leslie
Bondy, Philip K
Bordner, Jon D D
Bormann, Barbara-Jean Anne
Briscoe, Anne M
Bronner, Felix
Bronsky, Albert J
Buck, Marion Gilmour
Cadman, Edwin Clarence
Canellakis, Evangelo S
Canellakis, Zoe Nakos
Carboni, Joan M
Celesk, Roger A
Cinti, Dominick Louis
Coleman, Joseph Emory
Crain, Richard Cullen
Crawford, Richard Bradway
Dannies, Priscilla Shaw
Das, Dipak K
Deutscher, Murray Paul
De Zeeuw, John Robert
Dingman, Douglas Wayne
Dix, Douglas Edward
Doeg, Kenneth Albert
Dooley, Joseph Francis
Dorsky, David Isaac
Edwards, Lawrence Jay
Eisenstadt, Jerome Melvin
Engelman, Donald Max
Epstein, Paul Mark
Eustice, David Christopher
Fenton, Wayne Alexander
Fiore, Joseph Vincent
Flavell, Richard Anthony
Forbush, Bliss, III
Fordham, Joseph Raymond
Forenza, Salvatore
Forget, Bernard G
Fruton, Joseph Stewart
Geiger, Edwin Otto
Geiger, Jon Ross
Gerritsen, Mary Ellen
Gordon, Malcolm Wofsy
Gorrell, Thomas Earl
Gortner, Ross Aiken, Jr
Grindley, Nigel David Forster
Gross, Ian
Gum, Ernest Kemp, Jr
Gunther, Jay Kenneth
Haas, Mark
Haley, Edward Everett
Hanson, Kenneth Ralph
Harding, Matthew William
Havir, Evelyn A
Heywood, Stuart Mackenzie
Hinman, Richard Leslie
Hobbs, Donald Clifford
Hook, Derek John
Howard, Phillenore Drummond
Howard, Robert Eugene
Huszar, Gabor

Hutchinson, Franklin
Infante, Anthony A
Janis, Ronald Allen
Jungas, Robert Leando
Keirns, James Jeffery
Kelleher, William Joseph
Khairallah, Edward A
Kind, Charles Albert
Kiron, Ravi
Koe, B Kenneth
Konigsbacher, Kurt S
Konigsberg, William Henry
Krause, Leonard Anthony
Kream, Barbara Elizabeth
Lande, Saul
Leadbetter, Edward Renton
Lee, Henry C
Lee, Thomas W
Lees, Thomas Masson
Lengyel, Peter
Lentz, Thomas Lawrence
Lerner, Aaron Bunsen
Lewis, Jonathan Joseph
Lipsky, Seymour Richard
Loomis, Stephen Henry
Lukens, Lewis Nelson
Lustig, Bernard
McCorkle, George Maston
McGregor, Donald Neil
MacNintch, John Edwin
Marchesi, Vincent T
Marks, Paul A
Matovcik, Lisa M
Mattes, William Bustin
Mayol, Robert Francis
Miller, Wilbur Hobart
Moore, Peter Bartlett
Mycek, Mary J
Myles, Diana Gold
Niblack, John Franklin
Noll, Clifford Raymond, Jr
Norton, Louis Arthur
Notation, Albert David
Novoa, William Brewster
Oates, Peter Joseph
O'Looney, Patricia Anne
Osborn, Mary Jane
Ozols, Juris
Paul, Jeddeo
Pawelek, John Mason
Pazoles, Christopher James
Pereira, Joseph
Peterson, Richard Burnett
Pfeiffer, Steven Eugene
Poincelot, Raymond Paul, Jr
Primakoff, Paul
Prusoff, William Herman
Pudelkiewicz, Walter Joseph
Putterman, Gerald Joseph
Rachinsky, Michael Richard
Radding, Charles Meyer
Rajendran, Vazhaikkurichi M
Ray, Verne A
Reazin, George Harvey, Jr
Rebuffe-Scrive, Marielle Francoise
Ressler, Charlotte
Retsema, James Allan
Richards, Frank Frederick
Roman, Laura M
Rosenbaum, Joel L
Rosenberg, Philip
Rossomando, Edward Frederick
Roth, Jay Sanford
Rothfield, Lawrence I
Rudnick, Gary
Rupp, W Dean
Sardinas, Joseph Louis
Schenkman, John Boris
Schmir, Gaston L
Schneider, E Gayle
Schwartz, Pauline Mary
Schwinck, Ilse
Seligson, David
Setlow, Peter
Simmonds, Sofia
Smilowitz, Henry Martin
Snider, Ray Michael
Spencer, Richard Paul
Steitz, Joan Argetsinger
Stowe, Bruce Bernot
Strittmatter, Philipp
Summers, William Cofield
Summers, Wilma Poos
Tallman, John Francis
Tanaka, Kay
Tanzer, Marvin Lawrence
Tappan, Donald Vester
Toralballa, Gloria C
Tourtellotte, Mark Eton
Turnipseed, Marvin Roy
Vasington, Frank D
Votaw, Robert Grimm
Ward, David Christian
Weissman, Sherman Morton
Wheeler, George Lawrence
Wilson, John Thomas
Wood, David Dudley
Woronick, Charles Louis
Wright, Herbert Fessenden
Yphantis, David Andrew
Zelitch, Israel

DELAWARE
Anton, David L

Billheimer, Jeffrey Thomas
Blomstrom, Dale Clifton
Bond, Elizabeth Dux
Boylen, Joyce Beatrice
Burns, Richard Charles
Campbell, Linzy Leon
Chen, Harry Wu-Shiong
Cheng, Yih-Shyun Edmond
Colman, Roberta F
Crippen, Raymond Charles
Davis, Leonard George
Dennis, Don
Dhurjati, Prasad S
Freerksen, Deborah Lynne (Chalmers)
Frey, William Adrian
Ganfield, David Judd
Gatenby, Anthony Arthur
Giles, Ralph E
Hayman, Selma
Herblin, William Fitts
Heytler, Peter George
Hodges, Charles Thomas
Holmes, Richard
Holsten, Richard David
Jackson, David Archer
Jefferies, Steven
Jenner, Edward L
Kerr, Janet Spence
Kinney, Anthony John
LaRossa, Robert Alan
Lichtner, Francis Thomas, Jr
Litchfield, William John
Loomis, Gary Lee
Lorimer, George Huntly
Marrs, Barry Lee
Miles, James Lowell
Morrissey, Bruce William
Myoda, Toshio Timothy
Perry, Kenneth W
Pierson, Keith Bernard
Saller, Charles Frederick
Salsbury, Robert Lawrence
Salzman, Steven Kerry
Sandberg, Robert Gustave
Schloss, John Vinton
Seetharam, Ramnath (Ram)
Sheppard, David E
Singleton, Rivers, Jr
Smith, David William
Smith, Jack Louis
Snyder, Jack Austin
Stevenson, Irone Edmund, Jr
Stopkie, Roger John
Strobach, Donald Roy
Sweetser, Philip Bliss
Tam, Sang William
Thompson, Jeffery Scott
Thorpe, Colin
Tolman, Chadwick Alma
Wermus, Gerald R
White, Harold Bancroft, III
Wriston, John Clarence, Jr
Yates, Richard Alan

DISTRICT OF COLUMBIA
Alving, Carl Richard
Argus, Mary Frances
Ashe, Warren (Kelly)
Bailey, John Martyn
Ball, William David
Barak, Eve Ida
Baron, Louis Sol
Bellin, Judith Schryver
Blecher, Melvin
Bridges, John Robert
Brooker, Gary
Brown, James Edward
Bundy, Bonita Marie
Calvert, Allen Fisher
Carroll, Alan G
Carson, Frederick Wallace
Cavanagh, Harrison Dwight
Cayle, Theodore
Cerveny, Thelma Jannette
Chaput, Raymond Leo
Chen, H R
Chiang, Peter K
Chirikjian, Jack G
Cocks, Gary Thomas
Cohn, Victor Hugo
Cowan, James W
Davidson, Eugene Abraham
Deutsch, Mike John
Doman, Elvira
Donaldson, Robert Paul
Edwards, Cecile Hoover
Ellis, Sydney
Fanning, George Richard
Field, Ruth Bisen
Finkelstein, James David
Fishbein, William Nichols
Fleming, Patrick John
Friedberg, Felix
Gallo, Linda Lou
Goldstein, Allan L
Gray, Irving
Hamosh, Margit
Hayden, George A
Henderson, Ellen Jane
Horner, William Harry
Irausquin, Hiltje
Jester, James Vincent
Jett-Tilton, Marti
Jordan, John Patrick

Kasbekar, Dinkar Kashinath
Khanna, Krishan L
Kimmel, Gary Lewis
Kornhauser, Andrija
Kumar, Soma
Lai, David Ying-lun
Lakshman, M Raj
Leto, Salvatore
Ligler, Frances Smith
Malinin, George I
Matyas, Gary Ralph
Matyas, Marsha Lakes
Mazel, Paul
Mehler, Alan Haskell
Merritt, William D
Miller, Linda Jean
Mitchell, Geraldine Vaughn
Morck, Timothy Anton
Nath, Jayasree
Naylor, Paul Henry
Nishioka, David Jitsuo
Olenick, John George
Papadopoulos, Nicholas M
Perfetti, Randolph B
Peters, Esther Caroline
Peterson, Malcolm Lee
Phillips, Terence Martyn
Prival, Michael Joseph
Quaife, Mary Louise
Reich, Melvin
Rennert, Owen M
Rhoads, Allen R
Richards, Roberta Lynne
Rosenberg, Robert Charles
Rothman, Sara Weinstein
Rubin, Martin Israel
Schneider, Bernard Arnold
Shank, Fred R
Shibko, Samuel Issac
Shukla, Kamal Kant
Singer, Maxine Frank
Smith, Robert Lawrence
Smith, Thomas Elijah
Smulson, Mark Elliott
Southerland, William M
Speidel, Edna W
Sridhar, Rajagopalan
Steers, Edward, Jr
Steinberg, Marcia Irene
Tarantino, Laura M(ary)
Todhunter, John Anthony
Tolbert, Margaret Ellen Mayo
Torrence, Paul Frederick
Valassi, Kyriake V
Vanderhoek, Jack Yehudi
Venkatesam, Malabi M
Wassef, Nabila M
Whitfield, Carolyn Dickson
Whittaker, Paul
Winter, William Phillips
Wolfe, Alan David
Wood, Garnett Elmer
Wright, Daniel Godwin
Wyngaarden, James Barnes
Yen-Koo, Helen C
Zimmer, Elizabeth Anne

FLORIDA
Abou-Khalil, Samir
Adair, Winston Lee, Jr
Albrecht, Stephan LaRowe
Allen, Charles Marshall, Jr
Baker, Stephen Phillip
Barber, Michael James
Bausher, Michael George
Beckhorn, Edward John
Bell, Paul Hadley
Bell, Robert Gale
Bellamy, Winthrop Dexter
Berry, Robert Eddy
Bieber, Theodore Emmanuel
Bleiweis, Arnold Sheldon
Bliznakov, Emile George
Bock, Fred G
Bosee, Roland Andrew
Bowes, George Ernest
Brown, Ross Duncan, Jr
Burchfield, Harry P
Buslig, Bela Stephen
Busse, Robert Franklyn
Carraway, Coralie Anne Carothers
Carraway, Kermit Lee
Castro, Alberto
Caswell, Anthony H
Chapman, Peter John
Childers, Steven Roger
Cline, Kenneth Charles
Coffey, Ronald Gibson
Cohen, Robert Jay
Coleman, Sylvia Ethel
Convertino, Victor Anthony
Cort, Winifred Mitchell
Coulson, Richard
Cousins, Robert John
Cowman, Richard Ammon
Crews, Fulton T
Cunningham, Glenn N
Dash, Harriman Harvey
Davenport, Thomas Lee
Dean, David Devereaux
DeKloet, Siwo R
Diamond, Steven Elliot
Dombro, Roy S
Downey, Kathleen Mary

Dunn, Ben Monroe
Dunn, William Arthur, Jr
Edelson, Jerome
Elliott, Paul Russell
Emerson, Geraldine Mariellen
Eoff, Kay M
Esser, Alfred F
Feaster, John Pipkin
Feinstein, Louis
Fisher, Waldo Reynolds
Fishman, Jack
Fontaine, Thomas Davis
Fried, Melvin
Frieden, Earl
Friedl, Frank Edward
Gander, John E
Ganguly, Rama
Gawron, Oscar
Gay, Don Douglas
Gibson, Joyce C
Giegel, Joseph Lester
Giner-Sorolla, Alfredo
Ginsler, Victor William
Gittelman, Donald Henry
Glaser, Luis
Goldberg, Melvin Leonard
Graham, W(alter) Donald
Graven, Stanley N
Green, Harry
Greenfield, Leonard Julian
Hahn, Elliot F
Halprin, Kenneth M
Hamilton, Franklin D
Hamilton, James Guthrie
Hanks, Robert William
Hargrave, Paul Allan
Hauswirth, William Walter
Hayashi, Teru
Holt, Thomas Manning
Homann, Peter H
Howell, Ralph Rodney
Hsia, Sung Lan
Hsu, Jeng Mein
Huber, Donald John
Huijing, Frans
Humphreys, Thomas Elder
Inana, George
Isaacks, Russell Ernest
Jensen, Roy A
Kapsalis, John George
Kelley, George Greene
Kilberg, Michael Steven
Kinsey, William Henderson
Knapp, Francis Marion
Koroly, Mary Jo
Kulwich, Roman
Lai, Patrick Kinglun
Lee, Ernest Y
Lewin, Alfred S
Light, Robley Jasper
Lim, Daniel V
Litman, Gary William
Litosch, Irene
Luer, Carl A
Lutz, Peter Louis
McLean, Mark Philip
McNary, Robert Reed
Makemson, John Christopher
Mallery, Charles Henry
Mans, Rusty Jay
Marzluff, William Frank, Jr
Mayer, Marion Sidney
Mende, Thomas Julius
Menzies, Robert Allen
Merdinger, Emanuel
Miller, Kent D
Muench, Karl Hugo
Mullins, John Thomas
Nakashima, Tadayoshi
Neary, Joseph Thomas
Neims, Allen Howard
Ness, Gene Charles
Noble, Nancy Lee
Nonoyama, Meihan
Nordby, Harold Edwin
Oberst, Fred William
O'Brien, Thomas W
Ott, Edgar Alton
Pita, Julio C
Polson, Charles David Allen
Pratt, Melanie M
Pressman, Berton Charles
Preston, James Faulkner, III
Previc, Edward Paul
Pruzansky, Jacob Julius
Purich, Daniel Lee
Reid, Parlane John
Riehm, John P
Roberts, R Michael
Roberts, Thomas L
Roeder, Martin
Romeo, John Thomas
Ross, Lynne Fischer
Rubin, Saul Howard
Ruegamer, William Raymond
Ryan, James Walter
Samis, Harvey Voorhees, Jr
Sarett, Herbert Paul
Sathe, Shridhar Krishna
Scarpace, Philip J
Schmidt, Robert Reinhart
Scott, Walter Alvin
Seibert, Florence B

Biochemistry (cont)

Shanmugam, Keelnatham
   Thirunavukkarasu
Shapiro, Jeffrey Paul
Shireman, Rachel Baker
Shirk, Paul David
Silhacek, Donald Le Roy
Silverman, David Norman
Silverstein, Herbert
Soldo, Anthony Thomas
Soliman, Magdi R I
Solomonson, Larry Paul
Soto, Aida R
Stahmann, Mark Arnold
Stearns, Thomas W
Stein, Abraham Morton
Stevens, Bruce Russell
Stewart, Ivan
Stone, Stanley S
Storrs, Eleanor Emerett
Sullivan, Lloyd John
Taylor, Barrie Frederick
Terranova, Andrew Charles
Tershakovec, George Andrew
Tocci, Paul M
Toporek, Milton
Traver, Janet Hope
Tsibris, John-Constantine Michael
Van Wart, Harold Edgar
Vickers, David Hyle
Voigt, Walter
Wallace, Robin A
Wecker, Lynn
Weissbach, Arthur
Wells, Gary Neil
Wheeler, Willis Boly
Whelan, William Joseph
White, Fred G
White, Roseann Spicola
Wilder, Violet Myrtle
Wodzinski, Rudy Joseph
Woessner, Jacob Frederick, Jr
Wood, William Otto
Woodside, Kenneth Hall
Young, David Michael
Yu, Simon Shyi-Jian
Zikakis, John Philip

GEORGIA

Abdel-Latif, Ata A
Abraham, Edathara Chacko
Adams, Robert Johnson
Agosin, Moises
Akhtak, Rashid Ahmed
Bailey, Gordon Burgess
Bernstein, Robert Steven
Black, Asa C, Jr
Black, Billy C, II
Black, Clanton Candler, Jr
Bonkovsky, Herbert Lloyd
Boutwell, Joseph Haskell
Brackett, Benjamin Gaylord
Bransome, Edwin D, Jr
Brewer, John Michael
Brooks, John Bill
Brubaker, Leonard Hathaway
Buccafusco, Jerry Joseph
Bustos-Valdes, Sergio Enrique
Calabrese, Ronald Lewis
Chen, Dillion Tong-ting
Cherniak, Robert
Chiu, Kirts C
Claybrook, James Russell
Coe, Elmon Lee
Collins, Delwood C
Cormier, Milton Joseph
Coryell, Margaret E
Cramer, Gisela Türck
Cramer, John Wesley
Curley, Winifred H
Dailey, Harry A
Danner, Dean Jay
Dembure, Philip Pito
Dervartanian, Daniel Vartan
Dirksen, Thomas Reed
Dixon, Dabney White
Doetsch, Paul William
Drummond, Margaret Crawford
Dull, Gerald G
Duncan, Robert Leon, Jr
Dure, Leon S, III
Edmondson, Dale Edward
Eriksson, Karl-Erik Lennart
Eriquez, Louis Anthony
Espelie, Karl Edward
Fales, Frank Weck
Fechheimer, Marcus
Finnerty, William Robert
Girardot, Jean Marie Denis
Glass, David Bankes
Glover, Claiborne V C, III
Gourse, Richard Lawrence
Green, John H
Greenberg, Jerrold
Hadd, Harry Earle
Hainline, Adrian, Jr
Hall, Dwight Hubert
Halper, Jaroslava
Harris, Henry Earl
Heise, John J
Hicks, Heraline Elaine
Hilton, James Lee
Hommes, Frits A

Howard, John Charles
Huisman, Titus Hendrik Jan
Iuvone, Paul Michael
Johnson, Joe
Jones, Dean Paul
Jones, George Henry
Kadis, Barney Morris
Karp, Warren B
Kiefer, Charles Randolph
Klee, Lucille Holljes
Kolbeck, Ralph Carl
Kraeling, Robert Russell
Kreitzman, Stephen Neil
Kuck, John Frederick Read, Jr
Kuo, Jyh-Fa
Kushner, Sidney Ralph
Leibach, Fredrick Hartmut
Lewis, Jasper Phelps
Ljungdahl, Lars Gerhard
Lyon, John Blakeslee, Jr
McCormick, Donald Bruce
Mackin, Robert Brian
MacMillan, Joseph Edward
McPherson, James C, Jr
McRorie, Robert Anderson
Madden, John Joseph
Martin, Roy Joseph, Jr
Mather, Jane H
May, Sheldon William
Meagher, Richard Brian
Mendicino, Joseph Frank
Miller, Lois Kathryn
Mills, John Blakely, III
Morin, Leo Gregory
Mortenson, Leonard Earl
Moss, Claude Wayne
Myers, Dirck V
Nelson, George Humphry
Noe, Bryan Dale
Ogle, Thomas Frank
Ove, Peter
Pandey, Kailash N
Papageorge, Evangeline Thomas
Paris, Doris Fort
Parrish, John W, Jr
Peck, Harry Dowd, Jr
Peifer, James J
Pine, Leo
Powers, James Cecil
Pratt, Lee Herbert
Pressey, Russell
Ragland, William Lauman, III
Ramachandran, Muthukrishnan
Reilly, Charles Conrad
Robertson, Alex F
Robinson, George Waller
Rogers, John Ernest
Sansing, Norman Glenn
Schadler, Daniel Leo
Schaefer, Gerald J
Schepartz, Abner Irwin
Schmidt, Gregory Wayne
Scott, David Frederick
Sgoutas, Demetrios Spiros
Shapira, Raymond
Shuster, Robert C
Sinor, Lyle Tolbot
Smith, David Fletcher
Sophianopoulos, Alkis John
Sridaran, Rajagopala
Stevens, Ann Rebecca
Stevens, Charles David
Suddath, Fred LeRoy, (Jr)
Thedford, Roosevelt
Travis, James
Tsang, Victor Chiu Wana
Underwood, Arthur Louis, Jr
Vegotsky, Allen
Wade, Adelbert Elton
Waitzman, Morton Benjamin
Wallace, Douglas Cecil
Wampler, John E
Warren, Stephen Theodore
Whitney, J(ohn) Barry, III
Wilhelmi, Alfred Ellis
Williams, Joy Elizabeth P
Williams, William Lawrence
Woodley, Charles Lamar
Woods, Wendell David
Ziegler, Harry Kirk

HAWAII

Ako, Harry Mu Kwong Ching
Allen, Richard Dean
Berger, Leslie Ralph
Bhagavan, Nadhipuram V
Burr, George Oswald
Chang, Franklin
Fok, Agnes Kwan
Gibbons, Barbara Hollingworth
Greenwood, Frederick C
Guillory, Richard John
Howton, David Ronald
Jackson, Mel Clinton
Krupp, David Alan
Loo, Yen-Hoong
McConnell, Bruce
McKay, Robert Harvey
Mandel, Morton
Maretzki, Andrew
Matsumoto, Hiromu
Moore, Paul Harris
Morton, Bruce Eldine
Phillips, John Howell, Jr

Putman, Edison Walker
Rechnitz, Garry Arthur
Scott, John Francis
Sherman, Martin
Stanley, Richard W
Stuart, William Dorsey
Sun, Samuel Sai-Ming
Tang, Chung-Shih
Van Reen, Robert
Vennesland, Birgit
Ward, Melvin A
Waslien, Carol Irene
Weems, Charles William
Yamamoto, Harry Y
Yasunobu, Kerry T

IDAHO

Armstrong, Marvin Douglas
Augustin, Jorg A L
Corsini, Dennis Lee
Crawford, Donald Lee
Crawford, Ronald L
Dalton, Jack L
Dreyfus, Pierre Marc
Dugan, Patrick R
Ellis, Robert William
Fontenelle, Lydia Julia
Hatcher, Herbert John
Keay, Leonard
LeTourneau, Duane John
McCune, Ronald William
Montoure, John Ernest
Oliver, David John
Ramagopal, Subbanaidu
Tollefson, Charles Ivar
Watson, Kenneth Fredrick
Wiese, Alvin Carl
Winkel, Cleve R
Winston, Vern

ILLINOIS

Abbott, William Ashton
Agarwal, Kan L
Aktipis, Stelios
Alberte, Randall Sheldon
Anderson, Byron
Anderson, David John
Anderson, Kenning M
Anderson, Louise Eleanor
Anderson, Robert Lewis
Applebury, Meredithe L
Argoudelis, Chris J
Armbruster, Frederick Carl
Arthur, Robert David
Aydelotte, Margaret Beesley
Baich, Annette
Baker, Harold Nordean
Baran, John Stanislaus
Barany, Michael
Baumann, Gerhard
Bechtel, Peter John
Becker, John Henry
Becker, Michael Allen
Beecher, Christopher W W
Bennett, Glenn Allen
Berman, Eleanor
Bernstein, Elaine Katz
Bernstein, Joel Edward
Berry, Robert Wayne
Beru, Nega
Betz, Robert F
Bezkorovainy, Anatoly
Bhattacharyya, Maryka Horsting
Bloomfield, Daniel Kermit
Blumenthal, Harold Jay
Bousquet, William F
Bowie, Lemuel James
Bradford, Marion McKinley
Braun, Donald Peter
Breen, Moira
Breillatt, Julian Paul, Jr
Brewer, Gregory J
Buetow, Dennis Edward
Burkwall, Morris Paton, Jr
Cabana, Veneracion Garganta
Calandra, Joseph Carl
Cammarata, Peter S
Campbell, Alfred Duncan
Casadaban, Malcolm John
Castignetti, Domenic
Ceithaml, Joseph James
Chambers, Donald A
Chatterton, Robert Treat, Jr
Chen, Wen Sherng
Cheung, Hou Tak
Christensen, Mary Lucas
Chung, Jiwhey
Clark, John Magruder, Jr
Cohen, Isaac
Cole, Edmond Ray
Collins, Vernon Kirkpatrick
Constantinou, Andreas I
Coolidge, Thomas B
Cork, Douglas J
Cronan, John Emerson, Jr
Czerlinski, George Heinrich
Damaskus, Charles William
Dasler, Waldemar
Davis, Carl Lee
Davis, Peyton Nelson
Dawson, Glyn
Decker, Richard H
DeFilippi, Louis J
De La Huerga, Jesus

DePinto, John A
DeSombre, Eugene Robert
Devries, Arthur Leland
Dixit, Saryu N
Dodge, Patrick William
Domanik, Richard Anthony
Doughty, Clyde Carl
Drucker, Harvey
Druse-Manteuffel, Mary Jeanne
Druyan, Mary Ellen
Dubin, Alvin
Dudkiewicz, Alan Bernard
Dumas, Lawrence Bernard
Dunaway, George Alton, Jr
Ebrey, Thomas G
Eddy, Dennis Eugene
Edwards, Harold Herbert
Ehrenpreis, Seymour
Emken, Edward Allen
Epstein, Wolfgang
Erve, Peter Raymond
Farnsworth, Wells Eugene
Feinstein, Robert Norman
Feldman, Fred
Fennewald, Michael Andrew
Ferrara, Louis W
Ferren, Larry Gene
Finnerty, James Lawrence
Flouret, George R
Foote, Carlton Dan
Ford, Susan Heim
Foster, Raymond Orrville
Fox, Sidney Walter
Frankfater, Allen
Freinkel, Norbert
Frenkel, Niza B
Friedman, Robert Bernard
Friedman, Yochanan
Friedmann, Herbert Claus
Fuchs, Elaine V
Gaballah, Saeed S
Gardner, Harold Wayne
Gassman, Merrill Loren
Gershbein, Leon Lee
Getz, Godfrey S
Giere, Frederic Arthur
Giometti, Carol Smith
Glaser, Janet H
Glaser, Michael
Glogovsky, Robert L
Goldberg, Erwin
Goldman, Allen S
Goldstone, Alfred D
Goldwasser, Eugene
Gratton, Enrico
Graves, Charles Norman
Griffin, Martin John
Gross, Martin
Gross, Thomas Lester
Gumport, Richard I
Gunsalus, Irwin Clyde
Hac, Lucile R
Hagar, Lowell Paul
Hager, Lowell Paul
Halfman, Clarke Joseph
Hampel, Arnold E
Hanlon, Mary Sue
Hanly, W Carey
Harrison, William Henry
Hass, George Michael
Hathaway, Robert J
Haugen, David Allen
Hauptmann, Randal Mark
Hawkins, Richard Albert
Hawrylewicz, Ervin J
Hayashi, James Akira
Helbert, James Raymond
Held, Irene Rita
Heller, Alfred
Henderson, Thomas Otis
Hiles, Richard Allen
Hiltibran, Robert Comegys
Himoe, Albert
Hjelle, Joseph Thomas
Hoerman, Kirk Conklin
Holbrook, Gabriel Peter
Holland, Louis Edward, II
Holleman, William H
Honig, George Raymond
Horwitz, Alan Fredrick
Hoskin, Francis Clifford George
Hospelhorn, Verne D
Hou, Ching-Tsang
Huffman, George Wallen
Ingle, Morton Blakeman
Iqbal, Zafar
Jackson, Richard W
Jaffe, Randal Craig
Javaid, Javaid Iqbal
Jeffay, Henry
Jiu, James
Johnston, Michael Adair
Jonas, Ana
Jungmann, Richard A
Kagan, Jacques
Kamath, Savitri Krishna
Kanabrocki, Eugene Ladislaus
Kantor, Harvey Sherwin
Kass, Leon Richard
Kassner, Richard J
Kathan, Ralph Herman
Katz, Adrian I
Kaufman, Stephen J
Kemp, Robert Grant

Kempner, David H
Kesler, Darrel J
Kim, Yung Dai
Kimura, James Hiroshi
Klopfenstein, William Elmer
Klubek, Brian Paul
Knight, Katherine Lathrop
Koch, Elizabeth Anne
Kokkinakis, Demetrius Michael
Kuczmarski, Edward R
Kuettner, Klaus E
Kumar, Sudhir
Kyncl, J Jaroslav
Lakshminarayanan, Krishnaiyer
Lamb, Robert Andrew
Lambert, Glenn Frederick
Lambert, Mary Pulliam
Lamberts, Burton Lee
Lange, Charles Ford
Lange, Yvonne
Larson, Bruce Linder
Layman, Donald Keith
Leven, Robert Maynard
Levin, Samuel Joseph
L'Heureux, Maurice Victor
Liang, Tehming
Liao, Shutsung
Lin, Reng-Lang
Loach, Paul A
Lobstein, Otto Ervin
Lockhart, Haines Boots
Lopatin, William
Lorand, Laszlo
Lucher, Lynne Annette
Lushbough, Channing Harden
Maass, Wolfgang Siegfried Gunther
McCandless, David Wayne
McDonald, Hugh Joseph
Mackal, Roy Paul
Madera-Orsini, Frank
Madsen, David Christy
Maier, George D
Mallia, Anantha Krishna
Marcotte, Patrick Allen
Margoliash, Emanuel
Markovitz, Alvin
Marr, James Joseph
Martin, Charles J
Matta, Michael Stanley
Mattenheimer, Hermann G W
Mehta, Rajendra G
Meredith, Stephen Charles
Miernyk, Jan Andrew
Miller, Charles G
Miller, James Edward
Miller, Robert Verne
Mistry, Sorab Pirozshah
Mohberg, Joyce
Molnar, Janos
Moore, Edwin Granville
Morley, Colin Godfrey Dennis
Moskal, Joseph Russell
Muck, George A
Myers, Ronald Berl
Nakagawa, Yasushi
Nakamoto, Tokumasa
Needleman, Saul Ben
Neet, Kenneth Edward
Neuhaus, Francis Clemens
Nolan, Chris
Norris, Frank Arthur
Northrop, Robert L
Novak, Robert Louis
Olsen, Kenneth Wayne
Ordal, George Winford
Ort, Donald Richard
Oshiro, Yuki
Ostrow, David Henry
Pai, Sadanand V
Pal, Dhiraj
Palmer, Warren K
Papaioannou, Stamatios E
Parvez, Zaheer
Pasterczyk, William Robert
Pekarek, Robert Sidney
Pepperberg, David Roy
Peraino, Carl
Perlman, Robert
Peterson, Rudolph Nicholas
Plate, Charles Alfred
Plate, Janet Margaret
Portis, Archie Ray, Jr
Potempa, Lawrence Albert
Preston, Robert Leslie
Pun, Pattle Pak-Toe
Rackis, Joseph John
Radloff, Harold David
Radzialowski, Frederick M
Rafelson, Max Emanuel, Jr
Rao, Gopal Subba
Rao, Mrinalini Chatta
Rasenick, Mark M
Rausch, David John
Ressler, Newton
Richardson, Arlan Gilbert
Richmond, Patricia Ann
Ringler, Ira
Robbins, Kenneth Carl
Robinson, James Lawrence
Rogalski-Wilk, Adrian Alice
Rotermund, Albert J, Jr
Rothfus, John Arden
Rothman-Denes, Lucia B
Rowe, William Bruce

Rownd, Robert Harvey
Royer, Garfield Paul
Rupprecht, Kevin Robert
Sargent, Malcolm Lee
Savidge, Jeffrey Lee
Scanu, Angelo M
Schenck, Jay Ruffner
Scherberg, Neal Harvey
Schlenk, Fritz
Schlueter, Robert J
Schnell, Gene Wheeler
Schultz, Richard Michael
Schumacher, Gebhard Friederich B
Schumer, William
Scott, Don
Seed, Randolph William
Sehgal, Lakshman R
Seidenfeld, Jerome
Sellers, Donald Roscoe
Shambaugh, George E, III
Shapiro, David Jordon
Shapiro, Stanley Kallick
Shaw, Paul Dale
Shen, Linus Liang-nene
Shepherd, Herndon Guinn
Sheppard, John Richard
Sievert, Herman William
Sigler, Paul Benjamin
Silver, Simon David
Simonson, Lloyd Grant
Sky-Peck, Howard H
Sligar, Stephen Gary
Slodki, Morey Eli
Smith, Steven Joel
Smouse, Thomas Hadley
Snyder, William Robert
Sorensen, Leif Boge
Spitzer, Robert Harry
Splittstoesser, Walter E
Stark, Benjamin Chapman
Steck, Theodore Lyle
Stein, Herman H
Steiner, Donald Frederick
Storti, Robert V
Straus, Werner
Strauss, Bernard S
Struthers, Barbara Joan Oft
Subbaiah, Papasani Venkata
Suzue, Ginzaburo
Switzer, Robert Lee
Sze, Paul Yi Ling
Tao, Mariano
Telser, Alvin Gilbert
Thompson, David Jerome
Thompson, Richard Edward
Thomson, John Ferguson
Thonar, Eugene Jean-Marie
Thorp, Frank Kedzie
Titchener, Edward Bradford
Toback, F(rederick) Gary
Tomita, Joseph Tsuneki
Tookey, Harvey Llewellyn
Tripathi, Satish Chandra
Tumbleson, M(yron) E(ugene)
Turnquist, Richard Lee
Urbas, Branko
Van de Kar, Louis David
Van Fossan, Donald Duane
Van Kley, Harold
Varricchio, Frederick
Vary, James Corydon
Vary, Patricia Susan
Veis, Arthur
Vodkin, Lila Ott
Wachtl, Carl
Wade, David Robert
Wagner, Gerald C
Walter, Robert John
Walter, Trevor John
Wang, Andrew H-J
Wang, Gary T
Wang, Hwa Lih
Wang, Li Chuan
Wawszkiewicz, Edward John
Weber, Gregorio
Webster, Dale Arroy
Weinstein, Hyman Gabriel
Weiss, Samuel Bernard
Weissmann, Bernard
Welker, Neil Ernest
Wells, Warren F
Westbrook, Edwin Monroe
Westley, John Leonard
Weyhenmeyer, James Alan
Whisler, Walter William
Wideburg, Norman Earl
Wiegand, Ronald Gay
Wilbraham, Antony Charles
Wilkinson, Brian James
Williams-Ashman, Howard Guy
Wilson, Curtis Marshall
Wilson, Donald Alan
Witmer, Heman John
Wittman, James Smythe, III
Wolf, Walter J
Womble, David Dale
Wong, Paul Wing-Kon
Wool, Ira Goodwin
Wu, Anna Fang
Wu, Tai Te
Wu-Wong, Jinshyun Ruth
Young, Jay Maitland
Yu, Fu-Li
Zahalsky, Arthur C

Zak, Radovan Hynek
Zaneveld, Lourens Jan Dirk
Zaroslinski, John F

INDIANA
Abdallah, Abdulmuniem Husein
Ali, Rida A
Allmann, David William
Anderson, David Bennett
Ashendel, Curtis Lloyd
Asteriadis, George Thomas, Jr
Axelrod, Bernard
Baker, A Leroy
Baldwin, William Walter
Barnhart, James William
Basu, Manju
Basu, Subhash Chandra
Bauer, Robert
Becker, Benjamin
Beeson, W Malcolm
Behrens, Otto Karl
Beranek, William, Jr
Bhatti, Waqar Hamid
Biggs, Homer Gates
Blair, Paul V
Blatt, Joel Martin
Bobbitt, Jesse LeRoy
Bonner, John Franklin, Jr
Borglum, Gerald Baltzer
Bosron, William F
Botero, J M
Bowman, Donald Edwin
Boyer, Ernest Wendell
Brandt, Karl Garet
Bretthauer, Roger K
Bridges, C David
Bromer, William Wallis
Bryan, William Phelan
Bumpus, John Arthur
Burleigh, Bruce Daniel, Jr
Butler, Larry G
Carmichael, Ralph Harry
Carrico, Robert Joseph
Castellino, Francis Joseph
Chance, Ronald E
Christner, James Edward
Clevenger, Sarah
Cole, Thomas A
Coolbaugh, Ronald Charles
Corrigan, John Joseph
Crane, Frederick Loring
Dantzig, Anne H
DeLong, Allyn F
Diller, Erold Ray
Dilley, Richard Alan
Dillon, John Joseph, III
Dixon, Jack Edward
Donoho, Alvin Leroy
Duman, John Girard
Dunkle, Larry D
Dunn, Peter Edward
Eble, John Nelson
Edenberg, Howard Joseph
Edmundowicz, John Michael
Elkin, Robert Glenn
Enders, George Leonhard, Jr
Evans, Michael Allen
Fayle, Harlan Downing
Filmer, David Lee
Fong, Shao-Ling
Fox, Owen Forrest
Frank, Bruce Hill
Free, Alfred Henry
Frolik, Charles Alan
Fuchs, Morton S
Fuller, Ray W
Gallo, Duane Gordon
Gantzer, Mary Lou
Gardner, David Arnold
Ghosh, Swapan Kumar
Gibson, David Mark
Gilham, Peter Thomas
Gledhill, Robert Hamor
Goetz, Frederick William, Jr
Goff, Charles W
Goh, Edward Hua Seng
Golab, Tomasz
Goldstein, David Joel
Gray, Peter Norman
Gunter, Claude Ray
Gunther, Gary Richard
Gurd, Frank Ross Newman
Guthrie, George Drake
Hagen, Richard Eugene
Hamilton, David Foster
Hancock, Deana Lori
Harper, Edwin T
Harris, Robert Allison
Hayes, Donald Charles
Hegeman, George D
Heinicke, Herbert Raymond
Heinstein, Peter
Hendershot, William Fred
Henry, Harry James
Hermodson, Mark Allen
Herr, Earl Binkley, Jr
Herrmann, Klaus Manfred
Hershberger, Charles Lee
Hidy, Phil Harter
Ho, Nancy Wang-Yang
Ho, Peter Peck Koh
Hudock, George Anthony
Ingraham, Joseph Sterling
Irwin, William Elliot

Jackson, Richard Lee
Jaehning, Judith A
Jenkins, Winborne Terry
Jilka, Robert Laurence
Johnson, Eric Richard
Kauffman, Raymond F
Kempson, Stephen Allan
Kim, Ki-Han
Kirksey, Avanelle
Kleinschmidt, Walter John
Kohlhaw, Gunter B
Laakso, John William
Ladisch, Michael R
Lantero, Oreste John, Jr
Larsen, Steven H
Laskowski, Michael, Jr
Li, Ting Kai
Light, Albert
Lin, Renee C
Lindstrom, Terry Donald
Litov, Richard Emil
Loesch-Fries, Loretta Sue
Loudon, Gordon Marcus
Low, Philip Stewart
Lumeng, Lawrence
Lutz, Wilson Boyd
McBride, William Joseph
McCann, Peter Paul
McDonald, James Lee, Jr
McMullen, James Robert
McRoberts, Milton R
Malacinksi, George M
Marquis, Norman Ronald
Marshall, James John
Mason, Norman Ronald
Matsumoto, Charles
Mertz, Edwin Theodore
Middendorf, Donald Floyd
Millar, Wayne Norval
Moorehead, Wells Rufus
Morre, D James
Muhler, Joseph Charles
Murphy, Patrick Joseph
Natarajan, Viswanathan
Niederpruem, Donald J
Nielsen, Niels Christian
Nordschow, Carleton Deane
Nowak, Thomas
Ockerse, Ralph
Olsen, Richard William
Oster, Mark Otho
Ostroy, Sanford Eugene
Osuch, Mary Ann V
Pace, Norman R
Parker, Herbert Edmund
Pettinga, Cornelius Wesley
Probst, Gerald William
Proksch, Gary J
Puski, Gabor
Putnam, Frank William
Quackenbush, Forrest Ward
Queener, Sherry Fream
Queener, Stephen Wyatt
Ragheb, Hussein S
Rand, Phillip Gordon
Raveed, Dan
Regnier, Frederick Eugene
Roach, Peter John
Rodwell, Victor William
San Pietro, Anthony
Santerre, Robert Frank
Saz, Howard Jay
Schick, Lloyd Alan
Schneider, Donald Louis
Schulz, Arthur R
Seely, James Ervin
Shaw, Walter Norman
Sherman, Louis Allen
Shields, James Edwin
Siakotos, Aristotle N
Sipe, Jerry Eugene
Skarstedt, Mark T(eofil)
Skjold, Arthur Christopher
Smith, Leonard Charles
Smith, Robert William
Somerville, Ronald Lamont
Steinrauf, Larry King
Stevenson, William Campbell
Stier, Theodore James Blanchard
Stiller, Mary Louise
Stillwell, William Harry
Story, Jon Alan
Subramanian, Sethuraman
Surzycki, Stefan Jan
Swain, Richard Russell
Szuhaj, Bernard F
Taylor, Harold Leland
Termine, John David
Thompson, Richard Michael
Tischfield, Jay Arnold
Tsai, Chia-Yin
Tunnicliff, Godfrey
Van Frank, Richard Mark
Vasko, Michael Richard
Vogelhut, Paul Otto
Wagner, Eugene Stephen
Waldman, Alan S
Waldman, Barbara Criscuolo
Wallander, Jerome F
Wegener, Warner Smith
Weiner, Henry
Weith, Herbert Lee
Wellso, Stanley Gordon
White, Harold Keith

Biochemistry (cont)

Wildfeuer, Marvin Emanuel
Williams, Gene R
Wong, David Taiwai
Wostmann, Bernard Stephan
Wright, Walter Eugene
Yan, Sau-Chi Betty
Yen, Terence Tsin Tsu
Yoder, John Menly
Young, Peter Chun Man
Zalkin, Howard
Zilz, Melvin Leonard
Zimmerman, Sarah E
Zygmunt, Walter A

IOWA

Anderson, Lloyd L
Arnone, Arthur Richard
Ascoli, Mario
Baron, Jeffrey
Benbow, Robert Michael
Bishop, Stephen Hurst
Bosch, Arthur James
Butler, John E
Cain, George D
Campbell, Kevin Peter
Cavalieri, Anthony Joseph, II
Cazin, John, Jr
Celander, Evelyn Faun
Chalkley, G Roger
Cheng, Frank Hsieh Fu
Christiansen, James Brackney
Clark, Robert A
Colilla, William
Conway, Thomas William
Cook, Robert Thomas
Cunningham, Bryce A
Dahm, Paul Adolph
Davis, Leodis
Donelson, John Everett
Downing, Donald Talbot
Dutton, Gary Roger
Enger, Merlin Duane
Fellows, Robert Ellis, Jr
Franklin, Robert Louis
Fromm, Herbert Jerome
Garbutt, John Thomas
Gibson, David Thomas
Goodridge, Alan G
Graves, Donald J
Greenberg, Everett Peter
Grigsby, William Redman
Gussin, Gary Nathaniel
Hammond, Earl Gullette
Horowitz, Jack
Horst, Ronald Lee
Johnson, Lawrence Alan
Kalnitsky, George
Keeney, Dennis Raymond
Kintanar, Agustin
Koerner, Theodore Alfred William, Jr
Koppelman, Ray
Lara-Braud, Carolyn Weathersbee
Lata, Gene Frederick
Lillehoj, Eivind B
Lim, Ramon (Khe Siong)
Mahoney, Joan Munroe
Makar, Adeeb Bassili
Markovetz, Allen John
Masat, Robert James
Maurer, Richard Allen
Menninger, John Robert
Metzler, David Everett
Montgomery, Rex
Morehouse, Alpha L
Norcia, Leonard Nicholas
Oliver, Denis Richard
Olson, James Allen
Padrini, Vittorio Arturo
Phillips, Marshall
Plapp, Bryce Vernon
Rebouche, Charles Joseph
Rebstock, Theodore Lynn
Rhead, William James
Robson, Richard Morris
Robyt, John F
Routh, Joseph Isaac
Schmerr, Mary Jo F
Schmidt, Thomas John
Schottelius, Dorothy Dickey
Shires, Thomas Kay
Southard, Wendell Homer
Spector, Arthur Abraham
Speer, Vaughn C
Stageman, Paul Jerome
Stanton, Thaddeus Brian
Stegink, Lewis D
Stellwagen, Earle C
Stewart, Mark Armstrong
Stewart, Mary E
Stinski, Mark Francis
Stromer, Marvin Henry
Swan, Patricia B
Tabatabai, Louisa Braal
Thomas, Byron Henry
Thomas, James Arthur
Tipton, Carl Lee
Wang, Wei-Yeh
Wertz, Philip Wesley
White, Bernard J
Williams, Phletus P

KANSAS

Arnold, Wilfred Niels
Baker, Joffre Bernard
Bechtel, Donald Bruce
Bednekoff, Alexander G
Bode, Vernon Cecil
Bradford, Lawrence Glenn
Brown, John Clifford
Burkhard, Raymond Kenneth
Calvet, James P
Carr, Daniel Oscar
Clegg, Robert Edward
Clevenger, Richard Lee
Cunningham, Franklin E
Davis, John Stewart
Davis, Lawrence Clark
Deyoe, Charles W
Dirks, Brinton Marlo
Ebner, Kurt E
Ericson, Alfred (Theodore)
Funderburgh, James Louis
Gegenheimer, Peter Albert
Grunewald, Katharine Klevesahl
Guikema, James Allen
Harris, Lewis Philip
Hedgcoth, Charlie, Jr
Helmkamp, George Merlin, Jr
Himes, Richard H
Hopkins, Theodore Louis
Houston, L L
Hsu, Howard Huai Ta
Hudson, Billy Gerald
Johnson, Donovan Earl
Kimmel, Joe Robert
Kitos, Paul Alan
Koeppe, Owen John
Kramer, Karl Joseph
Lanman, Robert Charles
Leavitt, Wendell William
MacGregor, Ronal Roy
Manning, Robert Thomas
Marchin, George Leonard
Michaelis, Elias K
Michaelis, Mary Louise
Mills, Russell Clarence
Mulford, Dwight James
Murphy, John Joseph
Newmark, Marjorie Zeiger
Noelken, Milton Edward
Nordin, Philip
Ochs, Raymond S
Parkinson, Andrew
Parrish, Donald Baker
Parrish, John Wesley, Jr
Peters, Ralph I
Pickrell, John A
Ramamurti, Krishnamurti
Rawitch, Allen Barry
Reeck, Gerald Russell
Reeves, Robert Donald
Rhodes, James B
Robertson, Donald Claus
Roche, Thomas Edward
Roufa, Donald Jay
Ruliffson, Willard Sloan
Sanders, Robert B
Scheid, Harold E
Schmidt, Robert W
Seib, Paul A
Shogren, Merle Dennis
Silverstein, Richard
Smith, Larry Dean
Stetler, Dean Allen
Suzuki, Tsuneo
Takemoto, Dolores Jean
Timberlake, Joseph William
Tsen, Cho Ching
Weaver, Robert F
Wilson, George Spencer
Wong, Peter P
Yarbrough, Lynwood R
Young, George Robert

KENTUCKY

Aleem, M I Hussain
Andersen, Roger Allen
Bass, Norman Herbert
Benz, Frederick W
Brown, John Wesley
Chan, Shung Kai
Cheniae, George Maurice
Cohn, David Valor
Dallam, Richard Duncan
Davidson, Jeffrey Neal
Dean, William L
Dickson, Robert Carl
Dunham, Valgene Loren
Farrar, William Wesley
Fell, Ronald Dean
Fonda, Margaret Lee
Galardy, Richard Edward
Geoghegan, Thomas Edward
Glauert, Howard Perry
Goodman, Norman L
Gray, Robert Dee
Haley, Boyd Eugene
Hammond, Ray Kenneth
Hanson, Barbara Ann
Hilton, Mary Anderson
Hu, Alfred Soy Lan
Humphreys, Wallace F
Kargl, Thomas E
Kasarskis, Edward Joseph

Kennedy, John Elmo, Jr
Knapp, Frederick Whiton
Kuc, Joseph
Lang, Calvin Allen
Lester, Robert Leonard
Levy, Robert Sigmund
Lowe, Richie Howard
McConnell, Kenneth Paul
McGeachin, Robert Lorimer
Maisch, Weldon Frederick
Mandelstam, Paul
Martin, Nancy Caroline
Noland, Jerre Lancaster
Otero, Raymond B
Packett, Leonard Vasco
Pavlik, Edward John
Perlin, Michael Howard
Petering, Harold George
Prough, Russell Allen
Ramp, Warren Kibby
Rao, Chalamalasetty Venkateswara
Rawls, John Marvin, Jr
Rees, Earl Douglas
Rhoads, Robert E
Richard, John P
Rosenthal, Gerald A
Rothwell, Frederick Mirvan
Schurr, Avital
Schwert, George William
Sisken, Jesse Ernest
Slagel, Donald E
Staat, Robert Henry
Steiner, Marion Rothberg
Taylor, John Fuller
Teller, David Norton
Toman, Frank R
Vanaman, Thomas Clark
Williams, Arthur Lee
Winer, Alfred D
Wittliff, James Lamar

LOUISIANA

Alam, Bassima Saleh
Alam, Jawed
Alam, Syed Qamar
Bernofsky, Carl
Bobbin, Richard Peter
Bryan, Sara E
Burch, Robert Emmett
Byers, Larry Douglas
Cauthen, Sally Eugenia
Chang, Simon H
Chen, Hoffman Hor-Fu
Claycomb, William Creighton
Clemetson, Charles Alan Blake
Cohen, William
Coulson, Roland Armstrong
Day, Donal Forest
Dessauer, Herbert Clay
Deutsch, Walter A
DiMaggio, Anthony, III
Dunn, Adrian John
Dupuy, Harold Paul
Ehrlich, Kenneth Craig
Ehrlich, Melanie
Gottlieb, A Arthur
Grimes, Sidney Ray, Jr
Griswold, Kenneth Edwin, Jr
Guthrie, John Daulton
Hamori, Eugene
Hart, Lewis Thomas
Hegsted, Maren
Henderson, Ralph Joseph, Jr
Herbert, Jack Durnin
Hill, Franklin D
Hill, James Milton
Hyde, Paul Martin
Ibanez, Manuel Luis
Jacks, Thomas Jerome
Jain, Sushil Kumar
Kastl, Peter Robert
Kennedy, Frank Scott
Kokatnur, Mohan Gundo
Kuck, James Chester
Laine, Roger Allan
Lartigue, Donald Joseph
Lee, Jordan Grey
Li, Su-Chen
Li, Yu-Teh
Lopez-Santolino, Alfredo
Lucas, Myron Cran
Mahajan, Damodar K
Marshall, Wayne Edward
Martin, Julia Mae
Marvel, John Thomas
Miceli, Michael Vincent
Mokrasch, Lewis Carl
Moore, Thomas Stephen, Jr
Morgan, William T
Newsome, David Anthony
Olmsted, Clinton Albert
Ory, Robert Louis
Pryor, William Austin
Radhakrishnamurthy, Bhandaru
Reed, Brent C
Robertson, William Van Bogaert
Rogers, Robert Larry
Rogers, Stearns Walter
Roskoski, Robert, Jr
Rudolph, Guilford George
Ruffin, Spaulding Merrick
Sarphie, Theodore G
Schoellmann, Guenther
Settoon, Patrick Delano

Shedlarski, Joseph George, Jr
Shepherd, Hurley Sidney
Smith, Ann
Smith, Robert Lewis
Spanier, Arthur M
Spitzer, Judy A
Springgate, Clark Franklin
Srinivasan, Sathanur Ramachandran
Srinivasan, Vadake Ram
Stanfield, Manie K
Steele, Richard Harold
Stjernholm, Rune Leonard
Swenson, David Harold
Terry, Maurice Ernest
Tou, Jen-sie Hsu
Tsuzuki, Junji
Upadhyay, Jagdish M
Vercellotti, John R
Vijayagopal, Parakat
Wang, Ting Chung
Welch, George Rickey
Woodring, J Porter
York, David Anthony
Younathan, Ezzat Saad

MAINE

Barton, Barbara Ann
Borei, Hans Georg
Brown, Gregory Neil
Cook, Richard Alfred
De Haas, Herman
Gabbert, Paul George
Goodman, Irving
Greenwood, Paul Gene
Hill, Marquita K
Howland, John LaFollette
Kandutsch, Andrew August
Kozak, Leslie P
McKerns, Kenneth (Wilshire)
Manyan, David Richard
Paigen, Beverly Joyce
Paigen, Kenneth
Radke, Frederick Herbert
Rhodes, William Gale
Ridgway, George Junior
Roxby, Robert
Settlemire, Carl Thomas
Sherblom, Anne P
Sidell, Bruce David
Steinhart, William Lee
Waymouth, Charity
Yonuschot, Gene R

MARYLAND

Aaronson, Stuart Alan
Ackerman, Eric J
Adams, James Miller
Adamson, Richard H
Addanki, Somasundaram
Adelstein, Robert Simon
Ades, Ibrahim Z
Adrouny, George Adour (Kuyumjian)
Ahluwalia, Gurpreet S
Aksamit, Robert Rosooe
Albers, Robert Wayne
Aldridge, Mary Hennen
Allison, Richard Gall
Alsmeyer, Richard Harvey
Alvares, Alvito Peter
Amende, Lynn Meridith
Amr, Sania
Amzel, L Mario
Anderson, Larry Douglas
Anderson, Lucy Macdonald
Anderson, Norman Leigh
Anderson, Richard Allen
Anfinsen, Christian Boehmer
Anhalt, Grant James
Ansher, Sherry Singer
Argraves, W Scott
Ashwell, G Gilbert
Attaway, David Henry
August, Joseph Thomas
Augustine, Patricia C
Austin, Faye Carol
Avigan, Joel
Bachur, Nicholas R, Sr
Baker, Carl Gwin
Baker, George Thomas, III
Balinsky, Doris
Ballentine, Robert
Barban, Stanley
Bareis, Donna Lynn
Barranger, John Arthur
Bashirelahi, Nasir
Baum, Bruce J
Beall, Robert Joseph
Beaven, Michael Anthony
Beaven, Vida Helms
Beeler, Troy James
Behar, Marjam Gojchlerner
Bellino, Francis Leonard
Benton, Allen William
Berger, Edward Alan
Berger, Robert Lewis
Berger, Shelby Louise
Berlin, Elliott
Berman, Howard Mitchell
Bessman, Maurice Jules
Bhathena, Sam Jehangirji
Bhatnagar, Gopal Mohan
Bieri, John Genther
Bigger, Cynthia Anita Hopwood
Bitman, Joel

Blackman, Marc Roy
Bleecker, Margit
Bloch, Robert Joseph
Blosser, Timothy Hobert
Blough, Herbert Allen
Blum, Stanley Walter
Blumberg, Peter Mitchell
Blumenthal, Herbert
Blumenthal, Robert Paul
Boldt, Roger Earl
Bollum, Frederick James
Bono, Vincent Horace, Jr
Boone, Charles Walter
Borkovec, Alexej B
Brady, Roscoe Owen
Brand, Ludwig
Breitman, Theodore Ronald
Brewer, H Bryan, Jr
Brink, Linda Holk
Broomfield, Clarence A
Bruns, Herbert Arnold
Bünger, Rolf
Burge, Wylie D
Bustin, Michael
Butcher, Henry Clay, IV
Butzow, James J
Cabib, Enrico
Cain, Dennis Francis
Carrico, Christine Kathryn
Cassman, Marvin
Catravas, George Nicholas
Catt, Kevin John
Chader, Gerald Joseph
Chakrabarti, Siba Gopal
Chan, Daniel Wan-Yui
Chandra, G Ram
Chang, Lucy Ming-Shih
Chang, Yung-Feng
Chappelle, Emmett W
Chassy, Bruce Matthew
Chatterjee, Subroto
Chaudhari, Anshumali
Chaykovsky, Michael
Chen, Chung-Ho
Chen, Hao-Chia
Chen, Winston Win-Hower
Cheng, Tu-chen
Chernick, Sidney Samuel
Chirigos, Michael Anthony
Chitwood, David Joseph
Cho-Chung, Yoon Sang
Chock, P Boon
Chrambach, Andreas C
Chuang, De-Maw
Cimbala, Michele
Coffey, Donald Straley
Cohen, Robert Martin
Colburn, Nancy Hall
Collins, John Henry
Colombani, Paul Michael
Colombini, Marco
Condliffe, Peter George
Constantopoulos, George
Coomes, Marguerite Wilton
Cornblath, Marvin
Costlow, Richard Dale
Cox, George Warren
Craig, Nessly Coile
Creighton, Donald John
Creveling, Cyrus Robbins
Crystal, Ronald George
Cushman, Samuel Wright
Cutler, Gordon Butler, Jr
Cysyk, Richard L
Dacre, Jack Craven
Daly, John William
Danielpour, David
Darby, Eleanor Muriel Kapp
Das, Saroj R
Dasch, Gregory Alan
Datta, Padma Rag
Davidian, Nancy McConnell
Davidson, Harold Michael
Dawid, Igor Bert
Dean, Ann
Dean, Donna Joyce
De Luca, Luigi Maria
Devreotes, Peter Nicholas
Dhyse, Frederick George
Dienel, Gerald Arthur
Dingman, Charles Wesley, II
Dintzis, Renee Zlochover
Domanski, Thaddeus John
Doukas, Harry Michael
Dudley, Peter Anthony
Dufau, Maria Luisa
Dunaway-Mariano, Debra
Dupont, Jacqueline (Louise)
Dwyer, Dennis Michael
Eaton, Barbra L
Edidin, Michael Aaron
Egan, William Michael
Eipper, Betty Anne
Eisenberg, Evan
Eisenberg, Frank, Jr
Elson, Hannah Friedman
Englund, Paul Theodore
Epstein, David Aaron
Evarts, Ritva Poukka
Fakunding, John Leonard
Falkenstein, Kathy Fay
Farrar, William L
Farrelly, James Gerard
Feldlaufer, Mark Francis

Ferguson, Frederick Palmer
Ferguson, James Joseph, Jr
Fernie, Bruce Frank
Filner, Barbara
Finlayson, John Sylvester
Fishman, Peter Harvey
Flavin, Martin
Folk, John Edward
Fornace, Albert J, Jr
Foulds, John Douglas
Fox, Barbara Saxton
Frank, Leonard Harold
Franklin, Renty Benjamin
Frattali, Victor Paul
Freire, Ernesto I
French, Richard Collins
Friedman, Leonard
Furfine, Charles Stuart
Gabay, Sabit
Gabriel, R Othmar
Gallo, Robert C
Gamble, James Lawder, Jr
Gantt, Ralph Raymond
Gardner, Jerry David
Gelboin, Harry Victor
Gellert, Martin Frank
Gerard, Gary Floyd
Gerin, John Louis
Gerlt, John Alan
Gerwin, Brenda Isen
Gherna, Robert Larry
Gidez, Lewis Irwin
Ginsburg, Ann
Ginsburg, Victor
Glassman, Harold Nelson
Glazer, Robert Irwin
Glenn, Barbara Peterson
Gold, Norman Irving
Goldenbaum, Paul Ernest
Goldman, David
Goldsmith, Paul Kenneth
Golumbic, Calvin
Gomatos, Peter John
Goor, Ronald Stephen
Gorden, Phillip
Gordon, Nathan
Graf, Lloyd Herbert
Graham, Dale Elliott
Gram, Theodore Edward
Granger, Donald Lee
Gravell, Maneth
Green, Marie Roder
Gross, Kenneth Charles
Grossman, Lawrence
Gryder, Rosa Meyersburg
Guchhait, Ras Bihari
Gueriguian, John Leo
Gupte, Sharmila Shaila
Gusovsky, Fabian
Guttman, Helene Augusta Nathan
Habig, William Henry
Hadidi, Ahmed Fahmy
Haggerty, James Francis
Hajiyani, Mehdi Hussain
Hall, Richard Leland
Hallfrisch, Judith
Hamer, Dean H
Handwerker, Thomas Samuel
Hanover, John Allan
Hansen, John Norman
Hansford, Richard Geoffrey
Harmon, Joan T
Harrison, Helen Coplan
Hascall, Vincent Charles, Jr
Hatanaka, Masakazu
Hayes, Joseph Edward, Jr
Hearing, Vincent Joseph, Jr
Hegyeli, Ruth I E J
Helmsen, Ralph John
Heming, Arthur Edward
Henderson, Louis E
Hendler, Richard Wallace
Henkart, Pierre
Herbert, Elton Warren, Jr
Herbst, Edward John
Herman, Eliot Mark
Herrett, Richard Allison
Herriott, Roger Moss
Hess, Helen Hope
Hickey, Robert Joseph
Hilmoe, Russell J(ulian)
Hobbs, Ann Snow
Hoffman, David J
Hollingdale, Michael Richard
Holloway, Caroline T
Housewright, Riley Dee
Howard, Barbara V
Howe, Juliette Coupain
Howley, Peter Maxwell
Huang, Charles Y
Huang, Kuo-Ping
Huang, Pien-Chien
Hubbard, Van Saxton
Hurd, Suzanne Sheldon
Ikeda, George J
Impraim, Chaka Cetewayo
Irwin, David
Jackson, Edward Milton
Jacobowitz, David
Jacobs, Abigail Conway
Jacobson, Kenneth Alan
Jacobus, William Edward
Jakoby, William Bernard
Jamieson, Graham Archibald

Jandorf, Bernard Joseph
Jerina, Donald M
Johns, David Garrett
Johnson, Carl Boone
Johnson, David Freeman
Johnson, George S
Johnson-Winegar, Anna
Jones, Theodore Charles
Kao, Kung-Ying Tang
Kaper, Jacobus M
Kaplan, Emanuel
Kapoor, Chiranjiv L
Karpel, Richard Leslie
Kauffman, Frederick C
Kaufman, Bernard Tobias
Kaufman, Elaine Elkins
Kaufman, Seymour
Kawalek, Joseph Casimir, Jr
Kearney, Philip C
Keister, Donald Lee
Keith, Jerry M
Kelley, John Francis, Jr
Kelly, Thomas J
Kennedy, Robert A
Ketley, Jeanne Nelson
Kibbey, Maura Christine
Kidwell, William Robert
Kies, Marian Wood
Kim, Sooja K
Kindt, Thomas James
Kingsbury, David Wilson
Kippenberger, Donald Justin
Kitzes, George
Kleinman, Hynda Karen
Knoblock, Edward C
Komoriya, Akira
Korn, Edward David
Kozarich, John Warren
Krakauer, Teresa
Krause, David
Kraybill, Herman Fink
Krichevsky, Micah I
Kroll, Martin Harris
Krutzsch, Henry C
Kuether, Carl Albert
Kuff, Edward Louis
Kumaroo, Kuziyilethu Krishnan
Kundig, Fredericka Dodyk
Kunos, George
Kwiterovich, Peter O, Jr
Lai, Chun-Yen
Lakowicz, Joseph Raymond
Lakshmanan, Florence Lazicki
Lambooy, John Peter
Lands, William Edward Mitchell
Landsman, David
Langlykke, Asger Funder
Leder, Irwin Gordon
Lee, Chi-Jen
Lee, Fang-Jen Scott
Lee, Theresa
Lee, Yuan Chuan
Lee, Yuen San
Leiter, Joseph
Leonard, Charles Brown, Jr
Leppla, Stephen Howard
Lesko, Stephen Albert
Levander, Orville Arvid
Levenbook, Leo
Levin, Judith Goldstein
Levy, Robert
Lewis, Marc Simon
Liang, Shu-Mei
Lijinsky, William
Lin, Diane Chang
Lin, Leewen
Lin, Michael C
Lin, Shin
Lippel, Kenneth
Lippincott-Schwartz, Jennifer
Liu, Leroy Fong
Liu, Teh-Yung
Liverman, James Leslie
London, Edythe D
Long, Cedric William
Longfellow, David G(odwin)
Loriaux, D Lynn
Lowensohn, Howard Stanley
Lymn, Richard Wesley
McCandliss, Russell John
McCarty, Richard Earl
McClure, Michael Edward
McCurdy, John Dennis
McDonald, Lee J
Maciag, Thomas Edward
McKenney, Keith Hollis
McLaughlin, Alan Charles
McMacken, Roger
McQuaid, Richard William
MacQuillan, Anthony M
Magnani, John Louis
Majchrowicz, Edward
Malech, Harry Lewis
Maloney, Peter Charles
Manen, Carol-Ann
Manganiello, Vincent Charles
Marcus, Carol Joyce
Margolis, Ronald Neil
Margolis, Sam Aaron
Margolis, Simeon
Margulies, Maurice
Marks, Edwin Potter
Marsho, Thomas V
Martin, Margaret Eileen

Martin, Robert G
Marwah, Joe
Mather, Ian Heywood
Mattern, Michael Ross
Matthews, Benjamin F
Mattick, Joseph Francis
Maurizi, Michael R
Max, Stephen Richard
Melancon, Mark J
Melera, Peter William
Mercado, Teresa I
Mertz, Walter
Metzger, Henry
Mickel, Hubert Sheldon
Middlebrook, John Leslie
Mihalyi, Elemer
Mildvan, Albert S
Miles, Edith Wilson
Minthorn, Martin Lloyd, Jr
Mockrin, Stephen Charles
Mohla, Suresh
Moment, Gairdner Bostwick
Montell, Craig
Morris, Stephen Jon
Moschel, Robert Carl
Moss, Bernard
Mudd, Stuart Harvey
Mueller, Helmut
Mufson, R Allan
Munn, John Irvin
Muralidharan, V B
Murano, Genesio
Murayama, Makio
Mushinski, J Frederic
Nair, Padmanabhan
Nakhasi, Hira Lal
Nandedkar, Arvindkumar Narhari
Nash, Howard Allen
Nayak, Ramesh Kadbet
Nelson, John Howard
Nelson, Ralph Francis
Neufeld, Harold Alex
Newburgh, Robert Warren
Newcombe, David S
Newkirk, David Royal
Newrock, Kenneth Matthew
Newton, Sheila A
Nino, Hipolito V
Nirenberg, Marshall Warren
Noonan, Kenneth Daniel
Norman, Philip Sidney
Nossal, Nancy
O'Brien, John C
O'Brien, Paul J
O'Donnell PhD, James Francis
Oliver, Eugene Joseph
Oppenheim, Joost J
O'Rangers, John Joseph
Otani, Theodore Toshiro
Owens, Ida S
Pace, Judith G
Paddock, Jean K
Pagano, Richard Emil
Palmer, Winifred G
Papas, Takis S
Passaniti, Antonino
Passonneau, Janet Vivian
Patterson, Glenn Wayne
Pedersen, Peter L
Perdue, James F
Peterkofsky, Alan
Peterson, Elbert Axel
Petrella, Vance John
Phang, James Ming
Piatigorsky, Joram Paul
Pierce, Jack Vincent
Pitha-Rowe, Paula Marie
Pitlick, Frances Ann
Pogell, Burton M
Poirier, Miriam Christine Mohrhoff
Pollard, Harvey Bruce
Pollard, Thomas Dean
Pomerantz, Seymour Herbert
Poston, John Michael
Powers, Dennis A
Preusch, Peter Charles
Price, Alan Roger
Prouty, Richard Metcalf
Qasba, Pradmann K
Quarles, Richard Hudson
Rabinovitz, Marco
Rabussay, Dietmar Paul
Rakhit, Gopa
Ram, J Sri
Read-Connole, Elizabeth Lee
Rechcigl, Miloslav, Jr
Reed, Warren Douglas
Rehak, Matthew Joseph
Reiser, Sheldon
Repaske, Roy
Resnik, Robert Alan
Rew, Robert Sherrard
Reynolds, Kevin A
Reynolds, Robert David
Rhee, Sue Goo
Rice, Jerry Mercer
Rice, Nancy Reed
Richert, Nancy Dembeck
Rider, Agatha Ann
Rivera, Americo, Jr
Roberts, Anita Bauer
Robinson, Cecil Howard
Robinson-White, Audrey Jean
Rochovansky, Olga Maria

Biochemistry (cont)

Rodkey, Frederick Lee
Roeder, Lois M
Roesijadi, Guritno
Rogers, Michael Joseph
Romanowski, Robert David
Roscoe, Henry George
Roseman, Saul
Rosenthal, Nathan Raymond
Roth, Thomas Frederic
Rotherham, Jean
Rubin, Robert Jay
Russell, James T
Sabol, Steven Layne
Sachs, David Howard
Sacktor, Bertram
St John, Judith Brook
Salem, Norman, Jr
Samelson, Lawrence Elliot
Sampugna, Joseph
Sanslone, William Robert
Sarin, Prem S
Sarngadharan, Mangalasseril G
Saslaw, Leonard David
Saunders, James Allen
Scheibel, Leonard William
Schepartz, Saul Alexander
Scherbenske, M James
Schiaffino, Silvio Stephen
Schiffmann, Elliot
Schlom, Jeffrey
Schnaar, Ronald Lee
Schneider, Donald Leonard
Schneider, John H
Schoenberg, Daniel Robert
Schoene, Norberta Wachter
Schrum, Mary Irene Knoller
Schwartz, Joan Poyner
Schwartz, Martin
Scocca, John Joseph
Seliger, Howard Harold
Seto, Belinda P L
Seydel, Frank David
Shamsuddin, Abulkalam Mohammad
Sharma, Dinesh C
Sharma, Opendra K
Shellenberger, Thomas E
Shih, Thomas Y
Shiver, John W
Silverman, Robert Eliot
Silverman, Robert Hugh
Simic, Michael G
Simons, Samuel Stoney, Jr
Simpson, Robert Todd
Sitkovsky, Michail V
Sjoblad, Roy David
Skidmore, Wesley Dean
Slein, Milton Wilbur
Slife, Charles W
Smith, Charlotte Damron
Smith, Lewis Wilbert
Smith, Sharron Williams
Smith, William Owen
Snyder, Stephen Laurie
Soares, Joseph Henry, Jr
Sobel, Mark E
Sobocinski, Philip Zygmund
Sogn, John Allen
Sokoloff, Louis
Song, Byoung-Joon
Sowers, Arthur Edward
Sporn, Michael Benjamin
Stadtman, Earl Reece
Stadtman, Thressa Campbell
Stanchfield, James Ernest
Steinert, Peter Malcolm
Stoloff, Leonard
Stoolmiller, Allen Charles
Stoskopf, Michael Kerry
Straat, Patricia Ann
Stratmeyer, Melvin Edward
Striker, G E
Surrey, Kenneth
Sze, Heven
Tabakoff, Boris
Tabor, Celia White
Tabor, Herbert
Talbot, Bernard
Tamminga, Carol Ann
Taniuchi, Hiroshi
Taylor, Martha Loeb
Tester, Cecil Fred
Tietze, Frank
Tildon, J Tyson
Toliver, Adolphus P
Topper, Yale Jerome
Triantaphyllopoulos, Eugenie
Tschudy, Donald P
Tso, Tien Chioh
Turner, R James
Tyree, Bernadette
Ulsamer, Andrew George, Jr
Underwood, Barbara Ann
Usdin, Vera Rudin
Vance, Hugh Gordon
Varma, Shambhu D
Vaughan, Martha
Veech, Richard L
Vijay, Inder Krishan
Villet, Ruxton Herrer
Vincent, Phillip G
Vonderhaar, Barbara Kay
Vydelingum, Nadarajen Ameerdanaden

Wachter, Ralph Franklin
Wagner, David Darley
Walker, Richard Ives
Walter, Donald K
Wannemacher, Robert, Jr
Waravdekar, Vaman Shivram
Warner, Huber Richard
Watkins, Clyde Andrew
Weinbach, Eugene Clayton
Weiner, Myron
Weinstein, Constance de Courcy
Weinstein, Howard
Weir, Edward Earl, II
Weirich, Gunter Friedrich
Weiss, Bernard
Weiss, Joseph Francis
Westphal, Heiner J
White, Edward Austin
Whitehurst, Virgil Edwards
Wickerhauser, Milan
Wiech, Norbert Leonard
Wiggert, Barbara Norene
Williams, Jimmy Calvin
Willingham, Allan King
Wittenberger, Charles Louis
Wolf, Richard Edward, Jr
Wolff, Jan
Wolffe, Alan Paul
Woodman, Peter William
Wray, Granville Wayne
Wray, Virginia Lee Pollan
Wu, Henry Chi-Ping
Wu, Roy Shih-Shyong
Yamada, Kenneth Manao
Yamamoto, Richard Susumu
Youle, Richard James
Young, Ronald Jerome
Yu, Mei-ying Wong
Zacharius, Robert Marvin
Zielke, H Ronald
Zimmerman, Daniel Hill
Zimmerman, Steven B

MASSACHUSETTS

Abeles, Robert Heinz
Ajami, Alfred Michel
Albert, Mary Roberts Forbes (Day)
Alper, Chester Allan
Alpers, Joseph Benjamin
Anderson, Julius Horne, Jr
Antoniades, Harry Nicholas
Appel, Michael Clayton
Arnaout, M Amin
Arvan, Peter
Atkinson-Templeman, Kristine Hofgren
Auld, David Stuart
Auskaps, Aina Marija
Babayan, Vigen Khachig
Backman, Keith Cameron
Bade, Maria Leipelt
Bagshaw, Joseph Charles
Baril, Earl Francis
Barngrover, Debra Anne
Beach, Eliot Frederick
Beck, William Samson
Begg, David A
Bernfeld, Peter
Bernfield, Merton Ronald
Biswas, Chitra
Bloch, Konrad Emil
Blout, Elkan Rogers
Brandts, John Frederick
Brawerman, George
Breakefield, Xandra Owens
Brecher, Peter I
Brennessel, Barbara A
Brown, Beverly Ann
Brown, Gene Monte
Brown, Neal Curtis
Browne, Douglas Townsend
Buchanan, John Machlin
Bunn, Howard Franklin
Burkhardt, Alan Elmer
Burstein, Sumner
Canale-Parola, Ercole
Cantley, Lewis Clayton
Carter, Edward Albert
Caspi, Eliahu
Cheetham, Ronald D
Chishti, Athar Husain
Chivian, Eric Seth
Chlapowski, Francis Joseph
Chou, Iih-Nan
Chu, Shu-Heh W
Cintron, Charles
Clemson, Harry C
Codington, John F
Cohen, Carl M
Cohen, Nadine Dale
Cohen, Seymour Stanley
Collier, Robert John
Collins, Carolyn Jane
Comer, M Margaret
Cooper, Geoffrey Mitchell
Cornell, Neal William
Corwin, Laurence Martin
Coyne, Mary Downey
Crusberg, Theodore Clifford
Cymerman, Allen
Cynkin, Morris Abraham
Czech, Michael Paul
D'Amore, Patricia Ann
Daniel, Peter Francis
Davidson, Betty

Davidson, Samuel James
Demain, Arnold Lester
Dennen, David W
Dennis, Patricia Ann
Desrosiers, Ronald Charles
Devlin, Robert Martin
DeWitt, William
Dice, J Fred
Dills, William L, Jr
Dobson, James Gordon, Jr
Doty, Paul Mead
Drysdale, James Wallace
Ducibella, Tom
Duhamel, Raymond C
Edsall, John Tileston
Ehlers, Mario Ralf Werner
Ehrlich, H Paul
Ellingboe, James
Emanuel, Rodica L
Essigmann, John Martin
Fairbanks, Grant
Fasman, Gerald David
Feig, Larry Allen
Feldberg, Ross Sheldon
Fillios, Louis Charles
Fine, Richard Eliot
Flatt, Jean-Pierre
Foster, John McGaw
Fournier, Maurille Joseph, Jr
Fox, Irving Harvey
Fox, Thomas Oren
Francesconi, Ralph P
Franzblau, Carl
Fraser, Thomas Hunter
Frederick, Sue Ellen
Fritsch, Edward Francis
Fuller, Rufus Clinton
Furbish, Francis Scott
Furie, Bruce
Garg, Hari G
Gaudette, Leo Eward
Gawienowski, Anthony Michael
Gefter, Malcolm Lawrence
George, Harvey
Gergely, John
Geyer, Robert Pershing
Gill, David Michael
Goldberg, Alfred L
Goldberg, Irvin H(yman)
Goldin, Stanley Michael
Goldmacher, Victor S
Goldman, Peter
Goldsby, Richard Allen
Goldstein, Lawrence S B
Goldstein, Richard Neal
Gonnella, Patricia Anne
Good, Carl M, III
Goodman, Howard Michael
Gorfien, Harold
Green, David J
Greenfield, Seymour
Gudas, Lorraine J
Guidotti, Guido
Hadjian, Richard Albert
Hagopian, Miasnig
Halkerston, Ian D K
Hamilton, James Arthur
Hamilton, Paul Barnard
Harbison, G Richard
Harris, S Richard
Harris, Wayne G
Hartman, Iclal Sirel
Hartman, Standish Chard
Hassan, William Ephriam, Jr
Hastings, John Woodland
Hauser, George
Hay, Donald Ian
Hecht, Norman B
Hegsted, David Mark
Heinrich, Gerhard
Hele, Priscilla
Henneberry, Richard Christopher
Hepler, Peter Klock
Herrmann, Robert Lawrence
Hicks, Sonja Elaine
Hinrichs, Katrin
Hirschberg, Carlos Benjamin
Hojnacki, Jerome Louis
Holick, Michael Francis
Hollocher, Thomas Clyde, Jr
Holmquist, Barton
Hong, Jen Shiang
Hu, Jing-Shan
Huxley, Hugh E
Hwang, San-Bao
Hynes, Richard Olding
Ingram, Vernon Martin
Ip, Stephen H
Irving, James Tutin
Irwin, Carol Lee
Irwin, Louis Neal
Isselbacher, Kurt Julius
Jacobson, Bruce Shell
Jameson, James Larry
Jeanloz, Roger William
Jencks, William Platt
Johnson, Corinne Lessig
Jungalwala, Firoze Bamanshaw
Kadkade, Prakash Gopal
Kagan, Herbert Marcus
Kalckar, Herman Moritz
Kaminskas, Edvardas
Karnovsky, Manfred L
Kashket, Eva Ruth

Kashket, Shelby
Kennedy, Eugene P
Kidder, George Wallace, Jr
Killick, Kathleen Ann
Kilpatrick, Daniel Lee
King, Stephen Murray
Kisliuk, Roy Louis
Kitz, Richard J
Knipe, David Mahan
Kobayashi, Yutaka
Kolodner, Richard David
Koul, Omanand
Krane, Stephen Martin
Kravitz, Edward Arthur
Krinsky, Norman Irving
Kubin, Rosa
Kuemmerle, Nancy Benton Stevens
Kupchik, Herbert Z
Leary, John Dennis
LeBaron, Francis Newton
Ledbetter, Mary Lee Stewart
Lees, Marjorie Berman
Lees, Robert S
Lehman, William Jeffrey
Lehrer, Sherwin Samuel
Li, Jeanne B
Liberatore, Frederick Anthony
Lin, Chi-Wei
Lin, Edmund Chi Chien
Lionetti, Fabian Joseph
Lipinski, Boguslaw
Liss, Maurice
Little, Henry Nelson
Lodish, Harvey Franklin
Loscalzo, Joseph
Lowenstein, John Martin
Lu, Renne Chen
Luna, Elizabeth J
Macchi, I Alden
McCluer, Robert Hampton
McDonagh, Jan M
MacDonald, Alex Bruce
McGowan, Joan A
MacKerell, Alexander Donald, Jr
McMaster, Paul D
McReynolds, Larry Austin
Madras, Bertha K
Mager, Milton
Maione, Theodore E
Maniscalco, Ignatius Anthony
Manning, Brenda Dale
Mariani, Henry A
Marquez, Ernest Domingo
Marten, James Frederick
Martz, Eric
Mautner, Henry George
Meedel, Thomas Huyck
Melchior, Donald L
Mohr, Scott Chalmers
Moner, John George
Moolten, Frederick London
Muni, Indu A
Munro, Hamish N
Murphy, John R
Muus, Jytte Marie
Neer, Eva Julia
Nelson, Donald John
Nichols, George, Jr
Nordin, John Hoffman
Nussbaum, Alexander Leopold
O'Brien, Richard Desmond
Offner, Gwynneth Davies
Ofner, Peter
Orkin, Stuart H
Orlando, Joseph Alexander
Orme-Johnson, Nanette Roberts
Orme-Johnson, William H
Ouellette, Andre J
Paine, Clair Maynard
Papastoitsis, Gregory
Pappenheimer, Alwin Max, Jr
Pardee, Arthur Beck
Park, James Theodore
Pathak, Madhukar
Paul, Benoy Bhushan
Paz, Mercedes Aurora
Pecci, Joseph
Petsko, Gregory Anthony
Pharo, Richard Levers
Poccia, Dominic Louis
Porter, William L
Powers Lee, Susan Glenn
Putney, Scott David
Raghavan, Srinivasa
Reif-Lehrer, Liane
Reinhold, Vernon Nye
Rendina, George
Reyero, Cristina
Reyes, Victor E
Rheinwald, James George
Rice, Robert Vernon
Richardson, Charles Clifton
Richmond, Martha Ellis
Richter, Erwin (William)
Rieder, Sidney Victor
Riley, Monica
Rio, Donald C
Riordan, James F
Rivera, Ezequiel Ramirez
Robbins, Phillips Wesley
Roberts, Mary Fedarko
Robinson, Trevor
Rogers, William Irvine
Rosemberg, Eugenia

Ross, Alonzo Harvey
Ryan, Kenneth John
Sacks, David B
Saide, Judith Dana
Sakura, John David
Sanadi, D Rao
Sarkar, Nilima
Sarkar, Satyapriya
Schimmel, Paul Reinhard
Schleif, Robert Ferber
Schmid, Karl
Schwarting, Gerald Allen
Schwartz, Edith Richmond
Searcy, Dennis Grant
Seaver, Sally S
Seidel, John Charles
Seyfried, Thomas Neil
Shaeffer, Joseph Robert
Shashoua, Victor E
Shemin, David
Shing, Yuen Wan
Shipley, Reginald A
Short, John Lawson
Silbert, Jeremiah Eli
Sinex, Francis Marott
Sizer, Irwin Whiting
Slakey, Linda Louise
Slovacek, Rudolf Edward
Smith, Barbara D
Smith, Edgar Eugene
Smith, Joseph Donald
Smith, Nathan Lewis, III
Smith, Wendy Anne
Spiegelman, Bruce M
Spiro, Mary Jane
Spiro, Robert Gunter
Sreter, Frank A
Stare, Fredrick J
Stark, James Cornelius
Stein, Gary S
Stein, Janet Lee Swinehart
Stern, Arthur Irving
Stern, Ivan J
Stiles, Charles Dean
Stollar, Bernard David
Stossel, Thomas Peter
Strauss, Phyllis R
Struhl, Kevin
Sullivan, William Daniel
Surgenor, Douglas MacNevin
Sweadner, Kathleen Joan
Szent-Gyorgyi, Andrew Gabriel
Szlyk, Patricia Carol
Szoka, Paula Rachel
Szostak, Jack William
Takeuchi, Kiyoshi Hiro
Talamo, Barbara Lisann
Tarlov, Alvin Richard
Tashjian, Armen H, Jr
Taunton-Rigby, Alison
Taylor, Allen
Taylor, James A
Terner, Charles
Therriault, Donald G
Toole, Bryan Patrick
Troxler, Robert Fulton
Tullis, James Lyman
Urry, Lisa Andrea
Vallee, Bert L
Vallee, Richard Bert
Van Vunakis, Helen
Villee, Claude Alvin, Jr
Vitale, Joseph John
Wacker, Warren Ernest Clyde
Wald, George
Waloga, Geraldine
Walsh, Christopher Thomas
Wang, Chih-Lueh Albert
Wang, James C
Webb, Andrew Clive
Weinberg, Crispin Bernard
Weinstein, Mark Joel
Westhead, Edward William, Jr
Westheimer, Frank Henry
Wharton, David Carrie
Widmaier, Eric Paul
Wildasin, Harry Lewis
Williamson, Patrick Leslie
Witman, George Bodo, III
Wohlrab, Hartmut
Wong, Shan Shekyuk
Woodcock, Christopher Leonard Frank
Wotiz, Herbert Henry
Wun, Chun Kwun
Yesair, David Wayne
Young, Delano Victor
Young, Richard Lawrence
Young, Vernon Robert
Zieske, James David
Zimmermann, Robert Alan
Ziomek, Carol A

MICHIGAN
Abraham, Irene
Adelman, Richard Charles
Adler, John Henry
Agranoff, Bernard William
Akil, Huda
Alcala, Jose Ramon
Aminoff, David
Anderson, Richard Lee
Andrews, Arthur George
Armant, D Randall
Armstrong, Robert Lee

Bach, Michael Klaus
Ballou, David Penfield
Bandurski, Robert Stanley
Banerjee, Surath Kumar
Barbour, Stephen D
Barraco, Robin Anthony
Barry, Roger Donald
Bauriedel, Wallace Robert
Beher, William Tyers
Benjamins, Joyce Ann
Bennink, Maurice Ray
Ben-Yoseph, Yoav
Bergen, Werner Gerhard
Berk, Richard Samuel
Bernstein, Isadore A
Beyer, Robert Edward
Bieber, Loran Lamoine
Block, Walter David
Bole, Giles G
Boyer, Rodney Frederick
Branson, Dean Russell
Braselton, Webb Emmett, Jr
Braughler, John Mark
Brawn, Mary Karen
Breznak, John Allen
Brooks, Samuel Carroll
Burnett, Jean Bullard
Byerrum, Richard Uglow
Campbell, Kenneth Lyle
Caraway, Wendell Thomas
Carrick, Lee
Carter-Su, Christin
Caspers, Mary Lou
Cerra, Robert Frank
Chang, Albert Yen
Chatterjee, Bandana
Chaudhry, G Rasul
Chaudry, Irshad Hussain
Chou, Kuo Chen
Christensen, A(lbert) Kent
Christman, Judith Kershaw
Clarke, Steven Donald
Clewell, Don Bert
Cody, Wayne Livingston
Cole, George Christopher
Cole, Versa V
Conover, James H
Coon, Minor J
Cotter, Richard
Cuatrecasas, Pedro
Cullum, Malford Eugene
Dabich, Danica
Datta, Prasanta
Dazzo, Frank Bryan
Dekker, Eugene Earl
Distler, Jack, Jr
Dolin, Morton Irwin
Doscher, Marilyn Scott
Duell, Elizabeth Ann
Eberts, Floyd S, Jr
Eble, Thomas Eugene
Einspahr, Howard Martin
Elhammer, Ake P
Erickson, Laurence A
Evans, David R
Fairley, James Lafayette, Jr
Ferguson-Miller, Shelagh Mary
Ferrell, William James
Fischer, George A
Fleisher, Lynn Dale
Fluck, Michele Marguerite
Foote, Joel Lindsley
Foy, Robert Bastian
Frank, Robert Neil
Freeman, Arthur Scott
Friar, Robert Edsel
Friedland, Melvyn
Frohman, Charles Edward
Gee, Robert William
Gelehrter, Thomas David
Ghering, Mary Virgil
Giblin, Frank Joseph
Gilbertson, Terry Joel
Goodwin, Jesse Francis
Gorman, Robert Roland
Goyings, Lloyd Samuel
Grady, Joseph Edward
Grant, Rhoda
Gray, Gary D
Greenberg, Goodwin Robert
Griffith, Thomas
Grossman, Lawrence I
Hajra, Amiya Kumar
Hauxwell, Ronald Earl
Heffner, Thomas G
Heinrikson, Robert L
Hinman, Jack Wiley
Hirata, Fusao
Hoch, Frederic Louis
Hoeksema, Walter David
Holland, Jack Calvin
Hollenberg, Paul Frederick
Holmlund, Chester Eric
Hultquist, Donald Elliott
Hupe, Donald John
Jackson, Craig Merton
Johnson, Garland A
Johnson, Robert Michael
Joseph, Ramon R
Jourdian, George William
Kahn, A Clark
Kawooya, John Kasajja
Kent, Claudia
Kessel, David Harry

Ketchum, Paul Abbott
Kezdy, Ferenc J
Kimura, Tokuji
Kindel, Paul Kurt
King, Charles Miller
Kithier, Karel
Kleinsmith, Lewis Joel
Klohs, Wayne D
Kominek, Leo Aloysius
Kratochvil, Clyde Harding
Krawetz, Stephen Andrew
Krzeminski, Leo Francis
Kupiecki, Floyd Peter
Lahti, Robert A
Lamport, Derek Thomas Anthony
Landefeld, Thomas Dale
Lee, Chuan-Pu
Leng, Marguerite Lambert
Leopold, Wilbur Richard, III
Li, Li-Hsieng
Lightbody, James James
Lillevik, Hans Andreas
Lomax, Margaret Irene
Loor, Rueyming
Louis, Lawrence Hua-Hsien
Ludwig, Martha Louise
Luecke, Richard William
Lueking, Donald Robert
Luk, Gordon David
McCall, Keith Bradley
McCann, Daisy S
McCauley, Roy Barnard
McConnell, David Graham
McCoy, Lowell Eugene
MacDonald, Roderick Patterson
McGroarty, Estelle Josephine
McLean, John Robert
Makinen, Kauko K
Malkin, Leonard Isadore
Marcelo, Cynthia Luz
Marks, Bernard Herman
Martens, Margaret Elizabeth
Martin, Joseph Patrick, Jr
Massey, Vincent
Mather, Edward Chantry
Matthews, Rowena Green
Maxon, William Densmore
Medzihradsky, Fedor
Meisler, Miriam Horowitz
Meyer, Edward Dell
Meyerhoff, Mark Elliot
Miller, Susan Mary
Miller, Thomas Lee
Miller, William Louis
Mitchell, Robert Alexander
Moos, Walter Hamilton
Morris, Allan J
Mostafapour, M Kazem
Moudgil, Virinder Kumar
Nadler, Kenneth David
Nag, Asish Chandra
Nairn, Roderick
Neely, Brock Wesley
Neff, Alven William
Neidhardt, Frederick Carl
Neubig, Richard Robert
Niekamp, Carl William
Njus, David Lars
Nooden, Larry Donald
Nordby, Gordon Lee
Novak, Raymond Francis
Olson, Steven T
Orten, James M
Ovshinsky, Iris M(iroy)
Oxender, Dale LaVern
Palmer, Kenneth Charles
Parcells, Alan Jerome
Parker, Charles J, Jr
Payne, Anita H
Pecoraro, Vincent L
Petty, Howard Raymond
Petzold, Edgar
Philipsen, Judith Lynne Johnson
Podila, Gopi Krishna
Powers, Wendell Holmes
Preiss, Jack
Punch, James Darrell
Purkhiser, E Dale
Pyke, Thomas Richard
Radin, Eric Leon
Radin, Norman Samuel
Rainey, John Marion, Jr
Ram, Jeffrey L
Rapundalo, Stephen T
Raz, Avraham
Reddy, Venkat N
Reitz, Richard Henry
Renis, Harold E
Revzin, Arnold
Reynolds, Orland Bruce
Riggs, Thomas Rowland
Riley, Michael Verity
Rosen, Barry Philip
Rosenspire, Allen Jay
Rozhin, Jurij
Rust, Steven Ronald
Sadoff, Harold Lloyd
Saltiel, Alan Robert
Saper, Mark A
Sardesai, Vishwanath M
Sarkar, Fazlul Hoque
Satoh, Paul Shigemi
Schacht, Jochen (Heinrich)
Schwartz, Stanley Allen

Scott, Ronald McLean
Sell, John Edward
Sellinger, Otto Zivko
Shappirio, David Gordon
Shichi, Hitoshi
Shipman, Charles, Jr
Shonk, Carl Ellsworth
Shore, Joseph D
Smith, Denise Myrtle
Smith, Eugene Joseph
Smith, Robert James
Smith, William Lee
Somerville, Christopher Roland
Speck, John Clarence, Jr
Spilman, Charles Hadley
Stach, Robert William
Stanley, John Pearson
Stenesh, Jochanan
Stewart, Sheila Frances
Stucki, William Paul
Suelter, Clarence Henry
Sun, Frank F
Swanson, Curtis James
Sweeley, Charles Crawford
Syner, Frank N
Taggart, R(obert) Thomas
Tannen, Richard L
Tashian, Richard Earl
Tchen, Tche Tsing
Thoene, Jess Gilbert
Tolbert, Nathan Edward
Trachtenberg, Michael Carl
Turley, June Williams
Van Dyke, Russell Austin
Vatsis, Kostas Petros
Venta, Patrick John
Vinogradov, Serge
Von Korff, Richard Walter
Walker, Gustav Adolphus
Walter, Richard Webb, Jr
Walz, Daniel Albert
Wang, Henry
Wang, John L
Warner, Donald Theodore
Warrington, Terrell L
Watenpaugh, Keith Donald
Weinhold, Paul Allen
Wells, William Wood
West, Bruce David
Weymouth, Patricia Perkins
Whitfield, George Buckmaster, Jr
Wickrema Sinha, Asoka J
Williams, Charles Haddon, Jr
Wilson, Golder North
Wilson, John Edward
Winkler, Barry Steven
Winter, Harry Clark
Wolk, Coleman Peter
Wovcha, Merle G
Yamazaki, Russell Kazuo
Yanari, Sam Satomi
Yocum, Charles Fredrick
Yurewicz, Edward Charles
Zannoni, Vincent G
Zealey, Marion Edward
Zemlicka, Jiri
Zile, Maija Helene
Zimmerman, Norman

MINNESOTA
Abul-Hajj, Jusuf J
Adolph, Kenneth William
Ahmed, Khalil
Anderson, Dwight Lyman
Anderson, John Seymour
Anderson, Paul M
Arneson, Richard Michael
Bach, Marilyn Lee
Barden, Roland Eugene
Baumann, Wolfgang J
Beggs, William H
Berg, Marie Hirsch
Berg, Steven Paul
Berry, James Frederick
Blomquist, Charles Howard
Bodley, James William
Breene, William Michael
Brockman, Howard Lyle, Jr
Brown, Rhoderick Edmiston, Jr
Brummond, Dewey Otto
Brunelle, Thomas E
Burton, Alice Jean
Carr, Charles William
Chandan, Ramesh Chandra
Chipault, Jacques Robert
Clapper, David Lee
Clayton, Robert Allen
Cohen, Harold P
Coleman, Patrick Louis
Coon, Craig Nelson
Cornelius, Steven Gregory
Dagley, Stanley
Daravingas, George Vasilios
De Master, Eugene Glenn
Dempsey, Mary Elizabeth
Derr, Robert Frederick
Dickie, John Peter
Dunshee, Bryant R
Durst, Jack Rowland
Eaton, John Wallace
Edstrom, Ronald Dwight
Ellefson, Ralph Donald
Ellis, Lynda Betty
Evans, Gary William

Biochemistry (cont)

Frantz, Ivan DeRay, Jr
Frey, William Howard, II
Fuchs, James Allen
Ganguli, Mukul Chandra
Gannon, Mary Carol
Gebhard, Roger Lee
Gilboe, Daniel Pierre
Glass, Robert Louis
Goldberg, Nelson D
Gorman, Colum A
Gray, Ernest David
Greenberg, Leonard Jason
Gutmann, Helmut Rudolph
Hadler, Herbert Isaac
Hanson, Richard Steven
Hanson, Russell Floyd
Hendrickson, Herman Stewart, II
Henry, Ronald Alton
Hogenkamp, Henricus Petrus C
Holman, Ralph Theodore
Homann, H Robert
Howard, James Bryant
Hurley, William Charles
Jensen, James Le Roy
Jiang, Nai-Siang
Johnson, Susan Bissette
Jones, James Donald
Joos, Richard W
Katz, Morris Howard
Kirkwood, Samuel
Koerner, James Frederick
Koval, Thomas Michael
Lee, Nancy Zee-Nee Ma
Leung, Benjamin Shuet-Kin
Liener, Irvin Ernest
Lin, Sping
Lipscomb, John DeWald
Loh, Horace H
Longley, Robert W(illiam)
Lovrien, Rex Eugene
Lyon, Richard Hale
McCall, John Temple
Magee, Paul Terry
Malewicz, Barbara Maria
Mannering, Gilbert James
Markhart, Albert Henry, III
Mason, Harold L
Miller, Richard Lynn
Mullen, Joseph David
Neft, Nivaed
Nelsestuen, Gary Lee
Nicol, Susan Elizabeth
Ober, Robert Elwood
Oegema, Theodore Richard, Jr
Orf, John W
Owen, Charles Archibald, Jr
Penniston, John Thomas
Pfeiffer, Douglas Robert
Plagemann, Peter Guenter Wilhelm
Pour-El, Akiva
Powis, Garth
Prendergast, Franklin G
Prohaska, Joseph Robert
Ralston, Douglas Edmund
Rapp, Waldean G
Rathbun, William B
Rodgers, Nelson Earl
Rogers, Palmer, Jr
Roon, Robert Jack
Rosevear, John William
Ruschmeyer, Orlando R
Ryan, Robert J
Salisbury, Jeffrey L
Salo, Wilmar Lawrence
Salstrom, John Stuart
Schlenk, Hermann
Schmid, Harald Heinrich Otto
Schmidt, Jane Ann
Schwartz, Harold Leon
Shier, Wayne Thomas
Singer, Leon
Smith, Quenton Terrill
Somers, Perrie Daniel
Sparnins, Velta L
Spelsberg, Thomas Coonan
Spencer, Richard L
Sperber, William H
Stankovich, Marian Theresa
Stearns, Eugene Marion, Jr
Stein, Paul John
Stern, Marshall Dana
Stoesz, James Darrel
Straw, Thomas Eugene
Sullivan, Betty J
Surdy, Ted E
Swaiman, Kenneth F
Szurszewski, Joseph Henry
Tan, Agnes W H
Tautvydas, Kestutis Jonas
Thenen, Shirley Warnock
Thorpe, Neal Owen
Tindall, Donald J
Tuominen, Francis William
Ungar, Frank
Van Pilsum, John Franklin
Verma, Anil Kumar
Williams, Edmond Brady
Wilson, Michael John
Wood, John Martin
Woodward, Clare K
Yasmineh, Walid Gabriel

Adams, Junius Greene, III
Angel, Charles
Biesiot, Patricia Marie
Cannon, Jerry Wayne
Dilworth, Benjamin Conroy
Eftink, Maurice R
Futrell, Mary Feltner
Graves, David E
Haining, Joseph Leo
Hardy, James D
Hedin, Paul A
Heitz, James Robert
Ho, Ing Kang
Hoagland, Robert Edward
Kellogg, Thomas Floyd
Kennedy, Maurice Venson
Koch, Robert B
Kodavanti, Prasada Rao S
Larsen, James Bouton
Long, Billy Wayne
McNaughton, James Larry
Martin, Billy Joe
Mehrotra, Bam Deo
Morrison, John Coulter
Olson, Mark Obed Jerome
Ousterhout, Lawrence Elwyn
Outlaw, Henry Earl
Read, Virginia Hall
Sadana, Ajit
Salin, Marvin Leonard
Slobin, Lawrence I
Stojanovic, Borislav Jovan
Sulya, Louis Leon
Van Aller, Robert Thomas
Verlangieri, Anthony Joseph
Wahba, Albert J
White, Harold Birts, Jr
Wills, Gene David
Wilson, Robert Paul
Woodley, Charles Leon

MISSOURI
Ackermann, Philip Gulick
Adams, Steven Paul
Agrawal, Harish C
Ahmed, Mahmoud S
Alexander, Stephen
Alkjaersig, Norma Kirstine (Mrs A P Fletcher)
Allen, Robert Wade
Alpers, David Hershel
Apirion, David
Armbrecht, Harvey James
Arneson, Dora Williams
Ayyagari, L Rao
Bachman, Gerald Lee
Badr, Mostafa
Bajaj, S Paul
Banaszak, Leonard Jerome
Barnes, Wayne Morris
Bartkoski, Michael John, Jr
Berenbom, Max
Blake, Robert L
Blatz, Paul E
Bolla, Robert Irving
Bond, Guy Hugh
Brot, Frederick Elliot
Brown, Barbara Illingworth
Brown, Karen Kay (Kilker)
Brown, Olen Ray
Bruns, Lester George
Burch, Helen Bulbrook
Burke, William Joseph
Burton, Robert Main
Campbell, Benedict James
Carey, Paul L
Cenedella, Richard J
Chiappinelli, Vincent A
Chilson, Oscar P
Clare, Stewart
Cockrell, Ronald Spencer
Connolly, John W
Cornell, Creighton N
Coscia, Carmine James
Coudron, Thomas A
Counts, David Francis
Crouch, Eduard Clyde
Cullen, Susan Elizabeth
David, John Dewood
Davis, James Wendell
De Blas, Angel Luis
Dorner, Robert Wilhelm
Ehrenthal, Irving
Eissenberg, Joel Carter
ElAttar, Tawfik Mohammed Ali
Eliceiri, George Louis
Elliott, William H
Elson, Elliot
Engle, Michael Jean
Evans, James William
Feather, Milton S
Feder, Joseph
Fernandez-Pol, Jose Alberto
Fisher, Harvey Franklin
Fitch, Coy Dean
Folk, William Robert
Fox, J Eugene
Franz, John Matthias
Frieden, Carl
Gale, Nord Loran
Garner, George Bernard
Gehrke, Charles William
Geisbuhler, Timothy Paul

Geller, David Melville
Gerhardt, Klaus Otto
Gledhill, William Emerson
Gold, Alvin Hirsh
Gordon, Annette Waters
Gorski, Jeffrey Paul
Gossling, Jennifer
Grady, Harold James
Grant, Gregory Alan
Green, Eric Douglas
Green, Maurice
Greene, Daryle E
Grogan, Donald E
Hamilton, James Wilburn
Hancock, Anthony John
Harakas, N(icholas) Konstantinos
Hart, Kathleen Therese
Heard, John Thibaut, Jr
Herzig, Geoffrey Peter
Hicks, Patricia Fain
Hirschberg, Rona L
Ho, David Tuan-hua
Hooper, Nigel
Horwitt, Max Kenneth
House, William Burtner
Howard, Susan Carol Pearcy
Huang, Jung San
Hunter, Francis Edmund, Jr
Jackson, John Edward
Jacob, Gary Steven
Jaworski, Ernest George
Jeffrey, John J
Jellinek, Max
Johnston, Marilyn Frances Meyers
Kane, James F
Kaplan, Arnold
Katz, Edward
Katzman, Philip Aaron
Kessler, Gerald
Khalifah, Raja Gabriel
Kim, Hyun Dju
Kim, Yee Sik
Kishore, Ganesh M
Kobayashi, George S
Koeltzow, Donald Earl
Kornfeld, Rosalind Hauk
Krivi, Gwen Grabowski
Larson, Russell L
Lie, Wen-Rong
Longmore, William Joseph
Lowry, Oliver Howe
McDaniel, Michael Lynn
McDonald, John Alexander
McMaster, Marvin Clayton, Jr
MacQuarrie, Ronald Anthony
Madl, Ronald
Marshall, Garland Ross
Marshall, Richard Oliver
Martin, Arlene Patricia
Martinez-Carrion, Marino
Mathews, F(rancis) Scott
Mayo, Joseph William
Mehrle, Paul Martin, Jr
Melechen, Norman Edward
Merlie, John Paul
Miller, Herman T
Miller, Lowell D
Moore, Blake William
Morris, Roy Owen
Morton, Phillip A
Moscatelli, Ezio Anthony
Munns, Theodore Willard
Murray, Patrick Robert
Nelson, Curtis Jerome
Nicholas, Harold Joseph
Nickols, G Allen
Noteboom, William Duane
O'Dell, Boyd Lee
Ortwerth, Beryl John
Padron, Jorge Louis
Pike, Linda Joy
Polakoski, Kenneth Leo
Porter, Clark Alfred
Pueppke, Steven Glenn
Quay, Wilbur Brooks
Raskas, Heschel Joshua
Rice, Charles Moen, III
Riddle, Donald Lee
Robbins, Ernest Aleck
Robins, Eli
Roodman, Stanford Trent
Rosenthal, Harold Leslie
Rovetto, Michael Julien
Sadler, J(asper) Evan
Salkoff, Lawrence Benjamin
Sanes, Joshua Richard
Schlesinger, Milton J
Schuler, Martin N
Schulze, Irene Theresa
Seltzer, Jo Louise
Shelton, Damon Charles
Sherman, William Reese
Shukla, Shivendra Dutt
Siehr, Donald Joseph
Silbert, David Frederick
Singh, Harbhajan
Spilburg, Curtis Allen
Staatz, William D
Starling, Jane Ann
Stern, Michele Suchard
Strauss, Arnold Wilbur
Sun, Albert Yung-Kwang
Sunde, Roger Allan
Sweet, Frederick

Taranto, Michael Vincent
Teaford, Margaret Elaine
Thach, Robert Edwards
Toom, Paul Marvin
Tucker, Robert Gene
Vandepopuliere, Joseph Marcel
Warren, James C
Weber, Morton M
Wells, Frank Edward
Welply, Joseph Kevin
Weppelman, Roger Michael
White, Arnold Allen
Whitley, James R
Wiest, Walter Gibson
Willard, Mark Benjamin
Wixom, Robert Llewellyn
Wosilait, Walter Daniel
Yadav, Kamaleshwari Prasad
Yu, Shiu Yeh
Yunger, Libby Marie
Zahler, Warren Leigh
Zech, Arthur Conrad
Zenser, Terry Vernon
Zepp, Edwin Andrew

Blake, Tom
Dratz, Edward Alexander
El-Negoumy, Abdul Monem
Ericsson, Ronald James
Fevold, Harry Richard
Garon, Claude Francis
Hapner, Kenneth D
Hill, Walter Ensign
Jackson, Larry Lee
Jesaitis, Algirdas Joseph
Mell, Galen P
Myers, Lyle Leslie
Robbins, John Edward
Rogers, Samuel John
Roush, Allan Herbert
Stallknecht, Gilbert Franklin
Thompson, Holly Ann
Wright, Barbara Evelyn

Adrian, Thomas E
Barak, Anthony Joseph
Baumstark, John Spann
Birt, Diane Feickert
Boernke, William E
Borchers, Raymond (Lester)
Brakke, Myron Kendall
Bresnick, Edward
Carson, Steven Douglas
Carver, Michael Joseph
Casey, Carol A
Cavalieri, Ercole Luigi
Chollet, Raymond
Cox, George Stanley
Curtis, Gary Lynn
Daly, Joseph Michael
Dam, Richard
Dubes, George Richard
Elthon, Thomas Eugene
Fishkin, Arthur Frederic
Gambal, David
Gessert, Carl F
Goldsmith, Dale Preston Joel
Grandjean, Carter Jules
Grotjan, Harvey Edward, Jr
Gupta, Naba K
Harman, Denham
Heidrick, Margaret Louise
Hill, Robert Mathew
Hofert, John Frederick
Johnson, John Arnold
Johnston, Robert Benjamin
Kimberling, William J
Klucas, Robert Vernon
Knoche, Herman William
Lane, Leslie Carl
Lee, Jean Chor-Yin Wong
McClurg, James Edward
Mackenzie, Charles Westlake, III
Mahowald, Theodore Augustus
Matschiner, John Thomas
Murrin, Leonard Charles
Murthy, Veeraraghavan Krishna
Nickerson, Kenneth Warwick
O'Leary, Marion Hugh
Partridge, James Enoch
Phalen, William Edmund
Phares, Cleveland Kirk
Raha, Chitta Ranjan
Ramaley, Robert Folk
Rogan, Eleanor Groeniger
Rongone, Edward Laurel
Ryan, Wayne L
Scholar, Eric M
Schwartzbach, Steven Donald
Sethi, V Sagar
Shahani, Khem Motumal
Stevens, Sue Cassell
Sullivan, Thomas Wesley
Swanson, Jack Lee
Tuma, Dean J
Vidaver, George Alexander
Wagner, Frederick William
Watt, Dean Day
Wells, Ibert Clifton

Blomquist, Gary James

Buxton, Iain Laurie Offord
Dreiling, Charles Ernest
Halarnkar, Premjit P
Harrington, Rodney E
Heisler, Charles Rankin
Hylin, John Walter
Lewis, Roger Allen
Pardini, Ronald Shields
Peck, Ernest James, Jr
Reitz, Ronald Charles
Schooley, David Allan
Starich, Gale Hanson
Welch, William Henry, Jr
Winicov, Ilga
Woodin, Terry Stern

NEW HAMPSHIRE
Brinck-Johnsen, Truls
Chang, Ta-Yuan
Cole, Charles N
Copenhaver, John Harrison, Jr
Fanger, Michael Walter
Gray, Clarke Thomas
Gross, Robert Henry
Harbury, Henry Alexander
Hatch, Frederick Tasker
Hockett, Robert C(asad)
Hubbard, Colin D
Jackson, Robert C
Kensler, Charles Joseph
Kremzner, Leon T
Lienhard, Gustav E
Minoca, Subhash C
Noda, Lafayette Hachiro
Ou, Lo-Chang
Sloboda, Roger D
Smith, Edith Lucile
Smith, Samuel Cooper
Stewart, James Anthony
Teeri, Arthur Eino
Trumpower, Bernard Lee
Zolton, Raymond Peter

NEW JERSEY
Abramsky, Tessa
Ahn, Ho-Sam
Alper, Carl
Anderson, David Wesley
Anderson, Ethel Irene
Arena, Joseph P
Avigad, Gad
Bahr, James Theodore
Bailin, Gary
Balazs, Endre Alexander
Barkalow, Fern J
Barton, Beverly E
Basu, Mitali
Bauman, John W, Jr
Beadling, Leslie Craig
Bellhorn, Margaret Burns
Bennun, Alfred
Benson, Charles Everett
Berg, Richard A
Bernstein, Eugene H
Bex, Frederick James
Bhagavan, Hemmige
Bieber, Mark Allan
Bird, John William Clyde
Blackburn, Gary Ray
Borman, Aleck
Botelho, Lynne H Parker
Boublik, Miloslav
Boyd, John Edward
Brandriss, Marjorie C
Brin, Myron
Brostrom, Charles Otto
Brostrom, Margaret Ann
Brot, Nathan
Brown, Harry Darrow
Browning, Edward T
Bruno, Charles Frank
Buono, Frederick J
Burg, Richard William
Burns, John J
Burrier, Robert E
Bush, Karen Jean
Buyske, Donald Albert
Cagan, Robert H
Caldwell, Carlyle Gordon
Calvin, Harold I
Caporale, Lynn Helena
Carman, George Martin
Carroll, James Joseph
Carter, James Evan
Cascieri, Margaret Anne
Cayen, Mitchell Ness
Cevallos, William Hernan
Champe, Pamela Chambers
Chao, Yu-Sheng
Chase, Theodore, Jr
Chasin, Mark
Cheng, Kang
Chiang, Bin-Yea
Cho, Kon Ho
Choi, Sook Y
Choi, Ye-Chin
Christakos, Sylvia
Clark, Irwin
Clark, Mary Jane
Coffey, John William
Cohen, Herman
Colby, Richard H
Collett, Edward
Comai-Fuerherm, Karen

Conney, Allan Howard
Cornell, James S
Coscarelli, Waldimero
Coyle, Catherine Louise
Crane, Robert Kellogg
Cryer, Dennis Robert
Cunningham, Earlene Brown
Dalrymple, Ronald Howell
Davis, Brian K
Denhardt, David Tilton
Denton, Arnold Eugene
DePamphilis, Melvin Louis
Ding, Victor Ding-Hai
Dion, Arnold Silva
Doebber, Thomas Winfield
Dougherty, Harry W
Dreyfuss, Jacques
Dubin, Donald T
Dumm, Mary Elizabeth
Eades, Charles Hubert, Jr
Easterday, Richard Lee
Egan, Robert Wheeler
Ellis, Daniel B(enson)
Ennis, Herbert Leo
Erenrich, Evelyn Schwartz
Faber, Marcel D
Fagan, Paul V
Farnsworth, Patricia Nordstrom
Feigelson, Muriel
Feigenbaum, Abraham Samuel
Feldman, Susan C
Fenichel, Richard Lee
Ferguson, Karen Anne
Fiedler-Nagy, Christa
Field, Arthur Kirk
Finkelstein, Paul
Firschein, Hilliard E
Fitzgerald, Paula Marie Dean
Fletcher, Martin J
Flick, Christopher E
Flint, Sarah Jane
Flora, Robert Montgomery
Foley, James E
Frances, Saul
Frank, Oscar
Frantz, Beryl May
Frascella, Daniel W
Free, Charles Alfred
Frenkel, Gerald Daniel
Fresco, Jacques Robert
Freund, Thomas Steven
Fu, Shou-Cheng Joseph
Gaffar, Abdul
Gaffney, Barbara Lundy
Gage, L Patrick
Galdes, Alphonse
Gallopo, Andrew Robert
Gaylor, James Leroy
Gaynor, John James
Georgopapadakou, Nafsika Eleni
Ghai, Geetha R
Ghosh, Arati
Gibian, Morton J
Gibson, Joyce Correy
Gilfillan, Alasdair Mitchell
Gillum, Amanda McKee
Gilvarg, Charles
Ginsberg, Barry Howard
Goldman, Emanuel
Goldsmith, Laura Tobi
Grattan, Jerome Francis
Greasham, Randolph Louis
Gyure, William Louis
Hager, Mary Hastings
Hall, James Conrad
Hansen, Hans John
Hanson, Ronald Lee
Harris, Don Navarro
Hartig, Paul Richard
Harvey, Richard Alexander
Heaslip, Richard Joseph
Heikkila, Richard Elmer
Herring, Harold Keith
Hill, Helene Zimmermann
Hirschberg, Erich
Hitchcock-DeGregori, Sarah Ellen
Hochstadt, Joy
Hoffman, Allan Jordan
Huang, Mou-Tuan
Huff, Jesse William
Huger, Francis P
Humayun, Mir Z
Humes, John Leroy
Inamine, Edward S(eiyu)
Ivatt, Raymond John
Jacob, Theodore August
Jakubowski, Hieronim Zbigniew
Jass, Herman Earl
Johnson, Dewey, Jr
Jones, Benjamin Lewis
Kaczorowski, Gregory John
Kadin, Harold
Kaiserman, Howard Bruce
Kamiyama, Mikio
Kao, John Y
Katchen, Bernard
Katzen, Howard M
Keane, Kenneth William
Kenkare, Divaker B
Klessig, Daniel Frederick
Knight, Wilson Blaine
Kreutner, William
Krishnan, Gopal
Kroll, Emanuel

Ku, Edmond Chiu-Choon
Kumar, Suriender
Kumbaraci-Jones, Nuran Melek
Kuntzman, Ronald Grover
Lada, Arnold
Laden, Karl
Lampen, J Oliver
Landau, Matthew Paul
Lane, Alexander Z
Laskin, Allen I
Laughlin, Alice
LaVoie, Edmond J
Lea, Michael Anthony
Lechevalier, Mary P
Ledeen, Robert
Lee, Lih-Syng
Leitz, Victoria Mary
Lenard, John
Lester, David
Leveille, Gilbert Antonio
Senda, Mototaka
Levin, Wayne
Levy, Robert I
Lewis, Katherine
Li, Shu-Tung
Lin, Grace Woan-Jung
Lin, Tsau-Yen
Linemeyer, David L
Linsky, Cary Bruce
Liu, Alice Yee-Chang
Liu, Wen Chih
Lloyd, Norman Edward
Lovell, Richard Arlington
Lu, Anthony Y H
McGuinness, Eugene T
McGuire, John L
Maclin, Ernest
Malathi, Parameswara
Mandel, Lewis Richard
Manne, Veeraswamy
Manowitz, Paul
Margolis, Frank L
Marquez, Joseph A
Marra, Michael Dominick
Martin, Charles Everett
Marx, Joseph Vincent
Massover, William H
Matthijssen, Charles
Meienhofer, Johannes Arnold
Mellin, Theodore Nelson
Messing, Joachim W
Meyer, Karl
Mezick, James Andrew
Miller, Douglas Kenneth
Miner, Karen Mills
Mochan, Eugene
Monestier, Marc
Moore, James Billy, Jr
Morck, Roland Anton
Moreland, Suzanne
Morris, James Albert
Mosley, Stephen T
Mowles, Thomas Francis
Mueller, Stephen Neil
Mullinix, Kathleen Patricia
Murthy, Vadiraja Venkatesa
Nager, Urs Felix
Naimark, George Modell
Nash, Harold Anthony
Nemecek, Georgina Marie
Neri, Anthony
Newmark, Harold Leon
Nichols, Joseph
Nicolau, Gabriela
Niederman, Robert Aaron
Novick, William Joseph, Jr
Ofengand, Edward James
Oien, Helen Grossbeck
OSullivan, Joseph
Ozer, Harvey Leon
Pachter, Jonathan Alan
Paisley, Nancy Sandelin
Palmere, Raymond M
Pamer, Treva Louise
Pan, Yu-Ching E
Panagides, John
Pang, David C
Papastephanou, Constantin
Paterniti, James R, Jr
Peake, Clinton J
Peets, Edwin Arnold
Pestka, Sidney
Petrack, Barbara Kepes
Pietruszko, Regina
Pitts, Barry James Roger
Pleasants, Elsie W
Plescia, Otto John
Pobiner, Bonnie Fay
Poretz, Ronald David
Prince, Roger Charles
Raetz, Christian Rudolf Hubert
Reckel, Rudolph P
Reeves, John Paul
Ren, Peter
Reuben, Roberta C
Robbins, James Clifford
Rodwell, John Dennis
Rogers, Dexter
Rossow, Peter William
Rothman, James Edward
Rothrock, John William
Rowin, Gerald L
Rowley, George Richard
Rubanyi, Gabor Michael
Rumsey, William LeRoy

Russell, Thomas J, Jr
Ryzlak, Maria Teresa
Saldarini, Ronald John
Sanders, Marilyn Magdanz
Saunders, Robert Norman
Scannell, James Parnell
Schaffner, Carl Paul
Schaich, Karen Marie
Scheiner, Donald M
Schlesinger, David H
Schreiner, Ceinwen Ann
Schulman, Marvin David
Schwartz, Herbert
Schwartz, Jeffrey Lee
Schwarz, Hans Jakob
Schwenzer, Kathryn Sarah
Seitz, Eugene W
Sekula, Bernard Charles
Seligmann, Bruce Edward
Senda, Mototaka
Sessa, Grazia L
Shapiro, Bennett Michaels
Shapiro, Herman Simon
Shapiro, Irwin Louis
Shapiro, Stanley Seymour
Shatkin, Aaron Jeffrey
Sheppard, Herbert
Sherman, Merry Rubin
Sherman, Michael Ian
Sherr, Stanley I
Siegel, Marvin I
Sills, Matthew A
Simmons, Jean Elizabeth Margaret
Sipos, Tibor
Slater, Eve Elizabeth
Smith, Marian Jose
Smith, Sidney R, Jr
Snyder, Robert
Sofer, William Howard
Sohler, Arthur
Solis-Gaffar, Maria Corazon
Spano, Francis A
Spiegel, Herbert Eli
Staub, Herbert Warren
Steinberg, Malcolm Saul
Steiner, Kurt Edric
Sterbenz, Francis Joseph
Stern, William
Stiefel, Edward
Stier, Elizabeth Fleming
Stone, Charles Dean
Stone, Edward John
Sullivan, Andrew Jackson
Sutherland, Donald James
Svokos, Steve George
Swenson, Theresa Lynn
Swislocki, Norbert Ira
Szymanski, Chester Dominic
Tate, Robert Lee, III
Teipel, John William
Thomas, Kenneth Alfred, Jr
Tischio, John Patrick
Tkacz, Jan S
Tobkes, Nancy J
Tornqvist, Erik Gustav Markus
Triscari, Joseph
Trotta, Paul P
Tsong, Yun Yen
Udenfriend, Sidney
Umbreit, Wayne William
Unrau, David George
Vagelos, P Roy
Vander Wende, Christina
Vasconcelos, Aurea C
Vellekamp, Gary John
Verma, Devi C
Viebrock, Frederick William
Vinson, Leonard J
Vogel, Wolfgang Hellmut
Wagman, Gerald Howard
Wainio, Walter W
Wang, Bosco Shang
Wang-Iverson, Patsy
Wase, Arthur William
Wasserman, Paul Michael
Wasserman, Bruce P
Watkins, Paul Donald
Watson, Richard White, Jr
Webb, David R, Jr
Weissbach, Herbert
Welton, Ann Frances
Wilson, Alan C
Wilson, Robert G
Winters, Harvey
Wolff, Donald John
Wong, Keith Kam-Kin
Wong, Rosie Bick-Har
Wood, Alexander W
Worthy, Thomas E
Wu, Joseph Woo-Tien
Yang, Chung Shu
Zilinskas, Barbara Ann
Zuckerman, Leo

NEW MEXICO
Anderson, William Loyd
Atencio, Alonzo C
Baker, Thomas Irving
Barton, Larry Lumir
Belinsky, Steven Alan
Brandvold, Donald Keith
Buss, William Charles
Carroll, Arthur Paul

Biochemistry (cont)

Casillas, Edmund Rene
Coffey, John Joseph
Darnall, Dennis W
Griffin, Travis Barton
Gurd, Ruth Sights
Gurley, Lawrence Ray
Hageman, James Howard
Herman, Ceil Ann
Hildebrand, Carl Edgar
Hoard, Donald Ellsworth
Jenness, Robert
Kelly, Gregory
Kemp, John Daniel
Kraemer, Paul Michael
Kuehn, Glenn Dean
Loftfield, Robert Berner
McNamara, Mary Colleen
Omdahl, John L
Pastuszyn, Andrzej
Ratliff, Robert L
Reyes, Philip
Rivett, Robert Wyman
Roberson, Robert H
Roberts, Peter Morse
Sae, Andy S W
Saponara, Arthur G
Saunders, George Cherdron
Scallen, Terence
Schoenfeld, Robert George
Smoake, James Alvin
Standefer, Jimmy Clayton
Strniste, Gary F
Trotter, John Allen
Turner, Ralph B
Vander Jagt, David Lee
Vogel, Kathryn Giebler
Walters, Ronald Arlen
Wild, Gaynor (Clarke)
Woodfin, Beulah Marie

NEW YORK

Aaronson, Sheldon
Abell, Liese Lewis
Abood, Leo George
Abramson, Morris Barnet
Abreu, Sergio Luis
Acs, George
Aisen, Philip
Albanese, Anthony August
Albrecht, Alberta Marie
Alderfer, James Landes
Alexander, George Jay
Alexander, Renee R
Allen, Howard Joseph
Allen, Sydney Henry George
Almon, Richard Reiling
Altman, Kurt Ison
Ambron, Richard Thomas
Amiraian, Kenneth
Anchel, Marjorie Wolff
Andersen, Jon Alan
Anderson, Mary Elizabeth
Andrews, John Parray
Archibald, Reginald MacGregor
Arion, William Joseph
Aronson, John Noel
Aronson, Robert Bernard
Arvan, Dean Andrew
Asch, Harold Lawrence
Asimov, Isaac
Assoian, Richard Kenneth
Astill, Bernard Douglas
Atkinson, Paul H
Atlas, Steven Alan
Auclair, Walter
Augenlicht, Leonard Harold
Augustyn, Joan Mary
Awad, Atif B
Bachvaroff, Radoslav J
Badoyannis, Helen Litman
Bagchi, Sakti Prasad
Baglioni, Corrado
Baird, Malcolm Barry
Balis, Moses Earl
Baltimore, David
Bambara, Robert Anthony
Banay-Schwartz, Miriam
Banerjee, Debendranath
Banerjee, Ranjit
Barany, Francis
Barber, Eugene Douglas
Bard, Enzo
Barker, Robert
Barnett, Ronald E
Basilico, Claudio
Batt, William
Baum, George
Bazinet, George Frederick
Beach, David H
Beck, David Paul
Beck, Jeanne Crawford
Bedard, Donna Lee
Beeler, Donald A
Bellvé, Anthony Rex
Belman, Sidney
Belsky, Melvin Myron
Benesch, Ruth Erica
Bensadoun, Andre
Benuck, Myron
Benzo, Camillo Anthony
Berezney, Ronald
Berl, Soll

Bernacki, Ralph J
Bernardis, Lee L
Betheil, Joseph Jay
Beyer, Carl Fredrick
Bhargava, Madhu Mittra
Bhattacharya, Malaya
Bhattacharya-Chatterjee, Malaya
Birecka, Helena M
Birken, Steven
Bishop, Charles (William)
Black, Virginia H
Blank, Martin
Bleidner, William Egidius
Bloch, Alexander
Bloch, Eric
Blosser, James Carlisle
Blume, Arthur Joel
Blumenfeld, Olga O
Blumenstein, Michael
Bonney, Robert John
Bora, Sunder S
Borenfreund, Ellen
Borgese, Thomas A
Bourne, Philip Eric
Bowman, Roger Holmes
Bowman, William Henry
Brachfeld, Norman
Bradlow, Herbert Leon
Braunstein, Joseph David
Bray, Bonnie Anderson
Brenner, Mortimer W
Breslow, Jan Leslie
Brewer, Curtis Fred
Briehl, Robin Walt
Broadway, Roxanne Meyer
Brockerhoff, Hans
Brody, Marcia
Brooks, Robert R
Brown, Oliver Monroe
Brownie, Alexander C
Bryan, Ashley Monroe
Bryan, John Kent
Buckley, Edward Harland
Burger, Richard Melton
Burke, William Thomas, Jr
Bushkin, Yuri
Calame, Kathryn Lee
Calhoun, David H
Campbell, Bruce (Nelson), Jr
Campbell, Thomas Colin
Cano, Francis Robert
Carter, Timothy Howard
Catapane, Edward John
Catterall, James F
Center, Sharon Anne
Cerami, Anthony
Chan, Arthur Wing Kay
Chan, Peter Sinchun
Chan, Phillip C
Chan, Samuel H P
Chang, Pei Kung (Philip)
Chang, Ta-Min
Chanley, Jacob David
Chargaff, Erwin
Charney, Martha R
Chasalow, Fred I
Chasin, Lawrence Allen
Chatterjee, Nando Kumar
Chatterjee, Sunil Kumar
Chauhan, Ved P S
Chen, Ching-Ling Chu
Chen, Shu-Ten
Chesley, Leon Carey
Chheda, Girish B
Chiao, Jen Wei
Chou, Ting-Chao
Christy, Nicholas Pierson
Chu, Tsann Ming
Chutkow, Jerry Grant
Cirillo, Vincent Paul
Clark, Virginia Lee
Clarke, Donald Dudley
Clarkson, Allen Boykin, Jr
Claus, Thomas Harrison
Clesceri, Lenore Stanke
Coderre, Jeffrey Albert
Cohen, Gerald
Coleman, Peter Stephen
Colman, David Russell
Conway de Macario, Everly
Cooper, Arthur Joseph L
Corpe, William Albert
Corradino, Robert Anthony
Costa, Max
Cote, Lucien Joseph
Cowger, Marilyn L
Cross, George Alan Martin
Cross, Richard Lester
Cunningham, Bruce Arthur
Cunningham, Richard Preston
Cunninham-Rundles, Charlotte
Daly, Marie Maynard
Damadian, Raymond
Dancis, Joseph
Danishefsky, Isidore
Darzynkiewicz, Zbigniew Dzierzykraj
Datta, Ranajit Kumar
Davenport, Lesley
Davidow, Bernard
Davies, Kelvin J A
De Duve, Christian Rene
De Luca, Chester
De Renzo, Edward Clarence
Deshmukh, Diwakar Shankar

Desnick, Robert John
Desplan, Claude
DeTitta, George Thomas
Detmers, Patricia Anne
Detwiler, Thomas C
D'Eustachio, Peter
Dickerman, Herbert W
DiMauro, Salvatore
Diwan, Joyce Johnson
Dixon, Robert Louis
Doering, Charles Henry
Dolan, Jo Alene
Donner, David Bruce
Doudney, Charles Owen
Downs, Frederick Jon
Doyle, Darrell Joseph
Drickamer, Kurt
Dubroff, Lewis Michael
Dudock, Bernard S
Duff, Ronald George
Eberhard, Anatol
Edds, Kenneth Tiffany
Edelman, Gerald Maurice
Edelman, Isidore Samuel
Ehrke, Mary Jane
Eidinoff, Maxwell Leigh
Elliott, Rosemary Waite
Ellsworth, Robert King
Elsbach, Peter
Elwood, John Clint
Elwyn, David Hunter
Elzinga, Marshall
Emmons, Scott W
Enea, Vincenzo
England, Sasha
Erlanger, Bernard Ferdinand
Erlichman, Jack
Esders, Theodore Walter
Etlinger, Joseph David
Ettinger, Murray J
Eudy, William Wayne
Evans, Mary Jo
Everhart, Donald Lee
Ewart, Mervyn H
Fasco, Michael John
Feigelson, Philip
Feinman, Richard David
Fenton, John William, II
Ferguson, Earl Wilson
Ferguson, John Barclay
Ferrari, Richard Alan
Ferraro, John J
Ferris, James Peter
Fewkes, Robert Charles Joseph
Fiala, Emerich Silvio
Fine, Albert Samuel
Fischbarg, Jorge
Fisher, Paul Andrew
Fishman, Myer M
Fitzgerald, Patrick James
Flanigan, Everett
Fleischmajer, Raul
Fletcher, Paul Wayne
Florini, James Ralph
Fondy, Thomas Paul
Fopeano, John Vincent, Jr
Forest, Charlene Lynn
Fox, Jack Jay
Fox, Thomas David
Frank, David Stanley
Frankel, Arthur Irving
Fredrick, Jerome Frederick
Free, Stephen J
Freed, Simon
Freedberg, Irwin Mark
Freedman, Aaron David
Freedman, Jeffrey Carl
Freedman, Lewis Simon
Frenkel, Krystyna
Freundlich, Martin
Furchgott, Robert Francis
Galivan, John H
Gall, William Einar
Garrels, James I
Garrick, Laura Morris
Garrick, Michael D
Garte, Seymour Jay
Gates, Robert Leroy
Gaver, Robert Calvin
Gelbard, Alan Stewart
George, Elmer, Jr
Gershengorn, Marvin Carl
Gershon, Herman
Ghebrehiwet, Berhane
Gibson, Audrey Jane
Gibson, Kenneth David
Gizis, Evangelos John
Glassman, Armand Barry
Glenn, Joseph Leonard
Goebel, Walther Frederick
Goff, Stephen Payne
Gogel, Germaine E
Gold, Allen Morton
Goldberg, Allan Roy
Goldberger, Robert Frank
Goldman, James Eliot
Goldman, Stephen Shepard
Goldsmith, Lowell Alan
Goldstein, Fred Bernard
Goldstein, Gilbert
Goldstein, Jack
Goldstein, Menek
Golub, Lorne Malcolm
Golubow, Julius

Goodhue, Charles Thomas
Goodman, DeWitt Stetten
Gordon, Portia Beverly
Grauer, Amelie L
Green, Saul
Greenberg, Jacob
Greenberger, Lee M
Greengard, Olga
Greengard, Paul
Gregory, John Delafield
Grieninger, Gerd
Griffith, Owen Wendell
Griffiths, Joan Martha
Grubman, Marvin J
Grumet, Martin
Grunberger, Dezider
Gurel, Okan
Gurpide, Erlio
Gutcho, Sidney J
Guttenplan, Joseph B
Haas, Gerhard Julius
Habicht, Jean-Pierre
Hackler, Lonnie Ross
Haines, Thomas Henry
Haldar, Dipak
Haley, Nancy Jean
Hamill, Robert W
Hanafusa, Hidesaburo
Hankes, Lawrence Valentine
Hardy, Ralph Wilbur Frederick
Harris, Alex L
Harris-Warrick, Ronald Morgan
Hartenstein, Roy
Haschemeyer, Audrey Elizabeth Veazie
Hatcher, Victor Bernard
Hausmann, Ernest
Haymovits, Asher
Haywood, Anne Mowbray
Hehre, Edward James
Heidelberger, Michael
Heintz, Roger Lewis
Held, William Allen
Hendrickson, Wayne Arthur
Henrikson, Katherine Pointer
Henshaw, Edgar Cummings
Heppel, Leon Alma
Herz, Fritz
Hess, George Paul
Higashiyama, Tadayoshi
Hilborn, David Alan
Hilf, Russell
Hill, Doyle Eugene
Hind, Geoffrey
Hinkle, David Currier
Hinkle, Patricia M
Hinkle, Peter Currier
Hipp, Sally Sloan
Hirschman, Albert
Hoberman, Henry Don
Hoch, George Edward
Hof, Liselotte Bertha
Hoffman, Philip
Hoffmann, Dietrich
Hogg, James Felter
Hohmann, Philip George
Holland, Mary Jean Carey
Hollander, Joshua
Horecker, Bernard Leonard
Horowitz, Martin I
Horwitz, Susan Band
Houck, David R
Hough, Jane Linscott
Hoyte, Robert Mikell
Hrazdina, Geza
Hsu, Ming-Ta
Hsu, Robert Ying
Huang, Cheng-Chun
Huang, Sylvia Lee
Huberman, Joel Anthony
Hudecki, Michael Stephen
Hughes, Walter Lee, Jr
Hui, Koon-Sea
Hui, Sek Wen
Iglewski, Barbara Hotham
Ikonne, Justus Uzoma
Iodice, Arthur Alfonso
Ip, Clement Cheung-Yung
Ip, Margot Morris
Isaacs, Charles Edward
Itter, Stuart
Jack, Robert Cecil Milton
Jackanicz, Theodore Michael
Jacobs, Laurence Stanton
Jacobs, Richard L
Jacobs, Ross
Jacobson, Herbert (Irving)
Jainchill, Jerome
Jean-Baptiste, Emile
Jensen, Elwood Vernon
Joh, Tong Hyub
Johnson, Alan J
Jung, Chan Yong
Juo, Pei-Show
Kabat, Elvin Abraham
Kaminsky, Laurence Samuel
Kaplan, Barry Hubert
Kaplan, John Ervin
Karlin, Arthur
Karpatkin, Simon
Karr, James Presby
Katsoyannis, Panayotis G
Katz, Eugene Richard
Kazarinoff, Michael N
Keese, Charles Richard

Keithly, Janet Sue
Kelleher, Raymond Joseph, Jr
Keller, Elizabeth Beach
Kenyon, Alan J
Kern, Harold L
Kesner, Leo
Khan, Anwar Ahmad
Khan, Nasim A
Kieras, Fred J
Kim, Giho
Kim, Young Joo
Kim, Young Tai
Kimmich, George Arthur
Kincl, Fred Allan
King, Tsoo E
Kirdani, Rashad Y
Kirschenbaum, Donald (Monroe)
Kish, Valerie Mayo
Kishore, Gollamudi Sitaram
Kiyasu, John Yutaka
Klingman, Jack Dennis
Knaak, James Bruce
Kobilinsky, Lawrence
Kochwa, Shaul
Koehn, Paul V
Kohler, Constance Anne
Koizumi, Kiyomi
Koretz, Jane Faith
Koritz, Seymour Benjamin
Korytnyk, Walter
Kosman, Daniel Jacob
Kowalski, David Francis
Kraft, Patricia Lynn
Krakow, Joseph S
Kramer, Fred Russell
Krasna, Alvin Isaac
Krasner, Joseph
Krasnow, Frances
Kream, Jacob
Kress, Lawrence Francis
Kronman, Martin Jesse
Krulwich, Terry Ann
Kuchinskas, Edward Joseph
Kull, Fredrick J
Lacks, Sanford
Lahita, Robert George
Lai, Fong M
Lajtha, Abel
Lamberg, Stanley Lawrence
Lamster, Ira Barry
Lang, Helga M (Sister Therese)
Lapin, Evelyn P
Largis, Elwood Eugene
LaRue, Thomas A
Last, Robert L
Lau, Joseph T Y
Lavallee, David Kenneth
Lazaroff, Norman
Le, Heng-Chun
Lee, Ching-Li
Lee, Teh Hsun
Lee, Wei-Li S
Lee-Huang, Sylvia
Lees, Helen
Leff, Judith
Lerner, Leon Maurice
Levitz, Mortimer
Levy, Hans Richard
Levy, Harvey Merrill
Lewis, Bertha Ann
Li, Lu Ku
Libby, Paul Robert
Lieberman, Seymour
Liem, Ronald Kian Hong
Lin, Mow Shiah
Lindahl, Lasse Allan
Linder, Regina
Lipke, Peter Nathan
Lis, John Thomas
Listowsky, Irving
Liu, Houng-Zung
Lloyd, Kenneth Oliver
Lockwood, Dean H
Loeb, Marilyn Rosenthal
Lonberg-Holm, Knud Karl
London, Morris
Loullis, Costas Christou
Low, Barbara Wharton
Lucas, John J
Lugay, Joaquin Castro
Luine, Victoria Nall
Lusty, Carol Jean
McCabe, R Tyler
McClintock, David K
MacColl, Robert
McCormick, J Robert D
McCormick, Paulette Jean
McEwen, Bruce Sherman
McGowan, Eleanor Brookens
McGuire, John Joseph
McKeehan, Wallace Lee
Maddaiah, Vaddanahally Thimmaiah
Madden, Robert E
Madison, James Thomas
Magliulo, Anthony Rudolph
Maio, Joseph James
Maitra, Umadas
Makman, Maynard Harlan
Malbon, Craig Curtis
Maley, Frank
Maley, Gladys Feldott
Malhotra, Ashwani
Malik, Mazhar N
Mandl, Ines

Manning, James Matthew
Mannino, Raphael James
Mansour, Moustafa M
Marcu, Kenneth Brian
Marenus, Kenneth D
Marfey, Sviatopolk Peter
Margolin, Paul
Margolis, Renee Kleimann
Margolis, Richard Urdangen
Margossian, Sarkis S
Marinetti, Guido V
Marks, Neville
Markus, Gabor
Marmur, Julius
Marsh, Walton Howard
Martin, Constance R
Martin, David Lee
Martin, Kumiko Oizumi
Martonosi, Anthony
Matsul, Sei-Ichi
Maturo, Joseph Martin, III
Maxfield, Frederick Rowland
Mayhew, Eric George
Mazumder, Rajarshi
Mazur, Abraham
Meister, Alton
Meltzer, Herbert Lewis
Merrick, Joseph M
Merrifield, Robert Bruce
Messing, Ralph Allan
Meyer, Haruko
Meyers, Marian Bennett
Miller, Herbert Kenneth
Miller, Leon Lee
Miller, Terry Lynn
Mitacek, Eugene Jaroslav
Mittag, Thomas Waldemar
Mizejewski, Gerald Jude
Model, Peter
Moline, Sheldon Walter
Monder, Carl
Mondy, Nell Irene
Monheit, Alan G
Moorehead, Thomas J
Moos, Carl
Moos, Gilbert Ellsworth
Morin, John Edward
Morrill, Gene A
Morrison, John B
Morrison, Sidonie A
Mortlock, Robert Paul
Mosbach, Erwin Heinz
Moscatelli, David Anthony
Muller, Miklós
Muller-Eberhard, Ursula
Munk, Vladimir
Munson, Benjamin Ray
Mussell, Harry W
Nandi, Jyotirmoy
Napolitano, Raymond L
Neal, Michael William
Neeman, Moshe
Nemerson, Yale
Nettleton, Donald Edward, Jr
Neubort, Shimon
Nguyen-Huu, Chi M
Nisselbaum, Jerome Seymour
Norton, William Thompson
Norvitch, Mary Ellen
Notides, Angelo C
Obendorf, Ralph Louis
O'Connor, John Francis
Oesper, Peter
Olmsted, Joanna Belle
Olsen, Kathie Lynn
Olson, Robert Eugene
Ong, Eng-Bee
Oratz, Murray
Oronsky, Arnold Lewis
Palmer, Eugene Charles
Pan, Samuel Cheng
Papsidero, Lawrence D
Pardee, Joel David
Parham, Margaret Payne
Parker, Frank S
Pasternak, Gavril William
Patterson, Ernest Leonard
Pavlos, John
Pearlmutter, Anne Frances
Peerschke, Ellinor Irmgard Barbara
Peisach, Jack
Pellicer, Angel
Peruzzotti, George Peter
Peters, Theodore, Jr
Peverly, John Howard
Philipp, Manfred (Hans)
Pierre, Leon L
Pine, Martin J
Plager, John Everett
Plummer, Thomas H, Jr
Pogo, A Oscar
Poindexter, Jeanne Stove
Polatnick, Jerome
Pollard, Jeffrey William
Popenoe, Edwin Alonzo
Portnoy, Debbi
Posner, Aaron Sidney
Praissman, Melvin
Price, Frederick William
Price, Peter Michael
Pruslin, Fred Howard
Pullarkat, Raju Krishnan
Pullman, Maynard Edward
Pumo, Dorothy Ellen

Puszkin, Elena Getner
Pye, Orrea F
Quaroni, Andrea
Quigley, James P
Racker, Efraim
Rambaut, Paul Christopher
Rapport, Maurice M
Rathnam, Premila
Ratner, Sarah
Rattazzi, Mario Cristiano
Reddy, Bandaru Sivarama
Redman, Colvin Manuel
Reed, Roberta Gable
Reeke, George Norman, Jr
Reichberg, Samuel Bringeissen
Reichert, Leo E, Jr
Reid, Lola Cynthia McAdams
Reith, Maarten E A
Richie, John Peter, Jr
Rieder, Conly LeRoy
Rizack, Martin A
Rizzuto, Anthony B
Roberts, James Lewis
Roberts, Jeffrey Warren
Roberts, Joseph
Roberts, Richard Norman
Robinson, Alix Ida
Roeder, Robert Gayle
Roel, Lawrence Edmund
Roepe, Paul David
Roffman, Steven
Rogerson, Allen Collingwood
Rosano, Thomas Gerard
Rosen, John Friesner
Rosen, Ora Mendelsohn
Rosenfeld, Louis
Rosenfeld, Martin Herbert
Rosenthal, Arthur Frederick
Rosner, William
Ross, John Brandon Alexander
Rossman, Toby Gale
Roth, Jerome Allan
Rothstein, Morton
Rothstein, Rodney Joel
Rowe, Arthur W(ilson)
Roy, Harry
Rubin, Charles Stuart
Rubin, Ronald Philip
Russell, Charlotte Sananes
Ryan, Thomas John
Sabatini, David Domingo
Sabban, Esther Louise
Sack, Robert A
Sacks, William
Saez, Juan Carlos
Salerno, John Charles
Salvo, Joseph J
Samuels, Stanley
Sanford, Karl John
Sankar, D V Siva
Santiago, Noemi
Sarcione, Edward James
Sarma, Raghupathy
Sarma, Ramaswamy Harihara
Sassa, Shigeru
Satir, Birgit H
Sauer, Leonard A
Saxena, Brij B
Sayegh, Joseph Frieh
Schechter, Nisson
Schenkein, Isaac
Scher, William
Scheuer, James
Schildkraut, Carl Louis
Schlessinger, Joseph
Schliselfeld, Louis Harold
Schneider, Allan Stanford
Schor, Joseph Martin
Schramm, Vern Lee
Schreiner, Heinz Rupert
Schubert, Edward Thomas
Schuel, Herbert
Schulman, LaDonne Heaton
Schwartz, Gerald Peter
Schwartz, Ira
Schwartz, Morton K
Scott, Walter Neil
Scott, William Addison, III
Segal, Alvin
Segal, Harold Lewis
Seifter, Sam
Shafer, Stephen Joel
Shaffer, Jacquelin Bruning
Shafit-Zagardo, Bridget
Shalita, Alan Remi
Shalloway, David Irwin
Shapiro, Robert
Sharpless, Nansie Sue
Sheid, Bertrum
Shields, Dennis
Short, Sarah Harvey
Siebert, Karl Joseph
Siegelman, Harold William
Siekevitz, Philip
Siever, Larry Joseph
Silverman, Morris
Silverstein, Emanuel
Silverstein, Samuel Charles
Simon, Eric Jacob
Simon, Martha Nichols
Simon, Sanford Ralph
Simplicio, Jon
Simpson, Melvin Vernon
Singh, Malathy

Skipski, Vlaidimir P(avlovich)
Slaunwhite, Wilson Roy, Jr
Sloan, Donald Leroy, Jr
Slocum, Harry Kim
Smith, Richard Alan
Snoke, Roy Eugene
Sobel, Jael Sabina
Soeiro, Ruy
Soffer, Richard Luber
Sonenberg, Martin
Sowinski, Raymond
Spaulding, Stephen Waasa
Spector, Abraham
Spector, Leonard B
Sprinson, David Benjamin
Spritz, Norton
Squinto, Stephen P
Squires, Catherine L
Squires, Richard Felt
Srinivasan, P R
Srivastava, Bejai Inder Sahai
Stahl, William J
Stanley, Evan Richard
Stanley, Pamela Mary
Stasiw, Roman Orest
Steffen, Daniel G
Stein, Theodore Anthony
Steinman, Charles Robert
Steinman, Howard Mark
Stempien, Martin F, Jr
Sternglanz, Rolf
Stöhrer, Gerhard
Stracher, Alfred
Strand, James Cameron
Straus, David Bradley
Sturman, John Andrew
Sulkowski, Eugene
Sullivan, David Thomas
Sun, Alexander Shihkaung
Sun, Tung-Tien
Sutherland, Betsy Middleton
Swank, Richard Tilghman
Szebenyi, Doletha M E
Szer, Wlodzimierz
Tabachnick, Milton
Tabor, John Malcolm
Tallan, Harris H
Tam, James Pingkwan
Tamir, Hadassah
Tamm, Igor
Tanenbaum, Stuart William
Tan-Wilson, Anna L(i)
Tate, Suresh S
Taub, Mary L
Teebor, George William
Teply, Lester Joseph
Terminiello, Louis
Testa, Raymond Thomas
Thau, Rosemarie B Zischka
Thiessen, Reinhardt, Jr
Thomas, John Owen
Thompson, John Fanning
Thysen, Benjamin
Titeler, Milt
Tomashefsky, Philip
Tomasz, Alexander
Tomasz, Maria
Tometsko, Andrew M
Tooney, Nancy Marion
Toth, Eugene J
Traber, Maret G
Treble, Donald Harold
Trimble, Robert Bogue
Troll, Walter
Tropp, Burton E
Tseng, Linda
Tunis, Marvin
Turinsky, Jiri
Ueno, Hiroshi
Ulm, Edgar H
Unterman, Ronald David
Valinsky, Jay E
Van Buren, Jerome Paul
Van Campen, Darrell R
Venter, J Craig
Vladutiu, Georgirene Dietrich
Vratsanos, Spyros M
Wagh, Premanand Vinayak
Wainfan, Elsie
Walker, Theresa Anne
Walton, Daniel C
Wang, Dalton T
Wang, Jui Hsin
Wang, Tung Yue
Wapnir, Raul A
Warren, William A
Wasserman, Robert Harold
Wasserman, William John
Watanabe, Kyoichi A
Weber, Peter B
Weinfeld, Herbert
Weinstock, Irwin Morton
Weiss, Benjamin
Weiss, Irma Tuck
Wellner, Daniel
Wellner, Vaira Pamiljans
Wenner, Charles Earl
Westerfeld, Wilfred Wiedey
Weyter, Frederick William
Wheatley, Victor Richard
White, Richard John
Wiberg, John Samuel
Wiesner, Rakoma
Wilk, Sherwin

Biochemistry (cont)

Williams, Harold Henderson
Williams, William Joseph
Wilson, David Buckingham
Wilson, Karl A
Winkler, Norman Walter
Wittenberg, Beatrice A
Wittenberg, Jonathan B
Wolf, Robert Lawrence
Wolf, Walter Alan
Wolff, John Shearer, III
Wolfner, Mariana Federica
Wolgemuth, Debra Joanne
Wolin, Meyer Jerome
Wong, Patrick Yui-Kwong
Wood, Henry Nelson
Woodard, Helen Quincy
Wootton, John Francis
Wright, Lemuel Dary
Wu, Joseph M
Wu, Ray J
Wulff, Daniel Lewis
Yang, Song-Yu
Yeagle, Philip L
Yeh, Kwo-Yih
Yeh, Samuel D J
Yen, Andrew
You, Kwan-sa
Young, John Ding-E
Young, Roger Grierson
Yue, Robert Hon-Sang
Yunghans, Wayne N
Zahnd, Hugo
Zakrzewski, Sigmund Felix
Zanetti, Nina Clare
Zapisek, William Francis
Zengel, Janice Marie
Ziegler, Frederick Dixon
Zigman, Seymour
Zitomer, Richard Stephen
Zorzoli, Anita

NORTH CAROLINA

Abdullah, Munir
Agris, Paul F
Albro, Phillip William
Allen, Nina Strömgren
Anderson, Marshall W
Anderson, Richmond K
Armstrong, Frank Bradley, Jr
Aurand, Leonard William
Baccanari, David Patrick
Bales, Connie Watkins
Banes, Albert Joseph
Barakat, Hisham A
Barnes, Donald Wesley
Barrett, J(ames) Carl
Barrow, Emily Mildred Stacy
Bartholow, Lester C
Bates, William K
Bell, Fred E
Bell, Robert Maurice
Berberian, Paul Anthony
Berkut, Michael Kalen
Birnbaum, Linda Silber
Bloom, Kerry Steven
Bond, James Anthony
Boone, Lawrence Rudolph, III
Booth, Raymond George
Bott, Kenneth F
Bray, John Thomas
Bushey, Dean Franklin
Buttke, Thomas Martin
Cahill, Charles L
Camp, Pamela Jean
Caplow, Michael
Carter, Charles Edward
Carter, Charles Williams, Jr
Caterson, Bruce
Chae, Chi-Bom
Chae, Kun
Chaney, Stephen Gifford
Chang, Kwen-Jen
Cheng, Pi-Wan
Cheng, Yung-Chi
Chilton, Mary-Dell Matchett
Clare, Debra A
Clark, Martin Ralph
Clendenon, Nancy Ruth
Coffey, James Cecil, Jr
Coke, James Logan
Coleman, Mary Sue
Cooke, Anson Richard
Cory, Joseph G
Cresswell, Peter
Cross, Robert Edward
Crounse, Robert Griffith
Cunningham, Carol Clem
Currie, William Deems
Curtis, Susan Julia
Daniel, Larry W
Dawson, Jeffrey Robert
DeJong, Donald Warren
De Miranda, Paulo
Devereux, Theodora Reyling
Devlin, Robert B
Dobrogosz, Walter Jerome
Dohm, Gerald Lynis
Douglas, Michael Gilbert
Duch, David S
Duncan, Gordon Duke
Dunnick, June K
Eddy, Edward Mitchell

Elder, Joe Allen
Eldridge, John Charles
Eling, Thomas Edward
Elion, Gertrude Belle
Ellis, Fred Wilson
Endow, Sharyn Anne
Erickson, Bruce Wayne
Erickson, Harold Paul
Errede, Beverly Jean
Faust, Robert Gilbert
Fernandes, Daniel James
Ferone, Robert
Ferris, Robert Monsour
Findlay, John W A
Fine, Jo-David
Forman, Donald T
Fowler, Elizabeth
Fridovich, Irwin
Fried, Howard Mark
Frisell, Wilhelm Richard
Fyfe, James Arthur
Garlich, Jimmy Dale
Goldstein, Joyce Allene
Goz, Barry
Greene, Ronald C
Groves, William Ernest
Gruber, Kenneth Allen
Hall, Iris Beryl Haddon
Hamilton, Pat Brooks
Hamner, Charles Edward, Jr
Harrison, John Henry, IV
Hartz, John William
Hassan, Hosni Moustafa
Hatch, Gary Ephraim
Heck, Henry d'Arcy
Hemperly, John Jacob
Hershfield, Michael Steven
Hill, Charles Horace, Jr
Hill, Robert Lee
Hitchings, George Herbert
Holbrook, David James, Jr
Holm, Robert E
Holmes, Edward Warren
Hook, Gary Edward Raumati
Horton, Horace Robert
Hulcher, Frank H
Ito, Takeru
Jacobson, Kenneth Allan
Jetten, Anton Marinus
Johnson, Jean Louise
Johnson, Ronald Sanders
Johnston, Harry Henry
Joklik, Wolfgang Karl
Jones, Evan Earl
Jones, Mary Ellen
Jordan, Thomas L
Kahn, Joseph Stephan
Kamin, Henry
Kasperek, George James
Kaufman, Bernard
Kaufman, David Gordon
Keene, Jack Donald
Kelly, Susan Jean
King, Ann Christie
King, Jonathan Stanton
King, Kendall Willard
Kirshner, Norman
Korach, Kenneth Steven
Koritnik, Donald Raymond
Krasny, Harvey Charles
Kredich, Nicholas M
Krenitsky, Thomas Anthony
Kucera, Louis Stephen
Kwock, Lester
Lack, Leon
LaFon, Stephen Woodrow
Lee, Martin Jerome
Lee, Si Duk
Li, Steven Shoei-Lung
Lipscomb, Elizabeth Lois
Lister, Mark David
Longmuir, Ian Stewart
Louie, Dexter Stephen
Lucier, George W
Lumb, Roger H
Lundblad, Roger Lauren
Lundeen, Carl Victor, Jr
McCarty, Kenneth Scott
MacDonald, James Cameron
McIlwain, David Lee
McKee, David John
McLachlan, John Alan
Main, Alexander Russell
Marks, Richard Henry Lee
Maroni, Donna F
Massaro, Edward Joseph
Mechanic, Gerald
Meissner, Gerhard
Melnick, Ronald L
Metzgar, Richard Stanley
Miller, David S
Miller, Joseph Edwin
Miller, Richard Lee
Miller, William Laubach
Minard, Frederick Nelson
Modrich, Paul L
Moreland, Donald Edwin
Morell, Pierre
Munson, Paul Lewis
Mushak, Paul
Nayfeh, Shihadeh Nasri
Naylor, Aubrey Willard
Neal, Robert A
Nemeroff, Charles Barnet

Newton, Jack W
Nichol, Charles Adam
O'Callaghan, James Patrick
O'Neal, Thomas Denny
Osbahr, Albert J, Jr
Parks, Leo Wilburn
Pattee, Harold Edward
Pearlman, William Henry
Pearlstein, Robert David
Pekala, Phillip H
Pellizzari, Edo Domenico
Penniall, Ralph
Pennington, Sammy Noel
Petes, Thomas Douglas
Pillsbury, Harold C
Pizzo, Salvatore Vincent
Poole, Doris Theodore
Posner, Herbert S
Powers, Daniel D
Quinn, Richard Paul
Rajagopalan, K V
Ramagopal, Mudumbi Vijaya
Reinhard, John Frederick, Jr
Remy, Charles Nicholas
Reynolds, Jacqueline Ann
Richardson, Stephen H
Rodbell, Martin
Roer, Robert David
Ross, Richard Henry, Jr
Rudel, Lawrence L
Sahyoun, Naji Elias
St Clair, Richard William
Sancar, Gwendolyn Boles
Sanders, Benjamin Elbert
Sanders, Ronald L
Sassaman, Anne Phillips
Scarborough, Gene Allen
Schmidt, Stephen Paul
Schneider, Howard Albert
Selkirk, James Kirkwood
Shaw, Helen Lester Anderson
Shih, Jason Chia-Hsing
Siegel, Lewis Melvin
Sizemore, Ronald Kelly
Smith, Harold Carter
Smith, Peter Blaise
Smith, Susan T
Spector, Thomas
Spicer, Leonard Dale
Spiker, Steven L
Spremulli, Gertrude H
Steege, Deborah Anderson
Stiles, Gary L
Stone, Henry Otto, Jr
Strittmatter, Cornelius Frederick
Sullivan, James Bolling
Summer, George Kendrick
Sundberg, David K
Sung, Michael Tse Li
Suzuki, Kunihiko
Swaisgood, Harold Everett
Switzer, Boyd Ray
Sylvia, Avis Latham
Tanford, Charles
Teng, Ching Sung
Theil, Elizabeth
Thomas, William Albert
Thompson, William Francis, III
Thurman, Ronald Glenn
Toews, Arrel Dwayne
Tourian, Ara Yervant
Tove, Samuel B
Tove, Shirley Ruth
Trainer, John Ezra, Jr
Traut, Thomas Wolfgang
Tucker, Charles Leroy, Jr
Twarog, Robert
Van Loon, Edward John
Van Wagtendonk, Willem Johan
Volk, Richard James
Wachsman, Joseph T
Wagner, Robert H
Waite, Moseley
Wall, Monroe Eliot
Wallace, James William, Jr
Wallen, Cynthia Anne
Ward, James B
Waters, Michael Dee
Watts, John Albert, Jr
Webb, Neil Broyles
Webster, Robert Edward
Weybrew, Joseph Arthur
Wheat, Robert Wayne
White, Helen Lyng
White, J Courtland
White, James Rushton
Whitehurst, Garnett Brooks
Wilson, John Eric
Wilson, William Ewing
Wise, Edmund Merriman, Jr
Wiseman, Jeffrey Stewart
Wittels, Benjamin
Wolfenden, Richard Vance
Wykle, Robert Lee
Zimmerman, Thomas Paul

NORTH DAKOTA

Bakke, Jerome E
Banasik, Orville James
Buckner, James Stewart
Cornatzer, William Eugene
Duerre, John A
Frear, Donald Stuart
Graf, George

Hunt, Janet R
Jacobs, Francis Albin
Johnson, Phyllis Elaine
Johnson, W Thomas
Klosterman, Harold J
Knull, Harvey Robert
Lambeth, David Odus
Lamoureux, Gerald Lee
Lukaski, Henry Charles
Myron, Duane R
Nelson, Dennis Raymond
Nielsen, Forrest Harold
Nordlie, Robert Conrad
Park, Chung Sun
Paulson, Gaylord D
Ray, Paul Dean
Roseland, Craig R
Rosenberg, Harry
Schnell, Robert Craig
Sheridan, Mark Alexander
Suttle, Jeffrey Charles
Vasey, Edfred H
Waller, James R
Zimmerman, Don Charles

OHIO

Alben, James O
Allred, John B
Alter, Gerald M
Anderson, Robert L
Askari, Amir
Astrachan, Lazarus
Austern, Barry M
Ball, William James, Jr
Banerjee, Amiya Kumar
Banerjee, Sipra
Barber, George Alfred
Baroudy, Bahige M
Batra, Prem Parkash
Becker, Carter Miles
Behnke, William David
Behr, Stephen Richard
Behrman, Edward Joseph
Berger, Melvin
Berliner, Lawrence J
Biagini, Raymond E
Blohm, Thomas Robert
Blumenthal, Kenneth Michael
Boczar, Barbara Ann
Boggs, Robert Wayne
Boyle, Michael Dermot
Bozian, Richard C
Brecher, Arthur Seymour
Brierley, Gerald Philip
Brooks, James Reed
Brunden, Kurt Russell
Buck, Stephen Henderson
Bumpus, Francis Merlin
Buzard, James Albert
Camiener, Gerald Walter
Caplan, Arnold I
Caston, J Douglas
Chase, John William
Chen, I-Wen
Ching, Melvin Chung Hing
Clark, Burr, Jr
Clark, Eloise Elizabeth
Clark, Leland Charles, Jr
Clemans, George Burtis
Cooper, Cecil
Cooper, Dale A
Coots, Robert Herman
Cornwell, David George
Couri, Daniel
Craine, Elliott Maurice
Crandall, Dana Irving
Darrow, Robert A
Davis, Pamela Bowes
Day, Richard Allen
De Fiebre, Conrad William
Denko, Charles W
Deodhar, Sharad Dinkar
Deters, Donald W
Devor, Arthur William
Dewar, Norman Ellison
Dickman, John Theodore
DiCorleto, Paul Eugene
Dollwet, Helmar Hermann Adolf
Dorer, Frederic Edmund
Dorman, Robert Vincent
Doskotch, Raymond Walter
Doyle, Richard Robert
DuBrul, Ernest
Ehrhart, L Allen
Eichel, Herman Joseph
Ekvall, Shirley W
Evans, Helen Harrington
Evans, William R
Faber, Lee Edward
Fanger, Bradford Otto
Feller, Dannis R
Fleischman, Darrell Eugene
Flick, Parke Kinsey
Frajola, Walter Joseph
Francis, Marion David
Frea, James Irving
Freed, James Melvin
Freisheim, James Harold
Fritz, Herbert Ira
Ganis, Frank Michael Gangarosa
Ganschow, Roger Elmer
Giannini, A James
Gingery, Roy Evans
Goldthwait, David Atwater

Griffin, Charles Campbell
Griffin, Richard Norman
Groce, John Wesley
Gross, Elizabeth Louise
Grossman, Charles Jerome
Gruenstein, Eric Ian
Haas, Erwin
Hagerman, Dwain Douglas
Hamilton, Thomas Alan
Handwerger, Stuart
Hanson, Richard W
Harmony, Judith A K
Hartman, Warren Emery
Hasselgren, Per-Olof J
Heckman, Carol A
Henningfield, Mary Frances
Hering, Thomas M
Hickenbottom, John Powell
Hieber, Thomas Eugene
Highsmith, Robert F
Hiremath, Shivanand T
Hirschmann, Hans
Hitt, John Burton
Hoff, Kenneth Michael
Hohnadel, David Charles
Homan, Ruth Elizabeth
Homer, George Mohn
Hopfer, Ulrich
Horowitz, Myer George
Horrocks, Lloyd Allen
Horseman, Nelson Douglas
Hoss, Wayne Paul
Hug, George
Hunter, James Edward
Hutterer, Ferenc
Imondi, Anthony Rocco
Ives, David Homer
Jackson, Richard Lee
Jacobs-Lorena, Marcelo
Janusz, Michael John
Jaworski, Jan Guy
Jentoft, Joyce Eileen
Joffe, Frederick M
Johnson, Carl Lynn
Johnson, J David
Johnson, Lee Frederick
Johnson, Thomas Raymond
Johnston, John O'Neal
Jones, Stephen Wallace
Jordan, Freddie L
Kaneshiro, Edna Sayomi
Kariya, Takashi
Kean, Edward Louis
Keller, Stephen Jay
Klein, LeRoy
Kmetec, Emil Philip
Kolenbrander, Harold Mark
Koo, Peter H
Koop, Dennis Ray
Kowal, Jerome
Krampitz, Lester Orville
Kranias, Evangelia Galani
Krueger, Robert Carl
Ku, Han San
Lamborg, Marvin
Landau, Bernard Robert
Lane, Lois Kay
Lane, Richard Durelle
Laughlin, Ethelreda R
Lavik, Paul Sophus
Leffak, Ira Michael
Leis, Jonathan Peter
Lemon, Peter Willian Reginald
Lepp, Cyrus Andrew
Lessard, James Louis
Lovenberg, Walter McKay
Lukin, Marvin
McClure, Jerry Weldon
McCorquodale, Donald James
MacGee, Joseph
McIntyre, Russell Theodore
MacKenzie, Robert Douglas
McQuarrie, Irvine Gray
McQuate, John Truman
Malemud, Charles J
Mallin, Morton Lewis
Mamrack, Mark Donovan
Manning, Maurice
Mao, Simon Jen-Tan
Marzluf, George A
Means, Gary Edward
Mellgren, Ronald Lee
Merola, A John
Messineo, Luigi
Meyer, Ralph Roger
Mieyal, John Joseph
Miller, Robert Harold
Milo, George Edward
Milsted, Amy
Misono, Kunio Shiraishi
Modrzakowski, Malcolm Charles
Moore, Richard Owen
Morgan, Alan Raymond
Moulton, Bruce Carl
Mukhtar, Hasan
Mukkada, Antony Job
Mulhausen, Hedy Ann
Murray, Finnie Ardrey, Jr
Myers, Ronald Elwood
Nagy, Bela Ferenc
Neiderhiser, Dewey Harold
Newman, Howard Abraham Ira
Nuenke, Richard Harold
Nutter, William Ermal

Nygaard, Oddvar Frithjof
Ogle, James D
Okerholm, Richard Arthur
Oleinick, Nancy Landy
Ottolenghi, Abramo Cesare
Pappas, Peter William
Patel, Mulchand Shambhubhai
Pensky, Jack
Perry, George
Prairie, Richard Lane
Pruss, Rebecca M
Pynadath, Thomas I
Rawat, Arun Kumar
Rawson, James Rulon Young
Ray, Richard Schell
Reddy, Padala Vykuntha
Reimann, Erwin M
Rieske, John Samuel
Ritter, Edmond Jean
Ritzert, Roger William
Rodman, Harvey Meyer
Roehrig, Karla Louise
Rosenberry, Terrone Lee
Rosenfeld, Robert Samson
Rossmiller, John David
Rottman, Fritz M
Rouslin, William
Rowe, John James
Rudney, Harry
Russell, Paul Telford
Rutishauser, Urs Stephen
Rynbrandt, Donald Jay
Sable, Henry Zodoc
Saffran, Judith
Saffran, Murray
Sakami, Warwick
Samols, David R
Sawicki, Stanley George
Scarpa, Antonio
Schaff, John Franklin
Schanbacher, Floyd Leon
Schlender, Keith K
Schroeder, Friedhelm
Schumm, Dorothy Elaine
Schwartz, Arnold
Serif, George Samuel
Servaites, Jerome Casimer
Serve, Munson Paul
Shamberger, Raymond J
Shannon, Barry Thomas
Shertzer, Howard Grant
Singer, Sanford Sandy
Skau, Kenneth Anthony
Skeggs, Leonard Tucker, Jr
Slonim, Arnold Robert
Smeby, Robert Rudolph
Smith, Daniel John
Smith, Kenneth Thomas
Snell, Junius Fielding
Sperelakis, Nick
Sprecher, Howard W
Stauffer, Clyde E
Stevenson, J(oseph) Ross
Stoner, Clinton Dale
Stoner, Gary David
Sundaralingam, Muttaiya
Sunkara, Sai Prasad
Swift, Terrence James
Tabita, F Robert
Tannenbaum, Carl Martin
Tejwani, Gopi Assudomal
Tennent, David Maddux
Tepperman, Katherine Gail
Thomas, Craig Eugene
Thomas, William Eric
Thomson, William Alexander Brown
Tomei, L David
Trela, John Michael
Trewyn, Ronald William
Tsai, Chun-Che
Tsai, Ming-Daw
Varnes, Marie Elizabeth
Versteegh, Larry Robert
Vester, John William
Vignos, Paul Joseph, Jr
Voss, Anne Coble
Wagner, Thomas Edwards
Walenga, Ronald W
Walker, Thomas Eugene
Wallick, Earl Taylor
Walz, Frederick George
Wang, Taitzer
Webb, Thomas Evan
Wei, Robert
Weisman, Robert A
Widder, James Stone
Wideman, Cyrilla Helen
Willard, James Matthew
Williams, Marshall Vance
Williams, Wallace Terry
Wilson, James Albert
Winget, Gary Douglas
Wood, Harland G
Woodworth, Mary Esther
Wright, George Joseph
Wuu, Ting-Chi
Yates, Allan James
Yiamouyiannis, John Andrew
Ziller, Stephen A, Jr
Zimmerman, Ernest Frederick
Zull, James E

OKLAHOMA
Abbott, Donald Clayton

Alaupovic, Petar
Badgett, Allen A
Banschbach, Martin Wayne
Begovac, Paul C
Bingham, Robert J
Birckbichler, Paul Joseph
Blair, James Bryan
Bradford, Reagan Howard
Briggs, Thomas
Briscoe, William Travis
Broyles, Robert Herman
Burgus, Roger Cecil
Carpenter, Mary Pitynski
Carubelli, Raoul
Chandler, Albert Morrell
Chaturvedi, Arvind Kumar
Ciereszko, Leon Stanley, Sr
Clarke, Margaret Burnett
Coleman, Ronald Leon
Comp, Philip Cinnamon
Cox, Andrew Chadwick
Croy, Lavoy I
Cunningham, Madeleine White
Dale, George Leslie
Delaney, Robert
Dell'Orco, Robert T
Dillwith, Jack W
Duncan, Michael Robert
Eddington, Carl Lee
Esmon, Charles Thomas
Essenberg, Margaret Kottke
Essenberg, Richard Charles
Ferretti, Joseph Jerome
Fish, Wayne William
Gardner, Charles Olda, Jr
Gardner, Richard Lynn
Gholson, Robert Karl
Harmon, H James
Hartsuck, Jean Ann
Hopkins, Thomas R (Tim)
Hurst, Jerry G
Johnson, B Connor
Johnson, Ronald Roy
Kampschmidt, Ralph Fred
Ketring, Darold L
Kirkham, William R
Kizer, Donald Earl
Klein, Ronald Don
Koeppe, Roger Erdman
Koh, Eunsook Tak
Larsen, Earl George
Leach, Franklin Rollin
Lee, Diana Mang
Lu, Christopher D
McCallum, Roderick Eugene
McCay, Paul Baker
Matsumoto, Hiroyuki
Messmer, Dennis A
Mills, John Norman
Mitchell, Earl Douglass, Jr
Morrissey, James Henry
Murphy, Marjory Beth
Niedbalski, Joseph S
Odell, George Van, Jr
Ontko, Joseph Andrew
Passey, Richard Boyd
Poyer, Joe Lee
Reinhart, Gregory Duncan
Reinke, Lester Allen
Roe, Bruce Allan
Sachdev, Goverdhan Pal
Sanny, Charles Gordon
Schindler, Charles Alvin
Schubert, Karel Ralph
Seeney, Charles Earl
Sherwood, John L
Shipley, John D L
Short, Everett C, Jr
Silverman, Philip Michael
Sims, Peter Jay
Smith, Eddie Carol
Soodsma, James Franklin
Tang, Jordan J N
Trachewsky, Daniel
Waller, George Rozier, Jr
Wang, Chi-Sun
Weatherby, Gerald Duncan
Wender, Simon Harold
Yi, Cho Kwang
Yu, Chang-An
Yunice, Andy Aniece

OREGON
Anderson, Sonia R
Arnold, Roy Gary
Arp, Daniel James
Badenhop, Arthur Fredrick
Baisted, Derek John
Bartos, Dagmar
Bartos, Frantisek
Battaile, Julian
Beatty, Clarissa Hager
Beaudreau, George Stanley
Becker, Robert Richard
Bennett, Robert M
Bentley, J Peter
Black, John Alexander
Bonhorst, Carl W
Brandt, Howard Allen
Brodie, Ann Elizabeth
Brookes, Victor Jack
Brownell, Philip Harry
Bundy, Hallie Flowers
Carlisle, Kay Susan
Civelli, Oliver

Claycomb, Cecil Keith
Clinton, Gail M
Craig, Albert Morrison (Morrie)
Daley, Laurence Stephen
Debons, Albert Frank
Deeney, Anne O'Connell
Dietz, Thomas John
Ellinwood, William Edward
Evans, Harold J
Fang, Ta-Yun
Freed, Virgil Haven
Gabler, Walter Louis
Gamble, Wilbert
Gatewood, Dean Charles
Gillett, James Warren
Gold, Michael Howard
Hare, James Frederic
Hawkinson, Stuart Winfield
Hearon, William Montgomery
Holm, Harvey William
Hoskins, Dale Douglas
Jones, Richard Theodore
Kabat, David
Keevil, Thomas Alan
Kilgour, Gordon Leslie
Kittinger, George William
Lis, Adam W
Lis, Elaine Walker
Litt, Michael
Lochner, Janis Elizabeth
Loomis, Walter David
MacDonald, Donald Laurie
McDuffie, Norton G(raham), Jr
McNulty, Wilbur Palmer
Malbica, Joseph Orazio
Malencik, Dean A
Mathews, Christopher King
Mela-Riker, Leena Marja
Miller, Lorraine Theresa
Mooney, Larry Albert
Palmes, Edward Dannelly
Rasmussen, Lois E Little
Reed, Donald James
Rigas, Demetrios A
Roselli, Charles Eugene
Russell, Peter James
Sanders-Loehr, Joann
Scott, Eion George
Shearer, Thomas Robert
Skala, James Herbert
Soderling, Thomas Richard
Somero, George Nicholls
Spencer, Elaine
Stouffer, Richard Lee
Swanson, J Robert
Szalecki, Wojciech Josef
Terwilliger, Nora Barclay
Tinsley, Ian James
Todd, Wilbert R
Trione, Edward John
Ugarte, Eduardo
Van Bruggen, John Timothy
Van Eikeren, Paul
Weiser, Conrad John
Wolfe, Raymond Grover, Jr
Young, J Lowell

PENNSYLVANIA
Abrams, Richard
Abrams, William R
Adachi, Kazuhiko
Alexander, James King
Alhadeff, Jack Abraham
Allen, Arthur
Alper, Robert
Andrews, Peter Walter
Angstadt, Carol Newborg
Arnott, Marilyn Sue
Aronson, Nathan Ned, Jr
Asakura, Toshio
Auerbach, Victor Hugo
Avadhani, Narayan G
Axelrod, Abraham Edward
Ayers, Arthur Raymond
Baggott, James Patrick
Bailey, David George
Baker, Alan Paul
Baker, Wilber Winston
Baldridge, Robert Crary
Balin, Arthur Kirsner
Basford, Robert Eugene
Bashey, Reza Ismail
Bates, Margaret Westbrook
Baum, Robert Harold
Baumgarten, Werner
Bayer, Margret Helene Janssen
Beining, Paul R
Bentley, Ronald
Bernlohr, Robert William
Bertland, Alexander U
Bertolini, Donald R
Bhaduri, Saumya
Bhavanandan, Veer P
Biaglow, John E
Biebuyck, Julien Francois
Black, Robert Corl
Blackburn, Michael N
Boden, Guenther
Bodine, Peter Van Nest
Bonner, Walter Daniel, Jr
Bradley, Matthews Ogden
Brennan, Thomas Michael
Brinigar, William Seymour, Jr
Brown, William E

Biochemistry (cont)

Buck, Clayton Arthur
Burch, Mary Kappel
Burdick, Donald
Burke, James Patrick
Butler, Thomas Michael
Butler, William Barkley
Campbell, Iain Malcolm
Campo, Robert D
Castellano, Salvatore, Mario
Castle, John Edwards
Castric, Peter Allen
Cerbulis, Janis
Chacko, George Kutty
Chaiken, Irwin M
Chance, Britton
Chen, Shih-Fong
Ch'ih, John Juwei
Chowrashi, Prokash K
Chun, Edward Hing Loy
Chung, Albert Edward
Clagett, Carl Owen
Clark, Charles Christopher
Cohen, Leonard Harvey
Cohen, Pinya
Cohn, Mildred
Cohn, Robert M
Colman, Robert W
Colony-Cokely, Pamela
Connors, William Matthew
Conover, Thomas Ellsworth
Cooper, Joseph E
Cooperman, Barry S
Cordes, Eugene H
Cortner, Jean A
Coss, Ronald Allen
Craven, Patricia A
Creasey, William Alfred
Cristofalo, Vincent Joseph
Crosby, Lon Owen
Crowell, Richard Lane
Dalal, Fram Rustom
D'Alisa, Rose M
Damle, Suresh B
Daniel, James L
Das, Manjusri
Davies, Helen Jean Conrad
Davies, Robert Ernest
De La Haba, Gabriel Luis
Della-Fera, Mary Anne
Delluva, Adelaide Marie
Del Vecchio, Vito Gerard
Devlin, Thomas McKeown
Dewey, Virginia Caroline
Diamond, Leila
Di Cuollo, C John
Diven, Warren Field
DiVincenzo, George D
Donnelly, Thomas Edward, Jr
Dowd, Susan Ramseyer
Doyne, Thomas Harry
Drake, Billy Blandin
Draus, Frank John
Dreisbach, Joseph Herman
Dresden, Carlton F
Dressler, Hans
Duck, William N, Jr
Duker, Nahum Johanan
Dulka, Joseph John
Dutton, P Leslie
Eagon, Patricia K
Edmonds, Mary P
Edwards, John R
Ellingson, John S
Ellis, Demetrius
English, Leigh Howard
Erecinska, Maria
Esfahani, Mojtaba
Evans, Audrey Elizabeth
Farrell, Harold Maron, Jr
Farren, Ann Louise
Feingold, David Sidney
Feinstein, Sheldon Israel
Fenderson, Bruce Andrew
Fenton, Marilyn Ruth
Ferrell, Robert Edward
Finn, Frances M
Fisher, Edward Allen
Flaks, Joel George
Flanagan, Thomas Leo
Fluck, Eugene Richards
Fox, Jay B, Jr
Franklin, Samuel Gregg
Franzen, James
Fraser, Nigel William
Fritz, Paul John
Furth, John J
Garfinkle, Barry David
Gealt, Michael Alan
Ghosh, Amal Kumar
Gilbertson, John R
Glaid, Andrew Joseph, III
Glick, Jane Mills
Godfrey, John Carl
Godfrey, Susan Sturgis
Goff, Christopher Godfrey
Golder, Richard Harry
Goldfine, Howard
Golub, Ellis Eckstein
Goodgal, Sol Howard
Gould, Anne Bramlee
Gould, Robert James
Grant, Norman Howard

Grebner, Eugene Ernest
Gregory, Francis Joseph
Grindel, Joseph Michael
Grunwald, Gerald B
Gurin, Samuel
Gustine, David Lawrence
Hackney, David Daniel
Haimovich, Beatrice
Hamilton, Robert Houston
Hammel, Jay Morris
Hammerstedt, Roy H
Harding, Roy Woodrow, Jr
Hardison, Ross Cameron
Harmon, George Andrew
Hartline, Richard
Hartzler, Eva Ruth
Hass, Louis F
Hassell, John Robert
Haugaard, Niels
Hayashi, Teruo Terry
Heimer, Ralph
Henderson, George Richard
Henderson, Linda Shlatz
Hendry, Richard Allan
Henry, Susan Armstrong
Herbert, Michael
Herlyn, Dorothee Maria
Hess, Eugene Lyle
Hickey, Richard James
Higgins, Michael Lee
Higgins, Terry Jay
Higman, Henry Booth
Hill, Charles Whitacre
Ho, Chien
Hoek, Joannes (Jan) Bernardus
Hoffee, Patricia Anne
Hofmann, Klaus Heinrich
Holmes, William Leighton
Hoober, John Kenneth
Hopper, Sarah Priestly
Howe, Chin Chen
Hudson, Alan P
Hung, Paul P
Hurst, William Jeffrey
Idell-Wenger, Jane Arlene
Isom, Harriet C
Iyengar, Raja M
Jackson, Ethel Noland
Jacob, Samson T
Jacobs, Mark
Jacobsohn, Gert Max
Jacobsohn, Myra K
Jacobson, Lewis A
Jarett, Leonard
Jen-Jacobson, Linda
Jezyk, Peter Franklin
Jim, Kam Fook
Jimenez, Sergio
Johnston, James Bennett
Jones, David Hartley
Jorns, Marilyn Schuman
Kaji, Akira
Kaji, Hideko (Katayama)
Kalf, George Frederick
Kallen, Roland Gilbert
Kao, Race Li-Chan
Karol, Meryl Helene
Kasvinsky, Peter John
Katz, Solomon H
Kayne, Fredrick Jay
Kayne, Marlene Steinmetz
Kefalides, Nicholas Alexander
Kennett, Roger H
Khatami, Mahin
Khoury, George
Kim, Sangduk
Kirby, Edward Paul
Kirtley, Mary Elizabeth
Kleinman, Roberta Wilma
Knauff, Raymond Eugene
Kohn, Michael Charles
Korchak, Helen Marie
Koszalka, Thomas R
Kozinski, Andrzei
Krah, David Lee
Kritchevsky, David
Lampson, George Peter
La Noue, Kathryn F
Lapidus, Milton
Layne, Porter Preston
Leboy, Phoebe Starfield
Lee, David K H
Lee, Shaw-Guang Lin
Lee, Ted C K
Lehman, Ernest Dale
Leibel, Wayne Stephan
Leon, Shalom A
Lerner, Leonard Joseph
Lewbart, Marvin Louis
Lieberman, Hillel
Liebman, Paul Arno
Lien, Eric Louis
Lindstrom, Jon Martin
Little, Brian Woods
Litwack, Gerald
Logan, David Alexander
Lotlikar, Prabhakar Dattaram
Lowe-Krentz, Linda Jean
Maass, Alfred Roland
McAleer, William Joseph
McCarl, Richard Lawrence
McCarthy, William John
McClure, William Robert
MacDonald, James Scott

Machlowitz, Roy Alan
Mack, Lawrence Lloyd
McManus, Ivy Rosabelle
McMorris, F Arthur
Magee, Wayne Edward
Malamud, Daniel F
Manolson, Morris Frank
Manson, Lionel Arnold
Mao, James Chieh Hsia
Marcus, Abraham
Markham, George Douglas
Marks, Dawn Beatty
Marsh, Julian Bunsick
Mathur, Carolyn Frances
Mears, James Austin
Merkel, Joseph Robert
Metrione, Robert M
Mifflin, Theodore Edward
Miller, Elizabeth Eshelman
Miller, Gail Lorenz
Miller, James Eugene
Mong, Seymour
Monson, Frederick Carlton
Morgan, Howard E
Morris, Sidney Machen, Jr
Mortimore, Glenn Edward
Moss, Melvin Lane
Mullin, James Michael
Mumma, Ralph O
Murer, Erik Homann
Murphy, Robert Francis
Muth, Eric Anthony
Na, George Chao
Nambi, Ponnal
Neil, Gary Lawrence
Nemer, Martin Joseph
Nes, William Robert
Neubauer, Russell Howard
Nishikawa, Alfred Hirotoshi
Niu, Mann Chiang
Novak, Josef Frantisek
O'Connor, John Dennis
Oesterling, Myrna Jane
Ohnishi, Tomoko
Ohnishi, Tsuyoshi
O'Neill, John Joseph
Opas, Evan E
Opella, Stanley Joseph
Orr, Nancy Hoffner
Paik, Woon Ki
Pairent, Frederick William
Panos, Charles
Passananti, Gaetano Thomas
Patterson, Elizabeth Knight
Paul, Harbhajan Singh
Pazur, John Howard
Peebles, Craig Lewis
Pegg, Anthony Edward
Perlis, Irwin Bernard
Persky, Harold
Pessen, Helmut
Phillips, Allen Thurman
Pieringer, Ronald Arthur
Plaut, Gerhard Wolfgang Eugen
Pleasure, David
Pollack, Robert Leon
Porter, Curt Culwell
Porter, Ronald Dean
Pratt, Elizabeth Ann
Press, Linda Seghers
Prockop, Darwin J
Pruett, Patricia Onderdonk
Punnett, Thomas R
Pye, Edward Kendall
Rabinowitz, Joseph Loshak
Ramachandran, Subramania
Rao, Kalipatnapu Narasimha
Ray, Eva K
Reed, Ruth Elizabeth
Reinhart, Michael P
Ricciardi, Robert Paul
Robb, Richard John
Rodan, Gideon Alfred
Rose, Irwin Allan
Rose, Zelda B
Rosenblatt, Michael
Rosenbloom, Joel
Rosenblum, Irwin Yale
Rosenfeld, Leonard M
Ross, Alta Catharine
Roth, Stephen
Rothblat, George H
Rottenberg, Hagai
Rovera, Giovanni
Rutman, Robert Jesse
Salama, Guy
Salganicoff, Leon
Sampson, Phyllis Marie
Sands, Howard
Savage, Carl Richard, Jr
Schepartz, Bernard
Scher, Charles D
Schlegel, Robert Allen
Schleyer, Heinz
Schmaier, Alvin Harold
Schultz, Richard Morris
Schwartz, Elias
Scott, Dwight Baker McNair
Searls, Robert L
Seery, Virginia Lee
Segal, Stanton
Shank, Richard Paul
Shannon, Jack Corum
Shapiro, Irving Meyer

Shaw, Leslie M J
Shiau, Yih-Fu
Shiman, Ross
Shockman, Gerald David
Sidie, James Michael
Siegel, Richard C
Silver, Melvin Joel
Singer, Alan G
Sink, John Davis
Skogerson, Lawrence Eugene
Smith, Colleen Mary
Smyrniotis, Pauline Zoe
Snipes, Charles Andrew
Sorof, Sam
Soslau, Gerald
Sprince, Herbert
Stambaugh, Richard L
Stewart, Barbara Yost
Stiller, Richard L
Stinson, Edgar Erwin
Strauss, Jerome Frank, III
Strauss, Robert R
Suhadolnik, Robert J
Sun, James Dean
Sung, Cheng-Po
Surmacz, Cynthia Ann
Swaney, John Brewster
Sylvester, James Edward
Szymanski, Edward Stanley
Tatum, Charles Maris
Taylor, William Daniel
Thayer, Donald Wayne
Thayer, William
Theiner, Micha
Thimann, Kenneth Vivian
Thompson, Marvin P
Tint, Howard
Tobes, Michael Charles
Tocco, Dominick Joseph
Tomarelli, Rudolph Michael
Tong, Winton
Touchstone, Joseph Cary
Treece, Jack Milan
Tristram-Nagle, Stephanie Ann
Tu, Chen-Pei David
Tuan, Rocky Sung-Chi
Tulenko, Thomas Norman
Tuszynski, George P
Tutwiler, Gene Floyd
Uitto, Jouni Jorma
Upton, G Virginia
Vanderkooi, Jane M
Vanderlinde, Raymond E
Van Inwegen, Richard Glen
Van Rossum, George Donald Victor
Vaughan, James Roland
Vergona, Kathleen Anne Dobrosielski
Villafranca, Joseph John
Vlasuk, George P
Voet, Judith Greenwald
Vogt, Molly Thomas
Wainer, Arthur
Wallace, Herbert William
Wampler, D Eugene
Warme, Paul Kenneth
Warren, Leonard
Watt, Robert M
Waxman, Lloyd H
Weber, Annemarie
Weinbaum, George
Weinberg, Eric S
Weinhouse, Sidney
Weinryb, Ira
Weiss, Sidney
Westley, John William
Whitfield, Carol F(aye)
Widnell, Christopher Courtenay
Williamson, John Richard
Wilson, David F
Winkler, James David
Winsten, Seymour
Winters, Mary Ann
Woychik, John Henry
Yellin, Tobias O
Yonetani, Takashi
Young, Franklin
Yushok, Wasley Donald
Zelson, Philip Richard
Zemaitis, Michael Alan
Zurawski, Vincent Richard, Jr
Zweidler, Alfred

RHODE ISLAND

Beale, Samuel I
Biggins, John
Brautigan, David L
Brown, Phyllis R
Cha, Sungman
Coleman, John Russell
Constantinides, Spiros Minas
Crabtree, Gerald Winston
Crowley, James Patrick
Dahlberg, Albert Edward
Dain, Joel A
Davis, Robert Paul
Donovan, Gerald Alton
Eil, Charles
Fausto, Nelson
Flanagan, Thomas Raymond
Hai, Chi-Ming
Hammen, Carl Schlee
Hegre, Carman Stanford
Kresina, Thomas Francis
Landy, Arthur

Lederberg, Seymour
Lusk, Joan Edith
Malcolm, Alexander Russell
Martin, Horace F
Miech, Ralph Patrick
O'Leary, Gerard Paul, Jr
Parks, Robert Emmett, Jr
Plotz, Richard Douglas
Purvis, John L
Rao, Girimaji J Sathyanarayana
Rosenstein, Barry Sheldon
Rothman, Frank George
Sapolsky, Asher Isadore
Shaikh, Zahir Ahmad
Steiner, Manfred
Stoeckler, Johanna D
Tremblay, George Charles
Turner, Michael D
Von Riesen, Daniel Dean
Weltman, Joel Kenneth
Young, Robert M
Zackroff, Robert V

SOUTH CAROLINA
Allen, Donald Orrie
Arnaud, Philippe
Baggett, Billy
Banik, Narendra Lal
Baynes, John William
Berger, Franklin Gordon
Bishop, Muriel Boyd
Camper, Nyal Dwight
Chao, Julie
Cowgill, Robert Warren
Davis, Craig Wilson
Davis, Leroy
Dodds, Alvin Franklin
Doig, Marion Tilton, III
Dunlap, Robert Bruce
Finlay, Mary Fleming
Fisher, Ronald Richard
Fowler, Stanley D
Gadsden, Richard Hamilton, Sr
Greenfield, Seymour
Gross, Stephen Richard
Hays, Ruth Lanier
Henricks, Donald Maurice
Hollis, Bruce Warren
Holstein, Arthur G
Hood, Samuel Lowry
Jonsson, Haldor Turner, Jr
Kistler, Wilson Stephen, Jr
Knapp, Daniel Roger
Knight, Anne Bradley
Kowalczyk, Jeanne Stuart
Krall, Albert Raymond
Lazarchick, John
Ledford, Barry Edward
Lin, Tu
McCord, William Mellen
McDonald, John Kennely
Maheshwari, Kewal Krishnan
Miller, Ronald Lee
Mitchell, Jack Harris, Jr
Morrow, William Scot
Orcutt, Donald Adelbert
Oswald, Edward Odell
Paynter, Malcolm James Benjamin
Powell, Gary Lee
Priest, David Gerard
Rawalay, Surjan Singh
Ritchie, Kim
Rittenbury, Max Sanford
Rohlfing, Duane L
Sawyer, Roger Holmes
Schmidt, Gilbert Carl
Schwabe, Christian
Shively, Jessup MacLean
Singh, Inderjit
Sodetz, James M
Spain, James Dorris, Jr
Stidworthy, George H
Stillway, Lewis William
Stratton, Lewis Palmer
Stutzenberger, Fred John
Swanson, Arnold Arthur
Taylor, Harold Allison, Jr
Turk, Donald Earle
Wells, John Arthur
Wheeler, Darrell Deane
Wilson, Gregory Bruce
Wilson, Steven Paul
Woodard, Geoffrey Dean Leroy
Wuthier, Roy Edward
Yoch, Duane Charles
Zemp, John Workman
Zimmerman, James Kenneth
Zucker, Robert Martin

SOUTH DAKOTA
Brady, Frank Owen
Brandwein, Bernard Jay
Cook, David Edgar
Dwivedi, Chandradhar
Ferguson, Michael William
Hills, Loran C
Krueger, Keatha Kathrine
Langworthy, Thomas Allan
Lindahl, Ronald
Lindahl, Ronald Gunnar
Neuhaus, Otto Wilhelm
Olson, Oscar Edward
Peanasky, Robert Joseph
Prescott, Lansing M

Small, Gary D
Thomas, John Alva
Westfall, Helen Naomi
Whitehead, Eugene Irving

TENNESSEE
Ahmed, Nahed K
Airee, Shakti Kumar
Andersen, Richard Nicolaj
Anderson, Ted L
Andrews, John Stevens, Jr
Anjaneyulu, P S R
Appleman, James R
Baxter, John Edwards
Ben-Porat, Tamar
Berney, Stuart Alan
Blakley, Raymond L
Blevins, Raymond Dean
Bond, Andrew
Bradham, Laurence Stobo
Brent, Thomas Peter
Briggs, Robert Chester
Broquist, Harry Pearson
Brown, Fountaine Christine
Brown, James Walker, Jr
Bryant, Robert Emory
Bucovaz, Edsel Tony
Bunick, Gerard John
Burk, Raymond Franklin, Jr
Burr, William Wesley, Jr
Burt, Alvin Miller, III
Burtis, Carl A, Jr
Byers, Lawrence Wallace
Carlson, Gerald Michael
Carpenter, Graham Frederick
Champney, William Scott
Chen, James Pai-Fun
Chen, Thomas Tien
Cheung, Wai Yiu
Christensen, John
Chytil, Frank
Cohen, Stanley
Cohn, Waldo E
Coniglio, John Giglio
Cook, George A
Cook, Robert James
Corbin, Jack David
Costlow, Mark Enoch
Cox, Ray
Croom, Henrietta Brown
Cullen, Marion Permilla
Cunningham, Leon William
Danzo, Benjamin Joseph
Darby, William Jefferson
Das, Salil Kumar
Desiderio, Dominic Morse, (Jr)
Di Pietro, David Louis
Dockter, Michael Edward
Duhl, David M
Dulaney, John Thornton
Epler, James L
Ernst-Fonberg, Marylou
Exton, John Howard
Fain, John Nicholas
Farkas, Walter Robert
Faulkner, Willard Riley
Fink, Robert David
Fleischer, Becca Catherine
Fleischer, Sidney
Francis, Sharron H
Fried, Victor A
Friedman, Daniel Lester
Fujimura, Robert
Garbers, David Lorn
Gates, Ronald Eugene
Geller, Arthur Michael
Gettins, Peter
Gillespie, Elizabeth
Godfrey, Paul Russell
Gotterer, Gerald S
Goulding, Charles Edwin, Jr
Granner, Daryl Kitley
Guengerich, Frederick Peter
Guyer, Cheryl Ann
Hardman, Joel G
Harshman, Sidney
Hartman, Frederick Cooper
Hash, John H
Hayes, Robert M
Heimberg, Murray
Henke, Randolph Ray
High, Edward Garfield
Hilker, Doris M
Hill, George Carver
Hill, Robert James
Hingerty, Brian Edward
Howell, Elizabeth E
Huang, Leaf
Inagami, Tadashi
Inman, Franklin Pope, Jr
Irving, Charles Clayton
Jacobson, Karl Bruce
James, Jesse
Jefferson, William Emmett, Jr
Jennings, Lisa Helen Kyle
Jones, Peter D
Joshi, Jayant Gopal
Katze, Jon R
Kenney, Francis T
Kitabchi, Abbas E
Kitchen, Hyram
Kono, Tetsuro
Kotb, Malak Y
Kraus, Lorraine Marquardt

Kretchmar, Arthur Lockwood
Kuiken, Kenneth (Alfred)
Laboda, Henry M
Landon, Erwin Jacob
Larimer, Frank William
Lee, Kai-Lin
Lee, Ten Ching
Lerner, Joseph
Lin, Kuang-Tzu Davis
Lyman, Beverly Ann
McCarthy, John F
McCoy, Sue
McDonald, Ted Painter
Mani, I
Mann, George Vernon
Marchok, Ann Catherine
Marnett, Lawrence Joseph
Martindale, William Earl
Maxwell, Richard Elmore
Mayberry, William Roy
Mayer, Steven Edward
Meeks, Robert G
Mithcell, William Marvin
Mitra, Sankar
Montie, Thomas C
Monty, Kenneth James
Morrell, George
Morrison, Martin
Moses, Henry A
Nair, C(hellappan) Rajagopalan
Niyogi, Salil Kumar
O'Connor, Timothy Edmond
Ohi, Seigo
Olins, Ada Levy
Olins, Donald Edward
Ong, David Eugene
Osheroff, Neil
Pabst, Michael John
Pal, Bimal Chandra
Papas, Andreas Michael
Park, Charles Rawlinson
Park, Jane Harting
Puett, J David
Reed, Peter William
Regen, David Marvin
Rogers, Beverly Jane
Sachan, Dileep Singh
Savage, Dwayne Cecil
Savage, Jane Ramsdell
Schwarz, Otto John
Senogles, Susan Elizabeth
Seyer, Jerome Michael
Shugart, Lee Raleigh
Sloane, Nathan Howard
Smith, John Thurmond
Snapper, James Robert
Snyder, Fred Leonard
Staros, James Vaughan
Stevens, Audrey L
Stevens, Stanley Edward, Jr
Stone, William Lawrence
Stulberg, Melvin Philip
Thakar, Jay H
Thomas, Edwin Lee
Thomason, Donald Brent
Totter, John Randolph
Touster, Oscar
Trupin, Joel Sunrise
Uziel, Mayo
Volkin, Elliot
Vroman, Hugh Egmont
Wagner, Conrad
Warnock, Laken Guinn
Waters, Larry Charles
Watterson, D Martin
Wicks, Wesley Doane
Wigler, Paul William
Wilcox, Henry G
Womack, Frances C
Wood, John Lewis
Woychik, Richard P
Yang, Wen-Kuang
Yarbro, Claude Lee, Jr
Zee, Paulus

TEXAS
Abell, Creed Wills
Adlakha, Ramesh Chander
Aggarwal, Bharat Bhushan
Allen, Julius Cadden
Anderson, John Arthur
Anderson, Robert E
Andia-Waltenbaugh, Ana Maria
Angelides, Kimon Jerry
Ansari, Guhlam Ahmad Shakeel
Arenaz, Pablo
Arlinghaus, Ralph B
Asimakis, Gregory K
Atassi, Zouhair
Atkinson, Mark Arthur Leonard
Awapara, Jorge
Awasthi, Yogesh C
Babitch, Joseph Aaron
Baldwin, Thomas Oakley
Baptist, James (Noel)
Barker, Kenneth Leroy
Barnes, Eugene Miller, Jr
Barnes, Larry D
Barnett, Don(ald) R(ay)
Barr, Charles Richard
Bartel, Allen Hawley
Beaudet, Arthur L
Beckingham, Kathleen Mary
Becvar, James Edgar

Beerstecher, Ernest, Jr
Behal, Francis Joseph
Bellion, Edward
Benedict, C R
Benedict, Chauncey
Bertrand, Helen Anne
Birnbaumer, Lutz
Boltralik, John Joseph
Bottino, Nestor Rodolfo
Bowles, William Howard
Brady, Scott T
Brand, Jerry Jay
Brattain, Michael Gene
Brewer, Franklin Douglas
Bryan, Joseph
Brysk, Miriam Mason
Buchanan, Christine Elizabeth
Burks, James Kenneth
Busch, Harris
Butcher, Reginald William
Butow, Ronald A
Camp, Bennie Joe
Campbell, James Wayne
Campbell, William Jackson
Caprioli, Richard Michael
Carney, Darrell Howard
Carr, Bruce R
Carson, Daniel Douglas
Chan, Lawrence Chin Bong
Chan, Lee-Nien Lillian
Chan, Pui-Kwong
Chang, Ching Hsong
Chinn, Herman Isaac
Chiou, George Chung-Yih
Clark, Dale Allen
Cohen, Allen Barry
Cottam, Gene Larry
Couch, James Russell
Crass, Maurice Frederick, III
Creger, Clarence R
Criscuolo, Dominic
Dahm, Karl Heinz
Davis, Alvie Lee
Davis, Claude Geoffrey
Davis, Virginia Eischen
Dawson, Earl B
Dedman, John Raymond
Deisenhofer, Johann
Deloach, John Rooker
DeMoss, John A
Denney, Richard Max
Dennis, Joe
DeShazo, Mary Lynn Davison
Dieckert, Julius Walter
Di Ferrante, Daniela Tavella
Di Ferrante, Nicola Mario
Dobson, Harold Lawrence
Doctor, Vasant Manilal
Doebbler, Gerald Francis
Douglas, Tommy Charles
Dresden, Marc Henri
Dubbs, Del Rose M
Ducis, Ilze
Dunn, Floyd Warren
Dusenbery, Ruth Lillian
Dzidic, Ismet
Earhart, Charles Franklin, Jr
Edmundson, Allen B
Eichberg, Joseph
Eidels, Leon
Entman, Mark Lawrence
Eppright, Margaret
Estabrook, Ronald (Winfield)
Evans, Claudia T
Everse, Johannes
Fair, Daryl S
Fischer, Susan Marie
Fleischmann, William Robert, Jr
Foster, Donald Myers
Fox, Jack Lawrence
Franzl, Robert E
Frenkel, Rene A
Friend, Patric Lee
Furlong, Norman Burr, Jr
Gan, Jose Cajilig
Ganther, Howard Edward
Garber, Alan J
Garcia, Hector D
Gatlin, Delbert Monroe, III
Geoghegan, William David
Georgiou, George
Gilbert, Brian E
Goldschmidt, Millicent
Goldstein, Joseph Leonard
Goodman, Joel Mitchell
Gordon, Wayne Lecky
Gorelic, Lester Sylvan
Gottlieb, Paul David
Gotto, Antonio Marion, Jr
Gracy, Robert Wayne
Griffin, James Emmett
Grimes, L Nichols
Grinnell, Frederick
Grundy, Scott Montgomery
Guarino, Armand John
Guentzel, M Neal
Guidry, Marion Antoine
Guirard, Beverly Marie
Gunn, John Martyn
Guo, Yan-Shi
Guynn, Robert William
Haber, Bernard
Hall, Frank Foy
Hall, Timothy Couzens

Biochemistry (cont)

Hanahan, Donald James
Hardcastle, James Edward
Hardesty, Boyd A
Harding, Winfred Mood
Harper, Michael John Kennedy
Harris, Ben Gerald
Harris, Edward David
Harris, Stephen Eubank
Hedges, Dorothea Huseby
Henning, Susan June
Hentges, David John
Hersh, Louis Barry
Hewett'Emmett, David
Hewitt, Roger R
Hillar, Marian
Hillis, David Mark
Hodgins, Daniel Stephen
Hogan, Michael Edward
Holoubek, Viktor
Horowitz, Paul Martin
Houston, Forrest Gish
Howard, Harriette Ella Pierce
Hsueh, Andie M
Huang, Bessie Pei-Hsi
Huang, Charles T L
Hurlbert, Robert Boston
Ippen-Ihler, Karin Ann
Irvin, James Duard
Jacobson, Elaine Louise
Jacobson, Myron Kenneth
James, Harold Lee
Johnson, Kenneth Maurice, Jr
Johnston, John Marshall
Jorgensen, George Norman
Joshi, Vasudev Chhotalal
Jurtshuk, Peter, Jr
Kaman, Robert Lawrence
Kasschau, Margaret Ramsey
Kaufmann, Anthony J
Kelley, Jim Lee
Kiel, Johnathan Lloyd
Kimball, Aubrey Pierce
King, Richard Joe
Kit, Saul
Kitto, George Barrie
Klebe, Robert John
Koenig, Virgil Leroy
Koeppe, David Edward
Konkel, David Anthony
Kramer, Gisela A
Krieg, Daniel R
Kubena, Leon Franklin
Kuramitsu, Howard Kikuo
Kurosky, Alexander
Lacko, Andras Gyorgy
Lagowski, Jeanne Mund
Lam, Kwok-Wai
Landmann, Wendell August
Langdon, Robert Godwin
Lansford, Edwin Myers, Jr
Laseter, John Luther
Lawrence, Addison Lee
Lawrence, Richard Aubrey
Lazzari, Eugene Paul
Lee, James C
Lee, John Chung
Lever, Julia Elizabeth
Lewis, Donald Everett
Lewis, Douglas Scott
Lingle, Sarah Elizabeth
Little, Gwynne H
Longenecker, John Bender
Lorenzetti, Ole J
Lospalluto, Joseph John
Lotan, Reuben
Ludueña, Richard Froilan
McCarthy, John Lawrence, Jr
McCord, Tommy Joe
McDonald, George Gordon
Mace, Kenneth Dean
McGarry, John Denis
MacLeod, Michael Christopher
McMillin, Jeanie
Magill, Jane Mary (Oakes)
Margolin, Solomon B
Masaracchia, Ruthann A
Mason, James Ian
Mason, Morton Freeman
Masoro, Edward Joseph
Massey, John Boyd
Masters, Bettie Sue Siler
Matthews, Kathleen Shive
Means, Anthony R
Medina, Miguel Angel
Meistrich, Marvin Lawrence
Mendel, Julius Louis
Mendelson, Carole Ruth
Mersmann, Harry John
Meyer, Franz
Michael, Lloyd Hal
Middleditch, Brian Stanley
Milewich, Leon
Miller, Edward Godfrey, Jr
Miller, Joyce Mary
Miller, Sanford Arthur
Mills, Gordon Candee
Mills, William Ronald
Milner, Alice N
Miner, James William
Mize, Charles Edward
Modak, Arvind T
Moore, Erin Colleen

Mooz, Elizabeth Dodd
Morgan, Page Wesley
Morrisett, Joel David
Mukherjee, Amal
Murgola, Emanuel J
Nakajima, Motowo
Nall, Barry T
Nelson, Thomas Evar
Newman, Robert Alwin
Nicholson, Wayne Lowell
Nishimura, Jonathan Sei
Noall, Matthew Wilcox
Nordyke, Ellis Larrimore
Norton, Scott J
O'Donovan, Gerard Anthony
Olson, Merle Stratte
Oro, Juan
Overturf, Merrill L
Palmer, Graham
Papaconstantinou, John
Patel, Nutankumar T
Patsch, Wolfgang
Perkins, John Phillip
Peterson, David Oscar
Peterson, Julian Arnold
Pettit, Flora Hunter
Pirtle, Robert M
Platsoucas, Chris Dimitrios
Plunkett, William Kingsbury
Poduslo, Shirley Ellen
Poenie, Martin Francis
Poffenbarger, Phillip Lynn
Poulsen, Lawrence Leroy
Powell, Bernard Lawrence
Pownall, Henry Joseph
Prager, Morton David
Prescott, John Mack
Purdy, Robert H
Quiocho, Florante A
Rainwater, David Luther
Randerath, Kurt
Rassin, David Keith
Raushel, Frank Michael
Ravel, Joanne Macow
Reagor, John Charles
Reber, Elwood Frank
Reed, Lester James
Reiser, Raymond
Ridgway, Helen Jane
Riehl, Robert Michael
Riggs, Austen Fox, II
Riser, Mary Elizabeth
Ritter, Nadine Marie
Ritzi, Earl Michael
Roberts, Susan Jean
Robinson, Neal Clark
Ro-Choi, Tae Suk
Roels, Oswald A
Rogers, Gary Allen
Roller, Herbert Alfred
Root, Elizabeth Jean
Rose, Kathleen Mary
Rosen, Jeffrey Mark
Ross, Doris Laune
Ross, Elliott M
Roux, Stanley Joseph
Rudolph, Frederick Byron
Sahasrabuddhe, Chintaman Gopal
Sallee, Verney Lee
Salomon, Lothar L
Sampson, Herschel Wayne
Samson, Willis Kendrick
Sato, Clifford Shinichi
Schaffer, Barbara Noyes
Schlamowitz, Max
Schneider, Dennis Ray
Schonbrunn, Agnes
Schram, Alfred C
Schroeder, Hartmut Richard
Schroepfer, George John, Jr
Scouten, William Henry
Seifert, William Edgar, Jr
Serwer, Philip
Shain, Sydney A
Shapiro, David M
Shaw, Emil Gilbert
Shaw, Robert Wayne
Sherrill, Bette Cecile Benham
Shetlar, Marvin Roy
Shive, William
Shum, Wan-Kyng Liu
Siciliano, Michael J
Siler-Khodr, Theresa M
Simpson, Evan Rutherford
Simpson, John Wayne
Sirbasku, David Andrew
Skelley, Dean Sutherland
Skinner, Charles Gordon
Skow, Loren Curtis
Smith, Edward Russell
Smith, Leland Leroy
Smith, Louis C
Smith, Michael Lew
Smith, Russell Lamar
Snell, Esmond Emerson
Snell, William J
Snyder, Gary Dean
Sordahl, Louis A
Spallholz, Julian Ernest
Sparkman, Dennis Raymond
Srere, Paul Arnold
Sridhara, S
Srivastava, Satish Kumar
Stancel, George Michael

Starcher, Barry Chapin
Starr, Jason Leonard
Stephens, Robert Lawrence
Stewart, James Ray
Stith, William Joseph
Stone, William Harold
Stoops, James King
Stouffer, John Emerson
Strinden, Sarah Taylor
Stroman, David Womack
Stull, James Travis
Sutton, Harry Eldon
Taegtmeyer, Heinrich
Taurog, Alvin
Taylor, Alan Neil
Taylor, Robert Dalton
Tcholakian, Robert Kevork
Temeyer, Kevin Bruce
Thompson, Edward Ivins Bradbridge
Thompson, Ernest Aubrey, Jr
Thompson, Guy A, Jr
Tong, Alex W
Towne, Jack C
Tsai, Ming-Jer
Tsin, Andrew Tsang Cheung
Tsutsui, Ethel Ashworth
Tu, Shiao-chun
Uyeda, Kosaku
Vanderzant, Erma Schumacher
Van Dreal, Paul Arthur
Van Eys, Jan
Vela, Gerard Roland
Vogel, James John
Vonder Haar, Raymond A
Wagner, Martin James
Wakil, Salih J
Walborg, Earl Fredrick, Jr
Walker, James Benjamin
Walter, Charles Frank
Wang, Kuan
Wang, Yeu-Ming Alexander
Ward, Darrell N
Washington, Arthur Clover
Webb, Bill D
Weigel, Paul H(enry)
Wentland, Stephen Henry
Werbin, Harold
Wheeler, Michael Hugh
Whelly, Sandra Marie
Whitenberg, David Calvin
Whittle, John Antony
Wiggans, Donald Sherman
Wilkes, Stella H
Williams, Charles Herbert
Williams, Ralph Edward
Williams, William Thomas
Willms, Charles Ronald
Wilson, John H
Wilson, Richard Hansel
Wohlman, Alan
Wold, Finn
Wolf, Don Paul
Wolinsky, Ira
Wood, Randall Dudley
Wood, Robert Charles
Worthen, Howard George
Wright, Woodring Erik
Wu, Ming-Chi
Yee, John Alan
Yeoman, Lynn Chalmers
Yorio, Thomas
Younes, Muazza A
Young, Ryland F
Ziegler, Daniel
Ziegler, Miriam Mary

UTAH

Aird, Steven Douglas
Allred, Keith Reid
Andersen, William Ralph
Anderson, Anne Joyce
Ash, Kenneth Owen
Aust, Steven Douglas
Beck, Jay Vern
Bennett, Jesse Harland
Boeker, Elizabeth Anne
Bradshaw, William S
Bryson, Melvin Joseph
Burnham, Bruce Franklin
Burton, Sheril Dale
Caldwell, Karin D
Casjens, Sherwood Reid
Choules, George Lew
Chuang, Hanson Yii-Kuan
Clark, C Elmer
Dickman, Sherman Russell
Ellis, LeGrande Clark
Ely, Kathryn R
Emery, Thomas Fred
Evans, Robert John
Franklin, Michael R(oger)
Franklin, Naomi C
Galster, William Allen
Gortatowski, Melvin Jerome
Gubler, Clark Johnson
Henderson, Lavell Merl
Heninger, Richard Wilford
Herrick, Glenn Arthur
Hill, John Mayes, Jr
Hutchings, Brian Lamar
Janatova, Jarmila
Johnson, LaVell R
Johnson, Ralph M, Jr
Klein, Sigrid Marta

Krieger, Carl Henry
Kuby, Stephen A
Kuehl, LeRoy Robert
Lancaster, Jack R, Jr
Lawson, Larry Dale
Linker, Alfred
Locy, Robert Donald
Mangelson, Farrin Leon
Mangum, John Harvey
Miller, Gene Walker
Mohammad, Syed Fazal
Nelson, Don Harry
Nielson, Eldon Denzel
Petryka, Zbyslaw Jan
Ramachandran, Chittoor Krishna
Rasmussen, Kathleen Goertz
Richards, Oliver Christopher
Rilling, Hans Christopher
Schweizer, Martin Paul
Seeley, Schuyler Drannan
Simmons, Daniel L
Simmons, John Robert
Sipe, David Michael
Stokes, Barry Owen
Straight, Richard Coleman
Sweat, Floyd Walter
Swensen, Albert Donald
Takemoto, Jon Yutaka
Torres, Anthony R
Vassel, Bruno
Velick, Sidney Frederick
Vernon, Leo Preston
Weber, Darrell J
West, Charles Donald
Williams, Roger Richard
Winder, William W
Winge, Dennis R
Wright, Donald N
Yoshikami, Doju

VERMONT

Ariano, Marjorie Ann
Carew, Lyndon Belmont, Jr
Chiu, Jen-Fu
Clemmons, Jackson Joshua Walter
Currier, William Wesley
Foote, Murray Wilbur
Hall, Ross Hume
Hartnett, John (Conrad)
Hayes, John William
Kelleher, Philip Conboy
Kelley, Jason
Kilpatrick, Charles William
Lamden, Merton Philip
Long, George Louis
Mann, Kenneth Gerard
Melville, Donald Burton
Meyer, William Laros
Otter, Timothy
Racusen, David
Rittenhouse, Susan E
Sims, Ethan Allen Hitchcock
Sinclair, Peter Robert
Sjogren, Robert Erik
Thanassi, John Walter
Tyzbir, Robert S
Wallace, Susan Scholes
Watters, Christopher Deffner
Weed, Lawrence Leonard
Weller, David Lloyd
Woodcock-Mitchell, Janet Louise
Woodworth, Robert Cummings

VIRGINIA

Abbott, Lynn De Forrest, Jr
Alphin, Reevis Stancil
Alscher, Ruth
Amero, Sally Ann
Anderson, Bruce Murray
Anderson, Carl William (Bill)
Appleton, Martin David
Baldwin, Robert Russel
Banks, William Louis, Jr
Bender, Patrick Kevin
Benzinger, Rolf Hans
Boatman, Sandra
Bonar, Robert Addison
Bond, Judith
Bradbeer, Clive
Brandt, Richard Bernard
Brockman, Robert W
Brown, Barry Lee
Brown, Elise Ann Brandenburger
Brown, Jay Clark
Brown, Loretta Ann Port
Bruns, David Eugene
Bunce, George Edwin
Calvert, Richard John
Caponio, Joseph Francis
Carson, Eugene Watson, Jr
Cavender, Finis Lynn
Chen, Jiann-Shin
Chin, Byong Han
Chlebowski, Jan F
Claus, George William
Cohen, I Kelman
Collins, James Malcolm
Contreras, Thomas Jose
Coursen, Bradner Wood
Creutz, Carl Eugene
Dalton, Harry P
Davison, John Philip
Deemer, Marie Nieft
De Long, Chester Wallace

DeLorenzo, Robert John
Dementi, Brian Armstead
DeVries, George H
Dewey, William Leo
Diegelmann, Robert Frederick
Dunn, John Thornton
Ebel, Richard E
Eden, Francine Claire
Elford, Howard Lee
Engel, Ruben William
Evans, Herbert John
Fager, Lei Yen
Feher, Joseph John
Fishbein, Lawrence
Franson, Richard Carl
Fugate, Kearby Joe
Garrett, Reginald Hooker
Garrison, Norman Eugene
Gaugler, Robert Walter
Gear, Adrian R L
Gilles, Kenneth Albert
Gould, David Huntington
Graff, Gustav
Gregory, Eugene Michael
Grider, John Raymond
Grisham, Charles Milton
Grogan, William McLean
Gruemer, Hanns-Dieter
Hager, Chester Bradley
Hanig, Joseph Peter
Hatzios, Kriton Kleanthis
Hayes, Dora Kruse
Hecht, Sidney Michael
Hess, John Lloyd
Higgins, Edwin Stanley
Hilu, Khidir Wanni
Holloway, Peter William
Hotta, Shoichi Steven
Huang, Ching-hsien
Huang, Laura Chi
Hutcheson, Eldridge Tilmon, III
Jennings, Allen Lee
Kadner, Robert Joseph
Kalimi, Mohammed Yahya
Kelly, Robert Frank
Kimbrough, Theo Daniel, Jr
King, Betty Louise
Kline, Edward Samuel
Kupke, Donald Walter
Kutchai, Howard C
Larner, Joseph
Liberti, Joseph Pollara
Litman, Burton Joseph
McGilvery, Robert Warren
MacLeod, Robert Meredith
Maggio, Bruno
Mangan, George Francis, Jr
Mavis, Richard David
Merlino, Glenn T
Misra, Hara Prasad
Moore, Cyril L
Narasimhachari, Nedathur
Newman, Jack Huff
Niehaus, Walter G, Jr
Nolin, Janet M
Ogilvie, James William, Jr
Olson, Lee Charles
Olson, William Arthur
O'Neal, Charles Harold
Owens, Kenneth
Pallansch, Michael J
Pinkston, Margaret Fountain
Place, Janet Dobbins
Polan, Carl E
Price, Byron Frederick
Price, Steven
Raizen, Carol Eileen
Rickett, Frederic Lawrence
Roberts, Catherine Harrison
Rodricks, Joseph Victor
Rogers, Kenneth Scipio
Rosenthal, Miriam Dick
Rosinski, Joanne
Rothberg, Simon
Rowe, Mark J
Rutherford, Charles
Rutter, Henry Alouis, Jr
Sando, Julianne J
Saunders, Joseph Francis
Schellenberg, Karl A
Schirch, Laverne Gene
Shelton, Keith Ray
Sitz, Thomas O
Sobel, Robert Edward
Spangenberg, Dorothy Breslin
Spearing, Cecilia W
Stansly, Philip Gerald
Stepka, William
Stewart, Kent Kallam
Storrie, Brian
Stout, Ernest Ray
Swell, Leon
Swope, Fred C
Teekell, Roger Alton
Thompson, Thomas Edward
Topham, Richard Walton
Treadwell, Carleton Raymond
Treadwell, George Edward, Jr
Trelawny, Gilbert Sterling
Van Tuyle, Glenn Charles
Venditti, John M
Vermeulen, Carl William
Villar-Palasi, Carlos
Voige, William Huntley

Walker, Charles R
White, Howard Dwaine
Wilkinson, Christopher Foster
Willett, James Delos
Woo, Yin-tak
Yip, George
Young, Nelson Forsaith
Young, William W, Jr
Yu, Robert Kuan-Jen

WASHINGTON
Adman, Elinor Thomson
Adman, Raymond Lance
Ammann, Harriet Maria
Andersen, Niels Hjorth
Anderson, Larry Ernest
Aronoff, Samuel
Ash, Roy Phillip
Avner, Ellis David
Bankson, Daniel Duke
Barron, Edward J
Barrueto, Richard Benigno
Bendich, Arnold Jay
Blake, James J
Bohnert, Janice Lee
Bornstein, Paul
Branca, Andrew A
Brekke, Clark Joseph
Brosemer, Ronald Webster
Brown, George Willard, Jr
Butler, Lillian Ida
Capp, Grayson L
Carpenter, Carolyn Virus
Carter, William G
Cataldo, Dominic Anthony
Cheevers, William Phillip
Chen, Shi-Han
Christensen, Gerald M
Churchill, Lynn
Clapshaw, Patric Arnold
Crampton, George H
Croteau, Rodney
Dailey, Frank Allen
Dale, Beverly A
Davie, Earl W
Felton, Samuel Page
Finkelstein, David B
Fischer, Edmond H
Fletcher, Dean Charles
Floss, Heinz G
Forrey, Arden W
Foster, Robert Joe
Fournier, R E Keith
Fraenkel-Conrat, Jane E
Frankart, William A
Gaines, Robert D
Glomset, John A
Goheen, Steven Charles
Gordon, Milton Paul
Griswold, Michael David
Hakomori, Sen-Itiroh
Hall, Benjamin Downs
Hall, Stanton Harris
Harding, Joseph Warren, Jr
Hauschka, Stephen D
Hawkins, Richard L
Hazelbauer, Gerald Lee
Herriott, Jon R
Ho, Lydia Su-yong
Horbett, Thomas Alan
Hu, Shiu-Lok
Kachmar, John Frederick
Keller, Patricia J
Kelly, Jeffrey John
Ketchie, Delmer O
Klebanoff, Seymour J
Kromann, Rodney P
Kutsky, Roman Joseph
Labbe, Robert Ferdinand
Lawrence, John McCune
Leid, R Wes
Leung, Frederick C
Lightfoot, Donald Richard
Loeb, Lawrence Arthur
Loewus, Frank A
Loewus, Mary W
Lygre, David Gerald
McAnally, John Sackett
McDonough, Leslie Marvin
McFadden, Bruce Alden
McMahon, Daniel Stanton
Magnuson, James Andrew
Magnuson, Nancy Susanne
Mahlum, Daniel Dennis
Mahoney, Walter C
Malins, Donald Clive
Maloff, Bruce L(arrie)
Margolis, Robert Lewis
Mirkes, Philip Edmund
Monsen, Elaine R
Morris, David Robert
Narayanan, A Sampath
Nardella, Francis Anthony
Neurath, Hans
Neve, Richard Anthony
Noyes, Claudia Margaret
Owen, Stanley Paul
Pall, Martin L
Paquette, Thomas Leroy
Parson, William Wood
Patt, Leonard Merton
Pomeranz, Yeshajahu
Powers, Joseph Robert
Prody, Gerry Ann

Randall, Linda Lea
Reeves, Raymond
Reh, Thomas Andrew
Reid, Brian Robert
Richmond, Virginia
Riederer-Henderson, Mary Ann
Ritter, Preston Peck Otto
Rohrschneider, Larry Ray
Ronzio, Robert A
Rosen, Henry
Sage, Helene E
Schnell, Jerome Vincent
Schrader, Lawrence Edwin
Shepherd, Linda Jean
Shiota, Tetsuo
Shultz, Terry D
Sieker, Larry Charles
Smerdon, Michael John
Smith, Elizabeth Knapp
Smith, Sam Corry
Spackman, Darrel H
Spence, Kemet Dean
Stahl, William Louis
Stenkamp, Ronald Eugene
Stevens, Vernon Lewis
Stevens, Vincent Leroy
Storm, Daniel Ralph
Su, Judy Ya-Hwa Lin
Thomas, John M
Thompson, Roy Charles, Jr
Uribe, Ernest Gilbert
Utter, Fred Madison
Varanasi, Usha
Vaughan, Burton Eugene
Walsh, Kenneth Andrew
Wedemeyer, Gary Alvin
White, Fredric Paul
Widmaier, Robert George
Wiley, William Rodney
Wyrick, Ronald Earl
Yancey, Paul Herbert
Yount, Ralph Granville
Yu, Ming-Ho
Zaugg, Waldo S

WEST VIRGINIA
Beattie, Diana Scott
Blaydes, David Fairchild
Brooks, James Lee
Butcher, Fred Ray
Calhoon, Donald Alan
Campbell, Clyde Del
Canady, William James
Capstack, Ernest
Diehl, John Edwin
Foster, Joyce Geraldine
Guyer, Kenneth Eugene, Jr
Hansard, Samuel L, II
Harris, Charles Lawrence
Jagannathan, Singanallur N
Kaczmarczyk, Walter J
Kletzien, Rolf Frederick
Konat, Gregory W
Krause, Reginald Frederick
Malin, Howard Gerald
Markiw, Roman Teodor
Martin, William Gilbert
Mashburn, Louise Tull
Mashburn, Thompson Arthur, Jr
Mecca, Christyna Emma
Moffa, David Joseph
Ong, Tong-man
Quinlan, Dennis Charles
Rafter, Gale William
Reasor, Mark Jae
Reichenbecher, Vernon Edgar, Jr
Roberts, Joseph Linton
Tryfiates, George P
Van Dyke, Knox
Williams, Leah Ann
Wirtz, George H

WISCONSIN
Adler, Julius
Amasino, Richard M
Anand, Amarjit Singh
Anderson, Laurens
Ankel, Helmut K
Attie, Alan D
Balish, Edward
Bavisotto, Vincent
Becker, Wayne Marvin
Beinert, Helmut
Bekersky, Ihor
Benevenga, Norlin Jay
Bergdoll, Merlin Scott
Bergtrom, Gerald
Bernstein, Bradley Alan
Boutwell, Roswell Knight
Brill, Winston J
Broderick, Glen Allen
Brown, Raymond Russell
Brown, William Henry
Buchanan-Davidson, Dorothy Jean
Burger, Warren Clark
Burgess, Ann Baker
Burgess, Richard Ray
Burris, Robert Harza
Calbert, Harold Edward
Calvanico, Nickolas Joseph
Campbell, Harold Alexander
Cassens, Robert G
Chakraburtty, Kalpana
Chen, Chong Maw

Chen, Franklin M
Chen, Shao Lin
Cherayil, George Devassia
Cleland, William Wallace
Cohen, Philip Pacy
Colás, Antonio E
Collins, Mary Lynne Perille
Courtright, James Ben
Cox, Michael Matthew
Dahlberg, James Eric
Deese, Dawson Charles
DeLuca, Hector Floyd
Derse, Phillip H
Dimond, Randall Lloyd
Dodson, Vernon N
Ehle, Fred Robert
Elliott, Bernard Burton
Fahien, Leonard A
Fahl, William Edwin
Feinberg, Benjamin Allen
Fischbach, Fritz Albert
Fredricks, Walter William
Frey, Perry Allen
Ghazarian, Jacob G
Gilboe, David Dougherty
Girotti, Albert William
Gleiter, Melvin Earl
Goldberg, Burton David
Goldberger, Amy
Goodfriend, Theodore L
Gorsica, Henry Jan
Gorski, Jack
Greaser, Marion Lewis
Greenspan, Daniel S
Haas, Michael John
Hager, Steven Ralph
Harkin, John McLay
Harkness, Donald R
Harper, Alfred Edwin
Heath, Timothy Douglas
Hoekstra, William George
Hokin, Lowell Edward
Hornemann, Ulfert
Horseman, Nelson Douglas
Hosler, Charles Frederick, Jr
Huss, Ronald John
Hussa, Robert Oscar
Jacobson, Gunnard Kenneth
Joel, Cliffe David
John, Kavanakuvhiy V
Johnson, Morris Alfred
Jones, Berne Lee
Kaesberg, Paul Joseph
Kahan, Lawrence
Kaltenbach, John Paul
Karavolas, Harry J
Kasper, Charles Boyer
Keegstra, Kenneth G
Kitchell, James Frederick
Klaassen, Dwight Homer
Klebba, Phillip E
Kochan, Robert George
Kornguth, Steven E
Kramer, Elizabeth
Kushnaryov, Vladimir Michael
Lai, Ching-San
Lardy, Henry Arnold
Lemann, Jacob, Jr
Littlewood, Barbara Shaffer
Ludden, Paul W
MacDonald, Michael J
McFarland, James Thomas
McIntosh, Elaine Nelson
Mahl, Mearl Carl
Martin, Thomas Fabian John
Mertz, Janet Elaine
Metzenberg, Robert Lee
Miller, James Alexander
Miziorko, Henry Michael
Mosesson, Michael W
Mosher, Deane Fremont, Jr
Moskowitz, Gerard Jay
Mueller, Gerald Conrad
Munroe, Stephen Horner
Myers, Charles R
Nagodawithana, Tilak Walter
Nelson, David Lee
Oaks, John Adams
Oberley, Terry De Wayne
Olson, Earl Burdette, Jr
Ordman, Alfred Bram
Pan, David
Pariza, Michael Willard
Perlman, Richard
Perry, Billy Wayne
Pitot, Henry C, III
Poland, Alan P
Porter, John Willard
Potter, Van Rensselaer
Pscheidt, Gordon Robert
Qureshi, Nilofer
Raines, Ronald T
Rao, Ghanta Nageswara
Reed, George Henry
Ritter, Karla Schwensen
Sasse, Edward Alexander
Schantz, Edward Joseph
Scheusner, Dale Lee
Schnoes, Heinrich Konstantin
Schulz, Leslie Olmstead
Sealy, Roger Clive
Selman, Bruce R
Shrago, Earl
Siegel, Frank Leonard

Biochemistry (cont)

Siegel, Jack Morton
Slautterback, David Buell
Smith, Milton Reynolds
Steele, Robert Darryl
Stone, William Ellis
Stratman, Frederick William
Sussman, Michael R
Suttie, John Weston
Swick, Robert Winfield
Sytsma, Kenneth Jay
Takayama, Kuni
Taketa, Fumito
Tewksbury, Duane Allan
Tews, Jean Kring
Tormey, Douglass Cole
Trautman, Jack Carl
Tsao, Francis Hsiang-Chian
Turkington, Roger W
Twining, Sally Shinew
Unsworth, Brian Russell
Verma, Ajit K
Wall, Joseph Sennen
Weaver, Robert Hinchman
Wege, Ann Christene
Weil, Michael Ray
Wejksnora, Peter James
Wells, Barbara Duryea
Wenzel, Frederick J
Wilken, David Richard
Williams, Jeffrey Walter
Witt, Patricia L
Woldegiorgis, Gebretateos

WYOMING
Asplund, Russell Owen
Bosshardt, David Kirn
Bulla, Lee Austin, Jr
Caldwell, Daniel R
George, Robert Porter
Isaak, Dale Darwin
Ji, Inhae
Ji, Tae H(wa)
Kaiser, Ivan Irvin
Lewis, Randolph Vance
McColloch, Robert James
Nunamaker, Richard Allan
Petersen, Nancy Sue
Sullivan, Brian Patrick
Villemez, Clarence Louis, Jr
Walton, Thomas Edward

PUERTO RICO
Banerjee, Dipak Kumar
El-Khatib, Shukri M
Sandza, Joseph Gerard
Toro-Goyco, Efrain

ALBERTA
Baker, Glen Bryan
Basu, Tapan Kumar
Bleackley, Robert Christopher
Bridger, William Aitken
Campbell, James Nicoll
Cass, Carol E
Clandinin, Michael Thomas
Collier, Herbert Bruce
Cossins, Edwin Albert
Davis, Norman Rodger
Denford, Keith Eugene
Dixon, Gordon H
Eggert, Frank Michael
Francis, Mike McD
Gaucher, George Maurice
Gooding, Ronald Harry
Goren, Howard Joseph
Hart, David Arthur
Henderson, Joseph Franklin
Hollenberg, Morley Donald
Huber, Rueben Eugene
Iatrou, Kostas
Jacobson, Ada Leah
Jensen, Susan Elaine
Kadis, Vincent William
Kaneda, Toshi
Kaplan, Jacob Gordin
Kay, Cyril Max
Kuntz, Garland Parke Paul
McElhaney, Ronald Nelson
Madiyalakan, Ragupathy
Madsen, Neil Bernard
Meintzer, Roger Bruce
Nash, David
Paetkau, Verner Henry
Paranchych, William
Paterson, Alan Robb Phillips
Rattner, Jerome Bernard
Roberts, David Wilfred Alan
Rorstad, Otto Peder
Russell, James Christopher
Schneider, Wolfgang Johann
Schroder, David John
Scott, Paul G
Singh, Bhagirath
Smillie, Lawrence Bruce
Sohal, Parmjit S
Spencer, Mary Stapleton
Stevenson, Kenneth James
Stinson, Robert Anthony
Suresh, Mavanur Rangarajan
Tamaoki, Taiki
Vance, Dennis E
Van De Sande, Johan Hubert

Walsh, Michael Patrick
Wang, Jerry Hsueh-Ching
Watanabe, Mamoru
Weiner, Joel Hirsch
Zalik, Saul

BRITISH COLUMBIA
Aebersold, Ruedi H
Autor, Anne Pomeroy
Bragg, Philip Dell
Buckley, James Thomas
Burton, Albert Frederick
Bushnell, Gordon William
Butler, Gordon Cecil
Candido, Edward Peter Mario
Cullis, Pieter Rutter
Dennis, Patrick P
Desai, Indrajit Dayalji
Dorchester, John Edmund Carleton
Emerman, Joanne Tannis
Finlay, Barton Brett
Friz, Carl T
Frohlich, Jiri J
Garg, Arun K
Godolphin, William
Green, Beverley R
Hancock, Robert Ernest William
Haunerland, Norbert Heinrich
Hochachka, Peter William
Honda, Barry Marvin
Kasinsky, Harold Edward
Lee, Melvin
Li-Chan, Eunice
McBride, Barry Clarke
Majak, Walter
Matheson, Alastair Taylor
Mauk, Arthur Grant
Mauk, Marcia Rokus
Meheriuk, Michael
Mendoza, Celso Enriquez
Molday, Robert S
Nichols, Jack Loran
O'Donnell, Vincent Joseph
Olive, Peggy Louise
Redfield, Rosemary Jeanne
Rennie, Paul Steven
Richards, James Frederick
Richards, William Reese
Sadowski, Ivan J
Salari, Hassan
Shrimpton, Douglas Malcolm
Singh, Raj Kumari
Skala, Josef Petr
Taylor, Iain Edgar Park
Tener, Gordon Malcolm
Thurlbeck, William Michael
Towers, George Hugh Neil
Underhill, Edward Wesley
Unrau, Abraham Martin
Vincent, Steven Robert
Waygood, Ernest Roy
Weeks, Gerald
Zbarsky, Sidney Howard

MANITOBA
Biswas, Shib D
Blanchaer, Marcel Corneille
Bodnaryk, Robert Peter
Borsa, Joseph
Burton, David Norman
Choy, Patrick C
Clayton, James Wallace
Dakshinamurti, Krishnamurti
Davie, James Ronald
Hamilton, Ian Robert
Hill, Robert D
Hougen, Frithjof W
Kanfer, Julian Norman
Kovacs, Miklos I P
Kruger, James Edward
Kunz, Bernard Alexander
Lukow, Odean Michelin
Marquardt, Ronald Ralph
Matsuo, Robert R
Perkins, Harold Jackson
Pritchard, Ernest Thackeray
Rosenmann, Edward A
Singh, Harwant
Standing, Kenneth Graham
Stevens, Frits Christiaan
Suzuki, Isamu
Wrogemann, Klaus
Yamada, Esther V
Yurkowski, Michael

NEW BRUNSWICK
Bush, Roy Sidney
Cashion, Peter Joseph
Fraser, Alan Richard
Gauthier, Didier
Haya, Katsuji
McQueen, Ralph Edward

NEWFOUNDLAND
Barnsley, Eric Arthur
Brosnan, Margaret Eileen
Burness, Alfred Thomas Henry
Davidson, William Scott
Feltham, Lewellyn Allister Woodrow
Hoekman, Theodore Bernard
Idler, David Richard
Ke, Paul Jenn
Keough, Kevin Michael William
Mookerjea, Sailen

Orr, James Cameron
Senciall, Ian Robert
Shaw, Derek Humphrey

NOVA SCOTIA
Addison, Richard Frederick
Beveridge, James MacDonald
  Richardson
Bidwell, Roger Grafton Shelford
Blair, Alan Huntley
Breckenridge, Carl
Brown, Robert George
Chambers, Robert Warner
Cohen, William David
Dolphin, Peter James
Gray, Michael William
Helleiner, Christopher Walter
Holzbecher, Jiri
Kimmins, Warwick Charles
Laycock, Maurice Vivian
McFarlane, Ellen Sandra
MacRae, Herbert F
MacRae, Thomas Henry
Mezei, Catherine
Palmer, Frederick B St Clair
Savard, Francis Gerald Kenneth
Vandermeulen, John Henri
Wainwright, Stanley D
White, Thomas David

ONTARIO
Alexander, James Craig
Allan, Robert K
Ananthanarayanan, Vettaikkoru S
Annett, Robert Gordon
Anwar, Rashid Ahmad
Arif, Basil Mumtaz
Armstrong, John Briggs
Atkinson, Burr Gervais
Bacchetti, Silvia
Bag, Jnanankur
Baines, A D
Barclay, Barry James
Barran, Leslie Rohit
Baxter, Robert MacCallum
Bayley, Henry Shaw
Bayley, Stanley Thomas
Beare-Rogers, Joyce Louise
Becking, George C
Begin-Heick, Nicole
Bennick, Anders
Benoiton, Normand Leo
Bettger, William Joseph
Bewley, John Derek
Bhavnani, Bhagu R
Birmingham, Brendan Charles
Birnboim, Hyman Chaim
Blais, Burton W
Bols, Niels Christian
Brash, John Law
Brown, Stewart Anglin
Brownstone, Yehoshua Shieky
Burnison, Bryan Kent
Cairns, William Louis
Callahan, John William
Camerman, Norman
Cameron, Ross G
Campbell, James
Cann, Malcolm Calvin
Canvin, David T
Carey, Paul Richard
Carroll, Kenneth Kitchener
Clarke, Anthony John
Cohen, Saul Louis
Collins, Frank William, Jr
Connell, George Edward
Cooke, T Derek V
Crowe, Arlene Joyce
Cummins, W(illiam) Raymond
Cunnane, Stephen C
Dales, Samuel
Davis, Alan
Davis, Frank
Deber, Charles Michael
De Bold, Adolfo J
Dedhar, Shoukat
Deeley, Roger Graham
Dennis, David Thomas
D'Iorio, Antoine
Donisch, Valentine
Dorrell, Douglas Gordon
Downer, Roger George Hamill
Dumbroff, Erwin Bernard
Endrenyi, Laszlo
Epand, Richard Mayer
Farber, Emmanuel
Ferrier, Barbara May
Fish, Eleanor N
Fitt, Peter Stanley
Fitz-James, Philip Chester
Flanagan, Peter Rutledge
Flynn, Thomas Geoffrey
Forsdyke, Donald Roy
Foster, Charles David Owen
Fraser, Ann Davina Elizabeth
Freeman, Karl Boruch
Fritz, Irving Bamdas
Fuleki, Tibor
Galsworthy, Peter Robert
Galsworthy, Sara B
Gardner, David R
Gentner, Norman Elwood
Ghosh, Hara Prasad
Glick, Bernard Robert

Gold, Marvin H
Goldberg, David Myer
Gornall, Allan Godfrey
Gould, William Douglas
Greenblatt, Jack Fred
Grinstein, Sergio
Gupta, Radhey Shyam
Hagen, Paul Beo
Hamilton, Robert Milton Gregory
Harfenist, Elizabeth Joyce
Harpur, Robert Peter
Heacock, Ronald A
Heagy, Fred Clark
Hew, Choy-Leong
Hickey, Donal Aloysius
Himms-Hagen, Jean
Hobkirk, Ronald
Hofmann, Theo
Hope, Hugh Johnson
Horner, Alan Alfred
Ianuzzo, C David
Ingles, Charles James
Israelstam, Gerald Frank
Jellinck, Peter Harry
Jenkins, Kenneth James William
Johnson, Willard Jesse
Jones, John Dewi
Kalab, Miloslav
Kaplan, Harvey
Kapur, Bhushan M
Kay, Ernest Robert MacKenzie
Keeley, Fred W
Khan, Abdul Waheed
Kisilevsky, Robert
Kluger, Ronald H
Kosaric, Naim
Kramer, John Karl Gerhard
Krupka, Richard M(orley)
Kuhns, William Joseph
Kuksis, Arnis
Kuroski-De Bold, Mercedes Lina
Kushner, Donn Jean
Lane, Byron George
Lau, Catherine Y
Lawford, George Ross
Layne, Donald Sainteval
Lepock, James Ronald
Letarte, Michelle Vinlaine
Liew, Choong-Chin
Lo, Theodore Ching-Yang
Logan, David Mackenzie
Lowden, J Alexander
McCalla, Dennis Robert
Mackie, George Alexander
MacLennan, David Herman
McMurray, William Colin Campbell
MacRae, Andrew Richard
Magee, William Lovel
Maguire, Robert James
Mak, William Wai-Nam
Malkin, Aaron
Mathur, Sukhdev Prashad
Mavrides, Charalampos
Mellors, Alan
Metuzals, Janis
Miller, Richard Wilson
Milligan, Larry Patrick
Mishra, Ram K
Moon, Thomas William
Morand, Peter
Murray, Robert Kincaid
Murray, William Douglas
Neelin, James Michael
Nesheim, Michael Ernest
Nicholls, Doris McEwen
Nicholls, Peter
Nikiforuk, Gordon
Nozzolillo, Constance
O'Brien, Peter J
Olson, Arthur Olaf
Packham, Marian Aitchison
Painter, Robert Hilton
Percy, Maire Ede
Perry, Malcolm Blythe
Polley, John Richard
Possmayer, Fred
Proulx, Pierre R
Rabin, Elijah Zephania
Rauser, Wilfried Ernst
Reed, Juta Kuttis
Riordan, John Richard
Rixon, Raymond Harwood
Robinson, Brian Howard
Rock, Gail Ann
Rogers, Charles Graham
Rolleston, Francis Stopford
Rosa, Nestor
Roslycky, Eugene Bohdan
Ryan, Michael T
Sanwal, Bishnu Dat
Sarkar, Bibudhendra
Sarma, Dittakavi S R
Schachter, H
Schneider, Henry
Scott, Fraser Wallace
Scrimgeour, Kenneth Gray
Shah, Bhagwan G
Sikorska, Marianna
Siu, Chi-Hung
Smiley, James Richard
Smith, David Burrard
Sole, Michael Joseph

Spencer, John Hedley
Sribney, Michael
Stainer, Dennis William
Stanacev, Nikola Ziva
Stavric, Stanislava
Steele, John Earle
Stephenson, Norman Robert
Stewart, Harold Brown
Storey, Kenneth Bruce
Strickland, Kenneth Percy
Suderman, Harold Julius
Szabo, Arthur Gustav
Taylor, Keith Edward
Taylor, Norman Fletcher
Tepperman, Barry Lorne
Thomas, Barry Holland
Thompson, John Eveleigh
Tinker, David Owen
Tobe, Stephen Solomon
Toews, Cornelius J
Trevithick, John Richard
Trick, Charles Gordon
Tsai, Chishiun S
Tsang, Charles Pak Wai
Tu, Chin Ming
Tustanoff, Eugene Reno
Vafopoulo, Xanthe
Van Huystee, Robert Bernard
Vardanis, Alexander
Veliky, Ivan Alois
Viswanatha, Thammaiah
Walker, Ian Gardner
Walker, Peter Roy
Warner, Alden Howard
White, Gordon Allan
Williams, John Peter
Williamson, Denis George
Wilson, David George
Wilson, Jack Harold
Winter, Peter
Wolfe, Bernard Martin
Wong, Jeffrey Tze-Fei
Yamazaki, Hiroshi
Yang, Man-chiu
Yip, Cecil Cheung-Ching
Younglai, Edward Victor
Ziegler, Peter
Zilkey, Bryan Frederick
Zitnak, Ambrose

PRINCE EDWARD ISLAND
Rigney, James Arthur
Suzuki, Michio

QUEBEC
Abshire, Claude James
Anderson, William Alan
Antakly, Tony
Barcelo, Raymond
Barden, Nicholas
Barta, Alice
Beaudoin, Adrien Robert
Bélanger, Luc
Bellabarba, Diego
Bensley, Edward Horton
Bilimoria, Minoo Hormasji
Blostein, Rhoda
Borgeat, Pierre
Boulet, Marcel
Braun, Peter Eric
Brown, Gregory Gaynor
Cedergren, Robert J
Chabot, Benoit
Chafouleas, James G
Clark, Michael Wayne
Cooper, David Gordon
Cousineau, Gilles H
Daigneault, Rejean
David, Jean
De Lamirande, Gaston
DeMedicis, E M J A
Devor, Kenneth Arthur
Dubé, François
DuBow, Michael Scott
Ducharme, Jacques R
Dupont, Claire Hammel
Dupuis, Gilles
Edward, Deirdre Waldron
Fazekas, Arpad Gyula
Ford-Hutchinson, Anthony W
Gagnon, Claude
Gariepy, Claude
Germinario, Ralph Joseph
Gianetto, Robert
Givner, Morris Lincoln
Godin, Claude
Goltzman, David
Graham, Angus Frederick
Gresser, Michael Joseph
Guderley, Helga Elizabeth
Hallenbeck, Patrick Clark
Hamet, Pavel
Hancock, Ronald Lee
Heisler, Seymour
Herscovics, Annette Antoinette
Hosein, Esau Abbas
Ibrahim, Ragai Kamel
Ingram, Jordan Miles
Isler, Henri Gustave
Itiaba, Kibe
Johnstone, Rose M
Kallai-Sanfacon, Mary-Ann
Kluepfel, Dieter
Labrie, Fernand

Lanthier, Andre
Lapointe, Jacques
Leduc, Gerard
Lehoux, Jean-Guy
Lemonde, Andre
Levy, Samuel Wolfe
MacKenzie, Robert Earl
Maclachlan, Gordon Alistair
Manjunath, Puttaswamy
Marceau, Normand Luc
Martineau, Ronald
Mathieu, Leo Gilles
Meighen, Edward Arthur
Momparler, Richard Lewis
Morais, Rejean
Moroz, Leonard Arthur
Murgita, Robert Anthony
Murthy, Mahadi Raghavandrarao Ven
Niven, Donald Ferries
Page, Michel
Palfree, Roger Grenville Eric
Pande, Shri Vardhan
Pappius, Hanna M
Pelletier, Omer
Phan, Chon-Ton
Picard, Gaston Arthur
Ponka, Premysl
Poole, Anthony Robin
Poole, Ronald John
Powell, William St John
Preiss, Benjamin
Roberge, Andree Groleau
Roberts, Kenneth David
Rochefort, Joseph Guy
Roughley, Peter James
Roy, Claude Charles
Roy, Robert Michael McGregor
Sandor, Thomas
Sarhan, Fathey
Schiffrin, Ernesto Luis
Schucher, Reuben
Schulman, Herbert Michael
Seidah, Nabil George
Sheinin, Rose
Shore, Gordon Charles
Shoubridge, Eric Alan
Sinclair, Ronald
Sirois, Pierre
Skup, Daniel
Slilaty, Steve N
Solomon, Samuel
Sourkes, Theodore Lionel
Srivastava, Ashok Kumar
Stephens, Heather R
Sygusch, Jurgen
Talbot, Pierre J
Tan, Ah-Ti Chu
Tan, Liat
Tanguay, Robert M
Taussig, Andrew
Tenenhouse, Alan M
Tenenhouse, Harriet Susie
Thirion, Jean Paul Joseph
Tonks, David Bayard
Van Gelder, Nico Michel
Van Lier, Johannes Ernestinus
Vezina, Claude
Villeneuve, Andre
Waithe, William Irwin
Wang, Eugenia
Wasserman, Aaron Reuben
Willemot, Claude
Wolfe, Leonhard Scott
Yousef, Ibrahim Mohmoud

SASKATCHEWAN
Angel, Joseph Francis
Bertrand, Helmut
Blair, Donald George Ralph
Boulton, Alan Arthur
Craig, Burton Mackay
Davis, Bruce Allan
Dawson, Peter Stephen Shevyn
Durden, David Alan
Faber, Albert John
Gear, James Richard
Grant, Donald R
Gupta, Vidya Sagar
Harold, Stephen
Howarth, Ronald Edward
Kalra, Jawahar
Khandelwal, Ramji Lal
Korsrud, Gary Olaf
McArthur, Charles Stewart
McGregor, Douglas Ian
McLennan, Barry Dean
Martin, Robert O
Patience, John Francis
Richardson, J(ohn) Steven
Robertson, Hugh Elburn
Shargool, Peter Douglas
Vella, Francis
Waygood, Edward Bruce
Wood, James Douglas
Yu, Peter Hao

OTHER COUNTRIES
Abelev, Garri Izrailevich
Ahn, Tae In
Alivisatos, Spyridon Gerasimos
    Anastasios
Alivisatos, Spyridon Gerasimos
    Anastasios
Amsterdam, Abraham

Argos, Patrick
Bachmann, Fedor W
Baddiley, James
Bader, Hermann
Baig, Mirza Mansoor
Baldwin, Robert William
Barald, Kate Francesca
Barel, Monique
Barouki, Robert
Bayev, Alexander A
Ben-Ze'ev, Avri
Binder, Bernd R
Boaz, David Paul
Boffa, Lidia C
Bogoch, Samuel
Brody, Edward Norman
Bronk, John Ramsey
Brown, Barker Hastings
Bunn, Clive Leighton
Buzina, Ratko
Cash, William Davis
Cerbon-Solorzano, Jorge
Chetsanga, Christopher J
Chipman, David Mayer
Choi, Yong Chun
Creighton, Thomas Edwin
Cromartie, Thomas Houston
De Haën, Christoph
Delmer, Deborah P
De Renobales, Mertxe
Dikstein, Shabtay
Diner, Bruce Aaron
Doerfler, Walter Hans
Donato, Rosario Francesco
Dorrington, Keith John
Dreosti, Ivor Eustace
Dutta-Roy, Asim Kanti
Dworkin, Mark Bruce
Edelman, Marvin
Edelstein, Stuart J
Eggers, Hans J
Eik-Nes, Kristen Borger
Engelborghs, Yves
Fahimi, Hossein Dariush
Falb, Richard D
Fawaz, George
Fujiwara, Keigi
Gal, Susannah
Garcia-Sainz, J Adolfo
Gendler, Sandra J
Ginis, Asterios Michael
Glick, David M
Goding, James Watson
Graber, Robert Philip
Gregory, Peter
Grisolia, Santiago
Haegele, Klaus D
Hanukoglu, Israel
Hayashi, Masao
Heinz, Erich
Helmreich, Ernst J M
Henderson, David Andrew
Hillcoat, Brian Leslie
Hirano, Toshio
Hiyama, Tetsuo
Hosokawa, Keiichi
Ikehara, Yukio
Ionescu, Lavinel G
Jaffe, Werner G
Jakob, Karl Michael
Jarvis, Simon Michael
Jost, Jean-Pierre
Ju, Jin Soon
Kang, Sungzong
Kato, Ikunoshin
Katz, Bernard
Kaye, Alvin Maurice
Kent, Stephen Brian Henry
Kobata, Akira
Koerber, Walter Ludwig
Kuhn, Klaus
Kumar, S Anand
Lawson, William Burrows
Lee, Joseph Chuen Kwun
Lee, Men Hui
Leitzmann, Claus
Lewin, Lawrence M
Liao, Ta-Hsiu
Livett, Bruce G
Lizardi, Paul Modesto
Lonsdale-Eccles, John David
Loppnow, Harald
Low, Chow-Eng
Low, Teresa Lingchun Kao
MacCarthy, Jean Juliet
Maruta, Hiroshi
Meins, Frederick, Jr
Merlini, Giampaolo
Mezquita, Cristobal
Mitchell, Peter
Mittal, Balraj
Mizuno, Shigeki
Modabber, Farrokh Z
Montgomery, Morris William
Muller-Eberhard, Hans Joachim
Muñoz, Maria De Lourdes
Nagata, Kazuhiro
Piña, Enrique
Querinjean, Pierre Joseph
Ray, Prasanta K
Razzell, Wilfred Edwin
Ribbons, Douglas William
Roy, Raman K
Salas, Pedro Jose I

Schmell, Eli David
Schwartz, Peter Larry
Seya, Tsukasa
Shaw, Elliott Nathan
Soberon, Guillermo
Spierer, Pierre
Stockell-Hartree, Anne
Stoppani, Andres Oscar Manuel
Strous, Ger J
Subramanian, Alap Raman
Sullivan-Kessler, Ann Clare
Sung, Chen-Yu
Tachibana, Takehiko
Taniyama, Tadayoshi
Teleb, Zakaria Ahmed
Thiery, Jean Paul
Thorson, John Wells
Toutant, Jean-Pierre
Tsolas, Orestes
Tsou, Chen-Lu
Varga, Janos M
Vina, Juan R
Webb, Cynthia Ann Glinert
Weil, Marvin Lee
Welsh, Richard Stanley
Wenger, Byron Sylvester
Willard-Gallo, Karen Elizabeth
Wu, Albert M
Wu, Cheng-Wen
Wu, Felicia Ying-Hsiueh
Yamamoto, Hiroshi
Yeon-Sook, Yun

Biology, General

ALABAMA
Appel, Arthur Gary
Beyers, Robert John
Boozer, Reuben Bryan
Carter, Charlotte
Chang, Chi Hsiung
Dagg, Charles Patrick
Dozier, Slater Mathew
Freeman, John A
Graham, Tom Maness
Hanson, Roger Wayne
Hook, Magnus Ao
Koulourides, Theodore I
McCullough, Herbert Alfred
Michalek, Suzanne M
Pearson, Allen Mobley
Perry, Nelson Allen
Poirier, Gary Raymond
Rhodes, Richard Kent
Sledge, Eugene Bondurant
Tucker, Charles Eugene
Turco, Jenifer
Wilkes, James C
Williams, Louis Gressett
Williams, Michael Ledell

ALASKA
Kudenov, Jerry David
Proenza, Luis Mariano
Rice, Stanley Donald
Schoen, John Warren
Shields, Gerald Francis
Shirley, Thomas Clifton
Strand, John A, III
Willson, Mary Frances

ARIZONA
Haase, Edward Francis
Hadley, Mac Eugene
Hildebrand, John Grant, III
Johnson, Mary Ida
Justice, Keith Evans
Keck, Konrad
Kurtz, Edwin Bernard, Jr
Lisonbee, Lorenzo Kenneth
McElhinney, Margaret M (Cocklin)
Nagle, Ray Burdell
Pittendrigh, Colin Stephenson
Pommerville, Jeffrey Carl
Rasmussen, David Irvin
Schaffer, William Morris
Schmidt, Justin Orvel
Slobodchikoff, Constantine Nicholas
Stahnke, Herbert Ludwig
Stini, William Arthur
Timmermann, Barbara Nawalany
Tischler, Marc Eliot
Topoff, Howard Ronald
Towill, Leslie Ruth
Verbeke, Judith Ann
Younggren, Newell A
Zegura, Stephen Luke

ARKANSAS
Buffaloe, Neal Dollison
Chacko, Rosy J
Clower, Dan Fredric
Demarest, Jeffrey R
Geren, Collis Ross
Karlin, Alvan A
McMillan, Harlan L
Scheving, Lawrence Einar
Shade, Elwood B

CALIFORNIA
Adomian, Gerald E
Aggeler, Judith
Amirkhanian, John David

Biology, General (cont)

Humphreys, Walter James
Iverson, Ray Mads
Jensen, Albert Christian
Keller, Laura R
Keppler, William J
Koevenig, James L
Larkin, Lynn Haydock
Lloyd, James Edward
Lutz, Peter Louis
McClure, Joseph A
MacDonald, Eve Lapeyrouse
Miskimen, Carmen Rivera
Morrill, John Barstow, Jr
Muller, Kenneth Joseph
Murison, Gerald Leonard
Nicosia, Santo Valerio
Norman, Eliane Meyer
Norstog, Knut Jonson
Ocampo-Friedmann, Roseli C
Potter, Lincoln Truslow
Raizada, Mohan K
Reiskind, Jonathan
Rokach, Joshua
Romrell, Lynn John
Roux, Kenneth H
Samuelson, Don Arthur
Schwassman, Horst Otto
Selman, Kelly
Sinclair, Thomas Russell
Smart, Grover Cleveland, Jr
Smith, Albert Carl
Stechschulte, Agnes Louise
Sumners, DeWitt L
Taylor, George Thomas
Taylor, J(ames) Herbert
Tschinkel, Walter Rheinhardt
Waggoner, Phillip Ray
Wagoner, Dale E
Westfall, Minter Jackson, Jr
Williams, Norris Hagan
Zikakis, John Philip

GEORGIA
Adkison, Daniel Lee
Andrews, Lucy Gordon
Black, Jessie Kate
Boudinot, Frank Douglas
Browne, John Mwalimu
Bryans, Trabue Daley
Butts, David
Calabrese, Ronald Lewis
Carter, Eloise B
Carter, James Richard
Chin, Edward
Coward, Stuart Jess
Dashek, William Vincent
DeHaan, Robert Lawrence
Duncan, Robert Leon, Jr
Duncan, Wilbur Howard
Evatt, Bruce Lee
Fetner, Robert Henry
Hamada, Spencer Hiroshi
Hancock, Kenneth Farrell
Hayes, John Thompson
Haynes, John Kermit
Heise, John J
Hunter, Roy, Jr
Kethley, Thomas William
Landt, James Frederick
Lindsay, David Taylor
McPherson, Robert Merrill
Miller-Stevens, Louise Teresa
Nabrit, Samuel Milton
Parrish, Fred Kenneth
Paulsen, Douglas F
Pendergrass, Levester
Pienaar, Leon Visser
Robinson, Margaret Chisolm
Srivastava, Prakash Narain
Stouffer, Richard Franklin
Sutton, William Wallace
White, Donald Henry
Willis, Judith Horwitz
Wood, John Grady
Wyatt, Robert Edward
Zinsmeister, Philip Price

HAWAII
Carr, Gerald Dwayne
Carson, Hampton Lawrence
De Feo, Vincent Joseph
Flickinger, Reed Adams
Hapai, Marlene Nachbar
Hsiao, Sidney Chihti
Humphreys, Tom Daniel
Kane, Robert Edward
Kaneshiro, Kenneth Yoshimitsu
Lee, Cheng-Sheng
Read, George Wesley
Siegel, Barbara Zenz
Ward, Melvin A
Yanagimachi, Ryuzo

IDAHO
Bratz, Robert Davis
Henderson, Douglass Miles
Mincher, Bruce J
Robbins, Robert Raymond
Wyllie, Gilbert Alexander

ILLINOIS
Anderson, David John
Augspurger, Carol Kathleen
Bachop, William Earl

Baron, David Alan
Barr, Susan Hartline
Beck, Sidney L
Beer, Alan E
Benkendorf, Carol Ann
Berry, James Frederick
Berry, Robert Wayne
Bieler, Rüdiger
Birkenholz, Dale Eugene
Bjorklund, Richard Guy
Briggs, Robert Wilbur
Carnes, Bruce Alfred
Carr, Virginia McMillan
Castignetti, Domenic
Chappell, Dorothy Field
Chen, Shepley S
Chung, Stephen
Constantinou, Andreas I
Cox, Thomas C
Dal Canto, Mauro Carlo
Daniel, Jon Cameron
Decker, Robert Scott
Disterhoft, John Francis
Drengler, Keith Allan
Fast, Dale Eugene
Feng, Albert Shih-Hung
Flinn, James Edwin
Foster, Raymond Orrville
Fox, Sidney Walter
Friedman, Yochanan
Gassman, Merrill Loren
Gerdy, James Robert
Gibbs, Daniel
Gillette, Rhanor
Gingle, Alan Raymond
Giometti, Carol Smith
Goldie, Mark
Gomes, Wayne Reginald
Gonzales, Federico
Goodheart, Clyde Raymond
Grdina, David John
Howe, Henry Franklin
Huberman, Eliezer
Irwin, Michael Edward
Jasper, Donald K
Josephs, Robert
Kasprow, Barbara Ann
Kaufman, Stephen J
Lamppa, Gayle K
Lande, Russell Scott
Lee, Merlin Raymond
Lucher, Lynne Annette
McKinley, Vicky L
MacLeod, Ellis Gilmore
Maiorana, Virginia Catherine
Melin, Brian Edward
Mitchell, John Laurin Amos
Mockford, Edward Lee
Muffley, Harry Chilton
Murphy, Richard Allan
Natalini, John Joseph
Norton, David L
Nyberg, Dennis Wayne
Overton, Jane Harper
Palincsar, Edward Emil
Pappas, George Demetrios
Parker, Nancy Johanne Rentner
Prahlad, Kadaba V
Pruett-Jones, Stephen Glen
Rolfe, Gary Lavelle
Rosales-Sharp, Maria Consolacion
Rotermund, Albert J, Jr
Roth, Robert Mark
Scharf, Arthur Alfred
Schennum, Wayne Edward
Siegel, Ivens Aaron
Simpson, Sidney Burgess, Jr
Sörensen, Paul Davidsen
Spiess, Luretta Davis
Spiroff, Boris E N
Stocum, David Leon
Straus, Helen Lorna Puttkammer
Straus, Werner
Sturtevant, Ruthann Patterson
Sze, Paul Yi Ling
Thiruvathukal, Kris V
Thonar, Eugene Jean-Marie
Towle, David Walter
Tripathi, Satish Chandra
Van Valen, Leigh Maiorana
Velardo, Joseph Thomas
Wade, Michael John
Yopp, John Herman
Zaki, Abd El-Moneim Emam
Zimmerman, Roger Paul

INDIANA
Baker, Edgar Gates Stanley
Bates, Charles Johnson
Borgens, Richard Ben
Broxmeyer, Hal Edward
Burkholder, Timothy Jay
Chiscon, J Alfred
Crowell, (Prince) Sears, Jr
Cushing, Bruce S
Delfin, Eliseo Dais
Dial, Norman Arnold
DiMicco, Joseph Anthony
Dolph, Gary Edward
Gallo, Duane Gordon
Gammon, James Robert
Gaudin, Anthony J
Goff, Charles W
Grabowski, Sandra Reynolds

Hammond, Charles Thomas
Hunt, Linda Margaret
Iten, Laurie Elaine
Jensen, Richard Jorg
Kammeraad, Adrian
Koch, Arthur Louis
Krogmann, David William
Levy, Morris
McGowan, Michael James
Malacinksi, George M
Maloney, Michael Stephen
Mescher, Anthony Louis
Miller, Richard William
Mitchell, Cary Arthur
Nolan, Val, Jr
O'Connor, Brian Lee
Olson, John Bennet
O'Neal, Lyman Henry
Ortoleva, Peter Joseph
Ostroy, Sanford Eugene
Peterson, Laverne E
Peterson, Richard George
Raff, Rudolf Albert
Rosenberg, Gary David
Ruesink, Albert William
Shoup, Jane Rearick
Simmons, Eric Leslie
Speece, Susan Phillips
Stabler, Timothy Allen
Stillwell, William Harry
Taparowsky, Elizabeth Jane
Tweedell, Kenyon Stanley
Vanable, Joseph William, Jr
Wermuth, Jerome Francis
Wilkin, Peter J
Witzmann, Frank A
Young, Frank Nelson, Jr

IOWA
Adams, Donald Robert
Allegre, Charles Frederick
Bamrick, John Francis
Bonfiglio, Michael
Buettner, Garry Richard
Christiansen, Kenneth Allen
Christiansen, Paul Arthur
Coleman, Richard Walter
Cruden, Robert William
DeBoer, Kenneth F
Denburg, Jeffrey Lewis
DiSpirito, Alan Angelo
Dolphin, Warren Dean
Eagleson, Gerald Wayne
Getting, Peter Alexander
Hadow, Harlo Herbert
Hallberg, Richard Lawrence
Hoffmann, Richard John
Horton, Robert, Jr
Isely, Duane
Kapler, Josheph Edward
Kirkland, Willis L
Laffoon, John
Larkin, Jeanne Holden
Milkman, Roger Dawson
Rogers, Rodney Albert
Soll, David Richard
Solursh, Michael
Te Paske, Everett Russell
Thompson, Sue Ann
Waziri, Rafiq
Williams, Norman Eugene

KANSAS
Ahshapanek, Don Colesto
Ashe, James S
Barnes, William Gartin
Beary, Dexter F
Bechtel, Donald Bruce
Bowen, Daniel Edward
Chowdhury, Mridula
Cink, Calvin Lee
Dahl, Nancy Ann
Dey, Sudhansu Kumar
Fortner, George William
Gaines, Michael Stephen
Halazon, George Christ
Haufler, Christopher Hardin
Johnston, Richard Fourness
Kammer, Ann Emma
Neely, Peter Munro
Pierson, David W
Prophet, Carl Wright
Rowe, Vernon Dodds
Spooner, Brian Sandford
Stewart, Richard Cummins
Weis, Jerry Samuel
Williams, Larry Gale

KENTUCKY
Bennett, Thomas Edward
Boyarsky, Lila Harriet
Calkins, John
Dixon, Wallace Clark, Jr
Ettensohn, Francis Robert
Ferner, John William
Fontaine, Julia Clare
Goldstein, Lester
Harmet, Kenneth Herman
Humphries, Asa Alan, Jr
Keefe, Thomas Leeven
McLaughlin, Barbara Jean
Norfleet, Morris L
Prior, David James
Sabharwal, Pritam Singh

Slagel, Donald E
Timmons, Thomas Joseph
Wagner, Charles Eugene
Wolfson, Alfred M

LOUISIANA
Avault, James W, Jr
Blackwell, Meredith
Borsari, Bruno
Borsari, Bruno
Dempesy, Colby Wilson
Ellgaard, Erik G
Etheridge, Albert Louis
Grodner, Mary Laslie
Heyn, Anton Nicolaas Johannes
Lesseps, Roland Joseph
Levitzky, Michael Gordon
Lonergan, Thomas A
McPherson, Alvadus Bradley
Makielski, Sarah Kimball
Misra, Raghunath P
Newsome, David Anthony
Owings, Addison Davis
Reese, William Dean
Shepherd, David Preston
Specian, Robert David
Spring, Jeffrey H
Sullivan, Victoria I
Thien, Leonard B
Webert, Henry S

MAINE
Bailey, Norman Sprague
Barker, Jane Ellen
Bell, Allen L
Bennett, Miriam Frances
Blake, Richard D
Champlin, Arthur Kingsley
Davisson, Muriel Trask
Kornfield, Irving Leslie
Labov, Jay Brian
Lewis, Alan James
Minkoff, Eli Cooperman
Mun, Alton M
Raisbeck, Barbara
Spirito, Carl Peter
Stevens, Leroy Carlton, Jr
Tyler, Mary Stott

MARYLAND
Aaronson, Stuart Alan
Akiyama, Steven Ken
Albuquerque, Edson Xavier
Andersen, Arnold E
Arnheiter, Heinz
Barker, Jeffery Lange
Bass, Eugene L
Beyer, W Nelson
Bloch, Robert Joseph
Blye, Richard Perry
Bodian, David
Britz, Steven J
Brookes, Neville
Brown, Donald D
Brown, Kenneth Stephen
Burgess, Wilson Hales
Butman, Bryan Timothy
Chappelle, Emmett W
Cheng, Sheue-yann
Chuang, De-Maw
Coomes, Marguerite Wilton
Crosby, Gayle Marcella
Dahl, Roy Dennis
Darden, Lindley
Dawid, Igor Bert
Delahunty, George
Deringer, Margaret K
Dixon, Dennis Michael
Donlon, Mildred A
Durum, Scott Kenneth
Eglitis, Martin Alexandris
Elson, Hannah Friedman
Enna, Salvatore Joseph
Fambrough, Douglas McIntosh
Feldmesser, Julius
Forman, Michele R
Freeman, Colette
Freiberg, Samuel Robert
Futcher, Anthony Graham
Gainer, Harold
Gallin, John I
Gantt, Elisabeth
Geis, Aelred Dean
Giacometti, Luigi
Gill, Douglas Edward
Glick, J Leslie
Gray, Paulette S
Griesbach, Robert James
Gulyas, Bela Janos
Gunn, Charles Robert
Hasselmeyer, Eileen Grace
Hathaway, Wilfred Bostock
Hausman, Steven J
Hay, Robert J
Hein, Rosemary Ruth
Heine, Ursula Ingrid
Hewitt, Arthur Tyl
Hiatt, James Lee
Highton, Richard
Horenstein, Evelyn Anne
Kalberer, John Theodore, Jr
Karklins, Olgerts Longins
Kayar, Susan Rennie
Kelley, Russell Victor

Biology, General (cont)

Kernaghan, Roy Peter
Klein, David C
Koslow, Stephen Hugh
Kundig, Fredericka Dodyk
Lane, H Clifford
Leather, Gerald Roger
Ledney, George David
Leonard, Warren J
LeRoith, Derek
Levin, Barbara Chernov
Lichtenfels, James Ralph
Lin, Shin
Liu, Leroy Fong
Lyons, Russette M
McCutchen, Charles Walter
Maddox, Yvonne T
Mahoney, Francis Joseph
Maier, Robert J
Martin, George Reilly
Mattson, Margaret Ellen
Moloney, John Bromley
Moment, Gairdner Bostwick
Moreira, Jorge Eduardo
Murphy, Douglas Blakeney
Oberdorfer, Michael Douglas
Odell, Lois Dorothea
Oka, Takami
Parakkal, Paul Fab
Peck, Carl Curtis
Perry, James Warner
Perry, Vernon P
Piatigorsky, Joram Paul
Pienta, Roman Joseph
Pilitt, Patricia Ann
Platt, Austin Pickard
Pollard, Harvey Bruce
Popper, Arthur N
Prabhakar, Bellur Subbanna
Reid, Clarice D
Rifkin, Erik
Rodrigues, Merlyn M
Roth, Thomas Frederic
Rotherham, Jean
Roti Roti, Joseph Lee
Scala, John Richard
Scully, Erik Paul
Shamsuddin, Abulkalam Mohammad
Shear, Charles Robert
Sieckmann, Donna G
Slife, Charles W
Sokolove, Phillip Gary
Sowers, Arthur Edward
Spector, Novera Herbert
Stanley, Steven Mitchell
Stratmeyer, Melvin Edward
Stringfellow, Frank
Strum, Judy May
Thorington, Richard Wainwright, Jr
Traystman, Richard J
Varma, Shambhu D
Vermeij, Geerat Jacobus
Vlahakis, George
Walker, Richard Ives
Weimer, John Thomas
Weinreich, Daniel
Wolf, Robert Oliver
Zelenka, Peggy Sue
Zornetzer, Steven F

MASSACHUSETTS

Ackerman, Steven J
Adams, David S
Ahlberg, Henry David
Albertini, David Fred
Ament, Alison Stone
Arnaout, M Amin
Bawa, Kamaljit S
Begg, David A
Berry, Vern Vincent
Bond, George Walter
Borgatti, Alfred Lawrence
Burggren, Warren William
Caulfield, John P
Cheetham, Ronald D
Christensen, Thomas Gash
Cochrane, David Earle
Cohen, Samuel H
Collier, Robert John
Cook, Robert Edward
Copeland, Donald Eugene
Copeland, Frederick Cleveland
Dacheux, Ramon F, II
Dainiak, Nicholas
D'Amore, Patricia Ann
D'Avanzo, Charlene
Davis, Elizabeth Allaway
Dick, Stanley
DiLiddo, Rebecca McBride
Dinsmore, Jonathan H
Dowling, John Elliott
Edgar, Robert Kent
Ehrlich, H Paul
Feder, William Adolph
Fimian, Walter Joseph, Jr
Fink, Rachel Deborah
Fitzgerald, Marie Anton
Freadman, Marvin Alan
Fulton, Chandler Montgomery
Furshpan, Edwin Jean
Garay, Leslie Andrew
Gibbs, Martin
Goldhor, Susan

Goldin, Stanley Michael
Golub, Samuel J(oseph)
Gonnella, Patricia Anne
Goodenough, Judith Elizabeth
Gould, Stephen Jay
Greeley, Frederick
Gross, Jerome
Haimes, Howard B
Hall, Jeffrey Connor
Hargis, Betty Jean
Hexter, William Michael
Hichar, Joseph Kenneth
Holle, Paul August
Horvitz, Howard Robert
Hubbard, Ruth
Inoue, Shinya
Irwin, Louis Neal
Kamien, Ethel N
Kelner, Albert
Kidder, George Wallace, Jr
Kimball, John Ward
Kunkel, Joseph George
Leavis, Paul Clifton
Levi, Herbert Walter
Levin, Andrew Eliot
Levins, Richard
Lewontin, Richard Charles
Liberatore, Frederick Anthony
Liss, Maurice
Longcope, Christopher
Lyerla, Jo Ann Harding
Lyman, Charles Peirson
McNeil, Paul L
McPherson, John M
Malecki, George Jerzy
Margulis, Lynn
Masland, Richard Harry
Matsumoto, Steven G
Matthews, Samuel Arthur
Maxam, Allan M
Mayr, Ernst
Millard, William James
Millette, Clarke Francis
Monette, Francis C
Muckenthaler, Florian August
Mulcare, Donald J
Mulrennan, Cecilia Agnes
Murphey, Rodney Keith
Naqvi, S Rehan Hasan
Powell, Jeanne Adele
Procaccini, Donald J
Putney, Scott David
Reardon, John Joseph
Redington, Charles Bahr
Ropes, John Warren
Rosemberg, Eugenia
Roth, John L, Jr
Ruben, Morris P
Schmitt, Francis Otto
Seaman, Edna
Seitz, Anna W
Shahrik, H Arto
Slavin, Ovid
Smith, Dennis Matthew
Spiegel, Judith E
Sprague, Isabelle Baird
Steele, John H
Stoffolano, John George, Jr
Tauber, Alfred Imre
Taubman, Martin Arnold
Thomas, Gail B
Todd, Neil Bowman
Townsend, Jane Kaltenbach
Treistman, Steven Neal
Vaillant, Henry Winchester
Walker, James Willard
Webb, Andrew Clive
West, Arthur James, II
Wetmore, Ralph Hartley
White, Alan Whitcomb
Widmaier, Eric Paul
Williams, Carroll Milton
Wilson, Edward Osborne
Wu, Jung-Tsung
Yerganian, George
Zetter, Bruce Robert
Zottoli, Steven Jaynes
Zweig, Ronald David

MICHIGAN

Allen, Edward David
Ansbacher, Rudi
Armant, D Randall
Bach, Shirley
Blinn, Lorena Virginia
Boving, Bent Giede
Brinkley, Linda Lee
Bromley, Stephen C
Brown, Donald Frederick Mackenzie
Bush, Guy L
Chang, Chia-Cheng
Cohen, Sanford Ned
Conover, James H
D'Amato, Constance Joan
Fetterolf, Carlos de la Mesa, Jr
Fisher, Leslie John
Frisancho, A Roberto
Garlick, Robert L
Gay, Helen
Hancock, James Findley, Jr
Hayward, James Lloyd
Hiscoe, Helen Brush
Hitchcock, Dorothy Jean
Houts, Larry Lee

Hudson, Roy Davage
Jackson, Matthew Paul
Koehler, Lawrence D
Laborde, Alice L
Leach, Karen Lynn
Lefrancois, Leo
Lewis, Ralph William
Lopushinsky, Theodore
Low, Bobbi Stiers
McCauley, Roy Barnard
McIntosh, Lee
Medlin, Julie Anne Jones
Miller, Donald Elbert
Miller, James Ray
Moore, William Samuel
Morawa, Arnold Peter
Moudgil, Virinder Kumar
Murnik, Mary Rengo
Narayan, Latha
Oakley, Bruce
Olexia, Paul Dale
Oliver, Daphna R
Peters, Joseph John
Racle, Fred Arnold
Rogers, Claude Marvin
Root-Bernstein, Robert Scott
San Clemente, Charles Leonard
Schneider, Michael J
Sloat, Barbara Furin
Snitgen, Donald Albert
Stebbins, William Cooper
Sussman, Alfred Sheppard
Taylor, John Dirk
Teeri, James Arthur
Thompson, Paul Woodard
Townsend, Samuel Franklin
Vandermeer, John H
Van Deventer, William Carlstead
Vatsis, Kostas Petros
Venta, Patrick John
Webber, Patrick John
Whitney, Marion Isabelle
Wiman, Fred Hawkins
Wood, Pauline J
Wu, Ching Kuei
Wynne, Michael James
Yates, Jon Arthur

MINNESOTA

Anderson, Beverly
Birney, Elmer Clea
Dewald, Gordon Wayne
Ellinger, Mark Stephen
Erickson, James Eldred
Frenzel, Louis Daniel, Jr
Heidcamp, William H
Kerr, Norman Story
Kerr, Sylvia Joann
McCann, Lester J
McCulloch, Joseph Howard
McKinnell, Robert Gilmore
Milliner, Eric Killmon
Mork, David Peter Sogn
Palm, John Daniel
Pusey, Anne E
Regal, Philip Joe
Reynhout, James Kenneth
Samuelson, Charles R
Schimpf, David Jeffrey
Sehe, Charles Theodore
Shapiro, Burton Leonard
Sherer, Glenn Keith
Smith, Quenton Terrill
Spangler, George Russell
Tihon, Claude
Tyce, Gertrude Mary
Widin, Katharine Douglas
Wittry, Esperance

MISSISSIPPI

Case, Steven Thomas
Corbett, James John
Cox, Prentiss Gwendolyn
Edney, Norris Allen
Helms, Thomas Joseph
Kurtz, Mark Edward
McClurkin, Iola Taylor
Rockhold, Robin William
Spann, Charles Henry
Stark, William Polson
Steen, James Southworth
Sutton, William Wallace
Thureson-Klein, Asa Kristina

MISSOURI

Bader, Robert Smith
Bagby, John R, Jr
Beckett, Jack Brown
Bischoff, Eric Richard
Bose, Subir Kumar
Burke, William Joseph
Chan, Albert Sun Chi
Crawford, Susan Young
De Buhr, Larry Eugene
Derby, Albert
Elliott, Dana Ray
Fahim, Mostafa Safwat
Fernandez, Hugo Lathrop
Finley, David Emanuel
Gerdes, Charles Frederick
Gossling, Jennifer
Green, Michael
Hagen, Gretchen
Hamilton, John Meacham

Hansen, Harry Louis
Hanson, Willis Dale
Hart, Richard Allen
Hayes, Alice Bourke
Ho, David Tuan-hua
Hodges, Glenn R(oss)
Huckabay, John Porter
Hughes, Stephen Edward
Jones, Allan W
Kirk, David Livingstone
Kirk, Marilyn M
Krogstad, Donald John
Krukowski, Marilyn
Lin, Hsiu-san
Lumb, Ethel Sue
Magill, Robert Earle
Nichols, Herbert Wayne
Oshima, Eugene Akio
Raven, Peter Hamilton
Sanes, Joshua Richard
Schaal, Barbara Anna
Schneiderman, Howard Allen
Slatopolsky, Eduardo
Smith, James Lee
Stalker, Harrison Dailey
Stevenson, Robert Thomas
Stiehl, Richard Borg
Stombaugh, Tom Atkins
Voorhees, Frank Ray
Wang, Bin Ching
Wartzok, Douglas
Wilkens, Lon Allan
Yonke, Thomas Richard

MONTANA

Cameron, David Glen
Carlson, Clinton E
Coe, John Emmons
Dorgan, William Joseph
Medora, Rustem Sohrab
Mitchell, Lawrence Gustave
Pickett, James M
Preece, Sherman Joy, Jr
Sheridan, Richard P

NEBRASKA

Ballinger, Royce Eugene
Banerjee, Mihir R
Bliese, John C W
Boernke, William E
Bolick, Margaret Ruth
Boohar, Richard Kenneth
Brooks, Merle Eugene
Buell, Katherine Mayhew
Casey, Carol A
Cavalieri, Ercole Luigi
Curtin, Charles Byron
Dalley, Arthur Frederick, II
Joern, Anthony
McCue, Robert Owen
Nair, Chandra Kunju
Platz, James Ernest

NEVADA

Hillyard, Stanley Donald
Kiefer, Barry Irwin
Lowenthal, Douglas H

NEW HAMPSHIRE

Bergman, Kenneth David
Dingman, Jane Van Zandt
Douple, Evan Barr
Ferm, Vergil Harkness
Foret, John Emil
Fozdar, Birendra Singh
Jackson, Herbert William
Manasek, Francis John
Milne, Lorus Johnson
Milne, Margery (Joan) (Greene)
Sloboda, Roger D
Stahl, Barbara Jaffe
Van Ummersen, Claire Ann
Walker, Charles Wayne

NEW JERSEY

Abbott, Joan
Adamo, Joseph Albert
Adler, Beatriz C
Arnold, Frederic G
Auletta, Carol Spence
Bacharach, Martin Max
Belmonte, Rocco George
Berman, Helen Miriam
Bolton, Laura Lee
Bonner, John Tyler
Borysko, Emil
Brown, Thomas Edward
Carlson, Richard P
Chen, James Che Wen
Chiang, Bin-Yea
Cohen, Alan Mathew
Cook, Marie Mildred
Cristini, Angela
Dashman, Theodore
Davis, Bill David
Dhruv, Rohini Arvind
Durand, James Blanchard
Fangboner, Raymond Franklin
Francoeur, Robert Thomas
Garone, John Edward
Gavurin, Lillian
Gepner, Ivan Alan
Gona, Ophelia Delaine
Hampton, Suzanne Harvey

Hartwick, Richard Allen
Hayat, M A
Hill, Helene Zimmermann
Hitchcock-DeGregori, Sarah Ellen
Hockel, Gregory Martin
Hsu, Linda
Hubbell, Stephen Philip
Jabbar, Gina Marie
Katz, Laurence Barry
Koepp, Leila H
Kuehn, Harold Herman
Kumbaraci-Jones, Nuran Melek
Ledeen, Robert
Lee, Hsin-Yi
Leone, Ida Alba
Levin, Barry Edward
Linsky, Cary Bruce
Liu, Alice Yee-Chang
Lonski, Joseph
Loveland, Robert Edward
Mayer, Thomas C
Meiss, Alfred Nelson
Mele, Frank Michael
Mosley, Stephen T
Nagle, James John
Neri, Anthony
Perrine, John W
Petry, Robert Kendrick
Proudfoot, Bernadette Agnes
Ritchie, David Malcolm
Rosenthal, Judith Wolder
Rossow, Peter William
Sabrosky, Curtis Williams
Salerno, Alphonse
Schreckenberg, Mary Gervasia
Sherman, Michael Ian
Siegel, Allan
Sinha, Arabinda Kumar
Stein, Donald Gerald
Trotta, Paul P
Vasconcelos, Aurea C
Venable, Patricia Lengel
Vrijenhoek, Robert Charles
Wallace, Edith Winchell
Weis, Judith Shulman
Willis, Jacalyn Giacalone
Zipf, Elizabeth M(argaret)

NEW MEXICO
Allred, Kelly Wayne
Anderson, Dean Mauritz
Baca, Oswald Gilbert
Barrera, Cecilio Richard
Degenhardt, William George
Dittmer, Howard James
Groblewski, Gerald Eugene
Hillsman, Matthew Jerome
Hubby, John L
Lechner, John Fred
McNamara, Mary Colleen
Malvin, Gary M

NEW YORK
Abbatiello, Michael James
Ablin, Richard J
Adler, Kraig (Kerr)
Agnello, Arthur Michael
Alcamo, I Edward
Allen, Howard Joseph
Alsop, David W
Angeletti, Ruth Hogue
Archimovich, Alexander S
Bablanian, Rostom
Ballantyne, Donald Lindsay
Banerjee, Ranjit
Bannister, Thomas Turpin
Basile, Dominick V
Bellvé, Anthony Rex
Berlinrood, Martin
Borowsky, Richard Lewis
Boyer, Barbara Conta
Boyer, John Frederick
Brenowitz, A Harry
Bretscher, Anthony P
Britten, Bryan Terrence
Brown, William Louis, Jr
Browner, Robert Herman
Bruns, Peter John
Brunson, John Taylor
Burke, William Thomas, Jr
Cardillo, Frances M
Catapane, Edward John
Chabot, Brian F
Chiao, Jen Wei
Cohen, Nicholas
Commoner, Barry
Coppock, Donald Leslie
Corpe, William Albert
Cottrell, Stephen F
Craddock, Elysse Margaret
Cramer, Eva Brown
D'Adamo, Amedeo Filiberto, Jr
Day, Ivana Podvalova
Desplan, Claude
Detmers, Patricia Anne
DiGaudio, Mary Rose
Doty, Stephen Bruce
Druger, Marvin
DuFrain, Russell Jerome
Durr, David P
Dykhuizen, Daniel Edward
Edds, Kenneth Tiffany
Edmunds, Leland Nicholas, Jr
Ellis, John Francis

Erk, Frank Chris
Faber, Donald S
Falkowski, Paul Gordon
Ferguson, John Barclay
Fischer, Richard Bernard
Fleagle, John G
Fortner, Joseph Gerald
Frair, Wayne
Friedland, Beatrice L
Gabriel, Mordecai Lionel
Garcia, Alfredo Mariano
Geisler, Grace
Girsch, Stephen John
Goldsmith, Eli David
Goode, Robert P
Gorovsky, Martin A
Griffiths, Raymond Bert
Grillo, Ramon S
Hagar, Silas Stanley
Hallahan, William Laskey
Hamburgh, Max
Hardy, Matthew Phillip
Hartung, John David
Hellstrom, Harold Richard
Helson, Lawrence
Henderson, Thomas E
Hershey, Linda Ann
Hiiemae, Karen Marion
Hillman, Dean Elof
Himes, Marion
Hirsch, Helmut V B
Hirshon, Jordon Barry
Hoffman, Roger Alan
Hoogstraal, Harry
Hooper, George Bates
Hotchkiss, Sharon K
Hoy, Ronald Raymond
Hrushesky, William John Michael
Huang, Chau-Ting
Huang, Chester Chen-Chiu
Hudecki, Michael Stephen
Ingalls, James Warren, Jr
Intres, Richard
Isenberg, George Raymond, Jr
Jacobs, Laurence Stanton
Jacques, Felix Anthony
Jahan-Parwar, Behrus
Kandel, Eric Richard
Katz, Gary Victor
Kelly, Richard Delmer
Kemphues, Kenneth J
Kerr, Marilyn Sue
Kessler, Dietrich
Keston, Albert S
Kim, Untae
Klotz, Richard Lawrence
Ko, Li-wen
Koenig, Edward
Korf, Richard Paul
Kramer, Alfred William, Jr
LaFountain, James Robert, Jr
Lamberg, Stanley Lawrence
Lapin, David Marvin
Laug, George Milton
Laychock, Suzanne Gale
Levin, Norman Lewis
Liguori, Vincent Robert
Llinas, Rodolfo
Lnenicka, Gregory Allen
Lockshin, Richard Ansel
Lossinsky, Albert S
McCann-Collier, Marjorie Dolores
McDaniel, Carl Nimitz
McDonnell, Mark Jeffery
McEwen, Bruce F
McGee-Russell, Samuel M
McKenna, Olivia Cleveland
Mandel, Irwin D
Mandl, Richard H
Marks, Neville
Mascarenhas, Joseph Peter
Mason, Larry Gordon
Meiselman, Newton
Meng, Heinz Karl
Miller, Charles William
Mirand, Edwin Albert
Mitchell, John Taylor
Mitra, Jyotirmay
Moore, Malcolm A S
Moor-Jankowski, J
Moran, Denis Joseph
Morrill, Gene A
Motzkin, Shirley M
Muschio, Henry M, Jr
Nasrallah, Mikhail Elia
Nathan, Henry C
Nicholson, J(ohn) Charles (Godfrey)
North, Robert J
Novacek, Michael John
Novak, Joseph Donald
Paine, Philip Lowell
Pattee, Howard Hunt, Jr
Pesce, Michael A
Phillips, Carleton Jaffrey
Pietraface, William John
Poindexter, Jeanne Stove
Possidente, Bernard Philip, Jr
Power, Richard B
Prestwidge, Kathleen Joyce
Privitera, Carmelo Anthony
Pumo, Dorothy Ellen
Puszkin, Saul
Quaroni, Andrea
Rabino, Isaac

Raventós-Suárez, Carmen Elvira
Reedy, John Joseph
Reilly, Margaret Anne
Reuss, Ronald Merl
Rhodes, Rondell H
Rich, Abby M
Richmond, Milo Eugene
Risley, Michael Samuel
Robbins, Edith Schultz
Rosenberg, Warren L
Rosi, David
Ryan, Simeon P
Saez, Juan Carlos
Salpeter, Miriam Mirl
Salthe, Stanley Norman
Satir, Birgit H
Sato, Gordon Hisashi
Scalia, Frank
Scharff, Matthew Daniel
Schwartz, Norman Martin
Selsky, Melvyn Ira
Shafit-Zagardo, Bridget
Shannon, Jerry A, Jr
Sharma, Sansar C
Sharpless, George Robert
Shaw, Spencer
Shechter, Yaakov
Sherman, Paul Willard
Siegel, Charles David
Siemann, Dietmar W
Simmons, Daryl Michael
Sklarew, Robert J
Sleeper, David Allanbrook
Slocum, Harry Kim
Smeriglio, Alfred John
Sokal, Robert Reuven
Southwick, Edward Earle
Spencer, Selden J
Springer, Alan David
Strand, Fleur Lillian
Styles, Twitty Junius
Sun, Tung-Tien
Tamm, Igor
Thompson, Robert Poole
Tokay, Elbert
Tomashefsky, Philip
Trimble, Mary Ellen
Unterman, Ronald David
Vasey, Carey Edward
Veith, Frank James
Vuilleumier, Francois
Walker, Philip Caleb
Wasserman, Marvin
Wasserman, William John
Weiss, Klaudiusz Robert
Weiss, Paul Alfred
Wells, Russell Frederick
White, James Edwin
Williams, Donald Benjamin
Witkin, Steven S
Witkus, Eleanor Ruth
Wolfner, Mariana Federica
Wysocki, Annette Bernadette
Ycas, Martynas
Yen, Andrew
Young, William Donald, Jr
Zitrin, Charlotte Marker
Zubay, Geoffrey

NORTH CAROLINA
Abramson, Jon Stuart
Adler, Kenneth B
Anderson, Charles Eugene
Bailey, Donald Etheridge
Banerjee, Umesh Chandra
Bates, William K
Berberian, Paul Anthony
Blau, William Stephen
Bode, Arthur Palfrey
Bond, James Anthony
Brody, Arnold R
Burrous, Stanley Emerson
Camp, Pamela Jean
Cheng, Pi-Wan
Daggy, Tom
Edwards, Nancy C
Endow, Sharyn Anne
Feduccia, John Alan
Fitzharris, Timothy Patrick
Glasser, Richard Lee
Glassman, Edward
Goodman, Major M
Gordon, Christopher John
Haning, Blanche Cournoyer
Harrison, Frederick Williams
Henson, Anna Miriam (Morgan)
Horton, James Heathman
Jenkins, Jimmy Raymond
Jenner, Charles Edwin
Jetten, Anton Marinus
Jolls, Claudia Lee
Kalmus, Gerhard Wolfgang
Kimmel, Donald Loraine, Jr
Kirk, Daniel Eddins
Langford, George Malcolm
Lauder, Jean Miles
Laurie, John Sewall
LeBaron, Homer McKay
Lehman, Harvey Eugene
Leise, Esther M
Lemasters, John J
Lytle, Charles Franklin
McCrady, Edward, III
Martin, Virginia Lorelle

Meyer, John Richard
Moore, Martha May
Mueller, Nancy Schneider
Mukunnemkeril, George Mathew
Nicklas, Robert Bruce
O'Rand, Michael Gene
Parker, Beulah Mae
Petters, Robert Michael
Pung, Oscar J
Rabb, Robert Lamar
Ramagopal, Mudumbi Vijaya
Randall, John Frank
Real, Leslie Allan
Reddish, Paul Sigman
Robinson, Kent
Roop, Richard Allan
Rustioni, Aldo
Sadler, Thomas William
Smith, Ned Allan
Sonstegard, Karen Sue
Spencer, Lorraine Barney
Stacy, David Lowell
Stiven, Alan Ernest
Stuart, Ann Elizabeth
Teng, Ching Sung
Terborgh, John J
Theil, Elizabeth
Thomas, Mary Beth
Triantaphyllou, Hedwig Hirschmann
Turner, James Eldridge
Tyndall, Jesse Parker
Vogel, Steven
Watts, John Albert, Jr
White, Richard Alan
Yongue, William Henry

NORTH DAKOTA
Bleier, William Joseph
Gladue, Brian Anthony
Khansari, David Nemat
Scoby, Donald Ray
Starks, Thomas Leroy

OHIO
Akeson, Richard Allan
Allenspach, Allan Leroy
Andreas, Barbara Kloha
Beal, Kathleen Grabaskas
Boczar, Barbara Ann
Brackenbury, Robert William
Brosnihan, K Bridget
Bullock, Ward Ervin, Jr
Caplan, Arnold I
Cerroni, Rose E
Chang, In-Kook
Chumlea, William Cameron
Cibula, Adam Burt
Clay, Mary Ellen
Costello, Walter James
Craft, Thomas Jacob, Sr
Crutcher, Keith A
Dasch, Clement Eugene
Davis, James Allen
Easterly, Nathan William
Egar, Margaret Wells
Epp, Leonard G
Fried, Joel Robert
Fry, Anne Evans
Graham, Shirley Ann
Gwatkin, Ralph Buchanan Lloyd
Hamre, Harold Thomas
Hartsough, Robert Ray
Haubrich, Robert Rice
Hitt, John Burton
Hoagstrom, Carl William
Holeski, Paul Michael
Houston, Willie Walter, Jr
Jacobs-Lorena, Marcelo
Jarroll, Edward Lee, Jr
Jegla, Dorothy Eldredge
Kaiserman-Abramof, Ita Rebeca
Karas, James Glynn
Keefe, John Richard
Kleiner, Susan Mala
Kolodziej, Bruno J
Koo, Peter H
Landis, Story Cleland
Lasek, Raymond J
Leonard, Billie Charles
Lesh-Laurie, Georgia Elizabeth
Lovejoy, Owen
Luckenbill-Edds, Louise
Mao, Simon Jen-Tan
Mikula, Bernard C
Mitchell, Rodger (David)
Moore, Randy
Murthy, Raman Chittaram
Norris, Bill Eugene
Peterson, Paul Constant
Pfohl, Ronald John
Riley, Charles Victor
Rosenberg, Howard C
Samols, David R
Sawicki, Stanley George
Schurr, Karl M
Scott, William James, Jr
Shaffer, Charles Franklin
Sherman, Thomas Fairchild
Sich, Jeffrey John
Stambrook, Peter J
Stokes, Bradford Taylor
Tassava, Roy A
Thomson, Dale S
Vardaris, Richard Miles

Biology, General (cont)

Voneida, Theodore J
Witters, Weldon L
Wong, Laurence Cheng Kong
Yoon, Jong Sik

OKLAHOMA

Bantle, John Albert
Bastian, Joseph
Bogenschutz, Robert Parks
Davis, Marjorie
Fisher, John Berton
Marek, Edmund Anthony
Miller, Helen Carter
Ogilvie, Marilyn Bailey
O'Neal, Steven George
Patterson, Manford Kenneth, Jr
Radke, William John
Rainbolt, Mary Louise
Shmaefsky, Brian Robert
Thornton, John William
Young, David Allen

OREGON

Arch, Stephen William
Barmack, Neal Herbert
Beatty, Joseph John
Benedict, Ellen Maring
Blaustein, Andrew Richard
Breakey, Donald Ray
Brownell, Philip Harry
Carlisle, Kay Susan
Dawson, Peter Sanford
Duncan, James Thayer
Ellinwood, William Edward
Fernald, Russell Dawson
Gerke, John Royal
Grant, Philip
Gwilliam, Gilbert Franklin
Harris, Patricia J
Hedberg, Karen K
Henny, Charles Joseph
Kimmel, Charles Brown
King, Charles Everett
McCollum, Gin
McNulty, Wilbur Palmer
Montagna, William
Orkney, G Dale
Owczarzak, Alfred
Phoenix, Charles Henry
Riffey, Meribeth M
Ruben, John Alex
Russell, Nancy Jeanne
Springer, Martha Edith
Walker, Kenneth Merriam
Warren, Charles Edward
Wessells, Norman Keith
Weston, James A
Willis, David Lee

PENNSYLVANIA

Akruk, Samir Rizk
Andrews, Peter Walter
Baker, Kenneth Melvin
Barr, Richard Arthur
Bean, Barry
Bergstresser, Kenneth A
Bertolini, Donald R
Binkley, Sue Ann
Birchem, Regina
Bongiovanni, Alfred Marius
Botelho, Stella Yates
Bradt, Patricia Thornton
Brunkard, Kathleen Marie
Burgess, David Ray
Campbell, Graham Le Mesurier
Campo, Robert D
Chacko, Samuel K
Chepenik, Kenneth Paul
Compher, Marvin Keen, Jr
Corey, Sharon Eva
Das, Manjusri
Davies, Ronald Edgar
Debroy, Chitrita
Desmond, Mary Elizabeth
Diamond, Leila
DiBerardino, Marie A
Donawick, William Joseph
Dropp, John Jerome
Ellis, Demetrius
Erickson, Ralph O
Everhart, Leighton Phreaner, Jr
Fansler, Bradford S
Feit, Ira (Nathan)
Fowler, H(oratio) Seymour
Fremont, Henry Neil
George, John Lothar
Gerstein, George Leonard
Gertz, Steven Michael
Giri, Lallan
Gottlieb, Frederick Jay
Greene, Robert Morris
Grobstein, Paul
Grove, Alvin Russell, Jr
Gundy, Samuel Charles
Halbrendt, John Marthon
Hassell, John Robert
Hazuda, Daria Jean
Herold, Richard Carl
Herzog, Karl A
Hillman, Ralph
Himes, Craig L
Hoffman, Albert Charles

Hoffman, Daniel Lewis
Hopper, James Ernest
Hull, Larry Allen
Hulse, Arthur Charles
Jacobsohn, Myra K
Janzen, Daniel Hunt
Jensen, Bruce David
Jim, Kam Fook
Karlsson, Ulf Lennart
Katoh, Arthur
Kauffman, Stuart Alan
Kayhart, Marion
Kennedy, Michael Craig
King, David Beeman
Krishtalka, Leonard
LaBelle, Edward Francis
Lanni, Frederick
Lash, James (Jay) W
Lee, John Cheung Han
Leighton, Charles Cutler
Levine, Elliot Myron
Lindstrom, Eugene Shipman
Logan, David Alexander
MacDermott, Richard J, Jr
McDevitt, David Stephen
McGary, Carl T
McNamara, Pamela Dee
Maksymowych, Roman
Malsberger, Richard Griffith
Mann, Stanley Joseph
Masteller, Edwin C
Maxson, Linda Ellen R
Meek, Edward Stanley
Meyer, R Peter
Miller, Richard Lee
Mintz, Beatrice
Miselis, Richard Robert
Moss, William Wayne
Murray, Marion
Muth, Eric Anthony
Nadelhaft, Irving
Neff, Stuart Edmund
Nei, Masatoshi
Norrbom, Allen Lee
Ode, Philip E
Olds-Clarke, Patricia Jean
Orth, Joanne M
Paoletti, Robert Anthony
Parker, Jon Irving
Peightel, William Edgar
Phoenix, Donald R
Pitkin, Ruthanne B
Pleasure, David
Pruss, Thaddeus P
Rappaport, Harry P
Roth, Stephen
Ruffolo, Robert Richard, Jr
Ryan, James Patrick
Sachs, Howard George
Salzberg, Brian Matthew
Samollow, Paul B
Sanger, Joseph William
Schmaier, Alvin Harold
Schuyler, Alfred Ernest
Schwartz, Jeffrey H
Settlemyer, Kenneth Theodore
Shostak, Stanley
Skutches, Charles L
Smith, Allen Anderson
Smith, John Bryan
Smith, W John
Snipes, Charles Andrew
Sollberger, Dwight Ellsworth
Sorensen, Ralph Albrecht
Stableford, Louis Tranter
Staetz, Charles Alan
Stere, Athleen Jacobs
Stewart, Barbara Yost
Studer, Rebecca Kathryn
Sussman, Maurice
Suyama, Yoshitaka
Taylor, D Lansing
Telfer, William Harrison
Thompson, Roger Kevin Russell
Tobin, Thomas Vincent
Travis, Robert Victor
Tristram-Nagle, Stephanie Ann
Uricchio, William Andrew
Walsh, Charles Joseph
Weinreb, Eva Lurie
Weiss, Kenneth Monrad
West, Keith P
Wheeler, Donald Alsop
Woodruff, Richard Ira
Yoho, Timothy Price

RHODE ISLAND

Ellis, Richard Akers
Fausto, Nelson
Glicksman, Arvin Sigmund
Goertemiller, Clarence C, Jr
Hufnagel, Linda Ann
Silver, Alene Freudenheim
Stern, Leo
Wilde, Charles Edward, Jr
Young, Robert M
Zackroff, Robert V
Zarcaro, Robert Michael

SOUTH CAROLINA

Bank, Harvey L
Browdy, Craig Lawrence
Congdon, Justin Daniel
Cromer, Jerry Haltiwanger

Dawson, Wallace Douglas, Jr
Dobbs, Harry Donald
Graef, Philip Edwin
Houk, Richard Duncan
Kistler, Wilson Stephen, Jr
Krebs, Julia Elizabeth
LaBar, Martin
Manley, Donald Gene
Meredith, Howard Voas
Mulvey, Margaret
Pielou, William P
Pollard, Arthur Joseph
Porcher, Richard Dwight
Rembert, David Hopkins, Jr
Rohlfing, Duane L
Schmidt, Roger Paul
Scott, Jaunita Simons
Shealy, Harry Everett, Jr
Singh, Inderjit
Smith, Michael Howard
Surver, William Merle, Jr
Swallow, Richard Louis
Vernberg, Winona B
Voit, Eberhard Otto
Wagner, Charles Kenyon
Wourms, John P
Zucker, Robert Martin

SOUTH DAKOTA

Froiland, Sven Gordon
Goldman, Max
Hodgson, Lynn Morrison
Lindahl, Ronald
Westfall, Helen Naomi

TENNESSEE

Alsop, Frederick Joseph, III
Biggers, Charles James
Briggs, Robert Chester
Campbell, James A
Cardoso, Sergio Steiner
Chandler, Robert Walter
Chen, Chung-Hsuan
Cohn, Sidney Arthur
Coons, Lewis Bennion
Cooper, Terrance G
Costlow, Mark Enoch
Donaldson, Donald Jay
Duhl, David M
Gallagher, Michael Terrance
Garth, Richard Edwin
Harris, John Wallace
Hasty, David Long
Hossler, Fred E
Jennings, Lisa Helen Kyle
Jeon, Kwang Wu
Kahlon, Prem Singh
Kathman, R Deedee
Kimball, Richard Fuller
McCauley, David Evan
MacFadden, Donald Lee
McGhee, Charles Robert
Mansbach, Charles M, II
Marchok, Ann Catherine
Mathis, Philip Monroe
Monaco, Paul J
Murti, Kuruganti Gopalakrishna
Neff, Robert Jack
Norden, Jeanette Jean
Norton, Virginia Marino
Nunnally, David Ambrose
Olsen, John Stuart
Orgebin-Crist, Marie-Claire
Owens, Charles Allen
Pui, Ching-Hon
Pulliam, James Augustus
Redman, Charles Edwin
Reed, Horace Beecher
Ripley, Thomas H
Roberts, Lee Knight
Rogers, Beverly Jane
Roop, Robert Dickinson
Samuels, Robert
Schilling, Edward Eugene
Shivers, Charles Alex
Smith, Arlo Irving
Yates, Harris Oliver

TEXAS

Anderson, Richard Gilpin Wood
Aufderheide, Karl John
Azizi, Sayed Ausim
Baptist, James (Noel)
Barnes, Eugene Miller, Jr
Bashour, Fouad A
Belk, Gene Denton
Benedict, Irvin J
Bhaskaran, Govindan
Bickham, John W
Buskirk, Ruth Elizabeth
Busteed, Robert Charles
Byrd, Earl William, Jr
Cameron, Guy Neil
Cate, Rodney Lee
Chan, Lee-Nien Lillian
Cole, Garry Thomas
Coulter, Murray W
Craik, Eva Lee
Crawford, Gladys P
DeShaw, James Richard
Di Ferrante, Daniela Tavella
Diggs, George Minor, Jr
Dillon, Lawrence Samuel
Dresden, Marc Henri

Dronamraju, Krishna Rao
Dunbar, Bonnie Sue
Emery, William Henry Perry
Eskew, Cletis Theodore
Faulkner, Russell Conklin, Jr
Forsyth, John Wiley
Franklin, Luther Edward
Frigyesi, Tamas L
Gary, Roland Thacher
Glasser, Stanley Richard
Gottesfeld, Zehava
Grant, Verne (Edwin)
Grimes, L Nichols
Grimm, Elizabeth Ann
Guentzel, M Neal
Hacker, Carl Sidney
Handler, Shirley Wolz
Harris, Kerry Francis Patrick
Harris, Venoia M
Hollyfield, Joe G
Howard, Harriette Ella Pierce
Hsu, Laura Hwei-Nien Ling
Hug, Verena
Johnson, Larry
Jordan, Chris Sullivan
Kalthoff, Klaus Otto
Knaff, David Barry
Koballa, Thomas Raymond, Jr
Kurzrock, Razelle
Larimer, James Lynn
Lee, James C
Lieb, Carl Sears
Litke, Larry Lavoe
Lockett, Clodovia
Mastromarino, Anthony John
Mecham, John Stephen
Meola, Roger Walker
Mukherjee, Amal
Muntz, Kathryn Howe
Murphy, Clifford Elyman
Myers, Jack Edgar
Nakajima, Motowo
Parkening, Terry Arthur
Parker, Roy Denver, Jr
Pinson, Ernest Alexander
Poduslo, Shirley Ellen
Pool, Thomas Burgess
Price, Harold James
Rajaraman, Srinivasan
Redburn, Dianna Ammons
Reese, Weldon Harold
Romsdahl, Marvin Magnus
Rosenfeld, Charles Richard
Rowell, Chester Morrison, Jr
Sant'Ambrogio, Giuseppe
Sauer, Helmut Wilhelm
Schafer, Rollie R
Schlueter, Edgar Albert
Schmidly, David James
Schneider, Barbara G
Schwartz, Colin John
Shur, Barry David
Simpson, Beryl Brintnall
Smatresk, Neal Joseph
Smith, Cornelia Marschall
Smith, William Russell
Sohal, Rajindar Singh
Spear, Irwin
Sullivan, Patricia Ann Nagengast
Sweet, Merrill Henry, II
Taylor, Albert Cecil
Taylor, D(orothy) Jane
Tcholakian, Robert Kevork
Telfair, Raymond Clark, II
Terry, Robert James
Tuff, Donald Wray
Van Bavel, Cornelius H M
Van Overbeek, Johannes
Waldbillig, Ronald Charles
Waymire, Jack Calvin
Weigel, Paul H(enry)
Wheeler, Michael Hugh
Wicksten, Mary Katherine
Wilson, Robert Eugene
Winkler, Matthew M
Wolf, Don Paul
Wolfenberger, Virginia Ann
Worthington, Richard Dane
Wright, David Anthony
Wright, Woodring Erik
Wu, Kenneth Kun-Yu
Yee, John Alan

UTAH

Bradshaw, William S
Burgess, Paul Richards
Cuellar, Orlando
Davidson, Diane West
Edmunds, George Francis, Jr
Ellis, LeGrande Clark
Galinsky, Raymond Ethan
Geddes, David Darwin
Hathaway, Ralph Robert
Johnston, Norman Paul
Palmblad, Ivan G
Schoenwolf, Gary Charles
Shearin, Nancy Louise
Straight, Richard Coleman
Stringham, Reed Millington, Jr
Vickery, Robert Kingston, Jr
Wagner, Frederic Hamilton
Warren, Reed Parley
Youssef, Nabil Naguib

VERMONT
Krupp, Patricia Powers
Landesman, Richard
Raper, Carlene Allen
Trombulak, Stephen Christopher
Woodworth, Robert Hugo

VIRGINIA
Alden, Raymond W, III
Baker, Jeffrey John Wheeler
Bender, Michael E
Bentz, Gregory Dean
Bieber, Samuel
Birdsong, Ray Stuart
Bodkin, Norlyn L
Bonar, Robert Addison
Brangan, Pamela J
Britt, Douglas Lee
Brown, Luther Park
Brubaker, Kenton Kaylor
Bull, Alice Louise
Burian, Richard M
Chinnici, Joseph (Frank) Peter
Churchill, Helen Mar
Daniel, Joseph Car, Jr
De Haan, Henry J(ohn)
Diehl, Fred A
Diwan, Bhalchandra Apparao
Fink, Linda Susan
Flint, Franklin Ford
Forbes, Michael Shepard
Friedman, Ruth T
Gaines, Gregory
Garrison, Norman Eugene
Grainger, Robert Michael
Griffin, Gary J
Grimm, James K
Gross, Paul Randolph
Hallinan, Edward Joseph
Hamilton, Howard Laverne
Heath, Everett
Holt, Perry Cecil
Hunt, Lindsay McLaurin, Jr
Jackson, Elizabeth Burger
Johnson, Thomas L
Jones, Melton Rodney
King, Betty Louise
Konigsberg, Irwin R
Lutes, Charlene McClanahan
Mandell, Alan
Marshall, Harold George
Mellinger, Clair
Moncrief, Nancy D
Moyer, Wayne A
Osgood, Christopher James
Owers, Noel Oscar
Patrick, Graham Abner
Pinschmidt, Mary Warren
Price, Steven
Quattropani, Steven L
Radice, Gary Paul
Reams, Willie Mathews, Jr
Richmond, Isabelle Louise
Rubel, Edwin W
Scheckler, Stephen Edward
Shear, William Albert
Spears, Joseph Faulconer
Stanley, Melissa Sue Millam
Stroup, Cynthia Roxane
Thompson, Jesse Clay, Jr
Turner, Bruce Jay
Williams, Robert Ellis
Wingard, Christopher Jon
Wiseman, Lawrence Linden

WASHINGTON
Anderson, Leigh
Bakken, Aimee Hayes
Benson, Keith Rodney
Binder, Marc David
Boersma, P Dee
Brookbank, John Warren
Cattolico, Rose Ann
Cloney, Richard Alan
Cohen, Arthur Leroy
Donaldson, Lauren Russell
Dube, Maurice Andrew
Eberhardt, Lester Lee
Ford, James
Fournier, R E Keith
Gaddum-Rosse, Penelope
Garber, Richard Lincoln
Griswold, Michael David
Guttman, Burton Samuel
Kundig, Werner
Kutsky, Roman Joseph
Leung, Frederick C
Lorenzen, Carl Julius
Luchtel, Daniel Lee
McMahon, Daniel Stanton
McMenamin, John William
Maguire, Yu Ping
Mills, Claudia Eileen
Mirkes, Philip Edmund
Monan, Gerald E
Nameroff, Mark A
Nathanson, Neil Marc
Orians, Gordon Howell
Prothero, John W
Robinovitch, Murray R
Roos, John Francis
Soltero, Raymond Arthur
Soule, Oscar Hommel
Stenchever, Morton Albert

Su, Judy Ya-Hwa Lin
Tamarin, Arnold
Tartar, Vance
Taub, Frieda B
Webber, Herbert H
Withner, Carl Leslie, Jr
Wood, James W

WEST VIRGINIA
Bayless, Laurence Emery
Burns, John Thomas
Castranova, Vincent
Mecca, Christyna Emma
Pauley, Thomas Kyle
Pritchett, William Henry
Shalaway, Scott D
Stephenson, Steven Lee

WISCONSIN
Akins, Virginia
Albrecht, Kenneth Adrian
Armstrong, George Michael
Balsano, Joseph Silvio
Bownds, M Deric
Centonze, Victoria E
Claflin, Tom O
Claude, Philippa
Cripps, Derek J
Duewer, Elizabeth Ann
Foote, Kenneth Gerald
Fossland, Robert Gerard
Gillis, John Ericsen
Goldberg, Burton David
Greenler, Robert George
Grunewald, Ralph
Hall, Marion Trufant
Hamerstrom, Frances
Hennen, Sally
Jordan, William R, III
Karrer, Kathleen Marie
Koval, Charles Francis
Kumaran, A Krishna
Kung, Ching
Lai, Ching-San
LaMarca, Michael James
Leonard, Thomas Joseph
Light, Douglas B
Mahadeva, Madhu Narayan
Mendelson, Carl Victor
Minock, Michael Edward
Norris, Dale Melvin, Jr
Oertel, Donata (Mrs Bill M Sugden)
Pierson, Edgar Franklin
Rapacz, Jan
Sattler, Carol Ann
Schatten, Gerald Phillip
Seeburger, George Harold
Sonneborn, David R
Staszak, David John
Stretton, Antony Oliver Ward
Strohm, Jerry Lee
Tormey, Douglass Cole
Uno, Hideo
Verma, Ajit K
Wenzel, Frederick J
Wikum, Douglas Arnold

WYOMING
George, Robert Porter
Nunamaker, Richard Allan

PUERTO RICO
Opava-Stitzer, Susan C
Rodriguez, Rocio del Pilar

ALBERTA
Ball, George Eugene
Bland, Brian Herbert
Browder, Leon Wilfred
Campenot, Robert Barry
Cavey, Michael John
Langridge, William Henry Russell
Pattie, Donald L
Prepas, Ellie E
Schultz, Gilbert Allan
Zalik, Sara E
Zammuto, Richard Michael

BRITISH COLUMBIA
Belton, Peter
Booth, Amanda Jane
Ellis, Derek V
Finnegan, Cyril Vincent
Freeman, Hugh J
Ganders, Fred Russell
Greer, Galen Leroy
Jasch, Laura Gwendolyn
Lee, Chi-Yu Gregory
Olive, Peggy Louise
Paul, Miles Richard
Shamoun, Simon Francis
Smith, Michael Joseph
Srivastava, Lalit Mohan
Sziklai, Oscar
Taylor, Iain Edgar Park
Webster, John Malcolm
Yorque, Ralf Richard

MANITOBA
Bodaly, Richard Andrew
Braekevelt, Charlie Roger
Hara, Toshiaki J
Huebner, Erwin
Huebner, Judith Dee

Novak, Marie Marta
Pepper, Evan Harold

NEW BRUNSWICK
Anderson, John Murray
Burt, Michael David Brunskill
Glebe, Brian Douglas
McKenzie, Joseph Addison
Thomas, Martin Lewis Hall

NEWFOUNDLAND
Cowan, Garry Ian McTaggart

NOVA SCOTIA
Dadswell, Michael John
Fentress, John Carroll
Ferrier, Gregory R
Hall, Brian Keith
McLaren, Ian Alexander
Vethamany, Victor Gladstone
Wassersug, Richard Joel

ONTARIO
Anderson, Terry Ross
Atkinson, Burr Gervais
Aubin, Jane E
Balon, Eugene Kornel
Bancroft, John Basil
Barrett, Spencer Charles Hilton
Barron, John Robert
Battle, Helen Irene
Baum, Bernard R
Brooks, Daniel Rusk
Carlone, Robert Leo
Catling, Paul Miles
Caveney, Stanley
Chambers, Ann Franklin
Cook, David Greenfield
Dedhar, Shoukat
Dodson, Edward O
Fallding, Margaret Hurlstone Hardy
Fisher, Kenneth Robert Stanley
Forer, Arthur H
Fritz, Irving Bamdas
Garfield, Robert Edward
Harding, Paul George Richard
Harmsen, Rudolf
Hickey, Donal Aloysius
Hutchinson, Thomas C
Israelstam, Gerald Frank
Keddy, Paul Anthony
Kerbel, Robert Stephen
Lean, David Robert Samuel
Liversage, Richard Albert
MacCrimmon, Hugh Ross
McCully, Margaret E
Malkin, Aaron
Marcus, George Jacob
Masui, Yoshio
Money, Kenneth Eric
Montgomerie, Robert Dennis
Murray, William Douglas
O'Day, Danton Harry
Ouellet, Henri
Ozburn, George W
Peck, Stewart Blaine
Philogene, Bernard J R
Phipps, James Bird
Possmayer, Fred
Prevec, Ludvik Anthony
Purko, John
Raaphorst, G Peter
Rand, Richard Peter
Rapport, David Joseph
Robertson, Raleigh John
Roth, René Romain
Ruckerbauer, Gerda Margareta
Shih, Chang-Tai
Skoryna, Stanley C
Small, Ernest
Somorjai, Rajmund Lewis
Sorger, George Joseph
Sprague, John Booty
Stavric, Bozidar
Teskey, Herbert Joseph
Threlkeld, Stephen Francis H
Trevors, Jack Thomas
Tung, Pierre S
Turk, Fateh (Frank) M
Vijay, Hari Mohan
Warner, Alden Howard

QUEBEC
Alavi, Misbahuddin Zafar
Bélanger, Luc
Bell, Graham Arthur Charlton
Benoit, Guy J C
Bussey, Arthur Howard
Chase, Ronald
Descarries, Laurent
Digby, Peter Saki Bassett
Dubé, François
Dugre, Robert
Gervais, Francine
Goltzman, David
Grad, Bernard Raymond
Guyda, Harvey John
Mailloux, Gerard
Maly, Edward J
Manjunath, Puttaswamy
Marceau, Normand Luc
Mulay, Shree
Pallotta, Dominick John
Roy, Robert Michael McGregor

Sattler, Rolf
Sergeant, David Ernest
Sikorska, Hanna
Sirois, Pierre
Skup, Daniel
Srivastava, Uma Shanker
Tannenbaum, Gloria Shaffer
Wolsky, Alexander

SASKATCHEWAN
Kartha, Kutty Krishnan
McLennan, Barry Dean
Schmutz, Josef Konrad
Scoles, Graham John
Spurr, David Tupper
Yu, Peter Hao

YUKON TERRITORY
Doermann, August Henry

OTHER COUNTRIES
Abelev, Garri Izrailevich
Anderson, Emory Dean
Barald, Kate Francesca
Braquet, Pierre G
Comfort, Alexander
Emmons, Lyman Randlett
Forbes, Milton L
Francke, Oscar F
Gould-Somero, Meredith
Gual, Carlos
Haight, John Richard
Hamilton, William Donald
Hammerman, Ira Saul
Holldobler, Berthold Karl
Jakowska, Sophie
Kahn, Albert
Karasaki, Shuichi
Klein, Jan
Kobata, Akira
Lewis, Arnold Leroy, II
Lin, Chin-Tarng
MacLean, William Plannette, III
Marikovsky, Yehuda
Milstein, Cesar
Muñoz, Maria De Lourdes
Nakagawa, Shizutoshi
Neher, Erwin
Nevo, Eviatar
Rand, Austin Stanley
Rickenberg, Howard V
Rosso, Pedro
Saito, Takuma
Schliwa, Manfred
Sleeter, Thomas David
Smokovitis, Athanassios A
Solter, Davor
Withers, Philip Carew

Biomathematics

ALABAMA
Hazelrig, Jane
Katholi, Charles Robinson
Mack, Timothy Patrick
Siler, William MacDowell
Turner, Malcolm Elijah, Jr

ALASKA
Fagen, Robert

ARIZONA
Arabyan, Ara
Gaud, William S
Vincent, Thomas Lange
Winfree, Arthur T

ARKANSAS
Dykman, Roscoe A

CALIFORNIA
Antoniak, Charles Edward
Baldwin, Ransom Leland
Blischke, Wallace Robert
Boulton, Roger Brett
Brokaw, Charles Jacob
Case, Ted Joseph
DeMaster, Douglas Paul
Deriso, Richard Bruce
Desharnais, Robert Anthony
DiStefano, Joseph John, III
Dixon, Wilfrid Joseph
Edelman, Jay Barry
Evans, John W
Feldman, Marcus William
Ferguson, William E
Fernandez, Alberto Antonio
Forsythe, Alan Barry
Francis, Robert Colgate
Holmquist, Walter Richard
Jenden, Donald James
Kaus, Peter Edward
Klinger, Allen
Landahl, Herbert Daniel
Lange, Kenneth L
Lein, Allen
Licko, Vojtech
Macey, Robert Irwin
Marsh, Donald Jay
Milstein, Jaime
Nunney, Leonard Peter
Oster, George F
Powell, Thomas Mabrey

Biomathematics (cont)

Rescigno, Aldo
Ritz-Gold, Caroline Joyce
Rose, Michael Robertson
Rotenberg, Manuel
Schnitzer, Jan Eugeneusz
Waterman, Michael S
Weiner, Daniel Lee
Whittemore, Alice S

COLORADO
Best, Jay Boyd
Osborn, Ronald George
Reiner, John Maximilian

CONNECTICUT
Bohning, Daryl Eugene
Holsinger, Kent Eugene
Lambrakis, Konstantine Christos
Russu, Irina Maria

DELAWARE
Abremski, Kenneth Edward

DISTRICT OF COLUMBIA
Rosenberg, Edith E
Tangney, John Francis
Thall, Peter Francis
Tucker, John Richard

FLORIDA
Barnard, Donald Roy
Bloom, Stephen Allen
Cropper, Wendell Parker
Haynes, Duncan Harold
Meltzer, Martin Isaac
Simberloff, Daniel S
Stamper, James Harris
Stetson, Robert Franklin
Travis, Joseph

GEORGIA
Longini, Ira Mann, Jr
Lyons, Nancy I
Pulliam, H Ronald

ILLINOIS
Barr, Lloyd
Carnes, Bruce Alfred
Charlesworth, Brian
Deysach, Lawrence George
Hausfater, Glenn
Jakobsson, Eric Gunnar, Sr
Jordan, Steven Lee
Levine, Michael W
Lloyd, Monte
Loehle, Craig S
Mittenthal, Jay Edward
Smeach, Stephen Charles
Trimble, John Leonard
Webber, Charles Lewis, Jr
Yokoyama, Shozo

INDIANA
Park, Richard Avery, IV
Parkhurst, David Frank
Stewart, Terry Sanford

KANSAS
Welch, Stephen Melwood

KENTUCKY
Sih, Andrew
Yaes, Robert Joel

MAINE
Dowse, Harold Burgess

MARYLAND
Blum, Harry
Corliss, John Burt
Covell, David Gene
FitzHugh, Richard
Ginevan, Michael Edward
Gorman, Cornelia M
Kroll, Martin Harris
Myers, Charles
Parker, Rodger D
Rinzel, John Matthew
Robinson, David Adair
Sastre, Antonio
Steinmetz, Michael Anthony
Thomas, Robert Glenn
Ulanowicz, Robert Edward

MASSACHUSETTS
Carpenter, Gail Alexandra
Caswell, Hal
Futrelle, Robert Peel
Levins, Richard
Levy, Elinor Miller
Shipley, Reginald A
Smith, Tim Denis
Wun, Chun Kwun

MICHIGAN
Bookstein, Fred Leon
Estabrook, George Frederick
Jacquez, John Alfred
Rosenspire, Allen Jay
Savageau, Michael Antonio

MINNESOTA
Andow, David A
Curtsinger, James Webb
Tilman, G David

MISSISSIPPI
Sadana, Ajit

MISSOURI
Luecke, Richard H
Wette, Reimut

MONTANA
MacNeil, Michael

NEBRASKA
Ueda, Clarence Tad

NEW JERSEY
Cronin, Jane Smiley
Hubbell, Stephen Philip
Quinn, James Allen
Saiger, George Lewis
Weiss, Stanley H

NEW MEXICO
Bell, George Irving
Beyer, William A
Chen, David J
Jett, James Hubert
Mewhinney, James Albert
Scott, Bobby Randolph

NEW YORK
Altshuler, Bernard
Andersen, Olaf Sparre
Blessing, John A
Blumenson, Leslie Eli
Chou, Ting-Chao
Clark, Alfred, Jr
Clark, Patricia Ann Andre
Cohen, Joel Ephraim
Durkin, Patrick Ralph
Emrich, Lawrence James
Ginzburg, Lev R
Greco, William Robert
Hastings, Harold Morris
Holland, Mary Jean Carey
Johnson, Lawrence Lloyd
Keller, Evelyn Fox
Kilpper, Robert William
Koretz, Jane Faith
Levin, Simon Asher
Millstein, Jeffrey Alan
Niklas, Karl Joseph
Okubo, Akira
Ringo, James Lewis
Schiff, Joel D
Schimmel, Herbert
Schoenly, Kenneth George
Sellers, Peter Hoadley
Silverman, Benjamin David
Slatkin, Daniel Nathan
Tanner, Martin Abba
Wittie, Larry Dawson

NORTH CAROLINA
Carter, Thomas Edward, Jr
Crawford-Brown, Douglas John
Crowder, Larry Bryant
Dillard, Margaret Bleick
Gold, Harvey Joseph
Kootsey, Joseph Mailen
Schaffer, Henry Elkin
Smith, Charles Eugene
Stinner, Ronald Edwin
Vaughan, Douglas Stanwood
Weir, Bruce Spencer
Woodbury, Max Atkin

OHIO
Boxenbaum, Harold George
Fraser, Alex Stewart
Gates, Michael Andrew
Geho, Walter Blair
Klapper, Michael H
MacKichan, Janis Jean
Stacy, Ralph Winston

OKLAHOMA
Buchanan, David Shane
Dugan, Kimiko Hatta
Nowack, William J
Wells, Harrington

OREGON
Craig, Albert Morrison (Morrie)
Fernald, Russell Dawson
Udovic, Daniel

PENNSYLVANIA
Altshuler, Martin David
Badler, Norman Ira
Chiang, Soong Tao
Cohen, Stanley
Erickson, Ralph O
Gibson, Raymond Edward
Karreman, George
Kushner, Harvey
Marks, Louis Sheppard
Rapp, Paul Ernest
Tallarida, Ronald Joseph

RHODE ISLAND
Costantino, Robert Francis
Hai, Chi-Ming

SOUTH CAROLINA
Kosinski, Robert Joseph
Voit, Eberhard Otto

TENNESSEE
Bernard, Selden Robert
Gutzke, William H N
Hallam, Thomas Guy
Ikenberry, Roy Dewayne
Rivas, Marian Lucy

TEXAS
Brown, Barry W
Crick, Rex Edward
Foster, Donald Myers
Fox, George Edward
Furlong, Norman Burr, Jr
Hanis, Craig L
Hurley, Laurence Harold
Jansson, Birger
Ladde, Gangaram Shivlingappa
Li, Wen-Hsiung
Marks, Gerald A
Saltzberg, Bernard
Schuhmann, Robert Ewald
Srinivasan, Ramachandra Srini
Thames, Howard Davis, Jr
Walter, Charles Frank
White, Robert Allen
Zimmerman, Stuart O

UTAH
Cole, Walter Eckle

VERMONT
Costanza, Michael Charles
Tyree, Melvin Thomas

VIRGINIA
Adams, James Milton
Chandler, Jerry LeRoy
Mikulecky, Donald C
Roth, Allan Charles
Tyson, John Jeanes

WASHINGTON
Anderson, James Jay
Bare, Barry Bruce
Barker, Morris Wayne
Dailey, Frank Allen
Kareiva, Peter Michael
Laevastu, Taivo
Moody, Michael Eugene
Moolgavkar, Suresh Hiraji

WEST VIRGINIA
Brown, Mark Wendell

WISCONSIN
Carpenter, Stephen Russell
Chover, Joshua
Curie-Cohen, Martin Michael
Holden, James Edward
Jungck, John Richard

WYOMING
Boyce, Mark S

ALBERTA
Beck, James S
Freedman, Herbert Irving
McGann, Locksley Earl
Prepas, Ellie E
Strobeck, Curtis
Voorhees, Burton Hamilton
Whitelaw, William Albert

BRITISH COLUMBIA
Hoffmann, Geoffrey William
LeBlond, Paul Henri

MANITOBA
Gordon, Richard

NEW BRUNSWICK
Van Groenewoud, Herman

NEWFOUNDLAND
Wroblewski, Joseph S

NOVA SCOTIA
Rosen, Robert
Silvert, William Lawrence
Wong, Alan Yau Kuen

ONTARIO
Barr, David Wallace
Endrenyi, Laszlo
Haynes, Robert Hall
Miller, Donald Richard
Norwich, Kenneth Howard
Robertson, George Wilber
Taylor, Peter D
Zuker, Michael

QUEBEC
Diksic, Mirko
Jolicoeur, Pierre
Legendre, Pierre
Mackey, Michael Charles

Outerbridge, John Stuart
Roberge, Fernand Adrien

SASKATCHEWAN
Lapp, Martin Stanley

OTHER COUNTRIES
Bodmer, Walter Fred
Haegele, Klaus D
Kang, Kewon
Lindenmayer, Aristid
May, Robert McCredie
Rio, María Esther
Stearns, Stephen Curtis
Wrede, Don Edward

Biometrics-Biostatistics

ALABAMA
Best, Troy Lee
Borton, Thomas Ernest
Hazelrig, Jane
Hoff, Charles Jay
Holt, William Robert
Johnson, Evert William
McGuire, John Albert
Watson, Jack Ellsworth
Wolff, Albert Eli

ALASKA
Dahlberg, Michael Lee
Fox, John Frederick
Heifetz, Jonathan
Olsen, James Calvin
Reynolds, James Blair
Snyder, George Richard

ARIZONA
Aickin, Mikel G
Allen, Stephen Gregory
Denny, John Leighton
Gaud, William S
Jacobowitz, Ronald
Moon, Thomas Edward
Nicolls, Ken E
Wilson, F(rank) Douglas

ARKANSAS
Brown, Connell Jean
Dunn, James Eldon
Gaylor, David William
Holson, Ralph Robert
Walls, Robert Clarence

CALIFORNIA
Abramson, Norman Jay
Ahumada, Albert Jil, Jr
Anderson, Dennis Elmo
Arthur, Susan Peterson
Bentley, Donald Lyon
Bernstein, Leslie
Bray, Richard Newton
Brown, Byron William, Jr
Burrill, Melinda Jane
Callahan-Compton, Joan Rea
Chang, Potter Chien-Tien
Chiang, Chin Long
Clark, Virginia
Cumberland, William Glen
Enright, James Thomas
Farris, David Allen
Forsythe, Alan Barry
Foster, Ken Wood
Franti, Charles Elmer
Garber, Morris Joseph
Gerrodette, Timothy
Gordon, Louis
Gray, Gregory Edward
Guthrie, Donald
Hill, Annlia Paganini
Hills, F Jackson
Hubert, Helen Betty
Jackson, Crawford Gardner, Jr
James, Kenneth Eugene
Jerison, Harry Jacob
Klauber, Melville Roberts
Kope, Robert Glenn
Kraemer, Helena Chmura
Kuzma, Jan Waldemar
Lachenbruch, Peter Anthony
Little, Thomas Morton
Lovich, Jeffrey Edward
Mackey, Bruce Ernest
Malone, Marvin Herbert
Marshall, Rosemarie
Massey, Frank Jones, Jr
Merchant, Roland Samuel, Sr
Metter, Gerald Edward
Mikel, Thomas Kelly, Jr
Moore, Dan Houston, II
Moriarty, David John
Moses, Lincoln Ellsworth
Mueller, Thomas Joseph
Murad, Turhon Allen
Parrish, Richard Henry
Rao, Jammalamadaka S
Ray, Rose Marie
Rice, Dorothy Pechman
Rocke, David M
Rotenberry, John Thomas
Shonick, William
Show, Ivan Tristan
Swan, Shanna Helen

Tarter, Michael E
Treloar, Alan Edward
Ury, Hans Konrad
Varady, John Carl
Wall, Francis Joseph
Ward, David Gene
Weiner, Daniel Lee
Wenzel, Zita Marta
Wiggins, Alvin Dennie
Wimer, Cynthia Crosby
Wyzga, Ronald Edward
Zippin, Calvin

COLORADO
Alpern, Herbert P
Angleton, George M
Bernstein, Stephen
Bowden, David Clark
Chase, Gerald Roy
Gough, Larry Phillips
Johnson, Thomas Eugene
Jones, Richard Hunn
Mielke, Paul W, Jr
Oldemeyer, John Lee
Porter, Kenneth Raymond

CONNECTICUT
Chan, Yick-Kwong
Holford, Theodore Richard
Holsinger, Kent Eugene
Honeyman, Merton Seymour
John, Hugo Herman
White, Colin
Woronick, Charles Louis

DELAWARE
Fellner, William Henry
Miller, Douglas Charles
Pell, Sidney
Schuler, Victor Joseph

DISTRICT OF COLUMBIA
Ahmed, Susan Wolofski
Ampy, Franklin
Chiazze, Leonard, Jr
Cohen, Michael Paul
Halperin, Max
Hayek, Lee-Ann Collins
Hurley, Frank Leo
Myers, Wayne Lawrence
Nayak, Tapan Kumar
Rao, Mamidanna S
Stroup, Cynthia Roxane
Thall, Peter Francis
Tyrrell, Henry Flansburg

FLORIDA
Banghart, Frank W
Barnard, Donald Roy
Brain, Carlos W
Chew, Victor
DeRousseau, C(arol) Jean
Gaunt, Stanley Newkirk
Greenberg, Richard Alvin
Hagans, James Albert
Hagenmaier, Robert Doller
Kastenbaum, Marvin Aaron
Littell, Ramon Clarence
Lyman, Gary Herbert
McCann, James Alwyn
McCoy, Earl Donald
Nichols, Robert Loring
Rathburn, Carlisle Baxter, Jr
Stamper, James Harris
Stroh, Robert Carl
Wilkerson, James Edward

GEORGIA
Bailey, Robert Leroy
Carter, Melvin Winsor
Cohen, Alonzo Clifford, Jr
Green, Willard Wynn
Hogue, Carol Jane Rowland
Hutcheson, Kermit
Karon, John Marshall
Kneib, Ronald Thomas
Kutner, Michael Henry
Longini, Ira Mann, Jr
Lyons, Nancy I
Mabry, John William
McLaurin, Wayne Jefferson
Mosteller, Robert Cobb
Nichols, Michael Charles
Pepper, William Donald
Pogue, Richard Ewert
Shure, Donald Joseph
Ware, Kenneth Dale

HAWAII
Chung, Chin Sik
Freed, Leonard Alan
Krupp, David Alan
Silva, James Anthony
Simon, Christine Mae

IDAHO
Everson, Dale O
Keller, Barry Lee

ILLINOIS
Best, William Robert
Carmer, Samuel Grant
Carnes, Bruce Alfred
Carrier, Steven Theodore

Chen, Edwin Hung-Teh
Corruccini, Robert Spencer
Deysach, Lawrence George
Dyer, Alan Richard
Fernando, Rohan Luigi
Flay, Brian Richard
Ghent, Arthur W
Grossman, Michael
Haenszel, William Manning
Katz, Alan Jeffrey
Kruse, Kipp Colby
Loehle, Craig S
Martin, Yvonne Connolly
Mattson, Dale Edward
Myers, Charles Christopher
Norusis, Marija Jurate
Parsons, Carl Michael
Sacks, Jerome
Schaeffer, David Joseph
Seif, Robert Dale
Shanks, Roger D
Sollberger, Arne Rudolph
Walker, William M
Zar, Jerrold Howard

INDIANA
Boston, Andrew Chester
Cerimele, Benito Joseph
Hui, Siu Lui
Martin, Truman Glen
Moser, John William, Jr
Nyquist, Wyman Ellsworth
Pachut, Joseph F(rancis)
Parkhurst, David Frank
Sampson, Charles Berlin
Samuels, Myra Lee
Stewart, Terry Sanford
Wolfrom, Glen Wallace

IOWA
Berger, Philip Jeffrey
Clark, William Richard
Cox, Charles Philip
Cox, David Frame
Fuller, Wayne Arthur
Hatfield, Jerry Lee
Hotchkiss, Donald K
Kempthorne, Oscar
Remington, Richard Delleraine
Wunder, William W

KANSAS
Fryer, Holly Claire
Kaufman, Glennis Ann
Lanman, Robert Charles
Schalles, Robert R
Slade, Norman Andrew
Stewart, William Henry
Wassom, Clyde E
Welch, Stephen Melwood

KENTUCKY
Hays, Virgil Wilford
O'Connor, Carol Alf
Stevenson, Robert Jan
Surwillo, Walter Wallace

LOUISIANA
Baldwin, Virgil Clark, Jr
Diem, John Edwin
Eggen, Douglas Ambrose
Koonce, Kenneth Lowell
Orlando, Anthony Michael
Owings, Addison Davis
Seigel, Richard Allyn
Webber, Larry Stanford
Weinberg, Roger
Wright, Vernon Lee

MAINE
Gilbert, James Robert

MARYLAND
Abbey, Helen
Anello, Charles
Blot, William James
Brant, Larry James
Brodsky, Allen
Byar, David Peery
Chen, Tar Timothy
Chiacchierini, Richard Philip
Cutler, Jeffrey Alan
Diamond, Earl Louis
Edwards, Brenda Kay
Elkin, William Futter
Ellenberg, Jonas Harold
Ellenberg, Susan Smith
Embody, Daniel Robert
Feinleib, Manning
Feldman, Jacob J
Fisher, Pearl Davidowitz
Garrison, Robert J
Ginevan, Michael Edward
Goldberg, Irving David
Greenhouse, Samuel William
Haynes, Suzanne G
Heath, Robert Gardner
Hebel, John Richard
Hill, James Leslie
Homer, Louis David
Horn, Susan Dadakis
Jablon, Seymour
Jessup, Gordon L, Jr
Johnson, Mary Frances

Kimball, Allyn Winthrop
Knatterud, Genell Lavonne
Knoke, James Dean
Knox, Robert Gaylord
Kramer, Morton
Lachin, John Marion, III
Langenberg, Patricia Warrington
Lao, Chang Sheng
Levy, Paul Samuel
Litt, Bertram D
Meisinger, John Joseph
Mellits, E David
Milton, Roy Charles
Mosimann, James Emile
Myers, Max H
Nichols, James Dale
Osborne, J Scott, III
Pickle, Linda Williams
Rastogi, Suresh Chandra
Rider, Rowland Vance
Rosenberg, Saul H
Ross, Alan
Royall, Richard Miles
Russek-Cohen, Estelle
Schneiderman, Marvin Arthur
Shaw, Richard Franklin
Simon, Richard Macy
Smith, Lewis Wilbert
Smith, Paul John
Steinberg, Seth Michael
Suomi, Stephen John
Tarone, Robert Ernest
Tonascia, James A
Walker, Michael Dirck
Wilson, P David

MASSACHUSETTS
Ash, Arlene Sandra
Finn, John Thomas
Gelber, Richard David
Gleason, Ray Edward
Gordon, Claire Catherine
Gray, Douglas Carmon
Haight, Thomas H
Helgesen, Robert Gordon
Hoaglin, David Caster
Kamen, Gary P
Kolakowski, Donald Louis
Kunkel, Joseph George
Lagakos, Stephen William
Lavin, Philip Todd
Lemeshow, Stanley Alan
Rand, William Medden
Ransil, Bernard J(erome)
Reich, James Harry
Schoenfeld, David Alan
Smith, Frederick Edward
Smith, Tim Denis
Snow, Beatrice Lee
Stanley, Kenneth Earl
Tritchler, David Lynn
Tsiatis, Anastasios A
Wyshak, Grace
Zelen, Marvin

MICHIGAN
Assenzo, Joseph Robert
Bennett, Carl Leroy
Bookstein, Fred Leon
Brunden, Marshall Nils
Chang, William Y B
Cornell, Richard Garth
Gill, John Leslie
Huitema, Bradley Eugene
Keen, Robert Eric
Kowalski, Charles Joseph
McCrimmon, Donald Alan, Jr
McGilliard, Lon Dee
Mao, Ivan Ling
Metzler, Carl Maust
Mohberg, Noel Ross
Moll, Russell Addison
Musch, David C(harles)
Nordby, Gordon Lee
Patterson, Richard L
Schork, Michael Anthony
Simmons, Gary Adair
Sing, Charles F
Sokol, Robert James
Stoline, Michael Ross
Taylor, Mitchell Kerry
Tilley, Barbara Claire
Ullman, Nelly Szabo
Winterstein, Scott Richard
Zemach, Rita

MINNESOTA
Bartsch, Glenn Emil
Boen, James Robert
Boylan, William J
Curtsinger, James Webb
Elveback, Lillian Rose
Gatewood, Lael Cranmer
Goldman, Anne Ipsen
Hansen, Leslie Bennett
Jacobs, David R, Jr
Johnson, Eugene A
Keenan, Kathleen Margaret
Kjelsberg, Marcus Olaf
Kvalseth, Tarald Oddvar
Le, Chap Than
Loewenson, Ruth Brandenburger
McHugh, Richard B
Martin, Margaret Pearl

O'Fallon, William M
Siniff, Donald Blair
Sowell, John Basil
Stucker, Robert Evan
Taylor, William F
Weckwerth, Vernon Ervin

MISSISSIPPI
Meade, James Horace, Jr
Meydrech, Edward Frank
Watson, Clarence Ellis, Jr

MISSOURI
Flora, Jairus Dale, Jr
Gaffey, William Robert
Harvey, Joseph Eldon
Hill, Jack Filson
Hintz, Richard Lee
Moulton, Robert Henry
Smith, Richard Jay
Smith, Robert Francis
Spitznagel, Edward Lawrence, Jr
Thomson, Gordon Merle
Wette, Reimut

MONTANA
Brown, James Kerr
Jurist, John Michael
Metzgar, Lee Hollis

NEBRASKA
Bennett, Gary Lee
Compton, William A
Gardner, Charles Olda
Kimberling, William J
Moore, Kenneth J
Mumm, Robert Franklin
Schutz, Wilfred M

NEVADA
Flueck, John A
Kinnison, Robert Ray
Leedy, Clark D

NEW HAMPSHIRE
Hill, Percy Holmes
Urban, Willard Edward, Jr

NEW JERSEY
Blickstein, Stuart I
Clymer, Arthur Benjamin
Egerton, John Richard
Givens, Samuel Virtue
Grover, Gary James
Holland, Paul William
Hsu, Chin-Fei
London, Mark David
McGhee, George Rufus, Jr
Mather, Robert Eugene
Mietlowski, William Leonard
Miller, Alex
Morin, Peter Jay
Ordille, Carol Maria
Roberts-Marcus, Helen Miriam
Rodda, Bruce Edward
Saiger, George Lewis
Sharma, Ran S
Shirk, Richard Jay
Walsh, Teresa Marie

NEW MEXICO
Finkner, Morris Dale
Southward, Glen Morris

NEW YORK
Allaway, Norman C
Blumenson, Leslie Eli
Calhoon, Robert Ellsworth
Carruthers, Raymond Ingalls
Cisne, John Luther
Dibennardo, Robert
Dudewicz, Edward John
DuFrain, Russell Jerome
Durkin, Patrick Ralph
Emrich, Lawrence James
Falk, Catherine T
Federer, Walter Theodore
Feldman, Joseph Gerald
Fleiss, Joseph L
Greco, William Robert
Hucke, Dorothy Marie
Levene, Howard
Levin, Bruce
Lininger, Lloyd Lesley
Locke, Charles Stephen
Marcus, Leslie F
Mendell, Nancy Role
Mike, Valerie
Neugebauer, Richard
Oates, Richard Patrick
Porter, William Frank
Priore, Roger L
Robinson, Harry
Rockwell, Robert Franklin
Rohlf, F James
Schey, Harry Moritz
Schmee, Josef
Schneiweiss, Jeannette W
Shah, Bhupendra K
Shore, Roy E
Siegel, Carole Ethel
Singh, Madho
Slack, Nelson Hosking
Smoller, Sylvia Wassertheil

Biometrics-Biostatistics (cont)

Sokal, Robert Reuven
Sommer, Charles John
Stein, Theodore Anthony
Tanner, Martin Abba
Turnbull, Bruce William
Varma, Andre A O
Weinstein, Abbott Samson
Whitby, Owen
Wood, Paul Mix
Yozawitz, Allan
Zielezny, Maria Anna

NORTH CAROLINA
Abernathy, James Ralph
Atchley, William Reid
Chester, Alexander Jeffrey
Coulter, Elizabeth Jackson
Dinse, Gregg Ernest
Edelman, David Anthony
George, Stephen L
Gold, Harvey Joseph
Guess, Harry Adelbert
Hartford, Winslow H
Hayne, Don William
Hicks, Robert Eugene
Hoel, David Gerhard
Lessler, Judith Thomasson
Marcus, Allan H
Margolin, Barry Herbert
Moll, Robert Harry
Quade, Dana Edward Anthony
Rawlings, John Oren
Reckhow, Kenneth Howland
Reice, Seth Robert
Schaaf, William Edward
Sen, Pranab Kumar
Starmer, C Frank
Stier, Howard Livingston
Swallow, William Hutchinson
Sweeny, Hale Caterson
Turnbull, Craig David
Vaughan, Douglas Stanwood
Weir, Bruce Spencer
Wells, Henry Bradley
Williams, Ann Houston

NORTH DAKOTA
Lieberman, Diana Dale
Wrenn, William J

OHIO
Badger, George Franklin
Buncher, Charles Ralph
Cacioppo, John T
Caldito, Gloria C
Deddens, James Albert
Garg, Mohan Lal
Garton, David Wendell
Gartside, Peter Stuart
Guo, Shumei
Heckman, Carol A
Hoy, Casey William
Khamis, Harry Joseph
Klein, John Peter
Lake, Robin Benjamin
Lima, John J
Liu, Ting-Ting Y
McGinnis, John Thurlow
MacLean, David Belmont
Mukherjee, Mukunda Dev
Neff, Raymond Kenneth
Powers, Jean D
Reiches, Nancy A
Runkle, James Reade
Rustagi, Jagdish S
Siervogel, Roger M
Slutzky, Gale David
Williams, George W

OKLAHOMA
Calvert, Jon Channing
Morrison, Robert Dean
Sanderson, George Albert
Schaefer, Carl Francis
Sonleitner, Frank Joseph
Verhalen, Laval
Weeks, David Lee

OREGON
Ary, Dennis
Beatty, Joseph John
Isaacson, Dennis Lee
Lu, Kuo Hwa
Matarazzo, Joseph Dominic
Petersen, Roger Gene
Phillips, Donald Lundahl
Rowe, Kenneth Eugene
Senner, John William
Urquhart, N Scott

PENNSYLVANIA
Arrington, Wendell S
Copenhaver, Thomas Wesley
Cronk, Christine Elizabeth
Emberton, Kenneth C
Fienberg, Stephen Elliott
Gertz, Steven Michael
Gleser, Leon Jay
Gould, A Lawrence
Hendrickson, John Alfred, Jr
Hensler, Gary Lee
Iglewicz, Boris

Kesner, Michael H
Kleban, Morton H
Klinger, Thomas Scott
Krall, John Morton
Kushner, Harvey
Li, Ching Chun
Lindsay, Bruce George
Lustbader, Edward David
McLaughlin, Francis X(avier)
Maksymowych, Roman
Mason, Thomas Joseph
Menduke, Hyman
Miller, G(erson) H(arry)
Morehouse, Chauncey Anderson
Pinski, Gabriel
Rockette, Howard Earl, Jr
Roush, William Burdette
Rubin, Leonard Sidney
Rucinska, Ewa J
Schneider, Bruce E
Schor, Stanley
Singer, Arthur Chester
Spielman, Richard Saul
Sullivan, Alfred Dewitt
Wysocki, Charles Joseph
Zubin, Joseph

RHODE ISLAND
Costantino, Robert Francis

SOUTH CAROLINA
Drane, John Wanzer
Gross, Alan John
Hazen, Terry Clyde
Hunt, Hurshell Harvey
Loadholt, Claude Boyd
Miller, Millage Clinton, III
Pinder, John Edgar, III
Rust, Philip Frederick
Woodard, Geoffrey Dean Leroy

TENNESSEE
Dupont, William Dudley
Federspiel, Charles Foster
Gosslee, David Gilbert
Groer, Peter Gerold
Mathis, Philip Monroe
Nanney, Lillian Bradley
Somes, Grant William
Tan, Wai-Yuan
Vander Zwaag, Roger
Van Winkle, Webster, Jr
West, Dennis R
Zeighami, Elaine Ann

TEXAS
Bratcher, Thomas Lester
Brokaw, Bryan Edward
Carroll, Raymond James
Deming, Stanley Norris
Downs, Thomas D
Formanowicz, Daniel Robert, Jr
Fucik, John Edward
Gates, Charles Edgar
Glasser, Jay Howard
Gruber, George J
Hallum, Cecil Ralph
Hawkins, C Morton
Hsi, Bartholomew P
Huffman, Ronald Dean
Johnston, Dennis Addington
Johnston, Walter Edward
Kirk, Ivan Wayne
Kirk, Roger E
Littell, Arthur Simpson
Meyer, Betty Michelson
Richardson, Richard Harvey
Riewe, Marvin Edmund
Ring, Dennis Randall
Stedman, James Murphey
Vaughn, William King
Wehrly, Thomas Edward
Whorton, Elbert Benjamin
Wilcox, Roberta Arlene
Willig, Michael Robert

UTAH
Dueser, Raymond D
Hunt, Steven Charles
Reading, James Cardon
Ryel, Lawrence Atwell
Sisson, Donald Victor
Turner, David L(ee)

VERMONT
Costanza, Michael Charles
Emerson, John David
Haugh, Larry Douglas

VIRGINIA
Bailey, Robert Clifton
Cady, Foster Bernard
Carter, Walter Hansbrough, Jr
Chase, Helen Christina (Matulic)
Choi, Sung Chil
Fashing, Norman James
Flora, Roger E
Frahm, Richard R
Gardenier, Turkan Kumbaraci
Henry, Neil Wylie
Holtzman, Golde Ivan
Kilpatrick, S James, Jr
Leczynski, Barbara Ann
Loesch, Joseph G

McGilliard, Michael Lon
Minton, Paul Dixon
Munson, Arvid W
Reynolds, Marion Rudolph, Jr
Taub, Stephan Robert
Weiss, William

WASHINGTON
Bare, Barry Bruce
Barker, Morris Wayne
Bax, Nicholas John
Benedetti, Jacqueline Kay
Breslow, Norman Edward
Chapman, Douglas George
Clark, William Greer
Conquest, Loveday Loyce
Crowley, John James
Dailey, Frank Allen
Diehr, Paula Hagedorn
Dixon, Kenneth Randall
Farnum, Peter
Feigl, Polly Catherine
Fisher, Lloyd D
Gilbert, Ethel Schaefer
Grizzle, James Ennis
Hatheway, William Howell
Kronmal, Richard Aaron
Landis, Wayne G
McCaughran, Donald Alistair
Moolgavkar, Suresh Hiraji
Perrin, Edward Burton
Richardson, Michael Lewellyn
Rustagi, Krishna Prasad
Thomas, John M
Thompson, Donovan Jerome
Towner, Richard Henry
Wahl, Patricia Walker
Weiss, Noel Scott

WEST VIRGINIA
Keller, Edward Clarence, Jr
Thayne, William V
Townsend, Edwin C

WISCONSIN
Carpenter, Stephen Russell
Casler, Michael Darwin
Klotz, Jerome Hamilton
Kowal, Robert Raymond
Long, Charles Alan
Owens, John Michael

WYOMING
Boyce, Mark S
Eddy, David Maxon
Mateer, Niall John

PUERTO RICO
Bangdiwala, Ishver Surchand

ALBERTA
Bella, Imre E
Hardin, Robert Toombs
Kalantar, Alfred Husayn
Yeh, Francis Cho-hao
Zammuto, Richard Michael

BRITISH COLUMBIA
Campbell, Alan
Hall, John Wilfred
Lauzier, Raymond B
Mitchell, Kenneth John
Pielou, Evelyn C
Routledge, Richard Donovan
Sterling, Theodor David
Todd, Mary Elizabeth

MANITOBA
Kingsley, Michael Charles Stephen
Scott, David Paul
Stephens, Newman Lloyd

ONTARIO
Balakrishnan, Narayanaswany
Bertell, Rosalie
Brown, Kenneth Stephen
Chambers, James Robert
Chapman, Judith-Anne Williams
DeBoer, Gerrit
Dunnett, Charles William
Endrenyi, Laszlo
Forbes, William Frederick
Gent, Michael
Goldsmith, Charles Harry
Ihssen, Peter Edowald
Robson, Douglas Sherman
Schneider, Bruce Alton
Wolynetz, Mark Stanley

QUEBEC
Bailar, John Christian, III
Corriveau, Armand Gerard
Jolicoeur, Pierre
Legendre, Pierre
Siemiatycki, Jack

SASKATCHEWAN
Spurr, David Tupper

OTHER COUNTRIES
Bernstein, David Maier
Fitzhugh, Henry Allen
Siu, Tsunpui Oswald

Biophysics

ALABAMA
Benos, Dale John
Cheung, Herbert Chiu-Ching
Cline, George Bruce
Downey, James Merritt
Elgavish, Ada S
Friedlander, Michael J
Friedman, Michael E
Golden, Michael Stanley
Goodall, Marcus Campbell
Halm, Dan Robert
Harvey, Stephen Craig
Hazelrig, Jane
Khaled, Mohammad Abu
Klip, Willem
Latimer, Paul Henry
Magargal, Wells Wrisley, II
Monti, John Anthony
Taylor, Aubrey Elmo
Tice, Thomas Robert
Urry, Dan Wesley
West, Seymour S

ALASKA
Fink, Thomas Robert

ARIZONA
Barr, Ronald Edward
Bier, Milan
Blankenship, Robert Eugene
Butler, Byron C
Connor, William Gorden
De Gaston, Alexis Neal
Diehn, Bodo
Epstein, L(udwig) Ivan
Erickson, Eric Herman, Jr
Holmes, William Farrar
Idso, Sherwood B
Kessler, John Otto
Kilkson, Rein
Lukas, Ronald John
Mardian, James K W
Moore, Ana Maria
Moore, Thomas Andrew
Mosher, Richard Arthur
Rupley, John Allen
Sanderson, Douglas G
Scott, Alwyn C
Siemankowski, Raymond Francis
Tollin, Gordon
Trautman, Rodes
Vermaas, Willem F J
Wyckoff, Ralph Walter Graystone

ARKANSAS
Baker, Max Leslie
Demarest, Jeffrey R
Everett, Wilbur Wayne
Hart, Ronald Wilson
Martin, Duncan Willis
Moss, Alfred Jefferson, Jr
Nagle, William Arthur
Steinmeier, Robert C

CALIFORNIA
Abbott, Bernard C
Amirkhanian, John David
Amis, Eric J
Baily, Norman Arthur
Baker, Richard Freligh
Bales, Barney Leroy
Baskin, Ronald J
Bearden, Alan Joyce
Berry, Christine Albachten
Berry, Edward Alan
Bhatnagar, Rajendra Sahai
Biedebach, Mark Conrad
Bjorkman, Pamela J
Blakely, Eleanor Alice
Boxer, Steven George
Brady, Allan Jordan
Breisch, Eric Alan
Bremermann, Hans J
Brokaw, Charles Jacob
Bruner, Leon James
Brunk, Clifford Franklin
Budinger, Thomas Francis
Burns, Victor Will
Buttlaire, Daniel Howard
Carrano, Anthony Vito
Chan, Sunney Ignatius
Chen, Yang-Jen
Chen, Yi-Der
Chou, Chung-Kwang
Chu, William Tongil
Clayton, Roderick Keener
Codrington, Robert Smith
Connelly, Clarence Morley
Cooke, Roger
Courtney, Kenneth Randall
Crandall, Walter Ellis
Curtis, Stanley Bartlett
Dalton, Francis Norbert
Davis, Thomas Pearse
Deamer, David Wilson, Jr
Deonier, Richard Charles
Dickinson, Wade
Dill, Kenneth Austin
Dismukes, Robert Key
Dobson, R Lowry
Dubuc, Paul U
Dudek, F Edward

Du Pont, Frances Marguerite
Duzgunes, Nejat
Edelman, Jay Barry
Edmonds, Peter Derek
Eiler, John Joseph
Eisenberg, David
Eisenman, George
Evans, Evan Cyfeiliog, III
Fain, Gordon Lee
Feher, George
Fink, Anthony Lawrence
Fletterick, Robert John
Fortes, George (Peter Alexander)
Frey, Terrence G
Gaffey, Cornelius Thomas
Garfin, David Edward
Gill, James Edward
Glaeser, Robert M
Glick, Harold Alan
Goerke, Jon
Gofman, John William
Goldman, Marvin
Grantz, David Arthur
Gray, Joe William
Griffith, Owen Malcolm
Groce, David Eiben
Hall, James Ewbank
Hammes, Gordon G
Hanawalt, Philip Courtland
Hanson, Carl Veith
Hathaway, Gary Michael
Hayes, Thomas L
Haymond, Herman Ralph
Hazelwood, Robert Nichols
Hearst, John Eugene
Heath, Robert Louis
Hegenauer, Jack C
Hemmingsen, Barbara Bruff
Hemmingsen, Edvard A(lfred)
Henderson, Sheri Dawn
Hill, Terrell Leslie
Holmes, Donald Eugene
Hong, Keelung
Hooker, Thomas M, Jr
Hopfield, John Joseph
Horwitz, Joseph
Hsieh, Jen-Shu
Iberall, Arthur Saul
Jackson, Meyer B
James, Thomas Larry
Jariwalla, Raxit Jayantilal
Jensen, Ronald Harry
Johnson, C Scott
Junge, Douglas
Kavenoff, Ruth
Keizer, Joel Edward
Kelly, Donald Horton
Kendig, Joan Johnston
Kenyon, George Lommel
Kiger, John Andrew, Jr
Kim, Sung-Hou
Kirby, Jon Allan
Kitting, Christopher Lee
Klein, Melvin Phillip
Kleinfeld, A M
Korenbrot, Juan Igal
Kortright, James McDougall
Lad, Pramod Madhusudan
Lanyi, Janos K
Latta, Harrison
Lecar, Harold
Lewis, Steven M
Licko, Vojtech
Lieber, Richard L
Lindgren, Frank Tycko
Live, David H
Lubliner, J(acob)
Lucas, Joe Nathan
Luxon, Bruce Arlie
Lyman, John Tompkins
Lyon, Irving
McConnell, Harden Marsden
Macey, Robert Irwin
Mackay, Ralph Stuart
McNamee, Mark G
Maestre, Marcos Francisco
Magde, Douglas
Maki, August Harold
Mandelkern, Mark Alan
Manning, JaRue Stanley
Marcus, Carol Silber
Marsh, Donald Jay
Matthews, David Allan
Matthews, Harry Roy
Meehan, Thomas (Dennis)
Mel, Howard Charles
Mendelsohn, Mortimer Lester
Mendelson, Robert Alexander, Jr
Milanovich, Fred Paul
Miledi, Ricardo
Miljanich, George Paul
Morales, Manuel Frank
Mortimer, Robert Keith
Mueller, Thomas Joseph
Nacht, Sergio
Nakamura, Robert Masao
Nelson, Gary Joe
Neville, James Ryan
Nichols, Alexander Vladimir
Norman, Amos
Nothnagel, Eugene Alfred
Nuccitelli, Richard Lee
O'Benar, John DeMarion
O'Konski, Chester Thomas

O'Leary, Dennis Patrick
Pacela, Allan F
Paolini, Paul Joseph
Papahadjopoulos, Demetrios Panayotis
Paselk, Richard Alan
Payne, Philip Warren
Phelps, Michael Edward
Phibbs, Roderic H
Pollycove, Myron
Quintanilha, Alexandre Tiedtke
Ray, Dan S
Reese, Robert Trafton
Rich, Terrell L
Richard, Christopher Alan
Richardson, Irvin Whaley
Ritz-Gold, Caroline Joyce
Robison, William Lewis
Ross, John
Rothman, Stephen Sutton
Ruby, Ronald Henry
Saifer, Mark Gary Pierce
Sammak, Paul J
Sanui, Hisashi
Sargent, Thornton William, III
Sauer, Kenneth
Schimmerling, Walter
Schmid, Carl William
Schmid, Peter
Schmid-Schoenbein, Geert W
Schmidt, Paul Gardner
Schulte, Alfons F
Shafer, Richard Howard
Shepanski, John Francis
Sheppard, Asher R
Shetlar, Martin David
Siegel, Edward
Silk, Margaret Wendy Kuhn
Simmons, Norman Stanley
Singer, Jerome Ralph
Sinsheimer, Robert Louis
Smith, Emil L
Smith, Helene Sheila
Smith, Kendric Charles
Smith, William R
Snowdowne, Kenneth William
Snyder, Robert Gene
Solomon, Edward I
Sondhaus, Charles Anderson
Spitzer, Nicholas Canaday
Stannard, J Newell
Stevens, Lewis Axtell
Strickland, Erasmus Hardin
Sullivan, Sean Michael
Thomas, Charles Allen, Jr
Thomas, Richard Sanborn
Thompson, Lawrence Hadley
Tinoco, Joan W H
Tobias, Cornelius Anthony
Tomich, John Matthew
Toy, Arthur John, Jr
Tsien, Richard Winyu
Van Atta, John R
Van Dilla, Marvin Albert
Vickery, Larry Edward
Wade, Michael James
Walter, Harry
Ward, John F
Welch, Graeme P
Wemmer, David Earl
White, Stephen Halley
Wilbur, David Wesley
Williams, Lawrence Ernest
Williams, Robley Cook
Winchell, Harry Saul
Winet, Howard
Withers, Hubert Rodney
Wong, Kin-Ping
Wyatt, Philip Joseph
Yang, Chui-Hsu (Tracy)
Yang, Jen Tsi
Yayanos, A Aristides
Yguerabide, Juan
Zimm, Bruno Hasbrouck
Zucker, Robert Stephen

COLORADO
Barisas, Bernard George, Jr
Beam, Kurt George, Jr
Best, LaVar
Cann, John Rusweiler
Carpenter, Donald Gilbert
Cech, Carol Martinson
Clark, Benton C
Cummings, Donald Joseph
Dahl, Adrian Hilman
Dewey, Thomas Gregory
Elkind, Mortimer M
Fotino, Mircea
Fox, Michael Henry
Furcinitti, Paul Stephen
Gamow, Rustem Igor
Gill, Stanley Jensen
Johnson, James Edward
Lett, John Terence
Levinson, Simon Rock
McHenry, Charles S
Mueller, Theodore Arnold
Naughton, Michael A
Olson, John Melvin
Pautler, Eugene L
Pettijohn, David E
Phillipson, Paul Edgar
Puck, Theodore Thomas
Sanderson, Richard James

Seibert, Michael
Solie, Thomas Norman
Todd, Paul Wilson
Uhlenbeck, Olke Cornelis
Woody, Robert Wayne
York, Sheldon Stafford

CONNECTICUT
Adler, Alan David
Agulian, Samuel Kevork
Allewell, Norma Mary
Armitage, Ian MacLeod
Aylor, Donald Earl
Boulpaep, Emile L
Braswell, Emory Harold
Bronner, Felix
Bursuker, Isia
Coleman, Joseph Emory
Crain, Richard Cullen
Engelman, Donald Max
Forbush, Bliss, III
Gagge, Adolf Pharo
Gent, Martin Paul Neville
Glasel, Jay Arthur
Golub, Efim I
Heller, John Herbert
Henderson, Edward George
Herbette, Leo G
Hoffman, Joseph Frederick
Howard-Flanders, Paul
Hutchinson, Franklin
Knox, James Russell, Jr
Lambrakis, Konstantine Christos
Loew, Leslie Max
Macnab, Robert Marshall
Markowitz, David
Moore, Peter Bartlett
Moretz, Roger C
Morowitz, Harold Joseph
Peterson, Cynthia Wyeth
Philo, John Sterner
Prestegard, James Harold
Prigodich, Richard Victor
Richards, Frederic Middlebrook
Rockwell, Sara Campbell
Russu, Irina Maria
Schor, Robert
Schulz, Robert J
Schuster, Todd Mervyn
Schwartz, Ilsa Roslow
Schwartz, Tobias Louis
Sha'afi, Ramadan Issa
Shulman, Robert Gerson
Slayman, Clifford L
Sturtevant, Julian Munson
Terry, Thomas Milton
Tourtellotte, Mark Eton
Waxman, Stephen George
Wolf, Elizabeth Anne
Wyckoff, Harold Winfield
Yphantis, David Andrew
Zubal, I George

DELAWARE
Bolton, Ellis Truesdale
Brenner, Stephen Louis
Dwivedi, David A
Ehrlich, Robert Stark
Hartzell, Charles Ross, III
Jain, Mahendra Kumar
LaRossa, Robert Alan
Lichtner, Francis Thomas, Jr
Moe, Gregory Robert
Nelson, Mark James
Reuben, Jacques
Roe, David Christopher
Sharnoff, Mark

DISTRICT OF COLUMBIA
Barr, Nathaniel Frank
Chang, Eddie Li
DeLevie, Robert
Goodenough, David John
Gray, Irving
Jennings, William Harney, Jr
Jones, Janice Lorraine
Karle, Isabella Lugoski
Kowalsky, Arthur
Ledley, Robert Steven
Malinin, George I
Misra, Prabhakar
Montrose, Charles Joseph
O'Leary, Timothy Joseph
Phillips, William Dale
Rall, William Frederick
Robertson, James Sydnor
Rosenberg, Edith E
Salu, Yehuda
Schnur, Joel Martin
Shukla, Kamal Kant
Sukow, Wayne William
Sulzman, Frank Michael
Trubatch, Sheldon L
Vaishnav, Ramesh
Ward, Keith Bolen, Jr
Wickman, Herbert Hollis
Wood, Robert Winfield

FLORIDA
Anderson, Peter Alexander Vallance
Barber, Michael James
Beidler, Lloyd M
Biersdorf, William Richard
Block, Ronald Edward

Blum, Alvin Seymour
Boyce, Richard P
Brandon, Frank Bayard
Buss, Daryl Dean
Caswell, Anthony H
Cohen, Robert Jay
Colahan, Patrick Timothy
Eagle, Donald Frohlichstein
Edwards, Charles
Esser, Alfred F
Finlayson, Birdwell
Gilmer, Penny Jane
Gordon, Kenneth Richard
Hayashi, Teru
Haynes, Duncan Harold
Herbert, Thomas James
Himel, Chester Mora
Huebner, Jay Stanley
Kasha, Michael
Kerrick, Wallace Glenn Lee
Landowne, David
Leif, Robert Cary
Lindsey, Bruce Gilbert
McConnell, Dennis Brooks
Magleby, Karl LeGrande
Mankin, Richard Wendell
Moulton, Grace Charbonnet
Mukherjee, Pritish
Muller, Kenneth Joseph
Pepinsky, Raymond
Pressman, Berton Charles
Purich, Daniel Lee
Rhodes, William Clifford
Roberts, Sanford B(ernard)
Rose, Birgit
Scarpace, Philip J
Schleyer, Walter Leo
Schoor, W Peter
Shah, Dinesh Ochhavlal
Sharp, John Turner
Snyder, Patricia Ann
Stevens, Bruce Russell
Szczepaniak, Krystyna
Thorhaug, Anitra L
Wickstrom, Eric
Williams, Theodore P
Young, David Michael
Zengel, Janet Elaine

GEORGIA
Abercrombie, Ronald Ford
Adams, Robert Johnson
De Felice, Louis John
DeHaan, Robert Lawrence
Dusenbery, David Brock
Edmondson, Dale Edward
Etzler, Frank M
Eubig, Casimir
Fechheimer, Marcus
Fox, Ronald Forrest
Godt, Robert Eugene
Green, Keith
Habte-Mariam, Yitbarek
Hall, Dwight Hubert
Joyner, Ronald Wayne
Lee, John William
Little, Robert Colby
Morgan, Karl Ziegler
Moriarty, C Michael
Netzel, Thomas Leonard
Nosek, Thomas Michael
Pohl, Douglas George
Pooler, John Preston
Sadun, Alberto Carlo
Scott, Robert Allen
Sophianopoulos, Alkis John
Sprawls, Perry, Jr
Stevenson, Dennis A
Wallace, Robert Henry
Wartell, Roger Martin
Wilson, William David
Yeargers, Edward Klingensmith
Yu, Nai-Teng

HAWAII
Cooke, Ian McLean

IDAHO
D'Aoust, Brian Gilbert
Moore, Richard
Tanner, John Eyer, Jr

ILLINOIS
Aktipis, Stelios
Andriacchi, Thomas Peter
Applebury, Meredithe L
Barany, Michael
Barr, Lloyd
Bond, Howard Edward
Borso, Charles S
Bosmann, Harold Bruce
Burt, Charles Tyler
Caspary, Donald M
Chan, Yun Lai
Chang, Chong-Hwan
Chignell, Derek Alan
Cohn, Gerald Edward
Cox, Thomas C
Crofts, Antony R
Czerlinski, George Heinrich
Dallos, Peter John
Dawson, M Joan
Debrunner, Peter Georg
DeVault, Don Charles

Biophysics (cont)

Dickinson, Helen Rose
Doi, Kunio
Domanik, Richard Anthony
Donnelly, Thomas Henry
Ducoff, Howard S
Dunn, Floyd
Dunn, Robert Bruce
Dutta, Pulak
Ebrey, Thomas G
Eder, Douglas Jules
Eisenberg, Robert S
Fields, Theodore
Fleming, Graham Richard
Frauenfelder, Hans (Emil)
Fung, Leslie Wo-Mei
Geil, Phillip H
Gennis, Robert Bennett
Gillette, Rhanor
Gingle, Alan Raymond
Gould, John Michael
Govindjee, M
Gratton, Enrico
Grdina, David John
Grossweiner, Leonard Irwin
Hagstrom, Ray Theodore
Harris, Lowell Dee
Helman, Sandy I
Hendee, William Richard
Hoffman, Brian Mark
Horwitz, Alan Fredrick
Hubbard, Lincoln Beals
Ingram, Forrest Duane
Jakobsson, Eric Gunnar, Sr
Johnson, Michael Evart
Josephs, Robert
Jursinic, Paul Andrew
Keiderling, Timothy Allen
Kidder, George Wallace, III
Kim, Yung Dai
Kimura, Mineo
Koushanpour, Esmail
Kubitschek, Herbert Ernest
Kuchnir, Franca Tabliabue
Kuczmarski, Edward R
Lange, Yvonne
Lanzl, Lawrence Herman
Lauffenburger, Douglas Alan
Lauterbur, Paul Christian
LeBreton, Pierre Robert
Levine, Michael W
Longworth, James W
MacHattie, Lorne Allister
Makinen, Marvin William
Miller, Donald Morton
Mittenthal, Jay Edward
Nakajima, Shigehiro
Nakajima, Yasuko
Narahashi, Toshio
Nilges, Mark J
O'Brien, William Daniel, Jr
Offner, Franklin Faller
Oldfield, Eric
Olsen, Kenneth Wayne
Oono, Yoshitsugu
Ovadia, Jacques
Page, Ernest
Peak, Meyrick James
Pepperberg, David Roy
Preston, Robert Leslie
Rakowski, Robert F
Reynolds, Larry Owen
Ries, Herman Elkan, Jr
Robinson, Thomas Frank
Rowland, Robert Edmund
Scharf, Arthur Alfred
Scheiner, Steve
Schiffer, Marianne Tsuk
Schlenker, Robert Alison
Sherman, Warren V
Shriver, John William
Singer, Irwin
Singh, Sant Parkash
Sleator, William Warner, Jr
Sligar, Stephen Gary
Spears, Kenneth George
Stevens, Fred Jay
Swartz, Harold M
Taylor, Edwin William
Thurnauer, Marion Charlotte
Uretz, Robert Benjamin
Wang, Andrew H-J
Wasielewski, Michael Roman
Weber, Gregorio
Weissman, Michael Benjamin
Westbrook, Edwin Monroe
White, William
Whitmarsh, Clifford John
Wilson, Hugh Reid
Wolynes, Peter Guy
Wraight, Colin Allen
Wynveen, Robert Allen

INDIANA

Allerhand, Adam
Alvager, Torsten Karl Erik
Ascarelli, Gianni
Bakken, George Stewart
Bina, Minou
Bockrath, Richard Charles, Jr
Bosron, William F
Bridges, C David
Bryan, William Phelan

Byrn, Stephen Robert
Cramer, William Anthony
Davis, Grayson Steven
Filmer, David Lee
Fong, Francis K
Frank, Bruce Hill
Friedman, Julius Jay
Haak, Richard Arlen
Havel, Henry Acken
Helrich, Carl Sanfrid, Jr
Hui, Chiu Shuen
Kelly-Fry, Elizabeth
Maass-Moreno, Roberto
Macchia, Donald Dean
Madden, Keith Patrick
Nowak, Thomas
Nussbaum, Elmer
Ortoleva, Peter Joseph
Pak, William Louis
Pearlstein, Robert Milton
Quay, John Ferguson
Randall, James Edwin
Schauf, Charles Lawrence
Sherman, Louis Allen
Strickholm, Alfred
Swez, John Adam
Van Frank, Richard Mark
Vogelhut, Paul Otto
Wolff, Ronald Keith

IOWA

Applequist, Jon Barr
Berg, Virginia Seymour
Campbell, Kevin Peter
Chandran, Krishnan Bala
Foss, John G(erald)
Getting, Peter Alexander
Greenberg, Everett Peter
Jennings, Michael Leon
Kawai, Masataka
Kintanar, Agustin
Mayfield, John Eric
Rougvie, Malcolm Arnold
Shen, Sheldon Shih-Ta
Six, Erich Walther
Struve, Walter Scott
Wu, Chun-Fang
Wunder, Charles C(ooper)

KANSAS

Barton, Janice Sweeny
Colen, Alan Hugh
Friesen, Benjamin S
Gordon, Michael Andrew
Guikema, James Allen
Hersh, Robert Tweed
Kimler, Bruce Franklin
Krishnan, Engil Kolaj
Manney, Thomas Richard
Michaelis, Elias K
Valenzeno, Dennis Paul
Welling, Daniel J
Yarbrough, Lynwood R

KENTUCKY

Bowen, Thomas Earle, Jr
Butterfield, David Allan
Christensen, Ralph C(hresten)
Coffey, Charles William, II
Coohill, Thomas Patrick
Engelberg, Joseph
Franke, Ernst Karl
Getchell, Thomas V
Hirsch, Henry Richard
Sayeg, Joseph A
Schilb, Theodore Paul
Syed, Ibrahim Bijli
Walker, Sheppard Matthew
Wyssbrod, Herman Robert

LOUISIANA

Azzam, Rasheed M A
Beuerman, Roger Wilmer
Eggen, Douglas Ambrose
Hamori, Eugene
Heyn, Anton Nicolaas Johannes
Hyman, Edward Sidney
Klyce, Stephen Downing
Levitzky, Michael Gordon
Luftig, Ronald Bernard
McAfee, Robert Dixon
Marino, A A
Miceli, Michael Vincent
Navar, Luis Gabriel
Taylor, Eric Robert
Welch, George Rickey
Wiechelman, Karen Janice

MAINE

Blake, Richard D
Hughes, William Taylor
Kirkpatrick, Francis Hubbard
Nelson, Clifford Vincent
Rhodes, William Gale

MARYLAND

Aamodt, Roger Louis
Ackerman, Eric J
Adelman, William Joseph, Jr
Amzel, L Mario
Andersen, Frank Alan
Andrews, Stephen Brian
Astumian, Raymond Dean
Battey, James F

Bax, Ad
Bellino, Francis Leonard
Berger, Robert Lewis
Berson, Alan
Berzofsky, Jay Arthur
Bicknell-Brown, Ellen
Blaustein, Mordecai P
Blumenthal, Robert Paul
Bockstahler, Larry Earl
Brinley, Floyd John, Jr
Brownell, William Edward
Burt, David Reed
Bush, C Allen
Cantor, Kenneth P
Carlson, Francis Dewey
Carroll, Raymond James
Cassman, Marvin
Charney, Elliot
Chen, Henry Lowe
Clark, Carl Cyrus
Coble, Anna Jane
Colombini, Marco
Copeland, Edmund Sargent
Coulter, Charles L
Cyr, W Howard
Davies, David R
Davies, Philip Winne
Devreotes, Peter Nicholas
Dintzis, Howard Marvin
Dintzis, Renee Zlochover
Eanes, Edward David
Eaton, William Allen
Edidin, Michael Aaron
Ehrenstein, Gerald
Eichhorn, Gunther Louis
Eipper, Betty Anne
Farrell, Richard Alfred
FitzHugh, Richard
Flower, Robert Walter
Foster, Margaret C
Freire, Ernesto I
Fried, Jerrold
Froehlich, Jeffrey Paul
Fuhr, Irvin
Gaffney, Betty Jean
Ganguly, Pankaj
Geduldig, Donald
Geisler, Fred Harden
Gilbert, Daniel Lee
Goldman, Arnold I
Goldman, Lawrence
Goode, Melvyn Dennis
Graham, Dale Elliott
Gunkel, Ralph D
Gupte, Sharmila Shaila
Gutierrez, Peter Luis
Hagins, William
Hansford, Richard Geoffrey
Harris, Andrew Leonard
Hartley, Robert William, Jr
Hendler, Richard Wallace
Hirsch, Roland Felix
Hirsh, Allen Gene
Ho, Henry Sokore
Hobbs, Ann Snow
Hofrichter, Harry James
Hollis, Donald Pierce
Houk, Albert Edward Hennessee
Howarth, John Lee
Huang, Charles Y
Hybl, Albert
Iwasa, Kunihiko
Jernigan, Robert Lee
Jost, Patricia Cowan
Kafka, Marian Stern
Kan, Lou Sing
Kaplan, Ann Esther
Kempner, Ellis Stanley
Ketley, Jeanne Nelson
Lakowicz, Joseph Raymond
Larrabee, Martin Glover
Lattman, Eaton Edward
Lederer, William Jonathan
Lee, Byungkook
Levinson, John Z
Levitan, Herbert
Lewis, Marc Simon
Lin, Diane Chang
Lin, Shin
Lipicky, Raymond John
Livengood, David Robert
Love, Warner Edwards
Lymangrover, John R
Lymn, Richard Wesley
Lytle, Carl David
McCally, Russell Lee
McLaughlin, Alan Charles
McPhie, Peter
Maddox, Yvonne T
Maslow, David E
Mildvan, Albert S
Miles, Edith Wilson
Mills, William Andy
Minton, Allen Paul
Moore, Richard Davis
Morris, Stephen Jon
Mullins, Lorin John
Myers, Lawrence Stanley, Jr
Nelson, Ralph Francis
Norvell, John Charles
Nossal, Ralph J
Padlan, Eduardo Agustin
Pagano, Richard Emil
Parsegian, Vozken Adrian

Podolsky, Richard James
Polinsky, Ronald John
Pollard, Thomas Dean
Pratt, Arnold Warburton
Puskin, Jerome Sanford
Rakhit, Gopa
Rall, Wilfrid
Rifkind, Joseph Moses
Rinzel, John Matthew
Robertson, Baldwin
Rodbard, David
Roti Roti, Joseph Lee
Saba, George Peter, II
Sastre, Antonio
Schambra, Philip Ellis
Shamoo, Adil E
Shin, Yong Ae Im
Shore, Moris Lawrence
Shropshire, Walter, Jr
Siatkowski, Ronald E
Sinclair, Warren Keith
Sinks, Lucius Frederick
Sjodin, Raymond Andrew
Smith, Elmer Robert
Sowers, Arthur Edward
Spero, Leonard
Steinert, Peter Malcolm
Steven, Alasdair C
Stevens, Walter Joseph
Straat, Patricia Ann
Swenberg, Charles Edward
Sze, Heven
Szu, Shousun Chen
Taylor, Lauriston Sale
Thomas, Robert Glenn
Torchia, Dennis Anthony
Ts'o, Paul On Pong
Tung, Ming Sung
Turner, R James
Vogl, Thomas Paul
Williams, Robert Jackson
Wolff, John B
Wolffe, Alan Paul
Yasbin, Ronald Eliott
Yockey, Hubert Palmer
Yu, Leepo Cheng

MASSACHUSETTS

Adelstein, Stanely James
Adler, Alice Joan
Anderson, Olivo Margaret
Auer, Henry Ernest
Auld, David Stuart
Bansil, Rama
Becker, David Stewart
Berg, Howard Curtis
Bethune, John Lemuel
Breckenridge, John Robert
Browne, Douglas Townsend
Brownell, Gordon Lee
Burke, John Michael
Cameron, John Stanley
Campbell, Mary Kathryn
Cantley, Lewis Clayton
Carey, Martin Conrad
Castronovo, Frank Paul, Jr
Catsimpoolas, Nicholas
Clapp, Roger Edge
Cohen, Carolyn
Correia, John Arthur
Crusberg, Theodore Clifford
Curtis, Joseph C
DeRosier, David J
Doane, Marshall Gordon
Downer, Nancy Wuerth
Epp, Edward Rudolph
Epstein, Herman Theodore
Fairbanks, Grant
Falchuk, Kenneth H
Fasman, Gerald David
Fine, Samuel
Ford, Norman Cornell, Jr
Fossel, Eric Thor
Frishkopf, Lawrence Samuel
George, Stephen Anthony
Glimcher, Melvin Jacob
Goldman, David Eliot
Gross, David John
Hamilton, James Arthur
Hardy, William Lyle
Harrison, Stephen Coplan
Hartline, Peter Haldan
Hastings, John Woodland
Herzfeld, Judith
Herzlinger, George Arthur
Hoffman, Allen Herbert
Holmquist, Barton
Hoop, Bernard, Jr
Hwang, San-Bao
Inoue, Shinya
Jaffe, Lionel F
Jordan, Peter C H
Karasz, Frank Erwin
Kase, Kenneth Raymond
King, John Gordon
Klibanov, Alexander M
Koehler, Andreas Martin
Koltun, Walter Lang
Kropf, Allen
Kuo, Lawrence C
Laing, Ronald Albert
Lamola, Angelo Anthony
Langley, Kenneth Hall
Lees, Sidney

LeFevre, Paul Green
Lehman, William Jeffrey
Lele, Padmakar Pratap
Lerman, Leonard Solomon
Levy, Elinor Miller
Loretz, Thomas J
Loscalzo, Joseph
MacKerell, Alexander Donald, Jr
McMahon, Thomas Arthur
MacNichol, Edward Ford, Jr
Makowski, Lee
Marrero, Hector G
Marx, Kenneth Allan
Melchior, Donald L
Miller, Keith Wyatt
Morgan, Kathleen Greive
Neuringer, Leo J
Orme-Johnson, Nanette Roberts
Orme-Johnson, William H
Papaefthymiou, Georgia Christou
Petsko, Gregory Anthony
Phillies, George David Joseph
Platt, John Rader
Plocke, Donald J
Rich, Alexander
Rigney, David Roth
Rose, Robert M(ichael)
Rosen, Philip
Rosenblith, Walter Alter
Rothschild, Kenneth J
Sandler, Sheldon Samuel
Scheuchenzuber, H Joseph
Scheuplein, Robert J
Schimmel, Paul Reinhard
Sekuler, Robert W
Shapiro, Jacob
Shipley, George Graham
Simons, Elizabeth Reiman
Singer, Joshua J
Solomon, Arthur Kaskel
Stephens, Raymond Edward
Sternick, Edward Selby
Stossel, Thomas Peter
Tanaka, Toyoichi
Taylor, Robert E
Tosteson, Daniel Charles
Tripathy, Sukant K
Umans, Robert Scott
Vallee, Bert L
Verma, Surendra P
Voigt, Herbert Frederick
Waloga, Geraldine
Wang, Chih-Lueh Albert
Weaver, James Cowles
Webb, Robert Howard
Webster, Edward William
Wiley, Don Craig
Woodhull-McNeal, Ann P
Yen, Peter Kai Jen
Zaner, Ken Scott
Zimmermann, Robert Alan

MICHIGAN
Axelrod, Daniel
Babcock, Gerald Thomas
Blanchard, Fred Ayres
Carson, Paul Langford
Carter-Su, Christin
Chapman, Robert Earl, Jr
Chou, Kuo Chen
Conrad, Michael
Cowlishaw, John David
Crippen, Gordon Marvin
Dambach, George Ernest
Ferguson-Miller, Shelagh Mary
Garlick, Robert L
Gates, David Murray
Goldstein, Albert
Halvorson, Herbert Russell
Hodgson, Voigt R
Horvath, William John
Hubbard, Robert Phillip
Jacobson, Baruch S
Kimura, Tokuji
Ko, Howard Wha Kee
Krimm, Samuel
Levich, Calman
Liboff, Abraham R
Lindemann, Charles Benard
McConnell, David Graham
McGrath, John Joseph
McGroarty, Estelle Josephine
Maggiora, Gerald M
Matthews, Rowena Green
Mizukami, Hiroshi
Neubig, Richard Robert
Njus, David Lars
Novak, Raymond Francis
Oncley, John Lawrence
Orton, Colin George
Parkinson, William Charles
Petty, Howard Raymond
Phan, Sem Hin
Pollack, Gerald Leslie
Rimai, Lajos
Rogers, William Leslie
Rohrer, Douglas C
Rosenspire, Allen Jay
Sands, Richard Hamilton
Saper, Mark A
Schullery, Stephen Edmund
Sevilla, Michael Douglas
Shichi, Hitoshi
Stephenson, Robert Storer

Tepley, Norman
Tien, H Ti
Trachtenberg, Michael Carl
Valeriote, Frederick Augustus
Webb, Paul
Wiley, John W
Wolterink, Lester Floyd
Zand, Robert

MINNESOTA
Ackerman, Eugene
Adiarte, Arthur Lardizabal
Adolph, Kenneth William
Baumann, Wolfgang J
Bloomfield, Victor Alfred
Brown, Rhoderick Edmiston, Jr
Carter, John Vernon
Donaldson, Sue Karen
Evans, Douglas Fennell
Forro, Frederick, Jr
Garrity, Michael K
Goldstein, Stuart Frederick
Haaland, John Edward
Haddad, Louis Charles
Khan, Faiz Mohammad
Kluetz, Michael David
Loken, Merle Kenneth
McCullough, Edwin Charles
Meyer, Frank Henry
O'Connor, Michael Kieran
Pfeiffer, Douglas Robert
Polnaszek, Carl Francis
Schmitt, Otto Herbert
Sowell, John Basil
Stoesz, James Darrel
Szurszewski, Joseph Henry
Taylor, Stuart Robert
Thomas, David Dale
Tsong, Tian Yow
Ugurbil, Kamil
Yapel, Anthony Francis, Jr

MISSISSIPPI
Chaires, Jonathan Bradford
Coleman, Thomas George
Dwyer, Terry M
Eftink, Maurice R
Graves, David E
Pandey, Ras Bihari

MISSOURI
Ackerman, Joseph John Henry
Beetham, Karen Lorraine
Burgess, James Harland
Cochran, Andrew Aaron
Eldredge, Donald Herbert
Ghiron, Camillo A
Henson, Bob Londes
Highstein, Stephen Morris
Keane, John Francis, Jr
Larson, Kenneth Blaine
Menz, Leo Joseph
Northrip, John Willard
Pickard, William Freeman
Pueppke, Steven Glenn
Rowe, Elizabeth Snow
Schmitz, Kenneth Stanley
Shear, David Ben
Ter-Pogossian, Michel Mathew
Thomas, George Joseph, Jr
Vaughan, William Mace
Wartzok, Douglas
Wong, Tuck Chuen

MONTANA
Callis, Patrik Robert
Hill, Walter Ensign
Jesaitis, Algirdas Joseph
Jurist, John Michael
Ribi, Edgar

NEBRASKA
Chakkalakal, Dennis Abraham
Hahn, George LeRoy
Parkhurst, Lawrence John
Salhany, James Mitchell
Schneiderman, Martin Howard
Song, Pill-Soon
Verma, Shashi Bhushan

NEVADA
Barnes, George
Harrington, Rodney E

NEW HAMPSHIRE
Blanchard, Robert Osborn
Dennison, David Severin
Douple, Evan Barr
Gross, Robert Henry
Kegeles, Gerson
Kidder, John Newell
Musiek, Frank Edward
Oldfield, Daniel G
Richmond, Robert Chaffee
Stanton, Bruce Alan
Vournakis, John Nicholas

NEW JERSEY
Ashkin, Arthur
Becker, Joseph Whitney
Bender, Max
Bennun, Alfred
Blumberg, William Emil
Boltz, Robert Charles, Jr

Breslow, Esther M G
Brudner, Harvey Jerome
Chandler, Louis
Connor, John Arthur
DeBari, Vincent A
DeBari, Vincent A
Duran, Walter Nunez
Eisenberger, Peter Michael
Fitzgerald, Paula Marie Dean
Fordham, William David
Fresco, Jacques Robert
Friedman, Kenneth Joseph
Gagliardi, L John
Gagna, Claude Eugene
Hartig, Paul Richard
Hearn, Ruby Puryear
Hill, David G
Horn, Leif
Isaacson, Allen
Jelinski, Lynn W
Ji, Sungchul
Jordan, Frank
Kaczorowski, Gregory John
Lee, Lih-Syng
McCaslin, Darrell
Madison, Vincent Stewart
Manning, Gerald Stuart
Massover, William H
Mendelsohn, Richard
Mims, William B
Pethica, Brian Anthony
Pilla, Arthur Anthony
Poe, Martin
Prince, Roger Charles
Reuben, John Philip
Reynolds, George Thomas
Rousseau, Denis Lawrence
Rumsey, William LeRoy
Siano, Donald Bruce
Siegel, Jeffry A
Steinberg, Malcolm Saul
Stephenson, Elizabeth Weiss
Strauss, George
Teipel, John William
Tobkes, Nancy J
Tomaselli, Vincent Paul
Treu, Jesse Isaiah
Trotta, Paul P
Tzanakou, M Evangelia
Wangemann, Robert Theodore
Wassarman, Paul Michael
Williams, Myra Nicol
Witz, Gisela
Zoltan, Bart Joseph

NEW MEXICO
Bartholdi, Marty Frank
Beckel, Charles Leroy
Bell, George Irving
Berardo, Peter Antonio
Burks, Christian
Coppa, Nicholas V
Cram, Leighton Scott
Crissman, Harry Allen
Diel, Joseph Henry
Fletcher, Edward Royce
Galey, William Raleigh
Goldstein, Byron Bernard
Gurd, Ruth Sights
Hanson, Kenneth Merrill
Heller, Leon
Hildebrand, Carl Edgar
Holm, Dale M
Jett, James Hubert
Kelsey, Charles Andrew
Matwiyoff, Nicholas Alexander
Omdahl, John L
Partridge, L Donald
Perelson, Alan Stuart
Raju, Mudundi Ramakrishna
Salzman, Gary Clyde
Scott, Bobby Randolph
Shera, E Brooks
Stapp, John Paul

NEW YORK
Akers, Charles Kenton
Al-Awqati, Qais
Alfano, Robert R
Almog, Rami
Anbar, Michael
Andersen, Olaf Sparre
Anderson, Lowell Leonard
Bahary, William S
Baier, Robert Edward
Baker, Robert George
Barish, Robert John
Bean, Charles Palmer
Becker, Robert O
Bell, Duncan Hadley
Bello, Jake
Benham, Craig John
Bernhard, William Allen
Bettelheim, Frederick A
Bickmore, John Tarry
Bigler, Rodney Errol
Bird, Richard Putnam
Birge, Robert Richards
Bisson, Mary A
Bittman, Robert
Blank, Martin
Blei, Ira
Bond, Victor Potter
Bookchin, Robert M

Bottomley, Paul Arthur
Brand, John S
Brandt, Philip Williams
Bray, James William
Brehm, Lawrence Paul
Breneman, Edwin Jay
Brenner, Henry Clifton
Briehl, Robin Walt
Brink, Frank, Jr
Brody, Marcia
Brody, Seymour Steven
Brown, Rodney Duvall, III
Bryant, Robert George
Bunch, Phillip Carter
Bushinsky, David Allen
Candia, Oscar A
Carlson, Roy Douglas
Carstensen, Edwin L(orenz)
Cerny, Laurence Charles
Chamberlain, Charles Craig
Chappell, Richard Lee
Cohen, Beverly Singer
Cole, Patricia Ellen
Coleman, James R
Coleman, Peter Stephen
Cooper, Philip Harlan
Cowburn, David
Dahl, John Robert
Damadian, Raymond
Davenport, Lesley
Day, Loren A
De Lisi, Charles
DeTitta, George Thomas
Dicello, John Francis, Jr
Diem, Max
Diwan, Joyce Johnson
Doetschman, David Charles
Dorset, Douglas Lewis
Dreizen, Paul
Duax, William Leo
Duffey, Michael Eugene
Eberle, Helen I
Eberstein, Arthur
Eickelberg, W Warren B
Eisenstadt, Maurice
Eisinger, Josef
Elbaum, Danek
Ely, Thomas Sharpless
Fabry, Mary E Riepe
Fabry, Thomas Lester
Fajer, Jack
Featherstone, John Douglas Bernard
Feigenson, Gerald William
Feldman, Isaac
Fenstermacher, Joseph Don
Fischbarg, Jorge
Fisher, Vincent J
Fishman, Jerry Haskel
Forgacs, Gabor
Foster, Kenneth William
Freedman, Jeffrey Carl
Friedberg, Carl E
Friedman, Helen Lowenthal
Friedman, Joel Mitchel
Gao, Jiali
Gardella, Joseph Augustus, Jr
Gardner, Daniel
Geacintov, Nicholas
Gershman, Lewis C
Gerstman, Hubert Louis
Giaever, Ivar
Gibson, Quentin Howieson
Gogel, Germaine E
Green, Michael Enoch
Griffin, Jane Flanigen
Gross, Leo
Grumet, Martin
Gunter, Karlene Klages
Gunter, Thomas E, Jr
Gupta, Raj K
Gurpide, Erlio
Gutstein, William H
Haas, David Jean
Haines, Thomas Henry
Hamblen, David Philip
Hanson, Louise I Karle
Harbison, Gerard Stanislaus
Harper, Richard Allan
Harris, Jack Kenyon
Harris-Warrick, Ronald Morgan
Harth, Erich Martin
Hayon, Elie M
Hirschman, Shalom Zarach
Ho, John Ting-Sum
Holowka, David Allan
Hoory, Shlomo
Horowicz, Paul
Huang, Sylvia Lee
Hui, Sek Wen
Hurley, Ian Robert
Jacobsen, Chris J
Johnson, William Harding
Jung, Chan Yong
Kalogeropoulos, Theodore E
Kaplan, Ehud
Karlin, Arthur
Kass, Robert S
Keese, Charles Richard
Keng, Peter C
Kim, Carl Stephen
Kimmich, George Arthur
Kirchberger, Madeleine
Knight, Bruce Winton, (Jr)
Knowles, Richard James Robert

Biophysics (cont)

RHODE ISLAND
Andrews, Howard Lucius
Chapman, Kent M
Estrup, Faiza Fawaz
Fisher, Harold Wilbur
Hai, Chi-Ming
Polk, C(harles)
Stewart, Peter Arthur

SOUTH CAROLINA
Bank, Harvey L
Bronk, Burt V
Carlton, William Herbert
Larcom, Lyndon Lyle
Sanders, Samuel Marshall, Jr
Wheeler, Darrell Deane
Wheeler, Darrell Deane
Wise, William Curtis
Zucker, Robert Martin

SOUTH DAKOTA
Hastings, David Frank
Thompson, John Darrell

TENNESSEE
Adams, Paul Louis
Anjaneyulu, P S R
Auxier, John A
Barrett, Terence William
Bhattacharyya, Mohit Lal
Bunick, Gerard John
Cunningham, Leon William
Darden, Edgar Bascomb
Dockter, Michael Edward
Dudney, Charles Sherman
Edwards, Harold Henry
Erickson, Jon Jay
Fleischer, Sidney
Frederiksen, Dixie Ward
Georghiou, Solon
Gettins, Peter
Goulding, Charles Edwin, Jr
Greenbaum, Elias
Gutzke, William H N
Hingerty, Brian Edward
Huang, Leaf
Incardona, Antonino L
Johnson, Carroll Kenneth
Macleod, Robert M
Mullaney, Paul F
Olins, Donald Edward
Pulliam, James Augustus
Randolph, Malcolm Logan
Riggsby, William Stuart
Senogles, Susan Elizabeth
Staros, James Vaughan
Stone, William Lawrence
Thomason, Donald Brent
Tibbetts, Clark
Venable, John Heinz, Jr
Wikswo, John Peter, Jr
Williams, Robley Cook, Jr
Wondergem, Robert

TEXAS
Adams, Emory Temple, Jr
Alexander, William Carter
Allen, Robert Charles
Andresen, Michael Christian
Angelides, Kimon Jerry
Ansevin, Allen Thornburg
Atkinson, Mark Arthur Leonard
Baker, Robert David
Barnes, Charles M
Bartel, Allen Hawley
Beall, Paula Thornton
Becker, Ralph Sherman
Berry, Michael James
Brady, Scott T
Brodwick, Malcolm Stephen
Brown, Arthur Morton
Butler, Bruce David
Chang, Donald Choy
Czerwinski, Edmund William
Dowben, Robert Morris
Downey, H Fred
Eakin, Richard Timothy
Ellis, Kenneth Joseph
Espey, Lawrence Lee
Fishman, Harvey Morton
Fox, Jack Lawrence
Garay, Andrew Steven
Gierasch, Lila Mary
Ginsburg, Nathan
Glantz, Raymon M
Gray, Donald Melvin
Hackert, Marvin LeRoy
Hademenos, James George
Hazlewood, Carlton Frank
Hendrickson, Constance McRight
Herbert, Morley Allen
Hewitt, Roger R
Hogan, Michael Edward
Hopkins, Robert Charles
Huang, Huey Wen
Hudspeth, Albert James
Jagger, John
Kalthoff, Klaus Otto
King, Richard Joe
Krohmer, Jack Stewart
Lang, Dimitrij Adolf
Leuchtag, H Richard
Livingston, Linda
Lott, James Robert

McCammon, James Andrew
McCarter, Roger John Moore
Macfarlane, Ronald Duncan
McSherry, Diana Hartridge
Mahendroo, Prem P
Massey, John Boyd
Masters, Bettie Sue Siler
Meistrich, Marvin Lawrence
Meyn, Raymond Everett, Jr
Morrisett, Joel David
Murphy, Paul Henry
Nachtwey, David Stuart
Nakajima, Motowo
Ong, Poen Sing
Parker, Cleofus Varren, Jr
Powell, Michael Robert
Robberson, Donald Lewis
Robinson, Neal Clark
Rorschach, Harold Emil, Jr
Roux, Stanley Joseph
Rupert, Claud Stanley
Scheie, Paul Olaf
Schrank, Auline Raymond
Serwer, Philip
Shalek, Robert James
Shaw, Robert Wayne
Skolnick, Malcolm Harris
Smith, Thomas Caldwell
Taylor, Morris Chapman
Tilbury, Roy Sidney
Trkula, David
Tu, Shiao-chun
Waggener, Robert Glenn
Waldbillig, Ronald Charles
Walmsley, Judith Abrams
Wills, Nancy Kay
Wong, Brendan So
Wright, Ann Elizabeth

UTAH
Ailion, David Charles
Billings, R Gail
Brown, Harold Mack, Jr
Burton, Frederick Glenn
Earley, Charles Willard
Ely, Kathryn R
Gardner, Reed McArthur
Greenfield, Harvey Stanley
Lancaster, Jack R, Jr
Lloyd, Ray Dix
Mohammad, Syed Fazal
Piette, Lawrence Hector
Powers, Linda Sue
Shearin, Nancy Louise
Sipe, David Michael
Spikes, John Daniel
Tuckett, Robert P
Wood, O Lew
Woodbury, John Walter
Yoshikami, Doju

VERMONT
Lipson, Richard L
Nyborg, Wesley LeMars
Otter, Timothy
Parsons, Rodney Lawrence
Sachs, Thomas Dudley
Spearing, Ann Marie
Tritton, Thomas Richard
Tyree, Melvin Thomas
Wallace, Susan Scholes
Webb, George Dayton

VIRGINIA
Averill, Bruce Alan
Biltonen, Rodney Lincoln
Brill, Arthur Sylvan
Chandross, Ronald Jay
Chlebowski, Jan F
Clarke, Alexander Mallory
Cleary, Stephen Francis
Creutz, Carl Eugene
Dominey, Raymond Nelson
Eldridge, Charles A
Feher, Joseph John
Flinn, Jane Margaret
Ford, George Dudley
Gabriel, Barbra L
Garrison, James C
Goldman, Israel David
Grogan, William McLean
Hackett, John Taylor
Ham, William Taylor, Jr
Hamilton, Thomas Charles
Hoy, Gilbert Richard
Hutchinson, Thomas Eugene
Johnson, Michael L
Keefe, William Edward
Kretsinger, Robert
Krueger, Peter George
Kutchai, Howard C
Liberti, Joseph Pollara
Litman, Burton Joseph
Looney, William Boyd
Maggio, Bruno
Martin, James Henry, III
Mikulecky, Donald C
Moore, Cyril L
Nielsen, Peter Tryon
Noble, Julian Victor
Ridgway, Ellis Branson
Ritter, Rogers C
Roper, L(eon) David
Rozzell, Thomas Clifton

Sahu, Saura Chandra
Saltz, Joel Haskin
Schwab, Walter Edwin
Szabo, Gabor
Terner, James
Tolles, Walter Edwin
Wingard, Christopher Jon
Wyeth, Newell Convers

WASHINGTON
Almers, Wolfhard
Barlow, Clyde Howard
Bassingthwaighte, James B
Bernard, Gary Dale
Binder, Marc David
Braby, Leslie Alan
Callis, James Bertram
Cellarius, Richard Andrew
Detwiler, Peter Benton
Dunker, Alan Keith
Fisher, Darrell R
Glass, James Clifford
Gordon, Albert McCague
Herriott, Jon R
Hille, Bertil
Jensen, Lyle Howard
Johnson, John Richard
Kathren, Ronald Laurence
Kaufman, William Carl, Jr
Kehl, Theodore H
Magnuson, James Andrew
Nutter, Robert Leland
Palmer, Harvey Earl
Peterson-Kennedy, Sydney Ellen
Prothero, John W
Schurr, John Michael
Sikov, Melvin Richard
Stern, Edward Abraham
Stirling, Charles E
Tam, Patrick Yui-Chiu
Tenforde, Thomas SeBastian
Towe, Arnold Lester
Windsor, Maurice William
Wootton, Peter
Yantis, Phillip Alexander
Zimbrick, John David

WEST VIRGINIA
Douglass, Kenneth Harmon
Franz, Gunter Norbert
Kartha, Mukund K
Wallace, William Edward, Jr

WISCONSIN
Anderegg, John William
Beeman, William Waldron
Blattner, Frederick Russell
Cameron, John Roderick
DeLuca, Paul Michael, Jr
Dennis, Warren Howard
Fischbach, Fritz Albert
Greenebaum, Ben
Holden, James Edward
Hyde, James Stewart
Kincaid, James Robert
Kroger, Larry A
Lai, Ching-San
Markley, John Lute
Martin, Thomas Fabian John
Miller, Joyce Fiddick
Nickles, Robert Jerome
Norman, John Matthew
Patrick, Michael Heath
Raines, Ronald T
Record, M Thomas, Jr
Reed, George Henry
Ritter, Mark Alfred
Romans-Hess, Alice Yvonne
Scheele, Robert Blain
Sealy, Roger Clive
Sorenson, James Alfred
Wells, Barbara Duryea
Westler, William Milo
Worley, John David
Yu, Hyuk

WYOMING
Smith, William Kirby

PUERTO RICO
Copson, David Arthur
Orkand, Richard K
Vargas, Fernando Figueroa

ALBERTA
Bazett-Jones, David Paul
Beck, James S
Challice, Cyril Eugene
Codding, Penelope Wixson
Hoffer, J(oaquin) A(ndres)
Kisman, Kenneth Edwin
Kumar, Shrawan
McElhaney, Ronald Nelson
McGann, Locksley Earl
Poznansky, Mark Joab
Roche, Rodney Sylvester
Rowlands, Stanley
Smith, Richard Sidney
Stein, Richard Bernard
Sykes, Brian Douglas
Umezawa, Hiroomi
Van De Sande, Johan Hubert
Walsh, Michael Patrick
Weiner, Joel Hirsch

BRITISH COLUMBIA
Belton, Peter
Cushley, Robert John
Durand, Ralph Edward
Eaves, Allen Charles Edward
Friedmann, Gerhart B
Gosline, John M
Green, Beverley R
Hoffmann, Geoffrey William
Lam, Gabriel Kit Ying
Mauk, Arthur Grant
Palaty, Vladimir
Skarsgard, Lloyd Donald
Stephens-Newsham, Lloyd G
Withers, Stephen George

MANITOBA
Barnard, John Wesley
Bihler, Ivan
Forrest, Bruce James
Froese, Gerd
Polimeni, Philip Iniziato
Scott, David Paul
Standing, Kenneth Graham

NEW BRUNSWICK
Edwards, Merrill Arthur
Sharp, Allan Roy

NOVA SCOTIA
Luner, Stephen Jay
McDonald, Terence Francis
Verpoorte, Jacob A
Wong, Alan Yau Kuen

ONTARIO
Ananthanarayanan, Vettaikkoru S
Baker, Robert G
Bayley, Stanley Thomas
Butler, Keith Winston
Cairns, William Louis
Canham, Peter Bennett
Carey, Paul Richard
Casal, Hector Luis
Cradduck, Trevor David
Cummins, W(illiam) Raymond
Cunningham, John (Robert)
El-Sharkawy, Taher Youssef
Epand, Richard Mayer
Farrar, John Keith
Fenster, Aaron
Ferrier, Jack Moreland
Finlay, Joseph Bryan
Forer, Arthur H
Gardner, David R
Goldsack, Douglas Eugene
Greenblatt, Jack Fred
Greenstock, Clive Lewis
Grinstein, Sergio
Groom, Alan Clifford
Hallett, Frederick Ross
Hawton, Margaret H
Haynes, Robert Hall
Hinke, Joseph Anthony Michael
Hunt, John Wilfred
Jacobson, Stuart Lee
Jeffrey, Kenneth Robert
Johns, Harold E
Kennedy, James Cecil
Kruuv, Jack
Kuehn, Lorne Allan
Langille, Brian Lowell
Lepock, James Ronald
Leung, Philip Man Kit
Macdonald, Peter Moore
MacLennan, David Herman
Mantsch, Henry H
Miller, Richard Graham
Millman, Barry Mackenzie
Morris, Catherine Elizabeth
Morton, Richard Alan
Nicholls, Peter
Noolandi, Jaan
Norwich, Kenneth Howard
Osborne, Richard Vincent
Ottensmeyer, Frank Peter
Paul, William
Pfalzner, Paul Michael
Raaphorst, G Peter
Racey, Thomas James
Rainbow, Andrew James
Rand, Richard Peter
Rauth, Andrew Michael
Rawlinson, John Alan
Roach, Margot Ruth
Rotenberg, A Daniel
Rothstein, Aser
Sherebrin, Marvin Harold
Siminovitch, Louis
Smith, Ian Cormack Palmer
Song, Seh-Hoon
Stinson, Robert Henry
Szabo, Arthur Gustav
Tannock, Ian Frederick
Taylor, C P(atrick) S(tirling)
Till, James Edgar
Tong, Bok Yin
Trainor, Lynne E H
Tung, Pierre S

QUEBEC
Cohen, Montague
Dugas, Hermann
Frojmovic, Maurice Mony

Biophysics (cont)

Gjedde, Albert
Goldsmith, Harry L
Hallenbeck, Patrick Clark
Lamarche, François
Leblanc, Roger M
Lehnert, Shirley Margaret
Lennox, Robert Bruce
Mackey, Michael Charles
Marceau, Normand Luc
Payet, Marcel Daniel
Pezolet, Michel
Podgorsak, Ervin B
Poole, Ronald John
Roberge, Fernand Adrien
Roy, Guy
Ruiz-Petrich, Elena
Schanne, Otto F
Schiller, Peter Wilhelm
Seidah, Nabil George
Seufert, Wolf D
Slilaty, Steve N
Sygusch, Jurgen
Van Calsteren, Marie-Rose

SASKATCHEWAN
Scheltgen, Elmer

OTHER COUNTRIES
Araki, Masasuke
Argos, Patrick
Azbel, Mark
Berns, Donald Sheldon
Brown, Larry Robert
Cerbon-Solorzano, Jorge
Creighton, Thomas Edwin
Curtis, Adam Sebastian G
Deranleau, David A
De Troyer, Andre Jules
Diner, Bruce Aaron
Engelborghs, Yves
Falk, Gertrude
Friedland, Stephen Sholom
Fujiwara, Keigi
Gontier, Jean Roger
Guerrero, Ariel Heriberto
Hall, Theodore (Alvin)
Hammerman, Ira Saul
Heinz, Erich
Hiyama, Tetsuo
Kang, Sungzong
Kuhn, Klaus
Lewis, Aaron
Liao, Ta-Hsiu
Lieh, William Robert
Maas, Peter
Marikovsky, Yehuda
Mehler, Ernest Louis
Michel, Hartmut
Mittal, Balraj
Monos, Emil
Nagata, Kazuhiro
Rettori, Ovidio
Rinehart, Frank Palmer
Salas, Pedro Jose I
Sargent, David Fisher
Scott, Mary Jean
Stoeber, Werner
Thorson, John Wells
Trentham, David R
Tsernoglou, Demetrius
Tsou, Chen-Lu
Van Driessche, Willy
Vartsky, David
Westerhof, Nicolaas
Whittembury, Guillermo
Wrede, Don Edward
Wu, Cheng-Wen
Wu, Felicia Ying-Hsiueh

Botany-Phytopathology

ALABAMA
Backman, Paul Anthony
Bhatnagar, Yogendra Mohan
Bowen, William R
Brewer, Jesse Wayne
Clark, Edward Maurice
Cochis, Thomas
Curl, Elroy Arvel
Darden, William H, Jr
Deason, Temd R
Diener, Urban Lowell
Estes, Edna E
Eyster, Henry Clyde
Freeman, John A
Freeman, John Daniel
Hagler, Thomas Benjamin
Haynes, Robert Ralph
Hocking, George Macdonald
Huntley, Jimmy Charles
Jacobsen, Barry James
Jones, Daniel David
Jones, Jeanette
Latham, Archie J
Lelong, Michel Georges
Lyle, James Albert
McCullough, Herbert Alfred
McGuire, Robert Frank
Marshall, Norton Little
Marvel, Mason E
Olson, Richard Louis

Peterson, Curtis Morris
Rosene, Walter, Jr
Sharma, Govind C
Shepherd, Raymond Lee
Steward, Frederick Campion
Thomas, Joab Langston
Truelove, Bryan
Wilson, Eugene M

ALASKA
Juday, Glenn Patrick
Laursen, Gary A
Logsdon, Charles Eldon
Mitchell, William Warren

ARIZONA
Ahlgren, Isabel Fulton
Alcorn, Stanley Marcus
Aronson, Jerome Melville
Baker, Gladys Elizabeth
Blinn, Dean Ward
Bloss, Homer Earl
Caldwell, Roger Lee
Canright, James Edward
Clark, William Dennis
Crosswhite, Carol D
Crosswhite, Frank Samuel
Davis, Owen Kent
Gilbertson, Robert Lee
Gries, George Alexander
Haase, Edward Francis
Hine, Richard Bates
Hoshaw, Robert William
Huff, Albert Keith
Jackson, Ernest Baker
Johnson, Raymond Roy
Kernkamp, Milton F
Klopatek, Jeffrey Matthew
Landrum, Leslie Roger
Leatherman, Anna D
Leathers, Chester Ray
Mason, Charles Thomas, Jr
Mogensen, Hans Lloyd
Nelson, Merritt Richard
Nigh, Edward Leroy, Jr
Pinkava, Donald John
Reeder, John Raymond
Rominger, James McDonald
Russell, Thomas Edward
Schoenwetter, James
Schuster, Rudolf Mathias
Sommerfeld, Milton R
Stanghellini, Michael Eugene
Timmermann, Barbara Nawalany
Van Etten, Hans D
Verbeke, Judith Ann
Wagle, Robert Fay
Wilson, F(rank) Douglas

ARKANSAS
Adams, Randall Henry
Dale, James Lowell
Franke, Robert G
Goode, Monroe Jack
Hardin, Hilliard Frances
Harris, William M
Hutchison, James A
Johnson, George Thomas
Jones, John Paul
Logan, Lowell Alvin
Meyer, Richard Lee
Riggs, Robert D
Sinclair, Clarence Bruce
Slack, Derald Allen
Smith, Edwin Burnell
Steinkraus, Donald Curtiss
Sutherland, John Bruce, IV
Te Beest, David Orien
Templeton, George Earl
Timmermann, Dan, Jr
Tucker, Gary Edward
Walters, Hubert Jack

CALIFORNIA
Adaskaveg, James Elliott
Aly, Raza
Anderson, Dennis Elmo
Anthony, Margery Stuart
Anzalone, Louis, Jr
Ashworth, Lee Jackson, Jr
Baad, Michael Francis
Baalman, Robert J
Baker, Herbert George
Baldwin, James Gordon
Barbour, Michael G
Bega, Robert V
Benjamin, Richard Keith
Bennett, James Peter
Benseler, Rolf Wilhelm
Bianchi, Donald Ernest
Bierzychudek, Paulette F
Bizzoco, Richard Lawremce Weiss
Bowerman, Earl Harry
Brown, Richard McPike
Bruening, George
Buddenhagen, Ivan William
Burleigh, James Reynolds
Butler, Edward Eugene
Butler, John Earl
Calavan, Edmond Clair
Calpouzos, Lucas
Cande, W Zacheus
Caplin, Samuel Milton
Capon, Brian

Carlquist, Sherwin
Cheadle, Vernon Irvin
Christianson, Michael Lee
Collins, O'Neil Ray
Constance, Lincoln
Cooke, Ron Charles
Cox, Hiden Toy
Cunningham, Virgil Dwayne
Cutler, Hugh Carson
DeMason, Darleen Audrey
Dempster, Lauramay Tinsley
Derr, William Frederick
Desjardins, Paul Roy
DeVay, James Edson
Dimitman, Jerome Eugene
Dixon, Peter Stanley
Dodds, James Allan
Doyle, William T
Duffus, James Edward
Duncan, Thomas Osler
Duniway, John Mason
Ebert, Wesley W
Eckard, Kathleen
Eckert, Joseph Webster
Ediger, Robert I
Einset, John William
Elias, Thomas S
Ellstrand, Norman Carl
Emboden, William Allen, Jr
Endo, Robert Minoru
English, William Harley
Erspamer, Jack Laverne
Esau, Katherine
Falk, Richard H
Ferguson, Noel Moore
Finch, Harry C
Ford, Donald Hoskins
French, Alexander Murdoch
Furumoto, Warren Akira
Game, John Charles
Garber, Richard Hammerle
Gardner, Wayne Scott
Gasser, Charles Scott
Gerik, James Stephen
Gifford, Ernest Milton
Gill, Harmohindar Singh
Goff, Lynda June
Goodman, Victor Herke
Goss, James Arthur
Gotelli, David M
Green, Norman Edward
Green, Paul Barnett
Grieve, Catherine Macy
Grillos, Steve John
Grogan, Raymond Gerald
Gumpf, David John
Hancock, Joseph Griscom, Jr
Harrington, Dalton
Harris, Hubert Andrew
Harvey, John Marshall
Heckard, Lawrence Ray
Henrickson, James Solberg
Hildebrand, Donald Clair
Hildreth, Robert Claire
Hill, Ray Allen
Hills, F Jackson
Hirsch, Ann Mary
Hoefert, Lynn Lucretia
Hogan, Christopher James
Horton, James Charles
Howard, Dexter Herbert
Ingram, John (William), Jr
Kado, Clarence Isao
Kantz, Paul Thomas, Jr
Kaplan, Donald Robert
Kaplan, Hesh J
Karle, Harry P
Kasapligil, Baki
Kavaljian, Lee Gregory
Keeley, Sterling Carter
Keen, Noel Thomas
Keil, David John
Kendrick, James Blair, Jr
Kimsey, Lynn Siri
Kinloch, Bohun Baker, Jr
Kjeldsen, Chris Kelvin
Kliejunas, John Thomas
Koonce, Andrea Lavender
Kowalski, Richard J
Kurtzman, Ralph Harold, Jr
Kyhos, Donald William
Laemmlen, Franklin
Laetsch, Watson McMillan
Lang, Norma Jean
Largent, David Lee
Lathrop, Earl Wesley
Laude, Horton Meyer
Lear, Bert
Levin, Geoffrey Arthur
Lewin, Ralph Arnold
Lewis, Frank Harlan
Lindsay, George Edmund
Long, Sharon Rugel
Lord, Elizabeth Mary
Lukens, Raymond James
McCain, Arthur Hamilton
McClintock, Elizabeth
MacDonald, James Douglas
McHale, John T
McNeal, Dale William, Jr
Madore, Monica Agnes
Mallory, Thomas E
Mathias, Mildred Esther
Menge, John Arthur

Meredith, Farris Ray
Mircetich, Srecko M
Mitchell, Norman L
Mooring, John Stuart
Moran, Reid (Venable)
Mortenson, Theodore Hampton
Moseley, Maynard Fowle
Muller, Cornelius Herman
Munnecke, Donald Edwin
Murphy, Terence Martin
Muth, Gilbert Jerome
Neher, Robert Trostle
Nelson, Klayton Edward
Nelson, Richard D(ouglas)
Neushul, Michael, Jr
Nichols, Carl William
Norris, Daniel Howard
Nyland, George
O'Neill, Thomas Brendan
Ornduff, Robert
Panopoulos, Nickolas John
Parker, Virgil Thomas
Parmeter, John Richard, Jr
Parnell, Dennis Richard
Paulus, Albert
Philbrick, Ralph
Phillips, Douglas J
Pitelka, Louis Frank
Pitts, Wanna Dene
Platt Aloia, Kathryn Ann
Pray, Thomas Richard
Prezelin, Barbara Berntsen
Price, Mary Vaughan
Purcell, Alexander Holmes, III
Quail, Peter Hugh
Quibell, Charles Fox
Raabe, Robert Donald
Rader, William Ernest
Radewald, John Dale
Raju, Namboori Bhaskara
Ramsey, Richard Harold
Rappaport, Lawrence
Rasmussen, Robert A
Reeve, Marian E
Reynolds, Don Rupert
Rosen, Howard
Rosenberg, Dan Yale
Rudd, Velva Elaine
Saltveit, Mikal Endre, Jr
Scharpf, Robert F
Schnathorst, William Charles
Schneider, Henry
Schroth, Milton Neil
Schwegmann, Jack Carl
Scora, Rainer Walter
Shapiro, Arthur Maurice
Sherman, Robert James
Sidhu, Gurmel Singh
Silk, Margaret Wendy Kuhn
Silva, Paul Claude
Simpson, Robert Blake
Smith, Alan Reid
Smith, James Payne, Jr
Sparling, Shirley
Spurr, Arthur Richard
Stebbins, George Ledyard
Sterling, Clarence
Stern, Kingsley Rowland
Stoner, Martin Franklin
Sung, Zinmay Renee
Swatek, Frank Edward
Tavares, Isabelle Irene
Teviotdale, Beth Luise
Thaw, Richard Franklin
Thiers, Harry Delbert
Thomas, John Hunter
Thomas, Walter Dill, Jr
Thomson, William Walter
Tiffney, Bruce Haynes
Tilton, Varien Russell
Tisserat, Brent Howard
Travis, Robert LeRoy
Tucker, John Maurice
Uyemoto, Jerry Kazumitsu
Van Gundy, Seymour Dean
Vasek, Frank Charles
Vinters, Harry Valdis
Vinyard, William Corwin
Vogl, Richard J
Vreeland, Valerie Jane
Walch, Henry Andrew, Jr
Walker, Dan B
Walker, Dennis Kendon
Walkington, David L
Wallace, Gary Dean
Waser, Nickolas Merritt
Watterson, Jon Craig
Weathers, Lewis Glen
Webster, Grady Linder
Webster, Robert K
Weiler, John Henry, Jr
Weinhold, Albert Raymond
Wells, Kenneth
Whaley, Julian Wendell
Whitney, Kenneth Dean
Wiedman, Harold W
Wiggins, Ira Loren
Wilson, Kenneth Allen
Worker, George F, Jr
Yoder, David Lee
Young, Donald Raymond
Yu, Grace Wei-Chi Hu
Zedler, Paul H(ugo)
Zentmyer, George Aubrey, Jr

Zscheile, Frederick Paul, Jr

COLORADO
Altman, Jack
Anderson, Roger Arthur
Bailey, Dana Kavanagh
Baker, R Ralph
Bauer, Penelope Jane Hanchey
Bock, Jane Haskett
Bragonier, Wendell Hughell
Costello, David Francis
Davidson, Darwin Ervin
Dunahay, Terri Goodman
Ferchau, Hugo Alfred
Feucht, James Roger
Grant, Michael Clarence
Hansmann, Eugene William
Harrison, Monty DeVerl
Hartman, Emily Lou
Hess, Dexter Winfield
Hinds, Thomas Edward
Kreutzer, William Alexander
Larsen, Arnold Lewis
Lindsey, Julia Page
Livingston, Clark Holcomb
Longpre, Edwin Keith
McIntyre, Gary A
Martin, Susan Scott
Mitchell, John Edwards
Nisbet, Jerry J
Reed, John Frederick
Reeves, Fontaine Brent, Jr
Ruppel, Earl George
Shaw, Robert Blaine
Shushan, Sam
Siemer, Eugene Glen
Steinkamp, Myrna Pratt
Weber, William Alfred
Wilken, Dieter H
Wingate, Frederick Huston

CONNECTICUT
Anagnostakis, Sandra Lee
Anderson, Gregory Joseph
Berlyn, Graeme Pierce
Bristol, Melvin Lee
Carluccio, Leeds Mario
Collins, Ralph Porter
Cooke, John Cooper
Duffy, Regina Maurice
Goodwin, Richard Hale
Gordon, Philip N
Greenblatt, Irwin M
Handel, Steven Neil
Hickey, Leo Joseph
Horsfall, James Gordon
Huang, Liang Hsiung
Koning, Ross E
Lier, Frank George
McClymont, John Wilbur
Miller, Russell Bensley
Rich, Saul
Schneider, Craig William
Smith, William Hulse
Stevenson, Harlan Quinn
Sussex, Ian Mitchell
Taylor, Gordon Stevens
Tomlinson, Harley
Trainor, Francis Rice
Wallner, William E
Walton, Gerald Steven
Webster, Terry R
Wulff, Barry Lee
Zavala, Maria Elena

DELAWARE
Bozarth, Gene Allen
Carroll, Robert Buck
Crossan, Donald Franklin
Delp, Charles Joseph
Dill, Norman Hudson
Gallagher, John Leslie
Hadley, Bruce Alan
Hodson, Robert Cleaves
Howard, Richard James
Lichtner, Francis Thomas, Jr
Matlack, Albert Shelton
Mitchell, William H
Morehart, Allen L
Pizzolato, Thompson Demetrio
Richards, Bert Lorin, Jr
Smith, Constance Meta

DISTRICT OF COLUMBIA
Barnes, John Maurice
Bellmer, Elizabeth Henry
Clutter, Mary Elizabeth
DeFilipps, Robert Anthony
DiMichele, William Anthony
Donaldson, Robert Paul
Dudley, Theodore R
Eyde, Richard Husted
Frederick, Lafayette
Goldberg, Aaron
Hale, Mason Ellsworth, Jr
Hines, Pamela Jean
Irving, Patricia Marie
Krauss, Robert Wallfar
Krigsvold, Dale Thomas
Lellinger, David Bruce
Littler, Mark Masterton
Meñez, Ernani Guingona
Meyer, Frederick Gustav
Mislivec, Philip Brian

Norris, James Newcome, IV
Robinson, Harold Ernest
Shetler, Stanwyn Gerald
Skog, Laurence Edgar
Smith, Richard S, Jr
Strauss, Michael S
Sze, Philip
Wasshausen, Dieter Carl
Werner, Patricia Ann Snyder
Wurdack, John J
Zimmer, Elizabeth Anne

FLORIDA
Agrios, George Nicholas
Aiello, Annette
Aldrich, Henry Carl
Anderson, Loran C
Austin, Daniel Frank
Bartz, Jerry A
Berger, Richard Donald
Blazquez Y Servin, Carlos Humberto
Bourne, Carol Elizabeth Mulligan
Burch, Derek George
Chase, Ann Renee
Childs, James Fielding Lewis
Cline, Kenneth Charles
Comstock, Jack Charles
Crall, James Monroe
Darby, John Feaster
Davis, Benjamin Harold
Davis, Michael Jay
Dawes, Clinton John
Dean, David Devereaux
Dean, Jack Lemuel
DeMort, Carole Lyle
Dickson, Donald Ward
DuCharme, Ernest Peter
Eilers, Frederick Irving
Ellison, Marlon L
Engelhard, Arthur William
Erdos, Gregory William
Essig, Frederick Burt
Ewart, R Bradley
Fisher, Jack Bernard
Foster, Virginia
Freeman, Thomas Edward
Friedmann, E(merich) Imre
Garnsey, Stephen Michael
Gilbert, Margaret Lois
Gilmer, Robert McCullough
Gottwald, Timothy R
Gray, Dennis John
Griffin, Dana Gove, III
Grimm, Robert Blair
Hartman, James Xavier
Hiebert, Ernest
Hopkins, Donald Lee
Janos, David Paul
Judd, Walter Stephen
Kimbrough, James W
Koevenig, James L
Kretschmer, Albert Emil, Jr
Kucharek, Thomas Albert
Lambe, Robert Carl
Langdon, Kenneth R
Lee, David Webster
Litz, Richard Earle
Luke, Herbert Hodges
McMillan, Robert Thomas, Jr
Magie, Robert Ogden
Meister, Charles William
Miller, Harvey Alfred
Miller, Kim Irving
Miskimen, Carmen Rivera
Mullin, Robert Spencer
Mullins, John Thomas
Mustard, Margaret Jean
Myers, Roy Maurice
Nemec, Stanley
Niblett, Charles Leslie
Norman, Eliane Meyer
Norris, Dean Rayburn
Payne, Willard William
Pring, Daryl Roger
Purdy, Laurence Henry
Ranzoni, Francis Verne
Ray, James Davis, Jr
Romeo, John Thomas
Roth, Frank J, Jr
Roubik, David Ward
Rumbach, William Ervin
Schenck, Norman Carl
Shanor, Leland
Shepard, James F
Simons, John Norton
Smith, Alan Paul
Smith, Harlan Eugene
Sonoda, Ronald Masahiro
Spalding, Donald Hood
Stall, Robert Eugene
Stern, William Louis
Stone, Margaret Hodgman
Stone, William Jack Hanson
Sweet, Herman Royden
Thayer, Paul Loyd
Thompson, Neal Philip
Thorhaug, Anitra L
Tryon, Rolla Milton, Jr
Vasil, Indra Kumar
Vasil, Vimla
Volin, Raymond Bradford
Ward, Daniel Bertram
Weingartner, David Peter
Whiteside, Jack Oliver

Whittier, Henry O
Williams, Norris Hagan
Williams, Tom Vare
Zettler, Francis William

GEORGIA
Ahearn, Donald G
Ajello, Libero
Andrews, Lucy Gordon
Bacon, Charles Wilson
Berry, Charles Richard
Boole, John Allen, Jr
Bozeman, John Russell
Brown, Ronald Harold
Burbank, Madeline Palmer
Carter, Eloise B
Carter, James Richard
Christenberry, George Andrew
Diboll, Alfred
Drapalik, Donald Joseph
Faircloth, Wayne Reynolds
Fuller, Melvin Stuart
Gitaitis, Ronald David
Hancock, Kenneth Farrell
Hanlin, Richard Thomas
Hurd, Maggie Patricianne
Hussey, Richard Sommers
James, Charles William
Johnson, Alva William
Jones, Samuel B, Jr
Kaufman, Leo
Kochert, Gary Dean
Kuhlman, Elmer George
Kuhn, Cedric W
LeNoir, William Cannon, Jr
McCarter, States Marion
Marx, Donald Henry
Minton, Norman A
Motsinger, Ralph E
Murdy, William Henry
Pendergrass, Levester
Porter, David
Powell, William Morton
Powers, Harry Robert, Jr
Ragsdale, Harvey Larimore
Robinson, Margaret Chisolm
Romanovicz, Dwight Keith
Roncadori, Ronald Wayne
Ruehle, John Leonard
Schadler, Daniel Leo
Snyder, Hugh Donald
Sommer, Harry Edward
Stouffer, Richard Franklin
Tinga, Jacob Hinnes
Wells, Homer Douglas
Whitehead, Marvin Delbert
Wyatt, Robert Edward
Wynn, Willard Kendall, Jr
Zimmer, David E

HAWAII
Abbott, Isabella Aiona
Alvarez, Anne Maino
Apt, Walter James
Aragaki, Minoru
Carr, Gerald Dwayne
Doty, Maxwell Stanford
Hemmes, Don E
Holtzmann, Oliver Vincent
Krauss, Beatrice Hilmer
Lamoureux, Charles Harrington
Newhouse, W Jan
Ooka, Jeri Jean
Patil, Suresh Siddheshwar
Rohrbach, Kenneth G
St John, Harold
Sakai, William Shigeru
Smith, Albert Charles
Sohmer, Seymour H
Sun, Samuel Sai-Ming
Theobald, William L
Trujillo, Eduardo E

IDAHO
Anderegg, Doyle Edward
Clary, Warren Powell
Davidson, Christopher
Davis, James Robert
Douglas, Dorothy Ann
Fenwick, Harry
Finley, Arthur Marion
Guthrie, James Warren
Helton, Audus Winzle
Holte, Karl E
LeTourneau, Duane John
Lindsay, Delbert W
McDonald, Geral Irving
Maloy, Otis Cleo, Jr
Morris, John Leonard
Moser, Paul E
Naskali, Richard John
Ohms, Richard Earl
Osgood, Charles Edgar
Packard, Patricia Lois
Partridge, Arthur Dean
Robbins, Robert Raymond
Romanko, Richard Robert
Simpson, William Roy
Sojka, Robert E
Tylutki, Edmund Eugene
Wiese, Maurice Victor

ILLINOIS
Anderson, Roger Clark

Armstrong, Joseph Everett
Bailey, Zeno Earl
Becker, Steven Allan
Bissing, Donald Ray
Bouck, G Benjamin
Cain, Jerome Richard
Carothers, Zane Bland
Chamberlain, Donald William
Chappell, Dorothy Field
Chuang, Tsan Iang
Crandall-Stotler, Barbara Jean
Crane, Joseph Leland
Dziadyk, Bohdan
Edwards, Dale Ivan
Engel, John Jay
Ford, Richard Earl
Foster, Robin Bradford
Gassman, Merrill Loren
Gessner, Robert V
Glassman, Sidney Frederick
Gottlieb, David
Gray, Lewis Richard
Green, Thomas L
Grosklags, James Henry
Hadley, Elmer Burton
Hanzely, Laszlo
Hartstirn, Walter
Henry, Robert David
Hesseltine, Clifford William
Himelick, Eugene Bryson
Hinchman, Ray Richard
Hoffman, Larry Ronald
Hooker, Arthur Lee
Huft, Michael John
Jedlinski, Henryk
Jones, Almut Gitter
Jump, John Austin
Kauffman, Harold
Keating, Richard Clark
Kessler, Kenneth J, Jr
Kurtzman, Cletus Paul
Liberta, Anthony E
Lim, Sung Man
Linn, Manson Bruce
Lippincott, Barbara Barnes
Lippincott, James Andrew
Lorenz, Ralph William
McCleary, James A
McPheeters, Kenneth Dale
Malek, Richard Barry
Marsh, Terrence George
Matten, Lawrence Charles
Middleton, Beth Ann
Miller, Charles Edward
Miller, Raymond Michael
Mohlenbrock, Robert H, Jr
Monoson, Herbert L
Murakishi, Harry Haruo
Murray, Mary Aileen
Myers, Ronald Berl
Nevling, Lorin Ives, Jr
O'Flaherty, Larrance Michael Arthur
Pappelis, Aristotel John
Paxton, Jack Dunmire
Perino, Janice Vinyard
Perry, Eugene Arthur
Plowman, Timothy
Rahn, Joan Elma
Richard, John L
Ries, Stephen Michael
Robertson, Kenneth Ray
Rouffa, Albert Stanley
Schemske, Douglas William
Schmid, Walter Egid
Schoeneweiss, Donald F
Scott, William Wallace
Seigler, David Stanley
Shurtleff, Malcolm C, Jr
Sinclair, James Burton
Singer, Rolf
Singh, Dilbagh
Smith, Richard Lawrence
Sörensen, Paul Davidsen
Spencer, Jack T
Stein-Taylor, Janet Ruth
Stidd, Benton Maurice
Stieber, Michael Thomas
Stotler, Raymond Eugene
Sundberg, Walter James
Taylor, Roy Lewis
Thornberry, Halbert Houston
Titman, Paul Wilson
Toliver, Michael Edward
Troll, Ralph
Troyer, Alvah Forrest
Tuveson, Robert Williams
Ugent, Donald
White, Donald Glenn
Whiteside, Wesley C
Wicklow, Donald Thomas
Wilcox, Wesley Crain
Yos, David Albert
Zimmerman, Craig Arthur

INDIANA
Abney, Thomas Scott
Adams, Preston
Bhattacharya, Pradeep Kumar
Bishop, Charles Franklin
Bracker, Charles E
Brenneman, James Alden
Brooks, Austin Edward
Bucholtz, Dennis Lee
Coolbaugh, Ronald Charles

Botany-Phytopathology (cont)

Crouch, Martha Louise
Davies, Harold William, (Jr)
Dunkle, Larry D
Eiser, Arthur L
Emerson, Frank Henry
Fitzgerald, Paul Jackson
Foard, Donald Edward
Foos, Kenneth Michael
Gastony, Gerald Joseph
Goonewardene, Hilary Felix
Green, Ralph J, Jr
Grove, Stanley Neal
Hammond, Charles Thomas
Heiser, Charles Bixler, Jr
Hennen, Joe Fleetwood
Huber, Don Morgan
Janutolo, Delano Blake
Jensen, Richard Jorg
Kenaga, Clare Burton
Kidd, Frank Alan
Laviolette, Francis A
Lister, Richard Malcolm
Low, Philip Stewart
McGrath, James J
Mahlberg, Paul Gordon
Marshall, James John
Nicholson, Ralph Lester
Orpurt, Philip Arvid
Pecknold, Paul Carson
Postlethwait, Samuel Noel
Roth, Jonathan Nicholas
Rothrock, Paul E
Savage, Earl John
Schoknecht, Jean Donze
Scott, Donald Howard
Shaner, Gregory Ellis
Simmons, Emory Guy
Stevenson, Forrest Frederick
Susalla, Anne A
Tansey, Michael Richard
Tseng, Charles C
Tuite, John F
Warren, Herman Lecil
Watson, Maxine Amanda
Webb, Mary Alice
Welch, Winona Hazel
Wellso, Stanley Gordon
Whitehead, Donald Reed
Williams, Edwin Bruce
Wilson, Kathryn Jay
Wilson, Kenneth Sheridan
Winternheimer, P Louis
Yates, Willard F, Jr
Youse, Howard Ray

IOWA
Borgman, Robert P
Cazin, John, Jr
Christiansen, Paul Arthur
Dunleavy, John M
Durkee, LaVerne H
Durkee, Lenore T
Eilers, Lawrence John
Embree, Robert William
Ford, Clark Fugier
Graham, Benjamin Franklin
Hodges, Clinton Frederick
Horner, Harry Theodore
Huffman, Donald Marion
Isely, Duane
Knaphus, George
Knutson, Roger M
Lersten, Nels R
Main, Stephen Paul
Martinson, Charlie Anton
Orr, Alan R
Pohl, Richard Walter
Poulter, Dolores Irma
Simons, Marr Dixon
Sjölund, Richard David
Tachibana, Hideo
Tiffany, Lois Hattery
Vakili, Nader Gholi
Walker, Waldo Sylvester
Whitson, Paul David
Wilkinson, Daniel R

KANSAS
Albrecht, Mary Lewnes
Barkley, Theodore Mitchell
Bechtel, Donald Bruce
Browder, Lewis Eugene
Dawson, James Thomas
Edmunds, Leon K
Haufler, Christopher Hardin
Hetrick, Barbara Ann
Ikenberry, Gilford John, Jr
Keeling, Richard Paire
Kramer, Charles Lawrence
Lane, Meredith Anne
Leisman, Gilbert Arthur
Lichtwardt, Robert William
McGregor, Ronald Leighton
Martin, Terry Joe
Merrill, Gary Lane Smith
Peterson, John Edward, Jr
Sauer, David Bruce
Schwenk, Fred Walter
Sperry, Theodore Melrose
Stuteville, Donald Lee
Thomasson, Joseph R
Tomanek, Gerald Wayne

Torres, Andrew M
Wells, Philip Vincent
Youngman, Arthur L

KENTUCKY
Bryant, William Stanley
Clark, Jimmy Dorral
Davis, William S
Diachun, Stephen
Dillard, Gary Eugene
Eversmeyer, Harold Edwin
Fuller, Marian Jane
Harris, Denny Olan
Hayden, Mary Victoria
Herron, James Watt
Jenkins, Jeff Harlin
Keefe, Thomas Leeven
King, Joe Mack
Nicely, Kenneth Aubrey
Niffenegger, Daniel Arvid
Perlin, Michael Howard
Pirone, Thomas Pascal
Shepherd, Robert James
Shugars, Jonas P
Siegel, Malcolm Richard
Smiley, Jones Hazelwood
Smith, Dale Metz
Stroube, William Hugh
Thieret, John William
Thompson, Ralph Luther
Wheeler, Harry Ernest
Wiedeman, Varley Earl
Williams, Albert Simpson
Winstead, Joe Everett
Wolfson, Alfred M

LOUISIANA
Allen, Arthur (Silsby)
Berggren, Gerard Thomas, Jr
Birchfield, Wray
Black, Lowell Lynn
Bond, William Payton
Chapman, Russell Leonard
Christian, James A
Clark, Christopher Alan
Damann, Kenneth Eugene, Jr
Darwin, Steven Peter
DePoe, Charles Edward
Erbe, Lawrence Wayne
Hardberger, Florian Max
Kalinsky, Robert George
Kenknight, Glenn
Kirk, Ben Truett
Koike, Hideo
Lieux, Meredith Hoag
Lindberg, George Donald
Longstreth, David J
Lowy, Bernard
Lutes, Dallas D
Lynch, Steven Paul
MacKenzie, David Robert
Morris, Everett Franklin
Saigo, Roy Hirofumi
Schneidau, John Donald, Jr
Shepherd, Hurley Sidney
Sims, Asa C, Jr
Snow, Johnnie Park
Steib, Rene J
Terry, Maurice Ernest
Thomas, Roy Dale
Tucker, Shirley Cotter
Urbatsch, Lowell Edward
Utley, John Foster, III
Vail, Sidney Lee
Walker, Harrell Lynn
Webert, Henry S
Welden, Arthur Luna
White, James Clarence
Williams, George, Jr

MAINE
Campana, Richard John
Gelinas, Douglas Alfred
Hehre, Edward James, Jr
Lewis, Alan James
Manzer, Franklin Edward
Neubauer, Benedict Francis
Richards, Charles Davis
Thomas, Robert James
Wave, Herbert Edwin

MARYLAND
Altevogt, Raymond Fred
Anderson, Mauritz Gunnar
Barclay, Arthur S
Barksdale, Thomas Henry
Batra, Lekh Raj
Bean, George A
Benjamin, Chester Ray
Bereston, Eugene Sydney
Bonde, Morris Reiner
Bromfield, Kenneth Raymond
Broome, Carmen Rose
Bruns, Herbert Arnold
Corbett, M Kenneth
Damsteegt, Vernon Dale
Davis, Robert Edward
Deahl, Kenneth Luvere
Devine, Thomas Edward
Diener, Theodor Otto
Dixon, Dennis Michael
Dowler, William Minor
Duke, James A
Farr, David Frederick

Fulkerson, John Frederick
Futcher, Anthony Graham
Gehris, Clarence Winfred
Goth, Robert W
Graham, Joseph Harry
Griesbach, Robert James
Gross, Kenneth Charles
Gunn, Charles Robert
Harding, Wallace Charles, Jr
Heggestad, Howard Edwin
Herman, Eliot Mark
Hiatt, Caspar Wistar, III
Hodgson, Richard Holmes
Imle, Ernest Paul
Jong, Shung-Chang
Kahn, Robert Phillip
Kantzes, James (George)
Karlander, Edward P
Kennedy, Robert A
Kingsolver, Charles H
Kirkbride, Joseph Harold, Jr
Kulik, Martin Michael
Lentz, Paul Lewis
Lewis, Jack A
Lockard, J David
Lumsden, Robert Douglas
Maas, John Lewis
Meiners, Jack Pearson
Melching, J Stanley
Merz, William George
Miksche, Jerome Phillip
Moline, Harold Emil
Morgan, Omar Drennan, Jr
Motta, Jerome J
Ostazeski, Stanley A
Palm, Mary Egdahl
Papavizas, George Constantine
Perdue, Robert Edward, Jr
Perry, James Warner
Phillips, William George
Price, Samuel
Reveal, James L
Rissler, Jane Francina
Rossman, Amy Yarnell
Sayre, Richard Martin
Simpson, Marion Emma
Sisler, Hugh Delane
Smith, Andrew George
Smith, William Owen
Stavely, Joseph Rennie
Terrell, Edward Everett
Uecker, Francis August
Warmbrodt, Robert Dale
Waterworth, Howard E
Weaver, Leslie O
Windler, Donald Richard
Zacharius, Robert Marvin

MASSACHUSETTS
Ahmadjian, Vernon
Ashton, Peter Shaw
Barke, Harvey Ellis
Bawa, Kamaljit S
Bertin, Robert Ian
Bigelow, Howard Elson
Boger, Edwin August, Sr
Bolotin, Moshe
Bowley, Donovan Robin
Branton, Daniel
Brennan, James Robert
Burk, Carl John
Camp, Russell R
Caruso, Frank Lawrence
Cheetham, Ronald D
Creighton, Harriet Baldwin
Davis, Edward Lyon
Dent, Thomas Curtis
DeWolf, Gordon Parker, Jr
DiLiddo, Rebecca McBride
Forman, Richard T T
Frederick, Sue Ellen
Freeberg, John Arthur
Godfrey, Paul Jeffrey
Golubic, Stephan
Gruber, Peter Johannes
Haight, Thomas H
Hare, Joan Conway
Haskell, David Andrew
Hewitson, Walter Milton
Hilferty, Frank Joseph
Hoffmann, George Robert
Holmes, Francis W(illiam)
Howard, Richard Alden
Johansen, Hans William
Jost, Dana Nelson
Kaplan, Lawrence
Kissmeyer-Nielsen, Erik
Klekowski, Edward Joseph, Jr
Lovejoy, David Arnold
Madore, Bernadette
Manning, William Joseph
Melan, Melissa A
Mish, Lawrence Bronislaw
Mount, Mark Samuel
Nickerson, Norton Hart
Nixon, Charles William
Page, Joanna R Ziegler
Pfister, Donald Henry
Primack, Richard Bart
Raup, Hugh Miller
Reese, Elwyn Thomas
Rollins, Reed Clark
Roth, John L, Jr
Sayre, Geneva

Scheirer, Daniel Charles
Schofield, Edmund Acton, Jr
Schubert, Bernice Giduz
Schultes, Richard Evans
Shapiro, Seymour
Signer, Ethan Royal
Spence, Willard Lewis
Spongberg, Stephen Alan
Stein, Diana B
Stevens, Peter Francis
Thomas, Aubrey Stephen, Jr
Tomlinson, Philip Barry
Troll, Joseph
Wetmore, Ralph Hartley
Wilce, Robert Thayer
Wilson, Brayton F
Wood, Carroll E, Jr
Woodwell, George Masters
Zimmermann, Martin Huldrych

MICHIGAN
Abou-El-Seoud, Mohamed Osman
Andersen, Axel Langvad
Anderson, William Russell
Andresen, Norman A
Barnes, Burton Verne
Bath, James Edmond
Beaman, John Homer
Beck, Charles Beverley
Belcher, Robert Orange
Beneke, Everett Smith
Bird, George W
Bloom, Miriam
Bowers, Maynard C
Bowers, Robert Charles
Burton, Clyde Leaon
Cantino, Edward Charles
Crum, Howard Alvin
Dazzo, Frank Bryan
Erbisch, Frederic H
Estabrook, George Frederick
Fogel, Robert Dale
Freeman, Dwight Carl
Gilbert, William James
Glime, Janice Mildred
Halloin, John McDonell
Hampton, Raymond Earl
Hart, Lynn Patrick
Hohn, Matthew Henry
Hollensen, Raymond Hans
Holt, Imy Vincent
Hooker, William James
Hurst, Elaine H
Jones, Alan Lee
Kaufman, Peter Bishop
Klos, Edward John
Knobloch, Irving William
Krueger, Robert John
Krupka, Lawrence Ronald
LaCroix, Joseph Donald
Lacy, Melvyn Leroy
Lin, Chang Kwei
Lockwood, John LeBaron
Lowry, Robert James
McMeekin, Dorothy
Meyer, Ronald Warren
Murphy, Peter George
Olexia, Paul Dale
Pippen, Richard Wayne
Potter, Howard Spencer
Raikhel, Natasha V
Ramsdell, Donald Charles
Reznicek, Anton Albert
Rippon, John Willard
Rogers, Alvin Lee
Rogers, Claude Marvin
Rusch, Wilbert H, Sr
Saettler, Alfred William
Sakai, Ann K
Scheffer, Robert Paul
Schlichting, Harold Eugene, Jr
Schmitter, Ruth Elizabeth
Shaffer, Robert Lynn
Shontz, John Paul
Smith, Eugene William
Stein, Howard Jay
Stephens, Christine Taylor
Stephenson, Stephen Neil
Stowell, Ewell Addison
Tarapchak, Stephen J
Taylor, William Randolph
Toczek, Donald Richard
Vande Berg, Warren James
Van Faasen, Paul
Vargas, Joseph Martin, Jr
Volz, Paul Albert
Voss, Edward Groesbeck
Wagner, Florence Signaigo
Wagner, Warren Herbert, Jr
Wells, James Ray
Wilson, Ronald Wayne
Wujek, Daniel Everett
Wynne, Michael James
Yu, Shih-An

MINNESOTA
Abbott, Rose Marie Savelkoul
Anderson, Neil Albert
Banttari, Ernest E
Bissonnette, Howard Louis
Blanchette, Robert Anthony
Burton, Daniel Frederick
Burton, Verona Devine
Bushnell, William Rodgers

Bushong, Jerold Ward
Carlson, John Bernard
Charvat, Iris
Clapp, Thomas Wright
Ezell, Wayland Lee
French, David W
Gordon, Donald
Grewe, Alfred H, Jr
Groth, James Vernon
Hansen, Harold Westberg
Hill, Eddie P
Jefferson, Carol Annette
Johnson, Herbert Gordon
Kennedy, Bill Wade
Kommedahl, Thor
Krupa, Sagar
Larsen, Philip O
Lawrence, Donald Buermann
Leonard, Kurt John
Levine, Allen Stuart
Lockhart, Benham Edward
Loeffler, Robert J
McCulloch, Joseph Howard
McLaughlin, James L
Markhart, Albert Henry, III
Mason, Charles Perry
Mirocha, Chester Joseph
Monson, Paul Herman
Nyvall, Robert Frederick
O'Rourke, Richard Clair
Ownbey, Gerald Bruce
Percich, James Angelo
Rines, Howard Wayne
Roberts, Glenn Dale
Roelfs, Alan Paul
Romig, Robert William McClelland
Rowell, John Bartlett
Schafer, John Francis
Schimpf, David Jeffrey
Sentz, James Curtis
Severin, Charles Hilarion
Silflow, Carolyn Dorothy
Silverman, William Bernard
Singer, Susan Rundell
Skilling, Darroll Dean
Skjegstad, Kenneth
Stewart, Elwin Lynn
Stienstra, Ward Curtis
Tolbert, Robert John
Wetmore, Clifford Major
Widin, Katharine Douglas
Wilcoxson, Roy Dell

MISSISSIPPI
Ammon, Vernon Dale
Batson, William Edward, Jr
Cibula, William Ganley
Cliburn, Joseph William
Davis, Robert Gene
Delouche, James Curtis
Eleuterius, Lionel Numa
Filer, Theodore H, Jr
Graves, Clinton Hannibal, Jr
Hare, Mary Louise Eckles
Huneycutt, Maeburn Bruce
Keeling, Bobbie Lee
McGuire, James Marcus
Pitre, Henry Nolle, Jr
Pullen, Thomas Marion
Raj, Baldev
Rosenkranz, Eugen Emil
Sciumbato, Gabriel Lon
Sherman, Harry Logan
Singh, Jaswant
Spencer, James Alphus
Stewart, Robert Archie, II
Trevathan, Larry Eugene
Watson, James Ray, Jr
Wooten, Jean W

MISSOURI
Althaus, Ralph Elwood
Anderson, Robert Gordon
Apirion, David
Bell, Max Ewart
Bond, Lora
Brown, Merton F
Calvert, Oscar Hugh
Castaner, David
Croat, Thomas Bernard
Crosby, Marshall Robert
Cumbie, Billy Glenn
D'Arcy, William Gerald
Davidse, Gerrit
De Buhr, Larry Eugene
Dunn, David Baxter
Dwyer, John Duncan
Ewan, Joseph (Andorfer)
Finley, David Emanuel
Forero, Enrique
Fraley, Robert Thomas
Gentry, Alwyn Howard
Goldblatt, Peter
Goodman, Robert Norman
Gowans, Charles Shields
Graham, James Carl
Hagen, Gretchen
Hanks, David L
Hayes, Alice Bourke
Huckabay, John Porter
Kobayashi, George S
Kullberg, Russell Gordon
Lewis, Walter Hepworth
Lindhorst, Taylor Erwin

Lissant, Ellen Kern
Magill, Robert Earle
Maniotis, James
Millikan, Daniel Franklin, Jr
Moore, James Frederick, Jr
Morin, Nancy Ruth
Moser, Stephen Adcock
Neely, Robert Dan
Nichols, Herbert Wayne
Pallardy, Stephen Gerard
Ponder, Felix, Jr
Raven, Peter Hamilton
Redfearn, Paul Leslie, Jr
Rhodes, Russell G
Sehgal, Om Parkash
Stevens, Warren Douglas
Suhovecky, Albert J
Tal, William
Wagenknecht, Burdette Lewis
Wallin, Jack Robb
Weber, Wallace Rudolph
Wood, Joseph M
Wyllie, Thomas Dean

MONTANA
Carlson, Clinton E
Carroll, Thomas William
Chessin, Meyer
Elliott, Eugene Willis
McCoy, Thomas Joseph
Mathre, Donald Eugene
Rumely, John Hamilton
Sands, David Chandler
Sawyer, Paul Thompson
Sharp, Eugene Lester
Solberg, Richard Allen
Strobel, Gary A

NEBRASKA
Becker, Donald A
Bolick, Margaret Ruth
Boosalis, Michael Gus
Brooks, Merle Eugene
Egan, Robert Shaw
Gauger, Wendell Lee
Ikenberry, Richard W
Kaplan, Sanford Sandy
Kaul, Robert Bruce
Kerr, Eric Donald
Langenberg, Willem G
Maier, Charles Robert
Morris, Thomas Jack
Partridge, James Enoch
Peterson, Glenn Walter
Steadman, James Robert
Sutherland, David M
Swift, Lloyd Harrison
Weihing, John Lawson
Wysong, David Serge

NEVADA
Bohmont, Dale W
Fox, Carl Alan
Lupan, David Martin
Mozingo, Hugh Nelson
Niles, Wesley E
Thyr, Billy Dale
Went, Fritz

NEW HAMPSHIRE
Blanchard, Robert Osborn
Croasdale, Hannah Thompson
Crow, Garrett Eugene
DeMaggio, Augustus Edward
Fagerberg, Wayne Robert
Fralick, Richard Allston
Jones, Richard Conrad
Mathieson, Arthur C
Minoca, Subhash C
Norton, Robert James
Rich, Avery Edmund
Rock, Barrett Nelson
Schreiber, Richard William
Shigo, Alex Lloyd
Shortle, Walter Charles
Wilson, Carl Louis

NEW JERSEY
Babcock, Philip Arnold
Brown, Thomas Edward
Cappellini, Raymond Adolph
Ceponis, Michael John
Chen, James Che Wen
Chen, Tseh-An
Davis, Spencer Harwood, Jr
Day, Peter Rodney
Deems, Robert Eugene
Fairbrothers, David Earl
Froyd, James Donald
Gaynor, John James
Greenfield, Sydney Stanley
Halisky, Philip Michael
Jacobs, William Paul
Kasper, Andrew E, Jr
Klessig, Daniel Frederick
Kuehn, Harold Herman
Kuhnen, Sybil Marie
Lewis, Gwynne David
Ling, Hubert
Maiello, John Michael
Markle, George Michael
Myers, Ronald Fenner
Palser, Barbara Frances
Podleckis, Edward Vidas

Quinn, James Allen
Quinn, James Amos
Reid, Hay Bruce, Jr
Schwalb, Marvin N
Springer, John Kenneth
Star, Aura E
Watts, Daniel Jay
Weber, Paul Van Vranken
Younkin, Stuart G
Zuck, Robert Karl

NEW MEXICO
Allred, Kelly Wayne
Bohrer, Vorsila Laurene
Booth, John Austin
Carter, Jack Lee
Dittmer, Howard James
Dunford, Max Patterson
Grover, Herbert David
Heiner, Terry Charles
Holloway, Richard George
Hooks, Ronald Fred
Hsi, David Ching Heng
Kidd, David Eugene
Leyendecker, Philip Jordon
Lindsey, Donald L
Martin, William Clarence
Peterson, Roger Shipp
Potter, Loren David
Risser, Paul Gillan
Spellenberg, Richard (William)
Todsen, Thomas Kamp

NEW YORK
Aist, James Robert
Alves, Leo Manuel
Andrus, Richard Edward
App, Alva A
Arneson, Phil Alan
Banks, Harlan Parker
Basile, Dominick V
Bates, David Martin
Bedford, Barbara Lynn
Beer, Steven Vincent
Bentley, Barbara Lee
Biondo, Frank X
Blasdell, Robert Ferris
Bobear, Jean B
Boothroyd, Carl William
Braun, Alvin Joseph
Brett, Betty Lou Hilton
Buck, William R
Carroll, Robert Baker
Castello, John Donald
Chadha, Kailash Chandra
Chase, Sherret Spaulding
Chock, Jan Sun-Lum
Churchill, Algernon Coolidge
Cox, Donald David
Cronquist, Arthur John
Davis, David
De Laubenfels, David John
Dietert, Margaret Flowers
Di Sanzo, Carmine Pasqualino
Dolan, Desmond Daniel
Dress, William John
Dudock, Bernard S
Dumont, Kent P
Earle, Elizabeth Deutsch
Erb, Kenneth
Forest, Charlene Lynn
Fry, William Earl
Goldstein, Solomon
Gordon, Morris Aaron
Green, John Irving
Greller, Andrew M
Griffin, David H
Haber, Alan Howard
Hager, Richard Arnold
Haines, John Haldor
Hammill, Terrence Michael
Harman, Gary Elvan
Heidrick, Lee E
Heusser, Calvin John
Hibben, Craig Rittenhouse
Hirshon, Jordon Barry
Ho, Hon Hing
Hoham, Ronald William
Holmgren, Noel Herman
Holmgren, Patricia Kern
Howard, Harold Henry
Howell, Stephen Herbert
Hudler, George William
Humber, Richard Alan
Hunter, James Edward
Irwin, Howard Samuel
Jones, Clive Gareth
Kiviat, Erik
Kohut, Robert John
Korf, Richard Paul
Koyama, Tetsuo
Larson, Donald Alfred
Latorella, A Henry
Law, David Martin
Ledbetter, Myron C
Leopold, Donald Joseph
Lorbeer, James W
Lowe, Josiah L(incoln)
Luteyn, James Leonard
McDonnell, Mark Jeffery
Mai, William Frederick
Maple, William Thomas
Marengo, Norman Payson
Marsh, Leland C

Meiselman, Newton
Metzner, Jerome
Mickel, John Thomas
Miles, Philip Giltner
Miller, Norton George
Miller, Pauline Monz
Mitchell, Richard Shepard
Mitra, Jyotirmay
Mori, Scott Alan
Mussell, Harry W
Nolan, James Robert
Pickett, Steward T A
Prescott, Henry Emil, Jr
Putala, Eugene Charles
Rana, Mohammad A
Randall, Eric A
Reddy, Kalluru Jayarami
Richardson, F C
Robinson, Albert Dean
Robinson, Alix Ida
Robinson, Beatrice Letterman
Rochow, William Frantz
Rogerson, Clark Thomas
Rosenthal, Stanley Arthur
Salkin, Ira Fred
Schumacher, George John
Selsky, Melvyn Ira
Settle, Wilbur Jewell
Shechter, Yaakov
Sherf, Arden Frederick
Sheviak, Charles John
Silva-Hutner, Margarita
Simone, Leo Daniel
Sirois, David Leon
Slack, Steven Allen
Smith, Ora
Stalter, Richard
Staples, Richard Cromwell
Steere, William Campbell
Sweeney, Robert Anderson
Tepper, Herbert Bernard
Thurston, Herbert David
Titus, John Elliott
Toenniessen, Gary Herbert
Torgeson, Dewayne Clinton
Uhl, Charles Harrison
Wang, Chun-Juan Kao
Webber, Edgar Ernest
Weinstein, Leonard Harlan
Wilson, Jack Belmont
Yoder, Olen Curtis
Zabel, Robert Alger
Zaitlin, Milton
Zitter, Thomas Andrew

NORTH CAROLINA
Acedo, Gregoria N
Antognini, Joe
Averre, Charles Wilson, III
Aycock, Robert
Baranski, Michael Joseph
Barker, Kenneth Reece
Bateman, Durward F
Beard, Luther Stanford
Bell, Clyde Ritchie
Benson, David Michael
Beute, Marvin Kenneth
Billings, William Dwight
Bland, Charles E
Blum, Udo
Butterfield, Earle James
Callahan, Kemper Leroy
Camp, Pamela Jean
Campbell, Peter Hallock
Carpenter, Irvin Watson, Jr
Chopra, Baldeo K
Couch, John Nathaniel
Cowling, Ellis Brevier
Culberson, William Louis
Cutter, Lois Jotter
Dickison, William Campbell
Downs, Robert Jack
DuBay, Denis Thomas
Duncan, Harry Ernest
Echandi, Eddie
Ellis, Don Edwin
Fantz, Paul Richard
Flagg, Raymond Osbourn
Fulcher, William Ernest
Garner, Jasper Henry Barkdoll
Gensel, Patricia Gabbey
Gooding, Guy V, Jr
Haning, Blanche Cournoyer
Hardin, James Walker
Heagle, Allen Streeter
Hebert, Teddy T
Henderson, Nannette Smith
Hilger, Anthony Edward
Hommersand, Max Hoyt
Horton, James Heathman
Huang, Jeng-Sheng
Hunt, Kenneth Whitten
Jenkins, Samuel Forest, Jr
Johnson, Terry Walter, Jr
Jolls, Claudia Lee
Kapraun, Donald Frederick
Khan, Sekender Ali
Koch, William Julian
Kohlmeyer, Jan Justus
Krochmal, Arnold
Lapp, Neil Arden
Lucas, George Blanchard
Lucas, Leon Thomas
McDonald, James Clifton

Botany-Phytopathology (cont)

McVaugh, Rogers
Main, Charles Edward
Mangelsdorf, Paul Christoph
Manly, Jethro Oates
Massey, Jimmy R
Matthews, James Francis
Mayfield, John Emory
Merritt, James Francis
Miller, Carol Raymond
Mott, Ralph Lionel
Mowbray, Thomas Bruce
Olive, Lindsay Shepherd
Padgett, David Emerson
Peet, Robert Krug
Philpott, Jane
Powell, Nathaniel Thomas
Radford, Albert Ernest
Reinert, Richard Allyn
Ross, John Paul
Searles, Richard Brownlee
Shafer, Steven Ray
Shoemaker, Paul Beck
Sieren, David Joseph
Sievert, Richard Carl
Spencer, Lorraine Barney
Spurr, Harvey Wesley, Jr
Stone, Donald Eugene
Strider, David Lewis
Swab, Janice Coffey
Tweedy, Billy Gene
Van Dyke, Cecil Gerald
Walkinshaw, Charles Howard, Jr
Wallace, James William, Jr
Ward, John Everett, Jr
Welch, Aaron Waddington
Wells, Charles Van
Whitford, Larry Alston
Wilbur, Robert Lynch
Winstead, Nash Nicks
Wyatt, Raymond L
Yarnell, Richard Asa
Yeats, Frederick Tinsley
Zublena, Joseph Peter

NORTH DAKOTA
Barker, William T
Esslinger, Theodore Lee
Freeman, Myron L
Freeman, Thomas Patrick
Hosford, Robert Morgan, Jr
Kiesling, Richard Lorin
LaDuke, John Carl
Lamey, Howard Arthur
Seiler, Gerald Joseph
Starks, Thomas Leroy
Suttle, Jeffrey Charles

OHIO
Alldridge, Norman Alfred
Andreas, Barbara Kloha
Baker, Dwight Dee
Benzing, David H
Boerner, Ralph E J
Byrne, John Maxwell
Cantino, Philip Douglas
Cavender, James C
Chitaley, Shyamala D
Chuey, Carl F
Collins, Gary Brent
Cooke, William Bridge
Cooperrider, Tom Smith
Coplin, David Louis
Crawford, Daniel John
Curtis, Charles R
Deal, Don Robert
Dean, Donald Stewart
De Jong, Diederik Cornelis Dignus
Delanglade, Ronald Allan
Easterly, Nathan William
Ellett, Clayton Wayne
Eshbaugh, William Hardy
Farley, James D
Fisher, T Richard
Floyd, Gary Leon
Fulford, Margaret Hannah
Garraway, Michael Oliver
Giesy, Robert
Gilbert, Gareth E
Gingery, Roy Evans
Gochenaur, Sally Elizabeth
Gorchov, David Louis
Gordon, Donald Theile
Graffius, James Herbert
Graham, Alan Keith
Graham, Shirley Ann
Gray, William Dudley
Hauser, Edward J P
Heimsch, Charles W
Herr, Leonard Jay
Hillis, Llewellya
Hobbs, Clinton Howard
Hoitink, Harry A J
Ichida, Allan A
Janson, Blair F
Jensen, William August
Kaufmann, Maurice John
King, Charles C
Kolattukudy, P E
Laufersweiler, Joseph Daniel
Laushman, Roger H
Leben, Curt (Charles)
Lilly, Percy Lane

Lloyd, Robert Michael
Long, Terrill Jewett
Loucks, Orie Lipton
Louie, Raymond
Lowe, Rex Loren
Macior, Lazarus Walter
Mapes, Gene Kathleen
Mason, David Lamont
Mattox, Karl
Meyer, Bernard Sandler
Moore, Randy
Mueller, Sabina Gertrude
Mulroy, Juliana Catherine
Partyka, Robert Edward
Pollack, J Dennis
Popham, Richard Allen
Powell, Charles Carleton, Jr
Rhodes, Landon Harrison
Romans, Robert Charles
Rowe, Randall Charles
Rudolph, Emanuel David
Runkle, James Reade
Sack, Fred David
Schmitt, John Arvid, Jr
Schmitthenner, August Fredrick
Schreiber, Lawrence
Seymour, Roland Lee
Silvius, John Edward
Smith, Calvin Albert
Smoot, Edith L
Snider, Jerry Allen
Snyder, Gary Wayne
Stoner, Clinton Dale
Stoutamire, Warren Petrie
Stuckey, Ronald Lewis
Stuessy, Tod Falor
Tulecke, Walt
Weidensaul, T Craig
Weishaupt, Clara Gertrude
Wickstrom, Conrad Eugene
Williams, Lansing Earl
Wilson, Kenneth Glade
Wilson, Thomas Kendrick

OKLAHOMA
Barnes, George Lewis
Bleckmann, Charles Allen
Boke, Norman Hill
Campbell, Thomas Hodgen
Conway, Kenneth Edward
Essenberg, Margaret Kottke
Estes, James Russell
Gardner, Charles Olda, Jr
Gough, Francis Jacob
Gregory, Garold Fay
Larsh, Howard William
Levetin Avery, Estelle
Littlefield, Larry James
Love, Harry Schroeder, Jr
Melouk, Hassan A
Richardson, Paul Ernest
Russell, Scott D
Sherwood, John L
Studlar, Susan Moyle
Sturgeon, Roy V, Jr
Taylor, Constance Elaine Southern
Taylor, R(aymond) John
Tyrl, Ronald Jay
Yi, Cho Kwang
Young, David Allen

OREGON
Aho, Paul E
Allen, Thomas Cort, Jr
Baker, Kenneth Frank
Brandt, William Henry
Brehm, Bertram George, Jr
Calvin, Clyde Lacey
Cameron, H Ronald
Carroll, George C
Chambers, Kenton Lee
Converse, Richard Hugo
Corden, Malcolm Ernest
Daley, Laurence Stephen
Denison, William Clark
Erdman, Kimball S
Florance, Edwin R
Goheen, Austin Clement
Hampton, Richard Owen
Hansen, Everett Mathew
Hardison, John Robert
Higgins, Paul Daniel
Horner, Chester Ellsworth
Johnson, John Morris
Lang, Frank Alexander
Leach, Charles Morley
Linderman, Robert G
Lippert, Byron E
Meints, Russel H
Milbrath, Gene McCoy
Miller, Paul William
Mills, Dallice Ivan
Minore, Don
Molina, Randolph John
Moore, Larry Wallace
Neiland, Bonita J
Nelson, Earl Edward
Novak, Robert Otto
Orkney, G Dale
Phinney, Harry Kenyon
Powelson, Robert Loran
Quatrano, Ralph Stephen
Ream, Lloyd Walter, Jr
Retallack, Gregory John

Schmidt, Clifford LeRoy
Sherwood-Pike, Martha Allen
Spears, Brian Merle
Spotts, Robert Allen
Tepfer, Sanford Samuel
Trappe, James Martin
Wagner, David Henry
Wagner, Orvin Edson
Welty, Ronald Earle
Wimber, Donald Edward

PENNSYLVANIA
Allison, William Hugh
Archibald, Patricia Ann
Ayers, Arthur Raymond
Banerjee, Sushanta Kumar
Barnes, William Shelley
Bartuska, Doris G
Bayer, Margret Helene Janssen
Biebel, Paul Joseph
Birchem, Regina
Black, Robert Corl
Bloom, James R
Bowen, Paul Ross
Bradt, Patricia Thornton
Brunkard, Kathleen Marie
Bryner, Charles Leslie
Carley, Harold Edwin
Cavaliere, Alphonse Ralph
Cole, Herbert, Jr
Dahl, A(nthony) Orville
DeFigio, Daniel A
DeMott, Howard Ephraim
Dobbins, David Ross
Ehrlich, Mary Ann
Erickson, Ralph O
Fett, William Frederick
Gaither, Thomas Walter
Grossman, Herbert H
Gustine, David Lawrence
Hickey, Kenneth Dyer
Hillson, Charles James
Hoffmaster, Donald Edeburn
Hunter, Barry B
Jacobsen, Terry Dale
Keener, Carl Samuel
Kelley, William Russell
Kiger, Robert William
Kneebone, Leon Russell
Leath, Kenneth T
Lukezic, Felix Lee
Maksymowych, Roman
Mears, James Austin
Mikesell, Jan Erwin
Miller, Helena Agnes
Miller, Robert Ernest
Mingrone, Louis V
Montgomery, James Douglas
Moorman, Gary William
Nelson, Paul Edward
Nelson, Richard Robert
Oswald, John Wieland
Overlease, William R
Parks, James C
Pell, Eva Joy
Pickering, Jerry L
Poethig, Richard Scott
Pottmeyer, Judith Ann
Pritchard, Hayden N
Pursell, Ronald A
Roia, Frank Costa, Jr
Royse, Daniel Joseph
Salch, Richard K
Salvin, Samuel Bernard
Schaeffer, Robert L, Jr
Schein, Richard David
Schipper, Arthur Louis, Jr
Schisler, Lee Charles
Schrock, Gould Frederick
Schuyler, Alfred Ernest
Settlemyer, Kenneth Theodore
Sherwood, Robert Tinsley
Sinden, James Whaples
Smith, Bruce Barton
Snow, Jean Anthony
Stone, Benjamin Clemens
Therrien, Chester Dale
Thimann, Kenneth Vivian
Uscavage, Joseph Peter
Utech, Frederick Herbert
Verhoek, Susan Elizabeth
Wheeler, Donald Alsop
Wuest, Paul J

RHODE ISLAND
Beckman, Carl Harry
Church, George Lyle
Dyer, Hubert Jerome
Goos, Roger Delmon
Hammen, Susan Lum
Harlin, Marilyn Miler
Hartmann, George Charles
Hauke, Richard Louis
Heywood, Peter
Howard, Frank Leslie
Hull, Richard James
Jackson, Noel
Mueller, Walter Carl
Palmatier, Elmer Arthur
Schmitt, Johanna
Swift, Dorothy Garrison

SOUTH CAROLINA
Alexander, Paul Marion

Baxter, Luther Willis, Jr
Blackmon, Cyril Wells
Cowley, Gerald Taylor
Dickerson, Ottie J
Dukes, Philip Duskin
Guram, Malkiat Singh
Houk, Richard Duncan
Husband, David Dwight
Kingsland, Graydon Chapman
Miller, Robert Walker, Jr
Pollard, Arthur Joseph
Porcher, Richard Dwight
Powell, Robert W, Jr
Rodgers, Charles Leland
Schoulties, Calvin Lee
Sharitz, Rebecca Reyburn
Shealy, Harry Everett, Jr
Shive, John Benjamine, Jr
Stewart, Shelton E
Strobel, James Walter
Witcher, Wesley

SOUTH DAKOTA
Brashier, Clyde Kenneth
Buchenau, George William
Hart, Charles Richard
Hodgson, Lynn Morrison
Horton, Maurice Lee
Jensen, Stanley George
Myers, Gerald Andy
Tatina, Robert Edward
Van Bruggen, Theodore

TENNESSEE
Artist, Russell (Charles)
Ballal, S K
Bell, Sandra Lucille
Browne, Edward Tankard, Jr
Caponetti, James Dante
Channell, Robert Bennie
Cox, Edmond Rudolph, Jr
Ellis, William Haynes
Gunasekaran, Muthukumaran
Heilman, Alan Smith
Herndon, Walter Roger
Hilty, James Willard
Hughes, Karen Woodbury
Hunter, Gordon Eugene
Jennings, Lisa Helen Kyle
Johnson, Leander Floyd
Jones, Larry Hudson
Karve, Mohan Dattatreya
Kral, Robert
Murrell, James Thomas, Jr
Nall, Ray(mond) W(illett)
Olsen, John Stuart
Pulido, Miguel L
Quarterman, Elsie
Ramseur, George Shuford
Rosing, Wayne C
Salk, Martha Scheer
Schilling, Edward Eugene
Sharp, Aaron John
Shriner, David Sylva
Staub, Robert J
Trigiano, Robert Nicholas
Van Horn, Gene Stanley
Vredeveld, Nicholas Gene
Walne, Patricia Lee
Whittier, Dean Page
Wolf, Frederick Taylor

TEXAS
Amato, Vincent Alfred
Arnott, Howard Joseph
Arp, Gerald Kench
Averett, John E
Berry, Robert Wade
Bischoff, Harry William
Boutton, Thomas William
Bragg, Louis Hairston
Brand, Jerry Jay
Brown, Richard Malcolm, Jr
Browning, John Artie
Bryant, Vaughn Motley, Jr
Burgess, Jack D
Camp, Earl D
Cole, Garry Thomas
Cox, Elenor R
Diggs, George Minor, Jr
Elliot, Arthur McAuley
Ellzey, Joanne Tontz
Fearing, Olin S
Fowler, Norma Lee
Fryxell, Paul Arnold
Fucik, John Edward
Fulton, Joseph Patton
Grant, Verne (Edwin)
Harris, Kerry Francis Patrick
Hatch, Stephan LaVor
Higgins, Larry Charles
Hobbs, Clifford Dean
Hoff, Victor John
Hollis, John Percy, Jr
Hotchkiss, Arland Tillotson
Hunter, Richard Edmund
Johnston, La Verne Albert
Johnston, Marshall Conring
Judd, Frank Wayne
Keller, Harold Willard
Kimber, Clarissa Therese
Kroschewsky, Julius Richard
La Claire, John Willard, II
Land, Geoffrey Allison

Lee, Addison Earl
Loeblich, Alfred Richard, III
Lonard, Robert (Irvin)
Lyda, Stuart D
McCracken, Michael Dwayne
McFeeley, James Calvin
McGinnis, Michael Randy
McGrath, William Thomas
Maguire, Bassett
Mahler, William Fred
Manis, Archie L
Mims, Charles Wayne
Mueller, Dale M J
Neill, Robert Lee
Ortega, Jacobo
Pilcher, Benjamin Lee
Powell, Michael A
Prior, Paul Verdayne
Proctor, Vernon Willard
Roach, Archibald Wilson Kilbourne
Rosberg, David William
Schneider, Edward Lee
Shane, John Denis
Sleeth, Bailey
Smith, Don Wiley
Smith, Gerald Ray
Smith, Roberta Hawkins
Sorensen, Lazern Otto
Stanford, Jack Wayne
Starr, Richard Cawthon
Staten, Raymond Dale
Stewart, Robert Blaylock
Sweet, Charles Edward
Szaniszlo, Paul Joseph
Taylor, Charles Arthur, Jr
Taylor, Robert Lee
Thomas, Ruth Beatrice
Thurston, Earle Laurence
Toler, Robert William
Turner, Billie Lee
Venketeswaran, S
Waddell, Henry Thomas
Wadsworth, Dallas Fremont
Ward, Calvin Herbert
Williams, Kenneth Bock
Williges, George Goudie
Willingham, Francis Fries, Jr
Wilson, Hugh Daniel
Wilson, Robert Eugene
Worthington, Richard Dane
Wright, Robert Anderson

UTAH
Barkworth, Mary Elizabeth
Belnap, Jayne
Blauer, Aaron Clyde
Bollinger, William Hugh
Bozniak, Eugene George
Campbell, William Frank
Cannon, Orson Silver
Cox, Paul Alan
Dalton, Patrick Daly
Epstein, William Warren
Griffin, Gerald D
Hansen, Afton M
Harper, Kimball T
Harrison, Bertrand Fereday
Harrison, H Keith
Hess, Wilford Moser Bill
Hoffmann, James Allen
Holmgren, Arthur Herman
Howarth, Alan Jack
McKell, Cyrus Milo
Mumford, David Louis
Shaw, Richard Joshua
Shultz, Leila McReynolds
Smith, Marvin Artell
Takemoto, Jon Yutaka
Treshow, Michael
Van Alfen, Neal K
Vest, Hyrum Grant, Jr
Weber, Darrell J
Whitney, Elvin Dale
Wiens, Delbert
Wood, Benjamin W
Wullstein, Leroy Hugh

VERMONT
Barrington, David Stanley
Cook, Philip W
Dodge, Carroll William
Gregory, Robert Aaron
Hyde, Beal Baker
Jervis, Robert Alfred
Klein, Deana Tarson
Ullrich, Robert Carl
Vogelmann, Hubert Walter

VIRGINIA
Al-Doory, Yousef
Berliner, Martha D
Bodkin, Norlyn L
Bonner, Robert Dubois
Bradley, Ted Ray
Breil, David A
Chen, Jiann-Shin
Couch, Houston Brown
Coursen, Bradner Wood
Decker, Robert Dean
Dodd, John Durrance
Drake, Charles Roy
Fuller, Stephen William
Garrison, Hazel Jeanne
Gates, James Edward

Gough, Stephen Bradford
Gupton, Oscar Wilmot
Hall, Gustav Wesley
Hatzios, Kriton Kleanthis
Hill, Lynn Michael
Hufford, Terry Lee
Johnson, Miles F
Jones, Johnnye M
Lacy, George Holcombe
Lawrey, James Donald
Little, Elbert Luther, Jr
McKinsey, Richard Davis
Marshall, Harold George
Miller, Lawrence Ingram
Miller, Orson K, Jr
Milton, Nancy Melissa
Moody, Arnold Ralph
Moore, Laurence Dale
Owens, Vivian Ann
Parker, Bruce C
Paterson, Robert Andrew
Phipps, Patrick Michael
Place, Janet Dobbins
Porter, Daniel Morris
Porter, Duncan MacNair
Ramsey, Gwynn W
Riopel, James L
Roane, Curtis Woodard
Roane, Martha Kotila
Roshal, Jay Yehudie
Rosinski, Joanne
Runk, Benjamin Franklin Dewees
Scott, Joseph Lee
Scott, Marvin Wade
Silberhorn, Gene Michael
Sumrall, H Glenn
Tenney, Wilton R
Tolin, Sue Ann
Treadwell, George Edward, Jr
Vaughan, Michael Ray
Ware, Donna Marie Eggers
Wass, Marvin Leroy
Wells, Elizabeth Fortson
Willis, Lloyd L, II
Wills, Wirt Henry
Wilson, Charles Maye
Wilson, Coyt Taylor
Winstead, Janet

WASHINGTON
Ammirati, Joseph Frank, Jr
Anderson, Edward Frederick
Athow, Kirk Leland
Bliss, Lawrence Carroll
Booth, Beatrice Crosby
Bristow, Peter Richard
Bruehl, George William
Carr, Robert Leroy
Chastagner, Gary A
Clark, Raymond Loyd
Cobb, William Thompson
Cook, Robert James
Covey, Ronald Perrin, Jr
Del Moral, Roger
Denton, Melinda Fay
Dietz, Sherl M
Drum, Ryan William
Duran, Ruben
Edwards, Gerald Elmo
Foley, Dean Carroll
Fridlund, Paul Russell
Gabrielson, Richard Lewis
Gilmartin, Amy Jean
Gould, Charles Jay
Gross, Dennis Charles
Gurusiddaiah, Sarangamat
Hadwiger, Lee A
Halperin, Walter
Haskins, Edward Frederick
Hecht, Adolph
Hendrix, John Walter
Hewitt, William Boright
Hindman, Joseph Lee
Hudson, Peggy R
Johnson, Dennis Allen
Johnson, Richard Evan
Kraft, John M
Kruckeberg, Arthur Rice
Leopold, Estella (Bergere)
Lippert, Laverne Francis
Parish, Curtis Lee
Phillips, Ronald Carl
Post, Douglas Manners
Rayburn, William Reed
Rickard, William Howard, Jr
Rogers, Jack David
Schrader, Lawrence Edwin
Shaw, Charles Gardner
Silbernagel, Matt Joseph
Smith, Samuel H
Staley, John M
Taylor, Ronald
Thornton, Melvin LeRoy
Turner, William Junior
Waaland, Joseph Robert
Walker, Richard Battson
Ward, George Henry
Whisler, Howard Clinton
Withner, Carl Leslie, Jr

WEST VIRGINIA
Barrat, Joseph George
Bell, Carl F
Biggs, Alan Richard

Binder, Franklin Lewis
Chapman, Carl Joseph
Clarkson, Roy Burdette
Clovis, Jesse Franklin
Core, Earl Lemley
Elkins, John Rush
Elliston, John E
Gain, Ronald Ellsworth
Gillespie, William Harry
Guthrie, Roland L
Hindal, Dale Frank
Larson, Gary Eugene
Mills, Howard Leonard
Nunley, Robert Gray
Plymale, Edward Lewis
Pore, Robert Scott
Smith, Glenn Edward
Sorenson, William George
Stephenson, Steven Lee
Van Der Zwet, Tom

WISCONSIN
Andrews, John Herrick
Arnholt, Philip John
Berbee, John Gerard
Blum, John Leo
Boone, Donald Milford
Bostrack, Jack M
Bowers, Frank Dana
Crone, Lawrence John
Croxdale, Judith Gerow
DeGroot, Rodney Charles
De Zoeten, Gustaaf A
Dibben, Martyn James
Duewer, Elizabeth Ann
Durbin, Richard Duane
Ellingboe, Albert Harlan
Evert, Ray Franklin
Fay, Marcus J
Follstad, Merle Norman
Freckmann, Robert W
Fulton, Robert Watt
Gasiorkiewicz, Eugene Constantine
Geeseman, Gordon E
Gerloff, Gerald Carl
Graham, Linda Kay Edwards
Grau, Craig Robert
Grittinger, Thomas Foster
Hagedorn, Donald James
Hanson, Earle William
Harriman, Neil Arthur
Hillier, Richard David
Iltis, Hugh Hellmut
Jordan, William R, III
Jowett, David
Kitzke, Eugene David
Koch, Rudy G
Kurup, Viswanath Parameswar
Larsen, Michael John
Larson, Philip Rodney
McDonough, Eugene Stowell
Maravolo, Nicholas Charles
Maxwell, Douglas Paul
Michaelson, Merle Edward
Millington, William Frank
Moore, John Duain
Nair, Gangadharan V M
Nelson, Allen Charles
Newsome, Richard Duane
Palm, Elmer Thurman
Parker, Alan Douglas
Rice, Marion McBurney
Salamun, Peter Joseph
Sequeira, Luis
Sharkey, Thomas D
Simon, Philipp William
Smalley, Eugene Byron
Smith, Stanley Galen
Stearns, Forest
Stevenson, Walter Roe
Sussman, Michael R
Sytsma, Kenneth Jay
Tews, Leonard L
Thiesfeld, Virgil Arthur
Tibbitts, Theodore William
Tiefel, Ralph Maurice
Unbehaun, Laraine Marie
Unger, James William
Wade, Earl Kenneth
Waller, Donald Macgregor
Warner, James Howard
Weber, Albert Vincent
Williams, Paul Hugh
Worf, Gayle L

WYOMING
Bohnenblust, Kenneth E
Christensen, Martha

PUERTO RICO
Rodriguez, Rocio del Pilar

ALBERTA
Bakshi, Trilochan Singh
Bird, Charles Durham
Campbell, John Duncan
Cormack, Robert George Hall
Cuny, Robert Michael
Degenhardt, Keith Jacob
Gorham, Paul Raymond
Harper, Frank Richard
Hickman, Michael
Hiratsuka, Yasuyuki
Neish, Gordon Arthur

Nelson, Gordon Albert
Parkinson, Dennis
Skoropad, William Peter
Stewart, Wilson Nichols
Suresh, Mavanur Rangarajan
Tamaoki, Taiki
Wilson, Donald Benjamin

BRITISH COLUMBIA
Bandoni, Robert Joseph
Belland, René Jean
Bigelow, Margaret Elizabeth Barr
Bisalputra, Thana
Bohm, Bruce Arthur
Buchanan, Ronald James
Copeman, Robert James
De Boer, Solke Harmen
Dueck, John
Fisher, Francis John Fulton
Fushtey, Stephen George
Gandera, Fred Russell
Green, Beverley R
Hansen, Anton Juergen
Harvey, Michael John
Hebda, Richard Joseph
Hughes, Gilbert C
Kuijt, Job
Maze, Jack Reiser
Newroth, Peter Russell
Owens, John N
Paden, John Wilburn
Pepin, Herbert Spencer
Pocock, Stanley Albert John
Rahe, James Edward
Schofield, Wilfred Borden
Shaw, Michael
Stace-Smith, Richard
Stock, John Joseph
Towers, George Hugh Neil
Wall, Ronald Eugene
Warrington, Patrick Douglas
Weintraub, Marvin
Whitney, Harvey Stuart
Wright, Norman Samuel

MANITOBA
Atkinson, Thomas Grisedale
Bernier, Claude
Chong, James York
Dever, Donald Andrew
Dugle, Janet Mary Rogge
Harder, Donald Edwald
Johnson, Karen Louise
Kerber, Erich Rudolph
Martens, John William
Mills, John T
Nielsen, Jens Juergen
Olsen, Orvil Alva
Rohringer, Roland
Samborski, Daniel James

NEW BRUNSWICK
Lakshminarayana, J S S
Magasi, Laszlo P
Taylor, Andrew Ronald Argo
Whitney, Norman John

NEWFOUNDLAND
Hampson, Michael Chisnall
Scott, Peter John

NOVA SCOTIA
Greenidge, Kenneth Norman Haynes
Hall, Ivan Victor
McFadden, Lorne Austin
McLachlan, Jack (Lamont)
Ross, Robert Gordon
Vander Kloet, Sam Peter
Van der Meer, John Peter

ONTARIO
Alex, Jack Franklin
Allen, Wayne Robert
Anderson, Terry Ross
Argus, George William
Badenhuizen, Nicolaas Pieter
Bannan, Marvin William
Barr, Donald John Stoddart
Barrett, Spencer Charles Hilton
Basham, Jack T
Baum, Bernard R
Benedict, Winfred Gerald
Bonn, William Gordon
Boyer, Michael George
Busch, Lloyd Victor
Catling, Paul Miles
Chaly, Nathalie
Clark, Robert Vernon
Cody, William James
Cook, Frankland Shaw
Corlett, Michael Philip
Cruise, James E
Dalpe, Yolande
Davidson, Thomas Ralph
Duthie, Hamish
Eckenwalder, James Emory
Elfving, Donald Carl
Fahselt, Dianne
Fedak, George
Gayed, Sobhy Kamel
Gerrath, Joseph Fredrick
Ginns, James Herbert
Good, Harold Marquis
Greyson, Richard Irving

Botany-Phytopathology (cont)

Grodzinski, Bernard
Harvais, Gaetan Hugues
Heath, Ian Brent
Heath, Michele Christine
Hellebust, Johan Arnvid
Higgins, Verna Jessie
Hughes, Stanley John
Illman, William Irwin
Jarzen, David MacArthur
Jeglum, John Karl
Johnson, Peter Wade
Jones, Roger
Kendrick, Bryce
Kevan, Peter Graham
Krug, John Christian
Lott, John Norman Arthur
McAndrews, John Henry
McKeen, Colin Douglas
McKeen, Wilbert Ezekiel
McKenzie, Allister Roy
McNeill, John
Manocha, Manmohan Singh
Marks, Charles Frank
Morton, John Kenneth
Northover, John
Nozzolillo, Constance
Olthof, Theodore Hendrikus Antonius
Paliwal, Yogesh Chandra
Patrick, Zenon Alexander
Peterson, Robert Lawrence
Posluszny, Usher
Pringle, James Scott
Redhead, Scott Alan
Ritchie, James Cunningham
Savile, Douglas Barton Osborne
Seaman, William Lloyd
Shoemaker, Robert Alan
Singh, Rama Shankar
Sinha, Ramesh Chandra
Soper, James Herbert
Sutton, John Clifford
Thompson, Hazen Spencer
Townshend, John Linden
Traquair, James Alvin
Tu, Jui-Chang
Van Huystee, Robert Bernard
Ward, Edmund William Beswick
Warwick, Suzanne Irene
Winterhalder, Keith
Zilkey, Bryan Frederick

PRINCE EDWARD ISLAND

Hanic, Louis A
MacQuarrie, Ian Gregor
Thompson, Leith Stanley
Willis, Carl Bertram

QUEBEC

Bessette, France Marie
Brouillet, Luc
Carlson, Lester William
Charest, Pierre M
Coulombe, Louis Joseph
Estey, Ralph Howard
Gagnon, Camilien Joseph Xavier
Gibbs, Sarah Preble
Goldstein, Melvin E
Langford, Arthur Nicol
Legault, Albert
Morisset, Pierre
Richard, Claude
Sackston, Waldemar E
Sattler, Rolf
Vieth, Joachim
Watson, Alan Kemball
Willemot, Claude

SASKATCHEWAN

Gruen, Hans Edmund
Harms, Vernon Lee
Haskins, Reginald Hinton
Lapp, Martin Stanley
Looman, Jan
Mortensen, Knud
Sawhney, Vipen Kumar
Smith, Jeffrey Drew
Steeves, Taylor Armstrong
Tinline, Robert Davies

OTHER COUNTRIES

Berg, Arthur R
Bulmer, Glenn Stuart
Bunting, George Sydney, Jr
Chiarappa, Luigi
Cowan, Richard Sumner
De Renobales, Mertxe
Eiten, George
Erke, Keith Howard
Gibbs, R Darnley
Gottlieb, Otto Richard
Gressel, Jonathan Ben
Huisingh, Donald
Jensen, Lawrence Craig-Winston
Koch, Stephen Douglas
Kurobane, Itsuo
Lieth, Helmut Heinrich Friedrich
Lindenmayer, Aristid
Lowden, Richard Max
MacCarthy, Jean Juliet
Meins, Frederick, Jr
Muchovej, James John
Nakayama, Takao
Pickett-Heaps, Jeremy David
Prance, Ghillean T
Saari, Eugene E
Smith, Douglas Roane
Steer, Martin William
Steyermark, Julian Alfred
Stover, Robert Harry

Cytology

ALABAMA

Blalock, James Edwin
Bowen, William R
Dietz, Robert Austin
Furuto, Donald K
Gray, Bruce William
Hajduk, Stephen Louis
Moore, Bobby Graham
Sapp, Walter J
Sapra, Val T
Steward, Frederick Campion
Watson, Jack Ellsworth
Wilborn, Walter Harrison
Williams, John Watkins, III
Wyss, James Michael

ARIZONA

Aposhian, Hurair Vasken
Capco, David G
Chandler, Douglas Edwin
Doane, Winifred Walsh
Endrizzi, John Edwin
Ferris, Wayne Robert
Gerner, Eugene Willard
Goll, Darrel Eugene
Grim, J(ohn)Norman
Harkins, Kristi R
Holmgren, Paul
Kischer, Clayton Ward
Pinkava, Donald John
Pommerville, Jeffrey Carl
Shimizu, Nobuyoshi
Trelease, Richard Norman

ARKANSAS

Casciano, Daniel Anthony
Chacko, Rosy J
Dippell, Ruth Virginia
Evans, William L
Johnson, Bob Duell
Townsend, James Willis

CALIFORNIA

Adinolfi, Anthony M
Alfert, Max
Amoore, John Ernest
Andrus, William DeWitt, Jr
Antipa, Gregory Alexis
Arcadi, John Albert
Armstrong, Peter Brownell
Barber, Albert Alcide
Barker, Mary Elizabeth
Bartholomew, James Collins
Basbaum, Carol Beth
Beers, William Howard
Bernard, George W
Berns, Michael W
Bernstein, Emil Oscar
Bils, Robert F
Bissell, Mina Jahan
Bizzoco, Richard Lawremce Weiss
Blanks, Janet Marie
Bloom, Floyd Elliott
Bok, P Dean
Boyles, Janet
Breisch, Eric Alan
Brodsky, Frances M(artha)
Bryant, Susan Victoria
Burki, Henry John
Burns, Victor Will
Burnside, Mary Beth
Burwen, Susan Jo
Cailleau, Relda
Cande, W Zacheus
Cantor, Marvin H
Cascarano, Joseph
Clark, William R
Clothier, Galen Edward
Coffino, Philip
Cohen, Morris
Cohen, Natalie Shulman
Connell, Carolyn Joanne
Cooper, Kenneth Willard
Cota-Robles, Eugene H
Cotman, Carl Wayne
Cronshaw, James
Cunningham, Dennis Dean
Daniel, Ronald Scott
Das, Nirmal Kanti
De Francesco, Laura
Deitch, Arline D
Demaree, Richard Spottswood, Jr
Dirksen, Ellen Roter
Dodge, Alice Hribal
Dunnebacke-Dixon, Thelma Hudson
Dvorak, Jan
Eiserling, Frederick A
Estilai, Ali
Falk, Richard H
Fisher, Knute Adrian
Fisher, Steven Kay
Flashman, Stuart Milton
Fosket, Donald Elston

Frenster, John H
Friend, Daniel S
Fristrom, Dianne
Gard, David Lynn
Gifford, Ernest Milton
Ginsberg, Mark Howard
Glabe, Charles G
Goff, Lynda June
Golder, Thomas Keith
Golub, Edward S
Gordon, Manuel Joe
Grey, Robert Dean
Griffith, Donal Louis
Gruber, Helen Elizabeth
Hackett, Nora Reed
Hackney, Robert Ward
Haimo, Leah T
Hamkalo, Barbara Ann
Hanson, James Charles
Harris, John Wayne
Harris, Morgan
Herschman, Harvey R
Hess, Frederick Dan
Hessinger, David Alwyn
Hill, Ray Allen
Hogan, Christopher James
Ignarro, Louis Joseph
Jeffery, William Richard
Jenkins, Burton Charles
Jensen, Ronald Harry
Johnston, George Robert
Jones, Gary Edward
Jordan, Mary Ann
King, Barry Frederick
King, Eileen Brenneman
Kirschner, Marc Wallace
Klevecz, Robert Raymond
Kluss, Byron Curtis
Kornberg, Thomas B
Lake, James Albert
Lang, Dennis Robert
LaVail, Matthew Maurice
Lewiston, Norman James
Long, John Arthur
Lucas, Joe Nathan
Lyke, Edward Bonsteel
McGaughey, Charles Gilbert
McHale, John T
Mak, Linda Louise
Mayall, Brian Holden
Nelson-Rees, Walter Anthony
Nemere, Ilka M
Niklowitz, Werner Johannes
Ohno, Susumu
Ohnuki, Yasushi
Okada, Tadashi A
Palade, George E
Parry, Gordon
Parsons, John Arthur
Pendse, Pratapsinha C
Perkins, David Dexter
Perryman, Elizabeth Kay
Pickett, Patricia Booth
Plopper, Charles George
Pollock, Edward G
Price, Maureen G
Pryer, Nancy Kathryn
Raju, Namboori Bhaskara
Ralph, Peter
Revel, Jean Paul
Ribak, Charles Eric
Rick, Charles Madeira, Jr
Rosen, Steven David
Sanui, Hisashi
Schechter, Joel Ernest
Schmidt, Barbara A
Schmucker, Douglas Lees
Schooley, Caroline Naus
Schraer, Harald
Schraer, Rosemary
Sedlak, Bonnie Joy
Sekhon, Sant Singh
Sercarz, Eli
Shih, Ching-Yuan G
Simpson, Larry P
Smith, Martyn Thomas
Snow, Mikel Henry
Speicher, Benjamin Robert
Stanbridge, Eric John
Stephens, Robert James
Stern, Herbert
Stewart, Juan Godsil
Stoeckenius, Walther
Strahs, Kenneth Robert
Strobel, Edward
Stryer, Lubert
Szabo, Arlene Slogoff
Szego, Clara Marian
Talbot, Prudence
Thomson, William Walter
Tokuyasu, Kiyoteru
Vacquier, Victor Dimitri
Vande Berg, Jerry Stanley
Vanderlaan, Martin
Weigle, William O
Weinstock, Alfred
Whitney, Kenneth Dean
Williams, Mary Carol
Wissig, Steven
Wolff, Sheldon
Wood, Richard Lyman
Zieg, Roger Grant

COLORADO

Angell, Robert Walker
Bonneville, Mary Agnes
Davis, Roger Alan
Fifkova, Eva
Fotino, Mircea
Fox, Michael Henry
Giddings, Thomas H, Jr
Gorthy, Willis Charles
Hahn, William Eugene
Ham, Richard George
Henson, Peter Mitchell
Horak, Donald L
Klymkowsky, Michael W
McIntosh, John Richard
Maylie-Pfenninger, M F
Prescott, David Marshall
Sanderson, Richard James
Snyder, Judith Armstrong
Stack, Stephen M
Staehelin, Lucas Andrew
Stein, Gretchen Herpel
Stone, Gordon Emory
Tsuchiya, Takumi
Webber, Mukta Mala (Maini)

CONNECTICUT

Anderson, Gregory Joseph
Ashley, Terry Fay
Beitch, Irwin
Blackburn, Daniel Glenn
Chang, Amy Y
Child, Frank Malcolm
Constantine-Paton, Martha
Cooperstein, Sherwin Jerome
Dembitzer, Herbert
Grasso, Joseph Anthony
Greenberg, Jay R
Hand, Arthur Ralph
Hogan, James C
Jamieson, James Douglas
Kent, John Franklin
Koerting, Lola Elisabeth
Levin, Martin Allen
Milici, Anthony J
Moellmann, Gisela E Bielitz
Mundkur, Balaji
Myles, Diana Gold
Patton, Curtis LeVerne
Pochron, Sharon
Rakic, Pasko
Rasmussen, Howard
Roman, Laura M
Rosenbaum, Joel L
Simpson, Tracy L
Stevenson, Harlan Quinn
Trinkaus, John Philip
Vitkauskas, Grace
Wachtel, Allen W
Woods, John Whitcomb

DELAWARE

Borgaonkar, Digamber Shankarrao
Chen, Harry Wu-Shiong
Giaquinta, Robert T
Howard, Richard James
Ruben, Regina Lansing
Wagner, Roger Curtis

DISTRICT OF COLUMBIA

Albert, Ernest Narinder
Baldwin, Kate M
Ball, William David
Chapman, George Bunker
Chiarodo, Andrew
Fleming, Patrick John
Forman, David S
Griffin, Joe Lee
Hunter, Jehu Callis
Jones, Janice Lorraine
Kapur, Shakti Prakash
Leak, Lee Virn
Mullins, James Michael
Schiff, Stefan Otto
Weintraub, Robert Louis
Wrathall, Jean Rew

FLORIDA

Barkalow, Derek Talbot
Bourguignon, Lilly Y W
Braunschweiger, Paul G
Broschat, Kay O
Bunge, Mary Bartlett
Bunge, Richard Paul
Carraway, Coralie Anne Carothers
Chambers, Edward Lucas
Chegini, Nasser
Chen, L T
Cohen, Glenn Milton
Coleman, Sylvia Ethel
Dawes, Clinton John
Deats, Edith Potter
Dunn, William Arthur, Jr
Gabridge, Michael Gregory
Gennaro, Joseph Francis
Goldberg, Walter M
Hayashi, Teru
Hinkley, Robert Edwin, Jr
Hofer, Kurt Gabriel
Hurst, Josephine M
Keller, Thomas C S
Krishan, Awtar
Leif, Robert Cary
Luykx, Peter (van Oosterzee)

MacDonald, Eve Lapeyrouse
Murison, Gerald Leonard
Nicosia, Santo Valerio
Polson, Charles David Allen
Rose, Birgit
Ross, Michael H
Ryan, Una Scully
Schank, Stanley Cox
Scott, Walter Alvin
Selman, Kelly
Smith, David Spencer
Warmke, Harry Earl
Warren, Richard Joseph
Warren, Robert Holmes
Weber, James Edward

GEORGIA
Adkison, Claudia R
Andrews, Lucy Gordon
Bates, Harold Brennan, Jr
Bryan, John Henry Donald
Byrd, J Rogers
Davis, Herbert L, Jr
Fechheimer, Marcus
Fritz, Michael E
Gelfant, Seymour
Gulati, Adarsh Kumar
Hamada, Spencer Hiroshi
Haynes, John Kermit
Herman, Chester Joseph
Howard, Eugene Frank
Jones, Betty Ruth
McKinney, Ralph Vincent, Jr
Merkle, Roberta K
Ove, Peter
Palevitz, Barry Allan
Parker, Curtis Lloyd
Patterson, Rosalyn Mitchell
Paulsen, Douglas F
Price, Paul Jay
Priest, Robert Eugene
Ritter, Hope Thomas Martin, Jr
Romanovicz, Dwight Keith
Satya-Prakash, K L
Schuster, George Sheah
Sharkey, Margaret Mary
Stevens, Ann Rebecca
Welter, Dave Allen

HAWAII
Ahearn, Gregory Allen
Allen, Richard Dean
Fok, Agnes Kwan
Hassold, Terry Jon
Hemmes, Don E
Kleinfeld, Ruth Grafman
Sagawa, Yoneo
Yang, Hong-Yi

IDAHO
D'Aoust, Brian Gilbert
Myers, James Robert
Rourke, Arthur W

ILLINOIS
Albach, Richard Allen
Albrecht-Buehler, Guenter Wilhelm
Atherton, Blair T
Aydelotte, Margaret Beesley
Ayres, Kathleen N
Bartles, James Richard
Becker, Robert Paul
Bibbo, Marluce
Chang, Kwang-Poo
Choe, Byung-Kil
Cole, Madison Brooks, Jr
Cordes, William Charles
Crang, Richard Francis Earl
Daniel, Jon Cameron
Daniels, Edward William
Decker, Robert Scott
Doyle, William Lewis
Dybas, Linda Kathryn
Farbman, Albert Irving
Garber, Edward David
Giordano, Tony
Goldman, Robert David
Grdina, David John
Griesbach, Robert Anthony
Hadley, Henry Hultman
Hanzely, Laszlo
Humphreys, Susie Hunt
Ingram, Forrest Duane
Khodadad, Jena Khadem
Kuczmarski, Edward R
Kulfinski, Frank Benjamin
LaVelle, Faith Wilson
Lerner, Jules
Leven, Robert Maynard
Levy, Michael R
Loizzi, Robert Francis
Ma, Te Hsiu
McNulty, John Alexander
Martin, Terence Edwin
Menco, Bernard
Millhouse, Edward W, Jr
Morgan, Juliet
Nakajima, Yasuko
Nardi, James Benjamin
Nielsen, Peter James
Novales, Ronald Richards
O'Morchoe, Patricia Jean
Overton, Jane Harper
Plewa, Michael Jacob

Polet, Herman
Rafferty, Nancy S
Rotermund, Albert J, Jr
Rowley, Janet D
Schleicher, Joseph Bernard
Schreiber, Hans
Seed, Thomas Michael
Sheppard, John Richard
Spear, Brian Blackburn
Straus, Werner
Swift, Hewson Hoyt
Telser, Alvin Gilbert
Trout, Jerome Joseph
Verhage, Harold Glenn
Weber, David Frederick
Willey, Ruth Lippitt
Wolosewick, John J
Zaliskö, Edward John

INDIANA
Asai, David J
Ashendel, Curtis Lloyd
BeMiller, Paraskevi Mavridis
Bracker, Charles E
Chernoff, Ellen Ann Goldman
Chiscon, Martha Oakley
Chrisman, Charles Larry
Glover, David Val
Goff, Charles W
Grove, Stanley Neal
Gunther, Gary Richard
Hammond, Charles Thomas
Heiman, Mark Louis
Maloney, Michael Stephen
Moskowitz, Merwin
Palmer, Catherine Gardella
Phadke, Kalindi
Rai, Karamjit Singh
St John, Philip Alan
Schoknecht, Jean Donze
Sheffer, Richard Douglas
Sinclair, John Henry
Tischfield, Jay Arnold
Togasaki, Robert K
Tweedell, Kenyon Stanley
Williams, Daniel Charles
Winicur, Sandra
Zilz, Melvin Leonard

IOWA
Carlson, Wayne R
Durkee, Lenore T
Hallberg, Richard Lawrence
Kodama, Robert Makoto
Longo, Frank Joseph
Maynard, Jerry Allen
Meetz, Gerald David
Orr, Alan R
Outka, Darryll E
Runyan, William Scottie
Sandra, Alexander
Shaw, Gaylord Edward
Stromer, Marvin Henry
Sullivan, Charles Henry
Swanson, Harold Dueker
Tomes, Dwight Travis
Welshons, William John

KANSAS
Baker, Joffre Bernard
Burchill, Brower Rene
Burton, Paul Ray
Dentler, William Lee, Jr
Funderburgh, James Louis
Haufler, Christopher Hardin
Kimler, Bruce Franklin
Klein, Robert Melvin
Poisner, Alan Mark
Sarras, Michael P, Jr
Spooner, Brian Sandford
Westfall, Jane Anne

KENTUCKY
Hoffman, Eugene James
Humphries, Asa Alan, Jr
Pavlik, Edward John
Sisken, Jesse Ernest
Traurig, Harold H
Wolfson, Alfred M

LOUISIANA
Allen, Emory Raworth
Baum, Lawrence Stephen
Cowden, Ronald Reed
Dyer, Robert Frank
Fuseler, John William
Gallaher, William Richard
Hurley, Maureen
Jacks, Thomas Jerome
Jeter, James Rolater, Jr
Kasten, Frederick K
Lanners, H Norbert
Lin, James C H
Lumsden, Richard
Nickerson, Stephen Clark
Page, Clayton R, III
Pisano, Joseph Carmen
Ramsey, Paul Roger
Specian, Robert David
Tsuzuki, Junji
Wakeman, John Marshall
Weidner, Earl
Yates, Robert Doyle

MAINE
Bell, Allen L
Cook, James Richard
Davisson, Muriel Trask
Greenwood, Paul Gene
Hunter, Susan Julia
Leiter, Edward Henry
Roberts, Franklin Lewis
Terry, Robert Lee
Waymouth, Charity

MARYLAND
Aamodt, Roger Louis
Adelman, Mark Robert
Adler, Ruben
Attallah, A M
Barry, Sue-ning C
Bell, Mary
Berger, Edward Alan
Bodammer, Joel Edward
Bowers, Mary Blair
Bradlaw, June A
Broadwell, Richard Dow
Brown, Joshua Robert Calloway
Buck, John Bonner
Buck, Raymond Wilbur, Jr
Chang, Yao Teh
Chen, T R
Chu, Elizabeth Wann
Coe, Gerald Edwin
Cohen, Maimon Moses
Colburn, Nancy Hall
Craig, Nessly Coile
Cushman, Samuel Wright
Daniels, Mathew Paul
Di Mario, Patrick Joseph
Dwivedi, Radhey Shyam
Evans, Virginia John
Finch, Robert Allen
Frank, Martin
Friedman, Marc Mitchell
Frost, John Kingsbury
Gall, Joseph Grafton
Gallo, Robert C
Galloway, Sheila Margaret
Garfield, Sanford Allen
Gobel, Stephen
Goldman, Arnold I
Goldsmith, Paul Kenneth
Goode, Melvyn Dennis
Greenhouse, Gerald Alan
Griesbach, Robert James
Gwynn, Edgar Percival
Hall, William Thomas
Hamburger, Anne W
Hanover, John Allan
Harris, Rudolph
Hascall, Gretchen Katharine
Hay, Robert J
Hegyeli, Ruth I E J
Heine, Ursula Ingrid
Horenstein, Evelyn Anne
Hunt, Lois Turpin
Imberski, Richard Bernard
Iype, Pullolickal Thomas
Jackson, Edward Milton
Kang, Yuan-Hsu
Kibbey, Maura Christine
Kidwell, William Robert
Kirby, Paul Edward
Kloetzel, John Arthur
Korn, Edward David
Krakauer, Teresa
Leach, William Matthew
Lee, Edward Hsien-Chi
Lin, Diane Chang
Linden, Carol D
Lippincott-Schwartz, Jennifer
Lowensohn, Howard Stanley
McAtee, Lloyd Thomas
Malmgren, Richard Axel
Mather, Ian Heywood
Maupin, Pamela
Meszler, Richard M
Moreira, Jorge Eduardo
Murli, Hemalatha
Murphy, Douglas Blakeney
Musson, Robert A
Nauman, Robert Karl
Nayak, Ramesh Kadbet
Neale, Elaine Anne
Newton, Sheila A
Noonan, Kenneth Daniel
Pagano, Richard Emil
Phelps, Patricia C
Pinto Da Silva, Pedro Goncalves
Pitlick, Frances Ann
Platz, Robert Dole
Pluznik, Dov Herbert
Podskalny, Judith Mary
Pollard, Thomas Dean
Price, Samuel
Read-Connole, Elizabeth Lee
Reissig, Magdalena
Robey, Pamela Gehron
Robinson, David Mason
Robinson-White, Audrey Jean
Robison, Wilbur Gerald, Jr
Roth, Thomas Frederic
Schneider, Walter Carl
Schuetz, Allen W
Shelton, Emma
Silverman, David J
Small, Eugene Beach

Smith, Gilbert Howlett
Spatz, Maria
Sun, Nai Chau
Taylor, William George
Tjio, Joe Hin
Trump, Benjamin Franklin
Vigil, Eugene Leon
Vincent, Monroe Mortimer
Warmbrodt, Robert Dale
Weihing, Robert Ralph
Wenk, Martin Lester
Wergin, William Peter
Willingham, Mark C
Wollman, Seymour Horace
Wolniak, Stephen M
Wray, Granville Wayne
Yip, Rick Ka Sun
Zirkin, Barry Ronald

MASSACHUSETTS
Albert, Mary Roberts Forbes (Day)
Belt, Warner Duane
Berlowitz Tarrant, Laurence
Billingham, Rupert Everett
Branton, Daniel
Chlapowski, Francis Joseph
Chodosh, Sanford
Cintron, Charles
Cohen, Carl M
Cohen, Samuel H
Cornman, Ivor
Curtis, Joseph C
D'Amore, Patricia Ann
Davidson, Samuel James
Deegan, Linda Ann
Dice, J Fred
Dorey, Cheryl Kathleen
Ducibella, Tom
Eisenbarth, George Stephen
Erikson, Raymond Leo
Fine, Richard Eliot
Fink, Rachel Deborah
Frederick, Sue Ellen
Fulton, Chandler Montgomery
Gauthier, Geraldine Florence
Gerbi, Susan Alexandra
Gimbrone, Michael Anthony, Jr
Goldsby, Richard Allen
Goodenough, Daniel Adino
Gruber, Peter Johannes
Haimes, Howard B
Harrison, Bettina Hall
Haudenschild, Christian C
Hausman, Robert Edward
Hay, Elizabeth Dexter
Hepler, Peter Klock
Herman, Ira Marc
Ito, Susumu
Kennedy, Ann Randtke
Kirchanski, Stefan J
Krane, Stanley Garson
Lawrence, Jeanne Bentley
Ledbetter, Mary Lee Stewart
Lehman, William Jeffrey
Liss, Robert H
Margulis, Lynn
Moner, John George
Neutra, Marian R
Owen, Frances Lee
Padykula, Helen Ann
Pardue, Mary Lou
Paul, David Louis
Pederson, Thoru Judd
Penman, Sheldon
Pino, Richard M
Poccia, Dominic Louis
Pryor, Marilyn Ann Zirk
Reincke, Ursula
Remillard, Stephen Philip
Reyero, Cristina
Rivera, Ezequiel Ramirez
Rollins, Reed Clark
Ryser, Hugues Jean-Paul
Schneeberger, Eveline E
Schwartz, Martin Alexander
Scordilis, Stylianos Panagiotis
Seals, Jonathan Roger
Shahrik, H Arto
Skobe, Ziedonis
Sluder, Greenfield
Smith, Dennis Matthew
Sonnenschein, Carlos
Stephens, Raymond Edward
Stossel, Thomas Peter
Strauss, Phyllis R
Sullivan, Susan Jean
Swanson, Carl Pontius
Ting, Yu-Chen
Toole, Bryan Patrick
Trinkaus-Randall, Vickery E
Ulrich, Frank
Williamson, Patrick Leslie
Witman, George Bodo, III
Woodcock, Christopher Leonard Frank
Young, Delano Victor
Zeldin, Michael Hermen
Zimmerman, William Frederick

MICHIGAN
Adams, Jack Donald
Aggarwal, Surinder K
Al Saadi, A Amir
Al-Saadi, Abdul A
Andresen, Norman A

Cytology (cont)

Armant, D Randall
Bagchi, Mihir
Bhuyan, Bijoy Kumar
Bollinger, Robert Otto
Bouma, Hessel, III
Bowers, Maynard C
Boxer, Laurence A
Christensen, A(lbert) Kent
Cronkite, Donald Lee
Erbisch, Frederic H
Ernst, Stephen Arnold
Essner, Edward Stanley
Falls, William McKenzie
Floyd, Alton David
Gordon, Sheldon Robert
Grant, Rhoda
Gray, Robert Howard
Han, Seong S
Harding, Clifford Vincent, Jr
Heidemann, Steven Richard
Jacobs, Charles Warren
Kim, Sun-Kee
Kleinsmith, Lewis Joel
Levine, Laurence
Lindemann, Charles Benard
Motwani, Nalini M
Nag, Asish Chandra
Nasjleti, Carlos Eduardo
Pence, Leland Hadley
Pourcho, Roberta Grace
Pysh, Joseph John
Reddan, John R
Rizki, Tahir Mirza
Schmitter, Ruth Elizabeth
Sicko-Goad, Linda May
Sink, Kenneth C, Jr
Tosney, Kathryn W
Unakar, Nalin J
Walker, Glenn Kenneth
Wiener, Joseph

MINNESOTA
Bauer, Gustav Eric
Beitz, Alvin James
Blanchette, Robert Anthony
Blumenfeld, Martin
Cahoon, Sister Mary Odile
Downing, Stephen Ward
Elde, Robert Philip
Forbes, Donna Jean
Godec, Ciril J
Gooch, Van Douglas
Heidcamp, William H
Johnson, Ross Glenn
Linck, Richard Wayne
Malewicz, Barbara Maria
Nelson, John Daniel
Oetting, William Starr
Palm, Sally Louise
Phillips, Ronald Lewis
Rosenberg, Murray David
Rottmann, Warren Leonard
Severson, Arlen Raynold
Shier, Wayne Thomas
Silverman, William Bernard
Stadelmann, Eduard Joseph
Thompson, Edward William
Weber, Alvin Francis

MISSISSIPPI
Hare, Mary Louise Eckles
Howse, Harold Darrow
Klein, Richard Lester
Martin, Billy Joe
St Amand, Wilbrod
Wise, Dwayne Allison

MISSOURI
Allen, Robert Wade
Bell, Rondal E
Beringer, Theodore Michael
Berlin, Jerry D
Bischoff, Eric Richard
Brown, David Hazzard
Burdick, Allan Bernard
Cook, Nathan Howard
Feeney-Burns, Mary Lynette
Goldstein, Milton Norman
Goodenough, Ursula Wiltshire
Gordon, Albert Raye
Gustafson, John Perry
Harakas, N(icholas) Konstantinos
Kikudome, Gary Yoshinori
Kimber, Gordon
Kort, Margaret Alexander
McQuade, Henry Alonzo
Melnykovych, George
Menton, David Norman
Montague, Michael James
Rifas, Leonard
Sears, Ernest Robert
Sheridan, Judson Dean
Stahl, Philip Damien
Tai, William
Taylor, John Joseph
Tolmach, L(eonard) J(oseph)
Tyrer, Harry Wakeley

MONTANA
Bilderback, David Earl
Dorgan, William Joseph
Thompson, Holly Ann

NEBRASKA
Bieber, Raymond W
Crouse, David Austin
Eisen, James David
Fawcett, James Davidson
McCue, Robert Owen
Rhode, Solon Lafayette, III
Rodriguez-Sierra, Jorge F
Sharp, John Graham
Sheehan, John Francis

NEVADA
Nauman, Charles Hartley
Stratton, Clifford James
Vig, Baldev K

NEW HAMPSHIRE
Carpenter, Stanley John
Fagerberg, Wayne Robert
Handler, Evelyn Erika
Manasek, Francis John
Oldfield, Daniel G
Rodgers, Frank Gerald
Schreiber, Richard William
Sokol, Hilda Weyl
Stanton, Bruce Alan

NEW JERSEY
Agathos, Spiros Nicholas
Avers, Charlotte J
Bayne, Ellen Kahn
Berendsen, Peter Barney
Blackburn, Gary Ray
Boltz, Robert Charles, Jr
Ciosek, Carl Peter, Jr
Colby, Richard H
Collier, Marjorie McCann
Dubin, Donald T
Eastwood, Abraham Bagot
Fiedler-Nagy, Christa
Goggins, Jean A
Gray, Harry Edward
Greene, Arthur E
Gupta, Godaveri Rawat
Hansen, Hans John
Hyndman, Arnold Gene
Kelly, Robert P
Khavkin, Theodor
Kleinschuster, Stephen J, III
Kmetz, John Michael
Laemle, Lois K
Lutz, Richard Arthur
McGarrity, Gerard John
Mack, James Patrick
Malamed, Sasha
Massover, William H
Miller, Douglas Kenneth
Mueller, Stephen Neil
Mullins, Deborra E
Orsi, Ernest Vinicio
Penningroth, Stephen Meader
Prensky, Wolf
Rohrs, Harold Clark
Rossow, Peter William
Schroff, Peter David
Schuh, Joseph Edward
Seiden, David
Singer, Irwin I
Singer, Robert Mark
Steinberg, Malcolm Saul
Studzinski, George P
Tesoriero, John Vincent
Turner, Matthew X
Vena, Joseph Augustus
Wille, John Jacob, Jr
Wilson, Frank Joseph

NEW MEXICO
Bartholdi, Marty Frank
Bear, David George
Bourne, Earl Whitfield
Crissman, Harry Allen
Deaven, Larry Lee
Dunford, Max Patterson
Johnson, Neil Francis
Nelson, Mary Anne
Oliver, Janet Mary
Saland, Linda C
Stricker, Stephev Alexander
Tobey, Robert Allen
Trotter, John Allen
Vogel, Kathryn Giebler

NEW YORK
Aist, James Robert
Alsop, David W
Ambron, Richard Thomas
Anderson, O(rvil) Roger
Anversa, Piero
Asch, Bonnie Bradshaw
Baker, Raymond Milton
Barr, Charles E
Bartfeld, Harry
Bassett, Charles Andrew Loockerman
Beard, Margaret Elzada
Beckert, William Henry
Bender, Michael A
Berezney, Ronald
Bergeron, John Albert
Bleyman, Lea Kanner
Blobel, Gunter
Bogart, Bruce Ian
Borek, Carmia Ganz
Bregman, Allyn A(aron)

Brett, Betty Lou Hilton
Brown, William J
Brownscheidle, Carol Mary
Bushkin, Yuri
Carriere, Rita Margaret
Chiquoine, A Duncan
Cline, Sylvia Good
Cox, Dudley
Cramer, Eva Brown
Darnell, James Edwin, Jr
Darzynkiewicz, Zbigniew Dzierzykraj
De Duve, Christian Rene
De Lemos, Carmen Loretta
De Luca, Chester
DeLuca, Patrick John
Deschner, Eleanor Elizabeth
De Terra, Noël
Detmers, Patricia Anne
Drakontides, Anna Barbara
Drickamer, Kurt
Dziak, Rose Mary
Earle, Elizabeth Deutsch
Eckert, Barry S
Eckhardt, Ronald A
Edds, Kenneth Tiffany
Edmunds, Leland Nicholas, Jr
Engelhardt, Dean Lee
Etlinger, Joseph David
Factor, Jan Robert
Faust, Irving M
Fischman, Harlow Kenneth
Fisher, Paul Andrew
Forest, Charlene Lynn
Friedman, Eileen Anne
Gabrusewycz-Garcia, Natalia
Galinsky, Irving
Garner, James G
Gil, Joan
Godman, Gabriel C
Goldfischer, Sidney L
Gordon, Mildred Kobrin
Goyert, Sanna Mather
Gueft, Boris
Gupta, Sanjeer
Haines, Thomas Henry
Hammill, Terrence Michael
Hanley, Kevin Joseph
Hansen, John Theodore
Herman, Lawrence
Herz, Fritz
Himes, Marion
Hitzeman, Jean Walter
Hoch, Harvey C
Holtzman, Eric
Hotchkiss, Sharon K
Huang, Chester Chen-Chiu
Huberman, Joel Anthony
Humber, Richard Alan
Israel, Herbert William
Jensen, Thomas E
Johnson, Patricia R
Jones, C Robert
Kaplan, Joel Howard
Kaye, Nancy Weber
Koss, Leopold George
Kreibich, Gert
Krim, Mathilde
LaFountain, James Robert, Jr
Lavett, Diane Kathryn Juricek
Ledbetter, Myron C
Leith, ArDean
Liem, Ronald Kian Hong
Lin, Yue Jee
Lipkin, George
Lorch, Joan
Loud, Alden Vickery
Luck, David Jonathan Lewis
Lyser, Katherine May
McCann-Collier, Marjorie Dolores
McKeehan, Wallace Lee
McKeown-Longo, Paula Jean
McLean, Robert J
McMahon, Rita Mary
Magdoff-Fairchild, Beatrice
Manzo, Rene Paul
Mawe, Richard C
Mazurkiewicz, Joseph Edward
Merriam, Robert William
Metzner, Jerome
Miller, Michael E
Millis, Albert Jason Taylor
Mitra, Jyotirmay
Moriber, Louis G
Mukherjee, Asit B
Nichols, Kathleen Mary
Norin, Allen Joseph
North, Robert J
Notter, Mary Frances
Novikoff, Phyllis Marie
Nur, Uzi
O'Brien, James Francis
Ohr, Eleonore A
Olmsted, Joanna Belle
Orlic, Donald
Ornstein, Leonard
Padawer, Jacques
Paine, Philip Lowell
Panessa-Warren, Barbara Jean
Partin, John Calvin
Parysek, Linda M
Pawlowski, Philip John
Penney, David P
Phillips, David Mann
Pokorny, Kathryn Stein

Preisler, Harvey D
Puszkin, Saul
Quigley, James P
Rabinovitch, Michel Pinkus
Raventós-Suárez, Carmen Elvira
Rich, Abby M
Richter, Goetz Wilfried
Rieder, Conly LeRoy
Robbins, Edith Schultz
Rome, Doris Spector
Rosenberg, Warren L
Rosenbluth, Jack
Rothstein, Howard
Rothwell, Norman Vincent
Ruben, Robert Joel
Ruknudin, Abdul Majeed
Sabatini, David Domingo
Saks, Norman Martin
Schin, Kissu
Schuel, Herbert
Shafiq, Saiyid Ahmad
Sharp, William R
Sharpless, Thomas Kite
Shields, Dennis
Shopsis, Charles S
Sklarew, Robert J
Skoultchi, Arthur
Sloboda, Adolph Edward
Slocum, Harry Kim
Soeiro, Ruy
Soifer, David
Straubinger, Robert M
Sun, Alexander Shihkaung
Sutherland, Betsy Middleton
Swinton, David Charles
Szalay, Jeanne
Sztul, Elizabeth Sabina
Tannenbaum, Janet
Tilley, Shermaine Ann
Tomasz, Alexander
Tonna, Edgar Anthony
Trombetta, Louis David
Vanderberg, Jerome Philip
Van't Hof, Jack
Viceps-Madore, Dace I
Ward, Robert T
Warner, Jonathan Robert
Wasserman, Marvin
Waxman, Samuel
Weiss, Leonard
Weissmann, Gerald
Wenk, Eugene J
Wheeless, Leon Lum, Jr
Wolgemuth, Debra Joanne
Woods, Philip Sargent
Worgul, Basil Vladimir
Yang, Shung-Jun
Yeh, Kwo-Yih
Yunghans, Wayne N
Zaleski, Marek Bohdan
Zanetti, Nina Clare
Zeigel, Robert Francis
Zucker-Franklin, Dorothea
Zuzolo, Ralph C

NORTH CAROLINA
Allen, Charles Marvin
Banerjee, Umesh Chandra
Browne, Carole Lynn
Caruolo, Edward Vitangelo
Chan, Po Chuen
Crouse, Helen Virginia
Faust, Robert Gilbert
Fitzharris, Timothy Patrick
Fletcher, Donald James
Gilligan, Diana Mary
Hackenbrock, Charles Robert
Hall, Iris Beryl Haddon
Hanker, Jacob S
Harris, Albert Kenneth, Jr
Herman, Brian
Johnston, William Webb
Juliano, Rudolph Lawrence
Kalfayan, Laura Jean
Kalmus, Gerhard Wolfgang
Langford, George Malcolm
Lee, Greta Marlene
Maroni, Donna F
Maroni, Gustavo Primo
Matthews, James Francis
Mickey, George Henry
Misch, Donald William
Moses, Montrose James
Nicklas, Robert Bruce
Padilla, George M
Phillips, Lyle Llewellyn
Pukkila, Patricia Jean
Roberts, John Fredrick
Salmon, Edward Dickinson
Sanders, Aaron Perry
Sheetz, Michael Patrick
Stalker, Harold Thomas
Thomas, Mary Beth
Thomas, William Albert
Timothy, David Harry
Ting-Beall, Hie Ping
Triantaphyllou, Anastasios Christos
Yongue, William Henry

NORTH DAKOTA
Freeman, Thomas Patrick
LaDuke, John Carl
Riemann, John G

OHIO

Brummett, Anna Ruth
Budd, Geoffrey Colin
Cardeil, Robert Ridley, Jr
Chakraborty, Jyotsna (Joana)
Cohn, Norman Stanley
Cope, Frederick Oliver
Craft, Thomas Jacob, Sr
Culp, Lloyd Anthony
Decker, Jane M
DiCorleto, Paul Eugene
Dollinger, Elwood Johnson
Downey, Ronald J
Egar, Margaret Wells
Flaumenhaft, Eugene
Floyd, Gary Leon
Gesinski, Raymond Marion
Gilliam, James M
Gilloteaux, Jacques Jean-Marie A
Glaser, Ronald
Greider, Marie Helen
Gwatkin, Ralph Buchanan Lloyd
Hamlett, William Cornelius
Heckman, Carol A
Hink, Walter Fredric
Hitt, John Burton
Holderbaum, Daniel
Honda, Shigeru Irwin
Ip, Wallace
Kaetzel, Marcia Aldyth
Kaneshiro, Edna Sayomi
Keefe, John Richard
Klaunig, James E
Kreutzer, Richard D
Lane, Richard Durelle
Lim, David J
Macintyre, Malcolm Neil
Macintyre, Stephen S
Michaels, John Edward
Milsted, Amy
Moore, Fenton Daniel
Moore, Randy
Nasatir, Maimon
Paddock, Elton Farnham
Perry, George
Pribor, Donald B
Przybylski, Ronald J
Robinson, John Mitchell
Samols, David R
Schweltiz, Faye Dorothy
Snider, Jerry Allen
Strauch, Arthur Roger, III
Sunkara, Sai Prasad
Tandler, Bernard
Vaughn, Jack C
Watanabe, Michiko
Yemma, John Joseph
Zambernard, Joseph
Zull, James E

OKLAHOMA

Bantle, John Albert
Begovac, Paul C
Bell, Paul Burton, Jr
Birckbichler, Paul Joseph
Brasch, Klaus Rainer
Chung, Kyung Won
Clarke, Margaret Burnett
De Bault, Lawrence Edward
Dugan, Kimiko Hatta
Duncan, Michael Robert
Kollmorgen, G Mark
Lerner, Michael Paul
McClung, J Keith
Murphy, Marjory Beth
Papka, Raymond Edward
Ramon, Serafin
Rogers, Steffen Harold
Roszel, Jeffie Fisher
Russell, Scott D
Thornton, John William

OREGON

Bajer, Andrew
Bethea, Cynthia Louise
Brenner, Robert Murray
Conte, Frank Philip
Dooley, Douglas Charles
Fahrenbach, Wolf Henrich
Florance, Edwin R
Hard, Robert Paul
Harris, Patricia J
Johnson, John Morris
Montagna, William
Newman, Lester Joseph
Owczarzak, Alfred
Quinton-Cox, Robert
Terwilliger, Nora Barclay
Wimber, Donald Edward

PENNSYLVANIA

Abel, John H, Jr
Alperin, Richard Junius
Anderson, Neil Owen
Atkinson, Barbara
Baker, Wilber Winston
Baumrucker, Craig Richard
Bayer, Manfred Erich
Bean, Barry
Bennett, Henry Stanley
Billheimer, Foster E
Birchem, Regina
Black, Mark Morris
Brooks, John J

Brooks, John S J
Burholt, Dennis Robert
Butler, William Barkley
Campbell, Graham Le Mesurier
Cannizzaro, Linda A
Castor, LaRoy Northrop
Cauna, Nikolajs
Collins, Jimmy Harold
Colony-Cokely, Pamela
Cooke, Peter Hayman
Coss, Ronald Allen
DeCaro, Thomas F
DiBerardino, Marie A
Ehrlich, Howard George
Farber, Phillip Andrew
Flemister, Sarah C
Flesch, David C
Fluck, Richard Allen
Freed, Jerome James
Fussell, Catharine Pugh
Gabriel, Edward George
Gaffney, Edwin Vincent
Gersh, Eileen Sutton
Gilbert, Susan Pond
Gross, Dennis Michael
Grun, Paul
Hansburg, Daniel
Heald, Charles William
Hilfer, Saul Robert
Hopper, James Ernest
Horan, Paul Karl
Hymer, Wesley C
Jacobsen, Terry Dale
Jarvik, Jonathan Wallace
Kleinzeller, Arnost
Knobler, Robert Leonard
Knudsen, Karen Ann
Koprowska, Irena
Koros, Aurelia M Carissimo
Kvist, Tage Nielsen
Ladman, Aaron J(ulius)
La Noue, Kathryn F
Levine, Elliot Myron
Levine, Rhea Joy Cottler
Lieberman, Irving
Linask, Kersti Katrin
Lockwood, Arthur H
Lotze, Michael T
Lowery, Thomas J
McDonald, Barbara Brown
Majumdar, Shyamal K
Malamud, Daniel F
Michel, Kenneth Earl
Miller, Robert Christopher
Monson, Frederick Carlton
Moorhead, Paul Sidney
Moorman, Gary William
Mundell, Robert David
Munger, Bryce Leon
Nichols, Warren Wesley
Nyquist, Sally Elizabeth
O'Brien, Joan A
Oels, Helen C
Olds-Clarke, Patricia Jean
Orr, Nancy Hoffner
Paavola, Laurie Gail
Partanen, Carl Richard
Patterson, Elizabeth Knight
Peachey, Lee DeBorde
Piesco, Nicholas Peter
Porter, Keith Roberts
Pruett, Patricia Onderdonk
Quillin, Charles Robert
Roth, Stephen
Rubin, Walter
Sanger, Jean M
Sas, Daryl
Savage, Robert E
Schuit, Kenneth Edward
Schwartz, Arthur Gerald
Sheffield, Joel Benson
Starling, James Lyne
Storey, Bayard Thayer
Sundarraj, Nirmala
Surmacz, Cynthia Ann
Takats, Stephen Tibor
Taylor, D Lansing
Tchao, Ruy
Therrien, Chester Dale
Tilney, Lewis Gawtry
Trotter, Nancy Louisa
Vaidya, Akhil Babubhai
Vergona, Kathleen Anne Dobrosielski
Weinreb, Eva Lurie
Weisenberg, Richard Charles
Weston, John Colby
Widnell, Christopher Courtenay
Wolfe, Allan Frederick

RHODE ISLAND

Coleman, Annette Wilbois
Goertemiller, Clarence C, Jr
Heywood, Peter
Hufnagel, Linda Ann
Keogh, Richard Neil
Leduc, Elizabeth
Matsumoto, Lloyd H
Miller, Kenneth Raymond
Mottinger, John P
Ripley, Robert Clarence
Zackroff, Robert V

SOUTH CAROLINA

Borg, Thomas K

Cheng, Thomas Clement
Dickey, Joseph Freeman
Dougherty, William J
Fowler, Stanley D
King, Elizabeth Norfleet
Mironescu, Stefan Gheorghe Dan
Odor, Dorothy Louise
Sawyer, Roger Holmes
Simson, Jo Anne V
Wourms, John P

SOUTH DAKOTA

Chen, Chen Ho
Evenson, Donald Paul
Neufeld, Daniel Arthur
Prescott, Lansing M

TENNESSEE

Bibring, Thomas
Bogitsh, Burton Jerome
Carlson, James Gordon
Duck, Bobby Neal
Dumont, James Nicholas
Fuhr, Joseph Ernest
Gao, Kuixiong
Gartner, T Kent
Handel, Mary Ann
Hasty, David Long
Henson, James Wesley
Jeon, Kwang Wu
Kennedy, John Robert
Mazur, Peter
Monaco, Paul J
Monty, Kenneth James
Mrotek, James Joseph
Mullin, Beth Conway
Olins, Donald Edward
Olsen, John Stuart
Preston, Robert Julian
Rasch, Ellen M
Reger, James Frederick
Risby, Edward Louis
Rosing, Wayne C
Roth, Linwood Evans
Schultz, Terry Wayne
Shepherd, Virginia L
Skalko, Richard G(allant)
Swenson, Paul Arthur
Wells, Marion Robert

TEXAS

Adair, Gerald Michael
Ahearn, Michael John
Albrecht, Thomas Blair
Altenburg, Lewis Conrad
Anderson, Richard Gilpin Wood
Andia-Waltenbaugh, Ana Maria
Arnott, Howard Joseph
Arrighi, Frances Ellen
Aufderheide, Karl John
Barcellona, Wayne J
Bashaw, Elexis Cook
Beall, Paula Thornton
Beckingham, Kathleen Mary
Bickham, John W
Biesele, John Julius
Blystone, Robert Vernon
Brady, Richard Charles
Brady, Scott T
Brodie, Edmund Darrell, Jr
Burson, Byron Lynn
Bursztajn, Sherry
Burton, Alexis Lucien
Butler, James Keith
Cannon, Marvin Samuel
Cavazos, Lauro Fred
Chang, Jeffrey Peh-I
Cockerline, Alan Wesley
Cole, Garry Thomas
Dauwalder, Marianne
Davis, Claude Geoffrey
Davis, Frances Maria
Davis, Walter Lewis
Elder, Fred F B
Ellzey, Joanne Tontz
Emery, William Henry Perry
Espey, Lawrence Lee
Fearing, Olin S
Goldstein, Margaret Ann
Gomer, Richard Hans
Goodman, Joel Mitchell
Gratzner, Howard G
Greenberg, Stanly Donald
Grinnell, Frederick
Hagino, Nobuyoshi
Hittelman, Walter Nathan
Hoage, Terrell Rudolph
Hoff, Victor John
Howard, Harriette Ella Pierce
Huang, Bessie Pei-Hsi
Hudspeth, Albert James
Hunter, Jerry Don
Jackson, Raymond Carl
Jalal, Syed M
Karnaky, Karl John, Jr
Kasschau, Margaret Ramsey
Kidd, Harold J
La Claire, John Willard, II
Leffingwell, Thomas Pegg, Jr
Levinson, Charles
Linner, John Gunnar
Litke, Larry Lavoe
McGraw, John Leon, Jr
Maguire, Marjorie Paquette

Mayberry, Lillian Faye
Meistrich, Marvin Lawrence
Moore, Charleen Morizot
Moyer, Mary Pat Sutter
Nations, Claude
Neaves, William Barlow
Ordonez, Nelson Gonzalo
Pathak, Sen
Philpott, Charles William
Pinero, Gerald Joseph
Pool, Thomas Burgess
Powell, Michael A
Price, Harold James
Rao, Potu Narasimha
Rizzo, Peter Jacob
Roberts, Susan Jean
Root, Elizabeth Jean
Sampson, Herschel Wayne
Savage, Howard Edson
Schertz, Keith Francis
Seman, Gabriel
Shannon, Wilburn Allen, Jr
Shay, Jerry William
Sirbasku, David Andrew
Smith, James R
Smith, Willard Newell
Socher, Susan Helen
Stevens, Clark
Stubblefield, Travis Elton
Thurston, Earle Laurence
Tomasovic, Stephen Peter
Van Dreal, Paul Arthur
Venketeswaran, S
Vonder Haar, Raymond A
Williams, Robert K
Williams, William Thomas
Wise, Gary E
Woo, Savio L C
Yee, John Alan
Yin, Helen Lu
Younes, Muazaz A

UTAH

Bollinger, William Hugh
Capecchi, Mario Renato
Dewey, Douglas R
Ehrenfeld, Elvera
Halleck, Margaret S
Kaplan, Jerry
Miles, Charles P
Miller, Scott Cannon
Nabors, Charles J, Jr
O'Neill, Frank John
Sites, Jack Walter, Jr
Whitton, Leslie
Wolstenholme, David Robert
Youssef, Nabil Naguib

VERMONT

Folinas, Helen
Jauhar, Prem P
Meyer, Diane Hutchins
Stevens, Dean Finley
Sullivan, Thomas Donald
Wells, Joseph
Woodcock-Mitchell, Janet Louise

VIRGINIA

Bahns, Mary
Banker, Gary A
Bempong, Maxwell Alexander
Bender, Patrick Kevin
Bloodgood, Robert Alan
Breil, Sandra J
Brown, Jay Clark
Carson, Keith Alan
Cone, Clarence Donald, Jr
Deck, James David
DeVries, George H
Forbes, Michael Shepard
Goldstein, Lewis Charles
Gorbsky, Gary James
Grogan, William McLean
Heinemann, Richard Leslie
Herr, John Christian
Hoegerman, Stanton Fred
Jakoi, Emma Raff
Jones, Johnnye M
Laurie, Gordon William
Little, Charles Durwood, Jr
Merz, Timothy
Miller, Oscar Lee, Jr
Owens, Vivian Ann
Rebhun, Lionel Israel
Rodewald, Richard David
Rosinski, Joanne
Saacke, Richard George
Sandow, Bruce Arnold
Schwartz, Lawrence B
Scott, Joseph Lee
Spangenberg, Dorothy Breslin
Stetler, David Albert
Teichler-Zallen, Doris
Vinores, Stanley Anthony
Weber, Michael Joseph

WASHINGTON

Atkinson, Lenette Rogers
Bakken, Aimee Hayes
Baskin, Denis George
Bolender, Robert P
Brooks, Antone L
Byers, Breck Edward
Carr, Robert Leroy

Cytology (cont)

Drum, Ryan William
Erickson, John (Elmer)
Frederickson, Richard Gordon
Hampton, James C
Hecht, Adolph
Hendrickson, Anita Elizabeth
Horn, Diane
Hosick, Howard Lawrence
Humphrey, Donald Glen
Koehler, James K
Leik, Jean
McCollum, Gilbert Dewey, Jr
Margolis, Robert Lewis
Matlock, Daniel Budd
Pauw, Peter George
Prody, Gerry Ann
Quinn, LeBris Smith
Ross, Russell
Stanley, Hugh P
Stern, Irving B
Szubinska, Barbara
Waaland, Joseph Robert
Yao, Meng-Chao

WEST VIRGINIA
Nath, Joginder
Quinlan, Dennis Charles
Sutter, Richard P
Tolstead, William Lawrence

WISCONSIN
Allred, Lawrence Ervin (Joe)
Austin, Bert Peter
Bavister, Barry Douglas
Bitgood, J(ohn) James
Burke, Janice Marie
Dove, William Franklin
Estervig, David Nels
Goodman, Eugene Marvin
Graham, Linda Kay Edwards
Harb, Joseph Marshall
Karrer, Kathleen Marie
Kowal, Robert Raymond
Kushnaryov, Vladimir Michael
Lough, John William, Jr
Meisner, Lorraine Faxon
Millington, William Frank
Newcomb, Eldon Henry
Oaks, John Adams
Plaut, Walter (Sigmund)
Press, Newtol
Ris, Hans
Ruffolo, John Joseph, Jr
Schatten, Gerald Phillip
Schatten, Heide
Siegesmund, Kenneth A
Snyder, Virginia
Sobkowicz, Hanna Maria
Thet, Lyn Aung
Witt, Patricia L

WYOMING
Jenkins, Robert Allan
Smith-Sonneborn, Joan

PUERTO RICO
Candelas, Graciela C
Kozek, Wieslaw Joseph
Virkki, Niilo

ALBERTA
Cass, Carol E
Huang, Henry Hung-Chang
Kuspira, J
Larson, Ruby Ila
Longenecker, Bryan Michael
McGann, Locksley Earl
Malhotra, Sudarshan Kumar
Neuwirth, Maria
Rattner, Jerome Bernard
Shnitka, Theodor Khyam
Stevenson, Bruce R
Wagenaar, Emile B
Wolowyk, Michael Walter
Zalik, Sara E

BRITISH COLUMBIA
Aebersold, Ruedi H
Auersperg, Nelly
Berger, James Dennis
Brunette, Donald Maxwell
Burke, Robert D
Emerman, Joanne Tannis
Fankboner, Peter Vaughn
Molday, Robert S
Singh, Raj Kumari
Stich, Hans F

MANITOBA
Aung, Taing
Chong, James York
Evans, Laurie Edward
Hamerton, John Laurence
Harder, Donald Edwald
Kerber, Erich Rudolph
Szekely, Joseph George

NEW BRUNSWICK
Bonga, Jan Max
Yoo, Bong Yul

NEWFOUNDLAND
Bal, Arya Kumar

NOVA SCOTIA
Angelopoulos, Edith W
Chen, Lawrence Chien-Ming
Collins, Janet Valerie
Dickson, Douglas Howard
Kamra, Om Perkash
Kapoor, Brij M
Vandermeulen, John Henri
Wood, Eunice Marjorie

ONTARIO
Aye, Maung Tin
Brailovsky, Carlos Alberto
Britton, Donald MacPhail
Brown, David Lyle
Brown, William Roy
Carr, David Harvey
Caveney, Stanley
Chaly, Nathalie
Cheng, Hazel Pei-Ling
Cummins, Joseph E
Dales, Samuel
Davidson, Douglas
Dingle, Allan Douglas
Fedak, George
Forer, Arthur H
Gorczynski, Reginald Meiczyslaw
Govind, Choonilal Keshav
Habowsky, Joseph Edmund Johannes
Hattori, Toshiaki
Heath, Ian Brent
Heath, Michele Christine
Horgen, Paul Arthur
Johnson, Byron F
Joneja, Madan Gopal
Kadanka, Zdenek Karel
Kasha, Kenneth John
Kidder, Gerald Marshall
Leatherland, John F
Leppard, Gary Grant
Liao, Shuen-Kuei
Lu, Benjamin Chi-Ko
Lynn, Denis Heward
McKeen, Wilbert Ezekiel
Metuzals, Janis
Moens, Peter B
Morris, Gerald Patrick
Pelletier, R Marc
Rixon, Raymond Harwood
Rosenthal, Kenneth Lee
Ruthstein, Aser
Sarkar, Priyabrata
Sergovich, Frederick Raymond
Setterfield, George Ambrose
Shaver, Evelyn Louise
Shivers, Richard Ray
Soudek, Dushan Edward
Stanisz, Andrzej Maciej
Sun, Anthony Mein-Fang
Thompson, John Eveleigh
Tung, Pierre S
Wallen, Donald George
Whitfield, James F
Zimmerman, Arthur Maurice
Zimmerman, Selma Blau

QUEBEC
Beaudoin, Adrien Robert
Bendayan, Moise
Bennett, Gary Colin
Bergeron, John Joseph Marcel
Boothroyd, Eric Roger
Brawer, James Robin
Brouillet, Luc
Charest, Pierre M
Clark, Michael Wayne
Couillard, Pierre
Dhindsa, K S
Enesco, Hildegard Esper
Gibbs, Sarah Preble
Grant, William Frederick
Lafontaine, Jean-Gabriel
Meisels, Alexander
Parent, Andre
Pickett, Cecil Bruce
Reiswig, Henry Michael
Roy, Robert Michael McGregor
Schulman, Herbert Michael
Simard, Rene
Sinclair, Ronald
Stephens, Heather R
Wang, Eugenia
Wolsky, Maria de Issekutz

SASKATCHEWAN
Fowke, Lawrence Carroll
Newstead, James Duncan MacInnes
Oliphant, Lynn Wesley
Scoles, Graham John

OTHER COUNTRIES
Ahn, Tae In
Amsterdam, Abraham
Araki, Masasuke
Armato, Ubaldo
Avila, Jesus
Bar-Shavit, Zvi
Ben-Ze'ev, Avri
Britten, Edward James
Elsevier, Susan Maria
Fahimi, Hossein Dariush

Fujiwara, Keigi
Geiger, Benjamin
Guimaraes, Romeu Cardoso
Hayashi, Masao
Heiniger, Hans-Jorg
Hotta, Yasuo
Ikehara, Yukio
Ishikawa, Hiroshi
Jakob, Karl Michael
Jensen, Cynthia G
Jensen, Lawrence Craig-Winston
Jost, Jean-Pierre
Kaltsikes, Pantouses John
Kobata, Akira
La Chance, Leo Emery
Lane, Nancy Jane
Lee, Joseph Chuen Kwun
Leuchtenberger, Cecile
Makita, Takashi
Mittal, Balraj
Pickett-Heaps, Jeremy David
Saito, Takuma
Salas, Pedro Jose I
Schliwa, Manfred
Steer, Martin William
Strous, Ger J
Troutt, Louise Leotta
Wakayama, Yoshihiro
Yamada, Eichi

Ecology

ALABAMA
Bayne, David Roberge
Best, Troy Lee
Beyers, Robert John
Blanchard, Richard Lee
Boyd, Claude Elson
Boyer, William Davis
DeBrunner, Louis Earl
Dusi, Julian Luigi
Golden, Michael Stanley
Hays, Kirby Lee
Hill, William Francis, Jr
Hodgkins, Earl Joseph
Holler, Nicholas Robert
Huntley, Jimmy Charles
Lehman, Robert Harold
Mack, Timothy Patrick
Mettee, Maurice Ferdinand
Modlin, Richard Frank
Mullen, Gary Richard
Nelson, David Herman
Rajanna, Bettaiya
Regan, Gerald Thomas
Rogers, David T, Jr
Rosene, Walter, Jr
Rouse, David B
Scheiring, Joseph Frank
Suberkropp, Keller Francis
Tennessen, Kenneth Joseph
Wetzel, Robert George

ALASKA
Ahlgren, Molly O
Alexander, Vera
Bane, Gilbert Winfield
Behrend, Donald Fraser
Fox, John Frederick
Gard, Richard
Goering, John James
Guthrie, Russell Dale
Hanley, Thomas Andrew
Heifetz, Jonathan
Helle, John Harold (Jack)
Juday, Glenn Patrick
Klein, David Robert
Laursen, Gary A
MacLean, Stephen Frederick, Jr
Mathisen, Ole Alfred
Morrison, John Albert
Norton, David William
Reynolds, James Blair
Rogers, John Gilbert, Jr
Samson, Fred Burton
Schoen, John Warren
Shirley, Thomas Clifton
Smoker, William Williams
Snyder, George Richard
Stekoll, Michael Steven
Viereck, Leslie A
Werner, Richard Allen
West, George Curtiss
Willson, Mary Frances

ARIZONA
Ahlgren, Isabel Fulton
Beck, John R
Blinn, Dean Ward
Bradley, Michael Douglas
Brathovde, James Robert
Calder, William Alexander, III
Carothers, Steven Warren
Chew, Robert Marshall
Cole, Gerald Ainsworth
Collins, James Paul
Crosswhite, Carol D
Crosswhite, Frank Samuel
Davis, Owen Kent
Decker, John Peter
Faeth, Stanley Herman
Fisher, Stuart Gordon
Foster, Kennith Earl

Fritts, Harold Clark
Gaud, William S
Gerking, Shelby Delos
Grier, Charles Crocker
Haase, Edward Francis
Heed, William Battles
Hendrickson, John Roscoe
Hodges, Carl Norris
Huber, Roger Thomas
Hughes, Malcolm Kenneth
Istock, Conrad Alan
Johnsen, Thomas Norman, Jr
Johnson, Jack Donald
Justice, Keith Evans
Klopatek, Jeffrey Matthew
Krausman, Paul Richard
Leatherman, Anna D
McGinnies, William Grovenor
McPherson, Guy Randall
Martin, Paul Schultz
Martin, Samuel Clark
Montgomery, Willson Linn
Muma, Martin Hammond
Patten, Duncan Theunissen
Patton, David Roger
Price, Peter Wilfrid
Rosenzweig, Michael Leo
Schaffer, William Morris
Schmutz, Ervin Marcell
Schnell, Jay Heist
Severson, Keith Edward
Shaw, William Wesley
Smith, Andrew Thomas
Smith, Norman Sherrill
Sommerfeld, Milton R
Springfield, Harry Wayne
States, Jack Sterling
Szarek, Stanley Richard
Thomson, Donald A
Tschirley, Fred Harold
Turner, Raymond Marriner
Van Asdall, Willard
Van Riper, Charles, III
Wagner, Kenneth
Walsberg, Glenn Eric
Watson, Theo Franklin
Zwolinski, Malcolm John

ARKANSAS
Andrews, Theodore Francis
Bacon, Edmond James
Burleigh, Joseph Gaynor
Dale, Edward Everett, Jr
Hanebrink, Earl L
Harp, George Lemaul
Logan, Lowell Alvin
Plummer, Michael V(an)
Smith, Kimberly Gray
Steinkraus, Donald Curtiss
Tave, Douglas
Thornton, Kent W
Zeide, Boris

CALIFORNIA
Addicott, Fredrick Taylor
Ainley, David George
Anderson, Daniel William
Anderson, John Richard
Anderson, Steven Clement
Anspaugh, Lynn Richard
Anthony, Margery Stuart
Azam, Farooq
Baalman, Robert J
Baker, Herbert George
Bakus, Gerald Joseph
Baldy, Richard Wallace
Barbour, Michael G
Beatty, Kenneth Wilson
Becking, Rudolf (Willem)
Beidleman, Richard Gooch
Bennett, Albert Farrell
Bennett, Charles Franklin
Bennett, James Peter
Benseler, Rolf Wilhelm
Bertsch, Hans
Beuchat, Carol Ann
Bierzychudek, Paulette F
Bjorkman, Olle Erik
Blackmon, James B
Bohnsack, Kurt K
Botkin, Daniel Benjamin
Bottjer, David John
Bowers, Darl Eugene
Boyd, Milton John
Brant, Daniel (Hosmer)
Brattstrom, Bayard Holmes
Bray, Richard Newton
Brittan, Martin Ralph
Brooks, William Hamilton
Brusca, Richard Charles
Buchsbaum, Ralph
Burnett, Bryan Reeder
Cailliet, Gregor Michel
Campbell, Bruce Carleton
Carlucci, Angelo Francis
Carpenter, Frances Lynn
Case, Ted Joseph
Cech, Joseph Jerome, Jr
Chadwick, Nanette Elizabeth
Chapin, F Stuart, III
Chappell, Mark Allen
Chen, Carl W(an-Cheng)
Cheng, Lanna
Chesemore, David Lee

Christianson, Lee (Edward)
Coan, Eugene Victor
Cobb, Fields White, Jr
Cody, Martin Leonard
Cogswell, Howard Lyman
Cohen, Anne Carolyn Constant
Collier, Boyd David
Collier, Gerald
Colwell, Robert Knight
Connell, Joseph H
Conway, John Bell
Cooper, Charles F
Costa, Daniel Paul
Cothran, Warren Roderic
Cowles, David Lyle
Cox, Cathleen Ruth
Cox, George W
Coyer, James A
Craig, Robert Bruce
Cummings, John Patrick
Dasmann, Raymond Fredric
Davies, Robert Milton
Davis, Gary Everett
Dawson, Kerry J
Dayton, Paul K
DeMaster, Douglas Paul
Demond, Joan
Deriso, Richard Bruce
Desharnais, Robert Anthony
Dewey, Kathryn G
Dexter, Deborah Mary
Diamond, Jared Mason
Di Girolamo, Rudolph Gerard
Dingle, Richard Douglas Hugh
Dole, Jim
Doutt, Richard Leroy
Dowell, Robert Vernon
Drewes, Robert Clifton
Duffey, Sean Stephen
Ebert, Thomas A
Edmunds, Peter James
Ehrlich, Anne Howland
Ellstrand, Norman Carl
Endler, John Arthur
Eriksen, Clyde Hedman
Erman, Don Coutre
Estes, James Allen
Evans, Kenneth Jack
Farris, David Allen
Firle, Tomas E(rasmus)
Fisler, George Frederick
Fletcher, Donald Warren
Ford, Richard Fiske
Fox, Laurel R
Frankie, Gordon William
Fuhrman, Jed Alan
Fusaro, Craig Allen
Gates, Gerald Otis
Gerrodette, Timothy
Gill, Robert Wager
Gilmer, David Seeley
Given, Robert R
Glynn, Peter W
Goeden, Richard Dean
Gohr, Frank August
Goldman, Charles Remington
Golueke, Clarence George
Gotshall, Daniel Warren
Greenfield, Stanley Marshall
Greenhouse, Nathaniel Anthony
Griffin, James Richard
Grubbs, David Edward
Hanes, Ted L
Hanson, Joe A
Hardy, Cecil Ross
Hare, John Daniel, III
Harte, John
Hassler, Thomas J
Haverty, Michael Irving
Hazen, William Eugene
Hemingway, George Thomson
Hendler, Gordon Lee
Hendricks, Lawrence Joseph
Hespenheide, Henry August, III
Hewston, John G
Hinds, David Stewart
Hobson, Edmund Schofield
Hodges, Lance Thomas
Hodson, William Myron
Horn, Michael Hastings
Hoskins, Cortez William
Houpis, James Louis Joseph
Houston, Roy Seamands
Howard, Walter Egner
Hoy, Marjorie Ann
Hoyt, Donald Frank
Huffaker, Carl Barton
Hunt, George Lester, Jr
Hurlbert, Stuart Hartley
Iltis, Wilfred Gregor
Jacobsen, Nadine Klecha
Jain, Subodh K
Jennrich, Ellen Coutlee
Johnson, Albert W
Jones, Gilbert Fred
Jorgenson, Edsel Carpenter
Kassakhian, Garabet Haroutioun
Keeley, Jon E
Keil, David John
Kemp, Paul Raymond
Kenk, Roman
Kilgore, Bruce Moody
Kitting, Christopher Lee
Knight, Allen Warner

Koonce, Andrea Lavender
Kuris, Armand Michael
Kushner, Arthur S
Kutilek, Michael Joseph
Lane, Robert Sidney
Lang, Kenneth Lyle
Langenheim, Jean Harmon
Lansing, Neal F(isk), Jr
Legner, E Fred
Leigh, Thomas Francis
Levin, Geoffrey Arthur
Lewis, Anthony Wetzel
Lidicker, William Zander, Jr
Lindberg, David Robert
Lindburg, Donald Gilson
Loehlich, Helen Nina (Tappan)
Loomis, Robert Henry
Loomis, Robert Simpson
Loudon, Catherine
Love, Robert Merton
Lovich, Jeffrey Edward
Lowenstam, Heinz Adolf
Ludwig, John Howard
Lysyj, Ihor
McClenaghan, Leroy Ritter, Jr
McCullough, Dale Richard
McDonald, Philip Michael
Maciolek, John A
MacMillen, Richard Edward
Mahall, Bruce Elliott
Main, Robert Andrew
Major, Jack
Mattice, Jack Shafer
Maurer, Donald Leo
Melack, John Michael
Mikel, Thomas Kelly, Jr
Miller, Alan Charles
Mooney, Harold A
Moriarty, David John
Mortenson, Theodore Hampton
Mossman, Archie Stanton
Moyle, Peter Briggs
Mueller, Peter Klaus
Muller, Cornelius Herman
Mulligan, Timothy James
Mulroy, Thomas Wilkinson
Murdoch, William W
Murray, Steven Nelsen
Myers, Willard Glazier, Jr
Myrick, Albert Charles, Jr
Neher, Robert Trostle
Nichols, Frederic Hone
Niesen, Thomas Marvin
Nobel, Park S
Nunney, Leonard Peter
Nybakken, James W
Oechel, Walter C
Oglesby, Larry Calmer
Ohlendorf, Harry Max
Ornduff, Robert
Page, Robert Eugene, Jr
Parker, Virgil Thomas
Parsons, David Jerome
Patterson, Mark Robert
Pearcy, Robert Woodwell
Phaff, Herman Jan
Phillips, Edwin Allen
Pitelka, Louis Frank
Pitts, Wanna Dene
Porcella, Donald Burke
Potts, Donald Cameron
Powell, Jerry Alan
Powell, Thomas Mabrey
Price, Mary Vaughan
Purcell, Alexander Holmes, III
Ralph, C(lement) John
Rand, Patricia June
Randall, Janet Ann
Rateaver, Bargyla
Rau, Gregory Hudson
Rechnitzer, Andreas Buchwald
Reeve, Marian E
Reisen, William Kenneth
Richerson, Peter James
Risch, Stephen John
Roberts, Stephen Winston
Robilliard, Gordon Allan
Robison, William Lewis
Rockland, Louis B
Rose, Michael Robertson
Rotenberry, John Thomas
Rothe, Karolyn Regina
Roughgarden, Jonathan David
Rundel, Philip Wilson
Salt, George William
Sammak, Paul J
Sawyer, John Orvel, Jr
Schoener, Thomas William
Schultz, Arnold Max
Schwab, Ernest Roe
Scow, Kate Marie
Shapiro, Arthur Maurice
Sherman, Robert James
Show, Ivan Tristan
Silver, Mary Wilcox
Slack, Keith Vollmer
Smith, Paul Edward
Smith, Robert William
Smith, Thomas Patrick
Soltz, David Lee
Soule, Dorothy (Fisher)
Stamps, Judy Ann
Sternberg, Hilgard O'Reilly
Stewart, Brent Scott

Stewart, Glenn Raymond
Stewart, Joan Godsil
Straughan, Isdale (Dale) Margaret
Strecker, Robert Louis
Strong, Donald Raymond, Jr
Tegner, Mia Jean
Tenaza, Richard Reuben
Thornburgh, Dale A
Thum, Alan Bradley
Towner, Howard Frost
Tremor, John W
Tribbey, Bert Allen
Turner, Frederick Brown
Vail, Patrick Virgil
Van Wagtendonk, Jan Willem
Verner, Jared
Vilkitis, James Richard
Vitousek, Peter Morrison
Vitt, Laurie Joseph
Voeks, Robert Allen
Vogl, Richard J
Walter, Hartmut
Warner, Robert Ronald
Waser, Nickolas Merritt
Waters, William E
Watt, Kenneth Edmund Ferguson
Weathers, Wesley Wayne
Weintraub, Joel D
Wells, Patrick Harrington
Wenzel, Zita Marta
Westman, Walter Emil
Weston, Charles Richard
Weston, Henry Griggs, Jr
Wilcox, Bruce Alexander
Willemsen, Roger Wayne
Williams, Daniel Frank
Williams, Joseph Burton
Williams, Stanley Clark
Wilson, Kent Raymond
Wirtz, William Otis, II
Woodruff, David Scott
Yen, Teh Fu
Young, Donald Raymond
Young, John Chancellor
Zedler, Joy Buswell
Zedler, Paul H(ugo)
Zinkl, Joseph Grandjean
ZoBell, Claude E
Zuk, Marlene

COLORADO
Anderson, David R
Apley, Martyn Linn
Bailey, James Allen
Barney, Charles Wesley
Benson, Norman G
Bock, Carl E
Bock, Jane Haskett
Bond, Richard Randolph
Bonham, Charles D
Bowers, Marion Deane
Bradshaw, William Newman
Branson, Farrel Allen
Brown, Lewis Marvin
Buckner, David Lee
Carey, Cynthia
Carlson, Clarence Albert, Jr
Cassel, J(oseph) Frank(lin)
Cohen, Ronald R H
Cole, C Vernon
Compton, Thomas Lee
Cook, Robert Sewell
Costello, David Francis
Coughenour, Michael B
Crockett, Allen Bruce
Crouch, Glenn LeRoy
Cruz, Alexander
Davidson, Darwin Ervin
Driscoll, Richard Stark
Drummond, Boyce Alexander, III
Fall, Michael William
Forester, Richard Monroe
Francy, David Bruce
Gilbert, Douglas L
Goetz, Harold
Gough, Larry Phillips
Graul, Walter Dale
Green, Jeffrey Scott
Hansmann, Eugene William
Havlick, Spenser Woodworth
Hein, Dale Arthur
Herrmann, Scott Joseph
Hoffman, Dale A
Hutchinson, Gordon Lee
Kaufmann, Merrill R
Keammerer, Warren Roy
Keith, James Oliver
Kissling, Don Lester
Knopf, Fritz L
Krear, Harry Robert
LaBounty, James Francis, Sr
Lauenroth, William Karl
Lavelle, James W
Lehner, Philip Nelson
Lindauer, Ivo Eugene
Linhart, Yan Bohumil
McClellan, John Forbes
McGinnies, William Joseph
McLean, Robert George
Mahoney, Charles Lindbergh
Martin, Stephen George
Marzolf, George Richard
Moore, Russell Thomas
Oldemeyer, John Lee

Packard, Gary Claire
Peterson, Richard Charles
Porter, Kenneth Raymond
Redente, Edward Francis
Reed, Edward Brandt
Reidinger, Russell Frederick, Jr
Schmidt, John Lancaster
Seilheimer, Jack Arthur
Shaw, Robert Blaine
Shepperd, Wayne Delbert
Sinensky, Michael
Smith, Dwight Raymond
Southwick, Charles Henry
Starr, Robert I
Stucky, Richard K(eith)
Swanson, Vern Bernard
Terrell, Terry Lee Tickhill
Thompson, Daniel Quale
Tomback, Diana Francine
Tracy, C Richard
Van Horn, Donald H
Ward, James Vernon
Ward, Richard Theodore
Wasser, Clinton Howard
Whicker, Floyd Ward
Wiens, John Anthony
Wilson, David George
Woodmansee, Robert George
Wunder, Bruce Arnold

CONNECTICUT
Anderson, Gregory Joseph
Baillie, Priscilla Woods
Bongiorno, Salvatore F
Bormann, Frederick Herbert
Borst, Daryll C
Brewer, Robert Hyde
Buss, Leo William
Calabrese, Anthony
Carlton, James Theodore
Collins, Stephen
Craig, Catherine Lee
Damman, Antoni Willem Hermanus
Denyes, Helen Arliss
DeSanto, Robert Spilka
Egler, Frank Edwin
Feng, Sung Yen
Goodwin, Richard Hale
Gorski, Leon John
Graikoski, John T
Guttay, Andrew John Robert
Haakonsen, Harry Olav
Haffner, Rudolph Eric
Handel, Steven Neil
Henry, Charles Stuart
Holsinger, Kent Eugene
Hutchinson, George Evelyn
Jokinen, Eileen Hope
Katz, Max
Lier, Frank George
Lynch, George Robert
McClure, Mark Stephen
Maguder, Theodore Leo, Jr
Maier, Chris Thomas
Malone, Thomas Francis
Miller, Russell Bensley
Moehlman, Patricia des Roses
Niering, William Albert
Ostrander, Darl Reed
Remington, Charles Lee
Renfro, William Charles
Rettenmeyer, Carl William
Rich, Peter Hamilton
Shaw, Brenda Roberts
Siccama, Thomas G
Silander, John August, Jr
Singletary, Robert Lombard
Smith, David Martyn
Smith, Dwight Glenn
Streams, Frederick Arthur
Taigen, Theodore Lee
Voigt, Garth Kenneth
Welsh, Barbara Lathrop
Wheeler, Bernice Marion
Whitlatch, Robert Bruce
Wulff, Barry Lee

DELAWARE
Cornell, Howard Vernon
Curtis, Lawrence Andrew
Dill, Norman Hudson
Eisenberg, Robert Michael
Gaffney, Patrick M
Hough-Goldstein, Judith Anne
Hurd, Lawrence Edward
Kalkstein, Laurence Saul
Karlson, Ronald Henry
Lotrich, Victor Arthur
Mason, Charles Eugene
Matlack, Albert Shelton
Roth, Roland Ray
Schuler, Victor Joseph
Smith, David William
Targett, Timothy Erwin

DISTRICT OF COLUMBIA
Adey, Walter Hamilton
Barrows, Edward Myron
Blockstein, David Edward
Boots, Sharon G
Brooks, John Langdon
Buffington, John Douglas
Burris, John Edward
Burris, William Edmon

Ecology (cont)

Buzas, Martin A
Callahan, James Thomas
Challinor, David
Cox, Geraldine Anne Vang
DiMichele, William Anthony
Dudley, Theodore R
Evans, Gary R
Field, Ronald James
Goodland, Robert James A
Gould, Edwin
Hoffmann, Robert Shaw
Irving, Patricia Marie
Littler, Mark Masterton
Lovejoy, Thomas E
McDiarmid, Roy Wallace
Malcom, Shirley Mahaley
Merchant, Henry Clifton
Nayak, Tapan Kumar
Nickum, John Gerald
Paris, Oscar Hall
Parsons, John David
Peters, Esther Caroline
Phelps, Harriette Longacre
Policansky, David J
Reichman, Omer James
Reisa, James Joseph, Jr
Rogers, John Patrick
Roper, Clyde Forrest Eugene
Roskoski, Joann Pearl
Silver, Francis
Strauss, Michael S
Sze, Philip
Wauer, Roland H
Werner, Patricia Ann Snyder
Wilson, Don Ellis
Zucchetto, James John

FLORIDA

Abele, Lawrence Gordon
Adams, Ralph M
Alevizon, William
Alexander, Taylor Richard
Allen, Ted Tipton
Bennette, Jerry Mac
Betzer, Peter Robin
Bitton, Gabriel
Bloom, Stephen Allen
Branch, Lyn Clarke
Branham, Joseph Morhart
Brenner, Richard Joseph
Britt, N Wilson
Brower, Lincoln Pierson
Brown, Larry Nelson
Burch, Derek George
Burney, Curtis Michael
Calkins, Carrol Otto
Castner, James Lee
Christman, Steven Philip
Clark, Kerry Bruce
Cooper, Raymond David
Cowell, Bruce Craig
Cropper, Wendell Parker
Crump, Martha Lynn
Daubenmire, Rexford
Davis, John Armstrong
Deevey, Edward Smith, Jr
DeMort, Carole Lyle
Dinsmore, Bruce Heasley
Dragovich, Alexander
Dunkle, Sidney Warren
Dutary, Bedsy Elsira
Feinsinger, Peter
Fitzpatrick, John Weaver
Fleming, Theodore Harris
Forrester, Donald Jason
Frank, John Howard
Frei, Sister John Karen
Friedmann, E(merich) Imre
Giesel, James Theodore
Gilbert, Margaret Lois
Goldberg, Walter M
Gupta, Virendra K
Haag, Kim Hartzell
Hansen, Keith Leyton
Harris, Lawrence Dean
Harwell, Mark Alan
Hoffmann, Harrison Adolph
Hofstetter, Ronald Harold
Hollis, Mark Dexter
Hursh, Charles Raymond
Jackson, Daniel Francis
Jackson, Jeremy Bradford Cook
James, Frances Crews
Janos, David Paul
Jenkins, Dale Wilson
Johnson, F Clifford
Jones, John A(rthur)
Kale, Herbert William, II
Kaufman, Clemens Marcus
Kerr, John Polk
Kirkwood, James Benjamine
Kitchens, Wiley M
Knowlton, Nancy
Kruczynski, William Leonard
Lanciani, Carmine Andrew
Lawrence, Pauline Olive
Lee, David Webster
Lessios, Harilaos Angelou
Levey, Douglas J
Lillywhite, Harvey B
Long, Alan Jack
Lounibos, Leon Philip

Lowe, Jack Ira
Lowe, Ronald Edsel
McCoy, Earl Donald
McKenney, Charles Lynn, Jr
McSweeny, Edward Shearman
Mariscal, Richard North
Maturo, Frank J S, Jr
Mayer, Foster Lee, Jr
Means, D Bruce
Miskimen, George William
Moshiri, Gerald Alexander
Murison, Gerald Leonard
Nelson, Walter Garnet
Nichols, Robert Loring
Nordlie, Frank Gerald
Norval, Richard Andrew
Odum, John Conrad
Ogden, John Conrad
Osborne, Lance Smith
Parker, John Hilliard
Porter, Sanford Dee
Pritchard, Parmely Herbert
Putz, Francis Edward
Reid, Charles Phillip Patrick
Reid, George Kell
Reynolds, John Elliott, III
Rich, Earl Robert
Roubik, David Ward
Schaiberger, George Elmer
Schelske, Claire L
Sheng, Yea-Yi Peter
Simberloff, Daniel S
Smith, Alan Paul
Snedaker, Samuel Curry
Stout, Isaac Jack
Strohecker, Henry Frederick
Taylor, Walter Kingsley
Telford, Sam Rountree, Jr
Tingle, Frederic Carley
Trama, Francesco Biagio
Travis, Joseph
Virnstein, Robert W
Walker, Thomas Jefferson
Walsh, John Joseph
Weber, Neal Albert
Wheeler, George Carlos
Wheeler, Jeanette Norris
Whitcomb, Willard Hall
White, Arlynn Quinton, Jr
Wilkinson, Robert Cleveland, Jr
Wolda, Hindrik
Woodruff, Robert Eugene
Woods, Charles Arthur
Woolfenden, Glen Everett

GEORGIA

Andrews, Charles Lawrence
Bozeman, John Russell
Bratton, Susan Power
Brower, John Harold
Burbanck, Madeline Palmer
Burbanck, William Dudley
Burns, Lawrence Anthony
Carter, Eloise B
Carter, James Richard
Clark, James Samuel
Coleman, David Cowan
Cotter, David James
Crossley, DeRyee Ashton, Jr
Davenport, Leslie Bryan, Jr
Davis, Robert
Fritz, William J
Glasser, John Weakley
Golley, Frank Benjamin
Gordon, Robert Edward
Greear, Philip French-Carson
Hamrick, James Lewis
Hayes, John Thompson
Haynes, Ronnie J
Hewlett, John David
Jordan, Carl Frederick
Kneib, Ronald Thomas
Lassiter, Ray Roberts
Lipps, Emma Lewis
McKeever, Sturgis
McPherson, Robert Merrill
Manger, Martin C
Meentemeyer, Vernon George
Meyer, Judy Lynn
Monk, Carl Douglas
Montgomery, Daniel Michael
Murray, Joan Baird
Odum, Eugene Pleasants
Parker, Albert John
Patten, Bernard Clarence
Payne, Jerry Allen
Plummer, Gayther Lynn
Porter, James W
Pulliam, H Ronald
Quertermus, Carl John, Jr
Ragsdale, Harvey Larimore
Relyea, Kenneth George
Scott, Donald Charles
Sharp, Homer Franklin, Jr
Shure, Donald Joseph
Skeen, James Norman
Smith, Euclid O'Neal
Stanton, George Edwin
Stober, Quentin Jerome
Walls, Nancy Williams
White, Donald Henry
Wiebe, William John
Wiegert, Richard G
Wood, Kenneth George

Woodall, William Robert, Jr
Wyatt, Robert Edward

HAWAII

Bardach, John E
Berg, Carl John, Jr
Bridges, Kent Wentworth
Brock, Richard Eugene
Clarke, Thomas Arthur
Coles, Stephen Lee
DeMartini, Edward Emile
Eddinger, Charles Robert
Flanagan, Patrick William
Freed, Leonard Alan
Hadfield, Michael Gale
Hamilton, Lawrence Stanley
Hapai, Marlene Nachbar
Howarth, Francis Gard
Hunter, Cynthia L
Landry, Michael Raymond
Mueller-Dombois, Dieter
Parrish, James Davis
Reese, Ernst S
Rutherford, James Charles
Scheuer, Paul Josef
Simon, Christine Mae
Skillman, Robert Allen
Stone, Charles Porter
Tomich, Prosper Quentin
Vargas, Roger I
Winget, Robert Newell

IDAHO

Ables, Ernest D
Anderson, Jay Ennis
Bratz, Robert Davis
Clark, William Hilton
Douglas, Dorothy Ann
Goddard, Stephen
Hungerford, Kenneth Eugene
Johnson, Donald Ralph
Johnson, Frederic Duane
Jonas, Robert James
Keller, Barry Lee
Laundre, John William
McCaffrey, Joseph Peter
MacFarland, Craig George
Minshall, Gerry Wayne
O'Keeffe, Lawrence Eugene
Pearson, Lorentz Clarence
Peek, James Merrell
Scott, James Michael
Spillett, James Juan
Spomer, George Guy
Trost, Charles Henry
Wyllie, Gilbert Alexander
Yensen, Arthur Eric

ILLINOIS

Ames, Peter L
Anderson, Roger Clark
Ashby, William Clark
Augspurger, Carol Kathleen
Batzli, George Oliver
Berenbaum, May Roberta
Betz, Robert F
Bjorklund, Richard Guy
Bond, James Arthur, II
Brand, Raymond Howard
Brown, Sandra
Brugam, Richard Blair
Bryant, Marvin Pierce
Burger, George Vanderkarr
Burley, Nancy
Burr, Brooks Milo
Carnes, Bruce Alfred
Corbett, Gail Rushford
Culver, David Clair
Drickamer, Lee Charles
Dunsing, Marilyn Magdalene
Durham, Leonard
Dziadyk, Bohdan
Engel, John Jay
Engstrom, Norman Ardell
Epstein, Samuel Stanley
Fell, George Brady
Flynn, Robert James
Forman, G Lawrence
Foster, Robin Bradford
Freiburg, Richard Eighme
Getz, Lowell Lee
Ghent, Arthur W
Giometti, Carol Smith
Girard, G Tanner
Gorden, Robert Wayne
Greenberg, Bernard
Gustafson, Philip Felix
Hadley, Elmer Burton
Haynes, Robert C
Heaney, Lawrence R
Heltne, Paul Gregory
Herendeen, Robert Albert
Herricks, Edwin E
Hesketh, J D
Hinchman, Ray Richard
Howe, Henry Franklin
Hunt, Lawrence Barrie
Jahn, Lawrence A
Jones, Carl Joseph
Klubek, Brian Paul
Kruse, Kipp Colby
Kulfinski, Frank Benjamin
Lamp, Herbert F
Le Febvre, Eugene Allen

Levenson, James B
Lima, Gail M
Lloyd, Monte
Loehle, Craig S
Lorenz, Ralph William
McKinley, Vicky L
Marsh, Terrence George
Mathis, Billy John
Melin, Brian Edward
Mertz, David B
Meserve, Peter Lambert
Middleton, Beth Ann
Miller, Raymond Michael
Moll, Edward Owen
Munyer, Edward Arnold
Nixon, Charles Melville
Park, Thomas
Passman, Frederick Jay
Perino, Janice Vinyard
Phipps, Richard L
Pruett-Jones, Stephen Glen
Robertson, Philip Alan
Rolfe, Gary Lavelle
Rosenthal, Gerson Max, Jr
Savitz, Jan
Schemske, Douglas William
Schennum, Wayne Edward
Singer, Rolf
Singh, Sant Parkash
Southern, William Edward
Sprugel, George, Jr
Stahl, John Benton
Streets, David George
Stull, Elisabeth Ann
Thompson, Charles Frederick
Thompson, Vinton Newbold
Toliver, Michael Edward
Voigt, John Wilbur
Ware, George Henry
Weiler, William Alexander
Werner, William Ernest, Jr
Whitacre, David Martin
Whitley, Larry Stephen
Wicklow, Donald Thomas
Willey, Ruth Lippitt
Williams, Thomas Alan
Yambert, Paul Abt
Zar, Jerrold Howard
Zimmerman, Craig Arthur
Zusy, Dennis

INDIANA

Allen, Durward Leon
Bakken, George Stewart
Berry, James William
Bjerregaard, Richard S
Byrnes, William Richard
Daily, Fay Kenoyer
Doemel, William Naylor
Eberly, William Robert
Flanders, Robert Vern
Frey, David Grover
Gammon, James Robert
Hellenthal, Ronald Allen
Holt, Harvey Allen
Howe, Robert H L
Jackson, Marion T
Johnson, Michael David
Kapustka, Lawrence A
Kirkpatrick, Charles Milton
Kirkpatrick, Ralph Donald
Lindsey, Alton Anthony
Lodge, David Michael
McCafferty, William Patrick
McClure, Polley Ann
McIntosh, Robert Patrick
Miller, Richard William
Montague, Fredrick Howard, Jr
Mueller, Wayne Paul
Mumford, Russell Eugene
Nelson, Craig Eugene
Olson, John Bennet
O'Neil, Robert James
Paladino, Frank Vincent
Park, Richard Avery, IV
Parker, George Ralph
Pelton, John Forrester
Petty, Robert Owen
Porter, Clyde L, Jr
Randolph, James Collier
Rowland, William Joseph
Russo, Raymond Joseph
Schmelz, Damian Vincent
Schoknecht, Jean Donze
Smith, Charles Edward, Jr
Squiers, Edwin Richard
Vorst, James J
Watson, Maxine Amanda
Weeks, Harmon Patrick, Jr
Whitaker, John O, Jr
Whitehead, Donald Reed
Willard, Daniel Edward
Williams, Eliot Churchill
Wise, Charles Davidson

IOWA

Ackerman, Ralph Austin
Bachmann, Roger Werner
Berg, Virginia Seymour
Best, Louis Brown
Bovbjerg, Richard Viggo
Cavalieri, Anthony Joseph, II
Cawley, Edward T
Christiansen, Paul Arthur

Clark, William Richard
Coleman, Richard Walter
Cruden, Robert William
Danforth, William (Frank)
Delong, Karl Thomas
Eckblad, James Wilbur
Franklin, William Lloyd
Gilbert, William Henry, III
Glenn-Lewin, David Carl
Graham, Benjamin Franklin
Kaufmann, Gerald Wayne
Klaas, Erwin Eugene
Knutson, Roger M
Lyon, David Louis
McDonald, Donald Burt
Main, Stephen Paul
Merkley, Wayne Bingham
Poulter, Dolores Irma
Showers, William Broze, Jr
Vohs, Paul Anthony, Jr
Whitson, Paul David

KANSAS
Armitage, Kenneth Barclay
Ashe, James S
Bellah, Robert Glenn
Bowen, Daniel Edward
Boyd, Roger Lee
Cink, Calvin Lee
Clark, George Richmond, II
Curless, William Toole
Durst, Harold Everett
Ealy, Robert Phillip
Fitch, Henry Sheldon
Fleharty, Eugene
Hanson, Hugh
Hays, Horace Albennie
Hulbert, Lloyd Clair
Jackson, Sharon Wesley
Johnson, John Christopher, Jr
Johnston, Richard Fourness
Kaufman, Donald Wayne
Kaufman, Glennis Ann
Kelting, Ralph Walter
Klaassen, Harold Eugene
Neely, Peter Munro
O'Brien, William John
Owensby, Clenton Edgar
Pierson, David W
Platt, Dwight Rich
Prophet, Carl Wright
Robel, Robert Joseph
Slade, Norman Andrew
Smith, Christopher Carlisle
Spencer, Dwight Louis
Sperry, Theodore Melrose
Terman, Max R
Wolf, Thomas Michael
Zimmerman, John Lester

KENTUCKY
Atlas, Ronald M
Barnes, Richard N
Baskin, Jerry Mack
Boehms, Charles Nelson
Bryant, William Stanley
Crowley, Philip Haney
Cupp, Paul Vernon, Jr
Dillard, Gary Eugene
Ettensohn, Francis Robert
Ferner, John William
Harmet, Kenneth Herman
Hendrix, James William
Howell, Jerry Fonce
Martin, William Haywood, III
Muller, Robert Neil
Pass, Bobby Clifton
Prins, Rudolph
Robinson, Thane Sparks
Sih, Andrew
Stevenson, Robert Jan
Thompson, Marvin Pete
Thompson, Ralph Luther
Thorp, James Harrison, III
Timmons, Thomas Joseph
Westneat, David French, Jr
White, David Sanford
Wiedeman, Varley Earl
Wilder, Cleo Duke
Winstead, Joe Everett
Yeargan, Kenneth Vernon

LOUISIANA
Avent, Robert M
Bahr, Leonard M, Jr
Baldwin, Virgil Clark, Jr
Baumgardner, Ray K
Bounds, Harold C
Brown, Kenneth Michael
Carpenter, Stanley Barton
Chabreck, Robert Henry
Cordes, Carroll Lloyd
Curry, Mary Grace
Deaton, Lewis Edward
Defenbaugh, Richard Eugene
DePoe, Charles Edward
Dunigan, Edward P
Englande, Andrew Joseph
Fleeger, John Wayne
Gosselink, James G
Hackbarth, Winston (Philip)
Hamilton, Robert Bruce
Hastings, Robert Wayne
Hensley, Sess D

Homberger, Dominique Gabrielle
Hughes, Janice S
Jaeger, Robert Gordon
Johnston, James Baker
Kalinsky, Robert George
Kinn, Donald Norman
Knaus, Ronald Mallen
Loden, Michael Simpson
Longstreth, David J
Lorio, Peter Leonce, Jr
Lynch, Steven Paul
Moore, Walter Guy
Nasci, Roger Stanley
Platt, William Joshua, III
Poirrier, Michael Anthony
Ramsey, Paul Roger
Remsen, James Vanderbeek, Jr
Seigel, Richard Allyn
Smalley, Alfred Evans
Stiffey, Arthur V
Teate, James Lamar
Trapido, Harold
Vidrine, Malcolm Francis
Wascom, Earl Ray

MAINE
Borei, Hans Georg
Brainerd, John Whiting
Butler, Ronald George
Clough, Garrett Conde
Cooper, George Raymond
Coulter, Malcolm Wilford
Crawford, Hewlette Spencer, Jr
Davis, John Dunning
Dimond, John Barnet
Drury, William Holland
Gilbert, James Robert
Glanz, William Edward
Greenwood, Paul Gene
Griffin, Ralph Hawkins
Haines, Terry Alan
Hidu, Herbert
Larsen, Peter Foster
Les, Edwin Paul
Lewis, Alan James
Owen, Ray Bucklin, Jr
Parker, James Willard
Schwintzer, Christa Rose
Stickney, Alden Parkhurst
Tjepkema, John Dirk
Townsend, David Warren
Vadas, Robert Louis
Watling, Les
Welch, Walter Raynes
Wilcox, Louis Van Inwegen, Jr

MARYLAND
Adams, Lowell William
Allan, J David
Ballentine, Robert
Barbehenn, Kyle Ray
Barry, Ronald Everett, Jr
Batra, Suzanne Wellington Tubby
Beyer, W Nelson
Bledsoe, Lewis Jackson (Sam)
Borgia, Gerald
Boynton, Walter Raymond
Brandt, Stephen Bernard
Bresler, Jack Barry
Brush, Grace Somers
Bunce, James Arthur
Calhoun, John Bumpass
Cammen, Leon Matthew
Capone, Douglas George
Chen, T R
Clark, Donald Ray, Jr
Clark, Trevor H
Commito, John Angelo
Costanza, Robert
Coulombe, Harry N
Creighton, Phillip David
Cronin, Lewis Eugene
Datta, Padma Rag
D'Elia, Christopher Francis
Duke, James A
Eny, Desire M(arc)
Flittner, Glenn Arden
Flyger, Vagn Folkmann
Fowler, Bruce Andrew
Freed, Arthur Nelson
Fuller, Mark Roy
Galler, Sidney Roland
Gehris, Clarence Winfred
Gerring, Irving
Gerritsen, Jeroen
Gill, Douglas Edward
Gillespie, Walter Lee
Ginevan, Michael Edward
Gipson, Philip
Griswold, Bernard Lee
Hassett, Charles Clifford
Hill, Jane Virginia Foster
Hines, Anson Hemingway
Hocutt, Charles H
Hodgdon, Harry Edward
Holland, Marjorie Miriam
Hornor, Sally Graham
Houde, Edward Donald
Hull, James Clark
Inouye, David William
Kearney, Michael Sean
Kemp, William Michael
Knox, Robert Gaylord
Krywolap, George Nicholas

Kutz, Frederick Winfield
Lamp, William Owen
Leather, Gerald Roger
Leedy, Daniel Loney
McBryde, F(elix) Webster
McKaye, Kenneth Robert
McVey, James Paul
Malone, Thomas C
Martin, John A
Meredith, William G
Mihursky, Joseph Anthony
Moon, Milton Lewis
Morcock, Robert Edward
Morgan, Raymond P
Moseman, John Gustav
Mosher, James Arthur
Mountford, Kent
Nichols, James Dale
Novotny, Robert Thomas
Odell, Lois Dorothea
Osman, Richard William
Paul, Robert William, Jr
Platt, Austin Pickard
Rebach, Steve
Ritchie, Jerry Carlyle
Robertson, Andrew
Ross, Philip
Russek-Cohen, Estelle
Schneider, Eric Davis
Seliger, Howard Harold
Sellner, Kevin Gregory
Setzler-Hamilton, Eileen Marie
Sparling, Donald Wesley, Jr
Steinhauer, Allen Laurence
Teramura, Alan Hiroshi
Todd, Robin Grenville
Trauger, David Lee
Ulanowicz, Robert Edward
Ward, Fraser Prescott
Weisberg, Stephen Barry
Williams, Robert Jackson
Wise, David Haynes
Yanagihara, Richard

MASSACHUSETTS
Ament, Alison Stone
Art, Henry Warren
Babcock, William James Verner
Baird, Ronald C
Bawa, Kamaljit S
Bazzaz, Fakhri Al
Beckwitt, Richard David
Bertin, Robert Ian
Binford, Michael W
Canale-Parola, Ercole
Capuzzo, Judith M
Castro, Gonzalo
Caswell, Hal
Cheetham, Ronald D
Chew, Frances Sze-Ling
Chisholm, Sallie Watson
Clark, William Cummin
Clemens, Daniel Theodore
Coler, Robert A
Cook, Robert Edward
Curran, Harold Allen
D'Avanzo, Charlene
Duncan, Stewart
Edwards, Robert Lomas
Field, Hermann Haviland
Finn, John Thomas
Fitch, John Henry
Forman, Richard T T
Furth, David George
Gifford, Cameron Edward
Godfrey, Paul Jeffrey
Golubic, Stephan
Gregory, Constantine J
Gunner, Haim Bernard
Harbison, G Richard
Healy, William Ryder
Helgesen, Robert Gordon
Hobbie, John Eyres
Hoff, James Gaven
Jahoda, John C
Jearld, Ambrose, Jr
Johanningmeier, Arthur George
Jones, Gwilym Strong
Jost, Dana Nelson
Kazmaier, Harold Eugene
Kricher, John C
Kroodsma, Donald Eugene
Kunz, Thomas Henry
Larson, Joseph Stanley
Lee, Siu-Lam
Levy, Charles Kingsley
Lockhart, James Arthur
Lockwood, Linda Gail
Lovejoy, David Arnold
MacCoy, Clinton Viles
Marshall, Nelson
Mayo, Barbara Shuler
Moir, Ronald Brown, Jr
Moomaw, William Renken
Moore, Johnes Kittelle
Mulcahy, David Louis
Nickerson, Norton Hart
Pearce, Jack B
Pechenik, Jan A
Perry, Alfred Eugene
Peterson, Bruce Jon
Primack, Richard Bart
Reardon, John Joseph
Reimold, Robert J

Rhoads, Donald Cave
Richards, F Paul
Rock, Paul Bernard
Ropes, John Warren
Ross, Michael Ralph
Ruber, Ernest
Sargent, Theodore David
Schofield, Edmund Acton, Jr
Shaver, Gaius Robert
Sheldon, William Gulliver
Smith, Frederick Edward
Smith, Grahame J C
Smith, Susan May
Smith, Tim Denis
Snyder, Dana Paul
Tamarin, Robert Harvey
Taylor, James Kenneth
Teal, John Moline
Telford, Sam Rountree, III
Thomas, Aubrey Stephen, Jr
Traniello, James Francis Anthony
Urry, Lisa Andrea
Wasserman, Frederick E
White, Alan Whitcomb
Wigley, Roland L
Williams, Ernest Edward
Woodwell, George Masters
Wright, Richard T
Zweig, Ronald David

MICHIGAN
Bajema, Carl J
Batts, Henry Lewis, Jr
Beaver, Donald Loyd
Beeton, Alfred Merle
Belyea, Glenn Young
Benninghoff, William Shiffer
Bowen, Stephen Hartman
Bowers, Maynard C
Bowker, Richard George
Branson, Dean Russell
Brewer, Richard (Dean)
Brown, Edward Herriot, Jr
Brown, Robert Thorson
Burton, Thomas Maxie
Caldwell, Larry D
Cantlon, John Edward
Caswell, Herbert Hall, Jr
Catenhusen, John
Chang, William Y B
Cotner, James Bryan, Jr
Cowan, David Prime
Daniels, Stacy Leroy
Dapson, Richard W
Douglas, Matthew M
Driscoll, Egbert Gotzian
Engelmann, Manfred David
Estabrook, George Frederick
Evans, Francis Cope
Fogel, Robert Dale
Foster, Neal Robert
Freeman, Dwight Carl
Gebben, Alan Irwin
Giesy, John Paul, Jr
Gillingham, James Clark
Gleason, Gale R, Jr
Glime, Janice Mildred
Grafius, Edward John
Gross, Katherine Lynn
Hartman, Wilbur Lee
Haynes, Dean L
Hayward, James Lloyd
Hill, Richard William
Hill, Susan Douglas
Hiltunen, Jarl Kalervo
Hohn, Matthew Henry
Holland-Beeton, Ruth Elizabeth
Hough, Richard Anton
Hunter, Robert Douglas
Hurst, Elaine H
Janke, Robert A
Keen, Robert Eric
Kerfoot, Wilson Charles
Kevern, Niles Russell
Kilham, Peter
King, Darrell Lee
Kitchell, Jennifer Ann
Klug, Michael J
Knudson, Vernie Anton
Kraft, Kenneth J
Kurta, Allen
Kutkuhn, Joseph Henry
Lauff, George Howard
Lehman, John Theodore
Lopushinsky, Theodore
Low, Bobbi Stiers
McCrimmon, Donald Alan, Jr
McNabb, Clarence Duncan, Jr
MacTavish, John N
Manny, Bruce Andrew
Marsh, Frank Lewis
Martin, Michael McCulloch
Meier, Peter Gustav
Moll, Russell Addison
Murphy, Peter George
Niemi, Alfred Otto
Oldfield, Thomas Edward
Olexia, Paul Dale
Parejko, Ronald Anthony
Paul, Eldor Alvin
Peterson, Rolf Olin
Petrides, George Athan
Prince, Harold Hoopes
Prychodko, William Wasyl

Ecology (cont)

Rathcke, Beverly Jean
Riebesell, John F
Robertson, G Philip
Robinson, William Laughlin
Rollins-Page, Earl Arthur
Sakai, Ann K
Scavia, Donald
Shontz, John Paul
Siddiqui, Waheed Hasan
Simmons, Gary Adair
Smith, Bryce Everton
Solomon, Allen M
Stapp, William B
Stehr, Frederick William
Stephenson, Stephen Neil
Stoermer, Eugene F
Studier, Eugene H
Tarapchak, Stephen J
Taylor, Mitchell Kerry
Taylor, William Waller
Teeri, James Arthur
Thompson, Paul Woodard
Tiedje, James Michael
Trix, Phelps
Vanderploeg, Henry Alfred
Webber, Patrick John
Werner, Earl Edward
Wiman, Fred Hawkins
Winterstein, Scott Richard
Witter, John Allen
Yerkes, William D(ilworth), Jr
Zimmerman, Norman

MINNESOTA

Abbott, Robinson S, Jr
Andow, David A
Baker, Robert Charles
Bakuzis, Egolfs Voldemars
Bedford, William Brian
Brown, Bruce Antone
Buech, Richard Reed
Christian, Donald Paul
Collins, Hollie L
Cooper, James Alfred
Corbin, Kendall Wallace
Crawford, Ronald Lyle
Cushing, Edward John
Davis, Margaret Bryan
Downing, William Lawrence
Erickson, James Eldred
Foose, Thomas John
Ford, Norman Lee
Fredrickson, Arnold G(erhard)
Fremling, Calvin R
Frenzel, Louis Daniel, Jr
Frydendall, Merrill J
Gorham, Eville
Grigal, David F
Heinselman, Miron L
Henry, Mary Gerard
Heuschele, Ann
Holt, Charles Steele
Jannett, Frederick Joseph, Jr
Jefferson, Carol Annette
Jordan, Peter Albion
Klemer, Andrew Robert
Krogstad, Blanchard Orlando
Lawrence, Donald Buermann
Lonnes, Perry Bert
McConville, David Raymond
McNaught, Donald Curtis
Mech, Lucyan David
Megard, Robert O
Miller, William Eldon
Ordway, Ellen
Pemble, Richard Hoppe
Perry, James Alfred
Rudolf, Paul Otto
Schimpf, David Jeffrey
Schmid, William Dale
Shapiro, Joseph
Siniff, Donald Blair
Sowell, John Basil
Spangler, George Russell
Spigarelli, Steven Alan
Stanley, Patricia Mary
Swanson, Gustav Adolph
Tanner, Ward Dean, Jr
Tester, John Robert
Tilman, G David
VanAmburg, Gerald Leroy
Waters, Thomas Frank
Whiteside, Melbourne C
Windels, Carol Elizabeth
Wright, Herbert Edgar, Jr
Zischke, James Albert

MISSISSIPPI

Anderson, Gary
Barko, John William
Cibula, William Ganley
Cross, William Henley
Fye, Robert Eaton
Hardee, Dicky Dan
Hendricks, Donovan Edward
Jackson, Jerome Alan
Keiser, Edmund Davis, Jr
Kirby, Conrad Joseph, Jr
Mylroie, John Eglinton
Parker, William Skinker
Randolph, Kenneth Norris
Sadler, William Otho

Sherman, Harry Logan
Smith, James Winfred
Stewart, Robert Archie, II
Switzer, George Lester
Turner, Barbara Holman
Wakeley, James Stuart
Walley, Willis Wayne
Woodmansee, Robert Asbury

MISSOURI

Alderfer, Ronald Godshall
Aldrich, Richard John
Belshe, John Francis
Bethel, William Mack
Calabrese, Diane M
Carrel, James Elliott
Coles, Richard Warren
Davis, D Wayne
Dina, Stephen James
Drobney, Ronald DeLoss
Elder, William Hanna
Elliott, Dana Ray
Faaborg, John Raynor
Field, Christopher Bower
Finger, Terry Richard
Franson, Raymond Lee
Fredrickson, Leigh H
Fritzell, Erik Kenneth
Funk, John Leon
Gerhardt, H Carl, Jr
Gerrish, James Ramsay
Gersbacher, Willard Marion
Hanson, Willis Dale
Hawksley, Oscar
Hess, John Berger
Houghton, John M
Howell, Dillon Lee
Hunt, James Howell
Ingersol, Robert Harding
Jones, John Richard
Mueller, Irene Marian
Myers, Richard F
Pallardy, Stephen Gerard
Rawlings, Gary Don
Sander, Ivan Lee
Sexton, Owen J
Shaddy, James Henry
Shiflet, Thomas Neal
Stern, Daniel Henry
Stern, Michele Suchard
Styron, Clarence Edward, Jr
Tomasi, Thomas Edward
Wartzok, Douglas
Williams, Henry Warrington
Wilson, Stephen W
Wiltshire, Charles Thomas

MONTANA

Arno, Stephen Francis
Ball, Irvin Joseph
Behan, Mark Joseph
Collins, Don Desmond
Craighead, John J
Fritts, Steven Hugh
Gould, William Robert, III
Habeck, James Robert
Heitschmidt, Rodney Keith
Johnson, Howard Ernest
Johnson, Oscar Walter
Lotan, James
Lyon, Leonard Jack
McClelland, Bernard Riley
Mackie, Richard John
Manion, James J
Metzgar, Lee Hollis
Moore, Robert Emmett
Nielsen, Gerald Alan
Rounds, Burton Ward
Rumely, John Hamilton
Sheldon, Andrew Lee
Smith, Jean E
Stanford, Jack Arthur
Stark, Nellie May
Taber, Richard Douglas
Taylor, John Edgar
Wright, John Clifford

NEBRASKA

Ballinger, Royce Eugene
Becker, Donald A
Bragg, Thomas Braxton
Case, Ronald Mark
Geluso, Kenneth Nicholas
Grube, George Edward
Hergenrader, Gary Lee
Higley, Leon George
Joern, Anthony
Johnsgard, Paul Austin
Louda, Svata Mary
Lunt, Steele Ray
Nagel, Harold George
Peters, Edward James
Rowland, Neil Wilson
Schlesinger, Allen Brian
Waller, Steven Scobee
Walsh, Gary Lynn

NEVADA

Brussard, Peter Frans
Deacon, James Everett
Douglas, Charles Leigh
Fiero, George William, Jr
Fox, Carl Alan
Hylton, Alvin Roy

Kay, Fenton Ray
Marlow, Ronald William
Meeuwig, Richard O'Bannon
Murvosh, Chad M
O'Farrell, Michael John
O'Farrell, Thomas Paul
Oring, Lewis Warren
Paulson, Larry Jerome
Robertson, Joseph Henry
Rust, Richard W
Starkweather, Peter Lathrop
Taylor, George Evans, Jr
Tueller, Paul T
Vinyard, Gary Lee

NEW HAMPSHIRE

Baldwin, Henry Ives
Barry, William James
Borg, Robert Munson
Croker, Robert Arthur
Eggleston, Patrick Myron
Francq, Edward Nathaniel Lloyd
Goder, Harold Arthur
Haney, James Filmore
Holmes, Richard Turner
Kotila, Paul Myron
McDowell, William H
Mautz, William Ward
Safford, Lawrence Oliver
Sale, Peter Francis
Spencer, Larry T
Wurster-Hill, Doris Hadley

NEW JERSEY

Alexander, Richard Raymond
Bartha, Richard
Breslin, Alfred J(ohn)
Bruno, Stephen Francis
Burger, Joanna
Carter, Richard John
Casey, Timothy M
Cribben, Larry Dean
Cristini, Angela
Cromartie, William James, Jr
Crossner, Kenneth Alan
Crow, John H
Dauerman, Leonard
Dobi, John Steven
Ehrenfeld, David W
Gardiner, Lion Frederick
Good, Ralph Edward
Grassle, John Frederick
Grassle, Judith Payne
Hamby, Robert Jay
Hennings, George
Horn, Henry Stainken
Hunter, Joseph Vincent
Isquith, Irwin R
Keating, Kathleen Irwin
Koditschek, Leah K
Lafornara, Joseph Philip
Lechevalier, Mary P
Leck, Charles Frederick
Lo Pinto, Richard William
McCormick, Jon Michael
McGhee, George Rufus, Jr
McGuire, Terry Russell
Maiello, John Michael
Manganelli, Raymond M(ichael)
Merrill, Leland (Gilbert), Jr
Morgan, Mark Douglas
Morin, Peter Jay
Morrison, Douglas Wildes
Murray, Bertram George, Jr
Mytelka, Alan Ira
Nebel, Carl Walter
Pollock, Leland Wells
Power, Harry Waldo, III
Pramer, David
Psuty, Norbert Phillip
Quinn, James Amos
Sather, Bryant Thomas
Scherer, Robert C
Schultz, George Adam
Sebetich, Michael J
Shubeck, Paul Peter
Simpson, Robert Lee
Spiegel, Leonard Emile
Stearns, Donald Edison
Swan, Frederick Robbins, Jr
Swift, Fred Calvin
Willis, Jacalyn Giacalone

NEW MEXICO

Bahr, Thomas Gordon
Brown, James Hemphill
Cole, Richard Allen
Cooch, Frederick Graham
Crawford, Clifford Smeed
Damon, Edward G(eorge)
Davis, Charles A
Degenhardt, William George
Duszynski, Donald Walter
Findley, James Smith
Gennaro, Antonio Louis
Glock, Waldo Sumner
Gosz, James Roman
Grover, Herbert David
Holechek, Jerry Lee
Jones, Kirkland Lee
Lewis, James Chester
Liddell, Craig Mason
Martin, William Clarence
Mexal, John Gregory

Molles, Manuel Carl, Jr
Pieper, Rex Delane
Rea, Kenneth Harold
Schemnitz, Sanford David
Scott, Norman Jackson, Jr
Secor, Jack Behrent
Taylor, Robert Gay
Whitford, Walter George
Zimmerman, Dale A

NEW YORK

Able, Kenneth Paul
Alexander, Maurice Myron
Allen, Douglas Charles
Allen, J Frances
Andrle, Robert Francis
Arnold, Steven Lloyd
Babich, Harvey
Bard, Gily Epstein
Beason, Robert Curtis
Bedford, Barbara Lynn
Bell, Michael Allen
Bentley, Barbara Lee
Bernard, John Milford
Bothner, Richard Charles
Boylen, Charles William
Boynton, John E
Brocke, Rainer H
Brothers, Edward Bruce
Brown, Jerram L
Brown, Robert Zanes
Burgess, Robert Lewis
Caraco, Thomas Benjamin
Carruthers, Raymond Ingalls
Cassin, Joseph M
Cerwonka, Robert Henry
Chabot, Brian F
Christian, John Jermyn
Churchill, Algernon Coolidge
Clarke, Raymond Dennis
Cohen, Joel Ephraim
Cole, Jonathan Jay
Confer, John L
Cook, Robert Edward
Cox, Donald David
Crowell, Kenneth L
Daniels, Robert Artie
Dayton, Bruce R
De Laubenfels, David John
Dietert, Margaret Flowers
Dindal, Daniel Lee
Dooley, James Keith
Eickwort, George Campbell
Emlen, Stephen Thompson
Esser, Aristide Henri
Feeny, Paul Patrick
Fisher, Nicholas Seth
Forest, Herman Silva
Frankle, William Ernest
Fraser, Douglas Fyfe
French, Alan Raymond
Futuyma, Douglas Joel
Gallagher, Jane Chispa
George, Carl Joseph Winder
Gerard, Valrie Ann
Ginzburg, Lev R
Goodwin, Robert Earl
Green, John Irving
Greene, Kingsley L
Greenlaw, Jon Stanley
Greller, Andrew M
Habicht, Ernst Rollemann, Jr
Hairston, Nelson George, Jr
Hall, Charles Addison Smith
Hammond, H David
Harman, Willard Nelson
Harrison, Richard Gerald
Haugh, John Richard
Hauser, Richard Scott
Haynes, James Mitchell
Hendrey, George Rummens
Henshaw, Robert Eugene
Herreid, Clyde F, II
Hertz, Paul Eric
Hoham, Ronald William
Holway, James Gary
Horn, Edward Gustav
Howard, Harold Henry
Howarth, Robert W
Hughes, Patrick Richard
Jarvis, Richard S
Johnson, Robert Walter
Jones, Clive Gareth
Kamran, Mervyn Arthur
Kanzler, Walter Wilhelm
Keen, William Hubert
Kelly, John Russell
Kiviat, Erik
Klotz, Richard Lawrence
Kohut, Robert John
Kubersky, Edward Sidney
Langer, Arthur M
LaRow, Edward J
Lasker, Howard Robert
Lauer, Gerald J
Laurence, John A
Leopold, Donald Joseph
Levandowsky, Michael
Levin, Simon Asher
Levinton, Jeffrey Sheldon
Lieberman, Arthur Stuart
Likens, Gene Elden
Loggins, Donald Anthony
Lovett, Gary Martin

Luteyn, James Leonard
McCune, Amy Reed
McDonnell, Mark Jeffery
McNaughton, Samuel J
Madison, Dale Martin
Makarewicz, Joseph Chester
Maple, William Thomas
Mattfeld, George Francis
Maxwell, George Ralph, II
Millstein, Jeffrey Alan
Mitchell, Myron James
Mosher, John Ivan
Muller-Schwarze, Dietland
Naidu, Janakiram Ramaswamy
Nyrop, Jan Peter
O'Connor, Joel Sturges
Okubo, Akira
Pasby, Brian
Payne, Harrison H
Peckarsky, Barbara Lynn
Phillips, Arthur William, Jr
Phillips, Robert Rhodes
Pickett, Steward T A
Pierce, Madelene Evans
Pimentel, David
Poindexter, Jeanne Stove
Porter, William Frank
Pough, Frederick Harvey
Pough, Richard Hooper
Rachlin, Joseph Wolfe
Rana, Mohammad A
Raynal, Dudley Jones
Reid, Archibald, IV
Ringler, Neil Harrison
Robinson, Myron
Rockwell, Robert Franklin
Root, Richard Bruce
Rough, Gaylord Earl
Roze, Uldis
Saks, Norman Martin
Sass, Daniel B
Schaffner, William Robert
Schneider, Kathryn Claire (Johnson)
Schoenly, Kenneth George
Seeley, Thomas Dyer
Selleck, George Wilbur
Siegfried, Clifford Anton
Slack, Nancy G
Slobodkin, Lawrence Basil
Slusarczuk, George Marcelius Jaremias
Smith, Sharon Louise
Sohacki, Leonard Paul
Southwick, Edward Earle
Stalter, Richard
Steineck, Paul Lewis
Stewart, Kenton M
Stewart, Margaret McBride
Storr, John Frederick
Strayer, David Lowell
Sweeney, Robert Anderson
Tietjen, John H
Tillman, Robert Erwin
Titus, John Elliott
Tobach, Ethel
Tobiessen, Peter Laws
Toenniessen, Gary Herbert
VanDruff, Larry Wayne
Vawter, Alfred Thomas
Vuilleumier, Francois
Wali, Mohan Kishen
Washton, Nathan Seymour
Wecker, Stanley C
Weinstein, David Alan
Weinstein, Leonard Harlan
White, James Edwin
Winston, Judith Ellen
Wolf, Larry Louis
Wright, Margaret Ruth
Wurster, Charles F
Wysolmerski, Theresa
Zinder, Stephen Henry

NORTH CAROLINA
Aldridge, David William
Anderson, Roger Fabian
Anderson, Thomas Ernest
Baranski, Michael Joseph
Barden, Lawrence Samuel
Bellis, Vincent J, Jr
Billings, William Dwight
Blau, William Stephen
Blum, Udo
Bolen, Eric George
Boyce, Stephen Gaddy
Boyette, Joseph Greene
Brinson, Mark McClellan
Bruck, Robert Ian
Bryden, Robert Richmond
Butts, Jeffrey A
Carson, Johnny Lee
Chester, Alexander Jeffrey
Chestnut, Alphonse F
Christensen, Norman Leroy, Jr
Cooper, Arthur Wells
Copeland, Billy Joe
Crowder, Larry Bryant
Darling, Marilyn Stagner
Davis, Graham Johnson
Derrick, Finnis Ray
Dickerson, Willard Addison
Dimock, Ronald Vilroy, Jr
Doerr, Phillip David
DuBay, Denis Thomas
Elias, Robert William

Flint, Elizabeth Parker
Gross, Harry Douglass
Hackney, Courtney Thomas
Hairston, Nelson George
Haney, Alan William
Harper, James Douglas
Hoss, Donald Earl
Hunt, Kenneth Whitten
Jolls, Claudia Lee
Knight, Clifford Burnham
Kuenzler, Edward Julian
Lacey, Elizabeth Patterson
Lee, Si Duk
Livingstone, Daniel Archibald
Lutz, Paul E
McLeod, Michael John
Mason, Robert Edward
Menhinick, Edward Fulton
Miller, Robert James, II
Moore, Allen Murdoch
Mowbray, Thomas Bruce
Mozley, Samuel Clifford
Parnell, James Franklin
Peet, Robert Krug
Peterson, Charles Henry
Pittillo, Jack Daniel
Powell, Roger Allen
Quay, Thomas Lavelle
Real, Leslie Allan
Reice, Seth Robert
Richardson, Curtis John
Rittschof, Daniel
Robinson, Peter John
Romanow, Louise Rozak
Rublee, Parke Alstan
Ryals, George Lynwood, Jr
Schaaf, William Edward
Schlesinger, William Harrison
Searles, Richard Brownlee
Singer, Philip C
Smathers, Garrett Arthur
Spencer, Lorraine Barney
Stewart, Paul Alva
Stillwell, Harold Daniel
Stinner, Ronald Edwin
Stiven, Alan Ernest
Strain, Boyd Ray
Sutherland, John Patrick
Swank, Wayne T
Thayer, Gordon Wallace
Turner, Alvis Greely
Van Pelt, Arnold Francis, Jr
Ward, John Everett, Jr
Weigl, Peter Douglas
Weiss, Charles Manuel
Wentworth, Thomas Ralph
Wiley, Richard Haven, Jr
Williams, Ann Houston
Wolcott, Thomas Gordon
Wright, Stuart Joseph
Yarnell, Richard Asa
Yongue, William Henry
Zeiger, Errol

NORTH DAKOTA
Clambey, Gary Kenneth
Crawford, Richard Dwight
Greenwald, Stephen Mark
Krapu, Gary Lee
Lieberman, Diana Dale
Lieberman, Milton Eugene
Neel, Joe Kendall, Sr
Ralston, Robert D
Ries, Ronald Edward
Scoby, Donald Ray
Shubert, L Elliot
Wrenn, William J

OHIO
Alrutz, Robert Willard
Andreas, Barbara Kloha
Barrett, Gary Wayne
Baumann, Paul C
Beatley, Janice Carson
Bieri, Robert
Boerner, Ralph E J
Brown, James Harold
Burky, Albert John
Burtt, Edward Howland, Jr
Camp, Mark Jeffrey
Case, Denis Stephen
Chesnut, Thomas Lloyd
Chuey, Carl F
Claussen, Dennis Lee
Cline, Morris George
Cobbe, Thomas James
Colinvaux, Paul Alfred
Collins, Gary Brent
Cooke, George Dennis
Crites, John Lee
Culver, David Alan
Davis, Craig Brian
Deonier, D L
Dexter, Ralph Warren
Egloff, David Allen
Elfner, Lynn Edward
Federle, Thomas Walter
Fraleigh, Peter Charles
Gates, Michael Andrew
Goldstein, David Louis
Gorchov, David Louis
Hamilton, Ernest Scovell
Hauser, Edward J P
Heaslip, Margaret Barkley

Heath, Robert Thornton
Hille, Kenneth R
Hillis, Llewellya
Hoagstrom, Carl William
Hobbs, Horton Holcombe, III
Holeski, Paul Michael
Horn, David Jacobs
Jackson, William Bruce
King, Charles C
Klemm, Donald J
Knoke, John Keith
Kohn, Harold William
Kowal, Norman Edward
Laing, Charles Corbett
Laufersweiler, Joseph Daniel
Laushman, Roger H
Lewis, Michael Anthony
Long, Edward B
Loucks, Orie Lipton
McCall, Peter Law
McGinnis, John Thurlow
MacLean, David Belmont
McLean, Edward Bruce
Mahan, Harold Dean
Martin, Elden William
Miller, Arnold I
Miller, Michael Charles
Miller, Richard Samuel
Mitsch, William Joseph
Mulroy, Juliana Catherine
Olive, John H
Orcutt, Frederic Scott, Jr
Orr, Lowell Preston
Pearson, Paul Guy
Peterjohn, Glenn William
Peterle, Tony J
Phinney, George Jay
Riley, Charles Victor
Rubin, David Charles
Runkle, James Reade
Sanger, Jon Edward
Schroeder, Lauren Alfred
Shah, Kanti L
Snyder, Gary Wayne
Stansbery, David Honor
Stein, Carol B
Stein, Roy Allen
Stoffer, Richard Lawrence
Svendsen, Gerald Eugene
Swift, Michael Crane
Taylor, Douglas Hiram
Thibault, Roger Edward
Thompson, John Leslie
Uetz, George William
Ungar, Irwin A
Valentine, Barry Dean
Vaughn, Charles Melvin
Vessey, Stephen H
Wickstrom, Conrad Eugene
Williams, Patrick Kelly
Wilson, Mark Allan
Winner, Robert William
Wissing, Thomas Edward
Wistendahl, Warren Arthur

OKLAHOMA
Baird, Troy Alan
Black, Jeffrey Howard
Buck, Paul
Burks, Sterling Leon
Chapman, Brian Richard
Crockett, Jerry J
Dorris, Troy Clyde
Fox, Stanley Forrest
Francko, David Alex
Gray, Thomas Merrill
Grula, Mary Muedeking
Hornuff, Lothar Edward, Jr
Hutchison, Victor Hobbs
Korstad, John Edward
LeGrand, Frank Edward
Love, Harry Schroeder, Jr
Lynn, Robert Thomas
McPherson, James King
Madden, Michael Preston
Mares, Michael Allen
Matthews, William John
Miller, Helen Carter
Mock, Douglas Wayne
Namminga, Harold Eugene
Nighswonger, Paul Floyd
Rice, Elroy Leon
Shaw, James Harlan
Shmaefsky, Brian Robert
Sonleitner, Frank Joseph
Sturgeon, Edward Earl
Talent, Larry Gene
Taylor, Constance Elaine Southern
Thurman, Lloy Duane
Tyler, Jack D
Vestal, Bedford Mather
Vishniac, Helen Simpson
Wells, Harrington
Whitcomb, Carl Erwin
Wilhm, Jerry L
Yunice, Andy Aniece

OREGON
Anderson, Norman Herbert
Beatty, Joseph John
Betts, Burr Joseph
Bouck, Gerald R
Bradshaw, William Emmons
Buscemi, Philip Augustus

Callahan, Clarence Arthur
Carolin, Valentine Mott, Jr
Castenholz, Richard William
Chalk, David Eugene
Chilcote, William W
Church, Marshall Robbins
Cimberg, Robert Lawrence
Colvin, Dallas Verne
Cook, Stanton Arnold
Crawford, John Arthur
Cross, Stephen P
Dealy, John Edward
Denison, William Clark
De Witt, John William, Jr
Eddleman, Lee E
Farris, Richard Austin
Field, Katharine G
Fish, Joseph Leroy
Forbes, Richard Bryan
Frank, Peter Wolfgang
Garton, Ronald Ray
Gray, Jane
Hall, Frederick Columbus
Hansen, Everett Mathew
Hawke, Scott Dransfield
Hazel, Charles Richard
Henny, Charles Joseph
Hermann, Richard Karl
Higgins, Paul Daniel
Hixon, Mark A
Holzapfel, Christina Marie
Jarrell, Wesley Michael
Johnson, Michael Paul
Johnson, Samuel Edgar, II
Kogan, Marcos
Kumler, Marion Lawrence
Lackey, Robert T
Linn, DeVon Wayne
Liss, William John
Lotspeich, Frederick Benjamin
Lubchenco, Jane
Lyford, John H, Jr
McConnaughey, Bayard Harlow
McIntire, Charles David
Maloney, Thomas Edward
Malueg, Kenneth Wilbur
Marriage, Lowell Dean
Mason, Richard Randolph
Menge, Bruce Allan
Meslow, E Charles
Minore, Don
Molina, Randolph John
Murphy, Thomas A
Nebeker, Alan V
Neiland, Bonita J
Neilson, Ronald Price
Newton, Michael
Osgood, David William
Pearcy, William Gordon
Perry, David Anthony
Petersen, Richard Randolph
Peterson, Spencer Alan
Phillips, Donald Lundahl
Pizzimenti, John Joseph
Poulton, Charles Edgar
Powers, Charles F
Preston, Eric Miles
Rottink, Bruce Allan
Ryan, Roger Baker
Schmidt, Clifford LeRoy
Schrumpf, Barry James
Senner, John William
Sherr, Barry Frederick
Sherr, Evelyn Brown
Spears, Brian Merle
Stein, William Ivo
Stout, Benjamin Boreman
Tappeiner, John Cummings, II
Terraglio, Frank Peter
Thorson, Thomas Bertel
Tinnin, Robert Owen
Tubb, Richard Arnold
Udovic, Daniel
Volland, Leonard Allan
Wagner, David Henry
Waring, Richard H
Wick, William Quentin
Winjum, Jack Keith
Wood, Anne Michelle
Worrest, Robert Charles
Yamada, Sylvia Behrens
Zobel, Donald Bruce

PENNSYLVANIA
Abrahamson, Warren Gene, II
Adelson, Lionel Morton
Arnold, Dean Edward
Barnett, Leland Bruce
Barone, John B
Beach, Neil William
Bosshart, Robert Perry
Bott, Thomas Lee
Bradt, Patricia Thornton
Brenner, Frederic J
Cale, William Graham, Jr
Carey, Michael Dean
Casida, Lester Earl, Jr
Clark, Richard James
Coffman, William Page
Cruzan, John
Cummins, Kenneth William
Cunningham, Harry N, Jr
Denoncourt, Robert Francis
Dudzinski, Diane Marie

Ecology (cont)

Dunson, William Albert
Emberton, Kenneth C
Feuer, Robert Charles
Fisher, Robert L
Franz, Craig Joseph
Gilbert, Scott F
Good, Norma Frauendorf
Goulden, Clyde Edward
Graybill, Donald Lee
Guida, Vincent George
Haase, Bruce Lee
Hart, David Dickinson
Hartman, Richard Thomas
Hendrickson, John Alfred, Jr
Hendrix, Sherman Samuel
Hickey, Richard James
Hoffman, Daniel Lewis
Hoffmaster, Donald Edeburn
Hutnik, Russell James
Jacobs, George Joseph
Janzen, Daniel Hunt
Jayne, Benjamin A
Kilham, Susan Soltau
Kimmel, William Griffiths
Kirkland, Gordon Laidlaw, Jr
Kodrich, William Ralph
Koide, Roger Tai
Kuserk, Frank Thomas
Lawrence, Vinnedge Moore
Legge, Thomas Nelson
Levin, Michael H(oward)
McCrea, Kenneth Duncan
McDiffett, Wayne Francis
McNair, Dennis M
Marquis, David Alan
Medve, Richard J
Mellinger, Michael Vance
Miller, Kenneth Melvin
Montgomery, James Douglas
Moon, Thomas Charles
Moore, John Robert
Mueller, Charles Frederick
Ostrofsky, Milton Lewis
Overlease, William R
Parker, Jon Irving
Patil, Ganapati P
Pearson, David Leander
Pottmeyer, Judith Ann
Prezant, Robert Steven
Ratzlaff, Willis
Richardson, Jonathan L
Ricklefs, Robert Eric
Rosenzweig, William David
Rymon, Larry Maring
Salamon, Kenneth J
Samollow, Paul B
Saunders, William Bruce
Schultz, Jack C
Settlemyer, Kenneth Theodore
Sheldon, Joseph Kenneth
Sherman, John Walter
Shipman, Robert Dean
Shontz, Charles Jack
Snyder, Donald Benjamin
Snyder, Robert LeRoy
Stauffer, Jay Richard, Jr
Stephenson, Andrew George
Unz, Richard F(rederick)
Vannote, Robin L
Weiner, Jacob
Wert, Jonathan Maxwell, Jr
Wilhelm, Eugene J, Jr
Williams, Frederick McGee
Williamson, Craig Edward
Yahner, Richard Howard

RHODE ISLAND
Brown, James Henry, Jr
Golet, Francis Charles
Gould, Mark D
Hammen, Susan Lum
Husband, Thomas Paul
Hyland, Kerwin Ellsworth, Jr
Janis, Christine Marie
Jossi, Jack William
Lazell, James Draper
Mayer, Garry Franklin
Miller, Don Curtis
Nixon, Scott West
Oviatt, Candace Ann
Pearson, Philip Richardson, Jr
Pilson, Michael Edward Quinton
Schmitt, Johanna
Shoop, C Robert
Tarzwell, Clarence Matthew
Tarzwell, Clarence Matthew
Waage, Jonathan King

SOUTH CAROLINA
Abernathy, A(twell) Ray
Banus, Mario Douglas
Biernbaum, Charles Knox
Blood, Elizabeth Reid
Brisbin, I Lehr, Jr
Congdon, Justin Daniel
Coull, Bruce Charles
Dame, Richard Franklin
Davis, Luckett Vanderford
Dean, John Mark
Edwards, John C
Feller, Robert Jarman
Forsythe, Dennis Martin

Gibbons, J Whitfield
Harlow, Richard Fessenden
Hazen, Terry Clyde
Helms, Carl Wilbert
Jacobs, Jacqueline E
Kelly, Robert Withers
Kennamer, James Earl
Kosinski, Robert Joseph
Krebs, Julia Elizabeth
Lacher, Thomas Edward, Jr
Lincoln, David Erwin
McKellar, Henry Northington, Jr
McLeod, Kenneth William
Morris, James T
Mulvey, Margaret
Muska, Carl Frank
Newman, Michael Charles
Olson, John Bernard
Patton, Ernest Gibbes
Pinder, John Edgar, III
Pollard, Arthur Joseph
Rice, Theodore Roosevelt
Sharitz, Rebecca Reyburn
Shepard, Buford Merle
Smith, Michael Howard
Stancyk, Stephen Edward
Teska, William Reinhold
Turner, Jack Allen
Van Dolah, Robert Frederick
Wagner, Charles Kenyon
Wood, Gene Wayne
Woodin, Sarah Ann

SOUTH DAKOTA
Berry, Charles Richard, Jr
Diggins, Maureen Rita
Dunlap, Donald Gene
Einhellig, Frank Arnold
Froiland, Sven Gordon
Haertel, Lois Steben
Hall, Jerry Dexter
Hodgson, Lynn Morrison
Hoffman, George R
Hutcheson, Harvie Leon, Jr
Keenlyne, Kent Douglas
Kieckhefer, Robert William
Tatina, Robert Edward

TENNESSEE
Abernethy, Virginia Deane
Alsop, Frederick Joseph, III
Ambrose, Harrison William, III
Amundsen, Clifford C
Auerbach, Stanley Irving
Barnthouse, Lawrence Warner
Bradshaw, Aubrey Swift
Brawley, Susan Howard
Breeden, John Elbert
Brinkhurst, Ralph O
Brode, William Edward
Bunting, Dewey Lee, II
Cada, Glenn Francis
Campbell, James A
Chapman, Joe Alexander
Clarke, James Harold
Clebsch, Edward Ernst Cooper
Coutant, Charles Coe
Dale, Virginia House
DeAngelis, Donald Lee
Dearden, Boyd L
DeSelm, Henry Rawie
Dimmick, Ralph W
Echternacht, Arthur Charles
Eddlemon, Gerald Kirk
Elmore, James Lewis
Emanuel, William Robert
Francis, Chester Wayne
Garten, Charles Thomas, Jr
Gehrs, Carl William
Hallam, Thomas Guy
Hildebrand, Stephen George
James, Ted Ralph
Kathman, R Deedee
Kaye, Stephen Vincent
Kimmel, Bruce Lee
Kocher, David Charles
Kroodsma, Roger Lee
Leffler, Charles William
Lessman, Gary M
Loar, James M
McBrayer, James Franklin
McCarthy, John F
McCormick, J Frank
Martin, Robert Eugene
Miller, Neil Austin
Nall, Ray(mond) W(illett)
Olson, J(erry) S
O'Neill, Robert Vincent
Parchment, John Gerald
Payne, James
Polis, Gary Allan
Quarterman, Elsie
Reed, Robert Marshall
Reichle, David Edward
Roop, Robert Dickinson
Salk, Martha Scheer
Schneider, Gary
Scott, Arthur Floyd
Semlitsch, Raymond Donald
Sharma, Gopal Krishan
Sharp, Aaron John
Sharples, Frances Ellen
Staub, Robert J
Suter, Glenn Walter, II

Tanner, James Taylor
Tarpley, Wallace Armell
Tolbert, Virginia Rose
Van Hook, Robert Irving, Jr
Van Winkle, Webster, Jr
Voorhees, Larry Donald
Walker, Kenneth Russell
Wallace, Gary Oren
Watson, Annetta Paule
Weaver, George Thomas
Webb, J(ohn) Warren
Willard, William Kenneth
Wiser, Cyrus Wymer
Witherspoon, John Pinkney, Jr
Woods, Frank Wilson
Yarbro, Claude Lee, Jr

TEXAS
Adams, Clark Edward
Anderson, Richard Orr
Arp, Gerald Kench
Aumann, Glenn D
Beitinger, Thomas Lee
Belk, Gene Denton
Blankenship, Lytle Houston
Boutton, Thomas William
Briske, David D
Britton, Carlton M
Browning, John Artie
Bryant, Vaughn Motley, Jr
Buskirk, Ruth Elizabeth
Cameron, Guy Neil
Camp, Frank A, III
Chrzanowski, Thomas Henry
Clark, William Jesse
Coulson, Robert N
Darnell, Rezneat Milton
Delco, Exalton Alfonso, Jr
DeShaw, James Richard
Dickson, Kenneth Lynn
Diggs, George Minor, Jr
Dodd, Jimmie Dale
DuBar, Jules R
Dyksterhuis, Edsko Jerry
Easley, Stephen Phillip
Erdman, Howard E
Fitzpatrick, Lloyd Charles
Fonteyn, Paul John
Formanowicz, Daniel Robert, Jr
Foster, John Robert
Fowler, Norma Lee
Freeman, Charles Edward, Jr
Fulbright, Timothy Edward
Garner, Herschel Whitaker
Gehlbach, Frederick Renner
Gerard, Cleveland Joseph
Grumbles, Jim Bob
Guthrie, Rufus Kent
Hannan, Herbert Herrick
Harcombe, Paul Albin
Harris, Arthur Horne
Hellier, Thomas Robert, Jr
Herbst, Richard Peter
Humphries, James Edward, Jr
Inglis, Jack Morton
Ingold, Donald Alfred
Johnston, John Spencer
Jones, Clyde Joe
Judd, Frank Wayne
Keith, Donald Edwards
Kennedy, Joseph Patrick
Kibler, Kenneth G
Kroh, Glenn Clinton
Kunz, Sidney Edmund
Lacy, Julia Caroline
Landers, Roger Q, Jr
Lanza, Guy Robert
Lind, Owen Thomas
Lopez, Genaro
Lynch, Daniel Matthew
McCarley, Wardlow Howard
McCullough, Jack Dennis
McCully, Wayne Gunter
Maclean, Graeme Stanley
McMahon, Robert Francis, III
McMillan, Calvin
Maguire, Bassett, Jr
Mangan, Robert Lawrence
Mansfield, Clifton Tyler
Martin, Robert Frederick
Mathews, Nancy Ellen
Middleditch, Brian Stanley
Moldenhauer, Ralph Roy
Mueller, Dale M J
Neill, William Harold
Newman, George Allen
Nixon, Elray S
Oppenheimer, Carl Henry, Jr
Parker, Robert Hallett
Peacock, John Talmer
Pettit, Russell Dean
Philips, Billy Ulyses
Pianka, Eric R
Pierce, Jack Robert
Powell, Michael A
Ray, James P
Reeder, William Glase
Rennie, Thomas Howard
Ring, Dennis Randall
Rylander, Michael Kent
Schmidly, David James
Schneider, Dennis Ray
Schroder, Gene David
Schuster, Joseph L

Scudday, James Franklin
Shake, Roy Eugene
Silvey, J K Gwynn
Sissom, Stanley Lewis
Smeins, Fred E
Smith, Alan Lyle
Spurr, Stephen Hopkins
Sterner, Robert Warner
Stewart, Kenneth Wilson
Stransky, John Janos
Telfair, Raymond Clark, II
Tomson, Mason Butler
Tuttle, Merlin Devere
Van Auken, Oscar William
Vincent, Jerry William
Walker, Laurence Colton
Waller, William T
Wendt, Theodore Mil
Whiteside, Charles Hugh
Willig, Michael Robert
Wilson, Robert Eugene
Wohlschlag, Donald Eugene
Wolfenberger, Virginia Ann
Wood, Carl Eugene
Wu, Hsin-i
Zagata, Michael DeForest

UTAH
Allred, Dorald Mervin
Baer, James L
Balph, David Finley
Barnes, James Ray
Belnap, Jayne
Blaisdell, James Pershing
Blauer, Aaron Clyde
Bozniak, Eugene George
Brotherson, Jack DeVon
Buchanan, Hayle
Caldwell, Martyn Mathews
Chapman, Joseph Alan
Cox, Paul Alan
Dalton, Patrick Daly
Davidson, Diane West
Dobrowolski, James Phillip
Dueser, Raymond D
Ehleringer, James Russell
Ekdale, Allan Anton
Epstein, William Warren
Fisher, Richard Forrest
Flinders, Jerran T
Gessaman, James A
Harper, Kimball T
Hayward, Charles Lynn
Helm, William Thomas
Hirth, Harold Frederick
Hull, Alvin C, Jr
Johnson, Douglas Allan
Kadlec, John A
Lighton, John R B
Lloyd, Ray Dix
McKell, Cyrus Milo
MacMahon, James A
Malechek, John Charles
Mangum, Fredrick Anthony
Messina, Frank James
Miller, Raymond Woodruff
Mueggler, Walter Frank
Neuhold, John Mathew
Olsen, Peter Fredric
Palmblad, Ivan G
Porter, Richard Dee
Reid, William Harper
Richardson, Jay Wilson, Jr
Rushforth, Samuel Roberts
Shultz, Leila McReynolds
Sites, Jack Walter, Jr
Skujins, John Janis
Smith, Howard Duane
Taylor, Robert Joe
West, Neil Elliott
White, Clayton M
Wood, Timothy E
Yorks, Terence Preston

VERMONT
Barton, James Don, Jr
Bean, Daniel Joseph
Dritschilo, William
Flaccus, Edward
Forcier, Lawrence Kenneth
Heinrich, Bernd
Henson, Earl Bennette
Illick, J(ohn) Rowland
Jervis, Robert Alfred
McIntosh, Alan William
Potash, Milton
Schall, Joseph Julian
Spearing, Ann Marie
Westing, Arthur H

VIRGINIA
Andrews, Robin M
Barbaro, Ronald D
Barton, Alexander James
Bass, Michael Lawrence
Benfield, Ernest Frederick
Bishop, John Watson
Blem, Charles R
Bliss, Dorothy Crandall
Bodkin, Norlyn L
Briggs, Jeffrey L
Britt, Douglas Lee
Britton, Maxwell Edwin
Brooks, Garnett Ryland, Jr

Brown, Luther Park
Cairns, John, Jr
Cherry, Donald Stephen
Clark, Mary Eleanor
Cocking, W Dean
Cranford, Jack Allen
Dane, Charles Warren
Day, Frank Patterson, Jr
Decker, Robert Dean
Elwood, Jerry William
Ernst, Carl Henry
Eyman, Lyle Dean
Fashing, Norman James
Fink, Linda Susan
Fisher, Elwood
Fuller, Stephen William
Gaines, Gregory
Gangstad, Edward Otis
Gibson, James H
Giles, Robert H, Jr
Gottschalk, John Simison
Grove, Thurman Lee
Harris, William Franklin, III
Hinckley, Alden Dexter
Homsher, Paul John
Howe, Marshall Atherton
Hufford, Terry Lee
Hyer, Paul Vincent
Jahn, Laurence R
Jameson, Donald Albert
Jenkins, Robert Ellsworth, Jr
Jenkins, Robert Walls, Jr
Jenssen, Thomas Alan
Johnson, Philip L
Kelly, Mahlon George, Jr
Kelso, Donald Preston
Kirkpatrick, Roy Lee
Lawrey, James Donald
Lee, James A
Levy, Gerald Frank
Little, Elbert Luther, Jr
Logan, Jesse Alan
Mellinger, Clair
Milton, Nancy Melissa
Moore, David Jay
Musick, John A
Neves, Richard Joseph
Odum, William Eugene
Osborn, Kenneth Wakeman
Pienkowski, Robert Louis
Semtner, Paul Joseph
Short, Henry Laughton
Shugart, Herman Henry, Jr
Shuster, Carl Nathaniel, Jr
Sigafoos, Robert Sumner
Simmons, George Matthew, Jr
Talbot, Lee Merriam
Terman, Charles Richard
Tiwari, Surendra Nath
Underwood, Lawrence Statton
Vaughan, Michael Ray
Ware, Stewart Alexander
Webster, Jackson Ross
Wells, Elizabeth Fortson
West, David Armstrong
Wilbur, Henry Miles
Willis, Lloyd L, II
Wood, Leonard E(ugene)
Zieman, Joseph Crowe, Jr

WASHINGTON

Adamson, Lucile Frances
Agee, James Kent
Allen, Julia Natalia
Banko, Winston Edgar
Bax, Nicholas John
Becker, Clarence Dale
Berryman, Alan Andrew
Bliss, Lawrence Carroll
Boersma, P Dee
Boersma, P Dee
Booth, Beatrice Crosby
Braham, Howard Wallace
Bredahl, Edward Arlan
Brenchley, Gayle Anne
Carr, Robert Leroy
Catts, Elmer Paul
Cook, Paul Pakes, Jr
Curl, Herbert (Charles), Jr
Cushing, Colbert Ellis
Dauble, Dennis Deene
Dixon, Kenneth Randall
Drum, Ryan William
Edmondson, W Thomas
Eickstaedt, Lawrence Lee
Emlen, John Merritt
Fahnestock, George Reeder
Favorite, Felix
Fickeisen, Duane H
Fitzner, Richard Earl
Fleming, Richard Seaman
Fonda, Richard Weston
Franklin, Jerry Forest
Funk, William Henry
Gentry, Roger Lee
Gibson, Flash
Gilmartin, Amy Jean
Ginn, Thomas Clifford
Goebel, Carl Jerome
Gray, Robert H
Gunderson, Donald Raymond
Hansen, David Henry
Hanson, Wayne Carlyle
Harris, Grant Anderson

Hatheway, William Howell
Heinle, Donald Roger
Hicks, David L
Huey, Raymond Brunson
Johnson, Richard Evan
Kareiva, Peter Michael
Karlstrom, Ernest Leonard
Karr, James Richard
Kenagy, George James
King, James Roger
Kohn, Alan Jacobs
Landis, Wayne G
Lorenzen, Carl Julius
McAlister, William Bruce
McDonough, Leslie Marvin
Mack, Richard Norton
Manuwal, David Allen
Mason, David Thomas
Mearns, Alan John
Meeuse, Bastiaan J D
Mehringer, Peter Joseph, Jr
Mills, Claudia Eileen
Morgan, Kenneth Robb
Muller-Parker, Gisèle Thérèse
Murphy, Mary Eileen
Naiman, Robert Joseph
Nelson, Jack Raymond
Novotny, Anthony James
Orians, Gordon Howell
Page, Thomas Lee
Paine, Robert T
Parker, Richard Alan
Paulson, Dennis Roy
Pearson, Walter Howard
Price, Keith Robinson
Quay, Paul Douglas
Rayburn, William Reed
Rogers, Lee Edward
Schneider, David Edwin
Schultz, Vincent
Seymour, Allyn H
Simenstad, Charles Arthur
Sluss, Robert Reginald
Soule, Oscar Hommel
Sprugel, Douglas George
Staley, James Trotter
Summers, William Clarke
Swartzman, Gordon Leni
Swedberg, Kenneth C
Tanaka, Yasuomi
Taub, Frieda B
Taylor, Peter Berkley
Templeton, William Lees
Thom, Ronald Mark
Thompson, John N
Tsukada, Matsuo
Van Voris, Peter
Vogt, Kristiina Ann
Wiedemann, Alfred Max
Woelke, Charles Edward

WEST VIRGINIA

Bissonnette, Gary Kent
Brown, Mark Wendell
Carvell, Kenneth Llewellyn
Constantz, George Doran
Keller, Edward Clarence, Jr
Lang, Gerald Edward
Ludke, James Larry
Muul, Illar
Pauley, Thomas Kyle
Shalaway, Scott D
Shan, Robert Kuocheng
Smith, Robert Leo
Stephenson, Steven Lee

WISCONSIN

Adams, Michael Studebaker
Anderson, Raymond Kenneth
Beals, Edward Wesley
Carpenter, Stephen Russell
Copes, Frederick Albert
Cottam, Grant
Davidson, Donald William
Edgington, David Norman
Fitzgerald, George Patrick
Foote, Kenneth Gerald
Grittinger, Thomas Foster
Halgren, Lee A
Hall, Kent D
Hamerstrom, Frederick Nathan
Hardin, James William
Hickey, Joseph James
Hillier, Richard David
Hine, Ruth Louise
Hole, Francis Doan
Howell, Evelyn Anne
Iltis, Hugh Hellmut
Johnson, Wendel J
Kitchell, James Frederick
Kline, Virginia March
Krezoski, John R
Larsen, James Arthur
Larsen, Michael John
Long, Claudine Fern
McCabe, Robert Albert
McCown, Brent Howard
McDonald, Malcolm Edwin
McIntosh, Thomas Henry
Magnuson, John Joseph
Miller, David Hewitt
Minock, Michael Edward
Moermond, Timothy Creighton
Morgan, Michael Dean

Newsome, Richard Duane
Ogren, Herman August
Porter, Warren Paul
Richman, Sumner
Rongstad, Orrin James
Rosson, Reinhardt Arthur
Ruff, Robert LaVerne
Rusch, Donald Harold
Seale, Dianne B
Smith, Stanley Galen
Stearns, Forest
Temple, Stanley A
Waller, Donald Macgregor
Warner, James Howard
Weise, Charles Martin
White, Charley Monroe
Whitford, Philip Burton
Wikum, Douglas Arnold
Yasukawa, Ken
Young, Allen Marcus
Yuill, Thomas MacKay

WYOMING

Adams, John Collins
Anderson, Stanley H
Bergman, Harold Lee
Boyce, Mark S
Christensen, Martha
DePuit, Edward J
Diem, Kenneth Lee
Edwards, William Charles
Fisser, Herbert George
Hayden-Wing, Larry Dean
Holbrook, Frederick R
Kennington, Garth Stanford
Knight, Dennis Hal
Laycock, William Anthony
Lockwood, Jeffrey Alan
Parker, Michael
Reiners, William A
Reynolds, Richard Truman
Scott, Richard Walter
Stanton, Nancy Lea
Sullivan, Brian Patrick

PUERTO RICO

Bruck, David Lewis
Lewis, Allen Rogers
Wilson, Marcia Hammerquist

ALBERTA

Addicott, John Fredrick
Anderson, Paul Knight
Bidgood, Bryant Frederick
Boag, David Archibald
Byers, John Robert
Clifford, Hugh Fleming
Dancik, Bruce Paul
Davies, Ronald Wallace
Evans, William George
Freeman, Milton Malcolm Roland
Fuller, William Albert
Gallup, Donald Noel
Hickman, Michael
Holmes, John Carl
Huang, Henry Hung-Chang
Macpherson, Andrew Hall
McPherson, Harold James
Murie, Jan O
Pattie, Donald L
Pielou, Douglas Patrick
Schindler, David William
Scotter, George Wilby
Seghers, Benoni Hendrik
Swanson, Robert Harold
Vitt, Dale Hadley
Wein, Ross Wallace
Wong, Horne Richard
Zammuto, Richard Michael

BRITISH COLUMBIA

Annas, Richard Morris
Bell, Marcus Arthur Money
Belland, René Jean
Buchanan, Ronald James
Bunnell, Frederick Lindsley
Campbell, Alan
Chitty, Dennis Hubert
Cowan, Ian McTaggart
Dill, Lawrence Michael
Druehl, Louis D
Fisher, Francis John Fulton
Foreman, Ronald Eugene
Frazer, Bryan Douglas
Hartwick, Earl Brian
Healey, Michael Charles
Hebda, Richard Joseph
Jamieson, Glen Stewart
Kimmins, James Peter (Hamish)
Krajina, Vladimir Joseph
Lane, Edwin David
Lauzier, Raymond B
Levings, Colin David
Lewin, Victor
Mackauer, Manfred
McMullen, Robert David
Mathewes, Rolf Walter
Pielou, Evelyn C
Quinton, Dee Arlington
Sibert, John Rickard
Stockner, John G
Strang, Robert M
Thomson, Alan John
Tunnicliffe, Verena Julia

Turkington, Robert (Roy) Albert
Walters, Carl John
Warrington, Patrick Douglas
Wellington, William George
Zwickel, Fred Charles

MANITOBA

Aung, Taing
Brunskill, Gregg John
Brust, Reinhart A
Gee, John Henry
Guthrie, John Erskine
Hartland-Rowe, Richard C B
Huebner, Judith Dee
Iverson, Stuart Leroy
Johnson, Karen Louise
Johnson, Lionel
Keleher, J J
Kingsley, Michael Charles Stephen
Patalas, Kazimierz
Preston, William Burton
Pruitt, William O, Jr
Rosenberg, David Michael
Sinha, Ranendra Nath
Smith, Roger Francis Cooper
Staniforth, Richard John
Turnock, William James
Ward, Fredrick James
White, Noel David George
Zach, Reto

NEW BRUNSWICK

Baskerville, Gordon Lawson
Cook, Robert Harry
Eidt, Douglas Conrad
Eveleigh, Eldon Spencer
Harries, Hinrich
Logan, Alan
MacLean, David Andrew
Taylor, Andrew Ronald Argo
Van Groenewoud, Herman
Waiwood, Kenneth George

NEWFOUNDLAND

Davis, Charles (Carroll)
Haedrich, Richard L
Ni, I-Hsun

NOVA SCOTIA

Bowen, William Donald
Dadswell, Michael John
Doyle, Roger Whitney
Hardman, John Michael
Harrington, Fred Haddox
Longhurst, Alan R
Mann, Kenneth H
Miller, Robert Joseph
Newkirk, Gary Francis
Ogden, James Gordon, III
Silvert, William Lawrence
Specht, Harold Balfour
Stobo, Wayne Thomas
Wangersky, Peter John
Wiles, Michael

ONTARIO

Alex, Jack Franklin
Ambrose, John Daniel
Balon, Eugene Kornel
Banfield, Alexander William Francis
Barica, Jan M
Baxter, Robert MacCallum
Behan-Pelletier, Valerie Mary
Birmingham, Brendan Charles
Brown, Seward Ralph
Buckner, Charles Henry
Burger, Dionys
Burnison, Bryan Kent
Calder, Dale Ralph
Cameron, Duncan MacLean, Jr
Catling, Paul Miles
Cavers, Paul Brethen
Cheng, Hsien Hua
Chengalath, Rama
Coad, Brian William
Collins, Nicholas Clark
Danks, Hugh Victor
Darling, Donald Christopher
Dickman, Michael David
Edwards, Roy Lawrence
Emery, Alan Roy
Falls, James Bruce
Glooschenko, Walter Arthur
Graham, Terry Edward
Graham, William Muir
Green, Roger Harrison
Halfon, Efraim
Harcourt, Douglas George
Harvey, Harold H
Hofstra, Gerald
Hogan, Gary D
Hynes, Hugh Bernard Noel
Jeglum, John Karl
Jones, Philip Arthur
Jones, Roger
Kaushik, Narinder Kumar
Keddy, Paul Anthony
Kelso, John Richard Murray
Kershaw, Kenneth Andrew
Kevan, Peter Graham
Kobluk, David Ronald
Kovacs, Kit M
Kudo, Akira
Laing, John E

Ecology (cont)

Lavigne, David M
Lee, David Robert
LeRoux, Edgar Joseph
McAndrews, John Henry
M'Closkey, Robert Thomas
Mackay, Rosemary Joan
Maycock, Paul Frederick
Merriam, Howard Gray
Middleton, Alex Lewis Aitken
Montgomerie, Robert Dennis
Morris, Ralph Dennis
Morrison, Ian Kenneth
Mudroch, Alena
Munroe, Eugene Gordon
Pierce, Ronald Cecil
Powles, Percival Mount
Rapport, David Joseph
Regier, Henry Abraham
Risk, Michael John
Ritchie, James Cunningham
Robertson, Raleigh John
Roberts-Pichette, Patricia Ruth
Ryder, Richard Armitage
Sears, Markham Karli
Shih, Chang-Tai
Smith, David William
Smith, Donald Alan
Smol, John Paul
Solman, Victor Edward Frick
Sprules, William Gary
Stebelsky, Ihor
Stokes, Pamela Mary
Svoboda, Josef
Vollenweider, Richard A
Wallen, Donald George
Warwick, Suzanne Irene
Weatherley, Alan Harold
Weinberger, Pearl
Winterbottom, Richard
Winterhalder, Keith

QUEBEC

Beland, Pierre
Bell, Graham Arthur Charlton
Bider, John Roger
Bourget, Edwin
Brunel, Pierre
Cloutier, Conrad Francois
Dansereau, Pierre
Ferron, Jean H
Grandtner, Miroslav Marian
Grant, James William Angus
Hill, Stuart Baxter
Kalff, Jacob
Lacroix, Guy
Laflamme, Gaston
Langford, Arthur Nicol
Lechowicz, Martin John
Legendre, Louis
Legendre, Pierre
Leggett, William C
Levasseur, Maurice Edgar
MacLeod, Charles Franklyn
McNeil, Jeremy Nichol
Maly, Edward J
Morisset, Pierre
Percy, Jonathan Arthur
Peters, Robert Henry
Rau, Manfred Ernst
Reiswig, Henry Michael
Richard, Pierre Joseph Herve
Sanborne, Paul Michael
Sergeant, David Ernest
Sharma, Madan Lal
Shoubridge, Eric Alan
Stewart, Robin Kenny
Titman, Rodger Donaldson
Winget, Carl Henry

SASKATCHEWAN

Bothwell, Max Lewis
Coupland, Robert Thomas
Evans, Marlene Sandra
Forsyth, Douglas John
Hammer, Ulrich Theodore
Lehmkuhl, Dennis Merle
Maher, William J
Makowski, Roberte Marie Denise
Mitchell, George Joseph
Mukerji, Mukul Kumar
Nelson, Louise Mary
Oliphant, Lynn Wesley
Redmann, Robert Emanuel
Rowe, John Stanley
Sarjeant, William Antony Swithin
Schmutz, Josef Konrad
Secoy, Diane Marie
Waddington, John
Walther, Alina
Zilke, Samuel

OTHER COUNTRIES

Adames, Abdiel Jose
Biederman-Thorson, Marguerite Ann
Bray, John Roger
Briand, Frederic Jean-Paul
Budowski, Gerardo
Buttemer, William Ashley
Camiz, Sergio
Caperon, John
Chesson, Peter Leith
Cheung, Mo-Tsing Miranda

Craig, John Frank
Dahl, Arthur Lyon
Eiten, George
Erdtmann, Bernd Dietrich
Fowler, Scott Wellington
Gottlieb, Otto Richard
Holldobler, Berthold Karl
Holsworth, William Norton
Johannes, Robert Earl
Kessell, Stephen Robert
Koerner, Heinz
Komarkova, Vera
Koslow, Julian Anthony
Laurent, Pierre
Lee, Douglas Harry Kedgwin
Leigh, Egbert G, Jr
Lieth, Helmut Heinrich Friedrich
Marsh, James Alexander, Jr
May, Robert McCredie
Myres, Miles Timothy
Nelson, Stephen Glen
Nevo, Eviatar
O'Neill, Patricia Lynn
Paulay, Gustav
Peakall, David B
Rodriguez, Gilberto
Saether, Ole Anton
Stam, Jos
Stearns, Stephen Curtis
Tosi, Joseph Andrew, Jr
Ugolini, Fiorenzo Cesare
Westcott, Peter Walter
Winnett, George

Embryology

ALABAMA

Bacon, Arthur Lorenza
Freeman, John A
Hand, George Samuel, Jr
Hood, Ronald David
Keys, Charles Everel
Lubega, Seth Gasuza
McLaughlin, Ellen Winnie
Rhodes, Richard Kent

ARIZONA

Bagnara, Joseph Thomas
Doane, Winifred Walsh
Hendrix, Mary J C
Kischer, Clayton Ward
Maienschein, Jane Ann
Pogany, Gilbert Claude
Terry, Lucy Irene

ARKANSAS

Johnston, Perry Max
Klapper, Clarence Edward
Sheehan, Daniel Michael
Tank, Patrick Wayne

CALIFORNIA

Alberch, Pere
Anderson, Gary Bruce
Armstrong, Peter Brownell
Armstrong, Rosa Mae
Baird, John Jeffers
Benzer, Seymour
Bernard, George W
Brownell, Anna Gale
Bryant, Peter James
Calarco, Patricia G
Celniker, Susan Elizabeth
Chuong, Cheng-Ming
Corson, George Edwin, Jr
Davidson, Eric Harris
Dawson, John E
Dunaway, Marietta
Dunnebacke-Dixon, Thelma Hudson
Durrant, Barbara Susan
Epel, David
Fisher, Robin Scott
Glass, Laurel Ellen
Golbus, Mitchell S
Grey, Robert Dean
Hammar, Allan H
Heath, Harrison Duane
Hendrickx, Andrew George
Hunter, Alice S (Baker)
Jeffery, William Richard
Kalland, Gene Arnold
Keller, Raymond E
King, Barry Frederick
King, Robbins Sydney
Ko, Chien-Ping
Krejsa, Richard Joseph
Laham, Quentin Nadime
Lambert, Charles Calvin
Lengyel, Judith Ann
Levine, Michael Steven
Lipshitz, Howard David
Lopo, Alina C
Luckock, Arlene Suzanne
Mak, Linda Louise
Martin, Gail Roberta
Martins-Green, Manuela M
Maxson, Robert E, Jr
Menees, James H
Mertes, David H
Meyerowitz, Elliot Martin
Moore, Betty Clark
Moretti, Richard Leo

Neff, William Medina
Nuccitelli, Richard Lee
Overturf, Gary D
Pedersen, Roger Arnold
Plopper, Charles George
Pryer, Nancy Kathryn
Schweisthal, Michael Robert
Scott, Matthew P
Shook, Brenda Lee
Smith, L Dennis
Sparling, Mary Lee
Spitzer, Nicholas Canaday
Stephens, Lee Bishop, Jr
Stockdale, Frank Edward
Sung, Zinmay Renee
Thurmond, William
Towers, Bernard
Vaughn, James E, Jr
Vreeland, Valerie Jane
Webster, Barbara Donahue
Willhite, Calvin Campbell
Wilson, Doris Burda
Wilson, Wilfred J
Wilt, Fred H
Wood, Richard Lyman
Woods, Geraldine Pittman
Zernik, Joseph

COLORADO

Barrett, Dennis
Bekoff, Anne C
Finger, Thomas Emanuel
Ham, Richard George
Hanken, James
Maylie-Pfenninger, M F
Moury, John David
Seidel, George Elias, Jr
Torbit, Charles Allen, Jr

CONNECTICUT

Beitch, Irwin
Berry, Spencer Julian
Boell, Edgar John
Clark, Hugh
Cohen, Melvin Joseph
Constantine-Paton, Martha
Fell, Paul Erven
Fiore, Carl
Forbes, Thomas Rogers
Herrmann, Heinz
Infante, Anthony A
Jaffe, Laurinda A
Kent, John Franklin
Kollar, Edward James
Lee, Thomas W
Levin, Martin Allen
Meriney, Stephen D
Morest, Donald Kent
Pierro, Louis John
Rakic, Pasko
Roman, Laura M
Rossomando, Edward Frederick
Staugaard, Burton Christian
Stenn, Kurt S
Upholt, William Boyce

DELAWARE

Ruben, Regina Lansing
Song, Jiakun

DISTRICT OF COLUMBIA

Allan, Frank Duane
Avery, Gordon B
Ball, William David
Chan, Wai-Yee
Dimond, Marie Therese
Goeringer, Gerald Conrad
Hayes, Raymond L, Jr
Hines, Pamela Jean
Kimmel, Carole Anne
Nishioka, David Jitsuo
Rall, William Frederick

FLORIDA

Bray, Joan Lynne
Cameron, Don Frank
Chen, L T
Cohen, Glenn Milton
Colwin, Arthur Lentz
Colwin, Laura Hunter
Deats, Edith Potter
Goldberg, Stephen
Grabowski, Casimer Thaddeus
Hinsch, Gertrude Wilma
Hopper, Arthur Frederick
Koevenig, James L
Lessios, Harilaos Angelou
Muller, Kenneth Joseph
Pratt, Melanie M
Rice, Stanley Alan
Shireman, Rachel Baker
Siden, Edward Joel
Taylor, George Thomas
Tripp, John Rathbone
Vasil, Vimla

GEORGIA

DeHaan, Robert Lawrence
Gulati, Adarsh Kumar
Hicks, Heraline Elaine
Ivey, William Dixon
Khan, Iqbal M
McKenzie, John Ward
Patterson, Rosalyn Mitchell

Paulsen, Douglas F
Smeltzer, Richard Homer
Sohal, Gurkirpal Singh
Whitney, J(ohn) Barry, III
Zinsmeister, Philip Price

HAWAII

Arnold, John Miller
Dalton, Howard Clark
Hadfield, Michael Gale
Haley, Samuel Randolph
Hsiao, Sidney Chihti
Nelson, Marita Lee
Wyban, James A

IDAHO

Cloud, Joseph George
Fuller, Eugene George
Stephens, Trent Dee

ILLINOIS

Aydelotte, Margaret Beesley
Carr, Virginia McMillan
Criley, Bruce
Dinsmore, Charles Earle
Doering, Jeffrey Louis
Dudkiewicz, Alan Bernard
Durica, Thomas Edward
Dybas, Linda Kathryn
Farbman, Albert Irving
Foote, Florence Martindale
Gaik, Geraldine Catherine
Goldman, Allen S
LaVelle, Faith Wilson
Menco, Bernard
Mittenthal, Jay Edward
Moskal, Joseph Russell
Nardi, James Benjamin
Overton, Jane Harper
Pollack, Emanuel Davis
Rabuck, David Glenn
Rafferty, Keen Alexander, Jr
Rafferty, Nancy S
Schmidt, Anthony John
Seale, Raymond Ulric
Spiroff, Boris E N
Van Alten, Pierson Jay
Watterson, Ray Leighton
Williams, Thomas Alan
Zalisko, Edward John

INDIANA

BeMiller, Paraskevi Mavridis
Chernoff, Ellen Ann Goldman
Crouch, Martha Louise
Das, Gopal Dwarka
Davis, Grayson Steven
Denner, Melvin Walter
Hoversland, Roger Carl
Hunt, Linda Margaret
Kennedy, Duncan Tilly
Maloney, Michael Stephen
Mescher, Anthony Louis
Pietsch, Paul Andrew
Schaible, Robert Hilton
Tweedell, Kenyon Stanley

IOWA

Benbow, Robert Michael
Kieso, Robert Alfred
Kollros, Jerry John
Meetz, Gerald David
Mennega, Aaldert
Milkman, Roger Dawson
Rogers, Frances Arlene
Shen, Sheldon Shih-Ta
Solursh, Michael
Sullivan, Charles Henry
Welshons, William John

KANSAS

Bode, Vernon Cecil
Conrad, Gary Warren
Dey, Sudhansu Kumar
Frost-Mason, Sally Kay
Gattone, Vincent H, II
Grebe, Janice Durr
Lavia, Lynn Alan
Smalley, Katherine N
Wyttenbach, Charles Richard

KENTUCKY

Alexander, Lloyd Ephraim
Birge, Wesley Joe
Fowler, Ira
Humphries, Asa Alan, Jr
Just, John Josef
Matulionis, Daniel H
Smith, Stephen D

LOUISIANA

Clawson, Robert Charles
Cobb, Glenn Wayne
Cowden, Ronald Reed
Grodner, Mary Laslie
Kern, Clifford H, III
Lesseps, Roland Joseph
Newsome, David Anthony
Peebles, Edward McCrady
Vaupel, Martin Robert
Weber, Joseph T

MAINE

Bailey, Donald Wayne

Hoppe, Peter Christian
Rappaport, Raymond, Jr

MARYLAND
Ackerman, Eric J
Berger, Edward Alan
Brown, Kenneth Stephen
Daniels, Mathew Paul
Dean, Jurrien
Dosier, Larry Waddell
Ebert, James David
Eglitis, Martin Alexandris
Finch, Robert Allen
Gall, Joseph Grafton
Goode, Melvyn Dennis
Greenhouse, Gerald Alan
Gulyas, Bela Janos
Hanover, John Allan
Hausman, Steven J
Heck, Margaret Mathilde Sophie
Hiatt, James Lee
Imberski, Richard Bernard
Kaighn, Morris Edward
McKnight, Steven L
Newrock, Kenneth Matthew
Oberdorfer, Michael Douglas
Provine, Robert Raymond
Schindler, Joel Marvin
Schnaar, Ronald Lee
Shafer, W Sue
Stratmeyer, Melvin Edward
Strum, Judy May
Van Arsdel, William Campbell, III
White, Elizabeth Lloyd
Wolffe, Alan Paul
Wu, Roy Shih-Shyong
Zirkin, Barry Ronald

MASSACHUSETTS
Adler, Richard R
Ahlberg, Henry David
Albertini, David Fred
Begg, David A
Bernfield, Merton Ronald
Crain, William Rathbone, Jr
Dittmer, John Edward
Ducibella, Tom
El-Bermani, Al-Walid I
Erickson, Alan Eric
Ernst, Susan Gwenn
Eschenberg, Kathryn (Marcella)
Fink, Rachel Deborah
Furshpan, Edwin Jean
Gibbons, Michael Francis, Jr
Gonnella, Patricia Anne
Hausman, Robert Edward
Hay, Elizabeth Dexter
Hertig, Arthur Tremain
Hinrichs, Katrin
Hoar, Richard Morgan
Horvitz, Howard Robert
Hynes, Richard Olding
Jaenisch, Rudolf
Kafatos, Fotis C
Lazarte, Jaime Esteban
Meedel, Thomas Huyck
Miller, James Albert, Jr
Papaioannou, Virginia Eileen
Pardue, Mary Lou
Poccia, Dominic Louis
Rio, Donald C
Rollason, Grace Saunders
Ruderman, Joan V
Saunders, John Warren, Jr
Sorokin, Sergei Pitirimovitch
Spiegelman, Martha
Stein, Otto Ludwig
Takeuchi, Kiyoshi Hiro
Toole, Bryan Patrick
Townsend, Jane Kaltenbach
Urry, Lisa Andrea
Webb, Andrew Clive
Whittaker, J Richard
Wolf, Merrill Kenneth
Wright, Mary Lou
Wu, Jung-Tsung
Ziomek, Carol A

MICHIGAN
Abraham, Irene
Armant, D Randall
Atkinson, James William
Avery, James Knuckey
Beaudoin, Allan Roger
Buss, Jack Theodore
Cather, James Newton
Dunbar, Joseph C
Duwe, Arthur Edward
Easter, Stephen Sherman, Jr
Eichler, Victor B
Fisher, Don Lowell
Fritts-Williams, Mary Louise Monica
Froiland, Thomas Gordon
Goustin, Anton Scott
Heady, Judith E
Hill, Susan Douglas
Kemp, Norman Everett
Koehler, Lawrence D
Mathews, Willis Woodrow
Sacco, Anthony G
Smith, Richard Harding
Stinson, Al Worth
Tosney, Kathryn W

MINNESOTA
Ellinger, Mark Stephen
Sherer, Glenn Keith
Shoger, Ross L
Sicard, Raymond Edward
Singer, Susan Rundell
Sinha, Akhouri Achyutanand
Smail, James Richard
Smithberg, Morris
Theisen, Charles Thomas
Todt, William Lynn

MISSISSIPPI
Ball, Carroll Raybourne
Keiser, Edmund Davis, Jr
Lawler, Adrian Russell
Martin, Billy Joe
Roy, William Arthur

MISSOURI
Cheney, Clarissa M
Chiappinelli, Vincent A
David, John Dewood
Decker, John D
Ericson, Avis J
Friedman, Harvey Paul
Hamburger, Viktor
Moffatt, David John
Osdoby, Philip
Price, Joseph Levering
Sanes, Joshua Richard
Schreiweis, Donald Otto
Sharp, John Roland

MONTANA
Foresman, Kerry Ryan
Thompson, Holly Ann

NEBRASKA
Dossel, William Edward
Fawcett, James Davidson
Holyoke, Edward Augustus
Lund, Douglas E
McCue, Robert Owen
Schlesinger, Allen Brian
Smith, Paula Beth
Turpen, James Baxter

NEVADA
Tibbitts, Forrest Donald

NEW HAMPSHIRE
Ferm, Vergil Harkness
Spiegel, Evelyn Sclufer
Spiegel, Melvin

NEW JERSEY
Agnish, Narsingh Dev
Arthur, Alan Thorne
Cohen, Alan Mathew
Collier, Marjorie McCann
Essien, Francine B
Fangboner, Raymond Franklin
Francoeur, Robert Thomas
Frost, David
Halpern, Myron Herbert
Hampton, Suzanne Harvey
Hart, Nathan Hoult
Hollinshead, May B
Hyndman, Arnold Gene
Ivatt, Raymond John
Mitala, Joseph Jerrold
Neri, Anthony
Pai, Anna Chao
Saiff, Edward Ira
Shelden, Robert Merten
Silver, Lee Merrill
Steinberg, Malcolm Saul
Tesoriero, John Vincent
Trelstad, Robert Laurence
Weis, Peddrick

NEW MEXICO
Stricker, Stephev Alexander

NEW YORK
Angerer, Robert Clifford
Arny, Margaret Jane
Ash, William James
Asnes, Clara F
Auclair, Walter
Bachvarova, Rosemary Faulkner
Benzo, Camillo Anthony
Black, Virginia H
Blackler, Antonie W C
Bornslaeger, Elayne A
Boylan, Elizabeth S
Bradbury, Michael Wayne
Brownscheidle, Carol Mary
Cairns, John Mackay
Collier, Jack Reed
Crain, Stanley M
Crotty, William Joseph
Currie, Julia Ruth
Easton, Douglas P
Factor, Jan Robert
Foote, Robert Hutchinson
Fortune, Joanne Elizabeth
Fowler, James A
Frair, Wayne
Gates, Allen H(azen), Jr
Gershon, Michael David
Gordon, Jon W
Greengard, Olga

Grumet, Martin
Harris, Jack Kenyon
Holtfreter, Johannes Friedrich Karl
Hopkins, Betty Jo Henderson
Katz, Eugene Richard
Kaye, Nancy Weber
Kuehnert, Charles Carroll
Lemanski, Larry Frederick
Loy, Rebekah
Lyser, Katherine May
McCann-Collier, Marjorie Dolores
McCormick, Paulette Jean
McCourt, Robert Perry
McDaniel, Carl Nimitz
Miller, Richard Kermit
Miller, Sue Ann
Mitchell, John Taylor
Murphy, Michael Joseph
Newman, Stuart Alan
Nguyen-Huu, Chi M
Percus, Jerome K
Rasweiler, John Jacob, IV
Reid, Lola Cynthia McAdams
Reynolds, Wynetka Ann King
Rich, Abby M
Robbins, Edith Schultz
Roeder, Robert Gayle
Ruddell, Alanna
Schlafer, Donald H
Schuel, Herbert
Sobel, Jael Sabina
Spence, Alexander Perkins
Swartz, Gordon Elmer
Szabo, Piroska Ludwig
Tabor, John Malcolm
Topp, William Carl
Udin, Susan Boymel
Waelsch, Salome Gluecksohn
Wenk, Eugene J
Wu, Joseph M
Yeh, Kwo-Yih
Zanetti, Nina Clare

NORTH CAROLINA
Anderton, Laura Gaddes
Black, Betty Lynne
Counce, Sheila Jean
Craig, Syndey Pollock, III
Devlin, Robert B
Edwards, Nancy C
Fail, Patricia A
Garris, David Roy
Harris, Albert Kenneth, Jr
Johnston, Malcolm Campbell
Kalfayan, Laura Jean
Kalmus, Gerhard Wolfgang
Lawrence, Irvin E, Jr
Lehman, Harvey Eugene
Lieberman, Melvyn
McLachlan, John Alan
Markert, Clement Lawrence
Morrissey, Richard Edward
Sadler, Thomas William
Shafer, Thomas Howard
Smith, Bradley Richard
Sulik, Kathleen Kay
Thomas, Mary Beth
Thomas, William Albert
Welsch, Frank

NORTH DAKOTA
Owen, Alice Koning
Thompson, Michael Bruce

OHIO
Baker, Peter C
Beal, Kathleen Grabaskas
Birky, Carl William, Jr
Brummett, Anna Ruth
Caston, J Douglas
Chakraborty, Jyotsna (Joana)
Coleman, Marilyn A
Crutcher, Keith A
DuBrul, Ernest
Egar, Margaret Wells
Ferguson, Marion Lee
Hamlett, William Cornelius
Herschler, Michael Saul
Hilliard, Stephen Dale
Houston, Willie Walter, Jr
Jacobs-Lorena, Marcelo
Jaffee, Oscar Charles
Kriebel, Howard Burtt
Lane, Roger Lee
Lesh-Laurie, Georgia Elizabeth
Manson, Jeanne Marie
Perry, George
Raghavan, Valayamghat
Rutishauser, Urs Stephen
Saksena, Vishnu P
Tepperman, Katherine Gail
Thomson, Dale S
Vorhees, Charles V
Watanabe, Michiko
Yow, Francis Wagoner
Zimmerman, Ernest Frederick

OKLAHOMA
Bell, Paul Burton, Jr
Dugan, Kimiko Hatta
Grubb, Randall Barth
Gumbreck, Laurence Gable
Lhotka, John Francis

OREGON
Adams, Frank William
Cimberg, Robert Lawrence
Fish, Joseph Leroy
Grant, Philip
Gunberg, David Leo
Morris, John Edward
Postlethwait, John Harvey
Weston, James A

PENNSYLVANIA
Aronson, John Ferguson
Brent, Robert Leonard
Brinster, Ralph L
Cohen, Leonard Harvey
DiBerardino, Marie A
Fabian, Michael William
Fenderson, Bruce Andrew
Fluck, Richard Allen
Gibley, Charles W, Jr
Grobstein, Paul
Grunwald, Gerald B
Heyner, Susan
Hilfer, Saul Robert
Holtzer, Howard
Idzkowsky, Henry Joseph
Jensh, Ronald Paul
Kennedy, Michael Craig
Kochhar, Devendra M
Kvist, Tage Nielsen
Lash, James (Jay) W
Leibel, Wayne Stephan
Linask, Kersti Katrin
McLaughlin, Patricia J
Mezger-Freed, Liselotte
Niu, Mann Chiang
Oppenheimer, Jane Marion
Piesco, Nicholas Peter
Poethig, Richard Scott
Pollock, John Archie
Ramasastry, Sai Sudarshan
Ray, Eva K
Roth, Stephen
Searls, Robert L
Short, John Albert
Telfer, William Harrison
Thomson, Keith Stewart
Tuan, Rocky Sung-Chi
Vogel, Norman William
Weston, John Colby
Zaccaria, Robert Anthony
Zagon, Ian Stuart

RHODE ISLAND
Coleman, John Russell
Fausto-Sterling, Anne
Fish, William Arthur
Silver, Alene Freudenheim
Weisz, Paul B

SOUTH CAROLINA
Dickey, Joseph Freeman
Finlay, Mary Fleming
Herr, John Mervin, Jr
Odor, Dorothy Louise
Sawyer, Roger Holmes
Taber, Elsie

SOUTH DAKOTA
Haertel, John David
Johnson, Leland Gilbert
Naughten, John Charles
Neufeld, Daniel Arthur
Settles, Harry Emerson

TENNESSEE
Brawley, Susan Howard
Dumont, James Nicholas
Gao, Kuixiong
Hasty, David Long
Hoffman, Loren Harold
MacCabe, Jeffrey Allan
McFee, Alfred Frank
McGavock, Walter Donald
Mallette, John M
Popp, Raymond Arthur
Skalko, Richard G(allant)
Trigiano, Robert Nicholas
Wachtel, Stephen Shoel
Walton, Barbara Ann
Wiser, Cyrus Wymer

TEXAS
Blystone, Robert Vernon
Carney, Darrell Howard
Carson, Daniel Douglas
Collins, Russell Lewis
Dalterio, Susan Linda
DiMichele, Leonard Vincent
Faulkner, Russell Conklin, Jr
George, Fredrick William
Gomer, Richard Hans
Grimes, L Nichols
Houston, Marshall Lee
Jacobson, Antone Gardner
Kalthoff, Klaus Otto
Litke, Larry Lavoe
Martin, Edward Williford
Phillips, Roger Guy
Pierce, Jack Robert
Ritter, Nadine Marie
Roberts, Susan Jean
Sauer, Helmut Wilhelm
Schwalm, Fritz Ekkehardt

Embryology (cont)

Siler-Khodr, Theresa M
Wordinger, Robert James
Zwaan, Johan Thomas

UTAH
Jacobson, Marcus
Schoenwolf, Gary Charles
Seegmiller, Robert Earl
Wurst, Gloria Zettle

VERMONT
Evans, Hiram John
Glade, Richard William
Woodcock-Mitchell, Janet Louise

VIRGINIA
Black, Robert Earl Lee
Creager, Joan Guynn
DeSesso, John Michael
Devine, Charles Joseph, Jr
Diehl, Fred A
Harris, Thomas Mason
Jollie, William Pucette
Laurie, Gordon William
Little, Charles Durwood, Jr
Merlino, Glenn T
Moyer, Wayne A
Palisano, John Raymond
Potts, Malcolm
Radice, Gary Paul
Reynolds, John Dick
Sandow, Bruce Arnold
Wiseman, Lawrence Linden

WASHINGTON
Avner, Ellis David
Bakken, Aimee Hayes
Davies, Jack
Hauschka, Stephen D
Hendrickson, Anita Elizabeth
Hille, Merrill Burr
Holbrook, Karen Ann
Hosick, Howard Lawrence
Lowe, Janet Marie
Mills, Claudia Eileen
Quinn, LeBris Smith
Reh, Thomas Andrew
Sikov, Melvin Richard
Wakimoto, Barbara Toshiko
Yancey, Paul Herbert

WEST VIRGINIA
Butcher, Roy Lovell
Reyer, Randall William
Williams, Leah Ann

WISCONSIN
Auerbach, Robert
Bavister, Barry Douglas
Bersu, Edward Thorwald
Bolender, David Leslie
Fallon, John Francis
Gilbert-Barness, Enid F
Kaplan, Stanley
Keefer, Carol Lyndon
Kostreva, David Robert
Orsini, Margaret Ward (Giordano)
Snook, Theodore
Sobkowicz, Hanna Maria

WYOMING
Lillegraven, Jason Arthur
Petersen, Nancy Sue

PUERTO RICO
Luckett, Winter Patrick

ALBERTA
Campenot, Robert Barry
Cass, David D
Cavey, Michael John
Cuny, Robert Michael
Heming, Bruce Sword
Schneider, Wolfgang Johann
Sperber, Geoffrey Hilliard

BRITISH COLUMBIA
Algard, Franklin Thomas
Burke, Robert D
Diewert, Virginia M
Finnegan, Cyril Vincent
Todd, Mary Elizabeth

MANITOBA
Gordon, Richard
Laale, Hans W

NOVA SCOTIA
Dickson, Douglas Howard

ONTARIO
Balon, Eugene Kornel
Buchanan, George Dale
Butler, Richard Gordon
Chua, Kian Eng
Chui, David H K
Dingle, Allan Douglas
Etches, Robert J
Fallding, Margaret Hurlstone Hardy
Fisher, Kenneth Robert Stanley
Kidder, Gerald Marshall
Leibo, Stanley Paul

Rossant, Janet
Zimmerman, Selma Blau

QUEBEC
Briere, Normand
Dubé, François
Mulay, Shree
Trasler, Daphne Gay
Wolsky, Alexander

SASKATCHEWAN
Butler, Harry
Fedoroff, Sergey
Flood, Peter Frederick

OTHER COUNTRIES
Araki, Masasuke
Dworkin, Mark Bruce
Geiger, Benjamin
Helander, Herbert Dick Ferdinand
Kordan, Herbert Allen
Mizuno, Shigeki
Shore, Laurence Stuart
Solter, Davor
Szebenyi, Emil
Tachibana, Takehiko
Thiery, Jean Paul
Toutant, Jean-Pierre
Webb, Cynthia Ann Glinert
Wenger, Byron Sylvester

Endocrinology

ALABAMA
Berecek, Kathleen Helen
Garver, David L
Lutwak, Leo
Paxton, Ralph
Yarbrough, James David

ARIZONA
Hagedorn, Henry Howard
Justus, Jerry T
Mooradian, Arshag Dertad

ARKANSAS
Allaben, William Thomas
Ruwe, William David
Sheehan, Daniel Michael
Stallcup, Odie Talmadge

CALIFORNIA
Arnold, Arthur Palmer
Byerley, Lauri Olson
Chambers, Kathleen Camille
Curry, Donald Lawrence
Engelmann, Franz
Ezrin, Calvin
Fagin, Katherine Diane
Gibson, Thomas Richard
Grumbach, Melvin Malcolm
Harding, Boyd W
Hostetler, Karl Yoder
Johnson, Randolph Mellus
Kalland, Gene Arnold
Krauss, Ronald
Laird, Cleve Watrous
Lev-Ran, Arye
Licht, Paul
Lu, John Kuew-Hsiung
Moberg, Gary Philip
Mohan, Chandra
Nemere, Ilka M
Parlow, Albert Francis
Pavgi, Sushama
Perryman, Elizabeth Kay
Rosenfeld, George
Rosenfeld, Ron Gershon
Russell, Sharon May
Sawyer, Wilbur Henderson
Shelesnyak, Moses Chiam
Snow, George Edward
Szego, Clara Marian
Turgeon, Judith Lee
Vickery, Larry Edward
Walker, Ameae M
Weber, Heather R(oss) Wilson
Wechter, William Julius
Wilcox, Ronald Bruce
Yund, Mary Alice
Zernik, Joseph

COLORADO
Hanken, James
Hofeldt, Fred Dan
Kano-Sueoka, Tamiko
Nett, T M
Roberts, P Elaine

CONNECTICUT
Goldman, Bruce Dale
Matovcik, Lisa M
Pawelek, John Mason
Rebuffe-Scrive, Marielle Francoise
Tamborlane, William Valentine

DELAWARE
Saller, Charles Frederick
Tam, Sang William

DISTRICT OF COLUMBIA
Alleva, John J
Callaway, Clifford Wayne

Chan, Wai-Yee
Lippman, Marc Estes
Lumpkin, Michael Dirksen
Martin, Malcolm Mencer
Naylor, Paul Henry

FLORIDA
Chegini, Nasser
Deats, Edith Potter
Gennaro, Joseph Francis
Howard, Guy Allen
McLean, Mark Philip
Mintz, Daniel Harvey
Neary, Joseph Thomas
Phelps, Christopher Prine
Rao, Krothapalli Ranga
Rosenbloom, Arlan Lee
Scarpace, Philip J
Shapiro, Jeffrey Paul
Shirk, Paul David
Wallace, Robin A

GEORGIA
Barb, C Richard
Brann, Darrell Wayne
Hoffman, William Hubert
Hollowell, Joseph Gurney, Jr
Khan, Iqbal M
Mackin, Robert Brian
Muldoon, Thomas George
Noe, Bryan Dale
Ogle, Thomas Frank
Tyler, Jean Mary

HAWAII
Atkinson, Shannon K C
Lee, Cheng-Sheng

IDAHO
Mead, Rodney A

ILLINOIS
Bombeck, Charles Thomas, III
Cohen, Rochelle Sandra
Fang, Victor Shengkuen
Freinkel, Norbert
Goldman, Allen S
Horwitz, David Larry
Kasprow, Barbara Ann
Katzenellenbogen, Benita Schulman
Kesler, Darrel J
Kulkarni, Bidy
Kyncl, J Jaroslav
Loizzi, Robert Francis
Lorenzen, Janice R
McNulty, John Alexander
Meyers, Cal Yale
Morishima, Akira
Rabuck, David Glenn
Rubenstein, Arthur Harold
Schwartz, Neena Betty
Shambaugh, George E, III
Singh, Sant Parkash
Velardo, Joseph Thomas
Verhage, Harold Glenn
Walter, Robert John
Yelich, Michael Ralph
Zaneveld, Lourens Jan Dirk

INDIANA
Cherbas, Peter Thomas
Frolik, Charles Alan
Grant, Alan Leslie
Hancock, Deana Lori
Henry, David P, II
Hoversland, Roger Carl
Howard, David K
Moore, Ward Wilfred
Seely, James Ervin
Waldman, Barbara Criscuolo
Young, Peter Chun Man

IOWA
Ascoli, Mario
Ford, Stephen Paul
Stewart, Mary E
Trenkle, Allen H

KANSAS
Dey, Sudhansu Kumar
Greenwald, Gilbert Saul
Lavia, Lynn Alan
MacGregor, Ronal Roy
Rawitch, Allen Barry
Smalley, Katherine W
Stevenson, Jeffrey Smith
Voogt, James Leonard

KENTUCKY
Anderson, James Wingo
Miller, Ralph English
Nikitovitch-Winer, Miroslava B
Rao, Chalamalasetty Venkateswara

LOUISIANA
Hansel, William
Saphier, David
Spring, Jeffrey H

MAINE
Gaskins, H Rex
Musgrave, Stanley Dean

MARYLAND
Aldrich, Jeffrey Richard
Alleva, Frederic Remo
Ball, Gregory Francis
Bhathena, Sam Jehangirji
Cutler, Gordon Butler, Jr
Estienne, Mark Joseph
Fradkin, Judith Elaine
Freeman, Colette
Gann, Donald Stuart
Gill, John Russell, Jr
Glowa, John Robert
Grave, Gilman Drew
Gueriguian, John Leo
Hanson, Robert C
Hoeg, Jeffrey Michael
Howe, Juliette Coupain
Loriaux, D Lynn
Manganiello, Vincent Charles
Mills, James Louis
Nakhasi, Hira Lal
Nisula, Bruce Carl
Raina, Ashok K
Rosen, Saul W
Roth, Jesse
Silverman, Robert Eliot
Simons, Samuel Stoney, Jr
Stolz, Walter S
Striker, G E
Vydelingum, Nadarajen Ameerdanaden
Wollman, Seymour Horace
Zimbelman, Robert George
Zirkin, Barry Ronald

MASSACHUSETTS
Arvan, Peter
Callard, Gloria Vincz
Coyne, Mary Downey
Dayal, Yogeshwar
Emanuel, Rodica L
Flatt, Jean-Pierre
Frisch, Rose Epstein
Gawienowski, Anthony Michael
Gustafson, Alvar Walter
Kosasky, Harold Jack
McArthur, Janet W
Neer, Eva Julia
Orme-Johnson, Nanette Roberts
Reppert, Steve Marion
Rosemberg, Eugenia
Rosner, Anthony Leopold
Sacks, David B
Sawin, Clark Timothy
Seaver, Sally S
Snyder, Benjamin Willard
Strumwasser, Felix
Thorn, George W
Villa-Komaroff, Lydia
Villee, Claude Alvin, Jr
Villee, Dorothy Balzer
Widmaier, Eric Paul
Wright, Mary Lou

MICHIGAN
Al Saadi, A Amir
Barney, Christopher Carroll
Carter-Su, Christin
Chavin, Walter
Christensen, A(lbert) Kent
Dunbar, Joseph C
Floyd, John Claiborne, Jr
Friar, Robert Edsel
Horowitz, Samuel Boris
Hsu, Chen-Hsing
Ireland, James
Keyes, Paul Landis
Lew, Gloria Maria
Midgley, A Rees, Jr
Payne, Anita H
Rivera, Evelyn Margaret
Saltiel, Alan Robert
Subramanian, Marappa G
Tucker, Herbert Allen

MINNESOTA
Bauer, Gustav Eric
Brown, David Mitchell
Fallon, Ann Marie
Gilboe, Daniel Pierre
Goetz, Frederick Charles
Ryan, Robert J
Tindall, Donald J
Towle, Howard Colgate
Wilson, Michael John

MISSISSIPPI
Brown, Robert Dale

MISSOURI
Armbrecht, Harvey James
Blaine, Edward Homer
Buonomo, Frances Catherine
Nickols, G Allen
Rifas, Leonard
Tannenbaum, Michael Glen
Tomasi, Thomas Edward

MONTANA
Coe, John Emmons
Ericsson, Ronald James

NEBRASKA
Fawcett, James Davidson
Grotjan, Harvey Edward, Jr

Veomett, George Ector

NEVADA
Potter, Gilbert David

NEW HAMPSHIRE
Sower, Stacia Ann

NEW JERSEY
Bagnell, Carol A
Cascieri, Margaret Anne
Doebber, Thomas Winfield
Goldsmith, Laura Tobi
Gray, Harry Edward
Landau, Matthew Paul
Lenard, John
Mackway-Girardi, Ann Marie
Malamed, Sasha
Meyers, Kenneth Purcell
Pachter, Jonathan Alan
Rabii, Jamshid
Rhoten, William Blocher
Royce, Paul C
Ryzlak, Maria Teresa
Scanes, Colin G
Schlesinger, David H
Sherman, Merry Rubin
Triscari, Joseph
Wang, Bosco Shang

NEW MEXICO
Tombes, Averett S

NEW YORK
Archibald, Reginald MacGregor
Atlas, Steven Alan
Bambara, Robert Anthony
Bardin, Clyde Wayne
Beermann, Donald Harold
Benjamin, William B
Black, Virginia H
Bloch, Eric
Blumenthal, Rosalyn D
Bockman, Richard Steven
Bofinger, Diane P
Bora, Sunder S
Brown, Lawrence S, Jr
Chandran, V Ravi
Chasalow, Fred I
Christian, John Jermyn
Christy, Nicholas Pierson
Dickerman, Herbert W
Dorsey, Thomas Edward
Furlanetto, Richard W
Gapp, David Alger
Gershengorn, Marvin Carl
Ginsberg-Fellner, Fredda Vita
Greif, Roger Louis
Hardy, Matthew Phillip
Henrikson, Katherine Pointer
Hilf, Russell
Hinkle, Patricia M
Houck, David R
Jakway, Jacqueline Sinks
Johnston, Robert E
Kim, Untae
Koide, Samuel Saburo
Kraus, Shirley Ruth
Kream, Jacob
Kristal, Mark Bennett
Laychock, Suzanne Gale
McEwen, Bruce Sherman
Maddaiah, Vaddanahally Thimmaiah
Martin, Kumiko Oizumi
Michael, Sandra Dale
Nisselbaum, Jerome Seymour
Pi-Sunyer, F Xavier
Rathnam, Premila
Rothstein, Howard
Sauer, Leonard A
Schneider, Bruce Solomon
Southren, A Louis
Spitz, Irving Manfred
Strand, Fleur Lillian
Sundaram, Kalyan
Voorhess, Mary Louise

NORTH CAROLINA
Clemmons, David Roberts
Eddy, Edward Mitchell
Eldridge, John Charles
Fail, Patricia A
Hedlund, Laurence William
Korach, Kenneth Steven
Lobaugh, Bruce
Lund, Pauline Kay
Meyer, Ralph A, Jr
O'Steen, Wendall Keith
Sahyoun, Naji Elias
Surwit, Richard Samuel
Teng, Christina Wei-Tien Tu
Toverud, Svein Utheim
Tyrey, Lee
Van Wyk, Judson John

NORTH DAKOTA
Fivizzani, Albert John, Jr
Lukaski, Henry Charles
Sheridan, Mark Alexander

OHIO
Case, Denis Stephen
Fanger, Bradford Otto
Geho, Walter Blair

Handwerger, Stuart
Hiremath, Shivanand T
Horseman, Nelson Douglas
Jegla, Thomas Cyril
Johnston, John O'Neal
Saffran, Murray
Stevenson, J(oseph) Ross
Thornton, Janice Elaine

OKLAHOMA
Maton, Paul Nicholas
Trachewsky, Daniel
Watson, Gary Hunter

OREGON
Brenner, Robert Murray
Debons, Albert Frank
Greer, Monte Arnold
Mason, Robert Thomas
Ramaley, Judith Aitken
Roselli, Charles Eugene
Stormshak, Fredrick

PENNSYLVANIA
Baumrucker, Craig Richard
Bertolini, Donald R
Bodine, Peter Van Nest
Brooks, David Patrick
Finn, Frances M
Lerner, Leonard Joseph
Lien, Eric Louis
Nambi, Ponnal
Rosenblatt, Michael
Santen, Richard J
Stewart, Charles Newby
Vasilatos-Younken, Regina
Wickersham, Edward Walker

RHODE ISLAND
Flanagan, Thomas Raymond
Jackson, Ivor Michael David
Oxenkrug, Gregory Faiva

SOUTH CAROLINA
Bates, G William

TENNESSEE
Anderson, Ted L
Chen, Thomas Tien
Coulson, Patricia Bunker
Danzo, Benjamin Joseph
Heimberg, Murray
Herting, David Clair
Hoffman, Loren Harold
Kitabchi, Abbas E
Sander, Linda Dian

TEXAS
Bertrand, Helen Anne
Bryan, George Thomas
Burghardt, Robert Casey
Carson, Daniel Douglas
Chubb, Curtis Evans
Cooper, Cary Wayne
Dahm, Karl Heinz
George, Fredrick William
Guo, Yan-Shi
Hug, Verena
Kutteh, William Hanna
Lakoski, Joan Marie
Linner, John Gunnar
Norman, Reid Lynn
Ordonez, Nelson Gonzalo
Rosenfeld, Charles Richard
Roy, Arun K
Schonbrunn, Agnes
Smith, Edward Russell
Smith, Eric Morgan
Stubbs, Donald William
Taylor, D(orothy) Jane
Thompson, Edward Ivins Bradbridge

UTAH
Rasmussen, Kathleen Goertz
Wurst, Gloria Zettle

VIRGINIA
Adler, Robert Alan
Brown, Barry Lee
Edwards, Leslie Erroll
Liberti, Joseph Pollara
McCarty, Richard Charles
McNabb, F M Anne
Turner, Terry Tomo

WASHINGTON
Baskin, Denis George
Beck, Thomas W
Carlberg, Karen Ann
Fujimoto, Wilfred Y
Hawkins, Richard L
Horn, Diane
Kenney, Nancy Jane
Porte, Daniel, Jr
Riddiford, Lynn Moorhead
Smith, Elizabeth Knapp
Steiner, Robert Alan
Truman, James William
Wood, Francis C, Jr
Woods, Stephen Charles

WEST VIRGINIA
Burns, John Thomas
Butcher, Roy Lovell

Inskeep, Emmett Keith

WISCONSIN
Drucker, William D
Goldberger, Amy
Haas, Michael John
Hager, Steven Ralph
Piacsek, Bela Emery
Spieler, Richard Earl
Takahashi, Lorey K
Terasawa, EI
Warner, Eldon Dezelle

WYOMING
Ji, Inhae

PUERTO RICO
Banerjee, Dipak Kumar

ALBERTA
Mears, Gerald John
Peter, Richard Ector
Rorstad, Otto Peder

BRITISH COLUMBIA
Donaldson, Edward Mossop
Emerman, Joanne Tannis
Jones, David Robert
Kraintz, Leon
Vincent, Steven Robert

MANITOBA
Angel, Aubie
Eales, John Geoffrey
Faiman, Charles

NOVA SCOTIA
Savard, Francis Gerald Kenneth

ONTARIO
Ainsworth, Louis
Armstrong, David Thomas
Bhavnani, Bhagu R
Bols, Niels Christian
Etches, Robert J
Graham, Richard Charles Burwell
Hegele, Robert A
Jellinck, Peter Harry
Kutcher, Stanley Paul
Leatherland, John F
Little, James Alexander
Mak, William Wai-Nam
Malkin, Aaron
Reid, Robert Leslie
Steel, Colin Geoffrey Hendry
Tobe, Stephen Solomon
Vafopoulo, Xanthe
Volpe, Robert
Weick, Richard Fred
Wiebe, John Peter

QUEBEC
Adamkiewicz, Vincent Witold
Antakly, Tony
Bateman, Andrew
Benard, Bernard
Bendayan, Moise
Birmingham, Marion Krantz
Chafouleas, James G
Duguid, William Paris
Dussault, Jean H
Engel, Charles Robert
Farmer, Chantal
Gagnon, Claude
Garcia, Raul
Manjunath, Puttaswamy
Mulay, Shree
Peronnet, Franc04ois R R
Solomon, Samuel
Tannenbaum, Gloria Shaffer

OTHER COUNTRIES
Amsterdam, Abraham
Armato, Ubaldo
Barouki, Robert
Buttermer, William Ashley
Eik-Nes, Kristen Borger
Farid, Nadir R
Garcia-Sainz, J Adolfo
Gustafsson, Jan-Ake
Hanukoglu, Israel
Henderson, David Andrew
Ichikawa, Shuichi
Ishikawa, Hiroshi
Kaye, Alvin Maurice
Kuo, Ching-Ming
Markert, Claus O
Piña, Enrique
Racotta, Radu Gheorghe
Sabry, Ismail
Shimaoka, Katsutaro
Shore, Laurence Stuart
Stoppani, Andres Oscar Manuel
Vane, John Robert
Varga, Janos M

Entomology

ALABAMA
Adams, Curtis H
Appel, Arthur Gary
Barnes, William Wayne
Berger, Robert S

Blake, George Henry, Jr
Brewer, Jesse Wayne
Brown, Jack Stanley
Chambliss, Oyette Lavaughn
Clark, Wayne Elden
Cross, Earle Albright, Jr
Gaylor, Michael James
Gilliland, Floyd Ray, Jr
Hays, Kirby Lee
Johnson, William E, Jr
Kouskolekas, Costas Alexander
Mack, Timothy Patrick
Mullen, Gary Richard
Sanford, L G
Scheiring, Joseph Frank
Snoddy, Edward L
Tate, Laurence Gray
Tennessen, Kenneth Joseph
Williams, Michael Ledell

ALASKA
Steffan, Wallace Allan
Werner, Richard Allen

ARIZONA
Bariola, Louis Anthony
Barker, Roy Jean
Bartlett, Alan C
Bernays, Elizabeth Anna
Butler, George Daniel, Jr
Cazier, Mont Adelbert
Collins, Richard Cornelius
Crosswhite, Carol D
Doane, Charles Chesley
Erickson, Eric Herman, Jr
Faeth, Stanley Herman
Flint, Hollis Mitchell
Fritz, Roy Fredolin
George, Boyd Winston
Gerhardt, Paul Donald
Graham, Harry Morgan
Hadley, Neil F
Hagedorn, Henry Howard
Henneberry, Thomas James
Hewitt, George Berlyn
Hildebrand, John Grant, III
Huber, Roger Thomas
Johnson, Clarence Daniel
Kauffeld, Norbert M
Langston, Dave Thomas
Law, John Harold
Loper, Gerald Milton
Minch, Edwin Wilton
Moore, Leon
Muma, Martin Hammond
Nutting, William Leroy
Pinter, Paul James, Jr
Price, Peter Wilfrid
Schmidt, Justin Orvel
Smith, Robert Lloyd
Spangler, Hayward Gosse
Stahnke, Herbert Ludwig
Terry, Lucy Irene
Tuttle, Donald Monroe
Ware, George Whitaker, Jr
Watson, Theo Franklin
Werner, Floyd Gerald
Wick, James Roy
Zimmerman, James Roscoe

ARKANSAS
Adams, Randall Henry
Allen, Robert Thomas
Barton, Harvey Eugene
Burleigh, Joseph Gaynor
Clower, Dan Fredric
Couser, Raymond Dowell
Lancaster, Jessie Leonard, Jr
Musick, Gerald Joe
Phillips, Jacob Robinson
Steelman, Carrol Dayton
Steinkraus, Donald Curtiss
Stephen, Frederick Malcolm, Jr
Thompson, Lynne Charles
Watson, Robert Lee
Yearian, William C
Young, Seth Yarbrough, III

CALIFORNIA
Adams, Phillip A
Allen, William Westhead
Allison, William Earl
Anderson, John Richard
Andres, Lloyd A
Arnaud, Paul Henri, Jr
Bacon, Oscar Gray
Baker, Frederick Charles
Barnes, Martin McRae
Barr, Allan Ralph
Bellinger, Peter F
Benson, Robert Leland
Bishop, Jack Lynn
Blanc, Francis Louis
Bohart, Richard Mitchell
Bohnsack, Kurt K
Brown, Leland Ralph
Calavan, Edmond Clair
Caltagirone, Leopoldo Enrique
Campbell, Bruce Carleton
Carman, Glenn Elwin
Case, Ted Joseph
Chang, Ming-Houng (Albert)
Cheng, Lanna
Christensen, Howard Anthony

Entomology (cont)

Conway, John Bell
Corey, Robert Arden
Cothran, Warren Roderic
Cox, H C
Cunningham, Virgil Dwayne
Curtis, Charles Elliott
Dahlsten, Donald L
Daly, Howell Vann
DaMassa, Al John
Darst, Philip High
Davis, Edward E
Dingle, Richard Douglas Hugh
Doutt, Richard Leroy
Dowell, Robert Vernon
Duffey, Sean Stephen
Dunbar, Dennis Monroe
Edwards, J Gordon
Eldridge, Bruce Frederick
Eriksen, Clyde Hedman
Falcon, Louis A
Ferguson, William E
Fisher, Theodore William
Force, Don Clement
Frankie, Gordon William
Fristrom, Dianne
Furman, Deane Philip
Garcia, Richard
Garth, John Shrader
Gary, Norman Erwin
Gaston, Lyle Kenneth
Georghiou, George Paul
Gilmore, James Eugene
Goeden, Richard Dean
Golder, Thomas Keith
Gordh, Gordon
Gould, Douglas Jay
Grigarick, Albert Anthony, Jr
Hackwell, Glenn Alfred
Hagen, Kenneth Sverre
Hammock, Bruce Dupree
Hare, John Daniel, III
Haverty, Michael Irving
Hespenheide, Henry August, III
Hogue, Charles Leonard
Hopkins, Lemac
Hoy, James Benjamin
Huffaker, Carl Barton
Iltis, Wilfred Gregor
Jameson, Everett Williams, Jr
Jefferson, Roland Newton
Jeppson, Lee Ralph
Jorgenson, Edsel Carpenter
Judson, Charles LeRoy
Kadner, Carl George
Kaloostian, George H
Kavanaugh, David Henry
Kaya, Harry Kazuyoshi
Kimsey, Lynn Siri
Kissinger, David George
Kistner, David Harold
Koehler, Carlton Smith
Laidlaw, Harry Hyde, Jr
Lane, Robert Sidney
Lauck, David R
Leahy, Sister Mary Gerald
Legner, E Fred
Leigh, Thomas Francis
Lindgren, David Leonard
Linsley, Earle Gorton
Loeblich, Karen Elizabeth
Loomis, Richard Biggar
Loudon, Catherine
McClelland, George Anderson Hugh
McCluskey, Elwood Sturges
McDonald, Paul Thomas
McMurtry, James A
March, Ralph Burton
Marrone, Pamela Gail
Maxwell, Kenneth Eugene
Michelbacher, Abraham E
Millar, Jocelyn Grenville
Miller, Thomas Albert
Miura, Takeshi
Monroe, Ronald Eugene
Morse, Joseph Grant
Mussen, Eric Carnes
Nelson, Bernard Clinton
Nelson, Richard D(ouglas)
Nickel, Phillip Arnold
Nowell, Wesley Raymond
Oatman, Earl R
Oldfield, George Newton
Page, Robert Eugene, Jr
Partida, Gregory John, Jr
Pieper, Gustav Rene
Pinto, John Darwin
Pipa, Rudolph Louis
Powell, Jerry Alan
Purcell, Alexander Holmes, III
Rammer, Irwyn Alden
Reisen, William Kenneth
Reynolds, Harold Truman
Risch, Stephen John
Robertson, Jacqueline Lee
Rogoff, William Milton
Rust, Michael Keith
Ryckman, Raymond Edward
Schaefer, Charles Herbert
Schlinger, Evert Irving
Scott, Matthew P
Scudder, Harvey Israel
Sevacherian, Vahram

Shapiro, Arthur Maurice
Simkover, Harold George
Sleeper, Elbert Launee
Smith, Ray Fred
Smith, Richard Harrison
Soderstrom, Edwin Loren
Staal, Gerardus Benardus
Stelzer, Lorin Roy
Stern, Vernon Mark
Strong, Donald Raymond, Jr
Strong, Rudolph Greer
Summers, Charles Geddes
Sylvester, Edward Sanford
Tanada, Yoshinori
Terwedow, Henry Albert, Jr
Thompson, Robert Kruger
Thorp, Robbin Walker
Truxal, Fred Stone
Vail, Patrick Virgil
Wade, William Howard
Wasbauer, Marius Sheridan
Washino, Robert K
Waters, William E
Weber, Barbara C
Wells, Patrick Harrington
Williams, Carroll Burns, Jr
Wood, David Lee
Zavortink, Thomas James

COLORADO

Barnes, Ralph Craig
Beal, Richard Sidney, Jr
Becker, George Charles
Bowers, Marion Deane
Breed, Michael Dallam
Drummond, Boyce Alexander, III
Evans, Howard Ensign
Francy, David Bruce
Gholson, Larry Estie
Gregg, Robert Edmond
Gubler, Duane J
Harrison, Monty DeVerl
Harriss, Thomas T
Johnsen, Richard Emanuel
Lanham, Urless Norton
Linam, Jay H
Owen, William Bert
Polhemus, John Thomas
Roberts, P Elaine
Simpson, Robert Gene
Stevens, Robert E
Sublette, James Edward
Yun, Young Mok

CONNECTICUT

Aitken, Thomas Henry Gardiner
Anderson, John Fredric
Andreadis, Theodore George
Berry, Spencer Julian
Block, Bartley C
Connor, Lawrence John
Davis, Norman Thomas
DeCoursey, Russell Myles
Henry, Charles Stuart
Krinsky, William Lewis
Levine, Harvey Robert
McClure, Mark Stephen
Magnarelli, Louis Anthony
Maier, Chris Thomas
Miller, Russell Bensley
Nelson, Vernon A
Raghuvir, Nuggehalli Narayana
Remington, Charles Lee
Rettenmeyer, Carl William
Savos, Milton George
Schaefer, Carl W, II
Slater, James Alexander
Stephens, George Robert
Streams, Frederick Arthur
Wallis, Robert Charles
Wallner, William E
Weseloh, Ronald Mack
Wyman, Robert J

DELAWARE

Carnahan, James Elliot
Caron, Dewey Maurice
Cornell, Howard Vernon
Day, William H
Hough-Goldstein, Judith Anne
Kelsey, Lewis Preston
Koeppe, Mary Kolean
Krone, Lawrence James
Mason, Charles Eugene
Meade, Alston Bancroft
Miller, Douglas Charles
Weber, Richard Gerald
Wood, Thomas Kenneth

DISTRICT OF COLUMBIA

Allen, George E
Anderson, Donald Morgan
Baerg, David Carl
Barrows, Edward Myron
Brickey, Paris Manaford
Bridges, John Robert
Burns, John McLauren
Callahan, James Thomas
Cate, James Richard, Jr
Chapman, George Bunker
Chaput, Raymond Leo
Clarke, John Frederick Gates
Davis, Donald Ray
Flint, Oliver Simeon, Jr

Gagne, Raymond J
Goodwin, James Thomas
Gorham, John Richard
Grissell, Edward Eric Fowler
Harbach, Ralph Edward
Hodges, Ronald William
Jackson, Robert Dewey
Krombein, Karl Von Vorse
Lyon, Robert Lyndon
Marsh, Paul Malcolm
Mathis, Wayne Neilsen
Nickle, David Allan
Poole, Robert Wayne
Reed, William Doyle
Robbins, Robert Kanner
Robinson, Harold Ernest
Smith, David Rollins
Sollers-Riedel, Helen
Spangler, Paul Junior
Tompkins, George Jonathan
Ward, Ronald Anthony
White, Richard Earl

FLORIDA

Adlerz, Warren Clifford
Agee, Herndon Royce
Allen, Jon Charles
Arnett, Ross Harold, Jr
Baranowski, Richard Matthew
Barnard, Donald Roy
Bick, George Herman
Bidlingmayer, William Lester
Brenner, Richard Joseph
Britt, N Wilson
Brooks, Robert Franklin
Bullock, Robert Crossley
Butler, Jerry Frank
Calkins, Carrol Otto
Callahan, Philip Serna
Castner, James Lee
Chambers, Derrell Lynn
Clements, Burie Webster
CmejLa, Howard Edward
Couch, Terry Lee
Cromroy, Harvey Leonard
Dame, David Allan
Davis, Dean Frederick
Denmark, Harold Anderson
Downey, John Charles
Dunkle, Sidney Warren
Dutary, Bedsy Elsira
Emerson, Kary Cadmus
Fairchild, Graham Bell
Fatzinger, Carl Warren
Fisk, Frank Wilbur
Flowers, Ralph Wills
Frank, John Howard
Giblin-Davis, Robin Michael
Glancey, Burnett Michael
Goss, Gary Jack
Gupta, Virendra K
Haag, Kim Hartzell
Habeck, Dale Herbert
Haines, Robert Gordon
Hall, David Goodsell, IV
Hall, Donald William
Hamon, Avas Burdette
Heppner, John Bernhard
Hetrick, Lawrence Andrew
Himel, Chester Mora
Howard, Forrest William
Kerr, Stratton H
Kitzmiller, James Blaine
Koehler, Philip Gene
Kuitert, Louis Cornelius
Lawrence, Pauline Olive
Levy, Richard
Lofgren, Clifford Swanson
Lounibos, Leon Philip
Lowe, Ronald Edsel
McCoy, Clayton William
McCoy, Earl Donald
McGray, Robert James
McLaughlin, John Ross
Mankin, Richard Wendell
Matheny, Ellis Leroy, Jr
Mayer, Marion Sidney
Mayer, Richard Thomas
Mead, Frank Waldreth
Mendez, Eustorgio
Menzer, Robert Everett
Mitchell, Everett Royal
Mount, Gary Arthur
Mulrennan, John Andrew, Jr
Nation, James Lamar
Nayar, Jai Krishen
Nigg, Herbert Nicholas
Norval, Richard Andrew
Oberlander, Herbert
O'Meara, George Francis
Osborne, Lance Smith
Patterson, Richard Sheldon
Peters, William Lee
Pletsch, Donald James
Porter, Sanford Dee
Price, James Felix
Rathburn, Carlisle Baxter, Jr
Reiskind, Jonathan
Roberts, Donald Ray
Roberts, Richard Harris
Roubik, David Ward
Sanford, Malcolm Thomas
Schuster, David J
Seawright, Jack Arlyn

Shankland, Daniel Leslie
Shapiro, Jeffrey Paul
Silhacek, Donald Le Roy
Simonet, Donald Edward
Simons, John Norton
Smith, Carroll N
Smith, William Ward
Smittle, Burrell Joe
Sosa, Omelio, Jr
Sprenkel, Richard Keiser
Stange, Lionel Alvin
Strohecker, Henry Frederick
Terranova, Andrew Charles
Tingle, Frederic Carley
Townes, Henry Keith, Jr
Tsai, James Hsi-Cho
Tschinkel, Walter Rheinhardt
Vickers, David Hyle
Waddill, Van Hulen
Waites, Robert Ellsworth
Walker, Thomas Jefferson
Weber, Neal Albert
Weems, Howard Vincent, Jr
Weidhaas, Donald E
West-Eberhard, Mary J
Wheeler, George Carlos
Wheeler, Jeanette Norris
Whitcomb, Willard Hall
White, Albert Cornelius
Wilkinson, Robert Cleveland, Jr
Williams, David Francis
Williams, Norris Hagan
Wolda, Hindrik
Woodruff, Robert Eugene
Workman, Ralph Burns
Young, David Grier
Yu, Simon Shyi-Jian
Yunker, Conrad Erhardt
Zimmerman, John Harvey

GEORGIA

Allee, Marshall Craig
Amerson, Grady Malcolm
Arbogast, Richard Terrance
Atyeo, Warren Thomas
Bass, Max H(erman)
Blum, Murray Sheldon
Brady, Ullman Eugene, Jr
Brower, John Harold
Caldwell, Sloan Daniel
Chalfant, Richard Bruce
Chamberlain, Roy William
Clark, James Samuel
Collins, William Erle
Davis, Robert
Dutcher, James Dwight
Espelie, Karl Edward
Evans, Burton Robert
Franklin, Rudolph Thomas
French, Frank Elwood, Jr
Gilbert, Edward E
Gitaitis, Ronald David
Harris, Emmett Dewitt, Jr
Harrison, James Ostelle
Hermann, Henry Remley, Jr
Hibbs, Edwin Thompson
Highland, Henry Arthur
Hunter, Preston Eugene
Johnson, Donald Ross
Kappus, Karl Daniel
Keirans, James Edward
Lea, Arden Otterbein
Lewis, Wallace Joe
Lum, Patrick Tung Moon
Lynch, Robert Earl
McPherson, Robert Merrill
Matthews, Robert Wendell
Neel, William Wallace
Newhouse, Verne Frederic
Nord, John C
Nyczepir, Andrew Peter
Oliver, James Henry, Jr
Payne, Jerry Allen
Pratt, Harry Davis
Rice, Paul LaVerne
Rogers, Charlie Ellic
Shure, Donald Joseph
Smith, Eric Howard
Sparks, Alton Neal
Sudia, William Daniel
Taylor, Robert Tieche
Throne, James Edward
Tietjen, William Leighton
Todd, James Wyatt
Vincent, Leonard Stuart
Wallace, James Bruce
Weathersby, Augustus Burns
Whitesell, James Judd
Willis, Judith Horwitz
Wiseman, Billy Ray

HAWAII

Andersen, Dean Martin
Beardsley, John Wyman, Jr
Chang, Franklin
Duckworth, Walter Donald
Eschle, James Lee
Fujii, Jack K
Haramoto, Frank H
Hardy, D Elmo
Harris, Ernest James
Howarth, Francis Gard
Little, Harold Franklin
Medler, John Thomas

Mitchell, Wallace Clark
Namba, Ryoji
Ota, Asher Kenhachiro
Sherman, Martin
Simon, Christine Mae
Tamashiro, Minoru
Vargas, Roger I

IDAHO
Anderson, Robert Curtis
Baird, Craig Riska
Barr, William Frederick
Brusven, Merlyn Ardel
Clark, William Hilton
Gittins, Arthur Richard
Hostetter, Donald Lee
Klowden, Marc Jeffrey
McCaffrey, Joseph Peter
O'Keeffe, Lawrence Eugene
Osgood, Charles Edgar
Schell, Stewart Claude
Schenk, John Albright
Stark, Ronald William
Stiller, David
Stoltz, Robert Lewis
Waters, Norman Dale

ILLINOIS
Appleby, James E
Beland, Gary LaVern
Berenbaum, May Roberta
Bhatti, Rashid
Bouseman, John Keith
Chandran, Satish Raman
Chang, Kwang-Poo
Cibulsky, Robert John
Estep, Charles Blackburn
Flood, Brian Robert
French, Allen Lee
Friedman, Stanley
Funk, Richard Cullen
Gibbs, Daniel
Godfrey, George Lawrence
Goodrich, Michael Alan
Green, Thomas L
Greenberg, Bernard
Hamilton, Robert W
Horsfall, William Robert
Irwin, Michael Edward
Jones, Carl Joseph
Kethley, John Bryan
Khan, Mohammed Abdul Quddus
Khattab, Ghazi M A
LaBerge, Wallace E
Larsen, Joseph Reuben
Luckmann, William Henry
MacLeod, Ellis Gilmore
McPherson, John Edwin
Maddox, Joseph Vernard
Malek, Richard Barry
Melin, Brian Edward
Metcalf, Robert Lee
Meyer, Ronald Harmon
Mockford, Edward Lee
Moore, Stevenson, III
Nappi, Anthony Joseph
Nardi, James Benjamin
Nelson, Harry Gladstone
Riegel, Garland Tavner
Robertson, Hugh Mereth
Sedman, Yale S
Selander, Richard Brent
Sternburg, James Gordon
Thompson, Vinton Newbold
Toliver, Michael Edward
Unzicker, John Duane
Vandehey, Robert C
Wagner, John Alexander
Wagner, Vaughn Edwin
Waldbauer, Gilbert Peter
Watson, David Livingston
Webb, Donald Wayne
Wenzel, Rupert Leon
Willey, Robert Bruce
Wittig, Gertraude Christa

INDIANA
Broersma, Delmar B
Burkholder, Timothy Jay
Chio, E(ddie) Hang
Corrigan, John Joseph
Craig, George Brownlee, Jr
Delfin, Eliseo Dais
Denner, Melvin Walter
Dunn, Peter Edward
Edwards, Charles Richard
Esch, Harald Erich
Flanders, Robert Vern
Fuchs, Morton S
Fulton, Winston Cordell
Gallun, Robert Louis
Goonewardene, Hilary Felix
Gould, George Edwin
Grimstad, Paul Robert
Hellenthal, Ronald Allen
Hickey, William August
Hoefer, Raymond H
Hunt, Linda Margaret
Jantz, O K
Johnson, Michael David
Krekeler, Carl Herman
McCafferty, William Patrick
McGowan, Michael James
O'Neal, Lyman Henry

O'Neil, Robert James
Ortman, Eldon Emil
Osmun, John Vincent
Paschke, John Donald
Peterson, Lance George
Rai, Karamjit Singh
Russo, Raymond Joseph
Schuder, Donald Lloyd
Turpin, Frank Thomas
Ward, Gertrude Luckhardt
Wellso, Stanley Gordon
Wilson, Mark Curtis
Zimmack, Harold Lincoln

IOWA
Coats, Joel Robert
Dahm, Paul Adolph
Dicke, Ferdinand Frederick
Eiben, Galen J
Guthrie, Wilbur Dean
Hart, Elwood Roy
Krafsur, Elliot Scoville
Lewis, Robert Earl
Mertins, James Walter
Mutchmor, John A
Pedigo, Larry Preston
Rowley, Wayne A
Showers, William Broze, Jr
Stay, Barbara
Stockdale, Harold James
Stoltzfus, William Bryan
Wilson, Nixon Albert
Wilson, Richard Lee

KANSAS
Ashe, James S
Ashlock, Peter Dunning
Beer, Robert Edward
Blocker, Henry Derrick
Byers, George William
Dinkins, Reed Leon
Eddy, Thomas A
Elzinga, Richard John
Greene, Gerald L
Harvey, Thomas Larkin
Hatchett, Jimmy Howell
Hopkins, Theodore Louis
Kadoum, Ahmed Mohamed
Kramer, Karl Joseph
Lungstrom, Leon
Martin, Terry Joe
Michener, Charles Duncan
Mills, Robert Barney
Owen, Bernard Lawton
Peters, Leroy Lynn
Stockhammer, Karl Adolf
Thompson, Hugh Erwin
Von Rumker, Rosmarie
Walker, Neil Allan
Whitney, Wendell Keith
Wilde, Gerald Eldon
Woodruff, Laurence Clark

KENTUCKY
Christensen, Christian Martin
Covell, Charles VanOrden, Jr
Dahlman, Douglas Lee
Ferrell, Blaine Richard
Freytag, Paul Harold
Gregory, Wesley Wright, Jr
Knapp, Fred William
Nordin, Gerald LeRoy
Pass, Bobby Clifton
Raney, Harley Gene
Rodriguez, Juan Guadalupe
Scheibner, Rudolph A
White, David Sanford
Yeargan, Kenneth Vernon

LOUISIANA
Boudreaux, Henry Bruce
Chapin, Joan Beggs
Dakin, Matt Eitel
Fuxa, James Roderick
Graves, Jerry Brook
Hammond, Abner M, Jr
Harbo, John Russell
Hensley, Sess D
Khalaf, Kamel T
Kinn, Donald Norman
Lambremont, Edward Nelson
Long, William Henry
Miller, Albert
Nasci, Roger Stanley
Newsom, Leo Dale
Oliver, Abe D, Jr
Reagan, Thomas Eugene
Rinderer, Thomas Earl
Rolston, Lawrence H
Roussel, John S
Scott, Harold George
Spring, Jeffrey H
Stewart, T Bonner
Thurman-Swartzwelder, Ernestine H
Trapido, Harold
Tucker, Kenneth Wilburn
Vidrine, Malcolm Francis
Walker, John Robert
Woodring, J Porter
Yadav, Raghunath P

MAINE
Burbutis, Paul Philip
Dimond, John Barnet

Forsythe, Howard Yost, Jr
Houseweart, Mark Wayne
Jennings, Daniel Thomas
Knight, Fred Barrows
McDaniel, Ivan Noel
Raisbeck, Barbara
Sommerman, Kathryn Martha
Storch, Richard Harry
Wave, Herbert Edwin
Yonuschot, Gene R

MARYLAND
Adams, Jean Ruth
Adler, Victor Eugene
Aldrich, Jeffrey Richard
Armstrong, Earlene
Azad, Abdu F(arhang)
Bailey, Charles Lavon
Baker, Edward William
Barnett, Douglas Eldon
Barry, Cornelius
Batra, Suzanne Wellington Tubby
Bickley, William Elbert
Borgia, Gerald
Bram, Ralph A
Burton, George Joseph
Cantelo, William Wesley
Cleveland, Merrill L
Conner, George William
Coulson, Jack Richard
Crystal, Maxwell Melvin
Dasch, Gregory Alan
Davidson, John Angus
Feldlaufer, Mark Francis
Futcher, Anthony Graham
Gerberg, Eugene Jordan
Gonzales, Ciriaco Q
Graham, Charles Lee
Gwadz, Robert Walter
Hajjar, Nicolas Philippe
Hansford, Richard Geoffrey
Hanson, Frank Edwin
Harding, Wallace Charles, Jr
Harrison, Floyd Perry
Hassett, Charles Clifford
Herbert, Elton Warren, Jr
Inscoe, May Nilson
Jones, Tappey Hughes
Joseph, Stanley Robert
Kingsolver, John Mark
Klassen, Waldemar
Krestensen, Elroy R
Kutz, Frederick Winfield
Lamp, William Owen
Mallis, Arnold
Marks, Edwin Potter
Menn, Julius Joel
Messersmith, Donald Howard
Miller, Douglass Ross
Morgan, Neal O
Neal, John William, Jr
Nelson, James Harold
Nelson, Judd Owen
Nickle, William R
Parrish, Dale Wayne
Phillips, William George
Pilitt, Patricia Ann
Raina, Ashok K
Ramsay, Maynard Jack
Reddy, Gunda
Redfern, Robert Earl
Ridgway, Richard L
Riley, Robert C
Rozeboom, Lloyd Eugene
Schwartz, Paul Henry Jr
Shapiro, Martin
Sherman, Kenneth Eliot
Shimanuki, Hachiro
Sholdt, Lester Lance
Sonawane, Babasaheb R
Steinhauer, Allen Laurence
Stoetzel, Manya Brooke
Svoboda, James Arvid
Todd, Robin Grenville
Traub, Robert
Trosper, James Hamilton
Trpis, Milan
Vaughn, James L
Weirich, Gunter Friedrich
Wise, David Haynes
Wood, Francis Eugene
Yoder, Wayne Alva

MASSACHUSETTS
Barke, Harvey Ellis
Borgatti, Alfred Lawrence
Bowdan, Elizabeth Segal
Carde, Ring Richard Tomlinson
Carpenter, Frank Morton
Cohen, Samuel H
Dalgleish, Robert Campbell
Edman, John David
Ferro, David Newton
Furth, David George
Helgesen, Robert Gordon
Kafatos, Fotis C
Kelts, Larry Jim
Kunkel, Joseph George
Lee, Siu-Lam
Leonard, David E
Levi, Herbert Walter
Nutting, William Brown
Peters, Thomas Michael

Smith, Grahame J C
Spielman, Andrew
Stoffolano, John George, Jr
Traniello, James Francis Anthony
Wall, William James
Weaver, Nevin
Williams, Carroll Milton
Yin, Chih-Ming

MICHIGAN
Bath, James Edmond
Bird, George W
Bland, Roger Gladwin
Bratt, Albertus Dirk
Breznak, John Allen
Bush, Guy L
Cantrall, Irving James
Cook, David Russell
Cowan, David Prime
Douglas, Matthew M
Evans, David Arnold
Fischer, Roland Lee
Fleming, Richard Cornwell
Gamboa, George John
Grafius, Edward John
Guyer, Gordon Earl
Haynes, Dean L
Hoffman, Julius R
Hollingworth, Robert Michael
Hoopingarner, Roger A
Howe, Robert George
Howitt, Angus Joseph
Husband, Robert W
Kawooya, John Kasajja
McCall, Robert B
Meier, Peter Gustav
Merritt, Richard William
Moore, Thomas Edwin
Newson, Harold Don
Racke, Kenneth David
Rizzo, Donald Charles
Ruppel, Robert Frank
Scarborough, Charles Spurgeon
Scheel, Carl Alfred
Shappirio, David Gordon
Simmons, Gary Adair
Sloan, Norman F
Stehr, Frederick William
Toczek, Donald Richard
Wilson, Louis Frederick
Witter, John Allen

MINNESOTA
Andow, David A
Baker, Griffin Jonathan
Bartel, Monroe H
Batzer, Harold Otto
Borchers, Harold Allison
Chiang, Huai C
Christian, Paul Jackson
Cook, Edwin Francis
Cutkomp, Laurence Kremer
Fallon, Ann Marie
Fremling, Calvin R
Gundersen, Ralph Wilhelm
Hamrum, Charles Lowell
Harein, Phillip Keith
Jones, Richard Lamar
Krogstad, Blanchard Orlando
Kulman, Herbert Marvin
Kurtti, Timothy John
Lewis, Standley Eugene
Lorimer, Nancy L
Mauston, Glenn Warren
Miller, William Eldon
Moore, Glenn D
Moore, Joseph B
Noetzel, David Martin
Ordway, Ellen
Peck, John Hubert
Preiss, Frederick John
Price, Roger DeForrest
Radcliffe, Edward B
Ragsdale, David Willard
Rao, Gundu Hirisave Rama
Richards, Albert Glenn
Samuelson, Charles R
Sauer, Richard John
Wallace, Marion Brooks

MISSISSIPPI
Bailey, Jack Clinton
Collison, Clarence H
Combs, Robert L, Jr
Cross, William Henley
Davich, Theodore Bert
Dorough, H Wyman
Fye, Robert Eaton
Hardee, Dicky Dan
Hays, Donald Brooks
Helms, Thomas Joseph
Hendricks, Donovan Edward
Kincade, Robert Tyrus
Lago, Paul Keith
Lloyd, Edwin Phillips
McKibben, Gerald Hopkins
Mather, Bryant
Mitchell, Earl Bruce
Nebeker, Thomas Evan
Norment, Beverly Ray
Ouzts, Johnny Drew
Parencia, Charles R
Parrott, William Lamar
Pfrimmer, Theodore Roscoe

Kim, Ke Chung
Lawrence, Vinnedge Moore
McKinstry, Donald Michael
McNair, Dennis M
Manley, Thomas Roy
Marks, Louis Sheppard
Masteller, Edwin C
Montgomery, Ronald Eugene
Moss, William Wayne
Mumma, Ralph O
Norrbom, Allen Lee
Ode, Philip E
Pearson, David Leander
Pitts, Charles W
Presser, Bruce Douglas
Rao, Balakrishna Raghavendra
Roback, Selwyn
Roberts, Howard Radclyffe
Rutschky, Charles William
Schroeder, Mark Edwin
Schultz, Jack C
Sheldon, Joseph Kenneth
Simpson, Geddes Wilson
Smilowitz, Zane
Snetsinger, Robert J
Soboczenski, Edward John
Staetz, Charles Alan
Streu, Herbert Thomas
Swisher, Ely Martin
Wheeler, Alfred George, Jr
Williamson, Craig Edward
Wolfersberger, Michael Gregg
Wooldridge, David Paul
Yendol, William G
Yoho, Timothy Price
Yurkiewicz, William J

RHODE ISLAND
Hyland, Kerwin Ellsworth, Jr
Kerr, Theodore William, Jr
LeBrun, Roger Arthur
Reichart, Charles Valerian
Waage, Jonathan King

SOUTH CAROLINA
Adkins, Theodore Roosevelt, Jr
Alverson, David Roy
Creighton, Charlie Scattergood
Davis, Luckett Vanderford
DuRant, John Alexander, III
Elsey, Kent D
Fery, Richard Lee
Forsythe, Dennis Martin
Hays, Sidney Brooks
Henson, Joseph Lawrence
Johnson, Albert Wayne
King, Edwin Wallace
Manley, Donald Gene
Moore, Raymond F, Jr
Noblet, Raymond
Schalk, James Maximillian
Shepard, Buford Merle
Skelton, Thomas Eugene
Spooner, John D
Wimer, Larry Thomas

SOUTH DAKOTA
Kantack, Benjamin H
Kieckhefer, Robert William
Kirk, Vernon Miles
McDaniel, Burruss, Jr
Stoner, Warren Norton
Sutter, Gerald Rodney
Walgenbach, David D
Walstrom, Robert John

TENNESSEE
Bancroft, Harold Ramsey
Brown, Carl Dee
Caron, Richard Edward
Copeland, Thompson Preston
Gerhardt, Reid Richard
Haines, Thomas Walton
Hull, George, Jr
Kring, James Burton
Lambdin, Paris Lee
Lentz, Gary Lynn
Liles, James Neil
McGhee, Charles Robert
Nelson, Charles Henry
Patrick, Charles Russell
Reed, Horace Beecher
Shipp, Oliver Elmo
Smith, Omar Ewing, Jr
Southards, Carroll J
Tarpley, Wallace Armell
Tolbert, Virginia Rose
Wang, Richard Hsu-Shien
Watson, Annetta Paule
Webb, J(ohn) Warren
White, Jane Vicknair

TEXAS
Adkisson, Perry Lee
Applegate, Richard Lee
Bailey, Leo L
Bay, Darrell Edward
Benedict, John Howard, Jr
Bhaskaran, Govindan
Bowling, Clarence C
Brewer, Beverly Sparks
Brewer, Franklin Douglas
Brook, Ted Stephens
Brooks, Derl

Bull, Don Lee
Cogburn, Robert Ray
Collins, Anita Marguerite
Cook, Benjamin Jacob
Coulson, Robert N
Curtin, Thomas J
Dean, Herbert A
DeLoach, Culver Jackson, Jr
Drummond, Roger Otto
Duhrkopf, Richard Edward
Dulmage, Howard Taylor
Durden, Christopher John
Ewert, Adam
Fincher, George Truman
French, Jeptha Victor
Frisbie, Raymond Edward
Gibson, William Wallace
Gillaspy, James Edward
Gilstrap, Franklin Ephriam
Goodwin, William Jennings
Greenberg, Les Paul
Hall, Clarence Coney, Jr
Hanna, Ralph Lynn
Harden, Philip Howard
Harris, Kerry Francis Patrick
Harris, Marvin Kirk
Harris, Robert Lee
Hartberg, Warren Keith
Hilliard, John Roy, Jr
Hollingsworth, Joseph Pettus
Horner, Norman V
Keeley, Larry Lee
Klaus, Ewald Fred, Jr
Kunz, Sidney Edmund
Lopez, Genaro
Mangan, Robert Lawrence
Martin, Paul Bain
Meadows, Charles Milton
Meola, Roger Walker
Meola, Shirlee May
Morrison, Eston Odell
Nachman, Ronald James
Nettles, William Carl, Jr
Olson, Jimmy Karl
Osburn, Richard Lee
Parker, Roy Denver, Jr
Peck, William B
Reinert, James A
Richardson, Richard Harvey
Richerson, Jim Vernon
Ring, Dennis Randall
Robinson, James Vance
Rubink, William Louis
Scanlon, John Earl
Schaffner, Joseph Clarence
Scholl, Philip Jon
Schuster, Michael Frank
Shake, Roy Eugene
Slosser, Jeffrey Eric
Smith, James Willie, Jr
Standifer, Lonnie Nathaniel
Sterling, Winfield Lincoln
Stewart, Kenneth Wilson
Stone, Jay D
Sugden, Evan A
Summers, Max Duane
Tuff, Donald Wray
Van Cleave, Horace William
Vinson, S Bradleigh
Walker, J Knox
Watkins, Julian F, II
Williams, Ralph Edward
Williams, Robert K
Wolfenbarger, Dan A
Wright, James Elbert

UTAH
Allred, Dorald Mervin
Amman, Gene Doyle
Anderson, Russell D
Baumann, Richard William
Booth, Gary Melvon
Cole, Walter Eckle
Davidson, Diane West
Davis, Donald Walter
Edmunds, George Francis, Jr
Elbel, Robert E
Grundmann, Albert Wendell
Havertz, David S
Haws, Byron Austin
Hsiao, Ting Huan
Jorgensen, Clive D
Knowlton, George Franklin
Messina, Frank James
Mulaik, Stanley B
Nielsen, Lewis Thomas
Nielsen, Mervin William
Richardson, Jay Wilson, Jr
Stark, Harold Emil
Tipton, Vernon John
Whitehead, Armand T
Wood, Stephen Lane

VERMONT
Freeman, Jeffrey VanDuyne
Happ, George Movius
MacCollom, George Butterick
McLean, Donald Lewis

VIRGINIA
Bobb, Marvin Lester
Carico, James Edwin
Cochran, Donald Gordon
Dogger, James Russell

Eaton, John LeRoy
Eckerlin, Ralph Peter
Emsley, Michael Gordon
Fashing, Norman James
Fink, Linda Susan
Fletcher, Lowell W
Foote, Richard Herbert
Gray, Faith Harriet
Grimm, James K
Gurney, Ashley Buell
Haines, Kenneth A
Heikkenen, Herman John
Hinckley, Alden Dexter
Hofmaster, Richard Namon
Homsher, Paul John
Horsburgh, Robert Laurie
Jubb, Gerald Lombard, Jr
Kliewer, John Wallace
Knipling, Edward Fred
Kok, Loke-Tuck
Kosztarab, Michael
Logan, Jesse Alan
Manzelli, Manlio Arthur
Mullins, Donald Eugene
Payne, Thomas Lee
Pienkowski, Robert Louis
Roberts, James Ernest, Sr
Robinson, William H
Ross, Mary Harvey
Schultz, Peter Berthold
Semtner, Paul Joseph
Shaffer, Jay Charles
Smith, John Cole
Smythe, Richard Vincent
Spyhalski, Edward James
Stevens, Thomas McConnell
Turner, Ernest Craig, Jr
Weidhaas, John August, Jr

WASHINGTON
Akre, Roger David
Bay, Ernest C
Bernard, Gary Dale
Biever, Kenneth Duane
Branson, Terry Fred
Burditt, Arthur Kendall, Jr
Burts, Everett C
Butt, Billy Arthur
Carroll, Devine Patrick
Catts, Elmer Paul
Clement, Stephen LeRoy
Cone, Wyatt Wayne
Eighme, Lloyd Elwyn
Gara, Robert I
Getzin, Louis William
Gibson, Flash
Halfhill, John Eric
Hoyt, Stanley Charles
Kareiva, Peter Michael
Klostermeyer, Edward Charles
Klostermeyer, Lyle Edward
Kraft, Gerald F
Krysan, James Louis
McDonough, Leslie Marvin
Meeuse, Bastiaan J D
Milne, David Hall
Moffett, David Franklin, Jr
Moffitt, Harold Roger
Pike, Keith Schade
Poston, Freddie Lee, Jr
Riddiford, Lynn Moorhead
Rogers, Lee Edward
Senger, Clyde Merle
Shanks, Carl Harmon, Jr
Sluss, Robert Reginald
Smith, Stamford Dennis
Spence, Kemet Dean
Toba, H(achiro) Harold

WEST VIRGINIA
Adkins, Dean Aaron
Brown, Mark Wendell
Butler, Linda
Cole, Larry King
Sheppard, Roger Floyd

WISCONSIN
Amin, Omar M
Beck, Stanley Dwight
Boush, George Mallory
Burkholder, Wendell Eugene
Carlson, Stanley David
Chapman, R Keith
Coppel, Harry Charles
DeFoliart, Gene Ray
Dixon, John Charles
Drecktrah, Harold Gene
Fisher, Ellsworth Henry
Ghazarian, Jacob G
Giese, Ronald Lawrence
Gojmerac, Walter Louis
Grothaus, Roger Harry
Halgren, Lee A
Hilsenhoff, William Leroy
Kennedy, M(aldon) Keith
Koval, Charles Francis
Lichtenstein, E Paul
Norris, Dale Melvin, Jr
Owens, John Michael
Ritter, Karla Schwensen
Shenefelt, Roy David
Singer, George
Willis, Harold Lester
Wittrock, Darwin Donald

Entomology (cont)

Downer, Roger George Hamill
Downes, John Antony
Ellis, Clifford Roy
Fitz-James, Philip Chester
Friend, William George
George, John Allen
Hamilton, Kenneth Gavin Andrew
Harcourt, Douglas George
Harris, Charles Ronald
Henson, Walter Robert
Holland, George Pearson
House, Howard Leslie
Howden, Henry Fuller
Jaques, Robert Paul
Jones, Philip Arthur
Judd, William Wallace
Kevan, Peter Graham
Laing, John E
LeRoux, Edgar Joseph
Lindquist, Evert E
Locke, Michael
McAlpine, James Francis
McEwen, Freeman Lester
Masner, Lubomir
Munroe, Eugene Gordon
Nesbitt, Herbert Hugh John
Oliver, Donald Raymond
Ozburn, George W
Peck, Stewart Blaine
Peterson, Bobbie Vern (Robert)
Philogene, Bernard J R
Retnakaran, Arthur
Sears, Markham Karli
Sharkey, Michael Joseph
Sippell, William Lloyd
Smetana, Ales
Smith, Jonathan Jeremy Berkeley
Steele, John Earle
Teskey, Herbert Joseph
Tobe, Stephen Solomon
Tomlin, Alan David
Tung, Pierre S
Watson, Wynnfield Young
Wiggins, Glenn Blakely
Wyatt, Gerard Robert
Yoshimoto, Carl Masaru

PRINCE EDWARD ISLAND
Thompson, Leith Stanley

QUEBEC
Auclair, Jacques Lucien
Bigras, Francine Jeanne
Cloutier, Conrad Francois
Harper, Pierre (Peter) Paul
Hill, Stuart Baxter
Hilton, Donald Frederick James
Kevan, Douglas Keith McEwan
McFarlane, John Elwood
McNeil, Jeremy Nichol
Mailloux, Gerard
Mangat, Baldev Singh
O'Neil, Louis C
Pilon, Jean-Guy
Quednau, Franz Wolfgang
Rau, Manfred Ernst
Sanborne, Paul Michael
Sharma, Madan Lal
Stewart, Robin Kenny

SASKATCHEWAN
Doane, John Frederick
Ewen, Alwyn Bradley
Khachatourians, George G
Lehmkuhl, Dennis Merle
Makowski, Roberte Marie Denise
Mukerji, Mukul Kumar
Peschken, Diether Paul
Riegert, Paul William
Zacharuk, R Y

OTHER COUNTRIES
Adames, Abdiel Jose
Bailey, Donald Leroy
Birch, Martin Christopher
Brown, Anthony William Aldridge
Cromartie, Thomas Houston
De Renobales, Mertxe
Gratz, Norman G
Hanson, Paul Eliot
Holldobler, Berthold Karl
Hummel, Hans Eckhardt
La Chance, Leo Emery
Leeper, John Robert
Muniappan, Rangaswamy Naicker
Perry, Albert Solomon
Rentz, David Charles
Saether, Ole Anton
Saunders, Joseph Lloyd

Food Science & Technology

ALABAMA
Chastain, Marian Faulkner
Marion, James Edsel
Rymal, Kenneth Stuart
Singh, Bharat

ARIZONA
Alton, Alvin John
Hill, Warner Michael

Kaufman, C(harles) W(esley)
Kline, Ralph Willard
Nelson, Frank Eugene
Price, Ralph Lorin
Seperich, George Joseph
Sleeth, Rhule Bailey
Stiles, Philip Glenn
Urbain, Walter Mathias
Wrobel, Raymond Joseph

ARKANSAS
Kattan, Ahmed A
Lewis, Paul Kermith, Jr
Snyder, Harry E

CALIFORNIA
Amen, Ronald Joseph
Ashton, David Hugh
Avera, Fitzhugh Lee
Bentley, David R
Brant, Albert Wade
Butterworth, Thomas Austin
Caporaso, Fredric
Chou, Tsong-Wen
Clark, Walter Leighton, III
Collins, Edwin Bruce
Davé, Bhalchandra A
Dunkley, Walter Lewis
Feather, A(lan) L(ee)
Feeney, Robert Earl
Finkle, Bernard Joseph
Goldstein, Walter Elliott
Holmes, David G
Hugunin, Alan Godfrey
Huskey, Glen E
Ingle, James Davis
Ito, Keith A
Jaye, Murray Joseph
King, A Douglas, Jr
King, Alfred Douglas, Jr
Lei, Shau-Ping Laura
Levenson, Harold Samuel
Little, Angela C
Lu, Nancy Chao
Luh, Bor Shiun
Lukes, Thomas Mark
Lundeen, Glen Alfred
Mazelis, Mendel
Meade, Reginald Eson
Meggison, David Laurence
Michener, H(arold) David
Midura, Thaddeus
Moshy, Raymond Joseph
Nelson, David A
Noble, Ann Curtis
Pangborn, Rose Marie Valdes
Prater, Arthur Nickolaus
Prend, Joseph
Robertson, George Harcourt
Rockland, Louis B
Russell, Gerald Frederick
Saunders, Robert Montgomery
Sayre, Robert Newton
Schneeman, Barbara Olds
Schweigert, Bernard Sylvester
Schwimmer, Sigmund
Serbia, George William
Sheneman, Jack Marshall
Silberstein, Otmar Otto
Singleton, Vernon LeRoy
Spector, Sheldon Laurence
Stevenson, Kenneth Eugene
Sutton, Donald Dunsmore
Vaughn, Reese Haskell
Webster, John Robert
Wodicka, Virgil O

COLORADO
Church, Brooks Davis
Deane, Darrell Dwight
Harper, Judson M(orse)
Maga, Joseph Andrew
Pan, Huo-Ping
Schmidt, Glenn Roy
Skelton, Marilyn Mae
Smith, Gary Chester
Stone, Martha Barnes

CONNECTICUT
Agarwal, Vipin K
Bibeau, Thomas Clifford
Cornell, Alan
Dobry, Reuven
Fiore, Joseph Vincent
Fordham, Joseph Raymond
Hall, Kenneth Noble
Hart, William James, Jr
Hopper, Paul Frederick
Kinsman, Donald Markham
Piehl, Donald Herbert
Ranalli, Anthony William
Ross, Robert Edgar
Servadio, Gildo Joseph

DELAWARE
Knorr, Dietrich W
Link, Bernard Alvin
Mitchell, Donald Gilman

DISTRICT OF COLUMBIA
Bigelow, Sanford Walker
Gardner, Sherwin
Garner, Richard Gordon
Jacobson, Michael F

Kromer, Lawrence Frederick
Kuznesof, Paul Martin
Modderman, John Philip
Shank, Fred R
Weinberg, Myron Simon

FLORIDA
Ahmed, Esam Mahmoud
Bacus, James Nevill
Barber, Franklin Weston
Bates, Robert P(arker)
Beckhorn, Edward John
Brown, Ross Duncan, Jr
Brown, William Lewis
Burgess, Hovey Mann
Carbonell, Robert Joseph
Clark, Allen Varden
Cotton, Robert Henry
Dubravcic, Milan Frane
Fox, Kenneth Ian
Gill, William Joseph
Hale, Kirk Kermit, Jr
Hankinson, Denzel J
Harris, Natholyn Dalton
Jackel, Simon Samuel
Janky, Douglas Michael
Jezeski, James John
Kapsalis, John George
Koburger, John Alfred
Moore, Edwin Lewis
Nanz, Robert Augustus Rollins
Otwell, Walter Steven
Showalter, Robert Kenneth
Teixeira, Arthur Alves
Vedamuthu, Ebenezer Rajkumar
Whipple, Royson Newton

GEORGIA
Ayres, John Clifton
Beuchat, Larry Ray
Bhatia, Darshan Singh
Brown, William E
Bryan, Frank Leon
Cecil, Sam Reber
Dekazos, Elias Demetrios
Donahoo, Pat
Doyle, Michael Patrick
Ebert, Andrew Gabriel
Eitenmiller, Ronald Ray
Green, John H
Jen, Joseph Jwu-Shan
Koehler, Philip Edward
Malaspina, Alex
Nakayama, Tommy
Powers, John Joseph
Prothro, Johnnie W
Ray, Apurba Kanti
Reagan, James Oliver
Schuler, George Albert
Smit, Christian Jacobus Bester
Tybor, Philip Thomas
Worthington, Robert Earl
Young, Louis Lee
Zallen, Eugenia Malone

HAWAII
Chan, Harvey Thomas, Jr
Frank, Hilmer Aaron
Moser, Roy Edgar
Moy, James Hee
Nip, Wai Kit
Seifert, Josef
Waslien, Carol Irene
Yamamoto, Harry Y

IDAHO
Augustin, Jorg A L
Branen, Alfred Larry
Greene, Barbara E
Heimsch, Richard Charles
Muneta, Paul
Zaehringer, Mary Veronica

ILLINOIS
Akin, Cavit
Auerbach, Earl
Blaschek, Hans P
Bookwalter, George Norman
Bothast, Rodney Jacob
Burkwall, Morris Paton, Jr
Campbell, Michael Floyd
Chen, Wen Sherng
Cheryan, Munir
Chung, Jiwhey
Clark, John Peter, III
Cole, Morton S
Dean, Robert Waters
Dickerson, Roger William, Jr
Dordick, Isadore
Duxbury, Dean David
Elliott, James Gary
Erdman, John Wilson, Jr
Flowers, Russell Sherwood, Jr
Gabis, Damien Anthony
Glenister, Paul Robson
Halaby, George Anton
Harris, Ronald David
Kelley, Keith Wayne
Kemp, Albert Raymond
Kummerow, Fred August
Lebermann, Kenneth Wayne
Lechowich, Richard V
Lundquist, Burton Russell
Lushbough, Channing Harden

McKeith, Floyd Kenneth
Milner, Reid Thompson
Moeller, Theodore William
Murphy, Robert Emmett
Newton, Stephen Bruington
Osri, Stanley M(aurice)
Pavey, Robert Louis
Pedraja, Rafael R
Rackis, Joseph John
Rakosky, Joseph, Jr
Schanefelt, Robert Von
Scott, Don
Sehgal, Lakshman R
Sherman, Joseph E
Siedler, Arthur James
Smittle, Richard Baird
Smouse, Thomas Hadley
Soucie, William George
Steinberg, Marvin Phillip
Tompkin, Robert Bruce
Trelease, Richard Davis
Udani, Kanakkumar Harilal
Vosti, Donald Curtis
Wei, Lun-Shin
Welsh, Thomas Laurence
White, John Francis
Williams, Marion Porter
Wistreich, Hugo Eryk
Witter, Lloyd David

INDIANA
Babel, Frederick John
Bates, Charles Johnson
Chambers, James Vernon
Cook, David Allan
Cousin, Maribeth Anne
Crawford, Thomas Michael
Creinin, Howard Lee
Endres, Joseph George
Forrest, John Charles
Hagen, Richard Eugene
Henke, Mitchell C
Hoff, Johan Eduard
Irwin, William Elliot
Judge, Max David
Kittaka, Robert Shinnosuke
Liska, Bernard Joseph
Litov, Richard Emil
Marks, Jay Stewart
Morris, John F
Nelson, Philip Edwin
Ogilvy, Winston Stowell
Parmelee, Carlton Edwin
Puski, Gabor
Radanovics, Charles
Stadelman, William Jacob
Swaminathan, Balasubramanian
Tennyson, Richard Harvey
Wallander, Jerome F
Weaver, Connie Marie
Weiss, Ronald
Whistler, Roy Lester
Williams, Leamon Dale

IOWA
Chung, Ronald Aloysius
Hammond, Earl Gullette
Hartman, Paul Arthur
Kraft, Allen Abraham
LaGrange, William Somers
Marion, William W
Rohlf, Marvin Euguene
Sebranek, Joseph George
Walker, Homer Wayne

KANSAS
Allen, Deloran Matthew
Bowers, Jane Ann (Raymond)
Caul, Jean Frances
Cunningham, Franklin E
Dirks, Brinton Marlo
Farrell, Eugene Patrick
Fung, Daniel Yee Chak
Harbers, Carole Ann Z
Harrison, Dorothy Lucile
Hoover, William Jay
Kropf, Donald Harris
Krum, Jack Kern
Ponte, Joseph G, Jr
Ramamurti, Krishnamurti
Setser, Carole Sue
Smith, Meredith Ford
Vetter, James Louis

KENTUCKY
Bujake, John Edward, Jr
Harmet, Kenneth Herman
Langlois, Bruce Edward
Luzzio, Frederick Anthony
Maisch, Weldon Frederick

LOUISIANA
Barkate, John Albert
Broeg, Charles Burton
Day, Donal Forest
Frank, Arlen W(alker)
Grodner, Robert Maynard
James, William Holden
Kadan, Ranjit Singh
Liuzzo, Joseph Anthony
Rao, Ramachandra M R
St Angelo, Allen Joseph
Windhauser, Marlene M
Younathan, Margaret Tims

MAINE
Highlands, Matthew Edward
Slabyj, Bohdan M

MARYLAND
Alsmeyer, Richard Harvey
Anderson, Sue Ann Debes
Attaway, David Henry
Butman, Bryan Timothy
Dame, Charles
Denny, Cleve B
Devreotes, Peter Nicholas
Goodman, Sarah Anne
Graebert, Eric W
Hall, Richard Leland
Heath, James Lee
Hoffman, David J
Kim, Sooja K
King, Raymond Leroy
Kotula, Anthony W
Matthews, Ruth H
Milner, Max
Moats, William Alden
Molenda, John R
Porzio, Michael Anthony
Seifried, Adele Susan Corbin
Seifried, Harold Edwin
Twigg, Bernard Alvin
Vaughn, Moses William
Wabeck, Charles J
Wagner, David Darley
Westhoff, Dennis Charles
Whitehair, Leo A
Wiley, Robert Craig
Williams, Eleanor Ruth
Yeatman, John Newton

MASSACHUSETTS
Anderson, Edward Everett
Ayuso, Katharine
Babayan, Vigen Khachig
Breslau, Barry Richard
Buck, Ernest Mauro
Bustead, Ronald Lorima, Jr
Cardello, Armand Vincent
Clayton, J(oe) T(odd)
Clydesdale, Fergus Macdonald
Esselen, William B
Fagerson, Irving Seymour
Fram, Harvey
Francis, Frederick John
Goldblith, Samuel Abraham
Hultin, Herbert Oscar
Karel, Marcus
Kepper, Robert Edgar
King, Frederick Jessop
Kissmeyer-Nielsen, Erik
Lampi, Rauno Andrew
Levin, Robert E
Lundstrom, Ronald Charles
Maloney, John F
Mehrlich, Ferdinand Paul
Narayan, Krishnamurthi Ananth
Potter, Frank Elwood
Powers, Edmund Maurice
Ross, Edward William, Jr
Rowley, Durwood B
Salant, Abner Samuel
Sawyer, Frederick Miles
Short, John Lawson
Silverman, Gerald
Sinskey, Anthony J
Tan, Barrie
Tillotson, James E
Tuomy, Justin M(atthew)
Whitney, Lester F(rank)
Yamins, J(acob) L(ouis)
Zapsalis, Charles

MICHIGAN
Bartholmey, Sandra Jean
Brunner, Jay Robert
Carlotti, Ronald John
Cotter, Richard
Crevasse, Gary A
Dawson, Lawrence E
Einset, Eystein
Frodey, Ray Charles
Haines, William C
Harmon, Laurence George
Hedrick, Theodore Isaac
Hollingsworth, Cornelia Ann
Jackson, John Mathews
Markakis, Pericles
Marsh, Alice Garrett
Merkel, Robert Anthony
Price, James F
Purvis, George Allen
Schaller, Daryl Richard
Shelef, Leora Aya
Smith, Denise Myrtle
Stine, Charles Maxwell
Uebersax, Mark Alan
Wishnetsky, Theodore
Zabik, Mary Ellen

MINNESOTA
Allen, Charles Eugene
Anderson, Ray Harold
Bauman, Howard Eugene
Behnke, James Ralph
Berube, Robert
Breene, William Michael
Busta, Francis Fredrick

Caldwell, Elwood F
Chandan, Ramesh Chandra
Christianson, George
Cronk, Ted Clifford
Daravingas, George Vasilios
Davis, Eugenia Asimakopoulos
Dunshee, Bryant R
Durst, Jack Rowland
Epley, Richard Jess
Franzen, Kay Louise
Gallaher, Daniel David
Gillett, Tedford A
Gordon, Joan
Heller, Steven Nelson
Hurley, William Charles
Johnson, Dale Waldo
Katz, Morris Howard
Labuza, Theodore Peter
Levine, Allen Stuart
McKay, Larry Lee
Morris, Howard Arthur
Mullen, Joseph David
Noland, Wayland Evan
Opie, Joseph Wendell
Packard, Vernal Sidney, Jr
Pflug, Irving John
Rapp, Waldean G
Reineccius, Gary (Aubrey)
Reinhart, Richard D
Rodgers, Nelson Earl
Sapakie, Sidney Freidin
Sherck, Charles Keith
Slavin, Joanne Louise
Stoll, William Francis
Sullivan, Betty J
Tatini, Sita Ramayya
Thomas, Elmer Lawrence
Touba, Ali R
Vermilyea, Barry Lynn
Vickers, Zata Marie
Zottola, Edmund Anthony

MISSISSIPPI
Ford, Robert Sedgwick
Garrett, Ephraim Spencer, III
Regier, Lloyd Wesley

MISSOURI
Anderson, David W, Jr
Bough, Wayne Arnold
Fields, Marion Lee
Gnaedinger, Richard H
Hood, Larry Lee
Kluba, Richard Michael
Lo, Grace S
Marshall, Robert T
Mavis, James Osbert
Naumann, Hugh Donald
Newell, Jon Albert
Schwall, Donald V
Taranto, Michael Vincent
Titus, Dudley Seymour
Unklesbay, Nan F
Wells, Frank Edward

NEBRASKA
Aberle, Elton D
Beach, Betty Laura
Brazis, A(dolph) Richard
Bullerman, Lloyd Bernard
Froning, Glenn Wesley
Hanna, Milford A
Hartung, Theodore Eugene
Issenberg, Phillip
Shahani, Khem Motumal
Smith, Durward A
Taylor, Steve L

NEW HAMPSHIRE
Hayes, Kirby Maxwell

NEW JERSEY
Adler, Irwin L
Akerboom, Jack
Bakal, Abraham I
Bednarcyk, Norman Earle
Bedrosian, Karakian
Brogle, Richard Charles
Buck, Robert Edward
Canzonier, Walter J
Chu, Horn Dean
Dalrymple, Ronald Howell
Farkas, Daniel Frederick
Finley, John Westcott
Franceschini, Remo
Fulde, Roland Charles
Giddings, George Gosselin
Gilbert, Seymour George
Goodenough, Eugene Ross
Hagberg, Elroy Carl
Hayakawa, Kan-Ichi
Herring, Harold Keith
Hlavacek, Robert John
Hsu, Kenneth Hsuehchia
Johnson, Bobby Ray
Joiner, Robert Russell
Jones, Benjamin Lewis
Karmas, Endel
Katz, Ira
Lachance, Paul Albert
Litman, Irving Isaac
Lloyd, Norman Edward
Lund, Daryl B
McAnelly, John Kitchel

Miller, Albert Thomas
Miller, Gary A
Montville, Thomas Joseph
Morck, Roland Anton
Pleasants, Elsie W
Richberg, Carl George
Rosenfield, Daniel
Schaich, Karen Marie
Scharpf, Lewis George, Jr
Seitz, Eugene W
Sekula, Bernard Charles
Shirk, Richard Jay
Smith, Robert Ewing
Solberg, Myron
Staub, Herbert Warren
Stone, Charles Dean
Supran, Michael Kenneth
Wasserman, Bruce P
Wells, Phillip Richard
Wolin, Alan George

NEW MEXICO
O'Brien, Robert Thomas
Ray, Earl Elmer
Roubicek, Rudolf V

NEW YORK
Armbruster, Gertrude D
Axelson, Marta Lynne
Baker, Robert Carl
Beermann, Donald Harold
Borisenok, Walter A
Bourne, Malcolm Cornelius
Chang, Pei Kung (Philip)
Dibble, Marjorie Veit
Downing, Donald Leonard
Ehmann, Edward Paul
Filandro, Anthony Salvatore
Gizis, Evangelos John
Graham, Donald C W
Hang, Yong Deng
Hickernell, Gary L
Holland, Robert Francis
Ishler, Norman Hamilton
Jordan, William Kirby
Kahan, Sidney
Khan, Paul
Kosikowski, Frank Vincent
Kraft, Patricia Lynn
Kramer, Franklin
Kroenberg, Berndt
Ledford, Richard Allison
Livingston, G E
Lugay, Joaquin Castro
March, Richard Pell
Marshall, James Tilden, Jr
Mehta, Bipin Mohanlal
Miller, Dennis Dean
Parliment, Thomas H
Potter, Norman N
Prescott, Henry Emil, Jr
Ramstad, Paul Ellertson
Rao, Mentreddi Anandha
Regenstein, Joe Mac
Schiffmann, Robert F
Shipe, William Franklin
Smith, Ora
Szczesniak, Alina Surmacka
Turkki, Pirkko Reetta
Wellington, George Harvey
White, James Carrick
Zall, Robert Rouben

NORTH CAROLINA
Beam, John E
Dahiya, Raghunath S
DeJong, Donald Warren
DiMarco, G Robert
Fleming, Henry Pridgen
Gregory, Max Edwin
Hassan, Hosni Moustafa
Jones, Ivan Dunlavy
Jones, Victor Alan
Lineback, David R
May, Kenneth Nathaniel
Morse, Roy E
Neumann, Calvin Lee
Oblinger, James Leslie
Reed, Gerald
Roop, Richard Allan
Schiffman, Susan S
Swaisgood, Harold Everett
Tarver, Fred Russell, Jr
Thomas, Frank Bancroft
Walter, William Mood, Jr
Webb, Neil Broyles
Young, Clyde Thomas

NORTH DAKOTA
Donnelly, Brendan James
Stanislao, Bettie Chloe Carter

OHIO
Banwart, George J
Blaisdell, John Lewis
Cahill, Vern Richard
Chism, Grady William, III
Cornelius, Billy Dean
De Fiebre, Conrad William
Gallander, James Francis
Gould, Wilbur Alphonso
Grab, Eugene Granville, Jr
Hansen, Poul M T
Hartman, Warren Emery

Holzinger, Thomas Walter
Hunt, Fern Ensminger
Hunter, James Edward
Husaini, Saeed A
Jandacek, Ronald James
Jaynes, John Alva
Joffe, Frederick M
Kristoffersen, Thorvald
Larkin, Edward P
Litchfield, John Hyland
McComis, William T(homas)
Neer, Keith Lowell
Parrett, Ned Albert
Sabharwal, Kulbir
Sharma, Shri C
Sinnhuber, Russell Otto
Stauffer, Clyde E
Wehrle, Louis, Jr
Wilcox, Joseph Clifford
Youngquist, R(udolph) William
Ziller, Stephen A, Jr

OKLAHOMA
Berry, Joe Gene
Gilliland, Stanley Eugene
Guenther, John James
Ray, Frederick Kalb

OREGON
Anglemier, Allen Francis
Arnold, Roy Gary
Badenhop, Arthur Fredrick
Crawford, David Lee
Dutson, Thayne R
Elliker, Paul R
Kennick, Walter Herbert
Kifer, Paul Edgar
Libbey, Leonard Morton
Mickelberry, William Charles
Morgan, Bruce Henry
Morita, Toshiko N
Pearson, Albert Marchant
Scanlan, Richard Anthony
Wehr, Herbert Michael
Woodburn, Margy Jeanette
Wrolstad, Ronald Earl
Yang, Hoya Y

PENNSYLVANIA
Barnes, William Shelley
Beelman, Robert B
Birdsall, Marion Ivens
Brody, Aaron Leo
Buchanan, Robert Lester
Craig, James Clifford, Jr
Dimick, Paul Slayton
Fugger, Joseph
Hills, Claude Hibbard
Johnson, Littleton Wales
Johnson, Ogden Carl
Keeney, Philip G
Kroger, Manfred
MacNeil, Joseph H
Medina, Marjorie B
Miller, Arthur James
Miller, Kenneth Melvin
Mottur, George Preston
Palumbo, Samuel Anthony
Ramachandran, N
Roseman, Arnold S(aul)
Sapers, Gerald M
Sawicki, John Edward
Segall, Stanley
Seligson, Frances Hess
Thompson, Donald B
Wohlpart, Kenneth Joseph
Ziegler, John Henry, Jr

RHODE ISLAND
Chichester, Clinton Oscar
Constantinides, Spiros Minas
Cosgrove, Clifford James
Dymsza, Henry A
Josephson, Edward Samuel
Rand, Arthur Gorham, Jr
Simpson, Kenneth L

SOUTH CAROLINA
Godfrey, W Lynn
Hollis, Bruce Warren
Surak, John Godfrey

SOUTH DAKOTA
Costello, William James

TENNESSEE
Bass, Mary Anna
Collins, Jimmie Lee
Das, Salil Kumar
Davidson, Philip Michael
Draughon, Frances Ann
Jaynes, Hugh Oliver
Perry, Margaret Nutt

TEXAS
Aguilera, Jose Miguel
Armstrong, George Glaucus, Jr
Arnott, Howard Joseph
Bourland, Charles Thomas
Brittin, Dorothy Helen Clark
Burns, Edward Eugene
Carpenter, Zerle Leon
Cross, Hiram Russell
Czajka-Narins, Dorice M

Food Science & Technology (cont)

Dill, Charles William
Goodwin, Tommy Lee
Hsueh, Andie M
Hunnell, John Wesley
Johnston, Melvin Roscoe
Kendall, John Hugh
King, General Tye
LaBree, Theodore Robert
Landmann, Wendell August
Lin, Sherman S
Lusas, Edmund W
Matz, Samuel Adam
Mehta, Rajen
Mellor, David Bridgwood
Orts, Frank A
Rooney, Lloyd William
Sistrunk, William Allen
Smith, Malcolm Crawford, Jr
Sweat, Vincent Eugene
Vanderzant, Carl
Varsel, Charles John

UTAH
Ames, Robert A
Chase, Richard Lyle
Huber, Clayton Shirl
Larsen, Lloyd Don
Mendelhall, Von Thatcher
Salunkhe, Dattajeerao K
Yorks, Terence Preston

VERMONT
Mitchell, William Alexander

VIRGINIA
Burnette, Mahlon Admire, III
Collins, William F
Cooler, Frederick William
Fellers, David Anthony
Flick, George Joseph, Jr
Hausermann, Max
Keenan, Thomas William
Khan, Mahmood Ahmed
Lopez, Anthony
Mulvaney, Thomas Richard
Palmer, James Kenneth
Pao, Eleanor M
Phillips, Jean Allen
Rippen, Thomas Edward
Stern, Joseph Aaron
Swope, Fred C
Symons, Hugh William
Wesley, Roy Lewis
Yip, George

WASHINGTON
Brekke, Clark Joseph
Drake, Stephen Ralph
Eklund, Melvin Wesley
Freund, Peter Richard
Hard, Margaret McGregor
Hoskins, Frederick Hall
Ingalsbe, David Weeden
Maguire, Yu Ping
Matches, Jack Ronald
Meeder, Jeanne Elizabeth
Nagel, Charles William
Pigott, George M
Swanson, Barry Grant
Wekell, Marleen Marie

WISCONSIN
Amundson, Clyde Howard
Bard, John C
Bartholomew, Darrell Thomas
Bernstein, Bradley Alan
Bernstein, Sheldon
Bradley, Robert Lester, Jr
Buege, Dennis Richard
Chu, Fun Sun
Curwen, David
Deese, Dawson Charles
Duke, Stanley Houston
Dutton, Herbert Jasper
Fennema, Owen Richard
Greaser, Marion Lewis
Grindrod, Paul (Edward)
Hartzell, Thomas H
Kainski, Mercedes H
Kauffman, Robert Giller
Lindsay, Robert Clarence
McDivitt, Maxine Estelle
Marsh, Benjamin Bruce
Marth, Elmer Herman
Maurer, Arthur James
Moskowitz, Gerard Jay
Nagodawithana, Tilak Walter
Nelson, John Howard
Olson, Norman Fredrick
Price, Walter Van Vranken
Ryan, Dale Scott
Snudden, Birdell Harry
Steinke, Paul Karl Willi
Suess, Gene Guy
Trautman, Jack Carl
Wall, Joseph Sennen
Wallenfeldt, Evert
Warner, H Jack
Weber, Frank E
Wege, Ann Christene

Winder, William Charles

WYOMING
Field, Ray A

ALBERTA
Doornenbal, Hubert
Hadziyev, Dimitri
Hawrysh, Zenia Jean
Jackson, Harold
Jeremiah, Lester Earl
Stiles, Michael Edgecombe

BRITISH COLUMBIA
Duncan, Douglas Wallace
Earl, Allan Edwin
Kitson, John Aidan
Leichter, Joseph
Li-Chan, Eunice
Powrie, William Duncan

MANITOBA
Borsa, Joseph
Cenkowski, Stefan
Singh, Harwant
Vaisey-Genser, Florence Marion
White, Noel David George

NEWFOUNDLAND
Ke, Paul Jenn

NOVA SCOTIA
Ackman, Robert George
Tung, Marvin Arthur

ONTARIO
Blais, Burton W
Chapman, Ross Alexander
Clark, David Sedgefield
Davidson, Charles Mackenzie
Diosady, Levente Laszlo
Ferrier, Leslie Kenneth
Fuleki, Tibor
Lawford, George Ross
Lentz, Claude Peter
Loughheed, Thomas Crossley
Patel, Girishchandra Babubhai
Rayman, Mohamad Khalil
Sahasrabudhe, Madhu R
Stanley, David Warwick
Stavric, Bozidar
Thompson, Lilian Umale
Timbers, Gordon Ernest
Todd, Ewen Cameron David
Usborne, William Ronald
Van Den Berg, L
Veliky, Ivan Alois
Wood, Peter John
Zubeckis, Edgar

QUEBEC
Britten, Michel
Champagne, Claude P
David, Jean
Gagnon, Marcel
Goulet, Jacques
Idziak, Edmund Stefan
Kuhnlein, Harriet V
Moreau, Jean Raymond
Riel, Rene Rosaire
Sood, Vijay Kumar
Vezina, Claude

SASKATCHEWAN
Ingledew, William Michael
Youngs, Clarence George

OTHER COUNTRIES
Borgstrom, Georg Arne
Brown, Delos D
Fairweather-Tait, Susan Jane
Filadelfi-Keszi, Mary Ann Stephanie
Ginis, Asterios Michael
Kaffezakis, John George
Montgomery, Morris William
Rio, María Esther
Segal, Mark
Valencia, Mauro Eduardo
Yang, Ho Seung

Genetics

ALABAMA
Barker, Peter Eugene
Briles, Connally Oran
Briles, David Elwood
Carter, Charlotte
Chambliss, Oyette Lavaughn
Clement, William Madison, Jr
Cosper, Paula
Denton, Tom Eugene
Fattig, W Donald
Finley, Sara Crews
Finley, Wayne House
Fuller, Gerald M
Garver, David L
Goggans, James F
Hardman, John Kemper
Hoff, Charles Jay
Hunter, Eric
Lavelle, George Cartwright
Lemke, Paul Arenz
Lubega, Seth Gasuza

McCombs, Candace Cragen
McDaniel, Gayner Raiford
McGuire, John Albert
Norton, Joseph Daniel
Olson, Richard Louis
Rhodes, Richard Kent
Roozen, Kenneth James
Sayers, Earl Roger
Shepherd, Raymond Lee
Singh, Jarnail
Smith-Somerville, Harriett Elizabeth
Thompson, Jerry Nelson
Watson, Jack Ellsworth
Williams, John Watkins, III

ALASKA
Gharrett, Anthony John
Helle, John Harold (Jack)
Shields, Gerald Francis
Snyder, George Richard

ARIZONA
Allen, Stephen Gregory
Bartlett, Alan C
Bemis, William Putnam
Bernstein, Harris
Cassidy, Suzanne Bletterman
Church, Kathleen
Day, Arden Dexter
Doane, Winifred Walsh
Dukepoo, Frank Charles
Endrizzi, John Edwin
English, Darrel Starr
Goldstein, Elliott Stuart
Harris, Robert Martin
Heed, William Battles
Istock, Conrad Alan
Ito, Junetsu
Johnson, Herbert Windal
Kidwell, Margaret Gale
Little, John Wesley
McDaniel, Robert Gene
MacDonald, Marnie L
Mendelson, Neil Harland
Mount, David William Alexander
Muramoto, Hiroshi
Rasmussen, David Irvin
Rubis, David Daniel
Shimizu, Nobuyoshi
Thompson, Anson Ellis
Turcotte, Edgar Lewis
Ward, Oscar Gardien
Ward, Samuel
Wilson, F(rank) Douglas
Wilson, Frank Douglas
Woolf, Charles Martin
Zegura, Stephen Luke

ARKANSAS
Andrews, Luther David
Boulware, Ralph Frederick
Bourland, Freddie Marshall
Brown, A Hayden, Jr
Brown, Connell Jean
Clayton, Frances Elizabeth
Collins, Frederick Clinton
Collister, Earl Harold
Dippell, Ruth Virginia
Epstein, Joshua
Guest, William C
Gyles, Nicholas Roy
Haggard, Bruce Wayne
Hansen, Deborah Kay
Reddy, RamaKrishna Pashuvula
Sears, Jack Wood
Smith, Edwin Burnell
Tave, Douglas
Wolff, George Louis
York, John Owen

CALIFORNIA
Abbott, Ursula K
Abelson, John Norman
Abplanalp, Hans
Adams, Jane N
Allard, Robert Wayne
Allen, Marcia Katzman
Ames, Bruce Nathan
Ames, Giovanna Ferro-Luzzi
Amirkhanian, John David
Angell, Frederick Franklyn
Arai, Ken-Ichi
Ayala, Francisco Jose
Bailey-Serres, Julia
Baker, Robert Frank
Baldy, Marian Wendorf
Barratt, Raymond William
Baxter, John Darling
Beadle, George Wells
Beard, Benjamin H
Beckendorf, Steven K
Bell, A(udra) Earl
Belser, William Luther, Jr
Benzer, Seymour
Bergh, Berthold Orphie
Bernoco, Domenico
Bernstein, Sanford Irwin
Bertani, Giuseppe
Beyer, Edgar Herman
Blischke, Wallace Robert
Bliss, Fredrick Allen
Bonner, James (Fredrick)
Botstein, David
Bournias-Vardiabasis, Nicole

Bowen, Sarane Thompson
Bowling, Ann L
Bowman, Barry J
Boyd, James Brown
Boyer, Herbert Wayne
Bradford, G Eric
Brody, Stuart
Brown, Howard S
Brownell, Anna Gale
Bryant, Peter James
Buehring, Gertrude Case
Bullas, Leonard Raymond
Burrill, Melinda Jane
Busch, Robert Edward
Buth, Donald George
Calarco, Patricia G
Campbell, Allan McCulloch
Campbell, David Paul
Campbell, Judith Lynn
Cande, W Zacheus
Capron, Alexander Morgan
Carbon, John Anthony
Carpenter, Adelaide Trowbridge Clark
Catlin, B Wesley
Cavalli-Sforza, Luigi Luca
Celniker, Susan Elizabeth
Center, Elizabeth M
Chamberlin, Michael John
Charlang, Gisela Wohlrab
Chihara, Carol Joyce
Choy, Wai Nang
Christianson, Michael Lee
Chu, Irwin Y E
Clark, Alvin John
Clegg, Michael Tran
Cline, Thomas Warren
Close, Perry
Clough, Wendy Glasgow
Coffino, Philip
Cohen, Larry William
Cohen, Stanley Norman
Comings, David Edward
Cooper, Kenneth Willard
Corcoran, Mary Ritzel
Critchfield, William Burke
Cross, Ronald Allan
Daly, Kevin Richard
Davis, Brian Kent
Davis, Elmo Warren
Davis, Ronald Wayne
Davis, Rowland Hallowell
De Francesco, Laura
Dempsey, Wesley Hugh
Denis, Kathleen A
Desharnais, Robert Anthony
Dragon, Elizabeth Alice Oosterom
Dvorak, Jan
Ebert, Wesley W
Eckhart, Walter
Edgar, Robert Stuart
Ellstrand, Norman Carl
Emr, Scott David
Englesberg, Ellis
Ensign, Stewart Ellery
Epstein, Charles Joseph
Erickson, Jeanne Marie
Esposito, Michael Salvatore
Estilai, Ali
Evans, Ronald M
Falk, Darrel Ross
Farish, Donald James
Feldman, Jerry F
Fisher, Kathleen Mary Flynn
Fitch, Walter M
Flashman, Stuart Milton
Fluharty, Arvan Lawrence
Fogel, Seymour
Fradkin, Cheng-Mei Wang
Francke, Uta
Freeling, Michael
Friedmann, Theodore
Fristrom, Dianne
Fristrom, James W
Fujimoto, Atsuko Ono
Futch, David Gardner
Gall, Graham A E
Game, John Charles
Ganesan, Adayapalam T
Ganesan, Ann K
Gautsch, James Willard
Gelfand, David H
George, Sarah B(rewster)
Gill, Ayesha Elenin
Goetinck, Paul Firmin
Golbus, Mitchell S
Gordon, Manuel Joe
Gray, Joe William
Green, Melvin Martin
Grody, Wayne William
Grunstein, Michael
Gutman, George Andre
Guze, Carol (Konrad Lydon)
Hackett, Adeline J
Hanawalt, Philip Courtland
Hankinson, Oliver
Harding, James A
Harris, Mary Styles
Hedgecock, Dennis
Helinski, Donald Raymond
Herskowitz, Ira
Herzenberg, Leonard Arthur
Hill, Annlia Paganini
Hoch, James Alfred
Hogness, David Swenson

Hood, Leroy E
Hoopes, Laura Livingston Mays
Horowitz, Norman Harold
Hoy, Marjorie Ann
Hughes, Norman
Humaydan, Hasib Shaheen
Hyman, Bradley Clark
Jacobs, Patricia Ann
Jain, Subodh K
Jameson, David Lee
Jarvik, Lissy F
Jefferson, Margaret Correan
Jenkins, Burton Charles
Johnson, Ben Francis
Johnson, Bertil Lennart
Johnson, Carl William
Johnston, George Robert
Jones, Gary Edward
Kaiser, Armin Dale
Kaplan, William David
Kashyap, Tapeshwar S
Kedes, Laurence H
Kevles, Daniel Jerome
Kiger, John Andrew, Jr
King, Mary-Claire
Kinloch, Bohun Baker, Jr
Kirschbaum, Joel Bruce
Knudson, Gregory Blair
Kroman, Ronald Avron
Kustu, Sydney Govons
Laben, Robert Cochrane
Laidlaw, Harry Hyde, Jr
Landolph, Joseph Richard, II
Lange, Kenneth L
Latimer, Howard Leroy
Law, George Robert John
Leary, John Vincent
Lederberg, Esther Miriam
Ledig, F Thomas
Lefevre, George, Jr
Lehman, William Francis
Lengyel, Judith Ann
Levinson, Arthur David
Libby, William John, (Jr)
Lieb, Margaret
Lindsley, Dan Leslie, Jr
Lipshitz, Howard David
Long, Sharon Rugel
Loo, Melanie Wai Sue
Luckock, Arlene Suzanne
Lugo, Tracy Gross
McClelland, Michael
McClenaghan, Leroy Ritter, Jr
McDonald, Paul Thomas
McEwen, Joan Elizabeth
McFarlane, John Spencer
Mack, Thomas McCulloch
McLaughlin, Calvin Sturgis
Mangan, Jerrome
Marsh, James Lawrence
Martin, David William, Jr
Martin, George Steven
Martinek, George William
Maxwell, Joyce Bennett
Merchant, Sabeeha
Merriam, John Roger
Meyerowitz, Elliot Martin
Miller, Alexander
Mohandas, Thuluvancheri
Morse, Daniel E
Mortimer, Robert Keith
Nelson, Keith
Nelson-Rees, Walter Anthony
Neufeld, Elizabeth Fondal
Nunney, Leonard Peter
Ohno, Susumu
Ohnuki, Yasushi
Okada, Tadashi A
Owen, Ray David
Packman, Seymour
Page, Robert Eugene, Jr
Pedersen, Roger Arnold
Pendse, Pratapsinha C
Perkins, David Dexter
Phinney, Bernard Orrin
Pratt, David
Puhalla, John Edward
Puri, Yesh Paul
Qualset, Calvin Odell
Quiros, Carlos F
Rachmeler, Martin
Raju, Namboori Bhaskara
Ratty, Frank John, Jr
Rawal, Kanti M
Rearden, Carole Ann
Riblet, Roy Johnson
Rick, Charles Madeira, Jr
Rimoin, David (Lawrence)
Rinehart, Robert R
Roederer, Mario
Rose, Michael Robertson
Rosen, Howard
Rosenberg, Marvin J
Rosenthal, Allan Lawrence
Ross, Ian Kenneth
Rossi, John Joseph
Rotter, Jerome Israel
Roy-Burman, Pradip
Ryder, Edward Jonas
Ryder, Oliver A
Sacher, Robert Francis
Sadanaga, Kiyoshi
Saier, Milton H, Jr
St Lawrence, Patricia

Savitsky, Helen
Schekman, Randy W
Schlegel, Robert John
Schneider, Edward Lewis
Schulke, James Darrell
Scott, Matthew P
Seegmiller, Jarvis Edwin
Shapiro, Larry Jay
Shultz, Fred Townsend
Sidhu, Gurmel Singh
Siegel, Richard Weil
Silagi, Selma
Silverman, Michael Robert
Simmel, Edward Clemens
Simons, Robert W
Simpson, Robert Blake
Singh, Arjun
Sinibaldi, Ralph Michael
Slatkin, Montgomery (Wilson)
Slavkin, Harold Charles
Smith, Andrew Philip
Smith, Cassandra Lynn
Smith, Douglas Wemp
Smith, Helene Sheila
Smith, Kendric Charles
Smith, Steven Sidney
Snow, Sidney Richard
Sokoloff, Alexander
Soost, Robert Kenneth
Sparkes, Robert Stanley
Spence, Mary Anne
Spieler, Richard Arno
Spieth, Philip Theodore
Stallcup, Michael R
Stansfield, William D
Stormont, Clyde J
Straus, Daniel Steven
Strauss, Ellen Glowacki
Strickberger, Monroe Wolf
Strobel, Edward
Stumph, William Edward
Sung, Zinmay Renee
Tallman, John Gary
Taylor, Charles Ellett
Teuber, Larry Ross
Thomas, Heriberto Victor
Thompson, David J
Thompson, Lawrence Hadley
Thwaites, William Mueller
Touchberry, Robert Walton
Treat-Clemons, Lynda George
Van Elswyk, Marinus, Jr
Vogt, Peter Klaus
Vyas, Girish Narmadashankar
Wallace, Robert Bruce
Warren, Don Cameron
Waser, Nickolas Merritt
Wechsler, Steven Lewis
Wehr, Carl Timothy
Weinberg, Barbara Lee Huberman
Welch, James Edward
Wheelis, Mark Lewis
Whissell-Buechy, Dorothy Y E
Whitaker, Thomas Wallace
Wiebe, Michael Eugene
Wiktorowicz, John Edward
Wilcox, Bruce Alexander
Wilcox, Gary Lynn
Williams, Bobby Joe
Williams, Hibbard E
Wills, Christopher J
Wolff, Sheldon
Woodruff, David Scott
Yager, Janice L Winter
Ying, Kuang Lin
Yoshida, Akira
Yund, Mary Alice
Zamenhof, Patrice Joy
Zary, Keith Wilfred
Zimm, Georgianna Grevatt
Zipser, David
Zuccarelli, Anthony Joseph
Zuckekandl, Emile

COLORADO
Balbinder, Elias
Barrett, Dennis
Bock, Jane Haskett
Brandom, William Franklin
Brinks, James S
Clark, Roger William
Crumpacker, David Wilson
Cuany, Robin Louis
DeFries, John Clarence
Deitrich, Richard Adam
Dixon, Linda Kay
Eberhart, Steve A
Fechner, Gilbert Henry
Heim, Werner George
Hughes, Harrison Gilliatt
Johnson, Thomas Eugene
Jones, Carol A
Kano-Sueoka, Tamiko
Kao, Fa-Ten
Keim, Wayne Franklin
Ladd, Sheldon Lane
Liao, Martha
Linhart, Yan Bohumil
Mitton, Jeffry Bond
Morse, Helvise Glessner
Morse, Melvin Laurance
Nash, Donald Joseph
Nora, James Jackson
Ogg, James Elvis

Parma, David Hopkins
Patterson, David
Peeples, Earle Edward
Pettijohn, David E
Prescott, David Marshall
Puck, Mary Hill
Puck, Theodore Thomas
Quick, James S
Robinson, Arthur
Schanfield, Moses Samuel
Sinensky, Michael
Stein, Gretchen Herpel
Sueoka, Noboru
Taylor, Austin Laurence
Townsend, Charley E
Tsuchiya, Takumi
Vandenberg, Steven Gerritjan
Waldren, Charles Allen
Wehner, Jeanne M
Wenger, David Arthur
Wilson, Vincent L
Wood, William Barry, III

CONNECTICUT
Adelberg, Edward Allen
Amacher, David E
Anagnostakis, Sandra Lee
Ashley, Terry Fay
Bachmann, Barbara Joyce
Berg, Claire M
Berlyn, Mary Berry
Breg, William Roy
Chovnick, Arthur
Donady, J James
Eisenstadt, Jerome Melvin
Fenton, Wayne Alexander
Flavell, Richard Anthony
Fu, Wei-ning
Galbraith, Donald Barrett
Geiger, Jon Ross
Ginsburg, Benson Earl
Golub, Efim I
Gordon, Philip N
Greenblatt, Irwin M
Grindley, Nigel David Forster
Halaban, Ruth
Hanson, Earl Dorchester
Holsinger, Kent Eugene
Honeyman, Merton Seymour
Kavathas, Paula
Kidd, Kenneth Kay
Koerting, Lola Elisabeth
Kreizinger, Jean Dolloff
Liskay, Robert Michael
Low, Kenneth Brooks, Jr
Lynch, Carol Becker
McCorkle, George Maston
Martinez, Robert Manuel
Matheson, Dale Whitney
Mergen, Francois
Pawelek, John Mason
Pierro, Louis John
Poole, Andrew E
Poulson, Donald Frederick
Proctor, Alan Ray
Radding, Charles Meyer
Ray, Verne A
Ricciuti, Florence Christine
Rice, Frank J
Rishell, William Arthur
Rosenberg, Leon Emanuel
Ruddle, Francis Hugh
Rupp, W Dean
Sarkar, Siddhartha
Schacter, Bernice Zeldin
Schultz, R Jack
Schwinck, Ilse
Shapiro, Nathan
Silander, John August, Jr
Slayman, Carolyn Walch
Somes, Ralph Gilmore, Jr
Stevenson, Harlan Quinn
Summers, Wilma Poos
Truesdell, Susan Jane
Ward, David Christian
Wheeler, Bernice Marion

DELAWARE
Borgaonkar, Digamber Shankarrao
Eleuterio, Marianne Kingsbury
Francis, David W
Gaffney, Patrick M
Gatenby, Anthony Arthur
Gould, Adair Brasted
Hobgood, Richard Troy, Jr
Hodson, Robert Cleaves
Hoover, Dallas Gene
Kundt, John Fred
Lighty, Richard William
Martin-Deleon, Patricia Anastasia
Pearson, Mark Landell
Pene, Jacques Jean
Pierce, Edward Ronald
Sheppard, David E
White, Harold Bancroft, III

DISTRICT OF COLUMBIA
Ampy, Franklin
Baron, Louis Sol
Baumiller, Robert Cahill
Bergmann, Fred Heinz
Brosseau, George Emile, Jr
Bundy, Bonita Marie
Calvert, Allen Fisher

Chan, Wai-Yee
Chen, H R
DeGiovanni-Donnelly, Rosalie F
Dunkel, Virginia Catherine
Dutta, Sisir Kamal
Eastwood, Basil R
Falke, Ernest Victor
Gelmann, Edward P
Harriman, Philip Darling
Haskins, Caryl Parker
Hawkins, Morris, Jr
Hill, Richard Norman
Hill, Walter Ernest
Hines, Pamela Jean
Kinney, Terry B, Jr
Kumar, Ajit
Landman, Otto Ernest
Murray, Robert Fulton, Jr
Nasser, DeLill
Nightingale, Elena Ottolenghi
Niswander, Jerry David
Petricciani, John C
Poillon, William Neville
Policansky, David J
Prival, Michael Joseph
Ralls, Katherine Smith
Robbins, Robert John
Rothman, Sara Weinstein
Santamour, Frank Shalvey, Jr
Setlow, Valerie Petit
Simopoulos, Artemis Panageotis
Smith, David Allen
Thomas, Walter Ivan
Townsend, Alden Miller
Townsend, Alden Miller
Winter, William Phillips
Wrathall, Jean Rew
Zimmer, Elizabeth Anne

FLORIDA
Adams, Roger Omar
Baker, Richard H
Barnett, Ronald David
Baylis, John Robert, Jr
Binninger, David Michael
Bourne, Carol Elizabeth Mulligan
Buslig, Bela Stephen
Davis, Ralph Lanier
Dean, Charles Edgar
DeBusk, A Gib
DeKloet, Siwo R
Duggan, Dennis E
Edwardson, John Richard
Emmel, Thomas C
Ferl, Robert Joseph
Gaunt, Stanley Newkirk
Gilman, Lauren Cundiff
Goddard, Ray Everett
Greer, Sheldon
Hall, Harlan Glenn
Hauswirth, William Walter
Hecht, Frederick
Herskowitz, Irwin Herman
Hinson, Kuell
Huijing, Frans
Ingram, Lonnie O'Neal
Johnson, F Clifford
Jones, John Paul
Keppler, William J
Kitzmiller, James Blaine
Knauft, David A
Kossuth, Susan
Kuhn, David Truman
Lai, Patrick Kinglun
Laipis, Philip James
Lee, Doh-Yeel
Lessios, Harilaos Angelou
Lewin, Alfred S
Linden, Duane B
Long, Alan Jack
Luykx, Peter (van Oosterzee)
Marcus, Nancy Helen
Martin, Franklin Wayne
Miyamoto, Michael Masao
Moffa-White, Andrea Marie
Mortensen, John Alan
Moyer, Richard W
Muench, Karl Hugo
Norden, Allan James
Peacock, Hugh Anthony
Pfahler, Paul Leighton
Polson, Charles David Allen
Reid, Parlane John
Richmond, Rollin Charles
Rife, David Cecil
Roberts, Thomas L
Roess, William B
Rumbach, William Ervin
Schank, Stanley Cox
Scott, John Warner
Seawright, Jack Arlyn
Shanmugam, Keelnatham
     Thirunavukkarasu
Shirk, Paul David
Smith, Rex L
Stine, Gerald James
Stock, David Allen
Streilein, Jacob Wayne
Stroh, Robert Carl
Taylor, J(ames) Herbert
Teas, Howard Jones
Tedesco, Thomas Albert
Terranova, Andrew Charles
Thomason, David Morton

Genetics (cont)

Tracey, Martin Louis, Jr
Travis, Joseph
Voelly, Richard Walter
Wagoner, Dale E
Walker, Thomas Jefferson
Warren, Richard Joseph
White, Timothy Lee
Wilfret, Gary Joe

GEORGIA

Ambrose, John Augustine
Anderson, Wyatt W
Andrews, Lucy Gordon
Arnold, Jonathan
Avise, John C
Bennett, Sara Neville
Bongarten, Bruce C
Brownell, George H
Bryan, John Henry Donald
Burk, Lawrence G
Byrd, J Rogers
Case, Mary Elizabeth
Chen, Andrew Tat-Leng
Clark, Flora Mae
Crenshaw, John Walden, Jr
Crouse, Gray F
Danner, Dean Jay
Davis, Herbert L, Jr
Dinus, Ronald John
Duncan, Ronny Rush
Dusenbery, David Brock
Elmer, William Arthur
Elsas, Louis Jacob, II
Falek, Arthur
Fleming, Attie Anderson
Gardner, Arthur Wendel
Giles, Norman Henry
Glover, Claiborne V C, III
Golden, Ben Roy
Gourse, Richard Lawrence
Green, Willard Wynn
Hall, Dwight Hubert
Hammons, Ray Otto
Hamrick, James Lewis
Hanna, Wayne William
Hollowell, Joseph Gurney, Jr
Holzman, Gerald Bruce
Hommes, Frits A
Howe, Henry Branch, Jr
Kraus, John Franklyn
Kushner, Sidney Ralph
Lee, Joshua Alexander
Lucchesi, John Charles
McCartney, Morley Gordon
Madden, John Joseph
Marks, Henry L
Meagher, Richard Brian
Miller, John David
Mixon, Aubrey Clifton
Neville, Walter Edward, Jr
Oakley, Godfrey Porter, Jr
Oliver, James Henry, Jr
Pandey, Kailash N
Papa, Kenneth E
Patterson, Rosalyn Mitchell
Ray, Charles, Jr
Reines, Daniel
Satya-Prakash, K L
Schierman, Louis W
Schmidt, Gregory Wayne
Sluder, Earl Ray
Summers, Anne O
Thompson, James Marion
Thompson, Peter Ervin
Timberlake, William Edward
Volpe, Erminio Peter
Wallace, Douglas Cecil
Warren, Stephen Theodore
Washburn, Kenneth W
Weaver, James B, Jr
Whitney, J(ohn) Barry, III
Widstrom, Neil Wayne
Woessner, Ronald Arthur
Woodley, Charles Lamar

HAWAII

Ahearn, Jayne Newton
Ashton, Geoffrey C
Brewbaker, James Lynn
Chung, Chin Sik
Dalton, Howard Clark
Folsome, Clair Edwin
Goodman, Madelene Joyce
Hassold, Terry Jon
Heinz, Don J
Hunt, John A
Malecha, Spencer R
Mi, Ming-Pi
Reimer, Diedrich
Rotar, Peter P
Sagawa, Yoneo
Scott, John Francis
Simon, Christine Mae
Stuart, William Dorsey

IDAHO

Christian, Ross Edgar
Dahmen, Jerome J
Farrell, Larry Don
Fins, Lauren
Forbes, Oliver Clifford
Hansen, Leon A

Myers, James Robert
Pavek, Joseph John
Pearson, Lorentz Clarence
Ramagopal, Subbanaidu
Stephens, Trent Dee
Tullis, James Earl
Winston, Vern

ILLINOIS

Alexander, Denton Eugene
Alexander, Nancy J
Allison, David C
Amarose, Anthony Philip
Ausich, Rodney L
Bailey, Zeno Earl
Baumgardner, Kandy Diane
Beck, Sidney L
Becker, Michael Allen
Bell, Clara G
Bennett, Cecil Jackson
Bernard, Richard Lawson
Bouck, Noel
Bowman, James E
Bradford, Laura Sample
Brewen, J Grant
Briggs, Robert Wilbur
Briles, Worthie Elwood
Brockman, Herman E
Casadaban, Malcolm John
Chakrabarty, Ananda Mohan
Charlesworth, Brian
Cole, Michael Allen
Cummings, Michael R
Daniel, William L
Davidson, Richard Laurence
DeWet, Jan M J
Doering, Jeffrey Louis
Dorus, Elizabeth
Dudley, John Wesley
Englert, Du Wayne Cleveland
Esposito, Rochelle E
Farnsworth, Marjorie Whyte
Fast, Dale Eugene
Fennewald, Michael Andrew
Fernando, Rohan Luigi
Fox, James David
Francis, Bettina Magnus
Frischer, Henri
Fuchs, Elaine V
Garber, Edward David
Gardner, Jeffery Fay
Garwood, Douglas Leon
Geer, Billy W
George, William Leo, Jr
Gerdy, James Robert
Giordano, Tony
Glaser, Janet H
Gorsic, Joseph
Grahn, Douglas
Gravett, Howard L
Griesbach, Robert Anthony
Grossman, Michael
Grundbacher, Frederick John
Hirsch, Jerry
Holland, Louis Edward, II
Hoo, Joe Jie
Hooker, Arthur Lee
Huberman, Eliezer
Hymowitz, Theodore
Jones, Carl Joseph
Juergensmeyer, Elizabeth B
Jugenheimer, Robert William
Jump, Lorin Keith
Kang, David Soosang
Katz, Alan Jeffrey
Kaufman, Thomas Charles
Keim, Barbara Howell
King, Robert Charles
King, Robert Willis
Koch, Elizabeth Anne
Laffler, Thomas G
Lambert, Robert John
Lamppa, Gayle K
Lande, Russell Scott
Laughnan, John Raphael
Leffler, Harry Rex
Lerner, Jules
Liebman, Susan Weiss
Ma, Te Hsiu
MacHattie, Lorne Allister
Mahowald, Anthony P
Marczynska, Barbara Mary
Markovitz, Alvin
Melvold, Roger Wayne
Mets, Laurens Jan
Miller, Darrell Alvin
Miller, Robert Verne
Morishima, Akira
Myers, Oval, Jr
Nadler, Henry Louis
Nagylaki, Thomas Andrew
Nair, Shankar P
Nanney, David Ledbetter
Nickell, Cecil D
Nyberg, Dennis Wayne
Panasenko, Sharon Muldoon
Patterson, Earl Byron
Plate, Charles Alfred
Plewa, Michael Jacob
Pratt, Charles Walter
Propst, Catherine Lamb
Prout, Timothy
Rasmusen, Benjamin Arthur
Rink, George

Robertson, Hugh Mereth
Robison, Norman Glenn
Roderick, William Rodney
Roizman, Bernard
Roth, Robert Mark
Rowley, Janet D
Rupprecht, Kevin Robert
Salisbury, Glenn Wade
Sargent, Malcolm Lee
Schemske, Douglas William
Schreiber, Hans
Shanks, Roger D
Shapiro, James Alan
Simon, Ellen McMurtrie
Smith, Patricia Anne
Spear, Brian Blackburn
Spiess, Eliot Bruce
Spofford, Janice Brogue
Steffensen, Dale Marriott
Strauss, Bernard S
Thompson, Vinton Newbold
Tripathi, Satish Chandra
Troyer, Alvah Forrest
Tuveson, Robert Williams
Vandehey, Robert C
Vary, Patricia Susan
Vodkin, Lila Ott
Vodkin, Michael Harold
Wade, Michael John
Weber, David Frederick
Whitt, Dixie Dailey
Whitt, Gregory Sidney
Widholm, Jack Milton
Yarger, James G
Yokoyama, Shozo
Yoon, Ji-Won

INDIANA

Anderson-Mauser, Linda Marie
Axtell, John David
Bard, Martin
Beineke, Walter Frank
Bender, Harvey Alan
Bixler, David
Bockrath, Richard Charles, Jr
Bohren, Bernard Benjamin
Bonner, James Jose
Bosron, William F
Boston, Andrew Chester
Brush, F(ranklin) Robert
Campbell, Douglas Arthur
Castleberry, Ron M
Cherbas, Peter Thomas
Chiscon, J Alfred
Chrisman, Charles Larry
Christian, Joe Clark
Cole, Thomas A
Dantzig, Anne H
Edenberg, Howard Joseph
Engstrom, Lee Edward
Fischler, Drake Anthony
Gallun, Robert Louis
Garriott, Michael Lee
Gastony, Gerald Joseph
Gidda, Jaswant Singh
Glover, David Val
Goff, Charles W
Goldstein, David Joel
Hegeman, George D
Hershberger, Charles Lee
Hickey, William August
Hodes, Marion Edward
Hudock, George Anthony
Jaehning, Judith A
Janick, Jules
Karn, Robert Cameron
Kinzel, Jerry J
Larsen, Steven H
Levinthal, Mark
McCann, Peter Paul
McClure, Polley Ann
Martin, Truman Glen
Mertens, Thomas Robert
Nicholson-Guthrie, Catherine Shirley
Nielsen, Niels Christian
Nyquist, Wyman Ellsworth
Ohm, Herbert Willis
Oliver, Montague
Pak, William Louis
Palmer, Catherine Gardella
Patterson, Fred La Vern
Polley, Lowell David
Potter, Rosario H Yap
Queener, Stephen Wyatt
Rai, Karamjit Singh
Reed, Terry Eugene
Rhoades, Marcus Morton
Roman, Ann
Santerre, Robert Frank
Schaible, Robert Hilton
Schwartz, Drew
Shaw, Margery Wayne
Sheffer, Richard Douglas
Simon, Edward Harvey
Skjold, Arthur Christopher
Somerville, Ronald Lamont
Spieth, John
Stewart, Terry Sanford
Tan, James Chien-Hua
Taylor, Milton William
Tessman, Irwin
Tigchelaar, Edward Clarence
Tischfield, Jay Arnold
Tomes, Mark Louis

Tsai, Chia-Yin
Waldman, Alan S
Wappner, Rebecca Sue
Watson, Maxine Amanda
Weaver, David Dawson
Wilcox, Frank H
Wilson, Gary August
Yen, Terence Tsin Tsu
Zuckerman, Steven H

IOWA

Atherly, Alan G
Bamrick, John Francis
Benbow, Robert Michael
Berger, Philip Jeffrey
Brown, William Lacy
Carlson, Wayne R
Christian, Lauren L
Dalton, Lonnie Gene
Duvick, Donald Nelson
Ford, Clark Fugier
Frankel, Joseph
Gendel, Steven Michael
Grant, David Miller
Gussin, Gary Nathaniel
Hall, Richard Brian
Hegmann, Joseph Paul
Hoffmann, Richard John
Imsande, John
Kieso, Robert Alfred
Krafsur, Elliot Scoville
Lucas, Gene Allan
McNeill, Michael John
Marshall, William Emmett
Mayfield, John Eric
Meeker, David Lynn
Menninger, John Robert
Milkman, Roger Dawson
Miller, Wilmer Jay
Moore, Jay Winston
Nanda, Dave Kumar
Palmer, Reid G
Pattee, Peter A
Peterson, Peter Andrew
Rhead, William James
Robertson, Donald Sage
Russell, Wilbert Ambrick
Stieler, Carol Mae
Sunshine, Melvin Gilbert
Thorne, John Carl
Tomes, Dwight Travis
Walker, Jean Tweedy
Wang, Wei-Yeh
Welshons, William John
Wu, Chun-Fang
Wunder, William W

KANSAS

Bassi, Sukh D
Bode, Vernon Cecil
Clayberg, Carl Dudley
Craig, James Verne
Denell, Robin Ernest
Epp, Melvin David
Gill, Bikram Singh
Goldberg, Ivan D
Grebe, Janice Durr
Haufler, Christopher Hardin
Heiner, Robert E
Jackson, Sharon Wesley
Kaufman, Glennis Ann
Kinsey, John Aaron, Jr
Leslie, John Franklin
Liang, George H
Pittenger, Thad Heckle, Jr
Roufa, Donald Jay
Schalles, Robert R
Schlager, Gunther
Shankel, Delbert Merrill
Stetler, Dean Allen
Van Haverbeke, David F
Wassom, Clyde E
Weir, John Arnold
Wheat, John David
Wolf, Thomas Michael
Wolfe, Herbert Glenn

KENTUCKY

Collins, Glenn Burton
Cotter, William Bryan, Jr
Davidson, Jeffrey Neal
Dickson, Robert Carl
Goodwill, Robert
Martin, Nancy Caroline
Perlin, Michael Howard
Poneleit, Charles Gustav
Rawls, John Marvin, Jr
Sheen, Shuh-Ji
Sisken, Jesse Ernest
Stuart, James Glen
Williams, Arthur Lee
Wolfson, Alfred M
Yungbluth, Thomas Alan

LOUISIANA

Adams, John Clyde
Bennett, Joan Wennstrom
Bhatnager, Deepak
Chambers, Doyle
Christian, James A
Ellgaard, Erik G
French, Wilbur Lile
Griswold, Kenneth Edwin, Jr
Harlan, Jack Rodney

Hayes, Donald H
Humes, Paul Edwin
Kern, Clifford H, III
Kloepfer, Henry Warner
Koonce, Kenneth Lowell
Lee, William Roscoe
Lin, James C H
Lucas, Myron Cran
Mizell, Merle
Ramsey, Paul Roger
Rinderer, Thomas Earl
Shepherd, Hurley Sidney
Thurmon, Theodore Francis
Tipton, Kenneth Warren
Tucker, Kenneth Wilburn
Warters, Mary

MAINE
Bailey, Donald Wayne
Bernstein, Seldon Edwin
Blake, Richard D
Chai, Chen Kang
Champlin, Arthur Kingsley
Davisson, Muriel Trask
Doolittle, Donald Preston
Dowse, Harold Burgess
Eicher, Eva Mae
Fox, Richard Romaine
Gaskins, H Rex
Guidi, John Neil
Harris, Paul Chappell
Harrison, David Ellsworth
Hoppe, Peter Christian
Jerkofsky, Maryann
Kaliss, Nathan
LaMarche, Paul H
Les, Edwin Paul
Mobraaten, Larry Edward
Paigen, Beverly Joyce
Paigen, Kenneth
Ringo, John Moyer
Roberts, Franklin Lewis
Roderick, Thomas Huston
Russell, Elizabeth Shull
Sanford, Barbara Hendrick
Snell, George Davis
Steinhart, William Lee
Terry, Robert Lee

MARYLAND
Adhya, Sankar L
Altevogt, Raymond Fred
Amato, R Stephen S
Anderson, W French
Aycock, Marvin Kenneth, Jr
Barnett, Audrey
Barnhart, Benjamin J
Barranger, John Arthur
Battey, James F
Bias, Wilma B
Bigger, Cynthia Anita Hopwood
Borgia, Gerald
Bottino, Paul James
Boyer, Samuel H, IV
Bresler, Jack Barry
Breyere, Edward Joseph
Briggle, Leland Wilson
Brown, Kenneth Stephen
Buck, Raymond Wilbur, Jr
Camerini-Otero, Rafael Daniel
Chase, Gary Andrew
Chen, T R
Childs, Barton
Cifone, Maria Ann
Coe, Gerald Edwin
Cohen, Bernice Hirschhorn
Cohen, Maimon Moses
Conner, George William
Cowan, Elliot Paul
Cyr, W Howard
Dean, Ann
Dean, Jurrien
Devine, Thomas Edward
DiPaolo, Joseph Amedeo
Eldridge, Roswell
Elgin, James H, Jr
Falkenstein, Kathy Fay
Fearon, Douglas T
Felix, Jeanette S
Gahl, William A
Galletta, Gene John
Galloway, Sheila Margaret
Garges, Susan
Genys, John B
Gethmann, Richard Charles
Gilden, Raymond Victor
Goldman, David
Gorman, Cornelia M
Gottesman, Michael
Graham, Dale Elliott
Gray, David Bertsch
Gray, Paulette S
Greenberg, Judith Horovitz
Griesbach, Robert James
Guss, Maurice Louis
Gwynn, Edgar Percival
Hamer, Dean H
Hansen, Carl Tams
Hartman, Philip Emil
Hoeg, Jeffrey Michael
Hoffman, Harold A
Holbrook, Nikki J
Holtzman, Neil Anton
Hoyer, Leon William

Huang, Pien-Chien
Hudson, Lynn Diane
Imberski, Richard Bernard
James, Stephanie Lynn
Jones, Theodore Charles
Jordan, Elke
Kaiser-Kupfer, Muriel I
Kelly, Thomas J
Kernaghan, Roy Peter
Kety, Seymour S
Kimball, Paul Clark
Kindt, Thomas James
Klassen, Waldemar
Korper, Samuel
Kraemer, Kenneth H
Kwiterovich, Peter O, Jr
Kyle, Wendell H(enry)
Lacy, Ann Matthews
Landsman, David
Lerman, Michael Isaac
Levin, Barbara Chernov
Levine, Arthur Samuel
Lewis, Herman William
Lunney, Joan K
McBride, Orlando W
McKenney, Keith Hollis
McKeon, Catherine
McMacken, Roger
MacQuillan, Anthony M
Margulies, David Harvey
Marsh, David George
Martin, Robert G
Mayer, Vernon William, Jr
Melera, Peter William
Merril, Carl R
Migeon, Barbara Ruben
Mittal, Kamal Kant
Money, John William
Moseman, John Gustav
Murli, Hemalatha
Myrianthopoulos, Ntinos
Nash, Howard Allen
Naylor, Alfred F
Newrock, Kenneth Matthew
Nickerson, John Munro
O'Brien, Stephen James
Plato, Chris C
Potter, Michael
Powers, Dennis A
Price, Samuel
Remondini, David Joseph
Robbins, April Ruth
Robbins, Jay Howard
Robison, Wilbur Gerald, Jr
Rosner, Judah Leon
Russek-Cohen, Estelle
Sack, George H(enry), Jr
Scheinberg, Sam Louis
Schindler, Joel Marvin
Schuellein, Robert Joseph
Schulman, Joseph Daniel
Seydel, Frank David
Shaw, Richard Franklin
Shearn, Allen David
Silverman, Jeffrey Alan
Silverman, Robert Hugh
Smith, Gilbert Howlett
Smith, Hamilton Othanel
Smith-Gill, Sandra Joyce
Snope, Andrew John
Solomon, Joel Martin
Sprott, Richard Lawrence
Stonehill, Elliott H
Strathern, Jeffrey Neal
Sun, Nai Chau
Tester, Cecil Fred
Thorgeirsson, Snorri S
Tjio, Joe Hin
Uphoff, Delta Emma
Vlahakis, George
Voll, Mary Jane
Weiss, Bernard
Wolf, Richard Edward, Jr
Wu, Henry Chi-Ping
Yasbin, Ronald Eliott
Young, Bobby Gene

MASSACHUSETTS
Alper, Chester Allan
Atkinson-Templeman, Kristine Hofgren
Ausubel, Frederick Michael
Backman, Keith Cameron
Bawa, Kamaljit S
Beckwith, Jonathan Roger
Beckwitt, Richard David
Birchler, James Arthur
Bloom, Arthur David
Breakefield, Xandra Owens
Brown, Beverly Ann
Chin, William W
Clark, Arnold M
Comer, M Margaret
Cooper, Geoffrey Mitchell
Demain, Arnold Lester
Dick, Stanley
Dorf, Martin Edward
Dubey, Devendra P
Erbe, Richard W
Feig, Larry Allen
Fields, Bernard N
Fink, Gerald Ralph
Fox, Maurice Sanford
Fox, Thomas Oren
Fox, Thomas Walton

Fridovich-Keil, Judith Lisa
Fulton, Chandler Montgomery
Goldberg, Edward B
Goldstein, Lawrence S B
Good, Carl M, III
Haber, James Edward
Hall, Jeffrey Connor
Hattis, Dale B
Hecht, Norman B
Hexter, William Michael
Hoffmann, George Robert
Holmes, Helen Bequaert
Horvitz, Howard Robert
Hotchkiss, Rollin Douglas
Ives, Philip Truman
Jameson, James Larry
Kafatos, Fotis C
Kelner, Albert
Kelton, Diane Elizabeth
King, Jonathan (Alan)
Klekowski, Edward Joseph, Jr
Knipe, David Mahan
Kolakowski, Donald Louis
Koul, Omanand
Kunkel, Louis M
Lawrence, Jeanne Bentley
Ledbetter, Mary Lee Stewart
Leder, Philip
Lemontt, Jeffrey Fielding
Lerman, Leonard Solomon
Levy, Deborah Louise
Lewontin, Richard Charles
Li, Frederick P
Ludwin, Isadore
Lyerla, Jo Ann Harding
Lyerla, Timothy Arden
Ma, Nancy Shui-Fong
Madhavan, Kornath
Malamy, Michael Howard
Mange, Arthur P
Marcum, James Benton
Marinus, Martin Gerard
Merritt, Robert Buell
Miller, Judith Evelyn
Miller, Lynn
Mitchell, David Hillard
Moolten, Frederick London
Mulcahy, David Louis
Myers, Richard Hepworth
Nixon, Charles William
Papaioannou, Virginia Eileen
Pardue, Mary Lou
Petri, William Hugh
Reiner, Albey M
Rheinwald, James George
Riley, Monica
Rio, Donald C
Rollins, Reed Clark
Sager, Ruth
Schaefer, Ernst J
Schwaber, Jerrold
Seidman, Jonathan G
Seyfried, Thomas Neil
Sidman, Richard Leon
Siegel, Eli Charles
Signer, Ethan Royal
Snow, Beatrice Lee
Solbrig, Otto Thomas
Stein, Diana B
Struhl, Kevin
Swanson, Carl Pontius
Szoka, Paula Rachel
Szostak, Jack William
Tamarin, Robert Harvey
Tilley, Stephen George
Todd, Neil Bowman
Townes, Philip Leonard
Tsuang, Ming Tso
Tsung, Yean-Kai
Vankin, George Lawrence
Warner, Carol Miller
Weinstein, Alexander
Widmayer, Dorothea Jane
Williamson, Patrick Leslie
Witman, George Bodo, III
Wolf, Merrill Kenneth
Young, Delano Victor
Zannis, Vassilis I

MICHIGAN
Aaron, Charles Sidney
Abraham, Irene
Adams, Jack Dean
Adams, Julian Philip
Allen, Sally Lyman
Al Saadi, A Amir
Aminoff, David
Arking, Robert
Bach, Shirley
Bacon, Larry Dean
Barbour, Stephen D
Ben-Yoseph, Yoav
Bloom, Miriam
Bouma, Hessel, III
Bowers, Maynard C
Burnett, Jean Bullard
Bush, Guy L
Butterworth, Francis M
Caldwell, Jerry
Calvert, Jay Gregory
Chang, Chia-Cheng
Christman, Judith Kershaw
Chu, Ernest Hsiao-Ying
Chung, Shiau-Ta

Collins, Francis Sellers
Conover, James H
Cooper, Stephen
Corcos, Alain Francois
Cress, Charles Edwin
Crittenden, Lyman Butler
Cronkite, Donald Lee
Eisenstein, Barry I
Erickson, Robert Porter
Everson, Everett Henry
Ficsor, Gyula
Fleisher, Lynn Dale
Fluck, Michele Marguerite
Forsthoefel, Paulinus Frederick
Freeman, Dwight Carl
Freytag, Svend O
Friedman, Thomas Baer
Froiland, Thomas Gordon
Gay, Helen
Gelehrter, Thomas David
Gentile, James Michael
Gershowitz, Henry
Goustin, Anton Scott
Hackel, Emanuel
Hancock, James Findley, Jr
Hanover, James W
Harpstead, Dale D
Harris, James Edward
Heberlein, Gary T
Helling, Robert Bruce
Higgins, James Victor
Jacobs, Charles Warren
Janca, Frank Charles
Jones, Lily Ann
Karnosky, David Frank
Krawetz, Stephen Andrew
Levine, Myron
Lomax, Margaret Irene
McGilliard, Lon Dee
Martin, Joseph Patrick, Jr
Mayeda, Kazutoshi
Meisler, Miriam Horowitz
Miller, Curtis C
Miller, Dorothy Anne Smith
Miller, Orlando Jack
Moll, Patricia Peyser
Montgomery, Ilene Nowicki
Murnik, Mary Rengo
Nasjleti, Carlos Eduardo
Neel, James Van Gundia
Neidhardt, Frederick Carl
Padgett, George Arnold
Pelzer, Charles Francis
Phillip, Michael J
Rizki, Tahir Mirza
Robbins, Leonard Gilbert
Rossman, Elmer Chris
Ruby, John L
Schoenhard, Delbert E
Shontz, Nancy Nickerson
Siegel, Albert
Sing, Charles F
Sink, Kenneth C, Jr
Smith, David I
Smith, Paul Dennis
Somerville, Christopher Roland
Steiner, Erich E
Stone, Howard Anderson
Taggart, R(obert) Thomas
Tashian, Richard Earl
Theurer, Jessop Clair
Thoene, Jess Gilbert
Trosko, James Edward
Tse, Harley Y
Venta, Patrick John
Verley, Frank A
Wilson, Golder North
Wolk, Coleman Peter
Wolman, Sandra R
Wu, Ching Kuei

MINNESOTA
Anderson, Victor Elving
Barnes, Donald Kay
Beck, Barbara North
Berg, Robert W
Boylan, William J
Busch, Robert Henry
Caldecott, Richard S
Cervenka, Jaroslav
Comstock, Verne Edward
Condell, Yvonne C
Conlin, Bernard Joseph
Corbin, Kendall Wallace
Curtsinger, James Webb
Dapkus, David Conrad
David, Chelladurai S
Dearden, Douglas Morey
Desborough, Sharon Lee
Dewald, Gordon Wayne
Enfield, Franklin D
Engh, Helmer A, Jr
Fallon, Ann Marie
Fausch, Homer David
Forro, Frederick, Jr
Gengenbach, Burle Gene
Glass, Arthur Warren
Gordon, Hymie
Hansen, Leslie Bennett
Hanson, Daniel Ralph
Hedman, Stephen Clifford
Herforth, Robert S
Jessen, Carl Roger
Johnson, Freeman Keith

Genetics (cont)

Johnson, Theodore Reynold
Jokela, Jalmer John
King, Richard Allen
Kline, Bruce Clayton
Kowles, Richard Vincent
Lefebvre, Paul Alvin
Leonard, Kurt John
Lindgren, Alice Marilyn Lindell
Lofgren, James R
Lorimer, Nancy L
Lukasewycz, Omelan Alexander
Lykken, David Thoreson
McKay, Larry Lee
Magee, Paul Terry
Merrell, David John
Mock, James Joseph
Oetting, William Starr
Orf, John W
Palm, John Daniel
Phillips, Ronald Lewis
Rasmusson, Donald C
Reed, Elizabeth Wagner
Rehwaldt, Charles A
Reilly, Bernard Edward
Rines, Howard Wayne
Rodell, Charles Franklin
Rudolf, Paul Otto
Schacht, Lee Eastman
Shapiro, Burton Leonard
Shoffner, Robert Nurman
Silflow, Carolyn Dorothy
Singer, Susan Rundell
Snustad, Donald Peter
Snyder, Leon Allen
Spelsberg, Thomas Coonan
Spurrell, Francis Arthur
Stucker, Robert Evan
Stuthman, Deon Dean
Sulerud, Ralph L
Weibust, Robert Smith
Wettstein, Peter J
White, Donald Benjamin
Witkop, Carl Jacob, Jr
Woodward, Val Waddoups
Yunis, Jorge J

MISSISSIPPI

Adams, Junius Greene, III
Berry, Charles Dennis
Case, Steven Thomas
Creech, Roy G
Gupton, Creighton Lee
Jackson, John Fenwick
Jenkins, Johnie Norton
Kilen, Thomas Clarence
Peeler, Dudley F, Jr
Rutger, John Neil
St Amand, Wilbrod
Thomas, Charles Hill
Watson, Clarence Ellis, Jr
Wise, Dwayne Allison
Yarbrough, Karen Marguerite

MISSOURI

Alexander, Stephen
Anderson, Robert Glenn
Apirion, David
Ayyagari, L Rao
Beckett, Jack Brown
Brandt, E J
Brescia, Vincent Thomas
Brot, Frederick Elliot
Burdick, Allan Bernard
Cheney, Clarissa M
Cloninger, Claude Robert
Coe, Edward Harold, Jr
Curtiss, Roy, III
David, John Dewood
Eissenberg, Joel Carter
Fischhoff, David Allen
Folk, William Robert
Friedman, Lawrence David
Goodenough, Ursula Wiltshire
Gowans, Charles Shields
Gustafson, John Perry
Hansen, Ted Howard
Hartl, Daniel L
Heard, John Thibaut, Jr
Hess, John Berger
Hill, Jack Filson
Hillman, Richard Ephraim
Hoffman, Jacqueline Louise
Horsch, Robert Bruce
Hufham, James Birk
Johnson, Terrell Kent
Kane, James F
Kikudome, Gary Yoshinori
Kimber, Gordon
Klotz, John William
Levine, Robert Paul
Lie, Wen-Rong
Lower, William Russell
Lye, Robert J
McQuade, Henry Alonzo
Melechen, Norman Edward
Mohamed, Aly Hamed
Neuffer, Myron Gerald
Olson, Maynard Victor
Redei, Gyorgy Pal
Riddle, Donald Lee
Salkoff, Lawrence Benjamin
Sawyer, Stanley Arthur

Schneiderman, Howard Allen
Schreiweis, Donald Otto
Sears, Ernest Robert
Shreffler, Donald Cecil
Sly, William S
Stalker, Harrison Dailey
Stufflebeam, Charles Edward
Tai, William
Trinklein, David Herbert
Tritz, Gerald Joseph
Waite, Albert B
Wang, Richard J
Willard, Mark Benjamin

MONTANA

Allendorf, Frederick William
Blackwell, Robert Leighton
Blake, Tom
Burfening, Peter J
Burris, Martin Joe
Cameron, David Glen
Cantrell, John Leonard
Carlson, George A
Ericsson, Ronald James
Hovin, Arne William
Kress, Donnie Duane
McCoy, Thomas Joseph
McNeal, Francis H
MacNeil, Michael
Opitz, John Marius
Priest, Jean Lane Hirsch
Stimpfling, Jack Herman
Warren, Guylyn Rea
Welsh, James Ralph

NEBRASKA

Ahlschwede, William T
Baltensperger, David Dwight
Bennett, Gary Lee
Brumbaugh, John (Albert)
Carson, Steven Douglas
Compton, William A
Dickerson, Gordon Edwin
Dubes, George Richard
Eastin, John A
Eisen, James David
Gorz, Herman Jacob
Gregory, Keith Edward
Harris, Dewey Lynn
Haskins, Francis Arthur
Jenkins, Thomas Gordon
Kimberling, William J
Laster, Danny Bruce
Lund, Douglas E
Lunt, Steele Ray
Ross, William Max
Sanger, Warren Glenn
Schutz, Wilfred M
Smeltzer, Dale Gardner
Van Vleck, Lloyd Dale
Veomett, George Ector
Young, Lawrence Dale

NEVADA

Bailey, Curtiss Merkel
Brussard, Peter Frans
Vig, Baldev K
Winicov, Ilga

NEW HAMPSHIRE

Berger, Edward Michael
Bergman, Kenneth David
Chase, Herman Burleigh
Cole, Charles N
Collins, Walter Marshall
Funk, David Truman
Garrett, Peter Wayne
Green, Donald MacDonald
Gross, Robert Henry
Hatch, Frederick Tasker
Kiang, Yun-Tzu
Loy, James Brent
Minoca, Subhash C
Rogers, Owen Maurice
Soares, Eugene Robbins
Wurster-Hill, Doris Hadley

NEW JERSEY

Agnish, Narsingh Dev
Arthur, Alan Thorne
Axelrod, David E
Bacharach, Martin Max
Benson, Charles Everett
Berg, Richard A
Brandriss, Marjorie C
Byrne, Barbara Jean McManamy
Carlson, James H
Champe, Sewell Preston
Choi, Ye-Chin
Cox, Edward Charles
Cryer, Dennis Robert
Day, Peter Rodney
Day-Salvatore, Debra-Lynn
Dhruv, Rohini Arvind
Egger, M(aurice) David
Essien, Francine B
Fisher, Kenneth Walter
Flick, Christopher E
Frenkel, Gerald Daniel
Friedman, Ronald Marvin
Gavurin, Lillian
Gepner, Ivan Alan
Gillum, Amanda McKee
Goldman, Emanuel

Grassle, Judith Payne
Hill, Helene Zimmermann
Hochstadt, Joy
Hu, Ching-Yeh
Huber, Ivan
Humayun, Mir Z
Jaffe, Ernst Richard
Joslyn, Dennis Joseph
Kaback, David Brian
Kapp, Robert Wesley, Jr
Kirsch, Donald R
Klug, William Stephen
Krause, Eliot
Kurtz, Myra Berman
Lee, Ming-Liang
Leibowitz, Michael Jonathan
Levine, Arnold J
McGuire, Terry Russell
Madison, Caroline Rabb
Maloy, Joseph T
Martin, Charles Everett
Mather, Robert Eugene
Megna, John C(osimo)
Messing, Joachim W
Mezick, James Andrew
Middleton, Richard B
Moyer, Samuel Edward
Nagle, James John
Newton, (William) Austin
Ozer, Harvey Leon
Pai, Anna Chao
Passmore, Howard Clinton
Prensky, Wolf
Quinn, James Amos
Rosen, Dianne L
Rosenstraus, Maurice Jay
Schreiner, Ceinwen Ann
Schroff, Peter David
Senda, Mototaka
Shapiro, Herman Simon
Shin, Seung-il
Silhavy, Thomas J
Silver, Lee Merrill
Smouse, Peter Edgar
Sofer, William Howard
Stern, Elizabeth Kay
Stevens, Merwin Allen
Swenson, Theresa Lynn
Unowsky, Joel
Weinstein, David
Weisbrot, David R
Witkin, Evelyn Maisel
Worthy, Thomas E

NEW MEXICO

Baker, William Kaufman
Bartholdi, Marty Frank
Burks, Christian
Chen, David J
Deaven, Larry Lee
Dillon, Richard Thomas
Dunford, Max Patterson
Fisher, James Thomas
Gurd, Ruth Sights
Holland, Lewis
Hsi, David Ching Heng
Hubby, John L
Johnson, William Wayne
Kelly, Gregory
Kraemer, Paul Michael
McCuistion, Willis Lloyd
Melton, Billy Alexander, Jr
Nelson, Mary Anne
Phillips, Gregory Conrad
Shortess, David Keen
Strniste, Gary F
Thomassen, David George
Wagner, Robert Philip

NEW YORK

Abdelnoor, Alexander Michael
Amidon, Thomas Edward
Anderson, Ronald Eugene
Auerbach, Arleen D
Baglioni, Corrado
Baird, Malcolm Barry
Baker, Raymond Milton
Banerjee, Ranjit
Basilico, Claudio
Bazinet, George Frederick
Beam, Carl Adams
Beardsley, Robert Eugene
Belfort, Marlene
Bell, Robin Graham
Bender, Michael A
Benjaminson, Morris Aaron
Bernheimer, Harriet P
Bleyman, Lea Kanner
Bloom, Stephen Earl
Bopp, Lawrence Howard
Borowsky, Richard Lewis
Bradbury, Michael Wayne
Bregman, Allyn A(aron)
Breslow, Jan Leslie
Broker, Thomas Richard
Bruns, Peter John
Bukhari, Ahmad Iqbal
Bushkin, Yuri
Buxbaum, Joel N
Calame, Kathryn Lee
Calhoon, Robert Ellsworth
Carlson, Elof Axel
Carlson, Marian B
Caspari, Ernst Wolfgang

Catterall, James F
Chaganti, Raju Sreerama Kamalasana
Chalfie, Martin
Chang, Tien-ding
Chapman, Verne M
Chase, Sherret Spaulding
Chasin, Lawrence Allen
Chen, Ching-Ling Chu
Chepko-Sade, Bonita Diane
Clark, Virginia Lee
Cohen, Elias
Cole, Randall Knight
Craddock, Elysse Margaret
Cunningham, Richard Preston
Desnick, Robert John
Desplan, Claude
D'Eustachio, Peter
Dibennardo, Robert
Dietert, Rodney Reynolds
Dottin, Robert P
Doudney, Charles Owen
DuFrain, Russell Jerome
Dykhuizen, Daniel Edward
Ebert, Patricia Dorothy
Ehrman, Lee
Elliott, Rosemary Waite
Emmons, Scott W
Emsley, James Alan Burns
Enea, Vincenzo
Erk, Frank Chris
Erlenmeyer-Kimling, L
Everett, Herbert Lyman
Falk, Catherine T
Fantini, Amedeo Alexander
Ferguson, John Barclay
Fischman, Harlow Kenneth
Forest, Charlene Lynn
Fortune, Joanne Elizabeth
Fox, Thomas David
Frair, Wayne
Free, Stephen J
Galinsky, Irving
Gallagher, Jane Chispa
Garner, James G
Garrick, Laura Morris
Gates, Allen H(azen), Jr
German, James Lafayette, III
Gilbert, Fred
Ginzburg, Lev R
Gladstone, William Turnbull
Glass, Hiram Bentley
Goff, Stephen Payne
Goldberg, Allan Roy
Goldschmidt, Raul Max
Goldsmith, Lowell Alan
Goldstein, Fred Bernard
Gonzalez, Eulogio Raphael
Gordon, Jon W
Grimes, Gary Wayne
Grimwood, Brian Gene
Grodzicker, Terri Irene
Grossfield, Joseph
Guthrie, Robert
Hall, Barry Gordon
Hanafusa, Hidesaburo
Harford, Agnes Gayler
Harrison, Richard Gerald
Hartung, John David
Henderson, Charles R
Hershey, Alfred D
Himes, John Harter
Hinkle, David Currier
Hirschhorn, Kurt
Hotchkiss, Sharon K
Howard, Irmgard Matilda Keeler
Huberman, Joel Anthony
Hurst, Donald D
Hutt, Frederick Bruce
Jagiello, Georgiana Mary
Jainchill, Jerome
Jamback, Hugo Andrew, Jr
Jayme, David Woodward
Jhanwar, Suresh Chandra
Jimenez-Marin, Daniel
Johnsen, Roger Craig
Johnson, Edward Michael
Johnson, Lawrence Lloyd
Jolly, Clifford J
Kaelbling, Margot
Kallman, Klaus D
Kambysellis, Michael Panagiotis
Kascsak, Richard John
Katz, Eugene Richard
Kelleher, Raymond Joseph, Jr
Kelly, Sally Marie
Kemphues, Kenneth J
Kessin, Richard Harry
Khan, Nasim A
King, James Clement
Kish, Valerie Mayo
Klinger, Harold P
Koehn, Richard Karl
Konopka, Ronald J
Korf, Richard Paul
Kramer, Fred Russell
Kucherlapati, Raju Suryanarayana
Lacks, Sanford
LaFountain, James Robert, Jr
Lange, Christopher Stephen
Last, Robert L
Latorella, A Henry
Lavett, Diane Kathryn Juricek
Lawrence, Christopher William
Lazzarini, Robert A

Leary, James Francis
Lederberg, Joshua
Leifer, Zer
Leslie, Paul Willard
Levene, Howard
Levine, Louis
Levitan, Max
Lewis, Leslie Arthur
Lilly, Frank
Lin, Yue Jee
Lindahl, Lasse Allan
Lis, John Thomas
Litwin, Stephen David
Liu, Houng-Zung
Lugthart, Garrit John, Jr
Maas, Werner Karl
McClintock, Barbara
McCormick, Paulette Jean
Margolin, Paul
Marien, Daniel
Marsh, William Laurence
Maynard, Charles Alvin
Mehta, Bipin Mohanlal
Meyers, Marian Bennett
Michels, Corinne Anthony
Miller, Morton W
Mitra, Jyotirmay
Model, Peter
Monheit, Alan G
Moore, Carol Wood
Mukherjee, Asit B
Munro, Donald W, Jr
Murphy, Donal B
Muschio, Henry M, Jr
Nasrallah, Mikhail Elia
Nguyen-Huu, Chi M
Nitowsky, Harold Martin
Novick, Richard P
Nur, Uzi
Oltenacu, Elizabeth Allison Branford
O'Reilly, Richard
Ottman, Ruth
Pellicer, Angel
Pesetsky, Irwin
Pollard, Jeffrey William
Possidente, Bernard Philip, Jr
Prakash, Louise
Prakash, Satya
Price, Peter Michael
Pumo, Dorothy Ellen
Qazi, Qutubuddin H
Quinn, Cosmas Edward
Rainer, John David
Rajan, Thiruchandurai Viswanathan
Ramirez, Francisco
Rattazzi, Mario Cristiano
Raveche, Elizabeth Marie
Reddy, Kalluru Jayarami
Reilly, Marguerite
Reisch, Bruce Irving
Reissig, Jose Luis
Robinson, Albert Dean
Rockwell, Robert Franklin
Rogerson, Allen Collingwood
Roll, David E
Rossman, Toby Gale
Rotheim, Minna B
Rothstein, Rodney Joel
Rothwell, Norman Vincent
Rowley, Peter Templeton
Ruddell, Alanna
Rudner, Rivka
Rudy, Bernardo
Russel, Marjorie Ellen
Russell, George K(eith)
Salwen, Martin J
Sank, Diane
Scott, William Addison, III
Sechrist, Lynne Luan
Shafer, Stephen Joel
Shaffer, Jacquelin Bruning
Shafit-Zagardo, Bridget
Shechter, Yaakov
Sherman, Fred
Shows, Thomas Byron
Siegel, Irwin Michael
Silverstein, Emanuel
Silverstein, Saul Jay
Singh, Madho
Sirlin, Julio Leo
Sirotnak, Francis Michael
Smith, Harold Hill
Solish, George Irving
Sorrells, Mark Earl
Srb, Adrian Morris
Stanley, Pamela Mary
Steinberg, Bettie Murray
Stinson, Harry Theodore, Jr
Stockert, Elisabeth
Studier, Frederick William
Sullivan, David Thomas
Taub, Mary L
Thompson, Steven Risley
Tilley, Shermaine Ann
Tomashefsky, Philip
Tye, Bik-Kwoon
Valentine, Fredrick Arthur
Verma, Ram S
Vladutiu, Georgirene Dietrich
Waelsch, Salome Gluecksohn
Wallace, Donald Howard
Warburton, Dorothy
Wasserman, Marvin
Weitkamp, Lowell R

Weyter, Frederick William
Wiberg, John Samuel
Williamson, David Lee
Wilson, Dwight Elliott, Jr
Winchester, Robert J
Wishnick, Marcia M
Wolfner, Mariana Federica
Wolgemuth, Debra Joanne
Wulff, Daniel Lewis
Yang, Shung-Jun
Yoder, Olen Curtis
Young, Charles Stuart Hamish
Zahler, Stanley Arnold
Zaleski, Marek Bohdan
Zengel, Janice Marie
Zinder, Norton David
Zitomer, Richard Stephen

NORTH CAROLINA
Acedo, Gregoria N
Allen, Wendall E
Amos, Dennis Bernard
Anderton, Laura Gaddes
Atchley, William Reid
Barrett, J(ames) Carl
Barry, Edward Gail
Bates, William K
Bell, Juliette B
Bewley, Glenn Carl
Bishop, Jack Belmont
Bishop, Paul Edward
Bloom, Kerry Steven
Boone, Lawrence Rudolph, III
Bostian, Carey Hoyt
Bott, Kenneth F
Boyd, Jeffrey Allen
Boynton, John E
Burchall, James J
Butterworth, Byron Edwin
Carter, Thomas Edward, Jr
Chae, Chi-Bom
Chaplin, James Ferris
Claxton, Larry Davis
Cockerham, Columbus Clark
Cohen, Carl
Collins, William Kerr
Cook, Robert Edward
Cope, Will Allen
Corley, Ronald Bruce
Counce, Sheila Jean
Craig, Syndey Pollock, III
Crowl, Robert Harold
Curtis, Susan Julia
Daugherty, Patricia A
Davis, Daniel Layten
de Serres, Frederick Joseph
Devlin, Robert B
Dillard, Emmett Urcey
Drake, John W
Edwards, James Wesley
Eisen, Eugene J
Endow, Sharyn Anne
Errede, Beverly Jean
Farber, Rosann Alexander
Fowler, Elizabeth
Frelinger, Jeffrey
Gerstel, Dan Ulrich
Gilbert, Lawrence Irwin
Gillham, Nicholas Wright
Glazener, Edward Walker
Gooder, Harry
Goodman, Harold Orbeck
Goodman, Major M
Grosch, Daniel Swartwood
Gross, Samson Richard
Guild, Walter Rufus
Hanson, Warren Durward
Harris, Elizabeth Holder
Haughton, Geoffrey
Havenstein, Gerald B
Hebert, Teddy T
Heise, Eugene Royce
Hershfield, Michael Steven
Hicks, Robert Eugene
Hildreth, Philip Elwin
Holmes, Edward Warren
Horner, Theodore Wright
Hutchison, Clyde Allen, III
Jolls, Claudia Lee
Judd, Burke Haycock
Kalfayan, Laura Jean
Kellison, Robert Clay
Kerschner, Jean
Kredich, Nicholas M
Leamy, Larry Jackson
Legates, James Edward
Levings, Charles Sandford, III
Li, Steven Shoei-Lung
McDaniel, Benjamin Thomas
McKenzie, Wendell Herbert
Malling, Heinrich Valdemar
Mangelsdorf, Paul Christoph
Mann, Thurston (Jefferson)
Markert, Clement Lawrence
Maroni, Donna F
Maroni, Gustavo Primo
Mason, James Michael
Matzinger, Dale Frederick
Merritt, James Francis
Mickey, George Henry
Mitchell, Ann Denman
Modrich, Paul L
Moll, Robert Harry
Moore, Martha May

Namkoong, Gene
Newell, Nanette
Ostrowski, Ronald Stephen
Perry, Thomas Oliver
Petes, Thomas Douglas
Petters, Robert Michael
Phelps, Allen Warner
Phillips, Lyle Llewellyn
Pollitzer, William Sprott
Pukkila, Patricia Jean
Rawlings, John Oren
Robinson, Harold Frank
Robison, Odis Wayne
Sargent, Frank Dorrance
Saylor, LeRoy C
Scandalios, John George
Schaffer, Henry Elkin
Sheridan, William
Sidhu, Bhag Singh
Smith, Gary Joseph
Spencer, Lorraine Barney
Spiker, Steven L
Stone, Donald Eugene
Stuber, Charles William
Sulik, Kathleen Kay
Suzuki, Kunihiko
Swift, Michael
Thompson, William Francis, III
Tice, Raymond Richard
Timothy, David Harry
Tourian, Ara Yervant
Triantaphyllou, Anastasios Christos
Verghese, Margrith Wehrli
Voelker, Robert Allen
Walker, Cheryl Lyn
Ward, Frances Ellen
Webster, Robert Edward
Weeks, Leo
Weir, Bruce Spencer
Welsch, Frank
Whittinghill, Maurice
Williamson, John Hybert
Wilson, James Franklin
Wong, Fulton
Wright, Clarence Paul
Wynne, Johnny Calvin
Zobel, Bruce John

NORTH DAKOTA
Doney, Devon Lyle
Hein, David William
Helm, James Leroy
Joppa, Leonard Robert
McDonald, Ian Cameron Crawford
Riemann, John G
Seiler, Gerald Joseph
Sheridan, William Francis
Smith, Garry Austin
Whited, Dean Allen
Williams, Norman Dale

OHIO
Atkins, Charles Gilmore
Baroudy, Bahige M
Barr, Harry L
Beck, Doris Jean
Bhattacharjee, Jnanendra K
Birky, Carl William, Jr
Blumenthal, Robert Martin
Brown, William Paul
Burns, George W
Byard, Pamela Joy
Carver, Eugene Arthur
Chase, John William
Clise, Ronald Leo
Cohn, Norman Stanley
Cooper, Richard Lee
Cullis, Christopher Ashley
Dean, Donald Harry
Dickerman, Richard Curtis
Dickinson, Frank N
Doerder, F Paul
Dollinger, Elwood Johnson
Erway, Lawrence Clifton, Jr
Fechheimer, Nathan S
Fraser, Alex Stewart
Freed, James Melvin
Fuerst, Paul Anthony
Ganschow, Roger Elmer
Garton, David Wendell
Gates, Michael Andrew
Gesinski, Raymond Marion
Glatzer, Louis
Goldman, Stephen L
Gregg, Thomas G
Griffing, J Bruce
Gromko, Mark Hedges
Hayes, Thomas G
Herschler, Michael Saul
Hines, Harold C
Hinton, Claude Willey
Hogg, Robert W
House, Verl Lee
Huether, Carl Albert
Jollick, Joseph Darryl
Juberg, Richard Caldwell
Kalter, Harold
Keever, Carolyn Anne
Kontras, Stella B
Kreutzer, Richard D
Kriebel, Howard Burtt
Kurczynski, Thaddeus Walter
Lafever, Howard N

Laughner, William James, Jr
Laushman, Roger H
Loper, John C
McCune, Sylvia Ann
McDougall, Kenneth J
Macintyre, Malcolm Neil
McQuate, John Truman
Marzluf, George A
Mikula, Bernard C
Milsted, Amy
Morrow, Grant, III
Nebert, Daniel Walter
Nestor, Karl Elwood
Oakley, Berl Ray
Paddock, Elton Farnham
Powelson, Elizabeth Eugenie
Powers, Jean D
Rake, Adrian Vaughan
Rayle, Richard Eugene
Robinow, Meinhard
Rucknagel, Donald Louis
Schafer, Irwin Arnold
Schwer, Joseph Francis
Seiger, Marvin Barr
Siervogel, Roger M
Skavaril, Russell Vincent
Steinberg, Arthur Gerald
Sullivan, Robert Little
Treichel, Robin Stong
Trewyn, Ronald William
Washington, Willie James
Waterson, John R
Williams, Marshall Vance
Wilson, Kenneth Glade
Woodruff, Ronny Clifford
Woodworth, Mary Esther
Wrensch, Dana Louise
Yoon, Jong Sik
Young, Sydney Sze Yih
Zartman, David Lester

OKLAHOMA
Altmiller, Dale Henry
Badgett, Allen A
Brasch, Klaus Rainer
Bruneau, Leslie Herbert
Buchanan, David Shane
Carpenter, Nancy Jane
Dale, George Leslie
Edwards, Lewis Hiram
Essenberg, Richard Charles
Grubb, Randall Barth
Gumbreck, Laurence Gable
Hunger, Robert Marvin
Klein, Ronald Don
LeGrand, Frank Edward
Merkle, Owen George
Miller, Paul George
Miner, Gary David
Morrissey, James Henry
Murray, Jay Clarence
Powell, Jerrel B
Ramon, Serafin
Schaefer, Frederick Vail
Smith, Edward Lee
Taliaferro, Charles M
Thompson, James Neal, Jr
Tobin, Sara L
Verhalen, Laval
Wann, Elbert Van
Wells, Harrington
Young, Sharon Clairene

OREGON
Baer, Adela (Dee)
Bagby, Grover Carlton
Bernier, Paul Emile
Bigley, Robert Harry
Buist, Neil R M
Cameron, H Ronald
Cameron, James Wagner
Civelli, Oliver
Dawson, Peter Sanford
Duffield, Deborah Ann
Field, Katharine G
Fowler, Gregory L
Gold, Michael Howard
Haunold, Alfred
Hays, John Bruce
Ho, Iwan
Hoornbeek, Frank Kent
Hrubant, Henry Everett
Kabat, David
Kohler, Peter
Koler, Robert Donald
Kronstad, Warren Ervind
Mills, Dallice Ivan
Mohler, James Dawson
Mok, Machteld Cornelia
Newman, Lester Joseph
Novitski, Edward
Orkney, G Dale
Postlethwait, John Harvey
Prescott, Gerald H
Ream, Lloyd Walter, Jr
Roberts, Paul Alfred
Russell, Peter James
Senner, John William
Stahl, Franklin William
Thompson, Maxine Marie
Wagner, David Henry

PENNSYLVANIA
Alperin, Richard Junius

Genetics (cont)

Anderson, Thomas Foxen
Andrews, Peter Walter
Armstrong, Robert John
Arnott, Marilyn Sue
Avadhani, Narayan G
Barnes, William Shelley
Bean, Barry
Beasley, Andrew Bowie
Berg, Clyde C
Bowne, Samuel Winter, Jr
Brownstein, Barbara L
Buss, Edward George
Cannizzaro, Linda A
Carlton, Bruce Charles
Cherry, John Paul
Clark, Andrew Galen
Cleveland, Richard Warren
Cooper, Jane Elizabeth
Cortner, Jean A
Corwin, Harry O
Craig, Richard
Croce, Carlo Maria
Currier, Thomas Curtis
D'Agostino, Marie A
Debouck, Christine Marie
Deering, Reginald
Del Vecchio, Vito Gerard
DiBerardino, Marie A
Dudzinski, Diane Marie
Eckhardt, Robert Barry
Eckroat, Larry Raymond
Farber, Phillip Andrew
Feinstein, Sheldon Israel
Ferrell, Robert Edward
Finger, Irving
Freed, Jerome James
Friedman, Robert David
Fuscaldo, Anthony Alfred
Fuscaldo, Kathryn Elizabeth
Gabriel, Edward George
Gasser, David Lloyd
Gealt, Michael Alan
Gerhold, Henry Dietrich
Gersh, Eileen Sutton
Gilbert, Scott F
Glorioso, Joseph Charles, III
Goff, Christopher Godfrey
Gollin, Susanne Merle
Goodgal, Sol Howard
Goodwin, Kenneth
Gots, Joseph Simon
Gottlieb, Frederick Jay
Gottlieb, Karen Ann
Grebner, Eugene Ernest
Grun, Paul
Harding, Roy Woodrow, Jr
Hargrove, George Lynn
Harris, Harry
Haskins, Mark
Hedrick, Philip William
Henry, Susan Armstrong
Higgins, Terry Jay
Hill, Charles Whitacre
Hill, Richard Ray, Jr
Hite, Mark
Hoffman, Albert Charles
Hopper, Anita Klein
Hsu, Susan Hu
Infanger, Ann
Irr, Joseph David
Jackson, Ethel Noland
Jacobsen, Terry Dale
Jargiello, Patricia
Jarvik, Jonathan Wallace
Jenkins, John Bruner
Jervis, Herbert Hunter
Jezyk, Peter Franklin
Johnson, Melvin Walter, Jr
Johnston, James Bennett
Jones, Elizabeth W
Kaney, Anthony Rolland
Kayhart, Marion
Kennett, Roger H
Knowles, Barbara B
Knudson, Alfred George, Jr
Kozinski, Andrzei
Ladda, Roger Louis
Leibel, Wayne Stephan
Li, Ching Chun
McCarthy, Patrick Charles
McDonald, Daniel James
McFeely, Richard Aubrey
McKinley, Carolyn May
McMorris, F Arthur
McNamara, Pamela Dee
Majumdar, S K
Majumdar, Shyamal K
Mann, Stanley Joseph
Manolson, Morris Frank
Marks, Louis Sheppard
Marshall, Harold Gene
Maxson, Linda Ellen R
Mezger-Freed, Liselotte
Michel, Kenneth Earl
Miller, Robert Christopher
Mode, Charles J
Moorhead, Paul Sidney
Morrison, William Joseph
Morrow, Terry Oran
Mullin, James Michael
Nass, Margit M K
Nei, Masatoshi

Nichols, Warren Wesley
Ohlsson-Wilhelm, Betsy Mae
Oka, Seishi William
Olds-Clarke, Patricia Jean
Opas, Evan E
Orr, Nancy Hoffner
Partanen, Carl Richard
Patterson, Donald Floyd
Peebles, Craig Lewis
Person, Stanley R
Poethig, Richard Scott
Porter, Ronald Dean
Punnett, Hope Handler
Pursell, Mary Helen
Ricciardi, Robert Paul
Rose, Raymond Wesley, Jr
Royse, Daniel Joseph
Rudkin, George Thomas
Russell, Richard Lawson
Samollow, Paul B
Sanders, Mary Elizabeth
Schmickel, Roy David
Schultz, Jane Schwartz
Schwartz, Elias
Shannon, Jack Corum
Sideropoulos, Aris S
Silvers, Willys Kent
Specht, Lawrence W
Spielman, Richard Saul
Staetz, Charles Alan
Steiner, Kim Carlyle
Stolc, Viktor
Suyama, Yoshitaka
Sylvester, James Edward
Takats, Stephen Tibor
Tartof, Kenneth D
Taylor-Mayer, Rhoda E
Tevethia, Mary Judith (Robinson)
Tonzetich, John
Tu, Chen-Pei David
Turner, J Howard
Turoczi, Lester J
Weinberg, Eric S
Weiss, Kenneth Monrad
Wheeler, Donald Alsop
Wright, James Everett, Jr
Wurst, Glen Gilbert
Wysocki, Charles Joseph
Yang, Da-Ping
Zink, Gilbert Leroy

RHODE ISLAND

Coleman, John Russell
Costantino, Robert Francis
Fausto-Sterling, Anne
Fischer, Glenn Albert
Gonsalves, Neil Ignatius
Hagy, George Washington
Holstein, Thomas James
Landy, Arthur
Lederberg, Seymour
Malcolm, Alexander Russell
Mottinger, John P
Quevedo, Walter Cole, Jr
Rao, Girimaji J Sathyanarayana
Rosenstein, Barry Sheldon
Rothman, Frank George
Schmitt, Johanna
Zarcaro, Robert Michael
Zimmering, Stanley

SOUTH CAROLINA

Arnaud, Philippe
Byrd, Wilbert Preston
Chapman, Stephen R
Davis, Leroy
Dawson, Wallace Douglas, Jr
Ely, Berten E, III
Finlay, Mary Fleming
Glick, Bruce
Godley, Willie Cecil
Graham, William Doyce, Jr
Jones, Alfred
Karam, Jim Daniel
Labanick, George Michael
Lincoln, David Erwin
McClain, Eugene Fredrick
Maheshwari, Kewal Krishnan
Mulvey, Margaret
Olson, John Bernard
Pollard, Arthur Joseph
Rupert, Earlene Atchison
Rust, Philip Frederick
Sawyer, Roger Holmes
Shipe, Emerson Russell
Surver, William Merle, Jr
Taylor, Harold Allison, Jr
Wang, An-Chuan
Wilson, Gregory Bruce
Yardley, Darrell Gene

SOUTH DAKOTA

Cholick, Fred Andrew
Kahler, Alex L
Lindahl, Ronald
Lunden, Allyn Oscar
Morgan, Walter Clifford
West, Thomas Patrick

TENNESSEE

Ball, Mary Uhrich
Barnett, Paul Edward
Bell, Sandra Lucille
Biggers, Charles James

Bryant, Robert Emory
Champney, William Scott
Chi, David Shyh-Wei
Conger, Bob Vernon
Dev, Vaithilingam Gangathara
Duck, Bobby Neal
Epler, James L
Garth, Richard Edwin
Generoso, Walderico Malinawan
Harris, John Wallace
Henke, Randolph Ray
Hickok, Leslie George
Hochman, Benjamin
Howe, Martha Morgan
Hughes, Karen Woodbury
Jones, Larry Hudson
Kahlon, Prem Singh
Larimer, Frank William
Lozzio, Carmen Bertucci
McCauley, David Evan
McFee, Alfred Frank
Mathis, Philip Monroe
Mosig, Gisela
Niyogi, Salil Kumar
Ohi, Seigo
Owens, Charles Allen
Popp, Raymond Arthur
Preston, Robert Julian
Pui, Ching-Hon
Ramey, Harmon Hobson, Jr
Regan, James Dale
Rinchik, Eugene M
Rivas, Marian Lucy
Russell, Liane Brauch
Russell, William Lawson
Schlarbaum, Scott E
Sega, Gary Andrew
Selby, Paul Bruce
Shirley, Herschel Vincent, Jr
Sotomayor, Rene Eduardo
Summitt, Robert L
Taft, Kingsley Arter, Jr
Thor, Eyvind
Tibbetts, Clark
Wachtel, Stephen Shoel
West, Dennis R
Womack, Frances C
Wood, Henderson Kingsberry
Woychik, Richard P
Yang, Wen-Kuang
Young, Lawrence Dale

TEXAS

Adair, Gerald Michael
Alexander, Mary Louise
Allen, Archie C
Alperin, Jack Bernard
Altenburg, Lewis Conrad
Anderson, David Eugene
Arenaz, Pablo
Artzt, Karen
Aufderheide, Karl John
Barnett, Don(ald) R(ay)
Bashaw, Elexis Cook
Bawdon, Roger Everett
Beaudet, Arthur L
Bennett, Dorothea
Bickham, John W
Bowman, Barbara Hyde
Brokaw, Bryan Edward
Brown, Michael S
Burdette, Walter James
Burson, Byron Lynn
Calub, Alfonso deGuzman
Caskey, Charles Thomas
Chakraborty, Ranajit
Chalmers, John Harvey, Jr
Chan, Lawrence Chin Bong
Chan, Teh-Sheng
Chang, Ching Hsong
Christadoss, Premkumar
Clowes, Royston Courtenay
Collier, Jesse Wilton
Collins, Anita Marguerite
Coulter, Murray W
Cox, Rody Powell
Creel, Gordon C
Darlington, Gretchen Ann Jolly
Dewees, Andre Aaron
Doe, Frank Joseph
Dorn, Gordon Lee
Douglas, Tommy Charles
Dronamraju, Krishna Rao
Duhrkopf, Richard Edward
Dusenbery, Ruth Lillian
Dyke, Bennett
Earhart, Charles Franklin, Jr
Elder, Fred F B
Erdman, Howard E
Evans, Raeford G
Fanguy, Roy Charles
Fowler, Norma Lee
Frederiksen, Richard Allan
Fresquez, Catalina Lourdes
Fuerst, Robert
Gardner, Florence Harrod
George, Fredrick William
Gilmore, Earl C
Girvin, Eb Carl
Gold, John Rush
Goldstein, Joseph Leonard
Gottlieb, Paul David
Gratzner, Howard G
Greenbaum, Ira Fred

Greenberg, Frank
Griffin, James Emmett
Handler, Shirley Wolz
Hanis, Craig L
Hart, Gary Elwood
Hartberg, Warren Keith
Hecht, Ralph Martin
Hewett'Emmett, David
Hillis, David Mark
Hittelman, Walter Nathan
Hixson, James Elmer
Hoff, Victor John
Hogan, Michael Edward
Hsie, Abraham Wuhsiung
Huang, Bessie Pei-Hsi
Ippen-Ihler, Karin Ann
Jackson, Raymond Carl
Jacob, Horace S
Jalal, Syed M
Johnston, John Spencer
Kidd, Harold J
Kieffer, Nat
Klaus, Ewald Fred, Jr
Klebe, Robert John
Kohel, Russell James
Konkel, David Anthony
Kraig, Ellen
Kramer, Nicholas William
Kroschewsky, Julius Richard
Krueger, Willie Frederick
Kumar, Vinay
Kurosky, Alexander
Kurzrock, Razelle
LaBrie, David Andre
LeBlanc, Donald Joseph
Ledley, Fred David
Lee, James C
Lee, Shih-Shun
Lester, Larry James
Li, Wen-Hsiung
Lingle, Sarah Elizabeth
Long, Walter K
McBride, Raymond Andrew
MacCluer, Jean Walters
McCrady, William B
McDaniel, Milton Edward
McDonald, Lynn Dale
McIlrath, William Oliver
McNutt, Clarence Wallace
Magill, Clint William
Magill, Jane Mary (Oakes)
Maguire, Marjorie Paquette
Margolin, Solomon B
Mathews, Nancy Ellen
Maunder, A Bruce
May, Sterling Randolph
Miller, Julian Creighton, Jr
Moody, Eric Edward Marshall
Moore, Charleen Morizot
Morrow, Kenneth John, Jr
Murgola, Emanuel J
Nguyen, Henry Thien
Nicholson, Wayne Lowell
Norwood, James S
O'Brien, William E
Osburn, Richard Lee
Parker, Nick Charles
Pathak, Sen
Perlman, Philip Stewart
Peterson, David Oscar
Prasad, Naresh
Prasad, Rupi
Pratt, David R
Price, Harold James
Rainwater, David Luther
Rao, Potu Narasimha
Redshaw, Peggy Ann
Richardson, Richard Harvey
Riser, Mary Elizabeth
Robberson, Donald Lewis
Rose, Kathleen Mary
Rosenblum, Eugene David
Roth, Jack A
Sanders, Bobby Gene
Schertz, Keith Francis
Schroeter, Gilbert Loren
Schull, William Jackson
Session, John Joe
Shay, Jerry William
Siciliano, Michael J
Skow, Loren Curtis
Smith, Gerald Ray
Smith, James Douglas
Snider, Philip Joseph
Srivastava, Satish Kumar
Stewart, Charles Ranous
Stone, William Harold
Stroman, David Womack
Strong, Louise Connally
Summers, Max Duane
Sutton, Harry Eldon
Temeyer, Kevin Bruce
Templeton, Joe Wayne
Thompson, Tommy Earl
Tierce, John Forrest
Trentin, John Joseph
Van Buijtenen, Johannes Petrus
VandeBerg, John Lee
Voigt, Paul Warren
Walter, Ronald Bruce
Ward, Jonathan Bishop, Jr
Wheeler, Marshall Ralph
Williams, Robert Sanders
Willig, Michael Robert

Wilson, Carol Maggart
Wilson, John H
Womack, James E
Woo, Savio L C
Wright, David Anthony
Wright, Stephen E
Wright, Woodring Erik
Yamauchi, Toshio
Young, Ryland F
Zwaan, Johan Thomas

UTAH
Albrechtsen, Rulon S
Andersen, William Ralph
Arave, Clive W
Asay, Kay Harris
Belcher, Bascom Anthony
Bowman, James Talton
Davern, Cedric I
Dewey, Douglas R
Dewey, Wade G
Dickinson, William Joseph
Farmer, James Lee
Franklin, Naomi C
Gardner, Eldon John
Georgopoulos, Constantine Panos
Gesteland, Raymond Frederick
Gurney, Elizabeth Tucker Guice
Hansen, Afton M
Hunt, Steven Charles
Jeffery, Duane Eldro
Knapp, Gayle
Lanner, Ronald Martin
Lark, Karl Gordon
Larsen, Lloyd Don
Locy, Robert Donald
McArthur, Eldon Durant
Mullen, Richard Joseph
Okun, Lawrence M
Park, Robert Lynn
Parkinson, John Stansfield
Shumway, Lewis Kay
Simmons, Daniel L
Simmons, John Robert
Sites, Jack Walter, Jr
Skolnick, Mark Henry
Stutz, Howard Coombs
Thomas, James H
Williams, Roger Richard
Wurst, Gloria Zettle

VERMONT
Albertini, Richard Joseph
Boraker, David Kenneth
Jauhar, Prem P
Kilpatrick, Charles William
Lalley, Peter Austin
Moody, Paul Amos
Novotny, Charles
Raper, Carlene Allen
Saul, George Brandon, II
Ullrich, Robert Carl
Wilkinson, Ronald Craig
Young, William Johnson, II

VIRGINIA
Annan, Murvel Eugene
Barber, John Clark
Bempong, Maxwell Alexander
Bender, Patrick Kevin
Benepal, Parshotam S
Benzinger, Rolf Hans
Berger, Beverly Jane
Bradley, Sterling Gaylen
Brown, Loretta Ann Port
Brown, Russell Vedder
Burian, Richard M
Buss, Glenn Richard
Chandler, Jerry LeRoy
Chinnici, Joseph (Frank) Peter
De Haan, Henry J(ohn)
Esen, Asim
Falkinham, Joseph Oliver, III
Frahm, Richard R
Gaines, James Abner
Garrett, Reginald Hooker
Grant, Bruce S
Heinemann, Richard Leslie
Hill, Jim T
Hilu, Khidir Wanni
Hoegerman, Stanton Fred
Hohenboken, William Daniel
Howard-Peebles, Patricia Nell
Howes, Cecil Edgar
Huskey, Robert John
Jarvis, Floyd Eldridge, Jr
Jones, Joyce Howell
Jones, Melton Rodney
Jones, William F
Kadner, Robert Joseph
Kelly, Thaddeus Elliott
Kilpatrick, S James, Jr
Kolsrud, Gretchen Schabtach
Krugman, Stanley Liebert
Lacy, George Holcombe
Leighton, Alvah Theodore, Jr
Lutes, Charlene McClanahan
McGilliard, Michael Lon
Marlowe, Thomas Johnson
Mauer, Irving
Merz, Timothy
Miller, Oscar Lee, Jr
Muir, William A
Nance, Walter Elmore

Osgood, Christopher James
Raizen, Carol Eileen
Sherald, Allen Franklin
Siegel, Paul Benjamin
Stevens, Robert Edward
Swiger, Louis Andre
Taub, Stephan Robert
Teichler-Zallen, Doris
Townsend, J(oel) Ives
Trout, William Edgar, III
Vinson, William Ellis
Wallace, Bruce
West, David Armstrong
White, John Marvin
Wolf, Barry
Wright, Theodore Robert Fairbank

WASHINGTON
Ahmed, Saiyed I
Becker, Walter Alvin
Bendich, Arnold Jay
Bogyo, Thomas P
Brooks, Antone L
Byers, Breck Edward
Carr, Robert Leroy
Cheevers, William Phillip
Chen, Shi-Han
Cosman, David John
Daniels, Jess Donald
Erickson, John (Elmer)
Fangman, Walton L
Felsenstein, Joseph
Fialkow, Philip Jack
Fournier, R E Keith
Gartler, Stanley Michael
Giblett, Eloise Rosalie
Grootes-Reuvecamp, Grada Alijda
Hall, Benjamin Downs
Hartwell, Leland Harrison
Hawthorne, Donald Clair
Hazelbauer, Gerald Lee
Hecht, Adolph
Heston, Leonard L
Hillers, Joe Karl
Humphrey, Donald Glen
Hutchison, Nancy Jean
Karp, Laurence Edward
Kleinhofs, Andris
Konzak, Calvin Francis
Kraft, Joan Creech
Kutter, Elizabeth Martin
Laird, Charles David
Landis, Wayne G
Lane, Wallace
Lasure, Linda Lee
Leng, Earl Reece
Lightfoot, Donald Richard
McClary, Cecil Fay
MacKay, Vivian Louise
Martin, Mark Wayne
Matlock, Daniel Budd
Motulsky, Arno Gunther
Nilan, Robert Arthur
Nute, Peter Eric
Osborne, Richard Hazelet
Pall, Martin L
Peterson, Clarence James, Jr
Pious, Donald A
Prieur, David John
Roman, Herschel Lewis
Salk, Darrell John
Sandler, Laurence Marvin
Schroeder, Alice Louise
Sibley, Carol Hopkins
Stadler, David Ross
Stamatoyannopoulos, George
Stenchever, Morton Albert
Stettler, Reinhard Friederich
Stonecypher, Roy W
Taylor, Ronald
Thelen, Thomas Harvey
Towner, Richard Henry
Utter, Fred Madison
Vigfusson, Norman V
Wakimoto, Barbara Toshiko
Yao, Meng-Chao
Young, Francis Allan

WEST VIRGINIA
Cech, Franklin Charles
Chisler, John Adam
Headings, Verle Emery
Kaczmarczyk, Walter J
Kidder, Harold Edward
Latterell, Richard L
Ong, Teong-man
Reichenbecher, Vernon Edgar, Jr
Thayne, William V
Ulrich, Valentin
Yelton, David Baetz

WISCONSIN
Abrahamson, Seymour
Adler, Julius
Anderson, Robert Philip
Azen, Edwin Allan
Bennett, Kenneth A
Bergtrom, Gerald
Bersu, Edward Thorwald
Bitgood, J(ohn) James
Burgess, Ann Baker
Casler, Michael Darwin
Chambliss, Glenn Hilton
Chapman, Arthur Barclay

Courtright, James Ben
Crow, James Franklin
Curie-Cohen, Martin Michael
Curtis, Robin Livingstone
Datta, Surinder P
Dimond, Randall Lloyd
Dove, William Franklin
Einspahr, Dean William
Ellingboe, Albert Harlan
Forsberg, Robert Arnold
Gabelman, Warren Henry
Geeseman, Gordon E
Gilbert-Barness, Enid F
Greenspan, Daniel S
Haas, Michael John
Hall, Marion Trufant
Higgs, Roger L
Hornemann, Ulfert
Hougas, Robert Wayne
Huss, Ronald John
Ihrke, Charles Albert
Jacobs, Lois Jean
Jacobson, Gunnard Kenneth
Jungck, John Richard
Kaplan, Stanley
Kermicle, Jerry Lee
Kung, Ching
Lim, Johng Ki
Littlewood, Barbara Shaffer
Long, Sally Yates
McDonough, Eugene Stowell
Meisner, Lorraine Faxon
Mertz, Janet Elaine
Miller, Paul Dean
Nagodawithana, Tilak Walter
Nelson, Oliver Evans, Jr
Pan, David
Phillips, Ruth Brosi
Porter, John Willard
Randerson, Sherman
Rapacz, Jan
Reznikoff, William Stanton
Rimm, Alfred A
Seelke, Ralph Walter
Sendelbach, Anton G
Simon, Philipp William
Smith, Oliver Hugh
Smith, Richard R
Sondel, Paul Mark
Spritz, Richard Andrew
Strohm, Jerry Lee
Susman, Millard
Sytsma, Kenneth Jay
Szybalski, Waclaw
Temin, Rayla Greenberg
Tracy, William Francis
Wejksnora, Peter James
Williams, Paul Hugh

WYOMING
Bear, Phyllis Dorothy
Petersen, Nancy Sue
Tabachnick, Walter J

PUERTO RICO
Bruck, David Lewis
Squire, Richard Douglas
Virkki, Niilo

ALBERTA
Ahmed, Asad
Andrews, John Edwin
Berg, Roy Torgny
Biederman, Brian Maurice
Bowen, Peter
Church, Robert Bertram
Dancik, Bruce Paul
Dixon, Gordon H
Francis, Mike McD
Fredeen, Howard T
Gooding, Ronald Harry
Hodgetts, Ross Birnie
Iatrou, Kostas
Kuspira, J
Langridge, William Henry Russell
Larson, Ruby Ila
Muendel, Hans-Henning
Nakamura, Kazuo
Nash, David
Roberts, David Wilfred Alan
Rudd, Noreen L
Sanderson, Kenneth Edwin
Scheinberg, Eliyahu
Strobeck, Curtis
Von Borstel, Robert Carsten
Wagenaar, Emile B
Wegmann, Thomas George
Weijer, Jan
Yeh, Francis Cho-hao

BRITISH COLUMBIA
Applegarth, Derek A
Baird, Patricia A
Druehl, Louis D
Ganders, Fred Russell
Griffiths, Anthony J F
Hill, Arthur Thomas
Holl, Frederick Brian
Holm, David George
Hunt, Richard Stanley
Illingworth, Keith
MacDiarmid, William Donald
Meagher, Michael Desmond
Miller, James Reginald

Person, Clayton Oscar
Redfield, Rosemary Jeanne
Sadowski, Ivan J
Salari, Hassan
Stich, Hans F
Styles, Ernest Derek
Suzuki, David Takayoshi
Sziklai, Oscar
Tener, Gordon Malcolm
Utkhede, Rajeshwar Shamrao

MANITOBA
Aung, Taing
Ayles, George Burton
Campbell, Allan Barrie
Clayton, James Wallace
Dyck, Peter Leonard
Evans, Laurie Edward
Giesbrecht, John
Hamerton, John Laurence
Helgason, Sigurdur Bjorn
Kerber, Erich Rudolph
Kondra, Peter Alexander
Kunz, Bernard Alexander
Larter, Edward Nathan
Lewis, Marion Jean
McAlpine, Phyllis Jean
Wrogemann, Klaus

NEW BRUNSWICK
Bonga, Jan Max
Fowler, Donald Paige
Young, Donald Alcoe

NEWFOUNDLAND
Davidson, William Scott
Scott, Peter John

NOVA SCOTIA
Chiasson, Leo Patrick
Crober, Donald Curtis
Doyle, Roger Whitney
Gray, Michael William
Haley, Leslie Ernest
Kamra, Om Perkash
Newkirk, Gary Francis
Van der Meer, John Peter
Wainwright, Lillian K (Schneider)
Welch, J Philip
Zouros, Eleftherios

ONTARIO
Anstey, Thomas Herbert
Arnison, Paul Grenville
Axelrad, Arthur Aaron
Bacchetti, Silvia
Battle, Helen Irene
Behme, Ronald John
Britton, Donald MacPhail
Burnside, Edward Blair
Cade, William Henry
Campbell, Kenneth Wilford
Carmody, George R
Carstens, Eric Bruce
Chaly, Nathalie
Chambers, James Robert
Childers, Walter Robert
Chui, David H K
Cinader, Bernhard
Cooke, Fred
Cox, Diane Wilson
Cummins, Joseph E
Davidson, Ronald G
Deeley, Roger Graham
Dodson, Edward O
Duncan, William Raymond
Fedak, George
Fejer, Stephen Oscar
Fiser, Paul S(tanley)
Galsworthy, Peter Robert
Galsworthy, Sara B
Gowe, Robb Shelton
Grunder, Allan Angus
Gupta, Radhey Shyam
Harney, Patricia Marie
Haynes, Robert Hall
Heddle, John A M
Hegele, Robert A
Hickey, Donal Aloysius
Ho, Keh Ming
Hutton, Elaine Myrtle
Ihssen, Peter Edowald
Joneja, Madan Gopal
Kadanka, Zdenek Karel
Kalow, Werner
Kang, C Yong
Kasha, Kenneth John
Kasupski, George Joseph
Kidder, Gerald Marshall
Krepinsky, Jiri J
Larsen, Ellen Wynne
Lin, Ching Y
Lu, Benjamin Chi-Ko
MacLennan, David Herman
Moens, Peter B
Nasim, Anwar
Newcombe, Howard Borden
Percy, Maire Ede
Petras, Michael Luke
Rajhathy, Tibor
Reed, Thomas Edward
Reid, Robert Leslie
Robinson, Brian Howard
Rossant, Janet

Genetics (cont)

Sampson, Dexter Reid
Saunders, Shelley Rae
Seligy, Verner Leslie
Sengar, Dharmendra Pal Singh
Sergovich, Frederick Raymond
Simpson, Nancy E
Singal, Dharam Parkash
Singh, Rama Shankar
Sinha, Raj P
Smiley, James Richard
Soltan, Hubert Constantine
Sorger, George Joseph
Soudek, Dushan Edward
Takahashi, Francois Iwao
Tallan, Irwin
Thompson, Margaret A Wilson
Threlkeld, Stephen Francis H
Trevors, Jack Thomas
Uchida, Irene Ayako
Walden, David Burton
Yeatman, Christopher William

PRINCE EDWARD ISLAND
Choo, Thin-Meiw

QUEBEC
Anderson, William Alan
Brown, Douglas Fletcher
Buckland, Roger Basil
Cedergren, Robert J
Chabot, Benoit
Chafouleas, James G
Chiang, Morgan S
Chung, Young Sup
Clark, Michael Wayne
Comeau, Andre I
Corriveau, Armand Gerard
DuBow, Michael Scott
Fahmy, Mohamed Hamed
Fraser, Frank Clarke
Gervais, Francine
Goodfriend, Lawrence
Grant, William Frederick
Green, David M(artin)
Guttmann, Ronald David
Hallenbeck, Patrick Clark
Hamelin, Claude
Kafer, Etta (Mrs E R Boothroyd)
Langford, Arthur Nicol
Lapointe, Jacques
Messing, Karen
Mukerjee, Barid
Nishioka, Yutaka
Palmour, Roberta Martin
Pinsky, Leonard
Preus, Marilyn Ione
Rosenblatt, David Sidney
Roy, Robert Michael McGregor
Sarhan, Fathey
Scriver, Charles Robert
Shoubridge, Eric Alan
Sikorska, Hanna
Southin, John L
Stanners, Clifford Paul
Stevenson, Mary M
Tanguay, Robert M
Thirion, Jean Paul Joseph
Trasler, Daphne Gay
Vezina, Claude
Walker, Mary Clare
Wolsky, Alexander

SASKATCHEWAN
Baker, Robert John
Bertrand, Helmut
Crawford, Roy Douglas
Goplen, Bernard Peter
Harvey, Bryan Laurence
Khachatourians, George G
Kraay, Gerrit Jacob
Mortensen, Knud
Rank, Gerald Henry
Rossnagel, Brian Gordon
Scheltgen, Elmer
Scoles, Graham John
Shokeir, Mohamed Hassan Kamel
Spurr, David Tupper
Vella, Francis
Williams, Charles Melville

YUKON TERRITORY
Doermann, August Henry

OTHER COUNTRIES
Agarwal, Shyam S
Arber, Werner
Bodmer, Walter Fred
Bortolozzi, Jehud
Britten, Edward James
Brody, Edward Norman
Bunn, Clive Leighton
Cantrell, Ronald P
Carballo-Quiros, Alfredo
Chomchalow, Narong
Collins, Vincent Peter
Curtis, Byrd Collins
Da Cunha, Antonio Brito
Doerfler, Walter Hans
Elsevier, Susan Maria
Fitzhugh, Henry Allen
Goding, James Watson
Goodman, Fred

Hickman, Charles Garner
Hotta, Yasuo
Jakob, Karl Michael
Kaltsikes, Pantouses John
Kang, Kewon
Kerr, Warwick Estevam
Kimoto, Masao
Klein, Jan
La Chance, Leo Emery
Leigh, Egbert G, Jr
Meins, Frederick, Jr
Mojica-a, Tobias
Morton, Newton Ennis
Nevo, Eviatar
Nienstaedt, Hans
Ou, Jonathan Tsien-hsiong
Pavan, Crodowaldo
Radford, Alan
Salzano, Francisco Mauro
Spencer, John Francis Theodore
Spierer, Pierre
Stearns, Stephen Curtis
Stodolsky, Marvin
Subramanian, Alap Raman
Tan, Y H
Watanabe, Takeshi
Yakura, Hidetaka
Yamamoto, Hiroshi

Hydrobiology

ALABAMA
Bayne, David Roberge
Beyers, Robert John
Brown, Jack Stanley
Folkerts, George William
Rouse, David B

ALASKA
Alexander, Vera
Warner, Mark Clayson

ARIZONA
Blinn, Dean Ward

ARKANSAS
Jones, Joe Maxey
Oliver, Kelly Hoyet, Jr

CALIFORNIA
Conklin, Douglas Edgar
Eriksen, Clyde Hedman
Fuhrman, Jed Alan
Horne, Alexander John
Kenk, Roman
Knight, Allen Warner
Landolfi, Nicholas F
McCosker, John E
Maciolek, John A
Mercer, Edward King
Slack, Keith Vollmer

COLORADO
Cohen, Ronald R H
Havlick, Spenser Woodworth
LaBounty, James Francis, Sr
Ward, James Vernon
Windell, John Thomas

CONNECTICUT
Jokinen, Eileen Hope
Katz, Max
Rho, Jinnque
Whitworth, Walter Richard

DELAWARE
Curtis, Lawrence Andrew
Smith, David William

DISTRICT OF COLUMBIA
Hart, Charles Willard, Jr
Sze, Philip

FLORIDA
Belanger, Thomas V
Bowes, George Ernest
Brezonik, Patrick Lee
Cardeilhac, Paul T
Dinsmore, Bruce Heasley
Haag, Kim Hartzell
Haller, William T
Kerr, John Polk
Nordlie, Frank Gerald

GEORGIA
Greeson, Phillip Edward
Meyer, Judy Lynn
Nichols, Michael Charles
Stirewalt, Harvey Lee
Stoneburner, Daniel Lee
Tietjen, William Leighton
Wallace, James Bruce
Winger, Parley Vernon

HAWAII
Fast, Arlo Wade

ILLINOIS
Brigham, Warren Ulrich
Brugam, Richard Blair
Haynes, Robert C
Jahn, Lawrence A
Lee, Ming T

Sparks, Richard Edward
Stull, Elisabeth Ann

INDIANA
Chamberlain, William Maynard
Davies, Harold William, (Jr)
Frey, David Grover
Gammon, James Robert
Lodge, David Michael
McCowen, Max Creager
Spacie, Anne

IOWA
Bovbjerg, Richard Viggo
McDonald, Donald Burt
Merkley, Wayne Bingham

KANSAS
Boles, Robert Joe

KENTUCKY
Stevenson, Robert Jan
Timmons, Thomas Joseph

LOUISIANA
Brown, Kenneth Michael
Moore, Walter Guy
Poirrier, Michael Anthony
Vidrine, Malcolm Francis

MARYLAND
Gerritsen, Jeroen
Pancella, John Raymond
Paul, Robert William, Jr
Picciolo, Grace Lee

MASSACHUSETTS
Binford, Michael W
Dacey, John W H
Guillard, Robert Russell Louis

MICHIGAN
Andresen, Norman A
Bowen, Stephen Hartman
Burton, Thomas Maxie
Chang, William Y B
Douglas, Matthew M
Foster, Neal Robert
Holland-Beeton, Ruth Elizabeth
Hough, Richard Anton
Kevern, Niles Russell
Kovalak, William Paul
Scavia, Donald
Stoermer, Eugene F
Wujek, Daniel Everett

MINNESOTA
Davis, Margaret Bryan
Henry, Mary Gerard
Holt, Charles Steele
Klemer, Andrew Robert
McConville, David Raymond
Perry, James Alfred
Ruschmeyer, Orlando R
Straw, Thomas Eugene
Welter, Alphonse Nicholas

MISSISSIPPI
Knight, Luther Augustus, Jr

MISSOURI
Hanson, Willis Dale
Stern, Daniel Henry
Stern, Michele Suchard
Welply, Joseph Kevin

MONTANA
Schumacher, Robert E

NEVADA
Deacon, James Everett

NEW HAMPSHIRE
Gilbert, John Jouett

NEW JERSEY
Dorfman, Donald
Keating, Kathleen Irwin
Morgan, Mark Douglas
Riemer, Donald Neil
Rosengren, John
Weis, Judith Shulman

NEW MEXICO
Molles, Manuel Carl, Jr

NEW YORK
Boylen, Charles William
Collins, Carol Desormeau
Forest, Herman Silva
Godshalk, Gordon Lamar
Hairston, Nelson George, Jr
Hoham, Ronald William
Klotz, Richard Lawrence
Oglesby, Ray Thurmond
Osterberg, Donald Mouretz
Rhee, G-Yull
Slobodkin, Lawrence Basil
Smith, Alden Ernest
Strayer, David Lowell

NORTH CAROLINA
Brinson, Mark McClellan
Derrick, Finnis Ray

Kamykowski, Daniel

OHIO
Baumann, Paul C
Colinvaux, Paul Alfred
Cooke, George Dennis
Culver, David Alan
Federle, Thomas Walter
Hummon, William Dale
Laushman, Roger H
Olive, John H
Staker, Robert D
Stansbery, David Honor
Stein, Carol B
Wickstrom, Conrad Eugene

OKLAHOMA
Francko, David Alex
Hawxby, Keith William
Korstad, John Edward
Matthews, William John

OREGON
Buscemi, Philip Augustus
Castenholz, Richard William
Chapman, Gary Adair
Church, Marshall Robbins
Garton, Ronald Ray
Malueg, Kenneth Wilbur

PENNSYLVANIA
Arnold, Dean Edward
Bott, Thomas Lee
Bouchard, Raymond William
Coffman, William Page
Haase, Bruce Lee
Hart, David Dickinson
Johnston, Richard Boles, Jr
Kilham, Susan Soltau
Kimmel, William Griffiths
Kleinman, Robert L P
Mickle, Ann Marie
Prezant, Robert Steven
Williamson, Craig Edward

SOUTH CAROLINA
Abernathy, A(twell) Ray
Spain, James Dorris, Jr
Wixson, Bobby Guinn

TENNESSEE
Brinkhurst, Ralph O
Bunting, Dewey Lee, II
Cada, Glenn Francis
Elmore, James Lewis

TEXAS
Anderson, Richard Orr
Clark, William Jesse
Dickson, Kenneth Lynn
Foster, John Robert
Hellier, Thomas Robert, Jr
Lind, Owen Thomas
Neill, William Harold
Sterner, Robert Warner
Waller, William T

UTAH
Quinn, Barry George

VIRGINIA
Benfield, Ernest Frederick
Bishop, John Watson
Brehmer, Morris Leroy
Buikema, Arthur L, Jr
Elwood, Jerry William
Gough, Stephen Bradford
Grove, Thurman Lee
Hendricks, Albert Cates
Stevens, Robert Edward
Webster, Jackson Ross
Zubkoff, Paul L(eon)

WASHINGTON
Bisson, Peter Andre
Damkaer, David Martin
Smith, Stamford Dennis
Soltero, Raymond Arthur

WEST VIRGINIA
Adkins, Dean Aaron
Bissonnette, Gary Kent
Jenkins, Charles Robert
Weaks, Thomas Elton

WISCONSIN
Brooks, Arthur S
Carpenter, Stephen Russell
Graham, Linda Kay Edwards
Krezoski, John R
Magnuson, John Joseph
Mortimer, Clifford Hiley
Remsen, Charles C, III
Schmitz, William Robert
Seale, Dianne B
Verch, Richard Lee

WYOMING
Dolan, Joseph Edward

ALBERTA
Gorham, Paul Raymond
Wallis, Peter Malcolm

BRITISH COLUMBIA
Geen, Glen Howard
Nordin, Richard Nels

MANITOBA
Flannagan, John Fullan
Hamilton, Robert Duncan
Hecky, Robert Eugene
Staniforth, Richard John

NEWFOUNDLAND
Davis, Charles (Carroll)

NOVA SCOTIA
Chen, Lawrence Chien-Ming

ONTARIO
Barica, Jan M
Chengalath, Rama
Coad, Brian William
Colby, Peter J
Collins, Nicholas Clark
Dickman, Michael David
Duthie, Hamish
Fernando, Constantine Herbert
Fraser, James Millan
Harrison, Arthur Desmond
MacCrimmon, Hugh Ross
McQueen, Donald James
Oliver, Donald Raymond
Smol, John Paul
Vollenweider, Richard A
Wiggins, Glenn Blakely

QUEBEC
De la Noue, Joel Jean-Louis
Kalff, Jacob

OTHER COUNTRIES
Davis, William Jackson
Koerner, Heinz
Ling, Chung-Mei
Stearns, Stephen Curtis

Immunology

ALABAMA
Abrahamson, Dale Raymond
Acton, Ronald Terry
Alford, Charles Aaron, Jr
Bennett, Joe Claude
Blalock, J Edwin
Briles, Connally Oran
Briles, David Elwood
Carrozza, John Henry
Chapatwala, Kirit D
Cooper, Max Dale
Egan, Marianne Louise
Elson, Charles O, III
Furuto, Donald K
Gathings, William Edward
Gillespie, George Yancey
Hajduk, Stephen Louis
Hazlegrove, Leven S
Hiramoto, Raymond Natsuo
Hunter, Eric
Kayes, Stephen Geoffrey
Kearney, John F
Kilpatrick, John Michael
Klesius, Phillip Harry
Lamon, Eddie William
Lauerman, Lloyd Herman, Jr
Lausch, Robert Nagle
Lidin, Bodil Inger Maria
Lobuglio, Albert Francis
McCarthy, Dennis Joseph
McCombs, Candace Cragen
McGhee, Jerry Roger
Mestecky, Jiri
Michalek, Suzanne M
Montgomery, John R
Niemann, Marilyn Anne
Panangala, Victor S
Patterson, Loyd Thomas
Peterson, Raymond Dale August
Pillion, Dennis Joseph
Ranney, Richard Raymond
Rohrer, James William
Russell, Michael W(illiam)
Schrohenloher, Ralph Edward
Singh, Shiva Pujan
Smith, Paul Clay
Turco, Jenifer
Waiia, Amrik Singh
Winters, Alvin L

ALASKA
Leon, Kenneth Allen
Wennerstrom, David E

ARIZONA
Aden, David Paul
Archer, Stanley J
Baier, Joseph George
Burkholder, Peter M
Cawley, Leo Patrick
Criswell, Bennie Sue
DeLuca, Dominick
Halonen, Marilyn Jean
Hendrix, Mary J C
Hersh, Evan Manuel
Janssen, Robert (James) J
Jeter, Wayburn Stewart

Kay, Marguerite M B
Lopez, Maria del Carmen
Lukas, Ronald John
Manak, Rita C
Marchalonis, John Jacob
Meinke, Geraldine Chciuk
Nagle, Ray Burdell
Northey, William T
Pinnas, Jacob Louis
Roholt, Oliver A, Jr
Wallraff, Evelyn Bartels
Watson, Ronald Ross
Woods, Alexander Hamilton

ARKANSAS
Alquire, Mary Slanina (Hirsch)
Barnett, John Brian
Leone, Charles Abner
Roberts, Dean Winn, Jr
Soderberg, Lee Stephen Freeman
Thoma, John Anthony

CALIFORNIA
Abel, Carlos Alberto
Aladjem, Frederick
Allison, Anthony Clifford
Altieri, Dario Carlo
Amkraut, Alfred A
Anthony, Ronald Lewis
Arai, Ken-Ichi
Arquilla, Edward R
Ascher, Michael S
Banovitz, Jay Bernard
Bartl, Paul
Bastian, John F
Beall, Gildon Noel
Bechtol, Kathleen B
Becker, Martin Joseph
Benjamini, Eliezer
Berek, Jonathan S
Bhattacharya, Prabir
Bishop, John Michael
Bishop, Nancy Horschel
Bjorkman, Pamela J
Black, Kirby Samuel
Blair, Phyllis Beebe
Blakeslee, Dennis L(auren)
Block, Jerome Bernard
Bluestein, Harry Gilbert
Bokoch, Gary M
Bonavida, Benjamin
Bond, Martha W(illis)
Bozdech, Marek Jiri
Bramhall, John Shepherd
Brandon, David Lawrence
Bremermann, Hans J
Brostoff, Steven Warren
Bruckner, David Alan
Burnham, Thomas K
Byers, Vera Steinberger
Carlson, James Reynold
Carson, Dennis A
Castles, James Joseph, Jr
Cernosek, Stanley Frank, Jr
Chang, Chin-Hai
Chang, Jennie C C
Chen, Benjamin P P
Chenoweth, Dennis Edwin
Chesnut, Robert W
Childress, Evelyn Tutt
Chisari, Francis Vincent
Church, Joseph August
Clark, William R
Clement, Loran T
Cooper, Edwin Lowell
Cooper, Neil R
Costea, Nicolas V
Cowan, Keith Morris
Cremer, Natalie E
Curnutte, John Tolliver, III
Dafforn, Geoffrey Alan
Dasch, James R
Davenport, Calvin Armstrong
David, Gary Samuel
Davies, Huw M
Davis, William Ellsmore, Jr
Delk, Ann Stevens
DeLustro, Frank Anthony
Del Villano, Bert Charles
Denis, Kathleen A
Dennert, Gunther
Dixon, Frank James
Dixon, James Francis Peter
Dreyer, William J
Duffey, Paul Stephen
Dunnebacke-Dixon, Thelma Hudson
Dutton, Richard W
Duzgunes, Nejat
Eardley, Diane Douglas
Eaton, Monroe Davis
Edgington, Thomas S
Effros, Rita B
Engvall, Eva Susanna
Ernst, David N
Etzler, Marilynn Edith
Evans, Ernest Edward, Jr
Fabrikant, Irene Berger
Fahey, John Leslie
Fathman, C Garrison
Feingold, Ben F
Feldman, Joseph David
Felgner, Philip Louis
Fey, George

Fidler, John M
Forghani-Abkenari, Bagher
Frick, Oscar L
Friou, George Jacob
Gadol, Nancy
Gale, Robert Peter
Garfin, David Edward
Garratty, George
Garvey, Justine Spring
Gascoigne, Nicholas Robert John
Gee, Adrian Philip
Gelfand, David H
Gigli, Irma
Gilbreath, Michael Joye
Ginsberg, Mark Howard
Ginsberg, Theodore
Giorgi, Janis V(incent)
Gitnick, Gary L
Glassock, Richard James
Glembotski, Christopher Charles
Glovsky, M Michael
Golde, David William
Golub, Edward S
Golub, Sidney Harris
Good, Anne Haines
Goodman, Joel Warren
Goodman, Michael Gordon
Goodman, Richard E
Gottlieb, Michael Stuart
Green, Donald Eugene
Green, Douglas R
Greene, Elias Louis
Greenspan, John Simon
Grey, Howard M
Guenther, Donna Marie
Gupta, Rishab Kumar
Gupta, Sudhir
Gutman, George Andre
Haddad, Zack H
Haen, Peter John
Hanes, Deanne Meredith
Harris, John Wayne
Herzenberg, Leonard Arthur
Heuer, Ann Elizabeth
Hilgard, Henry Rohrs
Hill, Sharon Wilkins
Hoch, Sallie O'Neil
Hofmann, Bo
Hoke, Carolyn Kay
Holman, Halsted Reid
Holmberg, Charles Arthur
Hood, Leroy E
Horwitz, Marcus Aaron
Howard, Russell John
Hubbard, William Jack
Huff, Dennis Karl
Ishizaka, Kimishige
Ishizaka, Teruko
Jackson, Anne Louise
Jacob, Chaim O
Johansson, Mats W
Jones, Patricia Pearce
Jordan, Stanley Clark
Kang, Chang-Yuil
Kan-Mitchell, June
Kaslow, Arthur L
Katz, David Harvey
Kelley, Darshan Singh
Kermani-Arab, Vali
Kibler, Ruthann
Kiprov, Dobri D
Klaustermeyer, William Berner
Klinman, Norman Ralph
Kohler, Heinz
Kornfeld, Lottie
Koshland, Marian Elliott
Koths, Kirston Edward
Lad, Pramod Madhusudan
Ladisch, Stephan
Landolfi, Nicholas F
Landy, Maurice
Lehrer, Harris Irving
Lembach, Kenneth James
Lerner, Richard Alan
Levinson, Warren E
Lewis, Susanna Maxwell
Linscott, William Dean
Litman, David Jay
Lou, Kingdon
Lucas, David Owen
Luzzio, Anthony Joseph
McCarthy, Robert Elmer
McCormick, Robert T
McDevitt, Hugh O'Neill
MacKenzie, Malcolm R
McLaughlin-Taylor, Elizabeth
McMillan, Robert
Maino, Vernon Carter
Makinodan, Takashi
Makker, Sudesh Paul
Mann, Lewis Theodore, Jr
Marangos, Paul Jerome
Marquardt, Diana
Martin, David William, Jr
Masouredis, Serafeim Panogiotis
Mathies, Margaret Jean
Matsumura, Kenneth N
Mayron, Lewis Walter
Merchant, Bruce
Mescher, Matthew F
Michaeli, Dov
Milburn, Gary L
Miller, Alexander
Miller, John Johnston, III

Mitchell, Malcolm Stuart
Mochizuki, Diane Yukiko
Morrison, Sherie Leaver
Morrow, William John Woodroofe
Müller-Sieburg, Christa E
Myhre, Byron Arnold
Nagy, Stephen Mears, Jr
Nakamura, Robert Motoharu
Nitecki, Danute Emilija
Oh, Chan Soo
Olsen, Charles Edward
Osebold, John William
Outzen, Henry Clair, Jr
Owen, Ray David
Pallavicini, Maria Georgina
Parker, John William
Parkman, Robertson
Payne, Rose Marise
Pellegrino, Michele A
Perez, Hector Daniel
Peter, James Bernard
Peterlin, Boris Matija
Phelps, Leroy Nash
Pischel, Ken Donald
Pitesky, Isadore
Poggenburg, John Kenneth, Jr
Pollack, William
Prince, Harry E
Quismorio, Francisco P, Jr
Ralph, Peter
Rearden, Carole Ann
Reese, Robert Trafton
Reeve, Peter
Reisfeld, Ralph Alfred
Remington, Jack Samuel
Riblet, Roy Johnson
Rokeach, Luis Alberto
Rosenau, Werner
Rosenberg, Leon T
Rosenquist, Grace Link
Rosenthal, Sol Roy
Rounds, Donald Edwin
Saifer, Mark Gary Pierce
Salk, Jonas Edward
Schiller, Neal Leander
Schoenholz, Walter Kurt
Schonfeld, Steven Emanuel
Schulkind, Martin Lewis
Sears, Duane William
Seeger, Robert Charles
Senyk, George
Sercarz, Eli
Sharma, Somesh Datt
Shen, Wei-Chiang
Sherman, Linda Arlene
Shively, John Ernest
Shulman, Ira Andrew
Silverman, Paul Hyman
Singer, Paul A
Singh, Iqbal
Smith, Richard Scott
Smithwick, Elizabeth Mary
Snow, John Thomas
Solomon, George Freeman
Spiegelberg, Hans L
Spitler, Lynn E
Stevens, Ronald Henry
Stiehm, E Richard
Stormont, Clyde J
Stringfellow, Dale Alan
Strober, Samuel
Stroynowski, Iwona T
Suffin, Stephen Chester
Sweet, Benjamin Hersh
Szego, Clara Marian
Tachibana, Dora K
Takasugi, Mitsuo
Tamerius, John
Taylor, Clive Roy
Tempelis, Constantine H
Terasaki, Paul Ichiro
Terres, Geronimo
Theofilopoulos, Argyrios N
Thoman, Marilyn Louise
Thueson, David Orel
Trowbridge, Ian Stuart
Ullman, Edwin Fisher
Usinger, William R
Uyeda, Charles Tsuneo
Vannier, Wilton Emile
Van Winkle, Michael George
Vaughan, John Heath
Vedros, Neylan Anthony
Vickrey, Herta Miller
Victoria, Edward Jess, Jr
Vogt, Peter Klaus
Vredevoe, Donna Lou
Vyas, Girish Narmadashankar
Wall, Thomas Randolph
Walter, Harry
Ware, Carl F
Wasserman, Stephen I
Webb, David Ritchie
Wedberg, Stanley Edward
Weigle, William O
Weimer, Henry Eben
Weiser, Russel Shively
Weliky, Norman
Wiebe, Michael Eugene
Williams, Terry Wayne
Wilson, Darcy Benoit
Winkelhake, Jeffrey Lee
Witte, Owen Neil
Wofsy, Leon

Immunology (cont)

Wright, Richard Kenneth
Wu, William Gay
Yamamoto, Richard
Young, Janis Dillaha
Yu, David Tak Yan
Zeleznick, Lowell D
Ziccardi, Robert John
Zighelboim, Jacob
Zimmerman, Theodore Samuel
Zlotnik, Albert
Zvaifler, Nathan J

COLORADO

Arend, William Phelps
Barisas, Bernard George, Jr
Benjamin, Stephen Alfred
Berkelhammer, Jane
Brooks, Bradford O
Brooks Springs, Suzanne Beth
Bunn, Paul A, Jr
Cambier, John C
Campbell, Priscilla Ann
Chai, Hyman
Claman, Henry Neumann
Cockerell, Gary Lee
Crowle, Alfred John
De Martini, James Charles
Freed, John Howard
Gelfand, Erwin William
Giclas, Patricia C
Glode, Leonard Michael
Goren, Mayer Bear
Grieve, Robert B
Hager, Jean Carol
Harbeck, Ronald Joseph
Hayward, Anthony R
Henson, Peter Mitchell
Hudson, Bruce William
Irons, Richard Davis
Jacobs, Barbara B
Jones, Carol A
Jordan, Russell Thomas
Kirkpatrick, Charles Harvey
Kubo, Ralph Teruo
Larson, Kenneth Allen
Laudenslager, Mark LeRoy
Lee, Lela A
Leung, Donald Yap Man
Levinson, Simon Rock
McCalmon, Robert T, Jr
McMannis, John D
Marrack, Philippa Charlotte
Moorhead, John Wilbur
Repine, John E
Riches, David William Henry
Rodriguez, Eugene
Schanfield, Moses Samuel
Schooley, Robert T
Smith, Ralph E
Spaulding, Harry Samuel, Jr
Talmage, David Wilson
Tan, Eng M
Tengerdy, Robert Paul
Thorpe, Bert Duane
Trapani, Ignatius Louis
Urban, Richard William
Vasil, Michael Lawrence

CONNECTICUT

Ariyan, Stephan
Askenase, Philip William
Baumgarten, Alexander
Becker, Elmer Lewis
Booss, John
Bormann, Barbara-Jean Anne
Braun, Phyllis C
Bursuker, Isia
Cone, Robert Edward
Cornell-Bell, Ann Hall
Doherty, Niall Stephen
Edberg, Stephen Charles
Epstein, Paul Mark
Faanes, Ronald
Flavell, Richard Anthony
Fong, Jack Sun-Chik
Fredericksen, Tommy L
Gomez-Cambronero, Julian
Grattan, James Alex
Greiner, Dale L
Harding, Matthew William
Heim, Lyle Raymond
Horowitz, Mark Charles
Jacoby, Robert Ottinger
Janeway, Charles Alderson, Jr
Kantor, Fred Stuart
Kavathas, Paula
Kiron, Ravi
Korn, Joseph Howard
Krause, Peter James
Lee, Sin Hang
Letts, Lindsay Gordon
Littman, Bruce H
Loose, Leland David
Martin, R Russell
Mayol, Robert Francis
Mittler, Robert S
Newborg, Michael Foxx
Niblack, John Franklin
Otterness, Ivan George
Pratt, Diane McMahon
Primakoff, Paul
Rishell, William Arthur

Rocklin, Ross E
Ross, Martin Russell
Ruddle, Nancy Hartman
Schacter, Bernice Zeldin
Silbart, Lawrence K
Smith, Brian Richard
Snider, Ray Michael
Spitalny, George Leonard
Tourtellotte, Mark Eton
Wood, David Dudley
Woronick, Charles Louis
Yang, Tsu-Ju (Thomas)

DELAWARE

Dohms, John Edward
Fahey, James R
Frey, William Adrian
Fries, Cara Rosendale
Ganfield, David Judd
Ginnard, Charles Raymond
Gorzynski, Timothy James
Karol, Robin A
Krell, Robert Donald
Linna, Timo Juhani
Mackin, William Michael
Ruben, Regina Lansing
Tripp, Marenes Robert
Tulip, Thomas Hunt

DISTRICT OF COLUMBIA

Alving, Carl Richard
Bundy, Bonita Marie
Calderone, Richard Arthur
Cocks, Gary Thomas
Diggs, Carter Lee
Gravely, Sally Margaret
Herscowitz, Herbert Bernard
Jester, James Vincent
Jett-Tilton, Marti
Johnson, Armead
Katz, Paul K
Kind, Phyllis Dawn
Leto, Salvatore
Ligler, Frances Smith
Lumpkin, Michael Dirksen
Lyon, Jeffrey A
McCarty, Gale A
Matyas, Gary Ralph
Miller, Linda Jean
Pearson, Gary Richard
Phillips, Terence Martyn
Richards, Roberta Lynne
Roane, Philip Ransom, Jr
Royal, George Calvin, Jr
Taylor, Diane Wallace
Tompkins, George Jonathan
Turner, Willie
Vogel, Carl-Wilhelm E
Wassef, Nabila M
White, Sandra L
Wright, Daniel Godwin
Zatz, Marion M

FLORIDA

Adams, Roger Omar
Avner, Barry P
Bernstein, Stanley H
Bliznakov, Emile George
Blomberg, Bonnie B
Brenner, Richard Joseph
Chambliss, Keith Wayne
Claflin, Alice J
Clark, William Burton, IV
Crandall, Richard B
Crews, Fulton T
Day, Noorbibi Kassam
Dunn, William Arthur, Jr
Elfenbein, Gerald Jay
Esser, Alfred F
Fletcher, Mary Ann
Fogel, Bernard J
Friedman, Herman
Gengozian, Nazareth
Gennaro, Joseph Francis
Gifford, George Edwin
Goihman-Yahr, Mauricio
Goodman, Howard Charles
Groupe, Vincent
Hadden, John Winthrop
Hoffmann, Edward Marker
Johnson, Howard Marcellus
Kiefer, David John
Kimura, Arthur K
Klein, Paul Alvin
Lai, Patrick Kinglun
Lawman, Michael John Patrick
Legler, Donald Wayne
Lin, Tsue-Ming
Litman, Gary William
Lockey, Richard Funk
Luer, Carl A
McArthur, William P
McGuigan, James E
Mahan, Suman Mulakhraj
Panush, Richard Sheldon
Paradise, Lois Jean
Pearl, Gary Steven
Peck, Ammon Broughton
Roux, Kenneth H
Rowlands, David T, Jr
Schultz, Duane Robert
Scornik, Juan Carlos
Siden, Edward Joel
Sinkovics, Joseph G

Small, Parker Adams, Jr
Smith, Dennis Eugene
Specter, Steven Carl
Stock, David Allen
Stone, Stanley S
Streilein, Jacob Wayne
Sweeney, Michael Joseph
Tamplin, Mark Lewis
Thorn, Richard Mark
Thuning-Roberson, Claire Ann
White, Roseann Spicola
Williams, Ralph C, Jr
Zarco, Romeo Morales

GEORGIA

Alonso, Kenneth B
Bennett, Sara Neville
Butcher, Brian T
Campbell, Gary Homer
Cavallaro, Joseph John
Chen, Dillion Tong-ting
Cherry, William Bailey
Damian, Raymond T
Daugharty, Harry
Duncan, Robert Leon, Jr
Evans, Donald Lee
Feeley, John Cornelius
Garver, Frederick Albert
Goldman, John Abner
Gooding, Linda R
Grayzel, Arthur I
Gulati, Adarsh Kumar
Hall, Charles Thomas
Holt, Peter Stephen
Hunter, Robert L
Kagan, Irving George
Kiefer, Charles Randolph
Knudsen, Richard Carl
Koerner, T J
Krauss, Jonathan Seth
McKinney, Roger Minor
Madden, John Joseph
Maddison, Shirley Eunice
Margolis, Harold Stephen
Martin, Linda Spencer
Mathews, Henry Mabbett
Murphy, Frederick A
Navalkar, Ram G
Nguyen-Dinh, Phuc
Pine, Leo
Pratt, Lee Herbert
Preblud, Stephen Robert
Ragland, William Lauman, III
Reese, Andy Clare
Reimer, Charles Blaisdell
Reiss, Errol
Rodey, Glenn Eugene
Roesel, Catherine Elizabeth
Roper, Maryann
Sawyer, Richard Trevor
Schierman, Louis W
Sell, Kenneth W
Sinor, Lyle Tolbot
Smith, David Fletcher
Stouffer, Richard Franklin
Stulting, Robert Doyle, Jr
Sulzer, Alexander Jackson
Tsang, Victor Chiu Wana
Tunstall, Lucille Hawkins
Urso, Paul
Vogler, Larry B
Walls, Kenneth W
Ziegler, Harry Kirk

HAWAII

Benedict, Albert Alfred
Desowitz, Robert
Hokama, Yoshitsugi
Person, Donald Ames
Roberts, Robert Russell
Stuart, William Dorsey
Vann, Douglas Carroll

IDAHO

Congleton, James Lee

ILLINOIS

Andersen, Burton
Anderson, Byron
Arnason, Barry Gilbert Wyatt
Arroyave, Carlos Mariano
Baram, Peter
Baum, Linda Louise
Bell, Clara G
Bhatti, Rashid
Bluestone, Jeffrey A
Brackett, Robert Giles
Braun, Donald Peter
Brennan, Patricia Conlon
Briles, Worthie Elwood
Cabana, Veneracion Garganta
Callaghan, Owen Hugh
Chang, Chong-Hwan
Cheung, Hou Tak
Chudwin, David S
Cohen, Edward P
Conta, Barbara Saunders
Dal Canto, Mauro Carlo
Dau, Peter Caine
Davie, Joseph Myrten
Dray, Sheldon
Dudkiewicz, Alan Bernard
Dybas, Linda Kathryn
Ekstedt, Richard Dean

Emeson, Eugene Edward
Falk, Lawrence A, Jr
Feldbush, Thomas Lee
Fenters, James Dean
Fitch, Frank Wesley
Friedman, Yochanan
Gewurz, Henry
Goldberg, Erwin
Gotoff, Samuel P
Grundbacher, Frederick John
Halfman, Clarke Joseph
Hanly, W Carey
Hanson, Lyle Eugene
Harris, Jules Eli
Hirata, Arthur Atsunobu
Holzer, Timothy J
Horng, Wayne J
James, Karen K(anke)
Jones, Carl Joseph
Jones, Marjorie Ann
Kim, Byung Suk
Kim, Yoon Berm
Kim, Yung Dai
Knight, Katherine Lathrop
Kraft, Sumner Charles
Kulkarni, Anant Sadashiv
Landay, Alan Lee
Lange, Charles Ford
Lazda, Velta Abuls
McConnachie, Peter Ross
McLeod, Rima W
Marczynska, Barbara Mary
Margoliash, Emanuel
Markowitz, Abraham Sam
Mathews, Herbert Lester
Melvold, Roger Wayne
Miller, Stephen Douglas
Moskal, Joseph Russell
Moticka, Edward James
Murrell, Kenneth Darwin
Myers, Walter Loy
Nevalainen, David Eric
Ostrow, David Henry
Pachman, Lauren M
Parvez, Zaheer
Plate, Janet Margaret
Potempa, Lawrence Albert
Prasad, Rameshwar
Pun, Pattle Pak-Toe
Ratajczak, Helen Vosskuhler
Rich, Kenneth C
Richman, David Paul
Robertson, Abel Alfred Lazzarini, Jr
Rogalski-Wilk, Adrian Alice
Rowe, William Bruce
Rowley, Donald Adams
Sabet, Tawfik Younis
Sandberg, Philip A
Schlager, Seymour I
Schreiber, Hans
Segre, Diego
Segre, Mariangela Bertani
Sharon, Nehama
Shipchandler, Mohammed Tyebji
Shulman, Stanford Taylor
Simonson, Lloyd Grant
Sinclair, Thomas Frederick
Spear, Patricia Gail
Springer, Georg F
Stevens, Fred Jay
Storb, Ursula
Tartof, David
Thompson, Kenneth David
Tomita, Joseph Tsuneki
Treadway, William Jack, Jr
Turner, Donald W
Van Alten, Pierson Jay
Van Epps, Dennis Eugene
Venkataraman, M
Voss, Edward William, Jr
Waltenbaugh, Carl
Walter, Robert John
Wass, John Alfred
Weisz-Carrington, Paul
Weyhenmeyer, James Alan
Wolf, Raoul
Wu-Wong, Jinshyun Ruth
Yokoyama, Mitsuo
Young, Jay Maitland
Zahalsky, Arthur C
Zeiss, Chester Raymond

INDIANA

Aldo-Benson, Marlene Ann
Ashendel, Curtis Lloyd
Basu, Manju
Bauer, Dietrich Charles
Bick, Peter Hamilton
Boguslaski, Robert Charles
Brahmi, Zacharie
Diffley, Peter
Dunn, Peter Edward
Edwards, Joshua Leroy
Freeman, Max James
Gale, Charles
Ghosh, Swapan Kumar
Gunther, Gary Richard
Hicks, Edward James
Ho, Peter Peck Koh
Hoversland, Roger Carl
Howard, David K
Hullinger, Ronald Loral
Ingraham, Joseph Sterling
Janski, Alvin Michael

Jenski, Laura Jean
Kenny, Michael Thomas
Koppel, Gary Allen
Kreps, David Paul
Lopez, Carlos
McGowan, Michael James
McIntyre, John A
Miller (Gilbert), Carol Ann
Mocharla, Raman
Morter, Raymond Lione
Osuch, Mary Ann V
Petersen, Bruce H
Pfeifer, Richard Wallace
Roales, Robert R
Schmidtke, Jon Robert
Simon, Edward Harvey
Taylor, Milton William
Wagner, Morris
Wilde, Charles Edward, III
Yoder, John Menly
Zimmerman, Sarah E

IOWA
Ashman, Robert F
Ballas, Zuhari K
Boney, William Arthur, Jr
Butler, John E
Butler, John Edward
Cheng, Frank Hsieh Fu
Clark, Robert A
Dailey, Morris Owen
Fitzgerald, Gerald Roland
Ford, Stephen Paul
Goeken, Nancy Ellen
Hartman, Paul Arthur
Hoffmann, Louis Gerhard
Hom, Ben Lin
Hsu, Hsi Fan
Hsu, Shu Ying Li
Lamont, Susan Joy
Lubaroff, David Martin
Meetz, Gerald David
Miller, Wilmer Jay
O'Berry, Phillip Aaron
Quinn, Loyd Yost
Rebers, Paul Armand
Richerson, Hal Bates
Schmerr, Mary Jo F
Steinmuller, David
Stieler, Carol Mae
Targowski, Stanislaw P
Thurston, John Robert
Weinstock, Joel Vincent
Welter, C Joseph

KANSAS
Abdou, Nabih I
Bailie, Wayne E
Bradford, Lawrence Glenn
Brown, John Clifford
Bryan, Christopher F
Draper, Laurence Rene
Fortner, George William
Greene, Nathan Doyle
Haas, Herbert Frank
Halsey, John Frederick
Krishnan, Engil Kolaj
McElree, Helen
Miller, Glendon Richard
Minocha, Harish C
Morrison, David Campbell
Parmely, Michael J
Ringle, David Allan
Stewart, Richard Cummins
Suzuki, Tsuneo
Sweet, George H
Wolf, Thomas Michael
Wood, Gary Warren

KENTUCKY
Ambrose, Charles T
Espinosa, Enrique
Higginbotham, Robert David
Justus, David Eldon
Kaplan, Alan Marc
Morgan, Marjorie Susan
Neefe, John R
Ruchman, Isaac
Sonnenfeld, Gerald
Thompson, John S
Wacker, Waldon Burdette
Wallace, John Howard

LOUISIANA
Bagby, Gregory John
Cannon, Donald Charles
Choi, Yong Sung
Domingue, Gerald James
Epps, Anna Cherrie
Espinoza, Luis Rolan
Etheredge, Edward Ezekiel
Franklin, Rudolph Michael
Fulginiti, Vincent Anthony
Gaumer, Herman Richard
Gebhardt, Bryan Matthew
Gormus, Bobby Joe
Gottlieb, A Arthur
Hastings, Robert Clyde
Hodgin, Ezra Clay
Hornung, Maria
Hyslop, Newton Everett, Jr
Ivens, Mary Sue
Klei, Thomas Ray
Kohler, Peter Francis

Krahenbuhl, James Lee
Kuan, Shia Shiong
Lanners, H Norbert
Lehrer, Samuel Bruce
McDonald, John C
Magin, Ralph Walter
Michalski, Joseph Potter
Millikan, Larry Edward
Misra, Raghunath P
Neucere, Joseph Navin
Nickerson, Stephen Clark
O'Neil, Carol Elliot
Pisano, Joseph Carmen
Salvaggio, John Edmond
Saphier, David
Shaffer, Morris Frank
Sheiness, Diana Kay
Silberman, Ronald
Sizemore, Robert Carlen
Sorensen, Ricardo U
Strauss, Arthur Joseph Louis
Sullivan, Karen A
Thompson, James Jarrard
Wolf, Robert E

MAINE
Donahoe, John Philip
Eskelund, Kenneth H
Evans, Robert
Fink, Mary Alexander
Founds, Henry William
Gaskins, H Rex
Harrison, David Ellsworth
Kaliss, Nathan
Moody, Charles Edward, Jr
Ng, Ah-Kau
Novotny, James Frank
Renn, Donald Walter
Sanford, Barbara Hendrick
Shultz, Leonard Donald
Snell, George Davis
Weiss, John Jay

MARYLAND
Abe, Ryo
Adkinson, Newton Franklin, Jr
Adler, William
Alexander, Nancy J
Amzel, L Mario
Anderson, Arthur O
Anderson, Robert Simpers
Anhalt, Grant James
Appella, Ettore
Asofsky, Richard Marcy
Augustine, Patricia C
Austin, Faye Carol
Azad, Abdu F(arhang)
Baer, Harold
Baker, Phillip John
Bala, Shukal
Barker, Lewellys Franklin
Beaven, Michael Anthony
Beckner, Suzanne K
Bennett, William Ernest
Berkower, Ira
Berman, Howard Mitchell
Berne, Bernard H
Berzofsky, Jay Arthur
Beschorner, W E
Bhatnagar, Gopal Mohan
Birx, Deborah L
Biswas, Robin Michael
Blaese, R Michael
Bloom, Eda Terri
Borsos, Tibor
Braatz, James Anthony
Breyere, Edward Joseph
Brink, Linda Holk
Brook, Itzhak
Bustin, Michael
Butman, Bryan Timothy
Carski, Theodore Robert
Caspi, Rachel R
Chaparas, Sotiros D
Chen, Joseph Ke-Chou
Chused, Thomas Morton
Cohen, Sheldon Gilbert
Cole, Gerald Alan
Coligan, John E
Colombani, Paul Michael
Cowan, Elliot Paul
Cox, George Warren
Craig, Susan Walker
Creasia, Donald Anthony
Crystal, Ronald George
Dannenberg, Arthur Milton, Jr
Das, Saroj R
Dasch, Gregory Alan
Davidson, Wendy Fay
Dean, Ann
De la Cruz, Vidal F
DeVries, Yuan Lin
Dickler, Howard B
Dintzis, Renee Zlochover
Djurickovic, Draginja Branko
Dodd, Roger Yates
Donlon, Mildred A
Durum, Scott Kenneth
Dwyer, Dennis Michael
Eckels, Kenneth Henry
Edidin, Michael Aaron
Ennist, David L
Epstein, Suzanne Louise
Esposito, Vito Michael

Evans, Charles Hawes
Falkler, William Alexander, Jr
Fast, Patricia E
Fearon, Douglas T
Feinman, Susan (Birnbaum)
Fernie, Bruce Frank
Fex, Jorgen
Fields, Kay Louise
Finkelman, Fred Douglass
Fleisher, Thomas A
Fox, Barbara Saxton
Frank, Michael M
Fraser, Blair Allen
Froehlich, Luz
Fuson, Roger Baker
Gainer, Joseph Henry
Gasbarre, Louis Charles
Gearhart, Patricia Johanna
Germain, Ronald N
Gerrard, Theresa Lee
Gery, Igal
Gilden, Raymond Victor
Glaudemans, Cornelis P
Goldman, Daniel Ware
Goldsmith, Paul Kenneth
Goldstein, Robert Arnold
Goodman, Sarah Anne
Gordon, Lance Kenneth
Granger, Donald Lee
Gravell, Maneth
Green, Ira
Gress, Ronald E
Griffin, Diane Edmund
Habig, William Henry
Hackett, Joseph Leo
Hale, Martha L
Hammer, Carl Helman
Hampar, Berge
Hanks, John Harold
Hanna, Edgar Ethelbert, Jr
Hanna, Michael G, Jr
Hardegree, Mary Carolyn
Harvarth, Liana
Harvath, Liana
Haspel, Martin Victor
Hausman, Steven J
Hearing, Vincent Joseph, Jr
Hellman, Alfred
Henkart, Pierre
Hess, Allan Duane
Hewetson, John Francis
Hoffeld, J Terrell
Hoffman, Thomas
Hollingdale, Michael Richard
Hook, William Arthur
Hoyer, Leon William
Jaffe, Elaine Sarkin
Jakab, George Joseph
James, Stephanie Lynn
Johnson, Leslye
Johnson-Winegar, Anna
Kagey-Sobotka, Anne
Keegan, Achsah D
Kenny, James Joseph
Kenyon, Richard H
Kickler, Thomas Steven
Krakauer, Henry
Krakauer, Teresa
Kramer, Norman Clifford
Krause, David
Krause, Richard Michael
Kruisbeek, Ada M
Lee, Chi-Jen
Leef, James Lewis
Liang, Shu-Mei
Lichtenstein, Lawrence M
Lillehoj, Hyun Soon
Lippincott-Schwartz, Jennifer
Lockshin, Michael Dan
Longo, Dan L
Lunney, Joan K
McCarron, Richard M
McCartney-Francis, Nancy L
McClintock, Patrick Ralph
McCurdy, John Dennis
McFarland, Henry F
McFarlin, Dale Elroy
MacGlashan, Donald W, Jr
Mackler, Bruce F
MacVittie, Thomas Joseph
Mage, Michael Gordon
Mage, Rose G
Mageau, Richard Paul
Malech, Harry Lewis
Malmgren, Richard Axel
Manclark, Charles Robert
Mardiney, Michael Ralph, Jr
Maret, S Melissa
Margulies, David Harvey
Marquardt, Warren William
Marsh, David George
Maurer, Bruce Anthony
Meltzer, Monte Sean
Mergenhagen, Stephan Edward
Metcalfe, Dean Darrel
Metzger, Henry
Mittal, Kamal Kant
Mond, James Jacob
Monjan, Andrew Arthur
Morris, Joseph Anthony
Morse, Herbert Carpenter, III
Mufson, R A
Mushinski, J Frederic
Nacy, Carol Anne

Nauman, Robert Karl
Nelson, David L
Neta, Ruth
Nordin, Albert Andrew
Norman, Philip Sidney
Notkins, Abner Louis
O'Beirne, Andrew Jon
Oppenheim, Joost J
O'Rangers, John Joseph
Order, Stanley Elias
Oroszian, Stephen
Ortaldo, John R
Ostrand-Rosenberg, Suzanne T
Ottesen, Eric Albert
Ozato, Keiko
Packard, Beverly Sue
Padarathsingh, Martin Lancelot
Palestine, Alan G
Paul, William Erwin
Peters, Clarence J
Picciolo, Grace Lee
Plaut, Marshall
Plotz, Paul Hunter
Pluznik, Dov Herbert
Pohl, Lance Rudy
Powers, Kendall Gardner
Prabhakar, Bellur Subbanna
Prendergast, Robert Anthony
Prince, Gregory Antone
Quinnan, Gerald Vincent, Jr
Ram, J Sri
Read-Connole, Elizabeth Lee
Robbins, John B
Rogers, Michael Joseph
Rommel, Frederick Allen
Rose, Noel Richard
Rosenstein, Robert William
Rossio, Jeffrey L
Russell, Philip King
Sachs, David Howard
Samelson, Lawrence Elliot
Sandberg, Ann Linnea
Santos, George Wesley
Saxinger, W(illiam) Carl
Schleimer, Robert P
Schlom, Jeffrey
Schluederberg, Ann Elizabeth Snider
Schricker, Robert Lee
Segal, David Miller
Seto, Belinda P L
Shaw, Stephen
Shevach, Ethan Menahem
Shih, James Waikuo
Shin, Moon L
Shiver, John W
Sieckmann, Donna G
Siegel, Jay Philip
Silverstein, Arthur M
Singer, Alfred
Siraganian, Reuben Paul
Sitkovsky, Michail V
Smith, Phillip Doyle
Smith-Gill, Sandra Joyce
Snader, Kenneth Means
Sogn, John Allen
Solomon, Joel Martin
Spero, Leonard
Stamper, Hugh Blair
Stein, Kathryn E
Steinberg, Alfred David
Stevenson, Henry C
Stobo, John David
Stone, Sanford Herbert
Strober, Warren
Stylos, William A
Suskind, Sigmund Richard
Sztein, Marcelo Benjamin
Taniuchi, Hiroshi
Taylor, Christopher E
Ting, Chou-Chik
Tobie, John Edwin
Tosato, Giovanna
Trapani, Robert-John
Turkeltaub, Paul Charles
Valentine, Martin Douglas
Vincent, Monroe Mortimer
Vogel, Stefanie N
Wahl, Sharon Knudson
Waldmann, Thomas A
Walker, Richard Ives
Walters, Curla Sybil
Weislow, Owen Stuart
Weiss, Joseph Francis
Whittum-Hudson, Judith Anne
Wilder, Ronald Lynn
Winchurch, Richard Albert
Winkelstein, Jerry A
Wong, Dennis Mun
Wong, Donald Tai On
Wunderlich, John R
Yarchoan, Robert
Youle, Richard James
Young, Howard Alan
Yu, Mei-ying Wong
Zbar, Berton
Zimmerman, Daniel Hill

MASSACHUSETTS
Abromson-Leeman, Sara R
Ackerman, Steven J
Ahmed, A Razzaque
Allansmith, Mathea R
Alper, Chester Allan
Anderson, Deborah Jean

Immunology (cont)

Andres, Giuseppe A
Andre-Schwartz, Janine
Argyris, Bertie
Arnaout, M Amin
Atkinson-Templeman, Kristine Hofgren
Austen, K(arl) Frank
Barlozzari, Teresa
Benacerraf, Baruj
Billingham, Rupert Everett
Bing, David H
Blanchard, Gordon Carlton
Bloch, Kurt Julius
Bogden, Arthur Eugene
Borysenko, Myrin
Brennessel, Barbara A
Brown, Beverly Ann
Cantor, Harvey
Carter, Edward Albert
Carvalho, Angelina C A
Cathou, Renata Egone
Chase, Arleen Ruth
Chishti, Athar Husain
Coleman, Robert Marshall
Colvin, Robert B
Crusberg, Theodore Clifford
Czop, Joyce K
Datta, Syamal K
David, John R
Dittmer, John Edward
Dorf, Martin Edward
Dubey, Devendra P
Ducibella, Tom
Dvorak, Ann Marie-Tompkins
Eisen, Herman Nathaniel
Esber, Henry Jemil
Essex, Myron
Finberg, Robert William
Geha, Raif S
Gerety, Robert John
Girard, Kenneth Francis
Goldmacher, Victor S
Greenstein, Julia L
Haber, Edgar
Hadjian, Richard Albert
Hansen, David Elliott
Hargis, Betty Jean
Harrison, Bettina Hall
Hartner, William Christopher
Herrmann, Steven H
Hirsch, Martin Stanley
Ho, May-Kin
Hochman, Paula S
Huber, Brigitte T
Humphreys, Robert Edward
Hwang, San-Bao
Janicki, Bernard William
Jung, Lawrence Kwok Leung
Kaplan, Melvin Hyman
Kelley, Vicki E
Khaw, Ban-An
Kimball, John Ward
Klempner, Mark Steven
Knight, Glenn B
Kupchik, Herbert Z
Lampson, Lois Alterman
Lees, Marjorie Berman
Leskowitz, Sidney
Levine, Lawrence
Levy, Elinor Miller
Liberatore, Frederick Anthony
Litt, Mortimer
Lundstrom, Ronald Charles
MacDonald, Alex Bruce
McIntire, Kenneth Robert
Madoff, Morton A
Malkiel, Saul
Margolies, Michael N
Martz, Eric
Miller-Graziano, Carol L
Millette, Clarke Francis
Mole, John Edwin
Monaco, Anthony Peter
Monette, Francis C
Mordes, John Peter
Mulder, Carel
Nagler-Anderson, Cathryn
Nisonoff, Alfred
Owen, Frances Lee
Parker, David Charles
Peri, Barbara Anne
Pier, Gerald Bryan
Plaut, Andrew George
Pober, Jordan S
Raulet, David Henri
Reif, Arnold E
Reisert, Patricia
Reiss, Carol S
Remold, Heinz G
Reyero, Cristina
Reyes, Victor E
Rodrick, Mary Lofy
Ross, Alonzo Harvey
Rule, Allyn H
Russell, Paul Snowden
Saravis, Calvin
Schlossman, Stuart Franklin
Schneeberger, Eveline E
Schur, Peter Henry
Schwaber, Jerrold
Schwartz, Anthony
Seaver, Sally S
Seidman, Jonathan G

Sheffer, Albert L
Sher, Stephanie Ellsworth
Smith, Daniel James
Smith, Nathan Lewis, III
Steiner, Lisa Amelia
Stelos, Peter
Stollar, Bernard David
Stossel, Thomas Peter
Strauss, Phyllis R
Strom, Terry Barton
Strominger, Jack L
Sullivan, David Anthony
Sunshine, Geoffrey H
Sy, Man-Sun
Tauber, Alfred Imre
Taubman, Martin Arnold
Terry, William David
Thomas, Gail B
Thomas, Peter
Tsung, Yean-Kai
Tyler, Kenneth Laurence
Vanderburg, Charles R
Walker, Bruce David
Warner, Carol Miller
Weetall, Howard H
Winn, Henry Joseph
Wolff, Sheldon Malcolm
Yarmush, Martin Leon

MICHIGAN

Bach, Michael Klaus
Bacon, Larry Dean
Barksdale, Charles Madsen
Berger, Ann Elizabeth
Boros, Dov Lewis
Caldwell, Jerry
Callewaert, Denis Marc
Campbell, Wilbur Harold
Carey, Thomas E
Carrick, Lee
Chang, Timothy Scott
Chatterjee, Bandana
Cohen, Flossie
Cole, George Christopher
Conder, George Anthony
Distler, Jack, Jr
Dore-Duffy, Paula
Duwe, Arthur Edward
Fiedler, Virginia Carol
Fraker, Pamela Jean
Goodman, Morris
Gray, Gary D
Hackel, Emanuel
Hakim, Margaret Heath
Heppner, Gloria Hill
Hirata, Fusao
Holden, Howard T
Jensen, James Burt
Johnson, Herbert Gardner
Kaplan, Joseph
Kauffman, Carol A
Kaufman, Donald Barry
Kithier, Karel
Kong, Yi-Chi Mei
Lefford, Maurice J
Leon, Myron A
Lerman, Stephen Paul
Lightbody, James James
Liu, Stephen C Y
Loor, Rueyming
Lopatin, Dennis Edward
McCann, Daisy S
Miller, Fred R
Miller, Harold Charles
Montgomery, Ilene Nowicki
Montgomery, Paul Charles
Mulks, Martha Huard
Nairn, Roderick
Narayan, Latha
Ownby, Dennis Randall
Patterson, Ronald James
Pence, Leland Hadley
Petty, Howard Raymond
Phan, Sem Hin
Poulik, Miroslav Dave
Puddington, Lynn
Rauch, Helene Coben
Root-Bernstein, Robert Scott
Rosenspire, Allen Jay
Rowe, Nathaniel H
Sacco, Anthony G
San Clemente, Charles Leonard
Saper, Mark A
Sarkar, Fazlul Hoque
Schwartz, Stanley Allen
Sell, John Edward
Simon, Michael Richard
Sloat, Barbara Furin
Smith, Jerry Warren
Smith, Richard Harding
Smith, Robert James
Sundick, Roy
Swanborg, Robert Harry
Taggart, R(obert) Thomas
Thomas, David Warren
Toledo-Pereyra, Luis Horacio
Tracey, Daniel Edward
Truden, Judith Lucille
Tse, Harley Y
Ward, Peter A
Watts, Jeffrey Lynn
Wei, Wei-Zen
Weiner, Lawrence Myron
Whalen, Joseph Wilson

White, Gordon Justice
Whitehouse, Frank, Jr
Wilhelm, Rudolf Ernst
Williams, Jeffrey F
Yanari, Sam Satomi
Yates, Jon Arthur
Yurewicz, Edward Charles

MINNESOTA

Abraham, Robert Thomas
Adams, Ernest Clarence
Azar, Miguel M
Bach, Marilyn Lee
Beck, Barbara North
Carter, Orwin Lee
Clark, Connie
Cunningham, Julie Margaret
Dalmasso, Agustin Pascual
David, Chelladurai S
Filipovich, Alexandra H
Gleich, Gerald J
Greenberg, Leonard Jason
Hallgren, Helen M
Homburger, Henry A
Jemmerson, Ronald Renomer Weaver
Jeska, Edward Lawrence
Johnson, Arthur Gilbert
Johnson, Russell Clarence
Johnson, Theodore Reynold
Kersey, John H
Kyle, Robert Arthur
Leibson, Paul Joseph
Lennon, Vanda Alice
Lindgren, Alice Marilyn Lindell
Lukasewycz, Omelan Alexander
McKean, David Jesse
McPherson, Thomas C(oatsworth)
Markowitz, Harold
Meyers, Paul
Mitchell, Roger Harold
Muscoplat, Charles Craig
Nelson, Robert D
Page, Arthur R
Rice, Thomas Kenneth
Ritts, Roy Ellot, Jr
Sharma, Jagdev Mittra
Shope, Richard Edwin, Jr
Song, Chang Won
Stromberg, Bert Edwin, Jr
Taswell, Howard Filmore
Vallera, Daniel A

MISSISSIPPI

Clem, Lester William
Cruse, Julius Major, Jr
Cuchens, Marvin A
Lewis, Robert Edwin, Jr
Sindelar, Robert D
Steen, James Southworth
Watson, Edna Sue

MISSOURI

Addinger, Hans Karl
Allen, Robert Wade
Atkinson, John Patterson
Barrett, James Thomas
Bartkoski, Michael John, Jr
Bell, Jeffrey
Bellone, Clifford John
Birch, Ruth Ellen
Braciale, Thomas Joseph, Jr
Braciale, Vivian Lam
Brown, Eric
Brown, Karen Kay (Kilker)
Buening, Gerald Matthew
Buonomo, Frances Catherine
Catalona, William John
Codd, John Edward
Colten, Harvey Radin
Cullen, Susan Elizabeth
De Blas, Angel Luis
Fernandez-Pol, Jose Alberto
Ferris, Deam Hunter
Friedman, Harvey Paul
Granoff, Dan Martin
Green, Theodore James
Hansen, Ted Howard
Harakas, N(icholas) Konstantinos
Herzig, Geoffrey Peter
Howard, Susan Carol Pearcy
Johnston, Marilyn Frances Meyers
Kulczycki, Anthony, Jr
Lafrenz, David E
Lie, Wen-Rong
Little, John Russell, Jr
Lyle, Leon Richards
Miles, Donald Orval
Miller, Herman T
Misfeldt, Michael Lee
Mohanakumar, Thalachallour
Munns, Theodore Willard
Murray, Patrick Robert
Nahm, Moon H
Nunez, William J, III
Olson, Gerald Allen
Parker, Charles W
Pierce, Carl William
Polmar, Stephen Holland
Premachandra, Bhartur N
Rifas, Leonard
Roodman, Stanford Trent
Strunk, Robert Charles
Tumosa, Nina Jean
Twining, Linda Carol

White, Gordon Justice
Unanue, Emil R
Viamontes, George Ignacio
Wang, Richard J
Willard, Mark Benjamin

MONTANA

Braune, Maximillian O
Cantrell, John Leonard
Chesbro, Bruce W
Coe, John Emmons
Ewald, Sandra J
Jesaitis, Algirdas Joseph
Jutila, John W
Lodmell, Donald Louis
Munoz, John Joaquin
Myers, Lyle Leslie
Portis, John L
Reed, Norman D
Rudbach, Jon Anthony
Speer, Clarence Arvon

NEBRASKA

Carson, Steven Douglas
Chaperon, Edward Alfred
Crouse, David Austin
Curtis, Gary Lynn
Dyer, John Kaye
Gerber, Jay Dean
Ghosh, Chitta Ranjan
Heidrick, Margaret Louise
Kay, H David
Klassen, Lynell W
Kobayashi, Roger Hideo
O'Brien, Richard Lee
Ryan, Wayne L
Sharp, John Graham
Smith, Paula Beth
Turpen, James Baxter
Waldman, Robert H

NEVADA

Henry, Claudia
Kozel, Thomas Randall

NEW HAMPSHIRE

Adler, Frank Leo
Adler, Louise Tale
Ball, Edward D
Fahey, John Vincent
Fanger, Michael Walter
Green, William Robert
Kaiser, C William
Pistole, Thomas Gordon
Smith, Kendall A

NEW JERSEY

Anderson, David Wesley
Babu, Uma Mahesh
Barton, Beverly E
Basu, Mitali
Becker, Joseph Whitney
Bendich, Adrianne
Boltz, Robert Charles, Jr
Brockman, John A, Jr
Brunda, Michael J
Capetola, Robert Joseph
Caporale, Lynn Helena
Carlson, Richard P
Chang, Joseph Yoon
Crane, Mark
Das, Kiron Moy
Davies, Philip
DeBari, Vincent A
Devlin, Richard Gerald, Jr
Dhruv, Rohini Arvind
Egan, Robert Wheeler
Ende, Norman
Fiedler-Nagy, Christa
Field, Arthur Kirk
Gaffar, Abdul
Gately, Maurice Kent
Gilfillan, Alasdair Mitchell
Gillette, Ronald William
Gilman, Steven Christopher
Goldenberg, David Milton
Goldstein, Gideon
Graham, Henry Alexander, Jr
Green, Erika Ana
Green, Gerald
Greenstein, Teddy
Grimes, David
Gross, Peter A
Guerrero, Jorge
Gupta, Ayodhya P
Gupta, Godaveri Rawat
Gyan, Nanik D
Heaslip, Richard Joseph
Heveran, John Edward
Hirsch, Robert L
Humes, John Leroy
Kamiyama, Mikio
Kawanishi, Hidenori
Khavkin, Theodor
Kleinschuster, Stephen J, III
Koepp, Leila H
Krishnan, Gopal
Letizia, Gabriel Joseph
McKearn, Thomas Joseph
Mark, David Fu-Chi
Miller, Douglas Kenneth
Miller, Sally Ann
Miner, Karen Mills
Monestier, Marc
Moore, Vernon Leon

Oleske, James Matthew
Panagides, John
Passmore, Howard Clinton
Peterson, Arthur Carl
Plescia, Otto John
Ponzio, Nicholas Michael
Poretz, Ronald David
Raychaudhuri, Anilbaran
Reckel, Rudolph P
Rivkin, Israel
Rodwell, John Dennis
Rosenstraus, Maurice Jay
Saha, Anil
Schenkel, Robert H
Schleifer, Steven Jay
Schlesinger, R(obert) Walter
Schroff, Peter David
Sehgal, Surendra N
Sellgmann, Bruce Edward
Senda, Mototaka
Sethi, Jitender K
Shakarjian, Michael Peter
Shin, Seung-il
Sigal, Nolan H
Sikder, Santosh K
Singer, Irwin I
Smith, Sidney R, Jr
Solinger, Alan M
Stolen, Joanne Siu
Taylor, Bernard Franklin
Teipel, John William
Tenoso, Harold John
Tierno, Philip M, Jr
Tripodi, Daniel
Turner, Matthew X
VanMiddlesworth, Frank L
Wang, Bosco Shang
Wang-Iverson, Patsy
Webb, David R, Jr
Welton, Ann Frances
Whiteley, Phyllis Ellen
Woehler, Michael Edward
Wojnar, Robert John
Wong, Rosie Bick-Har
Zuckerman, Leo

NEW MEXICO
Anderson, William Loyd
Bell, George Irving
Bice, David Earl
Crago, Sylvia S
Davis, Larry Ernest
Edwards, Bruce S
Mold, Carolyn
Perelson, Alan Stuart
Rhodes, Buck Austin
Saland, Linda C
Saunders, George Cherdron
Shopp, George Milton, Jr
Smith, David Marshall
Sopori, Mohan L
Tokuda, Sei

NEW YORK
Abdelnoor, Alexander Michael
Ablin, Richard J
Abraham, George N
Al-Askari, Salah
Allen, Howard Joseph
Allen, Peter Zachary
Ames, Ira Harold
Amiraian, Kenneth
Arny, Margaret Jane
Ascensao, Joao L
Bachrach, Howard L
Bachvaroff, Radoslav J
Baird, Barbara A
Baltimore, David
Bankert, Richard Burton
Bartal, Arie H
Bartfeld, Harry
Basch, Ross S
Baum, George
Baum, John
Bauman, Norman
Bazinet, George Frederick
Bealmear, Patricia Maria
Bell, Robin Graham
Beutner, Ernst Herman
Beyer, Carl Fredrick
Bhattacharya, Malaya
Bhattacharya-Chatterjee, Malaya
Bielat, Kenneth L
Birken, Steven
Bloom, Barry R
Blumenthal, Rosalyn D
Bovbjerg, Dana H
Bowen, William H
Brandriss, Michael W
Braunstein, Joseph David
Brentjens, Jan R
Brophy, Mary O'Reilly
Bush, Maurice E
Bushkin, Yuri
Butler, Vincent Paul, Jr
Buxbaum, Joel N
Cabrera, Edelberto Jose
Calame, Kathryn Lee
Calvelli, Theresa A
Campbell, Samuel Gordon
Cerini, Constantino Peter
Chase, Merrill Wallace
Chase, Randolph Montieth, Jr
Chen, Priscilla B(urtis)

Chess, Leonard
Chiao, Jen Wei
Chisholm, David R
Christian, Charles L
Chu, Tsann Ming
Clarkson, Allen Boykin, Jr
Cohen, Elias
Cohen, Martin William
Cohen, Nicholas
Cohn, Deirdre Arline
Coico, Richard F
Collins, Arlene Rycombel
Collins, Frank Miles
Contento, Isobel
Conway de Macario, Everly
Cooper, Norman S
Cowell, James Leo
Craig, John Philip
Cramer, Eva Brown
Cravitz, Leo
Cunninham-Rundles, Charlotte
Curley, Dennis M
D'Alesandro, Philip Anthony
Day, Eugene Davis
Dean, David A
Dean, Jack Hugh
DeForest, Peter Rupert
Despommier, Dickson
D'Eustachio, Peter
Dietert, Rodney Reynolds
Divgi, Chaitanya R
Duff, Ronald George
Dupont, Bo
Durkin, Helen Germaine
Durr, Friedrich (E)
Dworetzky, Murray
Edelson, Paul J
Efthymiou, Constantine John
Ehrke, Mary Jane
Ellner, Paul Daniel
Erlanger, Bernard Ferdinand
Evans, Mary Jo
Evans, Richard Todd
Everhart, Donald Lee
Farah, Fuad Salim
Feit, Carl
Felten, David L
Fenton, John William, II
Ferrone, Soldano
Fisher, Clark Alan
Flanagan, Thomas Donald
Fleisher, Martin
Fleit, Howard B
Fondy, Thomas Paul
Fontana, Vincent J
Fotino, Marilena
Frair, Wayne
Frangione, Blas
Friedman, Eli A
Fritz, Katherine Elizabeth
Fuji, Hiroshi
Furmanski, Philip
Gabrielsen, Ann Emily
Gall, William Einar
Genco, Robert J
Ghebrehiwet, Berhane
Gibbs, David Lee
Gibofsky, Allan
Ginsberg-Fellner, Fredda Vita
Glassman, Armand Barry
Godfrey, H(enry) P(hilip)
Goldsmith, Lowell Alan
Gotschlich, Emil C
Grimwood, Brian Gene
Grob, David
Gutcho, Sidney J
Habicht, Gail Sorem
Hammerling, Ulrich
Han, Tin
Hawrylko, Eugenia Anna
Heidelberger, Michael
Henley, Walter L
Herr, Harry Wallace
Hirschhorn, Kurt
Hirschhorn, Rochelle
Holowka, David Allan
Holzman, Robert Stephen
Houghton, Alan N
Howard, Irmgard Matilda Keeler
Hsu, Konrad Chang
Iglewski, Barbara Hotham
Isaacs, Charles Edward
Isenberg, Henry David
Joel, Darrel Dean
Johnson, Lawrence Lloyd
Johnston, Dean
Josephson, Alan S
Kalsow, Carolyn Marie
Kaplan, Joel Howard
Kascsak, Richard John
Kenyon, Alan J
Khan, Abdul Jamil
Kim, Charles Wesley
Kim, Untae
Kim, Young Tai
Kimberly, Robert Parker
Kishore, Gollamudi Sitaram
Kite, Joseph Hiram, Jr
Kobilinsky, Lawrence
Kochwa, Shaul
Kowolenko, Michael D
Kratzel, Robert Jeffrey
Krown, Susan E
Lahita, Robert George

Lamberson, Harold Vincent, Jr
Lambert, Reginald Max
Lattime, Edmund Charles
Lau, Joseph T Y
Lawrence, David A
Leary, James Francis
Leddy, John Plunkett
Lee, Ching-Li
Levine, Bernard Benjamin
Lewis, Robert Miller
Lilly, Frank
Lipson, Steven Mark
Litwin, Stephen David
Lloyd, Kenneth Olof
Loeb, Marilyn Rosenthal
Lord, Edith M
Lossinsky, Albert S
Lutton, John D
Macario, Alberto J L
McClintock, David K
MacGillivray, Margaret Hilda
McNamara, Thomas Francis
Macphail, Stuart
Macris, Nicholas T
Magill, Thomas Pleines
Manjula, Belur N
Mannino, Raphael James
Manski, Wladyslaw J
Mansour, Moustafa M
Marboe, Charles Chostner
Marcu, Kenneth Brian
Marsh, William Laurence
Mayers, George Louis
Meruelo, Daniel
Michael, Sandra Dale
Miller, Frederick
Miller, Michael E
Millis, Albert Jason Taylor
Mizejewski, Gerald Jude
Mohn, James Frederic
Mohos, Steven Charles
Moline, Sheldon Walter
Mongini, Patricia Katherine Ann
Moor-Jankowski, J
Morgan, Donald O'Quinn
Moroson, Harold
Morse, Jane H
Morse, Stephen Scott
Mueller, August P
Murphy, Donal B
Muschel, Louis Henry
Nathenson, Stanley G
Neiders, Mirdza Erika
Neimark, Harold Carl
Neumann, Norbert Paul
Nisselbaum, Jerome Seymour
Noble, Bernice
Norcross, Neil Linwood
Norin, Allen Joseph
North, Robert J
Norvitch, Mary Ellen
Nussenzweig, Victor
Oettgen, Herbert Friedrich
Ogra, Pearay L
Old, Lloyd John
O'Reilly, Richard
Ovary, Zoltan
Packman, Charles Henry
Papsidero, Lawrence D
Pelus, Louis Martin
Perlman, Ely
Perper, Robert J
Phillips-Quagliata, Julia Molyneux
Pierce, Carol S
Pitt, Jane
Pollara, Bernard
Pruslin, Fred Howard
Pure, Ellen
Quaroni, Andrea
Quimby, Fred William
Rafla, Sameer
Raine, Cedric Stuart
Rajan, Thiruchandurai Viswanathan
Rao, Yalamanchili A K
Rapaport, Felix Theodosius
Rapp, Dorothy Glaves
Rappaport, Irving
Raveche, Elizabeth Marie
Raventós-Suárez, Carmen Elvira
Reddy, Mohan Muthireval
Reid, Lola Cynthia McAdams
Reimers, Thomas John
Rosenfeld, Richard Ernest
Rosenstreich, David Leon
Sanford, Karl John
Santiago, Noemi
Sayegh, Joseph Frieh
Scharff, Matthew Daniel
Schauf, Victoria
Schiffman, Gerald
Schoenfeld, Cy
Scott, David William
Sehgal, Pravinkumar B
Seiden, Philip Edward
Senitzer, David
Senterfit, Laurence Benfred
Seon, Ben Kuk
Shapiro, Caren Knight
Shayegani, Mehdi
Shulman, Sidney
Siegal, Frederick Paul
Silverstein, Samuel Charles
Siminoff, Paul
Simon, Sidney

Siskind, Gregory William
Slovin, Susan Faith
Socha, Wladyslaw Wojciech
Sorensen, Craig Michael
Spar, Irving Leo
Spitzer, Roger Earl
Stark, Dennis Michael
Stevens, Roy White
Stewart, Carleton C
Stolfi, Robert Louis
Stoner, Richard Dean
Stutman, Osias
Sultzer, Barnet Martin
Sussdorf, Dieter Hans
Tabor, John Malcolm
Taubman, Sheldon Bailey
Theis, Gail Goodman
Thorbecke, Geertruida Jeanette
Tilley, Shermaine Ann
Tomasi, Thomas B, Jr
Utermohlen, Virginia
Uzgiris, Egidijus E
Valentine, Fred Townsend
Van Oss, Carel J
Venter, J Craig
Vilcek, Jan Tomas
Vladutiu, Adrian O
Von Strandtmann, Maximillian
Waksman, Byron Halstead
Waltzer, Wayne C
Weksler, Marc Edward
Wicher, Victoria
Williams, Curtis Alvin, Jr
Winchester, Robert J
Wing, Edward Joseph
Winter, Alexander J
Witkin, Steven S
Woodruff, Judith J
Wright, Samuel D
Yang, Shung-Jun
Yang, Song-Yu
Young, John Ding-E
Zaleski, Marek Bohdan
Zauderer, Maurice
Zimmerman, Barry
Zolla-Pazner, Susan Beth

NORTH CAROLINA
Abramson, Jon Stuart
Adams, Dolph Oliver
Amos, Dennis Bernard
Barnes, Donald Wesley
Bast, Robert Clinton, Jr
Bolognesi, Dani Paul
Buttke, Thomas Martin
Carter, Philip Brian
Caruolo, Edward Vitangelo
Caterson, Bruce
Ciancolo, George J
Clinton, Bruce Allan
Cohen, Carl
Cohen, Harvey Jay
Cohen, Myron S
Collins, Jeffrey Jay
Corley, Ronald Bruce
Cresswell, Peter
Cronenberger, Jo Helen
Dawson, Jeffrey Robert
Durack, David Tulloch
Edens, Frank Wesley
Eisenberg, Robert
Erickson, Bruce Wayne
Falletta, John Matthew
Fine, Jo-David
Folds, James Donald
Fowler, Elizabeth
Frelinger, Jeffrey
Gardner, Edward, Jr
Golden, Carole Ann
Gusdon, John Paul
Hall, Russell P
Haskill, John Stephen
Haughton, Geoffrey
Heise, Eugene Royce
Hershfield, Michael Steven
Hoffman, Donald Richard
Hunt, William B, Jr
Hutt, Randy
Issitt, Peter David
Jacobs, Diane Margaret
Kariman, Khalil
Kataria, Yash P
Keene, Jack Donald
Kirkbride, L(ouis) D(ale)
Klapper, David G
Kostyu, Donna D
Kuhn, Raymond Eugene
Lee, David Charles
Luster, Michael I
McReynolds, Richard A
Mason, James Michael
Metzgar, Richard Stanley
Mitchell, Thomas Greenfield
Mizel, Steven B
Mueller, Nancy Schneider
Newell, Nanette
Newman, Simon Louis
Noga, Edward Joseph
O'Callaghan, James Patrick
Palczuk, Nicholas C
Paluch, Edward Peter
Pizzo, Salvatore Vincent
Quinn, Richard Paul
Richardson, Stephen H

Immunology (cont)

Rosse, Wendell Franklyn
Sage, Harvey J
Sassaman, Anne Phillips
Schwab, John Harris
Seigler, Hilliard Foster
Silverman, Myron Simeon
Singer, Kay Hiemstra
Smith, A Mason
Snyderman, Ralph
Strausbauch, Paul Henry
Stuhlmiller, Gary Michael
Thomas, Francis T
Thomas, Judith M
Thomas, William Albert
Volkman, Alvin
Ward, Frances Ellen
Winfield, John Buckner
Wolberg, Gerald
Yount, William J
Zimmerman, Thomas Paul

NORTH DAKOTA

Khansari, David Nemat
Rosenberg, Harry
Wikel, Stephen Kenneth

OHIO

Alexander, James Wesley
Bajpai, Praphulla K
Ball, Edward James
Ball, William James, Jr
Barriga, Omar Oscar
Barth, Rolf Frederick
Battisto, Jack Richard
Baxter, William D
Berger, Melvin
Bernstein, I Leonard
Biagini, Raymond E
Bigley, Nancy Jane
Blanton, Ronald Edward
Boyle, Michael Dermot
Brenner, Lorry Jack
Bullock, Ward Ervin, Jr
Cathcart, Martha K
Chakraborty, Jyotsna (Joana)
Chang, Jae Chan
Chorpenning, Frank Winslow
Daniel, Thomas Mallon
Decker, Lucile Ellen
Finke, James Harold
Frey, James R
Grossman, Charles Jerome
Gupta, Manjula K
Haas, Erwin
Hamilton, Thomas Alan
Harmony, Judith A K
Henningfield, Mary Frances
Herman, Jerome Herbert
Hess, Evelyn V
Hinman, Channing L
Horton, John Edward
Houchens, David Paul
Jackson, Robert Henry
Jamasbi, Roudabeh J
Janusz, Michael John
Jentoft, Joyce Eileen
Kapral, Frank Albert
Karp, Richard Dale
Kearns, Robert J
Keever, Carolyn Anne
Kier, Ann B
Kochan, Ivan
Koo, Peter H
Kreier, Julius Peter
Krueger, Robert George
Lamm, Michael Emanuel
Lane, Richard Durelle
Lang, Raymond W
Lepp, Cyrus Andrew
Mahmoud, Adel A F
Malemud, Charles J
Mao, Simon Jen-Tan
Megel, Herbert
Merritt, Katharine
Michael, Jacob Gabriel
Modrzakowski, Malcolm Charles
Morris, Randal Edward
Muckerheide, Annette
Murray, Finnie Ardrey, Jr
Nordlund, James John
Olsen, Richard George
Ooi, Boon Seng
Orosz, Charles George
Pesce, Amadeo J
Proffitt, Max Rowland
Rheins, Lawrence A
Rheins, Melvin S
Rhodes, Judith Carol
Robinson, Michael K
Rodman, Harvey Meyer
Rutishauser, Urs Stephen
Saif, Linda Jean
St Pierre, Ronald Leslie
Schanbacher, Floyd Leon
Shaffer, Charles Franklin
Shannon, Barry Thomas
Sheridan, John Francis
Sich, Jeffrey John
Smith, Kenneth Larry
Stavitsky, Abram Benjamin
Stevenson, J(oseph) Ross
Stevenson, John Ray

Stotts, Jane
Tarr, Melinda Jean
Tomei, L David
Treichel, Robin Stong
Tykocinski, Mark L
Wei, Robert
West, Clark Darwin
Whitacre, Caroline C
Widder, James Stone
Wolfe, Seth August, Jr
Wolos, Jeffrey Alan
Yen, Belinda R S
Zwilling, Bruce Stephen

OKLAHOMA

Bell, Paul Burton, Jr
Blackstock, Rebecca
Cain, William Aaron
Comp, Philip Cinnamon
Confer, Anthony Wayne
Cunningham, Madeleine White
Epstein, Robert B
Esmon, Charles Thomas
Graves, Donald C
Hall, Leo Terry
Hall, Nancy K
Harley, John Barker
Hyde, Richard Moorehead
Kincade, Paul W
Kollmorgen, G Mark
Leu, Richard William
McCallum, Roderick Eugene
Patnode, Robert Arthur
Sanborn, Mark Robert
Sherwood, John L
Teague, Perry Owen
Veltri, Robert William
Wolgamott, Gary

OREGON

Bagby, Grover Carlton
Bardana, Emil John, Jr
Barry, Ronald A
Bartos, Frantisek
Bayne, Christopher Jeffrey
Bennett, Robert M
Chilgren, John Douglas
Clark, David Thurmond
Fang, Ta-Yun
Koller, Loren D
Leslie, Gerrie Allen
Lochner, Janis Elizabeth
McClard, Ronald Wayne
Malley, Arthur
Norman, Douglas James
Pirofsky, Bernard
Rittenberg, Marvin Barry
Riviere, George Robert
Ruben, Laurens Norman
Schaeffer, Morris

PENNSYLVANIA

Andrews, Peter Walter
Atchison, Robert Wayne
Atkins, Paul C
Ayers, Arthur Raymond
Badger, Alison Mary
Bagasra, Omar
Baker, Alan Paul
Barker, Clyde Frederick
Becker, Anthony J, Jr
Berney, Steven
Bertolini, Donald R
Brooks, John J
Brooks, John S J
Brown, John M
Burdash, Nicholas Michael
Caligiuri, Lawrence Anthony
Calkins, Catherine E
Callahan, Hugh James
Cancro, Michael P
Caso, Louis Victor
Cebra, John Joseph
Ceglowski, Walter Stanley
Cohen, Pinya
Cohen, Stanley
Creech, Hugh John
Crowell, Richard Lane
D'Alisa, Rose M
Dalton, Barbara J
Das, Manjusri
Davis, Hugh Michael
Debouck, Christine Marie
DeHoratius, Raphael Joseph
Donnelly, John James, III
Douglas, Steven Daniel
Dryden, Richard Lee
Duquesnoy, Rene J
Eisenstein, Toby K
Farrell, Jay Phillip
Fenderson, Bruce Andrew
Fenton, Marilyn Ruth
Fiedel, Barry Allen
Fireman, Philip
Fleetwood, Mildred Kaiser
Flick, John A
Gasser, David Lloyd
Gerhard, Walter Ulrich
Gilbert, Scott F
Gill, Thomas James, III
Glorioso, Joseph Charles, III
Greenstein, Jeffrey Ian
Gregory, Francis Joseph
Harris, Tzvee N

Harrison, Aline Margaret
Herberman, Ronald Bruce
Herlyn, Dorothee Maria
Higgins, Terry Jay
Hsu, Susan Hu
Jegasothy, Brian V
Jensen, Bruce David
Jimenez, Sergio
Joseph, Jeymohan
Kamoun, Malek
Karakawa, Walter Wataru
Karol, Meryl Helene
Karush, Fred
Kennett, Roger H
Khatami, Mahin
Knobler, Robert Leonard
Knowles, Barbara B
Koffler, David
Korngold, Robert
Koros, Aurelia M Carissimo
Krah, David Lee
Kreider, John Wesley
Lacouture, Peter George
Lee, John C
Lee, John Cheung Han
Liberti, Paul A
Lindstrom, Jon Martin
Live, Israel
Long, Walter Kyle, Jr
Lotze, Michael T
Loughman, Barbara Ellen Evers
Lubeck, Michael D
McAlack, Robert Francis
McCarthy, Patrick Charles
McKenna, William Gillies
Maguire, Henry C, Jr
Manson, Lionel Arnold
Mashaly, Magdi Mohamed
Mastrangelo, Michael Joseph
Mastro, Andrea M
Maurer, Paul Herbert
Maxson, Linda Ellen R
Merryman, Carmen F
Miller, Elizabeth Eshelman
Milligan, Wilbert Harvey, III
Millman, Irving
Misra, Dhirendra N
Mitchell, Kenneth Frank
Mong, Seymour
Morahan, Page Smith
Murasko, Donna Marie
Nathanson, Neal
Natuk, Robert James
Neubauer, Russell Howard
Nowotny, Alois Henry Andre
Ogburn, Clifton Alfred
Ohlsson-Wilhelm, Betsy Mae
Opas, Evan E
Owen, Charles Scott
Pearson, David D
Pepe, Frank Albert
Phillips, S Michael
Proctor, Julian Warrilow
Raikow, Radmila Boruvka
Rest, Richard Franklin
Ritter, Carl A
Robb, Richard John
Rockey, John Henry
Roesing, Timothy George
Romano, Paula Josephine
Rosenberg, Jonathan S
Rozmiarek, Harry
Rubin, Benjamin Arnold
Rubin, Donald Howard
Salerno, Ronald Anthony
Salvin, Samuel Bernard
Sanders, Martin E
Sands, Howard
Schlegel, Robert Allen
Schmaier, Alvin Harold
Schreiber, Alan D
Schultz, Jane Schwartz
Schwartzman, Robert M
Siegel, Richard C
Simmons, Richard Lawrence
Six, Howard R
Smith, Jackson Bruce
Southam, Chester Milton
Steplewski, Zenon
Tachovsky, Thomas Gregory
Taichman, Norton Stanley
Tax, Anne
Tevethia, Satvir S
Thomas, Paul Milton
Thomas, William J
Tint, Howard
Vessey, Adele Ruth
Watt, Robert M
Weber, Wilfried T
Weidanz, William P
Wheelock, Earle Frederick
Whiteside, Theresa L
Winkelstein, Alan
Yurchenco, John Alfonso
Zarkower, Arian
Zink, Gilbert Leroy
Zmijewski, Chester Michael
Zurawski, Vincent Richard, Jr
Zurier, Robert B

RHODE ISLAND

Bailey, Garland Howard
Biron, Christine A
Blazar, Beverly A

Christenson, Lisa
DiLeone, Gilbert Robert
Dolyak, Frank
Knopf, Paul M
Kresina, Thomas Francis
Laux, David Charles
McMaster, Philip Robert Bache
Shank, Peter R
Wanebo, Harold
Weltman, Joel Kenneth
Zawadzki, Zbigniew Apolinary

SOUTH CAROLINA

Arnaud, Philippe
Boackle, Robert J
Bowers, William E
Chao, Julie
Cheng, Thomas Clement
Feller, Robert Jarman
Fudenberg, H Hugh
Galbraith, Robert Michael
Ghaffar, Abdul
Glick, Bruce
Goust, Jean Michel
Graber, Charles David
Hazen, Terry Clyde
Holper, Jacob Charles
Lazarchick, John
LeRoy, Edward Carwile
Lill, Patsy Henry
Mackaness, George Bellamy
Maheshwari, Kewal Krishnan
Mathur, Subbi
Sigel, M(ola) Michael
Virella, Gabriel T
Wang, An-Chuan
White, Edward
Willoughby, William Franklin
Wilson, Gregory Bruce

SOUTH DAKOTA

Hildreth, Michael B
Howard, Ronald M
Lynn, Raymond J

TENNESSEE

Altemeier, William Arthur, III
Briggs, Robert Chester
Brown, Alan R
Chandler, Robert Walter
Chen, James Pai-Fun
Chi, David Shyh-Wei
Chiang, Thomas M
Clark, Winston Craig
Colley, Daniel George
Cooper, Terrance G
Di Sabato, Giovanni
Draughon, Frances Ann
Duhl, David M
Fishman, Marvin
Fuson, Ernest Wayne
Gallagher, Michael Terrance
Gao, Kuixiong
Gillespie, Elizabeth
Griffin, Guy David
Gupta, Ramesh C
Harwood, Thomas Riegel
Herrod, Henry Grady
Ichiki, Albert Tatsuo
Inman, Franklin Pope, Jr
Isa, Abdallah Mohammad
Jennings, Lisa Helen Kyle
Johnson, Rother Rodenious
Jurand, Jerry George
Kennel, Stephen John
Kotb, Malak Y
Metzger, Dennis W
Meyers-Elliott, Roberta Hart
Ourth, Donald Dean
Perkins, Eugene Hafen
Portner, Allen
Rinchik, Eugene M
Roberts, Audrey Nadine
Roberts, Lee Knight
Rouse, Barry Tyrrell
South, Mary Ann
Stout, Robert Daniel
Sun, Deming
Townes, Alexander Sloan
Turner, Ella Victoria
Vredeveld, Nicholas Gene
Wachtel, Stephen Shoel
Walker, William Stanley
Webster, Robert G
Wust, Carl John
Yoo, Tai-June

TEXAS

Aggarwal, Bharat Bhushan
Allen, Robert Charles
Alvarez-Gonzalez, Rafael
Atassi, Zouhair
Baughn, Robert Elroy
Bjorndahl, Jay Mark
Boley, Robert B
Bowen, James Milton
Bursztajn, Sherry
Capra, J Donald
Christadoss, Premkumar
Cohen, Allen Barry
Cohen, Geraldine H
Collisson, Ellen Whited
Daniels, Jerry Claude
Davis, Claude Geoffrey

Denney, Richard Max
Douglas, Tommy Charles
Dreesman, Gordon Ronald
Dusenbery, Ruth Lillian
Eichberg, Jorg Wilhelm
Evans, John Edward
Everse, Johannes
Fahlberg, Willson Joel
Fanguy, Roy Charles
Faust, Charles Harry, Jr
Fernandes, Gabriel
Fidler, Isaiah J
Fisher, William Francis
Fleischmann, William Robert, Jr
Forman, James
Foster, Terry Lynn
Franzl, Robert E
Gardner, Florence Harrod
Geoghegan, William David
Gillard, Baiba Kurins
Goldman, Armond Samuel
Goldschmidt, Millicent
Gottlieb, Paul David
Grant, John Andrew, Jr
Grimm, Elizabeth Ann
Gutterman, Jordan U
Guy, Leona Ruth
Heggers, John Paul
Hejtmancik, Kelly Erwin
Hollinger, F(rederick) Blaine
Jasin, Hugo E
Jerrells, Thomas Ray
Johnson, Alice Ruffin
Jordon, Robert Earl
Joys, Terence Michael
Kahan, Barry D
Kemp, Walter Michael
Kettman, John Rutherford, Jr
Kiel, Johnathan Lloyd
Killion, Jerald Jay
Klimpel, Gary R
Kniker, William Theodore
Koppenheffer, Thomas Lynn
Kraig, Ellen
Kripke, Margaret Louise (Cook)
Krueger, Gerhard R F
Kumar, Vinay
Kurosky, Alexander
Kutteh, William Hanna
Lanford, Robert Eldon
Laughter, Arline H
Lieberman, Michael Merril
Loan, Raymond Wallace
Lopez-Berestein, Gabriel
McBride, Raymond Andrew
McConnell, Stewart
McKenna, John Morgan
McMurray, David N
Maleckar, James R
Mandy, William John
Marcus, Donald M
Marshall, Gailen D, Jr
Mavligit, Giora M
Measel, John William
Morrow, Kenneth John, Jr
Nakajima, Motowo
Nash, Donald Robert
Niederkorn, Jerry Young
Norris, Steven James
Oishi, Noboru
Ordonez, Nelson Gonzalo
Paque, Ronald E
Patsch, Wolfgang
Paull, Barry Richard
Pellis, Neal Robert
Periman, Phillip
Pinckard, Robert Neal
Platsoucas, Chris Dimitrios
Poduslo, Shirley Ellen
Poenie, Martin Francis
Powell, Bernard Lawrence
Prager, Morton David
Rael, Eppie David
Rainwater, David Luther
Rajaraman, Srinivasan
Ranney, David Francis
Rich, Robert Regier
Rich, Susan Marie Solliday
Ritzi, Earl Michael
Robertson, Stella M
Rodkey, Leo Scott
Rossen, Roger Downey
Roth, Jack A
Rudolph, Frederick Byron
Samson, Willis Kendrick
Sanders, Bobby Gene
Sanford, Barbara Ann
Schlamowitz, Max
Schmalstieg, Frank Crawford
Schmidt, Jerome P
Schneider, Sandra Lee
Sell, Stewart
Session, John Joe
Shadduck, John Allen
Shearer, William Thomas
Shillitoe, Edward John
Smiley, James Donald
Smith, Eric Morgan
Smith, Eric Morgan
Sparkman, Dennis Raymond
Spellman, Craig William
Stone, William Harold
Streckfuss, Joseph Larry
Taylor, D(orothy) Jane

Templeton, Joe Wayne
Thomas, James Ward
Tizard, Ian Rodney
Tom, Baldwin Heng
Tong, Alex W
Trentin, John Joseph
Twomey, Jeremiah John
Valdivieso, Dario
Veit, Bruce Clinton
Villacorte, Guillermo Vilar
Wagner, Gerald Gale
Walker, David Hughes
Winters, Wendell Delos
Wolinsky, Jerry Saul
Yang, Ovid Y H
Yeoman, Lynn Chalmers

UTAH
Araneo, Barbara Ann
Barnett, Bill B
Click, Robert Edward
Daynes, Raymond Austin
DeWitt, Charles Wayne, Jr
Donaldson, David Miller
Ely, Kathryn R
Fujinami, Robert S
Hammond, Mary Elizabeth Hale
Hayes, Sheldon P
Healey, Mark Calvin
Lancaster, Jack R, Jr
Marcus, Stanley
Mohammad, Syed Fazal
Muna, Nadeem Mitri
Nelson, Jerry Rees
North, James A
Pincus, Seth Henry
Rasmussen, Kathleen Goertz
Sidwell, Robert William
Singh, Vijendra Kumar
Sipe, David Michael
Spendlove, Rex S
Straight, Richard Coleman
Torres, Anthony R
Warren, Reed Parley
Wiley, Bill Beauford

VERMONT
Albertini, Richard Joseph
Ariano, Marjorie Ann
Boraker, David Kenneth
Cooper, Sheldon Mark
Falk, Leslie Alan
Kelley, Jason

VIRGINIA
Affronti, Lewis Francis
Barta, Ota
Bradley, Sterling Gaylen
Brown, Barry Lee
Brown, Russell Vedder
Burger, Carol J
DeVries, George H
Elgert, Klaus Dieter
Engelhard, Victor H
Escobar, Mario R
Esen, Asim
Eyre, Peter
Fu, Shu Man
Hager, Chester Bradley
Hard, Richard C, Jr
Herr, John Christian
Hsu, Hsiu-Sheng
Kelley, John Michael
Leech, Stephen H
Lobel, Steven A
Lubiniecki, Anthony Stanley
McWright, Cornelius Glen
Mullinax, Perry Franklin
Munson, Albert Enoch
Myrvik, Quentin N
Nagarkatti, Mitzi
Nagarkatti, Prakash S
Normansell, David E
Place, Janet Dobbins
Sando, Julianne J
Smibert, Robert Merrall, II
Smith, Wade Kilgore
Szakal, Andras Kalman
Taylor, Ronald Paul
Tew, John Garn
Tucker, Anne Nichols
Wilkins, Tracy Dale
Wright, George Green
Wright, George Leonard, Jr

WASHINGTON
Altman, Leonard Charles
Baker, Patricia J
Banks, Keith L
Barnes, Glover William
Bean, Michael Arthur
Clapshaw, Patric Arnold
Cosman, David John
Couser, William Griffith
Dailey, Frank Allen
Davis, William C
Eidinger, David
Ely, John Thomas Anderson
Farr, Andrew Grant
Frankart, William A
Graves, Scott Stoll
Hakomori, Sen-Itiroh
Harlan, John Marshall
Hellstrom, Ingegerd Elisabet

Hellstrom, Karl Erik Lennart
Henney, Christopher Scot
Horn, Diane
Houck, John Candee
Hsu, Chin Shung
Hu, Shiu-Lok
Kenny, George Edward
Lang, Bruce Z
Leid, R Wes
Lynch, David H
McGuire, Travis Clinton
McIvor, Keith L
Magnuson, James Andrew
Magnuson, Nancy Susanne
Maguire, Yu Ping
Maliszewski, Charles R
Mannik, Mart
Mason, Herman Charles
Morrissey, Philip John
Nardella, Francis Anthony
Nelson, Karen Ann
Pauley, Gilbert Buckhannan
Perlmutter, Roger M
Pious, Donald A
Pollack, Sylvia Byrne
Raff, Howard V
Rosse, Cornelius
Shultz, Terry D
Sibley, Carol Hopkins
Urdal, David L
Wedgwood, Ralph Josiah Patrick
Weir, Michael Ross

WEST VIRGINIA
Anderson, Douglas Poole
Burrell, Robert
Landreth, Kenneth S
Lewis, Daniel Moore
Mengoli, Henry Francis
Olenchock, Stephen Anthony
Reichenbecher, Vernon Edgar, Jr

WISCONSIN
Abramoff, Peter
Allen, Elizabeth Morei
Auerbach, Robert
Blattner, Frederick Russell
Blazkovec, Andrew A
Borden, Ernest Carleton
Brown, Raymond Russell
Burgess, Richard Ray
Calvanico, Nickolas Joseph
Chusid, Michael Joseph
Cook, Mark Eric
Czuprynski, Charles Joseph
Datta, Surinder P
Dimond, Randall Lloyd
Fink, Jordan Norman
Fredricks, Walter William
Goldberger, Amy
Grossberg, Sidney Edward
Heath, Timothy Douglas
Hinsdill, Ronald D
Hinshaw, Virginia Snyder
Hinze, Harry Clifford
Hirsch, Samuel Roger
Hong, Richard
Jensen, Richard Harvey
Kahan, Lawrence
Kim, Zaezeung
Klebba, Phillip E
Koethe, Susan M
Long, Claudine Fern
Manning, Dean David
Mansfield, John Michael
Marx, James John, Jr
Paulnock, Donna Marie
Rapacz, Jan
Schultz, Ronald David
Sidky, Younan Abdel Malik
Smith, Donald Ward
Sondel, Paul Mark
Tomar, Russell H
Truitt, Robert Lindell
Tufte, Marilyn Jean
Twining, Sally Shinew
Yoshino, Timothy Phillip

WYOMING
Belden, Everett Lee
Isaak, Dale Darwin
Pier, Allan Clark

PUERTO RICO
Bolaños, Benjamin
Garcia-Castro, Ivette
Habeeb, Ahmed Fathi Sayed Ahmed
Hillyer, George Vanzandt
Kozek, Wieslaw Joseph

ALBERTA
Baron, Robert Walter
Befus, A(lbert) Dean
Bleackley, Robert Christopher
Dossetor, John Beamish
Eggert, Frank Michael
Hart, David Arthur
Lee, Kwok-Choy
Longenecker, Bryan Michael
McPherson, Thomas Alexander
Madiyalakan, Ragupathy
Paetkau, Verner Henry
Paul, Leendert Cornelis
Rice, Wendell Alfred

Singh, Bhagirath
Stemke, Gerald W
Suresh, Mavanur Rangarajan
Wegmann, Thomas George

BRITISH COLUMBIA
Aebersold, Ruedi H
Agnew, Robert Morson
Bower, Susan Mae
Chase, William Henry
De Boer, Solke Harmen
Ekramoddoullah, Abul Kalam M
Ferguson, Alexander Cunningham
Hancock, Robert Ernest William
Hoffmann, Geoffrey William
Kelly, Michael Thomas
McBride, Barry Clarke
Matheson, David Stewart
Osoba, David
Salinas, Fernando A
Silver, Hulbert Keyes Belford
Teh, Hung-Sia

MANITOBA
Berczi, Istvan
Chow, Donna Arlene
Froese, Arnold
Garther, John G
Greenberg, Arnold Harvey
Kepron, Wayne
Lewis, Marion Jean
Paraskevas, Frixos
Sabbadini, Edris Rinaldo
Sehon, Alec

NEW BRUNSWICK
Bagnall, Richard Herbert

NEWFOUNDLAND
Hawkins, David Geoffrey
Michalak, Thomas Ireneusz

NOVA SCOTIA
Carr, Ronald Irving
Dolphin, Peter James
Jones, John Verrier
MacSween, Joseph Michael

ONTARIO
Baker, Michael Allen
Barber, Brian Harold
Barr, Ronald Duncan
Beaulieu, J A
Bishop, Claude Titus
Blais, Burton W
Broder, Irvin
Brown, William Roy
Campbell, James B
Chadwick, June Stephens
Chaly, Nathalie
Chechik, Boris
Cinader, Bernhard
Clark, David A
Cooke, T Derek V
Crookston, Marie Cutbush
Dales, Samuel
Delovitch, Terry L
Dent, Peter Boris
De Veber, Leverett L
Dosch, Hans-Michael
Dubiski, Stanislaw
Duncan, John Robert
Duncan, William Raymond
Elce, John Shackleton
Fish, Eleanor N
Forsdyke, Donald Roy
Freedman, Melvin Harris
Galsworthy, Sara B
Gauldie, Jack
Gorczynski, Reginald Meiczyslaw
Grinstein, Sergio
Gupta, Krishana Chandara
Hozumi, Nobumichi
Hozumi, Nobumichi
Inman, Robert Davies
Kaushik, Azad Kumar
Kennedy, James Cecil
Kerbel, Robert Stephen
Krepinsky, Jiri J
Kuroski-De Bold, Mercedes Lina
Lang, Gerhard Herbert
Lau, Catherine Y
Letarte, Michelle Vinlaine
Lingwood, Clifford Alan
Logan, James Edward
McDermott, Mark Rundle
Mahdy, Mohamed Sabet
Mak, William Wai-Nam
Medzon, Edward Lionel
Mellors, Alan
Metzgar, Don P
Miller, Richard Graham
Minta, Joe Oduro
Nielsen, Klaus H B
Nisbet-Brown, Eric Robert
Phillips, Robert Allan
Pickering, Richard Joseph
Pope, Barbara L
Pross, Hugh Frederick
Pruzanski, Waldemar
Ranadive, Narendranath Santuram
Rice, Christine E
Richter, Maxwell
Rosenthal, Kenneth Lee

Immunology (cont)

Ruckerbauer, Gerda Margareta
Samagh, Bakhshish Singh
Sauder, Daniel Nathan
Sengar, Dharmendra Pal Singh
Shewen, Patricia Ellen
Sinclair, Nicholas Roderick
Singal, Dharam Parkash
Singhal, Sharwan Kumar
Stanisz, Andrzej Maciej
Stemshorn, Barry William
Stewart, Thomas Henry McKenzie
Stiller, Calvin R
Strejan, Gill Henric
Szewczuk, Myron Ross
Tryphonas, Helen
Tsang, Charles Pak Wai
Underdown, Brian James
Vadas, Peter
Vijay, Hari Mohan
Wilkie, Bruce Nicholson
Wilson, Robert James

QUEBEC
Adamkiewicz, Vincent Witold
Antel, Jack Perry
Borgeat, Pierre
Cantor, Ena D
Delespesse, Guy Joseph
Ford-Hutchinson, Anthony W
Fournier, Michel
Freedman, Samuel Orkin
Gervais, Francine
Gold, Phil
Gordon, Julius
Guttmann, Ronald David
Lamoureux, Gilles
Lapp, Wayne Stanley
Martineau, Ronald
Menezes, Jose Piedade Caetano Agnelo
Miller, Sandra Carol
Moroz, Leonard Arthur
Murgita, Robert Anthony
Osmond, Dennis Gordon
Page, Michel
Palfree, Roger Grenville Eric
Poole, Anthony Robin
Potworowski, Edouard François
Rola-Pleszczynski, Marek
Schulz, Jan Ivan
Shuster, Joseph
Siboo, Russell
Sikorska, Hanna
Sirois, Pierre
Skamene, Emil
Stevenson, Mary M
Tanner, Charles E
Thomson, David M P
Trudel, Michel D
Waithe, William Irwin
Walker, Mary Clare

SASKATCHEWAN
Babiuk, Lorne A
Blakley, Barry Raymond
McLennan, Barry Dean

OTHER COUNTRIES
Aalund, Ole
Abelev, Garri Izrailevich
Acres, Robert Bruce
Agarwal, Shyam S
Arala-Chaves, Mario Passalaqua
Baldwin, Robert William
Barald, Kate Francesca
Barel, Monique
Binder, Bernd R
Bodmer, Walter Fred
Bojalil, Luis Felipe
Bolhuis, Reinder L H
Borek, Felix
Borel, Yves
Braquet, Pierre G
Brondz, Boris Davidovich
Campbell, Virginia Wiley
Dausset, Jean Baptiste Gabriel
Dianzani, Ferdinando
Dorrington, Keith John
Dyrberg, Thomas Peter
Eggers, Hans J
English, Leonard Stanley
Farid, Nadir R
Fernández-Cruz, Eduardo P
Franklin, Richard Morris
Fujimoto, Shigeyoshi
Fujiwara, Keigi
Gelzer, Justus
Gemsa, Diethard
Ghiara, Paolo
Goding, James Watson
Gorski, Andrzej
Gowans, James L
Ha, Tai-You
Hansson, Goran K
Häyry, Pekka J
Heiniger, Hans-Jorg
Hirano, Toshio
Hirokawa, Katsuiki
Hsu, Clement C S
Ide, Hiroyuki
Izui, Shozo
Jerne, Niels Kaj
Kerjaschki, Dontscho

Kessel, Rosslyn William Ian
Kimoto, Masao
Kobata, Akira
Kuhn, Klaus
Lamoyi, Edmundo
Lee, Men Hui
Levy, David Alfred
Lin, Chin-Tarng
Loppnow, Harald
Low, Chow-Eng
Low, Teresa Lingchun Kao
Lumb, Judith Rae H
MacDonald, Thomas Thornton
Marikovsky, Yehuda
Markert, Claus O
Maruta, Hiroshi
Maruyama, Koshi
Merlini, Giampaolo
Michaelsen, Terje E
Minowada, Jun
Muller-Eberhard, Hans Joachim
Muñoz, Maria De Lourdes
Pan, In-Chang
Prowse, Stephen James
Quastel, Michael Reuben
Querinjean, Pierre Joseph
Ray, Prasanta K
Revillard, Jean-Pierre Rémy
Romagnani, Sergio
Santoro, Ferrucio Fontes
Saravia, Nancy G
Schena, Francesco Paolo
Seya, Tsukasa
Solter, Davor
Stingl, Georg
Tachibana, Takehiko
Takai, Yasuyuki
Taniyama, Tadayoshi
Van Furth, Ralph
Van Loveren, Henk
Varga, Janos M
Wang, Soo Ray
Watanabe, Takeshi
Webb, Cynthia Ann Glinert
Weil, Marvin Lee
Weiss, David Walter
Willard-Gallo, Karen Elizabeth
Wu, Albert M
Yakura, Hidetaka
Yamamoto, Hiroshi
Yeon-Sook, Yun
Yoshida, Takeshi

Microbiology

ALABAMA
Acton, Ronald Terry
Alford, Charles Aaron, Jr
Ammerman, Gale Richard
Backman, Paul Anthony
Ball, Laurence Andrew
Bhatnagar, Yogendra Mohan
Blalock, James Edwin
Brady, Yolanda J
Briles, David Elwood
Brown, Alfred Ellis
Carrozza, John Henry
Carter, Charlotte
Cassell, Gail Houston
Chapatwala, Kirit D
Cody, Reynolds M
Coggin, Joseph Hiram
Compans, Richard W
Davis, Norman Duane
Dimopoullos, George Takis
Edgar, Samuel Allen
Egan, Marianne Louise
Estes, Edna E
Francis, Robert Dorl
Giambrone, Joseph James
Gillespie, George Yancey
Gottlieb, Sheldon F
Green, Margaret
Higgins, N Patrick
Hill, William Francis, Jr
Hunter, Eric
Hunter, Katherine Morton
Jones, Jeanette
Kearney, John F
Klesius, Phillip Harry
Lauerman, Lloyd Herman, Jr
Lavelle, George Cartwright
Lemke, Paul Arenz
Lidin, Bodil Inger Maria
McCaskey, Thomas Andrew
McGhee, Jerry Roger
Michalek, Suzanne M
Pass, Robert Floyd
Patterson, Loyd Thomas
Plumb, John Alfred
Ranney, Richard Raymond
Rivers, Douglas Bernard
Russell, Michael W(illiam)
Sakai, Ted Tetsuo
Schulze, Karl Ludwig
Schutzbach, John Stephen
Shannon, William Michael
Siddique, Irtaza H
Singh, Shiva Pujan
Smith, Paul Clay
Smith-Somerville, Harriett Elizabeth
Stagno, Sergio Bruno
Suberkropp, Keller Francis

Suling, William John
Thomas, Julian Edward, Sr
Turco, Jenifer
Wertz, Gail T Williams
Wilkoff, Lee Joseph
Winkler, Herbert H
Winters, Alvin L
Wood, David Oliver
Worley, S D

ALASKA
Brown, Edward James
Button, Don K
Lee, Jong Sun
Wennerstrom, David E

ARIZONA
Alcorn, Stanley Marcus
Appelgren, Walter Phon
Archer, Stanley J
Bachman, Marvin Charles
Baker, Gladys Elizabeth
Chalquest, Richard Ross
Gerba, Charles Peter
Hewlett, Martinez Joseph
Hill, Warner Michael
Janssen, Robert (James) J
Jeter, Wayburn Stewart
Johnson, Roy Melvin
Koffler, Henry
Mandle, Robert Joseph
Mare, Cornelius John
Meinke, William John
Mendelson, Neil Harland
Mosher, Richard Arthur
Moulder, James William
Nelson, Frank Eugene
Pommerville, Jeffrey Carl
Schmidt, Jean M
Shull, James Jay
Sinclair, Norval A
Sinski, James Thomas
Spizizen, John
Vincent, Walter Sampson
Wallraff, Evelyn Bartels
Yall, Irving

ARKANSAS
Abernathy, Robert Shields
Barnett, John Brian
Barron, Almen Leo
Bower, Raymond Kenneth
Bowling, Robert Edward
Champlin, William G
Elbein, Alan D
Ferguson, Dale Vernon
Hardin, Hilliard Frances
Heddleston, Kenneth Luther
Heflich, Robert Henry
Hinck, Lawrence Wilson
Howell, Robert T
Paulissen, Leo John
Pynes, Gene Dale
Skeeles, John Kirkpatrick
Soderberg, Lee Stephen Freeman
Steinkraus, Donald Curtiss
Sutherland, John Bruce, IV
Weeks, Robert Joe
Young, Seth Yarbrough, III

CALIFORNIA
Adams, Jane N
Adaskaveg, James Elliott
Allen, Charles Freeman
Allison, Anthony Clifford
Aly, Raza
Anthony, Ronald Lewis
Ashton, David Hugh
Ayengar, Padmasini (Mrs Frederick Aladjem)
Balch, William E
Bartnicki-Garcia, Salomon
Baxter, William Leroy
Beaman, Blaine Lee
Belisle, Barbara Wolfanger
Benenson, Abram Salmon
Berk, Arnold J
Berry, Edward Alan
Bertani, Giuseppe
Bertani, Lillian Elizabeth
Bibel, Debra Jan
Bishop, John Michael
Bishop, Nancy Horschel
Bissett, Marjorie Louise
Bittle, James Long
Bochner, Barry Ronald
Borchardt, Kenneth
Botzler, Richard George
Bovell, Carlton Rowland
Bowman, Barry J
Boyer, Herbert Wayne
Bradshaw, Lawrence Jack
Braemer, Allen C
Brosbe, Edwin Allan
Brownell, Anna Gale
Bruckner, David Alan
Brunell, Philip Alfred
Brunke, Karen J
Buchanan, Bob Branch
Buehring, Gertrude Case
Bulich, Anthony Andrew
Byatt, Pamela Hilda
Cabasso, Victor Jack
Calavan, Edmond Clair

Campbell, Judith Lynn
Canawati, Hanna N
Capers, Evelyn Lorraine
Carbon, John Anthony
Cardiff, Robert Darrell
Caren, Linda Davis
Carlberg, David Marvin
Carlson, James Reynold
Catlin, B Wesley
Chamberlin, Michael John
Chan, Sham-Yuen
Chang, George Washington
Chang, Jennie C C
Chang, Robert Shihman
Charlang, Gisela Wohlrab
Chatigny, Mark A
Chen, Anthony Bartheolomew
Chen, Benjamin P P
Childress, Evelyn Tutt
Chou, Tsong-Wen
Clark, Alvin John
Clough, Wendy Glasgow
Cohen, Morris
Cohen, Robert Elliot
Cohen, Stanley Norman
Collin, William Kent
Collins, Edwin Bruce
Connor, James D
Cooper, Robert Chauncey
Cota-Robles, Eugene H
Craig, James Morrison
Cremer, Natalie E
Crocker, Thomas Timothy
Dahms, Arthur Stephen
DaMassa, Al John
Daniels, Jeffrey Irwin
Dasgupta, Asim
Davé, Bhalchandra A
Davenport, Calvin Armstrong
Davis, Charles Edward
Davis, Rowland Hallowell
Del Villano, Bert Charles
Di Girolamo, Rudolph Gerard
Doi, Roy Hiroshi
Dragon, Elizabeth Alice Oosterom
Duffey, Paul Stephen
Dulbecco, Renato
Duniway, John Mason
Duzgunes, Nejat
Eckhart, Walter
Eiserling, Frederick A
Eldredge, Kelly Husbands
Emmons, Richard William
Englesberg, Ellis
Eppstein, Deborah Anne
Erickson, Jeanne Marie
Evans, Ernest Edward, Jr
Fabrikant, Irene Berger
Falkow, Stanley
Farmer, Walter Joseph
Finegold, Sydney Martin
Flashman, Stuart Milton
Forghani-Abkenari, Bagher
Fox, Eugene N
Froman, Seymour
Fuhrman, Jed Alan
Fukuyama, Thomas T
Fulco, Armand J
Fung, Henry C, Jr
Funk, Glenn Albert
Gaertner, Frank Herbert
Ganesan, Ann K
Geiduschek, Ernest Peter
Gilbreath, Michael Joye
Giorgi, Janis V(incent)
Gitnick, Gary L
Glembotski, Christopher Charles
Goehler, Brigitte Hanna
Gold, William
Goodman, Nelson
Goodman, Richard E
Gordon, Irving
Granger, Gale A
Green, Melvin Howard
Green, Richard H
Greene, Elias Louis
Grilione, Patricia Louise
Gumpf, David John
Gunsalus, Robert Philip
Gupta, Rishab Kumar
Gutman, George Andre
Hackett, Adeline J
Hadley, William Keith
Hagen, Charles Alfred
Haglund, John Richard
Haight, Roger Dean
Halde, Carlyn Jean
Hammar, Allan H
Hankinson, Oliver
Hanna, Lavelle
Hanson, Carl Veith
Hatfield, G Wesley
Hayflick, Leonard
Haygood, Margo Genevieve
Heckly, Robert Joseph
Hedrick, Ronald Paul
Heinrich, Milton Rollin
Hemmingsen, Barbara Bruff
Henderson, Gary Borgar
Herskowitz, Ira
Heuer, Ann Elizabeth
Hill, Ray Allen
Hirsh, Dwight Charles, III
Hirst, George Keble

Ho, Yuk Lin
Hoch, Sallie O'Neil
Hoeprich, Paul Daniel
Hofmann, Bo
Hoke, Carolyn Kay
Holland, John Joseph
Holmberg, Charles Arthur
Horvath, Kalman
Howard, Russell John
Huff, Dennis Karl
Hunderfund, Richard C
Hungate, Robert Edward
Hunter, Tony
Huskey, Glen E
Hyman, Bradley Clark
Hyman, Richard W
Imagawa, David Tadashi
Ingraham, John Lyman
Jackson, Andrew Otis
Jackson, James Oliver
Janda, John Michael
Jariwalla, Raxit Jayantilal
Jawetz, Ernest
Jaye, Murray Joseph
Johansson, Karl Richard
Jones, Theodore Harold Douglas
Kallman, Burton Jay
Kang, Chang-Yuil
Kang, Kenneth S
Kang, Tae Wha
Kaplan, Milton Temkin
Kavenoff, Ruth
Kennedy, Elhart James
Kettering, James David
Kibler, Ruthann
Kim, Juhee
King, A Douglas, Jr
King, Alfred Douglas, Jr
Klein, Harold Paul
Klement, Vaclav
Knudson, Gregory Blair
Koshland, Daniel Edward, Jr
Kozloff, Lloyd M
Kuhn, Daisy Angelika
Kunkee, Ralph Edward
Kutner, Leon Jay
Lai, Michael Ming-Chiao
Lane, Robert Sidney
Lanyi, Janos K
Lascelles, June
Lechtman, Max D
Lederberg, Esther Miriam
Lester, William Lewis
Levinson, Warren E
Levintow, Leon
Lewis, William Perry
Lindberg, Lois Helen
Liu, Chi-Li
Louie, Robert Eugene
Lynch, Richard Vance, III
McCarthy, Robert Elmer
McEwen, Joan Elizabeth
Mah, Robert A
Mankau, Reinhold
Manning, JaRue Stanley
Marr, Allen Gerald
Marrone, Pamela Gail
Martin, George Steven
Masover, Gerald K
Mathies, Margaret Jean
Matthews, Thomas Robert
Merigan, Thomas Charles, Jr
Merriam, Esther Virginia
Metcalf, Robert Harker
Michener, H(arold) David
Midura, Thaddeus
Mielenz, Jonathan Richard
Miller, Martin Wesley
Miller, Sol
Miranda, Quirinus Ronnie
Mitchell, John Alexander
Moore, Andrew Brooks
Morenzoni, Richard Anthony
Müller-Sieburg, Christa E
Mussen, Eric Carnes
Nahhas, Fuad Michael
Nassos, Patricia Saima
Nayak, Debi Prosad
Neilands, John Brian
Nematollahi, Jay
Newton, William Edward
Nichol, Francis Richard, Jr
Nichols, Barbara Ann
Nierlich, Donald P
Nikaido, Hiroshi
Nomura, Masayasu
Obijeski, John Francis
Oh, Jang Ok
Olson, Betty H
O'Neill, Thomas Brendan
Opfell, John Burton
Osebold, John William
Oshiro, Lyndon Satoru
Overby, Lacy Rasco
Padgett, Billie Lou
Pappagianis, Demosthenes
Penhoet, Edward Etienne
Perrault, Jacques
Peterson, George Harold
Phaff, Herman Jan
Phelps, Leroy Nash
Pitts, Robert Gary
Porter, David Dixon
Pratt, David

Pritchett, Randall Floyd
Prusiner, Stanley Ben
Puhvel, Sirje Madli
Purcell, Alexander Holmes, III
Quilligan, James Joseph, Jr
Rachmeler, Martin
Raj, Harkisan D
Redfield, William David
Reeve, Peter
Remington, Jack Samuel
Riggs, John L
Rivier, Catherine L
Roantree, Robert Joseph
Rogoff, Martin Harold
Rosenberg, Steven Loren
Ross, Ian Kenneth
Rowland, Ivan W
Ruby, Edward George
Rudnick, Albert
Russell, Ruth Lois
Sacks, Lawrence Edgar
Saier, Milton H, Jr
Samuel, Charles Edward
Saperstein, Sidney
Sapico, Francisco L
Schachter, Julius
Schaffer, Frederick Leland
Schein, Arnold Harold
Schekman, Randy W
Scherrer, Rene
Schiller, Neal Leander
Schmid, Peter
Schmidt, Nathalie Joan
Schonfeld, Steven Emanuel
Schwerdt, Carlton Everett
Scow, Kate Marie
Semancik, Joseph Stephen
Senyk, George
Seto, Joseph Tobey
Shapiro, Lucille
Sheneman, Jack Marshall
Shifrine, Moshe
Shum, Archie Chue
Silliker, John Harold
Simpson, Robert Blake
Smith, Cassandra Lynn
Smith, Douglas Wemp
Smith, Michael R
Soli, Giorgio
Soller, Arthur
Spahn, Gerard Joseph
Spanis, Curt William
Spotts, Charles Russell
Springer, Wayne Richard
Stallcup, Michael R
Stanbridge, Eric John
Starr, Mortimer Paul
Steenbergen, James Franklin
Stephens, William Leonard
Stevens, Jack Gerald
Stevens, Ronald Henry
Stevenson, Kenneth Eugene
Steward, John P
Stohlman, Stephen Arnold
Stoller, Benjamin Boris
Strauss, Ellen Glowacki
Strauss, James Henry
Stringfellow, Dale Alan
Stroynowski, Iwona T
Suffin, Stephen Chester
Sutton, Donald Dunsmore
Swatek, Frank Edward
Sweet, Benjamin Hersh
Sy, Jose
Sypherd, Paul Starr
Tachibana, Dora K
Tanada, Yoshinori
Taylor, Barry L
Taylor, John Marston
Tenenbaum, Saul
Theis, Jerold Howard
Thorner, Jeremy William
Thrupp, Lauri David
Tidwell, William Lee
Tomlinson, Geraldine Ann
Treagan, Lucy
Umanzio, Carl Beeman
Utterback, Nyle Gene
Uyeda, Charles Tsuneo
Vail, Patrick Virgil
VanBruggen, Ariena H C
Varmus, Harold Elliot
Vaughn, Reese Haskell
Vedros, Neylan Anthony
Vice, John Leonard
Vickrey, Herta Miller
Vilker, Vincent Lee
Villarreal, Luis Perez
Vogt, Peter Klaus
Volcani, Benjamin Elazari
Vold, Barbara Schneider
Volz, Michael George
Vredevoe, Donna Lou
Vyas, Girish Narmadashankar
Wagner, Edward Knapp
Waitz, Jay Allan
Wall, Thomas Randolph
Wallace, Robert Bruce
Wayne, Lawrence Gershon
Webb, David Ritchie
Wechsler, Steven Lewis
Wedberg, Stanley Edward
Wehrle, Paul F
Weinberg, Barbara Lee Huberman

Whang, Sukoo Jack
Wheelis, Mark Lewis
White, Maurice Leopold
White, Thomas James
Wiebe, Michael Eugene
Wilcox, Gary Lynn
Wilhelm, Alan Roy
Williams, Terry Wayne
Wistreich, George A
Wolf, Beverly
Wolochow, Hyman
Wood, Willis Avery
Woodhour, Allen F
Woolfolk, Clifford Allen
Wu, Jia-Hsi
Wu, William Gay
Wyatt, Philip Joseph
Yamamoto, Richard
Yayanos, A Aristides
York, Charles James
York, George Kenneth, II
Yu, David Tak Yan
Zamenhof, Stephen
Zee, Yuan Chung
Zlotnik, Albert
Zuccarelli, Anthony Joseph

COLORADO
Andrews, Ken J
Avens, John Stewart
Barber, Thomas Lynwood
Berens, Randolph Lee
Blair, Carol Dean
Bowles, Jean Alyce
Brennan, Patrick Joseph
Brooks, Bradford O
Bunn, Paul A, Jr
Calisher, Charles Henry
Cambier, John C
Chow, Tsu Ling
Church, Brooks Davis
Crowle, Alfred John
Danna, Kathleen Janet
De Martini, James Charles
Dobersen, Michael J
Francy, David Bruce
Frerman, Frank Edward
Goren, Mayer Bear
Grant, Dale Walter
Grieve, Robert B
Gubler, Duane J
Hager, Jean Carol
Harold, Franklin Marcel
Harold, Ruth L
Ivory, Thomas Martin, III
Janes, Donald Wallace
Johnson, Donal Dabell
Jordan, Russell Thomas
Keller, Frederick Albert, Jr
Klein, Donald Albert
Kloppel, Thomas Mathew
Kubo, Ralph Teruo
Linden, James Carl
McClatchy, Joseph Kenneth
Mika, Leonard Aloysius
Norstadt, Fred A
Ogg, James Elvis
Pizer, Lewis Ivan
Poyton, Robert Oliver
Prentice, Neville
Ranu, Rajinder S
Rich, Marvin A
Roberts, Walden Kay
Shulls, Wells Alexander
Smith, Ralph E
Talmage, David Wilson
Taylor, Austin Laurence
Tengerdy, Robert Paul
Trent, Dennis W
Ulrich, John August
Updegraff, David Maule
Urban, Richard William
Vasil, Michael Lawrence
Wyss, Orville

CONNECTICUT
Aaslestad, Halvor Gunerius
Aitken, Thomas Henry Gardiner
Armstrong, Martine Yvonne Katherine
Benz, Edward John
Berg, Claire M
Black, Francis Lee
Blogoslawski, Walter
Bluhm, Leslie
Booss, John
Borgia, Julian Frank
Braun, Phyllis C
Buck, John David
Burke, Carroll N
Cameron, J A
Celesk, Roger A
Coleman, William H
Coykendall, Alan Littlefield
Cummings, Dennis Paul
Curnen, Edward Charles, Jr
Dingman, Douglas Wayne
Dorsky, David Isaac
Downs, Wilbur George
Edberg, Stephen Charles
Eustice, David Christopher
Firshein, William
Forenza, Salvatore
Freedman, Michael Lewis
Fried, John H

Geiger, Edwin Otto
Geiger, Jon Ross
Golub, Efim I
Gorrell, Thomas Earl
Heim, Lyle Raymond
Hsiung, Gueh Djen
Huang, Liang Hsiung
Kemp, Gordon Arthur
Kirber, Maria Wiener
Kiron, Ravi
Krause, Peter James
LaMondia, James A
Leadbetter, Edward Renton
Lee, Sin Hang
Lentz, Thomas Lawrence
Loose, Leland David
Low, Kenneth Brooks, Jr
McGregor, Donald Neil
McKhann, Charles Fremont
Marcus, Philip Irving
Martin, R Russell
Mazzone, Horace M
Miller, I George, Jr
Natarajan, Kottayam Viswanathan
Nielsen, Peter Adams
Norton, Louis Arthur
Ornston, Leo Nicholas
Pazoles, Christopher James
Rauscher, Frank Joseph, Jr
Ray, Verne A
Repak, Arthur Jack
Retsema, James Allan
Rho, Jinnque
Romano, Antonio Harold
Ross, Martin Russell
Rothfield, Lawrence I
Routien, John Broderick
Sardinas, Joseph Louis
Sekellick, Margaret Jean
Setlow, Peter
Shope, Robert Ellis
Sieckhaus, John Francis
Slayman, Clifford L
Terry, Thomas Milton
Treffers, Henry Peter
Truesdell, Susan Jane
Ukeles, Ravenna
Verses, Christ James
Ward, David Christian
Wernau, William Charles
Wyckoff, Delaphine Grace Rosa

DELAWARE
Anderson, Jeffrey John
Benton, William J
Campbell, Linzy Leon
Carberry, Judith B
Cheng, Yih-Shyun Edmond
De Courcy, Samuel Joseph, Jr
Dexter, Stephen C
Dohms, John Edward
Eleuterio, Marianne Kingsbury
Enquist, Lynn William
Fries, Cara Rosendale
Gray, John Edward
Herson, Diane S
Hoover, Dallas Gene
Howard, Richard James
Jackson, David Archer
Korant, Bruce David
Lockart, Royce Zeno, Jr
Maciag, William John, Jr
Marrs, Barry Lee
Myoda, Toshio Timothy
Pene, Jacques Jean
Rosenberger, John Knox
Sariaslani, Sima
Singleton, Rivers, Jr
Smith, David William
Stopkie, Roger John
Torregrossa, Robert Emile
Yates, Richard Alan
Yin, Fay Hoh
Zajac, Barbara Ann
Zajac, Ihor

DISTRICT OF COLUMBIA
Albright, Julia W
Andrews, Wallace Henry
Ashe, Warren (Kelly)
Barak, Eve Ida
Baron, Louis Sol
Bellanti, Joseph A
Berman, Sanford
Bernard, Dane Thomas
Berry, Vinod K
Brandt, Carl David
Bundy, Bonita Marie
Calderone, Richard Arthur
Cayle, Theodore
Chapman, George Bunker
Cocks, Gary Thomas
Cutchins, Ernest Charles
DeCicco, Benedict Thomas
Dunkel, Virginia Catherine
Frederick, Lafayette
Gemski, Peter
Gibbs, Clarence Joseph, Jr
Gravely, Sally Margaret
Halstead, Thora Waters
Hartley, James William
Heineken, Frederick George
Henderson, Ellen Jane
Herscowitz, Herbert Bernard

Microbiology (cont)

Hill, Walter Ernest
Hollinshead, Ariel Cahill
Hugh, Rudolph
Kingsbury, David Thomas
Krigsvold, Dale Thomas
Kumar, Ajit
Madden, David Larry
Malveaux, Floyd J
Mandel, Benjamin
Matyas, Gary Ralph
Meyers, Wayne Marvin
Miller, William Robert
Nasser, DeLill
Neihof, Rex A
O'Hern, Elizabeth Moot
Okrend, Harold
Olenick, John George
Olson, James Gordon
Oyewole, Saundra Herndon
Parker, Richard H
Parrott, Robert Harold
Pearson, Gary Richard
Prival, Michael Joseph
Rabin, Robert
Reich, Melvin
Roane, Philip Ransom, Jr
Roskoski, Joann Pearl
Sabin, Albert B(ruce)
Salzman, Norman Post
Saz, Arthur Kenneth
Sever, John Louis
Silver, Sylvia
Sreevalsan, Thazepadath
Stokes, Gerald V
Tompkins, George Jonathan
Turner, Willie
Twedt, Robert Madsen
Venkatesam, Malabi M
Wassef, Nabila M
White, Sandra L
Woodman, Daniel Ralph
Young, Frank E

FLORIDA

Aldrich, Henry Carl
Baer, Herman
Balkwill, David Lee
Beckhorn, Edward John
Bellamy, Winthrop Dexter
Bergs, Victor Visvaldis
Betz, John Vianney
Bitton, Gabriel
Bleiweis, Arnold Sheldon
Bliznakov, Emile George
Block, Seymour Stanton
Blomberg, Bonnie B
Bourne, Carol Elizabeth Mulligan
Brandon, Frank Bayard
Brown, Ross Duncan, Jr
Brown, Thomas Allen
Brown, William Lewis
Burney, Curtis Michael
Carlson, Harve J
Carvajal, Fernando
Chapman, Peter John
Chester, Brent
Clark, William Burton, IV
Cleary, Timothy Joseph
Cohen, Marc Singman
Coleman, Sylvia Ethel
Cort, Winifred Mitchell
Cowman, Richard Ammon
Dardiri, Ahmed Hamed
Dierks, Richard Ernest
Dutary, Bedsy Elsira
Eilers, Frederick Irving
Ellis, Edwin M
Farrow, Wendall Moore
Fitzgerald, Robert James
Fox, Kenneth Ian
Frankel, Jack William
Frediani, Harold Arthur
Friedman, Herman
Friedmann, E(merich) Imre
Gabridge, Michael Gregory
Gaskin, Jack Michael
Gennaro, Robert Nash
Gifford, George Edwin
Godzeski, Carl William
Goihman-Yahr, Mauricio
Gottwald, Timothy R
Groupe, Vincent
Halbert, Seymour Putterman
Halkias, Demetrios
Hartman, James Xavier
Hauswirth, William Walter
Heinis, Julius Leo
Helmstetter, Charles E
Hiebert, Ernest
Houts, Garnette Edwin
Ingram, Lonnie O'Neal
Jenkin, Howard M
Jensen, Roy A
Jezeski, James John
Kiefer, David John
Klein, Paul Alvin
Koburger, John Alfred
Lai, Patrick Kinglun
Leonard, Frederic Adams
Lim, Daniel V
Lin, Tsue-Ming
Lopez, Diana Montes De Oca

McArthur, William P
McClung, Norvel Malcolm
McGray, Robert James
Makemson, John Christopher
Mans, Rusty Jay
Martinez, Octavio Vincent
Moscovici, Carlo
Moyer, Richard W
Nakashima, Tadayoshi
Nonoyama, Meihan
Ocampo-Friedmann, Roseli C
Paradise, Lois Jean
Peralta, Pauline Huntington
Prebluda, Harry Jacob
Preston, James Faulkner, III
Previc, Edward Paul
Purcifull, Dan Elwood
Putnam, Hugh D
Radhakrishnan, Chittur Venkitasubhan
Ringel, Samuel Morris
Robb, James Arthur
Roth, Frank J, Jr
Rubin, Harvey Louis
Sadler, George D
Sallman, Bennett
Sandoval, Howard Kenneth
Schuytema, Eunice Chambers
Scudder, Walter Tredwell
Shands, Joseph Walter, Jr
Silver, Warren Seymour
Sinkovics, Joseph G
Smith, Kenneth Leroy
Specter, Steven Carl
Spencer, John Lawrence
Stechschulte, Agnes Louise
Stock, David Allen
Streitfeld, Murray Mark
Sweet, Herman Royden
Tamplin, Mark Lewis
Taylor, Barrie Frederick
Tsai, James Hsi-Cho
Verkade, Stephen Dunning
White, Franklin Henry
White, Roseann Spicola
Wilson, William D
Wodzinski, Rudy Joseph

GEORGIA

Ahearn, Donald G
Anderson, David Prewitt
Balows, Albert
Barbaree, James Martin
Bard, Raymond Camillo
Batra, Gopal Krishan
Beard, Charles Walter
Bennett, Sara Neville
Best, Gary Keith
Beuchat, Larry Ray
Blank, Carl Herbert
Boring, John Rutledge, III
Brooks, John Bill
Brown, William E
Brownell, George H
Brugh, Max, Jr
Bryans, Trabue Daley
Burnett, George Wesley
Cassel, William Alwein
Cavallaro, Joseph John
Cherry, William Bailey
Cohen, Jay O
Cole, John Rufus, Jr
Cox, Nelson Anthony
Crouse, Gray F
Dailey, Harry A
Dobrovolny, Charles George
Dowdle, Walter R
Dowell, Vulus Raymond, Jr
Doyle, Michael Patrick
Duncan, Robert Leon, Jr
Eagon, Robert Garfield
Eriksson, Karl-Erik Lennart
Eriquez, Louis Anthony
Evans, Donald Lee
Ewing, William Howell
Farmer, John James, III
Fincher, Edward Lester
Finnerty, William Robert
Garver, Frederick Albert
Gitaitis, Ronald David
Good, Robert Campbell
Gourse, Richard Lawrence
Gratzek, John B
Green, John H
Hall, Charles Thomas
Harrell, William Knox
Hatheway, Charles Louis
Hedden, Kenneth Forsythe
Herrmann, Kenneth L
Hierholzer, John Charles
Hill, Edward Orson
Holt, Peter Stephen
Hubbard, Jerry S
Hunt, Dale E
Ibrahim, Adly N
Jones, Gilda Lynn
Jones, Rena Talley
Jones, Wilbur Douglas, Jr
Kadis, Solomon
Kaplan, William
Kellogg, Douglas Sheldon, Jr
Kerr, Thomas James
Kleckner, Albert Louis
Kleven, Stanley H
Krishnamurti, Pullabhotla V

Kubica, George P
LaMotte, Louis Cossitt, Jr
Lehner, Andreas Friedrich
Ljungdahl, Lars Gerhard
Luck, John Virgil
Lukert, Phil Dean
McClure, Harold Monroe
McDade, Joseph Edward
Margolis, Harold Stephen
Martin, William Randolph
Mathews, Henry Mabbett
Miller, Lois Kathryn
Morse, Stephen Allen
Moss, Claude Wayne
Murphy, Frederick A
Nahmias, Andre Joseph
Nakano, James Hiroto
Navalkar, Ram G
Paris, Doris Fort
Payne, William Jackson
Peck, Harry Dowd, Jr
Pine, Leo
Preblud, Stephen Robert
Pusey, P Lawrence
Reinhardt, Donald Joseph
Reiss, Errol
Rogers, John Ernest
Roth, Ivan Lambert
Sawyer, Richard Trevor
Schuler, George Albert
Schuster, George Sheah
Shotts, Emmett Booker, Jr
Shuster, Robert C
Snyder, Hugh Donald
Spitznagel, John Keith
Stulting, Robert Doyle, Jr
Sudia, William Daniel
Suggs, Morris Talmage, Jr
Sulzer, Alexander Jackson
Summers, Anne O
Taylor, Gerald C
Thomason, Berenice Miller
Tornabene, Thomas Guy
Tunstall, Lucille Hawkins
Tzianabos, Theodore
Vogel, Ralph A
Volkmann, Keith Robert
Wallace, Douglas Cecil
White, Lendell Aaron
Whitehead, Marvin Delbert
Wiebe, William John
Wilkinson, Hazel Wiley
Wolven Garrett, Anne M
Woodley, Charles Lamar
Wyatt, Roger Dale
Ziegler, Harry Kirk
Ziffer, Jack

HAWAII

Allen, Richard Dean
Berger, Leslie Ralph
Contois, David Ely
Flanagan, Patrick William
Fok, Agnes Kwan
Frank, Hilmer Aaron
Fujioka, Roger Sadao
Furusawa, Eiichi
Hall, John Bradley
Higa, Harry Hiroshi
Klemmer, Howard Wesley
Loh, Philip Choo-Seng
Mandel, Morton
Marchette, Nyven John
Patterson, Gregory Matthew Leon
Person, Donald Ames
Roberts, Robert Russell
Scott, John Francis
Taussig, Steven J
Vennesland, Birgit
Ward, Melvin A

IDAHO

Crawford, Donald Lee
Crawford, Ronald L
Dugan, Patrick R
Farrell, Larry Don
Frank, Floyd William
Hatcher, Herbert John
Heimsch, Richard Charles
Hostetter, Donald Lee
Keay, Leonard
Laurence, Kenneth Allen
Lingg, Al J
McCune, Mary Joan Huxley
Winston, Vern

ILLINOIS

Akin, Cavit
Albach, Richard Allen
Alexander, Nancy J
Armbruster, Frederick Carl
Bahn, Arthur Nathaniel
Baram, Peter
Bartell, Marvin H
Bhatti, Rashid
Bird, Thomas Joseph
Blaschek, Hans P
Blumenthal, Harold Jay
Bothast, Rodney Jacob
Braendle, Donald Harold
Braun, Donald Peter
Brennan, Patricia Conlon
Brewer, Gregory J
Brockman, Herman E

Buchholz Shaw, Donna Marie
Burmeister, Harland Reno
Cabana, Veneracion Garganta
Came, Paul E
Campbell, Michael Floyd
Casadaban, Malcolm John
Castignetti, Domenic
Chakrabarty, Ananda Mohan
Chang, Kwang-Poo
Chen, Shepley S
Cheng, Shu-Sing
Cheung, Hou Tak
Choe, Byung-Kil
Christensen, Mary Lucas
Chu, Daniel Tim-Wo
Chudwin, David S
Cibulsky, Robert John
Cohen, Sidney
Cooper, Morris Davidson
Corbett, Jules John
Cork, Douglas J
Crane, Anatole
Crouch, Norman Albert
Davis, Edwin Nathan
DePinto, John A
Dhaliwal, Amrik S
Doering, Jeffrey Louis
Drucker, Harvey
Duncan, James Lowell
Eachus, Alan Campbell
Ekstedt, Richard Dean
Engelbrecht, R(ichard) S(tevens)
Falk, Lawrence A, Jr
Farrand, Stephen Kendall
Fast, Dale Eugene
Fennewald, Michael Andrew
Fenters, James Dean
Flowers, Russell Sherwood, Jr
Ford, Richard Earl
Frampton, Elon Wilson
Frank, James Richard
Freeman, Maynard Lloyd
Gabis, Damien Anthony
Galsky, Alan Gary
Gessner, Robert V
Girolami, Roland Louis
Glaser, Janet H
Goldin, Milton
Goldman, Manuel
Gorden, Robert Wayne
Goss, George Robert
Gould, John Michael
Hac, Lucile R
Hagar, Lowell Paul
Halpern, Bernard
Hanson, Lyle Eugene
Herrmann, Ernest Carl, Jr
Holland, Louis Edward, II
Holleman, William H
Holzer, Timothy J
Hsu, Wen-Tah
Huberman, Eliezer
Ingle, Morton Blakeman
Jackson, George G
Jackson, Robert W
Jayaswal, Radheshyam K
Jonah, Margaret Martin
Jones, Marjorie Ann
Kaizer, Herbert
Kallio, Reino Emil
Kaneshiro, Tsuneo
Kantor, Harvey Sherwin
Kaye, Saul
Keudell, Kenneth Carson
Khodadad, Jena Khadem
Khoobyarian, Newton
Kim, Yoon Berm
King, Robert Willis
Konisky, Jordan
Kubitschek, Herbert Ernest
Kuehner, Calvin Charles
Laffler, Thomas G
Lakshminarayanan, Krishnaiyer
Lamb, Robert Andrew
Landau, William
Landay, Alan Lee
Lechowich, Richard V
Levy, Allan Henry
Lippincott, Barbara Barnes
Lippincott, James Andrew
Lucher, Lynne Annette
Markovitz, Alvin
Martin, Scott Elmore
Mateles, Richard I
Mathews, Herbert Lester
Matsumura, Philip
Melin, Brian Edward
Meyer, Richard Charles
Miller, Charles G
Miller, Raymond Michael
Miller, Robert Verne
Miller, Stephen Douglas
Moon, Robert John
Morello, Josephine A
Murphy, Mary Nadine
Murphy, Richard Allan
Nath, K Rajinder
Nath, Nrapendra
Neuhaus, Francis Clemens
Newton, Stephen Bruington
Northrop, Robert L
Orland, Frank J
Ostrow, David Henry
Pal, Dhiraj

Papoutsakis, Eleftherios Terry
Passman, Frederick Jay
Peak, Meyrick James
Pedraja, Rafael R
Pekarek, Robert Sidney
Perry, Dennis
Perry, Eugene Arthur
Peterson, David Allan
Plate, Charles Alfred
Plewa, Michael Jacob
Prakasam, Tata B S
Pratt, Charles Walter
Pumper, Robert William
Pun, Pattle Pak-Toe
Rabinovich, Sergio Rospigliosi
Rakosky, Joseph, Jr
Randall, Eileen Louise
Regunathan, Perialwar
Reilly, Christopher Aloysius, Jr
Rittmann, Bruce Edward
Roizman, Bernard
Rowan, Dighton Francis
Roy, Dipak
Rundell, Mary Kathleen
Rupprecht, Kevin Robert
Ryan, Jon Michael
Salyers, Abigail Ann
Scherr, George Harry
Schlager, Seymour I
Schleicher, Joseph Bernard
Schlenk, Fritz
Schnell, Gene Wheeler
Schwartz, Robert David
Seed, Thomas Michael
Segal, William
Shapiro, James Alan
Shapiro, Stanley Kallick
Shechmeister, Isaac Leo
Shen, Linus Liang-nene
Sherman, Joseph E
Shipkowitz, Nathan L
Simon, Ellen McMurtrie
Simon, Selwyn
Simonson, Lloyd Grant
Sinclair, Thomas Frederick
Singer, Samuel
Smittle, Richard Baird
Snyder, William Robert
Spear, Patricia Gail
Starzyk, Marvin John
Stevens, Joseph Alfred
Strano, Alfonso J
Strauss, Bernard S
Sutton, Lewis McMechan
Taylor, Welton Ivan
Thompson, Kenneth David
Tomita, Joseph Tsuneki
Tompkin, Robert Bruce
Tripathy, Deoki Nandan
Turner, Donald W
Vary, Patricia Susan
Walter, Trevor John
Wawszkiewicz, Edward John
Weary, Marlys E
Welker, Neil Ernest
Wetegrove, Robert Lloyd
Whitt, Dixie Dailey
Wideburg, Norman Earl
Widra, Abe
Wilkinson, Brian James
Witmer, Heman John
Witter, Lloyd David
Wolfe, Ralph Stoner
Wolff, Robert John
Yackel, Walter Carl
Yamashiroya, Herbert Mitsugi
Yoon, Ji-Won
Yotis, William William

INDIANA
Allen, Norris Elliott
Argot, Jeanne
Aronson, Arthur Ian
Asteriadis, George Thomas, Jr
Babel, Frederick John
Baldwin, William Walter
Becker, Benjamin
Begue, William John
Bockrath, Richard Charles, Jr
Boisvenue, Rudolph Joseph
Boyd, Frederick Mervin
Boyer, Ernest Wendell
Bozarth, Robert F
Bracker, Charles E
Brehm, Sylvia Patience
Bretz, Harold Walter
Burnstein, Theodore
Caltrider, Paul Gene
Counter, Frederick T, Jr
Cousin, Maribeth Anne
Crawford, James Gordon
Daily, Fay Kenoyer
Daily, William Allen
Day, Lawrence Eugene
Diffley, Peter
Doemel, William Naylor
Edenberg, Howard Joseph
Enders, George Leonhard, Jr
Foglesong, Mark Allen
Folkerts, Thomas Mason
Foos, Kenneth Michael
Gale, Charles
Gavin, John Joseph
Gest, Howard

Gordee, Robert Stouffer
Gustafson, Donald Pink
Haak, Richard Arlen
Haelterman, Edward Omer
Hanson, Robert Jack
Hegeman, George D
Hendrickson, Donald Allen
Hershberger, Charles Lee
Hicks, Garland Fisher, Jr
Hodson, Phillip Harvey
Hoehn, Marvin Martin
Huber, Floyd Milton
Ingraham, Joseph Sterling
Jaehning, Judith A
Kapustka, Lawrence A
Keith, Paula Myers
Kenny, Michael Thomas
Kinsel, Norma Ann
Kinzel, Jerry J
Kirkham, Wayne Wolpert
Kirsch, Edwin Joseph
Kittaka, Robert Shinnosuke
Klausmeier, Robert Edward
Konetzka, Walter Anthony
Konopka, Allan Eugene
Kreps, David Paul
Larsen, Steven H
Lees, Norman Douglas
Levine, Alvin Saul
Levinthal, Mark
Lister, Richard Malcolm
Lively, David Harryman
Loesch-Fries, Loretta Sue
Lopez, Carlos
Lovett, James Satterthwaite
McCann, Peter Paul
McClung, Leland Swint
McMullen, James Robert
Mallett, Gordon Edward
Mason, Earl James
Matsuoka, Tats
Matsuoka, Tats
Miescher, Guido
Millar, Wayne Norval
Miller (Gilbert), Carol Ann
Miller, Chris H
Minton, Sherman Anthony
Mocharla, Raman
Morse, Erskine Vance
Moskowitz, Merwin
Niederpruem, Donald J
Niss, Hamilton Frederick
Oster, Mark Otho
Paschke, John Donald
Pike, Loy Dean
Pittenger, Robert Carlton
Pleasants, Julian Randolph
Pollard, Morris
Queener, Stephen Wyatt
Ragheb, Hussein S
Roales, Robert R
Rogers, Howell Wade
Saz, Howard Jay
Schloemer, Robert Henry
Schoknecht, Jean Donze
Sherman, Louis Allen
Simon, Edward Harvey
Sipe, Jerry Eugene
Smith, Gary Lee
Smith, Robert William
Sojka, Gary Allan
Stone, Robert Louis
Summers, William Allen, Sr
Swaminathan, Balasubramanian
Taylor, Milton William
Tessman, Irwin
Turner, Jan Ross
Wegener, Warner Smith
Weinberg, Eugene David
White, David
Whitney, John Glen
Williams, Daniel Charles
Williams, Robert Dee
Wilson, Gary August
Zimmerman, Sarah E
Zygmunt, Walter A

IOWA
Allison, Milton James
Bryner, John Henry
Cazin, John, Jr
Coleman, Richard Walter
Collins, Richard Francis
Crawford, Irving Pope
De Jong, Peter J
DiSpirito, Alan Angelo
Durand, Donald P
Dworschack, Robert George
Fitzgerald, Gerald Roland
Frederick, Lloyd Randall
Frey, Merwin Lester
Gibson, David Thomas
Goss, Robert Charles
Greenberg, Everett Peter
Hanson, Austin Moe
Harp, James A
Hartman, Paul Arthur
Hausler, William John, Jr
Hughes, David Edward
Johnson, William
Kemeny, Lorant
Kieso, Robert Alfred
Kohn, Frank S
LaGrange, William Somers

Lambert, George
McClurkin, Arlan Wilbur
Markovetz, Allen John
Marshall, William Emmett
Martinson, Charlie Anton
Mayfield, John Eric
Mengeling, William Lloyd
Menninger, John Robert
Newcomb, Harvey Russell
Nicholson, Donald Paul
O'Berry, Phillip Aaron
Packer, R(aymond) Allen
Paul, Prem Sagar
Pirtle, Eugene Claude
Quinn, Loyd Yost
Rodriguez, Jose Enrique
Ross, Richard Francis
Sabiston, Charles Barker, Jr
Shaw, Gaylord Edward
Smith, Claire Leroy
Songer, Joseph Richard
Stahly, Donald Paul
Stalheim, Ole H Viking
Stanton, Thaddeus Brian
Stinski, Mark Francis
Sweat, Robert Lee
Targowski, Stanislaw P
Thompson, Sue Ann
Tjostem, John Leander
Van Der Maaten, Martin Junior
Van Eck, Edward Arthur
Walker, Homer Wayne
Welter, C Joseph
Williams, Fred Devoe
Williams, Phletus P
Wood, Richard Lee

KANSAS
Akagi, James Masuji
Amelunxen, Remi Edward
Bailie, Wayne E
Barnes, William Gartin
Bechtel, Donald Bruce
Behbehani, Abbas M
Bode, Vernon Cecil
Bradford, Lawrence Glenn
Buller, Clarence S
Carlson, Roger Harold
Cho, Cheng T
Coles, Embert Harvey, Jr
Consigli, Richard Albert
Decedue, Charles Joseph
Dirks, Brinton Marlo
Duncan, Bettie
Dykstra, Mark Allan
Field, Marvin Frederick
Fina, Louis R
Fung, Daniel Yee Chak
Furtado, Dolores
Gegenheimer, Peter Albert
Goldberg, Ivan D
Guikema, James Allen
Haas, Herbert Frank
Ham, George Eldon
Hetrick, Barbara Ann
Jensen, Thorkil
Johnson, Donovan Earl
Johnson, Terry Charles
Langston, Clarence Walter
Leslie, John Franklin
Lindsey, Norma Jack
Liu, Chien
MacFarlane, John O'Donnell
McMillen, Janis Kay
Marchin, George Leonard
Mayeux, Jerry Vincent
Miller, Glendon Richard
Morrison, David Campbell
Paretsky, David
Peterson, John Edward, Jr
Ramamurti, Krishnamurti
Robertson, Donald Claus
Sarachek, Alvin
Shankel, Delbert Merrill
Stetler, Dean Allen
Stewart, Richard Cummins
Su, Robert Tzyh-Chuan
Takemoto, Dolores Jean
Urban, James Edward
Werder, Alvar Arvid
Wong, Peter P
Wood, Gary Warren

KENTUCKY
Allen, George Perry
Andreasen, Arthur Albinus
Atlas, Ronald M
Bryans, John Thomas
Budde, Mary Laurence
Clark, Evelyn Genevieve
Dalzell, Robert Clinton
Dickson, Robert Carl
Elliott, Larry P
Goodman, Norman L
Hammond, Ray Kenneth
Harris, Delbert Linn
Harris, Denny Olan
Hendrix, James William
Issel, Charles John
Johnston, Paul Bruns
Keller, Kenneth F
Langlois, Bruce Edward
Lillich, Thomas Tyler
Liu, Pinghui Victor

McCollum, William Howard
Maisch, Weldon Frederick
Otero, Raymond B
Overman, Timothy Lloyd
Rhoads, Robert E
Rodriguez, Lorraine Ditzler
Rothwell, Frederick Mirvan
Ruchman, Isaac
Skean, James Dan
Sonnenfeld, Gerald
Squires, Robert Wright
Staat, Robert Henry
Steiner, Sheldon
Streips, Uldis Normunds
Vanaman, Thomas Clark
Vittetoe, Marie Clare
Wacker, Waldon Burdette
Wallace, John Howard
Weber, Frederick, Jr
Williams, Arthur Lee
Wiseman, Ralph Franklin
Zimmer, Stephen George

LOUISIANA
Akers, Thomas Gilbert
Alam, Jawed
Barkate, John Albert
Braymer, Hugh Douglas
Bryant, James Berry, Jr
Burns, Kenneth Franklin
Centifanto, Ysolina M
Corstvet, Richard E
Courtney, Richard James
Coward, Joe Edwin
Day, Donal Forest
Deas, Jane Ellen
Domer, Judith E
Dommert, Arthur Roland
Dunigan, Edward P
Ehrlich, Kenneth Craig
Ehrlich, Melanie
Gallaher, William Richard
Gaumer, Herman Richard
Gerone, Peter John
Gormus, Bobby Joe
Gray, Raymond Francis
Greer, Donald Lee
Guidry, Dieu-Donne Joseph
Hart, Lewis Thomas
Hedrick, Harold Gilman
Hurley, Maureen
Huxsoll, David Leslie
Hyman, Edward Sidney
Hyslop, Newton Everett, Jr
Ivens, Mary Sue
Jamison, Richard Melvin
Johnson, Emmett John
Kadan, Ranjit Singh
Koike, Hideo
Kotcher, Emil
Larkin, John Montague
Lembeck, William Jacobs
Luftig, Ronald Bernard
Meyers, Samuel Philip
Monsour, Victor
O'Callaghan, Dennis John
O'Callaghan, Richard J
Palmgren, Muriel Signe
Pierce, William Arthur, Jr
Shaffer, Morris Frank
Sheiness, Diana Kay
Siebeling, Ronald Jon
Silberman, Ronald
Sizemore, Robert Carlen
Socolofsky, Marion David
Soike, Kenneth Fieroe
Spence, Hilda Adele
Srinivasan, Vadake Ram
Starrett, Richmond Mullins
Stevenson, L Harold
Stiffey, Arthur V
Storz, Johannes
Thompson, James Jarrard
Tsuzuki, Junji
Upadhyay, Jagdish M
Weidner, Earl
Wilson, John Thomas

MAINE
Bain, William Murray
Curtis, Paul Robinson
De Siervo, August Joseph
Donahoe, John Philip
Founds, Henry William
Gezon, Horace Martin
Jerkofsky, Maryann
Lycette, R(ichard) (Milton)
Moody, Charles Edward, Jr
Nicholson, Bruce Lee
Norton, Cynthia Friend
Novotny, James Frank
Ridgway, George Junior
Terry, Robert Lee

MARYLAND
Abramson, I Jerome
Albano, Marianita Madamba
Albrecht, Paul
Allen, William Peter
Anderson, Arthur O
Anderson, Mauritz Gunnar
Armstrong, Earlene
Aurelian, Laure
Austin, Faye Carol

Microbiology (cont)

Azad, Abdu F(arhang)
Baker, Phillip John
Bala, Shukal
Barile, Michael Frederick
Bassin, Robert Harris
Bausum, Howard Thomas
Beemon, Karen Louise
Berger, Edward Alan
Berne, Bernard H
Berry, Bradford William
Biswal, Nilambar
Biswas, Robin Michael
Bloom, Eda Terri
Blough, Herbert Allen
Blumberg, Peter Mitchell
Boyd, Virginia Ann Lewis
Bozeman, F Marilyn
Bradlaw, June A
Brandt, Walter Edmund
Brook, Itzhak
Brown, Paul Wheeler
Burge, Wylie D
Burton, George Joseph
Butman, Bryan Timothy
Calabi, Ornella
Capone, Douglas George
Carski, Theodore Robert
Chang, Kenneth Shueh-Shen
Chang, Yung-Feng
Chaparas, Sotiros D
Charache, Patricia
Chen, Joseph Ke-Chou
Choppin, Purnell Whittington
Clark, Trevor H
Cole, Gerald Alan
Colwell, Rita R
Cook, Thomas M
Costlow, Richard Dale
Dalrymple, Joel McKeith
Daniels, William Fowler, Sr
Das, Naba Kishore
Dasch, Gregory Alan
Dashiell, Thomas Ronald
DeFrank, Joseph J
DeLamater, Edward Doane
Denny, Cleve B
Diener, Theodor Otto
Dodd, Roger Yates
Donlon, Mildred A
Eaves, George Newton
Eckels, Kenneth Henry
Edelman, Robert
Elin, Ronald John
Epstein, David Aaron
Esposito, Vito Michael
Evans, George Leonard
Feinman, Susan (Birnbaum)
Fernie, Bruce Frank
Fine, Donald Lee
Fischinger, Peter John
Fiset, Paul
Fish, Donald C
FitzGerald, David J P
Fitzgerald, Edward Aloysius
Formal, Samuel Bernard
Foulds, John Douglas
Frasch, Carl Edward
Friedman, Robert Morris
Froehlich, Luz
Fulkerson, John Frederick
Gajdusek, Daniel Carleton
Galasso, George John
Ganaway, James Rives
Garges, Susan
Gerin, John Louis
Gibby, Irvin Welch
Goldenbaum, Paul Ernest
Goldstein, Robert Arnold
Gomatos, Peter John
Gonda, Matthew Allen
Goodman, Sarah Anne
Goodman, Sarah Anne
Gordon, Lance Kenneth
Gordon, Milton
Gougé, Susan Cornelia Jones
Granger, Donald Lee
Gravell, Maneth
Griffin, Diane Edmund
Gruber, Jack
Grunberg, Neil Everett
Guss, Maurice Louis
Guttman, Helene Augusta Nathan
Habig, William Henry
Hackett, Joseph Leo
Hadidi, Ahmed Fahmy
Hale, Martha L
Hamilton, Bruce King
Hampar, Berge
Hanna, Edgar Ethelbert, Jr
Haspel, Martin Victor
Hatanaka, Masakazu
Hawley, Robert John
Hearn, Henry James, Jr
Hellman, Alfred
Henry, Timothy James
Hetrick, Frank M
Hewetson, John Francis
Heydrick, Fred Painter
Hill, James Carroll
Hochstein, Herbert Donald
Hoffmann, Conrad Edmund
Hoggan, Malcolm David

Holmes, Kathryn Voelker
Holmes, Randall Kent
Hook, William Arthur
Hopps, Hope Elizabeth Byrne
Hornor, Sally Graham
Housewright, Riley Dee
Howell, David McBrier
Howley, Peter Maxwell
Hoyer, Bill Henriksen
Huebner, Robert Joseph
Hunt, Lois Turpin
Hurlburt, Evelyn McClelland
Jakoby, William Bernard
James, Stephanie Lynn
Jaouni, Katherine Cook
Jemski, Joseph Victor
Johnson, Roger W
Johnson-Winegar, Anna
Joseph, J Mehsen
Joseph, Sammy William
Kaighn, Morris Edward
Kaper, James Bennett
Kelly, Thomas J
Kenyon, Richard H
Kern, Jerome
Kimball, Paul Clark
Kingsbury, David Wilson
Kingsbury, Elizabeth W
Kramer, Tim R
Krause, Richard Michael
Krichevsky, Micah I
Kwon-Chung, Kyung Joo
La Montagne, John Ring
Landon, John Campbell
Langlykke, Asger Funder
Ledney, George David
Lee, Chi-Jen
Lee, Fang-Jen Scott
Lee, Theresa
Leppla, Stephen Howard
Levin, Judith Goldstein
Levin, Morris A
Levy, Hilton Bertram
Libonati, Joseph Peter
Long, Cedric William
Lovett, Paul Scott
Lucas, John Paul
McCarron, Richard M
McGowan, John Joseph
McKenney, Keith Hollis
McNicol, Lore Anne
MacQuillan, Anthony M
MacVittie, Thomas Joseph
Mageau, Richard Paul
Maloney, Peter Charles
Manaker, Robert Anthony
Manclark, Charles Robert
Margulies, Maurice
Marquardt, Warren William
Martin, Robert G
Maurer, Bruce Anthony
Mayer, Vernon William, Jr
Mergenhagen, Stephan Edward
Merz, William George
Minah, Glenn Ernest
Moloney, John Bromley
Monath, Thomas P
Morgan, Herbert R
Morris, Joseph Anthony
Moss, Bernard
Murphy, Patrick Aidan
Murray, Gary Joseph
Narayan, Opendra
Nash, Howard Allen
Nauman, Robert Karl
Neva, Franklin Allen
Nomura, Shigeko
Notkins, Abner Louis
Nutter, John
O'Beirne, Andrew Jon
Oliver, Eugene Joseph
Papas, Takis S
Parkman, Paul Douglas
Pavlovskis, Olgerts Raimonds
Payton, Cecil Warren
Pearson, John William
Phillips, Grace Briggs
Pienta, Roman Joseph
Pilchard, Edwin Ivan
Pledger, Richard Alfred
Pollok, Nicholas Lewis, III
Poston, John Michael
Powers, Kendall Gardner
Prabhakar, Bellur Subbanna
Preble, Olivia Toby
Purcell, Robert Harry
Puziss, Milton
Quinnan, Gerald Vincent, Jr
Rader, Ronald Alan
Rafjako, Robert Richard
Ranhand, Jon M
Read-Connole, Elizabeth Lee
Reissig, Magdalena
Rhim, Johng Sik
Rice, Nancy Reed
Rizzo, Anthony Augustine
Rochovansky, Olga Maria
Rommel, Frederick Allen
Rose, Noel Richard
Rosenstein, Robert William
Rosner, Judah Leon
Rossio, Jeffrey L
Russell, Philip King
Sack, Richard Bradley

Sarma, Padman S
Saxinger, W(illiam) Carl
Schaub, Stephen Alexander
Schluederberg, Ann Elizabeth Snider
Schneider, Walter Carl
Schricker, Robert Lee
Schultz, Warren Walter
Schwab, Bernard
Shaffer, Charles Henry, Jr
Shah, Keerti V
Shapiro, Martin
Shelokov, Alexis
Sherman, Kenneth Eliot
Shih, James Waikuo
Shih, Thomas Y
Sibal, Louis Richard
Sieckmann, Donna G
Silverman, David J
Silverman, Robert Hugh
Simmon, Vincent Fowler
Simons, Daniel J
Sjoblad, Roy David
Slyter, Leonard L
Small, Eugene Beach
Smith, Andrew George
Smith, Dorothy Gordon
Snyder, Merrill J
Speedie, Marilyn Kay
Stadtman, Thressa Campbell
Stamper, Hugh Blair
Steinman, Harry Gordon
Steinman, Irvin David
Stevenson, Robert Edwin
Strickland, George Thomas
Stromberg, Kurt
Stunkard, Jim A
Sydiskis, Robert Joseph
Sztein, Marcelo Benjamin
Taube, Sheila Efron
Taylor, David Neely
Tepper, Byron Seymour
Theodore, Theodore Spiros
Tingle, Marjorie Anne
Tonik, Ellis J
Tully, Joseph George
Tyler, Bonnie Moreland
Vadlamudi, Sri Krishna
Vera, Harriette Dryden
Vincent, Phillip G
Voll, Mary Jane
Wachter, Ralph Franklin
Wahl, Sharon Knudson
Walker, Richard Ives
Walter, Eugene LeRoy, Jr
Walters, Curla Sybil
Weiner, Ronald Martin
Weislow, Owen Stuart
Weiss, Emilio
Westhoff, Dennis Charles
Whitehurst, Virgil Edwards
Williams, Jimmy Calvin
Winchurch, Richard Albert
Wisseman, Charles Louis, Jr
Wolf, Richard Edward, Jr
Wolff, David A
Wong, Dennis Mun
Woodward, Theodore Englar
Yanagihara, Richard
Yasbin, Ronald Eliott
Young, Bobby Gene
Young, Howard Alan
Young, Viola Mae (Horvath)
Zebovitz, Eugene
Zierdt, Charles Henry
Zimmerman, Eugene Munro

MASSACHUSETTS
Allen, Mary A Mennes
Asato, Yukio
Baker, Edgar Eugene, Jr
Berry, Vern Vincent
Black, Paul H
Blacklow, Neil Richard
Blaustein, Ernest Herman
Boger, Edwin August, Sr
Borgatti, Alfred Lawrence
Brennessel, Barbara A
Broitman, Selwyn Arthur
Canale-Parola, Ercole
Chang, Te Wen
Chou, Iih-Nan
Coffin, John Miller
Cohen, Joel Ralph
Cole, Edward Anthony
Collier, Robert John
Collins, Carolyn Jane
Comer, M Margaret
Conti, Samuel F
Cooney, Charles Leland
Cooney, Joseph Jude
Cooper, Geoffrey Mitchell
Craven, Donald Edward
Crisley, Francis Daniel
Cukor, George
Cynkin, Morris Abraham
Davis, Bernard David
Deitz, William Harris
Demain, Arnold Lester
Dennen, David W
DePalma, Philip Anthony
Derow, Matthew Arnold
Desrosiers, Ronald Charles
Dowell, Clifton Enders
Esber, Henry Jemil

Essex, Myron
Eubanks, Elizabeth Ruberta
Fairbanks, Grant
Farina, Joseph Peter
Fields, Bernard N
Foley, George Edward
Fraenkel, Dan Gabriel
Fritsch, Edward Francis
Fuller, Rufus Clinton
Gabliks, Jānis
Gerety, Robert John
Gibbons, Ronald J
Gill, David Michael
Goldberg, Edward B
Goldstein, Richard Neal
Greenfield, Seymour
Halvorson, Harlyn Odell
Hanshaw, James Barry
Hastings, John Woodland
Hawiger, Jack Jacek
Henneberry, Richard Christopher
Hirsch, Martin Stanley
Hotchkiss, Rollin Douglas
Jaenisch, Rudolf
Janicki, Bernard William
Jannasch, Holger Windekilde
Jensen, Erling
Johnson, Corinne Lessig
Jordan, Harold Vernon
Jost, Dana Nelson
Kaplan, David Lee
Kelley, William S
Kelner, Albert
Kepper, Robert Edgar
Kibrick, Sidney
Kieff, Elliott Dan
Klempner, Mark Steven
Knipe, David Mahan
Kuemmerle, Nancy Benton Stevens
Kundsin, Ruth Blumfeld
Labbe, Ronald Gilbert
Lessie, Thomas Guy
Levin, Robert E
Levy, Stuart B
Lingappa, Banadakoppa Thimmappa
Lingappa, Yamuna
Litsky, Bertha Yanis
Lodish, Harvey Franklin
Mabuchi, Katsuhide
McCabe, William R
McCormick, Neil Glenn
Magasanik, Boris
Mandels, Mary Hickox
Manning, William Joseph
Mata, Leonardo J
Mathews-Roth, Micheline Mary
Medearis, Donald N, Jr
Meedel, Thomas Huyck
Milani, Victor John
Miller, Judith Evelyn
Mitchell, Ralph
Mount, Mark Samuel
Mulder, Carel
Murphy, John R
Nichols, Roger Loyd
Orme-Johnson, William H
Park, James Theodore
Powers, Edmund Maurice
Previte, Joseph James
Prindle, Bryce
Putney, Scott David
Reiner, Albey M
Reyes, Victor E
Robinton, Elizabeth Dorothy
Rogers, Morris Ralph
Rosenberg, Fred A
Rowley, Durwood B
Rustigian, Robert
Sarkar, Nilima
Schaechter, Moselio
Schaffer, Priscilla Ann
Sharp, Phillip Allen
Silverman, Gerald
Sinskey, Anthony J
Sonenshein, Abraham Lincoln
Struhl, Kevin
Suit, Joan C
Sussman, Raquel Rotman
Szostak, Jack William
Thayer, Philip Standish
Thorne, Curtis Blaine
Tipper, Donald John
Tyler, Kenneth Laurence
Tyrrell, Elizabeth Ann
Walker, Robert W
Wallace, David H
Wilder, Martin Stuart
Williams, Ray Clayton
Wirsen, Carl O, Jr
Wun, Chun Kwun
Zetter, Bruce Robert

MICHIGAN
Adler, Richard
Anderson, Richard Lee
Bach, John Alfred
Baker, John Christopher
Baublis, Joseph V
Beardmore, William Boone
Bender, Robert Algerd
Beneke, Everett Smith
Benham, Ross Stephen
Berk, Richard Samuel
Berlin, Byron Sanford

Boros, Dov Lewis
Brawn, Mary Karen
Breznak, John Allen
Britt, Eugene Maurice
Brockman, Ellis R
Brockman, William Warner
Brown, William John
Brubaker, Robert Robinson
Burgoyne, George Harvey
Calvert, Jay Gregory
Carrick, Lee
Chandra, Purna
Chang, Timothy Scott
Chatterjee, Bandana
Chaudhry, G Rasul
Clewell, Don Bert
Coburn, Joel Thomas
Cochran, Kenneth William
Cohen, Michael Alan
Cole, George Christopher
Conder, George Anthony
Cooper, Stephen
Corner, Thomas Richard
Cotner, James Bryan, Jr
Cunningham, Charles Henry
Dazzo, Frank Bryan
Denison, Frank Willis, Jr
Diglio, Clement Anthony
Domagala, John Michael
Dore-Duffy, Paula
Eisenberg, Robert C
Erlandson, Arvid Leonard
Fadly, Aly Mahmoud
Fisher, Linda E
Fitzgerald, Robert Hannon, Jr
Fogel, Robert Dale
Friedman, Stephen Burt
Fulbright, Dennis Wayne
Garvin, Donald Frank
Gerhardt, Philipp
Grady, Joseph Edward
Haines, Richard Francis
Haines, William C
Hanka, Ladislav James
Hari, V
Harmon, Laurence George
Heberlein, Gary T
Heifetz, Carl Louis
Heppner, Gloria Hill
Hoeksema, Walter David
Holt, John Gilbert
Jackson, Matthew Paul
Jacobs, Charles Warren
Jay, James Monroe
Jeffries, Charles Dean
Jones, Garth Wicks
Jourdian, George William
Kauffman, Carol A
Ketchum, Paul Abbott
Klug, Michael J
Knight, Paul R
Kominek, Leo Aloysius
Kong, Yi-Chi Mei
Krabbenhoft, Kenneth Louis
Lampky, James Robert
Lefford, Maurice J
Levine, Myron
Levine, Seymour
Liu, Stephen C Y
Lomax, Margaret Irene
McCauley, Roy Barnard
McGroarty, Estelle Josephine
Mack, Walter Noel
Marshall, Vincent dePaul
Mason, Donald Joseph
Meyer, Edward Dell
Miller, Brinton Marshall
Miller, Harold Charles
Miller, Thomas Lee
Mitchell, John Richard
Molinari, John A
Montgomery, Paul Charles
Mostafapour, M Kazem
Mulks, Martha Huard
Murphy, William Henry, Jr
Mychajlonka, Myron
Narayan, Latha
Nazerian, Keyvan
Neidhardt, Frederick Carl
Olsen, Ronald H
Osborn, June Elaine
Parejko, Ronald Anthony
Patterson, Maria Jevitz
Peabody, Frank Robert
Pence, Leland Hadley
Peterson, Ward Davis, Jr
Podila, Gopi Krishna
Preiss, Jack
Puddington, Lynn
Punch, James Darrell
Pyke, Thomas Richard
Rauch, Helene Coben
Renis, Harold E
Reusser, Fritz
Righthand, Vera Fay
Robertson, G Philip
Robertson, John Harvey
Rogers, Alvin Lee
Rowe, Nathaniel H
Sadoff, Harold Lloyd
San Clemente, Charles Leonard
Sarkar, Fazlul Hoque
Savageau, Michael Antonio
Schuurmans, David Meinte

Schwartz, Stanley Allen
Sebek, Oldrich Karel
Serafini, Angela
Shelef, Leora Aya
Shipman, Charles, Jr
Smith, Eugene William
Smith, Jerry Warren
Sokolski, Walter Thomas
Steel, Robert
Stone, Howard Anderson
Trapp, Allan Laverne
Truden, Judith Lucille
Tsuji, Kiyoshi
Tung, Fred Fu
Uffen, Robert L
Walker, Gustav Adolphus
Walter, Richard Webb, Jr
Wang, Henry
Watts, Jeffrey Lynn
Wei, Wei-Zen
Weiner, Lawrence Myron
Wentworth, Berttina Brown
Whalen, Joseph Wilson
Whitehouse, Frank, Jr
Witz, Dennis Fredrick
Wolk, Coleman Peter
Yancey, Robert John, Jr
Yang, Gene Ching-Hua
Yokoyama, Melvin T
Zealey, Marion Edward
Zeikus, J Gregory

MINNESOTA
Anderson, Dwight Lyman
Bach, Marilyn Lee
Balfour, Henry H, Jr
Beggs, William H
Berube, Robert
Blazevic, Donna Jean
Burton, Alice Jean
Bushong, Jerold Ward
Busta, Francis Fredrick
Cleary, Paul Patrick
Crawford, Ronald Lyle
Dworkin, Martin
Faras, Anthony James
Fitzgerald, Thomas James
Fredrickson, Arnold G(erhard)
Gerding, Dale Nicholas
Gilboe, Daniel Pierre
Goldstein, Stuart Frederick
Grant, David James William
Haase, Ashley Thomson
Hanson, Richard Steven
Hapke, Bern
Henry, Ronald Alton
Hooper, Alan Bacon
Hu, Wei-Shou
Johnson, Russell Clarence
Johnson, Theodore Reynold
Karlson, Alfred Gustav
Katz, Morris Howard
Kerr, Sylvia Joann
Kline, Bruce Clayton
Loken, Keith I
Lyon, Richard Hale
McDuff, Charles Robert
McKay, Larry Lee
Magee, Paul Terry
Manning, Patrick James
Meyers, Paul
Miller, Richard Lynn
Nash, Peter
Nelson, Robert D
Oetting, William Starr
Oxborrow, Gordon Squires
Percich, James Angelo
Pflug, Irving John
Plagemann, Peter Guenter Wilhelm
Pomeroy, Benjamin Sherwood
Prince, James T
Ragsdale, David Willard
Reilly, Bernard Edward
Ritts, Roy Ellot, Jr
Roberts, Glenn Dale
Robertsen, John Alan
Rodgers, Nelson Earl
Rogers, Palmer, Jr
Rohlfing, Stephen Roy
Roon, Robert Jack
Sabath, Leon David
Schachtele, Charles Francis
Schrank, Gordon Dabney
Schuman, Leonard Michael
Sharma, Jagdev Mittra
Shope, Richard Edwin, Jr
Silverman, William Bernard
Sperber, William H
Stanley, Patricia Mary
Straw, Thomas Eugene
Tatini, Sita Ramayya
Thenen, Shirley Warnock
Tilman, G David
Tsien, Hsienchyang
Vallera, Daniel A
Vermilyea, Barry Lynn
Wagenaar, Raphael Omer
Watson, Dennis Wallace
Widin, Katharine Douglas
Ziegler, Richard James

MISSISSIPPI
Arceneaux, Joseph Lincoln
Bardsley, Charles Edward

Brown, Lewis Raymond
Brundage, William Gregory
Byers, Benjamin Rowe
Cook, David Wilson
Davis, Robert Gene
Downer, Donald Newson
Gafford, Lanelle Guyton
Garrett, Ephraim Spencer, III
Gentry, Glenn Aden
Grogan, James Bigbee
Lewis, Robert Edwin, Jr
Magee, Lyman Abbott
Mickelson, John Clair
Peterson, Harold LeRoy
Randall, Charles Chandler
Sadana, Ajit
Snazelle, Theodore Edward
Stojanovic, Borislav Jovan
Uzodinma, John E
Wahba, Albert J
Ward, Herbert Bailey
White, Charles Henry
Williamson, John S
Yarbrough, Karen Marguerite

MISSOURI
Adldinger, Hans Karl
Anderson, David W, Jr
Arens, Max Quirin
Bajpai, Rakesh Kumar
Barrett, James Thomas
Bell, Jeffrey
Blenden, Donald C
Bogosian, Gregg
Braciale, Thomas Joseph, Jr
Braciale, Vivian Lam
Branson, Dorothy Swingle
Brescia, Vincent Thomas
Brown, Karen Kay (Kilker)
Brown, Olen Ray
Burchard, Jeanette
DiFate, Victor George
Eissenberg, Joel Carter
Eltz, Robert Walter
Elvin-Lewis, Memory P F
Engley, Frank B, Jr
Fales, William Harold
Ferris, Deam Hunter
Fields, Marion Lee
Finkelstein, Richard Alan
Fischhoff, David Allen
Fleischman, Julian B
Folk, William Robert
Fraley, Robert Thomas
Franson, Raymond Lee
Gale, Nord Loran
Garrison, Robert Gene
Gledhill, William Emerson
Goldberg, Herbert Sam
Gossling, Jennifer
Granoff, Dan Martin
Green, Theodore James
Gustafson, Mark Edward
Hallas, Laurence Edward
Hamilton, Thomas Reid
Harrington, Glenn William
Harrison, Arthur Pennoyer, Jr
Hirschberg, Rona L
Hodges, Glenn R(oss)
Irgens, Roar L
Jacob, Gary Steven
Kane, James F
Kaplan, Arnold
Katz, Edward
Kishore, Ganesh M
Koplow, Jane
Kos, Edward Stanley
Laskowski, Leonard Francis, Jr
Lin, Hsiu-san
McCune, Emmett L
McIntosh, Arthur Herbert Cranstoun
Marshall, Robert T
Medoff, Gerald
Merlie, John Paul
Miles, Donald Orval
Misfeldt, Michael Lee
Moser, Stephen Adcock
Mueckler, Mike Max
Murray, Patrick Robert
Naumann, Hugh Donald
Nunez, William J, III
Olson, Gerald Allen
Olson, Lloyd Clarence
Parisi, Joseph Thomas
Pueppke, Steven Glenn
Rice, Charles Moen, III
Riddle, Donald Lee
Riggs, Hammond Greenwald, Jr
Rogolsky, Marvin
Rosenquist, Bruce David
Sargentini, Neil Joseph
Savage, George Roland
Scherr, David DeLano
Schlesinger, Milton J
Schlessinger, David
Schultz, Irwin
Schulze, Irene Theresa
Shieh, Kenneth Kuang-Zen
Solorzano, Robert Francis
Sonnenwirth, Alexander Coleman
Spurrier, Elmer R
Symington, Janey
Tritz, Gerald Joseph
Wang, Richard J

Weber, Morton M
Wenner, Herbert Allan
Winicov, Murray William
Yaverbaum, Sidney

MONTANA
Anacker, Robert Leroy
Bacon, Marion
Bond, Clifford Walter
Braune, Maximillian O
Combie, Joan D
Cox, Herald Rea
Cutler, Jimmy Edward
Faust, Richard Ahlvers
Firehammer, Burton Deforest
Fiscus, Alvin G
Garon, Claude Francis
Jutila, John W
Koostra, Walter L
McFeters, Gordon Alwyn
Munoz, John Joaquin
Nakamura, Mitsuru J
Nelson, Nels M
Newman, Franklin Scott
Parker, John C
Rudbach, Jon Anthony
Speer, Clarence Arvon
Stoenner, Herbert George
Swanson, John L
Taylor, John Jacob
Temple, Kenneth Loren
Thomas, Leo Alvon
Warren, Guylyn Rea

NEBRASKA
Baumstark, John Spann
Booth, Sheldon James
Brazis, A(dolph) Richard
Brown, Albert Loren
Bullerman, Lloyd Bernard
Chaperon, Edward Alfred
Davis, Eldon Vernon
Doran, John Walsh
Dubes, George Richard
Dyer, John Kaye
Ghosh, Chitta Ranjan
Ikenberry, Richard W
Kelling, Clayton Lynn
Kobayashi, Roger Hideo
Kolar, Joseph Robert, Jr
Lane, Leslie Carl
McFadden, Harry Webber, Jr
Miller, Norman Gustav
Morris, Thomas Jack
Nickerson, Kenneth Warwick
Ramaley, Robert Folk
Rhode, Solon Lafayette, III
Sanders, Christine Culp
Sanders, W Eugene, Jr
Schwartzbach, Steven Donald
Severin, Matthew Joseph
Torres-Medina, Alfonso
Underdahl, Norman Russell
Van Etten, James L
Veomett, George Ector
Wallen, Stanley Eugene
Weber, Allen Thomas
White, Roberta Jean

NEVADA
DiSalvo, Arthur F
Henry, Claudia
Kozel, Thomas Randall
Winicov, Ilga

NEW HAMPSHIRE
Adler, Louise Tale
Bent, Donald Frederick
Blakemore, Richard Peter
Blanchard, Robert Osborn
Brand, Karl Gerhard
Chesbro, William Ronald
Gray, Clarke Thomas
Green, William Robert
Hines, Mark Edward
Jones, Galen Everts
Lubin, Martin
Pfefferkorn, Elmer Roy, Jr
Pistole, Thomas Gordon
Rodgers, Frank Gerald
Smith, Edith Lucile
Smith, Mary Bunting
Tuttle, Robert Lewis
Zsigray, Robert Michael

NEW JERSEY
Adamo, Joseph Albert
Agathos, Spiros Nicholas
Anderson, David Wesley
Axelrod, David E
Bardell, David
Barkalow, Fern J
Barton, Beverly E
Baughn, Charles (Otto), Jr
Benson, Charles Everett
Bernstein, Eugene H
Bonner, Daniel Patrick
Bontempo, John A
Brandriss, Marjorie C
Buono, Frederick J
Burg, Richard William
Byrne, Barbara Jean McManamy
Byrne, Kevin M
Carman, George Martin

Microbiology (cont)

Carter, James Evan
Champe, Pamela Chambers
Charney, William
Choi, Ye-Chin
Cino, Paul Michael
Cook, Elizabeth Anne
Coscarelli, Waldimero
Daoust, Donald Roger
DePamphilis, Melvin Louis
Dhruv, Rohini Arvind
Dondershine, Frank Haskin
Dulaney, Eugene Lambert
Eagar, Robert Gouldman, Jr
Eaglesham, Allan Robert James
Eigen, Edward
Ellison, Solon Arthur
Eng, Robert H K
Ennis, Herbert Leo
Evans, Ralph H, Jr
Eveleigh, Douglas Edward
Feighner, Scott Dennis
Feldman, Lawrence A
Fernandes, Prabhavathi Bhat
Finstein, Melvin S
Flick, Christopher E
Frances, Saul
Gaffar, Abdul
Gale, George Osborne
Gately, Maurice Kent
Genetelli, Emil J
Ghosh, Arati
Ghosh, Bijan K
Gillum, Amanda McKee
Gilman, Steven Christopher
Girardi, Anthony Joseph
Goldemberg, Robert Lewis
Goldman, Emanuel
Green, Erika Ana
Greenstein, Teddy
Gross, Peter A
Gullo, Vincent Philip
Heck, James Virgil
Hendlin, David
Henry, Sydney Mark
Hirsch, Robert L
Hochstadt, Joy
Humayun, Mir Z
Jacks, Thomas Mauro
Jakubowski, Hieronim Zbigniew
Jansons, Vilma Karina
Johnson, Layne Mark
Jones, Benjamin Lewis
Kaminski, Zigmund Charles
Keller, John Randall
Kirchoff, William F
Kirsch, Donald R
Klosek, Richard C
Koepp, Leila H
Kraskin, Kenneth Stanford
Kuchler, Robert Joseph
Kuehn, Harold Herman
Kurtz, Myra Berman
Lampen, J Oliver
Lasfargues, Etienne Yves
Laskin, Allen I
Lechevalier, Hubert Arthur
Lechevalier, Mary P
Lenard, John
Letizia, Gabriel Joseph
Leung, Albert Yuk-Sing
Levin, Joseph David
Leyson, Jose Florante Justiniane
Ling, Hubert
Louria, Donald Bruce
McAnelly, John Kitchel
McDaniel, Lloyd Everett
McGarrity, Gerard John
Martin, James Franklin
Masurekar, Prakash Sharatchandra
Matthijssen, Charles
Mayernik, John Joseph
Megna, John C(osimo)
Meyers, Edward
Middleton, Richard B
Midlige, Frederick Horstmann, Jr
Miraglia, Gennaro J
Montville, Thomas Joseph
Neri, Anthony
Ofengand, Edward James
Orsi, Ernest Vinicio
Osborne, Frank Harold
OSullivan, Joseph
Ozer, Harvey Leon
Padhi, Sally Bulpitt
Parmegiani, Raulo
Perritt, Alexander M
Pestka, Sidney
Pianotti, Roland Salvatore
Pobiner, Bonnie Fay
Pramer, David
Prince, Herbert N
Raska, Karel Frantisek, Jr
Reilly, Hilda Christine
Rosen, Marvin
Rosen, William Edward
Rowin, Gerald L
Rupp, Frank Adolph
Sall, Theodore
Schaffner, Carl Paul
Scheiner, Donald M
Schlesinger, R(obert) Walter
Schulman, Marvin David

Schwalb, Marvin N
Sehgal, Surendra N
Seitz, Eugene W
Sekula, Bernard Charles
Shaw, Eugene
Shearer, Marcia Cathrine (Epple)
Shovlin, Francis Edward
Sikder, Santosh K
Simpson, Robert Wayne
Sipos, Tibor
Sjolander, Newell Oscar
Skalka, Anna Marie
Sohler, Arthur
Solberg, Myron
Stapley, Edward Olley
Sterbenz, Francis Joseph
Stockton, John Richard
Stollar, Victor
Stoudt, Thomas Henry
Strohl, William Allen
Tate, Robert Lee, III
Taylor, Bernard Franklin
Tenoso, Harold John
Tierno, Philip M, Jr
Tkacz, Jan S
Trivedi, Nayan B
Unowsky, Joel
Vallese, Frank M
Voos, Jane Rhein
Walton, Robert Bruce
Wang, Bosco Shang
Wang-Iverson, Patsy
Wasserman, Bruce P
Watkins, Paul Donald
Watrel, Warren George
Watson, Richard White, Jr
Wellerson, Ralph, Jr
Werth, Jean Marie
Winters, Harvey
Woehler, Michael Edward
Woodruff, Harold Boyd
Zimmerman, Sheldon Bernard

NEW MEXICO
Allison, David Coulter
Baca, Oswald Gilbert
Baker, Thomas Irving
Barrera, Cecilio Richard
Barton, Larry Lumir
Botsford, James L
Cords, Carl Ernest, Jr
Davis, Larry Ernest
Gregg, Charles Thornton
Kraemer, Paul Michael
Liddell, Craig Mason
McCarthy, Charlotte Marie
McLaren, Leroy Clarence
Martignoni, Mauro Emilio
Matchett, William H
Nelson, Mary Anne
O'Brien, Robert Thomas
Radloff, Roger James
Rypka, Eugene Weston
Sivinski, Jacek Stefan
Taylor, Robert Gay
Tokuda, Sei

NEW YORK
Abdelnoor, Alexander Michael
Ablin, Richard J
Abramowicz, Daniel Albert
Ajl, Samuel Jacob
Albrecht, Alberta Marie
Alcamo, I Edward
Amemiya, Kei
Amsterdam, Daniel
Andersen, Jon Alan
Andersen, Kenneth J
Anderson, Carl William
Anderson, Lucia Lewis
Appel, Max J
Arden, Sheldon Bruce
Atkinson, Paul H
Babich, Harvey
Bablanian, Rostom
Balduzzi, Piero
Barany, Francis
Bard, Enzo
Barksdale, Lane W
Bassett, Emmett W
Battley, Edwin Hall
Batzing, Barry Lewis
Bazinet, George Frederick
Beach, David H
Beardsley, Robert Eugene
Bedard, Donna Lee
Belly, Robert T
Benjaminson, Morris Aaron
Bergstrom, Gary Carlton
Bernheimer, Alan Weyl
Bernheimer, Harriet P
Berns, Kenneth
Bielat, Kenneth L
Biondo, Frank X
Birken, Steven
Blank, Robert H
Bopp, Lawrence Howard
Bottone, Edward Joseph
Bowen, William H
Boylen, Charles William
Breese, Sydney Salisbury, Jr
Brody, Marcia
Brown, Eric Reeder
Bruner, Dorsey William

Bukhari, Ahmad Iqbal
Bukovsan, Laura A
Burger, Richard Melton
Calhoun, David H
Calnek, Bruce Wixson
Cano, Francis Robert
Carmichael, Leland E
Carp, Richard Irvin
Carter, Timothy Howard
Casals, Jordi
Castello, John Donald
Catterall, James F
Celis, Roberto T F
Cerini, Costantino Peter
Chadha, Kailash Chandra
Chase, Merrill Wallace
Chen, Ching-Ling Chu
Chen, Priscilla B(urtis)
Chiao, Jen Wei
Chisholm, David R
Chiulli, Angelo Joseph
Choman, Bohdan Russell
Christensen, James Roger
Cirillo, Vincent Paul
Clark, Virginia Lee
Clesceri, Lenore Stanke
Cohen, Elias
Cole, Jonathan Jay
Collins, Arlene Rycombel
Collins, Carol Desormeau
Collins, Frank Miles
Conway de Macario, Everly
Cooper, Norman S
Corpe, William Albert
Cowell, James Leo
Cox, Dudley
Craig, John Philip
Cramer, Eva Brown
Crook, Philip George
Cunningham, Richard Preston
Danielson, Irvin Sigwald
Dean, Jack Hugh
Deibel, Rudolf
DeLuca, Patrick John
DiGaudio, Mary Rose
DiLiello, Leo Ralph
Dondero, Norman Carl
Dougherty, Robert Malvin
Eaton, Norman Ray
Efthymiou, Constantine John
Ehrlich, Henry Lutz
Elander, Richard Paul
Ellner, Paul Daniel
Elsbach, Peter
Eudy, William Wayne
Evans, Mary Jo
Evans, Richard Todd
Fabricant, Catherine G
Ferguson, John Barclay
Fewkes, Robert Charles Joseph
Finn, R(obert) K(aul)
Fisher, Clark Alan
Flanagan, Thomas Donald
Forgacs, Joseph
Fox, Sally Ingersoll
Francis, Arokiasamy Joseph
Free, Stephen J
Frickey, Paul Henry
Friedman, Selwyn Marvin
Froelich, Ernest
Furmanski, Philip
Garay, Gustav John
Gehrig, Robert Frank
Gelbard, Alan Stewart
George, Elmer, Jr
Gibbs, David Lee
Gibson, Audrey Jane
Gilardi, Gerald Leland
Gillespie, James Howard
Gilmour, Marion Nyholm H
Ginsberg, Harold Samuel
Godfrey, H(enry) P(hilip)
Goff, Stephen Payne
Goldberg, Allan Roy
Goldschmidt, Raul Max
Gonsalves, Dennis
Goodhue, Charles Thomas
Gordon, Morris Aaron
Gorzynski, Eugene Arthur
Graham, Donald C W
Griffin, David H
Haas, Gerhard Julius
Hadley, Susan Jane
Hagar, Silas Stanley
Hall, William Myron, Jr
Halstead, Scott Barker
Hammill, Terrence Michael
Hare, John Donald
Harman, Gary Elvan
Hawkins, Linda Louise
Hechemy, Karim E
Hehre, Edward James
Held, Abraham Albert
Hipp, Sally Sloan
Ho, Hon Hing
Hoffman, Heiner
Hoffmann, Michael K
Horwitz, Marshall Sydney
Hotchin, John Elton
Hsu, Ming-Ta
Huang, Alice Shih-Hou
Humber, Richard Alan
Hunt, George Albert
Hurwitz, Jerard

Hutchison, Dorris Jeannette
Hutner, Seymour Herbert
Iglewski, Barbara Hotham
Iglewski, Wallace
Isaacs, Charles Edward
Isenberg, Henry David
Isseroff, Hadar
Jahiel, Rene
Jarolmen, Howard
Juo, Pei-Show
Kalsow, Carolyn Marie
Katz, Michael
Keithly, Janet Sue
Kele, Roger Alan
Keller, Dolores Elaine
Khan, Nasim A
Khan, Paul
Kilbourne, Edwin Dennis
Kim, Charles Wesley
Kim, Kwang Shin
Klein, Elena Buimovici
Klein, Richard Joseph
Korf, Richard Paul
Kowolenko, Michael D
Kraft, William Gerald
Kratzel, Robert Jeffrey
Lacks, Sanford
Laffin, Robert James
Lahita, Robert George
Lamberson, Harold Vincent, Jr
Landsberger, Frank Robbert
Lasley, Betty Jean
Lazaroff, Norman
Ledford, Richard Allison
Lee, John Joseph
Leff, Judith
Lehman, John Michael
Leifer, Zer
Levandowsky, Michael
Lewis, Leslie Arthur
Liguori, Vincent Robert
Lilly, Frank
Lindahl, Lasse Allan
Linder, Regina
Lindsay, Harry Lee
Linke, Harald Arthur Bruno
Linkins, Arthur Edward
Linzer, Rosemary
Lipke, Peter Nathan
Lipson, Edward David
Lipson, Steven Mark
Loeb, Marilyn Rosenthal
Lossinsky, Albert S
Loughlin, James Francis
Lowe, Josiah L(incoln)
Lusty, Carol Jean
McAllister, William T
Macario, Alberto J L
McCormack, Grace
McCormick, J Robert D
McLaughlin, John J A(nthoany)
McNamara, Thomas Francis
McSharry, James John
Maestrone, Gianpaolo
Magill, Thomas Pleines
Mahoney, Robert Patrick
Maio, Joseph James
Maniloff, Jack
Manly, Kenneth Fred
Mannino, Raphael James
Manski, Wladyslaw J
Maramorosch, Karl
Marcu, Kenneth Brian
Margolin, Paul
Marquis, Robert E
Mashimo, Paul Akira
Maynard, Charles Alvin
Mehta, Bipin Mohanlal
Merrick, Joseph M
Meyer, Haruko
Michaud, Ronald Normand
Milgrom, Felix
Miller, Terry Lynn
Millian, Stephen Jerry
Mindich, Leonard Eugene
Misiek, Martin
Morin, John Edward
Morse, Stephen Scott
Mortlock, Robert Paul
Mundorff, Sheila Ann
Munk, Vladimir
Murphy, James Slater
Murphy, Timothy F
Muschio, Henry M, Jr
Nachbar, Martin Stephen
Napolitano, Raymond L
Neimark, Harold Carl
Nellis, Lois Fonda
Neu, Harold Conrad
Neurath, Alexander Robert
Noronha, Fernando M Oliveira
North, Robert J
Nuzzi, Robert
Oates, Richard Patrick
O'Brien, James Francis
Parham, Margaret Payne
Patterson, Ernest Leonard
Pavlova, Maria T
Peterson, Robert H F
Pfau, Charles Julius
Phillips, Arthur William, Jr
Pierce, Carol S
Pisano, Michael A
Pitt, Jane

Potter, Norman N
Powell, Sharon Kay
Price, Peter Michael
Prince, Alfred M
Pruslin, Fred Howard
Puttlitz, Donald Herbert
Quigley, James P
Reddy, Kalluru Jayarami
Reilly, Marguerite
Riha, William E, Jr
Rizzuto, Anthony B
Robinson, Alix Ida
Roeder, Robert Gayle
Rogerson, Allen Collingwood
Rosenfeld, Martin Herbert
Rosi, David
Rossman, Toby Gale
Roth, Frieda
Rotheim, Minna B
Russel, Marjorie Ellen
Salton, Milton Robert James
Santoro, Thomas
Sawyer, William D
Schaeffer, James Robert
Schauf, Victoria
Scher, William
Schloer, Gertrude M
Schuster, Frederick Lee
Scott, Fredric Winthrop
Scranton, Mary Isabelle
Sechrist, Lynne Luan
Sehgal, Pravinkumar B
Senitzer, David
Senterfit, Laurence Benfred
Shapiro, Caren Knight
Sharp, William R
Shayegani, Mehdi
Sheffy, Ben Edward
Shuler, Michael Louis
Silva-Hutner, Margarita
Silverstein, Samuel Charles
Siminoff, Paul
Simon, Robert David
Sinha, Asru Kumar
Slepecky, Ralph Andrew
Smart, Kathryn Marilyn
Smith, James Eldon
Spencer, Frank
Stamer, John Richard
Stanley, Pamela Mary
Stark, Egon
Starr, Theodore Jack
Steinberg, Bernard Albert
Steinberg, Bettie Murray
Steinkraus, Keith Hartley
Stevens, Roy Harris
Stevens, Roy White
Stotzky, Guenther
Stryker, Martin H
Su, Tah-Mun
Sudds, Richard Huyette, Jr
Sultzer, Barnet Martin
Surgalla, Michael Joseph
Swaney, Lois Mae
Taber, Harry Warren
Tamm, Igor
Tanenbaum, Stuart William
Tanzer, Charles
Taschdjian, Claire Louise
Tendler, Moses David
Testa, Raymond Thomas
Thacore, Harshad Rai
Toenniessen, Gary Herbert
Trowbridge, Richard Stuart
Udem, Stephen Alexander
Uzgiris, Egidijus E
Van Tassell, Morgan Howard
Vilcek, Jan Tomas
Vogel, Henry
Wassermann, Felix Emil
Weiner, Matei
Werber, Erna Alture
Wetmur, James Gerard
White, James Carrick
White, James Patrick
White, Richard John
Wiberg, John Samuel
Wicher, Konrad J
Wilhelm, James Maurice
Williams, Curtis Alvin, Jr
Wimmer, Eckard
Wing, Edward Joseph
Winter, Jeanette E
Wolff, John Shearer, III
Wolfson, Leonard Louis
Wolin, Meyer Jerome
Wulff, Daniel Lewis
Young, Charles Stuart Hamish
Zabel, Robert Alger
Zahler, Stanley Arnold
Zeigel, Robert Francis
Zero, Domenick Thomas
Zinder, Stephen Henry
Zolla-Pazner, Susan Beth
Zwarun, Andrew Alexander

NORTH CAROLINA
Allen, Wendall E
Arnold, Luther Bishop, Jr
Bachenheimer, Steven Larry
Banes, Albert Joseph
Barakat, Hisham A
Barker, James Cathey
Bates, William K

Bishop, Paul Edward
Bolognesi, Dani Paul
Boone, Lawrence Rudolph, III
Bott, Kenneth F
Brewer, Carl Robert
Brown, Talmage Thurman, Jr
Burchall, James J
Buttke, Thomas Martin
Carter, Philip Brian
Chapman, John Franklin, Jr
Cheng, Yung-Chi
Claxton, Larry Davis
Clinton, Bruce Allan
Coggins, Leroy
Collins, Jeffrey Jay
Colwell, William Maxwell
Connolly, Kevin Michael
Corley, Ronald Bruce
Craig, Syndey Pollock, III
Crawford, James Joseph L
Cromartie, William James
Dahiya, Raghunath S
Devereux, Theodora Reyling
Dobrogosz, Walter Jerome
Douros, John Drenkle
Drake, John W
Drexler, Henry
Durack, David Tulloch
Estevez, Enrique Gonzalo
Ferone, Robert
Fleming, Henry Pridgen
Folds, James Donald
Frisell, Wilhelm Richard
Fyfe, James Arthur
Gardner, Donald Eugene
Gilfillan, Robert Frederick
Golden, Carole Ann
Gonder, Eric Charles
Gooder, Harry
Hamilton, Pat Brooks
Harper, James Douglas
Hassan, Hosni Moustafa
Hill, Gale Bartholomew
Hirschberg, Nell
Hopfer, Roy L
Huang, Eng-Shang(Clark)
Hutt, Randy
Jacobson, Kenneth Allan
Jeffreys, Donald Bearss
Johnston, Harry Henry
Johnston, Robert Edward
Joklik, Wolfgang Karl
Katz, Samuel Lawrence
Keene, Jack Donald
Klapper, David G
Klein, Dolph
Knuckles, Joseph Lewis
Kucera, Louis Stephen
Lecce, James Giacomo
Luginbuhl, Geraldine Hobson
McNeill, John J
McReynolds, Richard A
Malling, Heinrich Valdemar
Manire, George Philip
Mitchell, Thomas Greenfield
Mizel, Steven B
Moody, Max Dale
Morrison, Ralph M
Muller, Uwe Richard
Newman, Simon Louis
Noga, Edward Joseph
Novitsky, James Alan
Oblinger, James Leslie
Osterhout, Suydam
Pagano, Joseph Stephen
Parks, Leo Wilburn
Perry, Jerome John
Pfaender, Frederic Karl
Phelps, Allen Warner
Phibbs, Paul Vester, Jr
Price, Kenneth Elbert
Pukkila, Patricia Jean
Richardson, Stephen H
Roop, Richard Allan
Rublee, Parke Alstan
Sandok, Paul Louis
Seed, John Richard
Shafer, Steven Ray
Shepard, Maurice Charles
Shih, Jason Chia-Hsing
Sizemore, Ronald Kelly
Sobsey, Mark David
Speck, Marvin Luther
Stone, Henry Otto, Jr
Tennant, Raymond Wallace
Turner, Alvis Greely
Van De Rijn, Ivo
Wachsman, Joseph T
Walkinshaw, Charles Howard, Jr
Ward, John Everett, Jr
Webb, Neil Broyles
Webster, Robert Edward
Wheat, Robert Wayne
Willett, Hilda Pope
Wilson, Harold Albert
Wilson, James Franklin
Wise, Ernest George
Witt, Donald James
Wolberg, Gerald
Wollum, Arthur George, II
Zeiger, Errol
Zwadyk, Peter, Jr

NORTH DAKOTA
Duerre, John A
Faust, Maria Anna
Fillipi, Gordon Michael
Fischer, Robert George
Kelleher, James Joseph
Marwin, Richard Martin
Shubert, L Elliot
Vennes, John Wesley
Waller, James R

OHIO
Astrachan, Lazarus
Atkins, Charles Gilmore
Baker, Dwight Dee
Bannan, Elmer Alexander
Banwart, George J
Baroudy, Bahige M
Battisto, Jack Richard
Berg, Gerald
Berman, Donald
Bertram, Timothy Allyn
Blumenthal, Robert Martin
Bonventre, Peter Frank
Bowman, Bernard Ulysses, Jr
Boyer, Jere Michael
Boyle, Michael Dermot
Brady, Robert James
Brent, Morgan McKenzie
Briner, William Watson
Brueske, Charles H
Bubel, Hans Curt
Burnham, Jeffrey C
Camiener, Gerald Walter
Carmichael, Wayne William
Chapple, Paul James
Chorpenning, Frank Winslow
Cornelius, Billy Dean
Cox, Charles Donald
Cox, Donald Cody
Cramblett, Henry G
Croft, Charles Clayton
Daniel, Thomas Mallon
Darrow, Robert A
De Fiebre, Conrad William
Dehority, Burk Allyn
Deters, Donald W
Dewar, Norman Ellison
Docherty, John Joseph
Evans, Helen Harrington
Federle, Thomas Walter
Flaumenhaft, Eugene
Frea, James Irving
Freimer, Earl Howard
Geldreich, Edwin E(mery)
Glaser, Ronald
Goldstein, Gerald
Graves, Anne Carol Finger
Hageage, George John, Jr
Hamparian, Vincent
Haynes, Ralph Edwards
Heuschele, Werner Paul
Hoff, John C
Hogg, Robert W
Holder, Ian Alan
Igel, Howard Joseph
Irmiter, Theodore Ferer
Jacobsen, Donald Weldon
Jalil, Mazhar
Jamasbi, Roudabeh J
Janusz, Michael John
Judge, Leo Francis, Jr
Kapral, Frank Albert
Kearns, Robert J
Kiggins, Edward M
Kirkland, Jerry J
Kohler, Erwin Miller
Kolodziej, Bruno J
Krampitz, Lester Orville
Kreier, Julius Peter
Krueger, Robert George
Krueger, Russell Francis
Kunin, Calvin Murry
Larkin, Edward P
Ledinko, Nada
Lichstein, Herman Carlton
Lingrel, Jerry B
Litchfield, John Hyland
Loper, John C
McFarland, Charles R
Maier, Siegfried
Mallin, Morton Lewis
Martin, James Harold
Martin, Scott McClung
Mayer, Gerald Douglas
Meredith, William Edward
Merola, A John
Merritt, Katharine
Meyer, Ralph Roger
Michael, William R
Mikolajcik, Emil Michael
Milo, George Edward
Modrzakowski, Malcolm Charles
Morris, Randal Edward
Morris Hooke, Anne
Muckerheide, Annette
Mukkada, Antony Job
Nankervis, George Arthur
Nunn, Dorothy Mae
Oakley, Berl Ray
Ockerman, Herbert W
Olsen, Richard George
Ottolenghi, Abramo Cesare
Pfister, Robert M

Pollack, J Dennis
Proffitt, Max Rowland
Rabin, Erwin R
Randles, Chester
Rawson, James Rulon Young
Rheins, Melvin S
Rhodes, Judith Carol
Rosen, Samuel
Rowe, John James
Safferman, Robert S
Saif, Linda Jean
Saif, Yehia M(ohamed)
Sawicki, Stanley George
Scarpino, Pasquale Valentine
Schiff, Gilbert Martin
Sheridan, John Francis
Sich, Jeffrey John
Smith, Kenneth Larry
Smith, Roger Dean
Somerson, Norman L
Staker, Robert D
Stanberry, Lawrence Raymond
Stephens, James Fred
Stevenson, John Ray
Stoner, Gary David
Stotts, Jane
Stukus, Philip Eugene
Sunkara, Sai Prasad
Tabita, F Robert
Thomas, Donald Charles
Treick, Ronald Walter
Trela, John Michael
Trewyn, Ronald William
Troller, John Arthur
Tuovinen, Olli Heikki
Turner, Thomas Bourne
Vestal, J Robie
Ward, Richard Leo
Washington, John A, II
Weber, George Russell
Wehrle, Louis, Jr
Weis, Dale Stern
Wickstrom, Conrad Eugene
Wiginton, Dan Allen
Wilcox, Joseph Clifford
Williams, Marshall Vance
Williamson, Clarence Kelly
Wise, Donald L
Wood, Harland G
Woodworth, Mary Esther
Yang, Tsanyen
Yohn, David Stewart

OKLAHOMA
Blackstock, Rebecca
Bryant, Rebecca Smith
Cain, William Aaron
Clark, James Bennett
Confer, Anthony Wayne
Conway, Kenneth Edward
Cozad, George Carmon
Cunningham, Madeleine White
Ferretti, Joseph Jerome
Flournoy, Dayl Jean
Fulton, Robert Wesley
Gee, Lynn LaMarr
Gilliland, Stanley Eugene
Graves, Donald C
Gumbreck, Laurence Gable
Hall, Leo Terry
Harmon, H James
Hatten, Betty Arlene
Hitzman, Donald Oliver
Hurley, James Edgar
Hyde, Richard Moorehead
Lancaster, John
Lerner, Michael Paul
Leu, Richard William
McCallum, Roderick Eugene
Messmer, Dennis A
Murphy, Juneann Wadsworth
Nordquist, Robert Ersel
Qadri, Syed M Hussain
Rhoades, Everett Ronald
Richardson, Lavon Preston
Sanborn, Mark Robert
Schaefer, Frederick Vail
Schindler, Charles Alvin
Schubert, Karel Ralph
Scott, Lawrence Vernon
Seideman, Walter E
Streebin, Leale E
Veltri, Robert William
Wegner, Gene H
Wolgamott, Gary
Woolsey, Marion Elmer
Yu, Linda

OREGON
Barry, Arthur Leland
Barry, Ronald A
Castenholz, Richard William
Clark, James Orie, II
Clinton, Gail M
Deeney, Anne O'Connell
Dietz, Thomas John
Elliker, Paul R
Elliott, Lloyd Floren
Florance, Edwin R
Fryer, John Louis
Gerke, John Royal
Hallum, Jules Verne
Holm, Harvey William
Kilbourn, Joan Priscilla Payne

Microbiology (cont)

Kim, Kenneth
Knittel, Martin Dean
Kuhnley, Lyle Carlton
Leong, Jo-Ann Ching
McConnaughey, Bayard Harlow
Mattson, Donald Eugene
Millette, Robert Loomis
Miner, J Ronald
Molina, Randolph John
Morgan, Bruce Henry
Morita, Richard Yukio
Morita, Toshiko N
Myrold, David Douglas
Nelson, Earl Edward
Oginsky, Evelyn Lenore
Pearson, George Denton
Rittenberg, Marvin Barry
Russell, Peter James
Sandine, William Ewald
Schaeffer, Morris
Seidler, Ramon John
Sherr, Barry Frederick
Sherr, Evelyn Brown
Siegel, Benjamin Vincent
Sistrom, William R
Starr, Patricia Rae
Taylor, Mary Lowell Branson
Taylor, Walter Herman, Jr
Ugarte, Eduardo
Van Dyke, Henry
Weaver, William Judson
Wedman, Elwood Edward
Wehr, Herbert Michael
Woodburn, Margy Jeanette

PENNSYLVANIA

Actor, Paul
Alexander, Aaron D
Alexander, James King
Anderson, Thomas Foxen
Atchison, Robert Wayne
Axler, David Allan
Baechler, Charles Albert
Bailey, Denis Mahlon
Baum, Robert Harold
Bayer, Manfred Erich
Bayer, Margret Helene Janssen
Bean, Barry
Beilstein, Henry Richard
Beining, Paul R
Benedict, Robert Curtis
Bering, Charles Lawrence
Bernlohr, Robert William
Bernstein, Alan
Bhaduri, Saumya
Bondi, Amedeo
Bott, Thomas Lee
Boutros, Susan Noblit
Brenchley, Jean Elnora
Brinton, Charles Chester, Jr
Brown, William E
Buchanan, Robert Lester
Caligiuri, Lawrence Anthony
Callahan, Hugh James
Casida, Lester Earl, Jr
Castric, Peter Allen
Cohen, Gary H
Cohen, Pinya
Coleman, Charles Mosby
Cronholm, Lois S
Crowell, Richard Lane
Cundy, Kenneth Raymond
Daneo-Moore, Lolita
Davies, Helen Jean Conrad
Davies, Warren Lewis
Debroy, Chitrita
Deering, Reginald
Deforest, Adamadia
DeMeio, Joseph Louis
Diamond, Leila
Dickerson, William H
Di Cuollo, C John
Dryden, Richard Lee
Dudzinski, Diane Marie
Eberhart, Robert J
Eisenstein, Toby K
Elliott, Arthur York
Fansler, Bradford S
Farber, Florence Eileen
Farber, Paul Alan
Feinstein, Sheldon Israel
Feit, Ira (Nathan)
Fletcher, Ronald D
Forbes, Martin
Fraser, Nigel William
Fuscaldo, Anthony Alfred
Gadebusch, Hans Henning
Gammon, Richard Anthony
Garfinkle, Barry David
Gealt, Michael Alan
Gerhard, Walter Ulrich
Glorioso, Joseph Charles, III
Godfrey, Susan Sturgis
Goldfine, Howard
Goldner, Herman
Gots, Joseph Simon
Gray, Alan
Gregory, Francis Joseph
Halleck, Frank Eugene
Hammel, Jay Morris
Hammond, Benjamin Franklin
Hankins, William Alfred

Harding, Roy Woodrow, Jr
Harmon, George Andrew
Harris, Susanna
Havas, Helga Francis
Henderson, Earl Erwin
Hendrix, Sherman Samuel
Henle, Gertrude
Higgins, Michael Lee
Hilleman, Maurice Ralph
Ho, Monto
Hobby, Gladys Lounsbury
Hoffee, Patricia Anne
Hummeler, Klaus
Hung, Paul P
Iralu, Vichazelhu
Isom, Harriet C
Jackson, Ethel Noland
Jacobson, Lewis A
Jarvik, Jonathan Wallace
Johnston, James Bennett
Joseph, Jeymohan
Jungkind, Donald Lee
Keenan, John Douglas
Kennedy, Harvey Edward
Khoury, George
Kirk, Billy Edward
Kissinger, John Calvin
Klein, Morton
Klens, Paul Frank
Kneebone, Leon Russell
Knobler, Robert Leonard
Koesterer, Martin George
Koprowski, Hilary
Kozak, Marilyn Sue
Krah, David Lee
Krawiec, Steven Stack
Kreider, John Wesley
Kuserk, Frank Thomas
Landau, Burton Joseph
Larson, Vivian M
Lashen, Edward S
Lawrence, William Chase
Layne, Porter Preston
Lee, Chin-Chiu
Lee, John Cheung Han
Liegey, Francis William
Live, Israel
Lobo, Francis X
Logan, David Alexander
Long, Carole Ann
Long, Walter Kyle, Jr
McAlack, Robert Francis
McCarthy, Frank John
McCarthy, William John
McCoy, John Philip, Jr
McKinstry, Donald Michael
McNamara, Pamela Dee
Magee, Wayne Edward
Majumdar, Shyamal K
Malamud, Daniel F
Mandel, John Herbert
Manson, Lionel Arnold
Mast, Morris G
Mathur, Carolyn Frances
Melamed, Sidney
Miller, James Eugene
Milligan, Wilbert Harvey, III
Millman, Irving
Minsavage, Edward Joseph
Miovic, Margaret Lancefield
Mizutani, Satoshi
Moorman, Gary William
Morahan, Page Smith
Morton, Harry E
Nathanson, Neal
Natuk, Robert James
Nelson, Paul Edward
Neubauer, Russell Howard
Neubeck, Clifford Edward
Osborne, William Wesley
Pagano, Joseph Frank
Pakman, Leonard Marvin
Palumbo, Samuel Anthony
Patrick, Ruth (Mrs Charles Hodge IV)
Patton, William Henry
Paucker, Kurt
Pepper, Rollin E
Phillips, Bruce A
Platt, David
Plotkin, Stanley Alan
Pootjes, Christine Fredricka
Porter, Ronald Dean
Poste, George Henry
Poupard, James Arthur
Pratt, Elizabeth Ann
Provost, Philip Joseph
Rapp, Fred
Ray, Eva K
Reich, Claude Virgil
Reisner, Gerald Seymour
Resconich, Emil Carl
Rest, Richard Franklin
Ricciardi, Robert Paul
Roesing, Timothy George
Roia, Frank Costa, Jr
Romano, Paula Josephine
Roosa, Robert Andrew
Rosan, Burton
Rosenkranz, Herbert S
Rosenzweig, William David
Rozmiarek, Harry
Rubin, Benjamin Arnold
Rubin, Donald Howard

Ruggieri, Michael Raymond
Sagik, Bernard Phillip
Salerno, Ronald Anthony
Santer, Melvin
Schaedler, Russell William
Scher, Charles D
Schiff, Paul L, Jr
Sheffield, Joel Benson
Shockman, Gerald David
Sideropoulos, Aris S
Singh, Balwant
Six, Howard R
Slifkin, Malcolm
Slotnick, Victor Bernard
Smith, Harry Logan, Jr
Smith, James Lee
Smith, Josephine Reist
Snow, Jean Anthony
Somkuti, George A
Southam, Chester Milton
Stempen, Henry
Steplewski, Zenon
Stere, Athleen Jacobs
Strauss, Robert R
Suyama, Yoshitaka
Taubler, James H
Tax, Anne
Taylor, Mary Marshall
Tevethia, Satvir S
Thayer, Donald Wayne
Thimann, Kenneth Vivian
Thomas, William J
Tint, Howard
Unz, Richard F(rederick)
Vaidya, Akhil Babubhai
Vaughan, James Roland
Vessey, Adele Ruth
Warren, George Harry
Watson, Joseph Alexander
Weidanz, William P
Weinbaum, George
Wilcox, Wesley C
Willett, Norman P
Wolf, Benjamin
Wolfersberger, Michael Gregg
Yamamoto, Nobuto
Youngner, Julius Stuart
Yurchenco, John Alfonso
Zaika, Laura Larysa
Zimmerer, Robert P
Zubrzycki, Leonard Joseph

RHODE ISLAND

Carpenter, Charles C J
Crowley, James Patrick
Goos, Roger Delmon
Howard, Frank Leslie
Howe, Calderon
Jones, Kenneth Wayne
Lederberg, Seymour
Lee, Tung-Ching
Miller, Robert Harold
O'Leary, Gerard Paul, Jr
Prager, Jan Clement
Rosenstein, Barry Sheldon
Shank, Peter R
Sieburth, John McNeill
Swift, Dorothy Garrison
Worthen, Leonard Robert
Yates, Vance Joseph

SOUTH CAROLINA

Barnett, Ortus Webb, Jr
Baxter, Ann Webster
Brown, Arnold
Davis, Leroy
Davis, Raymond F
Ely, Berten E, III
Galbraith, Robert Michael
Ghaffar, Abdul
Grady, Cecil Paul Leslie, Jr
Gustafson, Ralph Alan
Hayasaka, Steven S
Hazen, Terry Clyde
Henson, Joseph Lawrence
Higerd, Thomas Braden
Paynter, Malcolm James Benjamin
Reddick-Mitchum, Rhoda Anne
Richards, Gary Paul
Schmidt, Roger Paul
Shively, Jessup MacLean
Sigel, M(ola) Michael
Stutzenberger, Fred John
Vereen, Larry Edwin
Virella, Gabriel T
Wilkinson, Thomas Ross
Yoch, Duane Charles

SOUTH DAKOTA

Hildreth, Michael B
Langworthy, Thomas Allan
Lynn, Raymond J
Pengra, Robert Monroe
Prescott, Lansing M
Smith, Paul Francis
Stoner, Warren Norton
Sword, Christopher Patrick
Westby, Carl A
Westfall, Helen Naomi

TENNESSEE

Avis, Kenneth Edward
Beachey, Edwin Henry
Bean, William J, Jr

Beck, Raymond Warren
Becker, Jeffrey Marvin
Belew-Noah, Patricia W
Ben-Porat, Tamar
Brown, Arthur
Bryant, Robert Emory
Chandler, Robert Walter
Chi, David Shyh-Wei
Colley, Daniel George
Cook, Robert James
Cooper, Terrance G
Cox, Edmond Rudolph, Jr
Cypess, Raymond Harold
Davidson, Philip Michael
Draughon, Frances Ann
Farrington, Joseph Kirby
Freeman, Bob A
Hadden, Charles Thomas
Harshman, Sidney
Herting, David Clair
Hollis, Cecil George
Hougland, Arthur Eldon
Howe, Martha Morgan
Howell, Elizabeth E
Incardona, Antonino L
Isa, Abdallah Mohammad
Johnson, Charles William
Kaplan, Albert Sydney
Karve, Mohan Dattatreya
Karzon, David T
Katze, Jon R
Larimer, Frank William
Leach, Eddie Dillon
Mayberry, William Roy
Mithcell, William Marvin
Montie, Thomas C
Niyogi, Salil Kumar
Ourth, Donald Dean
Portner, Allen
Postlethwaite, Arnold Eugene
Rinchik, Eugene M
Roberts, Audrey Nadine
Robinson, John Price
Rouse, Barry Tyrrell
Ryden, Fred Ward
Stevens, Audrey L
Stevens, Stanley Edward, Jr
Todd, William McClintock
Trentham, Jimmy N
Vredeveld, Nicholas Gene
Walker, William Stanley
Webster, Robert G
White, David Cleaveland
Wiesmeyer, Herbert
Wilson, Benjamin James

TEXAS

Albrecht, Thomas Blair
Allen, Lois Brenda
Alvarez-Gonzalez, Rafael
Ascenzi, Joseph Michael
Atkinson, Mark Arthur Leonard
Baer, Richard
Baine, William Brennan
Baptist, James (Noel)
Baron, Samuel
Baseman, Joel Barry
Baughn, Robert Elroy
Bawdon, Roger Everett
Bellion, Edward
Black, Samuel Harold
Blouse, Louis E, Jr
Bose, Henry Robert, Jr
Bowen, James Milton
Brand, Jerry Jay
Brown, Lee Roy, Jr
Browning, John Artie
Brysk, Miriam Mason
Buchanan, Christine Elizabeth
Butel, Janet Susan
Castro, Gilbert Anthony
Cate, Thomas Randolph
Chan, James C
Chan, Teh-Sheng
Chen, Young Chang
Chrzanowski, Thomas Henry
Clowes, Royston Courtenay
Cole, Garry Thomas
Collisson, Ellen Whited
Cooper, Ronda Fern
Crandell, Robert Allen
Crawford, Gladys P
Croley, Thomas Edgar
Davis, Charles Patrick
Davis, James Royce
Dibble, John Thomas
Donnelly, Patricia Vryling
Dorn, Gordon Lee
Dreesman, Gordon Ronald
Dubbs, Del Rose M
Dubose, Robert Trafton
Dwyer, Lawrence Arthur
Earhart, Charles Franklin, Jr
East, James Lindsay
Eichenwald, Heinz Felix
Eidels, Leon
Eklund, Curtis Einar
Eldridge, David Wyatt
Ellzey, Joanne Tontz
Ennever, John Joseph
Eugster, A Konrad
Evans, John Edward
Fleenor, Marvin Bension
Fleischmann, William Robert, Jr

Foster, Billy Glen
Foster, Terry Lynn
Fox, George Edward
French, Jeptha Victor
Friend, Patric Lee
Fuerst, Robert
Gardner, Earl William, Jr
Gardner, Florence Harrod
Gardner, Frederick Albert
Georgiou, George
Gilbert, Brian E
Goldschmidt, Millicent
Goodman, Joel Mitchell
Grimm, Elizabeth Ann
Guentzel, M Neal
Guthrie, Rufus Kent
Harris, Elizabeth Forsyth
Harris, Kerry Francis Patrick
Haye, Keith R
Heberling, Richard Leon
Heck, Fred Carl
Heggers, John Paul
Hejtmancik, Kelly Erwin
Henney, Henry Russell, Jr
Henry, Clay Allen
Hentges, David John
Hillis, William Daniel, Sr
Holguin, Alfonso Hudson
Hollis, John Percy, Jr
Huber, Thomas Wayne
Humphrey, Ronald DeVere
Humphrey, Ronald Mack
Jorgensen, George Norman
Joys, Terence Michael
Jurtshuk, Peter, Jr
Kalter, Seymour Sanford
Kasel, Julius Albert
Kaufmann, Anthony J
Keyser, Peter D
Krueger, Gerhard R F
Kumar, Vinay
Kurosky, Alexander
Kutteh, William Hanna
LaBrie, David Andre
Lacko, Andras Gyorgy
Lansford, Edwin Myers, Jr
LeBlanc, Donald Joseph
Lee, William Thomas
Lees, George Edward
Lefkowitz, Stanley S
Leibowitz, Julian Lazar
Leininger, Harold Vernon
Lieberman, Michael Merril
Loan, Raymond Wallace
McBride, Mollie Elizabeth
McConnell, Stewart
McCracken, Alexander Walker
McDonald, William Charles
Mace, Kenneth Dean
Maleckar, James R
Mandel, Manley
Mattingly, Stephen Joseph
Mayberry, Lillian Faye
Mayor, Heather Donald
Measel, John William
Mehta, Rajen
Melnick, Joseph Louis
Miget, Russell John
Moyer, Mary Pat Sutter
Moyer, Rex Carlton
Murthy, Krishna K
Nash, Donald Robert
Nicholson, Wayne Lowell
O'Donovan, Gerard Anthony
Ohlenbusch, Robert Eugene
Olson, Leroy Justin
Oujesky, Helen Matusevich
Paque, Ronald E
Perez, John Carlos
Peterson, Johnny Wayne
Phillips, Guy Frank
Powell, Bernard Lawrence
Prasad, Rupi
Prashad, Nagindra
Quarles, John Monroe
Rael, Eppie David
Reeves, James Blanchette
Rich, Susan Marie Solliday
Riehl, Robert Michael
Riggs, Stuart
Ritzi, Earl Michael
Rolfe, Rial Dewitt
Rolston, Kenneth Vijaykumar Issac
Romeo, Tony
Rosenblum, Eugene David
Rudolph, Frederick Byron
Safe, Stephen Harvey
Sanford, Barbara Ann
Schlech, Barry Arthur
Schmidt, Jerome P
Schneider, Dennis Ray
Schwarz, John Robert
Seman, Gabriel
Serwer, Philip
Shadduck, John Allen
Shillitoe, Edward John
Simpson, Russell Bruce
Smith, Gerald Ray
Smith, Kendall O
Smith, Russell Lamar
Smith, William Russell
Snell, William J
Spellman, Craig William
Stenback, Wayne Albert

Stephens, Robert Lawrence
Stevens, Clark
Stewart, James Ray
Straus, David Conrad
Streckfuss, Joseph Larry
Stroman, David Womack
Summers, Max Duane
Szaniszlo, Paul Joseph
Taber, Willard Allen
Taylor, Robert Dalton
Thomas, Virginia Lynn
Toler, Robert William
Tong, Alex W
Trentin, John Joseph
Trkula, David
Valdivieso, Dario
Vanderzant, Carl
Vela, Gerard Roland
Villa, Vicente Domingo
Walker, David Hughes
Walker, James Roy
Wang, Augustine Weisheng
Weichlein, Russell George
Welch, Gordon E
Wendt, Theodore Mil
Whitford, Howard Wayne
Williams, Robert Pierce
Wilson, Robert Eugene
Winters, Wendell Delos
Wynne, Elmer Staten
Young, Ryland F
Zajic, James Edward
Zimmermann, Eugene Robert
Zuberer, David Alan

UTAH
Barnett, Bill B
Beck, Jay Vern
Belnap, Jayne
Blauer, Aaron Clyde
Bradshaw, Willard Henry
Brierley, Corale Louise
Brierley, James Alan
Burton, Sheril Dale
Cole, Barry Charles
Crane, George Thomas
Ehrenfeld, Elvera
Franklin, Naomi C
Fujinami, Robert S
Georgopoulos, Constantine Panos
Glasgow, Lowell Alan
Hill, Douglas Wayne
Jensen, Marcus Martin
Johnson, F Brent
Kern, Earl R
Klein, Sigrid Marta
Lancaster, Jack R, Jr
Lark, Cynthia Ann
Larsen, Austin Ellis
Larsen, Don Hyrum
Larsen, Lloyd Don
Lombardi, Paul Schoenfeld
Marcus, Stanley
Matsen, John Martin
Melton, Arthur Richard
Muna, Nadeem Mitri
Nelson, Jerry Rees
North, James A
O'Neill, Frank John
Post, Frederick Just
Prescott, Stephen M
Rees, Horace Benner, Jr
Roth, John R
Sagers, Richard Douglas
Sidwell, Robert William
Skujins, John Janis
Spendlove, Rex S
Stockland, Alan Eugene
Stokes, Barry Owen
Van Alfen, Neal K
Warren, Reed Parley
Welkie, George William
Wiley, Bill Beauford
Wood, Timothy E
Wullstein, Leroy Hugh

VERMONT
Baldwin, Jack Norman
Craighead, John Edward
Fives-Taylor, Paula Marie
Johnstone, Donald Boyes
Lachapelle, Rene Charles
Moehring, Joan Marquart
Moehring, Thomas John
Raper, Carlene Allen
Schaeffer, Warren Ira
Sjogren, Robert Erik
Ullrich, Robert Carl
Wallace, Susan Scholes
Weed, Lawrence Leonard

VIRGINIA
Affronti, Lewis Francis
Al-Doory, Yousef
Andrykovitch, George
Ayers, William Arthur
Barr, Fred S
Bates, Robert Clair
Bauerle, Ronald H
Benjamin, David Charles
Berliner, Martha D
Boardman, Gregory Dale
Boyle, John Joseph
Bradley, Sterling Gaylen

Brown, Jay Clark
Cabral, Guy Antony
Chalgren, Steve Dwayne
Chen, Jiann-Shin
Chung, Choong Wha
Clarke, Gary Anthony
Claus, George William
Coleman, Philip Hoxie
Cordes, Donald Ormond
Coursen, Bradner Wood
Cummins, Cecil Stratford
Dalton, Harry P
Domermuth, Charles Henry, Jr
Elwood, Jerry William
Escobar, Mario R
Evans, Herbert John
Formica, Joseph Victor
Fu, Shu Man
Fugate, Kearby Joe
Hempfling, Walter Pahl
Hench, Miles Ellsworth
Hoptman, Julian
Hsu, Hsiu-Sheng
Huang, H(sing) T(sung)
Johnson, James Carl
Johnson, John LeRoy
Kelley, John Michael
King, Betty Louise
Kirk, Paul Wheeler, Jr
Krieg, Noel Roger
Lamanna, Carl
Leech, Stephen H
Lemp, John Frederick, Jr
Lewis, Cornelius Crawford
Lobel, Steven A
Loria, Roger Moshe
Lubiniecki, Anthony Stanley
McCombs, Robert Matthew
McCowen, Sara Moss
McCuen, Robert William
McDonald, Gerald O
McVicar, John West
McWright, Cornelius Glen
Merchant, Donald Joseph
Moore, Walter Edward C
Munson, Albert Enoch
Myrvik, Quentin N
Nagarkatti, Mitzi
Nagarkatti, Prakash S
Neal, John Lloyd, Jr
Normansell, David E
Palisano, John Raymond
Pierson, Merle Dean
Raizen, Carol Eileen
Roberts, Catherine Harrison
Roshal, Jay Yehudie
Royt, Paulette Anne
Russell, Catherine Marie
Scheld, William Michael
Scott, Marvin Wade
Shadomy, Smith
Smibert, Robert Merrall, II
Somers, Kenneth Donald
Stern, Joseph Aaron
Stevens, Thomas McConnell
Sumrall, H Glenn
Tankersley, Robert Walker, Jr
Tew, John Garn
Tolin, Sue Ann
Toney, Marcellus E, Jr
Trelawny, Gilbert Sterling
Tucker, Anne Nichols
Vermeulen, Carl William
Volk, Wesley Aaron
Wagner, Robert Roderick
Ware, Lawrence Leslie, Jr
Weber, Michael Joseph
Wilkins, Judd Rice
Wilkins, Tracy Dale
Wright, George Green
Yousten, Allan A

WASHINGTON
Ahmed, Saiyed I
Baross, John Allen
Benedict, Robert Glenn
Bezdicek, David Fred
Boatman, Edwin S
Booth, Beatrice Crosby
Bourquin, Al Willis J
Cheevers, William Phillip
Cho, Byung-Ryul
Cohen, Arthur Leroy
Cooney, Marion Kathleen
Cosman, David John
Coyle, Marie Bridget
Davis, William C
Deming, Jody W
Duncan, James Byron
Eidinger, David
Eklund, Melvin Wesley
Evermann, James Frederick
Finkelstein, David B
Gold, Eli
Grable, Albert E
Graves, Scott Stoll
Groman, Neal Benjamin
Grootes-Reuvecamp, Grada Alijda
Gurusiddaiah, Sarangamat
Hawthorne, Donald Clair
Hu, Shiu-Lok
Johnstone, Donald Lee
Kenny, George Edward
Kirchheimer, Waldemar Franz

Kuo, Cho-Chou
Leid, R Wes
Lewin, Joyce Chismore
Lowe, Janet Marie
McDonald, John Stoner
McFadden, Bruce Alden
Magnuson, Nancy Susanne
Matches, Jack Ronald
Nakata, Herbert Minoru
Nelson, Karen Ann
Nester, Eugene William
Nutter, Robert Leland
Pacha, Robert Edward
Parish, Curtis Lee
Pierson, Beverly Kanda
Raff, Howard V
Rayburn, William Reed
Reeves, Raymond
Rohrschneider, Larry Ray
Sherris, John C
Smith, Louis De Spain
Spence, Kemet Dean
Stokes, Jacob Leo
Stuart, Kenneth Daniel
Van Hoosier, Gerald L, Jr
Wang, San-Pin
Wekell, Marleen Marie
Whisler, Howard Clinton
Whiteley, Helen Riaboff
Wiley, William Rodney

WEST VIRGINIA
Albertson, John Newman, Jr
Anderson, Douglas Poole
Binder, Franklin Lewis
Bissonnette, Gary Kent
Buckelew, Albert Rhoades, Jr
Calhoon, Donald Alan
Charon, Nyles William
Cook, Harold Andrew
Hahon, Nicholas
Lewis, Daniel Moore
Lim, James Khai-Jin
Moat, Albert Groombridge
Olenchock, Stephen Anthony
Ong, Tong-man
Pore, Robert Scott
Wolf, Kenneth Edward
Yelton, David Baetz

WISCONSIN
Armstrong, George Michael
Balish, Edward
Bernstein, Sheldon
Brawner, Thomas A
Brill, Winston J
Brock, Katherine Middleton
Brock, Thomas Dale
Burgess, Richard Ray
Burris, Robert Harza
Chambliss, Glenn Hilton
Clausz, John Clay
Cliver, Dean Otis
Collins, Mary Lynne Perille
Courtright, James Ben
Crabtree, Koby Takayashi
Czuprynski, Charles Joseph
DeMars, Robert Ivan
Dick, Elliot C
Ehle, Fred Robert
Friend, Milton
Fritz, Robert B
Gerberich, John Barnes
Golubjatnikov, Rjurik
Graham, Linda Kay Edwards
Grossberg, Sidney Edward
Hinshaw, Virginia Snyder
Hinze, Harry Clifford
Hoerl, Bryan G
Howard, Thomas Hyland
Huss, Ronald John
Jacobson, Gunnard Kenneth
Jameson, Patricia Madoline
Jeffries, Thomas William
Klebba, Phillip E
Kostenbader, Kenneth David, Jr
Kurup, Viswanath Parameswar
Kushnaryov, Vladimir Michael
Mahl, Mearl Carl
Mansfield, John Michael
Marsh, Richard Floyd
Marth, Elmer Herman
Moskowitz, Gerard Jay
Myers, Charles R
Nagodawithana, Tilak Walter
Nealson, Kenneth Henry
Nelson, Allen Charles
Nelson, Thomas Clifford
Olson, Norman Fredrick
Pariza, Michael Willard
Parker, Dorothy Lundquist
Peppler, Henry James
Remsen, Charles C, III
Rigney, Mary Margaret
Rosson, Reinhardt Arthur
Rude, Theodore Alfred
Rueckert, Roland R
Scheusner, Dale Lee
Schultz, Ronald David
Schwartz, Leander Joseph
Seelke, Ralph Walter
Sieber, Fritz
Skatrud, Thomas Joseph
Smith, Clyde Konrad

Microbiology (cont)

Smith, Donald Ward
Sonneborn, David R
Spalatin, Josip
Steeves, Richard Allison
Steinke, Paul Karl Willi
Szybalski, Waclaw
Takayama, Kuni
Taylor, Jerry Lynn
Temin, Howard Martin
Truitt, Robert Lindell
Walker, Duard Lee
Wege, Ann Christene
Wejksnora, Peter James
Wilcox, Kent Westbrook
Williams, Anna Maria
Williams, Jeffrey Walter
Yuill, Thomas MacKay
Zabransky, Ronald Joseph

WYOMING
Adams, John Collins
Belden, Everett Lee
Bulla, Lee Austin, Jr
Caldwell, Daniel R
George, Robert Porter
Isaak, Dale Darwin
Pier, Allan Clark
Walton, Thomas Edward

PUERTO RICO
Bolaños, Benjamin
Colon, Julio Ismael
Copson, David Arthur
Garcia-Castro, Ivette
Ramirez-Ronda, Carlos Hector
Torres-Blasini, Gladys

ALBERTA
Albritton, William Leonard
Athar, Mohammed Aqueel
Birdsell, Dale Carl
Campbell, James Nicoll
Cheng, Kuo-Joan
Colter, John Sparby
Costerton, J William F
Darcel, Colin Le Q
Dixon, John Michael Siddons
Francis, Mike McD
Fujita, Donald J
Gaucher, George Maurice
Greer, George Gordon
Jensen, Susan Elaine
Kadis, Vincent William
Kaneda, Toshi
Langridge, William Henry Russell
Larke, R(obert) P(eter) Bryce
McElhaney, Ronald Nelson
Marusyk, Raymond George
Paranchych, William
Parkinson, Dennis
Rennie, Robert John
Rice, Wendell Alfred
Schroder, David John
Scraba, Douglas G
Stiles, Michael Edgecombe
Wallis, Peter Malcolm
Weiner, Joel Hirsch
Westlake, Donald William Speck
Whitehouse, Ronald Leslie S
Yamamoto, Tatsuzo

BRITISH COLUMBIA
Agnew, Robert Morson
Albright, Lawrence John
Anderson, John Donald
Antia, Naval Jamshedji
Bell, Gordon Russell
Bower, Susan Mae
Chiko, Arthur Wesley
Child, Jeffrey James
Copeman, Robert James
Dempster, George
Dolman, Claude Ernest
Duncan, Douglas Wallace
Finlay, Barton Brett
Hancock, Robert Ernest William
Hawirko, Roma Zenovea
Kelly, Michael Thomas
Levy, Julia Gerwing
McBride, Barry Clarke
McCarter, John Alexander
McLean, Donald Millis
Majak, Walter
Miller, Robert Carmi, Jr
Redfield, Rosemary Jeanne
Sadowski, Ivan J
Stock, John Joseph
Townsley, Philip McNair
Tremaine, Jack H
Trust, Trevor John
Walden, C(ecil) Craig
Weeks, Gerald

MANITOBA
Borsa, Joseph
Burton, David Norman
Campbell, Norman E Ross
Gill, Clifford Cressey
Hamilton, Ian Robert
Hamilton, Robert Duncan
Hannan, Charles Kevin
Kunz, Bernard Alexander

Lukow, Odean Michelin
Maniar, Atish Chandra
Pepper, Evan Harold
Rollo, Ian McIntosh
Ronald, Allan Ross
Shiu, Robert P C
Suzuki, Isamu
Wallbank, Alfred Mills
Warner, Peter
Westdal, Paul Harold

NEW BRUNSWICK
Singh, Rudra Prasad

NEWFOUNDLAND
Barnsley, Eric Arthur
Burness, Alfred Thomas Henry
Davidson, William Scott
Nolan, Richard Arthur
Ozere, Rudolph L

NOVA SCOTIA
Easterbrook, Kenneth Brian
Kind, Leon Saul
MacRae, Thomas Henry
Mahony, David Edward
Rozee, Kenneth Roy
Shieh, Hang Shan
Stewart, James Edward
Vining, Leo Charles
White, Robert Lester
Wort, Arthur John

ONTARIO
Arif, Basil Mumtaz
Barnum, Donald Alfred
Barran, Leslie Rohit
Bather, Roy
Behme, Ronald John
Bouillant, Alain Marcel
Brailovsky, Carlos Alberto
Branton, Philip Edward
Burnison, Bryan Kent
Cairns, William Louis
Campbell, James B
Carstens, Eric Bruce
Chadwick, June Stephens
Chaudhary, Rabindra Kumar
Chernesky, Max Alexander
Clark, A Gavin
Clark, David Sedgefield
Clarke, Anthony John
Dales, Samuel
Dalpe, Yolande
Davidson, Charles Mackenzie
Deeley, Roger Graham
Derbyshire, John Brian
Doane, Frances Whitman
Duncan, I B R
Dutka, Bernard J
Farkas-Himsley, Hannah
Faulkner, Peter
Fish, Eleanor N
Fitt, Peter Stanley
Forsberg, Cecil Wallace
Franklin, Mervyn
Fraser, Ann Davina Elizabeth
Friesen, James Donald
Furesz, John
Galsworthy, Sara B
Gentner, Norman Elwood
Ghosh, Hara Prasad
Gochnauer, Thomas Alexander
Goepfert, John McDonnell
Gould, William Douglas
Graham, Frank Lawson
Gregory, Kenneth Fowler
Gupta, Krishana Chandara
Gyles, C L
Hauschild, Andreas H W
Heath, Ian Brent
Hendry, Anne Teresa
Higgins, Verna Jessie
Inman, Robert Davies
Inniss, William Edgar
Iyer, Rajul V
Johnson, Byron F
Johnson-Lussenburg, Christine Margaret
Kang, C Yong
Kasupski, George Joseph
Kaushik, Azad Kumar
Khan, Abdul Waheed
Lang, Gerhard Herbert
Lee, Peter E
Liu, Dickson Lee Shen
McCready, Ronald Glen Lang
McCurdy, Howard Douglas, Jr
Macdonald, John Barfoot
Mahdy, Mohamed Sabet
Mak, Stanley
Medzon, Edward Lionel
Metzgar, Don P
Milazzo, Francis Henry
Miller, John James
Moo-Young, Murray
Murray, William Douglas
Nielsen, Klaus H B
Paliwal, Yogesh Chandra
Patel, Girishchandra Babubhai
Pope, Barbara L
Pross, Hugh Frederick
Rao, Salem S
Rawls, William Edgar
Rayman, Mohamad Khalil

Rhodes, Andrew James
Richardson, Harold
Rigby, Charlotte Edith
Robinow, Carl Franz
Rogers, Charles Graham
Roslycky, Eugene Bohdan
Sabina, Leslie Robert
Sattar, Syed Abdus
Savan, Milton
Schneider, Henry
Seligy, Verner Leslie
Siminovitch, Louis
Smiley, James Richard
Sorger, George Joseph
Spence, Leslie Percival
Sprott, Gordon Dennis
Stainer, Dennis William
Stavric, Stanislava
Stemshorn, Barry William
Stevens, John Bagshaw
Stevenson, Ian Lawrie
Stewart, Robert Bruce
Szewczuk, Myron Ross
Todd, Ewen Cameron David
Trevors, Jack Thomas
Truscott, Robert Bruce
Tu, Chin Ming
Tu, Jui-Chang
Vas, Stephen Istvan
Veliky, Ivan Alois
Ward, Edmund William Beswick
Wayman, Morris
Wellman, Angela Myra
Wilkins, Peter Osborne
Wilson, Jack Harold
Winterhalder, Keith
Young, James Christopher F

QUEBEC
Abshire, Claude James
Ackermann, Hans Wolfgang
Anderson, William Alan
Belloncik, Serge
Benoit, Guy J C
Bilimoria, Minoo Hormasji
Bordeleau, Lucien Mario
Bourgaux, Pierre
Cantor, Ena D
Chabot, Benoit
Chagnon, Andre
Champagne, Claude P
Chan, Eddie Chin Sun
Chung, Young Sup
Comeau, Andre I
DeVoe, Irving Woodrow
DuBow, Michael Scott
Dugre, Robert
Gervais, Francine
Gibbs, Sarah Preble
Goulet, Jacques
Graham, Angus Frederick
Hallenbeck, Patrick Clark
Hamelin, Claude
Idziak, Edmund Stefan
Joncas, Jean Harry
Jurasek, Lubomir
Kluepfel, Dieter
Knowles, Roger
Laflamme, Gaston
Leduy, Anh
Lussier, Gilles L
MacLeod, Robert Angus
Martineau, Ronald
Mathieu, Leo Gilles
Menezes, Jose Piedade Caetano Agnelo
Murgita, Robert Anthony
Niven, Donald Ferries
Paice, Michael
Portelance, Vincent Damien
Richard, Claude
Roberge, Marcien Romeo
Sheinin, Rose
Siboo, Russell
Simard, Rene
Simard, Ronald E
Skup, Daniel
Slilaty, Steve N
Stanners, Clifford Paul
Talbot, Pierre J
Taussig, Andrew
Trepanier, Pierre
Trudel, Michel D
Vezina, Claude
Walker, Mary Clare
Weber, Joseph M

SASKATCHEWAN
Babiuk, Lorne A
Cullimore, Denis Roy
Haskins, Reginald Hinton
Ingledew, William Michael
Jones, Graham Alfred
Julien, Jean-Paul
Khachatourians, George G
Kurz, Wolfgang Gebhard Walter
Nelson, Louise Mary
Robertson, Hugh Elburn
Waygood, Edward Bruce

YUKON TERRITORY
Doermann, August Henry

OTHER COUNTRIES
Aalund, Ole

Arber, Werner
Baddiley, James
Bernal-Llanas, Enrique
Bishop, David Hugh Langler
Bojalil, Luis Felipe
Bolhuis, Reinder L H
Borlaug, Norman Ernest
Cerbon-Solorzano, Jorge
Coulston, Mary Lou
Dianzani, Ferdinando
Doerfler, Walter Hans
Dutta-Roy, Asim Kanti
Dyrberg, Thomas Peter
Eaton, Bryan Thomas
Eggers, Hans J
Feldman, Jose M
Franklin, Richard Morris
Gelzer, Justus
Gendler, Sandra J
Gjessing, Helen Witton
Graber, Robert Philip
Greenblatt, Charles Leonard
Greene, Velvl William
Gresser, Ion
Ha, Tai-You
Hiyama, Tetsuo
Hoadley, Alfred Warner
Hosokawa, Keiichi
Huang, Kun-Yen
Justines, Gustavo
Kaffezakis, John George
Kessel, Rosslyn William Ian
Kimoto, Masao
Koerber, Walter Ludwig
Kourany, Miguel
Loppnow, Harald
Marikovsky, Yehuda
Maruyama, Koshi
Mitruka, Brij Mohan
Mizuno, Shigeki
Mou, Duen-Gang
Nakagawa, Shizutoshi
Ou, Jonathan Tsien-hsiong
Pan, In-Chang
Racotta, Radu Gheorghe
Ray, Prasanta K
Razzell, Wilfred Edwin
Ribbons, Douglas William
Roy, Raman K
Scheid, Stephan Andreas
Shapiro, Stuart
Spencer, John Francis Theodore
Stingl, Georg
Stoppani, Andres Oscar Manuel
Subramanian, Alap Raman
Sykes, Richard Brook
Taniyama, Tadayoshi
Thormar, Halldor
Tsang, Joseph Chiao-Liang
Varga, Janos M
Weil, Marvin Lee
Weiss, David Walter
Yoshida, Takeshi

Molecular Biology

ALABAMA
Anderson, Peter Glennie
Aull, John Louis
Ball, Laurence Andrew
Bradley, James T
Chapatwala, Kirit D
Christian, Samuel Terry
Cook, William Joseph
Egan, Marianne Louise
Furuto, Donald K
Gay, Steffen
Giambrone, Joseph James
Gillespie, George Yancey
Hajduk, Stephen Louis
Hardman, John Kemper
Harvey, Stephen Craig
Higgins, N Patrick
Hunter, Eric
Jenkins, Ronald Lee
Lebowitz, Jacob
Lidin, Bodil Inger Maria
Oparil, Suzanne
Roozen, Kenneth James
Sani, Brahma Porinchu
Siegal, Gene Philip
Umeda, Patrick Kaichi
Wells, Robert Dale
Winters, Alvin L
Wood, David Oliver

ARIZONA
Bernstein, Carol
Bernstein, Harris
Bourque, Don Philippe
Chandler, Douglas Edwin
Doane, Winifred Walsh
Goldstein, Elliott Stuart
Goll, Darrel Eugene
Guerriero, Vincent, Jr
Hagedorn, Henry Howard
Hall, Jennifer Dean
Hansen, Jo Ann Brown
Harkins, Kristi R
Hewlett, Martinez Joseph
Kay, Marguerite M B
Krahl, Maurice Edward
Little, John Wesley

Mendelson, Neil Harland
Mooradian, Arshag Dertad
Mount, David William Alexander
Nagle, Ray Burdell
Pommerville, Jeffrey Carl
Rose, Seth David
Vermaas, Willem F J
Vincent, Walter Sampson
Ward, Samuel
Wright, Daniel Craig
Yohem, Karin Hummell

ARKANSAS
Cave, Mac Donald
Cornett, Lawrence Eugene
Heflich, Robert Henry
Leakey, Julian Edwin Arundell
Norris, James Scott
Poirier, Lionel Albert
Winter, Charles Gordon

CALIFORNIA
Abelson, John Norman
Alberts, Bruce Michael
Alousi, Adawia A
Ames, Giovanna Ferro-Luzzi
Anthony, Ronald Lewis
Arai, Ken-Ichi
Arnheim, Norman
Attardi, Giuseppe M
Bailey-Serres, Julia
Baker, Robert Frank
Balch, William E
Baluda, Marcel A
Bandman, Everett
Baxter, John Darling
Beckendorf, Steven K
Bekhor, Isaac
Benzer, Seymour
Berk, Arnold J
Bernstein, Sanford Irwin
Bhattacharya, Prabir
Bissell, Dwight Montgomery
Bjorkman, Pamela J
Blackburn, Elizabeth Helen
Bokoch, Gary M
Bonner, James (Fredrick)
Bower, Annette
Boyd, James Brown
Boyer, Paul Delos
Bradshaw, Ralph Alden
Bramhall, John Shepherd
Branscomb, Elbert Warren
Breidenbach, Rowland William
Briggs, Winslow Russell
Britten, Roy John
Bruening, George
Brunke, Karen J
Brutlag, Douglas Lee
Burgess, Teresa Lynn
Byerley, Lauri Olson
Calendar, Richard
Campagnoni, Anthony Thomas
Campbell, Judith Lynn
Cantor, Charles Robert
Carbon, John Anthony
Carrano, Anthony Vito
Celniker, Susan Elizabeth
Chamberlin, Michael John
Chan, Sham-Yuen
Chang, Chung-Nan
Chang, Jennie C C
Choudary, Prabhakara Velagapudi
Clark, Alvin John
Cline, Thomas Warren
Clough, Wendy Glasgow
Cohen, Edward Hirsch
Cohen, Larry William
Cohen, Robert Elliot
Conner, Brenda Jean
Cooper, James Burgess
Coyer, James A
Craik, Charles S
Crisp, Carl Eugene
Cronin, Michael John
Crooke, Stanley T
Curnutte, John Tolliver, III
Curry, Donald Lawrence
Dahms, Arthur Stephen
Daniell, Ellen
Davidson, Eric Harris
Davidson, Norman Ralph
Davies, Huw M
Davis, Craig H
Davis, Ronald Wayne
Davis, Rowland Hallowell
De Francesco, Laura
Denis, Kathleen A
Dennert, Gunther
Deonier, Richard Charles
Dickerson, Richard Earl
Dietz, George William, Jr
Donoghue, Daniel James
Downing, Michael Richard
Dragon, Elizabeth Alice Oosterom
Dreyer, William J
Duesberg, Peter H
Duffey, Paul Stephen
Dugaiczyk, Achilles
Dunaway, Marietta
Dunnebacke-Dixon, Thelma Hudson
Echols, Harrison
Ecker, David John
Eckhart, Walter

Edelman, Jay Barry
Eiserling, Frederick A
Emr, Scott David
Eng, Lawrence F
Englesberg, Ellis
Erickson, Jeanne Marie
Ernest, Michael Jeffrey
Feigal, Ellen G
Felgner, Philip Louis
Felton, James Steven
Fessler, John Hans
Flashman, Stuart Milton
Fleming, James Emmitt
Fraenkel-Conrat, Heinz Ludwig
Friedmann, Theodore
Fukuda, Minoru
Fulco, Armand J
Funk, Glenn Albert
Gaertner, Frank Herbert
Galas, David John
Game, John Charles
Ganesan, Adayapalam T
Ganesan, Ann K
Gascoigne, Nicholas Robert John
Gasser, Charles Scott
Gautsch, James Willard
Geiduschek, Ernest Peter
Gelfand, David H
Gerace, Larry R
Gibson, Thomas Richard
Gilula, Norton Bernard
Ginsberg, Theodore
Glaser, Donald Arthur
Glazer, Alexander Namiot
Golde, David William
Gray, Gary M
Grody, Wayne William
Grunstein, Michael
Gum, James Raymond, Jr
Gunsalus, Robert Philip
Guthrie, Christine
Gutman, George Andre
Hall, Raymond G, Jr
Hamkalo, Barbara Ann
Hanawalt, Philip Courtland
Hankinson, Oliver
Hanson, Carl Veith
Hathaway, Gary Michael
Haygood, Margo Genevieve
Heffernan, Laurel Grace
Henry, Helen L
Herschman, Harvey R
Hill, Ray Allen
Hirsch, Ann Mary
Ho, Begonia Y
Hoch, James Alfred
Hoch, Paul Edwin
Hoch, Sallie O'Neil
Hofmann, Bo
Hogan, Christopher James
Hoke, Glenn Dale
Hoopes, Laura Livingston Mays
Hosoda, Junko
Howard, Bruce David
Howard, Russell John
Huang, Anthony Hwoon Chung
Hunter, Tony
Hurd, Ralph Eugene
Hyman, Bradley Clark
Itakura, Keiichi
Jacob, Chaim O
Jacob, Mary
Jacobson, Ralph Allen
Janda, John Michael
Jardetzky, Oleg
Jariwalla, Raxit Jayantilal
Jeffery, William Richard
Johansson, Mats W
Johnson, Paul Hickok
Jones, Kenneth Charles
Kado, Clarence Isao
Kan, Yuet Wai
Kang, Tae Wha
Kan-Mitchell, June
Kasamatsu, Harumi
Katz, Louis
Kedes, Laurence H
Kelley, Darshan Singh
Kelly, Regis Baker
Kennedy, Mary Bernadette
Kim, Sung-Hou
King, A Douglas, Jr
Kirschbaum, Joel Bruce
Klevecz, Robert Raymond
Knudson, Gregory Blair
Kopito, Ron Rieger
Korn, David
Koshland, Daniel Edward, Jr
Koski, Raymond Allen
Koths, Kirston Edward
Kozloff, Lloyd M
Lai, Michael Ming-Chiao
Lake, James Albert
Lambert, Charles Calvin
Landolfi, Nicholas F
Landolph, Joseph Richard, II
Langridge, Robert
Larrick, James William
Last, Jerold Alan
Lee, Amy Shiu
Leffert, Hyam Lerner
Lei, Shau-Ping Laura
Leighton, Terrance J
Lengyel, Judith Ann

Leung, David Wai-Hung
Levine, Michael Steven
Levinson, Arthur David
Lewis, Susanna Maxwell
Lieb, Margaret
Lindh, Allan Goddard
Lipka, James J
Lipshitz, Howard David
Lipsick, Joseph Steven
Long, Sharon Rugel
Lopo, Alina C
Lowenstein, Jerold Marvin
Lusis, Aldons Jekabs
McClelland, Michael
McCullough, Richard Donald
McEwen, Joan Elizabeth
MacInnis, Austin J
Mackman, Nigel
MacLeod, Carol Louise
Mangan, Jerrome
Marshall, Charles Richard
Martin, David William, Jr
Martin, George Steven
Martins-Green, Manuela M
Martinson, Harold Gerhard
Matthews, Harry Roy
Maxson, Robert E, Jr
Meehan, Thomas (Dennis)
Melis, Anastasios
Merchant, Sabeeha
Meyerowitz, Elliot Martin
Miledi, Ricardo
Morse, Daniel E
Mosteller, Raymond Dee
Murdoch, Joseph Richard
Nagel, Glenn M
Napolitano, Leonard Michael, Jr
Neufeld, Berney Roy
Nierlich, Donald P
Nomura, Masayasu
Norman, Gary L
Oh, Chan Soo
Olson, Betty H
Ordahl, Charles Philip
Paik, Young-Ki
Pallavicini, Maria Georgina
Pauling, Edward Crellin
Pellegrini, Maria C
Perrault, Jacques
Perry, L Jeanne
Peterlin, Boris Matija
Philpott, Delbert E
Phleger, Charles Frederick
Pilgeram, Laurence Oscar
Pischel, Ken Donald
Pryer, Nancy Kathryn
Quail, Peter Hugh
Raj, Harkisan D
Ray, Dan S
Reeve, Peter
Reiness, Gary
Reitz, Richard Elmer
Reizer, Jonathan
Remsen, Joyce F
Rho, Joon H
Rice, Robert Hafling
Riggs, Arthur Dale
Robinson, William Sidney
Rokeach, Luis Alberto
Rome, Leonard H
Rosen, Howard
Rosenfeld, Ron Gershon
Rosenthal, Allan Lawrence
Roy-Burman, Pradip
Rubin, Gerald M
Ryder, Oliver A
Sadee, Wolfgang
Saifer, Mark Gary Pierce
Salser, Winston Albert
Samuel, Charles Edward
Sazama, Kathleen
Schachman, Howard Kapnek
Scheffler, Immo Erich
Schekman, Randy W
Scott, Matthew P
Sedat, John William
Seegmiller, Jarvis Edwin
Shapiro, Lucille
Shen, Che-Kun James
Shih, Jean Chen
Simons, Robert W
Simpson, Robert Blake
Singer, B
Sinibaldi, Ralph Michael
Sinsheimer, Robert Louis
Sjostrand, Fritiof S
Smith, Andrew Philip
Smith, Cassandra Lynn
Smith, Charles Allen
Smith, Douglas Wemp
Smith, Helene Sheila
Smith, Steven Sidney
Song, Moon K
Spray, Clive Robert
Stallcup, Michael R
Stanley, Wendell Meredith, Jr
Stellwagen, Robert Harwood
Stent, Gunther Siegmund
Stevens, Lewis Axtell
Strauss, James Henry
Stringfellow, Dale Alan
Strobel, Edward
Stroynowski, Iwona T
Stryer, Lubert

Stubbs, John Dorton
Stumph, William Edward
Subramani, Suresh
Sung, Zinmay Renee
Sussman, Howard H
Sy, Jose
Talhouk, Rabih Shakib
Taylor, Barry L
Thompson, Lawrence Hadley
Thorner, Jeremy William
Tilton, Varien Russell
Tisserat, Brent Howard
Tjian, Robert Tse Nan
Tobin, Allan Joshua
Traugh, Jolinda Ann
Traut, Robert Rush
Triche, Timothy J
Varmus, Harold Elliot
Vedvick, Thomas Scott
Verheyden, Julien P H
Vickery, Larry Edward
Villarreal, Luis Perez
Vogt, Peter Klaus
Volcani, Benjamin Elazari
Wade, Michael James
Wall, Thomas Randolph
Wallace, Robert Bruce
Warner, Robert Collett
Warshel, Arieh
Wasterlain, Claude Guy
Waterman, Michael S
Webb, David Ritchie
Weber, Heather R(oss) Wilson
Weinberg, Barbara Lee Huberman
Wettstein, Felix O
Wiebe, Michael Eugene
Wiktorowicz, John Edward
Wilcox, Gary Lynn
Williams, Robley Cook
Willson, Clyde D
Wong, Chi-Huey
Wood, William Irwin
Woodward, Dow Owen
Wu, Sing-Yung
Yamamoto, Keith Robert
Yamamoto, Richard
Yanofsky, Charles
Yu, Grace Wei-Chi Hu
Yund, Mary Alice
Zabin, Irving
Zarucki, Tanya Z
Zeichner-David, Margarita
Zernik, Joseph
Zieg, Roger Grant
Zlotnik, Albert
Zuccarelli, Anthony Joseph
Zuckekandl, Emile

COLORADO
Andrews, Ken J
Barrett, Dennis
Brown, Lewis Marvin
Cech, Thomas Robert
Clark, Roger William
Danna, Kathleen Janet
Dunahay, Terri Goodman
Dynan, William Shelley
Gerschenson, Lazaro E
Giclas, Patricia C
Glode, Leonard Michael
Hahn, William Eugene
Horwitz, Kathryn Bloch
Ishii, Douglas Nobuo
Johnson, Thomas Eugene
Kano-Sueoka, Tamiko
Liao, Martha
Lowndes, Joseph M
McHenry, Charles S
Maller, James Leighton
Mattoon, James Richard
Morse, Helvise Glessner
Nett, T M
Patterson, David
Pettijohn, David E
Poyton, Robert Oliver
Ranu, Rajinder S
Rash, John Edward
Roberts, P Elaine
Simon, Nancy Jane
Sinensky, Michael
Sneider, Thomas W
Stein, Gretchen Herpel
Stushnoff, Cecil
Taylor, Austin Laurence
Trent, Dennis W
Urban, Richard William
Vasil, Michael Lawrence
Waldren, Charles Allen
Wood, William Barry, III
Yarus, Michael J

CONNECTICUT
Aaslestad, Halvor Gunerius
Altman, Sidney
Amacher, David E
Bacopoulos, Nicholas G
Berg, Claire M
Bormann, Barbara-Jean Anne
Bothwell, Alfred Lester Meador
Braun, Phyllis C
Buck, Marion Gilmour
Carboni, Joan M
Deutscher, Murray Paul
Dingman, Douglas Wayne

Goodridge, Alan G
Gussin, Gary Nathaniel
Hallberg, Richard Lawrence
Horowitz, Jack
Makar, Adeeb Bassili
Maurer, Richard Allen
Mayfield, John Eric
Menninger, John Robert
Milkman, Roger Dawson
Schmerr, Mary Jo F
Stanton, Thaddeus Brian
Sullivan, Charles Henry
Thiermann, Alejandro Bories
Tomes, Dwight Travis
Trenkle, Allen H

KANSAS
Baker, Joffre Bernard
Besharse, Joseph Culp
Bradford, Lawrence Glenn
Calvet, James P
Davis, Lawrence Clark
Decedue, Charles Joseph
Denell, Robin Ernest
Funderburgh, James Louis
Gegenheimer, Peter Albert
Goldberg, Ivan D
Hedgcoth, Charlie, Jr
Johnson, Lowell Boyden
Johnson, Terry Charles
Leavitt, Wendell William
Leslie, John Franklin
Minocha, Harish C
Muthukrishnan, Subbaratnam
Ochs, Raymond S
Parkinson, Andrew
Rawitch, Allen Barry
Samson, Frederick Eugene, Jr
Singhal, Ram P
Smith, Larry Dean
Stetler, Dean Allen
Takemoto, Dolores Jean
Weaver, Robert F
Wong, Peter P
Yarbrough, Lynwood R

KENTUCKY
Davidson, Jeffrey Neal
Dickson, Robert Carl
Feese, Bennie Taylor
Geoghegan, Thomas Edward
Lesnaw, Judith Alice
Martin, Nancy Caroline
Perlin, Michael Howard
Pirone, Thomas Pascal
Rao, Chalamalasetty Venkateswara
Rawls, John Marvin, Jr
Sisken, Jesse Ernest
Slagel, Donald E
Toman, Frank R
Westneat, David French, Jr
Williams, Arthur Lee
Zimmer, Stephen George

LOUISIANA
Alam, Jawed
Anderson, William McDowell
Bhattacharyya, Ashim Kumar
Cardelli, James Allen
Cohen, J Craig
Ehrlich, Kenneth Craig
Ehrlich, Melanie
Grimes, Sidney Ray, Jr
Hackney, Cameron Ray
Haycock, John Winthrop
Ivens, Mary Sue
Moore, Thomas Stephen, Jr
Nasci, Roger Stanley
O'Callaghan, Richard J
Olmsted, Clinton Albert
Rothschild, Henry
Sarphie, Theodore G
Sheiness, Diana Kay
Shepherd, Hurley Sidney
Tsuzuki, Junji
Warren, Lionel Gustave
Weidner, Earl

MAINE
Gaskins, H Rex
McKerns, Kenneth (Wilshire)
Renn, Donald Walter
Roxby, Robert
Steinhart, William Lee
Welsch, Federico

MARYLAND
Aaronson, Stuart Alan
Abedin, Zain-ul
Ackerman, Eric J
Adelstein, Robert Simon
Ades, Ibrahim Z
Adhya, Sankar L
Anderson, Norman Leigh
Anderson, W French
August, Joseph Thomas
Austin, Faye Carol
Avigan, Joel
Azad, Abdu F(arhang)
Baker, George Thomas, III
Bala, Shukal
Beemon, Karen Louise
Beer, Michael
Bellino, Francis Leonard

Berger, Shelby Louise
Berkower, Ira
Berrettini, Wade Hayhurst
Bhatnagar, Gopal Mohan
Bigger, Cynthia Anita Hopwood
Bird, Robert Earl
Bloch, Robert Joseph
Blough, Herbert Allen
Bockstahler, Larry Earl
Bonner, Tom Ivan
Boyd, Virginia Ann Lewis
Burt, David Reed
Bustin, Michael
Camerini-Otero, Rafael Daniel
Carter, Barrie J
Cashel, Michael
Chang, Henry
Chassy, Bruce Matthew
Cho-Chung, Yoon Sang
Choppin, Purnell Whittington
Cleveland, Don W
Cohen, Gerson H
Cohen, Jack Sidney
Coleman, William Gilmore, Jr
Colombini, Marco
Cowan, Elliot Paul
Cox, George Warren
Craig, Nessly Coile
Crouch, Robert J
Crystal, Ronald George
Cutler, Richard Gail
Davies, David R
Dean, Ann
Dean, Donna Joyce
Dean, Jurrien
Deitzer, Gerald Francis
Devreotes, Peter Nicholas
Dintzis, Renee Zlochover
Doniger, Jay
Dwivedi, Radhey Shyam
Eckels, Kenneth Henry
Eglitis, Martin Alexandris
Eipper, Betty Anne
Elespuru, Rosalie K
Elson, Hannah Friedman
Ennist, David L
Epstein, David Aaron
Ewing, June Swift
Fedoroff, Nina V
Felsenfeld, Gary
Fields, Kay Louise
Fornace, Albert J, Jr
Foulds, John Douglas
Franklin, Renty Benjamin
Freese, Ernst
Gall, Joseph Grafton
Garges, Susan
Gartland, William Joseph
Gelboin, Harry Victor
Gellert, Martin Frank
Gerard, Gary Floyd
Goldberg, Michael Ian
Goldman, David
Goldsmith, Merrill E
Gonda, Matthew Allen
Goode, Melvyn Dennis
Gorman, Cornelia M
Graham, Dale Elliott
Gravell, Maneth
Greenhouse, Gerald Alan
Gusovsky, Fabian
Guss, Maurice Louis
Hadidi, Ahmed Fahmy
Hahn, Fred Ernst
Hairstone, Marcus A
Hanna, Edgar Ethelbert, Jr
Hansen, John Norman
Harrington, William Fields
Hatfield, Dolph Lee
Haun, Randy S
Hearing, Vincent Joseph, Jr
Henry, Timothy James
Herman, Eliot Mark
Hickey, Robert Joseph
Hoeg, Jeffrey Michael
Hollingdale, Michael Richard
Holloway, Caroline T
Howley, Peter Maxwell
Huang, Ru-Chih Chow
Hudson, Lynn Diane
Imberski, Richard Bernard
Impraim, Chaka Cetewayo
Irwin, David
Jernigan, Robert Lee
Jordan, Elke
Josephs, Steven F
Kaper, Jacobus M
Kaper, James Bennett
Karpel, Richard Leslie
Keister, Donald Lee
Keith, Jerry M
Kelly, Thomas J
Ketley, Jeanne Nelson
Kibbey, Maura Christine
Kimball, Paul Clark
Kleinman, Hynda Karen
Kraemer, Kenneth H
Krause, David
Kuff, Edward Louis
Kundig, Fredericka Dodyk
Kung, Shain-Dow
Kunos, George
Landsman, David
Lane, Malcolm Daniel

Lederer, William Jonathan
Lee, Chi-Jen
Lee, Fang-Jen Scott
Lee, Theresa
Leppla, Stephen Howard
Lerman, Michael Isaac
Levin, Barbara Chernov
Levin, Judith Goldstein
Levine, Arthur Samuel
Liang, Shu-Mei
Lillehoj, Hyun Soon
Linehan, William Marston
Lippincott-Schwartz, Jennifer
Liu, Leroy Fong
Liu, Teh-Yung
Longfellow, David G(odwin)
Luborsky, Samuel William
Lunney, Joan K
McCandliss, Russell John
McCarron, Richard M
McClure, Michael Edward
McCune, Susan K
McGowan, John Joseph
Maciag, Thomas Edward
McKenney, Keith Hollis
McKeon, Catherine
McLachlin, Jeanne Ruth
McMacken, Roger
McNicol, Lore Anne
Malech, Harry Lewis
Maloney, Peter Charles
Manganiello, Vincent Charles
Margolis, Sam Aaron
Margulies, David Harvey
Margulies, Maurice
Marks, Edwin Potter
Martin, Malcolm Alan
Martin, Robert G
Mattern, Michael Ross
Matthews, Benjamin F
Max, Edward E
Melera, Peter William
Merril, Carl R
Miles, Edith Wilson
Miles, Harry Todd
Mockrin, Stephen Charles
Montell, Craig
Mora, Peter T
Moss, Bernard
Mufson, R Allan
Mushinski, J Frederic
Nakhasi, Hira Lal
Nash, Howard Allen
Nathans, Daniel
Nayak, Ramesh Kadbet
Nienhuis, Arthur Wesley
Noguchi, Philip D
Nomura, Shigeko
Nossal, Nancy
Nowak, Thaddeus Stanley, Jr
Owens, Ida S
Pace, Judith G
Pastan, Ira Harry
Pedersen, Peter L
Peterson, Jane Louise
Phillips, Leo Augustus
Pitha-Rowe, Paula Marie
Rabussay, Dietmar Paul
Repaske, Roy
Rice, Nancy Reed
Rick, Paul David
Rommel, Frederick Allen
Rose, James A
Rosner, Judah Leon
Roth, Thomas Frederic
Sabol, Steven Layne
Sachs, David Howard
Sack, George H(enry), Jr
Safer, Brian
Samuel, Albert
Sanford, Katherine Koontz
Saunders, James Allen
Saxinger, W(illiam) Carl
Schindler, Joel Marvin
Schneider, Walter Carl
Schoenberg, Daniel Robert
Schultz, Warren Walter
Schwartz, Joan Poyner
Scocca, John Joseph
Seifried, Adele Susan Corbin
Seto, Belinda P L
Sexton, Thomas John
Sharma, Opendra K
Shih, Thomas Y
Shin, Yong Ae Im
Silverman, Robert Hugh
Simic, Michael G
Simons, Samuel Stoney, Jr
Smith, Gilbert Howlett
Sobel, Mark E
Sollner-Webb, Barbara Thea
Song, Byoung-Joon
Spero, Leonard
Stanchfield, James Ernest
Steinert, Peter Malcolm
Steven, Alasdair C
Stonehill, Elliott H
Strathern, Jeffrey Neal
Stromberg, Kurt
Suskind, Sigmund Richard
Szepesi, Bela
Tabakoff, Boris
Talbot, Bernard
Taniuchi, Hiroshi

Tester, Cecil Fred
Thorgeirsson, Snorri S
Vahey, Maryanne T
Vandenbergh, David John
Vande Woude, George
Venkatesan, S
Villet, Ruxton Herrer
Vonderhaar, Barbara Kay
Waldmann, Thomas A
Westphal, Heiner J
White, Elizabeth Lloyd
Whitfield, Harvey James, Jr
Wilder, Ronald Lynn
Williams, Noreen
Willingham, Mark C
Wolf, Richard Edward, Jr
Wolffe, Alan Paul
Woodstock, Lowell Willard
Wu, Henry Chi-Ping
Yarmolinsky, Michael Bezalel
Yasbin, Ronald Eliott
Young, Howard Alan
Young, Ronald Jerome
Zinmerman, Steven B

MASSACHUSETTS
Ahmed, A Razzaque
Backman, Keith Cameron
Bagshaw, Joseph Charles
Beckwitt, Richard David
Bernhard, Jeffrey David
Boedtker Doty, Helga
Brennessel, Barbara A
Brown, Beverly Ann
Carter, Edward Albert
Carvalho, Angelina C A
Chen, Anne Chi
Chikaraishi, Dona M
Chin, William W
Chishti, Athar Husain
Coffin, John Miller
Collins, Carolyn Jane
Collins, Tucker
Comer, M Margaret
Cooper, Geoffrey Mitchell
Coor, Thomas
Coyne, Mary Downey
Crain, William Rathbone, Jr
DeRosier, David J
Desrosiers, Ronald Charles
DeWitt, William
Dvorak, Ann Marie-Tompkins
Ehlers, Mario Ralf Werner
Emanuel, Rodica L
Ernst, Susan Gwenn
Feig, Larry Allen
Fenton, Matthew John
Fitzgerald, Marie Anton
Fournier, Maurille Joseph, Jr
Fox, Maurice Sanford
Fox, Thomas Oren
Fraser, Thomas Hunter
Fridovich-Keil, Judith Lisa
Fritsch, Edward Francis
Fulton, Chandler Montgomery
Furie, Bruce
Gefter, Malcolm Lawrence
Gilbert, Walter
Gill, David Michael
Godine, John Elliott
Goldberg, Alfred L
Goldberg, Edward B
Goldstein, Lawrence S B
Goldstein, Richard Neal
Gordon, Katherine
Guterman, Sonia Kosow
Hecht, Norman B
Heinrich, Gerhard
Hoffmann, George Robert
Holmquist, Barton
Horvitz, Howard Robert
Hotchkiss, Rollin Douglas
Hu, Jing-Shan
Hwang, San-Bao
Hynes, Richard Olding
Ingwall, Joanne S
Isselbacher, Kurt Julius
Jameson, James Larry
Kafatos, Fotis C
Kelley, William S
Kilpatrick, Daniel Lee
King, Jonathan (Alan)
Knipe, David Mahan
Kolodner, Richard David
Krane, Stanley Garson
Kuemmerle, Nancy Benton Stevens
Lai, Elaine Y
Lander, Arthur Douglas
Lawrence, Jeanne Bentley
Ledbetter, Mary Lee Stewart
Leder, Philip
Lehman, William Jeffrey
Lemontt, Jeffrey Fielding
Lerman, Leonard Solomon
Levy, Stuart B
Little, John Bertram
Losick, Richard Marc
Lou, Peter Louis
McNally, Elizabeth Mary
McReynolds, Larry Austin
Magasanik, Boris
Malamy, Michael Howard
Malt, Ronald A
Maniatis, Thomas Peter

Molecular Biology (cont)

Marotta, Charles Anthony
Marx, Kenneth Allan
Meedel, Thomas Huyck
Melan, Melissa A
Meselson, Matthew Stanley
Moolten, Frederick London
Mount, Mark Samuel
Mulder, Carel
Nadal-Ginard, Bernardo
Neer, Eva Julia
Offner, Gwynneth Davies
Orkin, Stuart H
Palubinskas, Felix Stanley
Papaioannou, Virginia Eileen
Pardee, Arthur Beck
Pardue, Mary Lou
Parker, David Charles
Pederson, Thoru Judd
Pero, Janice Gay
Petri, William Hugh
Poccia, Dominic Louis
Potts, John Thomas, Jr
Ptashne, Mark
Putney, Scott David
Rheinwald, James George
Rich, Alexander
Rigney, David Roth
Riley, Monica
Rio, Donald C
Ross, Alonzo Harvey
Ruderman, Joan V
Sarkar, Nilima
Sarkar, Satyapriya
Sass, Heinz
Schaechter, Moselio
Schleif, Robert Ferber
Schwaber, Jerrold
Seaver, Sally S
Seidman, Jonathan G
Shaeffer, Joseph Robert
Sharp, Phillip Allen
Shing, Yuen Wan
Signer, Ethan Royal
Sonenshein, Abraham Lincoln
Springer, Timothy Alan
Stein, Diana B
Stein, Gary S
Stein, Janet Lee Swinehart
Stephenson, Mary Louise
Strauss, Phyllis R
Struhl, Kevin
Szostak, Jack William
Taunton-Rigby, Alison
Thomas, Clayton Lay
Thomas, Peter
Tipper, Donald John
Tonegawa, Susumu
Torriani Gorini, Annamaria
Troxler, Robert Fulton
Urry, Lisa Andrea
Vanderburg, Charles R
Villa-Komaroff, Lydia
Wald, George
Wang, Chih-Lueh Albert
Wang, James C
Wang, Yu-Li
Webb, Andrew Clive
Weinberg, Robert A
Williamson, Patrick Leslie
Witman, George Bodo, III
Wright, Andrew
Young, Delano Victor
Zannis, Vassilis I
Zimmermann, Robert Alan

MICHIGAN
Abraham, Irene
Bender, Robert Algerd
Bergen, Werner Gerhard
Bernstein, Isadore A
Bouma, Hessel, III
Brawn, Mary Karen
Briggs, Josephine P
Brockman, William Warner
Burnett, Jean Bullard
Bush, Guy L
Butterworth, Francis M
Calvert, Jay Gregory
Chang, Chia-Cheng
Chatterjee, Bandana
Chaudhry, G Rasul
Chopra, Dharam Pal
Chou, Kuo Chen
Christman, Judith Kershaw
Datta, Prasanta
Douthit, Harry Anderson, Jr
Duell, Elizabeth Ann
Einspahr, Howard Martin
Eisenstein, Barry I
Fernandez-Madrid, Felix
Ficsor, Gyula
Fisher, Linda E
Fluck, Michele Marguerite
Freeman, Arthur Scott
Fulbright, Dennis Wayne
Garvin, Jeffrey Lawrence
Gelehrter, Thomas David
Gentile, James Michael
Goustin, Anton Scott
Grossman, Lawrence I
Hari, V
Heady, Judith E

Hirata, Fusao
Hupe, Donald John
Jackson, Matthew Paul
Janik, Borek
Jones, Lily Ann
Kim, Sun-Kee
Krawetz, Stephen Andrew
Lahti, Robert A
Landefeld, Thomas Dale
Levine, Myron
Lomax, Margaret Irene
Loor, Rueyming
Lutter, Leonard C
McCauley, Roy Barnard
McCormick, J Justin
McGroarty, Estelle Josephine
McIntosh, Lee
Maher, Veronica Mary
Meisler, Miriam Horowitz
Miller, Orlando Jack
Mizukami, Hiroshi
Montgomery, Ilene Nowicki
Morse, Philip Dexter, II
Moudgil, Virinder Kumar
Mulks, Martha Huard
Neidhardt, Frederick Carl
Payne, Anita H
Pecoraro, Vincent L
Pelzer, Charles Francis
Podila, Gopi Krishna
Preiss, Jack
Reusser, Fritz
Revzin, Arnold
Righthand, Vera Fay
Robbins, Leonard Gilbert
Rollins-Page, Earl Arthur
Saltiel, Alan Robert
Sarkar, Fazlul Hoque
Savageau, Michael Antonio
Schramm, John Gilbert
Siegel, Albert
Smith, David I
Smith, Paul Dennis
Snyder, Loren Russell
Somerville, Christopher Roland
Taggart, R(obert) Thomas
Tashian, Richard Earl
Trosko, James Edward
Tse, Harley Y
Uhler, Michael David
Varani, James
Venta, Patrick John
Whalen, Joseph Wilson

MINNESOTA
Adolph, Kenneth William
Anderson, Dwight Lyman
Blumenfeld, Martin
Cleary, Paul Patrick
Ellinger, Mark Stephen
Fallon, Ann Marie
Fass, David N
Forro, Frederick, Jr
Francis, Gary Stuart
Gengenbach, Burle Gene
Gerding, Dale Nicholas
Getz, Michael John
Haase, Ashley Thomson
Hedman, Stephen Clifford
Johnson, Russell Clarence
Johnson, Theodore Reynold
Kline, Bruce Clayton
Lefebvre, Paul Alvin
Lindgren, Alice Marilyn Lindell
Magee, Paul Terry
Malewicz, Barbara Maria
Miller, Richard Lynn
Nelson, John Daniel
Oetting, William Starr
Reilly, Bernard Edward
Rubenstein, Irwin
Salisbury, Jeffrey L
Salstrom, John Stuart
Schottel, Janet L
Sicard, Raymond Edward
Silflow, Carolyn Dorothy
Tautvydas, Kestutis Jonas
Tindall, Donald J
Todt, William Lynn
Toscano, William Agostino, Jr
Verma, Anil Kumar
Weatherbee, James A

MISSISSIPPI
Adams, Junius Greene, III
Biesiot, Patricia Marie
Boyle, John Aloysius, Jr
Case, Steven Thomas
Chaires, Jonathan Bradford
Gafford, Lanelle Guyton
Graves, David E
Olson, Mark Obed Jerome
Outlaw, Henry Earl
Slobin, Lawrence I
Wahba, Albert J
Williamson, John S

MISSOURI
Adams, Steven Paul
Adldinger, Hans Karl
Alexander, Stephen
Allen, Robert Wade
Apirion, David
Armbrecht, Harvey James

Ayyagari, L Rao
Barnes, Wayne Morris
Beachy, Roger Neil
Bell, Jeffrey
Bolla, Robert Irving
Bovy, Philippe R
Braciale, Vivian Lam
Brandt, E J
Brown, Karen Kay (Kilker)
Counts, David Francis
David, John Dewood
De Blas, Angel Luis
DiFate, Victor George
Drahos, David Joseph
Eissenberg, Joel Carter
Elgin, Sarah Carlisle Roberts
Eliceiri, George Louis
Fernandez-Pol, Jose Alberto
Ferris, Deam Hunter
Fischhoff, David Allen
Folk, William Robert
Fraley, Robert Thomas
Gordon, Jeffrey I
Grant, Gregory Alan
Green, Eric Douglas
Guilfoyle, Thomas J
Hagen, Gretchen
Hirschberg, Rona L
Holtzer, Marilyn Emerson
Hopkins, Johns Wilson
Horsch, Robert Bruce
Jaworski, Ernest George
Kane, James F
Kaplan, Arnold
Kennell, David Epperson
Koplow, Jane
Korsmeyer, Stanley Joel
Krivi, Gwen Grabowski
Lafrenz, David E
Lie, Wen-Rong
Lin, Hsiu-san
Lohman, Timothy Michael
Lozeron, Homer A
Melechen, Norman Edward
Merlie, John Paul
Miles, Charles Donald
Mortensen, Harley Eugene
Mueckler, Mike Max
Munns, Theodore Willard
Rice, Charles Moen, III
Riddle, Donald Lee
Sadler, J(asper) Evan
Salkoff, Lawrence Benjamin
Sargentini, Neil Joseph
Schlesinger, Sondra
Schlessinger, David
Schuck, James Michael
Staatz, William D
Strauss, Arnold Wilbur
Thach, Robert Edwards
Thomas, George Joseph, Jr
Wang, Richard J
Willard, Mark Benjamin
Wood, David Collier
Yarbro, John Williamson

MONTANA
Fevold, Harry Richard
Warren, Guylyn Rea

NEBRASKA
Boernke, William E
Brumbaugh, John (Albert)
Cox, George Stanley
Dubes, George Richard
Elthon, Thomas Eugene
Leuschen, M Patricia
Morris, Thomas Jack
Partridge, James Enoch
Schneiderman, Martin Howard
Schwartzbach, Steven Donald
Sharp, John Graham
Veomett, George Ector
Weeks, Donald Paul

NEVADA
Harrington, Rodney E
Winicov, Ilga

NEW HAMPSHIRE
Bergman, Kenneth David
Cole, Charles N
Gross, Robert Henry
Trumpower, Bernard Lee
Vournakis, John Nicholas

NEW JERSEY
Abou-Sabe, Morad A
Agathos, Spiros Nicholas
Alonso-Caplen, Firelli V
Axelrod, David
Babu, Uma Mahesh
Barkalow, Fern J
Barton, Beverly E
Basu, Mitali
Boublik, Miloslav
Brandriss, Marjorie C
Byrne, Barbara Jean McManamy
Byrne, Bruce Campbell
Caporale, Lynn Helena
Cascieri, Margaret Anne
Champe, Sewell Preston
Chizzonite, Richard A
Choi, Ye-Chin

Collett, Edward
Collier, Marjorie McCann
Cunningham, Earlene Brown
Daie, Jaleh
Day, Peter Rodney
Day-Salvatore, Debra-Lynn
Denhardt, David Tilton
DePamphilis, Melvin Louis
Egan, Robert Wheeler
Ennis, Herbert Leo
Essien, Francine B
Feighner, Scott Dennis
Fernandes, Prabhavathi Bhat
Fitzgerald, Paula Marie Dean
Flick, Christopher E
Flint, Sarah Jane
Frenkel, Gerald Daniel
Fresco, Jacques Robert
Friedman, Thea Marla
Gage, L Patrick
Gagna, Claude Eugene
Garro, Anthony Joseph
Gaynor, John James
Ghosh, Bijan K
Gillum, Amanda McKee
Gilman, Steven Christopher
Ginsberg, Barry Howard
Goldman, Emanuel
Goldsmith, Laura Tobi
Gray, Harry Edward
Hitchcock-DeGregori, Sarah Ellen
Hochstadt, Joy
Hu, Ching-Yeh
Humayun, Mir Z
Inouye, Masayori
Jakubowski, Hieronim Zbigniew
Johnson, Frank Harris
Kaback, David Brian
Kaiserman, Howard Bruce
Kamiyama, Mikio
Kirsch, Donald R
Klessig, Daniel Frederick
Kramer, Richard Allen
Krishnan, Gopal
Kumar, Suriender
Kurtz, Myra Berman
Laskin, Allen I
Leibowitz, Michael Jonathan
Lenard, John
Levine, Arnold J
Liu, Alice Yee-Chang
Lomedico, Peter T
McKearn, Thomas Joseph
Manne, Veeraswamy
Margolskee, Robert F
Mark, David Fu-Chi
Martin, Charles Everett
Massover, William H
Messing, Joachim W
Monestier, Marc
Mosley, Stephen T
Mullinix, Kathleen Patricia
Nelson, Nathan
Nemecek, Georgina Marie
Niederman, Robert Aaron
Ofengand, Edward James
Ozer, Harvey Leon
Pestka, Sidney
Pieczenik, George
Plescia, Otto John
Prince, Roger Charles
Reuben, Roberta C
Rhoten, William Blocher
Rodwell, John Dennis
Rosenstraus, Maurice Jay
Ryzlak, Maria Teresa
Sanders, Marilyn Magdanz
Schlesinger, R(obert) Walter
Schroff, Peter David
Senda, Mototaka
Shakarjian, Michael Peter
Shatkin, Aaron Jeffrey
Sherman, Michael Ian
Shin, Seung-il
Sikder, Santosh K
Silver, Lee Merrill
Sinha, Navin Kumar
Skalka, Anna Marie
Sofer, William Howard
Tesoriero, John Vincent
Thomas, Kenneth Alfred, Jr
Tilghman, Shirley Marie
Trotta, Paul F
Unowsky, Joel
Vasconcelos, Aurea C
Wang-Iverson, Patsy
Wasserman, Paul Michael
Wasserman, Bruce P
Weissbach, Herbert
Werth, Jean Marie

NEW MEXICO
Bartholdi, Marty Frank
Belinsky, Steven Alan
Burks, Christian
Buss, William Charles
Crissman, Harry Allen
Goad, Walter Benson, Jr
Kelly, Gregory
Kuehn, Glenn Dean
Lechner, John Fred
Nelson, Mary Anne
Strniste, Gary F
Thomassen, David George

NEW YORK

Abramowicz, Daniel Albert
Alexander, Renee R
Allfrey, Vincent George
Anderson, Carl William
Angerer, Lynne Musgrave
Angerer, Robert Clifford
Asch, Harold Lawrence
Assoian, Richard Kenneth
Atkinson, Paul H
Augenlicht, Leonard Harold
Axel, Richard
Bachrach, Howard L
Bachvaroff, Radoslav J
Bachvarova, Rosemary Faulkner
Badoyannis, Helen Litman
Baltimore, David
Bambara, Robert Anthony
Banerjee, Debendranath
Banerjee, Ranjit
Barany, Francis
Bard, Enzo
Bartus, Raymond T
Basilico, Claudio
Basilico, Claudio
Bauer, William R
Bedard, Donna Lee
Belfort, Marlene
Bender, Michael A
Benjamin, William B
Berezney, Ronald
Black, Lindsay MacLeod
Bloch, Alexander
Bofinger, Diane P
Bopp, Lawrence Howard
Bradbury, Michael Wayne
Bregman, Allyn A(aron)
Broadway, Roxanne Meyer
Broker, Thomas Richard
Brooks, Robert R
Brunner, Michael
Bruns, Peter John
Buckley, Edward Harland
Budd, Thomas Wayne
Bukhari, Ahmad Iqbal
Burger, Richard Melton
Buxbaum, Joel N
Calame, Kathryn Lee
Carlson, Marian B
Carter, Timothy Howard
Catterall, James F
Celis, Roberto T F
Chan, Samuel H P
Chapman, Verne M
Chatterjee, Nando Kumar
Chen, Ching-Ling Chu
Chovan, James Peter
Chow, Louise Tsi
Clark, Virginia Lee
Coderre, Jeffrey Albert
Cohen, Leonard A
Cole, Patricia Ellen
Colman, David Russell
Conway de Macario, Everly
Cooper, Norman S
Coppock, Donald Leslie
Craddock, Elysse Margaret
Cross, George Alan Martin
Cunningham, Richard Preston
D'Adamo, Amedeo Filiberto, Jr
Davies, Kelvin J A
Day, Loren A
De Leon, Victor M
Delihas, Nicholas
Desnick, Robert John
Desplan, Claude
D'Eustachio, Peter
Drickamer, Kurt
Drlica, Karl
Dubroff, Lewis Michael
Dudock, Bernard S
Duff, Ronald George
Dunn, John Patrick James
Easton, Douglas P
Eberle, Helen I
Eckhardt, Ronald A
Edelman, Gerald Maurice
Eisinger, Josef
Emmons, Scott W
Enea, Vincenzo
Engelhardt, Dean Lee
Evans, Irene M
Evans, Mary Jo
Figurski, David Henry
Finkelstein, Jacob Noah
Finlay, Thomas Hiram
Fischman, Harlow Kenneth
Fisher, Paul Andrew
Fishman, Jerry Haskel
Fox, Thomas David
Free, Stephen J
Freedberg, Irwin Mark
Freedman, Aaron David
Freundlich, Martin
Friedman, Selwyn Marvin
Garrels, James I
Garrick, Laura Morris
Garte, Seymour Jay
Gibson, Audrey Jane
Gibson, Kenneth David
Goff, Stephen Payne
Goldberg, Allan Roy
Goldschmidt, Raul Max
Goldstein, Mindy Sue

Gordon, Jon W
Grodzicker, Terri Irene
Grubman, Marvin J
Grumet, Martin
Gulati, Subhash Chander
Gupta, Sanjeev
Gurel, Okan
Gutcho, Sidney J
Haas, David Jean
Hall, Barry Gordon
Hanafusa, Hidesaburo
Hardy, Ralph Wilbur Frederick
Harford, Agnes Gayler
Hehre, Edward James
Held, William Allen
Henshaw, Edgar Cummings
Hilborn, David Alan
Hinkle, David Currier
Horiuchi, Kensuke
Hough, Paul Van Campen
Howard, Glenn Willard, Jr
Howell, Stephen Herbert
Hsu, Ming-Ta
Huang, Alice Shih-Hou
Huberman, Joel Anthony
Hurley, Ian Robert
Iglewski, Wallace
Isseroff, Hadar
Jacobson, Ann Beatrice
Jakes, Karen Sorkin
Janakidevi, K
Joshi, Sharad Gopal
Kascsak, Richard John
Katz, Eugene Richard
Kaye, Jerome Sidney
Kaye, Nancy Weber
Keng, Peter C
Kerwar, Suresh
Khan, Nasim A
Kish, Valerie Mayo
Kline, Larry Keith
Koide, Samuel Saburo
Kowalski, David Francis
Krakow, Joseph S
Kramer, Fred Russell
Krim, Mathilde
Lacks, Sanford
LaFountain, James Robert, Jr
Landau, Joseph Victor
Lange, Christopher Stephen
Lanks, Karl William
Last, Robert L
Lau, Joseph T Y
Lavett, Diane Kathryn Juricek
Lazzarini, Robert A
Leary, James Francis
Ledbetter, Myron C
Lee-Huang, Sylvia
Leifer, Zer
Leleiko, Neal Simon
Lemanski, Larry Frederick
Lewis, James Bryan
Lin, Fu Hai
Lindahl, Lasse Allan
Lipke, Peter Nathan
Lis, John Thomas
Lucas, John J
Lusty, Carol Jean
Lyman, Harvard
Macario, Alberto J L
McCabe, R Tyler
McCann-Collier, Marjorie Dolores
McCormick, Paulette Jean
McEwen, Bruce Sherman
McFall, Elizabeth
Madison, James Thomas
Maio, Joseph James
Maitra, Umadas
Maniloff, Jack
Manly, Kenneth Fred
Mannino, Raphael James
Marcu, Kenneth Brian
Marenus, Kenneth D
Margolin, Paul
Marians, Kenneth J
Marmur, Julius
Mascarenhas, Joseph Peter
Massie, Harold Raymond
Mazurkiewicz, Joseph Edward
Mehta, Bipin Mohanlal
Meruelo, Daniel
Michels, Corinne Anthony
Miller, David Lee
Millis, Albert Jason Taylor
Model, Peter
Morin, John Edward
Morse, Stephen Scott
Murphy, Timothy F
Neal, Michael William
Neubort, Shimon
Newman, Stuart Alan
Nguyen-Huu, Chi M
Norvitch, Mary Ellen
Novick, Richard P
Obendorf, Ralph Louis
Olmsted, Joanna Belle
Paine, Philip Lowell
Pelicci, Pier Guiseppe
Pellicer, Angel
Pierucci, Olga
Pietraface, William John
Pogo, A Oscar
Pogo, Beatriz G T
Polatnick, Jerome

Pollack, Robert Elliot
Powell, Sharon Kay
Prakash, Louise
Price, Frederick William
Price, Peter Michael
Pruslin, Fred Howard
Pumo, Dorothy Ellen
Quigley, Gary Joseph
Reddy, Kalluru Jayarami
Redman, Colvin Manuel
Reichert, Leo E, Jr
Reid, Lola Cynthia McAdams
Reissig, Jose Luis
Riley, Monica
Roberts, James Lewis
Roberts, Jeffrey Warren
Roberts, Richard John
Robinson, Alix Ida
Rodman, Toby C
Roeder, Robert Gayle
Roepe, Paul David
Rogerson, Allen Collingwood
Roll, David E
Rosenberg, Barbara Hatch
Rothstein, Rodney Joel
Roy, Harry
Ruddell, Alanna
Rudy, Bernardo
Russel, Marjorie Ellen
Sabatini, David Domingo
Sabban, Esther Louise
Saez, Juan Carlos
Salley, John Jones, Jr
Salvo, Joseph J
Sanford, Karl John
Sarma, Raghupathy
Sarma, Ramaswamy Harihara
Satir, Peter
Scher, William
Schlessinger, Joseph
Schloer, Gertrude M
Schneider, Bruce Solomon
Schoenborn, Benno P
Schoenfeld, Cy
Schulman, LaDonne Heaton
Seeman, Nadrian Charles
Sehgal, Pravinkumar B
Shafer, Stephen Joel
Shaffer, Jacquelin Bruning
Shafit-Zagardo, Bridget
Shafritz, David Andrew
Shahn, Ezra
Shalloway, David Irwin
Sherman, Fred
Shodell, Michael J
Silverstein, Emanuel
Silverstein, Samuel Charles
Silverstein, Saul Jay
Simon, Martha Nichols
Sirlin, Julio Leo
Smith, Issar
Sobel, Jael Sabina
Sobell, Henry Martinique
Soifer, David
Squinto, Stephen P
Srivastava, Bejai Inder Sahai
Stanley, Pamela Mary
Steinberg, Bettie Murray
Sternglanz, Rolf
Straus, David Bradley
Studier, Frederick William
Sullivan, David Thomas
Swaney, Lois Mae
Sweet, Robert Mahlon
Taber, Harry Warren
Tabor, John Malcolm
Taub, Mary L
Thomas, John Owen
Tilley, Shermaine Ann
Toenniessen, Gary Herbert
Tomasi, Thomas B, Jr
Trimble, Robert Bogue
Tsai, Jir Shiong
Tye, Bik-Kwoon
Unterman, Ronald David
Van't Hof, Jack
Vulliemoz, Yvonne
Wang, Tung Yue
Warner, Jonathan Robert
Wasserman, William John
West, Norman Reed
Wiberg, John Samuel
Wilhelm, James Maurice
Wilson, David Buckingham
Wimmer, Eckard
Wishnia, Arnold
Wolfner, Mariana Federica
Wolgemuth, Debra Joanne
Wood, Harry Alan
Woods, Philip Sargent
Wu, Joseph M
Wulff, Daniel Lewis
Yang, Song-Yu
Yoder, Olen Curtis
Young, Charles Stuart Hamish
Yunghans, Wayne N
Zahler, Stanley Arnold
Zaitlin, Milton
Zapisek, William Francis
Zengel, Janice Marie
Zitomer, Richard Stephen

NORTH CAROLINA

Abramson, Jon Stuart

Agris, Paul F
Allis, John W
Anderson, Marshall W
Bear, Richard Scott
Bell, Juliette B
Bell, Robert Maurice
Birnbaum, Linda Silber
Blackman, Carl F
Bloom, Kerry Steven
Boone, Lawrence Rudolph, III
Bott, Kenneth F
Boyd, Jeffrey Allen
Boynton, John E
Burchall, James J
Carter, Charles Williams, Jr
Chae, Chi-Bom
Chen, Michael Yu-Men
Cheng, Yung-Chi
Chilton, Mary-Dell Matchett
Clinton, Bruce Allan
Coleman, Mary Sue
Corley, Ronald Bruce
Craig, Syndey Pollock, III
Cunningham, Carol Clem
Curtis, Susan Julia
Deher, Kevin L
de Serres, Frederick Joseph
Devereux, Theodora Reyling
Devlin, Robert B
Dibner, Mark Douglas
Dieter, Michael Phillip
Douglas, Michael Gilbert
Eddy, Edward Mitchell
Edgell, Marshall Hall
Endow, Sharyn Anne
Erickson, Harold Paul
Errede, Beverly Jean
Farber, Rosann Alexander
Fernandes, Daniel James
Fowler, Elizabeth
Gruber, Kenneth Allen
Guild, Walter Rufus
Harris, Elizabeth Holder
Hassan, Hosni Moustafa
Heise, Eugene Royce
Hemperly, John Jacob
Huang, Eng-Shang(Clark)
Hutchison, Clyde Allen, III
Jetten, Anton Marinus
Johnson, Ronald Sanders
Judd, Burke Haycock
Kalfayan, Laura Jean
Kaufmann, William Karl
Keene, Jack Donald
King, Ann Christie
Kredich, Nicholas M
Kucera, Louis Stephen
Lapetina, Eduardo G
Lee, Greta Marlene
Lund, Pauline Kay
Lundblad, Roger Lauren
Markert, Clement Lawrence
Maroni, Gustavo Primo
Mason, James Michael
Matthysse, Ann Gale
Metzgar, Richard Stanley
Mitchell, Thomas Greenfield
Mott, Ralph Lionel
Negishi, Masahiko
Newell, Nanette
Pagano, Joseph Stephen
Petes, Thomas Douglas
Phelps, Allen Warner
Pizzo, Salvatore Vincent
Pukkila, Patricia Jean
Richardson, Jane S
Sahyoun, Naji Elias
Sancar, Aziz
Sancar, Gwendolyn Boles
Scandalios, John George
Shafer, Thomas Howard
Shih, Jason Chia-Hsing
Sonntag, William Edmund
Spiker, Steven L
Steege, Deborah Anderson
Stiles, Gary L
Stone, Henry Otto, Jr
Suk, William Alfred
Sung, Michael Tse Li
Teng, Christina Wei-Tien Tu
Theil, Elizabeth
Thompson, William Francis, III
Toews, Arrel Dwayne
Walker, Cheryl Lyn
Webster, Robert Edward
Wheeler, Kenneth Theodore, Jr
White, Glenn E
Wilson, Elizabeth Mary
Witt, Donald James
Wong, Fulton

NORTH DAKOTA

Duerre, John A
Park, Chung Sun

OHIO

Atkins, Charles Gilmore
Banerjee, Amiya Kumar
Banerjee, Sipra
Baroudy, Bahige M
Birky, Carl William, Jr
Blanton, Ronald Edward
Blumenthal, Robert Martin
Boczar, Barbara Ann

Molecular Biology (cont)

Brunden, Kurt Russell
Byers, Thomas Jones
Cardell, Robert Ridley, Jr
Cohn, Norman Stanley
Cooper, Dale A
Cope, Frederick Oliver
Coplin, David Louis
Cox, Donald Cody
Cullis, Christopher Ashley
Culp, Lloyd Anthony
D'Ambrosio, Steven M
Decker, Lucile Ellen
Deters, Donald W
DiCorleto, Paul Eugene
Doerder, F Paul
Drake, Richard Lee
DuBrul, Ernest
Evans, Helen Harrington
Evans, Thomas Edward
Fenoglio, Cecilia M
Flick, Parke Kinsey
Floyd, Gary Leon
Gesinski, Raymond Marion
Goldthwait, David Atwater
Hamilton, Thomas Alan
Handwerger, Stuart
Harpst, Jerry Adams
Hillis, Llewellya
Hiremath, Shivanand T
Horseman, Nelson Douglas
Hutton, John James, Jr
Jacobs-Lorena, Marcelo
Jegla, Thomas Cyril
Johnson, Lee Frederick
Johnson, Thomas Raymond
Kantor, George Joseph
Keller, Stephen Jay
Kolattukudy, P E
Kopchick, John J
Kriebel, Howard Burtt
Lane, Richard Durelle
Laughner, William James, Jr
Lechner, Joseph H
Leis, Jonathan Peter
Luck, Dennis Noel
McCorquodale, Donald James
McQuarrie, Irvine Gray
Mahmoud, Adel A F
Mamrack, Mark Donovan
Martin, Scott McClung
Meyer, Ralph Roger
Miller, Robert Harold
Milsted, Amy
Moffet, Robert Bruce
Nebert, Daniel Walter
Oakley, Berl Ray
Oleinick, Nancy Landy
Rake, Adrian Vaughan
Rhodes, Judith Carol
Rowe, John James
Samols, David R
Sawicki, Stanley George
Scarpa, Antonio
Scott, Roy Albert, III
Sen, Ganes C
Shannon, Barry Thomas
Shuster, Charles W
Slonczewski, Joan Lyn
Stanberry, Lawrence Raymond
Stukus, Philip Eugene
Sunkara, Sai Prasad
Swenson, Richard Paul
Tabita, F Robert
Tepperman, Katherine Gail
Thomas, Donald Charles
Thornton, Janice Elaine
Tomei, L David
Treichel, Robin Stong
Trewyn, Ronald William
Tykocinski, Mark L
Versteegh, Larry Robert
Waterson, John R
Wiginton, Dan Allen
Williams, Marshall Vance
Wilson, Kenneth Glade
Wise, Donald L
Woodruff, Ronny Clifford
Woodworth, Mary Esther

OKLAHOMA
Altmiller, Dale Henry
Brasch, Klaus Rainer
Broyles, Robert Herman
Chung, Kyung Won
Delaney, Robert
Epstein, Robert B
Essenberg, Richard Charles
Gardner, Charles Olda, Jr
Graves, Donald C
Hall, Leo Terry
Harmon, H James
Johnson, Arthur Edward
Klein, Ronald Don
Matsumoto, Hiroyuki
Melcher, Ulrich Karl
Morrissey, James Henry
Roe, Bruce Allan
Schaefer, Frederick Vail
Schubert, Karel Ralph
Silverman, Philip Michael
Trachewsky, Daniel
Watson, Gary Hunter

Wolgamott, Gary
Yi, Cho Kwang

OREGON
Bagby, Grover Carlton
Burgeson, Robert Eugene
Civelli, Oliver
Clinton, Gail M
Dietz, Thomas John
Dooley, Douglas Charles
Field, Katharine G
Gold, Michael Howard
Gould, Steven James
Hays, John Bruce
Ho, Iwan
Kohler, Peter
Matthews, Brian Wesley
Meints, Russel H
Millette, Robert Loomis
Novick, Aaron
Postlethwait, John Harvey
Ream, Lloyd Walter, Jr
Rittenberg, Marvin Barry
Robinson, Arthur B
Russell, Peter James
Schellman, John Anthony
Soderling, Thomas Richard
Thorsett, Grant Orel

PENNSYLVANIA
Abrams, William R
Anderson, Thomas Foxen
Ayers, Arthur Raymond
Aynardi, Martha Whitman
Baker, Alan Paul
Balin, Arthur Kirsner
Baserga, Renato
Bering, Charles Lawrence
Bernlohr, Robert William
Bhaduri, Saumya
Billheimer, Foster E
Bodine, Peter Van Nest
Brenchley, Jean Elnora
Brinigar, William Seymour, Jr
Brooks, John S J
Brownstein, Barbara L
Burnett, Roger Macdonald
Cannizzaro, Linda A
Chang, Frank N
Chung, Albert Edward
Cohen, Leonard Harvey
Cohen, Stanley
Cooperman, Barry S
Currier, Thomas Curtis
Cutler, Winnifred Berg
D'Agostino, Marie A
Das, Manjusri
Debouck, Christine Marie
Debroy, Chitrita
Deering, Reginald
Diorio, Alfred Frank
Diven, Warren Field
Donnelly, John James, III
Emberton, Kenneth C
Evensen, Kathleen Brown
Fansler, Bradford S
Farber, Florence Eileen
Feinstein, Sheldon Israel
Ferrone, Frank Anthony
Fisher, Edward Allen
Frankel, Fred Robert
Fraser, Nigel William
Freed, Jerome James
Gay, Carol Virginia Lovejoy
Gealt, Michael Alan
Gilbert, Scott F
Glorioso, Joseph Charles, III
Godfrey, Susan Sturgis
Goff, Christopher Godfrey
Goldstein, Norma Ornstein
Goodgal, Sol Howard
Graetzer, Reinhard
Grebner, Eugene Ernest
Grunstein, Michael M
Grunwald, Gerald B
Hardison, Ross Cameron
Hendrix, Roger Walden
Henry, Susan Armstrong
Hopper, Anita Klein
Howe, Chin Chen
Hudson, Alan P
Jackson, Ethel Noland
Jacobson, Lewis A
Jarvik, Jonathan Wallace
Johnson, Kenneth Allen
Johnston, James Bennett
Karush, Fred
Kayne, Marlene Steinmetz
Kennett, Roger H
Khatami, Mahin
Khoury, George
Ladda, Roger Louis
Lawrence, William Chase
Layne, Porter Preston
Lazo, John Stephen
Leboy, Phoebe Starfield
Leibel, Wayne Stephan
Lindstrom, Jon Martin
Llinas, Miguel
Lockwood, Arthur H
Logan, David Alexander
Long, Walter Kyle, Jr
Lotze, Michael T
Lu, Ponzy

McCarthy, Patrick Charles
McClure, William Robert
McKenna, William Gillies
McLaughlin, Patricia J
McMorris, F Arthur
McNamara, Pamela Dee
Manolson, Morris Frank
Manson, Lionel Arnold
Masker, Warren Edward
Mastro, Andrea M
Mifflin, Theodore Edward
Milcarek, Christine
Morris, Sidney Machen, Jr
Nass, Margit M K
Neubauer, Russell Howard
Niu, Mann Chiang
Oldham, James Warren
Opas, Evan E
Peebles, Craig Lewis
Perry, Robert Palese
Person, Stanley R
Phillips, Allen Thurman
Phillips, Stephen Lee
Pollock, John Archie
Porter, Ronald Dean
Poste, George Henry
Pratt, Elizabeth Ann
Press, Linda Seghers
Punnett, Hope Handler
Ramachandran, N
Ricciardi, Robert Paul
Ritter, Carl A
Robb, Richard John
Roesing, Timothy George
Rosenberg, Martin
Rosenblum, Irwin Yale
Rubin, Donald Howard
Russell, Alan James
Rutman, Robert Jesse
Salerno, Ronald Anthony
Sanger, Jean M
Sanger, Joseph William
Sato, Masahiko
Scher, Charles D
Schlegel, Robert Allen
Scholnick, Steven Bruce
Schwartz, Elias
Shockman, Gerald David
Sideropoulos, Aris S
Singer, Alan G
Stewart, Barbara Yost
Stolc, Viktor
Sussman, Maurice
Sylvester, James Edward
Taubler, James H
Tax, Anne
Taylor, William Daniel
Tevethia, Mary Judith (Robinson)
Tu, Chen-Pei David
Tuan, Rocky Sung-Chi
Turoczi, Lester J
Vaidya, Akhil Babubhai
Vlasuk, George P
Wachsberger, Phyllis Rachelle
Weinberg, Eric S
Wetzel, Ronald Burnell
Whitfield, Carol F(aye)
Zweidler, Alfred

RHODE ISLAND
Brautigan, David L
Cohen, Paul S(idney)
Coleman, John Russell
Dahlberg, Albert Edward
Eil, Charles
Estrup, Faiza Fawaz
Fausto, Nelson
Fisher, Harold Wilbur
Knopf, Paul M
Landy, Arthur
Lederberg, Seymour
O'Leary, Gerard Paul, Jr
Rosenstein, Barry Sheldon
Rotman, Boris
Sapolsky, Asher Isadore
Shank, Peter R
Ware, Vassie C
Zackroff, Robert V

SOUTH CAROLINA
Arnaud, Philippe
Barnett, Ortus Webb, Jr
Berger, Franklin Gordon
Brown, Arnold
Chao, Julie
Chao, Lee
Davis, Leroy
Ely, Berten E, III
Higerd, Thomas Braden
Kistler, Wilson Stephen, Jr
Larcom, Lyndon Lyle
Ledford, Barry Edward
Lin, Tu
Maheshwari, Kewal Krishnan
Paddock, Gary Vincent
Richards, Gary Paul
Wilson, Gregory Bruce
Yardley, Darrell Gene

SOUTH DAKOTA
Hildreth, Michael B
Lindahl, Ronald
West, Thomas Patrick
Westfall, Helen Naomi

TENNESSEE
Anderson, Ted L
Barnett, William Edgar
Bhorjee, Jaswant S
Billen, Daniel
Blaser, Martin Jack
Bryant, Robert Emory
Bunick, Gerard John
Clark, Winston Craig
Close, David Matzen
Cook, George A
Cooper, Terrance G
Duhl, David M
Fleischer, Sidney
Friedman, Daniel Lester
Fujimura, Robert
Granner, Daryl Kitley
Granoff, Allan
Griffin, Guy David
Guyer, Cheryl Ann
Hadden, Charles Thomas
Hazelton, Bonni Jane
Henke, Randolph Ray
Hingerty, Brian Edward
Howe, Martha Morgan
Howell, Elizabeth E
Incardona, Antonino L
Jarrett, Harry Wellington, III
Jeon, Kwang Wu
Katze, Jon R
Kitabchi, Abbas E
Larimer, Frank William
LeStourgeon, Wallace Meade
McGavock, Walter Donald
Mitchell, William Marvin
Mitra, Sankar
Monaco, Paul J
Mullin, Beth Conway
Niyogi, Salil Kumar
Ohi, Seigo
Olins, Ada Levy
Olins, Donald Edward
Portner, Allen
Purcell, William Paul
Rettenmier, Carl Wayne
Riggsby, William Stuart
Rinchik, Eugene M
Savage, Dwayne Cecil
Skinner, Dorothy M
Staros, James Vaughan
Stevens, Stanley Edward, Jr
Thomason, Donald Brent
Tibbetts, Clark
Touster, Oscar
Trigiano, Robert Nicholas
Trupin, Joel Sunrise
Watterson, D Martin
Williams, Robley Cook, Jr
Woychik, Richard P
Yang, Wen-Kuang

TEXAS
Adair, Gerald Michael
Aggarwal, Bharat Bhushan
Alhossainy, Effat M
Alvarez-Gonzalez, Rafael
Amoss, Max St Clair
Angelides, Kimon Jerry
Arenaz, Pablo
Arntzen, Charles Joel
Atkinson, Mark Arthur Leonard
Aufderheide, Karl John
Beaudet, Arthur L
Beckingham, Kathleen Mary
Bellion, Edward
Bennett, Dorothea
Bennett, George Nelson
Breen, Gail Anne Marie
Bremer, Hans
Brysk, Miriam Mason
Buchanan, Christine Elizabeth
Bursztajn, Sherry
Campbell, James Wayne
Carney, Darrell Howard
Carson, Daniel Douglas
Chadwick, Arthur Vorce
Chalmers, John Harvey, Jr
Chan, John Yeuk-hon
Chan, Lawrence Chin Bong
Chan, Lee-Nien Lillian
Chang, Donald Choy
Clowes, Royston Courtenay
Collisson, Ellen Whited
Coulter, Murray W
Crawford, Isaac Lyle
Davis, Claude Geoffrey
Dedman, John Raymond
Dempsey, Walter B
Denney, Richard Max
Donnelly, Patricia Vryling
Dowhan, William
Earhart, Charles Franklin, Jr
Epstein, Henry F
Evans, Claudia T
Evans, John Edward
Faust, Charles Harry, Jr
Fleischmann, William Robert, Jr
Forrest, Hugh Sommerville
Fox, George Edward
Fox, Jack Lawrence
Frederiksen, Richard Allan
Fresquez, Catalina Lourdes
Garber, Alan J
Garcia, Hector D

Garrard, William T
Georgiou, George
Gomer, Richard Hans
Goodman, Joel Mitchell
Gorelic, Lester Sylvan
Gottlieb, Paul David
Gratzner, Howard G
Griffin, James Emmett
Grimm, Elizabeth Ann
Hackert, Marvin LeRoy
Hagino, Nobuyoshi
Hall, Timothy Couzens
Hanis, Craig L
Harris, Stephen Eubank
Hecht, Ralph Martin
Heimbach, Richard Dean
Hewitt, Roger R
Hillar, Marian
Hillis, David Mark
Hixson, James Elmer
Holoubek, Viktor
Hopkins, Robert Charles
Howard, Harriette Ella Pierce
Huang, Bessie Pei-Hsi
Ihler, Garret Martin
Ippen-Ihler, Karin Ann
Jacobson, Elaine Louise
Jagger, John
James, Harold Lee
Joys, Terence Michael
Kit, Saul
Konkel, David Anthony
Kraig, Ellen
Kroschewsky, Julius Richard
Kurzrock, Razelle
Kutteh, William Hanna
Lang, Dimitrij Adolf
LeBlanc, Donald Joseph
Ledley, Fred David
Lee, Chong Sung
Leuchtag, H Richard
Lever, Julia Elizabeth
Lieberman, Michael Williams
Linner, John Gunnar
Lupski, James Richard
McBride, Raymond Andrew
McCammon, James Andrew
MacLeod, Michael Christopher
Mandel, Manley
Marsh, Robert Cecil
Masters, Bettie Sue Siler
Mayor, Heather Donald
Means, Anthony R
Meltz, Martin Lowell
Mills, William Ronald
Murgola, Emanuel J
Nguyen, Henry Thien
Nicholson, Wayne Lowell
O'Malley, Bert W
Patel, Nutankumar T
Perlman, Philip Stewart
Peterson, David Oscar
Pirtle, Robert M
Platsoucas, Chris Dimitrios
Poduslo, Shirley Ellen
Poenie, Martin Francis
Powell, Bernard Lawrence
Rajaraman, Srinivasan
Ritter, Nadine Marie
Ritzi, Earl Michael
Robberson, Donald Lewis
Romeo, Tony
Rose, Kathleen Mary
Rosen, Jeffrey Mark
Ross, Elliott M
Roth, Jack A
Roy, Arun K
Rupert, Claud Stanley
Sauer, Helmut Wilhelm
Saunders, Grady Franklin
Schaffer, Barbara Noyes
Schrader, William Thurber
Serwer, Philip
Shay, Jerry William
Smith, Russell Lamar
Snell, William J
Sparkman, Dennis Raymond
Stone, William Harold
Stroman, David Womack
Stubblefield, Travis Elton
Summers, Max Duane
Temeyer, Kevin Bruce
Thompson, Edward Ivins Bradbridge
Thompson, Ernest Aubrey, Jr
Tomasovic, Stephen Peter
Tong, Alex W
Trkula, David
Tsai, Ming-Jer
Vinson, S Bradleigh
Walmsley, Judith Abrams
Walter, Ronald Bruce
Warner, Marlene Ryan
Washington, Arthur Clover
Wild, James Robert
Williams, Robert Sanders
Wilson, Carol Maggart
Woo, Savio L C
Wood, Robert Charles
Wright, Stephen E
Wright, Woodring Erik
Yang, Funnei
Yeh, Lee-Chuan Caroline
Yeoman, Lynn Chalmers
Yielding, K Lemone

Young, Ryland F

UTAH
Aird, Steven Douglas
Anderson, Anne Joyce
Carroll, Dana
Farmer, James Lee
Franklin, Naomi C
Fujinami, Robert S
Georgopoulos, Constantine Panos
Gesteland, Raymond Frederick
Gurney, Theodore, Jr
Herrick, Glenn Arthur
Hirning, Lane D
Howarth, Alan Jack
Huang, Wai Mun
Kemp, John Wilmer
Knapp, Gayle
Lark, Karl Gordon
Larsen, Lloyd Don
Lighton, John R B
Okun, Lawrence M
Parkinson, John Stansfield
Pincus, Seth Henry
Rasmussen, Kathleen Goertz
Roth, John R
Simmons, Daniel L
Smith, Marvin Artell
Summers, Donald F
Takemoto, Jon Yutaka
Torres, Anthony R
Van Alfen, Neal K
Warters, Raymond Leon
Williams, Roger Richard
Wolstenholme, David Robert
Yoshikami, Doju

VERMONT
Albertini, Richard Joseph
Hoagland, Mahlon Bush
Kelley, Jason
Long, George Louis
Raper, Carlene Allen
Schaeffer, Warren Ira
Ullrich, Robert Carl
Wallace, Susan Scholes
Weller, David Lloyd
Woodcock-Mitchell, Janet Louise

VIRGINIA
Amero, Sally Ann
Auton, David Lee
Bates, Robert Clair
Bauerle, Ronald H
Bender, Patrick Kevin
Beyer, Ann L
Blatt, Jeremiah Lion
Brown, Jay Clark
Chen, Jiann-Shin
Chlebowski, Jan F
Collins, James Malcolm
DeLorenzo, Robert John
Eden, Francine Claire
Esen, Asim
Garrett, Reginald Hooker
Garrison, James C
Grogan, William McLean
Gross, Paul Randolph
Hamlin, Joyce Libby
Jakoi, Emma Raff
Kalimi, Mohammed Yahya
Kelley, John Michael
King, Betty Louise
Kretsinger, Robert
Lacy, George Holcombe
Laurie, Gordon William
Lemp, John Frederick, Jr
Marron, Michael Thomas
Merlino, Glenn T
Miller, Oscar Lee, Jr
Osheim, Yvonne Nelson
Roberts, Catherine Harrison
Rutherford, Charles
Stout, Ernest Ray
Tolin, Sue Ann
Tyson, John Jeanes
Vermeulen, Carl William
Wagner-Bartak, Claus Gunter

WASHINGTON
Adman, Elinor Thomson
Albers, John J
Bakken, Aimee Hayes
Bendich, Arnold Jay
Bornstein, Paul
Byers, Breck Edward
Camerman, Arthur
Cattolico, Rose Ann
Cheevers, William Phillip
Clapshaw, Patric Arnold
Cosman, David John
Deming, Jody W
Finkelstein, David B
Gallant, Jonathan A
Garber, Richard Lincoln
Graves, Scott Stoll
Gurusiddaiah, Sarangamat
Hall, Benjamin Downs
Hall, Stanton Harris
Hauschka, Stephen D
Haydock, Paul Vincent
Hazelbauer, Gerald Lee
Hosick, Howard Lawrence
Hu, Shiu-Lok

Kohlhepp, Sue Joanne
Kraft, Joan Creech
Kutter, Elizabeth Martin
McFadden, Bruce Alden
MacKay, Vivian Louise
Magnuson, Nancy Susanne
Margolis, Robert Lewis
Matlock, Daniel Budd
Narayanan, A Sampath
Nathanson, Neil Marc
Pious, Donald A
Prody, Gerry Ann
Randall, Linda Lea
Reeder, Ronald Howard
Reeves, Jerry John
Reeves, Raymond
Riddiford, Lynn Moorhead
Rohrschneider, Larry Ray
Sage, Helene E
Salk, Darrell John
Sibley, Carol Hopkins
Smith, Gerald Ralph
Steiner, Robert Alan
Stenkamp, Ronald Eugene
Stuart, Kenneth Daniel
Wakimoto, Barbara Toshiko
Weintraub, Harold M
Yancey, Paul Herbert
Yao, Meng-Chao
Young, Elton Theodore

WEST VIRGINIA
Kletzien, Rolf Frederick
Konat, Gregory W
Reichenbecher, Vernon Edgar, Jr
Sutter, Richard P
Yelton, David Baetz

WISCONSIN
Allred, Lawrence Ervin (Joe)
Amasino, Richard M
Anderson, Robert Philip
Armstrong, George Michael
Attie, Alan D
Becker, Wayne Marvin
Bergen, Lawrence
Bergtrom, Gerald
Blattner, Frederick Russell
Bock, Robert Manley
Borisy, Gary Guy
Brown, Raymond Russell
Burgess, Ann Baker
Burgess, Richard Ray
Calvanico, Nickolas Joseph
Chambliss, Glenn Hilton
Collins, Mary Lynne Perille
Courtright, James Ben
Cox, Michael Matthew
Cramer, Jane Harris
Dahlberg, James Eric
Dimond, Randall Lloyd
Dwyer-Hallquist, Patricia
Ellingboe, Albert Harlan
Fahl, William Edwin
Goldberger, Amy
Goodman, Robert Merwin
Greenspan, Daniel S
Haas, Michael John
Hokin, Lowell Edward
Hopwood, Larry Eugene
Hornemann, Ulfert
Horseman, Nelson Douglas
Hutton, James Robert
Inman, Ross
Jacobson, Gunnard Kenneth
John, Maliyakal Eappen
Johnson, Warren Victor
Kaesberg, Paul Joseph
Kahan, Lawrence
Karrer, Kathleen Marie
Keegstra, Kenneth G
Kermicle, Jerry Lee
Klebba, Phillip E
Kubinski, Henry A
Littlewood, Roland Kay
Markley, John Lute
Martin, Thomas Fabian John
Mertz, Janet Elaine
Moskowitz, Gerard Jay
Munroe, Stephen Horner
Nelson, David Lee
Norris, Dale Melvin, Jr
Pan, David
Rapacz, Jan
Record, M Thomas, Jr
Ross, Jeffrey
Scheele, Robert Blain
Seelke, Ralph Walter
Slaunterback, David Buell
Spritz, Richard Andrew
Sussman, Michael R
Sytsma, Kenneth Jay
Szybalski, Waclaw
Tufte, Marilyn Jean
Verma, Ajit K
Waring, Gail L
Wejksnora, Peter James
Wells, Barbara Duryea
Wilcox, Kent Westbrook
Williams, Jeffrey Walter
Wolter, Karl Erich

WYOMING
Ji, Inhae

Lewis, Randolph Vance
Petersen, Nancy Sue
Smith-Sonneborn, Joan

PUERTO RICO
Candelas, Graciela C

ALBERTA
Bazett-Jones, David Paul
Biederman, Brian Maurice
Church, Robert Bertram
Dancik, Bruce Paul
Dixon, Gordon H
Hart, David Arthur
Iatrou, Kostas
Jensen, Susan Elaine
Kaplan, Jacob Gordin
Langridge, William Henry Russell
Lederis, Karolis (Karl)
Lemieux, Raymond Urgel
Morgan, Antony Richard
Paetkau, Verner Henry
Rattner, Jerome Bernard
Schneider, Wolfgang Johann
Scraba, Douglas G
Singh, Bhagirath
Vance, Dennis E
Van De Sande, Johan Hubert
Weiner, Joel Hirsch

BRITISH COLUMBIA
Aebersold, Ruedi H
Astell, Caroline R
Burke, Robert D
Candido, Edward Peter Mario
Dennis, Patrick P
Druehl, Louis D
Ekramoddoullah, Abul Kalam M
Emerman, Joanne Tannis
Finlay, Barton Brett
Green, Beverley R
Griffiths, Anthony J F
Hancock, Robert Ernest William
Honda, Barry Marvin
Jasch, Laura Gwendolyn
Matheson, Alastair Taylor
Redfield, Rosemary Jeanne
Rennie, Paul Steven
Sadowski, Ivan J
Salari, Hassan
Shamoun, Simon Francis
Skala, Josef Petr
Smith, Michael Joseph
Tener, Gordon Malcolm
Weeks, Gerald

MANITOBA
Borsa, Joseph
Davie, James Ronald
Kunz, Bernard Alexander
Rosenmann, Edward A
Wrogemann, Klaus

NEW BRUNSWICK
Krause, Margarida Oliveira

NEWFOUNDLAND
Burness, Alfred Thomas Henry
Davidson, William Scott
Michalak, Thomas Ireneusz
Michalski, Chester James
Morris, Claude C

NOVA SCOTIA
Chambers, Robert Warner
Doolittle, Warren Ford, III
Gray, Michael William
MacRae, Thomas Henry
Vining, Leo Charles

ONTARIO
Abrahamson, Edwin William
Adams, Gabrielle H M
Arif, Basil Mumtaz
Asculai, Samuel Simon
Atkinson, Burr Gervais
Axelrad, Arthur Aaron
Bacchetti, Silvia
Barclay, Barry James
Barran, Leslie Rohit
Bayley, Stanley Thomas
Behki, Ram M
Birnbaum, George I
Birnboim, Hyman Chaim
Brown, Ian Ross
Brown, William Roy
Burke, Derek Clissold
Camerman, Norman
Carstens, Eric Bruce
Chambers, Ann Franklin
Chui, David H K
Cinader, Bernhard
Davis, Alan
Dedhar, Shoukat
Deeley, Roger Graham
Etches, Robert J
Fish, Eleanor N
Fitt, Peter Stanley
Forsdyke, Donald Roy
Fraser, Ann Davina Elizabeth
Freedman, Melvin Harris
Friesen, James Donald
Ghosh, Hara Prasad
Glick, Bernard Robert

Molecular Biology (cont)

Graham, Frank Lawson
Greenblatt, Jack Fred
Gupta, Radhey Shyam
Haynes, Robert Hall
Hegele, Robert A
Heikkila, John J
Hickey, Donal Aloysius
Hofmann, Theo
Horgen, Paul Arthur
Hozumi, Nobumichi
Huber, Carol (Saunderson)
Ingles, Charles James
Jeffrey, Kenneth Robert
Johnson-Lussenburg, Christine Margaret
Kang, C Yong
Kasha, Kenneth John
Kasupski, George Joseph
Kidder, Gerald Marshall
Krepinsky, Jiri J
Lau, Catherine Y
Liew, Choong-Chin
Logan, David Mackenzie
Lu, Benjamin Chi-Ko
Lynn, Denis Heward
Mackie, George Alexander
MacLennan, David Herman
Mak, William Wai-Nam
Metuzals, Janis
Mishra, Ram K
Narang, Saran A
Neelin, James Michael
Ottensmeyer, Frank Peter
Pearlman, Ronald E
Percy, Maire Ede
Przybylska, Maria
Rainbow, Andrew James
Rayman, Mohamad Khalil
Scott, Fraser Wallace
Seligy, Verner Leslie
Sells, Bruce Howard
Singal, Dharam Parkash
Singh, Rama Shankar
Smiley, James Richard
Sorger, George Joseph
Spencer, John Hedley
Stavric, Stanislava
Stiller, Calvin R
Straus, Neil Alexander
Sun, Anthony Mein-Fang
Trick, Charles Gordon
Tu, Jui-Chang
Walker, Ian Gardner
Walker, Peter Roy
Warner, Alden Howard
Warwick, Suzanne Irene
Wu, Tai Wing

QUEBEC
Anderson, William Alan
Antakly, Tony
Barden, Nicholas
Beaudoin, Adrien Robert
Bélanger, Luc
Bellabarba, Diego
Brown, Gregory Gaynor
Chabot, Benoit
Chafouleas, James G
Charest, Pierre M
Chung, Young Sup
Cousineau, Gilles H
DuBow, Michael Scott
Dunn, Robert James
Fournier, Michel
Green, David M(artin)
Kafer, Etta (Mrs E R Boothroyd)
Lapointe, Jacques
Lebel, Susan
Levasseur, Maurice Edgar
Murgita, Robert Anthony
Murthy, Mahadi Raghavandrarao Ven
Nishioka, Yutaka
Paice, Michael
Palfree, Roger Grenville Eric
Pickett, Cecil Bruce
Ponka, Premysl
Sarhan, Fathey
Seufert, Wolf D
Shoubridge, Eric Alan
Simard, Rene
Skup, Daniel
Slilaty, Steve N
Stanners, Clifford Paul
Sygusch, Jurgen
Talbot, Pierre J
Tanguay, Robert M
Trepanier, Pierre
Trudel, Michel D
Waithe, William Irwin
Wang, Eugenia

SASKATCHEWAN
Babiuk, Lorne A
Bertrand, Helmut
Blair, Donald George Ralph
Faber, Albert John
Quail, John Wilson
Robertson, Beverly Ellis

YUKON TERRITORY
Doermann, August Henry

OTHER COUNTRIES
Abelev, Garri Izrailevich
Ahn, Tae In
Amsterdam, Abraham
Anagnou, Nicholas P
Arber, Werner
Argos, Patrick
Armato, Ubaldo
Arnott, Struther
Bachmann, Fedor W
Baldwin, Robert William
Barel, Monique
Barouki, Robert
Bayev, Alexander A
Ben-Ze'ev, Avri
Bishop, David Hugh Langler
Boffa, Lidia C
Bolhuis, Reinder L H
Brody, Edward Norman
Bunn, Clive Leighton
Campbell, Virginia Wiley
Caro, Lucien G
Chetsanga, Christopher J
Choi, Yong Chun
Collins, Vincent Peter
Creighton, Thomas Edwin
Curtis, Adam Sebastian G
De Haën, Christoph
Doerfler, Walter Hans
Donato, Rosario Francesco
Dworkin, Mark Bruce
Dyrberg, Thomas Peter
Edelman, Marvin
Eggers, Hans J
Elsevier, Susan Maria
Engelborghs, Yves
Fermi, Giulio
Fernández-Cruz, Eduardo P
Franklin, Richard Morris
Ganten, Detlev
Garcia-Sainz, J Adolfo
Geiger, Benjamin
Gendler, Sandra J
Ghiara, Paolo
Goding, James Watson
Greene, Lewis Joel
Grisolia, Santiago
Gruss, Peter H
Guimaraes, Romeu Cardoso
Haggis, Geoffrey Harvey
Hansson, Goran K
Hanukoglu, Israel
Hashimoto, Paulo Hitonari
Hayashi, Masao
Henderson, David Andrew
Hirano, Toshio
Hosokawa, Keiichi
Ikehara, Yukio
Ishikawa, Hiroshi
Jakob, Karl Michael
Kaye, Alvin Maurice
Kimoto, Masao
Klug, Aaron
Kuhn, Klaus
Kuriyama, Kinya
Lamoyi, Edmundo
Liao, Ta-Hsiu
Ling, Chung-Mei
Lizardi, Paul Modesto
Loppnow, Harald
Maruta, Hiroshi
Meins, Frederick, Jr
Mezquita, Cristobal
Michaels, Allan
Michaelsen, Terje E
Mittal, Balraj
Mizuno, Shigeki
Nagata, Kazuhiro
Richmond, Robert H
Roy, Raman K
Santoro, Ferrucio Fontes
Sargent, David Fisher
Schena, Francesco Paolo
Schmell, Eli David
Sheldrick, Peter
Spencer, John Francis Theodore
Spierer, Pierre
Stark, George Robert
Stebbing, Nowell
Steer, Martin William
Stingl, Georg
Stodolsky, Marvin
Strous, Ger J
Subramanian, Alap Raman
Tan, Y H
Taniyama, Tadayoshi
Teleb, Zakaria Ahmed
Thiery, Jean Paul
Toutant, Jean-Pierre
Tsolas, Orestes
Tsou, Chen-Lu
Wakayama, Yoshihiro
Watanabe, Takeshi
Willard-Gallo, Karen Elizabeth
Wyman, Jeffries

Neurosciences

ALABAMA
Berecek, Kathleen Helen
Bradley, Laurence A
Egan, Marianne Louise
Friedlander, Michael J

Garver, David L
Loop, Michael Stuart
Marchase, Richard Banfield
Oparil, Suzanne
Oyster, Clyde William
Williams, Raymond Crawford

ARIZONA
Hruby, Victor J
Mooradian, Arshag Dertad
Scott, Alwyn C
Triffet, Terry

ARKANSAS
Dykman, Roscoe A
Gilmore, Shirley Ann
Holson, Ralph Robert
Ruwe, William David
Skinner, Robert Dowell
Wessinger, William David

CALIFORNIA
Allman, John Morgan
Arnold, Arthur Palmer
Aserinsky, Eugene
Ashe, John Herman
Beckman, Alexander Lynn
Berger, Ralph Jacob
Bradshaw, Ralph Alden
Bremermann, Hans J
Brown, Warren Shelburne, Jr
Buchwald, Jennifer S
Campagnoni, Anthony Thomas
Chambers, Kathleen Camille
Chapman, Loring Frederick
Cohn, Major Lloyd
Davis, William Jackson
Ellman, George Leon
Eng, Lawrence F
Enoch, Jay Martin
Estrada, Norma Ruth
Fagin, Katherine Diane
Fain, Gordon Lee
Farley, Roger Dean
Fields, Howard Lincoln
Fisher, Robin Scott
Fluharty, Arvan Lawrence
Fromkin, Victoria A
Gehrmann, John Edward
Gibson, Thomas Richard
Glaser, Donald Arthur
Goldberg, Mark Arthur
Goodman, Corey Scott
Greenwood, Mary Rita Cooke
Grinnell, Alan Dale
Haines, Richard Foster
Hall, Zach Winter
Hammerschlag, Richard
Haun, Charles Kenneth
Hefti, Franz F
Hillyard, Steven Allen
Ho, Begonia Y
Horowitz, John M
Howd, Robert A
Johnson, Randolph Mellus
Junge, Douglas
Kelly, Regis Baker
Kennedy, Mary Bernadette
Kishimoto, Yasuo
Ko, Chien-Ping
Levine, Michael S
Lewis, Nina Alissa
Libet, Benjamin
Lindsley, Donald B
Lingappa, Jaisri Rao
McCaman, Richard Eugene
McClure, William Owen
McCullough, Richard Donald
McGaugh, James L
McGinnis, James F
Marg, Elwin
Marmor, Michael F
Martin, James Tillison
Miledi, Ricardo
Miller, Arthur Joseph
Mohan, Chandra
Morse, Daniel E
Mulloney, Brian
Patterson, Paul H
Peper, Erik
Perkel, Donald Howard
Ralston, Henry James, III
Ramachandran, Vilayanur Subramanian
Ribak, Charles Eric
Richard, Christopher Alan
Ridgway, Sam H
Rome, Leonard H
Ruesch, Jurgen
Russell, Sharon May
Sadee, Wolfgang
Sawyer, Charles Henry
Schwab, Ernest Roe
Segall, Paul Edward
Sheppard, Asher R
Shih, Jean Chen
Skrzypek, Josef
Snow, George Edward
Spitzer, Nicholas Canaday
Starr, Arnold
Stryker, Michael Paul
Terry, Robert Davis
Thompson, Jeffrey Michael
Thompson, Richard Frederick
Tobin, Allan Joshua

Tomich, John Matthew
Tschirgi, Robert Donald
Tyler, Christopher William
Vaughn, James E, Jr
Vinters, Harry Valdis
Wasterlain, Claude Guy
Westheimer, Gerald
Williston, John Stoddard
Wimer, Richard E
Yund, E William
Yuwiler, Arthur
Zucker, Robert Stephen

COLORADO
Bamburg, James Robert
Bekoff, Anne C
Best, LaVar
Betz, William J
Deitrich, Richard Adam
Dubin, Mark William
Eaton, Robert Charles
Finger, Thomas Emanuel
Ishii, Douglas Nobuo
Kater, Stanley B
Kulkosky, Paul Joseph
Mykles, Donald Lee
Roper, Stephen David

CONNECTICUT
Barry, Michael Anhalt
Berlind, Allan
Bohannon, Richard Wallace
Charney, Dennis S
Cohen, Lawrence Baruch
Cohen, Melvin Joseph
Davis, Michael
Denenberg, Victor Hugo
Glasel, Jay Arthur
Goldman, Bruce Dale
Herbette, Leo G
LaMotte, Carole Choate
Landmesser, Lynn Therese
Lentz, Thomas Lawrence
Marks, Lawrence Edward
Miller, William Henry
Morest, Donald Kent
Morrow, Jon S
Potashner, Steven Jay
Riker, Donald Kay
Ritchie, Brenda Rachel (Bigland)
Sachs, Benjamin David
Salafia, W(illiam) Ronald
Schwartz, Ilsa Roslow
Snider, Ray Michael
Tallman, John Francis
Taub, Arthur
Wallace, Robert B
Waxman, Stephen George
Welch, Annemarie S

DELAWARE
Brown, Barry Stephen
Carnahan, James Elliot
Granda, Allen Manuel
Gulick, Walter Lawrence
Holyoke, Caleb William, Jr
Saller, Charles Frederick
Song, Jiakun
Tam, Sang William

DISTRICT OF COLUMBIA
Alleva, John J
Bick, Katherine Livingstone
Chow, Ida
Cohn, Victor Hugo
Doctor, Bhupendra P
Finn, Edward J
Fishbein, William Nichols
Herman, Barbara Helen
Jacobsen, Frederick Marius
Louttit, Richard Talcott
Lumpkin, Michael Dirksen
Millar, David B
Perry, David Carter
Raslear, Thomas G
Sobotka, Thomas Joseph
Weinberger, Daniel R
Wylie, Richard Michael

FLORIDA
Ajmone-Marsan, Cosimo
Anderson, Peter Alexander Vallance
Barrett, Ellen Faye
Barrett, John Neil
Bennett, Gudrun Staub
Bunge, Mary Bartlett
Evoy, William (Harrington)
Gennaro, Joseph Francis
Gleeson, Richard Alan
Goldstein, Mark Kane
Griffith, Gail Susan Tucker
Hamasaki, Duco I
Heaton, Marieta Barrow
Knapp, Francis Marion
Lindsey, Bruce Gilbert
McKinney, Michael
Muller, Kenneth Joseph
Neary, Joseph Thomas
Phelps, Christopher Prine
Potter, Lincoln Truslow
Price, David Alan
Rowland, Neil Edward
Wecker, Lynn
Wilson, David Louis

GEORGIA
Barb, C Richard
Bates, Harold Brennan, Jr
Benoit, Peter Wells
Black, Asa C, Jr
Boothe, Ronald G
Bowen, John Metcalf
Brann, Darrell Wayne
Byrd, Larry Donald
Denson, Donald D
Doetsch, Gernot Siegmar
Edwards, Gaylen Lee
Fletcher, James Erving
Gibson, John Michael
Holland, Robert Campbell
Isaac, Walter
Iuvone, Paul Michael
Jackson, William James
Loring, David William
McDaniel, William Franklin
Mackin, Robert Brian
May, Sheldon William
Mulroy, Michael Joseph
Schaefer, Gerald J
Schweri, Margaret Mary
Sutin, Jerome
Wolf, Steven L
Zinsmeister, Philip Price

HAWAII
Bitterman, Morton Edward
Hardman, John M
Morton, Bruce Eldine

IDAHO
DeSantis, Mark Edward

ILLINOIS
Agarwal, Gyan C
Applebury, Meredithe L
Becker, Robert Paul
Bernstein, Joel Edward
Brewer, Gregory J
Brown, Meyer
Carr, Virginia McMillan
Cohen, Rochelle Sandra
Dal Canto, Mauro Carlo
Daley, Darryl Lee
Dallos, Peter John
Dodge, Patrick William
Ebrey, Thomas G
Farbman, Albert Irving
Fox, Sidney Walter
Gillette, Martha Ulbrick
Giordano, Tony
Goldberg, Jay M
Gottlieb, Gerald Lane
Greenough, William Tallant
Hoshiko, Michael S
Jensen, Robert Alan
Kebabian, John Willis
Kleinman, Kenneth Martin
Kochman, Ronald Lawrence
Lambert, Mary Pulliam
Lavelle, Arthur
Levine, Michael W
McKelvy, Jeffrey F
McNulty, John Alexander
Mafee, Mahmood Forootan
Menco, Bernard
Miller, Gregory Duane
Nakajima, Shigehiro
Narahashi, Toshio
Nardi, James Benjamin
Neet, Kenneth Edward
Offner, Franklin Faller
Phelps, Creighton Halstead
Pokorny, Joel
Pollack, Emanuel Davis
Rasenick, Mark M
Richman, David Paul
Robinson, Charles J
Satinoff, Evelyn
Schmidt, Robert Sherwood
Takahashi, Joseph S
Weyhenmeyer, James Alan
Wilson, Hugh Reid
Wu, Chau Hsiung

INDIANA
Aprison, Morris Herman
Chernoff, Ellen Ann Goldman
Devoe, Robert Donald
Echtenkamp, Stephen Frederick
Frommer, Gabriel Paul
Gidda, Jaswant Singh
Henry, David P, II
Juraska, Janice Marie
Kennedy, Duncan Tilly
Leander, John David
Mason, Norman Ronald
Nichols, David Earl
Nicholson-Guthrie, Catherine Shirley
Pak, William Louis
Pietsch, Paul Andrew
Rebec, George Vincent
Strickholm, Alfred
Thor, Karl Bruce
Wasserman, Gerald Steward

IOWA
Biller, Jose
Coulter, Joe Dan
Kollros, Jerry John

Myers, Glenn Alexander
Wu, Chun-Fang

KANSAS
Alper, Richard H
Imig, Thomas Jacob
Michaelis, Mary Louise
Norris, Patricia Ann
Pazdernik, Thomas Lowell
Samson, Frederick Eugene, Jr
Smalley, Katherine N
Voogt, James Leonard
Westfall, Jane Anne

KENTUCKY
Essig, Carl Fohl
Randall, David Clark
Schurr, Avital
Surwillo, Walter Wallace

LOUISIANA
Davis, George Diament
Dunn, Adrian John
Garcia, Meredith Mason
Johnsen, Peter Berghsey
Miceli, Michael Vincent
Nair, Pankajam K
Prasad, Chandan
Saphier, David
Seaman, Ronald L
Silverman, Harold
Smith, Diane Elizabeth
Strauss, Arthur Joseph Louis
York, David Anthony

MAINE
Dowse, Harold Burgess

MARYLAND
Alberts, Walter Watson
Anderson, David Everett
Andrews, Stephen Brian
Arora, Prince Kumar
Atrakchi, Aisar Hasan
Ball, Gregory Francis
Battey, James F
Bhathena, Sam Jehangirji
Bloch, Robert Joseph
Broadwell, Richard Dow
Brownell, William Edward
Cantor, David S
Cereghino, James Joseph
Chiueh, Chuang Chin
Cohen, Robert Martin
Creveling, Cyrus Robbins
Dingman, Charles Wesley, II
Dubner, Ronald
Elson, Hannah Friedman
Estienne, Mark Joseph
Fechter, Laurence David
Fex, Jorgen
Fields, Kay Louise
Garruto, Ralph Michael
Gilbert, Daniel Lee
Glowa, John Robert
Gold, Philip William
Goldman, David
Goodwin, Frederick King
Green, Martin David
Guilarte, Tomas R
Gusovsky, Fabian
Guttman, Helene Augusta Nathan
Hallett, Mark
Hanbauer, Ingeborg
Hanley, Daniel F, Jr
Harris, Andrew Leonard
Helke, Cinda Jane
Hempel, Franklin Glenn
Henkin, Robert I
Hill, James Leslie
Hoffman, Paul Ned
Hudson, Lynn Diane
Jacobowitz, David
Jacobson, Kenneth Alan
Johnson, Kenneth Olafur
Kaufman, Elaine Elkins
Kety, Seymour S
Klatzo, Igor
Kuenzel, Wayne John
Larrabee, Martin Glover
Leach, Berton Joe
Lerman, Michael Isaac
Leshner, Alan Irvin
Ludlow, Christy L
McCarron, Richard M
McCune, Susan K
MacLean, Paul Donald
Mains, Richard E
Mishkin, Mortimer
Montell, Craig
Moran, Jeffrey Chris
Muma, Nancy A
Nelson, Phillip Gillard
Nowak, Thaddeus Stanley, Jr
Nuite-Belleville, Jo Ann
Parsegian, Vozken Adrian
Pickar, David
Poggio, Gian Franco
Popper, Arthur N
Rall, Wilfrid
Rawlings, Samuel Craig
Reese, Thomas Sargent
Rennels, Marshall L
Rogawski, Michael Andrew

Roman, Gustavo Campos
Ruchkin, Daniel S
Russell, James T
Saavedra, Juan M
Salem, Norman, Jr
Schnaar, Ronald Lee
Selmanoff, Michael Kidd
Sheridan, Philip Henry
Stoolmiller, Allen Charles
Summers, Raymond
Vandenbergh, David John
Vogl, Thomas Paul
Yu, Mei-ying Wong
Zielke, H Ronald

MASSACHUSETTS
Anderson, Olivo Margaret
Anderson Olivo, Margaret
Berman, Marlene Oscar
Bird, Stephanie J
Callard, Gloria Vincz
Caplan, Louis Robert
DeBold, Joseph Francis
Dethier, Vincent Gaston
Eldred, William D
Emanuel, Rodica L
Epstein, Irving Robert
Gamzu, Elkan R
Graybiel, Ann M
Griffin, Donald R(edfield)
Gross, David John
Hall, Robert Dilwyn
Hauser, George
Hausman, Robert Edward
Henneberry, Richard Christopher
Herrup, Karl Franklin
Irwin, Louis Neal
Kamen, Gary P
Kanarek, Robin Beth
Kazemi, Homayoun
Koul, Omanand
Kravitz, Edward Arthur
Lee, Gloria
Lees, Marjorie Berman
Madras, Bertha K
Marder, Eve Esther
Masek, Bruce James
Masland, Richard Harry
Micheli, Lyle Joseph
Morris, Robert
Murphey, Rodney Keith
Neer, Eva Julia
Olivo, Richard Francis
O'Malley, Alice T
Palay, Sanford Louis
Papastoitsis, George
Reppert, Steve Marion
Ross, Alonzo Harvey
Schneider, Gerald Edward
Schomer, Donald Lee
Schwarting, Gerald Allen
Schwartz, William Joseph
Sidman, Richard Leon
Sihag, Ram K
Singer, Joshua J
Strumwasser, Felix
Sweadner, Kathleen Joan
Talamo, Barbara Lisann
Tyler, H Richard
Tyler, Kenneth Laurence
Villa-Komaroff, Lydia
Voigt, Herbert Frederick
Weinberg, Crispin Bernard
Wolf, Merrill Kenneth
Wyse, Gordon Arthur

MICHIGAN
Anderson, Thomas Edward
Barksdale, Charles Madsen
Barney, Christopher Carroll
Casey, Kenneth L(yman)
Diaz, Fernando G
Dore-Duffy, Paula
Easter, Stephen Sherman, Jr
Fernandez, Hector R C
Freeman, Arthur Scott
Gebber, Gerard L
Glover, Roy Andrew
Hall, Edward Dallas
Hatton, Glenn Irwin
Hawkins, Joseph Elmer
Heidemann, Steven Richard
Johnson, John Irwin, Jr
Lahti, Robert A
Lew, Gloria Maria
Nuttall, Alfred L
Richardson, Rudy James
Stach, Robert William
Stout, John Frederick
Tosney, Kathryn W
Tse, Harley Y
Tweedle, Charles David
Uhler, Michael David
Wiley, John W
Zand, Robert

MINNESOTA
Alsum, Donald James
Anderson, Victor Elving
Beitz, Alvin James
Berry, James Frederick
Cunningham, Julie Margaret
Drewes, Lester Richard
Dysken, Maurice William

Iadecola, Costantino
Koerner, James Frederick
Levine, Allen Stuart
Low, Walter Cheney
Messing, Rita Bailey
Newman, Eric Allan
Rathbun, William B
Ray, Charles Dean
Ruggero, Mario Alfredo
Soechting, John F
Stauffer, Edward Keith
Terzuolo, Carlo A
Westmoreland, Barbara Fenn

MISSISSIPPI
Alford, Geary Simmons
Jennings, David Phipps
Lynch, James Carlyle
Ma, Terence P
Peeler, Dudley F, Jr

MISSOURI
Agrawal, Harish C
Ahmed, Mahmoud S
Cohen, Adolph Irvin
Hughes, Stephen Edward
Merlie, John Paul
Mills, Steven Harlon
Natani, Kirmach
Pearlman, Alan L
Powers, William John
Rosenthal, Harold Leslie
Sanes, Joshua Richard
Stein, Paul S G
Tannenbaum, Michael Glen
Thach, William Thomas
Tumosa, Nina Jean
Wright, Dennis Charles
Young, Paul Andrew

NEBRASKA
Mann, Michael David
Murrin, Leonard Charles

NEVADA
Ort, Carol

NEW HAMPSHIRE
Nattie, Eugene Edward
Sower, Stacia Ann
Velez, Samuel Jose

NEW JERSEY
Allen, Richard Charles
Cascieri, Margaret Anne
Connor, John Arthur
Conway, Paul Gary
Egger, M(aurice) David
Fangboner, Raymond Franklin
Feldman, Susan C
Fryer, Rodney Ian
Gelperin, Alan
Gertner, Sheldon Bernard
Gray, Harry Edward
Gross, Charles Gordon
Hey, John Anthony
Hubbard, John W
Jabbar, Gina Marie
Laemle, Lois K
Malamed, Sasha
Manowitz, Paul
Margolskee, Robert F
Moyer, John Allen
Pachter, Jonathan Alan
Panagides, John
Rabii, Jamshid
Schlesinger, David H
Schreckenberg, Mary Gervasia
Solla, Sara A
Szewczak, Mark Russell
Tallal, Paula
Zbuzek, Vratislav

NEW MEXICO
Osbourn, Gordon Cecil
Wild, Gaynor (Clarke)

NEW YORK
Aschner, Michael
Badoyannis, Helen Litman
Barlow, Robert Brown, Jr
Beason, Robert Curtis
Begleiter, Henri
Bennett, Michael Vander Laan
Blosser, James Carlisle
Bolton, David C
Brody, Harold
Cabot, John Boit, II
Capranica, Robert R
Carpenter, David O
Chappell, Richard Lee
Chauhan, Ved P S
Chutkow, Jerry Grant
Cohan, Christopher Scott
Crain, Stanley M
Del Cerro, Manuel
Demeter, Steven
Diakow, Carol
Doty, Robert William
Drakontides, Anna Barbara
Edelman, Gerald Maurice
Engbretson, Gustav Alan
Feldman, Samuel M
Felten, David L

Neurosciences (cont)

Fenstermacher, Joseph Don
Fetcho, Joseph Robert
Gall, William Einar
Gardner, Daniel
Gardner, Esther Polinsky
Gershon, Michael David
Gibson, Gary Eugene
Gintautas, Jonas
Gogel, Germaine E
Goldman, James Eliot
Grafstein, Bernice
Greenberg, Danielle
Greenberger, Lee M
Grob, David
Hartung, John David
Hiller, Jacob Moses
Holland, Mary Jean Carey
Holloway, Ralph L, Jr
Holtzman, Eric
Hood, Donald C
Horel, James Alan
Hough, Lindsay B
Jacquet, Yasuko F
Jakway, Jacqueline Sinks
John, E Roy
Karpiak, Stephen Edward
Kimelberg, Harold Keith
Koizumi, Kiyomi
Kream, Jacob
Kristal, Mark Bennett
Kupfermann, Irving
Latimer, Clinton Narath
Leibovic, K Nicholas
Leibowitz, Sarah Fryer
Levinthal, Charles F
Lin, Fu Hai
Lnenicka, Gregory Allen
Low, Barbara Wharton
Lyser, Katherine May
McCabe, R Tyler
McEwen, Bruce Sherman
Mahadik, Sahebarao P
Martin, David Lee
Mazurkiewicz, Joseph Edward
Mendell, Lorne Michael
Mozell, Maxwell Mark
Murphy, Randall Bertrand
Noback, Charles Robert
Norton, William Thompson
Okamoto, Michiko
Pasik, Pedro
Pasik, Tauba
Pentney, Roberta Pierson
Podleski, Thomas Roger
Possidente, Bernard Philip, Jr
Powell, Sharon Kay
Preston, James Benson
Raine, Cedric Stuart
Reeke, George Norman, Jr
Reilly, Margaret Anne
Ringo, James Lewis
Roberts, James Lewis
Rubinson, Kalman
Rudy, Bernardo
Ruggiero, David A
Saez, Juan Carlos
Scott, Sheryl Ann
Seegal, Richard Field
Shapley, Robert M
Sladek, Celia Davis
Smith, Gerard Peter
Soifer, David
Solomon, Seymour
Sperling, George
Squinto, Stephen P
Strand, Fleur Lillian
Thomas, Garth Johnson
Tieman, Suzannah Bliss
Udin, Susan Boymel
Victor, Jonathan David
Walcott, Benjamin
Wallman, Joshua
West, Norman Reed
Whitaker-Azmitia, Patricia Mack
Williams, Curtis Alvin, Jr
Williamson, Samuel Johns
Wineburg, Elliot N
Yazulla, Stephen
Yoburn, Byron Crocker
York, James Lester
Zukin, Stephen R

NORTH CAROLINA

Bell, Mary Allison
Booth, Raymond George
Defoggi, Ernest
Edens, Frank Wesley
Farel, Paul Bertrand
Ferris, Robert Monsour
Gerhardt, Don John
Hedlund, Laurence William
Hemperly, John Jacob
Hicks, T Philip
Howard, James Lawrence
Leise, Esther M
Levitt, Melvin
Lewin, Anita Hana
Ludel, Jacqueline
Lund, Pauline Kay
McNamara, James O'Connell
MacPhail, Robert C
Morell, Pierre

Morgan, Kevin Thomas
O'Steen, Wendall Keith
Sahyoun, Naji Elias
Smith, Ned Allan
Somjen, George G
Sonntag, William Edmund
Stern, Warren C
Surwit, Richard Samuel
Suzuki, Kunihiko
Tilson, Hugh Arval
Toews, Arrel Dwayne
Tyrey, Lee
Weiner, Richard D
Williams, Redford Brown, Jr
Wilson, John Eric
Wong, Fulton

NORTH DAKOTA

Penland, James Granville

OHIO

Brunden, Kurt Russell
Buck, Stephen Henderson
Carr, Albert A
Clark, David Lee
Cras, Patrick
Cruce, William L R
Dorman, Robert Vincent
Gesteland, Robert Charles
Giannini, A James
Glenn, Loyd Lee
Godfrey, Donald Albert
Goodrich, Cecilie Ann
Heesch, Cheryl Miller
Jones, Stephen Wallace
Jordan, Freddie L
Kornacker, Karl
Krontiris-Litowitz, Johanna Kaye
Lacey, John I
LaManna, Joseph Charles
Landis, Dennis Michael Doyle
McQuarrie, Irvine Gray
Messer, William Sherwood, Jr
Mulick, James Anton
Myers, Ronald Elwood
Nasrallah, Henry A
Patterson, Michael Milton
Robbins, David O
Robbins, Norman
Schwarz, Marvin
Sheridan, John Francis
Stuesse, Sherry Lynn
Thomas, William Eric
Thornton, Janice Elaine
Waller, Hardress Jocelyn
Watanabe, Michiko
Wolfe, Seth August, Jr
Zigmond, Richard Eric

OKLAHOMA

Dryhurst, Glenn
Farber, Jay Paul
Holloway, Frank A
Jones, Sharon Lynn
Leonard, Janet Louise
Nowack, William J
Papka, Raymond Edward
Revzin, Alvin Morton

OREGON

Brown, Arthur Charles
Fahrenbach, Wolf Henrich
Gwilliam, Gilbert Franklin
Lewy, Alfred James
Pubols, Benjamin Henry, Jr
Roselli, Charles Eugene
Seil, Fredrick John
Soderling, Thomas Richard
Tunturi, Archie Robert

PENNSYLVANIA

Allen, Michael Thomas
Bridger, Wagner H
Brooks, David Patrick
Cameron, William Edward
Carpenter, Malcolm Breckenridge
Clark, T(helma) K
Connor, John D
Davies, Richard Oelbaum
De Groat, William C
Eisenman, Leonard Max
Ermentrout, George Bard
Erulkar, Solomon David
Fernstrom, John Dickson
Gerstein, George Leonard
Goldberg, Michael Ellis
Greenstein, Jeffrey Ian
Grobstein, Paul
Grunwald, Gerald B
Hand, Peter James
Hartman, Herman Bernard
Hill, Shirley Yarde
Hurvich, Leo Maurice
Jameson, Dorothea
Jaweed, Mazher
Joseph, Jeymohan
Josiassen, Richard Carlton
Kovachich, Gyula Bertalan
Kozak, Wlodzimierz M
Kozikowski, Alan Paul
Kvist, Tage Nielsen
Leavitt, Marc Laurence
Lindgren, Clark Allen
Lindstrom, Jon Martin

Lublin, Fred D
McLaughlin, Patricia J
McMorris, F Arthur
Newmark, Jonathan
Pollock, John Archie
Poplawsky, Alex James
Reynolds, Charles F
Ridenour, Marcella V
Roemer, Richard Arthur
Rubin, Leonard Sidney
Russell, Richard Lawson
Salmoiraghi, Gian Carlo
Saunders, James Charles
Senft, Joseph Philip
Smyth, Thomas, Jr
Sodicoff, Marvin
Sprague, James Mather
Stewart, Charles Newby
Storella, Robert J, Jr
Stricker, Edward Michael
Summy-Long, Joan Yvette
Whinnery, James Elliott
Wilberger, James Eldridge
Wysocki, Charles Joseph
Zagon, Ian Stuart
Zigmond, Michael Jonathan

RHODE ISLAND

Christenson, Lisa
Flanagan, Thomas Raymond
Greenblatt, Samuel Harold
Hufnagel, Linda Ann
Jackson, Ivor Michael David
Oxenkrug, Gregory Faiva
Sanberg, Paul Ronald

SOUTH CAROLINA

Krall, Albert Raymond

TENNESSEE

Bealer, Steven Lee
Burt, Alvin Miller, III
DeSaussure, Richard Laurens, jr
Desiderio, Dominic Morse, (Jr)
Dettbarn, Wolf Dietrich
Gao, Kuixiong
Kostrzewa, Richard Michael
Lawler, James E
Lubar, Joel F
Partridge, Lloyd Donald
Robertson, David
Sanders-Bush, Elaine
Senogles, Susan Elizabeth
Wikswo, John Peter, Jr

TEXAS

Alkadhi, Karim A
Andresen, Michael Christian
Bjorndahl, Jay Mark
Brady, Scott T
Bursztajn, Sherry
Chang, Donald Choy
Cooper, Cary Wayne
Dafny, Nachum
Denney, Richard Max
Dilsaver, Steven Charles
Ducis, Ilze
Eidelberg, Eduardo
Finitzo-Hieber, Terese
Fox, Donald A
Franzl, Robert E
Gerken, George Manz
Gonzalez-Lima, Francisco
Gorry, G Anthony
Hamilton, Charles R
Huffman, Ronald Dean
Johnson, Kenneth Maurice, Jr
Klemm, William Robert
Kozlowski, Gerald P
Lakoski, Joan Marie
Leuchtag, H Richard
Linner, John Gunnar
McAdoo, David J
Marks, Gerald A
Miikkulainen, Risto Pekka
Mikiten, Terry Michael
Norman, Reid Lynn
Pease, Paul Lorin
Pirch, James Herman
Ross, Elliott M
Rylander, Michael Kent
Saltzberg, Bernard
Schaffer, Barbara Noyes
Sherry, Clifford Joseph
Skolnick, Malcolm Harris
Smith, Edward Russell
Smith, Eric Morgan
Thompson, Wesley Jay
Vinson, David Berwick
Wilczynski, Walter
Wilson, L Britt
Wong, Brendan So
Yung, W K Alfred

UTAH

Black, Dean
Cheney, Carl D
Creel, Donnell Joseph
Fidone, Salvatore Joseph
Galster, William Allen
Jacobson, Marcus
Jarcho, Leonard Wallenstein
Mullen, Richard Joseph
Okun, Lawrence M

VERMONT

Freedman, Steven Leslie
Parsons, Rodney Lawrence

VIRGINIA

Banker, Gary A
Carson, Keith Alan
De Haan, Henry J(ohn)
Denbow, Donald Michael
Fine, Michael Lawrence
Friesen, Wolfgang Otto
Gold, Paul Ernest
Grant, John Wallace
Gray, Faith Harriet
Grider, John Raymond
Guth, Lloyd
Kurtzke, John F
McCarty, Richard Charles
Macdonald, Timothy Lee
Maggio, Bruno
Murray, Jeanne Morris
Schulze, Gene Edward
Stein, Barry Edward
Vinores, Stanley Anthony

WASHINGTON

Artru, Alan Arthur
Baksi, Samarendra Nath
Barnes, Charles Dee
Baskin, Denis George
Beck, Thomas W
Bernard, Gary Dale
Bowden, Douglas Mchose
Calvin, William Howard
Churchill, Lynn
Clapshaw, Patric Arnold
Hawkins, Richard L
Hendrickson, Anita Elizabeth
Kenney, Nancy Jane
Larsen, Lawrence Harold
Lovely, Richard Herbert
Mendelson, Martin
Nathanson, Neil Marc
Stahl, William Louis
Steiner, Robert Alan
Truman, James William
Whatmore, George Bernard
Woods, Stephen Charles
Young, Francis Allan

WEST VIRGINIA

Azzaro, Albert J
Culberson, James Lee
Gladfelter, Wilbert Eugene
Konat, Gregory W

WISCONSIN

Berman, Alvin Leonard
Bloom, Alan S
Claude, Philippa
Curtis, Robin Livingstone
Hamsher, Kerry de Sandoz
Hetzler, Bruce Edward
Hind, Joseph Edward
Hokin, Lowell Edward
Keesey, Richard E
Kostreva, David Robert
Light, Douglas B
Rhode, William Stanley
Sufit, Robert Louis
Takahashi, Lorey K
Tulunay-Keesey, Ulker
Welker, Wallace I
Wong-Riley, Margaret Tze Tung

PUERTO RICO

Banerjee, Dipak Kumar
De La Sierra, Angell O
Kicliter, Ernest Earl, Jr
Orkand, Richard K
Wolstenholme, Wayne W

ALBERTA

Kolb, Bryan Edward
Lukowiak, Kenneth Daniel
Roth, Sheldon H
Spencer, Andrew Nigel
Stell, William Kenyon

BRITISH COLUMBIA

Crockett, David James
Jones, David Robert
McGeer, Edith Graef
McGeer, Patrick L
Mackie, George Owen
Pate, Brian David
Phillips, Anthony George
Schwarz, Dietrich Walter Friedrich
Sinclair, John G
Steeves, John Douglas
Vincent, Steven Robert

MANITOBA

Dakshinamurti, Krishnamurti
Glavin, Gary Bertrun

NEW BRUNSWICK

Seabrook, William Davidson
Sivasubramanian, Pakkirisamy

NOVA SCOTIA

Connolly, John Francis
Howlett, Susan Ellen

ONTARIO
Barr, Murray Llewellyn
Beninger, Richard J
Bisby, Mark A
Brooks, Vernon Bernard
Brown, Ian Ross
Coscina, Donald Victor
De la Torre, Jack Carlos
Donald, Merlin Wilfred
Downer, Roger George Hamill
Govind, Choonilal Keshav
Hachinski, Vladimir C
Hallett, Peter Edward
Himms-Hagen, Jean
Howard, Ian Porteous
Hrdina, Pavel Dusan
Kalant, Harold
Kutcher, Stanley Paul
Leung, Lai-Wo Stan
Mazurkiewicz-Kwilecki, Irena Maria
Pace Asciak, Cecil
Percy, Maire Ede
Rollman, Gary Bernard
Seguin, Jerome Joseph
Smith, Jonathan Jeremy Berkeley
Stanisz, Andrzej Maciej
Vanderwolf, Cornelius Hendrik
Weaver, Lynne C
Weick, Richard Fred
Witelson, Sandra Freedman

QUEBEC
Anctil, Michel
Antel, Jack Perry
Birmingham, Marion Krantz
Capek, Radan
Chouinard, Guy
Clark, Michael Wayne
Cohen, Monroe W
Collier, Brian
Diksic, Mirko
Gjedde, Albert
Gloor, Pierre
Karpati, George
MacIntosh, Frank Campbell
Murthy, Mahadi Raghavandrarao Ven
Olivier, Andre
Outerbridge, John Stuart
Palmour, Roberta Martin
Pappius, Hanna M
Pasztor, Valerie Margaret
Robert, Suzanne
Steriade, Mircea
Tannenbaum, Gloria Shaffer
Young, Simon N

SASKATCHEWAN
Durden, David Alan
Irvine, Donald Grant
Johnson, Dennis Duane
Richardson, J(ohn) Steven
Wood, James Douglas

OTHER COUNTRIES
Araki, Masasuke
Brown, Charles Eric
Brust-Carmona, Hector
Cavonius, Carl Richard
Collins, Vincent Peter
Curtis, Adam Sebastian G
De Troyer, Andre Jules
Donato, Rosario Francesco
Ganchrow, Donald
Guerrero-Munoz, Frederico
Hashimoto, Paulo Hitonari
Hirokawa, Katsuiku
Hisada, Mituhiko
Holman, Richard Bruce
Hughes, Abbie Angharad
Kawamura, Hiroshi
Kuriyama, Kinya
Ladinsky, Herbert
Levi-Montalcini, Rita
Lieb, William Robert
Livett, Bruce G
Lu, Guo-Wei
Maruta, Hiroshi
Mølhave, Lars
Nabeshima, Toshitaka
Reuter, Harald
Sabry, Ismail
Sampson, Sanford Robert
Toutant, Jean-Pierre
Tung, Che-Se
Wakayama, Yoshihiro
Wenger, Byron Sylvester
Wettstein, Joseph G
Yamada, Eichi
Zapata, Patricio

Nutrition

ALABAMA
Chastain, Marian Faulkner
Clark, Alfred James
Kochakian, Charles Daniel
Menaker, Lewis
Otto, David A
Roland, David Alfred, Sr
Sani, Brahma Porinchu
Sauberlich, Howerde Edwin

ARIZONA
Brannon, Patsy M
Gordon, Richard Seymour
Kaufman, C(harles) W(esley)
Kemmerer, Arthur Russell
McCaughey, William Frank
McNamara, Donald J
Peng, Yeh-Shan
Sander, Eugene George
Stini, William Arthur

ARKANSAS
Greenman, David Lewis
Horan, Francis E
Lewis, Sherry M
Schieferstein, George Jacob

CALIFORNIA
Allerton, Samuel E
Amy, Nancy Klein
Baldwin, Ransom Leland
Barnard, R James
Betschart, Antoinette
Brandon, David Lawrence
Briggs, George McSpadden
Burri, Betty Jane
Calloway, Doris Howes
Canham, John Edward
Caporaso, Fredric
Carlisle, Edith M
Carlson, Don Marvin
Chang, Jason Washington
Chen, Tung-Shan
Christensen, Halvor Niels
Curry, Donald Lawrence
Dobbins, John Potter
Feeney, Robert Earl
Fleming, Sharon Elayne
Gietzen, Dorothy Winter
Gray, Gary M
Greenwood, Mary Rita Cooke
Harrison, Michael R
Hawkes, Wayne Christian
Hill, Fredric William
Jacob, Mary
Johnson, Herman Leonall
Kaslow, Arthur L
Keagy, Pamela M
Keen, Carl L
Kelley, Darshan Singh
Kretsch, Mary Josephine
Lönnerdal, Bo L
Lu, Nancy Chao
Mizuno, Nobuko S(himotori)
Nassos, Patricia Saima
Nelson, Gary Joe
Norman, Anthony Westcott
Painter, Ruth Coburn Robbins
Prater, Arthur Nickolaus
Rahlmann, Donald Frederick
Rodriquez, Mildred Shepherd
Rucker, Robert Blain
Russell, Gerald Frederick
Schweigert, Bernard Sylvester
Scott, Karen Christine
Slater, Grant Gay
Slavkin, Harold Charles
Smith, Christine H
Strother, Allen
Thomas, Heriberto Victor
Thomson, John Ansel Armstrong
Tinoco, Joan W H
Tsao, Constance S
Turnlund, Judith Rae
Yang, Meiling T
Zamenhof, Stephen

COLORADO
Allen, Kenneth G D
Grundleger, Melvin Lewis
Ham, Richard George
Hambidge, K Michael
Stifel, Fred B

CONNECTICUT
Bronner, Felix
Ferris, Ann M
Fordham, Joseph Raymond
Jensen, Robert Gordon
Khairallah, Edward A
Rebuffe-Scrive, Marielle Francoise
Ross, Donald Joseph

DELAWARE
Billheimer, Jeffrey Thomas
Cartwright, Aubrey Lee, Jr
Kerr, Janet Spence
Rasmussen, Arlette Irene
White, Harold Bancroft, III

DISTRICT OF COLUMBIA
Callaway, Clifford Wayne
Canary, John J(oseph)
Fox, Mattie Rae Spivey
Harris, Suzanne Straight
Jacobson, Michael F
Pennington, Jean A T
Pla, Gwendolyn W
Prosky, Leon
Shank, Fred R
Simopoulos, Artemis Panageotis
Sundaresan, Peruvemba Ramnathan
Whittaker, Paul
Young, John Karl

FLORIDA
Bauernfeind, Jacob (Jack) C(hristopher)
Borum, Peggy R
Cousins, Robert John
Nanz, Robert Augustus Rollins
Ott, Edgar Alton
Rowland, Neil Edward
Sathe, Shridhar Krishna

GEORGIA
Ambrose, John Augustine
Berdanier, Carolyn Dawson
Edwards, Hardy Malcolm, Jr
Girardot, Jean Marie Denis
McCormick, Donald Bruce
Mickelsen, Olaf
Mullen, Barbara J
Wang, Marian M
Warner, Harold
Woods, Wendell David

HAWAII
Hartung, G(eorge) Harley
Wood, Betty J

IDAHO
Scott, James Michael

ILLINOIS
Bhattacharyya, Maryka Horsting
Bookwalter, George Norman
Campbell, Michael Floyd
Clark, Jimmy Howard
Dasler, Waldemar
Druyan, Mary Ellen
Fahey, George Christopher, Jr
Hatfield, Efton Everett
Hawrylewicz, Ervin J
Heybach, John Peter
Janghorbani, Morteza
Jeffay, Henry
Larson, Bruce Linder
Layman, Donald Keith
Lerner, Louis L(eonard)
McCormick, David Loyd
Mallia, Anantha Krishna
Miller, Gregory Duane
Mobarhan, Sohrab
Robinson, James Lawrence
Shambaugh, George E, III
Sutton, Lewis McMechan
Thompson, David Jerome
Wittman, James Smythe, III

INDIANA
Akrabawi, Salim S
Biggs, Homer Gates
Cline, Tilford R
Cook, David Allan
Day, Harry Gilbert
Forsyth, Dale Marvin
Frolik, Charles Alan
Pleasants, Julian Randolph
Schulz, Arthur R

IOWA
Beitz, Donald Clarence
Goodridge, Alan G
Morriss, Frank Howard, Jr
Olson, James Allen
Rebouche, Charles Joseph
Roderuck, Charlotte Elizabeth
Thompson, Robert Gary
Young, Jerry Wesley

KANSAS
Harbers, Carole Ann Z
Krishnan, Engil Kolaj
Parrish, John Wesley, Jr
Ranhotra, Gurbachan Singh
Smith, Meredith Ford

KENTUCKY
Boling, James A
Chen, Linda Li-Yueh Huang
Chow, Ching Kuang
Fell, Ronald Dean
Glauert, Howard Perry
Kasarskis, Edward Joseph
Mercer, Leonard Preston, II
Petering, Harold George

LOUISIANA
Alam, Bassima Saleh
Alam, Syed Qamar
Anderson, William McDowell
Culley, Dudley Dean, Jr
Hegsted, Maren
Misra, Raghunath P
Pryor, William Austin
Ramsey, Paul Roger
Windhauser, Marlene M

MAINE
Cook, Richard Alfred

MARYLAND
Beecher, Gary Richard
Bhathena, Sam Jehangirji
Chakrabarti, Siba Gopal
Chang, Mei Ling (Wu)
Church, Lloyd Eugene
Combs, Gerald Fuson
De Luca, Luigi Maria

Edelman, Robert
Evarts, Ritva Poukka
Frattali, Victor Paul
Goor, Ronald Stephen
Gori, Gio Batta
Green, Martin David
Guilarte, Tomas R
Hanks, John Harold
Henkin, Robert I
Hornstein, Irwin
Howe, Juliette Coupain
Hubbard, Van Saxton
Kim, Sooja K
Knapka, Joseph J
Kwiterovich, Peter O, Jr
McKenna, Mary Catherine
Matthews, Ruth H
Micozzi, Marc S
Nakhasi, Hira Lal
Nowak, Thaddeus Stanley, Jr
Passwater, Richard Albert
Pilch, Susan Marie
Poston, John Michael
Preusch, Peter Charles
Read, Merrill Stafford
Rechcigl, Miloslav, Jr
Reynolds, Robert David
Salem, Norman, Jr
Sampugna, Joseph
Sanslone, William Robert
Sass, Neil Leslie
Smith, James Cecil
Taylor, Lauriston Sale
Taylor, Martha Loeb
Taylor, Phillip R
Trout, David Linn
Vanderslice, Joseph Thomas
Varma, Shambhu D
Vydelingum, Nadarajen Ameerdanaden
Yergey, Alfred L, III

MASSACHUSETTS
Auskaps, Aina Marija
Babayan, Vigen Khachig
Barngrover, Debra Anne
Bert, Mark Henry
Flatt, Jean-Pierre
Geyer, Robert Pershing
Grodberg, Marcus Gordon
Kanarek, Robin Beth
McGowan, Joan A
Narayan, Krishnamurthi Ananth
Nauss, Kathleen Minihan
Pober, Zalmon
Rand, William Medden
Roubenoff, Ronenn
Russell, Robert M
Sawyer, Frederick Miles
Schaefer, Ernst J
Shaw, James Headon
Stare, Fredrick J
Tan, Barrie
Taylor, Allen
Udall, John Nicholas, Jr
Zannis, Vassilis I

MICHIGAN
Cossack, Zafrallah Taha
Frisancho, A Roberto
Hunt, Charles E
Jen, Kai-Lin Catherine
McBean, Lois D
Marshall, Charles Wheeler
Saldanha, Leila Genevieve
Sardesai, Vishwanath M
Thomas, John William
Van Dyke, Russell Austin
Wishnetsky, Theodore

MINNESOTA
Barker, Norval Glen
Cleary, Margot Phoebe
Coon, Craig Nelson
Cornelius, Steven Gregory
Dixit, Padmakar Kashinath
Holman, Ralph Theodore
Johnson, Susan Bissette
Levine, Allen Stuart
Lukasewycz, Omelan Alexander
Savaiano, Dennis Alan
Slavin, Joanne Louise

MISSISSIPPI
Brown, Robert Dale
McNaughton, James Larry
Robinson, Edwin Hollis
Wilson, Robert Paul

MISSOURI
Chi, Myung Sun
Coudron, Thomas A
Dixit, Rakesh
Hopkins, Daniel T
Sargent, William Quirk
Shelton, Damon Charles
Sunde, Roger Allan
Tucker, Robert Gene
Vineyard, Billy Dale

NEBRASKA
Birt, Diane Feickert
Ferrell, Calvin L
Gessert, Carl F
Grandjean, Carter Jules

Nutrition (cont)

Lewis, Austin James
Yen, Jong-Tseng

NEW JERSEY
Colby, Richard H
Comai-Fuerherm, Karen
Dalrymple, Ronald Howell
Deetz, Lawrence Edward, III
Ellenbogen, Leon
Farkas, Daniel Frederick
Fu, Shou-Cheng Joseph
Griminger, Paul
Hager, Mary Hastings
Keating, Kathleen Irwin
Lin, Grace Woan-Jung
Machlin, Lawrence Judah
Mellies, Margot J
Mohammed, Kasheed
Morck, Roland Anton
Oser, Bernard Levussove
Pleasants, Elsie W
Rosenfield, Daniel
Staub, Herbert Warren
Stein, T Peter
Triscari, Joseph
Weissman, Paul Morton

NEW MEXICO
Del Valle, Francisco Rafael
Omdahl, John L

NEW YORK
Albrecht, Alberta Marie
Ames, Stanley Richard
Awad, Atif B
Bauman, Dale E
Bernardis, Lee L
Chan, Mabel M
Chisholm, David R
Cooperman, Jack M
Corradino, Robert Anthony
Cunninham-Rundles, Charlotte
De Luca, Chester
DiFrancesco, Loretta
Finberg, Laurence
Fusco, Carmen Louise
Garza, Cutberto
Geary, Norcross D
Goodman, DeWitt Stetten
Graham, Donald C W
Greenberg, Danielle
Haas, Jere Douglas
Haley, Nancy Jean
Hall, William Myron, Jr
Henshaw, Edgar Cummings
Heymsfield, Steven B
Howard, Irmgard Matilda Keeler
Isaacs, Charles Edward
Johnson, Patricia R
Kahan, Sidney
Khan, Paul
Kishore, Gollamudi Sitaram
Klavins, Janis Vilberts
Kuftinec, Mladen M
Leleiko, Neal Simon
Lengemann, Frederick William
Lossinsky, Albert S
McGuire, John Joseph
Morrison, Mary Alice
Napoli, Joseph Leonard
Patterson, Ernest Leonard
Rasmussen, Kathleen Maher
Richie, John Peter, Jr
Roel, Lawrence Edmund
Rosensweig, Norton S
Sauer, Leonard A
Shayegani, Mehdi
Singh, Malathy
Sparks, Charles Edward
Steffen, Daniel G
Traber, Maret G
Wapnir, Raul A
Wasserman, Robert Harold
Welch, Ross Maynard
Wu, Joseph M

NORTH CAROLINA
Bales, Connie Watkins
Bartholow, Lester C
Borgman, Robert F
Burge, Jean C
Catignani, George Louis
Gallagher, Margie Lee
Hayes, Johnnie Ray
Kamin, Henry
Louie, Dexter Stephen
Miller, Donald F
Morse, Roy E
Reinhard, John Frederick, Jr
Schiffman, Susan S
Shaw, Helen Lester Anderson
Swaisgood, Harold Everett

NORTH DAKOTA
Hunt, Janet R
Jacobs, Francis Albin
Milne, David Bayard
Nielsen, Forrest Harold
Stanislao, Bettie Chloe Carter

OHIO
Acosta, Phyllis Brown
Behr, Stephen Richard
Benton, Duane Allen
Birkhahn, Ronald H
Boggs, Robert Wayne
Cardell, Robert Ridley, Jr
Cooper, Dale A
Ekvall, Shirley W
Farrier, Noel John
Green, Ralph
Grooms, Thomas Albin
Hagerman, Larry M
Hecker, Art L
Hink, Walter Fredric
Jandacek, Ronald James
Lemon, Peter Willian Reginald
McCarthy, F D
Martin, Elden William
Miller, Robert Harold
Naito, Herbert K
Pfeifer, Gerard David
Reddy, Padala Vykuntha
Richardson, Keith Erwin
Robinow, Meinhard
Stevenson, John Ray
Varma, Raj Narayan
Voss, Anne Coble
Weber, George Russell
Williams, Wallace Terry

OKLAHOMA
Koh, Eunsook Tak
Metcoff, Jack
Patterson, Manford Kenneth, Jr
Rikans, Lora Elizabeth
Seideman, Walter E
Thurman, Lloy Duane

OREGON
Fang, Ta-Yun
Hackman, Robert Mark
Martin, Lloyd W
Swank, Roy Laver
Whanger, Philip Daniel
Wyatt, Carolyn Jane
Yearick, Elisabeth Stelle

PENNSYLVANIA
Albright, Fred Ronald
Barnes, William Shelley
Biebuyck, Julien Francois
Birdsall, Marion Ivens
Davies, Ronald Edgar
Della-Fera, Mary Anne
Dryden, Richard Lee
Fenton, Marilyn Ruth
Green, Michael H(enry)
Kare, Morley Richard
Klurfeld, David Michael
Leach, Roland Melville, Jr
Mottur, George Preston
Mullen, James L
Pike, Ruth Lillian
Pollack, Robert Leon
Ramachandran, N
Rao, Kalipatnapu Narasimha
Ross, Alta Catharine
Scholz, Richard W
Seligson, Frances Hess
Smith, John Edgar
Sprince, Herbert
Stewart, Charles Newby
Thompson, Donald B
Tristram-Nagle, Stephanie Ann
Tuan, Rocky Sung-Chi
Vasilatos-Younken, Regina
Wright, Helen S
Yushok, Wasley Donald

RHODE ISLAND
Dymsza, Henry A
Nippo, Murn Marcus
Rand, Arthur Gorham, Jr
Shaikh, Zahir Ahmad

SOUTH CAROLINA
Hollis, Bruce Warren
Maurice, D V
Turk, Donald Earle

SOUTH DAKOTA
Grove, John Amos

TENNESSEE
Benton, Charles Herbert
Cook, Robert James
Cullen, Marion Permilla
Herting, David Clair
Jones, Peter D
Ong, David Eugene
Papas, Andreas Michael
Perry, Margaret Nutt
Robbins, Kelly Roy
Sintes, Jorge Luis
Wagner, Conrad

TEXAS
Anthony, W Brady
Arnold, Watson Caufield
Bertrand, Helen Anne
Brittin, Dorothy Helen Clark
Carney, Darrell Howard
Chang, Ching Hsong

Dudrick, Stanley John
Fernandes, Gabriel
Freeland-Graves, Jeanne H
Fry, Peggy Crooke
Gatlin, Delbert Monroe, III
Harris, Edward David
Haskell, Betty Echternach
Hsueh, Andie M
Huber, Gary Louis
Jacobson, Elaine Louise
Klein, Gordon Leslie
Kubena, Leon Franklin
Landmann, Wendell August
Lane, Helen W
Lawrence, Addison Lee
Lewis, Douglas Scott
Liepa, George Uldis
Lifschitz, Meyer David
Madsen, Kenneth Olaf
Masoro, Edward Joseph
Mastromarino, Anthony John
Moyer, Mary Pat Sutter
Oberleas, Donald
Pond, Wilson Gideon
Rassin, David Keith
Reagor, John Charles
Root, Elizabeth Jean
Rudolph, Frederick Byron
Spallholz, Julian Ernest
Washington, Arthur Clover
Wood, Randall Dudley
Worthy, Graham Anthony James
Yeh, Lee-Chuan Caroline
Yu, Byung Pal

UTAH
Chan, Gary Mannerstedt
Graff, Darrell Jay
Hansen, Roger Gaurth
Johnston, Stephen Charles
Lawson, Larry Dale
Mahoney, Arthur W
Ramachandran, Chittoor Krishna
Stoddard, George Edward
Windham, Carol Thompson
Wyse, Bonita W

VERMONT
Bartel, Lavon L
Merrow-Hopp, Susan B
Tyzbir, Robert S

VIRGINIA
Brandt, Richard Bernard
Christiansen, Marjorie Miner
Dannenburg, Warren Nathaniel
Feher, Joseph John
Herbein, Joseph Henry, Jr
Kirkpatrick, Roy Lee
Pao, Eleanor M
Sheppard, Alan Jonathan
Webb, Ryland Edwin

WASHINGTON
Adamson, Lucile Frances
Bankson, Daniel Duke
Childs, Marian Tolbert
Drum, Ryan William
Felton, Samuel Page
Fletcher, Dean Charles
Hardy, Ronald W
Kachmar, John Frederick
Kutter, Elizabeth Martin
Lee, Donald Jack
Meeder, Jeanne Elizabeth
Murphy, Mary Eileen
Pubols, Merton Harold
Ronzio, Robert A
Sasser, Lyle Blaine
Shultz, Terry D
Swanson, Barry Grant
Yu, Ming-Ho

WEST VIRGINIA
Nomani, M Zafar
Schubert, John Rockwell

WISCONSIN
Armentano, Louis Ernst
Attie, Alan D
Baumann, Carl August
Brown, Raymond Russell
Deese, Dawson Charles
Harper, Alfred Edwin
Hoekstra, William George
Jackson, Marion Leroy
Keesey, Richard E
Kemnitz, Joseph William
Kochan, Robert George
McIntosh, Elaine Nelson
Marlett, Judith Ann
Ogden, Robert Verl
Olsen, Ward Alan
Pariza, Michael Willard
Roberts, Willard Lewis
Schulz, Leslie Olmstead
Steele, Robert Darryl
Tews, Jean Kring
Wege, Ann Christene
Woldegiorgis, Gebretateos

PUERTO RICO
El-Khatib, Shukri M

ALBERTA
Basu, Tapan Kumar
Hawrysh, Zenia Jean
Mathison, Gary W(ayne)
Sohal, Parmjit S

BRITISH COLUMBIA
Freeman, Hugh J

MANITOBA
Angel, Aubie
Ingalls, Jesse Ray
Marquardt, Ronald Ralph

NEWFOUNDLAND
Orr, Robin Denise Moore

NOVA SCOTIA
Castell, John Daniel
Lall, Santosh Prakash
Weld, Charles Beecher

ONTARIO
Alexander, James Craig
Beare-Rogers, Joyce Louise
Begin-Heick, Nicole
Bettger, William Joseph
Cinader, Bernhard
Dalpe, Yolande
Flanagan, Peter Rutledge
Fortin, Andre Francis
Hill, Eldon G
Holub, Bruce John
Mills, David Edward
Pace Asciak, Cecil
Shah, Bhagwan G
Stavric, Bozidar
Yeung, David Lawrence

QUEBEC
Donefer, Eugene
Kallai-Sanfacon, Mary-Ann
Kay, Ruth McPherson
Kuhnlein, Harriet V
Pelletier, Omer
Peronnet, Franc04ois R R
Srivastava, Uma Shanker
Young, Simon N

SASKATCHEWAN
Patience, John Francis

OTHER COUNTRIES
Bronk, John Ramsey
Brunser, Oscar
Crompton, David William Thomasson
Dutta-Roy, Asim Kanti
Fairweather-Tait, Susan Jane
Jaffe, Werner G
Ju, Jin Soon
Mela, David Jason
Palmer, Sushma Mahyera
Rio, María Esther
Santidrian, Santiago
Segal, Mark
Vina, Juan R

Physical Anthropology

ALABAMA
Hoff, Charles Jay
Wyss, James Michael

ALASKA
Milan, Frederick Arthur

ARIZONA
Birkby, Walter H
Bleibtreu, Hermann Karl
Harrison, Gail Grigsby
Merbs, Charles Francis
Olsen, Stanley John
Rittenbaugh, Cheryl K
Schoenwetter, James
Stini, William Arthur
Turner, Christy Gentry, II
Zegura, Stephen Luke

CALIFORNIA
Brodsky, Carroll M
Dolhinow, Phyllis Carol
Howell, Francis Clark
Jurmain, Robert Douglas
Lindburg, Donald Gilson
McHenry, Henry Malcolm
McLeod, Samuel Albert
Murad, Turhon Allen
Napton, Lewis Kyle
Rogers, Spencer Lee
Roll, Barbara Honeyman Heath
Sarich, Vincent M
Todd, Harry Flynn, Jr
Treloar, Alan Edward
Williams, Bobby Joe
Wood, Corinne Shear
Zihlman, Adrienne Louella

COLORADO
Brues, Alice Mossie
Charney, Michael
Greene, David Lee
Hackenberg, Robert Allan
Kelso, Alec John (Jack)

Moore, Lorna Grindlay
Schanfield, Moses Samuel
Schulter-Ellis, Frances Pierce

CONNECTICUT
Coe, Michael Douglas
Heller, John Herbert
Kidd, Kenneth Kay
Lamm, Foster Philip
Laughlin, William Sceva

DELAWARE
Leslie, Charles Miller

DISTRICT OF COLUMBIA
Bernor, Raymond Louis
Ericksen, Mary Frances
Ortner, Donald John
Ubelaker, Douglas Henry

FLORIDA
Armelagos, George John
DeRousseau, C(arol) Jean
Lieberman, Leslie Sue
Maples, William Ross
Wienker, Curtis Wakefield

GEORGIA
Smith, Euclid O'Neal

HAWAII
Goodman, Madelene Joyce
Hanna, Joel Michael
Pietrusewsky, Michael, Jr

ILLINOIS
Buikstra, Jane Ellen
Corruccini, Robert Spencer
Costa, Raymond Lincoln, Jr
Giles, Eugene
Hausfater, Glenn
Klepinger, Linda Lehman
Reed, Charles Allen
Rosenberger, Alfred L
Simon, Mark Robert
Singer, Ronald
Tuttle, Russell Howard

INDIANA
O'Connor, Brian Lee

IOWA
Andelson, Jonathan Gary
Ciochon, Russell Lynn
Dawson, David Lynn
Staley, Robert Newton

KANSAS
Finnegan, Michael

KENTUCKY
Hochstrasser, Donald Lee
Reid, Russell Martin
Savage, Steven Paul
Wiese, Helen Jean Coleman

MAINE
Stoudt, Howard Webster

MARYLAND
Bresler, Jack Barry
Garruto, Ralph Michael
Kerley, Ellis R
Micozzi, Marc S
Thorp, James Wilson
Yin, Frank Chi-Pong

MASSACHUSETTS
Gibbons, Michael Francis, Jr
Gordon, Claire Catherine
Howells, William White
Hunt, Edward Eyre
Kolakowski, Donald Louis
Seltzer, Carl Coleman
Swedlund, Alan Charles

MICHIGAN
Bajema, Carl J
Brace, C Loring
Garn, Stanley Marion
Jordan, Brigitte
Lasker, Gabriel (Ward)
Livingstone, Frank Brown
Reynolds, Herbert McGaughey
Snow, Loudell Fromme
Weiss, Mark Lawrence
Wolpoff, Milford Howell

MISSOURI
Gavan, James Anderson
Rasmussen, David Tab
Smith, Richard Jay

MONTANA
Smith, Charline Galloway

NEVADA
Brooks, Sheilagh Thompson

NEW YORK
Ascher, Robert
Chepko-Sade, Bonita Diane
Delson, Eric
Dibennardo, Robert

Dubroff, Lewis Michael
Dyson-Hudson, V Rada
Fleagle, John G
Geise, Marie Clabeaux
Gerber, Linda M
Gustav, Bonnie Lee
Haas, Jere Douglas
Himes, John Harter
Holloway, Ralph L, Jr
Jolly, Clifford J
Kennedy, Donald Alexander
Kennedy, Kenneth Adrian Raine
Kinzey, Warren Glenford
Leslie, Paul Willard
Little, Michael Alan
Mendel, Frank C
Pasamanick, Benjamin
Rightmire, George Philip
Sank, Diane
Shapiro, Harry Lionel
Sirianni, Joyce E
Sokal, Robert Reuven
Steegmann, Albert Theodore, Jr
Stern, Jack Tuteur, Jr
Susman, Randall Lee
Tattersall, Ian Michael
Taylor, James Vandigriff
Townsend, John Marshall
Winter, John Henry

NORTH CAROLINA
Cartmill, Matt
Daniel, Hal J
Holcomb, George Ruhle
O'Barr, William McAlston
Pollitzer, William Sprott
Robbins, Louise Marie
Simons, Elwyn LaVerne

OHIO
Blank, John Edward
Byard, Pamela Joy
Chumlea, William Cameron
Latimer, Bruce Millikin
McConville, John Theodore
Marras, William Steven
Oyen, Ordean James
Poirier, Frank Eugene
Saul, Frank Philip
Saul, Julie Mather
Sciulli, Paul William
Slutzky, Gale David

OKLAHOMA
Bell, Robert Eugene
Snow, Clyde Collins

OREGON
Moreno-Black, Geraldine S

PENNSYLVANIA
Baker, Paul Thornell
Blumberg, Baruch Samuel
Cronk, Christine Elizabeth
Eckhardt, Robert Barry
Friedlaender, Jonathan Scott
Gottlieb, Karen Ann
Harpending, Henry Cosad
Johnston, Francis E
Katz, Solomon H
Mann, Alan Eugene
Schwartz, Jeffrey H
Siegel, Michael Ian
Tinsman, James Herbert, Jr
Weiss, Kenneth Monrad
Zimmerman, Michael Raymond

SOUTH CAROLINA
Rathbun, Ted Allan

TENNESSEE
Bass, William Marvin, III
McNutt, Charles Harrison

TEXAS
Biggerstaff, Robert Huggins
Bramblett, Claud Allen
Bryant, Vaughn Motley, Jr
Coelho, Anthony Mendes, Jr
Easley, Stephen Phillip
Fry, Edward Irad
Gibson, Kathleen Rita
Hixson, James Elmer
Lamb, Neven P
Malina, Robert Marion
Novak, Ladislav Peter
Race, George Justice
Steele, David Gentry
Wetherington, Ronald K
Wohlschlag, Donald Eugene

UTAH
McCullough, John Martin

WASHINGTON
Chrisman, Noel Judson
Eck, Gerald Gilbert
Hurlich, Marshall Gerald
Lundy, John Kent
Newell-Morris, Laura
Nute, Peter Eric

WEST VIRGINIA
Fix, James D

Hilloowala, Rumy Ardeshir

WISCONSIN
Bennett, Kenneth A
Leutenegger, Walter
Tappen, Neil Campbell

WYOMING
Gill, George Wilhelm

BRITISH COLUMBIA
Booth, Amanda Jane

MANITOBA
De Pena, Joan Finkle
Kaufert, Joseph Mossman
Lewis, Marion Jean

NOVA SCOTIA
Walker, Joan Marion

ONTARIO
Anderson, James Edward
Beevis, David
Gaherty, Geoffrey George
Helmuth, Hermann Siegfried
Hunter, William Stuart
McFeat, Tom Farrar Scott
Saunders, Shelley Rae
Singh, Ripu Daman

QUEBEC
Kuhnlein, Harriet V

OTHER COUNTRIES
Goldstein, Marcus Solomon
Oxnard, Charles Ernest

Physiology, Animal

ALABAMA
Appel, Arthur Gary
Ardell, Jeffrey Laurence
Barker, Samuel Booth
Beasley, Philip Gene
Beckett, Sidney D
Berecek, Kathleen Helen
Bone, Leon Wilson
Boshell, Buris Raye
Bowie, Walter C
Carlo, Waldemar Alberto
Dixon, Earl, Jr
Downey, James Merritt
Dudley, Susan D
Elgavish, Ada S
Friedlander, Michael J
Gibbons, Ashton Frank Eleazer
Goldman, Ronald
Grubbs, Clinton Julian
Hageman, Gilbert Robert
Harding, Thomas Hague
Holloway, Clarke L
Jeffcoat, Marjorie K
Jenkins, Ronald Lee
Kochakian, Charles Daniel
Logic, Joseph Richard
Magargal, Wells Wrisley, II
Marple, Dennis Neil
Matalon, Sadis
Modlin, Richard Frank
Neill, Jimmy Dyke
Oparil, Suzanne
Paxton, Ralph
Pegram, George Vernon, Jr
Pittman, James Allen, Jr
Pritchett, John Franklyn
Redding, Richard William
Schnaper, Harold Warren
Schneyer, Charlotte A
Schoultz, Ture William
Sharma, Udhishtra Deva
Smith, Curtis R
Takahashi, Ellen Shizuko
Thomas, Julian Edward, Sr
Wit, Lawrence Carl
Yarbrough, James David

ALASKA
Babcock, Malin Marie
Feist, Dale Daniel
Miller, Lyster Keith
Paul, Augustus John, III
Proenza, Luis Mariano
Shirley, Thomas Clifton

ARIZONA
Bloedel, James R
Blouin, Leonard Thomas
Bowers, William Sigmond
Chandler, Douglas Edwin
Chen, Hsien-Jen James
Chiasson, Robert Breton
Clayton, John Wesley, Jr
Fowler, Dona Jane
Hadley, Mac Eugene
Hadley, Neil F
Hagedorn, Henry Howard
Harkins, Kristi R
Hazel, Jeffrey Ronald
Heine, Melvin Wayne
Hildebrand, John Grant, III
Hull, Hugh Boden
Johnson, Mary Ida

Kazal, Louis Anthony
Knudson, Ronald Joel
Krahl, Maurice Edward
Laird, Hugh Edward, II
Lei, David Kai Yui
McCauley, William John
McCuskey, Robert Scott
Nicolls, Ken E
Nugent, Charles Arter, Jr
Parsons, L Claire
Pickens, Peter E
Tarby, Theodore John
Tischler, Marc Eliot
Walsberg, Glenn Eric
Wegner, Thomas Norman
Winfree, Arthur T

ARKANSAS
Allaben, William Thomas
Amlaner, Charles Joseph, Jr
Bridgman, John Francis
Cockerham, Lorris G(ay)
Cornett, Lawrence Eugene
Daniels, L B
Demarest, Jeffrey R
Doyle, Lee Lee
Elders, Minnie Joycelyn
Garcia-Rill, Edgar E
Greenman, David Lewis
Harris, Grover Cleveland, Jr
Kellogg, David Wayne
Light, Kim Edward
McGilliard, A Dare
McMillan, Harlan L
Marvin, Horace Newell
Piper, Edgar L
Rayford, Phillip Leon
Reddy, RamaKrishna Pashuvula
Ruwe, William David
Soulsby, Michael Edward
Stallcup, Odie Talmadge
Vesely, David Lynn

CALIFORNIA
Adams, Thomas Edwards
Adams, William S
Adey, William Ross
Ahmad, Nazir
Alexander, Natalie
Alousi, Adawia A
Amirkhanian, John David
Anand, Rajen S(ingh)
Anderson, Gary Bruce
Antipa, Gregory Alexis
Arnaud, Claude Donald, Jr
Arp, Alissa Jan
Ashe, John Herman
Ashmore, Charles Robert
Austin, George M
Avila, Vernon Lee
Bacchus, Habeeb
Baker, Mary Ann
Baldwin, Ransom Leland
Barker, David Lowell
Barnes, Paul Richard
Barrett, Robert
Baylor, Denis Aristide
Beattie, Randall Chester
Beekman, Bruce Edward
Bennett, Albert Farrell
Bennett, Edward Leigh
Bentley, David R
Bercovitz, Arden Bryan
Berger, Ralph Jacob
Bern, Howard Alan
Bethune, John Edmund
Beuchat, Carol Ann
Bickford, Reginald G
Biglieri, Edward George
Binggeli, Richard Lee
Bloom, Floyd Elliott
Bolaffi, Janice Lerner
Book, Steven Arnold
Botstein, David
Bovell, Carlton Rowland
Bower, Annette
Brace, Robert Allen
Bradford, G Eric
Braunstein, Glenn David
Breisch, Eric Alan
Brown, Joan Heller
Brown, Marvin Ross
Brunton, Laurence
Buchwald, Jennifer S
Buchwald, Nathaniel Avrom
Bullock, Leslie Patricia
Bullock, Theodore Holmes
Burrill, Melinda Jane
Callantine, Merritt Reece
Carew, Thomas Edward
Carlsen, Richard Chester
Carson, Virginia Rosalie Gottschall
Carstens, Earl E
Cavalieri, Ralph R
Cech, Joseph Jerome, Jr
Chalupa, Leo M
Chang, Betty
Chang, Ernest Sun-Mei
Chappell, Mark Allen
Chopra, Inder Jit
Chow, Kao Liang
Clayton, Raymond Brazenor
Cohen, Natalie Shulman
Colvin, Harry Walter, Jr

Taylor, Malcolm Herbert

DISTRICT OF COLUMBIA
Ahluwalia, Balwant Singh
Allen, Robert Erwin
Barker, Winona Clinton
Beck, Lucille Bluso
Becker, Kenneth Louis
Blanquet, Richard Steven
Bowling, Lloyd Spencer, Sr
Cerveny, Thelma Jannette
Chaput, Raymond Leo
Crawford, Lester M
Crisp, Thomas Mitchell, Jr
Dimond, Marie Therese
Eagles, Douglas Alan
Forman, David S
Frey, Mary Anne Bassett
Gold, Armand Joel
Harmon, John W
Henry, Walter Lester, Jr
Holloway, James Ashley
Ison-Franklin, Eleanor Lutia
Jenkins, Mamie Leah Young
Jordan, Alexander Walker, III
Kenney, Richard Alec
Kot, Peter Aloysius
Lavine, Robert Alan
Lumpkin, Michael Dirksen
McEwen, Gerald Noah, Jr
Malveaux, Floyd J
Morck, Timothy Anton
Murphy, James John
Packer, Randall Kent
Pearlman, Ronald C
Pennington, Jean A T
Phillips, Robert Ward
Pointer, Richard Hamilton
Ramey, Estelle R
Rapisardi, Salvatore C
Rosenberg, Edith E
Shukla, Kamal Kant
Silva, Omega Logan
Sulzman, Frank Michael
Tearney, Russell James
Tidball, Charles Stanley
Tidball, M Elizabeth Peters
Umminger, Bruce Lynn
Wagner, Henry George
Weiss, Ira Paul
Whidden, Stanley John
Wilson, Don Ellis
Wong, Harry Yuen Chee
Wylie, Richard Michael
Wyngaarden, James Barnes
Young, John Karl

FLORIDA
Adams, Ralph M
Adams, Roger Omar
Baker, Carleton Harold
Barrett, Ellen Faye
Barrett, John Neil
Bassett, Arthur L
Bensen, Jack F
Besch, Emerson Louis
Biersdorf, William Richard
Blackwell, Harold Richard
Booth, Nicholas Henry
Boyd, Eleanor H
Bramante, Pietro Ottavio
Brown, Dawn LaRue
Bunch, Wilton Herbert
Bzoch, Kenneth R
Cassin, Sidney
Castro, Alberto
Chen, Chao Ling
Clendenin, Martha Anne
Convertino, Victor Anthony
Davenport, Paul W
Davidoff, Robert Alan
Dietz, John R
Dunn, William Arthur, Jr
Easton, Dexter Morgan
Ellias, Loretta Christine
Emerson, Geraldine Mariellen
Emmanuel, George
Freeman, Marc Edward
Freund, Gerhard
Friedl, Frank Edward
Galindo, Anibal H
Gaunt, Robert
Gordon, Kenneth Richard
Gorniak, Gerard Charles
Greenberg, Michael John
Gruber, Samuel Harvey
Haag, Kim Hartzell
Hackman, John Clement
Hammer, Lowell Clarke
Harrison, Robert J
Holt, Joseph Paynter, Sr
Hope, George Marion
Jaeger, Marc Jules
Joftes, David Lion
Kalra, Satya Paul
Kessler, Richard Howard
Knapp, Francis Marion
Krzanowski, Joseph John, Jr
Lawrence, Pauline Olive
Levey, Douglas J
Levy, Norman Stuart
Lillywhite, Harvey B
Lindsey, Bruce Gilbert
Lipner, Harry Joel

Loewenstein, Werner Randolph
Lutz, Peter Louis
McKenney, Charles Lynn, Jr
McKenzie, John Maxwell
McLean, Mark Philip
Magleby, Karl LeGrande
Merritt, Alfred M, II
Michie, David Doss
Miles, Richard David
Munson, John Bacon
Nichols, Wilmer Wayne
Nordlie, Frank Gerald
Oberlander, Herbert
Otis, Arthur Brooks
Ott, Edgar Alton
Palmore, William P
Pargman, David
Penhos, Juan Carlos
Pfeiffer, Eric A
Phelps, Christopher Prine
Pieper, Heinz Paul
Pollock, Michael L
Rao, Krothapalli Ranga
Rao, Papineni Seethapathi
Reynolds, David George
Robinson, Gerald Garland
Rodrick, Gary Eugene
Root, Allen William
Rowland, Neil Edward
Russell, Diane Haddock
Sackner, Marvin Arthur
Shankland, Daniel Leslie
Sharp, John Turner
Snow, Thomas Russell
Soliman, Karam Farag Attia
Stafford, Robert Oppen
Stainsby, Wendell Nicholls
Sturbaum, Barbara Ann
Sypert, George Walter
Thatcher, William Watters
Thomas, William Clark, Jr
Thompson, Ronald Halsey
Thuning-Roberson, Claire Ann
Tiffany, William James, III
Tobey, Frank Lindley, Jr
Vail, Edwin George
Vallowe, Henry Howard
Vander Meer, Robert Kenneth
Voigt, Walter
Walker, Don Wesley
Weber, James Edward
White, Arlynn Quinton, Jr
White, Arlynn Quinton, Jr
Wilcox, Christopher Stuart
Wilde, Walter Samuel
Wilkerson, James Edward
Wright, Paul Albert
Zengel, Janet Elaine

GEORGIA
Baldwin, Bernell Elwyn
Barb, C Richard
Bhalla, Vinod Kumar
Binnicker, Pamela Caroline
Black, John B
Bond, Gary Carl
Brann, Darrell Wayne
Bransome, Edwin D, Jr
Buccafusco, Jerry Joseph
Calabrese, Ronald Lewis
Catravas, John D
Comeau, Roger William
Costoff, Allen
Coulter, Dwight Bernard
Crim, Joe William
Dale, Edwin
Doetsch, Gernot Siegmar
Dusenbery, David Brock
Eaton, Douglas Charles
Edwards, Gaylen Lee
Ehrhart, Ina C
English, Arthur William
Franch, Robert H
Givens, James Robert
Hadd, Harry Earle
Hendrich, Chester Eugene
Hofman, Wendell Fey
Howarth, Birkett, Jr
Humphrey, Donald R
Hunt, William Daniel
Inge, Walter Herndon, Jr
Innes, David Lyn
Joyner, Ronald Wayne
Kraeling, Robert Russell
Leibach, Fredrick Hartmut
Leitch, Gordon James
Letbetter, William Dean
Little, Robert Colby
McDaniel, William Franklin
Mahesh, Virendra B
Martin, David Edward
Michael, Richard Phillip
Miller, David Arthur
Mills, Thomas Marshall
Mokler, Corwin Morris
Mulroy, Michael Joseph
Nadler, Ronald D
Neville, Walter Edward, Jr
Nosek, Thomas Michael
Ogle, Thomas Frank
Parks, John S
Parrish, John W, Jr
Pashley, David Henry
Phillips, Lawrence Stone

Porter, James W
Porterfield, Susan Payne
Rinard, Gilbert Allen
Smith, Anderson Dodd
Smith, Jesse Graham, Jr
Sridaran, Rajagopala
Stoney, Samuel David, Jr
Thompson, Frederick Nimrod, Jr
West, Charles Hutchison Keesor
Wood, John Grady

HAWAII
Ahearn, Gregory Allen
Atkinson, Shannon K C
Brick, Robert Wayne
Bryant-Greenwood, Gillian Doreen
De Feo, Vincent Joseph
Gillary, Howard L
Grau, Edward Gordon
Greenwood, Frederick C
Hall, John Bradley
Hapai, Marlene Nachbar
Hartline, Daniel Keffer
Hartung, G(eorge) Harley
Krupp, David Alan
Lee, Cheng-Sheng
Nelson, Marita Lee
Pang-Ching, Glenn K
Read, George Wesley
Wayman, Oliver
Weems, Charles William
Whittow, George Causey
Yount, David Eugene

IDAHO
Cloud, Joseph George
House, Edwin W
Kelley, Fenton Crosland
Mead, Rodney A

ILLINOIS
Agin, Daniel Pierre
Bahr, Janice M
Banerjee, Chandra Madhab
Barr, Lloyd
Barr, Susan Hartline
Bartell, Marvin H
Bartke, Andrzej
Bastian, James W
Baumann, Gerhard
Becker, Donald Eugene
Becker, John Henry
Berger, Sheldon
Berry, Robert Wayne
Bhattacharyya, Maryka Horsting
Bingel, Audrey Susanna
Bombeck, Charles Thomas, III
Brewer, Nathan Ronald
Bruce, David Stewart
Burhop, Kenneth Eugene
Buschmann, MaryBeth Tank
Buschmann, Robert J
Caspary, Donald M
Chatterton, Robert Treat, Jr
Chesky, Jeffrey Alan
Clayton-Hopkins, Judith Ann
Cline, William H, Jr
Cohen, David Harris
Cohen, Maynard
Cox, Thomas C
Cralley, John Clement
Crystal, George Jeffrey
Delcomyn, Fred
DeSombre, Eugene Robert
Disterhoft, John Francis
Dobrin, Philip Boone
Doemling, Donald Bernard
Dunn, Robert Bruce
Dziuk, Philip J
Elble, Rodger Jacob
Enroth-Cugell, Christina
Fanslow, Don J
Fay, Richard Rozzell
Feigen, Larry Philip
Feng, Albert Sih-Hung
Ferguson, James L
Fisher, Cletus G
Flouret, George R
Ford, Lincoln Edmond
Freinkel, Norbert
Gasdorf, Edgar Carl
Gewertz, Bruce Labe
Gibbs, Daniel
Gibori, Geula
Giere, Frederic Arthur
Gillette, Rhanor
Gold, Jay Joseph
Goldfarb, Roy David
Goldstick, Thomas Karl
Graves, Charles Norman
Green, Orville
Griffiths, Thomas Alan
Hansen, Timothy Ray
Harrison, Paul C
Hatfield, Efton Everett
Heybach, John Peter
Holmes, Kenneth Robert
Houk, James Charles
Hughes, John Russell
Jackson, Gary Loucks
Jaffe, Randal Craig
Janusek, Linda Witek
Jensen, Donald Reed
Jones, Stephen Bender

Kallio, Reino Emil
Katz, Adrian I
Katzenellenbogen, Benita Schulman
Kelley, Keith Wayne
Kesler, Darrel J
Klabunde, Richard Edwin
Konisky, Jordan
Kusano, Kiyoshi
Landau, Richard Louis
Layman, Donald Keith
Ledwitz-Rigby, Florence Ina
Lee, Chung
Levitsky, Lynne Lipton
Lindheimer, Marshall D
Litteria, Marilyn
Lodge, James Robert
McCormick, David Loyd
MacHattie, Lorne Allister
Manohar, Murli
Marczynski, Thaddeus John
Masserman, Jules Homan
Matsumura, Philip
Mayor, Gilbert Harold
Metzger, Boyd Ernest
Michael, Joel Allen
Moon, Richard C
Mullendore, James Myers
Mulvihill, Mary Lou Jolie
Nakajima, Shigehiro
Nakajima, Yasuko
Nicolette, John Anthony
Nielsen, Peter James
Novales, Ronald Richards
Nutting, Ehard Forrest
Omachi, Akira
Oscai, Lawrence B
Pappas, George Demetrios
Pederson, Vernon Clayton
Pepperberg, David Roy
Peterson, Barry Wayne
Peterson, Darryl Ronnie
Pindok, Marie Theresa
Pinto, Lawrence Henry
Polley, Edward Herman
Preston, Robert Leslie
Radwanska, Ewa
Rakowski, Robert F
Rao, Mrinalini Chatta
Rechtschaffen, Allan
Riddle, Wayne Allen
Robinson, Thomas Frank
Romack, Frank Eldon
Rosen, Arthur Leonard
Rosenfield, Robert Lee
Rotermund, Albert J, Jr
Routtenberg, Aryeh
Rovick, Allen Asher
Roys, Chester Crosby
Rymer, William Zev
Saper, Clifford B
Satinoff, Evelyn
Schwartz, Neena Betty
Sehgal, Lakshman R
Shearer, William McCague
Shriver, John William
Siegel, Jonathan Howard
Singer, Irwin
Smith, Douglas Calvin
Stearner, Sigrid Phyllis
Steger, Richard Warren
Steiner, Donald Frederick
Stickney, Janice Lee
Sweeney, Daryl Charles
Swiatek, Kenneth Robert
Takahashi, Joseph S
Tang, Pei Chin
Thompson, Phebe Kirsten
Thonar, Eugene Jean-Marie
Tone, James N
Towle, David Walter
Tse, Warren W
Turek, Fred William
Twardock, Arthur Robert
Verhage, Harold Glenn
Wade, David Robert
Wagner, William Charles
Wass, John Alfred
Webber, Charles Lewis, Jr
Webster, James Randolph, Jr
Wetzel, Allan Brooke
Wilber, Laura Ann
Wilson, Donald Alan
Winter, Robert John
Yelich, Michael Ralph
Zar, Jerrold Howard
Zehr, John E
Zemlin, Willard R

INDIANA
Anderson, David Bennett
Anderson, John Nicholas
Andrews, Frederick Newcomb
Babbs, Charles Frederick
Bakken, George Stewart
Begue, William John
Ben-Jonathan, Nira
Bohlen, Harold Glenn
Bottoms, Gerald Doyle
Brennan, David Michael
Bridges, C David
Brush, F(ranklin) Robert
Burkholder, Timothy Jay
Carter, James M
Chance, Ronald E

Physiology, Animal (cont)

Clemens, James Allen
Costill, David Lee
Devoe, Robert Donald
Duman, John Girard
Dunn, Peter Edward
Dustman, John Henry
Echtenkamp, Stephen Frederick
Epstein, Aubrey
Esch, Harald Erich
Everson, Ronald Ward
Fasola, Alfred Francis
Garriott, Michael Lee
Geddes, Leslie Alexander
Gidda, Jaswant Singh
Goetsch, Gerald D
Goetz, Frederick William, Jr
Grant, Alan Leslie
Gunther, Gary Richard
Hammel, Harold Theodore
Heiman, Mark Louis
Hill, Donald Louis
Holland, James Philip
Irwin, Glenn Ward, Jr
Johnston, Cyrus Conrad, Jr
Konetzka, Walter Anthony
Lesh, Thomas Allan
Macchia, Donald Dean
McClure, Polley Ann
Malven, Paul Vernon
Meyer, Frederick Richard
Neff, William Duwayne
Ochs, Sidney
Ostroy, Sanford Eugene
Outhouse, James Burton
Paladino, Frank Vincent
Paschall, Homer Donald
Peterson, Richard George
Pflanzer, Richard Gary
Powell, Richard Cinclair
Prange, Henry Davies
Reading, Rogers W
Renzi, Alfred Arthur
Roales, Robert R
Rothe, Carl Frederick
Rowland, David Lawrence
St John, Philip Alan
Sojka, Gary Allan
Stabler, Timothy Allen
Steer, Max David
Tacker, Willis Arnold, Jr
Tanner, George Albert
Thompson, Daniel James
Thor, Karl Bruce
Voelz, Michael H
Wagner, Wiltz Walker, Jr
Weinstein, Paul P
Williams, Daniel Charles
Yoder, John Menly
Zeller, Frank Jacob

IOWA

Ackerman, Ralph Austin
Anderson, Lloyd L
Bhalla, Ramesh C
Campbell, Kevin Peter
Carithers, Jeanine Rutherford
Conn, P Michael
Coulter, Joe Dan
Dellmann, H Dieter
Denburg, Jeffrey Lewis
Dunham, Jewett
Eagleson, Gerald Wayne
Fellows, Robert Ellis, Jr
Fishman, Irving Yale
Ford, Stephen Paul
Getting, Peter Alexander
Hardin, Carolyn Myrick
Heistad, Donald Dean
Hembrough, Frederick B
Henninger, Ann Louise
Hubel, Kenneth Andrew
Husted, Russell Forest
Kaplan, Murray Lee
Knott, John Russell
Masat, Robert James
Maurer, Richard Allen
Morrical, Daniel Gene
Morriss, Frank Howard, Jr
Perret, George (Edward)
Randic, Mirjana
Read, Charles H
Redmond, James Ronald
Riggs, Dixon L
Schmidt, Thomas John
Shaw, Gaylord Edward
Shen, Sheldon Shih-Ta
Simpson, Robert John
Spaziani, Eugene
Stratton, Donald Brendan
Swenson, Melvin John
Thompson, Sue Ann
Waziri, Rafiq
Whipp, Shannon Carl
Williams, Dean E
Wu, Chun-Fang

KANSAS

Baranczuk, Richard John
Berman, Nancy Elizabeth Johnson
Besharse, Joseph Culp
Bunag, Ruben David
Cheney, Paul David

Dahl, Nancy Ann
Davis, John Stewart
Dey, Sudhansu Kumar
Dunn, Jon D
Erickson, Howard Hugh
Fedde, Marion Roger
Ferraro, John Anthony
Fina, Louis R
Gattone, Vincent H, II
Goetzinger, Cornelius Peter
Gonzalez, Norberto Carlos
Hopkins, Theodore Louis
Johnson, Donald Charles
Kammer, Ann Emma
Keller, Leland Edward
Kramer, Karl Joseph
Leavitt, Wendell William
McCroskey, Robert Lee
Maher, Michael John
Meek, Joseph Chester, Jr
Michaelis, Elias K
Miller, William Eugene
Neufeld, Gaylen Jay
Orr, James Anthony
Parrish, John Wesley, Jr
Peters, Ralph I
Quadagno, David Michael
Quadri, Syed Kaleemullah
Rowe, Edward C
Rowe, Vernon Dodds
Smalley, Katherine N
Stevenson, Jeffrey Smith
Sullivan, Lawrence Paul
Tessel, Richard Earl
Vacca, Linda Lee
Valenzeno, Dennis Paul
Wilson, Fred E
Yochim, Jerome M

KENTUCKY

Barber, William J
Baur, John M
Bennett, Thomas Edward
Blake, James Neal
Boehms, Charles Nelson
Boyarsky, Louis Lester
Cohn, David Valor
Creek, Robert Omer
Fell, Ronald Dean
Ferrell, Blaine Richard
Frazier, Donald Tha
Gladden, Bruce
Green, William Warden
Harris, Patrick Donald
Hirsch, Henry Richard
Jones, Sanford L
Just, John Josef
Kargl, Thomas E
Legan, Sandra Jean
Loy, Robert Graves
McCook, Robert Devon
McGraw, Charles Patrick
McLaughlin, Barbara Jean
Miller, Ralph English
Moody, William Glenn
Ott, Cobern Erwin
Peterson, Roy Phillip
Prior, David James
Puckett, Hugh
Rao, Chalamalasetty Venkateswara
Smothers, James Llewellyn
Spamford, Bryant
Spoor, William Arthur
Traurig, Harold H
Urbscheit, Nancy Lee
Wead, William Badertscher
Williams, Walter Michael

LOUISIANA

Anderson, Mary Loucile
Arimura, Akira
Bagby, Gregory John
Bailey, R L
Barker, Hal B
Bartell, Clelmer Kay
Battarbee, Harold Douglas
Baum, Lawrence Stephen
Beadle, Ralph Eugene
Berlin, Charles I
Beuerman, Roger Wilmer
Brizzee, Kenneth Raymond
Caprio, John Theodore
Carmines, Pamela Kay
Christian, Frederick Ade
Coy, David Howard
Deaton, Lewis Edward
Dietz, Thomas Howard
Di Luzio, Nicholas Robert
Ely, Thomas Harrison
Fingerman, Milton
Gaar, Kermit Albert, Jr
Godke, Robert Alan
Happel, Leo Theodore, Jr
Harbo, John Russell
Harrison, Richard Miller
Hess, Melvin
Istre, Clifton O, Jr
Jenkins, William L
Johnsen, Peter Berghsey
Kastin, Abba J
Kreisman, Norman Richard
Lang, Charles H
Laurent, Sebastian Marc
Levitzky, Michael Gordon

Loewenstein, Joseph Edward
Lowe, Robert Franklin, Jr
McNamara, Dennis B
Mahajan, Damodar K
Miller, Harvey I
Nair, Pankajam K
O'Dell Smith, Roberta Maxine
Olmsted, Clinton Albert
Olson, Richard David
Peaslee, Margaret H
Porter, Johnny Ray
Robertson, George Leven
St Angelo, Allen Joseph
Saphier, David
Schally, Andrew Victor
Shepherd, David Preston
Silverman, Harold
Spring, Jeffrey H
Wakeman, John Marshall
Wallin, John David
Windhauser, Marlene M
Woodring, J Porter
York, David Anthony

MAINE

Bain, William Murray
Bayer, Robert Clark
Bernstein, Seldon Edwin
Dowse, Harold Burgess
Gainey, Louis Franklin, Jr
Harris, Paul Chappell
Harrison, David Ellsworth
Knoll, Henry Albert
Labov, Jay Brian
Norton, James Michael
Rand, Peter W
Schmidt-Nielsen, Bodil Mimi
Shumway, Sandra Elisabeth
Spirito, Carl Peter

MARYLAND

Allen, Willard M
Anderson, David Everett
Anderson, Larry Douglas
Anderson, Robert Simpers
Angelone, Luis
Augustine, Patricia C
Ballard, Kathryn Wise
Bareis, Donna Lynn
Barker, Jeffery Lange
Barraclough, Charles Arthur
Bass, Eugene L
Batson, David Banks
Bean, Barbara Louise
Beisel, William R
Benevento, Louis Anthony
Benson, Dennis Alan
Bergey, Gregory Kent
Berman, Michael Roy
Blaustein, Mordecai P
Bloom, Sherman
Bodammer, Joel Edward
Bodian, David
Bolt, Douglas John
Boucher, John H
Brownstein, Michael Jay
Buck, John Bonner
Bünger, Rolf
Burke, Robert Emmett
Burt, David Reed
Burton, Dennis Thorpe
Carlson, Drew E
Cecil, Helene Carter
Chapin, John Ladner
Chen, Hao-Chia
Chitwood, David Joseph
Choudary, Jasti Bhaskararao
Clarke, David Harrison
Colombini, Marco
Commissiong, John Wesley
Contrera, Joseph Fabian
Coulombe, Harry N
Cronin, Thomas Wells
Dahl, Roy Dennis
Dahlen, Roger W
Delahunty, George
De Monasterio, Francisco M
Dhindsa, Dharam Singh
Dubin, Norman H
Dubner, Ronald
Dubois, Andre T
Dudley, Peter Anthony
Dufau, Maria Luisa
Ehrlich, Walter
Eipper, Betty Anne
Engel, Bernard Theodore
Estienne, Mark Joseph
Ewing, Larry Larue
Fajer, Abram Bencjan
Fambrough, Douglas McIntosh
Fitch, Kenneth Leonard
Fitzgerald, Robert Schaefer
Foster, Giraud Vernam
Fowler, Arnold K
Frank, Martin
Franklin, Renty Benjamin
Freed, Arthur Nelson
Freed, Michael Abraham
Fuller, Mark Roy
Gainer, Harold
Gann, Donald Stuart
Gardner, Jerry David
Graff, Morris Morse
Graham, Charles Raymond, Jr

Greulich, Richard Curtice
Haddy, Francis John
Hansen, Barbara Caleen
Hansford, Richard Geoffrey
Hanson, Frank Edwin
Hanson, Robert C
Hardy, Lester B
Helke, Cinda Jane
Hempel, Franklin Glenn
Hill, Elwood Fayette
Hirshman, Carol A
Hobbs, Ann Snow
Homer, Louis David
Hopkins, Thomas Franklin
Hoversland, Arthur Stanley
Jasper, Robert Lawrence
Jenkins, Melvin Earl
Kafka, Marian Stern
Kayar, Susan Rennie
Kennedy, Thomas James, Jr
Khan, Mushtaq Ahmad
Kidd, Bernard Sean Langford
Kinnard, Matthew Anderson
Kowarski, A Avinoam
Kundig, Fredericka Dodyk
Lange, Gordon David
Larrabee, Martin Glover
Lazar-Wesley, Eliane M
Levitan, Herbert
Lin, Diane Chang
Livengood, David Robert
Loeb, Marcia Joan
Lowensohn, Howard Stanley
Lymangrover, John R
MacCanon, Donald Moore
McIndoe, Darrell W
McLaughlin, Alan Charles
MacVittie, Thomas Joseph
Marban, Eduardo
Margolis, Ronald Neil
Marx, Stephen John
Mattson, Margaret Ellen
Meszler, Richard M
Migeon, Claude Jean
Milnor, William Robert
Monjan, Andrew Arthur
Murray, Gary Joseph
Murray, George Cloyd
Myslinski, Norbert Raymond
Nelson, Phillip Gillard
Nelson, Ralph Francis
O'Rangers, John Joseph
Pancella, John Raymond
Paré, William Paul
Permutt, Solbert
Plato, Chris C
Poggio, Gian Franco
Powers, Dennis A
Provine, Robert Raymond
Raina, Ashok K
Rall, Wilfrid
Rattner, Barnett Alvin
Rechcigl, Miloslav, Jr
Reed, Randall R
Revoile, Sally Gates
Robbins, Jacob
Robinson, David Lee
Roesijadi, Guritno
Rose, John Charles
Roth, George Stanley
Sacktor, Bertram
Salans, Lester Barry
Saudek, Christopher D
Schuetz, Allen W
Scow, Robert Oliver
Sexton, Thomas John
Shapiro, Bert Irwin
Sherins, Richard J
Shimizu, Hiroshi
Smith, Lewis Wilbert
Smith, Thomas Graves, Jr
Spector, Novera Herbert
Spooner, Peter Michael
Steinmetz, Michael Anthony
Stewart, Doris Mae
Straile, William Edwin
Stringfellow, Frank
Symmes, David
Tabakoff, Boris
Tasaki, Ichiji
Taylor, Martha Loeb
Terris, James Murray
Theodore, Theodore Spiros
Traystman, Richard J
Trout, David Linn
Turner, R James
Tyler, Bonnie Moreland
Umberger, Ernest Joy
Van Arsdel, William Campbell, III
Vener, Kirt J
Watkins, Clyde Andrew
Watson, John Thomas
Weight, Forrest F
Weinreich, Daniel
Weintraub, Bruce Dale
Weisfeldt, Myron Lee
Whitehorn, William Victor
Wildt, David Edwin
Williams, Walter Ford
Wintercorn, Eleanor Stiegler
Wolff, Jan
Wollman, Seymour Horace
Wurtz, Robert Henry
Yellin, Herbert

Yoshinaga, Koji
Zimbelman, Robert George

MASSACHUSETTS
Adolph, Alan Robert
Alpers, Joseph Benjamin
Ames, Adelbert, III
Anderson, Olivo Margaret
Anderson, Rosalind Coogan
Anderson, Olivo Margaret
Askew, Eldon Wayne
Atema, Jelle
Axelrod, Lloyd
Banzett, Robert B
Barber, Saul Benjamin
Beitins, Inese Zinta
Bengele, Howard Henry
Berman, Marlene Oscar
Bird, Stephanie J
Black, Donald Leighton
Bougas, James Andrew
Bridges, Robert Stafford
Brink, John Jerome
Bruno, Merle Sanford
Burggren, Warren William
Butler, James Preston
Callard, Gloria Vincz
Cameron, John Stanley
Camougis, George
Capuzzo, Judith M
Carey, Francis G
Chan-Palay, Victoria
Chappel, Scott Carlton
Chattoraj, Sati Charan
Clemens, Daniel Theodore
Corkin, Suzanne Hammond
Coyne, Mary Downey
Crawford, John Douglas
Dacheux, Ramon F, II
Damassa, David Allen
Deegan, Linda Ann
Delaney, Patrick Francis, Jr
Denniston, Joseph Charles
Dobson, James Gordon, Jr
Douglas, Pamela Susan
Dowling, John Elliott
Duby, Robert T
Duffy, Frank Hopkins
Durkot, Michael John
Eisenbarth, George Stephen
Farina, Joseph Peter
Feig, Larry Allen
Furshpan, Edwin Jean
George, Stephen Anthony
Goldberg, Alfred L
Goodman, Henry Maurice
Graves, William Earl
Griffin, Donald R(edfield)
Grossberg, Stephen
Grossman, William
Guimond, Robert Wilfrid
Gustafson, Alvar Walter
Habener, Joel Francis
Hales, Charles A
Halkerston, Ian D K
Hall, Robert Dilwyn
Hartline, Peter Haldan
Hartner, William Christopher
Henneberry, Richard Christopher
Hichar, Joseph Kenneth
Hinrichs, Katrin
Hobson, John Allan
Hoffmann, Joan Carol
Hubel, David Hunter
Irving, James Tutin
Irwin, Louis Neal
Kahn, Carl Ronald
Kauer, John Stuart
Kazemi, Homayoun
Kennedy, Ann Randtke
Kiang, Nelson Yuan-Sheng
Kunkel, Joseph George
Kunz, Thomas Henry
Lambert, Helen Haynes
Landowne, Milton
LeFevre, Paul Green
Lessie, Thomas Guy
Lettvin, Jerome Y
Liss, Robert H
Longcope, Christopher
Lowenstein, Edward
Lynch, Harry James
McCarley, Robert William
Macchi, I Alden
McCormick, Stephen Daniel
McCracken, John Aitken
McGowan, Joan A
Maloof, Farahe
Mandels, Mary Hickox
Maran, Janice Wengerd
Marieb, Elaine Nicpon
Marks, Leon Joseph
Marsh, Richard L
Matsumoto, Steven G
Millard, William James
Miller, James Albert, Jr
Mitchell, David Hillard
Moore-Ede, Martin C
Mordes, John Peter
Morgan, James Philip
Morgane, Peter J
Morin, Walter Arthur
Moskowitz, Michael Arthur
Mountain, David Charles, Jr

Murphey, Rodney Keith
Naqvi, S Rehan Hasan
O'Malley, Alice T
Palmer, John Derry
Payne, Bertram R
Pfister, Richard Charles
Pober, Zalmon
Pontoppidan, Henning
Poon, Chi-Sang
Potter, David Dickinson
Prestwich, Kenneth Neal
Raviola d'Elia, Giuseppina E(nrica)
Remmel, Ronald Sylvester
Reppert, Steve Marion
Rheinwald, James George
Richardson, George S
Riggi, Stephen Joseph
Rigney, David Roth
Roberts, John Stephen
Ross, James Neil, Jr
Ryan, Kenneth John
Saide, Judith Dana
Saunders, Richard Henry, Jr
Schaub, Robert George
Schneeberger, Eveline E
Schomer, Donald Lee
Senior, Boris
Shepro, David
Singer, Joshua J
Skavenski, Alexander Anthony
Skrinar, Gary Stephen
Smith, Wendy Anne
Snyder, Benjamin Willard
Solomon, Jolane Baumgarten
Stein, Diana B
Talamo, Barbara Lisann
Tashjian, Armen H, Jr
Torda, Clara
Townsend, Jane Kaltenbach
Treistman, Steven Neal
Trinkaus-Randall, Vickery E
Villee, Dorothy Balzer
Waloga, Geraldine
Webb, Helen Marguerite
Weinberger, Steven Elliott
Weiss, Earle Burton
Wenger, Christian Bruce
Widmaier, Eric Paul
Wiegner, Allen W
Woodhull-McNeal, Ann P
Wyse, Gordon Arthur
Zapol, Warren Myron
Zeldin, Michael Hermen
Zottoli, Steven Jaynes

MICHIGAN
Alpern, Mathew
Anderson, Gordon Frederick
Aulerich, Richard J
Barman, Susan Marie
Barney, Christopher Carroll
Beck, Ronald Richard
Bergen, Werner Gerhard
Bernard, Rudy Andrew
Bohr, David Francis
Borer, Katarina Tomljenovic
Bowker, Richard George
Bradley, Robert Martin
Bredeck, Henry E
Brooks, Samuel Carroll
Campbell, Kenneth Lyle
Casey, Kenneth L(yman)
Chou, Ching-Chung
Courtney, Gladys (Atkins)
Coyle, Peter
Cronkite, Donald Lee
Cunningham, James Gordon
D'Amato, Constance Joan
Dambach, George Ernest
Davenport, Horace Willard
Davis, Lynne
Dunbar, Joseph C
Duncan, Gordon W
England, Barry Grant
Fernandez, Hector R C
Foster, Douglas Layne
Frank, Robert Neil
Friar, Robert Edsel
Froiland, Thomas Gordon
Fromm, David
Gala, Richard R
Garlick, Robert L
Gebber, Gerard L
Gossain, Ved Vyas
Goudsmit, Esther Marianne
Grant, Rhoda
Gratz, Ronald Karl
Green, Daniel G
Guthe, Karl Frederick
Hajra, Amiya Kumar
Hall, Edward Dallas
Halter, Jeffrey Brian
Hatton, Glenn Irwin
Hawkins, Joseph Elmer
Heisey, S(amuel) Richard
Henneman, Harold Albert
Hill, Richard William
Hoch, Frederic Louis
Hoffert, Jack Russell
Hsu, Chen-Hsing
Ireland, James
Johnson, Herbert Gardner
Karsch, Fred Joseph
Keyes, Paul Landis

Kimball, Frances Adrienne
Kostyo, Jack Lawrence
Kurta, Allen
Lancaster, Cleo
Lauderdale, James W, Jr
Lawson, David Michael
Lehman, Grace Church
Mather, Edward Chantry
Midgley, A Rees, Jr
Mistretta, Charlotte Mae
Mitchell, Jerald Andrew
Moore, Thomas Edwin
Morrow, Thomas John
Nachreiner, Raymond F
Nag, Asish Chandra
Neudeck, Lowell Donald
Nuttall, Alfred L
Oakley, Bruce
Penney, David George
Person, Steven John
Phillis, John Whitfield
Piercey, Montford F
Pittman, Robert Preston
Polin, Donald
Puro, Donald George
Rapundalo, Stephen T
Retzlaff, Ernest (Walter)
Reynolds, Orland Bruce
Riegle, Gail Daniel
Rillema, James Alan
Rivera, Evelyn Margaret
Robert, Andre
Robinson, Norman Edward
Rovner, David Richard
Rust, Steven Ronald
Sedensky, James Andrew
Shappirio, David Gordon
Sharf, Donald Jack
Sparks, Harvey Vise
Spielman, William Sloan
Spilman, Charles Hadley
Stebbins, William Cooper
Stephenson, Robert Bruce
Stephenson, Robert Storer
Stewart, Sheila Frances
Stromsta, Courtney Paul
Stucki, Jacob Calvin
Studier, Eugene H
Tannen, Richard L
Tigchelaar, Peter Vernon
Tosi, Oscar I
Tucker, Herbert Allen
Valenstein, Elliot Spiro
Vander, Arthur J
Webb, Paul
Webb, R Clinton
Wiley, John W
Wolterink, Lester Floyd

MINNESOTA
Alsum, Donald James
Barker, Laren Dee Stacy
Barnwell, Franklin Hershel
Bauer, Gustav Eric
Brimijoin, William Stephen
Clapper, David Lee
Di Salvo, Joseph
Doe, Richard P
Donaldson, Sue Karen
Duke, Gary Earl
Dziuk, Harold Edmund
Ebner, Timothy John
Eisenberg, Richard Martin
Fohlmeister, Jurgen Fritz
Forbes, Donna Jean
From, Arthur Harvey Leigh
Gallaher, Daniel David
Good, A L
Gorman, Colum A
Haller, Edwin Wolfgang
Hedgecock, Le Roy Darien
Heller, Lois Jane
Hooper, Alan Bacon
Hunter, Alan Graham
Iadecola, Costantino
Jacobson, Joan
Kallok, Michael John
Keshaviah, Prakash Ramnathpur
Klicka, John Kenneth
Knox, Charles Kenneth
Lambert, Edward Howard
Lennon, Vanda Alice
Leung, Benjamin Shuet-Kin
Low, Walter Cheney
McCaffrey, Thomas Vincent
McDermott, Richard P
Marx, George Donald
Mayberry, William Eugene
Mortimer, James Arthur
Ogburn, Phillip Nash
Oppenheimer, Jack Hans
Phillips, Richard Edward
Polzin, David J
Poppele, Richard E
Purple, Richard L
Robinette, Martin Smith
Ruggero, Mario Alfredo
Schmalz, Philip Frederick
Schwartz, Harold Leon
Sehe, Charles Theodore
Sicard, Raymond Edward
Smith, Thomas Jay
Szurszewski, Joseph Henry
Thompson, Edward William

Tindall, Donald J
Ungar, Frank
Weir, Edward Kenneth
Wheaton, Jonathan Edward
Wilson, Michael John

MISSISSIPPI
Anderson, Gary
Biesiot, Patricia Marie
Bradley, Doris P
Chambers, Janice E
Coleman, Thomas George
Fuquay, John Wade
Larsen, James Bouton
Long, Billy Wayne
Lynch, James Carlyle
Montani, Jean-Pierre
Nelson, Norman Crooks
Norman, Roger Atkinson, Jr
Read, Virginia Hall
Smith, Manis James, Jr

MISSOURI
Ackerman, Joseph John Henry
Ahmed, Mahmoud S
Armbrecht, Harvey James
Avioli, Louis
Baer, Robert W
Bergmann, Steven R
Blaine, Edward Homer
Breitenbach, Robert Peter
Buonomo, Frances Catherine
Burns, Thomas Wade
Burton, Harold
Carrel, James Elliott
Coles, Richard Warren
Collier, Robert Joseph
Coyer, Philip Exton
Dale, Homer Eldon
Davis, Hallowell
Daw, Nigel Warwick
Derby, Albert
Dixit, Rakesh
Feir, Dorothy Jean
Fernandez, Hugo Lathrop
Fondacaro, Joseph D
Fredrickson, John Murray
Geisbuhler, Timothy Paul
Gerhardt, H Carl, Jr
Gibbs, Finley P
Goetz, Kenneth Lee
Grunt, Jerome Alvin
Hammerman, Marc R
Highstein, Stephen Morris
Hillman, Richard Ephraim
Hunt, Carlton Cuyler
Jeffrey, John J
Johnson, J(oseph) Alan
Kipnis, David Morris
Kohlmeier, Ronald Harold
Liu, Maw-Shung
Longley, William Joseph
Martin, Charles Everett
Mercer, Robert William
Mills, Steven Harlon
Mitchell, Henry Andrew
Molnar, Charles Edwin
Morton, Phillip A
Nickols, G Allen
Nyquist-Battie, Cynthia
Packman, Paul Michael
Parker, Mary Langston
Paull, Willis K, Jr
Peck, William Arno
Premachandra, Bhartur N
Quay, Wilbur Brooks
Rash, Jay Justen
Rovainen, Carl (Marx)
Ruh, Mary Frances
St Omer, Vincent Victor
Salkoff, Lawrence Benjamin
Sargent, William Quirk
Savery, Harry P
Schonfeld, Gustav
Sharp, John Roland
Starling, Jane Ann
Stein, Paul S G
Tannenbaum, Michael Glen
Tolbert, Daniel Lee
Tomasi, Thomas Edward
Tumosa, Nina Jean
Vineyard, Billy Dale
Volkert, Wynn Arthur
Warren, James C
Welply, Joseph Kevin
Weppelman, Roger Michael
Whitten, Elmer Hammond
Wilkens, Lon Allan
York, Donald Harold

MONTANA
Bradbury, James Thomas
Chaney, Robert Bruce, Jr
Coe, John Emmons
Dratz, Edward Alexander
Fevold, Harry Richard
Foresman, Kerry Ryan
Hull, Maurice Walter
Kilgore, Delbert Lyle, Jr
Kirkpatrick, Jay Franklin
McFeters, Gordon Alwyn
McMillan, James Alexander
Parker, Charles D
Patent, Gregory Joseph

Physiology, Animal (cont)

Tietz, William John, Jr

NEBRASKA
Barnawell, Earl B
Clark, Francis John
Clemens, Edgar Thomas
Cornish, Kurtis George
DeGraw, William Allen
Ellingson, Robert James
Ellington, Earl Franklin
Farr, Lynne Adams
Ford, Johny Joe
Fougeron, Myron George
Geluso, Kenneth Nicholas
Grotjan, Harvey Edward, Jr
Jones, Timothy Arthur
Larson, Larry Lee
Laster, Danny Bruce
Littledike, Ernest Travis
Maurer, Ralph Rudolf
Pancoe, William Louis, Jr
Pekas, Jerome Charles
Roberts, Jane Carolyn
Rodriguez-Sierra, Jorge F
Tharp, Gerald D
Yen, Jong-Tseng

NEVADA
Anderson, Bernard A
Foote, Wilford Darrell
Garner, Duane LeRoy
Marlow, Ronald William
Nellor, John Ernest
Ort, Carol
Potter, Gilbert David
Sanders, Kenton M
Starkweather, Peter Lathrop
Turner, John K
Van Remoortere, Emile C

NEW HAMPSHIRE
Bogdonoff, Philip David, Jr
Brinck-Johnsen, Truls
Cahill, George Francis, Jr
Crandell, Walter Bain
Daubenspeck, John Andrew
Edwards, Brian Ronald
Galton, Valerie Anne
Gosselin, Robert Edmond
Hermann, Howard T
Munck, Allan Ulf
Musiek, Frank Edward
Nattie, Eugene Edward
St John, Walter McCoy
Sokol, Hilda Weyl
Sower, Stacia Ann
Stanton, Bruce Alan
Tokay, F Harry
Van Ummersen, Claire Ann

NEW JERSEY
Ahn, Ho-Sam
Arbeeny, Cynthia M
Arnold, Frederic G
Arthur, Alan Thorne
Bagnell, Carol A
Bex, Frederick James
Bird, John William Clyde
Boyle, Joseph, III
Brooks, Jerry R
Buchweitz, Ellen
Bullock, John
Carlson, Richard P
Cevallos, William Hernan
Chang, Joseph Yoon
Chinard, Francis Pierre
Collins, Elliott Joel
Conn, Hadley Lewis, Jr
Connor, John Arthur
Convey, Edward Michael
Cornell, James S
Curcio, Lawrence Nicholas
Curro, Frederick A
Daggs, Ray Gilbert
Davis, Brian K
Deetz, Lawrence Edward, III
Denhardt, David Tilton
Duran, Walter Nunez
Edelman, Norman H
Edinger, Henry Milton
Egan, Robert Wheeler
Ertel, Norman H
Farmanfarmaian, Allahverdi
Farnsworth, Patricia Nordstrom
Foley, James E
Goldsmith, Laura Tobi
Gona, Amos G
Gona, Ophelia Delaine
Goodman, Frank R
Grover, Gary James
Gruszczyk, Jerome Henry
Hafs, Harold David
Hahn, DoWon
Hall, James Conrad
Hall, Joseph L
Halmi, Nicholas Stephen
Hearney, Elaine Schmidt
Hendler, Edwin
Hockel, Gregory Martin
Hofmann, Frederick Gustave
Hruza, Zdenek
Hubbard, John W

Huber, Ivan
Julesz, Bela
Kapoor, Inder Prakash
Katz, George Maxim
Kirschner, Marvin Abraham
Kumbaraci-Jones, Nuran Melek
Lenz, Paul Heins
Levenstein, Irving
Levin, Barry Edward
Lisk, Robert Douglas
Lowndes, Herbert Edward
McArdle, Joseph John
Mellin, Theodore Nelson
Merrill, Gary Frank
Meyers, Kenneth Purcell
Mittler, James Carlton
Mohammed, Kasheed
Moreland, Suzanne
Mowles, Thomas Francis
Mueller, Stephen Neil
Neri, Rudolph Orazio
Newton, Paul Edward
Oddis, Leroy
Page, Charles Henry
Paterniti, James R, Jr
Perryman, James Harvey
Pitts, Barry James Roger
Ponessa, Joseph Thomas
Rabii, Jamshid
Royce, Paul C
Rubanyi, Gabor Michael
Rumsey, William LeRoy
Saini, Ravinder Kumar
Sanders, Jay W
Siegel, Allan
Solomon, Thomas Allan
Steele, Ronald Edward
Stein, Donald Gerald
Steiner, Kurt Edric
Steinetz, Bernard George, Jr
Stevenson, Nancy Roberta
Swenson, Theresa Lynn
Swislocki, Norbert Ira
Szewczak, Mark Russell
Virkar, Raghunath Atmaram
Wallace, Edith Winchell
Watnick, Arthur Saul
Weiss, Gerson
Wilbur, Robert Daniel
Zbuzek, Vlasta Kmentova
Zbuzek, Vratislav
Zulalian, Jack

NEW MEXICO
Bernstein, Marvin Harry
Botsford, James L
Buss, William Charles
Chick, Thomas Wesley
Duszynski, Donald Walter
Hallford, Dennis Murray
Herman, Ceil Ann
Luft, Ulrich Cameron
McNamara, Mary Colleen
Martinez, J Ricardo
Matchett, William H
Mauderly, Joe L
Mauro, Jack Anthony
Muggenburg, Bruce Al
Omdahl, John L
Partridge, L Donald
Priola, Donald Victor
Ratliff, Floyd
Ratner, Albert

NEW YORK
Adkins-Regan, Elizabeth Kocher
Adolph, Edward Frederick
Aldrich, Thomas K
Alexander, Robert Spence
Altszuler, Norman
Amassian, Vahe Eugene
Antzelevitch, Charles
April, Ernest W
Aronson, Ronald Stephen
Asanuma, Hiroshi
Atlas, Steven Alan
Baizer, Joan Susan
Baker, Robert George
Barac-Nieto, Mario
Beermann, Donald Harold
Bender, David Bowman
Bennett, Michael Vander Laan
Benzo, Camillo Anthony
Bergman, Emmett Norlin
Berlyne, Geoffrey Merton
Bickerman, Hylan A
Bienvenue, Gordon Raymond
Bisson, Mary A
Bito, Laszlo Z
Blair, Martha Longwell
Bleicher, Sheldon Joseph
Bookchin, Robert M
Borowsky, Betty Marian
Brachfeld, Norman
Broadway, Roxanne Meyer
Brown, Joel Edward
Brown, Oliver Monroe
Brown, Patricia Stocking
Brown, Stephen Clawson
Browner, Robert Herman
Brust, Manfred
Bukovsan, William
Burchfiel, James Lee
Bushinsky, David Allen

Butler, Robert Neil
Cabot, John Boit, II
Canfield, Robert E
Canfield, William H
Carlson, Albert Dewayne, Jr
Carr, Ronald E
Catapane, Edward John
Center, Sharon Anne
Centola, Grace Marie
Cerny, Frank J
Chan, Stephen
Chang, Chin-Chuan
Christy, Nicholas Pierson
Claus, Thomas Harrison
Clitheroe, H John
Cohan, Christopher Scott
Cohen, Avis Hope
Cohen, Bernard
Cohen, Morton Irving
Coleman, Paul David
Cooper, George William
Coppola, John Anthony
Corradino, Robert Anthony
Crain, Stanley M
Crandall, David L
Cushman, Paul, Jr
Davies, Kelvin J A
Davis, Paul Joseph
Diakow, Carol
Dickson, Stanley
Dobson, Alan
Doering, Charles Henry
Dow, Bruce MacGregor
Duffey, Michael Eugene
DuFrain, Russell Jerome
Durkovic, Russell George
Egan, Edmund Alfred
Eisenstein, Albert Bernard
Emmers, Raimond
Engbretson, Gustav Alan
Erasmus, Beth de Wet (Fleming)
Faber, Donald S
Farhi, Leon Elie
Federico, Olga Maria
Felig, Philip
Fenstermacher, Joseph Don
Ferguson, Earl Wilson
Fetcho, Joseph Robert
Finkelstein, Jacob Noah
Fletcher, Paul Wayne
Foote, Robert Hutchinson
Fortune, Joanne Elizabeth
Fourtner, Charles Russell
Fox, Kevin A
Frankel, Arthur Irving
Freed, Simon
French, Alan Raymond
Frumkes, Thomas Eugene
Gapp, David Alger
Gardner, Daniel
Gardner, Eliot Lawrence
Gardner, Esther Polinsky
Gasteiger, Edgar Lionel
Geary, Norcross D
Gershengorn, Marvin Carl
Gershon, Michael David
Gibson, Kenneth David
Gil, Joan
Goldman, Stephen Shepard
Gonzalez, Eulogio Raphael
Gootman, Norman Lerner
Gootman, Phyllis Myrna Adler
Greenberg, Danielle
Grota, Lee J
Habicht, Gail Sorem
Hainline, Louise
Halpern, Bruce Peter
Halpern, Mimi
Halpryn, Bruce
Han, Jaok
Harkavy, Allan Abraham
Harris, Charles Leon
Harris-Warrick, Ronald Morgan
Harth, Erich Martin
Hawkins, Robert Drake
Hayes, Carol J
Heath, Martha Ellen
Hechemy, Karim E
Higashiyama, Tadayoshi
Hillman, Dean Elof
Hirsch, Helmut V B
Hirsh, Kenneth Roy
Hochberg, Irving
Hoffman, Roger Alan
Holtzman, Eric
Holtzman, Seymour
Hood, Donald C
Horn, Eugene Harold
Horvath, Fred Ernest
Hotchkiss, Sharon K
Houpt, Katherine Albro
Houpt, Thomas Richard
Howland, Howard Chase
Hudecki, Michael Stephen
Hughes, Patrick Richard
Hyde, Richard Witherington
Isseroff, Hadar
Jacklet, Jon Willis
Jacobs, Laurence Stanton
Jacobson, Herbert (Irving)
Jahan-Parwar, Behrus
Jayme, David Woodward
Johnson, Patricia R
Johnston, Robert E

Kandel, Eric Richard
Kao, Chien Yuan
Kaplan, John Ervin
Kashin, Philip
Kass, Robert S
Kaufman, Albert Irving
Kelly, Dennis D
Kennedy, Margaret Wiener
Kent, Barbara
Kerlan, Joel Thomas
Khanna, Shyam Mohan
Kimmich, George Arthur
Kincl, Fred Allan
King, Thomas K C
Klocke, Robert Albert
Knigge, Karl Max
Koenig, Edward
Kohn, Michael
Koizumi, Kiyomi
Konopka, Ronald J
Krasney, John Andrew
Krey, Lewis Charles
Krulwich, Terry Ann
Kupfer, Sherman
Kupperman, Herbert Spencer
Kydd, David Mitchell
Lawton, Richard Woodruff
Lee, James B
Leeper, Robert Dwight
Lengemann, Frederick William
Levey, Harold Abram
Levitz, Mortimer
Levy, David Edward
Liebman, Frederick Melvin
Llinas, Rodolfo
Lnenicka, Gregory Allen
Loew, Ellis Roger
Loretz, Christopher Alan
Lutton, John D
MacGillivray, Margaret Hilda
McKenna, Olivia Cleveland
Mahler, Richard Joseph
Mahoney, Robert Patrick
Makous, Walter L
Malhotra, Ashwani
Malik, Asrar B
Mantel, Linda Habas
Maple, William Thomas
Martin, Constance R
Mason, Elliott Bernard
Medici, Paul T
Mehaffey, Leatham, III
Mellins, Robert B
Menguy, Rene
Messina, Edward Joseph
Michaelson, Solomon M
Miller, Myron
Miller, Richard Avery
Mindich, Leonard Eugene
Montefusco, Cheryl Marie
Morris, Marilyn Emily
Mueller, August P
Murphy, Michael Joseph
Nahas, Gabriel Georges
Nathan, Marc A
Nelson, Margaret Christina
Newell, Jonathan Clark
Nichols, Kathleen Mary
Nicholson, J(ohn) Charles (Godfrey)
Nickerson, Peter Ayers
Notides, Angelo C
Oberdörster, Günter
O'Connor, John Francis
Oesterreich, Roger Edward
Olsen, Kathie Lynn
Ornt, Daniel Burrows
Overweg, Norbert I A
Pandit, Hemchandra M
Pasik, Pedro
Pasik, Tauba
Peters, Theodore, Jr
Pfaff, Donald Wells
Pfaffmann, Carl
Pollard, Jeffrey William
Potts, Gordon Oliver
Pough, Frederick Harvey
Poulos, Dennis A
Primack, Marshall Philip
Purpura, Dominick Paul
Quinn, Cosmas Edward
Reilly, Margaret Anne
Reimers, Thomas John
Rifkind, Arleen B
Ringo, James Lewis
Robertson, Douglas Reed
Rockwood, William Philip
Roel, Lawrence Edmund
Rogers, Philip Virgilius
Rosen, John Friesner
Rosen, Ora Mendelsohn
Rosenberg, Warren L
Rosensweig, Norton S
Rosoff, Betty
Rubin, Ronald Philip
Ruknudin, Abdul Majeed
Sack, Robert A
Sakitt, Barbara
Salpeter, Miriam Mirl
Saxena, Brij B
Scalia, Frank
Scarpelli, Emile Michael
Schiff, Joel D
Schlafer, Donald H
Schlesinger, Richard B

Schmidt, John Thomas
Schoenfeld, Cy
Schreibman, Martin Paul
Schryver, Herbert Francis
Sellers, Alvin Ferner
Sharma, Sansar C
Shellabarger, Claire J
Shepherd, Julian Granville
Sherman, Frederick George
Sherman, Samuel Murray
Sherwood, Louis Maier
Slack, Nelson Hosking
Sonenberg, Martin
Sorrentino, Sandy, Jr
Southwick, Edward Earle
Sparks, Charles Edward
Spaulding, Stephen Waasa
Spray, David Conover
Stagg, Ronald M
Steffen, Daniel G
Stein, Theodore Anthony
Stephenson, John Leslie
Stoner, Larry Clinton
Stouter, Vincent Paul
Strand, Fleur Lillian
Strand, James Cameron
Sturr, Joseph Francis
Sundaram, Kalyan
Tamir, Hadassah
Tapper, Daniel Naphtali
Thau, Rosemarie B Zischka
Thornborough, John Randle
Tobach, Ethel
Tsan, Min-Fu
Udin, Susan Boymel
Van Hattum, Rolland James
Van Liew, Hugh Davenport
Van Tienhoven, Ari
Ventura, William Paul
Vogel, Henry
Vroman, Leo
Wallman, Joshua
Walters, Deborah K W
Waterhouse, Joseph Stallard
Weiss, Klaudiusz Robert
Weitzman, Elliot D
White, James Patrick
Wilson, Victor Joseph
Wit, Andrew Lewis
Witkovsky, Paul
Wittig, Kenneth Paul
Wulff, Verner John
Yao, Alice C
Yarns, Dale A
Yasumura, Seiichi
Yazulla, Stephen
Yellin, Edward L
Yu, Ts'ai-Fan
Zablow, Leonard
Zimmerman, Jay Alan

NORTH CAROLINA
Aldridge, David William
Anderson, Nels Carl, Jr
Anderson, Roger Fabian
Arendshorst, William John
Beaty, Orren, III
Beckman, David Lee
Bentzel, Carl Johan
Black, Betty Lynne
Bo, Walter John
Breese, George Richard
Brewer, Carl Robert
Brinn, Jack Elliott, Jr
Burden, Hubert White
Carroll, Robert Graham
Caruolo, Edward Vitangelo
Casseday, John Herbert
Castell, Donald O
Clark, Martin Ralph
Clemmons, David Roberts
Coffey, James Cecil, Jr
Corless, Joseph Michael James
Cornwell, John Calhoun
Coulter, Norman Arthur, Jr
Croom, Warren James, Jr
Daw, John Charles
Di Giulio, Richard Thomas
Dimmick, John Frederick
Dingledine, Raymond J
Dreyer, Duane Arthur
Dusseau, Jerry William
Edens, Frank Wesley
Eldridge, Frederic L
Eldridge, John Charles
Erickson, Robert Porter
Faber, James Edward
Fail, Patricia A
Fairchild, Homer Eaton
Fletcher, Donald James
Forward, Richard Blair, Jr
Furth, Eugene David
Garren, Henry Wilburn
Garris, David Roy
Gatten, Robert Edward, Jr
Glasser, Richard Lee
Glassman, Edward
Goetsch, Dennis Donald
Goode, Lemuel
Gordon, Christopher John
Gottschalk, Carl William
Green, Harold D
Greenfield, Joseph C, Jr
Gruber, Kenneth Allen

Hazzard, William Russell
Hedlund, Laurence William
Henson, Anna Miriam (Morgan)
Hicks, T Philip
Hodson, Charles Andrew
Javel, Eric
Johnson, Bryan Hugh
Kaufman, William
Kizer, John Stephen
Klitzman, Bruce
Korach, Kenneth Steven
Koritnik, Donald Raymond
Langford, George Malcolm
Lassiter, William Edmund
Lawson, Edward Earle
Lieberman, Melvyn
Little, William C
Lloyd, Charles Wait
Lobaugh, Bruce
Louis, Thomas Michael
Lund, Pauline Kay
McManus, Thomas (Joseph)
Malindzak, George S, Jr
Meissner, Gerhard
Meyer, Ralph A, Jr
Mikat, Eileen M
Miller, David S
Miller, William Laubach
Millstein, Lloyd Gilbert
Moore, John Wilson
Munson, Paul Lewis
Myers, Robert Durant
Nayfeh, Shihadeh Nasri
O'Brien, Deborah A
Ontjes, David A
Parker, John Curtis
Pasipoularides, Ares D
Perlmutt, Joseph Hertz
Pritchard, John B
Purves, Dale
Ramagopal, Mudumbi Vijaya
Reel, Jerry Royce
Robinson, Jerry Allen
Roer, Robert David
Rustioni, Aldo
Schanberg, Saul M
Schmidt, Stephen Paul
Schmidt-Nielsen, Knut
Shaw, Helen Lester Anderson
Simpson, Everett Coy
Siopes, Thomas David
Smith, Donald Eugene
Smith, Frank Houston
Smith, Ned Allan
Sonntag, William Edmund
Stacy, David Lowell
Stevens, Charles Edward
Strandhoy, Jack W
Stuart, Ann Elizabeth
Stumpf, Walter Erich
Thaxton, James Paul
Thubrikar, Mano J
Toverud, Svein Utheim
Turner, James Eldridge
Tyrey, Lee
Underwood, Herbert Arthur, Jr
Underwood, Louis Edwin
Vandemark, Noland Leroy
Vandenbergh, John Garry
Van Wyk, Judson John
Wallace, Andrew Grover
Watts, John Albert, Jr
Whitsel, Barry L
Williams, Redford Brown, Jr
Wise, George Herman
Wolcott, Thomas Gordon
Yarger, William E

NORTH DAKOTA
Albright, Bruce Calvin
Aschbacher, Peter William
Fivizzani, Albert John, Jr
Haunz, Edgar Alfred
Joshi, Madhusudan Shankarrao
Leopold, Roger Allen
Roseland, Craig R
Saari, Jack Theodore
Sheridan, Mark Alexander
Stinnett, Henry Orr
Struble, Craig Bruce
Zogg, Carl A

OHIO
Adams, Walter C
Akeson, Richard Allan
Arlian, Larry George
Baker, Saul Phillip
Barnes, Karen Louise
Behr, Stephen Richard
Berger, Kenneth Walter
Bettice, John Allen
Biagini, Raymond E
Billman, George Edward
Birkhahn, Ronald H
Black, Craig Patrick
Bohrman, Jeffrey Stephen
Bolls, Nathan J, Jr
Boyer, Jere Michael
Buck, Stephen Henderson
Butz, Andrew
Chester, Edward Howard
Ching, Melvin Chung Hing
Clanton, Thomas L
Clark, Kenneth Edward

Claussen, Dennis Lee
Coleman, Marilyn A
Cooke, Helen Joan
Cooper, Gary Pettus
Cornhill, John Fredrick
Costello, Walter James
Cruce, William L R
Curry, John Joseph, III
Decker, Clarence Ferdinand
DeVillez, Edward Joseph
Donnelly, Kenneth Gerald
Dutta, Hiran M
Faber, Lee Edward
Foreman, Darhl Lois
Freed, James Melvin
Fritz, Marc Anthony
Genuth, Saul M
Glaser, Roger Michael
Goldstein, David Louis
Greenstein, Julius S
Grossie, James Allen
Grossman, Charles Jerome
Gwinn, John Frederick
Hamlin, Robert Louis
Harder, John Dwight
Harrison, Donald C
Haxhiu, Musa A
Hebbard, Frederick Worthman
Heesch, Cheryl Miller
Henschel, Austin
Hill, Richard M
Hinman, Channing L
Hopfer, Ulrich
Jegla, Thomas Cyril
Johnson, Melvin Andrew
Johnston, John O'Neal
Jones, Stephen Wallace
Kaiserman-Abramof, Ita Rebeca
Katona, Peter Geza
Kern, Michael Don
Kogut, Maurice D
Kolenbrander, Harold Mark
Kolodziej, Bruno J
LaManna, Joseph Charles
Landis, Story Cleland
Lasek, Raymond J
Latshaw, J David
Lemon, Peter Willian Reginald
Linkenheimer, Wayne Henry
Malhotra, Om P
Martin, Elden William
Meserve, Lee Arthur
Metz, David A
Michal, Edwin Keith
Millard, Ronald Wesley
Moulton, Bruce Carl
Mudry, Karen Michele
Murray, Finnie Ardrey, Jr
Needham, Glen Ray
Nishikawara, Margaret T
Nodar, Richard (H)enry
Nussbaum, Noel Sidney
Orcutt, Frederic Scott, Jr
Orosz, Charles George
Paradise, Norman Francis
Patterson, Michael Milton
Pettegrew, Raleigh K
Pilati, Charles Francis
Pinheiro, Marilyn Lays
Quinn, David Lee
Rall, Jack Alan
Ramsey, James Marvin
Ray, Richard Schell
Replogle, Clyde R
Robbins, Norman
Robinson, Michael K
Rogers, Richard C
Rothchild, Irving
Sabourin, Thomas Donald
Saiduddin, Syed
Schoessler, John Paul
Sherman, Robert George
Shertzer, Howard Grant
Shipley, Michael Thomas
Smith, Charles Roger
Smith, Clifford James
Smith, David Varley
Soloff, Melvyn Stanley
Srivastava, Laxmi Shanker
Stacy, Ralph Winston
Stevenson, J(oseph) Ross
Stokes, Bradford Taylor
Stuesse, Sherry Lynn
Swift, Michael Crane
Tassava, Roy A
Teyler, Timothy James
Thomas, William Eric
Treick, Ronald Walter
Tzagournis, Manuel
Vardaris, Richard Miles
Vogel, Thomas Timothy
Voneida, Theodore J
Wagner, Thomas Edwards
Watson, John Ernest
Weiss, Harold Samuel
White, Robert J
Wiley, Ronald Lee
Willett, Lynn Brunson
Wilson, David Franklin
Wolin, Lee Roy
Zigmond, Richard Eric

OKLAHOMA
Baird, Troy Alan

Bastian, Joseph
Bourdeau, James Edward
Burgus, Roger Cecil
Chung, Kyung Won
Clark, James Bennett
Dormer, Kenneth John
Farber, Jay Paul
Faulkner, Kenneth Keith
Faulkner, Lloyd (Clarence)
Foreman, Robert Dale
Gass, George Hiram
Gumbreck, Laurence Gable
Gunn, Chesterfield Garvin, Jr
Haines, Howard Bodley
Houlihan, Rodney T
Hurst, Jerry G
Hutchison, Victor Hobbs
Jones, Sharon Lynn
Kinasewitz, Gary Theodore
King, Mary Margaret
Kling, Ozro Ray
Martin, Loren Gene
Massion, Walter Herbert
Meier, Charles Frederick, Jr
Miner, Gary David
Pento, Joseph Thomas
Peterson, Donald Frederick
Rich, Clayton
Robertson, William G
Sauer, John Robert
Schaefer, Carl Francis
Scherlag, Benjamin J
Shirley, Barbara Anne
Simpson, Ocleris C
Stith, Rex David
Thies, Roger E
Trachewsky, Daniel
Turman, Elbert Jerome
Wettemann, Robert Paul

OREGON
Arch, Stephen William
Bandick, Neal Raymond
Barmack, Neal Herbert
Bethea, Cynthia Louise
Bradshaw, William Emmons
Brookhart, John Mills
Brown, Arthur Charles
Brownell, Philip Harry
Carlisle, Kay Susan
Chilgren, John Douglas
Christensen, Ned Jay
Church, David Calvin
Coleman, Ralph Orval, Jr
Davis, Steven Lewis
Ellinwood, William Edward
Fernald, Russell Dawson
Grimm, Robert John
Gunberg, David Leo
Gwilliam, Gilbert Franklin
Harris, Nellie Robbins
Hillman, Stanley Severin
Hisaw, Frederick Lee, Jr
Keenan, Edward James
Kohler, Peter
Levine, Leonard
Lilly, David J
Ludlam, William Myrton
McCollum, Gin
Maksud, Michael George
Marrocco, Richard Thomas
Mason, Robert Thomas
Meikle, Mary B
Montagna, William
Moore, Frank Ludwig
Novy, Miles Joseph
Oginsky, Evelyn Lenore
Phoenix, Charles Henry
Pritchard, Austin Wyatt
Pubols, Lillian Menges
Ramaley, Judith Aitken
Rasmussen, Lois E Little
Richards, Richard Earl
Roberts, Michael Foster
Roselli, Charles Eugene
Roth, Niles
Soderwall, Arnold Larson
Spies, Harold Glen
Stormshak, Fredrick
Stouffer, Richard Lee
Swanson, Lloyd Vernon
Swanson, Robert E
Taylor, Mary Lowell Branson
Taylor, Walter Herman, Jr
Terwilliger, Nora Barclay
Thorson, Thomas Bertel
Wilson, Marlene Moore
Young, Norton Bruce
Zimmerman, Earl Abram

PENNSYLVANIA
Albertine, Kurt H
Angstadt, Robert B
Aston-Jones, Gary Stephen
Binkley, Sue Ann
Borle, Andre Bernard
Botelho, Stella Yates
Bradley, Stanley Edward
Brooks, David Patrick
Buckelew, Thomas Paul
Burgi, Ernest, Jr
Bursey, Charles Robert
Buskirk, Elsworth Robert
Cameron, William Edward

Physiology, Animal (cont)

Clark, James Michael
Colby, Howard David
Compher, Marvin Keen, Jr
Corbin, Alan
Cowan, Alan
Crutchfield, Floy Love
Davies, Richard Oelbaum
Davis, Robert Harry
DeCaro, Thomas F
De Groat, William C
DeGroat, William Chesney, Jr
Della-Fera, Mary Anne
DeMartinis, Frederick Daniel
Demers, Laurence Maurice
Di George, Angelo Mario
Ertel, Robert James
Fish, Frank Eliot
Flemister, Sarah C
Florant, Gregory Lester
Flynn, John Thomas
Frazier, James Lewis
Fritz, George Richard, Jr
Fromm, Gerhard Hermann
Fuchs, Franklin
Fuller, Ellen Oneil
Gerstein, George Leonard
Goodman, David Barry Poliakoff
Gregg, Christine M
Grosvenor, Clark Edward
Grunstein, Michael M
Guida, Vincent George
Hagen, Daniel Russell
Hale, Creighton J
Harris, Dorothy Virginia
Hartley, Harold V, Jr
Henderson, Linda Shlatz
Herlyn, Dorothee Maria
Hoffman, Eric Alfred
Holliday, Charles Walter
Hook, Jerry Bruce
Horn, Lyle William
Horner, George John
Hymer, Wesley C
Idzkowsky, Henry Joseph
Isom, Harriet C
Jacobs, George Joseph
Jacobson, Frank Henry
Jacoby, Henry I
Jones, Robert Leroy
Joos, Barbara
Kelsey, Ruben Clifford
Kennedy, Michael Craig
Kidawa, Anthony Stanley
Kinter, Lewis Boardman
Kivlighn, Salah Dean
Klinger, Thomas Scott
Kornfield, A(lfred) T(heodore)
Kozak, Wlodzimierz M
Kydd, George Herman
Lamperti, Albert A
Leach, Roland Melville, Jr
Lefer, Allan Mark
Levey, Gerald Saul
Liebman, Paul Arno
Lindgren, Clark Allen
London, William Thomas
McElligott, James George
McKinstry, Donald Michael
McNair, Dennis M
Martin, John Samuel
Martinez-Hernandez, Antonio
Mashaly, Magdi Mohamed
Mendelson, Emanuel Share
Miller, Kirk
Minsker, David Harry
Miselis, Richard Robert
Molt, James Teunis
Monson, Frederick Carlton
Mote, Michael Isnardi
Motoyama, Etsuro K
Mullin, James Michael
Murray, Marion
Nadelhaft, Irving
Orr, Nancy Hoffner
Owen, Oliver Elon
Palmer, Larry Alan
Panuska, Joseph Allan
Peacock, Samuel Moore, Jr
Persky, Harold
Pirone, Louis Anthony
Pitkin, Ruthanne B
Pitkow, Howard Spencer
Powell, William John, Jr
Pruett, Esther D R
Rao, Kalipatnapu Narasimha
Rapp, Paul Ernest
Redgate, Edward Stewart
Robertson, Robert James
Rodan, Gideon Alfred
Ross, Leonard Lester
Salzberg, Brian Matthew
Sandler, Rivka Black
Savage, Carl Richard, Jr
Scherer, Peter William
Schroeder, Allen C
Scott, Donald, Jr
Sekerka, Robert Floyd
Shank, Richard Paul
Shaw, Ralph Arthur
Short, John Albert
Sidie, James Michael
Siegfried, John Barton

Smith, M Susan
Snipes, Charles Andrew
Somlyo, Avril Virginia
Sperling, Mark Alexander
Steele, Craig William
Strauss, Jerome Frank, III
Stricker, Edward Michael
Studer, Rebecca Kathryn
Summy-Long, Joan Yvette
Surmacz, Cynthia Ann
Takashima, Shiro
Tanabe, Tsuneo Y
Taylor-Mayer, Rhoda E
Thomas, Steven P
Tobin, Thomas Vincent
Trelka, Dennis George
Trevino, Daniel Louis
Trippodo, Nick Charles
Tulenko, Thomas Norman
Tutwiler, Gene Floyd
Van Thiel, David H
Vargo, Steven William
Vasilatos-Younken, Regina
Vickers, Florence Foster
Vollmer, Regis Robert
Weiss, Michael Stephen
Weisz, Judith
Wickersham, Edward Walker
Wideman, Robert Frederick, Jr
Winchester, Richard Albert
Yip, Joseph W
Young, In Min
Yushok, Wasley Donald
Zabara, Jacob
Zaccaria, Robert Anthony
Zelis, Robert Felix
Zimmerman, Irwin David

RHODE ISLAND

Arnold, Mary B
Durfee, Wayne King
Ebner, Ford Francis
Elbaum, Charles
Engeland, William Charles
Hai, Chi-Ming
Hammen, Carl Schlee
Hammen, Susan Lum
Harrison, Robert William
Morris, David Julian
Pilson, Michael Edward Quinton
Richardson, Peter Damian

SOUTH CAROLINA

Abel, Francis Lee
Baxter, Ann Webster
Bell, Norman H
Birrenkott, Glen Peter, Jr
Blackburn, John Gill
Campbell, Gary Thomas
Cheng, Thomas Clement
Clark, George
Coleman, James Roland
Cordova-Salinas, Maria Asuncion
Dickey, Joseph Freeman
Finlay, Mary Fleming
Gibson, Mary Morton
Glick, Bruce
Hays, Sidney Brooks
Henricks, Donald Maurice
Katz, Sidney
McFarland, Kay Flowers
Ondo, Jerome G
Peterson, Ralph Edward
Powell, Harold
Weymouth, Richard J
Wilbur, Donald Lee
Wise, William Curtis

SOUTH DAKOTA

Goldman, Max
Heisinger, James Fredrick
Johnson, Jeffery Lee
Langworthy, Thomas Allan
Pengra, Robert Monroe
Roller, Michael Harris
Schlenker, Evelyn Heymann
Westby, Carl A
Zawada, Edward T, Jr
Zeigler, David Wayne

TENNESSEE

Asp, Carl W
Banerjee, Mukul Ranjan
Bealer, Steven Lee
Beard, James David
Biaggioni, Italo
Bogitsh, Burton Jerome
Brigham, Kenneth Larry
Camacho, Alvro Manuel
Carter, Clint Earl
Cherrington, Alan Douglas
Chytil, Frank
Cook, Robert James
Cragle, Raymond George
Davis, Kenneth Bruce, Jr
Davis, William James
Dooley, Elmo S
Durham, Ross M
Eiler, Hugo
Elberger, Andrea June
Fleischer, Sidney
Foreman, Charles William
Freeman, John A
Hoover, Donald Barry

Huggins, James Anthony
Ikenberry, Roy Dewayne
Law, Peter Koi
Lee, Kai-Lin
Leffler, Charles William
Levene, John Reuben
Levine, Jon Howard
Liddle, Grant Winder
Mahajan, Satish Chander
Mallette, John M
Mrotek, James Joseph
Norden, Jeanette Jean
Norton, Virginia Marino
Orgebin-Crist, Marie-Claire
Peterson, Darwin Wilson
Powers, J Bradley
Puett, J David
Reddy, Churku Mohan
Regen, David Marvin
Richmond, Chester Robert
Rieke, Garl Kalman
Sander, Linda Dian
Shirley, Herschel Vincent, Jr
Snapper, James Robert
Solomon, Solomon Sidney
Stone, Robert Edward, Jr
Taylor, Robert Emerald, Jr
Turner, Barbara Bush
Weinstein, Ira
Welch, Hugh Gordon
Wondergem, Robert

TEXAS

Achilles, Robert F
Alexander, William Carter
Alkadhi, Karim A
Allen, Edward Patrick
Alter, William A, III
Amoss, Max St Clair
Ashworth, Robert David
Azizi, Sayed Ausim
Banerji, Tapan Kumar
Barker, Kenneth Leroy
Barnes, George Edgar
Bellinger, Larry Lee
Bertrand, Helen Anne
Birnbaumer, Lutz
Blankenship, James Emery
Bockholt, Anton John
Bonner, Hugh Warren
Booth, Frank
Boudreau, James Charles
Breen, Gail Anne Marie
Brooks, Barbara Alice
Brown, J(ack) H(arold) U(pton)
Burks, James Kenneth
Burns, John Mitchell
Burton, Karen Poliner
Burton, Russell Rohan
Butler, Bruce David
Caffrey, James Louis
Cameron, James N
Campbell, Bonnalie Oetting
Castracane, V Daniel
Cate, Rodney Lee
Chang, Ching Hsong
Chowdhury, Ajit Kumar
Christensen, Burgess Nyles
Chubb, Curtis Evans
Clark, James Henry
Claus-Walker, Jacqueline Lucy
Clench, Mary Heimerdinger
Coats, Alfred Cornell
Cooper, Cary Wayne
Cooper, John C, Jr
Cranford, Jerry L
Crass, Maurice Frederick, III
Crawford, Isaac Lyle
Crews, David Pafford
Dafny, Nachum
Dalterio, Susan Linda
Danhof, Ivan Edward
Davis, James Royce
Dedman, John Raymond
DeFrance, Jon Fredric
DiMichele, Leonard Vincent
Downey, H Fred
Easley, Stephen Phillip
Eldridge, David Wyatt
Entman, Mark Lawrence
Espey, Lawrence Lee
Evans, James Warren
Falck, Frank James
Fawcett, Colvin Peter
Fisher, Frank M, Jr
Foster, Daniel W
Fox, Donald A
Frasher, Wallace G, Jr
Frazier, Loy William, Jr
Fremming, Benjamin DeWitt
Frigyesi, Tamas L
Gabel, Joseph C
Gaitz, Charles M
Gaugl, John F
Gerken, George Manz
German, Dwight Charles
Glantz, Raymon M
Glasser, Stanley Richard
Goldstein, Margaret Ann
Gonzalez-Lima, Francisco
Gordon, Wayne Lecky
Gottesfeld, Zehava
Gotto, Antonio Marion, Jr
Guest, Maurice Mason

Guo, Yan-Shi
Haensly, William Edward
Hancock, Michael B
Harper, Michael John Kennedy
Hartley, Craig Jay
Hartsfield, Sandee Morris
Hazelwood, Robert Leonard
Hazlett, David Richard
Hill, Ronald Stewart
Holman, Gerald Hall
Hotchkiss, Julane
Hudspeth, Albert James
Jauchem, James Robert
Johnson, Larry
Kasschau, Margaret Ramsey
Keating, Robert Joseph
Keele, Doman Kent
Kenny, Alexander Donovan
King, Richard Joe
Knobil, Ernst
Konecci, Eugene B
Korr, Irvin Morris
Kubena, Leon Franklin
Kunze, Diana Lee
Larimer, James Lynn
Lawrence, Addison Lee
Leach-Huntoon, Carolyn S
Lebovitz, Robert Mark
Lees, George Edward
Lewis, Douglas Scott
Lifschitz, Meyer David
Lloyd, James Armon
Lucas, Edgar Arthur
McCann, Samuel McDonald
McCrady, James David
McGrath, James Joseph
McMahon, Robert Francis, III
Madison, Leonard Lincoln
Mailman, David Sherwin
Margolin, Solomon B
Martin, Frederick N
Masoro, Edward Joseph
Mattingly, Stephen Joseph
Means, Anthony R
Meola, Roger Walker
Michael, Lloyd Hal
Mikiten, Terry Michael
Mitchell, Jere Holloway
Moldawer, Marc
Moore, Lee E
Morrow, Dean Huston
Moss, Robert L
Nathan, Richard D
Norman, Reid Lynn
Nusynowitz, Martin Lawrence
Ojeda, Sergio Raul
Olson, John S
O'Malley, Bert W
Owens, David William
Pak, Charles Y
Parkening, Terry Arthur
Parker, Nick Charles
Peake, Robert Lee
Pitts, Donald Graves
Poluhowich, John Jacob
Pond, Wilson Gideon
Powell, Michael Robert
Preslock, James Peter
Proppe, Duane W
Redburn, Dianna Ammons
Reeves, T Joseph
Reid, Michael Baron
Reiter, Russel Joseph
Richards, Joanne S
Riehl, Robert Michael
Ritter, Nadine Marie
Rock, Michael Keith
Roeser, Ross Joseph
Roffwarg, Howard Philip
Rosenfeld, Charles Richard
Ross, Griff Terry, Sr
Rudenberg, Frank Hermann
Samaan, Naguib A
Samson, Willis Kendrick
Sant'Ambrogio, Giuseppe
Schafer, Rollie R
Scheel, Konrad Wolfgang
Schindler, William Joseph
Schrader, William Thurber
Scurry, Murphy Townsend
Senseman, David Michael
Shain, Sydney A
Shum, Wan-Kyng Liu
Silverthorn, Dee Unglaub
Simmons, David J
Skinner, James Ernest
Smatresk, Neal Joseph
Smith, Keith Davis
Snyder, Gary Dean
Soriero, Alice Ann
Spence, Dale William
Sperling, Harry George
Stancel, George Michael
Stein, Jerry Michael
Storey, Arthur Thomas
Stubbs, Donald William
Suddick, Richard Phillips
Tate, Charlotte Anne
Taubert, Kathryn Anne
Tsin, Andrew Tsang Cheung
Vanatta, John Crothers, III
Varma, Surendra K
Vaughan, Mary Kathleen
Villa, Vicente Domingo

Vinson, S Bradleigh
Walsh, Scott Wesley
Ward, Walter Frederick
Warner, Marlene Ryan
Waymire, Jack Calvin
Wayner, Matthew John
Weems, William Arthur
White, Fred Newton
Whitmore, Donald Herbert, Jr
Whitsett, Johnson Mallory, II
Williams, Charles Herbert
Williams, Gary Lynn
Willis, William Darrell, Jr
Wilson, Everett D
Wilson, L Britt
Witzel, Donald Andrew
Wolfe, James Wallace
Wolfenberger, Virginia Ann
Wordinger, Robert James
Worthy, Graham Anthony James
Yorio, Thomas
Yu, Byung Pal

UTAH
Baranowski, Robert Louis
Bland, Richard David
Brown, Harold Mack, Jr
Burgess, Paul Richards
Conlee, Robert Keith
Ellis, LeGrande Clark
Eyzaguirre, Carlos
Foote, Warren Christopher
Galster, William Allen
Gessaman, James A
Grosser, Bernard Irving
Hansen, Afton M
Hsiao, Ting Huan
Jensen, David
Kesner, Raymond Pierre
Lighton, John R B
Lords, James Lafayette
McKenzie, Jess Mack
Mecham, Merlin J
Musick, James R
Nelson, Don Harry
Rhees, Reuben Ward
Rodin, Martha Kinscher
Ruhmann Wennhold, Ann Gertrude
Shininger, Terry Lynn
Shumway, Richard Phil
Tuckett, Robert P
Wilson, Dana E
Winder, William W
Wurst, Gloria Zettle
Yoshikami, Doju

VERMONT
Bartel, Lavon L
Evans, John N
Otter, Timothy
Webb, George Dayton
Welsh, George W, III

VIRGINIA
Adams, James Milton
Andrykovitch, George
Bass, Michael Lawrence
Berne, Robert Matthew
Bhatnagar, Ajay Sahai
Blackard, William Griffith
Blatt, Elizabeth Kempske
Block, Gene David
Bogdanove, Emanuel Mendel
Burr, Helen Gunderson
Calvert, Richard John
Clark, Mary Eleanor
Clarke, Gary Anthony
Colmano, Germille
Corwin, Jeffrey Todd
Costanzo, Linda Schupper
Dane, Charles Warren
Davis, Joel L
Denbow, Donald Michael
Dendinger, James Elmer
Duling, Brian R
Dunn, John Thornton
Edwards, Leslie Erroll
Fariss, Bruce Lindsay
Feher, Joseph John
Fenner-Crisp, Penelope Ann
Fine, Michael Lawrence
Friesen, Wolfgang Otto
Gray, Faith Harriet
Grider, John Raymond
Gwazdauskas, Francis Charles
Hackett, John Taylor
Hall, Richard Eugene
Hanig, Joseph Peter
Hanna, George R
Hardie, Edith L
Harper, Laura Jane
Heath, Alan Gard
Herbein, Joseph Henry, Jr
Hickman, Cleveland Pendleton, Jr
Hodgen, Gary Dean
Johnston, David Ware
Jones, Joyce Howell
Kirkpatrick, Roy Lee
Kramp, Robert Charles
Kripke, Bernard Robert
Kronfeld, David Schultz
Lloyd, John Willie, III
McCarty, Richard Charles
McCowen, Sara Moss

Mackowiak, Robert Carl
Manson, Nancy Hurt
Martin, James Henry, III
Mayer, David Jonathan
Mellon, DeForest, Jr
Nolan, Stanton Peelle
Owen, John Atkinson, Jr
Owens, Gary Keith
Paulsen, Elsa Proehl
Pittman, Roland Nathan
Prewitt, Russell Lawrence, Jr
Price, Steven
Rogol, Alan David
Roth, Allan Charles
Rubel, Edwin W
Rubio, Rafael
Ruhling, Robert Otto
Schwab, Walter Edwin
Scott, David Evans
Stein, Barry Edward
Stewart, Jennifer Keys
Swanson, Robert James
Szumski, Alfred John
Talbert, George Brayton
Teekell, Roger Alton
Vinores, Stanley Anthony
Wei, Enoch Ping
Wells, John Morgan, Jr
Williams, Gerald Albert
Wingard, Christopher Jon
Wolford, John Henry

WASHINGTON
Adkins, Ronald James
Ammann, Harriet Maria
Artru, Alan Arthur
Avner, Ellis David
Baksi, Samarendra Nath
Barnes, Charles Dee
Beck, Thomas W
Berger, Albert Jeffrey
Bernard, Gary Dale
Bierman, Edwin Lawrence
Binder, Marc David
Blinks, John Rogers
Bowden, Douglas Mchose
Bremner, William John
Brengelmann, George Leslie
Canfield, Robert Charles
Carlberg, Karen Ann
Crampton, George H
Dauble, Dennis Deene
Deyrup-Olsen, Ingrith Johnson
Dickhoff, Walton William
Dickson, William Morris
Ensinck, John William
Estergreen, Victor Liné
Gibson, Flash
Gollnick, Philip D
Goodner, Charles Joseph
Gordon, Albert McCague
Halbert, Sheridan A
Hampton, John Kyle, Jr
Harding, Joseph Warren, Jr
Heitkemper, Margaret M
Hicks, David L
Hildebrandt, Jacob
Hosick, Howard Lawrence
Huey, Raymond Brunson
Kahan, Linda Beryl
Kaufman, William Carl, Jr
Kenagy, George James
Kennedy, Thelma Temy
Kromann, Rodney P
Landau, Barbara Ruth
Leung, Frederick C
Maloff, Bruce L(arrie)
Martin, Arthur Wesley
Mendelson, Martin
Moffett, David Franklin, Jr
Morgan, Kenneth Robb
Murphy, Mary Eileen
Nathanson, Neil Marc
Palka, John Milan
Palmer, John M
Paulsen, Charles Alvin
Prinz, Patricia N
Prusch, Robert Daniel
Reeves, Jerry John
Reh, Thomas Andrew
Reid, Donald House
Riddiford, Lynn Moorhead
Rodieck, Robert William
Ruvalcaba, Rogelio H A
Santisteban, George Anthony
Schneider, David Edwin
Simpson, John Barclay
Smith, Lynwood S
Smith, Orville Auverne
Spencer, Merrill Parker
Steiner, Robert Alan
Taborsky, Gerald J, Jr
Teller, Davida Young
Toivola, Pertti Toivo Kalevi
Van Citters, Robert L
White, Fredric Paul
Whiteley, Helen Riaboff
Willows, Arthur Owen Dennis
Yancey, Paul Herbert
Youmans, William Barton

WEST VIRGINIA
Brown, David E
Brown, Paul B

Castranova, Vincent
Chambers, John William
Cochrane, Robert Lowe
Collins, William Edgar
Connors, John Michael
Culberson, James Lee
Dailey, Robert Arthur
Einzig, Stanley
Fisher, Martin Joseph
Gladfelter, Wilbert Eugene
Inskeep, Emmett Keith
Jones, John Evan
Kidder, Harold Edward
Overbeck, Henry West
Peterson, Ronald A
Reilly, Frank Daniel
Weber, Kenneth C

WISCONSIN
Anand, Amarjit Singh
Attie, Alan D
Bach-Y-Rita, Paul
Bavister, Barry Douglas
Benjamin, Robert Myles
Bergtrom, Gerald
Bisgard, Gerald Edwin
Bownds, M Deric
Bremel, Robert Duane
Chi, Che
Coon, Robert L
Cowley, Allen Wilson, Jr
Davis, Larry Dean
Dierschke, Donald Joe
Ehle, Fred Robert
Flax, Stephen Wayne
Folts, John D
Gilboe, David Dougherty
Goldberg, Alan Herbert
Gorski, Jack
Hager, Steven Ralph
Hall, Terence Robert
Hamilton, Lyle Howard
Harper, Alfred Edwin
Hetzler, Bruce Edward
Horseman, Nelson Douglas
Howard, Thomas Hyland
Keesey, Richard E
Kemnitz, Joseph William
Knoll, Jack
Kohler, Elaine Eloise Humphreys
Kostreva, David Robert
Lalley, Peter Michael
Lanphier, Edward Howell
Lombard, Julian H
Marsh, Benjamin Bruce
Minnich, John Edwin
Murray, Mary Patricia
Nagle, Francis J
Nichols, Roy Elwyn
Piacsek, Bela Emery
Plotka, Edward Dennis
Raff, Hershel
Reneau, John
Rhode, William Stanley
Ritchie, Betty Caraway
Rose, Jerzy Edwin
Rouse, Thomas C
Schlough, James Sherwyn
Sidky, Younan Abdel Malik
Smith, Dean Orren
Spieler, Richard Earl
Spurr, Gerald Baxter
Staszak, David John
Steele, Robert Darryl
Stekiel, William John
Stevens, Richard Joseph
Stowe, David F
Stretton, Antony Oliver Ward
Sullivan, John Joseph
Terasawa, EI
Vomachka, Archie Joel
Walgenbach-Telford, Susan Carol
Weil, Michael Ray
Wentworth, Bernard C
Will, James Arthur
Wolf, Richard Clarence
Woolsey, Clinton Nathan
Yin, Tom Chi Tien

WYOMING
Bowerman, Robert Francis
Dunn, Thomas Guy
Rose, James David
Smith, William Kirby

PUERTO RICO
Afanador, Arthur Joseph
Del Castillo, Jose
Haddock, Lillian
Opava-Stitzer, Susan C
Orkand, Richard K

ALBERTA
Bland, Brian Herbert
Campenot, Robert Barry
Cottle, Merva Kathryn Warren
Coulter, Glenn Hartman
Dawson, J W
Hoffer, J(oaquin) A(ndres)
Hohn, Emil Otto
Hutchison, Kenneth James
Jaques, Louis Barker
Jones, Richard Lee
LeBlanc, Francis Ernest

Lederis, Karolis (Karl)
Lorscheider, Fritz Louis
Lukowiak, Kenneth Daniel
Mackay, William Charles
McMahon, Brian Robert
Mears, Gerald John
Molnar, George D
Moore, Graham John
Neuwirth, Maria
Pang, Peter Kai To
Pearson, Keir Gordon
Peter, Richard Ector
Pittman, Quentin J
Russell, James Christopher
Smith, Richard Sidney
Spencer, Andrew Nigel
Stein, Richard Bernard
Stell, William Kenyon
Tyberg, John Victor
Walsh, Michael Patrick
Wang, Lawrence Chia-Huang
Whitelaw, William Albert
Wilkens, Jerrel L
Wilson, Frank B
Young, Bruce Arthur

BRITISH COLUMBIA
Bailey, Charles Basil Mansfield
Bruchovsky, Nicholas
Copp, Douglas Harold
Cynader, Max Sigmund
Donaldson, Edward Mossop
Fankboner, Peter Vaughn
Freeman, Hugh J
Gosline, John M
Haunerland, Norbert Heinrich
Hayward, John S
Hughes, Maryanne Robinson
Hughes, Maryanne Robinson
Jones, David Robert
Kraintz, Leon
Krishnamurti, Cuddalore Rajagopal
Milsom, William Kenneth
Nowaczynski, Wojciech
Rennie, Paul Steven
Schwarz, Dietrich Walter Friedrich
Semba, Kazue
Steeves, John Douglas
Vincent, Steven Robert

MANITOBA
Anthonisen, Nicholas R
Bihler, Ivan
Dyck, Gerald Wayne
Faiman, Charles
Friesen, Henry George
Giles, Michael Arthur
Greenway, Clive Victor
Hara, Toshiaki J
Huebner, Judith Dee
Nance, Dwight Maurice
Polimeni, Philip Iniziato
Singal, Pawan Kumar
Swierstra, Ernest Emke
Winter, Jeremy Stephen Drummond

NEW BRUNSWICK
Anderson, John Murray
Paim, Uno
Seabrook, William Davidson
Waiwood, Kenneth George

NEWFOUNDLAND
Brosnan, John Thomas
Fletcher, Garth L
Hillman, Donald Arthur
Kao, Ming-hsiung
Morris, Claude C
Rusted, Ian Edwin L H
Steele, Vladislava Julie
Virgo, Bruce Barton

NOVA SCOTIA
Dudar, John Douglas
Ferrier, Gregory R
Finley, John P
Horrobin, David Frederick
Leslie, Ronald Allan
McDonald, Terence Francis
Marshall, William Smithson
Rasmusson, Douglas Dean
Savard, Francis Gerald Kenneth
Szerb, John Conrad
Toews, Daniel Peter
Weld, Charles Beecher

ONTARIO
Abbott, Frank Sidney
Abrahams, Vivian Cecil
Ackermann, Uwe
Ainsworth, Louis
Alikhan, Muhammad Akhtar
Atwood, Harold Leslie
Bayliss, Colin Edward
Begin-Heick, Nicole
Biro, George P
Bisby, Mark A
Bols, Niels Christian
Bonkalo, Alexander
Broughton, Roger James
Buchanan, George Dale
Burton, John Heslop
Butler, Richard Gordon
Cafarelli, Enzo Donald

Lutt, Carl J
Lyon, Irving
McBlair, William
McClintic, Joseph Robert
Macey, Robert Irwin
McGuigan, Frank Joseph
Maffly, LeRoy Herrick
Martinson, Ida Marie
Meehan, John Patrick
Meerbaum, Samuel
Michael, Ernest Denzil, Jr
Mines, Allan Howard
Mitchell, Robert A
Mommaerts, Wilfried
Morey-Holton, Emily Rene
Morse, John Thomas
Morton, Martin Lewis
Muscatine, Leonard
Newsom, Bernard Dean
Nicoll, Charles S
Nolan, Janiece Simmons
Ogasawara, Frank X
Ogden, Thomas E
Omid, Ahmad
Oyama, Jiro
Pace, Nello
Paolini, Paul Joseph
Parlow, Albert Francis
Pearl, Ronald G
Pengelley, Eric T
Peper, Erik
Peterson, Charles Marquis
Pickwell, George Vincent
Pilgeram, Laurence Oscar
Pipes, Gayle Woody
Pollard, James Edward
Povzhitkov, Moysey Michael
Powanda, Michael Christopher
Power, Gordon D
Pratley, James Nicholas
Rabinowitz, Lawrence
Ralston, Henry James
Reid, Ian Andrew
Renkin, Eugene Marshall
Reynolds, Robert Williams
Rhode, Edward A, Jr
Ridley, Peter Tone
Robin, Eugene Debs
Rosenfeld, Sheldon
Ross, Gordon
Rothman, Stephen Sutton
Rubinstein, Eduardo Hector
Rudolph, Abraham Morris
Runion, Howell Irwin
Russell, Sharon May
Sammak, Paul J
Sandler, Harold
Sassaman, Clay Alan
Scheer, Bradley Titus
Schein, Arnold Harold
Schmid-Schoenbein, Geert W
Schneider, Charles L
Schnitzer, Jan Eugeneusz
Schooley, John C
Schwab, Robert G
Schweisthal, Michael Robert
Scobey, Robert P
Searle, Gilbert Leslie
Seegers, Walter Henry
Sellers, Alvin Louis
Setler, Paulette Elizabeth
Shaw, Jane E
Shellock, Frank G
Shimizu, C Susan
Shore, Virgie Guinn
Shvartz, Esar
Simmons, Daniel Harold
Simmons, Francis Blair
Sirota, Jonas H
Smith, Arthur Hamilton
Smith, Robert Elphin
Smulders, Anthony Peter
Sobin, Sidney S
Sonnenschein, Ralph Robert
Stabenfeldt, George H
Starr, Arnold
Staub, Norman Croft
Stein, Myron
Stein, Seymour Norman
Stiffler, Daniel F
Stolzenberg, Sidney Joseph
Strautz, Robert Lee
Sugar, Oscar
Swan, Harold James Charles
Swanson, George D
Talbot, Prudence
Taylor, Anna Newman
Testerman, John Kendrick
Thomas, Robert E
Thrasher, Terry Nicholas
Tierney, Donald Frank
Timiras, Paola Silvestri
Torre-Bueno, Jose Rollin
Trei, John Earl
Tremor, John W
Tullis, Richard Eugene
Turgeon, Judith Lee
Turley, Kevin
Van Harreveld, Anthonie
Vernikos, Joan
Vidoli, Vivian Ann
Vomhof, Daniel William
Wagner, Jeames Arthur
Ward, Susan A

Warnock, David Gene
Wasserman, Karlman
Wasterlain, Claude Guy
Weathers, Wesley Wayne
Weidner, William Jeffrey
Weissberg, Robert Murray
Wells, J Gordon
Wendel, Otto Theodore, Jr
Westfahl, Pamela Kay
Whipp, Brian James
White, Stephen Halley
Wilson, Archie Fredric
Wilson, Barry William
Woolley, Dorothy Elizabeth Schumann
Wright, Ernest Marshall
Wright, Richard Kenneth
Yates, Francis Eugene
Young, Ho Lee
Younoszai, Rafi
Yu, Jen
Zieg, Roger Grant
Zweifach, Benjamin William

COLORADO
Audesirk, Teresa Eck
Balke, Bruno
Battaglia, Frederick Camillo
Beam, Kurt George, Jr
Best, Jay Boyd
Betz, William J
Brockway, Alan Priest
Burke, Edmund R
Burke, Thomas Joseph
Carey, Cynthia
Filley, Giles Franklin
Fuller, Barbara Fink Brockway
Gray, John Stephens
Hand, Steven Craig
Harold, Franklin Marcel
Krausz, Stephen
Levinson, Simon Rock
McMurtry, Ivan Fredrick
Masken, James Frederick
Meschia, Giacomo
Mykles, Donald Lee
Neville, Margaret Cobb
Pautler, Eugene L
Pickett, Bill Wayne
Quissell, David Olin
Ralph, Charles Leland
Reeve, Ernest Basil
Rich, Royal Allen
Richards, Edmund A
Roberts, P Elaine
Roper, Stephen David
Sexton, Alan William
Snyder, Gregory Kirk
Takeda, Yasuhiko
Trapani, Ignatius Louis
Wilber, Charles Grady
Williams, Ronald Lee

CONNECTICUT
Agulian, Samuel Kevork
Anderson, Rebecca Jane
Aronson, Peter S
Bailie, Michael David
Baird, James Leroy, Jr
Behrman, Harold R
Berliner, Robert William
Borgia, Julian Frank
Boulpaep, Emile L
Boylan, John W
Briscoe, Anne M
Bronner, Felix
Chandler, William Knox
Clark, Nancy Barnes
Constantine, Jay Winfred
Cooperstein, Sherwin Jerome
Crean, Geraldine L
Dawson, Margaret Ann
Dumont, Allan E
Edwards, Lawrence Jay
Fiore, Carl
Fleming, James Stuart
Gagge, Adolf Pharo
Gee, J Bernard L
Giebisch, Gerhard Hans
Gwynn, Robert H
Haas, Mark
Hayslett, John P
Hirsh, Eva Maria Hauptmann
Hoffman, Joseph Frederick
Jaffe, Laurinda A
Jungas, Robert Leando
Katz, Arnold Martin
Koeppen, Bruce Michael
Landmesser, Lynn Therese
Lee, Thomas W
Lynch, George Robert
Mack, Gary W
Matovcik, Lisa M
Meriney, Stephen D
Naftolin, Frederick
Oates, Peter Joseph
Patterson, John Ward
Rajendran, Vazhaikkurichi M
Rawson, Robert Orrin
Renfro, J Larry
Riblet, Leslie Alfred
Ritchie, Joseph Murdoch
Roos, Henry
Schnitzler, Ronald Michael
Schwartz, Tobias Louis

Sha'afi, Ramadan Issa
Siegel, Norman Joseph
Sikand, Rajinder S
Slayman, Clifford L
Sobol, Bruce J
Stitt, John Thomas
Taylor, Kenneth J W
Turnipseed, Marvin Roy
Verses, Christ James
Waterman, Talbot H(owe)
Weissman, Sherman Morton
Woods, Joseph James
Wright, Hastings Kemper
Zaret, Barry L

DELAWARE
DiPasquale, Gene
Skopik, Steven D
South, Frank E
Turlapaty, Prasad
Wheeler, Allan Gordon

DISTRICT OF COLUMBIA
Atkins, James Lawrence
Bellamy, Ronald Frank
Burger, Edward James, Jr
Carpentier, Robert George
Cassidy, Marie Mullaney
Chaput, Raymond Leo
Coleman, Bernell
Freygang, Walter Henry, Jr
Geelhoed, Glenn William
Hamosh, Margit
Hamosh, Paul
Haskins, Caryl Parker
Hoskin, George Perry
Johnson, Thomas F
Jordan, Alexander Walker, III
Kasbekar, Dinkar Kashinath
Koering, Marilyn Jean
Lakshman, M Raj
Leto, Salvatore
Lilienfield, Lawrence Spencer
Lorber, Mortimer
Nardone, Roland Mario
Nath, Jayasree
Orlans(Morton), F Barbara
Pearce, Frederick James
Phelps, Harriette Longacre
Ramey, Estelle R
Ramwell, Peter William
Ronkin, R(aphael) R(ooser)
Schreiner, George E
Sinkford, Jeanne C
Starr, Matthew C
Sudia, Theodore William
Sulzman, Frank Michael
Verrier, Richard Leonard
Watkins, Don Wayne
Wong, Harry Yuen Chee
Yamamoto, William Shigeru
Yen-Koo, Helen C

FLORIDA
Abraham, William Michael
Abrams, Robert Marlow
Anderson, John Francis
Baiardi, John Charles
Balis, John Ulysses
Barkalow, Derek Talbot
Barrera, Frank
Battista, Sam P
Bazer, Fuller Warren
Beidler, Lloyd M
Bressler, Steven L
Brown, John Lott
Bundy, Roy Elton
Buss, Daryl Dean
Chaet, Alfred Bernard
Cody, D Thane
Cousins, Robert John
Davis, Darrell Lawrence
Dietz, John R
Drummond, Willa H
Dudley, Gary A
Ebbs, Jane Cotton
Edwards, Charles
Eitzman, Donald V
Ekberg, Donald Roy
Ellington, William Ross
Elmadjian, Fred
Epstein, Murray
Evans, David Hudson
Ferguson, John Carruthers
Fregly, Melvin James
Gerencser, George A
Gilmore, Joseph Patrick
Glennon, Joseph Anthony
Hajdu, Stephen
Handford, Stanley Wing
Hayashi, Teru
Hays, Elizabeth Teuber
Ho, Ren-jye
Howell, David Sanders
Hurst, Josephine M
Jacobson, Howard Newman
Katovich, Michael J
Kemph, John Patterson
Kerrick, Wallace Glenn Lee
Kessler, Richard Howard
Kissen, Abbott Theodore
Landowne, David
Lawrence, John M
Lawrence, Merle

Legler, Donald Wayne
Lemberg, Louis
Lotz, W Gregory
Mallery, Charles Henry
Mayrovitz, Harvey N
Natzke, Roger Paul
Nicolosi, Gregory Ralph
Nicosia, Santo Valerio
Nolasco, Jesus Bautista
Olsson, Ray Andrew
Penhos, Juan Carlos
Posner, Philip
Raizada, Mohan K
Rao, Krothapalli Ranga
Rao, Papineni Seethapathi
Reininger, Edward Joseph
Reynolds, Orr Esrey
Robbins, Ralph Compton
Rockstein, Morris
Roeder, Martin
Rose, Birgit
Rosenthal, Myron
Samis, Harvey Voorhees, Jr
Schwassman, Horst Otto
Shireman, Rachel Baker
Shiverick, Kathleen Thomas
Sick, Thomas J
Silverman, Sol Richard
Soliman, Magdi R I
Sparkman, Marjorie Frances
Speckmann, Elwood W
Stevens, Bruce Russell
Still, Eugene Updike
Sturbaum, Barbara Ann
Thatcher, William Watters
Tuttle, Ronald Ralph
Vogh, Betty Pohl
Wells, Charles Henry
Wiggers, Harold Carl
Wilcox, Christopher Stuart
Williams, Clyde Michael
Williams, Joseph Francis
Wilson, Henry R
Wood, Charles Evans
Woodside, Kenneth Hall

GEORGIA
Abercrombie, Ronald Ford
Ashley, Doyle Allen
Balcer-Brownstein, Josefine P
Blum, Murray Sheldon
Boyd, Louis Jefferson
Chew, Catherine S
Collins, Delwood C
Davidson, John Keay, III
De Felice, Louis John
DeHaan, Robert Lawrence
Edelhauser, Henry F
Eriksson, Karl-Erik Lennart
Geber, William Frederick
Ginsburg, Jack Martin
Godt, Robert Eugene
Green, Keith
Guinan, Mary Elizabeth
Gunn, Robert Burns
Hendrich, Chester Eugene
Hersey, Stephen J
Hockman, Charles Henry
Huber, Thomas Lee
Hunter, Frissell Roy
Innes, David Lyn
Jackson, Richard Thomas
Khan, Mohammad Iqbal
Kolbeck, Ralph Carl
Mann, David R
Manning, John W
Martin, David Edward
Matsumoto, Yorimi
Moriarty, C Michael
Nabrit, Samuel Milton
Nichols, J Wylie
Pashley, David Henry
Pooler, John Preston
Popovic, Vojin
Porterfield, Susan Payne
Reichard, Sherwood Marshall
Rinard, Gilbert Allen
Steinbeck, Klaus
Taylor, Robert Clement
Tuttle, Elbert P, Jr
Waitzman, Morton Benjamin
Weatherred, Jackie G
White, John Francis
Whitford, Gary M
Wiedmeier, Vernon Thomas
Willis, John Steele
Woods, Wendell David
Young, Leona Graff

HAWAII
Ahearn, Gregory Allen
Chang, Franklin
Druz, Walter S
Hammer, Ronald Page, Jr
Kodama, Arthur Masayoshi
Lin, Yu-Chong
Read, George Wesley
Reed, Stuart Arthur
Rogers, Terence Arthur
Segal, Earl
Smith, Richard Merrill

IDAHO
Bunde, Daryl E

Physiology, General (cont)

Christian, Ross Edgar
Dahmen, Jerome J
Ferguson, James Homer
Laskowski, Michael Bernard
McKean, Thomas Arthur
Seeley, Rod R

ILLINOIS

Abraira, Carlos
Al-Bazzaz, Faiq J
Albrecht, Ronald Frank
Aldred, J Phillip
Allgood, Joseph Patrick
Anderson, John Denton
Bacus, James William
Barany, Kate
Barr, Lloyd
Bartke, Andrzej
Becker, John Henry
Beecher, Christopher W W
Bell, Richard Dennis
Blivaiss, Ben Burton
Browning, Ronald Anthony
Buchholz, Robert Henry
Buetow, Dennis Edward
Burhop, Kenneth Eugene
Carlson, Gerald Eugene
Catchpole, Hubert Ralph
Chan, Yun Lai
Cralley, John Clement
Cromer, John A
Cureton, Thomas (Kirk), Jr
Curtis, Brian Albert
Dawe, Albert Rolke
Dawson, M Joan
DeClue, Jim W
Denbo, John Russell
Devries, Arthur Leland
Domroese, Kenneth Arthur
Ducoff, Howard S
Dunagan, Tommy Tolson
Dziuk, Philip J
Eagle, Edward
Ebert, Paul Allen
Ecker, Richard Eugene
Eder, Douglas Jules
Eisenberg, Brenda Russell
Eisenberg, Robert S
Ellert, Martha Schwandt
Ernest, J Terry
Filkins, James P
Fozzard, Harry A
Frehn, John
Friedlander, Ira Ray
Garthwaite, Susan Marie
Gartner, Lawrence Mitchel
Gibbons, Larry V
Gibori, Geula
Gillette, Martha Ulbrick
Glaviano, Vincent Valentino
Green, Jonathan P
Greenberg, Ruven
Grimm, Arthur F
Harris, Ruth B S
Harris, Stanley Cyril
Hassan, Aslam Sultan
Hast, Malcolm Howard
Hawkins, Richard Albert
Hawley, Philip Lines
Heath, James Edward
Hechter, Oscar Milton
Helman, Sandy I
Hertig, Bruce Allerton
Houk, James Charles
Humphrey, Gordon Laird
Hunter, William Sam
Hunzicker-Dunn, Mary
Hutchens, John Oliver
Hyatt, Robert Eliot
Jackson, Gary Loucks
Jacobs, H Kurt
Jakobsson, Eric Gunnar, Sr
Jung, Frederic Theodore
Kaplan, Harold M
Karczmar, Alexander George
Kelso, Albert Frederick
Khan, Mohammed Nasrullah
Kidder, George Wallace, III
Kim, Mi Ja
Klein, Diane M
Koblick, Daniel Cecil
Koushanpour, Esmail
Kyncl, J Jaroslav
Lax, Louis Carl
Leven, Robert Maynard
Lin, James Chih-I
Lindheimer, Marshall D
Lipsius, Stephen Lloyd
Lit, Alfred
Lodge, James Robert
Loeb, Jerod M
Loizzi, Robert Francis
Lopez-Majano, Vincent
Lorand, Laszlo
Luisada, Aldo Augusto
McCormack, Charles Elwin
Madsen, David Christy
Marotta, Sabath Fred
Mason, George Robert
Metz, John Thomas
Miller, Donald Morton
Morrison, Shaun Francis

Myers, James Hurley
Neidle, Enid Anne
Nelson, Douglas O
Nelson, Ralph A
Nelson, Thomas Eustis, Jr
Nickerson, John Lester
Nicolette, John Anthony
Nyhus, Lloyd Milton
Olson, Howard H
Palincsar, Edward Emil
Pederson, Vernon Clayton
Peiss, Clarence Norman
Perkins, William Eldredge
Perlman, Robert
Perry, Harold Tyner, Jr
Preston, Robert Leslie
Prosser, Clifford Ladd
Proudfit, Carol Marie
Rabuck, David Glenn
Rakowski, Robert F
Ratzlaff, Kermit O
Reiser, Peter Jacob
Ringham, Gary Lewis
Ripps, Harris
Sanderson, Glen Charles
Sayeed, Mohammed Mahmood
Schreider, Bruce David
Scommegna, Antonio
Simmons, William Howard
Singh, Sant Parkash
Skosey, John Lyle
Sleator, William Warner, Jr
Snarr, John Frederic
Solaro, R John
Sukowski, Ernest John
Tallitsch, Robert Boyde
Taylor, Ardell Nichols
Ten Eick, Robert Edwin
Thilenius, Otto G
Thomas, John Xenia, Jr
Thompson, Leif Harry
Toback, F(rederick) Gary
Tobin, Martin John
Triano, John Joseph
Truong, Xuan Thoai
Ts'ao, Chung-Hsin
Turnquist, Richard Lee
Van de Kar, Louis David
Vesselinovitch, Stan Dushan
Wessel, Hans U
Wilkinson, Harold L
Zak, Radovan Hynek
Zimmerman, Roger Paul

INDIANA

Alliston, Charles Walter
Asano, Tomoaki
BeMiller, Paraskevi Mavridis
Blasingham, Mary Cynthia
Bottoms, Gerald Doyle
Brett, William John
Christ, Daryl Dean
Dantzig, Anne H
Dungan, Kendrick Webb
Echtenkamp, Stephen Frederick
Elharrar, Victor
Elizondo, Reynaldo S
Friedman, Julius Jay
Fuller, Forst Donald
Geddes, LaNelle Evelyn
Grabowski, Sandra Reynolds
Greenspan, Kalman
Heath, Gordon Glenn
Henzlik, Raymond Eugene
Hertzler, Emanuel Cassel
Hinds, Marvin Harold
Howard, David K
Hui, Chiu Shuen
Hunt, Linda Margaret
Hurst, Robert Nelson
Hyslop, Paul A
Kempson, Stephen Allan
Knoebel, Leon Kenneth
Lane, Ardelle Catherine
Lin, Tsung-Min
Maass-Moreno, Roberto
Marquis, Norman Ronald
Mays, Charles Edwin
Meiss, Richard Alan
Nunn, Arthur Sherman, Jr
Oliver, Montague
Randall, Barbara Feucht
Randall, James Edwin
Randall, Walter Clark
Renzi, Alfred Arthur
Rhoades, Rodney A
Rinkema, Lynn Ellen
Schaeffer, John Frederick
Schauf, Charles Lawrence
Selkurt, Ewald Erdman
Shea, Philip Joseph
Silbaugh, Steven A
Singleton, Wayne Louis
Sisson, George Maynard
Stephenson, William Kay
Strickholm, Alfred
Suthers, Roderick Atkins
Tamar, Henry
Wilkin, Peter J
Witzmann, Frank A
Wolff, Ronald Keith

IOWA

Bergman, Ronald Arly

Bishop, Stephen Hurst
Christensen, James
Deavers, Daniel Ronald
Dunham, Jewett
Engen, Richard Lee
Folk, George Edgar, Jr
Gisolfi, Carl Vincent
Guernsey, Duane L
Guest, Mary Frances
Gurll, Nelson
Hardin, Carolyn Myrick
Hatch, Robert Harold
Horst, Ronald Lee
Huiatt, Ted W
Imig, Charles Joseph
Jennings, Michael Leon
Kirkland, Willis L
Kolder, Hansjoerg E
Masat, Robert James
Mennega, Aaldert
Norcia, Leonard Nicholas
Osborne, James William
Rulon, Russell Ross
Schottelius, Byron Arthur
Searle, Gordon Wentworth
Siebes, Maria
Spaziani, Eugene
Tade, William Howard
Welsh, Michael James
Woolley, Donald Grant
Wunder, Charles C(ooper)

KANSAS

Alper, Richard H
Balfour, William Mayo
Call, Edward Prior
Cheney, Paul David
Clancy, Richard L
Gillespie, Jerry Ray
Greenberger, Norton Jerald
Greenwald, Gilbert Saul
Hermreck, Arlo Scott
Michaelis, Mary Louise
Morrissey, J Edward
Noell, Werner K
Norris, Patricia Ann
Paulsen, Gary Melvin
Pickrell, John A
Ringle, David Allan
Shirer, Hampton Whiting
Smith, Joseph Emmitt
Thompson, Alan Morley
Trank, John W
Upson, Dan W
Urban, James Edward
Valenzeno, Dennis Paul
Yochim, Jerome M

KENTUCKY

Anderson, Gary Lee
Bailey, Donald Wycoff
Bennett, Thomas Edward
Bowen, Thomas Earle, Jr
Chappell, Guy Lee Monty
Crawford, Eugene Carson, Jr
Diana, John N
Dixon, Wallace Clark, Jr
Douglas, Robert Hazard
Fell, Ronald Dean
Frazier, Donald Tha
Gailey, Franklin Bryan
Getchell, Thomas V
Gillilan, Lois Adell
Gordon, Helmut Albert
Huang, Kee-Chang
Humphreys, Wallace F
Jimenez, Agnes E
Jokl, Ernst
Kraman, Steve Seth
Lee, Lu-Yuan
Legan, Sandra Jean
Moore, James Carlton
Musacchia, X J
Parkins, Frederick Milton
Passmore, John Charles
Peretz, Bertram
Peterson, Roy Phillip
Ramp, Warren Kibby
Randall, David Clark
Richardson, David Ray
Schilb, Theodore Paul
Thornton, Paul A
Walker, Sheppard Matthew
Wekstein, David Robert
Wiegman, David L
Wyssbrod, Herman Robert
Zechman, Frederick William, Jr

LOUISIANA

Anderson, William McDowell
Arimura, Akira
Beard, Elizabeth L
Bhattacharyya, Ashim Kumar
Cullen, John Knox
Davis, George Diament
Falcon, Carroll James
Flournoy, Robert Wilson
Franklin, Beryl Cletis
Gasser, Raymond Frank
George, Ronald Baylis
Granger, D Niel
Harper, Jon William
Harrison, Lura Ann
Hayes, Donald H

Heneghan, James Beyer
Hyman, Edward Sidney
Klyce, Stephen Downing
Kreider, Jack Leon
Liles, Samuel Lee
McAfee, Robert Dixon
McDonough, Kathleen H
McElroy, William Tyndell, Jr
Mayerson, Hymen Samuel
Meier, Albert Henry
Miller, William Wadd, III
Morrissette, Maurice Corlette
Nance, Francis Carter
Navar, Luis Gabriel
Newman, Wiley Clifford, Jr
Olmsted, Clinton Albert
Petri, William Henry, III
Porter, Charles Warren
Porter, Johnny Ray
Randall, Howard M
Ricks, Beverly Lee
Robinson, George Edward, Jr
Robinson, Roy Garland, Jr
Roheim, Paul Samuel
Roussel, Joseph Donald
Sloop, Charles Henry
Spanier, Arthur M
Spitzer, John J
Vela, Adan Richard
Wallin, John David
Weill, Hans

MAINE

Bennett, Miriam Frances
Bishop, David Wakefield
Bredenberg, Carl Eric
Chute, Robert Maurice
Gelinas, Douglas Alfred
Hiatt, Robert Burritt
Hogben, Charles Adrian Michael
Logan, Rowland Elizabeth
Lycette, R(ichard) (Milton)
Major, Charles Walter
Sidell, Bruce David
Twarog, Betty Mack

MARYLAND

Abbrecht, Peter H
Adelman, William Joseph, Jr
Ahrens, Richard August
Alexander, Nancy J
Altland, Paul Daniel
Anderson, James Howard
Anderson, Norman Gulack
Baker, George Thomas, III
Baker, Houston Richard
Balaban, Robert S
Baum, Siegmund Jacob
Beall, James Robert
Benjamin, Fred Berthold
Benton, Allen William
Biersner, Robert John
Blackman, Marc Roy
Bodammer, Joel Edward
Bolt, Douglas John
Bricker, Jerome Gough
Brock, Mary Anne
Bromberger-Barnea, Baruch (Berthold)
Brusilow, Saul W
Canonico, Peter Guy
Chaudhari, Anshumali
Choudary, Jasti Bhaskararao
Coble, Anna Jane
Covell, David Gene
Craig, Francis Northrop
Critz, Jerry B
Cullen, Joseph Warren
Cummings, Edmund George
Dalton, John Charles
Darby, Eleanor Muriel Kapp
Dasler, Adolph Richard
Delahunty, George
Eichholz, Alexander
Ewing, Larry Larue
Fajer, Abram Bencjan
Ferguson, Frederick Palmer
Fitzgerald, Robert Schaefer
Freas, William
Gamble, James Lawder, Jr
Georgopoulos, Apostolos P
Gilbert, Daniel Lee
Glaser, Edmund M
Gonzalez-Fernandez, Jose Maria
Greenbaum, Leon J, Jr
Greisman, Sheldon Edward
Griffo, Zora Jasincuk
Grollman, Sigmund
Hallfrisch, Judith
Hamilton, Clara Eddy
Hammer, John A
Hassett, Charles Clifford
Hegyeli, Ruth I E J
Hellman, Alfred
Hertz, Roy
Higgins, William Joseph
Himwich, Williamina (Elizabeth)
 Armstrong
Hoffman, David J
Horton, Richard Greenfield
Jackson, Michael J
Jaeger, James J
Johnson, Lawrence Arthur
Joy, Robert John Thomas
Jurf, Amin N

Kafka, Marian Stern
Kalberer, John Theodore, Jr
Kaplan, Ann Esther
Kenimer, James G
Ketchel, Melvin M
Kety, Seymour S
Kiley, James P
King, Theodore Matthew
Kinnamon, Kenneth Ellis
Koehler, Raymond Charles
Kunos, George
Lall, Abner Bishamber
Lederer, William Jonathan
Lee-Ham, Doo Young
LeRoith, Derek
Levinson, John Z
Levy, Alan C
Litten, Raye Z, III
Liu, Ching-Tong
Lo, Chu Shek
Lowy, R Joel
Lymn, Richard Wesley
McCormick, Kathleen Ann
Maciag, Thomas Edward
Maddox, Yvonne T
Maloney, Peter Charles
Marban, Eduardo
Marwah, Joe
Mercado, Teresa I
Meyer, Richard Arthur
Mitchell, Thomas George
Moran, Walter Harrison, Jr
Morgan, Raymond P
Mountcastle, Vernon Benjamin
Mullins, Lorin John
Murano, Genesio
Newball, Harold Harcourt
Orahovats, Peter Dimiter
Orloff, Jack
Paape, Max J
Packard, Barbara B K
Pamnani, Motilal Bhagwandas
Patanelli, Dolores J
Pierce, Sidney Kendrick
Pinter, Gabriel George
Podolsky, Richard James
Pursel, Vernon George
Quimby, Freeman Henry
Rall, Joseph Edward
Rapoport, Stanley I
Raub, William F
Rawlings, Samuel Craig
Redick, Thomas Ferguson
Rexroad, Caird Eugene, Jr
Reynafarje, Baltazar
Robinson, David Adair
Rose, John Charles
Rosenstein, Laurence S
Roth, Jan Jean
Rovelstad, Gordon Henry
Sagawa, Kiichi
Salcman, Michael
San Antonio, James Patrick
Sastre, Antonio
Scherbenske, M James
Schmidt, Edward Matthews
Schoenberg, Mark
Schweizer, Malvina
Scott, William Wallace
Scow, Robert Oliver
Shamoo, Adil E
Shields, Jimmie Lee
Shock, Nathan Wetherill
Siebens, Arthur Alexandre
Siegel, John H
Sokoloff, Louis
Solomon, Neil
Specht, Heinz
Striker, G E
Triantaphyllopoulos, Demetrios
Trusal, Lynn R
Varma, Shambhu D
Vollmer, Erwin Paul
Wade, James B
Waltrup, Paul John
Weinberg, Robert P
Wier, Withrow Gil
Williams, Robert Jackson
Yau, King-Wai
Zaharko, Daniel Samuel
Zierler, Kenneth

MASSACHUSETTS
Adams, Jack H(erbert)
Arvan, Peter
Barger, A(braham) Clifford
Barkman, Robert Cloyce
Bass, David Eli
Beasley, Debbie Sue
Berman, Herbert Joshua
Biggers, John Dennis
Bizzi, Emilio
Blake, Thomas R
Bloomquist, Eunice
Boisse, Norman Robert
Botticelli, Charles Robert
Brain, Joseph David
Brown, Robert Glenn
Bruce, John Irvin
Burggren, Warren William
Burse, Richard Luck
Chang, Min Chueh
Chidsey, Jane Louise
Clowes, George Henry Alexander, Jr

Cochrane, David Earle
Cole, Edward Anthony
Covino, Benjamin Gene
Cymerman, Allen
Czeisler, Charles Andrew
Dane, Benjamin
Dayal, Yogeshwar
Deen, William Murray
Epstein, Franklin Harold
Essig, Alvin
Fay, Fredric S
Freadman, Marvin Alan
Gaensler, Edward Arnold
Ganley, Oswald Harold
Gifford, Cameron Edward
Gissen, Aaron J
Goldberg, Alfred L
Goldman, David Eliot
Goldman, Ralph Frederick
Gonzalez, Richard Rafael
Goodman, Henry Maurice
Graves, William Earl
Green, Howard
Guimond, Robert Wilfrid
Hardy, William Lyle
Hastings, John Woodland
Hedley-Whyte, John
Heglund, Norman C
Hilden, Shirley Ann
Hodgson, Edward Shilling
Howe, George R
Ingram, Roland Harrison
Irwin, Richard Stephen
Ishikawa, Sadamu
Jackson, Benjamin T
Jaffe, Lionel F
Jain, Rakesh Kumar
Johnson, Elsie Ernest
Kamen, Gary P
Kaminer, Benjamin
Kayne, Herbert Lawrence
Kazemi, Homayoun
Keeney, Clifford Emerson
Kolka, Margaret A
Lazarte, Jaime Esteban
Leavis, Paul Clifton
Leeman, Susan Epstein
LeFevre, Marian E Willis
LeFevre, Paul Green
Li, Jeanne B
Libby, Peter
Lipinski, Boguslaw
Liscum, Laura
Little, John Bertram
Loewenfeld, Irene Elizabeth
Loring, Stephen H
Marcus, Elliot M
Mayer, Jean
Meldon, Jerry Harris
Milburn, Nancy Stafford
Millard, William James
Morgan, Kathleen Greive
Mulvey, Philip Francis, Jr
O'Connor, William Brian
Orlidge, Alicia
Pandolf, Kent Barry
Pappenheimer, John Richard
Patton, John F
Pearincott, Joseph V
Prestwich, Kenneth Neal
Previte, Joseph James
Procaccini, Donald J
Pryor, Marilyn Ann Zirk
Quine, Willard V
Reichlin, Seymour
Resnick, Oscar
Ricci, Benjamin
Roberts, John Lewis
Roberts, John Stephen
Rock, Paul Bernard
Romanoff, Elijah Bravman
Rosemberg, Eugenia
Scheid, Cheryl Russell
Schwartz, Bernard
Schwartz, John H
Scott, George Taylor
Shipley, Reginald A
Slechta, Robert Frank
Smith, Curtis Griffin
Snedecor, James George
Solomon, Jolane Baumgarten
Stephenson, Lou Ann
Szlyk, Patricia Carol
Taylor, Charles Richard
Taylor, Robert E
Tosteson, Daniel Charles
Ullrick, William Charles
Vogel, James Alan
Waggener, Thomas Barrow
Walkowiak, Edmund Francis
Walsh, John V
Wilgram, George Friederich
Wilson, Thomas Hastings
Zetter, Bruce Robert

MICHIGAN
Adams, Thomas
Aiken, James Wavell
Barnhart, Marion Isabel
Be Ment, Spencer L
Bernard, Rudy Andrew
Blackwell, Leo Herman
Bloom, Miriam
Bradley, Robert Martin

Briggs, Josephine P
Buhl, Allen Edwin
Burlington, Roy Frederick
Butterworth, Francis M
Carter-Su, Christin
Chimoskey, John Edward
Christiansen, Richard Louis
Churchill, Paul Clayton
Cohen, Leonard Arlin
Collins, Robert James
Cooper, Theodore
D'Alecy, Louis George
Dawson, David Charles
Dawson, William Ryan
Deuben, Roger R
Dillon, Patrick Francis
Dukelow, W Richard
England, Barry Grant
Ernst, Stephen Arnold
Faulkner, John A
Ferguson-Miller, Shelagh Mary
Fernandez-Madrid, Felix
Fernandez y Cossio, Hector Rafael
Foa, Piero Pio
Foster, Douglas Layne
Freedman, Robert Russell
Frohman, Charles Edward
Fromm, Paul Oliver
Gala, Richard R
Garvin, Jeffrey Lawrence
Gelderloos, Orin Glenn
Gerritsen, George Contant
Gordon, Sheldon Robert
Hansen-Smith, Feona May
Hatton, Glenn Irwin
Heffner, Thomas G
Henry, Raymond Leo
Horowitz, Samuel Boris
Howatt, William Frederick
Hsu, Chen-Hsing
Jackson, William F
Jacquez, John Alfred
Jochim, Kenneth Erwin
Johnson, Shirley Mae
Johnston, Raymond F
Kaldor, George
Kluger, Matthew Jay
Kostyo, Jack Lawrence
Kratochvil, Clyde Harding
Lauderdale, James W, Jr
Lewis, Benjamin Marzluff
Lindemann, Charles Benard
McCoy, Lowell Eugene
Malvin, Richard L
Marshall, Norman Barry
Meites, Joseph
Menge, Alan C
Meyer, Ronald Anthony
Mitchell, Jerald Andrew
Morrow, Thomas John
Nelson, Eldon Lane, Jr
Neudeck, Lowell Donald
Noe, Frances Elsie
Pax, Ralph A
Penney, David George
Piercey, Montford F
Ram, Jeffrey L
Reynolds, Orland Bruce
Riegle, Gail Daniel
Rillema, James Alan
Ringer, Robert Kosel
Rutledge, Lester T
Schneidkraut, Marlowe J
Schwartz, Jessica
Shansky, Michael Steven
Shepard, Robert Stanley
Sherman, James H
Silbergleit, Allen
Steiman, Henry Robert
Stockwell, Charles Warren
Stout, John Frederick
Swanson, Curtis James
Trachtenberg, Michael Carl
Van Harn, Gordon L
Van Huss, Wayne D
Walz, Daniel Albert
Weg, John Gerard
White, Timothy P
Whitten, Bertwell Kneeland
Williams, John Andrew
Williams, William James
Winbury, Martin M
Winkler, Barry Steven
Wittle, Lawrence Wayne
Wood, Jack Sheehan

MINNESOTA
Alsum, Donald James
Bacaner, Marvin Bernard
Bache, Robert James
Beck, Kenneth Charles
Benson, Katherine Alice
Carter, Earl Thomas
Delaney, John P
Donaldson, Sue Karen
Dousa, Thomas Patrick
Eaton, John Wallace
Frederick, Edward C
Gallant, Esther May
Ganguli, Mukul Chandra
Gebhard, Roger Lee
Godec, Ciril J
Graham, Edmund F
Graubard, Mark Aaron

Grim, Eugene
Gunst, Susan Jane
Haas, John Arthur
Haller, Edwin Wolfgang
Hancock, Peter Adrian
Housmans, Philippe Robert H P
Humphrey, Edward William
Jankus, Edward Francis
Johnson, Ivan M
Johnson, John Alexander
Johnson, Vincent Arnold
Kane, William J
Keys, Ancel (Benjamin)
Knox, Franklin G
Levitt, David George
Lifson, Nathan
Mabry, Paul Davis
Meyer, Maurice Wesley
Michels, Lester David
Purple, Richard L
Rehder, Kai
Robb, Richard A
Shepherd, John Thompson
Taylor, Stuart Robert
Wallace, Kendall B
Wangensteen, Ove Douglas
Welter, Alphonse Nicholas
Wood, Earl Howard
Zanjani, Esmail Dabaghchian

MISSISSIPPI
Ashburn, Allen David
Bearden, Henry Joe
Douglas, Ben Harold
Dzielak, David J
Granger, Joey Paul
Guyton, Arthur Clifton
Hall, John Edward
Hester, Robert Leslie
Jennings, David Phipps
Langford, Herbert Gaines
Manning, R Davis, Jr
Mehendale, Harihara Mahadeva
Pace, Henry Buford
Russell, Raymond Alvin
Turner, Manson Don
Venkataramiah, Amaraneni
Walker, James Frederick
Young, David Bruce

MISSOURI
Baile, Clifton Augustus, III
Biellier, Harold Victor
Blount, Don Houston
Bond, Guy Hugh
Boyarsky, Saul
Brown, Herbert Ensign
Chaplin, Susan Budd
Collier, Robert Joseph
Cook, Mary Rozella
Cornell, Creighton N
Coyer, Philip Exton
Csaky, Tihamer Zoltan
Dahms, Thomas Edward
Davis, James Othello
Eldredge, Donald Herbert
Ellsworth, Mary Litchfield
Fickess, Douglas Ricardo
Fleming, Warren R
Forker, E Lee
Gaddis, Monica Louise
Graham, Charles
Griggs, Douglas M, Jr
Hessler, Jack Ronald
Holloszy, John O
Howlett, Allyn C
Jen, Philip Hungsun
Johnson, Harold David
Jones, Allan W
Keane, John Francis, Jr
Kim, Hyun Dju
Krukowski, Marilyn
Lenhard, James M
Lind, Alexander R
Longley, William Joseph
Martin, Richard Harvey
Mehrle, Paul Martin, Jr
Merrick, Arthur West
Mock, Orin Bailey
Morrill, Callis Gary
Mueckler, Mike Max
Peissner, Lorraine C
Platner, Wesley Stanley
Polakoski, Kenneth Leo
Rich, Travis Dean
Roos, Albert
Rovetto, Michael Julien
Rubin, Maryiln Bernice
Ruh, Mary Frances
Russell, Robert Lee
Sage, Martin
Savery, Harry P
Schadt, James C
Senay, Leo Charles, Jr
Stahl, Philip Damien
Suga, Nobuo
Van Beaumont, Karel William
Waite, Albert B
Walsh, Raymond Robert
Wang, Bin Ching
Welling, Larry Wayne
Winter, Henry Frank, Jr
Zatzman, Marvin Leon
Zepp, Edwin Andrew

Physiology, General (cont)

Garren, Henry Wilburn
Garrett, Ruby Joyce Burriss
Glenn, Thomas M
Gutknecht, John William
Hayek, Dean Harrison
Hofmann, Lorenz M
Horres, Charles Russell, Jr
Humm, Douglas George
Hutchins, Phillip Michael
Jobsis, Frans Frederik
Kokas, Eszter B
Kootsey, Joseph Mailen
Kylstra, Johannes Arnold
Lazarus, Lawrence H
Lieberman, Edward Marvin
Lieberman, Melvyn
Longmuir, Ian Stewart
Louie, Dexter Stephen
Mandel, Lazaro J
Menhinick, Edward Fulton
Mikat, Eileen M
Miller, Augustus Taylor, Jr
Miller, Inglis J, Jr
Mills, Elliott
Morse, Roy E
Mouw, David Richard
Murdock, Harold Russell
Negro-Vilar, Andres F
Nusser, Wilford Lee
O'Neil, John J
Opdyke, David F
Pallotta, Barry S
Parker, John Curtis
Patterson, David Thomas
Perl, Edward Roy
Proffit, William R
Robinson, Jerry Allen
Robison, Odis Wayne
Rose, James C
Sar, Madhabananda
Sargent, Frank Dorrance
Schmidt-Nielsen, Knut
Shinkman, Paul G
Skinner, Newman Sheldon, Jr
Smith, Donald Eugene
Smith, Thomas Lowell
Somjen, George G
Swan, Algernon Gordon
Talmage, Roy Van Neste
Thurber, Robert Eugene
Townes, Mary McLean
Tucker, Vance Alan
Ulberg, Lester Curtiss
Vogel, Steven
Waugh, William Howard
Welsch, Frank
Wilbur, Karl Milton
Wooles, Wallace Ralph
Yonce, Lloyd Robert

NORTH DAKOTA
Akers, Thomas Kenny
Brumleve, Stanley John
Davison, Kenneth Lewis
Duerr, Frederick G
Gerst, Jeffery William
Joshi, Madhusudan Shankarrao
Lukaski, Henry Charles
Rose, Richard Carrol
Saari, Jack Theodore

OHIO
Adams, Walter C
Adragna, Norma C
Appert, Hubert Ernest
Bajpai, Praphulla K
Banks, Robert O
Banwell, John G
Barlow, George
Barr, Harry L
Bhattacharya, Amar Nath
Bozler, Emil
Brand, Paul Hyman
Bryant, Shirley Hills
Budd, Geoffrey Colin
Burns, Elizabeth Mary
Cerroni, Rose E
Chakraborty, Jyotsna (Joana)
Clark, David Lee
Connors, Alfred F, Jr
Cooke, Helen Joan
Corson, Samuel Abraham
Daroff, Robert Barry
Dujardin, Jean-Pierre L
Ely, Daniel Lee
Ferguson, Marion Lee
Ferrario, Carlos Maria
Foulkes, Ernest Charles
Freed, James Melvin
Fry, Donald Lewis
Garton, David Wendell
Geha, Alexander Salim
Gotshall, Robert William
Greenwald, Lewis
Grossie, James Allen
Grupp, Gunter
Grupp, Ingrid L
Hall, Philip Wells, III
Handwerger, Stuart
Hanson, Kenneth Marvin
Hassler, Craig Reinhold
Holtkamp, Dorsey Emil
Horseman, Nelson Douglas
Hoshiko, Tomuo

Howell, John N
Johnson, Dale Richard
Jones, Patricia H
Jones, Richard Dell
Jung, Dennis William
Khairallah, Philip Asad
Kirby, Albert Charles
Kleinerman, Jerome
Kline, Daniel Louis
Krontiris-Litowitz, Johanna Kaye
Kunz, Albert L
Lauf, Peter Kurt
Lessler, Milton A
Levy, Matthew Nathan
Lewis, Lena Armstrong
Lindley, Barry Drew
Linkenheimer, Wayne Henry
Lipsky, Joseph Albin
Lustiok, Sheldon Irving
McGrady, Angele Vial
Macht, Martin Benzyl
Maron, Michael Brent
Martin, Paul Joseph
Meserve, Lee Arthur
Metting, Patricia J
Miles, Daniel S
Mostardi, Richard Albert
Naito, Herbert K
Nathan, Paul
Neiman, Gary Scott
Nelson, Leonard
Newman, Howard Abraham Ira
Nielson, Read R
Noyes, David Holbrook
Nussbaum, Noel Sidney
Olson, Lynne E
Patil, Popat N
Paul, Lawrence Thomas
Paul, Richard Jerome
Peterjohn, Glenn William
Recknagel, Richard Otto
Robbins, David O
Rudy, Yoram
Saiduddin, Syed
Scarpa, Antonio
Senturia, Jerome B(asil)
Sernka, Thomas John
Slonim, Arnold Robert
Smith, Charles Welstead
Sperelakis, Nick
Stokes, Robert Mitchell
Stone, Kathleen Sexton
Strohl, Kingman P
Stuesse, Sherry Lynn
Sutherland, James McKenzie
Thomson, Dale S
Tomashefski, Joseph Francis
Travis, Randall Howard
Webb, Paul
Wideman, Cyrilla Helen
Wilson, James Albert
Wilson, Kenneth Glade
Witters, Weldon L
Wolgemuth, Richard Lee
Wood, Jackie Dale

OKLAHOMA
Armstrong, Robert Beall
Beames, Calvin G, Jr
Beesley, Robert Charles
Benyajati, Siribhinya
Blair, Robert William
Bogenschutz, Robert Parks
Cooper, Richard Grant
Cox, Beverley Lenore
DeLacerda, Fred G
Dowell, Russell Thomas
Harrison, Aix B
Higgins, E Arnold
Hinshaw, Lerner Brady
Houlihan, Rodney T
Hurst, Jerry G
Johnson, Becky Beard
Lovallo, William Robert
McDonald, Leslie Ernest
Maton, Paul Nicholas
Mobley, Bert A
Newcomer, Wilbur Stanley
Olson, Robert Leroy
Oyler, J Mack
Shirley, Barbara Anne
Thomas, Sarah Nell
Wickham, M Gary

OREGON
Anderson, Debra F
Bennett, Robert M
Carleton, Blondel Henry
Chilgren, John Douglas
Conte, Frank Philip
Crawshaw, Larry Ingram
Debons, Albert Frank
Dejmal, Roger Kent
Dow, Robert Stone
Faber, Jan Job
Fang, Ta-Yun
Hardt, Alfred Black
Hillman, Stanley Severin
Hohimer, A Roger
Hoskins, Dale Douglas
Johnson, Larry Reidar
Keyes, Jack Lynn
Kleinholz, Lewis Hermann
Levine, Leonard

Lucas, Oscar Nestor
Mela-Riker, Leena Marja
Peterson, Clare Gray
Rampone, Alfred Joseph
Resko, John A
Richards, Oscar White
Rudy, Paul Passmore, Jr
Shannon, James Augustine
Somero, George Nicholls
Stones, Robert C
Stouffer, Richard Lee
Tanz, Ralph
Van Hassel, Henry John
Weimar, Virginia Lee
Worrest, Robert Charles
Yatvin, Milton B
Zauner, Christian Walter

PENNSYLVANIA
Agersborg, Helmer Pareli Kjerschow, Jr
Agus, Zalman S
Ambromovage, Anne Marie
Angelakos, Evangelos Theodorou
Anthony, Adam
Armstrong, Clay M
Ashton, Francis T
Baechler, Charles Albert
Becker, Anthony J, Jr
Bianchi, Carmine Paul
Borle, Andre Bernard
Brobeck, John Raymond
Brooks, Frank Pickering
Butler, Thomas Michael
Civan, Mortimer M
Coburn, Ronald F
Cox, Robert Harold
Cristofalo, Vincent Joseph
Cutler, Winnifred Berg
Davies, Robert Ernest
DeBias, Domenic Anthony
Detweiler, David Kenneth
De Weer, Paul Joseph
Doghramji, Karl
Drees, John Allen
Driska, Steven P
Dunson, William Albert
Edwards, McIver Williamson, Jr
English, Leigh Howard
Fawley, John Philip
Fischer, Grace Mae
Fishman, Alfred Paul
Flaim, Kathryn Erskine
Flemister, Launcelot Johnson
Flemister, Sarah C
Flickinger, George Latimore, Jr
Fluck, Richard Allen
Forster, Robert E, II
Fox, Karl Richard
Freeman, Alan R
Friedman, M H
Fritz, George Richard, Jr
Fuchs, Franklin
Gay, Carol Virginia Lovejoy
Gee, William
Gellai, Miklos
Glauser, Elinor Mikelberg
Goldberg, Martin
Goldman, Yale E
Greene, Charlotte Helen
Grego, Nicholas John
Hammerstedt, Roy H
Hart, Robert Gerald
Hartman, Arthur Dalton
Hasler, Marilyn Jean
Hausberger, Franz X
Henderson, George Richard
Herbison, Gerald J
Hitner, Henry William
Holliday, Charles Walter
Hollis, Theodore M
Horn, Lyle William
Idell-Wenger, Jane Arlene
Jaweed, Mazher
Jefferson, Leonard Shelton
Johnson, Ernest Walter
Kamon, Eliezer
Kao, Race Li-Chan
Kare, Morley Richard
Killian, Gary Joseph
Kimbel, Philip
Knuttgen, Howard G
Lahiri, Sukhamay
Langan, William Bernard
Lentini, Eugene Anthony
Levin, Sidney Seamore
Levine, Elliot Myron
Lindgren, Clark Allen
Loeb, Alex Lewis
Loewy, Ariel Gideon
Lowery, Thomas J
Lynch, Peter Robin
Magno, Michael Gregory
Martin, John Samuel
Mashaly, Magdi Mohamed
Michelson, Eric L
Minsker, David Harry
Mir, Ghulam Nabi
Mitchell, Robert Bruce
Moore, E(arl) Neil
Morgan, Howard E
Mortimore, Glenn Edward
Neff, William H
Newell, Allen
Peachey, Lee DeBorde

Pegg, Anthony Edward
Pelleg, Amir
Piscitelli, Joseph
Pitt, Bruce R
Polgar, George
Post, Robert Lickely
Rannels, Donald Eugene, Jr
Reibel-Shinfeld, Diane Karen
Reinking, Larry Norman
Roberts, Shepherd (Knapp de Forest)
Robishaw, Janet D
Rosenfeld, Leonard M
Salama, Guy
Schauer, Richard C
Senft, Joseph Philip
Shaffer, Thomas Hillard
Shapiro, Herbert
Shiau, Yih-Fu
Sidic, James Michael
Siegfried, John Barton
Siegman, Marion Joyce
Sigg, Ernest Beat
Smith, M Susan
Spear, Joseph Francis
Tansy, Martin F
Taylor, Paul M
Thomas, Steven P
Tong, Winton
Torchiana, Mary Louise
Torres, Joseph Charles
Trainer, David Gibson
Upton, G Virginia
Vagnucci, Anthony Hillary
Van Inwegen, Richard Glen
Wardell, Joe Russell, Jr
Wasserman, Martin Allan
Waterhouse, Barry D
Watrous, James Joseph
Weber, Annemarie
Weiser, Philip Craig
Weisz, Judith
Wesson, Laurence Goddard, Jr
Whinnery, James Elliott
Whitfield, Carol F(aye)
Wickersham, Edward Walker
Winegrad, Saul
Wolf, Stewart George, Jr
Wolfersberger, Michael Gregg
Wolken, Jerome Jay
Woychik, John Henry
Zabara, Jacob
Zavodni, John J

RHODE ISLAND
Constantine, Herbert Patrick
Cserr, Helen F
Degnan, Kevin John
Dolyak, Frank
Galletti, Pierre Marie
Goldstein, Leon
Harrison, Robert William
Hill, Robert Benjamin
Jackson, Donald Cargill
Lehmkuhle, Stephen W
Marshall, Jean McElroy
Stewart, Peter Arthur

SOUTH CAROLINA
Allen, Thomas Charles
Bond, Robert Franklin
Cole, Benjamin Theodore
Colwell, John Amory
Cook, James Arthur
Cooper, George, IV
Cromer, Jerry Haltiwanger
Fairbanks, Gilbert Wayne
Frayser, Katherine Regina
Fredericks, Christopher M
Fulton, George P(earman)
Hays, Ruth Lanier
Hempling, Harold George
Horres, Alan Dixon
Hughes, Buddy Lee
Katz, Sidney
Kinard, Fredrick William
Knight, Anne Bradley
Leonard, Walter Raymond
McCutcheon, Ernest P
McNamee, James Emerson
Pivorun, Edward Broni
Runey, Gerald Luther
Sias, Fred R, Jr
Smiley, James Watson
Taber, Elsie
Tempel, George Edward
Watson, Philip Donald
Wheeler, Alfred Portius
Wheeler, Darrell Deane
William, James C, Jr
Wolf, Matthew Bernard

SOUTH DAKOTA
Diggins, Maureen Rita
Goodman, Barbara Eason
Hastings, David Frank
Hietbrink, Bernard Edward
Johnson, Leland Gilbert
Schlenker, Evelyn Heymann
Schramm, Mary Arthur
Slyter, Arthur Lowell

TENNESSEE
Ahokas, Robert A
Armstead, William M

Physiology, General (cont)

Arnold, William Archibald
Bagby, Roland Mohler
Banerjee, Mukul Ranjan
Caldwell, Robert William
Chen, Thomas Tien
Coburn, Corbett Benjamin, Jr
Cook, John Samuel
Corbin, Jack David
Coulson, Patricia Bunker
Davis, William James
Fitzgerald, Laurence Rockwell
Friesinger, Gottlieb Christian
Fry, Richard Jeremy Michael
Ginski, John Martin
Groer, Maureen
Hall, Hugh David
Hardman, Joel G
Harris, Thomas R(aymond)
Johnson, Leonard Roy
Joyner, William Lyman
Kant, Kenneth James
Kono, Tetsuro
Lawrence, William Homer
Leffler, Charles William
Liles, James Neil
MacFadden, Donald Lee
McGuinness, Owen P
Mahajan, Satish Chander
Manley, Emmett S
Manthey, Arthur Adolph
Meng, H C
Moses, Henry A
Neff, Robert Jack
Park, Charles Rawlinson
Rasch, Robert
Reynolds, Leslie Boush, Jr
Rhamy, Robert Keith
Sander, Linda Dian
Schneider, Edward Greyer
Share, Leonard
Stahlman, Mildred
Stiles, Robert Neal
Stinson, Joseph McLester
Thakar, Jay H
Thomason, Donald Brent
Van Middlesworth, Lester
Williams, Carole A
Wiser, Cyrus Wymer
Wondergem, Robert
Wood, Henderson Kingsberry
Wood, William Booth

TEXAS

Alkadhi, Karim A
Allen, Julius Cadden
Allen, Thomas Hunter
Andia-Waltenbaugh, Ana Maria
Angelides, Kimon Jerry
Armstrong, George Glaucus, Jr
Ashton, Juliet H
Baker, Lee Edward
Baker, Robert David
Baumgardner, F Wesley
Beard, James B
Bishop, Jack Garland
Bishop, Vernon S
Blomqvist, Carl Gunnar
Brewer, Franklin Douglas
Bristol, John Richard
Brodwick, Malcolm Stephen
Bronson, Franklin Herbert
Brown, Arthur Morton
Brown, Jack Harold Upton
Buderer, Melvin Charles
Bullard, Truman Robert
Burns, John Mitchell
Butcher, Reginald William
Butler, Bruce David
Byrne, John Howard
Campbell, Bonnalie Oetting
Carnes, David Lee, Jr
Castracane, V Daniel
Castro, Gilbert Anthony
Chang, Donald Choy
Chowdhury, Ajit Kumar
Clark, James Henry
Cooper, William Anderson
Dietlein, Lawrence Frederick
Dill, Russell Eugene
Dowben, Robert Morris
Eddy, Carlton Anthony
Fife, William Paul
Fishman, Harvey Morton
Fletcher, John Lynn
Franklin, Thomas Doyal, Jr
Frishman, Laura J(ean)
Glantz, Raymon M
Green, Gary Miller
Grundy, Scott Montgomery
Guentherman, Robert Henry
Guo, Yan-Shi
Gutierrez, Guillermo
Hall, Charles Eric
Hamilton, Terrell Hunter
Hatch, William James
Hazlewood, Carlton Frank
Herd, James Alan
Herlihy, Jeremiah Timothy
Hester, Richard Kelly
Hettinger, Deborah D R
Hightower, Nicholas Carr, Jr
Hill, Ronald Stewart

Hoage, Terrell Rudolph
Holly, Frank Joseph
Horger, Lewis Milton
Horne, Francis R
Hughes, Maysie J H
Humphrey, Ronald DeVere
Hupp, Eugene Wesley
James, Harold Lee
Johnson, John Marshall
Johnson, Robert Lee
Jordan, Paul H, Jr
Kallus, Frank Theodore
Kalu, Dike Ndukwe
Kasschau, Margaret Ramsey
Kellaway, Peter
Kitay, Julian I
Knobil, Ernst
Korr, Irvin Morris
Kraemer, Duane Carl
Kraus-Friedmann, Naomi
Krise, George Martin
Krock, Larry Paul
Kronenberg, Richard Samuel
Langston, Jimmy Byrd
Leach-Huntoon, Carolyn S
Lehmkuhl, L Don
Leroy, Robert Frederick
Lewis, Simon Andrew
Liepa, George Uldis
Lipton, James Matthew
Little, Perry L
Livingston, Linda
Loftus, Joseph P, Jr
Lott, James Robert
Lutherer, Lorenz O
McCarter, Roger John Moore
McCarthy, John Lawrence, Jr
McDonald, Harry Sawyer
McGrath, James Joseph
Maclean, Graeme Stanley
Masoro, Edward Joseph
Maxwell, Leo C
Meininger, Gerald A
Moldenhauer, Ralph Roy
Montgomery, Edward Harry
Mukherjee, Amal
Musgrave, F Story
Norris, William Elmore, Jr
Norwood, James S
O'Brien, Larry Joe
Ordway, George A
Orem, John
Peterson, Lysle Henry
Poenie, Martin Francis
Porter, John Charles
Pratt, David R
Preslock, James Peter
Raven, Peter Bernard
Reid, Michael Baron
Rogers, Walter Russell
Rosborough, John Paul
Russell, John McCandless
Sallee, Verney Lee
Sant'Ambrogio, Giuseppe
Schrank, Auline Raymond
Schuhmann, Robert Ewald
Schultz, Stanley George
Schwalm, Fritz Ekkehardt
Shade, Robert Eugene
Shepherd, Albert Pitt, Jr
Sikes, James Klingman
Siler-Khodr, Theresa M
Simmons, David J
Slack, Jim Marshall
Smatresk, Neal Joseph
Smith, Edwin Lee
Smith, Thomas Caldwell
Sordahl, Louis A
Sorensen, Elsie Mae (Boecker)
Soriero, Alice Ann
Srebro, Richard
Sybers, Harley D
Szilagyi, Julianna Elaine
Taurog, Alvin
Taylor, Alan Neil
Templeton, Gordon Huffine
Thompson, James Charles
Traber, Daniel Lee
Valentich, John David
Vanatta, John Crothers, III
Vick, Robert Lore
Wall, Malcolm Jefferson, Jr
Ward, Walter Frederick
Weisbrodt, Norman William
Weitlauf, Harry
White, Fred Newton
Widner, William Richard
Wildenthal, Kern
Williams, Fred Eugene
Williams, Robert Sanders
Wills, Nancy Kay
Wilson, Everett D
Wilson, Forest Ray, II
Wohlman, Alan
Wong, Brendan So
Woodward, Donald Jay
Wright, Kenneth C
Young, Margaret Claire

UTAH

Bloxham, Don Dee
Clark, C Elmer
Conlee, Robert Keith
Dixon, John Aldous

Galster, William Allen
Geddes, David Darwin
Graff, Darrell Jay
Heninger, Richard Wilford
Hibbs, John Burnham
Jaussi, August Wilhelm
Johnston, Stephen Charles
Kuida, Hiroshi
Parkin, James Lamar
Renzetti, Attilio D, Jr
Sanders, Raymond Thomas
Stringham, Reed Millington, Jr
Urry, Ronald Lee
Walker, John Lawrence, Jr
Warner, Homer R
Windham, Carol Thompson
Woodbury, John Walter

VERMONT

Alpert, Norman Roland
Bevan, Rosemary D
Chambers, Alfred Hayes
Davison, John (Amerpohl)
Foss, Donald C
Gennari, F John
Halpern, William
Hanson, John Sherwood
Heinrich, Bernd
Hendley, Edith Di Pasquale
Johnson, Robert Eugene
King, Patricia A
Low, Robert Burnham
McCrorey, Henry Lawrence
Mulieri, Berthann Scubon
Mulieri, Louis A
Nye, Robert Eugene, Jr
Parsons, Rodney Lawrence
Simmons, Kenneth Rogers

VIRGINIA

Baker, Donald Granville
Barrett, Richard John
Belardinelli, Luiz
Biber, Thomas U L
Bond, Judith
Bornmann, Robert Clare
Bowman, Edward Randolph
Briggs, Fred Norman
Brown, Barry Lee
Buikema, Arthur L, Jr
Cavender, Finis Lynn
Chevalier, Robert Louis
Conrad, Margaret C
Contreras, Thomas Jose
Costanzo, Richard Michael
Daniel, Joseph Car, Jr
De Long, Chester Wallace
Desjardins, Claude
Edwards, Leslie Erroll
Ford, George Dudley
Gewirtz, David A
Gourley, Desmond Robert Hugh
Greenfield, Wilbert
Hart, Jayne Thompson
Hinton, Barry Thomas
Hoptman, Julian
Howards, Stuart S
Howes, Cecil Edgar
Hundley, Louis Reams
Johnson, Joseph L
Kalimi, Mohammed Yahya
Kaul, Sanjiv
Kimbrough, Theo Daniel, Jr
Knight, James William
Kontos, Hermes A
Kory, Ross Conklin
Kutchai, Howard C
Leighton, Alvah Theodore, Jr
Lincicome, David Richard
McNabb, Roger Allen
Maio, Domenic Anthony
Makhlouf, Gabriel Michel
Mangum, Charlotte P
Menaker, Michael
Mengebier, William Louis
Mikulecky, Donald C
Murphy, Richard Alan
Osborne, Paul James
Peach, Michael Joe
Pinschmidt, Mary Warren
Pitts, Grover Cleveland
Poland, James Leroy
Price, Steven
Ridgway, Ellis Branson
Rochester, Dudley Fortescue
Saacke, Richard George
Scanlon, Patrick Francis
Sellers, Cletus Miller, Jr
Somlyo, Andrew Paul
Stillwell, Edgar Feldman
Suter, Daniel B
Szabo, Gabor
Szumski, Alfred John
Talbot, Richard Burritt
Tolles, Walter Edwin
Turner, Terry Tomo
Ulvedal, Frode
Van Krey, Harry P
Watlington, Charles Oscar
Wei, Enoch Ping
Wilson, John Drennan
Witorsch, Raphael Jay

WASHINGTON

Almers, Wolfhard
Anderson, Marjorie Elizabeth
Bassingthwaighte, James B
Bhansali, Praful V
Blinks, John Rogers
Bredahl, Edward Arlan
Burnell, James McIndoe
Campbell, Kenneth B
Catterall, William A
Cohen, Arthur Leroy
Detwiler, Peter Benton
Dickhoff, Walton William
Farner, Donald Sankey
Feigl, Eric O
Free, Michael John
Gale, Charles C, Jr
Gellert, Ronald J
Glomset, John A
Gordon, Albert McCague
Hanegan, James L
Hegyvary, Csaba
Heinle, Donald Roger
Hille, Bertil
Hornbein, Thomas F
Jackson, Kenneth Lee
Kaplan, Alex
Kastella, Kenneth George
Kirschner, Leonard Burton
Koch, Alan R
Koenig, Jane Quinn
Martin, Arthur Wesley
Matlock, Daniel Budd
Morrison, Peter Reed
Patton, Harry Dickson
Phillips, Richard Dean
Plisetskaya, Erika Michael
Pollack, Gerald H
Reid, Donald House
Rowell, Loring B
Scher, Allen Myron
Schoene, Robert B
Stien, Howard M
Stirling, Charles E
Su, Judy Ya-Hwa Lin
Taylor, Eugene M
Towe, Arnold Lester
Vaughan, Burton Eugene
Went, Hans Adriaan
Whatmore, George Bernard
White, Ronald Jerome
White, Thomas Taylor
Woods, Stephen Charles

WEST VIRGINIA

Aserinsky, Eugene
Aulick, Louis H
Boyd, James Edward
Castranova, Vincent
Cochrane, Robert Lowe
Dailey, Robert Arthur
Franz, Gunter Norbert
Hedge, George Albert
Irish, James McCredie, III
Kidder, Harold Edward
Larson, Gary Eugene
Lindsay, Hugh Alexander
Stauber, William Taliaferro

WISCONSIN

Bass, Paul
Benjamin, Hiram Bernard
Bittar, Evelyn Edward
Brittain, David B
Brooks, Jack Carlton
Brugge, John F
Carlson, Stanley David
Colás, Antonio E
Condon, Robert Edward
Cowley, Allen Wilson, Jr
Craig, Elizabeth Anne
Cvancara, Victor Alan
Daniels, Farrington, Jr
Dawson, Christopher A
Dennis, Warren Howard
Effros, Richard Matthew
Fahning, Melvyn Luverne
First, Neal L
Forster, Hubert Vincent
Gaspard, Kathryn Jane
Hall, Kent D
Harder, David Rae
Holden, James Edward
Kendrick, John Edsel
Kleinman, Jack G
Klitgaard, Howard Maynard
Kochan, Robert George
Lauson, Henry Dumke
Light, Douglas B
Lombard, Julian H
Macintyre, Bruce Alexander
Moss, Richard
Olsen, Ward Alan
Osborn, Jeffrey L
Pace, Marvin M
Plotka, Edward Dennis
Porth, Carol Mattson
Rankin, John
Rankin, John Horsley Grey
Reddan, William Gerald
Rickaby, David A
Rieselbach, Richard Edgar
Roman, Richard J
Sarna, Sushil K

Smith, Curtis Alan
Smith, James John
Smith, Luther Michael
Smith, Milton Reynolds
Snyder, Ann C
Staszak, David John
Stone, William Ellis
Stowe, David F
Terry, Leon Cass
Thibodeau, Gary A
Van Horn, Diane Lillian
Wen, Sung-Feng
Wolf, Richard Clarence
Woodson, Robert D

PUERTO RICO
De Mello, W Carlos
Fernández-Repollet, Emma D
García-Castro, Ivette
Santos-Martinez, Jesus
Vargas, Fernando Figueroa

ALBERTA
Baumber, John Scott
Beatty, David Delmar
Birdsell, Dale Carl
Cooper, Keith Edward
Doornenbal, Hubert
Famiglietti, Edward Virgil, Jr
Frank, George Barry
Hammond, Brian Ralph
Henderson, Ruth McClintock
Lauber, Jean Kautz
Lorscheider, Fritz Louis
Mears, Gerald John
Pang, Peter Kai To
Poznansky, Mark Joab
Schachter, Melville
Schneider, Wolfgang Johann
Thomas, Norman Randall
Veale, Warren Lorne

BRITISH COLUMBIA
Bates, David Vincent
Belton, Peter
Bressler, Bernard Harvey
Cottle, Walter Henry
Cramer, Carl Frederick
Cynader, Max Sigmund
Dehnel, Paul Augustus
Dorchester, John Edmund Carleton
Hahn, Peter
Hoar, William Stewart
Keeler, Ralph
Ledsome, John R
Leung, So Wah
Lioy, Franco
McLennan, Hugh
Noble, Robert Laing
Palaty, Vladimir
Perks, Anthony Manning
Phillips, John Edward
Quastel, D M J
Randall, David John
Sanders, Harvey David
Skala, Josef Petr
Thurlbeck, William Michael
Webber, William A
Wong, Norman L M

MANITOBA
Bodnaryk, Robert Peter
Carter, Stefan A
Chernick, Victor
Dandy, James William Trevor
Dhalla, Naranjan Singh
Faiman, Charles
Gaskell, Peter
Gerber, George Hilton
Kepron, Wayne
Lautt, Wilfred Wayne
Moorhouse, John A
Stephens, Newman Lloyd
Younes, Magdy K

NEW BRUNSWICK
Cook, Robert Harry
Cowan, F Brian M
Driedzic, William R
Saunders, Richard Lee

NEWFOUNDLAND
Mookerjea, Sailen
Nolan, Richard Arthur
Payton, Brian Wallace

NOVA SCOTIA
Armour, John Andrew
Dolphin, Peter James
Freeman, Harry Cleveland
Hatcher, James Donald
Howlett, Susan Ellen
Issekutz, Bela, Jr
Rautaharju, Pentti M
Wong, Alan Yau Kuen

ONTARIO
Ackles, Kenneth Norman
Andrew, George McCoubrey
Ashwin, James Guy
Atwood, Harold Leslie
Barclay, Jack Kenneth
Calaresu, Franco Romano
Campbell, Edward J Moran

Campbell, James
Cowan, John
Cummins, W(illiam) Raymond
Cunningham, David A
Dirks, John Herbert
Downie, Harry G
England, Sandra J
Ettinger, George Harold
Farrar, John Keith
Firstbrook, John Bradshaw
Fiser, Paul S(tanley)
Fritz, Irving Bamdas
Froese, Alison Barbara
Frost, Barrie James
George, John Caleekal
Girvin, John Patterson
Grayson, John
Grinstein, Sergio
Groom, Alan Clifford
Harding, Paul George Richard
Hoffman-Goetz, Laurie
Hoffstein, Victor
Jacobson, Stuart Lee
Jennings, Donald B
Johnston, Miles Gregory
Jones, Douglas L
Karmazyn, Morris
Kinson, Gordon A
Kooh, Sang Whay
Kraicer, Jacob
Lefcoe, Neville
Logothetopoulos, J
Lynn, Denis Heward
Machin, J
Martin, Julio Mario
Mercer, Paul Frederick
Mettrick, David Francis
Milazzo, Francis Henry
Money, Kenneth Eric
Moon, Thomas William
Morris, Catherine Elizabeth
Napke, Edward
Norwich, Kenneth Howard
O'Hea, Eugene Kevin
Phillipson, Eliot Asher
Philogene, Bernard J R
Raeside, James Inglis
Rakusan, Karel Josef
Rand, Richard Peter
Rappaport, Aron M
Roslycky, Eugene Bohdan
Roth, René Romain
Sen, Amar Kumar
Shah, Bhagwan G
Shephard, Roy Jesse
Sirek, Anna
Sirek, Otakar Victor
Sun, Anthony Mein-Fang
Talesnik, Jaime
Trevors, Jack Thomas
Veliky, Ivan Alois
Vranic, Mladen
Weaver, Lynne C
Younglai, Edward Victor
Zimmerman, Arthur Maurice

QUEBEC
Adamkiewicz, Vincent Witold
Anand-Srivastava, Madhu Bala
Barcelo, Raymond
Bergeron, Georges Albert
Bergeron, Michel
Billette, Jacques
Bolte, Edouard
Briere, Normand
Burgess, John Herbert
Capek, Radan
Castellucci, Vincent F
Collier, Brian
Cronin, Robert Francis Patrick
Cruess, Richard Leigh
De Champlain, Jacques
DeRoth, Laszlo
Digby, Peter Saki Bassett
Dufour, Jacques John
Dugal, Louis Paul
Dunnigan, Jacques
Frojmovic, Maurice Mony
Gagnon, Andre
Galeano, Cesar
Gjedde, Albert
Goltzman, David
Itiaba, Kibe
Kapoor, Narinder N
Larochelle, Jacques
Leblanc, Jacques Arthur
Lemonde, Andre
MacIntosh, Frank Campbell
Milic-Emili, Joseph
Mortola, Jacopo Prospero
Olivier, Andre
Payet, Marcel Daniel
Polosa, Canio
Ponka, Premysl
Potvin, Pierre
Quillen, Edmond W, Jr
Ruiz-Petrich, Elena
Schanne, Otto F
Tenenhouse, Harriet Susie
Van Gelder, Nico Michel
Wolsky, Alexander

SASKATCHEWAN
Murphy, Bruce Daniel

Sulakhe, Prakash Vinayak
Thornhill, James Arthur
Weisbart, Melvin
Williams, Charles Melville

OTHER COUNTRIES
Alivisatos, Spyridon Gerasimos
Anastasios
Binder, Bernd R
Bligh, John
Boullin, David John
Braquet, Pierre G
Chertok, Robert Joseph
De Alba Martinez, Jorge
Delahayes, Jean
De Troyer, Andre Jules
Dikstein, Shabtay
Fawaz, George
Fleckenstein, Albrecht
Garcia Ramos, Juan
Gatz, Randall Neal
Gontier, Jean Roger
Hall, Peter
Hall, Peter Francis
Hazeyama, Yuji
Heiniger, Hans-Jorg
Heinz, Erich
Hisada, Mituhiko
Josenhans, William T
Kawamura, Hiroshi
King, Dorothy Wei (Cheng)
Kordan, Herbert Allen
Kraicer, Peretz Freeman
Lu, Guo-Wei
McBroom, Marvin Jack
Machne, Xenia
McMillan, Joseph Patrick
Meli, Alberto L G
Mezquita, Cristobal
Monos, Emil
Radford, Edward Parish
Reuter, Harald
Rodahl, Kaare
Sampson, Sanford Robert
Sawyer, Richard Leander
Schönbaum, Eduard
Shiraki, Keizo
Smolander, Martti Juhani
Torún, Benjamin
Tung, Che-Se
Valencia, Mauro Eduardo
Van Driessche, Willy
Vane, John Robert
Veicsteinas, Arsenio
Watanabe, Yoshio
Weidmann, Silvio
Welty, Joseph D
Zapata, Patricio

Physiology, Plant

ALABAMA
Allen, Seward Ellery
Backman, Paul Anthony
Biswas, Prosanto K
Cherry, Joe H
Davis, Donald Echard
Elgavish, Ada S
Eyster, Henry Clyde
Gilliam, Charles Homer
Henderson, James Henry Meriwether
Jones, Daniel David
Jones, Jeanette
Landers, Kenneth Earl
O'Kelley, Joseph Charles
Peterson, Curtis Morris
Rajanna, Bettaiya
Rogers, Hugo H, Jr
Singh, Bharat
Truelove, Bryan
Weete, John Donald
Whatley, Booker Tillman

ALASKA
Stekoll, Michael Steven

ARIZONA
Allen, Stephen Gregory
Bartels, Paul George
Davis, Edwin Alden
Erickson, Eric Herman, Jr
Fowler, Dona Jane
Guinn, Gene
Harris, Leland
Hendrix, Donald Louis
Hull, Herbert Mitchell
Katterman, Frank Reinald Hugh
Kimball, Bruce Arnold
Larkins, Brian Allen
Lipke, William G
McDaniel, Robert Gene
Mellor, Robert Sydney
Morton, Howard LeRoy
Newman, David William
O'Leary, James William
Radin, John William
Tinus, Richard Willard
Towill, Leslie Ruth
Trelease, Richard Norman
Upchurch, Robert Phillip
Verbeke, Judith Ann
Wright, Daniel Craig

ARKANSAS
Einert, Alfred Erwin
Lane, Forrest Eugene
McMasters, Dennis Wayne
Morris, Justin Roy
Nickell, Louis G
Oosterhuis, Derrick M
Reed, Hazell
Roberson, Ward Bryce
Stutte, Charles A
Turnipseed, Glyn D
West, Charles Patrick
Wickliff, James Leroy

CALIFORNIA
Addicott, Fredrick Taylor
Anderson, Lars William James
Arditti, Joseph
Ashton, Floyd Milton
Bailey-Serres, Julia
Baldy, Richard Wallace
Bayer, David E
Beevers, Harry
Bell, Charles W
Benson, Andrew Alm
Bils, Robert F
Bird, Harold L(eslie), Jr
Blakely, Lawrence Mace
Blinks, Lawrence Rogers
Bolar, Marlin L
Bonner, Bruce Albert
Brandon, David Lawrence
Brecht, Patrick Ernest
Breidenbach, Rowland William
Briggs, Winslow Russell
Buchanan, Bob Branch
Campbell, Bruce Carleton
Campbell, Neil Allison
Castelfranco, Paul Alexander
Catlin, Peter Bostwick
Chrispeels, Maarten Jan
Coggins, Charles William, Jr
Conn, Eric Edward
Cooper, Geoffrey Kenneth
Cooper, James Burgess
Corcoran, Mary Ritzel
Corse, Joseph Walters
Cronshaw, James
Currier, Herbert Bashford
Dalton, Francis Norbert
Davies, Huw M
Downing, Michael Richard
Dugger, Willie Mack, Jr
Duniway, John Mason
Du Pont, Frances Marguerite
Durzan, Donald John
Eaks, Irving Leslie
Einset, John William
Elkin, Lynne Osman
Epstein, Emanuel
Erickson, Louis Carl
Estermann, Eva Frances
Finkle, Bernard Joseph
Finn, James Crampton, Jr
Fork, David Charles
Fosket, Donald Elston
French, Charles Stacy
Fuller, Glenn
Grantz, David Arthur
Gray, Reed Alden
Greene, Richard Wallace
Grieve, Catherine Macy
Grossenbacher, Karl A(lbert)
Haard, Norman F
Haxo, Francis Theodore
Heath, Robert Louis
Hess, Charles
Hess, Frederick Dan
Hewitt, Allan A
Hirsch, Ann Mary
Holm-Hansen, Osmund
Houpis, James Louis Joseph
Howe, George Franklin
Hsiao, Theodore Ching-Teh
Huang, Anthony Hwoon Chung
Huffaker, Ray C
Jacobson, Louis
Johnson, Kenneth Duane
Jones, Russell Lewis
Jordan, Lowell Stephen
Kader, Adel Abdel
Kemp, Paul Raymond
Kester, Dale Emmert
Ketellapper, Hendrik Jan
Kliewer, Walter Mark
Koehler, Don Edward
Kohl, Harry Charles, Jr
Kozlowski, Theodore Thomas
Labanauskas, Charles K
Labavitch, John Marcus
Laties, George Glushanok
Leonard, Robert Thomas
Levitt, Jacob
Lewis, Lowell N
Liebhardt, William C
Lincoln, Richard G
Lipton, Werner Jacob
Loe, Robert Wayne
Long, Sharon Rugel
Loomis, Robert Simpson
Lovatt, Carol Jean
Lovelace, C James
Lucas, William John
Lyons, James Martin

Physiology, Plant (cont)

Maas, Eugene Vernon
MacDonald, James Douglas
McNairn, Robert Blackwood
Madore, Monica Agnes
Mazelis, Mendel
Melis, Anastasios
Merchant, Sabeeha
Morris, Leonard Leslie
Murashige, Toshio
Neel, James William
Nevins, Donald James
Ngo, That Tjien
Nieman, Richard Hovey
Nobel, Park S
Norlyn, Jack David
Norris, Robert Francis
Nothnagel, Eugene Alfred
Oechel, Walter C
Park, Roderic Bruce
Percival, Frank William
Phinney, Bernard Orrin
Pierce, Wayne Stanley
Pratt, Harlan Kelley
Purves, William Kirkwood
Rappaport, Lawrence
Ray, Peter Martin
Rayle, David Lee
Rendig, Victor Vernon
Ritenour, Gary Lee
Roberts, Stephen Winston
Rogers, Bruce Joseph
Romani, Roger Joseph
Rubatzky, Vincent E
Ryugo, Kay
Sacher, Robert Francis
Sachs, Roy M
Saltveit, Mikal Endre, Jr
Sayre, Robert Newton
Schieferstein, Robert Harold
Shih, Ching-Yuan G
Shugarman, Peter Melvin
Silberstein, Otmar Otto
Sommer, Noel Frederick
Stehsel, Melvin Louis
Stemler, Alan James
Stocking, Clifford Ralph
Stone, Edward Curry
Taiz, Lincoln
Tanada, Takuma
Terry, Norman
Thornton, Robert Melvin
Tilton, Varien Russell
Ting, Irwin Peter
Tobin, Elaine Munsey
Uriu, Kiyoto
Vanderhoef, Larry Neil
Vinters, Harry Valdis
Wallace, Arthur
Wallace, Joan M
Weatherspoon, Charles Phillip
Weaver, Ellen Cleminshaw
Weaver, Robert John
Weber, John R
Weimberg, Ralph
Wiley, Lorraine
Wilkins, Harold
Willemsen, Roger Wayne
Wu, Jia-Hsi
Yamaguchi, Masatoshi
Yang, Shang Fa
Zscheile, Frederick Paul, Jr

COLORADO
Alford, Donald Kay
Barney, Charles Wesley
Basham, Charles W
Bonde, Erik Kauffmann
Brown, Lewis Marvin
Dever, John E, Jr
Eley, James H
Fly, Claude Lee
Gough, Larry Phillips
Hanan, Joe John
Hendrix, John Edwin
Holm, David Garth
Kaufmann, Merrill R
Linck, Albert John
Nabors, Murray Wayne
Pollock, Bruce McFarland
Ross, Cleon Walter
Schweizer, Edward E
Seibert, Michael
Storey, Richard Drake
Stushnoff, Cecil
Wilson, Alma McDonald
Workman, Milton

CONNECTICUT
Ahrens, John Frederick
Berlyn, Graeme Pierce
Crain, Richard Cullen
Galston, Arthur William
Geballe, Gordon Theodore
Goldsmith, Mary Helen Martin
Gordon, John C
Kennard, William Crawford
Koning, Ross E
Koontz, Harold Vivien
Parker, Johnson
Peterson, Richard Burnett
Reazin, George Harvey, Jr
Rother, Ana

Satter, Ruth
Slayman, Clifford L
Stowe, Bruce Bernot
Wargo, Philip Matthew
Warren, Richard Scott
Wetherell, Donald Francis
Zavala, Maria Elena
Zelitch, Israel

DELAWARE
Beyer, Elmo Monroe, Jr
Boyer, John Strickland
Bozarth, Gene Allen
Carroll, Robert Buck
Gatenby, Anthony Arthur
Giaquinta, Robert T
Green, Jerome
Hodson, Robert Cleaves
Knowles, Francis Charles
Lichtner, Francis Thomas, Jr
Lin, Willy
Long, James Delbert
Matlack, Albert Shelton
Reasons, Kent M
Riggleman, James Dale
Somers, George Fredrick, Jr
Wittenbach, Vernon Arie

DISTRICT OF COLUMBIA
Allen, James Ralston
Burns, Russell MacBain
Burris, John Edward
Chen, H R
Donaldson, Robert Paul
Gordon, William Ransome
Halstead, Thora Waters
Irving, Patricia Marie
Islam, Nurul
Keitt, George Wannamaker, Jr
Krauss, Robert Wallfar
Littler, Mark Masterton
Quebedeaux, Bruno, Jr
Rabson, Robert
Sulzman, Frank Michael
Tompkins, Daniel Reuben
Weintraub, Robert Louis
Wiggans, Samuel Claude

FLORIDA
Albrigo, Leo Gene
Ayers, Alvin Dearing
Bausher, Michael George
Bennette, Jerry Mac
Biggs, Robert Hilton
Boss, Manley Leon
Bowes, George Ernest
Bryan, Herbert Harris
Burdine, Howard William
Buslig, Bela Stephen
Calvert, David Victor
Campbell, Carl Walter
Cantliffe, Daniel James
Cooper, William Cecil
Cropper, Wendell Parker
Davenport, Thomas Lee
Dusky, Joan Agatha
Fitzpatrick, George
Fritz, George John
Gaskins, Murray Hendricks
Gay, Don Douglas
Gilreath, James Preston
Grunwald, Claus Hans
Hall, Chesley Barker
Haller, William T
Homann, Peter H
Huber, Donald John
Jackson, William Thomas
Karsten, Kenneth Stephen
Knauft, David A
Kossuth, Susan
Lee, David Webster
Mecklenburg, Roy Albert
Mullins, John Thomas
Orgell, Wallace Herman
Owens, Clarence Burgess
Parsons, Lawrence Reed
Phillips, Richard Lee
Reid, Charles Phillip Patrick
Roberts, Donald Ray
Ruelke, Otto Charles
Sager, John Clutton
Schroder, Vincent Nils
Silverman, David Norman
Sinclair, Thomas Russell
Smith, Paul Frederick
Smith, Richard Clark
Steward, Kerry Kalen
Stewart, Herbert
Sweet, Haven C
Thompson, Neal Philip
Vasil, Vimla
Verkade, Stephen Dunning
Weigel, Russell C(ornelius), Jr
Wells, Gary Neil
West, Sherlie Hill
Wheaton, Thomas Adair
White, Timothy Lee
Wilcox, Merrill
Williams, Robert Haworth
Wilson, William Curtis
Wiltbank, William Joseph
Yelenosky, George

GEORGIA
Burns, Robert Emmett
Cutler, Horace Garnett
Dashek, William Vincent
Dekazos, Elias Demetrios
Dinus, Ronald John
Donoho, Clive Wellington, Jr
Duncan, Ronny Rush
Eastin, Emory Ford
Espelie, Karl Edward
Hendrix, Floyd Fuller, Jr
Hilton, James Lee
Hilton, James Lee
Key, Joe Lynn
McLaurin, Wayne Jefferson
Michel, Burlyn Everett
Mixon, Aubrey Clifton
Nes, William David
Ohki, Kenneth
Palevitz, Barry Allan
Phatak, Sharad Chintaman
Pratt, Lee Herbert
Pressey, Russell
Reger, Bonnie Jane
Reilly, Charles Conrad
Robinson, Margaret Chisolm
Romanovicz, Dwight Keith
Sansing, Norman Glenn
Schmidt, Gregory Wayne
Smith, Albert Ernest, Jr
Smith, Morris Wade
Smittle, Doyle Allen
Sommer, Harry Edward
Sparks, Darrell
Vines, Herbert Max
Walker, Alma Toevs
Wilkinson, Robert Eugene
Wood, Bruce Wade

HAWAII
Akamine, Ernest Kisei
Cooil, Bruce James
De la Pena, Ramon Serrano
Demanche, Edna Louise
Kefford, Noel Price
Krauss, Beatrice Hilmer
Moore, Paul Harris
Patterson, Gregory Matthew Leon
Rauch, Fred D
Sakai, William Shigeru
Sanford, Wallace Gordon
Skolmen, Roger Godfrey
Sun, Samuel Sai-Ming
Tanabe, Michael John
Vennesland, Birgit

IDAHO
Bowmer, Richard Glenn
Dwelle, Robert Bruce
Kleinkopf, Gale Eugene
Kochan, Walter J
LeTourneau, Duane John
Oliver, David John
Ramagopal, Subbanaidu
Roberts, Lorin Watson
Sojka, Robert E
Spomer, George Guy
Thill, Donald Cecil

ILLINOIS
Alberte, Randall Sheldon
Ashby, William Clark
Ausich, Rodney L
Blair, Louis Curtis
Brown, Lindsay Dietrich
Chappell, Dorothy Field
Chen, Shepley S
Cordes, William Charles
Dickinson, David Budd
Dove, Lewis Dunbar
Edwards, Harold Herbert
Eskins, Kenneth
Galsky, Alan Gary
Gassman, Merrill Loren
Gould, John Michael
Govindjee, M
Hageman, Richard Harry
Hanson, John Bernard
Harper, James Eugene
Hauptmann, Randal Mark
Heichel, Gary Harold
Heller, Alfred
Hesketh, J D
Hinchman, Ray Richard
Holbrook, Gabriel Peter
Holt, Donald Alexander
Howell, Robert Wayne
Huck, Morris Glen
Leffler, Harry Rex
Lippincott, Barbara Barnes
Lippincott, James Andrew
Lorenz, Ralph William
Maass, Wolfgang Siegfried Gunther
McCracken, Derek Albert
McIlrath, Wayne Jackson
Melhado, L(ouisa) Lee
Meyer, Martin Marinus, Jr
Miernyk, Jan Andrew
Ogren, William Lewis
Ort, Donald Richard
Pappelis, Aristotel John
Player, Mary Anne
Portis, Archie Ray, Jr
Portz, Herbert Lester

Rinne, Robert W
Romberger, John Albert
Ruddat, Manfred
Skirvin, Robert Michael
Skok, John
Splittstoesser, Walter E
Spomer, Louis Arthur
Stoller, Edward W
Van Sambeek, Jerome William
Verduin, Jacob
Weber, Evelyn Joyce
Weidner, Terry Mohr
Widholm, Jack Milton
Williams, David James
Woolley, Joseph Tarbet
Yopp, John Herman

INDIANA
Alder, Edwin Francis
Barr, Rita
Bauman, Thomas Trost
Bhattacharya, Pradeep Kumar
Boyd, Frederick Mervin
Brenneman, James Alden
Bucholtz, Dennis Lee
Castleberry, Ron M
Chaney, William R
Coolbaugh, Ronald Charles
Crouch, Martha Louise
Dilley, Richard Alan
Dunkle, Larry D
Fischler, Drake Anthony
Gentile, Arthur Christopher
Gerwick, Ben Clifford, III
Hagen, Charles William, Jr
Hamilton, David Foster
Hodges, Thomas Kent
Housley, Thomas Lee
Kapustka, Lawrence A
Keck, Robert William
Manthey, John August
Miller, Carlos Oakley
Mitchell, Cary Arthur
Nichols, Kenneth E
Ockerse, Ralph
Polley, Lowell David
Raveed, Dan
Rhykerd, Charles Loren
Ruesink, Albert William
Schreiber, Marvin Mandel
Smith, Charles Edward, Jr
Stiller, Mary Louise
Togasaki, Robert K
Tsai, Chia-Yin
Waldrep, Thomas William
Watson, Maxine Amanda
Webb, Mary Alice
Williams, Gene R
Williams, James Lovon, Jr
Wright, William Leland

IOWA
Anderson, Irvin Charles
Berg, Virginia Seymour
Burris, Joseph Stephen
Cavalieri, Anthony Joseph, II
Chaplin, Michael H
Durkee, Lenore T
Ford, Clark Fugier
George, John Ronald
Imsande, John
LaMotte, Clifford Elton
Lillehoj, Eivind B
Muir, Robert Mathew
Pearce, Robert Brent
Shibles, Richard Marwood
Showers, William Broze, Jr
Sjölund, Richard David
Smith, Frederick George
Stewart, Cecil R
Svec, Leroy Vernon
Tjostem, John Leander
Yager, Robert Eugene

KANSAS
Borchert, Rolf
Coyne, Patrick Ivan
Davis, Lawrence Clark
Guikema, James Allen
Jennings, Paul Harry
Johnson, Lowell Boyden
Kirkham, M B
Murphy, John Joseph
Murphy, Larry S
Wiest, Steven Craig
Wong, Peter P

KENTUCKY
Bush, Lowell Palmer
Dunham, Valgene Loren
Dyar, James Joseph
Harmet, Kenneth Herman
Hiatt, Andrew Jackson
Kasperbauer, Michael J
Lacefield, Garry Dale
Lambert, Roger Gayle
Lasheen, Aly M
Leggett, James Everett
Lowe, Richie Howard
Morgan, Marjorie Susan
Roberts, Clarence Richard
Toman, Frank R
Wagner, George Joseph

LOUISIANA
Alam, Jawed
Baker, John Bee
Barber, John Threlfall
Barnett, James P
Benda, Gerd Thomas Alfred
Bhatnager, Deepak
Board, James Ellery
Carpenter, Stanley Barton
Carter, Mason Carlton
Faw, Wade Farris
Gosselink, James G
Hough, Walter Andrew
Jacks, Thomas Jerome
Longstreth, David J
Lorio, Peter Leonce, Jr
Martin, Freddie Anthony
Moore, Thomas Stephen, Jr
Musgrave, Mary Elizabeth
Nair, Pankajam K
Rogers, Robert Larry
Standifer, Leonides Calmet, Jr
Stevenson, Enola L
Terry, Maurice Ernest
Webert, Henry S
Yatsu, Lawrence Y

MAINE
Brown, Gregory Neil
Eggert, Franklin Paul
Schwintzer, Christa Rose
Thomas, Robert James
Tjepkema, John Dirk

MARYLAND
Baker, James Earl
Bandel, Vernon Allan
Barnett, Neal Mason
Britz, Steven J
Bruns, Herbert Arnold
Bunce, James Arthur
Butcher, Henry Clay, IV
Chappelle, Emmett W
Chitwood, David Joseph
Christiansen, Meryl Naeve
Christy, Alfred Lawrence
Cleland, Charles Frederick
Deitzer, Gerald Francis
Falkenstein, Kathy Fay
Filner, Barbara
Foy, Charles Daley
Freiberg, Samuel Robert
Gantt, Elisabeth
Gouin, Francis R
Gross, Kenneth Charles
Habermann, Helen M
Heggestad, Howard Edwin
Herman, Eliot Mark
Herrett, Richard Allison
Hill, Jane Virginia Foster
Hirsh, Allen Gene
Hruschka, Howard Wilbur
Hurtt, Woodland
Isensee, Allan Robert
Josephs, Melvin Jay
Kennedy, Robert A
Krizek, Donald Thomas
Leather, Gerald Roger
Lee, Edward Hsien-Chi
Liverman, James Leslie
Margulies, Maurice
Matthews, Benjamin F
Mulchi, Charles Lee
Owens, Lowell Davis
Patterson, Glenn Wayne
Racusen, Richard Harry
Ridley, Esther Joanne
St John, Judith Brook
Saunders, James Allen
Schwartz, Martin
Shaw, Warren Cleaton
Shear, Cornelius Barrett
Shropshire, Walter, Jr
Simpson, Marion Emma
Sloger, Charles
Smith, William Owen
Stanley, Ronald Alwin
Steffens, George Louis
Sterrett, John Paul
Sze, Heven
Teramura, Alan Hiroshi
Tester, Cecil Fred
Vigil, Eugene Leon
Wadleigh, Cecil Herbert
Wang, Chien Yi
Warmbrodt, Robert Dale
Watada, Alley E
Williams, Robert Jackson
Woodstock, Lowell Willard
Youle, Richard James
Zacharius, Robert Marvin
Zimmerman, Richard Hale

MASSACHUSETTS
Barker, Allen Vaughan
Bazzaz, Maarib Bakri
Bogorad, Lawrence
Craker, Lyle E
Dacey, John W H
Devlin, Robert Martin
DiLiddo, Rebecca McBride
Feinleib, Mary Ella (Harman)
Frederick, Sue Ellen
Freeberg, John Arthur

Gruber, Peter Johannes
Hepler, Peter Klock
Howe, Kenneth Jesse
Kadkade, Prakash Gopal
Kamien, Ethel N
Klein, Attila Otto
Lockhart, James Arthur
McLeod, Guy Collingwood
Melan, Melissa A
Offner, Gwynneth Davies
Orme-Johnson, William H
Papastoitsis, Gregory
Reid, Philip Dean
Rivera, Ezequiel Ramirez
Schiff, Jerome A
Stern, Arthur Irving
Tattar, Terry Alan
Thomas, Aubrey Stephen, Jr
Torrey, John Gordon
Whitaker, Ellis Hobart
Zimmermann, Martin Huldrych

MICHIGAN
Adams, Paul Allison
Adler, John Henry
Anderson, Richard Lee
Bukovac, Martin John
Campbell, Wilbur Harold
Christenson, Donald Robert
Dennis, Frank George, Jr
Dilley, David Ross
Good, Norman Everett
Halloin, John McDonell
Hare, Leonard N
Heberlein, Gary T
Henry, Egbert Winston
Hough, Richard Anton
Howell, Gordon Stanley, Jr
Ikuma, Hiroshi
Inselberg, Edgar
Isleib, Donald Richard
Johnston, Taylor Jimmie
Kaufman, Peter Bishop
Kende, Hans Janos
Kivilaan, Aleksander
Lang, Anton
McIntosh, Lee
Mullison, Wendell Roxby
Nadler, Kenneth David
Nooden, Larry Donald
Podila, Gopi Krishna
Preiss, Jack
Putnam, Alan R
Ries, Stanley K
Rossman, Elmer Chris
Schneider, Michael J
Somerville, Christopher Roland
Stein, Howard Jay
Vander Beek, Leo Cornelis
Wolt, Jeffrey Duaine
Yocum, Conrad Schatte
Zeevaart, Jan Adriaan Dingenis

MINNESOTA
Ahlgren, George E
Andersen, Robert Neils
Behrens, Richard
Brenner, Mark
Bushnell, William Rodgers
Choe, Hyung Tae
Frenkel, Albert W
Gengenbach, Burle Gene
Hackett, Wesley P
Jonas, Herbert
Klemer, Andrew Robert
Kraft, Donald J
Li, Pen H (Paul)
Pratt, Douglas Charles
Rehwaldt, Charles A
Silflow, Carolyn Dorothy
Singer, Susan Rundell
Smith, Lawrence Hubert
Soulen, Thomas Kay
Sowell, John Basil
Stadelmann, Eduard Joseph
Stadtherr, Richard James
Sucoff, Edward Ira
Sullivan, Timothy Paul
Tautvydas, Kestutis Jonas
Zeyen, Richard John

MISSISSIPPI
Creech, Roy G
Duke, Stephen Oscar
Elmore, Carroll Dennis
Hoagland, Robert Edward
Hodges, Harry Franklin
Hodges, John Deavours
Kurtz, Mark Edward
McChesney, James Dewey
McWhorter, Chester Gray
Roark, Bruce (Archibald)
Spiers, James Monroe
Wills, Gene David

MISSOURI
Blevins, Dale Glenn
Carpenter, Will Dockery
Donald, William Waldie
Field, Christopher Bower
Fox, J Eugene
Franson, Raymond Lee
George, Milon Fred
Hayes, Alice Bourke

Houghton, John M
Kohl, Daniel Howard
Lambeth, Victor Neal
Mertz, Dan
Miles, Charles Donald
Montague, Michael James
Nelson, Curtis Jerome
Pallardy, Stephen Gerard
Paul, Kamalendu Bikash
Pickard, Barbara Gillespie
Porter, Clark Alfred
Schumacher, Richard William
Sells, Gary Donnell
Stern, Michele Suchard
Symington, Janey
Varner, Joseph Elmer
Vogt, Albert R
Wochok, Zachary Stephen

MONTANA
Behan, Mark Joseph
Bilderback, David Earl
Carpenter, Bruce H
Chessin, Meyer
Mills, Ira Kelly
Sheridan, Richard P
Stallknecht, Gilbert Franklin
Stark, Nellie May

NEBRASKA
Chollet, Raymond
Clark, Ralph B
Daly, Joseph Michael
Davies, Eric
Eastin, John A
Elthon, Thomas Eugene
Higley, Leon George
Kinbacher, Edward John
Partridge, James Enoch
Read, Paul Eugene
Roeth, Frederick Warren
Rowland, Neil Wilson
Schwartzbach, Steven Donald
Specht, James Eugene

NEVADA
Fox, Carl Alan
Howland, Joseph E(mery)
Taylor, George Evans, Jr

NEW HAMPSHIRE
Eggleston, Patrick Myron
Laber, Larry Jackson
Minoca, Subhash C

NEW JERSEY
Bahr, James Theodore
Brown, Thomas Edward
Bruno, Stephen Francis
Carr, Richard J
Chen, Tsong Meng
Clark, Harold Eugene
Daie, Jaleh
Davis, Robert Foster, Jr
Durkin, Dominic J
Eaglesham, Allan Robert James
Gaynor, John James
Ghai, Geetha R
Gramlich, James Vandle
Grant, Neil George
Greenfield, Sydney Stanley
Gruenhagen, Richard Dale
Hu, Ching-Yeh
Marrese, Richard John
Nelson, Nathan
Price, C A
Reid, Hay Bruce, Jr
Vasconcelos, Aurea C

NEW MEXICO
Carroll, Arthur Paul
Cotter, Donald James
Essington, Edward Herbert
Fisher, James Thomas
Fowler, James Lowell
Gosz, James Roman
Johnson, Gordon Verle
Mexal, John Gregory
Phillips, Gregory Conrad
Shortess, David Keen
Throneberry, Glyn Ogle

NEW YORK
Alves, Leo Manuel
Ammirato, Philip Vincent
Amundson, Robert Gale
Bing, Arthur
Birecka, Helena M
Bisson, Mary A
Brody, Marcia
Buckley, Edward Harland
Budd, Thomas Wayne
Chabot, Brian F
Collins, Carol Desormeau
Creasy, Leroy L
Curtis, Otis Freeman, Jr
Davies, Peter John
Dietert, Margaret Flowers
Domozych, David S
Earle, Elizabeth Deutsch
Ecklund, Paul Richard
Edgerton, Louis James
Evans, Lance Saylor
Ewing, Elmer Ellis

Fisher, Nicholas Seth
Gallagher, Jane Chispa
Gerard, Valrie Ann
Gussin, Arnold E S
Haber, Alan Howard
Hammond, H David
Hillman, William Sermolino
Hind, Geoffrey
Jacobson, Jay Stanley
Jagendorf, Andre Tridon
Keng, Peter C
Khan, Anwar Ahmad
Klingensmith, Merle Joseph
Krikorian, Abraham D
Lakso, Alan Neil
LaRue, Thomas A
Law, David Martin
Leopold, Aldo Carl
Linkins, Arthur Edward
Lyman, Harvard
McCune, Delbert Charles
McDaniel, Carl Nimitz
MacLean, David Cameron
Mancinelli, Alberto L
Mantai, Kenneth Edward
Miller, John Henry
Obendorf, Ralph Louis
Pietraface, William John
Pool, Robert Morris
Posner, Herbert Bernard
Powell, Loyd Earl, Jr
Rana, Mohammad A
Rhee, G-Yull
Robinson, Terence Lee
Schaedle, Michail
Shrift, Alex
Siegelman, Harold William
Simon, Robert David
Sirois, David Leon
Spanswick, Roger Morgan
Steponkus, Peter Leo
Tan-Wilson, Anna L(i)
Titus, John Elliott
Truscott, Frederick Herbert
Weinstein, Leonard Harlan
Welch, Ross Maynard
Wilcox, Hugh Edward
Wilson, Karl A

NORTH CAROLINA
Allen, Nina Strömgren
Burns, Joseph Charles
Camp, Pamela Jean
Carter, Thomas Edward, Jr
Collins, Henry A
Cooke, Anson Richard
Corbin, Frederick Thomas
Cowett, Everett R
Danielson, Loran Leroy
Darling, Marilyn Stagner
Davis, Daniel Layten
Davis, Graham Johnson
De Hertogh, August Albert
DeJong, Donald Warren
Downs, Robert Jack
DuBay, Denis Thomas
Ellenson, James L
Fiscus, Edwin Lawson
Friedrich, James Wayne
Gross, Harry Douglass
Haun, Joseph Rhodes
Heck, Walter Webb
Hellmers, Henry
Holm, Robert E
Jaffe, Mordecai J
Jeffreys, Donald Bearss
Kamykowski, Daniel
Kramer, Paul Jackson
Long, Raymond Carl
Maier, Robert Hawthorne
Manning, David Treadway
Matthysse, Ann Gale
Miller, Conrad Henry
Miller, Joseph Edwin
Miller, Robert James, II
Moll, Robert Harry
Moreland, Donald Edwin
Morrison, Ralph M
Mott, Ralph Lionel
Naylor, Aubrey Willard
Nelson, Paul Victor
O'Neal, Thomas Denny
Pattee, Harold Edward
Peet, Mary Monnig
Ramus, Joseph S
Richardson, Curtis John
Rogerson, Asa Benjamin
Sanders, Douglas Charles
Scofield, Herbert Temple
Scott, Tom Keck
Sehgal, Prem P
Seltmann, Heinz
Shafer, Thomas Howard
Sheets, Thomas Jackson
Siedow, James N
Sisler, Edward C
Spiker, Steven L
Stier, Howard Livingston
Sunda, William George
Terborgh, John J
Thompson, William Francis, III
Troyer, James Richard
Volk, Richard James
Wallace, James William, Jr

Vanden Born, William Henry
Whitehouse, Ronald Leslie S
Zalik, Saul

BRITISH COLUMBIA
Ballantyne, David John
Fisher, Francis John Fulton
Lavender, Denis Peter
Looney, Norman E
Mitchell, Donald John
Norton, Colin Russell
Pollard, Douglas Frederick William
Prasad, Raghubir (Raj)
Runeckles, Victor Charles
Srivastava, Lalit Mohan
Stout, Darryl Glen
Taylor, Iain Edgar Park
Unrau, Abraham Martin
Vidaver, William Elliott
Warrington, Patrick Douglas
Waygood, Ernest Roy

MANITOBA
LaCroix, Lucien Joseph
Staniforth, Richard John

NEW BRUNSWICK
Bonga, Jan Max
Coleman, Warren Kent
Cumming, Bruce Gordon
Fensom, David Strathern
Little, Charles Harrison Anthony
Yoo, Bong Yul

NEWFOUNDLAND
Hampson, Michael Chisnall

NOVA SCOTIA
Bidwell, Roger Grafton Shelford
Blatt, Carl Roger
Chen, Lawrence Chien-Ming
Curry, George Montgomery
Kimmins, Warwick Charles
Laycock, Maurice Vivian

ONTARIO
Arnison, Paul Grenville
Barker, William George
Bewley, John Derek
Birmingham, Brendan Charles
Canvin, David T
Chang, Fa Yan
Colman, Brian
Cummins, W(illiam) Raymond
Czuba, Margaret
Dainty, Jack
Dumbroff, Erwin Bernard
Effer, W R
Elfving, Donald Carl
Farmer, Robert E, Jr
Ferrier, Jack Moreland
Fletcher, Ronald Austin
Forsyth, Frank Russell
Heath, Michele Christine
Higgins, Verna Jessie
Hofstra, Gerald
Hogan, Gary D
Hope, Hugh Johnson
Hopkins, William George
Horton, Roger Francis
Ingratta, Frank Jerry
Israelstam, Gerald Frank
Joy, Kenneth Wilfred
Lee, Tsung Ting
Leppard, Gary Grant
Liptay, Albert
Mack, Alexander Ross
Minshall, William Harold
Nozzolillo, Constance
Oaks, B Ann
Ormrod, Douglas Padraic
O'Sullivan, John
Peirson, David Robert
Pillay, Dathathry Trichinopoly Natraj
Proctor, John Thomas Arthur
Rauser, Wilfried Ernst
Riekels, Jerald Wayne
Rosa, Nestor
Stokes, Pamela Mary
Switzer, Clayton Macfie
Ursino, Donald Joseph
Webb, David Thomas
Weinberger, Pearl
White, Gordon Allan
Wightman, Frank

PRINCE EDWARD ISLAND
MacQuarrie, Ian Gregor
Suzuki, Michio

QUEBEC
Bigras, Francine Jeanne
Bolduc, Reginald J
Boll, William George
Corriveau, Armand Gerard
Furlan, Valentin
Gibbs, Sarah Preble
Girouard, Ronald Maurice
Levasseur, Maurice Edgar
Maclachlan, Gordon Alistair
Phan, Chon-Ton
Poole, Ronald John
St Pierre, Jean Claude
Sarhan, Fathey

Willemot, Claude

SASKATCHEWAN
Clarke, John Mills
Gusta, Lawrence V
Kartha, Kutty Krishnan
King, John
Kurz, Wolfgang Gebhard Walter
Quick, William Andrew
Tanino, Karen Kikumi
Turel, Franziska Lili Margarete
Walther, Alina

OTHER COUNTRIES
Delmer, Deborah P
Dreosti, Ivor Eustace
Edelman, Marvin
Gressel, Jonathan Ben
MacCarthy, Jean Juliet
Sadik, Sidki
Steer, Martin William
Vlitos, August John

Toxicology

ALABAMA
Hood, Ronald David
Liu, Ray Ho
Yarbrough, James David

ALASKA
Kennish, John M

ARIZONA
Halpert, James Robert
North-Root, Helen May

ARKANSAS
Allaben, William Thomas
Hansen, Deborah Kay
Hinson, Jack Allsbrook
Holson, Ralph Robert
Jackson, Carlton Darnell
Leakey, Julian Edwin Arundell
Paule, Merle Gale
Roberts, Dean Winn, Jr
Schieferstein, George Jacob
Sheehan, Daniel Michael
Soderberg, Lee Stephen Freeman
Wessinger, William David
Young, John Falkner

CALIFORNIA
Bailey, David Nelson
Baldwin, Robert Charles
Ballard, Ralph Campbell
Bendix, Selina (Weinbaum)
Corby, Donald G
Coughlin, James Robert
Crosby, Donald Gibson
Danse, Ilene H Raisfeld
Davis, Brian Kent
Davis, William Ellsmore, Jr
DePass, Linval R
Elsayed, Nabil M
Farber, Sergio Julio
Furst, Arthur
Goldman, Marvin
Hackney, Jack Dean
Hochstein, Paul Eugene
Hoke, Glenn Dale
Hollinger, Mannfred Alan
Howd, Robert A
Hutchin, Maxine E
Kassakhian, Garabet Haroutiun
Kelman, Bruce Jerry
Kland, Mathilde June
Koschier, Francis Joseph
Landolph, Joseph Richard, II
Last, Jerold Alan
Leung, Peter
Malone, Marvin Herbert
Newell, Gordon Wilfred
Ohlendorf, Harry Max
Parker, Kenneth D
Rabovsky, Jean
Sanders, Brenda Marie
Schneider, Meier
Smith, Martyn Thomas
Sunshine, Irving
Tune, Bruce Malcolm
Victery, Winona Whitwell
Way, Jon Leong
Wilson, Barry William
Witham, Clyde Lester
Witschi, Hanspeter R

COLORADO
Aldrich, Franklin Dalton
Brooks, Bradford O
Hall, Alan H
Hesterberg, Thomas William
Johnsen, Richard Emanuel
Sokol, Ronald Jay
Wilber, Charles Grady
Wilson, Vincent L
Wingeleth, Dale Clifford

CONNECTICUT
Amacher, David E
Berdick, Murray
Buck, Marion Gilmour
Cohen, Steven Donald

Mattes, William Bustin
Milzoff, Joel Robert
Saunders, Donald Roy
Schurig, John Eberhard
Snellings, William Moran
Stadnicki, Stanley Walter, Jr
Tyler, Tipton Ransom

DELAWARE
Malley, Linda Angevine
Pierson, Keith Bernard
Staples, Robert Edward

DISTRICT OF COLUMBIA
April, Robert Wayne
Bigelow, Sanford Walker
Bolger, P Michael
Cerveny, Thelma Jannette
Feeney, Gloria Comulada
Hathcock, John Nathan
Herman, Eugene H
Horakova, Zdenka
Lai, David Ying-lun
Lee, I P
McMahon, Timothy F
Nadolney, Carlton H
Page, Samuel William
Raslear, Thomas G
Rhoden, Richard Allan
Schwartz, Sorell Lee
Silver, Francis
Sobotka, Thomas Joseph
Thomas, Richard Dean
Todhunter, John Anthony
Whittaker, Paul

FLORIDA
Boxill, Gale Clark
Corbett, Michael Dennis
Frank, H Lee
Levy, Richard
Lu, Frank Chao
McKenney, Charles Lynn, Jr
Mayer, Foster Lee, Jr
Milton, Robert Mitchell
Murchison, Thomas Edgar
Porvaznik, Martin
Radomski, Jack London
Rao, Krothapalli Ranga
Richards, Norman Lee
Teaf, Christopher Morris

GEORGIA
Bates, Harold Brennan, Jr
Berg, George G
Catravas, John D
Curley, Winifred H
Denson, Donald D
Hargrove, Robert John
Holt, Peter Stephen
Lehner, Andreas Friedrich
McGrath, William Robert
White, Donald Henry
Winger, Parley Vernon
Wolven-Garrett, Anne M

HAWAII
Doerge, Daniel Robert
Seifert, Josef
Sherman, Martin

ILLINOIS
Benkendorf, Carol Ann
Bhattacharyya, Maryka Horsting
Brockman, Herman E
Feldbush, Thomas Lee
Foreman, Ronald Louis
Francis, Bettina Magnus
Garvin, Paul Joseph, Jr
Hansen, Larry George
Kaistha, Krishan K
Kaminski, Edward Jozef
Kokkinakis, Demetrius Michael
McConnachie, Peter Ross
McCormick, David Loyd
Miller, Gregory Duane
Narahashi, Toshio
Northup, Sharon Joan
Patterson, Douglas Reid
Reilly, Christopher Aloysius, Jr
Taylor, Julius David

INDIANA
Amundson, Merle E
Born, Gordon Stuart
Borowitz, Joseph Leo
Caplis, Michael E
Gehring, Perry James
Lindstrom, Terry Donald
Litov, Richard Emil
Maickel, Roger Philip
Meyers, Donald Bates
Paladino, Frank Vincent
Pfeifer, Richard Wallace
Thompson, Daniel James
Vodicnik, Mary Jo
Wheeler, Ralph John
Wolff, Ronald Keith

IOWA
Biermann, Janet Sybil
Coats, Joel Robert
McDonald, Donald Burt
Stahr, Henry Michael

KANSAS
Browne, Ronald K
Oehme, Frederick Wolfgang
Parkinson, Andrew
Pazdernik, Thomas Lowell
Schroeder, Robert Samuel

KENTUCKY
Chen, Theresa S
Glauert, Howard Perry
Gupta, Ramesh C
Hurst, Harrell Emerson
Jarboe, Charles Harry
Kasarskis, Edward Joseph
Myers, Steven Richard
Nerland, Donald Eugene
Volp, Robert Francis
Yokel, Robert Allen

LOUISIANA
Hughes, Janice S
Lee, Jordan Grey
Manno, Barbara Reynolds
Pryor, William Austin
Smith, Robert Leonard
Swenson, David Harold

MARYLAND
Alleva, Frederic Remo
Atrakchi, Aisar Hasan
Barbehenn, Elizabeth Kern
Barbour, Michael Thomas
Bass, Eugene L
Beyer, W Nelson
Bigger, Cynthia Anita Hopwood
Burton, Dennis Thorpe
Caplan, Yale Howard
Chatterjee, Subroto
Cifone, Maria Ann
Clark, Donald Ray, Jr
Cone, Edward Jackson
Creasia, Donald Anthony
Dacre, Jack Craven
Fechter, Laurence David
Finch, Robert Allen
Frazier, John Melvin
Friedman, Leonard
Glocklin, Vera Charlotte
Goldman, Dexter Stanley
Gori, Gio Batta
Green, Martin David
Gryder, Rosa Meyersburg
Habig, William Henry
Hajjar, Nicolas Philippe
Hall, Richard Leland
Hill, Elwood Fayette
Hsia, Mong Tseng Stephen
Jacobs, Abigail Conway
Joshi, Sewa Ram
Kokoski, Charles Joseph
Landauer, Michael Robert
Lee, Fang-Jen Scott
Lijinsky, William
Melancon, Mark J
Milman, Harry Abraham
Morcock, Robert Edward
Morton, Joseph James Pandozzi
Mufson, R Allan
Newcombe, David S
Nuite-Belleville, Jo Ann
Palmer, Winifred G
Prasanna, Hullahalli Rangaswamy
Putman, Donald Lee
Ragsdale, Nancy Nealy
Rattner, Barnett Alvin
Roesijadi, Guritno
Rubin, Robert Jay
Saffiotti, Umberto
Saslaw, Leonard David
Shih, Tsung-Ming Anthony
Steele, Vernon Eugene
Stoskopf, Michael Kerry
Vocci, Frank Joseph
Wannemacher, Robert, Jr
Whaley, Wilson Monroe
Whiting, John Dale, Jr
Wiemeyer, Stanley Norton
Yager, James Donald, Jr
Yang, Shen Kwei
Zeeman, Maurice George

MASSACHUSETTS
Adams, Richard A
Anderson, Rosalind Coogan
Busby, William Fisher, Jr
Chou, Iih-Nan
Edwards, Gordon Stuart
Goldmacher, Victor S
Graffeo, Anthony Philip
Hoar, Richard Morgan
Hoffmann, George Robert
Hollocher, Thomas Clyde, Jr
Robinson, William Edward
Sivak, Andrew
Slapikoff, Saul Abraham
Smith, Dennis Matthew
Smuts, Mary Elizabeth
Tan, Barrie
Wogan, Gerald Norman

MICHIGAN
Aaron, Charles Sidney
Aulerich, Richard J
Beaudoin, Allan Roger

Toxicology (cont)

Bedford, James William
Bernstein, Isadore A
Braselton, Webb Emmett, Jr
Butterworth, Francis M
Chang, Chia-Cheng
Chopra, Dharam Pal
Cochran, Kenneth William
Elliott, George Algimon
Frade, Peter Daniel
Garg, Bhagwan D
Giesy, John Paul, Jr
Hult, Richard Lee
Kamrin, Michael Arnold
Lindblad, William John
Mac, Michael John
MacDonald, John Robert
Moolenaar, Robert John
Passino, Dora R May
Piper, Walter Nelson
Richardson, Rudy James
Roth, Robert Andrew, Jr
Sinsheimer, Joseph Eugene
Sleight, Stuart Duane
Van Dyke, Russell Austin
Zimmerman, Norman

MINNESOTA
Goble, Frans Cleon
McNaught, Donald Curtis
Messing, Rita Bailey
Smith, Thomas Jay
Willard, Paul W

MISSISSIPPI
Benson, William Hazlehurst
Chambers, Janice E
Kodavanti, Prasada Rao S
Larsen, James Bouton
Mehendale, Harihara Mahadeva

MISSOURI
Badr, Mostafa
Brown, Olen Ray
Dixit, Rakesh
Howlett, Allyn C
Kuhns, John Farrell
Lau, Brad W C
Mehrle, Paul Martin, Jr
Mills, Steven Harlon
Schoettger, Richard A
Su, Kwei Lee
Yanders, Armon Frederick

NEBRASKA
Berndt, William O
Cohen, Samuel Monroe
Nipper, Henry Carmack
Schneider, Norman Richard
Scholes, Norman W

NEW HAMPSHIRE
Gosselin, Robert Edmond

NEW JERSEY
Amemiya, Kenjie
Arthur, Alan Thorne
Auletta, Carol Spence
Bogden, John Dennis
Cooper, Keith Raymond
Curcio, Lawrence Nicholas
Dairman, Wallace M
De Salva, Salvatore Joseph
Durham, Stephen K
Easterday, Otho Dunreath
Farmanfarmaian, Allahverdi
Frenkel, Gerald Daniel
Gertner, Sheldon Bernard
Ghosh, Arati
Hayes, Terence James
Jaeger, Rudolph John
Kapp, Robert Wesley, Jr
Katz, Laurence Barry
Lagunowich, Laura Andrews
Levin, Wayne
Lewis, Neil Jeffrey
Lin, Yi-Jong
Mehlman, Myron A
Rohovsky, Michael William
Rothstein, Edwin C(arl)
Scala, Robert Andrew
Schreiner, Ceinwen Ann
Segelman, Alvin Burton
Tierney, William John
Willson, John Ellis

NEW MEXICO
Bechtold, William Eric
Belinsky, Steven Alan
Corcoran, George Bartlett, III
Dahl, Alan Richard
Hadley, William Melvin
Henderson, Rogene Faulkner
Hoover, Mark Douglas
Johnson, Neil Francis
Lechner, John Fred
Mauderly, Joe L
Mewhinney, James Albert
Shopp, George Milton, Jr
Thomassen, David George
Thompson, Jill Charlotte

NEW YORK
Abraham, Rajender
Aschner, Michael
Astill, Bernard Douglas
Barber, Eugene Douglas
Bloom, Stephen Earl
Bogoroch, Rita
Bora, Sunder S
Borenfreund, Ellen
Boyd, Juanell N
Carpenter, David O
Castellion, Alan William
Chauhan, Ved P S
Cronkite, Eugene Pitcher
David, Oliver Joseph
Davies, Kelvin J A
Dixon, Robert Louis
Drobeck, Hans Peter
Durkin, Patrick Ralph
Ely, Thomas Sharpless
Ferin, Juraj
Finkelstein, Jacob Noah
Gates, Allen H(azen), Jr
Gordon, Harry William
Hoffman, Donald Bertrand
Kneip, Theodore Joseph
Kostyniak, Paul J
Kowolenko, Michael D
Krasavage, Walter Joseph
Latimer, Clinton Narath
Lin, George Hung-Yin
McCann, Michael Francis
Maines, Mahin D
Mantel, Linda Habas
Miller, Richard Kermit
Milmore, John Edward
Muessig, Paul Henry
Neal, Michael William
Oberdörster, Günter
O'Donoghue, John Lipomi
Pool, William Robert
Richie, John Peter, Jr
Salwen, Martin J
Smith, Thomas Harry Francis
Stark, Dennis Michael
Sundaram, Kalyan
Weinstein, Leonard Harlan
Weisburger, John Hans
Weiss, Bernard
Zedeck, Morris Samuel

NORTH CAROLINA
Adler, Kenneth B
Aldridge, David William
Anderson, Roger Fabian
Bishop, Jack Belmont
Bond, James Anthony
Boreiko, Craig John
Boyd, Jeffrey Allen
Carter, Charleata A
Chan, Po Chuen
Claxton, Larry Davis
Dauterman, Walter Carl
Dieter, Michael Phillip
Di Giulio, Richard Thomas
Dunnick, June K
Durham, William Fay
Eastin, William Clarence, Jr
Fail, Patricia A
Fouts, James Ralph
Gardner, Donald Eugene
Graham, Doyle Gene
Hart, Larry Glen
Heck, Henry d'Arcy
Heck, Jonathan Daniel
Hodgson, Ernest
Johnson, Franklin M
Krasny, Harvey Charles
Little, Patrick Joseph
McBay, Arthur John
MacPhail, Robert C
Massaro, Edward Joseph
Matthews, Hazel Benton, Jr
Mennear, John Hartley
Mitchell, Ann Denman
Morgan, Kevin Thomas
Morrissey, Richard Edward
Popp, James Alan
Riviere, Jim Edmond
St Clair, Mary Beth Genter
Schiller, Carol Masters
Schwetz, Bernard Anthony
Steinhagen, William Herrick
Suk, William Alfred
Sumner, Darrell Dean
Tilson, Hugh Arval
Toews, Arrel Dwayne
Tucker, Walter Eugene, Jr
Turner, Alvis Greely
Walters, Douglas Bruce
Waters, Michael Dee
Welsch, Frank

NORTH DAKOTA
Hein, David William
Schnell, Robert Craig

OHIO
Alden, Carl L
Banning, Jon Willroth
Baumann, Paul C
Benedict, James Harold
Bertram, Timothy Allyn
Biagini, Raymond E

Bingham, Eula
Brain, Devin King
Carmichael, Wayne William
Carpenter, Robert Leland
Carter, R Owen, Jr
Collins, William John
Coots, Robert Herman
Dodd, Darol Ennis
Dougherty, John A
Homer, George Mohn
Klemm, Donald J
Langley, Albert E
Lewis, Trent R
Milstein, Stanley Richard
Pesce, Amadeo J
Peterle, Tony J
Pfohl, Ronald John
Robinson, Michael K
Rynbrandt, Donald Jay
Staubus, Alfred Elsworth
Tabor, Marvin Wilson
Thomas, Craig Eugene
Tong, James Ying-Peh
Wong, Laurence Cheng Kong
Woodward, James Kenneth
Ziller, Stephen A, Jr

OKLAHOMA
Bantle, John Albert
Coleman, Nancy Pees
Crane, Charles Russell
Dubowski, Kurt M(ax)
Fletcher, John Samuel
Rikans, Lora Elizabeth
Rolf, Lester Leo, Jr

OREGON
Dost, Frank Norman
Garton, Ronald Ray
Henny, Charles Joseph
Koller, Loren D
Miller, Terry Lee

PENNSYLVANIA
Alarie, Yves
Alperin, Richard Junius
Blum, Lee M
Carpenter, Charles Patten
Coon, Julius Mosher
Deckert, Fred W
English, Leigh Howard
Fogleman, Ralph William
Forman, Debra L
Goldstein, Robin Sheryl
Hite, Mark
Hurt, Susan Schilt
Jaweed, Mazher
Johnson, Elmer Marshall
Karol, Meryl Helene
Kinter, Lewis Boardman
Lacouture, Peter George
Lazo, John Stephen
Majumdar, Shyamal K
Mattison, Donald Roger
Miller, Arthur James
Oldham, James Warren
Schwartz, Edward
Seligson, Frances Hess
Sherman, Larry Ray
Smith, Roger Bruce
Steele, Craig William
Sun, James Dean
Thoman, Charles James
Urbach, Frederick
Weil, Carrol S
Yankell, Samuel L
Zacchei, Anthony Gabriel
Zarkower, Arian
Zemaitis, Michael Alan

RHODE ISLAND
Brungs, William Aloysius
Yevich, Paul Peter

SOUTH CAROLINA
Woodard, Geoffrey Dean Leroy

SOUTH DAKOTA
Evenson, Donald Paul
Hamilton, Steven J

TENNESSEE
Briggs, Robert Chester
Dettbarn, Wolf Dietrich
Lawrence, William Homer
Lyman, Beverly Ann
Marnett, Lawrence Joseph
Shugart, Lee Raleigh
Skalko, Richard G(allant)
Suter, Glenn Walter, II
White, David Cleaveland

TEXAS
Ansari, Guhlam Ahmad Shakeel
Arenaz, Pablo
Bailey, Everett Murl, Jr
Bjorndahl, Jay Mark
Bost, Robert Orion
Cody, John T
Ganther, Howard Edward
Garcia, Hector D
Gingell, Ralph
Gothelf, Bernard
Hsie, Abraham Wuhsiung

Kim, Hyeong Lak
Kuhn, Janice Oseth
Lal, Harbans
McGrath, James Joseph
Meistrich, Marvin Lawrence
Meltz, Martin Lowell
Mitchell, Jerry R
Moreland, Ferrin Bates
Nechay, Bohdan Roman
Reagor, John Charles
Tindall, Donald R
Wallace, Jack E
Ward, Jonathan Bishop, Jr

UTAH
Cheney, Carl D
Moody, David Edward
Sharma, Raghubir Prasad

VERMONT
Schaeffer, Warren Ira
Sinclair, Peter Robert

VIRGINIA
Allison, Trenton B
Bass, Michael Lawrence
Batra, Karam Vir
Bleiberg, Marvin Jay
Borzelleca, Joseph Francis
Byrd, Daniel Madison, III
Cairns, John, Jr
Chinnici, Joseph (Frank) Peter
Donohue, Joyce Morrissey
Douglas, Jocelyn Fielding
Du, Julie (Yi-Fang) Tsai
Egle, John Lee, Jr
Gregory, Arthur Robert
Gross, Stanley Burton
Hess, John Lloyd
Hill, Jim T
Lipnick, Robert Louis
Locke, Krystyna Kopaczyk
Locke, Raymond Kenneth
McConnell, William Ray
Macdonald, Timothy Lee
Munson, Albert Enoch
Nagarkatti, Mitzi
Nagarkatti, Prakash S
Pages, Robert Alex
Reno, Frederick Edmund
Schulze, Gene Edward
Wilkinson, Christopher Foster

WASHINGTON
Dauble, Dennis Deene
Karagianes, Manuel Tom
Long, Edward R
Lovely, Richard Herbert
Mahlum, Daniel Dennis
Sasser, Lyle Blaine
Sikov, Melvin Richard
Wehner, Alfred Peter
Whelton, Bartlett David
Yu, Ming-Ho

WEST VIRGINIA
Castranova, Vincent
Rankin, Gary O'Neal

WISCONSIN
Balke, Nelson Edward
Brooks, Arthur S
Cook, Mark Eric
Fahl, William Edwin
Gallenberg, Loretta A
Gilbert-Barness, Enid F
Hake, Carl (Louis)
Kaplan, Stanley
Kitzke, Eugene David
Pariza, Michael Willard
Saryan, Leon Aram
Spieler, Richard Earl

WYOMING
Lockwood, Jeffrey Alan

ALBERTA
Cook, David Alastair
Coppock, Robert Walter
Hart, David Arthur
Madiyalakan, Ragupathy

BRITISH COLUMBIA
Autor, Anne Pomeroy
Banister, Eric Wilton
Bellward, Gail Dianne
Davison, Allan John
Lauzier, Raymond B
Law, Francis C P
Warrington, Patrick Douglas

MANITOBA
Giles, Michael Arthur
Glavin, Gary Bertrun
Szekely, Joseph George

NEW BRUNSWICK
Cook, Robert Harry
Haya, Katsuji

NEWFOUNDLAND
Virgo, Bruce Barton

NOVA SCOTIA
Gilgan, Michael Wilson
MacRae, Thomas Henry

ONTARIO
Burnison, Bryan Kent
Chappel, Clifford
Cherian, M George
Feuer, George
Goyer, Robert Andrew
Graham, Richard Charles Burwell
Harris, Charles Ronald
Hollbach, Natasha Coffin
Kalow, Werner
Kelso, John Richard Murray
Khanna, Jatinder Mohan
Liu, Dickson Lee Shen
McCalla, Dennis Robert
Robinson, Gerald Arthur
Roy, Marie Lessard
Smith, Lorraine Catherine
Trenholm, Harold Locksley

QUEBEC
Bensley, Edward Horton
Kafer, Etta (Mrs E R Boothroyd)

SASKATCHEWAN
Blakley, Barry Raymond
Huang, P M
Irvine, Donald Grant
Khachatourians, George G

OTHER COUNTRIES
Back, Kenneth Charles
Gustafsson, Jan-Ake
Ide, Hiroyuki
Jondorf, W Robert
Mølhave, Lars
Nabeshima, Toshitaka
Nakagawa, Shizutoshi
Orrenius, Sten
Stoppani, Andres Oscar Manuel
Sung, Chen-Yu
Van Loveren, Henk

Zoology

ALABAMA
Adams, Curtis H
Arrington, Richard, Jr
Bacon, Arthur Lorenza
Best, Troy Lee
Beyers, Robert John
Boozer, Reuben Bryan
Boschung, Herbert Theodore
Brown, Jack Stanley
Carter, Howard Payne
Current, William L
Davis, David Gale
Dixon, Carl Franklin
Dusi, Julian Luigi
Fitzpatrick, J(oseph) F(erris), Jr
Folkerts, George William
Freeman, John A
Gilliland, Floyd Ray, Jr
Hajduk, Stephen Louis
Hartley, Marshall Wendell
Hemphill, Andrew Frederick
Holler, Nicholas Robert
Holliman, Dan Clark
Jenkins, Ronald Lee
Loop, Michael Stuart
Mason, William Hickman
Mitchell, Joseph Christopher
Modlin, Richard Frank
Mount, Robert Hughes
Sanford, L G
Sharma, Udhishtra Deva
Shipp, Robert Lewis
Yokley, Paul, Jr

ALASKA
Fay, Francis Hollis
Feist, Dale Daniel
Hameedi, Mohammad Jawed
Kessel, Brina
Klein, David Robert
Kudenov, Jerry David
Peyton, Leonard James
Schoen, John Warren
Thomas, Gary Lee
Weeden, Robert Barton
Wennerstrom, David E
West, George Curtiss
Williamson, Francis Sidney Lanier

ARIZONA
Alvarado, Ronald Herbert
Anderson, Glenn Arthur
Bradshaw, Gordon Van Rensselaer
Clark, Clarence Floyd
Cockrum, Elmer Lendell
Cohen, Andrew Scott
Davidson, Elizabeth West
Davis, Russell Price
Faeth, Stanley Herman
Fouquette, Martin John, Jr
Goslow, George E, Jr
Grim, J(ohn)Norman
Hadley, Mac Eugene
Hendrickson, John Roscoe
Johnson, Oliver William

Krutzsch, Philip Henry
Landers, Earl James
Lowe, Charles Herbert, Jr
McClure, Michael Allen
Maienschein, Jane Ann
Mead, Albert Raymond
Mead, James Irving
Minckley, Wendell Lee
Montgomery, Willson Linn
Ohmart, Robert Dale
Patterson, Robert Allen
Rosenzweig, Michael Leo
Russell, Stephen Mims
Smith, Andrew Thomas
Spofford, Sally Hoyt
Spofford, Walter Richardson, II
Van Riper, Charles, III
Vaughan, Terry Alfred
Walsberg, Glenn Eric
Welch, Claude Alton
White, John Anderson
Wilkes, Stanley Northrup
Woodin, William Hartman, III
Zimmerman, James Roscoe
Zweifel, Richard George

ARKANSAS
Amlaner, Charles Joseph, Jr
Bailey, Claudia F
Baker, Frank Hamon
Beadles, John Kenneth
Bentley, Cleo L
Couser, Raymond Dowell
Dorris, Peggy Rae
Fribourgh, James H
Hanebrink, Earl L
Harp, George Lemaul
Heidt, Gary A
Kilambi, Raj Varad
Kraemer, Louise Russert
McDaniel, Van Rick
Morgans, Leland Foster
Plummer, Michael V(an)
Rakes, Jerry Max
Robison, Henry Welborn
Schmitz, Eugene H
Sealander, John Arthur, Jr
Smith, Kimberly Gray
Walker, James Martin

CALIFORNIA
Alberch, Pere
Alexander, Claude Gordon
Alvarino de Leira, Angeles
Anderson, Bertin W
Andrews, Fred Gordon
Andrus, William DeWitt, Jr
Anthony, Ronald Lewis
Antipa, Gregory Alexis
Armstrong, Peter Brownell
Arnold, John Ronald
Arora, Harbans Lall
Arp, Alissa Jan
Austin, Joseph Wells
Bair, Thomas De Pinna
Baker, Mary Ann
Baldwin, James Gordon
Balgooyen, Thomas Gerrit
Barlow, George Webber
Bartholomew, George Adelbert
Beatty, Kenneth Wilson
Beck, Albert J
Beeman, Robert D
Benes, Elinor Simson
Bennett, Albert Farrell
Bern, Howard Alan
Berrend, Robert E
Bertsch, Hans
Beuchat, Carol Ann
Bierzychudek, Paulette F
Black, John David
Bolaffi, Janice Lerner
Boolootian, Richard Andrew
Bowers, Darl Eugene
Bowers, Roger Raymond
Boyd, Milton John
Bradbury, Margaret G
Brand, Leonard Roy
Brattstrom, Bayard Holmes
Brittan, Martin Ralph
Broadbooks, Harold Eugene
Brownell, Robert Leo, Jr
Brusca, Gary J
Brusca, Richard Charles
Buchsbaum, Ralph
Buth, Donald George
Cailliet, Gregor Michel
Calarco, Patricia G
Callison, George
Campbell, James L
Carpenter, Roger Edwin
Case, Ted Joseph
Castro, Peter
Cech, Joseph Jerome, Jr
Chamberlain, Dilworth Woolley
Chang, Ernest Sun-Mei
Chappell, Mark Allen
Chen, Lo-Chai
Cheng, Lanna
Christianson, Lee (Edward)
Church, Ronald L
Cliff, Frank Samuel
Coan, Eugene Victor
Cody, Martin Leonard

Cogswell, Howard Lyman
Cohen, Anne Carolyn Constant
Cohen, Daniel Morris
Cohen, Nathan Wolf
Collias, Elsie Cole
Collias, Nicholas Elias
Collier, Gerald
Collins, Charles Thompson
Costa, Daniel Paul
Courchesne, Eric
Cowles, David Lyle
Cox, George W
Crane, Jules M, Jr
Crouch, James Ensign
DeMartini, John
Desharnais, Robert Anthony
DeWolfe, Barbara Blanchard Oakeson
Dowell, Armstrong Manly
Drewes, Robert Clifton
Dudek, F Edward
Durrant, Barbara Susan
Eakin, Richard Marshall
Ebeling, Alfred W
Edmunds, Peter James
Edwards, J Gordon
Egge, Alfred Severin
Endler, John Arthur
Enright, James Thomas
Eschmeyer, William Neil
Estes, James Allen
Etheridge, Richard Emmett
Evans, Kenneth Jack
Federici, Brian Anthony
Feldmeth, Carl Robert
Fell, Howard Barraclough
Ferguson, William E
Fierstine, Harry Lee
Filice, Francis P
Fiorindo, Robert Philip
Fisler, George Frederick
Flaim, Francis Richard
Fleminger, Abraham
Follett, Wilbur Irving
Fox, Laurel R
Fusaro, Craig Allen
Gabel, James Russel
Galt, Charles Parker, Jr
Garth, John Shrader
Gascoigne, Nicholas Robert John
George, Sarah B(rewster)
Ghiselin, Michael Tenant
Goldberg, Stephen Robert
Goldstein, Bernard
Gorman, George Charles
Gray, Constance Helen
Greene, Richard Wallace
Guthrie, Daniel Albert
Haas, Richard
Hackney, Robert Ward
Hammar, Allan H
Hand, Cadet Hammond, Jr
Hanson, William Roderick
Hardy, Cecil Ross
Harmon, Wallace Morrow
Harris, John Michael
Harris, Lester Earle, Jr
Hemingway, George Thomson
Hendler, Gordon Lee
Hermans, Colin Olmsted
Hessler, Robert Raymond
Hildebrand, Milton
Ho, Ju-Shey
Hobson, Edmund Schofield
Holland, Nicholas Drew
Hooper, Emmet Thurman, Jr
Horn, Michael Hastings
Houston, Roy Seamands
Howell, Thomas Raymond
Hoyt, Donald Frank
Hsueh, Aaron Jen Wang
Huckaby, David George
Hunsaker, Don, II
Iwamoto, Tomio
Jackson, Crawford Gardner, Jr
Jacobs, John Allen
James, Thomas William
Jameson, Everett Williams, Jr
Jennrich, Ellen Coutlee
Jessop, Nancy Meyer
Johnson, Eric Van
Johnson, Ned Keith
Jones, Ira
Jones, James Henry
Jones, Ronald McClung
Jorgenson, Edsel Carpenter
Keller, Raymond E
Kenk, Roman
Kenk, Vida Carmen
Kitting, Christopher Lee
Kooyman, Gerald Lee
Krejsa, Richard Joseph
Lambert, Charles Calvin
Lawrence, George Edwin
Lee, Sue Ying
Leitner, Philip
Lenhoff, Howard Maer
Leviton, Alan Edward
Lewis, Cynthia Lucille
Licht, Paul
Lidicker, William Zander, Jr
Lindberg, David Robert
Loher, Werner J
Longhurst, John Charles
Loomis, Richard Biggar

Loomis, Robert Henry
Lovich, Jeffrey Edward
Lownsbery, Benjamin Ferris
Luckock, Arlene Suzanne
Lyke, Edward Bonsteel
McCosker, John E
Mace, John Weldon
McFarland, William Norman
McLaughlin, Charles Albert
McLean, James H
McLean, Norman, Jr
McLeod, Samuel Albert
MacMillen, Richard Edward
Maggenti, Armand Richard
Main, Robert Andrew
Mankau, Sarojam Kurudamannil
Marler, Peter
Martin, Juel William
Mayhew, Wilbur Waldo
Mazia, Daniel
Mead, Giles Willis
Mercer, Edward King
Morin, James Gunnar
Morris, Robert Wharton
Morton, Martin Lewis
Mossman, Archie Stanton
Moyle, Peter Briggs
Mulligan, Timothy James
Mulloney, Brian
Murphy, Ted Daniel
Muscatine, Leonard
Myrick, Albert Charles, Jr
Nandi, Satyabrata
Nelson, Keith
Newberry, Andrew Todd
Nishioka, Richard Seiji
Noffsinger, Ella Mae
Norris, Kenneth Stafford
Nybakken, James W
Ohlendorf, Harry Max
Ono, Joyce Kazuyo
Orr, Robert Thomas
Padian, Kevin
Parrish, Richard Henry
Patton, James Lloyd
Pavgi, Sushama
Pearse, Vicki Buchsbaum
Perrin, William Fergus
Petersen, Bruce Wallace
Phleger, Charles Frederick
Pitelka, Dorothy Riggs
Pitelka, Frank Alois
Pitelka, Louis Frank
Platzer, Edward George
Plopper, Charles George
Plymale, Harry Hambleton
Pohlo, Ross
Poinar, George O, Jr
Potts, Donald Cameron
Power, Dennis Michael
Presch, William Frederick
Price, Mary Vaughan
Prothero, Donald Ross
Randall, Janet Ann
Raski, Dewey John
Reisen, William Kenneth
Reish, Donald James
Risch, Stephen John
Risser, Arthur Crane, Jr
Roche, Edward Towne
Roe, Pamela
Roest, Aryan Ingomar
Rogoff, William Milton
Rosenblatt, Richard Heinrich
Rudd, Robert L
Ruibal, Rodolfo
Saffo, Mary Beth
Scherba, Gerald Marron
Schooley, Caroline Naus
Sekhon, Sant Singh
Shellhammer, Howard Stephen
Sherman, Irwin William
Sibley, Charles Gald
Simpson, Larry P
Smith, Ralph Ingram
Smith, Thomas Patrick
Soltz, David Lee
Soule, John Dutcher
Spieler, Richard Arno
Spieth, Herman Theodore
Standing, Keith M
Stanton, Toni Lynn
Starrett, Andrew
Stebbins, Robert Cyril
Steele, Arnold Edward
Stephens, Grover Cleveland
Stewart, Glenn Raymond
Stiffler, Daniel F
Stohler, Rudolf
Stott, Kenhelm Welburn, Jr
Strecker, Robert Louis
Strohman, Richard Campbell
Swift, Camm Churchill
Taylor, Dwight Willard
Taylor, Leighton Robert, Jr
Thomas, Barry
Thomas, Bert O
Thueson, David Orel
Tisserat, Brent Howard
Tjioe, Djoe Tjhoo
Tomlinson, Jack Trish
Torch, Reuben
Torell, Donald Theodore
Trapp, Gene Robert

Zoology (cont)

Tribbey, Bert Allen
Triplett, Edward Lee
Turner, Frederick Brown
Udvardy, Miklos Dezso Ferenc
VanBlaricom, Glenn R
Vance, Velma Joyce
Vehrencamp, Sandra Lee
Viglierchio, David Richard
Vitt, Laurie Joseph
Wake, Marvalee H
Walters, Roland Dick
Warter, Stuart L
Washburn, Sherwood L
Waters, James Frederick
Way, Jon Leong
Welsh, James Francis
Wenner, Adrian Manley
Westervelt, Clinton Albert, Jr
Westfahl, Pamela Kay
Weston, Henry Griggs, Jr
Wickler, Steven John
Widmer, Elmer Andreas
Williams, Daniel Frank
Wilson, Allan Charles
Wilson, Barry William
Wirtz, William Otis, II
Wood, Forrest Glenn
Wood, Richard Lyman
Woodwick, Keith Harris
Wright, Richard Kenneth
Yarnall, John Lee
Young, Donald Raymond
Zieg, Roger Grant
Zimmer, Russel Leonard
Zuk, Marlene

COLORADO
Angell, Robert Walker
Apley, Martyn Linn
Armstrong, David M(ichael)
Audesirk, Teresa Eck
Bedinger, Charles Arthur, Jr
Belden, Don Alexander, Jr
Bock, Carl E
Bushnell, John Horace
Capen, Ronald L
Cassel, J(oseph) Frank(lin)
Cohen, Robert Roy
Cruz, Alexander
Enderson, James H
Erickson, James George
Fall, Michael William
Finger, Thomas Emanuel
Finley, Robert Byron, Jr
Fuller, Barbara Fink Brockway
Green, Jeffrey Scott
Hall, Joseph Glenn
Hanken, James
Hansen, Richard M
Harriss, Thomas T
Herrmann, Scott Joseph
Hibler, Charles Phillip
Janes, Donald Wallace
Jones, Richard Evan
Lehner, Philip Nelson
Linder, Allan David
McClellan, John Forbes
McLean, Robert George
Marquardt, William Charles
Mayer, William Vernon
Moury, John David
Nash, Donald Joseph
Packard, Gary Claire
Pennak, Robert William
Pettus, David
Plakke, Ronald Keith
Schmidt, Gerald D
Seilheimer, Jack Arthur
Smith, Hobart Muir
Spencer, Albert William
Sublette, James Edward
Voth, David Richard
Ward, Gerald Madison
Woolley, Tyler Anderson
Wunder, Bruce Arnold
Zeiner, Frederick Neyer

CONNECTICUT
Aylesworth, Thomas Gibbons
Bernard, Richard Fernand
Blackburn, Daniel Glenn
Booth, Charles E
Brewer, Robert Hyde
Brush, Alan Howard
Carlton, James Theodore
Chichester, Lyle Franklin
Clark, George Alfred, Jr
Cohen, Melvin Joseph
DeCoursey, Russell Myles
Fiore, Carl
Gable, Michael F
Grimm-Jorgensen, Yvonne
Hanks, James Elden
Hartman, Willard Daniel
Hutchinson, George Evelyn
James, Hugo A
Jokinen, Eileen Hope
Kent, John Franklin
Loomis, Stephen Henry
Lund, William Albert, Jr
Schake, Lowell Martin
Simpson, Tracy L

Singletary, Robert Lombard
Spiers, Donald Ellis
Steinmetz, Charles, Jr
Swain, Elisabeth Ramsay
Taigen, Theodore Lee
Thurberg, Frederick Peter
Wells, Kentwood David
Wheeler, Bernice Marion

DELAWARE
Boord, Robert Lennis
Carriker, Melbourne Romaine
Curtis, Lawrence Andrew
Ferguson, Thomas
Gaffney, Patrick M
Karlson, Ronald Henry
Koeppe, Mary Kolean
Pierson, Keith Bernard
Roth, Roland Ray
Song, Jiakun
Targett, Timothy Erwin
Tripp, Marenes Robert

DISTRICT OF COLUMBIA
Allen, George E
Banks, Richard Charles
Banks, William Michael
Banta, William Claude
Barnard, Jerry Laurens
Bayer, Frederick Merkle
Bellmer, Elizabeth Henry
Bernor, Raymond Louis
Bowman, Thomas Elliot
Burns, John McLauren
Bushman, John Branson
Cairns, Stephen Douglas
Campbell, Carlos Boyd Godfrey
Collette, Bruce Baden
Cooper, George Everett
Dickens, Charles Henderson
Dimond, Marie Therese
Domning, Daryl Paul
Fauchald, Kristian
Feeney, Gloria Comulada
Harbach, Ralph Edward
Higgins, Robert Price
Hobbs, Horton Holcombe, Jr
Hope, William Duane
Hoskin, George Perry
Kapur, Shakti Prakash
Knapp, Leslie W
Knowlton, Robert Earle
Lachner, Ernest Albert
Lovejoy, Thomas E
McDiarmid, Roy Wallace
McLaughlin, David
Manning, Raymond B
Mathis, Wayne Neilsen
Merchant, Henry Clifton
Nickle, David Allan
Nickum, John Gerald
Olson, Storrs Lovejoy
Paris, Oscar Hall
Pawson, David Leo
Perez-Farfante, Isabel Cristina
Pettibone, Marian Hope
Puglia, Charles Raymond
Ralls, Katherine Smith
Rehder, Harald Alfred
Reichman, Omer James
Rice, Mary Esther
Ripley, Sidney Dillon, II
Robinson, Michael Hill
Roper, Clyde Forrest Eugene
Springer, Victor Gruschka
Swinebroad, Jeff
Tyler, James Chase
Umminger, Bruce Lynn
Waller, Thomas Richard
Watson, George E, III
Wauer, Roland H
Weitzman, Stanley Howard
West, William Lionel
Williams, Austin Beatty
Williams, Jeffrey Taylor
Wilson, Don Ellis
Zug, George R

FLORIDA
Adams, Roger Omar
Allen, Ted Tipton
Anderson, Peter Alexander Vallance
Auffenberg, Walter
Austin, Oliver Luther, Jr
Bazer, Fuller Warren
Berner, Lewis
Bick, George Herman
Bortone, Stephen Anthony
Branch, Lyn Clarke
Britt, N Wilson
Brockmann, Helen Jane
Brodkorb, Pierce
Brown, Larry Nelson
Caldwell, David Keller
Caldwell, Melba Carstarphen
Carpenter, James Woodford
Castner, James Lee
Clark, Kerry Bruce
Colwin, Arthur Lentz
Colwin, Laura Hunter
Cooley, Nelson Reede
Courtenay, Walter Rowe, Jr
Courtney, Charles Hill
De Witt, Robert Merkle

Dickinson, Joshua Clifton, Jr
Dickson, Donald Ward
Dietz, John R
Dunkle, Sidney Warren
Eckelbarger, Kevin Jay
Ehrhart, Llewellyn McDowell
Elliott, Alfred Marlyn
Ellis, Leslie Lee, Jr
Evans, David Hudson
Feddern, Henry A
Ferguson, John Carruthers
Fitzpatrick, John Weaver
Fleming, Theodore Harris
Flowers, Ralph Wills
Forrester, Donald Jason
Fournie, John William
Giblin-Davis, Robin Michael
Gilbert, Carter Rowell
Gilbert, Perry Webster
Gleeson, Richard Alan
Goldberg, Stephen
Goldberg, Walter M
Gordon, Kenneth Richard
Gorniak, Gerard Charles
Grabowski, Casimer Thaddeus
Greenberg, Michael John
Groupe, Vincent
Gude, Richard Hunter
Gupta, Virendra K
Hardy, John William
Hays, Elizabeth Teuber
Heard, William Herman
Heckerman, Raymond Otto
Hinsch, Gertrude Wilma
Holling, Crawford Stanley
Kale, Herbert William, II
Kaufmann, John Henry
Kerr, John Polk
Kinloch, Robert Armstrong
Kirkwood, James Benjamine
Kruczynski, William Leonard
Lane, Charles Edward
Layne, James Nathaniel
Lessios, Harilaos Angelou
Levey, Douglas J
Levy, Richard
Lillywhite, Harvey B
McKenney, Charles Lynn, Jr
McSorley, Robert
McSweeny, Edward Shearman
Mariscal, Richard North
Maturo, Frank J S, Jr
Means, D Bruce
Mendez, Eustorgio
Meryman, Charles Dale
Michel, Harding B
Miyamoto, Michael Masao
Moore, Donald Richard
Moore, Joseph Curtis
Myrberg, Arthur August, Jr
Nordlie, Frank Gerald
O'Bannon, John Horatio
Odell, Daniel Keith
Overman, Amegda Jack
Owre, Oscar Theodore
Patterson, Richard Sheldon
Perry, Vernon G
Porter, James Armer, Jr
Porter, Sanford Dee
Povar, Morris Leon
Reddish, Robert Lee
Reynolds, John Elliott, III
Rhoades, Harlan Leon
Rice, Nolan Ernest
Rice, Stanley Alan
Richards, William Joseph
Riggs, Carl Daniel
Robins, Charles Richard
Robinson, Gerald Garland
Savage, Jay Mathers
Sayles, Everett Duane
Schwartz, Albert
Simon, Joseph Leslie
Smith, Frederick George Walton
Snelson, Franklin F, Jr
Stafford, Robert Oppen
Stevenson, Henry Miller
Stokes, Donald Eugene
Sturbaum, Barbara Ann
Tarjan, Armen Charles
Taylor, George Thomas
Taylor, Walter Kingsley
Telford, Sam Rountree, Jr
Thommes, Robert Charles
Thompson, Fred Gilbert
Thurber, Walter Arthur
Travis, Joseph
Tripp, John Rathbone
Wallbrunn, Henry Maurice
Webb, Sawney David
White, Arlynn Quinton, Jr
Wing, Elizabeth S
Witham, P(hilip) Ross
Wolff, Ronald Gilbert
Woodruff, Robert Eugene
Woods, Charles Arthur
Woolfenden, Glen Everett
Wyneken, Jeanette
Yerger, Ralph William
Yunker, Conrad Erhardt

GEORGIA
Adkison, Daniel Lee
Amerson, Grady Malcolm

Avise, John C
Bates, Harold Brennan, Jr
Briggs, John Carmon
Brown, Paul Lopez
Burbanck, William Dudley
Caldwell, Sloan Daniel
Comeau, Roger William
Costoff, Allen
Cutler, Horace Garnett
Davenport, Leslie Bryan, Jr
English, Arthur William
Fechheimer, Marcus
Fitt, William K
Green, Edwin Alfred
Hanson, William Lewis
Hicks, Heraline Elaine
Hunter, Frissell Roy
Husain, Ansar
Johnson, Alva William
Keeler, Clyde Edgar
Kneib, Ronald Thomas
Laerm, Joshua
McKeever, Sturgis
Mapp, Frederick Everett
Miller, George C
Murray, Joan Baird
Nadler, Ronald D
Odum, Eugene Pleasants
Patterson, Rosalyn Mitchell
Paulin, Jerome John
Porter, James W
Provost, Ernest Edmund
Relyea, Kenneth George
Saladin, Kenneth S
Scott, Donald Charles
Seerley, Robert Wayne
Sharp, Homer Franklin, Jr
Shibley, John Luke
Sridaran, Rajagopala
Stirewalt, Harvey Lee
Taylor, Robert Clement
Tietjen, William Leighton
Urban, Emil Karl
Urso, Paul
Utley, Philip Ray
White, Donald Henry
Willis, John Steele

HAWAII
Ahearn, Gregory Allen
Arnold, John Miller
Bailey-Brock, Julie Helen
Berg, Carl John, Jr
Berger, Andrew John
DeMartini, Edward Emile
Eddinger, Charles Robert
Eldredge, Lucius G
Freed, Leonard Alan
Grau, Edward Gordon
Greenfield, David Wayne
Hadfield, Michael Gale
Haley, Samuel Randolph
Hsiao, Sidney Chihti
Hunter, Cynthia L
Kamemoto, Fred Isamu
Kay, Elizabeth Alison
Kinzie, Robert Allen, III
Kondo, Yoshio
Koshi, James H
Little, Harold Franklin
Losey, George Spahr, Jr
Pyle, Robert Lawrence
Reed, Stuart Arthur
Reimer, Diedrich
Tinker, Spencer Wilkie
Tomich, Prosper Quentin
Vargas, Roger I
Wyban, James A

IDAHO
Berrett, Delwyn Green
Bunde, Daryl E
Cade, Thomas Joseph
Fritchman, Harry Kier, II
Goddard, Stephen
Hibbert, Larry Eugene
Kelley, Fenton Crosland
Mead, Rodney A
Reno, Harley W
Schell, Stewart Claude
Trost, Charles Henry
Wootton, Donald Merchant

ILLINOIS
Alger, Nelda Elizabeth
Andrews, Richard D
Axtell, Ralph William
Bartell, Marvin H
Beck, Sidney L
Beecher, William John
Bennett, Cecil Jackson
Berry, James Frederick
Bieler, Rüdiger
Binford, Laurence Charles
Blake, Emmet Reid
Bolt, John Ryan
Brandon, Ronald Arthur
Brown, Lauren Evans
Buhse, Howard Edward, Jr
Burhop, Kenneth Eugene
Burr, Brooks Milo
Carr, Tommy Russell
Cracraft, Joel Lester
Damaskus, Charles William

Dinsmore, Charles Earle
Dudkiewicz, Alan Bernard
Dunn, Robert Bruce
Dybas, Linda Kathryn
Edwards, Dale Ivan
Fanslow, Don J
Finkel, Asher Joseph
Fooden, Jack
Forman, G Lawrence
Foster, Merrill W
Foulkes, Robert Hugh
Franks, Edwin Clark
Freiburg, Richard Eighme
Funk, Richard Cullen
Garoian, George
Garrigus, Upson Stanley
Garthe, William A
George, William
Giere, Frederic Arthur
Girard, G Tanner
Goldberg, Robert Jack
Goodrich, Michael Alan
Graber, Richard Rex
Gramza, Anthony Francis
Green, Jonathan P
Griffiths, Thomas Alan
Hausler, Carl Louis
Heaney, Lawrence R
Heath, James Edward
Hetzel, Howard Roy
Heuberger, Glen (Louis)
Hoffmeister, Donald Frederick
Holmes, Edward Bruce
Hunt, Lawrence Barrie
Jablonski, David
Jensen, Donald Reed
Johnson-Wint, Barbara Paule
Jollie, Malcolm Thomas
Kanatzar, Charles Leplie
Ketterer, John Joseph
Kruidenier, Francis Jeremiah
Kruse, Kipp Colby
Larson, Ingemar W
Le Febvre, Eugene Allen
Lehmann, Wilma Helen
Levine, Norman Dion
Lima, Gail M
Lombard, Richard Eric
Lutsch, Edward F
Mackal, Roy Paul
Malek, Richard Barry
Marx, Hymen
Mehta, Rajendra G
Meseth, Earl Herbert
Moll, Edward Owen
Morgan, Juliet
Munyer, Edward Arnold
Myer, Donal Gene
Nadler, Charles Fenger
Naples, Virginia L
Natalini, John Joseph
Nielsen, Peter James
Page, Lawrence Merle
Peterson, Darryl Ronnie
Pruett-Jones, Stephen Glen
Rabb, George Bernard
Rabinowitch, Victor
Rafferty, Nancy S
Ratzlaff, Kermit O
Reed, Charles Allen
Ridgeway, Bill Tom
Robertson, Hugh Mereth
Rosenthal, Gerson Max, Jr
Rosenthal, Marcia White
Sanderson, Glen Charles
Sather, J Henry
Semrad, Joseph Edward
Shepherd, Benjamin Arthur
Shomay, David
Solem, G Alan
Sprugel, George, Jr
Stains, Howard James
Thiruvathukal, Kris V
Thomerson, Jamie E
Thompson, Charles Frederick
Thonar, Eugene Jean-Marie
Throckmorton, Lynn Hiram
Thurow, Gordon Ray
Traylor, Melvin Alvah, Jr
Troll, Ralph
Uzzell, Thomas
Vetter, Richard L
von Zellen, Bruce Walfred
Voris, Harold K
Waltenbaugh, Carl
Waring, George Houstoun, IV
Warnock, John Edward
Weigel, Robert David
Whitt, Dixie Dailey
Williams, Thomas Alan
Wittig, Gertraude Christa
Wolff, Robert John
Zalisko, Edward John

INDIANA

Bergeson, Glenn Bernard
Burkholder, Timothy Jay
Cable, Raymond Millard
Corrigan, John Joseph
Crowell, (Prince) Sears, Jr
Cushing, Bruce S
Durflinger, Elizabeth Ward
Dustman, John Henry
Eberly, William Robert

Ferris, John Mason
Ferris, Virginia Rogers
Frey, David Grover
Gaudin, Anthony J
Gidda, Jaswant Singh
Goetz, Frederick William, Jr
Harrington, Rodney B
Hopp, William Beecher
Iverson, John Burton
Jacobs, Merle Emmor
Johnson, Willis Hugh
Jones, Duvall Albert
Judge, Max David
Kelley, George W, Jr
Krekeler, Carl Herman
List, James Carl
Lodge, David Michael
McCafferty, William Patrick
Maloney, Michael Stephen
Mays, Charles Edwin
Meyer, Frederick Richard
Minton, Sherman Anthony
Nelson, Craig Eugene
Olson, John Bennet
O'Neal, Lyman Henry
Ott, Karen Jacobs
Pachut, Joseph F(rancis)
Perrill, Stephen Arthur
Platt, Thomas Reid
Preer, John Randolph, Jr
Roales, Robert R
Rowland, William Joseph
Santerre, Robert Frank
Sever, David Michael
Singleton, Wayne Louis
Stob, Martin
Tamar, Henry
Tihen, Joseph Anton
Tweedell, Kenyon Stanley
Webster, Jackson Dan
Wellso, Stanley Gordon
Werth, Robert Joseph
Williams, Eliot Churchill
Wise, Charles Davidson
Zinsmeister, William John

IOWA

Ackerman, Ralph Austin
Adams, Donald Robert
Beams, Harold William
Bowles, John Bedell
Buttrey, Benton Wilson
Ciochon, Russell Lynn
Cook, Kenneth Marlin
Dawson, David Lynn
Dolphin, Warren Dean
Dunham, Jewett
Goellner, Karl Eugene
Hicks, Ellis Arden
Jahn, J Russell
Kessel, Richard Glen
Klaas, Erwin Eugene
Kline, Edwin A
Krafsur, Elliot Scoville
Laffoon, John
Lewis, Charles J
Lewis, Robert Earl
Lyon, David Louis
Menzel, Bruce Willard
Mertins, James Walter
Norton, Don Carlos
Outka, Darryll E
Parrish, Frederick Charles, Jr
Reitan, Phillip Jennings
Robson, Richard Morris
Rogers, Frances Arlene
Rogers, Thomas Edwin
Rundell, Harold Lee
Solursh, Michael
Stevermer, Emmett J
Sullivan, Charles Henry
Ulrich, Merwyn Gene
Vohs, Paul Anthony, Jr
Welshons, William John
Wilson, Nixon Albert
Zmolek, William G

KANSAS

Allen, Deloran Matthew
Ashe, James S
Besharse, Joseph Culp
Boyd, Roger Lee
Boyer, Don Raymond
Burkholder, John Henry
Burton, Paul Ray
Choate, Jerry Ronald
Cink, Calvin Lee
Clark, George Richmond, II
Clarke, Robert Francis
Cross, Frank Bernard
Dehner, Eugene William
Duellman, William Edward
Ely, Charles Adelbert
Fautin, Daphne Gail
Fleharty, Eugene
Franzen, Dorothea Susanna
Good, Don L
Hays, Horace Albennie
Humphrey, Philip Strong
Hunter, Wanda Sanborn
Jenkinson, Marion Anne
Johnson, John Christopher, Jr
Johnston, Richard Fourness
Kaufman, Donald Wayne

Klemm, Robert David
Leavitt, Wendell William
McCrone, John David
Mengel, Robert Morrow
Michener, Charles Duncan
Morrissey, J Edward
Nelson, Victor Eugene
Neufeld, Gaylen Jay
Orr, James Anthony
Owen, Bernard Lawton
Platt, Dwight Rich
Prophet, Carl Wright
Stockhammer, Karl Adolf
Thompson, Max Clyde
Walker, Richard Francis
Westfall, Jane Anne
Wiley, E(dward) O(rlando), III
Wolfe, James Leonard

KENTUCKY

Barbour, Roger William
Birge, Wesley Joe
Branson, Branley Allan
Bryant, William Stanley
Budde, Mary Laurence
Cole, Evelyn
Crawford, Eugene Carson, Jr
Cupp, Paul Vernon, Jr
Davis, Wayne Harry
Derrickson, Charles M
Dutt, Ray Horn
Ely, Donald Gene
Ferner, John William
Ferrell, Blaine Richard
Hamon, J Hill
Hilton, Frederick Kelker
Howell, Jerry Fonce
Hoyt, Robert Dan
Just, John Josef
Kuehne, Robert Andrew
Lindsay, Dwight Marsee
Monroe, Burt Leavelle, Jr
Musacchia, X J
Oetinger, David Frederick
Prins, Rudolph
Rawls, John Marvin, Jr
Shadowen, Herbert Edwin
Sih, Andrew
Spoor, William Arthur
Stephens, Noel, Jr
Tamburro, Kathleen O'Connell
Thorp, James Harrison, III
Tucker, Ray Edwin
Uglem, Gary Lee
Westneat, David French, Jr
Whiteker, McElwyn D
Wilder, Cleo Duke
Williams, John C

LOUISIANA

Abram, James Baker, Jr
Avent, Robert M
Bamforth, Stuart Shoosmith
Bennett, Harry Jackson
Black, Joe Bernard
Boertje, Stanley
Bryan, Charles F
Christian, Frederick Ade
Collins, Richard Lapointe
Cordes, Carroll Lloyd
Corkum, Kenneth C
Davis, Billy J
Dessauer, Herbert Clay
Douglas, Neil Harrison
Dundee, Harold A
Etheridge, Albert Louis
Eyster, Marshall Blackwell
Fairbanks, Laurence Dee
Felder, Darryl Lambert
Fleeger, John Wayne
Gassie, Edward William
Goertz, John William
Green, Jeffrey David
Grodner, Robert Maynard
Hardberger, Florian Max
Hardy, Laurence McNeil
Harman, Walter James
Harrison, Richard Miller
Hastings, Robert Wayne
Homberger, Dominique Gabrielle
Hughes, Janice S
Kee, David Thomas
Kent, George Cantine, Jr
Khalaf, Kamel T
Kinn, Donald Norman
Kreider, Jack Leon
Lane, James Dale
Lanners, H Norbert
Loden, Michael Simpson
Meier, Albert Henry
Nair, Pankajam K
Norris, William Warren
Platt, William Joshua, III
Poirrier, Michael Anthony
Porter, Charles Warren
Remsen, James Vanderbeek, Jr
Ricks, Beverly Lee
Rossman, Douglas Athon
Seigel, Richard Allyn
Shaw, Richard Francis
Shepherd, David Preston
Silverman, Harold
Sizemore, Robert Carlen
Smalley, Alfred Evans

Stewart, T Bonner
Suttkus, Royal Dallas
Turner, Hugh Michael
Wakeman, John Marshall
Williams, Kenneth L

MAINE

Allen, Kenneth William
Bell, Allen L
Borei, Hans Georg
Butler, Ronald George
Dean, David
Dearborn, John Holmes
De Witt, Hugh Hamilton
Gainey, Louis Franklin, Jr
Glanz, William Edward
Greenwood, Paul Gene
Huntington, Charles Ellsworth
Larsen, Peter Foster
McCleave, James David
Martin, Robert Lawrence
Palmer, Ralph Simon
Parker, James Willard
Pettingill, Olin Sewall, Jr
Townsend, David Warren
Tyler, Seth
Watling, Les
Yuhas, Joseph George

MARYLAND

Adams, Lowell William
Ahl, Alwynelle S
Albright, Joseph Finley
Alleva, Frederic Remo
Altland, Paul Daniel
Altman, Philip Lawrence
Anderson, Lucy Macdonald
Ball, Gregory Francis
Barbour, Michael Thomas
Barnett, Audrey
Barry, Ronald Everett, Jr
Bass, Eugene L
Beyer, W Nelson
Brock, Mary Anne
Buck, John Bonner
Burton, Dennis Thorpe
Cammen, Leon Matthew
Cargo, David Garrett
Carney, William Patrick
Chace, Fenner Albert, Jr
Chen, T R
Clark, Eugenie
Coulombe, Harry N
Creighton, Phillip David
Cronin, Thomas Wells
Dalton, John Charles
Delahunty, George
Dwyer, Dennis Michael
Eckardt, Michael Jon
Endo, Burton Yoshiaki
Erickson, Howard Ralph
Fayer, Ronald
Feldmesser, Julius
Forester, Donald Charles
Freed, Arthur Nelson
Futcher, Anthony Graham
Gillespie, Walter Lee
Golden, Alva Morgan
Goldsmith, Paul Kenneth
Gorman, Cornelia M
Graham, Charles Raymond, Jr
Green, Marie Roder
Hairstone, Marcus A
Hamilton, Clara Eddy
Hansen, Ira Bowers
Highton, Richard
Hines, Anson Hemingway
Hocutt, Charles H
Hotton, Nicholas, III
Hudson, A Sue
Huheey, James Edward
Kayar, Susan Rennie
Kirby, Paul Edward
Kloetzel, John Arthur
Lall, Abner Bishamber
Landauer, Michael Robert
Leach, Berton Joe
Leedy, Daniel Loney
Lewis, Herman William
Linder, Harris Joseph
McAtee, Lloyd Thomas
McVey, James Paul
Marsden, Halsey M
Maslow, David E
Messersmith, Donald Howard
Meszler, Richard M
Michelson, Edward Harlan
Miller, Dorothea Starbuck
Moment, Gairdner Bostwick
Mosher, James Arthur
Munson, Donald Albert
Nichols, James Dale
Nickle, William R
Novotny, Robert Thomas
Oberdorfer, Michael Douglas
O'Grady, Richard Terence
Orejas-Miranda, Braulio
Osman, Richard William
Palmer, Timothy Trow
Pancella, John Raymond
Park, Lee Crandall
Phillips, William George
Piavis, George Walter
Pierson, Bernice Frances

Zoology (cont)

Pilitt, Patricia Ann
Popper, Arthur N
Potter, Jane Huntington
Powers, Kendall Gardner
Provenza, Dominic Vincent
Reddy, Gunda
Reichelderfer, Charles Franklin
Remondini, David Joseph
Rice, Clifford Paul
Robbins, Chandler Seymour
Roesijadi, Guritno
Rose, Kenneth David
Scully, Erik Paul
Shafer, W Sue
Simons, Daniel J
Small, Eugene Beach
Smith, F Harrell
Snyder, Jack Russell
Sparling, Donald Wesley, Jr
Steinmetz, Michael Anthony
Stewart, Doris Mae
Stringfellow, Frank
Tarshis, Irvin Barry
Trauger, David Lee
Van Arsdel, William Campbell, III
Vener, Kirt J
Weisberg, Stephen Barry
Wildt, David Edwin
Wiley, Martin Lee
Yoder, Wayne Alva

MASSACHUSETTS
Ahlberg, Henry David
Anderson, Olivo Margaret
Atkinson-Templeman, Kristine Hofgren
Barber, Saul Benjamin
Bertin, Robert Ian
Bond, George Walter
Booke, Henry Edward
Borton, Anthony
Boss, Kenneth Jay
Bowdan, Elizabeth Segal
Bridges, Robert Stafford
Broseghini, Albert L
Burggren, Warren William
Capuzzo, Judith M
Castro, Gonzalo
Chlapowski, Francis Joseph
Clemens, Daniel Theodore
Cornman, Ivor
Dasgupta, Arijit M
Deegan, Linda Ann
Duhamel, Raymond C
Duncan, Stewart
Eschenberg, Kathryn (Marcella)
Ezzell, Robert Marvin
Furth, David George
Gibbons, Michael Francis, Jr
Grant, William Chase, Jr
Grice, George Daniel, Jr
Griffin, Donald R(edfield)
Gustafson, Alvar Walter
Harbison, G Richard
Healy, William Ryder
Hichar, Joseph Kenneth
Hillman, Robert Edward
Hoff, James Gaven
Honigberg, Bronislaw Mark
Horner, B Elizabeth
Howe, Robert Johnston
Jahoda, John C
Jearld, Ambrose, Jr
Jones, Gwilym Strong
Klingener, David John
Kroodsma, Donald Eugene
Kunz, Thomas Henry
Laprade, Mary Hodge
Liem, Karel F
Lovejoy, David Arnold
McCormick, Stephen Daniel
McLaren, J Philip
Millard, William James
Moir, Ronald Brown, Jr
Morse, M Patricia
Muckenthaler, Florian August
Mulcare, Donald J
Naqvi, S Rehan Hasan
OHara, Robert J
Paracer, Surindar Mohan
Payne, Bertram R
Paynter, Raymond Andrew, Jr
Pearincott, Joseph V
Pechenik, Jan A
Perry, Alfred Eugene
Prescott, John Hernage
Prestwich, Kenneth Neal
Pyle, Robert Wendell
Rauch, Harold
Riser, Nathan Wendell
Robinson, William Edward
Ropes, John Warren
Sames, George L
Sanders, Howard L(awrence)
Sargent, Theodore David
Smith, Frederick Edward
Smith, Wendy Anne
Snow, Beatrice Lee
Snyder, Dana Paul
Telford, Sam Rountree, III
Tilley, Stephen George
Trinkaus-Randall, Vickery E
Turner, Ruth Dixon

Vankin, George Lawrence
Waters, Joseph Hemenway
Werntz, Henry Oscar
Wichterman, Ralph
Widmayer, Dorothea Jane
Williams, Ernest Edward
Williamson, Peter George
Wright, Mary Lou
Wyse, Gordon Arthur
Zinn, Donald Joseph
Zottoli, Robert

MICHIGAN
Aggarwal, Surinder K
Albright, Raymond Gerard
Alexander, Richard Dale
Atkinson, James William
Balaban, Martin
Band, Rudolph Neal
Blankespoor, Harvey Dale
Bowen, Stephen Hartman
Bowker, Richard George
Brady, Allen Roy
Brewer, Richard (Dean)
Buhl, Allen Edwin
Burch, John Bayard
Butsch, Robert Stearns
Calvert, Jay Gregory
Caswell, Herbert Hall, Jr
Cather, James Newton
Chobotar, Bill
Conder, George Anthony
Connelly, Thomas George
Cook, David Russell
Cooper, William E
Courtney, Gladys (Atkins)
Dillery, Dean George
Distler, Jack, Jr
Dulin, William E
Edgar, Arlan Lee
Eichler, Victor B
Engemann, Joseph George
Eyer, Lester Emery
Fennel, William Edward
Fink, William Lee
Fleming, Richard Cornwell
Frantz, William Lawrence
Gangwere, Stanley Kenneth
Gans, Carl
Gillingham, James Clark
Gordon, Sheldon Robert
Hensley, Marvin Max
Hill, Richard William
Hooper, Frank Fincher
Horowitz, Samuel Boris
Houts, Larry Lee
Hudson, Roy Davage
Hughes, Walter William, III
Kemp, Norman Everett
Kerfoot, Wilson Charles
King, John Arthur
Kluge, Arnold Girard
Kurta, Allen
Lagler, Karl Frank
Lamberts, Austin E
Lauff, George Howard
Lehman, Grace Church
Lehnert, James Patrick
Lew, Gloria Maria
Lunk, William Allan
McCrimmon, Donald Alan, Jr
McKitrick, Mary Caroline
Martin, Joseph Patrick, Jr
Merrill, Dorothy
Miller, Brinton Marshall
Miller, Robert Rush
Moore, Thomas Edwin
Myers, Philip
Nussbaum, Ronald Archie
Payne, Robert B
Pelzer, Charles Francis
Person, Steven John
Peters, Lewis
Porter, Thomas Wayne
Richards, Lawrence Phillips
Ritland, Richard Martin
Rivera, Evelyn Margaret
Rivera, Evelyn Margaret
Rusch, Wilbert H, Sr
Sacco, Anthony G
Saltiel, Alan Robert
Scarborough, Charles Spurgeon
Shaver, Robert John
Smith, Gerald Ray
Smith, Paul Dennis
Smith, R Jay
Speare, Edward Phelps
Stehr, Frederick William
Storer, Robert Winthrop
Tenbroek, Bernard John
Thompson, William Lay
Thoresen, Asa Clifford
Vatsis, Kostas Petros
Waffle, Elizabeth Lenora
Walker, Glenn Kenneth
Weiss, Mark Lawrence
Werner, Earl Edward
Werner, John Kirwin
Whiteman, Eldon Eugene
Winterstein, Scott Richard

MINNESOTA
Arthaud, Raymond Louis
Arthur, John W

Ballard, Neil Brian
Barker, S(hirley)Hugh
Barnwell, Franklin Hershel
Bartel, Monroe H
Birney, Elmer Clea
Brown, Rhoderick Edmiston, Jr
Collins, Hollie L
Corbin, Kendall Wallace
Dapkus, David Conrad
Dearden, Douglas Morey
Downing, William Lawrence
Dropkin, Victor Harry
Erickson, James Eldred
Foose, Thomas John
Ford, Norman Lee
Frederick, Edward C
Frydendall, Merrill J
Gilbertson, Donald Edmund
Goble, Frans Cleon
Grewe, Alfred H, Jr
Hanson, Lester Eugene
Hazard, Evan Brandao
Heist, Herbert Ernest
Herman, William S
Heuschele, Ann
Hofslund, Pershing Benard
Hoppe, David Matthew
Jannett, Frederick Joseph, Jr
Johnson, Ivan M
Johnson, Vincent Arnold
Koval, Thomas Michael
McCann, Lester J
Mech, Lucyan David
Murdock, Gordon Robert
Noetzel, David Martin
Noland, Wayland Evan
Olson, Magnus
Parmelee, David Freeland
Preiss, Frederick John
Pusey, Anne E
Rathbun, William B
Saccoman, Frank (Michael)
Shoger, Ross L
Sinha, Akhouri Achyutanand
Sulerud, Ralph L
Tordoff, Harrison Bruce
Toscano, William Agostino, Jr
Underhill, James Campbell
Wagenbach, Gary Edward
Warner, Dwain Willard
Weibust, Robert Smith
Williams, Steven Frank
Wyatt, Ellis Junior
Zischke, James Albert

MISSISSIPPI
Anderson, Gary
Boyd, Leroy Houston
Boykins, Ernest Aloysius, Jr
Cliburn, Joseph William
Dawson, Charles Eric
Gunter, Gordon
Helms, Thomas Joseph
Jackson, Jerome Alan
Keiser, Edmund Davis, Jr
Longest, William Douglas
Randolph, Kenneth Norris
Ross, Stephen Thomas
Spann, Charles Henry
Sutton, William Wallace
Tyson, Greta E
Venkataramiah, Amaraneni
Walley, Willis Wayne

MISSOURI
Aldridge, Robert David
Belshe, John Francis
Blaine, Edward Homer
Bolla, Robert Irving
Breitenbach, Robert Peter
Calabrese, Diane M
Clare, Stewart
Coles, Richard Warren
Cook, Nathan Howard
Davis, Jerry Collins
deRoos, Roger McLean
Elliott, Dana Ray
Fickess, Douglas Ricardo
Gerdes, Charles Frederick
Gerhardt, H Carl, Jr
Gersbacher, Willard Marion
Goodge, William Russell
Grobman, Arnold Brams
Harmon, Bud Gene
Hawksley, Oscar
Hazelwood, Donald Hill
Hughes, Stephen Edward
Ignoffo, Carlo Michael
Ingersol, Robert Harding
Luecke, Richard H
McCollum, Clifford Glenn
Metter, Dean Edward
Miles, Charles David
Mitchell, Henry Andrew
Mock, Orin Bailey
Momberg, Harold Leslie
Myers, Richard F
Orr, Orty Beam
Pflieger, William Leo
Rasmussen, David Tab
Russell, Robert Julian, Jr
Sage, Martin
Schreiweis, Donald Otto
Sexton, Owen J

Sharp, John Roland
Smith, Nathan Elbert
Smith, Richard Jay
Sorenson, Marion W
Taber, Charles Alec
Tannenbaum, Michael Glen
Train, Carl T
Twente, John W
Williams, Henry Warrington
Williamson, James Lawrence
Wilson, Stephen W

MONTANA
Bakken, Arnold
Ball, Irvin Joseph
Brunson, Royal Bruce
Dynes, J Robert
Foresman, Kerry Ryan
Harrington, Joseph D
Johnson, Oscar Walter
Kilgore, Delbert Lyle, Jr
Mitchell, Lawrence Gustave
O'Gara, Bartholomew Willis
Pfeiffer, Egbert Wheeler
Picton, Harold D
Sheldon, Andrew Lee
Smith, Ervin Paul
Tibbs, John Francisco
Weisel, George Ferdinand, Jr
Wright, Philip Lincoln

NEBRASKA
Adams, Charles Henry
Ballinger, Royce Eugene
Baumel, Julian Joseph
Bliese, John C W
Boernke, William E
Dalley, Arthur Frederick, II
Fritschen, Robert David
Geluso, Kenneth Nicholas
Genoways, Hugh Howard
Gunderson, Harvey Lorraine
Hergenrader, Gary Lee
Janovy, John, Jr
Johnsgard, Paul Austin
Kutler, Benton
Lynch, John Douglas
McGaugh, John Wesley
Peters, Edward James
Pritchard, Mary (Louise) Hanson
Sharpe, Roger Stanley

NEVADA
Douglas, Charles Leigh
Garner, Duane LeRoy
Kay, Fenton Ray
O'Farrell, Michael John
O'Farrell, Thomas Paul
Ryser, Fred A, Jr
Winokur, Robert Michael

NEW HAMPSHIRE
Bergman, Kenneth David
Bibeau, Armand A
Borror, Arthur Charles
Dingman, Jane Van Zandt
Eggleston, Patrick Myron
Francq, Edward Nathaniel Lloyd
Gilbert, John Jouett
Harrises, Antonio Efthemios
Hatt, Robert Torrens
Lavoie, Marcel Elphege
Roos, Thomas Bloom
Sasner, John Joseph, Jr
Spencer, Larry T
Stettenheim, Peter
Strout, Richard Goold
Walker, Warren Franklin, Jr
Wurster-Hill, Doris Hadley

NEW JERSEY
Agnish, Narsingh Dev
Ahn, Ho-Sam
Beer, Colin Gordon
Brattsten, Lena B
Brubaker, Paul Eugene, II
Campbell, William Cecil
Chizinsky, Walter
Collier, Marjorie McCann
Cook, Marie Mildred
Crane, Mark
Crossner, Kenneth Alan
Eggert, Robert Glenn
Feldman, Susan C
Ford, Daniel Morgan
Foster, William Burnham
Gardiner, Lion Edward
Gaumer, Albert Edwin Hellick
Gittleson, Stephen Mark
Grant, Peter Raymond
Grassle, Judith Payne
Griffo, James Vincent, Jr
Gupta, Godaveri Rawat
Hall, James Conrad
Halpern, Myron Herbert
Isquith, Irwin R
Jenkins, William Robert
Johnson, Leslie Kilham
Joslyn, Dennis Joseph
Kantor, Sidney
Kiley, Charles Walter
Koepp, Stephen John
Leck, Charles Frederick
Levine, Donald Martin

Lutz, Richard Arthur
McGhee, George Rufus, Jr
Madison, Caroline Rabb
Martin, Robert Allen
Massa, Tobias
Petriello, Richard P
Pollock, Leland Wells
Power, Harry Waldo, III
Sacks, Martin
Saiff, Edward Ira
Sellmer, George Park
Stearns, Donald Edison
Strausser, Helen R
Van Gelder, Richard George
Virkar, Raghunath Atmaram
Wallace, Edith Winchell
Wasserman, Aaron Osias
Weiss, Mitchell Joseph
Wilhoft, Daniel C
Willis, Jacalyn Giacalone

NEW MEXICO
Brown, James Hemphill
Conant, Roger
Cooch, Frederick Graham
Corliss, John Ozro
Crissman, Harry Allen
Degenhardt, William George
Duncan, Irma W
Findley, James Smith
Hayward, Bruce Jolliffe
Herman, Ceil Ann
Hubbard, John Patrick
Jones, Kirkland Lee
La Pointe, Joseph L
Ligon, James David
Martignoni, Mauro Emilio
Raitt, Ralph James, Jr
Scott, Norman Jackson, Jr
Stallcup, William Blackburn, Jr
Stricker, Stephev Alexander
Thaeler, Charles Schropp, Jr
Wolberg, Donald Lester
Zimmerman, Dale A

NEW YORK
Abbatiello, Michael James
Adler, Kraig (Kerr)
Alexander, A Allan
Anderson, Sydney
Andrle, Robert Francis
Baker-Cohen, Katherine France
Battin, William T
Beason, Robert Curtis
Behrens, Mildred Esther
Bell, Michael Allen
Bell, Robin Graham
Benton, Allen Haydon
Bernardis, Lee L
Blackler, Antonie W C
Blake, Robert Wesley
Bock, Walter Joseph
Borowsky, Richard Lewis
Bradbury, Michael Wayne
Brannigan, David
Breed, Helen Illick
Brett, Betty Lou Hilton
Brett, Carlton Elliot
Britten, Bryan Terrence
Brothers, Edward Bruce
Brown, Patricia Stocking
Brown, Stephen Clawson
Bruning, Donald Francis
Chamberlain, James Luther
Clark, Ralph M
Cooper, Arthur Joseph L
Crocker, Denton Winslow
Crowell, Robert Merrill
Cupp, Eddie Wayne
Cutler, Edward Bayler
Daniels, Robert Artie
De Gennaro, Louis D
Delson, Eric
Deubler, Earl Edward, Jr
Dowling, Herndon Glenn, (Jr)
Dudley, Patricia
Easton, Douglas P
Eaton, Stephen Woodman
Eisner, Thomas
Elrod, Joseph Harrison
Emerson, William Keith
Engbretson, Gustav Alan
Factor, Jan Robert
Fetcho, Joseph Robert
Finks, Robert Melvin
Finlay, Peter Stevenson
Fowler, James A
Frankel, Arthur Irving
French, Alan Raymond
Gering, Robert Lee
Goodwin, Robert Earl
Graham, William Joseph
Greenhall, Arthur Merwin
Greenlaw, Jon Stanley
Grove, Patricia A
Haefner, Paul Aloysius, Jr
Hainsworth, Fenwick Reed
Hamilton, William John, Jr
Hardy, Matthew Phillip
Haresign, Thomas
Harman, Willard Nelson
Harris, Charles Leon
Haugh, John Richard
Heath, Martha Ellen

Hecht, Max Knobler
Heidenthal, Gertrude Antoinette
Hertz, Paul Eric
Holtzman, Seymour
Holz, George Gilbert, Jr
Horst, G Roy
Howland, Howard Chase
Hutt, Frederick Bruce
Johansson, Tage Sigvard Kjell
Johnson, Lawrence Lloyd
Johnston, Robert E
Jolly, Clifford J
Kallen, Frank Clements
Kallman, Klaus D
Kanzler, Walter Wilhelm
Keen, William Hubert
Keithly, Janet Sue
Kirsteuer, Ernst
Klein, Harold George
Kolmes, Steven Albert
Koopman, Karl Friedrich
Krause, David Wilfred
Krieg, David Charles
Krishna, Kumar
Lackey, James Alden
Landry, Stuart Omer, Jr
Lanyon, Wesley Edwin
LaRow, Edward J
Lasker, Howard Robert
Laychock, Suzanne Gale
Lemanski, Larry Frederick
Leonard, Samuel Leeson
Levin, Norman Lewis
Lipke, Peter Nathan
LoGerfo, John J
Lorch, Joan
Loretz, Christopher Alan
McCourt, Robert Perry
McCune, Amy Reed
McDowell, Sam Booker
McGraw, James Carmichael
MacIntyre, Giles T
McIntyre, Judith Watland
Madison, Dale Martin
Mantel, Linda Habas
Maple, William Thomas
Maxwell, George Ralph, II
Miller, Sue Ann
Monaco, Lawrence Henry
Mueller, Justus Frederick
Muller-Schwarze, Dietland
Murphy, Michael Joseph
Myers, Charles William
Nelson, Sigurd Oscar, Jr
Nichols, Kathleen Mary
Nigrelli, Ross Franco
Nur, Uzi
Oaks, Emily Caywood Jordan
Organ, James Albert
Ortman, Robert A
Osterberg, Donald Mouretz
Payne, Harrison H
Peckarsky, Barbara Lynn
Peckham, Richard Stark
Perlmutter, Alfred
Phillips, Carleton Jaffrey
Phillips-Quagliata, Julia Molyneux
Platnick, Norman I
Pokorny, Kathryn Stein
Pough, Frederick Harvey
Pough, Richard Hooper
Rasweiler, John Jacob, IV
Rausch, James Peter
Reilly, Marguerite
Reisman, Howard Maurice
Rich, Abby M
Richmond, Milo Eugene
Rieder, Conly LeRoy
Rivest, Brian Roger
Roecker, Robert Maar
Rosi, David
Rough, Gaylord Earl
Rudzinska, Maria Anna
Ryan, Richard Alexander
Satir, Peter
Schneider, Kathryn Claire (Johnson)
Schreibman, Martin Paul
Schuster, Frederick Lee
Shaffer, Jacquelin Bruning
Short, Lester Le Roy
Siegfried, Clifford Anton
Slobodkin, Lawrence Basil
Smith, Clarence Lavett
Smith, Sharon Louise
Sohacki, Leonard Paul
Stewart, Kenton M
Stewart, Margaret McBride
Stouffer, James Ray
Stouter, Vincent Paul
Strayer, David Lowell
Summers, Robert Gentry, Jr
Swartz, Gordon Elmer
Swartzendruber, Donald Clair
Szabo, Piroska Ludwig
Tattersall, Ian Michael
Taub, Aaron M
Thompson, Steven Risley
Thornborough, John Randle
Titus, John Elliott
Tobach, Ethel
VanDruff, Larry Wayne
Vasey, Carey Edward
Vuilleumier, Francois
Ward, Robert T

Wells, John West
Wemyss, Courtney Titus, Jr
Wenk, Eugene J
Werner, Robert George
Whiting, Anne Margaret
Williams, George Christopher
Windsor, Donald Arthur
Winston, Judith Ellen
Wolf, Larry Louis
Wood, Raymond Arthur
Wright, Margaret Ruth
Wysolmerski, Theresa
Zimmerman, Jay Alan

NORTH CAROLINA
Aldridge, David William
Allen, Charles Marvin
Ankel-Simons, Friderun Annursel
Bailey, Joseph Randle
Baker, Elizabeth McIntosh
Barrick, Elliott Roy
Biggs, Walter Clark, Jr
Boliek, Irene
Bookhout, Cazlyn Green
Bradbury, Phyllis Clarke
Brown, Richard Dean
Bruce, Richard Conrad
Butts, Jeffrey A
Carter, Charleata A
Colvard, Dean Wallace
Costlow, John DeForest
Coyle, Frederick Alexander
Darling, Marilyn Stagner
Derrick, Finnis Ray
Dimock, Ronald Vilroy, Jr
Edwards, James Wesley
Feduccia, John Alan
Fehon, Jack Harold
Freeman, John Alderman
Gilbert, Lawrence Irwin
Gordon, Christopher John
Gregg, John Richard
Grosch, Daniel Swartwood
Guyselman, John Bruce
Hairston, Nelson George
Hampton, Carolyn Hutchins
Heckrotte, Carlton
Hedlund, Laurence William
Hendrickson, Herbert T
Hodson, Charles Andrew
Hylander, William Leroy
Jones, Claiborne Stribling
Kalmus, Gerhard Wolfgang
Klopfer, Peter Hubert
Koritnik, Donald Raymond
Lassiter, Charles Albert
Lay, Douglas M
Lee, Greta Marlene
Leise, Esther M
Lessler, Judith Thomasson
Lindquist, David Gregory
Link, Garnett William, Jr
Lutz, Paul E
Lynn, William Gardner
Lytle, Charles Franklin
McCoy, John J
McCrary, Anne Bowden
McDaniel, Susan Griffith
McDowell, Robert E, Jr
McLeod, Michael John
McMahan, Elizabeth Anne
Manooch, Charles Samuel, III
Meyer, John Richard
Miller, Grover Cleveland
Moore, Allen Murdoch
Moseley, Lynn Johnson
Mueller, Helmut Charles
Mueller, Nancy Schneider
Parnell, James Franklin
Peterson, Charles Henry
Poulton, Bruce R
Powell, Roger Allen
Radovsky, Frank Jay
Reice, Seth Robert
Reynolds, Joshua Paul
Rittschof, Daniel
Roberts, John Fredrick
Roer, Robert David
Schwartz, Frank Joseph
Sharer, Archibald Wilson
Simpson, Everett Coy
Smith, Ned Allan
Spuller, Robert L
Stafford, Darrel Wayne
Steinhagen, William Herrick
Stewart, Paul Alva
Stroud, Richard Hamilton
Sulik, Kathleen Kay
Sullivan, James Bolling
Sutcliffe, William Humphrey, Jr
Thomas, Mary Beth
Trainer, John Ezra, Jr
Triantaphyllou, Hedwig Hirschmann
Vogel, Steven
Wainwright, Stephen Andrew
Williams, Ann Houston
Wolk, Robert George
Yarbrough, Charles Gerald
Yongue, William Henry

NORTH DAKOTA
Cherian, Sebastian K
Duerr, Frederick G
Erickson, Duane Otto

Gerst, Jeffery William
Gladue, Brian Anthony
Grier, James William
Kannowski, Paul Bruno
Lieberman, Milton Eugene
Riemann, John G
Seabloom, Robert W
Struble, Craig Bruce
Wrenn, William J

OHIO
Alicino, Nicholas J
Baker, Peter C
Banning, Jon Willroth
Barbour, Clyde D
Barnes, Herbert M
Barrow, James Howell, Jr
Beal, Kathleen Grabaskas
Berra, Tim Martin
Bettice, John Allen
Birky, Carl William, Jr
Burns, Robert David
Burtt, Edward Howland, Jr
Case, Denis Stephen
Claussen, Dennis Lee
Clise, Ronald Leo
Cooke, Helen Joan
Crites, John Lee
Culver, David Alan
Daniel, Paul Mason
Delphia, John Maurice
Dexter, Ralph Warren
Dickerman, Richard Curtis
Elfner, Lynn Edward
Etges, Frank Joseph
Faber, Lee Edward
Ferguson, Marion Lee
Foreman, Darhl Lois
Fry, Anne Evans
Garton, David Wendell
Gates, Michael Andrew
Gatz, Arthur John, Jr
Gaunt, Abbot Stott
Gilloteaux, Jacques Jean-Marie A
Goldstein, David Louis
Gottschang, Jack Louis
Grassmick, Robert Alan
Greenstein, Julius S
Guttman, Sheldon
Hahnert, William Franklin
Hintz, Howard William
Hoagstrom, Carl William
Hubschman, Jerry Henry
Hummon, William Dale
Ingersoll, Edwin Marvin
Isler, Gene A
Jackson, Dale Latham
Klemm, Donald J
Kopchick, John J
Lane, Roger Lee
Lawrence, Robert G
Lesh-Laurie, Georgia Elizabeth
McLean, Edward Bruce
Mahan, Harold Dean
Martin, Scott McClung
Mayfield, Harold Ford
Miller, John Wesley, Jr
Moore, Nelson Jay
Moore, Randy
Myser, Willard C
Orcutt, Frederic Scott, Jr
Pannabecker, Richard Floyd
Parrish, Wayne
Patton, Wendell Keeler
Peterjohn, Glenn William
Peterle, Tony J
Pettegrew, Raleigh K
Pilati, Charles Francis
Plimpton, Rodney F, Jr
Przybylski, Ronald J
Putnam, Loren Smith
Rab, Paul Alexis
Rothenbuhler, Walter Christopher
Rovner, Jerome Sylvan
Rubin, David Charles
St John, Fraze Lee
Saksena, Vishnu P
Scott, John Paul
Sherman, Robert George
Shetlar, David John
Silverman, Jerald
Stansbery, David Honor
Stein, Carol B
Stein, Roy Allen
Stoffer, Richard Lawrence
Swift, Michael Crane
Tandler, Bernard
Thomas, Cecil Wayne
Trautman, Milton Bernhard
Uetz, George William
Valentine, Barry Dean
Vaughn, Charles Melvin
Venard, Carl Ernest
Weis, Dale Stern
White, Andrew Michael
Wilson, Richard Ferrin
Wood, Jackie Dale
Wood, Timothy Smedley

OKLAHOMA
Baird, Troy Alan
Black, Jeffrey Howard
Brown, Harley Procter
Brown, Marie Jenkins

Zoology (cont)

Carpenter, Charles Congden
Carter, William Alfred
Chapman, Brian Richard
Clemens, Howard Paul
Cox, Beverley Lenore
Curd, Milton Rayburn
Davis, Marjorie
Floyd, Robert A
Fox, Stanley Forrest
Glass, Bryan Pettigrew
Haines, Howard Bodley
Hellack, Jenna Jo
Henrickson, Robert Lee
Hill, Loren Gilbert
Hutchison, Victor Hobbs
Korstad, John Edward
Lindsay, Hague Leland, Jr
Mares, Michael Allen
Matthews, William John
Miller, Helen Carter
Miller, Rudolph J
Nelson, John Marvin, Jr
Ownby, Charlotte Ledbetter
Radke, William John
Reinke, Lester Allen
Russell, Charles Clayton
Schnell, Gary Dean
Seto, Frank
Shorter, Daniel Albert
Sonleitner, Frank Joseph
Talent, Larry Gene
Thornton, John William
Tyler, Jack D
Vestal, Bedford Mather
Vial, James Leslie
Walters, Lowell Eugene
Whitmore, Mary (Elizabeth) Rowe
Young, Sharon Clairene

OREGON
Beatty, Joseph John
Bethea, Cynthia Louise
Boddy, Dennis Warren
Bond, Carl Eldon
Callahan, Clarence Arthur
Carter, Richard Thomas
Cheeke, Peter Robert
Cimberg, Robert Lawrence
Coffey, Marvin Dale
Crawford, John Arthur
Cross, Stephen P
Doudoroff, Peter
Field, Katharine G
Forbes, Richard Bryan
Gonor, Jefferson John
Gunberg, David Leo
Gwilliam, Gilbert Franklin
Henny, Charles Joseph
Hillman, Stanley Severin
Hisaw, Frederick Lee, Jr
Hixon, Mark A
Jensen, Harold James
Kessel, Edward Luther
Lubchenco, Jane
Markle, Douglas Frank
Maser, Chris
Mix, Michael Cary
Neiland, Kenneth Alfred
Olson, Robert Eldon
Osgood, David William
Owczarzak, Alfred
Quast, Jay Charles
Roberts, Michael Foster
Simpson, Leonard
Spencer, Peter Simner
Storm, Robert MacLeod
Taylor, Jocelyn Mary
Thorson, Thomas Bertel
Warren, Charles Edward
Weaver, Morris Eugene
Wirtz, John Harold

PENNSYLVANIA
Alperin, Richard Junius
Anderson, David Robert
Barnes, Robert Drane
Barnett, Leland Bruce
Beach, Neil William
Bell, Edwin Lewis, II
Bellis, Edward David
Bergstresser, Kenneth A
Buckelew, Thomas Paul
Carpenter, Esther
Catalano, Raymond Anthony
Chase, Robert Silmon, Jr
Clark, Richard James
Cole, James Edward
Conner, Robert Louis
Cundall, David Langdon
Cunningham, Harry N, Jr
Dalby, Peter Lenn
Davis, George Morgan
Debouck, Christine Marie
Denoncourt, Robert Francis
Dollahon, Norman Richard
Dunson, William Albert
Fabian, Michael William
Feuer, Robert Charles
Fish, Frank Eliot
Foor, W Eugene
Franz, Craig Joseph
Fredrickson, Richard William

Gallati, Walter William
Gibley, Charles W, Jr
Gill, Frank Bennington
Grossman, Herbert H
Grosvenor, Clark Edward
Guida, Vincent George
Halbrendt, John Marthon
Hall, John Sylvester
Harclerode, Jack E
Hayes, Wilbur Frank
Henderson, Alex
Hendrix, Sherman Samuel
Hill, Frederick Conrad
Holliday, Charles Walter
Hulse, Arthur Charles
Jacobs, George Joseph
Jeffries, William Bowman
Joos, Barbara
Kamon, Eliezer
Karlson, Eskil Leannart
Kesner, Michael H
Kimmel, William Griffiths
Kirkland, Gordon Laidlaw, Jr
Klinger, Thomas Scott
Krishtalka, Leonard
Legge, Thomas Nelson
Linask, Kersti Katrin
McCoy, Clarence John, Jr
McDermott, John Joseph
Mann, Alan Eugene
Meinkoth, Norman August
Meyer, R Peter
Miller, Richard Lee
Neff, William H
O'Connor, John Dennis
Ogren, Robert Edward
Ostrovsky, David Saul
Otte, Daniel
Parkes, Kenneth Carroll
Pearson, David Leander
Pitkin, Ruthanne B
Porter, Keith Roberts
Prezant, Robert Steven
Raikow, Robert Jay
Reif, Charles Braddock
Robertson, Robert
Rockwell, Kenneth H
Rogers, William Edwin
Rosenberg, Gary
Ruffer, David G
Samollow, Paul B
Schlitter, Duane A
Selander, Robert Keith
Sherritt, Grant Wilson
Shostak, Stanley
Siegel, Michael Ian
Sillman, Emmanuel I
Snyder, Donald Benjamin
Stauffer, Jay Richard, Jr
Streu, Herbert Thomas
Taylor, D Lansing
Thomas, Roger David Keen
Thompson, Roger Kevin Russell
Thomson, Keith Stewart
Trainer, John Ezra, Sr
Turoczi, Lester J
Twiest, Gilbert Lee
Uricchio, William Andrew
Webb, Glenn R
West, Keith P
Williams, Russell Raymond
Williams, Timothy C
Wilson, Lowell L
Winkelmann, John Roland
Yahner, Richard Howard
Zaccaria, Robert Anthony
Zenisek, Cyril James

RHODE ISLAND
Dolyak, Frank
Hammen, Carl Schlee
Harrison, Robert William
Heppner, Frank Henry
Hyland, Kerwin Ellsworth, Jr
Jeffries, Harry Perry
Lazell, James Draper
LeBrun, Roger Arthur
Miller, Don Curtis

SOUTH CAROLINA
Anderson, William Dewey, Jr
Biernbaum, Charles Knox
Bildstein, Keith Louis
Brisbin, I Lehr, Jr
Browdy, Craig Lawrence
Carew, James L
Fincher, John Albert
Forsythe, Dennis Martin
Fox, Richard Shirley
Gauthreaux, Sidney Anthony
Gibbons, J Whitfield
Guram, Malkiat Singh
Harrison, Julian R, III
Helms, Carl Wilbert
Johnson, Albert Wayne
Johnson, Robert Karl
Kelly, Robert Withers
Labanick, George Michael
Lacher, Thomas Edward, Jr
Leonard, Walter Raymond
Lewis, Stephen Albert
Mulvey, Margaret
Newman, Michael Charles
Pivorun, Edward Broni

Roache, Lewie Calvin
Sandifer, Paul Alan
Schindler, James Edward
Scott, Jaunita Simons
Skelley, George Calvin, Jr
Spooner, John D
Stewart, Shelton E
Swallow, Richard Louis
Teska, William Reinhold
Watabe, Norimitsu
Wheeler, Alfred Portius

SOUTH DAKOTA
Dillon, Raymond Donald
Dunlap, Donald Gene
Haertel, John David
Hall, Jerry Dexter
Harrell, Byron Eugene
Martin, James Edward
Musson, Alfred Lyman
Parke, Wesley Wilkin
Romans, John Richard
Slyter, Arthur Lowell
Smolik, James Darrell
Tatina, Robert Edward
Wahlstrom, Richard Carl
Zeigler, David Wayne

TENNESSEE
Alsop, Frederick Joseph, III
Amy, Robert Lewis
Auerbach, Stanley Irving
Bernard, Ernest Charles
Brinkhurst, Ralph O
Brode, William Edward
Burnham, Kenneth Donald
Carlton, Robert Austin
Chance, Charles Jackson
Chandler, Clay Morris
Coutant, Charles Coe
Dennison, Clifford C
Dumont, James Nicholas
Dunn, Mary Catherine
Dyer, Melvin I
Echternacht, Arthur Charles
Eddlemon, Gerald Kirk
Elliott, Alice
Etnier, David Allen
Ford, Floyd Mallory
Foreman, Charles William
Freeman, John Richardson
Greenberg, Neil
Gutzke, William H N
Harrison, Robert Edwin
Hochman, Benjamin
Ikenberry, Roy Dewayne
James, Ted Ralph
Kathman, R Deedee
Kennedy, Michael Lynn
Lawson, James Everett
Loar, James M
McCarthy, John F
McGavock, Walter Donald
McGhee, Charles Robert
Millemann, Raymond Eagan
Murphy, George Graham
Nelson, Diane Roddy
Nunnally, David Ambrose
Parchment, John Gerald
Parmalee, Paul Woodburn
Payne, James
Polis, Gary Allan
Richardson, Don Orland
Riechert, Susan Elise
Schultz, Terry Wayne
Scott, Mack Tommie
Semlitsch, Raymond Donald
Sharples, Frances Ellen
Snyder, David Hilton
Staub, Robert J
Vogel, Howard H, Jr
Voorhees, Larry Donald
Wallace, Gary Oren
Welch, Hugh Gordon
Wilhelm, Walter Eugene
Wilson, James Lester
Yeatman, Harry Clay

TEXAS
Adams, Clark Edward
Aggarwal, Bharat Bhushan
Allen, Archie C
Applegate, Richard Lee
Arnold, Keith Alan
Baker, James Haskell
Baker, Rollin Harold
Banerji, Tapan Kumar
Barth, Robert Hood, Jr
Bassett, James Wilbur
Beasley, Clark Wayne
Belk, Gene Denton
Bickham, John W
Bischoff, Harry William
Bramblett, Claud Allen
Brodie, Edmund Darrell, Jr
Cameron, Ivan Lee
Carpenter, Zerle Leon
Chaney, Allan Harold
Chrapliwy, Peter Stanley
Clayton, Dale Leonard
Clench, Mary Heimerdinger
Cooper, William Anderson
Crawford, Isaac Lyle
Crews, David Pafford

Dalquest, Walter Woelberg
Davis, Gordon Wayne
Dixon, James Ray
Dunbar, Bonnie Sue
Faulkner, Russell Conklin, Jr
Flury, Alvin Godfrey
Formanowicz, Daniel Robert, Jr
Garner, Herschel Whitaker
Glenn, William Grant
Greding, Edward J, Jr
Greenbaum, Ira Fred
Greenberg, Les Paul
Hanlon, Roger Thomas
Hannan, Herbert Herrick
Harris, Arthur Horne
Harry, Harold William
Hillis, David Mark
Hubbs, Clark
Hudson, Frank Alden
Huggins, Sara Espe
Ingold, Donald Alfred
Jacobson, Antone Gardner
Johnston, John Spencer
Jones, Clyde Joe
Jones, J Knox, Jr
Judd, Frank Wayne
Kalthoff, Klaus Otto
Killebrew, Flavius Charles
Kirkpatrick, Mark Adams
Klaus, Ewald Fred, Jr
Kuntz, Robert Elroy
Lewis, Simon Andrew
Lieb, Carl Sears
Long, James Duncan
McCarley, Wardlow Howard
McDonald, Harry Sawyer
McEachran, John D
McMahon, Robert Francis, III
Martin, Floyd Douglas
Martin, Robert Frederick
Meacham, William Ross
Meade, Thomas Gerald
Mecham, John Stephen
Metcalf, Artie Lou
Murray, Harold Dixon
Neill, Robert Lee
Newman, George Allen
Nichols, James Ross
Owens, David William
Packchanian, Ardzroony (Arthur)
Park, Taisoo
Parker, Nick Charles
Parker, Robert Hallett
Pierce, Jack Robert
Pyburn, William F
Ramsey, Jed Junior
Rennie, Thomas Howard
Roberts, Larry Spurgeon
Robertson, Walter Volley
Roller, Herbert Alfred
Rose, Francis L
Rubink, William Louis
Rylander, Michael Kent
Sanders, Ottys E
Schlueter, Edgar Albert
Schmidly, David James
Schwalm, Fritz Ekkehardt
Scudday, James Franklin
Self, Hazzle Layfette
Shake, Roy Eugene
Sissom, Stanley Lewis
Smatresk, Neal Joseph
Smith, William H
Smith, William Russell
Steele, David Gentry
Strawn, Robert Kirk
Tamsitt, James Ray
Telfair, Raymond Clark, II
Terry, Robert James
Thames, Walter Hendrix, Jr
Throckmorton, Gaylord Scott
Tribble, Leland Floyd
Turco, Charles Paul
Tuttle, Merlin Devere
Vaughan, Mary Kathleen
Warner, Marlene Ryan
Webb, Robert G
Wheeler, Marshall Ralph
Wicksten, Mary Katherine
Wilson, William Thomas
Wohlschlag, Donald Eugene
Worthington, Richard Dane
Wursig, Bernd Gerhard
Yaden, Senkalong
Zimmerman, Earl Graves

UTAH
Aird, Steven Douglas
Bahler, Thomas Lee
Balph, Martha Hatch
Behle, William Harroun
Cuellar, Orlando
Dixon, Keith Lee
Evans, Frederick Read
Fitzgerald, Paul Ray
Frost, Herbert Hamilton
Gaufin, Arden Rupert
Graff, Darrell Jay
Griffin, Gerald D
Hansen, Afton M
Hulet, Clarence Veloid
Jeffery, Duane Eldro
Jensen, Emron Alfred
Johnston, Stephen Charles

Legler, John Marshall
Liddell, William David
MacMahon, James A
Messina, Frank James
Murphy, Joseph Robison
Negus, Norman Curtiss
Petty, William Clayton
Schoenwolf, Gary Charles
Sites, Jack Walter, Jr
Smith, Howard Duane
Smith, Nathan McKay
Van De Graaff, Kent Marshall
White, Clayton M

VERMONT
Bell, Ross Taylor
Chipman, Robert K
Detwyler, Robert
Gray, John Patrick
Hitchcock, Harold Bradford
Kilpatrick, Charles William
Moody, Paul Amos
Schall, Joseph Julian
Stevens, Dean Finley

VIRGINIA
Adkisson, Curtis Samuel
Alden, Raymond W, III
Aldrich, John Warren
Andrews, Jay Donald
Atkins, David Lynn
Barton, Alexander James
Birdsong, Ray Stuart
Blood, Benjamin Donald
Briggs, Jeffrey L
Brown, Luther Park
Burreson, Eugene M
Bush, Francis M
Byrd, Mitchell Agee
Carico, James Edwin
Chipley, Robert MacNeill
Corwin, Jeffrey Todd
Cranford, Jack Allen
Cruthers, Larry Randall
Davis, John Edward, Jr
Dent, James (Norman)
Diehl, Fred A
Eckerlin, Ralph Peter
Eller, Arthur L, Jr
Elwood, Jerry William
Emsley, Michael Gordon
Ernst, Carl Henry
Fenner-Crisp, Penelope Ann
Gaines, Gregory
Gourley, Eugene Vincent
Gray, Faith Harriet
Green, George G
Heath, Alan Gard
Hickman, Cleveland Pendleton, Jr
Hoffman, Richard Lawrence
Holsinger, John Robert
Holt, Perry Cecil
Howe, Marshall Atherton
Jahn, Laurence R
Johnson, Raymond Earl
Johnson, Rose Mary
Jopson, Harry Gorgas Michener
Kornegay, Ervin Thaddeus
Kornicker, Louis Sampson
McNabb, Roger Allen
Mangum, Charlotte P
Marcellini, Dale Leroy
Maroney, Samuel Patterson, Jr
Mehner, John Frederick
Merlino, Glenn T
Moncrief, Nancy D
Muncy, Robert Jess
Murray, Joseph James, Jr
Musick, John A
Neves, Richard Joseph
Opell, Brent Douglas
Osborne, Paul James
Pinschmidt, William Conrad, Jr
Pugh, Jean Elizabeth
Ray, G Carleton
Rebhun, Lionel Israel
Russell, Catherine Marie
Sandow, Bruce Arnold
Savitzky, Alan Howard
Shugart, Herman Henry, Jr
Shuster, Carl Nathaniel, Jr
Simpson, Margaret
Sonenshine, Daniel E
Spangenberg, Dorothy Breslin
Stanley, Melissa Sue Millam
Stevens, Robert Edward
Turner, Bruce Jay
Turney, Tully Hubert
Vaughan, Michael Ray
Wass, Marvin Leroy
Wells, Ouida Carolyn
West, Warwick Reed, Jr
Wilbur, Henry Miles
Wilson, John William, III
Wood, Robertson Harris Langley
Woolcott, William Starnold

WASHINGTON
Boersma, P Dee
Braham, Howard Wallace
Brenchley, Gayle Anne
Broad, Alfred Carter
Brown, Herbert Allen
Brown, Robert Harrison

Clark, Glen W
Dickhoff, Walton William
Drabek, Charles Martin
Dumas, Philip Conrad
Eastlick, Herbert Leonard
Edwards, John S
Farner, Donald Sankey
Ford, James
Gibson, Flash
Gilbert, Frederick Franklin
Gorbman, Aubrey
Harrington, Edward James
Hayes, Murray Lawrence
Henry, Dora Priaulx
Illg, Paul Louis
Johnson, Murray Leathers
Johnson, Phyllis Truth
Johnson, Richard Evan
Karlstrom, Ernest Leonard
Kohn, Alan Jacobs
Kozloff, Eugene Nicholas
Laird, Charles David
Landis, Wayne G
Larsen, John Herbert, Jr
Long, Edward R
Loughlin, Thomas Richard
Luchtel, Daniel Lee
McCloskey, Lawrence Richard
McLaughlin, Patsy Ann
Martin, Arthur Wesley
Martin, Dennis John
Miller, Elwood Morton
Mills, Claudia Eileen
Moffett, David Franklin, Jr
Morgan, Kenneth Robb
Murphy, Mary Eileen
Paine, Robert T
Paulson, Dennis Roy
Pietsch, Theodore Wells
Plisetskaya, Erika Michael
Reischman, Placidus George
Rempel, Arthur Gustav
Rice, Dale Warren
Ross, June Rosa Pitt
Scheffer, Victor B
Schroeder, Paul Clemens
Senger, Clyde Merle
Stien, Howard M
Turner, William Junior
Turner, William Junior
Wakimoto, Barbara Toshiko
White, Ronald Jerome
Yao, Meng-Chao

WEST VIRGINIA
Hall, George Arthur, Jr
Hall, John Edgar
Hurlbutt, Henry Winthrop
Pauley, Thomas Kyle
Reyer, Randall William
Seidel, Michael Edward
Shalaway, Scott D
Tarter, Donald Cain
Taylor, Ralph Wilson

WISCONSIN
Abramoff, Peter
Barrett, James Martin
Baylis, Jeffrey Rowe
Beck, Stanley Dwight
Bloom, Alan S
Brynildson, Oscar Marius
Crowe, David Burns
Emlen, John Thompson, Jr
Fraser, Lemuel Anderson
Gottfried, Bradley M
Grittinger, Thomas Foster
Guilford, Harry Garrett
Hasler, Arthur Davis
Hodgson, James Russell
Hoffman, William F
Horseman, Nelson Douglas
Johnson, Wendel J
Lai, Ching-San
Leutenegger, Walter
Light, Douglas B
Lombard, Julian H
Long, Charles Alan
Long, Claudine Fern
Lukens, Paul W, Jr
Mahadeva, Madhu Narayan
Mahmoud, Ibrahim Younis
Michaud, Ted C
Minock, Michael Edward
Norden, Carroll Raymond
North, Charles A
Passano, Leonard Magruder
Riegel, Ilse Leers
Ruffolo, John Joseph, Jr
Rusch, Donald Harold
Rust, Charles Chapin
Seale, Dianne B
Spieler, Richard Earl
Staszak, David John
Sweet, Gertrude Evans
Tappen, Neil Campbell
Vomachka, Archie Joel
Weil, Michael Ray
Weise, Charles Martin
Wilson, Richard Howard
Wittrock, Darwin Donald
Young, Howard Frederick

WYOMING
Botkin, Merwin P
Diem, Kenneth Lee
Dolan, Joseph Edward
George, Robert Porter
Kingston, Newton
Nunamaker, Richard Allan
Parker, Michael
Reynolds, Richard Truman
Tabachnick, Walter J

PUERTO RICO
Crabtree, David Melvin
Cutress, Charles Ernest
Rivero, Juan Arturo
Roman, Jesse
Squire, Richard Douglas
Thomas, (John) (Paul) Richard
Villella, John Baptist
Voltzow, Janice

ALBERTA
Cavey, Michael John
Clifford, Hugh Fleming
Currie, Philip John
Dixon, Elisabeth Ann
Geist, Valerius
Holmes, John Carl
Lauber, Jean Kautz
Macpherson, Andrew Hall
Mahrt, Jerome L
Nelson, Joseph Schieser
Neuwirth, Maria
Pittman, Quentin J
Rosenberg, Herbert Irving
Ross, Donald Murray
Russell, Anthony Patrick
Seghers, Benoni Hendrik
Spencer, Andrew Nigel
Steiner, Andre Louis
Stirling, Ian G
Strobeck, Curtis
Wang, Lawrence Chia-Huang
Wilson, Mark Vincent Hardman
Zammuto, Richard Michael

BRITISH COLUMBIA
Adams, James Russell
Bandy, Percy John
Booth, Amanda Jane
Bousfield, Edward Lloyd
Burke, Robert D
Cowan, Ian McTaggart
Dehnel, Paul Augustus
Dill, Lawrence Michael
Fields, William Gordon
Fontaine, Arthur Robert
Geen, Glen Howard
Gosline, John M
Jamieson, Glen Stewart
Jones, David Robert
Kasinsky, Harold Edward
Lewis, Alan Graham
Liley, Nicholas Robin
Lindsey, Casimir Charles
Mackie, George Owen
McLean, John Alexander
Marliave, Jeffrey Burton
Milsom, William Kenneth
Newman, Murray Arthur
Nordan, Harold Cecil
Nursall, John Ralph
Randall, David John
Stringam, Elwood Williams
Tunnicliffe, Verena Julia
Wilimovsky, Norman Joseph

MANITOBA
Ayles, George Burton
Braekevelt, Charlie Roger
Clayton, James Wallace
Dandy, James William Trevor
Hara, Toshiaki J
Huebner, Judith Dee
Iverson, Stuart Leroy
Laale, Hans W
Novak, Marie Marta
Preston, William Burton
Pruitt, William O, Jr
Smith, Roger Francis Cooper
White, Noel David George
Zach, Reto

NEW BRUNSWICK
Cowan, F Brian M
Eveleigh, Eldon Spencer
Scott, William Beverley

NEWFOUNDLAND
Aldrich, Frederick Allen
Bennett, Gordon Fraser
Davis, Charles (Carroll)
Kao, Ming-hsiung
Khan, Rasul Azim
Mercer, Malcolm Clarence
Morris, Claude C
Steele, Donald Harold
Steele, Vladislava Julie
Threlfall, William

NOVA SCOTIA
Bleakney, John Sherman
Brown, Richard George Bolney
Chapman, David MacLean

Dadswell, Michael John
Ferrier, Gregory R
Garside, Edward Thomas
Leslie, Ronald Allan
Rojo, Alfonso

ONTARIO
Atwood, Harold Leslie
Balon, Eugene Kornel
Banfield, Alexander William Francis
Barlow, Jon Charles
Barr, David Wallace
Beamish, Frederick William Henry
Berger, Jacques
Beverley-Burton, Mary
Bols, Niels Christian
Bond, Edwin Joshua
Buchanan-Smith, Jock Gordon
Butler, Richard Gordon
Cade, William Henry
Calder, Dale Ralph
Cameron, Duncan MacLean, Jr
Carlisle, David Brez
Casselman, John Malcolm
Chambers, Ann Franklin
Chant, Donald A
Chengalath, Rama
Chia, Fu-Shiang
Ciriello, John
Coad, Brian William
Cook, David Greenfield
Craigie, Edward Horne
Crossman, Edwin John
Darling, Donald Christopher
Desser, Sherwin S
Dodson, Edward O
Emery, Alan Roy
Fenton, M Brock
Freeman, Reino Samuel
George, John Caleekal
Godfrey, William Earl
Govind, Choonilal Keshav
Gray, David Robert
Harington, Charles Richard
Helmuth, Hermann Siegfried
Johansen, Peter Herman
Keenleyside, Miles Hugh Alston
Kelso, John Richard Murray
Kott, Edward
Kovacs, Kit M
Lai-Fook, Joan Elsa I-Ling
Lavigne, David M
Leatherland, John F
Lee, David Robert
Lingwood, Clifford Alan
Lynn, Denis Heward
McAllister, Donald Evan
McMillan, Donald Burley
Middleton, Alex Lewis Aitken
Morris, Catherine Elizabeth
Muir, Barry Sinclair
Munroe, Eugene Gordon
Olthof, Theodorus Hendrikus Antonius
Ouellet, Henri
Pang, Cho Yat
Parsons, Thomas Sturges
Petras, Michael Luke
Rising, James David
Robertson, Raleigh John
Ronald, Keith
Roots, Betty Ida
Rossant, Janet
Saleuddin, Abu S
Scott, David Maxwell
Seligy, Verner Leslie
Shih, Chang-Tai
Shivers, Richard Ray
Singal, Dharam Parkash
Singh, Ripu Daman
Smetana, Ales
Smith, Donald Alan
Smith, Jonathan Jeremy Berkeley
Smol, John Paul
Solman, Victor Edward Frick
Steele, John Earle
Stevens, Ernest Donald
Stone, John Bruce
Tomlin, Alan David
Watson, Wynnfield Young
Winterbottom, Richard
Wright, Kenneth A
Youson, John Harold

PRINCE EDWARD ISLAND
Drake, Edward Lawson

QUEBEC
Ali, Mohamed Ather
Anctil, Michel
Brunel, Pierre
Ferron, Jean H
Filteau, Gabriel
Grant, James William Angus
Green, David M(artin)
Guderley, Helga Elizabeth
Johnston, C Edward
Kapoor, Narinder N
McNeil, Raymond
Magnin, Etienne Nicolas
Mangat, Baldev Singh
Nishioka, Yutaka
Pirlot, Paul
Reiswig, Henry Michael
Sanborne, Paul Michael

Zoology (cont)

SASKATCHEWAN
Brandell, Bruce Reeves
Forsyth, Douglas John
Gilmour, Thomas Henry Johnstone
Makowski, Roberte Marie Denise
Mitchell, George Joseph
Oliphant, Lynn Wesley
Secoy, Diane Marie

OTHER COUNTRIES
Buttemer, William Ashley
Craig, John Frank
Haight, John Richard
Heatwole, Harold Franklin
Hisada, Mituhiko
Jahn, Ernesto
Laurent, Pierre
Lochhead, John Hutchison
Lorenz, Konrad Zacharias
McMillan, Joseph Patrick
Marsh, James Alexander, Jr
Myres, Miles Timothy
Nemenzo, Francisco
O'Neill, Patricia Lynn
Paulay, Gustav
Phillips, Allan Robert
Rich, Thomas Hewitt
Richmond, Robert H
Rodriguez, Gilberto
Rubinoff, Ira
Sabry, Ismail
Saether, Ole Anton
Schliwa, Manfred
Schmidt-Koenig, Klaus
Tinbergen, Nikolaas
Westcott, Peter Walter
Withers, Philip Carew

Other Biological Sciences

ALABAMA
Barfuss, Delon Willis
Bradley, James T
Lewis, Marian L Moore
McDonald, Jay M
Magargal, Wells Wrisley, II
Singh, Jarnail
Smith-Somerville, Harriett Elizabeth
Strada, Samuel Joseph
Wilson, G Dennis

ALASKA
Proenza, Luis Mariano
Rockwell, Julius, Jr

ARIZONA
Flint, Hollis Mitchell
George, M Colleen
Gerner, Eugene Willard
Gilkey, John Clark
Henshaw, Paul Stewart
Reitan, Ralph Meldahl
Taylor, Richard G

ARKANSAS
Baker, Max Leslie
Epstein, Joshua
Griffin, Edmond Eugene
Hart, Ronald Wilson
Leone, Charles Abner
Nagle, William Arthur
Sherman, Jerome Kalman

CALIFORNIA
Abramson, Stephan B
Ainsworth, Earl John
Alpen, Edward Lewis
Amerine, Maynard Andrew
Ashe, John Herman
Bagshaw, Malcolm A
Bailey, Ian L
Bechtol, Kathleen B
Belisle, Barbara Wolfanger
Bennett, John Francis
Bentley, David R
Billingham, John
Birren, James E(mmett)
Blakely, Eleanor Alice
Bonura, Thomas
Bradbury, Margaret G
Brown, J Martin
Bryant, Peter James
Burgess, Teresa Lynn
Campbell, Kenneth Eugene, Jr
Chao, Fu-chuan
Cobb, Fields White, Jr
Cohn, Stanton Harry
Cox, Cathleen Ruth
Cramer, Donald V
Crandall, Edward D
Crane, Jules M, Jr
Cross, Carroll Edward
Deen, Dennis Frank
Dement, William Charles
Dewey, William
Dillman, Robert O
Dobson, R Lowry
Epstein, John Howard
Euteneuer, Ursula Brigitte
Fox, Joan Elizabeth Bothwell
Fuller, Thomas Charles

Gish, Duane Tolbert
Glick, David
Go, Vay Liang
Goeddel, David V
Gold, Daniel P
Goldfine, Ira D
Goldstein, Norman N
Goldyne, Marc E
Gondos, Bernard
Hager, E Jant
Hamilton, Robert L, Jr
Hammel, Eugene A
Harris, John Wayne
Hoch, Paul Edwin
Hoffman, Robert M
Holley, Daniel Charles
Holmquist, Walter Richard
Howard, Walter Egner
Innerarity, Tom L
Jerison, Harry Jacob
Jordan, Mary Ann
Jukes, Thomas Hughes
Kallman, Robert Friend
Kohn, Fred R
Kraft, Lisbeth Martha
Kurohara, Samuel S
Kuroki, Gary W
Kutas, Marta
Lanier, Lewis L
Larabell, Carolyn A
Lawyer, Arthur L
Lee, Nancy L
Leon, Michael Allan
Lewis, Robert Allen
Li, Thomas M
Liburdy, Robert P
Lorence, Matthew C
McDonnel, Gerald M
McEwen, Joan Elizabeth
Marcus, Carol Silber
Marini, Mario A
Martins-Green, Manuela M
Matthes, Michael Taylor
Meiselman, Herbert Joel
Meizel, Stanley
Meredith, Carol N
Miledi, Ricardo
Miquel, Jaime
Mussen, Eric Carnes
Nichols, Barbara Ann
Nissenson, Robert A
Padian, Kevin
Painter, Robert Blair
Peterson, Andrew R
Phillips, Theodore Locke
Ralph, Peter
Rees, Rees Bynon
Reiness, Gary
Reiss, Diana
Ribak, Charles Eric
Rivier, Catherine L
Robles, Laura Jeanne
Rosenzweig, Mark Richard
Ryder, Oliver A
Schonewald-Cox, Christine Micheline
Seto, Joseph Tobey
Shapiro, Lucille
Smith, Charles G
Sondhaus, Charles Anderson
Springer, Wayne Richard
Squire, Larry Ryan
Tamerius, John
Urch, Umbert Anthony
Varon, Myron Izak
Vinters, Harry Valdis
Washburn, Sherwood L
Weil, Jon David
Weinstein, Irwin M
Weissman, Irving L
Wenzel, Bernice Martha
Wyrwicka, Wanda

COLORADO
Bedford, Joel S
Cook, James L
Elkind, Mortimer M
Gillette, Edward LeRoy
Johnson, Thomas Eugene
Johnson, Timothy John Albert
Kloppel, Thomas Mathew
Lett, John Terence
Liechty, Richard Dale
Parton, William Julian, Jr
Prasad, Kedar N
Runner, Meredith Noftzger
Stith, Bradley James
Verdeal, Kathey Marie
Voss, James Leo
Watkins, Linda Rothblum
Whicker, Floyd Ward

CONNECTICUT
Eustice, David Christopher
Farmer, Mary Winifred
Fischer, James Joseph
Klein, Norman W
Lurie, Alan Gordon
Moretz, Roger C
Morrow, Jon S
Pooley, Alan Setzler
Smilowitz, Henry Martin
Walton, Alan George

DISTRICT OF COLUMBIA
Ampy, Franklin
Barak, Eve Ida
Bednarek, Jana Marie
Carlson, William Dwight
Cuatrecasas, José
Field, Ruth Bisen
Green, Rayna Diane
Halstead, Thora Waters
Henderson, Ellen Jane
Hollinshead, Ariel Cahill
Lakshman, M Raj
Lavine, Robert Alan
McCarty, Gale A
Malone, Thomas
Malone, Thomas E
Milbert, Alfred Nicholas
Miles, Corbin I
Parasuraman, Raja
Rall, William Frederick
Robinette, Charles Dennis
Smith, David Allen
Tangney, John Francis
Vietmeyer, Noel Duncan
Wassef, Nabila M

FLORIDA
Achey, Phillip M
Armelagos, George John
Bloom, Stephen Allen
Bruner, Harry Davis
Cromroy, Harvey Leonard
Dribin, Lori
Dutary, Bedsy Elsira
Hope, George Marion
Keane, Robert W
Litosch, Irene
Lombardi, Max H
Patterson, Richard Sheldon
Raizada, Mohan K
Robitaille, Henry Arthur
Routh, Donald Kent
Smith, Lawton Harcourt
Stroud, Robert Michael
Teitelbaum, Philip
Thuning-Roberson, Claire Ann
West, Richard Lowell
Wheeler, George Carlos

GEORGIA
Ades, Edwin W
Adkison, Daniel Lee
Bryan, John Henry Donald
Cureton, Kirk J
Eriksson, Karl-Erik Lennart
Gould, Kenneth G
Isaac, Walter
Karow, Armand Monfort, Jr
Kaufman, Leo
Kaul, Pushkar N
Kubica, George P
Proctor, Charles Darnell
Reichard, Sherwood Marshall
Romanovicz, Dwight Keith
Tzianabos, Theodore
Urso, Paul
Zaia, John Anthony

HAWAII
Carlson, John Gregory
Fast, Arlo Wade
Fok, Agnes Kwan
Howarth, Francis Gard
Kaneshiro, Kenneth Yoshimitsu
Lee, Cheng-Sheng
Schmitt, Donald Peter
Townsley, Sidney Joseph
Walsh, William Arthur

ILLINOIS
Barr, Susan Hartline
Becker, John Henry
Bevan, William
Brues, Austin
Capone, James J
Coulson, Richard L
Dawson, M Joan
Desai, Parimal R
DiMuzio, Michael Thomas
Dudley, Horace Chester
Frenkel, Niza B
Fritz, Thomas Edward
Grahn, Douglas
Heaney, Lawrence R
Heller, Alfred
Holtzman, Richard Beves
Kaine, Brian Paul
King, David George
Klass, Michael R
Knospe, William H
Kundu, Samar K
Mathews, Martin Benjamin
Mednieks, Maija
Mewissen, Dieudonne Jean
Naples, Virginia L
Peak, Meyrick James
Rabb, George Bernard
Ratajczak, Helen Vosskuhler
Rechtschaffen, Allan
Reichmann, Manfred Eliezer
Rice, Thomas B
Rosner, Marsha R
Rundo, John
Saper, Clifford B

INDIANA
Chandrasekhar, S
Chernoff, Ellen Ann Goldman
Daily, Fay Kenoyer
Green, Morris
Guth, S(herman) Leon
Juraska, Janice Marie
Kenny, Michael Thomas
Rausch, Steven K
Wagner, Morris
Zinsmeister, William John

IOWA
Dutton, Gary Roger
Isely, Duane
LaBrecque, Douglas R
Laffoon, John
Osborne, James William
Riley, Edgar Francis, Jr
Smith, Norman B

KANSAS
Cho, Cheng T
Choate, Jerry Ronald
Frost-Mason, Sally Kay
Hays, Horace Albennie
Kimler, Bruce Franklin
Shaw, Edward Irwin
Uyeki, Edwin M

KENTUCKY
Bonner, Philip Hallinder
Calkins, John
Carter, Julia H
Crooks, Peter Anthony
Feola, Jose Maria
Friedler, Robert M
Giammara, Beverly L Turner Sites
Hirsch, Henry Richard
Kary, Christina Dolores
Maruyama, Yosh
Urano, Muneyasu

LOUISIANA
Anderson, William McDowell
Hughes, Janice S
Knaus, Ronald Mallen
Musgrave, Mary Elizabeth
Shaw, Richard Francis
Swenson, David Harold

MAINE
Founds, Henry William
Harrison, David Ellsworth
Lovett, Edmund J, III
Musgrave, Stanley Dean

MARYLAND
Anderson, Robert Simpers
Atwell, Constance Woodruff
Baker, Carl Gwin
Baker, George Thomas, III
Bank-Schlegel, Susan Pamela
Baum, Siegmund Jacob
Brook, Itzhak
Burton, Dennis Thorpe
Cereghino, James Joseph
Chen, Raymond F
Chou, Janice Y
Crampton, Theodore Henry Miller
Dermody, William Christian
Doubt, Thomas J
Driscoll, Bernard F
Earnshaw, William C
Eddy, Hubert Allen
Ederer, Fred
Fischman, Marian Weinbaum
Fozard, James Leonard
Gallin, John I
Geller, Ronald G
Gill, John Russell, Jr
Golding, Hana
Gonzalez, Frank J
Gorden, Phillip
Goroff, Diana K
Gray, Paulette S
Harrison, George H
Hawkins, Michael John
Hecht, Toby T
Hendrix, Thomas Russell
Henrich, Curtis J
Hochstein, Herbert Donald
Hodge, Frederick Allen
Hubbard, Ann Louise
James, Stephen P
Kammula, Raju G
Khachaturian, Zaven Setrak
Kimes, Brian Williams
Kinnamon, Kenneth Ellis
Kraemer, Kenneth H
Krusberg, Lorin Ronald
Kung, Hsiang-Fu
Lakatta, Edward G

Simpson, Sidney Burgess, Jr
Sinclair, James Burton
Sinclair, Thomas Frederick
Sitrin, Michael David
Smith, Douglas Calvin
Stefani, Stefano
Stehney, Andrew Frank
Swartz, Harold M
Thomson, John Ferguson
Zimmerman, Roger Paul

Larsen, Paul M
Leach, William Matthew
Leef, James Lewis
Levenson, Robert
Ley, Herbert L, Jr
Lipman, David J
Lowder, James N
Lowy, R Joel
McDonald, Lee J
McGowan, John Joseph
Margolis, Ronald Neil
Marino, Pamela J
Marzella, Louis
Meryman, Harold Thayer
Michelsen, Arve
Milman, Gregory
Mitsuya, Hiroaki
Morris, Stephen Jon
Neville, David Michael, Jr
O'Grady, Richard Terence
Oliver, Constance
Ozato, Keiko
Pagano, Richard Emil
Page, Norbert Paul
Peterson, Jane Louise
Pienta, Roman Joseph
Pilitt, Patricia Ann
Plato, Chris C
Riesz, Peter
Rifkind, Basil M
Robinson, David Mason
Sarin, Prem S
Saslaw, Leonard David
Shulman, N Raphael
Skidmore, Wesley Dean
Sobocinski, Philip Zygmund
Solomon, Howard Fred
Spector, Novera Herbert
Spiegel, Allen M
Stratmeyer, Melvin Edward
Top, Franklin Henry, Jr
Vande Woude, George
Wachholz, Bruce William
Weiss, Bernard
Weiss, Joseph Francis
Wright, William W
Zbar, Berton
Zoon, Kathryn C

MASSACHUSETTS
Adams, Richard A
Allen, Philip G
Barlow, John Sutton
Barone, Leesa M
Bell, Eugene
Bennett, Holly Vander Laan
Bynum, T E
Davis, William Edwin, Jr
Ennis, Francis A
Ezzell, Robert Marvin
Fimian, Walter Joseph, Jr
Geisterfer-Lowrance, Anja A T
Giffin, Emily Buchholtz
Greene, Reginald
Grigg, Peter
Hagen, Susan James
Hartwig, John
Hebben, Nancy
Henneman, Elwood
Hu, Jing-Shan
Huggins, Charles Edward
Jacobs, Jerome Barry
Jones, Rosemary
Kennedy, Ann Randtke
Kung, Patrick C
Landsberg, Lewis
Lee, John K
McIntire, Kenneth Robert
Marcus, Elliot M
Mulvey, Philip Francis, Jr
Naqvi, S Rehan Hasan
Ofner, Peter
Orlidge, Alicia
Paul, Benoy Bhushan
Reincke, Ursula
Sekuler, Robert W
Sommer Smith, Sally K
Stein, Janet Lee Swinehart
Swann, David A
Tannenbaum, Steven Robert
Tilley, Stephen George
Volkmann, Frances Cooper
Ziomek, Carol A

MICHIGAN
Allen, Edward David
Berent, Stanley
Bernstein, Maurice Harry
Brawn, Mary Karen
Bursian, Steven John
Chavin, Walter
Dajani, Adnan S
Dixon-Holland, Deborah Ellen
Drescher, Dennis G
Erickson, Laurence A
Fisher, Leslie John
Gordon, Sheldon Robert
Goustin, Anton Scott
Hansen-Smith, Feona May
Hirata, Fusao
Jensen, James Burt
Kaufman, Donald Barry
Kawooya, John Kasajja
Keyes, Paul Landis

Klomparens, Karen L
Kurachi, Kotoku
Laborde, Alice L
Ledbetter, Steven R
Lindsay, Robert Kendall
Lum, Lawrence
Lynne-Davies, Patricia
McKitrick, Mary Caroline
McManus, Myra Jean
Marcelo, Cynthia Luz
Miller, Josef Mayer
Motiff, James P
Moudgil, Virinder Kumar
Radin, Eric Leon
Raub, Thomas Jeffrey
Roberts, Joan Marie
Rutledge, Lester T
Simpson, William L
Sloat, Barbara Furin
Smith, David I
Solomon, Allen M
Wilks, John William
Zacks, James Lee

MINNESOTA
Bishop, Jonathan S
Brambl, Robert Morgan
Gores, Gregory J
Graham, Edmund F
Hamilton, David Whitman
Jenkins, Robert Brian
Kornblith, Carol Lee
La Russo, Nicholas F
Lindgren, Alice Marilyn Lindell
Mabry, Paul Davis
Mount, Donald I
Ordway, Ellen
Percich, James Angelo
Rohrbach, Michael Steven
Rosenberg, Andreas
Salisbury, Jeffrey L
Song, Chang Won
Swingle, Karl Frederick
Vetter, Richard J
Yellin, Absalom Moses

MISSISSIPPI
Alexander, William Nebel
Lloyd, Edwin Phillips
McClurkin, Iola Taylor

MISSOURI
Alexander, Stephen
Braciale, Vivian Lam
Burchard, Jeanette
Cook, Mary Rozella
Dilley, William G
Glenn, Kevin Challon
Harding, Clifford Vincent, III
Kuo, Scot Charles
Leimgruber, Richard M
Lenhard, James M
Mecham, Robert P
Medoff, Judith
Menz, Leo Joseph
Monzyk, Bruce Francis
Morton, Phillip A
Moulton, Robert Henry
Roloff, Marston Val
Shugart, Jack Isaac
Tolmach, L(eonard) J(oseph)

MONTANA
Bergman, Robert K
Rumely, John Hamilton

NEBRASKA
Dalrymple, Glenn Vogt
Kolar, Joseph Robert, Jr
Morris, Mary Rosalind

NEVADA
Hunter, Kenneth W, Jr

NEW HAMPSHIRE
Douple, Evan Barr
Fagerberg, Wayne Robert
Normandin, Robert F
Wurster-Hill, Doris Hadley

NEW JERSEY
Alfano, Michael Charles
Bex, Frederick James
Black, Ira B
Burger, Joanna
Carlson, Gerald M
DeForrest, Jack M
DeMaio, Donald Anthony
Dugan, Gary Edwin
Frances, Saul
Heaslip, Richard Joseph
Hirsch, Robert L
Jacobus, David Penman
Levine, O Robert
Lewis, Steven Craig
Manne, Veeraswamy
Mele, Frank Michael
Naro, Paul Anthony
Riley, David
Schrankel, Kenneth Reinhold
Sherrick, Carl Edwin
Snape, William J
Stern, William
Tiku, Moti

Tischio, John Patrick
Triemer, Richard Ernest

NEW MEXICO
Guilmette, Raymond Alfred
Lechner, John Fred
Lundgren, David L(ee)
Mewhinney, James Albert
Pena, Hugo Gabriel
Thomas, Kimberly W
Thomassen, David George
Trotter, John Allen
Walters, Ronald Arlen

NEW YORK
Adesnik, Milton
Altman, Kurt Ison
Asch, Bonnie Bradshaw
Bachrach, Howard L
Badoyannis, Helen Litman
Banerjee, Debendranath
Bateman, John Laurens
Baust, John G
Bell, Michael Allen
Berkowitz, Jesse M
Bernstein, Alvin Stanley
Beyer, Carl Fredrick
Black, Lindsay MacLeod
Bruce, Alan Kenneth
Bullough, Vern L
Carsten, Arland L
Casarett, Alison Provoost
Center, Sharon Anne
Chasalow, Fred I
Chou, Shyan-Yih
Cohen, Norman
Colman, David Russell
Concannon, Joseph N
Corradino, Robert Anthony
Cronkite, Eugene Pitcher
Crouch, Billy G
Deschner, Eleanor Elizabeth
DiCioccio, Richard A
DiGaudio, Mary Rose
Domozych, David S
Dougherty, Thomas John
Dounce, Alexander Latham
Duff, Ronald George
Erlij, David
Ferris, Steven Howard
Godson, Godfrey Nigel
Goldfarb, David S
Greene, Lloyd A
Grove, Patricia A
Hall, Eric John
Hammond, H David
Hanafusa, Hidesaburo
Haycock, Dean A
Heisermann, Gary J
Hopkins, Betty Jo Henderson
Horst, Ralph Kenneth
Janakidevi, K
Kaplan, Ehud
Kilpper, Robert William
Kim, Jae Ho
Lau, Joseph T Y
Law, Paul Arthur
Lawrence, Christopher William
Lazarow, Paul B
Lennarz, William J
McGarry, Michael P
McOsker, Charles C
Macris, Nicholas T
Maillie, Hugh David
Makous, Walter L
Masur, Sandra Kazahn
Meruelo, Daniel
Miller, Morton W
Moroson, Harold
Movshon, J Anthony
Norton, Larry
Olsen, Kathie Lynn
Osterberg, Donald Mouretz
Peerschke, Ellinor Irmgard Barbara
Pfaffmann, Carl
Pogo, Beatriz G T
Powell, Sharon Kay
Redmond, Billy Lee
Reichert, Leo E, Jr
Ritch, Robert
Rivenson, Abraham S
Rowe, Arthur W(ilson)
Sackeim, Harold A
Sanders, F Kingsley
Schmidt, John Thomas
Serafy, D Keith
Shellabarger, Claire J
Siddiqui, M A Q
Sobel, Jael Sabina
Sperling, George
Stross, Raymond George
Sutherland, Robert Melvin
Treser, Gerhard
Vladutiu, Georgirene Dietrich
Wiesel, Torsten Nils
Wigler, Michael H
Winston, Judith Ellen
Woodard, Helen Quincy
Wurster, Charles F
Wysocki, Annette Bernadette
Yang, Shung-Jun
Yazulla, Stephen

NORTH CAROLINA
Allen, Nina Strömgren
Barrett, J(ames) Carl
Branch, Clarence Joseph
Butts, Jeffrey A
Curnow, Randall T
Hall, Warren G
Horstman, Donald H
Hoss, Donald Earl
Jameson, Charles William
Jones(Stover), Betsy
Kovacs, Charles J
Kvalnes-Krick, Kalla L
Kwock, Lester
Massaro, Edward Joseph
Miwa, Gerald T
Morgan, Kevin Thomas
Nettesheim, Paul
O'Brien, Deborah A
Reed, William
Rotman, Harold H
Rublee, Parke Alstan
Sanders, Aaron Perry
Smith, Gary Keith
Wheeler, Kenneth Theodore, Jr
Wise, Ernest George

NORTH DAKOTA
Kelleher, James Joseph

OHIO
Boissy, Raymond E
Britton, Steven Loyal
Brunner, Robert Lee
Chapple, Paul James
Chen, I-Wen
Coots, Robert Herman
Fanaroff, Avroy A
Fucci, Donald James
Goldblatt, Peter Jerome
Grossman, Charles Jerome
Grund, Vernon Roger
Hiles, Maurice
Hilmas, Duane Eugene
Jegla, Dorothy Eldredge
Lacey, Beatrice Cates
Lee, Harold Hon-Kwong
Leis, Jonathan Peter
Lieberman, Michael A
McCarthy, F D
Mendenhall, Charles L
Nixon, Charles William
Nordlund, James John
Nygaard, Oddvar Frithjof
Olsen, Richard George
Oyen, Ordean James
Patterson, Michael Milton
Pribor, Donald B
Rein, Diane Carla
Thomas, Craig Eugene
Varnes, Marie Elizabeth
Yang, Tsanyen
Yates, Allan James

OKLAHOMA
Francko, David Alex
Friedberg, Wallace
Gimble, Jeffrey M
Holloway, Frank A
Mares, Michael Allen
Simpson, Ocleris C
Talent, Larry Gene
Thomas, Sarah Nell

OREGON
Bethea, Cynthia Louise
Brownell, Philip Harry
Burry, Kenneth A
Callahan, Clarence Arthur
Clinton, Gail M
Cross, Stephen P
Denison, William Clark
Koong, Ling-Jung
Liss, William John
Pizzimenti, John Joseph
Straube, Robert Leonard
Yatvin, Milton B

PENNSYLVANIA
Adelson, Lionel Morton
Agarwal, Jai Bhagwah
Baldino, Frank, Jr
Biaglow, John E
Boggs, Sallie Patton Slaughter
Brennan, James Thomas
Clark, T(helma) K
Conger, Alan Douglas
Courtney, Richard J
Cox, Malcolm
Dinges, David Francis
Edelstone, Daniel I
Emberton, Kenneth C
Ewing, David Leon
Foley, Thomas Preston, Jr
Forbes, Paul Donald
Fussell, Catharine Pugh
Glick, Mary Catherine
Gollin, Susanne Merle
Gombar, Charles T
Greene, Mark Irwin
Halbrendt, John Marthon
Hauptman, Stephen Phillip
Isom, Harriet C
Jacobs, George Joseph

Other Biological Sciences (cont)

Kim, Ke Chung
Ladman, Aaron J(ulius)
Leeper, Dennis Burton
Leon, Shalom A
Maloy, W Lee
Mason, James Russell
Maul, Gerd G
Maxson, Linda Ellen R
Moyer, Robert (Findley)
Mullin, James Michael
Murphy, Robert Francis
Norrbom, Allen Lee
Olds-Clarke, Patricia Jean
Olexa, Stephanie A
Pruett, Esther D R
Saunders, William Bruce
Sax, Martin
Vlasuk, George P
Watson, Joseph Alexander
Zigmond, Sally H

RHODE ISLAND
Biancani, Piero
Bibb, Harold David
Eil, Charles
McMillan, Paul N
Waage, Jonathan King

SOUTH CAROLINA
Appel, James B
Bank, Harvey L
Creek, Kim E
Ghaffar, Abdul
Mitchell, Hugh Bertron
Powell, Donald Ashmore
Powers, Edward Lawrence
Rembert, David Hopkins, Jr

TENNESSEE
Alsop, Frederick Joseph, III
Carlson, James Gordon
Clapp, Neal K
Coons, Lewis Bennion
Fry, Richard Jeremy Michael
Gibbs, Samuel Julian
Gutzke, William H N
Hadden, Charles Thomas
Hubner, Karl Franz
Isom, Billy Gene
Kuo, Chao-Ying
Limbrd, Lee Eberhardt
Mazur, Peter
Monaco, Paul J
Noonan, Thomas Robert
Reddick, Bradford Beverly
Rushton, Priscilla Strickland
Schlarbaum, Scott E
Sharp, Aaron John
Storer, John B
Sun, Deming

TEXAS
Alhossainy, Effat M
Aumann, Glenn D
Baldwin, William Russell
Baron, Samuel
Barranco, Sam Christopher, III
Boldt, David H
Byrne, John Howard
Christ, John Ernest
Cooper, Ronda Fern
Cox, Ann Bruger
Deisenhofer, Johann
Draper, Rockford Keith
Dung, H C
Ford, George Peter
Frishman, Laura J(ean)
Gardner, Florence Harrod
Gaulden, Mary Esther
Gyorkey, Ferenc
Haye, Keith R
Howe, William Edward
Jenkins, Vernon Kelly
Kennedy, Joseph Patrick
Klaus, Ewald Fred, Jr
Ku, Bernard Siu-Man
Kunau, Robert, Jr
Lachman, Lawrence B
La Claire, John Willard, II
Lapeyre, Jean-Numa
Lopez-Berestein, Gabriel
Meltz, Martin Lowell
Mitchell, Robert W
Moller, Peter C
Nachtwey, David Stuart
Oujesky, Helen Matusevich
Pennebaker, James W
Prasad, Naresh
Rodarte, Joseph Robert
Sauer, Helmut Wilhelm
Schonbrunn, Agnes
Soltes, Edward John
Steinberger, Anna
Thalmann, Robert H
Tomasovic, Stephen Peter
Ullrich, Robert Leo
Wiederhold, Michael L

UTAH
Barnett, Bill B

Dethlefsen, Lyle A
Mahoney, Arthur W
Mays, Charles William
Ramachandran, Chittoor Krishna
Stevens, Walter
Warters, Raymond Leon

VERMONT
Watters, Christopher Deffner

VIRGINIA
Bonvillian, John Doughty
Carson, Keith Alan
Corwin, Jeffrey Todd
Davis, Joel L
De Haan, Henry J(ohn)
Edwards, Ernest Preston
Fuller, Stephen William
Hilu, Khidir Wanni
Holsinger, John Robert
Jones, Johnnye M
Knauer, Thomas E
Looney, William Boyd
Moncrief, Nancy D
Nelson, Neal Stanley
Rozzell, Thomas Clifton
Senti, Frederic R(aymond)
Sirica, Alphonse Eugene
Stevens, Robert Edward
Townsend, Lawrence Willard
Wilson, John Drennan

WASHINGTON
Bair, William J
Bonham, Kelshaw
Branca, Andrew A
Christensen, Gerald M
Crill, Wayne Elmo
Dale, Beverly A
Edwards, John S
Greenberg, Philip D
Greenlee, Theodore K
Hungate, Frank Porter
McDougall, James K
Mackensie, Alan P
Martin, Carroll James
Moffett, Stacia Brandon
Monan, Gerald E
Rasey, Janet Sue
Sanders, Charles Leonard, Jr
Sieker, Larry Charles
Singer, Jack W
Teller, Davida Young
Thompson, Roy Charles, Jr
Wolf, Norman Sanford

WEST VIRGINIA
Reid, Robert Leslie

WISCONSIN
Bosnjak, Zeljko J
Clifton, Kelly Hardenbrook
Cummings, John Albert
Duewer, Elizabeth Ann
Gilbert-Barness, Enid F
Goldstein, Robert
Gould, Michael Nathan
Greiff, Donald
Grunewald, Ralph
Iltis, Hugh Hellmut
Kostenbader, Kenneth David, Jr
Kowal, Robert Raymond
Mandel, Neil
Norris, William Penrod
Schwartz, Bradford S
Sieber, Fritz
Tulunay-Keesey, Ulker

WYOMING
Doerges, John E
Humburg, Neil Edward

PUERTO RICO
Jordan-Molero, Jaime E
Squire, Richard Douglas
Voltzow, Janice

ALBERTA
Bidgood, Bryant Frederick
Church, Robert Bertram
Famiglietti, Edward Virgil, Jr
McGann, Locksley Earl
Ruth, Royal Francis
Sohal, Parmjit S
Wyse, David George

BRITISH COLUMBIA
Campbell, Alan
Freeman, Hugh J
Hoffmann, Geoffrey William
Sinclair, John G
Skarsgard, Lloyd Donald

MANITOBA
Chong, James York
Gerrard, Jonathan M
Gill, Clifford Cressey

NEW BRUNSWICK
Waiwood, Kenneth George

NOVA SCOTIA
Connolly, John Francis
Horackova, Magda

Stobo, Wayne Thomas

ONTARIO
Adamson, S Lee
Basu, Prasanta Kumar
Burba, John Vytautas
Bush, Raymond Sydney
Carstens, Eric Bruce
Chadwick, June Stephens
Clark, Gordon Murray
Connolly, Joseph Anthony
Cosmos, Ethel
Cutz, Ernest
Czuba, Margaret
Dalpe, Yolande
Elinson, Richard Paul
Forer, Arthur H
Hare, William Currie Douglas
Hill, Richard Peter
Inch, William Rodger
Kovacs, Kit M
Kruuv, Jack
Lau, Catherine Y
Leibo, Stanley Paul
Lingwood, Clifford Alan
McKneally, Martin F
Redhead, Scott Alan
Schneider, Bruce Alton
Tannock, Ian Frederick
Whitmore, Gordon Francis

QUEBEC
Antakly, Tony
Legendre, Pierre
Lehnert, Shirley Margaret
Verdy, Maurice

SASKATCHEWAN
Lapp, Martin Stanley
Ripley, Earle Allison

OTHER COUNTRIES
Acres, Robert Bruce
Avila, Jesus
Barald, Kate Francesca
Black, James
Brondz, Boris Davidovich
Campbell, Virginia Wiley
Farid, Nadir R
Ganchrow, Donald
Gendler, Sandra J
Hahn, Eric Walter
Jarvis, Simon Michael
Jondorf, W Robert
Kang, Kewon
Lin, Chin-Tarng
Lynn, Robert K
MacCarthy, Jean Juliet
Oxnard, Charles Ernest
Quastel, Michael Reuben
Shimaoka, Katsutaro

CHEMISTRY

Agricultural & Food Chemistry

ALABAMA
Ammerman, Gale Richard
Mortvedt, John Jacob
Rymal, Kenneth Stuart
Small, Robert James

ALASKA
Kennish, John M

ARIZONA
Emery, Donald F
Hendrix, Donald Louis
Hoffmann, Joseph John
Kline, Ralph Willard
Shaw, Ellsworth
Steffen, Albert Harry

ARKANSAS
Crandall, Philip Glen
Hood, Robert L
Horan, Francis E

CALIFORNIA
Baker, Frederick Charles
Bernhard, Richard Allan
Bolin, Harold R
Chen, Ming Sheng
Chen, Tung-Shan
Chou, Tsong-Wen
Cleary, Michael E
Coughlin, James Robert
Cramer, Archie Barrett
Crosby, Donald Gibson
Dutra, Ramiro Carvalho
Eisenberg, Sylvan
Fancher, Llewellyn W
Farmer, Walter Joseph
Feeney, Robert Earl
Fleming, Sharon Elayne
Fung, Steven
Goss, James Arthur
Haddon, William F, (Jr)
Hamaker, John Warren
Hammock, Bruce Dupree
Heil, John F, Jr
Henika, Richard Grant
Henrick, Clive Arthur

Hill, Fredric William
Hokama, Takeo
Ja, William Yin
Josephson, Ronald Victor
Kean, Chester Eugene
Khayat, Ali
Kinsella, John Edward
L'Annunziata, Michael Frank
Lee, David Louis
Leo, Albert Joseph
Lewis, Sheldon Noah
Liang, Yola Yueh-o
Lundeen, Glen Alfred
McNall, Lester R
Maier, V(incent) P(aul)
Masri, Merle Sid
Mayer, William Joseph
Mihailovski, Alexander
Millar, Jocelyn Grenville
Nielsen, Milo Alfred
Patton, Stuart
Petrowski, Gary E
Quistad, Gary Bennet
Rendig, Victor Vernon
Richardson, Thomas
Riffer, Richard
Rockland, Louis B
Roitman, James Nathaniel
Russell, Gerald Frederick
Rynd, James Arthur
Samuels, Robert Bireley
Sayre, Robert Newton
Schaefer, Charles Herbert
Singleton, Vernon LeRoy
Smith, Lloyd Muir
Spatz, David Mark
Swisher, Horton Edward
Thomson, John Ansel Armstrong
Thomson, Tom Radford
Thornton, John Irvin
Vreeland, Valerie Jane
Walker, Howard David
Walker, Howard George, Jr
Wauchope, Robert Donald
Weast, Clair Alexander
Williams, John Wesley
Winterlin, Wray Laverne
Wright, Terry L
Young, Donald C
Yuen, Wing
Zavarin, Eugene

COLORADO
Bower, Nathan Wayne
Cannon, William Nathaniel
Heberlein, Douglas Garavel
Kimoto, Walter Iwao
Lorenz, Klaus J
Mosier, Arvin Ray
Nachtigall, Guenter Willi
Norstadt, Fred A
Turner, James Howard
Verdeal, Kathey Marie

CONNECTICUT
Agarwal, Vipin K
Bibeau, Thomas Clifford
Hankin, Lester
Mattina, Mary Jane Incorvia
Pierce, James Benjamin
Putterman, Gerald Joseph
Relyea, Douglas Irving
Schaaf, Thomas Ken

DELAWARE
Anderson, Jeffrey John
Austin, Paul Rolland
Beestman, George Bernard
Dubas, Lawrence Francis
Finn, James Walter
Jelinek, Arthur Gilbert
Koeppe, Mary Kolean
May, Ralph Forrest
Metzger, James Douglas
Moberg, William Karl
Montague, Barbara Ann
Morrissey, Bruce William
Rasmussen, Arlette Irene
Roe, David Christopher
Saegebarth, Klaus Arthur
Sandell, Lionel Samuel
Taves, Milton Arthur
Thoroughgood, Carolyn A
Wommack, Joel Benjamin, Jr
Woods, Thomas Stephen

DISTRICT OF COLUMBIA
Baker, Charles Wesley
Flora, Lewis Franklin
Hammer, Charles F
Kuznesof, Paul Martin
Lombardo, Pasquale
Modderman, John Philip
Mog, David Michael
Page, Samuel William
Sarmiento, Rafael Apolinar
Schneider, Bernard Arnold
Staruszkiewicz, Walter Frank, Jr
Treichler, Ray

FLORIDA
Attaway, John Allen
Berry, Robert Eddy
Boehme, Werner Richard

Calvert, David Victor
Corbett, Michael Dennis
Cort, Winifred Mitchell
Gregory, Jesse Forrest, III
Grunwald, Claus Hans
Hagenmaier, Robert Doller
Hunter, George L K
Jackel, Simon Samuel
Jezeski, James John
Kilsheimer, John Robert
Knapp, Joseph Leonce, Jr
Matthews, Richard Finis
Moser, Robert E
Nagy, Steven
Nakashima, Tadayoshi
Newsom, William S, Jr
Prebluda, Harry Jacob
Ross, Lynne Fischer
Sadler, George D
Shuey, William Carpenter
Wolf, Benjamin
Yu, Simon Shyi-Jian

GEORGIA
Barton, Franklin Ellwood, II
Brown, David Smith
Chortyk, Orestes Timothy
Churchill, Frederick Charles
Clark, Benjamin Cates, Jr
Gaines, Tinsley Powell
Iacobucci, Guillermo Arturo
Jain, Anant Vir
Jen, Joseph Jwu-Shan
Leffingwell, John C
Lillard, Dorris Alton
Loewenstein, Morrison
Myers, Dirck V
Smit, Christian Jacobus Bester
Wilson, David Merl

HAWAII
Doerge, Daniel Robert
Hilton, H Wayne
Moritsugu, Toshio
Sherman, Martin
Young, Hong Yip

IDAHO
Hibbs, Robert A
Lehrer, William Peter, Jr
Muneta, Paul
Natale, Nicholas Robert

ILLINOIS
Abbott, Thomas Paul
Berenbaum, May Roberta
Bietz, Jerold Allen
Blair, Louis Curtis
Bloom, Jack
Bookwalter, George Norman
Burkwall, Morris Paton, Jr
Campbell, Alfred Duncan
Chang, Hsien-Hsin
Cheryan, Munir
Christianson, Donald Duane
Crovetti, Aldo Joseph
Donnelly, Thomas Henry
Fink, Kenneth Howard
Germino, Felix Joseph
Hahn, Richard Ray
Halaby, George Anton
Hamer, Martin
Hefley, Alta Jean
Heydanek, Menard George, Jr
Hill, Kenneth Richard
Hoepfinger, Lynn Morris
Honig, David Herman
Hynes, John Thomas
Inglett, George Everett
Kinghorn, Alan Douglas
Kite, Francis Ervin
Klein, Barbara P
Knake, Ellery Louis
Krishnamurthy, Ramanathapur
  Gundachar
Larson, Bruce Linder
Levin, Alfred A
Lundquist, Burton Russell
Maher, George Garrison
Moser, Kenneth Bruce
Mounts, Timothy Lee
Nielsen, Harald Christian
Nishida, Toshiro
Pai, Sadanand V
Parrill, Irwin Homer
Polen, Percy B
Rackis, Joseph John
Rankin, John Carter
Schaefer, Wilbur Carls
Schanefelt, Robert Von
Sessa, David Joseph
Sheldon, Victor Lawrence
Singh, Laxman
Soucie, William George
Standing, Charles Nicholas
Stanford, Marlene A
Tiemstra, Peter J
Traxler, James Theodore
Trelease, Richard Davis
Van der Kloot, Albert Peter
Wang, Pie-Yi
Weber, Evelyn Joyce
Whitney, Robert McLaughlin
Wilke, Robert Nielsen

Williams, Joseph Lee
Woolson, Edwin Albert
Yates, Shelly Gene

INDIANA
Alter, John Emanuel
Axelrod, Bernard
Beck, James Richard
Butler, Larry G
Corrigan, John Joseph
Glover, David Val
Hoff, Johan Eduard
Katz, Frances R
Kossoy, Aaron David
Michel, Karl Heinz
Pratt, Dan Edwin
Rogers, Richard Brewer
Walker, Jerry C

IOWA
Barthel, William Frederick
Coats, Joel Robert
Murphy, Patricia A
Rebstock, Theodore Lynn
Stahr, Henry Michael

KANSAS
Chung, Okkyung Kim
Hooker, Mark L
Hoseney, Russell Carl
Krum, Jack Kern
Lookhart, George LeRoy
Scheid, Harold E
Schroeder, Robert Samuel
Seitz, Larry Max
Shellenberger, John Alfred
Shogren, Merle Dennis
Stutz, Robert L
Tsen, Cho Ching
Von Rumker, Rosmarie

KENTUCKY
Colby, David Anthony
Hamilton-Kemp, Thomas Rogers
Knapp, Frederick Whiton
Luzzio, Frederick Anthony
Massik, Michael
Wong, John Lui

LOUISIANA
Bayer, Arthur Craig
Blouin, Florine Alice
Connick, William Joseph, Jr
Domelsmith, Linda Nell
Fischer, Nikolaus Hartmut
Gonzales, Elwood John
Liuzzo, Joseph Anthony
Marshall, Wayne Edward
Morris, Cletus Eugene
Mustafa, Shams
Palmgren, Muriel Signe
St Angelo, Allen Joseph
Settoon, Patrick Delano
Spanier, Arthur M
Vercellotti, John R

MAINE
Pierce, Arleen Cecilia

MARYLAND
Argauer, Robert John
Baker, Doris
Beroza, Morton
Brause, Allan R
Caret, Robert Laurent
Christy, Alfred Lawrence
Fulger, Charles V
Galetto, William George
Gilmore, Thomas Meyer
Hall, Richard Leland
Herner, Albert Erwin
Horwitz, William
Inscoe, May Nilson
Jacobson, Martin
Kaufman, Donald DeVere
Kleier, Daniel Anthony
Lee, Yuen San
Liang, Shu-Mei
Mandava, Naga Bhushan
Menn, Julius Joel
Moats, William Alden
Murphy, Elizabeth Wilcox
Nelson, Judd Owen
Porzio, Michael Anthony
Ragsdale, Nancy Nealy
Rice, Clifford Paul
Salwin, Harold
Schechter, Milton Seymour
Skidmore, Wesley Dean
Sphon, James Ambrose
Veitch, Fletcher Pearre, Jr
Wong, Noble Powell
Zeleny, Lawrence

MASSACHUSETTS
Angelini, Pio
Fagerson, Irving Seymour
Gorfien, Harold
Hagopian, Miasnig
Lapuck, Jack Lester
Learson, Robert Joseph
Lerner, Harry
Mabrouk, Ahmed Fahmy
Marquis, Judith Kathleen

Nawar, Wassef W
Potter, Frank Elwood
Salant, Abner Samuel
Taub, Irwin A(llen)

MICHIGAN
Arrington, Jack Phillip
Bjork, Carl Kenneth, Sr
Brunner, Jay Robert
Budde, Paul Bernard
Carlotti, Ronald John
Ferguson-Miller, Shelagh Mary
Frawley, Nile Nelson
Gilbertson, Terry Joel
Gray, Ian
Hightower, Kenneth Ralph
Hollingsworth, Cornelia Ann
Hollingworth, Robert Michael
Jensen, David James
Laks, Peter Edward
Leng, Marguerite Lambert
Mergentime, Max
Muentener, Donald Arthur
Pynnonen, Bruce W
Shelef, Leora Aya
Skarsaune, Sandra Kaye
Smith, Denise Myrtle
Walters, Stephen Milo

MINNESOTA
Anderson, George Robert
Atwell, William Alan
Chandan, Ramesh Chandra
Fulmer, Richard W
Goforth, Deretha Rainey
Lonergan, Dennis Arthur
Reinhart, Richard D
Roach, J Robert
Throckmorton, James Rodney
Van Valkenburg, Jeptha Wade, Jr

MISSISSIPPI
Alley, Earl Gifford
Cardwell, Joe Thomas
Collins, Johnnie B
Hedin, Paul A
Minyard, James Patrick

MISSOURI
Bailey, Milton (Edward)
Conn, James Frederick
Dutra, Gerard Anthony
Erickson, David R
Franz, John Edward
Freeman, Robert Clarence
Gash, Virgil Walter
Hammann, William Curl
Hardwick, William Aubrey, Jr
Holmes, Edward Lawson
Klein, Andrew John
Koenig, Karl E
Long, Lawrence William
Moedritzer, Kurt
Newell, Jon Albert
Perry, Paul
Pfander, William Harvey
Radke, Rodney Owen
Rogic, Milorad Mihailo
Schleppnik, Alfred Adolf
Schumacher, Richard William
Seymour, Keith Goldin
Sharp, Dexter Brian
Shaver, Kenneth John
Stalling, David Laurence
Stenseth, Raymond Eugene
Stout, Edward Irvin
Waggle, Doyle H

MONTANA
McGuire, Charles Francis
Richards, Geoffrey Norman

NEBRASKA
Issenberg, Phillip
McClurg, James Edward
Taylor, Steve L
Wallen, Stanley Eugene

NEW HAMPSHIRE
Feliciotti, Enio
Ikawa, Miyoshi

NEW JERSEY
Addor, Roger Williams
Barringer, Donald F, Jr
Borenstein, Benjamin
Brown, Dale Gordon
Carlucci, Frank Vito
Cavender, Patricia Lee
Cevasco, Albert Anthony
Chang, Stephen Szu Shiang
Chiang, Bin-Yea
Coggon, Philip
Crosby, Guy Alexander
Dahle, Leland Kenneth
Focella, Antonino
Fost, Dennis L
Gatterdam, Paul Esch
Giacobbe, Thomas Joseph
Giddings, George Gosselin
Gruenhagen, Richard Dale
Halfon, Marc
Harding, Maurice James Charles
Hatch, Charles Eldridge, III

Ho, Chi-Tang
Hsu, Kenneth Hsuehchia
Jacobson, Glen Arthur
Johnson, Bobby Ray
Joiner, Robert Russell
Klemann, Lawrence Paul
Labows, John Norbert, Jr
Ladner, David William
Lavanish, Jerome Michael
Leeder, Joseph Gorden
Lutz, Albert William
Lynch, Charles Andrew
Manley, Charles Howland
Maravetz, Lester L
Miller, Albert Thomas
Naipawer, Richard Edward
Pearce, David Archibald
Quinn, James Allen
Ramsey, Arthur Albert
Rein, Alan James
Rinehart, Jay Kent
Rosen, Joseph David
Schaich, Karen Marie
Scharpf, Lewis George, Jr
Seitz, Eugene W
Stults, Frederick Howard
Sullivan, Andrew Jackson
Tway, Patricia C
Vallese, Frank M
Vanden Heuvel, William John Adrian,
  III
Van Saun, William Arthur
Volpp, Gert Paul Justus
Waltking, Arthur Ernest
Wislocki, Peter G
Yoder, Donald Maurice

NEW MEXICO
Del Valle, Francisco Rafael
Roubicek, Rudolf V
Smith, Garmond Stanley

NEW YORK
Acree, Terry Edward
Aten, Carl Faust, Jr
Bourke, John Butts
Bourne, Malcolm Cornelius
Brenner, Mortimer W
Buchel, Johannes A
Cante, Charles John
Christenson, Philip A
Dowling, Joseph Francis
Durst, Richard Allen
Ehmann, Edward Paul
Kahan, Sidney
Lewis, Bertha Ann
Lisk, Donald J
Mackay, Donald Alexander Morgan
Mansfield, Kevin Thomas
Marov, Gaspar J
Mirviss, Stanley Burton
Prestwich, Glenn Downes
Reich, Ismar M(eyer)
Roelofs, Wendell L
Roth, Howard
Shallenberger, Robert Sands
Smith, Laura Lee Weisbrodt
Walter, Reginald Henry

NORTH CAROLINA
Anderson, Thomas Ernest
Apperson, Charles Hamilton
Bell, Jimmy Holt
Cooke, Anson Richard
Engle, Thomas William
Gemperline, Margaret Mary Cetera
Hermanson, Harvey Philip
Holm, Robert E
Jennings, Carl Anthony
Kuhr, Ronald John
Kurtz, A Peter
LeBaron, Homer McKay
Lineback, David R
McDaniel, Roger Lanier, Jr
Manning, David Treadway
Marable, Nina Louise
Maselli, John Anthony
Moates, Robert Franklin
Murphy, Robert T
Neumann, Calvin Lee
Patterson, Ronald Brinton
Perfetti, Patricia F
Reed, Gerald
Schwindeman, James Anthony
Senzel, Alan Joseph
Sheets, Thomas Jackson
Worsham, Arch Douglas

NORTH DAKOTA
D'Appolonia, Bert Luigi
Davison, Kenneth Lewis
Donnelly, Brendan James
Johnson, Phyllis Elaine
McDonald, Clarence Eugene
Stolzenberg, Gary Eric
Struble, Craig Bruce
Walsh, David Ervin

OHIO
Battershell, Robert Dean
Brown, David Francis
Buck, Keith Taylor
Cotton, Wyatt Daniel
Dolfini, Joseph E

Agricultural & Food Chemistry (cont)

Dollimore, David
Eisenhardt, William Anthony, Jr
Flautt, Thomas Joseph, Jr
Gird, Steven Richard
Greenlee, Kenneth William
Heckert, David Clinton
Jackson, Earl Rogers
Litchfield, John Hyland
Magee, Thomas Alexander
Marks, Alfred Finlay
Meinert, Walter Theodore
Mohlenkamp, Marvin Joseph, Jr
Myhre, David V
Ockerman, Herbert W
Peng, Andrew Chung Yen
Powers, Larry James
Schafer, Mary Louise
Schenz, Anne Filer
Schweitzer, Mark Glenn
Sevenants, Michael R
Strobel, Rudolf G K
Watson, Maurice E
Watson, Stanley Arthur
Yamazaki, William Toshi
Youngquist, R(udolph) William

OKLAHOMA
Abbott, Donald Clayton
Bingham, Robert J
Fish, Wayne William
Seideman, Walter E

OREGON
Arnold, Roy Gary
Fang, Sheng Chung
Laver, Murray Lane
Norris, Logan Allen
Paul, Pauline Constance
Wrolstad, Ronald Earl

PENNSYLVANIA
Albright, Fred Ronald
Carter, Linda G
Cerbulis, Janis
Cherry, John Paul
Di Cuollo, C John
Doner, Landis Willard
Dyott, Thomas Michael
Farrell, Harold Maron, Jr
Fishman, Marshall Lewis
Foglia, Thomas Anthony
Gluntz, Martin L
Hardesty, Patrick Thomas
Hess, Earl Hollinger
Hicks, Kevin B
Holsinger, Virginia Harris
Hood, Lamartine Frain
Hurst, William Jeffrey
Hurt, Susan Schilt
Johnson, Wayne Orrin
Kurtz, David Allan
Lopez, R C Gerald
Lyman, William Ray
McCurry, Patrick Matthew, Jr
Maerker, Gerhard
Marmer, William Nelson
Moreau, Robert Arthur
Mozersky, Samuel M
Neubeck, Clifford Edward
Newirth, Terry L
Ocone, Luke Ralph
Pfeffer, Philip Elliot
Rogerson, Thomas Dean
Rothman, Alan Michael
Satterlee, Lowell Duggan
Schwing, Gregory Wayne
Sheeley, Richard Moats
Strohm, Paul F
Strong, Frederick Carl, III
Talley, Eugene Alton
Thayer, Donald Wayne
Wasserman, Aaron E
Wolff, Ivan A
Yih, Roy Yangming
Zaika, Laura Larysa

RHODE ISLAND
Lee, Tung-Ching
Olney, Charles Edward
Rand, Arthur Gorham, Jr

SOUTH CAROLINA
Dellicolli, Humbert Thomas
Hunter, George William
Morr, Charles Vernon
Surak, John Godfrey

SOUTH DAKOTA
Halverson, Andrew Wayne
Parsons, John G

TENNESSEE
Buchanan, Russell
Campbell, Ada Marie
Davidson, Philip Michael
Demott, Bobby Joe
Shah, Jitendra J
Woods, Alvin Edwin

TEXAS
Beier, Ross Carlton
Chen, Anthony Hing
Elder, Vincent Allen
Hendrickson, Constance McRight
Kendall, John Hugh
Lin, Sherman S
Mehta, Rajen
Mills, William Ronald
Mitchell, Roy Ernest
Ray, Allen Cobble
Sand, Ralph E
Stoner, Graham Alexander
Varsel, Charles John
Will, Paul Arthur

UTAH
Ames, Robert A
Balandrin, Manuel F
Ernstrom, Carl Anthon
Johnson, John Hal
Richardson, Gary Haight
Segelman, Florence H
Skujins, John Janis
Wallace, Volney

VIRGINIA
Bishop, John Russell
Evans, Herbert John
Felton, Staley Lee
Ferguson, Robert Nicholas
Glasser, Wolfgang Gerhard
Harowitz, Charles Lichtenberg
Harvey, William Ross
Huang, H(sing) T(sung)
Ikeda, Robert Mitsuru
Marcy, Joseph Edwin
Palmer, James Kenneth
Phillips, Jean Allen
Rickett, Frederic Lawrence
Rippen, Thomas Edward
Sprinkle, Robert Shields, III

WASHINGTON
Chinn, Clarence Edward
Dorward-King, Elaine Jay
Ericsson, Lowell Harold
Gruger, Edward H, Jr
Hill, Herbert Henderson, Jr
Ingalsbe, David Weeden
Pigott, George M
Pomeranz, Yeshajahu
Powers, Joseph Robert
Rasco, Barbara A
Rigby, F Lloyd
Rubenthaler, Gordon Lawrence
Spinelli, John
Stansby, Maurice Earl
Stout, Virginia Falk
Sullivan, John W
Swanson, Barry Grant
Wekell, Marleen Marie
Worthington, Ralph Eric
Zimmermann, Charles Edward

WEST VIRGINIA
Barnes, Robert Keith
Bartley, William J
Foster, Joyce Geraldine

WISCONSIN
Balke, Nelson Edward
Ballantine, Larry Gene
Bartholomew, Darrell Thomas
Besch, Gordon Otto Carl
Boggs, Lawrence Allen
Broderick, Glen Allen
Bundy, Larry Gene
Chicoye, Etzer
Cooper, Elmer James
Hollenberg, David Henry
Lindsay, Robert Clarence
Olson, Norman Fredrick
Peterson, David Maurice
Priest, Matthew A
Sanderson, Gary Warner
Sebree, Bruce Randall
Springer, Edward L(ester)
Wall, Joseph Sennen
Woldegiorgis, Gebretateos

WYOMING
Miller, Glenn Joseph
Rohwer, Robert G

ALBERTA
Campbell, Stewart John
Hawrysh, Zenia Jean
Hill, Bernard Dale

BRITISH COLUMBIA
Anderson, John Ansel
Li-Chan, Eunice
Majak, Walter
Nakai, Shuryo

MANITOBA
Bendelow, Victor Martin
Bushuk, Walter
Eskin, Neason Akiva Michael
Kovacs, Miklos I P
Kruger, James Edward
Lukow, Odean Michelin
MacGregor, Alexander William

Noll, John Stephen
Singh, Harwant
Tipples, Keith H
Tkachuk, Russell

NOVA SCOTIA
Ackman, Robert George
Gilgan, Michael Wilson
Robinson, Arthur Robin

ONTARIO
Brewer, Arthur David
Campbell, James Alexander
Campbell, Kenneth Wilford
Collins, Frank William, Jr
Court, William Arthur
Cunningham, Hugh Meredith
DeMan, John Maria
Ferrier, Leslie Kenneth
Fortin, Andre Francis
Greenhalgh, Roy
Harris, Charles Ronald
Hay, George William
Khan, Shahamat Ullah
Lougheed, Thomas Crossley
Pare, J R Jocelyn
Ratnayake, Walisundera Mudiyanselage Nimal
Rubin, Leon Julius
Sarwar, Ghulam
Scott, Fraser Wallace
Scott, Peter Michael
Spencer, Elvins Yuill
Stavric, Bozidar
Thompson, Lilian Umale
Trenholm, Harold Locksley
Yaguchi, Makoto
Young, James Christopher F

QUEBEC
Bélanger, Jacqueline M R
Britten, Michel
Common, Robert Haddon
Panalaks, Thavil
Picard, Gaston Arthur
Sood, Vijay Kumar

SASKATCHEWAN
Grant, Donald R
Grover, Rajbans
Julien, Jean-Paul
Smith, Allan Edward
Sosulski, Frank Walter

OTHER COUNTRIES
Brown, Barker Hastings
Buckmire, Reginald Eugene
Buckwold, Sidney Joshua
Chan, Hak-Foon
Gregory, Peter
Kurobane, Itsuo
Lee, Men Hui
Malherbe, Roger F
Perry, Albert Solomon
Turner, Fred, Jr

Analytical Chemistry

ALABAMA
Arendale, William Frank
Askew, William Crews
Aull, John Louis
Baker, June Marshall
Barrett, William Jordan
Beeman, Curt Pletcher
Bertsch, Wolfgang
Campbell, Robert Terry
Daniel, Robert Eugene
Emerson, Merle T
Garber, Robert William
Garmon, Ronald Gene
Geppert, Gerard Allen
Hamer, Justin Charles
Harris, Albert Zeke
Ho, Mat H
Hung, George Wen-Chi
Johnson, Frank Junior
Lambert, James LeBeau
Liu, Ray Ho
Lloyd, Nelson Albert
McNutt, Ronald Clay
Makhija, Suraj Parkash
Miller, Herbert Crawford
Mountcastle, William R, Jr
Retief, Daniel Hugo
Rymal, Kenneth Stuart
Sides, Gary Donald
Small, Robert James
Smith, Frederick Paul
Strohl, John Henry
Tan, Boen Hie
Timkovich, Russell
Toren, Eric Clifford, Jr
Veazey, Thomas Mabry
Ward, Edward Hilson
Weinberg, David Samuel
Woodham, Donald W

ALASKA
Kennish, John M
LaPerriere, Jacqueline Doyle
Stolzberg, Richard Jay

ARIZONA
Andreen, Brian H
Blouin, Leonard Thomas
Boynton, William Vandegrift
Burke, Michael Francis
Casto, Clyde Christy
Denton, M Bonner
Fedder, Steven Lee
Fernando, Quintus
Freiser, Henry
Fuchs, Jacob
George, Thomas D
Gilbert, Don Dale
Hildebrandt, Wayne Arthur
Juvet, Richard Spalding, Jr
Kirby, Robert Emmet
Knowlton, Gregory Dean
Liu, Chui Hsun
Mosher, Richard Arthur
Moss, Rodney Dale
Muggli, Robert Zeno
Swinehart, Bruce Arden
Towle, Louis Wallace
Valenty, Steven Jeffrey
Whitcomb, Donald Leroy
Wright, Daniel Craig
Yates, Ann Marie

ARKANSAS
Bobbitt, Donald Robert
Bowman, Leo Henry
Fortner, Wendell Lee
Fu, Peter Pi-Cheng
Gosnell, Aubrey Brewer
Hood, Robert L
Howick, Lester Carl
Jimerson, George David
Komoroski, Richard Andrew
Korfmacher, Walter Averill
Ku, Victoria Feng
Mitchell, Richard Sibley
Nisbet, Alex Richard
Nix, Joe Franklin
Parks, Albert Fielding

CALIFORNIA
Abbott, Seth R
Abdel-Baset, Mahmoud B
Aberth, William H
Adinoff, Bernard
Amoore, John Ernest
Anson, Fred (Colvig)
Bailey, David Nelson
Baker, Frederick Charles
Barrall, Edward Martin, II
Beilby, Alvin Lester
Belsky, Theodore
Bentley, Kenton Earl
Bernhard, Richard Allan
Bilow, Norman
Boyle, Walter Gordon, Jr
Braun, Donald E
Bremner, Raymond Wilson
Brookman, David Joseph
Burlingame, Alma L
Burnham, Alan Kent
Burtner, Dale Charles
Bushey, Albert Henry
Buttrill, Sidney Eugene, Jr
Bystroff, Roman Ivan
Calkins, Russel Crosby
Camenzind, Mark J
Cape, John Anthony
Caputi, Roger William
Carle, Glenn Clifford
Carr, Edward Mark
Carr, Robert J(oseph)
Chaney, Charles Lester
Chang, Ming-Houng (Albert)
Chatfield, Dale Alton
Chodos, Arthur A
Christoffersen, Donald John
Ciaramitaro, David A
Clark, Dennis Richard
Clarkson, Jack E
Cleary, Michael E
Cocks, George Gosson
Colovos, George
Cotts, Patricia Metzger
Crawford, Richard Whittier
Creek, Jefferson Louis
Crichton, David
Crutchfield, Charlie
Current, Jerry Hall
Curry, Bo(stick) U
Czuha, Michael, Jr
Davison, John Blake
Day, Robert James
DeJongh, Don C
Delker, Gerald Lee
Deuble, John Lewis, Jr
Donnelly, Timothy C
Doyle, Walter M
Duffield, Jack Jay
Epstein, Barry D
Espoy, Henry Marti
Estill, Wesley Boyd
Evans, Charles Andrew, Jr
Farrington, Paul Stephen
Fassel, Velmer Arthur
Feldman, Fredric J
Fenimore, David Clarke
Fernandez, Alberto Antonio
Fetzer, John Charles

Fine, Dwight Albert
Fischer, Robert Blanchard
Fisher, Richard Paul
Fraser, James Mattison
Frye, Herschel Gordon
Fujikawa, Norma Sutton
Gallegos, Emilio Juan
Gantzel, Peter Kellogg
Gardiner, Kenneth William
Gaston, Lyle Kenneth
Gillen, Keith Thomas
Goldberg, Ira Barry
Gordon, Benjamin Edward
Gordon, Joseph Grover, II
Gordon, Robert Julian
Goss, James Arthur
Gouw, T(an) H(ok)
Grant, Patrick Michael
Green, Robert Bennett
Greig, Douglas Richard
Gump, Barry Hemphill
Haddon, William F, (Jr)
Hanneman, Walter W
Harrar, Jackson Elwood
Harris, Daniel Charles
Hartigan, Martin Joseph
Haugen, Gilbert R
Hausmann, Werner Karl
Hayes, Janan Mary
Helmer, John
Hem, John David
Hessel, Donald Wesley
Hirschfeld, Tomas Beno
Hoffmann, Ronald Lee
Holmes, Ivan Gregory
Hovanec, B(ernard) Michael
Hrubesh, Lawrence Wayne
Hubert, Jay Marvin
Hurd, Ralph Eugene
Ja, William Yin
Jacob, Peyton, III
Jacob, Robert Allen
Jain, Naresh C
Jakob, Fredi
Jankowski, Stanley John
Janota, Harvey Franklin
Jardine, Ian
Jayne, Jerrold Clarence
Jenden, Donald James
Jennings, Walter Goodrich
Jensen, James Leslie
Jentoft, Ralph Eugene, Jr
Johanson, Robert Gail
Johnson, LeRoy Franklin
Judd, Stanley H
Kabra, Pokar Mal
Kang, Tae Wha
Katekaru, James
Kehoe, Thomas J
Kennedy, John Harvey
Kim, Ki Hong
Kirkpatrick, James W
Kissel, Charles Louis
Kland, Mathilde June
Klein, David Henry
Klemm, Waldemar Arthur, Jr
Kong, Eric Siu-Wai
Kothny, Evaldo Luis
Koziol, Brian Joseph
Kuhlmann, Karl Frederick
Kuwahara, Steven Sadao
Landgraf, William Charles
Laub, Richard J
Lawless, James George
Lee, Alfred Tze-Hau
Liang, Yola Yueh-o
Liberman, Martin Henry
Liggett, Lawrence Melvin
Lin, Jiann-Tsyh
Lincoln, Kenneth Arnold
Lindner, Elek
Lord, Harry Chester, III
Lundin, Robert Enor
Lysyj, Ihor
McAuliffe, Clayton Doyle
McCauley, Gerald Brady
McGown, Evelyn L
Mach, Martin Henry
McIver, Robert Thomas, Jr
McKaveney, James P
McKinney, Ted Meredith
MacWilliams, Dalton Carson
Mallett, William Robert
Marantz, Laurence Boyd
Markowitz, Samuel Solomon
Marshall, Donald D
Mason, James Willard
Mason, William Burkett
Matsuyama, George
Medrud, Ronald Curtis
Meinhard, James Edgar
Miller, George E
Moody, Kenton J
Mooney, John Bernard
Moore, Gerald L
Morris, Jerry Lee, Jr
Mosen, Arthur Walter
Neptune, John Addison
Neufeld, Jerry Don
Newton, John Chester
Nilsson, William A
Nowak, Anthony Victor
Oh, Chan Soo
Okamura, Judy Paulette

Ozari, Yehuda
Palmer, Thomas Adolph
Pappatheodorou, Sofia
Parrish, William
Parsons, Michael L
Paselk, Richard Alan
Passchier, Arie Anton
Passell, Thomas Oliver
Pearson, Robert Melvin
Pease, Burton Frank
Penton, Zelda Eve
Perkins, Willis Drummond
Perone, Samuel Patrick
Perry, Dale Lynn
Pesek, Joseph Joel
Pierce, Matthew Lee
Pollack, Louis Rubin
Potter, John Clarkson
Preus, Martin William
Putz, Gerard Joseph
Pyper, James William
Que Hee, Shane Stephen
Rabenstein, Dallas Leroy
Raby, Bruce Alan
Rainis, Andrew
Ratcliff, Milton, Jr
Rhodes, David R
Rice, Elmer Harold
Rider, Benjamin Franklin
Rivier, Jean E F
Roberts, Julian Lee, Jr
Robinson, Merton Arnold
Rocklin, Roy David
Russ, Guston Price, III
Rynd, James Arthur
Sahbari, Javad Jabbari
Sanborn, Russell Hobart
Saperstein, David Dorn
Scattergood, Thomas W
Schaleger, Larry L
Schelar, Virginia Mae
Schwall, Richard Joseph
Seim, Henry Jerome
Selig, Walter S
Shields, Loran Donald
Short, Michael Arthur
Skoog, Douglas Arvid
Sloane, Howard J
Smith, Charles Aloysius
Smith, James Hart
Smith, Leverett Ralph
Solomon, Malcolm David
Stevens, William George
Stevenson, Robert Lovell
Strahl, Erwin Otto
Suffet, I H (Mel)
Sunshine, Irving
Suryaraman, Maruthuvakudi Gopalasastri
Sussman, Howard H
Teeter, Richard Malcolm
Thomas, Richard Sanborn
Thompson, John Michael
Thornton, John Irvin
Throop, Lewis John
Ting, Chih-Yuan Charles
Tsao, Constance S
Tseng, Chien Kuei
Tuffly, Bartholomew Louis
Ullman, Edwin Fisher
Ulrich, William Frederick
VanAntwerp, Craig Lewis
Vedvick, Thomas Scott
Vomhof, Daniel William
Walker, Joseph
Wallcave, Lawrence
Walz, Alvin Eugene
Warren, Paul Horton
Weiss, Fred Toby
Weiss, Harold Gilbert
Weiss, Roger Harvey
Wensley, Charles Gelen
West, Charles David
West, Donald Markham
Westover, Lemoyne Byron
Whatley, Thomas Alvah
Whiteker, Roy Archie
Whittemore, Irville Merrill
Wiesendanger, Hans Ulrich David
Wiley, Michael David
Wilkins, Charles Lee
Willis, William Van
Winterlin, Wray Laverne
Worden, Earl Freemont, Jr
Wright, John Marlin
Wu, Jiann-Long
Young, Donald Charles
Young, Ho Lee
Yuen, Wing
Zabin, Burton Allen
Zander, Andrew Thomas
Zellmer, David Louis
Ziegler, Carole L

COLORADO
Albert, Harrison Bernard
Barisas, Bernard George, Jr
Birks, John William
Bower, Nathan Wayne
Chao, Tsun Tien
Chisholm, James Joseph
Christie, Joseph Herman
Edwards, Kenneth Ward
Elliott, Cecil Michael
Erdmann, David E

Fennessey, Paul V
Fishman, Marvin Joseph
Fitzgerald, Jerry Mack
Fitzpatrick, Francis Anthony
Gillespie, John Paul
Herrmann, Scott Joseph
Hill, Walter Edward, Jr
Hitchcock, Eldon Titus
Hutto, Francis Baird, Jr
Janes, Donald Wallace
Jones, Berwyn E
Kinsinger, James A
Kniebes, Duane Van
Koval, Carl Anthony
MacCarthy, Patrick
Maciel, Gary Emmet
Mahan, Kent Ira
Mehs, Doreen Margaret
Micheli, Roger Paul
Morrison, Charles Freeman, Jr
Murphy, Robert Carl
Norton, Daniel Remsen
Rhoads, William Denham
Ritchey, John Michael
Rogerson, Peter Freeman
Schalge, Alvin Laverne
Sievers, Robert Eugene
Skogen, Haven Sherman
Skogerboe, Rodney K
Skougstad, Marvin Wilmer
Smith, Dwight Morrell
Smith, John Elvans
Stull, Dean P
Tackett, James Edwin, Jr
Thompson, Ronald G
Trusell, Fred Charles
Turner, James Howard
Van Vorous, Ted
Vejvoda, Edward
Voorhees, Kent Jay
Walton, Harold Frederic
Webb, John Day
Wildeman, Thomas Raymond
Williams, Jean Paul
Wingeleth, Dale Clifford
Woerner, Dale Earl

CONNECTICUT
Agarwal, Vipin K
Alpert, Nelson Leigh
Anderson, Carol Patricia
Andrade, Manuel
Auerbach, Michael Howard
Baldwin, Ronald Martin
Barney, James Earl, II
Buck, Marion Gilmour
Burnett, Robert Walter
Campbell, Bruce Henry
Chambers, William Edward
Chang, Ted T
Clarke, George
Cohen, Edward Morton
Cooper, James William
Csejka, David Andrew
Curley, James Edward
Edman, Walter W
Ettre, Leslie Stephen
Forcier, George Arthur
Goedert, Michel G
Golden, Gerald Seymour
Gordon, Philip N
Gray, Edward Theodore, Jr
Greenhouse, Steven Howard
Greenough, Ralph Clive
Groth, Joyce Lorraine
Henderson, David Edward
Hiskey, Clarence Francis
Humphreys, Robert William Riley
I, Ting-Po
Johnson, Bruce McDougall
Kavarnos, George James
Kluender, Harold Clinton
Krol, George J
Kuck, Julius Anson
Lamm, Foster Philip
Levy, Gabor Bela
MacDonald, John Chisholm
McMurray, Walter Joseph
McNally, John G
Mattina, Mary Jane Incorvia
Mayell, Jaspal Singh
Michel, Robert George
Miller, Jerry K
Murai, Kotaro
Page, Edgar J(oseph)
Park, George Bennet
Paul, Jeddeo
Reffner, John A
Rennhard, Hans Heinrich
Richter, G Paul
Rook, Harry Lorenz
Rosenberg, Ira Edward
Rusling, James Francis
Santacana-Nuet, Francisco
Sarneski, Joseph Edward
Savitzky, Abraham
Sease, John William
Shah, Shirish A
Shain, Irving
Shalvoy, Richard Barry
Shaw, Brenda Roberts
Stock, John Thomas
Stuart, James Davies
Suib, Steven L

Swanson, Donald Leroy
Tom, Glenn McPherson
Voorhies, John Davidson
Ziegler, William Arthur

DELAWARE
Abrahamson, Earl Arthur
Baaske, David Michael
Barth, Howard Gordon
Bauchwitz, Peter S
Benson, Richard Edward
Bly, Donald David
Brame, Edward Grant, Jr
Buchta, Raymond Charles
Carberry, Judith B
Chase, David Bruce
Chiu, Jen
Clarke, John Frederick Gates, Jr
Conner, Albert Z
Crecely, Roger William
Crippen, Raymond Charles
Dippel, William Alan
Durbin, Ronald Priestley
Duswalt, Allen Ainsworth, Jr
Dwivedi, Anil Mohan
Evans, Dennis Hyde
Finn, James Walter
Firment, Lawrence Edward
Fleming, Sydney Winn
Gardiner, John Alden
Gillow, Edward William
Gold, Harvey Saul
Haq, Mohammad Zamir-ul
Harlow, Richard Leslie
Hartzell, Charles Ross, III
Harvey, John, Jr
Hayman, Alan Conrad
Hobgood, Richard Troy, Jr
Hoegger, Erhard Fritz
Jack, John James
Johnson, David Russell
Johnson, Donald Richard
Kaiser, Mary Agnes
Ketterer, Paul Anthony
Kirkland, Joseph Jack
Kissa, Erik
Koeppe, Mary Kolean
Lam, Gilbert Nim-Car
Larsen, Barbara Seliger
Lee, Shung-Yan Luke
Levy, Paul F
McCurdy, Wallace Hutchinson, Jr
McEwen, Charles Nehemiah
Mathre, Owen Bertwell
Memeger, Wesley, Jr
Milian, Alwin S, Jr
Mitchell, John, Jr
Mollica, Joseph Anthony
Monagle, Daniel J
Moore, Charles B(ernard)
Moore, Ralph Bishop
Moser, Glenn Allen
Mowery, Richard Allen, Jr
Munson, Burnaby
Narvaez, Richard
Neal, Thomas Edward
Neff, Bruce Lyle
Nichols, James Randall
Paris, Jean Philip
Patterson, Gordon Derby, Jr
Pierson, Keith Bernard
Quarry, Mary Ann
Quon, Check Yuen
Reuben, Jacques
Rudat, Martin August
Salzman, Steven Kerry
Sammak, Emil George
Sauerbrunn, Robert Dewey
Scott, James Alan
Slade, Arthur Laird
Stockburger, George Joseph
Swann, Charles Paul
Sweetser, Philip Bliss
Tise, Frank P
Twelves, Robert Ralph
Ward, George A
Wermus, Gerald R
Wetlaufer, Donald Burton
Whitney, Charles Candee, Jr
Williams, Reed Chester
Yau, Wallace Wen-Chuan
Zinser, Edward John

DISTRICT OF COLUMBIA
Boots, Sharon G
Breen, Joseph John
Collat, Justin White
DeLevie, Robert
Frank, Richard Stephen
Frodyma, Michael Mitchell
Gunn, John William, Jr
Hammer, Charles F
Hoffman, Michael K
Jett-Tilton, Marti
Karle, Jean Marianne
Kruegel, Alice Virginia
Lerner, Pauline
McCown, John Joseph
Miller, David Jacob
Modderman, John Philip
Morris, Joseph Burton
Nesheim, Stanley
Nordquist, Paul Edgard Rudolph, Jr
O'Grady, William Edward

Analytical Chemistry (cont)

Paabo, Maya
Page, Samuel William
Pannu, Sardul S
Poon, Bing Toy
Preer, James Randolph
Rehwoldt, Robert E
Sarmiento, Rafael Apolinar
Schmidt, William Edward
Scott, Kenneth Richard
Shirk, James Siler
Staruszkiewicz, Walter Frank, Jr
Tanner, James Thomas
Turner, Anne Halligan
Wade, Clarence W R
Wainer, Irving William
Wang, Jin Tsai
Wilkinson, John Edwin
Wyatt, Jeffrey Renner

FLORIDA
Attaway, John Allen
Bates, Roger Gordon
Baumann, Arthur Nicholas
Bledsoe, James O, Jr
Bowman, Ray Douglas
Braman, Robert Steven
Brezonik, Patrick Lee
Carlson, David Arthur
Cohen, Martin Joseph
Coolidge, Edwin Channing
Davidson, Mark Rogers
Debnath, Sadhana
Diamond, Steven Elliot
Dinsmore, Howard Livingstone
Drake, Robert Firth
Dubravcic, Milan Frane
Foner, Samuel Newton
Francis, Stanley Arthur
Frediani, Harold Arthur
Fuller, Harold Wayne
Gay, Don Douglas
Gordon, Saul
Green, Floyd J
Gropp, Armin Henry
Hamill, James Junior
Hannan, Roy Barton, Jr
Hardman, Bruce Bertolette
Harrison, Willard Wayne
Hartley, Arnold Manchester
Heidner, Robert Hubbard
Hormats, Ellis Irving
Huber, Joseph William, III
Hunter, George L K
Jackson, George Frederick, III
Jackson, Harold Woodworth
Keirs, Russell John
Kelley, Myron Truman
King, Roy Warbrick
Krc, John, Jr
Kroger, Hanns H
Kurtz, George Wilbur
Laitinen, Herbert August
Loder, Edwin Robert
Long, Calvin H
McGee, William Walter
Mann, Charles Kenneth
Martin, Barbara Bursa
Meisels, Gerhard George
Miller, William Knight
Mounts, Richard Duane
Moye, Hugh Anson
Munk, Miner Nelson
Myers, Richard Lee
Normile, Hubert Clarence
Olsen, Eugene Donald
Paterson, Arthur Renwick
Piotrowicz, Stephen R
Poet, Raymond B
Roach, Don
Rogers, Lewis Henry
Ross, Lynne Fischer
Schilt, Alfred Ayars
Schmid, Gerhard Martin
Schulman, Stephen Gregory
Schultz, Franklin Alfred
Senftleber, Fred Carl
Shaw, Philip Eugene
Sneade, Barbara Herbert
Spiegelhalter, Roland Robert
Strunk, Duane H
Swartz, William Edward, Jr
Szonntagh, Eugene L(eslie)
Urone, Paul
Vickers, Thomas J
Wiebush, Joseph Roy
Willard, Thomas Maxwell
Winefordner, James D
Yost, Richard A

GEORGIA
Anderson, James Leroy
Baughman, George Larkins
Bayer, Charlene Warres
Black, Billy C, II
Blum, Murray Sheldon
Bottomley, Lawrence Andrew
Boudinot, Frank Douglas
Brewer, John Gilbert
Brooks, John Bill
Busch, Kenneth Louis
Cadwallader, Donald Elton
Chandler, James Harry, III

Chortyk, Orestes Timothy
Churchill, Frederick Charles
Clark, Benjamin Cates, Jr
Cunningham, Alice Jeanne
Day, Reuben Alexander, Jr
Doetsch, Paul William
Donaldson, William Twitty
Ellington, James Jackson
Flaschka, Hermenegild Arved
Freese, William P, II
Frierson, William Joe
Furse, Clare Taylor
Gaines, Tinsley Powell
Garrison, Arthur Wayne
Gary, Julia Thomas
Gollob, Lawrence
Guerrant, Gordon Owen
Gunter, Bobby J
Hicks, Donald Gail
Hodges, Linda Carol
Holzer, Gunther Ulrich
Hoover, Thomas Burdett
Husa, William John, Jr
Isaac, Robert A
Jackson, William Morrison
Jain, Anant Vir
James, Franklin Ward
James, Jeffrey
Koelsche, Charles L
Kohl, Paul Albert
Krajca, Kenneth Edward
Lehner, Andreas Friedrich
Lindauer, Maurice William
Love, Jimmy Dwane
McBride, Clifford Hoyt
McGuire, John Murray
Matthews, Edward Whitehouse
Nicolson, Paul Clement
O'Neal, Floyd Breland
Parker, Lloyd Robinson, Jr
Pyle, John Tillman
Rhoades, James Lawrence
Robinson, George Waller
Rogers, Lockhart Burgess
Schweri, Margaret Mary
Simmons, George Allen
Sommer, Harry Edward
Starr, Thomas Louis
Steele, Jack
Stratton, Cedric
Sturrock, Peter Earle
Thruston, Alfred Dorrah, Jr
Trawick, William George
Warner, Isiah Manuel
Wilson, David Merl
Woodford, James

HAWAII
Crane, Sheldon Cyr
Forster, William Owen
Hertlein, Fred, III
Kemp, Paul James
Malmstadt, Howard Vincent
Moritsugu, Toshio
Pecsok, Robert Louis
Roberts, Robert Russell
Sansone, Francis Joseph
Seifert, Josef

IDAHO
Arcand, George Myron
Delmastro, Ann Mary
Delmastro, Joseph Raymond
Farwell, Sherry Owen
Hibbs, Robert A
Lewis, Leroy Crawford
Mincher, Bruce J
Sill, Claude Woodrow
Sutton, John Curtis
Wright, Kenneth James

ILLINOIS
Adlof, Richard Otto
Afremow, Leonard Calvin
Altpeter, Lawrence L, Jr
Anderson, Arnold Lynn
Atoji, Masao
Babcock, Robert Frederick
Bailey, David Newton
Bath, Donald Alan
Bernetti, Raffaele
Bingham, Carleton Dille
Bloemer, William Louis
Brezinski, Darlene Rita
Brobst, Kenneth Martin
Brooks, Kenneth Conrad
Brubaker, Inara Mencis
Burns, Richard Price
Carnahan, Jon Winston
Caskey, Albert Leroy
Chadde, Frank Ernest
Chandler, Dean Wesley
Chao, Sherman S
Chipman, Gary Russell
Chou, Mei-In Melissa Liu
Connolly, James D
Cummings, Thomas Fulton
Davis, Andrew Morgan
DeFord, Donald Dale
Dickerson, Donald Robert
Dietz, Mark Louis
Domsky, Irving Isaac
Doody, Marijo
Erickson, Mitchell Drake

Eskins, Kenneth
Fairman, William Duane
Faulkner, Larry Ray
Ferrara, Louis W
Ferren, Larry Gene
Filson, Don P
Firsching, Ferdinand Henry
Fitch, Alanah
Ganchoff, John Christopher
Gibbons, Larry V
Goss, George Robert
Gowda, Netkal M Made
Gray, Linsley Shepard, Jr
Greene, John Philip
Grove, Ewart Lester
Guyon, John Carl
Haberfeld, Joseph Lennard
Halfman, Clarke Joseph
Haugen, David Allen
Hazdra, James Joseph
Hefley, Alta Jean
Hill, Kenneth Richard
Hillis, Mary Olive
Hime, William Gene
Hockman, Deborah C
Hoepfinger, Lynn Morris
Hollins, Robert Edward
Holtzman, Richard Beves
Homeier, Edwin H, Jr
Hughes, Benjamin G
Inskip, Ervin Basil
Jakubiec, Robert Joseph
Janghorbani, Morteza
Jarke, Frank Henry
Jaselskis, Bruno
Jiu, James
Johari, Om
Johnson, John Harold
Justen, Lewis Leo
Kaplan, Ephraim Henry
Kennedy, Albert Joseph
Khattab, Ghazi M A
Kieft, Richard Leonard
Kokkinakis, Demetrius Michael
Kornel, Ludwig
Krawetz, Arthur Altshuler
Lambert, Glenn Frederick
Lambert, Joseph B
Lane, William James
Lasley, Stephen Michael
Layloff, Thomas
Levenberg, Milton Irwin
Levi-Setti, Riccardo
Linde, Harry Wight
Linder, Louis Jacob
Lucchesi, Claude A
Lucci, Robert Dominick
McCrone, Walter C
Maclay, G Jordan
Marcus, Mark
Markunas, Peter Charles
Marquart, John R
Martin, Ronald LeRoy
Matulis, Raymond M
Melendres, Carlos Arciaga
Melford, Sara Steck
Metcalfe, Lincoln Douglas
Minear, Roger Allan
Moore, Carl Edward
Mounts, Timothy Lee
Nagy, Zoltan
Nealy, Carson Louis
Neas, Robert Edwin
Newcome, Marshall Millar
Newhart, M(ary) Joan
Nieman, Timothy Alan
Norman, Richard Daviess
Palmer, Curtis Allyn
Pankratz, Ronald Ernest
Pasterczyk, William Robert
Peck, Theodore Richard
Pesyna, Gail Marlane
Petro, Peter Paul, Jr
Pietri, Charles Edward
Piwoni, Marvin Dennis
Rein, James Earl
Rinehart, Kenneth Lloyd
Rohwedder, William Kenneth
Ruch, Rodney R
Scheeline, Alexander
Scholfield, Charles Rexel
Scholz, Robert George
Schroeer, Juergen Max
Sellers, Donald Roscoe
Sennello, Lawrence Thomas
Sessa, David Joseph
Shapiro, Rubin
Sheft, Irving
Sherren, Anne Terry
Sibbach, William Robert
Singh, Laxman
Steele, Ian McKay
Stetter, Joseph Robert
Stoffer, Robert Llewellyn
Stubblefield, Robert Douglas
Sutton, Lewis McMechan
Sytsma, Louis Frederick
Szpunar, Carole Bryda
Thomas, Elizabeth Wadsworth
Tomkins, Marion Louise
Trent, John Ellsworth
Vandeberg, John Thomas
Vander Velde, George
Van Duyne, Richard Palmer

Vanýsek, Petr
Washburn, William H
Wei, Lester Yeehow
Wenzel, Bruce Erickson
Whaley, Thomas Patrick
Wimer, David Carlisle
Winans, Randall Edward
Wingender, Ronald John
Woolson, Edwin Albert
Young, Austin Harry
Young, David W

INDIANA
Alter, John Emanuel
Amundson, Merle E
Amy, Jonathan Weekes
Bair, Edward Jay
Baird, William McKenzie
Bambenek, Mark A
Becker, Elizabeth Ann (White)
Bishara, Rafik Hanna
Boaz, Patricia Anne
Boguslaski, Robert Charles
Bottei, Rudolph Santo
Bowser, James Ralph
Burden, Stanley Lee, Jr
Chernoff, Donald Alan
Clark, Larry P
Contario, John Joseph
Day, Edgar William, Jr
DeLong, Allyn F
Demkovich, Paul Andrew
Doeden, Gerald Ennen
Dube, David Gregory
Dubin, Paul Lee
Dyer, Rolla McIntyre, Jr
Elkin, Robert Glenn
Ellefsen, Paul
Fox, Owen Forrest
Free, Helen M
Fricke, Gordon Hugh
Gainer, Frank Edward
Gallo, Duane Gordon
Gunter, Claude Ray
Guthrie, Frank Albert
Hayes, John Michael
Hemmes, Paul Richard
Heydegger, Helmut Roland
Hieftje, Gary Martin
Hites, Ronald Atlee
Hoefer, Raymond H
Hoff, Johan Eduard
Housmyer, Carl Leonidas
Huitink, Geraldine M
Hulbert, Matthew H
Jansing, Jo Ann
Kamat, Prashant V
Kennedy, Edward Earl
Kissinger, Peter Thomas
Kramer, William J
Kuzel, Norbert R
Lytle, Fred Edward
McLafferty, John J, Jr
McMasters, Donald L
Margerum, Dale William
Marsh, Max Martin
Mays, David Lee
Mazac, Charles James
Merritt, Lynne Lionel, Jr
Meyerson, Seymour
Michel, Karl Heinz
Miller, James Franklin
Morris, David Alexander Nathaniel
Nagel, Edgar Herbert
Niss, Hamilton Frederick
Occolowitz, John Lewis
Pacer, Richard A
Pardue, Harry L
Peters, Dennis Gail
Pribush, Robert A
Price, Howard Charles
Quinney, Paul Reed
Reilly, James Patrick
Reuland, Donald John
Rickard, Eugene Clark
Rivers, Paul Michael
Schaap, Ward Beecher
Schall, Elwyn DeLaurel
Shaw, Vernon Reed
Shockey, William Lee
Shultz, Clifford Glen
Siefker, Joseph Roy
Smith, David Lee
Smith, Jean Blair
Smucker, Arthur Allan
Stevenson, William Campbell
Stroube, William Bryan, Jr
Sutula, Chester Louis
Swanson, Lynn Allen
Tennyson, Richard Harvey
Thompson, Richard Michael
Timma, Donald Lee
Waldman, Barbara Criscuolo
Wallace, Gerald Wayne
Weaver, Michael John
Wernimont, Grant (Theodore)
Wheeler, Ralph John
Wolszon, John Donald
Yan, Sau-Chi Betty
Zimmerman, John F

IOWA
Baetz, Albert L
Binz, Carl Michael

Buchanan, Edward Bracy, Jr
Cotton, Therese Marie
Diehl, Harvey
Edelson, Martin Charles
Fritz, James Sherwood
Goetz, Charles Albert
Hammerstrom, Harold Elmore
Houk, Robert Samuel
Jacob, Fielden Emmitt
Keiser, Jeffrey E
Kniseley, Richard Newman
Koerner, Theodore Alfred William, Jr
Kracher, Alfred
McClelland, John Frederick
Malone, Diana
Markuszewski, Richard
Pflaum, Ronald Trenda
Schmidt, Reese Boise
Stahr, Henry Michael
Svec, Harry John
Swartz, James E
Watkins, Stanley Read
Weeks, Stephan John
Yeung, Edward Szeshing

KANSAS
Adams, Ralph Norman
Cole, Jerry Joe
Davis, Lawrence Clark
Decedue, Charles Joseph
Hammaker, Robert Michael
Hawley, Merle Dale
Hiebert, Allen G
Isenhour, Thomas Lee
Iwamoto, Reynold Toshiaki
Johnson, David Barton
Judson, Charles Morrill
Kuwana, Theodore
Lambert, Jack Leeper
Landis, Arthur Melvin
Lehman, Thomas Alan
Lindenbaum, Siegfried
Lookhart, George LeRoy
Macke, Gerald Fred
Marshall, Delbert Allan
Meloan, Clifton E
Pfluger, Clarence Eugene
Schorno, Karl Stanley
Seitz, Larry Max
Sellers, Douglas Edwin
Singhal, Ram P
Sunderman, Herbert D
Wilson, George Spencer
Wright, Charles Hubert
Youngstrom, Richard Earl

KENTUCKY
Byrn, Ernest Edward
Byrne, Francis Patrick
Clark, Howell R
Colby, David Anthony
Cooke, Samuel Leonard
Davidson, John Edwin
Hamilton-Kemp, Thomas Rogers
Hartman, David Robert
Helms, Boyce Dewayne
Hessley, Rita Kathleen
Holler, Floyd James
Hurst, Harrell Emerson
Johnson, Lawrence Robert
Keely, William Martin
Klingenberg, Joseph John
Lauterbach, John Harvey
McClellan, Bobby Ewing
McGinness, James Donald
Moorhead, Edward Darrell
O'Reilly, James Emil
Phillips, John Perrow
Ray, Jesse Paul
Reed, Kenneth Paul
Riley, John Thomas
Robertson, John David
Rosenberg, Alexander F
Salyer, Darnell
Schulz, William
Shank, Lowell William
Smith, Walter Thomas, Jr
Thio, Alan Poo-An
Wong, John Lui

LOUISIANA
Bailey, George William
Berg, Eugene Walter
Berni, Ralph John
Braun, Robert Denton
Brown, Lawrence E(ldon)
Carver, James Clark
Colgrove, Steven Gray
Connick, William Joseph, Jr
DeLeon, Ildefonso R
Drushel, Harry (Vernon)
Evilia, Ronald Frank
Fischer, Nikolaus Hartmut
Fitzpatrick, Jimmie Doile
Fontenot, Martin Mayance, Jr
Geissler, Paul Robert
Gibson, David Michael
Graves, Robert Joseph
Greco, Edward Carl
Ham, Russell Allen
Hankins, B(obby) E(ugene)
Hargis, Larry G
Imhoff, Donald Wilbur
Kuan, Shia Shiong

Lee, John Yuchu
Lin, Denis Chung Kam
Mark, Harold Wayne
Merrill, Howard Emerson
Montalvo, Joseph G, Jr
Morris, Cletus Eugene
Morris, Nancy Mitchell
Moseley, Patterson B
Mustafa, Shams
Oberding, Dennis George
Overton, Edward Beardslee
Peard, William John
Riddick, John Allen
Rimes, William John
Robinson, James William
Rommel, Marjorie Ann
Seidler, Rosemary Joan
Shih, Frederick F
Slaven, Robert Walter
Smith, Isaac Litton
Smith, Robert Leonard
Thomas, Samuel Gabriel
Vercellotti, John R
Vidaurreta, Luis E
Walters, Fred Henry
West, Philip William

MAINE
Cronn, Dagmar Rais
Machemer, Paul Ewers
Robbins, Wayne Brian
Whitten, Maurice Mason

MARYLAND
Adams, James Miller
Alexander, Thomas Goodwin
Argauer, Robert John
Armstrong, Daniel Wayne
Auel, RaeAnn Marie
Babrauskas, Vytenis
Baker, Doris
Barnes, Ira Lynus
Becker, Donald Arthur
Beecher, Gary Richard
Behar, Marjam Gojchlerner
Benson, Walter Roderick
Beroza, Morton
Block, Jacob
Brancato, David Joseph
Bryden, Wayne A
Buchanan, John Donald
Burrows, Elizabeth Parker
Callahan, John Joseph
Callahan, Mary Vincent
Caplan, Yale Howard
Cassidy, James Edward
Chakrabarti, Siba Gopal
Christensen, Richard G
Cincotta, Joseph John
Cone, Edward Jackson
Cotter, Robert James
Darneal, Robert Lee
Davis, Henry McRay
DeToma, Robert Paul
Devoe, James Rollo
Downing, Robert Gregory
Duignan, Michael Thomas
Eng, Leslie
Fassett, John David
Fauth, Mae Irene
Fenselau, Catherine Clarke
Ferretti, James Alfred
Feyns, Liviu Valentin
Fisher, Dale John
Freeman, David Haines
Gajan, Raymond Joseph
Garrett, Benjamin Caywood
Gilfrich, John Valentine
Goldin, Abraham Samuel
Gordon, Glen Everett
Grady, Lee Timothy
Graham, Joseph H
Groopman, John Davis
Hackert, Raymond L
Haenni, Edward Otto
Hartstein, Arthur M
Heinrich, Kurt Francis Joseph
Herner, Albert Erwin
Hertz, Harry Steven
Hillery, Paul Stuart
Hirsch, Roland Felix
Horwitz, William
Hsu, Chen C
Hsu, Stephen M
Issaq, Haleem Jeries
Jones, Donald Eugene
Jovancicevic, Vladimir
Kasler, Franz Johann
Kaufman, Samuel
Kawalek, Joseph Casimir, Jr
Keily, Hubert Joseph
Kelly, William Robert
Kirkendall, Thomas Dodge
Klein, Jerry Robert
Kline, Jerry Robert
Koch, Thomas Richard
Koch, William Frederick
Kroll, Martin Harris
Lepley, Arthur Ray
Lijinsky, William
Linder, Seymour Martin
Lusby, William Robert
McCandliss, Russell John
MacFadden, Kenneth Orville

McGuire, Francis Joseph
McKinney, Robert Wesley
Marinenko, George
Martinez, Richard Isaac
Masters, Larry William
May, Irving
May, Willie Eugene
Mead, Marshall Walter
Melancon, Mark J
Miller, C David
Miziolek, Andrzej Wladyslaw
Muschik, Gary Mathew
Muser, Marc
Musselman, Nelson Page
Newbury, Dale Elwood
Novak, Thaddeus John
O'Haver, Thomas Calvin
O'Rangers, John Joseph
Ostrofsky, Bernard
Paulsen, Paul
Peterson, John Ivan
Pomerantz, Irwin Herman
Prouty, Richard Metcalf
Rakhit, Gopa
Ranganathan, Brahmanpalli
    Narasimhamurthy
Rasberry, Stanley Dexter
Rice, Clifford Paul
Rice, James K
Risby, Terence Humphrey
Roller, Peter Paul
Rollins, Orville Woodrow
Rotherham, Jean
St John, Peter Alan
Salwin, Harold
Samuel, Aryeh Hermann
Schaffer, Robert
Schifreen, Richard Steven
Schwartz, Robert Saul
Sherman, Anthony Michael
Smardzewski, Richard Roman
Snow, Milton Leonard
Snow, Philip Anthony
Sphon, James Ambrose
Stuntz, Calvin Frederick
Taylor, John Keenan
Theimer, Edgar E
Thompson, Donald Leroy
Tompa, Albert S
Topping, Joseph John
Turk, Gregory Chester
Vance, Hugh Gordon
Vanderslice, Joseph Thomas
Veillon, Claude
Warshowsky, Benjamin
Wexler, Arthur Samuel
Whiting, John Dale, Jr
Willeboordse, Friso
Williams, Frederick Wallace
Wolf, Wayne Robert
Yergey, Alfred L, III
Young, Harold Henry
Yurow, Harvey Warren
Zacharius, Robert Marvin

MASSACHUSETTS
Adler, Norman
Ahearn, James Joseph, Jr
Ammlung, Richard Lee
Annino, Raymond
Atkins, Jaspard Harvey
Bankston, Donald Carl
Baratta, Edmond John
Barnes, Ramon M
Berera, Geetha Poonacha
Berlandi, Francis Joseph
Berry, Vern Vincent
Bessette, Russell Romulus
Bidlingmeyer, Brian Arthur
Billo, Edward Joseph
Boehm, Paul David
Bracco, Donato John
Brauner, Phyllis Ambler
Buono, John Arthur
Cembrola, Robert John
Chung, Frank H
Clarke, Michael J
Coleman, Geoffry N
Coppola, Elia Domenico
Cormier, Alan Dennis
Cosgrove, James Francis
Costello, Catherine E
Curran, David James
David, Donald J
Davies, Geoffrey
DiNardi, Salvatore Robert
Donoghue, John Timothy
Ellingboe, James
Ellis, David Wertz
Ezrin, Myer
Flagg, John Ferard
Fletcher, Kenneth Steele, III
Forman, Earl Julian
Frant, Martin S
Fritzsche, Alfred Keith
Gersh, Michael Elliot
Gershman, Louis Leo
Gerteisen, Thomas Jacob
Gibb, Thomas Robinson Pirie
Gilbert, Theodore William, Jr
Gilbert, Thomas Rexford
Girard, Francis Henry
Goodman, Philip
Gopikanth, M L

Gore, William Earl
Graffeo, Anthony Philip
Griffin, Paul Joseph
Hammond, Sally Katharine
Hanselman, Raymond Bush
Hayden, Thomas Day
Heyn, Arno Harry Albert
Hobbs, John Robert
Hochella, Norman Joseph
Hunt, Albert Melvin
Jankowski, Conrad M
Kaplan, David Lee
Karger, Barry Lloyd
Klanfer, Karl
Kliem, Peter O
Kocon, Richard William
Krull, Ira Stanley
Langmuir, Margaret Elizabeth Lang
Learson, Robert Joseph
Legg, Kenneth Deardorff
Lembo, Nicholas J
Leonard, Edward H
Lester, Joseph Eugene
Li, Kuang-Pang
Licht, Stuart Lawrence
Light, Truman S
Lingane, James Joseph
Little, James Noel
Livingston, Hugh Duncan
Lublin, Paul
Macaione, Domenic Paul
McGarry, Margaret
Malenfant, Arthur Lewis
Mattina, Charles Frederick
Merritt, Margaret Virginia
Mowery, Dwight Fay, Jr
Neue, Uwe Dieter
Olmez, Ilhan
Olver, John Walter
Peace, George Earl, Jr
Pinkus, Jack Leon
Pojasek, Robert B
Ram, Neil Marshall
Robbat, Albert, Jr
Robertson, Donald Hubert
Ross, James William
Rubin, Leon E
Rupich, Martin Walter
Serafin, Frank G
Shepardson, John U
Siegal, Bernard
Siggia, Sidney
Strong, Robert Stanley
Sudmeier, James Lee
Takman, Bertil Herbert
Tan, Barrie
Thomas, Martha Jane Bergin
Thorstensen, Thomas Clayton
Trummer, Steven
Turnquist, Carl Richard
Uden, Peter Christopher
Varco-Shea, Theresa Camille
Vouros, Paul
Wallace, David H
Wallace, Frederic Andrew
Wechter, Margaret Ann
Weigand, Willis Alan

MICHIGAN
Allison, John
Anderson, Charles Thomas
Armentrout, David Noel
Bauer, William Eugene
Benson, Edmund Walter
Bernius, Mark Thomas
Bone, Larry Irvin
Bowman, Phil Bryan
Braselton, Webb Emmett, Jr
Bredeweg, Robert Allen
Cadle, Steven Howard
Cantor, David Milton
Chance, Robert L
Chrepta, Stephen John
Chulski, Thomas
Coburn, Joel Thomas
Coleman, David Manley
Cope, Virgil W
Corrigan, Dennis Arthur
Crouch, Stanley Ross
Crow, Frank Warren
Crummett, Warren B
D'Itri, Frank M
Dragun, James
Dukes, Gary Rinehart
Elving, Philip Juliber
Enke, Christie George
Erlich, Ronald Harvey
Fawcett, Timothy Goss
Floutz, William Vaughn
Forist, Arlington Ardeane
Frade, Peter Daniel
Frawley, Nile Nelson
Gaarenstroom, Stephen William
Gill, Harold Hatfield
Goetz, Rudolph W
Goodwin, Jesse Francis
Gordus, Adon Alden
Gulick, Wilson M, Jr
Gump, J R
Hajratwala, Bhupendra R
Hamlin, William Earl
Hart, Donald John
Heeschen, Jerry Parker
Herrinton, Paul Matthew

Analytical Chemistry (cont)

Holcomb, Ira James
Holkeboer, Paul Edward
Hollenberg, Paul Frederick
Hood, Robin James
Howell, James Arnold
Hutchinson, Kenneth A
Ikuma, Hiroshi
Jackson, Larry Lynn
Jensen, David James
Johnson, Jack (Lamar)
Johnson, Ray Leland
Kaiser, David Gilbert
Kallos, George J
Kang, Uan Gen
Kelly, Nelson Allen
King, Stanley Shih-Tung
Kolat, Robert S
Kuo, Mingshang
Langvardt, Patrick William
Larson, John Grant
Leenheer, Mary Janeth
Lewis, Lynn Loraine
Lindblad, William John
Linowski, John Walter
Lipton, Michael Forrester
Loomis, Thomas Charles
Lorch, Steven Kalman
Lubman, David Mitchell
McEwen, David John
McLean, James Dennis
Maheswari, Shyam P
Mahle, Nels H
Marquardt, Roland Paul
Maskal, John
Merrill, Jerald Carl
Meyerhoff, Mark Elliot
Morris, Michael D
Mosher, Robert Eugene
Munson, James William
Murie, Richard A
Murphy, William R
Myers, Harvey Nathaniel
Nader, Bassam Salim
Nestrick, Terry John
Neumann, Fred William
Nevius, Timothy Alfred
Petzold, Edgar
Pfeiffer, Curtis Dudley
Poole, Colin Frank
Popov, Alexander Ivan
Potter, Noel Marshall
Price, Harold Anthony
Prostak, Arnold S
Przybytek, James Theodore
Rainey, Mary Louise
Reim, Robert E
Rengan, Krishnaswamy
Roberts, Charles Brockway
Robertson, John Harvey
Rorabacher, David Bruce
Rulfs, Charles Leslie
Schuetzle, Dennis
Seymour, Michael Dennis
Shimp, Neil Frederick
Siegl, Walter Otto
Sinsheimer, Joseph Eugene
Skelly, Norman Edward
Slywka, Gerald William Alexander
Small, Hamish
Smart, James Blair
Smith, Kenneth Edward
Smith, Ralph G
Smith, Richard Harding
Spang, Arthur William
Spurlock, Carola Henrich
Steinhaus, Ralph K
Stenger, Vernon Arthur
Struck, William Anthony
Suggs, William Terry
Swarin, Stephen John
Swartz, Grace Lynn
Swathirajan, S
Swovick, Melvin Joseph
Symonds, Robert B
Szutka, Anton
Taraszka, Anthony John
Thomas, Kenneth Eugene, III
Timnick, Andrew
Tou, James Chieh
Tsuji, Kiyoshi
Turley, June Williams
Vanderwielen, Adrianus Johannes
VanEffen, Richard Michael
Van Hall, Clayton Edward
Walters, Stephen Milo
Warren, H(erbert) Dale
Watson, Jack Throck
Weber, Dennis Joseph
Whitfield, Richard George
Wong, Peter Alexander
Zak, Bennie

MINNESOTA

Adams, Ernest Clarence
Atwell, William Alan
Baude, Frederic John
Bowers, Larry Donald
Brenner, Mark
Bydalek, Thomas Joseph
Carr, Peter William
Clapp, C(harles) Edward
DeRoos, Fred Lynn

Duerst, Richard William
Durnick, Thomas Jackson
Ellefson, Ralph Donald
Evans, John Fenton
Farm, Raymond John
Farnum, Sylvia A
Forrette, John Elmer
Freier, Herbert Edward
Fritsch, Carl Walter
Gyberg, Arlin Enoch
Haddad, Louis Charles
Hagen, Donald Frederick
Haile, Clarence Lee
Henney, Robert Charles
Holden, John B, Jr
Jensen, Richard Erling
Kariv-Miller, Essie
Katz, William
Kolthoff, Izaak Maurits
Kovacic, Joseph Edward
Latterell, Joseph J
Lodge, Timothy Patrick
Lonnes, Perry Bert
MacKellar, William John
McKenna, Jack F(ontaine)
McMullen, James Clinton
Marsh, Frederick Leon
Marshall, John Clifford
Mishmash, Harold Edward
Newmark, Richard Alan
Olsen, Rodney L
Oxborrow, Gordon Squires
Poe, Donald Patrick
Potts, Lawrence Walter
Ramette, Richard Wales
Rejto, Peter A
Thatcher, Walter Eugene
Tiers, George Van Dyke
Todd, Jerry William
Toren, Paul Edward
Trusk, Ambrose
Walters, John Philip

MISSISSIPPI

Alley, Earl Gifford
Carter, Fairie Lyn
Doumit, Carl James
ElSohly, Mahmoud Ahmed
Fawcett, Newton Creig
Hussey, Charles Logan
Kalasinsky, Victor Frank
Kopfler, Frederick Charles
Luker, William Dean
Minyard, James Patrick
Monts, David Lee
Tai, Han
White, June Broussard

MISSOURI

Ahmed, Moghisuddin
Arneson, Dora Williams
Beaver, Earl Richard
Bier, Dennis Martin
Brescia, Vincent Thomas
Cheng, Kuang Lu
Cole, Douglas L
Conkin, Robert A
Coria, Jose Conrado
Craver, Clara Diddle (Smith)
Dahl, William Edward
Drew, Henry D
Emery, Edward Mortimer
Field, Byron Dustin
Freeland, Max
Freeman, John Jerome
Geisman, Raymond August, Sr
Gerhardt, Klaus Otto
Gibbons, James Joseph
Glascock, Michael Dean
Gnaedinger, Richard H
Grayson, Michael A
Greenlief, Charles Michael
Grubbs, Charles Leslie
Guerry, Davenport, Jr
Gustafson, Mark Edward
Haggerty, William Joseph, Jr
Hallas, Laurence Edward
Hardtke, Fred Charles, Jr
Haynes, William Miller
Kaelble, Emmett Frank
Kaley, Robert George, II
Keller, Robert Ellis
Kenyon, Allen Stewart
Klein, Andrew John
Kluba, Richard Michael
Koirtyohann, Samuel Roy
Korotev, Randall Lee
Ladenson, Jack Herman
Larson, Wilbur John
Levy, Ram Leon
Lott, Peter F
Ludwig, Frederick John, Sr
Macias, Edward S
Malik, Joseph Martin
Mori, Erik Jun
Moulton, Robert Henry
Newell, Jon Albert
Ogilvie, James Louis
Pickett, Edward Ernest
Pierce, John Thomas
Plimmer, Jack Reynolds
Podosek, Frank A
Rath, Nigam Prasad
Roy, Rabindra (Nath)

Rueppel, Melvin Leslie
Schaeffer, Harold F(ranklin)
Sharp, Dexter Brian
Sherman, William Reese
Smith, David Warren
Smith, Ronald Gene
Stout, Edward Irvin
Sunde, Roger Allan
Talbott, Ted Delwyn
Thielmann, Vernon James
Tuthill, Samuel Miller
Wilcox, Harold Kendall
Wilkinson, Ralph Russell
Woodhouse, Edward John
Worley, Jimmy Weldon

MONTANA

Amend, John Robert
Layman, Wilbur A
Volborth, Alexis

NEBRASKA

Blickensderfer, Peter W
Carr, James David
Freidline, Charles Eugene
Gross, Michael Lawrence
Heath, Eugene Cartmill
Howell, Daniel Bunce
Issenberg, Phillip
Johar, J(ogindar) S(ingh)
Kaplan, Sanford Sandy
Kenkel, John V
Mattern, Paul Joseph
Mattes, Frederick Henry
Phelps, George Clayton
Solsky, Joseph Fay
Struempler, Arthur W

NEVADA

Billingham, Edward J, Jr
Eastwood, DeLyle
Fletcher, Aaron Nathaniel
Gerard, Jesse Thomas
Klainer, Stanley M
Miles, Maurice Jarvis
Pierson, William R
Rhees, Raymond Charles
Schooley, David Allan
Sovocool, G Wayne
Tanner, Roger Lee
Vincent, Harold Arthur

NEW HAMPSHIRE

Damour, Paul Lawrence
Fogleman, Wavell Wainwright
Hagan, William John, Jr
Hebert, Normand Claude
Illian, Carl Richard
Leipziger, Fredric Douglas
Merritt, Charles, Jr
Neil, Thomas C
Roberts, John Edwin
Seitz, William Rudolf
Stepenuck, Stephen Joseph, Jr

NEW JERSEY

Abdou, Hamed M
Achari, Raja Gopal
Aczel, Thomas
Adler, Seymour Jacob
Aldrich, Haven Scott
Altman, Lawrence Jay
Anthony, Linda J
Ardelt, Wojciech Joseph
Aronovic, Sanford Maxwell
Ashworth, Harry Arthur
Banick, William Michael, Jr
Barnes, Robert Lee
Bathala, Mohinder S
Beck, John Louis
Beispiel, Myron
Benz, Wolfgang
Bernath, Tibor
Bhattacharyya, Pranab K
Boczkowski, Ronald James
Bodin, Jerome Irwin
Bornstein, Alan Arnold
Bornstein, Michael
Borysko, Emil
Bose, Ajay Kumar
Bozzelli, Joseph William
Bradstreet, Raymond Bradford
Brand, William Wayne
Brody, Stuart Martin
Brofazi, Frederick R
Brown, John Angus
Burnett, Bruce Burton
Cardarelli, Joseph S
Carlucci, Frank Vito
Carter, James Evan
Chandrasekhar, Prasanna
Chaudhri, Safee U
Chidsey, Christopher E
Citron, Irvin Meyer
Cohen, Allen Irving
Cone, Conrad
Conley, Jack Michael
Coyle, Catherine Louise
Crane, Laura Jane
Cryer, Dennis Robert
Daly, Robert E
Dean, Donald E
DeCastro, Arthur
De Silva, John Arthur F

Diegnan, Glenn Alan
Downing, George V, Jr
Duerr, J Stephen
Dunham, John Malcolm
Dunn, William Howard
Eastman, David Willard
Edelson, Edward Harold
Edelstein, Harold
Egan, Richard Stephen
Egli, Peter
Ehrlich, Julian
Eider, Norman George
Erdmann, Duane John
Fairchild, Edward H
Feldman, Nicholas
Ferraro, Charles Frank
Fink, David Warren
Finston, Harmon Leo
Fix, Kathleen A
Flanders, Clifford Auten
Fong, Jones W
Foster, Walter H, Jr
Frankoski, Stanley P
Frantz, Beryl May
Funke, Phillip T
Ganapathy, Ramachandran
Geller, Milton
Gerecht, J Fred
Gierer, Paul L
Glass, John Richard
Goffman, Martin
Gosser, Leo Anthony
Granchi, Michael Patrick
Grandolfo, Marian Carmela
Grey, Peter
Gullo, Vincent Philip
Hackman, Martin Robert
Hagel, Robert B
Hall, Gene Stephen
Hall, Gretchen Randolph
Hammond, Willis Burdette
Hanley, Arnold V
Hansen, Holger Victor
Hansen, Ralph Holm
Hartkopf, Arleigh Van
Heacock, Craig S
Head, William Francis, Jr
Helrich, Kenneth
Heveran, John Edward
Hills, Stanley
Ho, Chi-Tang
Hobart, Everett W
Hoffman, Clark Samuel, Jr
Hoffman, Henry Tice, Jr
Hokanson, Gerard Clifford
Huger, Francis P
Huneke, James Thomas
Iorns, Terry Vern
Jackson, Thomas A J
Jacobs, Morton Howard
Jacobson, Harold
Jain, Nemichand B
Janssen, Richard William
Javick, Richard Anthony
Jemal, Mohammed
Jensen, Norman P
Johnson, Hilding Reynold
Joseph, John Mundancheril
Kabadi, Balachandra N
Kadin, Harold
Kallmann, Silve
Kaplan, Gerald
Katz, Sidney A
Kender, Donald Nicholas
Kern, Werner
Kikta, Edward Joseph, Jr
Kim, Benjamin K
Kirschbaum, Joel Jerome
Kline, Berry James
Kobrin, Robert Jay
Kohout, Frederick Charles, III
Kolis, Stanley Joseph
Kutsher, George Samuel
Labows, John Norbert, Jr
Lalancette, Roger A
Laughlin, Alice
Letterman, Herbert
Lewis, Arnold D
Liebowitz, Stephen Marc
Lofstrom, John Gustave
Lorenz, Patricia Ann
Love, L J Cline
MacDonald, Alexander, Jr
McGuire, David Kelty
McMahon, David Harold
McSharry, William Owen
Maldacker, Thomas Anton
Maloy, Joseph T
Medwick, Thomas
Melveger, Alvin Joseph
Mergens, William Joseph
Miller, James Monroe
Miller, Warren Victor
Milliman, George Elmer
Mitchell, James Winfield
Montana, Anthony J
Montgomery, Richard Millar
Moros, Stephen Andrew
Mowitz, Arnold Martin
Mukai, Cromwell Daisaku
Murthy, Vadiraja Venkatesa
Mussinan, Cynthia June
Nadkarni, Ramachandra Anand
Naik, Datta Vittal

Nicolau, Gabriela
Niedermayer, Alfred O
O'Connor, Joseph Michael
O'Connor, Matthew James
Okinaka, Yutaka
Opila, Robert L, Jr
Oyler, Alan Richard
Papastephanou, Constantin
Patrick, James Edward
Phillips, Wendell Francis
Pilkiewicz, Frank George
Platt, Thomas Boyne
Pluscec, Josip
Pobiner, Harvey
Rabel, Fredric M
Rau, Eric
Redden, Patricia Ann
Reents, William David, Jr
Reissmann, Thomas Lincoln
Riedhammer, Thomas M
Rigler, Neil Edward
Roberts, Ronald Frederick
Rodgers, Robert Stanleigh
Rolle, F Robert
Romano, Salvatore James
Rose, Ira Marvin
Rosen, Joseph David
Rothstein, Edwin C(arl)
Sattur, Theodore W
Searle, Norma Zizmer
Shinkai, Ichiro
Sibilia, John Philip
Sieh, David Henry
Sinclair, James Douglas
Singleton, Bert
Smyers, William Hays
Somkaite, Rozalija
Spohn, Ralph Joseph
Sterbenz, Francis Joseph
Stillman, John Edgar
Stober, Henry Carl
Szap, Peter Charles
Szyper, Mira
Taylor, John William
Tibbetts, Merrick Sawyer
Trewella, Jeffrey Charles
Triglia, Emil J
Tse, Francis Lai-Sing
Turi, Paul George
Turse, Richard S
Tway, Patricia C
Vallese, Frank M
Venturella, Vincent Steven
Vickroy, David Gill
Virgili, Luciano
Vogel, Veronica Lee
Waltking, Arthur Ernest
Ward, Laird Gordon Lindsay
Westerdahl, Carolyn Ann Lovejoy
Westerdahl, Raymond P
Whigan, Daisy B
Wilson, Mabel F
Wittick, James John
Wolf, Thomas
Woodruff, Hugh Boyd
Zaim, Semih
Zanzucchi, Peter John
Zenchelsky, Seymour Theodore

NEW MEXICO
Apel, Charles Turner
Balagna, John Paul
Beattie, Willard Horatio
Bechtold, William Eric
Bentley, Glenn E
Binder, Irwin
Bowen, Scott Michael
Burns, Frank Bernard
Campbell, George Melvin
Caton, Roy Dudley, Jr
Clark, Robert Paul
Cunningham, Paul Thomas
Essington, Edward Herbert
Ewing, Galen Wood
Fritz, Georgia T(homas)
George, Raymond S
Haaland, David Michael
Hakkila, Eero Arnold
Heaton, Richard Clawson
Hobart, David Edward
Jackson, Darryl Dean
Kelly, Clark Andrew
Kenna, Bernard Thomas
Kosiewicz, Stanley Timothy
Loughran, Edward Dan
Matlack, George Miller
Mead, Richard Wilson
Morales, Raul
Neary, Michael Paul
Nielsen, Stuart Dee
Niemczyk, Thomas M
Ohline, Robert Wayne
Ott, Donald George
Park, Su-Moon
Patterson, James Howard
Renschler, Clifford Lyle
Richardson, Albert Edward
Rogers, Raymond N
Sedlacek, William Adam
Sherman, Robert Howard
Smith, James Lewis
Smith, Maynard E
Smith, Wayne Howard
Spall, Walter Dale

Stanbro, William David
Trujillo, Patricio Eduardo
Vanderborgh, Nicholas Ernest
Walton, George
Wang, Joseph
Weissman, Suzanne Heisler
West, Mike Harold
Williams, Mary Carol
Yasuda, Stanley K

NEW YORK
Abraham, Jerrold L
Abruna, Hector D
Ahuja, Satinder
Aikens, David Andrew
Alliet, David F
Ambrose, Robert T
Anderson, Frank Wallace
Andrews, John Parray
Andrews, Mark Allen
Angelino, Norman J
Antonucci, Frank Ralph
Apai, Gustav Richard, II
Baden, Harry Christian
Beach, David H
Becker, Harry Carroll
Berger, Selman A
Billmeyer, Fred Wallace, Jr
Birke, Ronald Lewis
Bixler, John Wilson
Blackwell, Crist Scott
Boettger, Susan D
Bonvicino, Guido Eros
Bottger, Gary Lee
Brown, Neil Harry
Brown, Oliver Monroe
Bruckenstein, Stanley
Burlitch, James Michael
Bush, David Graves
Campion, James J
Cappel, C Robert
Capretta, Umberto
Cardenas, Raúl R, Jr
Carnahan, James Claude
Carpenter, Thomas J
Carson, Chester Carrol
Chang, Jack Che-Man
Christenson, Philip A
Conway, Walter Donald
Cooke, William Donald
Cooper, Aaron David
Corth, Richard
Cover, Richard Edward
Cratty, Leland Earl, Jr
Creasy, William Russel
D'Angelo, Gaetano
Daves, Glenn Doyle, Jr
Davis, Abram
DeForest, Peter Rupert
Dehm, Richard Lavern
DeStefano, Anthony Joseph
Dexter, Theodore Henry
Dietz, Edward Albert, Jr
Dietz, Russell Noel
Dinan, Frank J
DiNunzio, James E
Douglass, Pritchard Calkins
Dumoulin, Charles Lucian
Durst, Richard Allen
Dutta, Shib Prasad
Dwyer, Robert Francis
Eadon, George Albert
Edsberg, Robert Leslie
Fay, Homer
Fiala, Emerich Silvio
Fong, Godwin Wing-Kin
Froelich, Philip Nissen
Gardella, Joseph Augustus, Jr
Gaver, Robert Calvin
Gilbert, Jack Pittard
Glickstein, Joseph
Gluck, Ronald Monroe
Goddard, John Burnham
Goldman, James Allan
Gordon, Harry William
Greizerstein, Hebe Beatriz
Grossman, William Elderkin Leffingwell
Gustin, Vaughn Kenneth
Halpern, Mordecai Joseph
Hanna, Samir A
Hanson, Jonathan C
Hardy, Ralph Wilbur Frederick
Heininger, Clarence George, Jr
Heintz, Edward Allein
Hopke, Philip Karl
Ilmet, Ivor
Irsa, Adolph Peter
Jacobson, Jay Stanley
Jaffe, Marvin Richard
Janauer, Gilbert E
Jayme, David Woodward
Jespersen, Neil David
Johnson, Raymond Nils
Jones, Stanley Leslie
Julian, Donald Benjamin
Karliner, Jerrold
Karp, Stewart
Keesee, Robert George
Keller, Roy Alan
Kesner, Leo
Kho, Boen Tong
Killian, Carl Stanley
Kingston, Charles Richard
Kissel, Thomas Robert

Klingele, Harold Otto
Kneip, Theodore Joseph
Kozak, Gary S
Krey, Philip W
Kross, Robert David
Kuritzkes, Alexander Mark
Lam, Stanley K
Lane, Keith Aldrich
Launer, Philip Jules
Leff, Judith
Lerman, Steven I
Lessor, Arthur Eugene, Jr
Lessor, Edith Schroeder
Levine, Solomon Leon
Levinson, Steven R
Levy, Arthur Louis
Levy, George Charles
Lewin, Seymour Z
Liang, Charles C
Liao, Hsueh-Liang
Ligon, Woodfin Vaughan, Jr
Loach, Kenneth William
Locke, David Creighton
Lorenzo, George Albert
Loscalzo, Anne Grace
Lovecchio, Frank Vito
Luders, Richard Christian
Ma, Maw-Suen
Macero, Daniel Joseph
McGriff, Richard Bernard
McHugh, James Anthony, Jr
McLafferty, Fred Warren
McLennan, Scott Mellin
Mahony, John Daniel
Manche, Emanuel Peter
Margoshes, Marvin
Marov, Gaspar J
Marr, David Henry
Matthews, Dwight Earl
May, Joan Christine
Meloon, Daniel Thomas, Jr
Mennitt, Philip Gary
Merritt, Paul Eugene
Meyer, John Austin
Michiels, Leo Paul
Miller, Raymond Sumner
Molina, John Francis
Molloy, Andrew A
Morrison, George Harold
Mourning, Michael Charles
Mundorff, Sheila Ann
Murphy, Kenneth Robert
Mylroie, Victor L
Nathanson, Benjamin
Nelson, Kurt Herbert
Neubort, Shimon
Noel, Dale Leon
Norton, Elinor Frances
Oberholtzer, James Edward
O'Donnell, Raymond Thomas
O'Mara, Michael Martin
Oppenheimer, Larry Eric
Orna, Mary Virginia
Orzech, Chester Eugene, Jr
Osteryoung, Janet G
Osteryoung, Robert Allen
Padmanabhan, G R
Parker, Frank S
Pearse, George Ancell, Jr
Pittman, Kenneth Arthur
Pomeroy, Gordon Ashby
Pontius, Dieter J J
Potter, George Henry
Prigot, Melvin
Przybylowicz, Edwin P
Rand, Salvatore John
Reardon, Joseph Daniel
Redalieu, Elliot
Reddy, Thomas Bradley
Reinmuth, William Henry
Renwick, J Alan A
Resnik, Frank Edward
Reuter, Wilhad
Richtol, Herbert H
Roboz, John
Rogers, Donald Warren
Rosenthal, Donald
Rothchild, Robert
Roy, Ram Babu
Russell, Virginia Ann
Sage, Gloria W
Salotto, Anthony W
Sandifer, James Roy
Sauer, Charles William
Sayegh, Joseph Frieh
Schaefer, Robert William
Schmeltz, Irwin
Schneider, Frank L
Schottmiller, John Charles
Schucker, Gerald D
Schupp, Orion Edwin, III
Schwartz, Herbert Mark
Segatto, Peter Richard
Sharkey, John Bernard
Sharma, Minoti
Shilman, Avner
Silver, Herbert Graham
Simmons, Daryl Michael
Singh, Surjit
Sinha, Asru Kumar
Smith, Alan Jerrard
Solomon, Jerome Jay
Spielholtz, Gerald I
Spink, Charles Harlan

Stryker, Martin H
Su, Yao Sin
Swartz, James Lawrence
Sweet, Richard Clark
Takeuchi, Esther Sans
Talley, Charles Peter
Taub, Aaron M
Teitelbaum, Charles Leonard
Thumm, Byron Ashley
Tong, Stephen S C
Triplett, Kelly B
Underkofler, William Leland
Van Geet, Anthony Leendert
Von Bacho, Paul Stephan, Jr
Waehner, Kenneth Arthur
Wandass, Joseph Henry
Wapnir, Raul A
Weisler, Leonard
Werner, Thomas Clyde
West, Kenneth Calvin
Whitlock, L Ronald
Wiberley, Stephen Edward
Williams, Evan Thomas
Wilson, Claude E
Wood, David
Wright, Charles Joseph
Wu, Konrad T
Yeransian, James A
Youker, John
Yu, Ming Lun
Zuehlke, Carl William
Zuman, Petr

NORTH CAROLINA
Ackerman, Donald Godfrey, Jr
Adcock, Louis Henry
Alam, Mohammed Ashraful
Anderegg, Robert James
Bell, Jimmy Holt
Blackburn, Thomas Roy
Boss, Charles Ben
Bryan, Horace Alden
Buck, Richard Pierson
Burnett, John Nicholas
Bursey, Joan Tesarek
Bursey, Maurice Moyer
Chou, David Yuan Pin
Clapp, William Lee
Clements, John B(elton)
Cochran, George Thomas
Collier, Herman Edward, Jr
Cross, Robert Edward
Crumpler, Thomas Bigelow
Cundiff, Robert Hall
Decker, Clifford Earl, Jr
Dobbins, James Talmage, Jr
Elder, James Franklin, Jr
Forman, Donald T
Gangwal, Santosh Kumar
Ganz, Charles Robert
Gemperline, Margaret Mary Cetera
Gemperline, Paul Joseph
Gibson, Robert Harry
Gillikin, Jesse Edward, Jr
Gleit, Chester Eugene
Green, Charles Raymond
Hadzija, Bozena Wesley
Hanck, Kenneth William
Harrell, T Gibson
Harrison, Stanley L
Hass, James Ronald
Heck, Henry d'Arcy
Heckman, Robert Arthur
Heintzelman, Richard Wayne
Herman, Harvey Bruce
Hinze, Willie Lee
Hong, Donald David
Hurlbert, Bernard Stuart
Jackman, Donald Coe
Jezorek, John Robert
Johnson, Delwin Phelps
Johnson, J(ames) Donald
Jorgenson, James Wallace
Kersey, Robert Lee, Jr
Kirby, James Ray
Knecht, Laurance A
Lewis, Claude Irenius
Li, Chia-Yu
Linton, Richard William
Ljung, Harvey Arthur
Lochmuller, Charles Howard
Lueck, Charles Henry
Lunney, David Clyde
Ma, Tsu Sheng
McBay, Arthur John
McCarty, Billy Dean
McDaniel, Roger Lanier, Jr
Martin, G(uy) William, Jr
Mays, Rolland Lee
Murphy, Robert T
Murray, Royce Wilton
O'Connor, Lila Hunt
Olander, Donald Paul
Padmore, Joel M
Panek, Edward John
Parker, Carol Elaine Greenberg
Parks, Ross Lombard
Patterson, Ronald Brinton
Pellizzari, Edo Domenico
Perfetti, Thomas Albert
Pickett, John Harold
Ramaswamy, H N
Remington, Lloyd Dean
Rochow, Theodore George

Analytical Chemistry (cont)

Sawardeker, Jawahar Sazro
Schickedantz, Paul David
Scott, Donald Ray
Senzel, Alan Joseph
Shackelford, Walter McDonald
Steinhagen, William Herrick
Stejskal, Edward Otto
Strobel, Howard Austin
Stubblefield, Charles Bryan
Tomer, Kenneth Beamer
Tyndall, John Raymond
Upton, Ronald P
Van Atta, Robert Ernest
Volk, Richard James
Wani, Mansukhlal Chhaganlal
Whitehurst, Garnett Brooks
Willey, Joan DeWitt
Wilson, Nancy Keeler
Woosley, Royce Stanley
Yeowell, David Arthur
Youmans, Hubert Lafay

NORTH DAKOTA

Johnson, Arnold Richard, Jr
Johnson, Phyllis Elaine
Khalil, Shoukry Khalil Wahba
Metzger, James David
Stolzenberg, Gary Eric
Tallman, Dennis Earl
Vogel, Gerald Lee

OHIO

Allenson, Douglas Rogers
Ampulski, Robert Stanley
Anderson, Larry Bernard
Andresen, Brian Dean
Andria, George D
Attalla, Albert
Austern, Barry M
Beres, John Joseph
Bimber, Russell Morrow
Black, Arthur Herman
Brain, Devin King
Brandt, Manuel
Bromund, Richard Hayden
Bromund, Werner Hermann
Budde, William L
Budke, Clifford Charles
Buell, Glen R
Burg, William Robert
Burrows, Kerilyn Christine
Butkus, Antanas
Campbell, Donald R
Chambers, Lee Mason
Chang, Jung-Ching
Chen, Yih-Wen
Chong, Clyde Hok Heen
Choung, Hun Ryang
Christian, John B
Clowers, Churby Conrad, Jr
Cordell, Richard William
Coulter, Paul David
Couri, Daniel
Cox, James Allan
Crable, John Vincent
Cryberg, Richard Lee
Danielson, Neil David
Dehne, George Clark
Deviney, Marvin Lee, Jr
Diem, Hugh E(gbert)
Dobbelstein, Thomas Norman
Duff, Robert Hodge
Durand, Edward Allen
Eckstein, Yona
Eisentraut, Kent James
Fabricant, Barbara Louise
Fabris, Hubert
Ferguson, John Allen
Fike, Winston
Folk, Theodore Lamson
Fraser, Alex Stewart
Freeberg, Fred E
Garn, Paul Donald
Gerlach, Edward Rudolph
Gilpin, Roger Keith
Gird, Steven Richard
Gordon, Gilbert
Gordon, Sydney Michael
Gorse, Joseph
Grasselli, Jeanette Gecsy
Greenberg, Mark Shiel
Greene, Arthur Frederick, Jr
Greenlee, Kenneth William
Greinke, Ronald Alfred
Grieshammer, Lawrence Louis
Gurley, Thomas Wood
Gustafson, David Harold
Gustafson, Terry Lee
Hall, Ronald Henry
Harmon, Dale Joseph
Heineman, William Richard
Henry, William Mellinger
Hess, George G
Hibbits, James Oliver, Jr
Hilton, Ashley Stewart
Hively, Robert Arland
Hoffman, William Andrew, Jr
Holbert, Gene W(arwick)
Holtman, Mark Steven
Holubec, Zenowie Michael
Homan, Ruth Elizabeth
Howell, Norman Gary

Hubbard, Arthur T
Hutchison, William Marwick
Ikenberry, Luther Curtis
Jakobsen, Robert John
Jensen, Adolph Robert
Karweik, Dale Herbert
Katon, John Edward
Katovic, Vladimir
Kay, Peter Steven
Knowlton, David A
Koch, Ronald Joseph
Koehler, Mark E
Koknat, Friedrich Wilhelm
Krishen, Anoop
Krivis, Alan Frederick
Krochta, William G
Kuemmel, Donald Francis
Lamb, Robert Edward
Lang, James Frederick
Laning, Stephen Henry
Lattimer, Robert Phillips
Latz, Howard W
Leussing, Daniel, Jr
Liao, Shu-Chung
Link, William Edward
Lott, John Alfred
Lucas, Kenneth Ross
Lucke, William E
McCreery, Richard Louis
McFarland, Charles Warren
MacGee, Joseph
MacKichan, Janis Jean
Mark, Harry Berst, Jr
Marks, Alfred Finlay
Marshall, Alan George
Mateescu, Gheorghe D
Meal, Larie L
Megargle, Robert G
Miller, Theodore Lee
Misono, Kunio Shiraishi
Mitchum, Ronald Kem
Mumtaz, Mohammad Moizuddin
Myers, Ronald Eugene
Neveu, Darwin D
Nicholson, D Allan
Oertel, Richard Paul
Olson, Carter LeRoy
Olynyk, Paul
Pacey, Gilbert E
Pappas, Leonard Gust
Pappenhagen, James Meredith
Parker, Gordon Arthur
Patel, Siddharth Manilal
Pausch, Jerry Bliss
Pearson, Karl Herbert
Porter, Leo Earle
Riechel, Thomas Leslie
Rubinson, Judith Faye
Rynasiewicz, Joseph
Saltzman, Bernard Edwin
Sams, Richard Alvin
Saraceno, Anthony Joseph
Schroeder, Friedhelm
Schumacher, Roy Joseph
Schweitzer, Mark Glenn
Seabaugh, Pyrtle W
Shaer, Elias Hanna
Shaw, Elwood R
Silver, Gary Lee
Silverman, Herbert Philip
Smith, Francis White
Smith, Jerome Paul
Spindler, Donald Charles
Sporek, Karel Frantisek
Srinivasan, Vakula S
Stadler, Louis Benjamin
Steichen, Richard John
Stephens, Marvin Wayne
Stotz, Robert William
Sympson, Robert F
Tabor, Marvin Wilson
Tan, Henry S I
Taylor, Michael Lee
Thomas, Quentin Vivian
Thompson, Robert Quinton
Thomson, William Alexander Brown
Tiernan, Thomas Orville
Tong, James Ying-Peh
Trivisonno, Charles F(rancis)
Turnquist, Truman Dale
Tyler, Willard Philip
Wadelin, Coe William
Watson, Maurice E
Weeks, Thomas Joseph, Jr
Wenclawiak, Bernd Wilhelm
Westneat, David French
Wharton, H(arry) Whitney
Wilde, Bryan Edmund
Williams, Robert Calvin
Williams, Theodore Roosevelt
Wilson, Larry Eugene
Wohlfort, Sam Willis
Wright, George Joseph
Yanko, William Harry
Yellin, Wilbur
Zakriski, Paul Michael
Zaye, David F

OKLAHOMA

Allen, Marvin Carrol
Al-Shaieb, Zuhair Fouad
Anderson, Paul Dean
Battiste, David Ray
Beckett, James Reid

Brown, Kenneth Henry
Bruner, Ralph Clayburn
Cabbiness, Dale Keith
Carubelli, Raoul
Coffman, Harold H
Cowley, Thomas Gladman
DiFeo, Daniel Richard, Jr
Dryhurst, Glenn
Evens, F Monte
Fish, Wayne William
Frost, Jackie Gene
Greenwood, Gil Jay
Grigsby, Ronald Davis
Groten, Barney
Hamming, Mynard C
Harris, Ray Edgar
Leslie, Wallace Dean
Linder, Donald Ernst
Maddin, Charles Milford
Miller, John Walcott
Moczygemba, George A
Monn, Donald Edgar
Mottola, Horacio Antonio
Natowsky, Sheldon
Paxson, John Ralph
Puls, Robert W
Reinbold, Paul Earl
Robinson, Jack Landy
Shew, Delbert Craig
Shioyama, Tod Kay
Van De Steeg, Garet Edward
Varga, Louis P
Wharry, Stephen Mark

OREGON

Adams, Frank William
Binder, Bernhard
Brinkley, John Michael
Church, Larry B
Clark, James Orie, II
Elia, Victor John
Freund, Harry
Gerke, John Royal
Goodney, David Edgar
Hackleman, David E
Hawkes, Stephen J
Humphrey, J Richard
Ingle, James Davis, Jr
Kay, Michael Aaron
Ko, Hon-Chung
Long, James William
Matthes, Steven Allen
Mooney, Larry Albert
Perlich, Robert Willard
Piepmeier, Edward Harman
Roe, David Kelmer
Szalecki, Wojciech Josef

PENNSYLVANIA

Achey, Frederick Augustus
Adler, Irving Larry
Albright, Fred Ronald
Allen, Eugene (Murray)
Allen, Herbert E
Almond, Harold Russell, Jr
Angeloni, Francis M
Anspon, Harry Davis
Baker, Harold Weldon
Bandi, William R
Bandy, Alan Ray
Barford, Robert A
Barnett, Herald Alva
Barnhart, Barry B
Baum, Harry
Beitchman, Burton David
Bhatt, Padmamabh P
Boggs, William Emmerson
Bouis, Paul Andre
Box, Larry
Brackmann, Richard Theodore
Brenner, Gerald Stanley
Brooks, Marvin Alan
Campbell, Iain Malcolm
Carlson, Dana Peter
Caruso, Sebastian Charles
Cavanaugh, James Richard
Chang, Cheng Allen
Cheng, Cheng-Yin
Cheng, Hung-Yuan
Christie, Michael Allen
Clavan, Walter
Coetzee, Johannes Francois
Connelly, Carolyn Thomas
Cook, Ronald Frank
Cornell, Donald Gilmore
Danchik, Richard S
De Jong, Gary Joel
Derby, James Victor
Diorio, Alfred Frank
Doner, Landis Willard
Dulka, Joseph John
Einhorn, Philip A
Ewing, Andrew Graham
Farlee, Rodney Dale
Felty, Wayne Lee
Fiddler, Walter
Finegold, Harold
Finseth, Dennis Henry
Foglia, Thomas Anthony
Follweiler, Douglas MacArthur
Fox, Jay B, Jr
Frankel, Lawrence (Stephen)
Franz, David Alan
Freedman, Robert Wagner

Frohliger, John Owen
Giannovario, Joseph Anthony
Godfrey, Robert Allen
Golton, William Charles
Gottlieb, Irvin M
Grinstein, Reuben H
Grob, Robert Lee
Guthrie, Joseph D
Hafford, Bradford C
Hardesty, Patrick Thomas
Harrington, George William
Hedrick, Jack LeGrande
Hercules, David Michael
Herman, Richard Gerald
Herzog, Leonard Frederick, II
Hicks, Kevin B
Hinkel, Robert Dale
Ho, Floyd Fong-Lok
Hoberman, Alfred Elliott
Holifield, Charles Leslie
Honig, Richard Edward
Hubbard, Willard Dwight
Hughes, Michael Charles
Hurst, William Jeffrey
Hurwitz, Jan Krosst
Janicki, Casimir A
Johnston-Feller, Ruth M
Jordan, Joseph
Judd, Jane Harter
Jurs, Peter Christian
Kanzelmeyer, James Herbert
Karp, Howard
Kelly, Ernest L
Khan, Shakil Ahmad
Khosah, Robinson Panganai
Kieft, Lester
Kimlin, Mary Jayne
King, Richard Warren
Kleinman, Roberta Wilma
Kramer, Raymond Arthur
Krapf, George
Krzeminski, Stephen F
Kuebler, John Ralph, Jr
Kugler, George Charles
Kurtz, David Allan
Lacoste, Rene John
Larkin, Robert Hayden
Latshaw, David Rodney
Lemmon, Donald H
Leyon, Robert Edward
Liberti, Paul A
Locke, Harold Ogden
Louie, Ming
Lyman, William Ray
McCallum, Keith Stuart
MacDonald, Hubert C, Jr
McElroy, Mary Kieran
McKay, James Brian
Maerker, Gerhard
Mainier, Robert
Mair, Robert Dixon
Majors, Ronald E
Manka, Dan P
Markham, James J
Marmer, William Nelson
Maroulis, Peter James
Martin, Aaron J
Martin, John Robert
Melnick, Laben Morton
Mifflin, Theodore Edward
Minard, Robert David
Mozersky, Samuel M
Nagy, Dennis J
Nikelly, John G
Noceti, Richard Paul
Obbink, Russell C
Ohnesorge, William Edward
Osol, Arthur
Ottenstein, Daniel
Owens, Kevin Glenn
Pacer, John Charles
Parees, David Marc
Perry, Mary Hertzog
Peterson, James Oliver
Phifer, Lyle Hamilton
Pinschmidt, Robert Krantz, Jr
Plankey, Francis William, Jr
Preti, George
Prohaska, Charles Anton
Reiff, Harry Elmer
Retcofsky, Herbert L
Robinson, Douglas Walter
Rodell, Michael Byron
Rodgers, Sheridan Joseph
Rogers, Horace Elton
Romberger, Karl Arthur
Ross, Stephen T
Rothbart, Herbert Lawrence
Rothman, Alan Michael
Rutgers, Jay G
Ruud, Clayton Olaf
Sadtler, Philip
Sanders, Charles Irvine
Scheirer, James E
Schobert, Harold Harris
Schroeder, Thomas Dean
Schultz, Hyman
Schweighardt, Frank Kenneth
Shepherd, Rex E
Shergalis, William Anthony
Sherma, Joseph A
Sherman, Larry Ray
Shive, Donald Wayne
Siegel, Melvin Walter

Siegel, Richard C
Simon, Joseph Matthew
Smith, James Stanley
Smith, Stewart Edward
Snider, Albert Monroe, Jr
Spritzer, Michael Stephen
Stahl, John Wendell
State, Harold M
Stelting, Kathleen Marie
Stiller, Richard L
Stone, Herman
Straub, William Albert
Streuli, Carl Arthur
Strong, Frederick Carl, III
Stuper, Andrew John
Sylvester, James Edward
Syty, Augusta
Tackett, Stanford L.
Talley, Eugene Alton
Taylor, David Cobb
Taylor, Robert Morgan
Thornton, Donald Carlton
Toberman, Ralph Owen
Turner, William Richard
Umbreit, Gerald Ross
Urenovitch, Joseph Victor
Varma, Asha
Veening, Hans
Vickers, Stanley
Warren, Richard Joseph
Weber, Frank L
Weiss, Paul Storch
Westmoreland, David Gray
White, Edward Roderick
Williams, John Roderick
Winograd, Nicholas
Wittle, John Kenneth
Yohe, Thomas Lester
Young, Irving Gustav
Zacchei, Anthony Gabriel
Zaika, Laura Larysa
Zalipsky, Jerome Jaroslaw

RHODE ISLAND
Brown, Christopher W
Cruickshank, Alexander Middleton
Fasching, James Le Roy
Gearing, Juanita Newman
Huebert, Barry Joe
Kirschenbaum, Louis Jean
Kreiser, Ralph Rank
Martin, Horace F
Morris, George V
Moyerman, Robert Max
Quinn, James Gerard
Swift, Dorothy Garrison
Walsh, John Thomas
Zuehlke, Richard William

SOUTH CAROLINA
Asleson, Gary Lee
Baumann, Elizabeth Wilson
Bidleman, Terry Frank
Blood, Elizabeth Reid
Clayton, Fred Ralph, Jr
Cogswell, George Wallace
Davis, Joseph B
Davis, Raymond F
Deanhardt, Marshall Lynn
Edwards, John C
Elzerman, Alan William
Emrick, Edwin Roy
Farmer, Larry Bert
Fulda, Myron Oscar
Goode, Julia Pratt
Goode, Scott Roy
Hochel, Robert Charles
Hofstetter, Kenneth John
Holcomb, Herman Perry
Hopper, Michael James
Hunter, George William
Jacobs, William Donald
Kendall, David Nelson
Kinard, W Frank
King, Lee Curtis
Knapp, Daniel Roger
Lincoln, David Erwin
Lovins, Robert E
McFarren, Earl Francis
McGill, Julian Edward
Maier, Herbert Nathaniel
Malstrom, Robert Arthur
Morrow, William Scot
Oswald, Edward Odell
Peterson, Stephen Frank
Philp, Robert Herron, Jr
Pike, LeRoy
Rains, Theodore Conrad
Sargent, Roger N
Sheridan, Jane Connor
Stampf, Edward John, Jr
Wilhite, Elmer Lee
Wynn, James Elkanah

SOUTH DAKOTA
Hilderbrand, David Curtis
Looyenga, Robert William

TENNESSEE
Bhattacharya, Syamal Kanti
Blanck, Harvey F, Jr
Burtis, Carl A, Jr
Caflisch, Edward George
Caflisch, George Barrett

Cain, Carl, Jr
Chambers, James Q
Christie, Warner Howard
Commerford, John D
Cook, Kelsey Donald
Dean, John Aurie
Desiderio, Dominic Morse, (Jr)
Dillard, James William
Dilts, Robert Voorhees
Duncan, Budd Lee
Dyer, Frank Falkoner
Engel, Adolph James
Ferguson, Robert Lynn
Franklin, James Curry
Guerin, Michael Richard
Hahn, Richard Balser
Hall, Larry Cully
Hicks, Jackson Earl
Honaker, Carl Boggess
Horton, Charles Abell
Jarrett, Harry Wellington, III
Johnson, Eric Robert
Klatt, Leon Nicholas
Lynch, John August
Lyons, Harold
McDowell, William Jackson
McNeely, Robert Lewis
Mahlman, Harvey Arthur
Mamantov, Gleb
Maya, Leon
Miller, Francis Joseph
Morie, Gerald Prescott
Morrow, Roy Wayne
Mueller, Theodore Rolf
Nicely, Vincent Alvin
Nyssen, Gerard Allan
Osborne, Charles Edward
Otis, Marshall Voigt
Patterson, Truett Clifton
Payne, DeWitt Allen
Raasch, Lou Reinhart
Rice, Walter Wilburn
Ross, Harley Harris
Rutenberg, Aaron Charles
Scott, Dan Dryden
Scroggie, Lucy E
Shah, Jitendra J
Shults, Wilbur Dotry, II
Smith, David Huston
Sweetman, Brian Jack
Taylor, Ellison Hall
Taylor, Kirman
Todd, Peter Justin
Torrey, Rubye Prigmore
Uziel, Mayo
Vo-Dinh, Tuan
Weber, Charles William
Wehry, Earl L, Jr
Whetsel, Kermit Bazil
White, David Cleaveland
White, James Carl
Wignall, George Denis
Wiser, James Eldred
Wojciechowski, Norbert Joseph
Yoakum, Anna Margaret
Young, Jack Phillip

TEXAS
Adams, Martha Lovell
Adkins, John Earl, Jr
Allison, Jean Batchelor
Alsop, John Henry, III
Anselmo, Vincent C
Ayres, Gilbert Haven
Baker, Charles Taft
Banta, Marion Calvin
Bard, Allen Joseph
Bartsch, Richard Allen
Bastiaans, Glenn John
Batten, Charles Francis
Beck, Benny Lee
Beier, Ross Carlton
Benson, Royal H
Betso, Stephen Richard
Bhatia, Kishan
Blay, George Albert
Botto, Robert Irving
Braterman, Paul S
Bushey, Michelle Marie
Campbell, Dan Norvell
Caprioli, Richard Michael
Cates, Vernon E
Chamberlain, Nugent Francis
Chen, Edward Chuck-Ming
Cogswell, Howard Winwood, Jr
Cole, Larry Lee
Cruser, Stephen Alan
Cummiskey, Charles
Daugherty, Kenneth E
Deming, Stanley Norris
DiFoggio, Rocco
Drake, Edgar Nathaniel, II
Dunn, Danny Leroy
Edgerley, Dennis A
Elder, Vincent Allen
Elthon, Donald L
Emerson, David Edwin
Faris, Sam Russell
Floyd, Willis Waldo
Foster, Norman George
Fowler, Robert McSwain
Frazee, Jerry D
Fritsche, Herbert Ahart, Jr
Fuller, Martin Emil

Gerlach, John Louis
Ghowsi, Kiumars
Gibson, Everett Kay, Jr
Grant, Clarence Lewis
Grogan, Michael John
Gupta, Vishnu Das
Hall, Randall Clark
Harris, Edward Lyndol
Hart, Haskell Vincent
Harvey, Mack Creede
Hausler, Rudolf H
Holcombe, James Andrew
Horning, Evan Charles
Howard, Charles
Humphrey, Ray Eicken
Hunt, Richard Henry
Iddings, Frank Allen
Iskander, Felih Youssef
Johnson, Ralph Alton
Jones, Lawrence Ryman
Jones, Llewellyn Claiborne, Jr
Jungclaus, Gregory Alan
Kadish, Karl Mitchell
Karchmer, Jean Herschel
Kaye, Howard
Kenner, Charles Thomas
Keyworth, Donald Arthur
Lee, George H, II
Le Febvre, Edward Ellsworth
Lerner, Melvin
Lewis, Donald Richard
Liehr, Joachim G
Lin-Vien, Daimay
Luttrell, George Howard
McDougall, Robert I
Malloy, Thomas Bernard, Jr
Mansfield, Clifton Tyler
Markwell, Dick Robert
Martin, Charles R
Maute, Robert Lewis
Mehta, Rajen
Melton, James Ray
Melton, Marilyn Anders
Mendez, Victor Manuel
Middleditch, Brian Stanley
Millar, John David
Miller, James Richard
Moltzan, Herbert John
Moreland, Ferrin Bates
Mosier, Benjamin
Muhs, Merrill Arthur
Nelson, Ivory Vance
Newton, Robert Andrew
Patel, Bhagwandas Mavjibhai
Patton, Leo Wesley
Peurifoy, Paul Vastine
Poe, Richard D
Pogue, Randall F
Posey, Daniel Earl
Rajeshwar, Krishnan
Ray, Allen Cobble
Reed, John J R
Reinecke, Manfred Gordon
Rekers, Robert George
Rhodes, Robert Carl
Ricca, Paul Joseph
Rivera, William Henry
Robinson, J(ames) Michael
Robinson, Robert Eugene
Rodriguez, Charles F
Rollwitz, William Lloyd
Saleh, Farida Yousry
Sand, Ralph E
Schweikert, Emile Alfred
Scott, Robert Edward
Shelly, Dennis C
Shore, Fred L
Sikes, James Klingman
Snapp, Thomas Carter, Jr
Southern, Thomas Martin
Spallholz, Julian Ernest
Starks, Aubrie Neal, Jr
Stoner, Graham Alexander
Stubbeman, Robert Frank
Thompson, Richard John
Tibbals, Harry Fred, III
Van Dreal, Paul Arthur
Varsel, Charles John
Vien, Steve Hung
Walters, Lester James, Jr
Wang, Ting-Tai Helen
Wendel, Carlton Tyrus
Wendlandt, Wesley W
Wentworth, Wayne
Whiteside, Charles Hugh
Wilson, Ray Floyd
Windham, Ronnie Lynn
Winfrey, J C
Worthington, James Brian
Yerick, Roger Eugene
Zlatkis, Albert

UTAH
Adler, Robert Garber
Brauner, Kenneth Martin
Bunger, James Walter
Burdett, Lorenzo Worth
Burton, Frederick Glenn
Butler, Eliot Andrew
Comeford, John J
Eatough, Delbert J
Eyring, Edward M
Farnsworth, Paul Burton
Garrard, Verl Grady

Goates, Steven Rex
Grey, Gothard C
Hall, David Warren
Harris, Joel Mark
James, Helen Jane
Janata, Jiri
McKenzie, Jess Mack
Mangelson, Nolan Farrin
Myers, Marcus Norville
Parry, Edward Petterson
Phillips, Lee Revell
Spence, Jack Taylor
Tuddenham, W(illiam) Marvin
Walker, Donald I
Wallace, Volney
Woolley, Earl Madsen

VERMONT
Abajian, Paul G
Campbell, Donald Edward
Geiger, William Ebling, Jr
Keenan, Robert Gregory
Kellner, Stephan Maria Eduard
Pipenberg, Kenneth James
Pool, Edwin Lewis
Provost, Ronald Harold
Shearer, Charles M
Tremmel, Carl George

VIRGINIA
Allen, Ralph Orville, Jr
Anderson, Carl William (Bill)
Anderson, Samuel
Armstrong, Alfred Ringgold
Baedecker, Philip A
Bartschmid, Betty Rains
Beyad, Mohammed Hossain
Bondurant, Charles W, Jr
Brandt, Richard Bernard
Breder, Charles Vincent
Burton, Willard White
Cantu, Antonio Arnoldo
Chandler, Carl Davis, Jr
Clarke, Alan R
Dalton, Roger Wayne
Decorpo, James Joseph
Demas, James Nicholas
Dessy, Raymond Edwin
Dominey, Raymond Nelson
Dyer, Randolph H
Einolf, William Noel
Fabbi, Brent Peter
Feinstein, Hyman Israel
Ferguson, Robert Nicholas
Fischbach, Henry
Fonong, Tekum
Glanville, James Oliver
Greene, Virginia Carvel
Grossman, Jeffrey N
Grunder, Fred Irwin
Gushee, Beatrice Eleanor
Harvey, William Ross
Hassler, William Woods
Hausermann, Max
Hawkridge, Fred Martin
Howard, John William
Huggett, Robert James
Hunt, Donald F
Kuhn, William Frederick
Kumari, Durga
Lewis, Cornelius Crawford
Leyden, Donald E
Link, William B
Loda, Richard Thomas
Macko, Stephen Alexander
McNair, Harold Monroe
Mahoney, Bernard Launcelot, Jr
Marcy, Joseph Edwin
Martin, Albert Edwin
Mason, John Grove
Mefford, David Allen
Meites, Louis
Moody, John Robert
Morgan, Evan
Morgan, John Walter
Olin, Jacqueline S
Parsons, James Sidney
Powell, William Allan
Price, Byron Frederick
Schnepfe, Marian Moeller
Settle, Frank Alexander, Jr
Shaheen, Donald G
Simon, Frederick Otto
Sobol, Stanley Paul
Spies, Joseph Reuben
Sprinkle, Robert Shields, III
Stewart, Kent Kallam
Tiedemann, Albert William, Jr
Van Norman, John Donald
Van't Riet, Bartholomeus
Vilcins, Gunars
Watson, Duane Craig
Whidby, Jerry Frank
Wickham, James Edgar, Jr
Will, Fritz, III
Wood, George Marshall

WASHINGTON
Alberts, Gene S
Anderson, Donald Hervin
Ballou, Nathan Elmer
Barnes, Edwin Ellsworth
Behm, Roy
Berry, Keith O

Analytical Chemistry (cont)

Briggs, William Scott
Butler, Lillian Ida
Butts, William Cunningham
Callis, James Bertram
Campbell, Milton Hugh
Christian, Gary Dale
Cormack, James Frederick
Daniel, J(ack) Leland
Diebel, Robert Norman
Duncan, Leonard Clinton
Easty, Dwight Buchanan
Ericsson, Lowell Harold
Felicetta, Vincent Frank
Felton, Samuel Page
Fordyce, David Buchanan
Frank, Andrew Julian
Gahler, Arnold Robert
Godar, Edith Marie
Goheen, Steven Charles
Goldring, Lionel Solomon
Healy, Michael L
Hill, Herbert Henderson, Jr
Huestis, Laurence Dean
Ingalsbe, David Weeden
Jones, Jerry Lynn
Jones, Thomas Evan
Kalkwarf, Donald Riley
Kaye, James Herbert
King, Donald M
Koehmstedt, Paul Leon
Kowalski, Bruce Richard
Mehlhaff, Leon Curtis
Mopper, Kenneth
Noyes, Claudia Margaret
Pool, Karl Hallman
Radziemski, Leon Joseph
Rasco, Barbara A
Robinson, Rex Julian
Scott, Frederick Arthur
Serne, Roger Jeffrey
Shull, Charles Morell, Jr
Sklarew, Deborah S
Smith, Richard Dale
Smith, Robert Victor
Spinelli, John
Stevens, Vernon Lewis
Stout, Virginia Falk
Stromatt, Robert Weldon
Taylor, Murray East
Warner, John Scott
Warner, Ray Allen
Weyh, John Arthur
Zimmerman, Gary Alan

WEST VIRGINIA
Bhasin, Madan M
Chadwick, David Henry
Das, Kamalendu
Dawson, Thomas Larry
Dunphy, James Francis
Elliston, John E
Finklea, Harry Osborn
Fisher, John F
Glick, Charles Frey
Kalb, G William
Lamey, Steven Charles
Macnaughtan, Donald, Jr
Martin, John Perry, Jr
Ode, Richard Herman
Pribble, Mary Jo
Sandridge, Robert Lee
Smith, James Allbee
Sperati, Carleton Angelo

WISCONSIN
Alvarez, Robert
Baetke, Edward A
Bayer, Richard Eugene
Bekersky, Ihor
Berge, Douglas G
Beyerlein, Floyd Hilbert
Blaedel, Walter John
Broderick, Glen Allen
Clark, Harlan Eugene
Conigliaro, Peter James
Connors, Kenneth A
Constant, Marc Duncan
Doerr, Robert George
Drexler, Edward James
Eckert, Alfred Carl, Jr
Eggert, Arthur Arnold
Evenson, Merle Armin
Ewald, Fred Peterson, Jr
Feinberg, Benjamin Allen
Hansen, Robert Conrad
Hassinger, Mary Colleen
Hoffman, Norman Edwin
Hynek, Robert James
Kao, Wen-Hong
Kehres, Paul W(illiam)
Kincaid, James Robert
Knight, Homer Talcott
Landucci, Lawrence L
Lanterman, Elma
Larson, Wilbur S
Lemm, Arthur Warren
Liang, Shoudeng
Lindsay, Robert Clarence
Markley, John Lute
Parnell, Donald Ray
Phillips, John Spencer
Polcyn, Daniel Stephen

Priest, Matthew A
Propp, Jacob Henry
Qureshi, Nilofer
Rainville, David Paul
Ritter, Garry Lee
Rosenthal, Jeffrey
Scheppers, Gerald J
Schwab, Helmut
Scott, Lawrence William
Sommers, Raymond A
Stone, William Ellis
Taylor, James Welch
Taylor, Paul John
Trischan, Glenn M
West, Kevin J
West, Kevin James
Wiersma, James H
Wright, John Curtis
Wright, Steven Martin
Zinkel, Duane Forst

WYOMING
Archer, Vernon Shelby
Dorrence, Samuel Michael
Duvall, John Joseph
Hurtubise, Robert John
Netzel, Daniel Anthony
Petersen, Joseph Claine
Poulson, Richard Edwin
Sweet, David Paul

PUERTO RICO
ALZERRECA, ARNALDO
Carrasquillo, Arnaldo
Infante, Gabriel A
Rose, Stuart Alan
Stephens, William Powell

ALBERTA
Birss, Viola Ingrid
Cantwell, Frederick Francis
Codding, Edward George
D'Agostino, Paul A
Harris, Walter Edgar
Hill, Bernard Dale
Hogg, Alan Mitchell
Horlick, Gary
Kratochvil, Byron
Plambeck, James Alan
Yamdagni, Raghavendra
Yeager, Howard Lane

BRITISH COLUMBIA
Anderson, John Ansel
D'Auria, John Michael
Hocking, Martin Blake
Law, Francis C P
McAuley, Alexander
Oliver, Barry Gordon
Wade, Adrian Paul
Yunker, Mark Bernard

MANITOBA
Attas, Ely Michael
Bigelow, Charles C
Chow, Arthur
Kovacs, Miklos I P
Lange, Bruce Ainsworth
Muir, Derek Charles G
Saluja, Preet Pal Singh
Stewart, Reginald Bruce
Tkachuk, Russell
Vandergraaf, Tjalle T
Williams, Philip Carslake

NEW BRUNSWICK
Brewer, Douglas G
Grant, Douglas Hope
Mallet, Victorin Noel

NEWFOUNDLAND
Georghiou, Paris Elias
Longerich, Henry Perry
Newlands, Michael John
Stein, Allan Rudolph

NOVA SCOTIA
Ackman, Robert George
Addison, Richard Frederick
Aue, Walter Alois
Bridgeo, William Alphonsus
Chatt, Amares
Gilgan, Michael Wilson
Holzbecher, Jiri
Jamieson, William David
Odense, Paul Holger
Ramaley, Louis
Ryan, Douglas Earl
Stephens, Roger
Stiles, David A

ONTARIO
Afghan, Baderuddin Khan
Arsenault, Guy Pierre
Atkinson, George Francis
Bannard, Robert Alexander Brock
Baptista, Jose Antonio
Berman, Shier
Bhatnagar, Dinech C
Bioleau, Luc J R
Boorn, Andrew William
Chakrabarti, Chuni Lal
Chau, Alfred Shun-Yuen
Chiba, Mikio

Corsini, A
Court, William Arthur
Crocker, Iain Hay
Cross, Charles Kenneth
Currah, Jack Ellwood
Dick, James Gardiner
Dumbroff, Erwin Bernard
Dunford, Raymond A
Foster, M Glenys
French, J(ohn) Barry
Goulden, Peter Derrick
Graham, Ronald Powell
Greenhalgh, Roy
Greenstock, Clive Lewis
Guevremont, Roger M
Gulens, Janis
Harrison, Alexander George
Hay, George William
Heggie, Robert Murray
Hileman, Orville Edwin, Jr
Hill, Martha Adele
Hogan, Gary D
Holland, William John
Holloway, Clive Edward
Ihnat, Milan
Jardine, John McNair
Kaiser, Klaus L(eo) E(duard)
Karasek, Francis Warren
Kelly, Patrick Clarke
Logan, James Edward
Loughheed, Thomas Crossley
Lucas, Douglas M
Marshall, Heather
Martin, Trevor Ian
Maynes, Albion Donald
Meresz, Otto
Miller, Jack Martin
Onuska, Francis Ivan
Pang, Henrianna Yicksing
Pare, J R Jocelyn
Perrin, Carrol Hollingsworth
Ramachandran, Vangipuram S
Reffes, Howard Allen
Russell, Douglas Stewart
Sahasrabudhe, Madhu R
Sandler, Samuel
Schroeder, William Henry
Scott, Peter Michael
Sekerka, Ivan
Shah, Bhagwan G
Shearer, Duncan Allan
Singer, Eugen
Sowa, Walter
Stairs, Robert Ardagh
Stillman, Martin John
Thibert, Roger Joseph
Ulpian, Carla
Westaway, Kenneth C
Westland, Alan Duane
Wolkoff, Aaron Wilfred
Wood, Gordon Walter
Wright, Maurice Morgan
Yang, Paul Wang
Young, James Christopher F
Zakaib, Daniel D

QUEBEC
Allen, Lawrence Harvey
Bélanger, Jacqueline M R
Belanger, Patrice Charles
Borgeat, Pierre
Desjardins, Claude W
Dorris, Gilles Marcel
Douek, Maurice
Feng, Rong
Fresco, James Martin
Goldstein, Sandu M
Gurudata, Neville
Lennox, Robert Bruce
Lennox, Robert Bruce
Panalaks, Thavil
Pelletier, Omer
Powell, William St John
Power, Joan F
Purdy, William Crossley
Roy, Jean-Claude
Verschingel, Roger H C
Yates, Claire Hilliard
Zienius, Raymond Henry

SASKATCHEWAN
Cassidy, Richard Murray
Durden, David Alan
Korsrud, Gary Olaf
Larson, D Wayne

OTHER COUNTRIES
Andreae, Meinrat Otto
Balakrishnan, Narayana Swamy
Bess, Robert Carl
Brown, Charles Eric
Buckwold, Sidney Joshua
Chang, Shuya
Charola, Asuncion Elena
Cheung, Mo-Tsing Miranda
Collins, Carol Hollingworth
Collins, Kenneth Elmer
Edmonds, James W
Finlayson, James Bruce
Guerrero, Ariel Heriberto
Haegele, Klaus D
Hasty, Robert Armistead
Irgolic, Kurt Johann
Lewis, Arnold Leroy, II

Lingane, Peter James
Loeliger, David A
Lonsdale-Eccles, John David
Milner, David
Perry, John Arthur
Raveendran, Ekarath
Winnett, George

Biochemistry

ALABAMA
Bennett, Leonard Lee, Jr
Bertsch, Wolfgang
Breithaupt, Lea Joseph, Jr
Cheung, Herbert Chiu-Ching
Dorai-Raj, Diana Glover
Friedman, Michael E
Funkhouser, Jane D
Furuto, Donald K
Ho, Mat H
Kendrick, Aaron Baker
Krishna, N Rama
Legg, Ivan
McCarthy, Dennis Joseph
Miller, George Paul
Parks, Paul Franklin
Prince, Charles William
Pritchard, David Graham
Sakai, Ted Tetsuo
Secrist, John Adair, III
Tan, Boen Hie
Timkovich, Russell
Truelove, Bryan
Weete, John Donald
Wilken, Leon Otto, Jr

ALASKA
Fink, Thomas Robert
Genaux, Charles Thomas
Kennish, John M
Lane, Robert Harold

ARIZONA
Bates, Robert Wesley
Blankenship, Robert Eugene
Chvapil, Milos
Cronin, John Read
Diehn, Bodo
Foltz, Thomas Roberts, Jr
Gunderson, Hans Magelssen
Harris, Leland
Lohr, Dennis Evan
Morton, Howard LeRoy
Rose, Seth David
Sander, Eugene George
Sanderson, Douglas G
Smith, Cecil Randolph, Jr
Tollin, Gordon
Valenty, Vivian Briones
Yamamura, Henry Ichiro

ARKANSAS
Beranek, David T
Doyle, Miles Lawrence
Everett, Wilbur Wayne
Fu, Peter Pi-Cheng
Geren, Collis Ross
Goodwin, James Crawford
Jones, Joe Maxey
Koeppe, Roger E, II
Ku, Victoria Feng
Thoma, John Anthony
Winter, Charles Gordon

CALIFORNIA
Albert, Jerry David
Allison, William S
Aloia, Roland C
Amoore, John Ernest
Ansari, Ali
Atwood, Linda
Baker, Frederick Charles
Baldwin, Robert Lesh
Baptist, Victor Harry
Barton, Jacqueline K
Becker, David Morris
Bernardin, John Emile
Bigler, William Norman
Bird, Harold L(eslie), Jr
Blank, Gregory Scott
Bond, Martha W(illis)
Bovard, Freeman Carroll
Boxer, Steven George
Boyer, Paul Delos
Brostoff, Steven Warren
Bruice, Thomas Charles
Burri, Betty Jane
Cabot, Myles Clayton
Campagnoni, Anthony Thomas
Cantor, Charles Robert
Chan, Bock G
Chan, David S
Chan, Sunney Ignatius
Chang, George Washington
Chang, Yi-Han
Chao, Li-Pen
Cheng, Sze-Chuh
Chenoweth, Dennis Edwin
Choong, Hsia Shaw-Lwan
Choudary, Prabhakara Velagapudi
Christianson, Michael Lee
Chuong, Cheng-Ming
Codrington, Robert Smith

Cole, Roger David
Conover, Woodrow Wilson
Coyle, Bernard Andrew
Craik, Charles S
Cronin, Michael John
Dafforn, Geoffrey Alan
Dahms, Arthur Stephen
Davis, Ward Benjamin
Dawson, Marcia Ilton
DeLange, Robert J
Delk, Ann Stevens
Dennis, Edward A
Deonier, Richard Charles
Dickerson, Richard Earl
Diederich, Francois Nico
Dunaway, Marietta
Eiduson, Samuel
Einstein, Elizabeth Roboz
Eisenberg, David
Ernst, Roberta Dorothea
Feeney, Robert Earl
Felgner, Philip Louis
Fink, Anthony Lawrence
Fleischer, Everly B
Fowler, Audree Vernee
Freedland, Richard A
Fukuda, Minoru
Fuller, Glenn
Gagne, Robert Raymond
Galaway, Ronald Alvin
Garrison, Warren Manford
Geller, Edward
George-Nascimento, Carlos
Gerig, John Thomas
Gilleland, Martha Jane
Gish, Duane Tolbert
Glaser, Robert J
Glick, David
Grassetti, Davide Riccardo
Green, Donald Eugene
Grimes, Carol Jane Galles
Gum, James Raymond, Jr
Hajdu, Joseph
Hamilton, Carole Lois
Hansen, Robert John
Hearst, John Eugene
Heath, Robert Louis
Hill, Terrell Leslie
Hinkson, Jimmy Wilford
Ho, Begonia Y
Hoke, Glenn Dale
Hong, Keelung
Hooker, Thomas M, Jr
Horowitz, Sylvia Teich
Hugli, Tony Edward
Hurd, Ralph Eugene
Hutchin, Maxine E
Iacono, James M
Jacob, Peyton, III
Jacobson, Ralph Allen
James, Thomas Larry
Jardine, Ian
Jensen, Ronald Harry
Jewett, Sandra Lynne
Johnson, Herman Leonall
Jones, Richard Elmore
Kasarda, Donald David
Kavenoff, Ruth
Keelung, Hong
Kenyon, George Lommel
Khanna, Pyare Lal
King, A Douglas, Jr
Klinman, Judith Pollock
Kohler, Heinz
Kosow, David Phillip
Koziol, Brian Joseph
Kraut, Joseph
Kuhlmann, Karl Frederick
Kuwahara, Steven Sadao
Ladisch, Stephan
Langerman, Neal Richard
Lee, David Louis
Leighton, Terrance J
Levy, Daniel
Lin, Jiann-Tsyh
Linder, Maria C
Lindner, Elek
Lipka, James J
Lippmann, Wilbur
Loew, Gilda Harris
Lohrmann, Rolf
Lorance, Elmer Donald
McConnell, Harden Marsden
McGee, Lawrence Ray
McKenna, Charles Edward
Maggio, Edward Thomas
Maier, V(incent) P(aul)
Maki, August Harold
Mathewson, James H
Matthews, David Allan
Metzger, Robert P
Miller, Jon Philip
Miller, Sol
Minch, Michael Joseph
Mitoma, Chozo
Miyada, Don Shuso
Mosher, Carol Walker
Nagel, Glenn M
Nambiar, Krishnan P
Neilands, John Brian
Ngo, That Tjien
Ning, Robert Y
Nitecki, Danute Emilija
Osborn, Terry Wayne

Parsons, Stanley Monroe
Paselk, Richard Alan
Patton, Stuart
Payne, Philip Warren
Pellegrini, Maria C
Perry, L Jeanne
Peter, James Bernard
Peters, Marvin Arthur
Petersen, Gene
Pettit, David J
Pigiet, Vincent P
Poling, Stephen Michael
Riffer, Richard
Riley, Richard Fowble
Ritzmann, Ronald Fred
Roberts, Sidney
Robins, Roland Kenith
Roche, George William
Rosenberg, Lawson Lawrence
Rosenthal, Allan Lawrence
Rynd, James Arthur
Sanchez, Robert A
Sartoris, David John
Sauer, Kenneth
Sayre, Robert Newton
Schein, Arnold Harold
Schmid, Carl William
Schmidt, Paul Gardner
Schumaker, Verne Norman
Scogin, Ron Lynn
Sendroy, Julius, Jr
Shafer, Richard Howard
Shaffer, Patricia Marie
Shen, Wei-Chiang
Siegel, Brock Martin
Silberstein, Otmar Otto
Slattery, Charles Wilbur
Solomon, Edward I
Starr, Mortimer Paul
Stumph, William Edward
Sullivan, John Lawrence
Swinehart, James Herbert
Thomas, Charles Allen, Jr
Thorner, Jeremy William
Tinoco, Joan W H
Tomich, John Matthew
Trevor, Anthony John
Tutupalli, Lohit Venkateswara
Ullman, Edwin Fisher
Vandlen, Richard Lee
Varon, Silvio Salomone
Vedvick, Thomas Scott
Vinogradoff, Anna Patricia
Walters, Roland Dick
Ward, Raymond Leland
Weber, Bruce Howard
Weiss, Theodore Joel
Weissman, Norman
Westcott, Keith R
White, David Halbert
Wikman-Coffelt, Joan
Willhite, Calvin Campbell
Wisnieski, Bernadine Joann
Wolfgram, Frederick John
Wong, Chi-Huey
Wright, Ernest Marshall
Wu, Sing-Yung
Wynston, Leslie K
Yang, Jen Tsi
Young, Ho Lee
Zahnley, James Curry
Zimm, Bruno Hasbrouck

COLORADO
Allen, Kenneth G D
Barisas, Bernard George, Jr
Beck, Steven R
Cann, John Rusweiler
Cech, Carol Martinson
Cech, Thomas Robert
Collins, Allan Clifford
Dewey, Thomas Gregory
Dynan, William Shelley
Elliott, Cecil Michael
Fitzpatrick, Francis Anthony
Freed, John Howard
Gal, Joseph
Gill, Stanley Jensen
Levinson, Simon Rock
Lin, Leu-Fen Hou
Martin, Susan Scott
Miller, William Theodore
Pitts, Malcolm John
Repine, John E
Schonbeck, Niels Daniel
Stushnoff, Cecil
Taber, Richard Lawrence
Uhlenbeck, Olke Cornelis
Uhlenhopp, Elliott Lee
Woody, A-Young Moon
York, Sheldon Stafford

CONNECTICUT
Adler, Alan David
Allewell, Norma Mary
Armitage, Ian MacLeod
Beardsley, George Peter
Berube, Gene Roland
Boell, Edgar John
Branchini, Bruce Robert
Braswell, Emory Harold
Celmer, Walter Daniel
Chaturvedi, Rama Kant
Chello, Paul Larson

Crothers, Donald M
Doyle, Terrence William
Galston, Arthur William
Geiger, Edwin Otto
Gent, Martin Paul Neville
Godchaux, Walter, III
Gomez-Cambronero, Julian
Grattan, James Alex
Haake, Paul
Haakonsen, Harry Olav
Hook, Derek John
I, Ting-Po
Knox, James Russell, Jr
Lamm, Foster Philip
Langner, Ronald O
Lee, Henry C
Phillips, Alvah H
Philo, John Sterner
Prestegard, James Harold
Prigodich, Richard Victor
Putterman, Gerald Joseph
Richards, Frederic Middlebrook
Rooney, Seamus Augustine
Ross, Donald Joseph
Sartorelli, Alan Clayton
Schuster, Todd Mervyn
Skinner, H Catherine W
Sturtevant, Julian Munson
Taylor, Duncan Paul
Tishler, Max
Verdi, James L

DELAWARE
Anderson, Jeffrey John
Beer, John Joseph
Beyer, Elmo Monroe, Jr
Billheimer, Jeffrey Thomas
Burton, Lester Percy Joseph
Colman, Roberta F
Confalone, Pat N
Davis, Leonard George
DeGrado, William F
Domaille, Peter John
Duke, Jodie Lee, Jr
Ehrlich, Robert Stark
Freerksen, Deborah Lynne (Chalmers)
Galbraith, William
Hartzell, Charles Ross, III
Herron, Norman
Hodges, Charles Thomas
Johnson, Donald Richard
Kettner, Charles Adrian
Knowles, Francis Charles
Larsen, Barbara Seliger
Lerman, Charles Lew
Link, Bernard Alvin
Malley, Linda Angevine
Moe, Gregory Robert
Morrissey, Bruce William
Nelson, Mark James
Sariaslani, Sima
Schloss, John Vinton
Sharp, Jonathan Hawley
Smith, Jerry Howard
Taves, Milton Arthur
Thompson, Jeffery Scott
Thoroughgood, Carolyn A
Thorpe, Colin
Wetlaufer, Donald Burton
Whitney, Charles Candee, Jr

DISTRICT OF COLUMBIA
Adomaitis, Vytautas Albin
Baker, Charles Wesley
Bednarek, Jana Marie
Brown, James Edward
Doctor, Bhupendra P
El Khadem, Hassan S
Field, Ruth Bisen
Gaber, Bruce Paul
Garavelli, John Stephen
Hare, Peter Edgar
Holt, James Allen
Isbell, Horace Smith
Jenkins, Mamie Leah Young
Karle, Jean Marianne
Khanna, Krishan L
Kowalsky, Arthur
Lerner, Pauline
May, Leopold
Miles, Corbin I
Mog, David Michael
Mullin, Brian Robert
O'Leary, Timothy Joseph
Perros, Theodore Peter
Pierce, Dorothy Helen
Rabin, Robert
Rogers, Senta S(tephanie)
Rosenberg, Robert Charles
Smith, Thomas Elijah
Tallent, William Hugh
Walsh, Stephen G
Weintraub, Robert Louis
Welt, Isaac Davidson
Woolridge, Edward Daniel
Yang, David Chih-Hsin
Zaborsky, Oskar Rudolf

FLORIDA
Ahmad, Fazal
Bash, Paul Anthony
Blossey, Erich Carl
Bowman, Ray Douglas
Chun, Paul W

Cross, Timothy Albert
Davison, Clarke
Dean, David Devereaux
Diamond, Steven Elliot
Dobbins, Robert Joseph
Dunn, Ben Monroe
Eichler, Duane Curtis
Elfenbein, Gerald Jay
Ferl, Robert Joseph
Finger, Kenneth F
Frishman, Daniel
Fuller, Harold Wayne
Garrett, Edward Robert
Gilmer, Penny Jane
Goldstein, Arthur Murray
Grossman, Steven Harris
Hackney, John Franklin
Halpern, Ephriam Philip
Hanks, Robert William
Henson, Carl P
Holt, Thomas Manning
Huber, Joseph William, III
Lee, Henry Joung
Leibman, Kenneth Charles
Man, Eugene H
Marzluff, William Frank, Jr
Meloche, Henry Paul
Miles, Richard David
Nagy, Steven
Olsen, Eugene Donald
Owen, Terence Cunliffe
Price, David Alan
Pritham, Gordon Herman
Rill, Randolph Lynn
Schleyer, Walter Leo
Schoor, W Peter
Schultz, Duane Robert
Shiverick, Kathleen Thomas
Smith, Dennis Eugene
Stewart, Ivan
Templer, David Allen
Trefonas, Louis Marco
Vander Meer, Robert Kenneth
White, Frederick Howard, Jr
Wickstrom, Eric
Williams, Joseph Francis
Winkler, Bruce Conrad
Yu, Simon Shyi-Jian

GEORGIA
Abdel-Latif, Ata A
Anderson, James Leroy
Bain, James Arthur
Batra, Gopal Krishan
Dervartanian, Daniel Vartan
Etzler, Frank M
Evans, Donald Lee
Habte-Mariam, Yitbarek
Hodges, Linda Carol
Holzer, Gunther Ulrich
Hommes, Frits A
Iacobucci, Guillermo Arturo
Jain, Anant Vir
Kuck, John Frederick Read, Jr
Mahesh, Virendra B
May, Sheldon William
Mickelsen, Olaf
Moinuddin, Jessie Fischer
Morse, Stephen Allen
Mortenson, Leonard Earl
Nes, William David
Pohl, Douglas George
Powers, James Cecil
Ray, Apurba Kanti
Reiss, Errol
Rhoades, James Lawrence
Scott, Robert Allen
Shapira, Raymond
Steele, Jack
Torres, Lourdes Maria
Trawick, William George
Wilson, William David
Worthington, Robert Earl
Yu, Nai-Teng

HAWAII
Doerge, Daniel Robert
Loo, Yen-Hoong
Moritsugu, Toshio
Seifert, Josef
Tang, Chung-Shih
Vennesland, Birgit

IDAHO
Dreyfus, Pierre Marc
Elmore, John Jesse, Jr
Goettsch, Robert
Keay, Leonard
Lehrer, William Peter, Jr

ILLINOIS
Anderson, Robert Lewis
Bachrach, Joseph
Baich, Annette
Bietz, Jerold Allen
Bolen, David Wayne
Brubaker, George Randell
Burt, Charles Tyler
Cannon, John Burns
Chen, Wen Sherng
Chignell, Derek Alan
Cifonelli, Joseph Anthony
Dikeman, Roxane Norris
Doane, William M

Biochemistry (cont)

Donnelly, Thomas Henry
Drengler, Keith Allan
Druse-Manteuffel, Mary Jeanne
Dybel, Michael Wayne
Emken, Edward Allen
Erman, James Edwin
Eskins, Kenneth
Farnsworth, Norman R
Ferren, Larry Gene
Foster, George A, Jr
Friedman, Robert Bernard
Gardner, Harold Wayne
Gennis, Robert Bennett
Gould, John Michael
Griffin, Harold Lee
Grotefend, Alan Charles
Gumport, Richard I
Hagar, Lowell Paul
Harris, Donald Wayne
Haselkorn, Robert
Hathaway, Robert J
Held, Irene Rita
Henkin, Jack
Herting, Robert Leslie
Hodge, John Edward
Hoepfinger, Lynn Morris
Hynes, John Thomas
Iqbal, Zafar
Jones, Marjorie Ann
Kass, Guss Sigmund
Katzenellenbogen, Benita Schulman
Katzenellenbogen, John Albert
Klotz, Irving Myron
Koritala, Sanbasivaroa
Kornel, Ludwig
Kresheck, Gordon C
Kumar, Sudhir
Lambert, Mary Pulliam
Lehman, Dennis Dale
Leonard, Nelson Jordan
Lerner, Louis L(eonard)
Loach, Paul A
Lopatin, William
Makinen, Marvin William
Margoliash, Emanuel
Martinek, Robert George
Matthews, Clifford Norman
Medcalf, Darrell Gerald
Melhado, L(ouisa) Lee
Mimnaugh, Michael Neil
Molotsky, Hyman Max
Moore, Edwin Granville
Moser, Kenneth Bruce
Mota de Freitas, Duarte Emanuel
Needleman, Saul Ben
Neet, Kenneth Edward
Olsen, Kenneth Wayne
Ostrow, Jay Donald
Panasenko, Sharon Muldoon
Papaioannou, Stamatios E
Parker, Helen Meister
Parsons, Carl Michael
Paschall, Eugene F
Paulson, Glenn
Peterson, Melbert Eugene
Portnoy, Norman Abbye
Pryde, Everett Hilton
Reynolds, Rosalie Dean (Sibert)
Rinehart, Kenneth Lloyd
Rotenberg, Keith Saul
Rothfus, John Arden
Salamon, Ivan Istvan
Satterlee, James Donald
Sawinski, Vincent John
Seidman, Martin
Sennello, Lawrence Thomas
Sessa, David Joseph
Sherman, Warren V
Shriver, John William
Siedler, Arthur James
Silverman, Richard Bruce
Simmons, William Howard
Smittle, Richard Baird
Snyder, William Robert
Soucie, William George
Subbaiah, Papasani Venkata
Suslick, Kenneth Sanders
Szuchet, Sara
Tao, Mariano
Treadway, William Jack, Jr
Van Cleve, John Woodbridge
Van Fossan, Donald Duane
Van Kley, Harold
Van Lanen, Robert Jerome
Walhout, Justine Isabel Simon
Wargel, Robert Joseph
Weber, Evelyn Joyce
Westley, John Leonard
Williams, Michael
Young, Jay Maitland
Zeffren, Eugene

INDIANA

Alter, John Emanuel
Baird, William McKenzie
Beckman, Jean Catherine
BeMiller, James Noble
Boguslaski, Robert Charles
Boschmann, Erwin
Bosron, William F
Burck, Philip John
Burleigh, Bruce Daniel, Jr

Byrn, Stephen Robert
Caplis, Michael E
Coburn, Stephen Putnam
Cox, David Jackson
Davis, Eldred Jack
Eyring, Edward J
Fife, Wilmer Krafft
Fox, Owen Forrest
Gantzer, Mary Lou
Gin, Jerry Ben
Gorenstein, David George
Gunter, Claude Ray
Hamill, Robert L
Harpold, Michael Alan
Hodes, Marion Edward
Janski, Alvin Michael
Jenkins, Winborne Terry
Johnson, Eric Richard
Kinzel, Jerry J
Kleber, John William
Koller, Charles Richard
Kory, Mitchell
Kreiser, Thomas H(arry)
Leoschke, William Leroy
McBride, William Joseph
Michel, Karl Heinz
Nicholson, Ralph Lester
Patwardhan, Bhalchandra H
Rainey, Donald Paul
Ray, William Jackson, Jr
Richardson, John Paul
Roach, Peter John
Robertson, Donald Edwin
Rodia, Jacob Stephen
Roeske, Roger William
Ryder, Kenneth William, Jr
Serianni, Anthony Stephan
Slavik, Nelson Sigman
Smith, Gerald Floyd
Smucker, Arthur Allan
Szuhaj, Bernard F
Tunnicliff, Godfrey
Umbarger, H Edwin
Van Etten, Robert Lee
Weller, Lowell Ernest
Wendel, Samuel Reece
Wild, Gene Muriel
Wildfeuer, Marvin Emanuel
Wishinsky, Henry
Wright, Walter Eugene
Yan, Sau-Chi Betty

IOWA

Abadi, Djhanguir M
Baetz, Albert L
Biermann, Janet Sybil
Cotton, Therese Marie
Foss, John G(erald)
Goff, Harold Milton
Hampton, David Clark
Hanson, Austin Moe
Jennings, Michael Leon
Kostic, Nenad M
Lim, Ramon (Khe Siong)
Linhardt, Robert John
Marshall, William Emmett
Maruyama, George Masao
Masat, Robert James
Meints, Clifford Leroy
Murphy, Patricia A
Nair, Vasu
Newton, John Marshall
Rosazza, John N
Verkade, John George
Wubbels, Gene Gerald

KANSAS

Adams, Ralph Norman
Barton, Janice Sweeny
Bates, Lynn Shannon
Borchardt, Ronald T
Clarenburg, Rudolf
Colen, Alan Hugh
Decedue, Charles Joseph
Grunewald, Gary Lawrence
Hanzlik, Robert Paul
Johnson, David Barton
Kastner, Curtis Lynn
Kramer, Karl Joseph
Lookhart, George LeRoy
Mueller, Delbert Dean
Murdock, Archie Lee
Muthukrishnan, Subbaratnam
Nicholson, Larry Michael
Schowen, Richard Lyle
Seitz, Larry Max

KENTUCKY

Bass, Norman Herbert
Butterfield, David Allan
Colby, David Anthony
Cottrell, Ian William
Diedrich, Donald Frank
Hartman, David Robert
Morgan, Marjorie Susan
Porter, William Hudson
Rhoads, Robert E
Spatola, Arno F
Teller, David Norton
Volp, Robert Francis
Wong, John Lui

LOUISIANA

Alworth, William Lee

Barker, Louis Allen
Bazan, Nicolas Guillermo
Bernofsky, Carl
Biersmith, Edward L, III
Doomes, Earl
Dunn, Adrian John
Gandour, Richard David
George, William Jacob
Goldberg, Stanley Irwin
Gormus, Bobby Joe
Haycock, John Winthrop
Kadan, Ranjit Singh
Kuan, Shia Shiong
Liuzzo, Joseph Anthony
Maverick, Andrew William
Neucere, Joseph Navin
St Angelo, Allen Joseph
Shepherd, Raymond Edward
Shih, Frederick F
Taylor, Eric Robert
Weidig, Charles F
Wiechelman, Karen Janice

MAINE

Blake, Richard D
Guiseley, Kenneth B
Libby, Carol Baker
Page, David Sanborn
Pierce, Arleen Cecilia
Renn, Donald Walter
Sherblom, Anne P

MARYLAND

Abrell, John William
Albers, Robert Wayne
Andres, Scott Fitzgerald
Aszalos, Adorjan
Axelrod, Julius
Bax, Ad
Beckner, Suzanne K
Beecher, Gary Richard
Bennett, David Arthur
Berzofsky, Jay Arthur
Bicknell-Brown, Ellen
Bills, Donald Duane
Bis, Richard F
Bleecker, Margit
Braatz, James Anthony
Broomfield, Clarence A
Bruck, Stephen Desiderius
Bucci, Enrico
Bush, C Allen
Callahan, Mary Vincent
Cantoni, Giulio L
Cassidy, James Edward
Chang, Henry
Cheh, Albert Mei-Chu
Chen, Joseph Ke-Chou
Chmurny, Alan Bruce
Chmurny, Gwendolyn Neal
Chuang, De-Maw
Cohen, Jack Sidney
Cohen, Louis Arthur
Collins, John Henry
Copeland, Edmund Sargent
Cross, David Ralston
Daly, John William
Dighe, Shrikant Vishwanath
Eanes, Edward David
Eaton, William Allen
Eby, Denise
Eichhorn, Gunther Louis
Elin, Ronald John
Fales, Henry Marshall
Fenselau, Catherine Clarke
Fitzgerald, Glenna Gibbs (Cady)
Fox, Barbara Saxton
Fraser, Blair Allen
Freimuth, Henry Charles
Fuhr, Irvin
Gaffney, Betty Jean
Ganguly, Pankaj
Garland, Donita L
Garner, Daniel Dee
Gillette, James Robert
Glaudemans, Cornelis P
Goldberg, Michael Ian
Goldman, Dexter Stanley
Goldstein, Jorge Alberto
Hairstone, Marcus A
Hajjar, Nicolas Philippe
Hanover, John Allan
Heman-Ackah, Samuel Monie
Herner, Albert Erwin
Hertz, Harry Steven
Hickey, Robert Joseph
Hinton, Deborah M
Holmes, Kathryn Voelker
Hosmane, Ramachandra Sadashiv
Hubbard, Donald
Ingham, Kenneth Culver
Inman, John Keith
Irving, George Washington, Jr
Johnston, Laurance S
Josephs, Steven F
Jost, Patricia Cowan
Kador, Peter Fritz
Kaplan, Ann Esther
Karpel, Richard Leslie
Kayser, Robert Helmut
Kelsey, Morris Irwin
Kerley, Ellis R
Kleinman, Hynda Karen
Kohn, Leonard David

Krakauer, Henry
Lamy, Peter Paul
Landsman, David
Langone, John Joseph
Lee-Ham, Doo Young
Litterst, Charles Lawrence
Liu, Darrell T
Lo, Theresa Nong
London, Edythe D
Luborsky, Samuel William
Maciag, Thomas Edward
McKenna, Mary Catherine
McPhie, Peter
Marciani, Dante Juan
Marquardt, Warren William
Marquez, Victor Esteban
Martenson, Russell Eric
Melville, Robert S
Menn, Julius Joel
Mickel, Hubert Sheldon
Miles, Edith Wilson
Miller, Paul Scott
Miller, Steve P F
Minton, Allen Paul
Murray, Gary Joseph
Myers, Charles
Ness, Robert Kiracofe
Passwater, Richard Albert
Phillips, Leo Augustus
Pollack, Ralph Martin
Ponnamperuma, Cyril Andrew
Poston, John Michael
Prasanna, Hullahalli Rangaswamy
Quarles, Richard Hudson
Repaske, Roy
Rifkind, Joseph Moses
Roberts, Anita Bauer
Roberts, David Duncan
Robinson-White, Audrey Jean
Rosenstein, Robert William
Sampugna, Joseph
Saslaw, Leonard David
Sass, Neil Leslie
Sayer, Jane M
Schaffer, Robert
Schechter, Alan Neil
Schifreen, Richard Steven
Schneider, Walter Carl
Seifried, Adele Susan Corbin
Seifried, Harold Edwin
Sharrett, A Richey
Shore, Moris Lawrence
Skelly, Jerome Philip, Sr
Skidmore, Wesley Dean
Snader, Kenneth Means
Speedie, Marilyn Kay
Spooner, Peter Michael
Sporn, Eugene Milton
Steiner, Robert Frank
Surrey, Kenneth
Suskind, Sigmund Richard
Swann, Madeline Bruce
Taniuchi, Hiroshi
Titus, Elwood Owen
Townsend, Craig Arthur
Townsley, John D
Ts'o, Paul On Pong
Tso, Tien Chioh
Tung, Ming Sung
Veitch, Fletcher Pearre, Jr
Vijay, Inder Krishan
Walter, Eugene LeRoy, Jr
Watkins, Paul Alan
Weinstein, Howard
Welch, Arnold D(eMerritt)
Whiting, John Dale, Jr
Williams, Noreen
Williamson, Charles Elvin
Wilson, Samuel H
Wolff, John B
Wu, Roy Shih-Shyong
Wykes, Arthur Albert
Yeh, Lai-Su Lee
Yoho, Clayton W

MASSACHUSETTS

Adler, Alice Joan
Ahern, David George
Ajami, Alfred Michel
Askew, Eldon Wayne
Auer, Henry Ernest
Babayan, Vigen Khachig
Ball, Derek Harry
Berlowitz Tarrant, Laurence
Bing, David H
Browne, Douglas Townsend
Campbell, Mary Kathryn
Cantley, Lewis Clayton
Cathou, Renata Egone
Champion, Paul Morris
Chase, Arleen Ruth
Chipman, Wilmon B
Clarke, Michael J
Cohen, Saul G
Demain, Arnold Lester
Dickinson, Leonard Charles
DiLiddo, Rebecca McBride
Doctrow, Susan R
Dooley, David Marlin
Downer, Nancy Wuerth
Ehlers, Mario Ralf Werner
Fossel, Eric Thor
Foye, William Owen
Friedman, Orrie Max

Gallop, Paul Myron
Galper, Jonas Bernard
Girard, Francis Henry
Gounaris, Anne Demetra
Greenaway, Frederick Thomas
Grodberg, Marcus Gordon
Haber, Stephen B
Hansen, David Elliott
Harrison, Stephen Coplan
Hassan, William Ephriam, Jr
Hauser, George
Hellman, Kenneth P
Hixson, Susan Harvill
Hobey, William David
Hochella, Norman Joseph
Holick, Sally Ann
Jahngen, Edwin Georg Emil, Jr
Jungalwala, Firoze Bamanshaw
Kaplan, David Lee
King, Jonathan (Alan)
Klibanov, Alexander M
Kline, Toni Beth
Klotz, Lynn Charles
Knowles, Aileen Foung
Knowles, Jeremy Randall
Kobayashi, Kazumi
Kocon, Richard William
Koltun, Walter Lang
Kominz, David Richard
Kropf, Allen
Kuo, Lawrence C
Kupfer, David
Laursen, Richard Allan
Learson, Robert Joseph
Lees, Marjorie Berman
Libby, Peter
Lippard, Stephen J
Lowey, Susan
Lundstrom, Ronald Charles
MacKerell, Alexander Donald, Jr
Makowski, Lee
Marquis, Judith Kathleen
Marx, Kenneth Allan
Nauss, Kathleen Minihan
Nickerson, Richard G
O'Brien, Richard Desmond
Paskins-Hurlburt, Andrea Jeanne
Pinkus, Jack Leon
Plocke, Donald J
Redfield, Alfred Guillou
Robbat, Albert, Jr
Robinson, William Edward
Rosenkrantz, Harris
Ruelius, Hans Winfried
Scordilis, Stylianos Panagiotis
Shuster, Louis
Silver, Marc Stamm
Simons, Elizabeth Reiman
Stephens, Raymond Edward
Thomas, Peter
Timasheff, Serge Nicholas
Umans, Robert Scott
Vellaccio, Frank
Walker-Nasir, Evelyne
Warren, Christopher David
Whitesides, George McClelland
Williamson, Kenneth Lee
Wineman, Robert Judson
Winkelman, James W
Worden, Patricia Barron
Yarmush, Martin Leon
Young, Richard Lawrence
Zannis, Vassilis I

MICHIGAN
Akil, Huda
Babcock, Gerald Thomas
Barraco, Robin Anthony
Bredeck, Henry E
Brown, Ray Kent
Callewaert, Denis Marc
Campbell, Wilbur Harold
Caspers, Mary Lou
Chang, Chi Kwong
Cook, Paul Laverne
Coward, James Kenderdine
Craig, Winston John
Crippen, Gordon Marvin
Cullum, Malford Eugene
Deal, William Cecil, Jr
Dekker, Eugene Earl
Drach, John Charles
Dukes, Gary Rinehart
Edwards, Brian F
Eliezer, Naomi
Fernandez-Madrid, Felix
Fischer, Lawrence J
Forist, Arlington Ardeane
Garlick, Robert L
Hakim, Margaret Heath
Halvorson, Herbert Russell
Herzig, David Jacob
Holland, Jack Calvin
Hultquist, Donald Elliott
Jackson, Craig Merton
Janik, Borek
Jay, James Monroe
Kaiser, David Gilbert
Kang, Uan Gen
Kiechle, Frederick Leonard
Kuo, Mingshang
La Du, Bert Nichols, Jr
Laks, Peter Edward

Lawton, Richard G
Lindblad, William John
Lorch, Steven Kalman
McBean, Lois D
McCarville, Michael Edward
McCullough, Willard George
McIntosh, Lee
McNair, Ruth Davis
Marshall, Charles Wheeler
Matthews, Rowena Green
Miller, Susan Mary
Mizukami, Hiroshi
Moore, Kenneth Edwin
Mostafapour, M Kazem
Moyer, Carl Edward
O'Connell, Paul William
Ogilvie, Marvin Lee
Parikh, Jekishan R
Penner-Hahn, James Edward
Petzold, Edgar
Phan, Sem Hin
Randinitis, Edward J
Root-Bernstein, Robert Scott
Roth, Robert Andrew, Jr
Salmond, William Glover
Schacht, Jochen (Heinrich)
Schullery, Stephen Edmund
Sellinger, Otto Zivko
Servis, Robert Eugene
Shore, Joseph D
Showalter, Howard Daniel Hollis
Smart, James Blair
Smith, Richard Harding
Stach, Robert William
Thoene, Jess Gilbert
Tormanen, Calvin Douglas
Tung, Fred Fu
Uhler, Michael David
Vaughn, Clarence Benjamin
Watson, Jack Throck
Welsch, Clifford William, Jr
Wilson, John Edward
Woster, Patrick Michael
Yancey, Robert John, Jr
Yurewicz, Edward Charles
Zand, Robert

MINNESOTA
Adams, Ernest Clarence
Adiarte, Arthur Lardizabal
Ames, Matthew Martin
Barany, George
Baumann, Wolfgang J
Bloomfield, Victor Alfred
Bowers, Larry Donald
Brockman, Howard Lyle, Jr
Brown, Rhoderick Edmiston, Jr
Carter, John Vernon
Conard, Gordon Joseph
Crawford, Ronald Lyle
Csallany, Agnes Saari
Farnum, Sylvia A
Freier, Esther Fay
Fulmer, Richard W
Gray, Gary Ronald
Haddad, Louis Charles
Hartman, Boyd Kent
Jemmerson, Ronald Renomer Weaver
Kluetz, Michael David
Kyle, Robert Arthur
Lawrenz, Frances Patricia
Lin, Spring
Lovrien, Rex Eugene
Malewicz, Barbara Maria
Nicol, Susan Elizabeth
Perisho, Clarence R
Que, Lawrence, Jr
Rao, Gundu Hirisave Rama
Schmidt, Jane Ann
Swaiman, Kenneth F
Swanson, Anne Barrett
Toscano, William Agostino, Jr
Tsong, Tian Yow
Ugurbil, Kamil
Wei, Guang-Jong Jason

MISSISSIPPI
Boyle, John Aloysius, Jr
Case, Steven Thomas
Chaires, Jonathan Bradford
Fawcett, Newton Creig
Graves, David E
Ho, Ing Kang
Kopfler, Frederick Charles
McGinnis, Gary David
Peterson, Harold LeRoy
Williamson, John S

MISSOURI
Ackers, Gary K
Adams, Steven Paul
Black, Wayne Edward
Brown, Barbara Illingworth
Brown, David Henry
Burke, William Joseph
Byington, Keith H
Chiappinelli, Vincent A
Cicero, Theodore James
Coudron, Thomas A
Covey, Douglas Floyd
De Blas, Angel Luis
Delaware, Dana Lewis
Feder, Joseph
Ferrendelli, James Anthony

Fraley, Robert Thomas
Francis, Faith Ellen
Geison, Ronald Leon
Green, Vernon Albert
Hertelendy, Frank
Holtzer, Alfred Melvin
Hood, Larry Lee
Howard, Susan Carol Pearcy
Lau, Brad W C
Lipkin, David
Lohman, Timothy Michael
Long, Lawrence William
Lozeron, Homer A
McCormick, John Pauling
MacQuarrie, Ronald Anthony
Mayer, Dennis T
Mortensen, Harley Eugene
Moscatelli, Ezio Anthony
Newell, Jon Albert
Piepho, Robert Walter
Ramsey, Robert Bruce
Rogic, Milorad Mihailo
Rosenthal, Harold Leslie
Rowe, Elizabeth Snow
Schleppnik, Alfred Adolf
Schmitz, Kenneth Stanley
Schuck, James Michael
Shukla, Shivendra Dutt
Sikorski, James Alan
Slatopolsky, Eduardo
Smith, Carl Hugh
Sun, Albert Yung-Kwang
Vineyard, Billy Dale
Wheeler, James Donlan
Whittle, Philip Rodger
Wilkinson, Ralph Russell
Wood, David Collier
Yarbro, John Williamson
Yunger, Libby Marie

MONTANA
Parker, Keith Krom
Thompson, Holly Ann

NEBRASKA
Casey, Carol A
Ford, Johny Joe
Gold, Barry Irwin
Lockridge, Oksana Maslivec
MacDonald, Richard G
Nagel, Donald Lewis
Parkhurst, Lawrence John
Rasmussen, Russell Lee
Smith, David Hibbard
Yen, Jong-Tseng

NEVADA
Dreiling, Charles Ernest
Lightner, David A
Peck, Ernest James, Jr
Wolf, Edward Charles

NEW HAMPSHIRE
Chang, Ta-Yuan
Chasteen, Norman Dennis
Green, David Francis
Hagan, William John, Jr
Ikawa, Miyoshi
Maudsley, David V
Wetterhahn, Karen E

NEW JERSEY
Alberts, Alfred W
Ardelt, Wojciech Joseph
Babu, Uma Mahesh
Bagnell, Carol A
Bathala, Mohinder S
Becker, Joseph Whitney
Bernholz, William Francis
Bhattacharjee, Himangshu Ranjan
Boskey, Adele Ludin
Breslow, Esther M G
Bryant, Robert Wesley
Caldwell, Carlyle Gordon
Chaikin, Philip
Chiao, Yu-Chih
Chinard, Francis Pierre
Comai-Fuerherm, Karen
Coyle, Catherine Louise
Crane, Laura Jane
Cushman, David Wayne
Dashman, Theodore
DeBari, Vincent A
Deetz, Lawrence Edward, III
Denhardt, David Tilton
Desjardins, David Francis
Diegnan, Glenn Alan
Dunikoski, Leonard Karol, Jr
Durso, Donald Francis
Dvornik, Dushan Michael
Eagar, Robert Gouldman, Jr
Ehrlich, Julian
Ellenbogen, Leon
Erckel, Ruediger Josef
Erenrich, Eric Howard
Erenrich, Evelyn Schwartz
Felix, Arthur M
Ferguson, Karen Anne
Fernandes, Prabhavathi Bhat
Fordham, William David
Frederick, James R
Fu, Shou-Cheng Joseph
Gagna, Claude Eugene
Garfinkel, Harmon Mark

Georgopapadakou, Nafsika Eleni
Giddings, George Gosselin
Gidwani, Ram N
Gold, Barry Ira
Grosso, John A
Hagan, James Joyce
Hall, Stan Stanley
Harkins, Robert W
Harrison-Johnson, Yvonne E
Haslanger, Martin Frederick
Heacock, Craig S
Heimer, Edgar P
Hensens, Otto Derk
Ivatt, Raymond John
Jacobs, Morton Howard
Jacobson, Martin Michael
Jakubowski, Hieronim Zbigniew
Jones, Martha Ownbey
Jordan, Frank
Joseph, John Mundancheril
Kaczorowski, Gregory John
Kaiserman, Howard Bruce
Kamm, Jerome J
Kauzmann, Walter (Joseph)
Kirschbaum, Joel Jerome
Kolis, Stanley Joseph
Koonce, Samuel David
Kripalani, Kishin J
Kristol, David Sol
Kroon, Paulus Arie
Lam, Yiu-Kuen Tony
Lin, Chin-Chung
Lin, Grace Woan-Jung
McCaslin, Darrell
MacCoss, Malcolm
Mackerer, Carl Robert
Madison, Vincent Stewart
Manning, Gerald Stuart
Manowitz, Paul
Margolis, Frank L
Mark, David Fu-Chi
Mendelsohn, Richard
Montville, Thomas Joseph
Murthy, Vadiraja Venkatesa
Muschek, Lawrence David
Naipawer, Richard Edward
Nelson, Nathan
Ocain, Timothy Donald
Ofengand, Edward James
Olson, Wilma King
OSullivan, Joseph
Palmer, Keith Henry
Podleckis, Edward Vidas
Quinn, Gertrude Patricia
Roland, Dennis Michael
Rothberg, Lewis Josiah
Rowin, Gerald L
Saferstein, Richard
Schaich, Karen Marie
Scharpf, Lewis George, Jr
Schrier, Eugene Edwin
Schugar, Harvey
Schwenker, Robert Frederick, Jr
Segelman, Alvin Burton
Shefer, Sarah
Singh, Vishwa Nath
Smith, Elizabeth Melva
Snyder, Robert
Spiro, Thomas George
Stillman, John Edgar
Strauss, George
Strumeyer, David H
Stults, Frederick Howard
Symchowicz, Samson
Takahashi, Mark T
Tindall, Charles Gordon, Jr
Tolman, Richard Lee
Vandegrift, Vaughn
VanMiddlesworth, Frank L
Wassarman, Paul Michael
Weill, Carol Edwin
White, Ronald E
Wildnauer, Richard Harry
Wislocki, Peter G

NEW MEXICO
Birnbaum, Edward Robert
Darnall, Dennis W
Fee, James Arthur
Gurd, Ruth Sights
Guziec, Frank Stanley, Jr
Guziec, Lynn Erin
Jones, Peter Frank
Matwiyoff, Nicholas Alexander
Thompson, Jill Charlotte
Trujillo, Ralph Eusebio
Wild, Gaynor (Clarke)
Woodfin, Beulah Marie

NEW YORK
Abramowicz, Daniel Albert
Abramson, Morris Barnet
Andrews, John Parray
Angelino, Norman J
Bachrach, Howard L
Baird, Barbara A
Bardos, Thomas Joseph
Baum, Stuart J
Bednar, Rodney Allan
Bello, Jake
Benuck, Myron
Berl, Soll
Bettelheim, Frederick A
Bitcover, Ezra Harold

Ocone, Luke Ralph
Parker, Leslie
Perlis, Irwin Bernard
Phillips, Michael Canavan
Prescott, David Julius
Ritter, Carl A
Rose, George David
Rosenberg, Jerome Laib
Russell, Alan James
Sax, Martin
Schengrund, Cara-Lynne
Schray, Keith James
Selinsky, Barry Steven
Seybert, David Wayne
Shaw, Leslie M J
Sideropoulos, Aris S
Sitrin, Robert David
Sloviter, Henry Allan
Smith, John Bryan
Soriano, David S
Southwick, Philip Lee
Spangler, Martin Ord Lee
Stevens, Charles Le Roy
Storey, Bayard Thayer
Straub, Thomas Stuart
Sun, James Dean
Takashima, Shiro
Tarver, James H, Jr
Thoman, Charles James
Thornton, Edward Ralph
Thornton, Elizabeth K
Tomezsko, Edward Stephen John
Tu, Shu-i
Turner, Paul Jesse
Tzodikov, Nathan Robert
Vickers, Stanley
Villafranca, Joseph John
Voet, Donald Herman
Wedler, Frederick Charles Oliver, Jr
Wetzel, Ronald Burnell
Witham, Francis H
Witting, Lloyd Allen
Wolfersberger, Michael Gregg
Wolken, Jerome Jay
Zacchei, Anthony Gabriel

RHODE ISLAND
Brautigan, David L
Cane, David Earl
Hartman, Karl August
Lee, Tung-Ching

SOUTH CAROLINA
Amma, Elmer Louis
Banik, Narendra Lal
Dawson, John Harold
Evans, David Wesley
Greene, George C, III
Greenfield, Seymour
Hunter, George William
Kinard, Fredrick William
Lindenmayer, George Earl
Wells, John Arthur

SOUTH DAKOTA
Brady, Frank Owen
Cook, David Edgar
Evenson, Donald Paul
Marshall, Finley Dee
West, Thomas Patrick

TENNESSEE
Benton, Charles Herbert
Bhattacharya, Syamal Kanti
Egan, B Zane
Farrington, Joseph Kirby
Fridland, Arnold
Goertz, Grayce Edith
Greenbaum, Elias
Griffin, Guy David
Harris, Durward Smith
Hartman, Frederick Cooper
Hossler, Fred E
Incardona, Antonino L
Jarrett, Harry Wellington, III
Jernigan, Howard Maxwell, Jr
Lowe, James N
Morrison, Martin
Pittz, Eugene P
Reddick, Bradford Beverly
Roberts, DeWayne
Staros, James Vaughan
Thakar, Jay H
Turner, Rex Howell
Uziel, Mayo
Waddell, Thomas Groth
Williams, Robley Cook, Jr
Worsham, Lesa Marie Spacek

TEXAS
Adams, Emory Temple, Jr
Albrecht, Thomas Blair
Allen, Robert Charles
Alvarez-Gonzalez, Rafael
Arntzen, Charles Joel
Ball, M Isabel
Bates, George Winston
Besch, Paige Keith
Brown, Wanda Lois
Burzynski, Stanislaw Rajmund
Bushey, Michelle Marie
Cameron, Bruce Francis
Cate, Rodney Lee
Cody, John T

Cook, Paul Fabyan
Crookshank, Herman Robert
Czerwinski, Edmund William
Davis, Donald Robert
Dear, Robert E A
Dedman, John Raymond
Deloach, John Rooker
Dempsey, Walter B
Dwyer, Lawrence Arthur
Eakin, Richard Timothy
Everse, Johannes
Fleenor, Marvin Bension
Funkhouser, Edward Allen
Furlong, Norman Burr, Jr
Gierasch, Lila Mary
Gillard, Baiba Kurins
Gilman, Alfred G
Gray, Donald Melvin
Gray, Horace Benton, Jr
Gutsche, Carl David
Guynn, Robert William
Hackert, Marvin LeRoy
Heffley, James D
Hendrickson, Constance McRight
Ho, Beng Thong
Ho, Dah-Hsi
Ivie, Glen Wayne
Joshi, Vasudev Chhotalal
Kellems, Rodney E
Kelley, Jim Lee
Kraus-Friedmann, Naomi
Kroschewsky, Julius Richard
Ludden, Thomas Marcellus
Lupski, James Richard
Mabry, Tom Joe
McAdoo, David J
Macfarlane, Ronald Duncan
Massey, John Boyd
Masters, Bettie Sue Siler
Modak, Arvind T
Moreland, Ferrin Bates
Nachman, Ronald James
Nordyke, Ellis Larrimore
Olson, John S
Oray, Bedii
Pace, Carlos Nick
Parry, Ronald John
Peck, Merlin Larry
Piziak, Veronica Kelly
Ranney, David Francis
Rassin, David Keith
Raval, Dilip N
Ray, Allen Cobble
Renthal, Robert David
Robberson, Donald Lewis
Robinson, Neal Clark
Robison, George Alan
Rogers, Lorene Lane
Romeo, Tony
Sato, Clifford Shinichi
Saunders, Priscilla Prince
Sawyer, Donald Turner, Jr
Scott, Alastair Ian
Scouten, William Henry
Shaw, Robert Wayne
Sherry, Allan Dean
Skelley, Dean Sutherland
Slaga, Thomas Joseph
Stone, Irving Charles, Jr
Teng, Jon Ie
Waldbillig, Ronald Charles
Waterman, Michael Roberts
Weigel, Paul H(enry)
Wendt, Theodore Mil
Widger, William Russell
Wiggins, Richard Calvin
Williams, Charles Herbert
Williams, Mary Carr
Wilson, Carol Maggart
Wilson, Lon James
Woo, Savio L C
Wright, Stephen E
Ziegler, Miriam Mary

UTAH
Ames, Robert A
Barnett, Bill B
Gesteland, Raymond Frederick
Gibb, James Woolley
Johnston, Stephen Charles
Kemp, John Wilmer
Knapp, Gayle
Linker, Alfred
Moody, David Edward
Musick, James R
Petryka, Zbyslaw Jan
Pitt, Charles H
Prescott, Stephen M
Ramachandran, Chittoor Krishna
Robins, Morris Joseph
Rothenberg, Mortimer Abraham
Seeley, Schuyler Drannan
Smith, Marvin Artell
Walker, Edward Bell Mar
Windham, Carol Thompson

VERMONT
Crooks, George Chapman
Mitchell, William Alexander
Moyer, Walter Allen, Jr
Rinse, Jacobus
Tritton, Thomas Richard
Walters, Carol Price
Wurzburg, Otto Bernard

VIRGINIA
Allison, Trenton B
Averill, Bruce Alan
Barnett, Lewis Brinkley
Biltonen, Rodney Lincoln
Blackmore, Peter Fredrick
Blair, Barbara Ann
Brunelle, Richard Leon
Chandler, Jerry LeRoy
Collins, Galen Franklin
Cone, Clarence Donald, Jr
Dannenburg, Warren Nathaniel
Donohue, Joyce Morrissey
Easter, Donald Philips
Esen, Asim
Ferguson, Robert Nicholas
Fonong, Tekum
Gander, George William
Hall, Philip Layton
Hunt, Donald F
Jacobs, Maryce Mercedes
Macdonald, Timothy Lee
Martin, Robert Bruce
Monroe, Stuart Benton
Moore, William Earl
Olin, Jacqueline S
Pages, Robert Alex
Pierson, Merle Dean
Place, Janet Dobbins
Roberts, David Craig
Rodig, Oscar Rudolf
Sahu, Saura Chandra
Schroeder, Walter Adolph
Sirica, Alphonse Eugene
Smibert, Robert Merrall, II
Smith, Harold Linwood
Stucky, Gary Lee
Taylor, Ronald Paul
Ulvedal, Frode
Whitney, George Stephen
Willett, James Delos
Wishner, Lawrence Arndt
Yu, Robert Kuan-Jen

WASHINGTON
Anderson, Larry Ernest
Atkins, William M
Baillie, Thomas Allan
Bland, Jeffrey S
Bourquin, Al Willis J
Deming, Jody W
Duncan, James Byron
Ericsson, Lowell Harold
Gorman, Marvin
Hill, Herbert Henderson, Jr
Jensen, Lyle Howard
Lybrand, Terry Paul
Meeuse, Bastiaan J D
Palmiter, Richard
Patt, Leonard Merton
Peterson-Kennedy, Sydney Ellen
Phillips, Steven J
Read, David Hadley
Rees, Allan W
Reeves, Raymond
Roubal, William Theodore
Russo, Salvatore Franklin
Teller, David Chambers
Wekell, Marleen Marie
Whelton, Bartlett David
Zimmerman, Gary Alan

WEST VIRGINIA
Elkins, John Rush
Foster, Joyce Geraldine
Lotspeich, Frederick Jackson
Roberts, Joseph Linton

WISCONSIN
Anderson, Laurens
Antholine, William E
Bernstein, Sheldon
Bock, Robert Manley
Boggs, Lawrence Allen
Brown, Charles Eric
Cohen, Philip Pacy
Craig, Elizabeth Anne
Cushing, Merchant Leroy
Dwyer-Hallquist, Patricia
Hake, Carl (Louis)
Hoffman, Norman Edwin
Johns, Philip Timothy
Johnson, Morris Alfred
Johnson, Warren Victor
Jones, Berne Lee
Kirk, Kent T
Klaassen, Dwight Homer
Krahnke, Harold C
Markley, John Lute
Miller, Joyce Fiddick
Petering, David Harold
Raines, Ronald T
Record, M Thomas, Jr
Rich, Daniel Hulbert
Roberts, Ronald C
Romans-Hess, Alice Yvonne
Rosenberg, Robert Melvin
Sanderson, Gary Warner
Saryan, Leon Aram
Shaw, C Frank, III
Sih, Charles John
Skatrud, Thomas Joseph
Smith, Milton Reynolds
Soerens, Dave Allen

Twining, Sally Shinew
Ward, Kyle, Jr
Westler, William Milo
Worley, John David

WYOMING
Lewis, Randolph Vance

PUERTO RICO
Habeeb, Ahmed Fathi Sayed Ahmed
Infante, Gabriel A
Ramirez, J Roberto
Salas-Quintana, Salvador

ALBERTA
Bide, Richard W
Denford, Keith Eugene
Dunford, Hugh Brian
Hill, Bernard Dale
Hodges, Robert Stanley
Jack, Thomas Richard
Kotovych, George
Moore, Graham John
Roche, Rodney Sylvester
Stiles, Michael Edgecombe
Sykes, Brian Douglas
Van De Sande, Johan Hubert
Vederas, John Christopher
Wiebe, Leonard Irving

BRITISH COLUMBIA
Anderson, John Ansel
Cushley, Robert John
Dolphin, David Henry
Ekramoddoullah, Abul Kalam M
Haunerland, Norbert Heinrich
Irvine, George Norman
Mauk, Arthur Grant
Oehlschlager, Allan Cameron
Richards, William Reese
Smith, Michael
Withers, Stephen George

MANITOBA
Duckworth, Henry William (Harry)
Forrest, Bruce James
Hasinoff, Brian Brennen
Hougen, Frithjof W
Jamieson, James C
McDonald, Bruce Eugene
MacGregor, Elizabeth Ann
Rosenmann, Edward A
Stevens, Frits Christiaan
Vitti, Trieste Guido

NEWFOUNDLAND
Shaw, Derek Humphrey

NOVA SCOTIA
Dolphin, Peter James
Freeman, Harry Cleveland
Gilgan, Michael Wilson
Odense, Paul Holger
Verpoorte, Jacob A
Vining, Leo Charles
White, Robert Lester
White, Thomas David

ONTARIO
Brisbin, Doreen A
Callahan, John William
Caplan, Donald
Chua, Kian Eng
Clarke, Anthony John
Cupp, Calvin R
Currie, Violet Evadne
Deslauriers, Roxanne Marie Lorraine
Dorrell, Douglas Gordon
Epand, Richard Mayer
Ferrier, Leslie Kenneth
Feuer, George
Gibson, Rosalind Susan
Goldsack, Douglas Eugene
Guthrie, James Peter
Janzen, Edward George
Kates, Morris
Khanna, Jatinder Mohan
Liew, Choong-Chin
Logan, James Edward
Macdonald, Peter Moore
Mantsch, Henry H
Marks, Charles Frank
Rodrigo, Russell Godfrey
Romans, Robert Gordon
Sarwar, Ghulam
Spencer, John Hedley
Spenser, Ian Daniel
Stancer, Harvey C
Stillman, Martin John
Thibert, Roger Joseph
Trenholm, Harold Locksley
Tsang, Charles Pak Wai
Wiebe, John Peter
Wood, Peter John
Wu, Tai Wing
Yaguchi, Makoto
Yang, Paul Wang
Yeung, David Lawrence

QUEBEC
Bateman, Andrew
Belanger, Patrice Charles
Boulet, Marcel
Braun, Peter Eric

Datz, Sheldon
Feigerle, Charles Stephen
Klots, Cornelius E
Miller, John Cameron
Taylor, Ellison Hall
Trowbridge, Lee Douglas

TEXAS
Albright, John Grover
Bartsch, Richard Allen
Becker, Ralph Sherman
Brooks, Philip Russell
Cares, William Ronald
DeBerry, David Wayne
Dobson, Gerard Ramsden
Eaker, Charles William
Hayes, Edward Francis
Holwerda, Robert Alan
Jacobson, Barry Martin
Martin, Charles R
Newcomb, Martin
Schmalz, Thomas G
Stanbury, David McNeil
Tibbals, Harry Fred, III

UTAH
Goates, Steven Rex
Wahrhaftig, Austin Levy

VIRGINIA
Bogan, Denis John
DeGraff, Benjamin Anthony
Demas, James Nicholas
Desjardins, Steven G
Jordan, Wade H(ampton), Jr
Meites, Louis
Metcalf, David Halstead
Sanzone, George
Schiavelli, Melvyn David
Tyson, John Jeanes

WASHINGTON
Baughcum, Steven Lee
Duvall, George Evered
Friedrich, Donald Martin
Heller, Eric Johnson
Root, John Walter
Tabbutt, Frederick Dean
Windsor, Maurice William

WISCONSIN
Crim, F(orrest) Fleming
Megahed, Sid A
Moore, John Ward
Viswanathan, Ramaswami
Zimmerman, Howard Elliot

WYOMING
Jacobson, Irven Allan, Jr

PUERTO RICO
Infante, Gabriel A

ALBERTA
Birss, Viola Ingrid
Dunford, Hugh Brian
Freeman, Gordon Russel
Sanger, Alan Rodney
Tschuikow-Roux, Eugene

BRITISH COLUMBIA
Basco, N

MANITOBA
Hasinoff, Brian Brennen
McFarlane, Joanna

NEW BRUNSWICK
Reinsborough, Vincent Conrad

NOVA SCOTIA
Pacey, Philip Desmond
Roscoe, John Miner

ONTARIO
Anlauf, Kurt Guenther
Back, Robert Arthur
Bolton, James R
Brumer, Paul William
Buncel, Erwin
Carrington, Tucker
Dove, John Edward
Halfon, Efraim
Henry, Bryan Roger
Kresge, Alexander Jerry
LeRoy, Donald James
Liu, Wing-Ki
Penner, Glenn H
Sadowski, Chester M

QUEBEC
Fleming, Bruce Ingram
Srivastava, Uma Shanker

SASKATCHEWAN
Steer, Ronald Paul

OTHER COUNTRIES
Balakrishnan, Narayana Swamy
Bondybey, Vladimir E
Forst, Wendell
Kay, Kenneth George

Chemistry, General

ALABAMA
Askew, William Crews
Aull, John Louis
Baker, June Marshall
Baker, Rees Terence Keith
Blanchard, Richard Lee
Carpenter, Charles H
Cataldo, Charles Eugene
Dorai-Raj, Diana Glover
Ellenburg, Janus Yentsch
Fendley, Ted Wyatt
Frings, Christopher Stanton
Garner, Robert Henry
Gill, Piara Singh
Govil, Narendra Kumar
Hardin, Ian Russell
Hughes, Edwin R
Kennedy, Frank Metler
Lieberman, Robert
Lucas, William R(ay)
Marano, Gerald Alfred
Metzger, Robert Melville
Mishra, Satya Narayan
Pan, Chai-Fu
St Pierre, Thomas
Scheiner, Bernard James
Sheridan, Richard Collins
Shoemaker, John Daniel, Jr
Smith, Lewis Taylor
Stevens, Frank Joseph
Summerlin, Lee R
Tamburin, Henry John
Toffel, George Mathias
Walia, Amrik Singh
Warfield, Carol Larson
Watkins, Charles Lee
Whitehead, Fred
Yee, Tin Boo
Ziegler, Paul Fout

ALASKA
Lokken, Donald Arthur

ARIZONA
Allen, John Rybolt
Barr, Charles (Robert)
Brown, Paul Wayne
Burgus, Warren Harold
Chapman, Robert Dale
Connor, Ralph (Alexander)
Cox, Fred Ward, Jr
Delbecq, Charles Jarchow
Fancher, Otis Earl
Glaunsinger, William Stanley
Helbert, John N
Irvine, John Withers, Jr
Kierbow, Julie Van Note Parker
Kriege, Owen Hobbs
Larson, Lester Mikkel
Nordstrom, Brian Hoyt
Robinette, Hillary, Jr
Rogers, Alan Barde
Shafer, M(errill) W(ilbert)
Tam, Wing Yim
Thomas, Charles L(amar)
Urry, Wilbert Herbert
Wise, Edward Nelson

ARKANSAS
Geren, Collis Ross
Koeppe, Roger E, II
Ku, Victoria Feng
Kuroda, Paul Kazuo
Obenland, Clayton O
Poirier, Lionel Albert
Shook, Thomas Eugene

CALIFORNIA
Abrams, Irving M
Ahuja, Jagan N
Alire, Richard Marvin
Allen, Esther C(ampbell)
Alley, Starling Kessler, Jr
Altgelt, Klaus H
Altman, David
Anderson, Robert Emra
Andreas, John M(oore)
Andrews, Lawrence James
Ansari, Ali
Arias, Jose Manuel
Arnold, James Richard
Atkins, Don Carlos, Jr
Atkinson, Roger
Avera, Fitzhugh Lee
Barusch, Maurice R
Battles, Willis Ralph
Beardslee, Ronald Allen
Becker, Joseph F
Beckman, Arnold Orville
Behr, Inga
Beilby, Alvin Lester
Berman, Horace Aaron
Bishop, John William
Bishop, Keith C, III
Bittner, Harlan Fletcher
Bondar, Richard Jay Laurent
Bonner, Lyman Gaylord
Bonner, William Andrew
Bradbury, E Morton
Brauman, John I
Brink, David Liddell
Brouillard, Robert Ernest

Brown, Weldon Grant
Bruenner, Rolf Sylvester
Byall, Elliott Bruce
Campbell, George Washington, Jr
Chaikin, Saul William
Chou, Yungnien John
Clark, Leigh Bruce
Cliath, Mark Marshall
Condit, Paul Carr
Condit, Ralph Howell
Cooper, Wilson Wayne
Cowperthwaite, Michael
Coyner, Eugene Casper
Craik, Charles S
Csicsery, Sigmund Maria
Cubicciotti, Daniel David
Current, Steven P
Cusumano, James A
Cutforth, Howard Glen
Davies, Robert Milton
Deck, Joseph Francis
Deleray, Arthur Loyd
De Vries, John Edward
Domeniconi, Michael John
Drewes, Patricia Ann
Eastman, John W
Eikrem, Lynwood Olaf
Etzler, Dorr Homer
Evans, Allison Bickle
Fajans, Edgar W
Feiler, William A, Jr
Feng, Da-Fei
Fenton, Donald Mason
Feramisco, James Robert
Ferguson, Lloyd Noel
Fletcher, Peter C
Foehr, Edward Gotthard
Frank, Joan Patricia
Frazer, Jack Winfield
Furby, Neal Washburn
Gamble, Fred Ridley, Jr
Gardner, Marjorie Hyer
Gerstein, Melvin
Gilbert, Francis Evalo
Gold, Marvin B
Goldberg, Sabine Ruth
Gordon, Benjamin Edward
Greene, Paul E
Griffin, Thomas Scott
Griffing, John Malcolm
Groot, Cornelius
Haas, Peter Herbert
Haig, Pierre Vahe
Haimes, Florence Catherine
Hale, Ron L
Halstead, Bruce W
Hamlin, Kenneth Eldred, Jr
Hansch, Corwin Herman
Harding-Barlow, Ingeborg
Hardy, Edgar Erwin
Harrar, Jackson Elwood
Hathaway, Gary Michael
Havlicek, Stephen
Hawkes, Wayne Christian
Hayter, Roy G
Heinemann, Heinz
Helman, Edith Zak
Hendricks, Grant Walstein
Hicks, Harry Gross
Hickson, Donald Andrew
Hill, Russell John
Hinkson, Jimmy Wilford
Hochberg, Melvin
Holmquist, Walter Richard
Holzmann, Richard Thomas
Hooper, Catherine Evelyn
Horn, Christian Friedrich
Hydock, Joseph J
Hygh, Earl Hampton
Ja, William Yin
Janus, Alan Robert
Jennings, Harley Young, Jr
Johnson, LeRoy Franklin
Johnson, Richard D
Jones, Francis Tucker
Jones, Rebecca Anne
Jorgensen, Paul J
Joseph, Marjory L
Kabra, Pokar Mal
Kamp, David Allen
Kane, Stephen Shimmon
Kaska, William Charles
Katsumoto, Kiyoshi
Kauffman, George Bernard
Kaye, Wilbur (Irving)
Keilin, Bertram
Kelley, Michael J
Kempter, Charles Prentiss
Keyzer, Hendrik
Kharasch, Norman
King, William Mattern
Kirsch, Milton
Klein, August S
Koenig, Nathan Hart
Kohn, Gustave K
Kolbezen, Martin (Joseph)
Kray, Louis Robert
Krikorian, Oscar Harold
Krummenacher, Daniel
Kuo, Harng-Shen
Kurnick, Allen Abraham
La Ganga, Thomas S
Lawton, Emil Abraham
Lee, Vin-Jang

Lee, Young-Jin
LeFave, Gene M(arion)
Leider, Herman R
Lemmon, Richard Millington
Leyden, Richard Noel
Liddicoet, Thomas Herbert
Linder, Maria C
Liska, Kenneth J
Loomis, Albert Geyer
Lukens, Herbert Richard, Jr
LuValle, James Ellis
Maas, Keith Allan
McCloskey, Allen Lyle
Madix, Robert James
Mar, Raymond W
Marantz, Laurence Boyd
Marquis, Marilyn A
Megraw, Robert Ellis
Meiners, Henry C(ito)
Meinhard, James Edgar
Metzger, Robert P
Michael, Leslie William
Michels, Lloyd R
Miller, Glenn Harry
Miller, Stanley Lloyd
Moffitt, Robert Allan
Monson, Richard Stanley
Moore, C Bradley
Morgal, Paul Walter
Morgan, James John
Morris, Jerry Lee, Jr
Mosen, Arthur Walter
Murov, Steven Lee
Murov, Steven Lee
Mysels, Estella Katzenellenbogen
Navratil, James Dale
Naylor, Benjamin Franklin
Needles, Howard Lee
Neesby, Torben Emil
Neff, Loren Lee
Neher, Maynard Bruce
Neu, Ernest Ludwig
Newsam, John M
Newsom, Herbert Charles
Newton, Amos Sylvester
Nixon, Alan Charles
Nozaki, Kenzie
Oesterreicher, Hans
O'Konski, Chester Thomas
Orgel, Leslie E
Owen, Walter Wycliffe
Ozari, Yehuda
Pallos, Ferenc M
Pathania, Rajeshwar S
Pattabhiraman, Tammanur R
Pauling, Linus Carl
Pearson, Michael J
Pearson, Ralph Gottfrid
Peeters, Randall Louis
Perrino, Charles T
Petrucci, Ralph Herbert
Pettijohn, Richard Robert
Phipps, Peter Beverley Powell
Pileggi, Vincent Joseph
Pimentel, George Claude
Pokras, Harold Herbert
Pollack, Louis Rubin
Potter, John Clarkson
Ralls, Jack Warner
Recht, Howard Leonard
Reed, Russell, Jr
Rempel, Herman G
Rocklin, Albert Louis
Rogoff, William Milton
Rowland, F Sherwood
Rudy, Thomas Philip
Savedoff, Lydia Goodman
Schall, Roy Franklin, Jr
Schlatter, James Cameron
Schneider, Meier
Schumacher, Joseph Charles
Schweiker, George Christian
Searcy, A(lan) W(inn)
Sedat, John William
Sellas, James Thomas
Sharts, Clay Marcus
Simon, John Douglas
Simpson, William Tracy
Singer, Stanley
Sloane, Howard J
Slota, Peter John, Jr
Smith, Charles Francis, Jr
Smith, Elbert George
Snyder, Lloyd Robert
Spangler, John Allen
Spitze, LeRoy Alvin
Stevenson, David P(aul)
Stonebraker, Peter Michael
Stosick, Arthur James
Strouse, Charles Earl
Subramani, Suresh
Suess, Hans Eduard
Sweeney, William Alan
Thomas, Gerald Andrew
Tiedcke, Carl Heinrich Wilhelm
Ting, Shih-Fan
Toy, Madeline Shen
Traylor, Patricia Shizuko
Trogler, William C
Truesdell, Alfred Hemingway
Tsao, Makepeace Uho
Turner, James A
Turnlund, Judith Rae
Vagnini, Livio L

Chignell, Derek Alan
Cohen, Donald
Cohen, Harry
Colton, Frank Benjamin
Curtis, Veronica Anne
Damusis, Adolfas
De Pasquali, Giovanni
Dinerstein, Robert Alvin
Drickamer, Harry George
Dybalski, Jack Norbert
Elkins, Robert Hiatt
Epstein, Morton Batlan
Field, Kurt William
Fiess, Harold Alvin
Foster, Harold Marvin
Frame, Robert Roy
Friedland, Waldo Charles
Friedman, Bernard Samuel
Goran, Morris
Goss, George Robert
Green, Frank Orville
Greenberg, Elliott
Grosz, Oliver
Hadley, Elbert Hamilton
Hampel, Clifford Allen
Harper, Jon Jay
Harrington, Joseph Anthony
Harris, Samuel William
Hartlage, James Albert
Henning, Lester Allan
Henry, Robert Edwin
Holtzman, Richard Beves
Howsmon, John Arthur
Ichniowski, Thaddeus Casimir
Inskip, Ervin Basil
Iveson, Herbert Todd
Jason, Emil Fred
Joffe, Morris H
Kalinowski, Mathew Lawrence
Kaufman, Priscilla C
Kirkpatrick, Joel Lee
Kleiman, Morton
Konopinski, Virgil J
Lee, Anthony L
Lerner, Louis L(eonard)
Lester, George Ronald
Lillwitz, Lawrence Dale
McCollum, John David
Magnus, George
Mahajan, Om Prakash
Maltenfort, George Gunther
Marks, Tobin Jay
Martinek, Robert George
Mason, Donald Frank
Matson, Howard John
Maynert, Everett William
Miller, John Robert
Nash, Ralph Glen
Nelson, Arthur Kendall
Neuzil, Richard William
Newhart, M(ary) Joan
Palmer, Curtis Allyn
Parent, Joseph D(ominic)
Penrose, William Roy
Perlow, Mina Rea Jones
Pollitzer, Ernest Leo
Porsche, Jules D(ownes)
Princen, Lambertus Henricus
Pullukat, Thomas Joseph
Regunathan, Perialwar
Richmond, Patricia Ann
Rietz, Edward Gustave
Riley, Reed Farrar
Rosenberg, Saul Howard
Ryan, Julian Gilbert
Salutsky, Murrell Leon
Sawinski, Vincent John
Sehgal, Lakshman R
Sharma, Brahma Dutta
Shida, Mitsuzo
Silvernail, Walter Lawrence
Simmons, William Howard
Simon, Wilbur
Simon Walhout, Justine I
Slama, Francis J
Snow, Adolph Isaac
Sommers, Armiger Henry
Spikes, John Jefferson
Stavrolakis, J(ames) A(lexander)
Steele, Ian McKay
Steele, Sidney Russell
Stehney, Andrew Frank
Steindler, Martin Joseph
Stipanovic, Bozidar J
Stuart, Alfred Herbert
Stucker, Joseph Bernard
Studier, Martin Herman
Sutton, Russell Paul
Sytsma, Louis Frederick
Treadway, William Jack, Jr
Trimble, Russell Fay
Walhout, Justine Isabel Simon
Warne, Thomas Martin
Welsh, Thomas Laurence
Wing, Robert Edward
Woolson, Edwin Albert
Wysocki, Allen John
Yang, Nien-Chu
Yuster, Philip Harold

INDIANA
Abdulla, Riaz Fazal
Amy, Jonathan Weekes
Bodner, George Michael

Booe, J(ames) M(arvin)
Boyd, Frederick Mervin
Chisholm, Malcolm Harold
Christou, George
Das, Paritosh Kumar
Demkovich, Paul Andrew
Diamond, Sidney
Edgell, Walter Francis
Felger, Maurice Monroe
Fricke, Gordon Hugh
Gardner, David Arnold
Getzendaner, Milton Edmond
Janski, Alvin Michael
Johnston, Katharine Gentry
Kammann, Karl Philip, Jr
Kissinger, Peter Thomas
Kroon, James Lee
Martin, Jerome
Mead, Darwin James
Meyerson, Seymour
Montieth, Richard Voorhees
Moorehead, Wells Rufus
Natarajan, Viswanathan
Proksch, Gary J
Rapkin, Myron Colman
Ross, Alberta B
Scherer, George Allen
Schick, Lloyd Alan
Shultz, Clifford Glen
Shupe, Robert Eugene
Sieloff, Ronald F
Sowers, Edward Eugene
Swain, Richard Russell
Taylor, Harold Mellon
Timma, Donald Lee
Truce, William Everett
Walsh, Patrick Noel
Wang, Jin-Liang
Wehrmeister, Herbert Louis
Wilks, Louis Phillip
Williams, Leamon Dale
Yordy, John David

IOWA
Carr, Duane Tucker
Danzer, Laurence Alfred
Dible, William Trotter, Jr
Firstenberger, B(urnett) G(eorge)
Hanak, Joseph J
Meints, Clifford Leroy
Reuland, Robert John
Wubbels, Gene Gerald

KANSAS
Baranczuk, Richard John
Beadle, Buell Wesley
Beard, William Quinby, Jr
Beecham, Curtis Michael
Bellet, Eugene Marshall
Berg, J(ohn) Robert
Boyle, Kathryn Moyne Ward Dittemore
Bricker, Clark Eugene
Choguill, Harold Samuel
Christian, Robert Vernon, Jr
Curless, William Toole
Dreschhoff, Gisela Auguste-Marie
Finney, Karl Frederick
Glick, John Henry, Jr
Hefferren, John James
Hiebert, Allen G
Hirschmann, Robert P
Lanning, Francis Chowing
Lehman, Thomas Alan
McCormick, Bailie Jack
Reynolds, Charles Albert
Rumpel, Max Leonard
Schrenk, William George

KENTUCKY
Beck, Lloyd Willard
Doderer, George Charles
Helms, Boyce Dewayne
Hettinger, William Peter, Jr
Hunter, Norman W
Kary, Christina Dolores
Klingenberg, Joseph John
Knopf, Daniel Peter
Kovach, Stephen Michael
McDermott, Dana Paul
Morris, Edward C
Plucknett, William Kennedy
Wagner, William Frederick

LOUISIANA
Andrews, Bethlehem Kottes
Baird, William C, Jr
Beeler, Myrton Freeman
Broeg, Charles Burton
Bromberg, Milton Jay
Brown, Jerome Engel
Brown, Lawrence E(ldon)
Calamari, Timothy A, Jr
Cook, Shirl Eldon
Daigle, Donald J
Danti, August Gabriel
Dearth, James Dean
DeMonsabert, Winston Russel
Englert, Mary Elizabeth
Foster, Walter Edward
Gormus, Bobby Joe
Greco, Edward Carl
Guthrie, Donald Arthur
Halbert, Thomas Risher
Hamer, Jan

Harvey, Clarence Charles, (Jr)
Kennedy, Frank Scott
Kewish, Ralph Wallace
Laurent, Sebastian Marc
Mangham, Jesse Roger
Merrill, Howard Emerson
Moseley, Harry Edward
Plonsker, Larry
Reeves, Wilson Alvin
Reichle, Alfred Douglas
Reid, John David
Risinger, Gerald E
Roberts, Earl John
Roberts, Reginald Francis
Sachdev, Sham L
Siddall, Thomas Henry, III
Smith, Charles Hooper
Srinivasan, Sathanur Ramachandran
Stuntz, Gordon Frederick
Vingiello, Frank Anthony
Watson, Dennis Ronald
Yeadon, David Allou

MAINE
Bragdon, Robert Wright
Cartier, George Thomas
Robbins, Wayne Brian
Rundell, Clark Ace
Sottery, Theodore Walter
Stanley, Norman Francis

MARYLAND
Addanki, Somasundaram
Affens, Wilbur Allen
Barrack, Carroll Marlin
Baummer, J Charles, Jr
Berch, Julian
Berkowitz, Sidney
Black, Simon
Braude, George Leon
Callahan, Mary Vincent
Carr, Charles Jelleff
Caswell, Robert Little
Chan, Daniel Wan-Yui
Cheng, Sheue-yann
Cohen, Alex
Cohen, Howard Joseph
Collins, John Henry
Conrad, Edward Ezra
Crisler, Joseph Presley
Currie, Lloyd Arthur
Dingell, James V
Edinger, Stanley Evan
Ellis, Rex
Fields, Richard Joel
Fisher, Dale John
Fitch, Steven Joseph
Fowler, Emil Eugene
Frank, Victor Samuel
Frattali, Victor Paul
Freimuth, Henry Charles
Friedman, Fred K
Fristrom, Robert Maurice
Gann, Richard George
Ghoshtagore, Rathindra Nath
Gibian, Thomas George
Goldenson, Jerome
Goldheim, Samuel Lewis
Grant, David Graham
Guinn, Vincent Perry
Gupta, Gian Chand
Haffner, Richard William
Hanford, William Edward
Harris, Milton
Heath, George A(ugustine)
Heimerl, Joseph Mark
Hinton, Deborah M
Horney, Amos Grant
Hoster, Donald Paul
Houk, Albert Edward Hennessee
Howell, Barbara Fennema
Ibrahim, A Mahammad
James, John Cary
Jones, Owen Lloyd
Jones, Thomas Oswell
Kappe, David Syme
Keily, Hubert Joseph
Koch, William Frederick
Kolobielski, Marjan
Krynitsky, John Alexander
Landgrebe, Albert R
Lee, Theresa
Lepley, Arthur Ray
Levine, Alan Stewart
Levy, Robert
Lijinsky, William
McCandliss, Russell John
McCormick, Anna M
McMurdie, Howard Francis
Marans, Nelson Samuel
Mariano, Patrick S
Marquart, Ronald Gary
Marton, Joseph
Massie, Samuel Proctor
Mazumder, Bibhuti R
Melville, Robert S
Merz, Kenneth M(alcolm), Jr
Miller, Hugh Hunt
Milne, George William Anthony
Misra, Renuka
Montgomery, Stewart Robert
Mowry, David Thomas
Murray, Gary Joseph
Naibert, Zane Elvin

Nandedkar, Arvindkumar Narhari
Ondov, John Michael
Panayappan, Ramanathan
Peiser, Herbert Steffen
Phelan, Earl Walter
Pick, Robert Orville
Placious, Robert Charles
Powell, Francis X
Prasanna, Hullahalli Rangaswamy
Rice, Rip G
Richards, Joseph Dudley
Richman, Robert Michael
Rochlin, Phillip
Rosenblatt, David Hirsch
Rosenblum, Annette Tannenholz
Rostenbach, Royal E(dwin)
Roswell, David Frederick
Rush, Cecil Archer
Sanders, James Grady
Schattner, Robert I
Schifreen, Richard Steven
Shah, Shirish
Siatkowski, Ronald E
Sieckmann, Donna G
Silverman, Joseph
Skalny, Jan Peter
Smith, Betty F
Smith, Ieuan Trevor
Strauss, Simon Wolf
Takagi, Shozo
Taylor, Harold Nathaniel
Trimmer, Robert Whitfield
Trus, Benes L
Vennos, Mary Susannah
Vrieland, Gail Edwin
Ward, Joseph Richard
Watters, Robert Lisle
Wells, James Robert
Weser, Don Benton
White, Blanche Babette
Wiggin, Edwin Albert
Wolford, Richard Kenneth
Wright, William Wynn
Yeh, Kwan-Nan
Yeh, Lai-Su Lee
Young, Jay Alfred
Zamora, Antonio
Zdravkovich, Vera

MASSACHUSETTS
Backman, Keith Cameron
Bailey, Milton
Baratta, Edmond John
Bartlett, Paul Doughty
Bazinet, Maurice L
Berlandi, Francis Joseph
Berry, Vern Vincent
Bronstein, Irena Y
Brown, Harold Hubley
Bump, Charles Kilbourne
Burkhardt, Alan Elmer
Chang, Raymond
Charkoudian, John Charles
Chase, Fred Leroy
Clemson, Harry C
Clydesdale, Fergus Macdonald
Coleman, George W(illiam)
Cook, Michael Miller
Coulter, Lowell Vernon
Crandlemere, Robert Wayne
Dasgupta, Arijit M
Deck, Joseph Charles
Deutsch, Marshall Emanuel
Doshi, Anil G
Elfbaum, Stanley Goodman
Evans, David A
Fine, David H
Forchielli, Americo Lewis
Frankel, Richard Barry
Freeman, John Paul
Friedenstein, Hanna
Gibson, George
Giese, Roger Wallace
Giessen, Bill C(ormann)
Goldberg, Gershon Morton
Golubovic, Aleksandar
Gorfien, Harold
Green, Milton
Hadjian, Richard Albert
Hamilton, Charles William
Handy, Carleton Thomas
Hepler, Peter Klock
Hoffman, Donald Oliver
Holland, Andrew Brian
Hume, David Newton
Ives, Robert Southwick
Kaplan, Lawrence Jay
Kliem, Peter O
Korenstein, Ralph
Lambert, Ronald
Levy, Boris
Lin, Sin-Shong
McGrath, W Patrick
MacIver, Donald Stuart
Markland, William R
Mears, Whitney Harris
Merken, Melvin
Morbey, Graham Kenneth
Naidus, Harold
Olson, Arthur Russell
O'Malley, Robert Francis
Panto, Joseph Salvatore
Pappalardo, Romano Giuseppe
Parkinson, R(obert) E(dward)

Chemistry, General (cont)

Peirent, Robert John
Peng, Fred Ming-Sheng
Raphael, Thomas
Raymond, Samuel
Ritt, Paul Edward, Jr
Rogers, Howard Gardner
Ross, Sidney David
Schaffel, Gerson Samuel
Scholberg, Harold Milton
Shepp, Allan
Sigai, Andrew Gary
Sleeman, Richard Alexander
Smith, William Edward
Stein, Samuel H
Strimling, Walter Eugene
Suplinskas, Raymond Joseph
Swain, Charles Gardner
Swank, Thomas Francis
Taunton-Rigby, Alison
Taylor, Lloyd David
Teeter, Martha Mary
Trotz, Samuel Isaac
Vasilos, Thomas
Veidis, Mikelis Valdis
Wald, Fritz Veit
Waller, David Percival
Wechter, Margaret Ann
Weller, Paul Franklin
Wiener, Robert Newman
Wood, John Stanley
Wyman, John E
Zapsalis, Charles

MICHIGAN

Allenstein, Richard Van
Anders, Oswald Ulrich
Anderson, Melvin Lee
Anderson, Robert Hunt
Bartleson, John David
Blinn, Lorena Virginia
Boundy, Ray Harold
Broene, Herman Henry
Bruner, Leonard Bretz, Jr
Colingsworth, Donald Rudolph
Cook, Paul Laverne
Corrigan, Dennis Arthur
Davis, Ralph Anderson
Deck, Charles Francis
Dirkse, Thedford Preston
Doyle, Daryl Joseph
Duggan, Helen Ann
Eldis, George Thomas
Epstein, Emanuel
Fearon, Frederick William Gordon
Felmlee, William John
Fischer, George A
Fortuna, Edward Michael, Jr
Foy, Robert Bastian
Frawley, Nile Nelson
Frevel, Ludo Karl
Frisch, Kurt Charles
Gardner, Wayne Stanley
Gayer, Karl Herman
Gendernalik, Sue Aydelott
Goliber, Edward William
Grochoski, Gregory T
Gruen, Fred Martin
Hallada, Calvin James
Hartman, Robert John
Hartwell, George E, Jr
Heyman, Laurel Elaine
Hinkamp, James Benjamin
Jacko, Michael George
Johnson, James Leslie
Kangas, Donald Arne
Kardos, Otto
Kennelly, William J
Kirschner, Stanley
Ksycki, Mary Joecile
Kummer, Joseph T
LaBarge, Robert Gordon
Leddy, James Jerome
LeGrow, Gary Edward
Lentz, Charles Wesley
Lustgarten, Ronald Krisses
McGraw, Leslie Daniel
Maheswari, Shyam P
Mance, Andrew Mark
Meyer, Heinz Friedrich
Moissides-Hines, Lydia Elizabeth
Moser, Frank Hans
Mullins, John A
Nelson, John Arthur
Northup, Melvin Lee
Peery, Clifford Young
Pierce, James Kenneth
Polmanteer, Keith Earl
Rabold, Gary Paul
Reim, Robert E
Ross, John Franklin
Ruof, Clarence Herman
Sardesai, Vishwanath M
Schaap, A Paul
Schneider, Eric West
Sheets, Donald Guy
Shelef, Mordecai
Snyder, Dexter Dean
Solomon, David Eugene
Speier, John Leo, Jr
Stevens, Calvin Lee
Streiff, Anton Joseph
Strojny, Edwin Joseph

Swartz, Grace Lynn
Swathirajan, S
Tai, Julia Chow
Tasker, Clinton Waldorf
Terry, Samuel Matthew
Vanderkooi, William Nicholas
Wang, Simon S
Wang, Yar-Ming
Webb, Philip Gilbert
Willson, Philip James
Yeung, Patrick Pui-hang
Zabik, Matthew John

MINNESOTA

Agre, Courtland LeVerne
Baker, Michael Harry
Baldwin, Arthur Richard
Beebe, George Warren
Benham, Judith Laureen
Brasted, Robert Crocker
Briggs, David R(euben)
Brink, Norman George
Carlson, Robert Leonard
Childs, William Ves
Cowles, Edward J
Di Gangi, Frank Edward
Durnick, Thomas Jackson
Eisenreich, Steven John
Finholt, Albert Edward
Frank, William Charles
Fridinger, Tomas Lee
Haase, Ashley Thomson
Hagen, Donald Frederick
Harrison, Stuart Amos
Hatcher, John Burton
Holler, Albert Cochran
Iwasaki, Iwao
Janes, Donald Lucian
Jewsbury, Wilbur
Kleber, Eugene Victor
McKenna, Jack F(ontaine)
McMullen, James Clinton
Magnuson, Vincent Richard
Malzer, Gary Lee
Mayerle, James Joseph
Meehan, Edward Joseph
Miessler, Gary Lee
Nowlin, Duane Dale
Paterson, William Gordon
Pearlson, Wilbur H
Sahyun, Melville Richard Valde
Schwartz, A(lbert) Truman
Seibold, Carol Duke
Sherman, Patsy O'Connell
Sobieski, James Fulton
Swofford, Harold S, Jr
Throckmorton, James Rodney
Wertz, John Edward

MISSISSIPPI

Altenkirch, Robert Ames
Emerich, Donald Warren
McHenry, William Earl
Pinson, James Wesley
Russell, Joseph Louis
Wolverton, Billy Charles

MISSOURI

Bovy, Philippe R
Brew, William Barnard
Churchill, Ralph John
Craver, Clara Diddle (Smith)
Craver, John Kenneth
Crutchfield, Marvin Mack
Dyroff, David Ray
Eime, Lester Oscar
Elliott, Joseph Robert
Farnsworth, Marie
Festa, Roger Reginald
Fox, Dale Bennett
Hellerstein, Stanley
Hemphill, Louis
Hodges, Glenn R(oss)
Homeyer, August Henry
James, William Joseph
Kern, Roland James
Khalifah, Raja Gabriel
King, Perry, Jr
Kishore, Ganesh M
Kuhns, John Farrell
Lembke, Roger Roy
Lynch, Dan K
McConaghy, John Stead, Jr
McGinnes, Edgar Allan, Jr
Magruder, Willis Jackson
Mohrman, Harold W
Morse, Ronald Loyd
Nason, Howard King
Orban, Edward
Palmer, John Frank, Jr
Perry, Randolph, Jr
Richard, William Ralph, Jr
Russell, Robert Raymond
Russo, Michael Eugene
Scallet, Barrett Lerner
Sherman, William Reese
Singh, Harbhajan
Stone, Bobbie Dean
Teaford, Margaret Elaine
Waldron, Harold Francis
Walsh, Robert Jerome
Wang, Maw Shiu
Welch, Michael John
Wilcox, Harold Kendall

Winicov, Murray William
Yeager, John Frederick
Zienty, Ferdinand B

MONTANA

Baker, Graeme Levo
Fessenden, Ralph James
Gloege, George Herman
Goering, Kenneth Justin
Julian, Gordon Ray
Mundy, Bradford Philip
Parker, Keith Krom

NEBRASKA

Garey, Carroll Laverne
Griswold, Norman Ernest
Heaney, Robert Proulx
Mattes, Frederick Henry
Nair, Chandra Kunju
Quigley, Herbert Joseph, Jr
Rack, Edward Paul
Sturgeon, George Dennis
Wang, Chin Hsien

NEVADA

Dean, John Gilbert
Dickson, Frank Wilson
Fox, Neil Stewart
Nazy, John Robert

NEW HAMPSHIRE

Cotter, Robert James
Custer, Michael
Kippax, Donald
Roberts, John Edwin
Wolff, Nikolaus Emanuel

NEW JERSEY

Adler, Irwin L
Ager, John Winfrid
Allen, Leland Cullen
Alvarez, Vernon Leon
Andreatch, Anthony J
Antonsen, Donald Hans
Babson, Arthur Lawrence
Barile, George Conrad
Barker, Samuel Lamar
Barnes, Robert Lee
Bassett, Alton Herman
Berk, Bernard
Binder, Michael
Blank, Zvi
Bloch, Aaron Nixon
Bluestein, Claire
Borowsky, Harry Herbert
Bosniack, David S
Brennan, James A
Brewer, Glenn A, Jr
Brody, Stuart Martin
Brown, George Lincoln
Brown, Lawrence Milton
Brown, Stanley Monty
Bulusu, Suryanarayana
Butensky, Irwin
Campbell, Clement, Jr
Carroll, James Joseph
Carter, Richard John
Chang, Jun Hsin
Cheng, Kang
Cohen, Howard Melvin
Cohn, J Gunther
Condon, Francis Edward
Crane, Laura Jane
Crowell, Edwin Patrick
Cruz, Mamerto Manahan, Jr
Cullen, Glenn Wherry
Curry, Michael Joseph
Cutler, Frank Allen, Jr
Dean, Donald E
Deen, Harold E(ugene)
DeMartino, Ronald Nicholas
Di Carlo, Frederick Joseph
Drelich, Arthur (Herbert)
Eachus, Spencer William
Eigen, Edward
Eiszner, James Richard
Elden, Richard Edward
Erdmann, Duane John
Feldman, Nicholas
Firth, William Charles, Jr
Fischer, Robert Leigh
Fischer, Ronald Howard
Flexser, Leo Aaron
Freedman, Arthur Jacob
Frey, William Carl
Fung, Shun Chong
Gajewski, Fred John
Gambino, S(alvatore) Raymond
Gavini, Muralidhara B
Gawley, Irwin H, Jr
Geiger, Marion Braxton
Gerecht, J Fred
Gershon, Sol D
Gilbert, Richard Lapham, Jr
Good, Mary Lowe
Goodloe, Paul Miller, II
Grabowski, Edward Joseph John
Granito, Charles Edward
Gray, Russell Houston
Greenblatt, Martha
Grubman, Wallace Karl
Guthrie, Roger Thackston
Haft, Jacob I
Halfon, Marc

Hay, Peter Marsland
Heffron, Peter John
Herdklotz, John Key
Herrington, Kermit (Dale)
Herzog, Hershel Leon
Hollander, Max Leo
Huger, Francis P
Hughes, O Richard
Idol, James Daniel, Jr
Idson, Bernard
Ivashkiv, Eugene
Jaruzelski, John Janusz
Jeter, Hewitt Webb
Jones, Martha Ownbey
Kaplan, Michael
Karg, Gerhart
Katz, Sidney A
Kern, Werner
Kirshenbaum, Isidor
Knapp, Malcolm Hammond
Koonce, Samuel David
Kraus, Hubert Adolph
Kuebler, William Frank, Jr
Lafornara, Joseph Philip
Lam, Fuk Luen
LaPalme, Donald William
Laudise, Robert Alfred
Layng, Edwin Tower
Levinson, Sidney Bernard
Levy, Marilyn
Li, Shu-Tung
Liebman, Arnold Alvin
Lobunez, Walter
Los, Marinus
LoSurdo, Antonio
Louis, Kwok Toy
Luberoff, Benjamin Joseph
Luthy, Jakob Wilhelm
McGinnis, James Lee
McVey, Jeffrey King
Maleeny, Robert Timothy
Manganaro, James Lawrence
Manganelli, Raymond M(ichael)
Maso, Henry Frank
Mellberg, James Richard
Melveger, Alvin Joseph
Miale, Joseph Nicolas
Mitchell, Thomas Owen
Mohr, Richard Arnold
Most, Joseph Morris
Nadkarni, Ramachandra Anand
Naglieri, Anthony N
Navrotsky, Alexandra
Neiswender, David Daniel
Obeji, John T
Pader, Morton
Parisse, Anthony John
Pedersen, Charles John
Perrotta, James
Price, Alson K
Priesing, Charles Paul
Pugliese, Michael
Ramaprasad, K R (Ram)
Rankel, Lillian Ann
Rau, Eric
Rees, Richard Wilhelm A
Reichle, Walter Thomas
Rinehart, Jay Kent
Robbins, Clarence Ralph
Robinson, Edwin Allin
Rolston, Charles Hopkins
Ropp, Richard C
Rosenthal, Fritz
Roth, Shirley H
Roukes, Michael L
Saldick, Jerome
Salerno, Alphonse
Sausville, Joseph Winston
Schlossman, Mitchell Lloyd
Schneider, Joseph
Schneider, Paul
Schrier, Melvin Henry
Schubert, Rudolf
Schuler, Mathias John
Schwartz, Newton
Scott, William Edwin
Sekutowski, Dennis G
Shah, Atul A
Sherrick, Carl Edwin
Shulman, George
Sikder, Santosh K
Silvestri, Anthony John
Sincius, Joseph Anthony
Sivco, Deborah L
Soled, Stuart
Sonn, George Frank
Sorkin, Marshall
Spiegel, Herbert Eli
Spiegelman, Gerald Henry
Spiro, Thomas George
Staum, Muni M
Steiger, Fred Harold
Stempel, Arthur
Stern, Eric Wolfgang
Stingl, Hans Alfred
Stoy, William S
Stults, Frederick Howard
Subramanyam, Dilip Kumar
Suciu, George Dan
Sugam, Richard Jay
Sutman, Frank X
Swenson, Richard Waltner
Swenson, Theresa Lynn
Sykes, Donald Joseph

Szyper, Mira
Tatyrek, Alfred Frank
Taylor, Paul Duane
Thelin, Jack Horstmann
Thich, John Adong
Thomas, Robert Joseph
Thurman, Richard Gary
Torok, Andrew, Jr
Townley, Robert William
Tripathi, Uma Prasad
Turse, Richard S
Valdsaar, Herbert
Van de Castle, John F
Van Saun, William Arthur
Vellturo, Anthony Francis
Venuto, Paul B
Von, Isaiah
Voss, Kenneth Edwin
Vyas, Brijesh
Watt, William Russell
Wehrli, Pius Anton
Weisgerber, George Austin
White, John Francis
White, John Greville
Wilson, Robert G
Wood, Darwin Lewis
Woodbridge, Joseph Eliot
Young, Edmond Grove
Young, Lewis Brewster
Zaim, Semih
Zoss, Abraham Oscar

NEW MEXICO
Anderson, William Loyd
Attrep, Moses, Jr
Baker, Richard Dean
Balagna, John Paul
Baxman, Horace Roy
Bentley, Glenn E
Bravo, Justo Baladjay
Browne, Charles Idol
Bryant, Ernest Atherton
Caton, Roy Dudley, Jr
Cheavens, Thomas Henry
Cowan, George A
Cox, Lawrence Edward
Daniels, William Richard
Eller, Phillip Gary
Ford, George Pratt
Giorgi, Angelo Louis
Gray, Eoin Wedderburn
Groblewski, Gerald Eugene
Guenther, Arthur Henry
Hardy, Paul Wilson
Heldman, Julius David
Hersh, Sylvan David
Hobart, David Edward
Kahn, Milton
Kjeldgaard, Edwin Andreas
Kunz, Walter Ernest
Levine, Herman Saul
Luehr, Charles Poling
McKenzie, James Montgomery
Magnani, Nicholas J
Melgard, Rodney
Ott, Donald George
Patterson, James Howard
Powers, Dana Auburn
Richardson, Albert Edward
Sparks, Morgan
Standefer, Jimmy Clayton
Steinhaus, David Walter
Storm, Carlyle Bell
Stubbs, Morris Frank
Summers, Donald Balch
Thompson, Joseph Lippard
Wolfsberg, Kurt
Zack, Neil Richard

NEW YORK
Adler, George
Adler, Stephen Fred
Albrecht, Andreas Christopher
Allen, Augustine Oliver
Allen, Donald Stewart
Allen, Gary William
Allentoff, Norman
Alliet, David F
Armamento, Eduardo T
Armour, Eugene Arthur
Auerbach, Clemens
Bacon, Egbert King
Bacon, Robert Elwin
Bahl, Om Parkash
Bard, Charleton Cordery
Baron, Arthur L
Barr, Donald Eugene
Bednar, Rodney Allan
Bhattacharya-Chatterjee, Malaya
Bielski, Benon H J
Blatt, Sylvia
Blumenthal, Warren Barnett
Booms, Robert Edward
Bove, John L
Brauer, Joseph B(ertram)
Brink, Gilbert Oscar
Brody, Bernard B
Brown, Eric Reeder
Brown, Robert Getman
Brust, David Philip
Bunce, Stanley Chalmers
Burge, Robert Ernest, Jr
Burtt, Benjamin Pickering
Cabeen, Samuel Kirkland

Cappel, C Robert
Caso, Marguerite Miriam
Chang, Hai-Won
Chen, Cindy Chei-Jen
Chen, Stephen P K
Chilcote, Max Eli
Chodroff, Saul
Christensen, Edward Richards
Clark, Herbert Mottram
Clemens, Carl Frederick
Cofrancesco, Anthony J
Cohen, Jacob Isaac
Coll, Hans
Conant, Robert Henry
Condrate, Robert Adam
Coppens, Philip
Copulsky, William
Corbeels, Roger
Dabrowiak, James Chester
Daniels, John Maynard
Dessauer, John Hans
Dichter, Michael
Dille, Kenneth Leroy
Dixon, William Brightman
Drickamer, Kurt
Edgerton, Robert Flint
Edwards, Leila
Eibeck, Richard Elmer
Eirich, Frederick Roland
Feit, Irving N
Feitler, David
Feldman, Larry Howard
Fischer, Leewellyn C
Fix, Delbert Dale
Flank, William H
Fleming, James Charles
Follows, Alan Greaves
Forsyth, Paul Francis
Freeman, Richard Carl
Friberg, Stig E
Fromageot, Henri Pierre-Marcel
Garza, Cutberto
Gladstone, Matthew Theodore
Gogel, Germaine E
Gregor, Harry Paul
Griffiths, Joan Martha
Grina, Larry Dale
Grubman, Marvin J
Gurel, Demet
Gwilt, John Ruff
Habig, Robert L
Harbison, Gerard Stanislaus
Hastings, Julius Mitchell
Hellman, Henry Martin
Hensler, Joseph Raymond
Hersh, Leroy S
Holder, Charles Burt, Jr
Holtz, Carl Frederick
Holub, Fred F
Horvat, Robert Emil
Hunt, Heman Dowd
Hunt, Roy Edward
Jabbur, Ramzi Jibrail
Jaffe, Marvin Richard
Jelling, Murray
Johnson, Hollister, Jr
Joseph, Solomon
Julian, Donald Benjamin
Kablaoui, Mahmoud Shafiq
Keirns, Mary Hull
Keisch, Bernard
Kellogg, Lillian Marie
Kennard, Kenneth Clayton
Kishore, Gollamudi Sitaram
Klerer, Julius
Kofron, James Thomas, Jr
Kurz, Richard Karl
Lai, Yuan-Zong
Lam, Stanley K
Larson, Allan Bennett
Law, Paul Arthur
LeBlanc, Jerald Thomas
Lee, Ching-Li
Lerman, Sidney
Levin, Robert Aaron
Lincoln, Lewis Lauren
Lippe, Robert Lloyd
Loscalzo, Anne Grace
Lum, Kin K
Luther, Herbert George
McEvoy, Francis Joseph
Mackles, Leonard
McLaen, Donald Francis
Martell, Michael Joseph, Jr
Marton, Renata
Maurer, Gernant E
Mausner, Jack
Merkel, Paul Barrett
Merrigan, Joseph A
Miller, Melvin J
Model, Frank Steven
Morrison, John Agnew
Mourning, Michael Charles
Muenter, Annabel Adams
Mullen, Patricia Ann
Mumbach, Norbert R
Murphy, Cornelius Bernard
Murphy, Daniel Barker
Mylroie, Victor L
Nahabedian, Kevork Vartan
Naider, Fred R
Nash, Robert Arnold
Nelson, Lawrence Barclay
Nerken, Albert

Neuberger, Dan
Nolan, John Thomas, Jr
Ober, Christopher Kemper
Oberender, Frederick G
Oja, Tonis
Osborn, H(arland) James
Osterholtz, Frederick David
Padalino, Stephen John
Pedroza, Gregorio Cruz
Pemrick, Raymond Edward
Pepe, Joseph Philip
Pepkowitz, Leonard Paul
Perlman, Morris Leonard
Perlstein, Jerome Howard
Pesce, Michael A
Post, Howard William
Potter, George Henry
Purandare, Yeshwant K
Quinan, James Roger
Rimai, Donald Saul
Rocks, Lawrence
Roffman, Steven
Rosano, Henri Louis
Ross, Sydney
Rothermel, Joseph Jackson
Rothstein, Samuel
Saeva, Franklin Donald
Saletan, Leonard Timothy
Salvo, Joseph J
Scaife, Charles Walter John
Schaffrath, Robert Eben
Schallhorn, Charles H
Schick, Jerome David
Schindler, Hans
Scholes, Samuel Ray, Jr
Schroeter, Siegfried Hermann
Schwartz, Morton K
Seanor, Donald A
Seus, Edward J
Shapiro, Raymond E
Sharbaugh, Amandus Harry
Sharma, Minoti
Sheeran, Stanley Robert
Shen, Samuel Yi-Wen
Silleck, Clarence Frederick
Singh, Baldev
Slate, Floyd Owen
Slezak, Jane Ann
Smith, Gale Eugene
Sparberg, Esther Braun
Spingola, Frank
Spittler, Terry Dale
Springer, Dwight Sylvan
Srivastava, Suresh Chandra
Stachel, Johanna
Stasiw, Roman Orest
Sterrett, Frances Susan
Stiel, Leonard Irwin
Stoll, William Russell
Sturges, Stuart
Sullivan, Michael Francis
Sutherland, George Leslie
Sweeney, William Mortimer
Swinehart, James Stephen
Tassinari, Silvio John
Texter, John
Thomson, Gerald Edmund
Tischer, Thomas Norman
Toribara, Taft Yutaka
Tuite, Robert Joseph
Urban, Joseph
Van Norman, Gilden Ramon
Via, Francis Anthony
Vincent, George Paul
Vogel, Alfred Morris
Wadlinger, Robert Louis Peter
Whittingham, M(ichael) Stanley
Winstrom, Leon Oscar
Wolfson, Leonard Louis
Wyatt, Benjamin Woodrow
Yao, Jerry Shi Kuang
Yourtee, Lawrence Karn
Zakkay, Victor
Zimar, Frank
Zingaro, Joseph S
Zinman, Walter George
Zwick, Daan Marsh
Zwicker, Walter Karl

NORTH CAROLINA
Apellaniz, Joseph E P
Aycock, Benjamin Franklin
Bauermeister, Herman Otto
Beaumont, Ralph Harrison, Jr
Beck, Keith Russell
Berry, George William
Bockstahler, Theodore Edwin
Bonk, James F
Bradley, Harris Walton
Bright, Gordon Stanley
Casey, James Patrick
Chang, Hou-Min
Chou, David Yuan Pin
Christman, Russell Fabrique
Cialdella, Cataldo
Clemens, Donald Faull
Crutcher, Harold L
Culbreth, Judith Elizabeth
Dawe, Harold Joseph
Deal, Glenn W, Jr
Dixit, Ajit Suresh
Edwards, Ben E
Ellestad, Reuben B
Findley, William Robert

Fine, Jo-David
Flokstra, John Hilbert
Gallent, John Bryant
Gillespie, Arthur Samuel, Jr
Goldstein, Irving Solomon
Green, Charles Raymond
Groszos, Stephen Joseph
Hall, Seymour Gerald
Herrell, Astor Y
Hess, Daniel Nicholas
Hilliard, Roy C
Hoelzel, Charles Bernard
Hoffman, Doyt K, Jr
Hovis, Louis Samuel
Hurlbert, Bernard Stuart
Jennings, Carl Anthony
Johnson, Ronald Sanders
Jones, Daniel Silas, Jr
Jones, Samuel O'Brien
Kaufmann, Peter John
Knight, Samuel Bradley
Leidy, Ross Bennett
Loranger, William Farrand
McPhail, Andrew Tennent
Martens, Christopher Sargent
Morris, Leo Raymond
Moseman, Robert Fredrick
Mushak, Paul
Nowell, John William
Purchase, Earl Ralph
Senkus, Murray
Smith, Fred R, Jr
Smith, Susan T
Sookne, Arnold Maurice
Spremulli, Linda Lucy
Squibb, Samuel Dexter
Teague, Claude Edward, Jr
Tschanz, Christian
Tulagin, Vsevolod
Watts, Plato Hilton, Jr
Wehner, Philip
Wollin, Goesta
Woosley, Royce Stanley

NORTH DAKOTA
Abrahamson, Harmon Bruce
Bauer, Armand
Fleeker, James R
McCarthy, Gregory Joseph
Rathmann, Franz Heinrich

OHIO
Antler, Morton
Atam-Alibeckoff, Galib-Bey
Bailey, John Clark
Ball, David Ralph
Bayless, Philip Leighton
Behrmann, Eleanor Mitts
Botros, Raouf
Brady, Thomas E
Brown, Glenn Halstead
Burrows, Kerilyn Christine
Campbell, Donald R
Cardon, Samuel Zelig
Carter, Carolyn Sue
Cobb, Thomas Berry
Coe, Kenneth Loren
Cole, John Oliver
Connor, Daniel S
Cook, William R, Jr
Croxton, Frank Cutshaw
Damian, Carol G
Darling, Samuel Mills
DeWitt, Bernard James
Dorfman, Leon Monte
Dove, Ray Allen
Eaker, Charles Mayfield
Finn, John Martin
Fischer, Mark Benjamin
Fortman, John Joseph
Gage, Frederick Worthington
Garascia, Richard Joseph
Garrett, Alfred Benjamin
Gibbins, Betty Jane
Gibbons, Louis Charles
Gird, Steven Richard
Goetz, Richard W
Goodson, Alan Leslie
Graham, Paul Whitener Link
Grisaffe, Salvatore J
Guo, Hua
Hall, Ronald Henry
Hanes, Ronnie Michael
Hanks, Richard Donald
Harmony, Judith A K
Harris, Arlo Dean
Hart, David Joel
Hein, Richard William
Helminiak, Thaddeus Edmund
Hine, Jack
Hohnadel, David Charles
Holland, William Frederick
Innes, John Edwin
Ish, Carl Jackson
Johnston, Herbert Norris
Kawahara, Fred Katsumi
Kay, Peter Steven
Kershner, Carl John
Khosla, Mahesh C
Klopman, Gilles
Knecht, Walter Ludwig
Knipple, Warren Russell
Kohl, Fred John
Koknat, Friedrich Wilhelm

Chemistry, General (cont)

Kolopajlo, Lawrence Hugh
Kozikowski, Barbara Ann
Krohn, Albertine
Kwiatek, Jack
Lateef, Abdul Bari
Lehn, William Lee
Lepp, Cyrus Andrew
Lewis, James Edward
Lichtin, J Leon
Luoma, Ernie Victor
McCarty, Lewis Vernon
McCune, Homer Wallace
McGandy, Edward Lewis
McKinnis, Charles Leslie
McLoughlin, Daniel Joseph
Mayer, Ramona Ann
Meckstroth, Wilma Koenig
Meites, Samuel
Mettee, Howard Dawson
Michael, William R
Milberger, Ernest Carl
Miller, William Reynolds, Jr
Morton, Maurice
Muntean, Richard August
Muse, Joel, Jr
Myers, Elliot H
Nelson, Gilbert Harry
Newby, John R
Nygaard, Oddvar Frithjof
Oesterling, Thomas O
Olson, Walter T
Ouimet, Alfred J, Jr
Patel, Siddharth Manilal
Paynter, John, Jr
Peters, Lynn Randolph
Rau, Allen H
Renoll, Mary Wilhelmine
Rich, Ronald Lee
Roberts, Timothy R
Rubinson, Judith Faye
Rubinson, Kenneth A
Schaff, John Franklin
Schales, Otto
Schmidt, Donald L
Semon, Waldo Lonsbury
Shackle, Dale Richard
Shine, Daniel Phillip
Sievert, Carl Frank
Simpson, William Henry
Skees, Hugh Benedict
Snyder, Carl Edward
Snyder, Milton Jack
Spencer, Walter William
Spittler, Ernest George
Spoehr, Albert Frederick
Stevens, Henry Conrad
Stevenson, Don R
Sullenger, Don Bruce
Suter, Robert Winford
Sweet, Thomas Richard
Tamborski, Christ
Thekdi, Arvind C
Thomas, George B
Thomas, McCalip Joseph
Toeniskoetter, Richard Henry
Towns, Robert Lee Roy
Turnbull, Kenneth
Van Winkle, Quentin
Velenyi, Louis Joseph
Volk, Murray Edward
Walters, Martha I
Watters, James I
Weidner, Bruce Van Scoyoc
Wiedenheft, Charles John
Wilkes, Charles Eugene
Yang, Philip Yung-Chin
Yang, Tsanyen
Yocum, Ronald Harris

OKLAHOMA
Axe, William Nelson
Ball, John Sigler
Beaver, W Don
Bills, John Lawrence
Bresson, Clarence Richard
Coffman, Harold H
Cowley, Thomas Gladman
Daake, Richard Lynn
Dermer, Otis Clifford
Doss, Richard Courtland
Fenstermaker, Roger William
Ferguson, William Sidney
Frost, Jackie Gene
Hamm, Donald Ivan
Harrison, Hugh Thomas
Heasley, Gene
Hermann, John Alexander
Holtmyer, Marlin Dean
Lee, Diana Mang
Linder, Donald Ernst
Matson, Michael Steven
Matson, Ted P
Mills, King Louis, Jr
Motz, Kaye La Marr
Natowsky, Sheldon
Nesbitt, Stuart Stoner
Norell, John Reynolds
Nye, Mary Jo
O'Neal, Steven George
Passey, Richard Boyd
Pitchford, Armin Cloyst
Pritchard, James Edward

Purdue, Jack Olen
Richardson, Verlin Homer
Scott, Charles Edward
Sheshtawy, Adel A
Varga, Louis P
Washcheck, Paul Howard
Wheelock, Kenneth Steven
Williamson, William Burton
Wimmer, Donn Braden
Witt, Donald Reinhold

OREGON
Arp, Daniel James
Barnum, Dennis W
Bartlett, James Kenneth
Brandt, Howard Allen
Charlton, David Berry
Dittmer, Karl
Fincke, Margaret Louise
Fleetwood, Charles Wesley
Gerke, John Royal
Graham, Beardsley
Halko, Barbara Tomlonovic
Henry, Jack Leland
Jaeger, Charles Wayne
Keedy, Curtis Russell
Larsen, Marlin Lee
Lochner, Janis Elizabeth
Long, James William
MacVicar, Robert William
Meredith, Robert E(ugene)
Pankow, James Frederick
Scott, Peter Carlton
Shoemaker, David Powell
Terraglio, Frank Peter
Torley, Robert Edward
Trent, Walter Russell
Wang, Chih Hsing
Winniford, Robert Stanley

PENNSYLVANIA
Abrams, Ellis
Adams, Roy Melville
Andelman, Julian Barry
Barker, Franklin Brett
Bartish, Charles Michael Christopher
Batzar, Kenneth
Beavers, Ellington McHenry
Bell, Ian
Berkowitz, Harry Leo
Bissey, Luther Trauger
Bodamer, George Willoughby
Bogard, Andrew Dale
Bossard, Mary Jeanette
Bowman, Kenneth Aaron
Bowne, Samuel Winter, Jr
Brantley, Susan Louise
Buckles, Marjorie Fox
Buechler, Peter Robert
Buhle, Emmett Loren
Calkins, Charles Richard
Casassa, Ethel Zaiser
Castle, John Edwards
Catalano, Anthony William
Cheng, Cheng-Yin
Cheng, Hung-Yuan
Chenot, Charles Frederic
Clark, T(helma) K
Clarke, Duane Grookett
Clement, Gerald Edwin
Coleman, Charles Mosby
Connors, William Matthew
Conroy, James Strickler
Constantinides, Panayiotis Pericleous
Cowen, William Frank
Cupper, Robert Alton
Dalal, Fram Rustom
Deischer, Claude Knauss
Dicciani, Nancy Kay
Diethorn, Ward Samuel
Dunkelberger, Tobias Henry
Einhorn, Philip A
Ernst, Richard Edward
Eror, Nicholas George, Jr
Fearing, Ralph Burton
Feeman, James Frederic
Feldman, Kenneth Scott
Ficher, Miguel
Filachione, Edward Mario
Fisher, James Delbert
Fowkes, Frederick Mayhew
Frank, William Benson
Friedman, Arnold Carl
Fugger, Joseph
Ghosh, Mriganka M(ouli)
Gilbert, William Irwin
Gillman, Hyman David
Gottscho, Alfred M(orton)
Greenberg, Charles Bernard
Grundmann, Christoph Johann
Guild, Lloyd V
Gur, David
Hall, Gary R
Hall, W Keith
Hall, William Heinlen
Halpern, Benjamin David
Harkins, Thomas Regis
Harren, Richard Edward
Harrison, John William
Ho, Shih Ming
Hoberman, Alfred Elliott
Hochstrasser, Robin
Hosler, Peter
Huff, George Franklin

Hunter, James Bruce
Johnston, Gordon Robert
Karo, Wolf
Kauffman, Joel Mervin
Khatami, Mahin
Kirklin, Perry William
Kirsch, Francis William
Kitazawa, George
Kleinman, Roberta Wilma
Kochar, Harvinder K
Kuebler, John Ralph, Jr
Lacoste, Rene John
Lange, Barry Clifford
Larson, Kenneth Curtis
Lloyd, Thomas Blair
Lyman, William Ray
McCurry, Patrick Matthew, Jr
MacDiarmid, Alan Graham
McGovern, John Joseph
Martinez de Pinillos, Joaquin Victor
Matocha, Charles K
Meadows, Geoffrey Walsh
Melamed, Nathan T
Mendelsohn, Morris A
Merner, Richard Raymond
Metzger, Sidney Henry, Jr
Millard, Frederick William
Mirhej, Michael Edward
Misra, Sudhan Sekher
Mortimer, Charles Edgar
Moscony, John Joseph
Myers, Earl Eugene
Myers, Raymond Reever
Nath, Amar
Nemeth, Edward Joseph
Neubeck, Clifford Edward
Nuessle, Albert Christian
Peterman, Keith Eugene
Pierce, Roberta Marion
Rao, V Udaya S
Renfrew, Edgar Earl
Rogers, Ralph Loucks
Romovacek, George R
Rowe, Jay Elwood
Ruh, Edwin
Rushton, Brian Mandel
Saffer, Charles Martin, Jr
Sanders, Charles Irvine
Sankar, Suryanarayan G
Sax, Sylvan Maurice
Scharpf, William George
Schramm, Robert Frederick
Schrum, Robert Wallace
Sexsmith, Frederick Hamilton
Shapiro, Zalman Mordecai
Sheeran, Patrick Jerome
Sieger, John S(ylvester)
Sloat, Charles Allen
Smith, Carolyn Jean
Smith, Charles Lea
Smith, James S(terrett)
Smith, James Stanley
Squire, Edward Noonan
Stanley, Edward Livingston
Steiner, Russell Irwin
Stengle, William Bernard
Suyama, Yoshitaka
Suydam, Frederick Henry
Sykes, James Aubrey, Jr
Tosch, William Conrad
Trachtman, Mendel
Tulk, Alexander Stuart
Van Dolah, Robert Wayne
Varma, Asha
Vaux, James Edward, Jr
Vick, Gerald Kieth
Wang, Ke-Chin
Warme, Paul Kenneth
Weisz, Paul Burg
Weldes, Helmut H
Werkman, Joyce
Windisch, Rita M
Wissow, Lennard Jay
Wohleber, David Alan
Woodhouse, John Crawford
Wootten, Michael John
Wu, Ching-Yong
Zaika, Laura Larysa
Zimmt, Werner Siegfried

RHODE ISLAND
Brown, Phyllis R
Caroselli, Remus Francis
De Bethune, Andre Jacques
Di Pippo, Ascanio G
Gerritsen, Hendrik Jurjen
Griffiths, William C
MacKenzie, Scott, Jr

SOUTH CAROLINA
Adams, Daniel Otis
Amidon, Roger Welton
Asleson, Gary Lee
Aspland, John Richard
Bailey, Carl Williams, III
Ball, Frank Jervery
Baynes, John William
Bibler, Ned Eugene
Billica, Harry Robert
Bowman, Wilfred William
Buurman, Clarence Harold
Cathey, LeConte
Chase, Vernon Lindsay
Clayton, Fred Ralph, Jr

Culbertson, Edwin Charles
DesMarteau, Darryl D
Earl, Charles Riley
Ghaffar, Abdul
Glocker, Edwin Merriam
Goldstein, Herman Bernard
Hayes, David Wayne
Holcomb, Herman Perry
Hunter, George William
King, Lee Curtis
Kistler, Malathi K
LaFleur, Kermit Stillman
Macaulay, Andrew James
Metz, Clyde
Moncrief, John William
Montenyohl, Victor Irl
Otto, Wolfgang Karl Ferdinand
Park, Conrad B
Pecka, James Thomas
Russell, Allen Stevenson
Sax, Karl Jolivette
Scott, Jaunita Simons
Sello, Stephen
Spain, James Dorris, Jr
Spooner, George Hansford
Sproul, Gordon Duane
Weiss, Michael Karl
Wheat, Joseph Allen
Wise, John Thomas
Worsham, Walter Castine

SOUTH DAKOTA
Hanson, Milton Paul
Jensen, William Phelps
Johnson, Elmer Roger
Landborg, Richard John

TENNESSEE
Adams, Robert Edward
Baker, Philip Schaffner
Bloor, John E
Brokaw, George Young
Butler, Thomas Arthur
Clark, Richard Bennett
Clemens, Robert Jay
Coleman, Charles Franklin
Davis, Phillip Howard
Davis, Wallace, Jr
Egan, B Zane
Elowe, Louis N
Embree, Norris Dean
Gleason, Geoffrey Irving
Gloyer, Stewart Edward
Goodson, Louie Aubrey, Jr
Gupta, Brij Mohan
Harris, Warren Whitman
Hefferlin, Ray (Alden)
Henry, Jonathan Flake
Hibbs, Roger Franklin
Horton, Charles Abell
Hurst, G Samuel
Katz, Sidney
Keim, Christopher Peter
Kelly, Minton J
Lappin, Gerald R
Lyon, William Southern, Jr
McDaniel, Edgar Lamar, Jr
McDuffie, Bruce
Martin, James Cuthbert
Martin, Richard Blazo
Mattison, Louis Emil
Natelson, Samuel
Nunez, Loys Joseph
Nyssen, Gerard Allan
Pearson, Donald Emanual
Penner, Hellmut Philip
Perkins, James
Perry, Lloyd Holden
Pinkerton, Frank Henry
Poutsma, Marvin Lloyd
Rainey, Robert Hamric
Rudolph, Philip S
Schmitt, Charles Rudolph
Schreyer, James Marlin
Schulert, Arthur Robert
Simhan, Raj
Sizemore, Douglas Reece
Skinner, John Taylor
Stratton, Wilmer Joseph
Strehlow, Richard Alan
Stroud, Robert Wayne
Sublett, Robert L
Taylor, Kirman
Torrey, Rubye Prigmore
Van Wazer, John Robert

TEXAS
Adams, Charles Rex
Adams, Herman Ray
Adcock, Willis Alfred
Aguilo, Adolfo
Ahlberg, Dan Leander
Altmiller, Henry
Aquino, Dolores Catherine
Asperger, Robert George, Sr
Bachmann, John Henry, Jr
Baker, Samuel I
Barton, Derek Harold Richard
Battista, Orlando Aloysius
Bellavance, David Walter
Benner, Gerald Stokes
Bennett, Richard Henry
Bernal, Ivan
Bowerman, Ernest William

Broun, Thorowgood Taylor, Jr
Bungo, Michael William
Burkart, Leonard F
Burke, Roger E
Bush, Warren Van Ness
Carbajal, Bernard Gonzales, III
Carlson, Kenneth Theodore
Castrillon, Jose P A
Cavitt, Stanley Bruce
Cleek, Given Wood
Cody, John T
Cotton, Frank Albert
Crecelius, Robert Lee
Cronkright, Walter Allyn, Jr
Cruse, Robert Ridgely
Daugherty, Kenneth E
DeBerry, David Wayne
Dille, Roger McCormick
Dobrott, Robert D
Ducis, Ilze
Edwards, Gayle Dameron
Edwards, George
Ellzey, Marion Lawrence, Jr
Escue, Richard Byrd, Jr
Farhataziz, Mr
Ferraris, John Patrick
Fleenor, Marvin Bension
Floyd, Willis Waldo
Folkers, Karl August
Fritsche, Herbert Ahart, Jr
Gibson, Joseph Woodward
Goodwin, John Thomas, Jr
Grissom, David
Gryting, Harold Julian
Guenther, Frederick Oliver
Haberman, John Phillip
Hackerman, Norman
Hale, Cecil Harrison
Ham, Joe Strother
Hammond, James W
Harbordt, C(harles) Michael
Hardcastle, James Edward
Hausler, Rudolf H
Heller, Adam
Henery, James Daniel
Herbert, Stephen Aven
Hoefelmeyer, Albert Bernard
Horridge, Patricia Emily
Hosmane, Narayan Sadashiv
Hutchinson, Bennett B
Ivie, Glen Wayne
Jeanes, Jack Kenneth
Johnson, Mary Lynn Miller
Jones, Jesse W
Jordan, John William
Keenan, Roy W
Kelley, Jim Lee
Kelso, Edward Albert
King, Edward Frazier
Koestler, Robert Charles
Kreglewski, Alexander
Kulkarni, Padmakar Venkatrao
Ladde, Gangaram Shivlingappa
Laity, John Lawrence
Lawson, Jimmie Brown
Levine, Duane Gilbert
Levy, Leon Bruce
Lew, Chel Wing
Lupski, James Richard
Lyle, Robert Edward, Jr
McAdoo, David J
Macaluso, Anthony, Sr
McCown, Joseph Dana
Machacek, Oldrich
McMillin, Jeanie
McPherson, Clinton Marsud
Manning, Harold Edwin
Massingill, John Lee, Jr
Meerbott, William Keddie
Melton, Marilyn Anders
Merchant, Philip, Jr
Moffat, James
Morgan, Bryan Edward
Morgan, Leon Owen
Moser, James Howard
Mosier, Benjamin
Moy, Mamie Wong
Naae, Douglas Gene
Nielsen, Robert Peter
Norton, Scott J
O'Connor, Rod
Ostroff, Anton G
Otto, John B, Jr
Pannell, Richard Byron
Parker, Patrick LeGrand
Patil, Kashinath Ziparu
Phillips, Guy Frank
Reeves, Perry Clayton
Reisberg, Joseph
Rigdon, Orville Wayne
Roberts, Ammarette
Rodriguez, Charles F
Roehrig, Gerald Ralph
Rosenfeld, Daniel David
Safe, Stephen Harvey
Saleh, Farida Yousry
Sartor, Albin Francis, Jr
Schroepfer, George John, Jr
Sears, Raymond Eric John
Shilstone, James Maxwell, Jr
Sicilio, Fred
Simonsen, Stanley Harold
Skarlos, Leonidas
Skelley, Dean Sutherland

Stewart, Frank Edwin
Stranges, Anthony Nicholas
Stuckwisch, Clarence George
Suttle, Andrew Dillard, Jr
Tang, Yi-Noo
Tanner, Alan Roger
Tiedemann, Herman Henry
Towne, Jack C
Unruh, Jerry Dean
Varma, Rajender S
Wagner, Lawrence Carl
Walbrick, Johnny Mac
Wald, Milton M
Waters, John Albert
Weigel, Paul H(enry)
Wendt, Theodore Mil
Woller, William Henry
Wurth, Thomas Joseph
Yao, Joe

UTAH
Alexander, Guy B
Ash, Kenneth Owen
Borrowman, S Ralph
Cagle, Fredric William, Jr
Gardner, William H
Gates, Henry Stillman
Hansen, Lee Duane
Haslem, William Joshua
Hill, George Richard
Hunter, Byron Alexander
Langheinrich, Armin P(aul)
McCloskey, James Augustus, Jr
Miller, Richard Roy
Nauman, Edward Franklin
Oblad, Alex Golden
Poulter, Charles Dale
Stokes, Barry Owen
Street, Joseph Curtis
Stuart, David Marshall
Sudweeks, Walter Bentley
Thamer, B(urton) J(ohn)
Ursenbach, Wayne Octave

VERMONT
Gilbert, Arthur Donald
Kuffner, Roy Joseph
Lang, William Harry
Preuss, Albert F
Steele, Richard
Wooding, William Minor

VIRGINIA
Alexander, Allen Leander
Bailey, Susan Goodman
Balmforth, Dennis
Berry, Clark Green
Blair, Barbara Ann
Bliss, Laura
Blodgett, Robert Bell
Borum, Olin H
Casali, Liberty
Clark, Walter Ernest
Clifford, Alan Frank
Cole, James Webb, Jr
Cool, Raymond Dean
De Marco, Ronald Anthony
DeVore, Thomas Carroll
Dodgen, Durward F
Durrill, Preston Lee
Finley, Arlington Levart
Fischbach, Henry
Garretson, Harold H
Greinke, Everett D
Gruemer, Hanns-Dieter
Haas, Carol Kressler
Hecht, Sidney Michael
Hemley, John Julian
Holmes, Joseph Charles
Huggett, Clayton (McKenna)
Inskeep, George Esler
Jackson, Elizabeth Burger
Jenkins, Robert Walls, Jr
Keihn, Frederick George
Kuhn, William Frederick
Moroni, Eneo C
Morrell, William Egbert
Munday, J(ohn) C(lingman)
Murray, John Wolcott
Newman, Richard Holt
Pohlmann, Juergen Lothar Wolfgang
Rainer, Norman Barry
Rainis, Albert Edward
Ritz, Victor Henry
Roche, James Norman
Rothstein, Lewis Robert
Savory, John
Sayala, Chhaya
Scully, Frank E, Jr
Segura, Gonzalo, Jr
Sobel, Robert Edward
Spindel, William
Stolow, Nathan
Turner, Carlton Edgar
Twilley, Ian Charles
Urbanik, Arthur Ronald
Van Norman, John Donald
Wayland, Rosser Lee, Jr
Wei, Enoch Ping
Wilber, Joe Casley, Jr
Zakrzewski, Thomas Michael

WASHINGTON
Atkins, William M

Barton, Gerald Blackett
Bennett, Clifton Francis
Berry, Keith O
Black, Otis Deitz
Brouns, Richard John
Cady, George Hamilton
Carlson, Lewis John
Chinn, Clarence Edward
Coe, Richard Hanson
Crittenden, Alden La Rue
Diebel, Robert Norman
Evans, John Charles, Jr
Exarhos, Gregory James
Fairhall, Arthur William
Fournier, R E Keith
Gahler, Arnold Robert
Garland, John Kenneth
Goldschmid, Otto
Guion, Thomas Hyman
Hausenbuiller, Robert Lee
Ho, Lydia Su-yong
Hrutfiord, Bjorn F
Jones, Carl Trainer
Kaplan, Alex
Kelly, Jeffrey John
Kent, Ronald Allan
Killingsworth, Lawrence Madison
Kokta, Milan Rastislav
Lackey, Homer Baird
Liang, Shou Chu
Lo, Cheng Fan
Ludwig, Charles Heberle
Macklin, John Welton
Morris, Daniel Luzon
Nowotny, Kurt A
Patt, Leonard Merton
Pool, Karl Hallman
Riehl, Jerry A
Ryan, Clarence Augustine, Jr
Sarkanen, Kyosti Vilho
Schmidt, Eckart W
Shepherd, Linda Jean
Sheppard, John Clarence
Tazuma, James Junkichi
Thomas, Berwyn Brainerd
Tinker, John Frank
Van Tuyl, Harold Hutchison
West, Martin Luther
Wilson, Archie Spencer
Worthington, Ralph Eric
Zimbrick, John David

WEST VIRGINIA
Bartley, William J
Bhasin, Madan M
Chadwick, David Henry
Cunningham, Newlin Buchanan
Giza, Chester Anthony
Karr, Clarence, Jr
Longanbach, James Robert
MacPeek, Donald Lester
Manyik, Robert Michael
Matthews, Virgil Edison
Moffa, David Joseph
Osborn, Claiborn Lee
Sherman, Paul Dwight, Jr
Smith, Percy Leighton
Stansbury, Harry Adams, Jr
Walker, Wellington Epler
Wilson, Thomas Putnam

WISCONSIN
Adams, James William
Adrian, Alan Patrick
Anderson, Stephen William
Andrews, Oliver Augustus
Beach, George Winchester
Bournique, Raymond August
Brebrick, Robert Frank, Jr
Cherayil, George Devassia
Deutsch, Harold Francis
Doumas, Basil T
Downs, Martin Luther
Dunn, Stanley Austin
Feist, William Charles
Glasoe, Paul Kirkwold
Gloyer, Stewart Wayne
Hellman, Nison Norman
Hill, John William
Holland, Dewey G
Hossain, Shafi Ul
Huber, Calvin
Ihde, Aaron John
Isenberg, Irving Harry
Jenny, Neil Allan
Kao, Wen-Hong
Klopotek, David L
Laessig, Ronald Harold
Leutgoeb, Rosalia Aloisia
Long, Claudine Fern
Makela, Lloyd Edward
Millett, Merrill Albert
Moore, John Ward
Morton, Stephen Dean
Oehmke, Richard Wallace
Perry, Billy Wayne
Phillips, John Spencer
Punwar, Jalamsinh K
Reinders, Victor A
Rosenthal, Waldemar Arthur
Roth, Marie M
Sanyer, Necmi
Sasse, Edward Alexander
Savereide, Thomas J

Schopler, Harry A
Schwab, Helmut
Scott, Lawrence William
Shakhashiri, Bassam Zekin
Showalter, Donald Lee
Siegfried, Robert
Smith, Milton Reynolds
Sommers, Raymond A
Stevens, Michael Fred
Trytten, Roland Aaker
Weipert, Eugene Allen
West, Kevin J
West, Kevin James
Woldegiorgis, Gebretateos
Wright, Steven Martin
Young, Raymond A

WYOMING
Haines, William Emerson
Latham, DeWitt Robert
Noe, Lewis John
Robertson, Raymond E(liot)
Robinson, Wilbur Eugene
Seese, William Shober

PUERTO RICO
ALZERRECA, ARNALDO
Baus, Bernard V(illars)
Ramos, Lillian
Stephens, William Powell

ALBERTA
Bird, Gordon Winslow
Birss, Viola Ingrid
Davis, Stuart George
Hepler, Loren George
Holmes, Owen Gordon
Hyne, James Bissett
Krouse, Howard Roy
Rauk, Arvi
Strausz, Otto Peter
Wieser, Helmut

BRITISH COLUMBIA
Becker, Edward Samuel
Evans, Russell Stuart
Gerry, Michael Charles Lewis
Godard, Hugh P(hillips)
James, Douglas Garfield Limbrey
Keevil, Norman Bell
Polglase, William James
Rosell, Karl-Gunnar
Spitzer, Ralph
Walker, David Crosby

MANITOBA
Biswas, Shib D
Lemire, Robert James
Letkeman, Peter
Muir, Derek Charles G
Singh, Ajit
Tkachuk, Russell
Vandergraaf, Tjalle T

NEW BRUNSWICK
Mallet, Victorin Noel
Mehra, Mool Chand
Unger, Israel

NEWFOUNDLAND
Scott, John Marshall William

NOVA SCOTIA
Knop, Osvald
McAlduff, Edward J
McMillan, Alan F
Matthews, Frederick White
Wyman, Harold Robertson

ONTARIO
Boorn, Andrew William
Borr, Mitchell
Brownstein, Sydney Kenneth
Buckler, Ernest Jack
Burgess, William Howard
Burns, George
Dyne, Peter John
George, Albert El Deeb
Gillespie, Ronald James
Goring, David Arthur Ingham
Gough, Sidney Roger
Gould, William Douglas
Gulens, Janis
Heyding, Robert Donald
Hollbach, Natasha Coffin
Holloway, Clive Edward
Ingruber, Otto Vincent
Jacobs, Patrick W M
Jones, R Norman
Joubin, Franc Renault
Koningstein, Johannes A
Kudo, Akira
Lean, David Robert Samuel
Leonard, John Alex
Levere, Trevor Harvey
Limerick, Jack McKenzie, Sr
Logan, Charles Donald
McBryde, William Arthur Evelyn
McIntosh, Alexander Omar
MacKay, Donald Douglas
Moo-Young, Murray
Morley, Harold Victor
Mutton, Donald Barrett
Nicolle, Francois Marcel Andre

Chemistry, General (cont)

Prince, Alan Theodore
Puddington, Ira E(dwin)
Ramachandran, Vangipuram S
Roovers, J
Ross, Robert Anderson
Ryan, Michael T
Sadler, Arthur Graham
Sadowski, Chester M
Saha, Jadu Gopal
Sanderson, Henry Preston
Schroeder, William Henry
Shurvell, Herbert Francis
Usselman, Melvyn Charles
Van Loon, Jon Clement
Wade, Robert Simson
Wiles, David M
Zakaib, Daniel D

QUEBEC
Ayroud, Abdul-Mejid
Beique, Rene Alexandre
Belanger, Alain
Bolker, Henry Irving
Cooper, David Gordon
Goel, Krishan Narain
Gurudata, Neville
Harpp, David Noble
Lau, Cheuk Kun
Onyszchuk, Mario
Pelletier, Gerard Eugene
Rivest, Roland
Roy, Jean-Claude
Schucher, Reuben
Tan, Ah-Ti Chu
Tomlinson, George Herbert
Tonks, David Bayard
Yaffe, Leo

SASKATCHEWAN
Cassidy, Richard Murray
Johnson, Keith Edward
Spinks, John William Tranter
Stewart, John Wray Black
Sutherland, Ronald George
Woods, Robert James

OTHER COUNTRIES
Ache, Hans Joachim
Bergstrom, K Sune D
Caesar, Philip D
Cornforth, John Warcup
Crabbé, Pierre
De Haën, Christoph
Dreosti, Ivor Eustace
Fey, George Ting-Kuo
Gorin, Philip Albert James
Greichus, Yvonne A
Hasty, Robert Armistead
Huber, Robert
Metzger, Gershon
Parker, David H
Pearson, Arthur David
Phillips, David Colin
Salk, Sung-Ho Suck
Schallenberg, Elmer Edward
Sleeter, Thomas David
Steadman, Robert George
Stumm, Werner
Taylor, Barry Edward
Webb, Cynthia Ann Glinert
Wilkinson, Geoffrey
Winn, Edward Barriere
Wittig, Georg Friedrich Karl

Inorganic Chemistry

ALABAMA
Beal, James Burton, Jr
Chastain, Benjamin Burton
Colburn, Charles Buford
Copeland, David Anthony
Jackson, Margaret E
Johnson, Frederic Allan
Knockemus, Ward Wilbur
Krannich, Larry Kent
Legg, Ivan
Ludwick, Larry Martin
McKee, Michael Leland
McNutt, Ronald Clay
Makhija, Suraj Parkash
Marano, Gerald Alfred
Perry, William Daniel
Retief, Daniel Hugo
Teggins, John E
Van Artsdalen, Ervin Robert
Vigee, Gerald S
Ward, Edward Hilson

ALASKA
Lane, Robert Harold
Lokken, Donald Arthur

ARIZONA
Beaumont, Randolph Campbell
Birk, James Peter
Chapman, Robert Dale
DeKorte, John Martin
Eads, Ewin Alfred
Eichinger, Jack Waldo, Jr
Enemark, John Henry
Eyring, LeRoy

Feltham, Robert Dean
Heath, Roy Elmer
Jull, Anthony John Timothy
Keller, Philip Charles
Knowlton, Gregory Dean
Lichtenberger, Dennis Lee
Liu, Chui Hsun
Marks, Ronald Lee
O'Keeffe, Michael
Post, Roy G
Quill, Laurence Larkin
Robertson, Frederick Noel
Roche, Thomas Stephen
Rund, John Valentine
Smith, Rodger Chapman
Willson, Donald Bruce
Zielen, Albin John

ARKANSAS
Cordes, Arthur Wallace
Dodson, B C
Gwinup, Paul D
Hood, Robert L
Jimerson, George David
Johnson, Dale A
Palmer, Bryan D
Pearson, Robert Stanley
Ransford, George Henry
Schafer, Lothar
Schultz, Donald Raymond
Stuckey, John Edmund
Teague, Marion Warfield
Trigg, William Walker
Wiggins, James William

CALIFORNIA
Almond, Hy
Alvarez, Vincent Edward
Bailin, Lionel J
Baisden, Patricia Ann
Balch, Alan Lee
Bartlett, Neil
Barton, Jacqueline K
Bau, Robert
Beck, Roland Arthur
Bennett, Larry E
Bercaw, John Edward
Blair, George Richard
Boardman, William Walter, Jr
Bonner, Norman Andrew
Boone, James Lightholder
Bowen, Ruth Justice
Brown, William E(ric)
Burg, Anton Behme
Callahan, Kenneth Paul
Calvin, Melvin
Camenzind, Mark J
Canning, T(homas) F
Caputi, Roger William
Chock, Ernest Phaynan
Ciaramitaro, David A
Cihonski, John Leo
Cobble, James Wikle
Cohn, Kim
Collman, James Paddock
Connick, Robert Elwell
Constant, Clinton
Covey, William Danny
Coyle, Bernard Andrew
Crutchfield, Charlie
Cubicciotti, Daniel David
Current, Steven P
Czuha, Michael, Jr
Darnell, Alfred Jerome
Davis, Clyde Edward
Davison, John Blake
De Haan, Frank P
Delker, Gerald Lee
Dines, Martin Benjamin
Doedens, Robert John
Eaton, David Leo
Ecker, David John
Elson, Robert Emanuel
Evans, William John
Feher, Frank J
Feiler, William A, Jr
Feldman, Fredric J
Fine, Dwight Albert
Fleischauer, Paul Dell
Fleischer, Everly B
Ford, Peter Campbell
Frenzel, Lydia Ann Melcher
Gagne, Robert Raymond
Gerlach, John Norman
Ginell, William Seaman
Goetschel, Charles Thomas
Gold, Marvin B
Goldberg, Sabine Ruth
Goldwhite, Harold
Gordon, Joseph Grover, II
Gorman, Melville
Grant, Louis Russell, Jr
Grantham, Leroy Francis
Gray, Harry B
Greenstadt, Melvin
Grubbs, Robert Howard
Grunthaner, Frank John
Hall, Kenneth Lynn
Hamilton, Hobart Gordon, Jr
Hardcastle, Kenneth Irvin
Harris, Daniel Charles
Harris, Gordon McLeod
Hawthorne, Marion Frederick
Hempel, Judith Cato

Hendrickson, David Norman
Herman, Zelek Seymour
Hiller, Frederick W
Hodgson, Keith Owen
Hoffman, Charles John
Hoffmann, Michael Robert
Holzmann, Richard Thomas
Hornig, Howard Chester
Hubred, Gale L
Hulet, Ervin Kenneth
Jaecker, John Alvin
James, Dean B
Jayne, Jerrold Clarence
Johanson, Robert Gail
Jolly, William Lee
Jones, Patrick Ray
Kaesz, Herbert David
Katzin, Leonard Isaac
Kauffman, George Bernard
Kay, Eric
Kehoe, Thomas J
Kennedy, James Vern
Kieft, John A
King, William Robert, Jr
Klemm, Waldemar Arthur, Jr
Knobler, Carolyn Berk
Korst, William Lawrence
Kothny, Evaldo Luis
Kratzer, Reinhold
Kubota, Mitsuru
Labinger, Jay Alan
Landesman, Herbert
Landis, Vincent J
Lawton, Emil Abraham
Lee, Yat-Shir
Lewis, Nathan Saul
Lind, Carol Johnson
Lipka, James J
Liu, Sung-Tsuen
Lustig, Max
McBride, William Robert
McKaveney, James P
Malik, Jim Gorden
Malouf, George M
Markowitz, Samuel Solomon
Marshall, Donald D
Martz, Harry Edward, Jr
Masters, Burton Joseph
Matheson, Arthur Ralph
Maya, Walter
Mee, Jack Everett
Meyer, Carl Beat
Michlmayr, Manfred
Mode, Vincent Alan
Moe, George
Monchamp, Roch Robert
Mosen, Arthur Walter
Murbach, Earl Wesley
Nathan, Lawrence Charles
Nelson, Norvell John
Newkirk, Herbert William
Newsam, John M
Newton, William Edward
Ogimachi, Naomi Neil
Palmer, Thomas Adolph
Panzer, Richard Earl
Pappatheodorou, Sofia
Pearson, Michael J
Perry, Dale Lynn
Phipps, Peter Beverley Powell
Po, Henry N
Potter, Norman D
Potts, John Calvin
Raby, Bruce Alan
Rard, Joseph Antoine
Raymond, Kenneth Norman
Reed, Christopher Alan
Reinhardt, Richard Alan
Rhein, Robert Alden
Ring, Morey Abraham
Robinson, Paul Ronald
Rogers, Howard H
Rosenberg, Sanders David
Rossman, George Robert
Russell, John Blair
Russell, Thomas Paul
Rustad, Douglas Scott
Salot, Stuart Edwin
Sangster, Raymond Charles
Schack, Carl J
Schrauzer, Gerhard N
Sellas, James Thomas
Sharpless, K Barry
Shoemaker, Carlyle Edward
Siegel, Bernard
Silber, Herbert Bruce
Smart, James Conrad
Smith, Robert Alan
Solomon, Edward I
Sprague, Robert W
Steinman, Robert
Stucky, Galen Dean
Swinehart, James Herbert
Taube, Henry
Tillay, Eldrid Wayne
Tilley, T(erry) Don
Toney, Joe David
Ulrich, William Frederick
Valentine, Joan Selverstone
Van Alten, Lloyd
Vander Wall, Eugene
Van Hecke, Gerald Raymond
Vernon, Gregory Allen
Vogel, Roger Frederick

Wadley, Margil Warren
Wagner, Ross Irving
Waldo, Willis Henry
Warf, James Curren
Weaver, Henry D, Jr
Weber, Carl Joseph
Webster, Clyde Leroy, Jr
Weiss, Harold Gilbert
Weisz, Robert Stephen
Wickham, Donald G
Wilk, William David
Williams, Colin James
Willis, William Van
Winchell, Robert E
Wong, Hans Kuomin
Wong, Joe
Young, Donald C
Zabin, Burton Allen
Zink, Jeffrey Irve

COLORADO
Anderson, Oren P
Barry, Henry F
Blake, Daniel Melvin
Clough, Francis Bowman
Conner, Jack Michael
Cummins, Jack D
Daugherty, Ned Arthur
Dickerhoof, Dean W
Eaton, Gareth Richard
Eaton, Sandra Shaw
Erdmann, David E
Fields, Clark Leroy
Fleming, Michael Paul
Frye, James Sayler
Geller, Seymour
Gerteis, Robert Louis
Gottschall, W Carl
Haas, Frank C
Hadzeriga, Pablo
Hall, James Louis
Heberlein, Douglas Garavel
Hutto, Francis Baird, Jr
Hyatt, David Ernest
Keder, Wilbert Eugene
King, Edward Louis
Koval, Carl Anthony
Legal, Casimer Claudius, Jr
Lewis, Clifford Jackson
Macalady, Donald Lee
Mehs, Doreen Margaret
Miner, Frend John
Moody, David Coit, III
Muscatello, Anthony Curtis
Norman, Arlan Dale
Pierpont, Cortlandt Godwin
Pundsack, Frederick Leigh
Ritchey, John Michael
Sanderson, Robert Thomas
Schaeffer, Riley
Sievers, Robert Eugene
Splittgerber, George H
Thompson, Gary Haughton
Thompson, Ronald G
Watkins, Kay Orville
Yang, In Che

CONNECTICUT
Alfieri, Charles C
Bogucki, Raymond Francis
Carrano, Salvatore Andrew
Chamberland, Bertrand Leo
Crabtree, Robert H
Eppler, Richard A
Faller, John William
Galasso, Francis Salvatore
Golden, Gerald Seymour
Grayson, Martin
Haake, Paul
Horrigan, Philip Archibald
Kostiner, Edward S
Kozlowski, Adrienne Wickenden
Krause, Ronald Alfred
Kuck, Julius Anson
Leavitt, Julian Jacob
Lines, Ellwood LeRoy
McKeon, James Edward
Moeller, Carl William, Jr
Moore, Robert Earl
Moyer, Ralph Owen, Jr
Noack, Manfred Gerhard
Prigodich, Richard Victor
Radtke, Schrade Fred
Richter, G Paul
Roberts, Elliott John
Roscoe, John Stanley, Jr
Sarada, Thyagaraja
Sarneski, Joseph Edward
Schiessl, H(enry) W(illiam)
Shaw, Brenda Roberts
Suib, Steven L
Tanaka, John
Tom, Glenn McPherson
Yang, Darchun Billy

DELAWARE
Baker, Ralph Thomas
Bissot, Thomas Charles
Boughton, John Harland
Braun, Juergen Hans
Brill, Thomas Barton
Bulkowski, John Edmund
Burmeister, John Luther
Burton, Lester Percy Joseph

Calabrese, Joseph C
Chowdhry, Uma
Domaille, Peter John
Drinkard, William Charles, Jr
Falletta, Charles Edward
Foley, Henry Charles
Glaeser, Hans Hellmut
Hertzenberg, Elliot Paul
Hess, Richard William
Hoppenjans, Donald William
Ittel, Steven Dale
Kolski, Thaddeus L(eonard)
Kruse, Walter M
Ling, Harry Wilson
Longhi, Raymond
Luther, George William, III
Lyon, Donald Wilkinson
McClelland, Alan Lindsey
McGinnis, William Joseph
McLain, Stephan James
Mahler, Walter
Mead, Edward Jairus
Mighton, Charles Joseph
Miller, Joel Steven
Moran, Edward Francis, Jr
Moser, Glenn Allen
Parkinson, Bruce Alan
Phillips, Brian Ross
Potrafke, Earl Mark
Roe, David Christopher
Rogers, Donald B
Rosenberg, Richard Martin
Schulz, Wallace Wendell
Shannon, Robert Day
Sharp, Kenneth George
Shozda, Raymond John
Sopp, Samuel William
Theopold, Klaus Hellmut
Thompson, Jeffery Scott
Thorpe, Colin
Tolman, Chadwick Alma
Tomic, Ernst Alois
Tulip, Thomas Hunt
Wasfi, Sadiq Hassan
Welch, Raymond Lee

DISTRICT OF COLUMBIA
Baker, Louis Coombs Weller
Beach, Harry Lee, Jr
Bertsch, Charles Rudolph
Burnett, John Lambe
Butcher, Raymond John
Butter, Stephen Allan
Crum, John Kistler
Earley, Joseph Emmet
Erstfeld, Thomas Ewald
Henry, Richard Lynn
Hunt, John Baker
Klein, Philipp Hillel
Kuznesof, Paul Martin
Larsen, Lynn Alvin
McCormack, Mike
Marianelli, Robert Silvio
Nordquist, Paul Edgard Rudolph, Jr
Ondik, Helen Margaret
Pope, Michael Thor
Preer, James Randolph
Rowley, David Alton
Shirk, Amy Emiko
Stern, Kurt Heinz
Venezky, David Lester
Wang, Jin Tsai
Webb, Alan Wendell
White, David Gover

FLORIDA
Adiseshu, Setty Ravanappa
Ahmann, Donald H(enry)
Austin, Alfred Ells
Babich, Michael Wayne
Baumann, Arthur Nicholas
Betzer, Peter Robin
Birdwhistell, Ralph Kenton
Burbage, Joseph James
Callahan, James Louis
Carlson, Gordon Andrew
Choppin, Gregory Robert
Clark, Ronald Jene
Clausen, Chris Anthony
Davidovits, Joseph
DeLap, James Harve
Drago, Russell Stephen
Drake, Robert Firth
Everett, Kenneth Gary
Frazier, Stephen Earl
Goedken, Virgil Linus
Hare, Curtis R
Hellwege, Herbert Elmore
Hoffman, Lewis Charles
Iloff, Phillip Murray, Jr
Jungbauer, Mary Ann
Kirshenbaum, Abraham David
Kroger, Hanns H
Lee, Kah-Hock
McFarlin, Richard Francis
Martin, Barbara Bursa
Martin, Dean Frederick
Mellon, Edward Knox, Jr
Neithamer, Richard Walter
Palmer, Jay
Penner, Siegfried Edmund
Perumareddi, Jayarama Reddi
Rees, William Smith, Jr
Remmel, Randall James

Roe, William P(rice)
Ryschkewitsch, George Eugene
Sisler, Harry Hall
Stanko, Joseph Anthony
Stoufer, Robert Carl
Whitaker, Robert Dallas
Whitney, Ellsworth Dow
Willard, Thomas Maxwell
Worrell, Jay H

GEORGIA
Anderson, Bruce Martin
Ashby, Eugene Christopher
Barefield, Edward Kent
Bertrand, Joseph Aaron
Block, Toby Fran
Bottomley, Lawrence Andrew
Brandau, Betty Lee
Centofanti, Louis F
Crawford, Van Hale
Daane, Adrian Hill
DeLorenzo, Ronald Anthony
Gailey, Kenneth Durwood
Griffiths, James Edward
Henneike, Henry Fred
Hicks, Donald Gail
Hunt, Gary W
Hunt, Harold Russell, Jr
James, Jeffrey
Johnson, Ronald Carl
King, Robert Bruce
Koelsche, Charles L
Kutal, Charles Ronald
Lockhart, William Lafayette
Netherton, Lowell Edwin
Netzel, Thomas Leonard
Neumann, Henry Matthew
Richardson, Susan D
Royer, Donald Jack
Ruff, John K
Scott, Robert Allen
Sears, Curtis Thornton, Jr
Spencer, Jesse G
Steele, Jack
Stratton, Cedric
Thomas, Frank Harry
Torres, Lourdes Maria
Waggoner, William Horace
Whitten, Kenneth Wayne
Williams, George Nathaniel
Young, Raymond Hinchcliffe, Jr

HAWAII
Cramer, Roger Earl
Gilje, John
Hirayama, Chikara
Kemp, Paul James
Waugh, John Lodovick Thomson
Wrathall, Jay W

IDAHO
Arcand, George Myron
Baker, John David
Batey, Harry Hallsted, Jr
Benson, Ernest Phillip, Jr
Bills, Charles Wayne
Carter, Loren Sheldon
Grahn, Edgar Howard
Hammer, Robert Russell
Lewis, Leroy Crawford
Mincher, Bruce J
Nagel, Terry Marvin
Shreeve, Jean'ne Marie
Slansky, Cyril M
Tracy, Joseph Walter
Wright, Kenneth James

ILLINOIS
Adams, Max Dwain
Allred, Albert Louis
Appelman, Evan Hugh
Bailar, John Christian, Jr
Basile, Louis Joseph
Basolo, Fred
Beese, Ronald Elroy
Belford, R Linn
Benzinger, William Donald
Berntsen, Robert Andyv
Bertolacini, Ralph James
Broach, Robert William
Brooks, Kenneth Conrad
Brown, Theodore Lawrence
Brubaker, George Randell
Brubaker, Inara Mencis
Bunting, Roger Kent
Burdett, Jeremy Keith
Burns, Richard Price
Cafasso, Fred A
Carlin, Richard Lewis
Carnahan, Jon Winston
Carrado, Kathleen Anne
Chang, Chin Hsiung
Chao, Sherman S
Chellew, Norman Raymond
Coley, Ronald Frank
Court, Anita
Curtiss, Larry Alan
Danzig, Morris Juda
De Pasquali, Giovanni
Dietz, Mark Louis
Doody, Marijo
Ebert, Lawrence Burton
Finn, Patricia Ann
Fischer, Albert Karl

Freeman, Wade Austin
Funck, Larry Lehman
Ganchoff, John Christopher
Hakewill, Henry, Jr
Halpern, Jack
Hamerski, Julian Joseph
Hanson, John Elbert
Harris, Ronald L
Henry, Patrick M
Herlinger, Albert William
Hess, Wendell Wayne
Hoffman, Alan Bruce
Hoffman, Brian Mark
Hohnstedt, Leo Frank
Hollins, Robert Edward
Homeier, Edwin H, Jr
Hopkins, Paul Donald
Horwitz, Earl Philip
House, James Evan, Jr
Huang, Leo W
Hughes, Benjamin G
Huston, John Lewis
Hutchison, David Allan
Ibers, James Arthur
Jaffe, Philip Monlane
Jaselskis, Bruno
Karayannis, Nicholas M
Karraker, Robert Harreld
Keiter, Ellen Ann
Keiter, Richard Lee
Kemper, Kathleen Ann
Kennelly, Mary Marina
Kieft, Richard Leonard
Klemperer, Walter George
Kokalis, Soter George
Kuhajek, Eugene James
Leibowitz, Leonard
Liu, Chui Fan
Maroni, Victor August
Mason, W Roy, III
Melford, Sara Steck
Mendelsohn, Marshall H
Mota de Freitas, Duarte Emanuel
Nash, Kenneth Laverne
Nebgen, John William
Nicolae, George G
Nilges, Mark J
Partenheimer, Walter
Pavkovic, Stephen F
Perkins, Alfred J
Perlow, Mina Rea Jones
Poeppelmeier, Kenneth Reinhard
Pohlmann, Hans Peter
Poskozim, Paul Stanley
Postmus, Clarence, Jr
Rathke, Jerome William
Rauchfuss, Thomas Bigley
Reagan, William Joseph
Reidies, Arno H
Rogers, Robin Don
Schmulbach, Charles David
Schreiner, Felix
Schug, Kenneth
Schultz, Arthur Jay
Schulz, Charles Emil
Sheft, Irving
Shriver, Duward F
Stafford, Fred E
Stein, Lawrence
Steunenberg, Robert Keppel
Suslick, Kenneth Sanders
Thompson, Arthur Robert
Thompson, Martin Leroy
Treptow, Richard S
Trevorrow, Laverne Everett
Trimble, Russell Fay
Vaughn, Joe Warren
Weil, Thomas Andre
West, Douglas Xavier
Whaley, Thomas Patrick
White, Jesse Edmund
Williams, Jack Marvin
Woods, Mary
Zimmerman, Donald Nathan

INDIANA
Bottei, Rudolph Santo
Bowser, James Ralph
Brown, Herbert Charles
Case, Vernon Wesley
Chisholm, Malcolm Harold
Christou, George
Davenport, Derek Alfred
DeSantis, John Louis
Fehlner, Thomas Patrick
Ferraudi, Guillermo Jorge
Friedel, Arthur W
Garber, Lawrence L
George, James E
Green, Mark Alan
Hofman, Emil Thomas
Holowaty, Michael O
Joyner, Ralph Delmer
Lipschutz, Michael Elazar
McMillin, David Robert
Margerum, Dale William
Morrison, William Alfred
Phillips, John R
Pilger, Richard Christian, Jr
Pitha, John Joseph
Pribush, Richard A
Reuland, Donald John
Robinson, William Robert
Schaap, Ward Beecher

Scheidt, Walter Robert
Schwan, Theodore Carl
Siefker, Joseph Roy
Storhoff, Bruce Norman
Swartz, Marjorie Louise
Todd, Lee John
Voorhoeve, Rudolf Johannes Herman
Walter, Joseph L
Wentworth, Rupert A D
Wyma, Richard J
Zeldin, Martel

IOWA
Angelici, Robert Joe
Bakac, Andreja
Bennett, William Earl
Chang, James C
Corbett, John Dudley
Deskin, William Arna
Doyle, John Robert
Espenson, James Henry
Eyman, Darrell Paul
Glick, Milton Don
Goff, Harold Milton
Hammerstrom, Harold Elmore
Heino, Walden Leo
Hutton, Wilbert, Jr
King, Walter Bernard
Kostic, Nenad M
McCarley, Robert Eugene
Martin, Don Stanley, Jr
Mentone, Pat Francis
Messerle, Louis
Miller, Gordon James
Powell, Jack Edward
Quass, La Verne Carl
Rila, Charles Clinton
Swartz, James E
Woods, Joe Darst

KANSAS
Boyle, Kathryn Moyne Ward Dittemore
Busch, Daryle Hadley
Cohen, Sheldon H
Frazier, A Joel
Gimple, Glenn Edward
Greene, Frank T
Griswold, Ernest
Gusenius, Edwin Maurtiz
Hiebert, Allen G
Johnson, William Jacob
Klabunde, Kenneth John
Kleinberg, Jacob
Lambert, Jack Leeper
Landis, Arthur Melvin
McCormick, Bailie Jack
McElroy, Albert Dean
Mertes, Kristin Bowman
Potts, Melvin Lester
Purcell, Keith Frederick
Rumpel, Max Leonard
Sullivan, Mary Louise

KENTUCKY
Beall, Gary Wayne
Bettinger, Donald John
Boggess, Gary Wade
Buchanan, Robert Martin
Chamberlin, John MacMullen
Clark, Howell R
Crawford, Thomas H
Daly, John Matthew
Davidson, John Edwin
Henrickson, Charles Henry
Hunt, Richard Lee
Keely, William Martin
Kiser, Robert Wayne
Klingenberg, Joseph John
McDermott, Dana Paul
Magnuson, Winifred Lane
Niedenzu, Kurt
Powell, Howard B
Richter, Edward Eugene
Riley, John Thomas
Sarma, Atul C
Selegue, John Paul
Sweeny, Daniel Michael
Tharp, A G
Vance, Robert Floyd

LOUISIANA
Anex, Basil Gideon
Bains, Malkiat Singh
Berni, Ralph John
Boudreaux, Edward A
Brown, Leo Dale
Carmichael, J W, Jr
Day, Marion Clyde, Jr
Doomes, Earl
Field, Jack Everett
Frey, Frederick Wolff, Jr
Halbert, Thomas Risher
Ham, Russell Allen
Jonassen, Hans Boegh
Kennedy, Frank Scott
Lanoux, Sigred Boyd
Mague, Joel Tabor
Maverick, Andrew William
Mustafa, Shams
Overton, Edward Beardslee
Peard, William John
Raisen, Elliott
Robson, Harry Edwin
Schweizer, Albert Edward

Inorganic Chemistry (cont)

Seidler, Rosemary Joan
Selbin, Joel
Sen, Buddhadev
Siriwardane, Upali
Sistrunk, Thomas Olloise
Stanley, George Geoffrey
Stuntz, Gordon Frederick
Ter Haar, Gary L
Watkins, Steven F
Weidig, Charles F
Wiewiorowski, Tadeusz Karol

MAINE
Patterson, Howard Hugh
Rowan, Nancy Gordon
Russ, Charles Roger
Smith, Wayne Lee
Wheatland, David Alan

MARYLAND
Albinak, Marvin Joseph
Auel, RaeAnn Marie
Barnes, James Alford
Biddle, Richard Albert
Block, Jacob
Bonnell, David William
Bonsack, James Paul
Boucher, Laurence James
Boyd, Alfred Colton, Jr
Brinckman, Frederick Edward, Jr
Carter, John Paul
Chamberlain, David Leroy, Jr
Coleman, James Stafford
Collins, Alva LeRoy, Jr
Cooperstein, Raymond
Coyle, Thomas Davidson
Debye, Nordulf Wiking Gerud
Decraene, Denis Fredrick
Eanes, Edward David
Eichhorn, Gunther Louis
Flanders, Kathleen Corey
Freeman, Horatio Putnam
Ginther, Robert J
Griffo, Joseph Salvatore
Grim, Samuel Oram
Gryder, John William
Hastie, John William
Hsu, Chen C
Huheey, James Edward
Humphrey, S(idney) Bruce
Jaquith, Richard Herbert
Joedicke, Ingo Bernd
Johannesen, Rolf Bradford
Josephs, Steven F
Karlin, Kenneth Daniel
Kask, Uno
Kayser, Richard Francis
Koubek, Edward
Levy, Newton, Jr
Lussier, Roger Jean
McCawley, Frank X(avier)
Magee, John Storey, Jr
Maselli, James Michael
Mathew, Mathai
Murch, Robert Matthews
Muser, Marc
Pace, Judith G
Poli, Rinaldo
Pribula, Alan Joseph
Richman, Robert Michael
Ritter, Joseph John
Roethel, David Albert Hill
Rollins, Orville Woodrow
Shade, Joyce Elizabeth
Swann, Madeline Bruce
Tarshis, Irvin Barry
Thompson, Joseph Kyle
Tuve, Richard Larsen
Wason, Satish Kumar
Williams, Oren Francis
Zimmerman, John Gordon
Zuckerbrod, David

MASSACHUSETTS
Ammlung, Richard Lee
Archer, Ronald Dean
Auld, David Stuart
Baird, Donald Heston
Bates, Carl H
Beall, Herbert
Berka, Ladislav Henry
Bessette, Russell Romulus
Billo, Edward Joseph
Brauner, Phyllis Ambler
Burgess, Thomas Edward
Caslavska, Vera Barbara
Caslavsky, Jaroslav Ladislav
Chin, David
Clarke, Michael J
Cohen, Saul Mark
Coleman, William Fletcher
Comer, Joseph John
Condike, George Francis
Croft, William Joseph
Davies, Geoffrey
Deckert, Cheryl A
Donoghue, John Timothy
Dooley, David Marlin
Duecker, Heyman Clarke
Eddy, Robert Devereux
Ekman, Carl Frederick W
Feakes, Frank

Foxman, Bruce Mayer
George, John Warren
Goldberg, Gershon Morton
Greenaway, Frederick Thomas
Greenblatt, David J
Haas, Terry Evans
Harrill, Robert W
Higgins, Dorothy
Hoffman, Morton Z
Hollocher, Thomas Clyde, Jr
Holm, Richard H
Holmes, Robert Richard
Hooper, Robert John
Kaczmarczyk, Alexander
Knox, Kerro
Korenstein, Ralph
Kowalak, Albert Douglas
Kustin, Kenneth
Langer, Horst Günter
Linck, Robert George
Lippard, Stephen J
Loretz, Thomas J
McGarry, Margaret
Marganian, Vahe Mardiros
Menashi, Jameel
Milburn, Ronald McRae
Minne, Ronn N
Moore, Robert Edmund
Natansohn, Samuel
Palilla, Frank C
Powers, Donald Howard, Jr
Quinlan, Kenneth Paul
Reiff, William Michael
Reis, Arthur Henry, Jr
Ricci, John Silvio, Jr
Richason, George R, Jr
Seyferth, Dietmar
Shinn, Dennis Burton
Smith, Joseph Harold
Sosinsky, Barrie Alan
Stuhl, Louis Sheldon
Sullivan, Edward Augustine
Suplinskas, Raymond Joseph
Swank, Thomas Francis
Taub, Irwin A(llen)
Thomas, Martha Jane Bergin
Urry, Grant Wayne
Wang, Robert T
Weibrecht, Walter Eugene
Weigand, Willis Alan
Weller, Paul Franklin
Wilson, Linda S (Whatley)
Wrighton, Mark Stephen
Yang, Julie Chi-Sun
Zompa, Leverett Joseph

MICHIGAN
Anderson, Charles Thomas
Anderson, Melvin Lee
Baney, Ronald Howard
Bennett, Stephen Lawrence
Benson, Edmund Walter
Brault, Margaret A
Brenner, Alan
Brewer, Stephen Wiley, Jr
Brown, Donald John
Brubaker, Carl H, Jr
Carpenter, Michael Kevin
Chance, Robert L
Chandra, Grish
Collins, Ronald William
Cooke, Dean William
Coucouvanis, Dimitri N
Curtis, Myron David
Dawson, Gladys Quinty
Deck, Charles Francis
DeKock, Roger Lee
Dennis, Mary
Dukes, Gary Rinehart
Eastland, George Warren, Jr
Eick, Harry Arthur
Eliezer, Isaac
Endicott, John F
Evans, B J
Fisher, Galen Bruce
Flynn, James Patrick
Frey, John Erhart
Galloway, Gordon Lynn
Gaswick, Dennis C
Gump, J R
Hamilton, James Beclone
Hammer, Robert Nelson
Harmon, Kenneth Millard
Harris, Paul Jonathan
Hartwell, George E, Jr
Hutchison, James Robert
Jekel, Eugene Carl
Johnson, David Alfred
Johnson, Wayne Douglas
Kanamueller, Joseph M
Kanatzidis, Mercouri G
Kirschner, Stanley
Knop, Charles Philip
Kopperl, Sheldon Jerome
Kuczkowski, Robert Louis
Lane, George Ashel
Latham, Ross, Jr
Lehman, Duane Stanley
Lintvedt, Richard Lowell
Lipowitz, Jonathan
Lofquist, Marvin John
Luehrs, Dean C
McCollough, Fred, Jr
McLean, John A, Jr

Magnell, Kenneth Robert
Mance, Andrew Mark
Maskal, John
Mericola, Francis Carl
Moyer, John Raymond
Murchison, Craig Brian
Oei, Djong-Gie
Oliver, John Preston
Pecoraro, Vincent L
Penner-Hahn, James Edward
Peters, Till Justus Nathan
Pinnavaia, Thomas J
Popov, Alexander Ivan
Potts, Richard Allen
Roberts, Charles Brockway
Rorabacher, David Bruce
Rulfs, Charles Leslie
Rycheck, Mark Rule
Seefurth, Randall N
Skelcey, James Stanley
Spees, Steven Tremble, Jr
Spike, Clark Ghael
Stowe, Robert Allen
Swartz, Grace Lynn
Tamres, Milton
Taylor, Robert Craig
Taylor, Stephen Keith
Tsiginos, George Andrew
Turner, Almon George, Jr
Van Doorne, William
Vukasovich, Mark Samuel
Warren, H(erbert) Dale
Williams, Donald Howard
Wilson, Laurence Edward
Wymore, Charles Elmer
Yamauchi, Masanobu
Yesinowski, James Paul

MINNESOTA
Alton, Earl Robert
Baum, Burton Murry
Bolles, Theodore Frederick
Brasted, Robert Crocker
Britton, William Giering
Bydalek, Thomas Joseph
Carberry, Edward Andrew
Carlson, Robert Leonard
Carpenter, John Harold
Chamberlain, Craig Stanley
Conroy, Lawrence Edward
Dendinger, Richard Donald
Dinga, Gustav Paul
Dreyer, Kirt A
Ellis, John Emmett
Finholt, James E
Fleming, Peter B
Foss, Frederick William, Jr
Gabor, Thomas
Gebelt, Robert Eugene
Gladfelter, Wayne Lewis
Graham, Robert Leslie
Haas, Larry Alfred
Hill, Brian Kellogg
Howell, Peter Adam
Johnson, Bryce Vincent
King, Reatha Clark
McKenna, Jack F(ontaine)
Magnuson, Vincent Richard
Mathiason, Dennis R
Mayerle, James Joseph
Miessler, Gary Lee
Mueting, Ann Marie
Neuvar, Erwin W
Oberle, Thomas M
Owens, Kenneth Eugene
Pignolet, Louis H
Plovnick, Ross Harris
Potts, Lawrence Walter
Que, Lawrence, Jr
Reynolds, Warren Lind
Siedle, Allen R
Sorensen, David T
Stadtherr, Leon
Stein, Paul John
Tarr, Donald Arthur
Thomas, John Paul
Thompson, Larry Clark
Thompson, Mary E
Wolsey, Wayne C
Wood, John Martin

MISSISSIPPI
Ahmed, Ismail Yousef
Bishop, Allen David, Jr
Doumit, Carl James
Howell, John Emory
Klingen, Theodore James
McMahan, William H
Wertz, David Lee
White, June Broussard

MISSOURI
Anderson, Harold D
Bahn, Emil Lawrence, Jr
Barton, Lawrence
Bauman, John E, Jr
Bleeke, John Richard
Brammer, Lee
Connolly, John W
Corey, Eugene R
Corey, Joyce Yagla
Coria, Jose Conrado
Craddock, John Harvey
Crutchfield, Marvin Mack

Dean, Walter Keith
Deutsch, Edward Allen
Donovan, John Richard
Droll, Henry Andrew
Ebner, Jerry Rudolph
Felthouse, Timothy R
Forsberg, John Herbert
Forster, Denis
French, James Edwin
Gard, David Richard
Gibbons, James Joseph
Grindstaff, Wyman Keith
Groenweghe, Leo Carl Denis
Haymore, Barry Lant
Heitsch, Charles Weyand
Johnson, Mary Frances
King, Thomas Morgan
Lang, Gerhard Paul
Lannert, Kent Philip
Lawless, Edward William
Leifield, Robert Francis
Long, Gary John
McDonald, Hector O
Malone, Leo Jackson, Jr
Moedritzer, Kurt
Monzyk, Bruce Francis
Moore, Edward Lee
Murmann, Robert Kent
O'Brien, James Francis
Readnour, Jerry Michael
Richard, William Ralph, Jr
Riley, Dennis Patrick
Ross, Frederick Keith
Schlemper, Elmer Otto
Solodar, Arthur John
Staniforth, Robert Arthur
Stary, Frank Edward
Stearns, Robert Inman
Stone, Bobbie Dean
Switzer, Jay Alan
Tappmeyer, Wilbur Paul
Thompson, Richard Claude
Ucko, David A
Wehrmann, Ralph F(rederick)
Wilcox, Harold Kendall

MONTANA
Abbott, Edwin Hunt
Berglund, Donna Lou
Elliott, Eugene Willis
Emerson, Kenneth
Jennings, Paul W
Osterheld, Robert Keith
Pagenkopf, Gordon K
Thomas, Forrest Dean, II
Van Meter, Wayne Paul

NEBRASKA
Freidline, Charles Eugene
George, T Adrian
Griswold, Norman Ernest
Harris, Holly Ann
Harris, Robert Hutchison
Holtzclaw, Henry Fuller, Jr
Howell, Daniel Bunce
Mattes, Frederick Henry
Rasmussen, Russell Lee
Rieke, Reuben Dennis
Sterner, Carl D

NEVADA
Grenda, Stanley C
LeMay, Harold E, Jr
MacDonald, David J
Nelson, John Henry
Olander, James Alton
Rhees, Raymond Charles
Williams, Loring Rider

NEW HAMPSHIRE
Chasteen, Norman Dennis
Dodson, Vance Hayden, Jr
Fogleman, Wavell Wainwright
Haendler, Helmut Max
Hoag, Roland Boyden, Jr
Housecroft, Catherine Elizabeth
Hughes, Russell Profit
Mooney, Richard Warren
Roberts, John Edwin
Soderberg, Roger Hamilton
Wetterhahn, Karen E
Wong, Edward Hou

NEW JERSEY
Aykan, Kamran
Bamberger, Curt
Barnes, Robert Lee
Benedict, Joseph T
Biagetti, Richard Victor
Bocarsly, Andrew B
Bogden, John Dennis
Bornstein, Alan Arnold
Brill, Yvonne Claeys
Brous, Jack
Buckley, R Russ
Castor, William Stuart, Jr
Celiano, Alfred
Chandrasekhar, Prasanna
Chester, Arthur Warren
Chianelli, Russell Robert
Clarke, Lilian A
Coakley, Mary Peter
Cohen, Bernard
Coyle, Catherine Louise

Crerar, David Alexander
Cullen, Glenn Wherry
Dentai, Andrew G
Dess, Howard Melvin
Dreeben, Arthur B
Dzierzanowski, Frank John
Ehrenstorfer, Sieglinde K M
Erckel, Ruediger Josef
Falconer, Warren Edgar
Falk, Charles David
Farrauto, Robert Joseph
Finston, Harmon Leo
Fix, Kathleen A
Forbes, Charles Edward
Ginsberg, Alvin Paul
Good, Mary Lowe
Gore, Ernest Stanley
Grams, Gary Wallace
Granchi, Michael Patrick
Grandolfo, Marian Carmela
Greenblatt, Martha
Haden, Walter Linwood, Jr
Hall, Gretchen Randolph
Hall, Richard Eugene
Hayles, William Joseph
Hills, Stanley
Huchital, Daniel H
Humiec, Frank S, Jr
Huneke, James Thomas
Inniss, Daryl
Isied, Stephan Saleh
Jackson, Julius
Jacobson, Allan Joseph
Jacobson, Stephen Ernest
Johnson, Jack Wayne
Jonte, John Haworth
Kern, Werner
Kestner, Mark Otto
Klein, Richard M
Kluiber, Rudolph W
Kowalski, Stephen Wesley
Kramer, George Mortimer
Krueger, Paul Carlton
Kuehl, Guenter Hinrich
Larkin, William Albert
Larsen, Marilyn Ankeney
Le Duc, J-Adrien Maher
Libowitz, George Gotthart
Lipp, Steven Alan
Lynde, Richard Arthur
McCarroll, William Henry
McClure, Donald Stuart
Magder, Jules
Marezio, Massimo
Matsuguma, Harold Joseph
Miller, Arthur
Miller, Warren Victor
Morrow, Scott
Munday, Theodore F
Murray, Leo Thomas
Naik, Datta Vittal
Nelson, Gregory Victor
Nelson, Roger Edwin
Pinch, Harry Louis
Potenza, Joseph Anthony
Rabinovich, Eliezer M
Radimer, Kenneth John
Rankel, Lillian Ann
Reichle, Walter Thomas
Rivela, Louis John
Robbins, Murray
Ropp, Richard C
Rosan, Alan Mark
Rubino, Andrew M
Schnable, George Luther
Schneemeyer, Lynn F
Schoenberg, Leonard Norman
Schugar, Harvey
Sekutowski, Dennis G
Sherry, Howard S
Sinclair, William Robert
Slepetys, Richard Algimantas
Spaulding, Len Davis
Spiro, Thomas George
Stange, Hugo
Stiefel, Edward
Strange, Ronald Stephen
Suchow, Lawrence
Szyper, Mira
Tannenbaum, Stanley
Tauber, Arthur
Tecotzky, Melvin
Thich, John Adong
Tiethof, Jack Alan
Tripathi, Uma Prasad
Vandegrift, Vaughn
Vanderspurt, Thomas Henry
Van Uitert, LeGrand G(erard)
Varga, Gideon Michael, Jr
Vickroy, David Gill
Ward, Laird Gordon Lindsay
Weschler, Charles John
Wheaton, Gregory Alan
Whigan, Daisy B
White, Lawrence Keith
Wise, John James
Wreford, Stanley S
Yocom, Perry Niel

NEW MEXICO
Alexander, Martin Dale
Asprey, Larned Brown
Baca, Glenn
Behrens, Robert George

Birnbaum, Edward Robert
Bowen, Scott Michael
Bravo, Justo Baladjay
Dahl, Alan Richard
DeArmond, M Keith
DuBois, Frederick Williamson
George, Raymond S
Heaton, Richard Clawson
Hobart, David Edward
Jackson, Darryl Dean
Kelber, Jeffry Alan
Kubas, Gregory Joseph
Levy, Samuel C
Maier, William Bryan, II
Mason, Caroline Faith Vibert
Newton, Thomas William
Onstott, Edward Irvin
Paine, Robert Treat, Jr
Penneman, Robert Allen
Peterson, Charles Leslie
Peterson, Eugene James
Plymale, Donald Lee
Popp, Carl John
Powers, Dana Auburn
Quinn, Rod King
Ryan, Robert Reynolds
Stinecipher, Mary Margaret
Swanson, Basil Ian
Tapscott, Robert Edwin
Thomas, Kimberly W
Thompson, Major Curt
West, Mike Harold
Yalman, Richard George
Zack, Neil Richard

NEW YORK
Abruna, Hector D
Adin, Anthony
Allenbach, Charles Robert
Amero, Bernard Alan
Anbar, Michael
Anthony, Thomas Richard
Ashen, Philip
Atwood, Jim D
Bailey, Ronald Albert
Banks, Ephraim
Baum, Parker Bryant
Baum, Stuart J
Beachley, Orville Theodore, Jr
Birnbaum, Ernest Rodman
Bonner, Francis Truesdale
Brady, James Edward
Brunschwig, Bruce Samuel
Bryan, Philip Steven
Bullock, R Morris
Bulloff, Jack John
Burlitch, James Michael
Campisi, Louis Sebastian
Chamberlain, Phyllis Ione
Chan, Shu Fun
Churchill, Melvyn Rowen
Cohen, Irwin A
Cole, Roger M
Cook, Edward Hoopes, Jr
Corfield, Peter William Reginald
Cratty, Leland Earl, Jr
Crayton, Philip Hastings
Creutz, Carol
Cumming, James B
Dabrowiak, James Chester
Danielson, Paul Stephen
DeCrosta, Edward Francis, Jr
Deiters, Joan A
Dexter, Theodore Henry
D'Heurle, Francois Max
Dillard, Clyde Ruffin
DiSalvo, Francis Joseph, Jr
Donovan, Thomas Arnold
Dumbaugh, William Henry, Jr
Eachus, Raymond Stanley
Eberts, Robert Eugene
Eibeck, Richard Elmer
Eisenberg, Max
Eisenberg, Richard
Fauth, David Jonathan
Fay, Robert Clinton
Featherstone, John Douglas Bernard
Feldman, Isaac
Flanigen, Edith Marie
Flank, William H
Frost, Robert Edwin
Gambino, Richard Joseph
Gancy, Alan Brian
Geiger, David Kenneth
Gentile, Philip
Gerace, Paul Louis
Goddard, John Burnham
Goldberg, David Elliott
Goldberg, Stephen Zalmund
Gortsema, Frank Peter
Gritmon, Timothy F
Gysling, Henry J
Hahn, Richard Leonard
Haim, Albert
Haitko, Deborah Ann
Halliday, Robert William
Hares, George Bigelow
Hauth, Willard E(llsworth), III
Holland, Hans J
Holtzberg, Frederic
Hubbard, Richard Alexander, II
Hyde, Kenneth E
Interrante, Leonard V
Jaffe, Marvin Richard

Jones, William Davidson
Kallen, Thomas William
Keeler, Robert Adolph
Keister, Jerome Baird
Kilbourn, Barry T
Kirchner, Richard Martin
Kivlighn, Herbert Daniel, Jr
Klein, Donald Lee
Klinck, Ross Edward
Koch, Stephen Andrew
Koetzle, Thomas F
Kokoszka, Gerald Francis
Kotz, John Carl
Kozlowski, Theodore R
Kwitowski, Paul Thomas
Landon, Shayne J
Larach, Simon
Lashewycz-Rubycz, Romana A
Lavallee, David Kenneth
Lazarus, Marc Samuel
Lees, Alistair John
Lefkowitz, Stanley A
Liu, Chong Tan
Luckey, George William
McCarthy, Paul James
MacDowell, John Fraser
McGrath, John F
McLendon, George L
Madan, Stanley Krishen
Malerich, Charles
Matienzo, Luis J
May, Joan Christine
Mayr, Andreas
Meloon, Daniel Thomas, Jr
Meloon, David Rand
Michiels, Leo Paul
Milton, Kirby Mitchell
Model, Frank Steven
Morrow, Jack I
Morrow, Janet Ruth
Muller, Olaf
Mullhaupt, Joseph Timothy
Myers, Clifford Earl
Nancollas, George H
Patmore, Edwin Lee
Patton, Robert Lyle
Pence, Harry Edmond
Pizer, Richard David
Poncha, Rustom Pestonji
Popp, Gerhard
Porter, Richard Francis
Price, Harry James
Puttlitz, Karl Joseph
Raider, Stanley Irwin
Robinson, Martin Alvin
Rossi, Miriam
Rupp, John Jay
Russell, Virginia Ann
Scaife, Charles Walter John
Schmeckenbecher, Arnold F
Schottmiller, John Charles
Scott, Bruce Albert
Sheard, John Leo
Shelby, James Elbert
Sheridan, Peter Sterling
Shineman, Richard Shubert
Shoup, Robert D
Silver, Herbert Graham
Simplicio, Jon
Smith, Alan Jerrard
Stephen, Keith H
Strekas, Thomas C
Sturmer, David Michael
Sujishi, Sei
Sutin, Norman
Swanson, Alan Wayne
Syed, Ashfaquzzaman
Szalda, David Joseph
Thompson, David Allen
Thurner, Joseph John
Triplett, Kelly B
Tyagi, Suresh C
Vaska, Lauri
Villa, Juan Francisco
Vinal, Richard S
Vogel, Glenn Charles
Von Winbush, Samuel
Walsh, Edward Nelson
Walter, Paul Hermann Lawrence
Wamser, Christian Albert
Weick, Charles Frederick
Weiser, David W
Wexell, Dale Richard
Wiedemeier, Heribert
Wilkes, Glenn Richard
Wiseman, George Edward
Wyman, Donald Paul
Ziolo, Ronald F
Zipp, Arden Peter
Zubieta, Jon Andoni

NORTH CAROLINA
Allen, Harry Clay, Jr
Bereman, Robert Deane
Bottjer, William George
Bowkley, Herbert Louis
Bryan, Horace Alden
Buchanan, James Wesley
Burrus, Robert Tilden
Clemens, Donald Faull
Cochran, George Thomas
Collier, Francis Nash, Jr
Collier, Herman Edward, Jr
Dilts, Joseph Alstyne

DuBois, Thomas David
Epperson, Edward Roy
Farona, Michael F
Frieser, Rudolf Gruenspan
Garrou, Philip Ernest
Gillikin, Jesse Edward, Jr
Gray, Walter C(larke)
Grimley, Eugene Burhans, III
Hammaker, Geneva Sinquefield
Hatfield, William E
Hentz, Forrest Clyde, Jr
Hicks, Kenneth Ward
Horner, William Wesley
Huang, Jamin
Hugus, Z Zimmerman, Jr
Jackels, Susan Carol
Jackman, Donald Coe
Jezorek, John Robert
Little, William Frederick
Long, George Gilbert
McAllister, Warren Alexander
McDonald, Daniel Patrick
MacInnes, David Fenton, Jr
May, Walter Ruch
Meyer, Thomas J
Murray, Royce Wilton
Noftle, Ronald Edward
Palmer, Richard Alan
Pruett, Roy L
Ramaswamy, H N
Reisman, Arnold
Rhyne, Thomas Crowell
Richter, Harold Gene
Schreiner, Anton Franz
Sink, Donald Woodfin
Spielvogel, Bernard Franklin
Switzer, Mary Ellen Phelan
Templeton, Joseph Leslie
Walsh, Edward John
Wasson, John R
Weaver, Ervin Eugene
Whangbo, Myung Hwan
Wilson, Lauren R
Zumwalt, Lloyd Robert

NORTH DAKOTA
Abrahamson, Harmon Bruce
Boudjouk, Philip Raymond
Broberg, Joel Wilbur
Garvey, Roy George
Hoggard, Patrick Earle
Morris, Melvin L
Young, Clifton A

OHIO
Ackermann, Martin Nicholas
Alexander, John J
Attig, Thomas George
Bacon, Frank Rider
Bacon, William Edward
Barnett, Kenneth Wayne
Bendure, Robert J
Bernays, Peter Michael
Bimber, Russell Morrow
Blinn, Elliott L
Blum, Patricia Rae
Brown, William Anderson
Burow, Duane Frueh
Bursten, Bruce Edward
Buttlar, Rudolph O
Carfagno, Daniel Gaetano
Carrabine, John Anthony
Chen, Yih-Wen
Chong, Clyde Hok Heen
Cooley, William Edward
Cummings, Sue Carol
Curry, John D
Dahlgren, Richard Marc
Darlington, William Bruce
Davies, Julian Anthony
Dowell, Michael Brendan
Duffy, Norman Vincent, Jr
Durand, Edward Allen
Edwards, Jimmie Garvin
Elder, Richard Charles
Eley, Richard Robert
Erbacher, John Kornel
Everson, Howard E
Ferguson, John Allen
Fischer, Mark Benjamin
Fortman, John Joseph
Fox, Robert Kriegbaum
Fryxell, Robert Edward
Gage, Frederick Worthington
Gallagher, Patrick Kent
Gaus, Paul Louis
Ghose, Hirendra M
Gilbert, George Lewis
Given, Kerry Wade
Goel, Anil B
Gordon, Gilbert
Gould, Edwin Sheldon
Greene, Arthur Frederick, Jr
Harrington, Roy Victor
Harris, Arlo Dean
Hendricker, David George
Hill, Ann Gertrude
Hohman, William H
Holtman, Mark Steven
Houk, Clifford C
Huff, Norman Thomas
Hunt, Dominic Joseph
Hunt, Jerry Donald
Hunter, James Charles

Hosmane, Narayan Sadashiv
Howard, Charles
Huege, Fred Robert
Hurlburt, H(arvey) Zeh
Job, Robert Charles
Johnson, Malcolm Pratt
Kelly, Henry Curtis
King, Edward Frazier
Klassen, David Morris
Koehler, William Henry
Kust, Roger Nayland
Kyker, Gary Stephen
Lagowski, Joseph John
Levitt, Leonard Sidney
Longo, John M
Lucid, Michael Francis
Mango, Frank Donald
Margrave, John Lee
Martell, Arthur Earl
Martin, Donald Ray
Messenger, Joseph Umlah
Meyer, James Melvin
Meyer, W(illiam) Keith
Meyers, M Douglas
Mills, Jerry Lee
Mosier, Benjamin
Neilson, Robert Hugh
Niu, Joseph H Y
Ortego, James Dale
Pennington, David Eugene
Porter, Vernon Ray
Puligandla, Viswanadham
Quane, Denis Joseph
Riggs, Charles Lee
Rollmann, Louis Deane
Rosenthal, Michael R
Sager, Ray Stuart
Sams, Lewis Calhoun, Jr
Sawyer, Donald Turner, Jr
Schultz, Linda Dalquest
Seaton, Jacob Alif
Sicilio, Fred
Singleton, David Michael
Slinkard, William Earl
Stanbury, David McNeil
Steinfink, Hugo
Stranges, Anthony Nicholas
Theriot, Leroy James
Urdy, Charles Eugene
Walmsley, Judith Abrams
Wendlandt, Wesley W
Whitmire, Kenton Herbert
Williams, Rickey Jay
Wilson, Lon James
Wilson, Peggy Mayfield Dunlap
Wisian-Neilson, Patty Joan
Witmer, William Byron
Zingaro, Ralph Anthony

UTAH
Adler, Robert Garber
Bills, James LaVar
Breckenridge, William H
Cannon, John Francis
Ernst, Richard Dale
Eyring, Edward M
Izatt, Reed McNeil
Kelsey, Stephen Jorgensen
Kodama, Goji
Malouf, Emil Edward
Morse, Joseph Grant
Morse, Karen W
Nordmeyer, Francis R
Parry, Robert Walter
Ragsdale, Ronald O
Richmond, Thomas G
Sauer, Dennis Theodore
Spence, Jack Taylor
Stoker, Howard Stephen
Waddell, Kidd M
Welch, Garth Larry
Wilson, Byron J

VERMONT
Allen, Christopher Whitney
Jasinski, Jerry Peter
Keenan, Robert Gregory
Schuele, William John

VIRGINIA
Addamiano, Arrigo
Anderson, Samuel
Averill, Bruce Alan
Bailey, Susan Goodman
Beck, James Donald
Berg, John Richard
Calle, Luz Marina
Clifford, Alan Frank
Crissman, Judith Anne
DeFotis, Gary Constantine
Demas, James Nicholas
Dominey, Raymond Nelson
Evans, Howard Tasker, Jr
Finkelman, Robert Barry
Finley, Arlington Levart
Fisher, James Russell
Fox, William B
Galloway, James Neville
Garrett, Barry B
Glanville, James Oliver
Goehring, John Brown
Grimes, Russell Newell
Gushee, Beatrice Eleanor
Hazlehurst, David Anthony

Hufstedler, Robert Sloan
Joebstl, Johann Anton
Kirby, Andrew Fuller
Liss, Ivan Barry
Malin, John Michael
Martin, Robert Bruce
Matuszko, Anthony Joseph
Melson, Gordon Anthony
Merola, Joseph Salvatore
Metcalf, David Halstead
Morgan, Evan
Mosbo, John Alvin
Myers, William Howard
Naeser, Charles Rudolph
Pasko, Thomas Joseph, Jr
Porterfield, William Wendell
Quagliano, James Vincent
Richardson, Frederick S
Saalfeld, Fred Eric
Sacks, Lawrence J
St Clair, Anne King
Schnepfe, Marian Moeller
Schreiber, Henry Dale
Sebba, Felix
Smith, Jerry Joseph
Stagg, William Ray
Taylor, Larry Thomas
Thompson, David Wallace
Topich, Joseph
Tyree, Sheppard Young, Jr
Watt, William Joseph
Wickham, James Edgar, Jr
Workman, Marcus Orrin

WASHINGTON
Axtell, Darrell Dean
Ballou, Nathan Elmer
Barney, Gary Scott
Berry, Keith O
Brehm, William Frederick
Burger, Leland Leonard
Cady, George Hamilton
Chinn, Clarence Edward
Cox, John Layton
Crook, Joseph Raymond
Crouthamel, Carl Eugene
Dreyfuss, Robert George
Duncan, Leonard Clinton
Eddy, Lowell Perry
Ellis, David Allen
Frankart, William A
Gibbins, Sidney Gore
Gregory, Norman Wayne
Grieb, Merland William
Haight, Gilbert Pierce, Jr
Hill, Orville Farrow
Hunt, John Philip
Jones, Thomas Evan
Kent, Ronald Allan
Koehmstedt, Paul Leon
Koerker, Frederick William
Ladd, Kaye Victoria
Macklin, John Welton
Millard, George Buente
Moore, Raymond H
Morrey, John Rolph
Neal, John Alexander
Norman, Joe G, Jr
Nowotny, Kurt A
Nyman, Carl John
O'Brien, Thomas Doran
Partridge, Jerry Alvin
Peterson-Kennedy, Sydney Ellen
Ryan, Jack Lewis
Sheppard, John Clarence
Templeton, John Charles
Terada, Kazuji
Weyh, John Arthur
Wicholas, Mark L
Wilson, Archie Spencer
Yunker, Wayne Harry

WEST VIRGINIA
Babb, Daniel Paul
Bhasin, Madan M
Bond, Stephon Thomas
Chakrabarty, Manoj R
Das, Kamalendu
Das Sarma, Basudeb
Dean, Warren Edgell
Hall, James Lester
Jones, Wilber Clark
Nakon, Robert Steven
Petersen, Jeffrey Lee
Peterson, James Lowell
Pribble, Mary Jo
Smith, James Allbee
Swiger, Elizabeth Davis

WISCONSIN
Antholine, William E
Bayer, Richard Eugene
Becker, Edward Brooks
Campbell, Donald L
Colton, Ervin
Dahl, Lawrence Frederick
Discher, Clarence August
Elsbernd, Helen
Evans, James Stuart
Fleming, Suzanne M
Fraser, Margaret Shirley
Gaines, Donald Frank
Hansen, Robert Conrad
Haworth, Daniel Thomas

Jache, Albert William
Kistner, Clifford Richard
Larsen, Edwin Merritt
Larson, Wilbur S
Marking, Ralph H
Moore, John Ward
Phillips, John Spencer
Pietro, William Joseph
Post, Elroy Wayne
Radtke, Douglas Dean
Rainville, David Paul
Roubal, Ronald Keith
Scott, Earle Stanley
Shaw, C Frank, III
Siebring, Barteld Richard
Spencer, Brock
Strommen, Dennis Patrick
Treichel, Paul Morgan, Jr
Utke, Allen R
Watters, Kenneth Lynn
Wright, Steven Martin

WYOMING
Coates, Geoffrey Edward
Hodgson, Derek John
Howatson, John
Morgan, George L
Schuman, Gerald E
Sullivan, Brian Patrick

PUERTO RICO
Lamba, Ram Sarup
Muir, Mariel Meents

ALBERTA
Boorman, Philip Michael
Cavell, Ronald George
Cowie, Martin
Graham, William Arthur Grover
Sanger, Alan Rodney

BRITISH COLUMBIA
Bushnell, Gordon William
Chaklader, Asoke Chandra Das
Cullen, William Robert
Dixon, Keith R
Einstein, Frederick W B
Howard, John
James, Brian Robert
Kirk, Alexander David
Lam, Gabriel Kit Ying
McAuley, Alexander
Orvig, Chris
Sutton, Derek

MANITOBA
Baldwin, W George
Janzen, Alexander Frank
Lange, Bruce Ainsworth
Taylor, Peter
Westmore, John Brian

NEW BRUNSWICK
Bottomley, Frank
Brewer, Douglas G
Fraser, Alan Richard
Mehra, Mool Chand
Passmore, Jack
Whitla, William Alexander

NEWFOUNDLAND
Clase, Howard John
Gogan, Niall Joseph
Haines, Robert Ivor
Lucas, Colin Robert
Newlands, Michael John
Rayner-Canham, Geoffrey William

NOVA SCOTIA
Cameron, T Stanley
Clark, Howard Charles
Faught, John Brian
Knop, Osvald
Peach, Michael Edwin
Ryan, Douglas Earl
Whiteway, Stirling Giddings

ONTARIO
Ashbrook, Allan William
Baldwin, Howard Wesley
Bancroft, George Michael
Barrett, Peter Fowler
Bhatnagar, Dinech C
Bioleau, Luc J R
Boorn, Andrew William
Brown, Ian David
Cale, William Robert
Carty, Arthur John
Chakrabarti, Chuni Lal
Cocivera, Michael
Dean, Philip Arthur Woodworth
Drake, John Edward
Ettel, Victor Alexander
Fyfe, William Sefton
Gillespie, Ronald James
Greedan, John Edward
Haines, Roland Arthur
Hair, Michael L
Hartman, John Stephen
Hemmings, Raymond Thomas
Kaiser, Klaus L(eo) E(duard)
Kidd, Robert Garth
Lever, Alfred B P
Lister, Maurice Wolfenden

Lock, Colin James Lyne
Mackiw, Vladimir Nicolaus
Miller, Jack Martin
Norris, A R
Payne, Nicholas Charles
Poe, Anthony John
Puddephatt, Richard John
Rempel, Garry Llewellyn
Richardson, Mary Frances
Ritcey, Gordon M
Rumfeldt, Robert Clark
Sadana, Yoginder Nath
Senoff, Caesar V
Stairs, Robert Ardagh
Stillman, Martin John
Thompson, James Charlton
Tomlinson, Richard Howden
Tuck, Dennis George
Walker, Alan
Walker, Ian Munro
Westland, Alan Duane
Wiles, Donald Roy
Willis, Christopher John

QUEBEC
Bourgon, Marcel
Hogan, James Joseph
Jabalpurwala, Kaizer E
Lalancette, Jean-Marc
Langford, Cooper Harold, III
Lau, Cheuk Kun
Onyszchuk, Mario
Theophanides, Theophile

SASKATCHEWAN
Senior, John Brian
Strathdee, Graeme Gilroy
Waltz, William Lee

OTHER COUNTRIES
Bulkin, Bernard Joseph
Burford, Neil
Collins, Kenneth Elmer
El-Awady, Abbas Abbas
Grunwald, John J
Irgolic, Kurt Johann
Joy, George Cecil, III
Loeliger, David A
Lonsdale-Eccles, John David
McGuire, George
McRae, Wayne A
Schleyer, Paul von Ragué
Shor, Arthur Joseph
Truex, Timothy Jay
Walton, Wayne J A, Jr
Wang, Yu
Wilkins, Ralph G
Wilkinson, Geoffrey

Nuclear Chemistry

ALABAMA
Lecklitner, Myron Lynn

ARIZONA
Boynton, William Vandegrift
Jull, Anthony John Timothy
Stamm, Robert Franz
Whitehurst, Harry Bernard

ARKANSAS
Day, James Meikle
Palmer, Bryan D

CALIFORNIA
Baisden, Patricia Ann
Bolt, Robert O'Connor
Bonner, Norman Andrew
Camp, David Conrad
Carr, Robert J(oseph)
Chaubal, Madhukar Gajanan
Croft, Paul Douglas
Diamond, Richard Martin
Donovan, P F
Dougan, Arden Diane
Gardner, Donald Glenn
Goishi, Wataru
Grant, Patrick Michael
Hawkes, Wayne Christian
Hicks, Harry Gross
Hoff, Richard William
Hoffman, Darleane Christian
Holmes, Donald Eugene
Holmes, Ivan Gregory
Hulet, Ervin Kenneth
Ide, Roger Henry
King, William Robert, Jr
Knox, William Jordan
Kocol, Henry
Kruger, Paul
Lederer, C Michael
Leich, Douglas Albert
Leventhal, Leon
Levy, Harris Benjamin
Ling, Alan Campbell
Lugmair, Guenter Wilhelm
MacDonald, Norman Scott
Markowitz, Samuel Solomon
Marti, Kurt
Martz, Harry Edward, Jr
Masters, Burton Joseph
Metzger, Albert E
Michels, Lloyd R

Nuclear Chemistry (cont)

Miller, George E
Miskel, John Albert
Mode, Vincent Alan
Moody, Kenton J
Najafi, Ahmad
Nethaway, David Robert
Norman, Eric B
Nurmia, Matti Juhani
Pehl, Richard Henry
Phelps, Michael Edward
Poggenburg, John Kenneth, Jr
Poskanzer, Arthur M
Raby, Bruce Alan
Ruiz, Carl P
Russ, Guston Price, III
Sextro, Richard George
Singer, Solomon Elias
Smith, Charles Francis, Jr
Stewart, Donald Charles
Struble, Gordon Lee
Sugihara, Thomas Tamotsu
Swanson, David G, Jr
Tewes, Howard Allan
Van Hise, James R
Varteressian, K(eghan) A(rshavir)
Wasson, John Taylor
Wiesendanger, Hans Ulrich David
Wild, John Frederick
Wolf, Walter
Wu, Jiann-Long
Wunderly, Stephen Walker

COLORADO
Caruthers, Marvin Harry
Gottschall, W Carl
Harlan, Ronald A
Heberlein, Douglas Garavel
Keder, Wilbert Eugene
Lucas, George Bond
Martell, Edward A
Muscatello, Anthony Curtis
Vejvoda, Edward
Yang, In Che

CONNECTICUT
Ziegler, Theresa Frances

DELAWARE
Narvaez, Richard
Tulip, Thomas Hunt

DISTRICT OF COLUMBIA
Brown, James Edward
Burnett, John Lambe
Houck, Frank Scanland
McCormack, Mike
Rzeszotarski, Waclaw Janusz

FLORIDA
Hertel, George Robert
Ostlund, H Gote
Stang, Louis George
Wethington, John A(bner), Jr

GEORGIA
Fink, Richard Walter
Jackson, William Morrison
Kahn, Bernd
Menon, Manchery Prabhakara
Neumann, Henry Matthew
Reif, Donald John

IDAHO
Baker, John David
Gehrke, Robert James

ILLINOIS
Adams, Richard Melverne
Ahmad, Irshad
Beasley, Thomas Miles
Bingham, Carleton Dille
Burrell, Elliott Joseph, Jr
Coley, Ronald Frank
Flynn, Kevin Francis
Horwitz, Earl Philip
Johnson, Irving
Kaufman, Sheldon Bernard
Kennedy, Albert Joseph
Liaw, Jye Ren
Mason, Donald Frank
Meadows, James Wallace, Jr
Melhado, L(ouisa) Lee
Nash, Kenneth Laverne
Nealy, Carson Louis
Pietri, Charles Edward
Steinberg, Ellis Philip
Steindler, Martin Joseph

INDIANA
Aprahamian, Ani
Contario, John Joseph
Kolata, James John
Lipschutz, Michael Elazar
Reuland, Donald John
Shaw, Stanley Miner
Stroube, William Bryan, Jr
Viola, Victor E, Jr

IOWA
Hill, John Christian
Martin, Don Stanley, Jr
Voigt, Adolf F

KENTUCKY
Ehmann, William Donald
Robertson, John David
Yates, Steven Winfield

LOUISIANA
Keeley, Dean Francis

MARYLAND
Becker, Donald Arthur
Buchanan, John Donald
Colle, Ronald
Cooperstein, Raymond
Del Grosso, Vincent Alfred
Downing, Robert Gregory
Friedman, Abraham S(olomon)
Gause, Evelyn Pauline
Gordon, Glen Everett
Lamaze, George Paul
Leak, John Clay, Jr
Mignerey, Alice Cox
Poochikian, Guiragos K
Roche, Lidia Alicia
Roethel, David Albert Hill
Smith, Elmer Robert
Walters, William Ben
Weber, Clifford E
Wing, James

MASSACHUSETTS
Brenner, Daeg Scott
Davis, Michael Allan
Huang, Shaw-Guang
Lis, Steven Andrew
Livingston, Hugh Duncan
Olmez, Ilhan
Richason, George R, Jr
Stark, Ruth E

MICHIGAN
Anders, Oswald Ulrich
Griffin, Henry Claude
Griffioen, Roger Duane
Huang, Che C
Morrissey, David Joseph
Rengan, Krishnaswamy
Robbins, John Alan
Schneider, Eric West
Thomas, Kenneth Eugene, III
Thompson, Paul Woodard

MINNESOTA
Carpenter, John Harold

MISSOURI
Ames, Donald Paul
Black, Wayne Edward
Freeman, Robert Clarence
Glascock, Michael Dean
Lyle, Leon Richards
Macias, Edward S
Manuel, Oliver K
Troutner, David Elliott
Volkert, Wynn Arthur

NEBRASKA
Blotcky, Alan Jay
Borchert, Harold R

NEVADA
Pierson, William R

NEW JERSEY
Black, James Francis
Cadet, Gardy
Egli, Peter
Ganapathy, Ramachandran
Gavini, Muralidhara B
Hall, Gene Stephen
Kramer, George Mortimer
Nicholls, Gerald P
Pickering, Miles Gilbert
Siegel, Jeffry A

NEW MEXICO
Apt, Kenneth Ellis
Barr, Donald Westwood
Bild, Richard Wayne
Binder, Irwin
Bowen, Scott Michael
Bryant, Ernest Atherton
Butler, Gilbert W
Catlett, Duane Stewart
Erdal, Bruce Robert
Fowler, Malcolm McFarland
Gancarz, Alexander John
Giesler, Gregg Carl
Hakkila, Eero Arnold
Hobart, David Edward
Jackson, Darryl Dean
Jackson, James F
Jackson, Sydney Vern
Kiley, Leo Austin
Marlow, Keith Winton
Mason, Allen Smith
Molecke, Martin A
Norris, Andrew Edward
Orth, Charles Joseph
Parker, Winifred Ellis
Paxton, Hugh Campbell
Penneman, Robert Allen
Pillay, K K Sivasankara
Reedy, Robert Challenger
Richardson, Albert Edward

SEDLACEK
Sedlacek, William Adam
Thomas, Charles Carlisle, Jr
Thomas, Kimberly W
Vieira, David John
Wahl, Arthur Charles
Wilhelmy, Jerry Barnard
Williams, Robert Allen
Wolfsberg, Kurt
Yates, Mary Anne

NEW YORK
Alexander, John Macmillan, Jr
Armamento, Eduardo T
Chan, Shu Fun
Chrien, Robert Edward
Chu, Yung Yee
Conn, Paul Kohler
Cooper, Arthur Joseph L
Cumming, James B
Dahl, John Robert
Davis, Raymond, Jr
Debiak, Ted Walter
Duek, Eduardo Enrique
Finn, Ronald Dennet
Foreman, Bruce Milburn, Jr
Friedlander, Gerhart
Gill, Ronald Lee
Grover, James Robb
Hahn, Richard Leonard
Hartmann, Francis Xavier
Haustein, Peter Eugene
Hoff, Wilford J, Jr
Holden, Norman Edward
Hopke, Philip Karl
Huizenga, John Robert
Husain, Liaquat
Katcoff, Seymour
Komorek, Michael Joesph, Jr
Krey, Philip W
Ma, Maw-Suen
McHugh, James Anthony, Jr
MacKenzie, Donald Robertson
Matuszek, John Michael, Jr
Mausner, Leonard Franklin
Meyer, John Austin
Schroder, Wolf-Udo
Tassinari, Silvio John
Ward, Thomas Edmund
Williams, Evan Thomas
Wolf, Alfred Peter

NORTH CAROLINA
Arnold, Luther Bishop, Jr
Kalbach, Constance
Spicer, Leonard Dale
Wyrick, Steven Dale

OHIO
Bogar, Louis Charles
Brandhorst, Henry William, Jr
Chong, Clyde Hok Heen
Decker, Clarence Ferdinand
Hummon, William Dale
John, George
Madia, William J
Muntean, Richard August
Nathan, Richard Arnold
Parikh, Sarvabhaum Sohanlal
Rudy, Clifford R
Saha, Gopal Bandhu
Stout, Barbara Elizabeth
Yanko, William Harry

OKLAHOMA
Klehr, Edwin Henry

OREGON
Church, Larry B
DeMent, Jack (Andrew)
Goles, Gordon George
Kay, Michael Aaron
Loveland, Walter (David)
Schmitt, Roman A

PENNSYLVANIA
Bogard, Andrew Dale
Bouis, Paul Andre
Brown, Paul Edmund
Catchen, Gary Lee
Chubb, Walston
Heys, John Richard
Hogan, Aloysius Joseph, Jr
Jaffe, Eileen Karen
Jester, William A
Kahler, Albert Comstock, III
Kaplan, Morton
Karol, Paul J(ason)
Kay, Jack Garvin
Kohman, Truman Paul
Pacer, John Charles
Sturm, James Edward
Tzodikov, Nathan Robert

SOUTH CAROLINA
Allen, Joe Frank
Bibler, Ned Eugene
Boyd, George Edward
Fowler, John Rayford
Groh, Harold John
Hawkins, Richard Horace
Hochel, Robert Charles
Hofstetter, Kenneth John
Holcomb, Herman Perry
Kinard, W Frank

McDONELL
McDonell, William Robert
Moore, Willard S
Stone, John Austin
Wilhite, Elmer Lee

TENNESSEE
Bemis, Curtis Elliot, Jr
Bhattacharya, Syamal Kanti
Bigelow, John E(aly)
Campbell, David Owen
Crompton, Charles Edward
Dyer, Frank Falkoner
Ferguson, Robert Lynn
Haas, Paul Arnold
Hahn, Richard Balser
Halperin, Joseph
Hamilton, Joseph H, Jr
Johnson, Noah R
Lindemer, Terrence Bradford
McDowell, William Jackson
Moyer, Bruce A
Plasil, Franz
Ricci, Enzo
Young, Jack Phillip

TEXAS
Conway, Dwight Colbur
Edgerley, Dennis A
Farhataziz, Mr
Gnade, Bruce E
Haenni, David Richard
Hassell, Clinton Alton
Hunt, Richard Henry
Iskander, Felib Youssef
Lamb, James Francis
Ludwig, Allen Clarence, Sr
Patel, Bhagwandas Mavjibhai
Tilbury, Roy Sidney
Trehan, Rajender
Walters, Lester James, Jr
Wright, James Foley

UTAH
Mangelson, Nolan Farrin
Petryka, Zbyslaw Jan

VIRGINIA
Allen, Ralph Orville, Jr
Baedecker, Philip A
Baker, Paul, Jr
Chulick, Eugene Thomas
Clark, Donald Eldon
Davidson, Charles Nelson
Grossman, Jeffrey N
Morgan, John Walter
Morrow, Richard Joseph
Owens, Charles Wesley
Perry, Dennis Gordon
Sprinkle, Robert Shields, III
Wong, George Tin Fuk

WASHINGTON
Anderson, Harlan John
Ballou, Nathan Elmer
Brodzinski, Ronald Lee
Brown, Donald Jerould
Chinn, Clarence Edward
Daniel, J(ack) Leland
Fisher, Darrell R
Garland, John Kenneth
Hendrickson, Waldemar Forrsel
Hill, Orville Farrow
Kaye, James Herbert
Krohn, Kenneth Albert
Reeder, Paul Lorenz
Tingey, Garth Leroy
Wheelwright, Earl J
Wilbur, Daniel Scott
Wogman, Ned Allen

WISCONSIN
Runnalls, Nelva Earline Gross

PUERTO RICO
Souto Bachiller, Fernando Alberto

ALBERTA
Apps, Michael John

BRITISH COLUMBIA
D'Auria, John Michael

MANITOBA
Attas, Ely Michael
Iverson, Stuart Leroy
Sunder, Sham
Sze, Yu-Keung
Tait, John Charles
Wikjord, Alfred George

NOVA SCOTIA
Chatt, Amares
Holzbecher, Jiri

ONTARIO
Ball, Gordon Charles
Crocker, Iain Hay
Davies, John Arthur
Eastwood, Thomas Alexander
Hay, George William
Horn, Dag
Jardine, John McNair
Jervis, Robert E
Marshall, Heather

QUEBEC
Diksic, Mirko
Hogan, James Joseph
Yaffe, Leo

SASKATCHEWAN
Cassidy, Richard Murray

OTHER COUNTRIES
Chung, Chien
Collins, Kenneth Elmer
Yellin, Joseph

Organic Chemistry

ALABAMA
Atwood, Jerry Lee
Ayers, Orval Edwin
Beasley, James Gordon
Benington, Frederick
Byrd, James Dotson
Caine, Drury Sullivan, III
Cava, Michael Patrick
Chou, Libby Wang
Christian, Samuel Terry
Coburn, William Carl, Jr
Cooper, Emerson Amenhotep
Dean, David Lee
Dukatz, Ervin L, Jr
Elliott, Robert Daryl
Evanochko, William Thomas
Feazel, Charles Elmo, Jr
Greene, Joseph Lee, Jr
Hamer, Justin Charles
Hargis, J Howard
Hart, David R
Huskins, Chester Walker
Isbell, Raymond Eugene
Jackson, Thomas Gerald
Kearley, Francis Joseph, Jr
Kiely, Donald Edward
Lambert, James LeBeau
Leslie, Thomas M
Lewis, Dennis Osborne
Livant, Peter David
Ludwick, Adriane Gurak
McManus, Samuel Plyler
Miller, Nathan C
Montgomery, John Atterbury
Murray, Thomas Pinkney
Nair, Madhavan G
Parish, Edward James
Patterson, William Jerry
Piper, James Robert
Ponder, Billy Wayne
Retief, Daniel Hugo
Richmond, Charles William
Riley, Thomas N
Sani, Brahma Porinchu
Sayles, David Cyril
Secrist, John Adair, III
Shealy, Y(oder) Fulmer
Shevlin, Philip Bernard
Short, William Arthur
Small, Robert James
Squillacote, Michael Edward
Stephens, William D
Tamburin, Henry John
Temple, Carroll Glenn
Thomas, Hazel Jeanette
Thomas, Joseph Calvin
Thompson, Wynelle Doggett
Thorpe, Martha Campbell
Veazey, Thomas Mabry
Weinberg, David Samuel
Worley, S D
Youngblood, Bettye Sue

ALASKA
Collins, Jeffery Allen
Reppond, Kermit Dale

ARIZONA
Adamson, Albert S, Jr
Adickes, H Wayne
Allen, James Durwood
Anderson, Floyd Edmond
Barry, Arthur John
Barstow, Leon E
Bates, Robert Brown
Berry, James Wesley
Burgoyne, Edward Eynon
Caple, Gerald
Carter, Herbert Edmund
Chapman, Robert Dale
Christensen, Kenner Allen
Coe, Beresford
Deutschman, Archie John, Jr
Doubek, Dennis Lee
Eskelson, Cleamond D
Glass, Richard Steven
Gust, J Devens, Jr
Harris, Leland
Heaton, Charles Daniel
Heinle, Preston Joseph
Hinman, Charles Wiley
Hoyt, Earle B, Jr
Hruby, Victor J
Huffman, Robert Wesly
Jungermann, Eric
Kelley, Alec Ervin
Kelley, Maurice Gordon
Kircher, Henry Winfried

ARKANSAS
Black, Howard Charles
De Luca, Donald Carl
Fry, Arthur James
Goodwin, James Crawford
Goodwin, Thomas Elton
Ku, Victoria Feng
Manion, Jerald Monroe
Meyer, Walter Leslie
Nave, Paul Michael
Pienta, Norbert J
Ransford, George Henry
Setliff, Frank Lamar
Siegel, Samuel
Sifford, Dewey H
Smith, Edgar Dumont
Wright, Joe Carrol
Yang, Dominic Tsung-Che
York, John Lyndal

CALIFORNIA
Abdel-Baset, Mahmoud B
Abegg, Victor Paul
Abell, Jared
Abrash, Henry I
Adams, John Howard
Ainsworth, Cameron
Akawie, Richard Isidore
Albert, Anthony Harold
Alexander, Edward Cleve
Allen, Charles Freeman
Allen, William Merle
Anderson, Clyde Lee
Anderson, Curtis Benjamin
Andrews, Lawrence James
Anet, Frank Adrien Louis
Assony, Steven James
Athey, Robert Douglas, Jr
Atkins, Ronald Leroy
Aue, Donald Henry
Bacha, John D
Bacskai, Robert
Baizer, Manuel M
Baker, Don Robert
Ballinger, Peter Richard
Banasiak, Dennis Stephen
Banigan, Thomas Franklin, Jr
Barnum, Emmett Raymond
Baum, John William
Belloli, Robert Charles
Benitez, Allen
Benson, Harriet
Benson, Robert Franklin
Bergman, Robert George
Berman, Elliot
Bernasconi, Claude Francois
Bernhard, Richard Allan
Berteau, Peter Edmund
Biggerstaff, Warren Richard
Billig, Franklin A
Bishop, Keith C, III
Bissell, Eugene Richard
Blair, Charles Melvin, Jr
Block, Michael Joseph
Bluestone, Henry
Boger, Dale L
Bohm, Howard A
Bond, Frederick Thomas
Boschan, Robert Herschel
Bottini, Albert Thomas
Bozak, Richard Edward
Braithwaite, Charles Henry, Jr
Brault, Robert George
Brauman, John I
Brauman, Sharon K(ruse)
Brotherton, Robert John
Brown, Charles Allan
Brown, Stuart Houston
Bruice, Thomas Charles
Bublitz, Donald Edward
Buckles, Robert Edwin
Bunnett, Joseph Frederick
Bunton, Clifford A
Burke, Richard Lerda
Burlingame, Alma L
Burrous, Merwyn Lee
Byrd, Norman Robert
Calvin, Melvin
Camenzind, Mark J
Campbell, Thomas Cooper
Carty, Daniel T
Casanova, Joseph
Caserio, Marjorie C
Casey, John Edward, Jr
Cason, James, Jr
Castro, Albert Joseph
Castro, Anthony J
Castro, Charles E

Chambers, Robert Rood
Chang, Ding
Chang, Frederic Chewming
Chapman, Orville Lamar
Chau, Michael Ming-Kee
Chauffe, Leroy
Chiddix, Max Eugene
Chin, Hsiao-Ling M
Ching, Ta Yen
Cho, Arthur Kenji
Choong, Hsia Shaw-Lwan
Ciaramitaro, David A
Ciula, Richard Paul
Clark, David Ellsworth
Cleary, Michael E
Cleland, George Horace
Cobb, Emerson Gillmore
Cole, Thomas Ernest
Collman, James Paddock
Colwell, William Tracy
Cooper, Douglas Elhoff
Cooper, Geoffrey Kenneth
Correia, John Sidney
Craig, John Cymerman
Craig, John Horace
Cram, Donald James
Cross, Alexander Dennis
Csendes, Ernest
Currell, Douglas Leo
Current, Steven P
Dafforn, Geoffrey Alan
Dahl, Klaus Joachim
Dalman, Gary
DaMassa, Al John
Daub, Guido William
Dauben, William Garfield
David, Gary Samuel
Davis, Jay C
Dawson, Daniel Joseph
DeAcetis, William
DeJongh, Don C
Dennis, Edward A
Dervan, Peter Brendan
Detert, Francis Lawrence
Deutsch, Daniel Harold
Dev, Vasu
DeVenuto, Frank
Diaz, Arthur Fred
Diederich, Francois Nico
Diekman, John David
DiGiorgio, Joseph Brun
Dills, Charles E
Dines, Martin Benjamin
Djerassi, Carl
Docks, Edward Leon
Dougherty, Dennis A
Dreyer, David
Drisko, Richard Warren
Dugaiczyk, Achilles
Duggins, William Edgar
Durham, Lois Jean
Dutra, Ramiro Carvalho
Eastman, Richard Hallenbeck
Eck, David Lowell
Embree, Harland Dumond
Endres, Leland Sander
Engler, Edward Martin
Erdman, Timothy Robert
Fahey, Robert C
Fancher, Llewellyn W
Fang, Fabian Tien-Hwa
Faulkner, D John
Feher, Frank J
Felix, Raymond Anthony
Field, George Francis
Fife, Thomas Harley
Fischback, Bryant C
Fish, Richard Wayne
Fisher, Richard Paul
Flood, Thomas Charles
Fohlen, George Marcel
Foote, Christopher S
Forkey, David Medrick
Foscante, Raymond Eugene
Frankel, Edwin N
Freeman, Fillmore
Frey, Thomas G
Fried, John
Friedman, Lester
Friedman, Mendel
Friedrich, Edwin Carl
Frost, Kenneth Almeron, Jr
Fuhrman, Albert William
Fukuto, Tetsuo Roy
Fuller, Glenn
Fung, Steven
Gaffield, William
Gale, Laird Housel
Gall, Walter George
Gallegos, Emilio Juan
Gandler, Joseph Rubin
Garber, John Douglas
Gash, Kenneth Blaine
Giants, Thomas W
Gibbons, Loren Kenneth
Gish, Duane Tolbert
Glaser, Charles Barry
Glaser, Robert J
Glover, Leon Conrad, Jr
Goetschel, Charles Thomas
Gold, Marvin H
Goldish, Dorothy May (Bowman)
Goodman, Murray
Goodrich, Judson Earl

Gordon, Chester Duncan
Gordon, Robert Julian
Grant, Barbara Dianne
Green, David Claude
Greene, Charles Richard
Grethe, Guenter
Gross, Paul Hans
Grubbs, Edward
Grubbs, Robert Howard
Gupta, Amitava
Hajdu, Joseph
Hale, Ron L
Hall, Thomas Kenneth
Hamel, Edward E
Hamilton, Carole Lois
Hammond, James Alexander, Jr
Han, Yuri Wha-Yul
Hanneman, Walter W
Harris, Edwin Randall
Harris, Francis Laurie
Harris, William Merl
Hart, Herbert Dorlan
Havlicek, Stephen
Haxo, Henry Emile, Jr
Hearst, Peter Jacob
Heasley, Victor Lee
Heathcock, Clayton Howell
Heller, Jorge
Helmkamp, George Kenneth
Henderson, Robert Burr
Hendrickson, Yngve Gust
Henrick, Clive Arthur
Henry, David Weston
Henry, Ronald Andrew
Hershey, John William Baker
Hess, Patrick Henry
Hessel, Donald Wesley
Hill, Marion Elzie
Hobbs, M Floyd
Hoch, Paul Edwin
Hogen-Esch, Thieo Eltjo
Hokama, Takeo
Holty, David Webster
Holtz, David
Horowitz, Robert Miller
Houk, Kendall N
Howell, Edward Tillson
Hsu, Ming-Ta Sung
Hughes, John Lawrence
Hullar, Theodore Lee
Ingraham, Lloyd Lewis
Isensee, Robert William
Ja, William Yin
Jacob, Peyton, III
Jacobs, Thomas Lloyd
Jaffe, Annette Bronkesh
James, Philip Nickerson
Jensen, Harbo Peter
Jensen, James Leslie
Johanson, Robert Gail
Johnson, Harry William, Jr
Johnson, Kenneth
Johnson, LeRoy Franklin
Johnson, Oscar Hugo
Johnson, William K
Johnson, William Summer
Jones, Gordon Henry
Jules, Leonard Herbert
Jung, Michael Ernest
Juster, Norman Joel
Kakis, Frederic Jacob
Kamp, David Allen
Kaneko, Thomas Motomi
Kaplan, Phyllis Deen
Kaufman, Daniel
Kaye, Samuel
Keeffe, James Richard
Kellman, Raymond
Kenyon, George Lommel
Kepner, Richard Edwin
Ketcham, Roger
Khanna, Pyare Lal
Kilday, Warren D
Kim, Chung Sul (Sue)
King, John Mathews
Kissel, Charles Louis
Klager, Karl
Kland, Mathilde June
Klinedinst, Paul Edward, Jr
Klinman, Judith Pollock
Kolyer, John M
Kong, Eric Siu-Wai
Konigsberg, Moses
Korte, William David
Koshland, Daniel Edward, Jr
Kovacs (Nagy), Hanna
Kozlowski, Robert H
Krantz, Reinhold John
Kratzer, Reinhold
Krbechek, Leroy O
Krubiner, Alan Martin
Kryger, Roy George
Kumamoto, Junji
Kumli, Karl F
Kurihara, Norman Hiromu
Kurkov, Victor Peter
Kushner, Arthur S
Labinger, Jay Alan
Laine, Richard Mason
Lamb, Sandra Ina
Lambert, Frank Lewis
Landrum, Billy Frank
Langford, Robert Bruce
Larrabee, Richard Brian

Organic Chemistry (cont)

Lawson, Daniel David
Lee, Alfred Tze-Hau
Lee, David Louis
Lee, Stuart M(ilton)
Lee, William Wei
Lee, Young-Jin
Levy, M(orton) Frank
Lewis, Sheldon Noah
Leyden, Richard Noel
Liang, Yola Yueh-o
Licari, James John
Liggett, Lawrence Melvin
Lin, Jiann-Tsyh
Lind, Carol Johnson
Lindquist, Robert Nels
Ling, Nicholas Chi-Kwan
Lipson, Melvin Alan
Little, John Clayton
Little, Raymond Daniel
Lorensen, Lyman Edward
Low, Hans
Lowe, Warren
Lowy, Peter Herman
Luibrand, Richard Thomas
Lyon, Cameron Kirby
Mabey, William Ray
McCaleb, Kirtland Edward
McCloskey, Chester Martin
Mach, Martin Henry
MacKay, Kenneth Donald
McKenna, Charles Edward
McNall, Lester R
Manatt, Stanley L
Mansfield, Joseph Victor
Maricich, Tom John
Markhart, Albert H, Jr
Marmor, Solomon
Marshall, Henry Peter
Marsi, Kenneth Larue
Marten, David Franklin
Martin, Eugene Christopher
Matthews, Gary Joseph
Matuszak, Charles A
Maya, Walter
Mayer, William Joseph
Mayfield, Darwin Lyell
Mayo, Frank Rea
Meader, Arthur Lloyd, Jr
Meikle, Richard William
Merrill, Ronald Eugene
Metzner, Ernest Kurt
Meyer, Gregory Carl
Meyers, Robert Allen
Micheli, Robert Angelo
Midland, Michael Mark
Mihailovski, Alexander
Mill, Theodore
Millar, Jocelyn Grenville
Miller, Leroy Jesse
Miller, Robert Dennis
Miller, Russell Bryan
Minch, Michael Joseph
Mirza, John
Mitchell, Alexander Rebar
Moffatt, John Gilbert
Molyneux, Russell John
Montana, Andrew Frederick
Moore, Harold W
Mori, Raymond I
Morton, Thomas Hellman
Mosher, Harry Stone
Moye, Anthony Joseph
Muchowski, Joseph Martin
Munger, Charles Galloway
Murdoch, Joseph Richard
Murov, Steven Lee
Musker, Warren Kenneth
Musser, John H
Myhre, Philip C
Najafi, Ahmad
Nambiar, Krishnan P
Nebenzahl, Linda Levine
Neher, Maynard Bruce
Nelson, Norvell John
Nestor, John Joseph, Jr
Neuman, Robert C, Jr
Newey, Herbert Alfred
Newsom, Herbert Charles
Ngo, That Tjien
Nielsen, Arnold Thor
Nilsson, William A
Ning, Robert Y
Nitecki, Danute Emilija
Noble, Paul, Jr
Northington, Dewey Jackson, Jr
Novak, Bruce Michael
Noyce, Donald Sterling
Nozaki, Kenzie
Nugent, Maurice Joseph, Jr
Nussenbaum, Siegfried Fred
Nyberg, David Dolph
Nyquist, Harlan LeRoy
Oakes, John Morgan
Ogimachi, Naomi Neil
Oh, Chan Soo
Okamura, William H
Olah, George Andrew
Olah, Judith Agnes
Olivier, Kenneth Leo
Olsen, Carl John
Onopchenko, Anatoli
Osborne, David Wendell

Ottke, Robert Crittenden
Overman, Larry Eugene
Oyama, Vance I
Ozari, Yehuda
Paik, Young-Ki
Palchak, Robert Joseph Francis
Pallos, Ferenc M
Pandell, Alexander Jerry
Pappatheodorou, Sofia
Patterson, Dennis Bruce
Paulson, Donald Robert
Pellegrini, Maria C
Pennington, Frank Cook
Percival, Douglas Franklin
Perrin, Charles Lee
Perry, Robert Hood, Jr
Peters, Howard McDowell
Phillips, Donald David
Phillips, Lee Vern
Phillips, Robert Edward
Pine, Stanley H
Pinnell, Robert Peyton
Prakash, Surya G K
Preus, Martin William
Putz, Gerard Joseph
Quistad, Gary Bennet
Raley, John Howard
Rapoport, Henry
Ratcliff, Milton, Jr
Reardon, Edward Joseph, Jr
Recsei, Andrew A
Reed, Russell, Jr
Reese, Floyd Ernest
Reifschneider, Walter
Reinheimer, John David
Reist, Elmer Joseph
Replogle, Lanny Lee
Reyes, Zoila
Richardson, Wallace Lloyd
Richardson, William Harry
Rickborn, Bruce Frederick
Rife, William C
Riffer, Richard
Rinderknecht, Heinrich
Ripka, William Charles
Roberts, John D
Robins, Roland Kenith
Rocklin, Albert Louis
Rodemeyer, Stephen A
Rodgers, James Edward
Roos, Leo
Ross, David Samuel
Ross, Donald Lewis
Rossiter, Bryant William
Rowley, Gerald L(eroy)
Rubenstein, Kenneth E
Rudy, Thomas Philip
Rusay, Ronald Joseph
Russell, John George
Ryan, Patrick Walter
Rynd, James Arthur
Saltman, William Mose
Saltonstall, Clarence William, Jr
Sanchez, Robert A
Santi, Daniel V
Savitz, Maxine Lazarus
Schaleger, Larry L
Schaumberg, Gene David
Schore, Neil Eric
Schwartz, Joseph Robert
Schwarzer, Carl G
Seapy, Dave Glenn
Seiber, James N
Selassie, Cynthia R
Sellas, James Thomas
Selover, James Carroll
Selter, Gerald A
Selwitz, Charles Myron
Servis, Kenneth L
Seubold, Frank Henry, Jr
Shapiro, Bernard Lyon
Sharpless, K Barry
Sharts, Clay Marcus
Shelden, Harold Raymond, II
Shellhamer, Dale Francis
Shelton, John C
Shen, Kelvin Kei-Wei
Shen, Yvonne Feng
Siegel, Brock Martin
Siegfried, William
Silva, Ricardo
Simpson, John Ernest
Sims, James Joseph
Singaram, Bakthan
Singer, Lawrence Alan
Singh, Prithipal
Sipos, Frank
Skinner, Wilfred Aubrey, Jr
Skowronski, Raymund Paul
Smith, Kevin Malcolm
Smith, Leverett Ralph
Smith, Louis Charles
Smith, Robert Alan
Smith, Thomas Henry
Sobolev, Igor
Solar, Samuel Louis
Solomon, Malcolm David
Soloway, S Barney
Sommer, Leo Harry
Spatz, David Mark
Spenger, Robert E
Spray, Clive Robert
Stanley, William Lyons
Stanton, Garth Michael

Stehsel, Melvin Louis
Steinberg, Gunther
Steinberg, Howard
Steinman, Robert
Sterling, Robert Fillmore
Still, Gerald G
Stone, Herbert
Stout, Charles Allison
Stoutamire, Donald Wesley
Straus, Alan Edward
Streitwieser, Andrew, Jr
Suzuki, Shigeto
Swidler, Ronald
Taagepera, Mare
Taft, Robert Wheaton, Jr
Tanabe, Masato
Teach, Eugene Gordon
Teeter, Richard Malcolm
Teranishi, Roy
Tess, Roy William Henry
Thelen, Charles John
Thompson, Evan M
Thomson, Tom Radford
Thorsett, Eugene Deloy
Tilles, Harry
Ting, Chih-Yuan Charles
Tokes, Laszlo Gyula
Tomich, John Matthew
Tong, Yulan Chang
Toy, Madeline Shen
Traylor, Teddy G
Trost, Barry M
Trowbridge, Dale Brian
Tsao, Constance S
Turner, James A
Ullman, Edwin Fisher
Untch, Karl George
Urbach, Karl Frederic
VanAntwerp, Craig Lewis
Varteressian, K(egham) A(rshavir)
Verbiscar, Anthony James
Verheyden, Julien P H
Vinogradoff, Anna Patricia
Virnig, Michael Joseph
Vollhardt, K Peter C
Vollmar, Arnulf R
Von HUngen, Kern
Wade, Robert Harold
Waiss, Anthony C, Jr
Walba, Harold
Walker, Francis H
Walker, Joseph
Wall, Robert Gene
Wallace, Edwin Garfield
Wallace, Robert Allan
Waters, William Lincoln
Webb, William Paul
Wechter, William Julius
Weck, Friedrich Josef
Wedegaertner, Donald K
Wegner, Patrick Andrew
Weinshenker, Ned Martin
Weisgraber, Karl Heinrich
Welch, Frank Joseph
Weliky, Norman
Wender, Paul Anthony
Wenkert, Ernest
Wensley, Charles Gelen
Westover, James Donald
White, David Halbert
Wight, Hewitt Glenn
Wikman-Coffelt, Joan
Wildes, Peter Drury
Wiley, Michael David
Wilgus, Donovan Ray
Williams, John Wesley
Williams, Peter M
Willson, Carlton Grant
Willson, Clyde D
Wilson, Elwood Justin, Jr
Winterlin, Wray Laverne
Wolf, Monte William
Wolfe, James Frederick
Wong, Chi-Huey
Wong, Shi-Yin
Woo, Gar Lok
Woods, William George
Wright, John Marlin
Wudl, Fred
Wunderly, Stephen Walker
Yamada, Yoshikazu
Yamagishi, Frederick George
Yang, Meiling T
Yee, Tucker Tew
Youngman, Edward August
Zavarin, Eugene
Zon, Gerald
Zweifel, George

COLORADO

Bailey, David Tiffany
Barrett, Anthony Gerard Martin
Bean, Gerritt Post
Beel, John Addis
Bierbaum, Veronica Marie
Blake, Daniel Melvin
Cannon, William Nathaniel
Caruthers, Marvin Harry
Champion, William (Clare)
Cook, William Boyd
Coomes, Richard Merril
Cristol, Stanley Jerome
Daughenbaugh, Randall Jay
DeBruin, Kenneth Edward

De Puy, Charles Herbert
Dewey, Fred McAlpin
DoMinh, Thap
Druelinger, Melvin L
Duke, Roy Burt, Jr
Dyckes, Douglas Franz
Ellison, Gayfree Barney
Evans, David Lane
Fennessey, Paul V
Fleming, Michael Paul
Gabel, Richard Allen
Gentry, Willard Max, Jr
Gillespie, John Paul
Goren, Mayer Bear
Gramera, Robert Eugene
Guss, Cyrus Omar
Hammer, Charles Rankin
Hayes, Jeffrey Charles
Hegedus, Louis Stevenson
Hirs, Christophe Henri Werner
Hornback, Joseph Michael
Hultquist, Martin Everett
Hyatt, David Ernest
Jones, Harold Lester
Kice, John Lord
Koch, Tad H
Langworthy, William Clayton
Lehman, Joe Junior
Levin, Ronald Harold
Lindquist, Robert Marion
Lodewijk, Eric
Lucas, George Bond
Meek, John Sawyers
Meyers, Albert Irving
Micheli, Roger Paul
Miller, William Theodore
Murphy, Robert Carl
Nachtigall, Guenter Willi
Nightingale, Dorothy Virginia
Ozog, Francis Joseph
Parungo, Farn Pwu
Patton, James Winton
Peters, Kevin Scott
Pyler, Richard Ernst
Robertson, Jerold C
Rosenburg, Dale Weaver
Schoffstall, Allen M
Schreck, James O(tto)
Schroeder, Herbert August
Shackelford, Scott Addison
Shelton, Robert Wayne
Spraggins, Robert Lee
Stermitz, Frank
Stickler, William Carl
Stille, John Kenneth
Stribley, Rexford Carl
Tomasi, Gordon Ernest
Turner, James Howard
Voorhees, Kent Jay
Walba, David Mark
Wilkes, John Stuart
Ziebarth, Timothy Dean

CONNECTICUT

Alexanian, Vazken Arsen
Alfieri, Charles C
Amundsen, Lawrence Hardin
Andrews, Glenn Colton
Arif, Shoaib
Baccei, Louis Joseph
Bailey, William Francis
Baker, Leonard Morton
Baldwin, Ronald Martin
Banucci, Eugene George
Batorewicz, Wadim
Bentz, Alan P(aul)
Berson, Jerome Abraham
Bindra, Jasjit Singh
Bloom, Barry Malcolm
Bobko, Edward
Bollyky, L(aszlo) Joseph
Bongiorni, Domenic Frank
Borden, George Wayne
Bordner, Jon D B
Branchini, Bruce Robert
Brown, Keith Charles
Byrne, Kevin J
Calbo, Leonard Joseph
Card, Roger John
Chandler, Roger Eugene
Chang, Pauline (Wuai) Kimm
Chu, Hsien-Kun
Church, Robert Fitz (Randolph)
Corbett, John Frank
Covey, Irene Mabel
Covey, Rupert Alden
Crabtree, Robert H
Crast, Leonard Bruce, Jr
Crawford, Thomas Charles
Cronin, Timothy H
Cue, Berkeley Wendell, Jr
Czuba, Leonard J
Dailey, Joseph Patrick
Desio, Peter John
Dirlam, John Philip
Dominy, Beryl W
Dow, Robert Lee
Doyle, Terrence William
Drucker, Arnold
Durante, Anthony Joseph
Elder, John William
Epling, Gary Arnold
Essenfeld, Amy
Evans, Charles P

Faller, John William
Fenton, Wayne Alexander
Figdor, Sanford Kermit
Fine, Leonard W
Fiore, Joseph Vincent
Firestone, Raymond A
Franck, Richard W
Frauenglass, Elliott
Friedlander, Henry Z
Fry, Albert Joseph
Gavin, David Francis
Gewanter, Herman Louis
Giuffrida, Robert Eugene
Gorham, William Franklin
Grayson, Martin
Harbert, Charles A
Hautala, Richard Roy
Heeren, James Kenneth
Henderson, William Arthur, Jr
Hendrick, Michael Ezell
Henkel, James Gregory
Herz, Jack L
Hess, Hans-Jurgen Ernst
Hines, Paul Steward
Hinman, Richard Leslie
Holland, Gerald Fagan
Hopper, Paul Frederick
Horrocks, Robert H
Huang, Samuel J
Huffman, K(enneth) Robert
Humphrey, Bingham Johnson
Humphreys, Robert William Riley
Husband, Robert Murray
Ives, Jeffrey Lee
Jacobi, Peter Alan
Jorgensen, William L
Kaplow, Leonard Samuel
Katz, Leon
Kavarnos, George James
Kerridge, Kenneth A
Kluender, Harold Clinton
Knollmueller, Karl Otto
Koe, B Kenneth
Konigsbacher, Kurt S
Korst, James Joseph
Kuck, Julius Anson
Kuryla, William C
LaMattina, John Lawrence
Lande, Saul
Larson, Eric George
Leavitt, Julian Jacob
Leis, Donald George
Leonard, Robert F
Lin, Tai-Shun
Lines, Ellwood LeRoy
Lipinski, Christopher Andrew
Lipka, Benjamin
Loew, Leslie Max
Lombardino, Joseph George
Luh, Yuhshi
McArthur, Richard Edward
McBride, James Michael
McClenachan, Ellsworth C
McDonough, Everett Goodrich
McFarland, James William
McGregor, Donald Neil
McGuinness, James Anthony
McKeon, James Edward
McManus, James Michael
MacMullen, Clinton William
Malofsky, Bernard Miles
Marfat, Anthony
Matsuda, Ken
Meierhoefer, Alan W
Melvin, Lawrence Sherman, Jr
Meriwether, Lewis Smith
Meschino, Joseph Albert
Miller, Audrey
Miller, Wilbur Hobart
Moppett, Charles Edward
Mosby, William Lindsay
Murai, Kotaro
Nielsen, John Merle
Noland, James Sterling
O'Connell, Edmond J, Jr
O'Shea, Francis Xavier
Padbury, John James
Panicci, Ronald J
Panzer, Hans Peter
Patterson, Elizabeth Chambers
Pegolotti, James Alfred
Perry, Clark William
Philips, Judson Christopher
Phillips, Benjamin
Pierce, James Benjamin
Pike, Roscoe Adams
Pinson, Ellis Rex, Jr
Poindexter, Graham Stuart
Prokai, Bela
Proverb, Robert Joseph
Raghu, Sivaraman
Rauhut, Michael McKay
Rennhard, Hans Heinrich
Ressler, Charlotte
Rich, Richard Douglas
Rim, Yong Sung
Rosati, Robert Louis
Rose, James Stephenson
Rosenberg, Ira Edward
Rosenman, Irwin David
Rothenberg, Alan S
Ryan, Richard Patrick
Salce, Ludwig
Sarges, Reinhard

Saunders, Martin
Scardera, Michael
Schmir, Gaston L
Schnur, Rodney Caughren
Sciarini, Louis J(ohn)
Sedlak, John Andrew
Shine, Timothy D
Singh, Ajaib
Smith, Curtis Page
Smith, Douglas Stewart
Smith, Frederick Albert
Stevens, Malcolm Peter
Stott, Paul Edwin
Strunk, Richard John
Stuber, Fred A
Suh, John Taiyoung
Tate, Bryce Eugene
Thomas, Paul David
Tishler, Max
Turk, Amos
Ulrich, Henri
Valentine, Donald H, Jr
Vinick, Fredric James
Volkmann, Robert Alfred
Wasserman, Harry H
Wassmundt, Frederick William
Wehman, Anthony Theodore
Welch, Willard McKowan, Jr
Wharton, Peter Stanley
Whelan, William Paul, Jr
Wiberg, Kenneth Berle
Willner, David
Wright, Alan Carl
Wuskell, Joseph P
Wynn, Charles Martin, Sr
Wystrach, Vernon Paul
Yevich, Joseph Paul
Ziegler, Frederick Edward

DELAWARE

Ainbinder, Zarah
Alderson, Thomas
Aldrich, Paul E
Ali-Khan, Ausat
Anderson, Albert Gordon
Anderson, Burton Carl
Angerer, J(ohn) David
Ansul, Gerald R
Armitage, John Brian
Athey, Robert Jackson
Austin, Paul Rolland
Autenrieth, John Stork
Bankert, Ralph Allen
Bare, Thomas M
Barker, Harold Clinton
Barron, Eugene Roy
Bartels, George William, Jr
Bauchwitz, Peter S
Baum, Martin David
Baylor, Charles, Jr
Beare, Steven Douglas
Beer, John Joseph
Bellina, Russell Frank
Benson, Frederic Rupert
Bingham, Richard Charles
Bjornson, August Sven
Black, Donald K
Blomstrom, Dale Clifton
Boden, Herbert
Bowers, George Henry, III
Breazeale, Almut Frerichs
Brehm, Warren John
Bremer, Keith George
Breslow, David Samuel
Brodoway, Nicolas
Brown, Barry Stephen
Brown, Robert Raymond
Burg, Marion
Burton, Lester Percy Joseph
Buser, Kenneth Rene
Cairncross, Allan
Carboni, Rudolph A
Carlson, Bruce Arne
Carlson, Norman Arthur
Carnahan, James Elliot
Chambers, Vaughan Crandall, (Jr)
Cherkofsky, Saul Carl
Chorvat, Robert John
Citron, Joel David
Class, Jay Bernard
Clayton, Anthony Broxholme
Clement, Robert Alton
Cone, Michael McKay
Confalone, Pat N
Cook, Gordon Smith
Cords, Donald Philip
Corner, James Oliver
Coulson, Dale Robert
Cripps, Harry Norman
Crouse, Dale McClish
Crow, Edwin Lee
Daly, John Joseph, Jr
D'Amore, Michael Brian
Darby, Robert Albert
David, Israel A
DeDominicis, Alex John
DeGrado, William F
Delmonte, David William
Dishart, Kenneth Thomas
Diveley, William Russell
Dyer, Elizabeth
Eaton, David Fielder
Ehrich, F(elix) Frederick
Eleuterio, Herbert Sousa

Engelhardt, Vaughn Arthur
Espy, Herbert Hastings
Evans, Franklin James, Jr
Fawcett, Mark Stanley
Feinberg, Stewart Carl
Felten, John James
Feltzin, Joseph
Finizio, Michael
Finlay, Joseph Burton
Foldi, Andrew Peter
Ford, Thomas Aven
Foss, Robert Paul
Frankenburg, Peter Edgar
Franklin, Richard Crawford
Frey, William Adrian
Fuchs, Julius Jakob
Fukunaga, Tadamichi
Fullhart, Lawrence, Jr
Gano, Robert Daniel
Ganti, Venkat Rao
Garland, Charles E
Garrison, William Emmett, Jr
Gervay, Joseph Edmund
Gibbs, Hugh Harper
Gilbert, Walter Wilson
Gorski, Robert Alexander
Gosser, Lawrence Wayne
Grot, Walther Gustav Fredrich
Gruber, Wilhelm F
Guggenberger, Lloyd Joseph
Gumprecht, William Henry
Hanzel, Robert Stephen
Haq, Mohammad Zamir-ul
Harmuth, Charles Moore
Harris, George Christe
Harris, John Ferguson, Jr
Hartzler, Harris Dale
Hasty, Noel Marion, Jr
Haubein, Albert Howard
Hayek, Mason
Hayman, Alan Conrad
Hays, John Thomas
Heck, Richard Fred
Heldt, Walter Z
Herkes, Frank Edward
Hilfiker, Franklin Roberts
Hill, Frederick Burns, Jr
Hirsty, Sylvain Max
Hirwe, Ashalata Shyamsunder
Hoegger, Erhard Fritz
Hoehn, Harvey Herbert
Hoeschele, Guenther Kurt
Hoh, George Lok Kwong
Hoiness, David Eldon
Holfeld, Winfried Thomas
Holmes, David Willis
Holmquist, Howard Emil
Holyoke, Caleb William, Jr
Honsberg, Wolfgang
Hood, Horace Edward
Hoover, Fred Wayne
Hornberger, Carl Stanley, Jr
Howe, King Lau
Hummel, Donald George
Immediata, Tony Michael
Inskip, Harold Kirkwood
Jabloner, Harold
Jackson, Harold Leonard
Jaffe, Edward E
Jamison, Joel Dexter
Jelinek, Arthur Gilbert
Jenner, Edward L
Johnson, Alexander Lawrence
Kaplan, Ralph Benjamin
Kassal, Robert James
Keating, James T
Kegelman, Matthew Roland
Kettner, Charles Adrian
Kittila, Richard Sulo
Knipmeyer, Hubert Elmer
Kosak, John R
Krackov, Mark Harry
Kramer, Brian Dale
Krespan, Carl George
Kwart, Harold
Kwok, Wo Kong
Laganis, Evan Dean
Landerl, Harold Paul
Langsdorf, William Philip
Lann, Joseph Sidney
Lazaridis, Christina Nicholson
LeClaire, Claire Dean
Lee, Shung-Yan Luke
Leser, Ernst George
Levitt, George
Libbey, William Jerry
Linn, William Joseph
Linsay, Ernest Charles
Lipscomb, Robert DeWald
Logullo, Francis Mark
Loken, Halvar Young
Longhi, Raymond
Loomis, Gary Lee
Lorenz, Carl Edward
Lukach, Carl Andrew
McCormack, William Brewster
McCoy, V Eugene, Jr
McGonigal, William E
McInerney, Eugene F
McKay, Sandra J
McLain, Stephan James
Mallonee, James Edgar
Manos, Philip
Matlack, Albert Shelton

Maury, Lucien Garnett
Maynes, Gordon George
Mazur, Stephen
Medeiros, Robert Whippen
Mekler, Arlen B
Memeger, Wesley, Jr
Metzger, James Douglas
Mighton, Charles Joseph
Monagle, Daniel J
Moncure, Henry, Jr
Monroe, Bruce Malcolm
Monroe, Elizabeth McLeister
Moore, James Alexander
Moore, Ralph Bishop
Morgan, Robert Lee
Mori, Peter Taketoshi
Moses, Francis Guy
Moyer, Calvin Lyle
Mrowca, Joseph J
Munn, George Edward
Murray, Roger Kenneth, Jr
Nader, Allan E
Naylor, Marcus A, Jr
Nelson, Jerry Allen
Newby, William Edward
Niedzielski, Edmund Luke
Norman, Oscar Loris
Norton, Lilburn Lafayette
Norton, Richard Vail
Ojakaar, Leo
Pagano, Alfred Horton
Panar, Manuel
Pappas, Nicholas
Park, Chung Ho
Patterson, George Harold
Paulshock, Marvin
Pelosi, Lorenzo Fred
Pensak, David Alan
Percival, William Colony
Petersen, Wallace Christian
Plambeck, Louis, Jr
Porter, Hardin Kibbe
Potrafke, Earl Mark
Pretka, John E
Prosser, Robert M
Prosser, Thomas John
Pruckmayr, Gerfried
Putzig, Donald Edward
Quisenberry, Richard Keith
Raasch, Maynard Stanley
Rajagopalan, Parthasarathi
Ramler, Edward Otto
Rave, Terence William
Raynolds, Stuart
Read, Robert E
Reardon, Joseph Edward
Reid, Donald Eugene
Reilly, George Joseph
Remington, William Roscoe
Richardson, Graham McGavock
Richardson, Paul Noel
Rondestvedt, Christian Scriver, Jr
Saegebarth, Klaus Arthur
Sauers, Richard Frank
Sausen, George Neil
Schadt, Frank Leonard, III
Schappell, Frederick George
Scherer, Kirby Vaughn, Jr
Schmiegel, Walter Werner
Schreyer, Ralph Courtenay
Schroeder, Herman Elbert
Schrof, William Ernst John
Schwartz, Jerome Lawrence
Schweitzer, Carl Earle
Schweizer, Edward Ernest
Senkler, George Henry, Jr
Shealy, Otis Lester
Shozda, Raymond John
Simmons, Howard Ensign, Jr
Simms, John Alvin
Simpson, David Alexander
Singh, Gurdial
Skolnik, Herman
Sloan, Gilbert Jacob
Sloan, Martin Frank
Smart, Bruce Edmund
Smat, Robert Joseph
Smiley, Robert Arthur
Smith, Claiborne Davis
Smith, Jerry Howard
Sroog, Cyrus Efrem
Stahl, Roland Edgar
Stanton, William Alexander
Steller, Kenneth Eugene
Stockburger, George Joseph
Strobach, Donald Roy
Subramanian, Pallatheri Manackal
Sundelin, Kurt Gustav Ragnar
Swerlick, Isadore
Taber, Douglass Fleming
Tan, Henry Harry
Targett, Nancy McKeever
Tatum, William Earl
Taves, Milton Arthur
Taylor, Robert Burns, Jr
Temple, Stanley
Terss, Robert H
Thamm, Richard C, Jr
Thompson, Robert Gene
Thornton, Roger Lea
Tise, Frank P
Trainor, George L
Traumann, Klaus Friedrich
Trost, Henry Biggs

Organic Chemistry (cont)

Twelves, Robert Ralph
Un, Howard Ho-Wei
Van Gulick, Norman Martin
Venkatachalam, Taracad Krishnan
Wagner, Hans
Wallenberger, Frederick Theodore
Walter, Henry Clement
Waring, Derek Morris Holt
Wasserman, Edel
Wat, Edward Koon Wah
Weaver, Jeremiah William
Webers, Vincent Joseph
Webster, Owen Wright
Welch, Raymond Lee
Wiley, Douglas Walker
Willer, Rodney Lee
Williams, Ebenezer David, Jr
Witterholt, Vincent Gerard
Wojtkowski, Paul Walter
Wolfe, James Richard, Jr
Woods, Thomas Stephen
Wright, Everett James
Wright, Leon Wendell
Wuonola, Mark Arvid
Yamamoto, Yasushi Stephen
Young, Charles Albert

DISTRICT OF COLUMBIA

Adomaitis, Vytautas Albin
Alexander, Benjamin H
Barry, Guy Thomas
Bednarek, Jana Marie
Bikales, Norbert M
Brown, Alfred Edward
Brownlee, Paula Pimlott
Cantrell, Thomas Samuel
Caress, Edward Alan
Carson, Frederick Wallace
Carter, Mary Eddie
Chamot, Dennis
Cotruvo, Joseph Alfred
Crist, DeLanson Ross
El Khadem, Hassan S
Engler, Reto Arnold
Fearn, James Ernest
Feldman, Martin Robert
Filipescu, Nicolae
Fleming, Patrick John
Girard, James Emery
Gist, Lewis Alexander, Jr
Gould, Jack Richard
Graminski, Edmond Leonard
Hammer, Charles F
Hancock, Kenneth George
Hare, Peter Edgar
Hirzy, John William
Hoffman, Michael K
Hudrlik, Anne Marie
Hudrlik, Paul Frederick
Jenkins, Mamie Leah Young
Keller, Teddy Monroe
King, Michael M
Klayman, Daniel Leslie
Kovach, Eugene George
Kruegel, Alice Virginia
Lapporte, Seymour Jerome
Lourie, Alan David
Merrifield, D Bruce
Metz, Fred L
Mog, David Michael
Perfetti, Randolph B
Phillips, Don Irwin
Pierce, Dorothy Helen
Ponaras, Anthony A
Poon, Bing Toy
Reed, Joseph
Rhoads, Allen R
Rogers, Senta S(tephanie)
Roscher, Nina Matheny
Rubottom, George M
Rzeszotarski, Waclaw Janusz
Showell, John Sheldon
Spurlock, Langley Augustine
Sunderlin, Charles Eugene
Talbert, Preston Tidball
Torrence, Paul Frederick
Vanderhoek, Jack Yehudi
Wade, Clarence W R
Wainer, Irving William
Walton, Theodore Ross
Wheeler, James William, Jr
White, Philip Cleaver
Wise, Hugh Edward, Jr
Zaborsky, Oskar Rudolf

FLORIDA

Arntzen, Clyde Edward
Astrologes, Gary William
Attaway, John Allen
Baker, Earl Wayne
Barnes, Garrett Henry, Jr
Batich, Christopher David
Battiste, Merle Andrew
Beiler, Theodore Wiseman
Bieber, Theodore Immanuel
Bishop, Charles Anthony
Black, William Bruce
Bledsoe, James O, Jr
Boehme, Werner Richard
Bonnell, James Monroe
Bordenca, Carl
Brown, Henry Clay, III

Burney, Curtis Michael
Busby, Hubbard Taylor, Jr
Busse, Robert Franklyn
Butler, George Bergen
Callen, Joseph Edward
Cardenas, Carlos Guillermo
Carlson, David Arthur
Carnes, Joseph John
Chang, Clifford Wah Jun
Chapman, Richard David
Cioslowski, Jerzy
Clark, Samuel Friend
Cline, Edward Terry
Coke, C(hauncey) Eugene
Conrad, Walter Edmund
Coolidge, Edwin Channing
Corbett, Michael Dennis
Couch, Margaret Wheland
Curry, Thomas Harvey
Dalton, Philip Benjamin
Davis, Curry Beach
Denny, George Hutcheson
Derfer, John Mentzer
DeTar, DeLos Fletcher
Deyrup, James Alden
Dobinson, Frank
Dolbier, William Read, Jr
Dougherty, Ralph C
Dunn, Ben Monroe
Ebetino, Frank Frederick
Eldred, Nelson Richards
Elliott, John Raymond
Fernandez, Jack Eugene
Fisher, George Harold, Jr
Fishman, Jack
Fozzard, George Broward
Freeman, Kenneth Alfrey
Friedlander, Herbert Norman
Gawley, Robert Edgar
Giner-Sorolla, Alfredo
Gokel, George William
Gordon, Joseph R
Gough, Robert George
Gourse, Jerome Allen
Green, Floyd J
Guida, Wayne Charles
Gupton, John
Gurst, Jerome E
Hahn, Elliot F
Hammer, Richard Hartman
Hanley, James Richard, Jr
Hardman, Bruce Bertolette
Harvey, George Ranson
Heberling, Jack Waugh, Jr
Hechenbleikner, Ingenuin Albin
Helling, John Frederic
Herriott, Arthur W
Herz, Werner
Heying, Theodore Louis
Hicks, Elija Maxie, Jr
Hisey, Robert Warren
Hoeppner, Conrad Henry
Hoffman, Warren E
Holmer, Donald A
Houtman, Thomas, Jr
Huba, Francis
Huber, Joseph William, III
Hunt, William Cecil
Hunter, George L K
Ihndris, Raymond Will
Iloff, Phillip Murray, Jr
Ingwalson, Raymond Wesley
Jackson, George Richard
Jaeger, Herbert Karl
Jones, William Maurice
Joyce, Robert Michael
Jurch, George Richard, Jr
Kagan, Benjamin
Kane, Bernard James
Kane, Howard L
Katchman, Arthur
King, Roy Warbrick
Klacsmann, John Anthony
Kleinschmidt, Albert Willoughby
Kowkabany, George Norman
Krafft, Marie Elizabeth
Kreider, Henry Royer
Kunisi, Venkatasubban S
Kurchacova, Elva S
Kutner, Abraham
Lawrence, Franklin Isaac Latimer
Leffler, John Edward
Leffler, Marlin Templeton
Levis, William Walter, Jr
Levy, Joseph
Lewis, George Edwin
Light, Robley Jasper
Lillien, Irving
Lombardo, Anthony
McBride, Joseph James, Jr
McCall, Marvin Anthony
McMullen, Warren Anthony
Mandell, Leon
Mattson, Guy C
Miles, David H
Milligan, Barton
Mitch, Frank Allan
Moser, Robert E
Moskowitz, Mark Lewis
Murphy, James Gilbert
Nersasian, Arthur
Newkome, George Richard
Noller, David Conrad
Nordby, Harold Edwin

Nystrom, Robert Forrest
Oberst, Fred William
Offenhauer, Robert Dwight
Orloff, Harold David
Owen, Gwilym Emyr, Jr
Owen, Terence Cunliffe
Parcell, Robert Ford
Parkanyi, Cyril
Parker, Earl Elmer
Payton, Albert Levern
Penner, Siegfried Edmund
Peterson, Robert Hampton
Pinder, Albert Reginald
Plapinger, Robert Edwin
Pop, Emil
Porter, Lee Albert
Postelnek, William
Prahl, Helmut Ferdinand
Ramsey, Brian Gaines
Ranck, Ralph Oliver
Rees, William Smith, Jr
Rice, Frederick Anders Hudson
Richards, Marvin Sherrill
Rider, Don Keith
Rogan, John B
Rokach, Joshua
Ross, Fred Michael
Royals, Edwin Earl
Sager, William Frederick
Salsbury, Jason Melvin
Saltiel, Jack
Saunders, James Henry
Schneller, Stewart Wright
Schultz, Harry Pershing
Schwartz, Martin Alan
Sellers, John William
Shabica, Anthony Charles, Jr
Sharkey, William Henry
Shaw, Philip Eugene
Sheffer, Howard Eugene
Shepard, Edwin Reed
Sherman, Albert Herman
Sherman, Edward
Smart, William Donald
Smith, William Mayo
Snyder, Carl Henry
Sollman, Paul Benjamin
Soto, Aida R
Spatz, Sydney Martin
Spayd, Richard W
Spencer, John Lawrence
Spilker, Clarence William
Stahl, Joel S
Starke, Albert Carl, Jr
Steadman, Thomas Ree
Stolberg, Marvin Arnold
Stump, Eugene Curtis, Jr
Sullivan, Lloyd John
Sweet, Ronald Lancelot
Szmant, Herman Harry
Templer, David Allen
Travnicek, Edward Adolph
Van Handel, Emile
Walborsky, Harry M
Walter, Charles Robert, Jr
Wander, Joseph Day
Warren, Harold Hubbard
Watkins, Spencer Hunt
Webb, Robert Lee
Wiles, Robert Allan
Williams, Robert Hackney
Wise, Raleigh Warren
Witte, Michael
Wood, James Brent, III
Woodward, Fred Erskine
Zimmer, William Frederick, Jr
Zoltewicz, John A

GEORGIA

Allinger, Norman Louis
Allison, John P
Anderson, Gloria Long
Ashby, Eugene Christopher
Baarda, David Gene
Barbas, John Theophani
Barefield, Edward Kent
Barreras, Raymond Joseph
Barton, Franklin Ellwood, II
Bayer, Charlene Warres
Bergman, Elliot
Blanton, Charles DeWitt, Jr
Blomquist, Richard Frederick
Boxer, Robert Jacob
Boykin, David Withers, Jr
Bralley, James Alexander
Burgess, Edward Meredith
Chandler, James Harry, III
Chortyk, Orestes Timothy
Chu, Chung K
Churchill, Frederick Charles
Clark, Benjamin Cates, Jr
Clendinning, Robert Andrew
Cole, Thomas Winston, Jr
Counts, Wayne Boyd
Derrick, Mildred Elizabeth
Dimmel, Donald R
Dinwiddie, Joseph Gray, Jr
Dixon, Dabney White
Duvall, Harry Marean
Ellington, James Jackson
Ennor, Kenneth Stafford
Esslinger, William Glenn
Fay, Alice D Awtrey
Freese, William P, II

Gianturco, Maurizio
Girardot, Jean Marie Denis
Goldsmith, David Jonathan
Gollob, Lawrence
Grovenstein, Erling, Jr
Hagan, Charles Patrick
Hargrove, Robert John
Harris, Henry Earl
Heyd, Charles E
Hicks, Arthur M
Hill, Richard Keith
Hodges, Linda Carol
House, Herbert Otis
Howard, John Charles
Hung, William Mo-Wei
Iacobucci, Guillermo Arturo
Iannicelli, Joseph
Jen, Yun
Johnson, Robert William, Jr
Jones, Ronald Goldin
Jones, William Howry
Kellogg, Craig Kent
Knight, James Albert, Jr
Knopka, W N
Krajca, Kenneth Edward
Langston, James Horace
Lester, Charles Turner
Lewis, Silas Davis
Liebeskind, Lanny Steven
Lomax, Eddie
Love, Jimmy Dwane
McCarthy, Neil Justin, Jr
McKinney, Roger Minor
Mahesh, Virendra B
Manger, Martin C
Marascia, Frank Joseph
Mathews, Walter Kelly
Matthews, Edward Whitehouse
Menger, Fred M
Mickelsen, Olaf
Netherton, Lowell Edwin
Newton, Melvin Gary
Pacifici, James Grady
Padwa, Albert
Pelletier, S William
Perdue, Edward Michael
Pohl, Douglas George
Polk, Malcolm Benny
Powers, James Cecil
Proctor, Charles Darnell
Radford, Terence
Raut, Kamalakar Balkrishna
Robbins, Paul Edward
Rodriguez, Augusto
Roobol, Norman R
Ruenitz, Peter Carmichael
Sandri, Joseph Mario
Schmalz, Alfred Chandler
Sears, Curtis Thornton, Jr
Shapira, Raymond
Sheehan, Desmond
Sommers, Jay Richard
Spence, Gavin Gary
Spriggs, Alfred Samuel
Stammer, Charles Hugh
Stanfield, James Armond
Su, Helen Chien-Fan
Thompson, Bobby Blackburn
Tolbert, Laren Malcolm
Trumbull, Elmer Roy, Jr
Tucker, Willie George
Turbak, Albin Frank
Tyler, Jean Mary
Walker, John J
Walsh, David Allan
Wiesboeck, Robert A
Williams, George Nathaniel
Wingard, Robert Eugene, Jr
Yost, Robert Stanley
Young, Raymond Hinchcliffe, Jr
Zalkow, Leon Harry
Zepp, Richard Gardner
Zvejnieks, Andrejs

HAWAII

Antal, Michael Jerry, Jr
Doerge, Daniel Robert
Dority, Guy Hiram
Hiu, Dawes Nyukleu
Howton, David Ronald
Kiefer, Edgar Francis
Kloetzel, Milton Carl
Larson, Harold Olaf
Liu, Robert Shing-Hei
Moore, Richard E
Mower, Howard Frederick
Norton, Ted Raymond
Scheuer, Paul Josef

IDAHO

Banks, Richard C
Braun, Loren L
Bush, David Clair
Cooley, James Hollis
Dalton, Jack L
Imel, Arthur Madison
Isaacson, Eugene I
Matjeka, Edward Ray
Natale, Nicholas Robert
Raunio, Elmer Kauno
Ronald, Bruce Pender
Sarett, Lewis Hastings
Stevenson, Charles Edward
Sutton, John Curtis

Wiegand, Gayl
Winkel, Cleve R

ILLINOIS
Adlof, Richard Otto
Allen, John Kay
Alm, Robert M
Anderson, Arnold Lynn
Andrews, Eugene Raymond
Applequist, Douglas Einar
Arendsen, David Lloyd
Argoudelis, Chris J
Armstrong, P(aul) Douglas
Arnold, Richard Thomas
Ashley, Warren Cotton
Babler, James Harold
Bagby, Marvin Orville
Baldoni, Andrew Ateleo
Baran, John Stanislaus
Barker, George Ernest
Bauer, Ludwig
Baumgarten, Ronald J
Beak, Peter
Bechara, Ibrahim
Beck, Karl Maurice
Berger, Daniel Richard
Bergstrom, Clarence George
Berman, Lawrence Uretz
Bernetti, Raffaele
Bernhardt, Randal Jay
Beyler, Roger Eldon
Blaha, Eli William
Bordwell, Frederick George
Boshart, Gregory Lew
Bowles, William Allen
Boyd, Mary K
Brace, Neal Orin
Breitbeil, Fred W, III
Brooks, Dee W
Brown, George Earl
Bryant, Rhys
Buchanan, David Hamilton
Buntrock, Robert Edward
Burgauer, Paul David
Burney, Donald Eugene
Bussert, Jack Francis
Cammarata, Peter S
Carr, Lawrence John
Carrick, Wayne Lee
Cengel, John Anthony
Cerefice, Steven A
Chang, Hsien-Hsin
Chao, Tai Siang
Cheng, Shu-Sing
Chou, Mei-In Melissa Liu
Churchill, Constance Louise
Clark, Frank S
Clark, Stephen Darrough
Closs, Gerhard Ludwig
Coates, Robert Mercer
Cochrane, Chappelle Cecil
Cordell, Geoffrey Alan
Cosper, David Russell
Crovetti, Aldo Joseph
Crumrine, David Shafer
Cunico, Robert Frederick
Currie, Bruce LaMonte
Curtin, David Yarrow
Curtis, Veronica Anne
Danzig, Morris Juda
DeBoer, Edward Dale
Debus, Allen George
Dellaria, Joseph Fred, Jr
DeYoung, Edwin Lawson
Dixit, Saryu N
Djuric, Stevan Wakefield
Doane, William M
Drengler, Keith Allan
DuBois, Grant Edwin
Dybalski, Jack Norbert
Eachus, Alan Campbell
Eaton, Philip Eugene
Ebner, Herman George
Eby, Lawrence Thornton
Economy, James
Ellis, Jerry William
Empen, Joseph A
Erby, William Arthur
Erickson, Mitchell Drake
Evans, Robert John
Falk, John Carl
Fanta, George Frederick
Fanta, Paul Edward
Fassnacht, John Hartwell
Faubl, Hermann
Feinstein, Allen Irwin
Field, Kurt William
Fields, Ellis Kirby
Filler, Robert
Firkins, John Lionel
Foote, Carlton Dan
Foster, George A, Jr
Frame, Robert Roy
Frank, Forrest Jay
Frey, David Allen
Fried, Josef
Gaffney, Jeffrey Steven
Garland, Robert Bruce
Garmaise, David Lyon
Garven, Floyd Charles
Gasser, William
Gast, Lyle Everett
Gavlin, Gilbert
Gearien, James Edward

Gebauer, Peter Anthony
Gehring, Harvey Thomas
Gemmer, Robert Valentine
Gershbein, Leon Lee
Gibbs, James Albert
Ginger, Leonard George
Giori, Claudio
Glaser, Milton Arthur
Gleeson, James Newman
Glover, Allen Donald
Goeckner, Norbert Anthony
Golinkin, Herbert Sheldon
Goretta, Louis Alexander
Gorman, Susan B
Gray, Nancy M
Grotefend, Alan Charles
Grothaus, Clarence (Edward)
Gupta, Goutam
Gutberlet, Louis Charles
Hac, Lucile R
Hakewill, Henry, Jr
Hall, J Herbert
Hamer, Martin
Hansen, Donald Willis, Jr
Hansen, John Frederick
Harper, Jon Jay
Hartlage, James Albert
Harvey, Ronald Gilbert
Haugen, David Allen
Henkin, Jack
Henning, Lester Allan
Henry, Patrick M
Hill, Kenneth Richard
Hoeg, Donald Francis
Hofreiter, Bernard T
Hogan, Philip
Hopkins, Paul Donald
Huang, Leo W
Huang, Shu-Jen Wu
Huffman, George Wallen
Hughes, Robert David
Hutchcroft, Alan Charles
Hutchison, David Allan
Illingworth, George Ernest
James, David Eugene
Janssen, Jerry Frederick
Jezl, James Louis
Johnson, Calvin Keith
Johnson, Carl Edwin
Johnson, Dale Howard
Johnson, John Harold
Jones, David A, Jr
Jones, Marjorie Ann
Jones, Peter Hadley
Jungmann, Richard A
Kagan, Jacques
Karll, Robert E
Katzenellenbogen, John Albert
Kaufman, Priscilla C
Kennelly, Mary Marina
Kerdesky, Francis A J
Kerkman, Daniel Joseph
Kevill, Dennis Neil
Khattab, Ghazi M A
Kim, Ki Hwan
King, Lafayette Carroll
Kipp, James Edwin
Kirkpatrick, Joel Lee
Kissel, William John
Kleiman, Morton
Klein, Larry L
Klimstra, Paul D
Knaggs, Edward Andrew
Knobloch, James Otis
Knutson, Clarence Arthur, Jr
Kolb, Doris Kasey
Kolb, Kenneth Emil
Krajewski, John J
Kretchmer, Richard Allan
Kruse, Carl William
Kucera, Thomas J
Kuceski, Vincent Paul
Kuhajek, Eugene James
Kuhlmann, George Edward
Kurath, Paul
Kurz, Michael E
Lambert, Joseph B
Lan, Ming-Jye
Larson, Richard Allen
Lauterbach, Richard Thomas
Lee, Cheuk Man
Leiby, Robert William
Lemper, Anthony Louis
Leonard, Nelson Jordan
Letsinger, Robert Lewis
Levin, Alfred A
Lewis, Frederick D
Lewis, Morton
Lillwitz, Lawrence Dale
Lindbeck, Wendell Arthur
Lindberg, Steven Edward
Lira, Emil Patrick
Little, Randel Quincy, Jr
Loire, Norman Paul
Lovald, Roger Allen
Lowrie, Harman Smith
Lucas, Glennard Ralph
Lucci, Robert Dominick
Lynch, Darrel Luvene
Lynch, Don Murl
McAlpine, James Bruce
McLaughlin, Robert Lawrence
Magnus, George
Maher, George Garrison

Mallia, Anantha Krishna
Malloy, Thomas Patrick
Matson, Howard John
Matt, Joseph
Matta, Michael Stanley
Matthews, Clifford Norman
Mazur, Robert Henry
Meguerian, Garbis H
Melhado, L(ouisa) Lee
Meyer, Delbert Henry
Meyers, Cal Yale
Mihina, Joseph Stephen
Mikulec, Richard Andrew
Miller, Arnold Reed
Miller, Francis Marion
Miller, Sidney Israel
Mimnaugh, Michael Neil
Mock, William L
Mohan, Prem
Morello, Edwin Francis
Moriarty, Robert M
Moser, Kenneth Bruce
Mounts, Timothy Lee
Muffley, Harry Chilton
Murphy, Thomas Joseph
Narske, Richard Martin
Neumiller, Harry Jacob, Jr
Nichols, George Morrill
Noren, Gerry Karl
Novak, Robert William
Oono, Yoshitsugu
Osman, Elizabeth Mary
Otey, Felix Harold
Padrta, Frank George
Pai, Sadanand V
Paisley, David M
Parmerter, Stanley Marshall
Patinkin, Seymour Harold
Patrick, Timothy Benton
Paul, Iain C
Perkins, Edward George
Perun, Thomas John
Peterson, Melbert Eugene
Piatak, David Michael
Piehl, Frank John
Pines, Herman
Pirkle, William H
Poel, Russell J
Portis, Archie Ray, Jr
Portnoy, Norman Abbye
Potempa, Lawrence Albert
Poulos, Nicholas A
Prout, Franklin Sinclair
Pryde, Everett Hilton
Radford, Herschel Donald
Raheel, Mastura
Rakoff, Henry
Rao, Yedavalli Shyamsunder
Rathke, Jerome William
Rausch, David John
Rawlinson, David John
Rees, Thomas Charles
Reidies, Arno H
Reily, William Singer
Rentmeester, Kenneth R
Reynolds, Rosalie Dean (Sibert)
Richardson, Arlan Gilbert
Richmond, James M
Rinehart, Kenneth Lloyd
Robin, Burton Howard
Rocek, Jan
Roderick, William Rodney
Rorig, Kurt Joachim
Rosen, Bruce Irwin
Rosenberg, Saul Howard
Rosenbrook, William, Jr
Sacco, Louis Joseph, Jr
Sample, James Halverson
Sause, H William
Schaap, Luke Anthony
Schaeffer, David Joseph
Schmukler, Seymour
Scholfield, Charles Rexel
Schriesheim, Alan
Schuster, Gary Benjamin
Schwab, Arthur William
Scouten, Charles George
Seigler, David Stanley
Shiner, Edward Arnold
Shipchandler, Mohammed Tyebji
Shoffner, James Priest
Shone, Robert L
Short, Rolland William Phillip
Shulman, Sol
Silverman, Richard Bruce
Sinclair, Henry Beall
Smith, Gerard Vinton
Smith, Homer Alvin, Jr
Smith, Robert Johnson
Smith, Stanley Glen
Snyder, Gary James
Snyder, William Robert
Spangler, Charles William
Spitzer, William Carl
Stephens, James Regis
Stock, Leon M
Stowell, James Kent
Strand, Robert Charles
Sytsma, Louis Frederick
Szpunar, Carole Bryda
Thomas, Alford Mitchell
Tiefenthal, Harlan E
Tomomatsu, Hideo
Tortorello, Anthony Joseph

Traxler, James Theodore
Trevillyan, Alvin Earl
Turner, Fred Allen
Tyner, David Anson
Ulrey, Stephen Scott
Unglaube, James M
Urbas, Branko
Vander Burgh, Leonard F
Van Lanen, Robert Jerome
Van Strien, Richard Edward
Vasiliauskas, Edmund
Venerable, James Thomas
Verbanac, Frank
Voedisch, Robert W
Vosti, Donald Curtis
Walhout, Justine Isabel Simon
Walter, Robert Irving
Walton, Henry Miller
Wang, Gary T
Warne, Thomas Martin
Wasielewski, Michael Roman
Weatherbee, Carl
Webber, Gayle Milton
Weier, Richard Mathias
Wentworth, Gary
Weston, Arthur Walter
Weyna, Philip Leo
Wiedenmann, Lynn G
Wilke, Robert Nielsen
Wilt, James William
Wimer, David Carlisle
Winans, Randall Edward
Witczak, Zbigniew J
Witt, John, Jr
Wolf, Leslie Raymond
Wolf, Richard Eugene
Wolff, William Francis
Wotiz, John Henry
Wysocki, Allen John
Yamamoto, Diane M
Yokelson, Howard Bruce
Yonan, Edward E
Young, Austin Harry
Young, David W
Young, Harland Harry
Zaugg, Harold Elmer
Zeffren, Eugene
Zletz, Alex

INDIANA
Archer, Robert Allen
Arndt, Henry Clifford
Bailey, Thomas Daniel
Bakker, Gerald Robert
Barnett, Charles Jackson
Bays, James Philip
Beck, James Richard
Becker, Elizabeth Ann (White)
Beckman, Jean Catherine
Benjaminov, Benjamin S
Benkeser, Robert Anthony
Benton, Francis Lee
Biddlecom, William Gerard
Blickenstaff, Robert Theron
Boguslaski, Robert Charles
Bolton, Benjamin A
Borchert, Peter Jochen
Bosin, Talmage R
Bostick, Edgar E
Bottorff, Edmond Milton
Brandt, Karl Garet
Brannon, Donald Ray
Brewster, James Henry
Brinkmeyer, Raymond Samuel
Brooker, Robert Munro
Brown, Herbert Charles
Burkett, Howard (Benton)
Burow, Kenneth Wayne, Jr
Byrn, Stephen Robert
Campbell, Kenneth Nielsen
Carlson, Merle Winslow
Carmack, Marvin
Carnahan, Robert Edward
Cassady, John Mac
Chateauneuf, John Edward
Clark, Larry P
Cook, Addison Gilbert
Cook, Donald Jack
Cook, Kenneth Emery
Cooper, Robin D G
Corey, Paul Frederick
Crandall, Jack Kenneth
Cushman, Mark
Cutshall, Theodore Wayne
Danehy, James Philip
Das, Paritosh Kumar
Debono, Manuel
DeSantis, John Louis
Dominianni, Samuel James
Dorman, Douglas Earl
Dunn, Howard Eugene
Dykstra, Stanley John
Easton, Nelson Roy
Egly, Richard S(amuel)
Ellingson, Rudolph Conrad
Emmick, Thomas Lynn
Evanega, George R
Farkas, Eugene
Feigl, Dorothy M
Feuer, Henry
Fife, Wilmer Krafft
Flaugh, Michael Edward
Flynn, John Joseph, Jr
Fornefeld, Eugene Joseph

Organic Chemistry (cont)

Freeman, Jeremiah Patrick
Frump, John Adams
Gajewski, Joseph J
Gallucci, Robert Russell
Garbrecht, William Lee
Gerns, Fred Rudolph
Gilham, Peter Thomas
Goe, Gerald Lee
Goossens, John Charles
Grieco, Paul Anthony
Grutzner, John Brandon
Gutowski, Gerald Edward
Han, Choong Yol
Harper, Edwin T
Harper, Richard Waltz
Helquist, Paul M
Hennion, George Felix
Herron, David Kent
Hull, Clarence Joseph
Ingraham, Joseph Sterling
Jacobs, Martin John
Johnston, Katharine Gentry
Kammann, Karl Philip, Jr
Kaslow, Christian Edward
Katner, Allen Samuel
Kelly, Walter James
Kilsheimer, Sidney Arthur
Kirst, Herbert Andrew
Kjonaas, Richard A
Koller, Charles Richard
Koppel, Gary Allen
Kornblum, Nathan
Kraemer, John Francis
Kramer, William J
Kreider, Leonard Cale
Kress, Thomas Joseph
Lantero, Oreste John, Jr
Lavagnino, Edward Ralph
Leonard, Jack E
Lewis, Dennis Allen
Lipkowitz, Kenny Barry
Longroy, Allan Leroy
Lorentzen, Keith Eden
Loudon, Gordon Marcus
Lutz, Wilson Boyd
McFarland, John William
McIntyre, Thomas William
Majewski, Robert Francis
Marshall, Frederick J
Matsumoto, Ken
Miesel, John Louis
Mikolasek, Douglas Gene
Miller, Edward George
Miller, Roy Glenn
Mixan, Craig Edward
Montgomery, Lawrence Kernan
Morris, John F
Morrison, Harry
Myerholtz, Ralph W, Jr
Negishi, Ei-ichi
Nelson, Nils Keith
Nevin, Robert Stephen
O'Doherty, George Oliver-Plunkett
O'Donnell, Martin James
Pasto, Daniel Jerome
Patwardhan, Bhalchandra H
Porter, Herschel Donovan
Pyle, James L
Quinn, Michael H
Rand, Leon
Rivers, Paul Michael
Robertson, David Wayne
Robertson, Donald Edwin
Robey, Roger Lewis
Ross, Joseph Hansbro
Safdy, Max Errol
Salerni, Oreste Leroy
Sallay, Stephen
Schroeder, Juel Pierre
Schwan, Theodore Carl
Serianni, Anthony Stephan
Shankland, Rodney Veeder
Shiner, Vernon Jack, Jr
Siebert, John
Sieloff, Ronald F
Smith, Gerald Floyd
Smith, Lewis Oliver, Jr
Sneen, Richard Allen
Soper, Quentin Francis
Sousa, Lynn Robert
Sowers, Edward Eugene
Stamper, Martha C
Stehouwer, David Mark
Tennyson, Richard Harvey
Thibault, Thomas Delor
Thompson, Gerald Lee
Toomey, Joseph Edward
Trinler, William A
Trischler, Floyd D
Trott, Gene F
Trozzolo, Anthony Marion
Uloth, Robert Henry
Van Dyke, John William, Jr
Van Etten, Robert Lee
Van Heyningen, Earle Marvin
Vaughn, Thomas Hunt
Wang, Jin-Liang
Webber, J Alan
Weiner, Henry
Welch, Zara D
Weller, Lowell Ernest
Wendel, Samuel Reece

White, Harold Keith
Winslow, Alfred Edwards
Wiseman, Park Allen
Wolinsky, Joseph
Wright, Ian Glaisby
Wu, Yao Hua
Yordy, John David

IOWA

Ault, Addison
Barton, Thomas J
Bearce, Winfield Hutchinson
Burton, Donald Joseph
Carlson, Emil Herbert
Czarny, Michael Richard
Docken, Adrian (Merwin)
Doorenbos, Harold E
Downing, Donald Talbot
Franklin, Robert Louis
Geels, Edwin James
Graybill, Bruce Myron
Hampton, David Clark
Kercheval, James William
Klemm, Richard Andrew
Kraus, Kenneth Wayne
Larock, Richard Craig
Lindberg, James George
Linhardt, Robert John
McGrew, Leroy Albert
MacMillan, James G
Nair, Vasu
Neumann, Marguerite
Osuch, Carl
Phillips, Marshall
Rila, Charles Clinton
Russell, Glen Allan
Swartz, James E
Trahanovsky, Walter Samuel
Verkade, John George
Welch, Dean Earl
Wertz, Philip Wesley
Wiemer, David F
Wubbels, Gene Gerald

KANSAS

Baumgartner, George Julius
Beecham, Curtis Michael
Boyle, Kathryn Moyne Ward Dittemore
Breed, Laurence Woods
Briles, George Herbert
Burgstahler, Albert William
Cahoy, Roger Paul
Carlson, Robert Gideon
Chappelow, Cecil Clendis, Jr
Cheng, Chia-Chung
Choguill, Harold Samuel
Clevenger, Richard Lee
Crandall, Elbert Williams
Duncan, William Perry
Englund, Charles R
Fu, Yun-Lung
Fuhlhage, Donald Wayne
Givens, Richard Spencer
Glazier, Robert Henry
Grunewald, Gary Lawrence
Harwell, Kenneth Elzer
Hua, Duy Huu
Huyser, Earl Stanley
Johnson, David Barton
Johnson, John Webster, Jr
Klabunde, Kenneth John
Landgrebe, John A
Lenhert, Anne Gerhardt
Linn, Carl Barnes
McDonald, Richard Norman
Marx, Michael
Moyer, Melvin Isaac
Munson, H Randall, Jr
Paukstelis, Joseph V
Schowen, Richard Lyle
Schroeder, Robert Samuel
Seib, Paul A
Slater, Carl David
Stutz, Robert L
Talaty, Erach R
Wiley, James C, Jr

KENTUCKY

Bendall, Victor Ivor
Brown, Ellis Vincent
Flachskam, Robert Louis, Jr
Gibson, Dorothy Hinds
Glenn, Furman Eugene
Graver, Richard Byrd
Guthrie, Robert D
Hamilton-Kemp, Thomas Rogers
Hendon, Joseph C
Hessley, Rita Kathleen
Hussung, Karl Frederick
Huttenlocher, Dietrich F
Johnson, Robert Reiner
Kadaba, Pankaja Kooveli
Kargl, Thomas E
Kennedy, John Elmo, Jr
Kornet, Milton Joseph
Lloyd, William Gilbert
Lukes, Robert Michael
Masters, John Edward
Meier, Mark Stephan
Meisenheimer, John Long
Owen, David Allan
Patterson, John Miles
Price, Martin Burton
Reasoner, John W

Richard, John P
Richter, Edward Eugene
Sagar, William Clayton
Sanford, Robert Alois
Schulz, William
Scott, George William
Selegue, John Paul
Shoemaker, Gradus Lawrence
Slocum, Donald Warren
Smith, Stanford Lee
Smith, Walter Thomas, Jr
Spatola, Arno F
Taylor, Kenneth Grant
Thio, Alan Poo-An
Tucker, Irwin William
Wilson, Gordon, Jr
Wilson, Joseph William

LOUISIANA

Avonda, Frank Peter
Babb, Robert Massey
Baird, William C, Jr
Bauer, Beverly Ann
Bayer, Arthur Craig
Benitez, Francisco Manuel
Bertoniere, Noelie Rita
Biersmith, Edward L, III
Blouin, Florine Alice
Boozer, Charles (Eugene)
Boyer, Joseph Henry
Brackenridge, David Ross
Bromberg, Milton Jay
Byrd, David Shelton
Carpenter, Paul Gershom
Cartledge, Frank
Chen, Hoffman Hor-Fu
Clarke, Wilbur Bancroft
Colgrove, Steven Gray
Connick, William Joseph, Jr
Conrad, Franklin
Corkern, Walter Harold
Couvillion, John Lee
Daly, William Howard
Daul, George Cecil
Davenport, Tom Forest, Jr
Davis, Bryan Terence
Domelsmith, Linda Nell
Doomes, Earl
Edwards, Joseph D, Jr
Eisenbraun, Allan Alfred
Ellzey, Samuel Edward, Jr
Ensley, Harry Eugene
Everly, Charles Ray
Filbey, Allen Howard
Fischer, Nikolaus Hartmut
Fitzpatrick, Jimmie Doile
Frank, Arlen W(alker)
Franklin, William Elwood
Gallo, August Anthony
Gandour, Richard David
Gibson, David Michael
Goldberg, Stanley Irwin
Goldblatt, Leo Arthur
Griffin, Gary Walter
Guthrie, Donald Arthur
Hanson, Marvin Wayne
Harper, Robert John, Jr
Holt, Robert Louis
Hornbaker, Edwin Dale
Ieyoub, Kalil Phillip
Ihrman, Kryn George
Impastato, Fred John
Jackisch, Philip Frederick
Jacks, Thomas Jerome
Jacobus, Otha John
Kehoe, Lawrence Joseph
Kennedy, Frank Scott
Knapp, Gordon Grayson
Koenig, Paul Edward
Kordoski, Edward William
Ledford, Thomas Howard
Lee, Burnell
Lee, John Yuchu
McGraw, Gerald Wayne
Magin, Ralph Walter
Mangham, Jesse Roger
Mark, Harold Wayne
Marvel, John Thomas
Mazzeno, Laurence William
Mollere, Phillip David
Moore, Richard Newton
Morris, Cletus Eugene
Munchausen, Linda Lou
Mustafa, Shams
Neher, Clarence M
Nevill, William Albert
Parent, Richard Alfred
Pepperman, Armand Bennett, Jr
Petterson, Robert Carlyle
Pine, Lloyd A
Plonsker, Larry
Price, Leonard
Pryor, William Austin
Rabideau, Peter W
Reeves, Richard Edwin
Reich, Donald Arthur
Reinhardt, Robert Milton
Roberts, Donald Duane
Robinson, Gene Conrad
Rogers, Stearns Moore
Rosene, Robert Bernard
Schexnayder, Mary Anne
Schoellmann, Guenther
Sevenair, John P

Shin, Kju Hi
Shubkin, Ronald Lee
Slaven, Robert Walter
Smalley, Arnold Winfred
Smith, Grant Warren, II
Smith, Robert Leonard
Stanfield, Manie K
Stanonis, David Joseph
Starrett, Richmond Mullins
Steinmetz, Walter Edmund
Stocker, Jack H(ubert)
Stowell, John Charles
Sumrell, Gene
Theriot, Kevin Jude
Timberlake, Jack W
Traynham, James Gibson
Trisler, John Charles
Vail, Sidney Lee
Vercellotti, John R
Walia, Jasjit Singh
Walter, Thomas James
Walton, Warren Lewis
Weidig, Charles F
Welch, Clark Moore
Wells, Darthon Vernon
Wright, Oscar Lewis
Zietz, Joseph R, Jr

MAINE

Aft, Harvey
Bentley, Michael David
Bonner, Willard Hallam, Jr
Borror, Alan L
Bragdon, Robert Wright
Duncan, Charles Donald
Fort, Raymond Cornelius, Jr
Green, Brian
Hawthorne, Robert Montgomery, Jr
Jensen, Bruce L
Libby, Carol Baker
McCarty, John Edward
Mattor, John Alan
Mayo, Dana Walker
Mueller, George Peter
Newton, Thomas Allen
Pierce, Arleen Cecilia
Reid, Evans Burton
Rhodes, William Gale
Shepard, Robert Andrews
Spencer, Claude Franklin
Tedeschi, Robert James
Tewksbury, L Blaine
Welldon, Paul Burke
Winicov, Herbert
Wolfhagen, James Langdon

MARYLAND

Aaron, Herbert Samuel
Adolph, Horst Guenter
Aldridge, Mary Hennen
Ammon, Herman L
Armstrong, Daniel Wayne
Bailey, William John
Banks, Harold Douglas
Barnes, Charlie James
Bauknight, Charles William, Jr
Beisler, John Albert
Bellet, Richard Joseph
Berkowitz, Lewis Maurice
Berkowitz, Sidney
Beroza, Morton
Blum, Stanley Walter
Borkovec, Alexej B
Bosmajian, George
Buras, Edmund Maurice
Burrows, Elizabeth Parker
Bush, Richard Wayne
Cahnmann, Hans Julius
Campbell, Paul Gilbert
Caret, Robert Laurent
Catravas, George Nicholas
Cavagna, Giancarlo Antonio
Chamberlain, David Leroy, Jr
Chaykovsky, Michael
Chiu, Ching Ching
Chmurny, Alan Bruce
Clarke, Frederic B, III
Coates, Arthur Donwell
Cogliano, Joseph Albert
Cohen, Alex
Cohen, Louis Arthur
Cowan, Dwaine O
Cragg, Gordon Mitchell
Creighton, Donald John
Cullen, William Charles
Davis, George Thomas
De Matte, Michael L
Dighe, Shrikant Vishwanath
DiNunno, Cecil Malmberg
Doukas, Harry Michael
Dragun, Henry L
Dubois, Ronald Joseph
Eng, Leslie
Engle, Robert Rufus
Epstein, Joseph
Estrin, Norman Frederick
Falci, Kenneth Joseph
Fales, Henry Marshall
Fatiadi, Alexander Johann
Feinberg, Robert Samuel
Feldman, Alfred Philip
Fenselau, Allan Herman
Ferretti, Aldo
Feyns, Liviu Valentin

Fike, Harold Lester
Fisher, E(arl) Eugene
Florin, Roland Eric
Fox, Robert Bernard
Frank, Victor Samuel
Fraser, Blair Allen
Friess, Seymour Louis
Gabrielsen, Bjarne
Gaidis, James Michael
Gajan, Raymond Joseph
Garman, John Andrew
Gibson, James Donald
Goldstein, Jorge Alberto
Graham, Joseph H
Gray, Allan P
Guthrie, James Leverette
Hamilton, William Lander
Harmon, Alan Dale
Hartwell, Jonathan Lutton
Henery-Logan, Kenneth Robert
Herndon, James W
Herner, Albert Erwin
Herrick, Elbert Charles
Hertz, Harry Steven
Heumann, Karl Fredrich
Highet, Robert John
Hillery, Paul Stuart
Hillstrom, Warren W
Hoffsommer, John C
Homberg, Otto Albert
Horak, Vaclav
Hosmane, Ramachandra Sadashiv
Hoster, Donald Paul
Hsia, Mong Tseng Stephen
Humphrey, S(idney) Bruce
Hyatt, Asher Angel
Inscoe, May Nilson
Isaacs, Tami Yvette
Jachimowicz, Felek
Jacobson, Arthur E
Jacobson, Martin
James, John Cary
Jaouni, Taysir M
Jarvis, Bruce B
Jerina, Donald M
Judson, Horace Augustus
Kamlet, Mortimer Jacob
Karpetsky, Timothy Paul
Karten, Marvin J
Kayser, Richard Francis
Keefer, Larry Kay
Kehr, Clifton Leroy
Kleinspehn, George Gehret
Kreysa, Frank Joseph
Krutzsch, Henry C
Kutik, Leon
Leak, John Clay, Jr
Lednicer, Daniel
Levy, Joseph Benjamin
Lewis, Cameron David
Liebman, Joel Fredric
Linder, Seymour Martin
Longenecker, William Hilton
McElroy, Wilbur Renfrew
McGovern, Terrence Phillip
McGuire, Francis Joseph
Markey, Sanford Philip
Marquez, Victor Esteban
Martenson, Russell Eric
Massie, Samuel Proctor
May, Willie Eugene
Mazzocchi, Paul Henry
Michejda, Christopher Jan
Mills, Frank D
Misra, Renuka
Morgan, Charles Robert
Moschel, Robert Carl
Murch, Robert Matthews
Murr, Brown L, Jr
Muschik, Gary Mathew
Naff, Marion Benton
Neta, Pedatsur
Newkirk, David Royal
Nickon, Alex
Novak, Thaddeus John
Oliverio, Vincent Thomas
Paull, Kenneth Dywain
Pitha, Josef
Plimmer, Jack Reynolds
Pohland, Albert
Pollack, Ralph Martin
Pomerantz, Irwin Herman
Posner, Gary Herbert
Preusch, Peter Charles
Pummer, Walter John
Raksis, Joseph W
Reynolds, Kevin A
Rice, Kenner Cralle
Robinson, Cecil Howard
Rodewald, Lynn B
Rosenblatt, David Hirsch
Rosenblum, Annette Tannenholz
Rotherham, Jean
Rowell, Charles Frederick
Saroff, Harry Arthur
Sayer, Jane M
Schechter, Milton Seymour
Schiffmann, Elliot
Schroeder, Michael Allan
Schwartz, Anthony Max
Schwarz, Meyer
Sherman, Anthony Michael
Silversmith, Ernest Frank
Simmons, Thomas Carl

Simons, Samuel Stoney, Jr
Slife, Charles W
Smedley, William Michael
Smith, Betty F
Smith, William Owen
Snader, Kenneth Means
Spande, Thomas Frederick
Sphon, James Ambrose
Stahly, Eldon Everett
Steck, Edgar Alfred
Stein, Harvey Philip
Steinman, Harry Gordon
Strier, Murray Paul
Sweeting, Linda Marie
Tamminga, Carol Ann
Terry, Paul H
Thompson, Malcolm J
Timony, Peter Edward
Titus, Elwood Owen
Tolgyesi, Eva
Townsend, Craig Arthur
Trimmer, Robert Whitfield
Tsai, Lin
Valega, Thomas Michael
Veitch, Fletcher Pearre, Jr
Vestling, Martha Meredith
Vitullo, Victor Patrick
Vonderhaar, Barbara Kay
Waters, James Augustus
Waters, Rolland Mayden
Weiss, Richard Gerald
Weiss, Ulrich
Werber, Frank Xavier
Weser, Don Benton
Whaley, Wilson Monroe
White, Emil Henry
Willette, Robert Edmond
Winestock, Claire Hummel
Wingrove, Alan Smith
Witkop, Bernhard
Wood, Louis L
Woods, Charles William
Yagi, Haruhiko
Yarbrough, Arthur C, Jr
Yeh, Lai-Su Lee
Yoho, Clayton W
Zaczek, Norbert Marion
Zdravkovich, Vera
Zehner, Lee Randall
Ziffer, Herman

MASSACHUSETTS
Aberhart, Donald John
Ahearn, James Joseph, Jr
Ahern, David George
Allen, Malwina I
Andrulis, Peter Joseph, Jr
Anselme, Jean-Pierre L M
Atkinson, Edward Redmond
Auld, David Stuart
Bader, Henry
Bannister, William Warren
Basdekis, Costas H
Bearden, William Harlie
Becker, Ernest I
Bender, Howard Sanford
Bennett, Ovell Francis
Berchtold, Glenn Allen
Biemann, Klaus
Biletch, Harry
Bluhm, Aaron Leo
Boehm, Paul David
Bornstein, Joseph
Bragole, Robert Anthony
Browne, Douglas Townsend
Browne, Sheila Ewing
Buchi, George
Buckler, Sheldon A
Burson, Sherman Leroy, Jr
Byrnes, Eugene William
Carpino, Louis A
Caspi, Eliahu
Chang, Kuang-Chou
Chapin, Earl Cone
Cheema, Zafarullah K
Chipman, Wilmon B
Clagett, Donald Carl
Clapp, Richard Crowell
Coffin, Perley Andrews
Cohen, Merrill
Cohen, Saul G
Cohen, Saul Mark
Comeford, Lorrie Lynn
Cook, Michael Miller
Coplan, Myron J
Corey, Elias James
Crawford, Jean Veghte
Danheiser, Rick Lane
Davis, Robert Bernard
Demember, John Raymond
Dixon, Brian Gilbert
Doering, William von Eggers
Donoghue, John Timothy
Drumm, Manuel Felix
Dunny, Stanley
Ehret, Anne
Erickson, Karen Louise
Fetscher, Charles Arthur
Filer, Crist Nicholas
Finkelstein, Manuel
Fox, Daniel Wayne
Foye, William Owen
Frank, Jean Ann
Freedman, Harold Hersh

Garner, Albert Y
Gartner, Edward A
Georgian, Vlasios
Gerteisen, Thomas Jacob
Giarrusso, Frederick Frank
Giering, Warren Percival
Girard, Francis Henry
Goldberg, Gordon Morton
Gomez, Ildefonso Luis
Gorbunoff, Marina J
Gore, William Earl
Gounaris, Anne Demetra
Grabowski, Joseph J
Grasshoff, Jurgen Michael
Greene, Frederick Davis, II
Grunwald, Ernest Max
Haber, Stephen B
Haensel, Vladimir
Hagopian, Miasnig
Hall, George E
Handrick, George Richard
Hansen, David Elliott
Hearn, Michael Joseph
Hendrickson, James Briggs
Herz, Matthew Lawrence
Hixson, Stephen Sherwin
Hodes, William
Hodgdon, Russell Bates, Jr
Holick, Sally Ann
Holmquist, Barton
Hoover, M Frederick
Howell, David Moore
Hurd, Richard Nelson
Idelson, Martin
Ika, Prasad Venkata
Isaks, Martin
Israel, Stanley C
Jahngen, Edwin Georg Emil, Jr
Jaquiss, Donald B G
Jarret, Ronald Marcel
Jennings, Bojan Hamlin
Jones, Elmer Everett
Jones, Guilford, II
Kafrawy, Adel
Kantor, Simon William
Keehn, Philip Moses
Kelly, Thomas Ross
Kemp, Daniel Schaeffer
Kenley, Richard Alan
Keough, Allen Henry
Khorana, Har Gobind
Kishi, Yoshito
Klibanov, Alexander M
Knapczyk, Jerome Walter
Knapp, Robert Lester
Kramer, Charles Edwin
Krieger, Jeanne Kann
Krueger, Robert A
Krull, Ira Stanley
Lapkin, Milton
LaSala, Edward Francis
Laufer, Daniel A
Lee, Kang In
Lenz, Robert William
Le Quesne, Philip William
Lichtin, Norman Nahum
Lillya, Clifford Peter
Litant, Irving
Long, Alan K
Lowry, Nancy
Lowry, Thomas Hastings
Macaione, Domenic Paul
McEwen, William Edwin
McGrath, Michael Glennon
McMaster, Paul D
Macnair, Richard Nelson
McWhorter, Earl James
Magarian, Charles Aram
Maniscalco, Ignatius Anthony
Markgraf, J(ohn) Hodge
Masamune, Satoru
Medalia, Avrom Izak
Mehta, Avinash C
Mehta, Mahendra
Miliora, Maria Teresa
Miller, Bernard
Moser, William Ray
Moses, Ronald Elliot
Mowery, Dwight Fay, Jr
Nickerson, Richard G
Nussbaum, Alexander Leopold
Obermayer, Arthur S
Ofner, Peter
Olsen, Robert Thorvald
O'Rell, Dennis Dee
Orphanos, Demetrius George
Ort, Morris Richard
Parham, Marc Ellous
Pars, Harry George
Pavlik, James William
Pearson, Myrna Schmidt
Perkins, Janet Sanford
Peterson, Janet Brooks
Pian, Charles Hsueh Chien
Pike, Ronald Marston
Pinkus, Jack Leon
Piper, James Underhill
Powers, Donald Howard, Jr
Quin, Louis DuBose
Rademacher, Leo Edward
Rastetter, William Harry
Rausch, Marvin D
Razdan, Raj Kumar
Rebek, Julius, Jr

Refojo, Miguel Fernandez
Rivin, Donald
Roberts, Francis Donald
Rogers, Howard Gardner
Rosenberg, Joseph
Rosenblum, Myron
Rosenfeld, Stuart Michael
Rosowsky, Andre
Rossitto, Conrad
Rowe, Paul E
Rubinstein, Harry
Sahatjian, Ronald Alexander
Salamone, Joseph C
Santer, James Owen
Sardella, Dennis Joseph
Scanio, Charles John Vincent
Schlein, Herbert
Schrock, Richard Royce
Schuler, Robert Frederick
Seyferth, Dietmar
Shafer, Sheldon Jay
Sheehan, John Clark
Shuford, Richard Joseph
Simon, Myron Sydney
Slifkin, Sam Charles
Smith, William Edward
Snider, Barry B
Soffer, Milton David
Sosinsky, Barrie Alan
Sripada, Pavanaram Kameswara
Stark, James Cornelius
Stein, Samuel H
Stemniski, John Roman
Stillwell, Richard Newhall
Stolow, Robert David
Strem, Michael Edward
Stuhl, Louis Sheldon
Sullivan, Charles Irving
Sybert, Paul Dean
Takman, Bertil Herbert
Taylor, Lloyd David
Tewksbury, Charles Isaac
Thomas, George Richard
Thomas, Peter
Towns, Donald Lionel
Trachtenberg, Edward Norman
Tramondozzi, John Edmund
Truesdale, Larry Kenneth
Van der Burg, Sjirk
Vanelli, Ronald Edward
Viola, Alfred
Vogel, George
Vouros, Paul
Waller, David Percival
Walter, Henry Alexander
Wang, Chi-Hua
Wang, Nancy Yang
Warner, John Charles
Warner, Philip Mark
Watterson, Arthur C, Jr
Webster, Eleanor Rudd
Weigand, Willis Alan
Weininger, Stephen Joel
Wentworth, Stanley Earl
Westheimer, Frank Henry
Whitesides, George McClelland
Whitney, Robert Byron
Whitney, Thomas Allen
Wick, Emily Lippincott
Williams, John Russell
Williamson, Kenneth Lee
Wilson, Angus
Wirth, Joseph Glenn
Wotiz, Herbert Henry
Young, Richard L
Ziomek, Carol A

MICHIGAN
Anderson, Amos Robert
Anderson, Hugh Verity
Andrews, Arthur George
Arrington, Jack Phillip
Ashe, Arthur James, III
Atwell, William Henry
Bailey, Donald Leroy
Bajzer, William Xavier
Baker, Robert Henry
Bannister, Brian
Barry, Roger Donald
Beal, Philip Franklin, III
Belmont, Daniel Thomas
Berger, Mitchell Harvey
Berman, Ellen Myra
Berndt, Donald Carl
Blair, Etcyl Howell
Blankespoor, Ronald Lee
Blankley, Clifton John
Blatchford, John Kerslake
Blecker, Harry Herman
Bluestein, Bernard Richard
Bowman, Carlos Morales
Boyer, Rodney Frederick
Braidwood, Clinton Alexander
Bremmer, Bart J
Brieger, Gottfried
Brodasky, Thomas Francis
Brower, Frank M
Brubaker, Carl H, Jr
Brust, Harry Francis
Bundy, Gordon Leonard
Burgert, Bill E
Butler, Donald Eugene
Camp, Ronald Lee
Caron, E(dgar) Louis

Organic Chemistry (cont)

Dale, Wesley John
Darby, Joseph Raymond
Delaware, Dana Lewis
Dias, Jerry Ray
DiFate, Victor George
Dill, Dale Robert
Diuguid, Lincoln Isaiah
Dixon, Marvin Porter
Dub, Michael
Duchek, John Robert
Dutra, Gerard Anthony
Eime, Lester Oscar
Elliott, William H
Engel, James Francis
Faulk, Dennis Derwin
Freeman, Robert Clarence
Froemsdorf, Donald Hope
Gaertner, Van Russell
Garin, David L
Gash, Virgil Walter
Gaspar, Peter Paul
Gerhardt, Klaus Otto
Glaspie, Peyton Scott
Godt, Henry Charles, Jr
Gordon, Annette Waters
Griffiths, David Warren
Grubbs, Charles Leslie
Guthrie, David Burrell
Hammann, William Curl
Hancock, Anthony John
Hedrick, Ross Melvin
Heininger, S(amuel) Allen
Herber, John Frederick
Hobbs, Charles Floyd
Hochwalt, Carroll Alonzo
Holm, Myron James
Hortmann, Alfred Guenther
Howe, Robert Kenneth
Hyndman, Harry Lester
Jamieson, Norman Clark
Jason, Mark Edward
Juenge, Eric Carl
Kaiser, Edwin Michael
Kalota, Dennis Jerome
Keller, William John
Kidwell, Roger Lynn
Knox, Walter Robert
Koeltzow, Donald Earl
Koenig, Karl E
Krueger, Paul A
Ku, Audrey Yeh
Kurz, Joseph Louis
Lanson, Herman Jay
Lennon, Patrick James
Levy, Edward Robert
Li, Tao Ping
Liao, Tsung-Kai
Lipkin, David
Loeppky, Richard N
Loncrini, Donald Francis
Lovett, Eva G
Ludwig, Frederick John, Sr
Lusskin, Robert Miller
McConaghy, John Stead, Jr
McCormick, John Pauling
McCoy, Layton Leslie
McHugh, Kenneth Laurence
MacQuarrie, Ronald Anthony
Malik, Joseph Martin
Malzahn, Ray Andrew
Mange, Franklin Edwin
Markowski, Henry Joseph
Matzner, Edwin Arthur
Moedritzer, Kurt
Morris, Donald Eugene
Morse, Ronald Loyd
Mosher, Melvyn Wayne
Mottus, Edward Hugo
Mount, Ramon Albert
Munch, John Howard
Murray, Robert Wallace
Murrill, Evelyn A
Oftedahl, Marvin Loren
Owsley, Dennis Clark
Palmer, John Frank, Jr
Paulik, Frank Edward
Peacock, Val Edward
Perozzi, Edmund Frank
Perry, Paul
Phillips, Bruce Edwin
Pivonka, William
Plimmer, Jack Reynolds
Podrebarac, Eugene George
Popp, Frank Donald
Rabjohn, Norman
Raizman, Paula
Rath, Nigam Prasad
Ratts, Kenneth Wayne
Redmore, Derek
Richard, William Ralph, Jr
Richards, Charles Norman
Ridgway, Robert Worrell
Robinson, Donald Alonzo
Rogic, Milorad Mihailo
Rothrock, Thomas Stephenson
Rueppel, Melvin Leslie
Sabacky, M Jerome
Salivar, Charles Joseph
Sathe, Sharad Somnath
Schisla, Robert M
Schuck, James Michael
Schultz, John E
Schultz, Robert George
Schumacher, Ignatius

Searles, Scott, Jr
Sears, J Kern
Sharp, Dexter Brian
Sikorski, James Alan
Singleton, Tommy Clark
Slocombe, Robert Jackson
Smith, Lowell R
Smith, Ronald Gene
Snyder, Joseph Quincy
Solodar, Arthur John
Stary, Frank Edward
Stenseth, Raymond Eugene
Stephens, John Arnold
Stout, Edward Irvin
Sweet, Frederick
Swisher, Robert Donald
Talbott, Ted Delwyn
Tallman, Ralph Colton
Teng, James
Tonkyn, Richard George
Vineyard, Billy Dale
Wagenknecht, John Henry
Wasson, Richard Lee
Watkins, Darrell Dwight, Jr
Whittle, Philip Rodger
Wilbur, James Myers, Jr
Wildi, Bernard Sylvester
Williams, Byron Lee, Jr
Wilson, James Dennis
Winicov, Murray William
Winter, Rudolph Ernst Karl
Woodbrey, James C
Worley, Jimmy Weldon
Wulfman, David Swinton
Yu, Shiu Yeh
Zee-Cheng, Robert Kwang-Yuen
Zey, Robert L
Zienty, Ferdinand B

MONTANA
Craig, Arnold Charles
Jennings, Paul W
Juday, Richard Evans
Stewart, John Mathews

NEBRASKA
Baumgarten, Henry Ernest
Cavalieri, Ercole Luigi
Clark, Ronald David
Cromwell, Norman Henry
Demuth, John Robert
George, Anne Denise
Gold, Barry Irwin
Gross, Michael Lawrence
Johnson, John Arnold
Jones, Lee Bennett
Kaufman, Don Allen
Kingsbury, Charles Alvin
Klein, Francis Michael
Laursen, Paul Herbert
Linstromberg, Walter William
Looker, James Howard
Nagel, Donald Lewis
O'Leary, Marion Hugh
Raha, Chitta Ranjan
Rieke, Reuben Dennis
Roark, James L
Ruyle, William Vance
Shearer, Greg Otis
Shiue, Chyng-Yann
Smith, David Hibbard
Takemura, Kaz H(orace)
Warner, Charles D
Wheeler, Desmond Michael Sherlock
Wood, James Kenneth

NEVADA
Emerson, David Winthrop
Fox, Neil Stewart
Kemp, Kenneth Courtney
Lightner, David A
Nazy, John Robert
Nelson, John Henry
Poziomek, Edward John
Rose, Charles Buckley
Scott, Lawrence Tressler
Smith, Robert Bruce
Sovocool, G Wayne
Titus, Richard Lee

NEW HAMPSHIRE
Andersen, Kenneth K
Bachman, Paul Lauren
Balmer, Clifford Earl
Beck, Mae Lucille
Bowen, Douglas Malcomson
Coburn, Everett Robert
Cotter, Robert James
Creagh-Deyter, Linda T
Curphey, Thomas John
Detweiler, W Kenneth
Donaruma, L Guy
English, Jackson Pollard
Friedman, Harris Leonard
Gribble, Gordon W
Hagan, William John, Jr
Harris, Thomas David
Hughes, Russell Profit
Jones, Paul Raymond
Lavigne, Andre Andre
Lemal, David M
Morrison, James Daniel
Neil, Thomas C
Pratt, Richard J

Shafer, Paul Richard
Spencer, Thomas A
Torkelson, Arnold
Upham, Roy Herbert
Weisman, Gary Raymond
Woodbury, Richard Paul
Worman, James John

NEW JERSEY
Aaronoff, Burton Robert
Abou-Gharbia, Magid
Abramovici, Miron
Achari, Raja Gopal
Adams, Phillip
Addor, Roger Williams
Afonso, Adriano
Agnello, Eugene Joseph
Aguiar, Adam Martin
Alekman, Stanley L
Allen, Richard Charles
Altman, Lawrence Jay
Anderson, Lowell Ray
Anderson, Paul LeRoy
Anderson, Robert Christian
Andose, Joseph D
Andrade, John Robert
Angel, Henry Seymour
Angier, Robert Bruce
Ariyan, Zaven S
Armbruster, David Charles
Asato, Goro
Ashby, Bruce Allan
Ashcraft, Arnold Clifton, Jr
Atwater, Norman Willis
Auerbach, Andrew Bernard
Auerbach, Victor
Augustine, Robert Leo
Bach, Frederick L
Badin, Elmer John
Bagli, Jenanbux Framroz
Ballina, Rudolph August
Bamberger, Curt
Barclay, Robert, Jr
Barile, George Conrad
Barnabeo, Austin Emidio
Baron, Frank A
Barringer, Donald F, Jr
Batcho, Andrew David
Battisti, Angelo James
Bauer, Frederick William
Bauman, Robert Andrew
Bavley, Abraham
Baylouny, Raymond Anthony
Beattie, Thomas Robert
Beinfest, Sidney
Beispiel, Myron
Bellis, Harold E
Benning, Calvin James
Berg, Jeffrey Howard
Berkelhammer, Gerald
Bernath, Tibor
Bernholz, William Francis
Berry, William Lee
Bertelo, Christopher Anthony
Bertz, Steven Howard
Besso, Michael M
Bezwada, Rao Srinivasa
Bihovsky, Ron
Blake, Jules
Bloom, Allen
Bluestein, Allen Channing
Bluestein, Claire
Bohrer, James Calvin
Boikess, Robert S
Boothe, James Howard
Borah, Kripanath
Bornstein, Alan Arnold
Borowitz, Grace Burchman
Bose, Ajay Kumar
Bosniack, David S
Bouboulis, Constantine Joseph
Bowman, Lewis Wilmer
Bowman, Robert Mathews
Boyle, Richard James
Boyle, William Johnston, Jr
Bradstreet, Raymond Bradford
Bradway, Keith E
Brand, William Wayne
Brandman, Harold A
Brennan, James A
Brenner, Douglas Milton
Brill, William Franklin
Brons, Cornelius Hendrick
Brooks, Robert Alan
Brown, Dale Gordon
Brown, John Angus
Brown, Richard Emery
Buck, Carl John
Bullock, Francis Jeremiah
Bulusu, Suryanarayana
Burlant, William Jack
Burton, Gilbert W
Buyske, Donald Albert
Cardis, Angeline Baird
Carrock, Frederick E
Cavender, Patricia Lee
Cevasco, Albert Anthony
Chamberlin, Earl Martin
Chandross, Edwin A
Charbonneau, Larry Francis
Chaudhri, Safee U
Chen, Shuhchung Steve
Chiao, Wen Bin
Chibnik, Sheldon

Chien, Ping-Lu
Clarke, Frank Henderson
Coan, Stephen B
Cohen, Murray Samuel
Cohen, Noal
Colicelli, Elena Jeanmarie
Comai-Fuerherm, Karen
Conciatori, Anthony Bernard
Condon, Michael Edward
Cone, Conrad
Controulis, John
Cook, Alan Frederick
Cooke, Robert Sanderson
Coombs, Robert Victor
Cordon, Martin
Crosby, Guy Alexander
Cruickshank, P A
Cummins, Richard Williamson
Cutler, Frank Allen, Jr
Dahill, Robert T, Jr
Dain, Jeremy George
Dante, Mark F
Davies, Richard Edgar
Dawson, Arthur Donovan
Dayan, Jason Edward
Deem, Mary Lease
Degginger, Edward R
Delaney, Edward Joseph
Denney, Donald Berend
Derieg, Michael E
DeStevens, George
Diassi, Patrick Andrew
DiBella, Eugene Peter
Diegnan, Glenn Alan
Dien, Chi-Kang
Draper, Richard William
Dreby, Edwin Christian, III
Duhl-Emswiler, Barbara Ann
Dunkel, Morris
Durette, Philippe Lionel
Dursch, Friedrich
Dvornik, Dushan Michael
Eastman, David Willard
Eby, John Martin
Eck, John Clifford
Eckler, Paul Eugene
Edelson, Edward Harold
Edelstein, Harold
Egli, Peter
Ehrenfeld, Robert Louis
Elsenbaumer, Ronald Lee
Emert, Jack Isaac
Erckel, Ruediger Josef
Essery, John M
Evers, William John
Fagan, Paul V
Fahey, John Leonard
Fahrenholtz, Kenneth Earl
Fahrenholtz, Susan Roseno
Fand, Theodore Ira
Farina, Thomas Edward
Fath, Joseph
Feig, Gerald
Fein, Marvin Michael
Feldman, Martin Louis
Ferstandig, Louis Lloyd
Finkelstein, Jacob
Finley, Joseph Howard
Firth, William Charles, Jr
Fishman, Morris
Fitchett, Gilmer Trower
Fitzgerald, Patrick Henry
Florey, Klaus
Fobare, William Floyd
Fono, Andrew
Ford, Neville Finch
Forster, Warren Schumann
Fost, Dennis L
Francis, John Elsworth
Frankenfeld, John William
Franz, Curtis Allen
Frey, Sheldon Ellsworth
Fry, John Sedgwick
Fryer, Rodney Ian
Funke, Phillip T
Gaffney, Barbara Lundy
Gal, George
Gall, Martin
Gallopo, Andrew Robert
Gander, Robert Johns
Ganguly, Ashit K
Gans, Manfred
Garty, Kenneth Thomas
Garwood, William Everett
Genet, Rene P H
Genzer, Jerome Daniel
Gerber, Samuel Michael
Gerecht, J Fred
Gessler, Albert Murray
Giacobbe, Thomas Joseph
Gibian, Morton J
Gierer, Paul L
Gilbert, Allan Henry
Gilman, Norman Washburn
Girijavallabhan, Viyyoor Moopil
Gladstone, Harold Maurice
Glamkowski, Edward Joseph
Glickman, Samuel Arthur
Glidden, Richard Mills
Goeke, George Leonard
Gold, Elijah Herman
Goldstein, Albert
Goldstein, Theodore Philip
Gorbaty, Martin Leo

Organic Chemistry (cont)

Wilson, Harold Frederick
Winslow, Field Howard
Witkowski, Joseph Theodore
Wittcoff, A Harold
Wittekind, Raymond Richard
Witz, Gisela
Wohl, Ronald A
Wolf, Frank James
Wolf, Philip Frank
Wolff, Steven
Woodworth, Curtis Wilmer
Wu, Joseph Woo-Tien
Wu, Mu Tsu
Wu, Tse Cheng
Yale, Harry Louis
Young, Lewis Brewster
Young, Sanford Tyler
Zabriskie, John Lansing, Jr
Zager, Ronald
Zalay, Andrew W(illiam)
Zambito, Arthur Joseph
Ziegler, John Benjamin
Zinnes, Harold
Zoss, Abraham Oscar

NEW MEXICO
Amai, Robert Lin Sung
Auerbach, Irving
Benicewicz, Brian Chester
Brower, Kay Robert
Burns, Frank Bernard
Cahill, Paul A
Cheavens, Thomas Henry
Clark, Ronald Duane
Clough, Roger Lee
Coburn, Michael Doyle
Corcoran, George Bartlett, III
Corrigan, John Raymond
Cowan, John C
Curtice, Jay Stephen
Dahl, Alan Richard
Danen, Wayne C
Davis, Dennis Duval
Dorko, Ernest A
Evans, Latimer Richard
George, Raymond S
Guziec, Frank Stanley, Jr
Guziec, Lynn Erin
Hatch, Melvin (Jay)
Hoffman, Robert Vernon
Hollstein, Ulrich
Kelly, Gregory
Kjeldgaard, Edwin Andreas
Larson, Thomas E
Loftfield, Robert Berner
Lwowski, Walter Wilhelm Gustav
Marcy, Willard
Morton, John West, Jr
Nielsen, Stuart Dee
Ogilby, Peter Remsen
Ott, Donald George
Papadopoulos, Eleftherios Paul
Stinecipher, Mary Margaret
Wewerka, Eugene Michael
Whaley, Thomas Williams
Williams, Joel Mann, Jr
Zeigler, John Martin

NEW YORK
Acerbo, Samuel Nicholas
Ackerman, Hervey Winfield, Jr
Adduci, Jerry M
Addy, John Keith
Adler, George
Aftergut, Siegfried
Agosta, William Carleton
Alaimo, Robert J
Albrecht, Frederick Xavier
Albright, Jay Donald
Allen, Duff Shederic, Jr
Allen, Gary William
Allen, Lewis Edwin
Althuis, Thomas Henry
Altland, Henry Wolf
Altschul, Rolf
Anderson, Wayne Keith
Andrews, Mark Allen
Angelino, Norman J
Arcesi, Joseph A
Archie, William C, Jr
Arkell, Alfred
Armour, Eugene Arthur
Armstrong, William Lawrence
Arnowich, Beatrice
Aviram, Ari
Axelrad, George
Axenrod, Theodore
Ayral-Kaloustian, Semiramis
Badding, Victor George
Baechler, Raymond Dallas
Bak, David Arthur
Balassa, Leslie Ladislaus
Baldwin, John E
Banay-Schwartz, Miriam
Bank, Shelton
Barany, Francis
Barkey, Kenneth Thomas
Barnett, Ronald E
Baron, Arthur L
Barr, Donald Eugene
Barrett, Edward Joseph
Barringer, William Charles
Bass, Jon Dolf
Baum, George

Baumgarten, Reuben Lawrence
Beavers, Dorothy (Anne) Johnson
Beavers, Leo Earice
Beck, Curt Werner
Bell, Malcolm Rice
Bell, Thomas Wayne
Bennett, James Gordy, Jr
Bent, Richard Lincoln
Benz, George William
Berdahl, Donald Richard
Berger, S Edmund
Bergmark, William R
Berry, E Janet
Bieron, Joseph F
Billman, John Henry
Bitha, Panayota
Blackman, Samuel William
Block, Eric
Blood, Charles Allen, Jr
Bluestein, Ben Alfred
Boeckman, Robert K, Jr
Boettger, Susan D
Bolon, Donald A
Bonvicino, Guido Eros
Booth, Robert Edwin
Borch, Richard Frederic
Borders, Donald B
Borowitz, Irving Julius
Boudakian, Max Minas
Box, Vernon G S
Brabander, Herbert Joseph
Bradley, Arthur
Bradlow, Herbert Leon
Breslow, Ronald
Brewer, Curtis Fred
Brown, John Francis, Jr
Brunings, Karl John
Bullock, R Morris
Bunce, Stanley Chalmers
Burge, Robert Ernest, Jr
Burgmaier, George John
Burrows, Cynthia Jane
Cahn, Arno
Campbell, Bruce (Nelson), Jr
Campbell, Gerald Allan
Cappel, C Robert
Carabateas, Philip M
Carle, Kenneth Roberts
Carnahan, James Claude
Carpenter, Barry Keith
Carpenter, James William
Carpenter, Thomas J
Carr, Russell L K
Carroll, Robert Baker
Cathcart, John Almon
Ceprini, Mario Q
Chafetz, Harry
Chapman, Derek D
Charton, Marvin
Chen, Chin Hsin
Chen, Shaw-Horng
Childers, Robert Lee
Christensen, Larry Wayne
Christenson, Philip A
Christiansen, Robert George
Christman, David R
Chu, Joseph Yung-Chang
Clardy, Jon Christel
Clark, Charles Austin
Closson, William Deane
Coburn, Robert A
Cochran, John Charles
Cofrancesco, Anthony J
Cohen, Hyman L
Cohen, Sidney
Collins, Joseph Charles, Jr
Collum, David Boshart
Comer, William Timmey
Condit, Paul Brainard
Conway, Walter Donald
Cooper, Arthur Joseph L
Copelin, Harry B
Crivello, James V
Cunningham, Michael Paul
Curran, William Vincent
Cutler, Louise Marie
Cutler, Royal Anzly, Jr
Dagostino, Vincent F
D'Angelo, Gaetano
Daniel, Daniel S
Daniels, Ralph
Danishefsky, Isidore
Dannenberg, Joseph
Darlak, Robert
Daum, Sol Jacob
Daves, Glenn Doyle, Jr
Davis, Horace Raymond
Davis, Marvin Lester
Day, Jack Calvin
Dehn, Joseph William, Jr
Delton, Mary Helen
Denk, Ronald H
De Selms, Roy Charles
Dickinson, William Borden
Dieterich, David Allan
Dille, Kenneth Leroy
Dinan, Frank J
Dittmer, Donald Charles
Dixon, James Edward
Donner, David Bruce
Doubleday, Charles E, Jr
Dougherty, Charles Michael
Dupont, Paul Emile
Dutta, Shib Prasad

Eadon, George Albert
Earing, Mason Humphry
Eberhard, Anatol
Eckroth, David Raymond
Edelman, Robert
Ehrenson, Stanton Jay
Eikenberry, Jon Nathan
Eisch, John Joseph
Ellestad, George A
Elwood, James Kenneth
Engebrecht, Ronald Henry
Engel, Robert Ralph
Epstein, Joseph William
Erickson, Wayne Francis
Esse, Robert Carlyle
Estes, John H
Evans, Francis Eugene
Evert, Henry Earl
Factor, Arnold
Fanshawe, William Joseph
Farb, Edith
Fauth, David Jonathan
Fedor, Leo Richard
Feigenson, Gerald William
Feit, Irving N
Feitler, David
Felty, Evan J
Fernandez, Jose Martin
Ferraro, John J
Ferris, James Peter
Fields, Donald Lee
Fields, Thomas Lynn
Figueras, John
Filandro, Anthony Salvatore
Finkbeiner, Herman Lawrence
Finley, Kay Thomas
Finnegan, Richard Allen
Finzel, Rodney Brian
Flanigan, Everett
Ford, John Albert, Jr
Fowler, Frank Wilson
Fowler, Joanna S
Fox, Adrian Samuel
Fox, Charles Junius
Francis, Arokiasamy Joseph
Frazza, Everett Joseph
Frechet, Jean M J
Frenkel, Krystyna
Friedman, Paul
Friedman, Seymour K
Friedrich, Louis Elbert
Frye, Robert Bruce
Fuerst, Adolph
Fulmor, William
Gage, Clarke Lyman
Ganem, Bruce
Gao, Jiali
Gates, Marshall DeMotte, Jr
Geering, Emil John
George, Philip Donald
Georgiev, Vassil St
Gershon, Herman
Gettler, Joseph Daniel
Giannotti, Ralph Alfred
Gilman, Robert Edward
Gilmont, Ernest Rich
Ginos, James Zissis
Gioannini, Theresa Lee
Glaros, George Raymond
Gletsos, Constantine
Gold, Allen Morton
Goldberg, Joseph Louis
Goldsby, Arthur Raymond
Goldschmidt, Eric Nathan
Gordon, Irving
Gortler, Leon Bernard
Gray, D Anthony
Greco, Claude Vincent
Green, Mark M
Greizerstein, Walter
Griffith, Owen Wendell
Grina, Larry Dale
Grubman, Marvin J
Gruenbaum, William Tod
Gruett, Monte Deane
Grushkin, Bernard
Gugig, William
Gurel, Demet
Gurien, Harvey
Gysling, Henry J
Haase, Jan Raymond
Haberfield, Paul
Hahn, Roger C
Haitko, Deborah Ann
Halm, James Maurice
Hamilton, Lewis R
Hammer, Richard Benjamin
Hargreaves, Ronald Thomas
Harris, Robert L
Harrison, James Beckman
Haske, Bernard Joseph
Hathaway, Susan Jane
Hawks, George H, III
Hecht, Stephen Samuel
Heidelberger, Michael
Hellmuth, Walter Wilhelm
Henderson, David Rippey
Henderson, Ulysses Virgil, Jr
Henion, Richard S
Henzel, Richard Paul
Hepfinger, Norbert Francis
Herbrandson, Harry Fred
Herbstman, Sheldon
Herdle, Lloyd Emerson

Hickernell, Gary L
Hileman, Robert E
Hill, John Hamon Massey
Hillman, Manny
Hindersinn, Raymond Richard
Hirschberg, Albert I
Hlasta, Dennis John
Hlavka, Joseph John
Hoffman, Linda M
Hoffman, Theodore P
Hoffmann, Dietrich
Honig, Milton Leslie
Hough, Lindsay B
Howard, Philip Hall
Hoyte, Robert Mikell
Hull, Carl Max
Indictor, Norman
Inman, Charles Gordon
Isaacson, Henry Verschay
Ishler, Norman Hamilton
Izzo, Patrick Thomas
Jaffe, Fred
Jahn, Edwin Cornelius
James, Daniel Shaw
Jean, George Noel
Jelling, Murray
Jenkins, Philip Winder
Jesaitis, Raymond G
Jochsberger, Theodore
Joffee, Irving Brian
Johansson, Sune
Johns, William Francis
Johnson, Bruce Fletcher
Johnson, Francis
Johnson, Thomas Lynn
Jones, Gerald Walter
Jones, Stephen Thomas
Jones, William Davidson
Kalenda, Norman Wayne
Kalman, Thomas Ivan
Kapecki, Jon Alfred
Kaplan, Melvin
Kapuscinski, Jan
Karliner, Jerrold
Katz, Thomas Joseph
Kaufman, Frank B
Kaye, Irving Allan
Keister, Jerome Baird
Kelly, Kenneth William
Kende, Andrew S
Kerber, Robert Charles
Kestner, Melvin Michael
Khan, Jamil Akber
Kinnel, Robin Bryan
Kirdani, Rashad Y
Klanderman, Bruce Holmes
Klein, Bernard
Klemchuk, Peter Paul
Klijanowicz, James Edward
Klingele, Harold Otto
Klose, Thomas Richard
Knauer, Bruce Richard
Koch, Heinz Frank
Koeng, Fred R
Kohlbrenner, Philip John
Korytnyk, Walter
Kosak, Alvin Ira
Krabbenhoft, Herman Otto
Krause, Josef Gerald
Kresse, Jerome Thomas
Kreuzer, James Leon
Krishnamurthy, Sundaram
Krueger, James Elwood
Krueger, William E
Kudzin, Stanley Francis
Kuivila, Henry Gabriel
Kullnig, Rudolph K
Kumler, Philip L
Kupchik, Eugene John
Kuritzkes, Alexander Mark
Kuzmak, Joseph Milton
Labianca, Dominick A
Laganis, Deno
Lai, Yu-Chin
LaLonde, Robert Thomas
Landesberg, Joseph Marvin
Lang, Stanley Albert, Jr
Lansbury, Peter Thomas
Lau, Philip T S
Laubach, Gerald D
LeBlanc, Jerald Thomas
Lee, Lieng-Huang
Lee, May D-Ming (Lu)
Lee, Tzoong-Chyh
Lee, Ving Jick
Leibman, Lawrence Fred
Lengyel, Istvan
Le Noble, William Jacobus
Leone, Ronald Edmund
Lesher, George Yohe
Leubner, Gerhard Walter
Levine, Stephen Alan
Ligon, Woodfin Vaughan, Jr
Limburg, William W
Lin, Mow Shiah
Lobo, Angelo Peter
Loev, Bernard
Lok, Roger
Looker, Jerome J
Lorenz, Roman R
Lorenzo, George Albert
Lowrance, William Wilson, Jr
Lyons, Joseph F
MacAvoy, Thomas Coleman

Organic Chemistry (cont)

McCann, Michael Francis
McClure, Judson P
McConnell, James Francis
MacDonald, Donald MacKenzie
McGinn, Clifford
McGriff, Richard Bernard
Machiele, Delwyn Earl
McKelvie, Neil
McLaen, Donald Francis
MacLaury, Michael Risley
MacLeay, Ronald E
McMurry, John Edward
McNelis, Edward Joseph
Malan, Rodwick LaPur
Malpass, Dennis B
Mansfield, Kevin Thomas
Marcelli, Joseph F
March, Jerry
Marfey, Sviatopolk Peter
Mark, Robert Vincent
Markees, Diether Gaudenz
Markovitz, Mark
Martellock, Arthur Carl
Martin, William Butler, Jr
Maul, James Joseph
Mayr, Andreas
Meinwald, Jerrold
Meinwald, Yvonne Chu
Menapace, Lawrence William
Meyers, Marian Bennett
Mihajlov, Vsevolod S
Miller, David Lee
Miller, James Robert
Miller, Theodore Charles
Miller, William Taylor
Milton, Kirby Mitchell
Mirviss, Stanley Burton
Misra, Anand Lal
Mitacek, Eugene Jaroslav
Modic, Frank Joseph
Mooberry, Jared Ben
Moon, Sung
Moore, James Alfred
Moos, Gilbert Ellsworth
Moran, Juliette May
Morduchowitz, Abraham
Morrill, Terence Clark
Mourning, Michael Charles
Murdock, Keith Chadwick
Murphy, Daniel Barker
Mustacchi, Henry
Mylroie, Victor L
Nealey, Richard H
Neeman, Moshe
Neill, Alexander Bold
Nettleton, Donald Edward, Jr
Neumann, Stephen Michael
Neveu, Maurice C
Nicholson, Isadore
Norcross, Bruce Edward
Ober, Christopher Kemper
Oberender, Frederick G
O'Brien, Anne T
Ojima, Iwao
Olafsson, Patrick Gordon
Oliver, Gene Leech
O'Mara, Michael Martin
Orlando, Charles M
Osawa, Yoshio
Osborn, H(arland) James
Osterholtz, Frederick David
Oyama, Shigeo Ted
Pachter, Irwin Jacob
Parham, James Crowder, II
Parikh, Hemant Bhupendra
Parsons, Timothy F
Partch, Richard Earl
Patmore, Edwin Lee
Pavlisko, Joseph Anthony
Pavlos, John
Pedroza, Gregorio Cruz
Pellegrini, Frank C
Pelosi, Evelyn Tyminski
Pelosi, Stanford Salvatore, Jr
Pemrick, Raymond Edward
Penn, Lynn Sharon
Pepe, Joseph Philip
Personeus, Gordon Rowland
Petersen, Ingo Hans
Petersen, Kenneth C
Petropoulos, Constantine Chris
Phillips, Donald Kenney
Pickett, James Edward
Pier, Harold William
Piper, Douglas Edward
Plue, Arnold Frederick
Pollart, Dale Flavian
Pomeroy, Gordon Ashby
Ponticello, Ignazio Salvatore
Pontius, Dieter J J
Portlock, David Edward
Poslusny, Jerrold Neal
Post, Howard William
Potter, George Henry
Potts, Kevin T
Powers, John Clancey, Jr
Prestwich, Glenn Downes
Prigot, Melvin
Pucknat, John Godfrey
Purandare, Yeshwant K
Rauch, Emil Bruno
Ravindran, Nair Narayanan

Reckhow, Warren Addison
Redalieu, Elliot
Rees, William Wendell
Regan, Thomas Hartin
Rehfuss, Mary
Reich, Marvin Fred
Reidlinger, Anthony A
Renfroe, Harris Burt
Rhodes, Yorke Edward
Richards, Jack Lester
Riecke, Edgar Erick
Ritchie, Calvin Donald
Roberts, Wendell Lee
Robertson, Dale Norman
Rosen, Milton Jacques
Rosenfeld, Jerold Charles
Ross, Robert Edward
Rothchild, Robert
Rousseau, Viateur
Roy, Ram Babu
Rush, Kent Rodney
Russell, Charlotte Sananes
Rutkin, Philip
Ryan, Thomas John
Salomone, Ramon Angelo
Sanchez, Jose
Sandhu, Mohammad Akram
Sands, Richard Dayton
Sapino, Chester, Jr
Sauer, Charles William
Saunders, William Hundley, Jr
Sauter, Frederick Joseph
Savage, Dennis Jeffrey
Schaeffer, James Robert
Schaffrath, Robert Eben
Scheiner, Peter
Schleigh, William Robert
Schlessinger, Richard H
Schlicht, Raymond Charles
Schmeltz, Irwin
Schmitz, Henry
Schuerch, Conrad
Schulenberg, John William
Schultz, Arthur George
Schultz, William Clinton
Schuster, David Israel
Schwab, Linda Sue
Schwan, Thomas James
Schwartz, Leonard H
Scott, Francis Leslie
Searle, Roger
Searles, Arthur Langley
Segal, Alvin
Seltzer, Stanley
Semmelhack, Martin F
Seus, Edward J
Shaath, Nadim Ali
Shannon, Frederick Dale
Shapiro, Robert
Sharma, Moheswar
Sheppard, Chester Stephen
Shields, Joan Esther
Siegart, William Raymond
Siew, Ernest L
Silberman, Robert G
Silleck, Clarence Frederick
Silveira, Augustine, Jr
Sims, Rex J
Singh, Baldev
Sisti, Anthony Joseph
Siuta, Gerald Joseph
Sleezer, Paul David
Slusarczuk, George Marcelius Jaremias
Smith, Richard Frederick
Smith, Thomas Woods
Snyder, Harry Raymond, Jr
Sogah, Dotsevi Yao
Sojka, Stanley Anthony
Solo, Alan Jere
Soloway, Harold
Soloway, Saul
Sorkin, Howard
Sorter, Peter F
Sowa, John Robert
Spangler, Fred Walter
Sparapany, John Joseph
Spittler, Terry Dale
Sprung, Joseph Asher
Srinivasan, Rangaswamy
Staklis, Andris A
Staples, Jon T
Starkey, Frank David
Stecher, Emma Dietz
Steinberg, David H
Stephens, Lawrence James
Stern, Max Herman
Sticker, Robert Earl
Still, W Clark, Jr
Stöhrer, Gerhard
Stone, Joe Thomas
Stork, Gilbert (Josse)
Stradling, Samuel Stuart
Sturmer, David Michael
Su, Tah-Mun
Sullivan, Michael Francis
Surrey, Alexander Robert
Sutherland, George Leslie
Swicklik, Leonard Joseph
Swinehart, James Stephen
Syed, Ashfaquzzaman
Takeuchi, Esther Sans
Teasdale, William Brooks
Teegarden, David Morrison
Tesoro, Giuliana C

Thomas, Harold Todd
Thomas, Telfer Lawson
Timell, Tore Erik
Tokoli, Emery G
Tomasz, Maria
Tometsko, Andrew M
Towle, Jack Lewis
Tremelling, Michael, Jr
Troll, Walter
Trucker, Donald Edward
Trust, Ronald I
Tufariello, Joseph James
Tuites, Richard Clarence
Turek, William Norbert
Turnblom, Ernest Wayne
Turner, S Richard
Uebel, Jacob John
Ursino, Joseph Anthony
van Duuren, Benjamin Louis
Van Verth, James Edward
Verbit, Lawrence
Verter, Herbert Sigmund
Vogel, Philip Christian
Von Strandtmann, Maximillian
Vratsanos, Spyros M
Vullo, William Joseph
Wagner, Gerald Roy
Walker, Russell Wagner
Wallis, Thomas Gary
Walsh, Edward Nelson
Ward, Frank Kernan
Warren, James Donald
Wasacz, John Peter
Watanabe, Kyoichi A
Weetman, David G
Weil, Edward David
Weisler, Leonard
Weiss, Benjamin
Weiss, David Steven
Weiss, Irma Tuck
Weiss, Martin Joseph
Weltman, Clarence A
Wentland, Mark Philip
White, Dwain Montgomery
White, Ralph Lawrence, Jr
Whitten, David G
Wiegert, Philip E
Wilcox, Charles Frederick, Jr
Wilen, Samuel Henry
Wilson, Burton David
Wilson, Stephen Ross
Windholz, Thomas Bela
Winter, Margaret Castle
Winter, Roland Arthur Edwin
Wiseman, George Edward
Wissner, Allan
Wolf, Alfred Peter
Wolfarth, Eugene F
Wolfe, Roger Thomas
Wolinski, Leon Edward
Wolny, Friedrich Franz
Wood, David
Worrall, Winfield Scott
Wright, George Carlin
Wright, Robert W
Wyman, Donald Paul
Yaffe, Roberta
You, Kwan-sa
Yourtee, Lawrence Karn
Yu, Chia-Nien
Yunick, Robert P
Yuska, Henry
Zalay, Ethel Suzanne
Zalewski, Edmund Joseph
Zavitsas, Andreas Athanasios
Zawadzki, Joseph Francis
Zerwekh, Charles Ezra, Jr
Zieger, Herman Ernst
Zimmerman, Barry
Zimmerman, Mary Prislopski
Zuckerman, Samuel
Zuman, Petr

NORTH CAROLINA

Abbey, Kirk J
Alston, Peter Van
Anthes, John Allen
Arnett, Edward McCollin
Arnold, Allen Parker
Arnold, Luther Bishop, Jr
Ayers, Paul Wayne
Bair, Kenneth Walter
Barborak, James Carl
Bauermeister, Herman Otto
Bealor, Mark Dabney
Beauchamp, Lilia Marie
Bell, Jimmy Holt
Bockstahler, Theodore Edwin
Boswell, Donald Eugene
Bradsher, Charles Kilgo
Brieaddy, Lawrence Edward
Bristol, Douglas Walter
Brookhart, Maurice S
Bryan, Carl Eddington
Bumgardner, Carl Lee
Bursey, Joan Tesarek
Bushey, Dean Franklin
Carroll, F Ivy
Carroll, Felix Alvin, Jr
Cavallito, Chester John
Chaney, David Webb
Chignell, Colin Francis
Chilton, William Scott
Chopra, Naiter Mohan

Cialdella, Cataldo
Clark, Russell Norman
Cline, Warren Kent
Coke, James Logan
Cook, Clarence Edgar
Cook, Lawrence C
Cote, Philip Norman
Cross, Robert Edward
Cubberley, Adrian H
Culberson, Chicita Frances
Cunningham, Carol Clem
Danieley, Earl
Daumit, Gene Philip
Davis, Roman
Detty, Wendell Eugene
Dhami, Kewal Singh
Dixit, Ajit Suresh
D'Silva, Themistocles Damasceno Joaquim
Dudley, Kenneth Harrison
Dufresne, Richard Frederick
Durden, John Apling, Jr
Dye, William Thomson, Jr
Edwards, Ben E
Eliel, Ernest Ludwig
Engel, John Francis
Erickson, Bruce Wayne
Etter, Robert Miller
Evans, Evan Franklin
Fabricius, Dietrich M
Forrester, Sherri Rhoda
Frank, Robert Loeffler
Fraser-Reid, Bertram Oliver
Fredericksen, James Monroe
Freedman, Leon David
Freeman, Harold Stanley
Fu, Wallace Yamtak
Fytelson, Milton
Gains, Lawrence Howard
Ganz, Charles Robert
Ghirardelli, Robert George
Giles, Jesse Albion, III
Gillikin, Jesse Edward, Jr
Goldstein, Irving Solomon
Graham, Jack Raymond
Greenspan, Frank Philip
Harfenist, Morton
Heckman, Robert Arthur
Heintzelman, Richard Wayne
Henderson, Thomas Ranney
Herbst, Robert Max
Higgins, Robert H
Hinze, Willie Lee
Hiskey, Richard Grant
Hix, Homer Bennett
Hoelzel, Charles Bernard
Holtkamp, Freddy Henry
Huang, Jamin
Huber, W(ilson) Frederick
Hurwitz, Melvin David
Husk, George Ronald
Izydore, Robert Andrew
Jameson, Charles William
Jeffs, Peter W
Jennings, Carl Anthony
Jiang, Jack Bau-Chien
Johnson, Michael Ross
Jones, Rufus Sidney
Jones, William Jonas, Jr
Jung, James Moser
Kasperek, George James
Kelley, James Leroy
King, Henry Lee
Knight, David Bates
Kropp, Paul Joseph
Kupstas, Edward Eugene
Kuyper, Lee Frederick
Lamb, Robert Charles
Leach, James Moore
Lee, Yue-Wei
Leiserson, Lee
Levine, Samuel Gale
Levy, Jack Benjamin
Levy, Louis A
Lewin, Anita Hana
Lewis, Robert Glenn
Little, William Frederick
Loeffler, Larry James
Loeppert, Richard Henry
Long, Robert Allen
Lynn, John Wendell
Ma, Tsu Sheng
McDaniel, Roger Lanier, Jr
McKay, Jerry Bruce
Manning, Robert Joseph
Marco, Gino Joseph
Martin, Gary Edwin
Martin, Ned Harold
Martin, Richard Hugo
Martin, William Royall, Jr
Mathewes, David A
Mechanic, Gerald
Mehta, Nariman Bomanshaw
Miles, George Benjamin
Miles, Marion Lawrence
Miller, Harry Brown
Miller, Walter Peter
Moates, Robert Franklin
Morris, Gene Franklin
Morris, Leo Raymond
Morrison, Robert W, Jr
Myers, John Albert
Neale, Robert S
Neumann, Calvin Lee

Newell, Marjorie Pauline
Nielsen, Lawrence Arthur
Norris, Terry Orban
Odell, Norman Raymond
Osbahr, Albert J, Jr
Panek, Edward John
Partridge, John Joseph
Patterson, Ronald Brinton
Peck, David W
Perfetti, Thomas Albert
Phillips, Arthur Page
Pirrung, Michael Craig
Porter, Ned Allen
Powers, Edward James
Preiss, Donald Merle
Preston, Jack
Proffitt, Thomas Jefferson, Jr
Pruett, Roy L
Purrington, Suzanne T
Raleigh, James Arthur
Ranieri, Richard Leo
Rapoport, Lorence
Reese, Millard Griffin, Jr
Rehberg, Chessie Elmer
Reid, Jack Richard
Rideout, Janet Litster
Riemann, James Michael
Rivers, Jessie Markert
Rodgman, Alan
Roth, Roy William
Russell, Henry Franklin
Scharver, Jeffrey Douglas
Schultz, Frederick John
Schumacher, Joseph Nicholas
Schwindeman, James Anthony
Sherer, James Pressly
Soeder, Robert W
Sparacino, Charles Morgan
Spears, Alexander White, III
Sprague, Robert Hicks
Squibb, Samuel Dexter
Steiner, Werner Douglas
Sternbach, Daniel David
Swaringen, Roy Archibald, Jr
Szabo, Emery D
Teague, Harold Junior
Thompson, Crayton Beville
Tinsley, Samuel Weaver
Tomer, Kenneth Beamer
Tucker, William Preston
Tyndall, John Raymond
Waite, Moseley
Walsh, Thomas David
Wani, Mansukhlal Chhaganlal
Wehner, Philip
Wetmore, David Eugene
Whangbo, Myung Hwan
Wheeler, Thomas Neil
Wilder, Pelham, Jr
Wilson, Nancy Keeler
Wiseman, Jeffrey Stewart
Woosley, Royce Stanley
Wyman, George Martin
Wyrick, Steven Dale

NORTH DAKOTA
Bergstrom, Donald Eugene
Boudjouk, Philip Raymond
Donnelly, Brendan James
Jasperse, Craig Peter
Johnson, A William
Lambeth, David Odus
Olson, Edwin S
Pomonis, James George
Rudesill, James Turner
Severson, Roland George
Shelver, William H
Stenberg, Virgil Irvin
Sugihara, James Masanobu
Tanaka, Fred Shigeru
Wardner, Carl Arthur
Woolsey, Neil Franklin

OHIO
Ager, David John
Albright, James Andrew
Alford, Harvey Edwin
Ambelang, Joseph Carlyle
Andresen, Brian Dean
Andrist, Anson Harry
Attig, Thomas George
Austin, Robert Andrae
Averill, Seward Junior
Baclawski, Leona Marie
Baker, Frank Weir
Ball, David Ralph
Baltazzi, Evan S
Bann, Robert (Francis)
Barnett, Kenneth Wayne
Barrer, Daniel Edward
Battershell, Robert Dean
Bauer, Richard G
Baughman, Glenn Laverne
Baur, Fredric John, Jr
Behrmann, Eleanor Mitts
Bender, Daniel Frank
Bergomi, Angelo
Bernstein, Stanley Carl
Bertelson, Robert Calvin
Bertsch, Robert Joseph
Bethea, Tristram Walker, III
Bimber, Russell Morrow
Binkley, Roger Wendell
Blewett, Charles William

Blum, Patricia Rae
Bockhoff, Frank James
Bond, William Bradford
Borgnaes, Dan
Brady, Leonard Everett
Brady, Thomas E
Brain, Devin King
Brannen, William Thomas, Jr
Brittain, William Joseph
Broaddus, Charles D
Brodhag, Alex Edgar, Jr
Brody, Richard Simon
Bromund, Werner Hermann
Brueggemeier, Robert Wayne
Brundage, Donald Keith
Bruno, David Joseph
Buck, Keith Taylor
Buell, Glen R
Bufkin, Billy George
Bumpus, Francis Merlin
Burke, Morris
Burt, Gerald Dennis
Butler, John Mann
Campbell, Robert Wayne
Canfield, James Howard
Carlock, John Timothy
Carr, Albert A
Chamberlin, William Bricker, III
Chen, Yih-Wen
Clark, Alan Curtis
Clemans, George Burtis
Clement, William H
Cohen, Irwin
Coleman, John Franklin
Coleman, Lester Earl, (Jr)
Colvin, Howard Allen
Compton, Ell Dee
Conley, Robert T
Connor, Daniel S
Convery, Robert James
Cook, Wendell Sherwood
Coran, Aubert Y
Corbin, James Lee
Cosgrove, Stanley Leonard
Costanza, Albert James
Cotton, Wyatt Daniel
Crano, John Carl
Cryberg, Richard Lee
Culbertson, Billy Muriel
Curry, Howard Millard
Dalton, James Christopher
Dana, Robert Clark
Dannley, Ralph Lawrence
Davies, Julian Anthony
Davis, Edward Melvin
Davis, James Allen
Delcamp, Robert Mitchell
Delvigs, Peter
DeMarco, John Gregory
Denham, Joseph Milton
Deters, Donald W
Dial, William Richard
Di Biase, Stephen Augustine
Dietrich, Joseph Jacob
Dimond, Harold Lloyd
Dinbergs, Kornelius
Dolfini, Joseph E
Doran, Thomas J, Jr
Doyle, Richard Robert
Duddey, James E
Duke, June Temple
Dunn, Horton, Jr
Dunning, John Walcott
Edman, James Richard
Eichel, Herman Joseph
Eicher, John Harold
Eisenhardt, William Anthony, Jr
Elmer, Otto Charles
Emrick, Donald Day
Erman, William F
Essig, Henry J
Eveslage, Sylvester Lee
Fabris, Hubert
Fayter, Richard George, Jr
Fechter, Robert Bernard
Feld, William Adam
Feldman, Julian
Fendler, Eleanor Johnson
Fentiman, Allison Foulds, Jr
Fieldhouse, John W
Fineberg, Herbert
Finelli, Anthony Francis
Fishel, Derry Lee
Flechtner, Thomas Welch
Fleischman, Darrell Eugene
Flynn, Gary Alan
Foldvary, Elmer
Ford, Emory A
Fox, Bernard Lawrence
Fraenkel, Gideon
Frank, Charles Edward
Franks, Allen P
Franks, John Anthony, Jr
Frederick, Marvin Ray
Fry, James Leslie
Furry, Benjamin K
Gainer, Gordon Clements
Galloway, Ethan Charles
Gano, James Edward
Gaul, Richard Joseph
George, Paul John
Germann, Richard P(aul)
Gibson, Thomas William
Gifford, David Stevens

Gilbert, Eugene Charles
Gilde, Hans-Georg
Gillis, Bernard Thomas
Ginning, P(aul) R(oll)
Glasgow, David Gerald
Gleim, Clyde Edgar
Gloth, Richard Edward
Goebel, Charles Gale
Goel, Anil B
Goetz, Richard W
Goodson, Alan Leslie
Gordon, John Edward
Gosselink, Eugene Paul
Gould, Edwin Sheldon
Graef, Walter L
Gray, Don Norman
Greenbaum, Sheldon Boris
Greene, Hoke Smith
Greene, Janice L
Griffin, Claibourne Eugene, Jr
Grimm, Robert Arthur
Grose, Herschel Gene
Grotta, Henry Monroe
Grummitt, Oliver Joseph
Gustafson, David Harold
Hanes, Ronnie Michael
Harrington, Roy Victor
Harris, Richard Lee
Hart, David Joel
Hausch, Walter Richard
Hawbecker, Byron L
Haynes, LeRoy Wilbur
Hazy, Andrew Christopher
Heckert, David Clinton
Hein, Richard William
Helmick, Larry Scott
Hergenrother, William Lee
Herliczek, Siegfried H
Herold, Robert Johnston
Hess, George G
Hickman, Howard Minor
Hirschmann, Hans
Hoke, Donald I
Holubec, Zenowie Michael
Horgan, Stephen William
Horne, Samuel Emmett, Jr
Horton, Derek
Hosansky, Norman Leon
Houser, John J
Howald, Jeremiah Mark
Hunt, Jerry Donald
Huntsman, William Duane
Hutchison, Robert B
Ingham, Robert Kelly
Innes, John Edwin
Jacobs, Harvey
Jacobs, Richard Lee
James, Floyd Lamb
Jarvi, Esa Tero
Johnson, Robert Gudwin
Jones, Winton D, Jr
Kagen, Herbert Paul
Kane, James Joseph
Katz, Morton
Katzen, Raphael
Kay, Edward Leo
Kay, Peter Steven
Ketcha, Daniel Michael
Kinsman, Donald Vincent
Kinstle, Thomas Herbert
Knowlton, David A
Kofron, William G
Kolattukudy, P E
Korach, Malcolm
Kotnik, Louis John
Kreuz, John Anthony
Kriens, Richard Duane
Kubisen, Steven Joseph, Jr
Kupel, Richard E
Kurtz, David Williams
Kwiatek, Jack
Kyung, Jai Ho
Laali, Khosrow
Lang, James Frederick
Laughlin, Robert Gene
Lawson, David Francis
Layer, Robert Wesley
Letton, James Carey
Levin, Robert Harold
Lewis, Irwin C
Leyland, Harry Mours
Lipinsky, Edward Solomon
Litt, Morton Herbert
Livigni, Russell A
Loening, Kurt L
Logan, Ted Joe
Losekamp, Bernard Francis
Loughran, Gerard Andrew, Sr
Lucier, John J
Lukin, Marvin
McCain, George Howard
McCarty, Frederick Joseph
McFarland, Charles Warren
McLoughlin, Daniel Joseph
McMillan, Robert McKee
McRowe, Arthur Watkins
Maender, Otto William
Magee, Thomas Alexander
Mao, Simon Jen-Tan
Marshall, Richard Allen
Mathers, Alexander Pickens
Matlin, Albert R
Mattson, Raymond Harding
Mayor, Rowland Herbert

Meinert, Walter Theodore
Meinhardt, Norman Anthony
Meloy, Carl Ridge
Michaelis, Carl I
Michelman, John S
Milone, Charles Robert
Milstein, Stanley Richard
Mitchell, George Redmond, Jr
Molnar, Stephen P
Morgan, Alan Raymond
Mouk, Robert Watts
Mundell, Percy Meldrum
Murdoch, Arthur
Murphy, Walter Thomas
Muse, Joel, Jr
Naples, Felix John
Nardi, John Christopher
Nathan, Richard Arnold
Neckers, Douglas
Nee, Michael Wei-Kuo
Nelson, Wayne Franklin
Newman, Melvin Spencer
Nicholas, Paul Peter
Nicholson, D Allan
Nielsen, Donald R
Nobis, John Francis
Nordlander, John Eric
Oberster, Arthur Eugene
O'Connor, David Evans
Odioso, Raymond C
Orchin, Milton
Ostrum, G(eorge) Kenneth
Ouellette, Robert J
Oziomek, James
Pace, Henry Alexander
Palopoli, Frank Patrick
Paquette, Leo Armand
Parish, Darrell Joe
Parker, Richard Ghrist
Peet, Norton Paul
Percec, Virgil
Peters, Lynn Randolph
Peterson, Donald J
Petrarca, Anthony Edward
Pialet, Joseph William
Platau, Gerard Oscar
Platz, Matthew S
Powell, Warren Howard
Powers, Larry James
Price, John Avner
Pritchett, Ervin Garrison
Prudence, Robert Thomas
Pynadath, Thomas I
Raciszewski, Zbigniew
Rader, Charles Phillip
Ramey, Chester Eugene
Ramp, Floyd Lester
Rau, Allen H
Raymond, Maurice A
Reilly, Charles Bernard
Rhinesmith, Herbert Silas
Richardson, Alfred, Jr
Richardson, Kathleen Schueller
Rizzi, George Peter
Roberts, Durward Thomas, Jr
Robins, Richard Dean
Robinson, Kenneth Robert
Roha, Max Eugene
Rosen, Irving
Ryan, Anne Webster
Ryan, Joseph Dennis
Ryerson, George Douglas
Sadurski, Edward Alan
Sampson, Paul
Sarbach, Donald Victor
Sartoris, Nelson Edward
Scalzi, Francis Vincent
Scanlan, Mary Ellen
Scheve, Bernard Joseph
Schlossman, Irwin S
Schmidt, Francis Henry
Schmidt, Paul J
Schneider, Wolfgang W
Schnettler, Richard Anselm
Schollenberger, Charles Sundy
Schroll, Gene E
Schulze, William Eugene
Schumacher, Roy Joseph
Schwerzel, Robert Edward
Scozzie, James Anthony
Sebastian, John Francis
Sedor, Edward Andrew
Selden, George
Selker, Milton Leonard
Serve, Munson Paul
Shechter, Harold
Sheibley, Fred Easly
Shelton, James Reid
Sheppard, William James
Shertzer, Howard Grant
Siedschlag, Karl Glenn, Jr
Siefken, Mark William
Siklosi, Michael Peter
Sill, Arthur DeWitt
Sinner, Donald H
Smith, Daniel John
Smith, Douglas Alan
Spanninger, Philip Andrew
Spessard, Dwight Rinehart
Spialter, Leonard
Spitzer, Jeffrey Chandler
Sporek, Karel Frantisek
Staker, Donald David
Stansfield, Roger Ellis

Organic Chemistry (cont)

Stein, Daryl Lee
Sterken, Gordon Jay
Sternfeld, Marvin
Stevens, Henry Conrad
Stevenson, Don R
Stoldt, Stephen Howard
Surbey, Donald Lee
Swartzentruber, Paul Edwin
Swenton, John Stephen
Tamborski, Christ
Tarvin, Robert Floyd
Taschner, Michael J
Tate, David Paul
Taylor, Lynn Johnston
Taylor, Richard Timothy
Temple, Robert Dwight
Theis, Richard James
Theisen, Cynthia Theres
Thomas, Alexander Edward, III
Thomas, Lewis Edward
Thompson, James Edwin
Throckmorton, Peter E
Traynor, Lee
Trivisonno, Charles F(rancis)
Tsai, Chung-Chieh
Tsai, Ming-Daw
Turkel, Rickey M(artin)
Turnbull, Kenneth
Undeutsch, William Charles
Vogel, Paul William
Volk, Murray Edward
Von Ostwalden, Peter Weber
Wade, Peter Cawthorn
Wagner, Melvin Peter
Walker, Thomas Eugene
Walsh, James Aloysius
Washburn, Lee Cross
Weaver, William Michael
Webb, Thomas Howard
Webster, James Albert
Weinstein, Arthur Howard
Weintraub, Philip Marvin
Werner, Raymond Edmund
Westerman, Ira John
Wideman, Lawson Gibson
Williger, Ervin John
Wilson, Gustavus Edwin
Winkler, Robert Randolph
Wise, Richard Melvin
Witiak, Donald T
Woo, James T K
Wright, Robert L
Yanko, William Harry
Yocum, Ronald Harris
York, Owen, Jr
Youngs, Wiley Jay
Zaremsky, Baruch
Zilch, Karl T
Zimmer, Hans

OKLAHOMA

Alaupovic, Petar
Baldwin, Roger Allan
Battiste, David Ray
Beckett, James Reid
Berlin, Kenneth Darrell
Bost, Howard William
Brady, Donnie Gayle
Burlett, Donald James
Clapper, Thomas Wayne
Craig, Louis Elwood
Crain, Donald Lee
Dermer, Otis Clifford
Durr, Albert Matthew, Jr
Eby, Harold Hildenbrandt
Eddington, Carl Lee
Edmison, Marvin Tipton
Efner, Howard F
Eisenbraun, Edmund Julius
Fahey, Darryl Richard
Farrar, Ralph Coleman
Geibel, Jon Frederick
Greenwood, Gil Jay
Hartzfeld, Howard Alexander
Hertzler, Donald Vincent
Hill, James Wagy
Hodnett, Ernest Matelle
Hughes, William Bond
Johnson, Wallace Delmar
Kleinschmidt, Roger Frederick
Kucera, Clare H
Lehr, Roland E
McGuire, Stephen Edward
McGurk, Donald J
Magarian, Robert Armen
Matson, Michael Steven
Minor, John Threecivelous
Moore, Donald R
Motz, Kaye La Marr
Natowsky, Sheldon
Nicholas, Kenneth M
Nye, Mary Jo
Pober, Kenneth William
Rotenberg, Don Harris
Schiff, Sidney
Seeney, Charles Earl
Selman, Charles Melvin
Shotts, Adolph Calveran
Snider, Theodore Eugene
Tillman, Richard Milton
Trepka, William James
Washcheck, Paul Howard

Wharry, Stephen Mark
White, Harold McCoy
Wurth, Michael John
Yu, Chang-An
Zuech, Ernest A

OREGON

Autrey, Robert Luis
Badger, Rodney Allan
Becker, Robert Richard
Bloomfield, Jordan Jay
Boekelheide, Virgil Carl
Brinkley, John Michael
Christensen, Bert Einar
Civelli, Oliver
Cronyn, Marshall William
Currie, James Orr, Jr
Deinzer, Max Ludwig
Dolby, Lloyd Jay
Duncan, James Alan
Elia, Victor John
Finke, Richard Gerald
Freeman, Peter Kent
Gleicher, Gerald Jay
Gould, Steven James
Grettie, Donald Pomeroy
Hancock, John Edward Herbert
Haugland, Richard Paul
Hauser, Frank Marion
Hudak, Norman John
Jaeger, Charles Wayne
Keana, John F W
Keevil, Thomas Alan
Klemm, LeRoy Henry
Klopfenstein, Charles E
Knott, Donald Macmillan
Koenig, Thomas W
Laver, Murray Lane
Levinson, Alfred Stanley
Lutz, Raymond Paul
McCoy, Joseph Wesley
Marvell, Elliot Nelson
Merz, Paul Louis
Oatfield, Harold
Paudler, William W
Pawlowski, Norman E
Postl, Anton
Rose, Norman Carl
Sandifer, Ronda Margaret
Schink, Chester Albert
Shilling, Wilbur Leo
Szalecki, Wojciech Josef
Thies, Richard William
Van Eikeren, Paul
Wamser, Carl Christian
White, James David
Whittemore, Charles Alan

PENNSYLVANIA

Ackermann, Guenter Rolf
Adler, Irving Larry
Albright, Robert Lee
Almond, Harold Russell, Jr
Alvino, William Michael
Anspon, Harry Davis
Austin, Thomas Howard
Baird, Merton Denison
Bakule, Ronald David
Barie, Walter Peter, Jr
Barnes, David Kennedy
Barr, John Tilman
Bauer, William, Jr
Baumann, Jacob Bruce
Bayer, Horst Otto
Beck, Paul Edward
Becker, Robert Hugh
Beckley, Ronald Scott
Beitchman, Burton David
Bell, Ian
Bell, Stanley C
Bender, Paul Elliot
Benfey, Otto Theodor
Benkovic, Stephen J
Bennett, Richard Bond
Benson, Barrett Wendell
Berges, David Alan
Bergo, Conrad Hunter
Berkheimer, Henry Edward
Berliner, Ernst
Berliner, Frances (Bondhus)
Bidlack, Verne Claude, Jr
Birnbaum, Hermann
Bissing, Donald Eugene
Blank, Benjamin
Blommers, Elizabeth Ann
Bluhm, Harold Frederick
Bock, Mark Gary
Bockrath, Bradley Charles
Bohen, Joseph Michael
Bolgiano, Nicholas Charles
Bolhofer, William Alfred
Bondinell, William Edward
Bortnick, Newman Mayer
Bosso, Joseph Frank
Bouis, Paul Andre
Bower, George Myron
Bowman, Robert Samuel
Brady, Stephen Francis
Braunstein, David Michael
Breazeale, Robert David
Brennan, Michael Edward
Brenner, Gerald Stanley
Brittelli, David Ross
Brockington, James Wallace

Brownell, George L
Browning, Daniel Dwight
Brutcher, Frederick Vincent, Jr
Buhle, Emmett Loren
Bulbenko, George Fedir
Burke, James David
Burns, H Donald
Cairns, Theodore L
Cammarata, Arthur
Cann, Michael Charles
Canter, Neil M
Capaldi, Eugene Carmen
Caputo, Joseph Anthony
Carlin, Robert Burnell
Carlson, Dana Peter
Carlson, Glenn Richard
Carter, Linda G
Caruso, Sebastian Charles
Casey, Adria Catala
Castle, John Edwards
Cawley, John Joseph
Cenci, Harry Joseph
Chakrabarti, Paritosh M
Chang, Charles Hung
Chang, Ching-Jen
Chang, Wen-Hsuan (Wayne)
Childress, Scott Julius
Chiola, Vincent
Chong, Berni Patricia
Chong, Joshua Anthony
Chow, Alfred Wen-Jen
Christie, Michael Allen
Christie, Peter Allan
Clapp, Charles H
Clemens, David Henry
Clovis, James S
Cohen, Gordon Mark
Cohen, Theodore
Condo, Albert Carman, Jr
Connor, James Edward, Jr
Conte, John Salvatore
Cook, Richard Sherrard
Coraor, George Robert
Cramer, Richard (David)
Crosby, Alan Hubert
Cunningham, Howard
Cupper, Robert Alton
Curran, Dennis Patrick
Dalton, Augustine Ivanhoe, Jr
Dalton, David Robert
Damle, Suresh B
Damodaran, Kalyani Muniratnam
Davie, William Raymond
Davis, Brian Clifton
Davis, Franklin A
DeLaMater, George (Bearse)
Deno, Norman C
Denoon, Clarence England, Jr
Dessauer, Rolf
De Tommaso, Gabriel Louis
DeWald, Horace Albert
De Witt, Hobson Dewey
Dickstein, Jack
Dimmig, Daniel Ashton
Dixon, Joseph Ardiff
Dohany, Julius Eugene
Donald, Dennis Scott
Dowbenko, Rostyslaw
Dowd, Paul
Dowd, Susan Ramseyer
Downs, James Joseph
Dreibelbis, John Adam
Dresden, Carlton F
Dressler, Hans
Dromgold, Luther D
Dunlap, Lawrence H
Dunn, George Lawrence
Durand, Marc L
Dymicky, Michael
Elbling, Irving Nelson
Ellis, Jeffrey Raymond
Emling, Bertin Leo
Emmons, William David
Engelhardt, Edward Louis
Erb, Dennis J
Erikson, J Alden
Ertelt, Henry Robinson
Evans, Ben Edward
Fainberg, Arnold Harold
Faltynek, Robert Allen
Farcasiu, Dan
Feairheller, Stephen Henry
Fearing, Ralph Burton
Feely, Wayne E
Feeman, James Frederic
Fehnel, Edward Adam
Feller, Robert Livingston
Felley, Donald Louis
Fenoglio, Richard Andrew
Ferren, Richard Anthony
Field, Nathan David
Finegold, Harold
Fishel, John B
Fitzgerald, James Allen
Flitter, David
Foglia, Thomas Anthony
Follweiler, Joanne Schaaf
Foltz, George Edward
Ford, Michael Edward
Francis, Peter Schuyler
Franko-Filipasic, Borivoj Richard Simon
Freed, Meier Ezra
Freedman, Robert Wagner
Freeman, James Harrison

Freidinger, Roger Merlin
Freyermuth, Harlan Benjamin
Friedman, Sidney
Fusco, Gabriel Carmine
Gadek, Frank Joseph
Gallagher, George Arthur
Gardner, David Milton
Gelb, Leonard Louis
Geoffroy, Gregory Lynn
Gerhardt, George William
Giuliano, Robert Michael
Givens, Edwin Neil
Glassick, Charles Etzweiler
Glover, George Irvin
Gluntz, Martin L
Godfrey, John Carl
Goldman, Theodore Daniel
Gormley, William Thomas
Greenberg, Herman Samuel
Grey, Roger Allen
Grezlak, John Henry
Griffith, Michael Grey
Grillot, Gerald Francis
Grinstein, Reuben H
Gwynn, Bernard Henry
Hager, Robert B
Haggard, Richard Allan
Hamel, Coleman Rodney
Hammons, James Hutchinson
Hanauer, Richard
Harbour, Robert Myron
Harrison, Aline Margaret
Harrison, Ernest Augustus, Jr
Hart, William Forris
Hartline, Richard
Hartman, John Alan
Hartman, Kenneth Eugene
Hartranft, George Robert
Hauptschein, Murray
Hausser, Jack W
Hedenburg, John Frederick
Heiberger, Philip
Heindel, Ned Duane
Heine, Harold Warren
Heller, Morgan Silliman
Henderson, Richard Elliott Lee
Hergert, Herbert L
Herman, Frederick Louis
Herskovitz, Thomas
Hertler, Walter Raymond
Hess, Ronald Eugene
Heys, John Richard
Hirsch, Stephen Simeon
Hirschmann, Ralph Franz
Hofferth, Burt Frederick
Hoffman, Jacob Matthew, Jr
Holden, Kenneth George
Holy, Norman Lee
Hoops, Stephen C
Hotelling, Eric Bell
Howard, Edgar, Jr
Huffman, Allan Murray
Humer, Philip Wilson
Hummer, James Knight
Hunter, Wood E
Hutchins, MaryGail Kinzer
Hutchins, Robert Owen, Sr
Hutton, Thomas Watkins
Hutzler, John R
Hwang, Bruce You-Huei
Ingram, Alvin Richard
Irwin, Philip George
Jackman, Lloyd Miles
Jacobson, Richard Martin
Johnson, Leroy Dennis
Johnson, Richard William
Johnson, Wayne Orrin
Johnston, Gordon Robert
Johnston, Jean Vance
Jones, James Holden
Jones, Jennings Hinch
Joullie, Madeleine M
Kaplan, Fred
Karo, Wolf
Kauer, James Charles
Kauffman, Joel Mervin
Kennedy, Flynt
Kennedy, Robert Michael
Kerst, A(l) Fred
Khan, Shabbir Ahmed
Khosah, Robinson Panganai
Kim, Jean Bartholomew
Kingsbury, William Dennis
Kirch, Lawrence S
Klapproth, William Jacob, Jr
Kleinman, Roberta Wilma
Koch, Walter Theodore
Koob, Robert Philip
Kopchik, Richard Michael
Kopple, Kenneth D(avid)
Kowalski, Conrad John
Kozikowski, Alan Paul
Kress, Bernard Hiram
Kronberger, Karlheinz
Krow, Grant Reese
Kruse, Lawrence Ivan
Kukla, Michael Joseph
Kulp, Stuart S
Labosky, Peter, Jr
LaCount, Robert Bruce
Laemmle, Joseph Thomas
Lange, Barry Clifford
Langeland, William Enberg
Larsen, John W

Larson, Gerald Louis
Lavoie, Alvin Charles
Leake, William Walter
Leber, Phyllis Ann
Leeper, Robert Walz
Lenox, Ronald Sheaffer
Leonard, John Joseph
Leston, Gerd
Lever, Cyril, Jr
Levesque, Charles Louis
Levine, Robert
Lieberman, Hillel
Lindsey, William B
Lipsitz, Paul
Livengood, Samuel Miller
Lopez, R C Gerald
Lovett, John Robert
Lugar, Richard Charles
Lumma, William Carl, Jr
Luskin, Leo Samuel
Lyons, James Edward
McAleer, William Joseph
McCarthy, James Francis
McCarthy, John Randolph
McGrath, Thomas Frederick
Machleder, Warren Harvey
McKeever, Charles H
McKelvey, Donald Richard
Madhav, Ram
Magill, Joseph Henry
Mallory, Clelia Wood
Mallory, Frank Bryant
Malter, Margaret Quinn
Manka, Dan P
Mantell, Gerald Jerome
Mao, Chung-Ling
Marmer, William Nelson
Maryanoff, Bruce Eliot
Maryanoff, Cynthia Anne Milewski
MASCIANTONIO, PHILIP
Mascioli, Rocco Lawrence
Maslak, Przemyslaw Boleslaw
Matyjaszewski, Krzysztof
Mendelson, Wilford Lee
Merner, Richard Raymond
Messer, Wayne Ronald
Middleton, William Joseph
Milakofsky, Louis
Miller, Thomas Gore
Minard, Robert David
Mokotoff, Michael
Monsimer, Harold Gene
Montgomery, Ronald Eugene
Mooney, David Samuel
Moore, Gordon George
Morse, Lewis David
Mortimer, Charles Edgar
Mount, Lloyd Gordon
Muck, Darrel Lee
Murphy, Clarence John
Musser, David Musselman
Myron, Thomas L(eo)
Naegele, Edward Wister, Jr
Naples, John Otto
Nedwick, John Joseph
Neidig, Howard Anthony
Neil, Gary Lawrence
Nelson, George Leonard
Nemec, Joseph William
Newirth, Terry L
Newitt, Edward James
Niederhauser, Warren Dexter
Noceti, Richard Paul
Nodiff, Edward Albert
Novak, Ronald William
Novello, Frederick Charles
Noyes, Paul R
Nutt, Ruth Foelsche
Nyi, Kayson
Obbink, Russell C
Ocone, Luke Ralph
Olofson, Roy Arne
Opie, Thomas Ranson
Orphanides, Gus George
Orr, Robert S
Osei-Gyimah, Peter
Owens, Frederick Hammann
Packman, Albert M
Papanikolaou, Nicholas E
Parish, Roger Cook
Parker, Winfred Evans
Patsiga, Robert A
Patterson, Dennis Ray
Paulson, Mark Clements
Peck, Richard Merle
Peffer, John Roscoe
Pelczar, Francis A
Pellegrini, John P, Jr
Perchonock, Carl David
Perry, Mary Hertzog
Peterson, James Oliver
Petronio, Marco
Piccolini, Richard John
Pinschmidt, Robert Krantz, Jr
Plant, William J
Pohland, Hermann W
Ponticello, Gerald S
Poos, George Ireland
Potter, Neil H
Precopio, Frank Mario
Press, Jeffery Bruce
Press, Linda Seghers
Preti, George
Price, Edward Hector

Quinn, Edwin John
Rabiger, Dorothy June
Ramachandran, Subramania
Rapp, Robert Dietrich
Rauch, Stewart Emmart, Jr
Reider, Malcolm John
Reifenberg, Gerald H
Reiff, Harry Elmer
Reingold, Iver David
Reitz, Allen Bernard
Reitz, Robert Rex
Remar, Joseph Francis
Remy, David Carroll
Renfrew, Edgar Earl
Resconich, Samuel
Reuwer, Joseph Francis, Jr
Reynolds, Brian Edgar
Richey, Herman Glenn, Jr
Roberts, Bryan Wilson
Robinson, Donald Nellis
Rogers, Ralph Loucks
Rogerson, Thomas Dean
Rosenthal, Rudolph
Ross, Stephen T
Rowe, Jay Elwood
Rowland, Alex Thomas
Rubin, Nathan
Rush, James E
Russell, Peter Byrom
Russey, William Edward
Russo, Thomas Joseph
Saari, Walfred Spencer
Saggiomo, Andrew Joseph
Samanen, James Martin
Sasin, Richard
Sattsangi, Prem Das
Sawicki, John Edward
Sawyer, David W(illiam)
Scala, Luciano Carlo
Schaeffer, Lee Allen
Scharpf, William George
Schauble, J Herman
Schearer, William Richard
Schiff, Paul L, Jr
Schneider, Henry Joseph
Scholnick, Frank
Schreiber, Kurt Clark
Schuster, Ingeborg I M
Schwing, Gregory Wayne
Segall, Stanley
Seybert, David Wayne
Shalit, Harold
Shaw, John Thomas
Sheeley, Richard Moats
Shepard, Kenneth LeRoy
Sidler, Jack D
Signorino, Charles Anthony
Silbert, Leonard Stanton
Sinotte, Louis Paul
Sittenfield, Marcus
Skell, Philip S
Smart, G N Russell
Smith, Francis Xavier
Smith, Graham Monro
Smith, Howard Leroy
Smith, James Stanley
Smith, Perrin Gary
Smith, Robert Lawrence
Smith, William Novis, Jr
Sollott, Gilbert Paul
Solodar, Warren E
Solomon, M Michael
Somkuti, George A
Sonnet, Philip E
Soriano, David S
Southwick, Philip Lee
Spangler, Martin Ord Lee
Spencer, Ralph Donald
Spindt, Roderick Sidney
Staiger, Roger Powell
Staley, Stuart Warner
Statton, Gary Lewis
Stedman, Robert John
Steiner, Russell Irwin
Stevens, Travis Edward
Stevenson, Robert William
Stewart, Thomas
Stine, William R
Stinson, Edgar Erwin
Stone, Herman
Straub, Thomas Stuart
Strohm, Paul F
Sundeen, Joseph Edward
Suter, Stuart Ross
Swamer, Frederic Wurl
Swift, Graham
Takeshita, Tsuneichi
Tatum, Charles Maris
Tekel, Ralph
Teller, Daniel Myron
Thoman, Charles James
Thompson, Wayne Julius
Thornton, Edward Ralph
Thornton, Elizabeth K
Tomezsko, Edward Stephen John
Touchstone, Joseph Cary
Trimitsis, George B
Tse, Rose (Lou)
Turner, Andrew B
Turner, Robert Lawrence
Tzodikov, Nathan Robert
Ullyot, Glenn Edgar
Vanderwerff, William D
Van Dyke, Charles H

Van Gemert, Barry
Van Horn, Ruth Warner
Varga, Charles E
Varkey, Thankamma Eapen
Varma, Asha
Vaux, James Edward, Jr
Veber, Daniel Frank
Vernon, John Ashbridge
Vick, Gerald Kieth
Vickers, Stanley
Vladutz, George E
Vogel, Martin
Waddell, Walter Harvey
Wade, Charles Gary
Wade, Peter Allen
Wagner, Charles Roe
Wagner, Robert Edwin
Walker, Ruth Angelina
Waller, Francis Joseph
Walsh, Edward Joseph, Jr
Walters, Lee Rudyard
Ward, Richard Bernard
Warfel, David Ross
Warrick, Percy, Jr
Washburne, Stephen Shepard
Watterson, Kenneth Franklin
Weaver, Leo James
Weese, Richard Henry
Wei, Yen
Weiler, Ernest Dieter
Weinreb, Steven Martin
Weinstock, Joseph
Weiss, James Allyn
Weldes, Helmut H
Welsh, David Albert
Wempe, Lawrence Kyran
Wender, Irving
Werner, Ervin Robert, Jr
Westley, John William
White, Harry Joseph
Whitney, Joel Gayton
Wilkins, Cletus Walter, Jr
Wilkins, Raymond Leslie
Williams, Donald Robert
Williams, John Roderick
Williamson, Hugh A
Wineholt, Robert Leese
Winstead, Meldrum Barnett
Wintner, Claude Edward
Witkoski, Francis Clement
Witschard, Gilbert
Woodward, David Willcox
Wu, Ching-Yong
Wuchter, Richard B
Wunz, Paul Richard, Jr
Yeager, Sandra Ann
Yost, John Franklin
Younes, Usama E
Young, Thomas Edwin
Zacharias, David Edward
Zajacek, John George
Zanger, Murray
Zell, Howard Charles
Zellner, Carl Naeher
Zook, Harry David

RHODE ISLAND
Abell, Paul Irving
Cheer, Clair James
Clapp, Leallyn Burr
Di Pippo, Ascanio G
Forbes, Malcolm Holloway
Galkowski, Theodore Thaddeus
Goodman, Leon
Kroll, Harry
Laferriere, Arthur L
Lawler, Ronald George
MacKay, Francis Patrick
Magyar, Elaine Stedman
Magyar, James George
Morris, David Julian
Moyerman, Robert Max
Nace, Harold Russ
Ouderkirk, John Thomas
Parker, Kathlyn Ann
Perry, Edward Mahlon
Quinn, James Gerard
Rerick, Mark Newton
Rieger, Anne Lloyd
Rosen, William M
Saltzman, Martin D
Slater, Schuyler G
Stokes, William Moore
Suggs, John William
Tien, Rex Yuan
Turcotte, Joseph George
Von Riesen, Daniel Dean
Williams, John Collins, Jr
Williard, Paul Gregory

SOUTH CAROLINA
Abramovitch, Rudolph Abraham Haim
Bailey, Roy Horton, Jr
Ball, Frank Jervery
Ballentine, Alva Ray
Barkley, Lloyd Blair
Beam, Charles Fitzhugh, Jr
Berger, Richard S
Bishop, Muriel Boyd
Bliss, Arthur Dean
Bly, Robert Stewart
Buurman, Clarence Harold
Cantrill, James Egbert
Carter, Kenneth Nolon

Cavin, William Pinckney
Cogswell, George Wallace
Crounse, Nathan Norman
DeBrunner, Ralph Edward
Desai, Vinodrai Ranchhodji
Dieter, Richard Karl
Eastes, Frank Elisha
Evans, David Wesley
Falkehag, S Ingemar
Farmer, Larry Bert
Farrar, Grover Louis
Gillespie, Robert Howard
Glymph, Eakin Milton
Goldstein, Herman Bernard
Griffith, Elizabeth Ann Hall
Groce, William Henry (Bill)
Hardwicke, James Ernest, Jr
Hartz, Roy Eugene
Henderson, Richard Wayne
Hill, Arthur Joseph, Jr
Holsten, John Robert
Huffman, John William, Jr
Hutter, George Frederick
Hynes, John Barry
Jeremias, Charles George
Jones, Edward Stephen
Kelly, Robert James
Knapp, Daniel Roger
Knowles, Cecil Martin
Koli, Andrew Kaitan
Krantz, Karl Walter
Kubler, Donald Gene
Kuhn, Hans Heinrich
Latta, Bruce McKee
Leopold, Robert Summers
Machell, Greville
Maricq, John
Marshall, James Arthur
Martin, Tellis Alexander
Marullo, Nicasio Philip
Miley, John Wulbern
Odom, Homer Clyde, Jr
Peterson, Paul E
Porter, John J
Porter, Rick Anthony
Posey, Robert Giles
Rawalay, Surjan Singh
Robinson, Myrtle Tonne
Robinson, Robert Earl
Ropp, Gus Anderson
Ross, Stanley Elijah
Rothrock, George Moore
Rowlett, Russell Johnston, Jr
Sanderfer, Paul Otis
Schaefer, Frederic Charles
Schneider, William Paul
Scruggs, Jack G
Shah, Hamish V
Shalaby, Shalaby W
Sheehan, William C
Sick, Lowell Victor
Sorell, Henry P
Teague, Peyton Clark
Theuer, William John
Tour, James Mitchell
Von Rosenberg, Joseph Leslie, Jr
Wagner, William Sherwood
Wearn, Richard Benjamin
Weinstock, Leonard M
Weiss, James Owen
Weiss, Michael Karl
Wishman, Marvin

SOUTH DAKOTA
Hanson, Milton Paul
Kintner, Robert Roy
Scott, George Prescott
Shryock, Gerald Duane
Stoner, Marshall Robert
Wadsworth, William Steele, Jr

TENNESSEE
Adcock, James Luther
Alexandratos, Spiro
Anjaneyulu, P S R
Bahner, Carl Tabb
Baillargeon, Victor Paul
Ball, Frances Louise
Barton, Kenneth Ray
Benjamin, Ben Monte
Benton, Charles Herbert
Bentz, Ralph Wagner
Boone, James Ronald
Bowman, Newell Stedman
Buchman, Russell
Bucovaz, Edsel Tony
Caflisch, Edward George
Chitwood, James Leroy
Clemens, Robert Jay
Cleveland, James Perry
Cliffton, Michael Duane
Collins, Clair Joseph
Collins, Jerry Dale
Commerford, John D
Coover, Harry Wesley, Jr
Dale, John Irvin, III
Dean, Walter Lee
Dombroski, John Richard
Eastham, Jerome Fields
Elliott, Irvin Wesley
Fagerburg, David Richard
Fenyes, Joseph Gabriel Egon
Field, Lamar
Finch, Gaylord Kirkwood

Organic Chemistry (cont)

Foster, Charles Howard
Gagen, James Edwin
Galbraith, Harry Wilson
Germroth, Ted Calvin
Gilliom, Richard D
Gilow, Helmuth Martin
Gleason, Edward Hinsdale, Jr
Gloyer, Stewart Edward
Griffin, Guy David
Griscom, Richard William
Gross, Benjamin Harrison
Hanley, Wayne Stewart
Harding, Charles Enoch
Harris, Thomas Munson
Hasek, Robert Hall
Hayes, Robert M
Hess, Bernard Andes, Jr
Ho, Patience Ching-Ru
Hoffman, Kenneth Wayne
Holmes, Jerry Dell
Hudnall, Phillip Montgomery
Hutchinson, James Herbert, Jr
Hyatt, John Anthony
Irick, Gether, Jr
Jenkins, James William
Jones, Glenn Clark
Kabalka, George Walter
Kennedy, Robert Wilson
Kirchner, Frederick Karl
Knee, Terence Edward Creasey
Kreh, Donald Willard
Krutak, James John, Sr
Kuo, Chung-Ming
Lane, Charles A
Langford, Paul Brooks
Lasslo, Andrew
Lowe, James N
Lowe, James Urban, II
Lu, Mary Kwang-Ruey Chao
Lura, Richard Dean
McCarthy, John F
McCombs, Charles Allan
McConnell, Richard Leon
McDaniel, Edgar Lamar, Jr
Magid, Linda Lee Jenny
Magid, Ronald
Mani, Rama I
Marnett, Lawrence Joseph
Martin, James Cullen
Martin, James Cuthbert
Meen, Ronald Hugh
Miller, James L
Miller, Robert Witherspoon
Nagel, Fritz John
Nealy, David Lewis
Newland, Gordon Clay
O'Connor, Timothy Edmond
Pagni, Richard
Pal, Bimal Chandra
Parish, Harlie Albert, Jr
Pearson, Donald Emanual
Penner, Hellmut Philip
Pera, John Dominic
Peterson, William Roger
Pond, David Martin
Poutsma, Marvin Lloyd
Puerckhauer, Gerhard Wilhelm Richard
Raaen, Vernon F
Rash, Fred Howard
Robinson, Charles Nelson
Rutenberg, Aaron Charles
Sachleben, Richard Alan
Schreiber, Eric Christian
Shah, Jitendra J
Shirley, David Allen
Smith, Howard E
Snell, Robert L
Solomons, William Ebenezer
Spadafino, Leonard Peter
Stites, Joseph Gant, Jr
Sublett, Bobby Jones
Sweetman, Brian Jack
Swindell, Robert Thomas
Tarbell, Dean Stanley
Thiessen, William Ernest
Thweatt, John G
Todd, Peter Justin
Tuleen, David L
Tyczkowski, Edward Albert
Vachon, Raymond Normand
Van Sickle, Dale Elbert
Volpe, Angelo Anthony
Waddell, Thomas Groth
Wang, Richard Hsu-Shien
Warren, Mitchum Ellison, Jr
Watts, Exum DeVer
Webb, James L A
Wescott, Lyle DuMond, Jr
Wicker, Thomas Hamilton, Jr
Wilkin, Louis Alden
Willcott, Mark Robert, III
Witzeman, Jonathan Stewart
Wyse, B O

TEXAS

Adam, Klaus
Adamcik, Joe Alfred
Adams, Leon Milton
Albach, Roger Fred
Allen, Robert Paul
Allen, Robert Ray
Ansari, Guhlam Ahmad Shakeel

Applewhite, Thomas Hood
Arganbright, Robert Philip
Ashton, Joseph Benjamin
Aspelin, Gary B(ertil)
Aufdermarsh, Carl Albert, Jr
Bailey, Philip Sigmon
Bartsch, Richard Allen
Bauer, Ronald Sherman
Bauld, Nathan Louis
Beede, Charles Herbert
Beier, Ross Carlton
Belew, John Seymour
Bennett, Robert Putnam
Benson, Royal H
Bergbreiter, David Edward
Besozzi, Alfio Joseph
Bhatia, Kishan
Biehl, Edward Robert
Billups, W Edward
Birney, David Martin
Borchardt, John Keith
Bost, Robert Orion
Brader, Walter Howe, Jr
Brady, William Thomas
Brandenberger, Stanley George
Burke, Roger E
Bush, Warren Van Ness
Cabanes, William Ralph, Jr
Cameron, Margaret Davis
Carlton, Donald Morrill
Castrillon, Jose P A
Caswell, Lyman Ray
Cate, Rodney Lee
Cates, Lindley A
Cavender, James Vere, Jr
Ciufolini, Marco A
Cobb, Raymond Lynn
Cogdell, Thomas James
Cole, Larry Lee
Cole, Walter Earl
Condray, Ben Rogers
Cook, Paul Fabyan
Cox, James Reed, Jr
Cragoe, Edward Jethro, Jr
Crawford, James Dalton
Crawford, James Worthington
Croce, Louis J
Croft, Thomas Stone
Cross, John Parson
Cupples, Barrett L(eMoyne)
Cuscurida, Michael
Cywinski, Norbert Francis
Darensbourg, Donald Jude
Darensbourg, Marcetta York
Darlage, Larry James
Dawes, John Leslie
Dear, Robert E A
De La Mare, Harold Elison
Denekas, Milton Oliver
Dill, Charles William
Dillard, Robert Garing, Jr
Dimitroff, Edward
Doyle, Michael P
Dunn, Danny Leroy
Dusenbery, Ruth Lillian
Duty, Robert C
Dvoretzky, Isaac
Dwyer, Lawrence Arthur
Edmondson, Morris Stephen
Ekerdt, John Gilbert
Elliott, J Lell
Ellison, Robert Hardy
Engel, Paul Sanford
Etter, Raymond Lewis, Jr
Everse, Johannes
Fain, Robert C
Fan, Joyce Wang
Farrissey, William Joseph, Jr
Fawcett, Colvin Peter
Ferraris, John Patrick
Fitch, John William, III
Floyd, Joseph Calvin
Floyd, Willis Waldo
Fodor, George E
Fonken, Gerhard Joseph
Ford, George Peter
Fox, Marye Anne
Francis, William Connett
Franks, Neal Edward
Franzl, Robert E
Fuchs, Richard
Fukuyama, Tohru
Garrett, James M
Gerlach, John Louis
Giam, Choo-Seng
Gilbert, John Carl
Goldstein, Herbert Jay
Gordon, Wayne Lecky
Graves, Robert Earl
Gryting, Harold Julian
Guerrant, William Barnett, Jr
Guidry, Carlton Levon
Gum, Wilson Franklin, Jr
Gutsche, Carl David
Gwynn, Donald Eugene
Ham, George Edward
Hamilton, Janet V
Harding, Kenn E
Harlan, Horace David
Haynes, George Rufus
Headley, Allan Dave
Hedrich, Loren Wesley
Henrichs, Paul Mark
Herndon, William Cecil

Hickner, Richard Allan
Hobbs, Charles Clifton, Jr
Hoblit, Louis Douglas
Hogg, John Leslie
Holst, Edward Harland
Horeczy, Joseph Thomas
Houston, James Grey
Huang, Charles T L
Hunt, Richard Henry
Hurdis, Everett Cushing
Idoux, John Paul
Imhoff, Michael Andrew
Ingram, Sammy Walker, Jr
Isbell, Arthur Furman
Jacobson, Barry Martin
Jaszberenyi, Joseph C
Jeskey, Harold Alfred
Johnson, Fred Lowery, Jr
Johnson, James Elver
Johnson, Malcolm Pratt
Journeay, Glen Eugene
Kamego, Albert Amil
Kaye, Howard
Keith, Lawrence H
Kerr, Ralph Oliver
Kirk, James Curtis
Kmiecik, James Edward
Knapp, Roger Dale
Kochi, Jay Kazuo
Koons, Charles Bruce
Kraychy, Stephen
Kyba, Evan Peter
Lakshmanan, P R
Landolt, Robert George
Larkin, John Michael
Leftin, Harry Paul
Levitt, Leonard Sidney
Levy, Leon Bruce
Lewis, Edward Sheldon
Liehr, Joachim G
Lin, Tz-Hong
Lin-Vien, Daimay
Lloyd, Winston Dale
Lonzetta, Charles Michael
Lowrey, Charles Boyce
Lundeen, Allan Jay
Lyle, Gloria Gilbert
Lyle, Robert Edward, Jr
Mabry, Tom Joe
McCollum, Anthony Wayne
McCown, Joseph Dana
McCoy, David Ross
McCullough, James Douglas, Jr
McCullough, Thomas F
McGirk, Richard Heath
Makin, Earle Clement, Jr
Mango, Frank Donald
Mangold, Donald J
Manning, Harold Edwin
Marchand, Alan Philip
Marquis, Edward Thomas
Marshall, James Lawrence
Martin, Charles William
Martin, Stephen Frederick
Marx, John Norbert
Mason, Perry Shipley, Jr
Massingill, John Lee, Jr
May, James Aubrey, Jr
Meinschein, Warren G
Melville, Marjorie Harris
Mendez, Victor Manuel
Miller, Emery B
Miller, James Richard
Mills, Nancy Stewart
Moltzan, Herbert John
Monti, Stephen Arion
Muhs, Merrill Arthur
Musser, James Tuttle
Naae, Douglas Gene
Nachman, Ronald James
Nagyvary, Joseph
Nash, William Donald
Naylor, Carter Graham
Nelson, Don B
Newcomb, Martin
Newsom, Raymond A
Newton, Robert Andrew
Niu, Joseph H Y
Oakes, Billy Dean
O'Brien, Daniel H
O'Farrell, Charles Patrick
O'Neal, Hubert Ronald
Otken, Charles Clay
Padgett, Algie Ross
Patil, Kashinath Ziparu
Patton, Leo Wesley
Patton, Tad LeMarre
Pelley, Ralph L
Pinkus, A(lbin) G(eorge)
Piziak, Veronica Kelly
Plank, Don Allen
Plummer, Benjamin Frank
Poe, Richard D
Pomerantz, Martin
Proops, William Robert
Pullig, Tillman R(upert)
Ramanujam, V M Sadagopa
Rao, Pemmaraju Narasimha
Razniak, Stephen L
Reeder, Charles Edgar
Reeves, W Preston
Reid, James Cutler
Reinecke, Manfred Gordon
Rhodes, Robert Carl

Richter, Reinhard Hans
Rigdon, Orville Wayne
Roberts, Royston Murphy
Roberts, Thomas David
Robinson, Alfred Green
Robinson, J(ames) Michael
Roehrig, Gerald Ralph
Rosenquist, Edward P
Rowton, Richard Lee
Russell, Thomas Webb
Sample, Thomas Earl, Jr
Sand, Ralph E
Schimelpfenig, Clarence William
Schlameus, Herman Wade
Schnizer, Arthur Wallace
Schueler, Bruno Otto Gottfried
Scott, Alastair Ian
Shine, Henry Joseph
Shore, Fred L
Shriner, Ralph Lloyd
Shumate, Kenneth McClellan
Simpson, Billy Doyle
Singleton, David Michael
Skarlos, Leonidas
Skinner, Charles Gordon
Smith, Curtis William
Smith, Leland Leroy
Smith, William Burton
Smutny, Edgar Josef
Snapp, Thomas Carter, Jr
Soltes, Edward John
Sonntag, Norman Oscar Victor
Sonntag, Roy Windham
Soulen, Robert Lewis
Speranza, George Phillip
Steinhardt, Charles Kendall
Stephenson, Danny Lon
Strom, E(dwin) Thomas
Summers, William Allen, Jr
Sund, Eldon H
Supple, Jerome Henry
Svetlik, Joseph Frank
Tanner, Alan Roger
Ternay, Andrew Louis, Jr
Thyagarajan, B S
Trotter, John Wayne
Tweedie, Virgil Lee
Unruh, Jerry Dean
Vanderbilt, Jeffrey James
Van Dijk, Christiaan Pieter
Varma, Rajender S
Vasilakos, Nicholas Petrou
Venier, Clifford George
Vogelfanger, Elliot Aaron
Wagner, Frank S, Jr
Walbrick, Johnny Mac
Wang, Ting-Tai Helen
Washington, Arthur Clover
Wentland, Stephen Henry
Wheeler, Edward Norwood
Whitesell, James Keller
Whittle, John Antony
Williams, Charles Herbert
Winfrey, J C
Witt, Enrique Roberto
Wood, Randall Dudley
Woodyard, James Douglas
Workman, Wesley Ray
Wulfers, Thomas Frederick
Yager, Billy Joe
Yarian, Dean Robert
Yeakey, Ernest Leon
Zlatkis, Albert

UTAH

Aldous, Duane Leo
Allred, Evan Leigh
Alvord, Donald C
Beishline, Robert Raymond
Bentrude, Wesley George
Blackham, Angus Udell
Bradshaw, Jerald Sherwin
Broadbent, Hyrum Smith
Broom, Arthur Davis
Dehm, Henry Christopher
Elmslie, James Stewart
Epstein, William Warren
Foltz, Rodger L
Gardner, Pete D
Gladysz, John A
Hall, David Warren
Hawkins, Richard Thomas
Hinshaw, Jerald Clyde
Horton, Walter James
Hunter, Byron Alexander
Johnson, John Hal
Johnson, LaVell R
Nelson, Kay LeRoi
Olsen, Richard Kenneth
Paul, Edward Gray
Phillips, Lee Revell
Reynolds, Richard Johnson
Robins, Morris Joseph
Smith, Grant Gill
Stang, Peter John
Sudweeks, Walter Bentley
Thompson, Grant
Van Orden, Harris O
Walling, Cheves (Thomson)

VERMONT

Bushweller, Charles Hackett
Chamberlain, Malcolm
Gleason, Robert Willard

Harnest, Grant Hopkins
Jewett, John Gibson
Krapcho, Andrew Paul
Kuehne, Martin Eric
McCormack, John Joseph, Jr
Magnien, Ernest
Murray, James Gordon
Shearer, Charles M
White, William North
Woodworth, Robert Cummings

VIRGINIA
Abraham, Donald James
Atkins, Robert Charles
Augl, Joseph Michael
Baedecker, Mary Jo
Bass, Robert Gerald
Bell, Charles E, Jr
Bell, Harold Morton
Bertram, Leon Leroy
Beyad, Mohammed Hossain
Bhatnagar, Ajay Sahai
Boatman, Sandra
Breder, Charles Vincent
Burke, Hanna Suss
Burton, Willard White
Carey, Francis Arthur
Castagnoli, Neal, Jr
Chlebowski, Jan F
Chong, Shuang-Ling
Clark, Allen Keith
Clough, Stuart Chandler
Cox, Richard Harvey
Crowell, Thomas Irving
Dahlgard, Muriel Genevieve
Dardoufas, Kimon C
DeBardeleben, John F
De Camp, Wilson Hamilton
Deinet, Adolph Joseph
DeVries, George H
Dillard, John Gammons
Durrell, William S
Eareckson, William Milton, III
Easter, Donald Philips
Eby, Charles J
Edwards, William Brundige, III
Einolf, William Noel
Ellis, Leonard Culberth
Fishbein, Lawrence
Fisher, Charles Harold
Fletcher, Harry Huntington
Fonong, Tekum
Gager, Forrest Lee, Jr
Gal, Andrew Eugene
Gamble, Dean Franklin
Gerow, Clare William
Gibson, Gerald W
Gibson, Harry William
Gillespie, Jesse Samuel, Jr
Goller, Edwin John
Gould, David Huntington
Gratz, Roy Fred
Grivsky, Eugene Michael
Haas, Carol Kressler
Hall, Philip Layton
Hammer, Gary G
Hammond, George Simms
Hansrote, Charles Johnson, Jr
Harowitz, Charles Lichtenberg
Hartzler, Jon David
Harvey, William Ross
Hassler, William Woods
Hausermann, Max
Hazlett, Robert Neil
Heisey, Lowell Vernon
Henson, Paul D
Heuberger, Oscar
Hill, Carl McClellan
Hill, Trevor Bruce
Hoffman, Henry Allen, Jr
Hudlicky, Milos
Hudlicky, Tomas
Hudson, Frederick Mitchell
Hunt, Donald F
Ireland, Robert Ellsworth
Jelinek, Charles Frank
Jensen, Arnold William
Johnson, Fatima Nunes
Johnson, John Enoch
Jonas, John Joseph
Kallianos, Andrew George
Kauffman, Glenn Monroe
Kingston, David George Ian
Klingensmith, George Bruce
Kondo, Norman Shigeru
Konizer, George Burr
Kumari, Durga
Lambert, Rogers Franklin
Leake, Preston Hildebrand
Lefebvre, Yvon
Link, William B
Lipnick, Robert Louis
McClenon, John R
Macdonald, Timothy Lee
McEntee, Thomas Edwin
McGrath, James Edward
Majewski, Theodore E
Mateer, Richard Austin
Matuszko, Anthony Joseph
Melton, Thomas Mason
Mengenhauser, James Vernon
Merkel, Timothy Franklin
Meyer, Leo Francis
Monroe, Stuart Benton

Moore, Theron Langford
Moroni, Eneo C
Narasimhachari, Nedathur
O'Brien, Michael Harvey
Ogilvie, James William, Jr
Ogliaruso, Michael Anthony
Ottenbrite, Raphael Martin
Patrick, James Burns
Powell, Justin Christopher
Rainer, Norman Barry
Ramirez, Fausto
Roberts, David Craig
Rodig, Oscar Rudolf
Roscher, David Moore
Rosenberg, Murray David
Ross, Alexander
Roth, Ronald John
Rubin, Alan Barry
Sahli, Muhammad S
Sayala, Chhaya
Schiavelli, Melvyn David
Schirch, Laverne Gene
Schulz, Johann Christoph Friedrich
Seligman, Robert Bernard
Shanklin, James Robert, Jr
Shapiro, Robert Howard
Shen, Tsung Ying
Shillington, James Keith
Slayden, Suzanne Weems
Smith, Craig La Salle
Smith, James Doyle
Southwick, Everett West
Stalick, Wayne Myron
Starkovsky, Nicolas Alexis
Starnes, William Herbert, Jr
Stewart, Roberta A
Strong, Jerry Glenn
Stubbins, James Fiske
Stubblefield, Frank Milton
Sundberg, Richard J
Taylor, William Irving
Teng, Lina Chen
Thompson, David Wallace
Turer, Jack
Urbanik, Arthur Ronald
Uwaydah, Ibrahim Musa
Verell, Ruth Ann
Walker, Grayson Watkins
Wasti, Khizar
Wayland, Rosser Lee, Jr
Wetmore, Stanley Irwin, Jr
Whitney, George Stephen
Wilkinson, William Kenneth
Willett, James Delos
Williams, Roy Lee
Winters, Lawrence Joseph
Wolfe, James F
Woo, Yin-tak
Yost, William Lassiter

WASHINGTON
Alberts, Arnold A
Allan, George Graham
Andersen, Niels Hjorth
Anderson, Arthur G, Jr
Anderson, Charles Dean
Babad, Harry
Baillie, Thomas Allan
Barrueto, Richard Benigno
Beelik, Andrew
Beug, Michael William
Blake, James J
Bocksch, Robert Donald
Borden, Weston Thatcher
Briggs, William Scott
Chambers, James Richard
Conca, Romeo John
Cooke, Manning Patrick, Jr
Dann, John Robert
Fletcher, Thomas Lloyd
Frankart, William A
Franz, James Alan
Giddings, William Paul
Godar, Edith Marie
Goheen, David Wade
Hedges, John Ivan
Herrick, Franklin Willard
Hine, John Maynard
Hoblitt, Richard Patrick
Holzman, George
Hopkins, Paul Brink
Huestis, Laurence Dean
Jacoby, Lawrence John
Johnson, Carl Arnold
Johnson, Donald Curtis
Kreibich, Roland
Kriz, George Stanley
Lago, James
Lampman, Gary Marshall
Latourette, Harold Kenneth
Leonard, John Edward
Lepse, Paul Arnold
McDonough, Leslie Marvin
McMichael, Kirk Dugald
Matteson, Donald Stephen
Mayer, Julian Richard
Meyer, Rich Bakke, Jr
Mulvey, Dennis Michael
Nelson, Randall Bruce
Orth, George Otto, Jr
Patt, Leonard Merton
Pavia, Donald Lee
Pocker, Yeshayau
Raucher, Stanley

Read, David Hadley
Reintjes, Marten
Ronald, Robert Charles
Rowland, Stanley Paul
Schubert, Wolfgang Manfred
Sears, Karl David
Senear, Allen Eugene
Stacy, Gardner W
Steckler, Bernard Michael
Stout, Virginia Falk
Trager, William Frank
Varanasi, Usha
Vessel, Eugene David
Wade, Leroy Grover, Jr
Walecka, Jerrold Alberts
Warner, John Scott
Wasserman, William Jack
Wilbur, Daniel Scott
Wither, Ross Plummer
Wollwage, Paul Carl
Wong, Chun-Ming
Young, Robert Hayward
Zimmerman, Gary Alan

WEST VIRGINIA
Anderson, Gary Don
Barnes, Robert Keith
Bartley, William J
Campbell, Clyde Del
Capstack, Ernest
Chadwick, David Henry
Digman, Robert V
Douglass, James Edward
Doumaux, Arthur Roy, Jr
Draper, John Daniel
Driesch, Albert John
Dunphy, James Francis
Elkins, John Rush
Fodor, Gabor
Frostick, Frederick Charles, Jr
Giza, Chester Anthony
Giza, Yueh-Hua Chen
Harrison, Arnold Myron
Henry, Joseph Peter
Hess, Lawrence George
Hubbard, John Lewis
Hussey, Edward Walter
Johnson, George Frederick
Johnson, Richard Lawrence
Kaplan, Leonard
Knopf, Robert John
Knowles, Richard N
Krabacher, Bernard
Kurland, Jonathan Joshua
Lamey, Steven Charles
Leckonby, Roy Alan
Lemke, Thomas Franklin
McCain, James Herndon
MacDowell, Denis W H
MacPeek, Donald Lester
Markiw, Roman Teodor
Matthews, Virgil Edison
Minckler, Leon Sherwood, Jr
Moore, William Robert
Muth, Chester William
Ode, Richard Herman
Osborn, Claiborn Lee
Papa, Anthony Joseph
Peascoe, Warren Joseph
Pianfetti, John Andrew
Rankin, Gary O'Neal
Ream, Bernard Claude
Robson, John Howard
Ruoff, William (David)
Sandridge, Robert Lee
Schilling, Curtis Louis, Jr
Sherman, Paul Dwight, Jr
Smith, James Allbee
Speck, Rhoads McClellan
Steinle, Edmund Charles, Jr
Volker, Eugene Jeno
Walker, Wellington Epler
Webber, Thomas Gray

WISCONSIN
Anderson, Stephen William
Baetke, Edward A
Bender, Margaret McLean
Biester, John Louis
Boggs, Lawrence Allen
Boye, Frederick C
Brown, Kenneth Howard
Brown, William Henry
Burke, Steven Douglas
Calvanico, Nickolas Joseph
Casey, Charles P
Chenier, Philip John
Chitharanjan, D
Cleary, James William
Conigliaro, Peter James
Cremer, Sheldon E
Crimmins, Timothy Francis
Dalrymple, David Lawrence
Dickerson, Charlesworth Lee
Doerr, Robert George
D'Orazio, Vincent T
Dwyer, Sean G
Dwyer-Hallquist, Patricia
Farnsworth, Carl Leon
Feist, William Charles
Goering, Harlan Lowell
Greene, Charles Edwin
Hamm, Kenneth Lee
Harkin, John McLay

Hart, Phillip A
Hedden, Gregory Dexter
Helms, John F
Hill, Elgin Alexander
Hoffman, Norman Edwin
Hollenberg, David Henry
Horton, Joseph William
Isenberg, Norbert
Johnson, Hal G(ustav)
Kester, Dennis Earl
Kinstle, James Francis
Klink, Joel Richard
Klopotek, David L
Klundt, Irwin Lee
Kolb, Vera
Kovacic, Peter
Landucci, Lawrence L
Lee, Kathryn Adele Bunding
Leekley, Robert Mitchell
Lin, Stephen Y
Lokensgard, Jerrold Paul
Losin, Edward Thomas
McClenahan, William St Clair
McKelvey, Ronald Deane
Magnuson, Eugene Robert
Manly, Donald G
Mathews, Frederick John
Megahed, Sid A
Moore, Leonard Oro
Nelsen, Stephen Flanders
Nitz, Otto William Julius
Ochrymowycz, Leo Arthur
Olson, Melvin Martin
Pearl, Irwin Albert
Perlman, Kato (Katherine) Lenard
Post, Elroy Wayne
Posvic, Harvey Walter
Priest, Matthew A
Puhl, Richard James
Raines, Ronald T
Rainville, David Paul
Rausch, Gerald
Reich, Hans Jurgen
Reich, Ieva Lazdins
Reichenbacher, Paul H
Roberts, Richard W
Roth, Marie M
Rowe, John Westel
Savereide, Thomas J
Scamehorn, Richard Guy
Schnack, Larry G
Schnoes, Heinrich Konstantin
Schroeder, Leland Roy
Shaver, Roy Allen
Soerens, Dave Allen
Sosnovsky, George
Splies, Robert Glenn
Stackman, Robert W
Stevens, Michael Fred
Theine, Alice
Thompson, Norman Storm
Thurmaier, Roland Joseph
Tonnis, John A
Tsao, Francis Hsiang-Chian
Turner, Robert James
Vedejs, Edwin
Ward, Kyle, Jr
Weipert, Eugene Allen
Wendland, Ray Theodore
Westler, William Milo
Whyte, Donald Edward
Wilds, Alfred Lawrence
Zimmerman, Howard Elliot
Zinkel, Duane Forst

WYOMING
Coates, Geoffrey Edward
Decora, Andrew Wayne
Dorrence, Samuel Michael
Duvall, John Joseph
Jaeger, David Allen
Kelly, Floyd W, Jr
McDonald, Francis Raymond
Maurer, John Edward
Nelson, David Alan
Northcott, Jean
Petersen, Joseph Claine
Raulins, Nancy Rebecca
Rhoads, Sara Jane
Robertson, Raymond E(liot)
Seese, William Shober
Speight, James G

PUERTO RICO
Carrasquillo, Arnaldo
Eberhardt, Manfred Karl
Gonzalez De Alvarez, Genoveva
Lamba, Ram Sarup
Souto Bachiller, Fernando Alberto

ALBERTA
Ayer, William Alfred
Bachelor, Frank William
Blackburn, Edward Victor
Brown, Robert Stanley
Cook, David Alastair
Coutts, Ronald Thomson
Crawford, Robert James
Creighton, Stephen Mark
Dixon, Elisabeth Ann
Elofson, Richard Macleod
Holmes, R H Lavergne
Hooz, John
Jack, Thomas Richard

Organic Chemistry (cont)

Kopecky, Karl Rudolph
Lemieux, Raymond Urgel
Lown, James William
Micetich, Ronald George
Moschopedis, Speros E
Robertson, Ross Elmore
Sorensen, Theodore Strang
Tanner, Dennis David
Tavares, Donald Francis
Vance, Dennis E
Williams, Jack L R

BRITISH COLUMBIA
Antia, Naval Jamshedji
Ashford, Walter Rutledge
Bragg, Philip Dell
Chalmers, William
Chow, Yuan Lang
Cox, Lionel Audley
Dutton, Guy Gordon Studdy
Freter, Kurt Rudolf
Gardner, Joseph Arthur Frederick
Hayward, Lloyl Douglas
Hocking, Martin Blake
Howard, John
Kiehlmann, Eberhard
Kutney, James Peter
Manville, John Fieve
Mitchell, Reginald Harry
Pincock, Richard Earl
Rosenthal, Alex
Sams, John Robert, Jr
Scheffer, John R
Singh, Raj Kumari
Slessor, Keith Norman
Stewart, Ross
Struble, Dean L
Tener, Gordon Malcolm
Weiler, Lawrence Stanley
Withers, Stephen George

MANITOBA
Charlton, James Leslie
Chow, Donna Arlene
Hunter, Norman Robert
Janzen, Alexander Frank
McKinnon, David M
Wong, Chiu Ming

NEW BRUNSWICK
Adams, Kenneth Allen Harry
Barclay, Lawrence Ross Coates
Findlay, John A
Jankowski, Christopher K
Kelly, Ronald Burger
Strunz, G(eorge) M(artin)
Valenta, Zdenek

NEWFOUNDLAND
Anderson, Hugh John
Orr, James Cameron
Shahidi, Fereidoon
Stein, Allan Rudolph

NOVA SCOTIA
Ackman, Robert George
Arnold, Donald Robert
Arseneau, Donald Francis
Aue, Walter Alois
Bunbury, David Leslie
Grossert, James Stuart
Hooper, Donald Lloyd
Leffek, Kenneth Thomas
Lynch, Brian Maurice
McCulloch, Archibald Wilson
Ogilvie, Kelvin Kenneth
Piorko, Adam M
White, Robert Lester

ONTARIO
Akhtar, Mohammad Humayoun
Alper, Howard
ApSimon, John W
Arsenault, Guy Pierre
Baer, Hans Helmut
Baines, Kim Marie
Bannard, Robert Alexander Brock
Bell, Russell A
Bharucha, Keki Rustomji
Birnbaum, George I
Bohme, Diethard Kurt
Brook, Adrian Gibbs
Buchanan, Gerald Wallace
Bunce, Nigel James
Buncel, Erwin
Bunting, John William
Burke, Derek Clissold
Butler, Douglas Neve
Cann, Malcolm Calvin
Casal, Hector Luis
Childs, Ronald Frank
Court, William Arthur
Darby, Nicholas
Dolenko, Allan John
Duff, James McConnell
Dunn, John Robert
Durst, Tony
Eastham, Arthur Middleton
Edwards, Douglas Cameron
Edwards, Oliver Edward
Eisenhauer, Hugh Ross
Evans, Taylor Herbert

Fallis, Alexander Graham
Fitt, Peter Stanley
Fraser, Robert Rowntree
Gingras, Bernard Arthur
Gordon, Myra
Graham, Bruce Allan
Guthrie, James Peter
Hach, Vladimir
Harris, John William
Harrison, Alexander George
Harrison, William Ashley
Heacock, Ronald A
Heggie, Robert Murray
Holland, Herbert Leslie
Hopkins, Clarence Yardley
Howard, James Anthony
Hughes, David William
Ingold, Keith Usherwood
Janzen, Edward George
Jones, John Bryan
Jones, Maurice Harry
Kaiser, Klaus L(eo) E(duard)
Kazmaier, Peter Michael
King, James Frederick
Kluger, Ronald H
Kramer, John Karl Gerhard
Krantz, Allen
Krepinsky, Jiri J
Kresge, Alexander Jerry
Lange, Gordon Lloyd
Laughton, Paul MacDonell
Lautens, Mark
Lautenschlaeger, Friedrich Karl
Lee-Ruff, Edward
Leitch, Leonard Christie
Lewars, Errol George
Leznoff, Clifford Clark
McArthur, Colin Richard
McIntosh, John McLennan
McKague, Allan Bruce
MacLean, David Bailey
McLean, Stewart
Marks, Gerald Samuel
Martin, Trevor Ian
Massiah, Thomas Frederick
Mayo, De Paul
Meresz, Otto
Moir, Robert Young
Morand, Peter
Morita, Hirokazu
Narang, Saran A
Onuska, Francis Ivan
Prokipcak, Joseph Michael
Puddephatt, Richard John
Ratnayake, Walisundera Mudiyanselage Nimal
Rees, Alun Hywel
Reid, Sidney George
Reynolds-Warnhoff, Patricia
Ritcey, Gordon M
Rodrigo, Russell Godfrey
Rutherford, Kenneth Gerald
Sanderson, Edwin S
Schmid, George Henry
Scott, Andrew Edington
Scott, Peter Michael
Serdarevich, Bogdan
Siddiqui, Iqbal Rafat
Smith, James Graham
Snieckus, Victor A
Sowa, Walter
Spencer, Elvins Yuill
Still, Ian William James
Stothers, John Bailie
Strachan, William Michael John
Szabo, Arthur Gustav
Szarek, Walter Anthony
Tidwell, Thomas Tinsley
Tuck, Dennis George
Usselman, Melvyn Charles
Vijay, Hari Mohan
Warkentin, John
Warnhoff, Edgar William
Wayman, Morris
Weedon, Alan Charles
Werstiuk, Nick Henry
Westaway, Kenneth C
Wigfield, Donald Compston
Wightman, Robert Harlan
Winnik, Françoise Martine
Winnik, Mitchell Alan
Winthrop, Stanley Oscar
Wood, Gordon Walter
Woolford, Robert Graham
Yates, Keith
Yates, Peter
Young, James Christopher F
Ziegler, Peter

PRINCE EDWARD ISLAND
Palmer, Glenn Earl
Rigney, James Arthur

QUEBEC
Adamkiewicz, Vincent Witold
Atkinson, Joseph George
Belanger, Patrice Charles
Belleau, Bernard Roland
Brown, Gordon Manley
Burnell, Robert H
Cannone, Persephone
Chan, Tak-Hang
Chenevert, Robert (Bernard)
Chubb, Francis Learmonth

Clayton, David Walton
Colebrook, Lawrence David
Cook, Robert Douglas
Cooper, Paul David
Daessle, Claude
Darling, Graham Davidson
Davis, Martin Arnold
Deans, Sidney Alfred Vindin
DeMedicis, E M J A
Deslongchamps, Pierre
Diksic, Mirko
Doughty, Mark
Dugas, Hermann
Edward, John Thomas
Engel, Charles Robert
Favre, Henri Albert
Ferdinandi, Eckhardt Stevan
Fleming, Bruce Ingram
Fliszar, Sandor
Garneau, Francois Xavier
Gaudry, Roger
Graham, Aloysius
Gravel, Denis Fernand
Gurudata, Neville
Hamlet, Zacharias
Hanessian, Stephen
Hay, Allan Stuart
Heisler, Seymour
Heitner, Cyril
Joly, Louis Philippe
Just, George
Khalil, Michel
Lau, Cheuk Kun
Lennox, Robert Bruce
Lennox, Robert Bruce
Lessard, Jean
O'Meara, Desmond
Paice, Michael
Perlin, Arthur Saul
Perron, Yvon G
Portelance, Vincent Damien
Powell, William St John
Rakhit, Sumanas
Richer, Jean-Claude
Roughley, Peter James
Vincent, Donald Leslie
Wu, Chisung
Yeats, Ronald Bradshaw
Zamir, Lolita Ora

SASKATCHEWAN
Chandler, William David
Gear, James Richard
Grant, Donald R
Lee, Choi Chuck
Lee, Donald Garry
Macphee, Kenneth Erskine
Martin, Robert O
Mezey, Paul G
Pepper, James Morley
Slater, George P
Smith, Allan Edward
Steck, Warren Franklin
Tulloch, Alexander Patrick

OTHER COUNTRIES
Baddiley, James
Barnes, Roderick Arthur
Boaz, David Paul
Boffa, Lidia C
Booth, Gary Edwin
Boustany, Kamel
Brown, Peter
Carter, Robert Everett
Cavanaugh, Robert J
Chaffin, Tommy L
Chan, Hak-Foon
Choi, Yong Chun
Collins, Carol Hollingworth
Conover, Lloyd Hillyard
Cromartie, Thomas Houston
Edmonds, James W
Evleth, Earl Mansfield
Ghosh, Anil Chandra
Godefroi, Erik Fred
Goldstein, Melvin Joseph
Graber, Robert Philip
Grisar, Johann Martin
Gschwend, Heinz W
Guehler, Paul Frederick
Hajos, Zoltan George
Hamb, Fredrick Lynn
Hassner, Alfred
Hounshell, William Douglas
Hummel, Hans Eckhardt
Hutchison, John Joseph
Kaufmann, Esteban
Kellogg, Richard Morrison
Kent, Stephen Brian Henry
Kober, Ehrenfried H
Kosower, Edward Malcolm
Krakower, Gerald W
Kuttab, Simon Hanna
Levand, Oscar
Levin, Gideon
Little, John Stanley
Lonsdale-Eccles, John David
Low, Chow-Eng
Low, Teresa Lingchun Kao
McCullough, John James
Malherbe, Roger F
Mancera, Octavio
Meyers, Martin Bernard
Millar, John Robert

Millington, James E
Prelog, Vladimir
Rakita, Philip Erwin
Schleyer, Paul von Ragué
Siegel, Herbert
Snyder, William Richard
Solomons, Thomas William Graham
Spaght, Monroe Edward
Trentham, David R
Trofimov, Boris Alexandrovich
Von Stryk, Frederick George
Ward, John Edward

Pharmaceutical Chemistry

ALABAMA
Beasley, James Gordon
Darling, Charles Milton
Elliott, Robert Daryl
Hocking, George Macdonald
Montgomery, John Atterbury
Morin, Richard Dudley
Nair, Madhavan G
Parish, Edward James
Riley, Thomas N
Sakai, Ted Tetsuo
Shealy, Y(oder) Fulmer
Struck, Robert Frederick
Tan, Boen Hie
Temple, Carroll Glenn
Thomas, Hazel Jeanette
Vacik, James P
Wilken, Leon Otto, Jr
Wintter, John Ernest

ARIZONA
Clinton, Raymond Otto
Cole, Jack Robert
Dale, Jack Kyle
Granatek, Alphonse Peter
Hoffmann, Joseph John
Martin, Arnold R
Parks, Lloyd McClain
Reed, Fred DeWitt, Jr
Remers, William Alan
Shively, Charles Dean
Yalkowsky, Samuel Hyman

ARKANSAS
Lattin, Danny Lee
Mittelstaedt, Stanley George
Sorenson, John R J
Voldeng, Albert Nelson

CALIFORNIA
Ainsworth, Cameron
Benet, Leslie Z
Biles, John Alexander
Boger, Dale L
Brochmann-Hanssen, Einar
Burlingame, Alma L
Carter, Kenneth
Chan, Kenneth Kin-Hing
Chaubal, Madhukar Gajanan
Chinn, Leland Jew
Chou, Yungnien John
Colwell, William Tracy
Craig, John Cymerman
Craik, Charles S
Cureton, Glen Lee
Degraw, Joseph Irving, Jr
Dev, Vasu
Felgner, Philip Louis
Fried, John
Friedman, Alan E
Fries, David Samuel
Fung, Steven
Gerber, Bernard Robert
Gordon, Samuel Morris
Goyan, Jere Edwin
Green, Donald Eugene
Hamilton, J(ames) Hugh
Hamor, Glenn Herbert
Henry, David Weston
Hoener, Betty-ann
Hooker, Thomas M, Jr
Hostetler, Karl Yoder
Hughes, John Lawrence
Jacob, Peyton, III
Johnson, Howard (Laurence)
Jones, Richard Elmore
Kaplan, Phyllis Deen
Katz, Martin
Kenyon, George Lommel
Kertesz, Jean Constance
Klein, David Henry
Koda, Robert T
Kollman, Peter Andrew
Kong, Eric Siu-Wai
Krezanoski, Joseph Z
Kryger, Roy George
Lad, Pramod Madhusudan
Landgraf, William Charles
Lawless, Gregory Benedict
Lazo-Wasem, Edgar A
Lee, David Louis
Leo, Albert Joseph
Lien, Eric Jung-Chi
Lobl, Thomas Jay
McGee, Lawrence Ray
McKenna, Charles Edward
Matin, Shaikh Badarul
Matuszak, Alice Jean Boyer

Mayer, William Joseph
Meehan, Thomas (Dennis)
Mitchner, Hyman
Mufson, Daniel
Musser, John H
Najafi, Ahmad
Nematollahi, Jay
Nestor, John Joseph, Jr
Nikoui, Nik
Ning, Robert Y
Okamura, William H
Ortiz de Montellano, Paul Richard
Pearlman, Rodney
Petracek, Francis James
Pierschbacher, Michael Dean
Pigiet, Vincent P
Reifenrath, William Gerald
Ripka, William Charles
Roscoe, Charles William
Rucker, Robert Blain
Safir, Sidney Robert
Sargent, Thornton William, III
Schwarz, Thomas Werner
Selassie, Cynthia R
Shell, John Weldon
Shen, Wei-Chiang
Singaram, Bakthan
Smith, Terry Douglas
Sorby, Donald Lloyd
Tozer, Thomas Nelson
Untch, Karl George
Vollhardt, K Peter C
Weinkam, Robert Joseph
Wender, Paul Anthony
Williams, Lewis David
Wolf, Walter
Wright, Terry L
Wu, Jiann-Long
Zderic, John Anthony

COLORADO
Barrett, Anthony Gerard Martin
Fitzpatrick, Francis Anthony
Micheli, Roger Paul
Nachtigall, Guenter Willi
Ruth, James Allan
Thompson, John Alec
Wingeleth, Dale Clifford

CONNECTICUT
Baldwin, Ronald Martin
Bongiorni, Domenic Frank
Church, Robert Fitz (Randolph)
Cohen, Edward Morton
Crast, Leonard Bruce, Jr
Crenshaw, Ronnie Ray
Cue, Berkeley Wendell, Jr
Dines, Allen I
Dirlam, John Philip
Doyle, Terrence William
Epling, Gary Arnold
Forssen, Eric Anton
Gauthier, George James
Ghosh, Dipak K
Harbert, Charles A
Henkel, James Gregory
Hess, Hans-Jurgen Ernst
Heyd, Allen
Hite, Gilbert J
Holland, Gerald Fagan
Horhota, Stephen Thomas
Ives, Jeffrey Lee
Johnson, Bruce McDougall
Kanig, Joseph Louis
Kerridge, Kenneth A
Kluender, Harold Clinton
Kotick, Michael Paul
Kreighbaum, William Eugene
Kriesel, Douglas Clare
LaMattina, John Lawrence
Lawson, John Edward
Lin, Tai-Shun
Luh, Yuhshi
McLamore, William Merrill
McManus, James Michael
Makriyannis, Alexandros
Marfat, Anthony
Mayell, Jaspal Singh
Melvin, Lawrence Sherman, Jr
Milne, George McLean, Jr
Moppett, Charles Edward
Nieforth, Karl Allen
Proverb, Robert Joseph
Rosati, Robert Louis
Ryan, Richard Patrick
Sarges, Reinhard
Schnur, Rodney Caughren
Schut, Robert N
Shapiro, Warren Barry
Simonelli, Anthony Peter
Spitznagle, Larry Allen
Suh, John Taiyoung
Tamorria, Christopher Richard
Temple, Davis Littleton, Jr
Thomas, Paul David
Tishler, Max
Volkmann, Robert Alfred
Yevich, Joseph Paul

DELAWARE
Aldrich, Paul E
Aungst, Bruce J
Baaske, David Michael
Boswell, George A

Carr, John B
Chaudhari, Bipin Bhudharlal
Cherkofsky, Saul Carl
Chorvat, Robert John
Crow, Edwin Lee
Dubas, Lawrence Francis
Finizio, Michael
Gakenheimer, Walter Christian
Holyoke, Caleb William, Jr
Huang, Chin Pao
Jackson, Thomas Edwin
Johnson, Alexander Lawrence
Kettner, Charles Adrian
Lam, Gilbert Nim-Car
Lee, Shung-Yan Luke
Lerman, Charles Lew
Mighton, Charles Joseph
Mollica, Joseph Anthony
Park, Chung Ho
Quarry, Mary Ann
Rajagopalan, Parthasarathi
Rifino, Carl Biaggio
Sundelin, Kurt Gustav Ragnar
Turlapaty, Prasad
Wuonola, Mark Arvid

DISTRICT OF COLUMBIA
Boots, Sharon G
Klayman, Daniel Leslie
La Pidus, Jules Benjamin
Milbert, Alfred Nicholas
Miller, David Jacob
Poon, Bing Toy
Rzeszotarski, Waclaw Janusz
Scott, Kenneth Richard
Telang, Vasant G
Thomas, Richard Dean
Weber, John Donald
Zalucky, Theodore B

FLORIDA
Child, Ralph Grassing
Curlin, Lemuel Calvert
Ebert, William Robley
Fitzgerald, Thomas James
Garrett, Edward Robert
Goldberg, Eugene P
Gramling, Lea Gene
Halpern, Ephriam Philip
Hammer, Henry Felix
Hammer, Richard Hartman
Howes, John Francis
Lee, Henry Joung
Nakashima, Tadayoshi
Parcell, Robert Ford
Pop, Emil
Rokach, Joshua
Schwartz, Michael Averill
Templer, David Allen
Wheeler, Frank Carlisle
Yost, Richard A
Young, Michael David

GEORGIA
Blanton, Charles DeWitt, Jr
Boudinot, Frank Douglas
Chu, Chung K
Churchill, Frederick Charles
Doetsch, Paul William
Ellington, James Jackson
Honigberg, Irwin Leon
Israili, Zafar Hasan
Lopez, Antonio Vincent
Matthews, Hewitt William
Pelletier, S William
Powers, James Cecil
Proctor, Charles Darnell
Rhodes, Robert Allen
Ruenitz, Peter Carmichael
Stewart, James T
Thompson, Bobby Blackburn
Walsh, David Allan
Waters, Kenneth Lee
Woodford, James

IDAHO
Goettsch, Robert
Isaacson, Eugene I
Natale, Nicholas Robert

ILLINOIS
Alam, Abu Shafiul
Amsel, Lewis Paul
Arendsen, David Lloyd
Armstrong, P(aul) Douglas
Baird, Michael Jefferson
Bauer, Ludwig
Borgman, Robert John
Brooks, Dee W
Burgauer, Paul David
Cannon, John Burns
Chen, Wen Sherng
Collins, Paul Waddell
Currie, Bruce LaMonte
Debus, Allen George
Dellaria, Joseph Fred, Jr
DeRose, Anthony Francis
Djuric, Stevan Wakefield
Dunn, William Joseph
Fitzloff, John Frederick
Garland, Robert Bruce
Gearien, James Edward
Gray, Nancy M
Hansen, Donald Willis, Jr

Hardwidge, Edward Albert
Jenkins, William Wesley
Jiu, James
Johnson, Dale Howard
Jones, Ralph William
Joslin, Robert Scott
Karim, Aziz
Kerdesky, Francis A J
Kerkman, Daniel Joseph
Kim, Ki Hwan
Kipp, James Edwin
Klimstra, Paul D
Krimmel, C Peter
Lee, Cheuk Man
Lu, Matthias Chi-Hwa
Lucci, Robert Dominick
Martin, Yvonne Connolly
Mohan, Prem
Moon, Byong Hoon
Musa, Mahmoud Nimir
Nysted, Leonard Norman
Pariza, Richard James
Piatak, David Michael
Possley, Leroy Henry
Propst, Catherine Lamb
Ranade, Vinayak Vasudeo
Rao, Gopal Subba
Roderick, William Rodney
Rorig, Kurt Joachim
Roseman, Theodore Jonas
Rosenberg, Saul Howard
Seidenfeld, Jerome
Sennello, Lawrence Thomas
Shipchandler, Mohammed Tyebji
Shone, Robert L
Siegel, Sheldon
Silverman, Richard Bruce
Singiser, Robert Eugene
Smith, Homer Alvin, Jr
Stout, David Michael
Thomas, Elizabeth Wadsworth
Ulrey, Stephen Scott
Weston, Arthur Walter
Winn, Martin
Witczak, Zbigniew J
Woroch, Eugene Leo
Yamamoto, Diane M
Young, David W
Zimmer, Arthur James

INDIANA
Amundson, Merle E
Biddlecom, William Gerard
Boenigk, John William
Born, Gordon Stuart
Boyd, Donald Bradford
Campaigne, Ernest Edwin
Cassady, John Mac
Chang, Ching-Jer
Childers, Ray Fleetwood
Clark, Larry P
Conine, James William
Cooper, Robin D G
Corey, Paul Frederick
Cushman, Mark
Cwalina, Gustav Edward
DeLong, Allyn F
Dinner, Alan
Dorman, Douglas Earl
Dube, David Gregory
Farkas, Eugene
Fricke, Gordon Hugh
Grant, Ernest Walter
Green, Mark Alan
Gutowski, Gerald Edward
Han, Choong Yol
Harper, Richard Waltz
Havel, Henry Acken
Herron, David Kent
Hulbert, Matthew H
Indelicato, Joseph Michael
Kaufman, Karl Lincoln
Kessler, Wayne Vincent
Knevel, Adelbert Michael
Kornfeld, Edmund Carl
Kossoy, Aaron David
Lappas, Lewis Christopher
Larsen, Aubrey Arnold
McIntyre, Thomas William
McKeehan, Charles Wayne
McLaughlin, Jerry Loren
Marshall, Winston Stanley
Massey, Eddie H
Matsumoto, Ken
Mikolasek, Douglas Gene
Morrow, Duane Francis
Murphy, Charles Franklin
Nichols, David Earl
Nirschl, Joseph Peter
Patwardhan, Bhalchandra H
Pikal, Michael Jon
Pioch, Richard Paul
Rickard, Eugene Clark
Robertson, David Wayne
Robertson, Donald Edwin
Safdy, Max Errol
Salerni, Oreste Leroy
Shaw, Stanley Miner
Stiver, James Frederick
Stroube, William Bryan, Jr
Su, Kenneth Shyan-Ell
Thompkins, Leon
Wagner, Eugene Ross
Wallace, Gerald Wayne

Wheeler, William Joe

IOWA
Cannon, Joseph G
Linhardt, Robert John
Neumann, Marguerite
Parrott, Eugene Lee
Wiley, Robert A
Wurster, Dale Eric, Jr

KANSAS
Beecham, Curtis Michael
Borchardt, Ronald T
Cheng, Chia-Chung
Gillingham, James Morris
Grunewald, Gary Lawrence
Hanzlik, Robert Paul
Haslam, John Lee
Lindenbaum, Siegfried
Marx, Michael
Mertes, Mathias Peter
Munson, H Randall, Jr
Owensby, Clenton Edgar
Pazdernik, Thomas Lowell
Rork, Gerald Stephen
Rytting, Joseph Howard
Stella, Valentino John
Stutz, Robert L
Youngstrom, Richard Earl

KENTUCKY
Crooks, Peter Anthony
DeLuca, Patrick Phillip
Digenis, George A
Kadaba, Pankaja Kooveli
Kornet, Milton Joseph
Luts, Heino Alfred
Luzzio, Frederick Anthony
Parker, Martin Dale
Smith, Walter Thomas, Jr
Thio, Alan Poo-An

LOUISIANA
Agrawal, Krishna Chandra
Bauer, Dennis Paul
Daigle, Josephine Siragusa
Davenport, Tom Forest, Jr
Gallo, August Anthony
Hornbaker, Edwin Dale
Weidig, Charles F

MAINE
Bragdon, Robert Wright
Dice, John Raymond
Tedeschi, Robert James

MARYLAND
Alexander, Thomas Goodwin
Barnstein, Charles Hansen
Benson, Walter Roderick
Blum, Stanley Walter
Brancato, David Joseph
Callery, Patrick Stephen
Carr, Charles Jelleff
Carter, John Paul, Jr
Chang, Charles Yu-Chun
Chang, Henry
Chiu, Tony Man-Kuen
Cohen, Louis Arthur
Driscoll, John Stanford
Ellin, Robert Isadore
Feyns, Liviu Valentin
Gardner, Jerry David
Gelberg, Alan
Grady, Lee Timothy
Graham, Joseph H
Gray, Allan P
Hajiyani, Mehdi Hussain
Henery-Logan, Kenneth Robert
Hillery, Paul Stuart
Hollenbeck, Robert Gary
Horak, Vaclav
Hosmane, Ramachandra Sadashiv
Jacobson, Kenneth Alan
Kador, Peter Fritz
Kaiser, Carl
Keefer, Larry Kay
Keily, Hubert Joseph
Klein, Michael
Klioze, Oscar
Kochhar, Man Mohan
Kramer, Stanley Phillip
Kumkumian, Charles Simon
Leak, John Clay, Jr
Leslie, James
Levin, Nathan
Levine, Alan Stewart
Lo, Theresa Nong
McShefferty, John
Marquez, Victor Esteban
Massie, Samuel Proctor
Misra, Renuka
Nemec, Josef
Oxman, Michael Allan
Pitha, Josef
Poochikian, Guiragos K
Quinn, Frank Russell
Rapaka, Rao Sambasiva
Rice, Kenner Cralle
Roller, Peter Paul
Schattner, Robert I
Schwartz, Paul
Schwartz, Samuel Meyer
Scovill, John Paul

Pharmaceutical Chemistry (cont)

Sharma, Dinesh C
Sheinin, Eric Benjamin
Short, James Harold
Shroff, Arvin Pranlal
Smith, Edward
Snow, Philip Anthony
Theimer, Edgar E
Waters, James Augustus
Weiss, Peter Joseph
Whaley, Wilson Monroe
Whiting, John Dale, Jr
Willette, Robert Edmond
Wolters, Robert John
Woodman, Peter William
Wykes, Arthur Albert
Yeh, Shu-Yuan
Zenitz, Bernard Leon
Zenker, Nicolas

MASSACHUSETTS

Ajami, Alfred Michel
Auer, Henry Ernest
Bessette, Russell Romulus
Cantley, Lewis Clayton
Cheema, Zafarullah K
Davis, Michael Allan
Filer, Crist Nicholas
Foye, William Owen
Gerteisen, Thomas Jacob
Granchelli, Felix Edward
Kafrawy, Adel
Kline, Toni Beth
Kobayashi, Kazumi
LaSala, Edward Francis
Margulis, Thomas N
Mickles, James
Neumeyer, John L
Parham, Marc Ellous
Razdan, Raj Kumar
Schomer, Donald Lee
Siegal, Bernard
Smith, Pierre Frank
Snider, Barry B
Takman, Bertil Herbert
Tuckerman, Murray Moses
Warner, John Charles
Williams, David Allen
Williams, David Lloyd
Wright, George Edward
Yesair, David Wayne

MICHIGAN

Abramson, Hanley N
Albertson, Noel Frederick
Axen, Udo Friedrich
Barksdale, Charles Madsen
Barksdale, Charles Madsen
Bartlett, Janeth Marie
Belmont, Daniel Thomas
Berman, Ellen Myra
Capps, David Bridgman
Cody, Wayne Livingston
Counsell, Raymond Ernest
Coward, James Kenderdine
Creger, Paul LeRoy
Culbertson, Townley Payne
Deering, Carl F
Delia, Thomas J
Domagala, John Michael
Dorman, Linneaus Cuthbert
Drach, John Charles
Dunbar, Joseph Edward
Dunker, Melvin Frederick William
Elslager, Edward Faith
Gadwood, Robert Charles
Goel, Om Prakash
Gregg, David Henry
Harmon, Robert E
Hatzenbuhler, Douglas Albert
Heinrikson, Robert L
Herrinton, Paul Matthew
Heyd, William Ernst
Hiestand, Everett Nelson
Hoefle, Milton Louis
Huang, Che C
Jacoby, Ronald Lee
Jones, Eldon Melton
Joshi, Mukund Shankar
Kagan, Fred
Kaiser, David Gilbert
Kiely, John Steven
Klutchko, Sylvester
Koch, Melvin Vernon
Koshy, K Thomas
Laks, Peter Edward
Lintner, Carl John, Jr
Lipton, Michael Forrester
McCall, John Michael
McGovren, James Patrick
Magerlein, Barney John
Mathison, Ian William
Maxey, Brian William
Meyer, Robert F
Moffett, Robert Bruce
Moissides-Hines, Lydia Elizabeth
Moore, Alexander Mazyck
Morozowich, Walter
Murphy, William R
Nagler, Robert Carlton
Nelson, Norman Allan

Pfeiffer, Curtis Dudley
Philipsen, Judith Lynne Johnson
Reid, William Bradley
Schumann, Edward Lewis
Short, Franklin Willard
Sinkula, Anthony Arthur
Sinsheimer, Joseph Eugene
Stanley, John Pearson
Suggs, William Terry
Taraszka, Anthony John
Thomas, Richard Charles
Tinney, Francis John
Tobey, Stephen Winter
Topliss, John G
Townsend, Leroy B
Unangst, Paul Charles
Ursprung, Joseph John
Valvani, Shri Chand
Van Rheenen, Verlan H
Wagner, John Garnet
Weisbach, Jerry Arnold
Westland, Roger D(ean)
Wilkinson, Paul Kenneth
Woo, P(eter) W(ing) K(ee)
Wormser, Henry C
Woster, Patrick Michael
Yankee, Ernest Warren

MINNESOTA

Ames, Matthew Martin
Bunker, James Edward
Crooks, Stephen Lawrence
Erickson, Edward Herbert
Farber, Elliott
Grant, David James William
Hanna, Patrick E
Jochman, Richard Lee
Kolthoff, Izaak Maurits
Mikhail, Adel Ayad
Nagasawa, Herbert Tsukasa
Portoghese, Philip S
Reever, Richard Eugene
Rippie, Edward Grant
Robertson, Jerry Earl
Vince, Robert
Wade, James Joseph
Yapel, Anthony Francis, Jr

MISSISSIPPI

Baker, John Keith
Borne, Ronald Francis
Gilmore, William Franklin
Nobles, William Lewis
Sam, Joseph
Thompson, Alonzo Crawford
Waller, Coy Webster

MISSOURI

Asrar, Jawed
Baker, Joseph Willard
Banakar, Umesh Virupaksh
Bovy, Philippe R
Chafetz, Lester
Delaware, Dana Lewis
Dill, Dale Robert
Eibert, John, Jr
Godt, Henry Charles, Jr
Gustafson, Mark Edward
Harland, Ronald Scott
Johnson, Richard Dean
Kenyon, Allen Stewart
Krueger, Paul A
Lennon, Patrick James
Liao, Tsung-Kai
Nuessle, Noel Oliver
Podrebarac, Eugene George
Popp, Frank Donald
Raffelson, Harold
Rost, William Joseph
Sabacky, M Jerome
Sathe, Sharad Somnath
Schleppnik, Alfred Adolf
Schumacher, Ignatius
Sikorski, James Alan
Stenseth, Raymond Eugene
Tucker, Robert Gene
Volkert, Wynn Arthur
Wasson, Richard Lee
Wheeler, James Donlan
Woodhouse, Edward John
Zee-Cheng, Robert Kwang-Yuen
Zienty, Ferdinand B
Zuzack, John W

MONTANA

Pettinato, Frank Anthony

NEBRASKA

Gold, Barry Irwin
Harris, Lewis Eldon
Murray, Wallace Jasper
Nagel, Donald Lewis
Roche, Edward Browining
Ruyle, William Vance
Saski, Witold
Shiue, Chyng-Yann
Small, LaVerne Doreyn

NEW JERSEY

Abdou, Hamed M
Abou-Gharbia, Magid
Andrade, John Robert
Ardelt, Wojciech Joseph
Asano, Akira

Bailey, Leonard Charles
Beattie, Thomas Robert
Beck, John Louis
Berger, Joel Gilbert
Bernady, Karel Francis
Bihovsky, Ron
Blackburn, Dale Warren
Bodin, Jerome Irwin
Bollinger, Frederick W(illiam)
Borah, Kripanath
Bose, Ajay Kumar
Brine, Charles James
Brofazi, Frederick R
Bryan, Wilbur Lowell
Butensky, Irwin
Carlucci, Frank Vito
Carter, James Evan
Cavender, Patricia Lee
Chen, James L
Chertkoff, Marvin Joseph
Chien, Yie W
Clarke, Frank Henderson
Clayton, John Mark
Colaizzi, John Louis
Controulis, John
Cook, Alan Frederick
Crawley, Lantz Stephen
Cross, John Milton
Dean, Donald E
Delaney, Edward Joseph
Denton, John Joseph
De Silva, John Arthur F
Diamond, Julius
DiFazio, Louis T
Draper, Richard William
Dugan, Gary Edwin
Duhl-Emswiler, Barbara Ann
Dunn, William Howard
Durette, Philippe Lionel
Egli, Peter
Elder, John Philip
Erhardt, Paul William
Essery, John M
Fahey, John Leonard
Fahrenholtz, Kenneth Earl
Fand, Theodore Ira
Farng, Richard Kwang
Faust, Richard Edward
Fedrick, James Love
Finkelstein, Jacob
Focella, Antonino
Fong, Jones W
Fryer, Rodney Ian
Gall, Martin
Garland, William Arthur
Geller, Milton
Gershon, Sol D
Gimelli, Salvatore Paul
Girijavallabhan, Viyyoor Moopil
Gold, Daniel Howard
Goldemberg, Robert Lewis
Gordon, Eric Michael
Gordziel, Steven A
Gosser, Leo Anthony
Grier, Nathaniel
Groeger, Theodore Oskar
Gund, Peter Herman Lourie
Gyan, Nanik D
Hager, Douglas Francis
Halpern, Donald F
Hardtmann, Goetz E
Haslanger, Martin Frederick
Head, William Francis, Jr
Heck, James Virgil
Heilman, William Paul
Helsley, Grover Cleveland
Hirsch, Allen Frederick
Hoff, Dale Richard
Hoffman, Allan Jordan
Hofmann, Corris Mabelle
Hokanson, Gerard Clifford
Houlihan, William J
Howe, Eugene Everett
Howell, Charles Frederick
Infeld, Martin Howard
Ivashkiv, Eugene
Jacobs, Allen Leon
Jacobson, Harold
Jaffe, James Mark
Jain, Nemichand B
Jensen, Norman P
Jirkovsky, Ivo
Jones, Howard
Kabadi, Balachandra N
Kaczorowski, Gregory John
Kalm, Max John
Kaplan, Leonard Louis
Kasparek, Stanley Vaclav
Katz, Irwin Alan
Kees, Kenneth Lewis
Kim, Benjamin K
Klaubert, Dieter Heinz
Kline, Berry James
Koster, William Henry
Kreft, Anthony Frank, III
La Rocca, Joseph Paul
Lehr, Hanns H
Lewis, Neil Jeffrey
Lieberman, Herbert A
Liebowitz, Stephen Marc
Lin, Yi-Jong
Linn, Bruce Oscar
Lippmann, Irwin
Liu, Yu-Ying

Love, L J Cline
Lukas, George
McCaully, Ronald James
MacCoss, Malcolm
McDowell, Wilbur Benedict
McKearn, Thomas Joseph
Marlowe, Edward
Martin, Lawrence Leo
Mehta, Atul Mansukhbhai
Merkle, F Henry
Mezick, James Andrew
Michel, Gerd Wilhelm
Mikes, John Andrew
Miner, Karen Mills
Miner, Robert Scott, Jr
Moros, Stephen Andrew
Nessel, Robert J
Nicolau, Gabriela
Ocain, Timothy Donald
Olson, Gary Lee
Parisse, Anthony John
Perrotta, James
Pflug, Gerald Ralph
Pilkiewicz, Frank George
Puri, Surendra Kumar
Rasmusson, Gary Henry
Rees, Richard Wilhelm A
Reif, Van Dale
Riedhammer, Thomas M
Roland, Dennis Michael
Rosenthal, Murray William
Rovnyak, George Charles
Santilli, Arthur A
Schaffner, Carl Paul
Schnaare, Roger L
Sciarrone, Bartley John
Segelman, Alvin Burton
Semenuk, Nick Sarden
Sestanj, Kazimir
Shah, Atul A
Sieh, David Henry
Simonoff, Robert
Skotnicki, Jerauld S
Slade, Joel S
Slusarchyk, William Allen
Solis-Gaffar, Maria Corazon
Somkaite, Rozalija
Staum, Muni M
Stein, Reinhardt P
Steinberg, Eliot
Steinman, Martin
Stober, Henry Carl
Strike, Donald Peter
Taylor, John William
Tolman, Richard Lee
Tway, Patricia C
Vanden Heuvel, William John Adrian, III
Venturella, Vincent Steven
Vida, Julius
Villani, Frank John, Sr
Virgili, Luciano
Vlattas, Isidoros
Walser, Armin
Watthey, Jeffrey William Herbert
Watts, Daniel Jay
Weintraub, Howard Steven
Weintraub, Leonard
Weiss, Marvin
Whigan, Daisy B
White, Ronald E
Wilkinson, Raymond George
Willis, Carl Raeburn, Jr
Wilson, Armin Guschel
Witkowski, Joseph Theodore
Wohl, Ronald A
Wright, William Blythe, Jr
Wu, Mu Tsu
Yakubik, John
Ziegler, John Benjamin

NEW MEXICO

Garst, John Eric
Kelly, Clark Andrew
Saenz, Reynaldo V

NEW YORK

Ackerman, James Howard
Agnew-Marcelli, G(ladys) Marie
Alks, Vitauts
Allen, George Rodger, Jr
Anderson, Wayne Keith
Archer, Sydney
Balassa, Leslie Ladislaus
Bardos, Thomas Joseph
Barringer, William Charles
Bitha, Panayota
Bizios, Rena
Bonvicino, Guido Eros
Brown, Neil Harry
Brunings, Karl John
Bryan, Hugh D
Burger, Richard Melton
Carabateas, Philip M
Chandran, V Ravi
Charton, Marvin
Chheda, Girish B
Christman, David R
Clarke, Robert LaGrone
Coburn, Robert A
Cohen, Elliott
Cooper, Aaron David
Curran, William Vincent
Dabrowiak, James Chester

Daum, Sol Jacob
Dutta, Shib Prasad
Feit, Irving N
Fields, Thomas Lynn
Finn, Ronald Dennet
Fliedner, Leonard John, Jr
Fondy, Thomas Paul
Fung, Ho-Leung
Gaver, Robert Calvin
Georgiev, Vassil St
Gershon, Herman
Ginos, James Zissis
Goldschmidt, Eric Nathan
Gordon, Maxwell
Gottesman, Elihu
Gottstein, William J
Gringauz, Alex
Gurwara, Sweet K
Haberfield, Paul
Hanna, Samir A
Hlasta, Dennis John
Hong, Chung Il
Hoyte, Robert Mikell
Jarowski, Charles I
Johns, William Francis
Johnson, Robert Ed
Kalman, Thomas Ivan
Kapoor, Amrit Lal
Kapuscinski, Jan
Khan, Jamil Akber
Klein, Harvey Gerald
Klingele, Harold Otto
Kohlbrenner, Philip John
Korytnyk, Walter
Krueger, James Elwood
Kushner, Samuel
Laubach, Gerald D
Lee, May D-Ming (Lu)
Lesher, George Yohe
Loev, Bernard
Lorenz, Roman R
Mackles, Leonard
Meyer, Walter Edward
Miller, Theodore Charles
Misra, Anand Lal
Moore, Maurice Lee
Morris, Marilyn Emily
Murdock, Keith Chadwick
Napoli, Joseph Leonard
Nash, Robert Arnold
Neenan, John Patrick
Novotny, Jaroslav
O'Brien, Anne T
Ojima, Iwao
Orzech, Chester Eugene, Jr
Paikoff, Myron
Parham, James Crowder, II
Partch, Richard Earl
Portmann, Glenn Arthur
Redalieu, Elliot
Reich, Marvin Fred
Renfroe, Harris Burt
Rich, Arthur Gilbert
Rossi, Miriam
Rubin, Byron Herbert
Salley, John Jones, Jr
Sayegh, Joseph Frieh
Scott, Francis Leslie
Sellstedt, John H
Shaath, Nadim Ali
Sharma, Ram Ashrey
Siggins, James Ernest
Singh, Baldev
Siuta, Gerald Joseph
Smith, Alan Jerrard
Snyder, Harry Raymond, Jr
Solo, Alan Jere
Soloway, Harold
Spiegel, Allen J
Tam, James Pingkwan
Taub, Abraham
Triggle, David J
Trust, Ronald I
Upeslacis, Janis
Urdang, Arnold
Venkataraghavan, R
Von Strandtmann, Maximillian
Warner, Paul Longstreet, Jr
Warren, James Donald
Webb, William Gatewood
Weiss, Martin Joseph
Welles, Harry Leslie
Wentland, Mark Philip
White, Ralph Lawrence, Jr
Wolf, Alfred Peter
Wolff, Manfred Ernst
Wood, David
You, Kwan-sa
Yu, Andrew B C

NORTH CAROLINA

Beverung, Warren Neil, Jr
Bigham, Eric Cleveland
Boettner, Fred Easterday
Booth, Raymond George
Boyd, Robert Edward, Sr
Brieaddy, Lawrence Edward
Brown, Horace Dean
Carl, Philip Louis
Carr, Fred K
Cavallito, Chester John
Chae, Kun
Cocolas, George Harry
Cory, Michael

Davis, Roman
Dixit, Ajit Suresh
Doak, George Osmore
Durden, John Apling, Jr
Engle, Thomas William
Fernandes, Daniel James
Foernzler, Ernest Carl
Freeman, Harold Stanley
Gemperline, Paul Joseph
Hadzija, Bozena Wesley
Hager, George Philip, Jr
Higgins, Robert H
Hinze, Willie Lee
Hong, Donald David
Hooper, Irving R
Hurlbert, Bernard Stuart
Ishaq, Khalid Sulaiman
Izydore, Robert Andrew
Jiang, Jack Bau-Chien
Kelley, James Leroy
Kelsey, John Edward
Kuyper, Lee Frederick
Lee, Kuo-Hsiung
Lee, Yue-Wei
Long, Robert Allen
McDermed, John Dale
Martin, Gary Edwin
Mehta, Nariman Bomanshaw
Millen, Jane
Misek, Bernard
Nielsen, Lawrence Arthur
Piantadosi, Claude
Poust, Rolland Irvin
Price, Kenneth Elbert
Reid, Jack Richard
Rideout, Janet Litster
Schaeffer, Howard John
Semeniuk, Fred Theodor
Sigel, Carl William
Sinha, Birandra Kumar
Spielvogel, Bernard Franklin
Swarbrick, James
Wyrick, Steven Dale
Yeowell, David Arthur

NORTH DAKOTA

Ary, Thomas Edward
Bergstrom, Donald Eugene
Jasperse, Craig Peter
Magarian, Edward O
Shelver, William H
Vincent, Muriel C

OHIO

Billups, Norman Frederick
Bope, Frank Willis
Brueggemeier, Robert Wayne
Carr, Albert A
Cheng, Lawrence Kar-Hiu
Cook, Elton Straus
Didchenko, Rostislav
Dolfini, Joseph E
Dunbar, Philip Gordon
Eisenhardt, William Anthony, Jr
Erman, William F
Fall, Harry H
Feld, William Adam
Fentiman, Allison Foulds, Jr
Fleming, Robert Willerton
Frank, Sylvan Gerald
Freedman, Jules
Germann, Richard P(aul)
Gilpin, Roger Keith
Glasser, Arthur Charles
Graef, Walter L
Greenlee, Kenneth William
Holbert, Gene W(arwick)
Homan, Ruth Elizabeth
Horgan, Stephen William
Huber, Harold E
Jarvi, Esa Tero
Jordan, Freddie L
Letton, James Carey
Leyda, James Perkins
Lichtin, J Leon
Lutton, John Kazuo
McCarty, Frederick Joseph
McVean, Duncan Edward
Messer, William Sherwood, Jr
Miller, Duane Douglas
Milstein, Stanley Richard
Morgan, Stanley L
Myhre, David V
Norris, Paul Edmund
Parker, Roger A
Peet, Norton Paul
Powers, Larry James
Rau, Allen H
Sampson, Paul
Schnettler, Richard Anselm
Sharma, Rameshwar Kumar
Soloway, Albert Herman
Stadler, Louis Benjamin
Stansloski, Donald Wayne
Staubus, Alfred Elsworth
Stotz, Robert William
Tan, Henry S I
Taschner, Michael J
Wade, Peter Cawthorn
Warner, Ann Marie
Warner, Victor Duane
Washburn, Lee Cross
Webb, Norval Ellsworth, Jr
Weintraub, Herschel Jonathan R

Witiak, Donald T
Zoglio, Michael Anthony

OKLAHOMA

Berlin, Kenneth Darrell
Carubelli, Raoul
Dryhurst, Glenn
Harris, Loyd Ervin
Magarian, Robert Armen
Ortega, Gustavo Ramon
Prabhu, Vilas Anandrao
Ratto, Peter Angelo
Shough, Herbert Richard

OREGON

Block, John Harvey
Hauser, Frank Marion
Schultz, Harry Wayne
Van Eikeren, Paul

PENNSYLVANIA

Ackermann, Guenter Rolf
Almond, Harold Russell, Jr
Bailey, Denis Mahlon
Bhatt, Padmamabh P
Blank, Benjamin
Bock, Mark Gary
Borke, Mitchell Louis
Brady, Stephen Francis
Brenner, Gerald Stanley
Brittelli, David Ross
Casey, Adria Catala
Chang, Cheng Allen
Chen, Shih-Fong
Cheng, Hung-Yuan
Chow, Alfred Wen-Jen
Cooperman, Barry S
Deist, Robert Paul
Dempski, Robert E
Douglas, Bryce
Evans, Ben Edward
Faiferman, Isidore
Falkiewicz, Michael Joseph
Feldman, Joseph Aaron
Fong, Kei-Lai Lau
Freidinger, Roger Merlin
Galinsky, Alvin M
Gangjee, Aleem
Gardella, Libero Anthony
Gennaro, Alfonso Robert
Grebow, Peter Eric
Grim, Wayne Martin
Grindel, Joseph Michael
Gunther, Wolfgang Hans Heinrich
Haines, William Joseph
Harrison, Aline Margaret
Hartman, Kenneth Eugene
Heindel, Ned Duane
Herzog, Karl A
Hoffman, Jacob Matthew, Jr
Hoover, John Russel Eugene
Howard, Stephen Arthur
Hwang, Bruce You-Huei
Janicki, Casimir A
Kade, Charles Frederick, Jr
Kauffman, Joel Mervin
Kelly, Ernest L
King, Robert Edward
Kingsbury, William Dennis
Kowalski, Conrad John
Kowarski, Chana Rose
Kruse, Lawrence Ivan
Kwan, King Chiu
Lapidus, Milton
Larson, Gerald Louis
Linfield, Warner Max
Lumma, William Carl, Jr
Mackowiak, Elaine DeCusatis
Martin, Bruce Douglas
Maryanoff, Bruce Eliot
Maryanoff, Cynthia Anne Milewski
Melamed, Sidney
Melander, Wayne Russell
Mishra, Dinesh S
Mokotoff, Michael
Morse, Lewis David
Motsavage, Vincent Andrew
Nutt, Ruth Foelsche
ONeill, Joseph Lawrence
Packman, Albert M
Parish, Roger Cook
Poole, John William
Press, Jeffery Bruce
Prugh, John Drew
Rasmussen, Chris Royce
Reiff, Harry Elmer
Reitz, Allen Bernard
Reynolds, Brian Edgar
Russo, Emanuel Joseph
Saggiomo, Andrew Joseph
Santora, Norman Julian
Sarantakis, Dimitrios
Scheindlin, Stanley
Schwartz, Joseph Barry
Sheeley, Richard Moats
Shepard, Kenneth LeRoy
Sinotte, Louis Paul
Sklar, Stanley
Smith, Robert Lawrence
Southwick, Philip Lee
Studt, William Lyon
Sunshine, Warren Lewis
Sutton, Blaine Mote
Tice, Linwood Franklin

Ullyot, Glenn Edgar
Urenovitch, Joseph Victor
Wei, Peter Hsing-Lien
Weinstock, Joseph
Wetzel, Ronald Burnell
Williams, John Roderick
Wilson, James William
Wineholt, Robert Leese
Woltersdorf, Otto William, Jr
Wood, Kurt Arthur
Yeager, Sandra Ann
Yellin, Tobias O
Yu, Ruey Jiin
Zanowiak, Paul

RHODE ISLAND

Abushanab, Elie
Sapolsky, Asher Isadore
Smith, Charles Irvel
Turcotte, Joseph George

SOUTH CAROLINA

Bauguess, Carl Thomas, Jr
Beamer, Robert Lewis
Dodds, Alvin Franklin
Haskell, Theodore Herbert, Jr
Martin, Tellis Alexander
Sadik, Farid
Schmidt, Gilbert Carl
Scruggs, Jack G
Shah, Hamish V
Walter, Wilbert George
Wynn, James Elkanah

SOUTH DAKOTA

Bailey, Harold Stevens
Omodt, Gary Wilson

TENNESSEE

Avis, Kenneth Edward
Buchman, Russell
Gilliom, Richard D
Gollamudi, Ramachander
Handorf, Charles Russell
Lasslo, Andrew
Lyman, Beverly Ann
Otis, Marshall Voigt
Parish, Harlie James, Jr
Pinajian, John Joseph
Sastry, Bhamidipaty Venkata Rama
Shah, Jitendra J
Sheth, Bhogilal
Smith, Howard E
Solomons, William Ebenezer
Wells, Jack Nulk
Wigler, Paul William

TEXAS

Alam, Maktoob
Bapatla, Krishna M
Bikin, Henry
Boblitt, Robert LeRoy
Cates, Lindley A
Ciufolini, Marco A
Cragoe, Edward Jethro, Jr
Delgado, Jaime Nabor
Doluisio, James Thomas
Gobuty, Allan Harvey
Harrell, William Broomfield
Hedrich, Loren Wesley
Hogan, Michael Edward
Hurley, Laurence Harold
Iskander, Felib Youssef
Jaszberenyi, Joseph C
Jones, Lawrence Ryman
Kasi, Leela Peshkar
Lemke, Thomas Lee
Lin, Tz-Hong
Marchand, Alan Philip
Marshall, Gailen D, Jr
Monk, Clayborne Morris
Mosier, Benjamin
Quintana, Ronald Preston
Reinecke, Manfred Gordon
Robinson, J(ames) Michael
Senkowski, Bernard Zigmund
Stavchansky, Salomon Ayzenman
Szczesniak, Raymond Albin
Ternay, Andrew Louis, Jr
Webber, Marion George

UTAH

Aldous, Duane Leo
Anderson, Bradley Dale
Balandrin, Manuel F
Broadbent, Hyrum Smith
Broom, Arthur Davis
Fullerton, Dwight Story
Higuchi, William Iyeo
Kim, Sung Wan
Kopecek, Jindrich
Mason, Robert C
Petersen, Robert Virgil
Roll, David Byron

VERMONT

McCormack, John Joseph, Jr

VIRGINIA

Abraham, Donald James
Andrako, John
Boots, Marvin Robert
Burger, Alfred
Cale, Albert Duncan, Jr

Creek, Jefferson Louis
Croll, Ian Murray
Crosley, David Risdon
Csicsery, Sigmund Maria
Cummings, John Patrick
Current, Jerry Hall
Curtis, Earl Clifton, Jr
Czuha, Michael, Jr
Dalla Betta, Ralph A
Dalton, Larry Raymond
Darnell, Alfred Jerome
Davenport, John Eaton
Davies, David Huw
Davis, Hubert Greenidge
Dawson, Kenneth Adrian
Day, Robert J(ames)
Day, Robert James
Dea, Phoebe Kin-Kin
Deal, Bruce Elmer
Dee, Diana
De Haan, Frank P
DeHollander, William Roger
Detz, Clifford M
Deutsch, Daniel Harold
Diederich, Francois Nico
DiGiorgio, Joseph Brun
Dill, Kenneth Austin
Dills, Charles E
Dobyns, Leona Danette
Dodge, Richard Patrick
Dodson, Charles Leon, Jr
Dodson, Richard Wolford
Dolbear, Geoffrey Emerson
Dole, Malcolm
Domeniconi, Michael John
Donnelly, Timothy C
Donovan, John W
Dormant, Leon M
Dorough, Gus Downs, Jr
Dows, David Alan
Duff, Russell Earl
Duisman, Jack Arnold
Dulmage, William James
Dyck, Rudolph Henry
Eatough, Norman L
Eckstrom, Donald James
Eichelberger, Robert Leslie
Ellis, Elizabeth Carol
El-Sayed, Mostafa Amr
El-Shimi, Ahmed Fayez
Elson, Robert Emanuel
Endres, Leland Sander
Engler, Edward Martin
English, Gerald Alan
Epstein, Barry D
Epstein, Leo Francis
Ervin, Guy, Jr
Evans, William Harrington
Farber, Joseph
Farber, Milton
Farone, William Anthony
Fass, Richard A
Fassel, Velmer Arthur
Faulkner, Thomas Richard
Fayer, Michael David
Fergin, Richard Kenneth
Feuerstein, Seymour
Fife, Thomas Harley
Fineman, Morton A
Fink, William Henry
Fisher, Robert Amos, Jr
Fisk, George Ayrs
Fleischauer, Paul Dell
Fleming, Edward Homer, Jr
Fox, Michael Jean
Frank, Henry Sorg
Fraser, James Mattison
Frashier, Loyd Dola
Fratiello, Anthony
Freer, Stephan T
Freiling, Edward Clawson
French, Edward P(erry)
Fritz, James John
Fuller, Milton E
Gale, Laird Housel
Gallegos, Emilio Juan
Galperin, Irving
Gantzel, Peter Kellogg
Gardiner, Kenneth William
Gardner, Donald Glenn
Gardner, Phillip John
Garrison, Warren Manford
Gatlin, Lila L
Gay, Charles Francis
Gay, Richard Leslie
Gee, Allen
Gelbart, William M
Gerber, Bernard Robert
Ghandehari, Mohammad Hossein
Gibson, Thomas Alvin, Jr
Gillen, Keith Thomas
Gilmore, John T
Ginell, William Seaman
Givens, William Geary
Goddard, Terrence Patrick
Godycki, Ludwig Edward
Goerke, Jon
Gofman, John William
Goldberg, Ira Barry
Goldberg, Sabine Ruth
Goldish, Elihu
Goldsmith, Henry Arnold
Goldstein, Elisheva
Golub, Morton Allan

Gordon, Alvin S
Gordon, Joseph Grover, II
Graf, Peter Emil
Grantham, Leroy Francis
Green, John William
Green, Robert Bennett
Greenberg, Sidney Abraham
Greene, Charles Richard
Grieman, Frederick Joseph
Gross, Paul Hans
Grossman, Jack Joseph
Grubbs, Edward
Gruhn, Thomas Albin
Grunthaner, Frank John
Gunnink, Raymond
Gupta, Amitava
Gustavson, Marvin Ronald
Gwinn, William Dulaney
Haag, Robert Marlay
Hadley, Steven George
Haendler, Blanca Louise
Haffner, James Wilson
Hahn, Alice Ulhee
Hahn, Harold Thomas
Hall, James Timothy
Hall, Kenneth Lynn
Hammond, R Philip
Handy, Lyman Lee
Hansen, Carl Frederick
Hansford, Rowland Curtis
Hanson, Carl Veith
Hanson, Mervin Paul
Harkins, Carl Girvin
Harnsberger, Hugh Francis
Harris, David Owen
Harris, Gordon McLeod
Harris, Robert A
Harrison, George Conrad, Jr
Harrison, Jonas P
Haskell, Vernon Charles
Haugen, Gilbert R
Hauk, Peter
Hayes, Claude Q C
Hazi, Andrew Udvar
Heinemann, Heinz
Heldman, Morris J
Helgeson, Harold Charles
Hem, John David
Hemminger, John Charles
Herm, Ronald Richard
Hertzler, Barry Lee
Hickson, Donald Andrew
Hidy, George Martel
Hiemenz, Paul C
Hill, Russell John
Hiller, Frederick W
Hills, Graham William
Hiraoka, Hiroyuki
Hisatsune, Isamu Clarence
Hobbs, M Floyd
Hoffman, Charles John
Hoffman, Darleane Christian
Hoffmann, Ronald Lee
Hollahan, John Ronald
Hollander, Jack Marvin
Hollenberg, J Leland
Holmes, Donald Eugene
Holzmann, Richard Thomas
Hong, Keelung
Hoover, William Graham
Hornig, Howard Chester
Hornung, Erwin William
Horsley, John Anthony
Horsma, David August
Hostettler, John Davison
Houle, Frances Anne
Howe, John A
Hsia, Yu-Ping
Hsieh, Paul Yao Tong
Hubert, Jay Marvin
Huntress, Wesley Theodore, Jr
Hunziker, Heinrich Erwin
Hutchinson, Eric
Imhof, William Lowell
Ingraham, Lloyd Lewis
Innes, William Beveridge
Irani, Riyad Ray
Jackson, William Morgan
Jaecker, John Alvin
Jaffe, Annette Bronkesh
Jaffe, Sigmund
James, Dean B
James, Edward, Jr
Jeffries, Jay B
Jendresen, Malcolm Dan
Jensen, James Leslie
Jentoft, Ralph Eugene, Jr
Johnson, Carl Emil, Jr
Johnson, LeRoy Franklin
Johnson, Ralph E
Johnston, Harold Sledge
Johnston, Stewart Archibald
Jones, Dane Robert
Jones, Patrick Ray
Jones, Richard Elmore
Jones, Wesley Morris
Kaelble, David Hardie
Kakis, Frederic Jacob
Kallo, Robert Max
Kamen, Martin David
Kaplan, David Gilbert
Kaplan, Selig N(eil)
Kardonsky, Stanley
Karukstis, Kerry Kathleen

Katekaru, James
Katzin, Leonard Isaac
Kay, Eric
Kearns, David R
Keefer, Raymond Marsh
Keeling, Charles David
Keizer, Joel Edward
Kellerman, Martin
Kempter, Charles Prentiss
Kendig, Martin William
Keneshea, Francis Joseph
Kennedy, James Vern
Kennedy, John Harvey
Kenney, John T
Keys, Richard Taylor
Keyzer, Hendrik
Kibby, Charles Leonard
Kieffer, William Franklin
Kieft, John A
Kinderman, Edwin Max
King, James, Jr
King, William Robert, Jr
Kinney, Gilbert Ford
Kinoshita, Kimio
Kirtman, Bernard
Kivelson, Daniel
Kliger, David Saul
Klinman, Judith Pollock
Knipe, Richard Hubert
Knobler, Charles Martin
Koenig, Inge Rabes
Kohler, Bryan Earl
Kollman, Peter Andrew
Konowalow, Daniel Dimitri
Korst, William Lawrence
Kosiba, Walter Louis
Kozawa, Akiya
Kramer, Henry Herman
Krikorian, Oscar Harold
Kroger, F(erdinand) A(nne)
Kropp, John Leo
Kumamoto, Junji
Kuntz, Irwin Douglas, Jr
Kuppermann, Aron
Lam, Leo Kongsui
Landel, Robert Franklin
Lang, Neil Charles
Langer, Sidney
Langlois, Gordon Ellerby
Lapple, Charles E
Laub, Richard J
Laudenslager, James Bishop
Lauer, George
Lee, Edward Kyung Chai
Lee, Yuan Tseh
Leffler, Esther Barbara
Leiga, Algird George
Leonard, William J, Jr
Lepie, Albert Helmut
Lerner, Narcinda Reynolds
Lester, William Alexander, Jr
Levine, Charles (Arthur)
Levine, Howard Bernard
Levy, Ricardo Benjamin
Lewis, Nathan Saul
Lewis, Sheldon Noah
Liberman, Martin Henry
Lincoln, Kenneth Arnold
Lind, Charles Douglas
Lind, Maurice David
Linde, Peter Franz
Lindenberg, Katja
Lindner, Duane Lee
Lindner, Manfred
Lindquist, Robert Henry
Ling, Alan Campbell
Littauer, Ernest Lucius
Liu, Ming-Biann
Liu, Sung-Tsuen
Live, David H
Lo, George Albert
Long, Franklin A
Lord, Harry Chester, III
Lorenz, Max R(udolph)
Ludwig, Frank Arno
Lundin, Robert Enor
Luther, Marvin L
LuValle, James Ellis
Lyon, David N
Maas, Keith Allan
McCalley, Roderick Canfield
McCarty, Jon Gilbert
McClellan, Aubrey Lester
McConnell, Harden Marsden
McIver, Robert Thomas, Jr
McKenzie, Donald Edward
Mackenzie, John D(ouglas)
McKinney, Ted Meredith
MacLaren, Richard Oliver
McMillan, William George
MacWilliams, Dalton Carson
Magde, Douglas
Magee, John Lafayette
Malik, Jim Gorden
Mallory, Herbert Dean
Maloney, Kenneth Long
Mansfeld, Florian Berthold
Mantei, Kenneth Alan
Maple, M Brian
Marcus, Rudolph Arthur
Marcus, Rudolph Julius
Margerum, John David
Margolis, Jack Selig
Markowitz, Samuel Solomon

Marsden, Sullivan S(amuel), Jr
Marsh, Glenn Anthony
Marsh, Richard Edward
Marshall, Henry Peter
Marsi, Kenneth Larue
Martin, L(aurence) Robbin
Martin, Richard McKelvy
Martin, Stanley Buel
Marx, Paul Christian
Mason, Harold Frederick
Mathews, Larry Arthur
Mayer, Stanley Wallace
Mayo, Frank Rea
Meads, Philip F
Meares, Claude Francis
Medrud, Ronald Curtis
Menefee, Emory
Merrin, Seymour
Merten, Ulrich
Meschi, David John
Metiu, Horia I
Meyer, Paul
Miles, Melvin Henry
Miller, Arnold
Miller, Donald Gabriel
Miller, George E
Miller, James Angus
Miller, William Hughes
Millikan, Roger Conant
Minch, Michael Joseph
Mishuck, Eli
Mlodozeniec, Arthur Roman
Moe, George
Moerner, William Esco
Moore, Gordon Earle
Moretto, Luciano G
Morris, Jerry Lee, Jr
Muller, Rolf Hugo
Murphy, James Francis
Mustafa, Mohammed G
Myers, Benjamin Franklin, Jr
Myers, Rollie John, Jr
Mysels, Karol Joseph
Nadler, Melvin Philip
Namboodiri, Madassery Neelakantan
Nanes, Roger
Nash, Charles Presley
Nash, James Richard
Naylor, Benjamin Franklin
Nazarian, Girair Mihran
Nebenzahl, Linda Levine
Neti, Radhakrishna Murty
Neuman, Robert C, Jr
Newkirk, Herbert William
Newman, John Scott
Newman, Roger
Nichols, Ambrose Reuben, Jr
Nicholson, Daniel Elbert
Nicholson, Margie May
Nicol, Malcolm Foertner
Niday, James Barker
Nimer, Edward Lee
Nishibayashi, Masaru
Noble, Paul, Jr
Noring, Jon Everett
Norman, John Harris
Notley, Norman Thomas
Nugent, Leonard James
Nunn, Roland Cicero
Oakes, John Morgan
Obremski, Robert John
Offen, Henry William
O'Keefe, Dennis Robert
Okumura, Mitchio
O'Loane, James Kenneth
O'Neal, Harry E
Orcutt, John Arthur
Orttung, William Herbert
Ory, Horace Anthony
Otto, Roland John
Otvos, John William
Panzer, Richard Earl
Parks, Norris Jim
Passchier, Arie Anton
Passell, Thomas Oliver
Payton, Patrick Herbert
Pearson, Michael J
Pearson, Robert Melvin
Pecora, Robert
Peller, Leonard
Perlman, Isadore
Peterson, Donald Bruce
Peterson, Donald Lee
Petrucci, Ralph Herbert
Petruska, John Andrew
Phillips, Norman Edgar
Phillips, Roger Winston
Philpott, Michael Ronald
Pickett, Herbert McWilliams
Pines, Alexander
Pitt, Colin Geoffrey
Pitzer, Kenneth Sanborn
Plock, Richard James
Po, Henry N
Potter, Norman D
Potts, John Calvin
Prenzlow, Carl Frederick
Pritchard, Glyn O
Pritchett, Thomas Ronald
Prussin, Stanley Gerald
Pye, Earl Louis
Pyper, James William
Quinlan, John Edward
Rainis, Andrew

Physical Chemistry (cont)

Raleigh, Douglas Overholt
Raley, John Howard
Rard, Joseph Antoine
Rasmussen, John Oscar, Jr
Rast, Howard Eugene, Jr
Razouk, Rashad Elias
Recht, Howard Leonard
Reichman, Sandor
Reisler, Hanna
Reiss, Howard
Rentzepis, Peter M
Rescigno, Thomas Nicola
Rhee, Jay Jea-yong
Rhim, Won-Kyu
Rhodes, David R
Richards, L(orenzo) Willard
Richardson, Jeffery Howard
Richter, Herbert Peter
Ritzman, Robert L
Roberts, Edwin Kirk
Rocklin, Albert Louis
Rodemeyer, Stephen A
Rodgers, James Edward
Roeder, Stephen Bernhard Walter
Roos, Leo
Rosa, Eugene John
Rosenblatt, Gerd Matthew
Rosenblum, Stephen Saul
Rosenkrantz, Marcy Ellen
Rosenthal, Lois C
Ross, David Samuel
Ross, John
Roth, Walter
Rowland, Richard Lloyd
Roy, Prodyot
Rudolph, Raymond Neil
Ruiz, Carl P
Russell, Kenneth Homer
Russell, Thomas Paul
Rustad, Douglas Scott
Ryason, Porter Raymond
Salovey, Ronald
Sanborn, Russell Hobart
Sancier, Kenneth Martin
Saperstein, David Dorn
Savitz, Maxine Lazarus
Scharber, Samuel Robert, Jr
Scharpen, LeRoy Henry
Schatzki, Thomas Ferdinant
Scher, Herbert Benson
Scherer, James R
Schimitschek, Erhard Josef
Schleich, Thomas W
Schmidt, Hartland H
Schmidt, Raymond LeRoy
Schnepp, Otto
Schremp, Frederic William
Schwartz, Joseph Robert
Schwartz, Larry L
Schwartz, Robert Nelson
Scott, Gary Walter
Scott, Robert Lane
Seaborg, Glenn Theodore
Seki, Hajime
Senozan, Nail Mehmet
Servis, Kenneth L
Shaka, Athan James
Sharp, Terry Earl
Sheehan, William Francis
Shelby, Robert McKinnon
Shetlar, Martin David
Shin, Soo H
Shirley, David Arthur
Shoolery, James Nelson
Shreve, George Wilcox
Shudde, Rex Hawkins
Shykind, David
Siebert, Eleanor Dantzler
Siegel, Bernard
Siegel, Irving
Siegel, Seymour
Silva, Robert Joseph
Silverman, Jacob
Sime, Rodney J
Sime, Ruth Lewin
Simon, John Douglas
Simpson, Howard Douglas
Singh, Hakam
Sinton, Steven Williams
Skolnick, Jeffrey
Skowronski, Raymund Paul
Sly, William Glenn
Smith, James Hart
Smith, Louis Charles
Snyder, Robert Gene
Solarz, Richard William
Solomon, Edward I
Somorjai, Gabor Arpad
Sporer, Alfred Herbert
Sposito, Garrison
Sprokel, Gerard J
Staudhammer, Peter
Steele, Warren Cavanaugh
Steffgen, Frederick Williams
Steinberg, Gunther
Steinberg, Martin
Steinmetz, Wayne Edward
Stephens, Edgar Ray
Stephens, Frank Samuel
Stern, John Hanus
Stern, Richard Cecil
Stevens, William George

Stewart, Robert Daniel
Strauss, Herbert L
Streitwieser, Andrew, Jr
Stross, Fred Helmut
Struble, Gordon Lee
Sullivan, Richard Frederick
Suryaraman, Maruthuvakudi Gopalasastri
Sutton, David George
Swalen, Jerome Douglas
Swanson, David G, Jr
Sweeney, Michael Anthony
Switkes, Eugene
Syage, Jack A
Symons, Philip Charles
Szwarc, Michael
Taagepera, Mare
Taft, Robert Wheaton, Jr
Tang, Stephen Shien-Pu
Tannenbaum, Irving Robert
Templeton, David Henry
Tewes, Howard Allan
Theeuwes, Felix
Theodorou, Doros Nicolas
Thomas, John Richard
Thompson, Douglas Stuart
Thompson, Evan M
Thomson, Tom Radford
Tice, Russell L
Tinoco, Ignacio, Jr
Tinti, Dino S
Tobias, Charles W
Toensing, C(larence) H(erman)
Topol, Leo Eli
Trotter, Philip James
True, Nancy S
Trueblood, Kenneth Nyitray
Tschoegl, Nicholas William
Tuck, Leo Dallas
Tuffly, Bartholomew Louis
Tully, Frank Paul
Valencich, Trina J
Valletta, Robert M
Vander Wall, Eugene
Van Hecke, Gerald Raymond
Van Thiel, Mathias
Varteressian, K(egham) A(rshavir)
Vaughan, Philip Alfred
Vermeulen, Theodore (Cole)
Vilker, Vincent Lee
Viola, John Thomas
Vogel, Richard Clark
Vold, Marjorie Jean
Vold, Regitze Rosenorn
Volman, David H
Vroom, David Archie
Wade, Charles Gordon
Walz, Alvin Eugene
Wang, John Ling-Fai
Ward, John William
Warshel, Arieh
Weaver, Henry D, Jr
Weck, Friedrich Josef
Weed, H(omer) C(lyde)
Weinberg, William Henry
Weir, William David
Weiss, Harold Gilbert
Wemmer, David Earl
Wensley, Charles Gelen
Wessel, John Emmit
Wessling, Ritchie A
Whatley, Thomas Alvah
Wheeler, John C
Whetten, Robert Lloyd
Whittam, James Henry
Wiesendanger, Hans Ulrich David
Wildes, Peter Drury
Williams, Colin James
Williams, Richard Stanley
Willis, Grover C, Jr
Wilson, Kent Raymond
Winston, Harvey
Wolf, Kathleen A
Wolfsberg, Max
Wong, Dorothy Pan
Wong, Hans Kuomin
Wong, Joe
Wong, K(wee) C
Wood, David Wells
Woods, William George
Woodson, John Hodges
Worden, Earl Freemont, Jr
Wright, Terry L
Wunderly, Stephen Walker
Wydeven, Theodore
Yang, Meiling T
Yanick, Nicholas Samuel
Yannoni, Costantino Sheldon
Yates, Robert Edmunds
Yayanos, A Aristides
Yee, Rena
Yee, Rena
Yeh, Yin
Young, David A
Zare, Richard Neil
Zewail, Ahmed H
Zink, Jeffrey Irve
Zwerdling, Solomon

COLORADO

Anderson, Larry Gene
Bernstein, Elliot R
Bierbaum, Veronica Marie
Birks, John William

Blakeslee, A Eugene
Cadle, Richard Dunbar
Callanan, Jane Elizabeth
Calvert, Jack George
Chiou, Cary T(sair)
Clough, Francis Bowman
Conner, Jack Michael
Connolly, John Stephen
Czanderna, Alvin Warren
Daughenbaugh, Randall Jay
de Heer, Joseph
DePoorter, Gerald Leroy
Dewey, Thomas Gregory
DoMinh, Thap
Douglas, Larry Joe
Drullinger, Robert Eugene
Eberhart, James Gettins
Edwards, Harry Wallace
Edwards, Kenneth Ward
Ely, James Frank
Estler, Ron Carter
Fischer, William Henry
Frank, Arthur Jesse
Frye, James Sayler
Gibbons, David Louis
Giulianelli, James Louis
Greene, Kenneth Titsworth
Greyson, Jerome
Hanley, Howard James Mason
Howard, Carleton James
James, MarLynn Rees
Keder, Wilbert Eugene
Kennedy, George Hunt
King, Lowell Alvin
Koch, William George
Ladanyi, Branka Maria
Lee, Raymond Curtis
Leone, Stephen Robert
Liao, Martha
Lineberger, William Carl
Macalady, Donald Lee
Maciel, Gary Emmet
Mahan, Kent Ira
Mills, James Wilson
Milne, Thomas Anderson
Mohilner, David Morris
Montgomery, Robert L
Nozik, Arthur Jack
Oetting, Franklin Lee
Olson, George Gilbert
Overend, Ralph Phillips
Ozog, Francis Joseph
Parungo, Farn Pwu
Peters, Kevin Scott
Phillipson, Paul Edgar
Presley, Cecil Travis
Rainwater, James Carlton
Riter, John Randolph, Jr
Robertson, Jerold C
Schwenz, Richard William
Shackelford, Scott Addison
Sherman, David Michael
Smith, Dwight Morrell
Solie, Thomas Norman
Solomon, Susan
Spain, Ian L
Stoner, Allan Wilbur
Strickler, Stewart Jeffery
Thompson, Jay Haughton
Tregillus, Leonard Warren
Turner, James Howard
Vaughan, John Dixon
Watkins, Kay Orville
Watkins, Kenneth Walter
Webb, John Day
Weiner, Eugene Robert
Whatley, Alfred T
Wilkes, John Stuart
Williams, John T
Witters, Robert Dale
Woody, Robert Wayne

CONNECTICUT

Adler, Alan David
Arnold, John Richard
Auerbach, Michael Howard
Bandes, Herbert
Barber, William Austin
Barrante, James Richard
Berets, Donald Joseph
Boggio, Joseph E
Bohn, Robert K
Bongiorni, Domenic Frank
Brinen, Jacob Solomon
Brooks, Clyde S
Brown, Charles Thomas
Brown, Keith Charles
Brown, Oliver Leonard Inman
Burford, M Gilbert
Camp, Eldridge Kimbel
Capwell, Robert J
Cardinal, John Robert
Casper, John Matthew
Castonguay, Richard Norman
Chang, Ted T
Cheng, Francis Sheng-Hsiung
Chupka, William Andrew
Clarke, George A
Corning, Mary Elizabeth
Csejka, David Andrew
David, Carl Wolfgang
Davidson, Theodore
DeDecker, Hendrik Kamiel Johannes
DePhillips, Henry Alfred, Jr

Dobbs, Gregory Melville
Douville, Phillip Raoul
Emond, George T
Epling, Gary Arnold
Falkenstein, Gary Lee
Frey, Douglas D
Fuoss, Raymond Matthew
Gallivan, James Bernard
Garcia, Mario Leopoldo
Glasel, Jay Arthur
Goldstein, Marvin Sherwood
Gray, Thomas James
Haller, Gary Lee
Hautala, Richard Roy
Hekal, Ihab M
Herbette, Leo G
Herron, Michael Myrl
Herz, Jack L
Hinchen, John J(oseph)
Horhota, Stephen Thomas
Hou, Kenneth C
I, Ting-Po
Katz, Lewis
Kowert, Bruce Arthur
Krol, George J
Landsman, Douglas Anderson
Lasaga, Antonio C
Leniart, Daniel Stanley
Linke, William Finan
Lipsky, Seymour Richard
Lisman, Frederick Louis
Loutfy, Rafik Omar
Lowry, Eric G
Lumb, Ralph F
Lyons, Philip Augustine
McBride, James Michael
McCarroll, Bruce
McKeon, Mary Gertrude
Markham, Claire Agnes
Mellor, John
Meriwether, Lewis Smith
Michels, H(orace) Harvey
Moore, Robert Earl
Murai, Kotaro
Novick, Stewart Eugene
O'Connell, Edmond J, Jr
Orloff, Malcolm Kenneth
Orr, William Campbell
Papazian, Louis Arthur
Patterson, Andrew, Jr
Petersson, George A
Phillips, George Wygant, Jr
Piehl, Donald Herbert
Prestegard, James Harold
Prigodich, Richard Victor
Pringle, Wallace C, Jr
Richards, Frederic Middlebrook
Roberts, Elliott John
Roscoe, John Stanley, Jr
Rusling, James Francis
Sanders, William Albert
Sarada, Thyagaraja
Schmitt, Joseph Lawrence, Jr
Schuster, Todd Mervyn
Scott, Raymond Peter William
Sethi, Dhanwant S
Shaw, Montgomery Throop
Simonelli, Anthony Peter
Singh, Ajaib
Skewis, John David
Skutnik, Bolesh Joseph
Smellie, Robert Henderson, Jr
Smith, Frederick Albert
Smith, Sidney Ruven
Smith, Trudy Enzer
Smith, Wendell Vandervort
Steingiser, Samuel
Strazdins, Edward
Stuber, Fred A
Swanson, Donald Leroy
Tobin, Marvin Charles
Tucci, James Vincent
Vaisnys, Juozas Rimvydas
Valentine, Donald H, Jr
Voorhies, John Davidson
Wasser, Richard Barkman
Wellman, Russel Elmer
Wheeler, George Lawrence
Whipple, Earl Bennett
Wolfram, Leszek January
Yang, Darchun Billy
Young, George Jamison
Ziegler, Theresa Frances

DELAWARE

Abrams, Lloyd
Alms, Gregory Russell
Andersen, Donald Edward
Arrington, Charles Hammond, Jr
Barteau, Mark Alan
Barton, Randolph, Jr
Bauchwitz, Peter S
Beare, Steven Douglas
Becher, Paul
Bissot, Thomas Charles
Blaker, Robert Hockman
Bond, Arthur Chalmer
Boyd, Robert Henry
Braun, Juergen Hans
Brill, Thomas Barton
Bristowe, William Warren
Brown, Charles Julian, Jr
Buck, Warren Howard
Cella, Richard Joseph, Jr

Chang, Catherine Teh-Lin
Chase, David Bruce
Chowdhry, Uma
Chu, Sung Gun
Coleman, Marcia Lepri
Cone, Michael McKay
Cooney, John Leo
Culberson, Charles Henry
Dam, Rudy Johan
Davison, Robert Wilder
Day, Herman O'Neal, Jr
Dettre, Robert Harold
Domaille, Peter John
Drake, Arthur Edwin
Dybowski, Cecil Ray
Eaton, David Fielder
Ehrich, F(elix) Frederick
Elliott, Ralph Benjamin
English, Alan Dale
Espy, Herbert Hastings
Evans, Dennis Hyde
Falletta, Charles Edward
Ferguson, Raymond Craig
Firment, Lawrence Edward
Flattery, David Kevin
Fleming, Sydney Winn
Foley, Henry Charles
Foss, Robert Paul
Frensdorff, H Karl
Futrell, Jean H
Gierke, Timothy Dee
Gillow, Edward William
Glasgow, Louis Charles
Gold, Harvey Saul
Golike, Ralph Crosby
Gore, Wilbert Lee
Gorski, Robert Alexander
Gosser, Lawrence Wayne
Graham, Arthur H(ughes)
Grant, David Evans
Haugh, Eugene (Frederick)
Hayek, Mason
Heldt, Walter Z
Herglotz, Heribert Karl Josef
Hiller, Dale Murray
Hoh, George Lok Kwong
Hopkins, Elbert Erskine
Hoppenjans, Donald William
Howe, King Lau
Hsiao, Benjamin S
Huntsberger, James Robert
Ingersoll, Henry Gilbert
Ireland, Carol Beard
John, Andrew
Johnson, David Russell
Johnson, Robert Chandler
Johnson, Rulon Edward, Jr
Kegelman, Matthew Roland
Keidel, Frederick Andrew
Keller, Philip Joseph
Kissa, Erik
Kobsa, Henry
Koch, Theodore Augur
Kolber, Harry John
Kramer, Brian Dale
Kruse, Walter M
Krusic, Paul Joseph
Kwart, Harold
Langsdorf, William Philip
Larry, John Robert
Lauzon, Rodrigue Vincent
Leung, Chung Ngoc
Lund, John Turner
Lyons, Peter Francis
McCoy, V Eugene, Jr
McCullough, Roy Lynn
McGinnis, William Joseph
McKinney, Charles Dana, Jr
Malone, Creighton Paul
Mannis, Fred
Marshall, William Joseph
Mastrangelo, Sebastian Vito Rocco
Maury, Lucien Garnett
Meakin, Paul
Merrifield, Richard Ebert
Mills, George Alexander
Moore, Ralph Bishop
Morrissey, Bruce William
Moses, Francis Guy
Munson, Burnaby
Neff, Bruce Lyle
Newby, William Edward
Nichols, James Randall
Niedzielski, Edmund Luke
Noggle, Joseph Henry
Odiorne, Truman J
Ovenall, Derick William
Overman, Joseph DeWitt
Palmer, Alan Blakeslee
Panar, Manuel
Paris, Jean Philip
Pariser, Rudolph
Parkins, John Alexander
Parkinson, Bruce Alan
Parrish, Robert G
Paulson, Charles Maxwell, Jr
Perri, Joseph Mark
Phillips, Brian Ross
Pierce, Marion Armbruster
Pierce, Robert Henry Horace, Jr
Pieski, Edwin Thomas
Podlas, Thomas Joseph
Polejes, J(acob) D
Reddy, Gade Subbarami

Restaino, Alfred Joseph
Reuben, Jacques
Ridge, Douglas Poll
Roe, David Christopher
Ross, William D(aniel)
Rudat, Martin August
Saffer, Henry Walker
St John, Daniel Shelton
Sandell, Lionel Samuel
Sarner, Stanley Frederick
Schmude, Keith E
Senkler, George Henry, Jr
Shain, Albert Leopold
Sheer, M(axine) Lana
Shozda, Raymond John
Simpson, David Alexander
Slade, Arthur Laird
Sloan, Gilbert Jacob
Smart, Bruce Edmund
Smith, Kenneth McGregor
Smith, Ronald W
Sopp, Samuel William
Spurlin, Harold Morton
Staikos, Dimitri Nickolas
Staley, Ralph Horton
Starkweather, Howard Warner, Jr
Tabibian, Richard
Tanikella, Murty Sundara Sitarama
Thayer, Chester Arthur
Thompson, Robert Gene
Trumbore, Conrad Noble
Van Dyk, John William
Walsh, Robert Michael
Wang, Victor Kai-Kuo
Webb, Charles Alan
Weiher, James F
Wendt, Robert Charles
Wetlaufer, Donald Burton
Whitehouse, Bruce Alan
Wiley, Douglas Walker
Wilhite, Douglas Lee
Wolfe, William Ray, Jr
Wood, Robert Hemsley
Wopschall, Robert Harold
Yiannos, Peter N
Zimmerman, Joseph

DISTRICT OF COLUMBIA
Abelson, Philip Hauge
Ali, Mahamed Asgar
Anderson, Jack R(oger)
Barkin, Stanley M
Bates, Richard Doane, Jr
Bednarek, Jana Marie
Berkowitz, Joan B
Brice, Luther Kennedy
Burley, Gordon
Burnett, John Lambe
Butler, James Ehrich
Callanan, Margaret Joan
Chang, Eddie Li
Chapin, Douglas Scott
Chiu, Ying-Nan
Corsaro, Robert Dominic
Cramer, William Herbert
Crist, DeLanson Ross
DeLevie, Robert
Donahue, D Joseph
Earley, Joseph Emmet
Fearn, James Ernest
Graminski, Edmond Leonard
Gress, Mary Edith
Griscom, David Lawrence
Gutman, David
Halpern, Joshua Baruch
Hancock, Kenneth George
Harris, William Charles
Herr, Frank Leaman, Jr
Heydemann, Peter Ludwig Martin
Hoering, Thomas Carl
Hoffman, Michael K
Hsu, David Shiao-Yo
Hughes, Robert Edward
Jawed, Inam
Karle, Jerome
Kirchhoff, William Hayes
Klein, Philipp Hillel
Laufer, Allan Henry
Lutz, George John
McCormack, Mike
Mack, Julius L
McNeal, Robert Joseph
Maisch, William George
Martire, Daniel Edward
May, Leopold
Merrifield, D Bruce
Meyer, Richard Adlin
Mies, Frederick Henry
Munson, Ronald Alfred
Nelson, David Lynn
O'Grady, William Edward
Okabe, Hideo
O'Leary, Timothy Joseph
Pierce, Dorothy Helen
Poranski, Chester F, Jr
Price, Elton
Raizen, Senta Amon
Ramaker, David Ellis
Roller, Paul S
Roscher, Nina Matheny
Rothman, Sam
Rowley, David Alton
Schnur, Joel Martin
Shaub, Walter M

Shirk, James Siler
Simmons, Joe Denton
Smith, Leslie E
Stern, Kurt Heinz
Sterrett, Kay Fife
Sutter, John Ritter
Swisher, James H(owe)
Tanczos, Frank I
Tolles, William Marshall
Trzaskoma, Patricia Povilitis
Tsang, Tung
Turner, Noel Hinton
Webb, Alan Wendell
Wickman, Herbert Hollis
Wilson, Mathew Kent
Wodarczyk, Francis John
Yankwich, Peter Ewald
Zwolenik, James J

FLORIDA
Adisesh, Setty Ravanappa
Allen, Eric Raymond
Ambrose, John Russell
Austin, Alfred Ells
Babich, Michael Wayne
Bailey, Thomas L, III
Banter, John C
Bartlett, Rodney Joseph
Batich, Christopher David
Baum, J(ames) Clayton
Binford, Jesse Stone, Jr
Birdwhistell, Ralph Kenton
Bishop, Charles Anthony
Blank, Charles Anthony
Block, Ronald Edward
Boehnke, David Neal
Boonstra, Bram B
Brey, Wallace Siegfried, Jr
Bright, Willard Mead
Brucat, Philip John
Carraher, Charles Eugene, Jr
Chiotti, Premo
Choppin, Gregory Robert
Chu, Ting Li
Cioslowski, Jerzy
Clay, John Paul
Colgate, Samuel Oran
Criss, Cecil M
Dakin, Thomas Wendell
Davis, Jefferson Clark, Jr
Debnath, Sadhana
DeLap, James Harve
Diamond, Jacob J(oseph)
Dibeler, Vernon Hamilton
Dougherty, Ralph C
Drake, Robert Firth
Drost-Hansen, Walter
Duedall, Iver Warren
Edelson, David
Ellis, William Hobert
Ernsberger, Fred Martin
Eubank, William Roderick
Eyler, John Robert
Foner, Samuel Newton
Francis, Stanley Arthur
Fried, Vojtech
Friedlander, Herbert Norman
Fulton, Robert Lester
Garrett, Richard Edward
Glick, Richard Edwin
Glover, Roland Leigh
Goll, Robert John
Gropp, Armin Henry
Gruntfest, Irving James
Gurry, Robert Wilton
Hall, Elton Harold
Hanrahan, Robert Joseph
Hardman, Bruce Bertolette
Hare, Curtis R
Harrow, Lee Salem
Henkin, Hyman
Hertel, George Robert
Hirko, Ronald John
Hirt, Thomas J(ames)
Huba, Francis
Hudson, Alice Peterson
Hudson, Reggie Lester
Hull, Harry H
Jackson, George Frederick, III
Johnsen, Russell Harold
Johnston, Milton Dwynell, Jr
Julien, Hiram Paul
Jurch, George Richard, Jr
Kasha, Michael
Kirshenbaum, Abraham David
Kreider, Henry Royer
Leidheiser, Henry, Jr
Linder, Bruno
Llewellyn, J(ohn) Anthony
Maybury, Paul Calvin
Meisels, Gerhard George
Micha, David Allan
Millero, Frank Joseph, Jr
Mills, Alfred Preston
Milton, Robert Mitchell
Moody, Leroy Stephen
Morehouse, Clarence Kopperl
Muga, Marvin Luis
Muschlitz, Earle Eugene, Jr
Myers, Gardiner Hubbard
Myerson, Albert Leon
Palmer, Jay
Parkanyi, Cyril
Parker, John Hilliard

Parrish, Clyde Franklin
Person, Willis Bagley
Perumareddi, Jayarama Reddi
Pierce, James Bruce
Pitt, Donald Alfred
Ramsey, Brian Gaines
Rikvold, Per Arne
Rossini, Frederick Dominic
Roth, James Frank
Sadler, Monroe Scharff
Saffer, Alfred
Safron, Sanford Alan
Sakhnovsky, Alexander Alexandrovitch
Schmidt, Klaus H
Schulman, Stephen Gregory
Schultz, Franklin Alfred
Sheldon, John William
Snyder, Patricia Ann
Steadman, Thomas Ree
Stevens, Brian
Swartz, William Edward, Jr
Szczepaniak, Krystyna
Turner, Ralph Waldo
Urone, Paul
Vala, Martin Thorvald, Jr
Van Wart, Harold Edgar
Van Zee, Richard Jerry
Vergenz, Robert Allan
Weltner, William, Jr
Whitney, Ellsworth Dow
Wolsky, Sumner Paul
Woodward, Fred Erskine
Worrell, Jay H
Wu, Jin Zhong
Young, Frank Glynn
Zaukelies, David Aaron
Zerner, Michael Charles

GEORGIA
Anderson, Bruce Martin
Anderson, James Leroy
Barr, John Baldwin
Barton, Franklin Ellwood, II
Batten, George L, Jr
Block, Toby Fran
Borkman, Raymond Francis
Brandau, Betty Lee
Camp, Ronnie Wayne
Carreira, Lionel Andrade
Cheng, Wu-Chieh
Clever, Henry Lawrence
Colvin, Clair Ivan
Cooper, Charles Dewey
Cooper, Gerald Rice
Crider, Fretwell Goer
Cummings, Frank Edson
DeLorenzo, Ronald Anthony
Derrick, Mildred Elizabeth
Dever, David Francis
Eberhardt, William Henry
Etzler, Frank M
Felton, Ronald H
Friedman, Harold Bertrand
Garrett, Bowman Staples
Gayles, Joseph Nathan, Jr
Goldstein, Jacob Herman
Gole, James Leslie
Grenga, Helen E(va)
Griffiths, James Edward
Habte-Mariam, Yitbarek
Harmer, Don Stutler
Hart, Raymond Kenneth
Heric, Eugene LeRoy
Hopkins, Harry P, Jr
James, Jeffrey
Johnston, Francis J
Jones, Alan Richard
Karickhoff, Samuel Woodford
Kaufman, Myron Jay
Keller, Oswald Lewin
Kohl, Paul Albert
Kotliar, Abraham Morris
Kozak, John Joseph
Kromann, Paul Roger
Lee, John William
Lin, Ming Chang
Lindauer, Maurice William
Liu, Fred Wei Jun
Lokken, Stanley Jerome
MacMillan, Joseph Edward
Macur, George J
Manson, Steven Trent
Melton, Charles Estel
Menon, Manchery Prabhakara
Miller, George Alford
Moran, Thomas Francis
Nelson, Robert Norton
Netherton, Lowell Edwin
Netzel, Thomas Leonard
Nickerson, John David
Orofino, Thomas Allan
Pacifici, James Grady
Perdue, Edward Michael
Pierotti, Robert Amadeo
Pignocco, Arthur John
Ray, Apurba Kanti
Reif, Donald John
Rhoades, James Lawrence
Richardson, Susan D
Rouvray, Dennis Henry
Samuels, Robert Joel
Sanders, T H, Jr
Sherry, Peter Burum
Smith, Darwin Waldron

Physical Chemistry (cont)

Spauschus, Hans O
Spencer, Jesse G
Starr, Thomas Louis
Steele, Jack
Story, Troy Lee, Jr
Sturrock, Peter Earle
Teja, Amyn Sadruddin
Thomas, Tudor Lloyd
Tincher, Wayne Coleman
Torres, Lourdes Maria
Trawick, William George
Walker, William Comstock
Watts, Sherrill Glenn
Wilson, William David
Winnick, Jack
Zepp, Richard Gardner

HAWAII

Bopp, Thomas Theodore
Brantley, Lee Reed
Chu, Victor Fu Hua
Dorrance, William Henry
Fadley, Charles Sherwood
Frystak, Ronald Wayne
Hanna, Melvin Wesley
Hirayama, Chikara
Ihrig, Judson La Moure
Inskeep, Richard Guy
McDonald, Ray Locke
Mader, Charles Lavern
Muenow, David W
Seff, Karl
Sood, Satya P
Waugh, John Lodovick Thomson

IDAHO

Baker, John David
Berreth, Julius R
Carter, Loren Sheldon
Christian, Jerry Dale
Faler, Kenneth Turner
Hagen, Jack Ingvald
Hammer, Robert Russell
Harmon, J Frank
Heckler, George Earl
Lewis, Leroy Crawford
Miller, Richard Lloyd
Nagel, Terry Marvin
Ramshaw, John David
Renfrew, Malcolm MacKenzie
Schuman, Robert Paul
Schutte, William Calvin
Tallman, Richard Louis
Tanner, John Eyer, Jr
Wai, Chien Moo

ILLINOIS

Abraham, Bernard M
Adams, Richard Melverne
Albrecht, William Lloyd
Anysas, Jurgis Arvydas
Arnold, Richard Thomas
Atoji, Masao
Bader, Samuel David
Baker, Louis, Jr
Beese, Ronald Elroy
Begala, Arthur James
Belford, R Linn
Benzinger, William Donald
Berkowitz, Joseph
Berry, Richard Stephen
Blander, Milton
Bloemer, William Louis
Blumberg, Avrom Aaron
Brubaker, Inara Mencis
Budrys, Rimgaudas S
Burdett, Jeremy Keith
Burns, Richard Price
Burrell, Elliott Joseph, Jr
Burwell, Robert Lemmon, Jr
Cafasso, Fred A
Carlson, Keith Douglas
Carnahan, Jon Winston
Carnall, William Thomas
Cengel, John Anthony
Ceperley, David Matthew
Chandler, Dean Wesley
Chang, Chin Hsiung
Chen, Juh Wah
Clardy, LeRoy
Clarkson, Robert Breck
Cochrane, Hector
Coley, Ronald Frank
Coutts, John Wallace
Crespi, Henry Lewis
Crumrine, David Shafer
Cummings, Thomas Fulton
Cunningham, George Lewis, Jr
Cutnell, John Daniel
Dehmer, Joseph Leonard
Dehmer, Patricia Moore
DeVault, Don Charles
Diamond, Herbert
Dickinson, Helen Rose
Didwania, Hanuman Prasad
Dlott, Dana D
Dobbs, Frank W
Dobry, Alan (Mora)
Ebdon, David William
Ebert, Lawrence Burton
Ehrlich, Gert
Eissler, Robert L

Eliason, Morton A
El Saffar, Zuhair M
Faber, Roger Jack
Farewell, John P
Faulkner, Larry Ray
Feng, Paul Yen-Hsiung
Fields, Paul Robert
Filson, Don P
Firkins, John Lionel
Fischer, Albert Karl
Fitch, Alanah
Fleming, Graham Richard
Frederick, Kenneth Jacob
Freed, Karl F
Freeman, Maynard Lloyd
Gaffney, Jeffrey Steven
Gemmer, Robert Valentine
Gindler, James Edward
Gislason, Eric Arni
Gleeson, James Newman
Glogovsky, Robert L
Glonek, Thomas
Golinkin, Herbert Sheldon
Gomer, Robert
Gonzalez, Richard Donald
Goodman, Leonard Seymour
Gordon, Robert Jay
Gordon, Sheffield
Granick, Steve
Gray, Linsley Shepard, Jr
Green, David William
Griffin, Harold Lee
Grove, Ewart Lester
Gruen, Dieter Martin
Gutfreund, Kurt
Gutowsky, Herbert S(ander)
Hadley, Fred Judson
Hakala, Reino William
Hardwidge, Edward Albert
Henderson, George Asa
Henderson, Giles Lee
Hersh, Herbert N
Hinckley, Conrad Cutler
Hinks, David George
Hoeg, Donald Francis
Hoffman, Alan Bruce
Hoffman, Brian Mark
Holcomb, David Nelson
Hollins, Robert Edward
Homeier, Edwin H, Jr
Hopkins, Paul Donald
House, James Evan, Jr
Huang, Leo W
Huang, Shu-Jen Wu
Hummel, John Philip
Huston, John Lewis
Hutchison, Clyde Allen, Jr
Illingworth, George Ernest
Iton, Lennox Elroy
Jaffey, Arthur Harold
Jameson, A Keith
Johnson, Irving
Johnson, Marvin Francis Linton
Jonah, Charles D
Jonas, Jiri
Jones, Richard Evan, Jr
Jones, Thomas Hubbard
Justen, Lewis Leo
Kaplan, Sam H
Kasner, Fred E
Katz, Joseph J
Katz, Sidney
Kaufman, Sheldon Bernard
Keiderling, Timothy Allen
Keiter, Ellen Ann
Kevill, Dennis Neil
Kiefer, John Harold
Kipp, James Edwin
Klein, Max
Kleppa, Ole Jakob
Kooser, Robert Galen
Koster, David F
Kotin, Leonard
Krawetz, Arthur Altshuler
Krueger, Robert Harold
Kuhlmann, George Edward
Kung, Harold Hing Chuen
Lam, Daniel J
Lang, Robert Phillip
LaPlanche, Laurine A
Larsen, Robert Peter
Lauterbur, Paul Christian
LeBreton, Pierre Robert
Leibowitz, Leonard
Leland, Frances E(lbridge)
Lester, George Ronald
Levy, Donald Harris
Lewis, Frederick D
Light, John Caldwell
Lisy, James Michael
Longworth, James W
Lucchesi, Claude A
Luerssen, Frank W
Lustig, Stanley
McDonald, Hugh Joseph
Mahajan, Om Prakash
Makinen, Marvin William
Makowski, Mieczyslaw Paul
Maroni, Victor August
Marquart, John R
Marvin, Henry Howard, Jr
Masel, Richard Isaac
Mason, Donald Frank
Meadows, James Wallace, Jr

Meguerian, Garbis H
Meisel, Dan
Melendres, Carlos Arciaga
Mendelsohn, Marshall H
Metz, Florence Irene
Meyer, Edwin F
Michael, Joe Victor
Miller, John Robert
Mooring, Francis Paul
Nachtrieb, Norman Harry
Nagy, Zoltan
Nandi, Satyendra Prosad
Nash, Kenneth Laverne
Nebgen, John William
Nichols, George Morrill
Nieman, George Carroll
Nilges, Mark J
Nordine, Paul Clemens
Oka, Takeshi
O'Neal, Harry Roger
Ostrow, Jay Donald
Oxtoby, David W
Padrta, Frank George
Parks, Eric K
Parrill, Irwin Homer
Paul, Iain C
Pearlstein, Arne Jacob
Perkins, Alfred J
Person, Lucy Wu
Pertel, Richard
Phifer, Harold Edwin
Pohlmann, Hans Peter
Pollitzer, Ernest Leo
Poppe, Wassily
Pratt, Stephen Turnham
Rakowsky, Frederick William
Ratner, Mark A
Rayudu, Garimella V S
Reed, John Francis
Rice, Stuart Alan
Richards, R Ronald
Ries, Herman Elkan, Jr
Riley, Stephen James
Rocek, Jan
Rothman, Alan Bernard
Russell, Morley Egerton
Saboungi, Marie-Louise Jean
Satterlee, James Donald
Sauer, Myran Charles, Jr
Schatz, George Chappell
Scheiner, Steve
Schrage, Samuel
Schreiner, Felix
Scouten, Charles George
Secrest, Donald H
Sedlet, Jacob
Sharma, Brahma Dutta
Sherman, Warren V
Shriver, John William
Sibener, Steven Jay
Sinha, Shome Nath
Smith, Gerard Vinton
Snelson, Alan
Snyder, Gary James
Spangler, Charles William
Spears, Kenneth George
Sproul, William Dallas
Stair, Peter Curran
Stetter, Joseph Robert
Stout, John Willard
Streets, David George
Strehlow, Roger Albert
Sugarman, Nathan
Tetenbaum, Marvin
Thurnauer, Marion Charlotte
Trevorrow, Laverne Everett
Trifunac, Alexander Dimitrije
Turkevich, Anthony
Unger, Lloyd George
Vandeberg, John Thomas
Vander Burgh, Leonard F
Van Duyne, Richard Palmer
Van Lente, Kenneth Anthony
Vanýsek, Petr
Vaughn, Joe Warren
Veis, Arthur
Veleckis, Ewald
Viehland, Larry Alan
Walter, Robert Irving
Washington, Elmer L
Wasielewski, Michael Roman
Weisleder, David
Weitz, Eric
Wenzel, Bruce Erickson
Wesolowski, Wayne Edward
Wharton, Lennard
White, Jesse Edmund
Williams, Lesley Lattin
Wilson, Robert Steven
Wingender, Ronald John
Woehler, Scott Edwin
Woo, Lecon
Wozniak, Wayne Theodore
Wu, Ying Victor
Young, Austin Harry
Young, Charles Edward
Young, Linda
Zaromb, Solomon
Zimmerman, Donald Nathan
Zletz, Alex

INDIANA

Allerhand, Adam
Alter, John Emanuel

Bair, Edward Jay
Balajee, Shankverm R
Bangs, Leigh Buchanan
Bischoff, Robert Francis
Boaz, Patricia Anne
Bonham, Russell Aubrey
Booe, J(ames) M(arvin)
Borden, Kenneth Duane
Bostick, Edgar E
Briggs, Thomas N
Bryan, William Phelan
Byrn, Stephen Robert
Chateauneuf, John Edward
Chernoff, Donald Alan
Daly, Patrick Joseph
Das, Paritosh Kumar
Desai, Pramod D
Diestler, Dennis Jon
Droege, John Walter
Dubin, Paul Lee
Dykstra, Clifford Elliot
Ewing, George Edward
Fehlner, Thomas Patrick
Ferraudi, Guillermo Jorge
Fessenden, Richard Warren
Fidelle, Thomas Patrick
Flick, Cathy
Fong, Francis K
Fong, Shao-Ling
Genshaw, Marvin Alden
Grant, Edward R
Grimley, Robert Thomas
Hagstrom, Stanley Alan
Hall, David Alfred
Halpern, Arthur Merrill
Hamill, William Henry
Havel, Henry Acken
Hayes, Robert Green
Helman, William Phillip
Hemmes, Paul Richard
Heydegger, Helmut Roland
Hubble, Billy Ray
Hulbert, Matthew H
Hunt, Ann Hampton
Jones, Noel Duane
Kamat, Prashant V
Kelly, Edward Joseph
King, Gayle Nathaniel
Kinsey, Philip A
Kirsch, Joseph Lawrence, Jr
Kosman, Warren Melvin
Kossoy, Aaron David
Kuo, Charles C Y
Langer, Lawrence Marvin
Langhoff, Peter Wolfgang
Larter, Raima
Leonard, Jack E
Lorentzen, Keith Eden
McKinney, Paul Caylor
Madden, Keith Patrick
Malik, David Joseph
Mark, Earl Larry
Marsh, Max Martin
Mazac, Charles James
Meiser, John H
Montgomery, Lawrence Kernan
Moore, Walter John
Morris, David Alexander Nathaniel
Mozumder, Asokendu
Mueller, Charles Richard
Muller, Norbert
Ogren, Paul Joseph
Onwood, David P
Ostroy, Sanford Eugene
Parmenter, Charles Stedman
Patterson, Larry K
Pearlstein, Robert Milton
Peters, Dennis Gail
Pierce, Louis
Pikal, Michael Jon
Pilger, Richard Christian, Jr
Pimblott, Simon Martin
Pinkham, Chester Allen, III
Porile, Norbert Thomas
Reilly, James Patrick
Richardson, James Wyman
Ricketts, John Adrian
Schuler, Robert Hugo
Shiner, Vernon Jack, Jr
Shoup, Charles Samuel, Jr
Simms, Paul C
Smith, David Lee
Smith, Gerald Duane
Sousa, Lynn Robert
Steinrauf, Larry King
Stevenson, Kenneth Lee
Streib, William E
Strieder, William
Strong, Laurence Edward
Subramanian, Sethuraman
Sutula, Chester Louis
Tatum, James Patrick
Tensmeyer, Lowell George
Thomas, John Kerry
Toomey, Joseph Edward
Tripathi, Govakh Nath Ram
Trozzolo, Anthony Marion
Vest, Robert W(ilson)
Voorhoeve, Rudolf Johannes Herman
Wagner, Eugene Stephen
Wallace, Gerald Wayne
Weaver, Michael John
Williams, Edward James
Wyma, Richard J

IOWA

Allen, Susan Davis
Applequist, Jon Barr
Baenziger, Norman Charles
Baker, James LeRoy
Breuer, George Michael
Buettner, Garry Richard
Cater, Earle David
Chang, James C
Coffman, Robert Edgar
Cotton, Therese Marie
Danzer, Laurence Alfred
DePristo, Andrew Elliott
Erickson, Luther E
Franzen, Hugo Friedrich
Friedrich, Bruce H
Gerstein, Bernard Clemence
Glick, Milton Don
Gschneidner, Karl A(lbert), Jr
Hansen, Peter Jacob
Hansen, Robert Suttle
Hoekstra, John Junior
Jackson, Herbert Lewis
Jacob, Fielden Emmitt
Jacobson, Robert Andrew
Jordan, Truman H
Kintanar, Agustin
Lutz, Robert William
Maatman, Russell Wayne
Martin, Don Stanley, Jr
Martin, S W
Mentone, Pat Francis
Moriarty, John Lawrence, Jr
Mottley, Carolyn
Powell, Jack Edward
Rider, Paul Edward, Sr
Smith, John F(rancis)
Smith, Karen Ann
Struve, Walter Scott
Stwalley, William Calvin
Svec, Harry John
Swenson, Charles Allyn
Wilhelm, Harley A
Wurster, Dale Eric, Jr
Yeung, Edward Szeshing
Zemke, Warren T

KANSAS

Allen, Anneke S
Argersinger, William John, Jr
Copeland, James Lewis
Fujimoto, Gordon Takeo
Funderburgh, James Louis
Gilles, Paul Wilson
Greene, Frank T
Greenlief, Charles M
Hammaker, Robert Michael
Harmony, Marlin D
Hiebert, Allen G
Ho, James Chien Ming
Judson, Charles Morrill
Kruh, Robert Frank
Kuwana, Theodore
Latschar, Carl Ernest
Lehman, Thomas Alan
Lindenbaum, Siegfried
Lookhart, George LeRoy
McElroy, Albert Dean
Moser, Herbert Charles
Noelken, Milton Edward
Potts, Melvin Lester
Renich, Paul William
Rose, Wayne Burl
Rumpel, Max Leonard
Rytting, Joseph Howard
Setser, Donald W
Shearer, Edmund Cook
Sorensen, Christopher Michael
Talaty, Erach R
Wahlbeck, Phillip Glenn
White, R Milford

KENTUCKY

Armendarez, Peter X
Beall, Gary Wayne
Beyer, Louis Martin
Bujake, John Edward, Jr
Butterfield, David Allan
Byrn, Ernest Edward
Chamberlin, John MacMullen
Clouthier, Dennis James
Companion, Audrey (Lee)
Conley, Harry Lee, Jr
Davis, Burtron H
Draper, Arthur Lincoln
Du Pre, Donald Bates
Farrar, David Turner
Field, Jay Ernest
Goren, Alan Charles
Guthrie, Robert D
Henley, Melvin Brent, Jr
Hettinger, William Peter, Jr
Johnson, Lawrence Robert
Keely, William Martin
Klein, Elias
Levey, Gerrit
Loeb, Leopold
McDermott, Dana Paul
Mapes, William Henry
Mitchell, Maurice McClellan, Jr
Moorhead, Edward Darrell
Mueller, Sister Rita Marie
Pattengill, Merle Dean
Pearson, Earl F

Plucknett, William Kennedy
Porter, Richard A
Price, Martin Burton
Reucroft, Philip J
Sands, Donald Edgar
Sears, Paul Gregory
Smiley, Harry M
Taylor, Morris D
Thompson, Ralph J
Trapp, Charles Anthony
Wilkins, Curtis C
Williams, Donald Elmer

LOUISIANA

Anex, Basil Gideon
Arthur, Jett Clinton, Jr
Bains, Malkiat Singh
Barrett, Robert Earl
Benerito, Ruth Rogan
Berni, Ralph John
Bissell, Charles Lynn
Boudreaux, Edward A
Bursh, Talmage Poutau
Byrd, David Shelton
Carmichael, J W, Jr
Carpenter, Dewey Kenneth
Craig, James Porter, Jr
Crowder, Gene Autrey
Day, Marion Clyde, Jr
Fagley, Thomas Fisher
Fischer, David John
Franklin, William Elwood
French, Alfred Dexter
Hamori, Eugene
Hanlon, Thomas Lee
Jones, Daniel Elven
Jung, Hilda Ziifle
Keeley, Dean Francis
Kern, Ralph Donald, Jr
Kestner, Neil R
Klasinc, Leo
Kumar, Devendra
La Rochelle, John Hart
Looney, Ralph William
McGlynn, Sean Patrick
Mark, Harold Wayne
Miller, Kenneth Jay
Moseley, Harry Edward
Nauman, Robert Vincent
Nelson, Mary Lockett
Oliver, James Russell
Perkins, Richard Scott
Phillips, Travis J
Plessy, Boake Lucien
Pullen, Bailey Price
Raisen, Elliott
Ratchford, Robert James
Riddick, Frank Adams, Jr
Roberts, Donald Duane
Robinson, Press L
Robson, Harry Edwin
Runnels, Lynn Kelli
Schexnayder, Mary Anne
Schroeder, Rudolph Alrud
Schucker, Robert Charles
Scott, John Delmoth
Sen, Buddhadev
Smith, Martin Bristow
Thompson, Ronald Hobart
Vickroy, Virgil Vester, Jr
Von Bodungen, George Anthony
Wan, Peter J
Ward, Truman L
Wendt, Richard P
Wharton, James Henry
Williams, Hulen Brown
Wood, James Manley, Jr
Wright, Oscar Lewis
Zagar, Walter T

MAINE

Allen, Roger Baker
Batha, Howard Dean
Birkett, James Davis
Boyles, James Glenn
Butcher, Samuel Shipp
Duncan, Charles Donald
Dunlap, Robert D
Fort, Raymond Cornelius, Jr
Goodfriend, Paul Louis
Harris, Robert Laurence
Holden, James Richard
Kirkpatrick, Francis Hubbard
Lauterbach, George Ervin
Merrifield, Paul Elliott
Patterson, Howard Hugh
Rasaiah, Jayendran C
Shattuck, Thomas Wayne
Simard, Gerald Lionel
Smith, Wayne Lee
Sottery, Theodore Walter
Stauffer, Charles Henry
Stebbins, Richard Gilbert
Wilson, Thomas Lee

MARYLAND

Abramowitz, Stanley
Adrian, Frank John
Albers, Edwin Wolf
Aranow, Ruth Lee Horwitz
Astumian, Raymond Dean
Auel, RaeAnn Marie
Avery, William Hinckley
Bardo, Richard Dale

Barnes, Charlie James
Barnes, James Alford
Bax, Ad
Becker, Edwin Demuth
Benson, Richard C
Berl, Walter G(eorge)
Bernecker, Richard Rudolph
Bertocci, Ugo
Bestul, Alden Beecher
Bicknell-Brown, Ellen
Blankenship, Floyd Allen
Block, Jacob
Block, Stanley
Blumenthal, Robert Paul
Bonnell, David William
Bonsack, James Paul
Bowen, Kit Hansel
Brenner, Abner
Brown, Samuel Heffner
Brown, Walter Eric
Bryden, Wayne A
Callahan, John Joseph
Callahan, Mary Vincent
Campbell, William Joseph
Casassa, Michael Paul
Castellan, Gilbert William
Cavanagh, Richard Roy
Cave, William Thompson
Chang, Ren-Fang
Chang, Shu-Sing
Charney, Elliot
Chen, Ching-Nien
Chmurny, Gwendolyn Neal
Chow, Laurence Chung-Lung
Clark, Joseph E(dward)
Clark, Paul Enoch
Clarke, Frederic B, III
Coates, Arthur Donwell
Cody, Regina Jacqueline
Cohen, Gerson H
Cohen, Louis Arthur
Coleman, James Stafford
Colle, Ronald
Colwell, Jack Harold
Coplan, Michael Alan
Coriell, Sam Ray
Coursey, Bert Marcel
Cross, David Ralston
Cunniff, Patricia A
Currie, Lloyd Arthur
Dagdigian, Paul J
Darwent, Basil de Baskerville
Davis, George Thomas
Davis, Henry McRay
Debye, Nordulf Wiking Gerud
Dehn, James Theodore
Deitz, Victor Reuel
DeToma, Robert Paul
DeVoe, Howard Josselyn
Dickens, Brian
Doan, Arthur Sumner, Jr
Dragun, Henry L
Duignan, Michael Thomas
Dunning, Herbert Neal
Eanes, Edward David
Eden, Murray
Edinger, Stanley Evan
Egan, William Michael
Egelhoff, William Frederick, Jr
Eng, Leslie
Epstein, Joseph
Ewing, June Swift
Ferretti, James Alfred
Fifer, Robert Alan
Flynn, Joseph Henry
Forziati, Alphonse Frank
Fox, Robert Bernard
Fraser, Gerald Timothy
Freedman, Eli (Hansell)
Freeman, David Haines
Frohnsdorff, Geoffrey James Carl
Fulmer, Glenn Elton
Gann, Richard George
Garvin, David
Gevantman, Lewis Herman
Gilbert, Francis Charles
Goldberg, Robert Nathan
Goldenberg, Neal
Gornick, Fred
Gottlieb, Melvin Harvey
Grant, Warren Herbert
Gravatt, Claude Carrington, Jr
Green, John Arthur Savage
Greenhouse, Harold Mitchell
Greer, Sandra Charlene
Greer, William Louis
Griffo, Joseph Salvatore
Gryder, John William
Guruswamy, Vinodhini
Hadermann, Albert Felix
Haller, Wolfgang Karl
Hamer, Walter Jay
Hampson, Robert F, Jr
Hartman, Kenneth Owen
Hartstein, Arthur M
Hashmall, Joseph Alan
Hastie, John William
Hedman, Fritz Algot
Herrick, Claude Cummings
Herron, John Thomas
Hiatt, Caspar Wistar, III
Hillstrom, Warren W
Hirsh, Allen Gene
Hofrichter, Harry James

Hollandsworth, Clinton E
Hollies, Norman Robert Stanley
Hougen, Jon T
Houseman, Barton L
Howell, Barbara Fennema
Hsu, Chen C
Huang, Charles Y
Hunt, Paul Payson
Iwasa, Kunihiko
Jachimowicz, Felek
Jackson, Jo-Anne Alice
Jacox, Marilyn Esther
Jaffe, Harold
James, Stanley D
Jarvis, Neldon Lynn
Jernigan, Robert Lee
Jonas, Leonard Abraham
Jones, Owen Lloyd
Jovancicevic, Vladimir
Julienne, Paul Sebastian
Kamlet, Mortimer Jacob
Kan, Lou Sing
Kandel, Richard Joshua
Kappe, David Syme
Katz, Joseph L
Kaufman, Joyce J
Kaufman, Samuel
Ketley, Arthur Donald
Khare, Mohan
Kippenberger, Donald Justin
Kirkien-Rzeszotarski, Alicja M
Kleier, Daniel Anthony
Klein, Nathan
Koski, Walter S
Kossiakoff, Alexander
Kreis, Ronald W
Krisher, Lawrence Charles
Krumbein, Simeon Joseph
Kumins, Charles Arthur
Kurylo, Michael John, III
Kushner, Lawrence Maurice
Lafferty, Walter J
Land, William Everett
Larkin, David
Larson, Clarence Edward
Lee, Byungkook
Lee, Frederick Strube
Lepley, Arthur Ray
LeRoy, André François
Levin, Ira William
Levin, Irvin
Levy, Joseph Benjamin
Lide, David Reynolds, Jr
Linzer, Melvin
Loebenstein, William Vaille
Long, Charles Anthony
Luborsky, Samuel William
Lyons, John Winship
McDaniel, Carl Vance
McDonald, Jimmie Reed
McDowell, Hershel
Mackay, Raymond Arthur
McLaughlin, Alan Charles
McLaughlin, William Lowndes
McNesby, James Robert
Magee, John Storey, Jr
Maher, Philip Kenerick
Malarkey, Edward Cornelius
Malmberg, Marjorie Schooley
Marinenko, George
Marqusee, Jeffrey Alan
Marshall, Walter Lincoln
Martinez, Richard Isaac
May, Willie Eugene
Maycock, John Norman
Mazumder, Bibhuti R
Michejda, Christopher Jan
Miller, Gerald Ray
Miller, Melvin P
Miller, Raymond Earl
Miller, Walter E
Minton, Allen Paul
Misra, Dwarika Nath
Miziolek, Andrzej Wladyslaw
Monchick, Louis
Moore, John Hays
Mopsik, Frederick Israel
Moses, Saul
Munn, Robert James
Murr, Brown L, Jr
Myers, Lawrence Stanley, Jr
Nash, Murray L
Neta, Pedatsur
Novotny, Donald Bob
O'Hare, Patrick
Olson, William Bruce
Ondov, John Michael
Orlick, Charles Alex
Ostrofsky, Bernard
Parsegian, Vozken Adrian
Paule, Robert Charles
Pawlowski, Anthony T
Peters, Alan Winthrop
Petersen, Raymond Carl
Peterson, Miller Harrell
Peterson, Norman Cornelius
Pierce, Elliot Stearns
Piermarini, Gasper J
Plumlee, Karl Warren
Pratt, Thomas Herring, Jr
Price, Donna
Rader, Charles Allen
Rakhit, Gopa
Raksis, Joseph W

Physical Chemistry (cont)

Redmon, Michael James
Reilly, Hugh Thomas
Riesz, Peter
Rifkind, Joseph Moses
Roberts, Ralph
Robertson, Baldwin
Robey, Frank A
Robinson, Dean Wentworth
Roche, Lidia Alicia
Romans, James Bond
Rosenberg, Arnold Morry
Rosenblatt, David Hirsch
Rosenfield, Joan Samour
Rotariu, George Julian
Rowell, Charles Frederick
Ruby, Stanley
Rush, John Joseph
Ruth, John Moore
Rutner, Emile
Saba, William George
Salwin, Arthur Elliott
Samuel, Aryeh Hermann
Satkiewicz, Frank George
Scheer, Milton David
Schroeder, LeRoy William
Schroeder, Michael Allan
Schuldiner, Sigmund
Schultz, John Wilfred
Schwartz, Anthony Max
Sheinson, Ronald Swiren
Silverstone, Harris Julian
Silverton, James Vincent
Simic, Michael G
Skrabek, Emanuel Andrew
Smardzewski, Richard Roman
Snow, Milton Leonard
Sondergaard, Neal Albert
Spangler, Glenn Edward
Stein, Stephen Ellery
Stephenson, John Carter
Stimler, Suzanne Stokes
Strauss, Mary Jo
Strier, Murray Paul
Suenram, Richard Dee
Swann, Madeline Bruce
Sweeting, Linda Marie
Tannenbaum, Harvey
Taylor, John Keenan
Tevault, David Earl
Thirumalai, Devarajan
Thompson, Donald Leroy
Thompson, Joseph Kyle
Thompson, Robert John, Jr
Thompson, Warren Elwin
Tompa, Albert S
Tompkins, Robert Charles
Trus, Benes L
Tung, Ming Sung
Tuve, Richard Larsen
Urbach, Herman B
Vander Hart, David Lloyd
Vanderryn, Jack
Verdier, Peter Howard
Vincent, James Sidney
Walker, Ronald Elliot
Wang, Francis Wei-Yu
Wason, Satish Kumar
Way, Kermit R
Weber, Alfons
Weber, Leon
Weiner, John
Weiss, Charles, Jr
Weiss, Richard Gerald
Williams, Ellen D
Willis, Phyllida Mave
Wilmot, George Barwick
Wing, James
Wu, En Shinn
Wu, Yung-Chi
Yagi, Haruhiko
Yarkony, David R
Yeh, Herman Jia-Chain
Young, Jay Alfred
Zimmerman, John Gordon
Zimmerman, Steven B
Zisman, William Albert
Zuckerbrod, David
Zwanzig, Robert Walter

MASSACHUSETTS

Ackerman, Jerome Leonard
Adelman, Albert H
Adler, Alice Joan
Adler, Norman
Alberty, Robert Arnold
Allen, Malwina I
Annino, Raymond
Armington, Alton
Atkins, Jaspard Harvey
Baboian, Robert
Baglio, Joseph Anthony
Baird, Donald Heston
Bald, Kenneth Charles
Bannister, William Warren
Barr, James K
Bearden, William Harlie
Bell, Jerry Alan
Benson, Bruce Buzzell
Berka, Ladislav Henry
Bernard, Walter Joseph
Bethune, John Lemuel
Birstein, Seymour J

Boedtker Doty, Helga
Bonner, Francis Joseph
Bowers, Peter George
Bracco, Donato John
Brandts, John Frederick
Brecher, Charles
Bridgman, Wilbur Benjamin
Browne, Sheila Ewing
Burke, John Michael
Butler, James Newton
Caledonia, George Ernest
Cathou, Renata Egone
Cembrola, Robert John
Cerankowski, Leon Dennis
Chance, Kelly Van
Chang, Kuang-Chou
Chang, Raymond
Charkoudian, John Charles
Chien, James C W
Chiu, Tak-Ming
Chu, Nori Yaw-Chyuan
Chung, Frank H
Clough, Stuart Benjamin
Cobb, Carolus M
Coleman, William Fletcher
Comeford, Lorrie Lynn
Connors, Robert Edward
Cormier, Alan Dennis
Coulter, Lowell Vernon
Daley, Henry Owen, Jr
D'Amato, Richard John
Dampier, Frederick Walter
Davidovits, Paul
Davison, Peter Fitzgerald
DeCosta, Peter F(rancis)
Deutch, John Mark
Dewald, Robert Reinhold
Dickinson, Alan Charles
Dickinson, Leonard Charles
DiNardi, Salvatore Robert
Donoghue, John Timothy
Dorain, Paul Brendel
Druy, Mark Arnold
Ehret, Anne
Elliott, John Frank
Engelke, John Leland
Epstein, Irving Robert
Eschenroeder, Alan Quade
Feakes, Frank
Fennelly, Paul Francis
Field, Robert Warren
Fink, Richard David
Fleck, George Morrison
Flowers, Ralph Grant
Flynn, George P(atrick)
Freedman, Andrew
Friedman, Raymond
Friend, Cynthia Marie
Fritzsche, Alfred Keith
Galligan, John D(onald)
Gardner, Donald Murray
Garland, Carl Wesley
George, James Henry Bryn
Gersh, Michael Elliot
Giner, Jose Domingo
Glazman, Yuli M
Goodman, Philip
Gopikanth, M L
Gordon, George Selbie
Gordon, Roy Gerald
Grabowski, Joseph J
Green, Byron David
Greenaway, Frederick Thomas
Greif, Mortimer
Griesinger, David Hadley
Grunwald, Ernest Max
Haas, John William, Jr
Hagnauer, Gary Lee
Hall, Lowell Headley, II
Hamilton, James Arthur
Hanselman, Raymond Bush
Harrison, Anna Jane
Harvey, Walter William
Hayden, Thomas Day
Henchman, Michael J
Herschbach, Dudley Robert
Herz, Matthew Lawrence
Herzfeld, Judith
Hess, John Monroe Converse
Hickman, James Joseph
Hobbs, John Robert
Hodgdon, Russell Bates, Jr
Hoffman, Morton Z
Hollister, Charlotte Ann
Hopkins, Esther Arvilla Harrison
Hornig, Donald Frederick
Huffman, Robert Eugene
Ika, Prasad Venkata
Illinger, Joyce Lefever
Illinger, Karl Heinz
Ingle, George William
Isaks, Martin
Johnson, Keith Huber
Johnston, A Sidney
Jones, Guilford, II
Jost, Ernest
Kafalas, Peter
Karplus, Martin
Keat, Paul Powell
Kellner, Jordan David
Ketteringham, John M
Klemperer, William
Klinedinst, Keith Allen
Kolodny, Nancy Harrison

Kopf, Peter W
Kowalak, Albert Douglas
Kramer, Jerry Martin
Krieger, Jeanne Kann
Kushick, Joseph N
Kustin, Kenneth
Lamola, Angelo Anthony
Langmuir, Margaret Elizabeth Lang
Lebowitz, Elliot
Lerner, Harry
Lester, Joseph Eugene
Levine, Oscar
Lewis, Armand Francis
Licht, Stuart Lawrence
Lichtin, Norman Nahum
Lin, Jeong-Long
Lindsay, William Tenney, Jr
Linschitz, Henry
Lipscomb, William Nunn, Jr
Loehlin, James Herbert
Lowey, Susan
Lowry, Nancy
Luft, Ludwig
McFadden, David Lee
Mac Knight, William John
McRae, Wayne Alan
Maher, Galeb Hamid
Malin, Murray Edward
Manchester, Kenneth Edward
Mandl, Alexander Ernst
Mariani, Henry A
Marrero, Hector G
Meal, Janet Hawkins
Medalia, Avrom Izak
Menashi, Jameel
Mendelson, Robert Allen
Messer, Charles Edward
Mettler, John D(aniel), Jr
Meyer, Vincent D
Millard, Richard James
Miller, Keith Wyatt
Minden, Henry Thomas
Molina, Mario Jose
Morris, Robert Alan
Murad, Edmond
Nash, Leonard Kollender
Nelson, David Robert
Nelson, Keith A
Nemeth, Ronald Louis
Neue, Uwe Dieter
Norment, Hillyer Gavin
Oppenheim, Irwin
Pandolfe, William David
Papaefthymiou, Georgia Christou
Patrick, Richard Montgomery
Paulsen, Duane E
Paulson, John Frederick
Pearson, Edwin Forrest
Pemsler, J(oseph) Paul
Peri, John Bayard
Perkins, Janet Sanford
Person, James Carl
Phillies, George David Joseph
Plumb, Robert Charles
Powell, Arnet L
Prock, Alfred
Pyun, Chong Wha
Ragle, John Linn
Ramsley, Alvin Olsen
Ransil, Bernard J(erome)
Rauh, Robert David, Jr
Rawlins, Wilson Terry
Ricci, John Silvio, Jr
Rivin, Donald
Roberts, Mary Fedarko
Rock, Elizabeth Jane
Roebber, John Leonard
Rose, Timothy Laurence
Ross, James William
Rowe, Paul E
Rowell, Robert Lee
Rupich, Martin Walter
Scala, Alfred Anthony
Schempp, Ellory
Scheuplein, Robert J
Schneider, Maxyne Dorothy
Schonhorn, Harold
Schwartz, Lowell Melvin
Serafin, Frank G
Shimizu, Nobumichi
Silbey, Robert James
Simons, Harold Lee
Slater, Richard Craig
Smith, Joseph Harold
Soltzberg, Leonard Jay
Spitler, Mark Thomas
Steel, Colin
Stein, Samuel H
Steinfeld, Jeffrey Irwin
Stengle, Thomas Richard
Stewart, Gerald Walter
Stidham, Howard Donathan
Stockman, David Lyle
Swank, Thomas Francis
Taub, Irwin A(llen)
Tauer, Kenneth J
Taylor, Lloyd David
Taylor, Raymond L
Thomas, Martha Jane Bergin
Torre, Frank John
Towns, Donald Lionel
Tsuk, Andrew George
Turnbull, David
Tuttle, Thomas R, Jr

Tykodi, Ralph John
Uhlig, Herbert H(enry)
Ukleja, Paul Leonard Matthew
Umans, Robert Scott
Underwood, Donald Lee
Vanderslice, Thomas Aquinas
Van Hook, Andrew
Vanpee, Marcel
Vidulich, George A
Viggiano, Albert
Waack, Richard
Wallace, Frederic Andrew
Wang, Chih-Lueh Albert
Wang, Robert T
Waugh, John Stewart
Weaver, Edwin Snell
Weiss, Karl H
Wen, Wen-Yang
Westmoreland, Phillip R
Wilde, Anthony Flory
Wilemski, Gerald
Wilson, Edgar Bright
Wisdom, Norvell Edwin, Jr
Wrathall, Donald Prior
Wright, David Franklin
Wrighton, Mark Stephen
Yannas, Ioannis Vassilios
Yannoni, Nicholas
Yoshino, Kouichi

MICHIGAN

Abu-Isa, Ismat Ali
Adams. Wade J
Anders, Oswald Ulrich
Anderson, James E
Bartell, Lawrence Sims
Bates, John Bertram
Bauer, David Robert
Ben, Manuel
Bennett, Stephen Lawrence
Berry, Myron Garland
Bettman, Max
Blint, Richard Joseph
Bone, Larry Irvin
Brenner, Alan
Brodasky, Thomas Francis
Camp, Ronald Lee
Carpenter, Michael Kevin
Chao, Mou Shu
Cho, Byong Kwon
Clarke, Richard Penfield
Cogan, Harold Louis
Coleman, David Manley
Cook, Richard James
Cornilsen, Bahne Carl
Craig, Robert George
Cratin, Paul David
Crippen, Gordon Marvin
Curnutt, Jerry Lee
Dahlgren, George
Dahm, Donald B
Deal, Ralph Macgill
Diamond, Howard
Dickie, Ray Alexander
Diesen, Ronald W
Dirkse, Thedford Preston
Downey, Joseph Robert, Jr
Doyle, Daryl Joseph
Drake, Michael Cameron
Duchamp, David James
Dunn, Thomas M
Duvall, Jacque L
Dye, James Louis
Dzieciuch, Matthew Andrew
Eastland, George Warren, Jr
Ebbing, Darrell Delmar
Eick, Harry Arthur
Eliezer, Isaac
Eliezer, Naomi
Ellis, Thomas Stephen
Endicott, John F
Enke, Christie George
Filisko, Frank Edward
Fisher, Edward Richard
Fisher, Galen Bruce
Flynn, James Patrick
Foister, Robert Thomas
Forist, Arlington Ardeane
Francis, Anthony Huston
Francisco, Joseph Salvadore, Jr
Freyberger, Wilfred L(awson)
Gagnon, Eugene Gerald
Gandhi, Harendra Sakarlal
Garrett, David L, (Jr)
Gland, John Louis
Gorse, Robert August, Jr
Graves, Bruce Bannister
Greene, Bettye Washington
Griffin, Henry Claude
Gulick, Wilson M, Jr
Gunther, Ronald George
Habermann, Clarence E
Hahne, Rolf Mathieu August
Hajratwala, Bhupendra R
Halberstadt, Marcel Leon
Hansen, Robert Douglas
Harris, Stephen Joel
Harrison, James Francis
Hase, William Louis
Hatzenbuhler, Douglas Albert
Heeschen, Jerry Parker
Heller, Hanan Chonon
Herk, Leonard Frank
Hillig, Kurt Walter, II

Hoare, James Patrick
Hong, Kuochih
Hood, Robin James
Houser, Thomas J
Howe, Norman Elton, Jr
Huettner, David Joseph
Hutchison, James Robert
Iyengar, Doreswamy Raghavachar
Jacobs, Gerald Daniel
Japar, Steven Martin
Johnson, David Alfred
Johnson, Ray Leland
Johnson, Richard Allen
Julien, Larry Marlin
Kagel, Ronald Oliver
Kaiser, Edward William, Jr
Kana'an, Adli Sadeq
Kawooya, John Kasajja
Kelly, Nelson Allen
Kenney, Donald J
Killgoar, Paul Charles, Jr
King, Peter Foster
King, Stanley Shih-Tung
Klempner, Daniel
Komarmy, Julius Michael
Kopelman, Raoul
Kuczkowski, Robert Louis
Kunz, Albert Barry
Kuo, Mingshang
Lane, George Ashel
Larsen, Eric Russell
Larson, John Grant
Lee, Chuan-Pu
Lee, Wei-Ming
Leffert, Charles Benjamin
Leifer, Leslie
Leroi, George Edgar
Levine, Samuel
Li, Chi-Tang
Lindfors, Karl Russell
Linowski, John Walter
Lintvedt, Richard Lowell
Lowry, George Gordon
Lubman, David Mitchell
McDonald, Charles Joseph
McHarris, William Charles
McTague, John Paul
Mani, Inder
Mansfield, Marc L
Maskal, John
Mayer, William John
Meibuhr, Stuart Gene
Mendenhall, George David
Merrill, Jerald Carl
Miller, Philip Joseph
Miller, Steven Ralph
Moore, Carl
Morgan, Roger John
Morrissey, David Joseph
Moyer, John Raymond
Murchison, Craig Brian
Murchison, Pamela W
Mutch, George William
Nazri, Gholam-Abbas
Neely, Brock Wesley
Niekamp, Carl William
Nordman, Christer Eric
Novak, Raymond Francis
Oei, Djong-Gie
O'Neil, James R
Otto, Klaus
Passino, Dora R May
Penner-Hahn, James Edward
Peters, James
Petersen, Donald Ralph
Petrella, Ronald Vincent
Platt, Alan Edward
Plaush, Albert Charles
Polik, William Frederick
Powell, Ralph Robert
Ramamurthy, Amurthur C
Reck, Gene Paul
Revzin, Arnold
Rieke, James Kirk
Riley, Bernard Jerome
Robbins, Omer Ellsworth, Jr
Rogers, Jerry Dale
Rothe, Erhard William
Rothschild, Walter Gustav
Rowe, Leonard C
Russell, Joel W
Sandel, Vernon Ralph
Sarge, Theodore William
Sarkar, Nitis
Saunders, Frank Linwood
Schlick, Shulamith
Schmidt, Parbury Pollen
Schnitker, Jurgen H
Schullery, Stephen Edmund
Schwendeman, Richard Henry
Seefurth, Randall N
Sell, Jeffrey Alan
Sevilla, Michael Douglas
Sharma, Ram Autar
Sharp, Robert Richard
Sheetz, David P
Skelly, Norman Edward
Skochdopole, Richard E
Sloane, Christine Scheid
Sloane, Thompson Milton
Slomp, George
Small, Hamish
Smith, A(lbert) Lee
Smith, Arthur Gerald

Snyder, Dexter Dean
Soulen, John Richard
Spurgeon, William Marion
Stark, Forrest Otto
Stein, Dale Franklin
Stowe, Robert Allen
Summitt, W(illiam) Robert
Swathirajan, S
Swets, Don Eugene
Tai, Julia Chow
Tamres, Milton
Taylor, Kathleen C
Taylor, Robert Cooper
Thacker, Raymond
Tischer, Ragnar P(ascal)
Tomalia, Donald Andrew
Tou, James Chieh
Tuesday, Charles Sheffield
Ullman, Robert
Vanderwielen, Adrianus Johannes
Vukasovich, Mark Samuel
Wahl, Werner Henry
Wang, Simon S
Warrick, Earl Leathen
Warrington, Terrell L
Watenpaugh, Keith Donald
Weber, Dennis Joseph
Weiner, Steven Allan
Weissman, Eugene Y(ehuda)
Westrum, Edgar Francis, Jr
Whitfield, Richard George
Wicke, Brian Garfield
Willermet, Pierre Andre
Williams, Ronald Lloyde
Wims, Andrew Montgomery
Wong, Peter Alexander
Yee, Albert Fan
Yesinowski, James Paul
Zanini-Fisher, Margherita

MINNESOTA
Adams, Roger James
Agarwal, Som Prakash
Allen, Martin
Amata, Charles David
Anderson, George Robert
Anderson, Ronald Keith
Barany, George
Barden, Roland Eugene
Beetch, Ellsworth Benjamin
Bohon, Robert Lynn
Bolles, Theodore Frederick
Bonne, Ulrich
Britton, William Giering
Brom, Joseph March, Jr
Bryan, Thomas T
Buchholz, Allan C
Carr, Robert Wilson, Jr
Carter, Orwin Lee
Childs, William Ves
Crawford, Bryce (Low), Jr
Dahler, John S
Davis, Howard Ted
Dempsey, John Nicholas
Dendinger, Richard Donald
Dickson, Arthur Donald
Dierenfeldt, Karl Emil
Drew, Bruce Arthur
Dreyer, Kirt A
Duerst, Richard William
Durnick, Thomas Jackson
Erickson, John M
Erickson, Randall L
Errede, Louis A
Evans, Douglas Fennell
Evans, John Fenton
Farnum, Sylvia A
Fester, Keith Edward
Finholt, James E
French, William George
Gabor, Thomas
Gavin, Robert M, Jr
Gentry, William Ronald
Giese, Clayton
Goon, David James Wong
Graham, Robert Leslie
Grant, David James William
Grundmeier, Ernest Winston
Gyberg, Arlin Enoch
Haas, Larry Alfred
Hanson, Allen Louis
Hardgrove, George Lind, Jr
Harriss, Donald K
Hexter, Robert Maurice
Hogle, Donald Hugh
Howell, Peter Adam
Jones, Lester Tyler
Khalafalla, Sanaa E
King, Reatha Clark
Kolthoff, Izaak Maurits
Koob, Robert Duane
Kreevoy, Maurice M
Labuza, Theodore Peter
Lai, Juey Hong
Lesikar, Arnold Vincent
Lumry, Rufus Worth, II
McBrady, John J
McKenna, Jack F(ontaine)
McKeown, James John
Marsh, Frederick Leon
Miessler, Gary Lee
Montgomery, Peter Williams
Moore, Francis Bertram
Moscowitz, Albert

Nichol, James Charles
Olsen, Douglas Alfred
Oriani, Richard Anthony
Owens, Boone Bailey
Owens, Kenneth Eugene
Peterson, Lowell E
Pocius, Alphonsus Vytautas
Polnaszek, Carl Francis
Prager, Stephen
Ramette, Richard Wales
Robins, Janis
Robinson, Glen Moore, III
Roquitte, Bimal C
Roska, Fred James
Rutledge, Robert L
Ryan, John Peter
Salmon, Oliver Norton
Schmidt, Lanny D
Schwartz, A(lbert) Truman
Slowinski, Emil J, Jr
Smyrl, William Hiram
Sobieski, James Fulton
Stankovich, Marian Theresa
Stoesz, James Darrel
Strong, Judith Ann
Subach, Daniel James
Suryanarayanan, Raj Gopalan
Tamsky, Morgan Jerome
Thomas, John Paul
Thompson, Herbert Bradford
Thompson, Mary E
Tiers, George Van Dyke
Truhlar, Donald Gene
Tsai, Bilin Paula
Ugurbil, Kamil
White, Joe Wade
Yapel, Anthony Francis, Jr

MISSISSIPPI
Baker, John Keith
Bass, Henry Ellis
Bishop, Allen David, Jr
Boyle, John Aloysius, Jr
Brady, Ruth Mary
Brent, Charles Ray
Calloway, E Dean
Combs, Leon Lamar, III
Crawford, Crayton McCants
Duffey, Donald Creagh
Fisher, Thomas Henry
Kalasinsky, Victor Frank
Klingen, Theodore James
Legg, John Wallis
Little, Brenda Joyce
McCain, Douglas C
Miller, Donald Piguet
Monts, David Lee
Myers, Richard Showse
Pinson, James Wesley
Posey, Franz Adrian
Stefani, Andrew Peter
Weissman, William
Wertz, David Lee
West, Rose Gayle
Wilson, Wilbur William
Yun, Kwang-Sik

MISSOURI
Ackerman, Joseph John Henry
Ames, Donald Paul
Bahn, Emil Lawrence, Jr
Baiamonte, Vernon D
Bard, James Richard
Baughman, Russell George
Beaver, Earl Richard
Beistel, Donald W
Benjamin, Philip Palamoottil
Berndt, Alan Fredric
Bertrand, Gary Lane
Bornmann, John Arthur
Bradburn, Gregory Russell
Brown, Robert Eugene
Bundschuh, James Edward
Burke, James Joseph
Byrd, Willis Edward
Carpenter, James Franklin
Conradi, Mark Stephen
Crutchfield, Marvin Mack
Dahl, William Edward
Dias, Jerry Ray
Dixon, Marvin Porter
Frederick, Raymond H
Freeman, John Jerome
Gard, Janice Koles
Gaspar, Peter Paul
Gerfen, Charles Otto
Glaspie, Peyton Scott
Gleaves, John Thompson
Greenlief, Charles Michael
Griffith, Edward Jackson
Griffiths, David Warren
Hansen, Richard Lee
Harris, Harold H
Henis, Jay Myls Stuart
Hinkebein, John Arnold
Holloway, Thomas Thornton
Holsen, James N(oble)
Holtzer, Marilyn Emerson
James, William Joseph
Jason, Mark Edward
Jones, Marvin Thomas
Kelley, John Daniel
Killingbeck, Stanley
Kim, Hyunyong

King, Thomas Morgan
Koerner, William Elmer
Kuntz, Robert Roy
Kurz, James Eckhardt
Kurz, Joseph Louis
Larsen, David W
Lee, Emerson Howard
Lembke, Roger Roy
Lilenfeld, Harvey Victor
Lin, Tien-Sung Tom
Lipkin, David
Lissant, Kenneth Jordan
Loeppky, Richard N
Lohman, Timothy Michael
Long, Gary John
Lott, Peter F
Lund, Louis Harold
McDonald, Hector O
McGraw, Robert Leonard
Macias, Edward S
Marshall, Charles Edmund
Mosher, Melvyn Wayne
Munch, Ralph Howard
O'Brien, James Francis
Palamand, Suryanarayana Rao
Piper, Roger D
Raw, Cecil John Gough
Readnour, Jerry Michael
Rice, Bernard
Richards, Charles Norman
Riehl, James Patrick
Roach, Donald Vincent
Robaugh, David Allan
Robertson, Bobby Ken
Roy, Rabindra (Nath)
Sarantites, Demetrios George
Schaefer, Jacob Franklin
Schaeffer, Harold F(ranklin)
Schmitt, John Leigh
Seeley, Robert D
Sheets, Ralph Waldo
Spilburg, Curtis Allen
Stary, Frank Edward
Takano, Masaharu
Teicher, Harry
Thomas, George Joseph, Jr
Thomas, Timothy Farragut
Thompson, Clifton C
Tria, John Joseph, Jr
Troutner, David Elliott
Unland, Mark Leroy
Wagenknecht, John Henry
Weissman, Samuel Isaac
Welsh, William James
Whitefield, Philip Douglas
Wilkinson, Ralph Russell
Winicov, Murray William
Wolf, Clarence J
Wong, Tuck Chuen
Woodbrey, James C
Yaris, Robert
Yasuda, Hirotsugu
Zetlmeisl, Michael Joseph

MONTANA
Barnhart, David M
Bradley, Daniel Joseph
Callis, Patrik Robert
Caughlan, Charles Norris
Conn, Paul Joseph
Craig, Arnold Charles
Emerson, Kenneth
Graham, Raymond
Howald, Reed Anderson
Layman, Wilbur A
Osterheld, Robert Keith
Pagenkopf, Gordon K
Smyth, Charles Phelps
Wood, William Wayne
Woodbury, George Wallis, Jr
Yates, Leland Marshall
Zelezny, William Francis

NEBRASKA
Coleman, George Hunt
Eckhardt, Craig Jon
Garey, Carroll Laverne
Howell, Daniel Bunce
King, Delbert Leo
Kuecker, John Frank
Rack, Edward Paul
Scholz, John Joseph, Jr
Shearer, Greg Otis
Skopp, Joseph Michael
Snipp, Robert Leo
Sonderegger, Theo Brown
Sterner, Carl D
Swanson, Jack Lee
Swanson, James A
Vanderzee, Cecil Edward
Zebolsky, Donald Michael

NEVADA
Blincoe, Clifton (Robert)
Burkhart, Richard Delmar
Earl, Boyd L
Eastwood, DeLyle
Farley, John William
Fletcher, Aaron Nathaniel
Gerard, Jesse Thomas
Harrington, Rodney E
Kemp, Kenneth Courtney
Klainer, Stanley M
MacDonald, David J

Physical Chemistry (cont)

Miller, Eugene
Pierson, William R
Shin, Hyung Kyu
Smith, Ross W
Sovocool, G Wayne

NEW HAMPSHIRE
Amell, Alexander Renton
Bel Bruno, Joseph James
Braun, Charles Louis
Chasteen, Norman Dennis
Cleland, Robert Lindbergh
Creagh-Deyter, Linda T
Damour, Paul Lawrence
Ellis, Samuel Benjamin
Greenwald, Harold Leopold
Hornig, James Frederick
Hubbard, Colin D
Kegeles, Gerson
Kelemen, Denis George
Mooney, Richard Warren
Nadeau, Herbert Gerard
Pilar, Frank Louis
Rice, Dale Wilson
Stepenuck, Stephen Joseph, Jr
Stockmayer, Walter Hugo
Swift, Robinson Marden
Weisman, Gary Raymond

NEW JERSEY
Aldrich, Haven Scott
Alekman, Stanley L
Alvarez, Vernon Leon
Alyea, Hubert Newcombe
Amron, Irving
Ander, Paul
Andrews, Rodney Denlinger, Jr
Appleby, Alan
Ardelt, Wojciech Joseph
Armstrong, William David
Auborn, James John
Axtmann, Robert Clark
Bachman, Kenneth Charles
Badin, Elmer John
Bahr, Charles Chester
Baidins, Andrejs
Bair, Harvey Edward
Baker, William Oliver
Ban, Vladimir Sinisa
Barnard, William Sprague
Bartok, William
Baughman, Ray Henry
Behl, Wishvender K
Bender, Max
Bendure, Raymond Lee
Bergh, Arpad A
Bernasek, Steven Lynn
Bernholz, William Francis
Bhattacharjee, Himangshu Ranjan
Bhattacharyya, Pranab K
Biagetti, Richard Victor
Binder, Michael
Bird, George Richmond
Black, James Francis
Blanc, Joseph
Bloch, Aaron Nixon
Bloom, Allen
Bocarsly, Andrew B
Bonacci, John C
Boskey, Adele Ludin
Bozzelli, Joseph William
Brenner, Douglas Milton
Brous, Jack
Brown, Joe Ned, Jr
Brown, Stanley Monty
Brugger, John Edward
Brus, Louis Eugene
Buck, Thomas M
Buckley, R Russ
Bulas, Romuald
Burley, David Richard
Burwasser, Herman
Calcote, Hartwell Forrest
Cardillo, Mark J
Carlucci, Frank Vito
Carrock, Frederick E
Castor, William Stuart, Jr
Chandross, Edwin A
Chatterjee, Pronoy Kumar
Chaudhri, Safee U
Chessin, Hyman
Chiao, Yu-Chih
Chidsey, Christopher E
Chiu, Tin-Ho
Cho, Kon Ho
Church, John Armistead
Coakley, Mary Peter
Cohen, Abraham Bernard
Cohen, Edward David
Cohen, George Lester
Cotter, Martha Ann
Couchman, Peter Robert
Cruice, William James
Dahlstrom, Bertil Philip, Jr
Davis, Stanley Gannaway
Dean, Anthony Marion
Debenedetti, Pablo Gaston
De Korte, Aart
DiMasi, Gabriel Joseph
Dismukes, Gerard Charles
Douglass, D C
Downing, George V, Jr

Dreby, Edwin Christian, III
Dreeben, Arthur B
Dunbar, Phyllis Marguerite
Dunn, William Howard
Dzierzanowski, Frank John
Ehrenstorfer, Sieglinde K M
Ekstrom, Lincoln
Elder, John Philip
Erenrich, Eric Howard
Ermler, Walter Carl
Evans, Warren William
Falconer, Warren Edgar
Falk, Charles David
Farrington, Thomas Allan
Farrow, Leonilda Altman
Felder, William
Feldstein, Nathan
Ferrigno, Thomas Howard
Fetters, Lewis
Filas, Robert William
Fisanick, Georgia Jeanne
Fitzgerald, Patrick Henry
Fix, Kathleen A
Flato, Jud B
Forbes, Charles Edward
Forster, Eric Otto
Foster, Walter H, Jr
Foy, Walter Lawrence
Frankenthal, Robert Peter
Fraser, Donald Boyd
Freund, Robert Stanley
Fuller, Everett J
Fung, Shun Chong
Gagliardi, L John
Galdes, Alphonse
Gale, Paula Jane
Gans, Manfred
Garfinkel, Harmon Mark
Garik, Vladimir L
Gelfand, Jack Jacob
Gethner, Jon Steven
Gibbard, H Frank
Glarum, Sivert Herth
Glassman, Irvin
Gold, Daniel Howard
Goldblatt, Irwin Leonard
Goldstein, Martin
Goldstein, Robert Lawrence
Gollob, Fred
Goodkin, Jerome
Goodman, Lionel
Gore, Ernest Stanley
Gottscho, Richard Alan
Grasselli, Robert Karl
Green, Michael Philip
Grelecki, Chester
Griffith, Martin G
Haag, Werner O
Haden, Walter Linwood, Jr
Hager, Douglas Francis
Halaby, Sami Assad
Halgren, Thomas Arthur
Harendza-Harinxma, Alfred Josef
Harpell, Gary Allan
Harris, Alexander L
Harris, Leonce Everett
Hayles, William Joseph
Hays, Byron G
Head-Gordon, Martin Paul
Healey, Frank Henry
Heilweil, Israel Joel
Held, Robert Paul
Hen, John
Herber, Rolfe H
Herdklotz, John Key
Herzog, Gregory F
Hills, Stanley
Hobson, Melvin Clay, Jr
Hoffman, Henry Tice, Jr
Horowitz, Hugh H(arris)
House, Edward Holcombe
Hsu, Edward Ching-Sheng
Huang, John S
Huneke, James Thomas
Hung, John Hui-Hsiung
Hunger, Herbert Ferdinand
Hunter, Edwin Thomas
Iglesia, Enrique
Ilardi, Joseph Michael
Inniss, Daryl
Isaac, Peter Ashley Hammond
Izod, Thomas Paul John
Jacobson, Harold
Jain, Nemichand B
Jelinski, Lynn W
Joffe, Joseph
Jones, Francis Thomas
Jordan, Andrew Stephen
Kagann, Robert Howard
Kaiserman, Howard Bruce
Kaldor, Andrew
Kamath, Yashavanth Katapady
Kampas, Frank James
Kaplan, Martin L
Kaplan, Michael
Karg, Gerhart
Kaufman, Ernest D
Kaufmann, Kenneth James
Kauzmann, Walter (Joseph)
Kenkare, Divaker B
Kennedy, Anthony John
Kerr, George Thomson
Kimmel, Elias
Kimmel, Howard S

Kleiner, Walter Bernhard
Kohout, Frederick Charles, III
Konde, Anthony Joseph
Kornblum, Saul S
Kosel, George Eugene
Kowalski, Ludwik
Kramer, George Mortimer
Krenos, John Robert
Krueger, Paul Carlton
Kunin, Robert
Kunzler, John Eugene
Kydd, Paul Harriman
Lai, Kuo-Yann
Laity, Richard Warren
Lang, Frank Theodore
Lanzerotti, Mary Yvonne DeWolf
LaPalme, Donald William
Larsen, Marilyn Ankeney
Larson, Hugo R
Laurie, Victor William
Le Duc, J-Adrien Maher
Lee, Do-Jae
Levi, Elliott J
Levine, Richard S
Link, Gordon Littlepage
Loh, Roland Ru-loong
Lorenz, Patricia Ann
LoSurdo, Antonio
Lowell, A(rthur) I(rwin)
Lucchesi, Peter J
Lui, Yiu-Kwan
Lyon, Richard Kenneth
McAfee, Kenneth Bailey, Jr
McCall, David Warren
McCauley, James A
McCleary, Harold Russell
McClure, Donald Stuart
McCullough, John Price
MacFarlane, Robert, Jr
McKnight, Lee Graves
McMahon, Paul E
McNevin, Susan Clardy
Madey, Theodore Eugene
Magder, Jules
Malinowski, Edmund R
Marcus, Robert Boris
Marra, Dorothea Catherine
Maulding, Hawkins Valliant, Jr
Meiboom, Saul
Mendelsohn, Richard
Meyer, Albert William
Miller, Arthur
Miller, Barry
Miller, Floyd Laverne
Mohacsi, Erno
Monse, Ernst Ulrich
Montana, Anthony J
Mraw, Stephen Charles
Mucenieks, Paul Raimond
Mukai, Cromwell Daisaku
Mullen, Robert Terrence
Nace, Donald Miller
Naro, Paul Anthony
Naumann, Robert Alexander
Nelson, Roger Edwin
Newman, Stephen Alexander
Nilsen, Walter Grahn
O'Connor, Matthew James
Okinaka, Yutaka
Olechowski, Jerome Robert
Olmstead, William N(eergaard)
Olson, David Harold
Opila, Robert L, Jr
Ors, Jose Alberto
Osuch, Christopher Erion
Owens, Clifford
Owens, Frank James
Panish, Morton B
Panson, Gilbert Stephen
Parisi, George I
Parker, Richard C
Parreira, Helio Correa
Patel, Gordhanbhai Nathalal
Paul, Edward W
Perkins, Harolyn King
Pethica, Brian Anthony
Pierce, Robert Charles
Pierson, William Grant
Pilla, Arthur Anthony
Pinch, Harry Louis
Plescia, Otto John
Pobiner, Harvey
Poe, Martin
Poindexter, Edward Haviland
Polestak, Walter John S
Potenza, Joseph Anthony
Princen, Henricus Mattheus
Puar, Mohindar S
Pugliese, Michael
Rabitz, Herschel Albert
Raghavachari, Krishnan
Raich, Henry
Raines, Thaddeus Joseph
Ramachandran, Pallassana N
Randall, James Carlton, Jr
Raveche, Harold Joseph
Raynor, Susanne
Readdy, Arthur F, Jr
Rebick, Charles
Redden, Patricia Ann
Reents, William David, Jr
Regna, Peter P
Reilly, Eugene Patrick
Rein, Alan James

Rennert, Joseph
Richlin, Jack
Riedhammer, Thomas M
Rieger, Martin Max
Roberts, Ronald Frederick
Robertson, Nat Clifton
Robins, Jack
Rollino, John
Rome, Martin
Rosenberg, Allan (Herbert)
Ross, Lawrence James
Roth, Heinz Dieter
Rousseau, Denis Lawrence
Rowe, Carleton Norwood
Rubin, Herbert
Rubino, Andrew M
Sagal, Matthew Warren
Saldick, Jerome
Salomon, Mark
Sanderson, Benjamin S
Sandus, Oscar
Sayres, Alden R
Schlegel, James M
Schueler, Paul Edgar
Schwartz, Bertram
Scott, Eric James Young
Searle, Norma Zizmer
Shah, Atul A
Shallcross, Frank V(an Loon)
Shanefield, Daniel J
Shaw, Henry
Sherry, Howard S
Shewmaker, James Edward
Shombert, Donald James
Siano, Donald Bruce
Sibilia, John Philip
Sifniades, Stylianos
Sinclair, William Robert
Sinfelt, John Henry
Slagg, Norman
Slichter, William Pence
Smith, Donald Eugene
Smith, Eileen Patricia
Smith, George Byron
Smith, Harlan Millard
Smith, Richard Pearson
Smith, Terry Edward
Soos, Zoltan Geza
Speert, Arnold
Spiro, Thomas George
Stillinger, Frank Henry
Stivala, Salvatore Silvio
Strauss, George
Strauss, Ulrich Paul
Suchow, Lawrence
Szamosi, Janos
Szyper, Mira
Tanenbaum, Morris
Thornton, C G
Tiethof, Jack Alan
Tietjen, James Joseph
Toby, Sidney
Toome, Voldemar
Trambarulo, Ralph
Trewella, Jeffrey Charles
Trifan, Daniel Siegfried
Trumbore, Forrest Allen
Tsang, Sien Moo
Tsonopoulos, Constantine
Tucci, Edmond Raymond
Tully, John Charles
Tunc, Deger Cetin
Turkevich, John
Tway, Patricia C
Vandenberg, Joanna Maria
Vanderspurt, Thomas Henry
Van Hook, James Paul
Vasile, Michael Joseph
Virgili, Luciano
Vogel, Veronica Lee
Vyas, Brijesh
Wagner, Sigurd
Wang, Chih Chun
Warren, Craig Bishop
Weakliem, Herbert Alfred, Jr
Wei, Chung-Chen
Wenger, Franz
Wenzel, John Thompson
Weschler, Charles John
Westerdahl, Carolyn Ann Lovejoy
Westerdahl, Raymond P
Wheaton, Gregory Alan
White, Lawrence Keith
Wiesenfeld, Jay Martin
Williams, Dale Gordon
Williams, Richard
Wise, John James
Wolf, Philip Frank
Woodbridge, Joseph Eliot
Wu, Ellen Lem
Yablonovitch, Eli
Yamin, Michael
Yardley, James Thomas, III

NEW MEXICO
Ames, Lynford Lenhart
Asprey, Larned Brown
Auerbach, Irving
Baca, Glenn
Baker, Floyd B
Bard, Richard James
Bayhurst, Barbara P
Bechtold, William Eric
Behrens, Robert George

Birely, John H
Blais, Normand C
Brabson, George Dana, Jr
Brandvold, Donald Keith
Breshears, Wilbert Dale
Butler, Michael Alfred
Cady, Howard Hamilton
Campbell, George Melvin
Carlson, Gary Alden
Catlett, Duane Stewart
Clark, Robert Paul
Clifton, David Geyer
Coppa, Nicholas V
Cromer, Don Tiffany
Cross, Jon Byron
Cunningham, Paul Thomas
Curtice, Jay Stephen
Cuthrell, Robert Eugene
Danen, Wayne C
DeArmond, M Keith
Dinegar, Robert Hudson
Doddapaneni, Narayan
Dorko, Ernest A
Dropesky, Bruce Joseph
Duffy, Clarence John
Eckert, Juergen
Ellis, Walton P
Engelke, Raymond Pierce
Erickson, Kenneth Lynn
Erikson, Jay Arthur
Ewing, Galen Wood
Ewing, Gordon J
Eyster, Eugene Henderson
Fowler, Malcolm McFarland
Fries, James Andrew
Fries, Ralph Jay
Grisham, Genevieve Dwyer
Haaland, David Michael
Hammel, Edward Frederic
Hartley, Danny L
Heaton, Maria Malachowski
Hoehn, Martha Vaughan
Hughes, Robert Clark
Jennings, Charles Warren
Jensen, Reed Jerry
Johnston, Roger Glenn
Jones, Peter Frank
Keizer, Clifford Richard
Keller, Richard Alan
Kenna, Bernard Thomas
Kerley, Gerald Irwin
Krupka, Milton Clifford
Laquer, Henry L
Larson, Thomas E
Levine, Herman Saul
Levy, Samuel C
Lieberman, Morton Leonard
Lyman, John L
Lynch, Richard Wallace
McDowell, Harding Keith
McLaughlin, Donald Reed
Mann, Joseph Bird, (Jr)
Martin, James Ellis
Mason, Caroline Faith Vibert
Mills, Robert Leroy
Mitchell, David Wesley
Morosin, Bruno
Mulford, Robert Neal Ramsay
Nathan, Charles C(arb)
Newton, Thomas William
Niemczyk, Thomas M
Nogar, Nicholas Stephen
Northrop, David A
O'Brien, Harold Aloysious, Jr
Ogard, Allen E
Ogilby, Peter Remsen
Olson, Douglas Bernard
Olson, William Marvin
Onstott, Edward Irvin
Orth, Charles Joseph
Park, Su-Moon
Peden, Charles H F
Peek, H(arry) Milton
Peterson, Charles Leslie
Peterson, Dean Everett
Pryor, Richard J
Purser, Fred O
Quinn, Rod King
Rabideau, Sherman Webber
Ramsay, John Barada
Renschler, Clifford Lyle
Rice, James Kinsey
Robinson, C Paul
Romig, Alton Dale, Jr
Sasmor, Daniel Joseph
Sattizahn, James Edward, Jr
Schaefer, Dale Wesley
Schott, Garry Lee
Schufle, Joseph Albert
Sedlacek, William Adam
Seegmiller, David W
Shepard, Joseph William
Sherman, Robert Howard
Smith, Gordon Meade
Smith, James Lewis
Sorem, Michael Scott
Stanbro, William David
Stark, Walter Alfred, Jr
Streetman, John Robert
Streit, Gerald Edward
Sullivan, John Henry
Taber, Joseph John
Tapscott, Robert Edwin
Taylor, Gene Warren

Thompson, Joseph Lippard
Vier, Dwayne Trowbridge
Walker, Robert Bridges
Wallace, Terry Charles
Walters, Edward Albert
Walton, George
Walton, Roddy Burke
Wampler, Fred Benny
Ward, John William
Werkema, George Jan
Wewerka, Eugene Michael
Whan, Ruth Elaine
Williams, David Cary
Wolfsberg, Kurt
Wood, Gerry Odell
Wu, Adam Yu
Young, Ainslie Thomas, Jr
Zeltmann, Alfred Howard

NEW YORK
Abramowicz, Daniel Albert
Abruna, Hector D
Addy, John Keith
Adelstein, Peter Z
Adin, Anthony
Albrecht, Andreas Christopher
Albrecht, Frederick Xavier
Albrecht, William Melvin
Alexander, John Macmillan, Jr
Alfieri, Gaetano T
Allen, Augustine Oliver
Allenbach, Charles Robert
Altschul, Rolf
Amborski, Leonard Edward
Amer, Nabil Mahmoud
Amero, Bernard Alan
Anbar, Michael
Anderson, Frank Wallace
Anderson, Herbert Rudolph, Jr
Apai, Gustav Richard, II
Araujo, Roger Jerome
Archie, William C, Jr
Arendt, Ronald H
Arents, John (Stephen)
Armanini, Louis Anthony
Arnowich, Beatrice
Aronson, Seymour
Arquette, Gordon James
Aten, Carl Faust, Jr
Atwood, Gilbert Richard
Aven, Manuel
Aviram, Ari
Avouris, Phaedon
Axe, John Donald
Baer, Norbert Sebastian
Baetzold, Roger C
Bagchi, Pranab
Bahary, William S
Baier, Robert Edward
Baird, Barbara A
Bak, David Arthur
Bakhru, Hassaram
Baldwin, John E
Bank, Shelton
Barile, Raymond Conrad
Barnett, Ronald E
Barron, Saul
Bartholomew, Roger Frank
Bauer, Simon Harvey
Baum, Parker Bryant
Baumgarten, Reuben Lawrence
Beck, Paul W
Beer, Sylvan Zavi
Belfort, Georges
Bent, Brian E
Bettelheim, Frederick A
Beuhler, Robert James, Jr
Bevak, Joseph Perry
Bharadwaj, Prem Datta
Bigeleisen, Jacob
Birge, Robert Richards
Birke, Ronald Lewis
Bishop, Marvin
Blackwell, Crist Scott
Bloch, Aaron N
Blum, Samuel Emil
Bobka, Rudolph J
Bodi, Lewis Joseph
Bonner, Francis Truesdale
Booman, Keith Albert
Bottger, Gary Lee
Boyer, Donald Wayne
Brady, George W
Bragg, John Kendal
Bramwell, Fitzgerald Burton
Bray, Norman Francis
Briehl, Robin Walt
Brill, Robert H
Browall, Kenneth Walter
Brown, Eric Richard
Bruckenstein, Stanley
Brumbaugh, Donald Verwey
Brumberger, Harry
Brunschwig, Bruce Samuel
Bryant, Robert George
Buff, Frank Paul
Buhks, Ephraim
Buhsmer, Charles P
Bulloff, Jack John
Burtt, Benjamin Pickering
Cadenhead, David Allan
Campion, James J
Cante, Charles John
Carle, Kenneth Roberts

Carroll, Harvey Franklin
Carroll, Robert William
Chaffee, Eleanor
Chandler, Horace W
Chang, Chin-An
Chang, Joseph Yung
Chang, Shih-Yung
Chapman, Sally
Charton, Marvin
Chen, Chang-Hwei
Chen, Cindy Chei-Jen
Cherin, Paul
Chiang, Joseph Fei
Chiang, Yuen-Sheng
Chin, Der-Tau
Chou, Chung-Chi
Chou, Tzi Shan
Chu, Benjamin Peng-Nien
Chu, Joseph Yung-Chang
Chu, Yung Yee
Clark, Alfred
Clark, Patricia Ann
Cloney, Robert Dennis
Cohen, Stephen Robert
Cole, David Le Roy
Collier, Susan S
Conan, Robert James, Jr
Cook, Edward Hoopes, Jr
Cooke, Derry Douglas
Cool, Terrill A
Cooper, Walter
Coppola, Patrick Paul
Corth, Richard
Cratty, Leland Earl, Jr
Creasy, William Russel
Cunningham, Michael Paul
Curme, Henry Garrett
Dailey, Benjamin Peter
Daniel, Daniel S
Dannenberg, Joseph
Dannhauser, Walter
Darrow, Frank William
Davenport, Lesley
Davis, William Donald
DeCrosta, Edward Francis, Jr
Dehn, Joseph William, Jr
Demerjian, Kenneth Leo
DeStefano, Anthony Joseph
Deutsch, John Ludwig
Diem, Max
Dieterich, David Allan
Dill, Aloys John
Dingledy, David Peter
Disch, Raymond L
Dixon, William Brightman
Doetschman, David Charles
Doigan, Paul
Dondes, Seymour
Donoian, Haig Cadmus
Doremus, Robert Heward
Doubleday, Charles E, Jr
Drauglis, Edmund
Dumbaugh, William Henry, Jr
Dwyer, James Michael
Dwyer, Robert Francis
Eachus, Raymond Stanley
Eberts, Robert Eugene
Egan, James John
Ehrenson, Stanton Jay
Ehrlich, Paul
Ehrlich, Sanford Howard
Eidinoff, Maxwell Leigh
Eikenberry, Jon Nathan
Eisenberg, Max
Elder, Fred A
Ells, Victor Raymond
Elmore, Glenn Van Ness
Everhard, Martin Edward
Factor, Arnold
Fajer, Jack
Farb, Edith
Farrar, James Martin
Fay, Homer
Featherstone, John Douglas Bernard
Fehlner, Francis Paul
Feigenson, Gerald William
Feldberg, Stephen William
Feldman, Isaac
Feldman, Larry Howard
Felty, Evan J
Filbert, Augustus Myers
Finkbeiner, Herman Lawrence
Finley, Kay Thomas
Finzel, Rodney Brian
Fishman, Jerry Haskel
Flank, William H
Flannery, John B, Jr
Flom, Donald Gordon
Flynn, George William
Folan, Lorcan Michael
Fontijn, Arthur
Fox, William
Fraenkel, George Kessler
Freed, Jack H
Freedman, Jeffrey Carl
Friedlander, Gerhart
Friedman, Harold Leo
Friedman, Joel Mitchel
Friedman, Lewis
Friedman, Seymour K
Fuerst, Adolph
Gaines, George Loweree, Jr
Gancy, Alan Brian
Gans, Paul Jonathan

Gardner, William Lee
Garetz, Bruce Allen
Garvey, James F
Gaudioso, Stephen Lawrence
Gawer, Albert Henry
Geacintov, Nicholas
Gentner, Robert F
Gershinowitz, Harold
Gettler, Joseph Daniel
Ghezzo, Mario
Gill, Robert Anthony
Gill, Ronald Lee
Gillies, Charles Wesley
Gilman, Paul Brewster, Jr
Gilmour, Hugh Stewart Allen
Ginell, Robert
Gledhill, Ronald James
Goldfarb, Theodore D
Good, Robert James
Goodisman, Jerry
Goodman, Jerome
Goodman, Seymour
Gordon, Barry Maxwell
Gore, Robert Cummins
Gould, Robert Kinkade
Green, Michael Enoch
Greenbaum, Steven Garry
Greenspan, Joseph
Griffel, Maurice
Griffin, Jane Flanigen
Grover, James Robb
Grubb, Willard Thomas
Grushkin, Bernard
Gum, Mary Lou
Haas, David Jean
Haas, Werner E L
Hach, Edwin E, Jr
Hall, Richard Travis
Haller, Ivan
Halpern, Mordecai Joseph
Hansen, Donald Joseph
Hanson, David M
Hanson, Jonathan C
Hanson, Louise I Karle
Harbottle, Garman
Hargitay, Bartholomew
Harrison, Shirley Wanda
Harwood, Colin Frederick
Hayon, Elie M
Hecht, Charles Edward
Heininger, Clarence George, Jr
Helfgott, Cecil
Hendricks, Robert William
Herczog, Andrew
Herkstroeter, William G
Herley, Patrick James
Herman, Irving Philip
Herskovits, Theodore Tibor
Hertl, William
Herz, Arthur H
Heston, William May, Jr
Hickmott, Thomas Ward
Hill, Cliff Otis
Hill, Derek Leonard
Hillig, William Bruno
Hillman, Manny
Hodge, Ian Moir
Holden, Norman Edward
Holleran, Eugene Martin
Hollinger, Henry Boughton
Holmes, Curtis Frank
Holroyd, Richard Allan
Hopke, Philip Karl
Houston, Paul Lyon
Howell, James MacGregor
Howery, Darryl Gilmer
Hsu, Ming-Ta
Htoo, Maung Shwe
Huang, Suei-Rong
Hubbard, Richard Alexander, II
Huggins, Charles Marion
Huizenga, John Robert
Hull, Michael Neill
Hurley, Ian Robert
Iden, Charles R
Ilmet, Ivor
Imperial, George Romero
Innes, Kenneth Keith
Irani, N F
Irsa, Adolph Peter
Isaacs, Hugh Solomon
Ishida, Takanobu
Jach, Joseph
Janauer, Gilbert E
Janz, George John
Jellinek, Hans Helmut Gunter
Jochsberger, Theodore
Joffee, Irving Brian
Johansson, Sune
Johnson, Myrle F
Johnson, Philip M
Johnston, Don Richard
Jorgensen, Helmuth Erik Milo
Jorne, Jacob
Joshi, Bhairav Datt
Kaarsberg, Ernest Andersen
Kambour, Roger Peabody
Kapecki, Jon Alfred
Kasper, John Simon
Katcoff, Seymour
Kaufman, Frank B
Keeler, Robert Adolph
Keesee, Robert George
Keller, D Steven

Physical Chemistry (cont)

Keller, Douglas Vern, Jr
Kennelley, James A
Kerker, Milton
Khare, Bishun Narain
Kim, Sang Hyung
Kiphart, Kerry
Kiss, Klara
Kittelberger, John Stephen
Kivlighn, Herbert Daniel, Jr
Klerer, Julius
Klinck, Ross Edward
Koch, Heinz Frank
Koeppl, Gerald Walter
Koetzle, Thomas F
Kohrt, Carl Fredrick
Kokoszka, Gerald Francis
Korinek, George Jiri
Kozak, Gary S
Krakow, Burton
Kratohvil, Josip
Krause, Sonja
Kreilick, Robert W
Kronman, Martin Jesse
Kuki, Atsuo
Kullnig, Rudolph K
Kumbar, Mahadevappa M
Kuzmak, Joseph Milton
Kwei, Ti-Kang
Ladd, John Herbert
Lam, Kai Shue
Lane, Richard L
Lange, Christopher Stephen
LaPietra, Joseph Richard
Larach, Simon
Lasker, Sigmund E
Laurenzi, Bernard John
Lazarus, Marc Samuel
LeBlanc, Oliver Harris, Jr
Lebowitz, Jacob Mordecai
Lee, Lieng-Huang
Lees, Alistair John
Lefkowitz, Stanley A
LeGrand, Donald George
Leone, James A
Leubner, Ingo Herwig
Leung, Pak Sang
Levine, Ira Noel
Levine, Samuel W
Levkov, Jerome Stephen
Levy, Arthur Louis
Lewin, Seymour Z
Lewis, David Kenneth
Li, Hong
Liang, Charles C
Liang, Kai
Lichtman, Irwin A
Liebeskind, Herbert
Lindholm, Robert D
Lindsay, Derek Michael
Lipsig, Joseph
Loebl, Ernest Moshe
Lombardi, John Rocco
Long, Michael Edgar
Longhi, John
Lorenzen, Jerry Alan
Loring, Roger Frederic
Lounsbury, John Baldwin
Low, Manfred Josef Dominik
Luborsky, Fred Everett
Luckey, George William
Ludlum, Kenneth Hills
Luner, Philip
Macaluso, Pat
McBreen, James
McCarthy, Paul James
MacColl, Robert
McDonald, Robert Skillings
McElligott, Peter Edward
McGee, Thomas Howard
McGrath, John F
McHugh, James Anthony, Jr
McKee, Douglas William
McKelvey, John Murray
MacKenzie, Donald Robertson
McLaren, Eugene Herbert
Mahony, John Daniel
Malerich, Charles
Mancuso, Richard Vincent
Marchetti, Alfred Paul
Marino, Robert Anthony
Marion, Alexander Peter
Mark, Herman Francis
Martin, Francis W
Martin, Frederick Johnson
Martin, William Butler, Jr
Martini, Catherine Marie
Massa, Dennis Jon
Massa, Louis
Matijevic, Egon
May, John Walter
Meloon, David Rand
Mendel, John Richard
Mennitt, Philip Gary
Merkel, Paul Barrett
Metz, Donald J
Meyerson, Bernard Steele
Mezei, Mihaly
Michiels, Leo Paul
Mihajlov, Vsevolod S
Miller, Harold A
Miller, Richard J
Miller, Warren James

Milner, Clifford E
Mitacek, Paul, Jr
Mitchell, John Jacob
Mittal, Kashmiri Lal
Model, Frank Steven
Modic, Frank Joseph
Molina, John Francis
Monahan, Alan Richard
Montano, Pedro Antonio
Morawetz, Herbert
Morehead, Frederick Ferguson, Jr
Morrow, Jack I
Moskowitz, Jules Warren
Moynihan, Cornelius Timothy
Mucci, Joseph Francis
Muenter, Annabel Adams
Muenter, John Stuart
Mullen, Patricia Ann
Mullhaupt, Joseph Timothy
Murphy, Daniel Barker
Murphy, James A
Myers, Anne Boone
Myers, Clifford Earl
Myers, Drewfus Young, Jr
Nafie, Laurence Allen
Nancollas, George H
Nash, Robert Joseph
Nathanson, Benjamin
Needham, Charles D
Némethy, George
Neubert, Theodore John
Neugebauer, Constantine Aloysius
Niedrach, Leonard William
Norcross, Bruce Edward
O'Donnell, Raymond Thomas
Oppenheimer, Larry Eric
Orem, Michael William
Orna, Mary Virginia
Osborg, Hans
Osgood, Richard Magee, Jr
Osipow, Lloyd Irving
Oster, Gerald
Osterhoudt, Hans Walter
Ott, Henry C(arl)
Overton, James Ray
Oyama, Shigeo Ted
Ozarow, Vernon
Padmanabhan, G R
Padnos, Norman
Pan, Kee-Chuan
Panagiotopoulos, Athanassios Z
Paraszczak, Jurij Rostyslan
Parravano, Carlo
Pasfield, William Horton
Patsis, Angelos Vlasios
Patton, Elizabeth VanDyke
Pavlisko, Joseph Anthony
Pearson, James Murray
Penn, Lynn Sharon
Perry, Edmond S
Perry, Ernest John
Petrakis, Leonidas
Petro, Anthony James
Petrucci, Sergio
Philips, Laura Alma
Piersma, Bernard J
Pliskin, William Aaron
Plummer, William Allan
Pochan, John Michael
Pomeroy, Gordon Ashby
Poncha, Rustom Pestonji
Porter, Richard Francis
Porter, Richard Needham
Poslusny, Jerrold Neal
Posner, Aaron Sidney
Powers, Dale Robert
Powers, Robert William
Praissman, Melvin
Prasad, Paras Nath
Preiss, Ivor Louis
Preses, Jack Michael
Price, Harry James
Proskauer, Eric S
Rand, Salvatore John
Ransom, Bruce Davis
Rashkin, Jay Arthur
Raymonda, John Warren
Reardon, Joseph Daniel
Reddy, Thomas Bradley
Reeves, Robert R
Rein, Robert
Remsberg, Louis Philip, Jr
Resing, Henry Anton
Rhodin, Thor Nathaniel, Jr
Ricci, John Ettore
Richtol, Herbert H
Ristow, Bruce W
Roepe, Paul David
Rogers, Donald Warren
Romankiw, Lubomyr Taras
Ronn, Avigdor Meir
Rose, Philip I
Rosen, Milton Jacques
Rosenthal, Donald
Rosman, Howard
Rosoff, Morton
Ross, John Brandon Alexander
Rossignton, David Ralph
Ruoff, Arthur Louis
Rzad, Stefan Jacek
Sage, Gloria W
Sakano, Theodore K
Saleeb, Fouad Zaki
Salotto, Anthony W

Saltsburg, Howard Mortimer
Salzberg, Hugh William
Samworth, Eleanor A
Sandifer, James Roy
Satir, Birgit H
Saturno, Antony Fidelas
Saunders, William Hundley, Jr
Scaringe, Raymond Peter
Schick, Martin J
Schottmiller, John Charles
Schreurs, Jan W H
Schulman, Jerome M
Schwartz, Geraldine Cogin
Schwartz, Stephen Eugene
Schwarz, Harold A
Schwarz, William Merlin, Jr
Scott, Bruce Albert
Segal, Bernice G
Segatto, Peter Richard
Seltzer, Stanley
Shahin, Michael M
Sheats, George Frederic
Shelby, James Elbert
Shiao, Daniel Da-Fong
Shilman, Avner
Shoup, Robert D
Shultz, Allan R
Siew, Ernest L
Silver, Herbert Graham
Singh, Surjit
Skarulis, John Anthony
Skelly, David W
Smid, Johannes
Smith, Gale Eugene
Smith, George David
Smith, Kenneth Judson, Jr
Smith, Michael
Smith, Norman Obed
Snyder, Lawrence Clement
Snyder, Robert Lyman
Sojka, Stanley Anthony
Solomon, Frank I
Solomon, Jack
Solomon, Jerome Jay
Somasundaran, P(onisseril)
Sorkin, Howard
Sparapany, John Joseph
Spink, Charles Harlan
Stanton, Richard Edmund
Stauffer, Robert Eliot
Sterman, Melvin David
Sterman, Samuel
Stern, Silviu Alexander
Stone, Joe Thomas
Stookey, Stanley Donald
Strekas, Thomas C
Strong, Robert Lyman
Sturmer, David Michael
Suggitt, Robert Murray
Sun, Siao Fang
Sundberg, Michael William
Sutin, Norman
Sweet, Richard Clark
Swingley, Charles Stephen
Szebenyi, Doletha M E
Szekely, Andrew Geza
Takacs, Gerald Alan
Tan, Julia S
Tang, Ignatius Ning-Bang
Taylor, Dale Frederick
Testa, Anthony Carmine
Texter, John
Thomas, Harold Todd
Thomas, Robert
Tomkiewicz, Micha
Tong, Stephen S C
Transue, Laurence Frederick
Tremelling, Michael, Jr
Trevoy, Donald James
Tseng, Linda
Valentini, James Joseph
Van Geet, Anthony Leendert
Vinciguerra, Michael Joseph
Vincow, Gershon
Violante, Michael Robert
Vogel, Philip Christian
Vonnegut, Bernard
Von Winbush, Samuel
Wait, Samuel Charles, Jr
Waitkins, George Raymond
Walker, Russell Wagner
Walnut, Thomas Henry, Jr
Wandass, Joseph Henry
Ward, Anthony Thomas
Warden, Joseph Tallman
Weiss, David Steven
Weiss, Jerome
Weller, S(ol) W(illiam)
Weltman, Clarence A
Wentorf, Robert H, Jr
Weston, Ralph E, Jr
White, Michael George
Whitten, David G
Widom, Benjamin
Wieder, Grace Marilyn
Wiegert, Philip E
Wiesenfeld, John Richard
Wijnen, Joseph M H
Wilcox, Charles Frederick, Jr
Will, Frit Gustav
Williams, David James
Wilson, James M
Winstrom, Leon Oscar
Winter, William Thomas

Wiswall, Richard H, Jr
Wolfarth, Eugene F
Woodward, Arthur Eugene
Wotherspoon, Neil
Wright, John Fowler
Wright, Robert W
Wrobel, Joseph Jude
Wu, Konrad T
Yablonsky, Harvey Allen
Yeagle, Philip L
Yencha, Andrew Joseph
Youker, John
Young, Ralph Howard
Yudelson, Joseph Samuel
Zavitsas, Andreas Athanasios
Zeltmann, Eugene W
Zevos, Nicholas
Ziegler, James Francis
Ziemniak, Stephen Eric
Zollweg, John Allman
Zuman, Petr

NORTH CAROLINA

Abbey, Kathleen Mary Kyburz
Ahrens, Rolland William
Alam, Mohammed Ashraful
Allen, Harry Clay, Jr
Allis, John W
Alston, Peter Van
Ambrosiani, Vincent F
Ammondson, Clayton John
Andrady, Anthony Lakshman
Arnett, Edward McCollin
Arnold, Luther Bishop, Jr
Ayers, Caroline LeRoy
Ayers, Paul Wayne
Baer, Tomas
Blackburn, Thomas Roy
Blackman, Carl F
Boldridge, David William
Bosley, David Emerson
Bowen, Lawrence Hoffman
Browne, Colin Lanfear
Buchanan, James Wesley
Buck, Richard Pierson
Burrus, Robert Tilden
Bush, Stewart Fowler
Cahill, Charles L
Carroll, Felix Alvin, Jr
Cassen, Thomas Joseph
Chantry, William Amdor
Chesnut, Donald Blair
Chou, David Yuan Pin
Coots, Alonzo Freeman
Daniels, William Ward
Davis, Henry Mauzee
Davis, Thomas Wilders
Dayton, Benjamin Bonney
Dearman, Henry Hursell
Echols, Joseph Todd, Jr
Elleman, Thomas Smith
Ellenson, James L
Ellison, Alfred Harris
Frieser, Rudolf Gruenspan
Gable, Ralph William
Getzen, Forrest William
Gillikin, Jesse Edward, Jr
Gillooly, George Rice
Gleit, Chester Eugene
Goodwin, Frank Erik
Gross, Paul Magnus, Jr
Hall, Carol Klein
Hall, Edward Duncan
Hall, William Earl
Hamilton, Willard Charlson
Hartzog, James Victor
Haseley, Edward Albert
Hembree, George Hunt
Hentz, Forrest Clyde, Jr
Herbst, Eric
Hermans, Jan Joseph
Hicks, Kenneth Ward
Hill, Eric Stanley
Hobbs, Marcus Edwin
Hornack, Frederick Mathew
Howell, James Milton
Hutchins, John R(ichard), III
Irvine, James Bosworth
Jackels, Charles Frederick
Jackson, Earl Graves
Jarnagin, Richard Calvin
Johnson, Charles Sidney, Jr
Johnson, James Edwin
Keith, Charles Herbert
Keller, Wayne Hicks
Kirk, David Clark
Knecht, Laurance A
Kornfeil, Fred
Krigbaum, William Richard
Kuppers, James Richard
Lee, Robert E, Jr
Leiserson, Lee
Lentz, Barry R
Lewin, Anita Hana
Lin, Stephen Fang-Maw
Loeppert, Richard Henry
McDonald, Daniel Patrick
MacPhail, Richard Allyn
Manning, Robert Joseph
Martin, G(uy) William, Jr
Martin, Richard Hugo
May, Walter Ruch
Meelheim, Richard Young
Miller, Robert L

Miller, Roger Ervin
Montet, George Louis
Moreland, Charles Glen
Morrow, John Charles, III
Neece, George A
Neumann, Calvin Lee
Park, Kisoon
Peck, David W
Pedersen, Lee G
Poirier, Jacques Charles
Raleigh, James Arthur
Reisman, Arnold
Reynolds, John Hughes, IV
Rhyne, Thomas Crowell
Rice, Richard Eugene
Rochow, Theodore George
Samulski, Edward Thaddeus
Schuh, Merlyn Duane
Scott, Donald Ray
Simon, Dorothy Martin
Sims, Leslie Berl
Smith, Peter
Spears, Alexander White, III
Spicer, Leonard Dale
Stejskal, Edward Otto
Stevens, John Gehret
Strobel, Howard Austin
Sunda, William George
Swarbrick, James
Tanford, Charles
Thompson, Nancy Lynn
Turner, Derek T
Vail, Charles Brooks
Vanselow, Clarence Hugo
Wasson, John R
Watts, Plato Hilton, Jr
Wertz, Dennis William
Wilfong, Robert Edward
Wilson, Nancy Keeler
Yang, Weitao
Young, William Anthony
Zeldes, Henry
Zumwalt, Lloyd Robert

NORTH DAKOTA
Abrahamson, Harmon Bruce
Bierwagen, Gordon Paul
Gillispie, Gregory David
Gordon, Mark Stephen
Hoggard, Patrick Earle
Kulevsky, Norman
Treumann, William Borgen
Vogel, Gerald Lee
Willson, Warrack Grant

OHIO
Abel, Alan Wilson
Adams, Harold Elwood
Allenson, Douglas Rogers
Amjad, Zahid
Ampulski, Robert Stanley
Andrist, Anson Harry
Attalla, Albert
Bacon, Frank Rider
Baczek, Stanley Karl
Bailey, John Clark
Baker, Charles Edward
Baranwal, Krishna Chandra
Batt, Russell Howard
Battino, Rubin
Bauman, Richard Gilbert
Belles, Frank Edward
Bender, Daniel Frank
Beres, John Joseph
Bernays, Peter Michael
Bernstein, Stanley Carl
Bertrand, Rene Robert
Biefeld, Paul Franklin
Bierlein, James A(llison)
Binkley, Roger Wendell
Binning, Robert Christie
Birkhahn, Ronald H
Bittker, David Arthur
Blocher, John Milton, Jr
Blomgren, George Earl
Bockhoff, Frank James
Bond, William Bradford
Brodd, Ralph James
Broge, Robert Walter
Bulgrin, Vernon Carl
Burow, Duane Frueh
Cameron, David George
Cantrell, Joseph Sires
Carfagno, Daniel Gaetano
Carlton, Terry Scott
Carman, Charles Jerry
Carter, R Owen, Jr
Cassim, Joseph Yusuf Khan
Ceasar, Gerald P
Chambers, Lee Mason
Chang, Jung-Ching
Chen, Yih-Wen
Chrysochoos, John
Coffin, Frances Dunkle
Collins, Edward A
Courchene, William Leon
Covitch, Michael J
Craig, Norman Castleman
Culbertson, Billy Muriel
Cummin, Alfred Samuel
Dahlgren, Richard Marc
Dalton, James Christopher
Darlington, William Bruce
Davies, Julian Anthony

Dawson, Steven Michael
Day, Jesse Harold
Den Besten, Ivan Eugene
Deviney, Marvin Lee, Jr
De Vries, Adriaan
Dewhurst, Harold Ainslie
Dietrick, Harry Joseph
Dollimore, David
Donovan, Sandra Steranka
Dumke, Warren Lloyd
Dunbar, Robert Copeland
Dymerski, Paul Peter
Eckert, George Frank
Eckstein, Bernard Hans
Eckstein, Yona
Edwards, Jimmie Garvin
Eisenhardt, William Anthony, Jr
Eley, Richard Robert
Endres, Paul Frank
Erickson, David Edward
Everson, Howard E
Fendler, Eleanor Johnson
Firestone, Richard Francis
Flautt, Thomas Joseph, Jr
Flechtner, Thomas Welch
Fleming, Paul Daniel, III
Foos, Raymond Anthony
Ford, William Ellsworth
Fordyce, James Stuart
Foreman, Dennis Walden, Jr
Foster, Alfred Field
Fox, Robert Kriegbaum
Fraenkel, Gideon
Frank, Sylvan Gerald
Frederick, John Edgar
Fryxell, Robert Edward
Fullmer, June Zimmerman
Gage, Frederick Worthington
Gardner-Chavis, Ralph Alexander
Garn, Paul Donald
Garner, Harry Richard
Ghose, Hirendra M
Gluyas, Richard Edwin
Goldman, Stephen Allen
Gordon, John Edward
Gordon, Sydney Michael
Gray, John Augustus, III
Greenberg, Daniel
Greene, Hoke Smith
Grieshammer, Lawrence Louis
Griffin, John Leander
Griffith, Cecil Baker
Gruber, Gerald William
Gupta, Vijay Kumar
Haas, Trice Walter
Hargis, I Glen
Harpst, Jerry Adams
Hassell, John Allen
Hatfield, Marcus Rankin
Herrmann, Kenneth Walter
Hickok, Robert Lee
Hill, Ann Gertrude
Hirst, Robert Charles
Hobrock, Don Leroy
Howell, Norman Gary
Hubbard, Arthur T
Huddleston, George Richmond, Jr
Hunter, James Charles
Hyndman, John Robert
Isbrandt, Lester Reinhardt
Ishida, Hatsuo
Jackobs, John Joseph
Jacobs, Harvey
Jaffe, Hans
Jakobsen, Robert John
Jandacek, Ronald James
Jendrek, Eugene Francis, Jr
Jenkins, Robert George
Jentoft, Joyce Eileen
Johnson, Gordon Lee
Johnson, Harlan Bruce
Jones, Edward Grant
Joyce, Blaine R
Karipides, Anastas
Karl, David Joseph
Katon, John Edward
Kaufman, Miron
Keil, Robert Gerald
Kerr, Robert Lowell
Kershner, Carl John
Kircher, John Frederick
Koehler, Mark E
Kohl, Fred John
Kollen, Wendell James
Kozikowski, Barbara Ann
Krieger, Irvin Mitchell
Kruse, Ferdinand Hobert
Kubisen, Steven Joseph, Jr
Kuska, Henry (Anton)
Lattimer, Robert Phillips
Laughlin, Robert Gene
Lauver, Richard William
Lee, Min-Shiu
Lemmerman, Karl Edward
Lewis, Irwin C
Lewis, James Edward
Lewis, Richard Thomas
Liao, Shu-Chung
Lim, Edward C
Litt, Morton Herbert
Livigni, Russell A
Livingston, Daniel Isadore
Loening, Kurt L
Lonadier, Frank Dalton

Lucas, Kenneth Ross
Luebbe, Ray Henry, Jr
Luoma, John Robert Vincent
McFarland, Charles Warren
McLachlan, Dan, Jr
McMillan, Garnett Ramsay
McQuigg, Robert Duncan
Magnusson, Lawrence Bersell
Mahadeviah, Inally
Malkin, Irving
Mark, Harry Berst, Jr
Marker, Leon
Marshall, Alan George
Martinsons, Aleksandrs
Mast, Roy Clark
Mateescu, Gheorghe D
Matesich, Mary Andrew
Mathews, A L
Mathews, Collis Weldon
Mattice, Wayne Lee
Meal, Larie L
Meckstroth, Wilma Koenig
Meeks, Frank Robert
Metcalfe, Joseph Edward, III
Meyer, Norman James
Middaugh, Richard Lowe
Miller, James Roland
Miller, Terry Alan
Miller, William Reynolds, Jr
Mochel, Virgil Dale
Moon, Tag Young
Mortensen, Earl Miller
Moss, Robert Henry
Mosser, John Snavely
Muchow, Gordon Mark
Mulligan, Bernard
Myers, R(alph) Thomas
Nardi, John Christopher
Neckers, Douglas
Neff, Vernon Duane
Nelson, Wayne Franklin
Newman, David S
Nikodem, Robert Bruce
Oertel, Richard Paul
Ogle, Pearl Rexford, Jr
Oliver, Joel Day
Olszanski, Dennis John
O'Neill, Richard Thomas
Palik, Emil Samuel
Parson, John Morris
Petschek, Rolfe George
Phillips, David Berry
Pinkerton, A Alan
Place, Robert Daniel
Platz, Matthew S
Pobereskin, Meyer
Ponter, Anthony Barrie
Porter, Raymond P
Porter, Spencer Kellogg
Post, Robert Elliott
Potter, Ralph Miles
Powell, David Lee
Poynter, James William
Provder, Theodore
Prusaczyk, Joseph Edward
Ramp, Floyd Lester
Reardon, Joseph Patrick
Reed, Allan Hubert
Reed, William Robert
Reno, Martin A
Reuter, Robert A
Rich, Joseph William
Rich, Ronald Lee
Richter, Helen Wilkinson
Rightmire, Robert
Ritchey, William Michael
Ritter, Hartien Sharp
Rodgers, Michael A J
Roe, Ryong-Joon
Rogers, Charles Edwin
Ross, Robert Talman
Ruch, Richard Julius
Russell, Leonard Nelson
Saupe, Alfred (Otto)
Schenz, Anne Filer
Schenz, Timothy William
Schildcrout, Steven Michael
Schmidt, Francis Henry
Schneider, Wolfgang W
Schochet, Melvin Leo
Schoonmaker, Richard Clinton
Schumm, Brooke, Jr
Schwerzel, Robert Edward
Scott, Roy Albert, III
Sebastian, John Francis
Seiger, Harvey N
Selker, Milton Leonard
Selover, Theodore Britton, Jr
Seybold, Paul Grant
Shaer, Elias Hanna
Shaw, Wilfrid Garside
Sicka, Richard Walter
Siebert, Alan Roger
Sieglaff, Charles Lewis
Silver, Gary Lee
Silverman, Herbert Philip
Simha, Robert
Singer, Leonard Sidney
Smith, Warren Harvey
Snavely, Deanne Lynn
Snyder, Donald Lee
Sodd, Vincent J
Sokoloski, Theodore Daniel
Spencer, Harry Edwin

Sprague, Estel Dean
Srinivasan, Vakula S
Stansbrey, John Joseph
Sternlicht, Himan
Stoner, Elaine Carol Blatt
Streng, William Harold
Strobel, Rudolf G K
Sullenger, Don Bruce
Swift, Terrence James
Swofford, Robert Lewis
Sympson, Robert F
Taylor, William Johnson
Temple, Robert Dwight
Thomas, Lazarus Daniel
Thomas, Terence Michael
Tiernan, Thomas Orville
Tokuhiro, Tadashi
Tong, James Ying-Peh
Townley, Charles William
Tsai, Chun-Che
Tuan, Debbie Fu-Tai
Turnbull, Bruce Felton
Ulmer, Richard Clyde
Uys, Johannes Marthinus
Vallee, Richard Earl
Van Lier, Jan Antonius
Vassiliades, Anthony E
Verhoek, Frank Henry
Wagner, Melvin Peter
Walsh, James Aloysius
Westenbarger, Gene Arlan
Whitman, Charles Inkley
Whitman, Donald Ray
Wilde, Bryan Edmund
Williams, Elmer Lee
Williams, Josephine Louise
Williams, Robert Calvin
Willson, Karl Stuart
Wolken, George, Jr
Woltz, Frank Earl
Wu, Richard Li-Chuan
Yeager, Ernest Bill
Yellin, Wilbur
Zoglio, Michael Anthony
Zubler, Edward George

OKLAHOMA
Ackerson, Bruce J
Andersen, Terrell Neils
Atkinson, Gordon
Baldwin, Bernard Arthur
Brown, Kenneth Henry
Burchfield, Thomas Elwood
Burr, John Green
Canham, Richard Gordon
Chatenever, Alfred
Christian, Sherril Duane
Cook, Charles Falk
Cunningham, Clarence Marion
Devlin, Joseph Paul
Dill, Edward D
Dwiggins, Claudius William, Jr
Eastman, Alan D
Fenstermaker, Roger William
Finch, Jack Norman
Fogel, Norman
Frame, Harlan D
Frech, Roger
Freeman, Robert David
Gall, James William
Gregory, M Duane
Grigsby, Ronald Davis
Grotheer, Morris Paul
Growcock, Frederick Bruce
Guillory, Jack Paul
Hall, Colby D(ixon), Jr
Harris, Jesse Ray
Harrison, R(oland) H(enry)
Harwood, William H
Heckelsberg, Louis Fred
Hobbs, Anson Parker
Howard, Robert Ernest
Janzen, Jay
Johnson, Arthur Edward
Johnson, Marvin M
Johnson, Timothy Walter
Kemp, Marwin K
Lanning, William Clarence
Lorenz, Philip Boalt
Lyon, William Graham
McCoy, Raymond Duncan
McDaniel, Max Paul
Mains, Gilbert Joseph
Matson, Ted P
Meshri, Dayaldas Tanumal
Murphy, George Washington
Neely, Stanley Carrell
Neptune, William Everett
Nolan, George Junior
Noll, Leo Albert
Nye, Mary Jo
Owen, James Robert
Palmer, Bruce Robert
Parrott, Stephen Laurent
Purdie, Neil
Purdue, Jack Olen
Raff, Lionel M
Renner, Terrence Alan
Schaffer, Arnold Martin
Schmidt, Donald Dean
Schmidt, Thomas William
Shell, Francis Joseph
Shock, D'Arcy Adriance
Stratton, Charles Abner

Physical Chemistry (cont)

Sundberg, Kenneth Randall
Thomason, William Hugh
Thompson, Donald Leo
Van der Helm, Dick
Van De Steeg, Garet Edward
Vinatieri, James Edward
Warren, Kenneth Wayne
Whitfill, Donald Lee
Williamson, William Burton
Wright, Paul McCoy

OREGON

Arthur, John Read, Jr
Barnett, Gordon Dean
Chittick, Donald Ernest
Daniels, Malcolm
Decius, John Courtney
Dunne, Thomas Gregory
Durham, George Stone
Dyke, Thomas Robert
Engelking, Paul Craig
Evans, Glenn Thomas
Feldman, Milton H
Gatz, Carole R
Hackelman, David E
Hamby, Drannan Carson
Hannum, Steven Earl
Hard, Thomas Michael
Hardwick, John Lafayette
Haygarth, John Charles
Hedberg, Kenneth Wayne
Hermens, Richard Anthony
Horne, Frederick Herbert
Hower, Charles Oliver
Hurst, James Kendall
Keedy, Curtis Russell
Klopfenstein, Charles E
Ko, Hon-Chung
Loehr, Thomas Michael
Lonsdale, Harold Kenneth
Loveland, Walter (David)
Lutz, Raymond Paul
McClure, David Warren
Maze, Robert Craig
Mazo, Robert Marc
Mickelsen, John Raymond
Nafziger, Ralph Hamilton
Nibler, Joseph William
Nielsen, Lawrence Ernie
Norris, Thomas Hughes
Noyes, Richard Macy
Payton, Arthur David
Photinos, Panos John
Prentice, Jack L
Pytkowicz, Ricardo Marcos
Robinson, Arthur B
Roe, David Kelmer
Schellman, John Anthony
Seevers, Robert Edward
Skinner, Gordon Bannatyne
Sleight, Arthur William
Swanson, Lynwood Walter
Tenney, Agnes
Thomas, Thomas Darrah
Tyler, David Ralph
Wamser, Carl Christian
Warren, William Willard, Jr
White, Horace Frederick
Winniford, Robert Stanley

PENNSYLVANIA

Allara, David Lawrence
Altares, Timothy, Jr
Ambs, William Joseph
Anderson, James B
Anderson, Jay Martin
Anderson, John Howard
Ascah, Ralph Gordon
Aukrust, Egil
Austin, James Bliss
Aviles, Rafael G
Bacher, Frederick Addison
Backus, John King
Bakule, Ronald David
Bandy, Alan Ray
Banks, Grace Ann
Barford, Robert A
Barnartt, Sidney
Barnes, Mary Westergaard
Barnett, Herald Alva
Bass, Jonathan Langer
Basseches, Harold
Batdorf, Robert Ludwig
Baurer, Theodore
Bechtold, Max Frederick
Becker, Aaron Jay
Beichl, George John
Belitskus, David
Benkovic, Stephen J
Bent, Henry Albert
Bergo, Conrad Hunter
Berk, Lawrence B
Berkheimer, Henry Edward
Bernheim, Robert A
Berry, Robert Walter
Biloen, Paul
Birks, Neil
Bivens, Richard Lowell
Blaustein, Bernard Daniel
Bock, Charles Walter
Boggs, William Emmerson
Bohning, James Joel

Bonewitz, Robert Allen
Borkowski, Raymond P
Bowman, Kenneth Aaron
Bowman, Robert Samuel
Brendley, William H, Jr
Breyer, Arthur Charles
Bromberg, J Philip
Brown, Patrick Michael
Brown, Paul Edmund
Bugosh, John
Burns, Allan Fielding
Butera, Richard Anthony
Buzzelli, Edward S
Cammarata, Arthur
Caretto, Albert A, Jr
Carroll, Robert J
Cartier, Peter G
Casassa, Edward Francis
Castellano, Salvatore, Mario
Castleman, Albert Welford, Jr
Catalano, Anthony William
Catchen, Gary Lee
Cavanaugh, James Richard
Cawley, John Joseph
Chaiken, Robert Francis
Chakrabarti, Paritosh M
Chang, Cheng Allen
Chase, Grafton D
Chesick, John Polk
Chiu, Tai-Woo
Christoffersen, Ralph Earl
Chu, Vincent H(ao) K(wong)
Chuang, Strong Chieu-Hsiung
Chung, Tze-Chiang
Clayton, John Charles (Hastings)
Clovis, James S
Cochran, Charles Norman
Cohn, Mildred
Cole, Milton Walter
Conroy, James Strickler
Cook, Donald Bowker
Cooper, John Neale
Corneliussen, Roger DuWayne
Cornell, Donald Gilmore
Courtney, Welby Gillette
Craig, Raymond S
Dai, Hai-Lung
Davis, Brian Clifton
Davis, Burl Edward
DeBroy, Tarasankar
Dedinas, Jonas
Deno, Norman C
Dent, Anthony L
DiCarlo, Ernest Nicholas
Diehl, Renee Denise
Diorio, Alfred Frank
Dougherty, Eugene P
Downey, Bernard Joseph
Downs, James Joseph
Dreibelbis, John Adam
Duck, William N, Jr
Dunkelberger, Tobias Henry
Durante, Vincent Anthony
Duttaahmed, A
Dux, James Philip
Dymicky, Michael
Dzombak, William Charles
El-Aasser, Mohamed S
Ellison, Frank Oscar
Elson, Jesse
Epstein, Lawrence Melvin
Erb, Robert Allan
Fabish, Thomas John
Fainberg, Arnold Harold
Falcone, James Salvatore, Jr
Falkiewicz, Michael Joseph
Farlee, Rodney Dale
Farrington, Gregory Charles
Feighan, Maria Josita
Feller, Robert Livingston
Ferrell, D Thomas, Jr
Ferren, Richard Anthony
Fetsko, Jacqueline Marie
Finegold, Harold
Finseth, Dennis Henry
Fischer, John Edward
Fishman, Marshall Lewis
Fitts, Donald Dennis
Forchheimer, Otto Louis
Forsman, William C(omstock)
Fortnum, Donald Holly
Foster, Norman Francis
Foster, Perry Alanson, Jr
Francis, Peter Schuyler
Franco, Nicholas Benjamin
Frank, William Benson
Frankel, Lawrence (Stephen)
Frenklach, Michael Y
Fresia, Elmo James
Frick, Neil Huntington
Fritsch, Arnold Rudolph
Fuget, Charles Robert
Furrow, Stanley Donald
Garbarini, Victor C
Garber, Charles A
Gardner, David Milton
Garrett, Michael Benjamin
Garrett, Robert Roth
Garrison, Barbara Jane
Gaughan, Renata Rysnik
Gerasimowicz, Walter Vladimir
Gittler, Franz Ludwig
Glandt, Eduardo Daniel
Gold, Lewis Peter

Goodwin, James Gordon, Jr
Gordon, William Edwin
Grady, Harold Roy
Grant, Richard J
Greeley, Richard Stiles
Greenshields, John Bryce
Greifer, Aaron Philip
Greiner, Richard William
Grey, James Tracy, Jr
Griffiths, Robert Budington
Gulbransen, Earl Alfred
Hall, W Keith
Haltner, Arthur John
Hammons, James Hutchinson
Hansen, Gerald Delbert, Jr
Hardman, Carl Charles
Harju, Philip Herman
Hart, Maurice I, Jr
Hatch, Richard C
Hazeltine, James Ezra, Jr
Heald, Emerson Francis
Hertzberg, Martin
Hillner, Edward
Hiltz, Arnold Aubrey
Hochstrasser, Robin
Hofer, Lawrence John Edward
Hogan, Aloysius Joseph, Jr
Hollibaugh, William Calvert
Hollingsworth, Charles Alvin
Hudson, Robert McKim
Hutta, Paul John
Irwin, Philip George
Jackovitz, John Franklin
James, Laylin Knox, Jr
Janda, Kenneth Carl
Jenkins, Wilmer Atkinson, II
Jhon, Myung S
Johnson, Edwin Wallace
Johnson, Kenneth Allen
Jordan, Kenneth David
Kakar, Anand Swaroop
Kaplan, Morton
Katsanis, Eleftherios P
Kay, Jack Garvin
Kay, Robert Leo
Keller, Rudolf
Kelly, Ernest L
Kenson, Robert Earl
Kerst, A(l) Fred
Kifer, Edward W
Kim, Tai Kyung
Kimlin, Mary Jayne
King, James P
King, Richard Warren
Klein, Michael Lawrence
Kleinsteuber, Tilmann Christoph Werner
Klier, Kamil
Kliman, Harvey Louis
Koller, Robert Dene
Kraitchman, Jerome
Krzeminski, Stephen F
Kumosinski, Thomas Francis
Kunesh, Charles Joseph
Labes, Mortimer Milton
Lampe, Frederick Walter
Lancet, Michael Savage
Lange, Klaus Robert
Larsen, John W
Larsen, Russell D
Lassettre, Edwin Nichols
Lautenberger, William J
Leader, Gordon Robert
Lee, Pang-Kai
Leffler, Amos J
Leininger, Paul Miller
Leister, Harry M
Leonard, John Joseph
Leston, Gerd
Lewis, Bernard
Lewis, Charles William
Li, Norman Chung
Liberti, Paul A
Lloyd, Thomas Blair
Locke, Harold Ogden
Longo, Frederick R
Loprest, Frank James
Lovejoy, Roland William
Lowe, John Philip
Ludwig, Howard C
Ludwig, Oliver George
McCarthy, James Francis
McCarthy, Raymond Lawrence
McDowell, Maurice James
Mack, Lawrence Lloyd
MacKay, Colin Francis
McKelvey, Donald Richard
McKinney, David Scroggs
Maclay, William Nevin
Magill, Joseph Henry
Maguire, Mildred May
Mair, Robert Dixon
Mandelcorn, Lyon
Manes, Milton
Manley, Thomas Clinton
Marcelin, George
Martin, Aaron J
Martin, Edward Shaffer
Martinez de Pinillos, Joaquin Victor
MASCIANTONIO, PHILIP
Mason, Charles Morgan
Matyjaszewski, Krzysztof
Mayer, George Emil
Mayo, Ralph Elliott
Meeker, Thrygve Richard

Melander, Wayne Russell
Messer, Wayne Ronald
Micale, Fortunato Joseph
Michaels, Adlai Eldon
Miller, Alfred Charles
Miller, Eugene D
Miller, John George
Milz, Wendell Collins
Minn, Fredrick Louis
Mishra, Dinesh S
Mitchell, Donald J
Molinari, Robert James
Montjar, Monty Jack
Morris, Marlene Cook
Motsavage, Vincent Andrew
Mozersky, Samuel M
Muck, Darrel Lee
Mulay, Laxman Nilakantha
Music, John Farris
Myers, Howard
Na, George Chao
Nagy, Dennis J
Nath, Amar
Naylor, Robert Ernest, Jr
Neddenriep, Richard Joe
Neidig, Howard Anthony
Newitt, Edward James
Nixon, Eugene Ray
Norris, Wilfred Glen
Nylund, Robert E
Ochiai, Ei-Ichiro
Ohlberg, Stanley Miles
O'Mara, James Herbert
Opella, Stanley Joseph
Owens, Kevin Glenn
Pacer, John Charles
Palmer, Howard Benedict
Parchen, Frank Raymond, Jr
Parker, Winfred Evans
Patterson, Gary David
Perzak, Frank John
Pessen, Helmut
Peterson, Jack Kenneth
Pettit, Frederick S
Pez, Guido Peter
Piccolini, Richard John
Pickering, Howard W
Pierce, Percy Everett
Pierce, Roberta Marion
Pinschmidt, Robert Krantz, Jr
Plazek, Donald John
Poole, John Anthony
Porter, Sydney W, Jr
Prados, Ronald Anthony
Pratt, David W
Prohaska, Charles Anton
Radspinner, John Asa
Randall, William Carl
Recktenwald, Gerald William
Reeder, Ray R
Retcofsky, Herbert L
Reuwer, Joseph Francis, Jr
Richardson, Ralph J
Richey, Willis Dale
Riethof, Thomas Robert
Ristey, William J
Roesmer, Josef
Rogers, Horace Elton
Romberger, Karl Arthur
Roper, Gerald C
Rosenbaum, Eugene Joseph
Rosenblum, Charles
Rothbart, Herbert Lawrence
Rozelle, Ralph B
Ruppel, Thomas Conrad
Sallavanti, Robert Armando
Salomon, Robert Ephriam
Samuelson, H Vaughn
Schaeffer, William Dwight
Scheirer, James E
Schell, William R
Schettler, Paul Davis, Jr
Schiesser, Robert H
Schiffman, Louis F
Schott, Hans
Schreiber, Kurt Clark
Schuster, Ingeborg I M
Schweighardt, Frank Kenneth
Shapiro, Zalman Mordecai
Sheasley, William David
Shergalis, William Anthony
Shim, Benjamin Kin Chong
Shoaf, Mary La Salle
Signorino, Charles Anthony
Silbert, Leonard Stanton
Simmons, Gary Wayne
Simonaitis, Romualdas
Singleton, Jack Howard
Siska, Peter Emil
Sloat, Charles Allen
Smith, Allan Laslett
Smith, David
Smith, Stewart Edward
Sollott, Gilbert Paul
Solomon, Joseph Alvin
Spear, Jo-Walter
Spell, Aldenlee
Spencer, James Nelson
Spinnler, Joseph F
Staskiewicz, Bernard Alexander
Staut, Ronald
Steele, William A
Steiger, Roger Arthur
Stewart, Robert F

Stokes, Charles Sommers
Stryker, Lynden Joel
Sturm, James Edward
Swain, Howard Aldred, Jr
Swarts, Elwyn Lowell
Swift, Harold Eugene
Szabo, Miklos Tamas
Takeshita, Tsuneichi
Taylor, George Russell
Taylor, Robert Morgan
Taylor, William Daniel
Tenney, Albert Seward, III
Tepper, Frederick
Tevebaugh, Arthur David
Thibeault, Jack Claude
Thomas, H Ronald
Thompson, Peter Trueman
Thornton, Elizabeth K
Tomezsko, Edward Stephen John
Tosch, William Conrad
Trachtman, Mendel
Tubbs, Robert Kenneth
Turner, Paul Jesse
Vanderhoff, John W
Varga, Charles E
Vastola, Francis J
Venable, Emerson
Voet, Donald Herman
Voltz, Sterling Ernest
Wahnsiedler, Walter Edward
Walker, Robert Winn
Wallace, Paul Francis
Warner, John Christian
Warrick, Percy, Jr
Watterson, Kenneth Franklin
Wefers, Karl
Wei-Berk, Caroline
Weismann, Theodore James
Weiss, Paul Storch
Welch, Cletus Norman
Wells, Adoniram Judson
Wells, James Edward
Westmoreland, David Gray
White, David
White, Edward Roderick
White, James Russell
White, Malcolm Lunt
White, Norman Edward
Williams, Ardis Mae
Williams, James Earl, Jr
Wilson, John Randall
Winters, Earl D
Wismer, Robert Kingsley
Witkowski, Robert Edward
Wojcik, John F
Wolke, Robert Leslie
Wootten, Michael John
Work, James Leroy
Wriedt, Henry Anderson
Wunderlich, Francis J
Yang, Arthur Jing-Min
Yao, Shang Jeong
Yates, John Thomas, Jr
Yeh, George Chiayou
Young, Harrison Hurst, Jr
Young, Irving Gustav
Yuan, Edward Lung
Zanger, Murray
Zelley, Walter Gauntt
Zeroka, Daniel
Zettlemoyer, Albert Charles
Ziegler, George William, Jr
Zimmerman, George Landis
Zordan, Thomas Anthony

RHODE ISLAND
Baird, James Clyde
Boyko, Edward Raymond
Calo, Joseph Manuel
Carpenter, Gene Blakely
Cole, Robert Hugh
Diebold, Gerald Joseph
Doll, Jimmie Dave
Estrup, Peder Jan Z
Greene, Edward Forbes
Hartman, Karl August
Huebert, Barry Joe
Johnson, Gordon Carlton
Kirschenbaum, Louis Jean
Kreiser, Ralph Rank
Kroll, Harry
Lawler, Ronald George
Marzzacco, Charles Joseph
Mason, Edward Allen
Morris, George V
Ouderkirk, John Thomas
Risen, William Maurice, Jr
Sprowles, Jolyon Charles
Stratt, Richard Mark
Williams, John Collins, Jr
Zuehlke, Richard William

SOUTH CAROLINA
Ashy, Peter Jawad
Avgeropoulos, George N
Bach, Ricardo O
Bailey, Charles Edward
Bailey, Roy Horton, Jr
Ballentine, Alva Ray
Bly, Robert Stewart
Bonner, Oscar Davis
Boyd, George Edward
Breazeale, William Horace, Jr
Clark, Hugh Kidder

Davis, Joseph B
Dellicolli, Humbert Thomas
Diefendorf, Russell Judd
Durig, James Robert
Dusenbury, Joseph Hooker
Dworjanyn, Lee O(leh)
Falkehag, S Ingemar
Faust, John William, Jr
Force, Carlton Gregory
Gibbs, Ann
Gilkerson, William Richard
Griffith, Elizabeth Ann Hall
Groh, Harold John
Harbour, John Richard
Heberger, John M
Henderson, Richard Wayne
Hennelly, Edward Joseph
Herdklotz, Richard James
Hochel, Robert Charles
Hopper, Michael James
Hyder, Monte Lee
Jacober, William John
Jarvis, Christine Woodruff
Jordan, Kenneth Gary
Jumper, Charles Frederick
Kendall, David Nelson
Knight, Jere Donald
Knight, Lon Bishop, Jr
Latta, Bruce McKee
Likes, Carl James
Longfield, James Edgar
Lowry, Bright Anderson
Maier, Herbert Nathaniel
Malstrom, Robert Arthur
Mercer, Edward Everett
Mercier, Philip Laurent
Metz, Clyde
Myrick, Michael Lenn
Nanney, Thomas Ray
Orebaugh, Errol Glen
Otto, Wolfgang Karl Ferdinand
Owen, John Harding
Patterson, William Alexander
Perkins, William Clopton
Piper, John
Plodinec, Matthew John
Porter, John J
Sargent, Roger N
Savitsky, George Boris
Sharpe, Louis Haughton
Simon, Frederick Tyler
Simonsen, David Raymond
Smith, Paul Kent
Spencer, Harold Garth
Stearns, Edwin Ira
Stone, John Austin
Taylor, Peter Anthony
Tolbert, Thomas Warren
Varin, Roger Robert
Walters, John Philip
Woolsey, Gerald Bruce

SOUTH DAKOTA
Coker, Earl Howard, Jr
Estee, Charles Remington
Hecht, Harry George
Martin, Willard John
Rue, Rolland R
Scott, George Prescott
Spinar, Leo Harold

TENNESSEE
Adams, Robert Edward
Airee, Shakti Kumar
Allen, Vernon R
Anderson, Roger W(hiting)
Annis, Brian Kitfield
Baes, Charles Frederick, Jr
Bamberger, Carlos Enrique Leopoldo
Banna, Salim M
Barber, Eugene John
Bates, John Bryant
Baxter, John Edwards
Beahm, Edward Charles
Bedoit, William Clarence, Jr
Begun, George Murray
Bell, Jimmy Todd
Bentz, Ralph Wagner
Blanck, Harvey F, Jr
Bleier, Alan
Bond, Walter D
Boston, Charles Ray
Brau, Charles Allen
Braunstein, Jerry
Brooks, Alfred Austin, Jr
Brown, Gilbert Morris
Brown, Lloyd Leonard
Browning, Horace Lawrence, Jr
Bullock, Jonathan S, IV
Bunick, Gerard John
Burns, John Howard
Busey, Richard Hoover
Busing, William Richard
Caflisch, George Barrett
Cantor, Stanley
Carman, Howard Smith, Jr
Cathcart, John Varn
Cheng, Ching-Chi Chris
Clarke, Ann Neistadt
Clarke, James Harold
Coleman, Charles Franklin
Collins, Clair Joseph
Compere, Edgar Lattimore
Condon, James Benton

Crompton, Charles Edward
Daunt, Stephen Joseph
Davis, Jimmy Henry
Davis, Montie Grant
Davis, Phillip Howard
Deshpande, Krishnanath Bhaskar
Dillard, James William
Doody, John Edward
Dudney, Nancy Johnston
Duncan, Budd Lee
Egan, B Zane
Evans, Joe Smith
Fagerburg, David Richard
Farrar, Robert Lynn, Jr
Feigerle, Charles Stephen
Field, Frank Henry
Fletcher, William Henry
Fort, Tomlinson, Jr
Franceschetti, Donald Ralph
Frederiksen, Dixie Ward
Fuzek, John Frank
Gangwer, Thomas E
Germroth, Ted Calvin
Gibson, John Knight
Glasstone, Samuel
Gray, Allen G(ibbs)
Guenther, William Benton
Gustafson, Bruce LeRoy
Habenschuss, Anton
Hall, Larry Cully
Halperin, Joseph
Harding, Charles Enoch
Hayter, John Bingley
Hinds, Nancy Webb
Ho, Patience Ching-Ru
Hochanadel, Clarence Joseph
Hoffman, Everett John
Holmes, Howard Frank
Huang, Thomas Tao Shing
Ijams, Charles Carroll
Jenks, Glenn Herbert
Johnson, Elijah
Johnston, David Owen
Judish, John Paul
Judkins, Roddie Reagan
Ketchen, Eugene Earl
Kocher, David Charles
Kovac, Jeffrey Dean
Krause, Herbert Francis
Krause, Manfred Otto
Langford, Paul Brooks
Leed, Russell Ernest
Li, Ying Sing
Lietzke, Milton Henry
Lindemer, Terrence Bradford
Livingston, Ralph
Longmire, Martin Shelling
Lura, Richard Dean
McConnell, Richard Leon
McDowell, William Jackson
McGill, Robert Mayo
McGraw, Gary Earl
McGrory, Joseph Bennett
Magid, Linda Lee Jenny
Malinauskas, Anthony Peter
Marshall, William Leitch
Martin, Thomas Waring
Maya, Leon
Miceli, Angelo Sylvestro
Miller, John Cameron
Moore, George Edward
Moore, Louis Doyle, Jr
Mortimer, Robert George
Moyer, Bruce A
Naddy, Badie Ihrahim
Newby, Frank Armon, Jr
Nicely, Vincent Alvin
Noid, Donald William
Nyssen, Gerard Allan
Parkinson, William Walker, Jr
Patton, Hugh Wilson
Pearson, R(ay) L(eon)
Petersen, Harold, Jr
Peterson, Joseph Richard
Petke, Frederick David
Phillips, Paul J
Polavarapu, Prasad Leela
Pond, David Martin
Poutsma, Marvin Lloyd
Powell, George Louis
Quist, Arvin Sigvard
Rauscher, Grant K
Reynolds, Jefferson Wayne
Robbins, Gordon Daniel
Roesler, Joseph Frank
Rudolph, Philip S
Rummel, Robert Edwin
Rush, Richard Marion
Russin, Nicholas Charles
Rutenberg, Aaron Charles
Schaad, Lawrence Joseph
Silverman, Meyer David
Smith, George Pedro
Soldano, Benny A
Staats, Percy Anderson
Starr, Duane Frank
Stoughton, Raymond Woodford
Sweeton, Frederick Humphrey
Sworski, Thomas John
Tarpley, Anderson Ray, Jr
Taylor, Ellison Hall
Tellinghuisen, Joel Barton
Thiessen, William Ernest
Thoma, Roy E

Tirman, Alvin
Toth, Louis McKenna
Trowbridge, Lee Douglas
Van Hook, William Alexander
Van Sickle, Dale Elbert
Vasofsky, Richard William
Waggener, William Cole
Ward, James Andrew
Watson, Marshall Tredway
Williams, Thomas Ffrancon
Wilson, David J
Woods, Clifton, III
Young, Frederick Walter, Jr
Young, Howard Seth
Zuber, William Henry, Jr

TEXAS
Abrams, Albert
Addy, Tralance Obuama
Ahlberg, Dan Leander
Albright, John Grover
Alexander, Robert Benjamin
Ali, Yusuf
Allison, Jean Batchelor
Altmiller, Henry
Anderson, Richard Louis
Anselmo, Vincent C
Armstrong, Andrew Thurman
Ashby, Carl Toliver
Baker, Charles Taft
Bannerman, James Knox
Banta, Marion Calvin
Bard, Allen Joseph
Barieau, Robert (Eugene)
Barna, Gabriel George
Basila, Michael Robert
Batten, Charles Francis
Becker, Ralph Sherman
Benson, Herbert Linne, Jr
Berg, Mark Alan
Berry, Michael James
Birney, David Martin
Blount, Floyd Eugene
Blytas, George Constantin
Bockris, John O'Mara
Boggs, James Ernest
Brady, William Thomas
Broodo, Archie
Brooks, Philip Russell
Brown, James Michael
Brown, Robert Wade
Brusie, James Powers
Busch, Marianna Anderson
Callender, Wade Lee
Cares, William Ronald
Carlson, Douglas W
Caudle, Danny Dearl
Chao, Jing
Chen, Edward Chuck-Ming
Chen, Michael Chia-Chao
Clark, Ronald Keith
Clingman, William Herbert, Jr
Closmann, Philip Joseph
Cole, David F
Collins, Francis Allen
Collins, Russell Lewis
Conway, Dwight Colbur
Cook, Evin Lee
Cook, Paul Fabyan
Cooley, Stone Deavours
Crawford, James Worthington
Croce, Louis J
Cross, Virginia Rose
Cruser, Stephen Alan
Cummiskey, Charles
Curl, Robert Floyd
Cusachs, Louis Chopin
Custard, Herman Cecil
Davis, Donald Robert
Davis, Raymond E
Deal, Carl Hosea, Jr
DeBerry, David Wayne
Denison, Jack Thomas
Desiderato, Robert, Jr
Dimitroff, Edward
Dobson, Gerard Ramsden
Dorris, Kenneth Lee
Doyle, Michael P
Drake, Edgar Nathaniel, II
Duncan, Walter Marvin, Jr
Duty, Robert C
Dyer, Lawrence D
Dzidic, Ismet
Eaker, Charles William
Ekerdt, John Gilbert
Elliott, J Lell
Ellzey, Marion Lawrence, Jr
Elward-Berry, Julianne
Engel, Paul Sanford
Farhataziz, Mr
Faris, Sam Russell
Faubion, Billy Don
Floyd, Willis Waldo
Forest, Edward
Foster, Norman George
Foster, William Roderick
Fox, Marye Anne
Frank, Steven Neil
Franklin, Thomas Chester
Frazee, Jerry D
Fredin, Leif G R
Freund, Howard John
Friedman, Robert Harold
Fuller, Martin Emil

Engel, Thomas Walter
Evans, Thomas Walter
Ewing, J J
Exarhos, Gregory James
Filby, Royston Herbert
Fordyce, David Buchanan
Franz, James Alan
Frasco, David Lee
Friedrich, Donald Martin
Gammon, Richard Harriss
Gerhold, George A
Goldring, Lionel Solomon
Gouterman, Martin (Paul)
Greager, Oswald Herman
Gregory, Norman Wayne
Halsey, George Dawson, Jr
Hamm, Randall Earl
Heller, Eric Johnson
Hendrickson, Waldemar Forrsel
Hipps, Kerry W
Hunt, John Philip
Johnson, A(lfred) Burtron, Jr
Johnson, Wilbur Vance
Kalkwarf, Donald Riley
Kazanjian, Armen Roupen
Keaton, Clark M
Kells, Lyman Francis
Kells, Milton Carlisle
Kelsh, Dennis J
Kissin, G(erald) H(arvey)
Koehmstedt, Paul Leon
Krier, Carol Alnoth
Kriz, George Stanley
Krohn, Kenneth Albert
Kwiram, Alvin L
Ladd, Kaye Victoria
Lambert, Maurice C
Leitz, Fred John, Jr
Leonard, John Edward
Lim, Young Woon (Peter)
Lindenmeyer, Paul Henry
Lingren, Wesley Earl
McDowell, Robin Scott
McGuire, Joseph Clive
Maki, Arthur George, Jr
Miller, John Howard
Minor, James E(rnest)
Moore, Emmett Burris, Jr
Moore, Robert Lee
Morrey, John Rolph
Moseley, William David, Jr
Murdock, Gordon Alfred
Neuzil, Edward F
Nickerson, Robert Fletcher
Nightingale, Richard Edwin
Ohta, Masao
Pavia, Donald Lee
Peterson-Kennedy, Sydney Ellen
Pocker, Yeshayau
Poole, Donald Ray
Rabinovitch, B(enton) S(eymour)
Read, David Hadley
Reinhardt, William Parker
Roake, William Earl
Root, John Walter
Ryan, Jack Lewis
Schomaker, Verner
Schurr, John Michael
Shull, Charles Morell, Jr
Slutsky, Leon Judah
Smith, Francis Marion
Smith, Richard Dale
Spencer, Arthur Milton, Jr
Stuve, Eric Michael
Styris, David Lee
Tabbutt, Frederick Dean
Tang, Kwong-Tin
Thorsell, David Linden
Tingey, Garth Leroy
Tobiason, Frederick Lee
Vandenbosch, Robert
Wei, Pax Samuel Pin
Wheelwright, Earl J
Whitmer, John Charles
Willett, Roger
Windsor, Maurice William
Wogman, Ned Allen
Yan, Johnson Faa
Yates, Albert Carl
Young, Robert Hayward
Yunker, Wayne Harry
Zoller, William H

WEST VIRGINIA
Abel, William T
Bassett, David R
Bhasin, Madan M
Chakrabarty, Manoj R
Coleman, James Edward
Dalal, Nar S
Das, Kamalendu
Dawson, Thomas Larry
Dean, Warren Edgell
Derderian, Edmond Joseph
Finklea, Harry Osborn
Fisher, John F
Grindstaff, Teddy Hodge
Hager, Stanley Lee
Hall, George Arthur, Jr
Hall, James Lester
Hanrahan, Edward S
Harrison, Arnold Myron
Hickman, James Blake
Hoback, John Holland

Humphrey, George Louis
Krabacher, Bernard
Kurland, Jonathan Joshua
Leckonby, Roy Alan
Lim, James Khai-Jin
Marcinkowsky, Arthur Ernest
Martin, John Perry, Jr
Murrmann, Richard P
Naumann, Alfred Wayne
Petersen, Jeffrey Lee
Rollins, Ronald Roy
Seshadri, Kalkunte S
Showalter, Kenneth
Stewart, Robert Francis
Vucich, M(ichael) G(eorge)
Wallace, William James Lord
Williamson, Kenneth Dale
Woodland, William Charles

WISCONSIN
Alvarez, Robert
Anderson, Harold J
Bahe, Lowell W
Barr, Tery Lynn
Beatty, James Wayne, Jr
Bender, Paul J
Blair, James Edward
Bondeson, Stephen Ray
Brandt, Werner Wilfried
Brebrick, Robert Frank, Jr
Bryce, Hugh Glendinning
Bullock, Kathryn Rice
Cataldi, Horace A(nthony)
Chen, Franklin M
Ciriacks, John A(lfred)
Clark, Harlan Eugene
Cornwell, Charles Daniel
Coward, Nathan A
Crim, F(orrest) Fleming
Dahl, Lawrence Frederick
Discher, Clarence August
Draeger, Norman Arthur
Duwell, Ernest John
Dwyer, Sean G
Ehlert, Thomas Clarence
Evans, James Stuart
Farrar, Thomas C
Fenrick, Harold William
Greenler, Robert George
Grotz, Leonard Charles
Haeberli, Willy
Hall, Lyle Clarence
Hansen, Paul Vincent, Jr
Harriman, John E
Hassinger, Mary Colleen
Hauser, Edward Russell
Holt, Matthew Leslie
Hyde, James Stewart
Inman, Ross
Kao, Wen-Hong
Keulks, George William
Kittsley, Scott Loren
Kresch, Alan J
Lang, Conrad Marvin
Lee, Kathryn Adele Bunding
Liang, Shoudeng
Liebe, Donald Charles
Lo, Mike Mei-Kuo
Losin, Edward Thomas
McAllister, Jerome Watt
McCoy, Charles Ralph
Mallmann, Alexander James
Miller, Arild Justesen
Morton, Stephen Dana
Mukerjee, Pasupati
Myers, George E
Nakamoto, Kazuo
Norman, Jack C
Phillips, John Spencer
Pietro, William Joseph
Pollnow, Gilbert Frederick
Prall, Bruce Randall
Radtke, Douglas Dean
Record, M Thomas, Jr
Romary, John Kirk
Rosenberg, Robert Melvin
Roskos, Roland R
St Louis, Robert Vincent
Schrader, David Martin
Schwab, Helmut
Sealy, Roger Clive
Sheppard, Erwin
Spencer, Brock
Steiner, John F
Su, Lao-Sou
Tcheurekdjian, Noubar
Thomas, Howard Major
Tiedemann, William Harold
Trischan, Glenn M
Vanselow, Ralf W
Vaughan, Worth E
Viswanathan, Ramaswami
Wang, Pao-Kuan
Wendricks, Roland N
Westler, William Milo
Willard, John Ela
Williams, John Warren
Woods, Robert Claude
Wright, John Curtis
Yu, Hyuk
Zimmerman, Howard Elliot
Zografi, George

WYOMING
Decora, Andrew Wayne
Duvall, John Joseph
Edmiston, Clyde
Hiza, Michael John, Jr
Jacobson, Irven Allan, Jr
McColloch, Robert James
Meyer, Edmond Gerald
Netzel, Daniel Anthony
Northcott, Jean
Poulson, Richard Edwin
Ryan, Victor Albert

PUERTO RICO
Blum, Lesser
Bode, James Daniel
Muir, Mariel Meents
Schwartz, Abraham
Siegel, Georges G
Souto Bachiller, Fernando Alberto
Toro-Goyco, Efrain
Vargas, Fernando Figueroa

ALBERTA
Ali, Shahida Parvin
Armstrong, David Anthony
Bertie, John E
Birss, Viola Ingrid
Blades, Arthur Taylor
Brown, Robert Stanley
Codding, Penelope Wixson
Creighton, Stephen Mark
Davis, Stuart George
Dunford, Hugh Brian
Freeman, Gordon Russel
Gunning, Harry Emmet
Harvey, Ross Buschlen
Heimann, Robert B(ertram)
Henderson, John Frederick
Hepler, Loren George
Hodgson, Gordon Wesley
Jacobson, Ada Leah
Kalantar, Alfred Husayn
Kebarle, Paul
Khanna, Faqir Chand
Kotovych, George
Krueger, Peter J
Kuntz, Garland Parke Paul
Lown, James William
McClung, Ronald Edwin Dawson
McCurdy, Keith G
Martin, John Scott
Muir, Donald Ridley
Plambeck, James Alan
Robertson, Ross Elmore
Russell, James Christopher
Schulz, Karlo Francis
Sommerfeldt, Theron G
Sorensen, Theodore Strang
Strausz, Otto Peter
Tollefson, Eric Lars
Tschuikow-Roux, Eugene
Wieser, Helmut
Yamdagni, Raghavendra
Yeager, Howard Lane

BRITISH COLUMBIA
Balfour, Walter Joseph
Barrow, Gordon M
Bell, Thomas Norman
Benston, Margaret Lowe
Bree, Alan V
Brion, Christopher Edward
Burnell, Edwin Elliott
Church, Shepard Earll, (Jr)
Dingle, Thomas Walter
Dunell, Basil Anderson
Farmer, James Bernard
Forgacs, Otto Lionel
Frost, David Cregreen
Funt, B Lionel
Gerry, Michael Charles Lewis
Harrison, Lionel George
Hatton, John Victor
Hayward, Lloyl Douglas
Hooley, Joseph Gilbert
Kirk, Alexander David
Korteling, Ralph Garret
Leja, J(an)
McAuley, Alexander
McDowell, Charles Alexander
Malli, Gulzari Lal
Morrison, Stanley Roy
Murray, Francis E
O'Brien, Robert Neville
Ogryzlo, Elmer Alexander
Peters, E(rnest)
Porter, Gerald Bassett
Reid, Cyril
Rieckhoff, Klaus E
Sams, John Robert, Jr
Schreier, Hanspeter
Sherwood, A Gilbert
Spratley, Richard Denis
Stewart, Ross
Trotter, James
Voigt, Eva-Maria
Wells, Edward Joseph
Zwarich, Ronald James

MANITOBA
Bigelow, Charles C
Bock, Ernst
Burchill, Charles Eugene

Charlton, James Leslie
Forrest, Bruce James
Gesser, Hyman Davidson
Hutton, Harold M
Kartzmark, Elinor Mary
McFarlane, Joanna
MacGregor, Elizabeth Ann
Sagert, Norman Henry
Saluja, Preet Pal Singh
Schaefer, Theodore Peter
Secco, Anthony Silvio
Singh, Ajit
Sunder, Sham
Sze, Yu-Keung
Tait, John Charles
Tomlinson, Michael
Torgerson, David Franklyn
Trick, Gordon Staples
Vikis, Andreas Charalambous
Westmore, John Brian

NEW BRUNSWICK
Brooks, Wendell V F
Grant, Douglas Hope
Grein, Friedrich
LeBel, Roland Guy
Read, John Frederick
Reinsborough, Vincent Conrad
Semeluk, George Peter
Thakkar, Ajit Jamnadas

NEWFOUNDLAND
Shahidi, Fereidoon
Smith, Frank Roylance
Stein, Allan Rudolph
Tremaine, Peter Richard

NOVA SCOTIA
Arnold, Donald Robert
Bunbury, David Leslie
Cooke, Robert Clark
Davies, Donald Harry
Falk, Michael
Gillis, Hugh Andrew
Hayes, Kenneth Edward
Jamieson, William David
Jones, William Ernest
Kreuzer, Han Jurgen
Kusalik, Peter Gerard
Kwak, Jan C T
Leffek, Kenneth Thomas
Linton, Everett Percival
Lynch, Brian Maurice
MacDonald, John James
Pacey, Philip Desmond
Secco, Etalo Anthony
Wasylishen, Roderick Ernest
White, Mary Anne
Whiteway, Stirling Giddings

ONTARIO
Abrahamson, Edwin William
Adams, Richard James
Allnatt, Alan Richard
Amenomiya, Yoshimitsu
Ananthanarayanan, Vettaikkoru S
Anlauf, Kurt Guenther
Aziz, Philip Michael
Aziz, Ronald A
Back, Margaret Helen
Baker, Robert G
Bancroft, George Michael
Bardwell, Jennifer Ann
Barradas, Remigio Germano
Barrie, Leonard Arthur
Barton, Stuart Samuel
Benson, George Campbell
Bernath, Peter Francis
Bharucha, Nana R
Bidinosti, Dino Ronald
Bluhm, Terry Lee
Bohme, Diethard Kurt
Bolton, James R
Breitman, Leo
Brown, James Douglas
Brown, Richard Julian Challis
Brumer, Paul William
Bulani, Walter
Carey, Paul Richard
Carrington, Tucker
Casal, Hector Luis
Chan, Raymond Kai-Chow
Cherniak, Eugene Anthony
Childs, Ronald Frank
Cocivera, Michael
Conway, Brian Evans
Cox, Brian
Dacey, John Robert
Dawes, David Haddon
Dawson, Peter Henry
Dawson, Peter Thomas
Deckers, Jacques (Marie)
De Kimpe, Christian Robert
Diaz, Carlos Manuel
Dick, James Gardiner
Dickson, Lawrence William
Dignam, Michael John
Dixon, William Rossander
Dugan, Charles Hammond
Dunn, John Robert
Eastwood, Thomas Alexander
Eaton, Donald Rex
Elias, Lorne
Filseth, Stephen V

Physical Chemistry (cont)

Ford, Richard Westaway
Gilbert, John Barry
Glew, David Neville
Goodings, John Martin
Gordon, Robert Dixon
Gough, Sidney Roger
Graham, Michael John
Graydon, William Frederick
Greenstock, Clive Lewis
Guillet, James Edwin
Hackett, Peter Andrew
Hair, Michael L
Hammerli, Martin
Handa, Paul
Hanna, Geoffrey Chalmers
Harrison, Alexander George
Hawton, Larry David
Henry, Bryan Roger
Hines, William Grant
Hitchcock, Adam Percival
Holmes, John Leonard
Holtslander, William John
Howard, James Anthony
Hunter, Geoffrey
Ingold, Keith Usherwood
Ingraham, Thomas Robert
Ingruber, Otto Vincent
Irish, Donald Edward
Jackman, Thomas Edward
Jacobs, Patrick W M
James, Herbert I
Janzen, Edward George
Jones, Alister Vallance
Jones, Maurice Harry
Kapral, Raymond Edward
Kavassalis, Tom A
Kenney-Wallace, Geraldine Anne
Kilp, Toomas
King, Gerald Wilfrid
Klassen, Norman Victor
Klug, Dennis Dwayne
Koningstein, Johannes A
Kresge, Alexander Jerry
Kruus, Peeter
Kumaran, Mavinkal K
Kuriakose, Areekattuthazhayil
Laidler, Keith James
Laposa, Joseph David
Leaist, Derek Gordon
Lepock, James Ronald
LeRoy, Donald James
Le Surf, Joseph Eric
Lin, Che-Shung
Litvan, Gerard Gabriel
Lockwood, David John
Lorimer, John William
Lossing, Frederick Pettit
Lowden, J Alexander
Lower, Stephen K
Lu, Wei-Kao
Lundell, O Robert
McAdie, Henry George
MacArthur, John Duncan
McCourt, Frederick Richard Wayne
Macdonald, Peter Moore
McGarvey, Bruce Ritchie
McKenney, Donald Joseph
Mackiw, Vladimir Nicolaus
Maguire, Robert James
Mantsch, Henry H
March, Raymond Evans
Marchessault, Robert Henri
Mayo, De Paul
Meath, William John
Moffat, John Blain
Morrison, James Alexander
Morrow, Barry Albert
Moskovits, Martin
Moule, David
Murphy, William Frederick
Nesheim, Michael Ernest
Niki, Hiromi
Norris, A R
Oldham, Keith Bentley
Paraskevopoulos, George
Penner, Glenn H
Platford, Robert Frederick
Polanyi, John Charles
Preston, Keith Foncell
Pritchard, Huw Owen
Ramachandran, Vangipuram S
Rand, Richard Peter
Reddoch, Allan Harvey
Reeves, Leonard Wallace
Ripmeester, John Adrian
Roovers, J
Ross, Robert Anderson
Rumfeldt, Robert Clark
Rummery, Terrance Edward
Sadowski, Chester M
Sandler, Samuel
Sarkar, Bibudhendra
Sawicka, Barbara Danuta
Schiff, Harold Irvin
Schneider, William George
Sharp, James H
Sharp, John Arthur
Shigeishi, Ronald A
Shurvell, Herbert Francis
Siebrand, Willem
Singleton, Donald Lee
Smith, Donald Reed

Southam, Frederick William
Stairs, Robert Ardagh
Stewart, James Allen
Sukava, Armas John
Symons, Edward Allan
Szabo, Arthur Gustav
Tedmon, Craig Seward, Jr
Tinker, David Owen
Tomlinson, Richard Howden
Vincett, Paul Stamford
Wan, Jeffrey Kwok-Sing
Ware, William Romaine
Weir, Ronald Douglas
Werstiuk, Nick Henry
Westaway, Kenneth C
Whalley, Edward
Williams, Harry Leverne
Willis, Clive
Winnik, Françoise Martine
Wright, Maurice Morgan
Yates, Keith

PRINCE EDWARD ISLAND
Liu, Michael T H

QUEBEC
Alince, Bohumil
Allen, Lawrence Harvey
Allen, Sandra Lee
Atack, Douglas
Bourgon, Marcel
Cabana, Aldee
Claessens, Pierre
Colebrook, Lawrence David
Daoust, Hubert
Davis, Herbert John
Desnoyers, Jacques Edouard
Dorris, Gilles Marcel
Edward, John Thomas
Fliszar, Sandor
Gray, Derek Geoffrey
Grosser, Arthur Edward
Harrod, John Frank
Heitner, Cyril
Herman, Jan Aleksander
Holden, Harold William
Jabalpurwala, Kaizer E
Kimmerle, Frank
Lam, Vinh-Te
Lamarche, François
Langford, Cooper Harold, III
Lau, Cheuk Kun
Leblanc, Roger M
Leigh, Charles Henry
Lennox, Robert Bruce
Leonard, Jacques Walter
LeRoy, Rodney Lash
Ouellet, Ludovic
Pezolet, Michel
Power, Joan F
Robertson, Alexander Allen
Sacher, Edward
Salahub, Dennis Russell
Savoie, Rodrigue
Schreiber, Henry Peter
Sicotte, Yvon
Theophanides, Theophile
Thompson, Allan Lloyd
Van De Van, Theodorus Gertrudus Maria
Whitehead, Michael Anthony
Yong, R(aymond)

SASKATCHEWAN
Bardwell, John Alexander Eddie
Barton, Richard J
Chandler, William David
Eager, Richard Livingston
Hontzeas, S
Knight, Arthur Robert
Kybett, Brian David
Larson, D Wayne
Lee, Donald Garry
McCallum, Kenneth James
Rummens, F H A
Steer, Ronald Paul
Strathdee, Graeme Gilroy
VanCleave, Allan Bishop
Verrall, Ronald Ernest
Waltz, William Lee
Weil, John A

OTHER COUNTRIES
Antoniou, Antonios A
Arnell, John Carstairs
Balakrishnan, Narayana Swamy
Band, Yehuda Benzion
Bargon, Joachim
Barrett, Jerry Wayne
Bearman, Richard John
Belton, Geoffrey Richard
Berns, Donald Sheldon
Bess, Robert Carl
Bondybey, Vladimir E
Brown, Larry Robert
Bulkin, Bernard Joseph
Caron, Aimery Pierre
Carter, Robert Everett
Cartwright, Hugh Manning
Chang, Shuya
Charola, Asuncion Elena
Chen, Chen-Tung Arthur
Collins, Carol Hollingworth
Collins, Kenneth Elmer

Dahl, Alton
Delahay, Paul
Eland, John Hugh David
El-Awady, Abbas Abbas
El-Bayoumi, Mohamed Ashraf
Engelborghs, Yves
Ernst, Richard R
Forst, Wendell
Freeman, Raymond
Grunwald, John J
Hansen, Charles M
Heckel, Edgar
Ilten, David Frederick
Ionescu, Lavinel G
Kay, Kenneth George
Kendall, James Tyldesley
Klein, Ralph
Kordesch, Karl Victor
Kuist, Charles Howard
Lambrecht, Richard Merle
Levin, Gideon
Lingane, Peter James
McAllister, William Albert
McRae, Wayne A
Mulas, Pablo Marcelo
Nowotny, Hans
Peatman, William Burling
Pierce, Timothy Ellis
Qian, Renyuan
Ree, Alexius Taikyue
Safran, Samuel A
Samuelson, Bengt Ingemar
Schlag, Edward William
Spaght, Monroe Edward
Stoeber, Werner
Tanny, Gerald Brian
Trofimov, Boris Alexandrovich
Turrell, George Charles
Walker, Michael Stephen

Polymer Chemistry

ALABAMA
Ayers, Orval Edwin
Beindorff, Arthur Baker
Byrd, James Dotson
Dean, David Lee
Galbraith, Ruth Legg
Galil, Fahmy
Hall, David Michael
Kispert, Lowell Donald
Lewis, Danny Harve
Ludwick, Adriane Gurak
Patterson, William Jerry
Sayles, David Cyril
Tice, Thomas Robert
Walsh, William K

ARIZONA
Adickes, H Wayne
Allen, James Durwood
Burke, William James
Carr, Clide Isom
Drake, Stevens Stewart
Hall, Henry Kingston, Jr
Heinle, Preston Joseph
Helbert, John N
Hunter, William Leslie
Kelley, Maurice Joseph
Kleinschmidt, Eric Walker
Ramohalli, Kumar Nanjunda Rao
Robertson, Frederick Noel
Valenty, Steven Jeffrey
Valenty, Vivian Briones

ARKANSAS
Gosnell, Aubrey Brewer

CALIFORNIA
Adicoff, Arnold
Akawie, Richard Isidore
Aklonis, John Joseph
Amis, Eric J
Athey, Robert Douglas, Jr
Bacskai, Robert
Baker, Richard William
Banigan, Thomas Franklin, Jr
Barasch, Werner
Bateman, John Hugh
Beck, Henry Nelson
Bilow, Norman
Boschan, Robert Herschel
Brauman, Sharon K(ruse)
Byrd, Norman Robert
Cahill, Richard William
Carty, Daniel T
Casella, John Francis
Castro, Anthony J
Chang, Hao-Jan
Chatfield, Dale Alton
Chen, Timothy Shieh-Sheng
Chiang, Anne
Chou, Yungnien John
Chung, Chi Hsiang
Colodny, Paul Charles
Cook, Robert Crossland
Cope, Oswald James
Cotts, David Bryan
Cotts, Patricia Metzger
Dahl, Klaus Joachim
Dawson, Daniel Joseph
Dawson, Kenneth Adrian
Denn, Morton M(ace)

Dyball, Christopher John
Eichinger, Bruce Edward
Epstein, George
Even, William Roy, Jr
Feay, Darrell Charles
Fedors, Robert Francis
Feher, Frank J
Fohlen, George Marcel
Fuhrman, Albert William
Gall, Walter George
Gaynor, Joseph
Giants, Thomas W
Glover, Leon Conrad, Jr
Golub, Morton Allan
Grieve, Catherine Macy
Han, Yuri Wha-Yul
Hass, Robert Henry
Haxo, Henry Emile, Jr
Hertzler, Barry Lee
Hess, Patrick Henry
Hoffman, Dennis Mark
Hogen-Esch, Thieo Eltjo
Hokama, Takeo
Hsieh, You-Lo
Hsu, Ming-Ta Sung
Jacobs, Thomas Lloyd
Johanson, Robert Gail
Kalfayan, Sarkis Hagop
Kamp, David Allen
Kaneko, Thomas Motomi
Kasai, Paul Haruo
Kellman, Raymond
Kim, Chung Sul (Sue)
Kissel, Charles Louis
Kolyer, John M
Kong, Eric Siu-Wai
Kratzer, Reinhold
Kurkov, Victor Peter
Landis, Abraham L
Landrum, Billy Frank
Lawson, Daniel David
Lawton, Emil Abraham
Leeg, Kenton J(ames)
Leeper, Harold Murray
LeFave, Gene M(arion)
Legge, Norman Reginald
Lewis, Robert Allen
Lewis, Sheldon Noah
Liberman, Martin Henry
Live, David H
Lorensen, Lyman Edward
Lovelace, Alan Mathieson
MacWilliams, Dalton Carson
Margerum, John David
Markhart, Albert H, Jr
Marks, Burton Stewart
Martin, Eugene Christopher
Meader, Arthur Lloyd, Jr
Menefee, Emory
Miller, Leroy Jesse
Molau, Gunther Erich
Murdoch, Joseph Richard
Needles, Howard Lee
Neidlinger, Hermann H
Newey, Herbert Alfred
Nikoui, Nik
Notley, Norman Thomas
Novak, Bruce Michael
Nyberg, David Dolph
O'Loane, James Kenneth
Ozari, Yehuda
Paciorek, Kazimiera J L
Patterson, Dennis Bruce
Percival, Douglas Franklin
Philipson, Joseph
Plamthottam, Sebastian S
Potter, Norman D
Preus, Martin William
Rabolt, John Francis
Reardon, Edward Joseph, Jr
Regulski, Thomas Walter
Reiss, Howard
Rhein, Robert Alden
Richter, Herbert Peter
Riley, Robert Lee
Ruetman, Sven Helmuth
Russell, Thomas Paul
Ryan, Patrick Walter
Salovey, Ronald
Saltman, William Mose
Saltonstall, Clarence William, Jr
Schell, William John
Seiling, Alfred William
Sellas, James Thomas
Shinohara, Makoto
Shrontz, John William
Siegfried, William
Singh, Hakam
Skolnick, Jeffrey
Smith, Gary Eugene
Smith, Thor Lowe
Snyder, Robert Gene
Sovish, Richard Charles
Steinberg, Gunther
Sterling, Robert Fillmore
Sundberg, John Edwin
Taft, David Dakin
Tess, Roy William Henry
Theodorou, Doros Nicolas
Thomson, Tom Radford
Tilley, T(erry) Don
Toy, Madeline Shen
Tung, Lu Ho
Volksen, Willi

Wade, Charles Gordon
Wade, Robert Harold
Weber, Carl Joseph
Weber, William Palmer
Wensley, Charles Gelen
Wessling, Ritchie A
Wheeler, John C
Willson, Carlton Grant
Winkler, DeLoss Emmet
Wismer, Marco
Wolfe, James Frederick
Wood, David Wells
Woods, William George
Wrasidlo, Wolfgang Johann
Wudl, Fred
Yee, Rena
Yeh, Gene Horng C
Yoon, Do Yeung
Youngman, Edward August
Zavarin, Eugene
Zeronian, Sarkis Haig
Ziegler, Carole L
Zimm, Bruno Hasbrouck

COLORADO
Bachman, Bonnie Jean Wilson
DoMinh, Thap
Druelinger, Melvin L
Dunn, Richard Lee
Frank, Arthur Jesse
Hermann, Allen Max
Lucas, George Bond
Mackinney, Herbert William
Shefter, Eli
Stille, John Kenneth
Strauss, Eric L
Thompson, Ronald G
Voorhees, Kent Jay
Webb, John Day

CONNECTICUT
Arnold, John Richard
Auerbach, Michael Howard
Baccei, Louis Joseph
Baker, Leonard Morton
Banucci, Eugene George
Baum, Bernard
Berdick, Murray
Bergen, Robert Ludlum, Jr
Cardinal, John Robert
Castonguay, Richard Norman
Cooke, Theodore Frederic
Cornell, Robert Joseph
Coscia, Anthony Thomas
Dobay, Donald Gene
Drucker, Arnold
Easterbrook, Eliot Knights
Goldberg, A Jon
Gorham, William Franklin
Hauser, Martin
Hsu, Nelson Nae-Ching
Huang, Samuel J
Huffman, K(enneth) Robert
Humphreys, Robert William Riley
Hwa, Jesse Chia Hsi
Johnson, Julian Frank
Koberstein, Jeffrey Thomas
Koch, Stanley D
Korin, Amos
Lamm, Foster Philip
Larson, Eric George
Leavitt, Julian Jacob
Leonard, Robert F
Levine, Leon
Luh, Yuhshi
Malofsky, Bernard Miles
Meierhoefer, Alan W
Noland, James Sterling
O'Shea, Francis Xavier
Padbury, John James
Page, Edgar J(oseph)
Panzer, Hans Peter
Pappas, S Peter
Pellon, Joseph
Philips, Judson Christopher
Powers, Joseph
Proverb, Robert Joseph
Rie, John E
Roberts, George P
Schile, Richard Douglas
Schmitt, Joseph Michael
Schrage, F Eugene
Scola, Daniel Anthony
Sedlak, John Andrew
Shaw, Montgomery Throop
Shine, William Morton
Sprague, G(eorge) Sidney
Steingiser, Samuel
Stevens, Malcolm Peter
Strazdins, Edward
Stuber, Fred A
Suen, T(zeng) J(iueq)
Tanaka, John
Ulrich, Henri
Wasser, Richard Barkman
Wellman, Russel Elmer
Wheeler, Edward Stubbs
Whelan, William Paul, Jr
White, Leroy Albert
Wolfram, Leszek January
Wuskell, Joseph P
Wystrach, Vernon Paul
Yang, Darchun Billy

DELAWARE
Abernathy, Henry Herman
Adelman, Robert Leonard
Ahmed, Syed Mahmood
Akeley, David Francis
Alderson, Thomas
Ali-Khan, Ausat
Alms, Gregory Russell
Andersen, Donald Edward
Anderson, Burton Carl
Angelo, Rudolph J
Angerer, J(ohn) David
Armbrecht, Frank Maurice, Jr
Armitage, John Brian
Austin, Paul Rolland
Bair, Thomas Irvin
Baird, Richard Leroy
Bannister, Robert Grimshaw
Barker, Harold Clinton
Barney, Arthur Livingston
Barron, Eugene Roy
Barth, Howard Gordon
Bauchwitz, Peter S
Baylor, Charles, Jr
Bercaw, James Robert
Beresniewicz, Aleksander
Black, Carl (Ellsworth)
Boettcher, F Peter
Breazeale, Almut Frerichs
Brehm, Warren John
Breslow, David Samuel
Brown, Morton
Brown, Robert Raymond
Brown, Stewart Cliff
Buck, Warren Howard
Burg, Marion
Cairncross, Allan
Calkins, William Harold
Caywood, Stanley William, Jr
Cella, Richard Joseph, Jr
Chiu, Jen
Chu, Sung Gun
Citron, Joel David
Class, Jay Bernard
Cluff, Edward Fuller
Collette, John Wilfred
Cone, Michael McKay
Cooper, Terence Alfred
Corner, James Oliver
Craig, Alan Daniel
David, Israel A
Dawson, Robert Louis
DeBrunner, Marjorie R
DeDominicis, Alex John
Delmonte, David William
Di Giacomo, Armand
Elliott, John Habersham
English, Alan Dale
Feinberg, Stewart Carl
Ferguson, Raymond Craig
Finn, James Walter
Fleming, Richard Allan
Flexman, Edmund A, Jr
Fogiel, Adolf W
Foldi, Andrew Peter
Ford, Thomas Aven
Foss, Robert Paul
Frankenburg, Peter Edgar
Franta, William Alfred
Frensdorff, H Karl
Funck, Dennis Light
Garrison, William Emmett, Jr
Gay, Frank P
Geer, Richard P
Gold, Harvey Saul
Golike, Ralph Crosby
Goodman, Albert
Grot, Walther Gustav Fredrich
Guggenberger, Lloyd Joseph
Hardie, William George
Harmuth, Charles Moore
Harrell, Jerald Rice
Harris, John Ferguson, Jr
Hasty, Noel Marion, Jr
Hatchard, William Reginald
Hill, Frederick Burns, Jr
Hirsty, Sylvain Max
Hoegger, Erhard Fritz
Hoehn, Harvey Herbert
Hoeschele, Guenther Kurt
Honsberg, Wolfgang
Howard, Edward George, Jr
Howe, King Lau
Hsiao, Benjamin S
Hurford, Thomas Rowland
Immediata, Tony Michael
Ittel, Steven Dale
Jabloner, Harold
Jackson, Harold Leonard
John, Andrew
Kassal, Robert James
Katz, Manfred
Keating, James T
Keown, Robert William
Ketterer, Paul Anthony
Killian, Frederick Luther
Kitson, Robert Edward
Knipmeyer, Hubert Elmer
Kogon, Irving Charles
Kohan, Melvin Ira
Kolber, Harry John
Krespan, Carl George
Kwolek, Stephanie Louise
Landoll, Leo Michael

Lauzon, Rodrigue Vincent
Lazaridis, Christina Nicholson
Lewis, John Raymond
Libbey, William Jerry
Lin, Kuang-Farn
Lipscomb, Robert DeWald
Logothetis, Anestis Leonidas
Logullo, Francis Mark
Loken, Halvar Young
Longhi, Raymond
Longworth, Ruskin
Loomis, Gary Lee
Lukach, Carl Andrew
Lyons, Peter Francis
MacDonald, Robert Neal
McLain, Stephan James
McTigue, Frank Henry
Mahler, Walter
Maloney, Daniel Edwin
Manos, Philip
Markiewitz, Kenneth Helmut
Martin, Wayne Holderness
Maynes, Gordon George
Miller, Ivan Keith
Monagle, Daniel J
Moncure, Henry, Jr
Monroe, Bruce Malcolm
Moore, Ralph Bishop
Moser, Glenn Allen
Narvaez, Richard
Naylor, Marcus A, Jr
Neff, Bruce Lyle
Nelb, Gary William
Niedzielski, Edmund Luke
Norton, Hilton Lafayette
Ojakaar, Leo
Pailthorp, John Raymond
Pappas, Nicholas
Patterson, Gordon Derby, Jr
Pelosi, Lorenzo Fred
Pieski, Edwin Thomas
Plambeck, Louis, Jr
Podlas, Thomas Joseph
Polejes, J(acob) D
Putnam, Stearns Tyler
Quisenberry, Richard Keith
Rave, Terence William
Raynolds, Stuart
Reardon, Joseph Edward
Repka, Benjamin C
Restaino, Alfred Joseph
Rodriguez-Parada, Jose Manuel
Saegebarth, Klaus Arthur
Sammak, Emil George
Sauerbrunn, Robert Dewey
Schadt, Frank Leonard, III
Schaefgen, John Raymond
Scheiber, David Hitz
Schiewetz, D(on) B(oyd)
Schmiegel, Walter Werner
Schreyer, Ralph Courtenay
Schroeder, Herman Elbert
Schrof, William Ernst John
Schwartz, Jerome Lawrence
Shain, Albert Leopold
Shambelan, Charles
Shih, Hsiang
Simms, John Alvin
Slade, Arthur Laird
Smart, Bruce Edmund
Smook, Malcolm Andrew
Smoot, Charles Richard
Sroog, Cyrus Efrem
Starkweather, Howard Warner, Jr
Sundet, Sherman Archie
Sweet, Arthur Thomas, Jr
Tabibian, Richard
Takvorian, Kenneth Bedrose
Tang, Walter Kwei-Yuan
Tanikella, Murty Sundara Sitarama
Tanner, David
Tarney, Robert Edward
Tillson, Henry Charles
Tise, Frank P
Tuites, Donald Edgar
Un, Howard Ho-Wei
Van Dyk, John William
Van Gulick, Norman Martin
Van Trump, James Edmond
Vassallo, Donald Arthur
Vriesen, Calvin W
Wagner, Richard Lloyd
Wallenberger, Frederick Theodore
Walter, Henry Clement
Webb, Charles Alan
Webers, Vincent Joseph
Webster, Owen Wright
Wendt, Robert Charles
Whitehouse, Bruce Alan
Williams, Ebenezer David, Jr
Williams, Richard Anderson
Wilson, Frank Charles
Wolfe, James Richard, Jr
Wu, Souheng
Zimmerman, Joseph
Zinser, Edward John
Zussman, Melvin Paul

DISTRICT OF COLUMBIA
Barkin, Stanley M
Bikales, Norbert M
Bultman, John D
Campbell, Francis James
Hirzy, John William

Keller, Teddy Monroe
Metz, Fred L
Smith, Leslie E
Wade, Clarence W R
Walton, Theodore Ross
Wood, Lawrence Arnell

FLORIDA
Bach, Hartwig C
Beatty, Charles Lee
Black, William Bruce
Blossey, Erich Carl
Brams, Stewart L
Carraher, Charles Eugene, Jr
Chapman, Richard David
Clark, Harold Arthur
Coke, C(hauncey) Eugene
Dannenberg, E(li) M
Davidovits, Joseph
Dobinson, Frank
Dunbar, Richard Alan
Elliott, John Raymond
Feit, Eugene David
Friedlander, Herbert Norman
Frishman, Daniel
Goldberg, Eugene P
Goldstein, Arthur Murray
Gordon, Joseph R
Hammer, Clarence Frederick, Jr
Hanley, James Richard, Jr
Hicks, Elija Maxie, Jr
Hopkins, George C
Hsu, Tsong-Han
Kenney, James Franklin
Kline, Gordon Mabey
Kraiman, Eugene Alfred
Kutner, Abraham
Levy, Richard
Mandelkern, Leo
Mattson, Guy C
Menyhert, William Robert
Moskowitz, Mark Lewis
Nelson, Gordon Leigh
Nersasian, Arthur
O'Leary, Kevin Joseph
Owen, Gwilym Emyr, Jr
Parker, Earl Elmer
Pollard, Robert Eugene
Potts, James Edward
Raich, William Judd
Remmel, Randall James
Rider, Don Keith
Ridgway, James Stratman
Rogan, John B
Russell, Charles Addison
Russell, James
Sadler, George D
Saunders, James Henry
Schwier, Chris Edward
Seelbach, Charles William
Sharkey, William Henry
Slade, Philip Earl, Jr
Smith, William Mayo
Stahl, Joel S
Steadman, Thomas Ree
Templer, David Allen
Ting, Robert Yen-Ying
Ucci, Pompelio Angelo
Weddleton, Richard Francis
Willwerth, Lawrence James
Wittbecker, Emerson LaVerne
Zimmer, William Frederick, Jr

GEORGIA
Allison, John P
Baxter, Gene Francis
Burch, George Nelson Blair
Clendinning, Robert Andrew
Freese, William P, II
Gollob, Lawrence
Gupta, Rakesh Kumar
Habte-Mariam, Yitbarek
Knopka, W N
Krajca, Kenneth Edward
Legare, Richard J
McCarthy, Neil Justin, Jr
Marshall, Donald Irving
Morman, Michael T
Netherton, Lowell Edwin
Orofino, Thomas Allan
Pohl, Douglas George
Polk, Malcolm Benny
Ray, Apurba Kanti
Reif, Donald John
Rodriguez, Augusto
Rouvray, Dennis Henry
Samoilov, Sergey Michael
Samuels, Robert Joel
Stratton, Robert Alan
Tincher, Wayne Coleman
Tolbert, Laren Malcolm
Weber, Robert Emil
Wingard, Robert Eugene, Jr
Yost, Robert Stanley
Young, Raymond Hinchcliffe, Jr

HAWAII
Kemp, Paul James
Sood, Satya P

IDAHO
Lambuth, Alan Letcher
Renfrew, Malcolm MacKenzie

Polymer Chemistry (cont)

Herk, Leonard Frank
Heyman, Duane Allan
Hinkamp, Paul Eugene, (II)
Holubka, Joseph Walter
Howell, Bob A
Jackson, Larry Lynn
Jarvis, Lactance Aubrey
Jones, Frank Norton
Kang, Uan Gen
Keinath, Steven Ernest
Killgoar, Paul Charles, Jr
Kim, Yungki
Koster, Robert Allen
Kresta, Jiri Erik
Krimm, Samuel
Labana, Santokh Singh
Langley, Neal Roger
Larsen, Eric Russell
Lee, Do Ik
Lee, Wei-Ming
Lipowitz, Jonathan
McDonald, Charles Joseph
McKeever, L Dennis
McLean, John A, Jr
Malchick, Sherwin Paul
Mani, Inder
Mansfield, Marc L
Meier, Dale Joseph
Miller, Frederick Arnold
Miller, Robert Llewellyn
Miranda, Thomas Joseph
Moore, Carl
Moore, Eugene Roger
Morgan, Roger John
Narayan, Tv Lakshmi
Newman, Seymour
Nordstrom, J David
Overberger, Charles Gilbert
Patton, John Thomas, Jr
Peters, James
Platt, Alan Edward
Pynnonen, Bruce W
Rasmussen, Paul G
Reynard, Kennard Anthony
Ryntz, Rose A
Sarkar, Nitis
Saunders, Frank Linwood
Schlick, Shulamith
Schmidt, Donald L
Schneider, John Arthur
Skochdopole, Richard E
Smith, Arthur Gerald
Smith, Harry Andrew
Solc, Karel
Stark, Forrest Otto
Stevens, Violete L
Swarin, Stephen John
Ullman, Robert
Walker, Frederick
Walles, Wilhelm Egbert
Weiss, Philip
Weyenberg, Donald Richard
Willermet, Pierre Andre
Williams, Fredrick David
Williamson, Jerry Robert
Wims, Andrew Montgomery
Wyzgoski, Michael Gary
Yee, Albert Fan
Zand, Robert

MINNESOTA
Abere, Joseph Francis
Agre, Courtland LeVerne
Andrus, Milton Henry, Jr
Cook, Jack E
Coury, Arthur Joseph
Diedrich, James Loren
Erickson, Randall L
Esmay, Donald Levern
Farber, Elliott
Farm, Raymond John
Fick, Herbert John
Floyd, Don Edgar
Forester, Ralph H
Gobran, Ramsis
Goken, Garold Lee
Hammar, Walton James
Hendricks, James Owen
Hill, Brian Kellogg
Hogle, Donald Hugh
Holmen, Reynold Emanuel
Johnson, Bryce Vincent
Johnson, Lennart Ingemar
Johnston, Manley Roderick
Kovacic, Joseph Edward
Kropp, James Edward
Kuder, Robert Clarence
Lai, Juey Hong
Langager, Bruce Allen
Li, Wu-Shyong
Lockwood, Robert Greening
Lodge, Timothy Patrick
Lu, Shih-Lai
Lucast, Donald Hurrell
Miller, Wilmer Glenn
Penney, William Harry
Pletcher, Wayne Albert
Prager, Julianne Heller
Punderson, John Oliver
Rice, David E
Robins, Janis
Roska, Fred James
Sandberg, Carl Lorens
Sherman, Patsy O'Connell

Smith, Samuel
Snell, John B
Talbott, Richard Lloyd
Tamsky, Morgan Jerome
Thalacker, Victor Paul
Tiers, George Van Dyke
Wear, Robert Lee
Weiss, Douglas Eugene
Wen, Richard Yutze
Williams, Todd Robertson
Wollner, Thomas Edward
Wright, Charles Dean

MISSISSIPPI
Griffin, Anselm Clyde, III
Mathias, Lon Jay
Pandey, Ras Biharl
Pittman, Charles U, Jr
Seymour, Raymond B(enedict)
Thames, Shelby Freland

MISSOURI
Ames, Donald Paul
Anagnostopoulos, Constantine E
Asrar, Jawed
Bechtle, Gerald Francis
Brown, Harold Probert
Burke, James Joseph
Carter, Don E
Conradi, Mark Stephen
Craver, Clara Diddle (Smith)
Eime, Lester Oscar
Gadberry, Howard M(ilton)
Greenley, Robert Z
Grubbs, Charles Leslie
Harland, Ronald Scott
Holtzer, Marilyn Emerson
Keller, William John
Klein, Andrew John
Koenig, Karl E
Ku, Audrey Yeh
Kurz, James Eckhardt
Lennon, Patrick James
Levy, Ram Leon
Markowski, Henry Joseph
Millich, Frank
Munch, John Howard
Richard, William Ralph, Jr
Sears, J Kern
Slocombe, Robert Jackson
Snyder, Joseph Quincy
Sonnino, Carlo Benvenuto
Stary, Frank Edward
Stoffer, James Osber
Stout, Edward Irvin
Thach, Robert Edwards
Tonkyn, Richard George
Twardowski, Zbylut Jozef
Wilbur, James Myers, Jr
Wohl, Martin H
Woodbrey, James C
Yasuda, Hirotsugu

NEBRASKA
Garey, Carroll Laverne
Schadt, Randall James

NEVADA
Borders, Alvin Marshall
Herzlich, Harold Joel

NEW HAMPSHIRE
Cotter, Robert James
Donaruma, L Guy
Greenwald, Harold Leopold
Petrasek, Emil John
Webb, Richard Lansing

NEW JERSEY
Aharoni, Shaul Moshe
Akkapeddi, Murali Krishna
Amundson, Karl Raymond
Arendt, Volker Dietrich
Ariemma, Sidney
Armbruster, David Charles
Ashcraft, Arnold Clifton, Jr
Auerbach, Andrew Bernard
Auerbach, Victor
Bair, Harvey Edward
Baker, William Oliver
Barclay, Robert, Jr
Bardoliwalla, Dinshaw Framroze
Barnabeo, Austin Emidio
Baum, Gerald A(llan)
Bender, Howard L
Berardinelli, Frank Michael
Berenbaum, Morris Benjamin
Bertelo, Christopher Anthony
Besso, Michael M
Bezwada, Rao Srinivasa
Bhattacharjee, Himangshu Ranjan
Blank, Zvi
Bornstein, Alan Arnold
Bouboulis, Constantine Joseph
Bovey, Frank Alden
Bowden, Murrae John Stanley
Bowen, J Hartley, Jr
Brine, Charles James
Burton, Gilbert W
Cacella, Arthur Ferreira
Canterino, Peter John
Chan, Maureen Gillen
Chance, Ronald Richard
Chandrasekhar, Prasanna

Chandross, Edwin A
Chapas, Richard Bernard
Charbonneau, Larry Francis
Chen, Catherine S H
Chen, Shuhchung Steve
Chen, William Kwo-Wei
Chenicek, Albert George
Chiao, Wen Bin
Chiao, Yu-Chih
Chimes, Daniel
Chlanda, Frederick P
Church, John Armistead
Cipriani, Cipriano
Cohen, Abraham Bernard
Conger, Robert Perrigo
Cooke, Robert Sanderson
Curry, Michael Joseph
Dante, Mark F
Davies, Richard Edgar
DeBona, Bruce Todd
Deshpande, Achyut Bhalchandra
Douglass, D C
Dyer, John
Eastman, David Willard
Eby, John Martin
Elsenbaumer, Ronald Lee
Erenrich, Eric Howard
Evers, William L
Fahrenholtz, Susan Roseno
Farrington, Thomas Allan
Fetters, Lewis
Filas, Robert William
Firth, William Charles, Jr
Fishman, David H
Fong, Jones W
Forbes, Charles Edward
Gander, Robert Johns
Gardiner, John Brooke
Gaughan, Roger Grant
Gaylord, Norman Grant
Giacobbe, Thomas Joseph
Giddings, Sydney Arthur
Gladstone, Harold Maurice
Gold, Daniel Howard
Goldblatt, Irwin Leonard
Goldstein, Albert
Gorbaty, Martin Leo
Green, Joseph
Griskey, R(ichard) G(eorge)
Grubman, Wallace Karl
Guthrie, Roger Thackston
Hager, Douglas Francis
Hale, Warren Frederick
Hammond, Willis Burdette
Hancock, James William
Hansen, Ralph Holm
Hanson, Harry Thomas
Hanson, James Edward
Harpell, Gary Allan
Hartkopf, Arleigh Van
Hawkins, Walter Lincoln
Helfand, Eugene
Herdklotz, John Key
Hirsch, Arthur
Hoffman, Henry Tice, Jr
Hort, Eugene Victor
Hulyalkar, Ramchandra K
Hundert, Murray Bernard
Isaacson, Robert B
Jackson, Thomas A J
Jaffe, Michael
Jelinski, Lynn W
Jirgensons, Arnold
Johnston, Christian William
Johnston, John
Johnston, John Eric
Kaback, Stuart Mark
Kamath, Yashavanth Katapady
Kaplan, Michael
Karl, Curtis Lee
Karol, Frederick J
Keaveney, William Patrick
Kelley, Joseph Matthew
Keogh, Michael John
Kirshenbaum, Gerald Steven
Kreeger, Russell Lowell
Kresge, Edward Nathan
Kronenthal, Richard Leonard
Kruh, Daniel
Kuntz, Irving
Kveglis, Albert Andrew
Landau, Edward Frederick
Larkin, William Albert
Lasky, Jack Samuel
Leal, Joseph Rogers
Lee, Lester Tsung-Cheng
Lem, Kwok Wai
Levi, David Winterton
Levy, Alan
Li, Hsin Lang
Lipowski, Stanley Arthur
Loan, Leonard Donald
Login, Robert Bernard
Lohse, David John
Lorenz, Donald H
Lovinger, Andrew Joseph
Lowell, A(rthur) I(rwin)
Lundberg, Robert Dean
Lunn, Anthony Crowther
Lyding, Arthur R
McGary, Charles Wesley, Jr
McGinnis, James Lee
Magder, Jules
Manning, Gerald Stuart

Marder, Herman Lowell
Matsuoka, Shiro
Matthews, Demetreos Nestor
Matzner, Markus
Melveger, Alvin Joseph
Meyer, Victor Bernard
Mikes, John Andrew
Miller, Albert Thomas
Miskel, John Joseph, Jr
Montana, Anthony J
Moroni, Antonio
Muldrow, Charles Norment, Jr
Munger, Stanley H(iram)
Munsell, Monroe Wallwork
Nguyen, Hien Vu
Noshay, Allen
O'Malley, James Joseph
Osuch, Christopher Erion
Pacansky, Thomas John
Palermo, Felice Charles
Pappalardo, Leonard Thomas
Parisek, Charles Bruce
Perkel, Robert Jules
Perry, Donald Dunham
Pietrusza, Edward Walter
Prapas, Aristotle G
Pugliese, Michael
Quan, Xina Shu-Wen
Quarles, Richard Wingfield
Ray-Chaudhuri, Dilip K
Register, Richard Alan
Reich, Leo
Reich, Murray H
Reichmanis, Elsa
Reilly, Eugene Patrick
Riccobono, Paul Xavier
Riley, David Waegar
Roberts, William John
Robins, Jack
Roper, Robert
Rosen, Marvin
Rosenthal, Arnold Joseph
Rossi, Robert Daniel
Sacher, Alex
Sacks, William
Saferstein, Lowell G
Salvesen, Robert H
Saxon, Robert
Scanley, Clyde Stephen
Schmidle, Claude Joseph
Schmitt, George Joseph
Schonfeld, Edward
Schulz, Donald Norman
Schwab, Frederick Charles
Schwenker, Robert Frederick, Jr
Searle, Norma Zizmer
Sekutowski, Dennis G
Shah, Hasmukh N
Shaw, Richard Gregg
Shechter, Leon
Sibilia, John Philip
Sidi, Henri
Skeist, Irving
Skoultchi, Martin Milton
Slagel, Robert Clayton
Smith, Terry Edward
Smyers, William Hays
Song, Won-Ryul
Sonn, George Frank
Sorenson, Wayne Richard
Spitsbergen, James Clifford
Spurr, Orson Kirk, Jr
Steiger, Fred Harold
Stone, Edward
Strauss, George
Sugathan, Kanneth Kochappan
Sulzberg, Theodore
Szamosi, Janos
Szymanski, Chester Dominic
Takahashi, Akio
Tamarelli, Alan Wayne
Tatyrek, Alfred Frank
Thich, John Adong
Thompson, Larry Flack
Tonelli, Alan Edward
Tornqvist, Erik Gustav Markus
Townsend, Palmer W
Van de Castle, John F
Ver Strate, Gary William
Vincent, Gerald Glenn
Vukov, Rastko
Wang, Tsuey Tang
Wasserman, David
Watt, William Russell
Weber, Thomas Andrew
Weintraub, Lester
Whelan, John Michael
Wilson, Donald Richard
Winslow, Field Howard
Wissbrun, Kurt Falke
Wu, Tse Cheng
Yourtee, John Boteler
Zaim, Semih
Zoss, Abraham Oscar

NEW MEXICO
Arnold, Charles, Jr
Assink, Roger Alyn
Benicewicz, Brian Chester
Benziger, Theodore Michell
Cahill, Paul A
Dowell, Flonnie
Gillen, Kenneth Todd
Liepins, Raimond

Polymer Chemistry (cont)

Martin, James Ellis
Nielsen, Stuart Dee
Ogilby, Peter Remsen
Schaefer, Dale Wesley
Trujillo, Ralph Eusebio
Van Deusen, Richard L
Wicks, Zeno W, Jr
Williams, Joel Mann, Jr
Young, Ainslie Thomas, Jr
Zeigler, John Martin

NEW YORK

Adduci, Jerry M
Adelstein, Peter Z
Alliet, David F
Altscher, Siegfried
Amborski, Leonard Edward
Arcesi, Joseph A
Bahary, William S
Baron, Arthur L
Barr, Donald Eugene
Benham, Craig John
Benzinger, James Robert
Berdahl, Donald Richard
Bettelheim, Frederick A
Beyer, George Leidy
Bluestein, Ben Alfred
Brust, David Philip
Bryant, Robert George
Cabasso, Israel
Campbell, Gerald Allan
Campbell, Gregory August
Carnahan, James Claude
Cathcart, John Almon
Ceprini, Mario Q
Chadwick, George F
Chamberlin, Howard Allen
Chen, Cindy Chei-Jen
Chen, Shaw-Horng
Chen, Tsang Jan
Chow, Che Chung
Ciccarelli, Roger N
Dagostino, Vincent F
DeCrosta, Edward Francis, Jr
Dehn, Joseph William, Jr
Dichter, Michael
Doetschman, David Charles
Earing, Mason Humphry
Edelman, Robert
Feger, Claudius
Felty, Evan J
Fox, Adrian Samuel
Frechet, Jean M J
Frye, Robert Bruce
Fuerniss, Stephen Joseph
Gardella, Joseph Augustus, Jr
Gardner, Sylvia Alice
Gardner, William Howlett
George, Philip Donald
Giannelis, Emmanuel P
Gladstone, Matthew Theodore
Godsay, Madhu
Green, Mark M
Grina, Larry Dale
Gruenbaum, William Tod
Hargitay, Bartholomew
Harris, Robert L
Harrison, James Beckman
Helfgott, Cecil
Henry, Arnold William
Hepfinger, Norbert Francis
Highsmith, Ronald Earl
Hindersinn, Raymond Richard
Hornibrook, Walter John
Horowitz, Carl
Immergut, Edmund H(einz)
Imperial, George Romero
Indictor, Norman
Isaacson, Henry Verschay
Jelling, Murray
Johansson, Sune
Johnson, Bruce Fletcher
Julinao, Peter C
Kamath, Vasanth Rathnakar
Kaplan, Mark Steven
Kauder, Otto Samuel
Khan, Jamil Akber
Khanna, Ravi
Kim, Ki-Soo
Kiss, Klara
Klein, Gerald Wayne
Klijanowicz, James Edward
Kolb, Frederick J(ohn), Jr
Kosky, Philip George
Krause, Sonja
Kronstein, Max
Kross, Robert David
Kumler, Philip L
Kuzmak, Joseph Milton
Kwei, Ti-Kang
Labianca, Dominick A
Laganis, Deno
Lai, Yu-Chin
Lau, Philip T S
Lee, Lieng-Huang
Lewis, Bertha Ann
Lhila, Ramesh Chand
Limburg, William W
Lupinski, John Henry
Lynn, Merrill
MacLaury, Michael Risley
Malpass, Dennis D

Mansfield, Kevin Thomas
Markovitz, Mark
Martellock, Arthur Carl
Mason, John Hugh
Massa, Dennis Jon
Mathes, Kenneth Natt
Meinwald, Yvonne Chu
Mendel, John Richard
Mermelstein, Robert
Metz, Donald J
Meyer, John Austin
Mijovic, Jovan
Miller, William Taylor
Mirviss, Stanley Burton
Model, Frank Steven
Monahan, Alan Richard
Moore, James Alfred
Moos, Gilbert Ellsworth
Morduchowitz, Abraham
Myers, Drewfus Young, Jr
Nelson, Robert Andrew
Ober, Christopher Kemper
Odian, George G
O'Mara, Michael Martin
Oppenheimer, Larry Eric
Orlando, Charles M
Otocka, Edward Paul
Overton, James Ray
Palladino, William Joseph
Pavlisko, Joseph Anthony
Pearce, Eli M
Pearson, James Murray
Pemrick, Raymond Edward
Penn, Lynn Sharon
Petersen, Kenneth C
Petropoulos, Constantine Chris
Pillar, Walter Oscar
Pollart, Dale Flavian
Post, Howard William
Potter, George Henry
Pucknat, John Godfrey
Rebel, William J
Riecke, Edgar Erick
Roedel, George Frederick
Rosenfeld, Jerold Charles
Rubin, Isaac D
Sabia, Raffaele
Sandhu, Mohammad Akram
Sarko, Anatole
Schuerch, Conrad
Seltzer, Raymond
Shannon, Frederick Dale
Silleck, Clarence Frederick
Smid, Johannes
Smith, Donald Arthur
Smith, Thomas Woods
Sogah, Dotsevi Yao
Sorkin, Howard
Staples, Jon T
Stein, Richard James
Sterman, Melvin David
Sterman, Samuel
Stoner, George Green
Sutherland, Judith Elliott
Tan, Julia S
Teegarden, David Morrison
Tesoro, Giuliana C
Thomas, Harold Todd
Triplett, Kelly B
Tuites, Richard Clarence
Turner, S Richard
Van Norman, Gilden Ramon
Vogl, Otto
Waltcher, Irving
Ward, Frank Kernan
Weiner, Milton Lawrence
Weiss, Jonas
White, Dwain Montgomery
Winter, Roland Arthur Edwin
Winter, William Thomas
Wolinski, Leon Edward
Wolny, Friedrich Franz
Woodward, Arthur Eugene
Wyman, Donald Paul
Yudelson, Joseph Samuel
Yunick, Robert P
Zalewski, Edmund Joseph

NORTH CAROLINA

Abbey, Kathleen Mary Kyburz
Abbey, Kirk J
Allis, John W
Ambrose, Richard Joseph
Andrady, Anthony Lakshman
Bryan, Carl Eddington
Burrus, Robert Tilden
Cates, David Marshall
Cuculo, John Anthony
Dhami, Kewal Singh
Drechsel, Paul David
Evans, Evan Franklin
Farona, Michael F
Gilbert, Richard Dean
Goldstein, Irving Solomon
Groszos, Stephen Joseph
Hall, Edward Duncan
Hamner, William Frederick
Hersh, Solomon Philip
Holzwarth, George Michael
Hurwitz, Melvin David
Jones, Rufus Sidney
Jones, William Jonas, Jr
King, Henry Lee
Kury, Robert Peter

Ledbetter, Harvey Don
Loeppert, Richard Henry
McKay, Jerry Bruce
McPeters, Arnold Lawrence
Magat, Eugene Edward
Miller, Walter Peter
Murray, Royce Wilton
Nash, James Lewis, Jr
Powers, Edward James
Preston, Jack
Rapoport, Lorence
Reese, Cecil Everett
Rochow, Theodore George
Roth, Roy William
Samulski, Edward Thaddeus
Sargeant, Peter Barry
Smith, Peter
Squibb, Samuel Dexter
Stannett, Vivian Thomas
Stejskal, Edward Otto
Szabo, Emery D
Theil, Michael Herbert
Tucker, Paul Arthur
Turner, Derek T
Walsh, Edward John
Williams, Joel Lawson
Wooten, Willis Carl, Jr
Work, Robert Wyllie

NORTH DAKOTA

Adams, David George
Bierwagen, Gordon Paul

OHIO

Adams, Harold Elwood
Aggarwal, Sundar Lal
Ambler, Michael Ray
Ampulski, Robert Stanley
Baczek, Stanley Karl
Baer, Eric
Ball, Lawrence Ernest
Baranwal, Krishna Chandra
Barlow, Anthony
Barsky, Constance Kay
Bauer, Richard G
Baughman, Glenn Laverne
Bennett, Richard Thomas
Benton, Kenneth Curtis
Beres, John Joseph
Bertsch, Robert Joseph
Bethea, Tristram Walker, III
Bond, William Bradford
Bonner, David Calhoun
Borgnaes, Dan
Bradley, Ronald W
Brittain, William Joseph
Butler, John Mann
Calderon, Nissim
Campbell, Robert Wayne
Carman, Charles Jerry
Carpenter, William Graham
Carter, R Owen, Jr
Causa, Alfredo G
Choung, Hun Ryang
Cinadr, Bernard F(rank)
Coleman, John Franklin
Coleman, Lester Earl, (Jr)
Collins, Edward A
Conley, Robert T
Cooper, Jack Loring
Costanza, Albert James
Cunningham, Robert Elwin
Delvigs, Peter
Deviney, Marvin Lee, Jr
D'Ianni, James Donato
Diem, Hugh E(gbert)
Dinbergs, Kornelius
Divis, Roy Richard
Dodge, James Stanley
Dollimore, David
Duddey, James E
Duke, June Temple
Dunn, Horton, Jr
Edman, James Richard
Ells, Frederick Richard
England, Richard Jay
Erman, William F
Essig, Henry J
Evans, Robert Morton
Evers, Robert C
Ewall, Ralph Xavier
Fabris, Hubert
Fall, Harry H
Fechter, Robert Bernard
Feld, William Adam
Fielding-Russell, George Samuel
Finelli, Anthony Francis
Ford, Emory A
Forsyth, Thomas Henry
Franks, Allen P
Frederick, John Edgar
Gates, Raymond Dee
Gebelein, Charles G
George, Paul John
Gerace, Michael Joseph
Giffen, William Martin, Jr
Gippin, Morris
Given, Kerry Wade
Gleim, Clyde Edgar
Goel, Anil B
Goldman, Stephen Allen
Gray, Don Norman
Griffin, Richard Norman
Grimm, Robert Arthur

Gromelski, Stanley John, Jr
Gruber, Elbert Egidius
Hall, Judd Lewis
Hamed, Gary Ray
Harris, Frank Wayne
Harris, Richard Lee
Hartsough, Robert Ray
Harwood, Harold James
Hassell, John Allen
Hayes, Robert Arthur
Healy, James C
Heineman, William Richard
Hergenrother, William Lee
Herliczek, Siegfried H
Herold, Robert Johnston
Hess, George G
Hiles, Maurice
Hoke, Donald I
Hollis, William Frederick
Horne, Samuel Emmett, Jr
Hoyt, John Manson
Hurley, William Joseph
Hyndman, John Robert
Ishida, Hatsuo
Jabarin, Saleh Abd El Karim
Jamieson, Alexander MacRae
Johnston, Norman Wilson
Kanakkanatt, Antony
Katz, Morton
Keck, Max Hans
Kell, Robert M
Kelley, Frank Nicholas
Kennedy, Joseph Paul
Keplinger, Orin Clawson
Koehler, Mark E
Koenig, Jack L
Kollen, Wendell James
Kreuz, John Anthony
Kuo, Cheng-Yih
Lal, Joginder
Lawson, David Francis
Lehr, Marvin Harold
Leininger, Robert Irvin
Lewis, Irwin C
Li, George Su-Hsiang
Lipinsky, Edward Solomon
Litt, Morton Herbert
Livigni, Russell A
Livigni, Russell Anthony
Livingston, Daniel Isadore
Lohr, Delmar Frederick, Jr
Losekamp, Bernard Francis
Lucas, Kenneth Ross
McCain, George Howard
McDonel, Everett Timothy
McGinniss, Vincent Daniel
McIntyre, Donald
McIntyre, William Ernest, Jr
Mark, James Edward
Marshall, Richard Allen
Mattice, Wayne Lee
Meinhardt, Norman Anthony
Merritt, Robert Edward
Metanomski, Wladyslaw Val
Mikesell, Sharell Lee
Miller, James Roland
Miller, William Reynolds, Jr
Milone, Charles Robert
Mitchell, George Redmond, Jr
Morton, Maurice
Mosser, John Snavely
Mouk, Robert Watts
Murphy, Walter Thomas
Myers, Ronald Eugene
Nakajima, Nobuyuki
Niemann, Theodore Frank
Ofstead, Eilert A
Ostrum, G(eorge) Kenneth
Oziomek, James
Pappas, Leonard Gust
Parish, Darrell Joe
Percec, Virgil
Petschek, Rolfe George
Piirma, Irja
Pritchett, Ervin Garrison
Prudence, Robert Thomas
Prusaczyk, Joseph Edward
Purdon, James Ralph, Jr
Purvis, John Thomas
Pyle, James Johnston
Quirk, Roderic P
Raciszewski, Zbigniew
Rader, Charles Phillip
Ramp, Floyd Lester
Raymond, Maurice A
Reardon, Joseph Patrick
Reilly, Charles Bernard
Roberts, Durward Thomas, Jr
Robins, Richard Dean
Roe, Ryong-Joon
Rogers, Charles Edwin
Roha, Max Eugene
Rosen, Irving
Sarbach, Donald Victor
Scheve, Bernard Joseph
Schneider, Wolfgang W
Schollenberger, Charles Sundy
Schuele, Donald Edward
Scott, Kenneth Mark
Sedor, Edward Andrew
Shelton, James Reid
Siebert, Alan Roger
Siefken, Mark William
Sieglaff, Charles Lewis

Simha, Robert
Sinclair, Richard Glenn, II
Sircar, Anil Kumer
Sliemers, Francis Anthony, Jr
Sommer, John G
Standish, Norman Weston
Steichen, Richard John
Stephens, Howard L
Stevenson, Don R
Stickney, Palmer Blaine
Strauss, Carl Richard
Tarvin, Robert Floyd
Tate, David Paul
Taylor, Lynn Johnston
Tessier, Claire Adrienne
Theis, Richard James
Thomas, George B
Throckmorton, Morford C
Toth, William James
Traynor, Lee
Trewiler, Carl Edward
Tsai, Chung-Chieh
Uebele, Curtis Eugene
Updegrove, Louis B
Vanderlind, Merwyn Ray
Vassiliades, Anthony E
Voigt, Charles Frederick
Wagner, Melvin Peter
Warner, Walter Charles
Weeks, Thomas Joseph, Jr
Weinstein, Arthur Howard
Westerman, Ira John
Wiff, Donald Ray
Williams, Robert Calvin
Williamson, Frederick Dale
Williger, Ervin John
Yang, Philip Yung-Chin
Yocum, Ronald Harris
Youngs, Wiley Jay
Zaremsky, Baruch

OKLAHOMA
Bailey, F Wallace
Brady, Donnie Gayle
Dix, James Seward
Efner, Howard F
Fodor, Lawrence Martin
Ford, Warren Thomas
Geibel, Jon Frederick
Graves, Toby Robert
Groten, Barney
Hogan, John Paul
Holtmyer, Marlin Dean
Hsieh, Henry Lien
Johnson, Timothy Walter
Jones, Faber Benjamin
Kokesh, Fritz Carl
Linder, Donald Ernst
Lindstrom, Merlin Ray
Miller, John Walcott
Moczygemba, George A
Moore, Donald R
Natowsky, Sheldon
Pober, Kenneth William
Rotenberg, Don Harris
Schmidt, Donald Dean
Schwab, Peter Austin
Seeney, Charles Earl
Selman, Charles Melvin
Short, James N
Shue, Robert Sidney
Sonnenfeld, Richard John
Stacy, Carl J
Trepka, William James
Welch, Melvin Bruce
Wharry, Stephen Mark
Witt, Donald Reinhold
Zelinski, Robert Paul

OREGON
Merz, Paul Louis
Nielsen, Lawrence Ernie
Shilling, Wilbur Leo
Smith, Kelly L

PENNSYLVANIA
Abate, Kenneth
Allara, David Lawrence
Allcock, Harry R(ex)
Altares, Timothy, Jr
Alvino, William Michael
Anspon, Harry Davis
Arkles, Barry Charles
Bagley, George Everett
Bakule, Ronald David
Balaba, Willy Mukama
Barie, Walter Peter, Jr
Bartovics, Albert
Bartz, Warren F(rederick)
Bassner, Sherri Lynn
Batzar, Kenneth
Bauer, William, Jr
Bauman, Bernard D
Beckley, Ronald Scott
Berk, Lawrence B
Berkheimer, Henry Edward
Berry, Guy C
Bertozzi, Eugene R
Bhatt, Padmamabh P
Blommers, Elizabeth Ann
Bolgiano, Nicholas Charles
Bolstad, Luther
Bower, George Myron
Braunstein, David Michael

Breazeale, Robert David
Brendley, William H, Jr
Brennan, Michael Edward
Buck, Jean Coberg
Burch, Mary Kappel
Buzzell, John Gibson
Canter, Neil M
Carlson, Dana Peter
Cartier, Peter G
Castle, John Edwards
Cenci, Harry Joseph
Chang, Ching-Jen
Chapman, Toby Marshall
Chen-Tsai, Charlotte Hsiao-yu
Chiola, Vincent
Chong, Berni Patricia
Christie, Peter Allan
Chung, Tze-Chiang
Claiborne, C Clair
Cohen, Gordon Mark
Condo, Albert Carman, Jr
Conyne, Richard Francis
Cook, Donald Bowker
Corneliussen, Roger DuWayne
Cornell, John Alston
Das, Suryya Kumar
De Tommaso, Gabriel Louis
Dickstein, Jack
Dimmig, Daniel Ashton
Doak, Kenneth Worley
Dohany, Julius Eugene
Dougherty, Eugene P
Dunlap, Lawrence H
Ebert, Philip E
Edelman, Leonard Edward
Ehrhart, Wendell A
Ehrig, Raymond John
El-Aasser, Mohamed S
Ellis, Jeffrey Raymond
Erhan, Semih M
Fellmann, Robert Paul
Fishman, Marshall Lewis
Fitzgerald, James Allen
Fitzgerald, Maurice E
Ford, Michael Edward
Francis, Peter Schuyler
Freeman, James Harrison
Frick, Neil Huntington
Frost, Lawrence William
Garber, Charles A
Garrett, Robert Roth
Gaughan, Renata Rysnik
Gelb, Leonard Louis
Gillis, Marina N
Goddu, Robert Fenno
Goldman, Theodore Daniel
Golton, William Charles
Goodman, Alan Lawrence
Goodman, Donald
Graham, Roger Kenneth
Greiner, Richard William
Grezlak, John Henry
Grinstein, Reuben H
Gunther, Wolfgang Hans Heinrich
Gutbezahl, Boris
Hanauer, Richard
Harris, James Joseph
Hartman, Marvis Edgar
Hartranft, George Robert
Heiberger, Philip
Hertler, Walter Raymond
Hirsch, Stephen Simeon
Hutchins, MaryGail Kinzer
Ingram, Alvin Richard
Irwin, Philip George
Irwin, William Edward
Johnston-Feller, Ruth M
Jones, Roger Franklin
Judge, Joseph Malachi
Karo, Wolf
Kennedy, Flynt
Khosah, Robinson Panganai
Klapproth, William Jacob, Jr
Kleinschuster, Jacob John
Kline, Richard William
Koob, Robert Philip
Kopchik, Richard Michael
Korchak, Ernest I(an)
Koziar, Joseph Cleveland
Kronberger, Karlheinz
Lake, Robert D
Langsam, Michael
Lantos, P(eter) R(ichard)
Lavoie, Alvin Charles
Lee, Ying Kao
Lesko, Patricia Marie
Leston, Gerd
Liu, Andrew T C
Lloyd, Thomas Blair
Louie, Ming
Luck, Russell M
Luskin, Leo Samuel
Lynch, Thomas John
McDowell, Maurice James
Mahlman, Bert H
Mair, Robert Dixon
Mantell, Gerald Jerome
Mao, Chung-Ling
Matyjaszewski, Krzysztof
Meier, Joseph Francis
Melamed, Sidney
Molinari, Robert James
Morgan, Paul Winthrop
Morse, Lewis David

Mueller, Donald Scott
Murphy, Clarence John
Nannelli, Piero
Naples, John Otto
Natoli, John
Novak, Ronald William
Noyes, Paul R
O'Mara, James Herbert
Orphanides, Gus George
Osei-Gyimah, Peter
Patsiga, Robert A
Patterson, Gary David
Peffer, John Roscoe
Peterson, Jack Kenneth
Petrich, Robert Paul
Pinschmidt, Robert Krantz, Jr
Plant, William J
Prane, Joseph W(illiam)
Quinn, Edwin John
Redmond, John Peter
Reuwer, Joseph Francis, Jr
Ristey, William J
Robinson, Donald Nellis
Samuelson, H Vaughn
Santamaria, Vito William
Scala, Luciano Carlo
Schaller, Edward James
Schimmel, Karl Francis
Schmitz, John Vincent
Schott, Hans
Schultz, Ray Karl
Seiner, Jerome Allan
Sheasley, William David
Shim, Benjamin Kin Chong
Sias, Charles B
Slysh, Roman Stephan
Smith, James David Blackhall
Smith, Stewart Edward
Snider, Albert Monroe, Jr
Sollott, Gilbert Paul
Solomon, M Michael
Soriano, David S
Sperling, Leslie Howard
Statton, Gary Lewis
Stevens, Travis Edward
Stone, Herman
Stryker, Lynden Joel
Swift, Graham
Tennent, Howard Gordon
Thibeault, Jack Claude
Thomas, H Ronald
Tuemmler, William Bruce
Turner, Robert Lawrence
Tweedie, Adelbert Thomas
Vanderhoff, John W
Victorius, Claus
Vijayendran, Bheema R
Vogel, Martin
Vorchheimer, Norman
Walker, Augustus Chapman
Warfel, David Ross
Watson, William Martin, Jr
Weese, Richard Henry
Wei, Yen
Wei-Berk, Caroline
Weisfeld, Lewis Bernard
Welsh, David Albert
Wempe, Lawrence Kyran
Werner, Ervin Robert, Jr
White, Malcolm Lunt
Whiteman, John David
Wilkins, Cletus Walter, Jr
Williams, Donald Robert
Work, James Leroy
Work, William James
Yanai, Hideyasu Steve
Younes, Usama E
Zimmt, Werner Siegfried

RHODE ISLAND
Barnett, Stanley M(arvin)
Scott, Peter Hamilton

SOUTH CAROLINA
Beam, Charles Fitzhugh, Jr
Berger, Richard S
Billica, Harry Robert
Bliss, Arthur Dean
Culbertson, Edwin Charles
Desai, Vinodrai Ranchhodji
Drews, Michael James
Hartz, Roy Eugene
Hendrix, James Easton
Hon, David Nyok-Sai
Hopkins, Allen John
Hudgin, Donald Edward
King, Lee Curtis
Krantz, Karl Walter
Longtin, Bruce
Machell, Greville
Otto, Wolfgang Karl Ferdinand
Pike, LeRoy
Porter, Rick Anthony
Posey, Robert Giles
Robinson, Myrtle Tonne
Robinson, Robert Earl
Ross, Stanley Elijah
Rothrock, George Moore
Scruggs, Jack G
Spencer, Harold Garth
Terry, Stuart Lee
Tour, James Mitchell
Wagner, William Sherwood
Walters, John Philip

Weiss, James Owen
Wishman, Marvin

SOUTH DAKOTA
Hanson, Milton Paul

TENNESSEE
Alexandratos, Spiro
Allen, Vernon R
Barton, Kenneth Ray
Bedoit, William Clarence, Jr
Brown, George Marshall
Browning, Horace Lawrence, Jr
Caflisch, George Barrett
Chambers, Ralph Arnold
Clark, Edward Shannon
Clemens, Robert Jay
Cliffton, Michael Duane
Collins, Jerry Dale
Cook, Kelsey Donald
Coover, Harry Wesley, Jr
Davis, Burns
Dean, Walter Lee
Dieck, Ronald Lee
Dombroski, John Richard
Dorsey, George Francis
Fagerburg, David Richard
Fordham, James Lynn
Fort, Tomlinson, Jr
Germroth, Ted Calvin
Gilkey, Russell
Gleason, Edward Hinsdale, Jr
Gray, Theodore Flint, Jr
Harmer, David Edward
Hoffman, Kenneth Wayne
Hutchinson, James Herbert, Jr
Jackson, Winston Jerome, Jr
Jones, Glenn Clark
Kashdan, David Stuart
Knee, Terence Edward Creasey
Kovac, Jeffrey Dean
Krutak, James John, Sr
Kuo, Chung-Ming
Leonard, Edward Charles, Jr
McConnell, Richard Leon
McFarlane, Finley Eugene
McPherson, James Louis
Mayberry, Thomas Carlyle
Moore, Louis Doyle, Jr
Newland, Gordon Clay
Nicely, Vincent Alvin
Parkinson, William Walker, Jr
Phillips, Paul J
Raynolds, Peter Webb
Sublett, Bobby Jones
Tibbetts, Clark
Vachon, Raymond Normand
Volpe, Angelo Anthony
Wang, Richard Hsu-Shien
White, Alan Wayne
Wicker, Thomas Hamilton, Jr
Wildman, Gary Cecil
Witzeman, Jonathan Stewart
Wunderlich, Bernhard
Yau, Cheuk Chung

TEXAS
Aufdermarsh, Carl Albert, Jr
Baker, Charles Taft
Beede, Charles Herbert
Bergbreiter, David Edward
Betso, Stephen Richard
Borchardt, John Keith
Brostow, Witold Konrad
Brown, James Michael
Bruins, Paul F(astenau)
Cabanes, William Ralph, Jr
Cannon, Dickson Y
Cassidy, Patrick Edward
Cavender, James Vere, Jr
Cross, John Parson
Daues, Gregory W, Jr
De La Mare, Harold Elison
Edmondson, Morris Stephen
Ellison, Robert Hardy
Elward-Berry, Julianne
Engle, Damon Lawson
Etter, Raymond Lewis, Jr
Farrissey, William Joseph, Jr
Fish, John G
Franks, Neal Edward
Fuller, Martin Emil
Goldstein, Herbert Jay
Goodwin, John Thomas, Jr
Goodwyn, Jack Ray
Gray, Kenneth W
Gryting, Harold Julian
Gum, Wilson Franklin, Jr
Handlin, Dale L
Heilman, William Joseph
Henry, Arthur Charles
Hickner, Richard Allan
Hill, Robert William
Holmes, Larry A
Hosmane, Narayan Sadashiv
Jennings, Alfred Roy, Jr
Kalfoglou, George
Kaye, Howard
Kennelley, Kevin James
Keskkula, Henno
Kochhar, Rajindar Kumar
Kohn, Erwin
Kyker, Gary Stephen
Lakshmanan, P R

Polymer Chemistry (cont)

Lloyd, Douglas Roy
Lyle, Robert Edward, Jr
McCullough, James Douglas, Jr
McDougall, Robert I
McGirk, Richard Heath
Mack, Mark Philip
Makin, Earle Clement, Jr
Mangold, Donald J
Martin, Charles R
Martin, Charles William
May, James Aubrey, Jr
Meyer, James Melvin
Moltzan, Herbert John
Munk, Petr
Naae, Douglas Gene
Neeley, Charles Mack
Newton, Robert Andrew
Olson, Danford Harold
Olstowski, Franciszek
Park, Vernon Kee
Patton, Tad LeMarre
Paul, Donald Ross
Pinkus, A(lbin) G(eorge)
Pomerantz, Martin
Powell, Richard James
Rawls, Henry Ralph
Richter, Reinhard Hans
Roberts, Thomas David
Robinson, Alfred Green
Sample, Thomas Earl, Jr
Schimelpfenig, Clarence William
Smith, Curtis William
Spenadel, Lawrence
Speranza, George Phillip
Stahl, Glenn Allan
Strom, E(dwin) Thomas
Sward, Edward Lawrence, Jr
Vogelfanger, Elliot Aaron
Webber, Stephen Edward
Wisian-Neilson, Patty Joan

UTAH
Dehm, Henry Christopher
Dror, Michael
Elmslie, James Stewart
Kopecek, Jindrich
Lyman, Donald Joseph
Reynolds, Richard Johnson
Thompson, Grant

VERMONT
Allen, Christopher Whitney
Berens, Alan Robert
Murray, James Gordon
Weimann, Ludwig Jan

VIRGINIA
Armstrong, Robert G
Barber, Patrick George
Barker, Robert Henry
Bass, Robert Gerald
Bell, Vernon Lee, Jr
Beyad, Mohammed Hossain
Bondurant, Charles W, Jr
Breder, Charles Vincent
Durrell, William S
Eareckson, William Milton, III
Farago, John
Farmer, Barry Louis
Fisher, Charles Harold
Fletcher, Harry Huntington
French, David Milton
Gerow, Clare William
Gibson, Harry William
Glasser, Wolfgang Gerhard
Gratz, Roy Fred
Gulrich, Leslie William, Jr
Hahn, Walter Leopold
Harowitz, Charles Lichtenberg
Hartzler, Jon David
Hazlehurst, David Anthony
Hewett, James Veith
Hodge, James Dwight
Huggett, Clayton (McKenna)
Hussamy, Samir
Jablonski, Werner Louis
Jensen, Arnold William
Johnson, William Randolph, Jr
Johnston, Norman Joseph
Kranbuehl, David Edwin
Lilleleht, L(embit) U(no)
Lodoen, Gary Arthur
McGrath, James Edward
Majewski, Theodore E
Merkel, Timothy Franklin
Meyer, Leo Francis
Milford, George Noel, Jr
Monroe, Stuart Benton
Orwoll, Robert Arvid
Patterson, Earl E(dgar)
Powell, Justin Christopher
Rainer, Norman Barry
Roberts, David Craig
Sahli, Brenda Payne
Sahli, Muhammad S
St Clair, Anne King
St Clair, Terry Lee
Schuurmans, Hendrik J L
Sherbeck, L Adair
Squire, David R
Starnes, William Herbert, Jr
Stiehl, Roy Thomas, Jr

Stump, Billy Lee
Tichenor, Robert Lauren
Ticknor, Leland Bruce
Urbanik, Arthur Ronald
Van Ness, Kenneth E
Wallace, Thomas Patrick
Weedon, Gene Clyde
Wilkinson, Thomas Lloyd, Jr
Wilkinson, William Kenneth
Wynne, Kenneth Joseph

WASHINGTON
Ericsson, Lowell Harold
Eustis, William Henry
Exarhos, Gregory James
Felicetta, Vincent Frank
Goldring, Lionel Solomon
Hine, John Maynard
Kesting, Robert E
Lim, Young Woon (Peter)
Lindenmeyer, Paul Henry
McCarthy, Joseph L(ePage)
Mehlhaff, Leon Curtis
Peterson, James Macon
Ratner, Buddy Dennis
Rowland, Stanley Paul
Subramanian, Ravanasamudram Venkatachalam
Tobiason, Frederick Lee
Vessel, Eugene David
Wasserman, William Jack
Wollwage, Paul Carl
Wong, Chun-Ming
Zollars, Richard Lee

WEST VIRGINIA
Bailey, Frederick Eugene, Jr
Barnes, Robert Keith
Bassett, David R
Bryant, George Macon
Dawson, Thomas Larry
Dunphy, James Francis
Giza, Yueh-Hua Chen
Hager, Stanley Lee
Knopf, Robert John
Koleske, Joseph Victor
Matthews, Virgil Edison
Minckler, Leon Sherwood, Jr
Novak, Ernest Richard
Osborn, Claiborn Lee
Peascoe, Warren Joseph
Rieck, James Nelson
Robson, John Howard
Schilling, Curtis Louis, Jr
Smith, Joseph James
Sperati, Carleton Angelo
Winston, Anthony

WISCONSIN
Anderson, Stephen William
Atalla, Rajai Hanna
Berge, John Williston
Bringer, Robert Paul
Brown, Charles Eric
Brown, Kenneth Howard
Cataldi, Horace A(nthony)
Caulfield, Daniel Francis
Chenier, Philip John
Christiansen, Alfred W
Cleary, James William
Cooper, Stuart L
Denton, Denice Dee
Dickerson, Charlesworth Lee
Dwyer, Sean G
Feist, William Charles
Ferry, John Douglass
Fischer, Richard Martin, Jr
Fitch, Robert McLellan
Gross, James Richard
Holland, Dewey G
Kester, Dennis Earl
Kinstle, James Francis
Klopotek, David L
Kurath, Sheldon Frank
Lemm, Arthur Warren
Lin, Stephen Y
Morris, Marion Clyde
Randall, Francis James
Reichenbacher, Paul H
Schrag, John L
Sheppard, Erwin
Soerens, Dave Allen
Stackman, Robert W
Strause, Sterling Franklin
Svoboda, Glenn Richard
Thurmaier, Roland Joseph
Verbrugge, Calvin James
Ward, Kyle, Jr
Wilkie, Charles Arthur
Yu, Hyuk

WYOMING
Hiza, Michael John, Jr
Petersen, Joseph Claine

PUERTO RICO
ALZERRECA, ARNALDO

ALBERTA
Fisher, Harold M
Henderson, John Frederick
Williams, Michael C(harles)

BRITISH COLUMBIA
Davis, Gerald Gordon
Funt, B Lionel
Hardwicke, Norman Lawson

MANITOBA
MacGregor, Elizabeth Ann
Singh, Ajit
Sze, Yu-Keung

NEW BRUNSWICK
Grant, Douglas Hope

NOVA SCOTIA
Lynch, Brian Maurice
Piorko, Adam M

ONTARIO
Alexandru, Lupu
Ananthanarayanan, Vettaikkoru S
Asculai, Samuel Simon
Bluhm, Terry Lee
Borr, Mitchell
Brash, John Law
Breitman, Leo
Buckler, Ernest Jack
Cameron, Irvine R
Carlsson, David James
Dolenko, Allan John
Eliades, Theo I
Golemba, Frank John
Huang, Robert Y M
Kabayama, Michiomi Abraham
Kazmaier, Peter Michael
Kilp, Toomas
Lautens, Mark
McEwan, Ian Hugh
Marchessault, Robert Henri
Martin, Trevor Ian
Milkie, Terence H
Noolandi, Jaan
O'Driscoll, Kenneth F(rancis)
Penlidis, Alexander
Rempel, Garry Llewellyn
Rudin, Alfred
Russell, Kenneth Edwin
Sanderson, Edwin S
Sundararajan, Pudupadi Ranganathan
Whittington, Stuart Gordon
Williams, Harry Leverne
Winnik, Françoise Martine
Winnik, Mitchell Alan

QUEBEC
Alince, Bohumil
Allen, Lawrence Harvey
Bartnikas, Ray
Darling, Graham Davidson
Dorris, Gilles Marcel
Eisenberg, Adi
Feldman, Dorel
Fleming, Bruce Ingram
Gray, Derek Geoffrey
Harrod, John Frank
Hay, Allan Stuart
Heitner, Cyril
Kokta, Bohuslav Vaclav
Lamarche, François
Leonard, Jacques Walter
Lepoutre, Pierre
Manley, Rockliffe St John
Patterson, Donald Duke
Prud'Homme, Jacques
Prud'homme, Robert Emery
St Pierre, Leon Edward
Sanctuary, Bryan Clifford
Utracki, Lechoslaw Adam
Wright, Archibald Nelson
Wu, Chisung

SASKATCHEWAN
Macphee, Kenneth Erskine

OTHER COUNTRIES
Brown, Charles Eric
Brown, Larry Robert
Cais, Rudolf Edmund
Chang, Shuya
Davidson, Daniel Lee
Grunwald, John J
Hamb, Fredrick Lynn
Hutchison, John Joseph
Isaacs, Philip Klein
Kuist, Charles Howard
Lin, Otto Chui Chau
McRae, Wayne A
Malherbe, Roger F
Markovitz, Hershel
Matsuo, Keizo
Meier, James Archibald
Millar, John Robert
Neuse, Eberhard Wilhelm
Qian, Renyuan
Reed, Thomas Freeman
Reid, William John
Snyder, Harold Lee
Snyder, William Richard
Tanny, Gerald Brian
Trofimov, Boris Alexandrovich
Weise, Jurgen Karl

Quantum Chemistry

ALABAMA
McKee, Michael Leland

ARIZONA
Balasubramanian, Krishnan
Lichtenberger, Dennis Lee
Steimle, Timothy C
Wing, William Hinshaw

ARKANSAS
Pulay, Peter

CALIFORNIA
Acrivos, Juana Luisa Vivo
Bagus, Paul Saul
Calabrese, Philip G
Carter, Emily Ann
Fink, William Henry
Goddard, William Andrew, III
Goldstein, Elisheva
Herman, Frank
Herman, Zelek Seymour
Hirschfelder, Joseph Oakland
Horsley, John Anthony
Huestis, David Lee
Karo, Arnold Mitchell
Keeports, David
Kollman, Peter Andrew
Konowalow, Daniel Dimitri
Lester, William Alexander, Jr
Miller, William Hughes
Morrison, Harry Lee
Rosenkrantz, Marcy Ellen
Sahbari, Javad Jabbari
Sheehan, William Francis
Shull, Harrison
Simon, Barry Martin
Stephens, Jeffrey Alan
Taylor, Howard S
Van Hecke, Gerald Raymond
Wolf, Kathleen A
Wolfsberg, Max

COLORADO
Morgan, Wm Lowell
Sherman, David Michael
Stewart, James Joseph Patrick

CONNECTICUT
Bohn, Robert K
Clarke, George A
Petersson, George A
Smooke, Mitchell D

DELAWARE
Doren, Douglas James
Morgan, John Davis, III
Szalewicz, Krzysztof
Wilhite, Douglas Lee

DISTRICT OF COLUMBIA
Hsu, David Shiao-Yo
Kertesz, Miklos

FLORIDA
Bartlett, Rodney Joseph
Bash, Paul Anthony
Baum, J(ames) Clayton
Cioslowski, Jerzy
Dash, Harriman Harvey
Jones, Walter H(arrison)
Lowdin, Per-Olov
Monkhorst, Hendrik J
Pop, Emil
Rhodes, William Clifford
Snyder, Patricia Ann
Szczepaniak, Krystyna
Vergenz, Robert Allan
Zerner, Michael Charles

GEORGIA
Gole, James Leslie
Owen, Gene Scott

ILLINOIS
Bouman, Thomas David
Curtiss, Larry Alan
Goodman, Gordon Louis
O'Donnell, Terence J
Sachtler, Wolfgang Max Hugo
Sanders, Frank Clarence, Jr
Scheiner, Steve
Woehler, Scott Edwin

INDIANA
Bentley, John Joseph
Boyd, Donald Bradford
Kosman, Warren Melvin
Lipkowitz, Kenny Barry
Malik, David Joseph

IOWA
Kostic, Nenad M
Miller, Gordon James

KANSAS
Layton, E Miller, Jr
Purcell, Keith Frederick
Zandler, Melvin E

KENTUCKY
Clouthier, Dennis James
Plucknett, William Kennedy

LOUISIANA
Anex, Basil Gideon
Mollere, Phillip David
Scott, John Delmoth

MARYLAND
Coplan, Michael Alan
Dehn, James Theodore
Kaufman, Joyce J
Liebman, Joel Fredric
McDiarmid, Ruth
Peters, Alan Winthrop
Rosenfield, Joan Samour
Ruffa, Anthony Richard
Silverstone, Harris Julian
Stevens, Walter Joseph

MASSACHUSETTS
Chang, Edward Shi Tou
Chen, Maynard Ming-Liang
Harrison, Ralph Joseph
Hornig, Donald Frederick
Platt, John Rader
Weiss, Karl H
Winick, Jeremy Ross

MICHIGAN
Chang, Tai Yup
DeKock, Roger Lee
Ebbing, Darrell Delmar
Francisco, Joseph Salvadore, Jr
Polik, William Frederick
Schnitker, Jurgen H
Tai, Julia Chow
Weidman, Robert Stuart

MINNESOTA
Mead, Chester Alden

MISSOURI
Baughman, Russell George
Ching, Wai-Yim
Hansen, Richard Lee
Thompson, Clifton C

MONTANA
Callis, Patrik Robert

NEBRASKA
Eckhardt, Craig Jon

NEW HAMPSHIRE
Fogleman, Wavell Wainwright

NEW JERSEY
Adams, William Henry
Ermler, Walter Carl
Frishberg, Carol
Haddon, Robert Cort
Harris, Leonce Everett
Head-Gordon, Martin Paul
Inniss, Daryl
McClure, Donald Stuart
Phillips, James Charles
Raghavachari, Krishnan
Raynor, Susanne
Sonn, George Frank
Stein, Reinhardt P
Stillinger, Frank Henry
Upton, Thomas Hallworth
Vogel, Veronica Lee

NEW MEXICO
Alldredge, Gerald Palmer
Cahill, Paul A
Cartwright, David Chapman
Martin, Richard Lee
Pack, Russell T
Redondo, Antonio
Switendick, Alfred Carl
Walters, Edward Albert

NEW YORK
Avouris, Phaedon
Barnett, Michael Peter
Beri, Avinash Chandra
Birge, Robert Richards
Chang, Shih-Yung
Ehrenson, Stanton Jay
Ezra, Gregory Sion
Feldman, Isaac
Finzel, Rodney Brian
Gao, Jiali
Hanson, Louise I Karle
Hattman, Stanley
Ho, Paul Siu-Chung
Jain, Duli Chandra
Joshi, Bhairav Datt
Loebl, Ernest Moshe
Padnos, Norman
Percus, Jerome K
Peterson, Otis G
Rubloff, Gary W
Silverman, Benjamin David
Weinstein, Harel
White, Michael George

NORTH CAROLINA
Boldridge, David William
Caves, Thomas Courtney

Miller, Robert L
Navangul, Himanshoo Vishnu
Rabinowitz, James Robert
Yang, Weitao

OHIO
Bursten, Bruce Edward
Del Bene, Janet Elaine
Klopman, Gilles
Kurtz, David Williams
Madia, William J
Smith, Douglas Alan
Swofford, Robert Lewis
Taylor, William Johnson
Weintraub, Herschel Jonathan R

OKLAHOMA
Harwood, William H
Lauffer, Donald Eugene
Neely, Stanley Carrell
Zetik, Donald Frank

OREGON
Engelking, Paul Craig

PENNSYLVANIA
Brittelli, David Ross
Garrison, Barbara Jane
Jordan, Kenneth David
Ludwig, Oliver George
Ranck, John Philip
Smith, Graham Monro
Staley, Stuart Warner
Turner, Paul Jesse
Zeroka, Daniel

SOUTH CAROLINA
Avegeropoulos, G
Larcom, Lyndon Lyle

SOUTH DAKOTA
Duffey, George Henry

TENNESSEE
Crawford, Oakley H
Fagerburg, David Richard
Jellison, Gerald Earle, Jr
Nicely, Vincent Alvin
Painter, Gayle Stanford
Tellinghuisen, Joel Barton

TEXAS
Becker, Ralph Sherman
Birney, David Martin
Boggs, James Ernest
Conway, Dwight Colbur
Ford, George Peter
Graham, William Richard Montgomery
Hayes, Edward Francis
Marcotte, Ronald Edward
Pettitt, Bernard Montgomery
Rossky, Peter Jacob
Scuseria, Gustavo Enrique

UTAH
Wahrhaftig, Austin Levy

VIRGINIA
Desjardins, Steven G
Kelley, Ralph Edward
Phillips, Donald Herman
Piepho, Susan Brand
Smith, Bertram Bryan, Jr

WASHINGTON
Gouterman, Martin (Paul)
Moore, Emmett Burris, Jr
Moseley, William David, Jr
Norman, Joe G, Jr
Poshusta, Ronald D

WEST VIRGINIA
Maruca, Robert Eugene

WISCONSIN
Barr, Tery Lynn
Bondeson, Stephen Ray
England, Walter Bernard

ALBERTA
Ali, Shahida Parvin
Boeré, René Theodoor
Grabenstetter, James Emmett
Klobukowski, Mariusz Andrzej

BRITISH COLUMBIA
Dingle, Thomas Walter

NEW BRUNSWICK
Grein, Friedrich
Sichel, John Martin
Thakkar, Ajit Jamnadas

NOVA SCOTIA
Boyd, Russell Jaye

ONTARIO
Baird, Norman Colin
Davison, Sydney George
Kazmaier, Peter Michael
King, Gerald Wilfrid
Meath, William John
Paldus, Josef
Schlesinger, Mordechay

Somorjai, Rajmund Lewis

QUEBEC
Salahub, Dennis Russell
Whitehead, Michael Anthony

SASKATCHEWAN
Mezey, Paul G

OTHER COUNTRIES
Balakrishnan, Narayana Swamy
Bunge, Carlos Federico
Kay, Kenneth George
Keller, Jaime
Mehler, Ernest Louis

Spectroscopy

ALABAMA
Coburn, William Carl, Jr

ARIZONA
Blankenship, Robert Eugene
Chapman, Robert Dale
Fuchs, Jacob
McMillan, Paul Francis
Vermaas, Willem F J

CALIFORNIA
Allamandola, Louis John
Anet, Frank Adrien Louis
Cain, William F
Curry, Bo(stick) U
Fassel, Velmer Arthur
Fetzer, John Charles
Hovanec, B(ernard) Michael
Jeffries, Jay B
Kasai, Paul Haruo
Keeports, David
Knudtson, John Thomas
Lee, Roland Robert
Meinhard, James Edgar
Okumura, Mitchio
Parsons, Michael L
Perkins, Willis Drummond
Pertica, Alexander José
Sahbari, Javad Jabbari
Shaka, Athan James
Shykind, David
Stöhr, Joachim

COLORADO
Vejvoda, Edward
Webb, John Day

CONNECTICUT
Bentz, Alan P(aul)
Bohn, Robert K
Dobbs, Gregory Melville
Greenhouse, Steven Howard
Reffner, John A
Siegel, Norman Joseph
Ultee, Casper Jan

DELAWARE
Barteau, Mark Alan
Brame, Edward Grant, Jr
Crawford, Michael Karl
Firment, Lawrence Edward
Wiley, Douglas Walker

DISTRICT OF COLUMBIA
May, Leopold
Misra, Prabhakar
Poranski, Chester F, Jr

FLORIDA
Baum, J(ames) Clayton
Block, Ronald Edward
Brucat, Philip John
Weltner, William, Jr

GEORGIA
De Sa, Richard John
Love, Jimmy Dwane
Tincher, Wayne Coleman
Van Assendelft, Onno Willem

ILLINOIS
Adlof, Richard Otto
Cutnell, John Daniel
Erickson, Mitchell Drake
Gruen, Dieter Martin
Horwitz, Alan Fredrick
Jonas, Jiri
Kemper, Kathleen Ann
LaPlanche, Laurine A
Masel, Richard Isaac
Mota de Freitas, Duarte Emanuel
Sibbach, William Robert
Thompson, Arthur Robert
Woehler, Scott Edwin
Wozniak, Wayne Theodore
Young, Charles Edward

INDIANA
Chernoff, Donald Alan
Havel, Henry Acken
Hunt, Ann Hampton
Thomas, John Kerry
Tripathi, Govakh Nath Ram

IOWA
Edelson, Martin Charles
Kintanar, Agustin
Koerner, Theodore Alfred William, Jr
Stahr, Henry Michael
Struve, Walter Scott

KANSAS
Carper, William Robert
Hammaker, Robert Michael
Purcell, Keith Frederick

KENTUCKY
Smith, Stanford Lee

LOUISIANA
Anex, Basil Gideon
Kumar, Devendra
Maverick, Andrew William
Miceli, Michael Vincent

MARYLAND
Fales, Henry Marshall
Frazier, Claude Clinton, III
Houseman, Barton L
Lafferty, Walter J
McDiarmid, Ruth
Miller, Steve P F
Robinson, Dean Wentworth
Siatkowski, Ronald E
Suenram, Richard Dee
Thompson, Warren Elwin

MASSACHUSETTS
Aronson, James Ries
Chiu, Tak-Ming
Dasari, Ramachandra R
Druy, Mark Arnold
Greenaway, Frederick Thomas
Varco-Shea, Theresa Camille

MICHIGAN
Coburn, Joel Thomas
Francisco, Joseph Salvadore, Jr
Hatzenbuhler, Douglas Albert
Langhoff, Charles Anderson
Leroi, George Edgar
Nader, Bassam Salim
Polik, William Frederick
Yesinowski, James Paul

MINNESOTA
Newmark, Richard Alan
Potts, Lawrence Walter
Smith, David Philip

MISSOURI
Ames, Donald Paul
Coria, Jose Conrado
Grayson, Michael A
Greenlief, Charles Michael
Kaelble, Emmett Frank
Wong, Tuck Chuen

NEW HAMPSHIRE
Bel Bruno, Joseph James

NEW JERSEY
Bhattacharjee, Himangshu Ranjan
Brody, Stuart Martin
Brons, Cornelius Hendrick
Brus, Louis Eugene
Fairchild, Edward H
Gethner, Jon Steven
Goodman, Lionel
Harris, Alexander L
Harris, Leonce Everett
McClure, Donald Stuart
Williams, Thomas Henry

NEW MEXICO
Beattie, Willard Horatio
Eckert, Juergen
McCulla, William Harvey
Williams, Mary Carol

NEW YORK
Angell, Charles Leslie
Billmeyer, Fred Wallace, Jr
Birke, Ronald Lewis
Blackwell, Crist Scott
Dixon, William Brightman
Doetschman, David Charles
Dumoulin, Charles Lucian
Feigenson, Gerald William
Hurd, Jeffery L
Lashewycz-Rubycz, Romana A
Loring, Roger Frederic
MacColl, Robert
Marrone, Paul Vincent
Myers, Anne Boone
Orzech, Chester Eugene, Jr
Philips, Laura Alma
Siggins, James Ernest
Wu, Konrad T

NORTH CAROLINA
Noftle, Ronald Edward
Pickett, John Harold
Stejskal, Edward Otto
Wilson, Nancy Keeler

OHIO
Diem, Hugh E(gbert)

Spectroscopy (cont)

Ford, William Ellsworth
Gorse, Joseph
McLean, Larry Raymond
Stout, Barbara Elizabeth
Thompson, Robert Quinton
Wilson, Gustavus Edwin

OKLAHOMA
Johnson, Arthur Edward

OREGON
Ferrante, Michael John
Matthes, Steven Allen

PENNSYLVANIA
Barnett, Herald Alva
Bodine, Peter Van Nest
Chang, Charles Hung
Gold, Lewis Peter
Howard, Wilmont Frederick, Jr
Jordan, Kenneth David
Lemmon, Donald H
Miller, Foil Allan
Nelson, George Leonard
Pfeffer, Philip Elliot
Selinsky, Barry Steven
Weiss, Paul Storch

SOUTH CAROLINA
Myrick, Michael Lenn

TENNESSEE
Carman, Howard Smith, Jr
Close, David Matzen
Li, Ying Sing

TEXAS
Berg, Mark Alan
Busch, Marianna Anderson
Fredin, Leif G R
Gnade, Bruce E
Gray, Kenneth W
Jaszberenyi, Joseph C
Laane, Jaan
Patel, Bhagwandas Mavjibhai
Rabalais, John Wayne
Rawls, Henry Ralph
Redington, Richard Lee

UTAH
Lyman, Donald Joseph

VIRGINIA
Bodnar, Robert John
Drew, Russell Cooper
Graybeal, Jack Daniel
Metcalf, David Halstead
Piepho, Susan Brand
Roberts, Catherine Harrison

WASHINGTON
Baer, Donald Ray
Friedrich, Donald Martin
Styris, David Lee

WISCONSIN
Lemm, Arthur Warren
West, Kevin James
Wright, John Curtis

WYOMING
Netzel, Daniel Anthony

ALBERTA
Egerton, Raymond Frank

BRITISH COLUMBIA
Balfour, Walter Joseph
Mitchell, Reginald Harry

MANITOBA
Attas, Ely Michael
Ens, E(rich) Werner
McFarlane, Joanna

NOVA SCOTIA
Coxon, John Anthony
Hooper, Donald Lloyd
McAlduff, Edward J
Wasylishen, Roderick Ernest

ONTARIO
Baines, Kim Marie
Carey, Paul Richard
LeRoy, Robert James
Macdonald, Peter Moore
Pare, J R Jocelyn
Rodrigo, Russell Godfrey
Shurvell, Herbert Francis
Winnik, Françoise Martine

QUEBEC
Bélanger, Jacqueline M R
Feng, Rong
Van Calsteren, Marie-Rose

OTHER COUNTRIES
Bondybey, Vladimir E
Trentham, David R

Structural Chemistry

ALABAMA
Cook, William Joseph
McKee, Michael Leland
Sayles, David Cyril
Squillacote, Michael Edward
Tamburin, Henry John

ARIZONA
Hruby, Victor J
Kukolich, Stephen George
McMillan, Paul Francis
Stamm, Robert Franz
Steimle, Timothy C

ARKANSAS
Fry, Arthur James
Pulay, Peter

CALIFORNIA
Abdel-Baset, Mahmoud B
Bau, Robert
Burlingame, Alma L
Callahan, Kenneth Paul
Delker, Gerald Lee
Dickerson, Richard Earl
Dougherty, Dennis A
Fetzer, John Charles
George, Patricia Margaret
Giants, Thomas W
Glaser, Robert J
Haddon, William F, (Jr)
Hardcastle, Kenneth Irvin
Hodgson, Keith Owen
Hurd, Ralph Eugene
Jaecker, John Alvin
Kim, Sung-Hou
Kirby, Jon Allan
Knobler, Carolyn Berk
La Mar, Gerd Neustadter
Leo, Albert Joseph
Lind, Maurice David
Medrud, Ronald Curtis
Murdoch, Joseph Richard
Patterson, Dennis Bruce
Peterson, Selmer Wilfred
Prakash, Surya G K
Raymond, Kenneth Norman
Samson, Sten
Schaefer, William Palzer
Scherer, James R
Shoolery, James Nelson
Simpson, Howard Douglas
Sims, James Joseph
Stöhr, Joachim
Stucky, Galen Dean
Tokes, Laszlo Gyula
True, Nancy S
Trueblood, Kenneth Nyitray
Untch, Karl George
Wong, Joe

COLORADO
Anderson, Oren P
Geller, Seymour

CONNECTICUT
Armitage, Ian MacLeod
Bohn, Robert K
Chang, Ted T
Katz, Lewis
Knox, James Russell, Jr
Rennhard, Hans Heinrich
Suib, Steven L

DELAWARE
Brame, Edward Grant, Jr
Calabrese, Joseph C
DeGrado, William F
Domaille, Peter John
Ferguson, Raymond Craig
Fukunaga, Tadamichi
Harlow, Richard Leslie
Herron, Norman
Ittel, Steven Dale
Lerman, Charles Lew
Reuben, Jacques
Shannon, Robert Day
Van Trump, James Edmond
Wallenberger, Frederick Theodore
Willer, Rodney Lee

DISTRICT OF COLUMBIA
Baker, Louis Coombs Weller
Butcher, Raymond John
Gilardi, Richard Dean
Gress, Mary Edith
Karle, Isabella Lugoski
Karle, Jean Marianne
Kertesz, Miklos

FLORIDA
Babich, Michael Wayne
Copeland, Richard Franklin
DeLap, James Harve
Pepinsky, Raymond
Person, Willis Bagley
Snyder, Patricia Ann
Stanko, Joseph Anthony
Trefonas, Louis Marco

GEORGIA
Pelletier, S William
Richardson, Susan D
Suddath, Fred LeRoy, (Jr)
Wilson, William David

HAWAII
Seff, Karl

ILLINOIS
Bouman, Thomas David
Broach, Robert William
Chang, Chong-Hwan
Curtin, David Yarrow
Guggenheim, Stephen
Johnson, Paul Lorentz
Keiderling, Timothy Allen
Lambert, Joseph B
LaPlanche, Laurine A
Levenberg, Milton Irwin
Magnus, George
Makinen, Marvin William
Maroni, Victor August
Mueller, Melvin H(enry)
Olsen, Kenneth Wayne
Pluth, Joseph John
Poeppelmeier, Kenneth Reinhard
Reynolds, Rosalie Dean (Sibert)
Ries, Herman Elkan, Jr
Rogers, Robin Don
Schultz, Arthur Jay
Shriver, Duward F
Tuomi, Donald
Vandeberg, John Thomas
Wang, Andrew H-J
Williams, Jack Marvin

INDIANA
Christou, George
Dorman, Douglas Earl
Huffman, John Curtis
Hunt, Ann Hampton
Koch, Kay Frances
Lipkowitz, Kenny Barry
Robinson, William Robert
Scheidt, Walter Robert
Serianni, Anthony Stephan
Streib, William E
Toomey, Joseph Edward
Williams, Edward James

IOWA
Jacobson, Robert Andrew
Kraus, Kenneth Wayne
McCarley, Robert Eugene
Messerle, Louis
Ruedenberg, Klaus

KANSAS
Fateley, William Gene
Harmony, Marlin D
Landis, Arthur Melvin
Mertes, Kristin Bowman
Pfluger, Clarence Eugene

KENTUCKY
Brock, Carolyn Pratt
Buchanan, Robert Martin
McDermott, Dana Paul
Selegue, John Paul

LOUISIANA
Brown, Leo Dale
Fronczek, Frank Rolf
Gandour, Richard David
Lee, Burnell
Watkins, Steven F

MAINE
Rhodes, William Gale

MARYLAND
Amzel, L Mario
Block, Stanley
Bowen, Kit Hansel
Coulter, Charles L
D'Antonio, Peter
Fraser, Blair Allen
Frazer, Benjamin Chalmers
Jones, Tappey Hughes
Lovas, Francis John
Lusby, William Robert
Mathew, Mathai
Nemec, Josef
Revesz, Akos George
Siatkowski, Ronald E
Suenram, Richard Dee
Takagi, Shozo
Trus, Benes L

MASSACHUSETTS
Baglio, Joseph Anthony
Burns, Roger George
Costello, Catherine E
Dahm, Donald J
Eriks, Klaas
Foxman, Bruce Mayer
Fritzsche, Alfred Keith
Gore, William Earl
Haas, Terry Evans
Hayden, Thomas Day
Kishi, Yoshito
Krull, Ira Stanley
Lippard, Stephen J

Loehlin, James Herbert
Margulis, Thomas N
Reis, Arthur Henry, Jr
Williamson, Kenneth Lee
Wilson, Linda S (Whatley)

MICHIGAN
Adams, Wade J
Bartell, Lawrence Sims
Bettman, Max
Cody, Wayne Livingston
Cornilsen, Bahne Carl
Duchamp, David James
Eick, Harry Arthur
Einspahr, Howard Martin
Fawcett, Timothy Goss
Heeschen, Jerry Parker
Hillig, Kurt Walter, II
Hohnke, Dieter Karl
Jacobs, Gerald Daniel
King, Stanley Shih-Tung
Kirschner, Stanley
Oliver, John Preston
Ovshinsky, Stanford R(obert)
Penner-Hahn, James Edward
Petersen, Donald Ralph
Rohrer, Douglas C
Ryntz, Rose A
Saper, Mark A
Schlick, Shulamith
Tai, Julia Chow
Tanis, Steven Paul
Tulinsky, Alexander
Woo, P(eter) W(ing) K(ee)

MINNESOTA
Baumann, Wolfgang J
Britton, Doyle
Etter, Margaret Cairns
Gray, Gary Ronald
Mayerle, James Joseph
Pignolet, Louis H
Stebbings, William Lee
Thompson, Herbert Bradford

MISSOURI
Baughman, Russell George
Brammer, Lee
Felthouse, Timothy R
Freeman, John Jerome
Haymore, Barry Lant
Heitsch, Charles Weyand
Jason, Mark Edward
Munch, John Howard
Ramsey, Robert Bruce
Rath, Nigam Prasad
Ross, Frederick Keith
Talbott, Ted Delwyn
Thomas, George Joseph, Jr
Wong, Tuck Chuen
Yelon, William B

NEBRASKA
Eckhardt, Craig Jon
Harris, Holly Ann

NEVADA
Lightner, David A

NEW HAMPSHIRE
Haendler, Helmut Max
Housecroft, Catherine Elizabeth

NEW JERSEY
Bahr, Charles Chester
Becker, Joseph Whitney
Boskey, Adele Ludin
Cohen, Allen Irving
Dreeben, Arthur B
Egan, Richard Stephen
Elk, Seymour B
Greenblatt, Martha
Head, William Francis, Jr
Inniss, Daryl
Jacobson, Allan Joseph
Klaubert, Dieter Heinz
Kleinschuster, Stephen J, III
Lalancette, Roger A
Lam, Yiu-Kuen Tony
Lawton, Stephen Latham
Li, Shu-Tung
Liesch, Jerrold Michael
McCaslin, Darrell
Marezio, Massimo
Mislow, Kurt Martin
Nelson, Gregory Victor
Pilkiewicz, Frank George
Raghavachari, Krishnan
Schlesinger, David H
Schneemeyer, Lynn F
Sibilia, John Philip
Suchow, Lawrence
Thomas, Kenneth Alfred, Jr
Vandenberg, Joanna Maria
Venturella, Vincent Steven
Williams, Thomas Henry

NEW MEXICO
Cromer, Don Tiffany
Eckert, Juergen
Kubas, Gregory Joseph
Lawson, Andrew Cowper, II
Mills, Robert Leroy
Morosin, Bruno

Peterson, Eugene James
Ryan, Robert Reynolds
Von Dreele, Robert Bruce

NEW YORK
Agarwal, Ramesh Chandra
Bauer, Simon Harvey
Bednowitz, Allan Lloyd
Bell, Thomas Wayne
Blackwell, Crist Scott
Blessing, Robert Harry
Bourne, Philip Eric
Clardy, Jon Christel
Clemans, Stephen D
Cody, Vivian
Cox, David Ernest
Diem, Max
Dorset, Douglas Lewis
Finzel, Rodney Brian
Flanigen, Edith Marie
Gillies, Charles Wesley
Goldberg, Stephen Zalmund
Hanson, Jonathan C
Hanson, Louise I Karle
Herley, Patrick James
Hoard, James Lynn
Kilbourn, Barry T
Kirchner, Richard Martin
Koetzle, Thomas F
Lashewycz-Rubycz, Romana A
Lee, May D-Ming (Lu)
Levy, George Charles
Low, Barbara Wharton
Morrow, Janet Ruth
Myers, Clifford Earl
Quigley, Gary Joseph
Reeke, George Norman, Jr
Rossi, Miriam
Rudman, Reuben
Rupp, John Jay
Seeman, Nadrian Charles
Smith, Douglas Lee
Snyder, Robert Lyman
Steinberg, David H
Syed, Ashfaquzzaman
Szalda, David Joseph
Szebenyi, Doletha M E
Thomas, Robert
Williams, Graheme John Bramald
Winter, William Thomas
Ziolo, Ronald F

NORTH CAROLINA
Allen, Harry Clay, Jr
Baird, Herbert Wallace
Hanker, Jacob S
Higgins, Robert H
Martin, Gary Edwin
Posner, Herbert S
Templeton, Joseph Leslie
Wahl, George Henry, Jr

NORTH DAKOTA
Garvey, Roy George
Radonovich, Lewis Joseph

OHIO
Burow, Duane Frueh
Campbell, James Edward
Cohen, Irwin
Elder, Richard Charles
Gano, James Edward
Hunter, James Charles
Jendrek, Eugene Francis, Jr
Kinstle, Thomas Herbert
Koknat, Friedrich Wilhelm
Kroenke, William Joseph
Mateescu, Gheorghe D
Meek, Devon Walter
Misono, Kunio Shiraishi
Oertel, Richard Paul
Oliver, Joel Day
Olszanski, Dennis John
Pinkerton, A Alan
Schweitzer, Mark Glenn
Shore, Sheldon Gerald
Sill, Arthur DeWitt
Sullenger, Don Bruce
Tsai, Chun-Che
Wolff, Gunther Arthur
Youngs, Wiley Jay

OKLAHOMA
Daake, Richard Lynn
Wharry, Stephen Mark

OREGON
Brown, Bruce Willard
Shoemaker, Clara Brink
Shoemaker, David Powell
Sleight, Arthur William

PENNSYLVANIA
Brady, Stephen Francis
Brixner, Lothar Heinrich
Craven, Bryan Maxwell
Finegold, Harold
Hutchins, MaryGail Kinzer
Janda, Kenneth Carl
Kopple, Kenneth D(avid)
Morris, Marlene Cook
Nelson, George Leonard
Perrotta, Anthony Joseph
Pez, Guido Peter

Ristey, William J
Ross, Stephen T
Sax, Martin
Shepherd, Rex E
Staley, Stuart Warner
Thomas, H Ronald

RHODE ISLAND
Carpenter, Gene Blakely
Hartman, Karl August
Williard, Paul Gregory

SOUTH CAROLINA
Amma, Elmer Louis
Brown, Farrell Blenn
Kendall, David Nelson
Posey, Robert Giles
Sproul, Gordon Duane
Stampf, Edward John, Jr

TENNESSEE
Barnes, Craig Eliot
Brown, George Marshall
Habenschuss, Anton
Hingerty, Brian Edward
Howell, Elizabeth E
Lenhert, P Galen
Li, Ying Sing
Polavarapu, Prasad Leela
Quist, Arvin Sigvard

TEXAS
Boggs, James Ernest
Czerwinski, Edmund William
Davis, Michael I
Frisch, P Douglas
Hance, Robert Lee
Ivey, Robert Charles
Kevan, Larry
Knapp, Roger Dale
Malloy, Thomas Bernard, Jr
Meyer, Edgar F
Naae, Douglas Gene
Redington, Richard Lee
Steinfink, Hugo
Watson, William Harold, Jr
Whitmire, Kenton Herbert

UTAH
Allred, Evan Leigh
Goates, Steven Rex
Kodama, Goji
Powers, Linda Sue

VERMONT
Allen, Christopher Whitney
Jasinski, Jerry Peter

VIRGINIA
Barber, Patrick George
Bryan, Robert Finlay
Evans, Howard Tasker, Jr
Graybeal, Jack Daniel
Grimes, Russell Newell
Huddle, Benjamin Paul, Jr
Richardson, Frederick S

WASHINGTON
Adman, Elinor Thomson
Camerman, Arthur
Garland, John Kenneth
Hakomori, Sen-Itiroh
Lingafelter, Edward Clay, Jr
Lytle, Farrel Wayne
Macklin, John Welton
Maki, Arthur George, Jr
Moore, Emmett Burris, Jr
Reid, Brian Robert
Schomaker, Verner
Stenkamp, Ronald Eugene
Whitmer, John Charles
Wilson, Archie Spencer

WISCONSIN
Landucci, Lawrence L
Qureshi, Nilofer
Schnoes, Heinrich Konstantin
Spencer, Brock

WYOMING
Hodgson, Derek John
Howatson, John
Netzel, Daniel Anthony

ALBERTA
Bayliss, Peter
Boeré, René Theodoor
Codding, Penelope Wixson
Cowie, Martin
Krueger, Peter J
Tschuikow-Roux, Eugene
Wieser, Helmut

BRITISH COLUMBIA
Bushnell, Gordon William
Dahn, Jeffery Raymond
Gerry, Michael Charles Lewis

MANITOBA
Lange, Bruce Ainsworth
Schaefer, Theodore Peter
Secco, Anthony Silvio
Sunder, Sham
Wikjord, Alfred George

NEW BRUNSWICK
Whitla, William Alexander

NEWFOUNDLAND
Newlands, Michael John

NOVA SCOTIA
Cameron, T Stanley
McAlduff, Edward J
Wasylishen, Roderick Ernest

ONTARIO
Baer, Hans Helmut
Birnbaum, George I
Brown, Ian David
Buncel, Erwin
Butler, Douglas Neve
Carty, Arthur John
Casal, Hector Luis
Drake, John Edward
Hitchcock, Adam Percival
Huber, Carol (Saunderson)
Lock, Colin James Lyne
MacLean, David Bailey
Mantsch, Henry H
Payne, Nicholas Charles
Penner, Glenn H
Przybylska, Maria

QUEBEC
Cedergren, Robert J
Jordan, Byron Dale
Onyszchuk, Mario
Sygusch, Jurgen
Theophanides, Theophile

SASKATCHEWAN
Barton, Richard J
Quail, John Wilson
Robertson, Beverly Ellis

OTHER COUNTRIES
Burford, Neil
Hirshfeld, Fred Lurie
Nyburg, Stanley Cecil
Schleyer, Paul von Ragué

Synthetic Inorganic & Organometallic Chemistry

ALABAMA
Haak, Frederik Albertus
Marano, Gerald Alfred

ARIZONA
Bartocha, Bodo
Lichtenberger, Dennis Lee
Roche, Thomas Stephen
Valenty, Steven Jeffrey

CALIFORNIA
Allen, William Merle
Bau, Robert
Bond, Martha W(illis)
Callahan, Kenneth Paul
Chen, Timothy Shieh-Sheng
Cole, Thomas Ernest
Current, Steven P
Davison, John Blake
Evans, William John
Feher, Frank J
Feigelson, Robert Saul
Fisher, Richard Paul
Gagne, Robert Raymond
Glaser, Robert J
Goldwhite, Harold
Grieve, Catherine Macy
Harris, Daniel Charles
Holzmann, Richard Thomas
Irvine, Stuart James Curzon
Jaecker, John Alvin
Kratzer, Reinhold
Kurkov, Victor Peter
Labinger, Jay Alan
Lawton, Emil Abraham
Lipka, James J
Lustig, Max
Miller, Russell Bryan
Mitchell, Dennis Keith
Murphy, James Francis
Nelson, Norvell John
Novak, Bruce Michael
Nozaki, Kenzie
Pappatheodorou, Sofia
Patterson, Dennis Bruce
Perry, Dale Lynn
Rainis, Andrew
Raymond, Kenneth Norman
Rosenberg, Sanders David
Schaefer, William Palzer
Shelton, Robert Neal
Singaram, Bakthan
Smith, Robert Alan
Stucky, Galen Dean
Tilley, T(erry) Don
Vogel, Roger Frederick
Vollhardt, K Peter C
Weber, Carl Joseph
Wender, Paul Anthony
Wilson, Charles Oren

COLORADO
Anderson, Oren P
Barrett, Anthony Gerard Martin
Evans, David Lane
Fleming, Michael Paul
Hayes, Jeffrey Charles
Ritchey, John Michael
Thompson, Ronald G

CONNECTICUT
Crabtree, Robert H
Haas, Thomas J
Holland, Gerald Fagan
Krause, Ronald Alfred
McKeon, James Edward
Nielsen, John Merle
Suib, Steven L
Yang, Darchun Billy

DELAWARE
Armbrecht, Frank Maurice, Jr
Baker, Ralph Thomas
Breslow, David Samuel
Bulkowski, John Edmund
Calabrese, Joseph C
Foley, Henry Charles
Ford, Thomas Aven
Fukunaga, Tadamichi
Glaeser, Hans Hellmut
Herron, Norman
Hess, Richard William
Ittel, Steven Dale
Karel, Karin Johnson
Laganis, Evan Dean
Longhi, Raymond
McKinney, Ronald James
McLain, Stephan James
Mills, George Alexander
Moser, Glenn Allen
Repka, Benjamin C
Rheingold, Arnold L
Rondestvedt, Christian Scriver, Jr
Taber, Douglass Fleming
Theopold, Klaus Hellmut
Tomic, Ernst Alois
Tulip, Thomas Hunt

DISTRICT OF COLUMBIA
Butcher, Raymond John
Butter, Stephen Allan
Hudrlik, Paul Frederick
Nordquist, Paul Edgard Rudolph, Jr
Pope, Michael Thor
Turner, Anne Halligan

FLORIDA
Hahn, Elliot F
Jones, William Maurice
Newkome, George Richard
Rees, William Smith, Jr
Rochow, Eugene George
Sisler, Harry Hall
Stanko, Joseph Anthony
Stoufer, Robert Carl

GEORGIA
Allison, John P
Anderson, Bruce Martin
Barbas, John Theophani
Bottomley, Lawrence Andrew
Chandler, James Harry, III
James, Jeffrey

HAWAII
Cramer, Roger Earl
Seff, Karl

IDAHO
Baker, John David

ILLINOIS
Arzoumanidis, Gregory G
Bernhardt, Randal Jay
Broach, Robert William
Fischer, Albert Karl
Gembicki, Stanley Arthur
Gupta, Goutam
Halpern, Jack
Harris, Ronald L
Henry, Patrick M
Herlinger, Albert William
Hollins, Robert Edward
Hopper, Steven Phillip
Hutchison, David Allan
Keiter, Richard Lee
Kieft, Richard Leonard
Lehman, Dennis Dale
Mendelsohn, Marshall H
Menke, Andrew G
Nichols, George Morrill
Poeppelmeier, Kenneth Reinhard
Rathke, Jerome William
Rauchfuss, Thomas Bigley
Rogers, Robin Don
Rosen, Bruce Irwin
Schultz, Arthur Jay
Shapley, John Roger
Shriver, Duward F
Suslick, Kenneth Sanders
Trevillyan, Alvin Earl
Weil, Thomas Andre
Whaley, Thomas Patrick
Yokelson, Howard Bruce

Synthetic Inorganic & Organometallic Chemistry (cont)

INDIANA
Bays, James Philip
Biddlecom, William Gerard
Bowser, James Ralph
Chisholm, Malcolm Harold
Christou, George
Green, Mark Alan
Helquist, Paul M
Kreider, Leonard Cale
Miller, Roy Glenn
Negishi, Ei-ichi
Scheidt, Walter Robert
Todd, Lee John

IOWA
Angelici, Robert Joe
Eyman, Darrell Paul
Hampton, David Clark
Kostic, Nenad M
McCarley, Robert Eugene
Messerle, Louis
Miller, Gordon James

KANSAS
Landis, Arthur Melvin
McCormick, Bailie Jack
Purcell, Keith Frederick

KENTUCKY
Brill, Joseph Warren
Buchanan, Robert Martin
Gibson, Dorothy Hinds
Henrickson, Charles Henry
Meier, Mark Stephan
Owen, David Allan
Selegue, John Paul
Slocum, Donald Warren

LOUISIANA
Conrad, Franklin
Doomes, Earl
Ehrlich, Kenneth Craig
Fronczek, Frank Rolf
Gautreaux, Marcelian Francis
Halbert, Thomas Risher
Ham, Russell Allen
Harper, Robert John, Jr
Hornbaker, Edwin Dale
Lee, John Yuchu
Mague, Joel Tabor
Mangham, Jesse Roger
Maverick, Andrew William
Mollere, Phillip David
Plonsker, Larry
Robson, Harry Edwin
Schweizer, Albert Edward
Selbin, Joel
Shin, Kju Hi
Shubkin, Ronald Lee
Siriwardane, Upali
Stanley, George Geoffrey
Theriot, Kevin Jude
Welch, Clark Moore
Zaweski, Edward F

MAINE
Bragdon, Robert Wright

MARYLAND
Argauer, Robert John
Barnes, James Alford
Coyle, Thomas Davidson
Decraene, Denis Fredrick
Frazier, Claude Clinton, III
Gause, Evelyn Pauline
Grim, Samuel Oram
Joedicke, Ingo Bernd
Poli, Rinaldo
Ritter, Joseph John
Schwartz, Paul
Shade, Joyce Elizabeth
Simon, Robert Michael
Viswanathan, C T

MASSACHUSETTS
Andrulis, Peter Joseph, Jr
Archer, Ronald Dean
Baglio, Joseph Anthony
Becker, Ernest I
Ciappenelli, Donald John
Clarke, Michael J
Evans, Michael Douglas
Foxman, Bruce Mayer
Goldberg, Gershon Morton
Harrill, Robert W
Idelson, Martin
Kline, Toni Beth
Kramer, Charles Edwin
Lippard, Stephen J
Moser, William Ray
Ort, Morris Richard
Parham, Marc Ellous
Rupich, Martin Walter
Snider, Barry B
Stuhl, Louis Sheldon
Sullivan, Edward Augustine
Taunton-Rigby, Alison
Urry, Grant Wayne

MICHIGAN
Anderson, Amos Robert
Anderson, Melvin Lee
Atwell, William Henry
Bailey, Donald Leroy
Bajzer, William Xavier
Baney, Ronald Howard
Brenner, Alan
Chandra, Grish
Harris, Paul Jonathan
Hartwell, George E, Jr
Lipton, Michael Forrester
Mance, Andrew Mark
Moehs, Peter John
Moser, Frank Hans
Oliver, John Preston
Rasmussen, Paul G
Rycheck, Mark Rule
Salinger, Rudolf Michael
Smart, James Blair
Stark, Forrest Otto
Weyenberg, Donald Richard
Wymore, Charles Elmer

MINNESOTA
Esmay, Donald Levern
Gladfelter, Wayne Lewis
Johnson, Bryce Vincent
Miessler, Gary Lee
Mueting, Ann Marie
Pignolet, Louis H
Reich, Charles
Schmit, Joseph Lawrence

MISSISSIPPI
Howell, John Emory
Pittman, Charles U, Jr

MISSOURI
Bleeke, John Richard
Chan, Albert Sun Chi
Felthouse, Timothy R
Frederick, Raymond H
Haymore, Barry Lant
Heitsch, Charles Weyand
Krueger, Paul A
Lennon, Patrick James
Markowski, Henry Joseph
Moedritzer, Kurt
Morris, Donald Eugene
Rath, Nigam Prasad
Riley, Dennis Patrick
Rogic, Milorad Mihailo
Sabacky, M Jerome
Wulfman, David Swinton

NEBRASKA
George, T Adrian
Rieke, Reuben Dennis

NEW HAMPSHIRE
Danforth, Raymond Hewes
Haendler, Helmut Max
Housecroft, Catherine Elizabeth
Hughes, Russell Profit
Meloni, Edward George

NEW JERSEY
Battisti, Angelo James
Bertelo, Christopher Anthony
Bertz, Steven Howard
Boyle, Richard James
Cullen, Glenn Wherry
Dreeben, Arthur B
Giddings, Sydney Arthur
Gitlitz, Melvin H
Granchi, Michael Patrick
Guthrie, Roger Thackston
Iorns, Terry Vern
Jacobson, Allan Joseph
Jacobson, Stephen Ernest
Johnson, Jack Wayne
Kuehl, Guenter Hinrich
Larkin, William Albert
McGinnis, James Lee
Magder, Jules
Pedersen, Charles John
Prapas, Aristotle G
Ramsden, Hugh Edwin
Reichle, Walter Thomas
Rosan, Alan Mark
Rossi, Robert Daniel
Schwartz, Herbert
Sekutowski, Dennis G
Spohn, Ralph Joseph
Thich, John Adong
Tiethof, Jack Alan
Tucci, Edmond Raymond
Voss, Kenneth Edwin
Weissman, Paul Morton
Wheaton, Gregory Alan
Wreford, Stanley S
Yocom, Perry Niel

NEW MEXICO
Bowen, Scott Michael
Kubas, Gregory Joseph
Liepins, Raimond
Mullendore, A(rthur) W(ayne)
Peterson, Eugene James
Smith, Wayne Howard
West, Mike Harold
Zeigler, John Martin

NEW YORK
Amero, Bernard Alan
Andrews, Mark Allen
Atwood, Jim D
Beachley, Orville Theodore, Jr
Bullock, R Morris
Burlitch, James Michael
Carpenter, Barry Keith
Dabrowiak, James Chester
Daves, Glenn Doyle, Jr
Dehn, Joseph William, Jr
Eschbach, Charles Scott
Flanigen, Edith Marie
Gancy, Alan Brian
Gardner, Sylvia Alice
Giannelis, Emmanuel P
Highsmith, Ronald Earl
Interrante, Leonard V
Jones, William Davidson
Katz, Thomas Joseph
Kauder, Otto Samuel
Kerber, Robert Charles
Kinnel, Robin Bryan
Koch, Stephen Andrew
Kotz, John Carl
Landon, Shayne J
Lashewycz-Rubycz, Romana A
Lau, Philip T S
Lees, Alistair John
McKelvie, Neil
MacLaury, Michael Risley
Malpass, Dennis B
Matienzo, Luis J
Mayr, Andreas
Miller, William Taylor
Mirviss, Stanley Burton
Morrow, Janet Ruth
Muller, Olaf
Ojima, Iwao
Philipp, Manfred (Hans)
Post, Howard William
Rupp, John Jay
Simmons, Harry Dady, Jr
Sogah, Dotsevi Yao
Sullivan, Michael Francis
Triplett, Kelly B
Ziolo, Ronald F

NORTH CAROLINA
Barborak, James Carl
Bereman, Robert Deane
Brookhart, Maurice S
Clemens, Donald Faull
Doak, George Osmore
Farona, Michael F
Gains, Lawrence Howard
Garrou, Philip Ernest
Gillooly, George Rice
Groszos, Stephen Joseph
Jackels, Susan Carol
McDonald, Daniel Patrick
Noftle, Ronald Edward
O'Connor, Lila Hunt
Panek, Edward John
Pruett, Roy L
Schwindeman, James Anthony
Spielvogel, Bernard Franklin
Switzer, Mary Ellen Phelan
Templeton, Joseph Leslie
Wasson, John R
Wilson, Lauren R

NORTH DAKOTA
Abrahamson, Harmon Bruce
Bergstrom, Donald Eugene
Boudjouk, Philip Raymond
Garvey, Roy George
Jasperse, Craig Peter

OHIO
Ackermann, Martin Nicholas
Alexander, John J
Attig, Thomas George
Blum, Patricia Rae
Bradley, Ronald W
Chrysochoos, John
Conder, Harold Lee
Cummings, Sue Carol
Davies, Julian Anthony
Didchenko, Rostislav
Fechter, Robert Bernard
Feldman, Julian
Fischer, Mark Benjamin
Fraenkel, Gideon
Franks, John Anthony, Jr
Gibson, Thomas William
Goel, Anil B
Horne, Samuel Emmett, Jr
Jacobsen, Donald Weldon
Johnson, Gordon Lee
Kang, Jung Wong
Katovic, Vladimir
Kenney, Malcolm Edward
Koknat, Friedrich Wilhelm
Kroenke, William Joseph
Litt, Morton Herbert
McDonald, John William
Meek, Devon Walter
Myers, Ronald Eugene
Nee, Michael Wei-Kuo
Nicholson, D Allan
Olszanski, Dennis John
Pearson, Karl Herbert
Pinkerton, A Alan

SHORE (NEW YORK continued)
Shore, Sheldon Gerald
Tamborski, Christ
Tessier, Claire Adrienne
Throckmorton, Peter E
Turkel, Rickey M(artin)
Wenclawiak, Bernd Wilhelm
Williams, John Paul
Wojcicki, Andrew
Youngs, Wiley Jay

OKLAHOMA
Clemenst, James Lee
Daake, Richard Lynn
Efner, Howard F
Graves, Victoria
Lindahl, Charles Blighe
Nicholas, Kenneth M

OREGON
Halko, Barbara Tomlonovic
Humphrey, J Richard
Tyler, David Ralph
Yoke, John Thomas

PENNSYLVANIA
Adams, Roy Melville
Allcock, Harry R(ex)
Balaba, Willy Mukama
Bartish, Charles Michael Christopher
Bassner, Sherri Lynn
Collins, Terrence James
Cooper, Norman John
Dent, Anthony L
Durante, Vincent Anthony
Ford, Michael Edward
Grey, Roger Allen
Gunther, Wolfgang Hans Heinrich
Howard, Wilmont Frederick, Jr
Jordan, Robert Kenneth
Kennedy, Flynt
Knoth, Walter Henry, Jr
Kraihanzel, Charles S
Laemmle, Joseph Thomas
Larson, Gerald Louis
Lavoie, Alvin Charles
Mair, Robert Dixon
Maryanoff, Cynthia Anne Milewski
Mascioli, Rocco Lawrence
Matyjaszewski, Krzysztof
Mocella, Michael Thomas
Murphy, Clarence John
Ochiai, Ei-Ichiro
Pez, Guido Peter
Pierce, Roberta Marion
Richey, Herman Glenn, Jr
Rodgers, Sheridan Joseph
Sherman, Larry Ray
Smart, G N Russell
Smith, William Novis, Jr
Soboczenski, Edward John
Sollott, Gilbert Paul
Stehly, David Norvin
Steward, Omar Waddington
Urenovitch, Joseph Victor
Waller, Francis Joseph
Weisfeld, Lewis Bernard
Willeford, Bennett Rufus, Jr

SOUTH CAROLINA
Adams, Richard Darwin
Cogswell, George Wallace
Culbertson, Edwin Charles
Evans, David Wesley
Lee, Burtrand Insung
Posey, Robert Giles
Reger, Daniel Lewis
Robinson, Myrtle Tonne
Robinson, Robert Earl
Shah, Hamish V
Stampf, Edward John, Jr
Tour, James Mitchell

TENNESSEE
Adcock, James Luther
Barnes, Craig Eliot
Fenyes, Joseph Gabriel Egon
Gibson, John Knight
Okrasinski, Stanley John
Summers, James Thomas
White, Alan Wayne

TEXAS
Bergbreiter, David Edward
Braterman, Paul S
Chen, Edward Chuck-Ming
Ciufolini, Marco A
Edmondson, Morris Stephen
Ekerdt, John Gilbert
Evans, Wayne Errol
Frisch, P Douglas
Hosmane, Narayan Sadashiv
Johnson, Malcolm Pratt
Jones, Paul Ronald
Kochhar, Rajindar Kumar
Laane, Jaan
Marchand, Alan Philip
Mills, Nancy Stewart
Newcomb, Martin
Pannell, Keith Howard
Singleton, David Michael
Whitmire, Kenton Herbert
Wisian-Neilson, Patty Joan

UTAH
Ernst, Richard Dale
Kodama, Goji
Li, Shin-Hwa
Richmond, Thomas G

VERMONT
Allen, Christopher Whitney
Jasinski, Jerry Peter

VIRGINIA
Averill, Bruce Alan
Bailey, Susan Goodman
Grimes, Russell Newell
Lodoen, Gary Arthur
Merola, Joseph Salvatore
Myers, William Howard
St Clair, Anne King
Sayala, Chhaya
Topich, Joseph

WASHINGTON
Bennett, Clifton Francis
Smith, Victor Herbert

WEST VIRGINIA
Breneman, William C
Schilling, Curtis Louis, Jr
Smith, James Allbee

WISCONSIN
Baetke, Edward A
Jache, Albert William
Rainville, David Paul
Reich, Hans Jurgen
Treichel, Paul Morgan, Jr

WYOMING
Branthaver, Jan Franklin
Sullivan, Brian Patrick

ALBERTA
Boeré, René Theodoor
Cowie, Martin
Graham, William Arthur Grover
Sanger, Alan Rodney

BRITISH COLUMBIA
James, Brian Robert
McAuley, Alexander
Mitchell, Reginald Harry
Orvig, Chris
Pomeroy, Roland Kenneth

NEWFOUNDLAND
Lucas, Colin Robert

NOVA SCOTIA
Bridgeo, William Alphonsus
Piorko, Adam M

ONTARIO
Alper, Howard
Baines, Kim Marie
Barrett, Peter Fowler
Brook, Adrian Gibbs
Buncel, Erwin
Carty, Arthur John
Drake, John Edward
Eliades, Theo I
Hemmings, Raymond Thomas
Holloway, Clive Edward
Lautens, Mark
Miller, Jack Martin
Payne, Nicholas Charles
Rempel, Garry Llewellyn
Walker, Alan

QUEBEC
Butler, Ian Sydney
Harrod, John Frank
Onyszchuk, Mario
Shaver, Alan Garnet
Theophanides, Theophile

OTHER COUNTRIES
Burford, Neil
Irgolic, Kurt Johann
Lemanski, Michael Francis

Synthetic Organic & Natural Products Chemistry

ALABAMA
Andrews, Russell S, Jr
Breithaupt, Lea Joseph, Jr
Czepiel, Thomas P
Hall, David Michael
Kiely, Donald Edward
Leslie, Thomas M
Parish, Edward James
Riley, Thomas N
Secrist, John Adair, III
Struck, Robert Frederick
Whorton, Rayburn Harlen

ARIZONA
Doubek, Dennis Lee
Herald, Cherry Lou
Herald, Delbert Leon, Jr
Hoffmann, Joseph John
Howells, Thomas Alfred
Hruby, Victor J

Jungermann, Eric
Kelley, Maurice Joseph
Moore, Ana Maria
Smith, Cecil Randolph, Jr
Timmermann, Barbara Nawalany

ARKANSAS
Fu, Peter Pi-Cheng
Glendening, Norman Willard

CALIFORNIA
Acton, Edward McIntosh
Baizer, Manuel M
Banigan, Thomas Franklin, Jr
Bird, Harold L(eslie), Jr
Boger, Dale L
Bohm, Howard A
Chan, David S
Chaubal, Madhukar Gajanan
Chen, Ming Sheng
Cole, Thomas Ernest
Conn, Eric Edward
Cooke, Ron Charles
Cooper, Geoffrey Kenneth
Correia, John Sidney
Corse, Joseph Walters
Craig, John Cymerman
Crews, Phillip
Crum, James Davidson
Daub, Guido William
Dev, Vasu
Diederich, Francois Nico
Djerassi, Carl
Dougherty, Dennis A
Erdman, Timothy Robert
Flath, Robert Arthur
Fung, Steven
Gaffield, William
Gall, Walter George
Gardner, Howard Shafer
Gaston, Lyle Kenneth
Ginsberg, Theodore
Goss, James Arthur
Grethe, Guenter
Harris, William Merl
Heather, James Brian
Henrick, Clive Arthur
Houk, Kendall N
Jurd, Leonard
Koepfli, Joseph B
Kryger, Roy George
Lee, Chi-Hang
Lewis, Robert Allen
Lin, Robert I-San
Lorance, Elmer Donald
McGee, Lawrence Ray
Matuszak, Charles A
Micheli, Robert Angelo
Mihailovski, Alexander
Millar, Jocelyn Grenville
Miller, Russell Bryan
Molyneux, Russell John
Mosher, Harry Stone
Musser, John H
Nambiar, Krishnan P
Nestor, John Joseph, Jr
Nitecki, Danute Emilija
Ogimachi, Naomi Neil
Ortiz de Montellano, Paul Richard
Overman, Larry Eugene
Pavlath, Attila Endre
Pierschbacher, Michael Dean
Pitt, Colin Geoffrey
Prakash, Surya G K
Preus, Martin William
Reardon, Edward Joseph, Jr
Rivier, Jean E F
Robins, Roland Kenith
Roitman, James Nathaniel
Sanchez, Robert A
Seapy, Dave Glenn
Shoolery, James Nelson
Siegel, Brock Martin
Sims, James Joseph
Singaram, Bakthan
Singleton, Vernon LeRoy
Snow, John Thomas
Spatz, David Mark
Spray, Clive Robert
Swidler, Ronald
Untch, Karl George
Vinogradoff, Anna Patricia
Vollhardt, K Peter C
Waiss, Anthony C, Jr
Welch, Steven Charles
Wender, Paul Anthony
Williams, John Wesley
Willson, Carlton Grant
Wong, Chi-Huey
Wright, Terry L
Zavarin, Eugene

COLORADO
Allen, Larry Milton
Barrett, Anthony Gerard Martin
Bartlett, William Rosebrough
Caruthers, Marvin Harry
DoMinh, Thap
Druelinger, Melvin L
Evans, David Lane
Fleming, Michael Paul
Gabel, Richard Allen
Stermitz, Frank
Walba, David Mark

CONNECTICUT
Arif, Shoaib
Baldwin, Ronald Martin
Brown, Keith Charles
Buchanan, Ronald Leslie
Church, Robert Fitz (Randolph)
Dirlam, John Philip
Doshan, Harold David
Dow, Robert Lee
Doyle, Terrence William
Forenza, Salvatore
Forssen, Eric Anton
Gordon, Philip N
Holland, Gerald Fagan
Juby, Peter Frederick
Kavarnos, George James
Kluender, Harold Clinton
Kotick, Michael Paul
Kuck, Julius Anson
Luh, Yuhshi
Marfat, Anthony
Melvin, Lawrence Sherman, Jr
Perry, Clark William
Proverb, Robert Joseph
Rennhard, Hans Heinrich
Rother, Ana
Schaaf, Thomas Ken
Schile, Richard Douglas
Shapiro, Warren Barry
Sieckhaus, John Francis
Tishler, Max
Volkmann, Robert Alfred
Wasser, Richard Barkman
Wright, Herbert Fessenden
Wuskell, Joseph P
Ziegler, Theresa Frances

DELAWARE
Alderson, Thomas
Bare, Thomas M
Burton, Lester Percy Joseph
Chorvat, Robert John
Class, Jay Bernard
Confalone, Pat N
Crow, Edwin Lee
DeGrado, William F
Dubas, Lawrence Francis
Ehrich, F(elix) Frederick
Freerksen, Robert Wayne
Fukunaga, Tadamichi
Hayek, Mason
Holyoke, Caleb William, Jr
Jenner, Edward L
Keim, Gerald Inman
Kettner, Charles Adrian
Krespan, Carl George
Laganis, Evan Dean
Loomis, Gary Lee
Monagle, Daniel J
Pratt, Herbert T
Pretka, John E
Putnam, Stearns Tyler
Reap, James John
Rogers, Donald B
Rondestvedt, Christian Scriver, Jr
Scalfarotto, Robert Emil
Schlecht, Matthew Fred
Taber, Douglass Fleming
Targett, Nancy McKeever
Thayer, Chester Arthur
Twelves, Robert Ralph
Wang, Chia-Lin Jeffrey
Webster, Owen Wright
Willer, Rodney Lee
Wuonola, Mark Arvid

DISTRICT OF COLUMBIA
Baker, Charles Wesley
Herz, Josef Edward
Hudrlik, Anne Marie
Hudrlik, Paul Frederick
Klayman, Daniel Leslie
Page, Samuel William
Ponaras, Anthony A
Sarmiento, Rafael Apolinar
Senftle, Frank Edward
Torrence, Paul Frederick
Wheeler, James William, Jr

FLORIDA
Battiste, Merle Andrew
Bledsoe, James O, Jr
Blossey, Erich Carl
Derfer, John Mentzer
Herz, Werner
Hull, Harry H
Kagan, Benjamin
Krafft, Marie Elizabeth
Lee, Henry Joung
Miles, Delbert Howard
Moskowitz, Mark Lewis
Most, David S
Ohrn, Nils Yngve
Owen, Terence Cunliffe
Pop, Emil
Porter, Lee Albert
Rokach, Joshua
Romeo, John Thomas
Russell, James
Schwarting, Arthur Ernest
Shapiro, Jeffrey Paul
Sisler, Harry Hall
Spencer, John Lawrence
Vander Meer, Robert Kenneth

Williams, Norris Hagan

GEORGIA
Bayer, Charlene Warres
Chiu, Kirts C
Chortyk, Orestes Timothy
Chu, Chung K
Clark, Charles Kittredge
Espelie, Karl Edward
Hill, Richard Keith
Iacobucci, Guillermo Arturo
Kiefer, Charles Randolph
Knopka, W N
Leffingwell, John C
Malcolm, Earl Walter
Nes, William David
Pelletier, S William
Polk, Malcolm Benny
Schramm, Lee Clyde
Sweeny, James Gilbert
Tolbert, Laren Malcolm
Tyler, Jean Mary
Walker, William Comstock
Woodford, James

HAWAII
Patterson, Gregory Matthew Leon
Scheuer, Paul Josef
Tius, Marcus Antonius

IDAHO
Lambuth, Alan Letcher
Natale, Nicholas Robert

ILLINOIS
Abbott, Thomas Paul
Adlof, Richard Otto
Arendsen, David Lloyd
Armstrong, P(aul) Douglas
Babler, James Harold
Bernhardt, Randal Jay
Brooks, Dee W
Cengel, John Anthony
Chu, Daniel Tim-Wo
Cole, Wayne
Cordell, Geoffrey Alan
Curtis, Veronica Anne
DeBoer, Edward Dale
Dellaria, Joseph Fred, Jr
Djuric, Stevan Wakefield
Drengler, Keith Allan
Eachus, Alan Campbell
Elliott, William John
Eskins, Kenneth
Estes, Timothy King
Field, Kurt William
Ford, Susan Heim
Friedman, Robert Bernard
Garland, Robert Bruce
Giori, Claudio
Grotefend, Alan Charles
Hofreiter, Bernard T
Hsu, Charles Fu-Jen
Jiu, James
Johnson, John Harold
Katzenellenbogen, John Albert
Kerdesky, Francis A J
Kerkman, Daniel Joseph
Klein, Larry L
Klimstra, Paul D
Lewis, Morton
Lucci, Robert Dominick
Maher, George Garrison
Miller, Fredric N
Mohan, Prem
Pariza, Richard James
Perun, Thomas John
Piatak, David Michael
Powell, Richard Grant
Rakoff, Henry
Richmond, James M
Rinehart, Kenneth Lloyd
Rosenberg, Saul Howard
Rosenbrook, William, Jr
Shotwell, Odette Louise
Silverman, Richard Bruce
Tortorello, Anthony Joseph
Traxler, James Theodore
Wallen, Lowell Lawrence
Wang, Gary T
Weier, Richard Mathias
Weston, Arthur Walter
Witczak, Zbigniew J
Yokelson, Howard Bruce

INDIANA
Barnett, Charles Jackson
Becker, Elizabeth Ann (White)
Biddlecom, William Gerard
Buchanan, Russell Allen
Bucholtz, Dennis Lee
Carmack, Marvin
Chang, Ching-Jer
Cooper, Robin D G
Corey, Paul Frederick
Dreikorn, Barry Allen
Fidelle, Thomas Patrick
Fife, Wilmer Krafft
Gresham, Robert Marion
Gutowski, Gerald Edward
Hamill, Robert L
Han, Choong Yol
Harper, Richard Waltz
Helquist, Paul M

Synthetic Organic & Natural Products Chemistry (cont)

Kjonaas, Richard A
Kossoy, Aaron David
McLaughlin, Jerry Loren
Marconi, Gary G
Negishi, Ei-ichi
O'Doherty, George Oliver-Plunkett
Patwardhan, Bhalchandra H
Rapkin, Myron Colman
Robertson, Donald Edwin
Rowe, James Lincoln
Shiner, Vernon Jack, Jr
Stevenson, William Campbell
Thompson, Gerald Lee
Wagner, Eugene Ross
Wang, Jin-Liang
Wildfeuer, Marvin Emanuel
Yordy, John David

IOWA
Carew, David P
Downing, Donald Talbot
Hampton, David Clark
Kraus, George Andrew
Kraus, Kenneth Wayne
Linhardt, Robert John
Nair, Vasu
Wiemer, David F

KANSAS
Beecham, Curtis Michael
Borchardt, Ronald T
Burgstahler, Albert William
Cheng, Chia-Chung
Duncan, William Perry
Hua, Duy Huu
Johnson, David Barton
McCullough, Elizabeth Ann
Marx, Michael
Mitscher, Lester Allen
Wiley, James C, Jr

KENTUCKY
Crooks, Peter Anthony
Kennedy, John Elmo, Jr
Lauterbach, John Harvey
Luzzio, Frederick Anthony
McGinness, William George, III
Meier, Mark Stephan
Slocum, Donald Warren
Taylor, Kenneth Grant
Wong, John Lui

LOUISIANA
Bauer, Dennis Paul
Bayer, Arthur Craig
Connick, William Joseph, Jr
Ensley, Harry Eugene
Fischer, Nikolaus Hartmut
Frank, Arlen W(alker)
Gallo, August Anthony
Gonzales, Elwood John
Harper, Robert John, Jr
Shin, Kju Hi
Smith, Grant Warren, II
Somsen, Roger Alan
Starrett, Richmond Mullins
Stowell, John Charles
Swenson, David Harold
Theriot, Kevin Jude
Vail, Sidney Lee
Welch, Clark Moore
Zentner, Thomas Glenn

MAINE
Eames, Arnold C
Green, Brian
McKerns, Kenneth (Wilshire)
Pierce, Arleen Cecilia
Renn, Donald Walter
Spencer, Claude Franklin

MARYLAND
Aldrich, Jeffrey Richard
Attaway, David Henry
Bauknight, Charles William, Jr
Brady, Robert Frederick, Jr
Burrows, Elizabeth Parker
Carter, John Paul, Jr
Cavagna, Giancarlo Antonio
Chiu, Tony Man-Kuen
Christensen, Richard G
Cragg, Gordon Mitchell
Engle, Robert Rufus
Falci, Kenneth Joseph
Fales, Henry Marshall
Feldlaufer, Mark Francis
Feyns, Liviu Valentin
Galetto, William George
Guruswamy, Vinodhini
Henery-Logan, Kenneth Robert
Horak, Vaclav
Hosmane, Ramachandra Sadashiv
Inscoe, May Nilson
Jachimowicz, Felek
Jacobson, Kenneth Alan
Jacobson, Martin
James, John Cary
Jones, Tappey Hughes
Klein, Michael
Lusby, William Robert

McShefferty, John
Marquez, Victor Esteban
Massie, Samuel Proctor
Miller, Steve P F
Misra, Renuka
Nemec, Josef
Otani, Theodore Toshiro
Paull, Kenneth Dywain
Poochikian, Guiragos K
Reynolds, Kevin A
Rice, Kenner Cralle
Roller, Peter Paul
Saunders, James Allen
Schwartz, Paul
Scovill, John Paul
Snader, Kenneth Means
Vandegaer, Jan Edmond
Warthen, John David, Jr
Waters, James Augustus
Weser, Don Benton
Whaley, Wilson Monroe
Yoho, Clayton W
Zeiger, William Nathaniel

MASSACHUSETTS
Ahern, David George
Ajami, Alfred Michel
Bide, Martin John
Carpino, Louis A
Chase, Arleen Ruth
Clapp, Richard Crowell
Cobb, R M Karapetoff
Eby, Robert Newcomer
Giarrusso, Frederick Frank
Hendrickson, James Briggs
Hodgins, George Raymond
Holick, Michael Francis
Holick, Sally Ann
Hurd, Richard Nelson
Idelson, Martin
Kelley, Charles Joseph
Kishi, Yoshito
Klanfer, Karl
Krull, Ira Stanley
Leary, John Dennis
Naidus, Harold
O'Rell, Dennis Dee
Raffauf, Robert Francis
Rosenfeld, Stuart Michael
Ruelius, Hans Winfried
Snider, Barry B
Soffer, Milton David
Sripada, Pavanaram Kameswara
Stinchfield, Carleton Paul
Sullivan, Edward Augustine
Takman, Bertil Herbert
Warner, John Charles
Williams, John Russell

MICHIGAN
Aaron, Charles Sidney
Axen, Udo Friedrich
Belmont, Daniel Thomas
Berman, Ellen Myra
Blankley, Clifton John
Clark, Peter David
Culbertson, Townley Payne
Deering, Carl F
Domagala, John Michael
Elrod, David Wayne
Elslager, Edward Faith
Frade, Peter Daniel
Greminger, George King, Jr
Hart, Harold
Herrinton, Paul Matthew
Hester, Jackson Boling, Jr
Huang, Che C
Huber, Joel E
Kiely, John Steven
Krueger, Robert John
Kuo, Mingshang
Laks, Peter Edward
LeBel, Norman Albert
Lipton, Michael Forrester
Logue, Marshall Woford
Marshall, Charles Wheeler
Moos, Walter Hamilton
Nader, Bassam Salim
Nagler, Robert Carlton
Nestrick, Terry John
Otto, Charlotte A
Pearlman, Bruce A
Petersen, Quentin Richard
Pfeiffer, Curtis Dudley
Putnam, Alan R
Sacks, Clifford Eugene
Schramm, John Gilbert
Skorcz, Joseph Anthony
Suggs, William Terry
Tanis, Steven Paul
Thomas, Richard Charles
Verseput, Herman Ward
Woo, P(eter) W(ing) K(ee)
Wormser, Henry C
Wuts, Peter G M

MINNESOTA
Barany, George
Carlson, Janet Lynn
Cheng, Hwei-Hsien
Crooks, Stephen Lawrence
Gray, Gary Ronald
Katz, Morris Howard
Lucast, Donald Hurrell

Noland, Wayland Evan
Pletcher, Wayne Albert
Schultz, Robert John
Spessard, Gary Oliver
Throckmorton, James Rodney

MISSISSIPPI
Cibula, William Ganley
Hufford, Charles David
McChesney, James Dewey
Panetta, Charles Anthony
Sindelar, Robert D
Williamson, John S

MISSOURI
Ahmed, Moghisuddin
Anderson, David W, Jr
Asrar, Jawed
Baker, Joseph Willard
Cole, Douglas L
Coudron, Thomas A
Delaware, Dana Lewis
Durley, Richard Charles
Elliott, William H
Gustafson, Mark Edward
Koenig, Karl E
Ku, Audrey Yeh
McCormick, John Pauling
Plimmer, Jack Reynolds
Popp, Frank Donald
Schleppnik, Alfred Adolf
Sikorski, James Alan
Slocombe, Robert Jackson
Wasson, Richard Lee

MONTANA
Richards, Geoffrey Norman

NEBRASKA
Gold, Barry Irwin
Takemura, Kaz H(orace)
Wheeler, Desmond Michael Sherlock

NEVADA
Lightner, David A

NEW HAMPSHIRE
Danforth, Raymond Hewes
Gribble, Gordon W
Neil, Thomas C
Woodbury, Richard Paul

NEW JERSEY
Abou-Gharbia, Magid
Addor, Roger Williams
Allen, Richard Charles
Andrade, John Robert
Armbruster, David Charles
Augustine, Robert Leo
Bamberger, Curt
Batcho, Andrew David
Bathala, Mohinder S
Bernholz, William Francis
Bertz, Steven Howard
Bodanszky, Miklos
Bose, Ajay Kumar
Brand, William Wayne
Carter, Richard John
Cavender, Patricia Lee
Chan, Ka-Kong
Chiao, Wen Bin
Christensen, Burton Grant
Cohen, Noal
Crawley, Lantz Stephen
Delaney, Edward Joseph
Draper, Richard William
Duhl-Emswiler, Barbara Ann
Durette, Philippe Lionel
Evans, Ralph H, Jr
Evers, William John
Fahey, John Leonard
Fairchild, Edward H
Finkelstein, Jacob
Fobare, William Floyd
Focella, Antonino
Fong, Jones W
Fost, Dennis L
Gabriel, Robert
Garfinkel, Harmon Mark
Gordon, Eric Michael
Grabowski, Edward Joseph John
Grosso, John A
Hardtmann, Goetz E
Haslanger, Martin Frederick
Hatch, Charles Eldridge, III
Head, William Francis, Jr
Heck, James Virgil
Hensens, Otto Derk
Jensen, Norman P
Jirkovsky, Ivo
Joseph, John Mundancheril
Kalm, Max John
Kanojia, Ramesh Maganlal
Kasparek, Stanley Vaclav
Kees, Kenneth Lewis
Klaubert, Dieter Heinz
Klemann, Lawrence Paul
Kroll, Emanuel
Ladner, David William
Lam, Yiu-Kuen Tony
Lavanish, Jerome Michael
LaVoie, Edmond J
Liebman, Arnold Alvin
Liebowitz, Stephen Marc

Liesch, Jerrold Michael
Linn, Bruce Oscar
Little, Edwin Demetrius
Liu, Yu-Ying
Lorey, Frank William
MacConnell, John Griffith
MacCoss, Malcolm
Maehr, Hubert
Mallams, Alan Keith
Marmor, Robert Samuel
Martin, Lawrence Leo
Mathew, Chempolil Thomas
Mertel, Holly Edgar
Michel, Gerd Wilhelm
Mikes, John Andrew
Miller, Thomas William
Montgomery, Richard Millar
Naipawer, Richard Edward
Ocain, Timothy Donald
Olson, Gary Lee
Palmere, Raymond M
Parker, William Lawrence
Pedersen, Charles John
Putter, Irving
Ramsden, Hugh Edwin
Ramsey, Arthur Albert
Rasmusson, Gary Henry
Reider, Paul Joseph
Reilly, Eugene Patrick
Rinehart, Jay Kent
Roland, Dennis Michael
Rosegay, Avery
Rossi, Robert Daniel
Rowin, Gerald L
Schaffner, Carl Paul
Schwartz, Herbert
Schwarz, John Samuel Paul
Sekula, Bernard Charles
Sieh, David Henry
Skotnicki, Jerauld S
Slade, Joel S
Slagel, Robert Clayton
Slusarchyk, William Allen
Sodano, Charles Stanley
Stein, Reinhardt P
Strike, Donald Peter
Sugathan, Kenneth Kochappan
Tilley, Jefferson Wright
VanMiddlesworth, Frank L
Vlattas, Isidoros
Wanat, Stanley Frank
Watthey, Jeffrey William Herbert
Wildman, George Thomas
Wilkinson, Raymond George
Williams, Thomas Henry
Zalay, Andrew W(illiam)
Ziegler, John Benjamin

NEW MEXICO
Cheavens, Thomas Henry
Liepins, Raimond
Ott, Donald George
Smith, Wayne Howard
Zeigler, John Martin

NEW YORK
Agosta, William Carleton
Alks, Vitauts
Ashen, Philip
Bardos, Thomas Joseph
Baron, Arthur L
Bell, Thomas Wayne
Berdahl, Donald Richard
Bitha, Panayota
Bittman, Robert
Boeckman, Robert K, Jr
Boettger, Susan D
Borowitz, Irving Julius
Box, Vernon G S
Brust, David Philip
Buchel, Johannes A
Burrows, Cynthia Jane
Christenson, Philip A
Clarke, Robert LaGrone
Curran, William Vincent
Daves, Glenn Doyle, Jr
Dick, William Edwin, Jr
Dickinson, William Borden
Dolak, Terence Martin
Edasery, James P
Feeny, Paul Patrick
Frechet, Jean M J
Fuerst, Adolph
Ganem, Bruce
Gershon, Herman
Gletsos, Constantine
Goldman, Norman L
Grina, Larry Dale
Gurel, Demet
Haase, Jan Raymond
Hahn, Roger C
Hall, Frederick Keith
Happel, John
James, Daniel Shaw
Johnson, Bruce Fletcher
Jones, Clive Gareth
Kende, Andrew S
Khan, Jamil Akber
Klingele, Harold Otto
Korytnyk, Walter
Kronstein, Max
Lai, Yu-Chin
LaLonde, Robert Thomas
Landon, Shayne J

Lee, Ving Jick
Leopold, Bengt
Loev, Bernard
McCormick, J Robert D
McMurry, John Edward
Malpass, Dennis B
Mansfield, Kevin Thomas
Meinwald, Jerrold
Meinwald, Yvonne Chu
Misra, Anand Lal
Murdock, Keith Chadwick
Murphy, Daniel Barker
Nair, Ramachandran Mukundalayam
   Sivarama
Nakanishi, Koji
Ojima, Iwao
Oliver, Gene Leech
Petersen, Ingo Hans
Premuzic, Eugene T
Prestwich, Glenn Downes
Pucknat, John Godfrey
Reich, Marvin Fred
Rich, Arthur Gilbert
Rosi, David
Salley, John Jones, Jr
Schultz, Arthur George
Schulz, John Hampshire
Schwab, Linda Sue
Silverstein, Robert Milton
Singh, Baldev
Siuta, Gerald Joseph
Solo, Alan Jere
Soloway, Harold
Stark, John Howard
Steinberg, David H
Tobkes, Martin
Trust, Ronald I
Van Verth, James Edward
Weisler, Leonard
Weiss, Martin Joseph
Zimmerman, Mary Prislopski

NORTH CAROLINA
Barborak, James Carl
Bassett, Joseph Yarnall, Jr
Bell, Jimmy Holt
Beverung, Warren Neil, Jr
Bigham, Eric Cleveland
Chilton, William Scott
Crimmins, Michael Thomas
Culberson, Chicita Frances
Davis, Roman
Dixit, Ajit Suresh
Dufresne, Richard Frederick
Edwards, Ben E
Engle, Thomas William
Fleischer, Thomas B
Fraser-Reid, Bertram Oliver
Gains, Lawrence Howard
Hooper, Irving R
Huang, Jamin
Hudson, Albert Berry
Jiang, Jack Bau-Chien
Kelsey, John Edward
Landes, Chester Grey
Lee, Kuo-Hsiung
Long, Robert Allen
Manning, David Treadway
Martin, Gary Edwin
Martin, Ned Harold
Mehta, Nariman Bomanshaw
Price, Kenneth Elbert
Ranieri, Richard Leo
Rivers, Jessie Markert
Russell, Henry Franklin
Schickedantz, Paul David
Schwindeman, James Anthony
Wallace, James William, Jr
Wani, Mansukhlal Chhaganlal
Woosley, Royce Stanley
Wyrick, Steven Dale

NORTH DAKOTA
Bergstrom, Donald Eugene
Donnelly, Brendan James
Jasperse, Craig Peter
Khalil, Shoukry Khalil Wahba

OHIO
Ager, David John
Awad, Albert T
Belletire, John Lewis
Bertelson, Robert Calvin
Brain, Devin King
Brillhart, Donald D
Buck, Keith Taylor
Carmichael, Wayne William
Darling, Stephen Deziel
Davis, Gerald Titus
Delvigs, Peter
Dolfini, Joseph E
Erman, William F
Fall, Harry H
Fechter, Robert Bernard
Fentiman, Allison Foulds, Jr
Gibson, Thomas William
Greenlee, Kenneth William
Holbert, Gene W(arwick)
Horne, Samuel Emmett, Jr
Ketcha, Daniel Michael
Kinstle, Thomas Herbert
Kurtz, David Williams
Letton, James Carey
Lollar, Robert Miller

Meinert, Walter Theodore
Merritt, Robert Edward
Morgan, Alan Raymond
Ostrum, G(eorge) Kenneth
Peterson, Robert C
Sampson, Paul
Sheets, George Henkle
Smith, Douglas Alan
Stevenson, Don R
Strobel, Rudolf G K
Taschner, Michael J
Throckmorton, Peter E
Turkel, Rickey M(artin)
Turnbull, Kenneth
Voss, Anne Coble
Walker, Thomas Eugene
Williamson, Frederick Dale
Ziller, Stephen A, Jr

OKLAHOMA
Eisenbraun, Edmund Julius
Nicholas, Kenneth M
Schmitz, Francis John

OREGON
Badger, Rodney Allan
Bublitz, Walter John, Jr
Dolby, Lloyd Jay
Duncan, James Alan
Gould, Steven James
Hauser, Frank Marion
Hudak, Norman John
Kelsey, Rick Guy
Laver, Murray Lane
McClard, Ronald Wayne
Mason, Robert Thomas
Shilling, Wilbur Leo
Szalecki, Wojciech Josef

PENNSYLVANIA
Abate, Kenneth
Anspon, Harry Davis
Beavers, Ellington McHenry
Berges, David Alan
Bloomer, James L
Bock, Mark Gary
Box, Larry
Brady, Stephen Francis
Canter, Neil M
Carter, Linda G
Chang, Charles Hung
Crissman, Jack Kenneth, Jr
Curran, Dennis Patrick
Erb, Dennis J
Erhan, Semih M
Evans, Ben Edward
Ford, Michael Edward
Freidinger, Roger Merlin
Gibson, Raymond Edward
Giuliano, Robert Michael
Godfrey, John Carl
Grey, Roger Allen
Gunther, Wolfgang Hans Heinrich
Gustine, David Lawrence
Hamel, Coleman Rodney
Heys, John Richard
Hicks, Kevin B
Hoffman, Jacob Matthew, Jr
Holden, Kenneth George
Hurt, Susan Schilt
Johnson, Richard William
Johnson, Wayne Orrin
Kauffman, Joel Mervin
Kowalski, Conrad John
Krow, Grant Reese
Kruse, Lawrence Ivan
Lange, Barry Clifford
Lapidus, Milton
Larson, Gerald Louis
Larson, Kenneth Curtis
Lavoie, Alvin Charles
Lesko, Patricia Marie
Leston, Gerd
Linfield, Warner Max
Lopez, R C Gerald
McCurry, Patrick Matthew, Jr
Maryanoff, Cynthia Anne Milewski
Mokotoff, Michael
Monsimer, Harold Gene
Morse, Lewis David
Nelson, George Leonard
Noceti, Richard Paul
Nutt, Ruth Foelsche
Osei-Gyimah, Peter
Prentiss, William Case
Press, Jeffery Bruce
Reingold, Iver David
Reitz, Allen Bernard
Robinson, Charles Albert
Ross, Stephen T
Schiff, Paul L, Jr
Shamma, Maurice
Sheeley, Richard Moats
Shepard, Kenneth LeRoy
Singer, Alan G
Soriano, David S
Stevens, Travis Edward
Thelman, John Patrick
Thoman, Charles James
Thompson, Wayne Julius
Turner, Andrew B
Veber, Daniel Frank
Wade, Peter Allen
Waller, Francis Joseph

Weinstock, Joseph
Weiss, James Allyn
Wempe, Lawrence Kyran
Werny, Frank
Williams, John Roderick
Wineholt, Robert Leese
Zajac, Walter William, Jr

RHODE ISLAND
Cane, David Earl
Shimizu, Yuzuru
Slater, Schuyler G
Williard, Paul Gregory

SOUTH CAROLINA
Bailey, Carl Williams, III
Dieter, Richard Karl
Haskell, Theodore Herbert, Jr
Hon, David Nyok-Sai
Martin, Tellis Alexander
Sarjeant, Peter Thomson
Scruggs, Jack G
Shah, Hamish V
Smith, William Edmond
Sproul, Gordon Duane
Strauss, Roger William
Tour, James Mitchell
Ward, Benjamin F, Jr
Weinstock, Leonard M
Wise, John Thomas

SOUTH DAKOTA
Gaines, Jack Raymond
Lewis, David Edwin

TENNESSEE
Adcock, James Luther
Bennett, Word Brown, Jr
Buchman, Russell
Clemens, Robert Jay
Cliffton, Michael Duane
Commerford, John D
Germroth, Ted Calvin
Ho, Patience Ching-Ru
Kashdan, David Stuart
Kuo, Chung-Ming
McConnell, Richard Leon
Olsen, John Stuart
Penner, Hellmut Philip
Rash, Fred Howard
Sachleben, Richard Alan
Schell, Fred Martin
Smith, Howard E
Spadafino, Leonard Peter
Waddell, Thomas Groth
White, Alan Wayne
Yau, Cheuk Chung

TEXAS
Alam, Maktoob
Ball, M Isabel
Bartsch, Richard Allen
Beier, Ross Carlton
Bergbreiter, David Edward
Ciufolini, Marco A
Dahm, Karl Heinz
Ellison, Robert Hardy
Ervin, Hollis Edward
Gilbert, John Carl
Hillery, Herbert Vincent
Jaszberenyi, Joseph C
Lin, Tz-Hong
Lyle, Robert Edward, Jr
Marchand, Alan Philip
Nachman, Ronald James
Newcomb, Martin
Parry, Ronald John
Reinecke, Manfred Gordon
Roehrig, Gerald Ralph
Sand, Ralph E
Schimelpfenig, Clarence William
Scott, Alastair Ian
Skarlos, Leonidas
Sonntag, Norman Oscar Victor
Stallings, James Cameron
Stipanovic, Robert Douglas
Teng, Jon Ie
Tindall, Donald R
Volgenau, Lewis
Watson, William Harold, Jr
Wheeler, Michael Hugh

UTAH
Balandrin, Manuel F
Broadbent, Hyrum Smith
Epstein, William Warren
Keeler, Richard Fairbanks
Petryka, Zbyslaw Jan
Robins, Morris Joseph
Segelman, Florence H
Walker, Edward Bell Mar

VERMONT
Hussak, Robert Edward

VIRGINIA
Beyad, Mohammed Hossain
Daylor, Francis Lawrence, Jr
Edwards, William Brundige, III
Gager, Forrest Lee, Jr
Harris, Paul Robert
Harvey, William Ross
Hausermann, Max
Houghton, Kenneth Sinclair, Sr

Hudlicky, Milos
Hudson, Frederick Mitchell
Hussamy, Samir
Kingston, David George Ian
Kinsley, Homan Benjamin, Jr
Majewski, Theodore E
Rodig, Oscar Rudolf
Rosenberg, Murray David
Sprinkle, Robert Shields, III
Teng, Lina Chen
Whitney, George Stephen

WASHINGTON
Anderson, Charles Dean
Bennett, Clifton Francis
Hopkins, Paul Brink
Ingalshe, David Weeden
Maranville, Lawrence Frank
Mulvey, Dennis Michael
Nelson, Randall Bruce
Sarkanen, Kyosti Vilho
Tostevin, James Earle
Trotter, Patrick Casey
Wade, Leroy Grover, Jr
Wilbur, Daniel Scott
Wollwage, Paul Carl

WEST VIRGINIA
Giza, Chester Anthony
Hubbard, John Lewis
Hussey, Edward Walter
Koleske, Joseph Victor
Ode, Richard Herman

WISCONSIN
Bernardin, Leo J
Burke, Steven Douglas
Chenier, Philip John
Cook, James Minton
Dugal, Hardev Singh
Holland, Dewey G
Hollenberg, David Henry
Hornemann, Ulfert
Klopotek, David L
Kocurek, Michael Joseph
Lin, Stephen Y
Manning, James Harvey
Park, Robert William
Pearl, Irwin Albert
Ratliff, Francis Tenney
Sommers, Raymond A
Wollwage, John Carl
Zinkel, Duane Forst

WYOMING
Branthaver, Jan Franklin
Seese, William Shober

PUERTO RICO
ALZERRECA, ARNALDO
Carrasquillo, Arnaldo
Stephens, William Powell

ALBERTA
Bachelor, Frank William
Liu, Hsing-Jang
Moore, Graham John
Singh, Bhagirath

BRITISH COLUMBIA
Becker, Edward Samuel
Boehm, Robert Louis
Goodeve, Allan McCoy
Hocking, Martin Blake
Liao, Ping-huang
Manville, John Fieve
Paszner, Laszlo
Procter, Alan Robert
Shaw, A(lexander) J(ohn)
Underhill, Edward Wesley
Yunker, Mark Bernard

MANITOBA
Chow, Donna Arlene
Hunter, Norman Robert
McKinnon, David M

NEW BRUNSWICK
Strunz, G(eorge) M(artin)

NEWFOUNDLAND
Georghiou, Paris Elias

NOVA SCOTIA
Chandler, Reginald Frank
Piorko, Adam M
Vining, Leo Charles
White, Robert Lester

ONTARIO
Aspinall, Gerald Oliver
Baer, Hans Helmut
Baines, Kim Marie
Bannard, Robert Alexander Brock
Bersohn, Malcolm
Birnbaum, George I
Butler, Douglas Neve
Collins, Frank William, Jr
Cory, Robert Mackenzie
Court, William Arthur
Darby, Nicholas
Fallis, Alexander Graham
Greenhalgh, Roy
Gupta, Virendra Nath

Synthetic Organic & Natural Products Chemistry (cont)

Hay, George William
Histed, John Allan
Hopkins, Clarence Yardley
Lange, Gordon Lloyd
Lautens, Mark
Leznoff, Clifford Clark
Lyne, Leonard Murray, Sr
MacLean, David Bailey
Morand, Peter
Pare, J R Jocelyn
Ratnayake, Walisundera Mudiyanselage Nimal
Rodrigo, Russell Godfrey
Saxton, William Reginald
Sowa, Walter
Starratt, Alvin Neil
Stavric, Stanislava
Still, Ian William James
Stoessl, Albert
Szarek, Walter Anthony
Vijay, Hari Mohan
Warnhoff, Edgar William
Weedon, Alan Charles
Wightman, Robert Harlan
Yates, Peter

QUEBEC
Alince, Bohumil
Atkinson, Joseph George
Belanger, Patrice Charles
Burnell, Robert H
Chenevert, Robert (Bernard)
Darling, Graham Davidson
Engel, Charles Robert
Escher, Emanuel
Feldman, Dorel
Garneau, Francois Xavier
Goldstein, Sandu M
Hamlet, Zacharias
Kokta, Bohuslav Vaclav
Kubes, George Jiri
Perron, Yvon G
Regoli, Domenico
Scallan, Anthony Michael
Schiller, Peter Wilhelm
Van Calsteren, Marie-Rose
Yeats, Ronald Bradshaw

SASKATCHEWAN
Abrams, Suzanne Roberta
Gupta, Vidya Sagar
Mikle, Janos J

OTHER COUNTRIES
Barnes, Roderick Arthur
Goldstein, Melvin Joseph
Gottlieb, Otto Richard
Graber, Robert Philip
Hassner, Alfred
Hummel, Hans Eckhardt
Kent, Stephen Brian Henry
Kurobane, Itsuo
Kuttab, Simon Hanna
Malherbe, Roger F
Mancera, Octavio
Mors, Walter B
Nair, Vijay
Shapiro, Stuart
Trofimov, Boris Alexandrovich
Tsang, Joseph Chiao-Liang

Theoretical Chemistry

ALABAMA
Baird, James Kern
Visscher, Pieter Bernard

ARIZONA
Balasubramanian, Krishnan
Golden, Sidney
Lichtenberger, Dennis Lee
Nordstrom, Brian Hoyt
Salzman, William Ronald

ARKANSAS
Pulay, Peter

CALIFORNIA
Alder, Berni Julian
Allen, Thomas Lofton
Blumstein, Carl Joseph
Brown, Nancy J
Carter, Emily Ann
Chandler, David
Chen, Joseph Cheng Yih
Chen, Yi-Der
Cook, Robert Crossland
Creighton, John Rogers
Dawson, Kenneth Adrian
Dill, James David
Dows, David Alan
Fetzer, John Charles
Fink, William Henry
Goddard, William Andrew, III
Goldstein, Elisheva
Harris, Robert A
Hazi, Andrew Udvar
Hempel, Judith Cato
Herman, Zelek Seymour

Horsley, John Anthony
Houk, Kendall N
Hsiung, Chi-Hua Wu
Huestis, David Lee
Keizer, Joel Edward
Kollman, Peter Andrew
Komornicki, Andrew
Konowalow, Daniel Dimitri
Lester, William Alexander, Jr
Loew, Gilda Harris
McKoy, Basil Vincent
McQuarrie, Donald Allan
Marcus, Rudolph Arthur
Matlow, Sheldon Leo
Medrud, Ronald Curtis
Metiu, Horia I
Miller, James Angus
Miller, William Hughes
Morrison, Harry Lee
Oakes, John Morgan
Olafson, Barry D
Palke, William England
Payne, Philip Warren
Reiss, Howard
Ritchie, Adam Burke
Rosenkrantz, Marcy Ellen
Segal, Gerald A
Shuler, Kurt Egon
Shull, Harrison
Simon, John Douglas
Skolnick, Jeffrey
Stephens, Jeffrey Alan
Stephens, Philip J
Strauss, Herbert L
Syage, Jack A
Whaley, Katharine Birgitta
Wheeler, John C
Whetten, Robert Lloyd
Wilkins, Roger Lawrence
Winter, Nicholas Wilhelm
Wolf, Kathleen A
Wolfsberg, Max
Woodson, John Hodges
Wulfman, Carl E

COLORADO
Ely, James Frank
Fixman, Marshall
Ladanyi, Branka Maria
Liao, Martha
Mansfield, Roger Leo
Parson, Robert Paul
Sanderson, Robert Thomas
Stewart, James Joseph Patrick
Weliky, Irving
Williams, Denis R

CONNECTICUT
Anderson, Carol Patricia
Beveridge, David L
Jorgensen, William L
Kavarnos, George James
Keyes, Thomas Francis
Michels, H(orace) Harvey
Petersson, George A
Sanders, William Albert
Sinanoglu, Oktay

DELAWARE
Doren, Douglas James
Foss, Robert Paul
Leung, Chung Ngoc
Morgan, John Davis, III
Pretka, John E
Shipman, Lester Lynn
Stewart, Charles Winfield, Sr
Szalewicz, Krzysztof
Wasserman, Edel
Wasserman, Zelda Rakowitz
Wiley, Douglas Walker

DISTRICT OF COLUMBIA
Chiu, Lue-Yung Chow
Kern, Charles William
Kertesz, Miklos
Ramaker, David Ellis
Rubin, Robert Joshua

FLORIDA
Bartlett, Rodney Joseph
Bash, Paul Anthony
Baum, J(ames) Clayton
Boehnke, David Neal
Cioslowski, Jerzy
Dufty, James W
Linder, Bruno
Micha, David Allan
Person, Willis Bagley
Rhodes, William Clifford
Zerner, Michael Charles

GEORGIA
Barbas, John Theophani
Bowman, Joel Mark
Kozak, John Joseph
Rouvray, Dennis Henry
Schaefer, Henry Frederick, III
Uzer, Ahmet Turgay

IDAHO
Shirts, Randall Brent

ILLINOIS
Bouman, Thomas David

Curtiss, Larry Alan
Freed, Karl F
Gislason, Eric Arni
Goodman, Gordon Louis
House, James Evan, Jr
Kim, Ki Hwan
Lykos, Peter George
Martin, Yvonne Connolly
O'Donnell, Terence J
Pariza, Richard James
Prais, Michael Gene
Sabelli, Nora Hojvat
Schatz, George Chappell
Scheiner, Steve
Snyder, Gary James
Tyrrell, James
Wagner, Albert Fordyce
Wilson, Robert Steven
Woehler, Scott Edwin
Wolynes, Peter Guy

INDIANA
Aprison, Morris Herman
Bentley, John Joseph
Boyd, Donald Bradford
Davidson, Ernest Roy
Diestler, Dennis Jon
Dykstra, Clifford Elliot
Kosman, Warren Melvin
Langhoff, Peter Wolfgang
Lipkowitz, Kenny Barry
Malik, David Joseph
Mueller, Charles Richard
Nazaroff, George Vasily
Pimblott, Simon Martin
Schwartz, Maurice Edward
Strieder, William

IOWA
DePristo, Andrew Elliott
Randic, Milan
Rider, Paul Edward, Sr
Ruedenberg, Klaus
Sando, Kenneth Martin

KANSAS
Chu, Shih I

KENTUCKY
Mueller, Sister Rita Marie

LOUISIANA
Bauer, Beverly Ann
Domelsmith, Linda Nell
Flurry, Robert Luther, Jr
Klasinc, Leo
Lee, Burnell
Politzer, Peter Andrew
Stanley, George Geoffrey
Watkins, Steven F

MARYLAND
Alexander, Millard Henry
Astumian, Raymond Dean
Bunyan, Ellen Lackey Spotz
Greer, William Louis
Hunter, Lawrence Wilbert
Kaye, Jack Alan
Kleier, Daniel Anthony
Liebman, Joel Fredric
Marqusee, Jeffrey Alan
Silver, David Martin
Stuebing, Edward Willis
Szabo, Attila
Weeks, John David

MASSACHUSETTS
Alper, Joseph Seth
Brenner, Howard
Chen, Maynard Ming-Liang
Epstein, Irving Robert
Hendrickson, James Briggs
Herzfeld, Judith
Hobey, William David
Jordan, Peter C H
Kirby, Kate Page
MacKerell, Alexander Donald, Jr
Muthukumar, Murugappan
Oppenheim, Irwin
Pan, Yuh Kang
Phillies, George David Joseph
Phillips, Philip Wirth
Pyun, Chong Wha
Ruskai, Mary Beth
Suplinskas, Raymond Joseph
Wilemski, Gerald
Winick, Jeremy Ross

MICHIGAN
Bartell, Lawrence Sims
Blankley, Clifton John
Blinder, Seymour Michael
Blint, Richard Joseph
Cheney, B Vernon
Crippen, Gordon Marvin
Cukier, Robert Isaac
Harrison, James Francis
Hase, William Louis
Lohr, Lawrence Luther, Jr
Rohrer, Douglas C
Rothschild, Walter Gustav
Schnitker, Jurgen H
Turner, Almon George, Jr

MINNESOTA
Mead, Chester Alden
Olmsted, Richard Dale
Polnaszek, Carl Francis
Thompson, Herbert Bradford

MISSOURI
Biolsi, Louis, Jr
Hansen, Richard Lee
McGraw, Robert Leonard
Plummer, Patricia Lynne Moore
Riehl, James Patrick
Rouse, Robert Arthur
Welsh, William James

NEBRASKA
Harris, Holly Ann

NEVADA
Shin, Hyung Kyu

NEW HAMPSHIRE
Housecroft, Catherine Elizabeth

NEW JERSEY
Adams, William Henry
Andose, Joseph D
Chance, Ronald Richard
Cotter, Martha Ann
Ermler, Walter Carl
Fisanick, Georgia Jeanne
Getzin, Paula Mayer
Goodman, Lionel
Gund, Peter Herman Lourie
Hager, Douglas Francis
Hammond, Willis Burdette
Head-Gordon, Martin Paul
Head-Gordon, Teresa Lyn
Helfand, Eugene
Paul, Edward W
Phillips, James Charles
Raghavachari, Krishnan
Raynor, Susanne
Strange, Ronald Stephen
Szamosi, Janos
Tully, John Charles
Upton, Thomas Hallworth
Weber, Thomas Andrew

NEW MEXICO
Bellum, John Curtis
Coppa, Nicholas V
Erpenbeck, Jerome John
Harary, Frank
Hay, Philip Jeffrey
Martin, Richard Lee
Pack, Russell T
Peek, James Mack
Redondo, Antonio
Schneider, Barry I
Wadt, Willard Rogers
Walker, Robert Bridges
Woodruff, Susan Beatty

NEW YORK
Berne, Bruce J
Birge, Robert Richards
Bishop, Marvin
Box, Vernon G S
Buff, Frank Paul
Campbell, Edwin Stewart
Carpenter, Barry Keith
Chapman, Sally
Ehrenson, Stanton Jay
Ezra, Gregory Sion
Frisch, Harry Lloyd
George, Thomas Frederick
Ginell, Robert
Hecht, Charles Edward
Hoffmann, Roald
Holmes, Curtis Frank
Kuki, Atsuo
Loring, Roger Frederic
Lounsbury, John Baldwin
McKelvey, John Murray
Marchese, Francis Thomas
Messmer, Richard Paul
Mezei, Mihaly
Miller, Kenneth John
Muckenfuss, Charles
Némethy, George
Newton, Marshall Dickinson
Pechukas, Philip
Porter, Richard Needham
Radel, Stanley Robert
Rhodes, Yorke Edward
Sage, Martin Lee
Snyder, Lawrence Clement
Stell, George Roger
Sufrin, Janice Richman
Widom, Benjamin
Zhang, John Zeng Hui
Zielinski, Theresa Julia

NORTH CAROLINA
Anderson, Marshall W
Bernholc, Jerzy
Choi, Sang-il
Hegstrom, Roger Allen
Jackels, Charles Frederick
Jones(Stover), Betsy
MacPhail, Richard Allyn
Parr, Robert Ghormley
Poirier, Jacques Charles

Whangbo, Myung Hwan
Whitten, Jerry Lynn
Yang, Weitao

OHIO
Carlton, Terry Scott
Colvin, Howard Allen
De Rocco, Andrew Gabriel
Huff, Norman Thomas
Jaffe, Hans
McFarland, Charles Warren
Mortensen, Earl Miller
Pitzer, Russell Mosher
Seiger, Harvey N
Seybold, Paul Grant
Shavitt, Isaiah
Smith, Douglas Alan
Taylor, William Johnson
Thomas, Terence Michael
Tuan, Debbie Fu-Tai
Weintraub, Herschel Jonathan R
Whitman, Donald Ray
Wolken, George, Jr

OKLAHOMA
Harwood, William H
Neely, Stanley Carrell
Sundberg, Kenneth Randall

OREGON
Herrick, David Rawls
Horne, Frederick Herbert
McClure, David Warren
Mazo, Robert Marc

PENNSYLVANIA
Davis, Jon Preston
Dougherty, Eugene P
Dyott, Thomas Michael
Garrison, Barbara Jane
Hameka, Hendrik Frederik
Irwin, Philip George
Jordan, Kenneth David
Kim, Shoon Kyung
Klein, Michael Lawrence
Kohn, Michael Charles
Kraihanzel, Charles S
Ludwig, Oliver George
Pople, John Anthony
Smith, David
Smith, Graham Monro
Stuper, Andrew John
Thibault, Jack Claude
Turner, Paul Jesse
Wood, Kurt Arthur
Yuan, Jian-Min
Zeroka, Daniel

RHODE ISLAND
Mason, Edward Allen
Stratt, Richard Mark

SOUTH CAROLINA
Gimarc, Benjamin M
Plodinec, Matthew John

TENNESSEE
Clarke, James Harold
Crawford, Oakley H
Ewig, Carl Stephen
Gilliom, Richard D
Hefferlin, Ray (Alden)
Kovac, Jeffrey Dean
Mortimer, Robert George
Painter, Gayle Stanford
Polavarapu, Prasad Leela
Sworski, Thomas John

TEXAS
Auchmuty, Giles
Birney, David Martin
Boggs, James Ernest
Cantrell, Cyrus D, III
Eaker, Charles William
Ford, George Peter
Hayes, Edward Francis
Klein, Douglas J
Kouri, Donald Jack
McCammon, James Andrew
Matcha, Robert Louis
Parr, Christopher Alan
Pettitt, Bernard Montgomery
Rossky, Peter Jacob
Schmalz, Thomas G
Scuseria, Gustavo Enrique
Webber, Stephen Edward
White, Ronald Joseph
Wyatt, Robert Eugene

UTAH
Haymet, Anthony Douglas-John
Henderson, Douglas J

VIRGINIA
Delos, John Bernard
Farmer, Barry Louis
LaBudde, Robert Arthur
Marron, Michael Thomas
Reynolds, Peter James
Richardson, Frederick S
Shillady, Donald Douglas
Smith, Bertram Bryan, Jr
Trindle, Carl Otis

WASHINGTON
Carlson, Charles Merton
Currah, Walter E
Dunning, Thomas Harold, Jr
Heller, Eric Johnson
Lybrand, Terry Paul
Moseley, William David, Jr
Reinhardt, William Parker
Scribner, John David
Yates, Albert Carl

WISCONSIN
Bird, R(obert) Byron
Bondeson, Stephen Ray
Curtiss, Charles Francis
England, Walter Bernard
Pietro, William Joseph
Skinner, James Lauriston
Zimmerman, Howard Elliot

ALBERTA
Birss, Fraser William
Boeré, René Theodoor
Clarke, Bruce Leslie
Fraga, Serafin
Grabenstetter, James Emmett
Huzinaga, Sigeru
Klobukowski, Mariusz Andrzej
Laidlaw, William George
Paul, Reginald
Thorson, Walter Rollier

BRITISH COLUMBIA
Dingle, Thomas Walter
Snider, Robert Folinsbee

MANITOBA
Gordon, Richard

NEW BRUNSWICK
Grein, Friedrich
Sichel, John Martin
Thakkar, Ajit Jamnadas

NOVA SCOTIA
Boyd, Russell Jaye
Coxon, John Anthony
Kusalik, Peter Gerard
Pacey, Philip Desmond
Wasylishen, Roderick Ernest

ONTARIO
Allnatt, Alan Richard
Bader, Richard Frederick W
Bishop, David Michael
Brumer, Paul William
Colpa, Johannes Pieter
Desai, Rashmi C
Henry, Bryan Roger
Kavassalis, Tom A
LeRoy, Robert James
Liu, Wing-Ki
Meath, William John
Ostlund, Neil Sinclair
Paldus, Josef
Penner, Alvin Paul
Pritchard, Huw Owen
Siebrand, Willem
Snider, Neil Stanley
Valleau, John Philip
Whittington, Stuart Gordon
Wright, James Sherman

QUEBEC
Eu, Byung Chan
Sanctuary, Bryan Clifford
Whitehead, Michael Anthony

SASKATCHEWAN
Mezey, Paul G

OTHER COUNTRIES
Bruns, Roy Edward
Bunge, Carlos Federico
Burnell, Louis A
Evleth, Earl Mansfield
Forst, Wendell
Hirshfeld, Fred Lurie
Kay, Kenneth George
Keller, Jaime
Mehler, Ernest Louis
Schleyer, Paul von Ragué

Thermodynamics & Material Properties

ALABAMA
Byrd, James Dotson
Harrison, Benjamin Keith
McKannan, Eugene Charles
Martin, James Arthur
Miller, George Paul
Pan, Chai-Fu
Thompson, Raymond G

ALASKA
Merrill, Robert Clifford, Jr

ARIZONA
Lunine, Jonathan Irving
Rupley, John Allen
Venables, John Anthony
Withers, James C

ARKANSAS
Pienta, Norbert J
Richardson, Charles Bonner

CALIFORNIA
Alivisatos, A Paul
Altman, Robert Leon
Amis, Eric J
Athey, Robert Douglas, Jr
Balzhiser, Richard E(arl)
Baskes, Michael I
Bohlen, Steven Ralph
Breitmayer, Theodore
Brewer, Leo
Carniglia, Stephen C(harles)
Carter, Emily Ann
Chang, Elfreda Te-Hsin
Chen, Timothy Shieh-Sheng
Condit, Ralph Howell
Cotts, David Bryan
Cotts, Patricia Metzger
Cubicciotti, Daniel David
Dawson, Kenneth Adrian
Domeniconi, Michael John
Doyle, Fiona Mary
Duff, Russell Earl
Dunn, Bruce
Elson, Robert Emanuel
Even, William Roy, Jr
Feigelson, Robert Saul
Fisher, Robert Amos, Jr
Fox, Joseph M(ickle), III
Fuller, Milton E
Ginell, William Seaman
Gür, Turgut M
Hahn, Harold Thomas
Hardt, Alexander P
Hem, John David
Hopper, Robert William
Hsieh, You-Lo
Hultgren, Ralph Raymond
Irvine, Stuart James Curzon
Kay, Robert Eugene
Kidd, Richard Wayne
King, William Robert, Jr
Krikorian, Oscar Harold
Langer, Sidney
Langerman, Neal Richard
Laub, Richard J
Lepie, Albert Helmut
Lewis, Nathan Saul
Lindner, Duane Lee
Liu, Frederick F
Ludwig, Frank Arno
McCarty, Jon Gilbert
McKenzie, William F
Martin, L(aurence) Robbin
Mason, D(avid) M(alcolm)
Meiser, Michael David
Meschi, David John
Miller, Donald Gabriel
Morris, John William, Jr
Munir, Zuhair A
Myronuk, Donald Joseph
Neidlinger, Hermann H
Nordstrom, Darrell Kirk
Noring, Jon Everett
Norman, John Harris
Novak, Bruce Michael
O'Loane, James Kenneth
Pask, Joseph Adam
Perry, Dale Lynn
Potter, Norman D
Rard, Joseph Antoine
Ritz-Gold, Caroline Joyce
Robin, Allen Maurice
Rock, Peter Alfred
Rothman, Albert J(oel)
Shelly, John Richard
Shine, M Carl, Jr
Simpson, Howard Douglas
Skowronski, Raymund Paul
Smith, Charles Aloysius
Smith, Gordon Stuart
Smith, James Hart
Smith, Louis Charles
Spiegler, Kurt Samuel
Stern, John Hanus
Strauss, Herbert L
Theodorou, Doros Nicolas
Tilley, T(erry) Don
Urtiew, Paul Andrew
Van Hecke, Gerald Raymond
Van Thiel, Mathias
Vernon, Gregory Allen
Whatley, Thomas Alvah
Wheeler, John C
White, Jack Lee
Williams, Richard Stanley
Wise, Henry
Yee, Rena

COLORADO
Arp, Vincent D
Bachman, Bonnie Jean Wilson
Blake, Daniel Melvin
Callanan, Jane Elizabeth
Chiou, Cary T(sair)
Chum, Helena Li
DePoorter, Gerald Leroy
Edwards, Harry Wallace
Ely, James Frank
Geller, Seymour
Hanley, Howard James Mason

Levenson, Leonard L
Modreski, Peter John
Montgomery, Robert L
Moore, John Jeremy
Rainwater, James Carlton
Reed, Thomas Binnington
Roshko, Alexana
Strauss, Eric L
Vaughan, John Dixon
Zoller, Paul

CONNECTICUT
Cohen, Myron Leslie
Frey, Douglas D
Grossman, Leonard N(athan)
Koberstein, Jeffrey Thomas
Kunz, Harold Russell
Larson, Eric George
McBride, James Michael
Rie, John E
Sarada, Thyagaraja
Wellman, Russel Elmer
Wolfram, Leszek January

DELAWARE
Chiu, Jen
Chu, Sung Gun
English, Alan Dale
Mills, Patrick Leo, Sr
Ross, William D(aniel)
Sammak, Emil George
Shozda, Raymond John

DISTRICT OF COLUMBIA
Carter, Gesina C
Corsaro, Robert Dominic
Dragoo, Alan Lewis
Gaines, Alan McCulloch
McKinney, John Edward
Stern, Kurt Heinz

FLORIDA
Beatty, Charles Lee
Chao, Raul Edward
Dufty, James W
Person, Willis Bagley
Van Zee, Richard Jerry
Zimmer, Martin F

GEORGIA
Anderson, James Leroy
Anderson, Robert Lester
Derrick, Mildred Elizabeth
Etzler, Frank M
Heric, Eugene LeRoy
Teja, Amyn Sadruddin
Williams, Emmett Lewis

HAWAII
Ming, Li Chung

IDAHO
Christian, Jerry Dale
Jacobsen, Richard T
Miller, Richard Lloyd
Ramshaw, John David
Slansky, Cyril M
Tanner, John Eyer, Jr

ILLINOIS
Abraham, Bernard M
Ali, Naushad
Blanks, Robert F
Burns, Richard Price
Cafasso, Fred A
Chan, Sai-Kit
Curtiss, Larry Alan
Finn, Patricia Ann
Fischer, Albert Karl
Giori, Claudio
Gray, Linsley Shepard, Jr
Gruen, Dieter Martin
Gutfreund, Kurt
House, James Evan, Jr
Huang, Leo W
Hwang, Yu-Tang
Johnson, Irving
Jonas, Jiri
Kemper, Kathleen Ann
Krawetz, Arthur Altshuler
Leibowitz, Leonard
Maroni, Victor August
Mendelsohn, Marshall H
Nagy, Zoltan
Nordine, Paul Clemens
Perkins, Alfred J
Peters, James Empson
Poeppelmeier, Kenneth Reinhard
Primak, William L
Sachtler, Wolfgang Max Hugo
Sather, Norman F(redrick)
Savidge, Jeffrey Lee
Schiffman, Robert A
Stafford, Fred E
Stout, John Willard
Wilson, Robert Steven
Zaromb, Solomon

INDIANA
Desai, Pramod D
Droege, John Walter
Fidelle, Thomas Patrick
Grimley, Robert Thomas
Henke, Mitchell C

Ho, Patience Ching-Ru
Klots, Cornelius E
Lenhert, P Galen
Li, Ying Sing
Lindemer, Terrence Bradford
Reynolds, Jefferson Wayne
Sales, Brian Craig
Schnelle, K(arl) B(enjamin), Jr
Seaton, William Hafford
Thiessen, William Ernest
Trowbridge, Lee Douglas
Williams, Robin O('Dare)

TEXAS
Albright, John Grover
Baker, Charles Taft
Brostow, Witold Konrad
Caflisch, Robert Galen
Chao, Jing
Chu, Ching-Wu
Clingman, William Herbert, Jr
DeBerry, David Wayne
Eakin, Bertram E
Fredin, Leif G R
Gooding, James Leslie
Gray, Kenneth W
Hance, Robert Lee
Ignatiev, Alex
Mack, Russel Travis
Margrave, John Lee
Marsh, Kenneth Neil
Munk, Petr
Pogue, Randall F
Spencer, John Edward
Whalen, James William
Whitmire, Kenton Herbert
Wilhoit, Randolph Carroll
Wood, Scott Emerson

UTAH
Carnahan, Robert D
Eatough, Delbert J
Haymet, Anthony Douglas-John
Izatt, Reed McNeil
Lisanke, Robert John, Sr
Miller, Akeley
Oyler, Dee Edward
Rowley, Richard L
Stringfellow, Gerald B

VIRGINIA
Bodnar, Robert John
Dewey, J(ohn) L(yons)
Erickson, Wayne Douglas
Field, Paul Eugene
Graham, Kenneth Judson
Hudson, Frederick Mitchell
Huebner, John Stephen
Jordan, Wade H(ampton), Jr
Laszlo, Tibor S
McGee, Henry A(lexander), Jr
Nuckolls, Joe Allen
Orwoll, Robert Arvid
Phillips, Donald Herman
Pohlmann, Juergen Lothar Wolfgang
Porzel, Francis Bernard
Schreiber, Henry Dale
Smith, Bertram Bryan, Jr
Ticknor, Leland Bruce
Toulmin, Priestley
Yates, Charlie Lee

WASHINGTON
Collins, Gary Scott
Ghiorso, Mark Stefan
Gregory, Norman Wayne
Hinman, Chester Arthur
Lindenmeyer, Paul Henry
Rai, Dhanpat
Riehl, Jerry A
Smith, Richard Dale
Spencer, Arthur Milton, Jr
Stuve, Eric Michael
Wilson, Charles Norman
Yunker, Wayne Harry

WISCONSIN
Atalla, Rajai Hanna
Cooper, Reid F
Draeger, Norman Arthur
Pollnow, Gilbert Frederick
Rosenberg, Robert Melvin
Sather, Glenn A(rthur)
Spencer, Brock

WYOMING
Hiza, Michael John, Jr
Hoffman, Edward Jack

ALBERTA
Checkel, M David
Gee, Norman
Heimann, Robert B(ertram)
Tschuikow-Roux, Eugene
Yeager, Howard Lane

MANITOBA
Cenkowski, Stefan
McFarlane, Joanna
Wikjord, Alfred George

NOVA SCOTIA
Kreuzer, Han Jurgen
White, Mary Anne

ONTARIO
Bluhm, Terry Lee
Glew, David Neville
Goodings, John Martin
Handa, Paul
Hemmings, Raymond Thomas
Kavassalis, Tom A
Koffyberg, Francois Pierre
Kumaran, Mavinkal K
Leaist, Derek Gordon
Milkie, Terence H
Perlman, Maier
Tombalakian, Artin S
Weir, Ronald Douglas
Westland, Alan Duane

PRINCE EDWARD ISLAND
Madan, Mahendra Pratap

QUEBEC
Britten, Michel
Chan, Sek Kwan
Dealy, John Michael
Desnoyers, Jacques Edouard
Dorris, Gilles Marcel
Leigh, Charles Henry
Sanctuary, Bryan Clifford
Utracki, Lechoslaw Adam

SASKATCHEWAN
Kybett, Brian David

OTHER COUNTRIES
Gordon, Jeffrey Miles
Jacob, K Thomas
Kafri, Oded
Nyburg, Stanley Cecil

Other Chemistry

ALABAMA
Acton, Ronald Terry
Krishna, N Rama
O'Kane, Kevin Charles
Pan, Chai-Fu
Sakai, Ted Tetsuo
Sayles, David Cyril
Tamburin, Henry John

ARIZONA
Denton, M Bonner
Muggli, Robert Zeno
Sanderson, Douglas G

ARKANSAS
Fu, Peter Pi-Cheng
Sears, Derek William George

CALIFORNIA
Bartlett, Paul A
Casey, John Edward, Jr
Chang, Ming-Houng (Albert)
Delaney, Margaret Lois
Domeniconi, Michael John
Folsom, Theodore Robert
Gardner, Marjorie Hyer
Garfin, David Edward
Garrison, Warren Manford
George, Patricia Margaret
George-Nascimento, Carlos
Gerstl, Bruno
Harrison, Jonas P
Hughes, Robert Alan
Jacob, Robert Allen
Juster, Norman Joel
Kluksdahl, Harris Eudell
Lu, Hsieng S
Ma, Joseph T
Maack, Christopher A
Maas, Keith Allan
Metiu, Horia I
Miyada, Don Shuso
Munger, Charles Galloway
Murphy, James Francis
Occelli, Mario Lorenzo
Prakash, Surya G K
Rogers, Howard H
Scattergood, Thomas W
Stucky, Galen Dean
Swidler, Ronald
Tin-Wa, Maung
Trogler, William C
Winterlin, Wray Laverne
Wu, Jiann-Long
Zeronian, Sarkis Haig
Zwerdling, Solomon

COLORADO
Batchelder, Arthur Roland
Bower, Nathan Wayne
Noufi, Rommel
Nozik, Arthur Jack
Schroeder, Herbert August
Seo, Eddie Tatsu

CONNECTICUT
Baker, Bernard S
Century, Bernard
Haydu, Juan B
Howard, Phillenore Drummond
Marfat, Anthony
Markham, Claire Agnes
Perry, Clark William

Thomas, Arthur L
Ziegler, Theresa Frances

DELAWARE
Ahmed, Syed Mahmood
Bare, Thomas M
Barron, Eugene Roy
Bergna, Horacio Enrique
Cairncross, Allan
Ehrich, F(elix) Frederick
Gibson, Joseph W(hitton), Jr
Hodges, Charles Thomas
Pagano, Alfred Horton
Vassiliou, Eustathios
Wayrynen, Robert Ellis
Williams, Ebenezer David, Jr

DISTRICT OF COLUMBIA
Colton, Richard J
Dennis, William E
Field, Ruth Bisen
McCormack, Mike
Turner, Anne Halligan

FLORIDA
Cohen, Martin Joseph
Cross, Timothy Albert
Dannenberg, E(li) M
Echegoyen, Luis Alberto
Ellis, Harold Hal
Hass, Michael A
Jackson, George Richard
Jungbauer, Mary Ann
Moudgil, Brij Mohan
Safron, Sanford Alan
Urone, Paul
Woods, James E

GEORGIA
Allison, John P
Block, Toby Fran
Bowen, Paul Tyner
Brown, William E
Derrick, Mildred Elizabeth
Goldsmith, Edward
Knight, James Albert, Jr
Kohl, Paul Albert
Lin, Ming Chang
Menon, Manchery Prabhakara
Netzel, Thomas Leonard
Shapira, Raymond
Tan, Kim H
Trawick, William George
Van Assendelft, Onno Willem

HAWAII
Crane, Sheldon Cyr

ILLINOIS
Albertson, Clarence E
Brown, Meyer
Doody, Marijo
Filler, Robert
Gowda, Netkal M Made
Gray, Nancy M
Horwitz, Earl Philip
Kaistha, Krishan K
Kemper, Kathleen Ann
Mahajan, Om Prakash
Matthews, Clifford Norman
Mertz, William J
Mohan, Prem
Moll, William Francis, Jr
Moschandreas, Demetrios J
Plautz, Donald Melvin
Ryan, Clarence E, Jr
Savit, Joseph
Schrader, Lawrence Edwin
Schurr, Garmond Gaylord
Walhout, Justine Isabel Simon
Whittington, Wesley Herbert
Zobel, Henry Freeman

INDIANA
Blatt, Joel Martin
Chen, Victor John
Dreikorn, Barry Allen
Dubin, Paul Lee
Gantzer, Mary Lou
Hemmes, Paul Richard
Kamat, Prashant V
Reuland, Donald John
Wagner, Eugene Ross

KANSAS
Ramamurti, Krishnamurti
Wright, Donald C

KENTUCKY
Baumgart, Richard
Giammara, Berverly L Turner Sites
Lloyd, William Gilbert
Riley, John Thomas
Wilson, Thomas Lamont
Zembrodt, Anthony Raymond

LOUISIANA
Carver, James Clark
Graves, Robert Joseph
Greco, Edward Carl
Hankins, B(obby) E(ugene)
Hanlon, Thomas Lee
Jung, John Andrew, Jr
Kane, Gordon Philo

Mollere, Phillip David
Morris, Cletus Eugene
Smith, Grant Warren, II
Wendt, Richard P
Zaweski, Edward F

MAINE
Tedeschi, Robert James

MARYLAND
Aaron, Herbert Samuel
Alperin, Harvey Albert
Craig, Paul N
Crawley, Jacqueline N
Darneal, Robert Lee
Hart, Gerald Warren
Holloway, Caroline T
Kanarowski, S(tanley) M(artin)
Kirkpatrick, Diana (Rorabaugh) M
Koch, Thomas Richard
McKenna, Mary Catherine
Mackler, Bruce F
McNairy, Sidney A, Jr
Malghan, Subhaschandra Gangappa
Pratt, Thomas Herring, Jr
Roethel, David Albert Hill
Ryback, Ralph Simon
Saba, William George
Sharrow, Susan O'Hott
Shrake, Andrew
Tarshis, Irvin Barry
Tower, Donald Bayley
Weser, Don Benton

MASSACHUSETTS
Barish, Leo
Caslavska, Vera Barbara
Caslavsky, Jaroslav Ladislav
Chen, Francis W
Chiu, Tak-Ming
Cohen, Robert Edward
Gutoff, Edgar B(enjamin)
Hoffman, Richard Laird
Idelson, Martin
Liggero, Samuel Henry
Matsuda, Seigo
Moser, William Ray
Rivin, Donald
Roffler-Tarlov, Suzanne K
Schlaikjer, Carl Roger
Serafin, Frank G
Sripada, Pavanaram Kameswara

MICHIGAN
Brower, Frank M
Chau, Vincent
Corrigan, Dennis Arthur
Dubpernell, George
Graham, John
Jaglan, Prem S
Janik, Borek
Kiechle, Frederick Leonard
Kipouros, Georges John
Korcek, Stefan
Lentz, Charles Wesley
Lorch, Steven Kalman
Maasberg, Albert Thomas
Mani, Inder
Murphy, William R
Nazri, Gholam-Abbas
Palaszek, Mary De Paul
Reising, Richard F
Riley, Bernard Jerome
Ryntz, Rose A
Sard, Richard
Schneider, John Arthur
Schubert, Jack
Seefurth, Randall N
Sharma, Ram Autar
Stevenson, Richard Marshall
Strom, Robert Michael
Swovick, Melvin Joseph
Zak, Bennie
Zanini-Fisher, Margherita

MINNESOTA
Beebe, George Warren
Douglas, David Lewis
Farber, Elliott
Grasse, Peter Brunner
Hagen, Donald Frederick
Hogle, Donald Hugh
Horner, James William, Jr
Johnson, Rodney L
Schultz, Robert John
Thompson, Phillip Gerhard

MISSISSIPPI
Little, Brenda Joyce

MISSOURI
Crutchfield, Marvin Mack
Kalota, Dennis Jerome
Knox, Walter Robert
Monzyk, Bruce Francis
Shen, Chung Yu
Slocombe, Robert Jackson
Whitefield, Philip Douglas

NEBRASKA
Allington, Robert W
Kenkel, John V

Other Chemistry (cont)

NEVADA
Herzlich, Harold Joel

NEW HAMPSHIRE
Greenwald, Harold Leopold
Nadeau, Herbert Gerard
Richmond, Robert Chaffee

NEW JERSEY
Aldrich, Haven Scott
Armbruster, David Charles
Bhandarkar, Suhas D
Chandrasekhar, Prasanna
Crawley, Lantz Stephen
Dante, Mark F
Frankoski, Stanley P
Gabriel, Robert
Gierer, Paul L
Gray, Russell Houston
Izod, Thomas Paul John
Jaffe, Sol Samson
Kline, Charles Howard
Kuehl, Guenter Hinrich
Liao, Mei-June
Lynch, Charles Andrew
Montgomery, Richard Millar
Nassau, Kurt
Orlando, Carl
Priesing, Charles Paul
Ramsden, Hugh Edwin
Roper, Robert
Shacklette, Lawrence Wayne
Sheppard, Keith George
Sugathan, Kenneth Kochappan
Walradt, John Pierce
Weil, Rolf
Zager, Ronald

NEW MEXICO
Kelber, Jeffry Alan
Maggiore, Carl Jerome
Walton, George

NEW YORK
Acharya, Seetharama A
Ambrose, Robert T
Beck, Curt Werner
Bent, Brian E
Cheh, Huk Yuk
Danielson, Paul Stephen
Figueras, John
Gluck, Ronald Monroe
Goldschmidt, Eric Nathan
Gorfien, Stephen Frank
Hargitay, Bartholomew
Harmon, Frederick Robert
Hicks, Bruce Lathan
Homan, Clarke Gilbert
Inman, Charles Gordon
James, Daniel Shaw
Joseph, Solomon
Keller, D Steven
Kohrt, Carl Fredrick
Lee, Lieng-Huang
Levine, Solomon Leon
Levitt, Morton
Lorenzo, George Albert
Luckey, George William
Myers, Drewfus Young, Jr
Orem, Michael William
Pochan, John Michael
Pomeroy, Gordon Ashby
Rebel, William J
Rhodes, Yorke Edward
Rubin, Isaac D
Sage, Gloria W
Schuster, David Israel
Takeuchi, Esther Sans
Yip, Yum Keung

NORTH CAROLINA
Bereman, Robert Deane
Beverung, Warren Neil, Jr
Fabricius, Dietrich M
Freeman, Harold Stanley
Fytelson, Milton
Garrou, Philip Ernest
Hildebolt, William Morton
Irene, Eugene Arthur
Kurtz, A Peter
McPeters, Arnold Lawrence
Perfetti, Patricia F
Roth, Barbara
Walters, Douglas Bruce

OHIO
Bailey, John Clark
Barsky, Constance Kay
Didchenko, Rostislav
Dinbergs, Kornelius
Dodge, James Stanley
Eley, Richard Robert
Epstein, Arthur Joseph
Fischer, Mark Benjamin
Halasa, Adel F
Hartsough, Robert Ray
Isenberg, Allen (Charles)
Lippert, Bruce J
Manning, Maurice
Merten, Helmut L
Myers, Elliot H
Ostrum, G(eorge) Kenneth

Rynbrandt, Donald Jay
Selover, Theodore Britton, Jr
Seybold, Paul Grant
Shaw, Wilfrid Garside
Sibley, Lucy Roy
Tong, James Ying-Peh
Wagner, Melvin Peter
Warner, Ann Marie

OKLAHOMA
Kucera, Clare H
Letcher, John Henry, III
Linder, Donald Ernst
Palmer, Bruce Robert

OREGON
Autrey, Robert Luis
Biermann, Christopher James
Jaeger, Charles Wayne
Laver, Murray Lane

PENNSYLVANIA
Archer, F Joy
Berges, David Alan
Berkheimer, Henry Edward
Biloen, Paul
Burman, Sudhir
Butera, Richard Anthony
Chang, Charles Hung
Cornell, John Alston
Ebert, Philip E
Feely, Wayne E
Friedman, Herbert Alter
Garbarini, Victor C
Garrison, Barbara Jane
Gerasimowicz, Walter Vladimir
Goodwin, James Gordon, Jr
Kugler, George Charles
Lai, Por-Hsiung
Louie, Ming
Marcelin, George
Mayer, Theodore Jack
Mehta, Prakash V
Reider, Malcolm John
Sax, Martin
Schoen, Kurt L
Shafer, Jules Alan
Sitrin, Robert David
Stryker, Lynden Joel
Urenovitch, Joseph Victor
Watterson, Kenneth Franklin
White, Malcolm Lunt
Wiehe, William Albert

RHODE ISLAND
Kreiser, Ralph Rank

SOUTH CAROLINA
Elzerman, Alan William
Kinard, W Frank
King, Lee Curtis
Longtin, Bruce
Spencer, Harold Garth

TENNESSEE
Barton, Kenneth Ray
Bleier, Alan
Gupta, Ramesh C
Gustafson, Bruce LeRoy
Harmer, David Edward
Higgins, Irwin Raymond
Horton, Charles Abell
James, Alton Everette, Jr
Karve, Mohan Dattatreya
Quist, Arvin Sigvard

TEXAS
Allison, Jean Batchelor
Hausler, Rudolf H
Hendrickson, Constance McRight
Johnson, Fred Lowery, Jr
Mitchell, Roy Ernest
Prasad, Rupi
Schlameus, Herman Wade

UTAH
Aldous, Duane Leo
Clark, David Neil
Dehm, Henry Christopher
Miller, Jan Dean
Pons, Stanley

VERMONT
Tremmel, Carl George

VIRGINIA
Allen, Ralph Orville, Jr
Barber, Patrick George
Bolleter, William Theodore
Gager, Forrest Lee, Jr
Hussamy, Samir
Jaisinghani, Rajan A
Majewski, Theodore E
Sedriks, Aristide John
Sobel, Robert Edward
Taylor, Ronald Paul
Tichenor, Robert Lauren

WASHINGTON
Berry, Keith O
Bierman, Sidney Roy
Nelson, Sidney D
Styris, David Lee
Wollwage, Paul Carl

WEST VIRGINIA
Bartley, William J

WISCONSIN
Berge, John Williston
Bloch, Daniel R
Bullock, Kathryn Rice
Fitch, Robert McLellan
Lee, Kathryn Adele Bunding
Mukerjee, Pasupati
Pearl, Irwin Albert
Schwab, Helmut

ALBERTA
Fisher, Harold M
Gee, Norman
Taylor, Wesley Gordon
Wiebe, Leonard Irving

BRITISH COLUMBIA
Rosell, Karl-Gunnar
Strong, David F

NEW BRUNSWICK
Passmore, Jack

NEWFOUNDLAND
Shahidi, Fereidoon

NOVA SCOTIA
Coxon, John Anthony
Ellenberger, Herman Albert
Fraser, Albert Donald

ONTARIO
Bolton, James R
Eliades, Theo I
Kazmaier, Peter Michael
Oldham, Keith Bentley
Wu, Tai Wing

QUEBEC
Clayton, David Walton
Lessard, Jean
Vijh, Ashok Kumar
Whitehead, Michael Anthony

OTHER COUNTRIES
Avila, Jesus
Grunwald, John J
Laurence, Alfred Edward

COMPUTER SCIENCES

Computer Sciences, General

ALABAMA
Atkinston, Ronald A
Barnard, Anthony C L
Dean, Susan Thorpe
De Maine, Paul Alexander Desmond
Dorai-Raj, Diana Glover
Downey, James Merritt
Fiesler, Emile
Fullerton, Larry Wayne
Hazelrig, Jane
Hill, David Thomas
Irwin, John David
Leslie, Thomas M
Lowry, James Lee
McCarthy, Dennis Joseph
Pattillo, Robert Allen
Reilly, Kevin Denis
Schroer, Bernard J
Seidman, Stephen Benjamin
Sides, Gary Donald
Stone, Max Wendell
White, Ronald
Yang, Chao-Chih
Zalik, Richard Albert

ALASKA
Dahlberg, Michael Lee
Lando, Barbara Ann
Larson, Frederic Roger
Wheeler, Keith Wilson

ARIZONA
Andrews, Gregory Richard
Barckett, Joseph Anthony
Bartels, Peter H
Bitter, Gary G
Carruthers, Lucy Marston
Clark, Wilburn O
Dempster, William Fred
Epstein, L(udwig) Ivan
Feldstein, Alan
Findler, Nicholas Victor
Fink, James Brewster
Formeister, Richard B
Gaud, William S
Holmes, William Farrar
Johnson, Randall Arthur
Kakar, Rajesh Kumar
Kreidl, Tobias Joachim
Lewis, William E(rvin)
Lorents, Alden C
Lovell, Stuart Estes
Mackulak, Gerald Thomas
Peterson, James Douglas
Randall, Lawrence Kessler, Jr
Smith, Josef Riley
Triffet, Terry

Wu, Hofu
Yakowitz, Sidney J

ARKANSAS
Abedi, Farrokh
Asfahl, C Ray
Crisp, Robert M, Jr
Goforth, Ronald R
Mink, Lawrence Albright
Rossa, Robert Frank
Sedelow, Sally Yeates
Sims, Robert Alan
Starling, Albert Gregory
Wilkes, Joseph Wray

CALIFORNIA
Adams, Ralph Melvin
Agostini, Romain Camille
Amdahl, Gene M
Anderson, Dennis Elmo
Anderson, Edward P(arley)
Anderson, Roger E
Aoki, Masanao
Arnovick, George (Norman)
Asendorf, Robert Harry
Ashworth, Edwin Robert
Avizienis, Algirdas
Backus, John
Baker, James Addison
Banerjee, Utpal
Barrett, William Avon
Barsky, Brian Andrew
Beach, Sharon Sickel
Beaver, W(illiam) L(awrence)
Bell, C Gordon
Benson, Robert Franklin
Bergquist, James William
Bernier, Charles L(lewellyn)
Bernstein, Ralph
Bhanu, Bir
Bharadvaj, Bala Krishnan
Bic, Lubomir
Bingham, Harry H, Jr
Blattner, Meera McCuaig
Blum, Manuel
Bly, Sara A
Boltinghouse, Joseph C
Boyle, Walter Gordon, Jr
Bradley, Hugh Edward
Brainerd, Walter Scott
Brown, David Randolph
Brown, Harold David
Bussell, Bertram
Butler, Jon Terry
Cantor, David Geoffrey
Carlan, Audrey M
Carlson, F Roy, Jr
Carlyle, Jack Webster
Carr, John Weber, III
Carson, George Stephen
Carterette, Edward Calvin Hayes
Case, Lloyd Allen
Cassel, Russell N
Caswell, John N(orman)
Cautis, C Victor
Chakrabarti, Supriya
Chambers, Frank Warman
Chellappa, Ramalingam
Chen, Alice Tung-Hua
Cheriton, David Ross
Choudary, Prabhakara Velagapudi
Chow, Wen Mou
Christiansen, Jerald N
Chu, Kai-Ching
Chu, Wesley W(ei-chin)
Clark, Crosman Jay
Clarke, Wilton E L
Clegg, Frederick Wingfield
Clinnick, Mansfield
Cochran, Stephen G
Codd, Edgar Frank
Cogan, Adrian Ilie
Coleman, Charles Clyde
Conant, Curtis Terry
Cooke, Ron Charles
Cottrell, Roger Leslie Anderton
Cover, Thomas Merrill
Crandall, Ira Carlton
Crawford, Richard Whittier
Cuadra, Carlos A(lbert)
Cunningham, Robert Stephen
Dairiki, Ned Tsuneo
Dalrymple, Stephen Harris
Deaton, Edmund Ike
Denning, Peter James
Dennis, Martha Greenberg
Despain, Alvin M(arden)
Deutsch, Laurence Peter
Donahue, James Edward
Dudziak, Walter Francis
Dutt, Nikil D
Elkind, Jerome I
Elspas, B(ernard)
Engel, Robert David
English, William Kirk
Erickson, Stanley Arvid
Estabrook, Kent Gordon
Estrin, Gerald
Estrin, Thelma A
Farone, William Anthony
Fateman, Richard J
Feigenbaum, Edward A(lbert)
Feign, David
Feldman, Jerome A

Fernbach, Sidney
Fischler, Martin A(lvin)
Flamm, Daniel Lawrence
Flegal, Robert Melvin
Floyd, Robert W
Fox, Geoffrey Charles
Frederick, David Eugene
Freund, Roland Wilhelm
Friedlander, Carl B
Friedman, Alan E
Friedman, Emily Perlinski
Gehrmann, John Edward
Gelinas, Robert Joseph
Gerhardt, Mark S
Geschke, Charles Matthew
Ghosh, Sakti P
Gittleman, Arthur P
Gladney, Henry M
Glaser, Edward L(ewis)
Goddard, Charles K
Goldberg, Jacob
Goldberg, Robert N
Gott, Euyen
Gottlieb, Peter
Gould, William E
Graham, Martin H(arold)
Greenberger, Martin
Greenwood, James Robert
Greig, Douglas Richard
Gretsky, Neil E
Grinnell, Robin Roy
Gruenberger, Fred J(oseph)
Gunckel, Thomas L, II
Gwinn, William Dulaney
Hamlin, Griffith Askew, Jr
Hamming, Richard W
Hance, Anthony James
Hardy, John W, Jr
Harris, David R
Harrison, Michael A
Hart, Peter E(lliot)
Hartman, James Kern
Harvey, Everett H
Hayes, Robert Mayo
Hearn, Anthony Clem
Hecht, Herbert
Hecht, Myron J
Hellerman, Herbert
Herriot, John George
Hightower, James K
Hilker, Harry Van Der Veer, Jr
Hillam, Bruce Parks
Hinrichs, Karl
Hoffman, Howard Torrens
Holl, Manfred Matthias
Hollander, Gerhard Ludwig
Holoien, Martin O
Horowitz, Ellis
Hsu, John Y
Hu, Te Chiang
Huang, Sung-Cheng
Huberman, Bernardo Abel
Hughett, Paul William
Hurwitz, Alexander
Israel, Jay Elliot
Jackson, Albert S(mith)
Jacobs, Irwin Mark
Jacobson, Raymond E
James, Kenneth Eugene
James, Philip Nickerson
Janos, William Augustus
Jefferson, Thomas Hutton, Jr
Kaelble, David Hardie
Kahan, William M
Karin, Sidney
Karplus, Walter J
Katz, Louis
Keaton, Michael John
Keddy, James Richard
Kehl, William Brunner
Kendall, Burton Nathaniel
Kenyon, Richard R(eid)
Kershaw, David Stanley
Keshavan, H R
Klahr, Philip
Kleinrock, Leonard
Knuth, Donald Ervin
Kolsky, Harwood George
Korf, Richard E
Korpman, Ralph Andrew
Kraabel, John Stanford
Kreuzer, Lloyd Barton
Kubik, Robert N
Kull, Lorenz Anthony
Lamie, Edward Louis
Lamport, Leslie B
Lang, Martin T
Lange, Rolf
Larkin, K(enneth) T(rent)
Larson, Harry Thomas
Lathrop, Kaye Don
Lay, Thorne
Ledin, George, Jr
Leedham, Clive D(ouglas)
Levin, Roy
Levit, Robert Jules
Lewis, William Perry
Lichter, James Joseph
Lim, Hong Seh
Lindal, Gunnar F
Linz, Peter
Loebner, Egon Ezriel
Lu, Adolph
Luckham, David Comptom

Luehrmann, Arthur Willett, Jr
Lum, Vincent Yu-Sun
Luqi
Lynch, William C
McCann, Gilbert Donald, Jr
McCarthy, John
McCluskey, Edward Joseph
McCool, John Macalpine
McCreight, Edward M
MacDonald, Gordon James Fraser
MacMillian, Stuart A
McMurray, Loren Robert
McQuillen, Howard Raymond
Magness, T(om) A(lan)
Maillot, Patrick Gilles
Malcolm, Michael Alexander
Maloy, John Owen
Mantey, Patrick E(dward)
Marder, Stanley
Maron, Melvin Earl
Mattson, Richard Lewis
Mead, Carver Andress
Meads, Philip Francis, Jr
Meisling, Torben (Hans)
Meissinger, Hans F
Meissner, Loren Phillip
Melvin, Jonathan David
Meyers, Gene Howard
Mihailovski, Alexander
Miller, William Frederick
Mitchell, James George
Mitchell, John Clifford
Moler, Cleve B
Monard, Joyce Anne
Monier, Louis Marcel
Morgenstern, Matthew
Morris, Jerry Lee, Jr
Morse, Joseph Grant
Motteler, Zane Clinton
Mulligan, Geoffrey C
Muntz, Richard Robert
Myers, George Scott, Jr
Neher, Maynard Bruce
Nelson, Eldred (Carlyle)
Neumann, Peter G
Newsam, John M
Ng, Edward Wai-Kwok
Ng, Lawrence Chen-Yim
Nickolls, John Richard
Nico, William Raymond
Niday, James Barker
Nikora, Allen P
Nilsson, Nils John
Obremski, Robert John
O'Dell, Austin Almond, Jr
Oliver, Ron
Olney, Ross David
O'Malley, Thomas Francis
Paal, Frank F
Parker, Alice Cline
Parker, Donn Blanchard
Patel, Arvindkumar Motibhai
Paulson, Boyd Colton, Jr
Pease, Marshall Carleton, III
Peddicord, Richard G
Pennington, Ralph Hugh
Perry, Robert Nathaniel, III
Petrowski, Gary E
Pignataro, Augustus
Plock, Richard James
Portnoff, Michael Rodney
Presser, Leon
Preston, Glenn Wetherby
Prokop, Jan Stuart
Pursglove, Laurence Albert
Ralston, Elizabeth Wall
Ramamoorthy, Chittoor V
Ramey, Daniel Bruce
Randall, Charles McWilliams
Ratcliff, Milton, Jr
Rau, Bantwal Ramakrishna
Reed, Irving Stoy
Requa, Joseph Earl
Rhodes, Edward Joseph, Jr
Richardson, John Mead
Rickard, James Joseph
Rider, Ronald Edward
Ritchie, Robert Wells
Rittenberg, Alan
Roberts, Charles Sheldon
Robin, Allen Maurice
Rodabaugh, David Joseph
Rohm, C E Tapie, Jr
Rose, Gene Fuerst
Rosen, Irwin Gary
Rosler, Lawrence
Rothenberg, Stephen
Rubin, Izhak
Rubin, Robert Howard
Rulifson, Johns Frederick
Russell, Gerald Frederick
Russell, Stuart Jonathan
Sangiovanni-Vincentelli, Alberto Luigi
Savitch, Walter John
Scherer, James R
Schlesinger, Stewart Irwin
Schwartz, Morton Donald
Schwartz, Sanford Bernard
Schweikert, Daniel George
Scott, Karen Christine
Seitz, S Stanley
Sengupta, Sumedha
Sequin, Carlo Heinrich
Shar, Leonard E

Shaw, Charles Alden
Shaw, George, II
Shemer, Jack Evvard
Shenker, Scott Joseph
Sherman, John Edwin
Shoch, John F
Silvester, John Andrew
Simons, Roger Mayfield
Singleton, Richard Collom
Sinton, Steven Williams
Slaughter, John Brooks
Smith, Alan Jay
Smith, Douglas Wemp
Smith, George C
Smith, James Thomas
Smoliar, Stephen William
Solomon, Malcolm David
Sondak, Norman Edward
Sorrels, John David
Spinrad, Robert J(oseph)
Stauffer, Howard Boyer
Stewart, John Joseph, Jr
Stone, William Ross
Sturm, Walter Allan
Suffin, Stephen Chester
Summers, Audrey Lorraine
Sutherland, William Robert
Talbot, Raymond James, Jr
Tang, Hwa-Tsang
Tanner, Robert Michael
Taylor, Richard N
Taylor, Robert William
Thacher, Henry Clarke, Jr
Thompson, Evan M
Toepfer, Richard E, Jr
Trenholme, John Burgess
Turin, George L
Ullman, Jeffrey D(avid)
Ung, Man T
Unti, Theodore Wayne Joseph
Valach, Miroslav
Vaughan, J Rodney M
Vemuri, Venkateswararao
Vidal, Jacques J
Wagner, Raymond Lee
Waldinger, Richard J
Wallace, Graham Franklin
Walters, Richard Francis
Ware, W(illis) H(oward)
Webre, Neil Whitney
Weibell, Fred John
Wexler, Jonathan David
White, Ray Henry
Wilensky, Robert
Wilkov, Robert Spencer
Williams, Jack Rudolph
Williams, Robin
Wilmer, Michael Emory
Wilson, Edward L
Winograd, Terry Allen
Wolf, Robert Peter
Wood, Roger Charles
Wood, William Irwin
Woods, Roger David
Woolard, Henry W(aldo)
Wray, John L
Yates, Scott Raymond
Young, John W(esley), Jr
Zadeh, L(otfi) A
Zimmerman, C Duane

COLORADO
Asherin, Duane Arthur
Brasch, Frederick Martin, Jr
Brown, Austin Robert, Jr
Buzbee, Billy Lewis
Carvalho, Sergio Eduardo Rodrigues De
Clementson, Gerhardt C
Cook, Robert Neal
Couger, J Daniel
Dandapani, Ramaswami
Dipner, Randy W
Dorn, William S
Eichelberger, W(illiam) H
Fosdick, Lloyd D(udley)
Frick, Pieter A
Frost, H(arold) Bonnell
Gentry, Willard Max, Jr
Gibson, William Loane
Gomberg, Joan Susan
Hittelman, Allen M
Irons, Edgar T(owar)
Irwin, Robert Cook
Johnson, Janice Kay
Johnstone, James G(eorge)
Jordan, Harry Frederick
Kelley, Neil Davis
Knight, Douglas Wayne
Lackey, Laurence
Lameiro, Gerard Francis
Lynch, Robert Michael
Malaiya, Yashwant Kumar
Marshall, Charles F
Meyer, Harvey John
Montague, Patricia Tucker
Mueller, Robert Andrew
Nemeth, Evi
Paine, Richard Bradford
Platter, Sanford
Quick, James S
Rechard, Ottis William
Sargent, Howard Harrop, III
Sauer, Jon Robert
Schnabel, Robert B

Sime, David Gilbert
Stanger, Andrew L
Sugar, George R
Vogel, Richard E
Wackernagel, Hans Beat
Wessel, William Roy
Wiatrowski, Claude Allan
Wonsiewicz, Bud Caesar
Zeiger, H Paul

CONNECTICUT
Agresta, Joseph
Apte, Chidanand
Bertram, J(ohn) E(lwood)
Booth, Taylor Lockwood
Buckley, Jay Selleck, Jr
Durnett, Robert Walter
Capwell, Robert J
Chatt, Allen Barrett
Close, Richard Thomas
Collins, Sylva Heghinian
Cowles, William Warren
Darcey, Terrance Michael
Dobbs, Gregory Melville
Dominy, Beryl W
Engel, Gerald Lawrence
Feit, Sidnie Marilyn
Frankel, Martin Richard
Harrison, Irene R
Hudak, Paul Raymond
King, George, III
Lang, George E, Jr
Levien, Roger Eli
Lof, John L(ars) C(ole)
Lovell, Bernard Wentzel
McCartin, Brian James
McCoy, Rawley D
Mann, Richard A(rnold)
Maryanski, Fred J
Oates, Peter Joseph
Prabulos, Joseph J, Jr
Rezek, Geoffrey Robert
Ridgway, William C(ombs), III
Robbins, David Alvin
Sachs, Martin William
Scherr, Allan L
Schultz, Martin H
Sharlow, John Francis
Sholl, Howard Alfred
Simpson, Wilburn Dwain
Smith, Edgar Clarence, Jr
Stein, Alan H
Steingiser, Samuel
Stephenson, Kenneth Edward
Stuart, James Davies
Youssef, Mary Naguib

DELAWARE
Abrahamson, Earl Arthur
Amer, Paul David
Benson, Frederic Rupert
Blaker, Robert Hockman
Bowyer, Kern M(allory)
Brown, Barry Stephen
Carberry, Judith B
Caviness, Bobby Forrester
Chester, Daniel Leon
Foster, Susan J
Golden, Kelly Paul
Hays, John Thomas
Hirsch, Albert Edgar
Jansson, Peter Allan
Jones, Louise Hinrichsen
Milian, Alwin S, Jr
Monet, Marion C(renshall)
Montague, Barbara Ann
Murphy, Arthur Thomas
Pensak, David Alan
Quarry, Mary Ann
Saunders, B David
Schultz, John Lawrence
Seetharam, Ramnath (Ram)
Skolnik, Herman
Taylor, Robert Burns, Jr
Ulery, Dana Lynn
Weiher, James F
Whitehouse, Bruce Alan

DISTRICT OF COLUMBIA
Abdali, Syed Kamal
Aufenkamp, Darrel Don
Baker, Dennis John
Baruch, Jordan J(ay)
Bassler, Richard Albert
Bertaut, Edgard Francis
Caceres, Cesar A
Della Torre, Edward
Diemer, F(erdinand) P(eter)
Edwards, Willard
Gammon, William Howard
Garcia, Oscar Nicolas
Hammer, Carl
Hart, Richard Cullen
Hodgson, James B, Jr
Hoffman, Lance J
Holland, Charles Jordan
Killion, Lawrence Eugene
Kitchens, Thomas Adren
Ledley, Robert Steven
Lehmann, John R(ichard)
McLean, John Dickson
Meltzer, Arnold Charles
Moraff, Howard
Murphy, James John

Computer Sciences, General (cont)

Nelson, Larry Dean
Newman, Simon M(eier)
Olmer, Jane Chasnoff
Park, Chan Mo
Perry, John Stephen
Pyke, Thomas Nicholas, Jr
Rosenblum, Lawrence Jay
Shore, John Edward
Stone, Philip M
Tidball, Charles Stanley
Vandergraft, James Saul
Werdel, Judith Ann
Williams, Leland Hendry
Wilson, James Bruce
Yamamoto, William Shigeru
Youm, Youngil

FLORIDA
Abou-Khalil, Samir
Baker, Theodore Paul
Bash, Paul Anthony
Bauer, Christian Schmid, Jr
Berk, Toby Steven
Bingham, Richard S(tephen), Jr
Bonn, T(heodore) H(ertz)
Bowers, James Clark
Carroll, Dennis Patrick
Cengeloglu, Yilmaz
Chartrand, Robert Lee
Clark, Kerry Bruce
Clutterham, David Robert
Cohen, Howard Lionel
Conaway, Charles William
Copeland, Richard Franklin
Coulter, Neal Stanley
Couturier, Gordon W
Fine, Rana Arnold
Friedl, Frank Edward
Goldwyn, Roger M(artin)
Goodman, Richard Henry
Gotterer, Malcolm Harold
Hall, David Goodsell, IV
Hand, Thomas
Harrison, Thomas J
Howard, Bernard Eufinger
Jahoda, Gerald
Jensen, Clayton Everett
Kandel, Abraham
Levow, Roy Bruce
Lighterman, Mark S
Lillien, Irving
Llewellyn, J(ohn) Anthony
Lofquist, George W
McMillan, Donald Ernest
Maddox, Billy Hoyte
Malthaner, W(illiam) A(mond)
Mankin, Richard Wendell
Martin, Charles John
Miles, David H
Naini, Majid M
Navlakha, Jainendra K
Neal, J(ames) P(reston), III
Rabbit, Guy
Rice, Randall Glenn
Roth, Paul Frederick
Sherman, Edward
Shershin, Anthony Connors
Stang, Louis George
Storm, Leo Eugene
Su, Stanley Y W
Timmer, Kathleen Mae
Tomonto, James R
Tulenko, James Stanley
Veinott, Cyril G
Vitagliano, Vincent J
Ward, James Audley
Wille, Luc Theo
Williams, Willie Elbert
Winton, Charles Newton

GEORGIA
Adams, David B
Arbogast, Richard Terrance
Arnold, Jonathan
Badre, Albert Nasib
Barnwell, Thomas Osborn, Jr
Bass, William Thomas
Blanke, Jordan Matthew
Camp, Ronnie Wayne
Caras, Gus J(ohn)
Chen, Dillion Tong-ting
Daniel, Leonard Rupert
Gallaher, Lawrence Joseph
Ghate, Suhas Ramkrishna
Goldman, John Abner
Graham, Roger Neill
Gupta, Rakesh Kumar
Holtum, Alfred G(erard)
Honkanen, Pentti A
Howard, Sethanne
Hunt, Gary W
Husson, Samir S
Koerner, T J
Krol, Joseph
Owen, Gene Scott
Pogue, Richard Ewert
Riddle, Lawrence H
Slamecka, Vladimir
Sparks, Arthur Godwin
Stone, David Ross

Straley, Tina
Suddath, Fred LeRoy, (Jr)
Techo, Robert
Vicory, William Anthony
Worth, Roy Eugene
Young, Raymond Hinchcliffe, Jr

HAWAII
Carlson, John Gregory
Kinariwala, Bharat K
Pager, David
Peterson, William Wesley
Walker, Terry M
Watanabe, Daniel Seishi
Watanabe, Michael Satosi
Weldon, Edward J, Jr

IDAHO
East, Larry Verne
Hoagland, Gordon Wood
Lawson, John D
Moore, Kenneth Virgil
Mortensen, Glen Albert
Wang, Ya-Yen Lee

ILLINOIS
Ahuja, Narendra
Arab-Ismaili, Mohammad Sharif
Ashenhurst, Robert Lovett
Ausman, Robert K
Bagley, John D(aniel)
Baldwin, Jack Timothy
Banerjee, Prithviraj
Banerjee, Prithviraj
Bareiss, Erwin Hans
Belford, Geneva Grosz
Bertoncini, Peter Joseph
Bookstein, Abraham
Boyle, James Martin
Brown, Patricia Lynn
Brown, Richard Maurice
Carrier, Steven Theodore
Carroll, John Terrance
Chaszeyka, Michael A(ndrew)
Cohen, Gloria
Cook, David Robert
Cook, Joseph Marion
Croke, Edward John
Divilbiss, James Leroy
Evens, Martha Walton
Faiman, Michael
Field, Kurt William
Fields, Thomas Henry
Filson, Don P
Finnerty, James Lawrence
Fisher, Richard Gary
Flora, Robert Henry
Fourer, Robert
Freeman, Maynard Lloyd
Friedlander, Ira Ray
Gay, Ben Douglas
Goodman, Gordon Louis
Gray, Linsley Shepard, Jr
Hammond, William Marion
Hansen, William Anthony
Hanson, Floyd Bliss
Hasdal, John Allan
Hattemer, Jimmie Ray
Herskovits, Lily Eva
Hosken, William H
Jacobsen, Fred Marius
Johnson, Donald Elwood
Jordan, Steven Lee
Kamath, Krishna
Kaper, Hans G
Kasube, Herbert Emil
Kim, Ki Hwan
Kovacs, Eve Veronika
Kuck, David Jerome
Larson, Richard Gustavus
Lawrie, Duncan H
Lax, Louis Carl
Lee, Der-Tsai
Levenberg, Milton Irwin
Levy, Allan Henry
Liu, Chung Laung
Loui, Michael Conrad
McKnight, Richard D
Mittman, Benjamin
Murata, Tadao
Muroga, Saburo
Newhart, M(ary) Joan
Nieman, George Carroll
O'Donnell, Terence J
Palmore, Julian Ivanhoe, III
Parker, Ehi
Parr, Phyllis Graham
Pesyna, Gail Marlane
Polychronopoulos, Constantine Dimitrius
Potter, Meredith Woods
Powers, Michael Jerome
Prais, Michael Gene
Price, David Thomas
Pyle, K Roger
Rafelson, Max Emanuel, Jr
Raffenetti, Richard Charles
Ray, Sylvian Richard
Reed, Daniel A
Reetz, Harold Frank, Jr
Renneke, David Richard
Rickert, Neil William
Ring, James George
Robertson, James E(vans)
Sabelli, Nora Hojvat

Scanlon, Jack M
Schipma, Peter B
Sellers, Donald Roscoe
Slotnick, Daniel Leonid
Smarr, Larry Lee
Snyder, James Newton
Soare, Robert I
Steele, Ian McKay
Stenberg, Charles Gustave
Swanson, Don R
Swoyer, Vincent Harry
Tomkins, Marion Louise
Ts'o, Timothy On-To
Van Ness, James E(dward)
Wells, Jane Frances
Wilcox, Lee Roy
Williams, Martha E
Wojcik, Anthony Stephen
Yau, Stephen Sik-Sang
Yoh, John K
Yu, Clement Tak
Yule, Herbert Phillip

INDIANA
Anderson-Mauser, Linda Marie
Atallah, Mikhail Jibrayil
Brown, Buck F(erguson)
Carlson, James C
Carlson, Lee Arnold
Carroll, John T, III
Costello, Donald F
Criss, Darrell E
Dalphin, John Francis
Davis, Grayson Steven
DeLong, Allyn F
DeMillo, Richard A
Drufenbrock, Diane
Easton, Richard J
Elharrar, Victor
Fuelling, Clinton Paul
Gersting, John Marshall, Jr
Gersting, Judith Lee
Guckel, Gunter
Hamblen, John Wesley
Hanes, Harold
Henry, Eugene W
Houstis, Elias N
Huffman, John Curtis
Jamieson, Leah H
Kallman, Ralph Arthur
Kent, Earle Lewis
Kuzel, Norbert R
Laxer, Cary
LoSecco, John M
Ma, Cynthia Sanman
Mansfield, Maynard Joseph
Marconi, Gary G
Marshall, Francis J
Moser, John William, Jr
Patrick, Edward Alfred
Pollock, G Donald
Prosser, Franklin Pierce
Putnam, Thomas Milton
Rego, Vernon J
Rice, John Rischard
Robertson, Edward L
Rosen, Saul
Smith, Peter David
Smucker, Arthur Allan
Springer, George
Stanton, Charles Madison
Thomas, Robert Jay
Thompson, Howard Doyle
Thuente, David Joseph
Weber, Janet Crosby
Wise, David Stephen
Young, Frank Hood
Yovits, Marshall Clinton

IOWA
Alton, Donald Alvin
Brearley, Harrington C(ooper), Jr
Garfield, Alan J
Jacob, Richard L
Kawai, Masataka
Lambert, Robert J
Lutz, Robert William
Meints, Clifford Leroy
Rila, Charles Clinton
Small, Arnold McCollum, Jr
Yohe, James Michael
Zettel, Larry Joseph

KANSAS
Bulgren, William Gerald
Carpenter, Kenneth Halsey
Conrow, Kenneth
Crowther, Robert Hamblett
Grzymala-Busse, Jerzy Witold
Hagen, Lawrence J
Hooker, Mark L
Legler, Warren Karl
Phillips, John Richard
Schweppe, Earl Justin
Van Swaay, Maarten
Wallace, Victor Lew
Wright, Donald C

KENTUCKY
Alter, Ronald
Carpenter, Dwight William
Davidson, Jeffrey Neal
Davis, Chester L
Davis, Thomas Haydn

Finkel, Raphael Ari
Franke, Charles H
Lindauer, George Conrad
Livingood, Marvin D(uane)
McCord, Michael Campbell
Pearson, Earl F

LOUISIANA
Agarwal, Arun Kumar
Bauer, Beverly Ann
Bedell, Louis Robert
Benard, Mark
Bonnette, Della T
Chen, Peter Pin-Shan
Dominick, Wayne Dennis
Fuchs, Richard E(arl)
Gajendar, Nandigam
Henry, Herman Luther, Jr
Jones, Daniel Elven
Kak, Subhash Chandra
Keys, L Ken
Lefton, Lew Edward
Looney, Stephen Warwick
Ovunc, Bulent Ahmet
Roquemore, Leroy

MAINE
Carter, William Caswell

MARYLAND
Agrawala, Ashok Kumar
Andreadis, Tim Dimitri
Ashman, Michael Nathan
Atchison, William Franklin
Barbe, David Franklin
Basili, Victor Robert
Bateman, Barry Lynn
Behforooz, Ali
Belliveau, Louis J
Berger, Robert Lewis
Berger, William J
Blanc, Robert Peter
Blevins, Gilbert Sanders
Blue, James Lawrence
Blum, Joseph
Blum, Robert Allan
Boisvert, Ronald Fernand
Bourbakis, Niklaos G
Braatz, James Anthony
Bukowski, Richard William
Cantor, David S
Carlstead, Edward Meredith
Chapin, Edward William, Jr
Chen, Ching-Nien
Chen, Lily Ching-Chun
Cherniavsky, Ellen Abelson
Chi, Donald Nan-Hua
Chi, L K
Clark, Bill Pat
Clark, Joseph E(dward)
Cohen, Gerson H
Conklin, James Byron, Jr
Coriell, Kathleen Patricia
Cornyn, John Joseph
Cotton, Ira Walter
Criss, Thomas Benjamin
Das, Prasanta
Deaven, Dennis George
Dhyse, Frederick George
Dubois, Andre T
Eaton, Barbra L
Elliot, Joe Oliver
Feldman, Alfred Philip
Fleming, James Joseph
Franks, David A
Freitag, Harlow
Fried, Jerrold
Gabriele, Thomas L
Garvin, James Brian
Geckle, William Jude
Geduldig, Donald
Gelberg, Alan
Ginevan, Michael Edward
Glaser, Edmund M
Gleissner, Gene Heiden
Glennan, T(homas) Keith
Goldfinger, Andrew David
Goldstein, Charles M
Goldstein, Gordon D(avid)
Gordon, Daniel Israel
Graham, Edward Underwood
Grosman, Louis Hirsch
Gross, James Harrison
Gull, Cloyd Dake
Heller, Stephen Richard
Hendrickson, Adolph C(arl)
Hevner, Alan Raymond
Hodes, Louis
Hollis, Jan Michael
Horna, Otakar Anthony
Hsu, Chen C
Hybl, Albert
Ingram, Glenn R
Jefferson, David Kenoss
Jefferson, Donald Earl
Jenkins, Robert Edward
Kahn, Arthur B
Kanal, Laveen Nanik
Karr, Alan Francis
Keenan, Thomas Aquinas
King, Joseph Herbert
Kippenberger, Donald Justin
Kissman, Henry Marcel
Knox, Robert Gaylord

Kosaraju, S Rao
Kroll, Bernard Hilton
Kurzhals, Peter R(alph)
Kyle, Herbert Lee
Landsburg, Alexander Charles
Lawson, Mildred Wiker
Lepson, Benjamin
Lidtke, Doris Keefe
Ligomenides, Panos Aristides
Lindberg, Donald Allan Bror
Little, John Llewellyn
Little, Joyce Currie
Lozier, Daniel William
Lyon, Gordon Edward
McRae, Vincent Vernon
Maisel, Herbert
Marimont, Rosalind Brownstone
Marquart, Ronald Gary
Martin, John A
May, Everette Lee, Jr
Messina, Carla Gretchen
Miller, Raymond Edward
Miller, Robert Gerard
Mohler, William C
Moshman, Jack
Nau, Dana S
Nusbickel, Edward M, Jr
Oza, Dipak H
Penn, Howard Lewis
Pickle, Linda Williams
Prewitt, Judith Martha Shimansky
Quinn, Frank Russell
Raff, Samuel J
Raub, William F
Rich, Robert Peter
Robock, Alan
Rombach, Hans Dieter
Rubinstein, Lawrence Victor
Sachs, Lester Marvin
Saltman, Roy G
Sammet, Jean E
Sandusky, Harold William
Sayer, John Samuel
Schlesinger, Judith Diane
Schmidt, Edward Matthews
Schneck, Paul Bennett
Schneider, John H
Schneider, William C
Schrum, Mary Irene Knoller
Sewell, Winifred
Shah, Shirish
Sher, Alvin Harvey
Sidhu, Deepinder Pal
Siegel, Elliot Robert
Silbergeld, Mae Driscoll
Silverton, James Vincent
Smith, Mark Stephen
Snow, Milton Leonard
Solomon, Jay Murrie
Spornick, Lynna
Stadter, James Thomas
Steven, Alasdair C
Stickler, Mitchell Gene
Stolz, Walter S
Sullivan, Francis E
Taragin, Morton Frank
Thakor, Nitish Vyomesh
Thoma, George Ranjan
Tippett, James T
Tripathi, Satish K
Trotter, Gordon Trumbull
Valley, Sharon Louise
Weigman, Bernard J
White, Eugene L
Wilkins, Michael Gray
Wooster, Harold Abbott
Yedinak, Peter Demerton
Zamora, Antonio
Zimmermann, Mark Edward
Zink, Sandra
Zusman, Fred Selwyn

MASSACHUSETTS
Abbott, Robert Classie
Adrion, William Richards
Akers, Sheldon Buckingham, Jr
Allen, Frank Lluberas
Alter, Ralph
Arbib, Michael A
Ash, Michael Edward
Bailey, Duane W
Ball, John Allen
Bankston, Donald Carl
Barngrover, Debra Anne
Belady, Laszlo Antal
Bennett, C Leonard
Bhandarkar, Dileep Pandurang
Bolker, Ethan D
Brackett, John Washburn
Branscomb, Lewis McAdory
Brayer, Kenneth
Briney, Robert Edward
Bruce, Kim Barry
Buono, John Arthur
Buzen, Jeffrey Peter
Cameron, Alastair Graham Walter
Cappallo, Roger James
Celmaster, William Noah
Champine, George Allen
Chen, Chi-Hau
Chen, Ching-chih
Chen, Robert Chia-Hua
Chu, J Chuan
Clarke, Lori A

Cohen, Howard David
Connors, Philip Irving
Cooke, Richard A
Corbató, Fernando Jose
Croft, Bruce
Dechene, Lucy Irene
Dennis, Jack Bonnell
Dertouzos, Michael L
Donovan, John Joseph
Dowd, John P
Dudgeon, Dan Ed
Eckhouse, Richard Henry
Eckstein, Jonathan
Evans, Thomas George
Fano, Robert M(ario)
Feustel, Edward Alvin
Floyd, William Beckwith
Friedenstein, Hanna
Friedman, Joyce Barbara
Fuller, Richard H
Fuller, Samuel Henry
Gagliardi, Ugo Oscar
Giarrusso, Frederick Frank
Giuliano, Vincent E
Glorioso, Robert M
Gold, Bernard
Gordon, Kurtiss Jay
Gorgone, John
Graham, Robert Montrose
Grodzinsky, Alan J
Hafen, Elizabeth Susan Scott
Hahn, Robert S(impson)
Harrison, Ralph Joseph
Hayden, Thomas Day
Heitman, Richard Edgar
Hershberg, Philip I
Holst, William Frederick
Hope, Lawrence Latimer
Hyatt, Raymond R, Jr
Ingham, Kenneth R
Jekeli, Christopher
Kaman, Charles Henry
Klensin, John
Klinedinst, Keith Allen
Kocher, Bryan S
Krause, Irvin
Kupferberg, Lenn C
Lacoss, Richard Thaddee
Lampson, Butler Wright
Laning, J Halcombe
Lechner, Robert Joseph
Levesque, Allen Henry
Levine, Randolph Herbert
Licklider, Joseph Carl Robnett
Lipner, Steven Barnett
Liskov, Barbara H
Logcher, Robert Daniel
McKnight, Lee Warren
Maeks, Joel
Makhoul, John Ibrahim
Martin, Richard Hadley, Jr
Martz, Eric
Menzin, Margaret Schoenberg
Meyer, Albert Ronald
Miffitt, Donald Charles
Milder, Fredric Lloyd
Morton, Donald John
Moses, Joel
Mountain, David Charles, Jr
Nassi, Isaac Robert
Nitecki, Zbigniew
Paquette, Gerard Arthur
Prerau, David Stewart
Press, William Henry
Rabin, Michael O
Raffel, Jack I
Rao, Ramgopal P
Riseman, Edward M
Rivest, Ronald L
Robbat, Albert, Jr
Roberts, Louis W
Ross, Douglas Taylor
Rothmeier, Jeffrey
Rowell, Robert Lee
Saltzer, Jerome H(oward)
Salveter, Sharon Caroline
Scheff, Benson H(offman)
Serr, Frederick E
Shaffer, Harry Leonard
Shambroom, Wiliam David
Shawcross, William Edgerton
Shum, Annie Waiching
Singer, Howard Joseph
Slack, Warner Vincent
Sneddon, Leigh
Stewart, Lawrence Colm
Stiffler, Jack Justin
Stillwell, Richard Newhall
Storer, James E(dward)
Szonyi, Geza
Teng, Chojan
Theobald, Charles Edwin, Jr
Thornton, Richard D(ouglas)
Tompkins, Howard E(dward)
Townley, Judy Ann
Trefethen, Lloyd Nicholas
Tweed, David George
Van Meter, David
Walter, Charlton M
Wand, Mitchell
Wang, An
Waters, Richard C(abot)
Weinberg, I Jack
Winett, Joel M

Woods, William A
Yarbrough, Lynn Douglas
Zachary, Norman

MICHIGAN
Anderson, Robert Hunt
Atreya, Arvind
Baker, Robert Henry, Jr
Barnard, Robert D(ane), Jr
Bollinger, Robert Otto
Bowman, Carlos Morales
Briggs, Darinka Zigic
Caron, E(dgar) Louis
Cheydleur, Benjamin Frederic
Conrad, Michael
Conway, Lynn Ann
Coyle, Peter
Denman, Harry Harroun
Dershem, Herbert L
Dubes, Richard C
Eichhorn, J(acob)
Elshoff, James L(ester)
Farah, Badie Naiem
Finerman, A(aron)
Flanders, Harley
Fry, James Palmer
Furlong, Robert B
Galler, Bernard Aaron
Getty, Ward Douglas
Gjostein, Norman A
Goldsmith, Donald Leon
Hansen, John C
Hee, Christopher Edward
Hicks, Darrell Lee
Hillig, Kurt Walter, II
Holland, John Henry
Holland, Steven William
Howe, William Jeffrey
Johnson, Whitney Larsen
Kaiser, Christopher B
Kallander, John William
Laurance, Neal L
Lowther, John Lincoln
McLaughlin, Renate
Marshall, Dale Earnest
Meteer, James William
Moore, Alexander Mazyck
Nelson, James Donald
Patt, Yale Nance
Petersen, Donald Ralph
Petzold, Edgar
Phillips, Richard Lang
Pynnonen, Bruce W
Ramamurthy, Amurthur C
Rollinger, Charles N(icholas)
Rosen, Jeffrey Kenneth
Rosenbaum, Manuel
Rovner, David Richard
Schriber, Thomas J
Schuetzle, Dennis
Simpson, William Albert
Sokol, Robert James
Tihansky, Diane Rice
Tsao, Nai-Kuan
Tuchinsky, Philip Martin
Vishnubhotla, Sarma Raghunadha
Watts, Jeffrey Lynn
Weinshank, Donald Jerome
Westervelt, Franklin Herbert
Williams, John Albert
Windeknecht, Thomas George
Wolfson, Seymour J
Wu, Hai

MINNESOTA
Ackerman, Eugene
Anderson, Ronald E
Austin, Donald Murray
Cohen, Arnold A
Cornelius, Steven Gregory
Crooks, Stephen Lawrence
Drew, Bruce Alvin
Du, David Hung-Chang
Dueltgen, Ronald Rex
Ek, Alan Ryan
Ellis, Lynda Betty
Gatewood, Lael Cranmer
Geokezas, Meletios
Glewwe, Carl W(illiam)
Gupta, Satish Chander
Henderson, Donald Lee
Jacobs, David R, Jr
Krueger, Eugene Rex
Kumar, Vipin
Lin, Benjamin Ming-Ren
McQuarrie, Donald G
Marshall, John Clifford
Oelberg, Thomas Jonathon
Patton, Peter C(lyde)
Pong, Ting-Chuen
Robb, Richard A
Schneider, G Michael
Schuldt, Marcus Dale
Seebach, J Arthur, Jr
Slagle, James R
Slavin, Joanne Louise
Smith, Robert Elijah
Stebbings, William Lee
Steen, Lynn Arthur
Stein, Marvin L
Wetmore, Clifford Major
Wolfe, Barbara Blair
Yasko, Richard N

MISSISSIPPI
Britt, James Robert
Coleman, Thomas George
Colonias, John S
Miller, James Edward
Taylor-Cade, Ruth Ann

MISSOURI
Agarwal, Ramesh K
Arvidson, Raymond Ernst
Ball, William E(rnest)
Banakar, Umesh Virupaksh
Blackwell, Paul K, II
Bodine, Richard Shearon
Bosanquet, Louis Percival
Bradburn, Gregory Russell
Cohick, A Doyle, Jr
Cottrell, Roy
Cox, Jerome R(ockhold), Jr
Franklin, Mark A
Griffith, Virgil Vernon
Groenweghe, Leo Carl Denis
Heard, John Thibaut, Jr
Ho, Chung You (Peter)
Johnson, Douglas William
Jones, Ronald Dale
Krone, Lester H(erman), Jr
Lee, Ralph Edward
Lembke, Roger Roy
MacDonald, Carolyn Trott
Moore, James Thomas
Morgan, Nancy H
Morley, Robert Emmett, Jr
Neuman, Rosalind Joyce
Ramsey, Robert Bruce
Schmidt, Bruno (Francis)
Stalling, David Laurence
Stuth, Charles James
Talbott, Ted Delwyn
Taylor, Ralph Dale
Thomas, Lewis Jones, Jr
Tyrer, Harry Wakeley
Wagner, Joseph Edward
Walsh, Robert Jerome

MONTANA
Amend, John Robert
Banaugh, Robert Peter
Combie, Joan D
Pierre, Donald Arthur
Weaver, Donald K(essler), Jr

NEBRASKA
Clements, Gregory Leland
Downing, John Scott
Fairchild, Robert Wayne
Gale, Douglas Shannon, II
Keller, Roy Fred
Leung, Joseph Yuk-Tong
Longley, William Warren, Jr
Scudder, Jeffrey Eric
Seth, Sharad Chandra
Sharp, Edward A
Zebolsky, Donald Michael

NEVADA
Brady, Allen H
Juliussen, J Egil
Manhart, Robert (Audley)
Marsh, David Paul
Minor, John Threecivelous
Telford, James Wardrop

NEW HAMPSHIRE
Epstein, Harvey Irwin
Fultyn, Robert Victor
Klotz, Louis Herman
Kuo, Shan Sun
Kurtz, Thomas Eugene
McKeeman, William Marshall
Taffe, William John
Tourgee, Ronald Alan

NEW JERSEY
Afshar, Siroos K
Agalloco, James Paul
Allen, Robert B
Amarel, S(aul)
Amron, Irving
Anderson, Milton Merrill
Anderson, Terry Lee
Andrews, Ronald Allen
Baker, Brenda Sue
Barnes, Derek A
Barsky, James
Bayer, Douglas Leslie
Beaton, Albert E
Becker, Sheldon Theodore
Beckman, Frank Samuel
Bergeron, Robert F(rancis) (Terry), Jr
Bobeck, Andrew H
Boylan, Edward S
Brandmaier, Harold Edmund
Burkhardt, Kenneth J
Cai, Jin-yi
Chambers, John McKinley
Cheo, Li-hsiang Aria S
Coffman, Edward G, Jr
Cornell, W(arren) A(lvan)
Coughran, William Marvin, Jr
Crowley, Thomas Henry
Cutler, Frank Allen, Jr
Dante, Mark F
DeBaun, Robert Matthew

Computer Sciences, General (cont)

Desjardins, Raoul
Dickerson, L(oren) L(ester), Jr
DiMasi, Gabriel Joseph
Eastwood, Abraham Bagot
Eisner, Mark Joseph
Epstein, Marvin Phelps
Fix, Kathleen A
Forys, Leonard J
Fox, Phyllis
Freeman, Herbert
Fu, Hui-Hsing
Gabriel, Edwin Z
Gargaro, Anthony
Gear, Charles William
George, Kenneth Dudley
Glass, John Richard
Goffman, Martin
Goldstein, Philip
Golin, Stuart
Gossmann, Hans Joachim
Goyal, Suresh
Grandolfo, Marian Carmela
Green, Edwin James
Gross, Arthur Gerald
Grosso, Anthony J
Hackman, Martin Robert
Halaby, Sami Assad
Hall, Gene Stephen
Hall, Homer James
Hamilton, Charles W(allace)
Harris, James Ridout
Hill, David G
Holzmann, Gerard J
Huang, Jennming Stephen
Huber, Richard V
Iorns, Terry Vern
Jackson, Thomas A J
Jarvis, John Frederick
Johannes, Virgil Ivancich
Johnson, Stephen Curtis
Joseph, John Mundancheril
Jurkat, Martin Peter
Kaita, Robert
Karney, Charles Fielding Finch
Karol, Mark J
Kasparek, Stanley Vaclav
Kaufman, Linda
Kellgren, John
Khalifa, Ramzi A
Kirch, Murray R
Kowalski, Ludwik
Krzyzanowski, Paul
Kulikowski, Casimir A
LaMastro, Robert Anthony
Levine, Richard S
Levy, Leon Sholom
Lichtenberger, Gerald Burton
Lohse, David John
Loscher, Robert A
Love, L J Cline
McCaslin, Darrell
McDonald, H(enry) S(tanton)
McGuinness, Deborah Louise
McIlroy, M Douglas
MacLennan, Carol G
Mahoney, Michael S
Malbrock, Jane C
Mamelak, Joseph Simon
Manickam, Janardhan
Marlin, Robert Lewis
Marmor, Robert Samuel
Marsh, John L
Matey, James Regis
Mauzey, Peter T
Mehta, Atul Mansukhbhai
Miller, Robert Alan
Moroni, Antonio
Most, Joseph Morris
Myers, Jeffrey
Newman, Stephen Alexander
Newton, Paul Edward
Nicholls, Gerald P
Nonnenmann, Uwe
O'Gorman, James
Ott, Teunis Jan
Pearsall, Thomas Perine
Philipp, Ronald E
Pinson, Elliot N
Polonsky, Ivan Paul
Popper, Robert David
Raychaudhuri, Kamal Kumar
Riggs, Richard
Robins, Jack
Roland, Dennis Michael
Rollino, John
Rosenblum, Martin Jacob
Rubin, Arthur I(srael)
Sannuti, Peddapullaiah
Schmidt, Barnet Michael
Shaer, Norman Robert
Shine, Robert John
Shipley, Edward Nicholas
Shober, Robert Anthony
Shulman, Herbert Byron
Siegel, Andrew Francis
Siegel, Jeffry A
Smagorinsky, Joseph
Steben, John D
Steiglitz, Kenneth
Stevens, John G

Stevenson, Robert Louis
Stoddard, James H
Strange, Ronald Stephen
Tarjan, Robert Endre
Traub, Joseph Frederick
Turoff, Murray
Uhrig, Jerome Lee
Verma, Pramode Kumar
Wagner, Richard Carl
Walsh, Teresa Marie
Weinberger, Peter Jay
Weiss, C Dennis
Weiss, Stanley H
Whigan, Daisy B
White, Myron Edward
Winder, Robert Owen
Woo, Nam-Sung
Wright, Margaret Hagen
Yamin, Michael
Yen, You-Hsin Eugene
Zwass, Vladimir

NEW MEXICO
Adams, J Mack
Ahmed, Nasir
Barsis, Edwin Howard
Bell, Stoughton
Berardo, Peter Antonio
Bleyl, Robert Lingren
Campbell, Katherine Smith
Curtis, Wesley E
Davey, William Robert
Divett, Robert Thomas
Erdal, Bruce Robert
Erickson, Kenneth Lynn
Faber, Vance
Fronek, Donald Karel
Henderson, Dale Barlow
Johnston, John B(everley)
Jones, Merrill C(alvin)
Kensek, Ronald P
Korman, N(athaniel) I(rving)
Lawrence, Raymond Jeffery
Luger, George F
McDonald, Timothy Scott
Miller, Edmund K(enneth)
Morrison, Donald Ross
Northrup, Clyde John Marshall, Jr
Phister, Montgomery, Jr
Snell, Charles Murrell
Sprinkle, James Kent, Jr
Stanbro, William David
Stark, Richard Harlan
Sung, Andrew Hsi-Lin
Tapp, Charles Millard
Terrell, C(harles) W(illiam)
Thorne, Billy Joe
Trowbridge, George Cecil
Weiss, Paul B
Wells, Mark Brimhall
Wienke, Bruce Ray
Wilkins, Ronald Wayne
Willbanks, Emily West

NEW YORK
Allen, Frances Elizabeth
Allen, James Frederick
Anastassiou, Dimitris
Anderson, Albert Edward
Andrews, Mark Allen
Anshel, Michael
Archibald, Julius A, Jr
Asprey, Winifred Alice
Augenstein, Moshe
Bahl, Lalit R(ai)
Ball, George William
Bashkow, Theodore R(obert)
Batter, John F(rederic), Jr
Bednowitz, Allan Lloyd
Berger, Jay Manton
Berger, Toby
Bernstein, Herbert Jacob
Bienstock, Daniel
Birnbaum, David
Bishop, Marvin
Blau, Lawrence Martin
Bloch, Alan
Borden, Edward B
Brown, David T
Brown, Theodore D
Bunch, Phillip Carter
Burnham, Dwight Comber
Cabeen, Samuel Kirkland
Campbell, Graham Hays
Carey, Bernard Joseph
Chan, Tat-Hung
Chang, John H(si-Teh)
Chi, Benjamin E
Chovan, James Peter
Christensen, Robert Lee
Cocke, John
Cohen, Daniel Isaac Aryeh
Cohen, Irving Allan
Cohen, Irwin A
Cohen, Martin O
Cohn, Harvey
Collins, Carol Desormeau
Cornacchio, Joseph V(incent)
Coutchie, Pamela Ann
Curtis, Ronald S
D'Auria, Thomas A
De Lillo, Nicholas Joseph
DeSieno, Robert P
Diamond, Fred Irwin

Dimmler, D(ietrich) Gerd
Douglas, Craig Carl
Drossman, Melvyn Miles
Dumoulin, Charles Lucian
Dwyer, Harry, III
Eberlein, Patricia James
Eckert, Richard Raymond
Eydgahi, Ali Mohammadzadeh
Falk, Catherine T
Fama, Donald Francis
Farber, Andrew R
Federighi, Francis D
Feeser, Larry James
Feingold, Alex Jay
Ferentz, Melvin
Finlay, Thomas Hiram
Fischbach, Joseph W(inston)
Flannery, John B, Jr
Fleisher, Harold
Flores, Ivan
Foley, Kenneth John
Foreman, Bruce Milburn, Jr
Frank, Thomas Stolley
Frantz, Andrew Gibson
Frazer, W(illiam) Donald
Freiman, Charles
Friedberg, Carl E
Frieder, Gideon
Gabay, Jonathan Glenn
Gabelman, Irving J(acob)
Gause, Donald C
Geiss, Gunther R(ichard)
Gelernter, Herbert Leo
Ghozati, Seyed-Ali
Gilchrist, Bruce
Gilman, S(ister) John Frances
Gledhill, Ronald James
Glimm, James Gilbert
Goertzel, Gerald
Goldberg, Conrad Stewart
Golden, Robert K
Goldstein, Lawrence Howard
Golumbic, Martin Charles
Gonzalez, Robert Anthony
Goodman, Seymour
Gordon, Irving
Greenberg, Donald P
Gries, David
Gruener, William B
Gustavson, Fred Gehrung
Hainline, Louise
Hammer, Richard Benjamin
Hannay, David G
Hanson, Jonathan C
Harrison, Malcolm Charles
Hartmanis, Juris
Heidelberger, Philip
Heller, Jack
Heller, William R
Henle, Robert A
Heppa, Douglas Van
Herrington, Lee Pierce
Hershenov, Joseph
Hoenig, Alan
Hong, Se June
Hsiao, Mu-Yue
Hubbard, Richard Alexander, II
Hudak, Michael J
Huddleston, John Vincent
Impagliazzo, John
Jabbur, Ramzi Jibrail
Jain, Duli Chandra
Jesaitis, Raymond G
Johnson, Wallace E
Jones, Leonidas John
Joyner, William Henry, Jr
Kaplan, Ehud
Kaplan, Martin Charles
Kashkoush, Ismail I
Kim, Carl Stephen
Kinzly, Robert Edward
Kirchmayer, Leon K
Klerer, Melvin
Komiak, James Joseph
Kotchoubey, Andrew
Kozak, Gary S
Kwok, Thomas Yu-Kiu
Kyburg, Henry
Laderman, Julian David
Lamster, Hal B
Lazareth, Otto William, Jr
Lerner, Rita Guggenheim
Lewis, Philip M
Lidofsky, Leon Julian
Lubowsky, Jack
Macaluso, Pat
McGloin, Paul Arthur
McLaughlin, Harry Wright
McNaughton, Robert
Maguire, Gerald Quentin, Jr
Maissel, Leon I
Mansfield, Victor Neil
Marcuvitz, Nathan
Marr, Robert B
Mattar, Farres Phillip
Matthews, Dwight Earl
Meenan, Peter Michael
Mercer, Robert Leroy
Merten, Alan Gilbert
Miller, Pauline Monz
Mohlke, Byron Henry
Moniot, Robert Keith
Morrison, John B
Moyne, John Abel

Muench, Donald Leo
Muller, Otto Helmuth
Musser, David Rea
Nagel, Harry Leslie
Nagy, George
Nevison, Christopher H
Notaro, Anthony
Ortel, William Charles Gormley
Padalino, Stephen John
Pardee, Otway O'Meara
Parikh, Hemant Bhupendra
Parker, Ronald W
Pechacek, Terry Frank
Peled, Abraham
Pence, Harry Edmond
Pifko, Allan Bert
Polivka, Raymond Peter
Postman, Robert Derek
Preiser, Stanley
Priore, Roger L
Ralston, Anthony
Ramaley, James Francis
Rauscher, Tomlinson Gene
Reckhow, Warren Addison
Reilly, Edwin David, Jr
Richter, Donald
Rivkin, Maxcy
Robinson, Louis
Rogers, Edwin Henry
Rosberg, Zvi
Rosenfeld, Jack Lee
Rosenkrantz, Daniel J
Roth, John Paul
Rowlett, Roger Scott
Rudin, Bernard D
Rudolph, Luther Day
Rutledge, Joseph Dela
Sakoda, William John
Sargent, Robert George
Schmitt, Erich
Schneider, Fred Barry
Schoenfeld, Robert Louis
Schwartz, Jacob Theodore
Schwartz, Mischa
Selman, Alan L
Shapiro, Donald M
Shapiro, Stephen D
Shapiro, Stuart Charles
Shaw, David Elliot
Sherman, Philip Martin
Sherman, Seymour
Shilepsky, Arnold Charles
Shooman, Martin L
Shriver, Bruce Douglas
Shuey, R(ichard) L(yman)
Sibert, Elbert Ernest
Sklarew, Robert J
Smith, Bernard
Sobel, Kenneth Mark
Sorter, Peter F
Spiegel, Robert
Srihari, Sargur N
Stanfel, Larry Eugene
Star, Martin Leon
Stearns, Richard Edwin
Stein, David Morris
Stevens, William Y(eaton)
Storm, Edward Francis
Streeter, Donald N(elson)
Su, Stephen Y H
Sublette, Ivan H(ugh)
Summers, Phillip Dale
Tassinari, Silvio John
Tewarson, Reginald P
Tobin, Michael
Todd, Aaron Rodwell
Tracz, Will
Tycko, Daniel H
Uretsky, Myron
Vollmer, Frederick Wolfer
Voss, Richard Frederick
Wadell, Lyle H
Wald, Alvin Stanley
Walker, Edward John
Waltzer, Wayne C
Wang, Chu Ping
Wang, Hsin-Pang
Weinberger, Arnold
Weinstein, David Alan
Whitby, Owen
White, Gerald M(ilton)
White, William Wallace
Wilkinson, John Peter Darrell
Williams, George Harry
Winkler, Leonard P
Wolf, Carol Euwema
Wong, Chak-Kuen
Worthing, Jurgen
Yu, Kai-Bor

NORTH CAROLINA
Agrawal, Dharma Prakash
Allen, Charles Michael
Barr, Roger Coke
Biermann, Alan Wales
Bright, George Walter
Brooks, Frederick P(hillips), Jr
Calingaert, Peter
Carlson, William Theodore
Chen, Michael Yu-Men
Chen, Su-shing
Chou, Wushow
Christian, Wolfgang C
Cullati, Arthur G

Dinse, Gregg Ernest
Dotson, Marilyn Knight
Ellis-Menaghan, John J
Epstein, George
Fuchs, Henry
Gallie, Thomas Muir
Gemperline, Margaret Mary Cetera
Grau, Albert A
Greim, Barbara Ann
Groves, William Ernest
Gustafson, Carl Gustaf, Jr
Hadeen, Kenneth Doyle
Hall, Edward Duncan
Halton, John Henry
Hammond, William Edward
Hartzema, Abraham Gijsbert
Hildebrandt, Theodore Ware
Kerr, Sandria Neidus
Kootsey, Joseph Mailen
Krasny, Harvey Charles
Kurtz, A Peter
Lapicki, Gregory
Long, Andrew Fleming, Jr
Lucas, Carol N
McAllister, David Franklin
McDaniel, Roger Lanier, Jr
Mago, Gyula Antal
Malindzak, George S, Jr
Marco, Gino Joseph
Marinos, Pete Nick
Meyer, Carl Dean, Jr
Miller, Donald F
Murray, Francis Joseph
Nair, Sreekantan S
Neidinger, Richard Dean
Niccolai, Nilo Anthony
Norgaard, Nicholas J
Norris, Fletcher R
Nutter, James I(rving)
Pizer, Stephen M
Plemmons, Robert James
Ramm, Dietolf
Riemann, James Michael
Roberts, Stephen D
Sawyer, John Wesley
Shelly, James H
Snodgrass, Rex Jackson
Snyder, Wesley Edwin
Sowell, Katye Marie Oliver
Stanat, Donald Ford
Stead, William Wallace
Struve, William George
Sussenguth, Edward H
Teague, David Boyce
Trivedi, Kishor Shridharbhai
Trussell, Henry Joel
Utku, Senol
Wagner, Robert Alan
Weiss, Stephen Fredrick
Wetmore, David Eugene
Whipkey, Kenneth Lee
Whitehurst, Garnett Brooks
Wright, William V(aughn)
Wright, William Vale

NORTH DAKOTA
Lieberman, Milton Eugene
Magel, Kenneth I
Tareski, Val Gerard
Vasey, Edfred H
Winrich, Lonny B

OHIO
Adeli, Hojjat
Beverly, Robert Edward, III
Blake, James Elwood
Buoni, John J
Burden, Richard L
Canfield, James Howard
Chamis, Alice Yanosko
Coe, Kenneth Loren
Coleman, Marilyn A
Cooke, Charles C
Coulter, Paul David
Cruce, William L R
Damian, Carol G
Darby, Ralph Lewis
Davis, Henry Werner
Davis, Michael E
Ernst, George W
Feld, William Adam
Fleming, Paul Daniel, III
Foulk, Clinton Ross
Fraser, Alex Stewart
Goodson, Alan Leslie
Graham, Paul Whitener Link
Gregory, Thomas Bradford
Hartrum, Thomas Charles
Heines, Thomas Samuel
Hess, Evelyn V
Hill, Ann Gertrude
Hulley, Clair Montrose
Hura, Gurdeep Singh
James, Thomas Ray
Jehn, Lawrence A
Jendrek, Eugene Francis, Jr
Kerr, Douglas S
Keyes, Marion Alvah, IV
Kolopajlo, Lawrence Hugh
Kwatra, Subhash Chander
Lamont, Gary Byron
LaRue, Robert D(ean)
Lechner, Joseph H
Little, Richard Allen

Liu, Ming-Tsan
Lutton, John Kazuo
McDonough, James Francis
Massey, L(ester) G(eorge)
Minchak, Robert John
Mochel, Virgil Dale
Mulhausen, Hedy Ann
Muller, Mervin Edgar
Murdoch, Arthur
Nardi, John Christopher
Neaderhouser Purdy, Carla Cecilia
Neff, Raymond Kenneth
Nicholson, Victor Alvin
Niemann, Theodore Frank
Nikolai, Paul John
Ogden, William Frederick
O'Neill, Edward True
Orr, Charles Henry
Ouimet, Alfred J, Jr
Papachristou, Christos A
Petrarca, Anthony Edward
Platau, Gerard Oscar
Potoczny, Henry Basil
Purvis, John Thomas
Quam, David Lawrence
Rajsuman, Rochit
Randels, James Bennett
Ratliff, Priscilla N
Rattan, Kuldip Singh
Reilly, Charles Bernard
Restemeyer, William Edward
Rich, Ronald Lee
Ricord, Louis Chester
Ross, Charles Burton
Roth, Mark A
Rubin, Stanley G(erald)
Ryan, Anne Webster
Schaefer, Donald John
Shields, George Seamon
Slotterbeck-Baker, Oberta Ann
Snider, John William
Solow, Daniel
Spialter, Leonard
Sterling, Leon Samuel
Sterrett, Andrew
Stickney, Alan Craig
Stobaugh, Robert Earl
Stoner, Ronald Edward
Suter, Bruce Wilsey
Way, Frederick, III
Weisgerber, David Wendelin
Wigington, Ronald L
Wolfe, Robert Kenneth
Zaye, David F

OKLAHOMA
Anderson, Paul Dean
Bednar, Jonnie Bee
Brady, Barry Hugh Garnet
Calvert, Jon Channing
Cheung, John Yan-Poon
Crafton, Paul A(rthur)
Eby, Harold Hildenbrandt
Fisher, Donald D
Folk, Earl Donald
Hale, William Henry, Jr
Hedrick, George Ellwood, III
Lauffer, Donald Eugene
McClain, Gerald Ray
McGurk, Donald J
Minor, John Threecivelous
Monn, Donald Edgar
Naymik, Daniel Allan
Noll, Leo Albert
Page, Rector Lee
Reimer, Dennis D
Rutledge, Carl Thomas
Summers, Gregory Lawson
Testerman, Jack Duane
Tomaja, David Louis
Usher, W(illia)m Mack
Vasicek, Daniel J
Vinatieri, James Edward
Walker, Billy Kenneth
Wilcox, Lyle C(hester)
Winter, William Kenneth

OREGON
Alin, John Suemper
Beyer, Terry
Currie, James Orr, Jr
Davenport, Wilbur B(ayley), Jr
Demarest, Harold Hunt, Jr
Ecklund, Earl Frank, Jr
Feldman, Milton H
Freiling, Michael Joseph
Gilbert, Barrie
Hancock, John Edward Herbert
Holmes, Zoe Ann
Hopkins, Theodore Emo
Jones, Donlan F(rancis)
Kieburtz, Richard B(ruce)
Mooney, Larry Albert
Moursund, David G
Rudd, Walter Greyson
Struble, George W
Terwilliger, Don William
Tinney, William Frank
Udovic, Daniel
Wagner, Orvin Edson

PENNSYLVANIA
Abend, Kenneth
Aburdene, Maurice Felix

Adams, William Sanders
Aiken, Robert McLean
Alvino, William Michael
Anderson, Jay Martin
Angstadt, Carol Newborg
Auerbach, Isaac L
Aviles, Rafael G
Badler, Norman Ira
Bajcsy, Ruzena K
Bamberger, Judy
Banerji, Ranan Bihari
Barnhart, Barry B
Bass, Leonard Joel
Baumann, Dwight Maylon Billy
Bowlden, Henry James
Box, Larry
Buriok, Gerald Michael
Burness, James Hubert
Burnett, Roger Macdonald
Cain, James Thomas
Caldwell, Christopher Sterling
Carbonell, Jaime Guillermo
Chang, C Hsiung
Chaplin, Norman John
Clemons, Eric K
Coleman, Anna M
Copenhaver, Thomas Wesley
Crawley, James Winston, Jr
Cupper, Robert
Director, Stephen William
Dwyer, Thomas A
Dyott, Thomas Michael
Eckert, J Presper
Eddy, William Frost
Eisenstein, Bruce A
Emmons, Larrimore Browneller
Ertekin, Turgay
Fabrey, James Douglas
Farley, Belmont Greenlee
Farren, Ann Louise
Favret, A(ndrew) G(illigan)
Feng, Tse-yun
Frazier, James Lewis
Friedman, Frank L
Fugger, Joseph
Furst, Merrick Lee
Garfield, Eugene
Garfinkel, David
Gerasimowicz, Walter Vladimir
Gibbs, Norman Edgar
Goldwasser, Samuel M
Gorn, Saul
Gottlieb, Frederick Jay
Gould, William Allen
Grabowski, Thomas J
Gross, Michael R
Gundersen, Larry Edward
Habermann, Arie Nicolaas
Haun, Robert Dee, Jr
Hermans, Hans J
Herzfeld, Valerius E
Hoffman, Eric Alfred
Hoover, L Ronald
Hurwitz, Jan Krosst
Huthnance, Edward Dennis, Jr
Isett, Robert David
Janicki, Casimir A
Joshi, Aravind Krishna
Kahler, Albert Comstock, III
Kane, John Vincent, Jr
Kazahaya, Masahiro Matt
Kelley, Jay Hilary
Kent, Allen
Kirk, David Blackburn
Kleban, Morton H
Klotz, Eugene Arthur
Koffman, Elliot B
Korsh, James F
Kozik, Eugene
Kung, Hsiang-Tsung
Kusic, George Larry, Jr
Laird, Donald T(homas)
Leas, J(ohn) W(esley)
Leibholz, Stephen W
Licini, Jerome Carl
McAllister, Marialuisa N
McArthur, William George
McFarlane, Kenneth Walter
McLaughlin, Francis X(avier)
Mahaffy, John Harlan
Mason, Thomas Joseph
Mayer, Joerg Werner Peter
Meyer, John Sigmund
Miller, G(erson) H(arry)
Mitchell, Tom M
Moravec, Hans Peter
Murgie, Samuel A
Nair, K Aiyappan
Newell, Allen
Olsen, Glenn W
O'Neill, John Cornelius
Penman, Paul D
Plonsky, Andrew Walter
Pool, James C T
Pruett, John Robert
Prywes, Noah S(hmarya)
Ramani, Raja Venkat
Reddy, Raj
Reed, Joseph Raymond
Reigel, Earl William
Reynolds, John C
Ridener, Fred Louis, Jr
Robl, Robert F(redrick), Jr
Roemer, Richard Arthur

Rogers, Ralph Loucks
Rose, George David
Roskies, Ralph Zvi
Ross, Arthur Leonard
Rush, James E
Sher, Irving Harold
Shuck, John Winfield
Sieber, James Leo
Smith, Chester Martin, Jr
Sproull, Robert Fletcher
Sullivan, George Allen
Sutherland, Ivan Edward
Tobias, Russell Lawrence
Tulenko, Thomas Norman
Vladutz, George E
Wagh, Meghanad D
Wagner, Clifford Henry
Walchli, Harold E(dward)
Walker, David Kenneth
Weinberger, Edward Bertram
Wiley, Samuel J
Wilkins, Raymond Leslie
Wismer, Robert Kingsley
Wu, John Naichi
Zaphyr, Peter Anthony

RHODE ISLAND
Carney, Edward J
Dewhurst, Peter
Gabriel, Richard Francis
Hemmerle, William J
Krikorian, John Sarkis, Jr
Preparata, Franco Paolo
Santos, Eugene (Sy)
Savage, John Edmund
Strauss, Charles Michael
Vitter, Jeffrey Scott
Wegner, Peter

SOUTH CAROLINA
Brons, Kenneth Allyn
Cannon, Robert L
Chuang, Ming Chia
Dobbins, James Gregory Hall
Eastman, Caroline Merriam
Foster, Kent Ellsworth
Gamble, Robert Oscar
Gillette, Paul Crawford
Hammond, Joseph Langhorne, Jr
Klemm, James L
Leathrum, James Frederick
Luedeman, John Keith
Miler, George Gibbon, Jr
Nanney, Thomas Ray
Porter, Hayden Samuel, Jr
Porter, Rick Anthony
Rowlett, Russell Johnston, Jr
Suich, John Edward
Turner, Albert Joseph, Jr
Wheat, Joseph Allen
Wise, William Curtis

SOUTH DAKOTA
Koepsell, Paul L(oel)

TENNESSEE
Bailes, Gordon Lee
Ball, Raiford Mill
Blass, William Errol
Braunstein, Helen Mentcher
Burton, John Williams
Campbell, Warren Elwood
Carter, Harvey Pate
Chitwood, Howard
Ferguson, Robert Lynn
Fischer, Charlotte Froese
Fischer, Patrick Carl
Garzon, Max
Gray, William Harvey
Habenschuss, Anton
Holdeman, Jonas Tillman, Jr
Hopkins, Richard Allen
Hutcheson, Paul Henry
Kiech, Earl Lockett
Kirchner, Frederick Karl
Lane, Charles A
Lea, James Wesley, Jr
LeBlanc, Larry Joseph
Murphy, Brian Donal
Otis, Marshall Voigt
Payne, DeWitt Allen
Petersen, Harold, Jr
Pfuderer, Helen A
Pleasant, James Carroll
Poore, Jesse H, Jr
Ross, Clay Campbell, Jr
Roussin, Robert Warren
Sherman, Gordon R
Sizemore, Douglas Reece
Springer, John Mervin
Trubey, David Keith
Umholtz, Clyde Allan
Van Rij, Willem Idaniel
Voorhees, Larry Donald
Wignall, George Denis

TEXAS
Albertson, Harold D
Ali, Moonis
Alo, Richard Anthony
Atkinson, Edward Neely
Aucoin, Paschal Joseph, Jr
Benedict, George Frederick
Blakley, George Robert

Computer Sciences, General (cont)

Bowdon, Edward K(night), Sr
Brown, Barry W
Brown, Bradford S
Browne, James Clayton
Carter, Elmer Buzby
Cecil, David Rolf
Childs, S(elma) Bart
Crawford, Richard Haygood, Jr
Curtin, Richard B
Daniel, James Wilson
Darwin, James T, Jr
DeGroot, Doug
Dobrott, Robert D
Dodson, David Scott
Dolling, David Stanley
Drew, Dan Dale
Eckles, Wesley Webster, Jr
Edgerley, Dennis A
Ellwood, Brooks B
Engelhardt, Albert George
Feagin, Terry
Frailey, Dennis J(ohn)
Frick, John P
Giese, Robert Paul
Gillette, Kevin Keith
Gladwell, Ian
Gorry, G Anthony
Gray, Kenneth W
Hamill, Dennis W
Hanne, John R
Hardcastle, Donald Lee
Harmon, Glynn
Herbert, Morley Allen
Hernandez, Norma
Hixon, Sumner B
Huang, Jung-Chang
Jennings, Alfred Roy, Jr
Johnson, Lloyd N(ewhall)
Keller, Thomas W
Kim, Won
King, Willis Kwongtsu
Konkel, David Anthony
Konstam, Aaron Harry
Koplyay, Janos Bernath
Ku, Bernard Siu-Man
Lacewell, Ronald Dale
Lee, Ellen Szeto
Linn, John Charles
McCoy, John Harold
Mack, Russel Travis
McLeod, Raymond, Jr
McSherry, Diana Hartridge
Malek, Miroslaw
Mao, Shing
Marcotte, Ronald Edward
Markenscoff, Pauline
Matula, David William
Maxwell, Donald A
Monk, Clayborne Morris
Moore, J Strother
Moore, Kris
Nassersharif, Bahram
Newhouse, Albert
Nickel, James Alvin
Novak, Gordon Shaw, Jr
Parberry, Ian
Parker, Arthur L (Pete)
Parr, Christopher Alan
Perry, Robert Riley
Pervin, William Joseph
Peterson, Lynn Louise Meister
Pitts, Gerald Nelson
Pooch, Udo Walter
Poucher, William B
Poulsen, Lawrence Leroy
Pyle, Leonard Duane
Randlett, Herbert Eldridge, Jr
Ransom, C J
Richardson, Arthur Jerold
Rinehart, Wilbur Allan
Ring, Dennis Randall
Rodriguez, Carlos Eduardo
Schlessinger, Bernard S
Schuster, Eugene F
Schutz, Bob Ewald
Self, Glendon Danna
Shen, Vincent Y
Shieh, Leang-San
Shooter, Jack Allen
Silberschatz, Abraham
Simmons, Dick Bedford
Smith, Philip Wesley
Smith, Rolf C, Jr
Srinivasan, Ramachandra Srini
Stein, William Edward
Stuart, Joe Don
Sudborough, Ivan Hal
Szygenda, Stephen Anthony
Volz, Richard A
Wang, Yuan R
Warlick, Charles Henry
Weiser, Alan
Whorton, Elbert Benjamin
Wiggins, James Wendell
Wilcox, Roberta Arlene
Will, Peter Milne
Williams, Glen Nordyke
Wolter, Jan D(ithmar)
Wood, Craig Adams
Wright, James Foley
Wrotenbery, Paul Taylor
Yeh, Raymond T
Yorio, Thomas
Zagata, Michael DeForest

UTAH
Butterfield, Veloy Hansen, Jr
Carlson, Gary
Chabries, Douglas M
Cox, Benjamin Vincent
Gates, Henry Stillman
Hurst, Rex LeRoy
Jones, Merrell Robert
Keller, Robert Marion
Lewis, Lawrence Guy
Musick, James R
Park, Robert Lynn
Pope, Wendell LaVon
Rasmussen, V Philip, Jr
Riesenfeld, Richard F
Stockham, Thomas Greenway, Jr
Stokes, Gordon Ellis
Viavant, William Joseph
Walker, LeRoy Harold

VERMONT
Ballou, Donald Henry
Gray, Kenneth Stewart
Kuffner, Roy Joseph
Rankin, Joanna Marie
Wiitala, Stephen Allen

VIRGINIA
Abrams, Marshall D
Anderson, Walter L(eonard)
Artna-Cohen, Agda
Balint, Francis Joseph
Barnes, Bruce Herbert
Batson, Alan Percy
Boehm, Barry William
Burns, William, Jr
Catlin, Avery
Cerf, Vinton Gray
Cheng, George Chiwo
Chimenti, Frank A
Chou, Tsai-Chia Peter
Claus, Alfons Jozef
Cook, Robert Patterson
Davis, Ruth Margaret
Deal, George Edgar
Eldridge, Charles A
Elkins, Judith Molinar
Epstein, Samuel David
Farrell, Robert Michael
Feyock, Stefan
Fisher, Gordon McCrea
Forman, Richard Allan
Fox, Edward A
Friedman, Fred Jay
Gander, George William
Goncz, John Henry
Goranson, H T
Gray, F(estus) Gail
Hancock, V(ernon) Ray
Heath, Lenwood S
Heffron, W(alter) Gordon
Heterick, Robert Cary, Jr
Hodge, Bartow
Hoffman, Karla Leigh
Hopper, Grace Murray
Huddle, Benjamin Paul, Jr
Hudgins, Aubrey C, Jr
Irick, Paul Eugene
Ivanetich, Richard John
Jacob, Robert J(oseph) K(assel)
Jones, Anita Katherine
Kahn, Robert Elliot
Koch, Carl Fred
Kripke, Bernard Robert
Lanzano, Bernadine Clare
Leczynski, Barbara Ann
Lee, John A N
Lefebvre, Yvon
Lewis, Jesse C
McCutchen, Samuel P(roctor)
MacKinney, Arland Lee
McNeil, Phillip Eugene
Mata-Toledo, Ramon Alberto
Mathis, Robert Fletcher
Meyrowitz, Alan Lester
Michalski, Ryszard Spencer
Moritz, Barry Kyler
Mullen, Joseph Matthew
Murray, Jeanne Morris
Nance, Richard E
Neher, Dean Royce
Norris, Eugene Michael
Oliver, Paul Alfred
Perry, Dennis Gordon
Pratt, Terrence Wendall
Reeker, Larry Henry
Reisler, Donald Laurence
Rine, David C
Roberts, David Craig
Rosenberg, Murray David
Rosenfeld, Robert L
Roussos, Constantine
Schneider, Philip Allen David
Seiner, John Milton
Shetler, Antoinette (Toni)
Shoosmith, John Norman
Shrier, Stefan
Small, Timothy Michael
Sorrell, Gary Lee

Stevens, Donald Meade
Taggart, Keith Anthony
Talbot, Richard Burritt
Thiel, Thomas J
Tole, John Roy
Torio, Joyce Clarke
Tripp, John Stephen
Van Tilborg, André Marcel
Voigt, Robert Gary
Warfield, J(ohn) N(elson)
Weaver, Alfred Charles
Weinstein, Stanley Edwin
Whitaker, William Armstrong
Wist, Abund Ottokar
Wolf, Eric W
Wulf, William Allan
Zilczer, Janet Ann

WASHINGTON
Benda, Miroslav
Bierman, Sidney Roy
Boctor, Magdy
Bolme, Mark W
Britt, Patricia Marie
Cole, Charles Ralph
Cutlip, William Frederick
Drevdahl, Elmer R(andolph)
Edison, Larry Alvin
Edwards, James Mark
Engleman, Christian L
Farnum, Peter
Golde, Hellmut
Gravitz, Sidney I
Jenner, David Charles
Kehl, Theodore H
Klepper, John Richard
Kowalik, Janusz Szczesny
Marsaglia, George
Meeder, Jeanne Elizabeth
Newman, Paul Harold
Noe, Jerre D(onald)
Oster, Clarence Alfred
Parker, Richard Alan
Richardson, Michael Lewellyn
Shamash, Yacov A
Shirley, Frank Connard
Smilen, Lowell I
Somani, Arun Kumar
Stokes, Gerald Madison
Thomas, James
Thompson, Robert Harry
Tinker, John Frank
Underwood, Douglas Haines

WEST VIRGINIA
Atkins, John Marshall
Becker, Stanley Leonard
Brown, Paul B
Jefimenko, Oleg D
Larson, Gary Eugene
Leech, H(arry) William
McGovern, Terence Joseph
Muth, Wayne Allen

WISCONSIN
Chen, Steve S
Cohen, Bernard Allan
Cook, David Marsden
Davida, George I
Davidson, Charles H(enry)
Desautels, Edouard Joseph
Divjak, August A
Dyer, Charles Robert
Eggert, Arthur Arnold
Evans, James Stuart
Firebaugh, Morris W
Fitzwater, Donald (Robert)
Fossum, Timothy V
Hafemann, Dennis Reinhold
Heinen, James Albin
Hutchinson, George Keating
Landweber, Lawrence H
Levine, Leonard P
Littlewood, Roland Kay
Malkus, David Starr
Miller, Paul Dean
Moses, Gregory Allen
Northouse, Richard A
Pollnow, Gilbert Frederick
Ramakrishnan, Raghu
Rang, Edward Roy
Rhyner, Charles R
Schultz, David Harold
Stahl, Neil
Uhr, Leonard Merrick

WYOMING
Bauer, Henry Raymond, III
Gastl, George Clifford
Magee, Michael Jack
Rowland, John H
Smith, William Kirby

PUERTO RICO
Lewis, Brian Kreglow
Lewis, Brian Murray
Montes, Maria Eugenia
Wolstenholme, Wayne W

ALBERTA
Bennion, Douglas Wilford
Chang, Chi
Fu, Cheng-Tze
George, Ronald Edison

Harvey, Ross Buschlen
Honsaker, John Leonard
Plambeck, James Alan
Prepas, Ellie E
Ulagaraj, Munivandy Seydunganallur
White, Lee James

BRITISH COLUMBIA
Cercone, Nicholas Joseph
Chanson, Samuel T
Clark, Stanley Ross
Davis, Wayne Alton
Davison, Allan John
Daykin, Philip Norman
Dill, John C
Dower, Gordon Ewbank
Ehle, Byron Leonard
Gilmore, Paul Carl
Hafer, Louis James
Halabisky, Lorne Stanley
Harrop, Ronald
Kennedy, James M
MacDonald, John Lauchlin
Manning, Eric G(eorge)
Odeh, Robert Eugene
Sheehan, Bernard Stephen
Sterling, Theodor David
Totten, James Edward
Van Emden, Maarten Herman
Wade, Adrian Paul

MANITOBA
King, Peter Ramsay
Tait, John Charles
Williams, Hugh Cowie

NEW BRUNSWICK
Wasson, W(alter) Dana

NEWFOUNDLAND
Longerich, Henry Perry
Shieh, John Shunen
Wroblewski, Joseph S

NOVA SCOTIA
Kruse, Robert Leroy
McEwan, Alan Thomas
Matthews, Frederick White

ONTARIO
Argyropoulos, Stavros Andreas
Ball, Gordon Charles
Bartels, Richard Harold
Batra, Tilak Raj
Bauer, Michael Anthony
Beatty, John C
Beylerian, Nurel
Bioleau, Luc J R
Boorn, Andrew William
Boulton, Peter Irwin Paul
Calder, Peter N
Chu, Chun-Lung (George)
Clarke, James Newton
Coll, David C
Cowan, Donald D
Fabian, Robert John
Fairman, Frederick Walker
Gaherty, Geoffrey George
Gentleman, Jane Forer
Gentleman, William Morven
Gotlieb, C(alvin) C(arl)
Graham, James W
Guevremont, Roger M
Haynes, Robert Brian
Hull, Thomas Edward
Ionescu, Dan
Jardine, D(onald) A(ndrew)
Jorch, Harald Heinrich
Lasker, George Eric
Lawson, John Douglas
Linders, James Gus
Meads, Jon A
Molder, S(annu)
Morris, Lawrence Robert
Munro, James Ian
Raaphorst, G Peter
Read, Ronald Cedric
Redish, Kenneth Adair
Reiter, Raymond
Riordon, J(ohn) Spruce
Robertson, George Wilber
Rowe, Ronald Kerry
Slonim, Jacob
Steiner, George
Tavares, Stafford Emanuel
Thierrin, Gabriel
Thomas, Paul A V
Tsichritzis, Dennis
Vanstone, Scott Alexander
Waddington, Raymond
Watson, Jeffrey
West, Eric Neil
White, George Michael
Wilson, John Cleland
Wong, Johnny Wai-Nang
Wood, Derick

QUEBEC
Agarwal, Vinod Kumar
Bui, Tien Dai
Davies, Roger
Ferguson, Michael John
Hyder, Syed S
Jordan, Byron Dale

Lecours, Michel
Newborn, Monroe M
Opatrny, Jaroslav
Poussart, Denis
Shizgal, Harry M
Suen, Ching Yee
Vaucher, Jean G
Zucker, Steven Warren

SASKATCHEWAN
Gupta, Madan Mohan
Symes, Lawrence Richard

OTHER COUNTRIES
Arimoto, Suguru
Baggi, Denis Louis
Bellanger, Maurice G
Blackburn, Jacob Floyd
Brown, Richard Harland
Burkhardt, Walter H
Bush, George Clark
Cartwright, Hugh Manning
Chen, Tien Chi
Cheng, Kuang-Fu
Cryer, Colin Walker
De Renobales, Mertxe
Fujiwara, Hideo
Glover, Francis Nicholas
Gritzmann, Peter
Halkias, Christos
Hall, Peter
Humphrey, Albert S
Ilten, David Frederick
Kim, Wan H(ee)
Ku, Victor Chia-Tai
Levialdi, Stefano
Lieth, Helmut Heinrich Friedrich
Low, Chow-Eng
Muskat, Joseph Baruch
Nestvold, Elwood Olaf
Pau, Louis F
Paulish, Daniel John
Querinjean, Pierre Joseph
Raviv, Josef
Sargent, David Fisher
Scheiter, B Joseph Paul
Sekine, Yasuji
Shafi, Mohammad
Shamir, Adi
Siegel, Herbert
Siklossy, Laurent
Vamos, Tibor
Wilkes, Maurice V

Hardware Systems

ALABAMA
Jolley, Homer Richard
Stone, Max Wendell

ALASKA
Watkins, Brenton John

ARIZONA
Dybvig, Paul Henry
Haynes, Munro K
McDaniel, Terry Wayne
Shoemaker, Richard Lee
Woodfill, Marvin Carl

CALIFORNIA
Albert, Paul A(ndre)
Allen, Charles A
Bardin, Russell Keith
Bertin, Michael C
Bic, Lubomir
Birecki, Henryk
Bush, George Edward
Butler, Jon Terry
Chen, Tsai Hwa
Davis, Alan Lynn
Dutt, Nikil D
Eaton, John Kelly
Fasang, Patrick Pad
Flynn, Michael J
Friedlander, Carl B
Gesner, Bruce D
Hennessy, John LeRoy
Herzog, Gerald B(ernard)
Hu, Sung Chiao
Kalbach, John Frederick
Kirk, Donald Evan
Kodres, Uno Robert
Kreuzer, Lloyd Barton
Langdon, Glen George, Jr
Latta, Gordon
Lee, Kenneth
Lieber, Richard L
Lin, Wen-C(hun)
Loomis, Herschel Hare, Jr
Meagher, Donald Joseph
Melvin, Jonathan David
Meyer, Lhary
Michaelson, Jerry Dean
Mitra, Sanjit K
Monier, Louis Marcel
Morgenstern, Matthew
Motteler, Zane Clinton
Napolitano, Leonard Michael, Jr
Nikora, Allen P
Onton, Aare
Parker, Alice Cline
Phipps, Peter Beverley Powell

Prasanna Kumar, V K
Requa, Joseph Earl
Riegert, R(ichard) P(aul)
Roberts, Charles Sheldon
Schechtman, Barry H
Schwartz, Morton Donald
Shaka, Athan James
Shar, Leonard E
Silvester, John Andrew
Sterling, Warren Martin
Strohman, Rollin Dean
Sung, Chia-Hsiaing
Swartzlander, Earl Eugene, Jr
Thomas, Hubert Jon
Thompson, David A
Winzenread, Marvin Russell
Young, Frank

COLORADO
Allen, James Lamar
Bennett, W Scott
Brady, Douglas MacPherson
Cathey, Wade Thomas, Jr
Clark, Jon D
Etter, Delores Maria
Frick, Pieter A
Furcinitti, Paul Stephen
Johnson, Gearold Robert
Lee, Yung-Cheng
Malaiya, Yashwant Kumar
Mueller, Robert Andrew
Wiatrowski, Claude Allan

CONNECTICUT
Capwell, Robert J
Goldman, Ernest Harold
Hudak, Paul Raymond
McDonald, John Charles

DELAWARE
Amer, Paul David
Christie, Phillip
Hayman, Alan Conrad
Holob, Gary M
Moore, Charles B(ernard)
Scofield, Dillon Foster

DISTRICT OF COLUMBIA
Bloch, Erich
Curtin, Frank Michael
Friedman, Arthur Daniel
Garcia, Oscar Nicolas
Lehmann, John R(ichard)

FLORIDA
Bonn, T(heodore) H(ertz)
Cengeloglu, Yilmaz
Coulter, Neal Stanley
Ege, Raimund K
Hales, Everett Burton
Hoffman, Thomas R(ipton)
Marcovitz, Alan Bernard

GEORGIA
Alford, Cecil Orie
Bomar, Lucien Clay
Cohn, Charles Erwin
Currie, Nicholas Charles
Husson, Samir S
Techo, Robert

HAWAII
Kinariwala, Bharat K

ILLINOIS
Chien, Andrew Andai
Coplien, James O
Day, Paul Palmer
DeFanti, Thomas A
Domanik, Richard Anthony
Flora, Robert Henry
Fuchs, W Kent
Hajj, Ibrahim Nasri
Harris, Lowell Dee
Kaplan, Daniel Moshe
Kubitz, William John
Merkelo, Henri
Murphy, Gordon J
Padua, David A
Pardo, Richard Claude
Polychronopoulos, Constantine Dimitrius
Reed, Daniel A
Slotnick, Daniel Leonid
Stafford, John William
Wojcik, Anthony Stephen

INDIANA
Aprahamian, Ani
Brown, Buck F(erguson)
Esch, Harald Erich
Harbron, Thomas Richard
Hinkle, Charles N(elson)
Place, Ralph L
Siegel, Howard Jay
Stanley, Gerald R
Western, Arthur Boyd
Wise, David Stephen

IOWA
Stewart, Robert Murray, Jr

KANSAS
Smaltz, Jacob Jay

KENTUCKY
Villareal, Ramiro

MAINE
Carter, William Caswell

MARYLAND
Alperin, Harvey Albert
Berson, Alan
Bourbakis, Niklaos G
Brenner, Alfred Ephraim
Buell, Duncan Alan
Hammond, Charles E
Hevner, Alan Raymond
Meyer, Gerard G L
South, Hugh Miles
Susskind, Alfred K(riss)

MASSACHUSETTS
Adrion, William Richards
Alter, Ralph
Chang, Robin
Cotter, Douglas Adrian
Dennison, Byron Lee
Eckhouse, Richard Henry
Foster, Caxton Croxford
Gagliardi, Ugo Oscar
Guenther, R(ichard)
Hofstetter, Edward
Hopkins, Albert Lafayette, Jr
Lis, Steven Andrew
Menon, Premachandran R(ama)
Moss, J Eliot B
Raffel, Jack I
Rosenberg, Arnold Leonard
Simovici, Dan
Speliotis, Dennis Elias
Stewart, Lawrence Colm
Stiffler, Jack Justin
Strong, Robert Michael

MICHIGAN
Becher, William D(on)
Be Ment, Spencer L
Bickart, Theodore Albert
Brumm, Douglas B(ruce)
Horvath, Ralph S(teve)
Larrowe, Boyd T
Nevius, Timothy Alfred
Patt, Yale Nance
Scott, Norman R(oss)
Westervelt, Franklin Herbert

MINNESOTA
Du, David Hung-Chang
Franta, William Roy
Gilbert, Barry Kent
Kain, Richard Yerkes
Lo, David S(hih-Fang)
MacKellar, William John
Sahni, Sartaj Kumar
Thomborson, Clark D

MISSISSIPPI
Burke, Robert Wayne
Cress, Daniel Hugg
Elsherbeni, Atef Zakaria

MISSOURI
Atwood, Charles LeRoy
Baer, Robert W
Franklin, Mark A
McFarland, William D
Morgan, Nancy H
Morley, Robert Emmett, Jr
Moss, Randy Hays
Rouse, Robert Arthur
Taylor, Ralph Dale

NEBRASKA
Gale, Douglas Shannon, II

NEVADA
Anderson, Arthur George
Benfield, Charles W(illiam)

NEW HAMPSHIRE
Brender, Ronald Franklin
Crowell, Merton Howard

NEW JERSEY
Burkhardt, Kenneth J
Carr, William N
Crochiere, Ronald E
Duttweiler, Donald Lars
French, Larry J
Gittleson, Stephen Mark
Goldstein, Philip
Graf, Hans Peter
Haas, Zygmunt
Horn, David Nicholas
Horng, Shi-Jinn
Katz, Sheldon Lane
Kobrin, Robert Jay
Krzyzanowski, Paul
Kuo, Ying L
Lo, Arthur W(unien)
Mersten, Gerald Stuart
Michelson, Leslie Paul
Murthy, Srinivasa K R
Nelson, Gregory Victor
Ninke, William Herbert
Schlam, Elliott
Schmidt, Barnet Michael

Sears, Raymond Warrick, Jr
Shahbender, R(abah) A(bd-El-Rahman)
Shieh, Ching-Chyuan
Taylor, Harold Evans
Thompson, Kenneth Lane
Tsaliovich, Anatoly
Woo, Nam-Sung
Zwass, Vladimir

NEW MEXICO
Butler, Harold S
Giesler, Gregg Carl
Hardin, James T
Muir, Patrick Fred
Phister, Montgomery, Jr
Smith, William Conrad

NEW YORK
Albicki, Alexander
Balabanian, Norman
Bashkow, Theodore R(obert)
Bossen, Douglas C
Brennemann, Andrew E(rnest), Jr
Carleton, Herbert Ruck
Chi, Chao Shu
Chrien, Robert Edward
Clarke, Kenneth Kingsley
Cole, James A
Costello, Richard Gray
Csermely, Thomas J(ohn)
Das, Pankaj K
Debany, Warren Harding, Jr
Dwyer, Harry, III
Eichelberger, Edward B
Fiedler, Harold Joseph
Fordon, Wilfred Aaron
Frank, Thomas Stolley
Gabay, Jonathan Glenn
Gottlieb, Allan
Grobman, Warren David
Heller, William R
Keyes, Robert William
Kotchoubey, Andrew
Landauer, Rolf William
Lorenzen, Jerry Alan
Maguire, Gerald Quentin, Jr
Maissel, Leon I
Myers, Robert Anthony
Notaro, Anthony
Padegs, Andris
Peled, Abraham
Pingali, Keshav Kumar
Pomerene, James Herbert
Puttlitz, Karl Joseph
Rauscher, Tomlinson Gene
Rogers, Edwin Henry
Rosenfeld, Jack Lee
Rosenkrantz, Daniel J
Schamberger, Robert Dean
Shooman, Martin L
Spencer, David R
Stern, Miklos
Su, Stephen Y H
Subramanian, Mani M
Syed, Ashfaquzzaman
Tatarczuk, Joseph Richard
Torng, Hwa-Chung
Trexler, Frederick David
Tummala, Rao Ramamohana
Wittie, Larry Dawson
Zukowski, Charles Albert

NORTH CAROLINA
Adelberger, Rexford E
Chen, Su-shing
Coates, Clarence L(eroy), Jr
D'Arruda, Jose Joaquim
Franzon, Paul Damian
Gray, James P
Hutchins, William R(eagh)
Javel, Eric
Johnson, G Allan
Kanopoulos, Nick
Nagle, H Troy
Pearlstein, Robert David
Wright, William V(aughn)
Zahed, Hyder Ali

NORTH DAKOTA
Tallman, Dennis Earl

OHIO
Dehne, George Clark
Gilfert, James C(lare)
Hollingsworth, Ralph George
Hubin, Wilbert N
Ismail, Amin Rashid
James, Thomas Ray
Jendrek, Eugene Francis, Jr
Moon, Tag Young
Papachristou, Christos A
Rajsuman, Rochit
Ramamoorthy, Panapakkam A
Rose, Charles William
Ross, Charles Burton
Taylor, Barney Edsel
Webster, Allen E
Zheng, Yuan Fang

OKLAHOMA
Grigsby, Ronald Davis
Gunter, Deborah Ann
Vanderwiele, James Milton

Hardware Systems (cont)

OREGON
Kocher, Carl A
Schwartz, James William
Stewart, Bradley Clayton

PENNSYLVANIA
Aburdene, Maurice Felix
Arrington, Wendell S
Barbacci, Mario R
Druffel, Larry Edward
Fisher, Tom Lyons
Goutmann, Michel Marcel
Grassi, Vincent G
Hardesty, Patrick Thomas
Humphrey, Watts S
Klafter, Richard D(avid)
Kung, Hsiang-Tsung
Larky, Arthur I(rving)
Mitchell, Tom M
Mondal, Kalyan
Murphy, Bernard T
Pinkerton, John Edward
Preston, Kendall, Jr
Reigel, Earl William
Rhodes, Donald Frederick
Tzeng, Kenneth Kai-Ming
Vastola, Francis J
Vogt, William G(eorge)
Wagh, Meghanad D
Walker, David Kenneth
Warme, Paul Kenneth

RHODE ISLAND
Gabriel, Richard Francis

SOUTH CAROLINA
Bennet, Archie Wayne
Bennett, Archie Wayne
Croft, George Thomas
Goode, Scott Roy
Lambert, Jerry Roy

SOUTH DAKOTA
Riemenschneider, Albert Louis

TENNESSEE
Bishop, Asa Orin, Jr
Campbell, Warren Elwood
Rajan, Periasamy Karivaratha
Skinner, George T
Trivedi, Mohan Manubhai

TEXAS
Abraham, Jacob A
Brakefield, James Charles
Brantley, William Cain, Jr
Brock, James Harvey
Franke, Ernest A
Garner, Harvey L(ouis)
Gonzalez, Mario J
Jump, J Robert
Knapp, Roger Dale
Lewis, Donald Richard
Lipovski, Gerald John (Jack)
Malek, Miroslaw
Markenscoff, Pauline
Martin, Norman Marshall
Nather, Roy Edward
Nute, C Thomas
Perez, Ricardo
Pritchard, John Paul, Jr
Rhyne, V(ernon) Thomas
Silberschatz, Abraham
Smith, Reid Garfield
Vines, Darrell Lee
Volz, Richard A
Watt, Joseph T(ee), Jr

UTAH
Hollaar, Lee Allen
Smith, Kent Farrell

VIRGINIA
Anderson, Walter L(eonard)
Beale, Guy Otis
Cook, Robert Patterson
Crawford, Daniel J
Eldridge, Charles A
Fisher, Arthur Douglas
Garcia, Albert B
Guerber, Howard P(aul)
Hilger, James Eugene
Larson, Arvid Gunnar
Liceaga, Carlos Arturo
McGrath, Arthur Kevin
Mathis, Robert Fletcher
Rony, Peter R(oland)
Saltz, Joel Haskin
Toida, Shunichi
Waxman, Ronald
Webber, John Clinton

WASHINGTON
Preikschat, F(ritz) K(arl)
Seamans, David A(lvin)
Smith, Burton Jordan
Somani, Arun Kumar
Spillman, Richard Jay

WEST VIRGINIA
Tewksbury, Stuart K

WISCONSIN
Brookshear, James Glenn
Dietmeyer, Donald L
Dobson, David A
Harris, J Douglas
Hind, Joseph Edward
Jacobi, George (Thomas)
Levine, Leonard P
Moore, Edward Forrest
Rosenthal, Jeffrey
Viswanathan, Ramaswami

WYOMING
Howell, Robert Richard

BRITISH COLUMBIA
Hafer, Louis James

NEWFOUNDLAND
Rahman, Md Azizur

ONTARIO
Emraghy, Hoda Abdel-Kader
Jullien, Graham Arnold
Majithia, Jayanti
Mouftah, Hussein T
Thomas, Paul A V
Wilson, John D(ouglas)

QUEBEC
Agarwal, Vinod Kumar
Atwood, John William
De Mori, Renato
Desai, Bipin C
Eddy, Nelson Wallace
Marin, Miguel Angel

OTHER COUNTRIES
Burkhardt, Walter H
Fujiwara, Hideo
Goguen, Joseph A, Jr
Halkias, Christos
Kim, Myunghwan
Kuroyanagi, Noriyoshi
Lewis, Arnold Leroy, II
Maas, Peter

Information Science & Systems

ARIZONA
Lorents, Alden C

ARKANSAS
Alquire, Mary Slanina (Hirsch)
Sedelow, Sally Yeates
Sedelow, Walter Alfred, Jr

CALIFORNIA
Alper, Marshall Edward
Birnbaum, Joel S
Block, Richard B
Bradley, Hugh Edward
Chock, Margaret Irvine
Cooper, William S
Firdman, Henry Eric
Glaser, Myron B(arnard)
Grethe, Guenter
Groner, Gabriel F(rederick)
Hellerman, Herbert
Hsu, Charles Teh-Ching
Jackson, Durward P
Jacobson, Allan Stanley (Bud)
Kerfoot, Branch Price
Lyman, John (Henry)
McCarthy, John Lockhart
Maillot, Patrick Gilles
Mathis, Ronald Floyd
Miel, George J
Minckler, Tate Muldown
Mulligan, Geoffrey C
Rohm, C E Tapie, Jr
Rulifson, Johns Frederick
Selinger, Patricia Griffiths
Star, Jeffrey L
Strand, Timothy Carl
Williams, Donald Spencer
Zebroski, Edwin L

COLORADO
Call, Patrick Joseph
Reiter, Elmar Rudolf

CONNECTICUT
Maizell, Robert Edward
Newman, Robert Weidenthal
Rezek, Geoffrey Robert
Sittig, Dean Forrest
Stuck, Barton W

DELAWARE
Bartkus, Edward Peter
Lehr, Gary Fulton
Panar, Manuel
Skolnik, Herman
Talley, John Herbert

DISTRICT OF COLUMBIA
Garavelli, John Stephen
Poon, Bing Toy
Robbins, Robert John
Stein, Marjorie Leiter

FLORIDA
Couturier, Gordon W
Kinnie, Irvin Gray
Korwek, Alexander Donald
Lurie, Arnold Paul
Swart, William W
Zanakis, Stelios (Steve) H

GEORGIA
Aronson, Jay E
Blanke, Jordan Matthew
Kollig, Heinz Philipp
Zunde, Pranas

ILLINOIS
Beecher, Christopher W W
Bookstein, Abraham
Buntrock, Robert Edward
Evens, Martha Walton
Foster, George A, Jr
Hoffman, Gerald M
Wells, Jane Frances

INDIANA
Eberts, Ray Edward
Koch, Kay Frances
Robertson, Edward L
Wynne, Bayard Edmund

KENTUCKY
Cooke, Samuel Leonard
Tran, Long Trieu

LOUISIANA
Kraft, Donald Harris

MAINE
Guidi, John Neil

MARYLAND
Bagg, Thomas Campbell
Blum, Bruce I
Buell, Duncan Alan
Campbell, William J
Cotton, Ira Walter
Henderson, Madeline M Berry
Johnson, Frederick Carroll
Kornman, Brent D
Little, Joyce Currie
Munro, Ronald Gordon
Randolph, Lynwood Parker

MASSACHUSETTS
Barlas, Julie S
Brayer, Kenneth
Champine, George Allen
Dyer, Charles Austen
Gilfix, Edward Leon
Giuliano, Vincent E
McKnight, Lee Warren
Morris, Robert
Moss, J Eliot B
O'Neil, Patrick Eugene
Prerau, David Stewart
Ross, Douglas Taylor
Royston, Richard John
Wasserman, Jerry

MICHIGAN
Bowman, Carlos Morales
Coburn, Joel Thomas
Farah, Badie Naiem
Kulier, Charles Peter
Sawatari, Takeo
Shahabuddin, Syed
Westland, Roger D(ean)

MINNESOTA
Dybvig, Douglas Howard
Hoffmann, Thomas Russell
March, Salvatore T
Pong, Ting-Chuen

MISSISSIPPI
Jennings, David Phipps
Leese, John Albert

MISSOURI
Morgan, Nancy H

NEW HAMPSHIRE
Egan, John Frederick

NEW JERSEY
Abeles, Francine
Andrews, Ronald Allen
Badalamenti, Anthony Francis
Blair, Grant Clark
Brown, William Stanley
Cardarelli, Joseph S
Carney, Richard William James
Hang, Hsueh-Ming
Horn, David Nicholas
Huber, Melvin Lefever
Johnson, Layne Mark
Kaback, Stuart Mark
McKenna, James
Matula, Richard Allen
Mersten, Gerald Stuart
Popper, Robert David
Poucher, John Scott
Rosin, Robert Fisher
Shieh, Ching-Chyuan
Sinclair, Brett Jason

Verdu, Sergio
Williams, Myra Nicol
Woodruff, Hugh Boyd
Zwass, Vladimir

NEW YORK
Ballou, Donald Pollard
Chang, Ifay F
Coden, Michael H
Farb, Edith
Fine, Terrence Leon
Gabay, Jonathan Glenn
Gannett, E K
Goldschmidt, Eric Nathan
Golibersuch, David Clarence
Hannay, David G
Harris, Alex L
Howell, Mary Gertrude
Koplowitz, Jack
Lerman, Steven I
Lommel, J(ames) M(yles)
McGee, James Patrick
Mowshowitz, Abbe
Neeper, Ralph Arnold
Quaile, James Patrick
Salwen, Martin J
Shenolikar, Ashok Kumar
Shuey, R(ichard) L(yman)
Windsor, Donald Arthur

NORTH CAROLINA
Butz, Robert Frederick
Godschalk, David Robinson
Kurtz, A Peter
Smith, William Adams, Jr
Starmer, C Frank

NORTH DAKOTA
Magel, Kenneth I

OHIO
Chamis, Alice Yanosko
Dunn, Horton, Jr
Hollis, William Frederick
Kaufman, John Gilbert, Jr
Pausch, Jerry Bliss

OKLAHOMA
McDevitt, Daniel Bernard

OREGON
Beaulieu, John David

PENNSYLVANIA
Anandalingam, G
Bhagavatula, VijayaKumar
Di Cuollo, C John
Farlee, Rodney Dale
Franz, Edmund C(larence)
Froehlich, Fritz Edgar
Garfield, Eugene
Goutmann, Michel Marcel
Grover, Carole Lee
Ho, Thomas Inn Min
Kam, Moshe
Kennedy, Harvey Edward
Korfhage, Robert R
Palmer, Jon (Carl)
Skoner, Peter Raymond
Treu, Siegfried

SOUTH CAROLINA
Eastman, Caroline Merriam

SOUTH DAKOTA
Leslie, Jerome Russell

TENNESSEE
Gonzalez, Rafael C
Kanciruk, Paul
LeBlanc, Larry Joseph
Trubey, David Keith
Upadhyaya, Belle Raghavendra
Voorhees, Larry Donald

TEXAS
Baggett, Millicent (Penny)
Bishop, Robert H
Brock, James Harvey
Fix, Ronald Edward
Gorry, G Anthony
Hedstrom, John Richard
Hennessey, Audrey Kathleen
Jackson, Eugene Bernard
Keller, Thomas W
McLeod, Raymond, Jr
Randolph, Paul Herbert
Sobol, Marion Gross
Wilhoit, Randolph Carroll
Wood, Kristin Lee
Wyllys, Ronald Eugene
Yao, James T-P

VIRGINIA
Awad, Elias M
Deal, George Edgar
Fox, Edward A
Gregory, Arthur Robert
Hodge, Donald Ray
Hunt, V Daniel
Ivanetich, Richard John
Johnson, Charles Minor
Liceaga, Carlos Arturo
Morris, Cecil Arthur, Sr

Rine, David C
Shetler, Antoinette (Toni)

WASHINGTON
Mendelson, Martin
Shin, Suk-han

WISCONSIN
McClenahan, William St Clair
Ramakrishnan, Raghu

NEW BRUNSWICK
Faig, Wolfgang
Lee, Y C

NEWFOUNDLAND
Thomeier, Siegfried

ONTARIO
Bauer, Michael Anthony
Mouftah, Hussein T
Slonim, Jacob

QUEBEC
De Mori, Renato
Li, Zi-Cai
Morgera, Salvatore Domenic

OTHER COUNTRIES
Arimoto, Suguru
Garcia-Santesmases, Jose Miguel
Kafri, Oded
Kaiser, Wolfgang A
Levialdi, Stefano
Levialdi, Stefano
Malah, David

Intelligent Systems

ALABAMA
Carlo, Waldemar Alberto
Fiesler, Emile
Habtemariam, Tsegaye
Harding, Thomas Hague
Werkheiser, Arthur H, Jr

ARIZONA
Cochran, Jeffery Keith
Findler, Nicholas Victor
Iverson, A Evan
Meieran, Eugene Stuart
Ryan, Thomas Wilton
Schowengerdt, Robert Alan
Trautman, Rodes
Young, Kenneth Christie

ARKANSAS
Asfahl, C Ray
Sedelow, Sally Yeates
Sedelow, Walter Alfred, Jr
Wrobel, Joseph Stephen

CALIFORNIA
Abbott, Seth R
Agoston, Max Karl
Ahumada, Albert Jil, Jr
Akonteh, Benny Ambrose
Altes, Richard Alan
Antonsson, Erik Karl
Arens, Yigal
Beach, Sharon Sickel
Bekey, George A(lbert)
Bendat, Julius Samuel
Bhanu, Bir
Bic, Lubomir
Birnir, Björn
Bobrow, Daniel G
Brutlag, Douglas Lee
Calabrese, Philip G
Carr, John Weber, III
Chang, Tien-Lin
Chellappa, Ramalingam
Cooper, William S
Curry, Bo(stick) U
Curtis, Earl Clifton, Jr
Dev, Parvati
Dornfeld, David Alan
Duda, Richard Oswald
Eckhardt, Wilfried Otto
Elspas, B(ernard)
Erman, Lee Daniel
Firdman, Henry Eric
Fischler, Martin A(lvin)
Freedy, Amos
Friedlander, Carl B
Gesner, Bruce D
Goldberg, Robert N
Green, Claude Cordell
Groce, David Eiben
Gutfinger, Dan Eli
Hackwood, Susan
Hamlin, Griffith Askew, Jr
Herbst, Noel Martin
Hightower, James K
Holeman, Dennis Leigh
Holtzman, Samuel
Hopfield, John Joseph
Hudson, Cecil Ivan, Jr
Jacob, George Korathu
Jacobson, Alexander Donald
Kagiwada, Harriet Hatsune
Kaplan, Ronald M
Klahr, Philip

Kreutz-Delgado, Kenneth Keith
Leavy, Paul Matthew
Lin, Wen-C(hun)
Loebner, Egon Ezriel
McCarthy, John Michael
MacMillian, Stuart A
Manna, Zohar
Matsumura, Kenneth N
Moran, Thomas Patrick
Morgenstern, Matthew
Mueller, Thomas Joseph
Murphy, Roy Emerson
Nelson, David A
Nesbit, Richard Allison
Nielsen, Norman Russell
Painter, Jeffrey Farrar
Parker, Donn Blanchard
Parkison, Roger C
Patterson, Tim J
Pearl, Judea
Perkins, Walton A, III
Phillips, Veril LeRoy
Ralston, Elizabeth Wall
Requicha, Aristides A G
Rowe, Lawrence A
Rulifson, Johns Frederick
Russell, Stuart Jonathan
Sadler, Charles Robinson, Jr
Sakaguchi, Dianne Koster
Salisbury, Stanley R
Schoenfeld, Alan Henry
Shepanski, John Francis
Skrzypek, Josef
Smith, Ora E
Sridharan, Natesa S
Surko, Pamela Toni
Teague, Tommy Kay
Thomas, Graham Havens
Van Klaveren, Nico
Wagner, Carl E
Wagner, William Gerard
Wexler, Jonathan David
White, Ray Henry
Williams, Donald Spencer
Woodriff, Roger L
Zadeh, L(otfi) A

COLORADO
Augusteijn, Marijke Francina
Clark, Jon D
Glover, Fred William
Mueller, Robert Andrew
Phillips, Keith L
Reiter, Elmar Rudolf
Troxell, Wade Oakes
Veirs, Val Rhodes

CONNECTICUT
Apte, Chidanand
Dreyfus, Marc George
Hanson, Trevor Russell
Streit, Roy Leon

DELAWARE
Abremski, Kenneth Edward
Chester, Daniel Leon
Dhurjati, Prasad S
Lehr, Gary Fulton
Lund, John Turner
Owens, Aaron James
Shipman, Lester Lynn

DISTRICT OF COLUMBIA
Garcia, Oscar Nicolas
Hartman, Patrick James
Moraff, Howard
Tangney, John Francis

FLORIDA
Biegel, John E
Blatt, Joel Herman
Bressler, Steven L
Buoni, Frederick Buell
Cengeloglu, Yilmaz
Clarke, Thomas Lowe
Detweiler, Steven Lawrence
Duersch, Ralph R
Genaidy, Ashraf Mohamed
Hand, Thomas
Hoffman, Frederick
Kandel, Abraham
Kirmse, Dale William
Martsolf, J David
Navlakha, Jainendra K
Nevill, Gale E(rwin), Jr
Peart, Robert McDermand
Seireg, Ali A
Tebbe, Dennis Lee
Tulenko, James Stanley
Winton, Charles Newton
Young, Tzay Y

GEORGIA
Alford, Cecil Orie
Aronson, Jay E
Foley, James David
Gibson, John Michael
Lee, Kok-Meng
Nevins, Arthur James
Prince, M(orris) David

HAWAII
Itoga, Stephen Yukio

IDAHO
Girse, Robert Donald
Johnson, John Alan
Stock, Molly Wilford

ILLINOIS
Agrawal, Arun Kumar
Ahuja, Narendra
Boyle, James Martin
Catrambone, Joseph Anthony, Sr
Chang, Shi Kuo
Coley, Ronald Frank
Crocker, Diane Winston
DeFanti, Thomas A
Evens, Martha Walton
Gabriel, John R
Harrington, Joseph Anthony
Heller, Barbara Ruth
Henschen, Lawrence Joseph
Jain, Ravinder Kumar
Krug, Samuel Edward
Lamont, Patrick
McAlpin, John Harris
Mount, Bertha Lauritzen
Polychronopoulos, Constantine Dimitrius
Thisted, Ronald Aaron
Wallis, Walter Denis
Wilson, Howell Kenneth

INDIANA
Ersoy, Okan Kadri
Hanson, Andrew Jorgen
Jamieson, Leah H
Kak, Avinash Carl
Lehto, Mark R
Mitchell, Owen Robert
Nof, Shimon Y
Patterson, Larry K
Salvendy, Gavriel
Siegel, Howard Jay
Stampfli, Joseph
Uhran, John Joseph, Jr

IOWA
Jeyapalan, Kandiah
Myers, Glenn Alexander
Quinn, Loyd Yost

KANSAS
Farokhi, Saeed
Grzymala-Busse, Jerzy Witold
Lerner, David Evan
Roddis, Winifred Mary Kim
Shanmugan, K Sam

KENTUCKY
Gruver, William A
McGinness, William George, III

LOUISIANA
Gajendar, Nandigam
Kraft, Donald Harris

MAINE
Guidi, John Neil
Tucker, Allen B

MARYLAND
Albus, James S
Benokraitis, Vitalius
Blum, Bruce I
Blum, Harry
Bonavita, Nino Louis
Bourbakis, Niklaos O
Campbell, William J
Chang, Alfred Tieh-Chun
Chapin, Edward William, Jr
Corliss, John Burt
Darden, Lindley
Dixon, John Kent
Eisenman, Richard Leo
Finkelstein, Robert
Hevner, Alan Raymond
Hodge, David Charles
Jenkins, Robert Edward
Jones, George R
Keenan, Thomas Aquinas
Kornman, Brent D
Ligomenides, Panos Aristides
Minker, Jack
Nanzetta, Philip Newcomb
Nau, Dana S
Powell, Edward Gordon
Rosenfeld, Azriel
Schlesinger, Judith Diane
Sidhu, Deepinder Pal
Sigillito, Vincent George
Tripathi, Satish K
Vogl, Thomas Paul
Weaver, Christopher Scot
Wilkins, Michael Gray

MASSACHUSETTS
Abbott, Robert Classie
Becker, David Stewart
Brown, Cynthia Ann
Burke, Leonarda
Carpenter, Gail Alexandra
Collins, Allan Meakin
Doyle, Jon
Dudgeon, Dan Ed
Dyer, Charles Austen
Dym, Clive L
Evans, Thomas George

Frawley, William James
Futrelle, Robert Peel
Kolodzy, Paul John
McConnell, Robert Kendall
Madden, Stephen James, Jr
Nassi, Isaac Robert
Nesbeda, Paul
Pauker, Stephen Gary
Prerau, David Stewart
Riseman, Edward M
Roberts, Louis W
Royston, Richard John
Salveter, Sharon Caroline
Seibel, Frederick Truman
Smith, Raoul Normand
Strong, Robert Michael
Waters, Richard C(abot)
Woods, William A

MICHIGAN
Conrad, Michael
Elrod, David Wayne
Enke, Christie George
Farah, Badie Naiem
Grosky, William Irvin
Hassoun, Mohamad Hussein
Holland, John Henry
Holland, Steven William
Kochen, Manfred
Krawetz, Stephen Andrew
Ku, Albert B
Lindsay, Robert Kendall
Marko, Kenneth Andrew
Page, Carl Victor
Ramamurthy, Amurthur C
Rillings, James H
Rosenbaum, Manuel
Sawatari, Takeo
Sethi, Ishwar Krishan
Stivender, Donald Lewis
Stockman, George C
Stout, Quentin Fielden
Teichroew, Daniel
Tuchinsky, Philip Martin
Westervelt, Franklin Herbert

MINNESOTA
Bahn, Robert Carlton
Connelly, Donald Patrick
Gini, Maria Luigia
Joseph, Earl Clark, II
Knutson, Charles Dwaine
Kumar, Vipin
La Bonte, Anton Edward
Pierce, Keith Robert
Pong, Ting-Chuen
Sadjadi, Firooz Ahmadi
Spurrell, Francis Arthur

MISSISSIPPI
Burke, Robert Wayne
Shim, Jung P

MISSOURI
Atwood, Charles LeRoy
Ball, William E(rnest)
Bovopoulos, Andreas D
Franz, John Matthias
Magill, Robert Earle
Peterson, Gerald E
Rodin, Ervin Y
Sabharwal, Chaman Lal
Wagner, Robert G
Yohe, Cleon Russell

MONTANA
Latham, Don Jay
Wright, Alden Halbert

NEVADA
Minor, John Threecivelous

NEW HAMPSHIRE
Frantz, Daniel Raymond
Misra, Alok C

NEW JERSEY
Afshar, Siroos K
Amarel, S(aul)
Bernstein, Lawrence
Burkhardt, Kenneth J
Cooper, Paul W
Gabbe, John Daniel
Gale, William Arthur
Graf, Hans Peter
Grek, Boris
Hubbard, William Marshall
Jurkat, Martin Peter
Kalley, Gordon S
Kirch, Murray R
Kuhl, Frank Peter, Jr
Layzer, Arthur James
McGuinness, Deborah Louise
Most, Joseph Morris
Nonnenmann, Uwe
Schindler, Max J
Schlam, Elliott
Shieh, Ching-Chyuan
Stengel, Robert Frank
Tishby, Naftali Z
Verma, Pramode Kumar
Weissman, Paul Morton
Woodruff, Hugh Boyd

Intelligent Systems (cont)

NEW MEXICO
Castle, John Granville, Jr
Frank, Robert Morris
Gilbert, Alton Lee
Luger, George F
Niemczyk, Thomas M
Osbourn, Gordon Cecil
Otway, Harry John
Ross, Timothy Jack
Sung, Andrew Hsi-Lin

NEW YORK
Allen, James Frederick
Anderson, Peter Gordon
Campbell, Gregory August
Chi, Chao Shu
Comly, James B
Damerau, Frederick Jacob
Dimmler, D(ietrich) Gerd
Dwyer, Harry, III
Fenrich, Richard Karl
Ferguson, David Lawrence
Gabay, Jonathan Glenn
Genova, James John
Golibersuch, David Clarence
Griesmer, James Hugo
Hajela, Prabhat
Hastings, Harold Morris
Herrington, Lee Pierce
Hong, Se June
Hudak, Michael J
Kalogeropoulos, Theodore E
Karnaugh, Maurice
Kaufman, Sol
Klerer, Melvin
Klir, George Jiri
Kornberg, Fred
Kyburg, Henry
Li, Chou H(siung)
McGee, James Patrick
Moyne, John Abel
Nerode, Anil
Pavlidis, Theo
Rapaport, William Joseph
Rauscher, Tomlinson Gene
Sakoda, William John
Sanderson, Arthur Clark
Schwartz, Jacob Theodore
Seeley, Thomas Dyer
Shapiro, Stuart Charles
Shasha, Dennis E
Shore, Richard A
Strand, Richard Carl
Summers, Phillip Dale
Taylor, James Hugh
Torng, Hwa-Chung
Wallace, Aaron
Walters, Deborah K W
Wang, Shu Lung
Westbrook, J(ack) H(all)
Wittie, Larry Dawson
Wolf, Walter Alan
Zdan, William
Zic, Eric A

NORTH CAROLINA
Bell, Norman R(obert)
Chang, Jeffrey C F
Chen, Su-shing
Dixon, Norman Rex
Elder, James Franklin, Jr
Haynes, John Lenneis
Loveland, Donald William
Ras, Zbigniew Wieslaw
Scott, Donald Ray

NORTH DAKOTA
Winrich, Lonny B

OHIO
Adeli, Hojjat
Burghart, James H(enry)
Chamis, Alice Yanosko
Chandrasekaran, Balakrishnan
Chawla, Mangal Dass
Hollingsworth, Ralph George
Law, Eric W
Ramamoorthy, Panapakkam A
Schlipf, John Stewart
Servais, Ronald Albert
Slotterbeck-Baker, Oberta Ann
Slutzky, Gale David
Speicher, Carl Eugene
Sterling, Leon Samuel
Suter, Bruce Wilsey
Tamburino, Louis A
Zheng, Yuan Fang

OKLAHOMA
Hochhaus, Larry
Koller, Glenn R
Naymik, Daniel Allan
Paxson, John Ralph

OREGON
Beck, John Robert
Bregar, William S
Freiling, Michael Joseph
Lendaris, George G(regory)
Schmisseur, Wilson Edward
Stewart, Bradley Clayton
Udovic, Daniel

PENNSYLVANIA
Aiken, Robert McLean
Altschuler, Martin David
Andrews, Peter Bruce
Badler, Norman Ira
Banerji, Ranan Bihari
Barbacci, Mario R
Berliner, Hans Jack
Bowlden, Henry James
Buchanan, Bruce G
Carbonell, Jaime Guillermo
Chaplin, Norman John
Dax, Frank Robert
Farlee, Rodney Dale
Finin, Timothy Wilking
Forgy, Charles Lanny
Fuchs, Walter
Garfinkel, David
Garrett, Thomas Boyd
Glaser, Robert
Goldwasser, Samuel M
Hahn, Peter Mathias
Herman, Martin
Hirschman, Lynette
Kam, Moshe
Klafter, Richard D(avid)
Koffman, Elliot B
Li, C(hing) C(hung)
Marquis, David Alan
Meystel, Alexander Michael
Mitchell, Tom M
Moser, Gene Wendell
Mowrey, Gary Lee
Reigel, Earl William
Siegel, Melvin Walter
Simon, Herbert A
Stern, Richard Martin, Jr
Stover, James Anderson, Jr
Treu, Siegfried
Verhanovitz, Richard Frank
Vladutz, George E
Vogt, William G(eorge)
Werner, Gerhard
Zabriskie, Franklin Robert

RHODE ISLAND
Fasching, James Le Roy
Kirschenbaum, Susan S
McClure, Donald Ernest
Santos, Eugene (Sy)

SOUTH CAROLINA
Cannon, Robert L
Hammond, Joseph Langhorne, Jr
Lambert, Jerry Roy

SOUTH DAKOTA
Hodges, Neal Howard, II

TENNESSEE
Bailes, Gordon Lee
Bartell, Steven Michael
Bourne, John Ross
Campbell, Warren Elwood
Franklin, Stanley Phillip
Gonzalez, Rafael C
Kawamura, Kazuhiko
Rodgers, Billy Russell
Schell, Fred Martin
Siirola, Jeffrey John
Trivedi, Mohan Manubhai
Uhrig, Robert Eugene
Umholtz, Clyde Allan

TEXAS
Aldridge, Jack Paxton, III
Ali, Moonis
Alo, Richard Anthony
Barstow, David Robbins
Bendapudi, Kasi Visweswararao
DeGroot, Doug
Deming, Stanley Norris
Ewell, James John, Jr
Franke, Ernest A
Frazer, Marshall Everett
Hennessey, Audrey Kathleen
Konstam, Aaron Harry
Kuipers, Benjamin Jack
Levinson, Stuart Alan
Miikkulainen, Risto Pekka
Mikiten, Terry Michael
Novak, Gordon Shaw, Jr
Peterson, Lynn Louise Meister
Self, Glendon Danna
Smith, Reid Garfield
Smith, Rolf C, Jr
Stiller, Peter Frederick
Stuart, Joe Don
Styblinski, Maciej A
Walkup, John Frank
Wolter, Jan D(ithmar)
Yao, James T-P

VIRGINIA
Abbott, Dean William
Awad, Elias M
Butler, Charles Thomas
Bynum, William Lee
Chou, Tsai-Chia Peter
Farrell, Robert Michael
Fink, Lester Harold
Fisher, Arthur Douglas
Fox, Edward A
Holzmann, Ernest G(unther)

Jones, Terry Lee
Kelly, Michael David
Kenyon, Stephen C
Levis, Alexander Henry
Loatman, Robert Bruce
McGrath, Arthur Kevin
Mandelberg, Martin
Masterson, Kleber Sanlin, Jr
Mata-Toledo, Ramon Alberto
Meyrowitz, Alan Lester
Miller, Betty M (Tinklepaugh)
Orcutt, Bruce Call
Sebastian, Richard Lee
Settle, Frank Alexander, Jr
Toida, Shunichi

WASHINGTON
Fetter, William Allan
Haralick, Robert M
Kowalik, Janusz Szczesny
Liu, Chen-Ching
MacVicar-Whelan, Patrick James
Savol, Andrej Martin
Waltar, Alan Edward

WISCONSIN
Dyer, Charles Robert
Eggert, Arthur Arnold
Firebaugh, Morris W
Levine, Leonard P
Moore, Edward Forrest
Ramakrishnan, Raghu

WYOMING
Cowles, John Richard
Magee, Michael Jack

ALBERTA
Gourishankar, V (Gouri)
Marsland, T(homas) Anthony
Martin, John Scott

BRITISH COLUMBIA
Calvert, Thomas W(illiam)
Dumont, Guy Albert
Meech, John Athol
Rosenberg, Richard Stuart
Thomson, Alan John
Wade, Adrian Paul

NEWFOUNDLAND
Shieh, John Shunen

NOVA SCOTIA
Oliver, Leslie Howard

ONTARIO
Argyropoulos, Stavros Andreas
Bauer, Michael Anthony
Chou, Jordan Quan Ban
Cohn-Sfetcu, Sorin
Dastur, Ardeshir Rustom
Emraghy, Hoda Abdel-Kader
Geddes, Keith Oliver
Giles, Robin
Goodenough, David George
Henry, Roger P
Ionescu, Dan
MacRae, Andrew Richard
Ratz, H(erbert) C(harles)
Slonim, Jacob
Stillman, Martin John

QUEBEC
De Mori, Renato
Desai, Bipin C
Guttmann, Ronald David
Hay, Donald Robert
Levine, Martin David
Newborn, Monroe M
Patel, Rajnikant V
Suen, Ching Yee
Zucker, Steven Warren

OTHER COUNTRIES
Arimoto, Suguru
Garcia-Santesmases, Jose Miguel
Goguen, Joseph A, Jr
Kyle, Thomas Gail
Levialdi, Stefano
Martin, Jim Frank
Niv, Yehuda
Pau, Louis F
Singh, Madan Gopal
Siu, Tsunpui Oswald
Tzafestas, Spyros G
Vamos, Tibor
Woo, Kwang Bang

Software Systems

ALABAMA
Alexander, David Michael
Copeland, David Anthony
Davis, Carl George
Jolley, Homer Richard
Klip, Dorothea A
Klip, Dorothea A
O'Kane, Kevin Charles
Roe, James Maurice, Jr
Seidman, Stephen Benjamin
Sides, Gary Donald
Stone, Max Wendell

Yeh, Pu-Sen

ALASKA
Hulsey, J Leroy
Rogers, James Joseph
Sheridan, John Roger
Watkins, Brenton John

ARIZONA
Andrews, Gregory Richard
Arabyan, Ara
Cathey, Everett Henry
Contractor, Dinshaw N
Daniel, Sam Mordochai
Dybvig, Paul Henry
Foltz, Craig Billig
Frieden, Bernard Roy
Gall, Donald Alan
Hayes, Donald Scott
Kieffer, Hugh Hartman
Korn, Granino A(rthur)
Kreidl, Tobias Joachim
Lorents, Alden C
Milnes, Dale J
Palusinski, Olgierd Aleksander
Sargent, Murray, III
Slaughter, Charles D
Trautman, Rodes
Wymore, Albert Wayne
Zielen, Albin John

CALIFORNIA
Adinoff, Bernard
Agostini, Romain Camille
Agoston, Max Karl
Anderson, Dennis Elmo
Apple, Martin Allen
Armstrong, John William
Astrahan, Morton M
Backus, John
Bahn, Gilbert S(chuyler)
Bailey, Michael John
Banerjee, Utpal
Barany, Ronald
Bar-Cohen, Yoseph
Barry, James Dale
Barsky, Brian Andrew
Basinger, Richard Craig
Benson, Donald Charles
Bertin, Michael C
Bic, Lubomir
Blattner, Meera McCuaig
Blinn, James Frederick
Blumstein, Carl Joseph
Bobrow, Daniel G
Bowles, Kenneth Ludlam
Boyle, Walter Gordon, Jr
Brainerd, Walter Scott
Brutlag, Douglas Lee
Burg, John Parker
Burton, Donald Eugene
Cantin, Gilles
Carr, John Weber, III
Casten, Richard G
Chen, Alice Tung-Hua
Cheriton, David Ross
Cherlin, George Yale
Chern, Ming-Fen Myra
Creighton, John Rogers
Cuadra, Carlos A(lbert)
Cunningham, Mary Elizabeth
Cunningham, Robert Stephen
Dalrymple, Stephen Harris
Davis, Alan Lynn
Denning, Dorothy Elizabeth Robling
Deutsch, Laurence Peter
Dill, James David
Dixon, Wilfrid Joseph
Doolittle, Robert Frederick, II
Dutt, Nikil D
Dwarakanath, Manchagondanahalli H
Edelman, Jay Barry
Eisberg, Robert Martin
Elliott, Denis Anthony
Elspas, B(ernard)
Emerson, Thomas James
Erman, Lee Daniel
Faulkner, Thomas Richard
Ferrari, Domenico
Firdman, Henry Eric
Fletcher, John George
Forster, Julian
Forsythe, Alan Barry
Frederick, David Eugene
Freiling, Edward Clawson
Friedlander, Carl B
Gallegos, Emilio Juan
Gaposchkin, Peter John Arthur
Giannetti, Ronald A
Gladney, Henry M
Goheen, Lola Coleman
Goldberg, Robert N
Gott, Euyen
Gottlieb, Peter
Graham, Susan Lois
Gray, Paul
Green, Claude Cordell
Green, Sherry Merrill
Greenberger, Martin
Grethe, Guenter
Groner, Gabriel F(rederick)
Hanscom, Roger H
Hecht, Herbert
Hegenauer, Jack C

Hennessy, John LeRoy
Herbst, Noel Martin
Hershey, Allen Vincent
Hightower, James K
Hilliker, David Lee
Hodson, William Myron
Holtz, David
Hopponen, Jerry Dale
Huebsch, Ian O
Hughett, Paul William
Innes, William Beveridge
Irvine, Cynthia Emberson
Jacobson, Allan Stanley (Bud)
Johnson, Mark Scott
Judd, Stanley H
Kaelble, David Hardie
Kaliski, Martin Edward
Kevorkian, Aram K
Kimball, Ralph B
Kissinger, David George
Klahr, Philip
Koenig, Daniel Rene
Korpman, Ralph Andrew
Kreuzer, Lloyd Barton
Kritz-Silverstein, Donna
Kuan, Teh S
Kuhlmann, Karl Frederick
Laiken, Nora Dawn
Landgraf, William Charles
Lang, Martin T
Latta, Gordon
Ledin, George, Jr
Lee, Keun Myung
Levine, Howard Bernard
Lewis, Nina Alissa
Licht, Paul
Lim, Hong Seh
Lipeles, Martin
Lodato, Michael W
Luongo, Cesar Augusto
Lupash, Lawrence O
Luqi,
McCarthy, John Lockhart
McCarthy, Mary Anne
McKay, Dale Robert
MacMillian, Stuart A
Magness, T(om) A(lan)
Maker, Paul Donne
Malcolm, Michael Alexander
Marcus, Rudolph Julius
Martin, Gordon Eugene
Melkanoff, Michel Allan
Melvin, Jonathan David
Meyers, Gene Howard
Mitchell, John Clifford
Mitra, Sanjit K
Mode, Vincent Alan
Moran, Thomas Patrick
Morgenstern, Matthew
Motteler, Zane Clinton
Muchmore, Robert B(oyer)
Mulligan, Geoffrey C
Murphy, Roy Emerson
Nelson, Eldred (Carlyle)
Neumann, Peter G
Nico, William Raymond
Nielsen, Norman Russell
Nikora, Allen P
Oddson, John Keith
Olsen, Edward Tait
Painter, Jeffrey Farrar
Parkison, Roger C
Payne, Philip Warren
Perloff, David Steven
Persky, George
Phillips, Veril LeRoy
Phinney, Nanette
Prag, Arthur Barry
Pratt, Vaughan Ronald
Pridmore-Brown, David Clifford
Pye, Earl Louis
Ratcliff, Milton, Jr
Rawal, Kanti M
Requa, Joseph Earl
Requicha, Aristides A G
Roberts, Charles Sheldon
Rogers, Howard H
Rowe, Lawrence A
Rudge, William Edwin
Rulifson, Johns Frederick
Runge, Richard John
Saffren, Melvin Michael
Sakaguchi, Dianne Koster
Saunders, Robert M(allough)
Schlafly, Roger
Schreiner, Robert Nicolas, Jr
Schwall, Richard Joseph
Schwartz, Morton Donald
Selinger, Patricia Griffiths
Shaka, Athan James
Shar, Leonard E
Silvester, John Andrew
Simons, Barbara Bluestein
Sincovec, Richard Frank
Smith, Frederick T(ucker)
Smith, William R
Solomon, Malcolm David
Strickland, Erasmus Hardin
Strohman, Rollin Dean
Sturgis, Howard Ewing
Sung, Chia-Hsiaing
Suri, Ashok
Surko, Pamela Toni
Szentirmai, George

Teague, Tommy Kay
Teeter, Richard Malcolm
Tilson, Bret Ransom
Tobey, Arthur Robert
VanAntwerp, Craig Lewis
Van Klaveren, Nico
Wagner, William Gerard
Walker, Kelsey, Jr
Wallace, Graham Franklin
Ward, David Gene
Weaver, William Bruce
Weiner, Daniel Lee
Wilbarger, Edward Stanley, Jr
Winkelmann, Frederick Charles
Worden, Paul Wellman, Jr
Zuppero, Anthony Charles

COLORADO
Abshier, Curtis Brent
Allen, James Lamar
Brasch, Frederick Martin, Jr
Call, Patrick Joseph
Chugh, Ashok Kumar
Clark, Jon D
Eichelberger, W(illiam) H
Etter, Delores Maria
Gibson, William Loane
Glover, Fred William
Hittle, Douglas Carl
Johnson, Gearold Robert
Montague, Stephen
Moore, Carla Jean
Morrison, John Stuart
Mueller, Robert Andrew
Munro, Richard Harding
Pearson, John Richard
Platt, Kenneth Allan
Robertson, Jerold C
Schnabel, Robert B
Sloss, Peter William
Stewart, James Joseph Patrick
Waite, William McCastline
Walden, Jack M
Wiatrowski, Claude Allan

CONNECTICUT
Baumgarten, Alexander
Booth, Taylor Lockwood
Cheng, David H S
Collins, John Barrett
Darcey, Terrance Michael
Engel, Gerald Lawrence
Greenwood, Ivan Anderson
Hanson, Trevor Russell
Harrison, Irene R
Henkel, James Gregory
Hudak, Paul Raymond
Kawaters, Woody H
McDonald, John Charles
Martinez, Robert Manuel
Relyea, Douglas Irving
Rusling, James Francis
Savitzky, Abraham
Scherr, Allan L
Seitelman, Leon Harold
Zubal, I George

DELAWARE
Amer, Paul David
Edwards, David Owen
Lehr, Gary Fulton
Nielsen, Paul Herron
Scofield, Dillon Foster
Ulery, Dana Lynn
Van Dyk, John William

DISTRICT OF COLUMBIA
Baker, Dennis John
Cherniavsky, John Charles
Curtin, Frank Michael
Filliben, James John
Finn, Edward J
Hartzler, Alfred James
Kaplan, David Jeremy
Kaplan, George Harry
McLean, John Dickson
Olmer, Jane Chasnoff
Penhollow, John O
Sjogren, Robert W, Jr
Spohr, Daniel Arthur

FLORIDA
Aubel, Joseph Lee
Bellenot, Steven F
Bonn, T(heodore) H(ertz)
Callan, Edwin Joseph
Cengeloglu, Yilmaz
Clark, Kerry Bruce
Coulter, Neal Stanley
Ege, Raimund K
Fisher, Robert Charles
Garrett, James Richard
Hand, Thomas
Hosni, Yasser Ali
Kelley, Myron Truman
Kolhoff, M(arvin) J(oseph)
Levow, Roy Bruce
Lloyd, Laurance H(enry)
Navlakha, Jainendra K
Selfridge, Ralph Gordon
Storm, Leo Eugene
Travis, Russell Burton
Veinott, Cyril G
Winton, Charles Newton

Young, John William

GEORGIA
Blanke, Jordan Matthew
Brubaker, Leonard Hathaway
Cohn, Charles Erwin
De Sa, Richard John
Fletcher, James Erving
Foley, James David
Gallaher, Lawrence Joseph
Hudson, Sigmund Nyrop
Mersereau, Russell Manning
Owen, Gene Scott
Tolbert, Laren Malcolm
Tooke, William Raymond, Jr

HAWAII
Ching, Chauncey T K
Johnson, Carl Edward
Kinariwala, Bharat K
Lindsay, Kenneth Lawson
Walker, Terry M

IDAHO
Anderegg, Doyle Edward
Mortensen, Glen Albert
Potter, David Eric

ILLINOIS
Aagaard, James S(tuart)
Al-Khafaji, Amir Wadi Nasif
Bagley, John D(aniel)
Bartlett, J Frederick
Boggs, George Johnson
Bombeck, Charles Thomas, III
Bowles, Joseph Edward
Boyle, James Martin
Chien, Andrew Andai
Cohen, Stanley
Coplien, James O
Cornwell, Larry Wilmer
Cowell, Wayne Russell
Davis, A Douglas
Day, Paul Palmer
DeFanti, Thomas A
Doerner, Robert Carl
Flora, Robert Henry
Fortner, Brand I
Gabriel, John R
Gay, Ben Douglas
Grace, Thom P
Guenther, Peter T
Gupta, Udaiprakash I
Hagstrom, Ray Theodore
Heinicke, Peter Hart
Horve, Leslie A
Johnson, Donald Elwood
Johnson, Paul Lorentz
Kerkman, Daniel Joseph
Lamont, Patrick
Land, Robert H
Marr, William Wei-Yi
Miller, Robert Carl
Mount, Bertha Lauritzen
Murata, Tadao
Nezrick, Frank Albert
Orthwein, W(illiam) C(oe)
Paddock, Robert Alton
Padua, David A
Phelan, James Joseph
Polychronopoulos, Constantine Dimitrius
Powers, Michael Jerome
Raffenetti, Richard Charles
Reed, Daniel A
Rickert, Neil William
Stutte, Linda Gail
Thisted, Ronald Aaron
Thomas, Gerald H
Toppel, Bert Jack
Udler, Dmitry
Ward, Charles Eugene Willoughby
Wells, Jane Frances
Yang, Shi-Tien
Yu, Clement Tak
Yule, Herbert Phillip

INDIANA
Boyd, Donald Bradford
Bunker, Bruce Alan
Dalphin, John Francis
Dube, David Gregory
Elmore, David
Harbron, Thomas Richard
Hellenthal, Ronald Allen
Helman, William Phillip
Jamieson, Leah H
Kuzel, Norbert R
Purdom, Paul W, Jr
Rego, Vernon J
Reklaitis, Gintaras Victor
Robertson, Edward L
Siegel, Howard Jay
Uhran, John Joseph, Jr
Vigdor, Steven Elliot
Weber, Janet Crosby
Wright, Jeffery Regan
Young, Frank Hood

IOWA
Alexander, Roger Keith
Christian, Lauren L
Fleck, Arthur C
Garfield, Alan J
Mischke, Charles R(ussell)

Tannehill, John Charles

KANSAS
Culvahouse, Jack Wayne
Giri, Jagannath
Kurt, Carl Edward
Lerner, David Evan

KENTUCKY
Simpson, James Edward
Villareal, Ramiro

LOUISIANA
Andrus, Jan Frederick
Brans, Carl Henry
Chen, Peter Pin-Shan
Denny, William F
Gajendar, Nandigam
Klyce, Stephen Downing
Roquemore, Leroy
Young, Myron H(wai-Hsi)

MAINE
Gordon, Geoffrey Arthur
Guidi, John Neil
Snyder, Arnold Lee, Jr
Tucker, Allen B

MARYLAND
Abramson, Fredric David
Agresti, William W
Babcock, Anita Kathleen
Basili, Victor Robert
Benokraitis, Vitalius
Blum, Bruce I
Bourbakis, Niklaos G
Brant, Larry James
Brenner, Alfred Ephraim
Bush, David E
Butler, Louis Peter
Chapin, Edward William, Jr
Chi, Donald Nan-Hua
Clark, Gary Edwin
Cornyn, John Joseph
Deprit, Andre A(lbert) M(aurice)
Dosier, Larry Waddell
Douglas, Lloyd Evans
Franks, David A
Goldstein, Charles M
Goldstein, Larry Joel
Graves, Harvey W(ilbur), Jr
Hammond, Charles E
Hevner, Alan Raymond
Hiatt, James Lee
Huffington, Norris J(ackson), Jr
Huneycutt, James Ernest, Jr
Jenkins, Robert Edward
Johnson, David Lee
Kahaner, David Kenneth
Kahn, Arthur B
Keenan, Thomas Aquinas
Kornman, Brent D
Kowalski, Richard
Lakein, Richard Bruce
Leach, Ronold
Lidtke, Doris Keefe
McCarn, Davis Barton
MacDonald, William
Macon, Nathaniel
Marquart, Ronald Gary
May, Everette Lee, Jr
Naddor, Eliezer
Pamidi, Prabhakar Ramarao
Paull, Kenneth Dywain
Perry, Peter M
Pisacane, Vincent L
Remondini, David Joseph
Rombach, Hans Dieter
Salwin, Arthur Elliott
Sammet, Jean E
Schlesinger, Judith Diane
Sidhu, Deepinder Pal
Smith, Richard Lloyd
South, Hugh Miles
Stickler, Mitchell Gene
Tissue, Eric Bruce
Tompkins, Robert Charles
Tripathi, Satish K
Weir, Edward Earl, II
Wende, Charles David
Whitmore, Bradley Charles
Zelkowitz, Marvin Victor

MASSACHUSETTS
Abrahams, Paul W
Adrion, William Richards
Aronson, James Ries
Barlas, Julie S
Belady, Laszlo Antal
Bertera, James H
Bloomfield, David Peter
Bruce, Kim Barry
Celmaster, William Noah
Champine, George Allen
Chang, Robin
Chew, Frances Sze-Ling
Clarke, Lori A
Cohen, Howard David
Comba, Paul Gustavo
Cullinane, Thomas Paul
Dhar, Sachidulal
Dyer, Charles Austen
Eckhouse, Richard Henry
Eoll, John Gordon

Software Systems (cont)

Estin, Robert William
Evans, Thomas George
Fitzgerald, Marie Anton
Floyd, William Beckwith
Foster, Caxton Croxford
Gagliardi, Ugo Oscar
Giuliano, Vincent E
Graham, Robert Montrose
Hahn, Robert S(impson)
Halperin-Maya, Miriam Patricia
Heitman, Richard Edgar
Hofstetter, Edward
Hollister, Charlotte Ann
Janowitz, Melvin Fiva
Johnson, Corinne Lessig
King, William Connor
Knighten, Robert Lee
Lees, David Eric Berman
Lerner, Harry
Lin, Alice Lee Lan
Lipner, Steven Barnett
Long, Alan K
McConnell, Robert Kendall
MacDougall, Edward Bruce
Makowski, Lee
Meal, Janet Hawkins
Menon, Premachandran R(ama)
Mooers, Calvin Northrup
Moss, J Eliot B
Nassi, Isaac Robert
Newman, Kenneth Wilfred
O'Neil, Elizabeth Jean
O'Neil, Patrick Eugene
Person, James Carl
Pope, Mary E
Ransil, Bernard J(erome)
Reinschmidt, Kenneth F(rank)
Ross, Douglas Taylor
Ross, Edward William, Jr
Royston, Richard John
Rubin, Allen Gershon
Salveter, Sharon Caroline
Salzberg, Betty
Scheff, Benson H(offman)
Schomer, Donald Lee
Seibel, Frederick Truman
Shaffer, Harry Leonard
Shorb, Alan McKean
Simovici, Dan
Smith, Raoul Normand
Snow, Beatrice Lee
Stewart, Lawrence Colm
Strecker, William D
Theobald, Charles Edwin, Jr
Tompkins, Howard E(dward)
Toscano, William Michael
Tripathy, Sukant K
Vidale, Richard F(rancis)
Wand, Mitchell
Waters, Richard C(abot)
Waud, Douglas Russell
Whitney, Cynthia Kolb
Williams, James G

MICHIGAN
Ancker-Johnson, Betsy
Bergmann, Dietrich R(udolf)
Bickart, Theodore Albert
Bookstein, Fred Leon
Campbell, Wilbur Harold
Cantor, David Milton
Caron, E(dgar) Louis
Chen, Min-Shih
Chen, Yudong
Clauer, C Robert, Jr
Duchamp, David James
Eagen, Charles Frederick
Estabrook, George Frederick
Freedman, M(orris) David
Herman, John Edward
Holland, Steven William
Jacoby, Ronald Lee
Johnson, Harold Hunt
Killgoar, Paul Charles, Jr
Knoll, Alan Howard
Ku, Albert B
Lenker, Susan Stamm
Lustgarten, Ronald Krisses
Metzler, Carl Maust
Rajlich, Vaclav Thomas
Rosen, Jeffrey Kenneth
Skaff, Michael Samuel
Taylor, Mitchell Kerry
Teichroew, Daniel
Tihansky, Diane Rice
Tsao, Nai-Kuan
Tuchinsky, Philip Martin
Turley, June Williams
Wasielewski, Paul Francis
Wims, Andrew Montgomery

MINNESOTA
Conlin, Bernard Joseph
Douglas, William Hugh
Du, David Hung-Chang
Fan, David P
Franta, William Roy
Gini, Maria Luigia
Kain, Richard Yerkes
Magnuson, Vincent Richard
Marshall, John Clifford
Polnaszek, Carl Francis

Pong, Ting-Chuen
Riley, Donald Ray
Sanders, Richard Mark
Schneider, G Michael
Spurrell, Francis Arthur
Thomborson, Clark D
Thompson, Herbert Bradford
Zitney, Stephen Edward

MISSISSIPPI
Elsherbeni, Atef Zakaria
McNaughton, James Larry
Pearce, David Harry

MISSOURI
Atwood, Charles LeRoy
Baer, Robert W
Ball, William E(rnest)
Koederitz, Leonard Frederick
McFarland, William D
Magill, Robert Earle
Moffatt, David John
Morgan, Nancy H
Peterson, Gerald E
Rouse, Robert Arthur
Styron, Clarence Edward, Jr
Zobrist, George W(inston)

MONTANA
Wright, Alden Halbert

NEBRASKA
Jenkins, Thomas Gordon
Schadt, Randall James
Scudder, Jeffrey Eric

NEVADA
Miller, Eugene
Minor, John Threecivelous

NEW HAMPSHIRE
Brender, Ronald Franklin
Farag, Ihab Hanna
Frantz, Daniel Raymond
Linnell, Richard D(ean)
MacGillivray, Jeffrey Charles
Misra, Alok C
Tourgee, Ronald Alan

NEW JERSEY
Afshar, Siroos K
Andose, Joseph D
Barnes, Derek A
Becker, Richard Alan
Belanger, David Gerald
Bergeron, Robert F(rancis) (Terry), Jr
Bernhardt, Ernest C(arl)
Bernstein, Lawrence
Bhagat, Phiroz Maneck
Bouboulis, Constantine Joseph
Burkhardt, Kenneth J
Catanese, Carmen Anthony
Chambers, John McKinley
Crane, Roger L
Delaney, Edward Joseph
Dobkin, David Paul
Duttweiler, Donald Lars
Elder, John Philip
Feiner, Alexander
Fischell, David R
Foregger, Thomas H
French, Larry J
Gabbe, John Daniel
Gargaro, Anthony
Gewirtz, Allan
Ginsberg, Barry Howard
Gittleson, Stephen Mark
Golin, Stuart
Halemane, Thirumala Raya
Henrich, Christopher John
Horng, Shi-Jinn
Huber, Richard V
Hulse, Russell Alan
Jeck, Richard Kahr
Kauffman, Ellwood
Kelsey, Edward Joseph
Kennedy, Anthony John
Kim, Uing W
Kobrin, Robert Jay
Krzyzanowski, Paul
Lechleider, J W
McGuinness, Deborah Louise
Maleeny, Robert Timothy
Manganaro, James Lawrence
Marscher, William Donnelly
Michelson, Leslie Paul
Mohr, Richard Arnold
Musa, John D
Nelson, Gregory Victor
Nonnenmann, Uwe
Ordille, Carol Maria
Polhemus, Neil W
Popper, Robert David
Prekopa, Andras
Rich, Kenneth Eugene
Robins, Jack
Robrock, Richard Barker, II
Rosin, Robert Fisher
Russell, Frederick A(rthur)
Sams, Burnett Henry, III
Saniee, Iraj
Schindler, Max J
Schivell, John Francis
Scholz, Lawrence Charles

Sears, Raymond Warrick, Jr
Shaer, Norman Robert
Shipley, Edward Nicholas
Sinclair, Brett Jason
Smith, William Bridges
Strother, J(ohn) A(lan)
Thompson, Kenneth Lane
Van Wyk, Christopher John
Vernon, Russel
Weiss, C Dennis
Woo, Nam-Sung
Wright, Margaret Hagen
Zabusky, Norman J
Zwass, Vladimir

NEW MEXICO
Berardo, Peter Antonio
Bergeron, Kenneth Donald
Biggs, Albert Wayne
Boland, W Robert
Butler, Harold S
Campbell, John Raymond
Cobb, Loren
Diel, Joseph Henry
Emigh, Charles Robert
Forslund, David Wallace
Giesler, Gregg Carl
Godfrey, Thomas Nigel King
Hartman, James Keith
Howell, Jo Ann Shaw
Hurd, Jon Rickey
Jones, Eric Daniel
Lambert, Howard W
Lilley, John Richard
McKay, Michael Darrell
Marks, Thomas, Jr
Mason, Rodney Jackson
Miller, Alan R(obert)
Nevitt, Thomas D
Phister, Montgomery, Jr
Poore, Emery Ray Vaughn
Rypka, Eugene Weston
Sicilian, James Michael
Smith, Brian Thomas
Sung, Andrew Hsi-Lin
Williams, Robert Allen
Wood, John Herbert

NEW YORK
Agarwal, Ramesh Chandra
Anastassiou, Dimitris
Archibald, Julius A, Jr
Arin, Kemal
Augenstein, Moshe
Axelrod, Norman Nathan
Ball, George William
Barnett, Michael Peter
Bell, Duncan Hadley
Boeck, William Louis
Bourne, Philip Eric
Brinch-Hansen, Per
Brunelle, Eugene John, Jr
Bunch, Phillip Carter
Burstein, Samuel Z
Castillo, Jessica Maguila
Chan, Tat-Hung
Cline, Harvey Ellis
Cole, James A
Conway, Richard Walter
Cornacchio, Joseph V(incent)
Coutchie, Pamela Ann
Curtis, Ronald S
Dasgupta, Gautam
Davenport, Lesley
Diebold, John Brock
Douglas, Craig Carl
Duek, Eduardo Enrique
Emrich, Lawrence James
Erdos, Marianne E
Esch, Louis James
Fenrich, Richard Karl
Ferentz, Melvin
Figueras, John
Figueras, Patricia Ann McVeigh
Flaherty, Joseph E
Folts, Dwight David
Goldberg, Conrad Stewart
Gottlieb, Allan
Greco, William Robert
Gustavson, Fred Gehrung
Hannay, David G
Harrington, Steven Jay
Hartwig, Curtis P
Heppa, Douglas Van
Herbert, Marc L
Homer, Eugene D(aniel)
Huddleston, John Vincent
Ishler, Norman Hamilton
Jaffe, Morry
Jain, Duli Chandra
Janak, James Francis
Jones, Leonidas John
Kaltofen, Erich L
Karnaugh, Maurice
Klerer, Melvin
Koplowitz, Jack
Kotchoubey, Andrew
Lamster, Hal B
La Tourrette, James Thomas
Levine, Stephen Alan
Levy, George Charles
Lichtig, Leo Kenneth
Litke, John David
Liu, Chamond

Loach, Kenneth William
Macaluso, Pat
Maguire, Gerald Quentin, Jr
Merten, Alan Gilbert
Monmonier, Mark
Morse, William M
Mosteller, Henry Walter
Notaro, Anthony
Pifko, Allan Bert
Pingali, Keshav Kumar
Ramaley, James Francis
Rauscher, Tomlinson Gene
Robinson, Harry
Rohlf, F James
Rosenkrantz, Daniel J
Ruston, Henry
Saunders, Burt A
Sayre, David
Schwartz, Jacob Theodore
Sciore, Edward
Sevian, Walter Andrew
Sherman, Philip Martin
Shooman, Martin L
Sinclair, Douglas C
Spang, H Austin, III
Spencer, David R
Srihari, Sargur N
Stark, Henry
Stenzel, Wolfram G
Strand, Richard Carl
Subramanian, Mani M
Tatarczuk, Joseph Richard
Thomas, Telfer Lawson
Tuli, Jagdish Kumar
Tycko, Daniel H
Wilson, Jack Martin
Wittie, Larry Dawson
Worthing, Jurgen

NORTH CAROLINA
Chen, Su-shing
Conners, Carmen Keith
Culbreth, Judith Elizabeth
Elder, James Franklin, Jr
Gould, Christopher Robert
Gray, James P
Higgins, Robert H
Javel, Eric
Johnson, G Allan
Latch, Dana May
Niccolai, Nilo Anthony
Reif, John H
Simon, Sheridan Alan
Smith-Thomas, Barbara
Sussenguth, Edward H
Sutcliffe, William Humphrey, Jr
Wallingford, John Stuart
Wright, William V(aughn)
Yarger, William E

NORTH DAKOTA
Magel, Kenneth I
Winrich, Lonny B

OHIO
Adeli, Hojjat
Blumenthal, Robert Martin
Brown, John Boyer
Buoni, John J
Chiu, Victor
Christensen, James Henry
Collins, George Edwin
Craig, Richard Gary
Dehne, George Clark
Giamati, Charles C, Jr
Greene, David C
Hall, Franklin Robert
Hartrum, Thomas Charles
Hollingsworth, Ralph George
Holtman, Mark Steven
Huff, Norman Thomas
Hulley, Clair Montrose
Ismail, Amin Rashid
Johnson, Robert Oscar
Juang, Ling Ling
Kerr, Douglas S
Klopman, Gilles
Lake, Robin Benjamin
Lipinsky, Edward Solomon
McConnell, H(oward) M(arion)
Mamrak, Sandra Ann
Mills, Wendell Holmes, Jr
Neff, Raymond Kenneth
Oliver, Joel Day
Peck, Lyman Colt
Petrarca, Anthony Edward
Potoczny, Henry Basil
Rose, Charles William
Roth, Mark A
Rounsley, Robert R(ichard)
Shick, Philip E(dwin)
Slotterbeck-Baker, Oberta Ann
Stansloski, Donald Wayne
Sterling, Leon Samuel
Wang, Paul Shyh-Horng
Waren, Allan D(avid)
Weintraub, Herschel Jonathan R
Winslow, Leon E
Wohlfort, Sam Willis
Zweben, Stuart Harvey

OKLAHOMA
Albright, James Curtice
Anderson, Paul Dean

Fisher, Donald D
Frame, Harlan D
Graves, Toby Robert
Grigsby, Ronald Davis
Hamming, Mynard C
Jennemann, Vincent Francis
Martner, Samuel (Theodore)
Walker, Billy Kenneth
Weston, Kenneth Clayton
Zetik, Donald Frank

OREGON
Anderson, Tera Lougenia
Beck, John Robert
Ecklund, Earl Frank, Jr
Freiling, Michael Joseph
Hamlet, Richard Graham
Kerlick, George David
Schmisseur, Wilson Edward
Schwartz, James William
Shapiro, Leonard David
Stewart, Bradley Clayton
Urquhart, N Scott
Wickes, William Castles

PENNSYLVANIA
Aburdene, Maurice Felix
Altschuler, Martin David
Ambs, William Joseph
Arrington, Wendell S
Bamberger, Judy
Barbacci, Mario R
Beck, Robert Edward
Beizer, Boris
Bowlden, Henry James
Bradley, James Henry Stobart
Bronzini, Michael Stephen
Cendes, Zoltan Joseph
Chang, C Hsiung
Confer, Gregory Lee
Detwiler, John Stephen
Druffel, Larry Edward
Ellison, Frank Oscar
Ermentrout, George Bard
Ertekin, Turgay
Fan, Ningping
Fenves, Steven J(oseph)
Finin, Timothy Wilking
Goodman, Alan Lawrence
Gottlieb, Frederick Jay
Greenfield, Roy Jay
Harbison, S P
Ho, Thomas Inn Min
Humphrey, Watts S
Hurwitz, Jan Krosst
Hyatt, Robert Monroe
Kaslow, David Edward
Kelemen, Charles F
Kleban, Morton H
Kline, Richard William
Korsh, James F
Larky, Arthur I(rving)
Lee, Insup
Lustbader, Edward David
Marcus, Robert Troy
Martin, Aaron J
Mondal, Kalyan
Monsimer, Harold Gene
Murgie, Samuel A
Nash, David Henry George
Pierson, Ellery Merwin
Pirnot, Thomas Leonard
Rehfield, David Michael
Reigel, Earl William
Ross, Bradley Alfred
Ryan, Thomas Arthur, Jr
Satyanarayanan, Mahadev
Scranton, Bruce Edward
Shaw, Mary M
Sommer, Holger Thomas
Szepesi, Zoltan Paul John
Treu, Siegfried
Vastola, Francis J
Warme, Paul Kenneth
Watson, William Martin, Jr
Wiley, Samuel J
Wilkins, Raymond Leslie
Young, Frederick J(ohn)

RHODE ISLAND
Doeppner, Thomas Walter, Jr
Evans, David L
Gabriel, Richard Francis
Paster, Donald L(ee)
Santos, Eugene (Sy)
Vitter, Jeffrey Scott

SOUTH CAROLINA
Cannon, Robert L
Dworjanyn, Lee O(leh)
Goode, Scott Roy
Gregory, Michael Vladimir
Lambert, Jerry Roy
Schwartz, Arnold Edward

SOUTH DAKOTA
Reuter, William L(ee)

TENNESSEE
Bailes, Gordon Lee
Campbell, Warren Elwood
Dobosy, Ronald Joseph
Ewig, Carl Stephen
Gray, William Harvey

Harrison, Robert Edwin
Hayter, John Bingley
Iskander, Shafik Kamel
Lane, Eric Trent
Lenhert, P Galen
Martin, Harry Lee
Ordman, Edward Thorne
Patterson, Malcolm Robert
Petersen, Harold, Jr
Pribor, Hugo C
Raridon, Richard Jay
Reister, David B(ryan)
Rivas, Marian Lucy
Sayer, Royce Orlando
Schach, Stephen Ronald
Sworski, Thomas John
Trivedi, Mohan Manubhai
Trowbridge, Lee Douglas
Wheeler, Orville Eugene

TEXAS
Aberth, Oliver George
Abraham, Jacob A
Bare, Charles L
Barstow, David Robbins
Bessman, Joel David
Brakefield, James Charles
Brantley, William Cain, Jr
Cameron, Bruce Francis
Campbell, Charles Edgar
Chakravarty, Indranil
Chen, Peter
Clingman, William Herbert, Jr
Czerwinski, Edmund William
Danburg, Jerome Samuel
Dolling, David Stanley
Ewell, James John, Jr
French, Robert Leonard
Frenger, Paul F
Garner, Harvey L(ouis)
Gentle, James Eddie
Gilmore, William Steven
Girou, Michael L
Glaspie, Donald Lee
Gonzalez, Mario J
Haenni, David Richard
Hanne, John R
Hasling, Jill Freeman
Hedstrom, John Richard
Heinze, William Daniel
Hennessey, Audrey Kathleen
Kingsley, Henry A(delbert)
Knapp, Roger Dale
Kochhar, Rajindar Kumar
Konstam, Aaron Harry
Kuipers, Benjamin Jack
Leiss, Ernst L
Leventhal, Stephen Henry
Levinson, Stuart Alan
Lurix, Paul Leslie, Jr
Markenscoff, Pauline
Massell, Wulf F
Midgley, James Eardley
Mistree, Farrokh
Moore, Thomas Matthew
Mortada, Mohamed
Nachlinger, R Ray
Nather, Roy Edward
Novak, Gordon Shaw, Jr
Palmeira, Ricardo Antonio Ribeiro
Parberry, Ian
Peterson, Lynn Louise Meister
Potts, Mark John
Ransom, C J
Reeder, Charles Edgar
Rubin, Richard Mark
Shen, Chin-Wen
Shen, Vincent Y
Shilstone, James Maxwell, Jr
Silberschatz, Abraham
Smith, Michael Kavanagh
Smith, Reid Garfield
Statz, Joyce Ann
Tibbals, Harry Fred, III
Vinson, David Berwick
Volz, Richard A
Ward, Phillip Wayne
Woodward, Joe William

UTAH
Embley, David Wayne
Herrin, Charles Selby
Hoggan, Daniel Hunter
Hollaar, Lee Allen
Karren, Kenneth W
Kruger, Robert A(lan)
Lisanke, Robert John, Sr

VIRGINIA
Bingham, Billy Elias
Cook, Robert Patterson
Dendrou, Stergios
Fitts, Richard Earl
Friedman, Fred Jay
Garcia, Albert B
Gardenier, John Stark
Gough, Stephen Bradford
Gowdy, Robert Henry
Hall, Otis F
Hartung, Homer Arthur
Henry, Neil Wylie
Jacob, Robert J(oseph) K(assel)
Kelly, Michael David
Liceaga, Carlos Arturo

McGean, Thomas J
McGrath, Arthur Kevin
Mata-Toledo, Ramon Alberto
Mathis, Robert Fletcher
Morris, Cecil Arthur, Sr
Orcutt, Bruce Call
Pitts, Thomas Griffin
Pratt, Terrence Wendall
Reeker, Larry Henry
Reierson, James (Dutton)
Richardson, Henry Russell
Rine, David C
Sage, Andrew Patrick
Saltz, Joel Haskin
Shetler, Antoinette (Toni)
Van Tilborg, André Marcel
Walker, Robert Paul
Yang, Ta-Lun

WASHINGTON
Bolme, Mark W
Britt, Patricia Marie
Carter, John Lemuel, Jr
Clark, William Greer
Gilmartin, Amy Jean
Golde, Hellmut
Henry, Robert R
Hinthorne, James Roscoe
Larrabee, Allan Roger
Lazowska, Edward Delano
MacKichan, Barry Bruce
MacLaren, Malcolm Donald
Malone, Stephen D
Mobley, Curtis Dale
Nickerson, Robert Fletcher
Nijenhuis, Albert
Padilla, Andrew, Jr
Savol, Andrej Martin
Smith, Burton Jordan
Spillman, Richard Jay
Tripp, Leonard L
Walker, Sharon Leslie
Wither, Ross Plummer

WEST VIRGINIA
Douglass, Kenneth Harmon
Fisher, John F
Ghigo, Frank Dunnington

WISCONSIN
Bates, Douglas Martin
Chi, Che
Clark, Harlan Eugene
Dalrymple, David Lawrence
Davida, George I
Fossum, Timothy V
Gelatt, Charles Daniel, Jr
Hafemann, Dennis Reinhold
Harris, J Douglas
Heinen, James Albin
Hinze, Harry Clifford
Jacobi, George (Thomas)
Kao, Wen-Hong
Morin, Dornis Clinton
Ramakrishnan, Raghu
Rosenthal, Jeffrey
Solomon, Marvin H

WYOMING
Bauer, Henry Raymond, III
Eddy, David Maxon

PUERTO RICO
Peinado, Rolando E

ALBERTA
Klobukowski, Mariusz Andrzej

BRITISH COLUMBIA
Booth, Andrew Donald
Chanson, Samuel T
Dill, John C
Hafer, Louis James
Wade, Adrian Paul

NEW BRUNSWICK
Lee, Y C
Vanicek, Petr

NEWFOUNDLAND
Rahman, Md Azizur
Shieh, John Shunen

NOVA SCOTIA
El-Hawary, Mohamed El-Aref
Kruse, Robert Leroy
Oliver, Leslie Howard
Rautaharju, Pentti M

ONTARIO
Argyropoulos, Stavros Andreas
Bauer, Michael Anthony
Blackwell, Alan Trevor
Dastur, Ardeshir Rustom
Emraghy, Hoda Abdel-Kader
Farrar, John Keith
Gentleman, William Morven
Hughes, Charles Edward
Jarvis, Roger George
Jeejeebhoy, Khursheed Nowrojee
Nash, John Christopher
Ogata, Hisashi
Okashimo, Katsumi
Roe, Peter Hugh O'Neil

Salvadori, Antonio
Slonim, Jacob
Taylor, David James
Tompa, Frank William
Waddington, Raymond

QUEBEC
Alagar, Vangalur S
Atwood, John William
Beron, Patrick
De Mori, Renato
Desai, Bipin C
Fabrikant, Valery Isaak
Opatrny, Jaroslav
Seldin, Jonathan Paul
Tomas, Francisco

SASKATCHEWAN
Lapp, Martin Stanley
Symes, Lawrence Richard

OTHER COUNTRIES
Bywater, Anthony Colin
Czepyha, Chester George Reinhold
Davis, Edward Alex
Goguen, Joseph A, Jr
Humphrey, Albert S
Maas, Peter
Nievergelt, Jurg
Revesz, Zsolt
Shafi, Mohammad
Siu, Tsunpui Oswald
Thorson, John Wells

Theory

ALABAMA
Dean, Susan Thorpe
Fiesler, Emile
Seidman, Stephen Benjamin

ALASKA
Lando, Barbara Ann

ARIZONA
Maier, Robert S

CALIFORNIA
Backus, John
Bull, Everett L, Jr
Butler, Jon Terry
Drobot, Vladimir
Emerson, Thomas James
Fagin, Ronald
Feferman, Solomon
Fraser, Grant Adam
Fredman, Michael Lawrence
Ginsburg, Seymour
Greibach, Sheila Adele
Hightower, James K
Inselberg, Alfred
Karp, Richard M
Lawler, Eugene L(eighton)
Lomax, Harvard
Manna, Zohar
Meagher, Donald Joseph
Melkanoff, Michel Allan
Mitchell, John Clifford
Monier, Louis Marcel
Moschovakis, Yiannis N
Motteler, Zane Clinton
Murphy, Roy Emerson
Nico, William Raymond
O'Dünlaing, Colm Pádraig
Pearl, Judea
Pease, Marshall Carleton, III
Pixley, Alden F
Pratt, Vaughan Ronald
Simons, Barbara Bluestein
Spanier, Edwin Henry
Stockmeyer, Larry Joseph
Sturgis, Howard Ewing
Tilson, Bret Ransom
Weeks, Dennis Alan

COLORADO
Manvel, Bennet

CONNECTICUT
Engel, Gerald Lawrence
Hajela, Dan
Hudak, Paul Raymond
Nilson, Edwin Norman
Rao, Valluru Bhavanarayana

DELAWARE
Caviness, Bobby Forrester

DISTRICT OF COLUMBIA
Abdali, Syed Kamal
Cherniavsky, John Charles
Della Torre, Edward
Friedman, Arthur Daniel
Hammer, Charles F

FLORIDA
Cenzer, Douglas
Levitz, Hilbert
Navlakha, Jainendra K
Roy, Dev Kumar
Stark, William Richard
Stark, William Richard

Theory (cont)

GEORGIA
Belinfante, Johan G F
Pomerance, Carl
Winkler, Peter Mann

ILLINOIS
Chang, Shi Kuo
Grace, Thom P
Gray, John Walker
Gupta, Udaiprakash I
Kalai, Ehud
Larson, Richard Gustavus
Lee, Der-Tsai
Livingston, Marilyn Laurene
Loui, Michael Conrad
Packel, Edward Wesler
Reingold, Edward Martin
Tier, Charles
Wallis, Walter Denis
Wells, Jane Frances
West, Douglas Brent

INDIANA
Atallah, Mikhail Jibrayil
Drufenbrock, Diane
Frederickson, Greg N
Fuelling, Clinton Paul
Gersting, Judith Lee
Purdom, Paul W, Jr
Robertson, Edward L
Szpankowski, Wojciech
Wise, David Stephen
Young, Frank Hood

IOWA
Alton, Donald Alvin

KANSAS
Strecker, George Edison

KENTUCKY
Lewis, Forbes Downer

LOUISIANA
Denny, William F
Kak, Subhash Chandra
Roquemore, Leroy

MAINE
Tucker, Allen B

MARYLAND
Atanasoff, John Vincent
Edmundson, Harold Parkins
Jablonski, Daniel Gary
Kosaraju, S Rao
Miller, Raymond Edward
Nau, Dana S
Rombach, Hans Dieter
Sadowsky, John

MASSACHUSETTS
Adrion, William Richards
Albertson, Michael Owen
Brown, Cynthia Ann
Bruce, Kim Barry
Chang, Robin
D'Alarcao, Hugo T
Doyle, Jon
Dyer, Charles Austen
Elias, Peter
Faber, Richard Leon
Gacs, Peter
Gardner, Marianne Lepp
Gessel, Ira Martin
Homer, Steven Elliott
Joni, Saj-nicole A
Levin, Leonid A
Lipner, Steven Barnett
Menon, Premachandran R(ama)
Nassi, Isaac Robert
O'Rourke, Joseph
Ostrovsky, Rafail M
Rogers, Hartley, Jr
Rosenberg, Arnold Leonard
Ross, Douglas Taylor
Simovici, Dan
Stonier, Tom Ted
Wand, Mitchell

MICHIGAN
Bachelis, Gregory Frank
Hinman, Peter Greayer
Malm, Donald E G
Rajlich, Vaclav Thomas
Stout, Quentin Fielden
Tihansky, Diane Rice
Windeknecht, Thomas George

MINNESOTA
Pierce, Keith Robert
Pour-El, Marian Boykan
Sahni, Sartaj Kumar
Seebach, J Arthur, Jr
Thomborson, Clark D

MISSISSIPPI
Burke, Robert Wayne

MISSOURI
Bruening, James Theodore
Nichols, Robert Ted

Peterson, Gerald E

MONTANA
Wright, Alden Halbert

NEBRASKA
Magliveras, Spyros Simos

NEVADA
Minor, John Threecivelous

NEW HAMPSHIRE
Bent, Samuel W
Bogart, Kenneth Paul

NEW JERSEY
Abeles, Francine
Afshar, Siroos K
Allender, Eric Warren
Amarel, S(aul)
Boesch, Francis Theodore
Brown, William Stanley
Dobkin, David Paul
Garey, Michael Randolph
Holzmann, Gerard J
Honeyman, Peter
Kirch, Murray R
Lagarias, Jeffrey Clark
Steiger, William Lee
Tindell, Ralph S
Van Wyk, Christopher John

NEW MEXICO
Harary, Frank
Sung, Andrew Hsi-Lin

NEW YORK
Anderson, Peter Gordon
Ball, George William
Bencsath, Katalin A
Case, John William
Chan, Tat-Hung
Cohen, Daniel Isaac Aryeh
Constable, Robert L
Coppersmith, Don
De Lillo, Nicholas Joseph
Di Paola, Robert Arnold
Goldfarb, Donald
Goodman, Jacob Eli
Gross, Jonathan Light
Hannay, David G
Hopcroft, John E(dward)
Kaltofen, Erich L
Kershenbaum, Aaron
Klir, George Jiri
Mitchell, Joseph Shannon Baird
Nerode, Anil
Pan, Victor
Parikh, Rohit Jivanlal
Rosenkrantz, Daniel J
Sciore, Edward
Selman, Alan L
Shasha, Dennis E
Shore, Richard A
Shub, Michael I
Sobczak, Thomas Victor
Stearns, Richard Edwin
Van Slyke, Richard M
Vasquez, Alphonse Thomas
Wagner, Eric G

NORTH CAROLINA
Blanchet-Sadri, Francine
Latch, Dana May
Loveland, Donald William
McAllister, David Franklin
Ras, Zbigniew Wieslaw
Reif, John H
Smith-Thomas, Barbara

OHIO
Collins, George Edwin
Friedman, Harvey Martin
Hartrum, Thomas Charles
Johnson, Robert Oscar
Potoczny, Henry Basil
Rolletschek, Heinrich Franz
Roth, Mark A
Rothstein, Jerome
Schlipf, John Stewart
Wells, Charles Frederick

OKLAHOMA
Letcher, John Henry, III
Walker, Billy Kenneth

OREGON
Hamlet, Richard Graham

PENNSYLVANIA
Becker, Stephen Fraley
Cupper, Robert
Furst, Merrick Lee
Kelemen, Charles F
Korsh, James F
Meystel, Alexander Michael
Shamos, Michael Ian
Sieber, James Leo
Simpson, Stephen G
Tzeng, Kenneth Kai-Ming
Wagh, Meghanad D

RHODE ISLAND
Doeppner, Thomas Walter, Jr

Leifman, Lev Jacob
Preparata, Franco Paolo
Savage, John Edmund
Simons, Roger Alan
Vitter, Jeffrey Scott

SOUTH CAROLINA
Comer, Stephen Daniel
Hammond, Joseph Langhorne, Jr

TENNESSEE
Franklin, Stanley Phillip
Ordman, Edward Thorne

TEXAS
Abraham, Jacob A
Borm, Alfred Ervin
Fix, George Joseph
Malek, Miroslaw
Matula, David William
Parberry, Ian
Wang, Yuan R
Wolter, Jan D(ithmar)

UTAH
Campbell, Douglas Michael
Higgins, John Clayborn

VERMONT
Weed, Lawrence Leonard

VIRGINIA
Friedman, Fred Jay
Heath, Lenwood S
Mata-Toledo, Ramon Alberto
Pratt, Terrence Wendall
Rine, David C
Spresser, Diane Mar
Stockmeyer, Paul Kelly

WASHINGTON
Kowalik, Janusz Szczesny
Somani, Arun Kumar
Storwick, Robert Martin
Young, Paul Ruel

WISCONSIN
Bach, Eric
Kinsinger, Richard Estyn
Moore, Edward Forrest
Rosser, John Barkley
Solomon, Marvin H

WYOMING
Cowles, John Richard

PUERTO RICO
Peinado, Rolando E

ONTARIO
Brzozowski, Janusz Antoni
Colbourn, Charles Joseph
Cook, Stephen Arthur
Davison, J Leslie
Dixon, John Douglas
Hughes, Charles Edward
Kent, Clement F

QUEBEC
Leroux, Pierre
Newborn, Monroe M

OTHER COUNTRIES
Fujiwara, Hideo
Goguen, Joseph A, Jr
Lindenmayer, Aristid
Myerson, Gerald
Nievergelt, Jurg

Other Computer Sciences

ALABAMA
Dukatz, Ervin L, Jr
Knox, Francis Stratton, III

ARIZONA
Daniel, Sam Mordochai
El-Ghazaly, Samir M
Honnell, Pierre M(arcel)
Justice, Keith Evans
Lorents, Alden C
Soderblom, Laurence Albert
Zeigler, Bernard Philip

CALIFORNIA
Baba, Paul David
Barsky, Brian Andrew
Bernstein, Ralph
Croft, Paul Douglas
Crowell, John Marshall
Cumberland, William Glen
Denning, Dorothy Elizabeth Robling
Dorfman, Leslie Joseph
Engel, Jan Marcin
Gailar, Owen H
Henderson, Sheri Dawn
Hotz, Henry Palmer
Keshavan, H R
Kuczynski, Eugene Raymond
Langdon, Allan Bruce
Lathrop, Richard C(harles)
Ma, Joseph T
Mathews, M(ax) V(ernon)

Mohanty, Nirode C
Prasanna Kumar, V K
Requa, Joseph Earl
Sangiovanni-Vincentelli, Alberto Luigi
Shevel, Wilbert Lee
Spencer, James Eugene
Tadjeran, Hamid
Thompson, Timothy J
Tuel, William Gole, Jr

COLORADO
Burkley, Richard M
Horton, Clifford E(dward)
Lett, Gregory Scott
Olson, George Gilbert
Pearlman, Sholom
Ralston, Margarete A
Sloss, Peter William
Stanger, Andrew L

CONNECTICUT
Engel, Gerald Lawrence
Frankel, Martin Richard
Nelson, Roger Peter
Ortner, Mary Joanne
Sittig, Dean Forrest

DELAWARE
Klein, Michael Tully

DISTRICT OF COLUMBIA
April, Robert Wayne
Chapin, Douglas Scott
Goldin, Edwin
Hawkins, Morris, Jr
Olmer, Jane Chasnoff
Welt, Isaac Davidson

FLORIDA
Chartrand, Robert Lee
Hand, Thomas
Jungbauer, Mary Ann
Levow, Roy Bruce
Mount, Gary Arthur
Naini, Majid M
Peterson, Ernest A
Rikvold, Per Arne
Sigmon, Kermit Neal
Starke, Albert Carl, Jr
Ulug, Esin M
Wu, Jin Zhong

GEORGIA
Cohn, Charles Erwin
Foley, James David
Hill, David W(illiam)
Husson, Samir S
Prince, M(orris) David
Zunde, Pranas

ILLINOIS
Chang, Shi Kuo
Eggan, Lawrence Carl
Grace, Thom P
Harris, Lowell Dee
Kubitz, William John
O'Morchoe, Patricia Jean
Saied, Faisal
Skeel, Robert David
Toy, Wing N
Wilke, Robert Nielsen

INDIANA
Fischler, Drake Anthony
Harbron, Thomas Richard
Markee, Katherine Madigan
Penna, Michael Anthony
Rego, Vernon J
Spicer, Donald Z
Wasserman, Gerald Steward
Wise, David Stephen
Young, Frank Hood

IOWA
Coy, Daniel Charles
Garfield, Alan J

KANSAS
Gordon, Michael Andrew
Lenhert, Donald H

MAINE
Carter, William Caswell

MARYLAND
Adams, James Alan
Belliveau, Louis J
Boggs, Paul Thomas
Boisvert, Ronald Fernand
Craig, Paul N
Demmerle, Alan Michael
Draper, Richard Noel
Eisner, Howard
Ephremides, Anthony
Fowler, Howland Auchincloss
Gary, Robert
Heilprin, Laurence Bedford
Hobbs, Robert Wesley
Kascic, Michael Joseph, Jr
Lepley, Arthur Ray
Marotta, Charles Rocco
Marx, Egon
Newman, David Bruce, Jr
Pallett, David Stephen

Schlesinger, Judith Diane
Smith, James Alan
Stuelpnagel, John Clay
Sullivan, Francis E
Turnrose, Barry Edmund
Valley, Sharon Louise
White, Eugene L

MASSACHUSETTS
Berera, Geetha Poonacha
Berkovits, Shimshon
Brenner, John Francis
Chang, Ching Shung
Crossman, David W
Eldred, William D
Gagliardi, Ugo Oscar
Gardner, Marianne Lepp
Graham, Robert Montrose
Hartline, Peter Haldan
Heyerdahl, Eugene Gerhardt
Hoffman, David Allen
Mesirov, Jill Portner
Murch, Laurence Everett
Olivo, Richard Francis
Read, Philip Lloyd
Richter, Stephen L(awrence)
Royston, Richard John
St Mary, Donald Frank
Sherman, Thomas Oakley
Winfrey, Richard Cameron

MICHIGAN
Bowman, Carlos Morales
Field, David Anthony
LePage, Raoul
Noe, Frances Elsie
Tsao, Nai-Kuan
Verhey, Roger Frank

MINNESOTA
Franta, William Roy
Rosen, Judah Ben

MISSISSIPPI
Andrews, Gordon Louis

MISSOURI
Hoover, Loretta White
Wagner, Robert G

NEBRASKA
Walters, Randall Keith

NEW HAMPSHIRE
Black, Sydney D
Kurtz, Thomas Eugene

NEW JERSEY
Blair, Grant Clark
Burrill, Claude Wesley
Dobkin, David Paul
Franks, Richard Lee
Gear, Charles William
Goffman, Martin
Halpern, Donald F
Hut, Piet
Jacquin, Arnaud Eric
Kalley, Gordon S
Lucas, Henry C, Jr
McGuinness, Deborah Louise
Quinlan, Daniel A
Schrenk, George L
Sinclair, Brett Jason
Stanton, Robert E
Weinstein, Stephen B
Weiss, Alan

NEW MEXICO
Alldredge, Gerald Palmer
Amann, James Francis
Argyros, Ioannis Konstantinos
Burks, Christian
Finkner, Morris Dale
Fries, Ralph Jay
Hanson, Kenneth Merrill
Williams, Robert Allen

NEW YORK
Ames, Stanley Richard
Archibald, Julius A, Jr
Augenstein, Moshe
Blumenson, Leslie Eli
Boettger, Susan D
Borowitz, Irving Julius
Braun, Ludwig
Conant, Francis Paine
Dudewicz, Edward John
Duggin, Michael J
Eydgahi, Ali Mohammadzadeh
Hicks, Bruce Lathan
Jedruch, Jacek
Kaplan, Martin Charles
Kenett, Ron
Kennicott, Philip Ray
Knowles, Richard James Robert
Koenig, Michael Edward Davison
Kriss, Michael Allen
Leith, ArDean
Maguire, Gerald Quentin, Jr
Mitchell, Joan LaVerne
Pan, Victor
Pennington, Keith Samuel
Schwartz, Mischa
Sobczak, Thomas Victor

Thomas, Richard Alan
Wadell, Lyle H
Walters, Deborah K W
Weinberger, Arnold
Yoffa, Ellen June
Zic, Eric A

NORTH CAROLINA
Beeler, Joe R, Jr
Cullati, Arthur G
Fuchs, Henry
Holcomb, Charles Edward
Land, Ming Huey
Oblinger, Diana Gelene
Patrick, Merrell Lee
Ras, Zbigniew Wieslaw
Smith, Bradley Richard
Stead, William Wallace
Sussenguth, Edward H
Vaughan, Douglas Stanwood
Woodbury, Max Atkin

NORTH DAKOTA
Kemper, Gene Allen
Magel, Kenneth I
Uherka, David Jerome

OHIO
Becker, Lawrence Charles
Dickinson, Frank N
Easterday, Jack L(eroy)
Fitzmaurice, Nessan
Ghandakly, Adel Ahmad
Hassell, John Allen
Hern, Thomas Albert
Hollingsworth, Ralph George
Hura, Gurdeep Singh
Kertz, George J
Levin, Jerome Allen
Ockerman, Herbert W
Potoczny, Henry Basil
Siervogel, Roger M
Stacy, Ralph Winston

OKLAHOMA
Cheung, John Yan-Poon
Fisher, Donald D
McClain, Gerald Ray
Nofziger, David Lynn
Robinson, Enders Anthony

OREGON
Matthes, Steven Allen

PENNSYLVANIA
Baker, Ronald G
Bergmann, Ernest Eisenhardt
Chase, Gene Barry
Green, Michael H(enry)
Grim, Larry B
Grinberg, Eric L
Hinton, Raymond Price
Jeruchim, Michel Claude
Kresh, J Yasha
Lambert, Joseph Michael
Meystel, Alexander Michael
Moss, William Wayne
Santora, Norman Julian
Shar, Albert O
Smith, Graham Monro
Szyld, Daniel Benjamin
Treu, Siegfried

RHODE ISLAND
Savage, John Edmund

SOUTH CAROLINA
Balintfy, Joseph L
Davis, Joseph B

SOUTH DAKOTA
Helsdon, John H, Jr

TENNESSEE
Baillargeon, Victor Paul
Evans, Joe Smith
Garzon, Max
James, Alton Everette, Jr
Sorrells, Frank Douglas
Watson, Evelyn E

TEXAS
Brown, J(ack) H(arold) U(pton)
Crossley, Peter Anthony
Cusachs, Louis Chopin
Dyke, Bennett
Gemignani, Michael C
Gonzalez, Mario J
Mikiten, Terry Michael
Penz, P Andrew
Potter, Robert Joseph
Renka, Robert Joseph
Treat, Charles Herbert
Watt, Joseph T(ee), Jr
Whittlesey, John R B

UTAH
Cohen, Elaine
Gardner, Reed McArthur
Hoggan, Daniel Hunter
Mosher, Loren Cameron

VIRGINIA
Anderson, Walter L(eonard)

Ell, William M
Field, Paul Eugene
Hamilton, Thomas Charles
Jacob, Robert J(oseph) K(assel)
Kilpatrick, S James, Jr
Lai, Chintu (Vincent C)
Latta, John Neal
Moritz, Barry Kyler
Nance, Richard E
Roussos, Constantine
Shetler, Antoinette (Toni)
Wolf, Eric W

WASHINGTON
Bax, Nicholas John
Millham, Charles Blanchard
Robinson, Barry Wesley

WISCONSIN
Borkovitz, Henry S
Kresch, Alan J
Strikwerda, John Charles

BRITISH COLUMBIA
Benston, Margaret Lowe
Ward, Lawrence McCue

NOVA SCOTIA
Moriarty, Kevin Joseph

ONTARIO
Geddes, Keith Oliver
Gough, Sidney Roger
Jackson, Kenneth Ronald
Law, Cecil E
Waddington, Raymond
Wright, Joseph D

QUEBEC
Châtillon, Guy
Desai, Bipin C
Newborn, Monroe M
Opatrny, Jaroslav
Poussart, Denis

SASKATCHEWAN
Law, Alan Greenwell

OTHER COUNTRIES
Lehman, Meir M
Massey, James L
Paulish, Daniel John

ENGINEERING

Aeronautical & Astronautical Engineering

ALABAMA
Anderson, Bernard Jeffrey
Bailey, J Earl
Bailey, Wayne L
Brainerd, Jerome J(ames)
Brandon, Walter Wiley, Jr
Cochran, John Euell, Jr
Cost, Thomas Lee
Cutchins, Malcolm Armstrong
Dannenberg, Konrad K
De Grandpre, Jean Louis
Doughty, Julian O
Dunn, Anne Roberts
French, Kenneth Edward
Geissler, Ernst D(ietrich)
Haeussermann, Walter
Hermann, R(udolf)
Hollub, Raymond M(athew)
Hung, Ru J
Krause, Helmut G L
McAuley, Van Alfon
Martin, James Arthur
Moore, Ronald Lee
Morgan, Bernard S(tanley), Jr
Nola, Frank Joseph
Rees, Eberhard F M
Reese, Bruce Alan
Rey, William K(enneth)
Ritter, A(lfred)
Roe, James Maurice, Jr
Sforzini, Richard Henry
Smith, Steven Patrick Decland
Speer, Fridtjof Alfred
Turner, Charlie Daniel, Jr
Weeks, George Eliot
Willenberg, Harvey Jack
Williams, James C(lifford), III
Wu, Shi Tsan

ALASKA
Johnson, Ralph Sterling, Jr

ARIZONA
Ahmed, Saad Attia
Bluestein, Theodore
Butler, Blaine R(aymond), Jr
Chen, Chuan Fang
Christensen, H(arvey) D(evon)
Fung, K Y
Glick, Robert L
Hirleman, Edwin Daniel, Jr
Makepeace, Gershom Reynolds
Mardian, James K W
Matthews, James B
Metzger, Darryl E

Miller, Harry
O'Leary, Brian Todd
Parks, E(dwin) K(etchum)
Ramohalli, Kumar Nanjunda Rao
Rummel, Robert Wiland
Saric, William Samuel
Sears, William R(ees)
Singhal, Avinash Chandra
Vincent, Thomas Lange
Wallace, C(harles) E(dward)
West, Robert Elmer
Wygnanski, Israel Jerzy

ARKANSAS
Burchard, John Kenneth
Carr, Gerald Paul

CALIFORNIA
Abzug, M(alcolm) J
Acheson, Louis Kruzan, Jr
Alperin, Morton
Araki, Minoru S
Ardema, Mark D
Ashley, Holt
Atwood, John Leland
Baldwin, Barrett S, Jr
Ballhaus, William F(rancis)
Barlow, Edward J(oseph)
Batdorf, Samuel Burbridge
Becker, Randolph Armin
Bell, M(arion) W(etherbee) Jack
Bharadvaj, Bala Krishnan
Blackmon, James B
Blackwelder, Ron F
Blasingame, Benjamin P(aul)
Bleviss, Zegmund O(scar)
Blumenthal, Irwin S(imeon)
Breakwell, John
Bright, Peter Bowman
Broadwell, James E(ugene)
Bromberg, R(obert)
Burden, Harvey Worth
Caflisch, Russel Edward
Cannon, Robert H, Jr
Caren, Robert Poston
Carlson, Richard M
Carterette, Edward Calvin Hayes
Casani, John R
Castle, Karen G
Chamberlain, Robert Glenn
Chambers, Robert J
Chapman, Dean R(oden)
Chapman, Gary Theodore
Charwat, Andrew F(ranciszek)
Chen, Tsai Hwa
Chin, Jin H
Cho, Alfred Chih-Fang
Cipriano, Leonard Francis
Clauser, Francis Hettinger
Cohen, Clarence B(udd)
Cohen, Lawrence Mark
Coles, Donald (Earl)
Compton, Dale L(eonard)
Cooper, George Emery
Copper, John A(lan)
Currie, Malcolm R
Cutforth, Howard Glen
Damonte, John Batista
Davis, Robert A(rthur)
Dillenius, Marnix Fritz Eugen
Donovan, Allen F(rancis)
Dowdy, William Louis
Dulock, Victor A, Jr
Eggers, A(lfred) J(ohn), Jr
Elachi, Charles
Elliott, David Duncan
Elverum, Gerard William, Jr
Farber, Joseph
Feign, David
Feinstein, Charles David
Feldman, Nathaniel E
Field, Lester
Fleming, Alan Wayne
Forbes, Judith L
Forbrich, Carl A, Jr
Forsberg, Kevin
Frazier, Edward Nelson
French, Edward P(erry)
Friedmann, Peretz Peter
Fuhrman, Robert Alexander
Gaskell, Robert Weyand
Gat, Nahum
Gazley, Carl, Jr
Gedeon, Geza S(cholcz)
Gevarter, William Bradley
Green, Allen T
Grobecker, Alan J
Gunkel, Robert James
Hackett, Colin Edwin
Haines, Richard Foster
Hall, Charles Frederick
Haloulakos, Vassilios E
Hansen, Grant Lewis
Hansmann, Douglas R
Harter, George A
Hawkins, Willis M(oore)
Hedgepeth, John M(ills)
Heinemann, Edward H
Heppe, R Richard
Hermsen, Robert W
Herrmann, George
Hess, R(onald) A(ndrew)
Hickman, Roy Scott
Hoff, N(icholas) J(ohn)

Aeronautical & Astronautical Engineering (cont)

Hoffmann, Jon Arnold
Horner, Richard E(lmer)
Hsu, Chieh-Su
Hu, Steve Seng-Chiu
Hult, John Luther
Hyatt, Abraham
Ikawa, Hideo
Iorillo, Anthony J
Jackson, Durward P
Jacobs, Michael Moises
Jeffs, George W
Johnson, Clarence L(eonard)
Jones, Robert Thomas
Judge, Roger John Richard
Kane, Thomas R(eif)
Kaplan, Richard E
Karamcheti, K(rishnamurty)
Kayton, Myron
Kelly, Robert Edward
Kennel, John Maurice
King, Hartley H(ughes)
Kirby, Jon Allan
Klestadt, Bernard
Knuth, Eldon L(uverne)
Kosmatka, John Benedict
Kraabel, John Stanford
Kraus, Samuel
Kropp, John Leo
Kuan, Teh S
Kwok, Munson Arthur
Laderman, A(rnold) J(oseph)
Lai, Kai Sun
Lampert, Seymour
Landecker, Peter Bruce
Landsbaum, Ellis M(erle)
Larson, Edward William, Jr
Lathrop, Richard C(harles)
Layton, Thomas William
Lederer, Jerome F
Leibowitz, Lewis Phillip
Leite, Richard Joseph
Leitmann, G(eorge)
Libby, Paul A(ndrews)
Liepman, H(ans) P(eter)
Lillie, Charles Frederick
Lin, Shao-Chi
Lindley, Charles A(lexander)
Lindsey, Gerald Herbert
Ling, Rung Tai
Lissaman, Peter Barry Stuart
Liu, Chung-Yen
Lomax, Harvard
Lopina, Robert F(erguson)
Lovelace, Alan Mathieson
Lowi, Alvin, Jr
Lynch, David Dexter
MacCready, Paul Beattie, Jr
McDonald, William True
McKenzie, Robert Lawrence
McLaughlin, William Irving
Macomber, Thomas Wesson
McRuer, Duane Torrance
Madan, Ram Chand
Mager, Artur
Magness, T(om) A(lan)
Manganiello, Eugene J(oseph)
Marcus, Bruce David
Martin, Elmer Dale
Maslach, George James
Maxworthy, Tony
Mayer, Harris Louis
Meissinger, Hans F
Melese, Gilbert B(ernard)
Melkanoff, Michel Allan
Miatech, Gerald James
Miel, George J
Miller, Richard Keith
Mingori, Diamond Lewis
Mirels, Harold
Moe, Osborne Kenneth
Monkewitz, Peter Alexis
Morgan, Lucian L(loyd)
Morris, Brooks T(heron)
Munson, Albert G
Myers, Dale Dehaven
Nachtsheim, Philip Robert
Netzer, David Willis
Neu, John Ternay
Ng, Lawrence Chen-Yim
Nguyen, Caroline Phuongdung
Nicolaides, John Dudley
Nielan, Paul E
Nielsen, Helmer L(ouis)
Niles, Philip William Benjamin
Park, Chul
Parker, Donn Blanchard
Peeters, Randall Louis
Peterson, Victor Lowell
Phillips, Samuel C
Pierucci, Mauro
Platzer, Maximilian Franz
Plotkin, Allen
Pottsepp, L(embit)
Preston, Robert Arthur
Puckett, Allen Emerson
Purcell, Everett Wayne
Putt, John Ward
Raj, Pradeep
Randolph, James E
Rauch, Herbert Emil

Rauch, Lawrence L(ee)
Rauch, Richard Travis
Raymond, Arthur E(mmons)
Reardon, Frederick H(enry)
Rechtin, Eberhardt
Retelle, John Powers, Jr
Revell, James D(ewey)
Reynolds, Harry Lincoln
Rich, Ben R
Ring, Robert E
Riparbelli, Carlo
Rivers, William J(ones)
Roberts, Leonard
Rockwell, Robert Lawrence
Rodabaugh, David Joseph
Rosen, Alan
Rossen, Joel N(orman)
Rott, Nicholas
Rubin, Sheldon
Rudy, Thomas Philip
Safonov, Michael G
Sawyer, Robert Fennell
Schairer, Robert S(org)
Schalla, Charence August
Schamberg, Richard
Schaufele, Roger Donald
Schjelderup, Hassel Charles
Schmit, Lucien A(ndre), Jr
Schneider, Alan M(ichael)
Schurmeier, Harris McIntosh
Scott, Paul Brunson
Seide, Paul
Seiff, Alvin
Sellars, John R(andolph)
Shaffar, Scott William
Shiffman, Max
Shu, Frank H
Shvartz, Esar
Simkin, Donald Jules
Sirignano, William Alfonso
Smelt, Ronald
Smith, A(pollo) M(ilton) O(lin)
Smith, George C
Snyder, Charles Thomas
Solomon, George E
Sparling, Rebecca Hall
Spier, Edward Ellis
Spiro, Irving J
Spreiter, John R(obert)
Springer, George S
Stella, Paul M
Stephens, Charles W
Stevenson, Robin
Stewart, H(omer) J(oseph)
Stone, Howard N(ordas)
Stoolman, Leo
Synolakis, Costas Emmanuel
Syvertson, Clarence A
Tang, Homer H(o)
Tang, Stephen Shien-Pu
Thelen, Charles John
Thompson, W P(aul)
Throner, Guy Charles
Trubert, Marc
Tubbs, Eldred Frank
Turchan, Otto Charles
Unal, Aynur
Vajk, J(oseph) Peter
Vanderplaats, Garret Niel
Vincenti, Walter G(uido)
Vlay, George John
Waaland, Irving T
Wagner, Raymond Lee
Walker, Kelsey, Jr
Wang, Ji Ching
Warren, Walter R(aymond), Jr
Wazzan, A R Frank
Weissman, Paul Robert
Wertz, James Richard
White, Moreno J
Whittier, James S(pencer)
Williams, E(dgar) P
Withee, Wallace Walter
Wittry, John P(eter)
Wood, Carlos C
Wood, Robert M(cLane)
Woolard, Henry W(aldo)
Yaggy, Paul Francis
Yakura, James K
Yang, H(sun) T(iao)
Yeh, Paul Pao
Yildiz, Alaettin

COLORADO
Ballhaus, William Francis, Jr
Barber, Robert Edwin
Born, George Henry
Brown, Alison Kay
Bunting, Jackie Ondra
Chow, Chuen-Yen
Colvis, John Paris
Culp, Robert D(udley)
Daley, Daniel H
Fester, Dale A(rthur)
Freymuth, Peter
Gibson, William Loane
Goldburg, Arnold
Hearth, Donald Payne
Jansson, David Guild
Kantha, Lakshmi
Klenknecht, Kenneth S(amuel)
Kohlman, David L(eslie)
Lett, Gregory Scott
Loehrke, Richard Irwin

Mansfield, Roger Leo
Marshall, Charles F
Mitchell, Charles Elliott
Morgenthaler, George William
Paynter, Howard L
Peterson, Harry C(larence)
Regenbrecht, D(ouglas) E(dward)
Sanchini, Dominick J
Sandborn, Virgil A
Seebass, Alfred Richard, III
Siuru, William D, Jr
Uberoi, M(ahinder) S(ingh)
Wall, Edward Thomas
Westfall, Richard Merrill
Wilbur, Paul James
Winn, C Byron

CONNECTICUT
Barnett, Mark
Campbell, George S(tuart)
Chu, Boa-Teh
Cohen, Myron Leslie
Duvivier, Jean Fernand
Fink, Martin Ronald
Ftaclas, Christ
Jenney, David S
Kaman, Charles Huron
Kuhrt, Wesley A
Lambrakis, Konstantine Christos
McCoy, Rawley D
Nilson, Edwin Norman
O'Brien, Robert L
Ojalvo, Irving U
Parker, Jack Steele
Paterson, Robert W
Ritkin, Edward Thaddeus
Porter, Richard W(illiam)
Robinson, Donald W(allace), Jr
Schreier, Stefan
Shainin, Dorian
Sreenivasan, Katepalli Raju
Verdon, Joseph Michael

DELAWARE
Danberg, James E(dward)
Kerr, Arnold D
Schwartz, Leonard William

DISTRICT OF COLUMBIA
Bainum, Peter Montgomery
Beach, Harry Lee, Jr
Brueckner, Guenter Erich
Chen, Davidson Tah-Chuen
Deshpande, Mohan Dhondorao
Donivan, Frank Forbes, Jr
Duncan, Robert C
Ellis, William Rufus
Fan, Dah-Nien
Flax, Alexander H(enry)
French, Francis William
Gould, Phillip
Green, Richard James
Hammersmith, John L(eo)
Hazelrigg, George Arthur, Jr
Kelley, Albert J(oseph)
Korkegi, Robert Hani
Liebowitz, Harold
Oran, Elaine Surick
Pyke, Thomas Nicholas, Jr
Rose, Raymond Edward
Stever, H Guyford
Townsend, Marjorie Rhodes
Urban, Peter Anthony
Whang, Yun Chow
Zlotnick, Martin

FLORIDA
Anderson, Roland Carl
Aronson, M(oses)
Bigley, Harry Andrew, Jr
Chamberlain, John
Chow, Wen Lung
Claridge, Richard Allen
Clark, John F
Clarkson, Mark H(all)
Cloutier, James Robert
Coar, Richard J
Dalehite, Thomas H
Daniel, Donald Clifton
Davis, Duane M
Diaz, Nils Juan
Dreves, Robert G(eorge)
Eisenberg, Martin A(llan)
Fearn, Richard L(ee)
Fleddermann, Richard G(rayson)
Hammond, Charles Eugene
Harmon, G Lamar
Hawkins, Robert C
Hoover, John W(esley)
Jones, Walter H(arrison)
Leadon, Bernard M(atthew)
Lijewski, Lawrence Edward
Lund, Frederick H(enry)
McLafferty, George H(oagland), Jr
Millsaps, Knox
Mulholland, John Derral
Peterson, James Robert
Proctor, Charles Lafayette, II
Scaringe, Robert P
Schimmel, Walter Paul
Smith, Jerome Allan
Suciu, S(piridon) N
Sun, Chang-Tsan
Thomas, Garland Leon

Wolf, Daniel Star
Yong, Yan

GEORGIA
Adomian, George
Atluri, Satya N
Boyles, Richard Quinn
Cahill, Jones F(rancis)
Carlson, Robert L
Cassanova, Robert Anthony
Craig, James I
Gray, Robin B(ryant)
Hackett, James E
Hubbart, James E
Lewis, David S(loan), Jr
Pierce, G(eorge) Alvin
Price, Edward Warren
Rehfield, Lawrence Wilmer
Strahle, Warren C(harles)
Zinn, Ben T

HAWAII
Cheng, Ping
Stuiver, W(illem)

IDAHO
Law, John
North, Paul

ILLINOIS
Barthel, Harold O(scar)
Benjamin, Roland John
Brewster, Marcus Quinn
Bullard, Clark W
Chiu, Huei-Huang
Dutton, Jonathan Craig
Errede, Steven Michael
George, John Angelos
Hofer, Kenneth Emil
Holmes, L(awrence) B(ruce)
Hugelman, Rodney D(ale)
Malloy, Donald Jon
Matalon, Moshe
Morel, Thomas
Morkovin, Mark V(ladimir)
Ormsbee, Allen I(ves)
Palmore, Julian Ivanhoe, III
Prussing, John E(dward)
Sentman, Lee H(anley), III
Sivier, Kenneth R(obert)
Stafford, John William
Tipei, Nicolae
Uherka, Kenneth Leroy
White, Robert Allan

INDIANA
Alspaugh, Dale W(illiam)
Gad-el-Hak, Mohamed
Gustafson, Winthrop A(dolph)
Hancock, John Ogden
Herrick, Thomas J(efferson)
Hoffman, Joe Douglas
Kareem, Ahsan
Kentzer, Czeslaw P(awel)
Lykoudis, Paul S
Marshall, Francis J
Mueller, Thomas J
Osborn, J(ohn) R(obert)
Razdan, Mohan Kishen
Schmidt, David Kelso
Sen, Mihir
Skelton, Robert Eugene
Snare, Leroy Earl
Thompson, Howard Doyle
Yang, Henry T Y

IOWA
Cook, William John
Hall, Jerry Lee
Hsu, Cheng-Ting
Iversen, James D(elano)
Lunde, Barbara Kegerreis, (BK)
Ma, Benjamin Mingli
Northup, Larry L(ee)
Seversike, Leverne K
Sturges, Leroy D
Tannehill, John Charles

KANSAS
Cogley, Allen C
Cook, Everett L
Downing, David Royal
Farokhi, Saeed
Giri, Jagannath
Lan, Chuan-Tau Edward
Lenzen, K(enneth) H(arvey)
Muirhead, Vincent Uriel
Razak, Charles Kenneth
Rodgers, Edward J(ohn)
Roskam, Jan
Tiffany, Charles F
Wentz, William Henry, (Jr)
Zumwalt, Glen W(allace)

LOUISIANA
Callens, Earl Eugene, Jr
Chieri, P(ericle) A(driano)
Courter, R(obert) W(ayne)
Kythe, Prem Kishore
Miller, Percy Hugh
Tewell, Joseph Robert
Thornton, William Edgar
Wetzel, Albert John

MARYLAND
Abed, Eyad Husni
Antman, Stuart S
Augustine, Norman R
Bastress, E(rnest) Karl
Beattie, Donald A
Berger, William J
Bernard, Peter Simon
Bersch, Charles Frank
Billig, Frederick S(tucky)
Burns, Bruce Peter
Butler, Louis Peter
Caveny, Leonard Hugh
Cleveland, William Grover, Jr
Cremens, Walter Samuel
Cronvich, Lester Louis
Day, LeRoy E(dward)
Deprit, Andre A(lbert) M(aurice)
Diamante, John Matthew
Dunham, David Waring
Eades, James B(everly), Jr
Eaton, Alvin Ralph
Edelson, Burton Irving
Edwards, Alan M
Ely, Raymond Lloyd
Evans, Larry Gerald
Fabunmi, James Ayinde
Fansler, Kevin Spain
Fifer, Robert Alan
Fischell, Robert E
Fleig, Albert J, Jr
Floyd, J F R(abardy)
Gartrell, Charles Frederick
Gessow, Alfred
Goldstein, Charles M
Gordon, Gary Donald
Gull, Theodore Raymond
Gupta, Ashwani Kumar
Hazen, David Comstock
Hobbs, Robert Wesley
Hornbuckle, Franklin L
Huffington, Norris J(ackson), Jr
Johnson, David Simonds
Jones, Everett
Kalil, Ford
Kamrass, Murray
Kemelhor, Robert Elias
Kiebler, John W(illiam)
Kitchens, Clarence Wesley, Jr
Koh, Severino Legarda
Korobkin, Irving
Kurzhals, Peter R(alph)
Lankford, J(ohn) L(lewellyn)
Levine, Robert S(idney)
Liu, Han-Shou
Lull, David B
McCourt, A(ndrew) W(ahlert)
Mahle, Christoph E
Markley, Francis Landis
Marsten, Richard B(arry)
Martin, John J(oseph)
Mathieu, Richard D(etwiler)
Mayers, Jean
Mecherikunnel, Ann Pottanat
Morgan, Walter L(eroy)
Mumford, Willard R
Murphy, C(harles) H(enry), Jr
Obremski, Henry J(ohn)
Oza, Dipak H
Paddack, Stephen J(oseph)
Paik, Ho Jung
Pamidi, Prabhakar Ramarao
Phinney, Ralph E(dward)
Pieper, George Francis
Powers, John Orin
Presser, Cary
Promisel, Nathan E
Raabe, Herbert P(aul)
Reupke, William Albert
Riley, Claude Frank, Jr
Rivello, Robert Matthew
Rogers, David Freeman
Rosen, Milton W(illiam)
Rueger, Lauren J(ohn)
Saarlas, Maido
Schneider, William C
Schuman, William John, Jr
Scialdone, John Joseph
Scipio, L(ouis) Albert, II
Sevik, Maurice
Smith, Richard Lloyd
Stadter, James Thomas
Street, William G(eorge)
Sturek, Walter Beynon
Tai, Tsze Cheng
Trainor, James H
Tschunko, Hubert F A
Utgoff, Vadym V
Weber, Richard Rand
Whittle, Frank
Yin, Frank Chi-Pong
Zien, Tse-Fou

MASSACHUSETTS
Anderson, J(ohn) Edward
Ash, Michael Edward
Baron, Judson R(ichard)
Battin, Richard H(orace)
Bender, Welcome W(illiam)
Budiansky, Bernard
Burke, Shawn Edmund
Cordero, Julio
Counselman, Charles Claude, III
Covert, Eugene Edzards

Crawley, Edward Francis
Delvaille, John Paul
Detra, Ralph W(illiam)
Dow, Paul C(rowther), Jr
Duffy, Robert A
Dugundji, John
Eoll, John Gordon
Eschenroeder, Alan Quade
Fay, James A(lan)
Fraser, Donald C
Garvey, R(obert) Michael
Gold, Harris
Gregers-Hansen, Vilhelm
Greitzer, Edward Marc
Hoag, David Garratt
Hollister, Walter M(ark)
Hursh, John W(oodworth)
Kautz, Frederick Alton, II
Kemp, Nelson Harvey
Kerrebrock, Jack Leo
Krebs, James N
Kreifeldt, John Gene
Lees, Sidney
Lemnios, A(ndrew) Z
Leung, Woon F (Wallace)
Lien, Hwachii
Lightfoot, Ralph B(utterworth)
Mastronardi, Richard
Milo, Henry L(ouis), Jr
Morse, Francis
Murman, Earll Morton
Neumann, Gerhard
Oman, Charles McMaster
Pallone, Adrian Joseph
Pan, Coda H T
Papernik, Lazar
Petschek, Harry E
Pian, Carlson Chao-Ping
Pian, Theodore H(sueh) H(uang)
Reeves, Barry L(ucas)
Reilly, James Patrick
Reinecke, William Gerald
Schmidt, George Thomas
Shea, Joseph F(rancis)
Stickler, David Bruce
Sutton, Emmett Albert
Tong, Pin
Topping, Richard Francis
Trilling, Leon
Udelson, Daniel G(erald)
Vander Velde, W(allace) E(arl)
Ventres, Charles Samuel
Wachman, Harold Yehuda
Waldman, George D(ewey)
Widnall, Sheila Evans
Wilk, Leonard Stephen
Witmer, Emmett A(tlee)
Wong, Po Kee

MICHIGAN
Adamson, Thomas C(harles), Jr
Alexandridis, Alexander A
Amick, James L(ewis)
Bitondo, Domenic
Blaser, Dwight A
Chen, Francis Hap-Kwong
Eisley, Joe G(riffin)
Faeth, Gerard Michael
Galan, Louis
Greenwood, D(onald) T(heodore)
Gutzman, Philip Charles
Howell, Larry James
Liu, Vi-Cheng
Lund, Charles Edward
Narain, Amitabh
Nelson, Wilbur C
Nicholls, J(ames) A(rthur)
Phillips, Richard Lang
Powell, Kenneth Grant
Rumpf, R(obert) J(ohn)
Segel, L(eonard)
Sichel, Martin
Spreitzer, William Matthew
Vinh, Nguyen Xuan
Whicker, Donald
Willmarth, William W(alter)

MINNESOTA
Beavers, Gordon Stanley
Cunningham, Thomas B
Greene, Christopher Storm
Harvey, Charles Arthur
Holdhusen, James S(tafford)
Moran, John P
Patton, Peter C(lyde)
Reilly, Richard J
Scott, Charles James
Stolarik, Eugene
Stone, Charles Richard
Sutton, Matthew Albert
Wilson, Theodore A(lexander)

MISSISSIPPI
Bennett, Albert George, Jr
Cliett, Charles Buren
McNaughton, James Larry
Smith, Allie Maitland
Thompson, Joe Floyd

MISSOURI
Agarwal, Ramesh K
Anders, William A
Bogar, Thomas John
Bower, William Walter

Branahl, Erwin Fred
Dharani, Lokeswarappa R
Elias, Thomas Ittan
Flowers, Harold L(ee)
Graff, George Stephen
Hakkinen, Raimo Jaakko
Hohenemser, Kurt Heinrich
Holsen, James N(oble)
Jischke, Martin C(harles)
Kibens, Valdis
Kotansky, D(onald) R(ichard)
Koval, Leslie R(obert)
Kozlowski, Don Robert
Leutzinger, Rudolph L(eslie)
Major, Schwab Samuel, Jr
Mirowitz, L(eo) I(saak)
Nelson, Harlan Frederick
Oetting, Robert B(enfield)
Painter, James Howard
Perisho, Clarence H(oward)
Poynton, Joseph Patrick
Rinehart, Walter Arley
Roos, Frederick William
Sajben, Miklos
Spaid, Frank William
Tsoulfanidis, Nicholas
Ulrich, Benjamin H(arrison), Jr
Warder, Richard C, Jr
Weissenburger, Jason T
Yardley, John Finley

NEBRASKA
Lu, Pau-Chang

NEVADA
Miller, Eugene
Wells, William Raymond

NEW HAMPSHIRE
Brilliant, Howard Michael
Frey, Elmer Jacob
Kantrowitz, Arthur (Robert)
Morduchow, Morris
Peline, Val P

NEW JERSEY
Abraham, John
Blackmore, Denis Louis
Bogdonoff, Seymour (Moses)
Brill, Yvonne Claeys
Clymer, Arthur Benjamin
Curtiss, Howard C(rosby), Jr
Davis, William F
DePalma, Anthony Michael
Durbin, Enoch Job
Fishburne, Edward Stokes, III
Florio, Pasquale J, Jr
Graham, (Frank) Dunstan
Kam, Gar Lai
Keigler, John Edward
Kornhauser, Alain Lucien
Lam, Sau-Hai
Lewellen, William Stephen
Mikolajczak, Alojzy Antoni
Minter, Jerry Burnett
Orszag, Steven Alan
Perry, Erik David
Raj, Rishi S
Schmidlin, Albertus Ernest
Schnapf, Abraham
Scholz, Lawrence Charles
Shefer, Joshua
Sisto, Fernando
Stengel, Robert Frank
Strom, Gordon H(aakon)
Strother, J(ohn) A(lan)
Yu, Yi-Yuan

NEW MEXICO
Baty, Richard Samuel
Berrie, David William
Bertin, John Joseph
Chen, Er-Ping
Curry, Warren H(enry)
Durham, Franklin P(atton)
El-Genk, Mohamed Shafik
Garcia, Carlos E(rnesto)
Hartley, Danny L
Maydew, Randall C
Moulds, William Joseph
Savage, Charles Francis
Scharn, Herman Otto Friedrich
Touryan, Kenell James
Williamson, Walton E, Jr
Woods, Robert Octavius

NEW YORK
Agosta, Vito
Anderson, Roy E
Austin, Fred
Banerjee, Prasanta Kumar
Barrows, John Frederick
Belove, Charles
Bennett, Leon
Berman, Julian
Bonan, Eugene J
Boyer, Donald Wayne
Breuhaus, W(aldemar) O(tto)
Brower, William B, Jr
Burns, Joseph A
Busnaina, Ahmed A
Caughey, David Alan
Dosanjh, Darshan S(ingh)
Duffy, Robert E(dward)

Falk, Theodore J(ohn)
Flaherty, Joseph E
Fleisig, Ross
Foreman, Kenneth M
Frank, Kozi
Fried, Erwin
Gavin, Joseph Gleason, Jr
George, A(lbert) R(ichard)
George, William Kenneth, Jr
Gilmore, Arthur W
Givi, Peyman
Goldstein, Stanley P(hilip)
Gran, Richard J
Grey, Jerry
Grossman, Norman
Guman, William J(ohn)
Hagerup, Henrik J(ohan)
Hajela, Prabhat
Hedrick, Ira Grant
Hicks, Bruce Lathan
Higuchi, Hiroshi
Hussain, Moayyed A
Isom, Morris P
Jedruch, Jacek
Kapila, Ashwani Kumar
Kashkoush, Ismail I
Kellogg, Spencer, II
Kelly, Thomas Joseph
Kempner, Joseph
Klosner, Jerome M
LaGraff, John Erwin
Lane, Frank
Lessen, Martin
Libove, Charles
Loeffler, Albert L, Jr
Loewy, Robert G(ustav)
Longman, Richard Winston
Longobardo, Anna Kazanjian
Lumley, John L(eask)
Mack, Charles Edward, Jr
Marrone, Paul Vincent
Mead, Lawrence Myers
Metzger, Ernest Hugh
Milliken, W(illiam) F(ranklin), Jr
Nardo, Sebastian V(incent)
Pifko, Allan Bert
Rae, William J
Reismann, Herbert
Resler, E(dwin) L(ouis), Jr
Scala, Sinclaire M(aximilian)
Scheuing, Richard A(lbert)
Schriro, George R
Sforza, Pasquale M
Shen, S(han) F(u)
Shepherd, D(ennis) G(ranville)
Shepherd, Joseph Emmett
Sherman, Zachary
Shube, Eugene E
Smedfjeld, John B
Sobel, Kenneth Mark
Soures, John Michael
Taulbee, Dale B(ruce)
Torrance, Kenneth E(ric)
Treanor, Charles Edward
Vaicaitis, Rimas
Visich, Marian, Jr
Watkins, Charles B
Wehle, Louis Brandeis, Jr
Weingarten, Norman C
Whiteside, James Brooks
Yu, Chia-Ping

NORTH CAROLINA
DeJarnette, Fred Roark
Dowell, Earl Hugh
Dunn, Joseph Charles
Hale, Francis Joseph
Hassan, Hassan Ahmad
Neal, C Leon
Torquato, Salvatore

OHIO
Armstrong, Neil A
Ault, George Mervin
Badertscher, Robert F(rederick)
Bagby, Frederick L(air)
Bahr, Donald Walter
Bailey, Cecil Dewitt
Banda, Siva S
Barranger, John P
Bodonyi, Richard James
Breuer, Delmar W
Burggraf, Odus R
Butz, Donald Josef
Chuang, Henry Ning
Donovan, Leo F(rancis)
Douglas, Donald Wills, Jr
Drake, Michael L
Dunford, Edsel D
Dunn, Robert Garvin
Edse, Rudolph
Fiore, Anthony William
Fitzmaurice, Nessan
Foster, Michael Ralph
Franke, Milton Eugene
Gatewood, Buford Echols
Glasgow, John Charles
Golovin, Michael N
Gorla, Rama S R
Gorland, Sol H
Graham, Robert William
Greber, Isaac
Grigger, David John
Groetsch, Charles William

Aeronautical & Astronautical Engineering (cont)

Guderley, Karl Gottfried
Halford, Gary Ross
Hamed, Awatef A
Hankey, Wilbur Leason, Jr
Hart, David Charles
Herbert, Thorwald
Hsia, John S
Keith, Theo Gordon, Jr
Kerr, Robert Lowell
Krishnamurthy, Lakshminarayanan
Kroll, Robert J
Krysiak, Joseph Edward
Larsen, Harold Cecil
Lee, John D(avid)
Lee, Jon H(yunkoo)
Leissa, A(rthur) W(illiam)
Lenhart, Jack G
Lieblein, Seymour
Lundin, Bruce T(heodore)
McBride, J(ames) W(allace)
Macke, H(arry) Jerry
Mall, Shankar
Marek, Cecil John
Moeckel, W(olfgang) E(rnst)
Mohn, Walter Rosing
Mularz, Edward Julius
Olson, Walter T
Pearson, Jerome
Peterson, George P
Povinelli, Louis A
Prahl, Joseph Markel
Quam, David Lawrence
Ramalingam, Mysore Loganathan
Reshotko, Eli
Richardson, David Louis
Rowe, Brian H
Rubin, Stanley G(erald)
Salkind, Michael Jay
Schultz, Edwin Robert
Schweiger, Marvin I
Servais, Ronald Albert
Sheskin, Theodore Jerome
Simitses, George John
Smith, Leroy H, Jr
Stetson, Kenneth F(rancis)
Tabakoff, Widen
Turchi, Peter John
Van Fossen, Don B
Voisard, Walter Bryan
Von Eschen, Garvin L(eonard)
Von Ohain, Hans Joachim
Wennerstrom, Arthur J(ohn)
Willeke, Klaus
Wolaver, Lynn E(llsworth)

OKLAHOMA

Blick, Edward F(orrest)
Earls, James Roe
Emanuel, George
Ketcham, Bruce V(alentine)
Omer, Savas
Rasmussen, Maurice L
Stafford, Thomas P(atten)
Striz, Alfred Gerhard
Swaim, Robert Lee
Vasicek, Daniel J
Weston, Kenneth Clayton
Zigrang, Denis Joseph

OREGON

Plummer, James Walter

PENNSYLVANIA

Amon Parisi, Cristina Hortensia
Blythe, Philip Anthony
Brunner, Mathias J
Cernansky, Nicholas P
Charp, Solomon
Chigier, Norman
Dash, Sanford Mark
Daywitt, James Edward
Fabish, Thomas John
Griffin, Jerry Howard
Hahn, Hong Thomas
Henderson, Robert E
Hnatiuk, Bohdan T
Hull, John Laurence
Kaplan, Marshall Harvey
Karras, Thomas William
Lakshminarayana, B
Lancaster, Otis Ewing
Lipsky, Stephen E
Loman, James Mark
Long, Lyle Norman
McCormick, Barnes W(arnock), Jr
McLaughlin, Philip V(an Doren), Jr
Mazzitelli, Frederick R(occo)
Merz, Richard A
Nolan, Edward J
Odrey, Nicholas Gerald
Paolino, Michael A
Parkin, Blaine R(aphael)
Piasecki, Frank Nicholas
Pinckney, Robert L
Ramos, Juan Ignacio
Rodini, Benjamin Thomas, Jr
Ross, Arthur Leonard
Rumbarger, John H
Schulman, Marvin
Settles, Gary Stuart

Sharma, Mangalore Gokulanand
Steg, L(eo)
Wang, Albert Show-Dwo
Whirlow, Donald Kent
Williams, Max L(ea), Jr
Zweben, Carl Henry

RHODE ISLAND

Clarke, Joseph H(enry)
Hayward, John T(ucker)
Liu, J(oseph) T(su) C(hieh)
McEligot, Donald M(arinus)
Mellberg, Leonard Evert
Viets, Hermann

SOUTH CAROLINA

Chuang, Ming Chia
Tull, William J

SOUTH DAKOTA

Winker, James A(nthony)

TENNESSEE

Burton, Robert Lee
Collins, Frank Gibson
Goethert, B(ernhard) H(ermann)
Harris, Wesley Leroy
Harwell, Kenneth Edwin
Keefer, Dennis Ralph
Keshock, Edward G
Kraft, Edward Michael
McGregor, Wheeler Kesey, Jr
Potter, John Leith
Roth, J(ohn) Reece
Shahrokhi, Firouz
Sissom, Leighton E(sten)
Skinner, George T
Toline, Francis Raymond
Wheeler, Orville Eugene
Whitfield, Jack D
Wimberly, C Ray
Wu, Ying-Chu Lin (Susan)
Young, Robert L(yle)

TEXAS

Adams, Richard E
Anderson, Dale Arden
Anderson, Edward Everett
Ball, Kenneth Steven
Bishop, Robert H
Bland, William M, Jr
Bluford, Guion Stewart, Jr
Bond, Aleck C
Brock, James Harvey
Carey, Graham Francis
Chevalier, Howard L
Cohen, Aaron
Cronk, Alfred E(dward)
Dalley, Joseph W(inthrop)
Dietz, William C
Dolling, David Stanley
Epp, Chirold Delain
Faget, Maxime A(llan)
Fairchild, Jack
Feagin, Terry
Fenter, Felix West
Fowler, Wallace T(homas)
Gourdine, Meredith Charles
Haisler, Walter Ervin
Henize, Karl Gordon
Hesse, Walter J(ohn)
Hinrichs, Paul Rutland
Hull, David G(eorge)
Hussain, A K M Fazle
Jackson, Eugene Bernard
Johnson, Johnny Albert
Jones, Jerold W
Junkins, John Lee
Juricic, Davor
Koplyay, Janos Bernath
Kraft, Christopher Columbus, Jr
Kranz, Eugene Francis
Lamb, J(amie) Parker, Jr
Levine, Joseph H
Loftus, Joseph P, Jr
Lowy, Stanley H(oward)
Lu, Frank Kerping
Miele, Angelo
Miller, Gerald E
Modisette, Jerry L
Morris, Owen G
Musa, Samuel A
Nicks, Oran Wesley
Norton, David Jerry
Page, Robert Henry
Payne, Fred R(ay)
Pray, Donald George
Purser, Paul Emil
Rand, James Leland
Ransleben, Guido E(rnst), Jr
Rodenberger, Charles Alvard
Sadler, Stanley Gene
Schapery, Richard Allan
Schutz, Bob Ewald
Secrest, Bruce Gill
Spindler, Max
Stanovsky, Joseph Jerry
Stearman, Ronald Oran
Szebehely, Victor
Talent, David Leroy
Tapley, Byron D(ean)
Thomas, Richard Eugene
Vance, John Milton
Ward, Donald Thomas

Webb, Theodore Stratton, Jr
Westkaemper, John C(onrad)
Widmer, Robert H
Young, John W(atts)

UTAH

Clyde, Calvin G(eary)
Hill, Eric von Krumreig
Hoeppner, David William
Krejci, Robert Henry
Lisanke, Robert John, Sr
Long, David G
Polve, James Herschal
Strozier, James Kinard
Zeamer, Richard Jere

VERMONT

Hermance, C(larke) E(dson)

VIRGINIA

Abbott, Dean William
Bingham, Billy Elias
Bishara, Michael Nageeb
Bolender, Carroll H
Brooks, Thomas Furman
Campbell, James Franklin
Carlson, Harry William
Chang, George Chunyi
Cliff, Eugene M
Cooley, William C
Cooper, Earl Dana
Cortright, Edgar Maurice
Crawford, Daniel J
Cross, Ernest James, Jr
Crossfield, A Scott
Curtin, Robert H
Donaldson, Coleman DuPont
Duberg, John E(dward)
Elias, Antonio L
Erickson, Wayne Douglas
Everett, Warren S
Finke, Reinald Guy
Finkelstein, Jay Laurence
Flack, Ronald Dumont, Jr
Freck, Peter G
Freitag, Robert Frederick
Gilbert, Arthur Charles
Gilruth, Robert R(owe)
Goodwin, Francis E
Graham, Kenneth Judson
Greenberg, Arthur Bernard
Hardrath, Herbert Frank
Haviland, John Kenneth
Henderson, Charles B(rooke)
Hidalgo, Henry
Holt, Alan Craig
Hou, Gene Jean-Win
Houbolt, John C(ornelius)
Huston, Robert James
Joshi, Suresh Meghashyam
Juang, Jer-Nan
Kapania, Rakesh Kumar
Katzoff, Samuel
Kelley, Henry J
King, Merrill Kenneth
Lauver, Dean C
Lee, Fred C
Lichtenberg, Byron Kurt
Lobb, R(odmond) Kenneth
Lovell, Robert R(oland)
Ludwig, George H
Lutze, Frederick Henry, Jr
Manfredi, Arthur Frank, Jr
Marchman, James F(ranklin), III
Margrave, Thomas Ewing, Jr
Michael, William Herbert, Jr
Myers, Michael Kenneth
Ng, Wing-Fai
Perkins, Courtland D(avis)
Pruett, Charles David
Puster, Richard Lee
Raju, Ivatury Sanyasi
Roberts, William Woodruff, Jr
Rokni, Mohammad Ali
Roy, Gabriel D
Sabadell, Alberto Jose
Savage, John W(frederick)
Schetz, Joseph A
Schmidt, Louis Vincent
Schriever, Bernard Adolf
Schy, Albert Abe
Scott, John E(dward), Jr
Seelig, Jakob Williams
Seiner, John Milton
Shore, David
Simmonds, James G
Singh, Mahendra Pal
Sobieszczanski-Sobieski, Jaroslaw
Summers, George Donald
Talapatra, Dipak Chandra
Telionis, Demetri Pyrros
Theon, John Speridon
Thompson, David Walker
Tiwari, Surendra Nath
Tole, John Roy
Tsang, Wai Lin
Waesche, R(ichard) H(enley) Woodward
Welch, Jasper Arthur, Jr
Whitesides, John Lindsey, Jr
Whitlock, Charles Henry
Wilson, Herbert Alexander, Jr
Wolfhard, Hans Georg
Wood, Albert D(ouglas)
Yates, Edward Carson, Jr

Zakrzewski, Thomas Michael

WASHINGTON

Ahlstrom, Harlow G(arth)
Arenz, Robert James
Boctor, Magdy
Bollard, R(ichard) John H
Bruckner, Adam Peter
Buonamici, Rino
Christiansen, Walter Henry
Clark, Robert Newhall
Delisi, Donald Paul
Dixon, Robert Jerome, Jr
Dvorak, Frank Arthur
Fetter, William Allan
Fyfe, I(an) Millar
Gravitz, Sidney I
Hamilton, William Thorne
Hebeler, Henry K
Heilenday, Frank W (Tod)
Hertzberg, Abraham
Holtby, Kenneth F
Kennet, Haim
Kurosaka, Mitsuru
Lin, H(ua)
Lurie, Harold
McCarthy, Douglas Robert
McMurtrey, Lawrence J
Martin, George C(oleman)
Oates, Gordon Cedric
Pennell, Maynard L
Pilet, Stanford Christian
Quist, William Edward
Rae, William H, Jr
Robel, Gregory Frank
Rubbert, Paul Edward
Russell, David A
Schairer, G(eorge) S(wift)
Schmidt, Eckart W
Shamash, Yacov A
Shank, Maurice E(dwin)
Steiner, John Edward
Street, Robert Elliott
Sutter, Joseph F
Tam, Patrick Yui-Chiu
Taya, Minoru
Varanasi, Suryanarayana Rao
Walsh, John Breffni
Walton, Vincent Michael
Wilson, Thornton Arnold

WEST VIRGINIA

Fanucci, Jerome B(en)
Kuhlman, John Michael
Loth, John Lodewyk
Peters, Penn A
Smith, James Earl

WISCONSIN

Lovell, Edward George
Moore, Gary T
Wnuk, Michael Peter
Zelazo, Nathaniel Kachorek

WYOMING

Adams, Donald F
Pell, Kynric M(artin)
Werner, Frank D(avid)
Wheasler, Robert

ALBERTA

Marsden, D(avid) J(ohn)

BRITISH COLUMBIA

Miura, Robert Mitsuru
Modi, V J

NEWFOUNDLAND

Muggeridge, Derek Brian

NOVA SCOTIA

Cochkanoff, O(rest)
Eames, M(ichael) C(urtis)

ONTARIO

Altman, Samuel Pinover
Barron, Ronald Michael
Chan, Yat Yung
Chu, Wing Tin
Cockshutt, E(ric) P(hilip)
Cowper, George Richard
DeLeeuw, J H
Dunn, D(onald) W(illiam)
Elfstrom, Gary Macdonald
Etkin, Bernard
French, J(ohn) Barry
Gladwell, Graham M L
Glass, I(rvine) I(srael)
Hughes, Peter C(arlisle)
Jacobson, Stuart Lee
Kosko, Eryk
Krishnappa, Govindappa
Lapp, P(hilip) A(lexander)
Lawson, John Douglas
Lukasiewicz, Julius
Matar, Said E
Moffatt, William Craig
Orlik-Ruckemann, Kazimierz Jerzy
Pindera, Jerzy Tadeusz
Plett, Edelbert Gregory
Prince, Robert Harry
Reid, Lloyd Duff
Ribner, Herbert Spencer
Richarz, Werner Gunter

Rimrott, F(riedrich) P(aul) J(ohannes)
Rosen, Marc Allen
Stevinson, Harry Thompson
Sullivan, Philip Albert
Tyler, R(onald) A(nthony)
Wallace, William
Weick, Richard Fred

QUEBEC
Bhat, Rama B
Galanis, Nicolas
Habashi, Wagdi George
Luckert, H(ans) J(oachim)
Mavriplis, F
Newman, B(arry) G(eorge)
Saber, Aaron Jaan
Vatistas, Georgios H

SASKATCHEWAN
Fuller, Gerald Arthur
Walerian, Szyszkowski

OTHER COUNTRIES
Al-Zubaidy, Sarim Naji
Hama, Francis R(yosuke)
Idelsohn, Sergio Rodolfo
Ince, A Nejat
Miller, Rene H(arcourt)
Morino, Luigi
Nomura, Yasumasa
Sato, Gentei
Voutsas, Alexander Matthew
    (Voutsadakis)
Vrebalovich, Thomas

Agricultural Engineering

ALABAMA
Corley, Tom Edward
Flood, Clifford Arrington, Jr
Hill, David Thomas
Johnson, Loyd
Renoll, Elmo Smith
Rochester, Eugene Wallace
Schafer, Robert Louis
Taylor, James H(obert)
Turnquist, Paul Kenneth

ARIZONA
Bessey, Paul Mack
Buras, Nathan
Cannon, M(oody) Dale
Fangmeier, Delmar Dean
Foster, Kennith Earl
Frevert, Richard Keller
Gitlin, Harris Martlin
Jordan, Kenneth A(llan)
Myers, Lloyd E(ldridge)
Rasmussen, William Otto
Replogle, John A(sher)
Smerdon, Ernest Thomas
Woolhiser, David A(rthur)

ARKANSAS
Berry, Ivan Leroy
Bryan, Billy Bird
Pitts, Donald James
Rokeby, Thomas R(upert) C(ollinson)

CALIFORNIA
Akesson, Norman B(erndt)
Bainer, Roy
Bery, Mahendera K
Fridley, Robert B
Goss, John R(ay)
Grinnell, Robin Roy
Hermsmeier, Lee F
Hickok, Robert Baker
Johnson, Maurice Verner, Jr
Loper, Willard H(ewitt)
Matthews, Floyd V(ernon), Jr
Merriam, John L(afayette)
Mohsenin, Nuri N
Mueller, Charles Carsten
Neubauer, L(oren) W(enzel)
O'Brien, Michael
Robertson, George Harcourt
Saltveit, Mikal Endre, Jr
Strohman, Rollin Dean
Wallender, Wesley William

COLORADO
Corey, Arthur Thomas
DeCoursey, Donn G(ene)
Evans, Norman A(llen)
Frasier, Gary Wayne
Harper, Judson M(orse)
Heermann, Dale F(rank)
Jensen, Marvin E(li)
Mickelson, Rome H
Van Schilfgaarde, Jan

CONNECTICUT
Alcantara, Victor Franco
Aldrich, Robert A(dams)
Bibeau, Thomas Clifford

DELAWARE
Collins, Norman Edward, Jr
Dwivedy, Ramesh C
Ritter, William Frederick
Scarborough, Ernest N

DISTRICT OF COLUMBIA
Doering, Eugene J(ohnson)
Harris, Wesley Lamar

FLORIDA
Bagnall, Larry Owen
Clayton, Joe Edward
Coppock, Glenn E(dgar)
Felton, Kenneth E(ugene)
Fluck, Richard Conard
Freeman, George R(oland)
Harrison, Dalton S(idney)
Holtan, Heggie Nordahl
Isaacs, Gerald W
Nordstedt, Roger Arlo
Overman, Allen Ray
Peart, Robert McDermand
Sager, John Clutton
Shaw, Lawrance Neil
Stephens, John C(arnes)
Teixeira, Arthur Alves
Whitney, Jodie Doyle
Zachariah, Gerald L

GEORGIA
Burgoa, Benali
Butler, James Lee
Clark, Rex L
Drury, Liston Nathaniel
Ghate, Suhas Ramkrishna
Henderson, George Edwin
Hoogenboom, Gerrit
Lawrence, Kurt C
Monroe, Gordon Eugene
Nelson, S(tuart) O(wen)
Smith, Ralph E(dward)
Thomas, Adrian Wesley
Threadgill, Ernest Dale
White, Harold D(ouglas)

HAWAII
Bartosik, Alexander Michael
Liang, Tung
Wang, Jaw-Kai

IDAHO
Bloomsburg, George L
Bondurant, James A(llison)
Brakensiek, Donald Lloyd
Dixon, John E(lvin)
Fitzsimmons, Delbert Wayne
Halterman, Jerry J
Humpherys, Allan S(tratford)
Johnson, Clifton W
Kincaid, Dennis Campbell
Pair, Claude H(erman)
Schreiber, David Laurence
Trout, Thomas James
Williams, Larry G

ILLINOIS
Cheryan, Munir
Curtis, James O(wen)
Day, Donald Lee
Fonken, David W(alter)
Goering, Carroll Eugene
Hansen, Ralph W(aldo)
Hummel, John William
Hunt, Donnell Ray
Jones, Benjamin A(ngus), Jr
Lloyd, Monte
Muehling, Arthur J
Olver, Elwood Forrest
Paulsen, Marvin Russell
Pickett, Leroy Kenneth
Puckett, Hoyle Brooks
Rodda, Errol David
Shove, Gene C(lere)
Siemens, John Cornelius
Steinberg, Marvin Phillip
Wakefield, Ernest Henry
Walter, Gordon H
Wang, Pie-Yi
Yoerger, Roger R

INDIANA
Dale, Alvin C
Hinkle, Charles N(elson)
Huggins, Larry Francis
Monke, Edwin J
Richey, Clarence B(entley)

IOWA
Baker, James LeRoy
Beer, Craig
Buchele, W(esley) F(isher)
Colvin, Thomas Stuart
Curry, Norval H
Hull, Dale O
Ives, Norton C(onrad)
Johnson, Howard P
Johnson, Lawrence Alan
Lane, Orris John, Jr
Marley, Stephen J
Melvin, Stewart W
Meyer, Vernon M(ilo)

KANSAS
Clark, Stanley Joe
Hagen, Lawrence J
Johnson, William Howard
Larson, George H(erbert)
Lyles, Leon
Schaper, Laurence Teis

KENTUCKY
Barfield, Billy Joe
Frank, John L
Henson, W(iley) H(ix), Jr
Parker, B(laine) F(rank)
Ross, Ira Joseph
Smith, Edward M(anson)
Walker, John Neal
Wells, Larry Gene
Yoder, Elmon Eugene

LOUISIANA
Braud, Harry J
Brown, William Henry
Cochran, Billy Juan
Fouss, James L(awrence)
Parish, Richard Lee
Robbins, Jackie Wayne Darmon
Rogers, James Samuel
Singh, Vijay P
Thomas, Carl H(enry)

MAINE
Hunter, James H
Klinge, Albert Frederick
Rowe, Richard J(ay)

MARYLAND
Ahrens, M(iles) Conner
Altman, Landy B, Jr
Amerman, Carroll R(ichard)
Finney, Essex Eugene, Jr
Johnson, Arthur Thomas
Langley, Maurice N(athan)
Norris, Karl H(oward)
Ross, David Stanley
Salomonson, Vincent Victor
Taylor, Leonard S
Yaramanoglu, Melih
Yeck, Robert Gilbert

MASSACHUSETTS
Rosenau, John (Rudolph)

MICHIGAN
Bakker-Arkema, Frederik Wilte
Bickert, William George
Butchbaker, Allen F
Cargill, B(urton) F(loyd)
Edwards, Donald M(ervin)
Esmay, Merle L(inden)
Kidder, Ernest H(igley)
McColly, Howard F(ranklin)
Mackson, Chester John
Maley, Wayne A
Marshall, Dale Earnest
Merva, George E

MINNESOTA
Allred, E(van) R(ich)
Bates, Donald W(esley)
Filkke, Arnold M(aurice)
Foster, George Rainey
Gupta, Satish Chander
Larson, Curtis L(uverne)
Morey, Robert Vance
Onstad, Charles Arnold
Schertz, Cletus E

MISSISSIPPI
Colwick, Rex Floyd
Fox, William R(obert)
Hendricks, Donovan Edward
Matthes, Ralph Kenneth, Jr
Meyer, Lawrence Donald
Mutchler, Calvin Kendal
Robinson, A(ugust) R(obert)
Shuman, Fred Leon, Jr
Welch, George Burns

MISSOURI
Brooker, Donald Brown
Day, Cecil LeRoy
Frisby, James Curtis
Harris, Franklin Dee
Hjelmfelt, Allen T, Jr
Kishore, Ganesh M
McFate, Kenneth L(everne)
Meador, Neil Franklin
Saran, Chitaranjan
Wilson, Clyde Livingston

NEBRASKA
Hahn, George LeRoy
Hanna, Milford A
Kleis, Robert W(illiam)
Leviticus, Louis I
Olson, Emanuel A(ole)
Splinter, William Eldon
Thompson, Thomas Leigh
Vanderholm, Dale Henry
Von Bargen, Kenneth Louis
Wittmuss, Howard D(ale)

NEVADA
Shanklin, Milton D(avid)

NEW JERSEY
Besley, Harry E(lmer)
Johnson, Curtis Allen
Lund, Daryl B

KENTUCKY (cont.)
Spillman, Charles Kennard
Thierstein, Gerald E

Mears, David R
Mills, William T(errell)
Ponessa, Joseph Thomas
Singley, Mark E(ldridge)

NEW MEXICO
Abernathy, George Henry
Morin, George Cardinal Albert
Rebuck, Ernest C(harles)
Yonas, Gerold

NEW YORK
Aries, Robert Sancier
Cooke, James Robert
Furry, Ronald B(ay)
Guest, Richard W(illiam)
Gunkel, Wesley W
Huntington, David Hans
Longhouse, Alfred Delbert
Millier, William F(rederick)
Rehkugler, Gerald E(dwin)
Scott, Norman Roy
Virk, Kashmir Singh

NORTH CAROLINA
Abrams, Charlie Frank, Jr
Barker, James Cathey
Bowen, Henry Dittimus
Cleland, John Gregory
Hassan, Awatif E
Hassler, F(rancis) J(efferson)
Huang, Barney K(uo-Yen)
Humphries, Ervin G(rigg)
Johnson, William Hugh
Kriz, George James
Lalor, William Francis
Overcash, Michael Ray
Rohrbach, Roger P(hillip)
Safley, Lawson McKinney, Jr
Sanoff, Henry
Skaggs, Richard Wayne
Suggs, Charles Wilson
Whitaker, Thomas Burton
Wiser, Edward H(empstead)
Young, James H

NORTH DAKOTA
Bauer, Armand
French, Ernest W(ebster)
Pratt, George L(ewis)
Promersberger, William J
Witz, Richard L

OHIO
Bondurant, Byron L(ee)
Brazee, Ross D
Curry, R(obert) Bruce
Fox, Robert Dean
Hall, Glenn Eugene
Hamdy, Mohamed Yousry
Herum, Floyd L(yle)
Huber, Samuel G(eorge)
Johnson, Carlton E(gbert)
Keener, Harold Marion
Lal, Rattan
Nelson, Gordon L(eon)
Rausch, David Leon
Reeve, Ronald C(ropper)
Roller, Warren L(eon)
Schwab, Glenn O(rville)
Sharma, Shri C
Short, Ted H
Taiganides, E Paul

OKLAHOMA
Batchelder, David G(eorge)
Brusewitz, Gerald Henry
Clary, Bobby Leland
Crow, Franklin Romig
Haan, Charles Thomas
Inoue, Riichi
Porterfield, Jay G
Roth, Lawrence O(rval)
Simpson, Ocleris C
Thompson, David Russell
Whitney, Richard Wilbur

OREGON
Booster, Dean Emerson
Hansen, Hugh J
Hellickson, Martin Leon
Kirk, Dale E(arl)
Miner, J Ronald
Moore, James Allan

PENNSYLVANIA
Anderson, Paul Milton
Daum, Donald Richard
Kjelgaard, William L
Morrow, Charles T(erry)
Myers, Earl A(braham)
Persson, Sverker
Skromme, Lawrence H
Walton, Harold V(incent)
White, John W

SOUTH CAROLINA
Bunn, Joe M(illard)
Doty, Coy William
Drew, Leland Overbey
Fischer, James Roland
Lambert, Jerry Roy
Ligon, James T(eddie)
Roberts, Darrell Lynn

Agricultural Engineering (cont)

Snell, A(bsalom) W(est)
Webb, Byron Kenneth
White, Richard Kenneth
Williamson, Robert Elmore

SOUTH DAKOTA
Hellickson, Mylo A
Moe, Dennis L
Myers, Victor (Ira)

TENNESSEE
Baxter, Denver O(len)
Helweg, Otto Jennings
Henry, Zachary Adolphus
McDow, John J(ett)
Sewell, John I

TEXAS
Allison, Robert Dean
Arkin, Gerald Franklin
Boyer, Robert Elston
Dvoracek, Marvin John
Engler, Cady Roy
Fryrear, Donald W
Gavande, Sampat A
Grub, Walter
Hauser, Victor La Vern
Hiler, Edward Allan
Hollingsworth, Joseph Pettus
Kirk, Ivan Wayne
Kunze, Otto Robert
Miller, John Allen
Morrison, John Eddy, Jr
Musick, Jack T(hompson)
Nixon, Paul R(obert)
Reddell, Donald Lee
Stermer, Raymond A
Stout, Bill A(lvin)
Ulich, Willie Lee

UTAH
Allen, Richard Glen
Anderson, Bruce Holmes
Griffin, Richard E
James, David Winston
Keller, Jack
Olsen, Edwin Carl, III
Peralta, Richard Carl
Peterson, Dean F(reeman), Jr
Rasmussen, V Philip, Jr
Silver, Barnard Stewart
Skogerboe, Gaylord Vincent
Stringham, Glen Evan
Willardson, Lyman S(essions)

VERMONT
Bornstein, Joseph

VIRGINIA
Becker, Clarence F(rederick)
Corey, Gilbert
Hall, Carl W(illiam)
Haugh, C(larence) Gene
Hurst, Homer T(heodore)
Mason, J(ohn) Philip Hanson, Jr
Miller, Adolphus James
Schlegelmilch, Reuben Orville
Shanholtz, Vernon Odell
Wright, Farrin Scott

WASHINGTON
Campbell, John C(arl)
Cykler, John Freuler
Davis, Denny Cecil
Hermanson, Ronald Eldon
James, Larry George
King, Larry Gene
McCool, D(onald) K
Powell, Albert E
Saxton, Keith E

WEST VIRGINIA
Diener, Robert G
Peters, Penn A

WISCONSIN
Amundson, Clyde Howard
Barrington, Gordon P
Bruhn, H(jalmar) D(iehl)
Bubenzer, Gary Dean
Buelow, Frederick H(enry)
Converse, James Clarence
Cramer, Calvin O
Evans, James Ornette
Kinch, Donald M(iles)
Northouse, Richard A
Schuler, Ronald Theodore
Thompson, Paul DeVries

WYOMING
Neibling, William Howard
Pochop, Larry Otto
Smith, James Lee

PUERTO RICO
Goyal, Megh R

ALBERTA
Domier, Kenneth Walter
Sommerfeldt, Theron G

BRITISH COLUMBIA
Coulthard, Thomas Lionel
Liao, Ping-huang
Staley, L(eonard) M(aurice)

MANITOBA
Cenkowski, Stefan
Jayas, Digvir Singh
Laliberte, Garland E
Lapp, H(erbert) M(elbourne)
Muir, William Ernest
Townsend, James Skeoch

ONTARIO
Dickinson, William Trevor
Mittal, Gauri S
Wong, Jo Yung

QUEBEC
Alward, Ron E
Broughton, Robert Stephen
McKyes, Edward

OTHER COUNTRIES
Finkel, Herman J(acob)
Krishna, J Hari
Preston, Thomas Alexander
Ward, Gerald T(empleton)

Bioengineering & Biomedical Engineering

ALABAMA
Goldman, Jay
Harding, Thomas Hague
Knox, Francis Stratton, III
McCutcheon, Martin J
Melville, Joel George
Sheffield, L Thomas
Vacik, James P
Wu, Xizeng

ARIZONA
Ahmed, Saad Attia
Arabyan, Ara
Bahill, A(ndrew) Terry
Barr, Ronald Edward
Gall, Donald Alan
Gitlin, Harris Martlin
Gross, Joseph F
Guilbeau, Eric J
Holloway, G Allen, Jr
Kauffman, John W
McRae, Lorin Post
Mylrea, Kenneth C
Smith, Josef Riley
Stiles, Philip Glenn

ARKANSAS
Amlaner, Charles Joseph, Jr
Clausen, Edgar Clemens
Schmitt, Neil Martin
Wear, James Otto

CALIFORNIA
Akontein, Benny Ambrose
Allen, Charles William
Altes, Richard Alan
Alvi, Zahoor M
Amstutz, Harlan Cabot
Angell, James Browne
Anger, Hal Oscar
Astrahan, Melvin Alan
Bagby, John P(endleton)
Beljan, John Richard
Berger, Stanley A(llan)
Bery, Mahendera K
Black, Kirby Samuel
Blanch, Harvey Warren
Bliss, James C(harles)
Boulton, Roger Brett
Bradley, A Freeman
Brant, Albert Wade
Britt, Edward Joseph
Bruley, Duane Frederick
Butterworth, Thomas Austin
Canawati, Hanna N
Carew, Thomas Edward
Chan, David S
Chenoweth, Dennis Edwin
Chou, Chung-Kwang
Codrington, Robert Smith
Cohn, Theodore E
Cook, Albert Moore
Coombs, William, Jr
Covell, James Wachob
Cromwell, Leslie
Daughters, George T, II
Davey, Trevor B(lakely)
Dev, Parvati
DiStefano, Joseph John, III
Eckenhoff, James Benjamin
Edmonds, Peter Derek
El-Refai, Mahmoud F
Estrin, Thelma A
Feather, A(lan) L(ee)
Flanigan, William Francis, Jr
Fung, Yuan-Cheng B(ertaam)
Glantz, Stanton Arnold
Goldstein, Walter Elliott
Gough, David Arthur
Gray, Joe William
Green, Philip S

Grodins, Fred Sherman
Gutfinger, Dan Eli
Haas, Gustav Frederick
Haines, Richard Foster
Hansmann, Douglas R
Harder, James Albert
Hayes, Thomas L
Hemmingsen, Edvard A(lfred)
Hill, Bruce Colman
Huang, Sung-Cheng
Hughes, Everett C
Iberall, Arthur Saul
Ingels, Neil Barton, Jr
Intaglietta, Marcos
Jacob, George Korathu
Jones, Jerry Latham
Jones, Joie Pierce
Karuza, Sarunas Kazys
Kaune, William Tyler
Keller, Edward Lowell
Kelly, Donald Horton
Knight, Patricia Marie
Kohler, George Oscar
Leake, Donald L
Leifer, Larry J
Levy, Donald M(arc)
Lewis, Edwin Reynolds
Lewis, Steven M
Likuski, Robert Keith
Lim, H(enry) C(hol)
Lin, Robert I-San
Llaurado, Josep G
Logsdon, Donald Francis, Jr
Lyman, John (Henry)
McRuer, Duane Torrance
Madan, Ram Chand
Marmarelis, Vasilis Z
Marsh, Donald Jay
Matsumura, Kenneth N
Matthys, Eric Francois
Maurice, David Myer
Meerbaum, Samuel
Midura, Thaddeus
Miller, Sol
Montgomery, Leslie D
Nimni, Marcel Efraim
Nunan, Craig S(pencer)
Pacela, Allan F
Pangborn, Rose Marie Valdes
Pedersen, Leo Damborg
Pinto, John Gilbert
Pittman, Ray Calvin
Powell, Frank Ludwig, Jr
Ramey, Melvin Richard
Randle, Robert James
Rasor, Ned S(haurer)
Robertson, George Harcourt
Rockwell, Robert Lawrence
Rugge, Henry F
Sadler, Charles Robinson, Jr
Saunders, Frank Austin
Schmid-Schoenbein, Geert W
Schneider, Alan M(ichael)
Shackelford, James Floyd
Shumate, Starling Everett, II
Singer, Jerome Ralph
Siposs, George G
Skrzypek, Josef
Smith, Robert Elphin
Smith, Warren Drew
Specht, Donald Francis
Spencer, Cherrill Melanie
Stark, Lawrence
Steele, Charles Richard
Sternberg, Moshe
Stroeve, Pieter
Stuhmiller, James Hamilton
Susskind, Charles
Szeto, Andrew Y J
Terdiman, Joseph Franklin
Thompson, Noel Page
Throner, Guy Charles
Twieg, Donald Baker
Unal, Aynur
Vurek, Gerald G
Ward, David Gene
Waring, Worden
Waterland, Larry R
Wayland, J(ames) Harold
Webbon, Bruce Warren
Weibell, Fred John
Weinberg, Daniel I
Werblin, Frank Simon
White, Moreno J
Wiberg, Donald M
Wieland, Bruce Wendell
Wirta, Roy W(illiam)
Yamashiro, Stanley Motohiro
Yen, Teh Fu
Yildiz, Alaettin
Young, Frank
Zweifach, Benjamin William

COLORADO
Aunon, Jorge Ignacio
Chang, Peter Hon
Dunn, Richard Lee
Ellis, Donald Griffith
Gamow, Rustem Igor
Hale, John
Hinman, Norman Dean
Histand, Michael B(enjamin)
Morgan, R John
Morrison, Charles Freeman, Jr

Newkirk, John Burt
Painter, Kent
Pan, Huo-Ping
Pauley, James Donald
Scherer, Ronald Callaway
Sodal, Ingvar E
Tengerdy, Robert Paul
Vaseen, V(esper) Albert

CONNECTICUT
Berglund, Larry Glenn
Bronzino, Joseph Daniel
Cohen, Myron Leslie
Crawford, John Okerson
Darcey, Terrance Michael
Dreyfus, Marc George
Epstein, Mary A Farrell
Hlavacek, Robert Allen
Horváth, Csaba Gyula
Hou, Kenneth C
Krummel, William Marion
Lambrakis, Konstantine Christos
Marchand, Nathan
Nehorai, Arye
Orphanoudakis, Stelios Constantine
Ranalli, Anthony William
Reisman, Harold Bernard
Salafia, W(illiam) Ronald
Sittig, Dean Forrest
Solonche, David Joshua
Stadnicki, Stanley Walter, Jr
Tuteur, Franz Benjamin
Wernau, William Charles

DELAWARE
Allen, Robert Harry
Bischoff, Kenneth Bruce
Dhurjati, Prasad S
Dwivedy, Ramesh C
Gaudy, Anthony F, Jr
Jefferies, Steven
Kilkson, Henn
Kingsbury, Herbert B
Pressman, Norman Jules
Read, Robert E
Santare, Michael Harold
Torregrossa, Robert Emile
Willis, Frank Marsden

DISTRICT OF COLUMBIA
Deshpande, Mohan Dhondorao
Heineken, Frederick George
Heldman, Dennis Ray
Jones, Janice Lorraine
Ledley, Robert Steven
Ligler, Frances Smith
Moraff, Howard
Parasuraman, Raja
Pollack, Herbert
Salu, Yehuda
Schnur, Joel Martin
Vaishnav, Ramesh
Wilson, James Bruce
Youm, Youngil

FLORIDA
Aronson, M(oses)
Block, Seymour Stanton
Bramante, Pietro Ottavio
Bressler, Steven L
Coulter, Wallace H
Deutsch, S(id)
Eckstein, Eugene Charles
Enger, Carl Christian
Finney, Roy Pelham
Freeman, Neil Julian
Gabridge, Michael Gregory
Genaidy, Ashraf Mohamed
Gibbs, Charles Howard
Gittelman, Donald Henry
Goldberg, Eugene P
Goldstein, Mark Kane
Heller, Zindel Herbert
Henson, Carl P
Kline, Jacob
McMillan, Donald Ernest
Martin, Charles John
Mayrovitz, Harvey N
Phillips, Winfred M(arshall)
Piotrowski, George
Price, Joel McClendon
Purdy, Alan Harris
Ramsey, Maynard, III
Rice, Stephen Landon
Seireg, Ali A
Spurlock, Jack Marion
Teixeira, Arthur Alves
Walburn, Frederick J
Whipple, Royson Newton
Williamson, Donald Elwin

GEORGIA
Bhatia, Darshan Singh
Busch, Kenneth Louis
Cassanova, Robert Anthony
Colton, Jonathan Stuart
English, Arthur William
Gibson, John Michael
Giddens, Don P(eyton)
Hamada, Spencer Hiroshi
Ku, David Nelson
May, Sheldon William
Morman, Michael T
Nerem, Robert Michael

Searle, John Randolph
Threadgill, Ernest Dale
Toler, J C
Wang, Johnson Jenn-Hwa
Warner, Harold
Wolf, Steven L
Yoganathan, Ajit Prithiviraj
Ziffer, Jack

HAWAII
Flanagan, Patrick William
Koide, Frank T
Moser, Roy Edgar
Pruder, Gary David

ILLINOIS
Agarwal, Gyan C
Bacus, James William
Barenberg, Sumner
Birnholz, Jason Cordell
Borso, Charles S
Breillatt, Julian Paul, Jr
Brown, Sherman Daniel
Chen, Michael Ming
Childress, Dudley Stephen
Cork, Douglas J
Dallos, Peter John
Dickerson, Roger William, Jr
Ducoff, Howard S
Dunn, Floyd
Elble, Rodger Jacob
Feinberg, Barry N
Flinn, James Edwin
Francis, Howard Thomas
Frizzell, Leon Albert
Goldstick, Thomas Karl
Gottlieb, Gerald Lane
Gupta, Ramesh
Harris, Lowell Dee
Holmes, Kenneth Robert
Houk, James Charles
Hua, Ping
Hughes, John Russell
Jacobs, John Edward
Joung, John Jongin
Kaganov, Alan Lawrence
Kertesz, Andrew (Endre)
Lauffenburger, Douglas Alan
Lin, James Chih-I
Miller, Irving F(ranklin)
Mockros, Lyle F(red)
Moore, John Fitzallen
O'Brien, William Daniel, Jr
Papoutsakis, Eleftherios Terry
Punwani, Dharam Vir
Rawlings, Charles Adrian
Reynolds, Larry Owen
Robinson, Charles J
Rosen, Arthur Leonard
Rymer, William Zev
Sather, Norman F(redrick)
Schnell, Gene Wheeler
Sellers, Donald Roscoe
Standing, Charles Nicholas
Steinberg, Marvin Phillip
Triano, John Joseph
Trimble, John Leonard
Weil, Max Harry
Widera, Georg Ernst Otto
Zehr, John E
Zuber, B(ert) L

INDIANA
Abel, Larry Allen
Brooks, Austin Edward
DeWitt, David P
Fearnot, Neal Edward
Geddes, Leslie Alexander
Hinds, Marvin Harold
Hinkle, Charles N(elson)
Hulbert, Samuel Foster
Kessler, David Phillip
Koivo, Antti J
Laxer, Cary
Liska, Bernard Joseph
Marks, Jay Stewart
Mastrototaro, John Joseph
Peppas, Nikolaos Athanassiou
Potvin, Alfred Raoul
Queener, Stephen Wyatt
Rothe, Carl Frederick
Stiver, James Frederick
Tiederman, William Gregg, Jr
Voelz, Michael H
Vogelhut, Paul Otto
Wasserman, Gerald Steward

IOWA
Buck, James R
Carlson, David L
Collins, Steve Michael
Engen, Richard Lee
Lakes, Roderic Stephen
Liu, Young King
Myers, Glenn Alexander
Park, Joon B
Rethwisch, David Gerard
Siebes, Maria
Titze, Ingo Roland
Young, Donald F(redrick)

KANSAS
Childress, Charles Curtis
Cooke, Francis W

Erickson, Howard Hugh
Erickson, Larry Eugene
Farrell, Eugene Patrick
Smith, Raymond V(irgil)

KENTUCKY
Edwards, Richard Glenn
Fleischman, Marvin
Hanley, Thomas Richard
Harris, Patrick Donald
Knapp, Charles Francis
Lafferty, James Francis
Lai-Fook, Stephen J
McCook, Robert Devon
Randall, David Clark
Squires, Robert Wright
Yoon, Hyo Sub

LOUISIANA
Bellina, Joseph Henry
Bundy, Kirk Jon
Chen, Isaac I H
Day, Donal Forest
Ewing, Channing Lester
Huckabay, Houston Keller
Klyce, Stephen Downing
Moulder, Peter Vincent, Jr
Saha, Subrata
Solomonow, Moshe
Sterling, Arthur MacLean
Walker, Cedric Frank

MAINE
Bush, George F(ranklin)
Hodgkin, Brian Charles

MARYLAND
Abbrecht, Peter H
Ashman, Michael Nathan
Berson, Alan
Bhagat, Hitesh Rameshchandra
Bowen, Rafael Lee
Bungay, Peter M
Carlson, Drew E
Chadwick, Richard Simeon
Charles, Harry Krewson, Jr
Chen, Ching-Nien
Cohen, Gerald Stanley
Covell, David Gene
Daniels, William Fowler, Sr
DeLoatch, Eugene M
Douglas, Andrew Sholto
Eden, Murray
Fischell, Robert E
Fishman, Elliot Keith
Fleshman, James Wilson, Jr
Frommer, Peter Leslie
Frost, John Kingsbury
Geckle, William Jude
Geduldig, Donald
Goldstein, Moise Herbert, Jr
Goldstein, Seth Richard
Gordon, Stephen L
Gupta, Vaikunth N
Hambrecht, Frederick Terry
Hamilton, Bruce King
Hamilton, Leroy Leslie
Heetderks, William John
Heinz, John Michael
Johnson, Arthur Thomas
Jones, Everett
Khachaturian, Zaven Setrak
Langlykke, Asger Funder
LeRoy, André François
Massey, Joe Thomas
Massof, Robert W
Maughan, W Lowell
Meyer, Richard Arthur
Michelsen, Arve
Moreira, Antonio R
Morris, Alan
Neugroschl, Daniel
Newcomb, Robert Wayne
Perry, Vernon P
Popel, Aleksander S
Ruttimann, Urs E
Sagawa, Kiichi
Salcman, Michael
Schmidt, Edward Matthews
Sheppard, Norman F, Jr
Stromberg, Robert Remson
Swift, David Leslie
Thakor, Nitish Vyomesh
Ulbrecht, Jaromir Josef
Watson, John Thomas
Webb, George N
Weisfeldt, Myron Lee
Yang, Xiaowei
Yin, Frank Chi-Pong

MASSACHUSETTS
Alter, Ralph
Aronow, Saul
Barngrover, Debra Anne
Breslau, Barry Richard
Bustead, Ronald Lorima, Jr
Chang, Kuo Wei
Charm, Stanley E
Chesler, David Alan
Clayton, J(oe) T(odd)
Colton, Clark Kenneth
Cook, Nathan Henry
Cooney, Charles Leland
Cravalho, Ernest G

Cudworth, Allen L
Cuffin, B(enjamin) Neil
Dale, William
Deluca, Carlo J
Dennison, Byron Lee
Dewey, C(larence) Forbes, Jr
Doane, Marshall Gordon
Downer, Nancy Wuerth
Drinker, Philip A
Eckhouse, Richard Henry
El-Bermani, Al-Walid I
Feldman, Charles Lawrence
Fine, Samuel
Fishman, Philip M
Gaw, C Vernon
Gordy, Edwin
Grodzinsky, Alan J
Gross, David John
Harris, Wayne G
Hatch, Randolph Thomas
Haudenschild, Christian C
Herzlinger, George Arthur
Hoffman, Allen Herbert
Jain, Rakesh Kumar
Kamen, Gary P
Kamm, Roger Dale
Klibanov, Alexander M
Krasner, Jerome L
Kreifeldt, John Gene
Lampi, Rauno Andrew
Langer, Robert Samuel
Lederman, David Mordechai
Lees, Sidney
Lele, Padmakar Pratap
Lerner, Harry
Lodwick, Gwilym Savage
Maloney, John F
Mann, Robert W(ellesley)
Maran, Janice Wengerd
Mark, Roger G
Mastenbrook, S Martin, Jr
Meldon, Jerry Harris
Mikic, Bora
Mountain, David Charles, Jr
Oman, Charles McMaster
Pan, Coda H T
Paul, Igor
Pedersen, Peder Christian
Petschek, Harry E
Peura, Robert Allan
Poirier, Victor L
Poon, Chi-Sang
Pope, Mary E
Rose, Timothy Laurence
Sahatjian, Ronald Alexander
Schoen, Frederick J
Shapiro, Ascher H(erman)
Shapiro, Howard Maurice
Sheridan, Thomas Brown
Singh, Param Indar
Sodickson, Lester A
Tuomy, Justin M(atthew)
Voigt, Herbert Frederick
Wallace, David H
Weinberg, Crispin Bernard
White, Augustus Aaron, III
Whitney, Lester F(rank)
Wiegner, Allen W
Yarmush, Martin Leon
Young, Laurence Retman
Zamenhof, Robert G A

MICHIGAN
Anderson, David J
Anderson, Thomas Edward
Asgar, Kamal
Be Ment, Spencer L
Cain, Charles Alan
Canale, Raymond Patrick
Chaffin, Don B
Cox, Mary E
Cullum, Malford Eugene
Francis, Ray Llewellyn
Fyhrie, David Paul
Hillegas, William Joseph
Horvath, Ralph S(teve)
Kim, Changhyun
King, Albert Ignatius
McGrath, John Joseph
Meyerhoff, Mark Elliot
Midgley, A Rees, Jr
Nuttall, Alfred L
Nyquist, Gerald Warren
Radin, Eric Leon
Repa, Brian Stephen
Reynolds, Herbert McGaughey
Robertson, John Harvey
Schramm, John Gilbert
Schultz, Albert Barry
Soutas-Little, Robert William
Stein, Paul David
Viano, David Charles
Wang, Henry
Wiitanen, Wayne Alfred
Williams, William James
Wolterink, Lester Floyd
Yang, Wen Jei

MINNESOTA
Bakken, Earl E
Beeler, George W, Jr
Carim, Hatim Mohamed
Clapper, David Lee
Cornelissen Guillaume, Germaine G

Coury, Arthur Joseph
Erdman, Arthur Guy
Finkelstein, Stanley Michael
Fredrickson, Arnold G(erhard)
Gibbons, Donald Frank
Greenleaf, James Fowler
Hu, Wei-Shou
Jacobs, Carl Henry
Kallok, Michael John
Keshaviah, Prakash Ramnathpur
Kvalseth, Tarald Oddvar
Mikhail, Adel Ayad
Miller, Nicholas Carl
Mueller, Rolf Karl
Neft, Nivaed
Nicoloff, Demetre M
Olson, Walter Harold
Owens, Boone Bailey
Ray, Charles Dean
Schmitt, Otto Herbert
Soechting, John F
Stephanedes, Yorgos Jordan
Thornton, Arnold William
Timm, Gerald Wayne

MISSISSIPPI
Matthes, Ralph Kenneth, Jr
Montani, Jean-Pierre
Pearce, David Harry

MISSOURI
Allen, William Corwin
Bajpai, Rakesh Kumar
Graham, Charles
Hahn, Allen W
Landiss, Daniel Jay
Larson, Kenneth Blaine
McFarland, William D
Mavis, James Osbert
Morley, Robert Emmett, Jr
Saran, Chitaranjan
Schuder, John Claude
Thomas, Lewis Jones, Jr
Titus, Dudley Seymour
Twardowski, Zbylut Jozef
Tyrer, Harry Wakeley
Yasuda, Hirotsugu

MONTANA
Banaugh, Robert Peter
Jurist, John Michael

NEBRASKA
Chakkalakal, Dennis Abraham
Ellingson, Robert James
Haack, Donald C(arl)
Leviticus, Louis I

NEW HAMPSHIRE
Baumann, Hans D
Brous, Don W
Converse, Alvin O
Daubenspeck, John Andrew
Douple, Evan Barr
Hayes, Kirby Maxwell
Kantrowitz, Arthur (Robert)
Strohbehn, John Walter

NEW JERSEY
Agathos, Spiros Nicholas
Akerboom, Jack
Amory, David William
Badalamenti, Anthony Francis
Bernstein, Alan D
Black, H(arold) S(tephen)
Castellana, Frank Sebastian
Chen, Nai Y
Chu, Horn Dean
Clements, Wayne Irwin
Connor, John Arthur
Crump, Jesse Franklin
Davis, Thomas Arthur
Edelman, Norman H
Farkas, Daniel Frederick
Fischell, David R
Flam, Eric
Goggins, Jean A
Goyal, Suresh
Greenstein, Teddy
Hall, Joseph L
Halpern, Teodoro
Hart, Colin Patrick
Houston, Vern Lynn
Huang, Ching-Rong
Kaufman, Arnold
Keller, Kenneth H(arrison)
Konikoff, John Jacob
Kristol, David Sol
Langrana, Noshir A
Li, John Kong-Jiann
Liebig, William John
Litman, Irving Isaac
Loscher, Robert A
Lunn, Anthony Crowther
Maclin, Ernest
McNicholas, James J
Mallison, George Franklin
Masurekar, Prakash Sharatchandra
Mears, David R
Megna, John C(osimo)
Meyer, Andrew U
Mohr, Richard Arnold
Pilla, Arthur Anthony
Reisman, Stanley S

Bioengineering & Biomedical Engineering (cont)

Ronel, Samuel Hanan
Salkind, Alvin J
Selwyn, Donald
Semmlow, John Leonard
Sofer, Samir Salim
Stanton, Robert E
Stickle, Gene P
Szarka, Laszlo Joseph
Tojo, Kakuji
Trivedi, Nayan B
Tzanakou, M Evangelia
Weiss, Lawrence H(eisler)
Welkowitz, Walter
West, John M(aurice)
Williams, Maryon Johnston, Jr
Ziskin, Marvin Carl
Zoltan, Bart Joseph

NEW MEXICO
Doss, James Daniel
Fukushima, Eiichi
Gray, Edwin R
Johnston, Roger Glenn
Novak, James Lawrence
Roubicek, Rudolf V
Small, James Graydon
Welford, Norman Traviss
Wilkins, Ebtisam A M Seoudi

NEW YORK
Akers, Charles Kenton
Baier, Robert Edward
Belfort, Georges
Bell, Duncan Hadley
Bennett, Leon
Bizios, Rena
Blesser, William B
Borden, Edward B
Bottomley, Paul Arthur
Branch, Garland Marion, Jr
Brenner, Mortimer W
Brunski, John Beyer
Bungay, Henry Robert, III
Carstensen, Edwin L(orenz)
Cheney, Margaret
Clark, Alfred, Jr
Cluxton, David H
Cohen, Irving Allan
Cokelet, Giles R(oy)
Cowin, Stephen Corteen
Csermely, Thomas J(ohn)
Dahl, John Robert
Dobelle, William Harvey
Doremus, Robert Heward
Duffey, Michael Eugene
Falk, Theodore J(ohn)
Fewkes, Robert Charles Joseph
Fordon, Wilfred Aaron
Fordyce, Wayne Edgar
Frankel, Victor H
Franse, Renard
Gilmore, Robert Snee
Gokhale, Vishwas Vinayak
Goodhue, Charles Thomas
Greatbatch, Wilson
Greene, Peter Richard
Harris, Jack Kenyon
Hart, Howard Roscoe, Jr
Hovnanian, H(rair) Philip
Hrushesky, William John Michael
Khanna, Shyam Mohan
Kim, Young Joo
Kinnen, Edwin
Kletsky, Earl J(ustin)
Knowles, Richard James Robert
Kohn, Michael
Konnerth, Karl Louis
Lai, W(ei) Michael
Leary, James Francis
Lee, George C
Lee, Richard Shao-Lin
Lenchner, Nathaniel Herbert
Levine, Sumner Norton
Lewis, Edward R(obert)
Litz, Lawrence Marvin
LoGerfo, John J
Loughlin, James Francis
Lubowsky, Jack
Mates, Robert Edward
Mellins, Robert B
Meltzer, Richard Stuart
Meth, Irving Marvin
Moghadam, Omid A
Moghadam, Omid A
Morris, Thomas Wilde
Moss, Gerald
Moss, Melvin Lionel
Mow, Van C
Mozley, James Marshall, Jr
Murphy, Eugene F(rancis)
Newell, Jonathan Clark
Oldshue, J(ames) Y(oung)
Ostrander, Lee E
Palmer, Harvey John
Pierson, Richard Norris, Jr
Ranu, Harcharan Singh
Redington, Rowland Wells
Riffenburgh, Robert Harry
Roy, Rob
Ruskin, Asa Paul

Savic, Michael I
Scott, Peter Douglas
Shapiro, Paul Jonathon
Sharpless, Thomas Kite
Shoemaker, Christine Annette
Shuler, Michael Louis
Simmons, Harry Dady, Jr
Slezak, Jane Ann
Smith, Lowell Scott
Smith, Robert L
Sotirchos, Stratis V
Susskind, Herbert
Taylor, Kenneth Doyle
Tichauer, Erwin Rudolph
Tobin, Michael
Torre, Douglas Paul
Tycko, Daniel H
Vance, Miles Elliott
Viernstein, Lawrence J
Vogelman, Joseph H(erbert)
Wald, Alvin Stanley
Wang, Lawrence K
Watkins, Charles B
Weinbaum, Sheldon
Weinstein, Herbert
Weissman, Michael Herbert
Wertman, Louis
Wheeless, Leon Lum, Jr
Wiegert, Philip E
Yellin, Edward L
Zablow, Leonard
Zelman, Allen
Zwislocki, Jozef John

NORTH CAROLINA
Abrams, Charlie Frank, Jr
Barnes, Ralph W
Barr, Roger Coke
Clark, Howard Garmany
Coulter, Norman Arthur, Jr
Defoggi, Ernest
Floyd, Carey E, Jr
Gerhardt, Don John
Hammond, William Edward
Haynes, John Lenneis
Hochmuth, Robert Milo
Hooper, Irving R
Horres, Charles Russell, Jr
Hsiao, Henry Shih-Chan
Huang, Barney K(uo-Yen)
Huang, Eng-Shang(Clark)
Hutchins, Phillip Michael
Jaszczak, Ronald Jack
Johnson, Richard Noring
Kelley, Thomas F
Klitzman, Bruce
Kootsey, Joseph Mailen
Lawson, Dewey Tull
Leatherman, Nelson E(arle)
Lucas, Carol N
Lundblad, Roger Lauren
Malindzak, George S, Jr
Markert, Clement Lawrence
Mosberg, Arnold T
Nagle, H Troy
Pasipoularides, Ares D
Pearlstein, Robert David
Plonsey, Robert
Sandok, Paul Louis
Smith, Charles Eugene
Starmer, C Frank
Thubrikar, Mano J
Tsui, Benjamin Ming Wah
Weiner, Richard D
Wiley, Albert Lee, Jr
Wolbarsht, Myron Lee
Wolcott, Thomas Gordon

NORTH DAKOTA
Bares, William Anthony
Mathsen, Don Verden

OHIO
Andonian, Arsavir Takfor
Bar-Yishay, Ephraim
Bennett, G(ary) F
Berliner, Lawrence J
Bettice, John Allen
Brown, Stanley Alfred
Brunner, Gordon Francis
Chu, Mamerto Loarca
Clark, David Lee
Dell'Osso, Louis Frank
Engin, Ali Erkan
Fink, David Jordan
Frank, Thomas Paul
Friedman, Morton Harold
Glaser, Roger Michael
Greber, Isaac
Greenberg, David B(ernard)
Grood, Edward S
Hall, Ernest Lenard
Harmon, Leon David
Hartrum, Thomas Charles
Hinman, Channing L
Hughes, Kenneth E(ugene)
Kahn, Alan Richard
Katona, Peter Geza
Katz, J Lawrence
Katzen, Raphael
Kinzel, Gary Lee
Ko, Wen Hsiung
Kolesar, Edward S
Kwatra, Subhash Chander

LaManna, Joseph Charles
Lee, Sunggyu
Leininger, Robert Irvin
Levy, Matthew Nathan
Lioi, Anthony Pasquale
McComis, William T(homas)
Macklin, Martin
McMillin, Carl Richard
Marras, William Steven
Martin, Paul Joseph
Miraldi, Floro D
Mortimer, J(ohn) Thomas
Mudry, Karen Michele
Neuman, Michael R
Peckham, P Hunter
Phillips, Chandler Allen
Primiano, Frank Paul, Jr
Reddy, Narender Pabbathi
Reynolds, David B
Ricord, Louis Chester
Rudy, Yoram
Saidel, Gerald Maxwell
Slonim, Arnold Robert
Stacy, Ralph Winston
Stevenson, James Francis
Taylor, Bruce Cahill
Thomas, Cecil Wayne
Tuovinen, Olli Heikki
Von Gierke, Henning Edgar
Wasserman, Donald Eugene
Weed, Herman Roscoe
Wolaver, Lynn E(llsworth)
Wong, Anthony Sai-Hung
Yoon, Jong Sik

OKLAHOMA
Begovac, Paul C
Berbari, Edward J
Campbell, John Alexander
Collins, William Edward
Dormer, Kenneth John
Foutch, Gary Lynn
Love, Tom Jay, Jr
Neathery, Raymond Franklin
Snow, Clyde Collins
Stover, Enos Loy
Troelstra, Arne

OREGON
Ehrmantraut, Harry Charles
Fields, R(ance) Wayne
Smith, Kelly L
Yang, Hoya Y

PENNSYLVANIA
Alteveer, Robert Jan George
Altschuler, Martin David
Angstadt, Carol Newborg
Baish, James William
Batterman, Steven C(harles)
Berkowitz, David Andrew
Bilgutay, Nihat Mustafa
Boston, John Robert
Brighton, John Austin
Briller, Stanley A
Buchsbaum, Gershon
Cavanagh, Peter R
Ching, Stephen Wing-Fook
Cox, Robert Harold
Crosby, Lon Owen
Detwiler, John Stephen
Dinges, David Francis
Eisenstein, Bruce A
Fan, Ningping
Fromm, Eli
Garfinkel, David
Gee, William
Geselowitz, David B(eryl)
Giannovario, Joseph Anthony
Grossmann, Elihu D(avid)
Jaron, Dov
Johnson, Ogden Carl
Kaufman, William Morris
Klafter, Richard D(avid)
Kline, Donald Edgar
Kornfield, A(lfred) T(heodore)
Kozak, Wlodzimierz M
Kresh, J Yasha
Kwatny, Eugene Michael
Li, C(hing) C(hung)
Litt, Mitchell
Liu, Andrew T C
Longini, Richard Leon
Natoli, John
Newhouse, Vernon Leopold
Nobel, Joel J
Noordergraaf, Abraham
Onik, Gary M
Oswald, Thomas Harold
Parker, Jennifer Ware
Pelleg, Amir
Pennock, Bernard Eugene
Pollack, Solomon R
Purdy, David Lawrence
Rajagopal, K R
Reid, John Mitchell
Russell, Alan James
Samuelson, H Vaughn
Scherer, Peter William
Schulman, Marvin
Schwan, Herman Paul
Sciamanda, Robert Joseph
Sharma, Mangalore Gokulanand
Sheers, William Sadler

Shung, K Kirk
Sitrin, Robert David
Spicher, John L
Spooner, Robert Bruce
Stewart, George Hamill
Stone, Douglas Roy
Sutter, Philip Henry
Thomas, Donald H(arvey)
Ultman, James Stuart
Van der Werff, Terry Jay
Villafana, Theodore
Voloshin, Arkady S
Wall, Conrad, III
Wampler, D Eugene
Weisz, Paul Burg
Whinnery, James Elliott
Williams, Albert J, Jr
Wolken, Jerome Jay

RHODE ISLAND
Barnett, Stanley M(arvin)
Christenson, Lisa
Galletti, Pierre Marie
Hoffman, Joseph Ellsworth, Jr
Richardson, Peter Damian

SOUTH CAROLINA
Hargest, Thomas Sewell
Lam, Chan F
McCutcheon, Ernest P
McNamee, James Emerson
Moyle, David Douglas
Turner, John Lindsey
Voit, Eberhard Otto
Watson, Philip Donald
Zucker, Robert Martin

TENNESSEE
Bahner, Carl Tabb
Barach, John Paul
Campbell, William B(uford)
Collins, Jerry C
Davison, Brian Henry
Deivanayagan, Subramanian
Donaldson, Terrence Lee
Fordham, James Lynn
Goulding, Charles Edwin, Jr
Harris, Thomas R(aymond)
King, Paul Harvey
McLeod, William D
Overholser, Knowles Arthur
Partridge, Lloyd Donald
Roselli, Robert J
Turitto, Vincent Thomas
Wasserman, Jack F
Wikswo, John Peter, Jr

TEXAS
Addy, Tralance Obuama
Aggarwal, Shanti J
Aguilera, Jose Miguel
Alexander, William Carter
Alfrey, Clarence P, Jr
Armeniades, Constantine D
Baker, Lee Edward
David, Yadin B
Diller, Kenneth Ray
Dunham, James George
Dykstra, Jerald Paul
Eberhart, Robert Clyde
Eberle, Jon William
Eggers, Fred M
Fleenor, Marvin Bension
Fletcher, John Lynn
Frenger, Paul F
Georgiou, George
Gilmour - Stallsworth, Lisa K
Gutierrez, Guillermo
Hartley, Craig Jay
Hellums, Jesse David
Homsy, Charles Albert
Howard, Lorn Lambier
Johnston, Melvin Roscoe
Jones, William B
Journeay, Glen Eugene
Koppa, Rodger J
Krouskop, Thomas Alan
McIntire, Larry V(ern)
McIntyre, John Armin
Miller, Gerald E
Ong, Poen Sing
Patel, Anil S
Popovich, Robert Peter
Ramsey, Jerry Dwain
Rollwitz, William Lloyd
Schermerhorn, John W
Senatore, Ford Fortunato
Sheppard, Louis Clarke
Skolnick, Malcolm Harris
Soltes, Edward John
Srinivasan, Ramachandra Srini
Stokely, Ernest Mitchell
Summers, Richard Lee
Sweat, Vincent Eugene
Taylor, Robert Dalton
Throckmorton, Gaylord Scott
Thurston, George Butte
Von Maltzahn, Wolf W
White, Ronald Joseph
Wu, Hsin-i

UTAH
Andrade, Joseph D
Berggren, Michael J

Billings, R Gail
Daniels, Alma U(riah)
Gardner, Reed McArthur
Germane, Geoffrey James
Horch, Kenneth William
Janata, Jiri
Kim, Sung Wan
Kopecek, Jindrich
Kratochvil, Jiri
Lee, James Norman
Normann, Richard A
Olsen, Don B
Peterson, Stephen Craig
Powers, Linda Sue
Stokes, Barry Owen
Strozier, James Kinard
Trujillo, Edward Michael
Tuckett, Robert P
Warner, Homer R

VERMONT
Dean, Robert Charles, Jr
Ellis, David M
Kim, Sukyoung
Lipson, Richard L
Sachs, Thomas Dudley

VIRGINIA
Ackerman, Roy Alan
Adams, James Milton
Arp, Leon Joseph
Attinger, Ernst Otto
Beran, Robert Lynn
Cheng, George Chiwo
Chi, Michael
Clarke, Alexander Mallory
Diller, Thomas Eugene
Edlich, Richard French
Eppink, Richard Theodore
Evans, Francis Gaynor
Gabelnick, Henry Lewis
Gabriel, Barbra L
Gillies, George Thomas
Grant, John Wallace
Hutchinson, Thomas Eugene
Kim, Yong Il
Lee, Jen-Shih
Lemp, John Frederick, Jr
Lichtenberg, Byron Kurt
Llewellyn, Gerald Cecil
Mikulecky, Donald C
Moon, Peter Clayton
Myers, Donald Albin
Reswick, James Bigelow
Roth, Allan Charles
Severns, Matthew Lincoln
Sudarshan, T S
Summers, George Donald
Theodoridis, George Constantin
Tole, John Roy
Updike, Otis L(ee), Jr
Wilkins, Judd Rice
Wist, Abund Ottokar
Worsham, James Essex, Jr

WASHINGTON
Andersen, Jonny
Auth, David C
Babb, Albert L(eslie)
Bhansali, Praful V
Binder, Marc David
Canfield, Robert Charles
Cholvin, Neal R
Daly, Colin Henry
Forster, Fred Kurt
Fuchs, Albert Frederick
Halbert, Sheridan A
Heideger, William J(oseph)
Hoffman, Allan Sachs
Hsu, Chin Shung
Huntsman, Lee L
Jacobson, John Obert
Klepper, John Richard
Lago, James
Luft, John Herman
Marshall, Robert P(aul)
Newman, Paul Harold
Pollack, Gerald H
Ratner, Buddy Dennis
Reeves, Jerry John
Riederer-Henderson, Mary Ann
Ringo, John Alan
Robkin, Maurice
Rushmer, Robert Frazer
Staiff, Donald C
Stear, Edwin Byron
Tam, Patrick Yui-Chiu
Taylor, Eugene M
Yee, Sinclair Shee-Sing

WEST VIRGINIA
Albertson, John Newman, Jr
Crum, Edward Hibbert
Kumar, Alok
Shaeiwitz, Joseph Alan
Shuck, Lowell Zane

WISCONSIN
Balmer, Robert Theodore
Cataldi, Horace A(nthony)
Cohen, Bernard Allan
DeWerd, Larry Albert
Divjak, August A
Flax, Stephen Wayne

Geisler, C(hris) D(aniel)
Hinkes, Thomas Michael
Jeutter, Dean Curtis
Linehan, John Henry
Sances, Anthony, Jr
Schopler, Harry A
Sfat, Michael R(udolph)
Skatrud, Thomas Joseph
Smith, Michael James
Thompson, Paul DeVries
Tompkins, Willis Judson
Trautman, Jack Carl
Vanderheiden, Gregg
Warner, H Jack
Webster, John Goodwin
Whiffen, James Douglass
Zuperku, Edward John

WYOMING
Ferris, Clifford D
Steadman, John William

PUERTO RICO
Vargas, Fernando Figueroa

ALBERTA
Hoffer, J(oaquin) A(ndres)
Shysh, Alec
Williams, Michael C(harles)

BRITISH COLUMBIA
Dower, Gordon Ewbank
Mantle, J(ohn) B(ertram)
Milsum, John H
Morrison, James Barbour
Morrison, James Barbour
Shaw, A(lexander) J(ohn)

MANITOBA
Gordon, Richard

NEW BRUNSWICK
Scott, Robert Nelson

NOVA SCOTIA
Rautaharju, Pentti M

ONTARIO
Albisser, Anthony Michael
Bewtra, Jatinder Kumar
Brach, Eugene Jenő
Brash, John Law
Cairns, William Louis
Cobbold, R S C
Cooke, T Derek V
Diosady, Levente Laszlo
Fenster, Aaron
Finlay, Joseph Bryan
Ghista, Dhanjoo Noshir
Hill, Martha Adele
Hopps, John Alexander
Jacobson, Stuart Lee
James, David F
Joy, Michael Lawrence Grahame
Kennedy, James Cecil
Kosaric, Naim
Kunov, Hans
Landolt, Jack Peter
Leung, Lai-Wo Stan
McNeice, Gregory Malcolm
Mittal, Gauri S
Norwich, Kenneth Howard
Piekarski, Konstanty
Roach, Margot Ruth
Robinson, Campbell William
Sherebrin, Marvin Harold
Slutsky, Arthur
Sun, Anthony Mein-Fang
Wayman, Morris
White, Denis Naldrett
Winter, David Arthur
Wu, Tai Wing
Zingg, Walter
Zubeckis, Edgar

PRINCE EDWARD ISLAND
Amend, James Frederick
Lodge, Malcolm A

QUEBEC
Billette, Jacques
Boyarsky, Abraham Joseph
Chang, Thomas Ming Swi
Cooper, David Gordon
Davis, John F
Frojmovic, Maurice Mony
Gulrajani, Ramesh Mulchand
Kearney, Robert Edward
Leduy, Anh
Roberge, Fernand Adrien
Roy, Guy
Volesky, Bohumil
Zucker, Steven Warren

SASKATCHEWAN
Cullimore, Denis Roy
Gupta, Madan Mohan
Macdonald, Douglas Gordon
Pollak, Viktor A

OTHER COUNTRIES
Chang, H K
De Bruyne, Peter
Dunn, Irving John

Eik-Nes, Kristen Borger
Falb, Richard D
Hazeyama, Yuji
Huckaba, Charles Edwin
Lee, Choong Woong
Monos, Emil
O'Neill, Patricia Lynn
Shapiro, Stuart
Suga, Hiroyuki
Westerhof, Nicolaas
Woo, Kwang Bang
Yang, Ho Seung
Yoshida, Fumitake

Ceramics Engineering

ALABAMA
Gates, D(aniel) W(illiam)
Heystek, Hendrik
Pattillo, Robert Allen
Yee, Tin Boo

ARIZONA
Carpenter, Ray Warren
Hansen, Kent W(endrich)
Risbud, Subhash Hanamant
Savrun, Ender
Smyth, Jay Russell
Sundahl, Robert Charles, Jr
Uhlmann, Donald Robert
Willson, Donald Bruce
Withers, James C

CALIFORNIA
Arrowood, Roy Mitchell, Jr
Baba, Paul David
Baldwin, Chandler Milnes
Bieler, Barrie Hill
Bragg, Robert H(enry)
Brown, William E(ric)
Bunshah, Rointan F(ramroze)
Cao, Hengchu
Carniglia, Stephen C(harles)
Chung, Chi Hsiang
Cline, Carl F
Dejonghe, Lutgard C
Dunn, Bruce
Ervin, Guy, Jr
Evans, Anthony Glyn
Feigelson, Robert Saul
Ferreira, Laurence E
Fletcher, Peter C
Gulden, Terry Dale
Gür, Turgut M
Halden, Frank
Harvey, A(lexander)
Harvey, Frances J, II
Hoenig, Clarence L
Hopper, Robert William
Johnson, Sylvia Marian
Kim, Ki Hong
Knapp, William John
Kuan, Teh S
Kurtz, Peter, Jr
Lamm, A Uno
Leipold, Martin H(enry)
Lin, Charlie Yeongching
McCreight, Louis R(alph)
Macha, Milo
McKee, W(illiam) Dean, Jr
Meiser, Michael David
Munir, Zuhair A
Nacamu, Robert Larry
Paquette, David George
Pask, Joseph Adam
Ram, Michael Jay
Rothman, Albert J(oel)
Sachdev, Suresh
Savitz, Maxine Lazarus
Scott, Garland Elmo, Jr
Seiling, Alfred William
Shackelford, James Floyd
Sines, George, Jr
Spera, Frank John
Van Dreser, Merton Lawrence
Watson, James Frederic
White, Jack Lee
Yang, Jenn-Ming
Yeh, Hun Chiang
Yoon, Rick J
Yu, Chyang John

COLORADO
DePoorter, Gerald Leroy
Koenig, John Henry
Moore, John Jeremy
Roshko, Alexana
Speil, Sidney
Thurnauer, Hans

CONNECTICUT
Eppler, Richard A
Grossman, Leonard N(athan)
Hulse, Charles O
Lemkey, Frankin David
Nath, Dilip K
Newman, Robert Weidenthal
Prasad, Arun
Strife, James Richard
Young, Robert Wesley

DELAWARE
Arots, Joseph B(artholomew)

Burn, Ian
Chowdhry, Uma
Hartmann, Hans S(iegfried)
Scherer, George Walter
Schiroky, Gerhard H
Slack, Lyle Howard
Tomic, Ernst Alois
Urquhart, Andrew Willard
Williams, Richard Anderson

DISTRICT OF COLUMBIA
Aggarwal, Ishwar D
Gottschall, Robert James
Kramer, Bruce Michael
Murray, Peter
Rao, Jaganmohan Boppana Lakshmi
Schioler, Liselotte Jensen
Tokar, Michael
Van Echo, Andrew

FLORIDA
Birch, Raymond E(mbree)
Clark, David Edward
Hench, Larry Leroy
McCracken, Walter John
Mitoff, S(tephan) P(aul)
Moudgil, Brij Mohan
Owens, James Samuel
Ricker, Richard W(ilson)
Ucci, Pompelio Angelo
Whitney, Ellsworth Dow
Wygant, J(ames) F(rederic)

GEORGIA
Chandan, Harish Chandra
Chapman, Alan T
Lackey, Walter Jackson
Logan, Kathryn Vance
Moody, Willis E, Jr
Pentecost, Joseph L(uther)
Proctor, W(illiam) J(efferson), Jr
Sanders, T H, Jr
Schulz, David Arthur
Soora, Siva Shunmugam
Starr, Thomas Louis

IDAHO
Tallman, Richard Louis
Weyand, John David

ILLINOIS
Albertson, Clarence E
Berger, Richard Lee
Bergeron, Clifton George
Berkelhamer, Louis H(arry)
Bickelhaupt, R(oy) E(dward)
Bratschun, William R(udolph)
Brown, Sherman Daniel
Buchanan, Relva Chester
Bunting, Bruce Gordon
Crowley, Michael Summers
Fenske, George R
Fessler, Raymond R
Johnson, D(avid) Lynn
Mason, Thomas Oliver
Myles, Kevin Michael
Poeppel, Roger Brian
Rothman, Steven J

INDIANA
Byers, Stanley A
Kennel, William E(lmer)
Kuo, Charles C Y
Lyon, K(enneth) C(assingham)

IOWA
Berard, Michael F
McGee, Thomas Donald
Martin, S W
Park, Joon B
Patterson, John W(illiam)
Wilder, David Randolph

KANSAS
Bauleke, Maynard P(aul)
Cooke, Francis W
Frazier, A Joel
Grisafe, David Anthony

KENTUCKY
Brock, Louis Milton
Currie, Thomas Eswin
Patrick, Robert F(ranklin)
Sarma, Atul C

LOUISIANA
Eaton, Harvill Carlton

MARYLAND
Adams, Edward Franklin
Bhalla, Sushil K
Block, Stanley
Chuang, Tze-jer
Cooperstein, Raymond
Coyle, Thomas Davidson
Dolhert, Leonard Edward
Economos, Geo(rge)
Fishman, Steven Gerald
Freiman, Stephen Weil
Gillich, William John
Guruswamy, Vinodhini
Haller, Wolfgang Karl
Hovmand, Svend
Kacker, Raghu N

Ceramics Engineering (cont)

Kerkar, Awdhoot Vasant
Kerper, Matthew J(ulius)
Lundsager, C(hristian) Bent
Machlin, Irving
McIlvain, Jess Hall
Malghan, Subhaschandra Gangappa
Merz, Kenneth M(alcolm), Jr
Munro, Ronald Gordon
Nagle, Dennis Charles
O'Keefe, John Aloysius
Parker, Frederick John
Peterson, Norman L(ee)
Quadir, Tariq
Ricker, Richard Edmond
Simiu, Emil
Wiederhorn, Sheldon M
Willis, James Byron

MASSACHUSETTS

Aldrich, Ralph Edward
Alliegro, Richard Alan
Ault, N(eil) N(orman)
Bates, Carl H
Blum, John Bennett
Blum, Seymour L
Chu, Gordon P K
Coble, R(obert) L(ouis)
Elliott, John Frank
Haggerty, John S
Hopkins, George Robert
Kalonji, Gretchen
Keat, Paul Powell
Khattak, Chandra Prakash
Kim, Han Joong
Kumar, Kaplesh
Lee, D(on) William
Lu, Grant
McCauley, James Weymann
Mahlo, Edwin K(urt)
Messier, Donald Royal
Passmore, Edmund M
Pasto, Arvid Eric
Peters, Edward Tehle
Poirier, Victor L
Quackenbush, Carr Lane W
Quinn, George David
Reynolds, Charles C
Rhodes, William Holman
Schmidt, Werner H(ans)
Schroter, Stanislaw Gustaw
Schwartz, Thomas Alan
Thieme, Cornelis Leo Hans
Trostel, Louis J(acob), Jr
Tuller, Harry Louis
Tustison, Randal Wayne
Vasilos, Thomas
Viechnicki, Dennis J
Wuensch, Bernhardt J(ohn)

MICHIGAN

Baney, Ronald Howard
Berg, M(orris)
Brubaker, Burton D(ale)
Chandra, Grish
Cornilsen, Bahne Carl
Fawcett, Timothy Goss
Hinton, Jonathan Wayne
Hucke, Edward E
Humenik, Michael, Jr
Insley, Robert H(iteshew)
Jain, Kailash Chandra
Platts, Dennis Robert
Pynnonen, Bruce W
Russell, William Charles
Subramanian, K N
Tien, Tseng-Ying
Van Vlack, Lawrence H(all)
Whalen, Thomas J(ohn)
Willson, Philip James

MINNESOTA

Bailey, Joseph T(homas)
Beck, Warren R(andall)
Fleming, Peter B
Huffine, Coy L(ee)
McHenry, Kelly David
Mar, Henry Y B
Owens, Kenneth Eugene
Sowman, Harold G
Stokes, Robert James

MISSISSIPPI

Marion, Robert Howard
Wehr, Allan Gordon

MISSOURI

Anderson, Harlan U(rie)
Crookston, J(ames) A(damson)
Day, D(elbert) E(dwin)
French, James Edwin
Grant, Sheldon Kerry
Harmon, Robert Wayne
Heimann, Robert L
Hunter, Orville, Jr
Ownby, P(aul) Darrell
Ramey, Roy Richard
Sparlin, Don Merle

NEVADA

Bradt, Richard Carl
Eichbaum, Barlane Ronald

NEW HAMPSHIRE

Beasley, Wayne M(achon)
Rice, Dale Wilson

NEW JERSEY

Abendroth, Reinhard P(aul)
Allegretti, John E
Bachman, George S(trickler)
Berkman, Samuel
Bhandarkar, Suhas D
Blank, Stuart Lawrence
Charvat, F(edia) R(udolf)
Comeforo, Jay E(ugene)
Dentai, Andrew G
Horowitz, Harold S
Johnson, David W, Jr
Jones, John Taylor
Khan, Saad Akhtar
Kramer, Carolyn Margaret
Kurkjian, Charles R(obert)
LaMastro, Robert Anthony
Lemaire, Paul J
Ling, Hung Chi
Loh, Roland Ru-loong
McLaren, Malcolm G(rant)
Newman, Stephen Alexander
Rabinovich, Eliezer M
Rorabaugh, Donald T
Rosen, Carol Zwick
Shanefield, Daniel J
Speronello, Barry Keven
Tenzer, Rudolf Kurt
Tischler, Oscar
Voss, Kenneth Edwin
Wachtman, John Bryan, Jr
Wenzel, John Thompson
Yan, Man Fei
Yoon, Euijoon

NEW MEXICO

Beauchamp, Edwin Knight
Diver, Richard Boyer, Jr
Eagan, Robert John
Kreidl, Norbert J(oachim)
Land, Cecil E(lvin)
Loehman, Ronald Ernest
Magnani, Nicholas J
Michalske, Terry Arthur
Milewski, John Vincent
Mitchell, Terence Edward
Stoddard, Stephen D(avidson)
Wilcox, Paul Denton
Wilder, James Andrew, Jr

NEW YORK

Adams, P B
Beall, George Halsey
Borom, Marcus P(reston)
Boyd, David Charles
Britton, Marvin Gale
Brownell, Wayne E(rnest)
Brun, Milivoj Konstantin
Buhsmer, Charles P
Chyung, Kenneth
Crandall, William B
Danielson, Paul Stephen
Doremus, Robert Heward
Duke, David Allen
Féhette, H(owells Achille) Van Derck
Giannelis, Emmanuel P
Greskovich, Charles David
Grossman, David G
Guile, Donald Lloyd
Gupta, Krishna Murari
Hammer, Richard Benjamin
Hauth, Willard E(llsworth), III
Herman, Herbert
Hillig, William Bruno
Jacoby, William R(ichard)
Johnson, Curtis Alan
Kilbourn, Barry T
King, Alexander Harvey
Lachman, Irwin Morris
Lakatos, Andras Imre
Lane, Richard L
Lanford, William Armistead
MacDowell, John Fraser
McMurtry, Carl Hewes
Meiling, Gerald Stewart
Meloon, David Rand
Minnear, William Paul
Morfopoulos, Vassilis C(onstantinos) P
Mueller, Edward E(ugene)
Myers, Mark B
Nowick, A(rthur) S(tanley)
Ott, Walter Richard
Pasco, Robert William
Prindle, William Roscoe
Prochazka, Svante
Psioda, Joseph Adam
Pye, Lenwood D(avid)
Reed, James Stalford
Reich, Ismar M(eyer)
Rossington, David Ralph
Rothermel, Joseph Jackson
Schreiber, Edward
Shaw, Robert Reeves
Shelby, James Elbert
Smith, Gail Preston
Smith, Russell D
Snyder, Robert Lyman
Spriggs, Richard Moore
Thompson, David Allen
Thompson, David Fred

Thompson, Robert Alan
Tinklepaugh, J(ames) R(oot)
Wang, Xingwu
Ward, Thomas J(ulian)
Westbrook, J(ack) H(all)
Wilson, James M
Zimar, Frank

NORTH CAROLINA

Davis, Robert F(oster)
Gilooly, George Rice
Graff, William (Arthur)
Hamme, John Valentine
Hurt, John Calvin
Hutchins, John R(ichard), III
Landes, Chester Grey
Manning, Charles Richard, Jr
Palmour, Hayne, III
Stoops, R(obert) F(ranklin)

OHIO

Alam, M Khairul
Baker, Robert J(ethro), Jr
Beals, Robert J(ennings)
Bradley, Ronald W
Choudhary, Manoj Kumar
Cooper, Alfred R, Jr
Copp, Albert Nils
DiCarlo, James Anthony
Drummond, Charles Henry, III
Duckworth, Winston H(oward)
Duderstadt, Edward C(harles)
Dutta, Sunil
Flynn, Ronald Thomas
Graham, Henry Collins
Graham, John W
Graham, Paul Whitener Link
Harmer, Richard Sharpless
Heasley, James Henry
Heuer, Arthur Harold
Hill, Brian
Huff, Norman Thomas
Johnson, Howard B(eattie)
Kampe, Dennis James
Kawasaki, Edwin Pope
Koenig, Charles Jacob
Kreidler, Eric Russell
Krysiak, Joseph Edward
La Mers, Thomas Herbert
Land, Peter L
Lee, Haynes A
Lennon, John W(illiam)
McCoy, Robert Allyn
Marchant, David Dennis
Metzger, A(rthur) J(oseph)
Mitchell, Terence Edward
Mohn, Walter Rosing
Moore, Arthur William
Parthasarathy, Triplicane Asuri
Petersen, Frederick Adolph
Piper, Ervin L
Pirooz, Perry Parviz
Rai, Amarendra Kumar
Readey, Dennis W(illiam)
Rosa, Casimir Joseph
Rosenfield, Alan R(obert)
Ruh, Robert
Ryder, Robert J
St Pierre, P(hilippe) D(ouglas) S
Sara, Raymond Vincent
Semler, Charles Edward
Shonebarger, F(rancis) J(oseph)
Shook, William Beattie
Snyder, Milton Jack
Stambaugh, Edgel Pryce
Stewart, Daniel Robert
Stover, E(dward) R(oy)
Swift, Howard R(aymond)
Tallan, Norman M
Thomas, Robert Hayne
Tooley, F(ay) V(aNisle)
Venkatu, Doulatabad A
Wurst, John Charles

OKLAHOMA

Harris, Jesse Ray

PENNSYLVANIA

Baran, George Roman
Belitskus, David
Bhalla, Amar S
Biggers, James Virgil
Blachere, Jean R
Cook, Charles S
Cox, John E(dward)
Davis, R(obert) E(lliot)
Emlemdi, Hasan Bashir
Esmen, Nurtan A
Garg, Diwakar
Gebhardt, Joseph John
Goldman, Kenneth M(arvin)
Goodwin, Charles Arthur
Greenberg, Charles Bernard
Gruver, Robert Martin
Gupta, Tapan Kumar
Hall, Charles A(insley)
Hall, Gary R
Harmer, Martin Paul
Harrison, Don Edward
Heiligman, Harold A
Hogg, Richard
Hummel, F(loyd) A(llen)
Jain, Himanshu
Jang, Sei Joo

Kilp, Gerald R
Kirchner, H(enry) P(aul)
Koczak, Michael Julius
Kotyk, Michael
Lidman, William G
Lo, W(ing) C(heuk)
MacZura, George
Mattox, Douglas Milton
Mecholsky, John Joseph, Jr
Messier, Russell
Mistler, Richard Edward
Osborn, Elburt Franklin
Perrotta, Anthony Joseph
Proske, Joseph Walter
Ray, Siba Prasad
Rindone, Guy E(dward)
Ruh, Edwin
Shapiro, Zalman Mordecai
Sieger, John S(ylvester)
Slyh, John A(llen)
Smid, Robert John
Smothers, William Joseph
Smyth, Donald Morgan
Staut, Ronald
Steiger, Roger Arthur
Stetson, Harold W(ilbur)
Stubican, Vladimir S(tjepan)
Tressler, Richard Ernest
Waldman, Jeffrey
Wang, Ke-Chin
Wasylyk, John Stanley
Weber, Frank L
Williams, David Bernard
Williamson, William O(wen)
Wood, Susan
Worrell, Wayne L
Yoldas, Bulent Erturk

RHODE ISLAND

Rockett, Thomas John
Suresh, Subra

SOUTH CAROLINA

Bradley, Fennimore N
Coffeen, W(illiam) W(eber)
Haertling, Gene Henry
Lee, Burtrand Insung
Lefort, Henry G(erard)
Lewis, Gordon
Piper, John
Robinson, Gilbert C(hase)

TENNESSEE

Bleier, Alan
Bloom, Everett E
Campbell, William B(uford)
Dudney, Nancy Johnston
Eyerly, George B(rown)
Harms, William Otto
Holcombe, Cressie Earl, Jr
Klueh, Ronald Lloyd
Lotts, Adolphus Lloyd
Scott, J(ames) L(ouis)
Stradley, James Grant

TEXAS

Brown, Richard Martin
Cleek, Given Wood
Diller, Kenneth Ray
Hillery, Herbert Vincent
Johnson, Elwin L Pete
McNamara, Edward P(aul)
Parikh, N(iranjan) M
Rulon, Richard M
Stradley, Norman H(enry)
Thornton, Hubert Richard
Von Rosenberg, Dale Ursini
Warner, Bert Joseph

UTAH

Carnahan, Robert D
Horton, Ralph M
Hyatt, Edmond Preston
Rasmussen, Jewell J
Rigby, E(ugene) B(ertrand)
Weed, Grant B(arg)

VERMONT

Campbell, Donald Edward
Kim, Sukyoung

VIRGINIA

Brown, Jesse J, Jr
Clarke, Alan R
Desu, Seshu Babu
Diness, Arthur M(ichael)
Hasselman, Didericus Petrus Hermannus
Johnson, David Harley
Kahn, Manfred
Kerr, John M(artin)
Lynch, Eugene Darrel
Matthews, R(obert) B(ruce)
Ormsby, W(alter) Clayton
Puster, Richard Lee
Rice, Roy Warren
Schreiber, Henry Dale
Vojnovich, Theodore
Wilcox, Benjamin A

WASHINGTON

Anderson, Harlan John
Burns, Robert Ward
Chikalla, Thomas D(avid)
Ding, Jow-Lian

Einziger, Robert E
Fischbach, David Bibb
Hart, Patrick E(ugene)
Hinman, Chester Arthur
Kruger, Owen L
Miller, Alan Dale
Roberts, J T Adrian
Scott, William D(oane)
Smith, Richard Dale
Stang, Robert George
Weber, William J
Whittemore, O(sgood) J(ames), Jr
Wilson, Charles Norman
Woods, Keith Newell

WEST VIRGINIA
Olcott, Eugene L

WISCONSIN
Dunn, Stanley Austin
Fournelle, Raymond Albert
Johnson, J(ames) R(obert)
Lenling, William James
Rohatgi, Pradeep Kumar
Szpot, Bruce F

ALBERTA
Heimann, Robert B(ertram)

BRITISH COLUMBIA
Chaklader, Asoke Chandra Das

NOVA SCOTIA
Gow, K V
King, Hubert Wylam

ONTARIO
Allan, George B
Kuriakose, Areekattuthazhayil
Nicholson, Patrick Stephen
Prasad, S E
Runnalls, O(liver) John C(lyve)
Sadler, Arthur Graham

QUEBEC
Angers, Roch
Saint-Jacques, Robert G

OTHER COUNTRIES
Jacob, K Thomas
Kishi, Keiji
Montierth, Max Romney

Chemical Engineering

ALABAMA
Achorn, Frank P
April, Gary Charles
Arnold, David Walker
Barber, J(ames) C(orbett)
Black, James H(ay)
Blouin, Glenn M(organ)
Curry, James Eugene
Galil, Fahmy
Harrison, Benjamin Keith
Hart, David R
Hatcher, William Julian, Jr
Hignett, Travis P(orter)
Hill, David Thomas
Hsu, Andrew C T
Hubbuch, Theodore N(orbert)
Landis, E K
Lanewala, Mohammed A
McKinley, Marvin Dyal
McMinn, Curtis J
Meline, Robert S(ven)
Nieberlein, Vernon Adolph
Pelt, Roland J
Phillips, Alvin B(urt)
Raymond, Dale Rodney
Rhoades, Richard G
Rodriguez, Harold Vernon
Roos, C(harles) William
Schrodt, Verle N(ewton)
Stonecypher, Thomas E(dward)
Veazey, Thomas Mabry
Walsh, William K
White, Niles C
Yett, Fowler Redford

ALASKA
Merrill, Robert Clifford, Jr
Ostermann, Russell Dean

ARIZONA
Berman, Neil Sheldon
Catterall, William E(dward)
Clark, Ezekail Louis
Dorson, William John, Jr
Edwards, Richard M(odlin)
Gealer, Roy L(ee)
Henderson, James Monroe
Kaufman, C(harles) W(esley)
Kleinschmidt, Eric Walker
Langdon, William Mondeo
Mohan, J(oseph) C(harles), Jr
Patterson, Ian D(avid)
Pearlstein, Arne Jacob
Post, Roy G
Randolph, Alan Dean
Reed, R(obert) M(arion LaFollette)
Rehm, Thomas R(oger)
Reiser, Castle O

Rosler, Richard S(tephen)
Sater, Vernon E(ugene)
Scott, Ralph Asa, Jr
Snoddon, W(illiam) J(ohn)
Talbert, Norwood K(eith)
Utagikar, Ajit Purushottam
Wendt, Jost O L

ARKANSAS
Bocquet, Philip E(dmund)
Burchard, John Kenneth
Clausen, Edgar Clemens
Couper, James R(iley)
Day, James Meikle
Fortner, Wendell Lee
Gaddy, James Leoma
Havens, Jerry Arnold
Julil, William G
Oxford, C(harles) W(illiam)
Ransford, George Henry
Shelley, William J
Thatcher, C(harles) M(anson)
Welker, J(ohn) Reed

CALIFORNIA
Adelman, Barnet Reuben
Alexander, Earl L(ogan), Jr
Allen, David Thomas
Anderson, Robert Neil
Appleman, Gabriel
Arias, Jose Manuel
Armstrong, Don L
Aroyan, H(arry) J(ames)
Arslancan, Ahmet N
Atherton, Robert W
Augood, Derek Raymond
Aziz, Khalid
Bacastow, Robert Bruce
Bae, Jae Ho
Baerg, William
Bailey, James E(dwin)
Baldauf, Gunther H(erman)
Balzhiser, Richard E(arl)
Bareis, David W(illard)
Barr, Frank T(homas)
Barry, Michael Lee
Beaton, Roy Howard
Beek, John
Bell, Alexis T
Bengtson, Kermit (Bernard)
Benson, Dale B(ulen)
Berg, Clyde H O
Bernath, L(ouis)
Berriman, Lester P
Bery, Mahendera K
Blanch, Harvey Warren
Blue, E(manuel) M(orse)
Bogart, Marcel J(ean) P(aul)
Boulton, Roger Brett
Brace, Robert Allen
Braithwaite, Charles Henry, Jr
Breitmayer, Theodore
Brice, Donat B(ennes)
Bridge, Alan G
Brigham, William Everett
Bromley, Le Roy Alton
Brown, Robert Schenck
Bruley, Duane Frederick
Buck, F(rank) A(lan) Mackinnon
Busch, Joseph Sherman
Butler, G(erard)
Caenepeel, Christopher Leon
Canning, T(homas) F
Carley, James F(rench)
Cashin, Kenneth D(elbert)
Chandrasekaran, Santosh Kumar
Charlesworth, Robert K(oridon)
Che, Stanley Chia-Lin
Chesworth, Robert Hadden
Chin, Jin H
Chin, Yu-Ren
Chittenden, David H
Chompff, Alfred J(ohan)
Chu, Ju Chin
Chung, Raymond
Clazie, Ronald N(orris)
Colmenares, Carlos Adolfo
Constant, Clinton
Cook, Glenn Melvin
Cook, Paul M
Coons, Fred F(leming)
Cooper, Wilson Wayne
Crandall, Edward D
Crookston, Reid B
Culler, F(loyd) L(eRoy), Jr
Current, Jerry Hall
Dance, Eldred Leroy
Dautzenberg, Frits Mathia
Davis, Bruce W
Deleray, Arthur Loyd
Denn, Morton M(ace)
Detz, Clifford M
Di Zio, Steven F(rank)
Dobbins, John Potter
Duffin, John H
Durai-swamy, Kandaswamy
Durland, John R(oyden)
Eagle, Sam
Edwards, Arthur L
Elton, Edward Francis
Engleman, Victor Solomon
Evans, James William
Favorite, John R
Feiler, William A, Jr

Feinleib, Morris
Ferm, Richard L
Ferreira, Laurence E
Flagan, Richard Charles
Flamm, Daniel Lawrence
Fletcher, Thomas Harvey
Fok, Samuel S(hiu) M(ing)
Foss, Alan Stuart
Foster, E(lton) Gordon
Fox, Joseph M(ickle), III
Friedlander, Sheldon K
Frischmuth, Robert Wellington
Furukawa, David Hiroshi
Gardner, Howard Shafer
Garrett, Donald E(verett)
Garrett, L(uther) W(eaver), Jr
Gavalas, George R(ousetos)
Gay, Richard Leslie
Gilkeson, M(urray) Mack
Goddard, Joe Dean
Goldfrank, Max
Goldstein, Walter Elliott
Goren, Simon L
Gouw, T(an) H(ok)
Green, Stanley J(oseph)
Grens, Edward A(nthony), II
Grimaldi, John Vincent
Grossberg, Arnold Lewis
Gunness, R(obert) C(harles)
Hahn, Harold Thomas
Haley, Kenneth William
Hamai, James Y
Hamilton, J(ames) Hugh
Hammond, R Philip
Hanna, Owen Titus
Hanson, D(onald) N(orman)
Hardt, Alexander P
Haritatos, Nicholas John
Harrison, Jonas P
Hartley, Fred L
Hass, Robert Henry
Hausmann, Werner Karl
Heil, John F, Jr
Heil, John F, Jr
Hennig, Harvey
Hermsen, Robert W
Hertwig, Waldemar R
Hester, John Nelson
Hidy, George Martel
Higgins, Brian Gavin
Hill, Roger W(arren)
Hinds, Horace, Jr
Hipkin, Howard G(eorge)
Holm, L(eRoy) W(allace)
Hong, Ki C(hoong)
Hovorka, Robert Bartlett
Hsieh, Paul Yao Tong
Hubbell, D(ean) S(terling)
Hubred, Gale L
Huff, James Eli
Hugill, J(ohn) T(empleton)
Irving, James P
Isenberg, Lionel
Jacobs, Joseph John
Jacobson, Robert Leroy
Jaros, Stanley E(dward)
Jones, Jerry Latham
Judson, Burton Frederick
Kaellis, Joseph
Kafesjian, R(alph)
Kalvinskas, John J(oseph)
Kane, Daniel E(dwin)
Kaneko, Thomas Motomi
Keaton, Michael John
Kelley, Arnold E
Kendall, Robert McCutcheon
Kennedy, James Vern
Kent, Clifford Eugene
Kertamus, Norbert John
Kieschnick, W(illiam) F(rederick)
King, C(ary) Judson, III
Klipstein, David Hampton
Knight, Patricia Anne
Knuth, Eldon L(uverne)
Kohl, A(rthur) L(ionel)
Kouzel, Bernard
Kuehne, Donald Leroy
Lai, San-Cheng
Laity, David Sanford
Lambert, Walter Paul
Lapple, Charles E
Lavendel, Henry W
Leal, L Gary
Lehmann, A(ldo) Spencer
Lenoir, J(ohn) M
Leslie, James C
Leventhal, Leon
Levine, Charles (Arthur)
Levy, Ricardo Benjamin
Li, Pei-Ching
Lichtblau, Irwin Milton
Lieberman, Alvin
Lim, H(enry) C(hol)
Lind, Wilton H(oward)
Lindner, Elek
Lindsay, W(esley) N(ewton)
Liu, Ming-Biann
Lockhart, F(rank) J(ones)
Longwell, P(aul) A(lan)
Lovell, Robert Edmund
Lowi, Alvin, Jr
Lund, J Kenneth
Lunde, Kenneth E(van)
Lynn, Scott

McBride, Lyle E(rwin), Jr
McCarty, Jon Gilbert
McCoy, Benjamin J(oe)
McCune, Conwell Clayton
McKay, Richard A(lan)
McKisson, R(aleigh) L(lewellyn)
Maddex, Phillip J(oseph)
Madix, Robert James
Maimoni, Arturo
Margolis, Stephen Barry
Mason, D(avid) M(alcolm)
Mason, John L(atimer)
Mathews, Larry Arthur
Mavity, Victor T(homas), Jr
Mayer, Jerome F
Mears, David Elliott
Meiners, Henry C(ito)
Meldau, R(obert) F(rederick)
Metzler, Charles Virgil
Meyer, Brad Anthony
Michels, Lloyd R
Mikolaj, Paul G(eorge)
Milligan, Robert T(homas)
Minet, Ronald G(eorge)
Mitchell, Reginald Eugene
Moll, Albert James
Morari, Manfred
Morgal, Paul Walter
Morgan, Lucian L(loyd)
Morrison, Malcolm Cameron
Muller, Rolf Hugo
Munger, Charles Galloway
Myers, John E(arle)
Navratil, James Dale
Naworski, Joseph Sylvester, Jr
Nazaroff, William W
Nelson, David A
Newman, John Scott
Newsom, Herbert Charles
Nguyen, Caroline Phuongdung
Nicholson, W(illiam) J(oseph)
Nixon, Alan Charles
Nobe, Ken
Noring, Jon Everett
Oldenburg, C(harles) C(lifford)
Oldenkamp, Richard D(ouglas)
Oliver, Earl Davis
Olson, Austin C(arlen)
Opfell, John Burton
Ore, Fernando
O'Rear, Dennis John
Orr, Franklin M, Jr
Ouano, Augustus Ceniza
Palen, Joseph W
Pan, Bingham Y(ing) K(uei)
Pastell, Daniel L(ouis)
Pearce, Frank G
Pearl, W(esley) L(loyd)
Pearson, Dale Sheldon
Petersen, Eugene E(dward)
Phillips, John Richard
Piasecki, Leonard R(ichard)
Pigford, Thomas H(arrington)
Pings, C(ornelius) J(ohn)
Pitcher, Wayne Harold, Jr
Porter, Marcellus Clay
Prausnitz, John Michael
Quentin, George Heinz
Raben, Irwin A(bram)
Ramey, H(enry) J(ackson), Jr
Ravicz, Arthur Eugene
Remedios, E(dward) C(harles)
Remer, Donald Sherwood
Richter, George Neal
Rinker, Robert G(ene)
Ritter, R(obert) Brown
Robertson, George Harcourt
Robertson, Glenn D(avid), Jr
Robillard, Geoffrey
Robin, Allen Maurice
Robinson, Richard C(lark)
Romig, Robert P
Rose, Aaron
Ross, Philip Norman, Jr
Rossen, Joel N(orman)
Rothman, Albert J(oel)
Rubin, Barney
Ruskin, Arnold M(ilton)
Russell, Grant E(dwin)
Rutz, Lenard O(tto)
Sanborn, Charles E(van)
Sandall, Orville Cecil
Sanders, Charles F(ranklin), Jr
Sarem, Amir M Sam
Sawyer, Frederick George
Schlatter, James Cameron
Schlinger, W(arren) G(leason)
Schneider, George Ronald
Schneider, Kenneth John
Schumacher, William John
Scott, John W(alter)
Seborg, Dale Edward
Seinfeld, John H
Semrau, Konrad (Troxel)
Serbia, George William
Sesonske, Alexander
Shah, Manesh J(agmohan)
Shair, Fredrick H
Sherwood, Albert E(dward)
Shrontz, John William
Shuler, Patrick James
Shumate, Starling Everett, II
Siegfried, William
Simon, Robert H

Chemical Engineering (cont)

Simpson, Howard Douglas
Slivinsky, Sandra Harriet
Smith, A(llen) N(athan)
Smith, Joe M(auk)
Smith, Ralph Carlisle
Snyder, Nathan W(illiam)
Stephens, Douglas Robert
Stillman, Richard Ernest
Stroeve, Pieter
Sun, Yun-Chung
Sundberg, John Edwin
Swidler, Ronald
Switzer, Robert L
Theodorou, Doros Nicolas
Tobias, Charles W
Todd, Eric E(dward)
Trilling, Charles A(lexander)
Valle-Riestra, J(oseph) Frank
Van Klaveren, Nico
Van Vorst, William D
Varteressian, K(egham) A(rshavir)
Vause, Edwin H(amilton)
Vermeulen, Theodore (Cole)
Vilker, Vincent Lee
Walker, Jimmy Newton
Walsh, William J
Wang, Chiu-Sen
Waterland, Larry R
Wazzan, A R Frank
Weinberg, William Henry
Weir, Alexander, Jr
Westerman, Edwin J(ames)
Whitehead, Kenneth E
Whitney, Gina Marie
Whittam, James Henry
Wilde, D(ouglass) J(ames)
Wilke, Charles R
Williams, Alan K
Willson, Alan Neil, Jr
Winter, Olaf Hermann
Witham, Clyde Lester
Wolf, Irving W
Wong, James B(ok)
Wong, Morton Min
Wright, Roger M
Yang, Meiling T
Yasui, George
Yen, I-Kuen
Yuan, Shao-yuen
Yuan, Sidney Wei Kwun
Yum, Su Il
Ziegenhagen, Allyn James

COLORADO
Baldwin, Lionel V
Barrick, P(aul) L(atrell)
Barry, Henry F
Beck, Steven R
Calcaterra, Robert John
Chase, Curtis Alden, Jr
Christiansen, Robert M(ilton)
Clough, David Edwards
Colver, C(harles) Phillip
Danzberger, Alexander Harris
Davis, John Albert, Jr
Drayer, Dennis Eugene
Ely, James Frank
Falconer, John Lucien
Flynn, Thomas M(urray)
Gary, James H(ubert)
Giller, E(dward) B(onfoy)
Gilliland, John L(awrence), Jr
Gogarty, W(illiam) B(arney)
Golden, John O(rville)
Graue, Dennis Jerome
Grobner, Paul Josef
Harper, Judson M(orse)
Hauser, Ray Louis
Hawley, Robert W(illiam)
Herring, Richard Norman
Hinman, Norman Dean
Howerton, Murlin T(homas)
Iyer, Ravi
Jargon, Jerry Robert
Jha, Mahesh Chandra
Jones, Stanley C(ulver)
Kelchner, Burton L(ewis)
Keller, Frederick Albert, Jr
Kidnay, Arthur J
Klein, James H(enry)
Knight, Bruce L
Kompala, Dhinakar S
Koons, David Swarner
Krantz, William Bernard
Lauer, B(yron) E(lmer)
Linden, James Carl
Mahajan, Roop Lal
Noble, Richard Daniel
Olien, Neil Arnold
Parsons, Robert W(estwood)
Peters, Max S(tone)
Plummer, Mark Alan
Poettmann, Fred H(einz)
Ramirez, W Fred
Rothfeld, Leonard B(enjamin)
Sani, Robert L(e Roy)
Savory, Leonard E(rwin)
Schmidt, Alan Frederick
Severson, Donald E(verett)
Streib, W(illiam) C(harles)
Timblin, Lloyd O, Jr
Timmerhaus, K(laus) D(ieter)

VanderLinden, Carl R
Vestal, Charles Russell
West, Ronald E(mmett)
York, J(esse) Louis
Zahradnik, Raymond Louis
Zimmerman, Carle Clark, Jr

CONNECTICUT
Baker, Bernard S
Bell, James Paul
Bennett, Carroll O(sborn)
Bilhorn, John Merlyn
Bollyky, L(aszlo) Joseph
Borden, George Wayne
Bozzuto, Carl Richard
Bretton, R(andolph) H(enry)
Bunney, Benjamin Stephenson
Butensky, Martin Samuel
Cadogan, W(illiam) P(atrick)
Carpenter, Kent Heisley
Casberg, John Martin
Chan, Wing Cheng Raymond
Chave, Charles Trudeau
Cheatham, Robert Gary
Cole, Stephen H(ervey)
Corwin, H(arold) E(arl)
Coughlin, Robert William
Cutlip, Michael B
Denton, William Irwin
DiBenedetto, Anthony T
Dobay, Donald Gene
Dobry, Reuven
Edman, Walter W
Epperly, W Robert
Eppler, Richard A
Falkenstein, Gary Lee
Forman, J(oseph) Charles
Frey, Douglas D
Geitz, R(obert) C(harles)
George, Edward Thomas
Goldstein, Marvin Sherwood
Grove, Herbert D(uncan), Jr
Gupta, Dharam Vir
Haller, Gary Lee
Harding, R(onald) H(ugh)
Horváth, Csaba Gyula
Hou, Kenneth C
Howard, G(eorge) Michael
Hsu, Nelson Nae-Ching
Hyde, John Welford
I, Ting-Po
Kellner, Henry L(ouis)
Kesten, Arthur S(idney)
Klei, Herbert Edward, Jr
Koberstein, Jeffrey Thomas
Korin, Amos
Kroll, Charles L(ouis)
Kunz, Harold Russell
Kurose, George
Leinroth, Jean Paul, Jr
Lobo, Walter E(der)
Lunde, Peter J
McCardell, W(illiam) M(arkham)
McKibbins, Samuel Wayne
McLain, William Harvey
Marsh, Bertrand Duane
Mickley, Harold S(omers)
Mitchell, Robert L(ynne)
Murai, Kotaro
Pai, Venkatrao K
Prabulos, Joseph J, Jr
Ramakrishnan, Terizhandur S
Reed, Charles E(li)
Rie, John E
Rosner, Daniel E(dwin)
Sabnis, Suman T
Schetky, L(aurence) M(cDonald)
Schiessl, H(enry) W(illiam)
Schoen, Herbert M
Schoenbrunn, Erwin F(rederick)
Schrage, F Eugene
Shaw, Montgomery Throop
Simerl, L(inton) E(arl)
Sinha, Vinod T(arkeshwar)
Skrivan, J(oseph) F(rancis)
Slotnick, Herbert
Stutzman, Leroy F
Suen, T(zeng) J(iueq)
Taylor, Roy Jasper
Thomas, Arthur L
Traskos, Richard Thomas
Walker, Charles A(llen)
Weidenbaum, Sherman S
Weiss, Robert Alan
Welch, John F, Jr
Wernau, William Charles

DELAWARE
Ahmed, Syed Mahmood
Akell, Robert B(erry)
Albers, Robert Edward
Allan, A J Gordon
Anderson, Howard W(ayne)
Armbrecht, Frank Maurice, Jr
Arots, Joseph B(artholomew)
Babcock, Byron D(ale)
Babcock, Dale F
Balder, Jay Royal
Bannister, Robert Grimshaw
Barteau, Mark Alan
Bedingfield, Charles H(osmer)
Bischoff, Kenneth Bruce
Borden, James B
Bours, William A(lsop), III

Brandreth, Dale Alden
Buckley, Page Scott
Busche, Robert M(arion)
Bydal, Bruce A
Carter, Richard P(ence)
Cichelli, Mario T(homas)
Clark, H(arold) B(lack)
Clark, William B
Comings, Edward Walter
Conklin, R(oger) N(orton)
Corty, Claude
Cunning, Joe David
Davies, Neg
Dentel, Steven Keith
Dhurjati, Prasad S
Dietz, John W
Dodge, Donald W(illiam)
Doerner, William A(llen)
Drew, David A(bbott)
Eidman, Richard August Louis
Evans, Franklin James, Jr
Foley, Henry Charles
Foster, Henry D(orroh)
Friel, Daniel Denwood
Gates, Bruce C(lark)
Gibson, Joseph W(hitton), Jr
Glaeser, Hans Hellmut
Goffinet, Edward P(eter), Jr
Gonzalez, Raul A(lberto)
Gray, Joseph B(urnham)
Greenewalt, Crawford Hallock
Habibi, Kamran
Hackman, Elmer Ellsworth, III
Hecht, J(ames) L(ee)
Hochberg, Jerome
Holob, Gary M
Hoopes, John W(alker), Jr
Hughes, John I(ngram)
Hull, Donald R(obert)
Immediata, Tony Michael
Isakoff, Sheldon Erwin
Jefferies, Steven
Kaler, Eric William
Kallal, R(obert) J(ohn)
Kamack, H(arry) J(oseph)
Kemp, Harold Steen
Kilkson, Henn
King, Charles O(rrin)
Kiviat, Fred E
Klein, Michael Tully
Kouba, Delore Loren
Lamb, David E(rnest)
Lamb, William Bolitho
Leffew, Kenneth W
Littleton, H(arold) T(homas) J(ackson)
Lombardo, R(osario) J(oseph)
Long, John Reed
Luckring, R(ichard) M(ichael)
McCullough, Roy Lynn
McCune, Leroy K(iley)
McGeorge, A(rthur), Jr
McNeely, James Braden
Manogue, William H(enry)
Marsh, Dean Mitchell
Mehra, Vinodkumar S
Metzner, Arthur B(erthold)
Miller, Donald Nelson
Mills, Patrick Leo, Sr
Moch, Irving, Jr
Monet, Marion C(renshall)
Mongan, Edwin Lawrence, Jr
Muendel, Carl H(einrich)
Ogunnaike, Babatunde Ayodeji
Olson, Jon H
Pecorini, Hector A(ndrew)
Pell, Mel
Pierce, Robert Henry Horace, Jr
Polejes, J(acob) D
Pontius, E(ugene) C(ameron)
Rabe, Allen E
Rawling, Frank L(eslie), Jr
Reis, Paul G(eorge)
Rick, Christian E(dward)
Roberts, John Burnham
Rodriguez-Parada, Jose Manuel
Ross, William D(aniel)
Royer, Dennis Jack
Rush, Frank E(dward), Jr
Russell, T W Fraser
Ryder, D(avid) F(rank)
St John, Daniel Shelton
Sample, Paul E(dward)
Sandler, Stanley I
Sashihara, Thomas F(ujio)
Sauerbrunn, Robert Dewey
Saxton, Ronald L(uther)
Sayre, Clifford M(orrill), Jr
Schiewetz, D(on) B(oyd)
Schotte, William
Sears, Leo A
Secor, Robert M(iller)
Senecal, Vance E(van)
Shedrick, Carl F(ranklin)
Shine, Annette Dudek
Smith, Donald W(anamaker)
Smith, Ronald W
Stine, William H, Jr
Tallman, J(ohn) C(ornwell)
Tang, Walter Kwei-Yuan
Ten Eyck, Edward H(anlon), Jr
Thayer, Chester Arthur
Thomas, Lee W(ilson)
Tomkowit, Thaddeus W(alter)
Uy, William Cheng

Vick Roy, Thomas Rogers
Wagner, Martin Gerald
Walmsley, Peter N(ewton)
Wilkens, George A(lbert)
Winer, Richard
Zevnik, Francis C(lair)

DISTRICT OF COLUMBIA
Aggarwal, Ishwar D
Bell, Michael J(oseph)
Bowen, David Hywel Michael
Branscome, James R
Burka, Maria Karpati
Buzzelli, Donald Edward
Chapin, Douglas McCall
Fisher, Farley
Goodman, Eli I
Hill, Christopher Thomas
Kirkbride, Chalmer Gatlin
Lang, Peter Michael
Lih, Marshall Min-Shing
McBride, Gordon Williams
Mukherjee, Tapan Kumar
Parczewski, Krzysztof I(gnacy)
Park, Chan Mo
Ramsey, Jerry Warren
Richman, David M(artin)
Roller, Paul S
Rosen, Howard Neal
Rzeszotarski, Waclaw Janusz
Stevenson, F Dee
Topper, Leonard
Trice, Virgil Garnett, Jr
Wallace, Carl J
Zaborsky, Oskar Rudolf

FLORIDA
Albert, R(obert) E(yer)
Ariet, Mario
Benedict, Manson
Bergelin, Olaf P(reysz)
Beuther, Harold
Blase, Edwin W(illiam)
Boehme, Werner Richard
Bonnell, James Monroe
Brooks, John A(lbert)
Browning, Joe Leon
Chaddock, Richard E(astman)
Chamberlain, Donald F(rank)
Chao, Raul Edward
Cline, Kenneth Charles
Cornelius, E(dward) B(ernard)
Curry, Thomas Harvey
Dailey, Charles E(lmer), III
Danly, Donald Ernest
Davidson, Mark Rogers
Dubravcic, Milan Frane
Edelson, David
Elzinga, D(onald) Jack
Ermenc, Eugene D
Fahien, Ray W
Fausett, Donald Wright
Feagin, Roy C(hester)
Fricke, Arthur Lee
Friedland, Daniel
Garwood, Maurice F
Gillin, James
Giraudi, Carlo
Gittelman, Donald Henry
Goll, Robert John
Harrow, Lee Salem
Henry, John Patrick, Jr
Hill, Frank B(ruce)
Hirsch, Donald Earl
Horrigan, Robert V(incent)
Hosler, E(arl) Ramon
Hsu, Tsong-Han
Hyman, Seymour C
Johns, Lewis E(dward), Jr
Jurgensen, Delbert F(rederick), Jr
Kaghan, Walter S(eidel)
Katell, Sidney
Kircher, Morton S(ummer)
Kirmse, Dale William
Kowalczyk, Leon S(tanislaw)
Kraybill, Richard R(eist)
Kumar, Ganesh N
Leibson, Irving
List, Harvey L(awrence)
Lynn, R(alph) Emerson
McAfee, Jerry
Maleady, N(oel) R(ichard)
Martin, Ralph H(arding)
Mason, D(onald) R(omagne)
May, Frank Pierce
Miles, Harry V(ictor)
Milton, Robert Mitchell
Moudgil, Brij Mohan
Normile, Hubert Clarence
Osborne, Franklin Talmage
Parker, Howard Ashley, Jr
Roe, William P(rice)
Ross, Fred Michael
Scheibel, Edward G(eorge)
Schimmel, Walter Paul
Schwier, Chris Edward
Sellers, John William
Selund, Robert B(ernard)
Seugling, Earl William, Jr
Shair, Robert C
Spurlock, Jack Marion
Stahl, Joel S
Stana, Regis Richard
Stern, Milton

Strohmaier, A(lfred) J(ohn)
Summers, Hugh B(loomer), Jr
Szonntagh, Eugene L(eslie)
Teller, Aaron Joseph
Travnicek, Edward Adolph
Tyner, Mack
Walker, Robert D(ixon), Jr
Waller, Richard Conrad
Watkins, Charles H(enry)
Wiles, Robert Allan
Wilson, Robert E(lwood)
Wornick, Robert C(harles)
Zegel, William Case
Zimmer, Martin F

GEORGIA
Bebbington, W(illiam) P(earson)
Caras, Gus J(ohn)
Dooley, William Paul
Eckert, Charles Alan
Ernst, William Robert
Forney, Larry J
Grubb, H(omer) V(ernon)
Gupta, Rakesh Kumar
Harbin, William T(homas)
Hart, John Robert
Hicks, Harold E(ugene)
Hicks, William B(ruce)
Holley, Richard Howard
Holt, David Lowell
Huang, Denis K
Jones, Alan Richard
Knight, James Albert, Jr
Lewis, H(erbert) Clay
McDonough, Thomas Joseph
Marshall, Clair Addison
Matteson, Michael Jude
Morman, Michael T
Nerem, Robert Michael
Nichols, G(eorge) Starr
O'Neil, Daniel Joseph
Orr, Clyde, Jr
Peake, Thaddeus Andrew, III
Poehlein, Gary Wayne
Price, Charles R(onald)
Roberts, Ronnie Spencer
Rousseau, Ronald William
Schneider, Alfred
Schwartz, Robert John
Sommerfeld, Jude T
Soora, Siva Shunmugam
Swank, Robert Roy, Jr
Szostak, Rosemarie
Techo, Robert
Teja, Amyn Sadruddin
Toledo, Romeo Trance
Tooke, William Raymond, Jr
Wethern, James Douglas
White, Mark Gilmore
Winnick, Jack
Wiseman, William H(oward)
Yirak, Jack J(unior)
Yoganathan, Ajit Prithiviraj

HAWAII
Antal, Michael Jerry, Jr
Brantley, Lee Reed
Krock, Hans-Jürgen
Moy, James Hee
Takahashi, Patrick Kenji

IDAHO
Bower, J(ohn) R(oy), Jr
Edwards, Louis Laird, Jr
Felt, Rowland Earl
Hanson, George H(enry)
Hyde, William W
Jackson, M(elbourne) L(eslie)
Jobe, Lowell A(rthur)
Lawroski, Harry
Obenchain, Carl F(ranklin)
Paige, David M(arsh)
Scheldorf, Jay J(ohn)
Slansky, Cyril M
Warner, R(ichard) E(lmore)

ILLINOIS
Agarwal, Ashok Kumar
Alkire, Richard Collin
Alsberg, Henry
Alwitt, Robert S(amuel)
Ayen, Richard J(ohn)
Babcock, Lyndon Ross, (Jr)
Babu, Suresh Pandurangam
Baird, Michael Jefferson
Baldwin, Richard H(arold)
Bankoff, S(eymour) George
Bengtson, Harlan Holger
Bezella, Winfred August
Blanks, Robert F
Boehler, Robert A
Bowman, Walker H(ill)
Brazelton, William T(homas)
Brockmeier, Norman Frederick
Brown, Charles S(aville)
Brown, George Martin
Brown, Sherman Daniel
Brusenback, Robert A(llen)
Bryan, William L
Bryan, William L
Bukacek, Richard F
Burris, Leslie
Butt, John B(aecher)
Cengel, John Anthony

Chao, Tai Siang
Chen, Juh Wah
Cheryan, Munir
Chilenskas, Albert Andrew
Clark, John Peter, III
Clauson, W(arren) W(illiam)
Cohen, William C(harles)
Coleman, Lester F
Conn, Arthur L(eonard)
Crawford, Raymond Maxwell, Jr
Cronauer, Donald (Charles)
Davis, Edward Nathan
Devgun, Jas S
Didwania, Hanuman Prasad
Dranoff, Joshua S(imon)
DuTemple, Octave J
Ekman, Frank
Fiedelman, Howard W(illiam)
Fink, Joanne Krupey
Finn, Patricia Ann
Fisher, Robert Earl
Fitzpatrick, Joseph A
Flinn, James Edwin
Forgac, John Michael
Gabor, John Dewain
Gavlin, Gilbert
Gembicki, Stanley Arthur
Goldman, Arthur Joseph
Goldstick, Thomas Karl
Gordon, Gerald Arthur
Griffin, Edward L(awrence), Jr
Gupta, Ramesh
Hacker, David S(olomon)
Hagenbach, W(illiam) P(aul)
Hanratty, Thomas J(oseph)
Harris, Ronald David
Heaston, Robert Joseph
Heil, Richard Wendell
Herring, William M(ayo)
Hesketh, Howard E
Honea, Franklin Ivan
Hough, Eldred W
Huang, Shu-Jen Wu
Hundley, John Gower
Hwang, Yu-Tang
Irvin, Howard H
Ivins, Richard O(rville)
Jacobsen, Fred Marius
Johnson, Terry R(obert)
Jostlein, Hans
Joung, John Jongin
Junk, William A(rthur), Jr
Kamm, Gilbert G(eorge)
Kessler, Nathan
Khan, Amanullah Rashid
Kiefer, John Harold
King, Edward P(eter)
Koenig, James J(acob)
Konopinski, Virgil J
Kramer, John Michael
Kramer, Sheldon J
Kruse, Carl William
Kumar, Romesh
Kung, Harold Hing Chuen
Kyle, Martin Lawrence
Lambert, John B(oyd)
Lasley, Stephen Michael
Lauffenburger, Douglas Alan
Lawroski, Stephen
Lee, Bernard S
Lester, George Ronald
Lewandowski, Melvin A
Li, Norman N(ian-Tze)
Linden, Henry R(obert)
Long, George
McHenry, K(eith) W(elles), Jr
McKnight, Richard D
Mah, Richard S(ze) H(ao)
Mannella, Gene Gordon
Mansoori, G Ali
Margolis, Asher J(acob)
Marianowski, Leonard George
Martin, Thomas L(yle), Jr
Masel, Richard Isaac
Mason, David McArthur
Mason, Edward Archibald
Mateles, Richard I
Matschke, Donald Edward
Mauzy, Michael P
May, Walter Grant
Meter, Donald M(ervyn)
Miller, David
Miller, Irving F(ranklin)
Miller, Shelby A(lexander)
Mosby, James Francis
Mulcahey, Thomas P
Murad, Sohail
Myles, Kevin Michael
Nagy, Zoltan
Nordine, Paul Clemens
Onischak, Michael
Osri, Stanley M(aurice)
Oster, Eugene Arthur
Ottino, Julio Mario
Papoutsakis, Eleftherios Terry
Parent, Joseph D(ominic)
Pearlstein, Arne Jacob
Pecina, Richard W
Peterson, Kenneth C(arl)
Pierce, R(obert) Dean
Poulikakos, Dimos
Pritzlaff, August H(enry), Jr
Punwani, Dharam Vir

Pyrcioch, Eugene Joseph
Rakowski, Robert F
Ramaswami, Devabhaktuni
Rasche, John Frederick
Reeve, Aubrey C
Rest, David
Rittmann, Bruce Edward
Robinson, Ralph M(yer)
Sather, Norman F(redrick)
Savidge, Jeffrey Lee
Schmalzer, David Keith
Schoepfer, Arthur E(ric)
Schowalter, William Raymond
Seefeldt, Waldemar Bernhard
Sherrill, J(oseph) C(yril)
Shimotake, Hiroshi
Singh, Shyam N
Snow, Richard Huntley
Sohr, Robert Trueman
Solvik, R S(ven)
Srinivasaraghavan, Rengachari
Staats, William R
Stadtherr, Mark Allen
Standing, Charles Nicholas
Steinberg, Marvin Phillip
Stevens, William F(oster)
Strong, E(rwin) R(ayford)
Swanson, Bernet S(teven)
Swearingen, John Eldred
Szepe, Stephen
Thiele, Ernest
Thodos, George
Travelli, Armando
Troscinski, Edwin S
Tsai, Boh Chang
Tsaros, C(onstantine) L(ouis)
Udani, Kanakkumar Harilal
Wadden, Richard Albert
Wang, Pie-Yi
Wasan, Darsh T
Westwater, J(ames) W(illiam)
Wilson, Gerald Gene
Wilson, Thomas Edward
Witter, Lloyd David
Wong, Wang Mo
Yao, Neng-Ping
Zaromb, Solomon

INDIANA
Abegg, Carl F(rank)
Albright, Lyle F(rederick)
Andres, Ronald Paul
Banchero, J(ulius) T(homas)
Booe, J(ames) M(arvin)
Carberry, James John
Cartmell, Robert Root
Caskey, Jerry Allan
Chang, Chien-Wu
Chao, Kwang Chu
Cornell, Stephen Watson
Cosway, Harry F(rancis)
Delgass, W Nicholas
Eckert, Roger E(arl)
Egly, Richard S(amuel)
Emery, Alden H(ayes), Jr
Fagan, J(ohn) R(obert)
Flock, John William
Greenkorn, Robert A(lbert)
Hite, S(amuel) C(harles)
Houze, Richard Neal
Hubble, Billy Ray
Hurley, Robert Edward
Jeffries, Quentin Ray
Kesler, G(eorge) H(enry)
Kessler, David Phillip
Knowlton, Floyd M(arion)
Kohn, James P(aul)
Kuo, Charles C Y
Ladisch, Michael R
Liley, Peter Edward
Ling, Ting H(ung)
Liu, Tung
Machtinger, Lawrence Arnold
Mills, Norman Thomas
Moore, Noel E(dward)
Nirschl, Joseph Peter
Nolting, Henry Frederick
Okos, Martin Robert
Peppas, Nikolaos Athanassiou
Ranade, Madhukar G
Razdan, Mohan Kishen
Reklaitis, Gintaras Victor
Robins, Norman Alan
Rominger, Michael Collins
Schmitz, Roger A(nthony)
Squires, Robert George
Strieder, William
Tsao, George T
Varma, Arvind
Wang, Jin-Liang
Wankat, Phillip Charles
Williams, Theodore J(oseph)
Wood, Rodney David

IOWA
Abraham, William H
Berry, Clyde Marvin
Boylan, D(avid) R(ay)
Brown, Melvin Henry
Burkhart, Lawrence E
Burnet, George
Carmichael, Gregory Richard
Coy, Daniel Charles
Egger, Carl Thomas

Gaertner, Richard F(rancis)
Jolls, Kenneth Robert
Kammermeyer, Karl
Larson, Maurice A(llen)
Reilly, Peter John
Rethwisch, David Gerard
Schustek, George W(illiam), Jr
Seagrave, Richard C(harles)
Sheeler, John B(riggs)
Tate, R(oger) W(allace)
Ulrichson, Dean LeRoy
Vetter, Arthur Frederick
Wheelock, Thomas David

KANSAS
Akins, Richard G(lenn)
Bates, Herbert T(empleton)
Brewer, Jerome
Chung, Do Sup
Crowther, Robert Hamblett
Cupit, Charles R(ichard)
Erickson, Larry Eugene
Fan, Liang-Tseng
Faw, Richard E
Fujimoto, Gordon Takeo
Green, Don Wesley
Honstead, William Henry
Jacobs, Louis John
Kurata, F(red)
Kyle, Benjamin G(ayle)
McElroy, Albert Dean
Maloney, J(ames) O(Hara)
Mesler, R(ussell) B(ernard)
Mingle, John O(rville)
Rosson, H(arold) F(rank)
Ryan, James M
Sutherland, John B(ennett)
Swift, George W(illiam)
Willhite, Glen Paul
Young, David A(nthony)

KENTUCKY
Flegenheimer, Harold H(ansleo)
Fleischman, Marvin
Funk, James Ellis
Gerhard, Earl R(obert)
Hamrin, Charles E(dward), Jr
Hanley, Thomas Richard
Harper, Dean Owen
Hoertz, Charles David, Jr
Jensen, Randolph A(ugust)
Kohnhorst, Earl Eugene
Laukhuf, Walden Louis Shelburne
Livingood, Marvin D(uane)
Mapes, William Henry
Mayland, Bertrand Jesse
Mellinger, Gary Andreas
Moore, Howard Francis
Plank, Charles Andrews
Ray, Asit Kumar
Sanford, Robert Alois
Schrodt, James Thomas
Wells, William Lochridge

LOUISIANA
Arthur, Jett Clinton, Jr
Ashby, B(illy) B(ob)
Bailey, Raymond Victor
Barona, Narses
Bertolette, W(illiam) deB(enneville)
Biggs, R Dale
Bunch, David William
Callihan, Clayton D
Carpenter, C(lifford) LeRoy
Carpenter, Frank G(ilbert)
Carr, James W(oodford), Jr
Causey, Ardee
Chen, Hoffman Hor-Fu
Chew, Woodrow W(ilson)
Clavenna, LeRoy Russell
Coates, Jesse
Collier, John Robert
Cordiner, James B(eattie), Jr
Corripio, Armando Benito
Crosby, Robert H(owell), Jr
Davidson, Donald H(oward)
Davis, P(hilip) C
Dearth, James Dean
Decossas, Kenneth Miles
Fang, Cheng-Shen
Gautreaux, Marcelian Francis
Gluckstein, Martin E(dwin)
Goerner, Joseph Kofahl
Griffin, Gregory Lee
Hansen, Robert M(arius)
Harrison, Douglas P
Hill, Archibald G
Ho, Chong Cheong
Holloway, Frederic Ancrum Lord
Hu, John Nan-Hai
Huckabay, Houston Keller
Johnson, Adrian Earl, Jr
Karkalits, Olin Carroll, Jr
Killgore, Charles A
Kingrea, C(harles) L(eo)
Knoepfler, Nestor B(eyer)
Koltun, Stanley Phelps
Kordoski, Edward William
Lippman, Alfred, Jr
Loechelt, Cecil P(aul)
McCarthy, Danny W
McLaughlin, Edward
Malone, James W(illiam)
Martinez, John L(uis)

Chemical Engineering (cont)

Mayer, Francis X(avier)
Murrill, Paul W(hitfield)
Pike, Ralph W
Polack, Joseph A(lbert)
Pressburg, Bernard S(amuel)
Reneau, Daniel Dugan, Jr
Riley, Kenneth Lloyd
Robson, Harry Edwin
Rosene, Robert Bernard
Samuels, Martin E(lmer)
Schucker, Robert Charles
Shuck, Frank O
Sterling, Arthur MacLean
Stuckey, A(lto) Nelson, Jr
Sullivan, Samuel Lane, Jr
Thibodeaux, Louis J
Voorhies, Alexis, Jr
Wan, Peter J
Weaver, R(obert) E(dgar) C(oleman)
Wilkins, Bert, Jr

MAINE

Chase, Andrew J(ackson)
Genco, Joseph Michael
Gorham, John Francis
Hill, Marquita K
Mumme, Kenneth Irving
Orcutt, John C
Rather, James B, Jr
Robbins, Wayne Brian
Shipman, C(harles) William
Thompson, Edward Valentine
Zieminski, S(tefan) A(ntoni)

MARYLAND

Adams, Edward Franklin
Adams, James Miller
Anderson, W(endell) L
Barclay, James A(lexander)
Barney, Duane Lowell
Beckmann, Robert B(ader)
Bournia, Anthony
Braude, George Leon
Brodsky, Allen
Bungay, Peter M
Cadman, Theodore W
Camp, Albert T(alcott)
Carlon, Hugh Robert
Chamberlain, David Leroy, Jr
Chiu, Ching Ching
Choi, Kyu-Yong
Cleveland, William Grover, Jr
Collins, William Henry
Cooperstein, Raymond
Dang, Vi Duong
Dart, Jack Calhoon
Dashiell, Thomas Ronald
Donohue, Marc David
Ednie, Norman A(lex)
Farmer, William S(ilas), Jr
Fishman, Steven Gerald
Foresti, Roy J(oseph), Jr
Gamson, Bernard W(illiam)
Gessner, Adolf Wilhelm
Gomezplata, Albert
Graebert, Eric W
Greenfeld, Sidney H(oward)
Gupta, Ashwani Kumar
Hader, Rodney N(eal)
Hall, Kimball Parker
Haller, Elden D
Hamilton, Bruce King
Hammond, Allen Lee
Harper, C(harles) A(rthur)
Harris, B(enjamin) L(ouis)
Harris, Milton
Heath, George A(ugustine)
Hegedus, L Louis
Herrick, Elbert Charles
Hovmand, Svend
Hsu, Stephen M
Humphrey, S(idney) Bruce
Jashnani, Indru
Jovancicevic, Vladimir
Kahn, Walter K(urt)
Kanarowski, S(tanley) M(artin)
Katz, Herbert M(arvin)
Katz, Joseph L
Kayser, Richard Francis
Kerkar, Awdhoot Vasant
Khatib-Rahbar, Mohsen
Kirby, Ralph C(loudsberry)
Knoedler, Elmer L
Ku, Chia-Soon
Langlykke, Asger Funder
Lard, Edwin Webster
Levine, Robert S(idney)
Lull, David B
McAvoy, Thomas John
McElroy, Wilbur Renfrew
McMullen, Bryce H
Moreira, Antonio R
Morgan, Arthur I, Jr
Moses, Saul
Muller, Robert E
Nash, Murray L
O'Loughlin, Walter K(lein)
Perry, Charles William
Peters, Alan Winthrop
Plumlee, Karl Warren
Prentice, Geoffrey Allan
Presser, Cary

Preusch, Peter Charles
Quandt, Earl Raymond, Jr
Rabussay, Dietmar Paul
Ranade, Madhav (Arun) Bhaskar
Regan, Thomas M(ichael)
Reichard, H(arold) F(orrest)
Reilly, Hugh Thomas
Rice, James K
Rockwell, Theodore
Rodger, Walton A
Schlain, David
Schroeder, W(ilburn) Carroll
Semerjian, Hratch G
Sengers, Johanna M H Levelt
Shane, Presson S
Shannon, Larry J(oseph)
Sheppard, Norman F, Jr
Sherwin, Martin Barry
Smith, Elmer Robert
Smith, Theodore G
Smutz, Morton
Steinberg, Ronald T
Swift, David Leslie
Szego, George C(harles)
Tippens, Dorr E(ugene) F(elt)
Ulanowicz, Robert Edward
Ulbrecht, Jaromir Josef
Villet, Ruxton Herrer
Walton, Ray Daniel, Jr
Wang, Francis Wei-Yu
Weigand, William Adam
White, Edmund W(illiam)
Wilkins, Michael Gray
Wing, James
Winters, C(harles) E(rnest)
Wong, Kin Fai
Yates, Harold W(illiam)
Zafiriou, Evanghelos
Zdravkovich, Vera

MASSACHUSETTS

Agarwal, Jagdish Chandra
Altieri, A(ngelo) M(ichael)
Ammlung, Richard Lee
Andersen, L(aird) Bryce
Apelian, Diran
Baddour, Raymond F(rederick)
Blake, Thomas R
Blankschtein, Daniel
Bloomfield, David Peter
Bodman, Samuel Wright, III
Bonner, Francis Joseph
Botsaris, Gregory D(ionysios)
Brenner, Howard
Breslau, Barry Richard
Brown, Robert A
Brown, Robert Lee
Buonopane, Ralph A(nthony)
Carter, Leo F
Chang, Tai Ming
Chen, Chuan Ju
Chen, Ning Hsing
Chen, Shiou-Shan
Chen, Sung Jen
Chu, Nori Yaw-Chyuan
Churchill, Geoffrey Barker
Clopper, Herschel
Cohen, Robert Edward
Cole, Harvey M
Cooney, Charles Leland
Cooper, William Wailes
Dale, William
Dalzell, William Howard
Dauphiné, T(honet) C(harles)
Davé, Raju S
DeCosta, Peter F(rancis)
Deen, William Murray
De Filippi, R(ichard) P(aul)
De Noto, Thomas Gerald
Donovan, James
Douglas, J(ames) M(errill)
Drake, Elisabeth Mertz
Eby, Robert Newcomer
Ehrenfeld, John R(oos)
Eldridge, John W(illiam)
Evans, Lawrence B(oyd)
Fagerson, Irving Seymour
Fariss, Robert Hardy
Feakes, Frank
Fine, David H
Fisher, Harold Wallace
Fox, Daniel Wayne
Friedman, Raymond
George, James Henry Bryn
Goettler, Lloyd Arnold
Gold, Harris
Gopikanth, M L
Gutoff, Edgar B(enjamin)
Hausslein, Robert William
Heitman, Richard Edgar
Hoffman, Paul Roger
Hoffman, Richard Laird
Holmes, Douglas Burnham
Hottel, Hoyt C(larke)
Howard, Jack Benny
Jacobson, Murray M
Jain, Rakesh Kumar
Jalan, Vinod Motilal
Johnson, Stephen Allen
Kaiser, Robert
Kingston, George C
Kirk, R(obert) S(tewart)
Kittrell, James Raymond
Klanfer, Karl

Kogos, L(aurence)
Kolodzy, Paul John
Kreiling, William H(erman)
Kruse, Robert Louis
Kulwicki, Bernard Michael
Kusik, Charles Lembit
La Mantia, Charles R
Latanision, Ronald Michael
Laurence, Robert L(ionel)
Lee, Eric Kin-Lam
Leung, Woon F (Wallace)
Longwell, John Ploeger
Luft, Ludwig
Lurie, Robert M(andel)
Ma, Yi Hua
McFarland, Charles Manter
Magarian, Charles Aram
Magnanti, Thomas L
Malloy, John B
Meldon, Jerry Harris
Michaels, Alan Sherman
Miekka, Richard G(eorge)
Mir, Leon
Morbey, Graham Kenneth
Mueller, Robert Kirk
Myers, Robert Frederick
Neue, Uwe Dieter
Novack, Joseph
Oster, Bernd
Packard, Richard Darrell
Peng, Fred Ming-Sheng
Petrovic, Louis John
Prodany, Nicholas W
Ragone, David Vincent
Reid, Robert C(lark)
Sarofim, Adel Fares
Satterfield, Charles N(elson)
Schoenberg, Theodore
Schutte, A(ugust) H(enry)
Schwartzberg, Henry G
Selke, William A
Shinskey, Francis Gregway
Short, John Lawson
Shusman, T(evis)
Simon, Robert H(erbert) M(elvin)
Sioui, Richard Henry
Sladek, Karl Josef, III
Smith, Kenneth A
Snedeker, Robert A(udley)
Sonnichsen, Harold Marvin
Stephens, Richard Harry
Stevens, J(ames) I(rwin)
Stone, H Nathan
Sussman, M(artin) V(ictor)
Szekely, Julian
Tester, Jefferson William
Thieme, Cornelis Leo Hans
Thomas, Martha Jane Bergin
Thorstensen, Thomas Clayton
Towns, Donald Lionel
Van Wormer, Kenneth A(ugustus), Jr
Vartanian, Leo
Wagner, Robert E(arl)
Wallace, David H
Wang, Daniel I-Chyau
Wei, James
Weiss, Alvin H(arvey)
Westmoreland, Phillip R
Wilemski, Gerald
Williams, Glenn C(arber)
Winter, Horst Henning
Yarmush, Martin Leon
Zotos, John

MICHIGAN

Amin, Sanjay Indubhai
Anderson, Donald K(eith)
Atreya, Arvind
Atwell, William Henry
Ben, Manuel
Bens, Frederick Peter
Breuer, Max Everett
Briggs, Dale Edward
Buss, David R(ichard)
Camp, David Thomas
Caplan, John D(avid)
Carmouche, L(ouis) N(orman)
Carnahan, Brice
Cha, Dae Yang
Chiu, Thomas T
Cho, Byong Kwon
Clarke, Richard Penfield
Cooper, Carl (Major)
Crittenden, John Charles
Cunningham, Fay Lavere
Curl, Rane L(ocke)
Daniels, Stacy Leroy
Donahue, Francis M(artin)
Ebach, Earl A
Eichhorn, J(acob)
Eldred, Norman Orville
Fauver, Vernon A(rthur)
Fenn, H(oward) N(athan)
Finch, C(harles) R(ichard)
Flynn, John M(athew)
Fogler, Hugh Scott
Foister, Robert Thomas
Forsblad, Ingemar Bjorn
Francis, Ray Llewellyn
Gagnon, Eugene Gerald
Gandhi, Harendra Sakarlal
Ginn, Robert Ford
Goksel, Mehmet Adnan
Graves, Harold E(dward)

Hall, Richard Harold
Hand, James Henry
Hawley, Martin C
Heiss, John F
Herz, Richard Kimmel
Hill, Robert F
Hinsch, James E
Hutchinson, Kenneth A
Hyun, Kun Sup
Jariwala, Sharad Lallubhai
Joshi, Mukund Shankar
Kadlec, Robert Henry
Katz, Donald L(aVerne)
Kempe, Lloyd L(ute)
Kenaga, Duane Leroy
Klimisch, Richard L
Klimpel, Richard Robert
Korcek, Stefan
Kummer, Joseph T
Kummler, Ralph H
Larson, John Grant
Lee, Do Ik
Leng, Douglas E
Lenton, Philip A(lfred)
Leopold, Wilbur Richard, III
Lewis, John G(alen)
Maasberg, Albert Thomas
McKinstry, Karl Alexander
McMicking, James H(arvey)
McMillan, Michael Lathrop
Macriss, Robert A
Martin, Joseph J
Martz, Lyle E(rwin)
Merritt, R(obert) W(alter)
Mickelson, Richard W
Moll, Harold Wesbrook
Moore, Eugene Roger
Murchison, Craig Brian
Narain, Amitabh
Nelson, John Arthur
Nesbitt, Carl C
Nevius, Timothy Alfred
Otrhalek, Joseph V
Permoda, Arthur J
Rea, David Richard
Revzin, Arnold
Robbins, Lanny Arnold
Robinson, Robert George
Rothe, Erhard William
Rounds, Fred G
Russell, William Charles
Salinger, Rudolf Michael
Savage, Albert B
Scherger, Dale Albert
Schwing, Richard C
Smith, Frederick W(ilson)
Soh, Sung Kuk
Stevenson, Richard Marshall
Stynes, Stanley K
Tison, Richard Perry
Voelker, C(larence) E(lmer)
Voorhees, Howard R(obert)
Wang, Henry
Wang, Simon S
Wegst, W(alter) F(rederick)
Weissman, Eugene Y(ehuda)
Wilkes, James Oscroft
Wilkinson, Bruce W(endell)
Williams, Eugene H(ugh)
Williams, G(eorge) Brymer
Ying, Ramona Yun-Ching
Yost, John R(obarts), Jr
Young, David Caldwell
Young, Edwin H(arold)

MINNESOTA

Adams, Robert McLean
Anderson, Harvey L(awrence)
Aris, Rutherford
Baker, Michael Harry
Baria, Dorab Naoroze
Benner, Blair Richard
Boening, Paul Henrik
Bohon, Robert Lynn
Britz, Galen C
Brown, Robert Ordway
Carr, Robert Wilson, Jr
Cooley, Albert Marvin
David, Moses M
Dreier, William Matthews, Jr
Dreshfield, Arthur C(harles), Jr
Drew, Bruce Arthur
Erickson, Eugene E
Erwin, James V
Fletcher, Edward A(braham)
Fredrickson, Arnold G(erhard)
Geankoplis, Christie J(ohn)
Hann, G(eorge) C(harles)
Haun, J(ames) W(illiam)
Hu, Wei-Shou
Huffine, Coy L(ee)
Isbin, Herbert S(tanford)
Jensen, Timothy B(erg)
Johnson, Lennart Ingemar
Johnson, W(illiam) C
Kwong, Joseph N(eng) S(hun)
Lai, Juey Hong
Macosko, Christopher Ward
Manske, Wendell J(ames)
Miller, Nicholas Carl
Moison, Robert Leon
Nam, Sehyun
Owens, Boone Bailey
Patel, Kalyanji U

Pearson, John W(illiam)
Penney, William Harry
Podas, William M(orris)
Ranz, William E(dwin)
Reinhart, Richard D
Sapakie, Sidney Freidin
Schmidt, Lanny D
Schoenherr, Roman Uhrich
Schoon, David Jacob
Schulz, Norman F
Shor, Steven Michael
Smuk, John Michael
Strauss, H(oward) J(erome)
Thalacker, Victor Paul
Thangaraj, Sandy
Tirrell, Matthew Vincent
Werth, Richard George
Yang, Jih Hsin (Jason)
Zitney, Stephen Edward

MISSISSIPPI
Anderson, Frank A(bel)
Aven, Russell E(dward)
Cade, Ruth Ann
Cornell, David
Genetti, William Ernest
George, Clifford Eugene
Kuo, Chiang-Hai
Leybourne, A(llen) E(dward), III
Norman, Roger Atkinson, Jr
Persell, Ralph M(ountjoy)
Sadana, Ajit
Sukanek, Peter Charles
Wehr, Allan Gordon

MISSOURI
Adams, C(harles) Howard
Anderson, Fletcher N
Bajpai, Rakesh Kumar
Ball, William E(rnest)
Barker, George Edward
Bosanquet, Louis Percival
Bucknell, Roger W(inston)
Carter, Don E
Carter, Lee
Chae, Young C
Cheng, William J(en) P(u)
Chien, Henry H(ung-Yeh)
Cova, Dario R
Cowherd, Chatten, Jr
Crocker, Burton B(lair)
Crosser, Orrin Kingsbery
Crum, Glen F(rancis)
Deam, James Richard
De Chazal, L(ouis) E(dmond) Marc
Dmytryszyn, Myron
Donovan, John Richard
Dudukovic, Milorad
Fellinger, L(owell) L(ee)
Ferchaud, John B(artholomew)
Findley, Marshall E(wing)
Fowler, Frank Cavan
Gadberry, Howard M(ilton)
Geisman, Raymond August, Sr
Getty, Robert J(ohn)
Gibbs, Marvin E
Grice, Harvey H(oward)
Hadden, Stuart Tracey
Haney, Paul D
Harakas, N(icholas) Konstantinos
Harland, Ronald Scott
Harris, J(ohn) S(terling)
Heitsch, Charles Weyand
Hines, Anthony Loring
Holsen, James N(oble)
Howard, Walter B(urke)
Hudson, Robert B
Johnson, James W(inston)
Jones, Otha Clyde
Kardos, John Louis
Kim, Keun Young
Koederitz, Leonard Frederick
Kramer, Richard Melvyn
Kwentus, Gerald K(enneth)
Lannert, Kent Philip
Lapple, Walter C(hristian)
Lee, Roberto
Long, Lawrence William
Luebbers, Ralph H(enry)
Luecke, Richard H
McKelvey, James M(organ)
Marc de Chazal, L E
Marrero, Thomas Raphael
Mason, Norbert
Motard, R(odolphe) L(eo)
Nellums, Robert (Overman)
Null, Harold R
Otto, Robert Emil
Patterson, Gary Kent
Peterson, Idelle M(arietta)
Poling, Bruce Earl
Preckshot, G(eorge) W(illiam)
Prelas, Mark Antonio
Privott, Wilbur Joseph, Jr
Proctor, Stanley Irving, Jr
Retzloff, David George
Robertson, David G C
Rosen, Edward M(arshall)
Rosen, Stephen L(ouis)
Scallet, Barrett Lerner
Schott, Jeffrey Howard
Schroy, Jerry M
Schuler, Rudolph William
Sisler, Charles Carleton

Smith, Buford Don
Snyder, Joseph Quincy
Sonnino, Carlo Benvenuto
Sparks, Robert Edward
Stone, Bobbie Dean
Storvick, Truman S(ophus)
Strunk, Mailand Rainey
Sutterby, John Lloyd
Teng, James
Thompson, Dudley
Throdahl, Monte C(orden)
Vandegrift, Alfred Eugene
Volz, William K(urt)
Waggoner, Raymond C
Walsh, Robert Jerome
Wasson, Richard Lee
Weddell, David S(tover)
Wohl, Martin H
Wulfman, David Swinton
Young, Henry H(ans)

MONTANA
Berg, Lloyd
Characklis, William Gregory
Deibert, Max Curtis
McCandless, Frank Philip
Nickelson, Robert L(eland)
Orth, John C(arl)
Sears, John T
Twidwell, Larry G

NEBRASKA
Eastin, John A
Gilbert, Richard E(arle)
Scheller, W(illiam) A(lfred)
Tao, L(uh) C(heng)
Timm, Delmar C
Weber, James H(arold)

NEVADA
Cox, Neil D
Dobo, Emerick Joseph
Eichbaum, Barlane Ronald
Hendrix, James Lauris
Herzlich, Harold Joel
MacDonald, David J
Miller, E(ugene)
Miller, Eugene
Miller, Wayne L(eroy)
Prater, C(harles) D(wight)

NEW HAMPSHIRE
Converse, Alvin O
Dewey, Bradley, Jr
Fan, Stephen S(hu-Tu)
Farag, Ihab Hanna
Fricke, Alan Francis
Kippax, Donald
Mathur, Virendra Kumar
Secord, Robert N
Smith, Frank W(illiam)
Sundberg, Donald Charles
Sze, Morgan Chuan-Yuan
Ulrich, Gael Dennis

NEW JERSEY
Adams, James Mills
Adler, Irwin L
Agalloco, James Paul
Ahlert, Robert Christian
Albohn, Arthur R(aymond)
Andreatch, Anthony J
Aubrecht, Don A(lbert)
Bajars, Laimonis
Bakal, Abraham I
Bartok, William
Baumann, George P(ershing)
Beinfest, Sidney
Berkman, Samuel
Bevans, Rowland S(cott)
Bhagat, Phiroz Maneck
Bhandarkar, Suhas D
Bieber, Harold H
Bieber, Herman
Biesenberger, Joseph A
Bisio, Atillio L
Blank, Zvi
Blyler, Lee Landis, Jr
Bockelmann, John B(urggraf)
Bohrer, Thomas Carl
Bonacci, John C
Bouloucon, Peter
Bowen, J Hartley, Jr
Bragg, Leslie B(artlett)
Brazinsky, Irving
Bryant, Howard Sewall
Bull, Daniel Newell
Burke, Robert F
Carluccio, Frank
Cart, Eldred Nolen, Jr
Cecchi, Joseph Leonard
Cerkanowicz, Anthony Edward
Cevasco, Albert Anthony
Chappelear, David C(onrad)
Chen, Nai Y
Cheng, Shang I
Cheremisinoff, Paul N
Chien, Luther C
Chimes, Daniel
Choi, Byung Chang
Chu, Horn Dean
Cipriani, Cipriano
Ciprios, George
Cohen, Edward David

Cruice, William James
Czeropski, Robert S(tephen)
Davis, Thomas Arthur
Debenedetti, Pablo Gaston
Deem, William Brady
Deen, Harold E(ugene)
De Lancey, George Byers
Detrick, John K(ent)
DiMasi, Gabriel Joseph
DiSalvo, Walter A
Dittman, Frank W(illard)
Dunning, Ranald G(ardner)
Dwyer, Francis Gerard
Eck, John Clifford
Eddinger, Ralph Tracy
Elgin, Joseph C(lifton)
Engel, Lawrence J
Farkas, Daniel Frederick
Farrauto, Robert Joseph
Fass, Stephen M
Feder, David O
Feder, Raymond L
Fischer, Ronald Howard
Fogg, Edward T(hompson)
Foroulis, Z Andrew
Fowles, Patrick Ernest
Frohlich, Gerhard J
Fung, Shun Chong
Gabriel, Robert
Gagliardi, George N(icholas)
Gans, Manfred
Gartside, Robert N(ifong)
Giddings, Sydney Arthur
Glass, Werner
Gogos, Costas G
Golding, Brage
Goldstein, David
Graessley, William W(alter)
Gray, Charles A(ugustus)
Gray, Ralph Donald, Jr
Greenstein, Teddy
Gregory, Richard Alan, Jr
Grelecki, Chester
Griskey, R(ichard) G(eorge)
Grubman, Wallace Karl
Grumer, Eugene Lawrence
Guernsey, Edwin O(wens)
Haag, Werner O
Halik, Raymond R(ichard)
Hamilton, Charles W(allace)
Han, Byung Joon
Hanesian, Deran
Hawkins, A(lbert) W(illiam)
Heath, Carl E(rnest), Jr
Heath, D(onald) P
Heck, Ronald Marshall
Heney, Lysle Joseph, Jr
Higginson, George W, Jr
Hnatow, Miquel Alexander
Hochman, Jack M(artin)
Hoffman, Robert Frank
Hopkins, Charles B(everley), Jr
Horzepa, John Philip
Hsu, Kenneth Hsuehchia
Huang, Ching-Rong
Huang, Jennming Stephen
Hundert, Irwin
Huppert, Irwin Neil
Hwu, Mark Chung-Kong
Iezzi, Robert Aldo
Iglesia, Enrique
Jackson, Roy
Jacolev, Leon
Jagel, Kenneth I(rwin), Jr
Joffe, Joseph
Johnson, Douglas L
Johnson, Ernest F(rederick), (Jr)
Johnson, James M(elton)
Johnston, Christian William
Kandiner, Harold J(ack)
Kane, Ronald S(teven)
Kant, Fred H(ugo)
Kaplan, Joel Howard
Kapner, Robert S(idney)
Karoly, Gabriel
Kaufman, Arnold
Kaufmann, Thomas G(erald)
Kaup, Edgar George
Kavesh, Sheldon
Keller, Kenneth H(arrison)
Khan, Saad Akhtar
Kibbel, William H, Jr
Klenke, Edward Frederick, Jr
Kmak, Walter S(teven)
Kydd, Paul Harriman
Ladenheim, Harry
Lai, Kuo-Yann
Lederman, Peter B
Lee, Wei-Kuo
Lee, Wooyoung
Lessard, Richard R
Levinson, Sidney Bernard
Lewis, Theodore
Li, Ke Wen
Lifson, William E(ugene)
Lindroos, Arthur E(dward)
Liu, David H(o-Feng)
Loscher, Robert A
Lovinger, Andrew Joseph
Lowenstein, J(ack) G(ert)
Lund, Daryl B
McCaffrey, David Saxer, Jr
McCormick, John E
McGarvey, Francis X(avier)

McNicholas, James J
Manganaro, James Lawrence
Mantell, Charles L(etnam)
Mathis, James Forrest
Matsen, John M(orris)
Matsuoka, Shiro
Maxwell, Bryce
Megna, John C(osimo)
Metzger, Charles O
Midler, Michael, Jr
Miller, Robert H(enry)
Miner, Robert Scott, Jr
Mohr, Richard Arnold
Mraw, Stephen Charles
Mukherjee, Debi Prasad
Munger, Stanley H(iram)
Murphy, Thomas Daniel
Murthy, Andiappan Kumaresa Sundara
Mytelka, Alan Ira
Nace, Donald Miller
Newberger, Mark
Nguyen, Hien Vu
Nusim, Stanley Herbert
Park, Young D
Patel, Rutton Dinshaw
Paul, George T(ompkins)
Peltzman, Alan
Perlmutter, Arthur
Prapas, Aristotle G
Prizer, Charles J(ohn)
Prud'Homme, Robert Krafft
Quan, Xina Shu-Wen
Rankin, Sidney
Register, Richard Alan
Reilly, James William
Robertson, Jerry L(ewis)
Robinson, Jerome David
Roe, Kenneth A
Rollman, Walter F(uhrmann)
Rosensweig, Ronald E(llis)
Rupp, Walter H(oward)
Russel, William Bailey
Sacks, Martin Edward
Salkind, Alvin J
Salzarulo, Leonard Michael
Sartor, Anthony
Saville, D(udley) A(lbert)
Sawin, Steven P
Schmidle, Claude Joseph
Schmidt, John P
Seltzer, Edward
Sermolins, Maris Andris
Sestanj, Kazimir
Shah, Hasmukh N
Sharp, Hugh T
Shaw, Henry
Short, W(illiam) Leigh
Shrensel, J(ulius)
Shrier, Adam Louis
Silla, Harry
Silvestri, Anthony John
Skaperdas, George T(heodore)
Sofer, Samir Salim
Sonn, George Frank
Spence, Sydney P(ayton)
Speronello, Barry Keven
Steele, Lawrence Russell
Sturzenegger, August
Suciu, George Dan
Sundaresan, Sankaran
Swabb, Lawrence E(dward), Jr
Swanson, David Bernard
Sweed, Norman Harris
Tamarelli, Alan Wayne
Tan, Victor
Taylor, William Francis
Tegge, B(ruce) R(obert)
Tio, Cesario O
Tischler, Oscar
Todd, David Burton
Tolsma, Jacob
Toner, Richard K(enneth)
Topcik, Barry
Townsend, Palmer W
Tsao, Utah
Tsonopoulos, Constantine
Tunkel, Steven Joseph
Udani, Lalit Kumar Harilal
Vardi, Joseph
Vaughn, Robert Donald
Vieth, Wolf R(andolph)
Vriens, Gerard N(ichols)
Weekman, Vern W(illiam), Jr
Weil, Benjamin Henry
Weinstein, Norman J(acob)
Weiss, Lawrence H(eisler)
Wenis, Edward
West, John M(aurice)
Westerdahl, Raymond P
Whitwell, John C(olman)
Wiehe, Irwin Andrew
Wildman, George Thomas
Williams, Larry McClease
Wright, Stuart R(edmond)
Wynn, Clayton S(cott)
Yan, Tsoung-Yuan
Yarrington, Robert M
Yourtee, John Boteler
Yurchak, Sergei
Zonis, Irwin S(amuel)
Zoss, Abraham Oscar
Zudkevitch, David

Chemical Engineering (cont)

Grigger, David John
Grisaffe, Salvatore J
Gurklis, John A(nthony)
Haering, Edwin Raymond
Healy, James C
Hedley, William H(enby)
Heibel, John T(homas)
Heines, Thomas Samuel
Heinold, Robert H
Henderson, Courtland M
Hershey, Daniel
Hershey, Harry Chenault
Hoffman, F(rank) E(dward)
Huddleston, George Richmond, Jr
Hwang, Sun-Tak
Ishida, Hatsuo
Jenkins, Robert George
Jones, Millard Lawrence, Jr
Kao, Yuen-Koh
Katzen, Raphael
Kawasaki, Edwin Pope
Kay, Webster Bice
Kell, Robert M
Kenat, Thomas Arthur
Kendall, H(arold) B(enne)
Khang, Soon-Jai
Kim, Byung C
Knaebel, Kent Schofield
Knowlton, David A
Knox, Kenneth L
Koegle, John S(tuart)
Kotnik, Louis John
Krieble, James G(erhard)
Lacksonen, James W(alter)
Lahti, Leslie Erwin
Lee, Jon H(yunkoo)
Lee, Sunggyu
Lemlich, Robert
Lewis, George R(obert)
Licht, W(illiam), Jr
Lindstedt, P(aul) M
Liu, Chung-Chiun
Lyons, Carl J(ohn)
Martin, John B(ruce)
Martin, Wilfred Samuel
Massey, L(ester) G(eorge)
Measamer, S(chubert) G(ernt)
Meinecke, Eberhard A
Metanomski, Wladyslaw Val
Minges, Merrill Loren
Moon, George D(onald), Jr
Moore, Arthur William
Mosser, John Snavely
Mounce, Troy G(aspard)
Oxley, Joseph H(ubbard)
Ponter, Anthony Barrie
Price, John Avner
Purdon, James Ralph, Jr
Radhakrishnan, Egyaraman
Rains, Roger Keranen
Rau, Allen H
Reynes, Enrique G
Roberts, Robert William
Roberts, Timothy R
Roblin, John M
Romesberg, Floyd Eugene
Rosa, Casimir Joseph
Rounsley, Robert R(ichard)
Rutherford, William M(organ)
Saidel, Gerald Maxwell
Scaccia, Carl
Schlaudecker, George F(rederick)
Schmitt, George Frederick, Jr
Schooley, Arthur Thomas
Schumm, Brooke, Jr
Schwind, Roger Allen
Selden, George
Selover, Theodore Britton, Jr
Servais, Ronald Albert
Sharma, Shri C
Sheets, George Henkle
Shick, Philip E(dwin)
Sinner, Donald H
Skidmore, Duane R(ichard)
Smith, Edwin E(arle)
Sommer, John G
Springer, Allan Matthew
Stevenson, James Francis
Stockman, Charles H(enry)
Stoops, Charles E(mmet), Jr
Strop, Hans R
Stubblebine, Warren
Swager, William L(eon)
Sweeney, Thomas L(eonard)
Sylvester, Nicholas Dominic
Syverson, Aldrich
Szirmay, Leslie v
Throckmorton, Peter E
Throne, James Louis
Turner, Andrew
Updegrove, Louis B
Vahldiek, Fred W(illiam)
Vidaurri, Fernando C, Jr
Wallace, Michael Dwight
Warshay, Marvin
Weber, Lester George
Webster, Allen E
Weisman, Joel
White, James L(indsay)
Wiederhold, Edward W(illiam)
Wilkes, William Roy
Winn, Hugh
Wolfe, Robert Kenneth
Wotzak, Gregory Paul

Yang, Philip Yung-Chin
Yu, Thomas Huei-Chung
Zager, Stanley E(dward)
Zakin, Jacques L(ouis)

OKLAHOMA
Adams, Don
Albright, Melvin A
Aldag, Arthur William, Jr
Arnold, D(onald) S(mith)
Arnold, Philip Mills
Banks, Robert L(ouis)
Becraft, Lloyd G(rainger)
Bell, Kenneth J(ohn)
Bray, Bruce G(lenn)
Cerro, Ramon Luis
Chang, Harry Lo
Cheng, Paul J(ih) T(ien)
Chiou, Chii-Shyoung
Chung, Ting-Horng
Clark, George C(harles)
Clemenst, James Lee
Crynes, Billy Lee
Dale, Glenn H(ilburn)
Daniels, Raymond D(eWitt)
Dew, John N(orman)
Doane, Elliott P
Donaldson, Erle C
Erbar, John Harold
Foutch, Gary Lynn
Gall, James William
Gilliland, Joe E(dward)
Graves, Toby Robert
Green, Ray Charles
Harrison, R(oland) H(enry)
Hays, George E(dgar)
Hellwig, L(angley) R(oberts)
Herbolsheimer, Glenn
Hughes, Kenneth James
Irvin, Howard Brownlee
Johnson, Marvin M
Johnson, Paul H(ilton)
Joshi, Sadanand D
Koerner, E(rnest) L(ee)
Kriegel, Monroe W(erner)
Leder, Frederic
Lilley, David Grantham
Logan, R(ichard) S(utton)
McCarthy, William Chase
Madden, Michael Preston
Maddox, R(obert) N(ott)
Manning, Francis S(cott)
Marwil, S(tanley) J(ackson)
Philoon, Wallace C, Jr
Pollock, L(yle) W(illiam)
Richardson, Raymond C(harles)
Robinson, Robert L(ouis), Jr
Sattler, Robert E(dward)
Skinner, Joseph L
Sliepcevich, Cedomir M
Soodsma, James Franklin
Starling, Kenneth Earl
Stover, Dennis Eugene
Tham, Min Kwan
Thompson, Richard E(ugene)
Vanderveen, John Warren
Van De Steeg, Garet Edward
Warren, Kenneth Wayne
Warzel, L(awrence) A(lfred)
Weis, Robert E(dward)
West, John B(ernard)
Williamson, William Burton
Wood, Harold Sinclair
Wride, W(illiam) James
Zetik, Donald Frank

OREGON
Barkelew, Chandler H(arrison)
Haygarth, John Charles
Jansen, George, Jr
Knudsen, J(ames) G(eorge)
Landsberg, Arne
Levenspiel, Octave
McDuffie, Norton G(raham), Jr
Maze, Robert Craig
Meredith, Robert E(ugene)
Miner, J Ronald
Mrazek, Robert Vernon
Olsen, Richard Standal
Petrie, John M(atthew)
Smith, Kelly L
Tucker, W(illiam) Henry
Wicks, Charles E(dward)
Woods, W(allace) Kelly

PENNSYLVANIA
Adams, Roy Melville
Alexander, Stuart David
Amero, R(obert) C
Anderson, James B
Anderson, John Leonard
Archer, David Horace
Austin, Leonard G(eorge)
Aviles, Rafael G
Baker, Frank William
Banks, Robert R(ae)
Barton, Paul
Bauer, William, Jr
Bay, Theodosios (Ted)
Beck, William F(rank)
Benson, H(omer) E(dwin)
Berty, Jozsef M
Biegler, Lorenz Theodor
Bielecki, Edwin J(oseph)

Bienstock, D(aniel)
Biloen, Paul
Blazey, Leland Way
Bogash, R(ichard)
Brainard, Alan J
Braun, Walter G(ustav)
Breston, Joseph N(orbert)
Brey, R(obert) N(ewton)
Brian, P(ierre) L(eonce) Thibaut
Byers, R Lee
Camp, Frederick William
Caram, Hugo Simon
Carr, Norman L(oren)
Castellano, Salvatore, Mario
Champagne, Paul Ernest
Chappelow, Cecil Clendis, III
Chelemer, Harold
Chen, John Chun-Chien
Chen, Michael S K
Chen-Tsai, Charlotte Hsiao-yu
Chiang, S(hiao) H(ung)
Chun, Sun Woong
Churchill, Stuart W(inston)
Claitor, L(ilburn) Carroll
Clump, Curtis William
Cobb, James Temple, Jr
Coghlan, David B(uell)
Condo, Albert Carman, Jr
Corbo, Vincent James
Cost, J(oe) L(ewis)
Cotabish, Harry N(elson)
Coughanowr, D(onald) R(ay)
Craig, James Clifford, Jr
Crits, George J(ohn)
Dalton, Augustine Ivanhoe, Jr
Danner, Ronald Paul
Daubert, Thomas Edward
Davis, Burl Edward
Dean, Sheldon Williams, Jr
Dell, M(anuel) Benjamin
Dent, Anthony L
DeVries, Frederick William
Dicciani, Nancy Kay
Doelp, Louis C(onrad), Jr
Dohany, Julius Eugene
Donald, Dennis Scott
Dromgold, Luther D
Duda, J(ohn) L(arry)
Dunckhorst, F(austino) T
Eagleton, Lee C(handler)
Ebert, Philip E
El-Aasser, Mohamed S
Engel, Alfred J
Enick, Robert M
Erskine, Donald B
Fear, J(ames) Van Dyck
Feathers, William D
Feinman, J(erome)
Fisher, Sallie Ann
Foltz, Donald Richard
Forney, Albert J
Forney, R(obert) C(lyde)
Forsman, William C(omstock)
Foster, Perry Alanson, Jr
Franco, Nicholas Benjamin
Frattini, Paul L
Frenklach, Michael Y
Frumerman, Robert
Gander, Frederick W(illiam)
Garber, Charles A
Garg, Diwakar
Gee, Edwin Austin
Geist, Jacob M(yer)
Georgakis, Christos
Giffen, Robert H(enry)
Glandt, Eduardo Daniel
Goldthwait, R(ichard) G(raham)
Goodwin, James Gordon, Jr
Gordon, Lyle J
Gottschlich, Chad F
Gottscho, Alfred M(orton)
Grace, Harold P(adget)
Grassi, Vincent G
Graves, David J(ames)
Griffith, Michael Grey
Grossmann, Elihu D(avid)
Grossmann, Ignacio E
Guger, Charles Edmund, Jr
Gupta, Vijai Prakash
Gurol, Mirat D
Hager, Wayne R
Halpern, Benjamin David
Hanauer, Richard
Hartzog, David G(eorge)
Haynes, William P
Heiligman, Harold A
Hein, R(owland) F(rank)
Helfferich, Friedrich G
Henly, Robert Stuart
Herman, Frederick Louis
Herman, Richard Gerald
Hess, Dennis William
Hoerner, George M, Jr
Holder, Gerald D
Horn, Lyle William
Horst, Ralph L, Jr
Humphrey, Arthur E(arl)
Jarrett, Noel
Jensen, James Ejler
Jester, William A
Jhon, Myung S
Jones, Roger Franklin
Jordan, Robert Kenneth
Jost, Donald E

Kabel, Robert L(ynn)
Kakar, Anand Swaroop
Kanter, Ira E
Keairns, Dale Lee
Keith, Frederick W(alter), Jr
Keller, Rudolf
Kelly, Conrad Michael
King, William Emmett, Jr
Kitzes, Arnold S(tanley)
Kivnick, Arnold
Klaus, E(lmer) Erwin
Klaus, Ronald Louis
Kline, Richard William
Klinzing, George Engelbert
Klugherz, Peter D(avid)
Ko, Edmond Inq-Ming
Korchak, Ernest I(an)
Lancet, Michael Savage
Lantos, P(eter) R(ichard)
Lavin, J Gerard
Lawrence, Sigmund J(oseph)
Lawson, Neal D(evere)
Lee, Young Hie
Lerner, B(ernard) J
Levy, Edward Kenneth
Li, Kun
Lindt, Jan Thomas
Litt, Mitchell
Lloyd, Robert
Lloyd, Wallis A(llen)
Lokay, Joseph Donald
Lugar, Richard Charles
Luthy, Richard Godfrey
Luyben, William Landes
McCormick, Robert H(enry)
McDonnell, Leo F(rancis)
McKinley, Clyde
McNeil, Kenneth Martin
MacRae, Donald Richard
Manka, Dan P
Manning, R(obert) E(dward)
Marcelin, George
Martin, Jay Ronald
Matulevicius, Edward S(tephen)
Mendelsohn, Morris A
Meyer, Bernard Henry
Misra, Sudhan Sekher
Montgomery, Ronald Eugene
Moore, Robert Byron
Moore, Walter Calvin
Mount, Lloyd Gordon
Mutharasan, Rajakkannu
Myers, Alan Louis
Myers, John Adams
Myron, Thomas L(eo)
Natoli, John
Nedwick, John Joseph
Nemeth, Edward Joseph
Neufeld, Ronald David
Nichols, Duane Guy
Nolan, Edward J
Ochiai, Ei-Ichiro
Orphanides, Gus George
Pardee, William A(ugustus)
Park, Paul Heechung
Parker, Jennifer Ware
Parker, Robert Orion
Pasceri, Ralph Edward
Paxton, R(alph) R(obert)
Peel, James Edwin
Penn, William B
Perlmutter, Daniel D
Pessen, Helmut
Petrich, Robert Paul
Petronio, Marco
Petura, John C
Pinckney, Robert L
Pinkston, John Turner
Pommersheim, James Martin
Powers, Gary James
Prane, Joseph W(illiam)
Prieve, Dennis Charles
Pulsifer, Allen Huntington
Quinn, John A(lbert)
Racunas, Bernard J
Raimondi, Pietro
Ramezan, Massood
Redmount, Melvin B(ernard)
Reeder, Clyde
Regan, Raymond Wesley
Reiff, Harry Elmer
Retallick, William Bennett
Rice, William James
Rosenthal, Howard
Ross, Bradley Alfred
Rothfus, Robert R(andle)
Rudershausen, Charles Gerald
Ruppel, Thomas Conrad
Russell, Alan James
Sampson, Ronald N
Sanabor, Louis John
Sattler, Frank A(nton)
Sawicki, John Edward
Scattergood, Edgar Morris
Schell, George W(ashington)
Schiesser, W(illiam) E(dward)
Schiffman, Louis F
Schlesinger, Martin D(avid)
Schnaible, H(arold) W(illiam)
Schultz, J(erome) S(amson)
Schwartz, Albert B
Scigliano, J Michael
Seider, Warren David
Seiner, Jerome Allan

Chemical Engineering (cont)

Shellenberger, Donald J(ames)
Sheridan, John Joseph, III
Siddiqui, Habib
Sides, Paul Joseph
Sieger, John S(ylvester)
Singleton, Alan Herbert
Sinnett, Carl E(arl)
Sittenfield, Marcus
Smith, George C(unningham)
Snyder, William James
Solt, Paul E
Soung, Wen Y
Sprowls, Donald O(tte)
Stein, Fred P(aul)
Stephenson, Robert L
Stone, Douglas Roy
Struck, Robert T(heodore)
Sweeny, Robert F(rancis)
Sykes, James Aubrey, Jr
Tallmadge, J(ohn) A(llen), Jr
Thompson, Ralph Newell
Thompson, Sheldon Lee
Thygeson, John R(obert), Jr
Tierney, John W(illiam)
Tinkler, Jack D(onald)
Tobias, George S
Toor, H(erbert) L(awrence)
Trunzo, Floyd F(rank)
Turner, Howard S(inclair)
Ultman, James Stuart
Vance, William Harrison
Vannice, Merlin Albert
Vidt, Edward James
Vrentas, Christine Mary
Vrentas, James Spiro
Wahnsiedler, Walter Edward
Waldman, L(ouis) A(braham)
Walpert, George W
Wamsley, W(elcome) W(illard)
Warburton, Charles E, Jr
Weddell, George G(ray)
Weimer, Robert Fredrick
Weinberger, Charles Brian
Weisser, Eugene P
Weisz, Paul Burg
Wenzel, Leonard A(ndrew)
Werner, John Ellis
West, A(rnold) Sumner
Westerberg, Arthur William
Wettach, William
Wetzel, Roland H(erman)
White, Robert E(dward)
Wiggill, John Bentley
Wilder, Harry D(ouglas)
Wilkens, John Albert
Wilkens, Lucile Shanes
Wilson, Harry W(alton), Jr
Wilson, Lawrence Albert, Jr
Wu, Dao-Tsing
Yang, Wen-Ching
Yavorsky, Paul M(ichael)
Yerazunis, Stephen
Zamanzadeh, Mehrooz
Zeigler, A(lfred) G(eyer)
Ziering, Lance K
Zolotorofe, Donald Lee

RHODE ISLAND

Barnett, Stanley M(arvin)
Calo, Joseph Manuel
Graham, Ronald A(rthur)
Hutzler, Leroy, III
Knickle, Harold Norman
Liu, J(oseph) T(su) C(hieh)
Mason, Edward Allen
Ricklin, Saul
Rose, Vincent C(elmer)
Satas, Donatas
Sprowles, Jolyon Charles
Suuberg, Eric Michael
Thompson, A(lexander) Ralph
Van Winkle, Thomas Leo
Votta, Ferdinand, Jr

SOUTH CAROLINA

Alley, Forrest C
Bart, Roger
Baumgarten, Peter Karl
Beckwith, William Frederick
Boelter, Edwin D, Jr
Bradley, Robert Foster
Bryan, James Clarence
Buckham, James A(ndrew)
Clarke, James
Cogan, Jerry Albert, Jr
Corbitt, Maurice R(ay)
Darnell, W(illiam) H(eaden)
Davis, Milton W(ickers), Jr
Dworjanyn, Lee O(leh)
Gibbons, Joseph H(arrison)
Glocker, Edwin Merriam
Grady, Cecil Paul Leslie, Jr
Greene, George C, III
Groce, William Henry (Bill)
Haile, James Mitchell
Hale, Raymond Joseph
Harshman, Richard C(alvert)
Henderson, James B(rooke)
Herdklotz, Richard James
Hon, David Nyok-Sai
Hootman, Harry Edward
Hubbell, Douglas Osborne

Jennings, Alfred S(tonebraker)
Johnson, Ben S(lemmons), Jr
Krumrei, W(illiam) C(larence)
Lee, Craig Chun-Kuo
Longtin, Bruce
Melsheimer, Stephen Samuel
Morris, J(ames) William
Mullins, Joseph Chester
O'Neill, John H(enry), Jr
O'Rear, Steward William
Otto, Wolfgang Karl Ferdinand
Porter, Rick Anthony
Proctor, J(ohn) F(rancis)
Robinson, William Courtney, Jr
Schilson, Robert E(arl)
Squires, Paul Herman
Terry, Stuart Lee
Webster, D(onald) S(teele)

SOUTH DAKOTA

Sandvig, Robert L(eRoy)

TENNESSEE

Anderson, Roger W(hiting)
Beadell, Donald Albert
Beall, S(amuel) E, Jr
Bigelow, John E(aly)
Bloom, Sanford Gilbert
Bogue, Donald Chapman
Bond, Walter D
Bopp, C(harles) Dan
Bowden, Warren W(illiam)
Boyle, William Robert
Buck, George Sumner, Jr
Bullock, Jonathan S, IV
Burwell, Calvin C
Byers, Charles Harry
Cain, Carl, Jr
Claiborne, H(arry) C(lyde)
Cochran, Henry Douglas, Jr
Coleman, Charles Franklin
Crawford, Lloyd W(illiam)
Crouse, David J, Jr
Culberson, Oran L(ouis)
Davison, Brian Henry
DeVan, Jackson H
Donaldson, Terrence Lee
Duckworth, William C(apell)
Fort, Tomlinson, Jr
Frazier, George Clark, Jr
Frye, Alva L(eonard)
Fulkerson, William
Gano, Richard W
Garber, Harold Jerome
Godbee, H(erschel) W(illcox)
Gregory, Dale R(ogers)
Haas, Paul Arnold
Harris, Thomas R(aymond)
Henry, Jonathan Flake
Hightower, Jesse Robert
Hoffman, H(erbert) W(illiam)
Hsu, Hsien-Wen
James, V(irgil) Eugene
Janna, William Sied
Jasny, George R
Jensen, J(ohn) H(enry), Jr
Johnson, Homer F(ields), Jr
Judkins, Roddie Reagan
Kasten, Paul R(udolph)
Keller, Charles A(lbert)
Klein, Jerry Alan
Leuze, Rex Ernest
Levin, Seymour A(rthur)
McClung, Robert W
McNeese, Leonard Eugene
Mahlman, Harvey Arthur
Mailen, James Clifford
Malling, Gerald F
Murray, Lawrence P(atterson), Jr
Newberry, J(ames) R(aymond)
North, Edward D(avid)
Ornitz, Barry Louis
Overholser, Knowles Arthur
Parish, Trueman Davis
Parsly, Lewis F(uller), Jr
Perona, Joseph James
Phillips, Bobby Mal
Poese, Lester E
Polahar, Andrew Francis
Prados, John W(illiam)
Rodgers, Billy Russell
Rosenthal, Murray Wilford
Roth, John Austin
Russell, Ross F
Ryon, Allen Dale
Sanders, John P(aul), Sr
Schnelle, K(arl) B(enjamin), Jr
Scott, Charles D(avid)
Sears, Mildred Bradley
Seaton, William Hafford
Sheth, Atul C
Siirola, Jeffrey John
Silverman, Meyer David
Singh, Suman Priyadarshi Narain
Sissom, Leighton E(sten)
Sittel, Chester Nachand
Spruiell, Joseph E(arl)
Tanner, Robert Dennis
Thomas, David Glen
Threadgill, W(alter) D(ennis)
Turitto, Vincent Thomas
Umholtz, Clyde Allan
Van Horn, Wendell Earl
Vaughen, Victor C(ornelius) A(dolph)

Wang, Chien Bang
Watson, Jack Samuel
Whatley, Marvin E
Woodard, Kenneth Eugene, Jr
Yarbrough, David Wylie
Yee, William C
Zachariasen, K(arsten) A(ndreas)

TEXAS

Acciarri, Jerry A(nthony)
Akgerman, Aydin
Ali, Yusuf
Al-Saadoon, Faleh T
Ambrose, John E
Amundson, Neal R
Andrews, Robert V(incent)
Anthony, Rayford Gaines
Armeniades, Constantine D
Atkins, George T(yng)
Baeder, Donald L(ee)
Bannon, Robert Patrick
Barlow, Joel William
Barrere, Clem A, Jr
Bawa, Mohendra S
Beck, Curt B(uxton)
Bethea, Robert Morrison
Biles, W(illiam) R(oy)
Blum, Harold A(rthur)
Bodnar, Stephen J
Bradford, John R(oss)
Brennecke, Henry Martin
Brennecke, Llewellyn F(rancis)
Bridges, Charles H(enry)
Brock, James Rush
Broodo, Archie
Brooke, M(axey)
Brunsting, Elmer H(enry)
Bush, Warren Van Ness
Camero, Arthur Anthony
Campbell, William M
Cardner, David V
Cares, William Ronald
Carradine, William Radell, Jr
Casad, Burton M
Chen, Anthony Hing
Chen, Edward Chuck-Ming
Chew, Ju-Nam
Chu, Chieh
Chyu, Ming-Chien
Cier, H(arry) E(vans)
Claassen, E(dwin) J(ack), Jr
Claridge, E(lmond) L(owell)
Clark, Stanley Preston
Cleland, Franklin Andrew
Coldren, Clarke L(incoln)
Colten, Oscar A(aron)
Cook, Evin Lee
Cruse, Carl Max
Cummings, George H(erbert)
Cupples, Barrett L(eMoyne)
Davis, Samuel Henry, Jr
Davison, Richard Read
Davitt, Harry James, Jr
Deans, Harry A
Dillard, Robert Garing, Jr
Domask, W(illiam) G(erhard)
Dotterweich, Frank H(enry)
Doumas, A(rthur) C(onstantinos)
Dukler, A(braham) E(manuel)
Duncan, Dennis Andrew
Durbin, Leonel Damien
Dye, Robert F(ulton)
Dyson, Derek C(harlesworth)
Eakin, Bertram E
Economou, Demetre J
Edgar, Thomas Flynn
Edwards, Victor Henry
Eggers, Fred M
Ehlig-Economides, Christine Anna
Ekerdt, John Gilbert
Ellison, Bart T
Engler, Cady Roy
Espino, Ramon Luis
Eubank, Philip Toby
Eubanks, L(loyd) Stanley
Fair, James R(utherford)
Flumerfelt, Raymond W
Fontaine, Marc F(rancis)
Fox, George Edward
Frantz, Joseph Foster
Furgason, Robert Roy
Fyfe, Richard Ross
Gidley, J(ohn) L(ynn)
Gilliland, Harold Eugene
Gilmour - Stallsworth, Lisa K
Goodman, Abraham H(arrison)
Goring, Geoffrey E(dward)
Graham, Harold L(aVerne)
Gregory, A(lvin) R(ay)
Grieves, Robert Belanger
Griffin, John R(obert)
Grimsby, F(rank) Norman
Gwyn, J(ohn) E(dward)
Haase, Donald J(ames)
Hales, Hugh B(radley)
Hall, Kenneth Richard
Hammond, James W
Harris, William Birch
Heenan, William A
Heichelheim, Hubert Reed
Heinrich, R(aymond) L(awrence)
Hellums, Jesse David
Henley, Ernest J(ustus)
Hightower, Joe W(alter)

Hill, Arthur S
Himmelblau, David M(autner)
Hirsch, Robert Louis
Hite, J Roger
Hocott, C(laude) R(ichard)
Hoffman, T(errence) W(illiam)
Holland, Charles D(onald)
Holmes, Larry A
Honeywell, Wallace I(rving)
Hopper, Jack R
Hougen, Joel O(liver)
Huang, Wann Sheng
Humphrey, Jimmy Luther
Hurlburt, H(arvey) Zeh
Im, Un Kyung
Ivey, E(dwin) H(arry), Jr
Jelen, Frederic Charles
Jiles, Charles William
Johnson, Lloyd N(ewhall)
Kalfoglou, George
Kemp, L(ebbeus) C(ourtright), Jr
Kingsley, Henry A(delbert)
Klecka, Miroslav Ezidor
Klein, V(ernon) A(lfred)
Knipp, Ernest A, (Jr)
Kobayashi, Riki
Koch, Howard A(lexander)
Koepf, Ernest Henry
Koppel, Lowell B
Koros, William John
Krautz, Fred Gerhard
Kridel, Donald Joseph
Latour, Pierre Richard
Lawler, James Henry Lawrence
Lee, Sidney
Leland, Thomas W, Jr
Lescarboura, Jaime Aquiles
Levy, Robert Edward
Lewis, James Pettis
Lloyd, Douglas Roy
Lockwood, Frances Ellen
Lowell, Philip S(iverly)
Ludwig, Allen Clarence, Sr
Luss, Dan
Lyon, John B(ennett)
McCutchan, Roy T(homas)
McFedries, Robert, Jr
McIntire, Larry V(ern)
McIntire, Louis V(ictor)
McKee, Herbert C(harles)
McKetta, John J, Jr
McNiel, James S(amuel), Jr
Maddox, Larry A(llen)
Mai, Klaus L(udwig)
Marberry, James E(dward)
Marple, Stanley, Jr
Marsh, Kenneth Neil
Massey, Michael John
Mattax, Calvin Coolidge
Meiser, K(enneth) D(onaldson)
Messenger, Joseph Umlah
Miller, Clarence A(lphonso)
Miller, Richard Linn
Mitchell, James Emmett
Moltzan, Herbert John
Morris, Herbert Comstock
Morrison, Milton Edward
Morse, N(orman) L(ester)
Moser, James Howard
Mullikin, Richard V(ickers)
Nicolaisen, B(ernard) H(enry)
Ochiai, Shinya
O'Connell, Harry E(dward)
Okamoto, K Keith
Olson, John S
Padgett, Algie Ross
Park, Vernon Kee
Parker, Arthur L (Pete)
Parker, Harry W(illiam)
Patel, Mayur
Patil, Kashinath Ziparu
Paul, Donald Ross
Pennington, Robert Elija
Perkins, Thomas K(eeble)
Pitt, Woodrow Wilson, Jr
Poddar, Syamal K
Popovich, Robert Peter
Poska, F(orrest) L(ynn)
Powell, Richard James
Prengle, H(erman) W(illiam), Jr
Pritchett, P(hilip) W(alter)
Rai, Charanjit
Rajagopalan, Raj
Randlett, Herbert Eldridge, Jr
Rase, Howard F
Rhinehart, Robert Russell, II
Richardson, James T(homas)
Richardson, Joseph Gerald
Roberts, Louis Reed
Roeger, Anton, III
Russell, James N(elson), Jr
Schechter, Robert Samuel
Schrader, R(obert) J
Schrage, Robert W
Scinta, James
Senatore, Ford Fortunato
Serth, Robert William
Shanfield, Henry
Sheppard, Louis Clarke
Silberberg, I(rwin) Harold
Simon, Eric
Slattery, John C
Slusser, M(arion) L(iles)
Somerville, George R

Souby, Armand Max
Speed, Raymond A(ndrew)
Stark, John, Jr
Stewart, J(ames) R(ush), Jr
Stone, Herbert L(osson)
Strauss, Howard W(illiam)
Styring, Ralph E
Swearingen, Judson Sterling
Tao, Frank F
Teague, Abner F
Tetlow, Norman Jay
Thompson, William Horn
Tiller, F(rank) M(onterey)
Tock, Richard William
Trachtenberg, Isaac
Tully, Philip C(ochran)
Van Horn, Lloyd Dixon
Vasilakos, Nicholas Petrou
Volpe, P(eter) J, Jr
von Rosenberg, H(ermann) E(ugene)
Wagner, John Philip
Walker, Richard E
Walls, Hugh A(lan)
Warren, Francis A(lbert)
Weatherford, W(illiam) D(ewey), Jr
Wheeler, Joe Darr
White, Bernard Henry
White, Ralph E
Wiener, L(udwig) D(avid)
Wilde, Kenneth Alfred
Williams, Curtis Chandler, III
Willoughby, Sarah Margaret C(laypool)
Wilson, J(ames) W(oodrow)
Wissler, Eugene H(arley)
Wong, S(oon) Y(uck)
Woods, J(ohn) M(elville)
Woodward, Joe William
Worley, Frank L, Jr
Wurth, Thomas Joseph
Yaws, Carl Lloyd
Yeung, Reginald Sze-Chit
Zaczepinski, Sioma

UTAH
Barker, Dee H(eaton)
Bartholomew, Calvin Henry
Bennion, Douglas Noel
Budge, Wallace Don
Christiansen, E(rnest) B(ert)
Coates, Ralph L
Dahlstrom, Donald A(lbert)
De Nevers, Noel Howard
Fay, John Edward, II
Hanks, Richard W(ylie)
Horton, M Duane
Kelsey, Stephen Jorgensen
Larsen, James Victor
Li, Shin-Hwa
Lighty, JoAnn Slama
Miller, Jan Dean
Oblad, Alex Golden
Othmer, Hans George
Pope, Bill Jordan
Powers, John E(dward)
Rowley, Richard L
Ryan, Norman W(allace)
Salt, Dale L(ambourne)
Seader, J(unior) D(eVere)
Silver, Barnard Stewart
Smoot, Leon Douglas
Sohn, Hong Yong
Thirkill, John D
Trujillo, Edward Michael
Tyler, Austin Lamont
Zeamer, Richard Jere

VERMONT
Bhatt, Girish M
Dean, Robert Charles, Jr
Dritschilo, William
Jones, John F(rederick)
Long, R(obert) B(yron)
Long, Robert Byron
Vivian, J(ohnson) Edward

VIRGINIA
Aly, Abdel Fattah
Bartlett, John W(esley)
Beach, Robert L
Beyer, Gerhard H(arold)
Bickling, Charles Robert
Blum, Edward H(oward)
Brown, Winton
Chiou, Minshon Jebb
Clark, Donald Eldon
Conger, William Lawrence
Cox, Edwin, III
Cummings, Peter Thomas
Dardoufas, Kimon C
Davis, Bernard Eric
Dedrick, Robert L(yle)
DePoy, Phil Eugene
Dewey, J(ohn) L(yons)
Diller, Thomas Eugene
Dorsey, Clark L(awler), Jr
Durfee, Robert Lewis
Erickson, Wayne Douglas
Fein, Harvey L(ester)
Ferguson, Thomas Lee
Foerster, Edward L(eRoy), Sr
Gabelnick, Henry Lewis
Gaden, Elmer L(ewis), Jr
Gainer, John Lloyd
Gallini, John B(attista)

Glasser, Wolfgang Gerhard
Gordon, John S(tevens)
Gross, George C(onrad)
Halsey, Brenton S
Harris, Paul Robert
Hauxwell, Gerald Dean
Hemm, Robert Virgil
Herbst, Mark Joseph
Hinkle, Barton L(eslie)
Hollingsworth, David S
Hudson, J(ohn) L
Hur, J James
Hussamy, Samir
Jaisinghani, Rajan A
Jimeson, Robert M(acKay), Jr
Joebstl, Johann Anton
Jonnard, Aimison
Keeler, Roger Norris
King, Merrill Kenneth
Kirkendall, Ernest Oliver
Kirtley, Thomas L(loyd)
Knap, James E(li)
Lacy, W(illiam) J(ohn)
Lasser, Howard Gilbert
LeVan, Martin Douglas, Jr
Lilleleht, L(embit) U(no)
Ling, James Gi-Ming
Lowe, William Webb
MacClaren, Robert H
McGee, Henry A(lexander), Jr
McHale, Edward Thomas
McKenna, John Dennis
Marchello, Joseph M(aurice)
Markels, Michael, Jr
Mayfield, Lewis G
Mertes, Frank Peter, Jr
Mischke, Roland A(lan)
Munday, J(ohn) C(lingman)
O'Connell, John P
Oliver, Robert C(arl)
Patterson, Earl E(dgar)
Perry, John E(dward)
Puyear, Donald E(mpson)
Roberts, Irving
Rony, Peter R(oland)
Sabadell, Alberto Jose
Schaffner, Robert M(ichael)
Sebba, Felix
Setterstrom, Carl A(lbert)
Shockley, Gilbert R
Sikri, Atam P
Southworth, Raymond W(illiam)
Squires, Arthur Morton
Stoddard, C(arl) Kerby
Thomson, Richard N
Tinsley, Richard Sterling
Uhl, V(incent) W(illiam)
Walther, James Eugene
Washington, James M(acknight)
Wills, George B(ailey)
Winslow, Charles Ellis, Jr
Youngs, Robert Leland

WASHINGTON
Ackerman, C(arl) D(avid)
Amberg, Herman R(obert)
Austin, George T(homas)
Babb, Albert L(eslie)
Barker, James J(oseph)
Beck, Theodore R(ichard)
Berg, John Calvin
Berger, Albert Jeffrey
Bolme, Donald W(eston)
Bolme, Mark W
Bowdish, Frank W(illiam)
Bowen, J(ewell) Ray
Brehm, William Frederick
Burns, Robert Ward
Cook, John Carey
Damon, Robert A(rthur)
Davis, E(arl) James
Dickinson, Dean Richard
Divine, James R(obert)
Dobratz, Carroll J
Duncan, James Byron
Edde, Howard Jasper
Evans, Thomas F(rederick)
Falco, James William
Faletti, Duane W
Finlayson, Bruce Alan
Finnigan, J(erome) W(oodruff)
Fosberg, Theodore M(ichael)
Foster, Norman Charles
Fullam, Harold Thomas
Garlid, Kermit L(eroy)
Hales, Jeremy M(organ)
Hanby, John Estes, Jr
Hanthorn, Howard E(ugene)
Heideger, William J(oseph)
Hendrickson, John R(eese), Sr
Hendrickson, Waldemar Forrsel
Hirsch, Horst Eberhard
Johanson, L(ennart) N(oble)
Johnson, B(enjamin) M(artineau)
Kaser, J(ohn) D(onald)
Kerbecek, Arthur J(oseph), Jr
Koehmstedt, Paul Leon
Krieger, Barbara Brockett
Krier, Carol Alnoth
LaFontaine, Edward
Lago, James
Lindstrom, Duaine Gerald
McCarthy, Joseph L(ePage)

McKean, William Thomas, Jr
Mahalingam, R
Martini, William Rogers
Miller, Reid C
Moore, Raymond H
Padilla, Andrew, Jr
Pearson, Donald A
Pilat, Michael Joseph
Putnam, G(arth) L(ouis)
Rein, Robert G, Jr
Richardson, Gerald Laverne
Ricker, Neil Lawrence
Rieck, H(enry) G(eorge)
Roake, William Earl
Rohrmann, Charles A(lbert)
Romero, Jacob B
Rosenwald, Gary W
Sehmel, George Albert
Sleicher, Charles A
Stuve, Eric Michael
Thomson, William Joseph
Woodfield, F(rank) W(illiam), Jr
Worthington, Ralph Eric
Zollars, Richard Lee

WEST VIRGINIA
Augstkalns, Valdis Ansis
Bailie, Richard Colsten
Beaton, Daniel H(arper)
Beeson, Justin Leo
Biddle, John Wilbur
Breneman, William C
Chaty, John Culver
Clark, Nigel Norman
Cope, Charles S(amuel)
Crum, Edward Hibbert
De Hoff, George R(oland)
Gillmore, Donald W(ood)
Green, R(alph) V(ernon)
Guthrie, Hugh D
Hazelton, Russell Frank
Howell, John H(ancock)
Humphreys, Kenneth K
Keller, George E, II
Larsen, Howland Aikens
Mellow, Ernest W(esley)
Powell, G(eorge) M(atthews), III
Shaeiwitz, Joseph Alan
Tatomer, Harry Nicholas
Verma, Surendra Kumar
Wales, Charles E
Ware, Charles Harvey, Jr
Weiser, Robert B(ruce)
Wen, Chin-Yung
Wertman, William Thomas
Wilson, John Sheridan

WISCONSIN
Amundson, Clyde Howard
Atalla, Rajai Hanna
Bajikar, Sateesh S
Bird, R(obert) Byron
Bringer, Robert Paul
Carstensen, Jens T(huroe)
Chapman, Thomas Woodring
Christiansen, Alfred W
Ciriacks, John A(lfred)
Coberly, Camden Arthur
Cooper, Stuart L
Doshi, Mahendra R
Duffie, John A(twater)
Gehrke, Willard H
Gottschalk, Robert Neal
Grace, Thomas Michael
Grindrod, Paul (Edward)
Han, Shu-Tang
Hansen, Paul B(ernard)
Hanway, John E(dgar), Jr
Hill, Charles Graham, Jr
Holland, Dewey G
Hughes, Richard Roberts
Hunter, William G(ordon)
Hurley, James R(obert)
Johnston, Peter Ramsey
Katz, William J(acob)
Koutsky, James A
Kurath, Sheldon Frank
Lightfoot, E(dwin) N(iblock), Jr
Makela, Lloyd Edward
Manning, James Harvey
Megahed, Sid A
Parker, Peter Eddy
Powell, Hugh N
Ray, W(illis) Harmon
Reid, Robert Lelon
Rudd, D(ale) F(rederick)
Sather, Glenn A(rthur)
Schaffer, Erwin Lambert
Seale, Dianne B
Sebree, Bruce Randall
Sell, Nancy Jean
Sfat, Michael R(udolph)
Springer, Edward L(ester)
Stewart, W(arren) E(arl)
Tiedemann, William Harold
Whitney, Roy P(owell)
Williams, Duane Alwin
Wingert, Louis Eugene

WYOMING
Carpenter, Harry C(lifford)
Cooney, David Ogden
Ewing, Richard Edward
Francis, Warren C(harles)

Gunn, Robert Dewey
Haynes, Henry William, Jr
Hiza, Michael John, Jr
Hoffman, Edward Jack
Hutchinson, Harold Lee
Odasz, F(rancis) B(ernard), Jr
Raynes, Bertram C(hester)
Silver, Howard Findlay
Stinson, Donald Leo

PUERTO RICO
Baus, Bernard V(illars)
Rodriguez, Luis F
Souto Bachiller, Fernando Alberto

ALBERTA
Bowman, C(lement) W(illis)
Brown, R(onald) A(nderson) S(teven)
Butler, R(oger) M(oore)
Dalla Lana, I(vo) G(iovanni)
Drum, I(an) M(ondelet)
Fu, Cheng-Tze
Govier, G(eorge) W(heeler)
Gregory, Garry Allen
Gregory, John
Mohtadi, Farhang
Nandakumar, Krishnaswamy
Otto, Fred Douglas
Pan, Chuen Yong
Rhodes, E(dward)
Tollefson, Eric Lars
Wanke, Sieghard Ernst
Williams, Michael C(harles)
Wood, Reginald Kenneth
Yarborough, Lyman

BRITISH COLUMBIA
Dumont, Guy Albert
Epstein, Norman
Forsyth, J(ames) S(neddon)
Godard, Hugh P(hillips)
Grace, John Ross
Kerekes, Richard Joseph
Levelton, B(ruce) Harding
Lielmezs, Janis
Meech, John Athol
Meisen, Axel
Pinder, Kenneth Lyle
Shaw, A(lexander) J(ohn)
Thompson, D(onald) W(illiam)
Young, Roland S

MANITOBA
Bassim, Mohamad Nabil
Rosinger, Eva L J
Tennankore, Kannan Nagarajan

NEW BRUNSWICK
Kristmanson, Daniel D
LeBel, Roland Guy
Morris, David Rowland
Picot, Jules Jean Charles
Ruthven, Douglas M
Tory, Elmer Melvin

NOVA SCOTIA
Chen, B(ih) H(wa)
McMillan, Alan F

ONTARIO
Adamek, Eduard Georg
Bacon, David W
Baird, Malcolm Henry Inglis
Basmadjian, Diran
Becker, Henry A
Beeckmans, Jan Maria
Benedek, Andrew
Bergougnou, Maurice A(medee)
Besik, Ferdinand
Bodnar, Louis Eugene
Brash, John Law
Bulani, Walter
Burns, C(harles) M(ichael)
Cale, William Robert
Campbell, W(illiam) M(unro)
Carr, Jack A(lbert)
Chaffey, Charles Elswood
Charles, Michael Edward
Chase, John Donald
Chen, Erh-Chun
Cho, Sang Ha
Chrones, James
Clark, Reginald Harold
Cranford, William B(rett)
Crowe, Cameron Macmillan
De Marco, F(rank) A(nthony)
Diaz, Carlos Manuel
Diosady, Levente Laszlo
Downie, John
Duever, Thomas Albert
Dullien, Francis A L
Ettel, Victor Alexander
Fahidy, Thomas Z(oltan)
Feuerstein, Irwin
Furter, W(illiam) F(rederick)
Gall, Carl Evert
Goulden, Peter Derrick
Hamielec, Alvin Edward
Handa, Paul
Hatcher, S(tanley) Ronald
Hayduk, Walter
Holtslander, William John
Hudgins, Robert R(oss)
Hummel, Richard Line

Chemical Engineering (cont)

Jardine, John McNair
Kosaric, Naim
Laughlin, R(obert) G(ardiner) W(illis)
Leaist, Derek Gordon
Leask, R(aymond) A(lexander)
Lu, Benjamin C(hih) Y(eu)
Luus, R(ein)
Macdonald, Ian Francis
McLeod, Norman William
Mann, Ranveer S
Mann, Ronald Francis
Matsuura, Takeshi
Milkie, Terence H
Miller, Alistair Ian
Missen, R(onald) W(illiam)
Mitsoulis, Evan
Mittal, Gauri S
Moo-Young, Murray
Mueller, Gerald Sylvester
Munns, W(illiam) O
Mutton, Donald Barrett
Napier, Douglas Herbert
Nicolle, Francois Marcel Andre
Nirdosh, Inderjit
Osberg, G(ustav) L(awrence)
Pei, David Chung-Tze
Penlidis, Alexander
Piggott, Michael R(antell)
Pogorski, Louis August
Reilly, Park McKnight
Rempel, Garry Llewellyn
Robinson, Campbell William
Rosehart, Robert George
Round, G(eorge) F(rederick)
Rowzee, E(dwin) R(alph)
Rubin, Leon Julius
Sandler, Samuel
Schlesinger, Mordechay
Scott, Donald S(trong)
Shemilt, L(eslie) W(ebster)
Silveston, Peter Lewis
Singer, Eugen
Smith, James W(ilmer)
Smith, William Robert
Souhrada, Frank
Southam, Frederick William
Sowa, Walter
Spink, D(onald) R(ichard)
Spinner, Irving Herbert
Sullivan, John Leslie
Tombalakian, Artin S
Trass, O(lev)
Turner, Godfrey Alan
Tzoganakis, Costas
Van Wagner, Charles Edward
Vijayan, Sivaraman
Vlachopoulos, John A(postolos)
Wayman, Morris
Weir, Ronald Douglas
Williams, Norman S W
Wojciechowski, Bohdan Wieslaw
Woods, Donald Robert
Wright, Joseph D
Wynnyckyj, John Rostyslav

QUEBEC

Beron, Patrick
Carreau, Pierre
Cholette, A(lbert)
Cloutier, Leonce
Cooke, Norman E(dward)
Croctogino, Reinhold Hermann
Dealy, John Michael
Douglas, W J Murray
Feldman, Dorel
Gauvin, William H
Kamal, Musa Rasim
Klassen, J(ohn)
Kubes, George Jiri
Leduy, Anh
Lee, John Hak Shan
Mujumdar, Arun Sadashiv
Roy, Paul-H(enri)
Waid, Ted Henry
Wu, Chisung

SASKATCHEWAN

Bakhshi, Narendra Nath
Catania, Peter J
DeCoursey, W(illiam) J(ames)
Esmail, Mohamed Nabil
Macdonald, Douglas Gordon
Mikle, Janos J
Phillips, Kenneth Lloyd
Shook, Clifton Arnold
Tinker, Edward Brian

OTHER COUNTRIES

Chang, H K
Chang, Shuya
Chen, Fred Fen Chuan
Cullinan, Harry T(homas), Jr
Dunn, Irving John
Economides, Michael John
Esterson, Gerald L(ee)
Fey, George Ting-Kuo
Fok, Siu Yuen
Fu, Yuan C(hin)
Hamada, Mokhtar M
Hansen, Charles M
Howell, John Anthony
Huckaba, Charles Edwin

Knott, Robert F(ranklin)
Kocatas, Babur M(ehmet)
Liu, Ta-Jo
McRae, Wayne A
Mathews, Joseph F(ranklin)
Miyano, Kenjiro
Mou, Duen-Gang
Shor, Arthur Joseph
Spottiswood, David James
Stam, Jos
Van Weert, Gezinus
Williams, F(ord) Campbell
Yoshida, Fumitake

Civil Engineering

ALABAMA

Appleton, Joseph Hayne
Blakney, William G G
Carlton, Thomas A, Jr
Caudill, Reggie Jackson
Cost, Thomas Lee
Costes, Nicholas Constantine
Gilbert, John Andrew
Hardin, Edwin M(ilton)
Hassell, H(orace) Paul, Jr
Judkins, Joseph Faulcon, Jr
Kallsen, Henry Alvin
Kandhal, Prithvi Singh
Keith, Warren Gray
Killingsworth, R(oy) W(illiam)
Marek, Charles R(obert)
Melville, Joel George
Roberts, Freddy Lee
Segner, Edmund Peter, Jr
Turner, Daniel Shelton
Walters, James Vernon
Zehrt, William H(arold)

ALASKA

Alter, Amos Joseph
Bennett, F Lawrence
Hulsey, J Leroy
Johansen, Nils Ivar
Sweet, Larry Ross
Tilsworth, Timothy
Zarling, John P

ARIZONA

Betz, Mathew J(oseph)
Blackburn, Jack Bailey
Brinker, Russell Charles
Burbank, Nathan C, Jr
Carder, David Ross
Contractor, Dinshaw N
Gitlin, Harris Martlin
Govil, Sanjay
Havers, John Alan
Hill, Louis A, Jr
Ince, Simon
Johnston, Bruce (Gilbert)
Klock, John W
Kriegh, James Douglas
Laursen, E(mmett) M(orton)
Lundgren, Harry Richard
Maddock, Thomas, Jr
Matthias, Judson S
Morris, Gene Ray
Newlin, Charles W(illiam)
Newlin, Philip Blaine
Nordby, Gene M
Reich, Brian M
Renard, Kenneth G
Replogle, John A(sher)
Richard, Ralph Michael
Rouse, Hunter
Singhal, Avinash Chandra
Smerdon, Ernest Thomas
Sorooshian, Soroosh
Sultan, Hassan Ahmed
Tuma, Jan J
Upchurch, Jonathan Everett
Woolhiser, David A(rthur)

ARKANSAS

Bissett, J(ames) R(obert)
Boyd, Donald Edward
Fletcher, Alan G(ordon)
Ford, Miller Clell, Jr
Heiple, Loren Ray
Jellinger, Thomas Christian
Jong, Ing-Chang

CALIFORNIA

Abdel-Ghaffar, Ahmed Mansour
Agardy, Franklin J
Alexander, Ira H(enris)
Alexander, Robert L
Ang, Alfredo H(ua)-S(ing)
Aroni, Samuel
Atwood, John Leland
Baker, Warren J
Banks, Harvey Oren
Bea, Robert G
Bellport, Bernard Philip
Birkimer, Donald Leo
Blume, John A(ugust)
Brahma, Chandra Sekhar
Breckenridge, Robert A(rthur)
Bresler, B(oris)
Brock, Richard R
Brooks, Norman H(errick)
Bugg, Sterling L(owe)

Butler, Stanley S
Carlson, Roy W(ashington)
Cassidy, John J(oseph)
Castellan, Norman J
Chadwick, Wallace Lacy
Chang, Howard
Chelapati, Chunduri V(enkata)
Chen, Carl W(an-Cheng)
Cheney, James A
Chi, Cheng-Ching
Chiang, George C(hihming)
Chopra, Anil K
Chou, Larry I-Hui
Chu, Kuang-Han
Chuang, Kuen-Puo (Ken)
Clough, Ray William
Collins, James Ian
Cooke, James Barry
Cornell, C(arl) Allin
Crawford, Norman Holmes
Cross, Ralph Herbert, III
Daganzo, Carlos Francisco
Daily, James W(allace)
Davis, Harmer E
Degenkolb, Henry John
Dodge, E(ldon) R(aymond)
Dong, Richard Gene
Dong, Stanley B
Douglas, James
Dowell, Douglas C
Dracup, John Albert
Eberhart, H(oward) D(avis)
Eliassen, Rolf
Evans, T(homas) H(ayhurst)
Everts, Craig Hamilton
Felton, Lewis P(eter)
Ferrara, Thomas Ciro
Filippou, Filip C
Fondahl, John W(alker)
Forrest, James Benjamin
Franzini, Joseph B(ernard)
Gabriel, Lester H
Gabrielsen, Bernard L
Garbarini, Edgar Joseph
Gere, James Monroe
Gerwick, Ben Clifford, Jr
Glenn, Richard A(llen)
Goodman, Richard E
Gordon, Ruth Vida
Grant, Eugene Lodewick
Hackel, Lloyd Anthony
Hahne, Henry V
Ham, Lee Edward
Hamilton, Gordon Wayne
Hammond, David G
Hanna, George P, Jr
Harder, James Albert
Hayes, Thomas Jay, III
Healey, Anthony J
Henderson, D(elbert) W
Herrmann, George
Herrmann, Leonard R(alph)
Hesse, Christian August
Heuze, Francois E
Hickok, Robert Baker
Housner, George W(illiam)
Hung, You-Tsai Joseph
Hwang, Li-San
Idriss, Izzat M
Ingram, Gerald E(ugene)
Iselin, Donald G
Jacobs, Joseph Donovan
Jeng, Raymond Ing-Song
Jenkins, David Isaac
Jennings, Paul C(hristian)
Johnston, Roy G
Kennedy, Robert P
Kiely, John Roche
Kim, Young C
Kiremidjian, Anne Setian
Kostyrko, George Jurij
Krishnamoorthy, Govindarajalu
Krone, Ray B
Kruger, Paul
Lam, Tenny N(icolas)
Larock, Bruce E
Lau, John H
Leidersdorf, Craig B
Lemcoe, M M(arshall)
Leonhard, William E
Leps, Thomas MacMaster
Lin, Tung Yen
Linsley, Ray K(eyes), Jr
Love, Joseph E(ugene), Jr
Lubliner, J(acob)
Luk, King Sing
Luscher, Ulrich
Lysmer, John
McCammon, Lewis B(rown), Jr
McCarty, Perry L(ee)
May, Adolf D(arlington), Jr
Medwadowski, Stefan J
Merriam, John L(afayette)
Merritt, J(oshua) L(evering), Jr
Meyers, Bernard Leonard
Miller, Richard Keith
Mitchell, James K(enneth)
Monismith, Carl L(eroy)
Moore, William W
Morgali, James R
Morgan, James John
Morris, Brooks T(heron)
Morris, Henry Madison, Jr
Mow, Maurice

Mueller, Charles Carsten
Naar, Jacques
Napolitano, Leonard Michael, Jr
Narasimhan, Thiruppudaimarudhur N
Nelson, Richard Bartel
Nemat-Nasser, Siavouche
Nordell, William James
Nowatzki, Edward Alexander
Oglesby, Clarkson Hill
Ongerth, Henry J
Oswald, William J
Parker, Henry Whipple
Paulson, Boyd Colton, Jr
Pearson, Erman A
Penzien, Joseph
Perloff, David Steven
Pietrzak, Lawrence Michael
Pincus, Howard Jonah
Popov, E(gor) P(aul)
Post, J(ames) L(ewis)
Ramey, Melvin Richard
Ratner, Robert (Stephen)
Riggs, Louis William
Ritchie, Stephen G
Rogers, Gifford Eugene
Rosen, Alan
Rowland, Walter Francis
Rubin, Sheldon
Rubinstein, Moshe Fajwel
Rudavsky, Alexander Bohdan
Sarkaria, Gurmukh S
Schiff, Anshel J
Schmit, Lucien A(ndre), Jr
Scordelis, Alexander Costicas
Scott, Ronald F(raser)
Scott, Verne H(arry)
Seaman, Lynn
Seide, Paul
Shanteau, Robert Marshall
Sharpe, Roland Leonard
Shaw, Warren A(rthur)
Shen, Hsieh Wen
Shepherd, Robin
Sicular, George M
Singh, Rameshwar
Skalak, Richard
Spicher, Robert G
Steinbrugge, Karl V
Stenstrom, Michael Knudson
Stockdale, William K
Stratton, Frank E(dward)
Sundaram, Panchanatham N
Supersad, Jankie Nanan
Sutherland, Louis Carr
Sve, Charles
Synolakis, Costas Emmanuel
Taylor, Robert Leroy
Thon, J George
Todd, David Keith
Vanoni, Vito A(ugust)
Venuti, William J(oseph)
Wallender, Wesley William
Watters, Gary Z
Westmann, Russell A
Wiegel, Robert L
Wiggins, John H(enry), Jr
Wilson, Basil W(rigley)
Wilson, Edward L
Yeh, William Wen-Gong
Yen, Bing Cheng
Ying, William H
Yu, David U L
Zwoyer, Eugene

COLORADO

Bartlett, Paul Eugene
Chugh, Ashok Kumar
Criswell, Marvin Eugene
Dahl, Arthur Richard
Day, David Allen
DeCoursey, Donn G(ene)
Evans, Norman A(llen)
Faddick, Robert Raymond
Fead, John William Norman
Feldman, Arthur
Feng, Chuan C(hung) D(avid)
Flack, J(ohn) E(rnest)
Frangopol, Dan Mircea
Frasier, Gary Wayne
Garstka, Walter U(rban)
Gerstle, Kurt H
Gessler, Johannes
Goble, George G
Goodman, James R
Hall, Warren A(cker)
Hendricks, David Warren
Holtz, Wesley G
Hughes, William Carroll
Illangasekare, Tissa H
Jibson, Randall W
Johnson, Arnold I(van)
Johnstone, James G(eorge)
Ketchum, Milo S
Ko, Hon-Yim
Koelzer, Victor A
Linstedt, Kermit Daniel
McLean, Francis Glen
Maierhofer, Charles Richard
Mays, John Rushing
Medearis, Kenneth Gordon
Moody, Martin L(uther)
Morel-Seytoux, Hubert Jean
Osterberg, J(orj) O(scar)
Pak, Ronald Y S

Rautenstraus, R(oland) C(urt)
Richardson, Everett V
Romig, William D(avis)
Savage, William Zuger
Schuster, Robert Lee
Simons, Daryl B
Smith, Roger Elton
Strom, Oren Grant
Summers, Luis Henry
Sunada, Daniel K(atsuto)
Suprenant, Bruce A
Tabler, Ronald Dwight
Thompson, Erik G(rinde)
Tulin, Leonard George
Ward, John C(layton)
Weingart, Richard
Wu, Jonathan Tzong
Yang, Chih Ted

CONNECTICUT
Alcantara, Victor Franco
DeWolf, John T
Epstein, Howard I
Flannelly, William G
Frantz, Gregory Clayton
Hemond, Conrad J(oseph), Jr
Johnston, E(lwood) Russell, Jr
Lin, Jia Ding
Long, Richard Paul
Nadel, Norman A
Olster, Elliot Frederick
Panuzio, Frank L
Posey, C(hesley) J(ohnston)
Stephens, Jack E(dward)

DELAWARE
Cheng, Alexander H-D
Chesson, Eugene, Jr
Dalrymple, Robert Anthony
Dentel, Steven Keith
Goel, Kailash C(handra)
Hsiao, George Chia-Chu
Huang, Chin Pao
Hume, Harold Frederick
Jones, Russel C(ameron)
Kaliakin, Victor Nicholas
Kerr, Arnold D
Kikuchi, Shinya
Kobayashi, Nobuhisa
Koch, Carl Mark
Nicholls, Robert Lee
Svendsen, Ib Arne
Wu, Jin
Yang, C(heng) Y(i)

DISTRICT OF COLUMBIA
Adams, Francis L(ee)
Bell, Bruce Arnold
Blanchard, Bruce
Chong, Ken Pin
Chung, Riley M
Culver, Charles George
Deen, Thomas B
Eggenberger, Andrew Jon
Hampton, Delon
Johnson, George Patrick
Jones, Irving Wendell
Kao, Timothy Wu
McGinnis, David Franklin, Jr
Mermel, Thaddeus Walter
Odar, Fuat
Parks, Vincent Joseph
Roberts, Paul Osborne, Jr
Senich, Donald
Smith, Waldo E(dward)
Soteriades, Michael C(osmas)
Toridis, Theodore George
Vaishnav, Ramesh
Way, George H
Woo, Dah-Cheng
Yachnis, Michael

FLORIDA
Anderson, Melvin W(illiam)
Brumer, Milton
Carrier, W(illiam) David, III
Chiu, Tsao Yi
Christensen, Bent Aksel
Chryssafopoulos, Hanka Wanda Sobczak
Chryssafopoulos, Nicholas
Claridge, Richard Allen
Dantin, Elvin J, Sr
Dean, Robert George
Dietz, Jess Clay
Drucker, Daniel Charles
Edson, Charles Grant
Ege, Raimund K
Fogarty, William Joseph
Givens, Paul Edward
Goodson, James Brown, Jr
Grinter, Linton E(lias)
Harrenstien, Howard P(aul)
Harris, Lee Errol
Hartman, John Paul
Heaney, James Patrick
Huang, Y(en) T(i)
Huber, Wayne Charles
Jenkins, David R(ichard)
Kersten, Robert D(onovan)
Kuzmanovic, B(ogdan) O(gnjan)
Lambe, T(homas) William
Langfelder, Leonard Jay
Lin, Y(u) K(weng)
Mantell, M(urray) I(rwin)

Michejda, Oskar
Miller, Charles Leslie
Nemerow, Nelson L(eonard)
Nichols, James Carlile
Nunnally, Stephens Watson
Partheniades, Emmanuel
Patterson, Archibald Oscar
Richards, A(lvin) M(aurer)
Ruth, Byron E
Schaub, James H(amilton)
Schmertmann, John H(enry)
Sirkin, Alan N
Stevens, John A(lexander)
Surti, Vasant H
Viessman, Warren, Jr
Vitagliano, Vincent J
Yong, Yan
Zollo, Ronald Francis

GEORGIA
Amirtharajah, Appiah
Barksdale, Richard Dillon
Barnhart, Cynthia
Barnwell, Thomas Osborn, Jr
Bowen, Paul Tyner
Brumund, William Frank
Caseman, A(ustin) Bert
Chang, Chin Hao
Chiu, Kirts C
Circeo, Louis Joseph, Jr
Covault, Donald O
Craig, James I
Fitzgerald, J(ohn) Edmund
Fragaszy, Richard J
Fulton, Robert E(arle)
Galeano, Sergio F(rancis)
Goodno, Barry John
Kahn, Lawrence F
Lnenicka, William J(oseph)
Sangster, William M(cCoy)
Saunders, Fred Michael
Sowers, George F(rederick)
Spetnagel, Theodore John
Stanley, Luticious Bryan, Jr
Stelson, T(homas) E(ugene)
Thomas, Adrian Wesley
Thornton, William Aloysius
Tooles, Calvin W(arren)
Traina, Paul J(oseph)
Whitehurst, Eldridge Augustus

HAWAII
Chiu, Arthur Nang Lick
Cox, Richard Horton
Flachsbart, Peter George
Fok, Yu-Si
Gerritsen, Franciscus
Go, Mateo Lian Poa
Grace, Robert Archibald
Hamada, Harold Seichi
Hufschmidt, Maynard Michael
Krock, Hans-Jürgen
Matsuda, Fujio
Nielsen, N Norby
Ogata, Akio
Papacostas, Constantinos Symeon
Saxena, Narendra K
Shupe, John W(allace)
Taoka, George Takashi
Williams, John A(rthur)
Wu, I-Pai
Young, Reginald H F

IDAHO
Humpherys, Allan S(tratford)
Johnson, Clifton W
Kincaid, Dennis Campbell
Lottman, Robert P(owell)
Sargent, Charles
Schreiber, David Laurence
Swiger, William F
Trout, Thomas James
Wilbur, Lyman D

ILLINOIS
Adams, John Rodger
Al-Khafaji, Amir Wadi Nasif
Ardis, Colby V, Jr
Babcock, Lyndon Ross, (Jr)
Barenberg, Ernest J(ohn)
Bazant, Zdenek P(avel)
Bell, Charles Eugene, Jr
Berger, Richard Lee
Berry, Donald S(tilwell)
Bowles, Joseph Edward
Boyer, LeRoy T
Briscoe, John William
Brotherson, Donald E
Calderon, Alberto Pedro
Chakrabarti, Subrata K
Chugh, Yoginder Paul
Cording, Edward J
Corley, William Gene
Daniel, Isaac M
Davisson, M T
Dempsey, Barry J
Ditmars, John David
Dobrovolny, Jerry S(tanley)
Engelbrecht, R(ichard) S(tevens)
Fonken, David W(alter)
Fraenkel, Stephen Joseph
Fucik, Edward Montford
Gamble, William Leo
Garcia, Marcelo Horacio

Gemmell, Robert S(tinson)
Gerstner, Robert W(illiam)
Ghosh, Satyendra Kumar
Gnaedinger, John P(hillip)
Gurfinkel, German R
Hall, W(illiam) J(oel)
Hall, William Joel
Haltiwanger, John D(avid)
Hanna, Steven J(ohn)
Hanson, John M
Hay, William Walter
Heil, Richard Wendell
Hendron, Alfred J, Jr
Herrin, Moreland
Hofer, Kenneth Emil
Hognestad, Eivind
Holland, Eugene Paul
Holsen, Thomas Michael
Ireland, Herbert O(rin)
Keer, Leon M
Kesler, Clyde E(rvin)
Khachaturian, Narbey
Klieger, Paul
Korn, Alfred
Krizek, Raymond John
Lawrence, Frederick Van Buren, Jr
Liu, Wing Kam
Lopez, Leonard Anthony
Lue-Hing, Cecil
McKee, Keith Earl
Maxwell, William Hall Christie
Maynard, Theodore Roberts
Merkel, Frederick Karl
Mockros, Lyle F(red)
Mosborg, Robert J(ohn)
Muehling, Arthur J
Munse, William H(erman)
Murtha, Joseph P
Muvdi, Bichara B
Paaswell, Robert E(mil)
Paintal, Amreek Singh
Pfeffer, John T
Prakash, Anand
Regunathan, Perialwar
Rittmann, Bruce Edward
Robinson, Arthur R(ichard)
Rodda, Errol David
Roy, Dipak
Russell, Henry George
Sami, Sedat
Saxena, Surendra K
Schlesinger, Lee
Shaffer, Louis Richard
Shah, Surendra P
Siess, Chester P(aul)
Silver, Marshall Lawrence
Smedskjaer, Lars Christian
Sozen, M(ete) A(vni)
Stallmeyer, J(ames) E(dward)
Tang, Wilson H
Thompson, Marshall Ray
Valocchi, Albert Joseph
Walker, William Hamilton
Wenzel, Harry G, Jr
Wong, Kam Wu
Woods, Kenneth R
Yen, Ben Chie
Zenz, David R

INDIANA
Bell, J(ohn) M
Chen, Wai-Fah
Delleur, Jacques W(illiam)
Drnevich, Vincent Paul
Eckelman, Carl A
Fredrich, Augustine Joseph
Gaunt, John Thixton
Goetz, William H(arner)
Graves, Leroy D
Gray, William Guerin
Halpin, Daniel William
Houck, Mark Hedrich
Huang, Nai-Chien
Kareem, Ahsan
Leonards, G(erald) A(llen)
Lewis, A(lbert) D(ale) M(ilton)
Lobo, Cecil T(homas)
Lovell, Charles W(illiam), Jr
McEntyre, John G(erald)
McLaughlin, J(ohn) F(rancis)
Meyers, Vernon J
Michael, Harold Louis
Miles, Robert D(ouglas)
Rao, R(amachandra) A
Satterly, Gilbert T(hompson)
Scholer, Charles Frey
Scott, Marion B(oardman)
Sinha, Kumares C
Skibniewski, Miroslaw Jan
Taylor, Richard L
Tenney, Mark W
Van der Heijde, Paul Karel Maria
Waling, J(oseph) L(ee)
Warder, David Lee
Winslow, Douglas Nathaniel
Wood, Leonard E(ugene)
Wright, Jeffery Regan
Yang, Henry T Y
Yeh, Pai-T(ao)
Yoder, Eldon J

IOWA
Austin, Tom Al
Baumann, E(dward) Robert

Branson, Dan E(arle)
Brewer, Kenneth Alvin
Carstens, Robert L(owell)
Cleasby, John Leroy
Dague, Richard R(ay)
Ekberg, Carl E(dwin), Jr
Fung, Honpong
Georgakakos, Konstantine P
Handy, Richard L(incoln)
Hoover, James M(yron)
Hubbard, Philip G(amaliel)
Jeyapalan, Kandiah
Kane, Harrison
Kennedy, John Fisher
Klaiber, Fred Wayne
Lane, Orris John, Jr
Lee, Dah-Yinn
McCauley, Howard W
Maze, Thomas Harold
Morgan, Paul E(merson)
Nakato, Tatsuaki
Nixon, Wilfrid Austin
Oulman, Charles S
Patel, Virendra C
Sanders, W(allace) W(olfred), Jr
Sayre, William Whitaker
Sheeler, John B(riggs)

KANSAS
Cook, Everett L
Cooper, Peter B(ruce)
Darwin, David
Ellis, Harold Bernard
Huang, Chi-Lung Dominic
Kahn, Charles Howard
Kurt, Carl Edward
Lee, Joe
Lenzen, K(enneth) H(arvey)
McBean, Robert Parker
McCabe, Steven Lee
McKinney, Ross E(rwin)
Moore, Raymond Knox
Roddis, Winifred Mary Kim
Rolfe, Stanley Theodore
Schaper, Laurence Teis
Smith, Bob L(ee)
Smith, Robert Lee
Snell, Robert Ross
Surampalli, Rao Yadagiri
Swartz, Stuart Endsley
Wilhelm, William Jean
Willems, Nicholas
Yu, Yun-Sheng

KENTUCKY
Blythe, David K(nox)
Deen, Robert C(urba)
Gesund, Hans
Hardin, Bobby Ott
Hutchinson, John W(endle)
Kao, David Teh-Yu
Kinman, Riley Nelson
Lauderdale, Robert A(mis), Jr
Man, Chi-Sing
Paz, Mario Meir
Vaziri, Menouchehr
Wang, Shien Tsun
Warner, Richard Charles
Yoder, Elmon Eugene

LOUISIANA
Aguilar, Rodolfo J
Arman, Ara
Benedict, Barry Arden
Bruce, Robert Nolan, Jr
Collins, Michael Albert
Courtney, John Charles
Dalia, Frank J
Grimwood, Charles
Johnson, Lee H(arnie)
Kazmann, Raphael Gabriel
Lee, Griff C
Lemke, Calvin A(ubrey)
McGhee, Terence Joseph
Mayer, J(ohn) K(ing)
Ovunc, Bulent Ahmet
Painter, Jack T(imberlake)
Price, Bobby Earl
Saha, Subrata
Singh, Vijay P
Stopher, Peter Robert
Szabo, A(ugust) J(ohn)
Wilson, Joe Robert
Wintz, William A, Jr

MAINE
Gorrill, William R
Greenwood, George W(atkins)
Kleinschmidt, R Stevens

MARYLAND
Amirikian, Arsham
Austin, John H(enry)
Babrauskas, Vytenis
Basdekas, Nicholas Leonidas
Blevins, R(alph) W(allace)
Browzin, Boris S(ergeevich)
Brush, Lucien M(unson), Jr
Bryan, Edward H
Cattaneo, L(ouis) E(mile)
Chen, Benjamin Yun-Hai
Chuang, Tze-jer
Corn, Morton
Corotis, Ross Barry

Civil Engineering (cont)

Dallman, Paul Jerald
Davidson, Bruce M
Douglas, Andrew Sholto
Duane, David Bierlein
Ellingwood, Bruce Russell
Eny, Desire M(arc)
Frank, Carolyn
Gaum, Carl H
Geyer, John Charles
Green, Richard Stedman
Greenfeld, Sidney H(oward)
Harris, Leonard Andrew
Heins, Conrad P, Jr
Jabbour, Kahtan Nicolas
Linaweaver, Frank Pierce
Loxley, Thomas Edward
Mallory, Charles William
Michalowski, Radoslaw Lucas
Moran, David Dunstan
Morris, Alan
Noblesse, Francis
Ostrom, Thomas Ross
Rosenberg, Arnold Morry
Sabnis, Gajanan Mahadeo
Saville, Thorndike, Jr
Scanlan, Robert Harris
Schlimm, Gerard Henry
Schruben, John H
Shalowitz, Erwin Emmanuel
Simiu, Emil
Statt, Terry G
Steiner, Henry M
Theuer, Paul John
Tholen, Albert David
Wolman, Abel
Wright, Richard N(ewport)
Yaramanoglu, Melih
Yokel, Felix Y

MASSACHUSETTS
Aldrich, Harl P, Jr
Ben-Akiva, Moshe E
Berger, Bernard Ben
Blanc, Frederic C
Bras, Rafael Luis
Breuning, Siegfried M
Brocard, Dominique Nicolas
Brown, Linfield Cutter
Carver, Charles E(llsworth), Jr
Casagrande, Leo
Chalabi, A Fattah
Chang, Ching Shung
Chen, Ming M
Collura, John
De Neufville, Richard Lawrence
Dietz, Albert (George Henry)
Entekhabi, Dara
Fiering, Myron B
Firnkas, Sepp
Fitzgerald, Robert William
Garrelick, Joel Marc
Gartner, Nathan Hart
Gumpertz, Werner H(erbert)
Hansen, R(obert) J(oseph)
Hecker, George Ernst
Hirschfeld, Ronald Colman
Hope, Elizabeth Greeley
Kachinsky, Robert Joseph
Kelsey, Ronald A(lbert)
Keshavan, Krishnaswamiengar
Ladd, Charles Cushing, III
LeMessurier, William James
Liepins, Atis Aivars
Lipner, Steven Barnett
Logcher, Robert Daniel
Madsen, Ole Secher
Male, James William
Marks, David Hunter
Marston, George Andrews
Mason, William C
Mei, Chiang C(hung)
Miller, Melton M, Jr
Moavenzadeh, Fred
Murphy, Peter John
Ostendorf, David William
Platt, Milton M
Raab, Allen Robert
Reinschmidt, Kenneth F(rank)
Schwartz, Thomas Alan
Selig, Ernest Theodore
Sheffi, Yosef
Shuldiner, Paul W(illiam)
Stolzenbach, Keith Densmore
Sutcliffe, Samuel
Tarkoy, Peter J
Thomas, Harold A(llen), Jr
Thompson, Charles
Webster, Lee Alan
White, Merit P(enniman)
Whitman, Robert V(an Duyne)
Wilson, Nigel Henry Moir

MICHIGAN
Andersland, Orlando Baldwin
Armstrong, John Morrison
Baillod, Charles Robert
Beaufait, Frederick W
Berg, Glen V(irgil)
Bergmann, Dietrich R(udolf)
Borchardt, Jack A
Bradley, William Arthur
Bulkley, Jonathan William

Canale, Raymond Patrick
Cleveland, Donald Edward
Cutts, Charles E(ugene)
Datta, Tapan K
Hanson, Norman Walter
Hanson, Robert D(uane)
Huang, Eugene Yuching
Johnston, Clair C
Kaldjian, Movses J(eremy)
Khasnabis, Snehamay
Knoll, Alan Howard
Ku, Albert B
Lubkin, James Leigh
McCauley, Robert F(orrestelle)
Morman, Kenneth N
Paulson, James M(arvin)
Prasuhn, Alan Lee
Richart, F(rank) E(dwin), Jr
Saul, William Edward
Shkolnikov, Moisey B
Smith, Hadley J(ames)
Snyder, Virgil W(ard)
Watwood, Vernon Bell, Jr
Woods, Richard David
Wu, Kuang Ming
Wylie, Evan Benjamin

MINNESOTA
Allred, E(van) R(ich)
Arndt, Roger Edward Anthony
Boening, Paul Henrik
Bowers, C(harles) Edward
Galambos, Theodore V
Goodman, L(awrence) E(ugene)
Gulliver, John Stephen
Johnson, Gerald Winford
Kersten, Miles S(tokes)
Ling, Joseph Tso-Ti
Reynolds, James Harold
Silberman, Edward
Stephanedes, Yorgos Jordan
Sterling, Raymond Leslie
Straub, Conrad P(aul)

MISSISSIPPI
Abdulrahman, Mustafa Salih
Bond, Marvin Thomas
Bowie, Andrew J(ackson)
Brown, Fred(erick) R(aymond)
Corey, Marion Willson
Cunny, Robert William
DeLeeuw, Samuel Leonard
George, Kalankamary Pily
Hudson, Robert Y(oung)
Johnson, L(awrence) D(avid)
McCrae, John Leonidas
Mather, Bryant
Perry, Edward Belk
Priest, Melville S(tanton)
Robinson, A(ugust) R(obert)
Shindala, Adnan
Shockley, W(oodland) G(ray)
Wang, Sam S(hu) Y(i)
Woodburn, Russell

MISSOURI
Andrews, William Allen
Baldwin, James W(arren), Jr
Banerji, Shankha K
Barr, David John
Cheng, Franklin Yih
Cronin, James Lawrence, Jr
Douty, Richard T
Edgerley, Edward, Jr
Emanuel, Jack Howard
Fowler, Timothy John
Gevecker, Vernon A(rthur) C(harles)
Goran, Robert Charles
Gould, Phillip L
Guell, David Lee
Hatheway, Allen Wayne
Hauck, George F(rederick) W(olfgang)
Heagler, John B(ay), Jr
Hemphill, Louis
Hjelmfelt, Allen T, Jr
Hoffman, Jerry C
Jawad, Maan Hamid
Jester, Guy Earlscort
Liu, Henry
McCarthy, John F(rancis)
Mains, Robert M(arvin)
Mansur, Charles I(saiah)
Munger, Paul R
O'Connor, John Thomas
Prakash, Shamsher
Robinson, Thomas B
Rockaway, John D, Jr
Salmons, John Robert
Schmidt, Norbert Otto
Schwartz, Henry Gerard, Jr
Senne, Joseph Harold, Jr
Szabo, Barna Aladar
Yu, Wei Wen

MONTANA
Friel, Leroy Lawrence
Hunt, William A(lfred)
Peavy, Howard Sidney
Scheer, Alfred C(arl)
Videon, Fred F(rancis)
Walker, Leland J

NEBRASKA
Andersen, Dewey Richard

Hammer, Mark J(ohn)
Marlette, Ralph R(oy)
Phelps, George Clayton
Riveland, A(rvin) R(oy)
Swihart, G(erald) R(obert)
Tunnicliff, David George

NEVADA
Brandstetter, Albin
Daemen, Jaak J K
Epps, Jon Albert
Ghosh, Amitava
Guitjens, Johannes C
Helm, Donald Cairney
Merdinger, Charles J(ohn)
Nelson, R William
Nielsen, John Palmer
Orcutt, Richard D(atton)
Thornburn, Thomas H(ampton)

NEW HAMPSHIRE
Batchelder, Gerald M(yles)
Cox, Gordon F N
Klotz, Louis Herman
Long, Carl F(erdinand)
Sproul, Otis J
Stearns, S(tephen) Russell
Wilson, Donald Alfred

NEW JERSEY
Balaguru, Perumalsamy N
Becht, Charles, IV
Billington, David Perkins
Blumberg, Alan Fred
Borg, Sidney Fred
Bruno, Michael Stephen
Chae, Yong Suk
Chan, Paul C
Cheng, David H(ong)
Derucher, Kenneth Noel
Domeshek, S(ol)
Esrig, Melvin I
Garrelts, Jewell Milan
Genetelli, Emil J
Gennaro, Joseph J(ohn)
Granstrom, Marvin L(e Roy)
Haefeli, Robert J(ames)
Horn, Harry Moore
Iyer, Ram R
Kagan, Harvey Alexander
Killam, Everett Herbert
Kim, Uing W
Kraft, Walter H
Lestingi, Joseph Francis
McCarthy, Gerald T(imothy)
Majumdar, Dalim K
Mark, Robert
Monahan, Edward James
Nathan, Kurt
Nawy, Edward George
Pfafflin, James Reid
Pignataro, Louis J(ames)
Pincus, George
Prevost, Jean Herve
Schmid, Werner E(duard)
Severud, Fred N
Spillers, William R
Thomann, Robert V
Vanmarcke, Erik Hector
Wiesenfeld, Joel

NEW MEXICO
Baerwald, John E(dward)
Bleyl, Robert Lingren
Christensen, N(ephi) A(lbert)
Clough, Richard H(udson)
Colp, John Lewis
Dillon, Robert Morton
Fuka, Louis Richard
Gafford, William R(odgers)
Hall, Jerome William
Hernandez, John W(hitlock)
Howell, Gregory A
Hulme, Bernie Lee
Hulsbos, C(ornie) (Leonard)
Lunsford, Jesse V(ernon)
Maggard, Samuel P
Martinez, Jose E(leazar)
Peck, Ralph B(razelton)
Perret, William Riker
Ross, Timothy Jack
Tremba, Edward Louis
Triandafilidis, George Emmanuel
Varan, Cyrus O
Von Riesemann, Walter Arthur
Yamada, Tetsuji
Zimmerman, Roger M

NEW YORK
Abel, John Fredrick
Ahmad, Jameel
Alpern, Milton
Alvarez, Ronald Julian
Armenakas, Anthony Emanuel
Banerjee, Prasanta Kumar
Baron, Melvin L(eon)
Barry, B(enjamin) Austin
Batson, Gordon B
Bieniek, Maciej P
Binger, Wilson Valentine
Birnstiel, Charles
Brandt, G(eorge) Donald
Brennan, Paul Joseph
Brown, William Augustin

Cantilli, Edmund Joseph
Cataldo, Joseph C
Clemence, Samuel Patton
Cohen, Edward
Constantinou, Michalakis
Dasgupta, Gautam
Deleanu, Aristide Alexandru-Ion
Dempsey, John Patrick
Deresiewicz, Herbert
Dicker, Daniel
Di Maggio, Frank Louis
Dobry, Ricardo
Ezra, Arthur Abraham
Feeser, Larry James
Fisher, Gordon P(age)
Florman, Samuel C
Fogel, Charles M(orton)
Fox, George A
Gallagher, Richard Hugo
Gergely, Peter
Gossett, James Michael
Grigoriu, Mircea Dan
Haines, Daniel Webster
Harlow, H(enry) Gilbert
Hazen, Richard
Helander, Martin Erik Gustav
Hodge, Raymond
Huddleston, John Vincent
Irwin, Lynne Howard
Isada, Nelson M
Jordan, Mark H(enry)
Kaarsberg, Ernest Andersen
Kahn, Elliott H
Ketter, Robert L(ewis)
Kulhawy, Fred Howard
Lai, W(ei) Michael
Lawler, John Patrick
Lee, George C
Lee, Robert Bumjung
Libove, Charles
Liggett, James Alexander
Lowe, John, III
Lynn, Walter R(oyal)
McDonald, Donald
McGuire, William
Mahtab, M Ashraf
Mandel, James A
Mead, Lawrence Myers
Meredith, Dale Dean
Meyburg, Arnim Hans
Meyer, Christian
Mijovic, Jovan
Miller, George Maurice
Morabito, Bruno P
Neal, John Alva
Nilson, Arthur H(umphrey)
O'Connor, Donald J
O'Rourke, Thomas Denis
Palevsky, Gerald
Pao, Yih-Hsing
Pekoz, Teoman
Pfrang, Edward Oscar
Prawel, Sherwood Peter, Jr
Raamot, Tonis
Rihm, Alexander, Jr
Robertson, Leslie Earl
Rumer, Ralph R, Jr
Salvadori, M(ario) G(iorgio)
Shaw, Richard P(aul)
Sherman, Zachary
Slate, Floyd Owen
Soifer, Herman
Soong, Tsu-Teh
Spencer, James W(endell)
Steven, James R
Swanson, Robert Lawrence
Tatlow, Richard H(enry), III
Tedesko, Anton
Testa, Rene B(iaggio)
Turkstra, Carl J
Turnquist, Mark Alan
Velzy, Charles O
Vemula, Subba Rao
Wang, Lawrence K
Wang, Muhao S
Wang, Ping Chun
Weidlinger, Paul
White, Richard Norman
Zimmie, Thomas Frank

NORTH CAROLINA
Amein, Michael
Brown, Earl Ivan, II
Burton, Ralph Gaines
Chanlett, Emil T(heodore)
Clark, Charles Edward
Cribbins, Paul Day
Evett, Jack B(urnie)
Fadum, Ralph Eigil
Fisher, C(harles) Page, Jr
Galler, William Sylvan
Glenn, George R(embert)
Gupta, Ajaya Kumar
Horn, J(ohn) W(illiam)
Humenik, Frank James
Iverson, F Kenneth
Kashef, A(bdel-Aziz) I(smail)
King, L(ee) Ellis
Lamb, James C(hristian), III
Matzen, Vernon Charles
Medina, Miguel Angel, Jr
Miranda, Constancio F
Pas, Eric Ivan
Peirce, James Jeffrey

Petroski, Henry J
Sharkoff, Eugene Gibb
Smallwood, Charles, Jr
Smith, J C
Tung, Chi Chao
Utku, Bisulay Bereket
Wahls, Harvey E(dward)
Zia, Paul Z

NORTH DAKOTA
Fossum, Guilford O
Mason, Earl Sewell
Phillips, Monte Leroy

OHIO
Adeli, Hojjat
Archer, Lawrence H(arry)
Bakos, Jack David, Jr
Bishop, Paul Leslie
Bondurant, Byron L(ee)
Cernica, John N
Chan, Yupo
Chawla, Mangal Dass
Chen, Chao-Hsing Stanley
Chen, T(ien) Y(i)
Cook, John P(hilip)
Cosens, Kenneth W
Crum, Ralph G
Eye, John David
Fertis, Demeter G(eorge)
Fleck, William G(eorge)
Fok, Thomas Dso Yun
Fu, Kuan-Chen
Gallo, Frank J
Hobbs, Benjamin F
Howe, Robert T(heodore)
Hyland, John R(oth)
Kaneshige, Harry Masato
Kellerstrass, Ernst Junior
Koo, Benjamin
Korda, Peter E
Lai, Jai-Lue
Laushey, Louis M(cNeal)
Leis, Brian Norman
McDonough, James Francis
Minich, Marlin
Nemeth, Zoltan Anthony
Ojalvo, Morris
Olesen, Douglas Eugene
Palazotto, Anthony Nicholas
Pao, Richard H(sien) F(eng)
Papadakis, Constantine N
Preul, Herbert C
Rai, Iqbal Singh
Ricca, Vincent Thomas
Sandhu, Ranbir Singh
Shah, Kanti L
Simon, Andrew L
Stiefel, Robert Carl
Treiterer, Joseph
Venkayya, Vipperla
Wandmacher, Cornelius
Ward, Roscoe Fredrick
Wu, Tien Hsing

OKLAHOMA
Abdel-Hady, M(ohamed) A(hmed)
Blaisdell, Fred W(illiam)
Brady, Barry Hugh Garnet
Canter, Larry Wayne
Dawkins, William Paul
Dunn, Clark A(llan)
Earls, James Roe
Forney, Bill E
Haliburton, T(racey) Allan
Harp, Jimmy Frank
Manke, Phillip Gordon
Parcher, James V(ernon)
Ree, William O(scar)
Reid, George W(illard)
Sack, Ronald Leslie
Stover, Enos Loy
Streebin, Leale E
Summers, Gregory Lawson

OREGON
Boucher, Raymond
Burgess, Fred J
Erzurumlu, H Chik
Klingeman, Peter C
Kocaoglu, Dundar F
Lall, B Kent
Mueller, Wendelin Henry, III
Rad, Franz N
Schroeder, Warren Lee
Sollitt, Charles Kevin
Tewinkel, G Carper

PENNSYLVANIA
Abrams, Joel Ivan
Aron, Gert
Au, Tung
Barnoff, Robert Mark
Batterman, Steven C(harles)
Becker, Stanley J
Beedle, Lynn Simpson
Bielak, Jacobo
Bieniawski, Zdzislaw Tadeusz
Bjorhovde, Reidar
Bronzini, Michael Stephen
Brungraber, Robert J
Bullen, Allan Graham Robert
Cady, Philip Dale
Christiano, Paul P

Clark, John W(ood)
Condo, Albert Carman, Jr
Daniels, John Hartley
D'Appolonia, Elio
Davinroy, Thomas Bernard
Dejaiffe, Ernest
Driscoll, George C
Durkee, Jackson L
Elsworth, Derek
Errera, Samuel J(oseph)
Fenves, Steven J(oseph)
Fisher, John William
Fleming, John F
Goel, Ram Parkash
Gotolski, William H(enry)
Hamel, James V(ictor)
Hartmann, Alois J(oseph)
Haythornthwaite, Robert M(orphet)
Hendrickson, Chris Thompson
Higgins, Frederick B(enjamin), Jr
Hribar, John Anthony
Kim, Jai Bin
Knaster, Tatyana
Koliner, Ralph
LaGrega, Michael Denny
Lepore, John A(nthony)
Lin, Larry Y H
Loigman, Harold
Lorsch, Harold G
Lu, Le-Wu
Luthy, Richard Godfrey
McClure, Richard Mark
McDonnell, Archie Joseph
McLaughlin, Philip V(an Doren), Jr
McMichael, Francis Clay
McNamee, Bernard M
Mangelsdorf, Clark P
Mecholsky, John Joseph, Jr
Myers, Earl A(braham)
Nesbitt, John B
Ostapenko, A(lexis)
Pangborn, Robert Northrup
Plonsky, Andrew Walter
Popovics, Sandor
Quimpo, Rafael Gonzales
Rachford, Thomas Milton
Ramanathan, M
Reed, Joseph Raymond
Regan, Raymond Wesley
Rolf, Richard L(awrence)
Roll, Frederic
Romualdi, James P
Roy, Pradip Kumar
Rumpf, John L
Saigal, Sunil
Samples, William R(ead)
Sangrey, Dwight A
Schuster, James J(ohn)
Sooky, Attila A(rpad)
Swedlow, Jerold Lindsay
Thompson, Ansel Frederick, Jr
Untrauer, Raymond E(rnest)
Van Horn, D(avid) A(lan)
Viest, Ivan M
Viscomi, B(runo) Vincent
Voight, Barry
Vuchic, Vukan R
Weggel, John Richard
White, Elizabeth Loczi
Wright, Thomas Wilson
Yen, Ben-Tseng
Younkin, Larry Myrle
Zandi, Iraj

RHODE ISLAND
Clifton, Rodney James
Dafermos, Stella
McEwen, Everett E(dwin)
Poon, Calvin P C
Pretzer, C Andrew
Silva, Armand Joseph

SOUTH CAROLINA
Alexander, William Davidson, III
Anand, Subhash Chandra
Anderson, Thomas L(eonard)
Awadalla, Nabil G
Bridges, Donald Norris
Clark, J(ames) Edwin
Dysart, Benjamin Clay, III
Grady, Cecil Paul Leslie, Jr
Henry, Harold Robert
Jackson, John Elwin, Jr
Jennett, Joseph Charles
Ledbetter, W(illiam) B(url)
McCormac, Jack Clark
Schwartz, Arnold Edward
Sparks, Peter Robert

SOUTH DAKOTA
Andersen, John R(obert)
Johnson, Emory Emanuel
Koepsell, Paul L(oel)
Ramakrishnan, Venkataswamy
Wegman, Steven M
Zogorski, John Steward

TENNESSEE
Armentrout, Daryl Ralph
Bonner, William Paul
Brewington, Percy, Jr
Brown, Bevan W, Jr
Cooper, Alfred J(oseph)
Foster, Edwin Powell

Freitag, Dean R(ichard)
Gray, William Harvey
Grecco, William L
Green, Robert S(mith)
Harrawood, Paul
Helweg, Otto Jennings
Hensley, Marble John, Sr
Humphreys, Jack Bishop
Madhavan, Kunchithapatham
Morrison, Thomas Golden
Perry, Randy L
Rowe, Robert S(eaman)
Rozenberg, J(uda) E(ber)
Sapirie, S(amuel) R(alph)
Schilling, Charles H(enry)
Smith, Dallas Glen, Jr
Tschantz, Bruce A
Wagner, Aubrey Joseph
Wessenauer, Gabriel Otto
Wheeler, Orville Eugene

TEXAS
Andrews, John Frank
Armstrong, James Clyde
Austin, Walter J
Bartel, Herbert H(erman), Jr
Beale, Luther A(lton)
Beard, Leo Roy
Bendapudi, Kasi Visweswararao
Benson, Fred J(acob)
Betts, Austin Wortham
Bigham, Robert Eric
Boyer, Robert Elston
Breen, John E(dward)
Brown, Daniel Mason
Brown, Robert Wade
Burns, Ned Hamilton
Buth, Carl Eugene
Caffey, James E
Charbeneau, Randall Jay
Clark, Robert Alfred
Collipp, Bruce Garfield
Coyle, Harry Michael
DeHart, Robert C(harles)
Delflache, Andre P
Everard, Noel James
Focht, John Arnold, Jr
Furlong, Richard W
Gallaway, Bob Mitchel
George, James Francis
Geyling, F(ranz) Th(omas)
Gloyna, Earnest F(rederick)
Graff, William J(ohn), Jr
Grieves, Robert Belanger
Hadley, William Owen
Halff, Albert Henry
Hann, Roy William, Jr
Haynes, John J
Herbich, John Bronislaw
Hirsch, Teddy James
Holcomb, Robert M(arion)
Holliday, George Hayes
Holt, Edward C(hester), Jr
Houston, Clyde Erwin
Huang, Tseng
Hudson, William Ronald
Ivey, Don Louis
Jirsa, James O
Junkins, Jerry R
Kennedy, Thomas William
Kiesling, Ernst W(illie)
Koehn, Enno
Kohl, John C(layton)
Krahl, Nat W(etzel)
Lazar, Benjamin Edward
Li, Wen-Hsiung
Loehr, Raymond Charles
Lowery, Lee Leon, Jr
Lu, Frank Kerping
Ludwig, Allen Clarence, Sr
Lytton, Robert Leonard
McClelland, Bramlette
McCullough, Benjamin Franklin
Machemehl, Jerry Lee
Mak, King Kuen
Manning, Sherrell Dane
Mathewson, Christopher Colville
Matlock, Hudson
Maxwell, Donald A
Mays, Larry Wesley
Michie, Jarvis D
Mohraz, Bijan
Monsees, James E
Moore, Walter L(eon)
Morgan, Carl William
Morgan, James Richard
Naismith, James Pomeroy
Nicastro, David Harlan
Olson, Roy Edwin
O'Neill, Michael Wayne
Parate, Natthu Sonbaji
Pavlovich, Raymond Doran
Phillips, Joseph D
Pinnell, Charles
Price, William Charles
Qasim, Syed Reazul
Rao, Shankaranarayana
    Ramohallinanjunda
Reddell, Donald Lee
Reese, Lymon C(lifton)
Richardson, Clarence Wade
Rogers, Bruce G(eorge)
Rozendal, David Bernard
Samson, Charles Harold

Shah, Haresh C
Simonis, John Charles
Sims, James R(edding)
Slotta, Larry Stewart
Spindler, Max
Stanovsky, Joseph Jerry
Streltsova, Tatiana D
Stubbs, Norris
Tarquin, Anthony Joseph
Thompson, Louis Jean
Truitt, Marcus M(cCafferty)
Tucker, Richard Lee
Vallabhan, C V Girija
Vann, W(illiam) Pennington
Veletsos, A(nestis)
Wah, Thein
Walton, Charles Michael
Weyler, Michael E
White, N(ikolas) F(rederick)
Williams, Glen Nordyke
Wolf, Harold William
Wright, Stephen Gailord
Yao, James T-P

UTAH
Allen, Richard Glen
Anderson, Bruce Holmes
Anderson, Douglas I
Bagley, Jay M(errill)
Bishop, A(very) A(lvin)
Christiansen, J(erald) E(mmett)
Clyde, Calvin G(eary)
Enke, Glenn L
Firmage, D(avid) Allan
Flammer, Gordon H(ans)
Fuhriman, D(ean) K(enneth)
Ghosh, Sambhunath
Grenney, William James
Hansen, Vaughn Ernest
Hargreaves, George H(enry)
Hoggan, Daniel Hunter
Israelsen, C Earl
Jeppson, Roland W
Karren, Kenneth W
Keller, Jack
Merritt, LaVere Barrus
Nordquist, Edwin C(lyde)
Olsen, Edwin Carl, III
Peralta, Richard Carl
Peterson, Dean F(reeman), Jr
Pyper, Gordon R(ichardson)
Rich, Elliot
Riley, John Paul
Sidle, Roy Carl
Tullis, J Paul
Vyas, Ravindra Kantilal
Waddell, Kidd M
Yu, Jason C

VERMONT
Butler, Donald J(oseph)
Cassell, Eugene Alan
Olson, James Paul
Oppenlander, Joseph Clarence
Pinder, George Francis
Saxe, Harry Charles

VIRGINIA
Blakey, Lewis Horrigan
Boardman, Gregory Dale
Buchanan, Thomas Joseph
Carrigan, P H, Jr
Chang, George Chunyi
Chi, Michael
Clarke, Frederick James
Cole, Ralph I(ttleson)
Cox, William Edward
Cragwall, J(oseph) S(amuel), Jr
Curtin, Robert H
Dendrou, Stergios
Diercks, Frederick O(tto)
Dobyns, Samuel Witten
Douma, Jacob H
Doyle, Frederick Joseph
Drewry, William Alton
Duncan, James M
Echols, Charles E(rnest)
Eppink, Richard Theodore
Estes, Edward Richard
Everett, Warren S
Fearnsides, John Joseph
Ferguson, George E(rnest)
Galvin, Cyril Jerome, Jr
Goins, Truman
Gray, George A(lexander)
Hakala, William Walter
Hardrath, Herbert Frank
Heller, Robert A
Heterick, Robert Cary, Jr
Higgins, Thomas Ernest
Hobeika, Antoine George
Hoel, Lester A
Holzer, Siegfried Mathias
Houbolt, John C(ornelius)
Hudson, Charles Michael
Jennings, Richard Louis
Knapp, John Williams
Kreipke, Merrill Vincent
Kuesel, Thomas Robert
Kuhlthau, A(lden) R(obert)
Lai, Chintu (Vincent C)
Larew, H(iram) Gordon
Lee, John A N
Lewis, Russell M(acLean)

Civil Engineering (cont)

Liu, Tony Chen-Yeh
McCormick, Fred C(ampbell)
McEwen, Robert B
Moore, Joseph Herbert
Morris, David
Morris, John Woodland
Ni, Chen-Chou
Noor, Ahmed Khairy
Nordlund, R(aymond) L(ouis)
Pasko, Thomas Joseph, Jr
Philleo, Robert Eugene
Pilkey, Walter David
Pletta, Dan Henry
Reilly, Thomas E
Rhode, Alfred S
Rojiani, Kamal B
Scalzi, John Baptist
Schad, Theodore M(acNeeve)
Schaefer, Francis T
Seelig, Jakob Williams
Simpson, James R(ussell)
Singh, Mahendra Pal
Snyder, Franklin F
Tsai, Frank Y
Wagner, John Edward
Walker, Richard David
Walker, William R
Wang, Leon Ru-Liang
Whitlock, Charles Henry
Wieczorek, Gerald Francis
Worrall, Richard D
Yang, Ta-Lun
Zuk, William

WASHINGTON
Amberg, Herman R(obert)
Anderson, Arthur R(oland)
Anderson, Richard Gregory
Bacon, Vinton Walker
Bender, Donald Lee
Berg, Richard Harold
Bhagat, Surinder Kumar
Bogan, Richard Herbert
Brown, Colin Bertram
Burges, Stephen John
Carlson, Dale Arvid
Carpenter, James E(dwin)
Chace, Alden Buffington, Jr
Christiansen, John V
Colcord, J E
Cole, Jon A(rthur)
De Goes, Louis
Edde, Howard Jasper
Evans, Roger James
Glenne, Bard
Hawkins, Neil Middleton
Hennes, Robert G(raham)
Horwood, Edgar M(iller)
Jansen, Robert Bruce
Janssen, Allen S
Kent, Joseph C(han)
King, Larry Gene
Law, Albert G(iles)
Luck, Leon D(an)
Mylroie, Willa W
Nece, Ronald Elliott
Nielson, Lyman J(ulius)
Pilat, Michael Joseph
Robbins, Richard J
Roberson, John A(rthur)
Roeder, Charles William
Rossano, August Thomas
Sawhill, Roy Bond
Shoemaker, Roy H(opkins)
Skilling, John Bower
Skrinde, Rolf T
Sorensen, Harold C(harles)
Sylvester, Robert O(hrum)
Terrel, Ronald Lee
Tichy, Robert J
Vasarhelyi, Desi F
Wenk, Edward, Jr

WEST VIRGINIA
Conway, Richard A
Jenkins, Charles Robert
Kemp, Emory Leland
McCaskey, A(mbrose) E(verett)
Thornton, Stafford E

WISCONSIN
Appel, David W(oodhull)
Bartel, Fred F(rank)
Bauer, Kurt W
Berthouex, Paul Mac
Boyle, William C(harles)
Busby, Edward Oliver
Christensen, Erik Regnar
Crabtree, Koby Takayashi
Crandall, Lee W(alter)
Day, Harold J
Edil, Tuncer Berat
Faherty, Keith F
Freas, Alan D('Yarmett)
Hoopes, John A
Johnson, John E(dwin)
Karadi, Gabor
Katz, William J(acob)
Kiefer, Ralph W
Kipp, Raymond J
Klus, John P
Larkin, Lawrence A(lbert)

Lillesand, Thomas Martin
Murphy, William G(rove)
Naik, Tarun Ratilal
Polkowski, Lawrence B(enjamin)
Roderick, Gilbert Leroy
Salmon, Charles G(erald)
Schaffer, Erwin Lambert
Scherz, James Phillip
Shaikh, A(bdul) Fattah
Sherman, D(onald) R
Wolf, Paul R
Zahn, John J
Zanoni, Alphonse E(ligius)

WYOMING
Bellamy, John C
Dolan, Charles W
Hoyt, Philip M(unro)
Humenick, Michael John, II
Lamb, Donald R(oy)
Rechard, Paul A(lbert)
Sailor, Samuel
Smith, James Lee
Tung, Yeou-Koung
Wilson, Eugene Madison

PUERTO RICO
Del Valle, Luis A
Lluch, Jose Francisco
Santiago-Melendez, Miguel

ALBERTA
Bakker, Jaap Jelle
De Paiva, Henry Albert Rawdon
Ellyin, Fernand
Epstein, Marcelo
Glockner, Peter G
Hunt, Robert Nelson
Johnston, Colin Deane
Joshi, Ramesh Chandra
Kivisild, H(ans) R(obert)
Kulak, Geoffrey Luther
Loov, Robert Edmund
MacGregor, James Grierson
Morgenstern, N(orbert) R
Murray, David William
Nasser, Tourai
Rajaratnam, N(allamuthu)
Scott, J(ohn) D(onald)
Simmonds, Sidney Herbert
Thomson, Stanley

BRITISH COLUMBIA
Bell, H(arry) R(ich)
Cherry, S(heldon)
Johnson, Joe W
Lipson, Samuel L(loyd)
Mindess, Sidney
Ruus, E(ugen)
Sexsmith, Robert G

MANITOBA
Baracos, Andrew
Lansdown, A(llen) M(aurice)
Rizkalla, Sami H
Soliman, Afifi Hassan
Sparling, Arthur Bambridge

NEW BRUNSWICK
Davar, K(ersi) S
Faig, Wolfgang
Lin, Kwan-Chow
Wells, David Ernest

NEWFOUNDLAND
Sharp, James Jack

NOVA SCOTIA
Adams, Peter Frederick Gordon
Sastry, Vankamamidi VRN

ONTARIO
Batchelor, B(arrington) D(eVere)
Bewtra, Jatinder Kumar
Beylerian, Nurel
Byer, Philip Howard
Chang, Paul Peng-Cheng
Coakley, John Phillip
Davenport, Alan Garnett
Davis, Merritt McGregor
Edwards, H(erbert) M(artell)
Ellis, J S
Gracie, G(ordon)
Handa, V(irender) K(umar)
Hipel, Keith William
Hope, Brian Bradshaw
Humar, Jagmohan Lal
Huseyin, Koncay
Irwin, Peter Anthony
Jones, L(lewellyn) E(dward)
Kamphuis, J(ohn) William
Kennedy, John B
Kennedy, Russell Jordan
Kenney, T Cameron
Khan, Ata M
Kirk, Donald Wayne
Krishnappa, Govindappa
Krishnappan, Bommanna Gounder
Leutheusser, H(ans) J(oachim)
Litvan, Gerard Gabriel
McCorquodale, John Alexander
MacInnis, Cameron
McLaughlin, Wallace Alvin
McLeod, Norman William

McNeice, Gregory Malcolm
Matyas, E(lmer) Leslie
Monforton, Gerard Roland
Ng, Simon S F
Novak, Milos
Poucher, Mellor Proctor
Quigley, Robert Murvin
Raymond, Gerald Patrick
Roorda, John
Rowe, Ronald Kerry
Selvadurai, A P S
Sharan, Shailendra Kishore
Sherbourne, Archibald Norbert
Tan, Peter Ching-Yao
Thompson, John Carl
Timusk, John
Topper, T(imothy) H(amilton)
Wilson, Kenneth Charles
Wright, Douglas Tyndall
Wright, Peter Murrell

QUEBEC
Bourque, Paul N(orbert)
Brandenberger, Arthur J
Broughton, Robert Stephen
Brunelle, Paul-Edouard
Douglass, Matthew McCartney
Gallez, Bernard
Giroux, Yves M(arie)
Hanna, Adel
Harris, P(hilip) J(ohn)
Johns, Kenneth Charles
Joly, George W(ilfred)
Ladanyi, Branko
McKyes, Edward
Marsh, Cedric
Pekau, Oscar A
Poorooshasb, Hormozd Bahman
Redwood, R(ichard) G(eorge)
Stathopoulos, Theodore
Troitsky, Michael S(erge)

SASKATCHEWAN
Curtis, Fred Allen
Fuller, Gerald Arthur
Hosain, Mahbub Ul
Mollard, John D
Nasser, Karim Wade
Viraraghavan, Thiruvenkatachari

OTHER COUNTRIES
Ersoy, Ugur
Finkel, Herman J(acob)
Hansen, Torben Christen
Jones, P(hilip) H(arrhy)
Laura, Patricio Adolfo Antonio
Lohtia, Rajinder Paul
McNown, John S
Thurlimann, Bruno

Electrical Engineering

ALABAMA
Aldridge, Melvin D(ayne)
Boland, Joseph S(amuel), III
Deck, Howard Joseph
Dezenberg, George John
Dodd, Curtis Wilson
Dowdle, Joseph C(lyde)
Gilbert, Stephen Marc
Graf, E(dward) R(aymond)
Grigsby, Leonard Lee
Haeussermann, Walter
Halijak, Charles A(ugust)
Honnell, Martial A(lfred)
Houts, Ronald C(arl)
Irwin, John David
Jaeger, Richard Charles
Johnson, David Edsel
Johnson, Johnny R(ay)
Kowel, Stephen Thomas
Lovingood, Judson Allison
Lowry, James Lee
Lueg, Russell E
McAuley, Van Alfon
McDuff, Odis P(elham)
MacIntyre, John R(ichard)
Moore, Fletcher Brooks
Morley, Lloyd Albert
Mott, Harold
Nalley, Donald Woodrow
Oglesby, Sabert, Jr
Padulo, Louis
Polge, Robert J
Poularikas, Alexander D
Randall, Joseph Lindsay
Rekoff, M(ichael) G(eorge), Jr
Rouse, John Wilson, Jr
Thurstone, Robert Leon

ALASKA
Bates, Howard Francis
Kokjer, Kenneth Jordan
Sweet, Larry Ross

ARIZONA
Akers, Lex Alan
Andeen, Richard E
Bahill, A(ndrew) Terry
Balanis, Constantine A
Barnes, Howard Clarence
Bose, Anjan
Carlile, Robert Nichols

Choi, Kwan-Yiu Calvin
Clark, Wilburn O
Collins, F(red) R(obert)
DeMassa, Thomas A
Dereniak, Eustace Leonard
Dickieson, Alton C
Eberhard, Everett
El-Ghazaly, Samir M
Fahey, Walter John
Ferry, David K
Findler, Nicholas Victor
Fliegel, Frederick Martin
Fordemwalt, James Newton
Galloway, Kenneth Franklin
Gerhard, Glen Carl
Gilkeson, Robert Fairbairn
Glenner, E J
Graham, Le Roy Cullen
Haden, Clovis Roland
Hamilton, Douglas J(ames)
Handy, Robert M(axwell)
Haynes, Munro K
Hessemer, Robert A(ndrew), Jr
Hill, Fredrick J
Hoehn, A(lfred) J(oseph)
Hoenig, Stuart Alfred
Honnell, Pierre M(arcel)
Howard, William Gates, Jr
Huelsman, Lawrence Paul
Huffman, Tommie Ray
Hunt, Bobby Ray
Johnson, Vern Ray
Karady, George Gyorgy
Kaufman, Irving
Kelly, Richard W(alter)
Kerwin, William J(ames)
Lonsdale, Edward Middlebrook
McRae, Lorin Post
Metz, Donald C(harles)
Milnes, Dale J
Mulvey, James Patrick
Myers, Ronald G
Mylrea, Kenneth C
Nabours, Robert Eugene
Palais, Joseph C(yrus)
Palusinski, Olgierd Aleksander
Peterson, Harold A(lbert)
Porcello, Leonard J(oseph)
Prince, John Luther, III
Ray, Howard Eugene
Reagan, John Albert
Ren, Shang Yuan
Rugge, Raymond A(lbert)
Russell, Paul E(dgar)
Ryan, Thomas Wilton
Schroder, Dieter K
Schultz, Donald Gene
Stott, Brian
Szilagyi, Miklos Nicholas
Thompson, Truet B(radford)
Tice, Thomas E(arl)
Ulich, Bobby Lee
Wait, John V
Welch, H(omer) William
White, Nathaniel Miller
Williams, Theodore L
Witulski, Arthur Frank
Wolfe, William Louis, Jr
Zeigler, Bernard Philip
Zimmer, Carl R(ichard)
Ziolkowski, Richard Walter

ARKANSAS
Cardwell, David Michael
Kaupp, Verne H
McCloud, Hal Emerson, Jr
Mix, Dwight Franklin
Setian, Leo
Stephenson, Stanley E(lbert)
Yaz, Engin
Yeargan, Jerry Reese

CALIFORNIA
Abbott, Wilton R(obert)
Altes, Richard Alan
Amazeen, Paul Gerard
Ames, John Wendell
Ammann, E(ugene) O(tto)
Anderson, Edward P(arley)
Anderson, George William
Anderson, Paul Maurice
Anderson, Warren R(onald)
Anderson, William W
Andrews, Austin Michael, II
Angelakos, Diogenes James
Angell, James Browne
Armantrout, Guy Alan
Arnold, James S(loan)
Astrahan, Morton M
Baer, Walter S
Barnes, C(asper) W(illiam), Jr
Barnum, James Robert
Barrett, William Avon
Barrick, Donald Edward
Basin, M(ichael) A(bram)
Beaver, W(illiam) L(awrence)
Beckwith, Sterling
Bedrosian, E(dward)
Belcher, Melvin B
Bell, C Gordon
Bell, Charles Vester
Bentley, Kenton Earl
Berg, Myles Renver
Berk, Aristid D

Electrical Engineering (cont)

Powers, John Patrick
Prabhakar, Jagdish Chandra
Purl, O(liver) Thomas
Quate, Calvin F(orrest)
Rabinowitz, Mario
Rahmat-Samii, Yahya
Ramamoorthy, Chittoor V
Ramo, Simon
Rauch, Donald J(ohn)
Rauch, Herbert Emil
Reddy, Parvathareddy Balarami
Redfield, David
Redinbo, G Robert
Reed, Eugene D
Reed, Irving Stoy
Ricardi, Leon J
Rice, Dennis Keith
Riese, Russell L(loyd)
Roberts, Lawrence G
Robinson, Guner Suzek
Rode, Jonathan Pace
Roney, Robert K(enneth)
Rosen, C(harles) A(braham)
Rosenheim, D(onald) E(dwin)
Ross, Hugh Courtney
Rothauge, Charles Harry
Rowen, William H(oward)
Rubin, Izhak
Sack, E(dgar) A(lbert), Jr
Safonov, Michael G
Salihi, Jalal T(awfiq)
Salzer, John M(ichael)
Sangiovanni-Vincentelli, Alberto Luigi
Saunders, Robert M(allough)
Sawchuk, Alexander Andrew
Saxe, Raymond Frederick
Schaffner, Gerald
Scharfetter, D(onald) L
Scheibe, Paul Otto
Scheuch, Don Ralph
Schinzinger, Roland
Schmars, William Thomas
Scholtz, Robert A
Schorr, Herbert
Schott, Frederick W(illiam)
Schwartz, Morton Donald
Schwarz, Steven E
Sclar, Nathan
Sebald, Anthony Vincent
Segal, Alexander
Seib, David Henry
Seitz, S Stanley
Seling, Theodore Victor
Senge, George H
Sensiper, S(amuel)
Sepmeyer, L(udwig) W(illiam)
Shaffner, Richard Owen
Shaw, Elden K
Shaw, George, II
Shemer, Jack Evvard
Shevel, Wilbert Lee
Shimada, Katsunori
Shoch, John F
Siddiqee, Muhammad Waheeduddin
Siegman, A(nthony) E(dward)
Silverman, Leonard
Simon, Marvin Kenneth
Simonen, Thomas Charles
Simpson, Richard Allan
Skalnik, J(ohn) G(ordon)
Sklansky, J(ack)
Skrzypek, Josef
Slaton, Jack H(amilton)
Slocum, Richard William
Smith, P Gene
Smith, Ralph J(udson)
Snyder, William
Soohoo, Ronald Franklin
Specht, Donald Francis
Srour, Joseph Ralph
Stahley, William
Staprans, Armand
Steenson, Bernard O(wen)
Steier, William H(enry)
Steinberg, Richard
Steiner, James W(esley)
Stenzel, Reiner Ludwig
Stinson, Donald Cline
Stoll, Paul James
Stolte, Charles
Stout, Thomas Melville
Strand, Timothy Carl
Stubberud, Allen Roger
Sun, Cheng
Sung, Chia-Hsiaing
Suran, Jerome J
Sutherland, William Robert
Sweeney, Lawrence Earl, Jr
Szentirmai, George
Tabak, Mark David
Tahiliani, Vasu H
Tang, Denny Duan-Lee
Tanner, Robert Michael
Temes, Gabor Charles
Thaler, G(eorge) J(ulius)
Thomas, David Tipton
Thomas, Hubert Jon
Thomasian, Aram John
Thompson, David A
Tilles, Abe
Tilley, Brian John

Tomasetta, Louis Ralph
Tortonese, Marco
Travis, J(ohn) C(harles)
Trujillo, Stephen Michael
Tu, Charles Wuching
Tuel, William Gole, Jr
Turin, George L
Tuttle, David F(ears)
Tyler, George Leonard
Uebbing, John Julian
Vallerga, Bernard A
Van Atta, Lester Clare
Vance, Edward F(lavus)
Vane, Arthur B(ayard)
Varaiya, Pravin Pratap
Vartanian, Perry H(atch), Jr
Vassiliou, Marius Simon
Veigele, William John
Vemuri, Venkateswararao
Vickers, J(ohn) M(ichael) F(rank)
Vickers, Roger Spencer
Viglione, Sam S
Villard, Oswald G(arrison), Jr
Villeneuve, A(lfred) T(homas)
Viterbi, Andrew J
Vitols, Visvaldis Alberts
Vlay, George John
Wachowski, Hillard M(arion)
Wada, George
Wade, Glen
Walden, Robert Henry
Walter, Carlton H(arry)
Wang, Carl C T
Wang, Ji Ching
Wang, Jon Y
Wang, P(aul) K(eng) C(hieh)
Wang, Paul Keng Chieh
Ward, John Robert
Washburn, Harold W(illiams)
Waterman, Alan T(ower), Jr
Watkins, Dean Allen
Webber, Stanley Eugene
Weber, Charles L
Weckler, Gene Peter
Weinberg, Daniel I
Weissman, Ira
Welch, Lloyd Richard
Welsh, David Edward
Wertz, Harvey J
Wheeler, Harold A(lden)
Whinnery, John R(oy)
White, George Matthews
White, Robert Lee
White, Stanley A
Whitmer, Robert Morehouse
Williams, Jack Rudolph
Willson, Alan Neil, Jr
Wilson, Barbara Ann
Wilson, Perry Baker
Wilson, William John
Wolf, Jack Keil
Wolff, Milo Mitchell
Wong, Kwan Y
Wood, F(rederick) B(ernard)
Woodbury, Eric John
Wozencraft, John McReynolds
Wu, Felix F
Wu, Te-Kao
Wuebbles, Donald J
Yaggy, Paul Francis
Yamakawa, Kazuo Alan
Yao, Kung
Yeh, Cavour W
Yeh, Hen-Geul
Yeh, Paul Pao
Young, James Forrest
Young, James R(alph)
Young, Konrad Kwang-Leei
Yousif, Salah Mohammad
Yu, Chyang John
Zarem, Abe Mordecai
Zeidler, James Robert
Zeoli, G(ene) W(esley)
Zucker, Oved Shlomo Frank

COLORADO

Allen, James Lamar
Avery, Susan Kathryn
Baird, Jack R
Baird, Ramon Condie
Barnes, Frank Stephenson
Bennett, W Scott
Bloom, L(ouis) R(ichard)
Booton, Richard C(rittenden), Jr
Bradford, Phillips Verner
Brady, Douglas MacPherson
Brandauer, Carl M(artin)
Britton, Charles Cooper
Broome, Paul W(allace)
Brown, Alison Kay
Brubaker, Thomas Allen
Calvert, James Bowles
Carpenter, Donald Gilbert
Cathey, Wade Thomas, Jr
Chen, Di
Chernow, Fred
Clifford, Steven Francis
Dichtl, Rudolph John
Eichelberger, W(illiam) H
El-Kareh, Auguste Badih
Elkins, Lincoln J
Engen, Glenn Forrest
Enloe, Louis Henry
Etter, Delores Maria

Faucett, Robert E
Feucht, Donald Lee
Flavin, Michael Austin
Flock, Warren L(incoln)
Frost, H(arold) Bonnell
Fuchs, Ewald Franz
Fuller, Jackson Franklin
Gage, Donald S(hepard)
Gless, George E
Goldfarb, Ronald B
Hanna, William J(ohnson)
Hayes, Russell E
Henkel, Richard Luther
Hermann, Allen Max
Hess, Howard M
Hill, David Allan
Hjelme, Dag Roar
Horton, Clifford E(dward)
Hull, Joseph A
Hume, Wayne C
Iyer, Ravi
Johnk, Carl T(heodore) A(dolf)
Jordan, Harry Frederick
Kanda, Motohisa
Kazmerski, Lawrence L
Kemmerly, Jack E(llsworth)
Kremer, Russell Eugene
Krenz, Jerrold H(enry)
Lewin, Leonard
Lubell, Jerry Ira
Ma, Mark T
Malaiya, Yashwant Kumar
Maler, George J(oseph)
Maley, S(amuel) W(ayne)
Mathys, Peter
Mattson, Roy Henry
Maxwell, Lee M(edill)
May, William G(ambrill)
Mengel, J T
Minnick, Robert C
Morgan, Alvin H(anson)
Morrison, John Stuart
Nahman, Norris S(tanley)
Nesenbergs, Martin
Ondrejka, Arthur Richard
O'Neill, John Francis
Ralston, Margarete A
Randa, James P
Roberts, Richard A
Schiffmacher, E(dward) R(obert)
Schleif, Ferber Robert
Serafin, Robert Joseph
Sisson, Ray L
Stone, J(ack) L(ee)
Summers, Claude M
Thomas, Joe Ed
Thompson, John Stewart
Twombly, John C
Utlaut, William Frederick
Van Pelt, Richard W(arren)
Wainwright, Ray M
Waite, William McCastline
Walden, Jack M
Wall, Edward Thomas
Webb, James R
Webb, Richard C(larence)
Wells, Marion Alva
Wiatrowski, Claude Allan
Wilmsen, Carl William
Winn, C Byron
Wu, Min-Yen
Ziemer, Rodger Edmund

CONNECTICUT

Akkapeddi, Prasad Rao
Anderson, R(obert) M(orris), Jr
Apte, Chidanand
Balasinski, Artur
Bennett, William Ralph
Bird, Leslie V(aughn)
Cheng, David H S
Cooper, Franklin Seaney
Cowles, William Warren
Crawford, John Okerson
Cunningham, W(alter) J(ack)
Davenport, Lee Losee
Dubin, Fred S
Ellis, Lynn W
Fitchen, Franklin Charles
Flynn, T(homas) F(rancis)
Fox, Martin Dale
Gaidis, Michael Christopher
Galbiati, Louis J
Glomb, Walter L
Hajela, Dan
Izenour, George C(harles)
Javidi, Bahram
Johnson, Loering M
Kerby, Hoyle Ray
Kirwin, Gerald James
Knapp, Charles H
Krummel, William Marion
Lewis, T(homas) Skipwith
Liu, Qing-Huo
Lof, John L(ars) C(ole)
Lovell, Bernard Wentzel
McDonald, John Charles
Mack, Donald R(oy)
Marchand, Nathan
Melehy, Mahmoud Ahmed
Nehorai, Arye
Northrop, Robert Burr
O'Brien, Brian
Orphanoudakis, Stelios Constantine

Quazi, Azizul H(aque)
Rich, Leonard G
Ruckebusch, Guy Bernard
Sapega, A(ugust) E(dward)
Schultheiss, Peter M(ax)
Schwartz, Michael H
Schwarz, Frank
Seely, Samuel
Shirer, Donald Leroy
Sparacino, Robert R
Stoker, Warren C
Streit, Roy Leon
Troutman, Ronald R
Tuteur, Franz Benjamin
Volpe, Gerald T
Wearly, William L
Zweig, Felix

DELAWARE

Barnett, Allen M
Bigliano, Robert P(aul)
Böer, Karl Wolfgang
Bolgiano, Louis Paul, Jr
Bowyer, Kern M(allory)
Dell, Curtis G(eorge)
Hardy, Henry Benjamin, Jr
Hounshell, David A
Ih, Charles Chung-Sen
McNeely, James Braden
Maher, John Philip
Moore, Charles B(ernard)
Murphy, Arthur Thomas
Vance, Paul A(ndrew), Jr
Walsh, Robert R(eddington)

DISTRICT OF COLUMBIA

Adams, Francis L(ee)
Aein, Joseph Morris
Atwood, Donald J
Berman, Gerald Adrian
Campbell, Arthur B
Campbell, Francis James
Campillo, Anthony Joseph
Chapin, Douglas McCall
Chien, Yi-Tzuu
Coffey, Timothy
Della Torre, Edward
Diemer, F(erdinand) P(eter)
Du, Li-Jen
Ellis, William Rufus
English, William Joseph
Frank, Howard
Friebele, Edward Joseph
Friedman, Arthur Daniel
Gabriel, William Francis
Garcia, Oscar Nicolas
Grafton, Robert Bruce
Grayson, Lawrence P(eter)
Gruen, H(arold)
Harmuth, Henning Friedolf
Hazelrigg, George Arthur, Jr
Hedges, Harry G
Jackson, William David
Kalmus, Henry P(aul)
Koo, Kee P
Kyriakopoulos, Nicholas
Lee, Alfred M
Lehmann, John R(ichard)
McDuffie, George E(addy), Jr
Meister, Robert
Meltzer, Arnold Charles
Mermel, Thaddeus Walter
Montgomery, G(eorge) Franklin
Nelson, David Brian
Nguyen, Charles Cuong
Niedenfuhr, Francis W(illiam)
Parasuraman, Raja
Pei, Richard Yu-Sien
Pickholtz, Raymond L
Rao, Jaganmohan Boppana Lakshmi
Sanders, John D
Schildcrout, Michael
Schriever, Richard L
Schutzman, Elias
Sjogren, Jon Arne
Skolnik, Merrill I
Stephanakis, Stavros John
Temes, Clifford Lawrence
Thomas, Leonard William, Sr
Trunk, Gerard Vernon
Valenzuela, Gaspar Rodolfo
Wang, Yen Chu
Wong, E(ugene)
Yen, Nai-Chyuan
Youngblood, William Alfred

FLORIDA

Aaron, M Robert
Adair, James Edwin
Amey, William G(reenville)
Ashley, J(ames) Robert
Beal, Jim C(ampbell)
Bixby, W(illiam) Herbert
Blecher, F(ranklin) H(ugh)
Boll, H(arry) J(oseph)
Bullock, Thomas Edward, Jr
Burdick, Glenn Arthur
Caldwell, Melba Carstarphen
Callahan, Thomas William
Carroll, Dennis Patrick
Chen, Tsong-Ming
Chen, Wayne H(wa-Wei)
Childers, Donald Gene
Cloutier, James Robert

Couch, Leon Worthington, II
Couturier, Gordon W
Dacey, George Clement
Davis, Carl F
DeLoach, Bernard Collins, Jr
Donaldson, Merle Richard
Dougherty, John William
Duersch, Ralph R
Durgin, Harold L
Easton, Ivan G(eorge)
Einspruch, Norman G(erald)
Elgerd, O(lle) I(ngemar)
Fiedler, George J
Finlon, Francis P(aul)
Flick, Carl
Forsman, M(arion) E(dwin)
Fossum, Jerry G
Fry, Francis J(ohn)
Gonzalez, Guillermo
Goss, Charles Rapp, Jr
Gould, G(erald) G(eza)
Hammer, Jacob
Harney, Robert Charles
Harrison, Thomas J
Haviland, Robert P(aul)
Hebb, Maurice F, Jr
Henning, Rudolf E
Hoeppner, Conrad Henry
Hoffman, Thomas R(ipton)
Holley, Charles H
Hollis, J(ohn) Searcy
Holmes, D Brainerd
Hoyer-Ellefsen, Sigurd
Jain, Vijay Kumar
Johnson, E(well) Calvin
Johnson, J(oseph) Stuart
Jury, Eliahu I(braham)
Kennedy, David P
Kessler, William J(oseph)
Ketchledge, Raymond Waibel
Kline, Jacob
Kolhoff, M(arvin) J(oseph)
Kornblith, Lester
Kostopoulos, George
Krausche, Dolores Smoleny
Lade, Robert Walter
Lai, David Chin
Laschever, Norman Lewis
Li, Sheng-San
Lister, Charles Allan
Malocha, Donald C
Malthaner, W(illiam) A(mond)
Mandil, I Harry
Mathews, Bruce Eugene
Melvin, E(ugene) A(very)
Messenger, Roger Alan
Migliaro, Marco William
Miller, Hillard Craig
Naini, Majid M
Neal, J(ames) P(reston), III
Nuese, Charles J
O'Malley, John Richard
Packer, Lewis C
Palmer, Ralph Lee
Peebles, Peyton Z, Jr
Ramond, Pierre Michel
Revay, Andrew W, Jr
Riblet, Henry B
Ross, Gerald Fred
Roth, Paul Frederick
Sells, Jackson S(tuart)
Shaffer, Charles V(ernon)
Sias, Frederick Ralph
Smith, Jack R(eginald)
Smythe, Robert C
Snyder, R L
Stach, Joseph
Staudhammer, John
Stavis, Gus
Stodola, Edwin King
Swartz, William Edward, Jr
Tapia, M(oiez) A(hmedale)
Taylor, Fredrick James
Tebbe, Dennis Lee
Timoshenko, Gregory Stephen
Tou, Julius T(su) L(ieh)
Towner, O W
Traexler, John F
Tusting, Robert Frederick
Ulm, Lester, Jr
Ulug, Esin M
Uman, Martin A(llan)
Ungvichian, Vichate
Van Ligten, Raoul Fredrick
Veinott, Cyril G
Vemuri, Suryanarayana
Wade, Thomas Edward
Wallace, James D
Weaver, Lynn E(dward)
Weber, Harry P(itt)
Weller, Richard Irwin
Whitman, Lawrence C
Wiseman, Robert S
Wolff, Hanns H
Yacoub, Kamal
Yii, Roland
Yon, E(ugene) T
Young, Tzay Y

GEORGIA
Adomian, George
Alford, Cecil Orie
Baxter, Samuel G
Bourne, Henry C(lark), Jr

Bruder, Jospeh Albert
Callahan, Leslie G, Jr
Connelly, J(oseph) Alvin
Cook, James H, Jr
Currie, Charles H
Dees, J W
Derrick, Robert P
Dutton, John C
Ecker, Harry Allen
Esogbue, Augustine O
Evans, Thomas P
Evans, W E
Feeney, Robert K
Fielder, D(aniel) C(urtis)
Finn, D(avid) L(ester)
Galeano, Sergio F(rancis)
Gaylord, Thomas Keith
Grace, John Nelson
Gray, James S
Griffin, Clayton Houstoun
Harrison, Gordon R
Holtum, Alfred G(erard)
Honkanen, Pentti A
Hunt, William Daniel
Hutchins, Samuel F
Jokerst, Nan Marie
Joy, Edward Bennett
Kalb, John W
Kreutel, Randall William, Jr
Landgren, John Jeffrey
Lee, Kok-Meng
McClellan, James Harold
Martin, Roy A
Mersereau, Russell Manning
Moad, M(ohamed) F(ares)
Moss, Richard Wallace
Owens, William Richard
Paris, Demetrius T
Peatman, John B(urling)
Ray, Dale C(arney)
Rhodes, William Terrill
Rodrigue, George Pierre
Russell, John Lynn, Jr
Ryan, Charles Edward, Jr
Schafer, Ronald W
Su, Kendall L(ing-Chiao)
Thumann, Albert
Vail, Charles R(owe)
Wallace, John M, Jr
Wang, Johnson Jenn-Hwa
Webb, Roger P(aul)
Wolla, Maurice L(eRoy)
Yokelson, Bernard J(ulius)

HAWAII
Abramson, N(orman)
Gaarder, Newell Thomas
Granborg, Bertil Svante Mikael
Hwang, Hu Hsien
Kinariwala, Bharat K
Lin, Shu
McFee, Richard
Peterson, William Wesley
Roelofs, Thomas Harwood
Weaver, Paul Franklin
Yuen, Paul C(han)

IDAHO
Baily, Everett M
DeBow, W Brad
Demuth, Howard B
Gray, Earl E
Hall, J(ames) A(lexander)
Hyde, William W
Law, John
Mann, Paul
Meyer, Orville R
Mueller, Fred(erick) M(arion)
Parish, William R
Stuffle, Roy Eugene

ILLINOIS
Aagaard, James S(tuart)
Agarwal, Gyan C
Ahuja, Narendra
Armington, Ralph Elmer
Arzbaecher, Robert
Behnke, Wallace B, Jr
Bitzer, Donald L
Briley, Bruce Edwin
Brodwin, Morris E(llis)
Brown, David P(aul)
Brown, Julius
Burnham, Robert Danner
Burtness, Roger William
Chang, Herbert Yu-Pang
Chen, W(ai) K(ai)
Chew, Weng Cho
Christianson, Clinton Curtis
Cohn, Jona
Coleman, P(aul) D(are)
Conant, Roger C
Cooper, Duane H(erbert)
Coplien, James O
Cox, Dennis Dean
Demaeyer, Bruce R
Deschamps, Georges Armand
DeTemple, Thomas Albert
Dunn, Floyd
Dyson, John Douglas
Eden, James Gary
Eilers, Carl G
Emery, W(illis) L(aurens)
Engel, Joel S

Epstein, Max
Erwin, David B(ishop), Sr
Friedlander, Alan L
Frizzell, Leon Albert
Fuchs, W Kent
Gaddy, Oscar
Gallagher, David Alden
Gardner, Chester Stone
Gillette, Richard F
Godhwani, Arjun
Gottlieb, Gerald Lane
Hajj, Ibrahim Nasri
Hammond, William Marion
Hess, Karl
Himmelstein, Sydney
Hoffman, Clyde H(arris)
Holmes, Kenneth Robert
Hrbek, George W(illiam)
Hua, Ping
Huang, Thomas Shi-Tao
Hunsinger, Bill Jo
Jenkins, William Kenneth
Jones, J(ohn) L(loyd), Jr
Jordan, Edward C(onrad)
Kang, Sung-Mo Steve
Kim, Kyekyoon K(evin)
Knop, Charles M(ilton)
Krein, Philip Theodore
Kumar, Panganamala Ramana
Kuo, Benjamin Chung-I
Lawrie, Duncan H
Laxpati, Sharad R
Lipinski, Walter C(harles)
Liu, Chao-Han
Lo, Y(uen) T(ze)
Luplow, Wayne Charles
Maclay, G Jordan
Mast, P(lessa) Edward
Mayeda, Wataru
Mayes, Paul E(ugene)
Messinger, Henry Peter
Metze, Gernot
Miller, W(endell) E(arl)
Murata, Tadao
Murphy, Gordon J
Naylor, David L
Neuhalfen, Andrew J
Newell, Darrell E
Newton, John S
O'Brien, William Daniel, Jr
Pai, Anantha M
Pai, Mangalore Anantha
Parker, Ehi
Perkins, William Randolph
Phillips, Thomas James
Plonus, Martin
Poppelbaum, Wolfgang Johannes
Priemer, Roland
Pursley, Michael Bader
Ransom, Preston Lee
Ray, Sylvian Richard
Reichard, Grant Wesley
Rekasius, Zenonas V
Reynolds, Larry Owen
Robbins, R(oger) W(ellington)
Robinson, Charles J
Rodriguez, Jacinto
Rutledge, Robert B
Saeks, Richard E
Saltzberg, Theodore
Sanathanan, C(hathilingath) K
Sarwate, Dilip Vishwanath
Sauer, Peter William
Savic, Stanley D
Scanlon, Jack M
Siegel, Jonathan Howard
Smith, James G(ilbert)
Smith, Leslie Garrett
Sognefest, Peter William
Soria, Rodolfo M(aximiliano)
Stafford, John William
Stein, Herbert Joseph
Stinaff, Russell Dalton
Studtmann, George H
Swenson, G(eorge) W(arner), Jr
Taflove, Allen
Toppeto, Alphonse A
Toy, Wing N
Trick, Timothy Noel
Trimble, John Leonard
Turnbull, Robert James
Ulrich, Werner
Unlu, M Selim
Uslenghi, Piergiorgio L
Van Ness, James E(dward)
Van Valkenburg, M(ac) E(lwyn)
Venema, Harry J(ames)
Verdeyen, Joseph T
Wakefield, Ernest Henry
Wax, Nelson
Webb, Harold Donivan
Weber, Erwin Wilbur
Weinberg, Philip
Wey, Albert Chin-Tang
Yau, Stephen Sik-Sang
Yeh, K(ung) C(hie)
Zell, Blair Paul

INDIANA
Acker, Frank Earl
Atallah, Mikhail Jibrayil
Berry, William B(ernard)
Brown, Buck F(erguson)
Chen, Chin-Lin

Chittick, K(enneth) A
Cohn, David L(eslie)
Coyle, Edward John
Criss, Darrell E
Detraz, Orville F
Dunipace, Kenneth Robert
El-Abiad, Ahmed H(anafi)
Ersoy, Okan Kadri
Fearnot, Neal Edward
Friedlaender, Fritz J(osef)
Fukunaga, Keinosuke
Gabriel, Garabet J(acob)
Gallagher, Neal Charles
Gelopulos, Demosthenes Peter
Gunshor, Robert Lewis
Haas, Violet Bushwick
Hammann, John William
Hayt, William H(art), Jr
Henke, Mitchell C
Henry, Eugene W
Heydt, Gerald Thomas
Holmes, Alvin W(illiam)
Jamieson, Leah H
Jefferies, Michael John
Johnson, C(harles) Bruce
Kashyap, Rangasami Laksminarayana
Kendall, Perry E(ugene)
Kim, Young Duc
Koivo, Antti J
Krause, Paul Carl, Jr
Lafuse, Harry G
Landgrebe, David Allen
Lin, P(en) M(in)
Lindenlaub, John Charles
Liu, Ruey-Wen
Lundstrom, Mark Steven
McGillem, C(lare) D(uane)
Mahmoud, Aly Ahmed
Melloch, Michael Raymond
Michel, Anthony Nikolaus
Mitchell, Owen Robert
Mowle, Frederic J
Neudeck, Gerold W(alter)
Ogborn, Lawrence L
Pierret, Robert Francis
Pierson, Edward S
Rikoski, Richard Anthony
Rogers, Charles C
Ross, Irvine E
Sabbagh, Harold A(braham)
Schalliol, Willis Lee
Schwartz, Richard John
Shelley, Austin L(inn)
Shewan, William
Siegel, Howard Jay
Sinha, Akhouri Suresh Chandra
Skelton, Robert Eugene
Snow, John Thomas
Szpankowski, Wojciech
Thompson, Hannis Woodson, Jr
Uhran, John Joseph, Jr
Vocke, Merlyn C
Williams, Walter Jackson, Jr
Winton, Henry J
Wintz, P(aul) A

IOWA
Alton, Everett Donald
Andersland, Mark Steven
Basart, John Philip
Boast, W(arren) B(enefield)
Brearley, Harrington C(ooper), Jr
Carlson, David L
Chyung, Dong Hak
Coady, Larry B
Collins, Steve Michael
Dalal, Vikram
Fanslow, Glenn E
Fouad, Abdel Aziz
Geiger, Randall L
Getting, Peter Alexander
Hale, Harry W(illiam)
Henrici, Carl Ressler
Horton, Richard E
Hsieh, Hsung-Cheng
Hubbard, Philip G(amaliel)
Jones, Edwin C, Jr
Kopplin, J(ulius) O(tto)
Lamont, John W(illiam)
Lonngren, Karl E(rik)
Lunde, Barbara Kegerreis, (BK)
Malik, Norbert Richard
Mericle, Morris H
Myers, Glenn Alexander
Pierce, Roger J
Pohm, A(rthur) V(incent)
Potter, Allan G
Reddy, Sudhakar M
Robinson, John Paul
Smay, Terry A
Stephenson, David Town
Titze, Ingo Roland
Wilson, Robert D(owning)

KANSAS
Carpenter, Kenneth Halsey
Cottom, Melvin C(lyde)
Daugherty, Don G(ene)
Dellwig, Louis Field
Dwyer, Samuel J, III
Grzymala-Busse, Jerzy Witold
Holtzman, Julian Charles
Hummels, Donald Ray
Jakowatz, Charles V

Riseman, Edward M
Roadstrum, William H(enry)
Roberge, James Kerr
Robinson, D(enis) M(orrell)
Rochefort, John S
Rona, Mehmet
Rosenkranz, Philip William
Ross, Myron Jay
Rotman, Walter
Rubin, Milton D(avid)
Rudko, Robert I
Ruina, J(ack) P(hilip)
Sakshaug, Eugene C
Salah, Joseph E
Saltzer, Jerome H(oward)
Sarles, F(rederick) Williams
Schattenburg, Mark Lee
Schaubert, Daniel Harold
Scherer, Harold Nicholas, Jr
Schetzen, Martin
Schleif, Robert Ferber
Schmidt, George Thomas
Schoendorf, William H(arris)
Schonhoff, Thomas Arthur
Seifer, Arnold David
Seifert, William W(alther)
Sethares, James C(ostas)
Shaffer, Harry Leonard
Shapiro, Jeffrey Howard
Sherman, Herbert
Shrader, William Whitney
Siebert, W(illiam) M(cConway)
Silevich, Michael B
Simmons, Alan J(ay)
Singer, Joshua J
Smith, Henry I
Smith, Paul E, Jr
Smullin, Louis Dijour
Sneddon, Leigh
Spears, David Lewis
Spencer, Gordon Reed
Spergel, Philip
Spindel, Robert Charles
Staelin, David Hudson
Stevenson, David Michael
Strong, Robert Michael
Stuart, John W(arren)
Sun, Fang-Kuo
Susman, Leon
Swift, Calvin Thomas
Tancrell, Roger Henry
Tang, Ting-Wei
Teager, Herbert Martin
Temme, Donald H(enry)
Teng, Chojan
Theobald, Charles Edwin, Jr
Thornton, Richard D(ouglas)
Troxel, Donald Eugene
Tsandoulas, Gerasimos Nicholas
Tsang, Dean Zensh
Tsaur, Bor-Yeu
Tweed, David George
Uhlir, Arthur, Jr
Van Meter, David
Voigt, Herbert Frederick
Wang, Christine A
Ward, Harold Richard
Ward, James
Ward, John E(rwin)
Wedlock, Bruce D(aniels)
Weng, Lih-Jyh
Wetzstein, H(anns) J(uergen)
Wheeler, Ned Brent
White, David Calvin
Whitehouse, David R(empfer)
Wiederhold, Pieter Rijk
Wiegner, Allen W
Wiesner, J(erome) B
Wilk, Leonard Stephen
Williamson, Richard Cardinal
Wilson, Gerald Loomis
Wogrin, Conrad A(nthony)
Woll, Harry Jean
Wu, Pei-Rin
Yaghjian, Arthur David
Yngvesson, K Sigfrid
Zimmermann, H(enry) J(oseph)
Zornig, John Grant
Zraket, Charles Anthony

MICHIGAN
Agarwal, Paul D(haram)
Aleksoff, Carl Chris
Anderson, Walter T(heodore)
Barr, Robert Ortha, Jr
Beglau, David Alan
Be Ment, Spencer L
Bickart, Theodore Albert
Bird, Michael Wesley
Bockemuehl, R(obert) R(ussell)
Borcherts, Robert H
Brammer, Forest E(vert)
Brown, Richard K(emp)
Brown, Verne R
Brumm, Douglas B(ruce)
Calahan, Donald A
Carey, John Joseph
Caron, E(dgar) Louis
Chau, Hin Fai
Chen, Kan
Chen, Kun-Mu
Chen, Yudong
Chuang, Kuei
Clark, John R(ay)

Conway, Lynn Ann
Davidson, Edward S(teinberg)
Dow, W(illiam) G(ould)
Dubes, Richard C
Eesley, Gary
Enns, Mark K
Fienup, James R
Frank, Max
Freedman, M(orris) David
Gaerttner, Martin R
Gagnon, Eugene Gerald
Getty, Ward Douglas
Ghoneim, Youssef Ahmed
Haddad, George I
Hanysz, Eugene Arthur
Hassoun, Mohamad Hussein
Hemdal, John Frederick
Hiatt, Ralph Ellis
Hillig, Kurt Walter, II
Ho, Bong
Holland, Steven William
Huebner, George L(ee), Jr
Jacovides, Linos J
Jones, R(ichard) James
Kerr, William
Khargonekar, Pramod P
Klingler, Eugene H(erman)
Knoll, Alan Howard
Koenig, Herman E
Kreer, John B(elshaw)
Kulkarni, Anand K
LaHaie, Ivan Joseph
LaRocca, Anthony Joseph
Larrowe, Boyd T
Larrowe, Vernon L
Lewis, Robert Warren
Lomax, Ronald J(ames)
Luxon, James Thomas
McMahon, E(dward) Lawrence
Macnee, Alan B(reck)
Malaczynski, Gerard W
Meisel, Jerome
Miller, M(urray) H(enri)
Nagy, Andrew F
Nevius, Timothy Alfred
Nyquist, Dennis Paul
Ohlsson, Robert Louis
Park, Gerald L(eslie)
Pendleton, Wesley William
Peterson, Lauren Michael
Polis, Michael Philip
Reid, Richard J(ames)
Reitan, Daniel Kinseth
Ribbens, William B(ennett)
Rigas, Anthony L
Rigas, Harriett B
Rohde, Steve Mark
Roll, James Robert
Salam, Fathi M A
Sattinger, Irvin J(ack)
Schubring, Norman W(illiam)
Sebastian, Franklin W
Sellers, Ernest E(dwin)
Sharpe, Charles Bruce
Smith, Robert W
Soper, Jon Allen
Stocker, Donald V(ernon)
Tai, Chen-To
Ulaby, Fawwaz Tayssir
Upatnieks, Juris
Vander Kooi, Lambert Ray
Vaz, Nuno A
Von Tersch, Lawrence W
Wang, Charles C(hen-Ding)
Weil, Herschel
Weng, Tung Hsiang
Wennerberg, A(llan) L(orens)
Williams, William James
Witt, Howard Russell
Zanini-Fisher, Margherita

MINNESOTA
Albertson, Vernon D(uane)
Bailey, Fredric Nelson
Champlin, Keith S(chaffner)
Davidson, Donald
Dinneen, Gerald Paul
Dixon, John D(ouglas)
Du, David Hung-Chang
DuBois, John R(oger)
Geddes, John Joseph
Geokezas, Meletios
Gilbert, Barry Kent
Graves, Wayne H(aigh)
Greene, Christopher Storm
Gruber, Carl L(awrence)
Harvey, Charles Arthur
Haxby, B(ernard) V(an Loan)
Higgins, Robert Arthur
Hocker, George Benjamin
Holte, James Edward
Kain, Richard Yerkes
Kaveh, Mostafa
Keitel, Glenn H(oward)
Killpatrick, Joseph E
Kinney, Larry Lee
Kumar, K S P
La Bonte, Anton Edward
Lambert, Robert F
Lea, Wayne Adair
Lee, Gordon M(elvin)
Lee, T(homas) S(hao-Chung)
Lindemann, Wallace W(aldo)
Lo, David S(hih-Fang)

McLane, Robert Clayton
Mohan, Narendra
Mooers, Howard T(heodore)
Motchenbacher, Curtis D
Nelson, Gerald Duane
Olyphant, Murray, Jr
Pinckaers, B(althasar) Hubert
Riaz, M(ahmoud)
Robbins, William Perry
Sackett, W(illiam) T(ecumseh), Jr
Sadjadi, Firooz Ahmadi
Schmitz, Harold Gregory
Shepherd, W(illiam) G(erald)
Smith, David Philip
Stephanedes, Yorgos Jordan
Thiede, Edwin Carl
Timm, Gerald Wayne
Tran, Nang Tri
Tufte, Obert Norman
Wollenberg, Bruce Frederick

MISSISSIPPI
Ball, Billie Joe
Burke, Robert Wayne
Elsherbeni, Atef Zakaria
Ferguson, Mary Hobson
Fry, Thomas R
Gassaway, James D
Glisson, Allen Wilburn, Jr
Hanson, Donald Farness
Herring, John Wesley, Jr
Ingels, Franklin M(uranyi)
Jacob, Paul B(ernard), Jr
Kajfez, Darko
Keulegan, Garbis
Kishk, Ahmed A
McDaniel, Willie L(ee), Jr
Miller, David Burke
Pearce, David Harry
Ramsdale, Dan Jerry
Simrall, Harry C(harles) F(leming)
Smith, Charles Edward
Wier, David Dewey

MISSOURI
Adams, Gayle E(ldredge)
Bertnolli, Edward Clarence
Betten, J Robert
Bialecke, Edward P
Bovopoulos, Andreas D
Brown, Lloyd Robert
Carson, Ralph S(t Clair)
Carter, Robert L(eroy)
Combs, Robert Glade
Constant, Paul C, Jr
Cox, Jerome R(ockhold), Jr
Denman, Eugene D(ale)
Dreifke, Gerald E(dmond)
DuBroff, Richard Edward
Everhart, James G
Farmer, John William
Forster, William H(all)
Glaenzer, Richard H
Goldstein, Julius L
Graham, Robert Reavis
Gruber, B(ernard) A
Hancock, John C(oulter)
Harbourt, Cyrus Oscar
Harmon, Robert Wayne
Hicks, Patricia Fain
Hoft, Richard Gibson
Indeck, Ronald S
Kimball, Charles Newton
Kline, Raymond Milton
Koopman, Richard J(ohn) W(alter)
Landiss, Daniel Jay
Lemon, Leslie Roy
Lindsay, James Edward, Jr
McFarland, William D
McLaren, Robert Wayne
McMillan, Gregory Keith
McMillan, Gregory Keith
McPherson, George, Jr
Mak, Sioe Tho
Marshall, Stanley V(ernon)
Martin, Edwin J, Jr
Morley, Robert Emmett, Jr
Moss, Randy Hays
Muller, Marcel Wettstein
Nau, Robert H(enry)
Prelas, Mark Antonio
Richards, Earl Frederick
Rosenbaum, Fred J(erome)
Ross, Donald K(enneth)
Ross, Monte
Ryan, Carl Ray
Sherman, Byron Wesley
Shrauner, Barbara Abraham
Slivinsky, Charles R
Sparlin, Don Merle
Spielman, Barry
Stanek, Eldon Keith
Tudor, James R
Tyrer, Harry Wakeley
Waid, Rex A(dney)
Waidelich, D(onald) L(ong)
Wann, Donald Frederick
Winter, David F(erdinand)
Wolfe, Charles Morgan
Wu, Cheng-Hsiao
Zobrist, George W(inston)

MONTANA
Bennett, Byron J(irden)

Durnford, Robert F(red)
Gerez, Victor
Hanton, John Patrick
Knox, James L(ester)
Pierre, Donald Arthur

NEBRASKA
Bahar, Ezekiel
Bashara, N(icolas) M
Brogan, William L
Chung, Hui-Ying
Davis, Harold
Edison, Allen Ray
Hultquist, Paul F(redrick)
Lagerstrom, John E(mil)
Myers, Grant G
Nelson, Don Jerome
Olive, David Wolph
Robison, Wendall C(loyd)
Seth, Sharad Chandra
Soukup, Rodney Joseph
Voelker, Robert Heth
Woods, Joseph
Woollam, John Arthur

NEVADA
Betts, Attie L(ester)
Halacsy, Andrew A
Manhart, Robert (Audley)
Messenger, George Clement
Rawat, Banmali Singh
Tryon, John G(riggs)

NEW HAMPSHIRE
Adjemian, Haroutioon
Blesser, Barry Allen
Brous, Don W
Buffler, Charles Rogers
Carroll, Lee Francis
Clark, Ronald Rogers
Crane, Robert Kendall
Crowell, Merton Howard
Glanz, Filson H
Hough, Richard R(alston)
Hraba, John Burnett
Hutchinson, Charles E(dgar)
Juhola, Carl
Klarman, Karl J(oseph)
Laaspere, Thomas
Melvin, Donald Walter
Moskowitz, Ronald
Murdoch, Joseph B(ert)
Pokoski, John Leonard
Samaroo, Winston R
Scott, Ronald E
Tillinghast, John Avery
Von-Recklinghausen, Daniel R

NEW JERSEY
Ackenhusen, John Goodyear
Allen, Jonathan
Amitay, Noach
Anderson, G(eorge) M(ullen)
Anderson, Robert E(dwin)
Anderson, Stanley William
Andrekson, Peter A(vo)
Andrews, Frederick T, Jr
Ashcroft, Dale Leroy
Assadourian, Fred
Atal, Bishnu S
Baldwin, George L(ee)
Ball, William Henry Warren
Bean, John Condon
Beard, Richard B
Becken, Eugene D
Berger, Paul Raymond
Bernstein, Alan D
Bernstein, Lawrence
Bharj, Sarjit Singh
Bitzer, Richard Allen
Black, H(arold) S(tephen)
Blanck, Andrew R(ichard)
Boesch, Francis Theodore
Bond, Robert Harold
Bondy, Michael F
Boyd, Gary Delane
Brandinger, Jay J
Bridges, Thomas James
Brilliant, Martin Barry
Brown, Irmel Nelson
Brownell, R(ichard) M(iller)
Bucher, T(homas) T(albot) Nelson
Buchner, Morgan M(allory), Jr
Butler, Herbert I
Cammarata, John
Capasso, Federico
Carter, Ashley Hale
Castenschiold, Rene
Chen, Fang Shang
Cheo, Li-hsiang Aria S
Chernak, Jess
Cheung, Shiu Ming
Chirlian, Paul M(ichael)
Cho, Alfred Y
Chu, Ta-Shing
Cimini, Leonard Joseph, Jr
Clements, Wayne Irwin
Cohen, Barry George
Cohen, Edwin
Cohen, Leonard George
Coleman, Norman P, Jr
Cooper, Paul W
Cornell, W(arren) A(lvan)
Coughran, William Marvin, Jr

Electrical Engineering (cont)

Cox, Donald Clyde
Crane, Roger L
Crochiere, Ronald E
Curtice, Walter R
Curtis, Thomas Hasbrook
Damen, Theo C
Davey, James R
David, Edward E(mil), Jr
Davis, William F
DeMarinis, Bernard Daniel
Denes, Peter B
Denlinger, Edgar J
Denno, Khalil I
Denton, Richard T
Deri, Robert Joseph
Detig, Robert Henry
Dickerson, L(oren) L(ester), Jr
Dickinson, Bradley William
Dixon, Paul King
Dodabalapur, Ananth
Dorros, Irwin
Drucker, Harris
Duttweiler, Donald Lars
Eager, George S, Jr
Easton, Elmer C(harles)
Eldumiati, Ismail Ibrahim
Elmendorf, Charles Halsey, III
Estrada, Herbert
Feiner, Alexander
Fich, Sylvan
Field, George Robert
Fischer, Russell Jon
Fishbein, William
Fisher, Reed Edward
Flanagan, James L(oton)
Flory, Leslie E(arl)
Forys, Leonard J
Franks, Richard Lee
Friedland, Bernard
Fukui, H(atsuaki)
Gabriel, Edwin Z
Gelnovatch, V
Giuffrida, Thomas Salvatore
Gloge, Detlef Christoph
Goodfriend, Lewis S(tone)
Gorog, Istvan
Gummel, Hermann K
Haas, Zygmunt
Hall, Joseph L
Hang, Hsueh-Ming
Hartung, John
Haskell, Barry G
Hatkin, Leonard
Heffes, Harry
Heirman, Donald N
Hershenov, B(ernard)
Herskowitz, Gerald Joseph
Hilibrand, J(ack)
Himmel, Leon
Holman, Wayne J(ames), Jr
Horing, Sheldon
House, Arthur Stephen
Houston, Vern Lynn
Hsu, H(wei) P(iao)
Hsuan, Hulbert C S
Jacquin, Arnaud Eric
Jagerman, David Lewis
Jain, Sushil C
Jansen, Frank
Jassby, Daniel Lewis
Jayant, Nuggehally S
Joel, Amos Edward, Jr
Johannes, Virgil Ivancich
Johnston, Wilbur Dexter, Jr
Kadota, T Theodore
Kaiser, James F(rederick)
Kaiser, Peter
Karol, Mark J
Katz, Sheldon Lane
Keigler, John Edward
Keizer, Eugene O(rville)
Kessler, Frederick Melvyn
Kessler, Irving Jack
Kieburtz, R(obert) Bruce
King, B(ernard) G(eorge)
Klapper, Jacob
Kluver, J(ohan) W(ilhelm)
Koch, Thomas L
Kobayashi, Hisashi
Kovats, Andre
Krevsky, Seymour
Krzyzanowski, Paul
Kuhl, Frank Peter, Jr
Kuo, Ying L
Labianca, Frank Michael
Langseth, Rollin Edward
LaPierre, Walter A(lfred)
Lasky, Jack Samuel
Lechner, Bernard J
Lee, T(ien) P(ei)
Lemberg, Howard Lee
Leu, Ming C
Levine, Nathan
Li, John Kong-Jiann
Li, Tingye
Lichtenberger, Gerald Burton
Lin, Chinlon
Lin, Dong Liang
Lindberg, Craig Robert
Liu, Bede
Liu, S(hing) G(ong)
Lo, Arthur W(unien)

Lopresti, Philip V(incent)
Lucky, Robert W
Maa, Jer-Shen
McCumber, Dean Everett
McDermott, Kevin J
McDonald, H(enry) S(tanton)
McLane, George Francis
McNair, Irving M, Jr
Manz, August Frederick
Marcatili, Enrique A J
Mattes, Hans George
Mauzey, Peter T
Mayhew, Thomas R
Mayo, John S
Meola, Robert Ralph
Mersten, Gerald Stuart
Meyer, Andrew U
Miller, Stewart E(dward)
Minter, Jerry Burnett
Misra, Raj Pratap
Mitchell, Olga Mary Mracek
Myers, George Henry
Nelson, W(inston) L(owell)
Netravali, Arun Narayan
Newhouse, Russell C(onwell)
Nowogrodzki, M(arkus)
Ohm, E(dward) A(llen)
Olive, Joseph P
Olken, Melvin I
Olmstead, John Aaron
O'Neill, Eugene F
Ong, Nai-Phuan
Osifchin, Nicholas
Padalino, Joseph John
Parkins, Bowen Edward
Partovi, Afshin
Personick, Stewart David
Pfafflin, James Reid
Pinson, Elliot N
Poor, Harold Vincent
Powers, K(erns) H(arrington)
Quinlan, Daniel A
Rabiner, Lawrence Richard
Rainal, Attilio Joseph
Rastani, Kasra
Raybon, Gregory
Reisman, Stanley S
Riggs, Victoria G
Rips, E(rvine) M(ilton)
Rosenberg, Aaron E(dward)
Rosenstark, Sol(omon)
Rosenthal, Lee
Rosenthal, Louis Aaron
Ross, Ian M(unro)
Rothschild, Gilbert Robert
Rowe, H(arrison) E(dward)
Rubinstein, Charles B(enjamin)
Russell, Frederick A(rthur)
Sagal, Matthew Warren
Saleh, Adel Abdel Moneim
Schaefer, Jacob W
Schiff, Leonard Norman
Schink, F E
Schmidt, Barnet Michael
Scholz, Lawrence Charles
Schulke, Herbert Ardis, Jr
Schwartz, Stuart Carl
Schwenker, J(ohn) E(dwin)
Seman, George William
Shaer, Norman Robert
Shea, Timothy Edward
Shefer, Joshua
Shen, Tek-Ming
Sherman, Ronald
Sherrick, Carl Edwin
Shulman, Herbert Byron
Silver, Howard I(ra)
Slepian, David
Sloane, Neil James Alexander
Smith, Peter William E
Smith, William Bridges
Sohn, Kenneth S (Kyu Suk)
Sontag, Eduardo Daniel
Staebler, David Lloyd
Stampfl, Rudolf A
Standley, Robert Dean
Stevens, James Everell
Strano, Joseph J
Strauss, Walter
Suozzi, Joseph John
Taylor, George William
Thelin, Lowell Charles
Thomas, David Gilbert
Thomas, Gary Lee
Thomas, John B(owman)
Thouret, Wolfgang E(mery)
Tien, P(ing) K(ing)
Tow, James
Tunis, C(yril) J(ames)
Uhrig, Jerome Lee
Vahaviolos, Sotirios J
Valenzuela, Reinaldo A
Vallese, Lucio M(ario)
Verdu, Sergio
Verma, Pramode Kumar
Von Aulock, Wilhelm Heinrich
Walsh, Teresa Marie
Weinberg, Louis
Weisbecker, Henry B
Weissenburger, Don William
Whitman, Gerald Martin
Wier, Joseph M(arion)
Wilkinson, W(illiam) C(layton)
Williams, Maryon Johnston, Jr

Williams, Thomas R(ay)
Winje, Russell A
Winston, Joseph
Winthrop, Joel Albert
Wischmeyer, Carl R(iehle)
Wood, Obert Reeves, II
Wyndrum, Ralph William, Jr
Wyner, Aaron D
Xie, Ya-Hong
Yadvish, Robert D
Yen, You-Hsin Eugene
Yi-Yan, Alfredo
Young, William Rae
Yue, On-Ching

NEW MEXICO

Adler, Richard John
Alers, George A
Anderson, Richard Ernest
Baty, Richard Samuel
Baum, Carl E(dward)
Biggs, Albert Wayne
Bolie, Victor Wayne
Bradshaw, Martin Daniel
Chaffin, Roger James
Davis, L(loyd) Wayne
DeVries, Ronald Clifford
Diver, Richard Boyer, Jr
Dorato, Peter
Doss, James Daniel
Duncan, Richard H(enry)
Elliott, Dana Edgar
Erteza, Ahmed
Feiertag, Thomas Harold
Freeman, Bruce L, Jr
Fronek, Donald Karel
Giles, Michael Kent
Gover, James E
Grannemann, W(ayne) W(illis)
Griffin, Patrick J
Gwyn, Charles William
Hardin, James T
Harrison, Charles Wagner, Jr
Hood, Jerry A
Jameson, Robert A
Jamshidi, Mohammad Mo
Joppa, Richard M
Karni, Shlomo
Kazda, Louis F(rank)
Kelley, Gregory M
Knudsen, Harold Knud
Kruger, Richard Paul
Laquer, Henry L
Linford, Rulon Kesler
Lucky, George W(illiam)
Maez, Albert R
Milton, Osborne
Morgan, John Derald
Muir, Patrick Fred
Myers, David Richard
Newby, Neal D(ow)
Noel, Bruce William
Norris, Carroll Boyd, Jr
Novak, James Lawrence
Osinski, Marek Andrzej
Paul, Robert Hugh
Rachele, Henry
Sandmeier, Henry Armin
Schueler, Donald G(eorge)
Short, Wallace W(alter)
Stevens, Roger Templeton
Tallerico, Paul Joseph
Tapp, Charles Millard
Taylor, Javin Morse
Thompson, Wiley Ernest
Tsao, Jeffrey Yeenien
VanDevender, John Pace
Weber, Harry A
Wecksung, George William
Welber, Irwin
Whitman, Rollin Lawrence
Wiggins, Carl M
Williams, James D
Wing, Janet E (Sweedyk) Bendt
Yonas, Gerold

NEW YORK

Abetti, Pier Antonio
Adams, Arlon Taylor
Adler, Michael Stuart
Agarwal, Ramesh Chandra
Agrawal, Govind P(rasad)
Ahmed-Zaid, Said
Ahn, K(ie) Y(eung)
Albicki, Alexander
Alexiou, Arthur George
Alfano, Robert R
Almasi, George Stanley
Anagnostopoulos, Constantine N
Anantha, Narasipur Gundappa
Anastassiou, Dimitris
Anderson, John M(elvin)
Angevine, Oliver Lawrence
Ankrum, Paul Denzel
Apai, Gustav Mar, II
Bachman, Henry L(ee)
Baertsch, Richard D
Bahl, Lalit R(ai)
Balabanian, Norman
Ballantyne, Joseph M(errill)
Barthold, Lionel O
Baule, Gerhard M
Baum, Eleanor Kushel
Bechtel, Marley E(lden)

Belove, Charles
Berger, Toby
Bewley, Loyal V
Blahut, Richard E
Blanchard, Fletcher A(ugustus), Jr
Blazey, Richard N
Blesser, William B
Blewett, John P
Bocko, Mark Frederick
Bolgiano, Ralph, Jr
Bolle, Donald Martin
Bond, Norman T(uttle)
Boorstyn, Robert Roy
Borrego, Jose M
Bossen, Douglas C
Brennemann, Andrew E(rnest), Jr
Brenner, Egon
Bresler, Aaron D
Brock, Robert H, Jr
Brodsky, Marc Herbert
Broers, Alec N
Brower, Allen S(pencer)
Brown, Burton Primrose
Brown, Dale Marius
Buchta, John C(harles)
Butler, Walter John
Calabrese, Carmelo
Caprio, James R
Carlson, A Bruce
Cassedy, Edward S(pencer), Jr
Chan, Jack-Kang
Chang, Hsu
Chang, Ifay F
Chang, Sheldon S L
Chass, Jacob
Chen, Arthur Chih-Mei
Chen, Cheng-Lin
Cheo, Bernard Ru-Shao
Chestnut, H(arold)
Chow, Chao K
Chow, Tat-Sing Paul
Christiansen, Donald David
Chubb, Charles F(risbie), Jr
Clarke, Kenneth Kingsley
Close, Charles M(ollison)
Cohen, Gerald H(oward)
Cole, James A
Colin, Lawrence
Constable, James Harris
Cornacchio, Joseph V(incent)
Costello, Richard Gray
Cotellessa, Robert F(rancis)
Cottingham, James Garry
Craft, Harold Dumont, Jr
Craig, Edward J(oseph)
Dalman, G(isli) Conrad
Daskam, Edward, Jr
Debany, Warren Harding, Jr
De France, Joseph J
Dehn, Rudolph A(lbert)
Del Toro, Vincent
De Mello, F Paul
Demerdash, Nabeel A O
Dennard, Robert H
DeRusso, Paul M(adden)
Desrochers, Alan Alfred
Diament, Paul
Diamond, Fred Irwin
Dickey, Frank R(amsey), Jr
Difranco, Julius V
Dill, Frederick H(ayes), Jr
Dolin, Richard
Duggin, Michael J
Duncan, Charles Clifford
Eastman, Dean Eric
Eastman, Lester F(uess)
Eichelberger, Edward B
Eichmann, George
Eisenberg, Lawrence
Elbaum, Marek
Ellert, Frederick J
Erdos, Marianne E
Erickson, William Harry
Ettenberg, M(orris)
Eveleigh, Virgil W(illiam)
Everett, Woodrow W, III
Feiker, George E(dward), Jr
Feisel, Lyle Dean
Felsen, Leopold Benno
Feth, George C(larence)
Fiedler, Harold Joseph
Fink, Donald G(len)
Flores, Ivan
Florman, Monte
Fordon, Wilfred Aaron
Forsyth, Eric Boyland
Fowler, James A
Frederick, Dean Kimball
Frisch, I(van) T
Fuller, Lynn Fenton
Geiss, Gunther R(ichard)
George, Nicholas
Gerace, Paul Louis
Gerhardt, Lester A
Ghozati, Seyed-Ali
Gildersleeve, Richard E
Giordano, Anthony B(runo)
Gisser, David G
Givone, Donald Daniel
Glasford, Glenn M(ilton)
Goldstein, Lawrence Howard
Gonzalez, Andrew
Goodheart, Clarence F(rancis)
Gran, Richard J

Grant, Ian S
Gratian, J(oseph) Warren
Greenwood, Allan Nunns
Griffith, John E(mmett)
Groner, Carl Fred
Gronner, A(lfred) D(ouglas)
Gross, Victor
Grover, Paul L, Jr
Gruenberg, Harry
Gutmann, Ronald Jay
Haddad, Richard A
Hall, Richard Travis
Hammam, M Shawky
Hanson, William A
Happ, Harvey Heinz
Harley, Naomi Hallden
Harrington, Dean Butler
Harrington, Roger F(uller)
Harris, Bernard
Harris, Lawson P(arks)
Harvey, Alexander Louis
Hauspurg, Arthur
Hayter, Walter R, Jr
Hedman, Dale E
Hesse, M(ax) H(arry)
Hessel, Alexander
Honsinger, Vernon Bertram
Hopcroft, John E(dward)
Houde, Robert A
Hsiang, Thomas Y
Hsu, Charles Ching-Hsiang
Hu, Ming-Kuei
Huening, Walter C, Jr
Hunt, Donald F(ulper)
Hyman, Abraham
Indusi, Joseph Paul
Jacobsen, William E
Jatlow, J(acob) L(awrence)
Javid, Mansour
Jelinek, Frederick
Jocoy, Edward Henry
Johnson, Charles Richard, Jr
Johnson, I Birger
Jones, Leonidas John
Karatzas, Ioannis
Karnaugh, Maurice
Katz, Harold W(illiam)
Kear, Edward B, Jr
Keller, Arthur Charles
Kelley, Michael C
Kent, Gordon
Ketchen, Mark B
Kinnen, Edwin
Kirby, Maurice J(oseph)
Kirchgessner, Joseph L
Kirchmayer, Leon K
Kiszenick, Walter
Kletsky, Earl J(ustin)
Kliman, Gerald Burt
Koehler, Richard Frederick, Jr
Komiak, James Joseph
Koplowitz, Jack
Kopp, Richard E
Korwin-Pawlowski, Michael Lech
Kranc, George M(aximilian)
Krembs, G(eorge) M(ichael)
Kunhardt, Erich Enrique
Kwok, Hoi S
Kwok, Thomas Yu-Kiu
Laghari, Javaid Rosoolbux
La Tourrette, James Thomas
Lauber, Thornton Stuart
Lee, Robert E
Leland, Harold R(obert)
Le Page, Wilbur R(eed)
Levi, Enrico
Levine, Alfred Martin
Ley, B James
Linder, Clarence H
Lindley, Kenneth Eugene
Linke, Simpson
Litman, Bernard
Litynski, Daniel Mitchell
Logan, Joseph Skinner
Lounsbury, John Baldwin
Low, Paul R
Lurch, E(dward) Norman
Lurkis, Alexander
Ma, William Hsioh-Lien
McCary, Richard O
McDonald, John F(rancis)
MacDonald, Noel C
McGaughan, Henry S(tockwell)
McGrath, Eugene R
McIsaac, Paul R(owley)
McMurry, William
Madhu, Swaminathan
Mahrous, Haroun
Maling, George Croswell, Jr
Malone, Dennis P(hilip)
Maltz, Martin Sidney
Margolis, Stephen G(oodfriend)
Marsocci, Velio Arthur
Mathes, Kenneth Natt
Matick, Richard Edward
Mayer, James W(alter)
Meadows, Henry E(merson), Jr
Merriam, Charles Wolcott, III
Meth, Irving Marvin
Mgrdechian, Raffee
Middleton, David
Mihran, Theodore Gregory
Moghadam, Omid A
Morris, George Michael

Morris, Thomas Wilde
Morton, Roger Roy Adams
Mosteller, Henry Walter
Mundel, August B(aer)
Nagy, George
Nation, John
Nawrocky, Roman Jaroslaw
Nelson, John Keith
Nichols, Benjamin
Nisteruk, Chester Joseph
Oden, Peter Howland
Oh, Se Jeung
O'Kean, Herman C
Padegs, Andris
Pardee, Otway O'Meara
Parks, Thomas William
Patel, Sharad A
Paul, George, Jr
Pavlidis, Theo
Paxton, K Bradley
Peng, Song-Tsuen
Perini, Jose
Perry, Michael Paul
Peskin, Arnold Michael
Peters, Philip H
Piatkowski, Thomas Frank
Plotkin, Martin
Possin, George Edward
Pottle, Christopher
Powell, Noble R
Price, Frank DuBois
Raghuveer, Mysore R
Rappaport, Stephen S
Reinhart, Stanley E, Jr
Richardson, Robert Lloyd
Richman, Donald
Ringlee, Robert J
Robinson, Charles C(anfield)
Rodems, James D
Rootenberg, Jacob
Rose, Kenneth
Rosenfeld, Jack Lee
Rosenthal, Saul W
Ross, John B(ye)
Rothschild, David (Seymour)
Rubin, Bruce Joel
Rubin, Joel E(dward)
Rubloff, Gary W
Ruehli, Albert Emil
Rushing, Allen Joseph
Russo, Roy Lawrence
Ryerson, Joseph L
Sanderson, Arthur Clark
Sandler, Melvin
Sanford, Richard Selden
Sarachik, Philip E(ugene)
Sarjeant, Walter James
Schanker, Jacob Z
Schlesinger, S Perry
Schultz, William C(arl)
Schwartz, Mischa
Schwartz, Richard F(rederick)
Schwartzman, Leon
Schwarz, Ralph J
Sen, Amiya K
Sen, Ujjal
Shamis, Sidney S
Shapiro, Sidney
Sharbaugh, Amandus Harry
Shaw, David T
Shaw, Leonard G
Shelley, Edwin F(reeman)
Shen, Wu-Mian
Shenolikar, Ashok Kumar
Shinners, Stanley Marvin
Shmoys, Jerry
Shulman, Carl
Sie, Charles H
Silverman, Bernard
Simpson, Murray
Slade, Bernard Newton
Slaymaker, Frank Harris
Smilowitz, Bernard
Smith, Albert A, Jr
Smith, David R(ichard)
Smith, Edward J(oseph)
Smith, Joseph H, Jr
Smith, Robert L
Snelling, Christopher
Songster, Gerard F(rancis)
Spang, H Austin, III
Stancampiano, Charles Vincent
Staples, Basil George
Star, Joseph
Stern, Miklos
Strait, Bradley Justus
Stratford, R P
Stratton, Roy Franklin, Jr
Strauss, Leonard
Strohm, Warren B(ruce)
Stutt, Charles A(dolphus)
Su, Stephen Y H
Sublette, Ivan H(ugh)
Subramanian, Mani M
Suenaga, Masaki
Tamir, Theodor
Tang, Donald T(ao-Nan)
Tanguay, A(rmand) R(ene)
Taub, Herbert
Taub, Jesse J
Taylor, James Hugh
Taylor, Kenneth Doyle
Tehon, Stephen Whittier
Thau, Frederick E

Thomann, Gary C
Thorp, James Shelby
Timms, Robert J
Titlebaum, Edward Lawrence
Truxal, John G(roff)
Tschang, Pin-Seng
Tseng, Fung-I
Tsividis, Yannis P
Tuan, Hang-Sheng
Unger, S(tephen) H(erbert)
Van Der Voorn, Peter C
Van Wagner, Edward M
Varshney, Pramod Kumar
Viertl, John Ruediger Mader
Vrana, Norman M
Waag, Robert Charles
Wang, Thomas Nie-Chin
Wang, Wen I
Wang, Xingwu
Watson, P(ercy) Keith
Weinberg, Norman
Weissman, David E(verett)
Welsh, James P
Weng, Wu Tsung
Westover, Thomas A(rchie)
Whalen, James Joseph
Wheeless, Leon Lum, Jr
Wiley, Richard G
Wilson, Delano D
Wilson, James William Alexander
Wing, Omar
Winkler, Leonard P
Winsor, Lauriston P(earce)
Winter, Herbert
Witt, Christopher John
Wobschall, Darold C
Wong, Edward Chor-Cheung
Wood, Allen John
Woodard, David W
Woods, John William
Worthing, Jurgen
Wouk, Victor
Wysocki, Joseph J(ohn)
Yao, David D
Yasar, Tugrul
York, Raymond A
Youla, Dante C
Yu, Hwa Nien
Yu, Kai-Bor
Zeidenbergs, Girts
Zic, Eric A
Zimmerman, Stanley William
Zukowski, Charles Albert

NORTH CAROLINA
Agrawal, Dharma Prakash
Allen, Charles Michael
Baliga, Jayant
Barclay, W(illiam) J(ohn)
Bell, Norman R(obert)
Bellack, Jack H
Burton, Ralph Gaines
Caldwell, Samuel Craig
Casey, H Craig, Jr
Casey, Horace Craig, Jr
Chou, Wushow
Coleman, Robert J(oseph)
Dickson, LeRoy David
Douglas, Edward Curtis
Easter, William Taylor
Eckels, Arthur R(aymond)
Ellis-Menaghan, John J
Fair, Richard Barton
Flood, Walter A(loysius)
Franzon, Paul Damian
Gordy, Thomas D(aniel)
Grady, Perry Linwood
Hacker, Herbert, Jr
Harris, Roy H
Hauser, John Reid
Hester, Donald L
Hoadley, George B(urnham)
Huber, William Richard, III
Jones, Creighton Clinton
Kauffman, James Frank
Kim, Chang-Sik
Kirschbaum, H(erbert) S(pencer)
Kuhn, Matthew
Lamorte, Michael Francis
Leech, Robert Leland
Lovick, Robert Clyde
McAlpine, George Albert
Masnari, Nino A
Massoud, Hisham Z
Matthews, N(eely) F(orsyth) J(ones)
Mink, James Walter
Monteith, Larry King
Moore, Edward T(owson)
Nagle, H Troy
Nolte, Loren W
O'Neal, John Benjamin, Jr
Owen, Harry A(shton), Jr
Peterson, David West
Pilkington, Theo C(lyde)
Redmon, John King
Reenstra, Arthur Leonard
Rhodes, Donald R(obert)
Samuels, Robert Lynn
Sandok, Paul Louis
Simons, Mayrant, Jr
Smith, Charles Eugene
Snyder, Wesley Edwin
Stafford, Bruce H(ollen)
Steer, Michael Bernard

Stevenson, William D(amon), Jr
Struve, William George
Suttle, Jimmie Ray
Tischer, Frederick Joseph
Tommerdahl, James B
Trussell, Henry Joel
Vander Lugt, Anthony
Wayne, Burton Howard
Weber, Ernst
Williams, Leo, Jr
Wilson, Thomas G(eorge)
Wright, William V(aughn)
Wylie, C J

NORTH DAKOTA
Anderson, Edwin Myron
Bares, William Anthony
Krueger, Jack N
Tareski, Val Gerard
Thomforde, C(lifford) J(ohn)
Yuvarajan, Subbaraya

OHIO
Adams, William Lawreson
Banda, Siva S
Barranger, John P
Batcher, Kenneth Edward
Bhagat, Pramode Kumar
Biedenbender, Michael David
Black, John Wilson
Brandeberry, James E
Brown, Frank Markham
Buckingham, William Thomas
Burghart, James H(enry)
Burhans, Ralph W(ellman)
Campbell, Richard M
Ceasar, Gerald P
Chen, Chiou Shiun
Chen, Hollis C(hing)
Chipman, R(obert) A(very)
Colbert, Charles
Collin, Robert E(manuel)
Compton, Ralph Theodore, Jr
Cornetet, Wendell Hillis, Jr
Corum, James Frederic
Cowan, John D, Jr
Dakin, James Thomas
Damon, Edward K(ennan)
Dayton, James Anthony, Jr
D'Azzo, John Joachim
Dell'Osso, Louis Frank
Dorsey, Robert T
El-Naggar, Mohammed Ismail
Eltimsahy, Adel H
Emmerling, Edward J
Engelmann, Richard H(enry)
Eppers, William C
Ernst, George W
Essman, Joseph Edward
Ewing, Donald J(ames), Jr
Faiman, Robert N(eil)
Falk, Harold Charles
Farison, James Blair
Fenton, Robert E
Forestieri, Americo F
Garbacz, Robert J
Ghandakly, Adel Ahmad
Golovin, Michael N
Goradia, Chandra P
Grant, Michael P(eter)
Gruber, Sheldon
Grumbach, Robert S(tephen)
Hall, Ernest Lenard
Hancock, Harold E(dwin)
Harpster, Joseph
Hazen, Richard R(ay)
Hemami, Hooshang
Hill, Jack Douglas
Hodge, Daniel B
Hogan, James J(oseph)
Holmes, Roger Arnold
Hornfeck, Anthony J
Horton, Billy Mitchusson
Hsu, Hsiung
Huang, John Wen-Chih
Ismail, Amin Rashid
Jackson, Warren, Jr
Johnson, Ray O
Jordan, Howard Emerson
Kashkari, Chaman Nath
Katona, Peter Geza
Keister, James E
Klock, Harold F(rancis)
Ko, H(sien) C(hing)
Koozekanai, Said H
Kramer, Raymond Edward
Kramerich, George L
Kraus, John Daniel
Ksienski, A(haron)
Kwatra, Subhash Chander
Lamont, Gary Byron
Lawler, Martin Timothy
Lawton, John G
Lee, Kai-Fong
Lemke, Ronald Dennis
Levis, C(urt) A(lbert)
Lewis, Donald Edward
Long, Ronald K(illworth)
McConnell, H(oward) M(arion)
McFarland, Richard Herbert
McMaster, Robert Charles
McNair, Robert J(ohn), Jr
Manning, Robert M
Mathews, John David

Electrical Engineering (cont)

Mayhan, Robert J(oseph)
Menq, Chia-Hsiang
Mergler, H(arry) W(inston)
Middendorf, William H
Moffatt, David Lloyd
Mudry, Karen Michele
Munk, Benedikt Aage
Neuman, Michael R
Olson, Karl William
Osterbrock, Carl H
Pao, Yoh-Han
Papachristou, Christos A
Pathak, Prabhakar H
Peters, Leon, Jr
Puglielli, Vincent George
Qureshi, A H
Raible, Robert H(enry)
Rajsuman, Rochit
Ramamoorthy, Panapakkam A
Rattan, Kuldip Singh
Repperger, Daniel William
Restemeyer, William Edward
Richmond, J(ack) H(ubert)
Rose, Harold Wayne
Rose, Mitchell
Schultz, Ronald G(len)
Schutta, James Thomas
Sebo, Stephen Andrew
Shenk, William E(dwin)
Shenoi, Belle Anantha
Shirk, B(rian) Thomas
Sibila, Kenneth Francis
Smith, Neal A(ustin)
Soska, Geary Victor
Springer, Robert Harold
Steckl, Andrew Jules
Stewart, John
Strnat, Karl J
Stuart, Thomas Andrew
Thiele, Gary Allen
Thomas, Cecil Wayne
Thorbjornsen, Arthur Robert
Thurston, M(arlin) O(akes)
Tsui, James Bao-Yen
Vassell, Gregory S
Ware, Brendan J
Warren, Claude Earl
Webster, Allen E
Wee, William Go
Weed, Herman Roscoe
Weimer, F(rank) Carlin
Wigington, Ronald L
Yeh, Hsi-Han
Zavodney, Lawrence Dennis
Zheng, Yuan Fang

OKLAHOMA

Baker, James E
Basore, B(ennett) L(ee)
Batchman, Theodore E
Berbari, Edward J
Bilger, Hans Rudolf
Breipohl, Arthur M
Buffum, C(harles) Emery
Cheung, John Yan-Poon
Cole, Charles F(rederick), Jr
Cronenwett, William Treadwell
Cummins, Richard L
El-Ibiary, Mohamed Yousif
Engle, Jessie Ann
Gudehus, Donald Henry
Hughes, William (Lewis)
Kahng, Seun Kwon
Kuriger, William Louis
Lal, Manohar
Lee, Samuel C
Lingelbach, D(aniel) D(ee)
McCollom, Kenneth A(llen)
Mulholland, Robert J(oseph)
Ramakumar, Ramachandra Gupta
Richardson, J Mark
Smith, Gerald Lynn
Strattan, Robert Dean
Tuma, Gerald
Vanderwiele, James Milton
Walker, Gene B(ert)
Wang, Richard Jinchi
Williams, Thomas Henry Lee
Yarlagadda, Radha Krishna Rao
Zelby, Leon W

OREGON

Alexander, Gerald Corwin
Arthur, John Read, Jr
Berglund, Carl Neil
Blumhagen, Vern Allen
Casperson, Lee Wendel
Chartier, Vernon L
Collins, John A(ddison)
Davenport, Wilbur B(ayley), Jr
Ede, Alan Winthrop
Engelking, Paul Craig
Engle, John Franklin
Gens, Ralph S
Ginsburg, Charles P
Hansen, Hugh J
Hayes, Thomas B
Herzog, James Herman
Holmes, James Frederick
Isaacson, Dennis Lee
Johnston, George I
Jones, Donlan F(rancis)

Lendaris, George G(regory)
Magnusson, Philip C(ooper)
Matson, Leslie Emmet, Jr
Mohler, Ronald Rutt
Morgan, Merle L(oren)
Schaumann, Rolf
Schrumpf, Barry James
Short, Robert Allen
Skinner, Richard Emery
Starr, E(ugene) C(arl)
Stewart, Bradley Clayton
Stone, Solon Allen
Taylor, Carson William
Tinney, William Frank
Tripathi, Vijai Kumar
Wager, John Fisher
Yamaguchi, Tadanori
Yau, Leopoldo D

PENNSYLVANIA

Abend, Kenneth
Aburdene, Maurice Felix
Adams, Raymond F
Adams, William Sanders
Allen, Charles C(orletta)
Anderson, A Eugene
Arora, Vijay Kumar
Ashok, S
Aston, Richard
Barus, Carl
Batchelor, John W
Baxter, Ronald Dale
Bedrosian, Samuel D
Berger, James M
Berkowitz, Raymond S
Bilgutay, Nihat Mustafa
Bloor, W(illiam) Spencer
Bockosh, George R
Bordogna, Joseph
Bose, Nirmal K
Boston, John Robert
Brastins, Auseklis
Brennan, Thomas Michael
Brown, Homer E
Brown, John Lawrence, Jr
Buchsbaum, Gershon
Cain, James Thomas
Caplan, Aubrey G
Carpenter, Lynn Allen
Casasent, David P(aul)
Cendes, Zoltan Joseph
Charap, Stanley H
Charp, Solomon
Cohn, Nathan
Colclaser, Robert Gerald
Cookson, Alan Howard
Coren, Richard L
Cross, Leslie Eric
Dahlke, Walter Emil
Das, Mukunda B
Davis, Paul Cooper
Delansky, James F
Denshaw, Joseph Moreau
Detwiler, John Stephen
Dick, George W
Director, Stephen William
Dodds, Wellesley Jamison
Dordick, Herbert S(halom)
Dorny, C Nelson
Douglas, Joseph Francis
Drumwright, Thomas Franklin, Jr
Dumbri, Austin C
Dunn, Charles Nord
Eisenstein, Bruce A
Embree, M(ilton) L(uther)
Engheta, Nader
Etzweiler, George Arthur
Falconer, Thomas Hugh
Fan, Ningping
Feero, William E
Fegley, Kenneth A(llen)
Felker, Jean Howard
Feng, Tse-yun
Finin, Timothy Wilking
Fischer, John Edward
Frost, L(eslie) S(wift)
Furfari, Frank A
Gallagher, John M(ichael), Jr
Goff, Kenneth W(ade)
Goldwasser, Samuel M
Goodman, Robert M(endel)
Gorham, R C
Gray, H(arry) J(oshua)
Grimes, Dale M(ills)
Guyker, William C, Jr
Haber, Fred
Hamilton, Howard Britton
Hamilton, William Howard
Harder, Edwin L
Hargens, Charles William, III
Harrold, Ronald Thomas
Herczfeld, Peter Robert
Hielscher, Frank Henning
Hileman, Andrew R
Hoburg, James Frederick
Hockenberry, Terry Oliver
Hoelzeman, Ronald G
Hoover, L Ronald
Houser, Wesley G(rant)
Hutcheson, J(ohn) A(lister)
Hutzler, John R
Hyatt, Robert Monroe
Jaggard, Dwight Lincoln
Jang, Sei Joo

Jarem, John
Jaron, Dov
Jeruchim, Michel Claude
Johnson, Alfred Theodore, Jr
Jones, A(ndrew) R(oss)
Jones, Clifford Kenneth
Jordan, A(ngel) G(oni)
Jordan, Angel G
Kam, Moshe
Karakash, John J
Kassam, Saleem Abdulali
Kaufman, William
Kaufman, William Morris
Kilgore, Lee A
Kingrea, James I
Klafter, Richard D(avid)
Klein, Gerald I(rwin)
Koepfinger, J L
Koffman, Elliot B
Kritikos, Haralambos N
Ku, Y(u) H(siu)
Lakhtakia, Akhlesh
Larky, Arthur I(rving)
Lavi, Abrahim
Leas, J(ohn) W(esley)
Leibholz, Stephen W
Lex, R(owland) G(arber), Jr
Li, C(hing) C(hung)
Liberman, Irving
Lory, Henry James
Lowry, Lewis Roy, Jr
McAdam, Will
McCormick, Paul R(ichard)
McCracken, Leslie G(uy), Jr
McCrumb, J(ohn) D(oench)
McGarity, Arthur Edwin
McMurtry, George James
MacRae, Donald Richard
Magison, Ernest Carroll
Maltby, Frederick L(athrop)
Mastascusa, Edward John
Matcovich, Thomas J(ames)
Mathews, C A
Mathias, Robert A(ddison)
Meiksin, Zvi H(ans)
Metzner, John J(acob)
Mickle, Marlin Homer
Milnes, Arthur G(eorge)
Mitchell, John Douglas
Mitchell, Tom M
Molter, Lynne Ann
Mondal, Kalyan
Mouly, Raymond J
Murphy, John N
Nelson, Richard L(loyd)
Neuman, Charles P(aul)
Nisbet, John S(tirling)
Nygren, Stephen Fredrick
Olson, H(ilding) M
Pearce, Charles Walter
Peck, D Stewart
Penn, William B
Perks, Norman William
Peterson, Wilbur Carroll
Phares, Alain Joseph
Pittman, G(eorge) F(rank), Jr
Przybysz, John Xavier
Putman, Thomas Harold
Rabii, Sohrab
Riemersma, H(enry)
Robinson, James William
Robl, Robert F(redrick), Jr
Rohrer, Ronald A
Ross, Charles W(arren)
Ross, William J(ohn)
Rowe, Joseph E(verett)
Rowlands, R(ichard) O(wen)
Rubin, Herbert
Rubinoff, Morris
Salati, Octavio M(ario)
Schatz, Edward R(alph)
Schroeder, Alfred C(hristian)
Schwan, Herman Paul
Shen, D(avid) W(ei) C(hi)
Shibib, M Ayman
Shobert, Erle Irwin, II
Showers, Ralph M(orris)
Shung, K Kirk
Siewiorek, Daniel Paul
Simaan, Marwan
Smith, David Beach
Smith, Richard G(rant)
Smith, Warren LaVerne
Smock, Dale Owen
Smyth, Michael P(aul)
Spandorfer, Lester M
Spielvogel, Lawrence George
Stein, Jack J(oseph)
Stern, Richard Martin, Jr
Stone, Gregory Michael
Stone, Robert P(orter)
Sun, Hun H
Sutherland, Ivan Edward
Sutter, Philip Henry
Sverdrup, Edward F
Sze, Tsung Wei
Szedon, John R(obert)
Taylor, Harry Elmer
Thompson, Francis Tracy
Thompson, William, Jr
Tomiyasu, Kiyo
Trainer, Michael Norman
Travis, Irven

Tretiak, Oleh John
Tzeng, Kenneth Kai-Ming
Uher, Richard Anthony
Urkowitz, Harry
Van Der Spiegel, Jan
Van Gelder, Arthur
Van Raalte, John A
Van Valkenburg, Ernest Scofield
Varadan, Vasundara Venkatraman
Varadan, Vijay K
Vogt, William G(eorge)
Voshall, Roy Edward
Wagh, Meghanad D
Wahl, A(rthur) J
Warren, S Reid, Jr
Warshawsky, Jay
Weaver, L(elland) A(ustin) C(harles)
Weeks, L(oraine) H(ubert)
Wemple, Stuart H(arry)
Werth, John
Whitehead, Daniel L(ee)
Whitney, Eugene C
Williams, Albert J, Jr
Wolfe, Reuben Edward
Young, Frederick J(ohn)
Young, Raymond H(ykes)
Yu, Francis T S
Zebrowitz, S(tanley)
Zocholl, Stanley E

RHODE ISLAND

Bartram, James F(ranklin)
Cooper, David B
Daly, James C(affrey)
Davis, John Sheldon
Hazeltine, Barrett
Hoffman, Joseph Ellsworth, Jr
Jackson, Leland Brooks
King, Charles W(illis)
Kornhauser, Edward T(heodore)
Krikorian, John Sarkis, Jr
Kushner, Harold J(oseph)
Lengyel, G(abriel)
Middleton, Foster H(ugh)
Park, Robert H
Petrou, Panayiotis Polydorou
Polk, C(harles)
Prince, Mack J(ames)
Savage, John Edmund
Spence, John Edwin
Tufts, Donald Winston
Wolovich, William Anthony

SOUTH CAROLINA

Askins, Harold Williams Jr, Jr
Bennet, Archie Wayne
Bennett, Archie Wayne
Bourkoff, Etan
Chisman, James Allen
Dumin, David Joseph
Eccles, William J
Ernst, Edward W
Fellers, Rufus Gustavus
Garzon, Ruben Dario
Geary, Leo Charles
Gilliland, Bobby Eugene
Gowdy, John Norman
Halford, Jake Hallie
Hammond, Joseph Langhorne, Jr
Hogan, Joseph C(harles)
Johnson, Luther Elman
Lathrop, Jay Wallace
Long, Jim T(homas)
Luh, Johnson Yang-Seng
McCarter, James Thomas
Maricq, John
Martin, John Campbell
Pearson, Lonnie Wilson
Peeples, Johnston William
Scoggin, James F, Jr
Sias, Fred R, Jr
Simpson, John W(istar)
Snelsire, Robert W
Sudarshan, Tangali S
Thurston, James N(orton)
Yang, Tsute

SOUTH DAKOTA

Cannon, Patrick Joseph
Cox, Cyrus W(illiam)
Gowen, Richard J
Hanson, Jerry Lee
Kurtenbach, Aelred J(oseph)
Manning, M(elvin) L(ane)
Reuter, William L(ee)
Riemenschneider, Albert Louis
Ross, Keith Alan
Sander, Duane E
Smith, Paul Letton, Jr
Storry, Junis O(liver)

TENNESSEE

Adams, Raymond Kenneth
Benson, Robert Wilmer
Bishop, Asa Orin, Jr
Blalock, Theron Vaughn
Bose, Bimal K
Brodersen, Arthur James
Cadzow, James A(rchie)
Chowdhuri, Pritindra
Cook, George Edward
Dicker, Paul Edward
Essler, Warren O(rvel)
Gonzalez, Rafael C

Green, Walter L(uther)
Harter, James A(ndrew)
Hoffman, G(raham) W(alter)
Hung, James Chen
Jermann, William Howard
Johnson, L(ee) Ensign
Joseph, Roy D
Kawamura, Kazuhiko
Kennedy, Eldredge Johnson
Leffell, W(ill) O(tis)
Malkani, Mohan J
Mosko, Sigmund W
Muly, Emil Christopher, Jr
Murakami, Masanori
Nash, Robert T
Neff, Herbert Preston, Jr
Noonan, John Robert
Oakes, Lester C
Ornltz, Barry Louis
Pace, Marshall Osteen
Parrish, Edward Alton, Jr
Pearsall, S(amuel) H(aff)
Pierce, John Frank
Rajan, Periasamy Karivaratha
Rochelle, Robert W(hite)
Roth, J(ohn) Reece
Shockley, Thomas D(ewey), Jr
Terry, Fred Herbert
Tillman, J(ames) D(avid), Jr
Waller, John Wayne
Weaver, Charles Hadley
Wessenauer, Gabriel Otto
White, Edward John

TEXAS
Aggarwal, Jagdishkumar Keshoram
Aiken, James G
Allen, John Burton
Anderson, Dale Arden
Anderson, Wallace L(ee)
Bailey, James Stephen
Baker, Wilfred E(dmund)
Barton, James Brockman
Bean, Wendell Clebern
Becker, Michael Franklin
Benning, Carl J, Jr
Bhattacharyya, Shankar P
Biard, James R
Bishop, Robert H
Bowdon, Edward K(night), Sr
Bryant, Michael David
Burrus, Charles Sidney
Cantrell, Cyrus D, III
Cash, Floyd Lee
Caswell, Gregory K
Chakravarty, Indranil
Chang, Christopher Teh-Min
Chen, Mo-Shing
Chenoweth, R(obert) D(ean)
Cherrington, Blake Edward
Choate, William Clay
Collins, Francis Allen
Cooke, James Louis
Cory, William Eugene
Crossley, Peter Anthony
Crum, Floyd M(axilas)
Crumb, Stephen Franklin
David, Yadin B
Denison, John Scott
Dodd, Edward Elliott
Donaldson, W(illis) Lyle
Dougal, Arwin A(delbert)
Duesterhoeft, William Charles, Jr
Dunham, James George
Dupuis, Russell Dean
Dutcher, Clinton Harvey, Jr
Eknoyan, Ohannes
Ekstrom, Michael P
Engelhardt, Albert George
Fannin, Bob M(eredith)
Fink, Lyman R
Firestone, William L(ouis)
Fite, Lloyd Emery
Fitzer, Jack
Franke, Ernest A
Friberg, Emil Edwards
Friedrich, Otto Martin, Jr
Gordon, William E(dwin)
Green, Terry C
Griffith, M S
Grimes, Craig Alan
Gully, John Houston
Gupta, Someshwar C
Hagler, Marion O(tho)
Hanne, John R
Hatsell, Charles Proctor
Haupt, L(ewis) M(cDowell), Jr
Hayden, Edgar C(lay)
Hayre, Harbhajan Singh
Heizer, Kenneth W
Hixson, Elmer LaVerne
Hoehn, G(ustave) L(eo)
Hoffmann-Pinther, Peter Hugo
Hopkins, George H, Jr
Howard, Lorn Lambier
Huang, Jung-Chang
Johnson, Richard Leon
Jones, William B(enjamin), Jr
Jump, J Robert
Kesler, Oren Byrl
Kilby, Jack St Clair
King, Willis Kwongtsu
Koepsel, Wellington Wesley
Korges, Emerson

Kostelnicek, Richard J
Krezdorn, Roy R
Krishen, Kumar
Kristiansen, Magne
Ku, Bernard Siu-Man
Kunze, Otto Robert
LaGrone, Alfred H(all)
Leeds, J Venn, Jr
Lengel, Robert Charles
Leonard, William F
Lewis, Frank Leroy
Liang, Charles Shih-Tung
Ling, Hao
Long, Stuart A
MacFarlane, Duncan Leo
Malek, Miroslaw
Marcus, Steven Irl
Markeuscoff, Pauline
Masten, Michael K
Matzen, Walter T(heodore)
Mays, Robert, Jr
Musa, Samuel A
Narasimhan, Mandayam A
Neikirk, Dean P
Nevels, Robert Dudley
Newton, Richard Wayne
Novak, Gordon Shaw, Jr
Ong, Poen Sing
Opiela, Alexander David, Jr
Ott, Granville E
Overzet, Lawrence John
Owen, Thomas E(dwin)
Paskusz, Gerhard F
Patel, Anil S
Paterson, James Lenander
Patton, Alton DeWitt
Pearson, James Boyd, Jr
Peikari, Behrouz
Perez, Ricardo
Pinkston, John Turner, III
Portnoy, William M
Powers, Edward Joseph, Jr
Powers, Robert S(inclair), Jr
Pritchett, William Carr
Provencio, Jesus Roberto
Puckett, Russell Elwood
Raju, Satyanarayana G V
Renner, Darwin S(prathard)
Reynolds, Jack
Rhyne, V(ernon) Thomas
Ribe, M(arshall) L(ouis)
Rogers, Phil H
Rupf, John Albert, Jr
Russell, B Don
Salis, Andrew E
Sandberg, I(rwin) W(alter)
Satterwhite, Ramon S(tewart)
Schell, Robert Ray
Schoenberger, Michael
Schulz, Richard Burkart
Shah, Pradeep L
Shen, Liang Chi
Shepherd, Mark, Jr
Sheppard, Louis Clarke
Shieh, Leang-San
Shrime, George P
Simpson, Richard S
Singh, Vijay Pal
Smith, Harold W(ood)
Smith, Jack
Sobol, Harold
Spooner, M(orton) G(ailend)
Srinath, Mandyam Dhati
Srinivasan, Ramachandra Srini
Stockton, James Evan
Stone, Orville L
Straiton, Archie Waugh
Streetman, Ben Garland
Szygenda, Stephen Anthony
Tasch, Al Felix, Jr
Taylor, Herbert Lyndon
Thorn, Donald Childress
Tittel, Frank K(laus)
Trench, William Frederick
Victor, Joe Mayer
Vines, Darrell Lee
Volz, Richard A
Von Maltzahn, Wolf W
Wakeland, William Richard
Walkup, John Frank
Wang, Yuan R
Ward, Ronald Wayne
Watt, Joseph T(ee), Jr
Welch, Ashley James
Weldon, William Forrest
Widerquist, V(ernon) R(oberts)
Will, Peter Milne
Willman, Joseph F(rank)
Wilton, Donald Robert
Woodson, Herbert H(orace)
Zrudsky, Donald Richard

UTAH
Atwood, Kenneth W
Baker, Kay Dayne
Barlow, Mark Owens
Berrett, Paul O(rin)
Bowman, Lawrence Sieman
Chabries, Douglas M
Chadwick, Duane G(eorge)
Christensen, Douglas Allen
Clark, Clayton
Clegg, John C(ardwell)
Cole, Larry S

Comer, David J
Durney, Carl H(odson)
Eer Nisse, Errol P(eter)
Embry, Bertis L(loyd)
Evans, David C(annon)
Gandhi, Om P
Hammond, Seymour Blair
Harris, Ronney D
Hollaar, Lee Allen
Keller, Robert Marion
Long, David G
Losee, Ferril A
Rushforth, Craig Knewel
Skinner, Robert L
Smith, Kent Farrell
Stephenson, Robert E(ldon)
Stringfellow, Gerald B
Woodbury, Richard C

VERMONT
Anderson, R(ichard) L(ouis)
Clark, Donald Lyndon
Critchlow, D(ale)
Ellis, David M
Evering, Frederick Christian, Jr
Golden, Kenneth Ivan
Gray, Kenneth Stewart
Higgins, John J
Quinn, Robert M(ichael)
Rush, Stanley
Stapper, Charles Henri
Wallis, Clifford Merrill
Williams, Ronald Wendell

VIRGINIA
Abbott, Dean William
Abrams, Marshall D
Aller, James Curwood
Aylor, James Hiram
Baeumler, Howard William
Bahl, Inder Jit
Bass, Steven Craig
Batte, William Granville
Beale, Guy Otis
Beran, Robert Lynn
Blackwell, William A(llen)
Blasbalg, Herman
Bostian, Charles William
Brown, Gary S
Bruno, Ronald C
Bunting, Christopher David
Casazza, John Andrew
Chen, Pi-Fuay
Cibulka, Frank
Clark, Ralph Leigh
Claus, Richard Otto
Cook, Gerald
Curry, Thomas F(ortson)
Davis, William Arthur
Dean, William C(orner)
Delnore, Victor Eli
De Wolf, David Alter
Dharamsi, Amin N
Doles, John Henry, III
Duff, William G
Edvane, John
Egan, Raymond D(avis)
Eggleston, John Marshall
Ellersick, Fred W(illiam)
Entzminger, John N
Farrell, Robert Michael
Farrukh, Usamah Omar
Fath, George R
Fink, Lester Harold
Fitts, Richard Earl
Fripp, Archibald Linley
Fthenakis, Emanuel John
Garcia, Albert B
Gee, Sherman
Gibson, John E(gan)
Ginsberg, Murry B(enjamin)
Godbole, Sadashiva Shankar
Goodman, A(lvin) M(alcolm)
Gordon, John P(etersen)
Gray, F(estus) Gail
Guerber, Howard P(aul)
Hall, Albert C(arruthers)
Hammond, Marvin H, Jr
Harrington, Robert Joseph
Hasenfus, Harold J(oseph)
Heebner, David Richard
Heffron, W(alter) Gordon
Henry, James H(erbert)
Herbst, Mark Joseph
Holt, Charles A(sbury)
Holzmann, Ernest G(unther)
Hopkins, Mansell Herbert, Jr
Huber, Paul W(illiam)
Humphris, Robert R
Hwang, William Gaong
Inigo, Rafael Madrigal
Jacobs, Ira
Johnson, Preston Benton
Jones, James L
Jordan, Kenneth L(ouis) Jr
Joshi, Suresh Meghashyam
Juang, Jer-Nan
Jud, Henry G
Kahn, Manfred
Kahn, Robert Elliot
Keblawi, Feisal Said
Keiser, Bernhard E(dward)
Kelly, Peter Michael
Kelsey, Eugene Lloyd

Kenyon, Stephen C
Kim, Yong Il
Landes, Hugh S(tevenson)
Larson, Arvid Gunnar
Latta, John Neal
Lee, Fred C
Lerner, Norman Conrad
Lord, Norman W
McCutchen, Samuel P(roctor)
McGregor, Dennis Nicholas
McVey, Eugene Steven
McWright, Glen Martin
Madan, Rabinder Nath
Manriquez, Rolando Paredes
Mattauch, Robert Joseph
Mikesell, Jon L
Miller, Robert H
Moseley, Sherrard Thomas
Nashman, Alvin E
Nichols, Lee L(ochhead), Jr
Nunnally, Huey Neal
Palmer, James D(aniel)
Park, John Howard, Jr
Phelan, James Frederick
Poon, Ting-Chung
Powley, George R(einhold)
Pryor, C(abell) Nicholas, Jr
Pulvari, Charles F
Raab, Harry Frederick, Jr
Rader, Louis T(elemacus)
Reed, Joseph
Russell, James Madison, III
Sage, Andrew Patrick
Schlegelmilch, Reuben Orville
Schneider, Ralph Jacob
Severns, Matthew Lincoln
Shur, Michael
Siegel, Clifford M(yron)
Singleton, James L
Smith, B(lanchard) D(rake), Jr
Stanley, William Daniel
Stephenson, Frederick William
Stutzman, Warren Lee
Tabak, Daniel
Thompson, Anthony Richard
Toida, Shunichi
Tsang, Wai Lin
Van Landingham, Hugh F(och)
Violette, Joseph Lawrence Norman
Walker, Loren Haines
Walsh, Edward Joseph
Waxman, Ronald
Weik, Martin H, Jr
Wu, Yung-Kuang

WASHINGTON
Ackerlind, E(rik)
Andersen, Jonny
Auth, David C
Baker, Richard A
Bhansali, Praful V
Blackburn, John Lewis
Brownlee, Donald E
Clark, Robert Newhall
Craine, Lloyd Bernard
Damborg, Mark J(ohannes)
Daniels, Patricia D
Dow, Daniel G(ould)
Dowis, W J
Eckersley, Alfred
Edwards, James Mark
El-Sharkawi, Mohamed A
Flechsig, Alfred J, Jr
Golde, Hellmut
Haralick, Robert M
Heindsmann, T(heodore) E(dward)
Heisler, Rodney
Helms, Ward Julian
Hill, W(illiam) Ryland
Holden, Alistair David Craig
Hower, Glen L
Hsu, C(hih) C(hi)
Hsu, Chin Shung
Isaak, Robert D(eets)
Ishimaru, Akira
Kim, Jae Hoon
Koerber, George G(regory)
Kofoid, Melvin J(ulius)
Lauritzen, Peter O
Lawrence, Gene Arthur
Liu, Chen-Ching
Lytle, Dean Winton
McSpadden, William Robert
Masden, Glenn W(illiam)
Medith, James S
Metz, Peter Robert
Mosher, Clifford Coleman, III
Noe, Jerre D(onald)
Noges, Endrik
Nutley, Hugh
Olsen, Robert Gerner
Oman, Henry
Peden, Irene C(arswell)
Preikschat, Ekkehard
Preikschat, F(ritz) K(arl)
Reynolds, Donald Kelly
Richardson, Richard Laurel
Ringo, John Alan
Robel, Gregory Frank
Rogers, Joseph Wood
Savol, Andrej Martin
Schrader, David Hawley
Schweitzer, Edmund Oscar, III
Seamans, David A(lvin)

Electrical Engineering (cont)

Seliga, Thomas A
Shamash, Yacov A
Somani, Arun Kumar
Spillman, Richard Jay
Stevens, Richard F
Stewart, John L(awrence)
Swarm, H(oward) Myron
Szablya, John F(rancis)
Tsang, Leung
Venkata, Subrahmanyam Saraswati
Walsh, John Breffni
Waltar, Alan Edward
Walton, Vincent Michael
West, Herschel J
Wood, Francis Patrick
Yore, Eugene Elliott
Young, James Arthur, Jr

WEST VIRGINIA
Blackwell, Lyle Marvin
Breece, Harry T, III
Brown, Paul B
Cannon, Walton Wayne
Cooley, Wils LaHugh
Jones, Edwin C
Keener, E(verett) L(ee)
Nutter, Roy Sterling, Jr
Russell, James A(lvin), Jr
Smith, Nelson S(tuart), Jr
Spaniol, Craig
Whittington, Bernard W

WISCONSIN
Battocletti, Joseph H(enry)
Bernstein, Theodore
Beyer, James B
Birkemeier, William P
Boettcher, Harold P(aul)
Bollinger, John G(ustave)
Brauer, John Robert
Burrage, Lawrence Minott
Cerrina, Francesco
Christensen, Erik Regnar
Davidson, Charles H(enry)
Denton, Denice Dee
Dietmeyer, Donald L
Eriksson, Larry John
Farrow, John H
Gray, Paul Eugene
Greiner, Richard A(nton)
Heinen, James Albin
Higgins, Thomas James
Horgan, James D(onald)
Hummert, George Thomas
Hyzer, William Gordon
Ishii, T(homas) Koryu
Jacobi, George (Thomas)
Kerkman, Russel John
Koenig, Eldo C(lyde)
Lind, Robert Wayne
Lipo, Thomas Anthony
Long, Willis Franklin
Martinson, Edwin O(scar)
Moeller, Arthur Charles
Morin, Dornis Clinton
Nash, William Hart
Niederjohn, Russell James
Nondahl, Thomas Arthur
Northouse, Richard A
Novotny, Donald Wayne
Prager, Stewart Charles
Rideout, Vincent C(harles)
Rosenthal, Jeffrey
Saleh, Bahaa E A
Scharer, John Edward
Schmitz, Norbert Lewis
Schotz, Larry
Scidmore, Allan K
Seshadri, Sengadu Rangaswamy
Shohet, Juda Leon
Skiles, James J(ean)
Smith, Luther Michael
Staats, Gustav W(illiam)
Steber, George Rudolph
Stremler, Ferrel G
Thompson, Paul DeVries
Tompkins, Willis Judson
Vairavan, Kasivisvanathan
Vernon, Ronald J
Walter, Gilbert G
Webster, John Goodwin
Wu, Sherman H
Zheng, Xiaoci

WYOMING
Beach, R(upert) K(enneth)
Constantinides, Christos T(heodorou)
Hakes, Samuel D(uncan)
Long, F(rancis) M(ark)
Rhodine, Charles Norman
Spenner, Frank J(ohn)
Weeks, Richard W(illiam)

PUERTO RICO
Hussain, Malek Gholoum Malek
Mercado-Jimenez, Teodoro

ALBERTA
Berg, Gunnar Johannes
Cormack, George Douglas
Goud, Paul A
Gourishankar, V (Gouri)

Malik, Om Parkash
Nutakki, Dharma Rao
Stein, Richard Adolph
Streets, Rubert Burley, Jr
Trofimenkoff, Frederick N(icholas)
Ulagaraj, Munivandy Seydunganallur
White, Lee James

BRITISH COLUMBIA
Antoniou, A(ndreas)
Beddoes, M(ichael) P(eter)
Booth, Andrew Donald
Davies, Michael Shapland
Deen, Mohamed Jamal
Dumont, Guy Albert
Kerr, James S(anford) S(tephenson)
Moore, A(rthur) Donald
Shulman, David-Dima
Soudack, Avrum Chaim
Zielinski, Adam

MANITOBA
Bridges, E(rnest)
Hamid, Michael
Johnson, R(ichard) A(llan)
Kao, Kwan Chi
Kim, Hyong Kap
Shafai, Lotfollah
Swift, Glenn W(illiam)

NEW BRUNSWICK
Lewis, J(ohn) E(ugene)
Scott, Robert Nelson
Wasson, W(alter) Dana

NEWFOUNDLAND
Rahman, Md Azizur
Vetter, William J

NOVA SCOTIA
Baird, Charles Robert
Bhartia, Prakash
El-Hawary, Mohamed El-Aref
El-Masry, Ezz Ismail
Holbrook, George W

ONTARIO
Abdelmalek, Nabin N
Ahmed, Nasir Uddin
Aitken, G(eorge) J(ohn) M(urray)
Balmain, Keith George
Bandler, John William
Biringer, Paul P(eter)
Blake, Ian Fraser
Boggs, Steven A
Broughton, M(ervyn) B(lythe)
Campbell, C(olin) K(ydd)
Campling, C(harles) H(ugh) R(amsay)
Carr, Jan
Castle, George Samuel Peter
Celinski, Olgierd J(erzy) Z(dzislaw)
Chow, Yung Leonard
Coll, David C
Cook, Kenneth R
Copeland, Miles Alexander
Costache, George Ioan
Davidson, Grant E(dward)
Debuda, Rudolf G
Densley, John R
Dewan, Shashi B
Di Cenzo, Colin D
Dohoo, Roy McGregor
Ellis, Jack Barry
Endrenyi, Janos
Fahmy, Moustafa Mahmoud
Fairman, Frederick Walker
Falconer, David Duncan
Findlay, Raymond D(avid)
Georganas, Nicolas D
Goldenberg, Andrew Avi
Hackam, Reuben
Ham, James M(ilton)
Hamacher, V(incent) Carl
Hanff, Ernest Salo
Heasell, E(dwin) L(ovell)
Hoefer, Wolfgang Johannes Reinhard
Ionescu, Dan
Janischewskyj, W
Ji, Guangda Winston
Jullien, Graham Arnold
Kalra, S(urindra) N(ath)
Kavanagh, Robert John
Kitai, Reuven
Kruus, Jaan
Kundur, Prabha Shankar
Kunov, Hans
Landolt, Jack Peter
Leach, Barrie William
Lemyre, C(lement)
Ling, Daniel
McCausland, Ian
McGee, William F
McLane, Peter John
Majithia, Jayanti
Mark, Jon Wei
Mathur, R(adhey) M(ohan)
Morris, Lawrence Robert
Mouftah, Hussein T
Mungall, Allan George
Penstone, S(idney) Robert
Perlman, Maier
Plant, B J
Raney, Russell Keith
Ratz, H(erbert) C(harles)

Reeve, John
Renfrew, Robert Morrison
Reza, Fazlollah M
Riordon, J(ohn) Spruce
Ristic, Velimir Mihailo
Robertson, Stuart Donald Treadgold
Roe, Peter Hugh O'Neil
Sablatash, Mike
Salama, Clement Andre Tewfik
Selvakumar, Chettypalayam Ramanathan
Semlyen, Adam
Sinclair, George
Sinha, Naresh Kumar
Slemon, Gordon R(ichard)
Smith, Kenneth Carless
Stevinson, Harry Thompson
Stuchly, Maria Anna
Stuchly, Stanislaw S
Tavares, Stafford Emanuel
Thomas, Raye Edward
Tsao, Sai Hoi
Van Driel, Henry Martin
Venetsanopoulos, Anastasios Nicolaos
Vervoort, Gerardus
Vidyasagar, Mathukumalli
Vlach, Jiri
Wallingford, Errol E(lwood)
Watt, Lynn A(lexander) K(eeling)
Wei, L(ing) Y(un)
Wilson, John D(ouglas)
Wittke, Paul H
Wong, George Shoung-Koon

PRINCE EDWARD ISLAND
Lodge, Malcolm A

QUEBEC
Adler, Eric L
Agarwal, Vinod Kumar
Aktik, Cetin
Bartnikas, Ray
Belanger, Pierre Rolland
Bhattacharyya, Bibhuti Bhusan
Boulet, J Lionel
Boyarsky, Abraham Joseph
Crine, Jean-Pierre C
Dumas, Jean
Farnell, G(erald) W(illiam)
Giguere, Joseph Charles
Hayes, Jeremiah Francis
Jumarie, Guy Michael
Kahrizi, Mojtaba
Lecours, Michel
Lefebvre, Mario
Levine, Martin David
Malowany, Alfred Stephen
Maruvada, P Sarma
Morgera, Salvatore Domenic
Mukhedkar, Dinkar
Paquet, Jean Guy
Patel, Rajnikant V
Pavlasek, Tomas J(an) F(rantisek)
Phan, Cong Luan
Plotkin, Eugene Isaak
Ramachandran, Venkatanarayana D
Roberge, Fernand Adrien
Saint-Arnaud, Raymond
Silvester, Peter Peet
St-Jean, Guy
Swamy, Mayasandra Nanjundiah
   Srikanta
Zames, George
Zucker, Steven Warren

SASKATCHEWAN
Billinton, Roy
Blachford, Cameron W
Boyle, A(rchibald) R(aymond)
Gupta, Madan Mohan
Hirose, Akira
Pollak, Viktor A

OTHER COUNTRIES
Aleksandrov, Georgij Nikolaevich
Angelini, Arnaldo M
Annestrand, Stig A
Baggi, Denis Louis
Balk, Pieter
Beckhoff, Gerhard Franz
Boxman, Raymond Leon
Burckhardt, Christoph B
Callier, Frank Maria
Chang, Chun-Yen
Chu, Kwo Ray
Conforti, Evandro
Czepyha, Chester George Reinhold
Fettweis, Alfred Leo Maria
Forrer, Max P(aul)
Gauster, Wilhelm Friedrich
Giarola, Attilio Jose
Jain, Anil Kumar
Jespersen, Nils Vidar
Johnson, E(dward) O
Kagawa, Yukio
Kaiser, Wolfgang A
Kao, Charles K
Khatib, Hisham M
Kim, Myunghwan
Kim, Wan H(ee)
Kimblin, Clive William
Kimura, Hidenori
Kirby, Richard C(yril)
Kishi, Keiji
Ku, Victor Chia-Tai

Kurata, Mamoru
Kurokawa, Kaneyuki
Lee, Choong Woong
Lindell, Ismo Veikko
Luckey, Paul David, Jr
Lueder, Ernst H
Luo, Peilin
Malah, David
Mansour, Mohamed
Maqusi, Mohammad
Marom, Emanuel
Massey, James L
Misugi, Takahiko
Paulish, Daniel John
Poloujadoff, Michel E
Rhodes, Ian Burton
Saba, Shoichi
Sakurai, Yoshifumi
Sato, Gentei
Sekimoto, Tadahiro
Sekine, Yasuji
Sessler, Gerhard Martin
Shtrikman, Shmuel
Steen, Robert Frederick
Strintzis, Michael Gerassimos
Tarui, Yasuo
Thiel, Frank L(ouis)
Tzafestas, Spyros G
Udo, Tatsuo
Un, Chong Kwan
Watanabe, Akira
Winn, Edward Barriere
Woo, Kwang Bang
Yamamoto, Mitsuyoshi
Yanabu, Satoru
Young, Ian Theodore
Zeheb, Ezra

Electronics Engineering

ALABAMA
Dixon, Julian Thomas
Edlin, George Robert
Fullerton, Larry Wayne
Gilbert, Stephen Marc
Honnell, Martial A(lfred)
Houts, Ronald C(arl)
Jaeger, Richard Charles
Nola, Frank Joseph
Post, M(aurice) Dean
Thomas, Albert Lee, Jr
Watson, Raymond Coke, Jr
White, Ronald
Wolin, Samuel

ALASKA
Hallinan, Thomas James

ARIZONA
Bahill, A(ndrew) Terry
Balanis, Constantine A
Bao, Qingcheng
Bluestein, Theodore
Chandra, Abhijit
Craib, James F(rederick)
De Gaston, Alexis Neal
De La Moneda, Francisco Homero
Dougherty, John Joseph
Dybvig, Paul Henry
El-Ghazaly, Samir M
Falk, Thomas
Fordemwalt, James Newton
Formeister, Richard B
Galloway, Kenneth Franklin
Gerhard, Glen Carl
Harper, Francis Edward
Hickernell, Fred Slocum
Hoehn, A(lfred) J(oseph)
Hohl, Jakob Hans
Holmgren, Paul
Jones, Roger C(lyde)
Karady, George Gyorgy
Kerwin, William J(ames)
McRae, Lorin Post
Mylrea, Kenneth C
O'Keefe, Michael Adrian
Palmer, James McLean
Perper, Lloyd
Pokorny, Gerold E(rwin)
Rutledge, James Luther
Schroder, Dieter K
Scott, Alwyn C
Selvidge, Harner
Silvern, Leonard Charles
Szilagyi, Miklos Nicholas
Wasson, James Walter
Weller, Edward F(rank), Jr
Wing, William Hinshaw
Witulski, Arthur Frank
Zeigler, Bernard Philip
Zelenka, Jerry Stephen

ARKANSAS
Cardwell, David Michael
Gordon, Millard F(reeman)

CALIFORNIA
Abraham, W(ayne) G(ordon)
Agostini, Romain Camille
Allen, Charles A
Alves, Ronald V
Anderson, Warren R(onald)
Andres, John Milton

Angell, James Browne
Anger, Hal Oscar
Antoniak, Charles Edward
Astrahan, Morton M
Atchison, F(red) Stanley
Aukland, Jerry C
Bacon, Lyle C(holwell), Jr
Baker, Richard H(owell)
Bardin, Russell Keith
Barhydt, Hamilton
Begovich, Nicholas A(nthony)
Bennett, Alan Jerome
Berg, Myles Renver
Berger, Josef
Bertram, Sidney
Bhaumik, Mani Lal
Bishop, Harold (Oswald)
Blasingame, Benjamin P(aul)
Blumenthal, Irwin S(imeon)
Borst, David W(ellington)
Breitwieser, Charles J(ohn)
Brennan, Lawrence Edward
Brewer, George R
Brodie, Ivor
Bullis, William Murray
Burnett, Lowell Jay
Caldwell, J(oseph) J(efferson), Jr
Carlan, Alan J
Carlyle, Jack Webster
Cass, Thomas Robert
Chakrabarti, Supriya
Chan, Bertram Kim Cheong
Chang, Juang-Chi (Joseph)
Chang, Kai
Chang, Tien-Lin
Chen, Tsai Hwa
Chiang, Anne
Chin, Maw-Rong
Chodorow, Marvin
Choi, Won Kil
Christie, John McDougall
Clarke, Lucien Gill
Cogan, Adrian Ilie
Conant, Curtis Terry
Cone, Donald R(oy)
Cook, Albert Moore
Cormier, Reginald Albert
Crain, Cullen Malone
Crandall, Ira Carlton
Crapuchettes, Paul W(ythe)
Creighton, John Rogers
Culler, Donald Merrill
Cutler, Leonard Samuel
Daehler, Max, Jr
Damonte, John Batista
Davis, William R
DeGrasse, Robert W(oodman)
Demetrescu, M
De Micheli, Giovanni
Dempsey, Martin E(wald)
Deshpande, Shivajirao M
Dev, Parvati
Dixon, Thomas Patrick
Dyce, Rolf Buchanan
Eargle, John Morgan
Eden, Richard Carl
Eldon, Charles A
Ellingboe, J(ules) K
Erdley, Harold F(rederick)
Everhart, Thomas E(ugene)
Ewing, Gerald Dean
Fasang, Patrick Pad
Feldman, Nathaniel E
Feng, Joseph Shao-Ying
Ferrari, Domenico
Fialer, Philip A
Fisher, Leon Harold
Flanigan, William Francis, Jr
Fleming, Alan Wayne
Fonck, Eugene J
Frankel, Sidney
Franklin, Gene F(arthing)
Frescura, Bert Louis
Fried, Walter Rudolf
Gamo, Hideya
Gardner, Floyd M
Gibson, Atholl Allen Vear
Grebene, Alan B
Green, Allen T
Green, Philip S
Grinnell, Robin Roy
Grobecker, Alan J
Groner, Gabriel F(rederick)
Groner, Paul Stephen
Gunckel, Thomas L, II
Gupta, Madhu Sudan
Haas, Gustav Frederick
Hackett, Le Roy Huntington, Jr
Haeff, Andrew V(asily)
Hagstrom, Stig Bernt
Haller, Eugene Ernest
Hankins, George Thomas
Hansen, Grant Lewis
Hardy, Vernon E
Harris, James Stewart, Jr
Hartwick, Thomas Stanley
Havens, Byron L(uther)
Hearn, Robert Henderson
Hellerman, Herbert
Helmer, John
Henning, Harley Barry
Herman, E(lvin) E(ugene)
Hewlett, W(illiam) R(edington)
Hingorani, Narain G

Hinrichs, Karl
Hoagland, Jack Charles
Hoch, Orion L
Hodges, Dean T, Jr
Hoffman, James Tracy
Hooper, Catherine Evelyn
Hornak, Thomas
Hovanessian, Shahen Alexander
Hrubesh, Lawrence Wayne
Hsia, Yukun
Hu, Chenming
Huang, Chaofu
Huang, Tiao-yuan
Hult, John Luther
Hummel, Steven G
Hung, John Wenchung
Hyams, Henry C
Ikezi, Hiroyuki
Isaman, Francis
Jacob, George Korathu
Jacobson, Raymond E
Jansen, Michael
Jobs, Steven P
Johannessen, Jack
Johnson, David Leroy
Johnson, Robert A
Jones, E(dward) M(cClung) T(hompson)
Jones, Earle Douglas
Jordan, Gary Blake
Josias, Conrad S(eymour)
Kafka, Robert W(illiam)
Kagiwada, Reynold Shigeru
Kahn, Frederic Jay
Kaisel, S(tanley) F(rancis)
Kamins, Theodore I
Kamphoefner, Fred J(ohn)
Kanter, Helmut
Kao, Tai-Wu
Karp, Arthur
Karuza, Sarunas Kazys
Kayton, Myron
Kazan, Benjamin
Kearney, Joseph W(illiam)
Kelly, Kenneth C
Kerfoot, Branch Price
King, Harry J
King, Howard E
Kinnison, Gerald Lee
Kjar, Raymond Arthur
Knighten, James Leo
Koetsch, Philip
Konrad, Gerhard T(hies)
Kosai, Kenneth
Kozlowski, Lester Joseph
Krause, Lloyd O(scar)
Kroemer, Herbert
Kroger, Marlin G(lenn)
Lacy, Peter D(empsey)
Lafranchi, Edward Alvin
Lally, Philip M(arshall)
Lampert, Carl Matthew
Lao, Binneg Yanbing
Larson, Harry Thomas
Lau, John H
Lau, S S
Law, Hsiang-Yi David
Leadabrand, Ray Laurence
Leavy, Paul Matthew
Leedham, Clive D(ouglas)
Lehovec, Kurt
Leipold, Martin H(enry)
Lender, Adam
Lepoff, Jack H
Leung, Charles Cheung-Wan
Levenson, Marc David
Levy, Ralph
Lewis, John E
Li, Chia-Chuan
Lin, Wen-C(hun)
Lisman, Perry Hall
Liu, Chi-Sheng
Lonky, Martin Leonard
Lopina, Robert F(erguson)
Lord, Harold Wilbur
Love, Sydney Francis
Lubman, David
Lynch, Frank W
McCaldin, J(ames) O(eland)
McColl, Malcolm
McCurdy, Alan Hugh
McDonald, William True
Macha, Milo
Mack, Dick A
McKay, Dale Robert
Mackay, Ralph Stuart
McNutt, Michael John
McQuillen, Howard Raymond
McWhorter, Malcolm M(yers)
Main, William Francis
Maloney, Timothy James
Maloy, John Owen
Manes, Kennneth Rene
Margerum, Donald L(ee)
Marshall, J Howard, III
Marxheimer, Rene B
Mataré, Herbert Franz
Mathews, W(arren) E(dward)
Mayper, V(ictor), Jr
Meagher, Donald Joseph
Meggers, William F(rederick), Jr
Meng, Shien-Yi
Meyer, Lhary
Michaelson, Jerry Dean
Middlebrook, R(obert) D(avid)

Miller, Arnold
Mitra, Sanjit K
Moerner, William Esco
Molnar, Imre
Morris, Fred(erick) W(illiam)
Morrow, Charles Tabor
Muchmore, Robert B(oyer)
Muller, Richard S
Mykkanen, Donald L
Naqvi, Iqbal Mehdi
Nelson, Richard Burton
Newman, Roger
Ohlson, John E
O'Keefe, Dennis Robert
Oleesky, Samuel S(imon)
Page, D(errick) J(ohn)
Palmer, John Parker
Paoli, Thomas Lee
Parker, Alice Cline
Parode, L(owell) C(arr)
Patel, Arvindkumar Motibhai
Persky, George
Pierce, John Robinson
Porter, John E(dward)
Portnoff, Michael Rodney
Presnell, Ronald I
Preston, Glenn Wetherby
Ranftl, Robert M(atthew)
Rao, R(angaiya) A(swathanarayana)
Rauch, Lawrence L(ee)
Reiche, Ludwig P(ercy)
Requicha, Aristides A G
Rickard, John Terrell
Robinson, Lloyd Burdette
Rogers, Howard H
Ross, Hugh Courtney
Rubenson, J(oseph) G(eorge)
Rude, Paul A
Rynn, Nathan
Sachdev, Suresh
Sadler, Charles Robinson, Jr
Safonov, Michael G
Samulon, Henry A
Sangiovanni-Vincentelli, Alberto Luigi
Sawyer, David Erickson
Schechter, Joel Ernest
Scherr, Harvey Murray
Schweikert, Daniel George
Scott, Paul Brunson
Sensiper, S(amuel)
Sequin, Carlo Heinrich
Sewell, Curtis, Jr
Shank, Charles Vernon
Shar, Leonard E
Shaw, Charles Alden
Sheingold, Abraham
Shevel, Wilbert Lee
Shin, Soo H
Silverstein, Elliot Morton
Sinclair, Robert
Singleton, John Byrne
Slaton, Jack H(amilton)
Smith, George Foster
Smith, Warren Drew
Spieler, Helmuth
Spinrad, Robert J(oseph)
Spitzer, William George
Stack (Stachiewicz), B(ogdan) R(oman)
Stapelbroek, Maryn G
Staudhammer, Peter
Steenson, Bernard O(wen)
Stein, William Earl
Sterling, Warren Martin
Stern, Arthur Paul
Stewart, James A
Stewart, Richard William
Sturm, Walter Allan
Swartzlander, Earl Eugene, Jr
Szentirmai, George
Sziklai, George C(lifford)
Tatum, Freeman A(rthur)
Tauber, Richard Norman
Teja, Edward Ray
Thiene, Paul G(eorge)
Thomas, Hubert Jon
Tilles, Abe
Torgow, Eugene N
Trenholme, John Burgess
Trubert, Marc
Tuszynski, Alfons Alfred
Vane, Arthur B(ayard)
Vaughan, J Rodney M
Victor, Walter K
Viswanathan, Chand R
Vlay, George John
Voreades, Demetrios
Waldhauer, F D
Wang, Kang-Lung
Wang, Shyh
Ware, W(illis) H(oward)
Washburn, Jack
Waugh, John Blake-Steele
Webster, Emilia
White, Richard Manning
White, Stanley A
Wiedow, Carl Paul
Wilcox, Jaroslava Zitkova
Wilson, Robert Gray
Wilson, William John
Wilts, Charles H(arold)
Wintroub, Herbert Jack
Wistreich, George A
Wright, John Marlin
Yang, Chung Ching

Yeh, Hen-Geul
Yeh, J J
Yeh, Paul Pao
Yeh, Yea-Chuan Milton
Yeh, Yin
Young, C(ecil) G(eorge), Jr
Young, Frank
Young, John A
Young, Konrad Kwang-Leei
Young, Richard D
Yue, A(lfred) S(hui-Choh)
Zamboni, Luciano
Zarem, Abe Mordecai
Zhou, Simon Zheng
Zuleeg, Rainer

COLORADO
Abshier, Curtis Brent
Adams, James Russell
Allen, James Lamar
Altschuler, Helmut Martin
Barnes, James Allen
Bauer, Charles Edward
Becker, F(loyd) K(enneth)
Boyd, Malcolm R(obert)
Brady, Douglas MacPherson
Brasch, Frederick Martin, Jr
Bushnell, Robert Hempstead
Crawford, Myron Lloyd
Dick, Donald Edward
Dixon, Robert Clyde
Eichelberger, W(illiam) H
Ellis, Donald Griffith
Elmore, Kimberly Laurence
Emery, Keith Allen
Engstrom, Herbert Leonard
Estin, Arthur John
Frick, Pieter A
Gardner, Edward Eugene
Gray, James Edward
Gupta, Kuldip Chand
Hanson, Donald Wayne
Haydon, George William
Hill, David Allan
Kanda, Motohisa
Kindig, Neal B(ert)
Lally, Vincent Edward
Lawton, Robert Arthur
Lebiedzik, Jozef
Lee, Yung-Cheng
Lewin, Leonard
Lubell, Jerry Ira
McAllister, Robert Wallace
McManamon, Peter Michael
Mahajan, Roop Lal
Mandics, Peter Alexander
Mathys, Peter
May, William G(ambrill)
Moddel, Garret R
Nahman, Norris S(tanley)
Ralston, Margarete A
Reiff, Glenn Austin
Sargent, Howard Harrop, III
Schiffmacher, E(dward) R(obert)
She, Chiao-Yao
Staehelin, Lucas Andrew
Strauch, Richard G
Sugar, George R
Swanson, Lawrence Ray
Van Pelt, Richard W(arren)
Wellman, Dennis Lee
Wertz, Ronald Duane
Zhang, Bing-Rong

CONNECTICUT
Apte, Chidanand
Balasinski, Artur
Barker, Richard Clark
Best, Stanley Gordon
Brienza, Michael Joseph
Burkhard, Mahlon Daniel
Carter, G Clifford
Castonguay, Richard Norman
Chang, Richard Kounai
Downes, William A(rthur)
Foyt, Arthur George, Jr
Gaertner, Wolfgang Wilhelm
Glomb, Walter L
Leonard, Robert F
Leonberger, Frederick John
Levenstein, Harold
Maples, Louis Charles
Marchand, Nathan
Nehorai, Arye
Newman, Robert Weidenthal
Rau, Richard Raymond
Rydz, John S
Schwartz, Michael H
Schwarz, Frank
Shankar, Ramamurti
Stein, Richard Jay
Troutman, Ronald R
Tuteur, Franz Benjamin
Zubal, I George

DELAWARE
Bowyer, Kern M(allory)
Burgess, John S(tanley)
Flattery, David Kevin
Golden, Kelly Paul
Hegedus, Steven Scott
Hunsperger, Robert G(eorge)
Ih, Charles Chung-Sen
Jansson, Peter Allan

Electronics Engineering (cont)

Miller, John H, III
Nielsen, Paul Herron
Sawers, James Richard, Jr
Young, M(ilton) G(abriel)

DISTRICT OF COLUMBIA
Anderson, Gordon Wood
Bloch, Erich
Borgiotti, Giorgio V
Borsuk, Gerald M
Donahue, D Joseph
Galane, Irma B(ereston)
Gordon, William Bernard
Gruen, H(arold)
Hartzler, Alfred James
Jackson, William David
Jones, Howard St Claire, Jr
Jordan, Arthur Kent
Killion, Lawrence Eugene
Knudson, Alvin Richard
Kohl, Walter H(einrich)
Kyriakopoulos, Nicholas
Leak, Lee Virn
Lebow, Irwin L(eon)
Lehmann, John R(ichard)
McKay, Jack Alexander
McMahon, John Michael
Marcus, Michael Jay
Moraff, Howard
Ngai, Kia Ling
Nguyen, Charles Cuong
Norton, Mahlon H
Peake, Harold J(ackson)
Pickholtz, Raymond L
Richardson, Mary Elizabeth
Roitman, Peter
Rojas, Richard Raimond
Roosevelt, C(ornelius) V(an) S(chaak)
Sterrett, Kay Fife
Tangney, John Francis
Taylor, Archer S
Thomas, Leonard William, Sr
Townsend, Marjorie Rhodes
Webb, Denis Conrad
Weinberg, Donald Lewis
Young, Leo
Youngblood, William Alfred

FLORIDA
Abbaschian, G J
Adhav, Ratnakar Shankar
Blatt, Joel Herman
Bowers, James Clark
Chen, Tsong-Ming
Chu, Shirley Shan-Chi
Chu, Ting Li
Clark, John F
Cohen, Martin Joseph
Coulter, Wallace H
Couturier, Gordon W
Davidson, Mark Rogers
Doyon, Leonard Roger
Enger, Carl Christian
Everett, Woodrow Wilson, Jr
Feit, Eugene David
Fischlschweiger, Werner
Fossum, Jerry G
Gottesman, Stephen T
Hales, Everett Burton
Hamman, Donald Jay
Henning, Rudolf E
Hoeppner, Conrad Henry
Hoffman, Thomas R(ipton)
Holst, Gerald Carl
Howell, John Foss
Hoyer-Ellefsen, Sigurd
Huebner, Jay Stanley
Johnson, Raymond C, Jr
Keenan, Robert Kenneth
Kelso, John Morris
Ketchledge, Raymond Waibel
Kiewit, David Arnold
Kornblith, Lester
Kretzmer, Ernest R(udolf)
Lachs, Gerard
Lindholm, Fredrik Arthur
Lister, Charles Allan
Lund, Frederick H(enry)
Mertens, Lawrence E(dwin)
Mooney, John Bradford, Jr
Mulson, Joseph F
Oman, Robert Milton
Pichal, Henri Thomas, II
Revay, Andrew W, Jr
Rosenblum, Harold
Ryder, John Douglass
Schmidt, Klaus H
Shea, Richard Franklin
Sheppard, Albert Parker
Singh, Rama Shankar
Suski, Henry M(ieczyslaw)
Tusting, Robert Frederick
Wade, Thomas Edward
Wang, Ru-Tsang
Watson, J(ames) Kenneth
Wu, Jin Zhong
Young, Charles Gilbert
Ziernicki, Robert S

GEORGIA
Bodnar, Donald George
Bomar, Lucien Clay

Bruder, Jospeh Albert
Camp, Ronnie Wayne
Currie, Nicholas Charles
Gibson, John Michael
Grace, Donald J
Holtum, Alfred G(erard)
Hooper, John William
Jones, Betty Ruth
Joy, Edward Bennett
Long, Maurice W(ayne)
McMillan, Robert Walker
Moss, Richard Wallace
Paulin, Jerome John
Pence, Ira Wilson, Jr
Pidgeon, Rezin E, Jr
Prince, M(orris) David
Ringel, Steven Adam
Robertson, Douglas Welby
Roper, Robert George
Sayle, William, II
Singletary, Thomas Alexander
Stoneburner, Daniel Lee
Vail, Charles R(owe)
Wiltse, James Cornelius
Witt, Samuel N(ewton), Jr
Yokelson, Bernard J(ulius)

HAWAII
Bixler, Otto C
Crane, Sheldon Cyr
Holm-Kennedy, James William
Koide, Frank T
Zisk, Stanley Harris

IDAHO
DeBow, W Brad
Stuffle, Roy Eugene

ILLINOIS
Adler, Robert
Agrawal, Arun Kumar
Ahuja, Narendra
Baier, William H
Banerjee, Prithviraj
Borso, Charles S
Camras, Marvin
Chien, Andrew Andai
Coleman, James J
DeVault, Don Charles
Domanik, Richard Anthony
Dyson, John Douglas
Eilers, Carl G
Errede, Steven Michael
Feinberg, Barry N
Gabriel, John R
Gupta, Nand K
Gupta, Udaiprakash A
Hajj, Ibrahim Nasri
Hoffman, Clyde H(arris)
Holonyak, N(ick), Jr
Hua, Ping
Hupert, Julius Jan Marian
Jostlein, Hans
Kang, Sung-Mo Steve
Kaplan, Daniel Moshe
Kaplan, Sam H
Kim, Kyekyoon K(evin)
Kumar, Panganamala Ramana
Kustom, Robert L
Laxpati, Sharad R
Lin, James Chih-I
Luplow, Wayne Charles
Macrander, Albert Tiemen
Merkelo, Henri
Mikulski, James J(oseph)
Millhouse, Edward W, Jr
Murphy, Gordon J
Neuhalfen, Andrew J
Offner, Franklin Faller
O'Neil, John J
Parker, Ehi
Parzen, Philip
Pursley, Michael Bader
Raccah, Paul M(ordecai)
Rudnick, Stanley J
Smedskjaer, Lars Christian
Sobel, Alan
Staunton, John Joseph Jameson
Studtmann, George H
Taflove, Allen
Toppeto, Alphonse A
Unlu, M Selim
Venema, Harry J(ames)
Weinstein, Ronald S
Windhorn, Thomas H
Zell, Blair Paul
Zitter, Robert Nathan

INDIANA
Brown, Buck F(erguson)
Detraz, Orville R
Fearnot, Neal Edward
Hammann, John William
Johnson, Charles F
Neudeck, Gerold W(alter)
Ogborn, Lawrence L
Orr, William H(arold)
Richter, Ward Robert
Roth, Lawrence Max
Schwartz, Richard John
Stanley, Gerald R
Stein, Frank S
Thomas, Lucius Ponder
Vogelhut, Paul Otto

White, Samuel Grandford, Jr

IOWA
Haendel, Richard Stone
Jolls, Kenneth Robert
Lunde, Barbara Kegerreis, (BK)

KANSAS
Moore, Richard K(err)
Norris, Roy Howard
Richard, Patrick
Shanmugan, K Sam
Wright, Donald C

KENTUCKY
Cathey, Jimmie Joe
Frank, John L
George, Ted Mason
Kadaba, Prasad Krishna
Steele, Earl L(arsen)

LOUISIANA
Bedell, Louis Robert
Cullen, John Knox
Erath, Louis W
Grimwood, Charles
Jones, Daniel Elven
Keys, L Ken
Marshak, Alan Howard
Seaman, Ronald L
Trahan, Russell Edward, Jr

MAINE
Holshouser, Don F
Snyder, Arnold Lee, Jr

MARYLAND
Adams, Robert John
Atanasoff, John Vincent
Atia, Ali Ezz Eldin
Barbe, David Franklin
Barrack, Carroll Marlin
Beal, Robert Carl
Bennighof, R(aymond) H(oward)
Berman, Michael Roy
Bishop, Walton B
Bluzer, Nathan
Bromberger-Barnea, Baruch (Berthold)
Brown, Harry Benjamin
Bukowski, Richard William
Cacciamani, Eugene Richard, Jr
Campanella, Samuel Joseph
Carp, Gerald
Cohen, Stanley Alvin
Corak, William Sydney
Cornett, Richard Orin
Daly, John Anthony
Daniel, Charles Dwelle, Jr
DeLamater, Edward Doane
Demmerle, Alan Michael
Drzewiecki, Tadeusz Maria
Eisner, Howard
Evans, Alan G
Fink, Don Roger
Frommer, Peter Leslie
Gabriele, Thomas L
Gammell, Paul M
Garver, Robert Vernon
Ghoshtagore, Rathindra Nath
Goepel, Charles Albert
Golding, Leonard S
Goldstein, Gordon D(avid)
Granatstein, Victor Lawrence
Greenhouse, Harold Mitchell
Gupta, Ramesh K
Gupta, Vaikunth N
Halpin, Joseph John
Hammond, Charles E
Hekimian, Norris Carroll
Hellwig, Helmut Wilhelm
Henderson, Beauford Earl
Ho, Henry Sokore
Horna, Otakar Armour
Hornbuckle, Franklin L
Hyde, Geoffrey
Jensen, Arthur Seigfried
Kanal, Laveen Nanik
Kaplan, Alexander E
Kiebler, John W(illiam)
Kravitz, Lawrence C
Kundu, Mukul Ranjan
Larrabee, R(obert) D(ean)
Leydorf, Glenn E(dwin)
Ligomenides, Panos Aristides
Lin, Hung Chang
Lowney, Jeremiah Ralph
Maccabee, Bruce Sargent
McConoughey, Samuel R
McLean, Flynn Barry
Mahle, Christoph E
Marsten, Richard B(arry)
Masters, Robert Wayne
Mayo, Santos
Michaelis, Michael
Michelsen, Arve
Miller, Kenneth M, Sr
Mills, Thomas K
Minneman, Milton J(ay)
Moore, Robert Avery
Munson, J(ohn) C(hristian)
Neale, Elaine Anne
Newcomb, Robert Wayne
Newman, David Bruce, Jr
Nusbickel, Edward M, Jr

Nylen, Marie Ussing
Oettinger, Frank Frederic
Oscar, Irving S
Palmer, Charles Harvey
Pande, Krishna P
Parakkal, Paul Fab
Paul, Dilip Kumar
Pirraglia, Joseph A
Pollack, Louis
Potyraj, Paul Anthony
Potzick, James Edward
Preisman, Albert
Pullen, Keats A(bbott), Jr
Raabe, Herbert P(aul)
Raines, Jeremy Keith
Rathbun, Edwin Roy, Jr
Riley, Patrick Eugene
Rosenblum, Howard Edwin
Saba, William George
Sann, Klaus Heinrich
Schafft, Harry Arthur
Schmidt, Edward Matthews
Schneck, Paul Bennett
Shelton, Emma
Sheridan, Michael N
Sicotte, Raymond L
Singh, Amarjit
South, Hugh Miles
Stickler, Mitchell Gene
Sugai, Iwao
Susskind, Alfred K(riss)
Sztankay, Zoltan Geza
Thakor, Nitish Vyomesh
Thoma, George Ranjan
Timm, Raymond Stanley
Tippett, James T
Vann, William L(onnie)
Vergara, William Charles
Wainwright, Richard Adolph
Watters, Edward C(harles), Jr
Weaver, Christopher Scot
Weber, Richard Rand
Weinreb, Sander
Woolston, Daniel D
Yang, Xiaowei
Young, C(harles), Jr

MASSACHUSETTS
Abraham, L(eonard) G(ladstone), Jr
Adler, Richard B(rooks)
Adlerstein, Michael Gene
Alexander, Michael Norman
Allen, Thomas John
Alter, Ralph
Altshuler, Edward E
Aprille, Thomas Joseph, Jr
Ball, John Allen
Band, Hans Eduard
Barrington, A(lfred) E(ric)
Barton, David Knox
Bell, Richard Oman
Benton, Stephen Anthony
Blake, Carl
Blum, John Bennett
Bose, Amar G(opal)
Bowman, David F(rancis)
Brayer, Kenneth
Breton, J Raymond
Brookner, Eli
Bugnolo, Dimitri Spartaco
Bussgang, J(ulian) J(akob)
Campbell, William (Aloysius)
Cardiasmenos, Apostle George
Castro, Alfred A
Chang, Robin
Chase, Charles Elroy, Jr
Clark, Melville, Jr
Clifton, Brian John
Colby, George Vincent, Jr
Comer, Joseph John
Corey, Brian E
Counselman, Charles Claude, III
Cronson, Harry Marvin
Dale, Brian
Dallos, Andras
Davis, Charles Freeman, Jr
Decker, C(harles) David
Dixit, Sudhir S
Eaves, Reuben Elco
Fano, Robert M(ario)
Feist, Wolfgang Martin
Figwer, J(ozef) Jacek
Forney, G(eorge) David, Jr
Forrester, Jay W
Foster, Caxton Croxford
Fowler, Charles A(lbert)
Fraser, Donald C
Galvin, Aaron A
Garvey, R(obert) Michael
Gaw, C Vernon
Geary, John Charles
Germeshausen, Kenneth J(oseph)
Giordano, Arthur Anthony
Grabel, Arvin
Grass, Albert M(elvin)
Gray, Truman S(tretcher)
Gregers-Hansen, Vilhelm
Groginsky, Herbert Leonard
Gross, Fritz A
Guidice, Donald Anthony
Hart, Harold M(artin)
Haugsjaa, Paul O
Hempstead, Charles Francis
Hershberg, Philip I

Herwald, Seymour W(illis)
Higgins, W(illiam) F(rancis)
Hockney, Richard L
Holmstrom, Frank Ross
Holway, Lowell Hoyt, Jr
Hope, Lawrence Latimer
Hurlburt, Douglas Herendeen
Hursh, John W(oodworth)
Ilic, Marija
Ippen, Erich Peter
James, George Ellert
Kanter, Irving
Keicher, William Eugene
Kincaid, Thomas Gardiner
Kleis, John Dieffenbach
Kosowsky, David I
Kovaly, John J
Kovatch, George
Krasner, Jerome L
Kyhl, Robert Louis
Lanyon, Hubert Peter David
Levesque, Allen Henry
Linden, Kurt Joseph
Loretz, Thomas J
Mack, Richard Bruce
McKnight, Lee Warren
Manley, Harold J(ames)
Mavretic, Anton
Miffitt, Donald Charles
Milburn, Nancy Stafford
Montazer, G Hosein
Moore, Robert Edmund
Moss, J Eliot B
Mullen, James A
Mulukutla, Sarma Sreerama
Naka, F(umio) Robert
Navon, David H
Nesbeda, Paul
Nesline, Frederick William, Jr
Palubinskas, Felix Stanley
Pearlman, Michael R
Penndorf, Rudolf
Peura, Robert Allan
Poon, Chi-Sang
Price, Robert
Raffel, Jack I
Rediker, Robert Harmon
Reich, Herbert Joseph
Reif, L Rafael
Reintjes, J Francis
Renner, Gerard W
Roberge, James Kerr
Rochefort, John S
Rudenberg, H(ermann) Gunther
Ruze, John
Safford, Richard Whiley
Sage, Jay Peter
Schloemann, Ernst
Schreiber, William F(rancis)
Schwartz, Jack
Searle, Campbell L(each)
Senturia, Stephen David
Shen, Hao-Ming
Shepherd, Freeman Daniel, Jr
Simmons, Alan J(ay)
Skolnikoff, Eugene B
Smith, Alan Bradford
Sommer, Alfred Hermann
Statz, Hermann
Stevenson, David Michael
Stewart, Lawrence Colm
Stiffler, Jack Justin
Stiglitz, Irvin G
Stockman, Harry E
Strelzoff, Alan G
Strimling, Walter Eugene
Teng, Chojan
Thun, Rudolf Eduard
Tsandoulas, Gerasimos Nicholas
Tsang, Dean Zensh
Tsaur, Bor-Yeu
Tuller, Harry Louis
Ward, James
Wasserman, Jerry
Weiler, Margaret Horton
Wilensky, Samuel
Yannoni, Nicholas

MICHIGAN
Ancker-Johnson, Betsy
Baldwin, Keith Malcolm
Becher, William D(on)
Birdsall, Theodore G
Brown, William M(ilton)
Brumm, Douglas B(ruce)
Bryant, John H(arold)
Cheydleur, Benjamin Frederic
Crary, Selden Bronson
Dow, W(illiam) G(ould)
Eagen, Charles Frederick
Eldred, Norman Orville
Enke, Christie George
Essner, Edward Stanley
Frank, Max
Gaerttner, Martin R
Getty, Ward Douglas
Ghoneim, Youssef Ahmed
Giacoletto, L(awrence) J(oseph)
Gustafson, Herold Richard
Heremans, Joseph P
Horvath, Ralph S(teve)
Irani, Keki B
Jacovides, Linos J
Jain, Kailash Chandra

Johansen, Elmer L
Johnson, Wayne Jon
Jones, R(ichard) James
Khargonekar, Pramod P
Kim, Changhyun
Kulkarni, Anand K
Larrowe, Vernon L
Lo, Wayne
Lomax, Ronald J(ames)
MacGregor, Donald Max
Malaczynski, Gerard W
Meitzler, Allen Henry
Miller, M(urray) H(enri)
Nazerian, Keyvan
Ristenbatt, Marlin P
Rouze, Stanley Ruple
Salam, Fathi M A
Schlax, T(imothy) Roth
Schubring, Norman W(illiam)
Sengupta, Dipak L(al)
Sethi, Ishwar Krishan
Wennerberg, A(llan) L(orens)
Woodyard, James Robert
Yeh, Chai

MINNESOTA
Bailey, Fredric Nelson
Champlin, Keith S(chaffner)
Dubbe, Richard F
Follingstad, Henry George
Fritze, Curtis W(illiam)
Gallo, Charles Francis
George, Peter Kurt
Glewwe, Carl W(illiam)
Gruber, Carl L(awrence)
Haxby, B(ernard) V(an Loan)
Lin, Benjamin Ming-Ren
Lutes, Olin S
McGlauchlin, Laurence D(onald)
Mooers, Howard T(heodore)
Olyphant, Murray, Jr
Premanand, Visvanatha
Sadjadi, Firooz Ahmadi
Stebbings, William Lee
Sutton, Matthew Albert
Thiede, Edwin Carl
Tran, Nang Tri
Van Doeren, Richard Edgerly
Werth, Richard George
Wesenberg, Clarence L
Yasko, Richard N

MISSISSIPPI
Smith, Charles Edward

MISSOURI
Carson, Ralph S(t Clair)
Ellison, John Vogelsanger
Everett, Paul Marvin
Flowers, Harold L(ee)
Forster, William H(all)
Franklin, Mark A
Gordon, Albert Raye
Griffith, Virgil Vernon
Hill, Dale Eugene
Kozlowski, Don Robert
Landiss, Daniel Jay
Leopold, Daniel J
Manson, Donald Joseph
Martin, William L, Jr
Moss, Randy Hays
Pickard, William Freeman
Richards, Earl Frederick
Rosenbaum, Fred J(erome)
Ross, Monte
Ryan, Carl Ray
Taylor, Ralph Dale
Wagner, Robert G
White, Warren D
Zobrist, George W(inston)

MONTANA
Gerez, Victor
Latham, Don Jay
Pierre, Donald Arthur
Weaver, Donald K(essler), Jr

NEBRASKA
Bahar, Ezekiel
Lenz, Charles Eldon

NEVADA
Benfield, Charles W(illiam)
Johnson, Bruce Paul
Rawat, Banmali Singh
Telford, James Wardrop

NEW HAMPSHIRE
Egan, John Frederick
Frost, Albert D(enver)
Hanson, Per Roland
Jeffery, Lawrence R
Misra, Alok C
Samaroo, Winston R
Von-Recklinghausen, Daniel R
Winn, Alden L(ewis)
Wolff, Nikolaus Emanuel

NEW JERSEY
Allen, Jonathan
Amitay, Noach
Anandan, Munisamy
Anderson, Milton Merrill
Anderson, Robert E(dwin)

Avins, Jack
Ballato, Arthur
Bechtle, Daniel Wayne
Belohoubek, Erwin F
Berthold, Joseph Ernest
Bharj, Sarjit Singh
Black, H(arold) S(tephen)
Blanck, Andrew R(ichard)
Blicher, Adolph
Bloom, Stanley
Borkan, Harold
Bowers, Klaus Dieter
Brown, Alfred Bruce, Jr
Brown, Irmel Nelson
Brudner, Harvey Jerome
Bucher, T(homas) T(albot) Nelson
Buckley, R Russ
Buiting, Francis P
Candy, J(ames) C(harles)
Carnes, James Edward
Carr, William N
Castenschiold, Rene
Celler, George K
Chand, Naresh
Chang, Tao-Yuan
Chang, T(ao)-Y(uan)
Chave, Alan Dana
Cho, Alfred Y
Chu, Sung Nee George
Chung, Yun C
Chynoweth, Alan Gerald
Clements, Wayne Irwin
Coffman, Edward G, Jr
Cohen, Abraham Bernard
Cox, Donald Clyde
Crochiere, Ronald E
Daly, Daniel Francis, Jr
D'Asaro, L Arthur
Derkits, Gustav
Detig, Robert Henry
Dickerson, L(oren) L(ester), Jr
Dickey, E(dward) T(hompson)
Dingle, Raymond
Dixon, Samuel, Jr
Dodabalapur, Ananth
Dodington, Sven Henry Marriott
Dolny, Gary Mark
Enstrom, Ronald Edward
Farber, Herman
Feuer, Mark David
Finkel, Leonard
Fishbein, William
Fisher, Reed Edward
French, Larry J
Freund, Robert Stanley
Gabriel, Edwin Z
Gerber, Eduard A
Gibson, James John
Giordmaine, Joseph Anthony
Goldberg, Edwin A(llen)
Goldstein, Robert Lawrence
Golin, Stuart
Goodman, David Joel
Gossmann, Hans Joachim
Grimmell, William C
Grosso, Anthony J
Gualtieri, Devlin Michael
Guenzer, Charles S P
Hakim, Edward Bernard
Hang, Hsueh-Ming
Harris, James Ridout
Harvey, Floyd Kallum
Hasegawa, Ryusuke
Hatkin, Leonard
Hernqvist, Karl Gerhard
Herold, E(dward) W(illiam)
Hilibrand, J(ack)
Hollywood, John M(atthew)
Hoover, Charles Wilson, Jr
Horn, David Nicholas
Hubbard, William Marshall
Hughes, James Sinclair
Hwang, Cherng-Jia
Johannes, Virgil Ivancich
Johnson, Walter C(urtis)
Kaminow, Ivan Paul
Katzmann, Fred L
Keigler, John Edward
Keizer, Eugene O(rville)
Kesselman, Warren Arthur
Kieburtz, R(obert) Bruce
Kihn, Harry
Klapper, Jacob
Knausenberger, Wulf H
Kniazuk, Michael
Kogelnik, H W
Krevsky, Seymour
Kuo, Ying L
Lawson, James Robert
Lechner, Bernard J
Levi, Anthony Frederic John
Li, John Kong-Jiann
Lin, Chinlon
Link, Fred M
Lubowe, Anthony G(arner)
Lunsford, Ralph D
Luryi, Serge
Maa, Jer-Shen
McLane, George Francis
McNair, Irving M, Jr
Mac Rae, Alfred Urquhart
MacRae, Alfred Urquhart
Mangulis, Visvaldis
Martin, Thomas A(ddenbrook)

Mauzey, Peter T
Meixler, Lewis Donald
Mesner, Max H(utchinson)
Millman, Sidney
Minter, Jerry Burnett
Mogab, Cyril Joseph
Morgan, Dennis Raymond
Mueller, Charles W(illiam)
Murthy, Srinivasa K R
Natapoff, Marshall
Ninke, William Herbert
Onyshkevych, Lubomyr S
Orlando, Carl
Partovi, Afshin
Pei, Shin-Shem
Pei, Shin-Shem
Perlman, Barry Stuart
Rainal, Attilio Joseph
Rajchman, Jan A(leksander)
Ravindra, Nuggehalli Muthanna
Reingold, Irving
Riddle, George Herbert Needham
Robinson, Daniel E
Robrock, Richard Barker, II
Rulf, Benjamin
Russell, Frederick A(rthur)
Saltzberg, Burton R
Schlam, Elliott
Schnable, George Luther
Schneider, Sol
Schulke, Herbert Ardis, Jr
Schulte, Harry John, Jr
Schumer, Douglas B
Schwenker, J(ohn) E(dwin)
Schwering, Felix
Sears, Raymond Warrick, Jr
Segelken, John Maurice
Semmlow, John Leonard
Shahbender, R(abah) A(bd-El-Rahman)
Shumate, Paul William, Jr
Sipress, Jack M
Sterzer, Fred
Strother, J(ohn) A(lan)
Suhir, Ephraim
Taylor, Geoff W
Taylor, Harold Evans
Thornton, C G
Tishby, Naftali Z
Tompsett, Michael F
Tsaliovich, Anatoly
Tsao-Wu, Nelson Tsin
Uhrig, Jerome Lee
Ullrich, Felix Thomas
Van Etten, James P(aul)
Verma, Pramode Kumar
Wagner, Sigurd
Wang, Chao Chen
Wang, Chen-Show
Warters, William Dennis
White, Lawrence Keith
Wilkinson, W(illiam) C(layton)
Wittenberg, Albert M
Woo, Nam-Sung
Wullert, John R, II
Yen, You-Hsin Eugene
Yi-Yan, Alfredo
Yoon, Euijoon
Zaininger, Karl Heinz
Ziegler, Hans K(onrad)

NEW MEXICO
Anderson, Richard Ernest
Berrie, David William
Brock, Ernest George
Brueck, Steven Roy Julien
Burr, Alexander Fuller
Castle, John Granville, Jr
Cole, Edward Issac, Jr
Dawes, William Redin, Jr
Doss, James Daniel
England, Talmadge Ray
Gilbert, Alton Lee
Gourley, Paul Lee
Gover, James E
Gurbaxani, Shyam Hassomal
Haberstich, Albert
Hankins, Timothy Hamilton
Hardin, James T
Hines, William Curtis
Jones, James Jordan
Korman, N(athaniel) I(rving)
Land, Cecil E(lvin)
Linford, Rulon Kesler
McInerney, John Gerard
Maez, Albert R
Marker, Thomas F(ranklin)
Metzger, Daniel Schaffer
Miller, Edmund K(enneth)
Northrup, Clyde John Marshall, Jr
Overhage, Carl F J
Parker, Joseph R(ichard)
Potter, James Martin
Rach, Randolph Carl
Rigrod, William W
Roberts, Peter Morse
Schueler, Donald G(eorge)
Shafer, Robert E
Shipley, James Parish, Jr
Sinha, Dipen N
Smith, William Conrad
Swingle, Donald Morgan
Tapp, Charles Millard

Electronics Engineering (cont)

NEW YORK
Abetti, Pier Antonio
Ahmed-Zaid, Said
Albicki, Alexander
Anderson, Roy E
Anderson, Wayne Arthur
Arams, F(rank) R(obert)
Bachman, Henry L(ee)
Baertsch, Richard D
Ballantyne, Joseph M(errill)
Basavaiah, Suryadevara
Belove, Charles
Bertoni, Henry Louis
Bilderback, Donald Heywood
Bloch, Alan
Blumenthal, Ralph Herbert
Borrego, Jose M
Bradshaw, John Alden
Brennemann, Andrew E(rnest), Jr
Brinch-Hansen, Per
Buchta, John C(harles)
Busnaina, Ahmed A
Calabrese, Carmelo
Carlin, Herbert J
Chang, L(eroy) L(i-Gong)
Chi, Chao Shu
Chiang, Yuen-Sheng
Chow, Tat-Sing Paul
Christiansen, Donald David
Clarke, Kenneth Kingsley
Coden, Michael H
Coleman, John Howard
Comly, James B
Coppola, Patrick Paul
Cottingham, James Garry
Daijavad, Shahvokh
Damouth, David Earl
Das, Pankaj K
Debany, Warren Harding, Jr
De France, Joseph J
Derman, Samuel
Dermit, George
Diamond, Fred Irwin
Dimmler, D(ietrich) Gerd
DiStefano, Thomas Herman
Drossman, Melvyn Miles
Dwyer, Harry, III
Eisenberg, Lawrence
Engeler, William E
Erdos, Marianne E
Ewing, Joan Rose
Fang, Frank F
Feerst, Irwin
Feth, George C(larence)
Fey, Curt F
Freundlich, Martin M
Fried, Erwin
Gabelman, Irving J(acob)
Garretson, Craig Martin
Genova, James John
Gerster, Robert Arnold
Ghandhi, Sorab Khushro
Ghezzo, Mario
Golden, Robert K
Good, William E
Gran, Richard J
Gratian, J(oseph) Warren
Green, Paul E(liot), Jr
Grischkowsky, Daniel Richard
Grobman, Warren David
Grosewald, Peter
Grumet, Alex
Hafner, Theodore
Hahnel, Alwin
Hall, Robert Noel
Harnden, John D, Jr
Harris, Bernard
Hartwig, Curtis P
Hawkins, Gilbert Allan
Herman, Lawrence
Higinbotham, William Alfred
Hill, Cliff Otis
Hsiang, Thomas Y
Hunts, Barney Dean
Hyman, Abraham
Isaacson, Michael Saul
Jatlow, J(acob) L(awrence)
Jelinek, Frederick
Jorne, Jacob
Kelly, John Henry
Keonjian, Edward
Ketchen, Mark B
Keyes, Robert William
Kiang, Ying Chao
Kiehl, Richard Arthur
King, Marvin
Kornberg, Fred
Kornreich, Philipp G
Kritz, J(acob)
Kuper, J B Horner
Kwok, Thomas Yu-Kiu
Lafferty, James M(artin)
LaMuth, Henry Lewis
Laponsky, Alfred Baer
La Russa, Joseph Anthony
Lawrence, David Joseph
Lebenbaum, Matthew T(obriner)
Lee, Young-Hoon
Litchford, George B
Littauer, Raphael Max
Liu, Yung Sheng
Logue, J(oseph) C(arl)

Loughlin, Bernard D
Lubowsky, Jack
Ludwig, Gerald W
Lustig, Howard E(ric)
McGee-Russell, Samuel M
McMurry, William
Maldonado, Juan Ramon
Martens, Alexander E(ugene)
Mattar, Farres Phillip
Meindl, James D
Mercer, Kermit R
Meth, Irving Marvin
Metzger, Ernest Hugh
Mgrdechian, Raffee
Mihran, Theodore Gregory
Milkovic, Miran
Moe, George Wylbur
Moghadam, Omid A
Mogro-Campero, Antonio
Morrison, John B
Morse, William M
Mortenson, Kenneth Ernest
Moy, Dan
Nelson, David Elmer
Ning, Tak Hung
Oh, Byungdu
Okwit, Seymour
Oliner, Arthur A(aron)
Ortel, William Charles Gormley
Packard, Karle Sanborn, Jr
Palocz, Istvan
Paraszczak, Jurij Rostyslan
Parks, Harold George
Peled, Abraham
Platner, Edward D
Plotkin, Martin
Polychronakos, Venetios Alexander
Powell, Noble R
Queen, Daniel
Raghuveer, Mysore R
Redington, Rowland Wells
Rosen, Paul
Rosenthal, Saul W
Rowe, Irving
Ruddick, James John
Sackman, George Lawrence
Salzberg, Bernard
Sample, Steven Browning
Sante, Daniel P(aul)
Satya, Akella V S
Savic, Michael I
Schachter, Rozalie
Schamberger, Robert Dean
Schanker, Jacob Z
Schindler, Joe Paul
Schoenfeld, Robert Louis
Schuster, Frederick Lee
Schwartz, Mischa
Schwartzman, Leon
Schwarz, Ralph J
Sewell, Frank Anderson, Jr
Shaw, John Askew
Sheryll, Richard Perry
Shevack, Hilda N
Silverman, Gordon
Simpson, Murray
Singer, Barry M
Sorokin, Peter
Spencer, David R
Spiegel, Robert
Spielman, Harold S
Spiro, Julius
Srinivasan, G(urumkonda) R
Star, Joseph
Stenger, Richard J
Stringer, L(oren) F(rank)
Su, Stephen Y H
Temple, Victor Albert Keith
Terman, Lewis Madison
Trevoy, Donald James
Trexler, Frederick David
Tummala, Rao Ramamohana
Varshney, Pramod Kumar
Vogelman, Joseph H(erbert)
Wald, Alvin Stanley
Walter, William Trump
Webster, Harold Frank
Wei, Ching-Yeu
Whalen, James Joseph
Wheatley, W(illiam) A(rthur)
Wie, Chu Ryang
Wiesner, Leo
Witt, Christopher John
Woodard, David W
Woods, John William
Worthing, Jurgen
Wouk, Victor
Yang, Edward S
Zdan, William
Zukowski, Charles Albert

NORTH CAROLINA
Appleby, Robert H
Bailey, Kincheon Hubert, Jr
Bailey, Robert Brian
Baliga, Jayant
Brinn, Jack Elliott, Jr
Burger, Robert M
Casey, H Craig, Jr
Defoggi, Ernest
Dotson, Marilyn Knight
Easter, William Taylor
Franzon, Paul Damian
Gray, James P

Hackenbrock, Charles Robert
Hauser, John Reid
Haynes, John Lenneis
Huber, William Richard, III
Hutchins, William R(eagh)
Kanopoulos, Nick
McEnally, Terence Ernest, Jr
Massoud, Hisham Z
Moses, Montrose James
Nagle, H Troy
Russell, Lewis Keith
Sander, William August, III
Steer, Michael Bernard
Tischer, Frederick Joseph
Wakeman, Charles B
Walters, Mark David
Wooten, Frank Thomas
Wortman, Jimmie J(ack)

NORTH DAKOTA
Yuvarajan, Subbaraya

OHIO
Antler, Morton
Arnett, Jerry Butler
Barranger, John P
Berthold, John William, III
Bhagat, Pramode Kumar
Blessinger, Michael Anthony
Connolly, Denis Joseph
Cummings, Charles Arnold
Cummins, Stewart Edward
Dell'Osso, Louis Frank
De Lucia, Frank Charles
Deubner, Russell L(eigh)
Duff, J(ack) E(rrol)
Duff, Robert Hodge
Easterday, Jack L(eroy)
El-Naggar, Mohammed Ismail
Falkenbach, George J(oseph)
Fritsch, Klaus
Gilfert, James C(lare)
Hill, James Stewart
Hostetler, Jeptha Ray
Hunter, William Winslow, Jr
Ingram, David Christopher
Ismail, Amin Rashid
Keim, John Eugene
Keyes, Marion Alvah, IV
Kosel, Peter Bohdan
Kosmahl, Henry G
Laskowski, Edward L
Lawson, Kenneth Dare
Madden, Thomas Lee
Middleton, Arthur Everts
Miller, Carl Henry, Jr
Morrill, Charles D(uncker)
Neuman, Michael R
Papachristou, Christos A
Pinchak, Alfred Cyril
Ramamoorthy, Panapakkam A
Redlich, Robert Walter
Rooney, Victor Martin
Rose, Harold Wayne
Rothstein, Jerome
Schneider, John Matthew
Snider, John William
Suter, Bruce Wilsey
Thorbjornsen, Arthur Robert
Vlcek, Donald Henry
Wickersham, Charles Edward, Jr
Wigington, Ronald L
Williams, Richard Alvin
Wong, Anthony Sai-Hung

OKLAHOMA
Baker, James E
Berbari, Edward J
Brown, Graydon L
Buck, Richard F
Cook, Charles F(oster), Jr
Cronenwett, William Treadwell
Gardner, John Omen
Hadley, Charles F(ranklin)
Hairfield, Harrell D, Jr
Hawley, Paul F(rederick)
McDevitt, Daniel Bernard
Robinson, Enders Anthony
Taylor, William L
Vanderwiele, James Milton

OREGON
Bhattacharya, Pallab Kumar
Chitwood, Paul H(erbert)
Clark, James Orie, II
Collins, John A(ddison)
Ede, Alan Winthrop
Engelbrecht, Rudolf S
Engelmann, Reinhart Wolfgang H
Gilbert, Barrie
Graham, Beardsley
Hackleman, David E
Kim, Dae Mann
Morgan, Merle L(oren)
Owen, Sydney John Thomas
Paarsons, James Delbert
Plummer, James Walter
Rosenthal, Jenny Eugenie
Schaumann, Rolf
Schwartz, James William
Skinner, Richard Emery
Stringer, Gene Arthur
Wagner, Orvin Edson
Yamaguchi, Tadanori

Yau, Leopoldo D

PENNSYLVANIA
Abend, Kenneth
Abramson, Edward
Adams, Raymond F
Akruk, Samir Rizk
Alexander, Frank Creighton, Jr
Allen, Charles C(orletta)
Anouchi, Abraham Y
Arora, Vijay Kumar
Ashok, S
Barber, M(ark) R
Beier, Eugene William
Bhagavatula, VijayaKumar
Bilgutay, Nihat Mustafa
Bordogna, Joseph
Bowler, David Livingstone
Brackmann, Richard Theodore
Bradley, W(illiam) E(arle)
Brastins, Auseklis
Cendes, Zoltan Joseph
Chaplin, Norman John
Clapp, Richard Gardner
Confer, Gregory Lee
Cresswell, Michael William
Davis, Paul Cooper
Davis, R(obert) E(lliot)
Dick, George W
Director, Stephen William
Druffel, Larry Edward
Dumbri, Austin C
Eberhardt, Nikolai
Elder, H E
Emtage, Peter Roesch
Fan, Ningping
Farr, Kenneth E(dward)
Favret, A(ndrew) G(illigan)
Fisher, Tom Lyons
Fonash, Stephan J(oseph)
Foster, Norman Francis
Gewartowski, J(ames) W(alter)
Goldman, Robert Barnett
Goldwasser, Samuel M
Goutmann, Michel Marcel
Grace, Harold P(adget)
Grim, Larry B
Hahn, Peter Mathias
Harrold, Ronald Thomas
Haun, Robert Dee, Jr
Hess, Dennis William
Hinton, Raymond Price
Jaffe, Donald
Jang, Sei Joo
Jaron, Dov
Jenny, Hans K
Jordan, A(ngel) G(oni)
Jordan, Angel G
Kane, John Vincent, Jr
Kaufman, William
Kazahaya, Masahiro Matt
Kryder, Mark Howard
Langer, Dietrich Wilhelm
Laughlin, David Eugene
Leas, J(ohn) W(esley)
Lee, Chin-Chiu
Licini, Jerome Carl
Lipsky, Stephen E
Long, Alton Los, Jr
Lory, Henry James
Lu, Chih Yuan
Martin, Aaron J
Meiksin, Zvi H(ans)
Mitchell, John Douglas
Mondal, Kalyan
Mowrey, Gary Lee
Murphy, Bernard T
Nagel, G(eorge) W(ood)
Niewenhuis, Robert James
Okress, Ernest Carl
Oneal, Glen, Jr
O'Neill, John Cornelius
Ontell, Marcia
Peterson, James Oliver
Pinkerton, John Edward
Post, Irving Gilbert
Putman, Thomas Harold
Revesz, George
Riebman, Leon
Robl, Robert F(redrick), Jr
Roy, Pradip Kumar
Sabnis, Anant Govind
Schroeder, Alfred C(hristian)
Shackelford, Charles L(ewis)
Shibib, M Ayman
Showers, Ralph M(orris)
Shung, K Kirk
Sibul, Leon Henry
Silzars, Aris
Simaan, Marwan
Smits, Friedolf M
Stehle, Philip McLellan
Steyert, William Albert
Stinger, Henry J(oseph)
Stone, Douglas Roy
Stone, Robert P(orter)
Sturdevant, Eugene J
Szepesi, Zoltan Paul John
Thompson, Eric Douglas
Thompson, Fred C
Tomiyasu, Kiyo
Trotter, Nancy Louisa
Van Der Spiegel, Jan
Wagner, David Loren

White, George Rowland
Whitley, James Heyward
Zebrowitz, S(tanley)

RHODE ISLAND
Baker, Walter L(ouis)
Grossi, Mario Dario
Krikorian, John Sarkis, Jr
Kroenert, John Theodore
Petrou, Panayiotis Polydorou

SOUTH CAROLINA
Askins, Harold Williams Jr, Jr
Blevins, Maurice Everett
Bourkoff, Etan
Doherty, William Humphrey
Dumin, David Joseph
Faust, John William, Jr
Garzon, Ruben Dario
Harrison, James William, Jr
Hogan, Joseph C(harles)
Miler, George Gibbon, Jr
Miller, Lee Stephen
Pritchard, Dalton H
Sheppard, Emory Lamar

SOUTH DAKOTA
Nelson, Vernon Ronald
Reuter, William L(ee)
Ross, Keith Alan
Swiden, LaDell Ray

TENNESSEE
Ball, Frances Louise
Barr, Dennis Brannon
Bose, Bimal K
Bundy, Robert W(endel)
Davidson, Robert C
Douglass, Terry Dean
Eads, B G
Joy, David Charles
Mosko, Sigmund W
Ornitz, Barry Louis
Pearsall, S(amuel) H(aff)
Rajan, Periasamy Karivaratha
Rome, James Alan
Trivedi, Mohan Manubhai
Watson, Robert Lowrie
Williams, Charles Wesley

TEXAS
Abraham, Jacob A
Banerjee, Sanjay Kumar
Barton, James Brockman
Basham, Jerald F
Bate, Robert Thomas
Bhuva, Rohit L
Biard, James R
Brakefield, James Charles
Brantley, William Cain, Jr
Bronaugh, Edwin Lee
Brown, Glenn Lamar
Callas, Gerald
Carvajal, Fernando David
Chamberlain, Nugent Francis
Chang, Christopher Teh-Min
Chyu, Ming-Chien
Cole, David F
Collins, Dean Robert
Cory, William Eugene
Crossley, Peter Anthony
Curtin, Richard B
Davidson, David Lee
Dougal, Arwin A(delbert)
Dykstra, Jerald Paul
Eknoyan, Ohannes
Esquivel, Agerico Liwag
Frazer, Marshall Everett
Frensley, William Robert
Ghowsi, Kiumars
Godbey, John Kirby
Grissom, David
Hahn, Larry Alan
Harper, James George
Hartley, Craig Jay
Hasty, Turner Elilah
Hazen, Gary Alan
Hinrichs, Paul Rutland
Hopkins, George H, Jr
Jones, William B(enjamin), Jr
Kirk, Wiley Price
Krishen, Kumar
Kristiansen, Magne
Lawrence, Joseph D, Jr
Leamy, Harry John
Levine, Jules David
Linder, John Scott
Ling, Hao
McDavid, James Michael
Mao, Shing
Mitchell, John Charles
Overzet, Lawrence John
Parker, Cleofus Varren, Jr
Penn, Thomas Clifton
Prabhu, Vasant K
Pritchard, John Paul, Jr
Pritchett, William Carr
Rabson, Thomas A(evelyn)
Reed, Mark Arthur
Rhyne, V(ernon) Thomas
Riter, Stephen
Schroen, Walter
Schulz, Richard Burkart
Schwetman, Herbert Dewitt

Singh, Vijay Pal
Smith, Reid Garfield
Sobol, Harold
Stenback, Wayne Albert
Stone, Orville L
Strieter, Frederick John
Styblinski, Maciej A
Summers, Richard Lee
Taylor, Morris Chapman
Trachtenberg, Isaac
Truchard, James Joseph
Von Maltzahn, Wolf W
White, Charles F(loyd)
Yanagisawa, Samuel T

UTAH
Atwood, Kenneth W
Barlow, Mark Owens
Butterfield, Veloy Hansen, Jr
Clegg, John C(ardwell)
Cox, Benjamin Vincent
Grow, Richard W
Hollaar, Lee Allen
Jeffery, Rondo Nelden
Jolley, David Kent
Lee, James Norman
Li, Shin-Hwa
Smith, Kent Farrell

VERMONT
Adler, Eric
Gray, Kenneth Stewart
Halpern, William
Handelsman, Morris
Pires, Renato Guedes
Pricer, Wilbur David
Raab, Frederick Herbert
Roth, Wilfred
Stephenson, J(ohn) Gregg
Wallis, Clifford Merrill

VIRGINIA
Alcaraz, Ernest Charles
Allen, John L(oyd)
Ammerman, Charles R(oyden)
Anderson, Walter L(eonard)
Bahl, Inder Jit
Bailey, Marion Crawford
Beale, Guy Otis
Blume, Hans-Juergen Christian
Burns, William, Jr
Burton, Larry C(lark)
Cabral, Guy Antony
Chun, Myung K(i)
Cibulka, Frank
Cooper, Larry Russell
Cosby, Lynwood Anthony
Crawford, Daniel J
Curry, Thomas F(ortson)
Daniel, Kenneth Wayne
Davis, Harry I
Davis, William Arthur
Dessouky, Dessouky Ahmad
Desu, Seshu Babu
Duff, William G
Entzminger, John N
Florman, Edwin F(rank)
Flory, Thomas Reherd
Ganguly, Suman
Gee, Sherman
Gerry, Edward T
Ginsberg, Murry B(enjamin)
Goodman, A(lvin) M(alcolm)
Guerber, Howard P(aul)
Haar, Jack Luther
Hammond, Marvin H, Jr
Hasegawa, Ichiro
Heebner, David Richard
Hilger, James Eugene
Howard, Dean Denton
Hvatum, Hein
Inigo, Rafael Madrigal
James, W(ilbur) Gerald
Jones, James L
Joyner, Weyland Thomas, Jr
Kalab, Bruno Marie
Keblawi, Feisal Said
Kellett, Claud Marvin
Kennedy, Andrew John
Kerr, Anthony Robert
Koch, Carl Fred
Kooij, Theo
Larson, Arvid Gunnar
Lee, Fred C
Liceaga, Carlos Arturo
McCutchen, Samuel P(roctor)
McLucas, John L(uther)
McWright, Glen Martin
Mandelberg, Martin
Mayer, Cornell Henry
Moseley, Sherrard Thomas
Neil, George Randall
Nunnally, Huey Neal
Nyman, Thomas Harry
Patterson, James Douglas
Plait, Alan Oscar
Pollard, John Henry
Potter, Richard R(alph)
Pry, Robert Henry
Raab, Harry Frederick, Jr
Rakestraw, James William
Reed, Joseph
Richard, Robert H(enry)
Robinson, Arthur S

Rogers, Thomas F
Schlegelmilch, Reuben Orville
Schneider, Ralph Jacob
Schutt, Dale W
Seibel, Hugo Rudolf
Seiler, Steven Wing
Siegel, Clifford M(yron)
Smith, Carey Daniel
Smith, Sidney Taylor
Speakman, Edwin A(aron)
Spitzer, Cary Redford
Stephenson, Frederick William
Stermer, Robert L, Jr
Stone, Harris B(obby)
Tole, John Roy
Tripp, John Stephen
Tsang, Wai Lin
VanDerlaske, Dennis P
Weaver, Alfred Charles
Weinstein, Iram J
Wolicki, Eligius Anthony
Woods, Roy Alexander
Worthington, John Wilbur
Yang, Ta-Lun

WASHINGTON
Boatman, Edwin S
Craine, Lloyd Bernard
Daniel, J(ack) Leland
Darling, Robert Bruce
Dunham, Glen Curtis
Gillingham, Robert J
Gunderson, Leslie Charles
Heilenday, Frank W (Tod)
Heindsmann, T(heodore) E(dward)
Henry, Robert R
Hirsch, Horst Eberhard
Hoisington, David B(oysen)
Hueter, Theodor Friedrich
Kim, Jae Hoon
Lorenzen, Howard O(tto)
Morton, Randall Eugene
Nomura, Kaworu Carl
Peterson, Arnold (Per Gustaf)
Preikschat, F(ritz) K(arl)
Ringo, John Alan
Russell, James Torrance
Shamash, Yacov A
Smilen, Lowell I
Stanley, Hugh P
Tsang, Leung

WEST VIRGINIA
Gould, Henry Wadsworth
Osborn, Claiborn Lee

WISCONSIN
Anderson, James Gerard
Anway, Allen R
Bajikar, Sateesh S
Bomba, Steven James
Brauer, John Robert
Coggins, James Ray
Crawmer, Daryl E
Davis, Thomas William
Divjak, August A
Dunn, Stanley Austin
Flax, Stephen Wayne
Hind, Joseph Edward
Jeutter, Dean Curtis
Kerkman, Russel John
Nordman, James Emery
Schlicke, Heinz M
Thompson, Paul DeVries
Vanderheiden, Gregg
Van Horn, Diane Lillian
Webster, John Goodwin
Zelazo, Nathaniel Kachorek

WYOMING
Constantinides, Christos T(heodorou)

ALBERTA
Cormack, George Douglas
Haslett, James William
Hunt, Robert Nelson

BRITISH COLUMBIA
Hafer, Louis James
Jull, Edward Vincent
Menon, Thuppalay K
Pomeroy, Richard James
Sallos, Joseph
Tiedje, J Thomas
Zielinski, Adam

MANITOBA
Bridges, E(rnest)
Shafai, Lotfollah

NEWFOUNDLAND
Rahman, Md Azizur

NOVA SCOTIA
Baird, Charles Robert
Bhartia, Prakash
El-Masry, Ezz Ismail
Roger, William Alexander
Wood, Eunice Marjorie

ONTARIO
Bahadur, Birendra
Bandler, John William
Barclay, Alexander Primrose Hutcheson

Barrington, Nevitt H J
Bigham, Clifford Bruce
Brach, Eugene Jenõ
Brannen, Eric
Carr, Jan
Carroll, John Millar
Celinski, Olgierd J(erzy) Z(dzislaw)
Chamberlain, S(avvas) G(eorgiou)
Chow, Yung Leonard
Cohn-Sfetcu, Sorin
Costain, Cecil Clifford
Earnshaw, John W
Fahmy, Moustafa Mahmoud
Fewer, Darrell R(aymond)
Ganza, Kresimir Peter
Greenaway, Keith R(ogers)
Hayes, J Edmund
Hobson, John Peter
Holford, Richard Moore
Johnston, Ronald Harvey
Jull, George W(alter)
Jullien, Graham Arnold
Jurkus, Algirdas Petras
Keeler, John S(cott)
Kuiper-Goodman, Tine
Langille, Robert C(arden)
Lemyre, C(lement)
Lindsey, George Roy
McIntosh, Bruce Andrew
Mark, Jon Wei
Milkie, Terence H
Mouftah, Hussein T
Northwood, Derek Owen
Purbo, Onno Widodo
Ratz, H(erbert) C(harles)
Rauch, Sol
Ryan, Dave
Salama, Clement Andre Tewfik
Sedra, Adel S
Selvakumar, Chettypalayam Ramanathan
Shewchun, John
Siddall, Ernest
Slater, Keith
Smith, Kenneth Carless
Stewart, Thomas William Wallace
Tavares, Stafford Emanuel
Tsao, Sai Hoi
Vanier, Jacques
Zukotynski, Stefan

QUEBEC
Agarwal, Vinod Kumar
Anderson, Donald Arthur
Atwood, John William
Blostein, Maier Lionel
Champness, Clifford Harry
Cheng, Richard M H
Ferguson, Michael John
Ghosh, Asoke Kumar
Gregory, Brian Charles
Gulrajani, Ramesh Mulchand
Hugon, Jean S
Lafontaine, Jean-Gabriel
L'Archevêque, Real Viateur
Lecours, Michel
Lombos, Bela Anthony
Maciejko, Roman
Patel, Rajnikant V
Schwelb, Otto
Shkarofsky, Issie Peter
Van Vliet, Carolyne Marina

OTHER COUNTRIES
Bellanger, Maurice G
Boxman, Raymond Leon
Carassa, Francesco
Chang, Chun-Yen
Chang, Morris
Chase, Robert L(loyd)
Chu, Kwo Ray
Conforti, Evandro
Cullen, Alexander Lamb
De Bruyne, Peter
Esposito, Raffaele
Eykhoff, Pieter
Fettweis, Alfred Leo Maria
Fujiwara, Hideo
Gindsberg, Joseph
Granger, John Van Nuys
Haggis, Geoffrey Harvey
Halkias, Christos
Heiblum, Mordehai
Ince, A Nejat
Jespersen, Nils Vidar
Johnson, E(dward) O
Kaiser, Wolfgang A
Kao, Charles K
Kim, Wan H(ee)
Kishi, Keiji
Kurokawa, Kaneyuki
Kuroyanagi, Noriyoshi
Lahiri, Syamal Kumar
Lee, Choong Woong
Lehman, Meir M
Lewis, Arnold Leroy, II
Lindell, Ismo Veikko
Luo, Peilin
Malah, David
Maqusi, Mohammad
Marom, Emanuel
Mengali, Umberto
Okamura, Sogo
Onoe, Morio
Pau, Louis F

Electronics Engineering (cont)

Paulish, Daniel John
Rozzi, Tullio
Sakurai, Yoshifumi
Sato, Gentei
Struzak, Ryszard G
Stumpers, Frans Louis H M
Sugaya, Hiroshi
Tarui, Yasuo
Thiel, Frank L(ouis)
Toshio, Makimoto
Un, Chong Kwan
Unger, Hans-Georg
Ushioda, Sukekatsu
Vander Vorst, Andre
Van Overstraeten, Roger Joseph
Wallmark, J(ohn) Torkel
Wilkes, Maurice V
Zeheb, Ezra

Engineering, General

ALABAMA
Carden, Arnold Eugene
Dannenberg, Konrad K
Edlin, George Robert
Gilbert, Stephen Marc
Johnson, Clarence Eugene
Nola, Frank Joseph
Tidwell, Eugene Delbert
Vance, Ollie Lawrence

ALASKA
Sweet, Larry Ross

ARIZONA
Brown, Gordon S
Brown, Paul Wayne
Chance, C(layton) W(illiam)
Corcoran, William P
Dempster, William Fred
Edlin, Frank E
Govil, Sanjay
McGuirk, William Joseph
McKenney, Dean Brinton
Miklofsky, Haaren A(lbert)
Osborn, Donald Earl
Randall, Lawrence Kessler, Jr
Rice, Warren
Rowand, Will H
Thomson, Quentin Robert

ARKANSAS
Crisp, Robert M, Jr
Sutherland, G Russell
Yaz, Engin

CALIFORNIA
Abbott, Seth R
Adelman, Barnet Reuben
Agrawal, Suphal Prakash
Arnold, Frank R(obert)
Arnold, Walter Frank
Arslancan, Ahmet N
Attwood, David Thomas, Jr
Atwood, John Leland
Ayars, James Earl
Bahn, Gilbert S(chuyler)
Balakrishnan, A V
Bhandarkar, Mangalore Dilip
Blankenship, Victor D(ale)
Blink, James Allen
Boone, James Lightholder
Bourke, Robert Hathaway
Brahma, Chandra Sekhar
Brand, Donald A
Bridges, William B
Buchberg, Harry
Burnett, James R
Cannon, Robert H, Jr
Cardwell, William Thomas, Jr
Carroll, William J
Chambre, Paul L
Chesworth, Robert Hadden
Chivers, Hugh John
Chu, Chung-Yu Chester
Clarke, Lucien Gill
Cleland, Laurence Lynn
Clewell, Dayton Harris
Cohn, Seymour B(ernard)
Conly, John F
Crooke, Robert C
Dahlberg, Richard Craig
Davies, John Tudor
Davis, Frank W
Davis, W Kenneth
Denavit, Jacques
Desoer, Charles A(uguste)
Dhir, Vijay Kumar
Dixon, Harry S(terling)
Donnan, William W
Durbeck, Robert C(harles)
Dyer, James Lee
Elder, Rex Alfred
Fan, Chien
Fergin, Richard Kenneth
Ferrari, Domenico
Fitzpatrick, Gary Owen
Foster, John Stuart, Jr
Freedy, Amos
Garrick, B(ernell) John
Gasich, Welko E

Gay, Charles Francis
Gilmartin, Thomas Joseph
Gleghorn, G(eorge) J(ay), (Jr)
Grant, Eugene Lodewick
Gray, A(ugustine) H(eard), Jr
Gunderson, Norman O
Hadley, Jeffery A
Haloulakos, Vassilios E
Hamilton, J(ames) Hugh
Hammond, R Philip
Hansen, Richard (Thomas)
Hanson, Merle Edwin
Harker, J M
Harmen, Raymond A
Hesse, Christian August
Hingorani, Narain G
Hoffman, Myron A(rnold)
Homer, Paul Bruce
Howard, Ronald A(rthur)
Hubka, William Frank
Hughes, Thomas Joseph
Ingram, Gerald E(ugene)
Ishihara, Kohei
Johannessen, Jack
Johnson, Conor Deane
Johnson, Paul L
Johnson, Robert L
Jones, Clarence S
Kaplan, Hesh J
Kesselring, John Paul
Kimme, Ernest Godfrey
King, William Stanely
Krause, Ralph A(lvin)
Kuczynski, Eugene Raymond
Kulgein, Norman Gerald
Kvaas, T(horvald) Arthur
Larson, Harry Thomas
LeFave, Gene M(arion)
Leonard, A(nthony)
Li, Seung P(ing)
Lindvall, F(rederick) C(harles)
Lloyd, Edward C(harles)
Lynch, Frank W
McDonald, Henry C
McHuron, Clark Ernest
McManigal, Paul Gabriel Moulin
Maillot, Patrick Gilles
Milstein, Frederick
Mohamed, Farghalli Abdelrahman
Molinder, John Irving
Moorhouse, Douglas C
Muskat, Morris
Mykkanen, Donald L
Nash, James Richard
Neal, Ralph Bennett
Nelson, Arthur L(ee)
O'Brien, Morrough P
Olfe, D(aniel)
Oliver, Bernard M(ore)
O'Neill, R(ussell) R(ichard)
Ormsby, Robert B, Jr
Oshman, M Kenneth
Palter, N(orman) H(oward)
Parden, Robert James
Parker, N(orman) F(rancis)
Parme, Alfred L
Paulling, J(ohn) R(andolph), Jr
Platus, Daniel Herschel
Reyhner, Theodore O
Richardson, Neal A(llen)
Roe, Arnold
Rosenbloom, Arnold
Rosenstein, A(llen) B
Sackman, Jerome L(eo)
Sadeh, Willy Zeev
Salinas, David
Sarkaria, Gurmukh S
Schmitendorf, William E
Schonberg, Russell George
Smith, Lloyd P
Sommers, William P(aul)
Spear, Robert Clinton
Sperling, Jacob L
Spitzer, Irwin Asher
Standing, Marshall B
Staudhammer, Peter
Stearns, H(orace) Myrl
Stimmell, K G
Stone, Henry E
Sworder, David D
Tang, Stephen Shien-Pu
Thomas, Frank J(oseph)
Thomas, Mitchell
Thompson, Leland E
Thomson, William Tyrrell
Thorpe, Howard A(lan)
Tribus, Myron
Trorey, A(lan) W(ilson)
Vassiliou, Marius Simon
Waldhauer, F D
Welsh, David Edward
Wheaton, Elmer Paul
Whipple, Christopher George
Wilkinson, Eugene P Dennis
Wood, Carlos C
Wood, Edward C(halmers)
Woods, Ralph Arthur
Wooldridge, Dean E
Wyllie, Loring A
Yates, Alden P

COLORADO
Albert, Harrison Bernard
Beck, Betty Anne

Beckstead, Leo William
Brown, Alison Kay
Chamberlain, A(drian) R(amond)
Eyman, Earl Duane
Glover, Robert E(llsworth)
Jansson, David Guild
Krill, Arthur Melvin
MacGregor, Ronald John
McManamon, Peter Michael
Mercer, Robert Allen
Recht, Rodney F(rank)
Roberts, Howard C(reighton)
Stauffer, Jack B
Timblin, Lloyd O, Jr
Verschoor, J(ack) D(ahlstrom)
Wainwright, William Lloyd

CONNECTICUT
Brenholdt, Irving R
Burkhard, Mahlon Daniel
Drebus, Richard William
Dubin, Fred S
Elrod, Harold G(lenn), Jr
Frey, Carl
Gerber, H Joseph
Gordon, Robert Boyd
Hood, Edward E, Jr
Jordan, Donald J
Ketchman, Jeffrey
Shichman, D(aniel)
Shuey, Merlin Arthur
Sternberg, Robert Langley
Strough, Robert I(rving)
Suprynowicz, Vincent A

DELAWARE
Riewald, Paul Gordon
Uy, William Cheng
Yiannos, Peter N

DISTRICT OF COLUMBIA
Ausubel, Jesse Huntley
Chapin, Douglas McCall
Dame, Richard Edward
Frair, Karen Lee
Green, Richard James
Groves, Donald George
Hodge, John Dennis
Kaye, John
Larson, Thomas D
Oran, Elaine Surick
Paige, Hilliard W
Sanders, Robert Charles
Sevin, E(ugene)
Sexton, Ken
Thomas, Leonard William, Sr
Townsend, Marjorie Rhodes
Tucker, John Richard
Urban, Peter Anthony
Weiss, Leonard
Willenbrock, Frederick Karl

FLORIDA
Amon, Max
Beil, Robert J(unior)
Concordia, Charles
Cox, Parker Graham
Davison, Beaumont
Dean, Donald L(ee)
Dilpare, Armand Leon
Dowdell, Rodger B(irtwell)
Ewald, William Philip
Florea, Harold R(obert)
Gaffney, F(rancis) J(oseph)
Gold, Edward
Goldwyn, Roger M(artin)
Henning, Rudolf E
Hirsch, Donald Earl
Hollis, Mark Dexter
Kalman, Rudolf Emil
Kolhoff, M(arvin) J(oseph)
McCune, Francis K(imber)
Maeder, P(aul) F(ritz)
Milton, James E(dmund)
Price, Donald Ray
Sirkin, Alan N
Squires, Lombard
Stefanakos, Elias Kyriakos
Swain, Geoffrey W
Walsh, Edward Kyran
Wiebusch, Charles Fred
Zimmer, Martin F

GEORGIA
Antolovich, Stephen D
Cassanova, Robert Anthony
Cullison, William Lester
Dickert, Herman A(lonzo)
Ghate, Suhas Ramkrishna
Green, G W
Gupton, Guy Winfred, Jr
Stanley, Luticious Bryan, Jr

HAWAII
Gersch, Will
Hallanger, Lawrence William
Yee, Alfred A

IDAHO
Bondurant, James A(llison)
Byers, Roland O
Froes, Francis Herbert (Sam)
Martin, J(ames) G(eorge)
Williams, George Arthur

ILLINOIS
Ban, Stephen Dennis
Burks, G Edwin
Christianson, Clinton Curtis
Conry, Thomas Francis
Drickamer, Harry George
Firstman, Sidney I(rving)
Gose, Earl E(ugene)
Hamming, Kenneth W
Hull, John R
Johari, Om
Jones, Leonard Clive
Kenny, Andrew Augustine
Kliphardt, Raymond A(dolph)
Lischer, Ludwig F
Morin, Charles Raymond
Safdari, Yahya Bhai
Sanathanan, C(hathilingath) K
Schallert, William Francis
Schlesinger, Lee
Shabana, Ahmed Abdelraouf
Sohr, Robert Trueman
Staunton, John Joseph Jameson
Stukel, James Joseph
Tao, Liang Neng
Weber, J K Richard
Woods, Kenneth R
Wright, Maurice Arthur

INDIANA
Acker, Frank Earl
Atallah, Mikhail Jibrayil
Bogdanoff, John Lee
Burden, Stanley Lee, Jr
Jefferies, Michael John
Kennedy, J(ohn) R(obert)
Kronmiller, C W
Mercer, Walter Ronald
Nagle, Edward John
Odor, David Lee
Pierson, Edward S
Plants, Helen Lester
Sen, Mihir
Skibniewski, Miroslaw Jan
Vanderbilt, Vern C, Jr
Voelz, Michael H

IOWA
Boylan, D(avid) R(ay)
Braaten, Melvin Ole
Firstenberger, B(urnett) G(eorge)
Greer, Raymond T(homas)
Kortanek, Kenneth O
Mack, Michael J
Mather, Roger Frederick
Northup, Larry L(ee)
Siebes, Maria

KANSAS
Frohmberg, Richard P
Gowdy, Kenneth King
Joyner, Howard Sajon
Kirmser, P(hilip) G(eorge)
Lindholm, John C
Razak, Charles Kenneth
Robinson, M John
Rodgers, Edward J(ohn)
Smith, Howard Wesley
Spears, Sholto Marion
Williams, Wayne Watson

KENTUCKY
Beattie, Horace S
Cathey, Jimmie Joe
Gold, Harold Eugene
Johnson, Alan Arthur
Viswanadham, Ramamurthy K

LOUISIANA
Decossas, Kenneth Miles
Latorre, Robert George
Molaison, Henri J(ean)
Robert, Kearny Quinn, Jr
Sabbaghian, Mehdy
Thurman, Henry L, Jr

MAINE
Kinnaird, Richard Farrell
Myers, Basil R

MARYLAND
Adams, Edward Franklin
Adams, Laurence J
Bascunana, Jose Luis
Bukowski, Richard William
Burns, Bruce Peter
Burns, John Joseph, Jr
Castella, Frank Robert
Chiogioji, Melvin Hiroaki
Christensen, Richard G
Cohen, Stanley Alvin
Coulter, James B
Coutinho, John
Darr, J(ack) E(dwin)
Field, Herbert Cyre
Gottfried, Paul
Graves, Harvey W(ilbur), Jr
Hall, Kimball Parker
Harman, George Gibson, Jr
Heider, Shirley (Scott) A(mborn)
Huang, Thomas Tsung-Tse
Johnston, Thomas M(atkins)
Katsanis, D(avid) J(ohn)
Kemelhor, Robert Elias

Knazek, Richard Allan
Ku, Harry Hsien Hsiang
Loxley, Thomas Edward
Lundsager, C(hristian) Bent
Makofski, Robert Anthony
Neild, A(lton) Bayne
Nichols, Kenneth David
Rabinow, Jacob
Ranganathan, Brahmanpalli
    Narasimhamurthy
Rostenbach, Royal E(dwin)
Sallet, Dirse Wilkis
Schubauer, Galen B
Schwoerer, F(rank)
Smith, Joseph Collins
Talbott, Edwin M
Theuer, Paul John
Treglia, Thomas A(nthony)
Troutman, James Scott
Walsh, James Paul
Walter, Donald K
Wimenitz, Francis Nathaniel
Zuber, Novak

MASSACHUSETTS
Apelian, Diran
Bahr, Karl Edward
Berkey, Donald C
Bowen, H Kent
Bradner, Mead
Drew, Philip Garfield
Ehrenbeck, Raymond
Emanuel, Alexander Eigeles
Forrester, Jay W
Glaser, Peter E(dward)
Hockney, Richard L
Hubbard, M(alcolm) M(acGregor)
Keil, Alfred Adolf Heinrich
Kroner, Klaus E(rlendur)
Latham, Allen, Jr
Lees, David Eric Berman
McCune, William James, Jr
McDonald, Donald C
Mack, Michael E
Markey, Winston Roscoe
Mastronardi, Richard
Merrill, E(dward) W(ilson)
Osgood, Elmer Clayton
Phillips, Thomas Leonard
Pickering, Frank E
Pirri, Anthony Nicholas
Pulling, Nathaniel H(osler)
Quinn, George David
Richardson, Allyn (St Clair)
Roadstrum, William H(enry)
Robinson, D(enis) M(orrell)
Rogowski, A(ugustus) R(udolph)
Scherer, Harold Nicholas, Jr
Schwartz, Thomas Alan
Shepherd, James E
Sikina, Thomas
Sosnowski, Thomas Patrick
Sprague, Robert C
Taylor, Edward S(tory)
Wang, An
Webster, Lee Alan

MICHIGAN
Alexandridis, Alexander A
Ball, George A(ppleton)
Barr, Harry
Batrin, George Leslie
Eltinge, Lamont
Fancher, Paul S(trimple)
Fisher, Edward Richard
Gilbert, E(dward) O(tis)
Gilbert, Elmer G(rant)
Greene, Bruce Edgar
Hrovat, Davorin
Jaicks, Frederick G
Luxon, James Thomas
Luzzi, Theodore E, Jr
Musinski, Donald Louis
Olte, Andrejs
Powers, William Francis
Prostak, Arnold S
Ressler, Neil William
Riley, Bernard Jerome
Ryant, Charles J(oseph), Jr
Scherger, Dale Albert
Stansel, John Charles
Thomson, Robert Francis
Wagner, Harvey Arthur
Weber, Walter J, Jr
Westervelt, Franklin Herbert

MINNESOTA
Bateson, Robert Neil
Foster, George Rainey
Mahmoodi, Parviz
Rao, Prakash
Scriven, L E(dward), (II)
Shulman, Yechiel
Stadtherr, Leon

MISSISSIPPI
Eubanks, William Hunter
Happ, Stafford Coleman
Rosenhan, A Kirk
Taylor-Cade, Ruth Ann

MISSOURI
Deam, James Richard
Glauz, William Donald

McDonnell, Sanford
Major, Schwab Samuel, Jr
Mathiprakasam, Balakrishnan
Oglesby, David Berger
Peltier, Eugene J
Roberts, J(asper) Kent
Schweiker, Jerry W

MONTANA
Friel, Leroy Lawrence
Taylor, William Robert

NEBRASKA
Lenz, Charles Eldon
Pao, Y(en)-C(hing)
Smith, Durward A

NEVADA
Snaper, Alvin Allyn
Tryon, John G(riggs)

NEW HAMPSHIRE
Fletcher, William L
Klarman, Karl J(oseph)
Laaspere, Thomas
Mellor, Malcolm
Mumma, Albert G
Noble, Charles Carmin
Spherley, Charles W, Jr
Wallis, Graham B

NEW JERSEY
Biskeborn, Merle Chester
Bohacek, Peter Karl
Brill, Yvonne Claeys
Clements, Wayne Irwin
Coleman, Norman P, Jr
Courtney-Pratt, Jeofry Stuart
DiMasi, Gabriel Joseph
Domeshek, S(ol)
Epstein, Seymour
Feinblum, David Alan
Fisher, Reed Edward
Fox, A(rthur) Gardner
Jain, Sushil C
Kandoian, A(rmig) G(hevont)
Kintner, Edwin E
Kittredge, Clifford Proctor
Lechner, Bernard J
Le Mee, Jean M
Lepselter, Martin P
Maclin, Ernest
Mikolajczak, Alojzy Antoni
Negin, Michael
Newman, Joseph Herbert
Patel, Chandra Kumar Naranbhai
Peskin, Richard Leonard
Phadke, Madhav Shridhar
Puri, Narindra Nath
Rigassio, James Louis
St Onge, G H
Shashidhara, Nagalapur Sastry
Shinozuka, Masanobu
Steigelmann, William Henry
Stevens, William D
Weiss, C Dennis
Zoltan, Bart Joseph

NEW MEXICO
Broyles, Carter D
Cronenberg, August William
Diver, Richard Boyer, Jr
Egbert, Robert B(aldwin)
Finch, Thomas Wesley
Howard, William Jack
Icerman, Larry
Jankowski, Francis James
Perkins, Roger Bruce
Russell, John Henry
Savage, Charles Francis
Tenney, Gerold H
Vortman, L(uke) J(erome)
Welber, Irwin
Wylie, Kyral Francis

NEW YORK
Amero, Bernard Alan
Baron, Seymour
Baum, Eleanor Kushel
Bigelow, John Edward
Bradburd, Ervin M
Cheng, Yean Fu
Cornacchio, Joseph V(incent)
Curtis, Huntington W(oodman)
Evans, Wayne Russell
Fama, Donald Francis
Fitzroy, Nancy Deloye
Flannery, John B, Jr
Freund, Richard A
Friedman, Morton (Benjamin)
Goertzel, Gerald
Greene, Peter Richard
Grunes, Robert Lewis
Haag, Fred George
Haddad, Jerrier Abdo
Hsieh, Shih-Yung
Johnson, Herbert Harrison
Krishna, C R
Lee, Charles Northam
Levy, Robert S(amuel)
Logan, H(enry) L(eon)
McMurry, William
Madey, Robert W
Matkovich, Vlado Ivan

Metz, Edward
Morabito, Bruno P
Parsegian, V(ozcan) Lawrence
Penwell, Richard Carlton
Rosenberg, Paul
Schanker, Jacob Z
Schultz, Andrew, Jr
Shaffer, Bernard W(illiam)
Shames, Irving H
Shube, Eugene E
Simons, Gene R
Snyder, Robert L(eon)
Soifer, Herman
Theobald, John J(acob)
Udolf, Roy
Velzy, Charles O
Vosburgh, Kirby Gannett
Weber, Arthur Phineas
Wendt, Robert L(ouis)
Wilcock, Donald F(rederick)
Witt, Christopher John
Wood, Wallace D(ean)
Zic, Eric A
Ziegler, Robert C(harles)

NORTH CAROLINA
Baker, Charles Ray
Chin, Robert Allen
Clark, Charles Edward
Gerhardt, Don John
Hess, Daniel Nicholas
Land, Ming Huey
Lee, William States
Lord, Peter Reeves
Owen, Warren H
Royster, L(arry) H(erbert)
Sander, William August, III
Utku, Senol
Wilson, Geoffrey Leonard
Wollin, Goesta
Wright, William Vale

NORTH DAKOTA
Rieder, William G(ary)

OHIO
Arnett, Jerry Butler
Bruno, David Joseph
Butz, Donald Josef
Ciricillo, Samuel F
Cook, James Robert
Cooke, Charles C
Cosgrove, Stanley Leonard
Cummings, Charles Arnold
Damianov, Vladimir B
Dunn, Robert Garvin
Eckert, John S
El-Naggar, Mohammed Ismail
Erbacher, John Kornel
Grigger, David John
Henderson, Courtland M
Hulley, Clair Montrose
Jones, Trevor O
Kelly, Sidney J(ohn)
Kornylak, Andrew T
LaRue, Robert D(ean)
Lownie, H(arold) W(illiam), Jr
Lynch, T(homas) E(lwin)
McCauley, Roy B, Jr
McNichols, Roger J(effrey)
Manson, S(amuel) S(tanford)
Messick, Roger E
Middendorf, William H
Newby, John R
Pugh, John W(illiam)
Putnam, Abbott (Allen)
Ramalingam, Mysore Loganathan
Robe, Thurlow Richard
Rubin, Saul H
Seagle, Stan R
Soska, Geary Victor
Sutton, J(ames) L(owell)
Voigt, Charles Frederick
Voisard, Walter Bryan
White, Willis S, Jr
Wolfe, Robert Kenneth
Zavodney, Lawrence Dennis

OKLAHOMA
Baker, James E
Basore, B(ennett) L(ee)
Bowman, Mark (McKinley), Jr
DeLacerda, Fred G
Donahue, Hayden Hackney
Forney, Bill E
Goddin, C(lifton) S(ylvanus)
Hays, George E(dgar)
Norton, Joseph R(andolph)
Rice, Charles Edward
Scisson, Sidney Eugene

OREGON
Fields, R(ance) Wayne
Skinner, Richard Emery

PENNSYLVANIA
Anouchi, Abraham Y
Ayres, Robert U
Bandel, Hannskarl
Bissinger, Barnard Hinkle
Brown, Chester Harvey, Jr
Burtis, Theodore A
Carfagno, Salvatore P
Champagne, Paul Ernest

Charp, Solomon
Chou, Pei Chi (Peter)
Comenetz, George
Confer, Gregory Lee
Crane, Lee Stanley
Davis, R(obert) E(lliot)
Deb, Arun K
DebRoy, Tarasankar
Dorne, Arthur
Falconer, Thomas Hugh
Feero, William E
Forscher, Frederick
Gallagher, C(harles) E(dward)
Gebhart, Benjamin
Geldmacher, R(obert) C(arl)
Hager, Wayne R
Hayes, Miles V(an Valzah)
Haythornthwaite, Robert M(orphet)
Heiligman, Harold A
Hoburg, James Frederick
Howland, Frank L
Kamen, Edward Walter
Knaster, Tatyana
Koch, Ronald N
Kreh, E(dward) J(oseph), Jr
Langston, Charles Adam
Lauchle, Gerald Clyde
Lazaridis, Anastas
Marks, Peter J
Meredith, David Bruce
Merz, Richard A
Michael, Norman
Moose, Louis F
Mouly, Raymond J
Palmer, Rufus N(elson)
Pan, Yuan-Siang
Paolino, Michael A
Penney, Gaylord W
Peterson, Richard Walter
Preusch, Charles D
Prywes, Noah S(hmarya)
Schenk, H(arold) L(ouis), Jr
Schulman, Marvin
Shobert, Erle Irwin, II
Silverstein, Calvin C(arlton)
Soung, Wen Y
Spencer, Arthur Coe, II
Spengos, Aris C(onstantine)
Stone, M(orris) D
Takeuchi, Kenji
Teller, J C
Tobias, Philip E
Tuffey, Thomas J
Viest, Ivan M
Walchli, Harold E(dward)
Williams, David Bernard
Wolgemuth, Carl Hess
Wray, Porter R

RHODE ISLAND
Kowalski, Tadeusz
Lerner, Samuel
Morse, Theodore Frederick
Pearson, Allan Einar
Roessler, Barton
Spero, Caesar A(nthony), Jr
Symonds, Paul S(outhworth)

SOUTH CAROLINA
Bridges, Donald Norris
Chao, Yuh J (Bill)
Harrison, James William, Jr
Hogan, Joseph C(harles)
Jaco, Charles M, Jr
Kellerman, Karl F(rederic)
Russell, Allen Stevenson
Sandrapaty, Ramachandra Rao
Williamson, Robert Elmore

SOUTH DAKOTA
Schleusener, Richard A

TENNESSEE
Buchanan, George R(ichard)
Burwell, Calvin C
Eissenberg, David M(artin)
Foster, Edwin Powell
Frounfelker, Robert E
Gat, Uri
Janna, William Sied
Kimmons, George H
Kinser, Donald LeRoy
Lim, Alexander Te
Moulden, Trevor Holmes
Pandeya, Prakash N
Reister, David B(ryan)
Ross, Harley Harris
Sheth, Atul C
Simhan, Raj
Spruiell, Joseph E(arl)
Waters, Dean Allison
Wichner, Robert Paul
Wimberly, C Ray

TEXAS
Abramson, H(yman) Norman
Albertson, Harold D
Austin, T Louis, Jr
Biard, James R
Bovay, Harry Elmo, Jr
Boyer, Robert Elston
Bravenec, Edward V
Burkey, Ronald Steven
Chandler, James Michael

Engineering, General (cont)

Childs, S(elma) Bart
Chyu, Ming-Chien
Clunie, Thomas John
Cragon, Harvey George
Daigh, John D(avid)
Davis, Alfred, Jr
Dhudshia, Vallabh H
French, Robert Leonard
Friberg, Emil Edwards
Frick, John P
Glaspie, Donald Lee
Godwin, James Basil, Jr
Griffin, Richard B
Harper, James George
Hillaker, Harry J
Jonsson, John Erik
Kiel, Otis Gerald
Lawrence, Joseph D, Jr
MacFarlane, Duncan Leo
Meyer, W(illiam) Keith
Musgrave, Albert Wayne
Odeh, A(ziz) S(alim)
Patel, Mayur
Raju, Satyanarayana G V
Rodenberger, Charles Alvard
Sawyer, Ralph Stanley
Schuhmann, Robert Ewald
Schweppe, Joseph L(ouis)
Sjoberg, Sigurd A
Stucker, Harry T
Vance, John Milton
Ward, Donald Thomas
Widmer, Robert H
Yuan, Robert L

UTAH
Allen, Dell K
Jacobsen, Stephen C
Johnson, Robert R(oyce)
MacGregor, Douglas
Nelson, Mark Adams

VERMONT
Foote, Vernon Stuart, Jr
Howard, Robert T(urner)

VIRGINIA
Balwanz, William Walter
Bishara, Michael Nageeb
Clough, G Wayne
Devens, W(illiam) George
Duscha, Lloyd A
Ell, William M
Foster, Richard B(ergeron)
Fubini, Eugene G(hiron)
Garcia, Albert B
Goins, Truman
Gordon, Ronald Stanton
Hall, Carl W(illiam)
Hammond, Marvin H, Jr
Hauxwell, Gerald Dean
Heller, Agnes S
Leon, H(erman) I
Lerner, Norman Conrad
Levine, Martin
Lowry, Ralph A(ddison)
McNichols, Gerald Robert
Marschall, Albert Rhoades
Mehta, Gurmukh D
Meirovitch, L(eonard)
Nuckolls, Joe Allen
Raju, Ivatury Sanyasi
Rall, Lloyd L(ouis)
Shrier, Stefan
Sikri, Atam P
Starbird, Alfred D
Summers, George Donald
Thorp, Benjamin A
Webb, George Randolph
Wilkinson, Thomas Lloyd, Jr

WASHINGTON
Bell, Milo C
Chiang, Karl Kiu-Kao
Clevenger, William A
Condit, Philip M
Glosten, Lawrence R
Gunderson, Leslie Charles
Hirschfelder, John Joseph
Mast, Robert F
Pearson, Carl E
Stanton, K Neil
Withington, Holden W

WEST VIRGINIA
Clark, Nigel Norman
Muth, Wayne Allen
Strickland, Larry Dean

WISCONSIN
Berthouex, Paul Mac
Bishop, Charles Joseph
Bomba, Steven James
Crawmer, Daryl E
Hodges, Lawrence H
Leenhouts, Lillian S
Lenz, Arno T(homas)
McMurray, David Claude
Richardson, Bobbie L
Schaffer, Erwin Lambert
Teletzke, Gerald H(oward)
Zanoni, Alphonse E(ligius)

WYOMING
Canfield, Carl Rex, Jr
Dolan, Charles W

PUERTO RICO
Bonnet, Juan A, Jr

ALBERTA
Duby, John
Karim, Ghazi A
Micko, Michael M
Nasser, Tourai
Verschuren, Jacobus Petrus

BRITISH COLUMBIA
Newbury, Robert W(illiam)
Srivastava, Krishan

MANITOBA
Cahoon, John Raymond

NEW BRUNSWICK
Chrzanowski, Adam

NOVA SCOTIA
Nugent, Sherwin Thomas
Sastry, Vankamamidi VRN
Strasser, John Albert

ONTARIO
Balakrishnan, Narayanaswany
Baronet, Clifford Nelson
Biggs, Ronald C(larke)
Carr, Jan
Jones, L(lewellyn) E(dward)
Lapp, P(hilip) A(lexander)
Matar, Said E
Novak, Milos
Vijay, Mohan Madan

PRINCE EDWARD ISLAND
Lodge, Malcolm A

QUEBEC
Bakhiet, Atef
Bhat, Rama B
Cherna, John C(harles)
Hanna, Adel
Marsh, Cedric
Massoud, Monir Fouad
Pekau, Oscar A
Sankar, Seshadri
Shaw, Robert Fletcher
Thirion, Jean Paul Joseph
Thompson, Allan Lloyd
Todorovic, Petar N
Tsui, Y(aw) T(zong)
Verschingel, Roger H C

SASKATCHEWAN
Riemer, Paul

OTHER COUNTRIES
Eichenberger, Hans P
Jacob, K Thomas
Mylonas, Constantine
Saba, Shoichi
Von Hoerner, Sebastian

Engineering Mechanics

ALABAMA
Beckett, Royce E(lliott)
Caudill, Reggie Jackson
Cochran, John Euell, Jr
Costes, Nicholas Constantine
Cutchins, Malcolm Armstrong
Davis, Anthony Michael John
Davis, Carl George
Doughty, Julian O
Evces, Charles Richard
Gambrell, Samuel C, Jr
Gilbert, John Andrew
Hill, James Lafe
Jones, Stanley E
Jordan, William D(itmer)
Kattus, J Robert
Killingsworth, R(oy) W(illiam)
Melville, Joel George
Schutzenhofer, Luke A
Turner, Charlie Daniel, Jr
Vance, Ollie Lawrence

ALASKA
Johansen, Nils Ivar
Skudrzyk, Frank J

ARIZONA
Arabyan, Ara
Chandra, Abhijit
Chen, Stanley Shiao-Hsiung
Chenea, Paul F(ranklin)
Govil, Sanjay
Malvick, Allan J(ames)
Neff, Richmond C(lark)
Saric, William Samuel
Savrun, Ender
Shaw, Milton C(layton)
Singhal, Avinash Chandra
Ulich, Bobby Lee
Wallace, C(harles) E(dward)

ARKANSAS
Boyd, Donald Edward
Jong, Ing-Chang
Ma, Er-Chieh
Smith, E(astman)

CALIFORNIA
Abdel-Ghaffar, Ahmed Mansour
Abrahamson, George R(aymond)
Adler, William Fred
Aggarwal, H(ans) R(aj)
Ardema, Mark D
Arnold, Frank R(obert)
Atchley, Bill Lee
Batdorf, Samuel Burbridge
Beal, Thomas R
Bendisz, Kazimierz
Blythe, William Richard
Boltinghouse, Joseph C
Bonora, Anthony Charles
Caligiuri, Robert Domenic
Cantin, Gilles
Cao, Hengchu
Carleone, Joseph
Chelapati, Chunduri V(enkata)
Chi, Cheng-Ching
Chiang, George C(hihming)
Chou, Larry I-Hui
Christensen, Richard Monson
Chuang, Kuen-Puo (Ken)
Collins, Jon David
Colton, James Dale
Das, Mihir Kumar
Dashner, Peter Alan
DeBra, Daniel Brown
Dong, Richard Gene
Dost, Martin Hans-Ulrich
Dowell, Douglas C
Eggers, A(lfred) J(ohn), Jr
Fleming, Alan Wayne
Flora, Edward B(enjamin)
Forrest, James Benjamin
Forsberg, Kevin
Galef, Arnold E
Geminder, Robert
Gere, James Monroe
Gleghorn, G(eorge) J(ay), (Jr)
Goldsmith, Werner
Hackel, Lloyd Anthony
Hackett, Colin Edwin
Herrmann, George
Hoff, N(icholas) J(ohn)
Horne, Roland Nicholas
Hsu, Chieh-Su
Hutchinson, James R(ichard)
Ingram, Gerald E(ugene)
Ito, Y(asuo) Marvin
Jacobson, Marcus
Johnson, Conor Deane
Kaul, Maharaj Krishen
Keller, Joseph Bishop
Kelly, Robert Edward
Kosmatka, John Benedict
Kraus, Samuel
Kuczynski, Eugene Raymond
Kwok, Munson Arthur
Lampert, Seymour
Larson, Edward William, Jr
Lau, John H
Leal, L Gary
Leckie, Frederick Alexander
Logan, James Columbus
Lysmer, John
Ma, Fai
Madan, Ram Chand
Mal, Ajit Kumar
Meissinger, Hans F
Meriam, James Lathrop
Mikes, Peter
Miller, Richard Keith
Monkewitz, Peter Alexis
Mote, C(layton) D(aniel), Jr
Mow, C(hao) C(how)
Muhlbauer, Karlheinz Christoph
Murch, S(tanley) Allan
Napolitano, Leonard Michael, Jr
Nelson, Richard Bartel
Nemat-Nasser, Siavouche
Nordell, William James
Piersol, Allan Gerald
Pierucci, Mauro
Pincus, Howard Jonah
Pinto, John Gilbert
Plotkin, Allen
Randolph, James E
Rau, Charles Alfred, Jr
Rubin, Sheldon
Sachs, Donald Charles
Schmit, Lucien A(ndre), Jr
Schuette, Evan H(enry)
Seaman, Lynn
Seide, Paul
Spier, Edward Ellis
Spreiter, John R(obert)
Stevenson, Merlon Lynn
Stockel, Ivar H(oward)
Stout, Ray Bernard
Strickland, Gordon Edward, Jr
Suer, H(erbert) S
Szego, Peter A
Trubert, Marc
Twiss, Robert John
Unal, Aynur
Vanblarigan, Peter

Vanderplaats, Garret Niel
Vanyo, James Patrick
Wallerstein, David Vandermere
Wehausen, John Vrooman
Weingarten, Victor I
Weng, Tu-Lung
Whittier, James S(pencer)
Wilbarger, Edward Stanley, Jr
Yeung, Ronald Wai-Chun
Zickel, John

COLORADO
Bacon, Merle D
Burnett, Jerrold J
Burnham, Marvin William
Cermak, J(ack) E(dward)
Chugh, Ashok Kumar
Criswell, Marvin Eugene
Datta, Subhendu Kumar
Feldman, Arthur
Frangopol, Dan Mircea
Geers, Thomas L
Gerdeen, James C
Medearis, Kenneth Gordon
Peterson, Harry C(larence)
Plows, William Herbert
Recht, Rodney F(rank)
Thompson, Erik G(rinde)
Weidman, Patrick Dan

CONNECTICUT
Adams, Brent Larsen
Cheatham, Robert Gary
Chin, Charles L(ee) D(ong)
Crossley, F(rancis) R(endel) Erskine
DeWolf, John T
Fink, Martin Ronald
Ketchman, Jeffrey
Maewal, Akhilesh
Mann, Richard A(rnold)
Nilson, Edwin Norman
Schile, Richard Douglas
Solecki, Roman
Verdon, Joseph Michael

DELAWARE
Cheng, Alexander H-D
Faupel, Joseph H(erman)
Hirsch, Albert Edgar
Kaliakin, Victor Nicholas
Kerr, Arnold D
Kingsbury, Herbert B
Lucas, James M
Santare, Michael Harold

DISTRICT OF COLUMBIA
Bainum, Peter Montgomery
Belsheim, Robert Oscar
Chong, Ken Pin
Herron, Isom H
Jones, Douglas Linwood
Menton, Robert Thomas
Moll, Magnus
Nachman, Arje
Niedenfuhr, Francis W(illiam)
Parks, Vincent Joseph
Rosenthal, F(elix)
Vaishnav, Ramesh
Walther, Carl H(ugo)
Wolko, Howard Stephen
Youm, Youngil

FLORIDA
Block, David L
Boykin, William H(enry), Jr
Catz, Jerome
Chiu, Tsao Yi
DeHart, Arnold O'Dell
Dilpare, Armand Leon
Ebrahimi, Fereshteh
Eisenberg, Martin A(llan)
Griffith, John E(dward)
Halperin, Don A(kiba)
Hsieh, Chung Kuo
Jenkins, David R(ichard)
Lijewski, Lawrence Edward
Lin, Y(u) K(weng)
Martin, Charles John
Mataga, Peter Andrew
Millsaps, Knox
Neff, Thomas O'Neil
Nevill, Gale E(rwin), Jr
Oline, Larry Ward
Plass, Harold J(ohn), Jr
Ranov, Theodor
Reichard, Ronnal Paul
Sandor, George N(ason)
Sidebottom, Omar M(arion)
Spears, Richard Kent
Stevens, Karl Kent
Sun, Chang-Tsan
Villanueva, Jose
Ward, Leonard George
Wiseman, H(arry) A(lexander) B(enjamin)
Yong, Yan

GEORGIA
Chang, Chin Hao
Ginsberg, Jerry Hal
Gupta, Rakesh Kumar
Jacobs, R(oy) K(enneth)
King, Wilton W(ayt)
McGill, David John

Neitzel, George Paul
Orloff, David Ira
Raville, Milton E(dward)
Sangster, William M(cCoy)
Spetnagel, Theodore John
Thornton, William Aloysius
Wang, James Ting-Shun
Wempner, Gerald Arthur

HAWAII
Burgess, John C(argill)
Stuiver, W(illem)
Taoka, George Takashi

ILLINOIS
Achenbach, Jan Drewes
Adrian, Ronald John
Ban, Stephen Dennis
Bazant, Zdenek P(avel)
Benjamin, Roland John
Berry, Gregory Franklin
Chakrabarti, Subrata K
Cheng, Herbert S
Chugh, Yoginder Paul
Conry, Thomas Francis
Cusano, Cristino
Daniel, Isaac M
Dantzig, Jonathan A
Davis, Philip K
Dutton, Jonathan Craig
Edelstein, Warren Stanley
Erwin, Lewis
Hall, William Joel
Hofer, Kenneth Emil
Hoffman, Clyde H(arris)
Hsui, Albert Tong-Kwan
Keer, Leon M
Kenny, Andrew Augustine
Kesler, Clyde E(rvin)
Kramer, John Michael
Kumar, Sudhir
Liu, Wing Kam
Miller, Robert Earl
Mockros, Lyle F(red)
Moran, Thomas J
Morrow, JoDean
Napadensky, Hyla S
Orthwein, W(illiam) C(oe)
Phillips, James Woodward
Pickett, Leroy Kenneth
Reichard, Grant Wesley
Riahi, Daniel Nourollah
Robinson, Arthur R(ichard)
Ruhl, Roland Luther
Schnobrich, William Courtney
Schoeberle, Daniel F
Sciammarella, Caesar August
Shabana, Ahmed Abdelraouf
Shack, William John
Shield, Richard Thorpe
Socie, Darrell Frederick
Stafford, John William
Tang, Wilson H
Ting, Thomas C(hi) T(sai)
Tucker, Charles L
Walker, John Scott
Weeks, Richard William
White, Robert Allan
Widera, Georg Ernst Otto
Wiley, Jack Cleveland
Wilson, William Robert Dunwoody
Worley, Will J
Wright, Maurice Arthur
Yen, Ben Chie

INDIANA
Chen, Wai-Fah
Gad-el-Hak, Mohamed
Huang, Nai-Chien
Jones, James Darren
Kareem, Ahsan
Kobayashi, F(rancis) M(asao)
Lee, L(awrence) H(wa) N(i)
Mortimer, Kenneth
Sen, Mihir
Snow, John Thomas
Strandhagen, Adolf G(ustav)
Sun, Chin-Teh
Sweet, Arnold Lawrence

IOWA
Akers, Arthur
Arora, Jasbir Singh
Choi, Kyung Kook
Chwang, Allen Tse-Yung
Cook, William John
Hekker, Roeland M T
Huston, Jeffrey Charles
Lakes, Roderic Stephen
Landweber, Louis
McConnell, Kenneth G
Nariboli, Gundo A
Nixon, Wilfrid Austin
Peterson, Paul W(eber)
Riley, William F(ranklin)
Rim, Kwan
Rogge, Thomas Ray
Sturges, Leroy D
Thompson, Robert Bruce
Tsai, Y(u)-M(in)
Zachary, Loren William

KANSAS
Huang, Chi-Lung Dominic

Lenzen, K(enneth) H(arvey)
Lindholm, John C
McBean, Robert Parker
McCabe, Steven Lee
Yu, Yun-Sheng

KENTUCKY
Bakanowski, Stephen Michael
Beatty, Millard Fillmore, Jr
Brock, Louis Milton
Dillon, Oscar Wendell, Jr
Huttenlocher, Dietrich F
Leigh, Donald C
Lu, Wei-yang
Man, Chi-Sing
Miller, C Eugene
Shippy, David James
Wang, Shien Tsun

LOUISIANA
Hewitt, Hudy C, Jr
Hibbeler, Russell Charles
Rubinstein, Asher A

MARYLAND
Alwan, Abdul-Mehsin
Amirikian, Arsham
Antman, Stuart S
Basdekas, Nicholas Leonidas
Belliveau, Louis J
Berger, Bruce S
Bernard, Peter Simon
Burns, Bruce Peter
Butler, Thomas W(esley)
Camponeschi, Eugene Thomas, Jr
Chadwick, Richard Simeon
Chuang, Tze-jer
Cunniff, Patrick F
Dhir, Surendra Kumar
Eades, James B(everly), Jr
Eitzen, Donald Gene
Ellingwood, Bruce Russell
Everstine, Gordon Carl
Fabunmi, James Ayinde
Flynn, Paul D(avid)
Fong, Jeffery Tse-Wei
Haberman, William L(awrence)
Huffington, Norris J(ackson), Jr
Jabbour, Kahtan Nicolas
Kitchens, Clarence Wesley, Jr
Koh, Severino Legarda
Lange, Eugene Albert
Michalowski, Radoslaw Lucas
Mordfin, Leonard
Noblesse, Francis
Pamidi, Prabhakar Ramarao
Peppin, Richard J
Poulose, Pathickal K
Rich, Harry Louis
Sallet, Dirse Wilkis
Schuman, William John, Jr
Sharpe, William Norman, Jr
Simiu, Emil
Smith, Russell Aubrey
Stein, Robert Alfred
Tai, Tsze Cheng
Tsai, Lung-Wen
Walsh, James Paul
Wang, Shou-Ling

MASSACHUSETTS
Adams, George G
Anderson, J(ohn) Edward
Archer, Robert Raymond
Budiansky, Bernard
Castro, Alfred A
Chang, Ching Shung
Cook, Nathan Henry
Cooper, W(illiam) E(ugene)
Covert, Eugene Edzards
De Fazio, Thomas Luca
Dym, Clive L
Hoffman, Allen Herbert
Kachanov, Mark L
Lardner, Thomas Joseph
Lemnios, A(ndrew) Z
MacGregor, C(harles) W(inters)
Pan, Coda H T
Pang, Yuan
Raab, Allen Robert
Raju, Palanichamy Pillai
Rapperport, Eugene J
Reed, F(lood) Everett
Rice, James R
Rogers, Hartley, Jr
Rossettos, John N(icholas)
Sanders, J(ohn) Lyell, Jr
Selig, Ernest Theodore
Sharp, A(rnold) G(ideon)
Tong, Pin
Wong, Po Kee
Work, Clyde E(verette)

MICHIGAN
Alexandridis, Alexander A
Altiero, Nicholas James, Jr
Barber, James Richard
Bolander, Richard
Browne, Alan Lampe
Chen, Francis Hap-Kwong
Cloud, Gary Lee
Comninou, Maria
Dawson, D(onald) E(merson)
DeSilva, Carl Nevin

Ellis, Robert William
Evaldson, Rune L
Ezzat, Hazem Ahmed
Ghoneim, Youssef Ahmed
Greenwood, D(onald) T(heodore)
Herzog, Bertram
Hess, Robert L(awrence)
Hovanesian, Joseph Der
Howell, Larry James
Iacocca, Lee A
Im, Jang Hi
Kamal, Mounir Mark
Kammash, Terry
Karnopp, Bruce Harvey
Kline, Kenneth A(lan)
Ku, Albert B
Kubis, Joseph J(ohn)
Kullgren, Thomas Edward
Kurajian, George Masrob
Li, Chin-Hsiu
Lord, Harold Wesley
Lubkin, James Leigh
Lund, Charles Edward
Medick, Matthew A
Morman, Kenneth N
Narain, Amitabh
Narayanaswamy, Onbathiveli S
Nefske, Donald Joseph
Nyquist, Gerald Warren
Passerello, Chris Edward
Robbins, D(elmar) Hurley
Rosinski, Michael A
Sachs, Herbert K(onrad)
Saul, William Edward
Schmidt, Robert
Schultz, Albert Barry
Shkolnikov, Moisey B
Sikarskie, David L(awrence)
Smith, Hadley J(ames)
Snyder, Virgil W(ard)
Soutas-Little, Robert William
Subramanian, K N
Tuchinsky, Philip Martin
Whicker, Donald
Wolf, Joseph A(llen), Jr
Wolf, Louis W
Wu, Hai
Zobel, Edward C(harles)

MINNESOTA
Beavers, Gordon Stanley
Ericksen, Jerald LaVerne
Garrard, William Lash, Jr
Gulliver, John Stephen
Hsiao, Chih Chun
Kallok, Michael John
Luskin, Mitchell B
Patton, Peter C(lyde)
Roska, Fred James
Sterling, Raymond Leslie
Warner, William Hamer

MISSISSIPPI
Butler, Dwain Kent
Cade, Ruth Ann
Carnes, Walter Rosamond
DeLeeuw, Samuel Leonard
Houston, James Robert
Posey, Joe Wesley
Scott, Charley

MISSOURI
Baldwin, James W(arren), Jr
Batra, Romesh Chander
Blundell, James Kenneth
Cheng, Franklin Yih
Cunningham, Floyd Mitchell
Davis, Robert Lane
Dharani, Lokeswarappa R
Fowler, Timothy John
Haas, Charles John
Hansen, Peter Gardner
Higdon, Archie
Hornsey, Edward Eugene
Jawad, Maan Hamid
Koval, Leslie R(obert)
Leutzinger, Rudolph L(eslie)
Liu, Henry
Mains, Robert M(arvin)
Mathiprakasam, Balakrishnan
Mirowitz, L(eo) I(saak)
Nikolai, Robert Joseph
Sastry, Shankara M L
Weissenburger, Jason T

MONTANA
Banaugh, Robert Peter
Friel, Leroy Lawrence

NEBRASKA
Blackman, J(ames) S(amuel)
Foral, Ralph Francis
Haack, Donald C(arl)
James, Merlin Lehn
Martin, Charles Wayne
Pierce, Donald N(orman)
Smith, Gerald M(ax)
Young, Lyle E(ugene)

NEVADA
Ghosh, Amitava

NEW HAMPSHIRE
Den Hartog, J(acob) P(ieter)

Hibler, William David, III
Klotz, Louis Herman
Long, Carl F(erdinand)
Mindlin, Raymond D(avid)
Morduchow, Morris
Spehrley, Charles W, Jr

NEW JERSEY
Balaguru, Perumalsamy N
Becht, Charles, IV
Benaroya, Haym
Chae, Yong Suk
Clymer, Arthur Benjamin
Denno, Khalil I
Dill, Ellis Harold
Frauenthal, James Clay
Goyal, Suresh
Hart, Colin Patrick
Killam, Everett Herbert
Langrana, Noshir A
Lee, Peter Chung-Yi
Lee, Wei-Kuo
Lestingi, Joseph Francis
Lubowe, Anthony G(arner)
Mark, Robert
Marscher, William Donnelly
Miller, Edward
Pai, David H(sien)-C(hung)
Pan, Ko Chang
Philipp, Ronald E
Pope, Daniel Loring
Prevost, Jean Herve
Sarkar, Kamalaksha
Schmidlin, Albertus Ernest
Shinozuka, Masanobu
Sisto, Fernando
Spillers, William R
Thomas, Ralph Henry, Sr
Tsakonas, Stavros
Weil, Rolf
Wilson, Charles Elmer
Yadvish, Robert D
Yu, Yi-Yuan
Zabusky, Norman J

NEW MEXICO
Albrecht, Bohumil
Chen, Er-Ping
Clouser, William Sands
Davison, Lee Walker
Drumheller, Douglas Schaeffer
Ebrahimi, Nader Dabir
Fisher, Franklin E(ugene)
Fuka, Louis Richard
Herrmann, Walter
Ju, Frederick D
Lowe, Terry Curtis
McTigue, David Francis
Moulds, William Joseph
Reuter, Robert Carl, Jr
Sutherland, Herbert James
Von Riesemann, Walter Arthur

NEW YORK
Ahmad, Jameel
Austin, Fred
Banerjee, Prasanta Kumar
Begell, William
Bennett, Leon
Benson, Richard Carter
Birnstiel, Charles
Cheng, Yean Fu
Chiang, Fu-Pen
Constantinou, Michalakis
Cozzarelli, Francis A(nthony)
Dempsey, John Patrick
Deresiewicz, Herbert
Haines, Daniel Webster
Hajela, Prabhat
Higuchi, Hiroshi
Hoenig, Alan
Huddleston, John Vincent
Hussain, Moayyed A
Hutchings, William Frank
Hwang, S Steve
Irwin, Arthur S(amuel)
Jwo, Chin-Hung
Kapila, Ashwani Kumar
Kayani, Joseph Thomas
Klosner, Jerome M
Krempl, Erhard
Kuchar, Norman Russell
Lai, W(ei) Michael
Lee, Erastus Henry
Lee, Richard Shao-Lin
Lessen, Martin
Levy, Alan Joseph
Libove, Charles
Lin, Sung P
Mendel, John Richard
Menkes, Sherwood Bradford
Moon, Francis C
Morfopoulos, Vassilis C(onstantinos) P
Mowbray, Donald F
Nilson, Arthur H(umphrey)
Panlilio, Filadelfo
Pifko, Allan Bert
Questad, David Lee
Sachse, Wolfgang H
Shaw, Richard P(aul)
Srinivasan, Vijay
Testa, Rene B(iaggio)
Thomas, John Howard
Vafakos, William P(aul)

Engineering Mechanics (cont)

Vallance, Michael Alan
Vemula, Subba Rao
Vidosic, J(oseph) P(aul)
Wagner, John George
Ward, Lawrence W(aterman)
Weidlinger, Paul
Whiteside, James Brooks
Wilkinson, John Peter Darrell

NORTH CAROLINA
Anderson, John P
Batra, Subhash Kumar
DeHoff, Paul Henry, Jr
Dow, Thomas Alva
Gupta, Ajaya Kumar
Hamann, Donald Dale
Hassan, Awatif E
Havner, Kerry S(huford)
Humphries, Ervin G(rigg)
McDonald, P(atrick) H(ill), Jr
Maday, Clarence Joseph
Matzen, Vernon Charles
Neal, C Leon
Pasipoularides, Ares D
Petroski, Henry J
Schmiedeshoff, Frederick William
Sharkoff, Eugene Gibb
Snyder, Robert Douglas
Sorrell, Furman Y(ates), Jr
Utku, Senol

NORTH DAKOTA
Maurer, Karl Gustav
Rieder, William G(ary)

OHIO
Andonian, Arsavir Takfor
Bogner, Fred K
Campbell, James Edward
Chakko, Mathew K(anjhirathinkal)
Chamis, Christos Constantinos
Chen, Chao-Hsing Stanley
Chu, Mamerto Loarca
Clausen, William E(arle)
Cummings, Charles Arnold
Damianov, Vladimir B
Dickey, David S
Drake, Michael L
Fu, Li-Sheng William
Gabb, Timothy Paul
Gallo, Frank J
Gent, Alan Neville
Glasgow, John Charles
Gorla, Rama S R
Halford, Gary Ross
Ho, Fanghuai H(ubert)
Holden, Frank C(harles)
Horton, Billy Mitchusson
Hulbert, Lewis E(ugene)
Jain, Vinod Kumar
Kicher, Thomas Patrick
Korda, Peter E
Kroll, Robert J
Lee, Jon H(yunkoo)
Lee, June Key
Leis, Brian Norman
Leissa, A(rthur) W(illiam)
Lemke, Ronald Dennis
Lewandowski, John Joseph
Macke, H(arry) Jerry
Minich, Marlin
Nara, Harry R(aymond)
Ojalvo, Morris
Popelar, Carl H(arry)
Raftopoulos, Demetrios D
Rai, Iqbal Singh
Robe, Thurlow Richard
Robinson, David Nelson
Sierakowski, Robert L
Smith, Robert Emery
Stouffer, Donald Carl
Torvik, Peter J
Von Ohain, Hans Joachim
Wagoner, Robert H
Walter, Joseph David
Whitney, James Martin
Zavodney, Lawrence Dennis

OKLAHOMA
Appl, Franklin John
Bert, Charles Wesley
Blenkarn, Kenneth Ardley
Earls, James Roe
Neathery, Raymond Franklin
Nolte, Kenneth George
Norton, Joseph R(andolph)
Striz, Alfred Gerhard

OREGON
Calder, Clarence Andrew
Ethington, Robert Loren

PENNSYLVANIA
Barsom, John M
Batterman, Steven C(harles)
Bielak, Jacobo
Bjorhovde, Reidar
Bucci, Robert James
Carelli, Mario Domenico
Chan, Siu-Kee
Conway, Joseph C, Jr
Dash, Sanford Mark

DeSanto, Daniel Frank
DiTaranto, Rocco A
Duncombe, E(liot)
Elsworth, Derek
Goel, Ram Parkash
Grannemann, Glenn Niel
Griffin, Jerry Howard
Hahn, Hong Thomas
Hardy, H(enry) Reginald, Jr
Hayek, Sabih I
Haythornthwaite, Robert M(orphet)
Hettche, Leroy Raymond
Hu, L(ing) W(en)
Jolles, Mitchell Ira
Kinnavy, M(artin) G(erald)
Ko, Frank K
Koch, Ronald N
Lakhtakia, Akhlesh
Leinbach, Ralph C, Jr
Likins, Peter William
Lorsch, Harold G
McCoy, John J
McLaughlin, Philip V(an Doren), Jr
McNitt, Richard Paul
Manjoine, Michael J(oseph)
Meyer, Paul A
Morehouse, Chauncey Anderson
Neubert, Vernon H
Newman, John B(ullen)
Novak, Stephen Robert
Ochs, Stefan A(lbert)
Odrey, Nicholas Gerald
Ostapenko, A(lexis)
Pangborn, Robert Northrup
Pfennigwerth, Paul Leroy
Plonsky, Andrew Walter
Raimondi, Albert Anthony
Rajagopal, K R
Rodini, Benjamin Thomas, Jr
Rogers, H(arry) C(arton), Jr
Ross, Arthur Leonard
Rumbarger, John H
Ruud, Clayton Olaf
Saigal, Sunil
Scheetz, Howard A(nsel)
Sharma, Mangalore Gokulanand
Sinclair, Glenn Bruce
Soler, Alan I(srael)
Sonnemann, George
Steg, L(eo)
Sturdevant, Eugene J
Trumpler, Paul R(obert)
Tuba, I Stephen
Varadan, Vasundara Venkatraman
Varadan, Vijay K
Vierck, Robert K
Visser, Cornelis
Voloshin, Arkady S
Williams, Max L(ea), Jr
Wu, John Naichi
Yen, Ben-Tseng
Zamrik, Sam Yusuf
Zweben, Carl Henry

RHODE ISLAND
Asaro, Robert John
Clifton, Rodney James
Duffy, Jacques Wayne
Ferrante, W(illiam) R(obert)
Suresh, Subra

SOUTH CAROLINA
Anand, Subhash Chandra
Awadalla, Nabil G
Bauld, Nelson Robert, Jr
Castro, Walter Ernest
Chao, Yuh J (Bill)
Goree, James Gleason
Jackson, John Elwin, Jr
Smith, Wilbur S
Sparks, Peter Robert
Turner, John Lindsey
Uldrick, John Paul
Waugh, John David
Yau, Wen-Foo

TENNESSEE
Baker, Allen Jerome
Blass, Joseph J(ohn)
Bretz, Philip Eric
Buchanan, George R(ichard)
Bundy, Robert W(endel)
Carley, Thomas Gerald
Foster, Edwin Powell
Iskander, Shafik Kamel
Janna, William Sied
Keedy, Hugh F(orrest)
Landes, John D
Lee, Ching-Wen
Lee, Donald William
Pandeya, Prakash N
Pawel, Janet Elizabeth
Pih, Hui
Pugh, Claud Ervin
Shobe, L(ouis) Raymon
Slade, Edward Colin
Smith, Dallas Glen, Jr
Snyder, William Thomas
Stoneking, Jerry Edward
Uhrig, Robert Eugene
Wasserman, Jack F
Wheeler, Orville Eugene
Wimberly, C Ray

TEXAS
Anderson, Charles E, Jr
Bendapudi, Kasi Visweswararao
Brock, James Harvey
Bryant, Michael David
Burger, Christian P
Carey, Graham Francis
Carroll, Michael M
Casey, James
Chevalier, Howard L
Dalley, Joseph W(inthrop)
Diller, Kenneth Ray
Ebner, Stanley Gadd
Eichberger, Le Roy Carl
Fischer, Ferdinand Joseph
Francis, Philip Hamilton
Gaines, J H
Geyling, F(ranz) Th(omas)
Gladwell, Ian
Hussain, A K M Fazle
Johnson, Brann
Jones, William B
Jordan, Neal F(rancis)
Juricic, Davor
Kana, Daniel D(avid)
Kanninen, Melvin Fred
Kinra, Vikram Kumar
Lawrence, Kent L(ee)
Lutes, Loren Daniel
Lytton, Robert Leonard
Mack, Lawrence R(iedling)
Marshek, Kurt M
Monsees, James E
Moon, Tessie Jo
Morgan, James Richard
Muster, Douglas Frederick
Nachlinger, R Ray
Nordgren, Ronald P
Oden, John Tinsley
Patrick, Wesley Clare
Pray, Donald George
Rader, Dennis
Ripperger, Eugene Arman
Schneider, William Charles
Schruben, Johanna Stenzel
Simonis, John Charles
Solberg, Ruell Floyd, Jr
Stanovsky, Joseph Jerry
Stern, Morris
Stubbs, Norris
Tien, John Kai
Tsahalis, Demosthenes Theodoros
Vallabhan, C V Girija
Von Maltzahn, Wolf W
Waid, Margaret Cowsar
Walton, Jay R
Wang, Chao-Cheng
Ward, Donald Thomas
Watson, Hal, Jr
Wilcox, Marion Walter

UTAH
Barton, Cliff S
Hill, Eric von Krumreig
Hoeppner, David William
Karren, Kenneth W
Rotz, Christopher Alan
Silver, Barnard Stewart

VIRGINIA
Barton, Furman W(yche)
Beran, Robert Lynn
Berry, James G(ilbert)
Chang, George Chunyi
Chi, Michael
Chisholm, Douglas Blanchard
Cooper, Henry Franklyn, Jr
Davidson, John Richard
Davis, John Grady, Jr
Duke, John Christian, Jr
Edlich, Richard French
Eppink, Richard Theodore
Fisher, Samuel Sturm
Frederick, Daniel
Gunter, Edgar Jackson, Jr
Hasenfus, Harold J(oseph)
Heller, Robert A
Herakovich, Carl Thomas
Horgan, Cornelius Oliver
Hou, Gene Jean-Win
Hudson, Charles Michael
Jones, Robert Millard
Juang, Jer-Nan
Kapania, Rakesh Kumar
Landgraf, Ronald William
Maderspach, Victor
Maher, F(rancis) J(oseph)
Mook, Dean Tritschler
Moon, Peter Clayton
Newman, James Charles, Jr
Pletta, Dan Henry
Post, Daniel
Raju, Ivatury Sanyasi
Rosenfeld, Robert L
Senseny, Paul Edward
Singh, Mahendra Pal
Smith, C(harles) William
Stinchcomb, Wayne Webster
Sword, James Howard
Talapatra, Dipak Chandra
Thurston, Gaylen James
Tompkins, Stephen Stern
Tsai, Frank Y
Wood, Albert D(ouglas)

Yang, Ta-Lun
Yates, Edward Carson, Jr

WASHINGTON
Chalupnik, James Dvorak
Ding, Jow-Lian
Dusto, Arthur Ronald
Duvall, George Evered
Goodstein, Robert
Hamilton, C Howard
Hartz, Billy J
Johnson, Jay Allan
Levinson, Mark
Madden, James H(oward)
Merchant, Howard Carl
Merckx, Kenneth R(ing)
Reyhner, Theodore Alison
Sherrer, Robert E(ugene)
Sorensen, Harold C(harles)
Stock, David Earl
Sutherland, Earl C
Varanasi, Suryanarayana Rao
Wolak, Jan

WEST VIRGINIA
Beeson, Justin Leo
Shuck, Lowell Zane
Venable, Wallace Starr

WISCONSIN
Bodine, Robert Y(oung)
Burck, Larry Harold
Cheng, Shun
Cook, Robert D(avis)
Crandall, Lee W(alter)
Crawmer, Daryl E
Cutler, Verne Clifton
Daane, Robert A
Edil, Tuncer Berat
Engelstad, Roxann Louise
Huang, T(zu) C(huen)
Hung, James Y(un-Yann)
Johnson, John E(dwin)
Johnson, Millard Wallace, Jr
Larkin, Lawrence A(lbert)
Lovell, Edward George
Malkus, David Starr
Naik, Tarun Ratilal
Pillai, Thankappan A K
Richard, Terry Gordon
Rowlands, Robert Edward
Saemann, Jesse C(harles), Jr
Schaffer, Erwin Lambert
Sorensen, Arthur (Sherman), Jr
Wnuk, Michael Peter
Zahn, John J

WYOMING
Adams, Donald F
Dolan, Charles W

PUERTO RICO
Goyal, Megh R
Khan, Winston

ALBERTA
Epstein, Marcelo
Hunt, Robert Nelson
Singh, Mansa C

BRITISH COLUMBIA
Mantle, J(ohn) B(ertram)
Pomeroy, Richard James

MANITOBA
Thomas, Robert Spencer David
Wilms, Ernest Victor

NOVA SCOTIA
Cochkanoff, O(rest)
Sastry, Vankamamidi VRN

ONTARIO
Barron, Ronald Michael
Cowper, George Richard
Huseyin, Koncay
Pindera, Jerzy Tadeusz
Rimrott, F(riedrich) P(aul) J(ohannes)
Schwaighofer, Joseph
Selvadurai, A P S
Sharan, Shailendra Kishore
Shelson, W(illiam)
Sherbourne, Archibald Norbert
Siddell, Derreck
Thompson, John Carl
Wong, Jo Yung

QUEBEC
Bhat, Rama B
Dealy, John Michael
Fabrikant, Valery Isaak
Hoa, Suong Van
Johns, Kenneth Charles
McKyes, Edward
Marsh, Cedric
Pekau, Oscar A
Stathopoulos, Theodore
Vatistas, Georgios H

OTHER COUNTRIES
Al-Zubaidy, Sarim Naji
Bodner, Sol R(ubin)
Ersoy, Ugur
Hjertager, Bjorn Helge

Laura, Patricio Adolfo Antonio
Lee, Choung Mook
Revesz, Zsolt
Ridha, R(aouf) A
Sunahara, Yoshifumi
Thurlimann, Bruno

Engineering Physics

ALABAMA
Askew, Raymond Fike
Cushing, Kenneth Mayhew
Mookherji, Tripty Kumar
Regner, John LaVerne
Wilhold, Gilbert A

ALASKA
Beebee, John Christopher
Wentink, Tunis, Jr

ARIZONA
Dempster, William Fred
El-Ghazaly, Samir M
Falk, Thomas
Fung, K Y
Hight, Ralph Dale
Mardian, James K W
Osborn, Donald Earl
Perper, Lloyd
Peyghambarian, Nasser
Rasmussen, William Otto
Schowengerdt, Robert Alan
Silvern, Leonard Charles
Szilagyi, Miklos Nicholas
Utagikar, Ajit Purushottam
Welch, H(omer) William
Ziolkowski, Richard Walter

CALIFORNIA
Adler, William Fred
Ames, Lawrence Lowell
Andersen, Wilford Hoyt
Bardsley, James Norman
Barhydt, Hamilton
Batdorf, Samuel Burbridge
Berdahl, Paul Hilland
Blackmon, James B
Blankenship, Victor D(ale)
Block, Richard B
Bradner, Hugh
Cao, Hengchu
Chesnut, Walter G
Clemens, Bruce Montgomery
Cohen, Lawrence Mark
Collins, Carter Compton
Cooper, Alfred William Madison
Coward, David Hand
Crandall, Michael Grain
Deshpande, Shivajirao M
Duneer, Arthur Gustav, Jr
Eby, Frank Shilling
Edmonds, Peter Derek
Edwall, Dennis Dean
Egermeier, R(obert) P(aul)
Elverum, Gerard William, Jr
Eukel, Warren W(enzl)
Everhart, Thomas E(ugene)
Fillius, Walker
Forbes, Judith L
Frank, Alan M
Frederick, David Eugene
Freis, Robert P
Fultz, Brent T
Garrett, Steven Lurie
Gillette, Philip Roger
Gin, W(inston)
Ginsburgh, Irwin
Glick, Harold Alan
Gould, Harvey Allen
Grant, Eugene F(redrick)
Grove, Andrew S
Heer, Ewald
Hendricks, Charles D(urrell), Jr
Hendry, George Orr
Herb, John A
Hesselink, Lambertus
Hsia, Yukun
Huberman, Marshall Norman
Hult, John Luther
Jaklevic, Joseph Michael
Jory, Howard Roberts
Judge, Roger John Richard
Karp, Arthur
Kennel, John Maurice
Khan, Mahbub R
Kittel, Peter
Kupfer, John Carlton
Lagarias, John S(amuel)
Lai, Kai Sun
Lapp, M(arshall)
Lee, Kenneth
Leibowitz, Lewis Phillip
Lemke, James Underwood
Leung, Charles Cheung-Wan
Leung, Ka-Ngo
Likuski, Robert Keith
Liu, Frederick F
Liu, Jia-ming
Longley-Cook, Mark T
Loo, Billy Wei-Yu
Lowi, Alvin, Jr
Marcus, John Stanley
Martin, Gordon Eugene

Maserjian, Joseph
Matthews, Stephen M
Miller, Joseph
Miller, Roger Heering
Moir, Ralph Wayne
Moore, Edgar Tilden, Jr
Muller, Richard S
Naqvi, Iqbal Mehdi
Nelson, Raymond Adolph
Nooker, Eugene L(eRoy)
Novotny, Vlad Joseph
Nunan, Craig S(pencer)
O'Donnell, C(edric) F(inton)
Palmer, John Parker
Parks, Lewis Arthur
Perkins, Kenneth L(ee)
Pomraning, Gerald C
Rasor, Ned S(haurer)
Rathmann, Carl Erich
Reinheimer, Julian
Remley, Marlin Eugene
Reynolds, Harry Lincoln
Reynolds, Harry Lincoln
Rosenblum, Stephen Saul
Saito, Theodore T
Salisbury, Stanley R
Salsig, William Winter, Jr
Sarwinski, Raymond Edmund
Scott, Franklin Robert
Scott, Paul Brunson
Shaffner, Richard Owen
Sher, Arden
Silverstein, Elliot Morton
Simonen, Thomas Charles
Sinclair, Kenneth F(rancis)
Singleton, John Byrne
Snowden, William Edward
Spector, Clarence J(acob)
Spencer, Cherrill Melanie
Starr, Chauncey
Stewart, Richard William
Stirn, Richard J
Stockel, Ivar H(oward)
Terhune, Robert William
Thompson, David A
Thomson, James Alex L
Trapp, Robert F(rank)
Tricoles, Gus P
Urtiew, Paul Andrew
Vanblarigan, Peter
Vaughan, J Rodney M
Veigele, William John
Victor, Andrew C
Viswanathan, R
Wagner, Carl E
Waldner, Michael
Wang, Jon Y
Warburton, William Kurtz
Warren, Mashuri Laird
Watt, Robert Douglas
Webber, Donald Salyer
Webster, Emilia
Weddell, James Blount
Weiss, Jeffrey Martin
Wiley, William Charles
Winkelmann, Frederick Charles
Yap, Fung Yen
Yeh, Edward H Y
Young, James Forrest
Young, Louise Gray
Zucca, Ricardo
Zuleeg, Rainer
Zuppero, Anthony Charles

COLORADO
Augusteijn, Marijke Francina
Brown, Edmund Hosmer
Bushnell, Robert Hempstead
Cathey, Wade Thomas, Jr
Chisholm, James Joseph
Gardner, Edward Eugene
Hanley, Howard James Mason
Harrington, Robert D(ean)
Hilchie, Douglas Walter
Hoffman, John Raleigh
Jaye, Seymour
Kremer, Russell Eugene
McAllister, Robert Wallace
Morse, J(erome) G(ilbert)
Radebaugh, Ray
Schwiesow, Ronald Lee
Verschoor, J(ack) D(ahlstrom)
Wertz, Ronald Duane

CONNECTICUT
Adams, Brent Larsen
Antkiw, Stephen
Brown, Arthur A(ustin)
Burwell, Wayne Gregory
Dreyfus, Marc George
Erlandson, Paul M(cKillop)
Greenwood, Ivan Anderson
Holzman, George Robert
Hufnagel, Robert Ernest
Paolini, Francis Rudolph
Rau, Richard Raymond
Shah, Mirza Mohammed
Similon, Philippe Louis
Sreenivasan, Katepalli Raju
Tittman, Jay
Troutman, Ronald R
Wormser, Eric M

DELAWARE
Bierlein, John David
Christie, Phillip
Hardy, Henry Benjamin, Jr
Hirsch, Albert Edgar
Jansson, Peter Allan
Moore, Charles B(ernard)
Owens, Aaron James
Scofield, Dillon Foster

DISTRICT OF COLUMBIA
Borgiotti, Giorgio V
Nachman, Arje
Needels, Theodore S
Pikus, Irwin Mark
Sanders, Robert Charles
Schriever, Richard L
Spohr, Daniel Arthur

FLORIDA
Adhav, Ratnakar Shankar
Anghaie, Samim
Baird, Alfred Michael
Clarke, Thomas Lowe
Cox, John David
Fry, James N
Hansel, Paul G(eorge)
Harmon, G Lamar
Harney, Robert Charles
Henning, Rudolf E
Langford, David
Melich, Michael Edward
Pichal, Henri Thomas, II
Robertson, Harry S(troud)
Sah, Chih-Tang
Schimmel, Walter Paul
Tomonto, James R
Tusting, Robert Frederick
Vollmer, James
Young, Charles Gilbert

GEORGIA
Batten, George L, Jr
Currie, Nicholas Charles
Jokerst, Nan Marie
Lau, Jark Chong
Lorber, Herbert William
McMillan, Robert Walker

HAWAII
Craven, John P
Curtis, George Darwin
Frazer, L Neil

IDAHO
Carpenter, Stuart Gordon
Mott, Jack Edward
Spencer, Paul Roger
Woodall, David Monroe

ILLINOIS
Bjorklund, John A(lexander)
Cardman, Lawrence Santo
Cohn, Gerald Edward
Cooper, William Edward
Fast, Ronald Walter
Flora, Robert Henry
Freedman, Stuart Jay
Gallagher, David Alden
Hafele, Joseph Carl
Herzenberg, Caroline Stuart Littlejohn
Horve, Leslie A
Jostlein, Hans
Kang, Sung-Mo Steve
Kasper, Gerhard
Kim, Kyekyoon K(evin)
Kinnmark, Ingemar Per Erland
Kuchnir, Moyses
Kustom, Robert L
Macrander, Albert Tiemen
Mokadam, Raghunath G(anpatrao)
Morkoc, Hadis
Pardo, Richard Claude
Saxena, Satish Chandra
Shabana, Ahmed Abdelraouf
Shea, Michael Francis
Sloma, Leonard Vincent
Spradley, Joseph Leonard
Sproul, William Dallas
Till, Charles Edgar
Uherka, Kenneth Leroy
Venema, Harry J(ames)
Wolsky, Alan Martin
Yamada, Ryuji

INDIANA
Chang, Chien-Wu
Ersoy, Okan Kadri
Henke, Mitchell C
Ho, Cho-Yen
Orr, William H(arold)
Schwartz, Richard John
Snare, Leroy Earl
Voelz, Frederick

IOWA
Schaefer, Joseph Albert
Spencer, Donald Lee
Thompson, Robert Bruce

KANSAS
Gray, Tom J
Stockli, Martin P
Unz, Hillel

KENTUCKY
Frank, John L

LOUISIANA
Holden, William R(obert)
Weiss, Louis Charles
Witriol, Norman Martin

MAINE
Nicol, James

MARYLAND
Bartley, William Call
Birnbaum, George
Blessing, Gerald Vincent
Blum, Norman Allen
Castruccio, Peter Adalbert
Connerney, John E P
Connor, Joseph Gerard, Jr
Cook, Richard Kaufman
Dauber, Edwin George
Dhir, Surendra Kumar
Doiron, Theodore Danos
Ethridge, Noel Harold
Fabunmi, James Ayinde
Fong, Jeffery Tse-Wei
French, Judson Cull
Gartrell, Charles Frederick
Geckle, William Jude
Gibson, Harold F(loyd)
Gillich, William John
Gonano, John Roland
Gordon, Gary Donald
Hansen, Robert Jack
Hauser, Michael George
Herzfeld, Charles Maria
Hubbell, John Howard
Huttlin, George Anthony
Hyde, Geoffrey
Kalil, Ford
Larrabee, R(obert) D(ean)
Lull, David B
Lundholm, J(oseph) G(ideon), Jr
McCoubrey, Arthur Orlando
Malarkey, Edward Cornelius
Merkel, George
Metze, George M
Moulton, James Frank, Jr
Ostaff, William A(llen)
Pearl, John Christopher
Pierce, Harry Frederick
Pisacane, Vincent L
Preisman, Albert
Reader, Wayne Truman
Rueger, Lauren J(ohn)
Sank, Victor J
Scialdone, John Joseph
Simonis, George Jerome
Skiff, Frederick Norman
Stone, Albert Mordecai
Turner, Robert Davison
Waldo, George Van Pelt, Jr
Warner, Brent A
Weigman, Bernard J
Zien, Tse-Fou
Zirkind, Ralph

MASSACHUSETTS
Adler, Richard B(rooks)
Allen, Steven
Arganbright, Donald G
Baird, Albert Washington, III
Bechis, Kenneth Paul
Brown, Walter Redvers John
Burke, John Michael
Castro, Alfred A
Chandler, Charles H(orace)
Cordero, Julio
Dakss, Mark Ludmer
Esch, Robin E
Fan, John C C
Feinleib, Julius
Garvey, R(obert) Michael
Guenther, R(ichard)
Holt, Frederick Sheppard
Huguenin, George Richard
Jones, Robert Allan
Jones, William J
Kamentsky, Louis A
Kanter, Irving
Kleis, John Dieffenbach
Kniazzeh, Alfredo G(iovanni) F(rancesco)
Kolodziejski, Leslie Ann
Kovaly, John J
Lederman, David Mordechai
Lien, Hwachii
Litzenberger, Leonard Nelson
Mack, Michael E
Nowak, Welville B(erenson)
Oettinger, Peter Ernest
Orloff, Lawrence
Rapperport, Eugene J
Reilly, James Patrick
Robinson, D(enis) M(orrell)
Schneider, Robert Julius
Schwartz, Jack
Sethares, James C(ostas)
Shapiro, Edward K Edik
Smith, Alan Bradford
Sosnowski, Thomas Patrick
Speliotis, Dennis Elias
Weggel, Robert John
Weiss, Jerald Aubrey

Engineering Physics (cont)

Winer, Bette Marcia Tarmey
Yngvesson, K Sigfrid

MICHIGAN
Bolander, Richard
Brown, Steven Michael
Dow, W(illiam) G(ould)
Johnson, Wayne Jon
Joseph, Bernard William
Kim, Changhyun
Logothetis, Eleftherios Miltiadis
Marko, Kenneth Andrew
Miller, Carl Elmer
Morgan, Roger John
Piccirelli, Robert Anthony
Schumacher, Berthold Walter
Smith, Hadley J(ames)
Thomson, Robert Francis

MINNESOTA
Agarwal, Vijendra Kumar
George, Peter Kurt
Gerlach, Robert Louis
Hocker, George Benjamin
Luskin, Mitchell B
Yasko, Richard N

MISSOURI
Carter, Robert L(eroy)
Kibens, Valdis
Kunze, Jay Frederick
Lind, Arthur Charles
Lowe, Forrest Gilbert
McMillan, Gregory Keith
Northrip, John Willard

NEBRASKA
Rohde, Suzanne Louise

NEW HAMPSHIRE
Buffler, Charles Rogers
Crowell, Merton Howard
Linnell, Richard D(ean)
Spehrley, Charles W, Jr
Temple, Peter Lawrence
Yen, Yin-Chao

NEW JERSEY
Berkman, Samuel
Bharj, Sarjit Singh
Bjorkholm, John Ernst
Browne, Robert Glenn
Brucker, Edward Byerly
Catanese, Carmen Anthony
Deri, Robert Joseph
Donovan, Richard C
Dutta, Mitra
Fisher, Reed Edward
Hammer, Jacob Meyer
Jahn, Robert G(eorge)
Jansen, Frank
Jassby, Daniel Lewis
Kaita, Robert
Kam, Gar Lai
Kugel, Henry W
Kunzler, John Eugene
Lanzerotti, Louis John
Lechner, Bernard J
Leheny, Robert Francis
Liao, Paul Foo-Hung
Michaelis, Paul Charles
Miles, Richard Bryant
Pargellis, Andrew Nason
Rainal, Attilio Joseph
Rothleder, Stephen David
Schivell, John Francis
Schrenk, George L
Schulte, Harry John, Jr
Schumer, Douglas B
Shahbender, R(abah) A(bd-El-Rahman)
Shen, Tek-Ming
Taylor, Geoff W
Wachtell, George Peter
Weissenburger, Don William
Wiesenfeld, Jay Martin
Williams, Neal Thomas
Yadvish, Robert D
Zweig, Gilbert

NEW MEXICO
Alers, George A
Barsis, Edwin Howard
Beckner, Everet Hess
Benjamin, Robert Fredric
Bergeron, Kenneth Donald
Bowman, Allen Lee
Chamberlin, Edwin Phillip
Clark, Wallace Thomas, III
Davis, L(loyd) Wayne
Dingus, Ronald Shane
Drumheller, Douglas Schaeffer
Farmer, William Michael
Fuka, Louis Richard
Gorham, Elaine Deborah
Gurbaxani, Shyam Hassomal
Haberstich, Albert
Hartman, James Keith
Henderson, Dale Barlow
Holm, Dale M
Jameson, Robert A
Janney, Donald Herbert
McNally, James Henry

Metzger, Daniel Schaffer
Oyer, Alden Tremaine
Palmer, Byron Allen
Parker, Jerald Vawer
Parker, Joseph R(ichard)
Perret, William Riker
Schneider, Jacob David
Sze, Robert Chia-Ting
Touryan, Kenell James
Yarger, Frederick Lynn

NEW YORK
Baker, Alan Gardner
Beckwith, Steven Van Walter
Benumof, Reuben
Bickmore, John Tarry
Bilderback, Donald Heywood
Birnbaum, David
Bittner, John William
Blumenthal, Ralph Herbert
Brehm, Lawrence Paul
Brunelle, Eugene John, Jr
Buhrman, Robert Alan
Cady, K Bingham
Chappell, Richard Lee
Cheng, Yean Fu
Cheo, Bernard Ru-Shao
Cohen, Mitchell S(immons)
Cole, Robert
Constable, James Harris
Corelli, John C
Cottingham, James Garry
Das, Pankaj K
Dowben, Peter Arnold
Flom, Donald Gordon
Friedman, Jack P
Genin, Dennis Joseph
George, Nicholas
Gokhale, Vishwas Vinayak
Goldman, Vladimir J
Greene, Peter Richard
Grunwald, Hubert Peter
Harris, Donald R, Jr
Hartmann, George Cole
Hoisie, Adolfy
Hoover, Thomas Earl
Jwo, Chin-Hung
Kasprzak, Lucian A
Kaup, David James
Komiak, James Joseph
Kuchar, Norman Russell
Kunhardt, Erich Enrique
Lakatos, Andras Imre
LaMuth, Henry Lewis
Li, Chung-Hsiung
Liimatainen, T(oivo) M(atthew)
Lindstrom, Gary J
MacDonald, Noel C
McGuire, Stephen Craig
Mahler, David S
Malaviya, Bimal K
Maling, George Croswell, Jr
Marcus, Michael Alan
Moe, George Wylbur
Morris, George Michael
Mortenson, Kenneth Ernest
Nation, John
Nicolosi, Joseph Anthony
Osgood, Richard Magee, Jr
Parks, Harold George
Pernick, Benjamin J
Podowski, Michael Zbigniew
Rudinger, George
Sai-Halasz, George Anthony
Salzberg, Bernard
Shenolikar, Ashok Kumar
Siegel, Benjamin Morton
Simpson, Murray
Smith, Lowell Scott
Star, Joseph
Stenzel, Wolfram G
Stephenson, Thomas E(dgar)
Stevens, William Y(eaton)
Stover, Raymond Webster
Temple, Victor Albert Keith
Vosburgh, Kirby Gannett
Walker, Michael Stephen
Wallace, Aaron
Weber, Arthur Phineas
Weeks, William Thomas
Wei, Ching-Yeu
Wertheimer, Alan Lee
Wie, Chu Ryang
Wysocki, Joseph J(ohn)

NORTH CAROLINA
Dayton, Benjamin Bonney
Dow, Thomas Alva
Ellis-Menaghan, John J
Grainger, John Joseph
Hinesley, Carl Phillip
Irene, Eugene Arthur
Jones, Creighton Clinton
Lawless, Philip Austin
Nader, John S(haheen)

OHIO
Andrews, Merrill Leroy
Bennett, Foster Clyde
Bhattacharya, Rabi Sankar
Curtis, Lorenzo Jan
Dakin, James Thomas
Golovin, Michael N
Griffing, David Francis

Horton, Billy Mitchusson
Hunter, William Winslow, Jr
Johnson, Ray O
Keim, John Eugene
Krishnamurthy, Lakshminarayanan
Landstrom, D(onald) Karl
Mayers, Richard Ralph
Mielke, Robert L
Reiling, Gilbert Henry
Rich, Joseph William
Seldin, Emanuel Judah
Smith, James Edward, Jr
Smith, Robert Emery
Springer, Robert Harold
Thompson, Hugh Ansley
Turchi, Peter John

OKLAHOMA
Blais, Roger Nathaniel
Kuenhold, Kenneth Alan
Lal, Manohar
Ryan, Stewart Richard

OREGON
Matson, Leslie Emmet, Jr
Paarsons, James Delbert
Yau, Leopoldo D

PENNSYLVANIA
Arnold, James Norman
Artman, Joseph Oscar
Ashok, S
Berman, Martin
Bilaniuk, Oleksa-Myron
Böer, Karl Wolfgang
Brunner, Mathias J
Carelli, Mario Domenico
Chaiken, Robert Francis
Chapman, John Donald
Cook, Donald Bowker
Cookson, Alan Howard
Coon, Darryl Douglas
Deis, Daniel Wayne
Dick, George W
Emtage, Peter Roesch
Fonash, Stephan J(oseph)
Garbuny, Max
Harrold, Ronald Thomas
Hoburg, James Frederick
Hurwitz, Jan Krosst
Jones, Clifford Kenneth
Kahn, David
Kanter, Ira E
Kryder, Mark Howard
Mahajan, Subhash
Messier, Russell
Meyer, Paul A
Ochs, Stefan A(lbert)
Oder, Robin Roy
Penney, Gaylord W
Pinckney, Robert L
Pinkerton, John Edward
Roy, Pradip Kumar
Schenk, H(arold) L(ouis), Jr
Slade, Paul Graham
Spira, Joel Solon
Steinbruegge, Kenneth Brian
Stinger, Henry J(oseph)
Tingle, William Herbert
Trainer, Michael Norman
Uher, Richard Anthony
Van Roggen, Arend
Varadan, Vijay K
Whitley, James Heyward
Wilson, Walter R
Zelac, Ronald Edward
Zemel, Jay N(orman)

RHODE ISLAND
Avery, Donald Hills

SOUTH CAROLINA
Avegeropoulos, G
Croft, George Thomas
Harrison, James William, Jr

SOUTH DAKOTA
Cannon, Patrick Joseph

TENNESSEE
Allison, Stephen William
Blankenship, James Lynn
Burton, Robert Lee
Foster, Edwin Powell
Kiech, Earl Lockett
Lee, Donald William
LeVert, Francis E
Longmire, Martin Shelling
McGregor, Wheeler Kesey, Jr
Maienschein, Fred (Conrad)
Morgan, Ora Billy, Jr
Morrison, Thomas Golden
Pearsall, S(amuel) H(aff)
Roth, J(ohn) Reece
Schwenterly, Stanley William, III
Slade, Edward Colin
Walter, F John
Watson, Robert Lowrie
Wilcox, William Jenkins, Jr

TEXAS
Anderson, Charles E, Jr
Baker, Wilfred E(dmund)
Birchak, James Robert

Boers, Jack E
Boyer, Lester Leroy
Brannon, H Raymond, Jr
Chivian, Jay Simon
Dhudshia, Vallabh H
Dykstra, Jerald Paul
Eberhart, Robert Clyde
Engelhardt, Albert George
Griffin, Richard B
Hebert, Joel J
Hewett, Lionel Donnell
Ignatiev, Alex
Jordan, Neal F(rancis)
Keister, Jamieson Charles
Kristiansen, Magne
Lacy, Lewis L
Matzkanin, George Andrew
Min, Kwang-Shik
Moore, Thomas Matthew
Nimitz, Walter W von
Quisenberry, Karl Spangler, Jr
Ransom, C J
Richard, Richard Ray
Skinner, G(eorge) M(acGillivray)
Stone, Orville L
Tao, Frank F
Waggoner, James Arthur
Wehring, Bernard William

UTAH
Clegg, John C(ardwell)
Cohen, Richard M
Jeffery, Rondo Nelden
Jolley, David Kent
Ward, Roger Wilson
Zeamer, Richard Jere

VIRGINIA
Bahl, Inder Jit
Bailey, Marion Crawford
Burley, Carlton Edwin
Cambel, Ali B(ulent)
Church, Charles Henry
De Wolf, David Alter
Fthenakis, Emanuel John
Goncz, John Henry
Hasenfus, Harold J(oseph)
Henry, James H(erbert)
Herbst, Mark Joseph
Hwang, William Gaong
James, W(ilbur) Gerald
Johnson, Robert Edward
Kapron, Felix Paul
Krassner, Jerry
Kuhlmann-Wilsdorf, Doris
Moon, Peter Clayton
Neubauer, Werner George
Ni, Chen-Chou
Pettus, William Gower
Porzel, Francis Bernard
Ramey, Robert Lee
Reed, Joseph
Reiss, Keith Westcott
Sager, Earl Vincent
Schremp, Edward Jay
Siebentritt, Carl R, Jr
Smith, Sidney Taylor
Wolicki, Eligius Anthony

WASHINGTON
Bruckner, Adam Peter
Center, Robert E
Christiansen, Walter Henry
Connell, James Roger
Gupta, Yogendra M(ohan)
Kim, Jae Hoon
Nichols, Davis Betz
Robinson, Barry Wesley
Russell, David A
Russell, James Torrance
Soreide, David Christien
Stitch, Malcolm Lane
Stoebe, Thomas Gaines

WEST VIRGINIA
Littleton, John Edward
Spaniol, Craig

WISCONSIN
Bomba, Steven James
Boom, Roger Wright
Hyzer, William Gordon
Kinsinger, Richard Estyn
Morin, Dornis Clinton
Nordman, James Emery
Tonner, Brian P

PUERTO RICO
Hussain, Malek Gholoum Malek

ALBERTA
Cormack, George Douglas

BRITISH COLUMBIA
Jull, Edward Vincent
Mackenzie, George Henry
Menon, Thuppalay K
Palmer, James Frederick

MANITOBA
Johnson, R(ichard) A(llan)
Mathur, Maya Swarup
Wilkins, Brian John Samuel

NEW BRUNSWICK
Boorne, Ronald Albert
Wells, David Ernest

NOVA SCOTIA
Elliott, James Arthur
El-Masry, Ezz Ismail
King, Hubert Wylam
Roger, William Alexander

ONTARIO
Berezin, Alexander A
Boggs, Steven A
Brach, Eugene Jenõ
Brimacombe, Robert Kenneth
Chang, Jen-Shih
Chou, Jordan Quan Ban
Cohn-Sfetcu, Sorin
Critoph, E(ugene)
Davies, John Arthur
Gold, Lorne W
Harms, Archie A
Jackson, David Phillip
Johnston, G(ordon) W(illiam)
Kumaran, Mavinkal K
Lipshitz, Stanley Paul
Lynch, Gerard Francis
McGee, William F
McNamara, Allen Garnet
Napier, Douglas Herbert
Orlik-Ruckemann, Kazimierz Jerzy
Ottensmeyer, Frank Peter
Pindera, Jerzy Tadeusz
Purbo, Onno Widodo
Rauch, Sol
Rosen, Marc Allen
Selvakumar, Chettypalayam Ramanathan
Siddell, Derreck
Smith, Kenneth Carless
Sonnenberg, Hardy
Thompson, David Allan
Uffen, Robert James
Yevick, David Owen

QUEBEC
Aktik, Cetin
Bose, Tapan Kumar
Champness, Clifford Harry
Chan, Sek Kwan
Gregory, Brian Charles
Hetherington, Donald Wordsworth
Izquierdo, Ricardo
Maciejko, Roman
Meunier, Michel
Power, Joan F
Schwelb, Otto
Terreault, Bernard J E J
Vatistas, Georgios H
Wertheimer, Michael Robert

SASKATCHEWAN
Nikiforuk, P(eter) N

OTHER COUNTRIES
Bryngdahl, Olof
Gordon, Jeffrey Miles
Kimblin, Clive William
Mengali, Umberto
Sato, Gentei
Steffen, Juerg
Sugaya, Hiroshi
Unger, Hans-Georg

Fuel Technology & Petroleum Engineering

ALASKA
Merrill, Robert Clifford, Jr
Ostermann, Russell Dean

ARIZONA
Ramohalli, Kumar Nanjunda Rao

CALIFORNIA
Adicoff, Arnold
Bauer, Robert Forest
Bridge, Alan G
Buck, F(rank) A(lan) Mackinnon
Burnham, Alan Kent
Campbell, Thomas Cooper
Chesnut, Dwayne A(llen)
Chilingarian, George V(aros)
Compton, Leslie Ellwyn
Coppel, Claude Peter
Culler, F(loyd) L(eRoy), Jr
Dautzenberg, Frits Mathia
Davidson, Lynn Blair
Davis, Bruce W
Day, Robert J(ames)
Dickinson, Wade
Dolbear, Geoffrey Emerson
Dormant, Leon M
Durai-swamy, Kandaswamy
Engleman, Victor Solomon
Fox, Joseph M(ickle), III
Frost, Kenneth Almeron, Jr
Gat, Nahum
Goss, John R(ay)
Greene, Charles Richard
Handy, Lyman Lee
Hass, Robert Henry
Healey, Anthony J
Heinemann, Heinz

Hennig, Harvey
Hess, Patrick Henry
Holm, L(eRoy) W(allace)
Horne, Roland Nicholas
Irving, James P
Johnson, Carl Emil, Jr
Keaton, Michael John
Keller, James Lloyd
Kertamus, Norbert John
Klotz, James Allen
Kohl, A(rthur) L(ionel)
Krueger, Roland Frederick
Kuehne, Donald Leroy
Lewis, Arthur Edward
Low, Hans
Ludwig, Harvey F
McKinney, Ted Meredith
Mallett, William Robert
Maloney, Kenneth Long
Marsden, Sullivan S(amuel), Jr
Meiners, Henry C(ito)
Meldau, R(obert) F(rederick)
Meyers, Robert Allen
Muskat, Morris
Nelson, David A
Nguyen, Caroline Phuongdung
Nunn, Roland Cicero
Offen, George Richard
Orr, Franklin M, Jr
Penner, S(tanford) S(olomon)
Ratcliffe, Charles Thomas
Richter, George Neal
Rosenberg, Sanders David
Rothman, Albert J(oel)
Sanders, Charles F(ranklin), Jr
Schwartz, Daniel Evan
Shrontz, John William
Shuler, Patrick James
Silcox, William Henry
Steffgen, Frederick Williams
Stephens, Douglas Robert
Sullivan, Richard Frederick
Sundberg, John Edwin
Turchan, Otto Charles
Uhl, Arthur E(dward)
Van Wingen, N(ico)
Vermeulen, Theodore (Cole)
Williams, Alan K
Williams, Forman A(rthur)
Witherspoon, Paul A(dams), Jr

COLORADO
Allred, V(ictor) Dean
Bass, Daniel Materson, Jr
Breitenbach, E A
Burdge, David Newman
Callanan, Jane Elizabeth
Collins, Royal Eugene
Gibbons, David Louis
Goldburg, Arnold
Graue, Dennis Jerome
Jargon, Jerry Robert
Jha, Mahesh Chandra
Kazemi, Hossein
Keder, Wilbert Eugene
McClellan, William Alan
Mast, Richard F(redrick)
Milne, Thomas Anderson
Osterhoudt, Walter Jabez
Rothfeld, Leonard B(enjamin)

CONNECTICUT
Bozzuto, Carl Richard
Galstaun, Lionel Samuel
Goldstein, Marvin Sherwood
Gray, Thomas James
Hyde, John Welford
Ramakrishnan, Terizhandur S
Rosner, Daniel E(dwin)
Ruckebusch, Guy Bernard
Seery, Daniel J
Solomon, Peter R
Stephenson, Kenneth Edward
Thomas, Arthur L

DELAWARE
Cantwell, Edward N(orton), Jr
Cheng, Alexander H-D
Klein, Michael Tully
Mills, George Alexander

DISTRICT OF COLUMBIA
Arnold, William Howard, Jr
Hershey, Robert Lewis
Leonard, Joseph Thomas

FLORIDA
Chao, Raul Edward
Francis, Stanley Arthur
Givens, Paul Edward
Hirt, Thomas J(ames)
Normile, Hubert Clarence
Parker, Howard Ashley, Jr
Schneider, Richard Theodore
Seireg, Ali A

GEORGIA
Hughes, Richard V(an Voorhees)
Winnick, Jack

IDAHO
Hyde, William W
Reno, Harley W

ILLINOIS
Agarwal, Ashok Kumar
Assanis, Dennis N
Baird, Michael Jefferson
Bunting, Bruce Gordon
Clark, Stephen Darrough
Conn, Arthur L(eonard)
Gembicki, Stanley Arthur
Gray, Kenneth Eugene
Gupta, Ramesh
Herskovits, Lily Eva
Hough, Eldred W
Hutchings, L(eRoi) E(arl)
Johnson, Irving
Kamath, Krishna
Klass, Donald Leroy
Krawetz, Arthur Altshuler
Kruse, Carl William
Kukes, Simon G
Lester, George Ronald
Linden, Henry R(obert)
Moyer, H(allard) C(harles)
Myles, Kevin Michael
Nandi, Satyendra Prosad
Olmstead, William Edward
Punwani, Dharam Vir
Rose, Walter Deane
Rusinko, Frank, Jr
Schwartz, Michael Muni
Staats, William R
Wert, Charles Allen

INDIANA
Cartmell, Robert Root
Droege, John Walter
Shankland, Rodney Veeder
Stehouwer, David Mark
Voelz, Frederick

IOWA
Egger, Carl Thomas
Fisher, Ray W
Hall, Jerry Lee
Markuszewski, Richard

KANSAS
Crowther, Robert Hamblett
DeForest, Elbert M
Preston, Floyd W
Ryan, James M
Smith, Raymond V(irgil)
Underwood, James Ross, Jr

KENTUCKY
Lloyd, William Gilbert
Mitchell, Maurice McClellan, Jr
Von Schonfeldt, Hilmar Armin
Wells, William Lochridge

LOUISIANA
Bourgoyne, Adam T, Jr
Brown, Leo Dale
Gluckstein, Martin E(dwin)
Huckabay, Houston Keller
Ledford, Thomas Howard
McCarthy, Danny W
Reichle, Alfred Douglas
Rosene, Robert Bernard
Stephens, Maynard Moody

MARYLAND
Argauer, Robert John
Bascunana, Jose Luis
Bock, Arthur E(mil)
Coates, Arthur Donwell
Foresti, Roy J(oseph), Jr
Gupta, Ashwani Kumar
Hornbuckle, Franklin L
Kolobielski, Marjan
Krynitsky, John Alexander
McQuaid, Richard William
Mangen, Lawrence Raymond
Plumlee, Karl Warren
Presser, Cary
Sevik, Maurice
Shannon, Larry J(oseph)
Tuttle, Kenneth Lewis
White, Edmund W(illiam)

MASSACHUSETTS
Armington, Alton
Hals, Finn
Hottel, Hoyt C(larke)
Jalan, Vinod Motilal
Johnson, Stephen Allen
Kincaid, Thomas Gardiner
Leung, Woon F (Wallace)
Petrovic, Louis John
Rowell, Robert Lee
Tester, Jefferson William

MICHIGAN
Eldred, Norman Orville
Eltinge, Lamont
Everett, Robert Line
Heminghous, William Wayne, Sr
Henein, Naeim A
Kabel, Richard Harvey
Klimpel, Richard Robert
Korcek, Stefan
Piccirelli, Robert Anthony
Schramm, John Gilbert
Shelef, Mordecai
Stansel, John Charles

Willermet, Pierre Andre

MINNESOTA
Ayeni, Babatunde J

MISSISSIPPI
Cotton, Frank Ethridge, Jr

MISSOURI
Carter, Lee
Fowler, Frank Cavan
Koederitz, Leonard Frederick
Preckshot, G(eorge) W(illiam)

MONTANA
Bradley, Daniel Joseph

NEW JERSEY
Bachman, Kenneth Charles
Bhagat, Phiroz Maneck
Bieber, Herman
Calcote, Hartwell Forrest
Chen, Nai Y
Church, John Armistead
Denno, Khalil I
Eddinger, Ralph Tracy
Fung, Shun Chong
Gethner, Jon Steven
Glassman, Irvin
Gorbaty, Martin Leo
Graham, Margaret Helen
Grandolfo, Marian Carmela
Griffith, Martin G
Grumer, Eugene Lawrence
Haag, Werner O
Halik, Raymond R(ichard)
Heath, Carl E(rnest), Jr
Jacolev, Leon
Kang, Chia-Chen Chu
Koehl, William John, Jr
Lessard, Richard R
Mraw, Stephen Charles
Munsell, Monroe Wallwork
Panzer, Jerome
Rankel, Lillian Ann
Rebick, Charles
Schlosberg, Richard Henry
Smyers, William Hays
Suciu, George Dan
Tiethof, Jack Alan
Upton, Thomas Hallworth
Van Hook, James Paul
Venuto, Paul B
Weinstein, Norman J(acob)
Yan, Tsoung-Yuan
Yarrington, Robert M

NEW MEXICO
Castle, John Granville, Jr
Drumheller, Douglas Schaeffer
Egbert, Robert B(aldwin)
Finch, Thomas Wesley
Granoff, Barry
Lee, David Oi
Mead, Richard Wilson
Muir, Patrick Fred
Nuttall, Herbert Ericksen, Jr
Taber, Joseph John
Vanderborgh, Nicholas Ernest

NEW YORK
Agosta, Vito
Child, Edward T(aylor)
Corbeels, Roger
Davis, Marshall Earl
Fein, R(ichard) S(aul)
Givi, Peyman
Grohse, Edward William
Handman, Stanley E
Harris, Colin C(yril)
Hileman, Robert E
Hunt, Everett Clair
Kaarsberg, Ernest Andersen
Keller, D Steven
Kim, Young Joo
Kolaian, Jack H
Krishna, C R
Neveu, Maurice C
Othmer, Donald F(rederick)
Pollart, Dale Flavian
Reber, Raymond Andrew
Vallance, Michael Alan
Virk, Kashmir Singh
Wang, Lin-Shu
Weinstein, Herbert
Wiese, Warren M(elvin)

NORTH CAROLINA
Eckerlin, Herbert Martin
Gangwal, Santosh Kumar
Kranich, Wilmer Leroy
Winston, Hubert

NORTH DAKOTA
Hasan, Abu Rashid
Mathsen, Don Verden
Willson, Warrack Grant

OHIO
Dorer, Casper John
Dunn, Robert Garvin
Fan, Liang-Shih
Harmon, David Elmer, Jr
Hazard, Herbert Ray

Fuel Technology & Petroleum Engineering (cont)

Jenkins, Robert George
Kampe, Dennis James
Lee, Sunggyu
Massey, L(ester) G(eorge)
Metcalfe, Joseph Edward, III
Mourad, A George
Murphy, Michael John
Putnam, Abbott (Allen)
Reid, William T(homas)
Shick, Philip E(dwin)
Slider, H C (Slip)
Sprafka, Robert J
Standish, Norman Weston
Webb, Thomas Howard

OKLAHOMA

Allen, Thomas Oscar
Azar, Jamal J
Baldwin, Roger Allan
Beckett, James Reid
Blenkarn, Kenneth Ardley
Blick, Edward F(orrest)
Brown, Kermit Earl
Brown, Larry Patrick
Bryant, Rebecca Smith
Campbell, John Morgan, Sr
Chatenever, Alfred
Chung, Ting-Horng
Coffman, Harold H
Dauben, Dwight Lewis
Davis, Robert Elliott
Eastman, Alan D
Elkins, L(loyd) E(dwin)
Fast, C(larence) R(obert)
Ferrell, Howard H
Forgotson, James Morris, Jr
Gall, James William
Geffen, T(heodore) M(orton)
Growcock, Frederick Bruce
Guerrero, E T
Hall, Colby D(ixon), Jr
Hays, George E(dgar)
Heath, Larman Jefferson
Helander, Donald P(eter)
Howard, Robert Adrian
Hurn, R(ichard) W(ilson)
Johnston, Harlin Dee
Kleinschmidt, Roger Frederick
Knapp, Roy M
Lal, Manohar
Lilley, David Grantham
Menzie, Donald E
Pollock, L(yle) W(illiam)
Raghavan, Rajagopal
Root, Paul John
Shell, Francis Joseph
Sheth, Ketankumar K
Sloan, Norman Grady
Thompson, Richard E(ugene)
Tiab, Djebbar
Vinatieri, James Edward
West, John B(ernard)
Wimmer, Donn Braden
Zetik, Donald Frank

PENNSYLVANIA

Acheson, Willard Phillips
Amero, R(obert) C
Archer, David Horace
Barmby, David Stanley
Bissey, Luther Trauger
Blaustein, Bernard Daniel
Boodman, Norman S
Camp, Frederick William
Catchen, Gary Lee
Cernansky, Nicholas P
Champagne, Paul Ernest
Charmbury, H(erbert) Beecher
Chigier, Norman
Churchill, Stuart W(inston)
Cobb, James Temple, Jr
Conroy, James Strickler
Davis, Burl Edward
Del Bel, Elsio
Dell, M(anuel) Benjamin
Enick, Robert M
Ertekin, Turgay
Finseth, Dennis Henry
Fulton, Paul F(ranklin)
Given, Peter Hervey
Givens, Edwin Neil
Grumer, Joseph
Holder, Gerald D
Kanter, Ira E
Levy, Edward Kenneth
Lovell, Harold Lemuel
McCormick, Robert H(enry)
Meredith, David Bruce
Nichols, Duane Guy
Oder, Robin Roy
Palmer, Howard Benedict
Petura, John C
Podgers, Alexander Robert
Romovacek, George R
Schiffman, Louis F
Siddiqui, Habib
Sinnett, Carl E(arl)
Soung, Wen Y
Stahl, C(harles) D(rew)
Szabo, Miklos Tamas

Tosch, William Conrad
Vastola, Francis J
Vick, Gerald Kieth
Vidt, Edward James
Warren, J(oseph) E(mmet)
Wender, Irving
Whyte, Thaddeus E, Jr
Wilkens, John Albert
Zauderer, Bert

RHODE ISLAND

Suuberg, Eric Michael

SOUTH DAKOTA

Wegman, Steven M

TENNESSEE

Hodgson, J(effrey) William
Phung, Doan Lien
Sheth, Atul C
Singh, Suman Priyadarshi Narain

TEXAS

Abernathy, Bobby F
Abouhalkah, T A
Akgerman, Aydin
Ali, Yusuf
Al-Saadoon, Faleh T
Arnold, C(harles) W(illiam)
Baber, Burl B, Jr
Bare, Charles L
Barnes, Allen Lawrence
Bradley, H(oward) B(ishop)
Brunsting, Elmer H(enry)
Bush, Warren Van Ness
Bussian, Alfred Erich
Calhoun, John C, Jr
Carlile, Robert Elliot
Carroll, Michael M
Caudle, Ben Hall
Chew, Ju-Nam
Chien, Sze-Foo
Claridge, E(lmond) L(owell)
Clark, James Donald
Closmann, Philip Joseph
Coats, Keith Hal
Cocanower, R(obert) D(unlavy)
Collipp, Bruce Garfield
Crandall, John R
Crawford, Duane Austin
Crawford, Paul B(erlowitz)
Crenshaw, Paul L
Cross, Virginia Rose
Crouse, Philip Charles
Curtis, Lawrence B
Dellinger, Thomas Baynes
Dorfman, Myron Herbert
Dunlap, Henry Francis
Eakin, Bertram E
Eckles, Wesley Webster, Jr
Fontaine, Marc F(rancis)
Gilchrist, Ralph E(dward)
Gilliland, Harold Eugene
Gilmore, Robert Beattie
Godwin, James Basil, Jr
Gregory, A(lvin) R(ay)
Gruy, Henry Jones
Haberman, John Phillip
Harris, Martin H
Heinrich, R(aymond) L(awrence)
Heins, Robert W
Hirasaki, George J
Holliday, George Hayes
Hrkel, Edward James
Huang, Wann Sheng
Ivey, E(dwin) H(arry), Jr
Jahns, Hans O(tto)
Jennings, Alfred Roy, Jr
Johnson, Lloyd N(ewhall)
Johnson, Mary Lynn Miller
Jones, Marvin Richard
Jorden, James Roy
Kalfoglou, George
Kellerhals, Glen E
Kelton, Frank Caleb
Lee, William John
Lestz, Sidney J
Levine, J(oseph) S(amuel)
Levy, Robert Edward
Lurix, Paul Leslie, Jr
McLeod, Harry O'Neal, Jr
Massey, Michael John
Mattax, Calvin Coolidge
Matthews, Charles Sedwick
Melrose, James C
Monaghan, Patrick Henry
Moore, Thomas Francis
Morgan, Bryan Edward
Morse, R(ichard) A(rden)
Mortada, Mohamed
Murdoch, Bruce Thomas
Nabor, George W(illiam)
Nancarrow, Warren George
Neavel, Richard Charles
Nemeth, Laszlo K
Olson, Danford Harold
Parker, Harry W(illiam)
Patton, Charles C(lifford)
Peaceman, Donald W(illiam)
Perkins, Thomas K(eeble)
Poddar, Syamal K
Podio, Augusto L
Posey, Daniel Earl
Richardson, Jasper E

Richardson, Joseph Gerald
Roebuck, Isaac Field
Roeger, Anton, III
Rowe, Allen McGhee, Jr
Russell, Donald Glenn
Russell, Thomas Webb
Shen, Chin-Wen
Sifferman, Thomas Raymond
Sinclair, A Richard
Slater, George E(dward)
Smith, Joseph Patrick
Souby, Armand Max
Stalkup, Fred I(rving), Jr
Streltsova, Tatiana D
Tao, Frank F
Thompson, William Horn
Vasilakos, Nicholas Petrou
Vernon, Lonnie William
Von Gonten, (William) Douglas
Von Rosenberg, Dale Ursini
Wagner, John Philip
Weatherford, W(illiam) D(ewey), Jr
White, N(ikolas) F(rederick)
Whiting, R(obert) L(ouis)
Wieland, Denton R
Wilhoit, Randolph Carroll
Winegartner, Edgar Carl
Zaczepinski, Sioma

UTAH

Bishop, Jay Lyman
Bunger, James Walter
Hill, George Richard
Langheinrich, Armin P(aul)
Sohn, Hong Yong
Trujillo, Edward Michael
Tuddenham, W(illiam) Marvin
Wiser, Wendell H(aslam)

VERMONT

Lang, William Harry
Long, Robert Byron

VIRGINIA

Bayliss, John Temple
Gordon, John S(tevens)
Hamrick, Joseph Thomas
Harney, Brian Michael
Hazlett, Robert Neil
Henderson, Charles B(rooke)
Jimeson, Robert M(acKay), Jr
Pitts, Thomas Griffin
Sarkes, Louis A(nthony)
Wayland, Russell Gibson

WASHINGTON

Holzman, George
Johnson, A(lfred) Burtron, Jr
Newman, Darrell Francis
Rosenwald, Gary W
Spencer, Arthur Milton, Jr

WEST VIRGINIA

Abel, William T
Clark, Nigel Norman
Das, Kamalendu
Gillmore, Donald W(ood)
Gwinn, James E
Rieke, Herman Henry, III
Rollins, Ronald Roy
Shuck, Lowell Zane
Spaniol, Craig
Wasson, James A(llen)
Wertman, William Thomas
Wilson, John Sheridan

WYOMING

Biggs, Paul
Dorrence, Samuel Michael
Ewing, Richard Edward
Hoffman, Edward Jack
Huff, Kenneth O
Hutchinson, Harold Lee
Marchant, Leland Condo
Odasz, F(rancis) B(ernard), Jr
Stinson, Donald Leo
Weinbrandt, Richard M

ALBERTA

Bennion, Douglas Wilford
Bird, Gordon Winslow
Brown, R(onald) A(nderson) S(teven)
Butler, R(oger) M(oore)
Chakravorty, S(ailendra) K(umer)
Checkel, M David
Clementz, David Michael
Dranchuk, Peter Michael
Flock, Donald Louis
Gregory, John
Kisman, Kenneth Edwin
Mungan, Necmettin
Scott, J(ohn) D(onald)
Stacy, T(homas) D(onnie)

ONTARIO

Bergougnou, Maurice A(medee)
Eliades, Theo I
Kosaric, Naim
Macdonald, Ian Francis
Napier, Douglas Herbert
Plett, Edelbert Gregory
Robinson, Campbell William
Shelson, W(illiam)
Walsh, John Heritage

Wojciechowski, Bohdan Wieslaw
Wynnyckyj, John Rostyslav

SASKATCHEWAN

Faber, Albert John

OTHER COUNTRIES

Boustany, Kamel
Economides, Michael John
Fu, Yuan C(hin)
Hjertager, Bjorn Helge
Lemanski, Michael Francis
Lingane, Peter James

Industrial & Manufacturing Engineering

ALABAMA

Brown, Robert Alan
Clark, Charles Edward
Copeland, David Anthony
Cox, Julius Grady
Hool, James N
Kallsen, Henry Alvin
Smith, Leo Anthony
Unger, Vernon Edwin, Jr
Webster, Dennis Burton
Workman, Gary Lee
Wyskida, Richard Martin

ALASKA

Bennett, F Lawrence

ARIZONA

Askin, Ronald Gene
Bailey, James Edward
Beaumariage, Terrence Gilbert
Bedworth, David D
Chandra, Abhijit
Cochran, Jeffery Keith
Ferrell, William Russell
Fitch, W(illiam) Chester
Hoyt, Charles D, Jr
Keats, John Bert
Kleinschmidt, Eric Walker
Mackulak, Gerald Thomas
Macleod, Hugh Angus
Metz, Donald C(harles)
Montgomery, Douglas C(arter)
Moor, William C(hattle)
Settles, F Stan
Shaw, Milton C(layton)
Slaughter, Charles D
Thompson, Truet B(radford)
Van Arsdel, John Hedde
Wymore, Albert Wayne
Young, Hewitt H

ARKANSAS

Asfahl, C Ray
Crisp, Robert M, Jr
Hobson, Leland S(tanford)
Imhoff, John Leonard
McBryde, Vernon E(ugene)
Thomas, Walter E
Wilkes, Joseph Wray

CALIFORNIA

Agostini, Romain Camille
Anjard, Ronald P, Sr
Antonsson, Erik Karl
Arnwine, William Carrol
Arslancan, Ahmet N
Bernard, Douglas Alan
Bertin, Michael C
Block, Richard B
Bonora, Anthony Charles
Brandeau, Margaret Louise
Campbell, Bonita Jean
Chi, Cheng-Ching
Chu, Chung-Yu Chester
Cook, Charles J
Cooper, George Emery
Czuha, Michael, Jr
Das, Mihir Kumar
Deriso, Richard Bruce
Dornfeld, David Alan
Eldon, Charles A
Fleischer, Gerald A
Grant, Eugene Lodewick
Gray, Paul
Greene, Charles Richard
Grimaldi, John Vincent
Hahne, Henry V
Hammond, Martin L
Harrison, George Conrad, Jr
Hausman, Warren H
Hecht, Myron J
Herrmann, George
Inoue, Michael Shigeru
Ireson, W(illiam) Grant
Irvine, Stuart James Curzon
Jacobson, Albert H(erman), Jr
Jillie, Don W
Kaliski, Martin Edward
Laitin, Howard
Lave, Roy E(llis), Jr
Lichter, James Joseph
Lindsay, Glenn Frank
Luther, Lester Charles
Martin, George
Mohamed, Farghalli Abdelrahman
Morgan, Donald E(arle)

Nadler, Gerald
Newnan, Donald G(lenn)
Nguyen, Luu Thanh
Oakford, Robert Vernon
Philipson, Lloyd Lewis
Remer, Donald Sherwood
Robinson, Gordon Heath
Rowe, Lawrence A
Schwartz, Daniel M(ax)
Schwartzbart, Harry
Shanthikumar, Jeyaveerasingam George
Shephard, Ronald W(illiam)
Singleton, Henry E
Thompson, David A(lfred)
Throner, Guy Charles
Vomhof, Daniel William
Wallack, Paul Mark
Wildes, Peter Drury
Wright, Edward Kenneth
Wymer, Joseph Peter
Yamakawa, Kazuo Alan

COLORADO
Bauer, Charles Edward
Giffin, Walter C(harles)
Litwhiler, Daniel W
McLean, Francis Glen
Ostwald, Phillip F
Tischendorf, John Allen
Troxell, Wade Oakes
Vejvoda, Edward
Wall, Edward Thomas
Wertz, Ronald Duane
Williams, Robert J(ames)

CONNECTICUT
Card, Roger John
Corwin, H(arold) E(arl)
De Groot, Sybil Gramlich
Enell, John Warren
Goldstein, Marvin Sherwood
Gupta, Dharam Vir
Haag, Robert Edwin
Johnson, Stephen Thomas
Kleinfeld, Ira H
Rezek, Geoffrey Robert
Schwartz, Michael H
Shainin, Dorian
Stein, Richard Jay
Wingfield, Edward Christian

DELAWARE
Gay, Frank P
Kikuchi, Shinya
Long, John Reed
Stewart, Edward William

DISTRICT OF COLUMBIA
Donahue, D Joseph
Esterling, Donald M
Kramer, Bruce Michael
Kyriakopoulos, Nicholas
Meier, Wilbur L(eroy), Jr
Trice, Virgil Garnett, Jr
Urban, Peter Anthony
Waters, Robert Charles
White, John Austin, Jr

FLORIDA
Brooks, George H(enry)
Brown, John Lott
Carpenter, James L(inwood), Jr
Doering, Robert Distler
Doyon, Leonard Roger
Elzinga, D(onald) Jack
Francis, Richard Lane
Genaidy, Ashraf Mohamed
Givens, Paul Edward
Gold, Edward
Haug, Arthur John
Hosni, Yasser Ali
Hoyer-Ellefsen, Sigurd
Hsu, Tsong-Han
Kruesi, William R
Kunce, Henry Warren
Leavenworth, Richard S
Muth, Eginhard Joerg
Nagy, Steven
Nattress, John Andrew
Proctor, Charles Lafayette
Schrader, George Frederick
Shaw, Lawrance Neil
Swart, William W
Van Ligten, Raoul Fredrick
Ward, Leonard George
Whitehouse, Gary E
Wiener, Earl Louis
Wilcox, Donald Brooks
Zanakis, Stelios (Steve) H

GEORGIA
Aronson, Jay E
Banks, Jerry
Callahan, Leslie G, Jr
Feldman, Edwin B(arry)
Fyffe, David Eugene
Gambrell, Carroll B(lake), Jr
Hines, William W
Jarvis, John J
Johnson, Cecil Gray
Johnson, Lynwood Albert
Krol, Joseph
Lehrer, Robert N(athaniel)
Malcolm, Earl Walter

Pence, Ira Wilson, Jr
Porter, Alan Leslie
Rogers, Harvey Wilbur
Rogers, Nelson K
Sadosky, Thomas Lee
Thuesen, Gerald Jorgen
Vicory, William Anthony
Wadsworth, Harrison M(orton)
Wood, Robert E
Zunde, Pranas

HAWAII
Bartosik, Alexander Michael

IDAHO
Lambuth, Alan Letcher
Preszler, Alan Melvin

ILLINOIS
Bell, Charles Eugene, Jr
Berkelhamer, Louis H(arry)
Block, Stanley M(arlin)
Bornong, Bernard John
Brauer, Roger L
Cannon, Howard S(uckling)
Cook, Harry Edgar
Dessouky, Mohamed Ibrahim
Doyle, Lawrence Edward
Frey, Donald N(elson)
Glod, Edward Francis
Haugen, Robert Kenneth
Hurter, Arthur P
McFee, Donald Ray
McKee, Keith Earl
Miller, Floyd Glenn
Morrow, Richard M
Parker, Ehi
Phillips, Rohan Hilary
Pigage, Leo C(harles)
Poulikakos, Dimos
Pritzker, Robert A
Rubenstein, Albert Harold
Ruhl, Roland Luther
Siegal, Burton L
Spector, Leo Francis
Thompson, Charles William Nelson
Triano, John Joseph
Wilson, William Robert Dunwoody

INDIANA
Amrine, Harold Thomas
Barany, James W
Compton, Walter Dale
Eaton, Paul Bernard
Eberts, Ray Edward
English, Robert E
Gorman, Eugene Francis
Greene, James H
Gresham, Robert Marion
Hemmes, Paul Richard
Klein, Howard Joseph
Lehto, Mark R
Leimkuhler, Ferdinand F
Nagle, Edward John
Nof, Shimon Y
Ovens, William George
Pielet, Howard M
Pritsker, A(drian) Alan B(eryl)
Salvendy, Gavriel
Solberg, James J
Sweet, Arnold Lawrence
Wynne, Bayard Edmund

IOWA
Adams, S Keith
Buck, James R
Cowles, Harold Andrews
Hempstead, J(ean) C(harles)
Hillyard, Lawrence R(obertson)
Kusiak, Andrew
Lamp, George Emmett, Jr
Mack, Michael J
McRoberts, Keith L
Maze, Thomas Harold
Papadakis, Emmanuel Philippos
Patterson, John W(illiam)
Rocklin, Isadore J
Smith, Gerald Wavern
Vardeman, Stephen Bruce
Vaughn, Richard Clements

KANSAS
Connor, Sidney G
Eckhoff, Norman Dean
Graham, A Richard
Harnett, R Michael
Hollis, Raymond Howard
Konz, Stephan A
Kramer, Bradley Alan
Lambert, Brian Kerry
Malzahn, Don Edwin
Tiffany, Charles F
Tillman, Frank A

KENTUCKY
Cathey, Jimmie Joe
Evans, Gerald William
Kohnhorst, Earl Eugene
Leep, Herman Ross
Lu, Wei-yang
Tran, Long Trieu

LOUISIANA
Cospolich, James D

Fontenot, Martin Mayance, Jr
Henry, Herman Luther, Jr
Keys, L Ken
Kraft, Donald Harris
Mann, Lawrence, Jr
Morrow, Norman Louis
Ristroph, John Heard
Rubinstein, Asher A
Siebel, M(athias) P(aul) L

MAINE
Webster, Karl Smith

MARYLAND
Albus, James S
Berger, Harold
Doiron, Theodore Danos
Haffner, Richard William
Hovmand, Svend
Jackson, Richard H F
Katsanis, D(avid) J(ohn)
Kemelhor, Robert Elias
McLaughlin, William Lowndes
Nusbickel, Edward M, Jr
Steiner, Henry M
Tarrants, William Eugene
Van Cott, Harold Porter
Weiss, Roland George

MASSACHUSETTS
Addiss, Richard Robert, Jr
Adler, Ralph Peter Isaac
Bahr, Karl Edward
Bailey, Milton
Brandler, Philip
Brown, Robert Lee
Chandler, Charles H(orace)
Chryssostomidis, Chryssostomos
Cook, Nathan Henry
Cullinane, Thomas Paul
De Fazio, Thomas Luca
Edelman, Julian
Emmons, Howard W(ilson)
Gaw, C Vernon
Gilfix, Edward Leon
Graves, Robert John
Gupta, Surendra Mohan
Hahn, Robert S(impson)
Hulbert, Thomas Eugene
Krause, Irvin
Kroner, Klaus E(rlendur)
Leamon, Tom B
Magnanti, Thomas L
Meal, Harlan C
Nelson, J(ohn) Byron
Papageorgiou, John Constantine
Rising, Edward James
Scudder, Henry J, III
Story, Anne Winthrop
Van der Burg, Sjirk
Vartanian, Leo

MICHIGAN
Aswad, A Adnan
Batrin, George Leslie
Carson, Gordon B(loom)
Chaffin, Don B
Chalfant, Forest Earle
Chen, Kan
Chen, Yudong
Dilley, David Ross
Farah, Badie Naiem
Flinn, Richard A(loysius)
Frosch, Robert Alan
Hayes, Edward J(ames)
Kachhal, Swatantra Kumar
Knappenberger, Herbert Allan
Lamberson, Leonard Roy
Lin, Paul Kuang-Hsien
Lorenz, John Douglas
Maley, Wayne A
Marks, Craig
Pandit, Sudhakar Madhavrao
Parsons, John Thoren
Pollock, Stephen M
Rollinger, Charles N(icholas)
Rosinski, Michael A
Sahney, Vinod K
Schriber, Thomas J
Schumacher, Berthold Walter
Spurgeon, William Marion
Stevenson, Robin
Taraman, Khalil Showky
Thomson, Robert Francis
Weiner, Steven Allan
Weinstein, Jeremy Saul
Williams, Gwynne
Wolf, Franklin Kreamer
Wu, Hai

MINNESOTA
Hoffmann, Thomas Russell
Jacobson, Allen F
Kvalseth, Tarald Oddvar
McElrath, Gayle W(illiam)
Raley, Frank Austin
Robertsen, John Alan
Smith, Thomas Jay
Starr, Patrick Joseph
Verma, Deepak Kumar
Wade, Michael George
White, John S(pencer)

MISSISSIPPI
Cotton, Frank Ethridge, Jr
Johnson, Larry Ray
Oswalt, Jesse Harrell
Parker, Murl Wayne

MISSOURI
Blundell, James Kenneth
Brumbaugh, Philip
David, Larry Gene
Donovan, John Richard
Eastman, Robert M(erriam)
Gillespie, LaRoux King
Gillespie, LaRoux King
Klein, Cerry M
Miller, Owen Winston
O'Connor, John Thomas
Vandegrift, Alfred Eugene

MONTANA
Gibson, David F(rederic)
Ritchey, John Arthur

NEBRASKA
Hoffman, Richard Otto
Riley, Michael Waltermier
Schneider, Morris Henry
Schniederjans, Marc James
Von Bargen, Kenneth Louis

NEVADA
See, John Bruce

NEW HAMPSHIRE
Black, Sydney D
Carreker, R(oland) P(olk), Jr

NEW JERSEY
Albin, Susan Lee
Anandan, Munisamy
Beinfest, Sidney
Boucher, Thomas Owen
Browne, Robert Glenn
Clifford, Paul Clement
Davis, William F
Elsayed, Elsayed Abdelrazik
Geskin, Ernest S
Hoover, Charles Wilson, Jr
Huang, Jenmming Stephen
Hunter, John Stuart
Jain, Sushil C
Jarvis, John Frederick
Kalley, Gordon S
Kellgren, John
Khalifa, Ramzi A
Landwehr, James M
Leu, Ming C
Liebig, William John
Luxhoj, James Thomas
McDermott, Kevin J
Magliola-Zoch, Doris
Mihalisin, John Raymond
Pargellis, Andrew Nason
Pasteelnick, Louis Andre
Perrotta, James
Polhemus, Neil W
Procknow, Donald Eugene
Riley, David Waegar
Schoenfeld, Theodore Mark
Sermolins, Maris Andris
Snowman, Alfred
Thomas, Stanislaus S(tephen)
Vyas, Brijesh
Walsh, Teresa Marie
Whitt, Ward
Wildman, George Thomas
Yan, Tsoung-Yuan
Young, Li

NEW MEXICO
Benziger, Theodore Michell
George, Raymond S
Jamshidi, Mohammad Mo
Lee, David Oi
Maez, Albert R
Martin, Richard Alan
Sandstrom, Donald James
Thode, E(dward) F(rederick)

NEW YORK
Armamento, Eduardo T
Balassa, Leslie Ladislaus
Benson, Richard Carter
Benzinger, James Robert
Berg, Daniel
Bialas, Wayne Francis
Bienstock, Daniel
Crooke, James Stratton
Desrochers, Alan Alfred
Drury, Colin Gordon
Dudewicz, Edward John
Eldin, Hamed Kamal
Franse, Renard
Frazer, J(ohn) Ronald
Gellman, Charles
Gerster, Robert Arnold
Goldfarb, Donald
Gottesman, Elihu
Greenberg, Jacob
Haddock, Jorge
Happel, John
Hauser, Norbert
Heath, David Clay
Helander, Martin Erik Gustav

Industrial & Manufacturing Engineering (cont)

Hunt, Everett Clair
Jaffe, William J(ulian)
Karwan, Mark Henry
Kashkoush, Ismail I
Kenett, Ron
Lindstrom, Gary J
McGee, James Patrick
Maxwell, William L
Meenan, Peter Michael
Menkes, Sherwood Bradford
Morfopoulos, Vassilis C(onstantinos) P
Mundel, August B(aer)
Mustacchi, Henry
Nanda, Ravinder
Penney, Carl Murray
Rosenberg, Paul
Sargent, Robert George
Schultz, Andrew, Jr
Simons, Gene R
Solomon, Jack
Srinivasan, Vijay
Stankiewicz, Raymond
Summers, Phillip Dale
Swalm, Ralph Oehrle
Thomas, Warren H(afford)
Thompson, Robert Alan
Ullmann, John E
Voelcker, Herbert B(ernhard)
Wang, Hsin-Pang
Wilczynski, Janusz S
Wright, Roger Neal
Yasar, Tugrul
Young, William Donald, Jr

NORTH CAROLINA

Appleby, Robert H
Ayoub, Mahmoud A
Bernhard, Richard Harold
Bockstahler, Theodore Edwin
Dyrkacz, W William
Eberhardt, Allen Craig
Harder, John Jurgen
Hurt, Alfred B(urman), Jr
Kilpatrick, Kerry Edwards
King, L(ee) Ellis
Lalor, William Francis
Mallik, Arup Kumar
Morgan, Donald Lee
Nutter, James I(rving)
Roberts, Stephen D
Smith, William Adams, Jr
Stidham, Shaler, Jr
Tartt, Thomas Edward

NORTH DAKOTA

Meldrum, Alan Hayward
Stanislao, Joseph

OHIO

Beckham, Robert R(ound)
Bishop, Albert B
Carmichael, Donald C(harles)
Chen, Kuei-Lin
Chrissis, James W
Daschbach, James McCloskey
Duttweiler, Russell E
Erbacher, John Kornel
Fraker, John Richard
Fritsch, Charles A(nthony)
Gilby, Stephen Warner
Hulley, Clair Montrose
Hyndman, John Robert
Jain, Vinod Kumar
Kaye, Christopher J
Keim, John Eugene
Kishel, Chester Joseph
Klosterman, Albert Leonard
Kornylak, Andrew T
Lahoti, Goverdhan Das
Lee, Peter Wankyoon
Lumsdaine, Edward
McNichols, Roger J(effrey)
Marras, William Steven
Menq, Chia-Hsiang
Merchant, Mylon Eugene
Roberts, Timothy R
Rockwell, Thomas H
Schutta, James Thomas
Sheskin, Theodore Jerome
Shetterly, Donivan Max
Slutzky, Gale David
Smith, George Leonard, Jr
Soska, Geary Victor
Srivatsan, Tirumalai Srinivas
Ventresca, Carol
Williams, Robert Lloyd
Wolfe, Robert Kenneth

OKLAHOMA

Aly, Adel Ahmed
Case, Kenneth E(ugene)
Ferguson, Earl J
Ferguson, William Sidney
Foote, Bobbie Leon
Kapur, Kailash C
Mize, Joe H(enry)
Schuermann, Allen Clark, Jr
Shamblin, James E
Shapiro, Robert Allen
Terrell, Marvin Palmer

Turner, Wayne Connelly
Vanderwiele, James Milton

OREGON

Kocaoglu, Dundar F

PENNSYLVANIA

Barsom, John M
Bise, Christopher John
Bockosh, George R
Burnham, Donald C
Crane, Lee Stanley
Dalal, Harish Maneklal
Dorny, C Nelson
Furfari, Frank A
Garg, Diwakar
Georgakis, Christos
Gottfried, Byron S(tuart)
Groover, Mikell
Ham, Inyong
Holzman, Albert G(eorge)
Johnson, Littleton Wales
Johnston-Feller, Ruth M
Kane, George E
Keyt, Donald E
Kingrea, James I
Kipp, Egbert Mason
Kottcamp, Edward H, Jr
Kuhn, Howard A
Lidman, William G
McCune, Duncan Chalmers
McDonnell, Leo F(rancis)
Mayer, George Emil
Mouly, Raymond J
Niebel, B(enjamin) W(illard)
Odrey, Nicholas Gerald
Penman, Paul D
Preusch, Charles D
Rogers, H(arry) C(arton), Jr
Saluja, Jagdish Kumar
Scigliano, J Michael
Thuering, George Lewis
Ventura, Jose Antonio
Wagner, David Loren
Weindling, Joachim I(gnace)
Whyte, Thaddeus E, Jr
Wolfe, Harvey
Woodburn, Wilton A

RHODE ISLAND

Needleman, Alan
Olson, David Gardels

SOUTH CAROLINA

Chisman, James Allen
Graulty, Robert Thomas
Holme, Thomas T(imings)
Kimbler, Delbert Lee
Sandrapaty, Ramachandra Rao

SOUTH DAKOTA

Johnson, Marvin Melrose
Swiden, LaDell Ray

TENNESSEE

Bontadelli, James Albert
Cox, Ronald Baker
Deivanayagan, Subramanian
Gano, Richard W
Gilbreath, Sidney Gordon, III
Isley, James Don
Lundy, Ted Sadler
Lyle, Benjamin Franklin
Pearsall, S(amuel) H(aff)
Prairie, Michael L
Sissom, Leighton E(sten)
Slaughter, Gerald M
Smith, James Ross
Sundaram, R Meenakshi

TEXAS

Ayoub, Mohamed Mohamed
Beightler, Charles Sprague
Bennett, George Kemble
Blank, Leland T
Bravenec, Edward V
Brown, Bradford S
Caswell, Gregory K
CoVan, Jack Phillip
Dudek, R(ichard) A(lbert)
Edgar, Thomas Flynn
Fletcher, John Lynn
Francis, Philip Hamilton
Gates, David G(ordon)
Geyling, F(ranz) Th(omas)
Gupton, Paul Stephen
Hoffman, Herbert I(rving)
Hogg, Gary Lynn
Koehn, Enno
Koppa, Rodger J
Lewis, Frank Leroy
Lutz, Raymond
McNees, Roger Wayne
Marshek, Kurt M
Meier, France Arnett
Miller, Gerald E
Mills, John James
Moon, Tessie Jo
Ozan, M Turgut
Phillips, Don T
Poage, Scott T(abor)
Ramaswamy, Kizhanatham V
Ramsey, Jerry Dwain
Rhodes, Allen Franklin

Smith, Milton Louis
Stevens, Gladstone Taylor, Jr
Street, Robert Lewis
Tims, George B(arton), Jr
Vernon, Ralph Jackson
Wilhelm, Wilbert Edward
Wolter, Jan D(ithmar)
Wood, Kristin Lee
Yakin, Mustafa Zafer

UTAH

Baum, Sanford
Nelson, Mark Adams
Rotz, Christopher Alan
Turley, Richard Eyring

VERMONT

Gray, John Patrick

VIRGINIA

Agee, Marvin H
Arp, Leon Joseph
Betti, John A
Cetron, Marvin J
Chiou, Minshon Jebb
Ell, William M
Fabrycky, Wolter J
Healy, John Joseph
Hunt, V Daniel
Jorstad, John Leonard
Kinsley, Homan Benjamin, Jr
Ling, James Gi-Ming
Maloney, Kenneth Morgan
Moore, James Mendon
Owczarski, William A(nthony)
Siebentritt, Carl R, Jr
Sink, David Scott
Smith, Michael Claude
Wierwille, Walter W(erner)
Williges, Robert Carl

WASHINGTON

Anderson, Richard Gregory
Bethel, James Samuel
Campbell, John C(arl)
Drui, Albert Burnell
Drumheller, Kirk
Jorgensen, Jens Erik
Madden, James H(oward)
Maranville, Lawrence Frank
Newman, Paul Harold
Ramulu, Mamidala
Savol, Andrej Martin
Tenenbaum, Michael
Wolak, Jan
Woodfield, F(rank) W(illiam), Jr

WEST VIRGINIA

Breneman, William C
Stoll, Richard E
Tompkins, Curtis Johnston

WISCONSIN

Bisgaard, Søren
Bollinger, John G(ustave)
Brisley, Chester L(avoyen)
Chang, Tsong-How
DeVries, Marvin Frank
Farrow, John H
Gustafson, David Harold
Hunter, William G(ordon)
Jacobi, George (Thomas)
James, Charles Franklin, Jr
Jared, Alva Harden
Lentz, Mark Steven
Moy, William A(rthur)
Robinson, Stephen Michael
Samuel, Jay Morris
Saxena, Umesh
Schotz, Larry
Smith, Michael James
Steudel, Harold Jude
Suri, Rajan
Szpot, Bruce F
Zelazo, Nathaniel Kachorek

WYOMING

Odasz, F(rancis) B(ernard), Jr

PUERTO RICO

Rodriguez, Luis F

ALBERTA

Gregory, John
Silver, Edward Allan
Sprague, James Clyde

BRITISH COLUMBIA

Franz, Norman Charles
Mitten, Loring G(oodwin)

NEW BRUNSWICK

Pham-Gia, Thu

NOVA SCOTIA

Kujath, Marek Ryszard
Wilson, George Peter

ONTARIO

Beevis, David
Buzacott, J(ohn) A(lan)
Carroll, John Millar
Cohn, Stanley Howard
Dhillon, Balbir Singh

Fahmy, Moustafa Mahmoud
Fenton, Robert George
Jeswiet, Jacob
Schey, John Anthony
Steiner, George
Wong, George Shoung-Koon

QUEBEC

Bakhiet, Atef
Cherna, John C(harles)
Osman, M(ohamed) O M
Saber, Aaron Jaan
Sankar, Seshadri

OTHER COUNTRIES

De Vedia, Luis Alberto
Eaton, Robert James
Hamacher, Horst W
Jost, Hans Peter
Malik, Mazhar Ali Khan
Pau, Louis F
Raouf, Abdul
Sekine, Yasuji
Singh, Madan Gopal
Steffen, Juerg
Stephens, Kenneth S
Van Weert, Gezinus

Materials Science Engineering

ALABAMA

Agresti, David George
Andrews, Russell S, Jr
Beindorff, Arthur Baker
Biblis, Evangelos J
Cataldo, Charles Eugene
Dean, David Lee
Dukatz, Ervin L, Jr
Hall, David Michael
Hardin, Ian Russell
Haygreen, John G
Hubbell, Wayne Charles
Kattus, J Robert
Lemons, Jack Eugene
McKannan, Eugene Charles
Marek, Charles R(obert)
Naumann, Robert Jordan
Pattillo, Robert Allen
Penn, Benjamin Grant
Stefanescu, Doru Michael
Thompson, Raymond G
Warfield, Carol Larson
Willenberg, Harvey Jack
Workman, Gary Lee

ALASKA

Johnson, Ralph Sterling, Jr
Skudrzyk, Frank J

ARIZONA

Agnihotri, Ram K
Berg, Howard Martin
Burke, William James
Carpenter, Ray Warren
Chu, William How-Jen
Cowley, John Maxwell
Dodd, Charles Gardner
Eyring, LeRoy
Fairbanks, Harold V
Falco, Charles Maurice
George, Thomas D
Hight, Ralph Dale
Hiskey, J Brent
Hunter, William Leslie
Jacobson, Michael Ray
Kingery, W(illiam) D(avid)
Kleinschmidt, Eric Walker
Leeser, David O(scar)
Neilson, George Francis, Jr
Orton, George W
Osborn, Donald Earl
Ramohalli, Kumar Nanjunda Rao
Ren, Shang Yuan
Risbud, Subhash Hanamant
Saul, Robert H
Savrun, Ender
Schroder, Dieter K
Shafer, M(errill) W(ilbert)
Shaw, Milton C(layton)
Simmons, Paul C
Singhal, Avinash Chandra
Smyth, Jay Russell
Sundahl, Robert Charles, Jr
Swalin, Richard Arthur
Tsong, Ignatius Siu Tung
Uhlmann, Donald Robert
Wagner, James Bruce, Jr
Willson, Donald Bruce
Withers, James C

CALIFORNIA

Adicoff, Arnold
Adinoff, Bernard
Adler, William Fred
Agrawal, Suphal Prakash
Allard, Douglas Lee
Ang, Tjoan-Liem
Anjard, Ronald P, Sr
Ardell, Alan Jay
Arnold, Walter Frank
Arrowood, Roy Mitchell, Jr
Atkinson, Russell H
Baba, Paul David

Bajaj, Jagmohan
Baldwin, Chandler Milnes
Barasch, Werner
Bar-Cohen, Yoseph
Barnett, David M
Baskes, Michael I
Bateman, John Hugh
Bean, Ross Coleman
Bienenstock, Arthur Irwin
Bilow, Norman
Botez, Dan
Bowman, Robert Clark, Jr
Bragg, Robert H(enry)
Bube, Richard Howard
Buffington, F(rancis) S(tephan)
Bullis, William Murray
Bullock, Ronald Elvin
Burland, Donald Maxwell
Burmeister, Robert Alfred
Bush, Gary Graham
Caligiuri, Robert Domenic
Cao, Hengchu
Carley, James F(rench)
Carniglia, Stephen C(harles)
Carruthers, John Robert
Cass, Thomas Robert
Chang, Hao-Jan
Chang, Kaung-Jain
Chen, C(hih) W(en)
Chen, Tu
Chiang, Anne
Chiou, C(harles)
Chompff, Alfred J(ohan)
Chou, Yungnien John
Christensen, Richard Monson
Chung, Chi Hsiang
Chung, Dae Hyun
Clark, Lloyd Douglas
Clemens, Bruce Montgomery
Clements, Linda L
Coburn, John Wyllie
Cockrell, Robert Alexander
Coleman, Charles Clyde
Colodny, Paul Charles
Comley, Peter Nigel
Condit, Ralph Howell
Cope, Oswald James
Dejonghe, Lutgard C
Dempsey, Martin E(wald)
Denn, Morton M(ace)
Docks, Edward Leon
Dorward, Ralph C(larence)
Dowell, Armstrong Manly
Doyle, Fiona Mary
Dunn, Bruce
Durham, William Bryan
Dyball, Christopher John
Eaton, David Leo
Eden, Richard Carl
Edwall, Dennis Dean
Edwards, Eugene H
Eichinger, Bruce Edward
Epstein, George
Estill, Wesley Boyd
Evans, Anthony Glyn
Even, William Roy, Jr
Farrar, John
Feather, David Hoover
Feay, Darrell Charles
Fedors, Robert Francis
Feigelson, Robert Saul
Feng, Joseph Shao-Ying
Fishman, Norman
Flamm, Daniel Lawrence
Flinn, Paul Anthony
Foiles, Stephen Martin
Forbes, James Franklin
Forbes, Judith L
Fouquet, Julie
Freise, Earl J
Fultz, Brent T
Gay, Richard Leslie
Geballe, Theodore Henry
Gehman, Bruce Lawrence
Geiss, Roy Howard
Gentile, Anthony L
Ghosh, Amit Kumar
Gill, William D(elahaye)
Gilman, John Joseph
Goetzel, Claus G(uenter)
Goo, Edward Kwock Wai
Graper, Edward Bowen
Green, Allen T
Green, Harry Western, II
Gronsky, Ronald
Gulden, Terry Dale
Gür, Turgut M
Hackett, Le Roy Huntington, Jr
Haegel, Nancy M
Hai, Francis
Hall, James Timothy
Haller, Eugene Ernest
Hammond, Martin L
Harkins, Carl Girvin
Harris, Daniel Charles
Harris, James Stewart, Jr
Hartsough, Larry Dowd
Harvey, Frances J, II
Hauber, Janet Elaine
Haycock, Ernest William
Haynes, Charles W(illard)
Hempstead, Robert Douglas
Hertzler, Barry Lee
Hoover, William Ringo

Hopper, Robert William
Hsieh, You-Lo
Huang, Tiao-yuan
Huggins, Robert A(lan)
Hummel, Steven G
Hunt, Arlon Jason
Jaffee, Robert I(saac)
Jaklevic, Joseph Michael
James, Edward, Jr
James, Michael Royston
Johanson, William Richard
Johnson, Sylvia Marian
Johnson, William Lewis
Johnson, William Robert
Jones, Brian Herbert
Jones, Robin L(eslie)
Jorgensen, Paul J
Joseph, Marjory L
Kaae, James Lewis
Kalfayan, Sarkis Hagop
Kamins, Theodore I
Kamp, David Allen
Kashar, Lawrence Joseph
Kay, Robert Eugene
Kendig, Martin William
Khan, Mahbub R
King, Ray J(ohn)
Klement, William, Jr
Klemm, Waldemar Arthur, Jr
Knight, Patricia Marie
Koenig, Nathan Hart
Kolek, Robert Louis
Kolyer, John M
Kosmatka, John Benedict
Kramer, David
Krikorian, Esther
Krivanek, Ondrej Ladislav
Kroemer, Herbert
Kuan, Teh S
Kupfer, John Carlton
LaChance, Murdock Henry
Lamb, Walter Robert
Lampert, Carl Matthew
Landel, Robert Franklin
Langer, Sidney
Lau, S S
Lavernia, Enrique Jose
Lee, Kenneth
Lee, Stuart M(ilton)
Leeg, Kenton J(ames)
Leeper, Harold Murray
Lehovec, Kurt
Leipold, Martin H(enry)
Lemke, James Underwood
Leslie, James C
Leverton, Walter Frederick
Lewis, Nathan Saul
Li, Chia-Chuan
Lin, Charlie Yeongching
Logan, James Columbus
Lomas, Charles Gardner
Lovelace, Alan Mathieson
Lucas, Glenn E
Lugassy, Armand Amram
McCreight, Louis R(alph)
McHugh, Stuart Lawrence
McKaveney, James P
Mackenzie, John D(ouglas)
McNutt, Michael John
Maloney, Timothy James
Mansfeld, Florian Berthold
Mar, Raymond W
Marcus, John Stanley
Margerum, John David
Marshall, Grayson William, Jr
Marshall, Sally J
Martin, George
Martin, Gordon Eugene
Maserjian, Joseph
Masters, Burton Joseph
Matthys, Eric Francois
Mehta, Povindar Kumar
Meike, Annemarie
Meiser, Michael David
Merriam, Marshal F(redric)
Meyer, Stephen Frederick
Meyers, Bernard Leonard
Mikami, Harry M
Mikes, Peter
Miller, Edward
Milstein, Frederick
Mohamed, Farghalli Abdelrahman
Molau, Gunther Erich
Moon, Donald W(ayne)
Mooney, John Bernard
Morris, John William, Jr
Mukherjee, Amiya K
Munir, Zuhair A
Needles, Howard Lee
Nellis, William J
Nelson, Howard Gustave
Nelson, Norvell John
Nemeth, Joseph
Nguyen, Luu Thanh
Nishi, Yoshio
Nix, William Dale
Nooker, Eugene L(eRoy)
Norman, John Harris
Novotny, Vlad Joseph
Odette, G(eorge) Robert
Oldfield, William
O'Meara, David Lillis
Ono, Kanji
Paciorek, Kazimiera J L

Packer, Charles M
Paquette, David George
Parrish, William
Pask, Joseph Adam
Pathania, Rajeshwar S
Paton, Neil (Eric)
Pearson, Dale Sheldon
Peeters, Randall Louis
Perkins, Kenneth L(ee)
Petroff, Pierre Marc
Pianetta, Piero Antonio
Pollack, Louis Rubin
Pye, Earl Louis
Rabolt, John Francis
Rau, Charles Alfred, Jr
Redfield, David
Regulski, Thomas Walter
Richman, Roger H
Riegert, R(ichard) P(aul)
Riley, Robert Lee
Ritchie, Robert Oliver
Rohde, Richard Whitney
Romanowski, Christopher Andrew
Rosenberg, Sanders David
Roy, Prodyot
Rubin, Barney
Rubin, Louis
Rudee, Mervyn Lea
Ruskin, Arnold M(ilton)
Sachdev, Suresh
Sahbari, Javad Jabbari
Salovey, Ronald
Sashital, Sanat Ramanath
Schalla, Charence August
Schjelderup, Hassel Charles
Schwab, Michael
Schwartzbart, Harry
Sclar, Nathan
Seaman, Lynn
Searcy, A(lan) W(inn)
Sedgwick, Robert T
Seiling, Alfred William
Shackelford, James Floyd
Sheehan, James Elmer
Shelton, Robert Neal
Sher, Arden
Sherby, Oleg D(imitri)
Sherman, Arthur
Shin, Soo H
Shinohara, Makoto
Shockey, Donald Albert
Simnad, Massoud T
Sinclair, Robert
Skowronski, Raymund Paul
Slivinsky, Sandra Harriet
Smith, Charles Aloysius
Snowden, William Edward
Spier, Edward Ellis
Spitzer, William George
Sree Harsha, Karnamadakala S
Stapelbroek, Maryn G
Steinberg, Daniel J
Stephens, Douglas Robert
Stetson, Alvin Rae
Stetson, Robert F
Stevenson, David Austin
Stirn, Richard J
Stolte, Charles
Stout, Ray Bernard
Stringer, John
Stroeve, Pieter
Suits, James Carr
Swanson, David G, Jr
Swearengen, Jack Clayton
Tauber, Richard Norman
Tennant, William Emerson
Thomas, Gareth
Thompson, Larry Dean
Tiller, William Arthur
Tombrello, Thomas Anthony, Jr
Treves, David
Tsai, Ming-Jong
Tung, Lu Ho
Twiss, Robert John
Vanblarigan, Peter
Vassiliou, Marius Simon
Viswanathan, R
Volksen, Willi
Vreeland, Thad, Jr
Vroom, David Archie
Wagner, C(hristian) N(ikolaus) J(ohann)
Wang, Kang-Lung
Warburton, William Kurtz
Washburn, Jack
Washburn, Jack
Weber, Carl Joseph
Weber, Eicke Richard
Westmacott, Kenneth Harry
White, Jack Lee
White, Moreno J
Whitney, Gina Marie
Wiener, Sidney
Williams, Richard Stanley
Williamson, Robert Brady
Winkler, DeLoss Emmet
Wood, David S(hotwell)
Wood, David Wells
Wright, Charles Cathbert
Wudl, Fred
Yang, Jenn-Ming
Yasui, George
Yoon, Rick J
Young, Konrad Kwang-Leei
Yu, Chyang John

Yu, Karl Ka-Chung
Zebroski, Edwin L
Zuleeg, Rainer

COLORADO
Adams, James Russell
Arp, Vincent D
Bachman, Bonnie Jean Wilson
Bauer, Charles Edward
Beck, Betty Anne
Blake, Daniel Melvin
Bodig, Jozsef
Brady, Brian T
Ciszek, Ted F
Czanderna, Alvin Warren
Dunn, Richard Lee
Ellis, Donald Griffith
Feldman, Arthur
Gardner, Edward Eugene
Goldfarb, Ronald B
Grobner, Paul Josef
Hauser, Ray Louis
Kazmerski, Lawrence L
Krantz, William Bernard
Kremer, Russell Eugene
Lebiedzik, Jozef
Lefever, Robert Allen
Levenson, Leonard L
Mackinney, Herbert William
Mahajan, Roop Lal
Moddel, Garret R
Moore, John Jeremy
Noufi, Rommel
Olson, David LeRoy
Owen, Robert Barry
Pankove, Jacques I
Pitts, John Roland
Predecki, Paul K
Rauch, Gary Clark
Richards, Harold Rex
Roshko, Alexana
Schmidt, Alan Frederick
Schramm, Raymond Eugene
Shuler, Craig Edward
Simon, Nancy Jane
Strauss, Eric L
Timblin, Lloyd O, Jr
Troxell, Harry Emerson, Jr
VanderLinden, Carl R
Webster, Larry Dale
Yarar, Baki
Zoller, Paul

CONNECTICUT
Adams, Brent Larsen
Agulian, Samuel Kevork
Arnold, John Richard
Balasinski, Artur
Berry, Richard C(hisholm)
Blackwood, Andrew W
Bradley, Elihu F(rancis)
Canonico, Domenic Andrew
Capwell, Robert J
Castonguay, Richard Norman
Cheatham, Robert Gary
Cheng, Francis Sheng-Hsiung
Chin, Charles L(ee) D(ong)
Cooke, Theodore Frederic
Davidson, Theodore
Easterbrook, Eliot Knights
Emond, George T
Eppler, Richard A
Falkenstein, Gary Lee
Fletcher, Frederick Bennett
Gardner, Fred Marvin
Giamei, Anthony Francis
Goldberg, A Jon
Gray, Thomas James
Grossman, Leonard N(athan)
Gutierrez-Miravete, Ernesto
Haacke, Gottfried
Haag, Robert Edwin
Hauser, Martin
Horhota, Stephen Thomas
Hsu, Nelson Nae-Ching
Johnson, Julian Frank
Koberstein, Jeffrey Thomas
Landsman, Douglas Anderson
Langeland, Kaare
Layden, George Kavanaugh
Lemkey, Frankin David
Levin, S Benedict
Levine, Leon
Maizell, Robert Edward
Nath, Dilip K
Nielsen, John Merle
Olster, Elliot Frederick
Parthasarathi, Manavasi Narasimhan
Potter, Donald Irwin
Powers, Joseph
Reffner, John A
Rie, John E
Roberts, George P
Rosner, Daniel E(dwin)
Scola, Daniel Anthony
Shaw, Montgomery Throop
Shichman, D(aniel)
Sprague, G(eorge) Sidney
Strife, James Richard
Sutton, W(illard) H(olmes)
Thompson, Earl Ryan
Traskos, Richard Thomas
Wheeler, Edward Stubbs

Materials Science Engineering (cont)

DELAWARE

Akeley, David Francis
Bair, Thomas Irvin
Bierlein, John David
Bindloss, William
Birkenhauer, Robert Joseph
Black, Carl (Ellsworth)
Böer, Karl Wolfgang
Brown, Stewart Cliff
Cessna, Lawrence C, Jr
Chesson, Eugene, Jr
Chou, Tsu-Wei
Chowdhry, Uma
Cooper, Terence Alfred
Dawson, Robert Louis
Dexter, Stephen C
English, Alan Dale
Faller, James George
Ferriss, Donald P
Firment, Lawrence Edward
Flattery, David Kevin
Fleming, Richard Allan
Freed, Robert Lloyd
Funck, Dennis Light
Gibbs, Thomas W(atson)
Goodman, Albert
Gorrafa, Adly Abdel-Moniem
Grant, David Evans
Greenfield, Irwin G
Hall, Ian Wavell
Hardie, William George
Hardy, Henry Benjamin, Jr
Harrell, Jerald Rice
Hartmann, Hans S(iegfried)
Hershkowitz, Robert L
Holob, Gary M
Hsiao, Benjamin S
Hsu, William Yang-Hsing
Hurford, Thomas Rowland
Immediata, Tony Michael
Isakoff, Sheldon Erwin
Jolley, John Eric
Katz, Manfred
Keidel, Frederick Andrew
Kramer, John J(acob)
Kwolek, Stephanie Louise
Landoll, Leo Michael
Lewis, John Raymond
Lin, Kuang-Farn
Logothetis, Anestis Leonidas
Loken, Halvar Young
McCullough, Roy Lynn
McNeely, James Braden
McTigue, Frank Henry
Markiewitz, Kenneth Helmut
Meakin, John David
Nelb, Gary William
Onn, David Goodwin
Ostmann, Bernard George
Paulson, Charles Maxwell, Jr
Pontrelli, Gene J
Pretka, John E
Ransom, J(ohn) T(hompson)
Riewald, Paul Gordon
Sands, Seymour
Schaefgen, John Raymond
Schenker, Henry Hans
Scherer, George Walter
Schiroky, Gerhard H
Schodt, Kathleen Patricia
Schultz, Jerold M
Scofield, Dillon Foster
Shealy, Otis Lester
Shoaf, Charles Jefferson
Staley, Ralph Horton
Tabb, David Leo
Tomic, Ernst Alois
Tuites, Donald Edgar
Urquhart, Andrew Willard
Van Trump, James Edmond
Vassallo, Donald Arthur
Vriesen, Calvin W
Wagner, Richard Lloyd
Weeks, Gregory Paul
Whitehouse, Bruce Alan
Williams, Richard Anderson

DISTRICT OF COLUMBIA

Aggarwal, Ishwar D
Alic, John A
Anderson, Gordon Wood
Ayers, Jack Duane
Belsheim, Robert Oscar
Bikales, Norbert M
Campbell, Francis James
Crouch, Roger Keith
Darby, Joseph B(ranch), Jr
Dragoo, Alan Lewis
Duesbery, Michael Serge
Fanconi, Bruno Mario
Fisher, Farley
Forester, Donald Wayne
Gottschall, Robert James
Grabowski, Kenneth S
Groves, Donald George
Haworth, William Lancelot
Jawed, Inam
Kaznoff, Alexis I(van)
Kramer, Bruce Michael
Larsen-Basse, Jorn

Lashmore, David S
McNeil, Michael Brewer
Moon, Deug Woon
O'Grady, William Edward
Pande, Chandra Shekhar
Radcliffe, S Victor
Reynik, Robert John
Schioler, Liselotte Jensen
Schnur, Joel Martin
Smidt, Fred August, Jr
Smith, Leslie E
Sprague, James Alan
Swisher, James H(owe)
Taggart, G(eorge) Bruce
Van Echo, Andrew
Wilsey, Neal David
Wolf, Stanley Myron
Zwilsky, Klaus M(ax)

FLORIDA

Abbaschian, G J
Ahmann, Donald H(enry)
Ambrose, John Russell
Anghaie, Samim
Anusavice, Kenneth John
Batich, Christopher David
Beatty, Charles Lee
Biver, Carl John, Jr
Bonifazi, Stephen
Burns, Michael J
Callan, Edwin Joseph
Catz, Jerome
Chu, Ting Li
Clark, David Edward
Coke, C(hauncey) Eugene
Dannenberg, E(li) M
Davidovits, Joseph
Davidson, Mark Rogers
Drucker, Daniel Charles
Dunbar, Richard Alan
Ebrahimi, Fereshteh
Feit, Eugene David
Fullman, R(obert) L(ouis)
Goldberg, David C
Goldberg, Eugene P
Gould, Robert William
Hartley, Craig Sheridan
Hecht, Ralph J
Holloway, Paul Howard
Hopkins, George C
Huffman, Jacob Brainard
Hummel, Rolf Erich
Jones, Kevin Scott
Kiewit, David Arnold
Kumar, Ganesh N
Linz, Arthur
McAlpine, Kenneth Donald
McCracken, Walter John
McKoy, James Benjamin, Jr
McSwain, Richard Horace
Maleady, N(oel) R(ichard)
Mandelkern, Leo
Mason, D(onald) R(omagne)
Mataga, Peter Andrew
Mizell, Louis Richard
Mohammed, M Hamdi A
Moore, Joseph B
Moudgil, Brij Mohan
O'Leary, Kevin Joseph
Pepinsky, Raymond
Postelnek, William
Potts, James Edward
Raich, William Judd
Rees, William Smith, Jr
Reichard, Ronnal Paul
Rice, Stephen Landon
Ridgway, James Stratman
Rikvold, Per Arne
Russell, Charles Addison
Russell, James
Saunders, James Henry
Schwier, Chris Edward
Seelbach, Charles William
Seth, Brij B
Shane, Robert S
Smith, William Fortune
Spears, Richard Kent
Springborn, Robert Carl
Steadman, Thomas Ree
Sun, Chang-Tsan
Sundaram, Swaminatha
Swain, Geoffrey W
Swartz, William Edward, Jr
Ting, Robert Yen-Ying
Weddleton, Richard Francis
Whitney, Ellsworth Dow
Wille, Luc Theo
Willwerth, Lawrence James
Wiseman, Robert S
Yu, A Tobey
Zollo, Ronald Francis

GEORGIA

Antolovich, Stephen D
Atluri, Satya N
Bacon, Roger
Burch, George Nelson Blair
Chandan, Harish Chandra
Colton, Jonathan Stuart
Coyne, Edward James, Jr
D'Annessa, A(nthony) T(homas)
Dickert, Herman A(lonzo)
Fitzgerald, J(ohn) Edmund
Kalish, David

Kaufman, Stanley
Korda, Edward J(ohn)
Krochmal, Jerome J(acob)
Logan, Kathryn Vance
Lundberg, John L(auren)
McCarthy, Neil Justin, Jr
Meyers, Carolyn Winstead
Morman, Michael T
Rice, James Thomas
Ringel, Steven Adam
Rodrigue, George Pierre
Rodriguez, Augusto
Roobol, Norman R
Samoilov, Sergey Michael
Samuels, Robert Joel
Sanders, T H, Jr
Schulz, David Arthur
Simmons, George Allen
Soora, Siva Shunmugam
Spauschus, Hans O
Spetnagel, Theodore John
Stratton, Robert Alan
Tooke, William Raymond, Jr
Weber, Robert Emil
Wellons, Jesse Davis, III
Yen, William Maoshung
Yeske, Ronald A

HAWAII

Brantley, Lee Reed
Daniel, Thomas Henry
Hihara, Lloyd Hiromi

IDAHO

Brown, Harry Lester
Buescher, Brent J
Froes, Francis Herbert (Sam)
Korth, Gary E
Miller, Richard Lloyd
Place, Thomas Alan
Telschow, Kenneth Louis
Walters, Leon C
Weyand, John David

ILLINOIS

Albertson, Clarence E
Albrecht, Edward Daniel
Aldred, Anthony T
Allen, Charles W(illard)
Alwitt, Robert S(amuel)
Baitinger, William F, Jr
Barnett, Scott A
Bazant, Zdenek P(avel)
Berkelhamer, Louis H(arry)
Bernstein, I(rving) Melvin
Bradley, Steven Arthur
Bratschun, William R(udolph)
Bridgeford, Douglas Joseph
Brittain, John O(liver)
Broutman, Lawrence Jay
Buchanan, Relva Chester
Burnham, Robert Danner
Carr, Stephen Howard
Chan, Sai-Kit
Chang, Franklin Shih Chuan
Chang, Shu-Pei
Chiou, Wen-An
Chipman, Gary Russell
Chung, Yip-Wah
Cohen, J(erome) B(ernard)
Connolly, James D
Cook, Harry Edgar
Crawford, Roy Kent
Crowley, Michael Summers
Damusis, Adolfas
Daniel, Isaac M
Dantzig, Jonathan A
Danyluk, Steven
Dutta, Pulak
Eades, John Alwyn
Ebert, Lawrence Burton
Edelstein, Warren Stanley
Erck, Robert Alan
Erwin, Lewis
Falk, John Carl
Fenske, George R
Fine, Morris Eugene
Firestone, Ross Francis
Fraenkel, Stephen Joseph
Frey, Donald N(elson)
Gaylord, Richard J
Geil, Phillip H
Gembicki, Stanley Arthur
Granick, Steve
Greener, Evan H
Griffith, James H
Gruber, George E, Jr
Guggenheim, Stephen
Hardwidge, Edward Albert
Hass, Alvin A
Hersh, Herbert N
Hilliard, John E
Hofer, Kenneth Emil
Irvin, Howard H
Isbister, Roger John
Johnson, D(avid) Lynn
Joung, John Jongin
Kamm, Gilbert G(eorge)
Kaufmann, Elton Neil
Kelman, L(eRoy) R
Kim, Kyekyoon K(evin)
Knox, Jack Rowles
Kramer, John Michael
Lam, Nghi Quoc

Lautenschlager, Eugene Paul
Lauterbach, Richard Thomas
Linde, Ronald K(eith)
Luplow, Wayne Charles
McKee, Keith Earl
Macrander, Albert Tiemen
Marks, Laurence D
Mason, Thomas Oliver
Meshii, Masahiro
Metz, Florence Irene
Mirabella, Francis Michael, Jr
Mura, Toshio
Muzyczko, Thaddeus Marion
Neuhalfen, Andrew J
Nordine, Paul Clemens
Paschke, Edward Ernest
Pincus, Irving
Poeppel, Roger Brian
Primak, William L
Raccah, Paul M(ordecai)
Raheel, Mastura
Rehn, Lynn Eduard
Reichard, Grant Wesley
Rey, Charles Albert
Rostoker, William
Rothman, Steven J
Routbort, Jules Lazar
Rowe, Charles David
Rowland, Theodore Justin
Rusinko, Frank, Jr
Sandrik, James Leslie
Schiffman, Robert A
Seidman, David N(athaniel)
Shaneyfelt, Duane L
Shenoy, Gopal K
Shroff, Ramesh N
Siegel, Richard W(hite)
Sinha, Shome Nath
Socie, Darrell Frederick
Sproul, William Dallas
Stavrolakis, J(ames) A(lexander)
Stowell, James Kent
Stubbins, James Frederick
Su, Cheh-Jen
Tsai, Boh Chang
Tucker, Charles L
Tuomi, Donald
Voorhees, Peter Willis
Weber, J K Richard
Weeks, Richard William
Weertman, Julia Randall
Wert, Charles Allen
Wessels, Bruce Warren
Wiedersich, H(artmut)
Wimber, R(ay) Ted
Wirtz, Gerald Paul
Wright, Maurice Arthur

INDIANA

Bellina, Joseph James, Jr
Benford, Arthur E
Bhattacharya, Debanshu
Booe, J(ames) M(arvin)
Chang, Chien-Wu
Chernoff, Donald Alan
Dayananda, Mysore Ananthamurthy
Diamond, Sidney
Dolch, William Lee
DuBroff, William
Eckelman, Carl A
Fritzlen, Glenn A
Gallucci, Robert Russell
Grace, Richard E(dward)
Gresham, Robert Marion
Jacobs, Martin Irwin
Kazem, Sayyed M
Koller, Charles Richard
Kuczynski, George Czeslaw
Kuo, Charles C Y
Liberti, Frank Nunzio
Liedl, Gerald L(eRoy)
McAleece, Donald John
Orr, William H(arold)
Peppas, Nikolaos Athanassiou
Phillips, Ralph W
Price, Howard Charles
Ranade, Madhukar G
Sato, Hiroshi
Schalliol, Willis Lee
Solomon, Alvin Arnold
Taylor, Raymond Ellory
Vest, Robert W(ilson)
Winslow, Douglas Nathaniel

IOWA

Allen, Susan Davis
Beaudry, Bernard Joseph
Buck, Otto
Demirel, T(urgut)
Gaertner, Richard F(rancis)
Greer, Raymond T(homas)
Hauenstein, Jack David
Lakes, Roderic Stephen
Lane, Orris John, Jr
Larsen, William L(awrence)
McGee, Thomas Donald
Moriarty, John Lawrence, Jr
Nariboli, Gundo A
Nixon, Wilfrid Austin
Park, Joon B
Rethwisch, David Gerard
Rocklin, Isadore J
Rosauer, Elmer Augustine
Schaefer, Joseph Albert

Schmidt, Frederick Allen
Spitzig, William Andrew
Thompson, Robert Bruce
Trivedi, Rohit K
Wechsler, Monroe S(tanley)

KANSAS
Best, Cecil H(amilton)
Cooke, Francis W
Darwin, David
Frohmberg, Richard P
Grisafe, David Anthony
Grosskreutz, Joseph Charles
Huang, Chi-Lung Dominic
Rose, Kenneth E(ugene)
Scherrer, Joseph Henry

KENTUCKY
Coe, Gordon Randolph
DeLong, Lance Eric
Giammara, Berverly L Turner Sites
Gillis, Peter Paul
Goren, Alan Charles
Huffman, Gerald P
Johnson, Alan Arthur
Kuhn, William E(rik)
Lipscomb, Nathan Thornton
Lu, Wei-yang
McGinness, James Donald
Mellinger, Gary Andreas
Mihelich, John L
Reucroft, Philip J
Sarma, Atul C
Viswanadham, Ramamurthy K

LOUISIANA
Beckman, Joseph Alfred
Bedell, Louis Robert
Bundy, Kirk Jon
Calamari, Timothy A, Jr
Choong, Elvin T
Diwan, Ravinder Mohan
Eaton, Harvill Carlton
Fogg, Peter John
Haefner, A(lbert) J(ohn)
Kyame, George John
Laurent, Sebastian Marc
Li, Hsueh Ming
McMillin, Charles W
Naidu, Seetala V
Penton, Harold Roy, Jr
Reid, John David
Rubinstein, Asher A
Somsen, Roger Alan
Vashishta, Priya Darshan
Vigo, Tyrone Lawrence
Weiss, Louis Charles
Wickson, Edward James
Zentner, Thomas Glenn

MAINE
Shottafer, James Edward
Thompson, Edward Valentine

MARYLAND
Ahearn, John Stephen
Armstrong, R(onald) W(illiam)
Bennett, Lawrence Herman
Benson, Richard C
Berger, Harold
Bersch, Charles Frank
Block, Ira
Bowen, Rafael Lee
Brauer, Gerhard Max
Butler, Thomas W(esley)
Camponeschi, Eugene Thomas, Jr
Carter, John Paul
Carver, Gary Paul
Cezairliyan, Ared
Chang, Shu-Sing
Charles, Harry Krewson, Jr
Chien, Chia-Ling
Chimenti, Dale Everett
Chuang, Tze-jer
Clark, Alan Fred
Corak, William Sydney
Coriell, Sam Ray
Cremens, Walter Samuel
Davis, George Thomas
Davis, Guy Donald
Dolhert, Leonard Edward
Downing, Robert Gregory
Eiss, Abraham L(ouis)
Feldman, Albert
Ferington, Thomas Edwin
Fishman, Steven Gerald
Franke, Gene Louis
Frederick, William George DeMott
Frederikse, Hans Pieter Roetert
Frohnsdorff, Geoffrey James Carl
Fulmer, Glenn Elton
Gammell, Paul M
Ghoshtagore, Rathindra Nath
Gillich, William John
Goktepe, Omer Faruk
Green, John Arthur Savage
Green, Robert E(dward), Jr
Greenfeld, Sidney H(oward)
Greenhouse, Harold Mitchell
Haffner, Richard William
Han, Charles Chih-Chao
Harper, C(harles) A(rthur)
Hasson, Dennis Francis
Hoffman, John D

Hoguet, Robert Gerard
Horowitz, Emanuel
Hovmand, Svend
Hsu, Stephen M
Huffington, Norris J(ackson), Jr
Ibrahim, A Mahammad
Kanarowski, S(tanley) M(artin)
Kirkendall, Thomas Dodge
Kopanski, Joseph J
Kreider, Kenneth Gruber
Krumbein, Simeon Joseph
Kumar, K Sharvan
Kuriyama, Masao
Levin, Roger L(ee)
Loss, Frank J
Mabie, Curtis Parsons
McElroy, Wilbur Renfrew
McFadden, Geoffrey Bey
Machlin, Irving
Malghan, Subhaschandra Gangappa
Marcinkowski, M(arion) J(ohn)
Marton, Joseph
Masters, Larry William
Merz, Kenneth M(alcolm), Jr
Metze, George M
Miller, Gerald R
Moorjani, Kishin
Mordfin, Leonard
Murphey, Wayne K
Nagle, Dennis Charles
Nelson, Elton Glen
Neugroschl, Daniel
Newbury, Dale Elwood
Ostrofsky, Bernard
Parker, Frederick John
Paul, Dilip Kumar
Peterlin, Anton
Poehler, Theodore O
Poulose, Pathickal K
Promisel, Nathan E
Quadir, Tariq
Rasberry, Stanley Dexter
Raskin, Betty Lou
Reno, Robert Charles
Revesz, Akos George
Rice, James K
Ricker, Richard Edmond
Ritter, Joseph John
Rosenberg, Arnold Morry
Rosenstein, Alan Herbert
Ruby, Stanley
Schaefer, Robert J
Schruben, John H
Schwartz, Lyle H
Schwartz, M(urray) A(rthur)
Scialdone, John Joseph
Sharp, William Broom Alexander
Skalny, Jan Peter
Smith, Jack Carlton
Smith, John Henry
Spivak, Steven Mark
Steiner, Bruce
Stickley, C(arlisle) Martin
Stromberg, Robert Remson
Swerdlow, Max
Thomson, Robb M(ilton)
Thurber, Willis Robert
Van Houten, Robert
Venkatesan, Thirumalai
Wacker, George Adolf
Wagner, Herman Leon
Waltrup, Paul John
Wang, Francis Wei-Yu
Warshaw, Israel
Warshaw, Stanley I(rving)
Westwood, Albert Ronald Clifton
Wolock, Irvin
Yeh, Kwan-Nan
Yolken, Howard Thomas

MASSACHUSETTS
Abkowitz, Stanley
Ackerman, Jerome Leonard
Adler, Ralph Peter Isaac
Alexander, Michael Norman
Allen, Samuel Miller
Ammlung, Richard Lee
Antal, John Joseph
Apelian, Diran
Arganbright, Donald G
Argon, Ali Suphi
Argy, Dimitri
Aronin, Lewis Richard
Averbach, B(enjamin) L(ewis)
Baboian, Robert
Bates, Carl H
Berera, Geetha Poonacha
Bernal G, Enrique
Bever, Michael B(erliner)
Biederman, Ronald R
Blum, Seymour L
Blumstein, Rita Blattberg
Bonner, Francis Joseph
Bougas, James Andrew
Brack, Karl
Buono, John Arthur
Chen, Chuan Ju
Chin, David
Chu, Nori Yaw-Chyuan
Chung, Frank H
Churchill, Geoffrey Barker
Clopper, Herschel
Cohen, Merrill
Cohen, Morris

Collins, Aliki Karipidou
Cooke, Richard A
Cooper, W(illiam) E(ugene)
Cosgrove, James Francis
Cotten, George Richard
Dale, William
Davé, Raju S
David, Donald J
Deckert, Cheryl A
Delagi, Richard Gregory
Desper, Clyde Richard
Dietz, Albert (George Henry)
Dionne, Gerald Francis
Ditchek, Brian Michael
Druy, Mark Arnold
Eagar, Thomas W
Eschenroeder, Alan Quade
Evans, Michael Douglas
Fan, John C C
Fettes, Edward Mackay
Flemings, Merton Corson
Folweiler, Robert Cooper
Fox, Daniel Wayne
French, David N(ichols)
Frushour, Bruce George
Garrett, Paul Daniel
Gatos, H(arry) C(onstantine)
Giessen, Bill C(ormann)
Godrick, Joseph Adam
Goela, Jitendra Singh
Goettler, Lloyd Arnold
Golub, Samuel J(oseph)
Gopikanth, M L
Grant, Nicholas J(ohn)
Haas, Howard Clyde
Hagnauer, Gary Lee
Handy, Carleton Thomas
Harman, T(heodore) C(arter)
Harrison, Ralph Joseph
Haugsjaa, Paul O
Hiatt, Norman Arthur
Hill, Loren Wallace
Hoadley, Robert Bruce
Hopkins, George Robert
Hynes, Thomas Vincent
Ika, Prasad Venkata
Illinger, Joyce Lefever
Iseler, Gerald William
Israel, Stanley C
Isserow, Saul
Jalan, Vinod Motilal
Johnson, Walter Roland
Jones, Alan A
Kachanov, Mark L
Kalonji, Gretchen
Kantor, Simon William
Karasz, Frank Erwin
Karulkar, Pramod C
Kelsey, Ronald A(lbert)
Kennedy, Edward Francis
Khattak, Chandra Prakash
Klein, Richard Morris
Klinedinst, Keith Allen
Kolodziejski, Leslie Ann
Kula, Eric Bertil
Kumar, Kaplesh
Latanision, Ronald Michael
Lauer, Robert B
Lavin, Edward
Lee, D(on) William
Lee, Kang In
Lees, Wayne Lowry
Lement, Bernard S
Levitt, Albert P
Liau, Zong-Long
Lin, Alice Lee Lan
Linden, Kurt Joseph
Litant, Irving
Livingston, James Duane
Lu, Grant
McCauley, James Weymann
McGarry, Frederick J(erome)
MacGregor, C(harles) W(inters)
MacVicar, Margaret Love Agnes
Maher, Galeb Hamid
Malozemoff, Alexis P
Marinaccio, Paul J
Martin, Richard Hadley, Jr
Masi, James Vincent
Matsuda, Seigo
Meyer, Robert Bruce
Milgrom, Jack
Minot, Michael Jay
Mont, George Edward
Moore, Robert Edmund
Morbey, Graham Kenneth
Murayama, Takayuki
Natansohn, Samuel
Nemphos, Speros Peter
Normandin, Raymond O
Nowak, Welville B(erenson)
O'Connell, Richard John
Ogilvie, Robert Edward
O'Rell, Dennis Dee
Oster, Bernd
Owen, Walter S
Papernik, Lazar
Park, B J
Pasto, Arvid Eric
Penchina, Claude Michel
Peters, Edward Tehle
Platt, Milton M
Pope, Mary E
Powers, Donald Howard, Jr

Quackenbush, Carr Lane W
Quynn, Richard Grayson
Reif, L Rafael
Rice, James R
Ritter, John Earl, Jr
Rose, Robert M(ichael)
Russell, Kenneth Calvin
Sadoway, Donald Robert
Sagalyn, Paul Leon
Sahatjian, Ronald Alexander
Sanger, Gregory Marshall
Sastri, Suri A
Sawan, Samuel Paul
Schloemann, Ernst
Schoen, Frederick J
Schroter, Stanislaw Gustaw
Schuler, Alan Norman
Schwartz, Thomas Alan
Schwensfeir, Robert James, Jr
Servi, I(talo) S(olomon)
Shapiro, Edward K Edik
Sharp, A(rnold) G(ideon)
Shaughnessy, Thomas Patrick
Shu, Larry Steven
Silver, Frank Morris
Sioui, Richard Henry
Smith, Donald Ross
Sonnichsen, Harold Marvin
Spaepen, Frans
Staker, Michael Ray
Strauss, Alan Jay
Thieme, Cornelis Leo Hans
Thomas, Edwin Lorimer
Thun, Rudolf Eduard
Tsaur, Bor-Yeu
Tuler, Floyd Robert
Tuller, Harry Louis
Tustison, Randal Wayne
Tzeng, Wen-Shian Vincent
Udipi, Kishore
Uhlig, Herbert H(enry)
Vander Sande, John Bruce
Viechnicki, Dennis J
Wald, Fritz Veit
Wang, Christine A
Weiner, Louis I
Wentworth, Stanley Earl
Williams, David John
Williams, John Russell
Wineman, Robert Judson
Winter, Horst Henning
Yau, Chiou Ching
Zotos, John

MICHIGAN
Aaron, Howard Berton
Arends, Charles Bradford
Baxter, William John
Bierlein, John Carl
Bigelow, Wilbur Charles
Bilello, John Charles
Brown, Steven Michael
Camp, David Thomas
Clark, Peter David
Clarke, Roy
Corey, Clark L(awrence)
Cox, Mary E
Dearlove, Thomas John
Dreyfuss, Patricia
Eldis, George Thomas
Ellis, Thomas Stephen
Filisko, Frank Edward
Fisher, Galen Bruce
Flaim, Thomas Alfred
Foister, Robert Thomas
Fortuna, Edward Michael, Jr
Francis, Ray Llewellyn
Fuerst, Carlton Dwight
Gallagher, James A
Gardon, John Leslie
Garrett, David L, (Jr)
Gjostein, Norman A
Greene, Bruce Edgar
Hart, Robert G
Harwood, Julius J
Hayes, Edward J(ames)
Heckel, Richard W(ayne)
Heminghous, William Wayne, Sr
Heremans, Joseph P
Hillegas, William Joseph
Hohnke, Dieter Karl
Hosford, William Fuller, Jr
Hucke, Edward E
Im, Jang Hi
Jain, Kailash Chandra
Jarvis, Lactance Aubrey
Jones, Frank Norton
Jones, Frederick Goodwin
Keinath, Steven Ernest
Kenig, Marvin Jerry
Kipouros, Georges John
Klempner, Daniel
Kresta, Jiri Erik
Kulkarni, Anand K
Kullgren, Thomas Edward
Langley, Neal Roger
Le Beau, Stephen Edward
Little, Robert E(ugene)
Logothetis, Eleftherios Miltiadis
Lund, Anders Edward
McKeever, L Dennis
Mansfield, Marc L
Meier, Dale Joseph
Miller, Carl Elmer

Materials Science Engineering (cont)

Miller, Robert Llewellyn
Montgomery, Donald Joseph
Morgan, Roger John
Morman, Kenneth N
Mukherjee, Kalinath
Narayanaswamy, Onbathiveli S
Newman, Seymour
O'Brien, William Joseph
Pehlke, Robert Donald
Reynard, Kennard Anthony
Rhee, Seong Kwan
Robertson, Richard Earl
Rol, Pieter Klaas
Rosenberg, Richard Carl
Rowe, Leonard C
Rowland, Sattley Clark
Russell, William Charles
Schrenk, Walter John
Schubring, Norman W(illiam)
Sell, Jeffrey Alan
Sliker, Alan
Smith, Darrell Wayne
Smith, John Robert
Soh, Sung Kuk
Srolovitz, David Joseph
Subramanian, K N
Suchsland, Otto
Summitt, W(illiam) Robert
Sun, Bernard Ching-Huey
Tillitson, Edward Walter
Van Vlack, Lawrence H(all)
Vogel, F(erdinand), L(incoln)
Vukasovich, Mark Samuel
Wang, Simon S
Wang, Yar-Ming
Warpehoski, Martha Anna
Whalen, Thomas J(ohn)
Williams, Fredrick David
Williams, Gwynne
Wyzgoski, Michael Gary
Yee, Albert Fan
Yeh, Gregory Soh-Yu

MINNESOTA
Bryan, Thomas T
Burkstrand, James Michael
Chamberlain, Craig Stanley
Chelikowsky, James Robert
David, Moses M
Diedrich, James Loren
Esmay, Donald Levern
Fick, Herbert John
French, William George
George, Peter Kurt
Gerberich, William Warren
Gerlach, Robert Louis
Goken, Garold Lee
Goldenberg, Barbara Lou
Gruber, Carl L(awrence)
Hann, Robert A
Hendricks, James Owen
Hill, Brian Kellogg
Huffine, Coy L(ee)
Jacobs, Carl Henry
Johnson, Lennart Ingemar
Johnson, Robert F
Katz, William
Koepke, Barry George
Krueger, Walter L(awrence)
Kuder, Robert Clarence
Langager, Bruce Allen
McHenry, Kelly David
Macosko, Christopher Ward
Miller, Wilmer Glenn
Mularie, William Mack
Nam, Sehyun
Olyphant, Murray, Jr
Oriani, Richard Anthony
Punderson, John Oliver
Rao, Prakash
Ross, Stuart Thom
Sandberg, Carl Lorens
Shores, David Arthur
Stokes, Robert James
Thalacker, Victor Paul
Thornton, Arnold William
Tirrell, Matthew Vincent
Tran, Nang Tri
Verma, Deepak Kumar
Wright, Charles Dean

MISSISSIPPI
Barnes, Hoyt Michael
Johnson, L(awrence) D(avid)
Marion, Robert Howard
Mather, Bryant
Mather, Katharine Kniskern
Taylor, Fred William

MISSOURI
Baldwin, James W(arren), Jr
Burke, James Joseph
Carpenter, James Franklin
Carter, Robert L(eroy)
Creighton, Donald L
Darby, Joseph Raymond
Dharani, Lokeswarappa R
Farmer, John William
Gadberry, Howard M(ilton)
Gibbons, Patrick C(handler)

Goran, Robert Charles
Greenley, Robert Z
Hale, Edward Boyd
Harmon, Robert Wayne
Hill, Dale Eugene
Indeck, Ronald S
Kenyon, Allen Stewart
Kurz, James Eckhardt
Leopold, Daniel J
Lind, Arthur Charles
McGinnes, Edgar Allan, Jr
Millich, Frank
Monfore, Gervaise Edwin
Mori, Erik Jun
Niesse, John Edgar
Ownby, P(aul) Darrell
Ross, Frederick Keith
Sastry, Shankara M L
Sears, J Kern
Sonnino, Carlo Benvenuto
Sparlin, Don Merle
Switzer, Jay Alan
Walsh, Robert Jerome
Wehrmann, Ralph F(rederick)
Wohl, Martin H
Woodbrey, James C
Yasuda, Hirotsugu

MONTANA
Bradley, Daniel Joseph
Koch, Peter

NEBRASKA
Phelps, George Clayton
Pierce, Donald N(orman)
Rohde, Suzanne Louise
Tunnicliff, David George
Verma, Shashi Bhushan

NEVADA
Chandra, Dhanesh
Epps, Jon Albert
Skaggs, Robert L
Snaper, Alvin Allyn

NEW HAMPSHIRE
Backofen, Walter A
Comerford, Matthias F(rancis)
Cox, Gordon F N
Creagh-Deyter, Linda T
Ge, Weikun
Hill, John Ledyard
Queneau, Paul E(tienne)
Rice, Dale Wilson
Samaroo, Winston R
Schulson, Erland Maxwell
Webb, Richard Lansing

NEW JERSEY
Abrahams, M(arvin) S(idney)
Anandan, Munisamy
Anthony, Philip John
Arendt, Volker Dietrich
Ariemma, Sidney
Bagley, Brian G
Balaguru, Perumalsamy N
Ballato, Arthur
Bardoliwalla, Dinshaw Framroze
Bassett, Alton Herman
Batlogg, Bertram
Battisti, Angelo James
Baughman, Ray Henry
Baum, Gerald A(llan)
Berardinelli, Frank Michael
Berenbaum, Morris Benjamin
Berkman, Samuel
Berkowitz, Barry Jay
Besso, Michael M
Bevk, Joze
Bhandarkar, Suhas D
Biskeborn, Merle Chester
Biswas, Dipak R
Blanck, Andrew R(ichard)
Blank, Stuart Lawrence
Blyler, Lee Landis, Jr
Borysko, Emil
Bovey, Frank Alden
Bowden, Murrae John Stanley
Bowen, J Hartley, Jr
Bradway, Keith E
Brenner, Douglas
Broer, Matthijs Meno
Brown, Walter Lyons
Burrus, Charles Andrew, Jr
Buteau, L(eon) J
Cadet, Gardy
Canter, Nathan H
Cassola, Charles A(lfred)
Cecchi, Joseph Leonard
Celler, George K
Chan, Maureen Gillen
Chand, Naresh
Chang, Tao-Yuan
Chang, T(ao)-Y(uan)
Chatterjee, Pronoy Kumar
Chen, Catherine S H
Chen, Shuhchung Steve
Chen, William Kwo-Wei
Chenicek, Albert George
Chidsey, Christopher E
Chin, Gilbert Yukyu
Chlanda, Frederick P
Cho, Alfred Y
Chu, Sung Nee George

Chynoweth, Alan Gerald
Cohen, Abraham Bernard
Conger, Robert Perrigo
Connolly, John Charles
Couchman, Peter Robert
Davis, Lance A(lan)
DeBona, Bruce Todd
Derkits, Gustav
Derucher, Kenneth Noel
Dittman, Frank W(illard)
Dodabalapur, Ananth
Donovan, Richard C
Drelich, Arthur (Herbert)
Duclos, Steven J
Dyer, John
Eachus, Spencer William
Ettenberg, Michael
Fahrenholtz, Susan Roseno
Farrauto, Robert Joseph
Feldman, Leonard Cecil
Finch, Rogers B(urton)
Fischer, Traugott Erwin
Forster, Eric Otto
Frankenthal, Robert Peter
Gaylord, Norman Grant
Gibson, J Murray
Gillham, John K
Glass, Alastair Malcolm
Graedel, Thomas Eldon
Green, Martin Laurence
Griskey, R(ichard) G(eorge)
Gross, Leonard
Gualtieri, Devlin Michael
Hale, Warren Frederick
Hannon, Martin J
Hanson, James Edward
Harpell, Gary Allan
Hawkins, Walter Lincoln
Hobstetter, John Norman
Hudson, Steven David
Hughes, O Richard
Huneke, James Thomas
Iezzi, Robert Aldo
Iglesia, Enrique
Isaacson, Robert B
Jackson, Kenneth Arthur
Jaffe, Michael
Jansen, Frank
Jin, Sungho
Jirgensons, Arnold
Johnson, David W, Jr
Johnston, John Eric
Johnston, Wilbur Dexter, Jr
Kalnin, Ilmar L
Karol, Frederick J
Kaufman, Herman S
Kear, Bernard Henry
Keramidas, Vassilis George
Khan, Saad Akhtar
Kimerling, Lionel Cooper
Kirshenbaum, Gerald Steven
Klein, Imrich
Knausenberger, Wulf H
Koul, Maharaj Kishen
Kramer, Carolyn Margaret
Kresge, Edward Nathan
Kunzler, John Eugene
Kveglis, Albert Andrew
Lamelas, Francisco Javier
Landrock, Arthur Harold
Larkin, William Albert
Lee, Lester Tsung-Cheng
Lemaire, Paul J
Leupold, Herbert August
Levi, David Winterton
Levy, Alan
Liang, Keng-San
Liebermann, Howard Horst
Ling, Hung Chi
Loh, Roland Ru-loong
Louis, Kwok Toy
Lowell, A(rthur) I(rwin)
Lundberg, Robert Dean
Lunn, Anthony Crowther
Lunt, Harry Edward
Maa, Jer-Shen
MacChesney, John Burnette
McGregor, Walter
MacRae, Alfred Urquhart
Mandel, Andrea Sue
Manders, Peter William
Mihalisin, John Raymond
Miller, Bernard
Miskel, John Joseph, Jr
Mitchell, Olga Mary Mracek
Mogab, Cyril Joseph
Morris, Robert Craig
Moustakas, Theodore D
Muldrow, Charles Norment, Jr
Nakahara, Shohei
Nardone, John
Nassau, Kurt
Novick, David Theodore
Ohring, Milton
Ong, Chung-Jian Jerry
Ong, Nai-Phuan
Pacansky, Thomas John
Pae, K(ook) D(ong)
Panish, Morton B
Patel, Gordhanbhai Nathalal
Paul, George T(ompkins)
Pearsall, Thomas Perine
Pebly, Harry E
Phillips, Julia M

Phillips, William
Pinch, Harry Louis
Plymale, Charles E
Poate, John Milo
Polhemus, Neil W
Polk, Conrad Joseph
Prevorsek, Dusan Ciril
Quarles, Richard Wingfield
Rabinovich, Eliezer M
Ramanan, V R V
Ravindra, Nuggehalli Muthanna
Raybon, Gregory
Ray-Chaudhuri, Dilip K
Register, Richard Alan
Reich, Leo
Reich, Murray H
Rorabaugh, Donald T
Royce, Barrie Saunders Hart
Rutherford, John L(oftus)
Sacks, William
Saferstein, Lowell G
Sarkar, Kamalaksha
Saxon, Robert
Scharfstein, Lawrence Robert
Schlabach, T(om) D(aniel)
Schoenfeld, Theodore Mark
Schonfeld, Edward
Schwab, Frederick Charles
Segelken, John Maurice
Shallcross, Frank V(an Loon)
Shanefield, Daniel J
Shelton, James Churchill
Sheppard, Keith George
Sinclair, James Douglas
Sivco, Deborah L
Smith, Carl Hofland
Smith, Donald Eugene
Snowman, Alfred
Song, Won-Ryul
Sorenson, Wayne Richard
Sprague, Basil Sheldon
Subramanyam, Dilip Kumar
Suchow, Lawrence
Suhir, Ephraim
Takahashi, Akio
Tan, Victor
Tatyrek, Alfred Frank
Tauber, Arthur
Tenzer, Rudolf Kurt
Terenzi, Joseph F
Thomas, Stanislaus S(tephen)
Thompson, Darrell Robert
Thompson, Larry Flack
Tischler, Oscar
Tonelli, Alan Edward
Townsend, Palmer W
Van Uitert, LeGrand G(erard)
Ver Strate, Gary William
Von Neida, Allyn Robert
Wagner, Richard S(iegfried)
Wang, Chih Chun
Wang, Tsuey Tang
Wasserman, David
Weil, Rolf
Weiss, Lawrence H(eisler)
Weissmann, Gerd Friedrich Horst
Wernick, Jack H(arry)
Westerdahl, Carolyn Ann Lovejoy
White, Lawrence Keith
Wong, Yiu-Huen
Woodward, Ted K
Xie, Ya-Hong
Yadvish, Robert D
Yim, W(oongsoon) Michael
Yoon, Euijoon
Zipfel, Christie Lewis

NEW MEXICO
Adler, Richard John
Arnold, Charles, Jr
Assink, Roger Alyn
Beauchamp, Edwin Knight
Christensen, N(ephi) A(lbert)
Claytor, Thomas Nelson
Clinard, Frank Welch, Jr
Curro, John Gillette
Davidson, Keith V(ernon)
Dick, Jerry Joel
Diegle, Ronald Bruce
Doss, James Daniel
Dowell, Flonnie
Ericksen, Richard Harold
Farnum, Eugene Howard
Follstaedt, David Martin
Fries, Ralph Jay
Frost, Harold Maurice, III
Gerstle, Francis Peter, Jr
Gourley, Paul Lee
Hopkins, Alan Keith
Kenna, Bernard Thomas
Kocks, U(lrich) Fred
Kreidl, Norbert J(oachim)
Kurtz, Steven Ross
Laquer, Henry L
Liepins, Raimond
Lowe, Terry Curtis
Miller, Alan R(obert)
Milton, Osborne
Mitchell, Terence Edward
Moir, David Chandler
Morosin, Bruno
Mullendore, A(rthur) W(ayne)
Northrop, David A
Northrup, Clyde John Marshall, Jr

O'Rourke, John Alvin
Parkin, Don Merrill
Peterson, Charles Leslie
Picraux, Samuel Thomas
Pike, Gordon E
Pope, Larry Elmer
Romig, Alton Dale, Jr
Salzbrenner, Richard John
Sandstrom, Donald James
Sasmor, Daniel Joseph
Saxton, Harry James
Seeger, Philip Anthony
Seiffert, Stephen Lockhart
Skaggs, Samuel Robert
Sutherland, Herbert James
Switendick, Alfred Carl
Taylor, Gene Warren
Travis, James Roland
Tsao, Jeffrey Yeenien
Wallace, Terry Charles
Warren, William Ernest
Wawersik, Wolfgang R
Wilder, James Andrew, Jr

NEW YORK
Adelstein, Peter Z
Adler, Philip N(athan)
Albers, Francis C
Albrecht, William Melvin
Altemose, Vincent O
Altscher, Siegfried
Anderson, Wayne Arthur
Anthony, Thomas Richard
Aven, Manuel
Axelrod, Norman Nathan
Banholzer, William Frank
Barr, Donald Eugene
Bayer, Raymond George
Benzinger, James Robert
Berkes, John Stephan
Berry, B(rian) S(hepherd)
Beshers, Daniel N(ewson)
Bilderback, Donald Heywood
Blakely, J(ohn) M
Boesch, William J
Bolon, Roger B
Borom, Marcus P(reston)
Borrelli, Nicholas Francis
Bourdillon, Antony John
Briant, Clyde Leonard
Brodsky, Marc Herbert
Brun, Milivoj Konstantin
Budinski, Kenneth Gerard
Buhks, Ephraim
Burke, J(oseph) E(ldrid)
Burns, Stephen James
Carrier, Gerald Burton
Chadwick, George F
Chamberlin, Howard Allen
Chang, Chin-An
Chang, L(eroy) L(i-Gong)
Cheh, Huk Yuk
Chen, Inan
Chen, Tsang Jan
Chiang, Yuen-Sheng
Chow, Tat-Sing Paul
Chyung, Kenneth
Coden, Michael H
Conn, Paul Kohler
Corelli, John Charles
Cosgarea, Andrew, Jr
Cozzarelli, Francis A(nthony)
Crandall, William B
Culver, Richard S
Cunningham, Michael Paul
Davidson, Robert W
Dermit, George
De Zeeuw, Carl Henri
D'Heurle, Francois Max
Doremus, Robert Heward
Dowben, Peter Arnold
Dudley, Michael
Duquette, David J(oseph)
Erhardt, Peter Franklin
Fay, Homer
Feger, Claudius
Feige, Norman G
Felty, Evan J
Fields, Alfred E
Fleischer, Robert Louis
Flom, Donald Gordon
Fox, Adrian Samuel
Frank, Kozi
Fritz, K(enneth) E(arl)
Galginaitis, Simeon Vitis
Gambino, Richard Joseph
George, Philip Donald
Ghosh, Arup Kumar
Giannelis, Emmanuel P
Gilmore, Robert Snee
Gluck, Ronald Monroe
Goland, Allen Nathan
Good, Robert James
Gorbatsevich, Serge N
Gould, Robert Kinkade
Greenbaum, Steven Garry
Grunes, Robert Lewis
Guile, Donald Lloyd
Gupta, Devendra
Gupta, Krishna Murari
Hart, Edward Walter
Hartwig, Curtis P
Hauth, Willard E(llsworth), III
Headrick, Randall L

Herczog, Andrew
Herley, Patrick James
Herman, Herbert
Highsmith, Ronald Earl
Hillig, William Bruno
Hirsh, Merle Norman
Hobbs, Stanley Young
Holland, Hans J
Homan, Clarke Gilbert
Hornibrook, Walter John
Horowitz, Carl
Hovel, Harold John
Hudson, John B(alch)
Hurd, Jeffery L
Huston, Ernest Lee
Interrante, Leonard V
Isaacs, Hugh Solomon
Isaacs, Leslie Laszlo
Jackson, Melvin Robert
Johansson, Sune
Johnson, Curtis Alan
Johnson, Herbert Harrison
Jordan, Albert Gustav
Julinao, Peter C
Kamath, Vasanth Rathnakar
Kambour, Roger Peabody
Kamdar, Madhusudan H
Kammer, Paul A
Kasprzak, Lucian A
King, Alexander Harvey
Klein, Gerald Wayne
Kohrt, Carl Fredrick
Komanduri, Ranga
Kosky, Philip George
Kramer, Edward J(ohn)
Krempl, Erhard
Kuan, Tung-Sheng
Kwok, Thomas Yu-Kiu
Laghari, Javaid Rosoolbux
Lane, Richard L
Lanford, William Armistead
Lauer, James Lothar
Lavine, James Philip
Lawson, Charles Alden
Lay, Kenneth W(ilbur)
Lee, Daeyong
Lee, Minyoung
Leigh, Richard Woodward
Lever, Reginald Frank
Lewis, Edward R(obert)
Li, Che-Yu
Li, Hong
Li, J(ames) C(hen) M(in)
Light, Thomas Burwell
Lindberg, Vern Wilton
Liu, Hao-Wen
Liu, Yung Sheng
Logan, Joseph Skinner
Lorenzen, Jerry Alan
Luborsky, Fred Everett
Ludeke, Rudolf
Luthra, Krishan Lal
Lynn, Kelvin G
Lynn, Merrill
MacCrone, Robert K
MacDonald, Noel C
MacDowell, John Fraser
McElligott, Peter Edward
Machlin, E(ugene) S(olomon)
MacKenzie, Donald Robertson
McMurtry, Carl Hewes
Mahler, David S
Marwick, Alan David
Mason, John Hugh
Mathes, Kenneth Natt
Matienzo, Luis J
Matkovich, Vlado Ivan
Maurer, Gernant E
Mayer, James W(alter)
Meloon, David Rand
Metz, Edward
Meyerson, Bernard Steele
Mijovic, Jovan
Mogro-Campero, Antonio
Mondolfo, L(ucio) F(austo)
Moore, Robert Stephens
Mort, Joseph
Mount, Eldridge Milford, III
Moy, Dan
Moynihan, Cornelius Timothy
Mueller, Edward E(ugene)
Mulford, Robert Alan
Muller, Olaf
Murphy, Eugene F(rancis)
Muzyka, Donald Richard
Myers, Drewfus Young, Jr
Myers, Mark B
Naar, Raymond Zacharias
Nauman, Edward Bruce
Neal, John Alva
Nelson, John Keith
Nielsen, John P(hillip)
Nowick, A(rthur) S(tanley)
Ober, Christopher Kemper
Oh, Byungdu
O'Reilly, James Michael
Otocka, Edward Paul
Ozimek, Edward Joseph
Pai, Damodar Mangalore
Palladino, William Joseph
Passoja, Dann E
Pearce, Eli M
Pedroza, Gregorio Cruz
Pemrick, Raymond Edward

Penn, Lynn Sharon
Penwell, Richard Carlton
Persans, Peter D
Pillar, Walter Oscar
Pliskin, William Aaron
Prager, Martin
Price, Frank DuBois
Prindle, William Roscoe
Psioda, Joseph Adam
Puttlitz, Karl Joseph
Quesnel, David John
Questad, David Lee
Rasmussen, Don Henry
Reardon, Joseph Daniel
Robinson, Charles C(anfield)
Roedel, George Frederick
Romankiw, Lubomyr Taras
Rothstein, Samuel
Sachse, Wolfgang H
Sadagopan, Varadachari
Saha, Bijay S
Sass, Stephen L
Satya, Akella V S
Schiff, Eric Allan
Schroder, Klaus
Schulze, Walter Arthur
Schuster, David Martin
Schwartz, Leon Joseph
Seltzer, Raymond
Setchell, John Stanford, Jr
Shaw, Robert Reeves
Shelby, James Elbert
Shen, Wu-Mian
Sheryll, Richard Perry
Shih, Kwang Kuo
Shoup, Robert D
Siau, John Finn
Sie, Charles H
Silcox, John
Skelly, David W
Slate, Floyd Owen
Snyder, Robert L(eon)
Snyder, Robert Lyman
Solomon, Harvey Donald
Spriggs, Richard Moore
Srinivasan, G(urumakonda) R
Srinivasan, Makuteswaran
Srinivasan, Rangaswamy
Sternstein, Sanford Samuel
Stookey, Stanley Donald
Strasser, Alfred Anthony
Sullivan, Peter Kevin
Sutherland, Judith Elliott
Swanson, Alan Wayne
Sweet, Richard Clark
Thompson, David Fred
Tonelli, John P, Jr
Totta, Paul Anthony
Trevoy, Donald James
Tu, King-Ning
Tummala, Rao Ramamohana
Vallance, Michael Alan
Valyi, Emery I
Vanier, Peter Eugene
Vaughan, Michael Thomas
Vidali, Gian Franco
Von Bacho, Paul Stephan, Jr
Von Gutfeld, Robert J
Vook, Richard Werner
Wachtell, Richard L(loyd)
Wagner, John George
Waltcher, Irving
Walter, William Trump
Wang, Franklin Fu-Yen
Ward, Samuel Abner
Webb, Watt Wetmore
Wei, Ching-Yeu
Weiner, Milton Lawrence
Weiss, Volker
Welch, David O(tis)
Westbrook, J(ack) H(all)
Wetherhold, Robert Campbell
Whang, Sung H
Whiteside, James Brooks
Whittingham, M(ichael) Stanley
Wie, Chu Ryang
Wilcox, W(illiam) R(oss)
Wilson, James M
Wittmer, Marc F
Woodard, David W
Wosinski, John Francis
Wright, Roger Neal
Yavorsky, John Michael
Yudelson, Joseph Samuel
Yuh, Huoy-Jen
Zakraysek, Louis
Zeldin, Arkady N
Zhang, Qiming

NORTH CAROLINA
Andrady, Anthony Lakshman
Austin, William W(yatt)
Bailey, Robert Brian
Baldwin, Vaniah Harmer, Jr
Baliga, Jayant
Barnhardt, Robert Alexander
Batra, Subhash Kumar
Beck, Keith Russell
Beeler, Joe R, Jr
Bernholc, Jerzy
Browne, Colin Lanfear
Buchanan, David Royal
Burton, Ralph Gaines
Cates, David Marshall

Clarke, John Ross
Clator, Irvin Garrett
Cocks, Franklin H
Cole, Jerome Foster
Conrad, Hans
Cuculo, John Anthony
Daumit, Gene Philip
Davis, Robert F(oster)
Douglas, Robert Alden
Drechsel, Paul David
Dyrkacz, W William
El-Shiekh, Aly H
Fornes, Raymond Earl
Gillooly, George Rice
Goldstein, Irving Solomon
Goodwin, Frank Erik
Grady, Perry Linwood
Graham, Louis Atkins
Gupta, Bhupender Singh
Hall, Seymour Gerald
Hamby, Dame Scott
Hamner, William Frederick
Haseley, Edward Albert
Havner, Kerry S(huford)
Haynie, Fred Hollis
Hersh, Solomon Philip
Hren, John J(oseph)
Hurt, John Calvin
Irene, Eugene Arthur
Kirk, Wilber Wolfe
Koch, Carl Conrad
Kusy, Robert Peter
Linton, Richard William
Lord, Peter Reeves
McNeil, Laurie Elizabeth
Masnari, Nino A
Massoud, Hisham Z
Mohanty, Ganesh Prasad
Narayan, Jagdish
Nash, James Lewis, Jr
Olf, Heinz Gunther
Pearsall, George W(ilbur)
Powers, Edward James
Prater, John Thomas
Reeber, Robert Richard
Reisman, Arnold
Rozgonyi, George A
Shepard, Marion L(averne)
Tanquary, Albert Charles
Taylor, Duane Francis
Walsh, Edward John
Walters, Mark David
Williams, Joel Lawson
Wooten, Willis Carl, Jr

NORTH DAKOTA
Adams, David George
Bierwagen, Gordon Paul

OHIO
Adams, Harold Elwood
Aggarwal, Sundar Lal
Antler, Morton
Baer, Eric
Ball, Lawrence Ernest
Baloun, Calvin H(endricks)
Barkley, John R
Bement, A(rden) L(ee), Jr
Bethea, Tristram Walker, III
Bhattacharya, Rabi Sankar
Biedenbender, Michael David
Birle, John David
Blackwell, John
Bohm, Georg G A
Bonner, David Calhoun
Bouton, Thomas Chester
Brandon, Clement Edwin
Brantley, William Arthur
Brooman, Eric William
Brown, Stanley Alfred
Buckley, Donald Henry
Bufkin, Billy George
Burte, Harris M(erl)
Calderon, Nissim
Campbell, James Edward
Campbell, Robert Wayne
Carmichael, Donald C(harles)
Causa, Alfredo G
Chakko, Mathew K(anjhirathinkal)
Choudhary, Manoj Kumar
Christian, John B
Clark, William Alan Thomas
Conant, Floyd Sanford
Copp, Albert Nils
D'Ianni, James Donato
DiCarlo, James Anthony
Divis, Roy Richard
Drake, Michael L
Duckworth, Winston H(oward)
Dudek, Thomas Joseph
Duderstadt, Edward C(harles)
Durand, Edward Allen
Duttweiler, Russell E
Eckstein, Yona
Ells, Frederick Richard
England, Richard Jay
Falkenbach, George J(oseph)
Fisher, George A, Jr
Gabb, Timothy Paul
Gates, John E(dward)
Gates, Raymond Dee
Gebelein, Charles G
Gerace, Michael Joseph
Germann, Richard P(aul)

Materials Science Engineering (cont)

Ginning, P(aul) R(oll)
Golovin, Michael N
Graham, Paul Whitener Link
Green, Robert Patrick
Griffin, John Leander
Griffin, Richard Norman
Gromelski, Stanley John, Jr
Halford, Gary Ross
Hall, Judd Lewis
Hamed, Gary Ray
Haque, Azeez C
Harmer, Richard Sharpless
Harrington, Roy Victor
Harris, Frank Wayne
Hassler, Craig Reinhold
Heasley, James Henry
Henderson, Courtland M
Henry, Leonard Francis, III
Hickok, Robert Lee
Hoch, Michael
Hough, Ralph L
Hughes, Kenneth E(ugene)
Ingram, David Christopher
Ishida, Hatsuo
Jain, Vinod Kumar
Jayaraman, Narayanan
Jayne, Theodore D
Kampe, Dennis James
Kanakkanatt, Antony
Kanakkanatt, Sebastian Varghese
Katz, J Lawrence
Kaufman, John Gilbert, Jr
Kaye, Christopher J
Kennedy, Joseph Paul
Keplinger, Orin Clawson
Koch, Ronald Joseph
Koenig, Jack L
Kolesar, Edward S
Kollen, Wendell James
Koros, Peter J
Kot, Richard Anthony
Krause, Horatio Henry
Kreidler, Eric Russell
Kroenke, William Joseph
Krysiak, Joseph Edward
Lad, Robert Augustin
Lando, Jerome B
Landstrom, D(onald) Karl
Lee, Min-Shiu
Lee, Peter Wankyoon
Lehn, William Lee
Leis, Brian Norman
Lewandowski, John Joseph
Li, George Su-Hsiang
Livigni, Russell Anthony
Livingston, Daniel Isadore
Lohr, Delmar Frederick, Jr
McCoy, Robert Allyn
McDonel, Everett Timothy
McGinniss, Vincent Daniel
McIntyre, Donald
McMillin, Carl Richard
Mall, Shankar
Mark, James Edward
Markworth, Alan John
Meinecke, Eberhard A
Mikesell, Sharell Lee
Miller, William Reynolds, Jr
Mindlin, Harold
Minges, Merrill Loren
Miyoshi, Kazuhisa
Mobley, Carroll Edward
Mohn, Walter Rosing
Moore, Arthur William
Morral, F(acundo) R(olf)
Nakajima, Nobuyuki
Payer, Joe Howard
Perrin, James Stuart
Peterson, Robert C
Peterson, Timothy Lee
Piirma, Irja
Porter, Leo Earle
Proctor, David George
Pronko, Peter Paul
Prusas, Zenon C
Pugh, John W(illiam)
Purdon, James Ralph, Jr
Pyle, James Johnston
Rai, Amarendra Kumar
Ramalingam, Mysore Loganathan
Rastogi, Prabhat Kumar
Reardon, Joseph Patrick
Reuter, Robert A
Rexer, Joachim
Rigney, David Arthur
Roe, Ryong-Joon
Roehrig, Frederick Karl
Rosenfield, Alan R(obert)
Rupert, John Paul
St John, Douglas Francis
Salkind, Michael Jay
Sara, Raymond Vincent
Saraceno, Anthony Joseph
Sargent, Gordon Alfred
Sawyer, Baldwin
Schmidt, Donald L
Schmitt, George Frederick, Jr
Seldin, Emanuel Judah
Semler, Charles Edward
Servais, Ronald Albert

Sicka, Richard Walter
Sinclair, Richard Glenn, II
Snyder, Milton Jack
Sommer, John G
Srivatsan, Tirumalai Srinivas
Stover, E(dward) R(oy)
Strnat, Karl J
Summers, James William
Tallan, Norman M
Thomas, Joseph Francis, Jr
Uebele, Curtis Eugene
Uralil, Francis Stephen
Uys, Johannes Marthinus
Vassamillet, Lawrence Francois
Venkatu, Doulatabad A
Versic, Ronald James
Wagoner, Robert H
Warner, Walter Charles
Welsch, Gerhard Egon
Westermann, Fred Ernst
White, James L(indsay)
Wickersham, Charles Edward, Jr
Wilde, Bryan Edmund
Williams, James Case
Williams, Wendell Sterling
Wittebort, Jules I
Wlodek, Stanley T
Wolff, Gunther Arthur
Wong, Wai-Mai Tsang
Yang, Philip Yung-Chin
Yu, Thomas Huei-Chung

OKLAHOMA
Binstock, Martin H(arold)
Bruner, Ralph Clayburn
Cook, Charles F(oster), Jr
Growcock, Frederick Bruce
Hsieh, Henry Lien
Jones, Faber Benjamin
Kline, Ronald Alan
Lou, Alex Yih-Chung
Schwab, Peter Austin
Sheshtawy, Adel A
Shue, Robert Sidney
Striz, Alfred Gerhard
Thomason, William Hugh
Zelinski, Robert Paul

OREGON
Arthur, John Read, Jr
Barnett, Gordon Dean
Bhattacharya, Pallab Kumar
Daellenbach, Charles Byron
Devletian, Jack H
Dooley, George Joseph, III
Ethington, Robert Loren
Ford, Wayne Keith
Merz, Paul Louis
Nielsen, Lawrence Ernie
Owen, Sydney John Thomas
Paarsons, James Delbert
Roberts, C Sheldon
Seaman, Geoffrey Vincent F
Sleight, Arthur William
Wager, John Fisher
Wood, William Edwin
Yau, Leopoldo D

PENNSYLVANIA
Abate, Kenneth
Albert, Robert Lee
Alley, Richard Blaine
Alper, Allen Myron
Arkles, Barry Charles
Armor, John N
Ashok, S
Balaba, Willy Mukama
Baran, George Roman
Baratta, Anthony J
Barnes, Mary Westergaard
Barsom, John M
Bartovics, Albert
Bauer, C(harles) L(loyd)
Bauman, Bernard D
Becker, Aaron Jay
Behrens, Ernst Wilhelm
Belitskus, David
Bhalla, Amar S
Blankenhorn, Paul Richard
Bluhm, Harold Frederick
Böer, Karl Wolfgang
Bolstad, Luther
Boltax, Alvin
Bramfitt, Bruce Livingston
Brody, Harold D
Brown, Norman
Brungraber, Robert J
Bucci, Robert James
Bucher, John Henry
Buck, Jean Coberg
Chan, Siu-Kee
Chen-Tsai, Charlotte Hsiao-yu
Cheung, Peter Pak Lun
Chou, Y(e) T(sang)
Chung, Deborah Duen Ling
Claiborne, C Clair
Clark, J(ohn) B(everley)
Coleman, Michael Murray
Conyne, Richard Francis
Cookson, Alan Howard
Corneliussen, Roger DuWayne
Dalal, Harish Maneklal
Das, Suryya Kumar
Dax, Frank Robert

DebRoy, Tarasankar
Deeg, Emil W(olfgang)
De Luccia, John Jerry
Doak, Kenneth Worley
Doty, W(illiam) D'Orville
Edelman, Leonard Edward
Egami, Takeshi
Ehrhart, Wendell A
Ehrig, Raymond John
Emlemdi, Hasan Bashir
Erhan, Semih M
Eror, Nicholas George, Jr
Field, Nathan David
Fitzgerald, Maurice E
Fong, James T(se-Ming)
Forscher, Frederick
Frattini, Paul L
Frost, Lawrence William
Garbarini, Victor C
Garber, Charles A
Garg, Diwakar
Gebhardt, Joseph John
Giannovario, Joseph Anthony
Gillis, Marina N
Gittler, Franz Ludwig
Goddu, Robert Fenno
Goel, Ram Parkash
Goldman, Kenneth M(arvin)
Goldstein, Joseph I
Goodyear, William Frederick, Jr
Graham, Charles D(anne), Jr
Greenberg, Charles Bernard
Gruenwald, Geza
Gutbezahl, Boris
Hahn, Hong Thomas
Hall, Gary R
Harkins, Thomas Regis
Harmer, Martin Paul
Harrison, Ian Roland
Hartman, Marvis Edgar
Herzog, Leonard Frederick, II
Hess, Dennis William
Hettche, Leroy Raymond
Heybey, Otfried Willibald Georg
Hildeman, Gregory John
Hillner, Edward
Hofferth, Burt Frederick
Hopkins, Richard H(enry)
Houlihan, John Frank
Howell, Paul Raymond
Hu, L(ing) W(en)
Hu, William H(sun)
Hunsicker, Harold Yundt
Ikeda, Richard Masayoshi
Irwin, William Edward
Jaffe, Donald
Jain, Himanshu
Jang, Sei Joo
Johnson, Gerald Glenn, Jr
Johnston, William V
Jones, Roger Franklin
Jordan, A(ngel) G(oni)
Jordan, Robert Kenneth
Kakar, Anand Swaroop
Kang, Joohee
Katz, Lewis E
Khare, Ashok Kumar
Kitazawa, George
Klapproth, William Jacob, Jr
Kline, Donald Edgar
Klingsberg, Cyrus
Knaster, Tatyana
Knox, Bruce E
Ko, Frank K
Koczak, Michael Julius
Komarneni, Sridhar
Korostoff, Edward
Kottcamp, Edward H, Jr
Koziar, Joseph Cleveland
Kraft, R(alph) Wayne
Kraitchman, Jerome
Krishnaswamy, S V
Kuhn, Howard A
Lake, Robert D
Lakhtakia, Akhlesh
Langsam, Michael
Lannin, Jeffrey S
Lawley, Alan
Leonard, Laurence
Libsch, Joseph F(rancis)
Lindt, Jan Thomas
Long, Alton Los, Jr
Lord, Arthur E, Jr
Loria, Edward Albert
Louie, Ming
Love, Gordon Ross
Lu, Chih Yuan
Luck, Russell M
McMahon, Charles J, Jr
Macmillan, Norman Hillas
Magill, Joseph Henry
Mahajan, Subhash
Mahlman, Bert H
Manjoine, Michael J(oseph)
Mayer, George Emil
Mecholsky, John Joseph, Jr
Mehta, Sudhir
Meibohm, Edgar Paul Hubert
Meier, Joseph Francis
Messier, Russell
Michael, Norman
Moss, Herbert Irwin
Muan, Arnulf
Muldawer, Leonard

Niebel, B(enjamin) W(illard)
Novak, Stephen Robert
Nuessle, Albert Christian
Pangborn, Robert Northrup
Peffer, John Roscoe
Perzak, Frank John
Peterson, Richard Walter
Pfennigwerth, Paul Leroy
Pike, Ralph Edwin
Pinckney, Robert L
Plamondon, Joseph Edward
Plazek, Donald John
Pollack, Solomon R
Pope, David Peter
Popovics, Sandor
Prane, Joseph W(illiam)
Preusch, Charles D
Prout, James Harold
Racette, George William
Ray, Siba Prasad
Redmond, John Peter
Reynolds, Claude Lewis, Jr
Rogers, H(arry) C(arton), Jr
Romovacek, George R
Rowe, Anne Prine
Roy, Della M(artin)
Roy, Rustum
Ruud, Clayton Olaf
Ryba, Earle Richard
Sabol, George Paul
Sander, Louis Frank
Scheetz, Howard A(nsel)
Schimmel, Karl Francis
Schwerer, Frederick Carl
Seiner, Jerome Allan
Shabel, Barrie Steven
Shaler, Amos J(ohnson)
Sharma, Mangalore Gokulanand
Sias, Charles B
Siddiqui, Habib
Sieger, John S(ylvester)
Simkovich, George
Simmons, Richard Paul
Smith, Deane Kingsley, Jr
Smith, James David Blackhall
Smith, William Novis, Jr
Smyth, Donald Morgan
Snider, Albert Monroe, Jr
Spitznagel, John A
Stefanou, Harry
Steiger, Roger Arthur
Stengle, William Bernard
Stinger, Henry J(oseph)
Sutter, Philip Henry
Swartz, John Croucher
Sykes, James Aubrey, Jr
Talvacchio, John
Thomas, David Alden
Thomas, Donald E(arl)
Thompson, Anthony W
Thrower, Peter Albert
Treadwell, Kenneth Myron
Tressler, Richard Ernest
Van Der Spiegel, Jan
Van Raalte, John A
Varadan, Vasundara Venkatraman
Varadan, Vijay K
Varnerin, Lawrence J(ohn)
Varrese, Francis Raymond
Vedam, Kuppuswamy
Venable, Emerson
Verleur, Hans Willem
Wagner, J Robert
Waldman, Jeffrey
Waldman, L(ouis) A(braham)
Walker, Augustus Chapman
Walker, Philip L(eroy), Jr
Wallace, William Edward
Wei, Robert Peh-Ying
Wei, Yen
Wei-Berk, Caroline
Weiner, Robert Allen
Wells, Ralph Gordon
Werner, F(red) E(ugene)
Werny, Frank
White, William Blaine
Whyte, Thaddeus E, Jr
Wilder, Harry D(ouglas)
Williams, David Bernard
Wood, Susan
Work, William James
Worrell, Wayne L
Wynblatt, Paul P
Yang, Arthur Jing-Min
Zamanzadeh, Mehrooz
Zweben, Carl Henry

RHODE ISLAND
Asaro, Robert John
Avery, Donald Hills
Caroselli, Remus Francis
Clifton, Rodney James
Findley, William N(ichols)
Freund, Lambert Ben
Gurland, Joseph
Richman, Marc H(erbert)
Rockett, Thomas John
Suresh, Subra

SOUTH CAROLINA
Aspland, John Richard
Avegeropoulos, G
Awadalla, Nabil G
Caskey, George R, Jr

Chase, Vernon Lindsay
Diefendorf, Russell Judd
Drews, Michael James
Dumin, David Joseph
Faust, John William, Jr
Fowler, John Rayford
Goldstein, Herman Bernard
Goswami, Bhuvenesh C
Hay, Ian Leslie
Hendrix, James Easton
Hopkins, Allen John
Jarvis, Christine Woodruff
Kimmel, Robert Michael
LaFleur, Kermit Stillman
Lee, Burtrand Insung
McDonell, William Robert
Minford, James Dean
Mosley, Wilbur Clanton, Jr
Moyle, David Douglas
Rack, Henry Johann
Rootare, Hillar Muidar
Schwartz, Elmer G(eorge)
Sello, Stephen
Shalaby, Shalaby W
Stevens, James Levon
Sturcken, Edward Francis
Taras, Michael Andrew
Taylor, Peter Anthony
Terry, Stuart Lee
Tolbert, Thomas Warren
Wicks, George Gary
Young, Franklin Alden, Jr

SOUTH DAKOTA
Cannon, Patrick Joseph
Han, Kenneth N
Ross, Keith Alan

TENNESSEE
Armentrout, Daryl Ralph
Bardos, Denes I(stvan)
Behr, Eldon August
Besmann, Theodore Martin
Besser, John Edwin
Billington, Douglas S(heldon)
Bleier, Alan
Bretz, Philip Eric
Bryan, Robert H(owell)
Budai, John David
Burton, Robert Lee
Chambers, Ralph Arnold
Cunningham, John Edward
Danko, Joseph Christopher
Davis, Burns
Dorsey, George Francis
Ford, James Arthur
Gilkey, Russell
Goodson, Louie Aubrey, Jr
Goodwin, Gene M
Googin, John M
Gorbatkin, Steven M
Gray, Theodore Flint, Jr
Grossbeck, Martin Lester
Harms, William Otto
Holcombe, Cressie Earl, Jr
Horak, James Albert
Horton, Joseph Arno, Jr
Iskander, Shafik Kamel
Joy, David Charles
Judkins, Roddie Reagan
Kinser, Donald LeRoy
Klueh, Ronald Lloyd
Langley, Robert Archie
Lichter, Barry D(avid)
Liu, Chain T
Lowndes, Douglas H, Jr
Lundy, Ted Sadler
Mansur, Louis Kenneth
Mayberry, Thomas Carlyle
Newland, Gordon Clay
Noonan, John Robert
Pawel, Janet Elizabeth
Pearson, R(ay) L(eon)
Phillips, Paul J
Pugh, Claud Ervin
Roberto, James Blair
Sales, Brian Craig
Scherpereel, Donald E
Scott, Herbert Andrew
Scott, J(ames) L(ouis)
Simhan, Raj
Slaughter, Gerald M
Smith, Michael James
Sparks, Cullie J(ames), Jr
Spooner, Stephen
Spruiell, Joseph E(arl)
Stiegler, James O
Stroud, Robert Wayne
Swindeman, Robert W
Wachs, Alan Leonard
Wang, Chien Bang
Waters, Dean Allison
Weeks, Robert A
Wert, James J
Wildman, Gary Cecil
Williams, Robin O('Dare)
Wunderlich, Bernhard
Yoo, Man Hyong

TEXAS
Armeniades, Constantine D
Arthur, Marion Abrahams
Aufdermarsh, Carl Albert, Jr

Banerjee, Sanjay Kumar
Barton, John R
Bokros, J(ack) C(hester)
Bourell, David Lee
Bovay, Harry Elmo, Jr
Bravenec, Edward V
Brostow, Witold Konrad
Bruins, Paul F(astenau)
Burger, Christian P
Burkart, Leonard F
Chen, Michael Chia-Chao
Chopra, Dev Raj
Chu, Wei-Kan
Cole, David F
Daues, Gregory W, Jr
Daugherty, Kenneth E
Davidson, David Lee
Davison, Sol
Dhudshia, Vallabh H
Eberhart, Robert Clyde
Economou, Demetre J
Fish, John G
Forest, Edward
Frick, John P
Geyling, F(ranz) Th(omas)
Ghowsi, Kiumars
Glosser, Robert
Goodwyn, Jack Ray
Griffin, John R(obert)
Griffin, Richard B
Gruber, George J
Gully, John Houston
Gupton, Paul Stephen
Harper, James George
Hasty, Turner Elilah
Hausler, Rudolf H
Hill, Robert William
Holliday, George Hayes
Holmes, Larry A
Huege, Fred Robert
Ignatiev, Alex
Johnson, Elwin L Pete
Jones, William B
Kanninen, Melvin Fred
Kaye, Howard
Kirk, Wiley Price
Koros, William John
Lacy, Lewis L
Leamy, Harry John
Levine, Jules David
Lewis, James Pettis
Lian, Shawn
Lindholm, Ulric S
Lloyd, Douglas Roy
Lytton, Robert Leonard
McConnell, Duncan
McDaniel, Floyd Delbert, Sr
McDavid, James Michael
Machacek, Oldrich
Marcus, Harris L
Marlow, William Henry
Matzkanin, George Andrew
Mills, John James
Moore, Thomas Matthew
Moss, Simon Charles
Murr, Lawrence Eugene
Olstowski, Franciszek
Packman, Paul Frederick
Park, Vernon Kee
Patriarca, Peter
Paul, Donald Ross
Pavlovich, Raymond Doran
Perez, Ricardo
Petersen, Donald H
Porter, Vernon Ray
Porter, Wilbur Arthur
Powers, John Michael
Pray, Donald George
Raba, Carl Franz, Jr
Ralls, Kenneth M(ichael)
Rao, Shankaranarayana
        Ramohallinanjunda
Reed, Mark Arthur
Roberts, John Melville
Salama, Kamel
Sanchez, Isaac Cornelius
Schapery, Richard Allan
Sharma, Suresh C
Shaw, Don W
Shilstone, James Maxwell, Jr
Singh, Vijay Pal
Stahl, Glenn Allan
Stehling, Ferdinand Christian
Steinfink, Hugo
Teller, Cecil Martin, II
Thornton, Joseph Scott
Tien, John Kai
Trachtenberg, Isaac
Turner, William Danny
Wang, Paul Weily
Williamson, Luther Howard
Winegartner, Edgar Carl
Yao, Joe
Yuan, Robert L

UTAH
Anderson, Douglas I
Andrade, Joseph D
Benner, Robert E
Boyd, Richard Hays
Carnahan, Robert D
Cohen, Richard M
Cook, Melvin Alonzo
Daniels, Alma U(riah)

Hoeppner, David William
Horton, Ralph M
Kim, Sung Wan
Li, Shin-Hwa
Lisanke, Robert John, Sr
Nauman, Edward Franklin
Nelson, Mark Adams
Orava, R(aimo) Norman
Rotz, Christopher Alan
Sohn, Hong Yong
Stringfellow, Gerald B
Sudweeks, Walter Bentley
Talbot, Eugene L(eroy)
Taylor, Philip Craig
Woodbury, Richard C

VERMONT
Anderson, R(ichard) L(ouis)
Berens, Alan Robert
Bhatt, Girish M
Fox, Bradley Alan
Furukawa, Toshiharu
Howard, Robert T(urner)
Kim, Sukyoung
Pires, Renato Guedes
Von Turkovich, Branimir F(rancis)

VIRGINIA
Almeter, Frank M(urray)
Armstrong, Robert G
Barker, Robert Edward, Jr
Birnbaum, Leon S
Blurton, Keith F
Buckley, John Dennis
Burley, Carlton Edwin
Cantrell, John H(arris)
Chiou, Minshon Jebb
Clark, Donald Eldon
Cook, Desmond C
Courtney, Thomas Hugh
Diness, Arthur M(ichael)
Duke, John Christian, Jr
Farago, John
Farmer, Barry Louis
Frederick, Daniel
Fripp, Archibald Linley
Gangloff, Richard Paul
Goodman, A(lvin) M(alcolm)
Gordon, Ronald Stanton
Graham, Kenneth Judson
Hahn, Henry
Hanneman, Rodney E
Hasselman, Didericus Petrus Hermannus
Hendricks, Robert Wayne
Henneke, Edmund George, II
Hibbard, Walter Rollo, Jr
Hodge, James Dwight
Holt, William Henry
Houska, Charles Robert
Hove, John Edward
Hudson, Charles Michael
Hutchinson, Thomas Eugene
Ijaz, Lubna Razia
Jesser, William Augustus
Johnson, Robert Alan
Johnson, William Randolph, Jr
Johnston, Norman Joseph
Kinsley, Homan Benjamin, Jr
Kirkendall, Ernest Oliver
Kranbuehl, David Edwin
Kuhlmann-Wilsdorf, Doris
Landgraf, Ronald William
Lane, Joseph Robert
Lawless, Kenneth Robert
Lodoen, Gary Arthur
Long, Edward Richardson, Jr
Lowe, A(rthur) L(ee), Jr
Lytton, Jack L(ester)
Maahs, Howard Gordon
Maloney, Kenneth Morgan
Matthews, R(obert) B(ruce)
Mayer, George
Milford, George Noel, Jr
Moon, Peter Clayton
Newman, James Charles, Jr
Outlaw, Ronald Allen
Patterson, James Douglas
Philleo, Robert Eugene
Pletta, Dan Henry
Pohlmann, Juergen Lothar Wolfgang
Puster, Richard Lee
Reifsnider, Kenneth Leonard
Reynolds, Richard Alan
Rice, Roy Warren
Rijke, Arie Marie
Robertson, W(illiam) D(onald)
Rothwarf, Frederick
St Clair, Anne King
St Clair, Terry Lee
Schuurmans, Hendrik J L
Sedriks, Aristide John
Senseny, Paul Edward
Sherbeck, L Adair
Shur, Michael
Squire, David R
Steele, Lendell Eugene
Stein, Bland Allen
Stinchcomb, Wayne Webster
Sudarshan, T S
Sumner, Barbara Elaine
Talapatra, Dipak Chandra
Tompkins, Stephen Stern
Unnam, Jalaiah
Van Ness, Kenneth E

Van Reuth, Edward C
Walker, Thomas Carl
Wawner, Franklin Edward, Jr
Wayland, Rosser Lee, Jr
Wilcox, Benjamin A
Wilkes, Garth L
Wilsdorf, Doris Kuhlmann
Wilsdorf, Heinz G(erhard) F(riedrich)
Wood, George Marshall
Wynne, Kenneth Joseph

WASHINGTON
Baer, Donald Ray
Bates, J(unior) Lambert
Boyer, Rodney Raymond
Brimhall, J(ohn) L
Bryant, Ben S
Burns, Robert Ward
Collins, Gary Scott
Cordingly, Richard Henry
Dahl, Roy Edward
Daniel, J(ack) Leland
Ding, Jow-Lian
Doran, Donald George
Dunham, Glen Curtis
Duran, Servet A(hmet)
Einziger, Robert E
Ellis, Everett Lincoln
Eustis, William Henry
Evans, Thomas Walter
Fischbach, David Bibb
Gelles, David Stephen
Guion, Thomas Hyman
Hamilton, C Howard
Hannay, Norman Bruce
Hawkins, Neil Middleton
Hinman, Chester Arthur
Huang, Fan-Hsiung Frank
Johns, William E
Johnson, A(lfred) Burtron, Jr
Johnson, Jay Allan
Jones, Russell Howard
Karagianes, Manuel Tom
Keating, John Joseph
Kent, Ronald Allan
Knotek, Michael Louis
Leney, Lawrence
Lindenmeyer, Paul Henry
McGuire, Joseph Clive
Mahalingam, R
Maloney, Thomas M
Matlock, John Hudson
Megraw, Robert Arthur
Polonis, Douglas Hugh
Quist, William Edward
Ramulu, Mamidala
Roake, William Earl
Roberts, J T Adrian
Smith, Brad Keller
Stang, Robert George
Stoebe, Thomas Gaines
Subramanian, Ravanasamudram
        Venkatachalam
Taya, Minoru
Taylor, Murray East
Tichy, Robert J
Tinder, Richard F(ranchere)
Tingey, Garth Leroy
Trotter, Patrick Casey
Weber, William J
Wilson, Charles Norman
Woodfield, F(rank) W(illiam), Jr
Woods, Keith Newell

WEST VIRGINIA
Bryant, George Macon
Crist, John Benjamin
De Barbadillo, John Joseph
Hamilton, John Robert
Humphreys, Kenneth K
Knight, Alan Campbell
Koch, Christian Burdick
Koleske, Joseph Victor
Smith, Walton Ramsay
Sperati, Carleton Angelo
Winston, Anthony

WISCONSIN
Aita, Carolyn Rubin
Bajikar, Sateesh S
Baker, George Severt
Barr, Tery Lynn
Boom, Roger Wright
Brown, Charles Eric
Burck, Larry Harold
Cataldi, Horace A(nthony)
Caulfield, Daniel Francis
Christiansen, Alfred W
Cooper, Reid F
Cooper, Stuart L
Crawmer, Daryl E
Dasgupta, Rathindra
Denton, Denice Dee
Draeger, Norman Arthur
Dunn, Stanley Austin
Ferry, John Douglass
Fischer, Richard Martin, Jr
Fournelle, Raymond Albert
Gross, James Richard
Hinkes, Thomas Michael
Hirthe, Walter M(atthew)
Isebrands, Judson G
Isenberg, Irving Harry
Johnson, J(ames) R(obert)

Materials Science Engineering (cont)

Johnson, John E(dwin)
Kehres, Paul W(illiam)
Kinstle, James Francis
Kocurek, Michael Joseph
Lagally, Max Gunter
Lemm, Arthur Warren
Lenling, William James
Liang, Shoudeng
Morris, Marion Clyde
Neumann, Joachim Peter
Nordman, James Emery
Pearl, Irwin Albert
Pillai, Thankappan A K
Randall, Francis James
Rechtin, Michael David
Reichenbacher, Paul H
Richard, Terry Gordon
Rohatgi, Pradeep Kumar
Rowlands, Robert Edward
Sanyer, Necmi
Schwarz, Eckhard C A
Svoboda, Glenn Richard
Tonner, Brian P
Van den Akker, Johannes Archibald
Verbrugge, Calvin James
Wnuk, Michael Peter
Young, Raymond A
Zerbe, John Irwin
Zheng, Xiaoci

WYOMING
Adams, Donald F
Dolan, Charles W

PUERTO RICO
Schwartz, Abraham

ALBERTA
Egerton, Raymond Frank
Micko, Michael M
Vroom, Alan Heard

BRITISH COLUMBIA
Davis, Gerald Gordon
Evans, Russell Stuart
Franz, Norman Charles
Garner, Andrew
Hardwicke, Norman Lawson
Hatton, John Victor
Mindess, Sidney
Paszner, Laszlo
Tiedje, J Thomas
Tromans, D(esmond)

MANITOBA
Bassim, Mohamad Nabil
Cahoon, John Raymond
Chaturvedi, Mahesh Chandra
Dutton, Roger
Kao, Kwan Chi
Simpson, Leonard Angus
Wilkins, Brian John Samuel
Woo, Chung-Ho

NEW BRUNSWICK
Schneider, Marc H
Sebastian, Leslie Paul

NEWFOUNDLAND
Molgaard, Johannes

NOVA SCOTIA
Jones, Derek William
King, Hubert Wylam
Steinitz, Michael Otto
Strasser, John Albert

ONTARIO
Alexandru, Lupu
Aust, Karl T(homas)
Bahadur, Birendra
Berezin, Alexander A
Biggs, Ronald C(larke)
Bratina, Woymir John
Cameron, Irvine R
Carlsson, David James
Cocivera, Michael
Convey, John
Cox, B(rian)
Cox, Brian
Dawes, David Haddon
Densley, John R
Geach, George Alwyn
Golemba, Frank John
Hackam, Reuben
Hope, Brian Bradshaw
Howard-Lock, Helen Elaine
Huang, Robert Y M
Hunt, Charles Edmund Laurence
Hurd, Colin Michael
Ives, Michael Brian
Jackman, Thomas Edward
Ji, Guangda Winston
Jorch, Harald Heinrich
Kavassalis, Tom A
Krausz, Alexander Stephen
Litvan, Gerard Gabriel
Logan, Charles Donald
McEwan, Ian Hugh
McGeer, James Peter

Morris, Larry Arthur
Nicholson, Patrick Stephen
O'Driscoll, Kenneth F(rancis)
Pascual, Roberto
Patchett, Joseph Edmund
Piekarski, Konstanty
Piggott, Michael R(antell)
Pilliar, Robert Mathews
Pindera, Jerzy Tadeusz
Plumtree, A(lan)
Pundsack, Arnold L
Purbo, Onno Widodo
Purdy, Gary Rush
Ramachandran, Vangipuram S
Russell, Kenneth Edwin
Saxton, William Reginald
Siddell, Derreck
Slater, Keith
Smeltzer, Walter William
Smith, Dennis Clifford
Storey, Robert Samuel
Sundararajan, Pudupadi Ranganathan
Thompson, David Allan
Tyson, William Russell
Varin, Robert Andrzej
Vincett, Paul Stamford
Wallace, William
Weir, Ronald Douglas
Wiles, David M
Williams, Harry Leverne
Yan, Maxwell Menuhin
Zukotynski, Stefan

QUEBEC
Angers, Roch
Bartnikas, Ray
Champness, Clifford Harry
Crine, Jean-Pierre C
Dealy, John Michael
Edington, Jeffrey William
Feldman, Dorel
Hanna, Adel
Hay, Donald Robert
Hoa, Suong Van
Izquierdo, Ricardo
Jonas, John Joseph
Kahrizi, Mojtaba
Kokta, Bohuslav Vaclav
Koran, Zoltan
Lepoutre, Pierre
Lombos, Bela Anthony
Manley, Rockliffe St John
Marsh, Cedric
Masut, Remo Antonio
Meunier, Michel
Osman, M(ohamed) O M
Page, Derek Howard
Patterson, Donald Duke
Prud'homme, Robert Emery
Rigaud, Michel Jean
Saint-Jacques, Robert G
St Pierre, Leon Edward
Tardif, Henri Pierre
Utracki, Lechoslaw Adam
Van Neste, Andre
Wertheimer, Michael Robert

OTHER COUNTRIES
Balk, Pieter
Brown, Peter
Cais, Rudolf Edmund
Chang, Chun-Yen
Davidson, Daniel Lee
De Vedia, Luis Alberto
Hansen, Torben Christen
Hansson, Inge Lief
Hashin, Zvi
Isaacs, Philip Klein
Jacob, K Thomas
Katz, Gerald
Kishi, Keiji
Lahiri, Syamal Kumar
Lin, Otto Chui Chau
Markovitz, Hershel
Martin, Jim Frank
Meier, James Archibald
Neuse, Eberhard Wilhelm
Reed, Thomas Freeman
Reid, William John
Sakurai, Yoshifumi
Sessler, Gerhard Martin
Solari, Mario Jose Adolfo
Wasa, Kiyotaka
Weinstock, Harold

Mechanical Engineering

ALABAMA
Barfield, Robert F(redrick) (Bob)
Bussell, William Harrison
Caudill, Reggie Jackson
Crawford, Martin
Davis, Carl George
Doughty, Julian O
Dybczak, Z(bigniew) W(ladyslaw)
French, Kenneth Edward
Gilbert, John Andrew
Hung, Ru J
Jones, Edward O(scar), Jr
Jones, Jess Harold
Jordan, William D(itmer)
Liu, C(hang) K(eng)

Liu, Frank C
Maples, Glennon
Morris, James Allen
Pears, Coultas D
Schaetzle, Walter J(acob)
Schroer, Bernard J
Talbot, T(homas) F
Thompson, Byrd Thomas, Jr
Thompson, Kenneth O(rval)
Walker, William F(red)
Yeh, Pu-Sen
Zalik, Richard Albert

ALASKA
Gosink, Joan P
Zarling, John P

ARIZONA
Ahmed, Saad Attia
Arnell, Walter James
Backus, Charles E
Baroczy, Charles J(ohn)
Barr, Lawrence Dale
Beakley, George Carroll, Jr
Bean, James J(oseph)
Chandra, Abhijit
Chen, Chuan Fang
Christensen, H(arvey) D(evon)
Cooperrider, Neil Kenneth
Cunningham, Richard G(reenlaw)
Davidson, Joseph Killworth
Evans, Donovan Lee
Fernando, Harindra Joseph
Florschuetz, Leon W(alter)
Fry, Harold
Fung, K Y
Gatley, William Stuart
Glick, Robert L
Govil, Sanjay
Hepworth, H(arry) Kent
Hilliard, Ronnie Lewis
Hirleman, Edwin Daniel, Jr
Jahsman, William Edward
Jordan, Richard Charles
Limbert, Douglas A(lan)
McGuirk, William Joseph
Matsch, L(ee) A(llan)
Matthews, James B
Metzger, Darryl E
Miller, Harry
Osborn, Donald Earl
Pearlstein, Arne Jacob
Peck, Robert E
Perkins, Henry Crawford, Jr
Ragsdell, Kenneth Martin
Rogers, Willard L(ewis)
Russell, Paul E(dgar)
Saric, William Samuel
Shaw, Milton C(layton)
Slaughter, Charles D
So, Ronald Ming Cho
Spurlock, Benjamin Hill, Jr
Suriano, F(rancis) J(oseph)
Thomas, R E
Wood, Bruce
Wood, Byard Dean
Wu, Hofu
Wygnanski, Israel Jerzy

ARKANSAS
Akin, Jim Howard
Deaver, Franklin Kennedy
Gilbrech, Donald Albert
Gleason, James Gordon
Jong, Ing-Chang
Kedzie, Donald P
Krohn, John Leslie
Reis, Irvin L

CALIFORNIA
Abdel-Ghaffar, Ahmed Mansour
Acosta, Allan James
Akin, Lee Stanley
Allen, Charles William
Alvares, Norman J
Antonsson, Erik Karl
Ardema, Mark D
Aref, Hassan
Arnold, Frank R(obert)
Arthur, Paul D(avid)
Auksmann, Boris
Austin, Arthur Leroy
Bagby, John P(endleton)
Bailey, Michael John
Barkan, P(hilip)
Battenburg, Joseph R
Beadle, Charles Wilson
Beighley, Clair M(yron)
Bell, Robert Alan
Bendisz, Kazimierz
Berger, Stanley A(llan)
Bernsen, Sidney A
Bevill, Vincent (Darell)
Bhushan, Bharat
Blackketter, Dennis O
Blackwelder, Ron F
Blake, Alexander
Blasingame, Benjamin P(aul)
Blink, James Allen
Bolt, Robert O'Connor
Boltinghouse, Joseph C
Bonin, John H(enry)
Bonora, Anthony Charles
Bowman, Craig T

Brandt, H(arry)
Bray, A Philip
Bright, Peter Bowman
Brock, John E(dison)
Broido, Jeffrey Hale
Brown, Tony Ray
Bruch, John C(larence), Jr
Burden, Harvey Worth
Bushnell, James Judson
Caligiuri, Robert Domenic
Cantin, Gilles
Carr, Robert Charles
Cass, Glen R
Chang, Daniel P Y
Charley, Philip J(ames)
Chen, Charles Shin-Yang
Cheng, Edward Teh-Chang
Chenoweth, James Merl
Chi, Cheng-Ching
Chou, Larry I-Hui
Choudhury, P Roy
Christensen, Richard Monson
Chu, Chung-Yu Chester
Clauser, Francis Hettinger
Clothier, Robert Frederic
Cloud, William K
Cochran, David L(eo)
Colton, James Dale
Comley, Peter Nigel
Conant, Curtis Terry
Conn, Robert William
Cooper, Thomas Edward
Cornet, I(srael) I(saac)
Culick, Fred E(llsworth) C(low)
Damonte, John Batista
D'Ardenne, Walter H
Das, Mihir Kumar
Dau, Gary John
Davey, Trevor B(lakely)
Davidson, Ernest
Davis, Charles Packard
Dean, Richard A
Deckert, Curtis Kenneth
Dergarabedian, Paul
Derr, Ronald Louis
Dipprey, Duane F(loyd)
Dong, Richard Gene
Dornfeld, David Alan
Dowell, Douglas C
Dudley, Darle W
Duffield, Jack Jay
Durbeck, Robert C(harles)
Dutt, Nikil D
Eaton, John Kelly
Edelman, Walter E(ugene), Jr
Edwards, Donald K
Egermeier, R(obert) P(aul)
Ellion, M Edmund
Eustis, Robert H(enry)
Evans, Gregory Herbert
Falcone, Patricia Kuntz
Fan, Chien
Faulders, Charles R(aymond)
Feinstein, Charles David
Ferziger, Joel H(enry)
Finnie, I(ain)
Fischer, George K
Flagan, Richard Charles
Fletcher, Thomas Harvey
Flora, Edward B(enjamin)
Folsom, Richard G(ilman)
Freberg, C(arl) Roger
Frey, Chris(tian) M(iller)
Friedmann, Peretz Peter
Frisch, Joseph
Fuchs, H(enry) O(tten)
Fuhs, Allen E(ugene)
Fung, Sui-an
Garbaccio, Donald Howard
Garry, Frederick W
Gay, Richard Leslie
Gerpheide, John H
Gershun, Theodore Leonard
Giedt, W(arren) H(arding)
Glenn, Lewis Alan
Gojny, Frank
Goluba, Raymond William
Goodwine, James K, Jr
Gordon, Hayden S(amuel)
Gould, William Richard
Grassi, Raymond Charles
Greif, Ralph
Haberman, Charles Morris
Haener, Juan
Hansmann, Douglas R
Harder, James Albert
Harvey, A(lexander)
Harvey, Frances J, II
Hauber, Janet Elaine
Haughton, Kenneth E
Haynes, Charles W(illard)
Healey, Anthony J
Helfman, Howard N
Henning, Carl Douglas
Hickman, Roy Scott
Hines, Douglas P
Hirasuna, Alan Ryo
Hoagland, Jack Charles
Hoffmann, Jon Arnold
Holeman, Dennis Leigh
Holliman, Albert Louis
Horne, Roland Nicholas
Horning, Donald O(ury)
Hovingh, Jack

Howe, Everett D(umser)
Hoyt, Jack W(allace)
Hsu, Chieh-Su
Huang, Francis F
Hudson, Donald E(llis)
Hussain, Nihad A
Jakubowski, Gerald S
Johanson, Jerry Ray
Johnson, Conor Deane
Johnson, David Leroy
Johnston, James P(aul)
Jory, Howard Roberts
Kane, E(neas) D(illon)
Karnopp, Dean Charles
Katz, Robert
Kaufman, Boris
Kays, William Morrow
Kayton, Myron
Kelleher, Matthew D(ennis)
Keller, Joseph Edward, Jr
Kelly, Robert Edward
Kemper, John D(ustin)
Kennedy, Ian Manning
Kesselring, John Paul
King, Hartley H(ughes)
Kline, Stephen Jay
Klipstein, David Hampton
Knapp, Karl
Kobayashi, Shiro
Kosmatka, John Benedict
Kraabel, John Stanford
Kruger, Charles Herman, Jr
Laderman, A(rnold) J(oseph)
La Fleur, James Kemble
Landis, James Noble
Lang, Thomas G(lenn)
Lavernia, Enrique Jose
Lavine, Adrienne Gail
Lay, Thorne
Lee, Peter H Y
Leifer, Larry J
Leitmann, G(eorge)
Levy, Salomon
Lewis, Francis Hotchkiss, Jr
Lick, Wilbert James
Lindsey, Gerald Herbert
Liu, Chen Ya
London, A(lexander) L(ouis)
Low, Lawrence J(acob)
Lowi, Alvin, Jr
Lucas, John W
Lucas, Robert Gillem
Luongo, Cesar Augusto
Ma, Fai
McCarthy, John Michael
McClure, Eldon Ray
McKillop, Allan A
McLeod, John Hugh, Jr
McMillan, Oden J
Macomber, Thomas Wesson
Madan, Ram Chand
Madsen, Richard Alfred
Marcus, Bruce David
Margolis, Stephen Barry
Marner, Wilbur Joseph
Martin, George
Marto, Paul James
Masri, Sami F(aiz)
Massier, Paul Ferdinand
Matheny, James Donald
Matthys, Eric Francois
Maulbetsch, John Stewart
Maxworthy, Tony
Melese, Gilbert B(ernard)
Meriam, James Lathrop
Merilo, Mati
Meyer, Brad Anthony
Meyers, Marc Andre
Miklowitz, Julius
Miller, Richard Keith
Mills, Anthony Francis
Mitchner, Morton
Moffat, Robert J
Moon, Donald W(ayne)
Moore, John R(obert)
Morrison, Frank Albert, Jr
Mow, C(hao) C(how)
Myers, Blake
Myronuk, Donald Joseph
Nathenson, Manuel
Netzer, David Willis
Newton, R(obert) E(ugene)
Nguyen, Luu Thanh
Nielan, Paul E
Nielsen, Helmer L(ouis)
Niles, Philip William Benjamin
Nooker, Eugene L(eRoy)
Noring, Jon Everett
Nunn, Robert Harry
Nypan, Lester Jens
Oehlberg, Richard N
Offen, George Richard
O'Hern, Eugene A
O'Meara, David Lillis
Oppenheim, A(ntoni) K(azimierz)
Orcutt, John Arthur
Owens, William Leo
Pagni, Patrick John
Pearson, John
Pedersen, Knud B(orge)
Pefley, Richard K
Pellinen, Donald Gary
Perlman, T(heodore)
Petrone, Rocco A

Pinkel, B(enjamin)
Pinto, John Gilbert
Potter, Richard C(arter)
Pressman, Ada Irene
Pucci, Paul F(rancis)
Rabl, Veronika Ariana
Rathmann, Carl Erich
Rau, Charles Alfred, Jr
Reardon, Frederick H(enry)
Reynolds, William Craig
Ritchie, Robert Oliver
Rivers, William J(ones)
Robben, Franklin Arthur
Robison, D(elbert) E(arl)
Ross, Bernard
Roth, Bernard
Rubin, Sheldon
Russell, T(homas) L(ee)
Russell, William T(reloar)
Salerno, Louis Joseph
Salmassy, Omar K
Salsig, William Winter, Jr
Samson, Sten
Sanders, Charles F(ranklin), Jr
Sarpkaya, Turgut
Sawyer, Robert Fennell
Schalla, Charence August
Scharton, Terry Don
Schefer, Robert Wilfred
Schmid-Schoenbein, Geert W
Schreiner, Robert Nicolas, Jr
Schrock, Virgil E(dwin)
Schurman, Glenn August
Schurmeier, Harris McIntosh
Schwartz, Daniel M(ax)
Schwartzbart, Harry
Seban, Ralph A
Seide, Paul
Shaffar, Scott William
Shah, Ramesh Trikamlal
Shanthikumar, Jeyaveerasingam George
Shelly, John Richard
Shoup, Terry Emerson
Silcox, William Henry
Sirignano, William Alfonso
Smith, Donald Stanley
Smith, Ora E
Smith, William R
Snyder, Nathan W(illiam)
Spier, Edward Ellis
Spinks, John Lee
Spitzer, Irwin Asher
Springett, David Roy
Steidel, Robert F(rancis), Jr
Stockel, Ivar H(oward)
Stone, Robert K(emper)
Stonebraker, Peter Michael
Stout, Ray Bernard
Street, Robert L(ynnwood)
Stuhmiller, James Hamilton
Swearengen, Jack Clayton
Sweeney, Donald Wesley
Takahashi, Yasundo
Talbot, Lawrence
Tellep, Daniel M
Thomas, Floyd W, Jr
Thomas, Graham Havens
Thrasher, L(awrence) W(illiam)
Throner, Guy Charles
Trezek, George J
Trubert, Marc
Tsao, Ching H
Unt, Hillar
Uyehara, Otto A(rthur)
Vanblarigan, Peter
Vanderplaats, Garret Niel
Van Sant, James Hurley, Jr
Velkoff, Henry Rene
Wang, Ji Ching
Wasley, Richard J(unior)
Welsh, David Edward
Wieland, Bruce Wendell
Wilde, D(ouglass) J(ames)
Wildmann, Manfred
Williams, Harry Edwin
Wood, Allen D(oane)
Wray, John L
Yang, Chun Chuan
Yeung, Ronald Wai-Chun
Yez, Martin S(imon)
Yildiz, Alaettin
Youngdahl, Paul F
Yuan, Sidney Wei Kwun
Zeren, Richard William
Zickel, John
Zivi, Samuel M(eisner)

COLORADO
Barber, Robert Edwin
Burnham, Marvin William
Carlson, Lawrence Evan
Chelton, Dudley B(oyd)
Crawford, Richard H
Daily, John W
Durham, Michael Dean
Ellis, Donald Griffith
Frangopol, Dan Mircea
Geers, Thomas L
Haberstroh, Robert D
Hittle, Douglas Carl
Jansson, David Guild
Johnson, Gearold Robert
Kassoy, David R
Kaufman, Harold Richard

Kerr, Robert McDougall
Kober, C(arl) L(eopold)
Kreith, Frank
Krill, Arthur Melvin
Kulacki, Francis Alfred
Ladd, Conrad Mervyn
Lee, Yung-Cheng
Loehrke, Richard Irwin
Lundstrom, Louis C
Mahajan, Roop Lal
Marshall, Charles F
Melsheimer, Frank Murphy
Meroney, Robert N
Mitchell, Charles Elliott
Murphy, John Michael
Norwood, Richard E(llis)
Pak, Ronald Y S
Paynter, Howard L
Peterson, Harry C(larence)
Radebaugh, Ray
Recht, Rodney F(rank)
Regenbrecht, D(ouglas) E(dward)
Rennat, Harry O(laf)
Schmidt, Alan Frederick
Siuru, William D, Jr
Smith, Frederick Willis
Stauffer, Jack B
Steward, W(illis) G(ene)
Suh, Chung-Ha
Summers, Luis Henry
Swanson, Lawrence Ray
Thompson, Erik G(rinde)
Troxell, Wade Oakes
Tuttle, Elizabeth R
Van Pelt, Richard W(arren)
Verschoor, J(ack) D(ahlstrom)
Wilbur, Paul James
Zoller, Paul

CONNECTICUT
Arnoldi, Walter Edwin
Berglund, Larry Glenn
Best, Stanley Gordon
Bowley, Wallace William
Bozzuto, Carl Richard
Brancato, Leo J(ohn)
Brand, Ronald S(cott)
Burridge, Robert
Burwell, Wayne Gregory
Chave, Charles Trudeau
Chin, Charles L(ee) D(ong)
Chiu, Yih-Ping
Cohen, Myron Leslie
Coogan, Charles H(enry), Jr
Crawford, John Okerson
Crossley, F(rancis) R(endel) Erskine
Dabora, Eli K
Dubin, Fred S
Enell, John Warren
Fink, Martin Ronald
Franz, Anselm
Garrett, Richard E
Johnson, Stephen Thomas
Kazerounian, Kazem
Ketchman, Jeffrey
Keyes, David Elliot
Kunz, Harold Russell
Lemkey, Frankin David
McFadden, Peter W(illiam)
Mack, Donald R(oy)
Maewal, Akhilesh
Ojalvo, Irving U
Paterson, Robert W
Peracchio, Aldo Anthony
Pitkin, Edward Thaddeus
Robinson, Donald W(allace), Jr
Rosner, Daniel E(dwin)
Schreier, Stefan
Shah, Mirza Mohammed
Sheets, Herman E(rnest)
Shichman, D(aniel)
Shuey, Merlin Arthur
Smith, Melvin I
Smooke, Mitchell D
Sreenivasan, Katepalli Raju
Warner, Thomas Clark, Jr
Williams, John Ernest

DELAWARE
Advani, Suresh Gopaldas
Bydal, Bruce A
Cantwell, Edward N(orton), Jr
Chou, Tsu-Wei
Dahlen, R(olf) J(ohn)
Dentel, Steven Keith
Dexter, Stephen C
Kaliakin, Victor Nicholas
Kingsbury, Herbert B
McKee, David Edward
Moore, Ralph Leslie
Murphy, Arthur Thomas
Pipes, Robert Byron
Santare, Michael Harold
Schwartz, Leonard William
Washburn, Robert Latham
Willis, Frank Marsden
Zimmerman, John R(ichard)

DISTRICT OF COLUMBIA
Alic, John A
Baer, Robert Lloyd
Baz, Amr Mahmoud Sabry
Beach, Harry Lee, Jr
Boehler, Gabriel D(ominique)

Chen, Leslie H(ung)
Cox, J E
Deshpande, Mohan Dhondorao
Foa, J(oseph) V(ictor)
Frair, Karen Lee
Gallagher, William J(oseph)
Gould, Phillip
Hartman, Patrick James
Hazelrigg, George Arthur, Jr
Huber, Peter William
Jackson, William David
Jones, Douglas Linwood
Kaye, John
Kelly, George Eugene
Kelnhofer, William Joseph
Korkegi, Robert Hani
Kramer, Bruce Michael
Larsen-Basse, Jorn
Larson, Charles Fred
Menton, Robert Thomas
Ojalvo, Morris S(olomon)
Parks, Vincent Joseph
Pei, Richard Yu-Sien
Robertson, A(lexander) F(rancis)
Rockett, John A
Rosenthal, F(elix)
Sanders, Robert Charles
Walker, M Lucius, Jr
Warnick, Walter Lee
Way, George H
Wilmotte, Raymond M
Wolko, Howard Stephen
Youm, Youngil
Zucchetto, James John

FLORIDA
Adt, Robert (Roy)
Anghaie, Samim
Anusavice, Kenneth John
Berger, Phyllis Belous
Bigley, Harry Andrew, Jr
Billings, Charles Edgar
Bober, William
Bowman, Thomas Eugene
Buzyna, George
Case, Robert Oliver
Catz, Jerome
Chow, Wen Lung
Claridge, Richard Allen
Coar, Richard J
Collier, Melvin Lowell
Cross, Ralph Emerson
DeHart, Arnold O'Dell
Diaz, Nils Juan
Dilpare, Armand Leon
Doering, Robert Distler
Drucker, Daniel Charles
Eckstein, Eugene Charles
Edwards, Thomas Claude
Etherington, Harold
Farber, Erich A(lexander)
Foster, J Earl
Gaither, Robert Barker
Gencsoy, Hasan Tahsin
Hahn, C(harles) Archie, Jr
Hartley, Craig Sheridan
Hartman, John Paul
Hoeppner, Conrad Henry
Hosler, E(arl) Ramon
Hosni, Yasser Ali
Hoyer-Ellefsen, Sigurd
Hrones, John Anthony
Hsieh, Chung Kuo
Juvinall, Robert C
King, Blake
Kinney, Robert Bruce
Kranc, Stanley Charles
Langford, David
Lijewski, Lawrence Edward
Lundgren, Dale A(llen)
McCracken, Walter John
Mandil, I Harry
Miller, Robert Gerry
Mittleman, John
Nimmo, Bruce Glen
Oliver, Calvin C(leek)
Phillips, Winfred M(arshall)
Piotrowski, George
Proctor, Charles Lafayette, II
Psarouthakis, John
Reichard, Ronnal Paul
Reisbig, Ronald Luther
Reiter, William Frederick, Jr
Rice, Stephen Landon
Roan, Vernon P
Sandor, George N(ason)
Scaringe, Robert P
Schimmel, Walter Paul
Schmidtke, R(ichard) A(llen)
Scott, Linus Albert
Seireg, Ali A
Shaw, Lawrance Neil
Sheppard, Donald M(ax)
Sias, Frederick Ralph
Silverstein, Abe
Silvestri, George J, Jr
Suciu, S(piridon) N
Sun, Chang-Tsan
Sundaram, Swaminatha
Tam, Christopher K W
Teixeira, Arthur Alves
Traexler, John F
Ward, Leonard George
Witzell, O(tto) W(illiam)

Mechanical Engineering (cont)

Woll, Edward
Zimmerman, Norman H(erbert)
Zubko, L(eonard) M(artin)

GEORGIA
Abdel-Khalik, Said Ibrahim
Antolovich, Stephen D
Atluri, Satya N
Barnett, Samuel C(larence)
Black, William Z(achary)
Colton, Jonathan Stuart
Craig, James I
Desai, Prateen V
Dickerson, Stephen L(ang)
Durbetaki, Pandeli
Flandro, Gary A
Freeston, W Denney, Jr
Ghate, Suhas Ramkrishna
Ginsberg, Jerry Hal
Goglia, M(ario) J(oseph)
Hodgdon, F(rank) E(llis)
Kadaba, Prasanna V
Knight, Lee H, Jr
Krol, Joseph
Ku, David Nelson
Lau, Jark Chong
Lee, Kok-Meng
Matula, Richard A
May, Gerald Ware
Meyers, Carolyn Winstead
Neitzel, George Paul
Nerem, Robert Michael
Orloff, David Ira
Rouse, William Bradford
Salant, Richard Frank
Shakun, Wallace
Stalford, Harold Lenn
Stanley, Luticious Bryan, Jr
Staton, Rocker Theodore, Jr
Strahle, Warren C(harles)
Vachon, Reginald Irenee
Vicory, William Anthony
Wempner, Gerald Arthur
Wepfer, William J
Winer, Ward Otis
Wu, James Chen-Yuan
Zinn, Ben T

HAWAII
Antal, Michael Jerry, Jr
Bartosik, Alexander Michael
Cheng, Ping
Chou, James C S
Fand, Richard Meyer
Fox, Joel S
Hallanger, Lawrence William
Hihara, Lloyd Hiromi

IDAHO
Jacobsen, Richard T
Korth, Gary E
Larson, Jay Reinhold
Sohal, Manohar Singh
Stewart, Richard Byron
Trout, Thomas James

ILLINOIS
Addy, Alva LeRoy
Adrian, Ronald John
Anderson, Thomas Patrick
Assanis, Dennis N
Balcerzak, Marion John
Bell, Charles Eugene, Jr
Benjamin, Roland John
Berry, Gregory Franklin
Block, Stanley M(arlin)
Brauer, Roger L
Brewster, Marcus Quinn
Bullard, Clark W
Bunting, Bruce Gordon
Chakrabarti, Subrata K
Chao, B(ei) T(se)
Chaszeyka, Michael A(ndrew)
Chato, John C(lark)
Chen, Michael Ming
Chittenden, William A
Chiu, Huei-Huang
Choi, Stephen U S
Chung, Paul M(yungha)
Clausing, A(rthur) M(arvin)
Conry, Thomas Francis
Cusano, Cristino
Daniel, Isaac M
Dantzig, Jonathan A
Deen, James Robert
Deitrich, L(awrence) Walter
Dix, Rollin Cumming
Dolan, Thomas J(ames)
D'Souza, Anthony Frank
Dunning, Ernest Leon
Dutton, Jonathan Craig
Erwin, Lewis
Eshleman, Ronald L
Farhadieh, Rouyentan
Francis, John Elbert
Freedman, Steven I(rwin)
Frey, Donald N(elson)
Graae, Johan E A
Gupta, Krishna Chandra
Halleen, Robert M(arvin)
Hartenberg, Richard S(cheunemann)
Hartnett, James P(atrick)

Holmes, L(awrence) B(ruce)
Horve, Leslie A
Hull, William L(avaldin)
Jefferson, Thomas Bradley
Jones, Barclay G(eorge)
Kaganov, Alan Lawrence
Kalpakjian, Serope
Kasuba, Romualdas
Kenny, Andrew Augustine
Kern, Roy Fredrick
Konzo, Seichi
Korst, Helmut Hans
Kottas, Harry
Kovitz, Arthur A(braham)
Leach, James L(indsay)
Lemke, Donald G(eorge)
Levine, Michael W
Liu, Wing Kam
Lottes, P(aul) A(lbert)
McKee, Keith Earl
Manos, William P
Mapother, Dillon Edward
Marr, William Wei-Yi
Martinec, Emil Louis
Matalon, Moshe
Minkowycz, W J
Mokadam, Raghunath G(anpatrao)
Moran, Thomas J
Morel, Thomas
Morkovin, Mark V(ladimir)
Nachtman, Elliot Simon
O'Bryant, David Claude
Okamura, Kiyohisa
Orthwein, W(illiam) C(oe)
Pearlstein, Arne Jacob
Peters, James Empson
Phillips, Rohan Hilary
Pickett, Leroy Kenneth
Porter, Robert William
Poulikakos, Dimos
Reichard, Grant Wesley
Rettaliata, John Theodore
Riahi, Daniel Nourollah
Ruhl, Roland Luther
Salzenstein, Marvin A(braham)
Sather, Norman F(redrick)
Shabana, Ahmed Abdelraouf
Shack, William John
Siegal, Burton L
Singh, Shyam N
Soo, Shao-Lee
Spector, Leo Francis
Spotts, M(erhyle) F(ranklin)
Thomas, Anthony
Thompson, W(illiam) E(dgar)
Ting, Thomas C(hi) T(sai)
Tipei, Nicolae
Trigger, Kenneth James
Tucker, Charles L
Tuzson, John J(anos)
Uherka, Kenneth Leroy
Verderber, Joseph Anthony
Walker, John Scott
Wang, Pie-Yi
Warinner, Douglas Keith
Weber, Norman
Weeks, Richard William
Wessler, Max Alden
White, Robert Allan
Widera, Georg Ernst Otto
Wiley, Jack Cleveland
Wilson, William Robert Dunwoody
Wirtz, Gerald Paul
Woods, Kenneth R
Worek, William Martin
Worley, Will J
Wurm, Jaroslav
Yoerger, Roger R
Zell, Blair Paul

INDIANA
Barash, Moshe M
Bergdolt, Vollmar Edgar
Brach, Raymond M
Brown, Charles L(eonard)
Carey, Alfred W(illiam), Jr
Carroll, John T, III
Chiang, Donald C
Citron, Stephen J
Cohen, Raymond
Dalphin, John Francis
DeWitt, David P
Fox, Robert William
Gad-el-Hak, Mohamed
Gibson, James Darrell
Goldschmidt, Victor W
Gorman, Eugene Francis
Hall, A(llen) S(trickland), Jr
Hamilton, James F(rancis)
Hansen, Arthur G(ene)
Hartsaw, William O
Hawks, Keith Harold
Ho, Cho-Yen
Hoffman, Joe Douglas
Holowenko, A(lfred) R(ichard)
Hooper, Irvin P(latt)
Howland, George Russell
Huang, Nai-Chien
Huffman, John Curtis
Incropera, Frank P
Jefferies, Michael John
Jerger, E(dward) W
Jones, James Darren
Kareem, Ahsan

Kazem, Sayyed M
Kruger, Fred W
Laurendeau, Normand Maurice
L'Ecuyer, Mel R
Lehmann, Gilbert Mark
Liley, Peter Edward
McAleece, Donald John
McComas, Stuart T
McDonald, Alan T(aylor)
Mercer, Walter Ronald
Messal, Edward Emil
Modrey, Joseph
Osborn, J(ohn) R(obert)
Ovens, William George
Pearson, Joseph T(atem)
Phillips, James W
Pierson, Edward S
Quinn, B(ayard) E(lmer)
Quinn, C Jack
Ramadhyani, Satish
Raven, Francis Harvey
Razdan, Mohan Kishen
Schoenhals, Robert James
Sen, Mihir
Shahidi, Freydoon
Sieloff, Ronald F
Smith, Charles O(liver)
Soedel, Werner
Stevenson, Warren H
Szewczyk, Albin A
Taylor, Harold Leroy
Taylor, Raymond Ellory
Thompson, Howard Doyle
Tiederman, William Gregg, Jr
Tree, David R
Wark, Kenneth, Jr
Warner, Cecil F(rancis)
Widener, Edward Ladd, Sr
Yang, Henry T Y
Yang, Kwang-Tzu
Zoss, Leslie M(ilton)

IOWA
Akers, Arthur
Baumgarten, Joseph Russell
Beckermann, Christoph
Black, Henry M(ontgomery)
Chen, Ching-Jen
Choi, Kyung Kook
Chwang, Allen Tse-Yung
Cook, William John
Fellinger, Robert C(ecil)
Gurll, Nelson
Hall, Jerry Lee
Haug, Edward J, Jr
Junkhan, George H
Kusiak, Andrew
Lance, George M(ilward)
Landweber, Louis
Larsen, William L(awrence)
McConnell, Kenneth G
Mack, Michael J
Madsen, Donald H(oward)
Mischke, Charles R(ussell)
Myers, Thomas Wilmer
Northup, Larry L(ee)
Okiishi, Theodore Hisao
Peters, Leo Charles
Pletcher, Richard H
Rocklin, Isadore J
Scholz, Paul Drummond
Serovy, George K(aspar)
Stephens, Ralph Ivan
Tannehill, John Charles
Trummel, J(ohn) Merle

KANSAS
Annis, Jason Carl
Appl, Fredric Carl
Barr, B(illie) Griffith
Burmeister, Louis C
Cheng, Lester (Le-Chung)
Forman, George W
Giri, Jagannath
Gorton, Robert Lester
Gosman, Albert Louis
Gowdy, Kenneth King
Graham, A Richard
Huang, Chi-Lung Dominic
Hwang, Ching-Lai
Johnson, Richard T(errell)
Jovanovic, M(ilan) K(osta)
Kipp, Harold Lyman
Lindholm, John C
Lyles, Leon
McCabe, Steven Lee
Miller, Paul Leroy, Jr
Razak, Charles Kenneth
Robinson, M John
Smith, Raymond V(irgil)
Thompson, Joseph Garth
Turnquist, Ralph Otto
Walker, Hugh S(anders)
Wattson, Robert K(ean), Jr
Wentz, William Henry, (Jr)

KENTUCKY
Beatty, Millard Fillmore, Jr
Conley, Weld E
Cremers, Clifford J
Drake, Robert M, Jr
Eaton, Thomas Eldon
Edwards, Richard Glenn
Frank, John L

Funk, James Ellis
Gard, O(liver) W(illiam)
Gold, Harold Eugene
Hahn, Ottfried J
Leep, Herman Ross
Leigh, Donald C
Lu, Wei-yang
Marshall, Maurice K(eith)
Mason, Harry Louis
Shupe, Dean Stanley
Tran, Long Trieu
Wilhoit, James Cammack, Jr

LOUISIANA
Adams, Kenneth H
Barron, Randall F(ranklin)
Chieri, P(ericle) A(driano)
Cospolich, James D
Deddens, J(ames) C(arroll)
Eaton, Harvill Carlton
Hamilton, D(ewitt) C(linton), Jr
Hewitt, Hudy C, Jr
Lai, Ying-San
Lowther, James David
Munchmeyer, Frederick Clarke
Parish, Richard Lee
Peyronnin, Chester A(rthur), Jr
Rotty, Ralph M(cGee)
Russo, Edwin Price
Saha, Subrata
Siebel, M(athias) P(aul) L
Sogin, H(arold) H
Thigpen, J(oseph) J(ackson)
Thompson, Hugh Allison
Trammell, Grover J(ackson), Jr
Whitehouse, Gerald D(ean)
Whitehurst, Charles A(ugustus)

MAINE
Hill, Richard C(onrad)
Sucec, James
Webster, Karl Smith
Weinstein, Alvin Seymour
Wilbur, L(eslie) C(lifford)

MARYLAND
Adams, James Alan
Armstrong, R(onald) W(illiam)
Baker, Kenneth L(eroy)
Barr, William A(lexander)
Bascunana, Jose Luis
Baum, Howard Richard
Berger, Bruce S
Bernard, Peter Simon
Bock, Arthur E(mil)
Brown, Paul Joseph
Burns, Bruce Peter
Burns, Timothy John
Camponeschi, Eugene Thomas, Jr
Chen, Benjamin Yun-Hai
Cleveland, William Grover, Jr
Dailey, George
Dally, James William
Douglas, Andrew Sholto
Drzewiecki, Tadeusz Maria
Dugoff, Howard
Duncan, James Playford
Fabic, Stanislav
Fabunmi, James Ayinde
Flynn, Paul D(avid)
Fowell, Andrew John
Franke, Gene Louis
Goldstein, Seth Richard
Gupta, Ashwani Kumar
Hansen, Robert Jack
Hardis, Leonard
Hasson, Dennis Francis
Hawkins, William M(adison), Jr
Haynes, Emanuel
Heller, Edward Lincoln
Hill, James Edward
Hopenfield, Joram
Hsu, Stephen M
Hunter, Lawrence Wilbert
Irwin, George Rankin
Jabbour, Kahtan Nicolas
Jones, Everett
Kamrass, Murray
Khatri, Hiralal C
Kiang, Robert L
Kitchens, Clarence Wesley, Jr
Koh, Severino Legarda
Kusuda, Tamami
Landsburg, Alexander Charles
Lange, Eugene Albert
Lopardo, Vincent Joseph
Loss, Frank J
Lu, Yeh-Pei
Lundsager, C(hristian) Bent
Marks, Colin H
Marlowe, Donald E(dward)
Martin, John J(oseph)
Mason, Henry Lea
Mintz, Fred
Mitler, Henri Emmanuel
Mordfin, Leonard
Morris, Alan
Nash, Jonathon Michael
Neild, A(lton) Bayne
Obremski, Henry J(ohn)
Pamidi, Prabhakar Ramarao
Peppin, Richard J
Perrone, Nicholas
Popel, Aleksander S

Poulose, Pathickal K
Presser, Cary
Quintiere, James G
Radermacher, Reinhard
Rich, Harry Louis
Riley, Claude Frank, Jr
Ritter, Joseph John
Rostamian, Rouben
Rushing, Frank C
Saarlas, Maido
Sallet, Dirse Wilkis
Sandusky, Harold William
Sayre, Clifford L(eRoy), Jr
Schleiter, Thomas Gerard
Schwartz, John T
Scialdone, John Joseph
Seigel, Arnold E(lliott)
Semerjian, Hratch G
Sevik, Maurice
Sharpe, William Norman, Jr
Smedley, William Michael
Smith, John Henry
Smith, Russell Aubrey
Smylie, Robert Edwin
Statt, Terry G
Strombotne, Richard L(amar)
Sturek, Walter Beynon
Tai, Tsze Cheng
Thiruvengadam, Alagu Pillai
Tsai, Lung-Wen
Tuttle, Kenneth Lewis
Urbach, Herman B
Wallace, James M
Walston, William H(oward), Jr
Weckesser, Louis Benjamin
Young, C(harles), Jr
Zmola, Paul C(arl)

MASSACHUSETTS

Adams, George G
Allen, Steven
Alpert, Ronald L
Ambs, Lawrence Lacy
Anderson, J(ohn) Edward
Anderson, John Edward
Argon, Ali Suphi
Astill, Kenneth Norman
Avila, Charles Francis
Bailey, Bruce M(onroe)
Blake, Thomas R
Bloomfield, David Peter
Bolz, R(ay) E(mil)
Borden, Roger R(ichmond)
Brosens, Pierre Joseph
Brown, Robert Lee
Burke, Shawn Edmund
Butterworth, George A M
Carson, John William
Chryssostomidis, Chryssostomos
Chu, Chauncey C
Cipolla, John William, Jr
Clark, Llewellyn Evans
Cook, Nathan Henry
Cooper, W(illiam) E(ugene)
Copley, Lawrence Gordon
Corey, Harold Scott
Crandall, Stephen H
Darkazalli, Ghazi
Daskin, W(alter)
De Fazio, Thomas Luca
Delagi, Richard Gregory
DiPippo, Ronald
Dittfach, John Harland
Dixon, John R
Dubowsky, Steven
Dunn, John Frederick, Jr
Ehrich, Fredric F(ranklin)
Fay, James A(lan)
Feldman, Charles Lawrence
Foster, Arthur R(owe)
Gaggioli, Richard A
Godrick, Joseph Adam
Goela, Jitendra Singh
Gold, Harris
Goss, William Paul
Greif, Robert
Greitzer, Edward Marc
Griffith, Peter
Hahn, Robert S(impson)
Hals, Finn
Haworth, D(onald) R(obert)
Heywood, John Benjamin
Hoffman, Allen Herbert
Hurt, William C, Jr
Iwasa, Yukikazu
Jakus, Karl
John, James Edward Albert
Johnson, Walter Roland
Kachanov, Mark L
Kamien, C Zelman
Katz, Israel
Kazimi, Mujid S
Kerney, Peter Joseph
Kirkpatrick, E(dward) T(homson)
Krause, Irvin
Kronauer, Richard Ernest
Lee, Harvey S
Lee, Shih-Ying
Lemnios, A(ndrew) Z
Leung, Woon F (Wallace)
Levitt, Albert P
Lindberg, Edward E
Loretz, Thomas J
McClintock, Frank A(mbrose)

McGowan, Jon Gerald
MacGregor, C(harles) W(inters)
Malkin, Stephen
Mann, Robert W(ellesley)
Manning, Jerome Edward
Mark, Melvin
Mastronardi, Richard
Milo, Henry L(ouis), Jr
Modak, Ashok Trimbak
Moore, Robert Edmund
Motherway, Joseph E
Murch, Laurence Everett
Nelson, Frederick Carl
Nye, Edwin (Packard)
Pan, Coda H T
Papernik, Lazar
Park, B J
Paul, Igor
Pellini, William S
Picha, Kenneth G(eorge)
Poirier, Victor L
Poli, Corrado
Pope, Joseph
Pope, Mary E
Probstein, Ronald F(ilmore)
Pulling, Nathaniel H(osler)
Rabinowicz, Ernest
Rapperport, Eugene J
Reed, F(lood) Everett
Reeves, Barry L(ucas)
Reichenbach, George Sheridan
Rios, Pedro Agustin
Rohsenow, Warren M(ax)
Rosenthal, Richard Alan
Rossettos, John N(icholas)
Schenck, Hilbert Van Nydeck, Jr
Scott, Kenneth Elsner
Selig, Ernest Theodore
Shapiro, Ascher H(erman)
Sharp, A(rnold) G(ideon)
Sheridan, Thomas Brown
Silvers, J(ohn) P(hillip)
Sioui, Richard Henry
Smith, Joseph LeConte, Jr
Sonin, Ain A(nts)
Staker, Michael Ray
Staszeky, Francis M
Stekly, Z J John
Stevens, Herbert H(owe), Jr
Suh, Nam Pyo
Sunderland, James Edward
Thompson, Charles
Tong, Pin
Toong, Tau-Yi
Topping, Richard Francis
Toscano, William Michael
Udelson, Daniel G(erald)
Ungar, Eric E(dward)
Vest, Charles Marstiller
Wadleigh, Kenneth R(obert)
Weinberg, Marc Steven
Wheildon, W(illiam) M(axwell), Jr
Whitney, Daniel Eugene
Williams, James Henry, Jr
Wilson, David Gordon
Winfrey, Richard Cameron
Wyler, John Stephen
Yener, Yaman
Zinsmeister, George Emil
Zotos, John
Zwiep, Donald N

MICHIGAN

Adams, William Eugene
Agnew, William G(eorge)
Akay, Adnan
Alexandridis, Alexander A
Amann, Charles A(lbert)
Ashcroft, Frederick H
Atkin, Rupert Lloyd
Atreya, Arvind
Barber, James Richard
Barnes, Gerald Joseph
Batrin, George Leslie
Beck, James V(ere)
Beltz, Charles R(obert)
Blaser, Dwight A
Boddy, David Edwin
Bolt, Jay A(rthur)
Brehob, W(ayne) M
Browne, Vance D'Armond
Caplan, John D(avid)
Chace, Milton A
Chen, Francis Hap-Kwong
Chen, Yudong
Chesebrough, Harry E
Cisler, Walker L(ee)
Clark, John A(lden)
Clark, Samuel Kelly
Cloud, Gary Lee
Cole, David Edward
Colucci, Joseph M(ichael)
Datsko, Joseph
Despres, Thomas A
Dhanak, Amritlal M(aganlal)
Edgerton, Robert Howard
Eltinge, Lamont
Everett, Robert Line
Ezzat, Hazem Ahmed
Felbeck, David K(niseley)
Fleischer, Henry
Foss, John F
Frutiger, Robert Lester
Fyhrie, David Paul

Galan, Louis
Gibson, Harold J(ames)
Gillespie, Thomas David
Graebel, William P(aul)
Gratch, Serge
Greene, Bruce Edgar
Habib, Izzeddin Salim
Hammitt, Frederick G(nichtel)
Haviland, Merrill L
Hayes, Edward J(ames)
Hays, Donald F(rank)
Henein, Naeim A
Hrovat, Davorin
Hunstad, Norman A(llen)
Im, Jang Hi
Ishihara, Teruo (Terry)
Johnson, Glen Eric
Johnson, John Harris
Jones, R(ichard) James
Kabel, Richard Harvey
Kauppila, Raymond William
Kehrl, Howard H
Keller, Robert B
Kerber, Ronald Lee
King, Albert Ignatius
Klomp, Edward
Krieger, Roger B
Kullgren, Thomas Edward
Kurajian, George Masrob
Lamberson, Leonard Roy
Lay, Joachim E(llery)
Lett, Philip W(ood), Jr
Li, Chin-Hsiu
Libertiny, George Zoltan
Little, Robert E(ugene)
Loofbourrow, Alan G
Lord, Harold Wesley
Lubkin, James Leigh
Ludema, Kenneth C
Lund, Charles Edward
McGrath, John Joseph
Martin, George H(enry)
Merte, Herman, Jr
Mirsky, W(illiam)
Misch, Herbert Louis
Narain, Amitabh
Nefske, Donald Joseph
Nyquist, Gerald Warren
Oh, Hilario Lim
Pandit, Sudhakar Madhavrao
Passerello, Chris Edward
Pearson, J(ohn) Raymond
Petersen, Donald E
Peterson, Richard Carl
Petrick, Ernest N(icholas)
Petrof, Robert Charles
Piccirelli, Robert Anthony
Potter, Merle C(larence)
Prasad, Biren
Quader, Ather Abdul
Rivard, Jerome G
Rodgers, John James
Rohde, Steve Mark
Roll, James Robert
Rollinger, Charles N(icholas)
Rosenberg, Richard Carl
Rosenberg, Ronald C(arl)
Rosinski, Michael A
Sagi, Charles J(oseph)
Schenk, Walter John
Schultz, Albert Barry
Schwerzler, Dennis David
Segel, L(eonard)
Shapton, William Robert
Shigley, Joseph E(dward)
Shipinski, John
Shkolnikov, Moisey B
Singh, Trilochan (Hardeep)
Smith, Gene E
Smith, Hadley J(ames)
Sonntag, Richard E
Soutas-Little, Robert William
Stivender, Donald Lewis
Suryanarayana, Narasipur Venkataram
Taplin, Lael Brent
Taraman, Khalil Showky
Taulbee, Carl D(onald)
Thompson, James Lowry
Tison, Richard Perry
Trachman, Edward G
Van Poolen, Lambert John
Van Wylen, Gordon J(ohn)
Wang, Dazong
Wedekind, Gilbert Leroy
Whicker, Donald
Willermet, Pierre Andre
Williams, Gwynne
Wolf, Joseph A(llen), Jr
Wu, Hai
Wu, Shien-Ming
Yang, Wen Jei
Yu, Mason K

MINNESOTA

Bailey, Fredric Nelson
Beavers, Gordon Stanley
Blackshear, Perry L(ynnfield), Jr
Bonne, Ulrich
Chase, Thomas Richard
Davidson, Donald
Erdman, Arthur Guy
Frohrib, Darrell A
Goldstein, R(ichard) J(ay)
Heberlein, Joachim Viktor Rudolf

Ibele, Warren Edward
Jacobs, Carl Henry
Johnson, Robert L(awrence)
Joseph, Daniel D
Joseph, Earl Clark, II
Kallok, Michael John
Keinath, Gerald E
Keshaviah, Prakash Ramnathpur
Kittelson, David Burnelle
Kleinhenz, William A
Knappe, Laverne F
Larson, Vincent H(ennix)
Liu, Benjamin Y H
McElrath, Gayle W(illiam)
Marple, Virgil Alan
Ogata, Katsuhiko
Olson, Reuben Magnus
Pfender, Emil
Raley, Frank Austin
Ramilingam, Subbiah
Riley, Donald Ray
Simon, Terrence William
Smith, David Philip
Sparrow, E(phraim) M(aurice)
Strait, John
Uban, Stephen A
Verma, Deepak Kumar
Whitby, Kenneth T(homas)

MISSISSIPPI

Altenkirch, Robert Ames
Bouchillon, Charles W
Brenkert, Karl, Jr
Brown, Joseph M
Carley, Charles Team, Jr
Fox, John Arthur
Jasper, Martin Theophilus
Posey, Joe Wesley
Powe, Ralph Elward
Rosenhan, A Kirk
Scott, Charley

MISSOURI

Armaly, Bassem Farid
Batra, Romesh Chander
Blundell, James Kenneth
Brooker, Donald Brown
Carter, Don E
Chen, Ta-Shen
Creighton, Donald L
Culp, Archie W
Dharani, Lokeswarappa R
Dreifke, Gerald E(dmond)
Duffield, Roger C
Elias, Thomas Ittan
Faucett, T(homas) R(ichard)
Flanigan, Virgil James
Fowler, Timothy John
Francis, Lyman L(eslie)
Gardner, Richard A
Gillespie, LaRoux King
Heimann, Robert L
Hines, Anthony Loring
Howell, Ronald Hunter
Husar, Rudolf Bertalan
Johnson, Kurt P
Kimel, William R(obert)
Kovacs, Sandor J, Jr
Koval, Leslie R(obert)
Laurenson, Robert Mark
Lehnhoff, Terry Franklin
Leutzinger, Rudolph L(eslie)
Look, Dwight Chester, Jr
Lowe, Forrest Gilbert
Loyalka, Sudarshan Kumar
Lysen, John C
Mathiprakasam, Balakrishnan
Miles, John B(ruce)
Oetting, Robert B(enfield)
Painter, James Howard
Perisho, Clarence H(oward)
Pringle, Oran A(llan)
Remington, C(harles) R(oy), Jr
Sauer, Harry John, Jr
Schott, Jeffrey Howard
Sutera, Salvatore P
Swenson, Donald Otis
Warder, Richard C, Jr
Weissenburger, Jason T
Young, Ross Darell

MONTANA

Albini, Frank Addison
Gerez, Victor
Herndon, Charles L(ee)
Scanlan, J(ack) A(ddison), Jr
Townes, Harry W(arren)

NEBRASKA

Brogan, William L
Lauck, Francis W
Leviticus, Louis I
Lu, Pau-Chang
McDougal, Robert Nelson
Newhouse, Keith N
Peters, Alexander Robert
Rohde, Suzanne Louise
Wittke, Dayton D
Wolford, James C
Woods, Joseph

NEVADA

Anderson, James T(reat)
Boehm, Robert Foty

Mechanical Engineering (cont)

Chandra, Dhanesh
Holdbrook, Ronald Gail
McKee, Robert B(ruce), Jr
Turner, Robert Harold
Wirtz, Richard Anthony

NEW HAMPSHIRE
Allmendinger, E Eugene
Baumann, Hans D
Black, Sydney D
Block, James A
Browne, W(illiam) H(ooper)
Colon, Frazier Paige
Feely, Frank Joseph, Jr
Fricke, Edwin Francis
Hill, Percy Holmes
Hoagland, Lawrence Clay
Jenike, Andrew W(itold)
Limbert, David Edwin
Liston, Ronald Argyle
Lunardini, Virgil J(oseph), Jr
Savage, Godfrey H
Spehrley, Charles W, Jr
Temple, Peter Lawrence

NEW JERSEY
Abraham, John
Becht, Charles, IV
Ben-Israel, Adi
Bergman, Robert
Berman, Irwin
Bhagat, Phiroz Maneck
Brady, Eugene F
Brandmaier, Harold Edmund
Briggs, David Griffith
Brunell, Karl
Bush, Charles Edward
Carlson, Harold C(arl) R(aymond)
Cerkanowicz, Anthony Edward
Chen, Rong Yaw
Cho, Soung Moo
Connolly, John Charles
Daman, Ernest L
Davis, William F
DePalma, Anthony Michael
Domeshek, S(ol)
Donovan, Richard C
Duerr, J Stephen
Durbin, Enoch Job
Ehrlich, I(ra) Robert
Eipper, Eugene B(retherton)
Fenster, Saul K
File, Joseph
Flaherty, Franklin Trimby, Jr
Florio, Pasquale J, Jr
Frenkiel, Richard H
Gabriel, Edwin Z
Geskin, Ernest S
Glassman, Irvin
Goyal, Suresh
Griskey, R(ichard) G(eorge)
Griswold, Edward Mansfield
Groeger, Theodore Oskar
Harford, James J
Hart, Colin Patrick
Hollenbach, Edwin
Hoover, Charles Wilson, Jr
Hrycak, Peter
Hsieh, Jui Sheng
Hung, Tonney H M
Imming, Harry S(tanley)
Iyer, Ram R
Jackson, Andrew
Jaluria, Yogesh
Jones, Charles
Kane, Ronald S(teven)
Khalifa, Ramzi A
Kirchner, Robert P
Kovats, Andre
Langrana, Noshir A
Lestingi, Joseph Francis
Leu, Ming C
Li, Hsin Lang
Lubowe, Anthony G(arner)
Luxhoj, James Thomas
McCullough, Robert William
McGregor, Walter
Magee, Richard Stephen
Mandel, Andrea Sue
Manson, Simon V(erril)
Marscher, William Donnelly
Matthews, E(dward) K
Maxwell, Bryce
Mellor, George Lincoln, Jr
Michaelis, Paul Charles
Miller, Edward
Moshey, Edward A
Mutter, William Hugh
Nardone, John
Orszag, Steven Alan
Pelan, Byron J
Percarpio, Edward P(eter)
Perry, Erik David
Philipp, Ronald E
Polaner, Jerome L(ester)
Prasad, Marehalli Gopalan
Pschunder, Ralph J(osef)
Raichel, Daniel R(ichter)
Raj, Rishi S
Roe, Kenneth A
Sachs, Harvey Maurice
Sarkar, Kamalaksha

Schlink, F(rederick) J(ohn)
Schmidlin, Albertus Ernest
Schnabel, George Joseph
Schnapf, Abraham
Segelken, John Maurice
Sernas, Valentinas A
Shah, Hasmukh N
Sisto, Fernando
Socolow, Robert H(arry)
Stamper, Eugene
Stengel, Robert Frank
Suhir, Ephraim
Temkin, Samuel
Thomas, Stanislaus S(tephen)
Thurston, Robert Norton
Weinzimmer, Fred
Wells, Herbert Arthur
Werner, Daniel Paul
Wesner, John William
Williams, Maryon Johnston, Jr
Wilson, Charles Elmer
Young, Li
Yu, Yi-Yuan
Zebib, Abdelfattah M G

NEW MEXICO
Bertin, John Joseph
Blackwell, Bennie Francis
Blejwas, Thomas Edward
Bryant, Lawrence E, Jr
Burchett, O Neill J
Burick, Richard Joseph
Clouser, William Sands
Cobble, Milan Houston
Diver, Richard Boyer, Jr
Dunn, Richard B
Ebbesen, Lynn Royce
Ebrahimi, Nader Dabir
El-Genk, Mohamed Shafik
Ellett, D Maxwell
Fisher, Franklin E(ugene)
Ford, C(larence) Quentin
Fuka, Louis Richard
Garcia, Carlos E(rnesto)
Gibbs, Terry Ralph
Gilbert, Joel Sterling
Godwin, Richard P
Harary, Frank
Hardin, James T
Hsu, Y(oun) C(hang)
Jones, Orval Elmer
Kashiwa, Bryan Andrew
Kelley, Gregory M
Kimbrell, Jack T(heodore)
Lam, Kin Leung
Lawrence, Raymond Jeffery
Lee, David Oi
Lowe, Terry Curtis
Metzger, Daniel Schaffer
Moulds, William Joseph
Mulholland, George
O'Rourke, Peter John
Philbin, Jeffrey Stephen
Powell, R(ay) B(edenkapp)
Reynolds, Stephen Edward
Shouman, A(hmad) R(aafat)
Stephans, Richard A
Straight, James William
Sullivan, J Al
Thurston, Rodney Sundbye
Touryan, Kenell James
Travis, John Richard
Von Riesemann, Walter Arthur
Wawersik, Wolfgang R
Wildin, Maurice W(ilbert)
Wolf, Helmut
Worstell, Hairston G(eorge)
Yeh, Hsu-Chi

NEW YORK
Adams, Donald E
Agosta, Vito
Aronson, Herbert
Banerjee, Prasanta Kumar
Barrows, John Frederick
Bartel, Donald L
Batter, John F(rederic), Jr
Baum, Richard T
Bayer, Raymond George
Begell, William
Bennett, Gerald William
Bennett, Leon
Benson, Richard Carter
Bergles, Arthur E(dward)
Birnstiel, Charles
Blesser, William B
Bolton, Joseph Aloysius
Booker, J(ohn) F
Brown, Dale H(enry)
Brownlow, Roger W
Busnaina, Ahmed A
Castelli, Vittorio
Cess, Robert D
Chandler, Jasper S(chell)
Chang, Shih-Yung
Cheng, Yean Fu
Chiang, Fu-Pen
Chinitz, Wallace
Clark, Alfred, Jr
Coffin, Louis F(ussell), Jr
Conta, Bart(holomew) J(oseph)
Corelli, John C
Crooke, James Stratton
Csermely, Thomas J(ohn)

Dahneke, Barton Eugene
De Boer, P(ieter) C(ornelis) Tobias
Deresiewicz, Herbert
Deshpande, Narayan V(aman)
Drucker, E(ugene) E(lias)
Eisenstadt, Raymond
Ellson, Robert A
Elston, Charles William
Endsley, L(ouis) E(ugene), Jr
Engel, Peter Andras
Ezra, Arthur Abraham
Femenia, Jose
Frank, Kozi
Freudenstein, Ferdinand
Fried, Erwin
Fritz, Robert J(acob)
Fuller, Dudley D(ean)
Gans, Roger Frederick
Garmezy, R(obert) H(arper)
Gaylord, Eber William
Gazda, I(rving) W(illiam)
George, A(lbert) R(ichard)
George, William Kenneth, Jr
Gibbon, Norman Charles
Ginsberg, Theodore
Givi, Peyman
Goldmann, Kurt
Goldspiel, Solomon
Granet, Irving
Greene, Peter Richard
Grisoli, John Joseph
Grossman, Perry L
Grunes, Robert Lewis
Gupta, Pradeep Kumar
Haag, Fred George
Haid, D(avid) A(ugustus)
Haines, Daniel Webster
Hall, Richard Travis
Handman, Stanley E
Harlam, Ruth (Mrs Fritz P Lehmann)
Harper, Kennard W(atson)
Harvey, Douglass Coate
Heimburg, Richard W
Hendrie, Joseph Mallam
Hetnarski, Richard Bozyslaw
Hoenig, Alan
Hollander, Milton B(ernard)
Hollenberg, Joel Warren
Holmes, Philip John
Hussain, Moayyed A
Hutchings, William Frank
Hwang, S Steve
Hwang, Shy-Shung
Imber, Murray
Irvine, Thomas Francis
Jedruch, Jacek
Jwo, Chin-Hung
Kapila, Ashwani Kumar
Karlekar, Bhalchandra Vasudeo
Kashkoush, Ismail I
Kayani, Joseph Thomas
Kear, Edward B, Jr
Kelly, Thomas Joseph
Kempner, Joseph
Kenyon, Richard A(lbert)
Kessler, Thomas J
Ketchum, Gardner M(ason)
Kramer, Aaron R
Krauter, Allan Irving
Krishna, C R
Kroeger, Peter G
Kuchar, Norman Russell
Kyanka, George Harry
Lahey, Richard Thomas, Jr
Landzberg, Abraham H(arold)
La Russa, Joseph Anthony
Laserson, Gregory
Laskaris, Trifon Evangelos
Lauber, Thornton Stuart
Lee, H(o) C(hong)
Lee, James Kelly
Lee, Richard Shao-Lin
Lessen, Martin
Levitsky, Myron
Lewis, Edward R(obert)
Li, Chung-Hsiung
Li, J(ames) C(hen) M(in)
Libove, Charles
Ling, F(rederick) F(ongsun)
Loeber, John Frederick
Loewen, Erwin G
Loewy, Robert G(ustav)
Longman, Richard Winston
Longobardo, Anna Kazanjian
Longobardo, Guy S
Lowen, Gerard G
Lowen, W(alter)
Lumley, John L(eask)
Lyman, Frederic A
McCalley, Robert B(ruce), Jr
Maclean, Walter M
Martens, Hinrich R
Mates, Robert Edward
Menkes, Sherwood Bradford
Mochel, Myron George
Montag, Mordechai
Moore, Franklin K(ingston)
Mosher, Melville Calvin
Mueller, W(heeler) K(ay)
Murphy, Eugene F(rancis)
Naar, Raymond Zacharias
Namer, Izak
Offenhartz, Edward
Osborn, Henry H(ooper)

Othmer, Donald F(rederick)
Pan, Huo-Hsi
Perkins, Kenneth Roy
Perkins, Richard W, Jr
Phelan, R(ichard) M(agruder)
Pita, Edward Gerald
Podowski, Michael Zbigniew
Purdy, D(avid) C(arl)
Quesnel, David John
Rae, William J
Reber, Raymond Andrew
Rice, John T(homas)
Rohatgi, Upendra Singh
Rothstein, Samuel
Sanders, W(illiam) Thomas
Scher, Robert Sander
Schroeder, John
Schwartzman, Leon
Sessler, John George
Sforza, Pasquale M
Shah, Ramesh Keshavlal
Shanebrook, J(ohn) Richard
Shapiro, Paul Jonathon
Shepherd, D(ennis) G(ranville)
Sherman, Zachary
Shube, Eugene E
Simon, Albert
Sneck, Henry James, (Jr)
Somerscales, Euan Francis Cuthbert
Somerset, James H
Srinivasan, Vijay
Staub, Fred W
Sternlicht, B(eno)
Sumner, Eric E
Tabi, Rifat
Tan, Chor-Weng
Thompson, Robert Alan
Thorsen, Richard Stanley
Torrance, Kenneth E(ric)
Tourin, Richard Harold
Vafakos, William P(aul)
Vassallo, Franklin A(llen)
Vaughan, David Sherwood
Velzy, Charles O
Vidosic, J(oseph) P(aul)
Virk, Kashmir Singh
Voelcker, Herbert B(ernhard)
Vohr, John H
Vopat, William A
Wagner, John George
Wallenfels, Miklos
Wang, Hsin-Pang
Wang, Kuo King
Wang, Lin-Shu
Watkins, Charles B
Wehe, Robert L(ouis)
Weinbaum, Sheldon
Wenig, Harold G(eorge)
Wetherhold, Robert Campbell
Whiteside, James Brooks
Wiese, Warren M(elvin)
Wiggins, Edwin George
Willmert, Kenneth Dale
Wilson, Merle R(obert)
Wolkoff, Harold
Wright, Roger Neal
Wulff, Wolfgang
Zimmer, G(eorge) A(rthur)
Zubaly, Robert B(arnes)
Zweig, John E

NORTH CAROLINA
Barrett, Rolin F(arrar)
Block, M Sabel
Boyers, Albert Sage
Brown, Theodore Cecil
Burton, Ralph A(shby)
Chaddock, Jack B
Chaplin, Frank S(prague)
Clark, Charles Edward
Cleland, John Gregory
Doolittle, J(esse) S(eymour)
Doster, Joseph Michael
Dow, Thomas Alva
Eberhardt, Allen Craig
Eckerlin, Herbert Martin
Elsevier, Ernest
El-Shiekh, Aly H
Garg, Devendra Prakash
Georgiadis, John G
Gupta, Ajaya Kumar
Harman, Charles M(organ)
Hill, Herbert Hewett
Hirshburg, Robert Ivan
Hochmuth, Robert Milo
Hurt, Alfred B(urman), Jr
Jennings, Burgess H(ill)
Jordan, Edward Daniel
Keltie, Richard Francis
Kim, Rhyn H(yun)
Kusy, Robert Peter
Lalor, William Francis
McDonald, P(atrick) H(ill), Jr
McElhaney, James Harry
Martin, David E(dwin)
Min, Tony C(harles)
Neal, C Leon
Ozisik, M Necati
Pearsall, George W(ilbur)
Petroski, Henry J
Reece, Joe Wilson
Smetana, Frederick Otto
Sorrell, Furman Y(ates), Jr
Stephenson, Harold Patty

Sud, Ish
Torquato, Salvatore
Zorowski, Carl F(rank)

NORTH DAKOTA
Li, Kam W(u)
Mathsen, Don Verden
Pfister, Philip Carl
Rieder, William G(ary)

OHIO
Adams, Otto Eugene, Jr
Alam, M Khairul
Anderson, William J
Andonian, Arsavir Takfor
Bagby, Frederick L(air)
Bahnfleth, Donald R
Bahniuk, Eugene
Bahr, Donald Walter
Barchfeld, Francis John
Beans, Elroy William
Berthold, John William, III
Bhagat, Pramode Kumar
Bisson, Edmond E(mile)
Bobo, Melvin
Bolz, Harold A(ugust)
Boulger, Francis W(illiam)
Brittain, Thomas M
Brown, Max Louis
Brush, John (Burke)
Buckley, Donald Henry
Chakko, Mathew K(anjirathinkal)
Chawla, Mangal Dass
Chu, Mamerto Loarca
Chuang, Henry Ning
Chung, Benjamin T
Ciricillo, Samuel F
Collins, Jack A(dam)
Damianov, Vladimir B
D'isa, Frank Angelo
Domholdt, Lowell Curtis
Eibling, James A(lexander)
Engdahl, Richard Bott
Ernst, Walter
Faulkner, Lynn L
Field, Michael
Franke, Milton Eugene
Freche, John C(harles)
Friar, Billy W(ade)
Frink, Donald W
Fritsch, Charles A(nthony)
Gabb, Timothy Paul
Gerhardt, Jon Stuart
Glaeser, William A(lfred)
Glasgow, John Charles
Glower, Donald D(uane)
Gorla, Rama S R
Gorland, Sol H
Graham, Robert William
Greber, Isaac
Greene, Neil E(dward)
Gregory, R(obert) Lee
Hall, Glenn Eugene
Hamed, Awatef A
Han, L(it) S(ien)
Hantman, Robert Gary
Harrisberger, Edgar Lee
Hazard, Herbert Ray
Heasley, James Henry
Heath, George L
Hemsworth, Martin C
Herbert, Thorwald
Ho, Fanghuai H(ubert)
Horton, Billy Mitchusson
Hursh, Robert W(illiam)
Huss, Harry O(tto)
Huston, Ronald L
Irey, Richard Kenneth
Jaikrishnan, Kadambi Rajgopal
Jain, Vinod Kumar
Janzow, Edward F(rank)
Jeffries, Neal Powell
Jehn, Lawrence A
Jeng, Duen-Ren
Jones, Charles Dingee
Kalivoda, Frank E, Jr
Keim, John Eugene
Keith, Theo Gordon, Jr
Kepler, Harold B(enton)
Kerka, William (Frank)
Kerr, Robert Lowell
Kinzel, Gary Lee
Klosterman, Albert Leonard
Komkov, Vadim
Kornylak, Andrew T
Krishnamurthy, Lakshminarayanan
Krysiak, Joseph Edward
Lahoti, Goverdhan Das
La Mers, Thomas Herbert
Landstrom, D(onald) Karl
Lawler, Martin Timothy
Lee, Peter Wankyoon
Leissa, A(rthur) W(illiam)
Lenhart, Jack G
Lewandowski, John Joseph
Lioi, Anthony Pasquale
Lumsdaine, Edward
Lumsdaine, Edward
Lundin, Bruce T(heodore)
Macke, H(arry) Jerry
Mall, Shankar
Marshall, Kenneth D
Mayers, Richard Ralph
Menq, Chia-Hsiang

Merchant, Mylon Eugene
Messinger, Richard C
Miller, William B
Miller, William Ralph
Morgan, William R(ichard)
Morse, Ivan E, Jr
Mularz, Edward Julius
Murphy, Michael John
Orndorff, Roy Lee, Jr
Ostrach, Simon
Peckham, P Hunter
Pinchak, Alfred Cyril
Powell, Gordon W
Prahl, Joseph Markel
Putnam, Abbott (Allen)
Ramalingam, Mysore Loganathan
Reddy, Narender Pabbathi
Reif, Thomas Henry
Rich, Joseph William
Richley, E(dward) A(nthony)
Robinson, Richard Alan
Rockstroh, Todd Jay
Rodman, Charles William
Sanders, John Claytor
Savage, Michael
Scavuzzo, Rudolph J, Jr
Schauer, John Joseph
Schleef, Daniel J
Schoch, Daniel Anthony
Sepsy, Charles Frank
Shine, Andrew J(oseph)
Siegel, Robert
Singh, Rajendra
Smith, Howard Edwin
Smith, L(eroy) H(arrington), Jr
Smith, Marion L(eroy)
Soni, Atmaram Harilal
Southam, Donald Lee
Srivatsan, Tirumalai Srinivas
Starkey, Walter L(eroy)
Sutton, George E
Tarantine, Frank J(ames)
Thompson, Hugh Ansley
Thorpe, James F(ranklin)
Tse, Francis S
Van Der Sluys, William
Voisard, Walter Bryan
Voorhees, John E
Waldron, Kenneth John
Webster, Allen E
Wennerstrom, Arthur J(ohn)
Whitacre, Gale R(obert)
White, John Joseph, III
Wilkinson, William H(adley)
Winiarz, Marek Leon
Zaman, Khairul B M Q
Zavodney, Lawrence Dennis

OKLAHOMA
Appl, Franklin John
Bennett, A(rthur) D(avid)
Bert, Charles Wesley
Blenkarn, Kenneth Ardley
Boggs, James H(arlow)
Chapel, Raymond Eugene
Cornelius, Archie J(unior)
Earls, James Roe
Egle, Davis Max
Emanuel, George
Fitch, Ernest Chester, Jr
Forney, Bill E
Gitzendanner, L G
Gollahalli, Subramanyam Ramappa
Guerrero, E T
Heath, Larman Jefferson
Hurn, R(ichard) W(ilson)
Inoue, Riichi
Joshi, Sadanand D
Khan, Akhtar Salamat
Lilley, David Grantham
Lowery, R(ichard) L
McQuiston, Faye C(lem)
Moretti, Peter M
Neathery, Raymond Franklin
Omer, Savas
Parker, Jerald D
Reid, Karl Nevelle, Jr
Riley, Richard King
Sheth, Ketankumar K
Striz, Alfred Gerhard
Upthegrove, W(illiam) R(eid)
Weston, Kenneth Clayton
Zigrang, Denis Joseph
Zirkle, Larry Don

OREGON
Boubel, Richard W
Calder, Clarence Andrew
Devletian, Jack H
Fanger, Carleton G(eorge)
Hansen, Hugh J
Larson, M(ilton) B(yrd)
LeTourneau, Budd W(ebster)
Light, Kenneth Freeman
Migliore, Herman James
Perry, Robert W(illiam)
Popovich, M(ilosh)
Reistad, Gordon Mackenzie
Savery, Clyde William
Thornburgh, George E(arl)
Welty, James Richard
Zaworski, R(obert) J(oseph)

PENNSYLVANIA
Amon Parisi, Cristina Hortensia
Antes, Harry W
Avitzur, Betzalel
Ayyaswamy, Portonovo S
Baish, James William
Bau, Haim Heinrich
Baumann, Dwight Maylon Billy
Benner, Russell Edward
Berkof, Richard Stanley
Berkowitz, David Andrew
Bridenbaugh, Peter R
Brighton, John Austin
Brown, Forbes Taylor
Brunner, Mathias J
Bucci, Robert James
Burnham, Donald C
Caporali, Ronald Van
Carlson, H(enning) Maurice
Catanach, Wallace M, Jr
Cernansky, Nicholas P
Chan, Siu-Kee
Chang, C Hsiung
Chigier, Norman
Chou, Pei Chi (Peter)
Chuang, Strong Chieu-Hsiung
Cohen, Ira M
Confer, Gregory Lee
Cook, Charles S
Curlee, Neil J(ames), Jr
Daum, Donald Richard
Devine, John Edward
Dillio, Charles Carmen
Dougall, Richard S(tephen)
Dunn, J(ohn) Howard
Everett, James Legrand, III
Falconer, Thomas Hugh
Farouk, Bakhtier
Fischer, Edward G(eorge)
Fong, James T(se-Ming)
Forstall, Walton, Jr
Fukushima, Toshiyuki
Fullerton, H(erbert) P(almer)
Geiger, Gene E(dward)
Goel, Ram Parkash
Goldberg, Joseph
Griffin, Jerry Howard
Gwiazda, S(tanley) J(ohn)
Hahn, Hong Thomas
Heinsohn, Robert J(ennings)
Henderson, Robert E
Herrick, Joseph Raymond
Hetzel, Theodore B(rinton)
Holl, J(ohn) William
Hughes, Arthur D(ouglas)
Hughes, William F(rank)
Hull, John Laurence
Hyatt, Robert Monroe
Ibrahim, Baky Badie
Jacobs, Harold Robert
Jensen, James Ejler
Juster, Allan
Karlovitz, Bela
Kaufman, Howard Norman
Keyser, David Richard
Keyt, Donald E
Kingrea, James I
Kinnavy, M(artin) G(erald)
Klaus, Ronald Louis
Koch, Ronald N
Kowal, George M
Kroon, R(einout) P(ieter)
Kuhn, Howard A
Lakshminarayana, B
Lavanchy, Andre C(hristian)
Lavin, J Gerard
Lawrence, Kenneth Bridge
Lazaridis, Anastas
Leidy, Blaine I(rvin)
Levy, Bernard, Jr
Levy, Edward Kenneth
Lior, Noam
Long, Lyle Norman
Lorsch, Harold G
Lucas, Robert Alan
McAssey, Edward V, Jr
McLaughlin, Philip V(an Doren), Jr
Macpherson, Alistair Kenneth
Mahaffy, John Harlan
Manjoine, Michael J(oseph)
Marston, Charles H
Matulevicius, Edward S(tephen)
May, Harold E(dward)
Mehta, Kishor Singh
Mendler, Oliver J(ohn)
Mercer, Samuel, Jr
Meredith, David Bruce
Merz, Richard A
Meyer, Wolfgang E(berhard)
Molyneux, John Ecob
Morrill, Bernard
Novak, Stephen Robert
Ochs, Stefan A(lbert)
Olson, Donald Richard
Orr, Leighton
Osterle, J(ohn) F(letcher)
Owczarek, J(erzy) A(ntoni)
Paolino, Michael A
Park, William H
Pfennigwerth, Paul Leroy
Podgers, Alexander Robert
Purdy, David Lawrence
Putman, Thomas Harold
Raimondi, Albert Anthony

Rajagopal, K R
Ramezan, Massood
Ramos, Juan Ignacio
Reethof, Gerhard
Remick, Forrest J(erome)
Roberts, Richard
Robertson, Donald
Rodini, Benjamin Thomas, Jr
Rohrer, Wesley M, Jr
Rose, Joseph Lawrence
Ross, Arthur Leonard
Rouleau, Wilfred T(homas)
Rubin, Edward S
Rumbarger, John H
Satz, Ronald Wayne
Schmidt, Alfred Otto
Schmidt, Frank W(illiam)
Schneider, Robert W(illiam)
Shearer, J(esse) Lowen
Silverstein, Calvin C(arlton)
Sinclair, Glenn Bruce
Solt, Paul E
Sommer, Holger Thomas
Spielvogel, Lawrence George
Stokey, W(illiam) F(armer)
Stradling, Lester J(ames), Jr
Sutherland, Ivan Edward
Szeri, Andras Zoltan
Tallian, Tibor E(ugene)
Thomas, Donald H(arvey)
Treadwell, Kenneth Myron
Trumpler, Paul R(obert)
Tsu, T(sung) C(hi)
Tuba, I Stephen
Wallace, Paul William
Wambold, James Charles
Webb, Ralph L
Weindling, Joachim I(gnace)
Weinstock, Manuel
Whitman, Alan M
Wiebe, Donald
Williams, Albert J, Jr
Williams, Max L(ea), Jr
Wilson, Myron Allen
Wolgemuth, Carl Hess
Woodburn, Wilton A
Wright, Dexter V(ail)
Wu, John Naichi
Yeh, Hsuan
Young, Frederick J(ohn)
Zaiser, James Norman
Zauderer, Bert
Zimmermann, F(rancis) J(ohn)

RHODE ISLAND
Boothroyd, Geoffrey
Bradbury, Donald
Brady, John F
Butler, Howard W(allace)
Clarke, Joseph H(enry)
Dahl, Norman C(hristian)
Dewhurst, Peter
Dobbins, Richard Andrew
Dunlap, Richard M(orris)
Fernandes, John Henry
Kestin, J(oseph)
Kim, Thomas Joon-Mock
Lessmann, Richard Carl
Liu, J(oseph) T(su) C(hieh)
McEligot, Donald M(arinus)
Nash, Charles Dudley, Jr
Ng, Kam Wing
Palm, William John
Richardson, Peter Damian
Sibulkin, Merwin
Spero, Caesar A(nthony), Jr
Test, Frederick L(aurent)
Thielsch, Helmut
Viets, Hermann

SOUTH CAROLINA
Awadalla, Nabil G
Bishop, Eugene H(arlan)
Byars, Edward F(ord)
Chao, Yuh J (Bill)
Christy, William O(liver)
Chuang, Ming Chia
Drane, John Wanzer
Elrod, Alvon Creighton
Gaines, Albert L(owery)
Garzon, Ruben Dario
Graulty, Robert Thomas
Grosh, Richard J(oseph)
Herdklotz, Richard James
Hester, Jarrett Charles
Hosny, Ahmed Nabil
Irwin, Lafayette K(ey)
Jackson, John Elwin, Jr
Jens, Wayne H(enry)
Joseph, J Walter, Jr
Lewis, Alexander D(odge)
McCarter, James Thomas
McMillan, Harry King
Overcamp, Thomas Joseph
Paul, Frank W(aters)
Progelhof, Richard Carl
Przirembel, Christian E G
Rudisill, Carl Sidney
Sandrapaty, Ramachandra Rao
Townes, George Anderson
Turner, John Lindsey
Uldrick, John Paul
Ward, David Aloysius
Westerman, William Joseph, II

Mechanical Engineering (cont)

SOUTH DAKOTA
Chiang, Chao-Wang
Scofield, Gordon L(loyd)

TENNESSEE
Attaway, Cecil Reid
Baker, Merl
Barnett, Paul Edward
Boudreau, William F(rancis)
Brown, Bevan W, Jr
Bundy, Robert W(endel)
Burton, Robert Lee
Cox, Ronald Baker
Deivanayagan, Subramanian
Edmondson, Andrew Joseph
Fraas, Arthur P(aul)
Frost, Walter
Gat, Uri
Gilbert, R(obert) J(ames)
Gray, William Harvey
Greenstreet, William (B) LaVon
Hodgson, J(effrey) William
Holland, Ray W(alter)
Houghton, James Richard
Iskander, Shafik Kamel
Isley, James Don
Janna, William Sied
Johnson, William S(tanley)
Kelso, Richard Miles
Kerr, Hugh Barkley
Keshock, Edward G
Kidd, George Joseph, Jr
Kinyon, Brice W(hitman)
Kress, Thomas Sylvester
Lim, Alexander Te
Martin, Harry Lee
Maxwell, Robert L(ouis)
Mellor, Arthur M(cLeod)
Milligan, Mancil W(ood)
Mullikin, H(arwood) F(ranklin)
Pandeya, Prakash N
Patterson, Malcolm Robert
Pih, Hui
Pitts, Donald Ross
Potter, John Leith
Pugh, Claud Ervin
Ray, John Delbert
Scherpereel, Donald E
Sissom, Leighton E(sten)
Slade, Edward Colin
Snyder, William Thomas
Sorrells, Frank Douglas
Stair, William K(enneth)
Sun, Pu-Ning
Sundaram, R Meenakshi
Swim, William B(axter)
Swindeman, Robert W
Swisher, George Monroe
Tunnell, William C(lotworthy)
Ventrice, Marie Busck
Wasserman, Jack F
Williamson, John W
Wimberly, C Ray
Witten, Alan Joel
Wright, William F(red)
Young, Robert L(yle)
Zuhr, Raymond Arthur

TEXAS
Addy, Tralance Obuama
Allen, Herbert
Anderson, Edward Everett
Baber, Burl B, Jr
Baker, Wilfred E(dmund)
Ball, Kenneth Steven
Barbin, Allen R
Barker, C(alvin) L R
Barnes, Allen Lawrence
Bergman, Theodore L
Bland, William M, Jr
Bourell, David Lee
Brown, Howard (Earl)
Brown, Otto G
Bryant, Michael David
Burger, Christian P
Caton, Jerald A
Chapman, Alan J(esse)
Cheatham, John B(ane), Jr
Chien, Sze-Foo
Chyu, Ming-Chien
Clark, L(yle) G(erald)
Cooper, John C, Jr
Crawford, Richard Haygood, Jr
Cunningham, Robert Ashley
Deily, Fredric H(arry)
Diller, Kenneth Ray
Dodge, Franklin Tiffany
Eberhart, Robert Clyde
Feltz, Donald Everett
Finch, Robert D
Fletcher, Leroy S(tevenson)
Flipse, John E
Flowers, Daniel F(ort)
Fogwell, Joseph Wray
Friberg, Emil Edwards
Glaspie, Donald Lee
Godwin, James Basil, Jr
Goodzeit, Carl Leonard
Graff, William J(ohn), Jr
Graiff, Leonard B(aldine)
Gully, John Houston
Gupton, Paul Stephen

Haji-Sheikh, Abdolhossein
Harkey, Jack W
Harris, William B(risbane) D(ick), Jr
Hebert, Joel J
Helmers, Donald Jacob
Hesse, Walter J(ohn)
Hoffman, Herbert I(rving)
Holman, Jack Philip
Hrkel, Edward James
Hussain, A K M Fazle
Jenkins, Lawrence E
Jones, Jerold W
Jones, Marvin Richard
Jordan, Duane Paul
Juricic, Davor
Kanninen, Melvin Fred
Kearns, Robert William
Kreisle, Leonardt F(erdinand)
Kunze, Otto Robert
Lamb, J(amie) Parker, Jr
Lawrence, James Harold, Jr
Lawrence, Kent L(ee)
Lengel, Robert Charles
Leonard, Robert Gresham
Lewis, James Pettis
Lienhard, John H(enry)
Lou, David Yeong-Suei
Lu, Frank Kerping
McBee, Frank W(ilkins), Jr
Mack, Russel Travis
Marshek, Kurt M
Martinez, Jose E(dwardo)
Mistree, Farrokh
Moon, Tessie Jo
Mouritsen, T(horvald) Edgar
Muster, Douglas Frederick
Nassersharif, Bahram
Ochiai, Shinya
Packman, Paul Frederick
Page, Robert Henry
Panton, R(onald) L(ee)
Perry, John Vivian, Jr
Peterman, David A
Plapp, John E(lmer)
Pope, Larry Debbs
Powell, Alan
Powers, Louis John
Purvis, Merton Brown
Rabins, Michael J
Rand, James Leland
Reynolds, Elbert Brunner, Jr
Rhodes, Allen Franklin
Richardson, Herbert Heath
Rodenberger, Charles Alvard
Russell, John Alvin
Rylander, Henry Grady, Jr
Sadler, Stanley Gene
Schmidt, P(hilip) S(tephen)
Short, Byron Elliott
Sifferman, Thomas Raymond
Simmang, C(lifford) M(ax)
Simonis, John Charles
Sinclair, A Richard
Solberg, Ruell Floyd, Jr
Springer, Karl Joseph
Staph, Horace E(ugene)
Strange, Lloyd K(eith)
Swope, Richard Dale
Teller, Cecil Martin, II
Tesar, Delbert
Tien, John Kai
Traver, Alfred Ellis
Treat, Charles Herbert
Turner, William Danny
Vance, John Milton
Vitkovits, John A(ndrew)
Vliet, Gary Clark
Waid, Margaret Cowsar
Wang, Chao-Cheng
Weldon, William Forrest
Wenzel, Alexander B
Westkaemper, John C(onrad)
Weyler, Michael E
Weynand, Edmund E
Wick, Robert S(enters)
Widmer, Robert H
Wilhelm, Wilbert Edward
Wilson, Claude Leonard
Winnie, Dayle David
Witte, Larry C(laude)
Wood, Charles D
Wood, Kristin Lee
Woolf, J(ack) R(oyce)
Young, Dana
Young, H(enry) Ben, Jr

UTAH
Andersen, Blaine Wright
Brown, W(ayne) S(amuel)
Cannon, John N(elson)
DeVries, K Lawrence
Free, Joseph Carl
Germane, Geoffrey James
Glenn, Alan Holton
Green, Sidney J
Heaton, Howard S(pring)
Hill, Eric von Krumreig
Hoeppner, David William
Holdredge, Russell M
Isaacson, Lavar King
Johnson, Arlo F
Kruger, Robert A(lan)
Limpert, Rudolf
MacGregor, Douglas

Moser, Alma P
Rotz, Christopher Alan
Sandquist, Gary Marlin
Silver, Barnard Stewart
Snow, Donald Ray
Strozier, James Kinard
Turley, Richard Eyring
Warner, Charles Y
Zeamer, Richard Jere

VERMONT
Dean, Robert Charles, Jr
Francis, Gerald Peter
Friihauf, Edward Joe
Hundal, Mahendra S(ingh)
Outwater, John O(gden)
Pope, Malcolm H
Tartaglia, Paul Edward
Tuthill, Arthur F(rederick)
Von Turkovich, Branimir F(rancis)
Wallace, Donald Macpherson, Jr

VIRGINIA
Allaire, Paul Eugene
Anderson, Willard Woodbury
Ash, Robert Lafayette
Bauer, Douglas Clifford
Beach, Robert L
Beard, James Taylor
Beran, Robert Lynn
Betti, John A
Bishara, Michael Nageeb
Brooks, Thomas Furman
Carpenter, James E(ugene)
Chappelle, Thomas W
Charvonia, David Alan
Comparin, Robert A(nton)
Cooley, William C
Cooper, Henry Franklyn, Jr
Davidson, John Richard
Diller, Thomas Eugene
Dutt, Gautam Shankar
Edyvane, John
Eiss, Norman Smith, Jr
Ell, William M
Flack, Ronald Dumont, Jr
Foster, Eugene L(ewis)
Fuller, Christopher Robert
Gorges, Heinz A(ugust)
Gouse, S William, Jr
Grant, John Wallace
Grunder, Hermann August
Gunter, Edgar Jackson, Jr
Hall, Carl W(illiam)
Hamrick, Joseph Thomas
Hasenfus, Harold J(oseph)
Hill, Floyd Isom
Hoffman, Kenneth Charles
Hou, Gene Jean-Win
Jacobs, Theodore Alan
Johnson, David Harley
Jones, J(ames) B(everly)
Kapania, Rakesh Kumar
Kauzlarich, James J(oseph)
Kramer, David Buckley
Layson, William M(cIntyre)
Levis, Alexander Henry
Long, Dale Donald
Mabie, Hamilton Horth
McGean, Thomas J
Manfredi, Arthur Frank, Jr
Mehta, Gurmukh D
Mindak, Robert Joseph
Moen, Walter B(oniface)
Moore, James W(allace) I
Moore, Robert E(dward)
Morgan, Charles D(avid)
Moses, Hal Lynwood
Ng, Wing-Fai
Ramberg, Steven Eric
Raney, J(ohn) P(hilip)
Reddy, Junuthula N
Reese, Ronald Malcolm
Roberts, A(lbert) S(idney), Jr
Rokni, Mohammad Ali
Roy, Gabriel D
Savage, William F(rederick)
Schonstedt, Erick O(scar)
Senseny, Paul Edward
Sexton, Michael Ray
Silveira, Milton Anthony
Smith, Carey Daniel
Snyder, Glenn J(acob)
Stiegler, T(heodore) Donald
Strawn, Oliver P(erry), Jr
Szeless, Adorjan Gyula
Talapatra, Dipak Chandra
Tompkins, Stephen Stern
Townsend, Miles Averill
Wang, C(hiao) J(en)
Whitelaw, R(obert) L(eslie)
Wilkinson, Thomas Lloyd, Jr
Wood, Albert D(ouglas)
Wood, Houston Gilleylen, III
Yates, Charlie Lee
Yates, Edward Carson, Jr
Young, Maurice Isaac

WASHINGTON
Arenz, Robert James
Beaubien, Stewart James
Bruckner, Adam Peter
Buonamici, Rino
Bush, S(pencer) H(arrison)

Childs, Morris E
Claudson, T(homas) T(ucker)
Crain, Richard Willson, Jr
Cross, Edward F
Currah, Walter E
Cykler, John Freuler
Day, Emmett E(lbert)
Depew, Creighton A
Ding, Jow-Lian
Drumheller, Kirk
Emery, Ashley F
Firey, Joseph Carl
Forster, Fred Kurt
Gajewski, W(alter) M(ichael)
Galle, Kurt R(obert)
Gessner, Frederick B(enedict)
Hale, Leonard Allen
Hamilton, C Howard
Jacobson, John Obert
Johnson, B(enjamin) M(artineau)
Jorgensen, Jens Erik
Kemper, Robert Schooley, Jr
Kippenhan, C(harles) J(acob)
Kobayashi, Albert S(atoshi)
LaFontaine, Edward
Leavenworth, Howard W, Jr
Levinson, Mark
McCormick, Norman Joseph
McFeron, D(ean) E(arl)
McMurtrey, Lawrence J
Madden, James H(oward)
Masden, Glenn W(illiam)
Matte, Joseph, III
Merchant, Howard Carl
Mills, B(lake) D(avid), Jr
Paynter, Gerald C(lyde)
Pratt, David Terry
Ramulu, Mamidala
Renkey, Edmund Joseph, Jr
Richmond, William D
Rieck, H(enry) G(eorge)
Robel, Stephen B(ernard)
Ross, Peter A
Sandwith, Colin John
Sengupta, Gautam
Shank, Maurice E(dwin)
Shoemaker, Roy H(opkins)
Smith, Thomas Hadwick
Spencer, Max M(arlin)
Stock, David Earl
Taggart, Raymond
Tam, Patrick Yui-Chiu
Taya, Minoru
Trent, Donald Stephen
Viggers, Robert F
Wolak, Jan
Yore, Eugene Elliott

WEST VIRGINIA
Chiang, Han-Shing
Clark, Nigel Norman
Kuhlman, John Michael
Kumar, Alok
Lantz, Thomas Lee
Nelson, Leonard C(arl)
Shuck, Lowell Zane
Smith, James Earl
Sneckenberger, John Edward
Steinhardt, Emil J
Vucich, M(ichael) G(eorge)
Wilson, John Sheridan

WISCONSIN
Balmer, Robert Theodore
Barrington, Gordon P
Beachley, Norman Henry
Beckman, William A
Bollinger, John G(ustave)
Borman, Gary Lee
Bradish, John Patrick
Bruhn, H(jalmar) D(iehl)
Burck, Larry Harold
Burstein, Sol
Chan, Shih Hung
DeVries, Marvin Frank
Doshi, Mahendra R
Dries, William Charles
El-Wakil, M(ohamed)
Engelstad, Roxann Louise
Fournelle, Raymond Albert
Frank, Arne
Garg, Arun
Gopal, Raj
Grogan, Paul J(oseph)
Henke, Russell W
Horntvedt, Earl W
Koeller, Ralph Carl
Kohli, Dilip
Landis, Fred
Lentz, Mark Steven
Linehan, John Henry
McMurray, David Claude
Mehta, N(avnit) C(hhaganlal)
Mitchell, John Wright
Myers, Glen E(verett)
Myers, Phillip S(amuel)
Nigro, Nicholas J
Obert, Edward Fredric
Pavelic, Vjekoslav
Reid, Robert Lelon
Riordan, H(ugh) E(rnest)
Rohatgi, Pradeep Kumar
Romer, I(rving) Carl, Jr
Rowlands, Robert Edward

Snyder, Warren Edward
Sprague, Clyde Howard
Steudel, Harold Jude
Stumpe, Warren Robert
Szpot, Bruce F
Tsao, Keh Cheng
Uicker, John Joseph, Jr
Ware, Chester Dawson
Weiss, Stanley
Wittnebert, Fred R
Zahn, John J
Zelazo, Nathaniel Kachorek

WYOMING
Adams, Donald F
Jacquot, Raymond G
Pell, Kynric M(artin)
Sutherland, Robert L(ouis)
Wheasler, Robert

PUERTO RICO
Sasscer, Donald S(tuart)
Soderstrom, Kenneth G(unnar)

ALBERTA
Bellow, Donald Grant
Checkel, M David
Dale, J(ames) D(ouglas)
Feingold, Adolph
Kentfield, John Alan Charles
Lock, G(erald) S(eymour) H(unter)
Mikulcik, E(dwin) C(harles)
Norrie, D(ouglas) H
Rodkiewicz, Czeslaw Mateusz
Sadler, G(erald) W(esley)

BRITISH COLUMBIA
Birkebak, Richard C(larence)
Hill, Philip G(raham)
Kerekes, Richard Joseph
Ko, Pak Lim
Leith, William Cumming
Mantle, J(ohn) B(ertram)
Milsum, John H
Morrison, James Barbour
Tabarrok, B(ehrouz)
Vitovec, Franz H
Wighton, John L(atta)

MANITOBA
Cenkowski, Stefan
Crosthwaite, John Leslie

NEWFOUNDLAND
Molgaard, Johannes

NOVA SCOTIA
Cochkanoff, O(rest)
Kujath, Marek Ryszard
Wilson, George Peter

ONTARIO
Baines, William Douglas
Baronet, Clifford Nelson
Brzustowski, Thomas Anthony
Chandrashekar, Muthu
Chou, Jordan Quan Ban
Cockshutt, E(ric) P(hilip)
Colborne, William George
Corneil, Ernest Ray
Currie, Iain George
Dalrymple, Desmond Grant
Dubey, Rajendra Narain
Emraghy, Hoda Abdel-Kader
Fenton, Robert George
Feraday, Melville Albert
Finlay, Joseph Bryan
Foreman, John E(dward) K(endall)
Gilbert, W(illiam) D(ouglas)
Gladwell, Graham M L
Goldenberg, Andrew Avi
Hancox, William Thomas
Henry, Roger P
Hooper, F(rank) C(lements)
Howard, Joseph H(erman) G(regg)
Jeswiet, Jacob
Jones, L(lewellyn) E(dward)
Judd, Ross Leonard
Kardos, Geza
Keffer, James F
Krausz, Alexander Stephen
Krishnappa, Govindappa
Ledwell, Thomas Austin
Lee, Benedict Huk Kun
Lee, Yung
Leutheusser, H(ans) J(oachim)
McCammond, David
McDonald, T(homas) W(illiam)
Makomaski, Andrzej Henryk
Marsters, Gerald F
Matar, Said E
Mercer, Alexander McDowell
Molder, S(annu)
Moroz, William James
Munro, Michael Brian
Nelles, John Sumner
North, Henry E(rick) T(uisku)
North, W(alter) Paul Tuisku
Orlik-Ruckemann, Kazimierz Jerzy
Perlman, Maier
Pick, Roy James
Pike, J(ohn) G(ibson)
Plett, Edelbert Gregory
Plumtree, A(lan)

Rice, W(illiam) B(othwell)
Rimrott, F(riedrich) P(aul) J(ohannes)
Rogers, James Terence, (Jr)
Rosen, Marc Allen
Rush, Charles Kenneth
Saravanamuttoo, Herbert Ian H
Schey, John Anthony
Shelson, W(illiam)
Southwell, P(eter) H(enry)
Tarasuk, John David
Tavoularis, Stavros
Varin, Robert Andrzej
Vijay, Mohan Madan
Ward, Charles Albert
Wong, Jo Yung

PRINCE EDWARD ISLAND
Lodge, Malcolm A

QUEBEC
Alward, Ron E
Berlyn, Robin Wilfrid
Bhat, Rama B
Bui, Tien Rung
Cheng, Richard M H
Croctogino, Reinhold Hermann
Galanis, Nicolas
Habashi, Wagdi George
Hoa, Suong Van
Kwok, Clyde Chi Kai
Lee, John Hak Shan
Lin, Sui
Mavriplis, F
Osman, M(ohamed) O M
Sankar, Seshadri
Sankar, Thiagas Sriram
Vatistas, Georgios H

SASKATCHEWAN
Esmail, Mohamed Nabil
Hedlin, Charles P(eter)
Koh, Eusebio Legarda
Le May, I(ain)
Ukrainetz, Paul Ruvim
Walerian, Szyszkowski

OTHER COUNTRIES
Al-Zubaidy, Sarim Naji
Bligh, Thomas Percival
Bodner, Sol R(ubin)
De Vedia, Luis Alberto
Duncan, John L(eask)
Eaton, Robert James
Gordon, Jeffrey Miles
Hjertager, Bjorn Helge
Idelsohn, Sergio Rodolfo
Jespersen, Nils Vidar
Jost, Hans Peter
Lee, Choung Mook
Nomura, Yasumasa
Pessen, David W
Revesz, Zsolt
Romero, Alejandro F
Sarmiento, Gustavo Sanchez
Solari, Mario Jose Adolfo
Tanner, Roger Ian
Taplin, David Michael Robert
Voutsas, Alexander Matthew
   (Voutsadakis)
Ward, Gerald T(empleton)
Wendland, Wolfgang Leopold
Yamamoto, Mitsuyoshi

Metallurgy & Physical Metallurgical Engineering

ALABAMA
Browning, J(ames) S(cott)
Cataldo, Charles Eugene
Chia, E Henry
Frye, John H, Jr
Geppert, Gerard Allen
Holland, James Read
Jemian, Wartan A(rmin)
Johnston, Mary Helen
Kattus, J Robert
Lucas, William R(ay)
Piwonka, Thomas Squire
Rampacek, Carl
Scheiner, Bernard James
Stanczyk, Martin Henry
Stefanescu, Doru Michael
Thompson, Raymond G
Wilcox, Roy Carl
Wilkins, Charles H(enry) T(ully)

ALASKA
Johnson, Ralph Sterling, Jr

ARIZONA
Arbiter, Nathaniel
Bhappu, Roshan B
Blickwede, D(onald) J(ohnson)
Carpenter, Ray Warren
Das, Subodh Kumar
Fairbanks, Harold V
Frame, John W
Hart, Lyman Herbert
Hendrickson, Lester Ellsworth
Hiskey, J Brent
Keating, Kenneth L(ee)
Kebler, Richard William
Kessler, Harold D

Kuhn, Martin Clifford
Leeser, David O(scar)
Levinson, David W
McPherson, Donald J(ames)
Meieran, Eugene Stuart
Murphy, Daniel John
Patterson, Vernon Howe
Peck, John F
Quadt, R(aymond) A(dolph)
Risbud, Subhash Hanamant
Savrun, Ender
Simmons, Paul C
Swalin, Richard Arthur
Taub, James M
Walker, Walter W(yrick)
Withers, James C
Zukas, Eugene G

CALIFORNIA
Agrawal, Suphal Prakash
Albert, Paul A(ndre)
Anjard, Ronald P, Sr
Armijo, Joseph Sam
Arrowood, Roy Mitchell, Jr
Baldwin, Chandler Milnes
Baldwin, David Hale
Bar-Cohen, Yoseph
Baskes, Michael I
Beebe, Robert Ray
Beerntsen, Donald J(oseph)
Bertossa, Robert C
Bhandarkar, Mangalore Dilip
Bonora, Anthony Charles
Boone, Donald H
Bragg, Robert H(enry)
Bray, Robert S
Brewer, Leo
Brimm, Eugene Oskar
Buhl, Robert Frank
Bunshah, Rointan F(ramroze)
Burke, Edmund C(harles)
Caligiuri, Robert Domenic
Cape, Arthur Tregoning (Tommy)
Chandler, Willis Thomas
Charley, Philip J(ames)
Chen, C(hih) W(en)
Chiou, C(harles)
Chung, Chi Hsiang
Clark, John R(obert)
Clemens, Bruce Montgomery
Clements, Linda L
Cline, Carl F
Comley, Peter Nigel
Crossley, Frank Alphonso
Dasher, John
Dejonghe, Lutgard C
Denney, Joseph M(yers)
Dorward, Ralph C(larence)
Douglass, David Leslie
Doyle, Fiona Mary
El Guindy, Mahmoud Ismail
Elsner, Norbert Bernard
Evans, Donald B
Evans, James William
Fellows, John A(lbert)
Feuerstein, Seymour
Finnegan, Walter Daniel
Forgeng, W(illiam) D(aniel)
Freise, Earl J
French, Edward P(erry)
Fuerstenau, D(ouglas) W(inston)
Fultz, Brent T
Gehman, Bruce Lawrence
Gerlach, John Norman
Ghosh, Amit Kumar
Goetzel, Claus G(uenter)
Goldberg, Alfred
Gordon, Barry Monroe
Gordon, Gerald M
Goubau, Wolfgang M
Gronsky, Ronald
Guttenplan, Jack David
Hackett, Le Roy Huntington, Jr
Haissig, Manfred
Hanafee, James Eugene
Harrison, J(ohn) D(avid)
Harvey, Frances J, II
Hauber, Janet Elaine
Haynes, Charles W(illard)
Helfrich, Wayne J(on)
Hill, Russell John
Hoover, William Ringo
Hubred, Gale L
Hultgren, Ralph Raymond
Hunnicutt, Richard P(earce)
Jaffe, Leonard David
Jaffee, Robert I(saac)
Johnson, William Robert
Jones, Brian Herbert
Jones, Robin L(eslie)
Kaneko, Thomas Motomi
Kashar, Lawrence Joseph
Kendig, Martin William
Kikuchi, Ryoichi
Kim, Ki Hong
Kneip, G(eorge) D(ewey), Jr
Koppenaal, Theodore J
Kramer, David
LaChance, Murdock Henry
Lampson, Francis Keith
Lavernia, Enrique Jose
Lewis, J(ack) R(ockley)
Li, Chia-Chuan

Lin, Charlie Yeongching
Littauer, Ernest Lucius
Loring, Blake M(arshall)
Lucas, Glenn E
Marshall, Sally J
Martin, George
Meyer, Brad Anthony
Meyers, Marc Andre
Mills, G(eorge) J(acob)
Mitchell, Michael Roger
Mohamed, Farghalli Abdelrahman
Mooz, William Ernst
Mukherjee, Amiya K
Murray, George T(homas)
Nachman, Joseph F(rank)
Oldfield, William
Paine, T(homas) O(tten)
Parker, Earl Randall
Pask, Joseph Adam
Pathania, Rajeshwar S
Paton, Neil (Eric)
Perkins, Roger A(llan)
Pritchett, Thomas Ronald
Rau, Charles Alfred, Jr
Raymond, Louis
Richter, Thomas A
Roberts, George A(dam)
Romanowski, Christopher Andrew
Rosenbaum, H(erman) S(olomon)
Rossin, A David
Schjelderup, Hassel Charles
Schuyten, John
Schwartzbart, Harry
Shackelford, James Floyd
Sherby, Oleg D(imitri)
Shine, M Carl, Jr
Shyne, J(ohn) C(ornelius)
Sinclair, Robert
Sines, George, Jr
Skinner, Loren Courtland, II
Slivinsky, Sandra Harriet
Sparling, Rebecca Hall
Steinberg, M(orris) A(lbert)
Stetson, Alvin Rae
Stewart, Robert Daniel
Strahl, Erwin Otto
Stringer, John
Sumsion, H(enry) T(heodore)
Syrett, Barry Christopher
Tanner, Lee Elliot
Tauber, Richard Norman
Teitel, Robert J(errell)
Thiers, Eugene Andres
Thomas, Gareth
Thompson, Larry Dean
Thornburg, David Devoe
Tietz, Thomas E(dwin)
Tiller, William Arthur
Tilles, Abe
Toensing, C(larence) H(erman)
Tomlinson, John Lashier
Vreeland, Thad, Jr
Wadsworth, Jeffrey
Wagner, C(hristian) N(ikolaus) J(ohann)
Waisman, Joseph L
Wang, John Ling-Fai
Watson, James Frederic
Westerman, Edwin J(ames)
Westmacott, Kenneth Harry
White, Jack Lee
Wong, Morton Min
Woods, Ralph Arthur
Yang, Charles Chin-Tze
Yang, Jenn-Ming
Yeh, Hun Chiang
Yoon, Rick J

COLORADO
Averill, William Allen
Beckstead, Leo William
Betterton, Jesse O, Jr
Bull, W(illiam) Rex
Carpenter, Steve Haycock
Copeland, William D
Edwards, Glen Robert
Fields, Davis S(tuart), Jr
Fischer, R(oland) B(arton)
Fraikor, Frederick John
Goth, John W
Grobner, Paul Josef
Haas, Frank C
Hager, John P(atrick)
Heiple, Clinton R
Holzwarth, James C
Jha, Mahesh Chandra
Krauss, George
Lewis, Clifford Jackson
Lewis, Homer Dick
McKinnell, W(illiam) P(arks), Jr
Michal, Eugene J(oseph)
Moore, John Jeremy
Mosher, Don R(aymond)
Mote, Jimmy Dale
Mueller, William M(artin)
Newkirk, John Burt
Olson, David LeRoy
Olson, George Gilbert
Rairden, John Ruel, III
Rauch, Gary Clark
Read, David Thomas
Reed, Richard P
Roshko, Alexana
Selle, James Edward
Simons, Sanford L(awrence)

Metallurgy & Physical Metallurgical Engineering (cont)

Vogt, Thomas Clarence, Jr
Wonsiewicz, Bud Caesar
Yarar, Baki
Yukawa, S(umio)

CONNECTICUT
Adams, Brent Larsen
Arnold, John Richard
Blackwood, Andrew W
Bradley, Elihu F(rancis)
Breinan, Edward Mark
Burghoff, Henry L
Canonico, Domenic Andrew
Cheatham, Robert Gary
Chernock, Warren Philip
Clark, C(harles) C(anfield)
Cole, Stephen H(ervey)
Devereux, Owen Francis
Donachie, Matthew J(ohn), Jr
Duhl, David N
Fletcher, Frederick Bennett
Galligan, James M
Gell, Maurice L
Giamei, Anthony Francis
Goldberg, A Jon
Gutierrez-Miravete, Ernesto
Haag, Robert Edwin
Hamjian, Harry J
Joesten, Raymond
Kattamis, Theodoulos Zenon
Lemkey, Frankin David
McEvily, Arthur Joseph, Jr
Mannweiler, Gordon B(annatyne)
Morral, John Eric
Parthasarathi, Manavasi Narasimhan
Pellier, Laurence (Delisle)
Popplewell, James Malcolm
Potter, Donald Irwin
Prasad, Arun
Pryor, Michael J
Radtke, Schrade Fred
Schetky, L(aurence) M(cDonald)
Shapiro, Eugene
Snow, David Baker
Sperry, Philip Roger
Strife, James Richard
Sullivan, Cornelius Patrick, Jr
Swan, D(avid)
Tangel, O(scar) F(rank)
Thomas, Alvin David, Jr
Williams, John Ernest
Winter, Joseph
Zackay, Victor Francis

DELAWARE
Birchenall, Charles Ernest
Cook, Leslie G(ladstone)
Dexter, Stephen C
Ferriss, Donald P
Gemmell, Gordon D(ouglas)
Gibbs, Thomas W(atson)
Graham, Arthur H(ughes)
Greenfield, Irwin G
Hall, Ian Wavell
Hentschel, Robert A(dolf) A(ndrew)
Kramer, John J(acob)
Lin, Kuang-Farn
Ransom, J(ohn) T(hompson)
Streicher, Michael A(lfred)
Urquhart, Andrew Willard

DISTRICT OF COLUMBIA
Ayers, Jack Duane
Beachem, Cedric D
Berry, Vinod K
Cullen, Thomas M
Dennis, William E
Gottschall, Robert James
Haworth, William Lancelot
Hirschhorn, Joel S(tephen)
Horton, Kenneth Edwin
Kaplan, Robert S
Kenahan, Charles Borromeo
Larsen-Basse, Jorn
Lashmore, David S
Meussner, R(ussell) A(llen)
Michel, David John
Murray, Peter
Pande, Chandra Shekhar
Radcliffe, S Victor
Rath, Bhakta Bhusan
Sullivan, Anna Mannevillette
Tokar, Michael
Van Echo, Andrew
Vardiman, Ronald G
Vold, Carl Leroy
Wolf, Stanley Myron
Zwilsky, Klaus M(ax)

FLORIDA
Abbaschian, G J
Anusavice, Kenneth John
Carney, D(ennis) J(oseph)
Caserio, Martin J
Dean, Walter Albert
DeHoff, Robert Thomas
Dudley, Richard H(arrison)
Ebrahimi, Fereshteh
Etherington, Harold

Fullman, R(obert) L(ouis)
Garwood, Maurice F
Gensamer, Maxwell
Gold, Edward
Goldberg, David C
Hartley, Craig Sheridan
Hartt, William Handy
Haseman, Joseph Fish
Hosni, Yasser Ali
Hummel, Rolf Erich
Lubker, Robert A(lfred)
MacDonnell, Wilfred Donald
McSwain, Richard Horace
Reed-Hill, Robert E(llis)
Reid, H(arry) F(rancis), Jr
Rhines, Frederick N(ims)
Roe, William P(rice)
Seth, Brij B
Smith, Edward S
Smith, William Fortune
Stern, Milton
Verink, Ellis D(aniel), Jr
Ward, Leonard George
Wyatt, James L(uther)

GEORGIA
Aseff, George V
Chandan, Harish Chandra
Cofer, Daniel Baxter
Coyne, Edward James, Jr
D'Annessa, A(nthony) T(homas)
Engel, Niels N(ikolaj)
Grenga, Helen E(va)
Hart, Raymond Kenneth
Hochman, Robert F(rancis)
Kalish, David
Korda, Edward J(ohn)
Logan, Kathryn Vance
Meyers, Carolyn Winstead
Sanders, T H, Jr
Underwood, Ervin E(dgar)
Williams, Emmett Lewis
Yeske, Ronald A
Young, C(larence) B(ernard) F(ehrler)

HAWAII
Hamaker, John C, Jr
Hihara, Lloyd Hiromi

IDAHO
Bartlett, R(obert) W(atkins)
Clifton, Donald F(rederic)
Emigh, G Donald
Griffith, W(illiam) A(lexander)
Keefer, Donald Walker
Korth, Gary E
Miller, Richard Lloyd
Place, Thomas Alan
Prisbrey, Keith A
Richardson, Lee S(pencer)
Tallman, Richard Louis
Walters, Leon C

ILLINOIS
Albrecht, Edward Daniel
Altstetter, Carl J
Askill, John
Barnett, Scott A
Battles, James E(verett)
Beck, Paul Adams
Bernstein, I(rving) Melvin
Birnbaum, H(oward) K(ent)
Bohl, Robert W
Bradley, Steven Arthur
Breen, Dale H
Breyer, Norman Nathan
Cannon, Howard S(uckling)
Chiswik, Haim H
Cohen, J(erome) B(ernard)
Cork, Douglas J
Danyluk, Steven
Domagala, Robert F
Erck, Robert Alan
Fenske, George R
Fessler, Raymond R
Fradin, Frank Yale
Frost, Brian R T
Gordon, P(aul)
Gruber, Eugene E, Jr
Kamm, Gilbert G(eorge)
Kelman, L(eRoy) R
Kern, Roy Fredrick
Keyser, N(aaman) H(enry)
Kittel, J Howard
Lam, Daniel J
Langenberg, F(rederick) C(harles)
Lawrence, Frederick Van Buren, Jr
Liu, Yung Yuan
Luerssen, Frank W
Metzger, Marvin
Mueller, Melvin H(enry)
Myles, Kevin Michael
Nachtman, Elliot Simon
Packer, Kenneth Frederick
Ripling, E(dward) J
Rostoker, William
Rothman, Steven J
Rowland, Theodore Justin
Sandoz, George
Schussler, M(ortimer)
Scott, Tom E
Shapiro, Stanley
Shenoy, Gopal K
Siegel, Richard W(hite)

Sinclair, George M(orton)
Sinha, Shome Nath
Smedskjaer, Lars Christian
Socie, Darrell Frederick
Speich, G(ilbert) R(obert)
Sproul, William Dallas
Sticha, Ernest August
Tuomi, Donald
Van Pelt, Richard H
Voorhees, Peter Willis
Walter, Gordon H
Wayman, C(larence) Marvin
Weeks, Richard William
Weertman, Julia Randall
Wensch, Glen W(illiam)
Westlake, Donald G(ilbert)
Wiedersich, H(artmut)
Wimber, R(ay) Ted
Wright, Maurice Arthur
Zambrow, J(ohn) L(ucian)

INDIANA
Balajee, Shankverm R
Bellina, Joseph James, Jr
Bhattacharya, Debanshu
Chang, Chien-Wu
Croat, John Joseph
Dayananda, Mysore Ananthamurthy
Drake, Justin R
Eaton, Paul Bernard
Fosnacht, Donald Ralph
Fritzlen, Glenn A
Ganguli, Amal(endu)
Gaskell, David R
Gorman, Eugene Francis
Grace, Richard E(dward)
Herman, Marvin
Holmes, Alvin W(illiam)
Hruska, Samuel Joseph
Kaye, Albert L(ouis)
Kelley, Thomas Neil
Klein, Howard Joseph
Lazaridis, Nassos A(thanasius)
Lee, Harvie Ho
Liedl, Gerald L(eRoy)
Miller, Albert Eugene
Mills, Norman Thomas
Ovens, William George
Pearson, Donald A
Peretti, Ettore A(lex)
Pielet, Howard M
Pops, Horace
Ranade, Madhukar G
Read, Robert H
Schumann, R(einhardt), Jr
Sherry, John M
Smith, Charles O(liver)
Tucker, Robert C, Jr
Wulpi, Donald James
Young, William Ben

IOWA
Beaudry, Bernard Joseph
Buck, Otto
Carlson, Oscar Norman
Gschneidner, Karl A(lbert), Jr
Kayser, Francis X
Larsen, William L(awrence)
Mather, Roger Frederick
Moriarty, John Lawrence, Jr
Patterson, John W(illiam)
Peterson, David T
Rocklin, Isadore J
Schmidt, Frederick Allen
Smith, John F(rancis)
Spitzig, William Andrew
Trivedi, Rohit K
Verhoeven, John Daniel
Wilder, David Randolph

KANSAS
Cooke, Francis W
Rose, Kenneth E(ugene)
Roth, Thomas Allan
Zerwekh, Robert Paul

KENTUCKY
Kuhn, William E(rik)
Mateer, Richard S(helby)
Mihelich, John L
Morris, James G
Stine, Philip Andrew
Viswanadham, Ramamurthy K

LOUISIANA
Bundy, Kirk Jon
Diwan, Ravinder Mohan
Kinchen, David G
Raman, Aravamudhan
Rubinstein, Asher A

MARYLAND
Armstrong, R(onald) W(illiam)
Boettinger, William James
Cahn, John W(erner)
Carter, John Paul
Coriell, Sam Ray
Cremens, Walter Samuel
Cuthill, John R(obert)
Das, Badri N(arayan)
Decraene, Denis Fredrick
Dieter, George E(llwood), Jr
Doan, Arthur Sumner, Jr
Early, James G(arland)

Economos, Geo(rge)
Fields, Richard Joel
Fishman, Steven Gerald
Fletcher, Stewart G(ailey)
Fraker, Anna Clyde
Franke, Gene Louis
Goode, Robert J
Hayden, H(oward) Wayne, Jr
Hayes, Earl T
Kirby, Ralph C(loudsberry)
Kramer, Irvin Raymond
Kreider, Kenneth Gruber
Kumar, K Sharvan
Lange, Eugene Albert
Lauriente, Mike
Low, John R(outh), Jr
McCawley, Frank X(avier)
Machlin, Irving
Manning, John Randolph
Moore, George A(ndrew)
Nash, Ralph Robert
Newbury, Dale Elwood
Nusbickel, Edward M, Jr
Peterson, Norman L(ee)
Poulose, Pathickal K
Pugh, E(dison) Neville
Quadir, Tariq
Ranganathan, Brahmanpalli Narasimhamurthy
Reuther, Theodore Carl, Jr
Ricker, Richard Edmond
Schaefer, Robert J
Sharp, William Broom Alexander
Smith, John Henry
Solow, Max
Wacker, George Adolf
Wang, Kung-Ping
Weber, Clifford E
Yakowitz, Harvey
Yolken, Howard Thomas
Zapffe, Carl Andrew

MASSACHUSETTS
Abkowitz, Stanley
Adler, Ralph Peter Isaac
Allen, Samuel Miller
Allen, Steven
Apelian, Diran
Argy, Dimitri
Aronin, Lewis Richard
Bachner, Frank Joseph
Balluffi, R(obert) W(eierter)
Bever, Michael B(erliner)
Biederman, Ronald R
Bolsaitis, Pedro
Bornemann, Alfred
Bruggeman, Gordon Arthur
Buffum, Donald C
Chalmers, Bruce
Chen, Sung Jen
Chu, Gordon P K
Cohen, Morris
Coleman, George W(illiam)
Cooke, Richard A
Cornie, James Allen
Coyne, James E
David, Donald J
Delagi, Richard Gregory
Ditchek, Brian Michael
Eagar, Thomas W
Elliott, John Frank
Evans, Michael Douglas
Flemings, Merton Corson
Freedman, George
French, David N(ichols)
French, Robert Dexter
Gangulee, Amitava
Garrett, Paul Daniel
Godrick, Joseph Adam
Grant, Nicholas J(ohn)
Gregory, Eric
Haggerty, John S
Harvey, Walter William
Horwedel, Charles Richard
Isserow, Saul
Johnson, Walter Roland
Kalonji, Gretchen
Karulkar, Pramod C
Khattak, Chandra Prakash
Kim, Han Joong
Kleis, John Dieffenbach
Krashes, David
Kula, Eric Bertil
Lander, Horace N(orman)
Latanision, Ronald Michael
Lement, Bernard S
Livingston, James Duane
McFarland, Charles Manter
MacGregor, C(harles) W(inters)
Malecki, George Jerzy
Marble, Earl R(obert), Jr
Matsuda, Seigo
Mettler, John D(aniel), Jr
Nowak, Welville B(erenson)
Old, Bruce S(cott)
Owen, Walter S
Parkinson, R(obert) E(dward)
Passmore, Edmund M
Pelloux, Regis M N
Petrovic, Louis John
Ragone, David Vincent
Rapperport, Eugene J
Reynolds, Charles C
Rhodes, William Holman

Sadoway, Donald Robert
Sastri, Suri A
Schoen, Frederick J
Schwensfeir, Robert James, Jr
Shen, Yuan-Shou
Smith, Cyril Stanley
Spaepen, Frans
Staker, Michael Ray
Strauss, Alan Jay
Sung, Changmo
Szekely, Julian
Thieme, Cornelis Leo Hans
Tustison, Randal Wayne
Tzeng, Wen-Shian Vincent
Wang, Chih-Chung
Zabielski, Chester V
Zotos, John

MICHIGAN
Aaron, Howard Berton
Albright, Darryl Louis
Beaman, Donald Robert
Beardmore, Peter
Ben, Manuel
Bens, Frederick Peter
Bilello, John Charles
Birkle, A(dolph) John
Bolling, Gustaf Fredric
Bond, Armand Paul
Carter, Giles Frederick
Cole, Gerald Sidney
Corey, Clark L(awrence)
Davies, Richard Glyn
Doane, Douglas V
Eldis, George Thomas
Flinn, Richard A(loysius)
Foreman, Robert Walter
Gibala, Ronald
Giszczak, Thaddeus
Gjostein, Norman A
Gluck, Jeremy V(ictor)
Habermann, Clarence E
Hagel, William C(arl)
Hallada, Calvin James
Harris, Kenneth
Harvey, Douglas J
Harwood, Julius J
Heldt, Lloyd A
Heminghous, William Wayne, Sr
Hendrickson, Alfred A
Hill, John Campbell
Hillegas, William Joseph
Hohnke, Dieter Karl
Hucke, Edward E
Hughel, Thomas J(osiah)
Humenik, Michael, Jr
Ivanso, Eugene V
Jain, Kailash Chandra
Jones, Frederick Goodwin
Keeler, Stuart P
Kipouros, Georges John
Kovacs, Bela Victor
Lahr, Gilbert M
Le Beau, Stephen Edward
Mikkola, Donald E(mil)
Mukherjee, Kalinath
Nagler, Charles Arthur
Nesbitt, Carl C
Pehlke, Robert Donald
Phillips, Charles W(illiam)
Quigg, Richard J
Rausch, Doyle W
Rhee, Seong Kwan
Riegger, Otto K
Robinson, George H(enry)
Rohrig, Ignatius A
Sard, Richard
Sharma, Ram Autar
Shoemaker, Robert Harold
Sinnott, M(aurice) J(oseph)
Smith, Darrell Wayne
Spencer, Andrew R
Sponseller, D(avid) L(ester)
Srolovitz, David Joseph
Stein, Dale Franklin
Stevenson, Robin
Stickels, Charles A
Subramanian, K N
Summitt, W(illiam) Robert
Thayer, Duane M
Thomson, Robert Francis
Tischer, Ragnar P(ascal)
Trojan, Paul K
Turcotte, William Arthur
Van Vlack, Lawrence H(all)
Wang, Yar-Ming
Whalen, Thomas J(ohn)
Williams, Gwynne
Winterbottom, W L
Young, Edwin H(arold)

MINNESOTA
Benner, Blair Richard
Bitsianes, Gust
Bleifuss, Rodney L
Dorenfeld, Adrian C
Fisher, Edward S
Geiger, Gordon Harold
Gerberich, William Warren
Hepworth, Malcolm T(homas)
Huffine, Coy L(ee)
Hultgren, Frank Alexander
Iwasaki, Iwao
Khalafalla, Sanaa E

Larson, Andrew Hessler
McHenry, Kelly David
Maciolek, Ralph Bartholomew
Nam, Sehyun
Nicholson, Morris E(mmons), Jr
Oriani, Richard Anthony
Premanand, Visvanatha
Ross, Stuart Thom
Sartell, Jack A(lbert)
Schmit, Joseph Lawrence
Schultz, Clifford W
Shah, Ishwarlal D
Shores, David Arthur
Staehle, Roger Washburne
Verma, Deepak Kumar

MISSISSIPPI
Wehr, Allan Gordon

MISSOURI
Askeland, Donald Raymond
Bolon, Albert Eugene
Brasunas, Anton Des
Fine, Morris M(ilton)
Hansen, John Paul
Kisslinger, Fred
Klein, Cerry M
Larson, Kenneth Blaine
Leighly, Hollis Philip, Jr
Morris, Arthur Edward
Neumeier, Leander Anthony
O'Keefe, Thomas Joseph
Robertson, David G C
Sastry, Shankara M L
Sonnino, Carlo Benvenuto
Weart, Harry W(aldron)
Wolf, Robert V(alentin)

MONTANA
Griffiths, Vernon
Twidwell, Larry G
Zucker, Gordon L(eslie)

NEBRASKA
Johnson, Donald L(ee)
Nelson, Russell C
Rohde, Suzanne Louise

NEVADA
Chandra, Dhanesh
Eichbaum, Barlane Ronald
Ferrell, Edward F(rancis)
Fuerstenau, M(aurice) C(lark)
See, John Bruce
Skaggs, Robert L

NEW HAMPSHIRE
Backofen, Walter A
Carreker, R(oland) P(olk), Jr
Colligan, George Austin
Dancy, Terence E(rnest)
Epremian, E(dward)
Gardner, Frank S(treeter)
Queneau, Paul E(tienne)
Schulson, Erland Maxwell

NEW JERSEY
Abendroth, Reinhard P(aul)
Albrethsen, A(drian) E(dysel)
Avery, Howard S
Bakish, Robert
Berkowitz, Barry Jay
Bevk, Joze
Boni, Robert Eugene
Brytczuk, Walter L(ucas)
Carlson, Harold C(arl) R(aymond)
Chapin, Henry J(acob)
Chatterji, Debajyoti
Chin, Gilbert Yukyu
Connolly, John Charles
Copson, Harry Rollason
Covert, Roger A(llen)
DeLuca, Robert D(avid)
Dittman, Frank W(illard)
Duerr, J Stephen
Eash, John T(rimble)
Feder, David O
Finke, Guenther Bruno
Foerster, George Stephen
Geskin, Ernest S
Green, Martin Laurence
Grupen, William Brightman
Hittinger, William C(harles)
Iezzi, Robert Aldo
Jin, Sungho
Jordan, Andrew Stephen
Kalish, Herbert S(aul)
Koch, Frederick Bayard
Koul, Maharaj Kishen
Kudman, Irwin
Kurtz, Anthony David
Lanam, Richard Delbert, Jr
Larson, Hugo R
Lemaire, Paul J
Levine, Richard S
Li, Hsin Lang
Liebermann, Howard Horst
Ling, Hung Chi
Lunsford, Ralph D
Lunt, Harry Edward
Manders, Peter William
Mathias, Joseph Simon
Mihalisin, John Raymond
Mitchell, Olga Mary Mracek

Morris, Robert Craig
Moyer, Kenneth Harold
Ohring, Milton
Opie, William R(obert)
Patel, Jamshed R(uttonshaw)
Peters, Dale Thompson
Polk, Conrad Joseph
Rorabaugh, Donald T
Rosen, Carol Zwick
Rosenberg, Robert
Rutherford, John L(oftus)
Scharfstein, Lawrence Robert
Schlabach, T(om) D(aniel)
Sheppard, Keith George
Snowman, Alfred
Subramanyam, Dilip Kumar
Tanenbaum, Morris
Taylor, Jean Ellen
Von Neida, Allyn Robert
Vyas, Brijesh
Weil, Rolf
Wernick, Jack H(arry)
Willens, Ronald Howard
Winsor, Frederick James
Yan, Man Fei
Yim, W(oongsoon) Michael
Yokelson, M(arshall) V(ictor)
Yoon, Euijoon

NEW MEXICO
Alers, George A
Armstrong, Philip E(dward)
Baker, Don H(obart), Jr
Behrens, Robert George
Catlett, Duane Stewart
Clouser, William Sands
Cost, James R(ichard)
Davidson, Keith V(ernon)
Dickinson, James M(illard)
Diegle, Ronald Bruce
Eckelmeyer, Kenneth Hall
Ericksen, Richard Harold
Fritschy, J(ohn) Melvin
George, Timothy Gordon
Hecker, Siegfried Stephen
Hopkins, Alan Keith
Lawson, Andrew Cowper, II
Lowe, Terry Curtis
Magnani, Nicholas J
Mead, Richard Wilson
Miller, Alan R(obert)
Milton, Osborne
Mitchell, David Wesley
Mitchell, Terence Edward
Mulford, Robert Neal Ramsay
Munson, Darrell E(ugene)
O'Rourke, John Alvin
Pope, Larry Elmer
Reiswig, Robert D(avid)
Romig, Alton Dale, Jr
Salzbrenner, Richard John
Sandstrom, Donald James
Tonks, Davis Loel
Van Den Avyle, James Albert
Yost, Frederick Gordon

NEW YORK
Adler, Philip N(athan)
Albers, Francis C
Albrecht, William Melvin
Aldrich, Robert George
Begley, James Andrew
Benz, M G
Beshers, Daniel N(ewson)
Bradley, William W(hitney)
Brauer, Joseph B(ertram)
Brock, Geoffrey E(dgar)
Bronnes, Robert L(ewis)
Brophy, Jere H(all)
Budinski, Kenneth Gerard
Burr, Arthur Albert
Carapella, S(am) C(harles), Jr
Castleman, L(ouis) S(amuel)
Cech, Robert E(dward)
Charles, R(ichard) J(oseph)
Chaudhari, Praveen
Chyung, Kenneth
Clayton, Clive Robert
Clum, James Avery
Cocca, M(ichael) A(nthony)
Cordovi, Marcel A(very)
Cosgarea, Andrew, Jr
Curran, Robert M
Death, Frank Stuart
Decker, L(ouis) H
D'Heurle, Francois Max
Duby, Paul F(rancois)
Duquette, David J(oseph)
Estes, John H
Fiedler, H(oward) C(harles)
Fischer, George J
Flom, Donald Gordon
Floreen, Stephen
Gagnebin, Albert Paul
Glicksman, Martin E
Goldspiel, Solomon
Grosewald, Peter
Grunes, Robert Lewis
Gupta, Devendra
Gupta, Krishna Murari
Gurinsky, David H(arris)
Handman, Stanley E
Harris, Colin C(yril)
Hausner, Henry H(erman)

Herman, Herbert
Homan, Clarke Gilbert
Huston, Ernest Lee
Ingvoldstad, D(onald) F(annon)
Jackson, Melvin Robert
Jordan, Albert Gustav
Judd, Gary
Kammer, Paul A
Kellogg, Herbert H(umphrey)
Kennelley, James A
Kim, Jonathan Jang-Ho
King, Alexander Harvey
Kotval, Pesho Sohrab
Kuan, Tung-Sheng
Le Maistre, Christopher William
Lencl, Fritz (Victor)
Li, Chou H(siung)
Lommel, J(ames) M(yles)
Lord, Arthur N(elson)
Luborsky, Fred Everett
Luthra, Krishan Lal
Lyman, W(ilkes) Stuart
McCarthy, John Joseph
Margolin, Harold
Marwick, Alan David
Maurer, Gernant E
Mayadas, A Frank
Minnear, William Paul
Mondolfo, L(ucio) F(austo)
Morfopoulos, Vassilis C(onstantinos) P
Morris, William Guy
Mulford, Robert Alan
Murakami, Masanori
Muzyka, Donald Richard
Nielsen, John P(hillip)
Nippes, Ernest F(rederick)
Oyler, Glenn W(alker)
Pasco, Robert William
Paul, David I
Pollock, D(aniel) D(avid)
Prager, Martin
Prater, T(homas) A(llen)
Preisler, Joseph J(ohn)
Prindle, William Roscoe
Psioda, Joseph Adam
Puttlitz, Karl Joseph
Pye, Lenwood D(avid)
Quesnel, David John
Reardon, Joseph Daniel
Rothstein, Samuel
Rowe, Raymond Grant
Saha, Bijay S
Satya, Akella V S
Scala, E(raldus)
Schadler, Harvey W(alter)
Schoefer, Ernest A(lexander)
Schulte, Robert Lawrence
Schuster, David Martin
Seigle, L(eslie) L(ouis)
Sims, Chester Thomas
Singh, Mrityunjay
Smith, G(eorge) V
Solomon, Harvey Donald
Srinivasan, Makuteswaran
Stockbridge, Christopher
Stoloff, Norman Stanley
Strasser, Alfred Anthony
Suenaga, Masaki
Sweet, Richard Clark
Szekely, Andrew Geza
Tavlarides, Lawrence Lasky
Taylor, Dale Frederick
Thompson, Charles Denison
Tobin, Albert George
Totta, Paul Anthony
Urban, Joseph
Van Den Sype, Jaak Stefaan
Vermilyea, D(avid) A(ugustus)
Wachtell, Richard L(loyd)
Weeks, John R(andel), IV
Weinig, Sheldon
Whang, Sung H
Woodford, David A(ubrey)
Wright, Roger Neal
Zakraysek, Louis

NORTH CAROLINA
Austin, William W(yatt)
Bailey, John Albert
Beiser, Carl A(dolph)
Cocks, Franklin H
Cox, John Jay, Jr
Dyrkacz, W William
Fahmy, Abdel Aziz
Goodwin, Frank Erik
Haynie, Fred Hollis
Hinesley, Carl Phillip
Holshouser, W(illiam) L(uther)
Kirk, Wilber Wolfe
Koch, Carl Conrad
Kozlik, Roland A
Kusy, Robert Peter
Moazed, K L
Pearsall, George W(ilbur)
Prater, John Thomas
Reeber, Robert Richard
Scattergood, Ronald O
Shepard, Marion L(averne)
Stadelmaier, H(ans) H(einrich)

OHIO
Adams, Robert R(oyston)
Ahmed, Shaffiq Uddin
Alam, M Khairul

Metallurgy & Physical Metallurgical Engineering (cont)

Ault, George Mervin
Babcock, Donald Eric
Baloun, Calvin H(endricks)
Bartlett, Edwin S(outhworth)
Bauer, James H(arry)
Beck, Franklin H(orne)
Bidwell, Lawrence Romaine
Bloom, Joseph Morris
Bomberger, H(oward) B(rubaker), Jr
Boulger, Francis W(illiam)
Buchheit, Richard D(ale)
Burte, Harris M(erl)
Buttner, F(rederick) H(oward)
Campbell, James Edward
Carlson, Robert G(ustav)
Carmichael, Donald C(harles)
Cary, Howard Bradford
Choudhary, Manoj Kumar
Clark, William Alan Thomas
Clauer, Allan Henry
Cooper, Thomas D(avid)
Copp, Albert Nils
Cordea, James Nicholas
Dawson, Steven Michael
Dembowski, Peter Vincent
Dickerson, R(onald) F(rank)
Dreshfield, Robert Lewis
Duttweiler, Russell E
Ebert, Lynn J
Fetters, Karl L(eroy)
Focke, Arthur E(ldridge)
Foster, Ellis L(ouis)
Freche, John C(harles)
Gabb, Timothy Paul
Gegel, Harold L(ouis)
Gelles, S(tanley) H(arold)
Gilby, Stephen Warner
Glaeser, William A(lfred)
Gonser, Bruce Winfred
Graham, John W
Griffith, Cecil Baker
Grisaffe, Salvatore J
Gurklis, John A(nthony)
Hall, Albert M(angold)
Harrison, Robert W
Heckard, David Custer
Hehemann, Robert F(rederick)
Henry, Donald J
Holden, Frank C(harles)
Hook, Rollin Earl
Howden, David Gordon
Ingram, David Christopher
Jackson, Curtis M(aitland)
Jacobs, J(ames) H(arrison)
Jayaraman, Narayanan
Kahles, John Frank
Kampe, Dennis James
Kawasaki, Edwin Pope
Klems, George J
Koros, Peter J
Koster, W(illiam) P(feiffer)
Kot, Richard Anthony
Krashin, Bernard R(obert)
Leber, Sam
Lee, Peter Wankyoon
Leontis, T(homas) E(rnest)
Lewandowski, John Joseph
Lipsitt, Harry A(llan)
Littmann, Martin F(rederick)
Lownie, H(arold) W(illiam), Jr
McCall, James Lodge
McCoy, Robert Allyn
Marschall, Charles W(alter)
Maykuth, D(aniel) J(ohn)
Mezey, Eugene Julius
Mitchell, Terence Edward
Mobley, Carroll Edward
Mohn, Walter Rosing
Morral, F(acundo) R(olf)
Myers, James R(ussell)
Newby, John R
Nikkel, Henry
Paine, Robert Madison
Pardue, William M
Parthasarathy, Triplicane Asuri
Payer, Joe Howard
Perlmutter, Isaac
Pierce, Cyril Marvin
Pool, Monte J
Powell, Gordon W
Pugh, John W(illiam)
Rai, Amarendra Kumar
Rapp, Robert Anthony
Rastogi, Prabhat Kumar
Ray, Alden E(arl)
Readey, Dennis W(illiam)
Rengstorff, George W(illard) P(epper)
Ricksecker, Ralph E
Rigney, David Arthur
Roehrig, Frederick Karl
Rosa, Casimir Joseph
Rosenfield, Alan R(obert)
Rowland, E(lbert) S(ands)
St Pierre, George R(oland)
St Pierre, P(hilippe) D(ouglas) S
Seagle, Stan R
Seltzer, Martin S
Shah, Bhupendra Umedchand
Shemenski, Robert Martin

Shewmon, Paul G(riffith)
Stover, E(dward) R(oy)
Thellmann, Edward L
Thomas, Joseph Francis, Jr
Thomas, Seth Richard
Trela, Edward
Troiano, A(lexander) R(obert)
Uebele, Curtis Eugene
Uys, Johannes Marthinus
Vassamillet, Lawrence Francois
Venkatu, Doulatabad A
Villar, James Walter
Voisard, Walter Bryan
Wagoner, Robert H
Wakelin, David Herbert
Wallace, John F(rancis)
Welsch, Gerhard Egon
Westerman, Arthur B(aer)
Westermann, Fred Ernst
Wickersham, Charles Edward, Jr
Wilde, Bryan Edmund
Williams, James Case
Willson, Karl Stuart
Wlodek, Stanley T
Zmeskal, Otto

OKLAHOMA
Binstock, Martin H(arold)
Block, Robert Jay
Bruner, Ralph Clayburn
Clemenst, James Lee
Daniels, Raymond D(eWitt)
Growcock, Frederick Bruce
Radd, F(rederick) J(ohn)
Richards, Kenneth Julian
Riggs, Olen Lonnie, Jr
Thomason, William Hugh

OREGON
Banning, Lloyd H(arold)
Beall, Robert Allan
Czyzewski, Harry
Daellenbach, Charles Byron
Dash, John
Devletian, Jack H
Dooley, George Joseph, III
Grange, Raymond A
Humphrey, J Richard
Nafziger, Ralph Hamilton
Olleman, Roger D(ean)
Paarsons, James Delbert
Roberts, C Sheldon
Schaefer, Seth Clarence
Thorne, John Kandelin

PENNSYLVANIA
Aaronson, Hubert Irving
Agarwala, Vinod Shanker
Aggen, George
Albert, Robert Lee
Antes, Harry W
Aplan, Frank F(ulton)
Aronson, Arthur H
Aspden, Robert George
Aukrust, Egil
Austin, James Bliss
Bajaj, Ram
Baran, George Roman
Barnartt, Sidney
Bauer, C(harles) L(loyd)
Bell, Gordon M(oreton)
Berman, David Alvin
Bhat, Gopal Krishna
Billman, Fred Richard
Birks, Neil
Bitler, William Reynolds
Bjorhovde, Reidar
Blayden, Lee Chandler
Boltax, Alvin
Bonewitz, Robert Allen
Bowman, Kenneth Aaron
Bramfitt, Bruce Livingston
Brenner, Sidney S(iegfried)
Bridenbaugh, Peter R
Brondyke, Kenneth J
Bryner, Joseph S(anson)
Bucher, John Henry
Burkart, Milton W(alter)
Bush, Glenn W
Butler, John F(rancis)
Caffrey, Robert E
Caton, Robert Luther
Chou, Y(e) T(sang)
Chubb, Walston
Chung, Deborah Duen Ling
Clark, J(ohn) B(everley)
Conard, George P(owell)
Corbett, Robert B(arnhart)
Daga, Raman Lall
Dax, Frank Robert
DebRoy, Tarasankar
Dell, M(anuel) Benjamin
DeLong, William T
De Luccia, John Jerry
Denny, J(ohn) P(almer)
Derge, G(erhard) (Julius)
Deurbrouck, Albert William
Dorschu, Karl E(dward)
Doty, W(illiam) D'Orville
Dukelow, Donald Allen
Dulis, Edward J(ohn)
Economy, George
Emerick, Harold B(urton)
Emlemdi, Hasan Bashir

Englehart, Edwin Thomas, Jr
Enrietto, Joseph Francis
Ferrari, Harry M
Fiore, Nicholas F
Fischer, Robert B
Fitterer, G(eorge) R(aymond)
Foltz, Donald Richard
Fonseca, Anthony Gutierre
Franz, Edmund C(larence)
Fricke, William G(eorge), Jr
Fruehan, Richard J
Fulton, J(ames) C(alvin)
Gill, C(harles) Burroughs
Girifalco, Louis A(nthony)
Goldman, Kenneth M(arvin)
Goldstein, Joseph I
Grabowski, Thomas J
Graham, Charles D(anne), Jr
Gregg, Roger Allen
Hahn, W(alter) C(harles), Jr
Harkness, Samuel Dacke
Hass, Kenneth Philip
Heger, James J
Hertzberg, Richard Warren
Hildeman, Gregory John
Hillner, Edward
Hogg, Richard
Hoke, John Henry
Hopkins, Richard H(enry)
Horne, G(erald) T(erence)
Howell, Paul Raymond
Hu, William H(sun)
Hull, Frederick Charles
Hunsicker, Harold Yundt
Jaffe, Donald
Jain, Himanshu
Jarrett, Noel
Jensen, Craig Leebens
Johnson, C Walter
Johnston, William V
Jordan, Robert Kenneth
Katz, Lewis E
Katz, Owen M
Keller, Rudolf
Khare, Ashok Kumar
Kilp, Gerald R
Koczak, Michael Julius
Korchynsky, M(ichael)
Kosco, John C(arroll)
Koss, Donald A
Kotyk, Michael
Kraft, R(alph) Wayne
Lai, Ralph Wei-Meen
Laird, Campbell
Langford, George
Laughlin, David Eugene
Lawley, Alan
Leinbach, Ralph C, Jr
Lena, Adolph J
Leonard, Laurence
Libsch, Joseph F(rancis)
Lidman, William G
Lifka, Bernard William
London, Gilbert J(ulius)
Loria, Edward Albert
Lustman, Benjamin
McCaughey, Joseph M
McGeady, Leon Joseph
McMahon, Charles J, Jr
MacRae, Donald Richard
Mahajan, Subhash
Manganello, S(amuel) J(ohn)
Manjoine, Michael J(oseph)
Marder, A R
Massalski, T(adeusz) B(ronislaw)
Mears, Dana Christopher
Mehrabian, Robert
Mehrkam, Quentin D
Meier, Gerald Herbert
Michalak, Joseph T(homas)
Misra, Sudhan Sekher
Mullendore, James Alan
Murphy, William J(ames)
Nilan, Thomas George
Novak, Stephen Robert
Orehotsky, John Lewis
Ostrowski, Edward Joseph
Pangborn, Robert Northrup
Pavlovic, Dusan M(ilos)
Paxton, H(arold) W(illiam)
Peiffer, Howard F
Pense, Alan Wiggins
Pettit, Frederick S
Pickering, Howard W
Pinnow, Kenneth Elmer
Pope, David Peter
Porter, Lew F(orster)
Preusch, Charles D
Proske, Joseph Walter
Radford, Kenneth Charles
Ray, Siba Prasad
Richmond, F(rancis) M(artin)
Rogers, H(arry) C(arton), Jr
Rossin, P(eter) C(harles)
Rowe, Anne Prine
Roy, Pradip Kumar
Ruud, Clayton Olaf
Ryba, Earle Richard
Sabol, George Paul
Sander, Louis Frank
Schaeffer, Gene Thomas
Schmidt, Richard
Shabel, Barrie Steven
Shellenberger, Donald J(ames)

Sheridan, John Joseph, III
Shields, Bruce Maclean
Simkovich, George
Smith, James S(terrett)
Spaeder, Carl Edward, Jr
Spitznagel, John A
Sprowls, Donald O(tte)
Staley, James T
Starr, C Dean
Stephenson, Edward T
Stephenson, Robert L
Stoehr, Robert Allen
Stout, Robert Daniel
Stumpf, H(arry) C(linch)
Tankins, Edwin S
Tarby, Stephen Kenneth
Taylor, A
Tepper, Frederick
Thompson, Anthony W
Thornburg, Donald Richard
Torok, Theodore Elwyn
Towner, R(aymond) J(ay)
Townsend, Herbert Earl, Jr
Turkdogan, Ethem Tugrul
Turnbull, G(ordon) Keith
Van Horn, Kent R(obertson)
Varrese, Francis Raymond
Waldman, Jeffrey
Weber, Frank L
Werner, F(red) E(ugene)
Werner, John Ellis
Whiteley, Roger L
Whitlow, Graham Anthony
Wiehe, William Albert
Williams, David Bernard
Wilson, Alexander D
Wood, John D(udley)
Wood, Susan
Woodburn, Wilton A
Woodruff, Kenneth Lee
Wriedt, Henry Anderson
Yeniscavich, William
Zamanzadeh, Mehrooz
Zuspan, G William

RHODE ISLAND
Asaro, Robert John
Avery, Donald Hills
Gurland, Joseph
Richman, Marc H(erbert)
Rockett, Thomas John
Roessler, Barton
Suresh, Subra

SOUTH CAROLINA
Lee, Craig Chun-Kuo
Linnert, George Edwin
McDonell, William Robert
Nevitt, Michael Vogt
Piper, John
Rack, Henry Johann
Rideout, Sheldon P
Robinson, William Courtney, Jr
Schilling, William Frederick
Spicer, Clifford W
Sturcken, Edward Francis
Weiss, Michael Karl
Wolf, James S

SOUTH DAKOTA
Han, Kenneth N
McHugh, Alexander E(dward)

TENNESSEE
Bayuzick, Robert J(ohn)
Borie, Bernard Simon
Bretz, Philip Eric
Brinkman, Charles R
Brooks, Charlie R(ay)
Bullock, Jonathan S, IV
Coleman, Charles Franklin
Danko, Joseph Christopher
David, Stanislaus Antony
DeVan, Jackson H
Flanagan, William F(rancis)
Ford, James Arthur
Glasser, Julian
Goodwin, Gene M
Gorbatkin, Steven M
Gray, Allen G(ibbs)
Gray, Robert J
Griess, John Christian, Jr
Grossbeck, Martin Lester
Harms, William Otto
Holcombe, Cressie Earl, Jr
Horak, James Albert
Horton, Joseph Arno, Jr
Jessen, Nicholas C, Jr
Klueh, Ronald Lloyd
Koger, John W
Lotts, Adolphus Lloyd
Lundin, Carl D
McElroy, David L(ouis)
McHargue, Carl J(ack)
Manly, William D
Miller, Ronald Eldon
Oliver, Bennie F(rank)
Pawel, Richard E
Slade, Edward Colin
Spooner, Stephen
Spruiell, Joseph E(arl)
Stansbury, E(le) E(ugene)
Stiegler, James O
Sump, Cord H(enry)

Sutherland, Charles William
Swindeman, Robert W
Vandermeer, R(oy) A
Weir, James Robert, Jr
Wert, James J
Williams, Robin O('Dare)
Yoakum, Anna Margaret
Yoo, Man Hyong
Yust, Charles S(imon)

TEXAS
Anderson, Robert Clark
Bokros, J(ack) C(hester)
Bourell, David Lee
Bravenec, Edward V
Brotzen, Franz R(ichard)
Caudle, Danny Dearl
Dobrott, Robert D
Eliezer, Zwy
Frick, John P
Gibson, Joseph Woodward
Gray, John Malcolm
Griffin, Richard B
Guard, Ray W(esley)
Gupton, Paul Stephen
Harper, James George
Harris, William J
Jerner, R Craig
Johnson, Robert M
Jones, William B
Kennelley, Kevin James
Koh, P(un) K(ien)
Lautzenheiser, Clarence Eric
Leverant, Gerald Robert
McLellan, Rex B
Marcus, Harris L
Mertens, Frederick Paul
Moore, Thomas Matthew
Murr, Lawrence Eugene
Parikh, N(iranjan) M
Patriarca, Peter
Patton, Charles C(lifford)
Prill, Arnold L
Ralls, Kenneth M(ichael)
Roberts, John Melville
Robinson, J(ames) Michael
Runke, Sidney Morris
Salama, Kamel
Scales, Stanley R
Schlitt, William Joseph, III
Spencer, Thomas H
Stark, J(ohn) P(aul), Jr
Staudenmayer, Ralph
Tems, Robin Douglas
Thomas, P(aul) D(aniel)
Tien, John Kai
Walsh, Kenneth Albert
Watkins, Maurice
Winegartner, Edgar Carl
Wiseman, Carl D

UTAH
Alexander, Guy B
Bishop, Jay Lyman
Bradford, Harold R(awsel)
Byrne, J(oseph) Gerald
Carnahan, Robert D
Dahlstrom, Donald A(lbert)
Daniels, Alma U(riah)
Healy, George W(illiam)
Horton, Ralph M
Last, Arthur W(illiam)
McKinney, William Alan
Malouf, Emil Edward
Michaelson, S(tanley) D(ay)
Miller, Jan Dean
Olson, Ferron Allred
Orava, R(aimo) Norman
Pitt, Charles H
Prater, John D
Sohn, Hong Yong
Spear, Carl D(avid)
Talbot, Eugene L(eroy)
Tuddenham, W(illiam) Marvin
Wadsworth, Milton E(lliot)

VERMONT
Fox, Bradley Alan
Frost, Paul D(avis)
Kim, Sukyoung

VIRGINIA
Almeter, Frank M(urray)
Cook, Desmond C
Courtney, Thomas Hugh
Davis, Bernard Eric
Floridis, Themistochles Philomilos
Gangloff, Richard Paul
Gillmor, R(obert) N(iles)
Hahn, Henry
Healy, John Joseph
Hendricks, Robert Wayne
Joebstl, Johann Anton
Jones, Thomas S
Jorstad, John Leonard
Kirkendall, Ernest Oliver
Kuhlmann-Wilsdorf, Doris
Landgraf, Ronald William
Lane, Joseph Robert
Leslie, William C(airns)
Levy, Sander Alvin
Lowe, A(rthur) L(ee), Jr
Lynd, Langtry Emmett
Lytton, Jack L(ester)

Mayer, George
Owczarski, William A(nthony)
Sansonetti, S John
Schwartz, Ira A(rthur)
Sedriks, Aristide John
Spangler, Grant Edward
Spencer, Chester W(allace)
Starke, Edgar Arlin, Jr
Stein, Bland Allen
Sudarshan, T S
Sullivan, Robert E(mmett)
Wilcox, Benjamin A
Zinkham, Robert Edward

WASHINGTON
Archbold, Thomas Frank
Basmajian, John Aram
Bowdish, Frank W(illiam)
Boyer, Rodney Raymond
Brehm, William Frederick
Brimhall, J(ohn) L
Bush, S(pencer) H(arrison)
Chikalla, Thomas D(avid)
Claudson, T(homas) T(ucker)
DeMoney, Fred William
Duran, Servet A(hmet)
Ekenes, J Martin
Evans, Ersel Arthur
Finlay, Walter L(eonard)
Fischbach, David Bibb
Gelles, David Stephen
Hamilton, C Howard
Hirsch, Horst Eberhard
Hirth, John P(rice)
Holt, Richard E(dwin)
Huang, Fan-Hsiung Frank
Jones, Russell Howard
Kissin, G(erald) H(arvey)
Kruger, Owen L
Leavenworth, Howard W, Jr
Leggett, Robert Dean
Marshall, Robert P(aul)
Masson, D(ouglas) Bruce
Matlock, John Hudson
Minor, James E(rnest)
Peterson, Warren Stanley
Polonis, Douglas Hugh
Quist, William Edward
Rao, Yalamanchili Krishna
Roberts, Earl C(hampion)
Roberts, J T Adrian
Sheely, W(allace) F(ranklyn)
Stang, Robert George
Sutherland, Earl C
Sweet, John W
Tenenbaum, Michael
Westerman, Richard Earl
Wick, O(swald) J
Woods, Keith Newell

WEST VIRGINIA
De Barbadillo, John Joseph
Eiselstein, Herbert Louis
Lemke, Thomas Franklin
Meloy, Thomas Phillips
Olcott, Eugene L
Stoll, Richard E

WISCONSIN
Baker, George Severt
Blumenthal, Robert N
Burck, Larry Harold
Chang, Y Austin
Curtis, Ralph Wendell
Daykin, Robert P
DeVries, Marvin Frank
Dodd, Richard Arthur
Dries, William Charles
Fournelle, Raymond Albert
Heine, Richard W
Hirthe, Walter M(atthew)
Lenling, William James
Loper, Carl R(ichard), Jr
Megahed, Sid A
Michael, Arthur B
Moll, Richard A
Neumann, Joachim Peter
Ramsey, Paul W
Richard, Terry Gordon
Rohatgi, Pradeep Kumar
Samuel, Jay Morris
Schultz, Jay Ward
Wasson, L(oerwood) C(harles)
Weiss, Stanley
Wnuk, Michael Peter
Worzala, F(rank) John

WYOMING
Jacquot, Raymond G

ALBERTA
Bradford, Samuel Arthur
Kuntz, Garland Parke Paul
Wayman, Michael Lash

BRITISH COLUMBIA
Alden, Thomas H(yde)
Barer, Ralph David
Brown, L(aurence) C(laydon)
Garner, Andrew
Hames, F(rederick) A(rthur)
Leja, J(an)
Meech, John Athol
Mular, A(ndrew) L(ouis)

Peters, B(runo) Frank
Peters, E(rnest)
Poling, George Wesley
Tromans, D(esmond)
Vitovec, Franz H
Wynne-Edwards, Hugh Robert

MANITOBA
Chaturvedi, Mahesh Chandra
Dutton, Roger
Rosinger, Herbert Eugene
Wilkins, Brian John Samuel

NOVA SCOTIA
Gow, K V
King, Hubert Wylam

ONTARIO
Agar, G(ordon) E(dward)
Alfred, Louis Charles Roland
Argyropoulos, Stavros Andreas
Ashbrook, Allan William
Atkinson, James T N
Aust, Karl T(homas)
Awadalla, Farouk Tawfik
Bennett, W Donald
Bibby, Malcolm
Bratina, Woymir John
Buhr, Robert K
Cameron, John
Caplan, Donald
Carpenter, Graham John Charles
Cox, Brian
Craig, George Black
Cupp, Calvin R
Diaz, Carlos Manuel
Ells, Charles Edward
Embury, John David
Ettel, Victor Alexander
Franklin, Ursula Martius
Gow, William Alexander
Hamilton, Bruce M
Hansson, Carolyn M
Heffernan, Gerald R
Holt, Richard Thomas
Hunt, Charles Edmund Laurence
Illis, Alexander
Ives, Michael Brian
Keys, John David
Kirkaldy, J(ohn) S(amuel)
Lakshmanan, Vaikuntam Iyer
Lu, Wei-Kao
McGeer, James Peter
Mackay, W(illiam) B(rydon) F(raser)
Morris, Larry Arthur
Northwood, Derek Owen
Orton, John Paul
Pascual, Roberto
Piercy, George Robert
Plumtree, A(lan)
Purdy, Gary Rush
Roberts, W(illiam) Neil
Runnalls, O(liver) John C(lyve)
Schey, John Anthony
Sekerka, Ivan
Siddell, Derreck
Smeltzer, Walter William
Spink, D(onald) R(ichard)
Tedmon, Craig Seward, Jr
Thornhill, Philip G
Toguri, James M
Varin, Robert Andrzej
Wallace, William
Walsh, John Heritage
Watt, Daniel Frank
Winegard, William Charles
Wynnyckyj, John Rostyslav
Youdelis, W(illiam) V(incent)

QUEBEC
Ajersch, Frank
Claisse, Fernand
Galibois, Andre
Habashi, Fathi
Hoa, Suong Van
Jonas, John Joseph
Lavigne, Maurice J
McQueen, Hugh J
Perron, Pierre Omer
Saint-Jacques, Robert G
Van Neste, Andre
Zepfel, William F

SASKATCHEWAN
Barton, Richard J
Le May, I(ain)

OTHER COUNTRIES
Belton, Geoffrey Richard
Boxman, Raymond Leon
Daniel, John Sagar
De Vedia, Luis Alberto
Guberman, H(erbert) D(avid)
Hansson, Inge Lief
Idelsohn, Sergio Rodolfo
Revesz, Zsolt
Rivier, Nicolas Yves
Sarmiento, Gustavo Sanchez
Solari, Mario Jose Adolfo
Spottiswood, David James
Taplin, David Michael Robert
Van Weert, Gezinus

Mining Engineering

ALABAMA
Ahrenholz, H(erman) William
Morley, Lloyd Albert
Park, Duk-Won

ALASKA
Cook, Donald J
Johansen, Nils Ivar
Rao, Pemmasani Dharma
Skudrzyk, Frank J
Wolff, Ernest N

ARIZONA
Coursen, David Linn
Hart, Lyman Herbert
Hiskey, J Brent

CALIFORNIA
Austin, Carl Fulton
Beebe, Robert Ray
Boynton, W(illiam) W(entworth)
Cahn, David Stephen
Conger, Harry M
Havard, John Francis
Hesse, Christian August
Heuze, Francois E
Hong, Ki C(hoong)
Kam, James Ting-Kong
Norman, L(ewis) A(rthur), Jr
Post, J(ames) L(ewis)
Pray, Ralph Emerson
Spokes, Ernest M(elvern)
Wallerstein, David Vandermere

COLORADO
Ayler, Maynard Franklin
Chugh, Ashok Kumar
Clark, George Bromley
Gentry, Donald William
Grosvenor, Niles E(arl)
Howard, Thomas E(dward)
Hustrulid, William A
Michal, Eugene J(oseph)
Panek, Louis A(nthony)
Ponder, Herman
Porter, Darrell Dean
Westfall, Richard Merrill

CONNECTICUT
Milliken, Frank R

DISTRICT OF COLUMBIA
Roosevelt, C(ornelius) V(an) S(chaak)
Yancik, Joseph J

FLORIDA
Garnar, Thomas E(dward), Jr
Goudarzi, Gus (Hossein)
Poundstone, William N

GEORGIA
Moody, Willis E, Jr

HAWAII
Cheng, Ping
Cruickshank, Michael James

IDAHO
Bartlett, R(obert) W(atkins)
Blake, Wilson
Froes, Francis Herbert (Sam)
Hoskins, John Richard

ILLINOIS
Chugh, Yoginder Paul
Eadie, George Robert
Heil, Richard Wendell

INDIANA
Lazaridis, Nassos A(thanasius)

KANSAS
Annis, Jason Carl

KENTUCKY
Leonard, Joseph William
Saperstein, Lee W(aldo)
Wright, Fred(erick) D(unstan)

LOUISIANA
Kazmann, Raphael Gabriel

MARYLAND
Malghan, Subhaschandra Gangappa
Morgan, John D(avis)
Wang, Kung-Ping

MASSACHUSETTS
Tarkoy, Peter J

MICHIGAN
Freyberger, Wilfred L(awson)
Greuer, Rudolf E A
Klimpel, Richard Robert
Sarkar, Nitis
Snyder, Virgil W(ard)

MINNESOTA
Atchison, Thomas Calvin, Jr
Benner, Blair Richard
Dorenfeld, Adrian C
Fairhurst, C(harles)

Mining Engineering (cont)

Johnson, Thys B(rentwood)
Schultz, Clifford W
Yardley, Donald H

MISSISSIPPI
Mather, Katharine Kniskern

MISSOURI
Barr, David John
Carmichael, Ronald L(ad)
Haas, Charles John
Hatheway, Allen Wayne
Scott, James J
Summers, David Archibold

MONTANA
Stout, Koehler

NEVADA
Akhtar, Salim
Daemen, Jaak J K
Ghosh, Amitava
Schilling, John H(arold)
Smith, Ross W

NEW JERSEY
Albrethsen, A(drian) E(dysel)
Avery, Howard S
Hassialis, Menelaos D(imitri)
Jin, Sungho
Maxim, Leslie Daniel

NEW MEXICO
Baker, Don H(obart), Jr
Chen, Er-Ping
Griswold, George B
Sutherland, Herbert James
Wawersik, Wolfgang R

NEW YORK
Boshkov, Stefan
Gambs, Gerard Charles
Harris, Colin C(yril)
Litz, Lawrence Marvin
Mahtab, M Ashraf
Malozemoff, Plato
Mika, Thomas Stephen
Wane, Malcolm T(rafford)
Yegulalp, Tuncel Mustafa

NORTH CAROLINA
Hamme, John Valentine

OHIO
Babcock, Donald Eric
Copp, Albert Nils
Konya, Calvin Joseph
Ponter, Anthony Barrie

OKLAHOMA
Binstock, Martin H(arold)
Brady, Barry Hugh Garnet
Bray, Bruce G(lenn)
Stover, Dennis Eugene

OREGON
Daellenbach, Charles Byron

PENNSYLVANIA
Aplan, Frank F(ulton)
Bell, Gordon M(oreton)
Bieniawski, Zdzislaw Tadeusz
Bise, Christopher John
Bockosh, George R
Core, Jesse F
Deurbrouck, Albert William
du Breuil, Felix L(emaigre)
Falkie, Thomas Victor
Fonseca, Anthony Gutierre
Grannemann, Glenn Niel
Hertzberg, Martin
Kelley, Jay Hilary
Kleinman, Robert L P
L'Esperance, Robert Louis
Lovell, Harold Lemuel
Mangelsdorf, Clark P
Mitchell, Donald W(illiam)
Mowrey, Gary Lee
Murphy, John N
Phelps, Lee Barry
Ramani, Raja Venkat
Rumbarger, John H
Skoner, Peter Raymond

SOUTH CAROLINA
McCormick, Robert Becker

SOUTH DAKOTA
Han, Kenneth N

TEXAS
Bigham, Robert Eric
Camp, Frank A, III
Heins, Robert W
Hoskins, Earl R, Jr
Mathewson, Christopher Colville
Monsees, James E
Parate, Natthu Sonbaji
Patrick, Wesley Clare
Russell, James E(dward)
Sahinen, Winston Martin
Schlitt, William Joseph, III

Shen, Chin-Wen
Winegartner, Edgar Carl

UTAH
Hucka, Vladimir Joseph
Michaelson, S(tanley) D(ay)
Miller, Jan Dean
Spedden, H Rush
Zavodni, Zavis Marian

VIRGINIA
Cooley, William C
Faulkner, Gavin John
Foreman, William Edwin
Karfakis, Mario George
Karmis, Michael E
Lucas, J B
Topuz, Ertugrul S
Torgersen, Paul E
Wayland, Russell Gibson

WASHINGTON
Bowdish, Frank W(illiam)
Drevdahl, Elmer R(andolph)
Finlay, Walter L(eonard)

WEST VIRGINIA
Adler, Lawrence
Chiang, Han-Shing
Humphreys, Kenneth K
Khair, Abdul Wahab
Peng, Syd Syh-Deng
Rollins, Ronald Roy

WISCONSIN
Dinter Brown, Ludmila

ALBERTA
Hoekstra, Pieter
Patching, Thomas
Scott, J(ohn) D(onald)

BRITISH COLUMBIA
Hollister, Victor F
Meech, John Athol
Mular, A(ndrew) L(ouis)

ONTARIO
Baird, Malcolm Henry Inglis
Calder, Peter N
Diaz, Carlos Manuel
Graham, A(lbert) Ronald
Simpson, Frank
Udd, John Eaman

QUEBEC
Barbery, Gilles A
Poorooshasb, Hormozd Bahman

OTHER COUNTRIES
Heerema, Ruurd Herre
Pegler, Alwynne Vernon
Spottiswood, David James
Wolfe, John A(llen)

Nuclear Engineering

ALABAMA
Hollis, Daniel Lester, Jr
Morris, James Allen
Perry, Nelson Allen
Willenberg, Harvey Jack
Williams, Louis Gressett

ARIZONA
Ganapol, Barry Douglas
Hetrick, David LeRoy
Koch, Leonard John
McKlveen, John William
Nelson, George William
Post, Roy G
Risbud, Subhash Hanamant
Seale, Robert L(ewis)

ARKANSAS
Krohn, John Leslie

CALIFORNIA
Auerbach, Jerome Martin
Balzhiser, Richard E(arl)
Bareis, David W(illard)
Beaton, Roy Howard
Bernath, L(ouis)
Bernsen, Sidney A
Blink, James Allen
Brehm, Richard Lee
Breitmayer, Theodore
Broido, Jeffrey Hale
Budnitz, Robert Jay
Bullock, Ronald Elvin
Busch, Joseph Sherman
Bush, Gary Graham
Carver, John Guill
Chan, Bertram Kim Cheong
Cheng, Edward Teh-Chang
Chesworth, Robert Hadden
Childs, Wylie J
Cohen, Karl (Paley)
Conn, Robert William
Culler, F(loyd) L(eRoy), Jr
Dahlberg, Richard Craig
D'Ardenne, Walter H
Dau, Gary John

Dean, Richard A
Deleray, Arthur Loyd
De Micheli, Giovanni
Dowdy, William Louis
Dunning, John Ray, Jr
Eggen, Donald T(ripp)
Ehrlich, Richard
Erdmann, Robert Charles
Esselman, W(alter) H(enry)
Fenech, Henri J
Forbes, Judith L
Forsen, Harold K(ay)
Freis, Robert P
Fullmer, George Clinton
Gershun, Theodore Leonard
Goodjohn, Albert J
Goodman, Julius
Gritton, Eugene Charles
Grossman, Lawrence M(orton)
Gyorey, Geza Leslie
Haake, Eugene Vincent
Hallet, Raymon William, Jr
Hammond, R Philip
Harris, Sigmund Paul
Hecht, Myron J
Heckman, Richard Ainsworth
Hendry, George Orr
Hill, Ernest E(lwood)
Hines, Douglas P
Holdren, John Paul
Hovingh, Jack
Howe, John P(erry)
Howerton, Robert James
Judge, Frank D
Kaplan, Selig N(eil)
Karin, Sidney
Kastenberg, William Edward
Keller, Joseph Edward, Jr
Kendrick, Hugh
Kim, Jinchoon
Koenig, Daniel Rene
Kull, Lorenz Anthony
Kupfer, John Carlton
Landsbaum, Ellis M(erle)
Langer, Sidney
Leung, Ka-Ngo
Levenson, Milton
Leverett, M(iles) C(orrington)
Lewis, Kenneth D
Loewenstein, Walter B
Long, Alexander B
Lucas, Glenn E
Lucas, Robert Gillem
Lurie, Norman A(lan)
McCreless, Thomas Griswold
Marto, Paul James
Martz, Harry Edward, Jr
Melese, Gilbert B(ernard)
Merilo, Mati
Merker, Milton
Miatech, Gerald James
Michels, Lloyd R
Moir, Ralph Wayne
Monard, Joyce Anne
Montague, L David
Moore, Richard Arthur
Morewitz, Harry Alan
Naymark, Sherman
O'Dell, Austin Almond, Jr
Odette, G(eorge) Robert
Oehlberg, Richard N
Olander, D(onald) R(aymond)
Orcutt, John Arthur
Orphan, Victor John
Papay, Lawrence T
Parks, Lewis Arthur
Partanen, Jouni Pekka
Passell, Thomas Oliver
Pathania, Rajeshwar S
Pedersen, Knud B(orge)
Perkins, Sterrett Theodore
Phillips, Thomas Joseph
Pinkel, B(enjamin)
Plebuch, Richard Karl
Pollack, Louis Rubin
Pomraning, Gerald C
Porter, Gary Dean
Profio, A(medeus) Edward
Remley, Marlin Eugene
Rickard, Corwin Lloyd
Rossin, A David
Rumble, Edmund Taylor, III
Sampson, Henry T
Saxe, Raymond Frederick
Schalla, Charence August
Schlaug, Robert Noel
Schrock, Virgil E(dwin)
Scott, Franklin Robert
Sesonske, Alexander
Sher, Rudolph
Siegel, Sidney
Simnad, Massoud T
Skavdahl, R(ichard) E(arl)
Slocum, Richard William
Smathers, James Burton
Snyder, Nathan W(illiam)
Stuhmiller, James Hamilton
Swanson, David G, Jr
Talley, Wilson K(inter)
Taylor, John Joseph
Theofanous, Theofanis George
Trippe, Anthony Philip
Turchan, Otto Charles
Uthe, P(aul) M(ichael), Jr

Vagelatos, Nicholas
Wagner, Carl E
Wazzan, A R Frank
Weber, Hans Josef
Williams, Alan K
Wolfe, Bertram
Wray, John L
Zebroski, Edwin L
Zimmerman, Elmer Leroy

COLORADO
Burnett, Jerrold J
Harlan, Ronald A
Jha, Mahesh Chandra
Ladd, Conrad Mervyn
Lewis, Homer Dick
Lubell, Jerry Ira
Meroney, Robert N
Reichardt, John William
Smith, Richard Cecil
Stauffer, Jack B
Wilbur, Paul James

CONNECTICUT
Chernock, Warren Philip
Cobern, Martin E
Hood, Edward E, Jr
Johnson, Loering M
Jones, Richard Bradley
Korin, Amos
Krisst, Raymond John
Lichtenberger, Harold V
Prabulos, Joseph J, Jr
Rohan, Paul E(dward)
Storrs, Charles Lysander

DELAWARE
Stewart, John Woods
Vick Roy, Thomas Rogers

DISTRICT OF COLUMBIA
Baer, Robert Lloyd
Campbell, Francis James
Chapin, Douglas McCall
Ferguson, George Alonzo
Florig, Henry Keith
Goodman, Eli I
Grabowski, Kenneth S
Killion, Lawrence Eugene
Lang, Peter Michael
Marcus, Gail Halpern
Michel, David John
Myers, Peter Briggs
Odar, Fuat
Rosen, Sol
Sanders, Robert Charles
Senich, Donald
Shon, Frederick John
Stone, Philip M
Thomas, Cecil Owen, Jr
Tokar, Michael
Trice, Virgil Garnett, Jr
Van Echo, Andrew

FLORIDA
Anghaie, Samim
Becker, Martin
Benedict, Manson
Buoni, Frederick Buell
Carroll, Edward Elmer
Cox, John David
Dalton, G(eorge) Ronald
Davis, Duane M
Diaz, Nils Juan
Dunford, James Marshall
Etherington, Harold
Hankel, Ralph D
Hill, Frank B(ruce)
Humphries, Jack Thomas
Hungerford, Herbert Eugene
Jacobs, Alan M(artin)
Kornblith, Lester
Langford, David
Mandil, I Harry
Mione, Anthony J
Rosenthal, Henry Bernard
Shea, Richard Franklin
Suciu, S(piridon) N
Tomonto, James R
Tulenko, James Stanley
Veziroglu, T Nejat
West, John M
Wethington, John A(bner), Jr

GEORGIA
Abdel-Khalik, Said Ibrahim
Circeo, Louis Joseph, Jr
Clement, Joseph D(ale)
Davis, Monte V
Eichholz, Geoffrey G(unther)
Graham, Walter Waverly, III
Griffin, Clayton Houstoun
Holt, David Lowell
Jessen, Nicholas C, Sr
Kallfelz, John Michael
Karam, Ratib A(braham)
Morgan, Karl Ziegler
Russell, John Lynn, Jr
Rust, James Harold
Schneider, Alfred
Schutt, Paul Frederick
Stacey, Weston Monroe, Jr
Swank, Robert Roy, Jr
Visner, Sidney

Wilkins, J Ernest, Jr

HAWAII
Russo, Anthony R

IDAHO
Bickel, John Henry
Bills, Charles Wayne
Buescher, Brent J
Carpenter, Stuart Gordon
Chapman, Ray LaVar
DeBow, W Brad
Gasidlo, Joseph Michael
Gehrke, Robert James
Hanson, George H(enry)
Harper, Henry Amos, Jr
Hyde, William W
Jobe, Lowell A(rthur)
Johnson, Stanley O(wen)
Kaiser, Richard Edward
Larson, Jay Reinhold
Lawroski, Harry
Leyse, Carl F(erdinand)
Long, John Kelley
McFarlane, Harold Finley
Majumdar, Debaprasad (Debu)
Meyer, Orville R
Moore, Kenneth Virgil
Mortensen, Glen Albert
Obenchain, Carl F(ranklin)
Olson, Willard Orvin
Reno, Harley W
Rice, Charles Merton
Sohal, Manohar Singh
Stevenson, Charles Edward
Tanner, John Eyer, Jr
Wilson, Albert E
Wood, Richard Ellet
Woodall, David Monroe

ILLINOIS
Albrecht, Edward Daniel
Axford, Roy Arthur
Bezella, Winfred August
Bhattacharyya, Samit Kumar
Braid, Thomas Hamilton
Carpenter, John Marland
Cember, Herman
Chang, Yoon Il
Chilton, A(rthur) B(ounds)
Crawford, Raymond Maxwell, Jr
Deen, James Robert
Deitrich, L(awrence) Walter
De Volpi, Alexander
Dickerman, Charles Edward
Doerner, Robert Carl
DuTemple, Octave J
Errede, Steven Michael
Fenske, George R
Fink, Charles Lloyd
Flynn, Kevin Francis
Goldman, Arthur Joseph
Gruber, Eugene E, Jr
Guenther, Peter T
Hang, Daniel F
Haugen, Robert Kenneth
Holtzman, Richard Beves
Horve, Leslie A
Ivins, Richard O(rville)
Jones, Barclay G(eorge)
Kanter, Manuel Allen
Kelman, L(eRoy) R
Kittel, J Howard
Lawroski, Stephen
Lellouche, Gerald S
LeSage, Leo G
Liaw, Jye Ren
Lipinski, Walter C(harles)
Liu, Yung Yuan
Macherey, Robert E
McKnight, Richard D
Malloy, Donald Jon
Marchaterre, John Frederick
Marr, William Wei-Yi
Massel, Gary Alan
Meneghetti, David
Miley, G(eorge) H(unter)
Monson, Harry O
Moran, Thomas J
Pizzica, Philip Andrew
Ramaswami, Devabhaktuni
Redman, William Charles
Rose, David
Seefeldt, Waldemar Bernhard
Sha, William T
Shaftman, David Harry
Snelgrove, James Lewis
Steindler, Martin Joseph
Stubbins, James Frederick
Toppel, Bert Jack
Travelli, Armando
Uherka, Kenneth Leroy
Wensch, Glen W(illiam)
Woodruff, William Lee
Yang, Shi-Tien
Youngdahl, Carl Kerr

INDIANA
Clikeman, Franklyn Miles
Lucey, John William
Lykoudis, Paul S
Ott, Karl O
Tucker, Robert C, Jr
Wellman, Henry Nelson

IOWA
Edelson, Martin Charles
Hendrickson, Richard A(llan)
Rohach, Alfred F(ranklin)

KANSAS
Eckhoff, Norman Dean
Faw, Richard E
Mingle, John O(rville)
Robinson, M John
Shultis, J Kenneth
Weller, Lawrence Allenby

KENTUCKY
Davis, Thomas Haydn
Eaton, Thomas Eldon
Lindauer, George Conrad
Ulrich, Aaron Jack

LOUISIANA
Courtney, John Charles
Deddens, J(ames) C(arroll)
Goodman, Alan Leonard
Hibbeler, Russell Charles
Lambremont, Edward Nelson
Trinko, Joseph Richard, Jr
Young, Myron H(wai-Hsi)

MAINE
Wilbur, L(eslie) C(lifford)

MARYLAND
Adamantiades, Achilles G
Almenas, Kazys Kestutis
Andreadis, Tim Dimitri
Barrett, Richard John
Bournia, Anthony
Buck, John Henry
Bustard, Thomas Stratton
Carter, Robert Emerson
Congel, Frank Joseph
Coursey, Bert Marcel
Dean, Stephen Odell
Duffey, Dick
Eisenhauer, Charles Martin
Eiss, Abraham L(ouis)
Englehart, Richard W(ilson)
Fabic, Stanislav
Farmer, William S(ilas), Jr
Fitzgerald, Duane Glenn
Goktepe, Omer Faruk
Graves, Harvey W(ilbur), Jr
Grotenhuis, Marshall
Hawkins, William M(adison), Jr
Huddleston, Charles Martin
Jabbour, Kahtan Nicolas
Kazi, Abdul Halim
Khatib-Rahbar, Mohsen
Loftness, Robert L(eland)
Loss, Frank J
McLeod, Norman Barrie
Mallory, Charles William
Marotta, Charles Rocco
Miller, Hugh Hunt
Mills, William Andy
Munno, Frank J
Oktay, Erol
Osborne, Morris Floyd
Papadopoulos, Konstantinos Dennis
Reupke, William Albert
Reuther, Theodore Carl, Jr
Riel, Gordon Kienzle
Rockwell, Theodore
Roush, Marvin Leroy
Sallet, Dirse Wilkis
Schleiter, Thomas Gerard
Schroeder, Frank, Jr
Schwoerer, F(rank)
Shalowitz, Erwin Emmanuel
Song, Yo Taik
Trombka, Jacob Israel
Tropf, Cheryl Griffiths
Van Houten, Robert
Walton, Ray Daniel, Jr
Weber, Clifford E
Wiggins, Peter F
Yockey, Hubert Palmer
Zmola, Paul C(arl)

MASSACHUSETTS
Abernathy, Frederick Henry
Bowman, H(arry) Frederick
Brown, Gilbert J
Carnesale, Albert
Connors, Philip Irving
Cooper, W(illiam) E(ugene)
Delagi, Richard Gregory
Driscoll, Michael John
Eoll, John Gordon
Gyftopoulos, Elias P(anayiotis)
Hansen, Kent F(orrest)
Hopkins, George Robert
Huffman, Fred Norman
Kautz, Frederick Alton, II
Kazimi, Mujid S
Kelsey, Ronald A(lbert)
Landis, John W
Lanning, David D(ayton)
Latanision, Ronald Michael
Rasmussen, Norman Carl
Russell, Kenneth Calvin
Sigmar, Dieter Joseph
Silvers, J(ohn) P(hillip)
Stevens, J(ames) I(rwin)

Yener, Yaman
Yip, Sidney

MICHIGAN
Akcasu, Ahmet Ziyaeddin
Dow, W(illiam) G(ould)
Duderstadt, James
Eldred, Norman Orville
Gomberg, Henry J(acob)
Gutzman, Philip Charles
Haidler, William B(ernard)
Hammitt, Frederick G(nichtel)
Hill, Robert F
Kammash, Terry
Kerr, William
Kikuchi, Chihiro
Killelea, Joseph R(ichard)
Knoll, Glenn F
Lee, John Chaeseung
McCarthy, Walter J, Jr
Mayer, Frederick Joseph
Miller, Herman Lunden
Shure, Fred C(harles)
Summerfield, George Clark
Vincent, Dietrich H(ermann)

MINNESOTA
Fisher, Edward S
Isbin, Herbert S(tanford)
Mohan, Narendra

MISSISSIPPI
Paulk, John Irvine

MISSOURI
Bolon, Albert Eugene
Carter, Robert L(eroy)
Donovan, John Richard
Edwards, Doyle Ray
Glascock, Michael Dean
Lowe, Forrest Gilbert
Loyalka, Sudarshan Kumar
Prelas, Mark Antonio
Richards, Earl Frederick
Tsoulfanidis, Nicholas
Werner, Samuel Alfred

MONTANA
Cohn, Paul Daniel

NEBRASKA
Blotcky, Alan Jay
Borchert, Harold R
Wittke, Dayton D

NEVADA
Brandstetter, Albin
Miller, Wayne L(eroy)

NEW HAMPSHIRE
Fricke, Edwin Francis
Price, Glenn Albert

NEW JERSEY
Adams, James Mills
Bartnoff, Shepard
Blackshaw, George Lansing
Bush, Charles Edward
Cho, Soung Moo
Denno, Khalil I
Fu, Hui-Hsing
George, Kenneth Dudley
Gordon, R(obert)
Halik, Raymond R(ichard)
Hendrickson, Tom A(llen)
Jensen, Betty Klainminc
Kane, Ronald S(teven)
Kovats, Andre
Levine, Jerry David
Long, Robert Leroy
Loscher, Robert A
McNicholas, James J
Mark, Robert
Reisman, Otto
Rothleder, Stephen David
Shaw, Henry
Shober, Robert Anthony
Towner, Harry H
Vann, Joseph M
Vernon, Russel
Wachtell, George Peter
Weinzimmer, Fred

NEW MEXICO
Baldwin, George C
Bergeron, Kenneth Donald
Booth, Lawrence A(shby)
Booth, Thomas Edward
Boudreau, Jay Edmond
Bryant, Lawrence E, Jr
Burick, Richard Joseph
Caldwell, John Thomas
Chambers, William Hyland
Chen, Er-Ping
Dowdy, Edward Joseph
Dudziak, Donald John
El-Genk, Mohamed Shafik
England, Talmadge Ray
Gauster, Wilhelm Belrupt
Gibbs, Terry Ralph
Gover, James E
Griffin, Patrick J
Guevara, Francisco A(ntonio)
Haight, Robert Cameron

Hoover, Mark Douglas
Jankowski, Francis James
Kelley, Gregory M
Kensek, Ronald P
Kirk, William Leroy
Lam, Kin Leung
Lanter, Robert Jackson
Lee, David Mallin
MacFarlane, Donald Robert
Marlow, Keith Winton
Mazarakis, Michael Gerassimos
Menlove, Howard Olsen
Miller, Warren Fletcher, Jr
Milton, Osborne
Muir, Douglas William
Nickel, George H(erman)
Nuttall, Herbert Ericksen, Jr
O'Dell, Ralph Douglas
Philbin, Jeffrey Stephen
Rahal, Leo James
Renken, James Howard
Sandmeier, Henry Armin
Sicilian, James Michael
Snyder, Albert W
Sprinkle, James Kent, Jr
Stephans, Richard A
Stevenson, Michael Gail
Tanaka, Nobuyuki
Terrell, C(harles) W(illiam)
Thomas, Charles Carlisle, Jr
Travis, John Richard
Valentine, James K
Vigil, John Carlos
Von Riesemann, Walter Arthur
Watson, Clayton Wilbur
Williams, James Marvin
Williams, Michael D
Williamson, Kenneth Donald, Jr
Yarnell, John Leonard

NEW YORK
Anthony, Donald Joseph
Bari, Robert Allan
Baron, Seymour
Baum, John W
Block, Robert Charles
Carew, John Francis
Chan, Shu Fun
Chappell, Richard Lee
Chen, Inan
Clark, David Delano
Cokinos, Dimitrios
Coppersmith, Frederick Martin
Diamond, David J(oseph)
Feinberg, Robert Jacob
Fishbone, Leslie Gary
Fried, Erwin
Fullwood, Ralph Roy
Goldmann, Kurt
Gralnick, Samuel Louis
Haag, Fred George
Hammer, David Andrew
Hanson, Albert L
Harris, Donald R, Jr
Hartmann, Francis Xavier
Hendrie, Joseph Mallam
Higinbotham, William Alfred
Hockenbury, Robert Wesley
Hoisie, Adolfy
Inhaber, Herbert
Ishida, Takanobu
Jedruch, Jacek
Jensen, Betty Klainminc
Kato, Walter Yoneo
Kayani, Joseph Thomas
King, Alexander Harvey
Kitchen, Sumner Wendell
Knight, Bruce Winton, (Jr)
Krieger, Gary Lawrence
Kroeger, Peter G
Lahey, Richard Thomas, Jr
Lazareth, Otto William, Jr
Lee, Richard Shao-Lin
Levine, Melvin Mordecai
Lidofsky, Leon Julian
Lin, Mow Shiah
Lubitz, Cecil Robert
McGuire, Stephen Craig
Marwick, Alan David
Meyer, Andrew
Ott, Henry C(arl)
Palmedo, Philip F
Perkins, Kenneth Roy
Podowski, Michael Zbigniew
Powell, James R, Jr
Simon, Albert
Stephenson, Thomas E(dgar)
Strasser, Alfred Anthony
Sturges, Stuart
Susskind, Herbert
Von Berg, Robert L(ee)
Vortuba, Jan
Ward, Thomas J(ulian)
Weber, Arthur Phineas
Weng, Wu Tsung
White, Frederick Andrew
Winsche, Warren Edgar
Wulff, Wolfgang

NORTH CAROLINA
Burton, Ralph Gaines
Doster, Joseph Michael
Fahmy, Abdel Aziz
Gardner, Robin P(ierce)

Nuclear Engineering (cont)

Jordan, Edward Daniel
MacMillan, J(ohn) H(enry)
Murray, Raymond L(eRoy)
Siewert, Charles Edward
Squire, Alexander
Turinsky, Paul Josef
Verghese, Kuruvilla
Watson, James E, Jr
Wiley, Albert Lee, Jr
Zaalouk, Mohamed Gamal
Zumwalt, Lloyd Robert

OHIO

Anno, James Nelson
Bailey, Robert E
Basham, Samuel Jerome, Jr
Bement, A(rden) L(ee), Jr
Beverly, Robert Edward, III
Bhattacharya, Ashok Kumar
Bridgman, Charles James
Carbiener, Wayne Alan
Denning, Richard Smith
Hofmann, Peter L(udwig)
Janzow, Edward F(rank)
John, George
Kaplan, Arthur Lewis
Kelly, Kevin Anthony
Lundin, Bruce T(heodore)
Miller, Don Wilson
Nakamura, Shoichiro
Pardue, William M
Robinson, Richard Alan
Rudy, Clifford R
Sanford, Edward Richard
Schlosser, Philip A
Shapiro, Alvin
Stockman, Charles H(enry)
Vanderburg, Vance Dilks
Weisman, Joel
Woodard, Ralph Emerson

OKLAHOMA

Bray, Bruce G(lenn)
Howard, Robert Adrian
Zigrang, Denis Joseph

OREGON

Chezem, Curtis Gordon
Hornyik, Karl
Kolar, Oscar Clinton
LeTourneau, Budd W(ebster)
Ringle, John Clayton
Ruby, Lawrence
Starr, E(ugene) C(arl)
Woods, W(allace) Kelly

PENNSYLVANIA

Baratta, Anthony J
Boltax, Alvin
Boyer, Vincent S
Brey, R(obert) N(ewton)
Carelli, Mario Domenico
Carfagno, Salvatore P
Catchen, Gary Lee
Chapin, David Lambert
Chelemer, Harold
Chi, John Wen Hua
Chubb, Walston
Cohen, Paul
Curlee, Neil J(ames), Jr
Diethorn, Ward Samuel
Doub, William Blake
Fischer, John Edward
Foderaro, Anthony Harolde
Forscher, Frederick
Freeman, Louis Barton
Giffen, Robert H(enry)
Hardy, Judson, Jr
Harkness, Samuel Dacke
Hogan, Aloysius Joseph, Jr
Houlihan, John Frank
Impink, Albert J(oseph), Jr
Jacobi, W(illiam) M(allett)
Jensen, James Ejler
Jester, William A
Johnston, William V
Kane, John Vincent, Jr
Kitzes, Arnold S(tanley)
Klevans, Edward Harris
Kline, Donald Edgar
Knief, Ronald Allen
Kortier, William E
Kowal, George M
Levine, Samuel Harold
Lippincott, Ezra Parvin
Lochstet, William A
Loman, James Mark
Martucci, John A
Mezger, Fritz Walter William
Michael, Norman
Palladino, Nunzio J(oseph)
Palusamy, Sam Subbu
Pfennigwerth, Paul Leroy
Redfield, John A(lden)
Remick, Forrest J(erome)
Saluja, Jagdish Kumar
Sarram, Mehdi
Shure, Kalman
Siddiqui, Habib
Silverstein, Calvin C(arlton)
Smid, Robert John
Steinbruegge, Kenneth Brian

Stern, Theodore
Tang, Y(u) S(un)
Treadwell, Kenneth Myron
Vance, William Harrison
Waldman, L(ouis) A(braham)
Weddell, George G(ray)
Witzig, Warren Frank
Yao, Shi Chune
Yeniscavich, William

RHODE ISLAND

Rose, Vincent C(elmer)

SOUTH CAROLINA

Ahlfeld, Charles Edward
Baumann, Norman Paul
Benjamin, Richard Walter
Bradley, Robert Foster
Bridges, Donald Norris
Chao, Yuh J (Bill)
Church, John Phillips
Clark, Hugh Kidder
Crandall, John Lou
Dworjanyn, Lee O(leh)
Fjeld, Robert Alan
Gaines, Albert L(owery)
Godfrey, W Lynn
Graulty, Robert Thomas
Graves, William Ewing
Gregory, Michael Vladimir
Groh, Harold John
Hahn, Walter I
Hootman, Harry Edward
Johnson, Ben S(lemmons), Jr
Macaulay, Andrew James
McCrosson, F Joseph
Schwartz, Elmer G(eorge)
Townes, George Anderson
Williamson, Thomas Garnett

TENNESSEE

Ackermann, Norbert Joseph, Jr
Armentrout, Daryl Ralph
Auxier, John A
Baer, Thomas Strickland
Besmann, Theodore Martin
Bigelow, John E(aly)
Booth, Ray S
Brewington, Percy, Jr
Bryan, Robert H(owell)
Burwell, Calvin C
Callihan, Dixon
Claiborne, H(arry) C(lyde)
Cole, Thomas Earle
Cope, David Franklin
Cottrell, William Barber
Crume, Enyeart Charles, Jr
Danko, Joseph Christopher
DeVan, Jackson H
Fontana, Mario H
Garber, Harold Jerome
Gat, Uri
Godbee, H(erschel) W(illcox)
Gray, Allen G(ibbs)
Grossbeck, Martin Lester
Haas, Paul Arnold
Harmer, David Edward
Haywood, Frederick F
Horak, James Albert
Johnson, Elizabeth Briggs
Judkins, Roddie Reagan
Kasten, Paul R(udolph)
Kerlin, Thomas W
Keshock, Edward G
Kress, Thomas Sylvester
Larkin, William (Joseph)
LeVert, Francis E
Lotts, Adolphus Lloyd
Lyon, Richard Norton
MacPherson, Herbert Grenfell
Maerker, Richard Erwin
Maienschein, Fred (Conrad)
Mansur, Louis Kenneth
Martin, Harry Lee
Mihalczo, John Thomas
Morgan, Marvin Thomas
Morgan, Ora Billy, Jr
Mott, Julian Edward
Mullikin, H(arwood) F(ranklin)
Murakami, Masanori
Nestor, C William, Jr
Oakes, Thomas Wyatt
Partain, Clarence Leon
Pasqua, Pietro F
Phung, Doan Lien
Roussin, Robert Warren
Row, Thomas Henry
Sanders, John P(aul), Sr
Sheth, Atul C
Stone, Robert Sidney
Stradley, James Grant
Toline, Francis Raymond
Trubey, David Keith
Uhrig, Robert Eugene
Upadhyaya, Belle Raghavendra
Wachspress, Eugene Leon
Webster, Curtis Cleveland
Weinberg, Alvin Martin

TEXAS

Betts, Austin Wortham
Bland, William M, Jr
Cochran, Robert Glenn
Dodson, David Scott

Feltz, Donald Everett
Hirsch, Robert Louis
Iddings, Frank Allen
Keister, Jamieson Charles
Klein, Dale Edward
Koen, Billy Vaughn
Levine, Joseph H
Moon, Tessie Jo
Murdoch, Bruce Thomas
Nassersharif, Bahram
Poston, John Ware
Rubin, Richard Mark
Shieh, Paulinus Shee-Shan
Simonis, John Charles
Wainerdi, Richard E(lliott)
Watson, Jerry M
Wehring, Bernard William
Wells, Michael Byron

UTAH

Kruger, Robert A(lan)
Lloyd, Ray Dix
Rogers, Vern Child
Sandquist, Gary Marlin
Thamer, B(urton) J(ohn)
Turley, Richard Eyring

VIRGINIA

Almeter, Frank M(urray)
Ball, Russell Martin
Bartlett, John W(esley)
Bauer, Douglas Clifford
Bingham, Billy Elias
DePoy, Phil Eugene
Deuster, Ralph W(illiam)
Finke, Reinald Guy
Finkelstein, Jay Laurence
Fitzgibbons, J(ohn) D(avid)
Foy, C Allan
Hanauer, Stephen H(enry)
Hauxwell, Gerald Dean
Johnson, W Reed
Kelly, James L(eslie)
Kinder, Thomas Hartley
Kurstedt, Harold Albert, Jr
Matthews, R(obert) B(ruce)
Meem, J(ames), Lawrence, Jr
Mertes, Frank Peter, Jr
Mikesell, Jon L
Mock, John E(dwin)
Mulder, Robert Udo
Neil, George Randall
Pettus, William Gower
Raab, Harry Frederick, Jr
Roberts, A(lbert) S(idney), Jr
Savage, William F(rederick)
Siebentritt, Carl R, Jr
Steele, Lendell Eugene
Townsend, Lawrence Willard
Whitelaw, R(obert) L(eslie)
Zweifel, Paul Frederick

WASHINGTON

Albrecht, Robert William
Anderson, Harlan John
Astley, Eugene Roy
Babb, Albert L(eslie)
Barker, James J(oseph)
Bowdish, Frank W(illiam)
Bunch, Wilbur Lyle
Buonamici, Rino
Bush, S(pencer) H(arrison)
Campbell, Milton Hugh
Carter, John Lemuel, Jr
Carter, Leland LaVelle
Clayton, Eugene Duane
Dahl, Roy Edward
Davies, Craig Edward
Evans, Ersel Arthur
Evans, Thomas Walter
Fisher, Darrell R
Gajewski, W(alter) M(ichael)
Garlid, Kermit L(eroy)
Gold, Raymond
Gore, Bryan Frank
Hendrickson, Waldemar Forrsel
Hill, Orville Farrow
Hopkins, Horace H, Jr
Huang, Fan-Hsiung Frank
Johnson, A(lfred) B(urton), Jr
Keating, John Joseph
Lewis, Milton
Lindstrom, Duaine Gerald
Little, Winston Woodard, Jr
Lloyd, Raymond Clare
Lurie, Harold
McCormick, Norman Joseph
McGuire, Joseph Clive
McSpadden, William Robert
Marshall, Robert P(aul)
Matte, Joseph, III
Morton, Randall Eugene
Newman, Darrell Francis
Nguyen, Dong Huu
Nicholson, Richard Benjamin
Reynolds, Roger Smith
Rieck, H(enry) G(eorge)
Roake, William Earl
Roberts, J T Adrian
Robkin, Maurice
Shen, Peter Ko-Chun
Shoemaker, Roy H(opkins)
Spinrad, Bernard Israel
Timmons, Darrol Holt

Trent, Donald Stephen
Waltar, Alan Edward
Weber, William J
Wedlick, Harold Lee
Wood, Donald Eugene
Woodfield, F(rank) W(illiam), Jr
Woodruff, Gene L(owry)
Woods, Keith Newell
Yoshikawa, Herbert Hiroshi

WEST VIRGINIA

Spaniol, Craig
Wilson, John Sheridan

WISCONSIN

Barschall, Henry Herman
Bevelacqua, Joseph John
Boom, Roger Wright
Burstein, Sol
Carbon, Max W(illiam)
Chan, Shih Hung
Christensen, Erik Regnar
Falk, Edward D
Kulcinski, Gerald La Vern
Linehan, John Henry
Moses, Gregory Allen
Van Sciver, Steven W

PUERTO RICO

Bonnet, Juan A, Jr
Rodriguez, Luis F
Sasscer, Donald S(tuart)

BRITISH COLUMBIA

Palmer, James Frederick

MANITOBA

Barclay, Francis Walter
Crosthwaite, John Leslie
McDonnell, Francis Nicholas
Rosinger, Herbert Eugene
Sargent, Frederick Peter
Smith, Roger M
Woo, Chung-Ho

ONTARIO

Andrews, Douglas Guy
Bethlendy, George
Biggs, Ronald C(larke)
Bigham, Clifford Bruce
Cox, B(rian)
Cox, Brian
Craig, Donald Spence
Dastur, Ardeshir Rustom
Dickson, Lawrence William
Furter, W(illiam) F(rederick)
Garland, William James
Gow, William Alexander
Graham, John
Hancox, William Thomas
Harms, Archie A
Hastings, Ian James
Holtslander, William John
Jackson, David Phillip
Nelles, John Sumner
Nirdosh, Inderjit
Robertson, J(ohn) A(rchibald) L(aw)
Rosen, Marc Allen
Siddall, Ernest
Vijay, Mohan Madan

QUEBEC

L'Archevêque, Real Viateur
Paskievici, Wladimir

SASKATCHEWAN

Amundrud, Donald Lorne

OTHER COUNTRIES

Angelini, Arnaldo M
Callier, Frank Maria
Chung, Chien
Dawson, Frank G(ates), Jr
Gauster, Wilhelm Friedrich
Mathews, Donald R(ichard)
Mulas, Pablo Marcelo
Solari, Mario Jose Adolfo
Williams, Michael Maurice Rudolph

Operations Research

ALABAMA

Brown, Robert Alan
Cox, Julius Grady
Griffin, Marvin A
Haak, Frederik Albertus
Regner, John LaVerne
Smith, Ernest Lee, Jr
Unger, Vernon Edwin, Jr
White, Charles Raymond
Wyskida, Richard Martin

ALASKA

Bennett, F Lawrence

ARIZONA

Askin, Ronald Gene
Buras, Nathan
Cochran, Jeffery Keith
Harris, DeVerle Porter
Lewis, William E(rvin)
Lord, William B
Mackulak, Gerald Thomas

Neuts, Marcel Fernand
Sorooshian, Soroosh
Szidarovszky, Ferenc
Szilagyi, Miklos Nicholas

ARKANSAS
Carr, Gerald Paul
Taha, Hamdy Abdelaziz

CALIFORNIA
Ancker, Clinton J(ames), Jr
Anderson, George William
Arnwine, William Carrol
Arrow, Kenneth J
Arthur, William Brian
Aster, Robert Wesley
Baer, Adrian Donald
Basinger, Richard Craig
Beckwith, Richard Edward
Bergman, Ray E(ldon)
Bhandarkar, Mangalore Dilip
Blumberg, Mark Stuart
Bolt, Robert O'Connor
Borsting, Jack Raymond
Bradley, Hugh Edward
Brandeau, Margaret Louise
Bratt, Bengt Erik
Bryson, Marion Ritchie
Capp, Walter B(ernard)
Chamberlain, Robert Glenn
Channel, Lawrence Edwin
Coile, Russell Cleven
Cottle, Richard W
Cretin, Shan
Cruz, Jose B(ejar), Jr
Daganzo, Carlos Francisco
Davidson, Lynn Blair
Dean, Burton Victor
Dresch, Francis William
Dreyfus, Stuart Ernest
Eaves, Burchet Curtis
Erickson, Stanley Arvid
Feinstein, Charles David
Forrest, Robert Neagle
Freiling, Edward Clawson
Gafarian, Antranig Vaughn
Gillette, Philip Roger
Goddard, Terrence Patrick
Gray, Paul
Greenberg, Bernard
Groce, David Eiben
Harrison, Don Edward, Jr
Hartman, James Kern
Holtzman, Samuel
Hong, Ki C(hoong)
Horne, Roland Nicholas
Howard, Gilbert Thoreau
Hudson, Cecil Ivan, Jr
Ingram, Gerald E(ugene)
Jensen, Paul Edward T
Jewell, William S(ylvester)
Keeney, Ralph Lyons
Kelley, Charles Thomas, Jr
Kendall, Burton Nathaniel
Koenigsberg, Ernest
Kreutz-Delgado, Kenneth Keith
Laitin, Howard
Lalchandani, Atam Prakash
Lam, Tenny N(icolas)
Lambert, Walter Paul
Lave, Roy E(llis), Jr
Lawler, Eugene L(eighton)
Lieberman, Gerald J
Lim, Hong Seh
Lindsay, Glenn Frank
Lucas, William Franklin
McMasters, Alan Wayne
Manne, Alan S
Mendelssohn, Roy
Miles, Ralph Fraley, Jr
Murray, George R(aymond), Jr
Nahmias, Steven
Oliver, Robert Marquam
Oren, Shmuel Shimon
Ott, Wayne R
Paulson, Boyd Colton, Jr
Philipson, Lloyd Lewis
Pietrzak, Lawrence Michael
Richards, Francis Russell
Rockower, Edward Brandt
Rowntree, Robert Fredric
Schamberg, Richard
Scholtz, Robert A
Schrady, David Alan
Shanteau, Robert Marshall
Shanthikumar, Jeyaveerasingam George
Shedler, Gerald Stuart
Silvester, John Andrew
Stauffer, Howard Boyer
Steingold, Harold
Stinson, Perri June
Sweeney, James Lee
Torbett, Emerson Arlin
Trauring, Mitchell
Trueman, Richard E(lias)
Vanderplaats, Garret Niel
Veinott, Arthur Fales, Jr
Wallerstein, David Vandermere
Whipple, Christopher George
Wilde, D(ouglass) J(ames)
Wong, Derek
Wright, Edward Kenneth
Wu, Felix F
Yu, Oliver Shukiang

Zimmerman, Elmer Leroy
Zweig, Hans Jacob

COLORADO
Clementson, Gerhardt C
Hall, Warren A(cker)
Lazzarini, Albert John
Litwhiler, Daniel W
Montague, Stephen
Morrison, John Stuart
Osborn, Ronald George
Underwood, Robert Gordon
Wackernagel, Hans Beat

CONNECTICUT
DeDecker, Hendrik Kamiel Johannes
Duvivier, Jean Fernand
Friedman, Don Gene
Hajela, Dan
Kleinfeld, Ira H
Newman, Robert Weidenthal
Powell, Bruce Allan
Terry, Herbert
Williams, David G(erald)

DELAWARE
Elterich, G Joachim
Kikuchi, Shinya
Kiviat, Fred E
Stark, Robert M

DISTRICT OF COLUMBIA
Baz, Amr Mahmoud Sabry
Briscoe, John
Gardner, Sherwin
Goor, Charles G
Hartman, Patrick James
Kaplan, David Jeremy
Kaye, John
Klemm, Rebecca Jane
Knight, John C (Ian)
Machol, Robert E
Marlow, William Henry
Meier, Wilbur L(eroy), Jr
Needels, Theodore S
Niedenfuhr, Francis W(illiam)
Park, Chan Mo
Seglie, Ernest Augustus
Singpurwalla, Nozer Drabsha
Soland, Richard Martin
Waters, Robert Charles
White, John Austin, Jr

FLORIDA
Bennet, George Kemble, Jr
Buoni, Frederick Buell
Cloutier, James Robert
Doyon, Leonard Roger
Duersch, Ralph R
Elzinga, D(onald) Jack
Francis, Richard Lane
Hanson, Morgan A
Hertz, David Bendel
Jones, Walter H(arrison)
Leavenworth, Richard S
Lloyd, Laurance H(enry)
Lund, Frederick H(enry)
Moder, Joseph J(ohn)
Muth, Eginhard Joerg
Neale, William McC(ormick), Jr
Olmstead, Paul Smith
Peart, Robert McDermand
Roth, Paul Frederick
Sivazlian, Boghos D
Stana, Regis Richard
Swart, William W
Whitehouse, Gary E

GEORGIA
Aronson, Jay E
Banks, Jerry
Barnhart, Cynthia
Callahan, Leslie G, Jr
Circeo, Louis Joseph, Jr
Cohen, Alonzo Clifford, Jr
Dickerson, Stephen L(ang)
Esogbue, Augustine O
Fyffe, David Eugene
Grum, Allen Frederick
Jarvis, John J
Jeroslow, Robert G
Johnson, Lynwood Albert
Longini, Ira Mann, Jr
Nemhauser, George L
Shakun, Wallace
Thomas, Michael E(dward)
Ware, Glenn Oren

HAWAII
Anderson, Gerald M
Cornforth, Clarence Michael
Lawson, Joel S(mith)
Liang, Tung
Papacostas, Constantinos Symeon
Wylly, Alexander

IDAHO
Bickel, John Henry
Stuffle, Roy Eugene

ILLINOIS
Caywood, Thomas E
Chitgopekar, Sharad Shankarrao
Chugh, Yoginder Paul

Conry, Thomas Francis
Cooper, Mary Weis
Cornwell, Larry Wilmer
Dessouky, Mohamed Ibrahim
Firstman, Sidney I(rving)
Fourer, Robert
Gillette, Richard F
Hassan, Mohammad Zia
Hoffman, Gerald M
Hurter, Arthur P
Kalai, Ehud
Kumar, Panganamala Ramana
Liebman, Jon C(harles)
Liebman, Judith Stenzel
McAlpin, John Harris
Maltz, Michael D
Miller, Robert Carl
Orden, Alex
Phelan, James Joseph
Phillips, Rohan Hilary
Smith, Spencer B
Steinberg, David Israel
Thomopoulos, Nick Ted
West, Douglas Brent

INDIANA
Cerimele, Benito Joseph
Coyle, Edward John
Hockstra, Dale Jon
Houck, Mark Hedrich
Leimkuhler, Ferdinand F
Morin, Thomas Lee
Nof, Shimon Y
Puri, Prem Singh
Reklaitis, Gintaras Victor
Skibniewski, Miroslaw Jan
Solberg, James J
Sweet, Arnold Lawrence
Szpankowski, Wojciech
Thuente, David Joseph
Wright, Jeffery Regan

IOWA
Andersland, Mark Steven
Kortanek, Kenneth O
Kusiak, Andrew
Liittschwager, John M(ilton)
McRoberts, Keith L
Sposito, Vincent Anthony

KANSAS
Grosh, Doris Lloyd
Harnett, R Michael
Hearne, Horace Clark, Jr
Kramer, Bradley Alan
Lee, E(ugene) Stanley
Lee, Joe
Yu, Yun-Sheng

KENTUCKY
Evans, Gerald William
Feibes, Walter
Seeger, Charles Ronald
Vaziri, Menouchehr
Warner, Richard Charles

LOUISIANA
Kraft, Donald Harris
McCarthy, Danny W
Morrow, Norman Louis
Naidu, Seetala V
Ristroph, John Heard
Spaht, Carlos G, II

MAINE
Wilms, Hugo John, Jr

MARYLAND
Adler, Sanford Charles
Ball, Michael Owen
Barrack, Carroll Marlin
Barrett, Richard John
Bateman, Barry Lynn
Belliveau, Louis J
Chacko, George Kuttickal
Chiogioji, Melvin Hiroaki
Costanza, Robert
Cushen, Walter Edward
Daly, John Anthony
Das, Prasanta
Dauber, Edwin George
Egner, Donald Otto
Eisner, Howard
Ellingwood, Bruce Russell
Flagle, Charles D(enhard)
Flum, Robert S(amuel), Sr
Frank, Carolyn
Fromovitz, Stan
Garrett, Benjamin Caywood
Garver, Robert Vernon
Gottfried, Paul
Grubbs, Frank Ephraim
Harris, Carl Matthew
Honig, John Gerhart
Hsieh, Richard Kuochi
Johnson, Russell Dee, Jr
Kamrass, Murray
Karr, Alan Francis
Katcher, David Abraham
Korobkin, Irving
Kushner, Harvey D(avid)
Levine, Eugene
Lundegard, Robert James
More, Kenneth Riddell

Moshman, Jack
Naddor, Eliezer
Offutt, William Franklin
Poltorak, Andrew Stephen
ReVelle, Charles S
Samuel, Aryeh Hermann
Steiner, Henry M
Temperley, Judith Kantack
Troutman, James Scott
Tullier, Peter Marshall, Jr
Voss, Paul Joseph
Walton, Ray Daniel, Jr
Wilson, John Phillips

MASSACHUSETTS
Bahr, Karl Edward
Brandler, Philip
Cline, James Edward
Cullinane, Thomas Paul
DeCosta, Peter F(rancis)
De Neufville, Richard Lawrence
Drake, Alvin William
Eckstein, Jonathan
Gartner, Nathan Hart
Giglio, Richard John
Graves, Robert John
Gupta, Surendra Mohan
Heitman, Richard Edgar
Hertweck, Gerald
Ilic, Marija
Kreifeldt, John Gene
Little, John Dutton Conant
Magnanti, Thomas L
Martland, Carl Douglas
Meal, Harlan C
Meal, Janet Hawkins
Papageorgiou, John Constantine
Reinschmidt, Kenneth F(rank)
Russo, Richard F
Safford, Richard Whiley
Sheffi, Yosef
Smith, Stephen Allen
Sun, Fang-Kuo
Thorpe, Rodney Warren
Tobin, Roger Lee
Wolfe, Harry Bernard
Zachary, Norman

MICHIGAN
Bergmann, Dietrich R(udolf)
Bonder, Seth
Chakravarthy, Srinivasaraghavan
Dittmann, John Paul
Horvath, William John
Johnson, Glen Eric
Kachhal, Swatantra Kumar
Khasnabis, Snehamay
Klimpel, Richard Robert
Kochen, Manfred
Kshirsagar, Anant Madhav
Lamberson, Leonard Roy
Magerlein, James Michael
Pandit, Sudhakar Madhavrao
Polis, Michael Philip
Pollock, Stephen M
Roll, James Robert
Sahney, Vinod K
Schriber, Thomas J
Simard, Albert Joseph
Taraman, Khalil Showky
Weiner, Steven Allan
Weinstein, Jeremy Saul
Wolf, Franklin Kreamer
Wolman, Eric

MINNESOTA
Ayeni, Babatunde J
Carstens, Allan Matlock
Hoffmann, Thomas Russell
Johnson, Thys B(rentwood)
Joseph, Earl Clark, II
Knutson, Charles Dwaine
Richards, James L
Sahni, Sartaj Kumar
Starr, Patrick Joseph
White, John S(pencer)
Wollenberg, Bruce Frederick

MISSISSIPPI
Rosenhan, A Kirk

MISSOURI
Bovopoulos, Andreas D
Glauz, William Donald
Klein, Cerry M
Krone, Lester H(erman), Jr
Murden, W(illiam) P(aul), Jr
Taibleson, Mitchell H

MONTANA
Jameson, William J, Jr

NEBRASKA
Ellwein, Leon Burnell

NEW HAMPSHIRE
Frey, Elmer Jacob
Jeffery, Lawrence R
Jolly, Stuart Martin
Stancl, Donald Lee

NEW JERSEY
Ahsanullah, Mohammad
Albin, Susan Lee

Operations Research (cont)

Ben-Israel, Adi
Boucher, Thomas Owen
Chasalow, Ivan G
Cinlar, Erhan
Cornell, W(arren) A(lvan)
Cullen, Daniel Edward
Eckler, Albert Ross
Elsayed, Elsayed Abdelrazik
Fask, Alan S
Forys, Leonard J
Garber, H(irsh) Newton
Gourary, Barry Sholom
Grumer, Eugene Lawrence
Haas, Zygmunt
Hatkin, Leonard
Hunter, John Stuart
Iyer, Ram R
Kaplan, Joel Howard
Kobayashi, Hisashi
Luxhoj, James Thomas
McCallum, Charles John, Jr
Maxim, Leslie Daniel
Ott, Teunis Jan
Polhemus, Neil W
Ramaswami, Vaidyanathan
Rothkopf, Michael H
Saniee, Iraj
Sarakwash, Michael
Schneider, Alfred Marcel
Segal, Moshe
Segers, Richard George
Sengupta, Bhaskar
Sherman, Ronald
Thomas, Ronald Emerson
Thomas, Stanislaus S(tephen)
Tsaliovich, Anatoly
Turoff, Murray
Whitt, Ward
Wood, Eric F

NEW MEXICO
Berrie, David William
Ebrahimi, Nader Dabir
Flanagan, Robert Joseph
Jamshidi, Mohammad Mo
Johnson, Mark Edward
Martz, Harry Franklin, Jr
Walvekar, Arun Govind

NEW YORK
Balachandran, Kashi Ramamurthi
Barnett, Michael Peter
Benson, Richard Carter
Bialas, Wayne Francis
Brown, Theodore D
Burke, Paul J
Burros, Raymond Herbert
Conway, Richard Walter
Derman, Cyrus
Dudewicz, Edward John
Fishbone, Leslie Gary
Goel, Amrit Lal
Harris, Colin C(yril)
Heidelberger, Philip
Helly, Walter S
Inhaber, Herbert
Johnson, Ellis Lane
Kabak, Irwin William
Karwan, Mark Henry
Kaufman, Sol
Kershenbaum, Aaron
Kingston, Paul L
Klein, Morton
Koenig, Michael Edward Davison
Leibowitz, Martin Albert
Lyons, Joseph Paul
Mitchell, Joseph Shannon Baird
Mosteller, Henry Walter
Owen, Joel
Palmedo, Philip F
Richter, Donald
Sargent, Robert George
Schweitzer, Paul Jerome
Shaw, Leonard G
Shoemaker, Christine Annette
Stanfel, Larry Eugene
Stedinger, Jery Russell
Stein, David Morris
Swann, Dale William
Thomas, Warren H(afford)
Todd, Michael Jeremy
Turnbull, Bruce William
Turnquist, Mark Alan
Van Slyke, Richard M
Vidosic, J(oseph) P(aul)
White, William Wallace
Wood, Paul Mix
Yablon, Marvin
Yao, David D
Yegulalp, Tuncel Mustafa

NORTH CAROLINA
Bernhard, Richard Harold
Elmaghraby, Salah Eldin
Galler, William Sylvan
Gold, Harvey Joseph
Hodgson, Thom Joel
Honeycutt, Thomas Lynn
Kilpatrick, Kerry Edwards
Martin, LeRoy Brown, Jr
Morris, Peter Alan
Papadopoulos, Alex Spero

Peterson, Elmor Lee
Roth, Walter John
Stidham, Shaler, Jr
Tolle, Jon Wright

NORTH DAKOTA
Hare, Robert Ritzinger, Jr

OHIO
Bishop, Albert B
Chan, Yupo
Chen, Kuei-Lin
Chrissis, James W
Clark, Gordon Meredith
Daschbach, James McCloskey
Duckstein, Lucien
Earhart, Richard Wilmot
Emmons, Hamilton
Flowers, Allan Dale
Gephart, Landis Stephen
Gleim, Clyde Edgar
Hurt, James Joseph
Jones, Chester George
Kantor, Paul B
Kowal, Norman Edward
McNair, Robert J(ohn), Jr
Martino, Joseph Paul
O'Neill, Edward True
Replogle, Clyde R
Sheskin, Theodore Jerome
Shock, Robert Charles
Solow, Daniel
Theusch, Colleen Joan
Ventresca, Carol
Waren, Allan D(avid)
Weisman, Joel
Williams, Robert Lloyd
Yu, Greta

OKLAHOMA
Aly, Adel Ahmed
Foote, Bobbie Leon
Johnston, Harlin Dee
Knapp, Roy M
Mize, Joe H(enry)
Schuermann, Allen Clark, Jr
Terrell, Marvin Palmer
Turner, Wayne Connelly

OREGON
Boyd, Dean Weldon
Kocaoglu, Dundar F

PENNSYLVANIA
Anandalingam, G
Arora, Vijay Kumar
Blumstein, Alfred
Bolmarcich, Joseph John
Bronzini, Michael Stephen
Clapp, Richard Gardner
Dorny, C Nelson
Farquhar, Peter Henry
Gottfried, Byron S(tuart)
Grossmann, Ignacio E
Hildebrand, David Kent
Holzman, Albert G(eorge)
Kaslow, David Edward
Korsh, James F
Kozik, Eugene
Leibholz, Stephen W
Levine, Samuel Harold
Longini, Richard Leon
Love, Carl G(eorge)
McCune, Duncan Chalmers
McGarity, Arthur Edwin
McLaughlin, Francis X(avier)
Mazumdar, Mainak
Mazzitelli, Frederick R(occo)
Odrey, Nicholas Gerald
Pierskalla, William P
Popovics, Sandor
Rosenshine, Matthew
Roush, William Burdette
Satz, Ronald Wayne
Sinclair, Glenn Bruce
Stephenson, Robert L
Stover, James Anderson, Jr
Ventura, Jose Antonio
Weindling, Joachim I(gnace)
Weir, William Thomas
Wilson, Myron Allen
Wolfe, Philip
Zaphyr, Peter Anthony
Zindler, Richard Eugene

RHODE ISLAND
Olson, David Gardels
Robertshaw, Joseph Earl

SOUTH CAROLINA
Drane, John Wanzer
Dysart, Benjamin Clay, III
Kimbler, Delbert Lee

TENNESSEE
Bartell, Steven Michael
LeBlanc, Larry Joseph
Lyle, Benjamin Franklin
Reister, David B(ryan)
Sherman, Gordon R
Siirola, Jeffrey John
Smith, James Ross
Sundaram, R Meenakshi

TEXAS
Bare, Charles L
Beightler, Charles Sprague
Bennett, George Kemble
Buckalew, Louis Walter
Disney, Ralph L(ynde)
Edgar, Thomas Flynn
Feldman, Richard Martin
Giese, Robert Paul
Greaney, William A
Harbaugh, Allan Wilson
Herman, Robert
Hogg, Gary Lynn
Innis, George Seth
Jensen, Paul Allen
Klingman, Darwin Dee
Lesso, William George
Lytton, Robert Leonard
Mak, King Kuen
Meier, France Arnett
Mistree, Farrokh
Modisette, Jerry L
Moore, Bill C
Ozan, M Turgut
Page, Thornton Leigh
Patton, Alton DeWitt
Poage, Scott T(abor)
Rhinehart, Robert Russell, II
Schrage, Robert W
Self, Glendon Danna
Sielken, Robert Lewis, Jr
Walton, Charles Michael
Wang, Yuan R
Wilhelm, Wilbert Edward
Woods, J(ohn) M(elville)
Yakin, Mustafa Zafer

UTAH
Anderson, Douglas I
Baum, Sanford
Davey, Gerald Leland
Kruger, Robert A(lan)
Peralta, Richard Carl

VERMONT
Brown, Robert Goodell

VIRGINIA
Barrois, Bertrand C
Bothwell, Frank Edgar
Brown, Gerald Leonard
Burkhalter, Barton R
Chappelle, Thomas W
Cummings, Peter Thomas
DePoy, Phil Eugene
Dockery, John T
Dorsey, Clark L(awler), Jr
Eldridge, Charles A
Englund, John Arthur
Epstein, Samuel David
Farrell, Robert Michael
Finkelstein, Jay Laurence
Freck, Peter G
Gale, Harold Walter
Gardenier, John Stark
Golub, Abraham
Greenberg, Irwin
Gross, Donald
Grotte, Jeffrey Harlow
Haering, George
Henry, James H(erbert)
Ho, S(hui)
Hobeika, Antoine George
Hodge, Donald Ray
Hoffman, Karla Leigh
Ivanetich, Richard John
Jimeson, Robert M(acKay), Jr
Johnson, Robert Ward
Kaplan, Robert Lewis
Kapos, Ervin
Karns, Charles W(esley)
Keblawi, Feisal Said
Kneece, Roland Royce, Jr
Kydes, Andy Steve
Lerner, Norman Conrad
Levis, Alexander Henry
Ling, James Gi-Ming
Lowry, Philip Holt
McNichols, Gerald Robert
Malcolm, Janet May
Masterson, Kleber Sanlin, Jr
May, Donald Curtis, Jr
Narula, Subhash Chander
Neuendorffer, Joseph Alfred
Overholt, John Lough
Pokrant, Marvin Arthur
Ratches, James Arthur
Rehm, Allan Stanley
Reynolds, Marion Rudolph, Jr
Rhode, Alfred S
Rosenbaum, David Mark
Rowe, William David
Sage, Andrew Patrick
Seelig, Jakob Williams
Shier, Douglas Robert
Smith, Michael Claude
Spivack, Harvey Marvin
Tabak, Daniel
Weinstein, Iram J
Wilkinson, William Lyle

WASHINGTON
Chaney, Robin W
Chiang, Karl Kiu-Kao

Lloyd, Raymond Clare
Pilet, Stanford Christian
Pina, Eduardo Isidorio
Ruby, Michael Gordon
Storwick, Robert Martin
Swartzman, Gordon Leni
Walsh, John Breffni

WEST VIRGINIA
Elmes, Gregory Arthur
Smith, Walton Ramsay
Stoll, Richard E
Tompkins, Curtis Johnston

WISCONSIN
Chang, Tsong-How
Moy, William A(rthur)
Myers, Vernon W
Robinson, Stephen Michael
Schmidt, John Richard
Schotz, Larry
Suri, Rajan

WYOMING
Eddy, David Maxon
Tung, Yeou-Koung

ALBERTA
Chan, Wah Chun
Elliott, Robert James
Fu, Cheng-Tze
Hamza, Mohamed Hamed
Silver, Edward Allan

BRITISH COLUMBIA
Chanson, Samuel T
Dumont, Guy Albert
Lambe, Thomas Anthony
Thomson, Alan John

NOVA SCOTIA
Eames, M(ichael) C(urtis)
El-Hawary, Mohamed El-Aref

ONTARIO
Bandler, John William
Buzacott, J(ohn) A(lan)
Byer, Philip Howard
Colbourn, Charles Joseph
Handa, V(irender) K(umar)
Hilldrup, David J
Hipel, Keith William
Hopkins, Nigel John
Hudson, John Leslie
Landolt, Jack Peter
Law, Cecil E
Magazine, Michael Jay
Mark, Jon Wei
Penner, Alvin Paul
Posner, Morton Jacob
Shelson, W(illiam)

QUEBEC
Ferguson, Michael John
Loulou, Richard Jacques

OTHER COUNTRIES
Davis, Edward Alex
Dunn, Irving John
Malik, Mazhar Ali Khan
Niv, Yehuda
Thissen, Wil A
Yazicigil, Hasan

Polymer Engineering

ALABAMA
McKannan, Eugene Charles
Tice, Thomas Robert

ARIZONA
Uhlmann, Donald Robert

CALIFORNIA
Carley, James F(rench)
Chatfield, Dale Alton
Chompff, Alfred J(ohan)
Clements, Linda L
Fishman, Norman
Giants, Thomas W
Lin, Charlie Yeongching
Matthys, Eric Francois
Nguyen, Luu Thanh
Saltman, William Mose
Seiling, Alfred William
Smith, Thor Lowe
Witham, Clyde Lester

COLORADO
Bachman, Bonnie Jean Wilson
Pitts, Malcolm John

CONNECTICUT
Casberg, John Martin
Gupta, Dharam Vir
Hsu, Nelson Nae-Ching
Prasad, Arun
Scola, Daniel Anthony

DELAWARE
Hardie, William George
Hsiao, Benjamin S
Kassal, Robert James

Kohan, Melvin Ira
Leffew, Kenneth W
Norton, Lilburn Lafayette
Ogunnaike, Babatunde Ayodeji
Rodriguez-Parada, Jose Manuel
Van Trump, James Edmond
Vick Roy, Thomas Rogers
Wagner, Martin Gerald

FLORIDA
Owen, Gwilym Emyr, Jr
Schwier, Chris Edward

GEORGIA
Colton, Jonathan Stuart

ILLINOIS
Bartimes, George F
Crist, Buckley, Jr
Fessler, Raymond R
Lustig, Stanley
Ottino, Julio Mario
Tucker, Charles L

INDIANA
Benford, Arthur E
Eckert, Roger E(arl)

IOWA
Park, Joon B

KANSAS
Macke, Gerald Fred
Young, David A(nthony)

LOUISIANA
Huckabay, Houston Keller

MARYLAND
Choi, Kyu-Yong
Ibrahim, A Mahammad
Kanarowski, S(tanley) M(artin)
Kerkar, Awdhoot Vasant
Lundsager, C(hristian) Bent
Neugroschl, Daniel

MASSACHUSETTS
Chen, Chuan Ju
Deanin, Rudolph D
Doshi, Anil G
Goettler, Lloyd Arnold
Kingston, George C
Manning, Monis Joseph
Mendelson, Robert Allen
Park, B J
Van der Burg, Sjirk
Wilde, Anthony Flory
Williams, John Russell

MICHIGAN
Filisko, Frank Edward
Foister, Robert Thomas
Heminghous, William Wayne, Sr

MINNESOTA
Nam, Sehyun
Thornton, Arnold William

MISSISSIPPI
Seymour, Raymond B(enedict)

MISSOURI
Harland, Ronald Scott
Lind, Arthur Charles

NEVADA
Herzlich, Harold Joel

NEW JERSEY
Bair, Harvey Edward
Filas, Robert William
Giddings, Sydney Arthur
Gregory, Richard Alan, Jr
Han, Byung Joon
Hansen, Ralph Holm
Hudson, Steven David
Khan, Saad Akhtar
Landrock, Arthur Harold
Lee, Wei-Kuo
Lem, Kwok Wai
Register, Richard Alan
Segelken, John Maurice
Shah, Hasmukh N
Tamarelli, Alan Wayne
Weiss, Lawrence H(eisler)
Yourtee, John Boteler

NEW MEXICO
Stampfer, Joseph Frederick

NEW YORK
Campbell, Gregory August
Feger, Claudius
Kosky, Philip George
Lhila, Ramesh Chand
Mount, Eldridge Milford, III
Nauman, Edward Bruce
Parikh, Hemant Bhupendra
Pasco, Robert William
Shen, Wu-Mian

NORTH CAROLINA
Aneja, Viney P
Hurt, Alfred B(urman), Jr

Szabo, Emery D

OHIO
Ahmad, Shamim
Bohm, Georg G A
Bradley, Ronald W
Drake, Michael L
Eby, Ronald Kraft
Gent, Alan Neville
Grimm, Robert Arthur
Jamieson, Alexander MacRae
Kaye, Christopher J
Leis, Brian Norman
McCoy, Robert Allyn
Orndorff, Roy Lee, Jr
Roberts, Timothy R
Standish, Norman Weston
Stevenson, James Francis
Thompson, Hugh Ansley
Tsai, Chung-Chieh

OKLAHOMA
Efner, Howard F
Rotenberg, Don Harris

PENNSYLVANIA
Balaba, Willy Mukama
Chen-Tsai, Charlotte Hsiao-yu
Claiborne, C Clair
Dohany, Julius Eugene
Ellis, Jeffrey Raymond
Frattini, Paul L
Georgakis, Christos
Grace, Harold P(adget)
Grassi, Vincent G
Hull, John Laurence
Lindsey, Roland Gray
Lynch, Thomas John
Prane, Joseph W(illiam)
Vrentas, Christine Mary

RHODE ISLAND
Avery, Donald Hills
Rockett, Thomas John

SOUTH CAROLINA
Wishman, Marvin

TENNESSEE
Bundy, Robert W(endel)
Griswold, Phillip Dwight
Phillips, Paul J

TEXAS
Kennelley, Kevin James
Krautz, Fred Gerhard
Paul, Donald Ross
Pritchett, P(hilip) W(alter)
Scinta, James

UTAH
Lyman, Donald Joseph

VERMONT
Bhatt, Girish M

VIRGINIA
Chiou, Minshon Jebb
Kinsley, Homan Benjamin, Jr
Wilkinson, Thomas Lloyd, Jr

WASHINGTON
Johns, William E

WEST VIRGINIA
De Hoff, George R(oland)

WISCONSIN
Bajikar, Sateesh S
Cooper, Stuart L
Denton, Denice Dee
Parnell, Donald Ray

ONTARIO
Duever, Thomas Albert
Mitsoulis, Evan
Slater, Keith
Tzoganakis, Costas

Sanitary & Environmental Engineering

ALABAMA
Anderson, Bernard Jeffrey
Cushing, Kenneth Mayhew
Hill, David Thomas
Leonard, Kathleen Mary
Marchant, Guillaume Henri (Wim), Jr
Perry, Nelson Allen
Walters, James Vernon
Wehner, Donald C
Whittle, George Patterson

ALASKA
Brown, Edward James
Tilsworth, Timothy

ARIZONA
Burke, James L(ee)
Gerba, Charles Peter
Householder, Michael K
Klock, John W
McKlveen, John William

Orton, George W
Poole, H K
Sears, Donald Richard

ARKANSAS
Alquire, Mary Slanina (Hirsch)
Fortner, Wendell Lee
Parker, David Garland
Snow, Donald L(oesch)

CALIFORNIA
Agardy, Franklin J
Akesson, Norman B(erndt)
Allen, George Herbert
Bery, Mahendera K
Bourke, Robert Hathaway
Brooks, Norman H(errick)
Cass, Glen R
Chang, Daniel P Y
Chen, Carl W(an-Cheng)
Chen, Kenneth Yat-Yi
Chu, Chung-Yu Chester
Church, Richard Lee
Cooper, Robert Chauncey
Cota, Harold Maurice
Diaz, Luis Florentino
Elimelech, Menachem
Everett, Lorne Gordon
Flagan, Richard Charles
Foxworthy, James E(rnest)
Furtado, Victor Cunha
Ghassemi, Masood
Hanna, George P, Jr
Heckman, Richard Ainsworth
Johnson, Kenneth Delford
Jones, Jerry Latham
Jones, Rebecca Anne
Kalinske, A(nton) A(dam)
Kam, James Ting-Kong
Kerri, Kenneth D
Lagarias, John S(amuel)
Lambert, Walter Paul
Lee, William Wai-Lim
Lim, H(enry) C(hol)
Longley-Cook, Mark T
Ludwig, Harvey F
Luscher, Ulrich
Lysyj, Ihor
Nazaroff, William W
Olson, Austin C(arlen)
Olson, Betty H
Ott, Wayne R
Perrine, R(ichard) L(eroy)
Porcella, Donald Burke
Robeck, Gordon G
Rogers, Gifford Eugene
Rudavsky, Alexander Bohdan
Scherfig, Jan W
Slote, Lawrence
Spicher, Robert G
Sproull, Wayne Treber
Stenstrom, Michael Knudson
Sutton, Donald Dunsmore
Tchobanoglous, George
Thomas, Jerome Francis
Waterland, Larry R
Wayne, Lowell Grant
Wilbarger, Edward Stanley, Jr
Wilson, Katherine Woods
Zison, Stanley Warren

COLORADO
Arnott, Robert A
Bradshaw, William Newman
Church, Brooks Davis
Durham, Michael Dean
Hittle, Douglas Carl
Jorden, Roger M
Lee, George Fred
Masse, Arthur N
Sabo, Julius Jay
Ward, Robert Carl
West, Ronald E(mmett)
Yarar, Baki

CONNECTICUT
Alcantara, Victor Franco
Bollyky, L(aszlo) Joseph
Bozzuto, Carl Richard
Dobay, Donald Gene
Widmer, Wilbur James

DELAWARE
Carberry, Judith B
Dam, Rudy Johan
Dentel, Steven Keith
Dwivedy, Ramesh C
Falk, Lloyd L(eopold)
Goel, Kailash C(handra)
Habibi, Kamran
Huang, Chin Pao
Hume, Harold Frederick
Jacobs, Emmett S
Koch, Carl Mark
Smith, Donald W(anamaker)

DISTRICT OF COLUMBIA
Bell, Bruce Arnold
Briscoe, John
Brown, Norman Louis
Cotruvo, Joseph Alfred
Hirschhorn, Joel S(tephen)
Middleton, John T(ylor)
Myers, Charles Edwin

Olem, Harvey
Shuster, Kenneth Ashton
Silver, Francis
Walker, John Martin
Westfield, James D

FLORIDA
Allen, Eric Raymond
Bevis, Herbert A(nderson)
Billings, Charles Edgar
Bolch, William Emmett, Jr
Dietz, Jess Clay
Echelberger, Wayne F, Jr
Hiser, Homer Wendell
Jackson, Daniel Francis
Lundgren, Dale A(llen)
Nichols, James Carlile
Nordstedt, Roger Arlo
Proctor, Charles Lafayette, II
Sirkin, Alan N
Stephens, N(olan) Thomas
Sullivan, James Haddon, Jr
Symons, George E(dgar)
Teller, Aaron Joseph
Wanielista, Martin Paul
Zegel, William Case

GEORGIA
Amirtharajah, Appiah
Bowen, Paul Tyner
Chiu, Kirts C
Galeano, Sergio F(rancis)
Hedden, Kenneth Forsythe
Hill, David W(illiam)
Rogers, Harvey Wilbur
Sanders, Walter MacDonald, III
Saunders, Fred Michael
Schliessmann, D(onald) J(oseph)
Stanley, Luticious Bryan, Jr

HAWAII
Krock, Hans-Jürgen
Lau, L(eung Ku) Stephen
Papacostas, Constantinos Symeon
Young, Reginald H F

IDAHO
Engen, Ivar A
Schreiber, David Laurence
Wallace, Alfred Thomas

ILLINOIS
Babcock, Lyndon Ross, (Jr)
Bickelhaupt, R(oy) E(dward)
Clark, John Peter, III
Ditmars, John David
Ekman, Frank
Ewing, Ben B
Hagenbach, W(illiam) P(aul)
Heil, Richard Wendell
Hermann, Edward Robert
Herricks, Edwin E
Holsen, Thomas Michael
Hsu, Deh Yuan
Jain, Ravinder Kumar
Kokoropoulos, Panos
Liebman, Jon C(harles)
Lue-Hing, Cecil
McFee, Donald Ray
Mauzy, Michael P
Minear, Roger Allan
Muehling, Arthur J
Nesselson, Eugene J(oseph)
Noll, Kenneth E(ugene)
Patterson, James W(illiam)
Prakasam, Tata B S
Reeve, Aubrey C
Regunathan, Perialwar
Rittmann, Bruce Edward
Roy, Dipak
Schaller, Robin Edward
Snoeyink, Vernon Leroy
Snow, Richard Huntley
Sohr, Robert Trueman
Srinivasaraghavan, Rengachari
Wadden, Richard Albert
Wang, Wun-Cheng W(oodrow)
Wilson, Thomas Edward
Yen, Ben Chie
Zenz, David R

INDIANA
Etzel, James Edward
Houck, Mark Hedrich
Howe, Robert H L
Lin, Ping-Wha
Shultz, Clifford Glen
Taylor, Richard L
Tenney, Mark W

IOWA
Austin, Tom Al
Baumann, E(dward) Robert
Cleasby, John Leroy
Dague, Richard R(ay)
Egger, Carl Thomas
Paulson, Wayne Lee
Schnoor, Jerald L

KANSAS
Cohen, Jules Bernard
Metzler, Dwight F
Surampalli, Rao Yadagiri

Sanitary & Environmental Engineering (cont)

KENTUCKY
Edwards, Richard Glenn
Fleischman, Marvin
Foree, Edward G(olden)
Jensen, Randolph A(ugust)
Tucker, Irwin William
Warner, Richard Charles

LOUISIANA
Collins, Michael Albert
Fontenot, Martin Mayance, Jr
Grimwood, Charles
Singh, Vijay P
Sterling, Arthur MacLean
Szabo, A(ugust) J(ohn)
Wrobel, William Eugene

MAINE
Rock, Chet A

MARYLAND
Bascunana, Jose Luis
Baummer, J Charles, Jr
Bryan, Edward H
Burrows, William Dickinson
Cookson, John T(homas), Jr
Dallman, Paul Jerald
Gilmore, Thomas Meyer
Goldberg, Irving David
Kawata, Kazuyoshi
LeRoy, André François
Linaweaver, Frank Pierce
Linder, Seymour Martin
O'Melia, Charles R(ichard)
ReVelle, Charles S
Rice, James K
Seagle, Edgar Franklin
Shipman, Harold R
Statt, Terry G
Tuttle, Kenneth Lewis
Varma, Man Mohan
Walter, Donald K
Weinberger, Leon Walter

MASSACHUSETTS
Blanc, Frederic C
Cudworth, Allen L
Ehrenfeld, John R(oos)
Eklund, Karl E
Eldred, Kenneth M
Fay, James A(lan)
Fennelly, Paul Francis
Golay, Michael W
Gutierrez, Leonard Villapando, Jr
Hanes, N Bruce
Harleman, Donald R(obert) F(ergusson)
Hart, Frederick Lewis
Holmes, Douglas Burnham
Kachinsky, Robert Joseph
Keshavan, Krishnaswamiengar
Male, James William
Marks, David Hunter
Moeller, D(ade) W(illiam)
Murphy, Brian Logan
Newman, Bernard
Ostendorf, David William
Pojasek, Robert B
Ram, Neil Marshall
Reimold, Robert J
Teal, John Moline
Tester, Jefferson William
Thorstensen, Thomas Clayton

MICHIGAN
Abriola, Linda Marie
Crittenden, John Charles
Daniels, Stacy Leroy
Domke, Charles J
Huebner, George J
Morman, Kenneth N
Ryant, Charles J(oseph), Jr
Scavia, Donald
Scherger, Dale Albert
Siddiqui, Waheed Hasan
Teichroew, Daniel

MINNESOTA
Baria, Dorab Naoroze
Boening, Paul Henrik
Eisenreich, Steven John
Gulliver, John Stephen
Haun, J(ames) W(illiam)
Johnson, Walter K
Ling, Joseph Tso-Ti
Reynolds, James Harold
Shen, Lester Sheng-wei
Stefan, Heinz G
Susag, Russell H(arry)
Uban, Stephen A

MISSISSIPPI
Bond, Marvin Thomas
Sherrard, Joseph Holmes
Wolverton, Billy Charles

MISSOURI
Bajpai, Rakesh Kumar
Banerji, Shankha K
Hatheway, Allen Wayne
Hoogheem, Thomas John

Huang, Ju-Chang
Koederitz, Leonard Frederick
Modesitt, Donald Ernest
Robinson, Thomas B
Ryckman, DeVere Wellington
Schroy, Jerry M
Schwartz, Henry Gerard, Jr
Sievers, Dennis Morline

MONTANA
Deibert, Max Curtis
Peavy, Howard Sidney

NEW HAMPSHIRE
Ballestero, Thomas P
Frost, Terrence Parker
Jackson, Herbert William

NEW JERSEY
Blank, Zvi
Blumberg, Alan Fred
Bruno, Michael Stephen
Brytczuk, Walter L(ucas)
Cohen, Irving David
Czeropski, Robert S(tephen)
Dobi, John Steven
Haefeli, Robert J(ames)
Iyer, Ram R
Lessard, Richard R
Mallison, George Franklin
Manganaro, James Lawrence
Mikes, John Andrew
Pfafflin, James Reid
Skovronek, Herbert Samuel
Toro, Richard Frank
Weinstein, Norman J(acob)

NEW MEXICO
Goodgame, Thomas H
Peil, Kelly M
Sivinski, Jacek Stefan

NEW YORK
Agosta, Vito
Ahmad, Jameel
Aulenbach, Donald Bruce
Bays, Jackson Darrell
Belfort, Georges
Bove, John L
Cataldo, Joseph C
Chadwick, George F
Clesceri, Nicholas Louis
Dick, Richard Irwin
Freeman, Eva Carol
Gale, Stephen Bruce
Gellman, Isaiah
Gerster, Robert Arnold
Gossett, James Michael
Grohse, Edward William
Hirtzel, Cynthia S
Hovey, Harry Henry, Jr
Jeris, John S(tratis)
Kiphart, Kerry
Koppelman, Lee Edward
Loucks, Daniel Peter
Lynn, Walter R(oyal)
Miller, George Maurice
Molof, Alan H(artley)
Murphy, Kenneth Robert
Oldshue, J(ames) Y(oung)
Palevsky, Gerald
Robinson, Myron
Shen, Thomas T
Shoemaker, Christine Annette
Soifer, Herman
Star, Joseph
Stedinger, Jery Russell
Tavlarides, Lawrence Lasky
Velzy, Charles O
Wagner, Peter Ewing
Wang, Lawrence K
Wang, Muhao S
Ziegler, Edward N
Ziegler, Robert C(harles)
Zimmie, Thomas Frank

NORTH CAROLINA
Ballard, Lewis Franklin
Barker, James Cathey
Bell, Carlos G(oodwin), Jr
Clark, Charles Edward
Ensor, David Samuel
Fox, Donald Lee
Lamb, James C(hristian), III
Larsen, Ralph Irving
Lawless, Philip Austin
Medina, Miguel Angel, Jr
Okun, D(aniel) A(lexander)
Peirce, James Jeffrey
Reckhow, Kenneth Howland
Reist, Parker Cramer
Saeman, W(alter) C(arl)
Senzel, Alan Joseph
Singer, Philip C
Stern, A(rthur) C(ecil)
Trenholm, Andrew Rutledge
Winston, Hubert

OHIO
Armstrong, James Anthony
Ayer, Howard Earle
Bennett, G(ary) F
Bishop, Paul Leslie
Bondurant, Byron L(ee)

Bostian, Harry (Edward)
Clark, C Scott
Geldreich, Edwin E(mery)
Gruber, Charles W
Gurklis, John A(nthony)
Hannah, Sidney Allison
Hyland, John R(oth)
Jain, Jain S
Kellerstrass, Ernst Junior
Morand, James M(cHuron)
Moulton, Edward Q(uentin)
Noss, Richard Robert
Olesen, Douglas Eugene
Papadakis, Constantine N
Preul, Herbert C
Prober, Richard
Saltzman, Bernard Edwin
Scarpino, Pasquale Valentine
Smith, James Edward, Jr
Springer, Allan Matthew
Stephan, David George
Stiefel, Robert Carl
Sykes, Robert Martin
Taiganides, E Paul
Williams, Robert Mack

OKLAHOMA
De Vries, Richard N
Foutch, Gary Lynn
Graves, Toby Robert
Kincannon, Donny Frank
Koerner, E(rnest) L(ee)
Law, James Pierce, Jr
Madden, Michael Preston
Middlebrooks, Eddie Joe
Stover, Enos Loy
Turner, Wayne Connelly

OREGON
Hsiung, Andrew K
Huntzicker, James John
Miner, J Ronald
Moore, James Allan
Schaumburg, Frank David
Tucker, W(illiam) Henry
Wohlers, Henry Carl

PENNSYLVANIA
Allen, Herbert E
Anandalingam, G
Atkins, Patrick Riley
Bhatla, Manmohan N (Bart)
Blayden, Lee Chandler
Cernansky, Nicholas P
Chang, C Hsiung
Cooper, Joseph E
Corbin, Michael H
Davidson, Cliff Ian
Dicciani, Nancy Kay
Esmen, Nurtan A
Ghosh, Mriganka M(ouli)
Gilardi, Edward Francis
Gill, C(harles) Burroughs
Grossmann, Elihu D(avid)
Gurol, Mirat D
Johnson, Glenn M
Keenan, John Douglas
Kugelman, Irwin Jay
LaGrega, Michael Denny
Lawrence, Alonzo William
Loigman, Harold
Luthy, Richard Godfrey
McGarity, Arthur Edwin
McMichael, Francis Clay
Manka, Dan P
Neufeld, Ronald David
Pena, Jorge Augusto
Petura, John C
Pipes, Wesley O
Pohland, Frederick G
Regan, Raymond Wesley
Rosenstock, Paul Daniel
Sagik, Bernard Phillip
Samples, William R(ead)
Sawicki, John Edward
Schoenberger, Robert J(oseph)
Shaler, Amos J(ohnson)
Shapiro, Maurice A
Sooky, Attila A(rpad)
Sorber, Charles Arthur
Spear, Jo-Walter
Stephenson, Robert L
Thompson, Ansel Frederick, Jr
Touhill, Charles Joseph
Unz, Richard F(rederick)
Weston, Roy Francis
Yohe, Thomas Lester

RHODE ISLAND
Hutzler, Leroy, III
Poon, Calvin P C

SOUTH CAROLINA
Dysart, Benjamin Clay, III
Elzerman, Alan William
Grady, Cecil Paul Leslie, Jr
Jennett, Joseph Charles
Keinath, Thomas M
Overcamp, Thomas Joseph
Patterson, C(leo) Maurice
Rich, Linvil G(ene)
Townes, George Anderson
White, Richard Kenneth
Wixson, Bobby Guinn

SOUTH DAKOTA
Han, Kenneth N
Hodges, Neal Howard, II
Zogorski, John Steward

TENNESSEE
Clarke, Ann Neistadt
Cowser, Kenneth Emery
Eckenfelder, William Wesley, Jr
Farrington, Joseph Kirby
Gallagher, Michael Terrance
Gano, Richard W
Godbee, H(erschel) W(illcox)
Hensley, Marble John, Sr
Jones, Larry Warner
Kelso, Richard Miles
Oakes, Thomas Wyatt
Parker, Frank L
Polahar, Andrew Francis
Roesler, Joseph Frank
Schnelle, K(arl) B(enjamin), Jr
Singh, Suman Priyadarshi Narain
Smith, John W(arren)
Smith, Michael James
Thackston, Edward Lee

TEXAS
Akgerman, Aydin
Andrews, John Frank
Armstrong, Neal Earl
Azad, Hardam Singh
Bodre, Robert Joseph
Busch, Arthur Winston
Caffey, James E
Camp, Frank A, III
Cawley, William Arthur
Charbeneau, Randall Jay
Davis, Ernst Michael
Eaton, David J
Fix, Ronald Edward
Gavande, Sampat A
Giammona, Charles P, Jr
Gilmour - Stallsworth, Lisa K
Gloyna, Earnest F(rederick)
Halff, Albert Henry
Harbordt, C(harles) Michael
Howe, Richard Samuel
McKee, Herbert C(harles)
Malina, Joseph F(rancis), Jr
Myrick, Henry Nugent
Naismith, James Pomeroy
Parate, Natthu Sonbaji
Qasim, Syed Reazul
Reynolds, Tom Davidson
Sherrill, Bette Cecile Benham
Stanford, Geoffrey
Stubbeman, Robert Frank
Symons, James M(artin)
Tarquin, Anthony Joseph
Wagner, John Philip
Wolf, Harold William

UTAH
Ghosh, Sambhunath
Lee, Jeffrey Stephen
Merritt, LaVere Barrus
Pyper, Gordon R(ichardson)
Rogers, Vern Child

VERMONT
Hussak, Robert Edward

VIRGINIA
Beach, Robert L
Boardman, Gregory Dale
Chi, Michael
Drewry, William Alton
Echols, Charles E(rnest)
Fang, Ching Seng
Higgins, Thomas Ernest
Kornicker, William Alan
Lassiter, William Stone
McKenna, John Dennis
Randall, Clifford W(endell)
Reilly, Thomas E
Shea, Timothy Guy
Simpson, James R(ussell)
Spofford, Walter O, Jr
Trout, Dennis Alan
Young, Earle F(rancis), Jr

WASHINGTON
Berg, Richard Harold
Bhagat, Surinder Kumar
Bogan, Richard Herbert
Carlson, Dale Arvid
Cormack, James Frederick
Darley, Ellis Fleck
Duncan, James Byron
Edde, Howard Jasper
Emery, Richard Meyer
Glenne, Bard
Jones, Kay H
LaFontaine, Edward
Mahalingam, R
Mar, Brian W(ayne)
Ruby, Michael Gordon
Skrinde, Rolf T
Sylvester, Robert O(hrum)

WEST VIRGINIA
Conway, Richard A
McCall, Robert G
McCaskey, A(mbrose) E(verett)

WISCONSIN
Arthur, Robert M
Boyle, William C(harles)
Christensen, Erik Regnar
Converse, James Clarence
Hanway, John E(dgar), Jr
Jayne, Jack Edgar
Kipp, Raymond J
Lentz, Mark Steven
Zanoni, Alphonse E(ligius)

WYOMING
Champlin, Robert L

PUERTO RICO
Del Valle, Luis A

BRITISH COLUMBIA
Liao, Ping-huang
Walden, C(ecil) Craig

NEW BRUNSWICK
Lin, Kwan-Chow

ONTARIO
Barabas, Silvio
Besik, Ferdinand
Bewtra, Jatinder Kumar
Boadway, John Douglas
Byer, Philip Howard
Heinke, Gerhard William
McBean, Edward A
Murphy, Keith Lawson
Nicolle, Francois Marcel Andre
Slawson, Peter (Robert)

QUEBEC
Beron, Patrick
De la Noue, Joel Jean-Louis
Leduy, Anh
Volesky, Bohumil

SASKATCHEWAN
Cullimore, Denis Roy
Curtis, Fred Allen
Viraraghavan, Thiruvenkatachari

OTHER COUNTRIES
Grigoropoulos, Sotirios G(regory)
Hoadley, Alfred Warner
Jokl, Miloslav Vladimir
Nebiker, John Herbert
Shuval, Hillel Isaiah
Stumm, Werner
Yao, Kuan Mu

Systems Design & Systems Science

ALABAMA
Brown, Robert Alan
Fullerton, Larry Wayne
Goldman, Jay
Haeussermann, Walter
McAuley, Van Alfon
Martin, James Arthur
Rekoff, M(ichael) G(eorge), Jr

ALASKA
Rogers, James Joseph
Skudrzyk, Frank J

ARIZONA
Askin, Ronald Gene
Bahill, A(ndrew) Terry
Bartocha, Bodo
Bluestein, Theodore
Burke, James L(ee)
Clark, Wilburn O
Cochran, Jeffery Keith
Dahl, Randy Lynn
Dempster, William Fred
Falk, Thomas
Ferrell, William Russell
Formeister, Richard B
Mackulak, Gerald Thomas
Miller, Harry
Milnes, Dale J
Ramler, W(arren) J(oseph)
Ryan, Thomas Wilton
Silvern, Leonard Charles
Slaughter, Charles D
Sorooshian, Soroosh
Szidarovszky, Ferenc
Wasson, James Walter
Welch, H(omer) William
Wymore, Albert Wayne
Zeigler, Bernard Philip
Zelenka, Jerry Stephen

ARKANSAS
Carr, Gerald Paul
Parsch, Lucas Dean
Sedelow, Walter Alfred, Jr

CALIFORNIA
Acheson, Louis Kruzan, Jr
Akonteh, Benny Ambrose
Alspach, Daniel Lee
Amara, Roy C
Ames, John Wendell
Anderson, George William
Antonsson, Erik Karl

Austin, Arthur Leroy
Baggerly, Leo L
Barr, Kevin Patrick
Berg, Myles Renver
Bertram, Sidney
Beyster, J Robert
Blackmon, James B
Bleviss, Zegmund O(scar)
Bloxham, Laurence Hastings
Bradley, Hugh Edward
Bradner, Hugh
Bryson, Arthur E(arl), Jr
Bush, George Edward
Caligiuri, Joseph Frank
Carlyle, Jack Webster
Carson, George Stephen
Carver, John Guill
Cashen, John F
Caswell, John N(orman)
Chamberlain, Robert Glenn
Channel, Lawrence Edwin
Chen, Kao
Chen, Tsai Hwa
Chester, Arthur Noble
Christiansen, Jerald N
Clark, Crosman Jay
Clemens, Jon K(aufmann)
Coleman, Paul Jerome, Jr
Cruz, Jose B(ejar), Jr
Culler, Donald Merrill
Davis, William Henke
Deckert, Curtis Kenneth
Depp, Joseph George
Dornfeld, David Alan
Dost, Martin Hans-Ulrich
Doyle, Walter M
Drugan, James Richard
Duffield, Jack Jay
Dulock, Victor A, Jr
Dunn, Donald A(llen)
Eden, Richard Carl
Egermeier, R(obert) P(aul)
Eggers, A(lfred) J(ohn), Jr
Elliott, David Duncan
Elverum, Gerard William, Jr
Faulkner, Thomas Richard
Feign, David
Firdman, Henry Eric
Fleischer, Gerald A
Fleming, Alan Wayne
Forsberg, Kevin
Fox, Joseph M(ickle), III
Franklin, Gene F(arthing)
Friedman, George J(erry)
Furst, Ulrich Richard
Gillette, Philip Roger
Gilmartin, Thomas Joseph
Gleghorn, G(eorge) J(ay), (Jr)
Goddard, Earl G(ascoigne)
Goheen, Lola Coleman
Grace, O Donn
Grant, Eugene F(redrick)
Gray, Paul
Green, Allen T
Gritton, Eugene Charles
Grossman, Jack Joseph
Gustavson, Marvin Ronald
Haake, Eugene Vincent
Hackett, Le Roy Huntington, Jr
Hackwood, Susan
Hamilton, Carole Lois
Hanson, Joe A
Hartwick, Thomas Stanley
Hecht, Herbert
Hecht, Myron J
Heer, Ewald
Hendrix, C(harles) E(dmund)
Hewer, Gary Arthur
Hilker, Harry Van Der Veer, Jr
Hoagland, Jack Charles
Hoffman, Howard Torrens
Hoffman, James Tracy
Holder, Harold Douglas
Holeman, Dennis Leigh
Holl, Richard Jacob
Hollander, Gerhard Ludwig
Holsztynski, Wlodzimierz
Holtzman, Samuel
Horn, Kenneth Porter
Hsu, Chieh-Su
Hult, John Luther
Hwang, Kai
Iberall, Arthur Saul
Ibrahim, Medhat Ahmed Helmy
Isenberg, Lionel
Jackson, Albert S(mith)
Jackson, Durward P
Jacobs, Michael Moises
Jacobson, Albert H(erman), Jr
Jaffe, Leonard David
Johannessen, Jack
Johnson, Reynold B
Kagiwada, Harriet Hatsune
Kaliski, Martin Edward
Kalvinskas, John J(oseph)
Kayton, Myron
Kelly, Kenneth C
Kendall, Burton Nathaniel
Kennel, John Maurice
Kerfoot, Branch Price
Kinnison, Gerald Lee
Klestadt, Bernard
Krause, Lloyd O(scar)
Kreutz-Delgado, Kenneth Keith

Kubik, Robert N
Kuczynski, Eugene Raymond
Laitin, Howard
Lampert, Seymour
Landecker, Peter Bruce
Langdon, Glen George, Jr
Langland, Barbara June
Larson, Harry Thomas
Laub, Alan John
Layton, Thomas William
Leavy, Paul Matthew
Leibowitz, Lewis Phillip
Leite, Richard Joseph
Leitmann, G(eorge)
Levan, Nhan
Lientz, Bennet Price
Lillie, Charles Frederick
Lindley, Charles A(lexander)
Lipeles, Martin
Lisman, Perry Hall
Liu, Frederick F
Longmuir, Alan Gordon
Lovell, Robert Edmund
Luenberger, David Gilbert
Lyman, John (Henry)
McBride, Lyle E(rwin), Jr
McDonald, William True
McQuillen, Howard Raymond
McRuer, Duane Torrance
Magness, T(om) A(lan)
Maloy, John Owen
Mantey, Patrick E(dward)
Marmarelis, Vasilis Z
Mathews, W(arren) E(dward)
Mayer, Harris Louis
Meissinger, Hans F
Meyer, Lhary
Michaelson, Jerry Dean
Mikes, Peter
Miles, Ralph Fraley, Jr
Mingori, Diamond Lewis
Mitra, Sanjit K
Mitsutomi, T(akashi)
Molinder, John Irving
Molnar, Imre
Moran, Thomas Patrick
Morrow, Charles Tabor
Morse, Joseph Grant
Mulligan, Geoffrey C
Muntz, Richard Robert
Mykkanen, Donald L
Nadler, Gerald
Nesbit, Richard Allison
Newton, Arthur R
Nickolls, John Richard
Nielsen, Norman Russell
Nunan, Craig S(pencer)
Nunn, Robert Harry
Omura, Jimmy Kazuhiro
Owen, Paul Howard
Park, Edward C(ahill), Jr
Parker, Alice Cline
Paulson, Boyd Colton, Jr
Pearl, Judea
Pedersen, Leo Damborg
Pelka, David Gerard
Philipson, Lloyd Lewis
Pietrzak, Lawrence Michael
Pinson, John C(arver)
Pinto, John Gilbert
Pister, Karl S(tark)
Poggio, Andrew John
Pressman, Ada Irene
Pye, Earl Louis
Randolph, James E
Ratner, Robert (Stephen)
Rau, Bantwal Ramakrishna
Rauch, Herbert Emil
Rechnitzer, Andreas Buchwald
Rechtin, Eberhardt
Reddy, Parvathareddy Balarami
Rickard, John Terrell
Roberson, Robert Errol
Roney, Robert K(enneth)
Rubin, Izhak
Ruskin, Arnold M(ilton)
Safonov, Michael G
Sakaguchi, Dianne Koster
Samson, Sten
Schneider, Alan M(ichael)
Schreiner, Robert Nicolas, Jr
Scudder, Harvey Israel
Seitz, S Stanley
Senge, George H
Sensiper, S(amuel)
Serdengecti, Sedat
Shepherd, Robin
Shoch, John F
Siljak, Dragoslav (D)
Silverstein, Elliot Morton
Slaton, Jack H(amilton)
Smith, Frederick T(ucker)
Smith, Otto J(oseph) M(itchell)
Sondak, Norman Edward
Speyer, Jason L
Spinrad, Robert J(oseph)
Spitzer, Irwin Asher
Sridharan, Natesa S
Steingold, Harold
Stenstrom, Michael Knudson
Stephanou, Stephen Emmanuel
Stern, Albert Victor
Stevenson, Robin
Stewart, Richard William

Sturm, Walter Allan
Sung, Chia-Hsiaing
Surko, Pamela Toni
Swanson, Eric Richmond
Talbot, Raymond James, Jr
Thompson, W P(aul)
Titus, Harold
Toepfer, Richard E, Jr
Tooley, Richard Douglas
Twieg, Donald Baker
Tyler, Christopher William
Wagner, Raymond Lee
Walker, Warren Elliott
Wallerstein, David Vandermere
Wang, H E Frank
Wang, Ji Ching
Ware, W(illis) H(oward)
Warren, Mashuri Laird
Weber, Charles L
Weinberg, Daniel I
Weiss, Herbert Klemm
Weissenberger, Stein
Wertz, James Richard
White, Moreno J
Wiberg, Donald M
Wilde, D(ouglass) J(ames)
Williams, Donald Spencer
Wray, John L
Wright, Edward Kenneth
Wu, Felix F
Yeh, Hen-Geul
Young, Richard D
Zander, Andrew Thomas
Zebroski, Edwin L
Zuppero, Anthony Charles

COLORADO
Barber, Robert Edwin
Brown, Alison Kay
Carpenter, Donald Gilbert
Frangopol, Dan Mircea
Frick, Pieter A
Gupta, Kuldip Chand
Jansson, David Guild
Johnson, Gearold Robert
Kelley, Neil Davis
Koldewyn, William A
Lazzarini, Albert John
Lund, Donald S
Mandics, Peter Alexander
Marshall, Charles F
Morrison, John Stuart
Munro, Richard Harding
Tischendorf, John Allen
Wackernagel, Hans Beat
Wall, Edward Thomas
Wertz, Ronald Duane

CONNECTICUT
Cooper, James William
Dubin, Fred S
Goldman, Ernest Harold
Halpern, Howard S
Hayes, Richard J
Howard, Robert Eugene
Johnson, Loering M
Ketchman, Jeffrey
Krummel, William Marion
Levenstein, Harold
Levien, Roger Eli
Mack, Donald R(oy)
Maewal, Akhilesh
Narendra, Kumpati S
Newman, R(obert) W(eidenthal)
O'Brien, Robert L
Porter, Richard W(illiam)
Poultney, Sherman King
Rezek, Geoffrey Robert
Shah, Mirza Mohammed
Simpson, Wilburn Dwain
Telfair, William Boys

DELAWARE
Arots, Joseph B(artholomew)
Bartkus, Edward Peter
Hayman, Alan Conrad
Kikuchi, Shinya
Leffew, Kenneth W
Murphy, Arthur Thomas
Nicholls, Robert Lee
Vick Roy, Thomas Rogers

DISTRICT OF COLUMBIA
Baker, Dennis John
Baz, Amr Mahmoud Sabry
Bloch, Erich
Burka, Maria Karpati
Donivan, Frank Forbes, Jr
Johnson, George Patrick
Kelley, Albert J(oseph)
Kelly, Henry Charles
Kroeger, Richard Alan
Kyriakopoulos, Nicholas
Marcus, Michael Jay
Meier, Wilbur L(eroy), Jr
Moll, Magnus
Penhollow, John O
Pyke, Thomas Nicholas, Jr
Turner, Kenneth Clyde
Zucchetto, James John

FLORIDA
Anderson, Roland Carl
Ariet, Mario

Systems Design & Systems Science (cont)

Bonn, T(heodore) H(ertz)
Boykin, William H(enry), Jr
Braswell, Robert Neil
Brooks, George H(enry)
Calhoun, Ralph Vernon, Jr
Capehart, Barney Lee
Chartrand, Robert Lee
Cox, John David
Dilpare, Armand Leon
Dougherty, John William
Duersch, Ralph R
Everett, Woodrow Wilson, Jr
Flick, Carl
Hammer, Jacob
Hampp, Edward Gottlieb
Harmon, G Lamar
Haviland, Robert P(aul)
Howard, Bernard Eufinger
Kinnie, Irvin Gray
Kirmse, Dale William
Klock, Benny LeRoy
Korwek, Alexander Donald
Kunce, Henry Warren
Lund, Frederick H(enry)
Mahig, Joseph
Melich, Michael Edward
Mertens, Lawrence E(dwin)
Mitoff, S(tephan) P(aul)
Neale, William McC(ormick), Jr
Nevill, Gale E(rwin), Jr
Peterson, James Robert
Proctor, Charles Lafayette
Rice, Stephen Landon
Roth, Paul Frederick
Sager, John Clutton
Scaringe, Robert P
Smerage, Glen H
Spurlock, Jack Marion
Stahl, Joel S
Storm, Leo Eugene
Suciu, S(piridon) N
Swain, Ralph Warner
Travis, Russell Burton
Tulenko, James Stanley
Tusting, Robert Frederick
Ulug, Esin M

GEORGIA
Braden, Charles Hosea
Callahan, Leslie G, Jr
Dickerson, Stephen L(ang)
Dutton, John C
Esogbue, Augustine O
Fletcher, James Erving
Hines, William W
Johnson, Cecil Gray
Krol, Joseph
Landgren, John Jeffrey
Lee, Kok-Meng
Lehrer, Robert N(athaniel)
Lorber, Herbert William
Nessmith, Josh T(homas), Jr
Pence, Ira Wilson, Jr
Rouse, William Bradford
Stalford, Harold Lenn
Vachon, Reginald Irenee
White, Harold D(ouglas)
Zunde, Pranas

HAWAII
Johnson, Carl Edward
Sagle, Arthur A

IDAHO
DeBow, W Brad
Jobe, Lowell A(rthur)
Potter, David Eric

ILLINOIS
Agarwal, Gyan C
Agrawal, Arun Kumar
Berry, Gregory Franklin
Coplien, James O
Dessouky, Mohamed Ibrahim
Finn, Patricia Ann
Frey, Donald N(elson)
Graupe, Daniel
Guenther, Peter T
Gupta, Nand K
Huck, Morris Glen
Hurter, Arthur P
Jacobsen, Fred Marius
Jordan, Bernard William, Jr
Kang, Sung-Mo Steve
Kuchnir, Moyses
Kumar, Panganamala Ramana
Luplow, Wayne Charles
Mikulski, James J(oseph)
Miller, Floyd Glenn
Moore, John Fitzallen
Murphy, Gordon J
Okamura, Kiyohisa
O'Neill, William D(ennis)
Phelan, James Joseph
Pursley, Michael Bader
Ruhl, Roland Luther
Siegal, Burton L
Thompson, Charles William Nelson
Toy, William W
Worley, Will J

Wurm, Jaroslav
Zell, Blair Paul

INDIANA
Carlson, James C
Coyle, Edward John
Fearnot, Neal Edward
Houck, Mark Hedrich
Kendall, Perry E(ugene)
Melsa, James Louis
Nof, Shimon Y
Reklaitis, Gintaras Victor
Rominger, Michael Collins
Sain, Michael K(ent)
Schmidt, David Kelso
Sinha, Kumares C
Williams, Theodore J(oseph)
Wolber, William George
Wood, Frederick Starr
Wright, Jeffery Regan
Wynne, Bayard Edmund

IOWA
Akers, Arthur
Chyung, Dong Hak
Haendel, Richard Stone
Hekker, Roeland M T
Kortanek, Kenneth O
Kusiak, Andrew
Lardner, James F
Mischke, Charles R(ussell)
Ong, John Nathan, Jr
Robinson, John Paul

KANSAS
Gray, Tom J
Harnett, R Michael
Kramer, Bradley Alan
Lee, E(ugene) Stanley
Mingle, John O(rville)
Robinson, M John
Sawan, Mahmoud Edwin
Smaltz, Jacob Jay
Thompson, Joseph Garth
Welch, Stephen Melwood

KENTUCKY
Chenoweth, Darrel Lee
Cooke, Samuel Leonard
Evans, Gerald William
Gold, Harold Eugene
Gruver, William A
Livingood, Marvin D(uane)
Merker, Stephen Louis
Squires, Robert Wright
Vaziri, Menouchehr

LOUISIANA
Cospolich, James D
Latorre, Robert George
McCarthy, Danny W
Ristroph, John Heard
Tewell, Joseph Robert

MAINE
Senders, John W

MARYLAND
Abed, Eyad Husni
Atia, Ali Ezz Eldin
Battin, William James
Blevins, Gilbert Sanders
Bugenhagen, Thomas Gordon
Butler, Louis Peter
Campbell, William J
Chi, Donald Nan-Hua
Choi, Kyu-Yong
Cohen, Stanley Alvin
Cotton, Ira Walter
Das, Prasanta
Drzewiecki, Tadeusz Maria
Eaton, Alvin Ralph
Eckerman, Jerome
Eisner, Howard
Ely, Raymond Lloyd
Evans, Alan G
Fischel, David
Fishman, Elliot Keith
Fleig, Albert J, Jr
Frank, Carolyn
Freeman, Ernest Robert
Gabriele, Thomas L
Gasparovic, Richard Francis
Gillis, James Thompson
Gleissner, Gene Heiden
Goepel, Charles Albert
Goldfinger, Andrew David
Gottfried, Paul
Graham, Edward Underwood
Grant, Conrad Joseph
Hall, Arthur David, III
Hammond, Charles E
Johnson, David Simonds
Kanal, Laveen Nanik
Kiebler, John W(illiam)
Kissell, Kenneth Eugene
Kossiakoff, Alexander
Ligomenides, Panos Aristides
Mahle, Christoph E
Makofski, Robert Anthony
Malarkey, Edward Cornelius
Martin, John A
Meyer, Gerard G L
Michelsen, Arve

Minneman, Milton J(ay)
Morgan, Walter L(eroy)
Mumford, Willard R
Paul, Dilip Kumar
Pisacane, Vincent L
Powell, Edward Gordon
Resnick, Joel B
Reupke, William Albert
Riley, Patrick Eugene
Rueger, Lauren J(ohn)
Saltman, Roy G
Sayer, John Samuel
Schneider, William C
Schruben, John H
Sharma, Dinesh C
Sleight, Thomas Perry
Thompson, Robert John, Jr
Tropf, William Jacob
Turner, Robert
Turner, Robert Davison
Vogelberger, Peter John, Jr
Voss, Paul Joseph
Weber, Richard Rand
Weiss, Roland George
Yaramanoglu, Melih
Zafiriou, Evanghelos

MASSACHUSETTS
Anderson, J(ohn) Edward
Aprille, Thomas Joseph, Jr
Bennett, C Leonard
Bloomfield, David Peter
Boodman, David Morris
Brayer, Kenneth
Brookner, Eli
Bussgang, J(ulian) J(akob)
Caron, Paul R(onald)
Clarke, Lori A
Colby, George Vincent, Jr
Cotter, Douglas Adrian
Cullinane, Thomas Paul
De Fazio, Thomas Luca
De Neufville, Richard Lawrence
Douglas, J(ames) M(errill)
Dow, Paul C(rowther), Jr
Dubowsky, Steven
Duffy, Robert A
Eaves, Reuben Elco
Faflick, Carl E(dward)
Finn, John Thomas
Forrester, Jay W
Gartner, Nathan Hart
Gelb, Arthur
Gilfix, Edward Leon
Gupta, Surendra Mohan
Hals, Finn
Hearn, David Russell
Hempstead, Charles Francis
Herwald, Seymour W(illis)
Ho, Yu-Chi
Holst, William Frederick
Hopkins, Albert Lafayette, Jr
Horowitz, Larry Lowell
Hursh, John W(oodworth)
Ilic, Marija
Keicher, William Eugene
King, William Connor
Kovaly, John J
Kovatch, George
Kreifeldt, John Gene
Lacoss, Richard Thaddee
Lees, Sidney
Lerner, Moisey
Levison, William H(enry)
Luft, Ludwig
Luther, Holger Martin
McBee, W(arren) D(ale)
Martland, Carl Douglas
Mastronardi, Richard
Mavretic, Anton
Miranda, Henry A, Jr
Mulukutla, Sarma Sreerama
Nelson, J(ohn) Byron
Nesbeda, Paul
Ogar, George W(illiam)
Paynter, Henry M(artyn)
Reinschmidt, Kenneth F(rank)
Richter, Stephen L(awrence)
Rubin, Milton D(avid)
Russell, George Albert
Safford, Richard Whiley
Sasiela, Richard
Schetzen, Martin
Schmidt, George Thomas
Schwartz, Jack
Seamans, R(obert) C(hanning), Jr
Seibel, Frederick Truman
Seifer, Arnold David
Shea, Joseph F(rancis)
Sheridan, Thomas Brown
Shorb, Alan McKean
Sorenson, Harold Wayne
Story, Anne Winthrop
Strelzoff, Alan G
Sun, Fang-Kuo
Thun, Rudolf Eduard
Tompkins, Howard E(dward)
Tweed, David George
Walter, Charlton M
Wetzstein, H(anns) J(uergen)
Wheeler, Ned Brent
Winer, Bette Marcia Tarmey
Wohl, Joseph G

MICHIGAN
Asik, Joseph R
Aswad, A Adnan
Ayres, Robert Allen
Barnard, Robert D(ane), Jr
Barr, Robert Ortha, Jr
Bergmann, Dietrich R(udolf)
Beutler, Frederick J(oseph)
Borcherts, Robert H
Camp, David Thomas
Chen, Kan
Fleischer, Henry
Freedman, M(orris) David
Ghoneim, Youssef Ahmed
Gjostein, Norman A
Hamilton, William Eugene, Jr
Haynes, Dean L
Hrovat, Davorin
Jones, R(ichard) James
Khargonekar, Pramod P
Khasnabis, Snehamay
Larrowe, Boyd T
Lett, Philip W(ood), Jr
McClamroch, N Harris
Mordukhovich, Boris S
Peterson, Lauren Michael
Petrick, Ernest N(icholas)
Polis, Michael Philip
Powers, William Francis
Rillings, James H
Rohde, Steve Mark
Roll, James Robert
Root, William L(ucas)
Rosenberg, Ronald C(arl)
Sahney, Vinod K
Salam, Fathi M A
Savageau, Michael Antonio
Schlax, T(imothy) Roth
Schriber, Thomas J
Scott, Norman R(oss)
Sethi, Ishwar Krishan
Shapton, William Robert
Simard, Albert Joseph
Stivender, Donald Lewis
Teichroew, Daniel
Thompson, James Lowry
Windeknecht, Thomas George

MINNESOTA
Anderson, Willard Eugene
Cunningham, Thomas B
DiBona, Peter James
Erdman, Arthur Guy
Glewwe, Carl W(illiam)
Greene, Christopher Storm
Johnson, Howard Arthur, Sr
Knutson, Charles Dwaine
Kumar, K S P
Kvalseth, Tarald Oddvar
Lee, E(rnest) Bruce
Starr, Patrick Joseph
Stephanedes, Yorgos Jordan
Thiede, Edwin Carl
Wollenberg, Bruce Frederick

MISSISSIPPI
Miller, James Edward

MISSOURI
Atwood, Charles LeRoy
Babcock, Daniel Lawrence
Blundell, James Kenneth
Bovopoulos, Andreas D
Creighton, Donald L
Deam, James Richard
Franklin, Mark A
Glauz, William Donald
Griffith, Virgil Vernon
Katz, Israel Norman
Lemon, Leslie Roy
Linder, Solomon Leon
McMillan, Gregory Keith
Richards, Earl Frederick
Zaborszky, John
Zobrist, George W(inston)

MONTANA
Gerez, Victor
Pierre, Donald Arthur

NEBRASKA
Ellwein, Leon Burnell
Lenz, Charles Eldon
Von Bargen, Kenneth Louis

NEW HAMPSHIRE
Frey, Elmer Jacob
Jeffery, Lawrence R
Jolly, Stuart Martin
Klarman, Karl J(oseph)
Limbert, David Edwin
Linnell, Richard D(ean)
Taft, Charles Kirkland

NEW JERSEY
Amarel, S(aul)
Amitay, Noach
An, Linda Huang
Anderson, Milton Merrill
Andrews, Ronald Allen
Belanger, David Gerald
Benaroya, Haym
Black, H(arold) S(tephen)
Boucher, Thomas Owen

Brenner, Douglas
Brown, Irmel Nelson
Browne, Robert Glenn
Canzonier, Walter J
Cardarelli, Joseph S
Castenschiold, Rene
Cheung, Shiu Ming
Clymer, Arthur Benjamin
Curtis, Thomas Hasbrook
DeMarinis, Bernard Daniel
Dobkin, David Paul
Dorros, Irwin
Falconer, Warren Edgar
Feiner, Alexander
Fretwell, Lyman Jefferson, Jr
Gabriel, Oscar V
Grauman, Joseph Uri
Haas, Zygmunt
Harris, James Ridout
Hsu, Jay C
Huang, Jennming Stephen
Huber, Richard V
Hung, Tonney H M
Johannes, Virgil Ivancich
Kalley, Gordon S
Kaplan, Joel Howard
Kesselman, Warren Arthur
Kobayashi, Hisashi
Lawson, James Robert
Leu, Ming C
Lucas, Henry C, Jr
McCumber, Dean Everett
Maclin, Ernest
McNicholas, James J
Mangulis, Visvaldis
Mauzey, Peter T
Meyer, Andrew U
Murthy, Srinivasa K R
Poor, Harold Vincent
Poucher, John Scott
Prekopa, Andras
Raychaudhuri, Kamal Kumar
Rich, Kenneth Eugene
Robrock, Richard Barker, II
Rosenblum, Martin Jacob
Rosenthal, Lee
Sachs, Harvey Maurice
Saniee, Iraj
Sannuti, Peddapullaiah
Scales, William Webb
Schaefer, Jacob W
Schnapf, Abraham
Scholz, Lawrence Charles
Sears, Raymond Warrick, Jr
Semmlow, John Leonard
Shefer, Joshua
Shen, Tek-Ming
Sherman, Ronald
Shieh, Ching-Chyuan
Shipley, Edward Nicholas
Shirk, Richard Jay
Shulman, Herbert Byron
Sipress, Jack M
Smith, William Bridges
Sontag, Eduardo Daniel
Steben, John D
Stengel, Robert Frank
Strother, J(ohn) A(lan)
Tzanakou, M Evangelia
Uhrig, Jerome Lee
Weiss, C Dennis
Westerman, Howard Robert
Young, Li

NEW MEXICO
Baty, Richard Samuel
Clark, Wallace Thomas, III
Coulter, Claude Alton
Curtis, Wesley E
Dudziak, Donald John
Ebrahimi, Nader Dabir
Gilbert, Alton Lee
Hakkila, Eero Arnold
Holmes, John Thomas
Jackson, Sydney Vern
Lee, David Mallin
Lundergan, Charles Donald
Novak, James Lawrence
Otway, Harry John
Shipley, James Parish, Jr
Sitney, Lawrence Raymond
Snider, Donald Edward
Swingle, Donald Morgan
Welber, Irwin
Williamson, Kenneth Donald, Jr
Woods, Robert Octavius

NEW YORK
Ahmed-Zaid, Said
Anastassiou, Dimitris
Anderson, Roy E
Axelrod, Norman Nathan
Balabanian, Norman
Bonan, Eugene J
Brownlow, Roger W
Burros, Raymond Herbert
Carlson, A Bruce
Carruthers, Raymond Ingalls
Chang, John H(si-Teh)
Chen, Chi-Tsong
Christiansen, Donald David
Clarke, Kenneth Kingsley
Cohen, Edwin
Costello, Richard Gray

Dahm, Douglas Barrett
Damouth, David Earl
Desrochers, Alan Alfred
Difranco, Julius V
Dimmler, D(ietrich) Gerd
Fine, Terrence Leon
Franse, Renard
Geiss, Gunther R(ichard)
Goel, Narendra Swarup
Gran, Richard J
Haas, David Jean
Hajela, Prabhat
Hanson, William A
Hauser, Norbert
Hetzel, Richard Ernest
Higinbotham, William Alfred
Howland, Howard Chase
Hutchings, William Frank
Jones, Leonidas John
Karnaugh, Maurice
Kershenbaum, Aaron
Kinzly, Robert Edward
Kirby, Maurice J(oseph)
Kirchmayer, Leon K
Klir, George Jiri
Koenig, Michael Edward Davison
Kopp, Richard E
La Russa, Joseph Anthony
Leigh, Richard Woodward
Levine, Stephen Alan
Ling, Huei
Litchford, George B
Logue, J(oseph) C(arl)
Longman, Richard Winston
Longobardo, Anna Kazanjian
Lowen, W(alter)
Lustig, Howard E(ric)
Maltz, Martin Sidney
Mead, Lawrence Myers
Meenan, Peter Michael
Menkes, Sherwood Bradford
Metzger, Ernest Hugh
Monmonier, Mark
Mosteller, Henry Walter
Notaro, Anthony
Nyrop, Jan Peter
Palmedo, Philip F
Papoulis, Athanasios
Pomerene, James Herbert
Raghuveer, Mysore R
Rivkin, Maxcy
Rogers, Edwin Henry
Rushing, Allen Joseph
Sarachik, Philip E(ugene)
Sargent, Robert George
Saridis, George N
Sevian, Walter Andrew
Shelley, Edwin F(reeman)
Shen, Chi-Neng
Shenolikar, Ashok Kumar
Sherman, Philip Martin
Sheryll, Richard Perry
Shevack, Hilda N
Shooman, Martin L
Shuey, R(ichard) L(yman)
Sinclair, Douglas C
Smith, Bernard
Sobczak, Thomas Victor
Sobel, Kenneth Mark
Spang, H Austin, III
Spencer, David R
Stein, David Morris
Sternlicht, B(eno)
Stone, Harold S
Subramanian, Mani M
Summers, Phillip Dale
Swanton, Walter F(rederick)
Swartz, James Lawrence
Tatarczuk, Joseph Richard
Taylor, James Hugh
Thompson, Robert Alan
Thorndike, Alan Moulton
Tiemann, Jerome J
Torng, Hwa-Chung
Tummala, Rao Ramamohana
Van Slyke, Richard M
Varshney, Pramod Kumar
Victor, Jonathan David
Wallace, Aaron
Wang, Shu Lung
Weingarten, Norman C
White, William Wallace
Winter, Herbert
Witt, Christopher John
Wong, Edward Chor-Cheung
Yablon, Marvin
Yao, David D
Yip, Kwok Leung
Youla, Dante C
Zdan, William

NORTH CAROLINA
Allen, Charles Michael
Bernhard, Richard Harold
Coates, Clarence L(eroy), Jr
Coulter, Norman Arthur, Jr
Defoggi, Ernest
Eberhardt, Allen Craig
Eckerlin, Herbert Martin
Franzon, Paul Damian
Gangwal, Santosh Kumar
Garg, Devendra Prakash
Gerhardt, Don John
Hutchins, William R(eagh)

Kanopoulos, Nick
Marinos, Pete Nick
Nutter, James I(rving)
Smith, William Adams, Jr
Struve, William George
Zaalouk, Mohamed Gamal

NORTH DAKOTA
Stanislao, Bettie Chloe Carter
Yuvarajan, Subbaraya

OHIO
Ash, Raymond H(ouston)
Bernard, John Wilfrid
Bhagat, Pramode Kumar
Bishop, Albert B
Burghart, James H(enry)
Chen, Chiou Shiun
Christensen, James Henry
Ciricillo, Samuel F
Cummings, Charles Arnold
Damianov, Vladimir B
D'Azzo, John Joachim
Easterday, Jack L(eroy)
El-Naggar, Mohammed Ismail
Faymon, Karl A(lois)
Fontana, Robert E
Fritsch, Charles A(nthony)
Ghandakly, Adel Ahmad
Giamati, Charles C, Jr
Gorland, Sol H
Grant, Michael P(eter)
Hajek, Otomar
Ho, Fanghuai H(ubert)
Hobbs, Benjamin F
Hoy, Casey William
Huang, John Wen-Chih
Jayne, Theodore D
Kantor, Paul B
Kaya, Azmi
Khalimsky, Efim D
Klosterman, Albert Leonard
Komkov, Vadim
Kramerich, George L
Kwatra, Subhash Chander
Lamont, Gary Byron
Laskowski, Edward L
Lee, Sunggyu
Lefkowitz, Irving
Lenhart, Jack G
Liu, Ming-Tsan
McGandy, Edward Lewis
McNichols, Roger J(effrey)
Menq, Chia-Hsiang
Miller, Richard Allen
Quam, David Lawrence
Rose, Charles William
Schultz, Edwin Powell
Shetterly, Donivan Max
Shick, Philip E(dwin)
Sprafka, Robert J
Steele, Jack Ellwood
Vassell, Gregory S
Wallace, Michael Dwight
Wyman, Bostwick Frampton

OKLAHOMA
Aly, Adel Ahmed
Baker, James E
Fagan, John Edward
McDevitt, Daniel Bernard
Martner, Samuel (Theodore)
Mize, Joe H(enry)
Richardson, J Mark
Schuermann, Allen Clark, Jr
Terrell, Marvin Palmer
Wilcox, Lyle C(hester)

OREGON
Boyd, Dean Weldon
Linstone, Harold A

PENNSYLVANIA
Abend, Kenneth
Allen, Charles C(orletta)
Anandalingam, G
Anouchi, Abraham Y
Baumann, Dwight Maylon Billy
Bedrosian, Samuel D
Bhagavatula, VijayaKumar
Biegler, Lorenz Theodor
Bise, Christopher John
Bordogna, Joseph
Brackmann, Richard Theodore
Bradley, James Henry Stobart
Brunner, Mathias J
Butkiewicz, Edward Thomas
Carbonell, Jaime Guillermo
Charp, Solomon
Clapp, Richard Gardner
Cohn, Nathan
Dax, Frank Robert
Dorny, C Nelson
Eisenberg, Lawrence
Eisenstein, Bruce A
Feng, Tse-yun
Frumerman, Robert
Fuchs, Walter
Garfinkel, David
Georgakis, Christos
Goldman, Robert Barnett
Goutmann, Michel Marcel
Grannemann, Glenn Niel
Griffin, Jerry Howard

Grossmann, Ignacio E
Hendrickson, Chris Thompson
Homsey, Robert John
Johnson, Stanley Harris
Kam, Moshe
Kaslow, David Edward
Kazahaya, Masahiro Matt
Kraft, R(alph) Wayne
Krendel, Ezra Simon
Ku, Y(u) H(siu)
Kusic, George Larry, Jr
Leibholz, Stephen W
Likins, Peter William
Longini, Richard Leon
McGarity, Arthur Edwin
Murgie, Samuel A
Nichols, Duane Guy
Popovics, Sandor
Preston, Kendall, Jr
Putman, Thomas Harold
Satz, Ronald Wayne
Scigliano, J Michael
Scranton, Bruce Edward
Sherwood, Bruce Arne
Sibul, Leon Henry
Siegel, Melvin Walter
Smith, David Beach
Stone, Gregory Michael
Stover, James Anderson, Jr
Thomas, Donald H(arvey)
Tierney, John W(illiam)
Tobias, Philip E
Uher, Richard Anthony
Verhanovitz, Richard Frank
Vogt, William G(eorge)
Walker, David Kenneth
Weinstock, W
Weir, William Thomas
Wilson, Myron Allen
Zygmont, Anthony J

RHODE ISLAND
Bartram, James F(ranklin)
Ehrlich, Stanley L(eonard)
Hazeltine, Barrett
Palm, William John
Petrou, Panayiotis Polydorou
Rerick, Mark Newton
Robertshaw, Joseph Earl

SOUTH CAROLINA
Balintfy, Joseph L
Blevins, Maurice Everett
Gaines, Albert L(owery)
Klemm, James L
Miler, George Gibbon, Jr

SOUTH DAKOTA
Hodges, Neal Howard, II
Oliver, Thomas Kilbury
Reuter, William L(ee)
Riemenschneider, Albert Louis

TENNESSEE
Arlinghaus, Heinrich Franz
Bartell, Steven Michael
Bontadelli, James Albert
Cox, Ronald Baker
Cox, Ronald Baker
Dabbs, John Wilson Thomas
Emanuel, William Robert
Haaland, Carsten Meyer
Hill, Hubert Mack
House, Robert W(illiam)
James, Alton Everette, Jr
Lane, Eric Trent
Martin, Harry Lee
Siirola, Jeffrey John
Upadhyaya, Belle Raghavendra

TEXAS
Bennett, George Kemble
Bhuva, Rohit L
Bishop, Robert H
Blackwell, Charles C, Jr
Blank, Leland T
Chakravarty, Indranil
Collins, Francis Allen
Cory, William Eugene
Curtin, Richard B
Dhudshia, Vallabh H
Dodd, Edward Elliott
Edgar, Thomas Flynn
Ewell, James John, Jr
Fletcher, John Lynn
French, Robert Leonard
Garner, Harvey L(ouis)
Hamill, Dennis W
Hennessey, Audrey Kathleen
Kesler, Oren Byrl
Koehn, Enno
Ladde, Gangaram Shivlingappa
Lawrence, Joseph D, Jr
Levine, Jules David
Linder, John Scott
Lipovski, Gerald John (Jack)
Marshek, Kurt M
Martin, Norman Marshall
Matlock, Gibb B
Maxwell, Donald A
Miller, Maurice Max
Mistree, Farrokh
Moser, James Howard
Mouritsen, T(horvald) Edgar

Systems Design & Systems Science (cont)

Nute, C Thomas
Ochiai, Shinya
Ostrofsky, Benjamin
Paddison, Fletcher C
Parker, Arthur L (Pete)
Patel, Mayur
Patton, Alton DeWitt
Phillips, Don T
Potter, Robert Joseph
Randolph, Paul Herbert
Rodenberger, Charles Alvard
Samson, Charles Harold
Secrest, Everett Leigh
Summers, Richard Lee
Tibbals, Harry Fred, III
VantHull, LORIN L
Wakeland, William Richard
Walkup, John Frank
Ward, Phillip Wayne
White, Charles F(loyd)
Widmer, Robert H
Wilhelm, Wilbert Edward
Wood, Kristin Lee
Yakin, Mustafa Zafer

UTAH
Baum, Sanford
Butterfield, Veloy Hansen, Jr
Clegg, John C(ardwell)
Cox, Benjamin Vincent
Hoggan, Daniel Hunter
Jolley, David Kent
Krejci, Robert Henry
MacGregor, Douglas
Sandquist, Gary Marlin
Turley, Richard Eyring
Yorks, Terence Preston

VERMONT
Ferris-Prabhu, Albert Victor Michael

VIRGINIA
Abbott, Dean William
Agee, Marvin H
Ammerman, Charles R(oyden)
Attinger, Ernst Otto
Ball, Joseph Anthony
Barmby, John G(lennon)
Beam, Walter R(aleigh)
Bingham, Billy Elias
Browell, Edward Vern
Burkhalter, Barton R
Charvonia, David Alan
Cheng, George Chiwo
Childress, Otis Steele, Jr
Cook, Charles William
Coram, Donald Sidney
Crawford, Daniel J
Daniel, Kenneth Wayne
Davis, Harry I
Doles, John Henry, III
Edyvane, John
Entzminger, John N
Epstein, Samuel David
Fabrycky, Wolter J
Fearnsides, John Joseph
Fink, Lester Harold
Finkelstein, Jay Laurence
Fletcher, James Chipman
Fthenakis, Emanuel John
Gamble, Thomas Dean
Gerlach, A(lbert) A(ugust)
Gibson, John E(gan)
Godbole, Sadashiva Shankar
Gordon, John S(tevens)
Greinke, Everett D
Guerber, Howard P(aul)
Hammond, Marvin H, Jr
Heffron, W(alter) Gordon
Hobeika, Antoine George
Hodge, Donald Ray
Holzmann, Ernest G(unther)
Hou, Gene Jean-Win
Hunt, V Daniel
Hwang, John Dzen
James, W(ilbur) Gerald
Johnson, Charles Minor
Joshi, Suresh Meghashyam
Kahn, Robert Elliot
Kahne, Stephen James
Kaplan, Robert Lewis
Keblawi, Feisal Said
Kuhlthau, A(lden) R(obert)
Larson, Arvid Gunnar
Leon, H(erman) I
Levis, Alexander Henry
Lord, Norman W
Ludwig, George H
McGean, Thomas J
McGregor, Dennis Nicholas
McKenna, John Dennis
McRee, G(riffith) J(ohn), Jr
Madan, Rabinder Nath
Manfredi, Arthur Frank, Jr
Marsh, Howard Stephen
Mehta, Gurmukh D
Murray, Jeanne Morris
Myers, Donald Albin
Nyman, Thomas Harry
Perry, Dennis Gordon

Reierson, James (Dutton)
Richards, Paul Bland
Rowe, William David
Sage, Andrew Patrick
Sawyer, James W
Schlegelmilch, Reuben Orville
Schwartz, Jay W(illiam)
Schy, Albert Abe
Seward, William Davis
Smith, Michael Claude
Sobieszczanski-Sobieski, Jaroslaw
Sorrell, Gary Lee
Thompson, Anthony Richard
Tsang, Wai Lin
Walker, Robert Paul
Warfield, J(ohn) N(elson)
Waxman, Ronald
Webber, John Clinton
Weinstein, Iram J
Wells, John Morgan, Jr
Wessel, Paul Roger
Whitelaw, R(obert) L(eslie)
Woeste, Frank Edward
Wolf, Eric W
Worthington, John Wilbur

WASHINGTON
Blackburn, John Lewis
Driver, Garth Edward
Gravitz, Sidney I
Hueter, Theodor Friedrich
MacVicar-Whelan, Patrick James
Madden, James H(oward)
Mar, Brian W(ayne)
Millham, Charles Blanchard
Preikschat, F(ritz) K(arl)
Ricker, Neil Lawrence
Stear, Edwin Byron
Tripp, Leonard L
Yore, Eugene Elliott

WEST VIRGINIA
Tewksbury, Stuart K

WISCONSIN
Bishop, Charles Joseph
Bomba, Steven James
Crandall, Lee W(alter)
Dobson, David A
Eriksson, Larry John
Goodson, Raymond Eugene
Hanway, John E(dgar), Jr
Heinen, James Albin
Henke, Russell W
Northouse, Richard A
Sanborn, I B
Schultz, Hilbert Kenneth
Suri, Rajan
Szpot, Bruce F

WYOMING
Constantinides, Christos T(heodorou)

PUERTO RICO
Hussain, Malek Gholoum Malek

ALBERTA
Chan, Wah Chun
Hamza, Mohamed Hamed

BRITISH COLUMBIA
Chanson, Samuel T
Pomeroy, Richard James

MANITOBA
Johnson, R(ichard) A(llan)

NEW BRUNSWICK
Lee, Y C

NEWFOUNDLAND
Shieh, John Shunen

NOVA SCOTIA
Baird, Charles Robert
Eames, M(ichael) C(urtis)
Kujath, Marek Ryszard

ONTARIO
Ahmed, Nasir Uddin
Altman, Samuel Pinover
Bandler, John William
Barrington, Nevitt H J
Beevis, David
Byer, Philip Howard
Chou, Jordan Quan Ban
Cohn, Stanley Howard
Coll, David C
Corneil, Ernest Ray
Davison, E(dward) J(oseph)
Fairman, Frederick Walker
Falconer, David Duncan
Georganas, Nicolas D
Goldenberg, Andrew Avi
Halfon, Efraim
Henry, Roger P
Hipel, Keith William
Huseyin, Koncay
Law, Cecil E
McClymont, Kenneth
Ratz, H(erbert) C(harles)
Rauch, Sol
Renfrew, Robert Morrison
Reza, Fazlollah M

Riordon, J(ohn) Spruce
Roe, Peter Hugh O'Neil
Ryan, Dave
Siddall, Ernest
Smith, Harold William
Wilson, John D(ouglas)
Wong, Jo Yung
Wonham, W Murray
Wright, Douglas Tyndall
Wright, Joseph D

QUEBEC
Bui, Tien Rung
Cheng, Richard M H
Davidson, Colin Henry
Ferguson, Michael John
Jumarie, Guy Michael
Lecours, Michel
Marin, Miguel Angel
Patel, Rajnikant V
Plotkin, Eugene Isaak
Sankar, Seshadri

SASKATCHEWAN
Curtis, Fred Allen
Fuller, Gerald Arthur
Gupta, Madan Mohan
Koh, Eusebio Legarda
Walerian, Szyszkowski

OTHER COUNTRIES
Arimoto, Masao
Bryngdahl, Olof
Czepyha, Chester George Reinhold
Esterson, Gerald L(ee)
Eykhoff, Pieter
Fettweis, Alfred Leo Maria
Fitzhugh, Henry Allen
Ince, A Nejat
Jespersen, Nils Vidar
Kimura, Hidenori
Kuroyanagi, Noriyoshi
Lehman, Meir M
Lindquist, Anders Gunnar
Mansour, Mohamed
Sekine, Yasuji
Singh, Madan Gopal
Sunahara, Yoshifumi
Thissen, Wil A
Tzafestas, Spyros G
Willems, Jan C
Yazicigil, Hasan
Zeheb, Ezra

Other Engineering

ALABAMA
Clark, Charles Edward
Hart, David R
Johnson, Evert William
Park, Duk-Won
Smoot, George Fitzgerald

ALASKA
Gedney, Larry Daniel

ARIZONA
Barrett, Harrison Hooker
Hart, Lyman Herbert
Helbert, John N
Makepeace, Gershom Reynolds
Mulvey, James Patrick
Reid, Ralph R
Young, Hewitt H

ARKANSAS
Bond, Robert Levi
Cardwell, David Michael
Carr, Gerald Paul
Glendening, Norman Willard
Sims, Robert Alan

CALIFORNIA
Agbabian, Mihran S
Agrawal, Suphal Prakash
Anger, Hal Oscar
Aster, Robert Wesley
Auksmann, Boris
Aziz, Khalid
Baba, Paul David
Booth, Newell Ormond
Braithwaite, Charles Henry, Jr
Byer, Robert L
Byram, George Wayne
Cook, Charles J
Cook, Neville G W
Crutchfield, Charlie
Damos, Diane Lynn
Davis, Harmer E
Elings, Virgil Bruce
Ewing, Gerald Dean
Fischer, George K
Forster, Julian
Fung, Yuan-Cheng B(ertaam)
Gearhart, Roger A
Graham, Martin H(arold)
Grier, Herbert E
Hackwood, Susan
Hayes, Thomas Jay, III
Hingorani, Narain G
John, Joseph
Kaye, Wilbur (Irving)
Kennedy, Robert P

Kivenson, Gilbert
Kurz, Richard J
Lamm, A Uno
Laub, Alan John
Lee, Hua
Leidersdorf, Craig B
Linnell, Robert Hartley
Longley-Cook, Mark T
Ma, Joseph T
McIntyre, Marie Claire
Maxfield, Bruce Wright
Meyer, Lhary
Munger, Charles Galloway
Nooker, Eugene L(eRoy)
O'Donnell, Ashton Jay
Paquette, Robert George
Rechnitzer, Andreas Buchwald
Remer, Donald Sherwood
Robinson, Lloyd Burdette
Robinson, Merton Arnold
Schade, Henry A(drian)
Schmidt, Raymond LeRoy
Sharpe, Roland Leonard
Shelly, John Richard
Shemdin, Omar H
Siljak, Dragoslav (D)
Smith, Levering
Smith, Warren James
Subramanya, Shiva
Suer, H(erbert) S
Swanson, Eric Richmond
Swartzlander, Earl Eugene, Jr
Sweeney, James Lee
Swerling, Peter
Thiers, Eugene Andres
Thomas, Graham Havens
Van Leer, John Cloud
Vomhof, Daniel William
Waldhauer, F D
Wallerstein, Edward Perry
Walsh, Don
Wenzel, James Gottlieb
Williamson, Robert Brady

COLORADO
Cohen, Robert
Holtz, Wesley G
Hotchkiss, William Rouse
Huisjen, Martin Albert
Keller, Frederick Albert, Jr
Kelley, Neil Davis
Knollenberg, Robert George
Kompala, Dhinakar S
Lightsey, Paul Alden
Nesenbergs, Martin
Peacock, Roy Norman
Plows, William Herbert
Puckorius, Paul Ronald
Schmidt, Alan Frederick
Strom, Oren Grant
Weber, James R

CONNECTICUT
Engelberger, Joseph Frederick
Hill, David Lawrence
Long, Richard Paul
Mancheski, Frederick J
Pooley, Alan Setzler
Powers, Joseph
Sheets, Herman E(rnest)
Shuey, Merlin Arthur
Tepas, Donald Irving
Weinstein, Curt David

DELAWARE
Arots, Joseph B(artholomew)
Dalrymple, Robert Anthony
Hanzel, Robert Stephen
Kobayashi, Nobuhisa
Levy, Paul F
Moore, Ralph Leslie
Sawers, James Richard, Jr
Sheer, M(axine) Lana

DISTRICT OF COLUMBIA
Baruch, Jordan J(ay)
Coulter, Henry W
Hartman, Patrick James
Johnson, Bruce
Lee, Alfred M
O'Fallon, Nancy McCumber
Parasuraman, Raja

FLORIDA
Arnold, Henry Albert
Beatty, Charles Lee
Braman, Robert Steven
Cassidy, William F
Daubin, Scott C
Davidson, James Blaine
Dunford, James Marshall
Freeman, Neil Julian
Koff, Bernard Louis
McAllister, Raymond Francis
Reynolds, Samuel D, Jr
Rona, Donna C
Rosenthal, Henry Bernard
Smith, Malcolm (Kinmonth)
Williamson, Donald Elwin
Wylde, Ronald James
Zborowski, Andrew

GEORGIA
Rogers, Nelson K

Woods, Wendell David

HAWAII
Bretschneider, Charles Leroy
Gerritsen, Franciscus
Hallanger, Lawrence William
Spielvogel, Lester Q

IDAHO
Engen, Ivar A
Spencer, Paul Roger

ILLINOIS
Brauer, Roger L
Cheng, Herbert S
Dolgoff, Abraham
Savic, Stanley D
Saxena, Satish Chandra
Toy, William W
Wise, Evan Michael
Wurm, Jaroslav

INDIANA
Ladisch, Michael R
McAleece, Donald John
Nolting, Henry Frederick
Skibniewski, Miroslaw Jan

IOWA
Nixon, Wilfrid Austin

KANSAS
Dean, Thomas Scott
Norris, Roy Howard
Razak, Charles Kenneth
Willhite, Glen Paul

LOUISIANA
Greco, Edward Carl
Kinchen, David G
Lee, Griff C

MAINE
Gibson, Richard C(ushing)
Norton, Karl Kenneth

MARYLAND
Acheson, Donald Theodore
Babrauskas, Vytenis
Belanger, Brian Charles
Brenner, Fivel Cecil
Cremens, Walter Samuel
Dallman, Paul Jerald
Doctor, Norman J(oseph)
Fowell, Andrew John
Franke, Gene Louis
Frey, Henry Richard
Gheorghiu, Paul
Gilford, Leon
Hall, Kimball Parker
Hyde, Geoffrey
Kacker, Raghu N
Kalil, Ford
Lange, Eugene Albert
Levine, Robert S(idney)
McConoughey, Samuel R
McIlvain, Jess Hall
Morgan, William Bruce
Navarro, Joseph Anthony
Papian, William Nathaniel
Potzick, James Edward
Quintiere, James G
Rao, Gopalakrishna M
Rochlin, Phillip
Rotariu, George Julian
Schulkin, Morris
Sikora, Jerome Paul
Stein, Robert Alfred
Sutter, David Franklin
Walker, William J, Jr
Weaver, Christopher Scot
Wilkins, Michael Gray
Wing, John Faxon
Wolde-Tinsae, Amde M

MASSACHUSETTS
Abkowitz, Martin A(aron)
Allen, Steven
Brandt, Bruce Losure
Chryssostomidis, Chryssostomos
Clarke, Edward Nielsen
Coor, Thomas
Copley, Lawrence Gordon
DiNardi, Salvatore Robert
Field, Hermann Haviland
Fitzgerald, Robert William
Friedman, Raymond
Hertweck, Gerald
Krasner, Jerome L
LeMessurier, William James
Lemons, Thomas M
Levin, Robert Edmond
MacMillan, Douglas Clark
Masubuchi, Koichi
Newman, J Nicholas
Park, B J
Pearlman, Michael R
Pope, Joseph
Read, Philip Lloyd
Shrader, William Whitney
Spergel, Philip
Wilkinson, Robert Haydn
Williams, Albert James, III

MICHIGAN
Ashcroft, Frederick H
Batrin, George Leslie
Beck, Robert Frederick
Benford, Harry (Bell)
Datta, Tapan K
Evans, Leonard
Fancher, Paul S(trimple)
Fortuna, Edward Michael, Jr
Greene, Bruce Edgar
Hammitt, Frederick G(nichtel)
Hickling, Robert
Nuttall, Alfred L
Prasad, Biren
Reising, Richard F
Risdon, Thomas Joseph
Rodgers, John James
Saul, William Edward
Taraman, Khalil Showky
Turcotte, William Arthur
Yagle, Raymond A(rthur)

MINNESOTA
Harvey, Charles Arthur
Mooers, Howard T(heodore)
Sterling, Raymond Leslie

MISSISSIPPI
Aughenbaugh, Nolan B(laine)
Bryant, Glenn D(onald)
Cade, Ruth Ann
Rosenhan, A Kirk

MISSOURI
Kurtz, Vincent E
Monzyk, Bruce Francis
Morgan, Robert P
Natani, Kirmach
Poynton, Joseph Patrick
Saran, Chitaranjan
Schaefer, Joseph Thomas
Walker, James Harris
Wulfman, David Swinton

NEVADA
Benfield, Charles W(illiam)

NEW HAMPSHIRE
Allen, Robert C(hadbourne)
Allmendinger, E Eugene
Crowell, Merton Howard
Kasper, Joseph F, Jr
Wilson, Donald Alfred

NEW JERSEY
Abrahams, Elihu
Biagetti, Richard Victor
Chan, Chun Kin
DeMarinis, Bernard Daniel
Dodington, Sven Henry Marriott
Endo, Youichi
Harvey, Floyd Kallum
Heirman, Donald N
Kiessling, George Anthony
Killam, Everett Herbert
Kluver, J(ohan) W(ilhelm)
Lund, Daryl B
McGregor, Walter
Mandel, Andrea Sue
Manz, August Frederick
Miner, Robert Scott, Jr
Newman, Joseph Herbert
Ponessa, Joseph Thomas
Quinlan, Daniel A
Robinson, Daniel E
Rothschild, Gilbert Robert
Sams, Burnett Henry, III
Savitsky, Daniel
Schoenberg, Leonard Norman
Sharp, Hugh T
Stiles, Lynn F, Jr
Wadlow, David
Warren, Clifford A
Wentworth, John W

NEW MEXICO
Cropper, Walter V
Davis, Alvin Herbert
Edeskuty, F(rederick) J(ames)
Elliott, Dana Edgar
Finch, Thomas Wesley
Holmes, John Thomas
Keepin, George Robert, Jr
Moulds, William Joseph
Neeper, Donald Andrew
Savage, Charles Francis
Stephans, Richard A
Strong, Ronald Dean
Sugerman, Leonard Richard
Valentine, James K

NEW YORK
Alexiou, Arthur George
Barone, Louis J(oseph)
Bayer, Raymond George
Bechhofer, Robert E(ric)
Bourdillon, Antony John
Brown, Neil Harry
Cohen, Leonard A
Drozdowicz, Zbigniew Marian
Fein, R(ichard) S(aul)
Femenia, Jose
Fisch, Oscar
Goodwin, Arthur VanKleek

Gould, James P
Greenspan, Joseph
Herman, Herbert
Hetzel, Richard Ernest
Hoover, Thomas Earl
Kahn, Elliott H
Kammer, Paul A
Klein, Harvey Gerald
Lakatos, Andras Imre
Lang, Martin
Le Maistre, Christopher William
Leone, James A
Lewis, Alan Laird
Lindstrom, Gary J
Lurkis, Alexander
Malaviya, Bimal K
Mathes, Kenneth Natt
Meth, Irving Marvin
Meyburg, Arnim Hans
Meyer, Robert Walter
Milkovic, Miran
Neeper, Ralph Arnold
Nicholson, Thomas Dominic
Palladino, William Joseph
Pennington, Keith Samuel
Pernick, Benjamin J
Pikarsky, Milton
Radeka, Veljko
Raghuveer, Mysore R
Rikmenspoel, Robert
Rothman, Herbert B
Schlee, Frank Herman
Sinclair, Douglas C
Staples, Basil George
Thomas, Richard Alan
Tobin, Michael
Vogel, Alfred Morris
Welsh, James P
Wiese, Warren M(elvin)
Wiggins, Edwin George
Wotherspoon, Neil
Youla, Dante C
Zubaly, Robert B(arnes)

NORTH CAROLINA
Ayoub, Mahmoud A
Ballard, Lewis Franklin
Buchanan, David Royal
Chin, Robert Allen
Ciccolella, Joseph A
Downs, Robert Jack
Fisher, C(harles) Page, Jr
Grainger, John Joseph
Kirkbride, L(ouis) D(ale)
Nader, John S(haheen)
Sherer, James Pressly
Utku, Bisulay Bereket

OHIO
Bisson, Edmond E(mile)
Butz, Donald Josef
Gallo, Frank J
Gregory, Thomas Bradford
Grimm, Robert Arthur
Houpis, Constantine H
Keyes, Marion Alvah, IV
Mielke, Robert L
Nash, David Byer
Schultz, Thomas J
Singh, Rajendra

OKLAHOMA
Forney, Bill E
Hendrickson, John Robert
McCoy, Raymond Duncan
Stover, Dennis Eugene
Summers, Jerry C
Winter, William Kenneth
Zigrang, Denis Joseph

OREGON
Devletian, Jack H
Duncan, Walter E(dwin)
Landers, John Herbert, Jr

PENNSYLVANIA
Bergmann, Ernest Eisenhardt
Bjorhovde, Reidar
Cavanagh, Peter R
Cohen, Anna (Foner)
Drumwright, Thomas Franklin, Jr
Ebert, Philip E
Hager, Wayne R
Hertzberg, Martin
Howe, Richard Hildreth
Ibrahim, Baky Badie
Ko, Frank K
Kraus, C Raymond
Krishnaswamy, S V
Levin, Michael H(oward)
Magaziner, Henry Jonas
Magison, Ernest Carroll
Mehta, Kishor Singh
Mouly, Raymond J
Oliver, Morris Albert
Ostapenko, A(lexis)
Rohrer, Ronald A
Sheppard, Walter Lee, Jr
Staley, James T
Sterk, Andrew A
Stroud, Jackson Swavely
Van Roggen, Arend
Wagner, J Robert
Werkman, Joyce

Westine, Peter Sven
White, Eugene Wilbert
Wiehe, William Albert
Zimmermann, F(rancis) J(ohn)

RHODE ISLAND
Ellison, William Theodore
Kirschenbaum, Susan S
Silva, Armand Joseph
Spaulding, Malcolm Lindhurst
Sternling, Charles V
Tyce, Robert Charles

SOUTH CAROLINA
Roddis, Louis Harry, Jr
Singleton, Chloe Joi
Sudarshan, Tangali S

SOUTH DAKOTA
Oliver, Thomas Kilbury
Sengupta, Sailes Kumar

TENNESSEE
Ackermann, Norbert Joseph, Jr
Baer, Thomas Strickland
Devgan, Satinderpaul Singh
Goans, Ronald Earl
Godbee, H(erschel) W(illcox)
Hensley, Marble John, Sr
Isley, James Don
Kopp, Manfred Kurt
Kress, Thomas Sylvester
Mueller, Theodore Rolf
Ornitz, Barry Louis
Pitcher, DePuyster Gilbert
Prairie, Michael L
Raudorf, Thomas Walter
Rayside, John Stuart
Slaughter, Gerald M

TEXAS
Abrams, Albert
Bigham, Robert Eric
Briggs, Edward M
Buckalew, Louis Walter
Collipp, Bruce Garfield
Crenshaw, Paul L
Eggers, Fred M
Ehlig-Economides, Christine Anna
Geer, Ronald L
Gourdine, Meredith Charles
Halff, Albert Henry
Hanson, Bernold M
Hinterberger, Henry
Hopkins, Robert Charles
Johnstone, C(harles) Wilkin
Levine, Jules David
Machemehl, Jerry Lee
McNiel, James S(amuel), Jr
Messenger, Joseph Umlah
Moore, Walter P, Jr
Naismith, James Pomeroy
Packman, Paul Frederick
Page, Robert Henry
Powell, Alan
Scinta, James
Senatore, Ford Fortunato
Smith, Richard Thomas
Spencer, Thomas H
Wagner, John Philip
Whiting, Allen R

UTAH
Bishop, Jay Lyman
Israelsen, C Earl
MacGregor, Douglas

VERMONT
Arns, Robert George
Smith, Mark K
Warfield, George

VIRGINIA
Arthur, Richard J(ardine)
Cambel, Ali B(ulent)
Campbell, Edward J
Claus, Alfons Jozef
Dalton, Roger Wayne
Gooding, Robert C
Haimes, Yacov Y
Harrington, Robert Joseph
King, Merrill Kenneth
Lai, Chintu (Vincent C)
Leopold, Reuven
Levy, Sander Alvin
Moore, Joseph Herbert
Ng, Wing-Fai
Phillips, William H
Reese, Ronald Malcolm
Rojiani, Kamal B
Schetz, Joseph A
Searle, Willard F, Jr
Ventre, Francis Thomas
Whitcomb, Richard T
Whitelaw, R(obert) L(eslie)
Woeste, Frank Edward
Yeager, Paul Ray

WASHINGTON
Adams, William Mansfield
Bolme, Donald W(eston)
Butler, Don
Hebeler, Henry K
Katz, Yale H

Other Engineering (cont)

Pigott, George M
Rasco, Barbara A
Sandstrom, Wayne Mark
Sutherland, Earl C
Zimmerer, Robert W

WEST VIRGINIA
Humphreys, Kenneth K
Lantz, Thomas Lee
Wales, Charles E

WISCONSIN
Falk, Edward D
Liebe, Donald Charles
Moore, Gary T
Sebree, Bruce Randall
Smith, Luther Michael
Vanderheiden, Gregg

ALBERTA
Fisher, Harold M
Vroom, Alan Heard

BRITISH COLUMBIA
Bell, H(arry) R(ich)
Thyer, Norman Harold

MANITOBA
Jayas, Digvir Singh

NEW BRUNSWICK
Faig, Wolfgang
Vanicek, Petr

NEWFOUNDLAND
Bajzak, Denes
Muggeridge, Derek Brian

NOVA SCOTIA
Eames, M(ichael) C(urtis)

ONTARIO
Atkinson, George Francis
Carr, Jan
Etkin, Bernard
Goldenberg, Andrew Avi
Gracie, G(ordon)
Jones, Alun Richard
Kamphuis, J(ohn) William
Keeler, John S(cott)
Mittal, Gauri S
Pajari, George Edward
Stirling, Andrew John

QUEBEC
Barbery, Gilles A
Ghosh, Sanjib Kumar

OTHER COUNTRIES
Al-Zubaidy, Sarim Naji
Bayev, Alexander A
Eisner, Edward
Eykhoff, Pieter
Farquhar, Gale Burton
Guberman, H(erbert) D(avid)
Kimura, Ken-ichi
Ku, Anthony Chia-Hung
Lohtia, Rajinder Paul
Naudascher, Eduard
Richards, Adrian F
Tzafestas, Spyros G

ENVIRONMENTAL, EARTH & MARINE SCIENCES

Atmospheric Dynamics

ALABAMA
Fowlis, William Webster

ARIZONA
Dickinson, Robert Earl
Fernando, Harindra Joseph

CALIFORNIA
Baboolal, Lal B
Chan, Kwoklong Roland
Chang, Chih-Pei
Chiu, Yam-Tsi
Chu, Pe-Cheng
Covey, Curtis Charles
Haltiner, George Joseph
Ingersoll, Andrew Perry
Kilb, Ralph Wolfgang
Marcus, Philip Stephen
Markowski, Gregory Ray
Molenkamp, Charles Richard
Schubert, Gerald
Seiff, Alvin
Simon, Richard L
Smith, Gary Richard
Street, Robert L(ynnwood)
Tag, Paul Mark
Wagner, Kit Kern
Weare, Bryan C
Williams, Roger Terry
Wurtele, Morton Gaither

COLORADO
Avery, Susan Kathryn

Balsley, Ben Burton
Baumhefner, David Paul
Diaz, Henry Frank
Dirks, Richard Allen
Elmore, Kimberly Laurence
Gage, Kenneth Seaver
Gille, John Charles
Gilman, Peter A
Haurwitz, Bernhard
Johnson, Richard Harlan
Kantha, Lakshmi
Kasahara, Akira
Kennedy, Patrick James
LeMone, Margaret Anne
Lenschow, Donald Henry
McWilliams, James Cyrus
Richmond, Arthur Dean
Snyder, Howard Arthur
Thaler, Eric Ronald
Van Zandt, Thomas Edward
Wessel, William Roy

CONNECTICUT
Brooks, Douglas Lee
Saltzman, Barry

DELAWARE
St John, Daniel Shelton

DISTRICT OF COLUMBIA
Deen, Thomas B
Kalnay, Eugenia
Kousky, Vernon E
Rao, Desiraju Bhavanarayana
Wagner, Andrew James

FLORIDA
Barcilon, Albert I
Bleck, Rainer
Chew, Frank
Clarke, Allan James
Mankin, Richard Wendell
Minzner, Raymond Arthur
Mooers, Christopher Northrup Kennard
Olson, Donald B
Pfeffer, Richard Lawrence
Ruscher, Paul H

GEORGIA
Alyea, Fred Nelson
Justus, Carl Gerald
Roper, Robert George

HAWAII
Wang, Bin

ILLINOIS
Adrian, Ronald John
Coulter, Richard Lincoln
Fultz, Dave
Kuo, Hsiao-Lan
Liu, Chao-Han
Man-Kin, Mak
Moschandreas, Demetrios J
Platzman, George William

INDIANA
Agee, Ernest M(ason)
Haines, Patrick A
Smith, Phillip J
Snow, John Thomas
Sun, Wen-Yin

IOWA
Hall, Jerry Lee

KANSAS
Braaten, David A

LOUISIANA
Rouse, Lawrence James, Jr

MAINE
Gordon, Geoffrey Arthur

MARYLAND
Adler, Robert Frederick
Baer, Ferdinand
Benton, George Stock
Conrath, Barney Jay
Diamante, John Matthew
Flasar, F Michael
Holloway, John Leith, Jr
Kitaigorodskii, Sergei Alexander
Pirraglia, Joseph A
Reeves, Robert William
Rice, Clifford Paul
Robock, Alan
Rodenhuis, David Roy
Scala, John Richard
Schemm, Charles Edward
Strobel, Darrell Fred

MASSACHUSETTS
Champion, Kenneth Stanley Warner
Denig, William Francis
Lindzen, Richard Siegmund
Nehrkorn, Thomas
Pedlosky, Joseph
Rosen, Richard David
Salah, Joseph E
Salstein, David A
Shapiro, Ralph
Spiegel, Stanley Lawrence

Stone, Peter Hunter

MICHIGAN
Banks, Peter Morgan
Boyd, John Philip
Lalas, Demetrius P

MINNESOTA
Pitcher, Eric John

MISSOURI
Lemon, Leslie Roy
Moore, James Thomas

NEBRASKA
Verma, Shashi Bhushan
Wehrbein, William Mead

NEVADA
Telford, James Wardrop

NEW HAMPSHIRE
Clark, Ronald Rogers

NEW JERSEY
Cheremisinoff, Paul N
Hayashi, Yoshikazu
Held, Isaac Meyer
Lau, Ngar-Cheung
Lewellen, William Stephen
Lipps, Frank B
Mahlman, Jerry David
Manabe, Syukuro
Mellor, George Lincoln, Jr
Oort, Abraham H
Ross, Bruce Brian
Smagorinsky, Joseph
Wetherald, Richard Tryon
Williams, Gareth Pierce

NEW MEXICO
Argo, Paul Emmett
Bleiweiss, Max Phillip
Campbell, John Raymond
Malone, Robert Charles
Orgill, Montie M
Rachele, Henry
Rokop, Donald J
Yamada, Tetsuji

NEW YORK
Bosart, Lance Frank
Cane, Mark A
Gross, Stanley H
Halitsky, James
Hameed, Sultan
Rao, Samohineeveesu Trivikrama
Rind, David Harold
Tenenbaum, Joel
Wang, Lin-Shu

NORTH CAROLINA
Bach, Walther Debele, Jr
Saucier, Walter Joseph
Vukovich, Fred Matthew

NORTH DAKOTA
Grainger, Cedric Anthony

OHIO
Burggraf, Odus R
Rayner, John Norman
Tuan, Tai-Fu

OKLAHOMA
Coffman, Harold H
Davies-Jones, Robert Peter
Hane, Carl Edward
Lilly, Douglas Keith
Sasaki, Yoshi Kazu

OREGON
Dews, Edmund

PENNSYLVANIA
Albrecht, Bruce Allen
Dutton, John Altnow
Wyngaard, John C

SOUTH CAROLINA
Larsen, Miguel Folkmar

TENNESSEE
Dobosy, Ronald Joseph
Schnelle, K(arl) B(enjamin), Jr
Skinner, George T

TEXAS
Brundidge, Kenneth Cloud
Chang, Ping
Curtis, Doris Malkin
Das, Phanindramohan
Few, Arthur Allen
Gillette, Kevin Keith
Johnson, Francis Severin
Maas, Stephen Joseph

UTAH
Geisler, John Edmund
Liou, Kuo-Nan
Long, David G
Paegle, Julia Nogues

VIRGINIA
Curtin, Brian Thomas
Grose, William Lyman
Spence, Thomas Wayne

WASHINGTON
Baldwin, Mark Phillip
Brown, Robert Alan
Clark, Kenneth Courtright
Connell, James Roger
Delisi, Donald Paul
Dorward-King, Elaine Jay
Harrison, Don Edmunds
Hernandez, Gonzalo J
Holton, James R
Overland, James Edward
Rhines, Peter Broomell
Sarachik, Edward S
Tung, Ka-Kit

WISCONSIN
Horn, Lyle Henry
Houghton, David Drew

WYOMING
Marwitz, John D

ONTARIO
Barrie, Leonard Arthur
Cho, Han-Ru
Peltier, William Richard
Shepherd, Gordon Greeley
Wiebe, H Allan

QUEBEC
Derome, Jacques Florian
Mysak, Lawrence Alexander

OTHER COUNTRIES
Subrahmanyam, D
Tolstoy, Ivan

Atmospheric Chemistry & Physics

ALABAMA
Abbas, Mian Mohammad
Anderson, Bernard Jeffrey
Davis, John Moulton
Helminger, Paul Andrew
Kassner, James Lyle, Jr
Meagher, James Francis
Stephens, James Briscoe
Stephens, Timothy Lee
Williamson, Ashley Deas

ALASKA
Jayaweera, Kolf
Ohtake, Takeshi
Shaw, Glenn Edmond
Wentink, Tunis, Jr

ARIZONA
Brault, James William
Herbert, Floyd Leigh
Hood, Lonnie Lamar
Jackson, Ray Dean
Krider, Edmund Philip
Layton, Richard Gary
Lee, Richard Norman
Lunine, Jonathan Irving
Nordstrom, Brian Hoyt
O'Brien, Keran
Pierce, Austin Keith
Shemansky, Donald Eugene
Twomey, Sean Andrew
Young, Kenneth Christie

ARKANSAS
Berger, Jerry Eugene
Korfmacher, Walter Averill
Richardson, Charles Bonner
Steele, Kenneth F

CALIFORNIA
Allen, Charles Freeman
Anderson, Kinsey Amor
Andre, Michael Paul
Appleby, John Frederick
Ashburn, Edward V
Atkinson, Roger
Baboolal, Lal B
Baker, Neal Kenton
Beer, Reinhard
Bonner, Norman Andrew
Borucki, William Joseph
Byers, Horace Robert
Cahill, Thomas A
Cain, William F
Canfield, Richard Charles
Cass, Glen R
Chahine, Moustafa Toufic
Chakrabarti, Supriya
Chan, Kwoklong Roland
Chiu, Yam-Tsi
Christensen, Andrew Brent
Cicerone, Ralph John
Collis, Ronald Thomas
Colovos, George
Cooper, Alfred William Madison
Coulson, Kinsell Leroy
Cruikshank, Dale Paul
Daisey, Joan M

Dowling, Jerome M
Eatough, Norman L
Ellis, Elizabeth Carol
Farber, Robert James
Farmer, Crofton Bernard
Feynman, Joan
Friedlander, Sheldon K
Gibbons, Mathew Gerald
Gilmore, Forrest Richard
Gordon, Robert Julian
Grossman, Jack Joseph
Hall, Freeman Franklin, Jr
Hamlin, Daniel Allen
Hansen, Anthony David Anders
Hardy, Kenneth Reginald
Harris, Kent Karren
Hill, Robert Dickson
Hoffmann, Michael Robert
Jacobs, Michael Moises
Jaklevic, Joseph Michael
Judge, Roger John Richard
Kalathil, James Sakaria
Kilb, Ralph Wolfgang
Landman, Donald Alan
Lange, Rolf
Larmore, Lewis
Leite, Richard Joseph
Lieberman, Alvin
Lipeles, Martin
Loos, Hendricus G
Mahadevan, Parameswar
Manalis, Melvyn S
Martin, L(aurence) Robbin
Mathews, Larry Arthur
Menzies, Robert Thomas
Moe, Mildred Minasian
Moe, Osborne Kenneth
Molenkamp, Charles Richard
Myers, Benjamin Franklin, Jr
Nero, Anthony V, Jr
Okumura, Mitchio
Old, Thomas Eugene
Paige, David A
Pang, Kevin Dit Kwan
Park, Chul
Park, Chung Gun
Penner, Joyce Elaine
Peterson, Kendall Robert
Pitts, James Ninde, Jr
Poppoff, Ilia George
Prag, Arthur Barry
Pueschel, Rudolf Franz
Randall, Charles McWilliams
Richards, L(orenzo) Willard
Rognlien, Thomas Dale
Rosen, Leonard Craig
Rowland, F Sherwood
Salanave, Leon Edward
Schwall, Richard Joseph
Sentman, Davis Daniel
Sharp, Richard Dana
Simon, Richard L
Smith, Charles Francis, Jr
Straus, Joe Melvin
Swenson, Gary Russell
Tag, Paul Mark
Tanenbaum, Basil Samuel
Thorne, Richard Mansergh
Toon, Owen Brian
Topol, Leo Eli
Turco, Richard Peter
Underwood, James Henry
Vedder, James Forrest
Vernazza, Jorge Enrique
Voss, Henry David
Walton, John Joseph
Whitten, Robert Craig, Jr
Winer, Arthur Melvyn
Wuebbles, Donald J
Young, Louise Gray

COLORADO
Amme, Robert Clyde
Anderson, Larry Gene
Balsley, Ben Burton
Boatman, Joe Francis
Bushnell, Robert Hempstead
Caracena, Fernando
Carbone, Richard Edward
Chappell, Charles Franklin
Clark, Wallace Lee
Cotton, William Reuben
Cox, Stephen Kent
Curtis, George William
Dana, Robert Watson
Davies, Kenneth
Derr, Vernon Ellsworth
Durham, Michael Dean
Dye, James Eugene
Eddy, John Allen
Edwards, Harry Wallace
Foote, Garvin Brant
Frisch, Alfred Shelby
Fritz, Richard Blair
Gage, Kenneth Seaver
Gille, John Charles
Goldan, Paul David
Hofmann, David John
Holzer, Thomas Edward
Hord, Charles W
Howard, Carleton James
Johnson, Richard Harlan
Joselyn, Jo Ann Cram
Kennedy, Patrick James

Kiehl, Jeffrey Theodore
Kleis, William Delong
Knight, Charles Alfred
Lenschow, Donald Henry
Lincoln, Jeannette Virginia
Little, Charles Gordon
Lodge, James Piatt, Jr
Mankin, William Gray
Martell, Edward A
Middleton, Paulette Bauer
Morgan, Wm Lowell
Murcray, David Guy
Murcray, Frank James
Neff, William David
Peterson, James T
Peterson, Vern Leroy
Pike, Julian M
Post, Madison John
Querfeld, Charles William
Richmond, Arthur Dean
Roberts, Walter Orr
Rosinski, Jan
Rufenach, Clifford L
Rush, Charles Merle
Ruttenberg, Stanley
Schonbeck, Niels Daniel
Schwiesow, Ronald Lee
Sinclair, Peter C
Solomon, Susan
Stedman, Donald Hugh
Thomas, Gary E
Van Valin, Charles Carroll
Wellman, Dennis Lee
Westwater, Edgeworth Rupert

CONNECTICUT
Cipriano, Ramon John
Poultney, Sherman King
Robinson, George David

DELAWARE
Glasgow, Louis Charles
Pierrard, John Martin

DISTRICT OF COLUMBIA
Apruzese, John Patrick
Carrigan, Richard Alfred
Fitzgerald, James W
Hawkins, Gerald Stanley
Ohring, George
Picone, J Michael
Rosenberg, Norman J
Summers, Michael Earl
Tarpley, Jerald Dan
Uman, Myron F
Weinreb, Michael Philip

FLORIDA
Allen, Eric Raymond
Block, David L
Heaps, Melvin George
Hill, Frank B(ruce)
Killinger, Dennis K
Llewellyn, Ralph A
Minzner, Raymond Arthur
Myerson, Albert Leon
Ostlund, H Gote
Piotrowicz, Stephen R
Reynolds, George William, Jr
Ruscher, Paul H
Winchester, John W
Windsor, John Golay, Jr

GEORGIA
Chameides, William Lloyd
Justus, Carl Gerald
Patterson, Edward Matthew
Wine, Paul Harris

HAWAII
Fullerton, Charles Michael

IDAHO
Farwell, Sherry Owen
Mills, James Ignatius

ILLINOIS
Beard, Kenneth Van Kirke
Birchfield, Gene Edward
Braham, Roscoe Riley, Jr
Cook, David Robert
Coulter, Richard Lincoln
Holt, Ben Dance
Kraakevik, James Henry
Murphy, Thomas Joseph
Semonin, Richard Gerard
Sheft, Irving
Wesely, Marvin Larry

INDIANA
Pribush, Robert A

IOWA
Carmichael, Gregory Richard
Yarger, Douglas Neal

KENTUCKY
Ray, Asit Kumar

LOUISIANA
Broeg, Charles Burton
Klasinc, Leo

MAINE
Blau, Henry Hess, Jr
Cronn, Dagmar Rais

MARYLAND
Aikin, Arthur Coldren
Bass, Arnold Marvin
Bauer, Ernest
Benesch, William Milton
Brace, Larry Harold
Brasunas, John Charles
Chandra, Sushil
Einaudi, Franco
Ellingson, Robert George
Epstein, Gabriel Leo
Feldman, Paul Donald
Fifer, Robert Alan
Goldberg, Richard Aran
Gordon, Glen Everett
Hanel, Rudolf A
Hedin, Alan Edgar
Herrero, Federico Antonio
Hess, Wilmot Norton
Hicks, Bruce Boundy
Hudson, Robert Douglas
Joiner, R(eginald) Gracen
Julienne, Paul Sebastian
Kaye, Jack Alan
Kostiuk, Theodor
Krueger, Arlin James
Kurylo, Michael John, III
Kyle, Herbert Lee
Larson, Reginald Einar
Lufkin, Daniel Harlow
McPeters, Richard Douglas
Martinez, Richard Isaac
Neupert, Werner Martin
Omidvar, Kazem
Potemra, Thomas Andrew
Prabhakara, Cuddapah
Quinn, Thomas Patrick
Rosenberg, Theodore Jay
Rosenfield, Joan Samour
Shettle, Eric Payson
Silver, David Martin
Stewart, Richard Willis
Stief, Louis J
Strobel, Darrell Fred
Thompson, Owen Edward
Tilford, Shelby G
Tuttle, Kenneth Lewis
Weller, Charles Stagg, Jr
Wiscombe, Warren Jackman

MASSACHUSETTS
Aarons, Jules
Birstein, Seymour J
Carlson, Herbert Christian, Jr
Champion, Kenneth Stanley Warner
Engelman, Arthur
Fitzgerald, Donald Ray
Gallagher, Charles Clifton
Glover, Kenneth Merle
Hedlund, Donald A
Huffman, Robert Eugene
Kaplan, Lewis David
Kolb, Charles Eugene, Jr
Levine, Randolph Herbert
McClatchey, Robert Alan
Markson, Ralph Joseph
Molina, Mario Jose
Morris, Robert Alan
Nehrkorn, Thomas
Newell, Reginald Edward
Paulsen, Duane E
Peace, George Earl, Jr
Penndorf, Rudolf
Person, James Carl
Picard, Richard Henry
Ratner, Michael Ira
Robertson, David C
Rothman, Laurence Sidney
Sagalyn, Rita C
Stakutis, Vincent John
Swider, William, Jr
Toman, Kurt
Tomljanovich, Nicholas Matthew
Traub, Wesley Arthur
Victor, George A
Viggiano, Albert
Volz, Frederic Ernst
Whitney, Cynthia Kolb
Winick, Jeremy Ross
Withbroe, George Lund
Wofsy, Steven Charles
Yamartino, Robert J

MICHIGAN
Atreya, Sushil Kumar
Burek, Anthony John
Chang, Tai Yup
Dingle, Albert Nelson
Donahue, Thomas Michael
Drayson, Sydney Roland
Francisco, Joseph Salvadore, Jr
Gorse, Robert August, Jr
Japar, Steven Martin
Kelly, Nelson Allen
Klimisch, Richard L
Kuhn, William R
Nagy, Andrew F
Reck, Ruth Annette
Rogers, Jerry Dale
Sharp, William Edward, III

Sloane, Christine Scheid
Wolff, George Thomas

MINNESOTA
Freier, George David
Mauersberger, Konrad

MISSISSIPPI
Rundel, Robert Dean

MISSOURI
Bragg, Susan Lynn
Carstens, John C
DuBroff, Richard Edward
Hagen, Donald E
Macias, Edward S
Nelson, George Driver
Pallmann, Albert J
Plummer, Patricia Lynne Moore
Podzimek, Josef
Vaughan, William Mace
Whitefield, Philip Douglas

MONTANA
Latham, Don Jay
Peavy, Howard Sidney

NEBRASKA
Billesbach, David P
Verma, Shashi Bhushan

NEVADA
Hallett, John
Hudson, James Gary
Long, Alexis Boris
Lowenthal, Douglas H
Pierson, William R
Pitter, Richard Leon
Potter, John Fred
Scott, William Taussig
Tanner, Roger Lee
Taylor, George Evans, Jr
Telford, James Wardrop

NEW HAMPSHIRE
Hollweg, Joseph Vincent
Taffe, William John

NEW JERSEY
Black, James Francis
Brenner, Douglas Milton
Graedel, Thomas Eldon
Jeck, Richard Kahr
Lau, Ngar-Cheung
Lioy, Paul James
McAfee, Kenneth Bailey, Jr
Mahlman, Jerry David
Manabe, Syukuro
Martin, John David
Ramaswamy, Venkatachalam
Weschler, Charles John

NEW MEXICO
Bleiweiss, Max Phillip
Brook, Marx
Chylek, Petr
Cunningham, Paul Thomas
Fowler, Malcolm McFarland
Goldman, S Robert
Guthals, Paul Robert
Jones, James Jordan
Karl, Robert Raymond, Jr
Maier, William Bryan, II
Mason, Allen Smith
Miller, August
Morse, Fred A
Nogar, Nicholas Stephen
Orgill, Montie M
Peek, H(arry) Milton
Popp, Carl John
Porch, William Morgan
Sedlacek, William Adam
Shively, Frank Thomas
Snider, Donald Edward
Streit, Gerald Edward
Tisone, Gary C
Wilkening, Marvin H
Winn, William Paul
Yamada, Tetsuji
Zinn, John

NEW YORK
Akers, Charles Kenton
Blanchard, Duncan Cromwell
Boeck, William Louis
Cess, Robert D
De Zafra, Robert Lee
Farley, Donald T, Jr
Feit, Julius
Geehern, Margaret Kennedy
Greenebaum, Michael
Hameed, Sultan
Hindman, Edward Evans
Hirsh, Merle Norman
Hopke, Philip Karl
Hopke, Philip Karl
Janik, Gerald S
Jiusto, James E
Jordan, David M
Keesee, Robert George
Kim, Jai Soo
Kumai, Motoi
Lovett, Gary Martin
McLaren, Eugene Herbert

Newton, Chester Whittier
Parton, William Julian, Jr
Peterson, James T
Peterson, Vern Leroy
Pielke, Roger Alvin
Purdom, James Francis Whitehurst
Reiter, Elmar Rudolf
Riehl, Herbert
Rottman, Gary James
Schneider, Stephen Henry
Schwiesow, Ronald Lee
Serafin, Robert Joseph
Thaler, Eric Ronald
Thompson, Starley Lee
Trenberth, Kevin Edward
Veal, Donald L
Vonder Haar, Thomas Henry
Washington, Warren Morton
Wellman, Dennis Lee
Williams, Merlin Charles
Wooldridge, Gene Lysle

CONNECTICUT
Aylor, Donald Earl
Cipriano, Ramon John
Friedman, Don Gene
Herron, Michael Myrl
Jackson, C(harles) Ian
Kawaters, Woody H
Monahan, Edward Charles
Newman, Steven Barry
Pandolfo, Joseph P
Paskausky, David Frank
Russo, John A, Jr
Waggoner, Paul Edward

DELAWARE
Brazel, Anthony James
Kalkstein, Laurence Saul
Mather, John Russell

DISTRICT OF COLUMBIA
Ausubel, Jesse Huntley
Bergman, Kenneth Harris
Bierly, Eugene Wendell
Brodrick, Harold James, Jr
Brueckner, Guenter Erich
De Leonibus, Pasquale S
Ellis, William Rufus
Epstein, Edward Selig
Fein, Jay Sheldon
Greenfield, Richard Sherman
Hecht, Alan David
Kalnay, Eugenia
Lehman, Richard Lawrence
Long, Paul Eastwood, Jr
McGinnis, David Franklin, Jr
Ohring, George
Perry, John Stephen
Quiroz, Roderick S
Rosenberg, Norman J
Summers, Michael Earl
Tarpley, Jerald Dan
Wagner, Andrew James
Ward, John Henry
White, Robert M

FLORIDA
Baum, Werner A
Brandli, Henry William
Clark, John F
Dubach, Leland L
Emiliani, Cesare
Fuelberg, Henry Ernest
Gerber, John Francis
Gleeson, Thomas Alexander
Haines, Donald Arthur
Harris, D Lee
Heaps, Melvin George
Hiser, Homer Wendell
Jamrich, John Xavier
Jensen, Clayton Everett
Jordan, Charles Lemuel
Kelso, John Morris
Krishnamurti, Tiruvalam N
La Seur, Noel Edwin
Lhermitte, Roger M
Long, Robert Radcliffe
Lundgren, Dale A(llen)
McFadden, James Douglas
Mach, William Howard
Martsolf, J David
Mecklenburg, Roy Albert
Minzner, Raymond Arthur
Neuberger, Hans Hermann
Newstein, Herman
O'Brien, James J
Pike, Arthur Clausen
Ray, Peter Sawin
Rosenthal, Stanley Lawrence
Ruscher, Paul H
Sheets, Robert Chester
Sinclair, Thomas Russell
Stephens, Jesse Jerald
Stuart, David W
Weinberg, Jerry L

GEORGIA
Arbogast, Richard Terrance
Blanton, Jackson Orin
Cooper, Charles Dewey
Evans, William Buell
Grams, Gerald William
Justus, Carl Gerald

Kiang, Chia Szu
Meentemeyer, Vernon George
Parker, Albert John
Peake, Thaddeus Andrew, III
Roper, Robert George
Taylor, Howard Edward

HAWAII
Chiu, Wan-Cheng
Chuan, Raymond Lu-Po
Murakami, Takio
Sadler, James C
Wang, Bin
Wyrtki, Klaus

IDAHO
Wright, James Louis

ILLINOIS
Ackerman, Bernice
Ardis, Colby V, Jr
Beard, Kenneth Van Kirke
Changnon, Stanley A, Jr
Changnon, Stanley Alcide, Jr
Cook, David Robert
Coulter, Richard Lincoln
Dailey, Paul William, Jr
Dordick, Isadore
Fisher, Perry Wright
Frenzen, Paul
Fujita, Tetsuya T
Handler, Paul
Muehling, Arthur J
Muller, Dietrich
Sasamori, Takashi
Semonin, Richard Gerard
Shannon, Jack Dee
Smith, Leslie Garrett
Staver, Allen Ernest
Stout, Glenn Emanuel
Webb, Harold Donivan
Wendland, Wayne Marcel

INDIANA
Gommel, William Raymond
Newman, James Edward
Smith, Phillip J
Vincent, Dayton George

IOWA
Georgakakos, Konstantine P
Hatfield, Jerry Lee
Shaw, Robert Harold
Waite, Paul J

KANSAS
Bark, Laurence Dean
Braaten, David A
Eagleman, Joe R
Hagen, Lawrence J
Welch, Stephen Melwood

KENTUCKY
Crawford, Nicholas Charles
Stewart, Robert Earl

LOUISIANA
Dessauer, Herbert Clay
Glenn, Alfred Hill
Hsu, Shih-Ang
Muller, Robert Albert

MAINE
Cronn, Dagmar Rais
Dethier, Bernard Emile
Gordon, Geoffrey Arthur
Snow, Joseph William
Snyder, Arnold Lee, Jr

MARYLAND
Abbey, Robert Fred, Jr
Acheson, Donald Theodore
Adler, Robert Frederick
Atlas, David
Baer, Ferdinand
Baer, Ledolph
Bandeen, William Reid
Barrientos, Celso Saquitan
Bauer, Ernest
Benton, George Stock
Berning, Warren Walt
Carlon, Hugh Robert
Carlstead, Edward Meredith
Cavalieri, Donald Joseph
Chang, Alfred Tieh-Chun
Comiso, Josefino Cacas
Conrath, Barney Jay
Cressman, George Parmley
Deaven, Dennis George
Ellingson, Robert George
Elliott, William Paul
Endal, Andrew Samson
Faller, Alan Judson
Friday, Elbert W, Jr
Fritz, Sigmund
Gilliland, Ronald Lynn
Glahn, Harry Robert
Goldberg, Richard Aran
Greenstone, Reynold
Gruber, Arnold
Gutman, George Garik
Hasler, Arthur Frederick
Heffter, Jerome L
Heppner, James P

Heymsfield, Gerald M
Holland, Joshua Zalman
Hornstein, John Stanley
Huang, Joseph Chi Kan
Hudlow, Michael Dale
Jacobs, Woodrow Cooper
Johnson, David Simonds
Johnson, Harry McClure
Kaiser, Jack Allen Charles
King, Michael Dumont
Kyle, Herbert Lee
Lavoie, Ronald Leonard
Leight, Walter Gilbert
McBryde, F(elix) Webster
McGovern, Wayne Ernest
Maier, Eugene Jacob Rudolph
Matthews, David LeSueur
Means, Lynn L
Mecherikunnel, Ann Pottanat
Meyer, James Henry
Miller, John Frederick
Mirabito, John A
Mogil, H Michael
Parkinson, Claire Lucille
Price, John Charles
Pritchard, Donald William
Rango, Albert
Rao, P Krishna
Rasmusson, Eugene Martin
Reeves, Robert William
Renne, David Smith
Robock, Alan
Rosenfield, Joan Samour
Rubin, Morton Joseph
Ruff, Irwin S
Salomonson, Vincent Victor
Schmugge, Thomas Joseph
Seguin, Ward Raymond
Shettle, Eric Payson
Simpson, Joanne
Stackpole, John Duke
Strong, Alan Earl
Tepper, Morris
Thompson, Owen Edward
Tracton, Martin Steven
Uhart, Michael Scott
Vernekar, Anandu Devarao
Vestal, Claude Kendrick
Wang, Ting-I
Young, George Anthony
Zankel, Kenneth L

MASSACHUSETTS
Anstey, Robert L
Austin, James Murdoch
Austin, Pauline Morrow
Barnes, Arnold Appleton, Jr
Bass, Arthur
Beckerle, John C
Bell, Barbara
Bradley, Raymond Stuart
Brooks, Edward Morgan
Carlson, Herbert Christian, Jr
Chisholm, Donald Alexander
Clough, Shepard Anthony
Conover, John Hoagland
Cunningham, Robert M
Darling, Eugene Merrill, Jr
Delvaille, John Paul
Entekhabi, Dara
Fougere, Paul Francis
Gray, Douglas Carmon
Hallgren, Richard E
Hanna, Steven Rogers
Harris, Miles Fitzgerald
Hummel, John Richard
Kaplan, Lewis David
Kornfield, Jack I
Lai, Shu Tim
Lorenz, Edward Norton
McClatchey, Robert Alan
McMenamin, Mark Allan
Minsinger, William Elliot
Myers, Robert Frederick
Nehrkorn, Thomas
Newell, Reginald Edward
Norment, Hillyer Gavin
Olmez, Ilhan
Penndorf, Rudolf
Ramakrishna, Kilaparti
Rork, Eugene Wallace
Salah, Joseph E
Sanders, Frederick
Schneider, Edwin Kahn
Sellers, Francis Bachman
Spengler, Kenneth C
Staelin, David Hudson
Stergis, Christos George
Stump, Robert
Twitchell, Paul F
Wheeler, Ned Brent
Wilson, F Wesley, Jr

MICHIGAN
Atreya, Sushil Kumar
Dingle, Albert Nelson
Gates, David Murray
Howe, George Marvel
Kuhn, William R
Kummler, Ralph H
Mason, Conrad Jerome
Menard, Albert Robert, III
Portman, Donald James
Ruffner, James Alan

Sloane, Christine Scheid
Wolff, George Thomas
Yerg, Donald G

MINNESOTA
Anderson, Frances Jean
Baker, Donald Gardner
Belmont, Arthur David
Dahlberg, Duane Arlen
Lawrence, Donald Buermann
McClure, Benjamin Thompson

MISSISSIPPI
Cross, Ralph Donald
Leese, John Albert

MISSOURI
Cowherd, Chatten, Jr
Darkow, Grant Lyle
Decker, Wayne Leroy
Johnson, Douglas William
Kung, Ernest Chen-Tsun
Lemon, Leslie Roy
Lin, Yeong-Jer
McNulty, Richard Paul
Moore, James Thomas
Podzimek, Josef
Rao, Gandikota V
Schaefer, Joseph Thomas

MONTANA
Caprio, Joseph Michael

NEBRASKA
Blad, Blaine L
Hahn, George LeRoy
Lawson, Merlin Paul
Wehrbein, William Mead

NEVADA
Flueck, John A
Gilbert, Dewayne Everett
Hoffer, Thomas Edward
Hudson, James Gary
Klieforth, Harold Ernest
Krenkel, Peter Ashton
Leipper, Dale F
Murino, Clifford John
Pitter, Richard Leon
Randerson, Darryl

NEW HAMPSHIRE
Bernstein, Abram Bernard
Bilello, Michael Anthony
Crane, Robert Kendall
Federer, C Anthony
Jastrow, Robert
Perovich, Donald Kole
Snively, Leslie O

NEW JERSEY
Becken, Eugene D
Broccoli, Anthony Joseph
Bryan, Kirk, (Jr)
Fels, Stephen Brook
Hamilton, Kevin
Havens, Abram Vaughn
House, Edward Holcombe
Hsu, Chin-Fei
Jeck, Richard Kahr
Kurihara, Yoshio
Lau, Ngar-Cheung
Letcher, David Wayne
Lipps, Frank B
McCabe, Gregory James, Jr
Manabe, Syukuro
Miyakoda, Kikuro
Ramaswamy, Venkatachalam
Spar, Jerome
Spelman, Michael John
Stouffer, Ronald Jay
Wetherald, Richard Tryon

NEW MEXICO
Barr, Sumner
Breiland, John Gustavson
Chylek, Petr
Cionco, Ronald Martin
Clements, William Earl
Davis, Cecil Gilbert
Gerstl, Siegfried Adolf Wilhelm
Hoard, Donald Ellsworth
Kelly, Robert Emmett
Mason, Allen Smith
Mokler, Brian Victor
Novlan, David John
Orgill, Montie M
Raymond, David James
Reed, Jack Wilson
Reifsnyder, William Edward
Schery, Stephen Dale
Yamada, Tetsuji

NEW YORK
Blum, Samuel Emil
Cess, Robert D
Chermack, Eugene E A
Czapski, Ulrich Hans
Demerjian, Kenneth Leo
Dubins, Mortimer Ira
Egan, Walter George
Geer, Ira W
Geller, Marvin Alan
Gross, Stanley H

Atmospheric Sciences, General (cont)

Hagfors, Tor
Halitsky, James
Hameed, Sultan
Havens, James Meryle
Herrington, Lee Pierce
Hindman, Edward Evans
Hubbard, John Edward
Jiusto, James E
Krey, Philip W
Laghari, Javaid Rosoolbux
Millman, George Harold
Murphy, Kenneth Robert
Myer, Glenn Evans
Neumann, Paul Gerhard
Oglesby, Ray Thurmond
Pack, Albert Boyd
Pierson, Willard James, Jr
Ram, Michael
Rampino, Michael Robert
Rao, Samohineeveesu Trivikrama
Rind, David Harold
Sato, Makiko
Simmons, Harry Dady, Jr
Stolov, Harold L
Travis, Larry Dean
Varanasi, Prasad
Vickers, William W

NORTH CAROLINA
Alpert, Leo
Arya, Satya Pal Singh
Binkowski, Francis Stanley
Brotak, Edward Allen
Ching, Jason Kwock Sung
Crutcher, Harold L
Decker, Clifford Earl, Jr
Dickerson, Willard Addison
Droessler, Earl George
Dubach, Harold William
Fox, Donald Lee
Hadeen, Kenneth Doyle
Haggard, William Henry
Haynie, Fred Hollis
Heck, Walter Webb
Karl, Thomas Richard
Kirk, Wilber Wolfe
Pooler, Francis, Jr
Ramage, Colin Stokes
Robinson, Peter John
Rodney, Earnest Abram
Saucier, Walter Joseph
Saxena, Vinod Kumar
Sethuraman, S
Shreffler, Jack Henry
Steila, Donald
Swift, Lloyd Wesley, Jr
Wollin, Goesta

NORTH DAKOTA
Enz, John Walter
Grainger, Cedric Anthony

OHIO
Arnfield, Anthony John
Brazee, Ross D
Cholak, Jacob
Colinvaux, Paul Alfred
James, Philip Benjamin
Rayner, John Norman
Sticksel, Philip Rice
Weiss, Harold Samuel
Willeke, Klaus

OKLAHOMA
Bluestein, Howard Bruce
Brock, Fred Vincent
Brown, Rodger Alan
Burgess, Donald Wayne
Carr, Meg Brady
Doswell, Charles Arthur, III
Doviak, Richard J
Eddy, George Amos
Hane, Carl Edward
Kessler, Edwin, 3rd
Kimpel, James Froome
Taylor, William L

OREGON
Coakley, James Alexander, Jr
Decker, Fred William
Gates, William Lawrence
Murphy, Allan Hunt
Neilson, Ronald Price
Niem, Wendy Adams
Paulson, Clayton Arvid
Peterson, Ernest W
Quinn, William Hewes
Wolf, Marvin Abraham

PENNSYLVANIA
Albrecht, Bruce Allen
Baurer, Theodore
Blackadar, Alfred Kimball
Cahir, John Joseph
Carlson, Toby Nahum
Curry, Judith Ann
DeWalle, David Russell
Guinan, Edward F
Hosler, Charles Luther, Jr
Jaggard, Dwight Lincoln

Justham, Stephen Alton
Kreitzberg, Carl William
Laws, Kenneth Lee
Lee, John Denis
McFarland, Mack
Mitchell, John Douglas
Mitchell, John Murray, Jr
Olivero, John Joseph, Jr
Russo, Joseph Martin
Sion, Edward Michael
Spalding, George Robert
Tang, Chung-Muh
Webster, Peter John

RHODE ISLAND
Godshall, Fredric Allen
Merrill, John T
Webb, Thompson, III

SOUTH CAROLINA
Crawford, Todd V
Davis, Francis Kaye
Duncan, Lewis Mannan
Gentry, Robert Cecil
Hofstetter, Kenneth John
Ulbrich, Carlton Wilbur
Weber, Allen Howard

SOUTH DAKOTA
Helsdon, John H, Jr
Orville, Harold Duvall
Schleusener, Richard A
Winker, James A(nthony)

TENNESSEE
Amundsen, Clifford C
Baldocchi, Dennis D
Blasing, Terence Jack
Gifford, Franklin Andrew, Jr
Holdeman, Jonas Tillman, Jr
Hosker, Rayford Peter, Jr
Kohland, William Francis
Murphy, Brian Donal

TEXAS
Brooks, David Arthur
Brundidge, Kenneth Cloud
Chamberlain, Joseph Wyan
Clark, Robert Alfred
Djuric, Dusan
Driscoll, Dennis Michael
Few, Arthur Allen
Franceschini, Guy Arthur
Frank, Neil LaVerne
Freeman, John Clinton, Jr
Goldman, Joseph L
Greene, Donald Miller
Griffiths, John Frederick
Haragan, Donald Robert
Harrison, Wilks Douglas
Hasling, Jill Freeman
Hill, Thomas Westfall
Ledley, Tamara Shapiro
McDaniel, Willard Rich
Miksad, Richard Walter
Monahan, Hubert Harvey
Moser, James Howard
Moyer, Vance Edwards
Pitts, David Eugene
Runnels, Robert Clayton
Scoggins, James R
Steiner, Jean Louise
Thompson, Aylmer Henry
Tinsley, Brian Alfred
Wagner, Norman Keith
Webb, Willis Lee

UTAH
Allen, Richard Glen
Astling, Elford George
Baker, Kay Dayne
Cramer, Harrison Emery
Dickson, Don Robert
Snellman, Leonard W

VERMONT
Illick, J(ohn) Rowland

VIRGINIA
Beal, Stuart Kirkham
Benton, Duane Marshall
Bishop, Joseph Michael
Browell, Edward Vern
Chang, Chieh Chien
Cooley, Duane Stuart
Deepak, Adarsh
Farrukh, Usamah Omar
Ganguly, Suman
Garstang, Michael
Gilman, Donald Lawrence
Hart, Michael H
Jess, Edward Orland
McCormick, Michael Patrick
Melfi, Leonard Theodore, Jr
Murino, Vincent S
O'Neill, Thomas Hall Robinson
Parry, Hubert Dean
Peck, Eugene Lincoln
Rigby, Malcolm
Russell, James Madison, III
Siegel, Clifford M(yron)
Spilhaus, Athelstan Frederick
Swissler, Thomas James
Theon, John Speridon

Tolson, Robert Heath
Trizna, Dennis Benedict
Trout, Dennis Alan
White, Fred D(onald)

WASHINGTON
Badgley, Franklin Ilsley
Baldwin, Mark Phillip
Barchet, William Richard
Businger, Joost Alois
Campbell, Gaylon Sanford
Craig, Richard Ansel
Craine, Lloyd Bernard
Droppo, James Garnet, Jr
Elderkin, Charles Edwin
Fleagle, Robert Guthrie
Fremouw, Edward Joseph
Fritschen, Leo J
Fuquay, James Jenkins
Hadlock, Ronald K
Hartmann, Dennis Lee
Hilst, Glenn Rudolph
Jones, Kay H
Katz, Yale H
Laevastu, Taivo
Lansinger, John Marcus
Leovy, Conway B
Mobley, Curtis Dale
Muench, Robin Davie
Radke, Lawrence Frederick
Reed, Richard John
Roden, Gunnar Ivo
Seliga, Thomas A
Tung, Ka-Kit
Wallace, John Michael
Weiss, Richard Raymond

WEST VIRGINIA
Covey, Winton Guy, Jr

WISCONSIN
Anderson, Charles Edward
Bretherton, Francis P
Bryson, Reid Allen
Haig, Thomas O
Hastenrath, Stefan Ludwig
Hedden, Gregory Dexter
Horn, Lyle Henry
Houghton, David Drew
Johnson, Donald R
Kuhn, Peter Mouat
Martin, David William
Miller, David Hewitt
Mitchell, Val Leonard
Moran, Joseph Michael
Nealson, Kenneth Henry
Ragotzkie, Robert Austin
Reinsel, Gregory Charles
Sharkey, Thomas D
Sikdar, Dhirendra N
Stearns, Charles R
Suchman, David
Suomi, Verner Edward
Tanner, Champ Bean
Wahl, Eberhard Wilhelm
Wang, Pao-Kuan
Young, John A

WYOMING
Marwitz, John D

PUERTO RICO
McDowell, Dawson Clayborn

ALBERTA
Fletcher, Roy Jackson
Giovinetto, Mario Bartolome
Goyer, Guy Gaston
Hage, Keith Donald

BRITISH COLUMBIA
Davis, Roderick Leigh
Humphries, Robert Gordon
Mathewes, Rolf Walter
Oke, Timothy Richard
Thyer, Norman Harold
Walker, Edward Robert
Wilson, Richard Garth

NEW BRUNSWICK
Hawkes, Robert Lewis

NOVA SCOTIA
Elliott, James Arthur

ONTARIO
Anlauf, Kurt Guenther
Baier, Wolfgang
Barrie, Leonard Arthur
Bhartendu, Srivastava
Blevis, Bertram Charles
Boville, Byron Walter
Cho, Han-Ru
Donelan, Mark Anthony
Ganza, Kresimir Peter
Gillespie, Terry James
Godson, Warren Lehman
Iribarne, Julio Victor
McBean, Edward A
McConnell, John Charles
MacHattie, Leslie Blake
McTaggart-Cowan, Patrick Duncan
Moorcroft, Donald Ross
Moroz, William James

Munn, Robert Edward
Robertson, George Wilber
Sanderson, Henry Preston
Strachan, William Michael John
Thomas, Morley Keith
Wiebe, H Allan

QUEBEC
Langleben, Manuel Phillip
Leighton, Henry George
Paulin, Gaston (Ludger)
Phan, Cong Luan
Rogers, Roddy R

SASKATCHEWAN
Ripley, Earle Allison

OTHER COUNTRIES
Adem, Julian
Blackburn, Jacob Floyd
Box, Michael Allister
Bray, John Roger
Bridgman, Howard Allen
Kyle, Thomas Gail
Levin, Zev
Lloyd, Christopher Raymond
Sakamoto, Clarence M
Stoeber, Werner
Williams, Michael Maurice Rudolph
Wilson, Cynthia
Zdunkowski, Wilford G

Earth Sciences, General

ALABAMA
Canis, Wayne F
Costes, Nicholas Constantine

ALASKA
Brown, Neal B

ARIZONA
Baker, Victor Richard
Baron, William Robert
Bates, Charles Carpenter
Chapman, Clark Russell
Chronic, Halka
Damon, Paul Edward
Dean, Jeffrey Stewart
Dodge, Theodore A
Dorn, Ronald I
Fellows, Larry Dean
Hartmann, William K
Hutchinson, Charles F
Jones, David Lloyd
Kamilli, Robert Joseph
Loomis, Frederick B
Lucchitta, Baerbel Koesters
Mead, James Irving
Moke, Charles Burdette
Parrish, Judith Totman
Wallace, Terry Charles, Jr
Wierenga, Peter J

ARKANSAS
Sears, Derek William George

CALIFORNIA
Addicott, Warren O
Austin, Steven Arthur
Bailey, Roy Alden
Bakun, William Henry
Barbat, William Franklin
Berg, Henry Clay
Bernstein, Ralph
Bertoldi, Gilbert LeRoy
Bohlen, Steven Ralph
Brahma, Chandra Sekhar
Campbell, Kenneth Eugene, Jr
Carlson, Carl E
Carr, Michael H
Carrigan, Charles Roger
Carsola, Alfred James
Castle, Robert O
Chesnut, Dwayne A(llen)
Clague, David A
Cloud, Preston
Covey, Curtis Charles
Denton, Joan E
Drake, Robert E
Galehouse, Jon Scott
Gottlieb, Peter
Grantz, Arthur
Hanan, Barry Benton
Harris, John Michael
Heusinkveld, Myron Ellis
Hirschfeld, Sue Ellen
Hodges, Carroll Ann
Holzer, Alfred
Huber, Norman King
Isherwood, William Frank
Iyer, Hariharaiyer Mahadeva
Kasameyer, Paul William
Kennedy, Burton Mack
Koenig, James Bennett
Ku, Teh-Lung
Lay, Thorne
Leps, Thomas MacMaster
Lier, John
Lippitt, Louis
Lupton, John Edward
McHugh, Stuart Lawrence
Mack, Seymour

Mahood, Gail Ann
Mooney, Walter D
Moore, James Gregory
Morton, Douglas M
Muffler, Leroy John Patrick
Nelson, Robert M
Nemat-Nasser, Siavouche
Nielson, Jane Ellen
Nur, Amos M
Orwig, Eugene R, Jr
Paige, David A
Phoenix, David A
Piel, Kenneth Martin
Pierce, William G
Reynolds, Michael David
Rogers, Thomas Hardin
Rossbacher, Lisa Ann
Servos, Kurt
Shreve, Ronald Lee
Silver, Eli Alfred
Smith, Richard Avery
Spence, Harlan Ernest
Sternberg, Hilgard O'Reilly
Stolper, Edward Manin
Stout, Ray Bernard
Sundaram, Panchanatham N
Switzer, Paul
Sylvester, Arthur Gibbs
Tewes, Howard Allan
Tilling, Robert Ingersoll
Tombrello, Thomas Anthony, Jr
Wadsworth, William Bingham
Welch, Robin Ivor
Wells, Ronald Allen
Werth, Glenn Conrad
Williams, Alan Evan
Zumberge, James Herbert

COLORADO
Adams, Samuel S
Armbrustmacher, Theodore J
Ayler, Maynard Franklin
Barber, George Arthur
Barron, Eric James
Blair, Robert G
Bryant, Bruce Hazelton
Bryant, Donald G
Burroughs, Richard Lee
Butler, Arthur P
Coates, Donald Allen
Cole, James Channing
Crockett, Allen Bruce
Crone, Anthony Joseph
Driscoll, Richard Stark
Ellingson, Jack A
Goetz, Alexander Franklin Hermann
Hay, William Winn
Henrickson, Eiler Leonard
Herrmann, Raymond
Hills, Francis Allan
Hittelman, Allen M
Holcombe, Troy Leon
Hotchkiss, William Rouse
Illangasekare, Tissa H
MacCarthy, Patrick
Malde, Harold Edwin
Menzer, Fred J
Morrison, Roger Barron
Myers, Donald Arthur
Naeser, Charles Wilbur
Naeser, Nancy Dearien
Palmer, Allison Ralph
Pickett, George R
Pierce, Kenneth Lee
Porter, Kenneth Raymond
Raup, Robert Bruce, Jr
Rold, John W
Rowley, Peter DeWitt
Schleicher, David Lawrence
Sheridan, Douglas Maynard
Sidder, Gary Brian
Tatsumoto, Mitsunobu
Toy, Terrence J
Walker, Theodore Roscoe
Warme, John Edward

CONNECTICUT
Anderson, Carol Patricia
Buss, Leo William
Chriss, Terry Michael
Clebnik, Sherman Michael
Friedman, Don Gene
Jackson, C(harles) Ian
Patton, Peter C
Tolderlund, Douglas Stanley

DELAWARE
Benson, Richard Norman
Brazel, Anthony James
Keidel, Frederick Andrew

DISTRICT OF COLUMBIA
Adey, Walter Hamilton
Baker, D(onald) James
Carlson, Richard Walter
Chung, Riley M
Corell, Robert Walden
Davis, Robert E
Doumani, George Alexander
Geelhoed, Glenn William
Haq, Bilal U
Matson, Michael
Melickian, Gary Edward
Ostenso, Ned Allen

Prewitt, Charles Thompson
Simkin, Thomas Edward
Towe, Kenneth McCarn
Wilson, Frederick Albert

FLORIDA
Andregg, Charles Harold
Clarke, Allan James
Compton, John S
Lovejoy, Donald Walker
Maurrasse, Florentin Jean-Marie Robert
Pirkle, Fredric Lee
Sanford, Malcolm Thomas
Tanner, William Francis, Jr
Tull, James Franklin
Winters, Stephen Samuel

GEORGIA
Clark, Benjamin Cates, Jr
Dod, Bruce Douglas
Hayes, Willis B
Tan, Kim H
Whitney, James Arthur

HAWAII
Garcia, Michael Omar
Khan, Mohammad Asad
Moberly, Ralph M
Peterson, Frank Lynn
Zisk, Stanley Harris

IDAHO
Knutson, Carroll Field
Ulliman, Joseph James

ILLINOIS
Altpeter, Lawrence L, Jr
Anderson, Richard Charles
Berry, Gregory Franklin
Chao, Sherman S
Chew, Weng Cho
Flynn, John Joseph
Fraunfelter, George H
Gealy, William James
Leighton, Morris Wellman
Mann, Christian John
Masters, John Michael
Maynard, Theodore Roberts
Peck, Theodore Richard
Poulikakos, Dimos
Rodolfo, Kelvin S
Sepkoski, J(oseph) J(ohn), Jr
Trask, Charles Brian

INDIANA
Biggs, Maurice Earl
Hamburger, Michael Wile
Kim, Yeong Ell
Martin, Charles Wellington, Jr
Park, Richard Avery, IV
Rosenberg, Gary David
Votaw, Robert Barnett

IOWA
Eaton, Gordon Pryor

KANSAS
Brown, Donald M
Clark, George Richmond, II
Moore, Richard K(err)
Welch, Jerome E
White, Stephen Edward

KENTUCKY
Blackburn, John Robert
Ettensohn, Francis Robert
Hoffman, Wayne Larry
Singh, Raman J

LOUISIANA
DeLaune, Ronald D
Rogers, James Edwin
Seglund, James Arnold
Walker, Harley Jesse

MARYLAND
Barker, John L, Jr
Brown, Michael
Chang, Alfred Tieh-Chun
Chery, Donald Luke, Jr
Cohen, Steven Charles
Crampton, Janet Wert
Debuchananne, George D
Gadsby, Dwight Maxon
Garvin, James Brian
Gloersen, Per
Gutman, George Garik
Hartle, Richard Eastham
Koblinsky, Chester John
Kolstad, George Andrew
Leatherman, Stephen Parker
Leedy, Daniel Loney
McCammon, Helen Mary
Mecherikunnel, Ann Pottanat
Mielke, James Edward
Mogil, H Michael
Parkinson, Claire Lucille
Phair, George
Price, John Charles
Raff, Samuel J
Salomonson, Vincent Victor
Schulkin, Morris
Stifel, Peter Beekman
Wells, Eddie N

Wolff, Roger Glen
Wolman, Markley Gordon

MASSACHUSETTS
Ballard, Robert D
Bradley, Raymond Stuart
Greenstein, Benjamin Joel
Hager, Bradford Hoadley
Hepburn, John Christopher
Hunt, John Meacham
Lin, Jian
McNutt, Marcia Kemper
Molnar, Peter Hale
Nunes, Paul Donald
Oldale, Robert Nicholas
Ridge, John Charles
Shimizu, Nobumichi
Takahashi, Kozo
Tarkoy, Peter J
Terzaghi, Ruth Doggett
Williamson, Peter George

MICHIGAN
Blinn, Lorena Virginia
Lefebvre, Richard Harold
MacMahan, Horace Arthur, Jr
Menninga, Clarence
Ronca, Luciano Bruno
Rusch, Wilbert H, Sr
Symonds, Robert B

MINNESOTA
Beck, Warren R(andall)
Grew, Priscilla Croswell Perkins
Hey, Richard N
Hudleston, Peter John
Parker, Ronald Bruce
Rapp, George Robert, Jr

MISSISSIPPI
Bruce, John Goodall, Jr
Meylan, Maurice Andre
Perry, Edward Belk
Saucier, Roger Thomas

MISSOURI
Arvidson, Raymond Ernst
Ashworth, William Bruce, Jr
Snowden, Jesse O
Unklesbay, Athel Glyde

MONTANA
Hyndman, Donald William

NEBRASKA
Doran, John Walsh

NEVADA
Gifford, Gerald F
Helm, Donald Cairney
Krenkel, Peter Ashton
Price, Jonathan Greenway
Raines, Gary L
Trexler, Dennis Thomas

NEW HAMPSHIRE
Richter, Dorothy Anne

NEW JERSEY
Agron, Sam Lazrus
Hanson, James Edward
Hordon, Robert M
Lindberg, Craig Robert
McCabe, Gregory James, Jr
Manabe, Syukuro
Monahan, Edward James
Strausberg, Sanford I
Weisberg, Joseph Simpson

NEW MEXICO
Arion, Douglas
Bish, David Lee
Broxton, David Edward
Colp, John Lewis
Downs, William Fredrick
Gancarz, Alexander John
Gerstl, Siegfried Adolf Wilhelm
Grambling, Jeffrey A
Kuellmer, Frederick John
Love, David Waxham
McTigue, David Francis
Manley, Kim
Shomaker, John Wayne
Trauger, Frederick Dale
Wawersik, Wolfgang R
Wells, Stephen Gene

NEW YORK
Bhattacharji, Somdev
Bond, Gerard C
Brett, Carlton Elliot
Bundy, Francis P
De Laubenfels, David John
Donnelly, Thomas Wallace
Egan, Walter George
Fleischer, Robert Louis
Hayes, Dennis E
Henderson, Floyd M
Henderson, Thomas E
Jacob, Klaus H
Kaufman, John H
Kent, Dennis V
Kubik, Peter W
Kumai, Motoi

Lindsley, Donald Hale
Mullins, Henry Thomas
Rampino, Michael Robert
Rokitka, Mary Anne
Romey, William Dowden
Trexler, Frederick David
Van Burkalow, Anastasia

NORTH CAROLINA
Cannon, Ralph S, Jr
Howard, Arthur David
Overcash, Michael Ray
Parsons, Willard H
Read, Floyd M
Ross, Thomas Edward
Stuart, Alfred Wright
Towell, William Earnest

NORTH DAKOTA
Scoby, Donald Ray

OHIO
Bladh, Katherine Laing
Bladh, Kenneth W
Friberg, LaVerne Marvin
Fuller, James Osborn
Harris-Noel, Ann (F H) Graetsch
Kovach, Jack
Mayer, Victor James
Nash, David Byer
Noble, Allen G
Savin, Samuel Marvin
Smoot, Edith L
Unrug, Raphael
Wojtal, Steven Francis

OKLAHOMA
Henderson, Frederick Bradley, III
Klehr, Edwin Henry
Martner, Samuel (Theodore)
Root, Paul John
Thomsen, Leon
Visher, Glenn S
Williams, Thomas Henry Lee

OREGON
Allison, Ira Shimmin
Drake, Ellen Tan
Lawrence, Robert D
McKinney, William Mark
Quinn, William Hewes
Young, J Lowell

PENNSYLVANIA
Beutner, Edward C
Brantley, Susan Louise
Briggs, Reginald Peter
Buis, Patricia Frances
Cassidy, William Arthur
Crawford, Maria Luisa Buse
Cuff, David J
Elsworth, Derek
Fritz, Lawrence William
Justham, Stephen Alton
Kasting, James Fraser
Krishtalka, Leonard
Luce, Robert James
Meisler, Harold
Pfefferkorn, Hermann Wilhelm
Rizza, Paul Frederick
Rohl, Arthur N
Schultz, Lane D
Shultz, Charles H
Simonds, John Ormsbee
Tuffey, Thomas J
Wegweiser, Arthur E

RHODE ISLAND
Cain, J(ames) Allan
Pieters, Carle M

SOUTH CAROLINA
Hawkins, Richard Horace
James, L(aurence) Allan

TENNESSEE
Bell, Persa Raymond
Fullerton, Ralph O
Gentry, Robert Vance
Johnson, Robert W, Jr
Jumper, Sidney Roberts
McSween, Harry Y, Jr
Miller, Calvin F
Siesser, William Gary
Tamura, Tsuneo
Wagner, Aubrey Joseph
Yau, Cheuk Chung

TEXAS
Berkhout, Aart W J
Brown, Glenn Lamar
Burton, Robert Clyde
Chatelain, Edward Ellis
Chronic, John
Cook, Earl Ferguson
De Ford, Ronald K
Dorfman, Myron Herbert
Eckelmann, Walter R
Eifler, Gus Kearney, Jr
Elsik, William Clinton
Greene, Donald Miller
Halpern, Martin
Harrison, Wilks Douglas
Hudson, Robert Frank

Earth Sciences, General (cont)

Jahns, Hans O(tto)
Jamieson-Magathan, Esther R
Jennings, Alfred Roy, Jr
Johnston, David Hervey
Jones, Ernest Austin, Jr
Judd, Frank Wayne
Keeney, Joe
LaBree, Theodore Robert
Lafon, Guy Michel
Lambert, Paul Wayne
Ledley, Tamara Shapiro
Lofgren, Gary Ernest
Murphy, Daniel L
Murray, Grover Elmer
Nestell, Merlynd Keith
Nicks, Oran Wesley
Nussmann, David George
Owens, Edward Henry
Pearson, Daniel Bester, III
Phinney, William Charles
Reaser, Donald Frederick
Robertson, James Douglas
Rowe, Marvin W
Schink, David R
Self, Stephen
Whitford-Stark, James Leslie
Woodruff, Charles Marsh, Jr
Woods, Raymond D

UTAH
Case, James B(oyce)
Christiansen, Eric H
Freethey, Geoffrey Ward
Gates, Joseph Spencer
Vaughn, Danny Mack

VERMONT
Doll, Charles George
Hussak, Robert Edward
Ratté, Charles A

VIRGINIA
Bowker, David Edwin
Burstyn, Harold Lewis
Campbell, Russell Harper
Corwin, Gilbert
Doyle, Frederick Joseph
Finkelman, Robert Barry
Gushee, Beatrice Eleanor
Harding, Duane Douglas
Hearn, Bernard Carter, Jr
James, Odette Bricmont
Johnson, Robert Ward
Leo, Gerhard William
Lipin, Bruce Reed
Macko, Stephen Alexander
Martens, James Hart Curry
Mikesell, Jon L
Miller, William Lawrence
Milton, Nancy Melissa
Molnia, Bruce Franklin
Mulder, Robert Udo
Newton, Elisabeth G
Odum, William Eugene
Outerbridge, William Fulwood
Padovani, Elaine Reeves
Radoski, Henry Robert
Said, Rushdi
Sato, Motoaki
Sayala, Dasharatham (Dash)
Spitzer, Cary Redford
Stebbins, Robert H
Suiter, Marilyn J
Wayland, Russell Gibson
Welch, Jasper Arthur, Jr

WASHINGTON
Ames, Lloyd Leroy, Jr
Armstrong, Frank Clarkson Felix
Barnes, Ross Owen
Cowan, Darrel Sidney
Dzurisin, Daniel
Edwards, James Mark
Heilman, Paul E
Hodges, Floyd Norman
Holcomb, Robin Terry
Malone, Stephen D
Riley, Robert Gene
Roberts, George E(dward)
Sanford, Thomas Bayes
Shin, Suk-han

WEST VIRGINIA
Khair, Abdul Wahab

WISCONSIN
Borowiecki, Barbara Zakrzewska
Bryson, Reid Allen
Fowler, Gerald Allan
Gernant, Robert Everett
Horn, Lyle Henry
Moran, Joseph Michael
Paull, Rachel Krebs
Read, William F

WYOMING
Huff, Kenneth O
Mateer, Niall John

ALBERTA
Charlesworth, Henry A K
Coen, Gerald Marvin

Goodarzi, Fariborz
Hunt, C Warren
McMillan, Neil John
Mossop, Grant Dilworth
Nesbitt, Bruce Edward
Rogerson, Robert James

BRITISH COLUMBIA
Davis, Roderick Leigh
Wheeler, John Oliver

MANITOBA
Welsted, John Edward
Whitaker, Sidney Hopkins
Wilkins, Brian John Samuel

NEW BRUNSWICK
Allen, Clifford Marsden
Ferguson, Laing

NOVA SCOTIA
Zentilli, Marcos

ONTARIO
Bell, Richard Thomas
Bevier, Mary Lou
Brown, Richard L
Cowan, W R
Fawcett, James Jeffrey
Flint, Jean-Jacques
Fox, Joseph S
Gussow, W(illia)m C(arruthers)
Jackson, Togwell Alexander
Long, Darrel Graham Francis
MacCrimmon, Hugh Ross
Macqueen, Roger Webb
Niki, Hiromi
Percival, John A
Prest, Victor Kent
Rowe, Ronald Kerry
Veizer, Ján
Wahl, William G

QUEBEC
Brandenberger, Arthur J
Chagnon, Jean Yves
David, Peter P
Dubois, Jean-Marie M
Ladanyi, Branko
Yong, R(aymond)

SASKATCHEWAN
Nisbet, Euan G
St Arnaud, Roland Joseph
Slater, George P
Strathdee, Graeme Gilroy

OTHER COUNTRIES
Alonso, Ricardo N
Dengo, Gabriel
Economides, Michael John
Lloyd, Christopher Raymond
Lueder, Ernst H
Rich, Thomas Hewitt
Ugolini, Fiorenzo Cesare

Environmental Sciences, General

ALABAMA
Baker, June Marshall
Campbell, Robert Terry
Cushing, Kenneth Mayhew
Garber, Robert William
Hung, Ru J
McCarl, Henry Newton
McDonald, Jack Raymond
Marano, Gerald Alfred
Marchant, Guillaume Henri (Wim), Jr
Meagher, James Francis
Olson, Willard Paul
Raymond, Dale Rodney
Reynolds, George Warren
Rivers, Douglas Bernard
Rogers, Hugo H, Jr
Smith, Wallace Britton
Vacik, James P
Wehner, Donald C
Workman, William Edward

ALASKA
Fahl, Charles Byron
Fink, Thomas Robert
Morrison, John Albert
Mowatt, Thomas C
Rice, Stanley Donald
Rockwell, Julius, Jr
Schindler, John Frederick
Strand, John A, III

ARIZONA
Avery, Charles Carrington
Bradley, Michael Douglas
Brathovde, James Robert
Day, Arden Dexter
Diehn, Bodo
Dorn, Ronald I
Dworkin, Judith Marcia
Fernando, Harindra Joseph
Gerba, Charles Peter
Haase, Edward Francis
Hodges, Carl Norris
Hutchinson, Charles F

Idso, Sherwood B
Klopatek, Jeffrey Matthew
McKlveen, John William
Myers, Lloyd E(ldridge)
Nordstrom, Brian Hoyt
Poole, H K
Robertson, Frederick Noel
Roth, Eldon Sherwood
Sears, Donald Richard
Snow, Eleanour Anne
Twomey, Sean Andrew
Wait, James Richard
Wierenga, Peter J
Zwolinski, Malcolm John

ARKANSAS
Anderson, Robbin Colyer
Andrews, Luther David
Burchard, John Kenneth
Chowdhury, Parimal
Day, James Meikle
Fortner, Wendell Lee
Goforth, Ronald R

CALIFORNIA
Anspaugh, Lynn Richard
Augood, Derek Raymond
Baez, Albert Vinicio
Bakus, Gerald Joseph
Barbour, Michael G
Baston, Janet Evelyn
Baugh, Ann Lawrence
Beck, Albert J
Bencala, Kenneth Edward
Bendix, Selina (Weinbaum)
Bennett, Charles Franklin
Bennett, James Peter
Benson, Andrew Alm
Bils, Robert F
Bishop, Keith C, III
Blume, Frederick Duane
Bubeck, Robert Clayton
Callahan-Compton, Joan Rea
Camenzind, Mark J
Carsola, Alfred James
Chamberlain, Dilworth Woolley
Chang, David Bing Jue
Chang, Shih-Ger
Chatigny, Mark A
Chen, Yang-Jen
Cliath, Mark Marshall
Cluff, Lloyd Sterling
Cole, Richard
Cooper, Wilson Wayne
Cosgrove, Gerald Edward
Crosby, Donald Gibson
Daniels, Jeffrey Irwin
Davis, Jay C
Davis, William Jackson
Dedrick, Kent Gentry
DeJongh, Don C
de Latour, Christopher
Dessel, Norman F
Deuble, John Lewis, Jr
Doner, Harvey Ervin
Dowdy, William Louis
Doyle, Fiona Mary
Edwards, J Gordon
Ehrlich, Anne Howland
Elimelech, Menachem
Elsberry, Russell Leonard
Farmer, Walter Joseph
Ferguson, William E
Fischback, Bryant C
Frank, Sidney Raymond
Freiling, Edward Clawson
Gaal, Robert A P
Gehri, Dennis Clark
Gordon, Robert Julian
Grantz, David Arthur
Greenfield, Stanley Marshall
Hamersma, J Warren
Hanna, George P, Jr
Hardebeck, Ellen Jean
Harris, Sigmund Paul
Harte, John
Hass, Robert Henry
Hazelwood, Robert Nichols
Hester, Norman Eric
Hodson, William Myron
Holdren, John Paul
Houpis, James Louis Joseph
Hovanec, B(ernard) Michael
Hrubesh, Lawrence Wayne
Hu, Steve Seng-Chiu
Hughes, Robert Alan
Hygh, Earl Hampton
Innes, William Beveridge
John, Walter
Johnson, Kenneth Delford
Jones, Rebecca Anne
Kallman, Burton Jay
Karuza, Sarunas Kazys
Kassakhian, Garabet Haroutioun
Kirch, Patrick Vinton
Kland, Mathilde June
Klein, David Henry
Knight, Allen Warner
Koenig, James Bennett
Kothny, Evaldo Luis
Ku, Teh-Lung
Kuehne, Donald Leroy
Lagarias, John S(amuel)
Laitin, Howard

Lamb, Sandra Ina
Langerman, Neal Richard
Lapple, Charles E
Lauer, George
Layton, David Warren
Leon, Henry A
Letey, John, Jr
Lick, Wilbert James
Lier, John
Lindberg, R(obert) G(ene)
Linnell, Robert Hartley
Lissaman, Peter Barry Stuart
Longley-Cook, Mark T
Ludwig, Claus Berthold
Lustig, Max
McAuliffe, Clayton Doyle
MacCready, Paul Beattie, Jr
MacKay, Kenneth Pierce, Jr
McKenzie, Donald Edward
McKone, Thomas Edward
Mahadevan, Parameswar
Malouf, George M
Martin, Michael
Mathews, Larry Arthur
Matthews, Stephen M
Melis, Anastasios
Mikel, Thomas Kelly, Jr
Mikkelsen, Duane Soren
Morse, John Thomas
Nagy, Kenneth Alex
Navratil, James Dale
Nero, Anthony V, Jr
Newsom, Bernard Dean
Nordstrom, Darrell Kirk
Oechel, Walter C
Offen, George Richard
Olson, Betty H
Orloff, Neil
Ott, Wayne R
Owen, Paul Howard
Paige, David A
Patterson, Clair Cameron
Perhac, Ralph Matthew
Perrine, R(ichard) L(eroy)
Pestrong, Raymond
Pierce, Matthew Lee
Piersol, Allan Gerald
Pipkin, Bernard Wallace
Pitts, James Ninde, Jr
Porcella, Donald Burke
Quan, William
Que Hee, Shane Stephen
Ragaini, Richard Charles
Ramspott, Lawrence Dewey
Rateaver, Bargyla
Remson, Irwin
Riffer, Richard
Riley, Robert Lee
Roberts, Stephen Winston
Robilliard, Gordon Allan
Robinson, Lewis Howe
Robinson, Paul Ronald
Robison, William Lewis
Rogers, Gifford Eugene
Rohde, Richard Whitney
Rosen, Alan
Rosen, Leonard Craig
Rossbacher, Lisa Ann
Rowland, F Sherwood
Russell, Philip Boyd
Sage, Orrin Grant, Jr
Salmon, Eli J
Sanders, Brenda Marie
Schaleger, Larry L
Schneider, Meier
Scow, Kate Marie
Segal, Alexander
Sextro, Richard George
Shaffar, Scott William
Sheppard, Asher R
Shinn, Joseph Hancock
Simon, Richard L
Smith, Charles Francis, Jr
Smith, Leverett Ralph
Spencer, Herbert W, III
Spicher, Robert G
Sprague, Robert W
Sproull, Wayne Treber
Star, Jeffrey L
Starr, Mortimer Paul
Stenstrom, Michael Knudson
Straughan, Isdale (Dale) Margaret
Suffet, I H (Mel)
Suman, Daniel Oscar
Switzer, Paul
Tewes, Howard Allan
Thomas, Jerome Francis
Thompson, Chester Ray
Thompson, Rosemary Ann
Thornton, John Irvin
Tilling, Robert Ingersoll
VanBlaricom, Glenn R
VanBruggen, Ariena H C
Vilkitis, James Richard
Voeks, Robert Allen
Walter, Hartmut
Warren, Mashuri Laird
Waterland, Larry R
Wayne, Lowell Grant
Weathers, Wesley Wayne
Weiss, Fred Toby
Wentworth, Carl M, Jr
Westman, Walter Emil
Wilcox, Bruce Alexander

Wilcox, Howard Albert
Wilson, Katherine Woods
Winer, Arthur Melvyn
Witham, Clyde Lester
Wuebbles, Donald J
Wyzga, Ronald Edward
Yamamoto, Sachio
Yen, I-Kuen
Ziegler, Carole L
Ziemer, Robert Ruhl

COLORADO
Arnott, Robert A
Boatman, Joe Francis
Boyd, Josephine Watson
Bradshaw, William Newman
Bush, Alfred Lerner
Chappell, Willard Ray
Chiou, Cary T(sair)
Christian, Wayne Gillespie
Cohen, Ronald R H
Colton, Roger Burnham
Craig, Roy Phillip
Crockett, Allen Bruce
Dabberdt, Walter F
Danzberger, Alexander Harris
Dewey, Fred McAlpin
Diaz, Henry Frank
Duray, John R
Eberhard, Wynn Lowell
Edwards, Harry Wallace
Fischer, William Henry
Fox, Douglas Gary
Havlick, Spenser Woodworth
Herrmann, Raymond
Hutchinson, Gordon Lee
Ivory, Thomas Martin, III
Johnson, Ross Byron
Kearney, Philip Daniel
Kemper, William Alexander
Krausz, Stephen
Langworthy, William Clayton
Legal, Casimer Claudius, Jr
McKown, Cora F
Martin, Stephen George
Meroney, Robert N
Mosier, Arvin Ray
Parker, H Dennison
Peterson, James T
Porter, Kenneth Raymond
Ratte, James C
Schneider, Stephen Henry
Smith, Dwight Raymond
Stauffer, Jack B
Steele, Timothy Doak
Thompson, Milton Avery
Thompson, Starley Lee
Timblin, Lloyd O, Jr
Toy, Terrence J
Tucker, Alan
Vaseen, V(esper) Albert
Waldren, Charles Allen
Weist, William Godfrey, Jr
Wellman, Dennis Lee
White, Gilbert Fowler
Wildeman, Thomas Raymond
York, J(esse) Louis

CONNECTICUT
Baillie, Priscilla Woods
Barske, Philip
Bentz, Alan P(aul)
Bollyky, L(aszlo) Joseph
Constant, Frank Woodbridge
Damman, Antoni Willem Hermanus
DeSanto, Robert Spilka
Dobay, Donald Gene
Egler, Frank Edwin
Friedlander, Henry Z
Friedman, Howard Stephen
Gauthier, George James
Geballe, Gordon Theodore
Groff, Donald William
Groth, Richard Henry
Haakonsen, Harry Olav
Haas, Thomas J
Hauser, Martin
Jackson, C(harles) Ian
Kawaters, Woody H
Koch, Richard Carl
Krause, Leonard Anthony
Mattina, Mary Jane Incorvia
Remington, Charles Lee
Renfro, William Charles
Rho, Jinnque
Rusling, James Francis
Seaman, Gregory G
Shaw, Brenda Roberts
Spiegel, Zane
Stuart, James Davies
Voorhies, John Davidson

DELAWARE
Brazel, Anthony James
Crippen, Raymond Charles
Fleming, Richard Allan
Geer, Richard P
Goel, Kailash C(handra)
Harvey, John, Jr
Hatchard, William Reginald
Huang, Chin Pao
Jacobs, Emmett S
Kalkstein, Laurence Saul
Krone, Lawrence James

Martin, Wayne Holderness
Mathre, Owen Bertwell
Mikell, William Gaillard
Milian, Alwin S, Jr
Pagano, Alfred Horton
Pierrard, John Martin
Potrafke, Earl Mark
Rocheleau, Robert
Spoljaric, Nenad
Willmott, Cort James

DISTRICT OF COLUMBIA
Adomaitis, Vytautas Albin
Alter, Harvey
April, Robert Wayne
Arcos, Joseph (Charles)
Argus, Mary Frances
Ballantine, David Stephen
Bannerman, Douglas George
Beard, Charles Irvin
Bell, Bruce Arnold
Bellin, Judith Schryver
Breen, Joseph John
Bushman, John Branson
Carrigan, Richard Alfred
Cochran, Thomas B
Cuatrecasas, José
DeCicco, Benedict Thomas
Doumani, George Alexander
El-Ashry, Mohamed T
Feeney, Gloria Comulada
Fisher, Farley
Goodland, Robert James A
Gould, Jack Richard
Hershey, Robert Lewis
Hirschhorn, Joel S(tephen)
Hirzy, John William
Keitt, George Wannamaker, Jr
Krinsley, Daniel B
MacArthur, Donald M
Meñez, Ernani Guingona
Middleton, John T(ylor)
Needels, Theodore S
Neiheisel, James
Neihof, Rex A
Olem, Harvey
Ostenso, Ned Allen
Paris, Oscar Hall
Preer, James Randolph
Richardson, Allan Charles Barbour
Rogers, Senta S(tephanie)
Roller, Paul S
Rubinoff, Roberta Wolff
Schneider, Bernard Arnold
Sexton, Ken
Shykind, Edwin B
Stroup, Cynthia Roxane
Sudia, Theodore William
Uman, Myron F
Wakelyn, Phillip Jeffrey
Walker, John Martin
Warnick, Walter Lee
Wilkinson, John Edwin
Zucchetto, James John

FLORIDA
Allen, Eric Raymond
Barile, Diane Dunmire
Belanger, Thomas V
Besch, Emerson Louis
Billings, Charles Edgar
Brezonik, Patrick Lee
Brooker, Hampton Ralph
Burrill, Robert Meredith
Carter, James Harrison, II
Chryssafopoulos, Hanka Wanda Sobczak
Convertino, Victor Anthony
Cooper, William James
Coppoc, William Joseph
Cropper, Wendell Parker
De Lorge, John Oldham
Dietz, Jess Clay
Donnalley, James R, Jr
Ekberg, Donald Roy
Fitzpatrick, George
Frazier, Stephen Earl
Frei, Sister John Karen
Fuller, Harold Wayne
Goldberg, Walter M
Goldstein, Harold William
Goss, Gary Jack
Hanley, James Richard, Jr
Harrold, Gordon Coleson
Haviland, Robert P(aul)
Herndon, Roy C
Huber, Wayne Charles
Jackson, Daniel Francis
Jenkins, Dale Wilson
Jones, John A(rthur)
Langford, David
Lim, Daniel V
Lundgren, Dale A(llen)
Mason, D(onald) R(omagne)
Menzer, Robert Everett
Meryman, Charles Dale
Miller, Harvey Alfred
Miskimen, George William
Myers, Richard Lee
Myerson, Albert Leon
Nelson, Gordon Leigh
Nelson, John William
Nichols, James Carlile
Parsons, Lawrence Reed
Posner, Gerald Seymour

Radomski, Jack London
Rao, Palakurthi Suresh Chandra
Rice, Stanley Alan
Richards, Norman Lee
Rona, Donna C
Ruscher, Paul H
Shair, Robert C
Silverman, David Norman
Sirkin, Alan N
Smith, William Mayo
Sneade, Barbara Herbert
Stamper, James Harris
Teller, Aaron Joseph
Tiffany, William James, III
Virnstein, Robert W
Wander, Joseph Day
White, Arlynn Quinton, Jr
White, Arlynn Quinton, Jr
Windsor, John Golay, Jr
Yost, Richard A

GEORGIA
Alberts, James Joseph
Amirtharajah, Appiah
Baarda, David Gene
Bailey, George William
Baughman, George Larkins
Bayer, Charlene Warres
Bozeman, John Russell
Brown, William E
Burns, Lawrence Anthony
Carver, Robert E
Chameides, William Lloyd
Chandler, James Harry, III
Cofer, Harland E, Jr
Ellington, James Jackson
Fineman, Manuel Nathan
Gaffney, Peter Edward
Garrison, Arthur Wayne
Ghuman, Gian Singh
Greear, Philip French-Carson
Hargrove, Robert John
Hedden, Kenneth Forsythe
Hill, David W(illiam)
Hilton, James A
Hoover, Thomas Burdett
Johnson, Barry Lee
Justus, Carl Gerald
Kollig, Heinz Philipp
Lee, Richard Fayao
Milham, Robert Carr
Nichols, Michael Charles
Odum, Eugene Pleasants
Peake, Thaddeus Andrew, III
Quisenberry, Dan Ray
Richardson, Susan D
Rodriguez, Augusto
Rogers, Harvey Wilbur
Schultz, Donald Paul
Shure, Donald Joseph
Spetnagel, Theodore John
Sutton, William Wallace
Swank, Robert Roy, Jr
Woodall, William Robert, Jr
Ziffer, Jack

HAWAII
Carpenter, Richard A
Coles, Stephen Lee
Cox, Doak Carey
Curtis, George Darwin
Flachsbart, Peter George
Frystak, Ronald Wayne
Fujioka, Roger Sadao
Halbig, Joseph Benjamin
Hufschmidt, Maynard Michael
Mackenzie, Frederick Theodore
Segal, Earl
Siegel, Sanford Marvin
Winget, Robert Newell

IDAHO
Clark, William Hilton
Ferguson, James Homer
Gehrke, Robert James
Gesell, Thomas Frederick
Hollenbaugh, Kenneth Malcolm
Knutson, Carroll Field
Lehrsch, Gary Allen
Reno, Harley W
Wright, Kenneth James

ILLINOIS
Akin, Cavit
Allen, John Kay
Babcock, Lyndon Ross, (Jr)
Banwart, Wayne Lee
Bath, Donald Alan
Baumgarten, Ronald J
Benioff, Paul
Bjorklund, Richard Guy
Carlson, Gerald Eugene
Casella, Alexander Joseph
Clark, Stephen Darrough
Doehler, Robert William
Erickson, Mitchell Drake
Flynn, Kevin Francis
Flynn, Robert James
Foster, John Webster
Frank, James Richard
Garvin, Paul Joseph, Jr
Gibbons, Larry V
Gilbert, Thomas Lewis
Girard, G Tanner

Glod, Edward Francis
Hagenbach, W(illiam) P(aul)
Hall, Stephen Kenneth
Hallenbeck, William Hackett
Harrison, Paul C
Harrison, Wyman
Herendeen, Robert Albert
Herricks, Edwin E
Hertig, Bruce Allerton
Hockman, Deborah C
Holsen, Thomas Michael
Hsui, Albert Tong-Kwan
Hutchcroft, Alan Charles
Jarke, Frank Henry
Johnson, John Harold
Kasper, Gerhard
Knake, Ellery Louis
Konopinski, Virgil J
Krug, Edward Charles
Lerman, Abraham
Loehle, Craig S
Marmer, Gary James
Michalski, Raymond J
Minear, Roger Allan
Murphy, Thomas Joseph
Nash, Kenneth Laverne
Nealy, Carson Louis
Paddock, Robert Alton
Paulson, Glenn
Penrose, William Roy
Piwoni, Marvin Dennis
Prakasam, Tata B S
Reilly, Christopher Aloysius, Jr
Rodolfo, Kelvin S
Roy, Dipak
Schaeffer, David Joseph
Schaller, Robin Edward
Seitz, Wesley Donald
Shen, Sin-Yan
Singh, Shyam N
Sohr, Robert Trueman
Southern, William Edward
Stearner, Sigrid Phyllis
Stetter, Joseph Robert
Stratton, James Forrest
Streets, David George
Stull, Elisabeth Ann
Sytsma, Louis Frederick
Vander Velde, George
Wadden, Richard Albert
Wahlgren, Morris A
Wang, Wun-Cheng W(oodrow)
Wesolowski, Wayne Edward
Wetegrove, Robert Lloyd
Wingender, Ronald John
Wolff, Robert John
Woods, Kenneth R
Zar, Jerrold Howard
Zenz, David R

INDIANA
Alliston, Charles Walter
Born, Gordon Stuart
Bumpus, John Arthur
Carlson, Merle Winslow
Cassidy, Harold Gomes
Chowdhury, Dipak Kumar
Eberly, William Robert
Hamelink, Jerry L
Hites, Ronald Atlee
Hubble, Billy Ray
Hurst, Robert Nelson
Landreth, Ronald Ray
Leap, Darrell Ivan
Leonard, Jack E
McCowen, Max Creager
McFee, William Warren
Mengel, David Bruce
Michaud, Howard H
Miller, Richard William
Mirsky, Arthur
Nelson, Craig Eugene
Parkhurst, David Frank
Pyle, James L
Speece, Susan Phillips
Squiers, Edwin Richard
Voelz, Frederick
Widener, Edward Ladd, Sr
Winkler, Erhard Mario

IOWA
Bremner, John McColl
Drake, Lon David
Graham, Benjamin Franklin
Karlen, Douglas Lawrence
Kniseley, Richard Newman
Markuszewski, Richard
Nakato, Tatsuaki
Schnoor, Jerald L
Walters, James Carter
Watkins, Stanley Read

KANSAS
Braaten, David A
McCullough, Elizabeth Ann
Pierson, David W
Platt, Dwight Rich
Schaper, Laurence Teis
Surampalli, Rao Yadagiri
Welch, Jerome E
Wright, Donald C

KENTUCKY
Birge, Wesley Joe

Environmental Sciences, General (cont)

Gauri, Kharaiti Lal
Hoffman, Wayne Larry
Jensen, Randolph A(ugust)
Kiefer, John David
Kupchella, Charles E
Livingood, Marvin D(uane)
Musacchia, X J
Sarma, Atul C
Seay, Thomas Nash
Sonnenfeld, Gerald
Stanonis, Francis Leo
Stevenson, Robert Jan
Volp, Robert Francis

LOUISIANA

Bahr, Leonard M, Jr
DeLaune, Ronald D
DeLeon, Ildefonso R
Ferrell, Ray Edward, Jr
Flournoy, Robert Wilson
Fontenot, Martin Mayance, Jr
Loden, Michael Simpson
Mills, Earl Ronald
Morrow, Norman Louis
Poirrier, Michael Anthony
Sartin, Austin Albert
Seigel, Richard Allyn
Thibodeaux, Louis J
Thompson, James Joseph
Wiewiorowski, Tadeusz Karol
Wilson, John Thomas
Wrobel, William Eugene

MAINE

Chute, Robert Maurice
Cronn, Dagmar Rais
Eastler, Thomas Edward
Howd, Frank Hawver
Robbins, Wayne Brian
Rock, Chet A
Snow, Joseph William
Tewhey, John David
Whitten, Maurice Mason

MARYLAND

Armstrong, Daniel Wayne
Auel, RaeAnn Marie
Barbour, Michael Thomas
Bauer, Ernest
Baummer, J Charles, Jr
Bausum, Howard Thomas
Becker, Donald Arthur
Berkson, Harold
Beroza, Morton
Brancato, David Joseph
Bregman, Jacob Israel
Caret, Robert Laurent
Chamberlain, David Leroy, Jr
Coble, Anna Jane
Colle, Ronald
Commito, John Angelo
Cookson, John T(homas), Jr
Costanza, Robert
Cothern, Charles Richard
Creasia, Donald Anthony
Dashiell, Thomas Ronald
Deitzer, Gerald Francis
Englehart, Richard W(ilson)
Fink, Don Roger
Foresti, Roy J(oseph), Jr
Fowler, Bruce Andrew
Galler, Sidney Roland
Gavis, Jerome
Gerritsen, Jeroen
Gessner, Adolf Wilhelm
Gevantman, Lewis Herman
Goktepe, Omer Faruk
Grotenhuis, Marshall
Gupta, Gian Chand
Guruswamy, Vinodhini
Heath, George A(ugustine)
Heath, Robert Gardner
Heggestad, Howard Edwin
Helz, George Rudolph
Herrett, Richard Allison
Hodge, Frederick Allen
Holland, Marjorie Miriam
Hopkins, Homer Thawley
Hubbard, Donald
Hunt, Charles Maxwell
Johnson, Charles C, Jr
Joiner, R(eginald) Gracen
Josephs, Melvin Jay
Katcher, David Abraham
Kelsey, Morris Irwin
Khare, Mohan
Kirkland, James T
Kline, Jerry Robert
Krizek, Donald Thomas
Krumbein, Simeon Joseph
Kurylo, Michael John, III
Kushner, Lawrence Maurice
Larson, Reginald Einar
Liggett, Walter Stewart, Jr
Linder, Seymour Martin
McBryde, F(elix) Webster
McCammon, Helen Mary
McVey, James Paul
Matta, Joseph Edward
May, Ira Philip

Melancon, Mark J
Mihursky, Joseph Anthony
Miller, Raymond Earl
Mogil, H Michael
Mountford, Kent
Mulchi, Charles Lee
Munasinghe, Mohan P
Muser, Marc
Myers, Lawrence Stanley, Jr
Nanzetta, Philip Newcomb
Newbury, Dale Elwood
Panayappan, Ramanathan
Plumlee, Karl Warren
Pomerantz, Irwin Herman
Reichel, William Louis
Rosenblatt, David Hirsch
Ross, Philip
Rubin, Robert Jay
Sansone, Eric Brandfon
Sarver, Emory William
Seifried, Harold Edwin
Sellner, Kevin Gregory
Shah, Shirish
Shannon, Larry J(oseph)
Shields, Jimmie Lee
Smith, Elmer Robert
Sonawane, Babasaheb R
Spangler, Glenn Edward
Swift, David Leslie
Talbot, Donald R(oy)
Teramura, Alan Hiroshi
Topping, Joseph John
Traystman, Richard J
Uhart, Michael Scott
Weir, Edward Earl, II
Weisberg, Stephen Barry
White, Harris Herman
Yakowitz, Harvey
Young, Alvin L
Young, George Anthony
Young, Jay Alfred
Zankel, Kenneth L
Zeeman, Maurice George

MASSACHUSETTS

Bass, Arthur
Berlandi, Francis Joseph
Bertin, Robert Ian
Bloom, Arthur David
Bowley, Donovan Robin
Busby, William Fisher, Jr
Butler, James Newton
Castro, Gonzalo
Chernosky, Edwin Jasper
Chisholm, Sallie Watson
Clark, William Cummin
Crandlemere, Robert Wayne
Dahm, Donald J
DeCosta, Peter F(rancis)
Dixon, Brian Gilbert
Ehrenfeld, John R(oos)
Eschenroeder, Alan Quade
Fay, James A(lan)
Feder, William Adolph
Fennelly, Paul Francis
Field, Hermann Haviland
Fitch, John Henry
Fix, Richard Conrad
Foster, Charles Henry Wheelwright
Genes, Andrew Nicholas
Golay, Michael W
Gold, Harris
Gregory, Constantine J
Hadlock, Charles Robert
Harvey, Walter William
Horne, Ralph Albert
Horzempa, Lewis Michael
Jankowski, Conrad M
Kaplan, David Lee
Kazmaier, Harold Eugene
Keast, David N(orris)
Lee, Kai Nien
Licht, Stuart Lawrence
Marini, Robert C
Marquis, Judith Kathleen
Moir, Ronald Brown, Jr
Murphy, Brian Logan
Nickerson, Norton Hart
Olmez, Ilhan
Ozonoff, David Michael
Pandolf, Kent Barry
Peace, George Earl, Jr
Platt, John Rader
Pojasek, Robert B
Primack, Richard Bart
Prodany, Nicholas W
Ram, Neil Marshall
Ramakrishna, Kilaparti
Redington, Charles Bahr
Reimold, Robert J
Richards, F Paul
Schempp, Ellory
Stevens, J(ames) I(rwin)
Stinchfield, Carleton Paul
Thomas, Aubrey Stephen, Jr
Trotz, Samuel Isaac
Wallace, Robert William
Wun, Chun Kwun

MICHIGAN

Ben, Manuel
Blanchard, Fred Ayres
Bone, Larry Irvin
Brewer, George E(ugene) F(rancis)

Burlington, Roy Frederick
Cadle, Steven Howard
Carpenter, Michael Kevin
Chang, Tai Yup
Cho, Byong Kwon
Chock, David Poileng
Cotner, James Bryan, Jr
Crummett, Warren B
Daniels, Stacy Leroy
D'Itri, Frank M
Domke, Charles J
Dzieciuch, Matthew Andrew
Farber, Hugh Arthur
Ficsor, Gyula
France, W(alter) DeWayne, Jr
Gates, David Murray
Gelderloos, Orin Glenn
Hahne, Rolf Mathieu August
Halberstadt, Marcel Leon
Hoerger, Fred Donald
Hough, Richard Anton
Howe, George Marvel
Jensen, David James
Johnson, Ray Leland
Kelly, Nelson Allen
Kerfoot, Wilson Charles
Kevern, Niles Russell
Klimisch, Richard L
Kullgren, Thomas Edward
Kummler, Ralph H
Kunkle, George Robert
Mac, Michael John
McClelland, Nina Irene
Mallinson, George Greisen
Maskal, John
Medlin, Julie Anne Jones
Miller, Robert Rush
Moolenaar, Robert John
Nestrick, Terry John
Northup, Melvin Lee
O'Shea, Timothy Allan
Otto, Klaus
Powitz, Robert W
Prostak, Arnold S
Racke, Kenneth David
Ramsey, John Charles
Reynolds, John Z
Robbins, John Alan
Rogers, Jerry Dale
Ronca, Luciano Bruno
Scherger, Dale Albert
Schuetzle, Dennis
Sloane, Christine Scheid
Solomon, Allen M
Stevens, Violete L
Stoermer, Eugene F
Stynes, Stanley K
Tomboulian, Paul
Tuesday, Charles Sheffield
Whitten, Bertwell Kneeland
Williams, Ronald Lloyde
Wolff, George Thomas
Wolt, Jeffrey Duaine
Wood, Jack Sheehan
Young, David Caldwell

MINNESOTA

Arthur, John W
Barber, Donald E
Berube, Robert
Bohon, Robert Lynn
Buchwald, Caryl Edward
Cheng, Hwei-Hsien
Clapp, C(harles) Edward
Eisenreich, Steven John
Gorham, Eville
Grew, Priscilla Croswell Perkins
Hagen, Donald Frederick
Haile, Clarence Lee
Henry, Mary Gerard
Lande, Sheldon Sidney
McNaught, Donald Curtis
Olness, Alan
Oxborrow, Gordon Squires
Peck, John Hubert
Rapp, George Robert, Jr
Reynolds, James Harold
Roessler, Charles Ervin
Sautter, Chester A
Stefan, Heinz G
Thompson, Fay Morgen
Ward, Wallace Dixon
Wetmore, Clifford Major

MISSISSIPPI

Davis, Robert Gene
Greever, Joe Carroll
Hendricks, Donovan Edward
Kodavanti, Prasada Rao S
Minyard, James Patrick
Overstreet, Robin Miles
Paulson, Oscar Lawrence
Saucier, Roger Thomas
Wolverton, Billy Charles

MISSOURI

Burst, John Frederick
Callis, Clayton Fowler
Carpenter, Will Dockery
Carson, Bonnie L Bachert
Churchill, Ralph John
Cowherd, Chatten, Jr
Crocker, Burton B(lair)
De Buhr, Larry Eugene

Dietrich, Martin Walter
Eigner, Joseph
Frye, Charles Isaac
Gledhill, William Emerson
Grubbs, Charles Leslie
Hallas, Laurence Edward
Hasan, Syed Eqbal
Henderson, Gray Stirling
Hoffman, Jerry C
Holloway, Thomas Thornton
Holroyd, Louis Vincent
Hoogheem, Thomas John
Howe, Robert Kenneth
Kaley, Robert George, II
Klein, Andrew John
Kramer, Richard Melvyn
Lawless, Edward William
Malik, Joseph Martin
Mieure, James Philip
Mori, Erik Jun
Raven, Peter Hamilton
Rawlings, Gary Don
Rueppel, Melvin Leslie
Saeger, Victor William
Schroy, Jerry M
Sheets, Ralph Waldo
Snowden, Jesse O
Swisher, Robert Donald
Vandegrift, Alfred Eugene
Vaughan, William Mace
Wilcox, Harold Kendall
Worley, Jimmy Weldon

MONTANA

Erickson, Ronald E
Knox, James L(ester)
Montagne, John M
Peavy, Howard Sidney
Temple, Kenneth Loren

NEBRASKA

Carson, Steven Douglas
Fairchild, Robert Wayne
Louda, Svata Mary
Pekas, Jerome Charles
Phelps, George Clayton
Raun, Earle Spangler
Walsh, Gary Lynn

NEVADA

Bishop, William P
Deacon, James Everett
Dickson, Frank Wilson
Flueck, John A
Jobst, Joel Edward
Kinnison, Robert Ray
Krenkel, Peter Ashton
Meier, Eugene Paul
Miles, Maurice Jarvis
Miller, Watkins Wilford
Nauman, Charles Hartley
Nichols, Chester Encell
Pitter, Richard Leon
Poziomek, Edward John
Trexler, Dennis Thomas
Yousef, Mohamed Khalil

NEW HAMPSHIRE

Bernstein, Abram Bernard
Bilello, Michael Anthony
Danforth, Raymond Hewes
Fogleman, Wavell Wainwright
Frost, Terrence Parker
Jackson, Herbert William
Petrasek, Emil John
Stepenuck, Stephen Joseph, Jr
Weber, James Harold

NEW JERSEY

Agron, Sam Lazrus
Barnes, Robert Lee
Bernath, Tibor
Borowsky, Harry Herbert
Brenner, Douglas Milton
Cheremisinoff, Paul N
Colby, Richard H
Cooper, Keith Raymond
Crerar, David Alexander
Dayan, Jason Edward
Delaney, Bernard T
De Planque, Gail
Dobi, John Steven
Falk, Charles David
Farrauto, Robert Joseph
Faust, Samuel Denton
Greenberg, Arthur
Hall, Herbert Joseph
Hall, Homer James
Holt, Vernon Emerson
Hrycak, Peter
Hundert, Irwin
Iglewicz, Raja
Jackson, Thomas A J
Jensen, Betty Klainmnc
Kahn, Donald Jay
Keating, Kathleen Irwin
Koehl, William John, Jr
Kukin, Ira
Leone, Ida Alba
Levine, Richard S
Lewellen, William Stephen
Lewis, Thomas Brinley
Ling, Hubert
Lioy, Paul James

London, Mark David
McCabe, Gregory James, Jr
McGuire, David Kelty
Makofske, William Joseph
Martin, John David
Menczel, Jehuda H
Morgan, Mark Douglas
Nicholls, Gerald P
Priesing, Charles Paul
Ramaswamy, Venkatachalam
Rankin, Sidney
Rau, Eric
Rigler, Neil Edward
Rothstein, Edwin C(arl)
Rupp, Walter H(oward)
Salvesen, Robert H
Schubert, Rudolf
Shaw, Henry
Skovronek, Herbert Samuel
Socolow, Robert H(arry)
Sugam, Richard Jay
Tamarelli, Alan Wayne
Tatyrek, Alfred Frank
Vallese, Frank M
Vanden Heuvel, William John Adrian, III
Van Pelt, Wesley Richard
Varga, Gideon Michael, Jr
Ward, Laird Gordon Lindsay
Watts, Daniel Jay
Weinstein, Norman J(acob)
Weisberg, Joseph Simpson
Wenzel, John Thompson
Wilson, Charles Elmer
Wolfe, Peter E
Wood, Eric F
Yan, Tsoung-Yuan

NEW MEXICO
Baker, Floyd B
Bhada, Rohinton(Ron) K
Bingham, Felton Wells
Brown, James Hemphill
Broxton, David Edward
Cheng, Yung-Sung
Clark, Ronald Duane
Close, Donald Alan
Darnall, Dennis W
Deal, Dwight Edward
Essington, Edward Herbert
Fries, James Andrew
Fritz, Georgia T(homas)
Hadley, William Melvin
Hansen, Wayne Richard
Henderson, Thomas Richard
Holloway, Richard George
Hoover, Mark Douglas
Jones, Kirkland Lee
Kenna, Bernard Thomas
Kuellmer, Frederick John
Laughlin, Alexander William
Lewis, James Vernon
Merritt, Melvin Leroy
Mokler, Brian Victor
Niemczyk, Thomas M
Palmer, Byron Allen
Phillips, Fred Melville
Polzer, Wilfred L
Popp, Carl John
Schery, Stephen Dale
Smith, James Lewis
Stanbro, William David
Strniste, Gary F
Tapscott, Robert Edwin
Williams, Mary Carol
Wilson, Lee
Wolberg, Donald Lester
Yeh, Hsu-Chi

NEW YORK
Abraham, Rajender
Altwicker, Elmar Robert
Amundson, Robert Gale
Andrus, Richard Edward
Babich, Harvey
Baer, Norbert Sebastian
Barber, Eugene Douglas
Bedford, Barbara Lynn
Berger, Selman A
Beyea, Jan Edgar
Boger, Phillip David
Bokuniewicz, Henry Joseph
Booman, Keith Albert
Boylen, Charles William
Bulloff, Jack John
Bush, David Graves
Carbone, Gabriel
Cardenas, Raúl R, Jr
Ceprini, Mario Q
Clarke, Raymond Dennis
Coch, Nicholas Kyros
Cohen, Beverly Singer
Cohen, Norman
Cole, Jonathan Jay
Conant, Francis Paine
Coppersmith, Frederick Martin
Dahneke, Barton Eugene
Dietz, Russell Noel
DiGaudio, Mary Rose
Dondero, Norman Carl
Dooley, James Keith
Edelstein, William Alan
Elder, Fred A
Fakundiny, Robert Harry

Feinberg, Robert Jacob
Fischman, Harlow Kenneth
Freudenthal, Hugo David
Garrell, Martin Henry
Gellman, Isaiah
Ginzburg, Lev R
Goldfarb, Theodore D
Gossett, James Michael
Gray, D Anthony
Grayson, Herbert G
Hang, Yong Deng
Harley, Naomi Hallden
Haynes, James Mitchell
Hecht, Stephen Samuel
Henderson, Floyd M
Henderson, Ulysses Virgil, Jr
Hennigan, Robert Dwyer
Henshaw, Robert Eugene
Hewitt, Philip Cooper
Hoffmann, Dietrich
Horvat, Robert Emil
Howarth, Robert W
Inhaber, Herbert
Jacobson, Jay Stanley
Jenkins, Philip Winder
Jensen, Betty Klainminc
Kennedy, Edwin Russell
Klanderman, Bruce Holmes
Kneip, Theodore Joseph
Kohut, Robert John
Konigsberg, Alvin Stuart
Koppelman, Lee Edward
Kozak, Gary S
Krey, Philip W
Krieger, Gary Lawrence
Kumai, Motoi
Lasday, Albert Henry
Levandowsky, Michael
Lieberman, Arthur Stuart
Locke, David Creighton
Lockhart, Haines Boots, Jr
Loggins, Donald Anthony
Lovett, Gary Martin
Ludlum, Kenneth Hills
Lundgren, Lawrence William, Jr
Ma, Maw-Suen
McCune, Delbert Charles
McNeil, Richard Jerome
Mellett, James Silvan
Millstein, Jeffrey Alan
Molloy, Andrew A
Nathanson, Benjamin
O'Brien, Anne T
Padnos, Norman
Palmer, James F
Pereira, Martin Rodrigues
Putman, George Wendell
Rae, Stephen
Rana, Mohammad A
Rifkind, Arleen B
Ring, James Walter
Robertson, William, IV
Robinson, Joseph Edward
Robinson, Myron
Sage, Gloria W
Salotto, Anthony W
Scala, Sinclaire M(aximilian)
Schulz, Helmut Wilhelm
Serafy, D Keith
Sheeran, Stanley Robert
Shen, Thomas T
Siegfried, Clifford Anton
Simons, Edward Louis
Smardon, Richard Clay
Smethie, William Massie, Jr
Smith, Thomas Harry Francis
Sternberg, Stephen Stanley
Sterrett, Frances Susan
Stewart, Donald Borden
Stotzky, Guenther
Taub, Aaron M
Weiss, Dennis
Weiss, Jonas
White, James Carrick
Williams, Evan Thomas
Winter, Margaret Castle
Wolff, Manfred Paul
Wolfson, Leonard Louis
Yang, John Yun-Wen

NORTH CAROLINA
Andrady, Anthony Lakshman
Andrews, Richard Nigel Lyon
Booth, Donald Campbell
Bruce, Richard Conrad
Bruck, Robert Ian
Bursey, Joan Tesarek
Carlson, William Theodore
Carpenter, Benjamin H(arrison)
Collins, Jeffrey Jay
Decker, Clifford Earl, Jr
Dement, John M
Elias, Robert William
Ellison, Alfred Harris
Ensor, David Samuel
Fischer, Janet Jordan
Flint, Elizabeth Parker
Fouts, James Ralph
Gardner, Donald Eugene
Gillespie, Arthur Samuel, Jr
Godschalk, David Robinson
Harris, Robert L, Jr
Hartford, Winslow H
Heck, Walter Webb

Hess, Daniel Nicholas
Howard, Arthur David
Karl, Thomas Richard
Knapp, Kenneth T
Knight, Clifford Burnham
Kohl, Jerome
Kuenzler, Edward Julian
Lamb, James C(hristian), III
Lao, Yan-Jeong
Lapp, Thomas William
Lawless, Philip Austin
LeBaron, Homer McKay
Lee, Robert E, Jr
Linton, Richard William
McKinney, James David
Maier, Robert Hawthorne
Malindzak, George S, Jr
Marcus, Allan H
O'Neal, Thomas Denny
Patterson, David Thomas
Peirce, James Jeffrey
Pellizzari, Edo Domenico
Pfaender, Frederic Karl
Purchase, Earl Ralph
Ross, Richard Henry, Jr
Ross, Thomas Edward
Sanoff, Henry
Scott, Donald Ray
Shafer, Steven Ray
Shendrikar, Arun D
Shreffler, Jack Henry
Shuman, Mark S
Stenger, William J(ames), Sr
Stern, A(rthur) C(ecil)
Stillwell, Harold Daniel
Sullivan, Arthur Lyon
Sumner, Darrell Dean
Swift, Lloyd Wesley, Jr
Textoris, Daniel Andrew
Trenholm, Andrew Rutledge
Turner, Alvis Greely
Volpe, Rosalind Ann
Walsh, Edward John
Walters, Douglas Bruce
Weaver, Ervin Eugene
Wilson, Lauren R

NORTH DAKOTA
Bickel, Edwin David

OHIO
Ayer, Howard Earle
Baasel, William David
Bendure, Robert J
Berman, Donald
Bimber, Russell Morrow
Bishop, Paul Leslie
Boerner, Ralph E J
Bohrman, Jeffrey Stephen
Burrows, Kerilyn Christine
Carfagno, Daniel Gaetano
Compton, Ell Dee
Disinger, John Franklin
Eckert, Donald James
Fan, Liang-Shih
Fearheller, William Russell, Jr
Fisher, Gerald Lionel
Forsyth, Jane Louise
Fortner, Rosanne White
Frank, Glenn William
Gieseke, James Arnold
Gordon, Gilbert
Gruber, Charles W
Hannah, Sidney Allison
Harris, Richard Lee
Hess, George G
Hill, James Stewart
Innes, John Edwin
Kagen, Herbert Paul
Kawahara, Fred Katsumi
Ku, Han San
Lollar, Robert Miller
Lustick, Sheldon Irving
McKenzie, Garry Donald
McQuigg, Robert Duncan
Manson, Jeanne Marie
Marks, Alfred Finlay
Matisoff, Gerald
Mitsch, William Joseph
Mukhtar, Hasan
Nobis, John Francis
Purvis, John Thomas
Ray, John Robert
Rubin, Alan J
Ruedisili, Lon Chester
Saraceno, Anthony Joseph
Scarpino, Pasquale Valentine
Shaw, Wilfrid Garside
Springer, Allan Matthew
Sticksel, Philip Rice
Stoldt, Stephen Howard
Sydnor, Thomas Davis
Taylor, Michael Lee
Thomas, Craig Eugene
Toeniskoetter, Richard Henry
Tuovinen, Olli Heikki
Utgard, Russell Oliver
Voigt, Charles Frederick
Wenclawiak, Bernd Wilhelm
Wohlfort, Sam Willis
Yanko, William Harry
Zimmerman, Ernest Frederick

OKLAHOMA
Bleckmann, Charles Allen
Bryant, Rebecca Smith
Coleman, Nancy Pees
Crockett, Jerry J
Dorris, Troy Clyde
Eby, Harold Hildenbrandt
Enfield, Carl George
Fish, Wayne William
Fletcher, John Samuel
Hounslow, Arthur William
Koerner, E(rnest) L(ee)
Law, James Pierce, Jr
Miller, Helen Carter
Miller, John Walcott
Monn, Donald Edgar
Mulholland, Robert J(oseph)
Pollock, L(yle) W(illiam)
Puls, Robert W
Runnels, John Hugh
Shmaefsky, Brian Robert
Smith, Samuel Joseph
Summers, Gregory Lawson
Thurman, Lloy Duane
Van De Steeg, Garet Edward
Warren, Kenneth Wayne

OREGON
Buscemi, Philip Augustus
Chapman, Gary Adair
Church, Marshall Robbins
Fish, William
Goodney, David Edgar
Higgins, Paul Daniel
Huntzicker, James John
Jarrell, Wesley Michael
Katen, Paul C
Neilson, Ronald Price
Pizzimenti, John Joseph
Reporter, Minocher C
Richards, Oscar White
Seyb, Leslie Philip
Stones, Robert C
Thorson, Thomas Bertel
Tingey, David Thomas
Warkentin, Benno Peter
Wohlers, Henry Carl
Worrest, Robert Charles

PENNSYLVANIA
Allen, Herbert E
Andelman, Julian Barry
Aplan, Frank F(ulton)
Arnold, Dean Edward
Baurer, Theodore
Birchem, Regina
Bott, Thomas Lee
Brady, Keith Bryan Craig
Browning, Daniel Dwight
Buis, Patricia Frances
Bushnell, Kent O
Caruso, Sebastian Charles
Casey, Adria Catala
Cheng, Cheng-Yin
Claiborne, C Clair
Clark, Roger Alan
Cohen, Anna (Foner)
Cohen, Bernard Leonard
Crits, George J(ohn)
Daniels-Severs, Anne Elizabeth
Denoncourt, Robert Francis
Einhorn, Philip A
Emrich, Grover Harry
Fletcher, Frank William
Fletcher, Ronald D
Franz, Craig Joseph
Gardner, Thomas William
Gertz, Steven Michael
Giffen, Robert H(enry)
Goodman, Donald
Goodspeed, Robert Marshall
Graybill, Donald Lee
Gurol, Mirat D
Heckscher, Stevens
Hogan, Aloysius Joseph, Jr
Keenan, John Douglas
Keller, Joseph Herbert
Khosah, Robinson Panganai
Kleinman, Robert L P
Komarneni, Sridhar
Ku, Peh Sun
Kurtz, David Allan
LaGrega, Michael Denny
Larkin, Robert Hayden
Levin, Michael H(oward)
Lochstet, William A
Lynch, Thomas John
McCarthy, Raymond Lawrence
MacDonald, Hubert C, Jr
McDonnell, Archie Joseph
McDonnell, Leo F(rancis)
McFarland, Mack
McGee, Charles E
McGovern, John Joseph
Maroulis, Peter James
Melamed, Sidney
Mellinger, Michael Vance
Meredith, David Bruce
Morgan, M(illett) Granger
Myron, Thomas L(eo)
Neufeld, Ronald David
Patil, Ganapati P
Pees, Samuel T(homas)
Pena, Jorge Augusto

Brown, Robert Melbourne
Buckner, Charles Henry
Carlisle, David Brez
Cedar, Frank James
Chua, Kian Eng
Coakley, John Phillip
Effer, W R
Fyles, John Gladstone
Gillham, Robert Winston
Glooschenko, Walter Arthur
Grant, Douglas Roderick
Green, Roger Harrison
Greenhalgh, Roy
Hare, Frederick Kenneth
Hill, Patrick Arthur
Hopton, Frederick James
Houston, Arthur Hillier
Ingraham, Thomas Robert
Jackson, Togwell Alexander
Jones, Philip Arthur
Jorgensen, Erik
Keddy, Paul Anthony
Kelso, John Richard Murray
Kesik, Andrzej B
Kramer, James Richard
Lawrence, John
Lee, David Robert
Liu, Dickson Lee Shen
McAdie, Henry George
McBean, Edward A
Mack, Alexander Ross
McKenney, Donald Joseph
Maguire, Robert James
Martin, Hans Carl
Matheson, De Loss H(ealy)
Meresz, Otto
Morley, Harold Victor
Moroz, William James
Morris, Larry Arthur
Nirdosh, Inderjit
Nriagu, Jerome Okonkwo
Osborne, Richard Vincent
Pajari, George Edward
Philogene, Bernard J R
Quigley, Robert Murvin
Roots, Ernest Frederick
Ruel, Maurice M J
Schroeder, William Henry
Singer, Eugen
Singh, Surinder Shah
Slawson, Peter (Robert)
Smol, John Paul
Stairs, Robert Ardagh
Stebelsky, Ihor
Strachan, William Michael John
Svoboda, Josef
Tu, Chin Ming
VandenHazel, Bessel J
Wiebe, H Allan
Wilkinson, Thomas Preston
Williams, David Trevor
Williams, George Ronald
Wolkoff, Aaron Wilfred
Yang, Paul Wang
Young, David Matheson

QUEBEC
Beland, Pierre
Bonn, Ferdinand J
Bouchard, Michel André
Chagnon, Jean Yves
Cooper, David Gordon
Desjardins, Claude W
Dubois, Jean-Marie M
Gangloff, Pierre
Kalff, Jacob
Lalancette, Jean-Marc
Paulin, Gaston (Ludger)
Shaw, Robert Fletcher
Yong, R(aymond)

SASKATCHEWAN
Cassidy, Richard Murray
Chew, Hemming
Forsyth, Douglas John
Huang, P M
Tanino, Karen Kikumi
Viraraghavan, Thiruvenkatachari
Waddington, John

OTHER COUNTRIES
Bennett, Burton George
Brydon, James Emerson
Carney, Gordon C
Chen, Chen-Tung Arthur
Coulston, Mary Lou
Dahl, Arthur Lyon
Davis, Edward Alex
Grigoropoulos, Sotirios G(regory)
Hutzinger, Otto
Jokl, Miloslav Vladimir
Lieth, Helmut Heinrich Friedrich
Mou, Duen-Gang
Raveendran, Ekarath
Richmond, Robert H
Stam, Jos
Stoeber, Werner
Tsuda, Roy Toshio
Winnett, George

Fuel Technology & Petroleum Engineering

ALABAMA
Mancini, Ernest Anthony

ALASKA
Ostermann, Russell Dean

CALIFORNIA
Abdel-Baset, Mahmoud B
Bahn, Gilbert S(chuyler)
Baugh, Ann Lawrence
Blumstein, Carl Joseph
Brown, Glenn A
Brownscombe, Eugene Russell
Compton, Leslie Ellwyn
Hill, Richard William
Lewis, Robert Allen
Long, H(ugh) M(ontgomery)
Nur, Amos M
Rainis, Andrew
Shrontz, John William
Silcox, William Henry
Sinton, Steven Williams
Sparling, Rebecca Hall
Uhl, Arthur E(dward)
Wright, Charles Cathbert

COLORADO
Fischer, William Henry
Giulianelli, James Louis
Montgomery, Robert L
Neel, Thomas H
Osmond, John C, Jr
Pickett, George R
Pitts, Malcolm John
Smith, Glenn S
Vogt, Thomas Clarence, Jr

CONNECTICUT
Bilhorn, John Merlyn
Cobern, Martin E
Epperly, W Robert
Gupta, Dharam Vir

DELAWARE
Cantwell, Edward N(orton), Jr

DISTRICT OF COLUMBIA
Brown, Norman Louis
Melickian, Gary Edward
Neihof, Rex A
Ramsey, Jerry Warren

FLORIDA
Nersasian, Arthur

GEORGIA
Woodall, William Robert, Jr

IDAHO
Maloof, Giles Wilson

ILLINOIS
Bickelhaupt, R(oy) E(dward)
Buchanan, David Hamilton
Gorody, Anthony Wagner
Harrington, Joseph Anthony
Herzenberg, Caroline Stuart Littlejohn
Kruse, Carl William
Kukes, Simon G
Marr, William Wei-Yi
Masel, Richard Isaac
Nealy, Carson Louis
Simon, Jack Aaron
Singh, Shyam N
Wilson, Gerald Gene

INDIANA
Shankland, Rodney Veeder
Stehouwer, David Mark

KENTUCKY
Cobb, James C
Huffman, Gerald P

LOUISIANA
Bourgoyne, Adam T, Jr
Forman, McLain J
Jones, Paul Hastings
Ledford, Thomas Howard
Sartin, Austin Albert
Schucker, Robert Charles

MARYLAND
Becker, Donald Arthur

MASSACHUSETTS
Hotchkiss, Henry
Johnson, Stephen Allen
Robbat, Albert, Jr
Tester, Jefferson William

MICHIGAN
Furey, Robert Lawrence
Henein, Naeim A
Thompson, Paul Woodard
Tuesday, Charles Sheffield

MISSISSIPPI
George, Clifford Eugene
Knight, Wilbur Hall

MISSOURI
Forero, Enrique
Goebel, Edwin DeWayne

NEVADA
Kukal, Gerald Courtney

NEW HAMPSHIRE
Frey, Elmer Jacob

NEW JERSEY
Blackburn, Gary Ray
Brenner, Douglas
Chen, Nai Y
Farrington, Thomas Allan
Hundert, Irwin
Jacolev, Leon
Mitchell, Thomas Owen
Salvesen, Robert H
Schutz, Donald Frank
Shrier, Adam Louis
Varga, Gideon Michael, Jr

NEW MEXICO
Bowsher, Arthur LeRoy
Cropper, Walter V
Hill, Mary Rae
Northrop, David A
Smith, James Lewis
Taber, Joseph John
Williams, Joel Mann, Jr

NEW YORK
Douglas, Craig Carl
Happel, John
Pannella, Giorgio
Sotirchos, Stratis V
Virk, Kashmir Singh

NORTH CAROLINA
O'Connor, Lila Hunt
Smith, Norman Cutler

NORTH DAKOTA
Harris, Steven H

OHIO
Bittker, David Arthur
Burow, Duane Frueh
Hall, Ronald Henry
Lewis, Irwin C
Mezey, Eugene Julius
Murphy, Michael John
Putnam, Abbott (Allen)
Stoldt, Stephen Howard

OKLAHOMA
Albright, James Curtice
Blais, Roger Nathaniel
Boade, Rodney Russett
Bryant, Rebecca Smith
Hurn, R(ichard) W(ilson)
Hyne, Norman John
Joshi, Sadanand D
Lorenz, Philip Boalt
Madden, Michael Preston
Menzie, Donald E
Pober, Kenneth William
Sloan, Norman Grady

PENNSYLVANIA
Aplan, Frank F(ulton)
Chigier, Norman
Durante, Vincent Anthony
Friedman, Sidney
Holder, Gerald D
Lovell, Harold Lemuel
Meyer, Wolfgang E(berhard)
Natoli, John
Noceti, Richard Paul
Nolan, Edward J
Retcofsky, Herbert L
Soung, Wen Y
Wicker, Robert Kirk
Wiggill, John Bentley

TENNESSEE
Guerin, Michael Richard
Rodgers, Billy Russell
Singh, Suman Priyadarshi Narain

TEXAS
Baker, Donald Roy
Barrow, Thomas D
Basan, Paul Bradley
Brown, James Michael
Campbell, William M
Closmann, Philip Joseph
Crawford, Duane Austin
Crouse, Philip Charles
Desai, Kantilal Panachand
DiFoggio, Rocco
Dorfman, Myron Herbert
Earlougher, Robert Charles, Jr
Font, Robert G
James, Bela Michael
Lacy, Lewis L
Leventhal, Stephen Henry
Levien, Louise
Lewis, James Pettis
Lowther, Frank Eugene
Massell, Wulf F
Maute, Robert Edgar
Meyer, W(illiam) Keith

Mortada, Mohamed
Murphy, Daniel L
Nabor, George W(illiam)
Nussmann, David George
Prats, Michael
Quisenberry, Karl Spangler, Jr
Randolph, Philip L
Rodriguez, Charles F
Roebuck, Isaac Field
Siegfried, Robert Wayne, II
Sifferman, Thomas Raymond
Souby, Armand Max
Stead, Frederick L
Thomas, Estes Centennial, III
Van Rensburg, Willem Cornellus Janse
Witterholt, Edward John

UTAH
Bartholomew, Calvin Henry
Comeford, John J

VERMONT
Long, Robert Byron

VIRGINIA
Bartis, James Thomas
Everett, Warren S
Gordon, John S(tevens)
Hudson, Frederick Mitchell
Sayala, Chhaya
Sikri, Atam P

WASHINGTON
LaFontaine, Edward
Spencer, Arthur Milton, Jr

WEST VIRGINIA
Strickland, Larry Dean

WISCONSIN
Steinhart, John Shannon

WYOMING
Branthaver, Jan Franklin
Meyer, Edmond Gerald
Miknis, Francis Paul
Poulson, Richard Edwin
Robinson, Wilbur Eugene

ALBERTA
Campbell, John Duncan
Collins, David Albert Charles
Feingold, Adolph
Goodarzi, Fariborz
Tollefson, Eric Lars
Wiggins, Ernest James

BRITISH COLUMBIA
Bustin, Robert Marc

NEWFOUNDLAND
King, Michael Stuart

NOVA SCOTIA
Lynch, Brian Maurice

ONTARIO
Gussow, W(illia)m C(arruthers)
Walsh, John Heritage
Williams, Peter J

QUEBEC
Leigh, Charles Henry

OTHER COUNTRIES
Raveendran, Ekarath

Geochemistry

ALABAMA
Chalokwu, Christopher Iloba
Cook, Robert Bigham, Jr
Griffin, Robert Alfred
Hughes, Travis Hubert
Isphording, Wayne Carter

ALASKA
Barsdate, Robert John
Hawkins, Daniel Ballou
Karl, Susan Margaret
Mowatt, Thomas C
Swanson, Samuel Edward

ARIZONA
Boynton, William Vandegrift
Burt, Donald McLain
Buseck, Peter R
Chase, Clement Grasham
Damon, Paul Edward
Drake, Michael Julian
Fedder, Steven Lee
Fisher, Frederick Stephen
Ganguly, Jibamitra
Graf, Donald Lee
Guilbert, John M
Haynes, Caleb Vance, Jr
Holloway, John Requa
Jull, Anthony John Timothy
Kamilli, Diana Chapman
Kamilli, Robert Joseph
Kieffer, Susan Werner
Knowlton, Gregory Dean
Larimer, John William

Geochemistry (cont)

Lewis, John Simpson
Long, Austin
McMillan, Paul Francis
Moore, Carleton Bryant
Nagy, Bartholomew Stephen
Norton, Denis Locklin
Robertson, Frederick Noel
Swindle, Timothy Dale
Titley, Spencer Rowe

ARKANSAS
Nix, Joe Franklin
Steele, Kenneth F

CALIFORNIA
Albee, Arden Leroy
Allan, John Ridgway
Arth, Joseph George, Jr
Austin, Steven Arthur
Bacastow, Robert Bruce
Bada, Jeffrey L
Bailey, Roy Alden
Barnes, Ivan
Bass, Manuel N
Beaty, David Wayne
Bischoff, James Louden
Boettcher, Arthur Lee
Bohlen, Steven Ralph
Brimhall, George H
Bubeck, Robert Clayton
Burnett, Donald Stacy
Burnham, Alan Kent
Burtner, Roger Lee
Cain, William F
Carmichael, Ian Stuart
Carpenter, Alden B
Carr, Michael H
Chapman, Robert Mills
Chodos, Arthur A
Chow, Tsaihwa James
Coe, Robert Stephen
Cohen, Lewis H
Conomos, Tasso John
Copelin, Edward Casimere
Craig, Harmon
Cruikshank, Dale Paul
Czamanske, Gerald Kent
Davidson, Jon Paul
Des Marais, David John
Einaudi, Marco Tullio
El Wardani, Sayed Aly
Emerson, Donald Orville
Epstein, Samuel
Ernst, Wallace Gary
Flegal, Arthur Russell, Jr
Floran, Robert John
Fournier, Robert Orville
Garlick, George Donald
Gerlach, John Norman
Goldberg, Edward D
Hanan, Barry Benton
Hein, James R
Helgeson, Harold Charles
Hem, John David
Housley, Robert Melvin
Hurst, Richard William
Irwin, James Joseph
Isherwood, Dana Joan
Jacobson, Stephen Richard
Jeanloz, Raymond
Kaplan, Issac R
Keeling, Charles David
Keith, Terry Eugene Clark
Kennedy, Burton Mack
Kennedy, Vance Clifford
Kharaka, Yousif Khoshu
Kothny, Evaldo Luis
Krauskopf, Konrad Bates
Ku, Teh-Lung
Kvenvolden, Keith Arthur
Lal, Devendra
Lanphere, Marvin Alder
Lind, Carol Johnson
Liou, Juhn G
Lugmair, Guenter Wilhelm
Lupton, John Edward
McAuliffe, Clayton Doyle
Macdougall, John Douglas
McKenzie, William F
Magoon, Leslie Blake, III
Mahood, Gail Ann
Majmundar, Hasmukhrai Hiralal
Manning, Craig Edward
Meike, Annemarie
Merrin, Seymour
Metzger, Albert E
Michalowski, Joseph Thomas
Moiseyev, Alexis N
Moore, James Gregory
Nelson, Robert M
Nielson, Jane Ellen
Niemeyer, Sidney
Nordstrom, Darrell Kirk
Papanastassiou, Dimitri A
Parks, George A(lbert)
Patterson, Clair Cameron
Perhac, Ralph Matthew
Pierce, Matthew Lee
Piper, David Zink
Pray, Ralph Emerson
Rau, Gregory Hudson
Redfield, William David

Reynolds, John Hamilton
Riese, Walter Charles Rusty
Russ, Guston Price, III
Saleeby, Jason Brian
Sarna-Wojcicki, Andrei M
Silver, Leon Theodore
Silverman, Sol Robert
Smalley, Robert Gordon
Smith, Robert Leland
Snetsinger, Kenneth George
Spera, Frank John
Sposito, Garrison
Stensrud, Howard Lewis
Stolper, Edward Manin
Suman, Daniel Oscar
Taylor, Hugh P, Jr
Thompson, John Michael
Tilling, Robert Ingersoll
Tilton, George Robert
Turco, Richard Peter
van de Kamp, Peter Cornelis
Warnke, Detlef Andreas
Warren, Paul Horton
Wasson, John Taylor
Webster, Clyde Leroy, Jr
Williams, Alan Evan
Williams, Peter M
Wyllie, Peter John
Ziegler, Carole L

COLORADO
Berger, Byron Roland
Bisque, Ramon Edward
Byers, Frank Milton, Jr
Cadigan, Robert Allen
Chaffee, Maurice A(hlborn)
Chao, Tsun Tien
Christie, Joseph Herman
Church, Stanley Eugene
Clark, Benton C
Dean, Walter E, Jr
Dembicki, Harry, Jr
Dubiel, Russell F
Edwards, Kenneth Ward
Erslev, Eric Allan
Ferris, Bernard Joe
Gerlach, Terrence Melvin
Griffitts, Wallace Rush
Heyl, Allen Van, Jr
Hill, Walter Edward, Jr
Hills, Francis Allan
Keighin, Charles William
Klusman, Ronald William
Langmuir, Donald
McCallum, Malcolm E
Mason, Malcolm
Miller, William Robert
Modreski, Peter John
Nash, J(ohn) Thomas
Olhoeft, Gary Roy
Palacas, James George
Patton, James Winton
Post, Edwin Van Horn
Rall, Raymond Wallace
Raup, Omer Beaver
Reitsema, Robert Harold
Romberger, Samuel B
Rouse, George Elverton
Runnells, Donald DeMar
Rye, Robert O
Sanford, Richard Frederick
Severson, Ronald Charles
Sherman, David Michael
Sidder, Gary Brian
Silberman, Miles Louis
Sommerfeld, Richard Arthur
Szabo, Barney Julius
Tatsumoto, Mitsunobu
Vine, James David
Wahl, Floyd Michael
Waples, Douglas Wendle
Wershaw, Robert Lawrence
Wildeman, Thomas Raymond
Yang, In Che
Young, Edward Joseph
Zartman, Robert Eugene

CONNECTICUT
Berner, Robert A
Herron, Michael Myrl
Lasaga, Antonio C
Megrue, George Henry
Rye, Danny Michael
Skinner, Brian John
Turekian, Karl Karekin

DELAWARE
Culberson, Charles Henry
Leavens, Peter Backus
Luther, George William, III
Sharp, Jonathan Hawley

DISTRICT OF COLUMBIA
Carlson, Richard Walter
Clarke, Roy Slayton, Jr
Dasch, Ernest Julius
Fleischer, Michael
Fredriksson, Kurt A
Fudali, Robert F
Gaines, Alan McCulloch
Hare, Peter Edgar
Herr, Frank Leaman, Jr
Hoering, Thomas Carl
Mao, Ho-Kwang

Mason, Brian Harold
Mysen, Bjorn Olav
Rumble, Douglas, III
Siegel, Frederic Richard
Snyder, John L
Sokoloff, Vladimir P
Usselman, Thomas Michael
Weill, Daniel Francis
Wright, Thomas Oscar

FLORIDA
Braids, Olin Capron
Brezonik, Patrick Lee
Burnett, William Craig
Compton, John S
Donoghue, Joseph F
Fanning, Kent Abram
Garrels, Robert Minard
Learned, Robert Eugene
Mueller, Paul Allen
Piotrowicz, Stephen R
Ragland, Paul C
Sackett, William Malcolm
Warburton, David Lewis

GEORGIA
Alberts, James Joseph
Bailey, George William
Brown, David Smith
Dod, Bruce Douglas
Hanson, Hiram Stanley
Holzer, Gunther Ulrich
Howard, James Hatten, III
Noakes, John Edward
Pollard, Charles Oscar, Jr
Wampler, Jesse Marion
Weaver, Charles Edward
Wenner, David Bruce
Whitney, James Arthur

HAWAII
Clark, Allen LeRoy
Erdman, John Gordon
Garcia, Michael Omar
Halbig, Joseph Benjamin
Mackenzie, Frederick Theodore
Manghnani, Murli Hukumal
Ming, Li Chung
Muenow, David W
Sansone, Francis Joseph

IDAHO
Buden, Rosemary V
Farwell, Sherry Owen
Harrison, William Earl
Knobel, LeRoy Lyle
Welch, Jane Marie

ILLINOIS
Anders, Edward
Anderson, Thomas Frank
Beasley, Thomas Miles
Bennett, Glenn Allen
Berg, Jonathan H
Bishop, Finley Charles
Chou, Chen-Lin
Clayton, Robert Norman
Corbett, Robert G
Davis, Andrew Morgan
Goldsmith, Julian Royce
Gorody, Anthony Wagner
Grossman, Lawrence
Guggenheim, Stephen
Hayatsu, Ryoichi
Holt, Ben Dance
Hower, John, Jr
Kirkpatrick, R James
Kirkpatrick, Robert James
Koster Van Groos, August Ferdinand
Krug, Edward Charles
Lerman, Abraham
Minear, Roger Allan
Moll, William Francis, Jr
Moore, Duane Milton
Palmer, Curtis Allyn
Perry, Eugene Carleton, Jr
Scouten, Charles George
Sood, Manmohan K
Stueber, Alan Michael

INDIANA
Elmore, David
Hayes, John Michael
Heydegger, Helmut Roland
Krothe, Noel C
Kullerud, Gunnar
Lipschutz, Michael Elazar
Meyer, Henry Oostenwald Albertijn
Shaffer, Nelson Ross
Shieh, Yuch-Ning
Towell, David Garrett
Wintsch, Robert P

IOWA
Kracher, Alfred
Lemish, John
Nordlie, Bert Edward
Seifert, Karl E

KANSAS
Angino, Ernest Edward
Bickford, Marion Eugene
Cullers, Robert Lee
Jones, Lois Marilyn

Van Schmus, William Randall
Welch, Jerome E
Whittemore, Donald Osgood
Zeller, Edward Jacob

KENTUCKY
Cobb, James C
Ehmann, William Donald
Thrailkill, John

LOUISIANA
Allen, Gary Curtiss
Ferrell, Ray Edward, Jr
Ham, Russell Allen
Overton, Edward Beardslee
Thibodeaux, Louis J
Wood, Elwyn Devere

MAINE
Howd, Frank Hawver
Mayer, Lawrence Michael
Murphy, Mary Ellen
Tewhey, John David

MARYLAND
Brown, Michael
Corliss, John Burt
Dunning, Herbert Neal
Everett, Ardell Gordon
Ferry, John Mott
French, Bevan Meredith
Helz, George Rudolph
Hinners, Noel W
Jones, Blair Francis
Kelly, William Robert
Krueger, Arlin James
Luth, William Clair
MacGregor, Ian Duncan
Marsh, Bruce David
May, Ira Philip
May, Irving
Mielke, James Edward
O'Keefe, John Aloysius
Riedel, Gerhardt Frederick
Thomas, Herman Hoit
Trombka, Jacob Israel
Walter, Louis S
Weidner, Jerry R
Wells, Eddie N
Winzer, Stephen Randolph

MASSACHUSETTS
Bates, Carl H
Bell, Peter M
Boehm, Paul David
Brownlow, Arthur Hume
Burns, Roger George
Cohen, Merrill
Dacey, John W H
Dennen, William Henry
Deuser, Werner Georg
Donohue, John J
Eby, G Nelson
Frey, Frederick August
Hart, Stanley Robert
Hepburn, John Christopher
Holland, Heinrich Dieter
Horzempa, Lewis Michael
Hunt, John Meacham
Jacobsen, Stein Bjornar
King, Kenneth, Jr
Leahy, Richard Gordon
Manheim, Frank T
Marshall, Donald James
Morse, Stearns Anthony
Naylor, Richard Stevens
Nunes, Paul Donald
Roedder, Edwin Woods
Sayles, Frederick Livermore
Selverstone, Jane Elizabeth
Shimizu, Nobumichi
Siever, Raymond
Teal, John Moline
Thompson, Geoffrey
Thompson, James Burleigh, Jr
Thompson, Mary Eleanor
Whelan, Jean King
Yuretich, Richard Francis

MICHIGAN
Barcelona, Michael Joseph
Cowgill, Ursula Moser
Essene, Eric J
Furlong, Robert B
Kilham, Peter
Leenheer, Mary Janeth
Meyers, Philip Alan
O'Neil, James R
Stonehouse, Harold Bertram
Symonds, Robert B
Vogel, Thomas A

MINNESOTA
Gorham, Eville
Murthy, Varanasi Rama
Parker, Ronald Bruce

MISSISSIPPI
Sundeen, Daniel Alvin

MISSOURI
Bolter, Ernst A
Crozaz, Ghislaine M
Haskin, Larry A

Johns, William Davis
Korotev, Randall Lee
Mantei, Erwin Joseph
Manuel, Oliver K
Podosek, Frank A

MONTANA
Alt, David D
Engel, Albert Edward John
Lange, Ian M
Volborth, Alexis

NEVADA
Christensen, Odin Dale
Cloke, Paul LeRoy
Dickson, Frank Wilson
Garside, Larry Joe
Hess, John Warren
Hsu, Liang-Chi
Kukal, Gerald Courtney
Lyons, William Berry
Nichols, Chester Encell
Price, Jonathan Greenway

NEW HAMPSHIRE
Gaudette, Henri Eugene
Hager, Jutta Lore
Hines, Mark Edward
McDowell, William H
Reynolds, Robert Coltart, Jr
Weber, James Harold

NEW JERSEY
Catanzaro, Edward John
Crerar, David Alexander
Feely, Herbert William
Ganapathy, Ramachandran
Gavini, Muralidhara B
Goldstein, Theodore Philip
Jonte, John Haworth
McVey, Jeffrey King
Mitchell, Thomas Owen
Puffer, John H
Schutz, Donald Frank
Sugam, Richard Jay
Takahashi, Taro

NEW MEXICO
Attrep, Moses, Jr
Baldridge, Warren Scott
Bild, Richard Wayne
Binder, Irwin
Bish, David Lee
Broadhead, Ronald Frigon
Brookins, Douglas Gridley
Broxton, David Edward
Bryant, Ernest Atherton
Cannon, Helen Leighton
Charles, Robert Wilson
Condie, Kent Carl
Downs, William Fredrick
Erdal, Bruce Robert
Ewing, Rodney Charles
Gancarz, Alexander John
Giordano, Thomas Henry
Grambling, Jeffrey A
Holley, Charles Elmer, Jr
Kuellmer, Frederick John
Laughlin, Alexander William
Meijer, Arend
Morris, David Albert
Norris, Andrew Edward
Norris, Andrew Edward
Northrop, David A
Ogard, Allen E
Papike, James Joseph
Phillips, Fred Melville
Polzer, Wilfred L
Popp, Carl John
Renault, Jacques Roland
Ristvet, Byron Leo
Rokop, Donald J
Shannon, Spencer Sweet, Jr
Taylor, G Jeffrey
Tremba, Edward Louis
Trujillo, Patricio Eduardo
Wolfsberg, Kurt

NEW YORK
Aller, Robert Curwood
Barnard, Walther M
Beall, George Halsey
Berkley, John Lee
Biscaye, Pierre Eginton
Boger, Phillip David
Broecker, Wallace
Brueckner, Hannes Kurt
Carl, James Dudley
Clemency, Charles V
DeLong, Stephen Edwin
Fisher, Nicholas Seth
Fountain, John Crothers
Friedman, Irving
Froelich, Philip Nissen
Gaffey, Michael James
Hanson, Gilbert N
Howarth, Robert W
Kay, Robert Woodbury
Kelly, John Russell
Lattimer, James Michael
Lawson, Charles Alden
Lewin, Seymour Z
Lindsley, Donald Hale
Longhi, John

Ma, Maw-Suen
McLaren, Eugene Herbert
McLennan, Scott Mellin
Miller, Donald Spencer
Moniot, Robert Keith
Premuzic, Eugene T
Putman, George Wendell
Reitan, Paul Hartman
Rutstein, Martin S
Scranton, Mary Isabelle
Smethie, William Massie, Jr
Speidel, David H
Thurber, David Lawrence
White, William Michael
Whitney, Philip Roy
Woodmansee, Donald Ernest

NORTH CAROLINA
Abbott, Richard Newton, Jr
Bray, John Thomas
Feiss, Paul Geoffrey
Fullagar, Paul David
Harris, William Burleigh
Livingstone, Daniel Archibald
Rogers, John James William
Willey, Joan DeWitt

NORTH DAKOTA
Sugihara, James Masanobu

OHIO
Banks, Philip Oren
Bladh, Katherine Laing
Bladh, Kenneth W
Carlson, Ernest Howard
Dahl, Peter Steffen
Dollimore, David
Faure, Gunter
Flynn, Ronald Thomas
Foland, Kenneth Austin
Gregor, Clunie Bryan
Kilinc, Attila Ishak
Koucky, Frank Louis, Jr
Kovach, Jack
Law, Eric W
Lo, Howard H
Matisoff, Gerald
Maynard, James Barry
Means, Jeffrey Lynn
Moody, Judith Barbara
Pushkar, Paul
Savin, Samuel Marvin
Stout, Barbara Elizabeth

OKLAHOMA
Al-Shaieb, Zuhair Fouad
Ames, Roger Lyman
Barker, Colin G
Collins, Arlee Gene
Enfield, Carl George
Ferguson, William Sidney
Fisher, John Berton
Gilbert, Murray Charles
Henderson, Frederick Bradley, III
Hitzman, Donald Oliver
Ho, Thomas Tong-Yun
Hounslow, Arthur William
London, David
Philp, Richard Paul
Powell, Benjamin Neff
Puls, Robert W
Smith, James Edward
Thompson, Robert Richard
Williams, Jack A

OREGON
Bennington, Kenneth Oliver
Field, Cyrus West
Fish, William
Goles, Gordon George
Goodney, David Edgar
Holser, William Thomas
Lotspeich, Frederick Benjamin
McBirney, Alexander Robert
Nafziger, Ralph Hamilton
Powell, James Lawrence

PENNSYLVANIA
Allen, Herbert E
Allen, Jack C, Jr
Allen, John Christopher
Arthur, Michael Allan
Barnes, Hubert Lloyd
Brady, Keith Bryan Craig
Brantley, Susan Louise
Buis, Patricia Frances
Cassidy, William Arthur
Clavan, Walter
Coe, Charles Gardner
Cohen, Alvin Jerome
Crawford, William Arthur
Deines, Peter
Eggler, David Hewitt
Given, Peter Hervey
Grubbs, Donald Keeble
Heald, Emerson Francis
Herman, Richard Gerald
Herzog, Leonard Frederick, II
Hodgson, Robert Arnold
Keith, MacKenzie Lawrence
Kerrick, Derrill M
Kohman, Truman Paul
Kump, Lee Robert
Liu, Mian

Lukert, Michael T
Mangelsdorf, Paul Christoph, Jr
Maroulis, Peter James
Muan, Arnulf
Niemitz, Jeffrey William
Ohmoto, Hiroshi
Osborn, Elburt Franklin
Rohl, Arthur N
Rose, Arthur William
Roy, Della M(artin)
Schmalz, Robert Fowler
Sclar, Charles Bertram
Seidell, Barbara Castens
Spear, Jo-Walter
Suhr, Norman Henry
Ulmer, Gene Carleton
White, William Blaine
Witkowski, Robert Edward
Yeh, Gour-Tsyh

RHODE ISLAND
Abell, Paul Irving
Duce, Robert Arthur
Felbeck, George Theodore, Jr
Gearing, Juanita Newman
Giletti, Bruno John
Hermes, O Don
Leinen, Margaret Sandra
Pilson, Michael Edward Quinton
Quinn, James Gerard

SOUTH CAROLINA
Clayton, Donald Delbert
Geidel, Gwendelyn
Moore, Willard S
Shervais, John Walter
Stakes, Debra Sue
Strom, Richard Nelsen
Walters, John Philip
Warner, Richard Dudley

TENNESSEE
Ault, Wayne Urban
Bohlmann, Edward Gustav
Emanuel, William Robert
Kopp, Otto Charles
Lindberg, Steven Eric
McSween, Harry Y, Jr
Marland, Gregg (Hinton)
Stow, Stephen Harrington
Taylor, Lawrence August

TEXAS
Adams, John Allan Stewart, Sr
Anthony, Elizabeth Youngblood
Arp, Gerald Kench
Baker, Donald Roy
Bence, Alfred Edward
Bernard, Bernie Boyd
Bishop, Richard Stearns
Bogard, Donald Dale
Botto, Robert Irving
Boutton, Thomas William
Brown, Robert Wade
Burkart, Burke
Casagrande, Daniel Joseph
Claypool, George Edwin
Clendening, John Albert
Cooper, James Erwin
Crick, Rex Edward
Dahl, Harry Martin
Eastwood, Raymond L
Eckelmann, Walter R
Elthon, Donald L
Ewing, Thomas Edward
Gibson, Everett Kay, Jr
Gooding, James Leslie
Haas, Herbert
Halpern, Martin
Henry, Christopher Duval
Ho, Clara Lin
Holdaway, Michael Jon
Ingerson, Fred Earl
Jungclaus, Gregory Alan
Kaiser, William Richard
Koepnick, Richard Borland
Kulla, Jean B
Lafon, Guy Michel
Land, Lynton S
Levien, Louise
Lewis, Donald Richard
Lindstrom, David John
Lindstrom, Marilyn Martin
Loeppert, Richard Henry, Jr
Lofgren, Gary Ernest
Long, Leon Eugene
Louden, L Richard
McDowell, Fred Wallace
Matthews, Martin David
Mitterer, Richard Max
Monaghan, Patrick Henry
Moore, Thomas Francis
Morse, John Wilbur
Nussmann, David George
Nyquist, Laurence Elwood
Orr, Wilson Lee
Pannell, Keith Howard
Parker, Patrick LeGrand
Parker, Robert Hallett
Pearson, Daniel Bester, III
Pearson, Frederick Joseph
Phinney, William Charles
Pogue, Randall F
Potts, Mark John

Presnall, Dean C
Roe, Glenn Dana
Rogers, Marion Alan
Schink, David R
Schumacher, Dietmar
Schwarzer, Theresa Flynn
Scott, Martha Richter
Self, Stephen
Smith, Michael A
Sommer, Sheldon E
Stormer, John Charles, Jr
Strom, E(dwin) Thomas
Von Rosenberg, Dale Ursini
Walters, Lester James, Jr
Warshauer, Steven Michael
Whelan, Thomas, III
Young, William Allen

UTAH
Christiansen, Eric H
Jurinak, Jerome Joseph
Kolesar, Peter Thomas
Nash, William Purcell
Parry, William Thomas
Phillips, Lee Revell
Wolgemuth, Kenneth Mark

VERMONT
Furukawa, Toshiharu
Westerman, David Scott

VIRGINIA
Allen, Ralph Orville, Jr
Baedecker, Mary Jo
Baedecker, Philip A
Bethke, Philip Martin
Bodnar, Robert John
Bricker, Owen P, III
Busenberg, Eurybiades
Craig, James Roland
Cunningham, Charles G
Dickinson, Stanley Key, Jr
Fabbi, Brent Peter
Finkelman, Robert Barry
Fisher, James Russell
Frye, Keith
Galloway, James Neville
Grossman, Jeffrey N
Grove, Thurman Lee
Hanna, William Jefferson
Hanshaw, Bruce Busser
Hearn, Bernard Carter, Jr
Hemingway, Bruce Sherman
Huang, Wen Hsing
Huebner, John Stephen
Huggett, Robert James
Kornicker, William Alan
Leo, Gerhard William
Levine, Joel S
Lipin, Bruce Reed
Macko, Stephen Alexander
Minkin, Jean Albert
Moore, Raymond Kenworthy
Morgan, John Walter
Mose, Douglas George
Padovani, Elaine Reeves
Pickering, Ranard Jackson
Rankin, Douglas Whiting
Ridge, John Drew
Roseboom, Eugene Holloway, Jr
Ross, Malcolm
Rubin, Meyer
Sato, Motoaki
Sayala, Dasharatham (Dash)
Schoen, Robert
Schreiber, Henry Dale
Sherwood, William Cullen
Simon, Frederick Otto
Smith, Richard Elbridge
Stewart, David Benjamin
Tanner, Allan Bain
Toulmin, Priestley
Walker, Alta Sharon
Wong, George Tin Fuk
Wood, Warren Wilbur

WASHINGTON
Ames, Lloyd Leroy, Jr
Bamford, Robert Wendell
Barnes, Ross Owen
Brownlee, Donald E
Chang, Yew Chun
Crecelius, Eric A
Ghiorso, Mark Stefan
Hedges, John Ivan
Hodges, Floyd Norman
Houck, John Candee
Jenne, Everett A
Laul, Jagdish Chander
Mopper, Kenneth
Olson, Edwin Andrew
Quay, Paul Douglas
Riley, Robert Gene
Serne, Roger Jeffrey
Sklarew, Deborah S
Stuiver, Minze
Walker, Sharon Leslie
Weis, Paul Lester
Wright, Judith

WEST VIRGINIA
Rauch, Henry William
Renton, John Johnston

Geochemistry (cont)

WISCONSIN
Bowser, Carl
Brown, Philip Edward
Hoffman, James Irvie
Krezoski, John R
Mudrey, Michael George, Jr
Myers, Charles R
Rosson, Reinhardt Arthur
Stenstrom, Richard Charles
West, Walter Scott

WYOMING
Branthaver, Jan Franklin
Howatson, John
Robinson, Wilbur Eugene

ALBERTA
Baadsgaard, Halfdan
Bird, Gordon Winslow
Clementz, David Michael
Cumming, George Leslie
Hitchon, Brian
Hunt, C Warren
Lambert, Richard St John
Nesbitt, Bruce Edward
Schmidt, Volkmar
Strausz, Otto Peter
Vitt, Dale Hadley

BRITISH COLUMBIA
Armstrong, Richard Lee
Arnold, Ralph Gunther
Clark, Lloyd Allen
Russell, Richard Doncaster
Sinclair, Alastair James

MANITOBA
Brunskill, Gregg John
Hecky, Robert Eugene
Vilks, Peter

NEW BRUNSWICK
Allen, Clifford Marsden

NEWFOUNDLAND
Haines, Robert Ivor
Longerich, Henry Perry
Welhan, John Andrew

NOVA SCOTIA
Buckley, Dale Eliot
Cooke, Robert Clark
Loring, Douglas Howard

ONTARIO
Anderson, Gregor Munro
Appleyard, Edward Clair
Baragar, William Robert Arthur
Bevier, Mary Lou
Brown, Seward Ralph
Cabri, Louis J
Carlisle, David Brez
Crocket, James Harvie
Dayal, Ramesh
De Kimpe, Christian Robert
Duke, John Murray
Edgar, Alan D
Fox, Joseph S
Fyfe, William Sefton
George, Albert El Deeb
Grieve, Richard Andrew
Haynes, Simon John
Jackson, Togwell Alexander
James, Richard Stephen
Kramer, James Richard
Kretz, Ralph
Krogh, Thomas Edvard
McNutt, Robert Harold
Macqueen, Roger Webb
MacRae, Neil D
Maxwell, John Alfred
Mudroch, Alena
Nriagu, Jerome Okonkwo
Pajari, George Edward
Pearce, Thomas Hulme
Roeder, Peter Ludwig
Schwarcz, Henry Philip
Scott, Steven Donald
Shaw, Denis Martin
Smith, Terence E
Turek, Andrew
Veizer, Ján
Walton, Alan
Weiler, Roland R

QUEBEC
Brown, Alex Cyril
Hesse, Reinhard
Hillaire-Marcel, Claude
Khalil, Michel
MacLean, Wallace H
Ouellet, Marcel
Power, Joan F
Webber, George Roger

SASKATCHEWAN
Chew, Hemming
Davies, John Clifford
Huang, P M

OTHER COUNTRIES
Andreae, Meinrat Otto
Charola, Asuncion Elena
Finlayson, James Bruce
Fritz, Peter
Hofmann, Albrecht Werner
Loeliger, David A
Schrayer, Grover J, Jr
Ugolini, Fiorenzo Cesare
Walton, Wayne J A, Jr

Geology, Applied, Economic & Engineering

ALABAMA
Bearce, Denny N
Carrington, Thomas Jack
Chalokwu, Christopher Iloba
Cook, Robert Bigham, Jr
Groshong, Richard Hughes, Jr
Hastings, Earl L
Hughes, Travis Hubert
Jones, Douglas Epps
LaMoreaux, Philip Elmer
McCarl, Henry Newton
McLaughlin, Kenneth Phelps
Mancini, Ernest Anthony
Marchant, Guillaume Henri (Wim), Jr
Moser, Paul H
Neathery, Thornton Lee
Simpson, Thomas A
Workman, William Edward

ALASKA
Brewer, Max C(lifton)
Calderwood, Keith Wright
Ferrians, Oscar John, Jr
Johansen, Nils Ivar
Mangus, Marvin D
Mowatt, Thomas C
Reed, Bruce Loring
Schmidt, Ruth A M
Triplehorn, Don Murray
Turner, Donald Lloyd
Weber, Florence Robinson
Wolff, Ernest N

ARIZONA
Agenbroad, Larry Delmar
Breed, Carol Sameth
Burt, Donald McLain
Cathey, Everett Henry
Childs, Orlo E
Cole, W Storrs
Cotera, Augustus S, Jr
Cree, Allan
Dapples, Edward Charles
DeCook, Kenneth James
Dickinson, William Richard
Dietz, Robert Sinclair
Elston, Donald (Parker)
Emery, Philip Anthony
Erskine, Christopher Forbes
Fellows, Larry Dean
Fink, James Brewster
Fisher, Frederick Stephen
Ganguly, Jibamitra
Gillis, James E, Jr
Greeley, Ronald
Guilbert, John M
Harris, DeVerle Porter
Hawkes, Herbert Edwin, Jr
Haynes, Caleb Vance, Jr
Heede, Burchard Heinrich
Johnsen, John Herbert
Kamilli, Robert Joseph
Kassander, Arno Richard, Jr
Kieffer, Susan Werner
Kiersch, George Alfred
Kremp, Gerhard Otto Wilhelm
Lacy, W(illard) C(arleton)
Lance, John Franklin
Loring, William Bacheller
Lundin, Robert Folke
McCauley, John Francis
Newman, William L
Norton, Denis Locklin
Ohle, Ernest Linwood
Peters, William Callier
Péwé, Troy Lewis
Radabaugh, Robert Eugene
Richardson, Randall Miller
Roddy, David John
Rush, Richard William
Schaber, Gerald Gene
Schreiber, Joseph Frederick, Jr
Shenkel, Claude W, Jr
Stump, Edmund
Sumner, John Stewart
Swann, Gordon Alfred
Thurston, William R
Titley, Spencer Rowe
Wait, James Richard
Willis, Clifford Leon
Wilson, Richard Fairfield

ARKANSAS
Jackson, Kern Chandler
Konig, Ronald H
MacDonald, Harold Carleton
Mack, Leslie Eugene
Manger, Walter Leroy
Oglesby, Gayle Arden

Spivey, Robert Charles
Steele, Kenneth F
Wagner, George Hoyt

CALIFORNIA
Abrams, Michael J
Adent, William A
Albee, Arden Leroy
Albers, John P
Alfors, John Theodore
Allan, John Ridgway
Allen, Clarence Roderic
Anderson, Charles Alfred
Arnal, Robert Emile
Ashley, Roger Parkmand
Austin, Carl Fulton
Avent, Jon C
Balthaser, Lawrence Harold
Barbat, William Franklin
Barron, John Arthur
Bass, Manuel N
Beaty, David Wayne
Berg, Henry Clay
Birman, Joseph Harold
Blackerby, Bruce Alfred
Blake, Milton Clark, Jr
Blom, Christian James
Bohannon, Robert Gary
Bonham, Lawrence Cook
Bonilla, Manuel George
Borax, Eugene
Borg, Iris Y P
Brabb, Earl Edward
Brahma, Chandra Sekhar
Bredehoeft, John Dallas
Brew, David Alan
Brimhall, George H
Brogan, George Edward
Brown, Glenn A
Bunch, Theodore Eugene
Carlisle, Donald
Case, James Edward
Castle, Robert O
Chapman, Robert Mills
Christensen, Mark Newell
Church, Clifford Carl
Churkin, Michael, Jr
Ciancanelli, Eugene Vincent
Clark, Bruce R
Clements, Thomas
Clifton, Hugh Edward
Cluff, Lloyd Sterling
Coash, John Russell
Coats, Robert Roy
Cogen, William Maurice
Coleman, Robert Griffin
Conrey, Bert L
Cook, Harry E, III
Cox, Dennis Purver
Creasey, Saville Cyrus
Curran, John Franklin
Dana, Stephen Winchester
Dewart, Gilbert
Donath, Fred Arthur
Douglas, Hugh
Dreyer, Robert Marx
Drugg, Warren Sowle
Easton, William Heyden
Eberlein, George Donald
Edmund, Rudolph William
Ehlig, Perry Lawrence
Ehrreich, Albert LeRoy
Einaudi, Marco Tullio
Elders, Wilfred Allan
Elliott, William J
Emerson, Donald Orville
Evans, James R
Evensen, James Millard
Evitt, William Robert
Fillippone, Walter R
Finney, Stanley Charles
Fisher, Richard Virgil
Floran, Robert John
Ford, Arthur B
Foster, Helen Laura
Foster, Robert John
Frautschy, Jeffery Dean
Frederickson, Arman Frederick
Friberg, James Frederick
Garrison, Robert Edward
Gastil, R Gordon
Gealy, Elizabeth Lee
Goodman, Richard E
Graham, Stephan Alan
Graves, Roy William, Jr
Gray, Cliffton Herschel, Jr
Greene, Robert Carl
Greenman, Norman
Griggs, Gary B
Gromme, Charles Sherman
Gryc, George
Gutstadt, Allan Morton
Guyton, James W
Hanscom, Roger H
Harbaugh, John Warvelle
Hart, Earl W
Harwood, David Smith
Havard, John Francis
Hawkins, James Wilbur, Jr
Hein, James R
Helin, Eleanor Kay
Henderson, Gerald Vernon
Henshaw, Paul Carrington
Herber, Lawrence Justin

Hesse, Christian August
Hietanen-Makela, Anna (Martta)
Higgins, James Woodrow
Hill, Donald Gardner
Hill, Mason Lowell
Hill, Merton Earle, III
Hirschfeld, Sue Ellen
Hodges, Carroll Ann
Hodges, Lance Thomas
Holwerda, James G
Holzer, Thomas Lequear
Hopson, Clifford Andrae
Hoskins, Cortez William
Hotz, Preston Enslow
Humphrey, Thomas Milton, Jr
Hurst, Richard William
Ingersoll, Raymond Vail
Inman, Douglas Lamar
Isherwood, William Frank
Jacobson, Stephen Richard
James, Laurence Beresford
Janke, Norman C
Jizba, Zdenek Vaclav
Johnson, Vard Hayes
Jones, David Lawrence
Kam, James Ting-Kong
Kaska, Harold Victor
Kastner, Miriam
Keith, Terry Eugene Clark
Kirby, James M
Kistler, Ronald Wayne
Klinger, Allen
Kopf, Rudolph William
Kruger, Fredrick Christian
Krummenacher, Daniel
LaFountain, Lester James, Jr
Lanphere, Marvin Alder
Leighton, Freeman Beach
Leith, Carlton James
Leps, Thomas MacMaster
Lewis, Arthur Edward
Light, Mitchell A
Lipman, Peter Waldman
Lippmann, Marcelo Julio
Lipps, Jere Henry
Loomis, Alden Albert
Lovering, Thomas Seward
Lumsden, William Watt, Jr
Lyon, Ronald James Pearson
McCrone, Alistair William
Macdonald, Kenneth Craig
McGeary, David F R
McKague, Herbert Lawrence
McLean, Hugh
McNitt, James R
Maddock, Marshall
Magoon, Leslie Blake, III
Majmundar, Hasmukhrai Hiralal
Majmundar, Hasmukhrai Hiralal
Mandra, York T
Martin, Roger Charles
Matthews, Robert A
Merifield, Paul M
Miatech, Gerald James
Moiseyev, Alexis N
Montgomery, Richard C
Moore, Henry J, II
Moran, William Rodes
Morris, Hal Tryon
Morris, William Joseph
Muessig, Siegfried Joseph
Muffler, Leroy John Patrick
Murphy, Michael A
Murray, Bruce C
Nash, Douglas E
Nelson, Carlton Hans
Nixon, Alan Charles
Norman, L(ewis) A(rthur), Jr
Norris, Robert Matheson
Nygreen, Paul W
Oakeshott, Gordon B
Olmsted, Franklin Howard
Olson, Jerry Chipman
Otte, Carel
Pampeyan, Earl H
Parker, Calvin Alfred
Patton, William Wallace, Jr
Peak, Wilferd Warner
Pestrong, Raymond
Peterson, Donald William
Phillips, Richard P
Phoenix, David A
Pike, Richard Joseph
Pincus, Howard Jonah
Pipkin, Bernard Wallace
Plafker, George
Pomeroy, John S
Pray, Ralph Emerson
Proctor, Richard James
Ptacek, Anton D
Radbruch-Hall, Dorothy Hill
Ramspott, Lawrence Dewey
Rex, Robert Walter
Richter, Raymond C
Rieg, Louis Eugene
Riese, Walter Charles Rusty
Rigsby, George Pierce
Riley, N Allen
Rodda, Peter Ulisse
Rogers, Thomas Hardin
Rose, Robert Leon
Rosenfeld, John L
Sabins, Floyd F
Sage, Orrin Grant, Jr

Saunders, Ronald Stephen
Schmidt, Richard Arthur
Schneiderman, Jill Stephanie
Schwartz, Daniel Evan
Scott, Edward W
Seed, Harry Bolton
Seiden, Hy
Seiders, Victor Mann
Singer, Donald Allen
Slosson, James E
Smalley, Robert Gordon
Smedes, Harry Wynn
Smith, Raymond James
Smith, Robert Leland
Smith-Evernden, Roberta Katherine
Snavely, Parke Detweiler, Jr
Stager, Harold Keith
Stanley, George M
Stensrud, Howard Lewis
Stewart, John Harris
Stout, Martin Lindy
Stuart-Alexander, Desiree Elizabeth
Sullivan, Raymond
Sullwold, Harold H
Sundaram, Panchanatham N
Swanberg, Chandler A
Swinney, Chauncey Melvin
Tabor, Rowland Whitney
Terry, Richard D
Testa, Stephen M
Theodore, Ted George
Thompson, George Albert
Tooker, Edwin Wilson
Townley, John Lewis, III
Towse, Donald Frederick
Travers, William Brailsford
Troxel, Bennie Wyatt
Truesdell, Alfred Hemingway
Turk, Leland Jan
Tyler, John Howard
Valentine, James William
van de Kamp, Peter Cornelis
Volbrecht, Stanley Gordon
Waggoner, Eugene B
Wahrhaftig, Clyde (Adolph)
Walawender, Michael John
Wallace, Robert Earl
Wasserburg, Gerald Joseph
Watson, Kenneth De Pencier
Weed, Elaine Greening Ames
Weiss, Lionel Edward
Wentworth, Carl M, Jr
White, Donald Edward
Williams, Alan Evan
Williams, Floyd James
Williams, John Wharton
Willis, David Edwin
Wilson, L Kenneth
Witherspoon, Paul A(dams), Jr
Wrucke, Chester Theodore, Jr
Wyllie, Peter John
Yeend, Warren Ernest
Zenger, Donald Henry
Ziony, Joseph Israel
Zoback, Mary Lou Chetlain
Zumberge, James Herbert

COLORADO
Adams, John Wagstaff
Ager, Thomas Alan
Aleinikoff, John Nicholas
Alpha, Andrew G
Applegate, James Keith
Arbenz, Johann Kaspar
Averitt, Paul
Ayler, Maynard Franklin
Bailey, R V
Ball, Douglas
Ball, Mahlon M
Barber, George Arthur
Barker, Fred
Barron, Eric James
Berger, Byron Roland
Berry, George Willard
Birkeland, Peter Wessel
Blair, Robert William
Bohor, Bruce Forbes
Braun, Lewis Timothy
Broin, Thayne Leo
Bromfield, Calvin Stanton
Bucknam, Robert Campbell
Bush, Alfred Lerner
Byers, Frank Milton, Jr
Cadigan, Robert Allen
Cady, Wallace Martin
Callender, Jonathan Ferris
Canney, Frank Cogswell
Cathcart, James B
Chaffee, Maurice A(hlborn)
Chamberlain, Theodore Klock
Chenoweth, William Lyman
Chico, Raymundo Jose
Choquette, Philip Wheeler
Clift, William Orrin
Cluff, Robert Murri
Coates, Donald Allen
Cobban, William Aubrey
Colton, Roger Burnham
Connor, Jon James
Craig, Dexter Hildreth
Crandell, Dwight Raymond
Cummings, David
Curry, James Kenneth
Curtis, Bruce Franklin

Cygan, Norbert Everett
Dahl, Arthur Richard
Daviess, Steven Norman
Desborough, George A
Dobrovolny, Ernest
Doehring, Donald O
Dolly, Edward Dawson
Dubiel, Russell F
Eckel, Edwin Butt
Edwards, John D
Eicher, Don Lauren
Elder, Curtis Harold
Everhart, Donald Lough
Fails, Thomas Glenn
Ferris, Bernard Joe
Ferris, Clinton S, Jr
Finch, Warren Irvin
Fleming, Robert William
Flores, Romeo M
Foster, Norman Holland
Frezon, Sherwood Earl
Friedman, Jules Daniel
Gerlach, Terrence Melvin
Goldich, Samuel Stephen
Graham, Charles Edward
Griffitts, Wallace Rush
Grose, Thomas Lucius Trowbridge
Grutt, Eugene Wadsworth, Jr
Hardaway, John E(vans)
Harms, John Conrad
Harrison, Jack Edward
Harshman, Elbert Nelson
Haun, John Daniel
Hayes, John Bernard
Hays, William Henry
Henrickson, Eiler Leonard
Heyl, Allen Van, Jr
Hittelman, Allen M
Hobbs, Samuel Warren
Howard, Thomas E(dward)
Hull, Joseph Poyer Deyo, Jr
Hutchinson, Richard William
Hutchinson, Robert Maskiell
Jacobson, Jimmy Joe
Jibson, Randall W
Johnson, Frederick Arthur, Jr
Johnson, Robert Britten
Johnson, Ross Byron
Johnson, Verner Carl
Johnstone, James G(eorge)
Jordan, James N
Kanizay, Stephen Peter
Keefer, William Richard
Keighin, Charles William
Kent, Harry Christison
Knight, Fred G
Kohls, Donald W
Kuntz, Mel Anton
Kupfer, Donald Harry
Lackey, Laurence
Landon, Robert E
Langmuir, Donald
Larson, Edwin E
Lee, Keenan
LeMasurier, Wesley Ernest
Leonard, B(enjamin) F(ranklin), (III)
Levings, William Stephen
Lewis, David Allen
Lindsey, David Allen
Lohman, Stanley William
Longley, W(illiam) Warren
Lowell, James Diller
Luft, Stanley Jeremie
McCallum, Malcolm E
McCubbin, Donald Gene
MacKenzie, David Brindley
MacLachlan, James Crawford
Mallory, William Wyman
Marvin, Richard F(rederick)
Mason, Malcolm
Mast, Richard F(redrick)
Maughan, Edwin Kelly
Maxwell, Charles Henry
Meade, Robert Heber, Jr
Meyer, Harvey John
Miesch, Alfred Thomas
Modreski, Peter John
Moench, Robert Hadley
Moore, David A
Moore, David Warren
Moore, Fred Edward
Mullineaux, Donal Ray
Murphy, John Francis
Mytton, James W
Naeser, Charles Wilbur
Nash, J(ohn) Thomas
Newman, Karl Robert
Nichols, Donald Ray
Olson, Richard Hubbell
Osmond, John C, Jr
Osterhoudt, Walter Jabez
Panek, Louis A(nthony)
Parker, John Marchbank
Pennington, Wayne David
Peterman, Zell Edwin
Peterson, Fred
Peterson, Richard Charles
Pilkington, Harold Dean
Pillmore, Charles Lee
Ponder, Herman
Post, Edwin Van Horn
Powell, James Daniel
Prather, Thomas Leigh
Pratt, Walden Penfield

Prichard, George Edwards
Quirke, Terence Thomas, Jr
Rall, Raymond Wallace
Raup, Omer Beaver
Reed, John Calvin, Jr
Robinson, Peter
Rold, John W
Romberger, Samuel B
Rooney, Lawrence Frederick
Rosenblum, Sam
Rouse, George Elverton
Rowley, Peter DeWitt
Russell, Phillip K
Sable, Edward George
Sanborn, Albert Francis
Sanford, Richard Frederick
Savage, James Zuger
Schmidt, Dwight Lyman
Schuster, Robert Lee
Scott, James Henry
Scott, Robert Blackburn
Seeland, David Arthur
Segerstrom, Kenneth
Shawe, Daniel Reeves
Sheridan, Douglas Maynard
Sidder, Gary Brian
Simpson, Howard Edwin
Sims, Paul Kibler
Skipp, Betty Ann
Sloss, Peter William
Smyth, Joseph Richard
Sommerfeld, Richard Arthur
Spencer, Charles Winthrop
Staatz, Mortimer Hay
Stark, Philip Herald
Strobell, John Dixon, Jr
Stucky, Richard K(eith)
Taylor, Michael E
Terriere, Robert T
Thomasson, Maurice Ray
Thompson, Thomas Luman
Thompson, Thomas Luther
Thompson, Tommy Burt
Thomsen, Harry Ludwig
Thorman, Charles Hadley
Tourtelot, Harry Allison
Trexler, David William
Tschanz, Charles McFarland
Turner, Daniel Stoughton
Turner, Mortimer Darling
Twenhofel, William Stephens
Tweto, Ogden
Van Zant, Kent Lee
Varnes, David Joseph
Vine, James David
Visher, Frank N
Wallace, Stewart Raynor
Warner, Lawrence Allen
Weimer, Robert J
Weist, William Godfrey, Jr
Weitz, Joseph Leonard
Wenrich, Karen Jane
Westphal, Warren Henry
Wilson, Elizabeth Allen
Wilson, John Coe
Wingate, Frederick Huston
Witkind, Irving Jerome
Wray, John Lee

CONNECTICUT
Ballmann, Donald Lawrence
Baskerville, Charles Alexander
Fox, G(eorge) Sidney
Frasche, Dean F
Groff, Donald William
Hyde, John Welford
Levin, S Benedict
MacFadyen, John Archibald, Jr
Miller, Franklin Stuart
Orville, Philip Moore
Skinner, Brian John
Spiegel, Zane
Waage, Karl Mensch
Winchell, Horace

DELAWARE
Ellis, Patricia Mench
Jordan, Robert R
Kraft, John Christian
Nicholls, Robert Lee
Pickett, Thomas Ernest
Stiles, A(lvin) B(arber)
Talley, John Herbert

DISTRICT OF COLUMBIA
Appleman, Daniel Everett
Benson, William Edward Barnes
Boyd, Francis R
Carroll, Gerald V
Duguid, James Otto
Dutro, John Thomas, Jr
Fiske, Richard Sewell
Irvine, T Neil
Jolly, Janice Laurene Willard
Jones, Gareth Hubert Stanley
Justus, Philip Stanley
Knox, Arthur Stewart
Maccini, John Andrew
Melickian, Gary Edward
Neiheisel, James
O'Donnell, Edward
Parham, Walter Edward
Pierce, Jack Warren
Whitmore, Frank Clifford, Jr

Zeizel, A(rthur) J(ohn)

FLORIDA
Blanchard, Frank Nelson
Chauvin, Robert S
Chawner, William Donald
Chryssafopoulos, Hanka Wanda Sobczak
Chryssafopoulos, Nicholas
Craig, Alan Knowlton
Deere, Don U(el)
DeVore, George Warren
Freeland, George Lockwood
Garnar, Thomas E(dward), Jr
Geraghty, James Joseph
Goudarzi, Gus (Hossein)
Guild, Philip White
Hartman, John Paul
Hoagland, Alan D
Joensuu, Oiva I
Kadey, Frederic L, Jr
Kirkemo, Harold
Kuhn, Truman Howard
Langfelder, Leonard Jay
Learned, Robert Eugene
Lemon, Roy Richard Henry
McClellan, Guerry Hamrick
Nichols, Robert Leslie
O'Hara, Norbert Wilhelm
Opdyke, Neil
Osmond, John Kenneth
Parker, Garald G, Sr
Patton, Thomas Hudson
Pirkle, Earl C
Pullen, Milton William, Jr
Randazzo, Anthony Frank
Smith, Douglas Lee
Snyder, Walter Stanley
Spangler, Daniel Patrick
Stehli, Francis Greenough
Travis, Russell Burton
Upchurch, Sam Bayliss
Weisbord, Norman Edward
Werner, Harry Emil
Wise, Sherwood Willing, Jr

GEORGIA
Allard, Gilles Olivier
Allen, Arthur T, Jr
Barksdale, Richard Dillon
Carpenter, Robert Halstead
Carver, Robert E
Cofer, Harland E, Jr
Darrell, James Harris, II
Ghuman, Gian Singh
Gonzales, Serge
Grant, Willard H
Hanson, Hiram Stanley
Henry, Vernon James, Jr
Herz, Norman
Howard, James Dolan
Hurst, Vernon James
Koch, George Schneider, Jr
Little, Robert Lewis
Ogren, David Ernest
Power, Walter Robert
Sanders, Richard Pat
Shrum, John W
Size, William Bachtrup
Smith, John M
Sowers, George F(rederick)
Sumartojo, Jojok
Weaver, Charles Edward
Wenner, David Bruce

HAWAII
Arkle, Thomas, Jr
Chave, Keith Ernest
Clark, Allen LeRoy
Cruickshank, Michael James
Divis, Allan Francis
Erdman, John Gordon
Fan, Pow-Foong
Helsley, Charles Everett
Kemp, Paul James
Lockwood, John Paul
Malahoff, Alexander
Pankiwskyj, Kost Andrij
Robertson, Forbes
Sinton, John Maynard

IDAHO
Bonnichsen, Bill
Buden, Rosemary V
Emigh, G Donald
Hall, William Bartlett
Hoggan, Roger D
Hollenbaugh, Kenneth Malcolm
Isaacson, Peter Edwin
Jones, Robert William
Knobel, LeRoy Lyle
Knutson, Carroll Field
Laval, William Norris
Miller, Maynard Malcolm
Reid, Rolland Ramsay
Siems, Peter Laurence
Waag, Charles Joseph
Warner, Mont Marcellus
Williams, George Arthur
Wilson, Monte Dale
Wood, Spencer Hoffman

ILLINOIS
Archbold, Norbert L
Baxter, James Watson

Geology, Applied, Economic & Engineering (cont)

Bergstrom, Robert Edward
Block, Douglas Alfred
Bradbury, James Clifford
Buschbach, Thomas Charles
Carozzi, Albert Victor
Carpenter, Philip John
Casella, Clarence J
Charlier, Roger Henri
Chiou, Wen-An
Collinson, Charles William
Corbett, Robert G
Crelling, John Crawford
Damberger, Heinz H
Davidson, Donald Miner, Jr
DeMott, Lawrence Lynch
Dolgoff, Abraham
DuMontelle, Paul Bertrand
Dutcher, Russell Richardson
Flemal, Ronald Charles
Foster, John Webster
Fryxell, Fritiof Melvin
Gealy, William James
Goldsmith, Julian Royce
Gorody, Anthony Wagner
Grim, Ralph Early
Gross, David Lee
Haddock, Gerald Hugh
Harrison, Wyman
Hart, Richard Royce
Hess, David Filbert
Hoffer, Abraham
Ireland, Herbert O(rin)
Kent, Lois Schoonover
Langenheim, Ralph Louis, Jr
Leighton, Morris Wellman
McKay, Edward Donald, III
Mann, Christian John
Masters, John Michael
Maynard, Theodore Roberts
Mikulic, Donald George
Norby, Rodney Dale
Okal, Emile Andre
Olsen, Edward John
Peppers, Russel A
Qutub, Musa Y
Rue, Edward Evans
Sandvik, Peter Olaf
Sasman, Robert T
Saxena, Surendra K
Schlanger, Seymour Oscar
Searight, Thomas Kay
Shabica, Charles Wright
Simon, Jack Aaron
Speed, Robert Clarke
Sundelius, Harold Wesley
Trask, Charles Brian
Utgaard, John Edward
Weiss, Malcolm Pickett
Wills, Donald L
Winans, Randall Edward

INDIANA

Biggs, Maurice Earl
Boneham, Roger Frederick
Carr, Donald Dean
Chowdhury, Dipak Kumar
Cleveland, John H
Davis, Michael Edward
Fairley, William Merle
Franks, Stephen Guest
Gray, Henry Hamilton
Horowitz, Alan Stanley
Judd, William Robert
Kane, Henry Edward
Levandowski, Donald William
Madison, James Allen
Mirsky, Arthur
Murray, Haydn Herbert
Owen, Donald Eugene
Patton, John Barratt
Ruhe, Robert Victory
Scholer, Charles Frey
Shaffer, Nelson Ross
Suttner, Lee Joseph
Totten, Stanley Martin
Vitaliano, Charles Joseph
Vitaliano, Dorothy Braunbeck
Votaw, Robert Barnett
West, Terry Ronald
Wier, Charles Eugene
Winkler, Erhard Mario

IOWA

Anderson, Wayne I
Biggs, Donald Lee
Carmichael, Robert Stewart
Demirel, T(urgut)
De Nault, Kenneth J
Dorheim, Fredrick Houge
Drake, Lon David
Foster, Charles Thomas, Jr
Hallberg, George Robert
Handy, Richard L(incoln)
Hansman, Robert H
Hendriks, Herbert Edward
Hershey, H Garland
Lemish, John
Lohnes, Robert Alan
Nordlie, Bert Edward
O'Brien, Dennis Craig

Vondra, Carl Frank

KANSAS

Avcin, Matthew John, Jr
Baars, Donald Lee
Berg, J(ohn) Robert
Brady, Lawrence Lee
Bridge, Thomas E
Davis, John Clements
Enos, Paul (Portenier)
Frazier, A Joel
Grant, Stanley Cameron
Gundersen, James Novotny
Hambleton, William Weldon
Jones, Lois Marilyn
Leonard, Roy J
Merriam, Daniel Francis
Merrill, William Meredith
Oros, Margaret Olava (Erickson)
Parizek, Eldon Joseph
Robison, Richard Ashby
Tasch, Paul
Walters, Robert F
Williams, Wayne Watson

KENTUCKY

Brant, Russell Alan
Brown, William Randall
Campbell, Lois Jeannette
Chesnut, Donald R, Jr
Clark, Armin Lee
Cobb, James C
Conkin, James E
Deen, Robert C(urba)
Drahovzal, James Alan
Ferm, John Charles
Gauri, Kharaiti Lal
Gildersleeve, Benjamin
Hagan, Wallace Woodrow
Haney, Donald C
Hunt, Graham Hugh
Kiefer, John David
MacQuown, William Charles, Jr
Philley, John Calvin
Seeger, Charles Ronald
Sendlein, Lyle V A
Smith, Gilbert Edwin
Stanonis, Francis Leo
Thrailkill, John
Whaley, Peter Walter

LOUISIANA

Arden, Daniel Douglas
Eisenstatt, Phillip
Fairchild, Joseph Virgil, Jr
Forman, McLain J
Franklin, George Joseph
Frey, Maurice G
Gagliano, Sherwood Moneer
Goldthwaite, Duncan
Hartman, James Austin
Herrmann, Leo Anthony
Johnson, Hamilton McKee
Jones, Paul Hastings
Kumar, Madhurendu B
Leutze, Willard Parker
MacDougall, Robert Douglas
McGehee, Richard Vernon
Moore, Clyde H, Jr
Roberts, Harry Heil
Seglund, James Arnold
Shaw, Nolan Gail
Skinner, Hubert Clayton
Stephens, Maynard Moody
Stephens, Raymond Weathers, Jr
Tabbert, Robert L
Utterback, Donald D
Van Den Bold, Willem Aaldert
Van Lopik, Jack Richard
Von Almen, William Frederick I
Weidie, Alfred Edward
Wilcox, Ronald Erwin
Young, Leonard M

MAINE

Eastler, Thomas Edward
Farnsworth, Roy Lothrop
Fisher, Irving Sanborn
Gates, Olcott
Guidotti, Charles V
Howd, Frank Hawver
Norton, Stephen Allen
Osberg, Philip Henry
Splettstoesser, John Frederick
Tewhey, John David
Trefethen, Joseph Muzzy

MARYLAND

Baker, Arthur A
Barker, John L, Jr
Beattie, Donald A
Brush, Lucien M(unson), Jr
Cleaves, Emery Taylor
Cohee, George Vincent
Conant, Louis Cowles
Doan, David Bentley
Dorr, John Van Nostrand, II
Everett, Ardell Gordon
Felsher, Murray
Gair, Jacob Eugene
Gaum, Carl H
Hamburger, Richard
Honkala, Fred Saul
Hubbert, Marion King

Kappel, Ellen Sue
Karklins, Olgerts Longins
Kinney, Douglas Merrill
Luth, William Clair
Mangen, Lawrence Raymond
May, Ira Philip
Mielke, James Edward
Miller, Ralph LeRoy
Milton, Charles
Nininger, Robert D
Ozol, Michael Arvid
Patterson, Charles Meade
Pavlides, Louis
Petrie, William Leo
Pettijohn, Francis John
Rozanski, George
Salisbury, John William, Jr
Segovia, Antonio
Singewald, Quentin Dreyer
Warren, Charles Reynolds
Weaver, Kenneth Newcomer
Wolff, Roger Glen
Yokel, Felix Y
Zall, Linda S
Zen, E-an

MASSACHUSETTS

Benziger, Charles P(hillip)
Berggren, William Alfred
Bromery, Randolph Wilson
Brophy, Gerald Patrick
Brown, George D, Jr
Brownlow, Arthur Hume
Bryan, Wilfred Bottriol
Burchfiel, Burrell Clark
Burns, Daniel Robert
Caldwell, Dabney Withers
Carr, Jerome Brian
Dennen, William Henry
Donohue, David Arthur Timothy
Donohue, John J
El-Baz, Farouk
Enzmann, Robert D
Farquhar, Oswald Cornell
Foose, Richard Martin
Foote, Freeman
Fox, William Templeton
Frimpter, Michael Howard
Furlong, Ira E
Genes, Andrew Nicholas
Gheith, Mohamed A
Hart, Stanley Robert
Hoff, Darrel Barton
Hope, Elizabeth Greeley
Hume, James David
Jacobsen, Stein Bjornar
Jenks, William Furness
McConnell, Robert Kendall
Milliman, John D
Motts, Ward Sundt
Patterson, Joseph Gilbert
Petersen, Ulrich
Pinson, William Hamet, Jr
Raab, Allen Robert
Reuss, Robert L
Robinson, Peter
Roy, David C
Schalk, Marshall
Schultz, John Russell
Skehan, James William
Socolow, Arthur A
Stalcup, Marvel C
Tarkoy, Peter J
Terzaghi, Ruth Doggett
Uchupi, Elazar
Walsh, Joseph Broughton
Wise, Donald U

MICHIGAN

Andersland, Orlando Baldwin
Catacosinos, Paul Anthony
Coyle, Peter
Elliott, Jane Elizabeth Inch
Fisher, James Harold
Hendrix, Thomas Eugene
Jackson, Philip Larkin
Kalliokoski, Jorma Osmo Kalervo
Kehew, Alan Everett
Kelly, William Crowley
Kesler, Stephen Edward
Knowles, David M
Kuenzi, W David
Kunkle, George Robert
McKee, Edith Merritt
MacTavish, John N
Mills, Madolia Massey
Moore, Wayne Elden
Ogden, Lawrence
Ronca, Luciano Bruno
Straw, William Thomas
Tharin, James Cotter
Trow, James
Vogel, Thomas A
Ward, Richard Floyd
Whitney, Marion Isabelle
Whitten, Eric Harold Timothy
Wicander, Edwin Reed

MINNESOTA

Bayer, Thomas Norton
Blake, Rolland Laws
Bleifuss, Rodney L
Bright, Robert C
Buchwald, Caryl Edward

Donovan, John Francis
Farnham, Paul Rex
Grew, Priscilla Croswell Perkins
Hoagberg, Rudolph Karl
Lepp, Henry
Marsden, Ralph Walter
Matsch, Charles Leo
Meyers, James Harlan
Morey, Glenn Bernhardt
Southwick, David Leroy
Sparling, Dale R
Swain, Frederick Morrill, Jr
Tuthill, Samuel James
Walton, Matt Savage
Weiblen, Paul Willard
Willard, Robert Jackson

MISSISSIPPI

Brunton, George Delbert
Butler, Dwain Kent
Coleman, Neil Lloyd
Egloff, Julius, III
Happ, Stafford Coleman
Henderson, Arnold Richard
Johnson, Henry Stanley, Jr
Knight, Wilbur Hall
Krinitsky, E(llis) L(ouis)
Laswell, Troy James
Lowrie, Allen
Malone, Philip Garcin
Mather, Bryant
Mather, Katharine Kniskern
Meylan, Maurice Andre
Paulson, Oscar Lawrence
Perry, Edward Belk
Reynolds, William Roger
Riggs, Karl A, Jr
Rucker, James Bivin
Russell, Ernest Everett
Shamburger, John Herbert

MISSOURI

Arvidson, Raymond Ernst
Barr, David John
Belt, Charles Banks, Jr
Bolter, Ernst A
Coveney, Raymond Martin, Jr
Danser, James W
Freeman, Thomas J
Gadberry, Howard M(ilton)
Hagni, Richard D
Hasan, Syed Eqbal
Hatheway, Allen Wayne
Hilpman, Paul Lorenz
Hopkins, M E
Houseknecht, David Wayne
Howe, Wallace Brady
Howell, Dillon Lee
Jester, Guy Earlscort
Keller, Walter David
Kisvarsanyi, Eva Bognar
Kisvarsanyi, Geza
Knox, Burnal Ray
Levin, Harold Leonard
Peck, Raymond Elliott
Prakash, Shamsher
Rockaway, John D, Jr
Ryan, Edward McNeill
Stitt, James Harry
Thomson, Kenneth Clair
Tibbs, Nicholas H
Viele, George Washington

MONTANA

Alt, David D
Balster, Clifford Arthur
Bartholomew, Mervin Jerome
Berg, Richard Blake
Chadwick, Robert Aull
Dresser, Hugh W
Earll, Fred Nelson
Engel, Albert Edward John
Fanshawe, John Richardson, II
Groff, Sidney Lavern
Lange, Ian M
Peterson, James Algert
Stout, Koehler
Temple, Kenneth Loren
Thompson, Gary Gene
Walker, Leland J
Weidman, Robert McMaster
Winston, Donald

NEBRASKA

Bentall, Ray
Carlson, Marvin Paul
Dreeszen, Vincent Harold
Meade, Grayson Eichelberger
Nettleton, Wiley Dennis
Shroder, John Ford, Jr
Skinner, Morris Fredrick
Stout, Thompson Mylan
Tanner, Lloyd George
Underwood, Lloyd B
Wayne, William John

NEVADA

Bonham, Harold F, Jr
Carr, James Russell
Christensen, Odin Dale
Daemen, Jaak J K
Dickson, Frank Wilson
Ellinwood, Howard Lyman
Felts, Wayne Moore

Fiero, George William, Jr
Flint, Delos Edward
Fuerstenau, M(aurice) C(lark)
Garside, Larry Joe
Ghosh, Amitava
Gustafson, Lewis Brigham
Helm, Donald Cairney
Hibbard, Malcolm Jackman
Horton, Robert Carlton
Jones, Michael Baxter
Kaufmann, Robert Frank
Kukal, Gerald Courtney
Larson, Lawrence T
Lintz, Joseph, Jr
Littleton, Robert T
Lucas, James Robert
Manning, John Craige
Nichols, Chester Encell
Paige, Russell Alston
Papke, Keith George
Payne, Anthony
Peck, John H
Price, Jonathan Greenway
Ritter, Dale Franklin
Robinson, Raymond Francis
Schilling, John H(arold)
Schweickert, Richard Allan
Slemmons, David Burton
Taranik, James Vladimir
Trexler, Dennis Thomas
Watters, Robert James
Wyman, Richard Vaughn

NEW HAMPSHIRE
Birnie, Richard Williams
Denny, Charles Storrow
Gaudette, Henri Eugene
Hager, Jutta Lore
Hoag, Roland Boyden, Jr
Holton, Adolphus
Johnson, Gary Dean
Klotz, Louis Herman
Lyons, John Bartholomew
McKim, Harlan L
Page, Lincoln Ridler
Richter, Dorothy Anne

NEW JERSEY
Agron, Sam Lazrus
Bonini, William E
Brons, Cornelius Hendrick
Dike, Paul Alexander
Faust, George Tobias
Gilliland, William Nathan
Grow, George Copernicus, Jr
Hargraves, Robert Bero
Harrison, Jack Lamar
Hedberg, Hollis Dow
Judson, Sheldon, Jr
Manspeizer, Warren
Mason, John Frederick
Metsger, Robert William
Meyer, Richard Fastabend
Monahan, Edward James
Morris, Robert Clarence
Navarra, John Gabriel
Suppe, John
Thomson, Alan
Widmer, Kemble
Yasso, Warren E

NEW MEXICO
Anderson, Roger Yates
Austin, George Stephen
Beckner, Everet Hess
Bejnar, Waldemere
Bieberman, Robert Arthur
Bish, David Lee
Broadhead, Ronald Frigon
Bullard, Edwin Roscoe, Jr
Chapin, Charles Edward
Clemons, Russell Edward
Cunningham, John E
Deal, Dwight Edward
Downs, William Fredrick
Duffy, Clarence John
Elston, Wolfgang Eugene
Erikson, Mary Jane
Finney, Joseph J
Flower, Rousseau Hayner
Geissman, John William
Glock, Waldo Sumner
Griswold, George B
Guthrie, George D, Jr
Harris, David Vernon
Havenor, Kay Charles
Hawley, John William
Heiken, Grant Harvey
Jacobsen, Lynn C
Kelley, Vincent Cooper
Kottlowski, Frank Edward
Lattman, Laurence Harold
Laughlin, Alexander William
Meijer, Arend
Morin, George Cardinal Albert
Morris, David Albert
Neal, James Thomas
Pierce, Robert Wesley
Rapaport, Irving
Renault, Jacques Roland
Riecker, Robert E
Robertson, James Magruder
Shannon, Spencer Sweet, Jr
Shomaker, John Wayne

Smith, Clay Taylor
Thompson, Samuel, III
Trauger, Frederick Dale
Tremba, Edward Louis
Weber, Robert Harrison
Wengerd, Sherman Alexander
Whetten, John T
Wilson, Lee

NEW YORK
Adams, Robert W
Benedict, Peter Carl
Bhattacharji, Somdev
Bird, John Malcolm
Boger, Phillip David
Boshkov, Stefan
Calkin, Parker E
Cazeau, Charles J
Clemence, Samuel Patton
Coates, Donald Robert
Coch, Nicholas Kyros
Dubins, Mortimer Ira
Dunn, James Robert
Elberty, William Turner, Jr
Fagan, John J
Fairbridge, Rhodes Whitmore
Fakundiny, Robert Harry
Fickies, Robert H
Gillett, Lawrence B
Gilman, Richard Atwood
Hatheway, Richard Brackett
Hawkins, William Max
Hay, Robert E
Hays, James D
Hewitt, Philip Cooper
Heyl, George Richard
Hutchison, David M
Isachsen, Yngvar William
Jacoby, Russell Stephen
Jicha, Henry Louis, Jr
Johnson, Frank Walker
Kulhawy, Fred Howard
Langway, Chester Charles, Jr
Liebe, Richard Milton
Loring, Arthur Paul
Lowe, Kurt Emil
Ludman, Allan
Lundgren, Lawrence William, Jr
Mahtab, M Ashraf
Manos, Constantine T
Mattson, Peter Humphrey
Maurer, Robert Eugene
Mercer, Kermit R
Morisawa, Marie
Muller, Ernest Hathaway
Muller, Otto Helmuth
Multer, H Gray
Nair, Ramachandran Mukundalayam Sivarama
Newell, Norman Dennis
O'Rourke, Thomas Denis
Philbrick, Shailer Shaw
Potter, Donald B
Prucha, John James
Putman, George Wendell
Reitan, Paul Hartman
Rhodes, Frank Harold Trevor
Roberson, Herman Ellis
Robinson, Joseph Edward
Rodriguez, Joaquin
Ross, Stewart Hamilton
Ryan, William B F
Sanders, John Essington
Sargent, Kenneth Albert
Sheldon, Richard P
Siemankowski, Francis Theodore
Sirkin, Leslie A
Sorauf, James E
Stewart, John Conyngham
Sturm, Edward
Sutton, Robert George
Teichert, Curt
Vollmer, Frederick Wolfer
Wagner, Gerald Roy
Wang, Kia K
Wohlford, Duane Dennis
Wolff, Manfred Paul
Young, Richard A

NORTH CAROLINA
Brown, Charles Quentin
Brown, Henry Seawell
Buie, Bennett Frank
Butler, James Robert
Callahan, John Edward
Chace, Frederic Mason
Chapman, John Judson
Feiss, Paul Geoffrey
Ferguson, Herman White
Forsythe, Randall D
Glenn, George R(embert)
Horstman, Arden William
Karson, Jeffrey Alan
Kashef, A(bdel-Aziz) I(smail)
Marlowe, James Irvin
Mauger, Richard L
Mock, Steven James
Neal, Donald Wade
Parker, John Mason, III
Riggs, Stanley R
Smith, Norman Cutler
Teague, Kefton Harding
Textoris, Daniel Andrew
White, William Alexander

NORTH DAKOTA
Bickel, Edwin David
Brophy, John Allen
Karner, Frank Richard
Phillips, Monte Leroy
Reid, John Reynolds, Jr

OHIO
Anderson, John Jerome
Banks, Philip Oren
Bates, Robert Latimer
Bergstrom, Stig Magnus
Bernhagen, Ralph John
Bever, James Edward
Bladh, Kenneth W
Bonnett, Richard Brian
Carlson, Ernest Howard
Carlton, Richard Walter
Caster, Kenneth Edward
Collins, Horace Rutter
Collinson, James W
Colombini, Victor Domenic
Coogan, Alan H
Cropp, Frederick William, III
Elliot, David Hawksley
Emerson, John Wilford
Everett, K R
Fay, Philip S
Fenner, Peter
Frank, Glenn William
Fuller, James Osborn
Gerrard, Thomas Aquinas
Hall, John Frederick
Harmon, David Elmer, Jr
Harris, C Earl
Harris-Noel, Ann (F H) Graetsch
Heimlich, Richard Allen
Herdendorf, Charles Edward, III
Hyland, John R(oth)
Kahle, Charles F
Khawaja, Ikram Ullah
Kneller, William Arthur
Kulander, Byron Rodney
La Rocque, Joseph Alfred Aurele
Laughon, Robert Bush
McKenzie, Garry Donald
Mancuso, Joseph J
Maynard, James Barry
Means, Jeffrey Lynn
Moody, Judith Barbara
Morris, Robert William
Nance, Richard Damian
Nash, David Byer
Noel, James A
Patchick, Paul Francis
Pope, John Keyler
Potter, Paul Edwin
Pride, Douglas Elbridge
Pryor, Wayne Arthur
Pushkar, Paul
Reinhart, Roy Herbert
Root, Samuel I
Schoonmaker, George Russell
Schumm, Brooke, Jr
Skinner, Walter Swart
Springer, George Henry
Steel, Warren G
Stone, John Grover, II
Summerson, Charles Henry
Turner, Andrew
Unrug, Raphael
Utgard, Russell Oliver
Villar, James Walter
Webb, Peter Noel
Weitz, John Hills
White, John Francis
White, Sidney Edward
Wiese, Robert George, Jr
Wu, Tien Hsing

OKLAHOMA
Barczak, Virgil J
Bennison, Allan P
Bergendahl, Maximilian Hilmar
Bradley, John Samuel
Busch, Daniel Adolph
Chenoweth, Philip Andrew
Dickey, Parke Atherton
Erickson, John William
Espach, Ralph H, Jr
Forgotson, James Morris
Friedman, Samuel Arthur
Goldstein, August, Jr
Granath, James Wilton
Green, Thom Henning
Hedlund, Richard Warren
Helander, Donald P(eter)
Henderson, Frederick Bradley, III
Hobson, John Peter, Jr
Holmes, Clifford Newton
Horn, Myron K
Huffman, George Garrett
Hyne, Norman John
Johnson, Kenneth Sutherland
Joshi, Sadanand D
Kent, Douglas Charles
Kleen, Harold J
Koller, Glenn R
Konkel, Philip M
London, David
Mankin, Charles John
Masters, Bruce Allen
Menzie, Donald E
Merrill, Robert Kimball

Meyers, William C
Moritz, Carl Albert
Mumma, Martin Dale
Naff, John Davis
Oliphant, Charles Winfield
Powell, Benjamin Neff
Ratliff, Larry E
Root, Paul John
Sanderson, George Albert
Scott, Robert W
Shelton, John Wayne
Stewart, Gary Franklin
Stone, John Elmer
Tillman, Roderick W
Toomey, Donald Francis
Wilson, Leonard Richard
Zimmer, Louis George

OREGON
Agnew, Allen Francis
Allen, John Eliot
Allison, Ira Shimmin
Baldwin, Ewart Merlin
Barnett, Stockton Gordon, III
Beaulieu, John David
Benson, Gilbert Thomas
Boggs, Sam, Jr
Boucot, Arthur James
Dole, Hollis Mathews
Field, Cyrus West
Griggs, Allan Bingham
Griswold, Daniel H(alsey)
Hook, John W
Hull, Donald Albert
Kays, M Allan
Lotspeich, Frederick Benjamin
Orr, William N
Palmer, Leonard A
Payne, Thomas Gibson
Purdom, William Berlin
Schroeder, Warren Lee
Staples, Lloyd William
Swanson, Douglas Neil
Taylor, Edward Morgan
Yeats, Robert Sheppard
Youngquist, Walter

PENNSYLVANIA
Adams, John Kendal
Allen, Jack C, Jr
Andrews, George William
Barnes, Hubert Lloyd
Biederman, Edwin Williams, Jr
Bopp, Frederick, III
Brady, Keith Bryan Craig
Briggs, Reginald Peter
Buckwalter, Tracy Vere, Jr
Burger, John Allan
Bushnell, Kent O
Clark, Roger Alan
Davis, Alan
Eggler, David Hewitt
Faas, Richard William
Fergusson, William Blake
Fletcher, Frank William
Flint, Norman Keith
Frantz, Wendelin R
Garber, Murray S
Gold, David Percy
Goodspeed, Robert Marshall
Gray, Ralph J
Greenberg, Seymour Samuel
Grossman, I G
Grubbs, Donald Keeble
Haefner, Richard Charles
Hamel, James V(ictor)
Harker, Robert Ian
Heck, Edward Timmel
Hoskins, Donald Martin
Howe, Richard Hildreth
Inners, Jon David
Jacobs, Alan Martin
Jenkins, George Robert
Kauffman, Marvin Earl
Kunasz, Ihor Andrew
L'Esperance, Robert Louis
Lidiak, Edward George
Lukert, Michael T
Mehta, Sudhir
Myer, George Henry
Nagell, Raymond H
Parizek, Richard Rudolph
Pees, Samuel T(homas)
Rose, Arthur William
Ryan, John Donald
Sangrey, Dwight A
Schneider, Michael Charles
Schultz, Lane D
Sclar, Charles Bertram
Short, Nicholas Martin
Sims, Samuel John
Skolnick, Herbert
Slingerland, Rudy Lynn
Smith, Arthur R
Stephenson, Robert Charles
Stingelin, Ronald Werner
Tomikel, John
Trexler, John Peter
Voight, Barry
Williams, Eugene G
Wright, Lauren Albert
Young, John Albion, Jr

Geology, Applied, Economic & Engineering (cont)

RHODE ISLAND
Cain, J(ames) Allan
Moore, George Emerson, Jr
Parmentier, Edgar M(arc)
Silva, Armand Joseph
Tullis, Julia Ann

SOUTH CAROLINA
Cohen, Arthur David
Colquhoun, Donald John
Ehrlich, Robert
Eyer, Jerome Arlan
Fisher, Stanley Parkins
Geidel, Gwendelyn
Gipson, Mack, Jr
Griffin, Villard Stuart, Jr
Harrington, John Wilbur
Haselton, George Montgomery
Hayes, Miles O
Lawrence, David Reed
McCormick, Robert Becker
Nairn, Alan Eben Mackenzie
Pirkle, William Arthur
Snipes, David Strange
Visvanathan, T R

SOUTH DAKOTA
Davis, Briant LeRoy
Gries, John Paul
Lisenbee, Alvis Lee
McGregor, Duncan J
Mickelson, John Chester
Norton, James Jennings
Parrish, David Keith
Redden, Jack A
Stevenson, Robert Evans
Tipton, Merlin J

TENNESSEE
Bergenback, Richard Edward
Brown, Bevan W, Jr
Byerly, Don Wayne
Deboo, Phili B
Deininger, Robert W
Delcourt, Paul Allen
Dodson, Chester Lee
Grant, Leland F(auntleroy)
Helton, Walter Lee
Hill, William T
Hopkins, Richard Allen
Johnson, Robert W, Jr
Klepser, Harry John
Kopp, Otto Charles
Lumsden, David Norman
Luther, Edward Turner
Maher, Stuart Wilder
Reesman, Arthur Lee
Stearns, Richard Gordon
Wilson, Robert Lake

TEXAS
Abernathy, Bobby F
Agatston, Robert Stephen
Ahr, Wayne Merrill
Albritton, Claude Carroll, Jr
Allen, Peter Martin
Al-Saadoon, Faleh T
Amsbury, David Leonard
Anderson, Duwayne Marlo
Anthony, Elizabeth Youngblood
Anthony, John Williams
Aronow, Saul
Arp, Gerald Kench
Baie, Lyle Fredrick
Ball, Stanton Mock
Bally, Albert Walter
Barnes, Virgil Everett
Barrow, Thomas D
Beaver, H H
Bence, Alfred Edward
Berg, Robert R
Berkhout, Aart W J
Bingler, Edward Charles
Bishop, Margaret S
Bishop, Richard Stearns
Blackwelder, Blake Winfield
Bloch, Salman
Boon, John Daniel, Jr
Boyer, Robert Ernst
Bramlette, William (Allen)
Bromberger, Samuel H
Brooks, James Elwood
Brown, Leonard Franklin, Jr
Buis, Otto J
Bullard, Fred Mason
Burgess, Jack D
Burk, Creighton
Burkart, Burke
Burton, Robert Clyde
Butler, John C
Caffey, James E
Campbell, Michael David
Castano, John Roman
Chakravarty, Indranil
Chuber, Stewart
Clabaugh, Stephen Edmund
Clark, Howard Charles, Jr
Clark, Kenneth Frederick
Claypool, George Edwin
Clendening, John Albert

Cochrum, Arthur L
Coffman, James Bruce
Cook, Earl Ferguson
Cook, Theodore Davis
Cordell, Robert James
Curtis, Doris Malkin
Dahl, Harry Martin
Dally, Jesse LeRoy
Daugherty, Franklin W
Davies, David K
Denison, Rodger Espy
Dodge, Charles Fremont
Dorfman, Myron Herbert
Doyle, Frank Lawrence
Duke, Michael SN
Dunlap, Henry Francis
Durham, Clarence Orson, Jr
Echols, Dorothy Jung
Eckles, Wesley Webster, Jr
Edwards, Marc Benjamin
Eifler, Gus Kearney, Jr
Elam, Jack Gordon
Ellison, Samuel Porter, Jr
Eveland, Harmon Edwin
Feazel, Charles Tibbals
Fett, John D
Finley, Robert James
Fisher, William Lawrence
Flawn, Peter Tyrrell
Folk, Stewart H
Font, Robert G
Fountain, Lewis Spencer
French, William Edwin
Frenzel, Hugh N
Frost, Stanley H
Gallagher, John Joseph, Jr
Galloway, William Edmond
Garbarini, George S
Gealy, John Robert
Gee, David Easton
Ghaly, Tharwat Shahata
Gibson, George R
Gore, Dorothy J
Gould, Howard Ross
Gregory, A(lvin) R(ay)
Gruy, Henry Jones
Gucwa, Paul Ramon
Gustavson, Thomas Carl
Guven, Necip
Haas, Merrill Wilber
Hadley, Kate Hill
Haeberle, Frederick Roland
Halbouty, Michel Thomas
Haller, Charles Regis
Hardin, George C, Jr
Harris, Rae Lawrence, Jr
Hayward, Oliver Thomas
Helwig, James A
Heroy, William Bayard, Jr
Hewitt, Charles Hayden
Heyman, Louis
Heymann, Dieter
Higgs, Donald Val
High, Lee Rawdon, Jr
Hills, John Moore
Hirsch, John Michele
Hoffer, Jerry M
Holden, Frederick Thompson
Hood, William Calvin
Horz, Friedrich
Howard, John Hall
Huffington, Roy Michael
Hurley, Neal Lilburn
Ingerson, Fred Earl
Jackson, Martin Patrick Arden
Jeffords, Russell MacGregor
Jennings, Albert Ray
Jodry, Richard L
Johnson, Carlton Robert
Johnson, Lloyd N(ewhall)
Jones, Ernest Austin, Jr
Jones, Eugene Laverne
Jones, Theodore Sidney
Kaiser, William Richard
Keenmon, Kendall Andrews
Kidwell, Albert Laws
King, Elbert Aubrey, Jr
Knoll, Kenneth Mark
Koenig, Karl Joseph
Koepnick, Richard Borland
Kulla, Jean B
Laird, Wilson Morrow
Land, Lynton S
Lane, Donald Wilson
Lawrence, Philip Linwood
LeBlanc, Arthur Edgar
Lehman, David Hershey
Lehmann, Elroy Paul
Levinson, Stuart Alan
Loetterle, Gerald John
Longacre, Susan Ann Burton
Louden, L Richard
Luttrell, Eric Martin
McCulloh, Thane H
McDaniel, Willard Rich
McFarlan, Edward, Jr
McGookey, Donald Paul
McIver, Norman L
Maddocks, Rosalie Frances
Massell, Wulf F
Mathewson, Christopher Colville
Matthews, Martin David
Matthews, William Henry, III
Mattis, Allen Francis

Mattox, Richard Benjamin
Matuszak, David Robert
Maxwell, John Crawford
Mazzullo, James Michael
Merrill, Glen Kenton
Meyer, Franz O
Milling, Marcus Eugene
Moehlman, Robert Stevens
Monsees, James E
Morrison, Donald Allen
Morton, Robert Alex
Mundt, Philip A
Murany, Ernest Elmer
Murphy, Daniel L
Murray, Grover Elmer
Murray, Joseph Buford
Mutis-Duplat, Emilio
Nanz, Robert Hamilton, Jr
Nugent, Robert Charles
Olson, Roy Edwin
Owen, Donald Edward
Oxley, Philip
Pampe, William R
Parker, Travis Jay
Patrick, Wesley Clare
Peterson, Hazel Agnes
Phillips, Joseph D
Pirie, Robert Gordon
Prescott, Basil Osborne
Price, William Charles
Pritchett, William Carr
Raba, Carl Franz, Jr
Rawson, Richard Ray
Reaser, Donald Frederick
Reeves, Robert Grier (Lefevre)
Reso, Anthony
Richardson, Joseph Gerald
Ries, Edward Richard
Riley, Charles Marshall
Roebuck, Isaac Field
Roehl, Perry Owen
Rogers, Marion Alan
Roper, Paul James
Rose, Peter R
Rose, William Dake
Rosen, Norman Charles
Ross, Charles Alexander
Sahinen, Winston Martin
St John, Bill
Saunders, Donald Frederick
Schramm, Martin William, Jr
Schroeder, Melvin Carroll
Schumacher, Dietmar
Schwarzer, Theresa Flynn
Scobey, Ellis Hurlbut
Shipman, Ross Lovelace
Slodowski, Thomas R
Smith, Charles Isaac
Smith, Jan G
Smith, Michael A
Sneider, Robert Morton
Snelson, Sigmund
Soderman, J William
Spencer, Alexander Burke
Sprunt, Eve Silver
Stanton, Robert James, Jr
Stead, Frederick L
Strickland, John Willis
Sullivan, Herbert J
Swanson, Donald Charles
Swanson, Eric Rice
Taylor, Ralph E
Tonking, William Harry
Trace, Robert Denny
Tyrrell, Willis W, Jr
Van Rensburg, Willem Cornelius Janse
Van Siclen, DeWitt Clinton
Walper, Jack Louis
Warshauer, Steven Michael
Watkins, Jackie Lloyd
Wegner, Robert Carl
Wermund, Edmund Gerald, Jr
White, David Archer
White, N(ikolas) F(rederick)
Wilde, Garner Lee
Woodruff, Charles Marsh, Jr
Young, William Allen

UTAH
Albee, Howard F
Alvord, Donald C
Bissell, Harold Joseph
Bullock, Kenneth C
Bushman, Jess Richard
Callaghan, Eugene
Christiansen, Eric H
Christiansen, Francis Wyman
Doelling, Hellmut Hans
Gates, Joseph Spencer
Godfrey, Andrew Elliott
Hamblin, William Kenneth
Hintze, Lehi Ferdinand
Hood, James Warren
Jensen, Mead LeRoy
Lohrengel, Carl Frederick, II
McCalpin, James P
McCleary, Jefferson Rand
McCoy, Roger Michael
Moore, Joseph Neal
Mosher, Loren Cameron
Moyle, Richard W
Neff, Thomas Rodney
Nielson, Dennis Lon
Pashley, Emil Frederick, Jr

Proctor, Paul Dean
Ross, Howard Persing
Smith, Robert Baer
Southard, Alvin Reid
Spotts, John Hugh
Vaughn, Danny Mack
Welsh, John Elliott, Sr
Whelan, James Arthur
Wiese, John Herbert
Willden, Charles Ronald
Zavodni, Zavis Marian

VERMONT
Illick, J(ohn) Rowland
Jaffe, Howard William
Klemme, Hugh Douglas
Ratté, Charles A
Stanley, Rolfe S
Stoiber, Richard Edwin
Warren, John Stanley

VIRGINIA
Ahlbrandt, Thomas Stuart
Andrews, Robert Sanborn
Bartlett, Charles Samuel, Jr
Barton, Paul Booth, Jr
Bennett, Gordon Daniel
Bergin, Marion Joseph
Bergquist, Harlan Richard
Bethke, Philip Martin
Bick, Kenneth F
Bird, Samuel Oscar, II
Bodenlos, Alfred John
Bowen, Zeddie Paul
Brobst, Donald Albert
Brown, Glen Francis
Burke, Kevin Charles
Carter, William Douglas
Chao, Edward Ching-Te
Clark, Sandra Helen Becker
Conley, James Franklin
Craig, James Roland
Cronin, Thomas Mark
Cunningham, Charles G
Davidson, David Francis
Doe, Bruce R
Dragonetti, John Joseph
Drew, Lawrence James
Eckelmann, Frank Donald
Eggleston, Jane R
Ellison, Robert L
Epstein, Jack Burton
Ericksen, George Edward
Eriksson, Kenneth Andrew
Ern, Ernest Henry
Fary, Raymond W, Jr
Froelich, Albert Joseph
Fryklund, Verne Charles, Jr
Gawarecki, Stephen Jerome
Gluskoter, Harold Jay
Goldsmith, Richard
Goodell, Horace Grant
Greenwood, William R
Grender, Gordon Conrad
Groat, Charles George
Hackett, Orwoll Milton
Hackman, Robert J(oseph)
Harbour, Jerry
Hart, Dabney Gardner
Heald, Pamela
Hearn, Bernard Carter, Jr
Hemley, John Julian
Herd, Darrell Gilbert
Hobbs, Charles Roderick Bruce, Jr
Hoover, Linn
Horton, James Wright, Jr
Hunt, Lee McCaa
Husted, John E
Inderbitzen, Anton Louis, Jr
Karfakis, Mario George
Kearns, Lance Edward
Konikow, Leonard Franklin
Kozak, Samuel J
Kreipke, Merrill Vincent
Krushensky, Richard D
Kydes, Andy Steve
Lesure, Frank Gardner
Lohman, Kenneth Elmo
Lowry, Wallace Dean
Lynd, Langtry Emmett
McCammon, Richard B
McGuire, Odell
Masters, Charles Day
Meissner, Charles Roebling, Jr
Milici, Robert Calvin
Miller, Betty M (Tinklepaugh)
Milton, Daniel Jeremy
Monroe, Watson Hiner
Newton, Elisabeth G
Noble, Edwin Austin
Outerbridge, William Fulwood
Ovenshine, A Thomas
Pasko, Thomas Joseph, Jr
Patterson, Sam H
Peck, Dallas Lynn
Pickering, Ranard Jackson
Ratcliffe, Nicholas Morley
Reilly, Thomas E
Reinemund, John Adam
Reinhardt, Juergen
Ridge, John Drew
Rioux, Robert Lester
Riva, Joseph Peter, Jr
Robbins, Eleanora Iberall

Robertson, Eugene Corley
Robinson, Edwin S
Roseboom, Eugene Holloway, Jr
Ryder, Robert Thomas
Said, Rushdi
Schmidt, Robert Gordon
Sears, Charles Edward
Shelkin, Barry David
Sinha, Akhaury Krishna
Smith, Richard Elbridge
Smith, William Lee
Spencer, Edgar Winston
Stebbins, Robert H
Sutter, John Frederick
Tolbert, Gene Edward
Toulmin, Priestley
Wagner, John Edward
Wayland, Russell Gibson
Weiss, Martin
Wieczorek, Gerald Francis
Wiesnet, Donald Richard
Wolff, Roger Glen
Wood, Leonard Alton
Wood, Leonard E(ugene)
Wright, Thomas L
Zadnik, Valentine Edward
Zeigler, John Milton

WASHINGTON
Anderson, Norman Roderick
Armstrong, Frank Clarkson Felix
Bamford, Robert Wendell
Banks, Norman Guy
Brewer, William Augustus
Brown, Donald Jerould
Brown, Randall Emory
Caggiano, Joseph Anthony, Jr
Carson, Robert James, III
Cheney, Eric Swenson
Christman, Robert Adam
Coombs, Howard A
Costa, John Emil
De Goes, Louis
Easterbrook, Don J
Ellis, Ross Courtland
Finlay, Walter L(eonard)
Foit, Franklin Frederick, Jr
Foley, Michael Glen
Fulmer, Charles V
Galster, Richard W
Hallet, Bernard
Henricksen, Thomas Alva
Hoblitt, Richard Patrick
Holmes, Mark Lawrence
Hopkins, William Stephen
James, Harold Lloyd
Karlstrom, Thor Nels Vincent
Kesler, Thomas L
Kiilsgaard, Thor H
Mallory, Virgil Standish
Miller, C Dan
Misch, Peter
Oles, Keith Floyd
Olson, Edwin Andrew
Pierson, Thomas Charles
Pollack, Jerome Marvin
Rau, Weldon Willis
Roberts, Ralph Jackson
Rosengreen, Theodore E
Scott, Kevin M
Scott, William Edward
Snook, James Ronald
Sorem, Ronald Keith
Steele, William Kenneth
Stewart, Richard John
Terrel, Ronald Lee
Threet, Richard Lowell
Vance, Joseph Alan
Waldron, Howard Hamilton (Hank)
Waters, Aaron Clement
Webster, Gary Dean
Weis, Paul Lester
Wodzicki, Antoni
Wolfe, Edward W

WEST VIRGINIA
Behling, Robert Edward
Chen, Ping-Fan
Donaldson, Alan C
Erwin, Robert Bruce
Gwinn, James E
Lessing, Peter
Ludlum, John Charles
Murphy, Allen Emerson
Rieke, Herman Henry, III
Ting, Francis Ta-Chuan

WISCONSIN
Broughton, William Albert
Brown, Bruce Elliot
Brown, Philip Edward
Byers, Charles Wesley
Cameron, Eugene Nathan
Dickas, Albert Binkley
Edil, Tuncer Berat
Emmons, Richard Conrad
Gates, Robert Maynard
Govin, Charles Thomas, Jr
LaBerge, Gene L
Lasca, Norman P, Jr
Meyer, Robert Paul
Mudrey, Michael George, Jr
Mursky, Gregory
Ostrom, Meredith Eggers

Palmquist, John Charles
Paull, Richard Allen
Peterson, Dallas Odell
Richason, Benjamin Franklin, Jr
Ross, Theodore William
Shea, James H
Stenstrom, Richard Charles
Stieglitz, Ronald Dennis
Tychsen, Paul C
Weege, Randall James
Weis, Leonard Walter
West, Walter Scott
Willis, Ronald Porter

WYOMING
Borgman, Leon Emry
Curry, William Hirst, III
Houston, Robert S
Huff, Kenneth O
Love, John David
Metzger, William John
Snoke, Arthur Wilmot

PUERTO RICO
Brooks, Harold Kelly

ALBERTA
Andrichuk, John Michael
Burwash, Ronald Allan
Campbell, Finley Alexander
Campbell, John Duncan
Cecile, Michael Peter
Cruden, David Milne
Folinsbee, Robert Edward
Fox, Frederick Glenn
Ghent, Edward Dale
Halferdahl, Laurence Bowes
Henderson, Gerald Gordon Lewis
Hills, Leonard Vincent
Hriskevich, Michael Edward
Hunt, C Warren
Klassen, Rudolph Waldemar
Levinson, Alfred Abraham
McMillan, Neil John
Masters, John Alan
Mellon, George Barry
Morgenstern, N(orbert) R
Neale, Ernest Richard Ward
Nesbitt, Bruce Edward
Norford, Brian Seeley
Norris, Donald Kring
Okulitch, Andrew Vladimir
Okulitch, Vladimir Joseph
Oliver, Thomas Albert
Ollerenshaw, Neil Campbell
Peirce, John Wentworth
Raasch, Gilbert O
Scafe, Donald William
Schmidt, Volkmar
Smith, Dorian Glen Whitney
Stelck, Charles Richard
Tankard, Anthony James
Wardlaw, Norman Claude
Young, Frederick Griffin

BRITISH COLUMBIA
Arnold, Ralph Gunther
Barnes, William Charles
Bustin, Robert Marc
Byrne, Peter M
Chase, Richard L
Clark, Lloyd Allen
Cooke, Herbert Basil Sutton
Danner, Wilbert Roosevelt
Gabrielse, Hubert
Gazin, Charles Lewis
Greenwood, Hugh J
Hughes, Richard David
Jackson, Lionel Eric, Jr
Lee, James William
Mathews, William Henry
Mothersill, John Sydney
Murray, James W
Payne, Myron William
Pocock, Stanley Albert John
Rice, Jack Morris
Robinson, Gershon Duvall
Roddick, James Archibald
Sinclair, Alastair James
Sutherland-Brown, Atholl
Tozer, E T
Warren, Harry Verney
Wheeler, John Oliver
Wynne-Edwards, Hugh Robert

MANITOBA
Anderson, Donald T
Baracos, Andrew
Lajtai, Emery Zoltan
McAllister, Arnold Lloyd
Teller, James Tobias
Turnock, A C
Whitaker, Sidney Hopkins
Wilson, Harry David Bruce

NEW BRUNSWICK
Moore, John Carman Gailey
Mossman, David John

NEWFOUNDLAND
King, Arthur Francis
King, Michael Stuart
McKillop, J(ohn) H
Williams, Harold

NOVA SCOTIA
Cormier, Randal
Greenblatt, Jayson Herschel
Hacquebard, Peter Albertus
Milligan, George Clinton
Riddell, John Evans
Shaw, William S
Williamson, Douglas Harris
Zentilli, Marcos

ONTARIO
Anderson, Francis David
Appleyard, Edward Clair
Armstrong, Herbert Stoker
Baird, David McCurdy
Baragar, William Robert Arthur
Beales, Francis William
Bell, Richard Thomas
Blackadar, Robert Gordon
Brummer, Johannes J
Burk, Cornelius Franklin, Jr
Calder, Peter N
Cameron, Robert Alan
Campbell, Ian Henry
Card, Kenneth D
Chandler, Frederick William
Chua, Kian Eng
Church, William Richard
Coakley, John Phillip
Cowan, W R
Crocket, James Harvie
Cumming, Leslie Merrill
Currie, John Bickell
Davies, James Frederick
Dence, Michael Robert
Donaldson, John Allan
Duke, John Murray
Emslie, Ronald Frank
Evans, James Eric Lloyd
Franklin, James McWillie
Frarey, Murray James
Fulton, Robert John
Fyles, John Gladstone
Gibb, Richard A
Goranson, Edwin Alexander
Grant, Douglas Roderick
Gussow, W(illia)m C(arruthers)
Harrison, James Merritt
Haynes, Simon John
Hill, Patrick Arthur
Hodder, Robert William
Hoffman, Douglas Weir
Hutchison, William Watt
Jenness, Stuart Edward
Jolliffe, Alfred Walton
Joubin, Franc Renault
Karrow, Paul Frederick
Kenney, T Cameron
Kissin, Stephen Alexander
Lee, Hulbert Austin
Leech, Geoffrey Bosdin
Lenz, Alfred C
Lumbers, Sydney Blake
McLaren, Digby Johns
McMurchy, Robert Connell
MacRae, Neil D
Menzies, John
Michener, Charles Edward
Middleton, Gerard Viner
Moore, John Marshall, Jr
Naldrett, Anthony James
Neilson, James Maxwell
Norris, Geoffrey
Palmer, H Currie
Peach, Peter Angus
Pearce, George William
Petruk, William
Poole, William Hope
Prince, Alan Theodore
Pye, Edgar George
Reesor, John Elgin
Roots, Ernest Frederick
Scott, John Stanley
Scott, Steven Donald
Smith, Charles Haddon
Sonnenfeld, Peter
Spooner, Ed Thornton Casswell
Stesky, Robert Michael
Symons, David Thorburn Arthur
Terasmae, Jaan
Thorpe, Ralph Irving
Tovell, Walter Massey
Van Loon, Jon Clement
Westermann, Gerd Ernst Gerold
Yole, Raymond William

QUEBEC
Beland, Jacques (Robert)
Blais, Roger A
Brown, Alex Cyril
Chagnon, Jean Yves
Chown, Edward Holton
Dionne, Jean-Claude
Doig, Ronald
Eakins, Peter Russell
Globensky, Yvon Raoul
Kerr, James William
Ladanyi, Branko
MacLean, Wallace H
Mamet, Bernard Leon
Martignole, Jacques
Martin, Robert Francois Churchill
Mountjoy, Eric W
Mukherji, Kalyan Kumar

Perrault, Guy
Poorooshasb, Hormozd Bahman
Sabourn, Robert Joseph Edmond
Seguin, Maurice Kaisholm
Webber, George Roger

SASKATCHEWAN
Christiansen, E A
Coleman, Leslie Charles
Davies, John Clifford
Kent, Donald Martin
Kupsch, Walter Oscar
Langford, Fred F
Mollard, John D
Vigrass, Laurence William

OTHER COUNTRIES
Alonso, Ricardo N
Arce, Jose Edgar
Conolly, John R
Cooke, Hermon Richard, Jr
Fritz, Peter
Govett, Gerald James
Hase, Donald Henry
Hoke, John Humphreys
Hsu, Kenneth Jinghwa
Kennedy, Michael John
Libby, Willard Gurnea
Lynts, George Willard
Rau, Jon Llewellyn
Sander, Nestor John
Schaetti, Henry Joachim
Von Huene, Roland
Walton, Wayne J A, Jr
Winkler, Virgil Dean
Wolfe, John A(llen)
Yazicigil, Hasan

Geology, Structural

ALABAMA
Groshong, Richard Hughes, Jr

ALASKA
Dillon, John Thomas
Karl, Susan Margaret
Weber, Florence Robinson

ARIZONA
Barnes, Charles Winfred
Chase, Clement Grasham
Coney, Peter James
Cree, Allan
Davis, George Herbert
Duffield, Wendell Arthur
Fink, Jonathan Harry
Lucchitta, Baerbel Koesters
Lucchitta, Ivo
McCullough, Edgar Joseph, Jr
Ragan, Donal Mackenzie
Sellin, H A
Shoemaker, Eugene Merle
Snow, Eleanour Anne
Titley, Spencer Rowe
Wallace, Terry Charles, Jr
Whorton, Chester D

ARKANSAS
Konig, Ronald H

CALIFORNIA
Alvarez, Walter
Arrowood, Roy Mitchell, Jr
Beard, Charles Noble
Beaty, David Wayne
Berkland, James Omer
Bird, Peter
Carr, Michael H
Chang, Kaung-Jain
Chapman, Robert Mills
Christiansen, Robert Lorenz
Christie, John McDougall
Ciancanelli, Eugene Vincent
Colburn, Ivan Paul
Compton, Robert Ross
Creely, Robert Scott
Crowell, John Chambers
Cruikshank, Dale Paul
Csejtey, Bela, Jr
Cserna, Eugene George
Davis, Gregory Arlen
Dennis, John Gordon
Donath, Fred Arthur
Durham, William Bryan
Farrington, William Benford
Foster, Helen Laura
Grantz, Arthur
Green, Harry Western, II
Hall, Clarence Albert, Jr
Hanan, Barry Benton
Heard, Hugh Corey
Hose, Richard K
Howard, Keith Arthur
Irwin, James Joseph
Janke, Norman C
Kirby, James M
Kopf, Rudolph William
Loney, Robert Ahlberg
Lyon, Ronald James Pearson
McKague, Herbert Lawrence
Mackevett, Edward M
Moores, Eldridge Morton
Nielson, Jane Ellen

Geology, Structural (cont)

Nur, Amos M
Oertel, Gerhard Friedrich
Olson, Jerry Chipman
Page, Benjamin Markham
Page, Robert Alan, Jr
Pierce, William G
Pincus, Howard Jonah
Rogers, Thomas Hardin
Rudnyk, Marian E
Saleeby, Jason Brian
Sarna-Wojcicki, Andrei M
Seff, Philip
Sieh, Kerry Edward
Silver, Eli Alfred
Smith, Robert Leland
Sylvester, Arthur Gibbs
Tabor, Rowland Whitney
Thompson, George Albert
Tobisch, Othmar Tardin
Twiss, Robert John
Vasco, Donald Wyman
Wadsworth, William Bingham
Wenk, Hans-Rudolf
Wentworth, Carl M, Jr
Wright, William Herbert, III

COLORADO
Anderson, Ernest R
Ayler, Maynard Franklin
Braddock, William A
Bryant, Bruce Hazelton
Cole, James Channing
Drewes, Harald D
Elder, Curtis Harold
Erslev, Eric Allan
Fails, Thomas Glenn
Friedman, Jules Daniel
Fuchs, Robert L
Grose, Thomas Lucius Trowbridge
Hamilton, Warren Bell
Henrickson, Eiler Leonard
Hustrulid, William A
Keefer, William Richard
Ketner, Keith B
Kupfer, Donald Harry
Link, Peter K
Lowell, James Diller
McCallum, Malcolm E
Machette, Michael N
Moore, David A
Moore, David Warren
Oriel, Steven S
Parker, John Marchbank
Pierce, Kenneth Lee
Pillmore, Charles Lee
Raup, Robert Bruce, Jr
Rold, John W
Rowley, Peter DeWitt
Sable, Edward George
Sanborn, Albert Francis
Sass, Louis Carl
Sheridan, Douglas Maynard
Silberling, Norman John
Skipp, Betty Ann
Swanson, Vernon E
Trimble, Donald E
Verbeek, Earl Raymond
Warner, Lawrence Allen
Wheeler, Russell Leonard
Zartman, Robert Eugene

CONNECTICUT
DeBoer, Jelle
Rodgers, John

DELAWARE
Ellis, Patricia Mench

DISTRICT OF COLUMBIA
Brocoum, Stephan John
Knox, Arthur Stewart
Nolan, Thomas Brennan
Rumble, Douglas, III
Wilson, Frederick Albert

FLORIDA
Draper, Grenville
Rosendahl, Bruce Ray
Tull, James Franklin

GEORGIA
Goodwin, Sidney S

HAWAII
Moore, Gregory Frank

IDAHO
Hall, William Bartlett

ILLINOIS
Buschbach, Thomas Charles
Klasner, John Samuel
Woodland, Bertram George

INDIANA
Chen, Wai-Fah
Chowdhury, Dipak Kumar
Dunning, Jeremy David
Hamburger, Michael Wile
Martin, Charles Wellington, Jr
Rigert, James Aloysius
Wintsch, Robert P

IOWA
Hoppin, Richard Arthur

KANSAS
Aber, James Sandusky
Blythe, Jack Gordon
Gries, John Charles
Underwood, James Ross, Jr

KENTUCKY
Brown, William Randall
Ewers, Ralph O
Hagan, Wallace Woodrow
Rast, Nicholas
Seeger, Charles Ronald
Von Schonfeldt, Hilmar Armin

LOUISIANA
Forman, McLain J
Frey, Maurice G
Kinsland, Gary Lynn
Kumar, Madhurendu B
Seglund, James Arnold
Wilcox, Ronald Erwin

MAINE
Hussey, Arthur M, II

MARYLAND
Allenby, Richard John, Jr
Baker, Arthur A
Beal, Robert Carl
Brown, Michael
Edwards, Jonathan, Jr
Fisher, George Wescott
Pavlides, Louis

MASSACHUSETTS
Ballard, Robert D
Bonnichsen, Martha Miller
Brace, William Francis
Burger, Henry Robert, III
Hepburn, John Christopher
Herda, Hans-Heinrich Wolfgang
Kamen-Kaye, Maurice
Kleinrock, Martin Charles
Lin, Jian
McGill, George Emmert
Selverstone, Jane Elizabeth
Takahashi, Kozo
Thompson, Margaret Douglas
Tucholke, Brian Edward

MICHIGAN
Frankforter, Weldon D
Huntoon, Jacqueline E
McKee, Edith Merritt
Van der Voo, Rob

MINNESOTA
Holst, Timothy Bailey
Hudleston, Peter John
Parker, Ronald Bruce
Southwick, David Leroy
Walton, Matt Savage

MISSISSIPPI
Egloff, Julius, III
Knight, Wilbur Hall
Morris, Gerald Brooks

MISSOURI
Arvidson, Raymond Ernst
Goebel, Edwin DeWayne
McKinnon, William Beall
Summers, David Archibold

MONTANA
Baker, David Warren
Hyndman, Donald William
Ruppel, Edward Thompson

NEBRASKA
Kaplan, Sanford Sandy
Nelson, Robert B

NEVADA
Felts, Wayne Moore
Garside, Larry Joe
Mifflin, Martin David
Peck, John H
Slemmons, David Burton
Wilbanks, John Randall

NEW HAMPSHIRE
Billings, Marland Pratt
Bothner, Wallace Arthur
Boudette, Eugene L
Richter, Dorothy Anne
Spink, Walter John

NEW JERSEY
Agron, Sam Lazrus
Hamil, Martha M
Metsger, Robert William

NEW MEXICO
Baldridge, Warren Scott
Budding, Antonius Jacob
Geissman, John William
Grambling, Jeffrey A
Love, David Waxham
Shomaker, John Wayne
Woodward, Lee Albert

NEW YORK
Bayly, M Brian
Bhattacharji, Somdev
Brock, Patrick Willet Grote
Brueckner, Hannes Kurt
Bursnall, John Treharne
Cassie, Robert MacGregor
Diebold, John Brock
Fakundiny, Robert Harry
Hayes, Dennis E
Heyl, George Richard
Isachsen, Yngvar William
Jacob, Klaus H
Karig, Daniel Edmund
Kidd, William Spencer Francis
King, John Stuart
Liebling, Richard Stephen
MacDonald, William David
Mattson, Peter Humphrey
Means, Winthrop D
Muller, Otto Helmuth
Prucha, John James
Seyfert, Carl K, Jr
Teichert, Curt
Thomas, John Jenks
Vandiver, Bradford B
Vollmer, Frederick Wolfer

NORTH CAROLINA
Butler, James Robert
Dennison, John Manley
Forsythe, Randall D
Johnson, J(ames) Donald
Karson, Jeffrey Alan
Lowry, Jean
Marlowe, James Irvin
Raymond, Loren Arthur
Smith, Norman Cutler

OHIO
Bever, James Edward
Burford, Arthur Edgar
Harmon, David Elmer, Jr
Kulander, Byron Rodney
Law, Eric W
Mahard, Richard H
Nance, Richard Damian
Richard, Benjamin H
Scotford, David Matteson
Skinner, William Robert
Unrug, Raphael
Wojtal, Steven Francis

OKLAHOMA
Ahern, Judson Lewis
Boade, Rodney Russett
Crosby, Gary Wayne
Espach, Ralph H, Jr
Granath, James Wilton
Meyerhoff, Arthur Augustus
Murray, Frederick Nelson
Sondergeld, Carl Henderson
Stearns, David Winrod
Williams, Lyman O

OREGON
Lawrence, Robert D
Moore, George William
Yeats, Robert Sheppard

PENNSYLVANIA
Alley, Richard Blaine
Clark, Roger Alan
Eggler, David Hewitt
Faill, Rodger Tanner
Haefner, Richard Charles
Hodgson, Robert Arnold
Langston, Charles Adam
Liu, Mian
Nickelsen, Richard Peter
Pees, Samuel T(homas)
Platt, Lucian B
Voight, Barry
Washburn, Robert Henry

RHODE ISLAND
McMaster, Robert Luscher
Moore, George Emerson, Jr
Tullis, Terry Edson

SOUTH CAROLINA
Secor, Donald Terry, Jr

SOUTH DAKOTA
Lisenbee, Alvis Lee
Parrish, David Keith

TENNESSEE
Brent, William B
Hatcher, Robert Dean, Jr
Johnson, Robert W, Jr

TEXAS
Ave Lallemant, Hans Gerhard
Berkhout, Aart W J
Bishop, Richard Stearns
Blackwelder, Blake Winfield
Buis, Otto J
Carter, Neville Louis
Cebull, Stanley Edward
Chatterjee, Sankar
Crouse, Philip Charles
Dalziel, Ian William Drummond
Dunn, David Evan

NEW YORK
Eifler, Gus Kearney, Jr
Ewing, Thomas Edward
Folk, Stewart H
Font, Robert G
Friedman, Melvin
Gallagher, John Joseph, Jr
Gee, David Easton
Green, Arthur R
Gruy, Henry Jones
Helwig, James A
Hilde, Thomas Wayne Clark
Hoge, Harry Porter
Huffington, Roy Michael
Ingerson, Fred Earl
Jackson, Martin Patrick Arden
Johnson, Brann
Lawrence, Philip Linwood
Logan, John Merle
Mosher, Sharon
Muehlberger, William Rudolf
Mutis-Duplat, Emilio
Nielsen, Kent Christopher
Norman, Carl Edgar
Pauken, Robert John
Reaser, Donald Frederick
Roper, Paul James
Russell, James E(dward)
St John, Bill
Stead, Frederick L
Swanson, Eric Rice
Van Siclen, DeWitt Clinton
Whitford-Stark, James Leslie
Wiltschko, David Vilander

UTAH
Fidlar, Marion M
Hardy, Clyde Thomas
McCleary, Jefferson Rand
Nielson, Dennis Lon
Proctor, Paul Dean

VERMONT
Doll, Charles George
Westerman, David Scott

VIRGINIA
Drake, Avery Ala, Jr
Eggleston, Jane R
Faust, Charles R
Gawarecki, Stephen Jerome
Goodwin, Bruce K
Harbour, Jerry
Hatch, Norman Lowrie, Jr
Herd, Darrell Gilbert
Horton, James Wright, Jr
McDowell, Robert Carter
Milici, Robert Calvin
Nelson, Arthur Edward
Offield, Terry Watson
Rankin, Douglas Whiting
Robertson, Eugene Corley
Russ, David Perry
Shanks, Wayne C, III
Stewart, David Benjamin
Weems, Robert Edwin

WASHINGTON
Armstrong, Frank Clarkson Felix
Beck, Myrl Emil, Jr
Cowan, Darrel Sidney
Holcomb, Robin Terry
Kesler, Thomas L
Lowes, Brian Edward
Smith, Brad Keller
Swanson, Donald Alan
Talbot, James Lawrence
Wahlstrom, Ernest E

WEST VIRGINIA
Gwinn, James E
Khair, Abdul Wahab
Peng, Syd Syh-Deng
Shumaker, Robert C

WISCONSIN
Craddock, J(ohn) Campbell
Dickas, Albert Binkley
Dutch, Steven Ian
Stenstrom, Richard Charles

WYOMING
Miller, Daniel Newton, Jr
Oman, Carl Henry
Snoke, Arthur Wilmot

ALBERTA
Cook, Frederick Ahrens
Cruden, David Milne
Cuny, Robert Michael
Okulitch, Andrew Vladimir
Tankard, Anthony James

BRITISH COLUMBIA
Armstrong, Richard Lee
Bustin, Robert Marc
Chase, Richard L
Monger, James William Heron
Wheeler, John Oliver

NEW BRUNSWICK
Allen, Clifford Marsden

NOVA SCOTIA
Keppie, John D

Milligan, George Clinton
Stevens, George Richard
Zentilli, Marcos

ONTARIO
Babcock, Elkanah Andrew
Bell, Richard Thomas
Fox, Joseph S
Grieve, Richard Andrew
Gussow, W(illia)m C(arruthers)
Haynes, Simon John
Kehlenbeck, Manfred Max
Percival, John A
Poole, William Hope
Price, Raymond Alex
Rousell, Don Herbert
Starkey, John
Udd, John Eaman

QUEBEC
Chown, Edward Holton
Laurin, Andre Frederic

SASKATCHEWAN
Stauffer, Mel R

OTHER COUNTRIES
Dengo, Gabriel
Dewey, John Frederick
Kennedy, Michael John
McIntyre, Donald B
Pegler, Alwynne Vernon
Sales, John Keith
Trümpy, D(aniel) Rudolf
Wolfe, John A(llen)
Wood, Dennis Stephenson

Geomorphology & Glaciology

ALASKA
Benson, Carl Sidney
Hamilton, Thomas Dudley
Hopkins, David Moody
Long, William Ellis
Osterkamp, Thomas Eugene
Ragle, Richard Harrison
Reger, Richard David
Weber, Florence Robinson
Weeks, Wilford Frank

ARIZONA
Baker, Victor Richard
Bergen, James David
Breed, Carol Sameth
Breed, William Joseph
Bull, William Benham
Chase, Clement Grasham
Childs, Orlo E
Dorn, Ronald I
Graf, William L
Greeley, Ronald
Haynes, Caleb Vance, Jr
Heede, Burchard Heinrich
Lucchitta, Baerbel Koesters
Lucchitta, Ivo
Péwé, Troy Lewis
Schoenwetter, James

CALIFORNIA
Brice, James Coble
Campbell, Catherine Chase
Carr, Michael H
Clark, Malcolm Mallory
Dewart, Gilbert
Forrest, James Benjamin
Foster, Helen Laura
Hart, Earl W
Higgins, Charles Graham
Huber, Norman King
Kamb, Walter Barclay
Keller, Edward Anthony
Kirch, Patrick Vinton
Leopold, Luna Bergere
Malin, Michael Charles
Nelson, Robert M
Nobles, Laurence Hewit
Page, Robert Alan, Jr
Rhodes, Dallas D
Rossbacher, Lisa Ann
Rudnyk, Marian E
Sharp, Robert Phillip
Shreve, Ronald Lee
Sternberg, Hilgard O'Reilly
Thorne, Robert Folger
Warnke, Detlef Andreas
Zumberge, James Herbert

COLORADO
Andrews, John Thomas
Behrendt, John Charles
Birkeland, Peter Wessel
Bradley, William Crane
Caine, T Nelson
Carrara, Paul Edward
Coates, Donald Allen
Cooley, Richard Lewis
Crone, Anthony Joseph
Doehring, Donald O
Emmett, William W
Fernald, Arthur Thomas
Friedman, Jules Daniel
Gomberg, Joan Susan
Hadley, Richard Frederick

Holcombe, Troy Leon
Jibson, Randall W
Krantz, William Bernard
Lackey, Laurence
Luft, Stanley Jeremie
Madole, Richard
Malde, Harold Edwin
Maughan, Edwin Kelly
Meier, Mark Frederick
Moore, David Warren
Morrison, Roger Barron
Nichols, Donald Ray
Pierce, Kenneth Lee
Schumm, Stanley Alfred
Strobell, John Dixon, Jr
Tabler, Ronald Dwight
Toy, Terrence J

CONNECTICUT
Baskerville, Charles Alexander
Clebnik, Sherman Michael
Hack, John Tilton
Patton, Peter C
Spiegel, Zane
Tolderlund, Douglas Stanley

DELAWARE
Kraft, John Christian

DISTRICT OF COLUMBIA
Melickian, Gary Edward
Vogt, Peter Richard
Zimbelman, James Ray

FLORIDA
Chryssafopoulos, Hanka Wanda Sobczak
Collins, Mary Elizabeth
Draper, Grenville
Goldthwait, Richard Parker
Lovejoy, Donald Walker
Parker, Garald G, Sr

IDAHO
Hall, William Bartlett
Megahan, Walter Franklin

ILLINOIS
Anderson, Richard Charles
Harris, Stanley Edwards, Jr
Johnson, William Hilton
McKay, Edward Donald, III
Masters, John Michael
Reams, Max Warren
Wallace, Ronald Gary

INDIANA
Bleuer, Ned Kermit
Melhorn, Wilton Newton
Miller, Victor Charles
Reshkin, Mark

IOWA
Handy, Richard L(incoln)
Hussey, Keith Morgan
Prior, Jean Cutler
Tuttle, Sherwood Dodge
Walters, James Carter

KANSAS
Aber, James Sandusky
Blythe, Jack Gordon
Dort, Wakefield, Jr

KENTUCKY
Ettensohn, Francis Robert

LOUISIANA
Coleman, James Malcolm
Walker, Harley Jesse

MAINE
Kellogg, Thomas B
Koons, Donaldson
Novak, Irwin Daniel

MARYLAND
Cleaves, Emery Taylor
Garvin, James Brian
Kearney, Michael Sean
Kravitz, Joseph Henry
Leatherman, Stephen Parker
Rango, Albert
Warren, Charles Reynolds
Wolman, Markley Gordon

MASSACHUSETTS
Bradley, Raymond Stuart
Colman, Steven Michael
Genes, Andrew Nicholas
Hartshorn, Joseph Harold
Kleinrock, Martin Charles
Newman, William Alexander
Newton, Robert Morgan
Ridge, John Charles
Southard, John Brelsford
Stearns, Charles Edward
Williams, Richard Sugden, Jr

MICHIGAN
Clark, James Alan
Eschman, Donald Frazier
Fekety, F Robert, Jr
Kehew, Alan Everett
Schmaltz, Lloyd John

Sommers, Lawrence M
Taylor, Lawrence Dow
Ten Brink, Norman Wayne

MINNESOTA
Cushing, Edward John
Hooke, Roger LeBaron
Hudleston, Peter John
Lawrence, Donald Buermann
Matsch, Charles Leo
Van Alstine, James Bruce

MISSISSIPPI
Happ, Stafford Coleman
Mylroie, John Eglinton
Otvos, Ervin George
Saucier, Roger Thomas

MISSOURI
Knox, Burnal Ray
McKinnon, William Beall
Miles, Randall Jay
Rockaway, John D, Jr
Schroeder, Walter Albert
Watson, Richard Allan

MONTANA
Montagne, John M

NEBRASKA
Diffendal, Robert Francis, Jr
Shroder, John Ford, Jr
Tanner, Lloyd George
Wayne, William John

NEVADA
Paige, Russell Alston
Slemmons, David Burton

NEW HAMPSHIRE
Ackley, Stephen Fred
Cox, Gordon F N
Hibler, William David, III
Mayewski, Paul Andrew

NEW JERSEY
Psuty, Norbert Phillip
Wolfe, Peter E
Yasso, Warren E

NEW MEXICO
Hawley, John William
Hill, Mary Rae
Love, David Waxham
Manley, Kim
Wells, Stephen Gene

NEW YORK
Abrahams, Athol Denis
Bloom, Arthur Leroy
Calkin, Parker E
Coch, Nicholas Kyros
Fleisher, Penrod Jay
Jacobs, Stanley S
Johnson, Kenneth George
Kidd, William Spencer Francis
Liebling, Richard Stephen
Lougeay, Ray Leonard
Morisawa, Marie
Street, James Stewart
Van Burkalow, Anastasia

NORTH CAROLINA
Dennison, John Manley
Howard, Arthur David
Mock, Steven James
Stephenson, Richard Allen
Stuart, Alfred Wright

NORTH DAKOTA
Clausen, Eric Neil
Reid, John Reynolds, Jr

OHIO
Andreas, Barbara Kloha
Bonnett, Richard Brian
Calhoun, Frank Gilbert
Craig, Richard Gary
Crowl, George Henry
Forsyth, Jane Louise
Laughon, Robert Bush
McKenzie, Garry Donald
Meinecke, Eberhard A
Nash, David Byer
Nave, Floyd Roger
Rich, Charles Clayton
Smith, Geoffrey W
Stewart, David Perry
Utgard, Russell Oliver
Whillans, Ian Morley

OKLAHOMA
Cannon, Philip Jan
Myers, Arthur John
Ross, Alex R
Salisbury, Neil Elliot

OREGON
Allison, Ira Shimmin
McKinney, William Mark
Niem, Wendy Adams

PENNSYLVANIA
Alley, Richard Blaine

Chapman, William Frank
Evenson, Edward B
Gardner, Thomas William
Hodgson, Robert Arnold
Potter, Noel, Jr
Sevon, William David, III

SOUTH CAROLINA
Haselton, George Montgomery
James, L(aurence) Allan
Pirkle, William Arthur

SOUTH DAKOTA
Rahn, Perry H

TENNESSEE
Lietzke, David Albert

TEXAS
Albritton, Claude Carroll, Jr
Allen, Peter Martin
Bohacs, Kevin Michael
Doyle, Frank Lawrence
Farrand, William Richard
Finley, Robert James
Gustavson, Thomas Carl
Harrison, Wilks Douglas
Johnson, Brann
Knoll, Kenneth Mark
McPherson, John G(ordon)
Melia, Michael Brendan
Owens, Edward Henry
Robertson, James Douglas
Rutford, Robert Hoxie
Soloyanis, Susan Constance
Woodruff, Charles Marsh, Jr

UTAH
Godfrey, Andrew Elliott
McCalpin, James P
Stokes, William Lee
Vaughn, Danny Mack

VERMONT
Larsen, Frederick Duane

VIRGINIA
Finkl, Charles William, II
Galvin, Cyril Jerome, Jr
Garner, Hessle Filmore
Gawarecki, Stephen Jerome
Goldsmith, Richard
Herd, Darrell Gilbert
Outerbridge, William Fulwood
Russ, David Perry
Weems, Robert Edwin
Wieczorek, Gerald Francis

WASHINGTON
Anderson, Norman Roderick
Caggiano, Joseph Anthony, Jr
Carson, Robert James, III
Costa, John Emil
De Goes, Louis
Dunne, Thomas
Easterbrook, Don J
Foley, Michael Glen
Grootes, Pieter Meiert
Hallet, Bernard
Hodge, Steven McNiven
Holcomb, Robin Terry
Kaatz, Martin Richard
Kiver, Eugene P
Miller, C Dan
Palmquist, Robert Clarence
Pessl, Fred, Jr
Pierson, Thomas Charles
Rosengreen, Theodore E
Scott, William Edward
Smith, J Dungan
Threet, Richard Lowell
Washburn, Albert Lincoln

WISCONSIN
Bentley, Charles Raymond
Borowiecki, Barbara Zakrzewska
Hole, Francis Doan
Knox, James Clarence
Lasca, Norman P, Jr
Richason, Benjamin Franklin, Jr
Schneider, Allan Frank
Stieglitz, Ronald Dennis

WYOMING
Mears, Brainerd, Jr

ALBERTA
Barendregt, Rene William
Cruden, David Milne
Giovinetto, Mario Bartolome
Honsaker, John Leonard
Klassen, Rudolph Waldemar
Rogerson, Robert James

BRITISH COLUMBIA
Armstrong, John Edward
Clarke, Garry K C
Foster, Harold Douglas
Jackson, Lionel Eric, Jr
Slaymaker, Herbert Olav

MANITOBA
Teller, James Tobias
Welsted, John Edward

Geomorphology & Glaciology (cont)

ONTARIO
Adams, William Alfred
Blake, Weston, Jr
Bunting, Brian Talbot
Davidson-Arnott, Robin G D
Dreimanis, Aleksis
Flint, Jean-Jacques
Greenwood, Brian
Johnson, Peter Graham
Karrow, Paul Frederick
Kesik, Andrzej B
LaValle, Placido Dominick
Menzies, John
Stalker, Archibald MacSween
St-Onge, Denis Alderic
Wilkinson, Thomas Preston

QUEBEC
Bonn, Ferdinand J
Bouchard, Michel André
Dubois, Jean-Marie M
Elson, John Albert
Gangloff, Pierre
Ladanyi, Branko
Langleben, Manuel Phillip

SASKATCHEWAN
Chew, Hemming
Langford, Fred F
Mollard, John D

OTHER COUNTRIES
Dury, George H
Hattersley-Smith, Geoffrey Francis
Jopling, Alan Victor
Van Andel, Tjeerd Hendrik

Geophysics

ALABAMA
Anderson, Christian Donald
Dalins, Ilmars
Smith, Robert Earl
Visscher, Pieter Bernard

ALASKA
Akasofu, Syun-Ichi
Belon, Albert Edward
Brewer, Max C(lifton)
Brown, Neal B
Davis, Thomas Neil
Degen, Vladimir
Gedney, Larry Daniel
Gosink, Joan P
Hallinan, Thomas James
Hunsucker, Robert Dudley
Rees, Manfred Hugh
Romick, Gerald J
Stenbaek-Nielsen, Hans C
Stone, David B
Watkins, Brenton John
Weeks, Wilford Frank
Wescott, Eugene Michael
Wilson, Charles R

ARIZONA
Bates, Charles Carpenter
Best, David M
Butler, Robert Franklin
Cathey, Everett Henry
Chase, Clement Grasham
Elston, Donald (Parker)
Fink, James Brewster
Fink, Jonathan Harry
Herbert, Floyd Leigh
Hood, Lonnie Lamar
Kieffer, Hugh Hartman
Kieffer, Susan Werner
Levy, Eugene Howard
Lunine, Jonathan Irving
Melosh, Henry Jay, IV
O'Leary, Brian Todd
Richardson, Randall Miller
Roddy, David John
Sass, John Harvey
Soderblom, Laurence Albert
Sonett, Charles Philip
Sumner, John Stewart
Wallace, Terry Charles, Jr
West, Robert Elmer

ARKANSAS
Leming, Charles William

CALIFORNIA
Abdel-Ghaffar, Ahmed Mansour
Ahrens, Thomas J
Akella, Jagannadham
Aki, Keiiti
Alexander, Ira H(enris)
Allen, Clarence Roderic
Anderson, Don Lynn
Anderson, Orson Lamar
Atwater, Tanya Maria
Bachman, Richard Thomas
Backus, George Edward
Bagby, John P(endleton)
Bakun, William Henry
Barnes, David Fitz

Becker, Alex
Berryman, James Garland
Biehler, Shawn
Bird, Peter
Bolt, Bruce A
Borchardt, Glenn (Arnold)
Borucki, William Joseph
Bradner, Hugh
Breiner, Sheldon
Brooke, John Percival
Brown, Maurice Vertner
Burg, John Parker
Burnett, Lowell Jay
Busse, Friedrich Hermann
Carrigan, Charles Roger
Case, James Edward
Cherry, Jesse Theodore
Christiansen, Robert Lorenz
Chung, Dae Hyun
Cicerone, Ralph John
Claerbout, Jon F
Cloud, William K
Coe, Robert Stephen
Coleman, Paul Jerome, Jr
Costantino, Marc Shaw
Cross, Ralph Herbert, III
Dalrymple, Gary Brent
Decker, Robert Wayne
Dewart, Gilbert
Dey, Abhijit
Doell, Richard Rayman
Donath, Fred Arthur
Donoho, David Leigh
Dorman, LeRoy Myron
Eaton, Jerry Paul
Eittreim, Stephen L
Esposito, Pasquale Bernard
Ewing, Clair Eugene
Fillippone, Walter R
Flatte, Stanley Martin
Fliegel, Henry Frederick
Focke, Alfred Bosworth
Frank, Sidney Raymond
Fraser-Smith, Antony Charles
Fuller, Michael D
Gaal, Robert A P
Gaskell, Robert Weyand
Gilbert, James Freeman
Gimlett, James I
Goins, Neal Rodney
Green, Harry Western, II
Grine, Donald Reaville
Griscom, Andrew
Grobecker, Alan J
Hall, Robert Earl
Hanks, Thomas Colgrove
Hannon, Willard James, Jr
Hardy, Rolland L(ee)
Harkrider, David Garrison
Harrison, John Christopher
Haubrich, Richard August
Healy, John H
Hearst, Joseph R
Heusinkveld, Myron Ellis
Hill, David Paul
Hill, Donald Gardner
Housley, Robert Melvin
Huebsch, Ian O
Inman, Douglas Lamar
Irwin, James Joseph
Isherwood, William Frank
Iyer, Hariharaiyer Mahadeva
Jachens, Robert C
Jackson, Bernard Vernon
Jackson, David Diether
Jeanloz, Raymond
Jones, Stanley Bennett
Joyner, William B
Kahle, Anne Bettine
Kanamori, Hiroo
Kaplan, Joseph
Kasameyer, Paul William
Kaula, William Mason
Kennedy, Burton Mack
Khurana, Krishan Kumar
King, Chi-Yu
Kirk, John Gallatin
Knopoff, Leon
Knudsen, William Claire
Lachenbruch, Arthur Herold
Latham, Gary V
Lay, Thorne
Lee, Tien-Chang
Lee, William Hung Kan
Leonard, Robert Stuart
Lin, Robert Peichung
Lin, Wunan
Lindh, Allan Goddard
Lippitt, Louis
Lipson, Joseph Issac
Luyendyk, Bruce Peter
MacDonald, Gordon James Fraser
Macdonald, Kenneth Craig
MacDoran, Peter Frank
McEvilly, Thomas V
McGarr, Arthur
McHugh, Stuart Lawrence
McIntyre, Marie Claire
McKenzie, William F
McNally, Karen Cook
McPherron, Robert Lloyd
Mal, Ajit Kumar
Martin, Terry Zachry
Meadows, Mark Allan

Moe, Osborne Kenneth
Mooney, Walter D
Morrison, Huntly Frank
Nakanishi, Keith Koji
Nason, Robert Dohrmann
Nathenson, Manuel
Nellis, William J
Northrop, John
Nur, Amos M
Nurmia, Matti Juhani
O'Keefe, John Dugan
Orcutt, John Arthur
Page, Robert Alan, Jr
Palluconi, Frank Don
Pang, Kevin Dit Kwan
Petersen, Carl Frank
Pfluke, John H
Phillips, Richard P
Poppoff, Ilia George
Porter, Lawrence Delpino
Portnoff, Michael Rodney
Pray, Ralph Emerson
Reynolds, John Hamilton
Reynolds, Ray Thomas
Rosenblatt, Daniel Bernard
Runge, Richard John
Russell, Christopher Thomas
Russell, Philip Boyd
Ryan, Jack A
Ryu, Jisoo Vinsky
St Amand, Pierre
Schock, Robert Norman
Scholl, James Francis
Schubert, Gerald
Schwing, Franklin Burton
Sentman, Davis Daniel
Shaw, Herbert Richard
Shor, George G, Jr
Shreve, Ronald Lee
Silver, Eli Alfred
Sims, John David
Spence, Harlan Ernest
Springer, Donald Lee
Stacey, John Sydney
Stephenson, Lee Palmer
Stevenson, David John
Stocker, Richard Louis
Sundaram, Panchanatham N
Swanberg, Chandler A
Swift, Charles Moore, Jr
Teng, Ta-Liang
Terrile, Richard John
Thatcher, Wayne Raymond
Thompson, George Albert
Thompson, Timothy J
Tooley, Richard Douglas
Tripp, R(ussell) Maurice
Tuman, Vladimir Shlimon
Tyler, George Leonard
Vacquier, Victor
Vanyo, James Patrick
Vasco, Donald Wyman
Vassiliou, Marius Simon
Verhoogen, John
Verosub, Kenneth Lee
Vickers, Roger Spencer
Walt, Martin
Ward, Peter Langdon
Warren, David Henry
Warshaw, Stephen I
Wasserburg, Gerald Joseph
Weiss, Jeffrey Martin
Wells, Ronald Allen
Wiggins, John H(enry), Jr
Williams, James Gerard
Willis, David Edwin
Wold, Richard John
Wood, Fergus James
Wright, Frederick Hamilton
Yoder, Charles Finney
Zoback, Mark D
Zoback, Mary Lou Chetlain
Zumberge, James Frederick

COLORADO
Abshier, Curtis Brent
Algermissen, Sylvester Theodore
Alldredge, Leroy Romney
Allen, Joe Haskell
Anderson, Edmund Hughes
Applegate, James Keith
Bailey, Dana Kavanagh
Balch, Alfred Hudson
Ball, Mahlon M
Behrendt, John Charles
Benton, Edward Rowell
Born, George Henry
Bufe, Charles Glenn
Campbell, Wallace Hall
Chang, Bunwoo Bertram
Chinnery, Michael Alistair
Cummings, David
Dana, Robert Watson
Diment, William Horace
Donovan, Terrence John
Engdahl, Eric Robert
Faller, James E
Friedman, Jules Daniel
Goetz, Alexander Franklin Hermann
Gomberg, Joan Susan
Grose, Thomas Lucius Trowbridge
Hansen, Richard Olaf
Hilt, Richard Leighton
Hittelman, Allen M

Holmer, Ralph Carrol
Hoover, Donald Brunton
Jacobson, Jimmy Joe
Johnson, Verner Carl
Kane, Martin Francis
Kisslinger, Carl
Kleinkopf, Merlin Dean
Lackey, Laurence
La Fehr, Thomas Robert
Lander, James French
Larner, Kenneth Lee
Larson, Edwin E
Leinbach, F Harold
Lincoln, Jeannette Virginia
Mason, Malcolm
Meyers, Herbert
Nabighian, Misac N
Obradovich, John Dinko
Olhoeft, Gary Roy
Osterhoudt, Walter Jabez
Pakiser, Louis Charles, Jr
Parker, John Marchbank
Pawlowicz, Edmund F
Pennington, Wayne David
Pickett, George R
Plows, William Herbert
Richmond, Arthur Dean
Romig, Phillip Richardson
Sanborn, Albert Francis
Sargent, Howard Harrop, III
Savage, William Zuger
Schiffmacher, E(dward) R(obert)
Scott, James Henry
Shedlock, Kaye M
Snyder, Howard Arthur
Speiser, Theodore Wesley
Spence, William J
Spencer, Joseph Walter
Thompson, Loren Edward, Jr
Thomsen, Harry Ludwig
Urban, Thomas Charles
Viksne, Andy
Watson, Kenneth
White, James Edward
Williams, David Lee
Wunderman, Richard Lloyd
Wyss, Max

CONNECTICUT
Aitken, Janet M
Burridge, Robert
Clark, Sydney P, Jr
DeBoer, Jelle
Dowling, John J
Friedman, Don Gene
Gordon, Robert Boyd
Kleinberg, Robert Leonard
Liu, Qing-Huo
O'Brien, Brian
Plona, Thomas Joseph
Schweitzer, Jeffrey Stewart
Tittman, Jay
Wiggins, Ralphe

DELAWARE
Hanson, Roy Eugene
Ness, Norman Frederick

DISTRICT OF COLUMBIA
Aldrich, Lyman Thomas
Alexander, Joseph Kunkle
Berkson, Jonathan Milton
Boss, Alan Paul
Chung, Riley M
Crary, Albert Paddock
Greenewalt, David
Hart, Pembroke J
Hays, James Fred
James, David Evan
Johnson, Leonard Evans
Linde, Alan Trevor
Mao, Ho-Kwang
Meier, Robert R
Penhollow, John O
Press, Frank
Reilly, Michael Hunt
Riemer, Robert Lee
Singer, S(iegfried) Fred
Smith, Waldo E(dward)
Usselman, Thomas Michael
Vogt, Peter Richard
Vogt, Peter Richard
Wetherill, George West
Zimbelman, James Ray

FLORIDA
Cain, Joseph Carter, III
Draper, Grenville
Harrison, Christopher George Alick
Helava, U(uno) V(ilho)
Jin, Rong-Sheng
Long, James Alvin
McLeroy, Edward Glenn
O'Hara, Norbert Wilhelm
Olson, Donald B
Opdyke, Neil
Orlin, Hyman
Parker, Garald G, Sr
Rona, Peter Arnold
Rosendahl, Bruce Ray
Sivjee, Gulamabas Gulamhusen

GEORGIA
Dainty, Anton Michael

Long, Leland Timothy
Lowell, Robert Paul
Whitney, James Arthur

HAWAII
Berg, Eduard
Daniel, Thomas Henry
Frazer, L Neil
Furumoto, Augustine S
Helsley, Charles Everett
Johnson, Carl Edward
Keating, Barbara Helen
Khan, Mohammad Asad
Malahoff, Alexander
Manghnani, Murli Hukumal
Ming, Li Chung
Moore, Gregory Frank
Raleigh, Cecil Baring
Rose, John Creighton
Saxena, Narendra K
Vitousek, Martin J
Walker, Daniel Alvin

IDAHO
Maloof, Giles Wilson

ILLINOIS
Carpenter, Philip John
Ervin, C Patrick
Fish, Ferol F, Jr
Foster, John Webster
Hoffer, Abraham
Hsui, Albert Tong-Kwan
Klasner, John Samuel
McGinnis, Lyle David
Okal, Emile Andre
Riahi, Daniel Nourollah
White, W(illiam) Arthur
Wong, Kam Wu

INDIANA
Biggs, Maurice Earl
Blakely, Robert Fraser
Chowdhury, Dipak Kumar
Christensen, Nikolas Ivan
Hamburger, Michael Wile
Hinze, William James
Kim, Yeong Ell
Kivioja, Lassi A
Madison, James Allen
Mead, Judson
Rudman, Albert J

IOWA
Carmichael, Robert Stewart
Hopkins, John Raymond
Jeyapalan, Kandiah

KANSAS
Hambleton, William Weldon
Steeples, Donald Wallace
Surampalli, Rao Yadagiri

KENTUCKY
Brock, Louis Milton
Seeger, Charles Ronald
Sendlein, Lyle V A

LOUISIANA
Bradley, Marshall Rice
Breeding, J Ernest, Jr
French, William Stanley
Higgs, Robert Hughes
Johnson, Hamilton McKee
Kinsland, Gary Lynn
Meriwether, John R
Pilger, Rex Herbert, Jr

MAINE
Wall, Robert Ecki

MARYLAND
Alers, Perry Baldwin
Allenby, Richard John, Jr
Baker, Leonard Samuel
Behannon, Kenneth Wayne
Bostrom, Carl Otto
Carter, William Eugene
Chovitz, Bernard H
Clark, Thomas Arvid
Cohen, Steven Charles
Comiso, Josefino Cacas
Connerney, John E P
Degnan, John James, III
Elsasser, Walter M
Fink, Don Roger
Fischer, Irene Kaminka
Flinn, Edward Ambrose
Garvin, James Brian
Gutman, George Garik
Hammond, Allen Lee
Heirtzler, James Ransom
Heppner, James P
Herrero, Federico Antonio
Hubbert, Marion King
Kappel, Ellen Sue
King, Joseph Herbert
Kohler, Max A
Krueger, Arlin James
Langel, Robert Allan
Lenoir, William Benjamin
Liu, Han-Shou
Loxley, Thomas Edward
Maier, Eugene Jacob Rudolph

Marsh, Bruce David
Martin, James Milton
Mead, Gilbert Dunbar
Meredith, Leslie Hugh
O'Keefe, John Aloysius
Olson, Peter Lee
Paik, Ho Jung
Pisacane, Vincent L
Robertson, Douglas Scott
Salisbury, John William, Jr
Smith, David Edmund
Spilhaus, Athelstan Frederick, Jr
Stone, Albert Mordecai
Svendsen, Kendall Lorraine
Taylor, Patrick Timothy
Thomas, Herman Hoit
Trainor, James H
Wells, Eddie N

MASSACHUSETTS
Beckerle, John C
Bell, Peter M
Birch, Albert Francis
Bogert, Bruce Plympton
Bowhill, Sidney Allan
Bowin, Carl Otto
Brecher, Aviva
Bromery, Randolph Wilson
Brown, Laurie Lizbeth
Brynjolfsson, Ari
Bunce, Elizabeth Thompson
Burger, Henry Robert, III
Burns, Daniel Robert
Carr, Jerome Brian
Cheng, Chuen Hon
Corey, Brian E
Counselman, Charles Claude, III
Dandekar, Balkrishna S
Denig, William Francis
Donohue, John J
Dziewonski, Adam Marian
Eckhardt, Donald Henry
Ewing, John I
Frisk, George Vladimir
Goody, Richard (Mead)
Guertin, Ralph Francis
Guidice, Donald Anthony
Hager, Bradford Hoadley
Hope, Elizabeth Greeley
Hoskins, Hartley
Jacobsen, Stein Bjornar
Jordan, Thomas Hillman
Kachanov, Mark L
Kerr, Donald M, Jr
King, Robert Wilson, Jr
Kleinrock, Martin Charles
Knecht, David Jordan
Lacoss, Richard Thaddee
Leblanc, Gabriel
Lin, Jian
Little, Sarah Alden
McConnell, Robert Kendall
McNutt, Marcia Kemper
Madden, Stephen James, Jr
Madden, Theodore Richard
Mahoney, William C
Meriwether, John Williams, Jr
Michael, Irving
Molnar, Peter Hale
O'Connell, Richard John
Parsignault, Daniel Raymond
Petschek, Harry E
Pike, Charles P
Rice, James R
Ridge, John Charles
Rooney, Thomas Peter
Rosen, Richard David
Rothwell, Paul L
Shaw, Peter Robert
Shuman, Bertram Marvin
Silverman, Sam M
Singer, Howard Joseph
Slowey, Jack William
Solomon, Sean Carl
Stephen, Ralph A
Toksoz, Mehmet Nafi
Tucholke, Brian Edward
Von Herzen, Richard P
Walsh, Joseph Broughton
Whitehead, John Andrews

MICHIGAN
Clark, James Alan
Clauer, C Robert, Jr
Cloud, Gary Lee
England, Anthony W
Frantti, Gordon Earl
Huntoon, Jacqueline E
Jackson, Philip Larkin
Jacobs, Stanley J
Nagy, Andrew F
Pollack, Henry Nathan
Smith, Thomas Jefferson
Trow, James
Van der Voo, Rob
Walker, James Callan Gray

MINNESOTA
Atchison, Thomas Calvin, Jr
Banerjee, Subir Kumar
Farnham, Paul Rex
Fogelson, David Eugene
Hey, Richard N
Mauersberger, Konrad

Yuen, David Alexander

MISSISSIPPI
Ballard, James Alan
Butler, Dwain Kent
Chang, Frank Keng
Cress, Daniel Hugg
Davis, Thomas Mooney
Geddes, Wilburt Hale
Lowrie, Allen
Malone, Philip Garcin
Morris, Gerald Brooks
Ramsdale, Dan Jerry
Sprague, Vance Glover, Jr
Sumner, Roger D

MISSOURI
DuBroff, Richard Edward
Heinrich, Ross Raymond
Herrmann, Robert Bernard
Hunt, Mahlon Seymour
McKinnon, William Beall
Mitchell, Brian James
Nuttli, Otto William
Rechtien, Richard Douglas
Rupert, Gerald Bruce
Stauder, William
Vincenz, Stanislaw Aleksander
White, Jon M

MONTANA
Banaugh, Robert Peter
Sill, William Robert
Wideman, Charles James

NEBRASKA
Goss, David

NEVADA
Anderson, John Gregg
Brune, James N
Hess, John Warren
Nichols, Chester Encell
Peck, John H
Ryall, Alan S, Jr
Slemmons, David Burton

NEW HAMPSHIRE
Ackley, Stephen Fred
Bothner, Wallace Arthur
Colbeck, Samuel C
Hager, Jutta Lore
Hibler, William David, III
Hoag, Roland Boyden, Jr
Perovich, Donald Kole
Simmons, Gene
Van Tassel, Roger A

NEW JERSEY
Bird, Harvey Harold
Bonini, William E
Chave, Alan Dana
Cox, Michael
Dahlen, Francis Anthony, Jr
Jeter, Hewitt Webb
Lanzerotti, Louis John
MacLennan, Carol G
Metsger, Robert William
Miyakoda, Kikuro
Morgan, William Jason
Orlanski, Isidoro
Phinney, Robert A
Sarmiento, Jorge Louis
Sheridan, Robert E
Suppe, John
Takahashi, Taro
Thiruvathukal, John Varkey
Thompson, William Baldwin
Tuleya, Robert E
Weissenburger, Don William

NEW MEXICO
Bullard, Edwin Roscoe, Jr
Close, Donald Alan
Derr, John Sebring
Deupree, Robert G
Dick, Richard Dean
Fehler, Michael C
Fugelso, Leif Erik
Geissman, John William
Gross, Gerardo Wolfgang
Holmes, Charles Robert
Huestis, Stephen Porter
Kuiper, Logan Keith
Lee, David Oi
Lysne, Peter C
McTigue, David Francis
Marker, Thomas F(ranklin)
Merritt, Melvin Leroy
Perret, William Riker
Phillips, John Lynch
Reiter, Marshall Allan
Rice, Robert Bruce
Riecker, Robert E
Roberts, Peter Morse
Sanders, Walter L
Sanford, Allan Robert
Searls, Craig Allen
Snell, Charles Murrell
Sun, James Ming-Shan
Tremba, Edward Louis
Weart, Wendell D

NEW YORK
Alsop, Leonard E
Bassett, William Akers
Bokuniewicz, Henry Joseph
Bolgiano, Ralph, Jr
Cathles, Lawrence MacLagan, III
Cisne, John Luther
Cozzarelli, Francis A(nthony)
Czapski, Ulrich Hans
Dickman, Steven Richard
Diebold, John Brock
Egan, Walter George
Fitzpatrick, Robert Charles
Fleischer, Robert Louis
Gilmore, Robert Snee
Hayes, Dennis E
Heyl, George Richard
Hodge, Dennis
Isacks, Bryan L
Jacob, Klaus H
Kaarsberg, Ernest Andersen
Kastens, Kim Anne
Katz, Samuel
Kaufman, Sidney
Kent, Dennis V
Kuo, John Tsung-fen
Langseth, Marcus G
Lawson, Charles Alden
Liebermann, Robert C
MacDonald, William David
Magorian, Thomas R
Meisel, David Dering
Meyer, Robert Jay
Millman, George Harold
Moniot, Robert Keith
Muller, Otto Helmuth
Oliver, Jack Ertle
Parks, Thomas William
Prucha, John James
Reitan, Paul Hartman
Remo, John Lucien
Revetta, Frank Alexander
Richards, Paul Granston
Rimai, Donald Saul
Rind, David Harold
Robinson, Joseph Edward
Ruddick, James John
Scholz, Christopher Henry
Schreiber, Edward
Senus, Walter Joseph
Sutton, George Harry
Sykes, Lynn Ray
Turcotte, Donald Lawson
Vaughan, Michael Thomas
Watt, James Peter
Weaver, John Scott
Woodmansee, Donald Ernest
Wu, Francis Taming

NORTH CAROLINA
Almy, Charles C, Jr
Johnson, Thomas Charles
Karson, Jeffrey Alan
Stavn, Robert Hans
Worzel, John Lamar

OHIO
Ahmad, Moid Uddin
Bossler, John David
Goad, Clyde Clarenton
Kellerstrass, Ernst Junior
Kulander, Byron Rodney
Kunze, Adolf Wilhelm Gerhard
Mourad, A George
Mueller, Ivan I
Noltimier, Hallan Costello
Rapp, Richard Henry
Richard, Benjamin H
Stierman, Donald John
Walter, Edward Joseph
Wolfe, Paul Jay

OKLAHOMA
Ahern, Judson Lewis
Brown, Graydon L
Byerly, Perry Edward
Chambers, Richard Lee
Crosby, Gary Wayne
Crowe, Christopher
Du Bois, Robert Lee
Elliott, Sheldon Ellwood
Frisillo, Albert Lawrence
Granath, James Wilton
Gutowski, Paul Ramsden
Hauge, Paul Stephen
Hill, Stephen James
Jennemann, Vincent Francis
Koller, Glenn R
Ratliff, Larry E
Robinson, Enders Anthony
Scales, John Alan
Sigal, Richard Frederick
Smith, James Edward
Sondergeld, Carl Henderson
Souder, Wallace William
Stoeckley, Thomas Robert
Stone, John Floyd
Thapar, Mangat Rai
Thomsen, Leon
Treitel, Sven

OREGON
Bodvarsson, Gunnar
Couch, Richard W

Geophysics (cont)

Demarest, Harold Hunt, Jr
Mather, Keith Benson
Trehu, Anne Martine

PENNSYLVANIA
Bell, Maurice Evan
Bennett, Lee Cotton, Jr
Biondi, Manfred Anthony
Greenfield, Roy Jay
Hodgson, Robert Arnold
Howell, Benjamin F, Jr
Langston, Charles Adam
Lavin, Peter Masland
Liu, Mian
Mowrey, Gary Lee
Olivero, John Joseph, Jr
Pilant, Walter L
Schmidt, Victor A
Simaan, Marwan
Strick, Ellis
Wuenschel, Paul Clarence

RHODE ISLAND
Hermance, John Francis
McMaster, Robert Luscher
Polk, C(harles)
Schultz, Peter Hewlett
Tullis, Terry Edson

SOUTH CAROLINA
Duncan, Lewis Mannan
Eyer, Jerome Arlan
Secor, Donald Terry, Jr

SOUTH DAKOTA
Davis, Briant LeRoy

TENNESSEE
Dorman, Henry James
Gentry, Robert Vance
Hopkins, Richard Allen
Johnson, Robert W, Jr
Thomson, Kerr Clive

TEXAS
Anderson, George Boine
Angona, Frank Anthony
Aucoin, Paschal Joseph, Jr
Backus, Milo M
Barrow, Thomas D
Baskir, Emanuel
Bean, Robert Jay
Behrens, Earl William
Berkhout, Aart W J
Bernard, Bernie Boyd
Blackwelder, Blake Winfield
Brown, Leonard Franklin, Jr
Bucy, J Fred
Burke, William Henry
Bussian, Alfred Erich
Carter, Neville Louis
Cheney, Monroe G
Chiburis, Edward Frank
Clark, Howard Charles, Jr
Clayton, Neal
Clough, John Wendell
Cook, Ernest Ewart
Currie, Robert Guinn
Cutler, Roger T
Dahm, Cornelius George
Damuth, John Erwin
Danburg, Jerome Samuel
Deaton, Bobby Charles
De Bremaecker, Jean-Claude
Deeming, Terence James
Delflache, Andre P
Dent, Brian Edward
Dobrin, Milton Burnett
Doser, Diane Irene
Dunlap, Henry Francis
Eastwood, Raymond L
Eaton, Jerome F
Eisner, Elmer
Ellwood, Brooks B
Fahlquist, Davis A
Feagin, Frank J
Fertl, Walter Hans
Fett, John D
Forester, Robert Donald
Fountain, Lewis Spencer
Galbraith, James Nelson, Jr
Gallagher, John Joseph, Jr
Gangi, Anthony Frank
Gardner, Gerald Henry Fraser
Gee, David Easton
Ginsburg, Merrill Stuart
Goforth, Thomas Tucker
Gregory, A(lvin) R(ay)
Hall, David Joseph
Handin, John Walter
Hanss, Robert Edward
Hardage, Bob Adrian
Hasbrook, Arthur F(erdinand)
Heinze, William Daniel
Heroy, William Bayard, Jr
Herrin, Eugene Thornton, Jr
Hilde, Thomas Wayne Clark
Hilterman, Fred John
Hoskins, Earl R, Jr
Huffington, Roy Michael
Hurley, Neal Lilburn
Johnson, Brann

Johnson, William W
Johnston, David Hervey
Jordan, Neal F(rancis)
Justice, James Horace
Keller, George Randy, Jr
Kern, John W
Kibler, Kenneth G
Kulla, Jean B
Landrum, Ralph Avery, Jr
Lawrence, Philip Linwood
Leiss, Ernst L
Levien, Louise
Levin, Franklyn Kussel
McCormack, Harold Robert
Massell, Wulf F
Mateker, Emil Joseph, Jr
Matumoto, Tosimatu
Mayne, William Harry
Miller, Max K
Moorhead, William Dean
Musgrave, Albert Wayne
Mut, Stuart Creighton
Neidell, Norman Samson
Nielsen, Kent Christopher
Nosal, Eugene Adam
Nugent, Robert Charles
Nussmann, David George
Opal, Chet Brian
Overmyer, Robert Franklin
Palmeira, Ricardo Antonio Ribeiro
Pan, Poh-Hsi
Peeples, Wayne Jacobson
Peters, Jack Warren
Peterson, Donald Neil
Phillips, Joseph D
Potts, Mark John
Price, William Charles
Pritchett, William Carr
Rader, Dennis
Ranganayaki, Rambabu Pothireddy
Reed, Dale Hardy
Reeves, Robert Grier (Lefevre)
Rezak, Richard
Rinehart, Wilbur Allan
Robertson, James Douglas
Rosenbaum, Joseph Hans
Roy, Robert Francis
Savit, Carl Hertz
Sbar, Marc Lewis
Schoenberger, Michael
Schramm, Martin William, Jr
Schutz, Bob Ewald
Sclater, John George
Self, Stephen
Seriff, Aaron Jay
Sheriff, Robert Edward
Shurbet, Deskin Hunt, Jr
Siegfried, Robert Wayne, II
Soloyanis, Susan Constance
Sorrells, Gordon Guthrey
Spencer, Terry Warren
Sprunt, Eve Silver
Srnka, Leonard James
Stoffa, Paul L
Sumner, John Randolph
Tajima, Toshiki
Talwani, Manik
Tixier, Maurice Pierre
Truxillo, Stanton George
Ward, Ronald Wayne
Ward-McLemore, Ethel
Watkins, Joel Smith, Jr
Watson, Robert Joseph
Wegner, Robert Carl
Wells, Frederick Joseph
Whittlesey, John R B
Widess, Moses B
Wiggins, James Wendell
Wilson, John Human
Wiltschko, David Vilander
Winbow, Graham Arthur
Witterholt, Edward John
Wu, Changsheng
Yarger, Harold Lee
Zemanek, Joseph, Jr
Zimmerman, Carol Jean

UTAH
Cook, Kenneth Lorimer
Foster, Robert H
Greene, Gordon William
Ross, Howard Persing
Smith, Robert Baer
Ward, Stanley Harry

VERMONT
Drake, Charles Lum

VIRGINIA
Bennett, Gordon Daniel
Berg, Joseph Wilbur, Jr
Blandford, Robert Roy
Bollinger, Gilbert A
Bowker, David Edwin
Chubb, Talbot Albert
Costain, John Kendall
Davis, Charles Mitchell, Jr
Delnore, Victor Eli
DeNoyer, John M
Doles, John Henry, III
Faust, Charles R
Gamble, Thomas Dean
Gawarecki, Stephen Jerome
Goncz, John Henry

Hamilton, Robert Morrison
Hanna, William F
Hartline, Beverly Karplus
Hemingway, Bruce Sherman
Hersey, John B
Hosterman, John W
Huebner, John Stephen
Johnson, Robert Edward
King, Elizabeth Raymond
Koch, Carl Fred
Lanzano, Paolo
Levine, Joel S
Michael, William Herbert, Jr
Milton, Nancy Melissa
Morrow, Richard Joseph
Orr, Marshall H
Radoski, Henry Robert
Robertson, Eugene Corley
Robinson, Edwin S
Romney, Carl Fredrick
Russ, David Perry
Sax, Robert Louis
Senseny, Paul Edward
Snoke, J Arthur
Spall, Henry Roger
Spinrad, Richard William
Stewart, David Benjamin
Tanner, Allan Bain
Ugincius, Peter
Unger, John Duey
Wallace, Raymond Howard, Jr
Wesson, Robert Laughlin
Williams, Owen Wingate

WASHINGTON
Adams, William Mansfield
Beck, Myrl Emil, Jr
Booker, John Ratcliffe
Bostrom, Robert Christian
Brown, Robert Alan
Bull, Colin Bruce Bradley
Caggiano, Joseph Anthony, Jr
Crosson, Robert Scott
Danes, Zdenko Frankenberger
Dehlinger, Peter
Dzurisin, Daniel
Fowles, George Richard
Heacock, Richard Ralph
Hoblitt, Richard Patrick
Hoch, Richmond Joel
Holmes, Mark Lawrence
Hussong, Donald MacGregor
Johnson, Harlan Paul
Kenney, James Francis
Koizumi, Carl Jan
Lansinger, John Marcus
Lister, Clive R B
Malone, Stephen D
Merrill, Ronald Thomas
Olsen, Kenneth Harold
Parks, George Kung
Quigley, Robert James
Raymond, Charles Forest
Schwarz, Sigmund D
Smith, Brad Keller
Smith, J Dungan
Smith, Stewart W
Steele, William Kenneth
Stottlemyre, James Arthur
Veress, Sandor A

WISCONSIN
Bentley, Charles Raymond
Clay, Clarence Samuel
Cooper, Reid F
Hammer, Sigmund Immanuel
Laudon, Thomas S
Meyer, Richard Ernst
Meyer, Robert Paul
Mortimer, Clifford Hiley
Mudrey, Michael George, Jr
Thurber, Clifford Hawes
Wang, Herbert Fan

WYOMING
Shive, Peter Northrop
Smithson, Scott Busby

ALBERTA
Argenal-Arauz, Roger
Cook, Frederick Ahrens
Cruden, David Milne
Cumming, George Leslie
Georgi, Daniel Taylan
Gough, Denis Ian
Hoekstra, Pieter
Jessop, Alan Michael
Kanasewich, Ernest Raymond
Lerbekmo, John Franklin
Peirce, John Wentworth
Rankin, David

BRITISH COLUMBIA
Barker, Alfred Stanley, Jr
Clowes, Ronald Martin
Dosso, Harry William
Ellis, Robert Malcolm
Galt, John (Alexander)
Horita, Robert Eiji
Hyndman, Roy D
Irving, Edward
Lewis, Trevor John
Mothersill, John Sydney
Nafe, John Elliott

Oldenburg, Douglas William
Russell, Richard Doncaster
Strangway, David W
Thomson, David James
Ulrych, Tadeusz Jan
Watanabe, Tomiya
Weaver, John Trevor
Weichert, Dieter Horst

MANITOBA
Hall, Donald Herbert

NEW BRUNSWICK
Burke, Kenneth B S
Hamilton, Angus Cameron
Vanicek, Petr

NEWFOUNDLAND
Deutsch, Ernst Robert
King, Michael Stuart
Rochester, Michael Grant
Wright, James Arthur

NOVA SCOTIA
Blanchard, Jonathan Ewart
Keen, Charlotte Elizabeth
Keen, Michael J
Ravindra, Ravi
Reynolds, Peter Herbert
Salisbury, Matthew Harold
Srivastava, Surat Prasad
Stevens, George Richard

ONTARIO
Berry, Michael John
Dence, Michael Robert
Drury, Malcolm John
Garland, George David
Gibb, Richard A
Gladwell, Graham M L
Grieve, Richard Andrew
Hare, Frederick Kenneth
Haworth, Richard Thomas
Hodgson, John Humphrey
Jones, Alister Vallance
Leith, Thomas Henry
McNamara, Allen Garnet
Manchee, Eric Best
Mansinha, Lalatendu
Mereu, Robert Frank
Morley, Lawrence Whitaker
Palmer, H Currie
Pearce, George William
Peltier, William Richard
Smylie, Douglas Edwin
Stesky, Robert Michael
Symons, David Thorburn Arthur
Thomas, Michael David
Uffen, Robert James
Watson, Michael Douglas
Weber, Jean Robert
Wesson, Paul Stephen
West, Gordon Fox
Wilson, John Tuzo
York, Derek H

QUEBEC
Crossley, David John
Doig, Ronald
Mukherji, Kalyan Kumar
Poorooshasb, Hormozd Bahman
Saull, Vincent Alexander
Seguin, Maurice Krisholm

SASKATCHEWAN
Nisbet, Euan G

OTHER COUNTRIES
Arce, Jose Edgar
Clement, William Glenn
Geller, Robert James
Hoke, John Humphreys
Horai, Ki-iti
Lomnitz, Cinna
Mason, Ronald George
Nestvold, Elwood Olaf
Pekeris, Chaim Leib
Tanis, James Iran
Taylor, Howard Lawrence
Tolstoy, Ivan
Vozoff, Keeva
Woodside, John Moffatt

Hydrology & Water Resources

ALABAMA
Bayne, David Roberge
Boyd, Claude Elson
Cook, Robert Bigham, Jr
LaMoreaux, Philip Elmer
Leonard, Kathleen Mary
Molz, Fred John, III
Moser, Paul H
Smoot, George Fitzgerald
Warman, James Clark

ALASKA
Brown, Edward James
Fahl, Charles Byron
Gosink, Joan P
Hawkins, Daniel Ballou
LaPerriere, Jacqueline Doyle
Long, William Ellis

Rogers, James Joseph
Slaughter, Charles Wesley

ARIZONA
Agenbroad, Larry Delmar
Avery, Charles Carrington
Baker, Malchus Brooks, Jr
Baker, Victor Richard
Bergen, James David
Bradley, Michael Douglas
Buras, Nathan
Callahan, Joseph Thomas
Cluff, Carwin Brent
Contractor, Dinshaw N
Davis, Edwin Alden
Davis, Stanley Nelson
DeCook, Kenneth James
Dickinson, Robert Earl
Dworkin, Judith Marcia
Emery, Philip Anthony
Erskine, Christopher Forbes
Evans, Daniel Donald
Fink, James Brewster
Gay, Lloyd Wesley
Graf, William L
Hawkins, Richard Holmes
Heede, Burchard Heinrich
Idso, Sherwood B
Ince, Simon
Lane, Leonard James
Limpert, Frederick Arthur
Lord, William B
Manera, Paul Allen
Montgomery, Errol Lee
Myers, Lloyd E(ldridge)
Neuman, Shlomo Peter
Osborn, Herbert B
Rasmussen, William Otto
Reich, Brian M
Renard, Kenneth G
Resnick, Sol Donald
Simpson, Eugene Sidney
Sorooshian, Soroosh
Sumner, John Stewart
Swenson, Frank Albert
Thames, John Long
Woolhiser, David A(rthur)
Zwolinski, Malcolm John

ARKANSAS
Mack, Leslie Eugene
Steele, Kenneth F

CALIFORNIA
Albert, Jerry David
Anderson, Henry Walter
Bean, Robert Taylor
Bencala, Kenneth Edward
Benes, Norman Stanley
Bertoldi, Gilbert LeRoy
Bredehoeft, John Dallas
Brice, James Coble
Brooks, Norman H(errick)
Bubeck, Robert Clayton
Butler, Stanley S
Carnahan, Chalon Lucius
Cassidy, John J(oseph)
Chen, Carl W(an-Cheng)
Chen, Cheng-lung
Chesnut, Dwayne A(llen)
Cooper, Charles F
Court, Arnold
Crawford, Norman Holmes
Dracup, John Albert
Everett, Lorne Gordon
Freeland, Forrest Dean, Jr
Getzen, Rufus Thomas
Griggs, Gary B
Hagan, Robert M(ower)
Ham, Lee Edward
Hanna, George P, Jr
Hassler, Thomas J
Humphrey, Thomas Milton, Jr
Huntley, David
Hwang, Li-San
Isherwood, William Frank
Janke, Norman C
Jeng, Raymond Ing-Song
Kam, James Ting-Kong
Kennedy, Vance Clifford
Kharaka, Yousif Khoshu
Lambert, Walter Paul
Lee, William Wai-Lim
Leps, Thomas MacMaster
Letey, John, Jr
Linsley, Ray K(eyes), Jr
Lippmann, Marcelo Julio
Lucas, Joe Nathan
Lysyj, Ihor
McKone, Thomas Edward
Manalis, Melvyn S
Morris, Henry Madison, Jr
Narasimhan, Thiruppudaimarudhur N
Nordstrom, Darrell Kirk
Oster, James Donald
Phoenix, David A
Poland, Joseph Fairfield
Remson, Irwin
Rogers, Gifford Eugene
Rudavsky, Alexander Bohdan
Saltonstall, Clarence William, Jr
Sarkaria, Gurmukh S
Scott, Verne H(arry)
Slack, Keith Vollmer

Snyder, Charles Theodore
Spieker, Andrew Maute
Sposito, Garrison
Sutton, Donald Dunsmore
Testa, Stephen M
Todd, David Keith
Turk, Leland Jan
Volz, Michael George
Yates, Scott Raymond
Ziemer, Robert Ruhl

COLORADO
Branson, Farrel Allen
Caine, T Nelson
Clebsch, Alfred, Jr
Cohen, Ronald R H
Cooley, Richard Lewis
Curtis, Bruce Franklin
DeCoursey, Donn G(ene)
Eisenlohr, W(illiam) S(tewart), Jr
Emmett, William W
Frasier, Gary Wayne
Garstka, Walter U(rban)
Griffith, Cecilia Girz
Hadley, Richard Frederick
Hardaway, John E(vans)
Herrmann, Raymond
Hoffer, Roger M(ilton)
Hotchkiss, William Rouse
Hughes, William Carroll
Illangasekare, Tissa H
Johnson, Arnold I(van)
Jones, Everett Bruce
Koelzer, Victor A
Langmuir, Donald
McLaughlin, Thad G
Meade, Robert Heber, Jr
Meiman, James R
Mickelson, Rome H
Miller, Glen A
Morel-Seytoux, Hubert Jean
Morrison, Roger Barron
Olhoeft, Gary Roy
Peckham, Alan Embree
Richardson, Everett V
Runnells, Donald DeMar
Sanford, Richard Frederick
Schumm, Stanley Alfred
Smith, Ralph Emerson
Smith, Roger Elton
Sommerfeld, Richard Arthur
Spence, William J
Steele, Timothy Doak
Striffler, William D
Tabler, Ronald Dwight
Toy, Terrence J
Urban, Thomas Charles
Visher, Frank N
Walker, Theodore Roscoe
Ward, Robert Carl
Weist, William Godfrey, Jr
Wershaw, Robert Lawrence
White, Gilbert Fowler
Yang, Chih Ted
Yang, In Che

CONNECTICUT
Bock, Paul
Fox, G(eorge) Sidney
Patton, Peter C
Spiegel, Zane

DELAWARE
Cheng, Alexander H-D
Habibi, Kamran
Mather, John Russell
Talley, John Herbert
Willmott, Cort James

DISTRICT OF COLUMBIA
Bell, Bruce Arnold
Blanchard, Bruce
Briscoe, John
Chung, Riley M
Duguid, James Otto
McGinnis, David Franklin, Jr
Mercer, James Wayne
Myers, Charles Edwin
Olem, Harvey
Smith, Waldo E(dward)
Stringfield, Victor Timothy
Terrell, Charles R
Walrafen, George Edouard
Warnick, Walter Lee
Woo, Dah-Cheng
Zeizel, A(rthur) J(ohn)

FLORIDA
Anderson, Melvin W(illiam)
Ballerini, Rocco
Barile, Diane Dunmire
Belanger, Thomas V
Bermes, Boris John
Chryssafopoulos, Nicholas
Conover, Clyde S(tuart)
Davis, John Armstrong
Dietz, Jess Clay
Hammond, Luther Carlisle
Huber, Wayne Charles
Jackson, Daniel Francis
Kantrowitz, Irwin H
Mace, Arnett C, Jr
Martsolf, J David
Nichols, James Carlile

O'Hara, Norbert Wilhelm
Orth, Paul Gerhardt
Parker, Garald G, Sr
Randazzo, Anthony Frank
Rao, Palakurthi Suresh Chandra
Sneade, Barbara Herbert
Stephens, John C(arnes)
Stewart, Mark Thurston
Tanner, William Francis, Jr
Upchurch, Sam Bayliss
Viessman, Warren, Jr
Warburton, David Lewis
Wozab, David Hyrum

GEORGIA
Barnwell, Thomas Osborn, Jr
Bowen, Paul Tyner
Burgoa, Benali
Carver, Robert E
Cofer, Harland E, Jr
Hewlett, John David
Hill, David W(illiam)
Johnston, Richard H
Lineback, Jerry Alvin
Meyer, Judy Lynn
Rogers, Harvey Wilbur
Snyder, Willard Monroe
Thomas, Adrian Wesley
Vorhis, Robert C
Wallace, James Robert

HAWAII
Ekern, Paul Chester
Fast, Arlo Wade
Hufschmidt, Maynard Michael
Lau, L(eung Ku) Stephen
Merriam, Robert Arnold
Peterson, Frank Lynn
Wu, I-Pai

IDAHO
Blackburn, Wilbert Howard
Brakensiek, Donald Lloyd
Clark, William Hilton
Crawford, Ronald L
Kincaid, Dennis Campbell
Knobel, LeRoy Lyle
Knutson, Carroll Field
Lehrsch, Gary Allen
Megahan, Walter Franklin
Schreiber, David Laurence
Williams, Roy Edward

ILLINOIS
Adams, John Rodger
Ardis, Colby V, Jr
Aubertin, Gerald Martin
Boast, Charles Warren
Carpenter, Philip John
Cartwright, Keros
Ditmars, John David
Foster, John Webster
Garcia, Marcelo Horacio
Gorody, Anthony Wagner
Heigold, Paul C
Kempton, John P(aul)
Krug, Edward Charles
McKay, Edward Donald, III
Marmer, Gary James
Mauzy, Michael P
Maxwell, William Hall Christie
Paddock, Robert Alton
Paintal, Amreek Singh
Piwoni, Marvin Dennis
Prakash, Anand
Sasman, Robert T
Semonin, Richard Gerard
Stout, Glenn Emanuel
Wendland, Wayne Marcel
Yen, Ben Chie

INDIANA
Fredrich, Augustine Joseph
Gray, William Guerin
Hellenthal, Ronald Allen
Krothe, Noel C
Leap, Darrell Ivan
McCafferty, William Patrick
Rao, R(amachandra) A
Shaffer, Nelson Ross
Spacie, Anne
Van der Heijde, Paul Karel Maria

IOWA
Austin, Tom Al
Georgakakos, Konstantine P
Hallberg, George Robert
Hershey, H Garland
Kirkham, Don
Koch, Donald Leroy
Nakato, Tatsuaki
Prior, Jean Cutler
Schnoor, Jerald L

KANSAS
Aber, James Sandusky
Beck, Henry V
Gordon, Ellis Davis
Grant, Stanley Cameron
Gries, John Charles
Metzler, Dwight F
Skidmore, Edward Lyman
White, Stephen Edward
Whittemore, Donald Osgood

KENTUCKY
Barfield, Billy Joe
Blackwell, Floyd Oris
Budde, Mary Laurence
Coltharp, George B
Crawford, Nicholas Charles
Kiefer, John David
Thrailkill, John
Warner, Richard Charles

LOUISIANA
Collins, Michael Albert
Ferrell, Ray Edward, Jr
Grimwood, Charles
Jones, Paul Hastings
Kazmann, Raphael Gabriel
Latorre, Robert George
Loden, Michael Simpson
Murray, Stephen Patrick
Nyman, Dale James
Singh, Vijay P
Wilson, John Thomas
Wrobel, William Eugene

MAINE
Kleinschmidt, R Stevens
Prescott, Glenn Carleton, Jr
Snow, Joseph William
Tewhey, John David

MARYLAND
Baummer, J Charles, Jr
Browzin, Boris S(ergeevich)
Chery, Donald Luke, Jr
Clarke, Frank Eldridge
Davis, George H
Everett, Ardell Gordon
Farnsworth, Richard Kent
Gaum, Carl H
Hudlow, Michael Dale
Jones, Blair Francis
Kohler, Max A
Linaweaver, Frank Pierce
Mack, Frederick K
May, Ira Philip
Miller, John Frederick
Munasinghe, Mohan P
Obremski, Henry J(ohn)
Papadopulos, Stavros Stefanu
Ragan, Robert Malcolm
Randall, John Douglas
Rango, Albert
ReVelle, Charles S
Salomonson, Vincent Victor
Schmugge, Thomas Joseph
Sternberg, Yaron Moshe
Watt, Mamadov Hame
Wolff, Roger Glen
Wolman, Markley Gordon
Yaramanoglu, Melih

MASSACHUSETTS
Bowley, Donovan Robin
Bras, Rafael Luis
Carr, Jerome Brian
Deegan, Linda Ann
Eagleson, Peter Sturges
Entekhabi, Dara
Frimpter, Michael Howard
Hillel, Daniel
Mader, Donald Lewis
Male, James William
Ostendorf, David William

MICHIGAN
Abriola, Linda Marie
Burton, Thomas Maxie
Clark, James Alan
D'Itri, Frank M
Holland-Beeton, Ruth Elizabeth
Kehew, Alan Everett
Kunkle, George Robert
Reynolds, John Z
Ronca, Luciano Bruno
Scavia, Donald
Tomboulian, Paul
Wylie, Evan Benjamin

MINNESOTA
Alexander, Emmit Calvin, Jr
Boening, Paul Henrik
Downing, William Lawrence
Foster, George Rainey
Gulliver, John Stephen
Gupta, Satish Chander
Klemer, Andrew Robert
Latterell, Joseph J
Linden, Dennis Robert
Lonergan, Dennis Arthur
McConville, David Raymond
McNaught, Donald Curtis
Perry, James Alfred
Pfannkuch, Hans Olaf
Silberman, Edward
Song, Charles Chieh-Shyang
Stefan, Heinz G
Tuthill, Samuel James
Walton, Matt Savage

MISSISSIPPI
Baker, Robert Andrew
Cross, Ralph Donald
Mutchler, Calvin Kendal
Priest, Melville S(tanton)

Dickinson, William Trevor
Elrick, David Emerson
Farvolden, Robert Norman
Flint, Jean-Jacques
Gillham, Robert Winston
Hughes, George Muggah
Jones, L(lewellyn) E(dward)
Kruus, Jaan
LaValle, Placido Dominick
Lee, David Robert
Lennox, Donald Haughton
McBean, Edward A
Meyboom, Peter
Moltyaner, Grigory
Schemenauer, Robert Stuart
Simpson, Frank
Tan, Chin Sheng
Wilkinson, Thomas Preston

QUEBEC
Beron, Patrick
Bouchard, Michel André
Ouellet, Marcel

SASKATCHEWAN
Chew, Hemming
Cullimore, Denis Roy
Fuller, Gerald Arthur
Mollard, John D
Tsang, Gee
Viraraghavan, Thiruvenkatachari

OTHER COUNTRIES
Briand, Frederic Jean-Paul
Krishna, J Hari
Raveendran, Ekarath
Ward, Gerald T(empleton)
Williams, Michael Maurice Rudolph
Yazicigil, Hasan

Marine Sciences, General

ALABAMA
Brady, Yolanda J
Modlin, Richard Frank
Williams, Louis Gressett

ALASKA
Ahlgren, Molly O
Dahlberg, Michael Lee
Hameedi, Mohammad Jawed
Mathisen, Ole Alfred
Olsen, James Calvin
Shirley, Thomas Clifton
Stekoll, Michael Steven
Weeks, Wilford Frank

ARIZONA
Parrish, Judith Totman

CALIFORNIA
Benson, Andrew Alm
Berry, William Benjamin Newell
Bottjer, David John
Bray, Richard Newton
Brumbaugh, Joe H
Chadwick, Nanette Elizabeth
Cheng, Lanna
Clague, David A
Cohen, Anne Carolyn Constant
Cohen, Daniel Morris
Cowles, David Lyle
Coyer, James A
Cummings, William Charles
Davis, Gary Everett
Davis, William Jackson
Dedrick, Kent Gentry
Delaney, Margaret Lois
Di Girolamo, Rudolph Gerard
Dill, Robert Floyd
Douglas, Robert G
Eittreim, Stephen L
Fiedler, Paul Charles
Field, A J
Fusaro, Craig Allen
Goff, Lynda June
Halpern, David
Ham, Lee Edward
Hanson, Joe A
Haygood, Margo Genevieve
Hemmingsen, Barbara Bruff
Hill, Merton Earle, III
Horn, Michael Hastings
Hunt, George Lester, Jr
Karl, Herman Adolf
Kitting, Christopher Lee
Kremer, Patricia McCarthy
Leidersdorf, Craig B
Lindner, Elek
Macdonald, Kenneth Craig
Mackenzie, Kenneth Victor
McLeod, Samuel Albert
Martin, Joel William
Morse, Daniel E
Mueller, James Lowell
Mulligan, Timothy James
Murray, Steven Nelsen
Muscatine, Leonard
Nybakken, James W
Oglesby, Larry Calmer
Orcutt, Harold George
Owen, Robert W, Jr
Parrish, Richard Henry

Patterson, Mark Robert
Pearse, Vicki Buchsbaum
Phleger, Charles Frederick
Potts, Donald Cameron
Rau, Gregory Hudson
Ripley, William Ellis
Robilliard, Gordon Allan
Roth, Ariel A
Sakagawa, Gary Toshio
Saleeby, Jason Brian
Sanders, Brenda Marie
Schmieder, Robert W
Schwing, Franklin Burton
Soule, Dorothy (Fisher)
Squires, Dale Edward
Stewart, Brent Scott
Stowe, Keith S
Straughan, Isdale (Dale) Margaret
Tegner, Mia Jean
Thierstein, Hans Rudolf
Thompson, Rosemary Ann
Yeung, Ronald Wai-Chun

COLORADO
Barron, Eric James
Brown, Lewis Marvin
Hay, William Winn
Kent, Harry Christison
Mykles, Donald Lee
Osterhoudt, Walter Jabez

CONNECTICUT
Baillie, Priscilla Woods
Booth, Charles E
Brooks, Douglas Lee
Farmer, Mary Winifred
Fell, Paul Erven
Haakonsen, Harry Olav
Katz, Max
Maples, Louis Charles
Monahan, Edward Charles
Tolderlund, Douglas Stanley

DELAWARE
Karlson, Ronald Henry
Kobayashi, Nobuhisa
Luther, George William, III
Maurmeyer, Evelyn Mary
Sayre, Clifford M(orrill), Jr
Schuler, Victor Joseph
Swann, Charles Paul
Targett, Nancy McKeever
Targett, Timothy Erwin
Taylor, Malcolm Herbert
Thoroughgood, Carolyn A

DISTRICT OF COLUMBIA
Adey, Walter Hamilton
Burris, John Edward
Cairns, Stephen Douglas
Diemer, F(erdinand) P(eter)
Feeney, Gloria Comulada
Gaffney, Paul G, II
Littler, Mark Masterton
Meñez, Ernani Guingona
Peters, Esther Caroline
Phelps, Harriette Longacre
Rubinoff, Roberta Wolff
Shykind, Edwin B
Sze, Philip
Terrell, Charles R
Williams, Austin Beatty

FLORIDA
Alevizon, William
Barkalow, Derek Talbot
Betzer, Peter Robin
Bloom, Stephen Allen
Courtenay, Walter Rowe, Jr
Craig, Alan Knowlton
Dodge, Richard E
Donoghue, Joseph F
Feddern, Henry A
Gifford, John A
Gleeson, Richard Alan
Griffith, Gail Susan Tucker
Hallock-Muller, Pamela
Joyce, Edwin A, Jr
Kerr, John Polk
Lutz, Peter Louis
McAllister, Raymond Francis
Marcus, Nancy Helen
Mariscal, Richard North
Maurrasse, Florentin Jean-Marie Robert
Miller, James Woodell
Nelson, Walter Garnet
Prince, Eric D
Reynolds, John Elliott, III
Rice, Stanley Alan
Richards, William Joseph
Stafford, Robert Oppen
Tamplin, Mark Lewis
Tappert, Frederick Drach
Taylor, Barrie Frederick
Teaf, Christopher Morris
Tiffany, William James, III
Virnstein, Robert W
White, Arlynn Quinton, Jr
White, Arlynn Quinton, Jr
Wiebusch, Charles Fred
Wise, Sherwood Willing, Jr
Wyneken, Jeanette
Zikakis, John Philip

GEORGIA
Adkison, Daniel Lee
Alberts, James Joseph
Blanton, Jackson Orin
Burns, Lawrence Anthony
Kneib, Ronald Thomas
Pomeroy, Lawrence Richards
Rogers, Peter H
Walls, Nancy Williams
Wiebe, William John

HAWAII
Bailey-Brock, Julie Helen
Boehlert, George Walter
Coles, Stephen Lee
Curtis, George Darwin
Eldredge, Lucius G
Garcia, Michael Omar
Helfrich, Philip
Hirota, Jed
Hunter, Cynthia L
Krock, Hans-Jürgen
Krupp, David Alan
Landry, Michael Raymond
Moberly, Ralph M
Moore, Gregory Frank
Wyban, James A
Wyrtki, Klaus
Zisk, Stanley Harris

ILLINOIS
Assanis, Dennis N
Bieler, Rüdiger
Dyer, William Gerald
Lima, Gail M
Towle, David Walter

INDIANA
Cushing, Bruce S
Hulbert, Matthew H
Spacie, Anne

IOWA
Landweber, Louis

KENTUCKY
Chesnut, Donald R, Jr

LOUISIANA
Bahr, Leonard M, Jr
Deaton, Lewis Edward
Defenbaugh, Richard Eugene
DeLaune, Ronald D
Mills, Earl Ronald
Shaw, Richard Francis
Van Lopik, Jack Richard
Walker, Harley Jesse
Weidner, Earl

MAINE
Borei, Hans Georg
Hawes, Robert Oscar
Larsen, Peter Foster
Marcotte, Brian Michael
Ridgway, George Junior
Shumway, Sandra Elisabeth
Townsend, David Warren
Welch, Walter Raynes

MARYLAND
Anderson, Robert Simpers
Baer, Ledolph
Berkson, Harold
Brandt, Stephen Bernard
Cammen, Leon Matthew
Capone, Douglas George
Carroll, Raymond James
Commito, John Angelo
Cronin, Thomas Wells
Fink, Don Roger
Helz, George Rudolph
Hines, Anson Hemingway
Kappel, Ellen Sue
Kayar, Susan Rennie
Kearney, Michael Sean
Koblinsky, Chester John
Kravitz, Joseph Henry
Leatherman, Stephen Parker
McCammon, Helen Mary
McVey, James Paul
Mielke, James Edward
Mihursky, Joseph Anthony
Moran, David Dunstan
Noblesse, Francis
Osman, Richard William
Powers, Dennis A
Pritchard, Donald William
Reader, Wayne Truman
Rich, Harry Louis
Riedel, Gerhardt Frederick
Sanders, James Grady
Stoskopf, Michael Kerry
Stubblefield, William Lynn
Szego, George C(harles)
Veitch, Fletcher Pearre, Jr
Weisberg, Stephen Barry
Wolfe, Douglas Arthur

MASSACHUSETTS
Atema, Jelle
Aubrey, David Glenn
Ballard, Robert D
Broadus, James Matthew
Bryan, Wilfred Bottrill

Butler, James Newton
Dacey, John W H
Dixon, Brian Gilbert
Jearld, Ambrose, Jr
Kleinrock, Martin Charles
Lambertsen, Richard H
Lin, Jian
Little, Sarah Alden
Lundstrom, Ronald Charles
McCoy, Floyd W, Jr
McLaren, J Philip
McNutt, Marcia Kemper
Marshall, Nelson
Mayo, Barbara Shuler
Parker, Henry Seabury, III
Pechenik, Jan A
Poag, Claude Wylie
Prescott, John Hernage
Reed, F(lood) Everett
Reimold, Robert J
Richards, F Paul
Robinson, William Edward
Stearns, Charles Edward
Stephen, Ralph A
Takeuchi, Kiyoshi Hiro
Wallace, Gordon Thomas
West, Arthur James, II
White, Alan Whitcomb
Wigley, Roland L
Wirsen, Carl O, Jr

MICHIGAN
Boyd, John Philip
Bratt, Albertus Dirk
Chang, William Y B
Lehman, John Theodore
Moll, Russell Addison

MINNESOTA
Hey, Richard N
Murdock, Gordon Robert

MISSISSIPPI
Anderson, Gary
Geddes, Wilburt Hale
Lawler, Adrian Russell
Leese, John Albert
Meylan, Maurice Andre
Morris, Gerald Brooks
Morris, Halcyon Ellen McNeil
Sprague, Vance Glover, Jr

MISSOURI
Kuhns, John Farrell
Sharp, John Roland
Snowden, Jesse O

NEVADA
Benfield, Charles W(illiam)
Deacon, James Everett

NEW HAMPSHIRE
Hines, Mark Edward
McDowell, William H
Weber, James Harold

NEW JERSEY
Brine, Charles James
Bruno, Michael Stephen
Canzonier, Walter J
Cooper, Keith Raymond
Farmanfarmaian, Allahverdi
Glaser, Keith Brian
Grassle, Judith Payne
London, Mark David
Pacheco, Anthony Louis
Psuty, Norbert Phillip
Sugam, Richard Jay

NEW MEXICO
Shopp, George Milton, Jr
Stricker, Stephev Alexander

NEW YORK
Biscaye, Pierre Eginton
Bokuniewicz, Henry Joseph
Boynton, John E
Clarke, Raymond Dennis
Cousteau, Jacques-Yves
Diebold, John Brock
Dugolinsky, Brent Kerns
Finks, Robert Melvin
Gerard, Valrie Ann
Hayes, Dennis E
Jacobs, Stanley S
Kaplan, Eugene Herbert
Karig, Daniel Edmund
Kastens, Kim Anne
Kent, Dennis V
Lasker, Howard Robert
Levinton, Jeffrey Sheldon
Multer, H Gray
Nuzzi, Robert
Quigley, James P
Rivest, Brian Roger
Scranton, Mary Isabelle
Serafy, D Keith
Smardon, Richard Clay
Smethie, William Massie, Jr
Smith, Sharon Louise
Webber, Edgar Ernest
Weyl, Peter K

Marine Sciences, General (cont)

NORTH CAROLINA
Chester, Alexander Jeffrey
Janowitz, Gerald S(aul)
Johnson, Thomas Charles
Karson, Jeffrey Alan
Kuenzler, Edward Julian
Lindquist, David Gregory
Link, Garnett William, Jr
Marlowe, James Irvin
Ramus, Joseph S
Rittschof, Daniel
Rulifson, Roger Allen
Schwartz, Frank Joseph
Searles, Richard Brownlee
Sizemore, Ronald Kelly
Sunda, William George
Thayer, Gordon Wallace
Williams, Ann Houston

NORTH DAKOTA
Faust, Maria Anna
Lieberman, Milton Eugene

OHIO
Boczar, Barbara Ann
Fortner, Rosanne White
Hamlett, William Cornelius
Hummon, William Dale
Hutchinson, Frederick Edward
Miller, Arnold I
Sabourin, Thomas Donald
Saksena, Vishnu P
Teeter, James Wallis

OKLAHOMA
Tillman, Roderick W
Visher, Glenn S

OREGON
Cimberg, Robert Lawrence
Dietz, Thomas John
Drake, Ellen Tan
Duffield, Deborah Ann
Lubchenco, Jane
Malouf, Robert Edward
Sherr, Barry Frederick
Sherr, Evelyn Brown
Terwilliger, Nora Barclay
Trehu, Anne Martine
Wood, Anne Michelle

PENNSYLVANIA
Adelson, Lionel Morton
Bowen, Vaughan Tabor
Carson, Bobb
Franz, Craig Joseph
Guida, Vincent George
Haase, Bruce Lee
Kump, Lee Robert
Miller, Richard Lee
Niemitz, Jeffrey William
Steele, Craig William
Taylor-Mayer, Rhoda E

RHODE ISLAND
Benoit, Richard J
Brungs, William Aloysius
Gould, Mark D
Leinen, Margaret Sandra
Miller, Don Curtis
Prell, Warren Lee

SOUTH CAROLINA
Bidleman, Terry Frank
Browdy, Craig Lawrence
Dame, Richard Franklin
Dean, John Mark
Feller, Robert Jarman
Higerd, Thomas Braden
McKellar, Henry Northington, Jr
Morris, James T
Smith, Theodore Isaac Jogues
Stakes, Debra Sue

SOUTH DAKOTA
Wiersma, Daniel

TENNESSEE
Bartell, Steven Michael
Brawley, Susan Howard
Brinkhurst, Ralph O
Simco, Bill Al

TEXAS
Aldrich, David Virgil
Baughn, Robert Elroy
Bohacs, Kevin Michael
Caillouet, Charles W, Jr
Chrzanowski, Thomas Henry
Damuth, John Erwin
DiMichele, Leonard Vincent
Foster, John Robert
Hanlon, Roger Thomas
Hasling, Jill Freeman
Hellier, Thomas Robert, Jr
La Claire, John Willard, II
McCloy, James Murl
Mazzullo, James Michael
Neill, William Harold
Parker, Robert Hallett

Pirie, Robert Gordon
Ray, James P
Rennie, Thomas Howard
Roberts, Susan Jean
Smith, Malcolm Crawford, Jr
Tindall, Donald R
Wohlschlag, Donald Eugene
Zimmerman, Carol Jean

UTAH
Waddell, Kidd M

VIRGINIA
Alden, Raymond W, III
Burreson, Eugene M
Curtin, Brian Thomas
Diaz, Robert James
Diehl, Fred A
Finkl, Charles William, II
Huggett, Robert James
Hurdle, Burton Garrison
Leung, Wing Hai
Loesch, Joseph G
Lynch, Maurice Patrick
McCormick-Ray, M Geraldine
Munday, John Clingman, Jr
Myers, Thomas DeWitt
Neves, Richard Joseph
Odum, William Eugene
Rippen, Thomas Edward
Shanks, Wayne C, III
Siapno, William David
Spinrad, Richard William
Weinstein, Alan Ira
Wingard, Christopher Jon
Wood, Robertson Harris Langley
Zieman, Joseph Crowe, Jr

WASHINGTON
Benson, Keith Rodney
Bernard, Eddie Nolan
Chace, Alden Buffington, Jr
Damkaer, David Martin
Hardy, John Thomas
Harry, George Yost
Hirschfelder, John Joseph
Long, Edward R
Mearns, Alan John
Miles, Edward Lancelot
Muench, Robin Davie
Muller-Parker, Gisèle Thérèse
Pearson, Walter Howard
Pietsch, Theodore Wells
Ross, June Rosa Pitt
Schwartz, Maurice Leo
Simenstad, Charles Arthur
Sorem, Ronald Keith
Swartzman, Gordon Leni
Templeton, William Lees
Thom, Ronald Mark

WEST VIRGINIA
Anderson, Douglas Poole

WISCONSIN
Amundson, Clyde Howard
Besch, Gordon Otto Carl
Brooks, Arthur S
Fowler, Gerald Allan
Gernant, Robert Everett
Hedden, Gregory Dexter
Horntvedt, Earl W
Krezoski, John R
Magnuson, John Joseph
Remsen, Charles C, III
Rosson, Reinhardt Arthur

PUERTO RICO
Crabtree, David Melvin
Voltzow, Janice

ALBERTA
Peirce, John Wentworth

BRITISH COLUMBIA
Chase, Richard L
Fankboner, Peter Vaughn
Hartwick, Earl Brian
Kelly, Michael Thomas
Mackie, George Owen
Tunnicliffe, Verena Julia
Yunker, Mark Bernard

NEW BRUNSWICK
Cook, Robert Harry
Frantsi, Christopher
Taylor, Andrew Ronald Argo
Wells, David Ernest

NEWFOUNDLAND
Haedrich, Richard L
Hiscott, Richard Nicholas
Khan, Rasul Azim
Muggeridge, Derek Brian
Ni, I-Hsun
Steele, Vladislava Julie

NOVA SCOTIA
Dadswell, Michael John
Jamieson, William David
Odense, Paul Holger
Sinclair, Michael Mackay
Stobo, Wayne Thomas
Vandermeulen, John Henri

ONTARIO
Calder, Dale Ralph
Geurts, Marie Anne H L
Green, Roger Harrison
LaValle, Placido Dominick
Leppard, Gary Grant
Moon, Thomas William
Ruel, Maurice M J
Smith, Lorraine Catherine
Winterbottom, Richard

QUEBEC
Beland, Pierre
Digby, Peter Saki Bassett
Filteau, Gabriel
Percy, Jonathan Arthur

SASKATCHEWAN
Nisbet, Euan G

OTHER COUNTRIES
Briand, Frederic Jean-Paul
Coulston, Mary Lou
Dahl, Arthur Lyon
Fowler, Scott Wellington
McCain, John Charles
Marsh, James Alexander, Jr
Paulay, Gustav
Richards, Adrian F
Rodriguez, Gilberto
South, Graham Robin

Mineralogy-Petrology

ALABAMA
Chalokwu, Christopher Iloba
Cook, Robert Bigham, Jr
Ehlers, Ernest George
Fang, Jen-Ho
Isphording, Wayne Carter
Prescott, Paul Ithel

ALASKA
Karl, Susan Margaret
Mowatt, Thomas C
Swanson, Samuel Edward

ARIZONA
Boyd, George Addison
Burt, Donald McLain
Buseck, Peter R
Duffield, Wendell Arthur
Fink, Jonathan Harry
Garske, David Herman
Gasparrini, Claudia
Guilbert, John M
Holloway, John Requa
Kamilli, Diana Chapman
Kamilli, Robert Joseph
Kieffer, Susan Werner
Loomis, Timothy Patrick
McMillan, Paul Francis
Page, Norman J
Snow, Eleanour Anne
Titley, Spencer Rowe
Wallace, Terry Charles, Jr
Williams, Sidney Arthur

ARKANSAS
Oglesby, Gayle Arden

CALIFORNIA
Albee, Arden Leroy
Allan, John Ridgway
Anderson, Orson Lamar
Austin, Steven Arthur
Beaty, David Wayne
Bieler, Barrie Hill
Boettcher, Arthur Lee
Bohlen, Steven Ralph
Boles, James Richard
Borg, Iris Y P
Brimhall, George H
Brooks, Elwood Ralph
Brown, Gordon Edgar, Jr
Burnett, John L
Burtner, Roger Lee
Carmichael, Ian Stuart
Chmura-Meyer, Carol A
Christiansen, Robert Lorenz
Clague, David A
Collins, Lorence Gene
Compton, Robert Ross
Cook, Harry E, III
Csejtey, Bela, Jr
Davidson, Jon Paul
Dodge, Franklin C W
Doner, Harvey Ervin
Dusel-Bacon, Cynthia
Ehlig, Perry Lawrence
Einaudi, Marco Tullio
Emerson, Donald Orville
Ernst, Wallace Gary
Floran, Robert John
Ford, Arthur B
Foster, Helen Laura
Fryer, Charles W
Gaal, Robert A P
Garlick, George Donald
Glassley, William Edward
Green, Harry Western, II
Hanan, Barry Benton
Harpster, Robert E

Harwood, David Smith
Hein, James R
Herber, Lawrence Justin
Hill, Hamilton Stanton
Hodges, Lance Thomas
Janke, Norman C
Jeanloz, Raymond
Kamb, Walter Barclay
Keith, Terry Eugene Clark
Liddicoat, Richard Thomas, Jr
Lin, Wunan
Lind, Carol Johnson
Liou, Juhn G
Loney, Robert Ahlberg
Longshore, John David
McFadden, Lucy-Ann Adams
McKague, Herbert Lawrence
McKenzie, William F
McLean, Hugh
Mahood, Gail Ann
Majmundar, Hasmukhrai Hiralal
Manning, Craig Edward
Mattinson, James Meikle
Meike, Annemarie
Mikami, Harry M
Moore, James Gregory
Moores, Eldridge Morton
Olson, Jerry Chipman
Pabst, Adolf
Parrish, William
Plummer, Charles Carlton
Ramspott, Lawrence Dewey
Rhoades, James David
Ross, Donald Clarence
Rossman, George Robert
Saleeby, Jason Brian
Sarna-Wojcicki, Andrei M
Schneiderman, Jill Stephanie
Silver, Leon Theodore
Smith, Robert Leland
Snetsinger, Kenneth George
Spera, Frank John
Stolper, Edward Manin
Strahl, Erwin Otto
Swinney, Chauncey Melvin
Sylvester, Arthur Gibbs
Tabor, Rowland Whitney
Taylor, Hugh P, Jr
Testa, Stephen M
Tilling, Robert Ingersoll
Vallier, Tracy L
Wadsworth, William Bingham
Walawender, Michael John
Warren, Paul Horton
Wenk, Hans-Rudolf
Williams, Alan Evan
Wilshire, Howard Gordon
Winchell, Robert E
Woyski, Margaret Skillman

COLORADO
Bohor, Bruce Forbes
Bryant, Bruce Hazelton
Cadigan, Robert Allen
Campbell, John Arthur
Cathcart, James B
Cole, James Channing
Desborough, George A
Dixon, Helen Roberta
Elder, Curtis Harold
Erslev, Eric Alan
Gerlach, Terrence Melvin
Griffitts, Wallace Rush
Grose, Thomas Lucius Trowbridge
Gude, Arthur James, 3rd
Hayes, John Bernard
Henrickson, Eiler Leonard
Hill, Walter Edward, Jr
Hills, Francis Allan
Jansen, George James
Keighin, Charles William
Kuntz, Mel Anton
Lee, Donald Edward
McCallum, Malcolm E
Modreski, Peter John
Munoz, James Loomis
Naeser, Nancy Dearien
Pilkington, Harold Dean
Ponder, Herman
Sanford, Richard Frederick
Sheppard, Richard A
Sheridan, Douglas Maynard
Sherman, David Michael
Sidder, Gary Brian
Smyth, Joseph Richard
Thompson, Tommy Burt
Verbeek, Earl Raymond
Wahl, Floyd Michael
Walker, Theodore Roscoe
Wallace, Chester Alan
Warner, Lawrence Allen
Wilcox, Ray Everett
Wunderman, Richard Lloyd
Young, Edward Joseph
Zartman, Robert Eugene

CONNECTICUT
Frueh, Alfred Joseph, Jr
Joesten, Raymond
Liese, Homer C
Philpotts, Anthony Robert
Piotrowski, Joseph Martin
Skinner, H Catherine W
Winchell, Horace

DELAWARE
Leavens, Peter Backus
Spoljaric, Nenad
Stiles, A(lvin) B(arber)
Thompson, Allan M

DISTRICT OF COLUMBIA
Finger, Larry W
Fleischer, Michael
Fudali, Robert F
Gaines, Alan McCulloch
Hays, James Fred
Hazen, Robert Miller
Jolly, Janice Laurene Willard
Mao, Ho-Kwang
Mysen, Bjorn Olav
Parham, Walter Edward
Prewitt, Charles Thompson
Rumble, Douglas, III
Simkin, Thomas Edward
Sukow, Wayne William
Usselman, Thomas Michael
Weill, Daniel Francis
Wilson, Frederick Albert
Yoder, Hatten Schuyler, Jr

FLORIDA
Compton, John S
Councill, Richard J
Draper, Grenville
Eades, James L
Gielisse, Peter Jacob Maria
Kadey, Frederic L, Jr
Lee, Harley Clyde
McClellan, Guerry Hamrick
Mueller, Paul Allen
Nagle, Frederick, Jr
Ragland, Paul C
Rosendahl, Bruce Ray
Travis, Russell Burton
Tull, James Franklin

GEORGIA
Cofer, Harland E, Jr
Dod, Bruce Douglas
Giardini, Armando Alfonzo
Gore, Pamela J(eanne) W(heeless)
Herz, Norman
Logan, Kathryn Vance
Pollard, Charles Oscar, Jr
Smith, John M
Sumartojo, Jojok
Weaver, Charles Edward
Weaver, Robert Michael
Whitney, James Arthur

HAWAII
Garcia, Michael Omar
Keil, Klaus
Ming, Li Chung
Scott, Edward Robert Dalton
Sinton, John Maynard
Wise, William Stewart

IDAHO
Bonnichsen, Bill
Buden, Rosemary V
Welch, Jane Marie

ILLINOIS
Ahmed, Wase U
Amos, Dewey Harold
Anderson, Alfred Titus, Jr
Baur, Werner Heinz
Baxter, James Watson
Berg, Jonathan H
Bishop, Finley Charles
Chapman, Carleton Abramson
Chiou, Wen-An
Doehler, Robert William
Forbes, Warren C
Glass, Herbert David
Guggenheim, Stephen
Haddock, Gerald Hugh
Harvey, Richard David
Hay, Richard Le Roy
Henderson, Donald Munro
Hess, David Filbert
Jones, Robert L
Kirchner, James Gary
Kirkpatrick, R James
Kirkpatrick, Robert James
Masters, John Michael
Moll, William Francis, Jr
Moore, Duane Milton
Moore, Paul Brian
Newton, Robert Chaffer
Odom, Ira Edgar
Reams, Max Warren
Sandberg, Philip A
Smith, Joseph Victor
Sood, Manmohan K
Speed, Robert Clarke
Steele, Ian McKay
Woodland, Bertram George

INDIANA
Brock, Kenneth Jack
Christensen, Nikolas Ivan
Franks, Stephen Guest
Madison, James Allen
Martin, Charles Wellington, Jr
Meyer, Henry Oostenwald Albertijn
Murray, Haydn Herbert

Roepke, Harlan Hugh
Shaffer, Nelson Ross
Shieh, Yuch-Ning
White, Joe Lloyd
Wintsch, Robert P

IOWA
De Nault, Kenneth J
Foster, Charles Thomas, Jr
Garvin, Paul Lawrence
McCormick, George R

KANSAS
Bickford, Marion Eugene
Twiss, Page Charles
Van Schmus, William Randall

KENTUCKY
Barnhisel, Richard I
Clark, Armin Lee
Stanonis, Francis Leo

LOUISIANA
Allen, Gary Curtiss
Byerly, Gary Ray
Fernandez, Louis Anthony
Ferrell, Ray Edward, Jr
Kinsland, Gary Lynn
McDowell, John Parmelee
Sartin, Austin Albert
Simmons, William Bruce, Jr
Wilcox, Ronald Erwin

MAINE
Grew, Edward Sturgis
Guidotti, Charles V
Morris, Horton Harold

MARYLAND
Brown, Michael
Chang, Luke Li-Yu
Darneal, Robert Lee
Ferry, John Mott
Fisher, George Wescott
Jones, Blair Francis
Kappel, Ellen Sue
MacGregor, Ian Duncan
Marsh, Bruce David
Mathew, Mathai
Parker, Frederick John
Pavlides, Louis
Roth, Robert S
Thomas, Herman Hoit
Veblen, David Rodli
Winzer, Stephen Randolph
Wittels, Mark C
Zen, E-an

MASSACHUSETTS
Aronson, James Ries
Bonnichsen, Martha Miller
Boutilier, Robert Francis
Brophy, Gerald Patrick
Brownlow, Arthur Hume
Burnham, Charles Wilson
Burns, Roger George
Dick, Henry Jonathan Biddle
Donohue, John J
Fairbairn, Harold Williams
Frondel, Clifford
Gheith, Mohamed A
Grove, Timothy Lynn
Hepburn, John Christopher
Hurlbut, Cornelius Searle, Jr
Jacobsen, Stein Bjornar
McCoy, Floyd W, Jr
Marvin, Ursula Bailey
Morse, Stearns Anthony
Naylor, Richard Stevens
Nunes, Paul Donald
Pasto, Arvid Eric
Peters, Edward Tehle
Robinson, Peter
Sand, Leonard B
Selverstone, Jane Elizabeth
Shimizu, Nobumichi
Siever, Raymond
Thompson, James Burleigh, Jr
Wobus, Reinhard Arthur

MICHIGAN
Dietrich, Richard Vincent
Essene, Eric J
Furlong, Robert B
Heinrich, Eberhardt William
Peacor, Donald Ralph
Ruotsala, Albert P
Sibley, Duncan Fawcett
Symonds, Robert B
Van Vlack, Lawrence H(all)
Vogel, Thomas A
Young, Davis Alan

MINNESOTA
Boardman, Shelby Jett
Grant, James Alexander
Green, John Chandler
Hatch, Robert Alchin
Rapp, George Robert, Jr
Walton, Matt Savage
Weiblen, Paul Willard
Zoltai, Tibor

MISSISSIPPI
Brunton, George Delbert
Collins, Johnnie B
Mather, Katharine Kniskern
Reynolds, William Roger
Sundeen, Daniel Alvin

MISSOURI
Frye, Charles Isaac
Hagni, Richard D
Himmelberg, Glen Ray
Johnson, Clayton Henry, Jr
Keller, Walter David
Kisvarsanyi, Eva Bognar
Lowell, Gary Richard
Maxwell, Dwight Thomas
Snowden, Jesse O
Thomson, Kenneth Clair

MONTANA
Hyndman, Donald William
Murray, Raymond Carl
Wehrenberg, John P

NEBRASKA
Treves, Samuel Blain

NEVADA
Chandra, Dhanesh
Christensen, Odin Dale
Felts, Wayne Moore
Garside, Larry Joe
Hsu, Liang-Chi
Kepper, John C
Kukal, Gerald Courtney
Larson, Lawrence T
Monroe, Eugene Alan
Pough, Frederick Harvey
Price, Jonathan Greenway
Smith, Eugene I(rwin)
Wilbanks, John Randall

NEW HAMPSHIRE
Reynolds, Robert Coltart, Jr
Richter, Dorothy Anne
Schneer, Cecil Jack

NEW JERSEY
Bundy, Wayne Miley
Carr, Michael John
Douglas, Lowell Arthur
Dowty, Eric
Faust, George Tobias
Hamil, Martha M
Hamilton, Charles Leroy
Hewins, Roger Herbert
Hollister, Lincoln Steffens
Puffer, John H
Vassiliou, Andreas H

NEW MEXICO
Baldridge, Warren Scott
Bish, David Lee
Broadhead, Ronald Frigon
Broxton, David Edward
Budding, Antonius Jacob
Charles, Robert Wilson
Condie, Kent Carl
Eichelberger, John Charles
Ewing, Rodney Charles
Finney, Joseph J
Geissman, John William
Grambling, Jeffrey A
Guthrie, George D, Jr
Klein, Cornelis
Kudo, Albert Masakiyo
Kuellmer, Frederick John
Meijer, Arend
Papike, James Joseph
Renault, Jacques Roland
Sun, James Ming-Shan
Taylor, G Jeffrey

NEW YORK
Baird, Gordon Cardwell
Barnard, Walther M
Bassett, William Akers
Berkley, John Lee
Boone, Gary M
Britton, Marvin Gale
Brock, Patrick Willet Grote
Bursnall, John Treharne
Clemency, Charles V
DeLong, Stephen Edwin
DeVries, Robert Charles
Dodd, Robert Taylor
Friedman, Gerald Manfred
Haas, Werner E L
Hanson, Gilbert N
Hatheway, Richard Brackett
Hawkins, William Max
Heyl, George Richard
Hodge, Dennis
Holland, Hans J
Jacoby, Russell Stephen
Kay, Suzanne Mahlburg
Langer, Arthur M
Lawson, Charles Alden
Leung, Irene Sheung-Ying
Liebling, Richard Stephen
Lindsley, Donald Hale
Longhi, John
Lowe, Kurt Emil
Ludman, Allan

Mattson, Peter Humphrey
Miyashiro, Akiho
Mumpton, Frederick Albert
Prinz, Martin
Prucha, John James
Reitan, Paul Hartman
Roberson, Herman Ellis
Rutstein, Martin S
Scharf, Walter
Schreiber, D Charlotte
Seyfert, Carl K, Jr
Sheridan, Michael Francis
Sturm, Edward
Teichert, Curt
Vandiver, Bradford B
Vaughan, Michael Thomas
Vollmer, Frederick Wolfer
Walker, David
Watt, James Peter
Weber, Julius
Whitney, Philip Roy
Wohlford, Duane Dennis
Wosinski, John Francis

NORTH CAROLINA
Abbott, Richard Newton, Jr
Butler, James Robert
Harris, William Burleigh
Howard, Clarence Edward
Neal, Donald Wade
Perkins, Ronald Dee
Raymond, Loren Arthur
Reeber, Robert Richard
Rogers, John James William
Textoris, Daniel Andrew

NORTH DAKOTA
Martin, DeWayne

OHIO
Bain, Roger J
Beard, William Clarence
Bever, James Edward
Bladh, Katherine Laing
Bladh, Kenneth W
Calhoun, Frank Gilbert
Carlson, Ernest Howard
Carlton, Richard Walter
Conrad, Malcolm Alvin
Cook, William R, Jr
Dahl, Peter Steffen
Elliot, David Hawksley
Foland, Kenneth Austin
Foreman, Dennis Walden, Jr
Friberg, LaVerne Marvin
Heimlich, Richard Allen
Hoekstra, Karl Egmond
Huff, Warren D
Koucky, Frank Louis, Jr
Laughon, Robert Bush
Law, Eric W
Lo, Howard H
Malcuit, Robert Joseph
Moody, Judith Barbara
Nance, Richard Damian
Schmidt, Mark Thomas
Scotford, David Matteson
Semler, Charles Edward
Skinner, William Robert
Tettenhorst, Rodney Tampa
Versic, Ronald James
Weitz, John Hills

OKLAHOMA
Al-Shaieb, Zuhair Fouad
Barczak, Virgil J
Blatt, Harvey
Gilbert, Murray Charles
Hounslow, Arthur William
London, David
Pittman, Edward D
Powell, Benjamin Neff
Tillman, Roderick W

OREGON
Borst, Roger Lee
Hammond, Paul Ellsworth
Kays, M Allan
McBirney, Alexander Robert
Powell, James Lawrence
Staples, Lloyd William
Taubeneck, William Harris
Taylor, Edward Morgan

PENNSYLVANIA
Allen, Jack C, Jr
Allen, John Christopher
Bates, Thomas Fulcher
Biederman, Edwin Williams, Jr
Buis, Patricia Frances
Carson, Bobb
Clavan, Walter
Cohen, Alvin Jerome
Crawford, Maria Luisa Buse
Crawford, William Arthur
Davis, Alan
Dunlap, Lawrence H
Eggler, David Hewitt
Erickson, Edwin Sylvester, Jr
Faill, Rodger Tanner
Gold, David Percy
Gray, Ralph J
Griffiths, John Cedric
Haefner, Richard Charles

Mineralogy-Petrology (cont)

Johnson, Leon Joseph
Kerrick, Derrill M
Komarneni, Sridhar
Liu, Mian
Muan, Arnulf
Myer, George Henry
Osborn, Elburt Franklin
Perrotta, Anthony Joseph
Rohl, Arthur N
Sclar, Charles Bertram
Shultz, Charles H
Simpson, Dale R
Smith, Deane Kingsley, Jr
Thornton, Charles Perkins
Vernon, William W
Walther, Frank H
White, Eugene Wilbert
Wiebe, Robert A
Williams, Eugene G
Williamson, William O(wen)

RHODE ISLAND
Giletti, Bruno John
Hermes, O Don
Hess, Paul C
Moore, George Emerson, Jr
Rutherford, Malcolm John
Sigurdsson, Haraldur
Yund, Richard Allen

SOUTH CAROLINA
Carew, James L
Shervais, John Walter
Stakes, Debra Sue
Strom, Richard Nelsen
Warner, Richard Dudley

SOUTH DAKOTA
Norton, James Jennings

TENNESSEE
Deininger, Robert W
Gavasci, Anna Teresa
Kohland, William Francis
Kopp, Otto Charles
McSween, Harry Y, Jr
Stow, Stephen Harrington
Tamura, Tsuneo
Taylor, Lawrence August

TEXAS
Ahr, Wayne Merrill
Anthony, Elizabeth Youngblood
Anthony, John Williams
Asquith, George Benjamin
Baker, Donald Roy
Barker, Daniel Stephen
Bence, Alfred Edward
Berkebile, Charles Alan
Butler, John C
Carman, Max Fleming, Jr
Cepeda, Joseph Cherubini
Chatelain, Edward Ellis
Clabaugh, Stephen Edmund
Clendening, John Albert
Dahl, Harry Martin
Dickey, John Sloan, Jr
Eastwood, Raymond L
Ehlmann, Arthur J
Elthon, Donald L
Ewing, Thomas Edward
Folk, Robert Louis
Ghaly, Tharwat Shahata
Guven, Necip
Hewitt, Charles Hayden
Heyman, Louis
Hixon, Sumner B
Holdaway, Michael Jon
Horz, Friedrich
Huege, Fred Robert
Ingerson, Fred Earl
Jackson, Martin Patrick Arden
Jonas, Edward Charles
Jones, Ernest Austin, Jr
Koepnick, Richard Borland
Kunze, George William
Levien, Louise
Lofgren, Gary Ernest
Longacre, Susan Ann Burton
McBride, Earle Francis
McCaleb, Stanley B
McConnell, Duncan
McPherson, John G(ordon)
Milford, Murray Hudson
Morrison, Donald Allen
Mosher, Sharon
Mutis-Duplat, Emilio
Namy, Jerome Nicholas
Ohashi, Yoshikazu
Phinney, William Charles
Presnall, Dean C
Smith, Douglas
Stevenson, Ralph Girard, Jr
Stormer, John Charles, Jr
Swanson, Eric Rice
Tieh, Thomas Ta-Pin
White, Stanton M
Whitford-Stark, James Leslie

UTAH
Christiansen, Eric H
Fiesinger, Donald William

Griffen, Dana Thomas
Morris, Elliot Cobia
Nash, William Purcell
Parry, William Thomas
Phillips, William Revell
Picard, M Dane

VERMONT
Doll, Charles George
Ratté, Charles A
Westerman, David Scott

VIRGINIA
Bethke, Philip Martin
Bloss, Fred Donald
Bodnar, Robert John
Brett, Robin
Calver, James Lewis
Corwin, Gilbert
Craig, James Roland
Cunningham, Charles G
Ericksen, George Edward
Ern, Ernest Henry
Evans, Howard Tasker, Jr
Finkelman, Robert Barry
Frye, Keith
Gibbs, Gerald V
Heald, Pamela
Hearn, Bernard Carter, Jr
Helz, Rosalind Tuthill
Hemingway, Bruce Sherman
Horton, James Wright, Jr
Huang, Wen Hsing
Huebner, John Stephen
James, Odette Bricmont
Kearns, Lance Edward
Lee, Kwang-Yuan
Leo, Gerhard William
Lipin, Bruce Reed
Lyons, Paul Christopher
McCartan, Lucy
Martens, James Hart Curry
Minkin, Jean Albert
Mitchell, Richard Scott
Moore, Raymond Kenworthy
Nelson, Bruce Warren
Nord, Gordon Ludwig, Jr
Padovani, Elaine Reeves
Rankin, Douglas Whiting
Ribbe, Paul Hubert
Robie, Richard Allen
Roseboom, Eugene Holloway, Jr
Ross, Malcolm
Stewart, David Benjamin
Toulmin, Priestley
Williams, Richard John
Wright, Thomas L
Zelazny, Lucian Walter
Zilczer, Janet Ann

WASHINGTON
Ames, Lloyd Leroy, Jr
Evans, Bernard William
Foit, Franklin Frederick, Jr
Ghiorso, Mark Stefan
Ghose, Subrata
Hinthorne, James Roscoe
Hodges, Floyd Norman
Huestis, Laurence Dean
Kesler, Thomas L
Kittrick, James Allen
Macklin, John Welton
Smith, Brad Keller
Sorem, Ronald Keith
Swanson, Donald Alan
Weis, Paul Lester

WEST VIRGINIA
Heald, Milton Tidd
Kalb, G William
Shumaker, Robert C

WISCONSIN
Bailey, Sturges Williams
Brown, Philip Edward
Cooper, Reid F
Medaris, L Gordon, Jr
Mudrey, Michael George, Jr
Woodard, Henry Herman, Jr

WYOMING
Houston, Robert S
Smithson, Scott Busby

ALBERTA
Bayliss, Peter
Burwash, Ronald Allan
Clementz, David Michael
Cook, Frederick Ahrens
Foscolos, Anthony E
Heimann, Robert B(ertram)
Lambert, Richard St John
Lerbekmo, John Franklin
Morton, Roger David
Nesbitt, Bruce Edward
Schmidt, Volkmar
Smith, Dorian Glen Whitney

BRITISH COLUMBIA
Armstrong, Richard Lee
Chase, Richard L
Greenwood, Hugh J
McTaggart, Kenneth C
Nuffield, Edward Wilfrid

Warren, Harry Verney

MANITOBA
Ferguson, Robert Bury
Hawthorne, Frank Christopher

NEW BRUNSWICK
Allen, Clifford Marsden
Mossman, David John

NEWFOUNDLAND
Hiscott, Richard Nicholas

NOVA SCOTIA
Schenk, Paul Edward
Zentilli, Marcos

ONTARIO
Appleyard, Edward Clair
Baragar, William Robert Arthur
Bevier, Mary Lou
Blackburn, William Howard
Cabri, Louis J
Corlett, Mabel Isobel
Davies, James Frederick
De Kimpe, Christian Robert
Duke, John Murray
Edgar, Alan D
Emslie, Ronald Frank
Gait, Robert Irwin
Gittins, John
Graham, A(lbert) Ronald
Grieve, Richard Andrew
Harris, Donald C
James, Richard Stephen
Jolly, Wayne Travis
Kissin, Stephen Alexander
Kodama, Hideomi
Lumbers, Sydney Blake
Mandarino, Joseph Anthony
Moore, John Marshall, Jr
Pajari, George Edward
Pearce, Thomas Hulme
Percival, John A
Quon, David Shi Haung
Roeder, Peter Ludwig
Rucklidge, John Christopher
Smith, Terence E
Starkey, John
Veizer, Ján
Wahl, William G
Wicks, Frederick John

QUEBEC
Chown, Edward Holton
Donnay, Joseph Desire Hubert
Feininger, Tomas
Laurent, Roger
Laurin, Andre Frederic
Laverdiere, Marc Richard
Ledoux, Robert Louis
Stevenson, Louise Stevens
Trzcienski, Walter Edward, Jr

SASKATCHEWAN
Davies, John Clifford
St Arnaud, Roland Joseph

OTHER COUNTRIES
Brydon, James Emerson
Dengo, Gabriel
Granata, Walter Harold, Jr
Kennedy, Michael John
Nickel, Ernest Henry
Walton, Wayne J A, Jr

Oceanography

ALABAMA
Davis, Anthony Michael John
Fowlis, William Webster
Reynolds, George Warren

ALASKA
Alexander, Vera
Burrell, David Colin
Elsner, Robert
Feder, Howard Mitchell
Goering, John James
Hameedi, Mohammad Jawed
Kelley, John Joseph, II
McRoy, C Peter
Mathisen, Ole Alfred
Matthews, John Brian
Naidu, Angi Satyanarayan
Nayudu, Y Rammohanroy
Paul, Augustus John, III
Reeburgh, William Scott
Royer, Thomas Clark
Sharma, Ghanshyam D
Shaw, David George
Smoker, William Williams
Wing, Bruce Larry
Wright, Frederick Fenning

ARIZONA
Agenbroad, Larry Delmar
Bates, Charles Carpenter
Dietz, Robert Sinclair
Fernando, Harindra Joseph

CALIFORNIA
Alldredge, Alice Louise

Arp, Alissa Jan
Arrhenius, Gustaf Olof Svante
Arthur, Robert Siple
Bacastow, Robert Bruce
Bakun, Andrew
Bascom, Willard
Beers, John R
Berger, Wolfgang Helmut
Bernstein, Robert Lee
Berry, Richard Warren
Bourke, Robert Hathaway
Bradshaw, John Stratlii
Brinton, Edward
Bubeck, Robert Clayton
Buffington, Edwin Conger
Burnett, Bryan Reeder
Carlson, Paul Roland
Carsola, Alfred James
Chamberlain, Dilworth Woolley
Chesworth, Robert Hadden
Childress, James J
Chow, Tsaihwa James
Chu, Pe-Cheng
Clark, Nathan Edward
Codispoti, Louis Anthony
Collins, Curtis Allan
Collins, James Ian
Conomos, Tasso John
Cox, Charles Shipley
Coyer, James A
Craig, Harmon
Crandell, George Frank
Cross, Ralph Herbert, III
Cummings, William Charles
Curray, Joseph Ross
Davis, Curtiss Owen
Davis, Gary Everett
Davis, Russ E
Dayton, Paul K
Demond, Joan
Di Girolamo, Rudolph Gerard
Dorman, Clive Edgar
Dugdale, Richard Cooper
Eittreim, Stephen L
Elsberry, Russell Leonard
El Wardani, Sayed Aly
Eppley, Richard Wayne
Everts, Craig Hamilton
Fast, Thomas Normand
Fiedler, Paul Charles
Field, Michael Ehrenhart
Fisher, Robert Lloyd
Flatte, Stanley Martin
Folsom, Theodore Robert
Ford, Richard Fiske
Foster, Michael Simmler
Foster, Theodore Dean
Fu, Lee Lueng
Fuhrman, Jed Alan
Fusaro, Craig Allen
Galt, Charles Parker, Jr
Gardner, James Vincent
Garwood, Roland William, Jr
Gast, James Avery
Geminder, Robert
Gerrodette, Timothy
Gibson, Carl H
Gordon, Malcolm Stephen
Gorsline, Donn Sherrin
Griggs, Gary B
Groves, Gordon William
Haderlie, Eugene Clinton
Hall, Robert Earl
Halpern, David
Hamilton, Edwin Lee
Hammond, Douglas Ellenwood
Haney, Robert Lee
Harrison, Florence Louise
Haury, Loren Richard
Hein, James R
Holl, Manfred Matthias
Holliday, Dale Vance
Holmes, Robert W
Honey, Richard Churchill
Hubred, Gale L
Hunt, Arlon Jason
Hunter, John Roe
Hurley, Robert Joseph
Hwang, Li-San
Inman, Douglas Lamar
Iyer, Hariharaiyer Mahadeva
Joseph, James
Joy, Joseph Wayne
Jung, Glenn Harold
Kaye, George Thomas
Kelley, James Charles
Kenyon, Kern Ellsworth
Knox, Robert Arthur
Kremer, James Nevin
Kremer, Patricia McCarthy
Ku, Teh-Lung
La Fond, Eugene Cecil
Laurs, Robert Michael
Leidersdorf, Craig B
Liu, Wingyuen Timothy
Loomis, Alden Albert
Lovett, Jack R
Lupton, John Edward
Lynch, Richard Vance, III
McBlair, William
McCarthy, Francis Davey
Macdonald, Kenneth Craig
McGeary, David F R
McGowan, John Arthur

McIntyre, Marie Claire
Mackenzie, Kenneth Victor
McLain, Douglas Robert
Martin, John Holland
Mathewson, James H
Medwin, Herman
Mikel, Thomas Kelly, Jr
Mohr, John Luther
Mulligan, Timothy James
Mullin, Michael Mahlon
Munk, Walter Heinrich
Newman, William Anderson
Nichols, Frederic Hone
Normark, William Raymond
North, Wheeler James
Nygreen, Paul W
Owen, Robert W, Jr
Paquette, Robert George
Parrish, Richard Henry
Pattabhiraman, Tammanur R
Patterson, Mark Robert
Patton, John Stuart
Pequegnat, Linda Lee Haithcock
Pequegnat, Willis Eugene
Phillips, David William
Phillips, Richard P
Phleger, Fred B
Pickwell, George Vincent
Pieper, Richard Edward
Pinkel, Robert
Potter, David Samuel
Potts, Donald Cameron
Powell, Thomas Mabrey
Rau, Gregory Hudson
Read, Robert G
Rechnitzer, Andreas Buchwald
Reid, Joseph Lee
Revelle, Roger (Randall Dougan)
Richards, Thomas L
Riedel, William Rex
Roach, David Michael
Rogers, David Peter
Roth, Ariel A
Ruby, Edward George
Saffo, Mary Beth
Schwing, Franklin Burton
Seckel, Gunter Rudolf
Seibel, Erwin
Semtner, Albert Julius, Jr
Seymour, Richard Jones
Show, Ivan Tristan
Silver, Eli Alfred
Silver, Mary Wilcox
Slack, Keith Vollmer
Smith, Raymond Calvin
Smith-Evernden, Roberta Katherine
Soule, Dorothy (Fisher)
Spies, Robert Bernard
Spiess, Fred Noel
Stephens, John Stewart, Jr
Stevenson, Robert Everett
Street, Robert L(ynnwood)
Suman, Daniel Oscar
Sund, Paul N
Tegner, Mia Jean
Terry, Richard D
Thierstein, Hans Rudolf
Thomas, William Hewitt
Thompson, Warren Charles
Trench, Robert Kent
Tsuchiya, Mizuki
VanBlaricom, Glenn R
Van Leer, John Cloud
Walsh, Don
Wang, Frank Feng Hui
Warnke, Detlef Andreas
Weare, Bryan C
Wilde, Pat
Wilkie, Donald W
Williams, Peter M
Wilson, Basil W(rigley)
Wimberley, Stanley
Wood, Fergus James
Yayanos, A Aristides
Zetler, Bernard David

COLORADO

Behrendt, John Charles
Born, George Henry
Chamberlain, Theodore Klock
Emery, William Jackson
Frisch, Alfred Shelby
Gossard, Earl Everett
Hay, William Winn
Holcombe, Troy Leon
Holland, William Robert
Kantha, Lakshmi
Kraus, Eric Bradshaw
Loughridge, Michael Samuel
McWilliams, James Cyrus
Moore, Carla Jean
Rufenach, Clifford L
Sawyer, Constance B
Shideler, Gerald Lee
Sloss, Peter William
Weihaupt, John George
Williams, David Lee

CONNECTICUT

Bohlen, Walter Franklin
Brooks, Albert Law
Carlton, James Theodore
Chriss, Terry Michael
Farmer, Mary Winifred

Fitzgerald, William Francis
Haffner, Rudolph Eric
Hanks, James Elden
Horne, Gregory Stuart
McGill, David A
Monahan, Edward Charles
Natarajan, Kottayam Viswanathan
Pandolfo, Joseph P
Paskausky, David Frank
Renfro, William Charles
Squires, Donald Fleming
Thurberg, Frederick Peter
Tolderlund, Douglas Stanley
Veronis, George
Welsh, Barbara Lathrop

DELAWARE

Culberson, Charles Henry
Daiber, Franklin Carl
Epifanio, Charles Edward
Garvine, Richard William
Glass, Billy Price
Karlson, Ronald Henry
Kobayashi, Nobuhisa
Luther, George William, III
Maurmeyer, Evelyn Mary
Miller, Douglas Charles
Price, Kent Sparks, Jr
Sharp, Jonathan Hawley
Svendsen, Ib Arne
Targett, Nancy McKeever
Targett, Timothy Erwin
Webster, Ferris
Wu, Jin

DISTRICT OF COLUMBIA

Andersen, Neil Richard
Baker, D(onald) James
Benson, Richard Hall
Berkson, Jonathan Milton
Bernor, Raymond Louis
Christofferson, Eric
De Leonibus, Pasquale S
Gross, M Grant, Jr
Heinrichs, Donald Frederick
Knauss, John Atkinson
Knowlton, Robert Earle
Manning, Raymond B
Meñez, Ernani Guingona
Mied, Richard Paul
Patterson, Steven Leroy
Pellenbarg, Robert Ernest
Penhale, Polly Ann
Rao, Desiraju Bhavanarayana
Roper, Clyde Forrest Eugene
Rosenblum, Lawrence Jay
Shen, Colin Yunkang
Shykind, Edwin B
Stanley, Daniel Jean
Valenzuela, Gaspar Rodolfo
Wilkerson, John Christopher

FLORIDA

Atwood, Donald Keith
Betzer, Peter Robin
Bullis, Harvey Raymond, Jr
Burnett, William Craig
Burney, Curtis Michael
Butler, Philip Alan
Byrd, Isaac Burlin
Byrne, Robert Howard
Carder, Kendall L
Chew, Frank
Clarke, Allan James
Compton, John S
Cooper, William James
Corcoran, Eugene Francis
Dean, Robert George
De Sylva, Donald Perrin
Dobkin, Sheldon
Doyle, Larry James
Dubbelday, Pieter Steven
Duedall, Iver Warren
Emiliani, Cesare
Fanning, Kent Abram
Fine, Rana Arnold
Freeland, George Lockwood
Gilbert, Perry Webster
Gordon, Howard R
Griffin, George Melvin, Jr
Hallock-Muller, Pamela
Hansen, Donald Vernon
Harris, D Lee
Harris, Lee Errol
Herrnkind, William Frank
Hsueh, Ya
Kruczynski, William Leonard
Lovejoy, Donald Walker
Ludeke, Carl Arthur
McAllister, Raymond Francis
McFadden, James Douglas
McLeroy, Edward Glenn
Mahadevan, Selvakumaran
Marcus, Nancy Helen
Mariscal, Richard North
Maul, George A
Menzel, Robert Winston
Mertens, Lawrence E(dwin)
Mitsui, Akira
Mooers, Christopher Northrup Kennard
Myrberg, Arthur August, Jr
Nakamura, Eugene Leroy
Nelson, Walter Garnet
Norris, Dean Rayburn

O'Brien, James J
Odum, Howard Thomas
Olson, Donald B
Ostlund, H Gote
Piotrowicz, Stephen R
Proni, John Roberto
Reichard, Ronnal Paul
Rona, Donna C
Smith, Ned Philip
Southam, John Ralph
Stewart, Harris Bates, Jr
Sturges, Wilton, III
Swain, Geoffrey W
Taylor, Barrie Frederick
Virnstein, Robert W
Voss, Gilbert Lincoln
Walsh, John Joseph
Wanless, Harold Rogers
Weatherly, Georges Lloyd
Weisberg, Robert H
Wiebusch, Charles Fred
Wiebush, Joseph Roy
Windsor, John Golay, Jr

GEORGIA

Blanton, Jackson Orin
Glasser, John Weakley
Hanson, Roger Brian
Harding, James Lombard
Hayes, Willis B
Henry, Vernon James, Jr
Menzel, David Washington
Meyer, Judy Lynn
Noakes, John Edward
Pomeroy, Lawrence Richards
Windom, Herbert Lynn

HAWAII

Abbott, Isabella Aiona
Bailey-Brock, Julie Helen
Bathen, Karl Hans
Berg, Carl John, Jr
Bienfang, Paul Kenneth
Boehlert, George Walter
Bretschneider, Charles Leroy
Clarke, Thomas Arthur
Cruickshank, Michael James
Daniel, Thomas Henry
Forster, William Owen
Gallagher, Brent S
Gopalakrishnan, Kakkala
Grigg, Richard Wyman
Keating, Barbara Helen
Kroenke, Loren William
Landry, Michael Raymond
Laws, Edward Allen
Loomis, Harold George
Mackenzie, Frederick Theodore
Mader, Charles Lavern
Magaard, Lorenz Carl
Peterson, Melvin Norman Adolph
Randall, John Ernest
Russo, Anthony R
Sansone, Francis Joseph
Spielvogel, Lester Q
Uchida, Richard Noboru
Wang, Bin
Woodcock, Alfred Herbert
Wyrtki, Klaus
Young, Richard Edward

ILLINOIS

Beasley, Thomas Miles
Charlier, Roger Henri
Chiou, Wen-An
Foster, Merrill W
Michelson, Irving
Platzman, George William
Rodolfo, Kelvin S
Shabica, Charles Wright
Walton, William Ralph

INDIANA

Christensen, Nikolas Ivan

IOWA

Hoffmann, Richard John
Walters, James Carter

LOUISIANA

Avent, Robert M
Bouma, Arnold Heiko
Breeding, J Ernest, Jr
Carney, Robert Spencer
Defenbaugh, Richard Eugene
Drummond, Kenneth Herbert
Farwell, Robert William
Harris, Alva H
Hastings, Robert Wayne
Hoese, Hinton Dickson
Huh, Oscar Karl
Korgen, Benjamin Jeffry
Murray, Stephen Patrick
Patrick, William H, Jr
Rouse, Lawrence James, Jr
Sen Gupta, Barun Kumar
Shaw, Richard Francis
Stiffey, Arthur V
Suhayda, Joseph Nicholas
Wiseman, William Joseph, Jr

MAINE

Dean, David
De Witt, Hugh Hamilton

Fink, Loyd Kenneth, Jr
Gilbert, James Robert
Gilmartin, Malvern
McAlice, Bernard John
Mayer, Lawrence Michael
Puri, Kewal Krishan
Revelante, Noelia
Schnitker, Detmar
Townsend, David Warren
Watling, Les
Welch, Walter Raynes
Yentsch, Charles Samuel

MARYLAND

Apel, John Ralph
Armstrong, Daniel Wayne
Baer, Ledolph
Barrientos, Celso Saquitan
Bjerkaas, Allan Wayne
Brandt, Stephen Bernard
Calder, John Archer
Callahan, Jeffrey Edwin
Cammen, Leon Matthew
Capone, Douglas George
Cavalieri, Donald Joseph
Chen, Benjamin Yun-Hai
Comiso, Josefino Cacas
Commito, John Angelo
Corliss, John Burt
Dantzler, Herman Lee, Jr
Del Grosso, Vincent Alfred
Diamante, John Matthew
Duane, David Bierlein
Elliott, William Paul
Etter, Paul Courtney
Faller, Alan Judson
Felsher, Murray
Fischer, Eugene Charles
Frey, Henry Richard
Fry, John Craig
Gereben, Istvan B
Gerritsen, Jeroen
Gloersen, Per
Hacker, Peter Wolfgang
Hanssen, George Lyle
Houde, Edward Donald
Huang, Joseph Chi Kan
Jacobs, Woodrow Cooper
Jefferson, Donald Earl
Johnson, Harry McClure
Kaiser, Jack Allen Charles
Kitaigorodskii, Sergei Alexander
Koblinsky, Chester John
Kravitz, Joseph Henry
Lambert, Richard Bowles, Jr
Larson, Reginald Einar
Legeckis, Richard Vytautas
Malone, Thomas C
Maynard, Nancy Gray
Mollo-Christensen, Erik Leonard
Monaldo, Francis Michael
Mountford, Kent
Osterberg, Charles Lamar
Park, Paul Kilho
Parkinson, Claire Lucille
Phillips, Owen M
Pritchard, Donald William
Rao, P Krishna
Rebach, Steve
Reeves, Robert William
Riedel, Gerhardt Frederick
Rivkin, Richard Bob
Sanders, James Grady
Saville, Thorndike, Jr
Schemm, Charles Edward
Schneider, Eric Davis
Schulkin, Morris
Sellner, Kevin Gregory
Sherman, Kenneth
Silver, David Martin
Sindermann, Carl James
Small, Eugene Beach
Snow, Milton Leonard
Spilhaus, Athelstan Frederick, Jr
Strong, Alan Earl
Taylor, Walter Rowland
Tenore, Kenneth Robert
Warsh, Catherine Evelyn
Weller, Charles Stagg, Jr
White, Harris Herman
Williams, Jerome
Williams, Robert Glenn
Wolfe, Douglas Arthur

MASSACHUSETTS

Aubrey, David Glenn
Backus, Richard Haven
Baird, Ronald C
Ballard, Robert D
Beardsley, Robert Cruce
Beckerle, John C
Beckwitt, Richard David
Boehm, Paul David
Bower, Amy Sue
Brink, Kenneth Harold
Bryden, Harry Leonard
Capuzzo, Judith M
Carr, Jerome Brian
Chisholm, Sallie Watson
Dillon, William Patrick
Dorman, Craig Emery
Emery, Kenneth Orris
Farrington, John William
Fofonoff, Nicholas Paul

Csanady, Gabriel Tibor
Curtin, Brian Thomas
Daniel, Kenneth Wayne
Delnore, Victor Eli
Dendrou, Stergios
Dolezalek, Hans
Dugan, John Philip
Dunstan, William Morgan
Edgar, Norman Terence
Fine, Michael Lawrence
Gaines, Gregory
Galvin, Cyril Jerome, Jr
Grabowski, Walter John
Grant, George C
Green, Edward Jewett
Hargis, William Jennings, Jr
Hicks, Steacy Dopp
Howard, William Eager, III
Hunt, Lee McCaa
Hurdle, Burton Garrison
Inderbitzen, Anton Louis, Jr
Johnson, George Leonard
Johnson, Ronald Ernest
Kelly, Mahlon George, Jr
Kinder, Thomas Hartley
Kirwan, Albert Dennis, Jr
Kornicker, William Alan
Kroll, John Ernest
Kuo, Albert Yi-Shuong
Lepple, Frederick Karl
Leung, Wing Hai
Lill, Gordon Grigsby
Ludwick, John Calvin, Jr
McGregor, Bonnie A
Macko, Stephen Alexander
Marshall, Harold George
Orth, Robert Joseph
Osborn, Kenneth Wakeman
Parvulescu, Antares
Perkins, Frank Overton
Pyle, Thomas Edward
Ramberg, Steven Eric
Raymond-Savage, Anne
Schwiderski, Ernst Walter
Snyder, J Edward, Jr
Spence, Thomas Wayne
Spilhaus, Athelstan Frederick
Spinrad, Richard William
Swift, Donald J P
Tatro, Peter Richard
Tipper, Ronald Charles
Trizna, Dennis Benedict
Webb, Kenneth L
Wells, John Morgan, Jr
Witting, James M
Wong, George Tin Fuk
Yates, Charlie Lee
Zieman, Joseph Crowe, Jr

WASHINGTON
Aagaard, Knut
Ahmed, Saiyed I
Alverson, Dayton L
Anderson, Donald Hervin
Anderson, James Jay
Aron, William Irwin
Baker, Edward Thomas
Banse, Karl
Barker, Morris Wayne
Barnes, Clifford Adrian
Barnes, Ross Owen
Bernard, Eddie Nolan
Best, Edgar Allan
Booth, Beatrice Crosby
Burns, Robert Earle
Buske, Norman L
Cannon, Glenn Albert
Carpenter, Roy
Chace, Alden Buffington, Jr
Chew, Kenneth Kendall
Coachman, Lawrence Keyes
Cokelet, Edward Davis
Collias, Eugene Evans
Creager, Joe Scott
Crecelius, Eric A
Curl, Herbert (Charles), Jr
Damkaer, David Martin
Dehlinger, Peter
Delisi, Donald Paul
Deming, Jody W
Duxbury, Alyn Crandall
Ebbesmeyer, Curtis Charles
Eickstaedt, Lawrence Lee
English, Thomas Saunders
Favorite, Felix
Feely, Richard Alan
Frost, Bruce Wesley
Glude, John Bryce
Gregg, Michael Charles
Gunderson, Donald Raymond
Hardy, John Thomas
Harlett, John Charles
Harrison, Don Edmunds
Hawkes, Joyce W
Healy, Michael L
Heath, G Ross
Hedges, John Ivan
Hickey, Barbara Mary
Holcomb, Robin Terry
Holmes, Mark Lawrence
Hood, Donald Wilbur
Jennings, Charles David
Johnson, Gregory Conrad
Jumars, Peter Alfred

Laevastu, Taivo
Lavelle, John William
Lingren, Wesley Earl
Long, Edward R
Lorenzen, Carl Julius
McManus, Dean Alvis
Martin, Seelye
Mobley, Curtis Dale
Mopper, Kenneth
Muench, Robin Davie
Muller-Parker, Gisèle Thérèse
Neve, Richard Anthony
Olsen, Kenneth Harold
Overland, James Edward
Pearson, Walter Howard
Perry, Mary Jane
Rattray, Maurice, Jr
Reed, Ronald Keith
Rhines, Peter Broomell
Richards, Francis Asbury
Rieck, H(enry) G(eorge)
Roden, Gunnar Ivo
Salo, Ernest Olavi
Sanford, Thomas Bayes
Sarachik, Edward S
Schwartz, Maurice Leo
Serne, Roger Jeffrey
Skov, Niels A
Smith, David Clement, IV
Smith, J Dungan
Sorem, Ronald Keith
Sparks, Albert Kirk
Sternberg, Richard Walter
Strathmann, Richard Ray
Taft, Bruce A
Taylor, Peter Berkley
Thom, Ronald Mark
Trent, Donald Stephen
Untersteiner, Norbert
Walker, Sharon Leslie
Wooster, Warren Scriver
Wright, Judith

WEST VIRGINIA
Calhoon, Donald Alan

WISCONSIN
Bentley, Charles Raymond
Difford, Winthrop Cecil
Edgington, David Norman
McCarthy, James Joseph
Meyer, Robert Paul
Mortimer, Clifford Hiley
Ragotzkie, Robert Austin

PUERTO RICO
Crabtree, David Melvin
Hernandez Avila, Manuel Luis
Lopez, Jose Manuel
Morelock, Jack

ALBERTA
Georgi, Daniel Taylan
Giovinetto, Mario Bartolome
Lane, Robert Kenneth
Scafe, Donald William

BRITISH COLUMBIA
Booth, Ian Jeremy
Burling, Ronald William
Clements, Reginald Montgomery
Hart, Josephine Frances Lavinia
Hughes, Blyth Alvin
Hyndman, Roy D
Krauel, David Paul
Lalli, Carol Marie
LeBlond, Paul Henri
Levings, Colin David
Lewis, Alan Graham
Lewis, Edward Lyn
Littlepage, Jack Leroy
Marko, John Robert
Maunsell, Charles Dudley
Murty, Tadepalli Satyanarayana
Parsons, Timothy Richard
Pickard, George Lawson
Pond, George Stephen
Skud, Bernard Einar
Stockner, John G
Tabata, Susumu
Taylor, Frank John Rupert (Max)
Thomson, Richard Edward
Tunnicliffe, Verena Julia
Waldichuk, Michael
Yunker, Mark Bernard

NEW BRUNSWICK
Kohler, Carl
Vanicek, Petr

NEWFOUNDLAND
Haedrich, Richard L
Templeman, Wilfred
Welhan, John Andrew
Wroblewski, Joseph S

NOVA SCOTIA
Boyd, Carl M
Bridgeo, William Alphonsus
Brown, Richard George Bolney
Cameron, Barry Winston
Castell, John Daniel
Cooke, Robert Clark
Dickie, Lloyd Merlin

Elliott, James Arthur
Gradstein, Felix Marcel
Keen, Michael J
King, Lewis H
Longhurst, Alan R
Mann, Kenneth H
Mills, Eric Leonard
Scarratt, David Johnson
Silvert, William Lawrence
Sinclair, Michael Mackay
Smith, Stuart D
Tan, Francis C
Trites, Ronald Wilmot

ONTARIO
Barica, Jan M
Campbell, Neil John
Carter, John C H
Chau, Yiu-Kee
Donelan, Mark Anthony
Drury, Malcolm John
Forrester, Warren David
Fyfe, William Sefton
Gaskin, David Edward
Glooschenko, Walter Arthur
Greenwood, Brian
Scott, Steven Donald
Shih, Chang-Tai
Trick, Charles Gordon
Walton, Alan

QUEBEC
Brunel, Pierre
Desjardins, Claude W
Dunbar, Maxwell John
Hesse, Reinhard
Ingram, Richard Grant
Khalil, Michel
Legendre, Louis
Leggett, William C
Levasseur, Maurice Edgar
Lewis, John Bradley
Mansfield, Arthur Walter
Mysak, Lawrence Alexander
Packard, Theodore Train
Picard, Gaston Arthur
Pounder, Elton Roy
Reiswig, Henry Michael

SASKATCHEWAN
Evans, Marlene Sandra

OTHER COUNTRIES
Andreae, Meinrat Otto
Aumento, Fabrizio
Caperon, John
Chen, Chen-Tung Arthur
Fowler, Scott Wellington
Frakes, Lawrence Austin
Koslow, Julian Anthony
Krause, Dale Curtiss
Mandelbaum, Hugo
Needler, George Treglohan
Rodriguez, Gilberto
Sleeter, Thomas David
Van Andel, Tjeerd Hendrik
Williams, Francis
Worthington, Lawrence Valentine

Paleontology

ALABAMA
Beals, Harold Oliver
Copeland, Charles Wesley, Jr
Hoover, Richard Brice
Huff, William J
Jones, Douglas Epps
Lamb, George Marion
Mancini, Ernest Anthony

ALASKA
Guthrie, Russell Dale

ARIZONA
Agenbroad, Larry Delmar
Beus, Stanley S
Canright, James Edward
Cohen, Andrew Scott
Colbert, Edwin Harris
Davis, Owen Kent
Flessa, Karl Walter
Hazel, Joseph Ernest
Kremp, Gerhard Otto Wilhelm
Lindsay, Everett Harold, Jr
Lundin, Robert Folke
Mead, James Irving
Nations, Jack Dale
Parrish, Judith Totman
White, John Anderson

ARKANSAS
Edson, James Edward, Jr
Manger, Walter Leroy
Thurmond, John Tydings

CALIFORNIA
Adam, David Peter
Addicott, Warren O
Allison, Carol Wagner
Allison, Richard C
Awramik, Stanley Michael
Axelrod, Daniel Isaac
Barnes, Lawrence Gayle

Barron, John Arthur
Berger, Wolfgang Helmut
Black, Craig C
Bottjer, David John
Brand, Leonard Roy
Buchheim, H Paul
Bukry, John David
Callison, George
Campbell, Catherine Chase
Campbell, Kenneth Eugene, Jr
Chmura-Meyer, Carol A
Church, Clifford Carl
Clemens, William Alvin
Cloud, Preston
Coffin, Harold Glen
Cohen, Anne Carolyn Constant
Cowen, Richard
Demond, Joan
Downs, Theodore
Drugg, Warren Sowle
Durham, John Wyatt
Easton, William Heyden
Estes, Richard
Etheridge, Richard Emmett
Evitt, William Robert
Finney, Stanley Charles
Fischer, Alfred George
Fry, Wayne Lyle
Gregory, Joseph Tracy
Hall, Donald D
Harris, John Michael
Hickman, Carole Stentz
Hill, Merton Earle, III
Howard, Hildegarde
Hutchison, John Howard
Ingle, James Chesney, Jr
Jacobson, Stephen Richard
Jakway, George Elmer
Jenkins, Floyd Albert
Jones, David Lawrence
Kaska, Harold Victor
Kennett, James Peter
Kern, John Philip
Kitts, David Burlingame
Kopf, Rudolph William
Krejsa, Richard Joseph
Lane, Bernard Owen
Laporte, Léo Frédéric
Lindberg, David Robert
Lipps, Jere Henry
Loeblich, Alfred Richard, Jr
Loeblich, Helen Nina (Tappan)
Lumsden, William Watt, Jr
McCrone, Alistair William
McLeod, Samuel Albert
Marshall, Charles Richard
Mintz, Leigh Wayne
Myrick, Albert Charles, Jr
Olson, Everett Claire
Padian, Kevin
Parker, Frances Lawrence
Piel, Kenneth Martin
Prothero, Donald Ross
Riedel, William Rex
Rudnyk, Marian E
Saul, Lou Ella Rankin
Savage, Donald Elvin
Schopf, James William
Schram, Frederick R
Smith, Judith Terry
Smith-Evernden, Roberta Katherine
Snyder, Charles Theodore
Stevens, Calvin H
Swift, Camm Churchill
Thierstein, Hans Rudolf
Tiffney, Bruce Haynes
Vaughn, Peter Paul
Warter, Janet Kirchner
Warter, Stuart L
Welles, Samuel Paul
Whistler, David Paul
Wiggins, Virgil Dale
Wilson, Edward Carl
Woodburne, Michael O

COLORADO
Ager, Thomas Alan
Bown, Thomas Michael
Cygan, Norbert Everett
Dubiel, Russell F
Fagerstrom, John Alfred
Forester, Richard Monroe
Guennel, Gottfried Kurt
Hanley, John Herbert
Harris, Judith Anne Van Couvering
Kauffman, Erle Galen
Kent, Bion H
Kent, Harry Christison
Kissling, Don Lester
Kosanke, Robert Max
Lewis, George Edward
Nichols, Douglas James
Nichols, Kathryn Marion
Palmer, Allison Ralph
Rall, Raymond Wallace
Robinson, Peter
Ross, Reuben James, Jr
Silberling, Norman John
Skipp, Betty Ann
Stucky, Richard K(eith)
Taylor, Michael E
Trexler, David William
Van Zant, Kent Lee

Hill, Frederick Conrad
Inners, Jon David
King, James Edward
Krishtalka, Leonard
Penny, John Sloyan
Pfefferkorn, Hermann Wilhelm
Rackoff, Jerome S
Ramsdell, Robert Cole
Rollins, Harold Bert
Rosenberg, Gary
Saunders, William Bruce
Spackman, William, Jr
Thayer, Charles Walter
Thomas, Roger David Keen
Thomson, Keith Stewart
Traverse, Alfred
Wegweiser, Arthur E

RHODE ISLAND
Imbrie, John
Janis, Christine Marie
Prell, Warren Lee
Webb, Thompson, III

SOUTH CAROLINA
Birkhead, Paul Kenneth
Carew, James L
Cohen, Arthur David
Lawrence, David Reed
MacKnight, Franklin Collester
Williams, Douglas Francis

SOUTH DAKOTA
Bjork, Philip R
Macdonald, James Reid
Martin, James Edward

TENNESSEE
Bordine, Burton W
Deboo, Phili B
Knox, Larry William
McLaughlin, Robert Everett
Siesser, William Gary

TEXAS
Abbott, William Harold
Beyer, Arthur Frederick
Bonem, Rena Mae
Chadwick, Arthur Vorce
Chatelain, Edward Ellis
Chatterjee, Sankar
Clarke, Robert Travis
Cornell, William Crowninshield
Crick, Rex Edward
Delevoryas, Theodore
DuBar, Jules R
Durden, Christopher John
Echols, Dorothy Jung
Echols, Joan
Elsik, William Clinton
Engelhardt, Donald Wayne
Feazel, Charles Tibbals
Fisher, William Lawrence
Frost, Stanley H
Gartner, Stefan, Jr
Gibson, Lee B
Gillette, David Duane
Gombos, Andrew Michael, Jr
Harris, Arthur Horne
James, Gideon T
Jamieson-Magathan, Esther R
Kocurko, Michael John
Lane, Harold Richard
Langston, Wann, Jr
LeBlanc, Arthur Edgar
LeMone, David V
Levinson, Stuart Alan
Lundelius, Ernest Luther, Jr
McAlester, Arcie Lee, Jr
McNulty, Charles Lee, Jr
Macurda, Donald Bradford, Jr
Melia, Michael Brendan
Merrill, Glen Kenton
Pampe, William R
Perkins, Bobby Frank
Pruitt, Basil Arthur, Jr
Reading, John Frank
Rohr, David M
Ross, Charles Alexander
Schultz, Gerald Edward
Schumacher, Dietmar
Shane, John Denis
Shaw, Alan Bosworth
Smith, Michael A
Sprinkle, James (Thomas)
Stanton, Robert James, Jr
Steele, David Gentry
Stewart, Robert Henry
Stover, Lewis Eugene
Sullivan, Herbert J
Vincent, Jerry William
Weart, Richard Claude
Wilde, Garner Lee
Yancey, Thomas Erwin
Young, Keith Preston
Zingula, Richard Paul

UTAH
Ash, Sidney Roy
Baer, James L
Ekdale, Allan Anton
Liddell, William David
Madsen, James Henry, Jr

Miller, Wade Elliott, II
Mosher, Loren Cameron
Moyle, Richard W
Rigby, J Keith
Roth, Peter Hans
Stokes, William Lee

VERMONT
Doll, Charles George
Trombulak, Stephen Christopher

VIRGINIA
Bird, Samuel Oscar, II
Bowen, Zeddie Paul
Cronin, Thomas Mark
Edwards, Lucy Elaine
Eggleston, Jane R
Ellison, Robert L
Gibson, Thomas George
Koch, Carl Fred
Lohman, Kenneth Elmo
Nelson, Clifford Melvin, Jr
Pojeta, John, Jr
Robbins, Eleanora Iberall
Said, Rushdi
Scheckler, Stephen Edward
Scolaro, Reginald Joseph
Sohl, Norman Frederick
Spencer, Randall Scott
Weems, Robert Edwin
Wilson, John William, III

WASHINGTON
Eck, Gerald Gilbert
Fields, Robert William
Gilmour, Ernest Henry
Hopkins, William Stephen
Kohn, Alan Jacobs
Lane, Wallace
Lowther, John Stewart
Mallory, Virgil Standish
Mehringer, Peter Joseph, Jr
Rau, Weldon Willis
Ross, June Rosa Pitt
Simmonds, Robert T
Tsukada, Matsuo
Webster, Gary Dean
Wright, Judith

WISCONSIN
Clark, David Leigh
Fowler, Gerald Allan
Gernant, Robert Everett
Long, Charles Alan
Maher, Louis James, Jr
Mendelson, Carl Victor
Paull, Rachel Krebs

WYOMING
Boyd, Donald Wilkin
Frerichs, William Edward
Lillegraven, Jason Arthur
Mateer, Niall John
Oman, Carl Henry

ALBERTA
Campbell, John Duncan
Chatterton, Brian Douglas Eyre
Currie, Philip John
Fox, Richard Carr
Naylor, Bruce Gordon
Norford, Brian Seeley
Okulitch, Vladimir Joseph
Osborn, Jeffrey Whitaker
Russell, Anthony Patrick
Stelck, Charles Richard
Wall, John Hallett
Wilson, Mark Vincent Hardman

BRITISH COLUMBIA
Hebda, Richard Joseph
Mathewes, Rolf Walter
Monger, James William Heron
Nelson, Samuel James
Pocock, Stanley Albert John
Rouse, Glenn Everett
Smith, Paul L

MANITOBA
Hecky, Robert Eugene

NEW BRUNSWICK
Ferguson, Laing
Logan, Alan

NEWFOUNDLAND
May, Sherry Jan

NOVA SCOTIA
Cameron, Barry Winston
Gradstein, Felix Marcel
Moore, Reginald George
Vilks, Gustavs
Zodrow, Erwin Lorenz

ONTARIO
Caldwell, William Glen Elliot
Churcher, Charles Stephen
Collins, Desmond H
Copeland, Murray John
Copper, Paul
Fritz, Madeleine Alberta
Fritz, William Harold
Harington, Charles Richard

Jarzen, David MacArthur
Kobluk, David Ronald
McLaren, Digby Johns
Norris, Geoffrey
Radforth, Norman William
Risk, Michael John
Russell, Dale A
Russell, Loris Shano
Terasmae, Jaan

QUEBEC
Beland, Pierre
Carroll, Robert Lynn
Clark, Thomas Henry
Hillaire-Marcel, Claude
Lesperance, Pierre J
Richard, Pierre Joseph Herve
Riva, John F
Stearn, Colin William

SASKATCHEWAN
Braun, Willi Karl
Sarjeant, William Antony Swithin

OTHER COUNTRIES
Alonso, Ricardo N
Archangelsky, Sergio
Erdtmann, Bernd Dietrich
Gordon, William Anthony
Lynts, George Willard
Paulay, Gustav
Rich, Thomas Hewitt
Sander, Nestor John
Thompson, Ida
Trümpy, D(aniel) Rudolf
Vickers-Rich, Patricia

Physical Geography

ALABAMA
Williams, Aaron, Jr

ARIZONA
Baron, William Robert
Berlin, Graydon Lennis
Breed, Carol Sameth
Childs, Orlo E
Crosswhite, Frank Samuel
Gillis, James E, Jr
Hutchinson, Charles F
Kremp, Gerhard Otto Wilhelm
Linehan, Urban Joseph
Manera, Paul Allen
Marcus, Melvin Gerald

CALIFORNIA
Beard, Charles Noble
Gordon, Burton LeRoy
Ives, John (Jack) David
Lier, John
McArthur, David Samuel
McElhoe, Forrest Lester, Jr
Pease, Robert Wright
Pike, Richard Joseph
Rees, John David
Sauer, Jonathan Deininger
Star, Jeffrey L
Sternberg, Hilgard O'Reilly
Stevens, Calvin H
Thorne, Robert Folger
Voeks, Robert Allen

COLORADO
Carrara, Paul Edward
Diaz, Henry Frank
Lehrer, Paul Lindner
Mason, Malcolm
Redente, Edward Francis
Steen-McIntyre, Virginia Carol

CONNECTICUT
Sakalowsky, Peter Paul, Jr

DELAWARE
Brazel, Anthony James
Kalkstein, Laurence Saul
Mather, John Russell
Willmott, Cort James

DISTRICT OF COLUMBIA
Abler, Ronald Francis
Matson, Michael
Strommen, Norton Duane

FLORIDA
Berry, Leonard
Burrill, Robert Meredith
Chauvin, Robert S
Craig, Alan Knowlton
Cross, Clark Irwin
Marcus, Robert Brown

GEORGIA
Meentemeyer, Vernon George
Parker, Albert John

HAWAII
Street, John Malcolm

ILLINOIS
Anderson, Richard Charles
Changnon, Stanley A, Jr
Charlier, Roger Henri

Espenshade, Edward Bowman, Jr
Gabler, Robert Earl
McMillan, R Bruce
Sharpe, David McCurry
Wendland, Wayne Marcel

INDIANA
Mausel, Paul Warner
Oliver, John Edward

IOWA
Akin, Wallace Elmus
Prior, Jean Cutler

KANSAS
White, Stephen Edward

KENTUCKY
Schwendeman, Joseph Raymond, Jr

LOUISIANA
Gagliano, Sherwood Moneer
Muller, Robert Albert
Walker, Harley Jesse
Webb, Robert MacHardy

MARYLAND
Campbell, William J
Chovitz, Bernard H
Fonaroff, Leonard Schuyler
Kearney, Michael Sean
McBryde, F(elix) Webster
Scala, John Richard
Vermeij, Geerat Jacobus

MASSACHUSETTS
Bradley, Raymond Stuart
Centorino, James Joseph
Dorman, Craig Emery
MacDougall, Edward Bruce

MICHIGAN
Aufdemberge, Theodore Paul
Erhart, Rainer R
Howe, George Marvel
Outcalt, Samuel Irvine
Sommers, Lawrence M

MISSISSIPPI
Cross, Ralph Donald
Quattrochi, Dale Anthony
Saucier, Roger Thomas

MISSOURI
Schroeder, Walter Albert
Stauffer, Truman Parker, Sr

NEBRASKA
Lawson, Merlin Paul

NEVADA
Mifflin, Martin David

NEW JERSEY
Hordon, Robert M
Psuty, Norbert Phillip
Wetherald, Richard Tryon

NEW MEXICO
Hill, Mary Rae
Holloway, Richard George

NEW YORK
Abrahams, Athol Denis
De Laubenfels, David John
Ebert, Charles H V
Henderson, Floyd M
Jarvis, Richard S
Lougeay, Ray Leonard
Metzger, Ernest Hugh
Monmonier, Mark
Palmer, James F
Van Burkalow, Anastasia
Vuilleumier, Francois

NORTH CAROLINA
Hidore, John J
Nunnally, Nelson Rudolph
Rees, John
Robinson, Peter John
Ross, Thomas Edward
Stillwell, Harold Daniel

OHIO
Arnfield, Anthony John
Basile, Robert Manlius
Crawford, Paul Vincent
Mahard, Richard H
Noble, Allen G
Ray, John Robert
Rayner, John Norman

OREGON
Johannessen, Carl L
McKinney, William Mark
Niem, Wendy Adams
Ripple, William John

PENNSYLVANIA
Justham, Stephen Alton
Sharer, Cyrus J
Wilhelm, Eugene J, Jr

Physical Geography (cont)

TENNESSEE
Fullerton, Ralph O
Jumper, Sidney Roberts

TEXAS
Butzer, Karl W(ilhelm)
Fix, Ronald Edward
Gore, Dorothy J
Humphries, James Edward, Jr
Kimber, Clarissa Therese
Lambert, Paul Wayne
McCloy, James Murl
Whitford-Stark, James Leslie

UTAH
Foster, Robert H
Godfrey, Andrew Elliott
Grey, Alan Hopwood
McCoy, Roger Michael
Stevens, Dale John
Vaughn, Danny Mack

VERMONT
Gade, Daniel W
VanderMeer, Canute

VIRGINIA
Pontius, Steven Kent

WASHINGTON
Kaatz, Martin Richard

WEST VIRGINIA
Calzonetti, Frank J
Elmes, Gregory Arthur
Murphy, Allen Emerson

WISCONSIN
Bhatia, Shyam Sunder
Knox, James Clarence
Richason, Benjamin Franklin, Jr

WYOMING
Reider, Richard Gary

ALBERTA
Klassen, Rudolph Waldemar
McPherson, Harold James
Rogerson, Robert James

BRITISH COLUMBIA
Barnes, Christopher Richard
Neilsen, Gerald Henry
Oke, Timothy Richard
Slaymaker, Herbert Olav

MANITOBA
Rounds, Richard Clifford
Welsted, John Edward

ONTARIO
Bunting, Brian Talbot
Greenwood, Brian
Kesik, Andrzej B
King, Roger Hatton
Menzies, John
Phillips, Brian Antony Morley
Reeds, Lloyd George
Sanderson, Marie Elizabeth
Wilkinson, Thomas Preston

QUEBEC
Bonn, Ferdinand J
Dionne, Jean-Claude
Gangloff, Pierre

OTHER COUNTRIES
Bridgman, Howard Allen
Francke, Oscar F
Rich, Thomas Hewitt

Stratigraphy-Sedimentation

ALABAMA
Copeland, Charles Wesley, Jr
Hooks, William Gary
Huff, William J
Lamb, George Marion
Mancini, Ernest Anthony

ALASKA
Karl, Susan Margaret
Naidu, Angi Satyanarayan

ARIZONA
Breed, Carol Sameth
Brennan, Daniel Joseph
Cohen, Andrew Scott
Davis, Owen Kent
Dickinson, William Richard
Kremp, Gerhard Otto Wilhelm
Krinsley, David
Lucchitta, Ivo
Nations, Jack Dale
Schreiber, Joseph Frederick, Jr

ARKANSAS
Edson, James Edward, Jr
Kehler, Philip Leroy
Vosburg, David Lee

CALIFORNIA
Abbott, Patrick Leon
Allan, John Ridgway
Alvarez, Walter
Austin, Steven Arthur
Bachman, Richard Thomas
Bassett, Henry Gordon
Bottjer, David John
Buchheim, H Paul
Burtner, Roger Lee
Campbell, Kenneth Eugene, Jr
Chapman, Robert Mills
Chmura-Meyer, Carol A
Colburn, Ivan Paul
Cook, Harry E, III
Fischer, Alfred George
Floran, Robert John
Ford, John Philip
Friberg, James Frederick
Galehouse, Jon Scott
Graham, Stephan Alan
Grantz, Arthur
Gutstadt, Allan Morton
Hill, Merton Earle, III
Hodges, Lance Thomas
Hose, Richard K
Hunter, Ralph Eugene
Ingersoll, Raymond Vail
Ingle, James Chesney, Jr
Kaska, Harold Victor
Kirby, James M
Kopf, Rudolph William
Laporte, Léo Frédéric
Lowe, Donald Ray
McCrone, Alistair William
McLean, Hugh
Mann, John Allen
Mintz, Leigh Wayne
Murphy, Michael A
Osborne, Robert Howard
Parker, Frances Lawrence
Peterson, Gary Lee
Prothero, Donald Ross
Riese, Walter Charles Rusty
Rogers, Thomas Hardin
Russell, Richard Dana
Sarna-Wojcicki, Andrei M
Schwartz, Daniel Evan
Sims, John David
Smith-Evernden, Roberta Katherine
Stevens, Calvin H
Tabor, Rowland Whitney
Thierstein, Hans Rudolf
Warnke, Detlef Andreas
Wentworth, Carl M, Jr

COLORADO
Bryant, Bruce Hazelton
Cadigan, Robert Allen
Campbell, John Arthur
Cathcart, James B
Clift, William Orrin
Crone, Anthony Joseph
Cross, Timothy Aureal
Dean, Walter E, Jr
Dubiel, Russell F
Elder, Curtis Harold
Ethridge, Frank Gulde
Fails, Thomas Glenn
Fassett, James Ernest
Flores, Romeo M
Hay, William Winn
Hayes, John Bernard
Holcombe, Troy Leon
Keefer, William Richard
Keighin, Charles William
Kent, Harry Christison
Ketner, Keith B
Lewis, George Edward
Link, Peter K
Luft, Stanley Jeremie
McCubbin, Donald Gene
MacKenzie, David Brindley
Maughan, Edwin Kelly
Moore, David A
Morrison, Roger Barron
Nichols, Kathryn Marion
Palmer, Allison Ralph
Parker, John Marchbank
Peterson, Fred
Pillmore, Charles Lee
Rall, Raymond Wallace
Raup, Robert Bruce, Jr
Ross, Reuben James, Jr
Sable, Edward George
Sanborn, Albert Francis
Shideler, Gerald Lee
Siemers-Blay, Charles T
Silberling, Norman John
Steen-McIntyre, Virginia Carol
Taylor, Michael E
Vine, James David
Walker, Theodore Roscoe
Wallace, Chester Alan
Warme, John Edward

CONNECTICUT
Chriss, Terry Michael
Drobnyk, John Wendel
Hickey, Leo Joseph
Horne, Gregory Stuart
Nadeau, Betty Kellett
Rodgers, John

DELAWARE
Ellis, Patricia Mench
Kraft, John Christian
Maurmeyer, Evelyn Mary
Talley, John Herbert
Thompson, Allan M

DISTRICT OF COLUMBIA
Boosman, Jaap Wim
Coates, Anthony George
Emry, Robert John
Haq, Bilal U
Hussain, Syed Taseer
Lindholm, Roy Charles
Neiheisel, James
Rumble, Douglas, III
Sando, William Jasper
Tracey, Joshua Irving, Jr
Wright, Thomas Oscar

FLORIDA
Councill, Richard J
Davis, Richard Albert, Jr
Donoghue, Joseph F
Freeland, George Lockwood
Gifford, John A
Ginsburg, Robert Nathan
Halley, Robert Bruce
Hallock-Muller, Pamela
Lovejoy, Donald Walker
McLeroy, Edward Glenn
Maurrasse, Florentin Jean-Marie Robert
Pirkle, Fredric Lee
Randazzo, Anthony Frank
Rosendahl, Bruce Ray
Tanner, William Francis, Jr
Wanless, Harold Rogers
Winters, Stephen Samuel

GEORGIA
Carver, Robert E
Cramer, Howard Ross
Fritz, William J
Gonzales, Serge
Gore, Pamela J(eanne) W(heeless)
Lineback, Jerry Alvin
Rich, Mark
Sumartojo, Jojok
Weaver, Charles Edward

HAWAII
Keating, Barbara Helen
Mackenzie, Frederick Theodore
Moberly, Ralph M
Moore, Gregory Frank

IDAHO
Isaacson, Peter Edwin
Ore, Henry Thomas
Spinosa, Claude

ILLINOIS
Baxter, James Watson
Block, Douglas Alfred
Buschbach, Thomas Charles
Doehler, Robert William
Flynn, John Joseph
Foster, Merrill W
Fraunfelter, George H
Gealy, William James
Hart, Richard Royce
Jablonski, David
Johnson, William Hilton
Klein, George deVries
Kluessendorf, Joanne
Kolata, Dennis Robert
Langenheim, Ralph Louis, Jr
Leighton, Morris Wellman
McKay, Edward Donald, III
Mann, Christian John
Mikulic, Donald George
Norby, Rodney Dale
Reams, Max Warren
Simon, Jack Aaron
Smith, Norman Dwight
Trask, Charles Brian
Ziegler, Alfred M

INDIANA
Dodd, James Robert
Droste, John Brown
Fraser, Gordon Simon
Hattin, Donald Edward
Mirsky, Arthur
Pachut, Joseph F(rancis)
Roepke, Harlan Hugh
Votaw, Robert Barnett

IOWA
Heckel, Philip Henry
Hinman, Eugene Edward
Kramer, Theodore Tivadar
Swett, Keene

KANSAS
Aber, James Sandusky
Avcin, Matthew John, Jr
Baars, Donald Lee
Blythe, Jack Gordon
Gerhard, Lee C
Jackson, Roscoe George, II
Twiss, Page Charles
Walton, Anthony Warrick

KENTUCKY
Chesnut, Donald R, Jr
Clark, Armin Lee
Helfrich, Charles Thomas
Kiefer, John David
Roberts, Thomas Glasdir
Singh, Raman J
Thomas, William Andrew

LOUISIANA
Braunstein, Jules
Craig, William Warren
Forman, McLain J
Horodyski, Robert Joseph
Huh, Oscar Karl
Kumar, Madhurendu B
Lock, Brian Edward
Meriwether, John R
Sartin, Austin Albert
Seglund, James Arnold
Vokes, Harold Ernest
Ward, William Cruse

MAINE
Kellogg, Thomas B
Rosenfeld, Melvin Arthur

MARYLAND
Jakab, George Joseph
Kravitz, Joseph Henry
Miller, Ralph LeRoy
Pavlides, Louis
Stifel, Peter Beekman

MASSACHUSETTS
Belt, Edward Scudder
Binford, Michael W
Greenstein, Benjamin Joel
Hubert, John Frederick
Knoll, Andrew Herbert
McCoy, Floyd W, Jr
McMenamin, Mark Allan
Pitrat, Charles William
Poag, Claude Wylie
Ridge, John Charles
Roth, John L, Jr
Schlee, John Stevens
Shrock, Robert Rakes
Southard, John Brelsford
Takahashi, Kozo
Tucholke, Brian Edward
White, Brian
Williamson, Peter George
Wilson, Philo Calhoun
Yuretich, Richard Francis

MICHIGAN
Briggs, Darinka Zigic
Catacosinos, Paul Anthony
Huntoon, Jacqueline E
McKee, Edith Merritt
Neal, William Joseph
Nordeng, Stephan C
Prouty, Chilton E
Wilkinson, Bruce H

MINNESOTA
Crews, Anita L
Davis, Margaret Bryan
Heller, Robert Leo
Morey, Glenn Bernhardt
Sloan, Robert Evan
Sparling, Dale R
Van Alstine, James Bruce

MISSISSIPPI
Brunner, Charlotte
Dunn, Dean Alan
Egloff, Julius, III
Fleischer, Peter
Happ, Stafford Coleman
Kelley, Patricia Hagelin
Laswell, Troy James
Lowrie, Allen
Otvos, Ervin George
Reynolds, William Roger

MISSOURI
Brill, Kenneth Gray, Jr
Frye, Charles Isaac
Gentile, Richard J
Goebel, Edwin DeWayne
Knox, Burnal Ray
Kurtz, Vincent E
Miller, James Frederick
Unklesbay, Athel Glyde

MONTANA
Moore, Johnnie Nathan

NEBRASKA
Carlson, Marvin Paul
Diffendal, Robert Francis, Jr
Kaplan, Sanford Sandy

NEVADA
Felts, Wayne Moore
Kepper, John C

NEW HAMPSHIRE
Hager, Jutta Lore
Spink, Walter John
Tischler, Herbert

NEW JERSEY
Metz, Robert
Morris, Robert Clarence
Olsson, Richard Keith
Van Houten, Franklyn Bosworth

NEW MEXICO
Bowsher, Arthur LeRoy
Broadhead, Ronald Frigon
Kottlowski, Frank Edward
Love, David Waxham
McTigue, David Francis
Manley, Kim
Pierce, Robert Wesley
Pierce, Robert William
Thompson, Samuel, III
Wells, Stephen Gene
Wengerd, Sherman Alexander
Wolberg, Donald Lester

NEW YORK
Abrahams, Athol Denis
Biscaye, Pierre Eginton
Bowman, James Floyd, II
Brett, Carlton Elliot
Calkin, Parker E
Cisne, John Luther
Coch, Nicholas Kyros
Dawson, James Clifford
Dugolinsky, Brent Kerns
Finks, Robert Melvin
Friedman, Gerald Manfred
Hand, Bryce Moyer
Johnson, Frank Walker
Kent, Dennis V
Kidd, William Spencer Francis
Krause, David Wilfred
Laub, Richard Steven
Lund, Richard
McLennan, Scott Mellin
Macomber, Richard Wiltz
Manos, Constantine T
Muskatt, Herman S
Newell, Norman Dennis
O'Brien, Neal Ray
Rampino, Michael Robert
Rickard, Lawrence Vroman
Sancetta, Constance Antonina
Savage, E Lynn
Scatterday, James Ware
Schreiber, B Charlotte
Tesmer, Irving Howard
Wang, Kia K
Wolff, Manfred Paul
Woodrow, Donald L

NORTH CAROLINA
Banerjee, Umesh Chandra
Briggs, Garrett
Cleary, William James
Dennison, John Manley
Harris, William Burleigh
Heron, S(tephen) Duncan, (Jr)
Ingram, Roy Lee
Johnson, Thomas Charles
Marlowe, James Irvin
Neal, Donald Wade
Textoris, Daniel Andrew
Webb, Fred, Jr
Welby, Charles William
Wheeler, Walter Hall
Zullo, Victor August

NORTH DAKOTA
Cvancara, Alan Milton

OHIO
Bain, Roger J
Boardman, Mark R
Bork, Kennard Baker
Burford, Arthur Edgar
Carlton, Richard Walter
Fuller, James Osborn
Harmon, David Elmer, Jr
Hatfield, Craig
Henniger, Bernard Robert
Kovach, Jack
McWilliams, Robert Gene
Mapes, Gene Kathleen
Martin, Wayne Dudley
Miller, Arnold I
Sturgeon, Myron Thomas
Wilson, Mark Allan

OKLAHOMA
Boellstorff, John David
Chambers, Richard Lee
Chenoweth, Philip Andrew
Espach, Ralph H, Jr
Friedman, Samuel Arthur
Granath, James Wilton
Huffman, George Garrett
Masters, Bruce Allen
Murray, Frederick Nelson
Pittman, Edward D
Powell, Benjamin Neff
Sanderson, George Albert
Smit, David Ernst
Suhm, Raymond Walter
Tillman, Roderick W
Toomey, Donald Francis
Visher, Glenn S

OREGON
Allison, Ira Shimmin
Beaulieu, John David
Fisk, Lanny Herbert
Gutschick, Raymond Charles
Niem, Alan Randolph
Niem, Wendy Adams
Retallack, Gregory John
Savage, Norman Michael

PENNSYLVANIA
Anderson, Edwin J
Carson, Bobb
Cotter, Edward
Donahue, Jack David
Faill, Rodger Tanner
Frantz, Wendelin R
Gardner, Thomas William
Goodwin, Peter Warren
Guber, Albert Lee
Hanson, Henry W A, III
Inners, Jon David
Jordan, William Malcolm
Lowright, Richard Henry
Pfefferkorn, Hermann Wilhelm
Ramsdell, Robert Cole
Saunders, William Bruce
Schmalz, Robert Fowler
Seidell, Barbara Castens
Slingerland, Rudy Lynn
Trexler, John Peter
Washburn, Robert Henry
Wegweiser, Arthur E

RHODE ISLAND
Burroughs, Richard
Head, James William, III
Leinen, Margaret Sandra
McMaster, Robert Luscher
Matthews, Robley Knight
Prell, Warren Lee

SOUTH CAROLINA
Birkhead, Paul Kenneth
Carew, James L
Cohen, Arthur David
Ehrlich, Robert
Gipson, Mack, Jr
Kanes, William H
Pirkle, William Arthur

SOUTH DAKODA
Martin, James Edward

TENNESSEE
Brent, William B
Brode, William Edward
Helton, Walter Lee
Klepser, Harry John
Siesser, William Gary
Walker, Kenneth Russell

TEXAS
Abbott, William Harold
Ahr, Wayne Merrill
Albritton, Claude Carroll, Jr
Amsbury, David Leonard
Baker, Donald Roy
Basan, Paul Bradley
Bebout, Don Gray
Behrens, Earl William
Blackwelder, Blake Winfield
Bohacs, Kevin Michael
Brooks, James Elwood
Brown, Leonard Franklin, Jr
Buis, Otto J
Burton, Robert Clyde
Chadwick, Arthur Vorce
Chatelain, Edward Ellis
Chatterjee, Sankar
Crick, Rex Edward
Curtis, Doris Malkin
Dahl, Harry Martin
Damuth, John Erwin
DuBar, Jules R
Echols, Dorothy Jung
Edwards, Marc Benjamin
Eifler, Gus Kearney, Jr
Elsik, William Clinton
Ewing, Thomas Edward
Feazel, Charles Tibbals
Finley, Robert James
Folk, Stewart H
Gee, David Easton
Gombos, Andrew Michael, Jr
Gruy, Henry Jones
Gustavson, Thomas Carl
Halpern, Martin
Heyman, Louis
Huffington, Roy Michael
Jackson, Martin Patrick Arden
Jones, James Ogden
Kaiser, William Richard
Kocurko, Michael John
Koepnick, Richard Borland
Lafon, Guy Michel
Lane, Harold Richard
Leblanc, Rufus Joseph
Longacre, Susan Ann Burton
McFarlan, Edward, Jr
McGrail, David Wayne
McPherson, John G(ordon)
Macurda, Donald Bradford, Jr
Matthews, Martin David

Mazzullo, James Michael
Melia, Michael Brendan
Meyer, Franz O
Milling, Marcus Eugene
Moiola, Richard James
Morton, Robert Alex
Murphy, Daniel L
Murray, Grover Elmer
Namy, Jerome Nicholas
North, William Gordon
Nugent, Robert Charles
Owen, Donald Edward
Owens, Edward Henry
Pirie, Robert Gordon
Reso, Anthony
Rezak, Richard
Rohr, David M
Roper, Paul James
Rose, Peter R
Rosen, Norman Charles
Ross, Charles Alexander
St John, Bill
Schumacher, Dietmar
Self, Stephen
Shane, John Denis
Shaw, Alan Bosworth
Smith, Michael A
Soloyanis, Susan Constance
Spencer, Alexander Burke
Stead, Frederick L
Stoudt, Emily Laws
Swanson, Donald Charles
Swanson, Eric Rice
Van Siclen, DeWitt Clinton
Warshauer, Steven Michael
Weart, Richard Claude
Wilde, Garner Lee
Wilson, James Lee
Wright, Ramil Carter
Yancey, Thomas Erwin
Young, Keith Preston

UTAH
Ash, Sidney Roy
Ekdale, Allan Anton
Fidlar, Marion M
Freethey, Geoffrey Ward
Kolesar, Peter Thomas
McCleary, Jefferson Rand
Oaks, Robert Quincy, Jr
Petersen, Morris Smith
Stokes, William Lee

VERMONT
Baldwin, Brewster

VIRGINIA
Ahlbrandt, Thomas Stuart
Darby, Dennis Arnold
Edwards, Lucy Elaine
Eggleston, Jane R
Eriksson, Kenneth Andrew
Finkl, Charles William, II
Galvin, Cyril Jerome, Jr
Hadley, Donald G
Harbour, Jerry
Lee, Kwang-Yuan
Lohman, Kenneth Elmo
McCartan, Lucy
Martens, James Hart Curry
Masters, Charles Day
Meissner, Charles Roebling, Jr
Milici, Robert Calvin
Miller, Betty M (Tinklepaugh)
Nelson, Bruce Warren
Nelson, Clifford Melvin, Jr
Outerbridge, William Fulwood
Reinhardt, Juergen
Russ, David Perry
Ryder, Robert Thomas
Said, Rushdi
Schwab, Frederic Lyon
Spencer, Randall Scott
Swift, Donald J P
Weems, Robert Edwin
Whisonant, Robert Clyde

WASHINGTON
Caggiano, Joseph Anthony, Jr
Fields, Robert William
Kesler, Thomas L
Miller, C Dan
Oles, Keith Floyd
Pierson, Thomas Charles
Rau, Weldon Willis
Scott, William Edward
Smith, J Dungan
Suczek, Christopher Anne
Webster, Gary Dean
Wright, Judith

WEST VIRGINIA
Khair, Abdul Wahab
Smosna, Richard Allan

WISCONSIN
Dott, Robert Henry, Jr
Fowler, Gerald Allan
Gernant, Robert Everett
Laudon, Thomas S
McKee, James W
Mendelson, Carl Victor
Paull, Rachel Krebs
Pray, Lloyd Charles

Schneider, Allan Frank
Stieglitz, Ronald Dennis
West, Walter Scott

WYOMING
Boyd, Donald Wilkin
Huff, Kenneth O
Lillegraven, Jason Arthur
Miller, Daniel Newton, Jr
Oman, Carl Henry

ALBERTA
Andrichuk, John Michael
Baillie, Andrew Dollar
Cecile, Michael Peter
Currie, Philip John
Lerbekmo, John Franklin
Mossop, Grant Dilworth
Schmidt, Volkmar
Stelck, Charles Richard
Tankard, Anthony James
Wilson, Mark Vincent Hardman

BRITISH COLUMBIA
Barnes, Christopher Richard
Hebda, Richard Joseph
Jackson, Lionel Eric, Jr
Monger, James William Heron
Nelson, Samuel James
Payne, Myron William
Smith, Paul L
Stott, Donald Franklin

MANITOBA
Teller, James Tobias

NEW BRUNSWICK
Ferguson, Laing

NEWFOUNDLAND
Harper, John David
Hiscott, Richard Nicholas
King, Arthur Francis

NOVA SCOTIA
Cameron, Barry Winston
Schenk, Paul Edward
Zodrow, Erwin Lorenz

ONTARIO
Bell, Richard Thomas
Caldwell, William Glen Elliot
Chandler, Frederick William
Coakley, John Phillip
Franklin, James McWillie
Geurts, Marie Anne H L
Greenwood, Brian
Jarzen, David MacArthur
Karrow, Paul Frederick
Lawson, David Edward
Long, Darrel Graham Francis
McLaren, Digby Johns
Macqueen, Roger Webb
Martini, Ireneo Peter
Menzies, John
Miall, Andrew D
Rukavina, Norman Andrew
Simpson, Frank
St-Onge, Denis Alderic
Walker, Roger Geoffrey
Yole, Raymond William
Young, Grant McAdam

QUEBEC
Bouchard, Michel André
Clark, Thomas Henry
Hesse, Reinhard
Lesperance, Pierre J
Mountjoy, Eric W
Stearn, Colin William

SASKATCHEWAN
Braun, Willi Karl
Hendry, Hugh Edward
Kent, Donald Martin
Nisbet, Euan G

OTHER COUNTRIES
Alonso, Ricardo N
Frakes, Lawrence Austin
Granata, Walter Harold, Jr
Kennedy, Michael John
Lloyd, Christopher Raymond
Trümpy, D(aniel) Rudolf
Van Andel, Tjeerd Hendrik
Vickers-Rich, Patricia

Other Environmental, Earth & Marine Sciences

ALABAMA
Daniel, Thomas W, Jr
Guarino, Anthony Michael
Hazlegrove, Leven S

ALASKA
Hamilton, Thomas Dudley
Hawkins, Daniel Ballou
Ragle, Richard Harrison
Reger, Richard David
Rockwell, Julius, Jr
Turner, Donald Lloyd

Bass, George F
Black, David Charles
Bullard, Fred Mason
Campbell, Charles Edgar
Cepeda, Joseph Cherubini
Denley, David R
Farrand, William Richard
Giese, Robert Paul
Haas, Herbert
Halpern, Martin
Hanson, Bernold M
Hay-Roe, Hugh
Henry, Christopher Duval
Kimber, Clarissa Therese
Leblanc, Rufus Joseph
Ledley, Tamara Shapiro
Murray, Grover Elmer
Overmyer, Robert Franklin
Pearson, Daniel Bester, III
Potter, Andrew Elwin, Jr
Roper, Paul James
Ross, Charles Alexander
Shelby, T H
Standford, Geoffrey Brian
Stotler, Raymond T
Taylor, Robert Dalton
Thill, Ronald E
Vail, Peter R

UTAH
Cramer, Harrison Emery
Freethey, Geoffrey Ward
Ghosh, Sambhunath
Haslem, William Joshua

VERMONT
Stoiber, Richard Edwin
VanderMeer, Canute

VIRGINIA
Beal, Stuart Kirkham
Benton, Duane Marshall
Calver, James Lewis
Delnore, Victor Eli
Glover, Lynn, III
Huggett, Robert James
Krushensky, Richard D
Lacy, W(illiam) J(ohn)
Masters, Charles Day
Mose, Douglas George
Murden, William Roland, Jr
Noble, Vincent Edward
Pontius, Steven Kent
Robbins, Eleanora Iberall
Spangenberg, Dorothy Breslin
Sutter, John Frederick
Theon, John Speridon
Tipper, Ronald Charles
Todd, Edward Payson
Tsai, Frank Y
Walker, Grayson Watkins
Weiss, Martin
Wright, Thomas L
Zimmerman, Peter David

WASHINGTON
Anderson, George Cameron
Coombs, Howard A
Fluharty, David Lincoln
Grootes, Pieter Meiert
Holmes, Mark Lawrence
Jenne, Everett A
Lipscomb, David M
Miller, C Dan
Porter, Stephen Cummings
Radke, Lawrence Frederick
Robkin, Maurice
Simenstad, Charles Arthur
Templeton, William Lees
Varanasi, Usha
Washburn, Albert Lincoln
Wilkening, Laurel Lynn

WISCONSIN
Lasca, Norman P, Jr
Lillesand, Thomas Martin
Maher, Louis James, Jr
Moore, Gary T
Mortimer, Clifford Hiley
Rhyner, Charles R

WYOMING
Barlow, James A, Jr
Doerges, John E
Hayden-Wing, Larry Dean
Howell, Robert Richard
Steidtmann, James R

ALBERTA
Goodarzi, Fariborz
Nasser, Tourai
Rutter, Nathaniel Westlund

BRITISH COLUMBIA
Armstrong, John Edward
Sutherland-Brown, Atholl

NEW BRUNSWICK
Lakshminarayana, J S S

NOVA SCOTIA
Bidwell, Roger Grafton Shelford
King, Lewis H
Reynolds, Peter Herbert

Stevens, George Richard

ONTARIO
Bend, John Richard
Cale, William Robert
Czuba, Margaret
Dawes, David Haddon
Dayal, Ramesh
Dence, Michael Robert
Donaldson, John Allan
Eade, Kenneth Edgar
Farquhar, Ronald McCunn
Gorman, William Alan
Hackett, Peter Andrew
Halliday, Ian
Hare, Frederick Kenneth
Jolly, Wayne Travis
McNutt, Robert Harold
Ruel, Maurice M J
Selvadurai, A P S
Simpson, Frank
Sly, Peter G
VandenHazel, Bessel J
Vollenweider, Richard A
Williams, Peter J

QUEBEC
Belanger, Pierre Andre
Dubois, Jean-Marie M
Elson, John Albert
Ghosh, Sanjib Kumar
Hillaire-Marcel, Claude
Khalil, Michel

SASKATCHEWAN
Hanley, Thomas ODonnell

OTHER COUNTRIES
Fowler, Scott Wellington
Hofmann, Albrecht Werner
Hughes, Charles James

MATHEMATICS
Algebra

ALABAMA
Brackin, Eddy Joe
Gibson, Robert Wilder
Kim, Ki Hang

ALASKA
Lando, Barbara Ann

ARIZONA
Jacobowitz, Ronald
Moore, Nadine Hanson
Pierce, Richard Scott
Toubassi, Elias Hanna
Velez, William Yslas

ARKANSAS
Rossa, Robert Frank
Schein, Boris M

CALIFORNIA
Akasaki, Takeo
Baker, Kirby Alan
Blattner, Robert J(ames)
Block, Richard Earl
Cannonito, Frank Benjamin
Carlson, David Hilding
Casey, John Edward, Jr
Celik, Hasan Ali
Cutler, Doyle O
Eklof, Paul Christian
Estes, Dennis Ray
Fraser, Grant Adam
Gaposchkin, Peter John Arthur
Gaskell, Robert Weyand
Gerstein, Larry J
Green, Sherry Merrill
Grigori, Artur
Guralnick, Robert Michael
Hales, Alfred Washington
Henkin, Leon (Albert)
Hopponen, Jerry Dale
Horn, Alfred
Johnsen, Eugene Carlyle
Lam, Tsit-Yuen
Lanski, Charles Philip
Lawson, Charles L
Luce, R(obert) Duncan
Marcus, Marvin
Mena, Roberto Abraham
Merris, Russell Lloyd
Mignone, Robert Joseph
Minc, Henryk
Montgomery, M Susan
Oppenheim, Joseph Harold
Phillips, Veril LeRoy
Pixley, Alden F
Pratt, Vaughan Ronald
Ralston, Elizabeth Wall
Ratliff, Louis Jackson, Jr
Rhodes, John Lewis
Rieffel, Marc A
Rodabaugh, David Joseph
Rosenberg, Alex
Small, Lance W
Smith, Velma Merriline
Smoke, William Henry
Sun, Hugo Sui-Hwan

Tang, Alfred Sho-Yu
Tang, Hwa-Tsang
Taussky, Olga
Teague, Tommy Kay
Tilson, Bret Ransom
Wiegmann, Norman Arthur
Yaqub, Adil Mohamed
Zelmanowitz, Julius Martin

COLORADO
Briggs, William Egbert
Lundgren, J Richard
Montague, Stephen
Roth, Richard Lewis
Struik, Ruth Rebekka

CONNECTICUT
Gürsey, Feza
Hager, Anthony Wood
Howe, Roger
Hurley, James Frederick
Mostov, George Daniel
Nilson, Edwin Norman
Reid, James Dolan
Spiegel, Eugene

DELAWARE
Caviness, Bobby Forrester
Ebert, Gary Lee
Saunders, B David

DISTRICT OF COLUMBIA
Rosenblatt, David
Sjogren, Jon Arne
Teller, John Roger
Thaler, Alvin Isaac

FLORIDA
Brewer, James W
Ferguson, Pamela A
Gilmer, Robert
Hoffman, Frederick
Magarian, Elizabeth Ann
Miller, Don Dalzell
Mott, Joe Leonard
Timmer, Kathleen Mae
Williams, Gareth
Yellen, Jay
Zerla, Fredric James

GEORGIA
Aust, Catherine Cowan
Belinfante, Johan G F
Bevis, Jean Harwell
Bompart, Billy Earl
Carlson, Jon Frederick
Falconer, Etta Zuber
Nail, Billy Ray
Neff, Mary Muskoff
Norwood, Gerald
Robinson, Daniel Alfred
Sharp, Thomas Joseph
Stone, David Ross
Straley, Tina
Watts, Sherrill Glenn

HAWAII
Nobusawa, Nobuo
Sagle, Arthur A

ILLINOIS
Bateman, Felice Davidson
Blair, William David
Blau, Harvey Isaac
Boyle, James Martin
Braunfeld, Peter George
Camburn, Marvin E
Clover, William John, Jr
Comerford, Leo P, Jr
Cutler, Janice Zemanek
Earnest, Andrew George
Evans, E Graham, Jr
Fossum, Robert Merle
Friedlander, Eric Mark
Glauberman, George Isaac
Haboush, William Joseph
Hsu, Nai-Chao
Janusz, Gerald Joseph
Kasube, Herbert Emil
Laible, Jon Morse
Lamont, Patrick
Larson, Richard Gustavus
Lazerson, Earl Edwin
Leibhardt, Edward
Otto, Albert Dean
Paley, Hiram
Parr, James Theodore
Price, David Thomas
Reznick, Bruce Arie
Stein, Michael Roger
Ullom, Stephen Virgil
Wagreich, Philip Donald
Wallis, Walter Denis
Walter, John Harris
Waterman, Peter Lewis
Weichsel, Paul M
Weiner, Louis Max
Wilcox, Lee Roy
Yau, Stephen Shing-Toung
Zelinsky, Daniel

INDIANA
Azumaya, Goro
Beehler, Jerry R

Drufenbrock, Diane
Edwards, Constance Carver
Hahn, Alexander J
Kronstein, Karl Martin
Kuczkowski, Joseph Edward
Machtinger, Lawrence Arnold
MacKenzie, Robert Earl
Murphy, Catherine Mary
Perlis, Sam
Phan, Kok-Wee
Pohl, Victoria Mary
Pollak, Barth
Shahidi, Freydoon
Sommese, Andrew John
Wong, Warren James

IOWA
Adelberg, Arnold M
Anderson, Daniel D
Barnes, Wilfred E
Dickson, Spencer E
Fuller, Kent Ralph
Hogben, Leslie
Johnson, Eugene W
Kim, Paik Kee
Kleinfeld, Erwin
Peake, Edmund James, Jr
Pigozzi, Don

KANSAS
Davis, Elwyn H
Emerson, Marion Preston
McCarthy, Paul Joseph
Shult, Ernest E

KENTUCKY
Beidleman, James C
Eaves, James Clifton
Elder, Harvey Lynn
Enochs, Edgar Earle
Franke, Charles H
Kearns, Thomas J
LeVan, Marijo O'Connor
Wallace, Kyle David

LOUISIANA
Benard, Mark
Cook, Virginia A
Cordes, Craig M
Fuchs, Laszlo
Heatherly, Henry Edward
Hildebrant, John A
McLean, R T
Ohm, Jack Elton
Spaht, Carlos G, II

MAINE
Lange, Gail Laura

MARYLAND
Andre, Peter P
Cohen, Joel M
Douglas, Lloyd Evans
Draper, Richard Noel
Dribin, Daniel Maccabaeus
Ehrlich, Gertrude
Federighi, Enrico Thomas
Goldhaber, Jacob Kopel
Good, Richard Albert
Helzer, Gerry A
Hutchinson, George Allen
Katz, Victor Joseph

MASSACHUSETTS
Altman, Allen Burchard
Auslander, Maurice
Bennett, Mary Katherine
Berkovits, Shimshon
Bernstein, Joseph N
Blackett, Donald Watson
Cattani, Eduardo
D'Alarcao, Hugo T
Dechene, Lucy Irene
Eisenbud, David
Fogarty, John Charde
Guterman, Martin Mayr
Hill, Victor Ernst, IV
Holland, Samuel S, Jr
Janowitz, Melvin Fiva
Joni, Saj-nicole A
Kass, Seymour
Kenney, Margaret June
Knighten, Robert Lee
Kundert, Esayas G
Lusztig, George
McQuarrie, Bruce Cale
Malone, Joseph James
Martindale, Wallace S
Morris, Robert
Newman, Kenneth Wilfred
Pollatsek, Harriet Suzanne
Reynolds, William Francis
Rudvalis, Arunas
Schafer, Richard Donald
Stanley, Richard Peter
Svanes, Torgny
Weiss, Edwin

MICHIGAN
Althoen, Steven Clark
Deering, Carl F
Feldman, Chester
Griess, Robert Louis, Jr
Heuvers, Konrad John

Blattner, Robert J(ames)
Bruckner, Andrew M
Caflisch, Russel Edward
Carson, George Stephen
Chang, Sun-Yung Alice
Cherlin, George Yale
Chernoff, Paul Robert
Coleman, Courtney (Stafford)
Concus, Paul
Crandall, Michael Grain
Curtis, Philip Chadsey, Jr
DePrima, Charles Raymond
Dickson, Lawrence John
Di Franco, Roland B
Dixon, Michael John
Donoghue, William F, Jr
Donoho, David Leigh
Drobot, Vladimir
Elderkin, Richard Howard
Fattorini, Hector Osvaldo
Feinstein, Charles David
Feldman, Jacob
Ferguson, Le Baron O
Finn, Robert
Fountain, Leonard Du Bois
Gamelin, Theodore W
Genensky, Samuel Milton
Gilbert, Richard Carl
Grabiner, Sandy
Greene, Robert Everist
Greenhall, Charles August
Gretsky, Neil E
Gruenberger, Fred J(oseph)
Halmos, Paul Richard
Helson, Henry
Helton, John W
Hench, David Le Roy
Herriot, John George
Hirsch, Morris William
Huddleston, Robert E
Hunt, Robert Weldon
James, Ralph L
James, Robert Clarke
Jewell, Nicholas Patrick
Juberg, Richard Kent
Keller, Herbert Bishop
Krabbe, Gregers Louis
Kurowski, Gary John
Lai, Kai Sun
Langford, Eric Siddon
Lapidus, Michel Laurent
Lashley, Gerald Ernest
Lesley, Frank David
Li, Peter Wai-Kwong
Lippman, Gary Edwin
Lu, Kau U
Luxemburg, Wilhelmus Anthonius
    Josephus
McCarthy, Mary Anne
Mignone, Robert Joseph
Miles, Frank Belsley
Nur, Hussain Sayid
O'Dúnlaing, Colm Pádraig
Oliger, Joseph Emmert
Osserman, Robert
Pinney, Edmund Joy
Radbill, John R(ussell)
Radnitz, Alan
Rao, Malempati Madhusudana
Reich, Simeon
Rieffel, Marc A
Riggle, Everett C
Rodabaugh, David Joseph
Rosen, Irwin Gary
Royden, Halsey Lawrence
Schechter, Martin
Schiffer, Menahem Max
Shapiro, Victor Lenard
Shiffman, Max
Simon, Barry Martin
Simons, Stephen
Thacher, Henry Clarke, Jr
Todd, John
Vasco, Donald Wyman
Vinokur, Marcel
Watson, Velvin Richard
Wehausen, John Vrooman
Weiss, Max Leslie
Widom, Harold
Wulbert, Daniel Eliot

COLORADO
Bleistein, Norman
Brown, Edmund Hosmer
Buzbee, Billy Lewis
Daly, James Edward
Darst, Richard B
Gibson, William Loane
Gill, John Paul, Jr
Hagler, James Neil
Jones, William B
Li, Hung Chiang
Manteuffel, Thomas Albert
Phillips, Keith L
Ramsay, Arlan (Bruce)
Shank, George Deane
Thaler, Eric Ronald
Thron, Wolfgang Joseph
Walter, Martin Edward

CONNECTICUT
Ahlberg, John Harold
Beals, Richard William
Blei, Ron Charles

Elliott, H(elen) Margaret
Giné-Masdéu, Evarist
Gosselin, Richard Pettengill
Howe, Roger
Leibowitz, Gerald Martin
Robbins, David Alvin
Schlesinger, Ernest Carl
Sidney, Stuart Jay

DELAWARE
Libera, Richard Joseph
Livingston, Albert Edward
Weinacht, Richard Jay

DISTRICT OF COLUMBIA
Chang, I-Lok
Espelie, (Mary) Solveig
Long, Paul Eastwood, Jr
Robinson, E Arthur, Jr
Ryff, John V
Vandergraft, James Saul

FLORIDA
Bellenot, Steven F
Cenzer, Douglas
Goodman, Richard Henry
Goodner, Dwight Benjamin
Isaacs, Godfrey Leonard
Ismail, Mourad E H
McArthur, Charles Wilson
McWilliams, Ralph David
Minzner, Raymond Arthur
Mukherjea, Arunava
Nagle, Richard Kent
Oberlin, Daniel Malcolm
Ratti, Jogindar Singh
Rubin, Richard Lee
Saff, Edward Barry
Squier, Donald Platte
Young, Eutiquio Chua

GEORGIA
Azoff, Edward Arthur
Batterson, Steven L
Bouldin, Richard H(indman)
Cash, Dewey Byron
Clark, Douglas Napier
Green, William Lohr
Herod, James Victor
Landgren, John Jeffrey
Meyer, Gunter Hubert
Osborn, James Maxwell
Riddle, Lawrence H
Ripy, Sara Louise
Shabazz, Abdulalim
Waltman, Paul Elvis
Worth, Roy Eugene
Youse, Bevan K

HAWAII
Iha, Franklin Takashi

IDAHO
Bobisud, Larry Eugene
Hausrath, Alan Richard
McClure, John Arthur

ILLINOIS
Albrecht, Felix Robert
Bank, Steven Barry
Barron, Emmanuel Nicholas
Bellow, Alexandra
Benzinger, Harold Edward, Jr
Berkson, Earl Robert
Berndt, Bruce Carl
Buntinas, Martin George
Burkholder, Donald Lyman
Calderon, Calixto Pedro
Carroll, Robert Wayne
Cox, Dennis Dean
Day, Mahlon Marsh
Diamond, Harold George
Driscoll, Richard James
Feinstein, Irwin K
Friedberg, Stephen Howard
Garder, Arthur
Gasper, George, Jr
Gupta, Chaitan Prakash
Ho, Chung-Wu
Ionescu Tulcea, Cassius
Jerome, Joseph Walter
Jones, Roger L
Kammler, David W
Lubin, Arthur Richard
McAlpin, John Harris
Nicholls, Peter John
Packel, Edward Wesler
Peck, Emily Mann
Peck, Newton Tenney
Peressini, Anthony L
Philipp, Walter V
Redmond, Donald Michael
Reznick, Bruce Arie
Rickert, Neil William
Robinson, Clark
Saylor, Paul Edward
Sherbert, Donald R
Waterman, Peter Lewis
Wenger, John C
Wetzel, John Edwin
Wheeler, Robert Francis
Yau, Stephen Shing-Toung
Zettl, Anton

INDIANA
Aliprantis, Charalambos Dionisios
Brothers, John Edwin
Chihara, Theodore Seio
Conte, Samuel D
Cowen, Carl Claudius, Jr
Drasin, David
Gautschi, Walter
Golomb, Michael
Kallman, Ralph Arthur
Kaminker, Jerome Alvin
Legg, David Alan
Lyons, Russell David
McKinney, Earl H
Penney, Richard Cole
Rhoades, Billy Eugene
Rothman, Neal Jules
Schober, Glenn E
Shahidi, Freydoon
Sommese, Andrew John
Specht, Edward John
Stampfli, Joseph
Stanton, Charles Madison
Stoll, Wilhelm
Synowiec, John A
Torchinsky, Alberto
Zachmanoglou, Eleftherios Charalambos
Zink, Robert Edwin

IOWA
Alexander, Roger Keith
Carlson, Bille Chandler
Dahiya, Rajbir Singh
Jorgensen, Palle E T
Peters, Justin
Ton-That, Tuong

KANSAS
Acker, Andrew French, III
Bajaj, Prem Nath
Brown, Robert Dillon
Burckel, Robert Bruce
Calys, Emanuel G
Eltze, Ervin Marvin
Himmelberg, Charles John
Linscheid, Harold Wilbert
Paschke, William Lindall
Ramm, Alexander G
Van Vleck, Fred Scott

KENTUCKY
Buckholtz, James Donnell
Howard, Henry Cobourn
King, Amy P
Shah, Swarupchand Mohanlal
Van Winter, Clasine

LOUISIANA
Agarwal, Arun Kumar
Anders, Edward B
Chan, Chiu Yeung
Conway, Edward Daire, III
Denny, William F
Dorroh, James Robert
Kalka, Morris
Knill, Ronald John
Kuo, Hui-Hsiung
Kythe, Prem Kishore
Lefton, Lew Edward
Liukkonen, John Robie
McGehee, Oscar Carruth
McKinney, Alfred Lee
Richardson, Leonard Frederick
Roques, Alban Joseph
Swetharanyam, Lalitha

MAINE
Balakrishnan, V K
Brace, John Wells

MARYLAND
Alexander, James Crew
Antman, Stuart S
Babuska, Ivo Milan
Benedetto, John
Berger, Alan Eric
Cawley, Robert
Cohen, Edgar A, Jr
Cohen, Joel M
Cook, Clarence Harlan
Diesen, Carl Edwin
Douglas, Lloyd Evans
Douglis, Avron
Ellis, Robert L
Goldberg, Seymour
Heins, Maurice Haskell
Horvath, John Michael
Hubbard, Bertie Earl
Hummel, James Alexander
Huneycutt, James Ernest, Jr
Johnson, David Lee
Johnson, Raymond Lewis
Kascic, Michael Joseph, Jr
Kirwan, William English
Kleppner, Adam
Krantz, Steven George
Lay, David Clark
Lepson, Benjamin
Liu, Tai-Ping
Lozier, Daniel William
Mahar, Thomas J
May, Everette Lee, Jr
Neri, Umberto
Newcomb, Robert Wayne

Raphael, Louise Arakelian
Rosenberg, Jonathan Micah
Saworotnow, Parfeny Pavolich
Shiffman, Bernard
Stephens, Arthur Brooke
Sullivan, Francis E
Wallace, Alton Smith
Warner, Charles Robert
Witzgall, Christoph Johann
Zabronsky, Herman
Zedek, Mishael

MASSACHUSETTS
Bailey, Duane W
Ben-Akiva, Moshe E
Berkey, Dennis Dale
Berkovits, Shimshon
Bernstein, Joseph N
Bilodeau, Gerald Gustave
Breton, J Raymond
Cole, Nancy
Cook, Thurlow Adrean
Cormack, Allan MacLeod
Dudley, Richard Mansfield
Filgo, Holland Cleveland, Jr
Giger, Adolf J
Grossberg, Stephen
Helgason, Sigurdur
Holland, Samuel S, Jr
Holmes, Richard Bruce
Horowitz, Larry Lowell
Kamowitz, Herbert M
Kon, Mark A
Lam, Kui Chuen
Melrose, Richard B
Quinto, Eric Todd
Ruskai, Mary Beth
St Mary, Donald Frank
Seeley, Robert T
Sherman, Thomas Oakley
Shuchat, Alan Howard
Siu, Yum-Tong
Warga, Jack

MICHIGAN
Anderson, Glen Douglas
Bachelis, Gregory Frank
Barnard, Robert D(ane), Jr
Bartle, Robert Gardner
Cesari, Lamberto
Dickson, Douglas Grassel
Duren, Peter Larkin
Embry-Wardrop, Mary Rodriguez
Fast, Henryk
Feldman, Chester
Hansen, Lowell John
Herzog, Fritz
Howard, Paul Edward
Johnson, George Philip
Lick, Don R
Loebl, Richard Ira
McLaughlin, Renate
Marik, Jan
Miklavcic, Milan
Mordukhovich, Boris S
Nelson, James Donald
Nyman, Melvin Andrew
Pearcy, Carl Mark, Jr
Piranian, George
Reade, Maxwell Ossian
Salehi, Habib
Simpson, William Albert
Skaff, Michael Samuel
Stoline, Michael Ross
Stout, Quentin Fielden
Vinh, Nguyen Xuan
Weil, Clifford Edward

MINNESOTA
Fabes, Eugene Barry
Gil de Lamadrid, Jesús
Gulliver, Robert David, II
Hejhal, Dennis Arnold
Hilding, Stephen R
Humke, Paul Daniel
Jodeit, Max A, Jr
Keynes, Harvey Bayard
Littman, Walter
McCarthy, Charles Alan
Mericle, R Bruce
Munro, William Delmar
Nitsche, Johannes Carl Christian
Olver, Peter John
Patton, Peter C(lyde)
Pour-El, Marian Boykan
Vessey, Theodore Alan

MISSISSIPPI
Causey, William McLain
Doblin, Stephen Alan
Mastin, Charles Wayne
Porter, James Franklin
Solomon, Jimmy Lloyd
Spikes, Paul Wenton

MISSOURI
Andalafte, Edward Ziegler
Baernstein, Albert, II
Connett, William C
Crownover, Richard McCranie
DeFacio, W Brian
Gillilan, James Horace
Haimo, Deborah Tepper
Hall, Leon Morris, Jr

Analysis & Functional Analysis (cont)

Buck, Robert Creighton
Carothers, Otto M
Engert, Martin
Forelli, Frank John
Hall, Robert Lester
Kurtz, Thomas Gordon
Levin, Jacob Joseph
Marden, Morris
Mullins, Robert Emmet
Nagel, Alexander
O'Malley, Richard John
Rabinowitz, Paul H
Rall, Louis Baker
Rudin, Walter
Shah, Ghulam M
Solomon, Donald W
Terwilliger, Paul M
Walter, Gilbert G
Ziegler, Michael Robert

WYOMING
Roth, Ben G

PUERTO RICO
Bobonis, Augusto

ALBERTA
Andersen, Kenneth F
Johnson, Dudley Paul
Riemenschneider, Sherman Delbert
Rod, David Lawrence

BRITISH COLUMBIA
Anderson, R F V
Boyd, David William
Bures, Donald John (Charles)
Cayford, Afton Herbert
Ehle, Byron Leonard
Granirer, Edmond
Leeming, David John
Pfaffenberger, William Elmer
Sion, Maurice
Srivastava, Hari Mohan
Srivastava, Rekha
Swanson, Charles Andrew
Ton, Bui An
Wray, Stephen Donald

MANITOBA
Finlayson, Henry C
Usmani, Riaz Ahmad
Wong, Roderick Sue-Cheun

NEW BRUNSWICK
Pham-Gia, Thu

NEWFOUNDLAND
Burry, John Henry William
Rees, Rolf Stephen
Singh, Sankatha Prasad
Thomeier, Siegfried

NOVA SCOTIA
Dunn, Kenneth Arthur
El-Hawary, Mohamed El-Aref
Fillmore, Peter Arthur
Grant, Douglass Lloyd
Tan, Kok-Keong

ONTARIO
Abdelmalek, Nabin N
Aczél, János D
Akhtar, Mohammad Humayoun
Arthur, James Greig
Atkinson, Harold Russell
Baker, John Alexander
Barbeau, Edward Joseph, Jr
Beesack, Paul Richard
Billigheimer, Claude Elias
Borwein, David
Bryan, Robert Neff
Caradus, Selwyn Ross
Chakravartty, Iswar C
Chang, Shao-chien
Choi, Man-Duen
Davis, Chandler
Duff, George Francis Denton
Eames, William
Elliott, George Arthur
Fleischer, Isidore
Giles, Robin
Graham, Colin C
Greiner, Peter Charles
Headley, Velmer Bentley
Henniger, James Perry
Howland, James Lucien
Husain, Taqdir
Kerman, R A
Langford, William Finlay
Macphail, Moray St John
Mercer, Alexander McDowell
Muldoon, Martin E
Nel, Louis Daniel
Prugovecki, Eduard
Rahman, Mizanur
Russell, Dennis C
Schubert, Cedric F
Stewart, James Drewry
Whitfield, John Howard Mervyn
Wigley, Neil Marchand

QUEBEC
Boyarsky, Abraham Joseph
Choksi, Jal R

Herschorn, Michael
Herz, Carl Samuel
Koosis, Paul
Lin, Paul C S
Schlomiuk, Dana
Taylor, John Christopher
Zaidman, Samuel

SASKATCHEWAN
Koh, Eusebio Legarda
Law, Alan Greenwell
Sato, Daihachiro

OTHER COUNTRIES
Artemiadis, Nicholas
Chesson, Peter Leith
Horwitz, Lawrence Paul
Kantorovitz, Shmuel
Korevaar, Jacob
Maltese, George J
Roseman, Joseph Jacob
Yang, Chung-Chun

Applied Mathematics

ALABAMA
Cochran, John Euell, Jr
Cook, Frederick Lee
Davis, Anthony Michael John
Fitzpatrick, Philip Matthew
Gilbert, Stephen Marc
Hatfield, John Dempsey
Holt, William Robert
Howell, James Levert
Johnson, Johnny R(ay)
Jones, Stanley E
McAuley, Van Alfon
McDonald, Jack Raymond
Yett, Fowler Redford
Zalik, Richard Albert

ALASKA
Beebee, John Christopher

ARIZONA
Ahmed, Saad Attia
Cushing, Jim Michael
Daniel, Sam Mordochai
Faris, William Guignard
Findler, Nicholas Victor
Fung, K Y
Hetrick, David LeRoy
Hoffman, William Charles
Iverson, A Evan
Kyrala, Ali
Lamb, George Lawrence, Jr
Lomen, David Orlando
Maier, Robert S
Melia, Fulvio
Perko, Lawrence Marion
Saric, William Samuel
Schowengerdt, Robert Alan
Szidarovszky, Ferenc
White, Simon David Manton
Ziolkowski, Richard Walter

ARKANSAS
Dunn, James Eldon
Keown, Ernest Ray
Perryman, John Keith
Yaz, Engin

CALIFORNIA
Ablow, Clarence Maurice
Amster, Harvey Jerome
Ancker, Clinton J(ames), Jr
Anderson, Charles Hammond
Anderson, Paul Maurice
Ardema, Mark D
Aref, Hassan
Arnold, Frank R(obert)
Aster, Robert Wesley
Auer, Lawrence H
Banerjee, Utpal
Banks, Dallas O
Bate, George Lee
Beaver, W(illiam) L(awrence)
Bendat, Julius Samuel
Berg, Paul Walter
Berger, Stanley A(llan)
Bergman, Ray E(ldon)
Bergquist, James William
Berryman, James Garland
Blum, Lenore Carol
Blum, Marvin
Bohachevsky, Ihor O
Bolt, Bruce A
Borrelli, Robert L
Brandeau, Margaret Louise
Brown, Harold David
Bruch, John C(larence), Jr
Bruley, Duane Frederick
Bucy, Richard Snowden
Bunch, James R
Burgin, George Hans
Burke, James Edward
Busenberg, Stavros Nicholas
Caflisch, Russel Edward
Cantin, Gilles
Carmona, Rene A
Casten, Richard G
Chamberlain, Robert Glenn
Chambre, Paul L

Chan, Bertram Kim Cheong
Chang, Tien-Lin
Chen, Yang-Jen
Cheng, Edward Teh-Chang
Cheng, Ralph T(a-Shun)
Cherlin, George Yale
Chou, Larry I-Hui
Chu, Kai-Ching
Chu, Pe-Cheng
Clark, Crosman Jay
Cohen, Donald Sussman
Cohen, Moses E
Concus, Paul
Cooke, Kenneth Lloyd
Corngold, Noel Robert David
Cruise, Donald Richard
Cumberbatch, Ellis
Dalrymple, Stephen Harris
Dashner, Peter Alan
Debreu, Gerard
De Figueiredo, Rui J P
DeFranco, Ronald James
De Micheli, Giovanni
Derenzo, Stephen Edward
Deriso, Richard Bruce
Dickson, Lawrence John
Di Franco, Roland B
Dost, Martin Hans-Ulrich
Eisemann, Kurt
Elderkin, Richard Howard
Eng, Genghmun
Fan, Hsin Ya
Fattorini, Hector Osvaldo
Faulkner, Frank David
Fernandez, Alberto Antonio
Filippenko, Vladimir I
Finn, Robert
Fisk, Robert Spencer
Fletcher, Thomas Harvey
Franklin, Joel Nicholas
Freund, Roland Wilhelm
Fried, Walter Rudolf
Friedman, George J(erry)
Fu, Lee Lueng
Galas, David John
Garfin, Louis
Garwood, Roland William, Jr
Gaskell, Robert Eugene
Gibson, James (Benjamin)
Gilbarg, David
Gill, Stephen Paschall
Glaser de Lugo, Frank
Glauz, Robert Doran
Goheen, Lola Coleman
Gold, Richard Robert
Golub, Gene H
Grace, O Donn
Grant, Eugene F(redrick)
Gretsky, Neil E
Gupta, Madhu Sudan
Haloulakos, Vassilios E
Hardy, Rolland L(ee)
Hausman, Arthur Herbert
Hedgepeth, John M(ills)
Heflinger, Lee Opert
Helton, F Joanne
Helton, John W
Hewer, Gary Arthur
Hirsch, Morris William
Ho, Hung-Ta
Holoien, Martin O
Holt, Maurice
Hovanessian, Shahen Alexander
Hu, Steve Seng-Chiu
Hughett, Paul William
Hunt, Robert Weldon
Ikawa, Hideo
Israel, Jay Elliot
Jacobson, Alexander Donald
Jefferson, Thomas Hutton, Jr
Johnsen, Eugene Carlyle
Jordan, James A, Jr
Juncosa, Mario Leon
Junge, Douglas
Kaellis, Joseph
Kalensher, Bernard Earl
Kaul, Maharaj Krishen
Keller, Edward Lee
Keller, Herbert Bishop
Keller, Joseph Bishop
Kevorkian, Aram K
Khurana, Krishan Kumar
Kidder, Ray Edward
Kimble, Gerald Wayne
Klotz, James Allen
Knobloch, Edgar
Knowles, James Kenyon
Koh, Robert Cy
Kraabel, John Stanford
Krener, Arthur James
Lang, Martin T
Lange, Charles Gene
Lapidus, Michel Laurent
Lathrop, Richard C(harles)
Lau, John H
Laub, Alan John
Lawson, Charles L
Lax, Melvin David
Layton, Thomas William
Levine, Harold
Lick, Wilbert James
Lindal, Gunnar F
Little, Thomas Morton
Lomax, Harvard

Lu, Kau U
Luce, R(obert) Duncan
Lupash, Lawrence O
Luxenberg, Harold Richard
Ma, Fai
McMurray, Loren Robert
Maillot, Patrick Gilles
Malmuth, Norman David
Marcus, Philip Stephen
Margulis, Stephen Barry
Maria, Narendra Lal
Mark, James Wai-Kee
Markowski, Gregory Ray
Marmarelis, Vasilis Z
Martin, Elmer Dale
Meadows, Mark Allan
Meissner, Loren Phillip
Metcalf, Frederic Thomas
Miel, George J
Miles, John Wilder
Milstein, Jaime
Mitchell, Reginald Eugene
Mitzner, Kenneth Martin
Moler, Cleve B
Mood, Alexander McFarlane
Moore, Douglas Houston
Moore, Edgar Tilden, Jr
Morgan, Donald E(arle)
Morizumi, S James
Mow, C(hao) C(how)
Muchmore, Robert B(oyer)
Nachbar, William
Nathanson, Weston Irwin
Nemat-Nasser, Siavouche
Neustadter, Siegfried Friedrich
Newell, Gordon Frank
Ng, Edward Wai-Kwok
Ng, Lawrence Chen-Yim
Oddson, John Keith
O'Dell, Austin Almond, Jr
Oliger, Joseph Emmert
Osher, Stanley Joel
Painter, Jeffrey Farrar
Parker, Donn Blanchard
Parlett, Beresford
Patel, Vithalbhai Ambalal
Phillips, Ralph Saul
Pierce, John Gregory
Plock, Richard James
Plotkin, Allen
Poggio, Andrew John
Pomraning, Gerald C
Porter, Lawrence Delpino
Portnoff, Michael Rodney
Pridmore-Brown, David Clifford
Prim, Robert Clay
Purcell, Everett Wayne
Rahmat-Samii, Yahya
Ramey, Daniel Bruce
Rao, Jammalamadaka S
Reissner, Eric
Richards, Francis Russell
Roberts, Leonard
Rocke, David M
Rockwell, Robert Lawrence
Rosen, Irwin Gary
Rosenblatt, Daniel Bernard
Rosenblatt, Murray
Runge, Richard John
Saffman, Philip Geoffrey
Salinas, David
Sangren, Ward Conrad
Schechter, Martin
Scheibe, Paul Otto
Schlafly, Roger
Schoenstadt, Arthur Loring
Scholl, James Francis
Serat, William Felkner
Shen, Chung Yi
Shi, Yun Yuan
Shiffman, Max
Shnider, Ruth Wolkow
Simons, Roger Mayfield
Sirignano, William Alfonso
Smith, Frederick T(ucker)
Spanier, Jerome
Stauffer, Howard Boyer
Synolakis, Costas Emmanuel
Szego, Peter A
Tadjeran, Hamid
Talley, Wilson K(inter)
Taub, Abraham Haskel
Taussky, Olga
Thorne, Charles Joseph
Tracy, Craig Arnold
Trenholme, John Burgess
Troesch, Beat Andreas
Tu, Yih-O
Vajk, J(oseph) Peter
Van Bibber, Karl Albert
Vaughan, J Rodney M
Wachowski, Hillard M(arion)
Walker, Kelsey, Jr
Wang, Paul Keng Chieh
Weber, Charles L
Weeks, Dennis Alan
Wells, Willard H
Whitham, Gerald Beresford
Wilcox, Jaroslava Zitkova
Willman, Warren Walton
Winter, Donald F
Wollmer, Richard Dietrich
Wood, Roger Charles
Woolard, Henry W(aldo)

Applied Mathematics (cont)

Wulfman, Carl E
Yates, Scott Raymond
Yee, Kane Shee-Gong
Yeh, Yea-Chuan Milton
Yeung, Ronald Wai-Chun
Ziegenhagen, Allyn James

COLORADO
Abshier, Curtis Brent
Bebernes, Jerrold William
Benton, Edward Rowell
Bleistein, Norman
Brady, Brian T
Colvis, John Paris
Cook, Robert Neal
Crawford, Myron Lloyd
Datta, Subhendu Kumar
Elmore, Kimberly Laurence
Esposito, Larry Wayne
Geers, Thomas L
Gomberg, Joan Susan
Greenberg, Herbert Julius
Hansen, Richard Olaf
Hector, David Lawrence
Heine, George Winfield, III
Hermes, Henry
Hittle, Douglas Carl
Hoots, Felix R
Hundhausen, Joan Rohrer
Irwin, Robert Cook
Jones, Richard Hunn
Kerr, Robert McDougall
Lett, Gregory Scott
Low, Boon-Chye
Lundgren, J Richard
Lynch, Robert Michael
McCormick, Stephen Fahrney
McKnight, Randy Sherwood
Morgenthaler, George William
Mueller, Raymond Karl
Nahman, Norris S(tanley)
Nesenbergs, Martin
Noble, Richard Daniel
Oneil, Stephen Vincent
Pak, Ronald Y S
Phillipson, Paul Edgar
Poore, Aubrey Bonner
Pruess, Steven Arthur
Pruess, Steven Arthur
Purdom, James Francis Whitehurst
Ralston, Margarete A
Sani, Robert L(e Roy)
Savage, William Zuger
Seebass, Alfred Richard, III
Shedlock, Kaye M
Slutz, Ralph Jeffery
Swanson, Lawrence Ray
Thaler, Eric Ronald
Underwood, Robert Gordon
Urban, Thomas Charles
Walden, Jack M
Walter, Martin Edward
Wessel, William Roy
West, Anita
Windholz, Walter M

CONNECTICUT
Agresta, Joseph
Ahlberg, John Harold
Barnett, Mark
Burridge, Robert
Chan, Wing Cheng Raymond
Crofts, Geoffrey
Goldstein, Rubin
Gutierrez-Miravete, Ernesto
Hajela, Dan
Harding, R(onald) H(ugh)
Harrison, Irene R
Keyes, David Elliot
Leutert, Werner Walter
Levien, Roger Eli
Liu, Qing-Huo
McCartin, Brian James
Mandelbrot, Benoit B
Masaitis, Ceslovas
Nehorai, Arye
Ojalvo, Irving U
Orphanoudakis, Stelios Constantine
Phillips, Peter Charles B
Ramakrishnan, Terizhandur S
Ruckebusch, Guy Bernard
Sachdeva, Baldev Krishan
Seitelman, Leon Harold
Sharlow, John Francis
Smooke, Mitchell D
Streit, Roy Leon
Stuck, Barton W
Verdon, Joseph Michael
Wachman, Murray

DELAWARE
Angell, Thomas Strong
Bydal, Bruce A
Caviness, Bobby Forrester
Cook-Ioannidis, Leslie Pamela
Hsiao, George Chia-Chu
Jones, Louise Hinrichsen
Kerr, Arnold D
Kleinman, Ralph Ellis
MacDonald, James
Mehl, James Bernard
Mills, Patrick Leo, Sr

Ogunnaike, Babatunde Ayodeji
Owens, Aaron James
St John, Daniel Shelton
Schwartz, Leonard William
Stakgold, Ivar
Taylor, Howard Milton, III
Trabant, Edward Arthur
Ulery, Dana Lynn
Weinacht, Richard Jay

DISTRICT OF COLUMBIA
Baer, Ralph Norman
Bainum, Peter Montgomery
Borgiotti, Giorgio V
Cohen, Michael Paul
English, William Joseph
Filliben, James John
Grossman, John Mark
Hall, William Spencer
Hammersmith, John L(eo)
Herron, Isom H
Holland, Charles Jordan
Kalnay, Eugenia
Kupperman, Robert Harris
Lee, David Allan
Lehman, Richard Lawrence
Marlow, William Henry
Menton, Robert Thomas
Nachman, Arje
Nelson, David Brian
Nelson, Larry Dean
Nguyen, Charles Cuong
Rockett, John A
Rojas, Richard Raimond
Sáenz, Albert William
Schweizer, Francois
Stokes, Arnold Paul
Valenzuela, Gaspar Rodolfo
Voytuk, James A
Weiss, Leonard
Yen, Nai-Chyuan

FLORIDA
Asner, Bernarad A, Jr
Beil, Robert J(unior)
Bezdek, James Christian
Bober, William
Chang, Lena
Chow, Wen Lung
Christiano, John G
Clark, Mary Elizabeth
Clarke, Allan James
Clarke, Thomas Lowe
Cloutier, James Robert
Daniel, Donald Clifton
Debnath, Lokenath
Fausett, Donald Wright
Fausett, Laurene van Camp
Fossum, Jerry G
Freeman, Neil Julian
Garrett, James Richard
Goldwyn, Roger M(artin)
Hager, William Ward
Hammer, Jacob
Howard, Bernard Eufinger
Howard, Louis Norberg
Hsieh, Chung Kuo
Hunter, Christopher
Ismail, Mourad E H
Jacobs, Alan M(artin)
Jacobs, Elliott Warren
Keesling, James Edgar
Kurzweg, Ulrich H(ermann)
Lacher, Robert Christopher
Lade, Robert Walter
Lee, Sung J
Leitner, Alfred
Lindholm, Fredrik Arthur
Loper, David Eric
McCormick, Clyde Truman
Martin, Charles John
Meacham, Robert Colegrove
Merrill, John Ellsworth
Millsaps, Knox
Nagle, Richard Kent
Piquette, Jean Conrad
Rautenstrauch, Carl Peter
Sandor, George N(ason)
Sigmon, Kermit Neal
Tam, Christopher K W
Todd, Hollis N
Tsokos, Chris Peter
Walsh, Edward Kyran
Williams, Carol Ann
Williams, Gareth
Young, John William

GEORGIA
Adomian, George
Atluri, Satya N
Boal, Jan List
Bodnar, Donald George
Clark, James Samuel
Ginsberg, Jerry Hal
Ho, Dar-Veig
Huberty, Carl J
Joyner, Ronald Wayne
Landgren, John Jeffrey
Longini, Ira Mann, Jr
Massey, Fredrick Alan
Meyer, Gunter Hubert
Mickens, Ronald Elbert
Neitzel, George Paul
Rice, Peter Milton

Sloan, Alan David
Smith, William Allen
Stalford, Harold Lenn
Wang, Johnson Jenn-Hwa
Warner, Isiah Manuel
Wilkins, J Ernest, Jr
Wiltse, James Cornelius

HAWAII
Antal, Michael Jerry, Jr
Coburn, Richard Karl
Frazer, L Neil
Loomis, Harold George
Sagle, Arthur A
Tuan, San Fu
Wang, Bin
Watanabe, Daniel Seishi

IDAHO
Bobisud, Larry Eugene
Girse, Robert Donald
Hausrath, Alan Richard
Hoagland, Gordon Wood
Johnson, John Alan
Knobel, LeRoy Lyle
Maloof, Giles Wilson
Moore, Kenneth Virgil
Mortensen, Glen Albert
Shaw, Charles Bergman, Jr

ILLINOIS
Albrecht, Felix Robert
Al-Khafaji, Amir Wadi Nasif
Ashenhurst, Robert Lovett
Barcilon, Victor
Bareiss, Erwin Hans
Barron, Emmanuel Nicholas
Barston, Eugene Myron
Bart, George Raymond
Benzinger, Harold Edward, Jr
Berger, Neil Everett
Berger, Steven Barry
Bernstein, Barry
Birnholz, Jason Cordell
Bookstein, Abraham
Buckmaster, John David
Carroll, Robert Wayne
Chen, Paul Ear
Chew, Weng Cho
Cook, Joseph Marion
Cowan, Jack David
Dupont, Todd F
Dupont, Todd F
Edelstein, Warren Stanley
Evans, William Paul
Fink, Charles Lloyd
Friedlander, Susan Jean
Golomski, William Arthur
Gruber, Eugene E, Jr
Hakala, Reino William
Hanson, Floyd Bliss
Hearn, Dwight D
Hsui, Albert Tong-Kwan
Hwang, Yu-Tang
Jerome, Joseph Walter
Jerrard, Richard Patterson
Kammler, David W
Kaper, Hans G
Kath, William Lawrence
Kinnmark, Ingemar Per Erland
Kramer, John Michael
Lebovitz, Norman Ronald
Liu, Wing Kam
McLinden, Lynn
Mann, Christian John
Mann, James Edward, Jr
Matalon, Moshe
Matkowsky, Bernard J
Maxwell, William Hall Christie
Miller, Arnold Reed
Moran, Thomas J
Muller, David Eugene
Mura, Toshio
Nagylaki, Thomas Andrew
Olmstead, William Edward
Ottino, Julio Mario
Palmore, Julian Ivanhoe, III
Papoutsakis, Eleftherios Terry
Pericak-Spector, Kathleen Anne
Prussing, John E(dward)
Ramanathan, Ganapathiagraharam V
Reynolds, Larry Owen
Riahi, Daniel Nourollah
Ritt, Robert King
Saari, Donald Gene
Saied, Faisal
Scott, Meckinley
Secrest, Donald H
Shaftman, David Harry
Skeel, Robert David
Slotnick, Daniel Leonid
Snyder, Herbert Howard
Tier, Charles
Tipei, Nicolae
Toy, William W
Travelli, Armando
Udler, Dmitry
Wacker, William Dennis
Wilson, Howell Kenneth
Yang, Shi-Tien
Zettl, Anton

INDIANA
Berkovitz, Leonard David

Carroll, John T, III
Chen, Wai-Fah
Citron, Stephen J
Criss, Darrell E
Ersoy, Okan Kadri
Fink, James Paul
Fuelling, Clinton Paul
Gersting, John Marshall, Jr
Golomb, Michael
Haas, Violet Bushwick
Haaser, Norman Bray
Hansen, Arthur G(ene)
Houstis, Elias N
Kentzer, Czeslaw P(awel)
Lee, Norman K
Lord, Gary Evans
Luke, Jon Christian
Lynch, Robert Emmett
Madan, Ved P
Mullikin, Thomas Wilson
Murphy, Catherine Mary
Ng, Bartholomew Sung-Hong
Reid, William Hill
Rice, John Rischard
Sweet, Arnold Lawrence
Synowiec, John A
Szpankowski, Wojciech
Thompson, Maynard
Thuente, David Joseph
Thurber, James Kent
Wernimont, Grant (Theodore)
Weston, Vaughan Hatherley

IOWA
Alexander, Roger Keith
Atkinson, Kendall E
Buck, James R
Carlson, Bille Chandler
Chen, Ching-Jen
Choi, Kyung Kook
Chwang, Allen Tse-Yung
Dahiya, Rajbir Singh
Haug, Edward J, Jr
Hethcote, Herbert Wayne
Horton, Robert, Jr
Jeyapalan, Kandiah
Jorgensen, Palle E T
Kawai, Masataka
Knowler, Lloyd A
Lambert, Robert J
Landweber, Louis
Luban, Marshall
Luecke, Glenn Richard
Miller, Richard Keith
Nariboli, Gundo A
Reckase, Mark Daniel
Seifert, George
Sheeler, John B(riggs)
Weiss, Harry Joseph

KANSAS
Acker, Andrew French, III
Davis, Elwyn H
Elcrat, Alan Ross
Kirmser, P(hilip) G(eorge)
Miller, Forest R
Phillips, John Richard
Ramm, Alexander G
Shultis, J Kenneth
Van Vleck, Fred Scott

KENTUCKY
Beatty, Millard Fillmore, Jr
Bowen, Ray M
Brock, Louis Milton
Davis, Chester L
Evans, Gerald William
Fairweather, Graeme
Gruver, William A
Howard, Henry Cobourn
Man, Chi-Sing
Ray, Asit Kumar
Russell, Marvin W
Schnare, Paul Stewart

LOUISIANA
Andrus, Jan Frederick
Bradley, Marshall Rice
Chan, Chiu Yeung
Chen, Isaac I H
Draayer, Jerry Paul
Foote, Joe Reeder
Gajendar, Nandigam
Heatherly, Henry Edward
Ho, Yew Kam
Kythe, Prem Kishore
Lefton, Lew Edward
Maxfield, John Edward
Puri, Pratap
Roquemore, Leroy
Temple, Austin Limiel, (Jr)
Tipler, Frank Jennings, III
Wenzel, Alan Richard
Yannitell, Daniel W

MAINE
Bowie, Oscar L
Farlow, Stanley Jerome
Lick, Dale W
McMillan, Brockway

MARYLAND
Abed, Eyad Husni
Alexander, James Crew

Andrew, Merle M
Antman, Stuart S
Atanasoff, John Vincent
Babcock, Anita Kathleen
Babuska, Ivo Milan
Baer, Ferdinand
Baum, Howard Richard
Benokraitis, Vitalius
Berger, Alan Eric
Berger, Bruce S
Bernard, Peter Simon
Bishop, Walton B
Blue, James Lawrence
Boggs, Paul Thomas
Bromberg, Eleazer
Burns, Timothy John
Cantor, David S
Cawley, Robert
Celmins, Aivars Karlis Richards
Chadwick, Richard Simeon
Chi, Donald Nan-Hua
Chi, L K
Ciment, Melvyn
Cohen, Edgar A, Jr
Colvin, Burton Houston
Covell, David Gene
Criss, John W
Criss, Thomas Benjamin
Davis, Frederic I
Eades, James B(everly), Jr
Ehrlich, Louis William
Ely, Raymond Lloyd
Everstine, Gordon Carl
Fletcher, John Edward
Fong, Jeffery Tse-Wei
Gates, Sylvester J, Jr
Gillis, James Thompson
Goldstein, Charles M
Grebogi, Celso
Hammond, Allen Lee
Herrmann, Robert Arthur
Hitchcock, Daniel Augustus
Hollenbeck, Robert Gary
Horak, Martin George
Horn, Roger Alan
Hubbell, John Howard
Hummel, James Alexander
Jackson, Richard H F
Johnson, Frederick Carroll
Kahaner, David Kenneth
Kahn, Arthur B
Kattakuzhy, George Chacko
Kellogg, Royal Bruce
Kitchens, Clarence Wesley, Jr
Kuttler, James Robert
Lozier, Daniel William
Lynn, Yen-Mow
McFadden, Geoffrey Bey
Macon, Nathaniel
Mahar, Thomas J
Marimont, Rosalind Brownstone
Martin, Monroe Harnish
Maslen, Stephen Harold
Massell, Paul Barry
Miller, Raymond Edward
Morgan, Richard C
Nau, Dana S
Noblesse, Francis
Odle, John William
Olver, Frank William John
Osborn, John Edward
Oser, Hans Joerg
Pai, S(hih) I
Rand, Robert Collom
Raphael, Louise Arakelian
Reader, Wayne Truman
Rehm, Ronald George
Rinzel, John Matthew
Roberts, Richard Calvin
Rosenstock, Herbert Bernhard
Rostamian, Rouben
Sadowsky, John
Schmid, Lawrence Alfred
Sedney, R(aymond)
Seidman, Thomas Israel
Serfling, Robert Joseph
Shaw, Harry, Jr
Simmons, John Arthur
Skiff, Frederick Norman
Smith, James Alan
Solomon, Jay Murrie
Stadter, James Thomas
Sullivan, Francis E
Tissue, Eric Bruce
Tropf, Cheryl Griffiths
Turner, Robert Davison
Watt, Mamadov Hame
Weiss, George Herbert
Weiss, Michael David
Wells, William T
Wu, Ching-Sheng
Yang, Xiaowei
Yorke, James Alan
Young, Hobart Peyton
Zabronsky, Herman
Zien, Tse-Fou

MASSACHUSETTS
Abbott, Douglas E(ugene)
Anderson, Donald Gordon Marcus
Archer, Robert Raymond
Ash, Michael Edward
Berman, Robert Hiram
Bernstein, Joseph N

Blake, Thomas R
Bower, Amy Sue
Brockett, Roger Ware
Burke, Shawn Edmund
Burns, Daniel Robert
Butler, James Preston
Calabi, Lorenzo
Cappallo, Roger James
Carpenter, Richard M
Carrier, George Francis
Chase, David Marion
Cipolla, John William, Jr
Clough, Shepard Anthony
Crowell, Julian
Cuffin, B(enjamin) Neil
Davis, Paul William
Delvaille, John Paul
Drane, Charles Joseph, Jr
Dym, Clive L
Esch, Robin E
Fiering, Myron B
Finney, Ross Lee, III
Fishman, Philip M
Fougere, Paul Francis
Freund, Robert M
Granoff, Barry
Greenspan, Harvey Philip
Hadlock, Charles Robert
Herda, Hans-Heinrich Wolfgang
Ho, Yu-Chi
Holst, Jewel Magee
Holst, William Frederick
Holt, Frederick Sheppard
Holway, Lowell Hoyt, Jr
Hovorka, John
Kanter, Irving
Kitchin, John Francis
Klein, William
Kon, Mark A
Lechner, Robert Joseph
Lemnios, A(ndrew) Z
Lin, Chia Chiao
Lindzen, Richard Siegmund
Lo, Andrew W
McElroy, Michael Brendan
Martland, Carl Douglas
Maskin, Eric S
Meldon, Jerry Harris
Miller, William Brunner
Muller, Karl Frederick
Nehrkorn, Thomas
Nesbeda, Paul
Noiseux, Claude Francois
O'Connell, Richard John
Oettinger, Anthony Gervin
Ogilvie, T(homas) Francis
O'Neil, Elizabeth Jean
Ostrovsky, Rafail M
Palubinskas, Felix Stanley
Pang, Yuan
Papernik, Lazar
Quinto, Eric Todd
Raab, Allen Robert
Richter, Stephen L(awrence)
Ross, Edward William, Jr
Rossettos, John N(icholas)
Roth, Robert Steele
St Mary, Donald Frank
Salstein, David A
Schetzen, Martin
Seifer, Arnold David
Senechal, Marjorie Lee
Shannon, Claude Elwood
Shapiro, Ralph
Sherman, Thomas Oakley
Shu, Larry Steven
Spiegel, Stanley Lawrence
Stiffler, Jack Justin
Stone, Peter Hunter
Strang, Gilbert
Sun, Fang-Kuo
Taff, Laurence Gordon
Tait, Kevin S
Teng, Chojan
Thompson, Charles
Thorpe, Rodney Warren
Toomre, Alar
Toscano, William Michael
Trefethen, Lloyd Nicholas
Warga, Jack
Weinberg, I Jack
Witsenhausen, Hans S
Wong, Po Kee
Woods, William A
Wunsch, Abraham David
Yaghjian, Arthur David
Yamartino, Robert J

MICHIGAN
Abriola, Linda Marie
Akay, Adnan
Alavi, Yousef
Anderson, Howard Benjamin
Birdsall, Theodore G
Bolander, Richard
Boyd, John Philip
Brown, James Ward
Chang, Jhy-Jiun
Cho, Byong Kwon
Chock, David Poileng
Chow, Pao Liu
Crittenden, John Charles
Curl, Rane L(ocke)
Dawson, D(onald) E(merson)

Denman, Harry Harroun
Dolph, Charles Laurie
Evans, David Hunden
Field, David Anthony
Flanders, Harley
Fleming, Richard J
Grosky, William Irvin
Habib, Izzeddin Salim
Hee, Christopher Edward
Hicks, Durrell Lee
Hill, Bruce M
Holland, John Henry
Hrovat, Davorin
Huntoon, Jacqueline E
Johnson, Harold Hunt
Kincaid, Wilfred Macdonald
Kochen, Manfred
Kubis, Joseph J(ohn)
LaHaie, Ivan Joseph
Larsen, Edward William
Lubman, David Mitchell
Lund, Charles Edward
Miklavcic, Milan
Mordukhovich, Boris S
Murphy, Brian Boru
Nesbitt, Cecil James
Pearcy, Carl Mark, Jr
Petersen, Donald Ralph
Polis, Michael Philip
Powell, Kenneth Grant
Ramamurthy, Amurthur C
Root, William L(ucas)
Salam, Fathi M A
Schmidt, Robert
Schneider, Eric West
Scott, Richard Anthony
Simon, Carl Paul
Soutas-Little, Robert William
Spahn, Robert Joseph
Thompson, James Lowry
Ullman, Nelly Szabo
Wang, Chang-Yi
Wasserman, Arthur Gabriel
Wasserman, Robert H
Weil, Herschel
Wineman, Alan Stuart
Yang, Wei-Hsuin
Yen, David Hsien-Yao

MINNESOTA
Anderson, Willard Eugene
Aris, Rutherford
Austin, Donald Murray
Cassola, Robert Louis
Castore, Glen M
Cornelissen Guillaume, Germaine G
Drew, Bruce Arthur
Follingstad, Henry George
Gaal, Steven Alexander
Harvey, Charles Arthur
Infante, Ettore F
Lodge, Timothy Patrick
Lundberg, Gustave Harold
Luskin, Mitchell B
Miller, Willard, Jr
Nitsche, Johannes Carl Christian
Olver, Peter John
Poling, Craig
Robb, Richard A
Rosen, Judah Ben
Schuldt, Spencer Burt
Scriven, L E(dward), (II)
Warner, William Hamer
Weinberger, Hans Felix
Wolfe, Barbara Blair
Zitney, Stephen Edward

MISSISSIPPI
Cade, Ruth Ann
Davis, Thomas Mooney
Fay, Temple Harold
George, Clifford Eugene
Green, Albert Wise
Johnston, Russell Shayne
Kishk, Ahmed A
Koshel, Richard Donald
Rohde, Florence Virginia
Solomon, Jimmy Lloyd
Warsi, Zahir U A

MISSOURI
Agarwal, Ramesh K
Ahlbrandt, Calvin Dale
Beem, John Kelly
Bruening, James Theodore
Cowherd, Chatten, Jr
Edwards, Carol Abe
Elliott, David LeRoy
Henson, Bob Londes
Ho, Chung You (Peter)
Huddleston, Philip Lee
Katz, Israel Norman
Krone, Lester H(erman), Jr
MacDonald, Carolyn Trott
Ornstein, Wilhelm
Pettey, Dix Hayes
Petty, Clinton Myers
Plummer, Otho Raymond
Polowy, Henry
Retzloff, David George
Rodin, Ervin Y
Sabharwal, Chaman Lal
Shrauner, Barbara Abraham
Spielman, Barry

Tarn, Tzyh-Jong

MONTANA
Barrett, Louis Carl
Derrick, William Richard
Grossman, Stanley I
Hess, Lindsay LaRoy

NEBRASKA
Conway, James Joseph
Haeder, Paul Albert
Maloney, John P
Schniederjans, Marc James
Scudder, Jeffrey Eric
Surkan, Alvin John

NEVADA
Lindstrom, Fredrick Thomas
Rawat, Banmali Singh
Tompson, Robert Norman
Turner, Robert Harold
Wells, William Raymond

NEW HAMPSHIRE
Albert, Arthur Edward
Black, Sydney D
Buffler, Charles Rogers
Darlington, Sidney
Kuo, Shan Sun
Meeker, Loren David
Morduchow, Morris
Radlow, James

NEW JERSEY
Abraham, John
Ahluwalia, Daljit Singh
Anderson, William Niles, Jr
Andrushkiw, Roman Ihor
Ball, William Henry Warren
Ben-Israel, Adi
Bergeron, Robert F(rancis) (Terry), Jr
Blackmore, Denis Louis
Brandmaier, Harold Edmund
Caffarelli, Luis Angel
Castor, William Stuart, Jr
Chai, Winchung A
Cheo, Li-hsiang Aria S
Chien, Victor
Cho, Soung Moo
Cohen, Edwin
Coleman, Norman P, Jr
Coughran, William Marvin, Jr
Crane, Roger L
Crochiere, Ronald E
Elk, Seymour B
Flannery, Brian Paul
Fornberg, Bengt
Foschini, Gerard Joseph
Frauenthal, James Clay
Gaer, Marvin Charles
Goldberg, Vladislav V
Gossmann, Hans Joachim
Gross, Arthur Gerald
Holford, Richard L
Holland, Paul William
Houston, John Lynn
Hughes, James Sinclair
Kam, Gar Lai
Kaufman, Linda
Kim, Uing W
Kirch, Murray R
Koss, Valery Alexander
Kriegsmann, Gregory A
Labianca, Frank Michael
Levine, Lawrence Elliott
Lieb, Elliott Hershel
Lindberg, Craig Robert
Lioy, Paul James
Lucantoni, David Michael
Mamelak, Joseph Simon
Mangulis, Visvaldis
Mazo, James Emery
Morgan, Samuel Pope
Morrison, John Allan
Neal, Scotty Ray
Nguyen, Hien Vu
Ogden, Joan Mary
Orszag, Steven Alan
Rebhuhn, Deborah
Roberts, Fred Stephen
Rosenblum, Martin Jacob
Saltzberg, Burton R
Schryer, Norman Loren
Shober, Robert Anthony
Sinclair, J Cameron
Sisto, Fernando
Slepian, David
Snygg, John Morrow
Sontag, Eduardo Daniel
Stehney, Ann Kathryn
Steiger, William Lee
Stepleman, Robert Saul
Stevens, John G
Stickler, David Collier
Strauss, Walter
Suhir, Ephraim
Sundaresan, Sankaran
Sverdlove, Ronald
Thomas, Larry Emerson
Vann, Joseph M
Weiss, Alan
White, Benjamin Steven
Whitman, Gerald Martin
Williams, Gareth Pierce

Gorman, Arthur Daniel
Grassi, Vincent G
Grover, Carole Lee
Gurtin, Morton Edward
Hageman, Louis Alfred
Hall, Charles Allan
Hickman, Warren David
Hoburg, James Frederick
Howland, Frank L
Joshi, Thomas J
Kazakia, Jacob Yakovos
Kolodner, Ignace Izaak
Lakhtakia, Akhlesh
Lepore, John A(nthony)
Lokay, Joseph Donald
Lynn, Roger Yen Shen
McLain, David Kenneth
Mallett, Russell Lloyd
Merrill, Samuel, III
Michalik, Edmund Richard
Molyneux, John Ecob
Moore, Walter Calvin
Murgie, Samuel A
Nash, David Henry George
Neuman, Charles P(aul)
Nolan, Edward J
Owen, David R
Pierce, Allan Dale
Pool, James C T
Porsching, Thomas August
Rajagopal, K R
Ramos, Juan Ignacio
Rapp, Paul Ernest
Rehfield, David Michael
Rheinboldt, Werner Carl
Rhodes, Donald Frederick
Rivlin, Ronald Samuel
Rorres, Chris
Sawyers, Kenneth Norman
Sibul, Leon Henry
Simaan, Marwan
Sinclair, Glenn Bruce
Smith, Gerald Francis
Smith, Warren LaVerne
Spencer, Arthur Coe, II
Starling, James Lyne
Szyld, Daniel Benjamin
Thompson, Gerald Luther
Thompson, William, Jr
Tozier, John E
Troy, William Christopher
Trumpler, Paul R(obert)
Tuba, I Stephen
Vrentas, Christine Mary
Wagner, Clifford Henry
Wayne, Clarence Eugene
Weiner, Robert Allen
Woo, Tse-Chien
Woodburn, Wilton A
Wright, Thomas Wilson
Wu, John Naichi
Yang, Arthur Jing-Min
Yukich, Joseph E

RHODE ISLAND
Asaro, Robert John
Bisshopp, Frederic Edward
Dafermos, Constantine M
Driver, Rodney D
Falb, Peter L
Freiberger, Walter Frederick
Grenander, Ulf
Kirwan, Donald Frazier
Krikorian, John Sarkis, Jr
Ladas, Gerasimos
Liu, Pan-Tai
McClure, Donald Ernest
Pipkin, Allen Compere
Roxin, Emilio O
Strauss, Charles Michael
Verma, Ghasi Ram
Weiner, Jerome Harris

SOUTH CAROLINA
Felling, William E(dward)
Fennell, Robert E
Honeck, Henry Charles
Hunt, Hurshell Harvey
Jackson, John Elwin, Jr
Klemm, James L
Kostreva, Michael Martin
Proctor, Thomas Gilmer
Sheppard, Emory Lamar
Uldrick, John Paul
Voit, Eberhard Otto

SOUTH DAKOTA
Gaalswyk, Arie
Raab, Wallace Albert

TENNESSEE
Baker, Allen Jerome
Beauchamp, John J
Bigelow, John E(aly)
Bownds, John Marvin
Cline, Randall Eugene
Crooke, Philip Schuyler
Cross, James Thomas
Davies, Kenneth Thomas Reed
Demetriou, Charles Arthur
Dobosy, Ronald Joseph
Dongarra, Jack Joseph
Emanuel, William Robert
Fischer, Charlotte Froese

Hallam, Thomas Guy
Harris, Wesley Leroy
Haynes, Tony Eugene
Henry, Jonathan Flake
Holdeman, Jonas Tillman, Jr
Hopkins, Richard Allen
Hutcherson, Joseph William
Hutcheson, Paul Henry
Jordan, George Samuel
Kreft, Edmund Michael
Nowlin, Charles Henry
Pugh, Claud Ervin
Rajan, Perisamy Karivaratha
Reddy, Kapuluru Chandrasekhara
Rozema, Edward Ralph
Schaefer, Philip William
Schumaker, Larry L
Sizemore, Douglas Reece
Sorrells, Frank Douglas
Stone, Robert Sidney
Strauss, Alvin Manosh
Turitto, Vincent Thomas
Wachspress, Eugene Leon
Whittle, Charles Edward, Jr

TEXAS
Anderson, Charles E, Jr
Anderson, David H
Anderson, Ronald M
Aronofsky, Julius S
Auchmuty, Giles
Auchmuty, James Francis Giles
Bakelman, Ilya J(acob)
Ball, Kenneth Steven
Barton, James Brockman
Bennett, George Kemble
Bhattacharyya, Shankar P
Bigham, Robert Eric
Blackwell, Charles C, Jr
Boyer, Lester Leroy
Breig, Edward Louis
Brockett, Patrick Lee
Brown, Dennison Robert
Brown, Glenn Lamar
Brown, Robert Wade
Bryant, Michael David
Caflisch, Robert Galen
Carey, Graham Francis
Carroll, Michael M
Chang, Ping
Charbeneau, Randall Jay
Chen, Goong
Chen, Peter
Cheney, Elliott Ward, (Jr)
Clunie, Thomas John
Coats, Keith Hal
Corduneanu, Constantin C
Cowan, Russell (Walter)
Deeming, Terence James
Deeter, Charles Raymond
Dodson, David Scott
Doser, Diane Irene
Eargle, George Marvin
Edgerley, Dennis A
Ehlig-Economides, Christine Anna
Eubank, Randall Lester
Ewell, James John, Jr
Finch, Leiko Hatta
Fischer, Ferdinand Joseph
Fix, George Joseph
Frenzen, Christopher Lee
Gardner, Clifford S
Giese, Robert Paul
Gilmour - Stallsworth, Lisa K
Gladwell, Ian
Goldman, Joseph L
Golubitsky, Martin Aaron
Haberman, Richard
Hasling, Jill Freeman
Hazen, Gary Alan
Heelis, Roderick Antony
Hurlburt, H(arvey) Zeh
Johnson, Darell James
Johnson, Johnny Albert
Junkins, John Lee
Justice, James Horace
Keister, Jamieson Charles
Kesler, Oren Byrl
Knudsen, John R
Koplyay, Janos Bernath
Leventhal, Stephen Henry
Li, Wen-Hsiung
Long, Stuart A
Lurix, Paul Leslie, Jr
MacFarlane, Duncan Leo
McIntyre, Robert Gerald
Mauldin, Richard Daniel
Midgley, James Eardley
Miikkulainen, Risto Pekka
Miller, James Gilbert
Mitchell, A Richard
Moorhead, William Dean
Morris, William Lewis
Musa, Samuel A
Nabor, George W(illiam)
Nachlinger, R Ray
Narcowich, Francis Joseph
Nassersharif, Bahram
Naugle, Norman Wakefield
Nickel, James Alvin
Ochiai, Shinya
Olson, Danford Harold
Peterson, Lynn Louise Meister
Prats, Michael

Price, Harvey Simon
Rajagopalan, Raj
Renka, Robert Joseph
Roberts, Thomas M
Rosenbaum, Joseph Hans
Sadler, Stanley Gene
Sandberg, I(rwin) W(alter)
Sanders, Bobby Lee
Schruben, Johanna Stenzel
Schulz, Richard Burkart
Shieh, Leang-San
Smith, Philip Wesley
Srnka, Leonard James
Stone, Herbert L(osson)
Stubbs, Norris
Styblinski, Maciej A
Tao, Frank F
Trench, William Frederick
Tsahalis, Demosthenes Theodoros
Turner, Danny William
Von Rosenberg, Dale Ursini
Waid, Margaret Cowsar
Walling, Derald Dee
Walton, Jay R
Ward, Ronald Wayne
Weiser, Alan
White, Ronald Joseph
Wiggins, James Wendell
Williams, Glen Nordyke
Wiltschko, David Vilander
Witterholt, Edward John
Yakin, Mustafa Zafer

UTAH
Alfeld, Peter
Bates, Peter William
Coles, William Jeffrey
Davey, Gerald Leland
Fletcher, Harvey Junior
Kelsey, Stephen Jorgensen
Larsen, Kenneth Martin
Othmer, Hans George
Paegle, Julia Nogues
Sandquist, Gary Marlin
Skarda, R Vencil, Jr
Snow, Donald Ray
Stenger, Frank
Walker, Homer Franklin
Wilcox, Calvin Hayden
Windham, Michael Parks

VERMONT
Olinick, Michael
Oughstun, Kurt Edmund
Pinder, George Francis

VIRGINIA
Adam, John Anthony
Andersen, Carl Marius
Arnberg, Robert Lewis
Bailey, Marion Crawford
Bakhshi, Vidya Sagar
Bartelt, Martin William
Buchal, Robert Norman
Burns, John Allen
Collier, Manning Gary
Coram, Donald Sidney
Davidson, John Richard
Davis, William Arthur
Debney, George Charles, Jr
Dozier, Lewis Bryant
Elkins, Judith Molinar
Evans, David Arthur
Fthenakis, Emanuel John
Gale, Harold Walter
Gerlach, A(lbert) A(ugust)
Greenblatt, Seth Alan
Grotte, Jeffrey Harlow
Gunter, Edgar Jackson, Jr
Gunzburger, Max Donald
Hamilton, Thomas Charles
Hannsgen, Kenneth Bruce
Henry, Myron S
Herdman, Terry Lee
Hoffman, Karla Leigh
Horgan, Cornelius Oliver
Hunt, Leon Gibson
Hwang, John Dzen
Hwang, William Gaong
Johnson, Charles Royal
Joshi, Suresh Meghashyam
Juang, Jer-Nan
Kapania, Rakesh Kumar
Kay, Irvin (William)
Kroll, John Ernest
Kydes, Andy Steve
LaBudde, Robert Arthur
Lanzano, Paolo
Lasiecka, Irena
LeVan, Martin Douglas, Jr
Low, Emmet Francis, Jr
McNeil, Phillip Eugene
Mandelberg, Martin
Mansfield, Lois E
Mathis, Robert Fletcher
Monacella, Vincent Joseph
Morse, Burt Jules
Myers, Michael Kenneth
Nayfeh, Ali Hasan
Perry, Charles Rufus, Jr
Pettus, William Gower
Poon, Ting-Chung
Pruett, Charles David
Raychowdhury, Pratip Nath

Rehm, Allan Stanley
Renardy, Michael
Renardy, Yuriko
Roberts, William Woodruff, Jr
Saltz, Joel Haskin
Schmeelk, John Frank
Schwiderski, Ernst Walter
Sebastian, Richard Lee
Shoosmith, John Norman
Shrier, Stefan
Simmonds, James G
Smith, Sidney Taylor
Sobieszczanski-Sobieski, Jaroslaw
Stone, Lawrence David
Tompkins, Stephen Stern
Triggiani, Roberto
Tweed, John
Tyson, John Jeanes
Wohl, Philip R

WASHINGTON
Berger, Albert Jeffrey
Burkhart, Richard Henry
Cochran, James Alan
Cokelet, Edward Davis
Criminale, William Oliver, Jr
Falco, James William
Gibbs, Alan Gregory
Goldstein, Allen A
Harrison, Don Edmunds
Kevorkian, Jirair
Lessor, Delbert Leroy
Levinson, Mark
Liemohn, Harold Benjamin
McSpadden, William Robert
Marshall, Donald E
Moody, Michael Eugene
Nievergelt, Yves
Nijenhuis, Albert
O'Malley, Robert Edmund, Jr
Reinhardt, William Parker
Richardson, Richard Laurel
Rimbey, Peter Raymond
Robel, Gregory Frank
Roetman, Ernest Levane
Storwick, Robert Martin
Street, Robert Elliott
Trent, Donald Stephen
Tripp, Leonard L
Tung, Ka-Kit
Wan, Frederic Yui-Ming

WEST VIRGINIA
Brady, Dennis Patrick
Bryant, Robert William
Cavalier, John F
Squire, William
Trapp, George E, Jr

WISCONSIN
Balmer, Robert Theodore
Bronikowski, Thomas Andrew
Dickey, Ronald Wayne
Divjak, August A
Hickman, James Charles
Johnson, Millard Wallace, Jr
Karreman, Herman Felix
Malkus, David Starr
Mangasarian, Olvi Leon
Moore, Robert H
Pollnow, Gilbert Frederick
Rang, Edward Roy
Schultz, David Harold
Shah, Ghulam M
Shen, Mei-Chang
Skinner, Lindsay A
Stahl, Neil
Strikwerda, John Charles
Tiedemann, William Harold
Uhlenbrock, Dietrich A
Wang, Pao-Kuan

WYOMING
Ewing, Richard Edward
George, John Harold

PUERTO RICO
Khan, Winston

ALBERTA
Aggarwala, Bhagwan D
Blais, J A Rodrigue
Collins, David Albert Charles
Dhaliwal, Ranjit S
Epstein, Marcelo
Freedman, Herbert Irving
Grabenstetter, James Emmett
Hsu, In-Ding
Ludwig, Garry (Gerhard Adolf)
Majumdar, Samir Ranjan
Nandakumar, Krishnaswamy
Westbrook, David Rex

BRITISH COLUMBIA
Clark, Colin Whitcomb
Fahlman, Gregory Gaylord
Graham, George Alfred Cecil
Halabisky, Lorne Stanley
Haussmann, Ulrich Gunther
Hughes, Blyth Alvin
Lardner, Robin Willmott
Leimanis, Eugene
Ludwig, Donald A
Millar, Robert Fyfe

Applied Mathematics (cont)

Miura, Robert Mitsuru
Reed, William J
Seymour, Brian Richard
Shoemaker, Edward Milton
Singh, Manohar
Srivastava, Hari Mohan
Srivastava, Rekha

MANITOBA
Barclay, Francis Walter
Jayas, Digvir Singh
Thomas, Robert Spencer David
Usmani, Riaz Ahmad
Wong, Roderick Sue-Cheun
Zeiler, Frederick

NEW BRUNSWICK
Basilevsky, Alexander
Tory, Elmer Melvin

NEWFOUNDLAND
May, Sherry Jan
Rochester, Michael Grant

NOVA SCOTIA
Bhartia, Prakash
Kujath, Marek Ryszard
Mohammed, Auyuab

ONTARIO
Ahmed, Nasir Uddin
Baillie, Donald Chesley
Barrie, Leonard Arthur
Barron, Ronald Michael
Barton, Richard Donald
Billigheimer, Claude Elias
Blackwell, John Henry
Campbell, Louis Lorne
Campeanu, Radu Ioan
Carver, Michael Bruce
Chan, Yat Yung
Cowper, George Richard
Davison, Sydney George
Derzko, Nicholas Anthony
Duever, Thomas Albert
Duff, George Francis Denton
Dunn, D(onald) W(illiam)
Ehrman, Joachim Benedict
Fairman, Frederick Walker
Fraser, Peter Arthur
Georganas, Nicolas D
Gladwell, Graham M L
Hancox, William Thomas
Hogarth, Jacke Edwin
Horbatsch, Marko M
Howland, James Lucien
Huschilt, John
Iverson, Kenneth Eugene
Jackson, Kenneth Ronald
Jarvis, Roger George
Kesarwani, Roop Narain
Kirby, Bruce John
Langford, William Finlay
Lawson, John Douglas
Leach, Barrie William
Lipshitz, Stanley Paul
McGee, William F
Mandelis, Andreas
Mandl, Paul
Mercer, Alexander McDowell
Molder, S(annu)
Mufti, Izhar H
Nash, John Christopher
Naylor, Derek
Nerenberg, Morton Abraham
Norminton, Edward Joseph
Oldham, Keith Bentley
Pfalzner, Paul Michael
Ponzo, Peter James
Rahman, Mizanur
Ranger, Keith Brian
Ray, Ajit Kumar
Reza, Fazlollah M
Rimrott, F(riedrich) P(aul) J(ohannes)
Ross, Roderick Alexander
Rowe, Ronald Kerry
Selvadurai, A P S
Sherbourne, Archibald Norbert
Sida, Derek William
Smith, William Robert
Stauffer, Allan Daniel
Sullivan, Paul Joseph
Tzoganakis, Costas
Wainwright, John
Wesson, Paul Stephen
Wonham, W Murray

QUEBEC
Clarke, Francis
Davies, Roger
Fabrikant, Valery Isaak
Jumarie, Guy Michael
Lefebvre, Mario
Li, Zi-Cai
Luckert, H(ans) J(oachim)
Morgera, Salvatore Domenic
Mysak, Lawrence Alexander
Roth, Charles
Sankoff, David
Schlomiuk, Dana
Tam, Kwok Kuen
Wallace, Philip Russell

Zlobec, Sanjo

SASKATCHEWAN
Esmail, Mohamed Nabil
Koh, Eusebio Legarda
Law, Alan Greenwell
Symes, Lawrence Richard
Walerian, Szyszkowski

OTHER COUNTRIES
Adem, Julian
Aumann, Robert John
Balinski, Michel Louis
Bazley, Norman William
Blackburn, Jacob Floyd
Callier, Frank Maria
Dougalis, Vassilios
Hamacher, Horst W
Hochberg, Kenneth J
Hsieh, Din-Yu
Idelsohn, Sergio Rodolfo
Kimura, Hidenori
Kydoniefs, Anastasios D
Lapostolle, Pierre Marcel
Laura, Patricio Adolfo Antonio
Lindquist, Anders Gunnar
Liu, Ta-Jo
Lueder, Ernst H
Luo, Peilin
Malah, David
Maqusi, Mohammad
May, Robert McCredie
Nievergelt, Jurg
Novotny, Eva
Raviv, Josef
Roseman, Joseph Jacob
Sarmiento, Gustavo Sanchez
Segel, Lee Aaron
Stoneham, Richard George
Taylor, Howard Lawrence
Tolstoy, Ivan
Wendland, Wolfgang Leopold
Willems, Jan C
Yang, Chung-Chun
Zeheb, Ezra

Biomathematics

ALABAMA
Cook, Frederick Lee
Katholi, Charles Robinson
Siler, William MacDowell
Turner, Malcolm Elijah, Jr

ARIZONA
Aickin, Mikel G
Colburn, Wayne Alan
Jacobowitz, Ronald
Kessler, John Otto

CALIFORNIA
Altes, Richard Alan
Biles, Charles Morgan
Bremermann, Hans J
Busenberg, Stavros Nicholas
Carew, Thomas Edward
Cavalli-Sforza, Luigi Luca
Cooke, Kenneth Lloyd
Cooper, William S
Dixon, Wilfrid Joseph
Elderkin, Richard Howard
Evans, John W
Feldman, Marcus William
Francis, Robert Colgate
Greever, John
Hirsch, Morris William
Huang, Sung-Cheng
Keller, Joseph Bishop
Kope, Robert Glenn
Landahl, Herbert Daniel
Lein, Allen
Licko, Vojtech
Luxon, Bruce Arlie
Milstein, Jaime
Mitchell, John Alexander
Neustadter, Siegfried Friedrich
Oster, George F
Purdue, Peter
Rescigno, Aldo
Richardson, Irvin Whaley
Scheibe, Paul Otto
Slatkin, Montgomery (Wilson)
Whittemore, Alice S
Wiggins, Alvin Dennie
Williams, William Arnold

COLORADO
Briggs, William Egbert
Reiner, John Maximilian
Warren, Peter

CONNECTICUT
Bohning, Daryl Eugene
Collins, Sylva Heghinian
Darcey, Terrance Michael
McCartin, Brian James
Wachman, Murray

DISTRICT OF COLUMBIA
Heineken, Frederick George

FLORIDA
Jones, Albert Cleveland

Keesling, James Edgar
Lyman, Gary Herbert
White, Timothy Lee

GEORGIA
Mickens, Ronald Elbert
Waltman, Paul Elvis

IDAHO
Bobisud, Larry Eugene
Hausrath, Alan Richard

ILLINOIS
Calderon, Calixto Pedro
Cowan, Jack David
Deysach, Lawrence George
Elble, Rodger Jacob
Feinstein, Irwin K
Friedlander, Ira Ray
Lauffenburger, Douglas Alan
Miller, Arnold Reed
Rosenblatt, Karin Ann
Skeel, Robert David
Smeach, Stephen Charles

INDIANA
Cerimele, Benito Joseph
Madan, Ved P
Thompson, Maynard

IOWA
Cornette, James L
Hethcote, Herbert Wayne
Kirkham, Don

KANSAS
Mainster, Martin Aron
Stewart, William Henry

KENTUCKY
Hayden, Thomas Lee
Hirsch, Henry Richard
McMorris, Fred Raymond

LOUISIANA
Heatherly, Henry Edward

MAINE
Murphy, Grattan Patrick

MARYLAND
Alexander, James Crew
Eden, Murray
Feldman, Jacob J
Hiatt, Caspar Wistar, III
Jessup, Gordon L, Jr
Lachin, John Marion, III
Marimont, Rosalind Brownstone
Massell, Paul Barry
Mazur, Jacob
Serfling, Robert Joseph
Yorke, James Alan

MASSACHUSETTS
Butler, James Preston
Carpenter, Gail Alexandra
Fishman, Philip M
Hattis, Dale B
Janowitz, Melvin Fiva
Lagakos, Stephen William
Stanley, Kenneth Earl

MICHIGAN
Conrad, Michael
Harrison, Michael Jay
Metzler, Carl Maust
Nyman, Melvin Andrew
Phelps, Frederick Martin, IV
Simon, Carl Paul
Verter, Joel I

MINNESOTA
Ek, Alan Ryan

MISSOURI
Larson, Kenneth Blaine
Taibleson, Mitchell H

MONTANA
Derrick, William Richard

NEBRASKA
Jenkins, Thomas Gordon
Schutz, Wilfred M
Ueda, Clarence Tad

NEW JERSEY
Ahsanullah, Mohammad
Badalamenti, Anthony Francis
Cronin, Jane Smiley
Elk, Seymour B
Frauenthal, James Clay
Pilla, Arthur Anthony
Siegel, Andrew Francis
Sverdlove, Ronald

NEW MEXICO
Cobb, Loren
Diel, Joseph Henry
Gibbs, William Royal
Ortiz, Melchor, Jr
Perelson, Alan Stuart

NEW YORK
Altshuler, Bernard
Anderson, Allan George
Bell, Jonathan George
Benham, Craig John
Berresford, Geoffrey Case
Blessing, John A
Blumenson, Leslie Eli
Brennan, Murray F
Chandran, V Ravi
Hartnett, William Edward
Hood, Donald C
Isaacson, David
Keller, Evelyn Fox
Leibovic, K Nicholas
Levin, Simon Asher
Lubowsky, Jack
Némethy, George
Niklas, Karl Joseph
Norton, Larry
Percus, Jerome K
Priore, Roger L
Ruppert, David
Schimmel, Herbert
Shalloway, David Irwin
Simon, William
Yarmush, David Leon

NORTH CAROLINA
Bishir, John William
Faulkner, Gary Doyle
Floyd, Carey E, Jr
Franke, John Erwin
Johnson, Thomas
Lobaugh, Bruce
Swallow, William Hutchinson
Woodbury, Max Atkin

NORTH DAKOTA
Penland, James Granville

OHIO
Bhargava, Triloki Nath
Dickinson, Frank N
Klein, John Peter

OKLAHOMA
Mulholland, Robert J(oseph)
Weeks, David Lee

OREGON
Mullooly, John P
Murphy, Lea Frances

PENNSYLVANIA
Ermentrout, George Bard
Hodgson, Jonathan Peter Edward
Karreman, George
Merrill, Samuel, III
Rorres, Chris
Scherer, Peter William
Tallarida, Ronald Joseph
Van der Werff, Terry Jay

SOUTH CAROLINA
Luedeman, John Keith

TENNESSEE
Bernard, Selden Robert
Dale, Virginia House
DeAngelis, Donald Lee
Demetriou, Charles Arthur
Dixon, Edmond Dale
Hallam, Thomas Guy
Partain, Clarence Leon
Partridge, Lloyd Donald
Tan, Wai-Yuan
Zeighami, Elaine Ann

TEXAS
Atkinson, Edward Neely
Brakefield, James Charles
Brown, Barry W
Eisenfeld, Jerome
Gutierrez, Guillermo
Jansson, Birger
Roberts, Thomas M
Saltzberg, Bernard
Sterner, Robert Warner
Thames, Howard Davis, Jr
White, Robert Allen
White, Ronald Joseph
Zimmerman, Stuart O

UTAH
Coles, William Jeffrey
Othmer, Hans George

VIRGINIA
Grant, John Wallace
Hohenboken, William Daniel
Holtzman, Golde Ivan
Hutchinson, Thomas Eugene
Leczynski, Barbara Ann
Wohl, Philip R

WASHINGTON
Britt, Patricia Marie
Conquest, Loveday Loyce
Gunderson, Donald Raymond
Moody, Michael Eugene
Moolgavkar, Suresh Hiraji
Swartzman, Gordon Leni

WISCONSIN
Chover, Joshua
Kurtz, Thomas Gordon
Schultz, David Harold
Walter, Gilbert G

PUERTO RICO
Peinado, Rolando E

ALBERTA
Voorhees, Burton Hamilton

BRITISH COLUMBIA
Anderson, R F V
Hewgill, Denton Elwood
Ludwig, Donald A
Miura, Robert Mitsuru
Seymour, Brian Richard

NOVA SCOTIA
Rosen, Robert

ONTARIO
Bertell, Rosalie
Ferrier, Jack Moreland
Halfon, Efraim
Holford, Richard Moore
Miller, Donald Richard
Rapoport, Anatol
Smith, William Robert
Somorjai, Rajmund Lewis
Zuker, Michael

QUEBEC
Boyarsky, Abraham Joseph
Jolicoeur, Pierre
Lacroix, Norbert Hector Joseph
Sankoff, David

SASKATCHEWAN
Spurr, David Tupper

OTHER COUNTRIES
Hochberg, Kenneth J
Segel, Lee Aaron

Combinatorics & Finite Mathematics

ALABAMA
Seidman, Stephen Benjamin
Slater, Peter John

ALASKA
Beebee, John Christopher
Hulsey, J Leroy

ARIZONA
Leonard, John Lander
Palusinski, Olgierd Aleksander

ARKANSAS
Schein, Boris M

CALIFORNIA
Alexanderson, Gerald Lee
Block, Richard Earl
Brandeau, Margaret Louise
Butler, Jon Terry
Dean, Richard Albert
Ford, Lester Randolph, Jr
Fraser, Grant Adam
Fredman, Michael Lawrence
Ghosh, Subir
Hales, Alfred Washington
Harper, Lawrence Hueston
Johnsen, Eugene Carlyle
Kaellis, Joseph
Kevorkian, Aram K
Lawler, Eugene L(eighton)
Minc, Henryk
Nathanson, Weston Irwin
O'Dúnlaing, Colm Pádraig
Pease, Marshall Carleton, III
Rebman, Kenneth Ralph
Reid, Kenneth Brooks
Rhodes, John Lewis
Sallee, G Thomas
Senge, George H
Simons, Barbara Bluestein
Stockmeyer, Larry Joseph
Sun, Hugo Sui-Hwan
Tanner, Robert Michael
Weeks, Dennis Alan
Welch, Lloyd Richard
Williamson, Stanley Gill

COLORADO
Clements, George Francis
Irwin, Robert Cook
Lundgren, J Richard
Manvel, Bennet
Maybee, John Stanley
Nemeth, Evi
Phillips, Keith L

CONNECTICUT
Blei, Ron Charles
Fine, Benjamin
Howe, Roger
Reid, James Dolan
Spiegel, Eugene
Stein, Alan H

Wolk, Elliot Samuel

DELAWARE
Ebert, Gary Lee

DISTRICT OF COLUMBIA
Garavelli, John Stephen
Sjogren, Jon Arne
Stein, Marjorie Leiter

FLORIDA
Butson, Alton Thomas
Cenzer, Douglas
Drake, David Allyn
Ferguson, Pamela A
Hoffman, Frederick
Ismail, Mourad E H
Olmstead, Paul Smith
Pettofrezzo, Anthony J
Roy, Dev Kumar
Shershin, Anthony Connors

GEORGIA
Canfield, Earl Rodney
Duke, Richard Alter
Neff, Mary Muskoff
Rouvray, Dennis Henry
Straley, Tina
Winkler, Peter Mann

IDAHO
Girse, Robert Donald

ILLINOIS
Appel, Kenneth I
Cormier, Romae Joseph
Eggan, Lawrence Carl
Grace, Thom P
Gupta, Udaiprakash I
Kasube, Herbert Emil
Lee, Der-Tsai
Livingston, Marilyn Laurene
Otto, Albert Dean
Reingold, Edward Martin
Reznick, Bruce Arie
Vanden Eynden, Charles Lawrence
Wallis, Walter Denis
Weichsel, Paul M
West, Douglas Brent

INDIANA
Beineke, Lowell Wayne
Kallman, Ralph Arthur
Murphy, Catherine Mary
Peltier, Charles Francis
Penna, Michael Anthony
Pippert, Raymond Elmer

IOWA
Robinson, John Paul

KANSAS
Davis, Elwyn H
Galvin, Fred
Shult, Ernest E

KENTUCKY
McMorris, Fred Raymond

LOUISIANA
Berman, David Michael
Maxfield, Margaret Waugh

MAINE
Balakrishnan, V K

MARYLAND
Brant, Larry James
Franks, David A
Rukhin, Andrew Leo
Siegel, Martha J
Wierman, John C

MASSACHUSETTS
Albertson, Michael Owen
Bolker, Ethan D
Dechene, Lucy Irene
Freund, Robert M
Gardner, Marianne Lepp
Gessel, Ira Martin
Hoffer, Alan R
Hyatt, Raymond R, Jr
Janowitz, Melvin Fiva
Joni, Saj-nicole A
Kenney, Margaret June
Levin, Leonid A
O'Neil, Patrick Eugene
O'Rourke, Joseph
Ostrovsky, Rafail M
Rosenberg, Arnold Leonard
Simovici, Dan
Smith, Raoul Normand
Stanley, Richard Peter
Woods, William A

MICHIGAN
Alavi, Yousef
Bachelis, Gregory Frank
Heuvers, Konrad John
Howard, Paul Edward
Lick, Don R
Stout, Quentin Fielden
VanderJagt, Donald W
White, Arthur Thomas, II

MINNESOTA
Anderson, Sabra Sullivan
Gallian, Joseph A
Hutchinson, Joan Prince

NEBRASKA
Chouinard, Leo George, II
Kramer, Earl Sidney
Magliveras, Spyros Simos
Walters, Randall Keith

NEW HAMPSHIRE
Bogart, Kenneth Paul

NEW JERSEY
Allender, Eric Warren
Blickstein, Stuart I
Boesch, Francis Theodore
Dekker, Jacob Christoph Edmond
Faith, Carl Clifton
Foregger, Thomas H
Garey, Michael Randolph
Gewirtz, Allan
Holzmann, Gerard J
Kessler, Irving Jack
Lagarias, Jeffrey Clark
Lepowsky, James Ivan
Lesniak, Linda
Mills, William Harold
Naus, Joseph Irwin
Roberts, Fred Stephen
Saniee, Iraj
Tindell, Ralph S
Tucker, Albert William

NEW MEXICO
Bradshaw, Martin Daniel
Entringer, Roger Charles
Faber, Vance
Harary, Frank
Hulme, Bernie Lee
Simmons, Gustavus James

NEW YORK
Anderson, Allan George
Baruch, Marjory Jean
Bencsath, Katalin A
Billera, Louis Joseph
Bing, Kurt
Buckley, Fred Thomas
Cohen, Daniel Isaac Aryeh
Coppersmith, Don
Cusick, Thomas William
Ecker, Joseph George
Feuer, Richard Dennis
Gerst, Irving
Goodman, Jacob Eli
Graver, Jack Edward
Green, Alwin Clark
Griesmer, James Hugo
Gross, Jonathan Light
Hausner, Melvin
Johnson, Sandra Lee
Kingston, Paul L
Knee, David Isaac
Kwong, Yui-Hoi Harris
Leary, Francis Christian
Martin, George Edward
Mattson, Harold F, Jr
Melter, Robert Alan
Mitchell, Joseph Shannon Baird
Mowshowitz, Abbe
Nathanson, Melvyn Bernard
Powers, David Leusch
Ralston, Anthony
Sellers, Peter Hoadley
Tamari, Dov
Tucker, Thomas William
Vasquez, Alphonse Thomas
Watkins, Mark E
Wong, Edward Chor-Cheung
Zaslavsky, Thomas

NORTH CAROLINA
Blanchet-Sadri, Francine
Geissinger, Ladnor Dale
Hartwig, Robert Eduard
Hodel, Margaret Jones
Kay, David Clifford
Klerlein, Joseph Ballard
Reif, John H
Reiter, Harold Braun

OHIO
Bannai, Eichi
Barger, Samuel Floyd
Bhargava, Triloki Nath
Gagola, Stephen Michael, Jr
Hahn, Samuel Wilfred
Hales, Raleigh Stanton, Jr
Hsia, John S
Khalimsky, Efim D
Maneri, Carl C
Nicholson, Victor Alvin
Ray-Chaudhuri, Dwijendra Kumar
Silverman, Robert
Solow, Daniel

OKLAHOMA
Goodey, Paul Ronald

OREGON
Stalley, Robert Delmer

PENNSYLVANIA
Andrews, George Eyre
Assmus, Edward F(erdinand), Jr
Balas, Egon
Bressoud, David Marius
Korfhage, Robert R
Mullen, Gary Lee
Parsons, Torrence Douglas
Schattschneider, Doris Jean
Simpson, Stephen G
Thompson, Gerald Luther
Tzeng, Kenneth Kai-Ming
Ventura, Jose Antonio
Wolfe, Dorothy Wexler

RHODE ISLAND
Simons, Roger Alan

SOUTH CAROLINA
Jamison, Robert Edward
Kostreva, Michael Martin
Luedeman, John Keith
McNulty, George Frank
Ringeisen, Richard Delose
Saxena, Subhash C

TENNESSEE
Faudree, Ralph Jasper, Jr
Ordman, Edward Thorne
Plummer, Michael David
Schelp, Richard Herbert

TEXAS
Cecil, David Rolf
Girou, Michael L
Hartfiel, Darald Joe
Matula, David William
Parberry, Ian
Tarwater, Jan Dalton

UTAH
Gross, Fletcher
Skarda, R Vencil, Jr
Snow, Donald Ray
Toledo, Domingo

VERMONT
Foote, Richard Martin

VIRGINIA
Collier, Manning Gary
Gray, F(estus) Gail
Heath, Lenwood S
Hoffman, Karla Leigh
Johnson, Charles Royal
Lipman, Marc Joseph
Norris, Eugene Michael
Rosenbaum, David Mark
Roussos, Constantine
Shier, Douglas Robert
Spresser, Diane Mar
Stockmeyer, Paul Kelly
Ward, Harold Nathaniel

WASHINGTON
DeTemple, Duane William
Grunbaum, Branko
Hansen, Rodney Thor
Kallaher, Michael Joseph
Long, Calvin Thomas
Nijenhuis, Albert
Webb, William Albert

WEST VIRGINIA
Gould, Henry Wadsworth

WISCONSIN
Bach, Eric
Barkauskas, Anthony Edward
Crowe, Donald Warren
Fournelle, Thomas Albert
Harris, Bernard
Kaye, Norman Joseph
Moore, Edward Forrest
Solomon, Marvin H
Terwilliger, Paul M

WYOMING
Bridges, William G
Di Paola, Jane Walsh
Porter, A Duane

PUERTO RICO
Moreno, Oscar

ALBERTA
Moon, John Wesley
Rhemtulla, Akbar Hussein

BRITISH COLUMBIA
Alspach, Brian Roger
Boyd, David William
Matheson, David Stewart
Riddell, James

MANITOBA
Dowling, Diane Mary
Mendelsohn, Nathan Saul

NEWFOUNDLAND
Rees, Rolf Stephen

NOVA SCOTIA
Kruse, Robert Leroy

Combinatorics & Finite Mathematics (cont)

ONTARIO
Balakrishnan, Narayanaswany
Colbourn, Charles Joseph
Coxeter, Harold Scott MacDonald
Davison, J Leslie
Djokovic, Dragomir Z
Garner, Cyril Wilbur Luther
Lehman, Alfred Baker
Pullman, Norman J
Roe, Peter Hugh O'Neil
Schellenberg, Paul Jacob
Shank, Herbert S
Steiner, George
Vanstone, Scott Alexander
Whittington, Stuart Gordon

QUEBEC
Allen, Harold Don
Brown, William G
Leroux, Pierre
Moser, William O J
Sabidussi, Gert Otto

SASKATCHEWAN
Sato, Daihachiro

OTHER COUNTRIES
Brown, Richard Harland
Gritzmann, Peter
Hamacher, Horst W
Myerson, Gerald
Olive, Gloria

Geometry

ARIZONA
Groemer, Helmut (Johann)

CALIFORNIA
Agoston, Max Karl
Alexanderson, Gerald Lee
Amemiya, Frances (Louise) Campbell
Barnett, David
Barsky, Brian Andrew
Blattner, Robert J(ames)
Blinn, James Frederick
Chakerian, Gulbank Donald
Chan, Jean B
Chern, Shiing-Shen
Collier, James Bryan
Feldman, Louis A
Finn, Robert
Green, Mark Lee
Greene, Robert Everist
Krener, Arthur James
Labarre, Anthony E, Jr
Lewis, George McCormick
Li, Peter Wai-Kwong
Moore, John Douglas
Osserman, Robert
Rabin, Jeffrey Mark
Sallee, G Thomas
Schlafly, Roger
Wavrik, John J
Weinstein, Alan David
Williamson, Robert Emmett
Wolf, Joseph Albert
Wu, Hung-Hsi

CONNECTICUT
Howe, Roger
Mostov, George Daniel
Seitelman, Leon Harold

DELAWARE
Ebert, Gary Lee

FLORIDA
Magarian, Elizabeth Ann

GEORGIA
Batterson, Steven L
Sparks, Arthur Godwin

ILLINOIS
Albrecht, Felix Robert
Bishop, Richard Lawrence
Earnest, Andrew George
Eichten, Estia Joseph
Fossum, Robert Merle
Haboush, William Joseph
Heitsch, James Lawrence
Ho, Chung-Wu
Jordan, Steven Lee
Lay, Steven R
Leibhardt, Edward
Portnoy, Esther
Richman, David Paul
Wagreich, Philip Donald
Waterman, Peter Lewis
Wetzel, John Edwin
Yau, Stephen Shing-Toung

INDIANA
Ludwig, Hubert Joseph
Mast, Cecil B
Peltier, Charles Francis
Penna, Michael Anthony
Penney, Richard Cole

Pohl, Victoria Mary
Sommese, Andrew John
Specht, Edward John
Stanton, Charles Madison

IOWA
Adelberg, Arnold M
Johnson, Norman L
Jorgensen, Palle E T
Kleinfeld, Erwin
Ton-That, Tuong

KANSAS
Shult, Ernest E

KENTUCKY
Lowman, Bertha Pauline

MAINE
Murphy, Grattan Patrick

MARYLAND
Blum, Harry
Fischer, Irene Kaminka
Gray, Alfred
Maloney, Clifford Joseph
Meyerson, Mark Daniel
Reinhart, Bruce Lloyd
Rosenberg, Jonathan Micah
Shiffman, Bernard
Witzgall, Christoph Johann

MASSACHUSETTS
Altman, Allen Burchard
Bennett, Mary Katherine
Bolker, Ethan D
Cattani, Eduardo
Cecil, Thomas E
Halperin-Maya, Miriam Patricia
Helgason, Sigurdur
Hoffer, Alan R
Hoffman, David Allen
Mann, Larry N
Matsusaka, Teruhisa
Mattuck, Arthur Paul
Melrose, Richard B
O'Rourke, Joseph
Sacks, Jonathan
Senechal, Marjorie Lee
Struik, Dirk Jan
Tanimoto, Taffee Tadashi

MICHIGAN
Arlinghaus, Sandra Judith Lach
Bookstein, Fred Leon
Chen, Bang-Yen
Houh, Chorng Shi
Johnson, Harold Hunt
Jones, Harold Trainer
Whitmore, Edward Hugh

MINNESOTA
Gulliver, Robert David, II
Nitsche, Johannes Carl Christian
Pohl, William Francis
Sperber, Steven Irwin

MISSOURI
Andalafte, Edward Ziegler
Beem, John Kelly
Cantwell, John Christopher
Freese, Raymond William
Jenkins, James Allister
Jensen, Gary Richard
Mashhoon, Bahram
Petty, Clinton Myers
Valentine, Joseph Earl
Wilson, Edward Nathan

NEBRASKA
Gross, Mildred Lucile
Magliveras, Spyros Simos

NEW HAMPSHIRE
Arkowitz, Martin Arthur
Gordon, Carolyn Sue
Snapper, Ernst

NEW JERSEY
Elk, Seymour B
Faith, Carl Clifton
Faltings, Gerd
Goldberg, Vladislav V
Nickerson, Helen Kelsall
Papantonopoulou, Aigli Helen
Taylor, Jean Ellen
Van Wyk, Christopher John
Young, Li

NEW MEXICO
Zund, Joseph David

NEW YORK
Abbott, John S
Adhout, Shahla Marvizi
Anderson, Peter Gordon
Barber, Sherburne Frederick
Barshay, Jacob
Billera, Louis Joseph
Blanton, John David
Chavel, Isaac
Dodziuk, Jozef
Dubisch, Russell John
Feldman, Edgar A

Fialkow, Aaron David
Gerber, Leon E
Goodman, Jacob Eli
Green, Alwin Clark
Hausner, Melvin
Henderson, David Wilson
Iwaniec, Tadeusz
Johnson, Sandra Lee
Martin, George Edward
Mitchell, Joseph Shannon Baird
Nirenberg, Louis
Pinkham, Henry Charles
Posamentier, Alfred S
Range, R Michael
Seppala-Holtzman, David N
Srinivasan, Vijay
Trasher, Donald Watson
Tuller, Annita
Vasquez, Alphonse Thomas
Warner, Frederic Cooper
Zaslavsky, Thomas
Zaustinsky, Eugene Michael

NORTH CAROLINA
Eberlein, Patrick Barry
Fulp, Ronald Owen
Gardner, Robert B
Jambor, Paul Emil
Kay, David Clifford
King, Lunsford Richardson
Schell, Joseph Francis
Sowell, Katye Marie Oliver
Stitzinger, Ernest Lester
Stroud, Junius Brutus
York, James Wesley, Jr

OHIO
Berton, John Andrew
Ferrar, Joseph C
Houston, William Bernard, Jr
Klosterman, Albert Leonard
Kullman, David Elmer
Lee, Dong Hoon
Maneri, Carl C
Millman, Richard Steven
Sholander, Marlow
Silverman, Robert
Weber, Waldemar Carl
Yaqub, Jill Courtaney Donaldson
　　Spencer

OKLAHOMA
Aichele, Douglas B
Breen, Marilyn
Goodey, Paul Ronald
McCullough, Darryl J

OREGON
Bhattacharyya, Ramendra Kumar
Flaherty, Francis Joseph
Gilkey, Peter Belden
Koch, Richard Moncrief
Parks, Harold Raymond
Tewinkel, G Carper
Todd, Philip Hamish

PENNSYLVANIA
Baildon, John David
Ehrmann, Rita Mae
Grinberg, Eric L
Harbater, David
James, Donald Gordon
Kazdan, Jerry Lawrence
Schäffer, Juan Jorge
Schattschneider, Doris Jean
Warner, Frank Wilson, III
Waterhouse, William Charles

RHODE ISLAND
Datta, Dilip Kumar
Rudolph, Lee

SOUTH CAROLINA
Jamison, Robert Edward
Layton, William Isaac
Saxena, Subhash C
Wylie, Clarence Raymond, Jr

TENNESSEE
Dobbs, David Earl
Wachspress, Eugene Leon

TEXAS
Aberth, Oliver George
Bakelman, Ilya J(acob)
Golubitsky, Martin Aaron
McCown, Malcolm G
Stiller, Peter Frederick
Vaaler, Jeffrey David
Vick, George R
Whittlesey, John R B

UTAH
Carlson, James Andrew
Clemens, Charles Herbert
Milicic, Dragan

VIRGINIA
Gold, Sydell Perlmutter
Gowdy, Robert Henry
Parshall, Brian J

WASHINGTON
Blumenthal, Robert Allan

DeTemple, Duane William
Dubois, Donald Ward
Duchamp, Thomas Eugene
Firkins, John Frederick
Grunbaum, Branko
Kallaher, Michael Joseph
Nievergelt, Yves
Nijenhuis, Albert

WISCONSIN
Bleicher, Michael Nathaniel
Crowe, Donald Warren

WYOMING
Di Paola, Jane Walsh
Roth, Ben G

ALBERTA
Antonelli, Peter Louis
Honsaker, John Leonard

BRITISH COLUMBIA
Alspach, Brian Roger
Boyd, David William
Lancaster, George Maurice
Totten, James Edward

MANITOBA
Thomas, Robert Spencer David

NEW BRUNSWICK
Dekster, Boris Veniamin

ONTARIO
Coxeter, Harold Scott MacDonald
Djokovic, Dragomir Z
Garner, Cyril Wilbur Luther
Mann, Robert Bruce
Robinson, Gilbert de Beauregard
Scherk, Peter
Sherk, Frank Arthur
Whitfield, John Howard Mervyn

QUEBEC
Schlomiuk, Dana

SASKATCHEWAN
Kaul, S K
Sato, Daihachiro

OTHER COUNTRIES
Crapo, Henry Howland
Gritzmann, Peter
Hughes, Daniel Richard
Kambayashi, Tatsuji
Thomason, Robert Wayne
Whitman, Andrew Peter

Logic

ARIZONA
Smarandache, Florentin

ARKANSAS
Schein, Boris M

CALIFORNIA
Addison, John West, Jr
Arens, Yigal
Blum, Lenore Carol
Bull, Everett L, Jr
Calabrese, Philip G
Cannonito, Frank Benjamin
Chang, Chen Chung
Church, Alonzo
Cooper, William S
Deutsch, Laurence Peter
Eklof, Paul Christian
Enderton, Herbert Bruce
Fagin, Ronald
Feferman, Solomon
Fraser, Grant Adam
Halmos, Paul Richard
Halpern, James Daniel
Henkin, Leon (Albert)
Janos, Ludvik
Lender, Adam
McLaughlin, William Irving
Manaster, Alfred B
Mignone, Robert Joseph
Moschovakis, Joan Rand
Moschovakis, Yiannis N
O'Dúnlaing, Colm Pádraig
Pixley, Alden F
Pratt, Vaughan Ronald
Russell, Stuart Jonathan
Sridharan, Natesa S
Vaught, Robert L
Wiebe, Richard Penner
Wolf, Robert Stanley

COLORADO
Reinhardt, William Nelson

CONNECTICUT
Goldman, Ernest Harold
Lerman, Manuel
Sharlow, John Francis
Wood, Carol Saunders

DELAWARE
Chester, Daniel Leon

DISTRICT OF COLUMBIA
Cherniavsky, John Charles
Curtin, Frank Michael
McLean, John Dickson
Nelson, David
Rosenblatt, David

FLORIDA
Cenzer, Douglas
Levitz, Hilbert
Miller, Don Dalzell
Roy, Dev Kumar
Stark, William Richard

GEORGIA
Winkler, Peter Mann

ILLINOIS
Baldwin, John Theodore
Clover, William John, Jr
Cormier, Romae Joseph
Hay, Louise (Schmir)
Henson, C Ward
Jockusch, Carl Groos, Jr
Soare, Robert I
Takeuti, Gaisi
Wilson, Howell Kenneth
Wojcik, Anthony Stephen

INDIANA
Cunningham, Ellen M
Peltier, Charles Francis
Rubin, Jean E
Wheeler, William Hollis

IOWA
Alton, Donald Alvin
Nelson, George
Pigozzi, Don

KANSAS
Galvin, Fred
Roitman, Judy

KENTUCKY
Crawford, Robert R

MAINE
Pogorzelski, Henry Andrew

MARYLAND
Bernstein, Allen Richard
Chapin, Edward William, Jr
Dubin, Henry Charles
Feldman, Robert H L
Green, Judy
Herrmann, Robert Arthur
Hodes, Louis
Horna, Otakar Anthony
Kueker, David William
Minker, Jack
Owings, James Claggett, Jr
Smith, Mark Stephen
Suppe, Frederick (Roy)

MASSACHUSETTS
Barnes, Arnold Appleton, Jr
Bruce, Kim Barry
Gacs, Peter
Henle, James Marston
Homer, Steven Elliott
Levin, Leonid A
Marsh, William Ernest
Meyer, Albert Ronald
Rogers, Hartley, Jr
Sacks, Gerald Enoch
Shorb, Alan McKean
Wand, Mitchell

MICHIGAN
Grosky, William Irvin
Hinman, Peter Greayer
Howard, Paul Edward
Lenker, Susan Stamm
Rajlich, Vaclav Thomas
Stob, Michael Jay

MINNESOTA
Fuhrken, Gebhard
Gaal, Ilse Lisl Novak
Pierce, Keith Robert
Pour-El, Marian Boykan

MISSOURI
Andalafte, Edward Ziegler

MONTANA
Barrett, Louis Carl

NEW JERSEY
Abeles, Francine
Dekker, Jacob Christoph Edmond
Nonnenmann, Uwe
Osofsky, Barbara Langer
Rosenstein, Joseph Geoffrey
Van Wyk, Christopher John

NEW MEXICO
Diel, Joseph Henry
Rypka, Eugene Weston

NEW YORK
Allen, James Frederick
Bing, Kurt

Bowen, Kenneth Alan
Case, John William
Chan, Tat-Hung
Cogan, Edward J
Constable, Robert L
Davis, Martin (David)
De Lillo, Nicholas Joseph
Di Paola, Robert Arnold
Goodman, Nicolas Daniels
Hartnett, William Edward
Hechler, Stephen Herman
Hughes, Harold K
Kopperman, Ralph David
Lercher, Bruce L
Mendelson, Elliott
Morley, Michael Darwin
Nerode, Anil
Parikh, Rohit Jivanlal
Platek, Richard Alan
Rapaport, William Joseph
Rutledge, Joseph Dela
Saracino, Daniel Harrison
Selman, Alan L
Shapiro, Stuart Charles
Shore, Richard A
Sibert, Elbert Ernest
Wang, Hao
Williams, Scott Warner
Zuckerman, Martin Michael

NORTH CAROLINA
Blanchet-Sadri, Francine
Hodel, Richard Earl
Loveland, Donald William
Mitchell, William John
Ras, Zbigniew Wieslaw

OHIO
Applebaum, Charles H
Collins, George Edwin
Foreman, Matthew D
Friedman, Harvey Martin
McCloskey, Teresemarie
Mendenhall, Robert Vernon
Rolletschek, Heinrich Franz
Schlipf, John Stewart
Schneider, Hubert H
Swardson, Mary Anne

OREGON
Parks, Harold Raymond
Penk, Anna Michaelides

PENNSYLVANIA
Andrews, Peter Bruce
Asenjo, Florencio Gonzalez
Chase, Gene Barry
Dawson, John William, Jr
Finin, Timothy Wilking
Furst, Merrick Lee
Hirschman, Lynette
Jech, Thomas J
Paulos, John Allen
Rescher, Nicholas
Simpson, Stephen G

RHODE ISLAND
Simons, Roger Alan

SOUTH CAROLINA
Comer, Stephen Daniel
McNulty, George Frank

TENNESSEE
Dixon, Edmond Dale
Purisch, Steven Donald

TEXAS
DeBakey, Lois
DeBakey, Selma
Martin, Norman Marshall
Vobach, Arnold R

VERMONT
Kreider, Donald Lester

VIRGINIA
Loatman, Robert Bruce
Orcutt, Bruce Call
Schneider, Philip Allen David

WASHINGTON
Young, Paul Ruel

WISCONSIN
Carothers, Otto M
Fournelle, Thomas Albert
Keisler, Howard Jerome
Kleene, Stephen Cole
Solomon, Marvin H

WYOMING
Cowles, John Richard

BRITISH COLUMBIA
Mekler, Alan
Thomason, Steven Karl

NEWFOUNDLAND
May, Sherry Jan

ONTARIO
Chang, Shao-chien
Cook, Stephen Arthur

Fleischer, Isidore
Giles, Robin
Hughes, Charles Edward
Kent, Clement F
Olin, Philip
Tall, Franklin David

QUEBEC
Daigneault, Aubert
Hatcher, William S
Seldin, Jonathan Paul

OTHER COUNTRIES
Vamos, Tibor

Mathematical Statistics

ALABAMA
Bradley, Edwin Luther, Jr
Bradley, Laurence A
Chow, Bryant
Drake, Albert Estern
Essenwanger, Oskar M
Gibbons, Jean Dickinson
Holt, William Robert
Hurst, David Charles
Mishra, Satya Narayan
Trawinski, Benon John
Trawinski, Irene Patricia Monahan
Welker, Everett Linus
Wheeler, Ruric E

ALASKA
Van Veldhuizen, Philip Androcles

ARIZONA
Anderson, Mary Ruth
Barckett, Joseph Anthony
Denny, John Leighton
Falk, Thomas
Freund, John Ernst
Gaumnitz, Erwin Alfred
Heinle, Preston Joseph
Jacobowitz, Ronald
Kakar, Rajesh Kumar
Keats, John Bert
Kececioglu, D(imitri) B(asil)
Yakowitz, Sidney J
Young, Dennis Lee

ARKANSAS
Alquire, Mary Slanina (Hirsch)
Brown, A Hayden, Jr
Dunn, James Eldon
McBryde, Vernon E(ugene)
Smith, Robert Paul
Springer, Melvin Dale
Strub, Mike Robert
Walls, Robert Clarence

CALIFORNIA
Alder, Henry Ludwig
Anderson, Theodore Wilbur
Antoniak, Charles Edward
Apostol, Tom M
Arnwine, William Carrol
Arthur, Susan Peterson
Astrachan, Max
Baker, George Allen
Barlow, Richard Eugene
Barr, Donald R
Beaver, Robert John
Beckwith, Richard Edward
Bell, Charles Bernard, Jr
Bendisz, Kazimierz
Benson, Robert Franklin
Bentley, Donald Lyon
Bickel, Peter J
Blackwell, David (Harold)
Blischke, Wallace Robert
Book, Stephen Alan
Brillinger, David Ross
Burg, John Parker
Cahn, David Stephen
Carlyle, Jack Webster
Chen, Chung Wei
Chern, Ming-Fen Myra
Chernick, Michael Ross
Cisin, Ira Hubert
Cover, Thomas Merrill
Cumberland, William Glen
Curry, Bo(stick) U
David, Florence N
Dolby, James Louis
Donoho, David Leigh
Dresch, Francis William
Drugan, James Richard
Efron, Bradley
Elashoff, Janet Dixon
Esary, James Daniel
Ferguson, Thomas S
Forsythe, Alan Barry
Franti, Charles Elmer
Freedman, David A
Gavin, Lloyd Alvin
Geng, Shu
Ghosh, Sakti P
Ghosh, Subir
Gokhale, Dattaprabhakar V
Goodman, Julius
Gordon, Louis
Gordon, Milton Andrew
Gruber, Helen Elizabeth

Halmos, Paul Richard
Hartman, James Kern
Heitkamp, Norman Denis
Hewer, Gary Arthur
Hodges, Joseph Lawson, Jr
Hoel, Paul Gerhard
Hsu, Yu-Sheng
Iglehart, Donald Lee
Jacobs, Patricia Anne
Jayachandran, Toke
Jennrich, Robert I
Jewell, Nicholas Patrick
Johns, Milton Vernon, Jr
Johnson, Carl William
Kalensher, Bernard Earl
Kane, Daniel E(dwin)
Karlin, Samuel
Katz, Richard Whitmore
King, R Maurice, Jr
Krieger, Henry Alan
Kritz-Silverstein, Donna
Lai, Tze Leung
Lange, Kenneth L
Larson, Harold Joseph
Le Cam, Lucien Marie
Lehmann, Erich Leo
Lewis, Peter Adrian Walter
Lieberman, Gerald J
McCall, Chester Hayden, Jr
MacDonald, Gordon James Fraser
Maksoudian, Y Leon
Mann, Nancy Robbins
Massey, Frank Jones, Jr
Mensing, Richard Walter
Mickey, Max Ray, Jr
Milch, Paul R
Miller, Rupert Griel
Molnar, Imre
Moser, Joseph M
Myers, Verne Steele
Myhre-Hollerman, Janet M
Newcomb, Robert Lewis
Olkin, Ingram
Olshen, Richard Allen
Park, Chong Jin
Park, Heebok
Piersol, Allan Gerald
Press, S James
Ramey, Daniel Bruce
Rao, Jammalamadaka S
Ray, Rose Marie
Read, Robert Richard
Resnikoff, George Joseph
Richards, Francis Russell
Rocke, David M
Romano, Albert
Rosenblatt, Daniel Bernard
Rosenblatt, Murray
Scheuer, Ernest Martin
Schwall, Richard Joseph
Scott, Elizabeth Leonard
Sengupta, Sumedha
Sevacherian, Vahram
Shen, Nien-Tsu
Show, Ivan Tristan
Siegmund, David O
Singleton, Richard Collom
Smith, Robert William
Smith, Wayne Earl
Smoke, Mary E
Solomon, Herbert
Sprowls, Riley Clay
Stauffer, Howard Boyer
Stewart, Leland Taylor
Stone, Charles Joel
Suich, Ronald Charles
Suppes, Patrick
Swan, Shanna Helen
Switzer, Paul
Thacker, John Charles
Trumbo, Bruce Edward
Tysver, Joseph Bryce
Ury, Hans Konrad
Wang, Peter Cheng-Chao
Waterman, Michael S
Watkins, William
Weiner, Daniel Lee
Wiggins, Alvin Dennie
Wu, Sing-Chou
Wurtele, Zivia Syrkin
Yahiku, Paul Y
Zweig, Hans Jacob

COLORADO
Bardwell, George
Barnes, James Allen
Bowden, David Clark
Brockwell, Peter John
Chase, Gerald Roy
Crow, Edwin Louis
Eberhart, Steve A
Gentry, Willard Max, Jr
Graybill, Franklin A
Healy, William Carleton, Jr
Heine, George Winfield, III
Heiny, Robert Lowell
Houston, Samuel Robert
Hume, Merril Wayne
Kafadar, Karen
Li, Hung Chiang
Mielke, Paul W, Jr
Miesch, Alfred Thomas
Morgenthaler, George William
Peterson, Gary A

Mathematical Statistics (cont)

Remmenga, Elmer Edwin
Schmid, John Carolus
Shedlock, Kaye M
Srivastava, Jagdish Narain
Tischendorf, John Allen
Weiss, Irving
West, Anita
Williams, James Stanley

CONNECTICUT
Anscombe, Francis John
Fischer, Diana Bradbury
Frankel, Martin Richard
Gelfand, Alan Enoch
Giné-Masdéu, Evarist
Goldstein, Rubin
Hartigan, John A
Kleinfeld, Ira H
Levenstein, Harold
Maples, Louis Charles
Moore, Melita Holly
Mukhopadhyay, Nitis
Nelson, Peter Reid
Noether, Gottfried Emanuel
Page, Edgar J(oseph)
Phillips, Peter Charles B
Posten, Harry Owen
Rao, Valluru Bhavanarayana
Savage, I Richard
Singer, Burton Herbert
Streit, Roy Leon
Tuteur, Franz Benjamin
Vitale, Richard Albert
Woods, Jimmie Dale

DELAWARE
Eissner, Robert M
Fellner, William Henry
Lucas, James M
Marquardt, Donald Wesley
Ogunnaike, Babatunde Ayodeji
Snee, Ronald D
Taylor, Howard Milton, III

DISTRICT OF COLUMBIA
Abramson, Lee Richard
Bishop, Yvonne M M
Bowker, Albert Hosmer
Brignoli Gable, Carol
Bryant, Barbara Everitt
Cohen, Michael Paul
Cox, Lawrence Henry
Crosby, David S
Demming, W(illiam) Edwards
Filliben, James John
Gastwirth, Joseph L
Goldfield, Edwin David
Goor, Charles G
Halperin, Max
Hammer, Carl
Hayek, Lee-Ann Collins
Hedlund, James Howard
Jabine, Thomas Boyd
Klemm, Rebecca Jane
Korin, Basil Peter
Kullback, Solomon
Lilliefors, Hubert W
Mandel, Benjamin J
Mann, Charles Roy
Melnick, Daniel
Nayak, Tapan Kumar
Olmer, Jane Chasnoff
Robinette, Charles Dennis
Rosenblatt, David
Rosenblatt, Joan Raup
Spruill, Nancy Lyon
Straf, Miron L
Thall, Peter Francis
Wallis, W Allen
Wiener, Howard Lawrence

FLORIDA
Ballerini, Rocco
Basu, Debabrata
Bingham, Richard S(tephen), Jr
Brain, Carlos W
Byrkit, Donald Raymond
Chang, Lena
Chew, Victor
Coldwell, Robert Lynn
Cornell, John Andrew
Dutton, Arthur Morlan
Eichhorn, Heinrich Karl
Fausett, Donald Wright
Frishman, Fred
Haase, Richard Henry
Hall, David Goodsell, IV
Hollander, Myles
Huerta, Manuel Andres
Jamrich, John Xavier
Kirmse, Dale William
Lin, Pi-Erh
Littell, Ramon Clarence
Magarian, Elizabeth Ann
Martin, Frank Garland
Meeter, Duane Anthony
Mendenhall, William, III
Olmstead, Paul Smith
Ostle, Bernard
Parr, William Chris
Pirkle, Fredric Lee
Proschan, Frank

Randles, Ronald Herman
Rao, Pejaver Vishwamber
Roth, Raymond Edward
Sartain, Jerry Burton
Scheaffer, Richard Lewis
Sethuraman, Jayaram
Shapiro, Samuel S
Somerville, Paul Noble
Stamper, James Harris
Storm, Leo Eugene
Tebbe, Dennis Lee
Ungvichian, Vichate
Williams, Willie Elbert

GEORGIA
Bargmann, Rolf Erwin
Bhagia, Gobind Shewakram
Bradley, Ralph Allan
Caras, Gus J(ohn)
Carter, Melvin Winsor
Chen, Hubert Jan-Peing
Choi, Keewhan
Cohen, Alonzo Clifford, Jr
Coyne, Edward James, Jr
Koch, George Schneider, Jr
Kutner, Michael Henry
Lavender, DeWitt Earl
Lyons, Nancy I
Mabry, John William
Neter, John
Pepper, William Donald
Riddle, Lawrence H
Taylor, Robert Lee
Thompson, Shirley Williams
Thompson, William Oxley, II
Tong, Yung Liang
Wadsworth, Harrison M(orton)
Ware, Glenn Oren
Warner, Isiah Manuel

HAWAII
Flachsbart, Peter George
Miura, Carole K Masutani

IDAHO
Butler, Calvin Charles
Everson, Dale O
Juola, Robert C
Laughlin, James Stanley
Stuffle, Roy Eugene

ILLINOIS
Abrams, Israel Jacob
Bahadur, Raghu Raj
Bohrer, Robert Edward
Burkholder, Donald Lyman
Carrier, Steven Theodore
Chakrabarti, Subrata K
Chitgopekar, Sharad Shankarrao
Cox, Dennis Dean
DeBoer, Edward Dale
Dwass, Meyer
Dyson, John Douglas
Eaton, Paul Wentland
Edge, Orlyn P
Fox, James David
Friedberg, Stephen Howard
Georgakis, Constantine
Golomski, William Arthur
Harter, H(arman) Leon
Hedayat, A Samad
Heller, Barbara Ruth
Ionescu Tulcea, Cassius
Jacobsen, Fred Marius
Kalai, Ehud
Knight, Frank B
Krug, Samuel Edward
Kruskal, William Henry
Laud, Purushottam Waman
Li, Jane Chiao
Lin, Lawrence I-Kuei
Madansky, Albert
Marden, John Iglehart
Martinsek, Adam Thomas
Meier, Paul
Monrad, Ditlev
Pendergrass, Robert Nixon
Philipp, Walter V
Portnoy, Stephen Lane
Roberts, Harry Vivian
Roth, Arthur Jason
Sacks, Jerome
Sanathanan, Lalitha P
Sclove, Stanley Louis
Scott, Meckinley
Seefeldt, Waldemar Bernhard
Starks, Thomas Harold
Stigler, Stephen Mack
Stout, William F
Tamhane, Ajit Chintaman
Taneja, Vidya Sagar
Tarver, Mae-Goodwin
Thisted, Ronald Aaron
Tiao, George Ching-Hwuan
Wacker, William Dennis
Wallace, David Lee
Wasserman, Stanley
Wijsman, Robert Arthur
Wolter, Kirk Marcus
Yang, Mark Chao-Kuen

INDIANA
Ali, Mir Masoom
Anderson, Virgil Lee

Bhattacharya, Rabindra Nath
Bradley, Richard Crane, Jr
Costello, Donald F
Cote, Louis J
Coyle, Edward John
Eberts, Ray Edward
Eckert, Roger E(arl)
Hamblen, John Wesley
Harrington, Rodney B
Hicks, Charles Robert
King, Edgar Pearce
Kinney, John James
Klein, Howard Joseph
Ma, Cynthia Sanman
McCabe, George Paul, Jr
Moore, David Sheldon
Murphy, John Riffe
Norton, James Augustus, Jr
Obenchain, Robert Lincoln
Puri, Prem Singh
Rubin, Herman
Sampson, Charles Berlin
Samuels, Myra Lee
Studden, William John
Tan, James Chien-Hua
Townsend, Douglas Wayne

IOWA
Alberda, Willis John
Athreya, Krishna Balasundaram
Berger, Philip Jeffrey
Braaten, Melvin Ole
Bright, Harold Frederick
Canfield, Earle Lloyd
Christian, Lauren L
Cressie, Noel A C
Cryer, Jonathan D
David, Herbert Aron
Dykstra, Richard Lynn
Feldt, Leonard Samuel
Fuller, Wayne Arthur
Geiger, Randall L
Harville, David Arthur
Hickman, Roy D
Hinz, Paul Norman
Hogg, Robert Vincent, Jr
Huntsberger, David Vernon
Isaacson, Stanley Leonard
Kempthorne, Oscar
Knowler, Lloyd A
Lott, Fred Wilbur, Jr
Meeker, William Quackenbush, Jr
Pollak, Edward
Reckase, Mark Daniel
Robertson, Tim
Russo, Ralph P
Sherman, Dorothy Helen
Snider, Bill Carl F
Sukhatme, Balkrishna Vasudeo
Sukhatme, Shashikala Balkrishna
Vardeman, Stephen Bruce

KANSAS
Arteaga, Lucio
Bulgren, William Gerald
Chopra, Dharam-Vir
Eklund, Darrel Lee
Feyerherm, Arlin Martin
Fryer, Holly Claire
Grosh, Doris Lloyd
Higgins, James Jacob
Johnson, Dallas Eugene
Kemp, Kenneth E
Paulie, M Catherine Therese
Platt, Ronald Dean
Price, Griffith Baley
Shanmugan, K Sam
Stewart, William Henry
Thomas, Harold Lee
White, Stephen Edward

KENTUCKY
Allen, David Mitchell
Anderson, Richard L
Bhapkar, Vasant Prabhakar
Feibes, Walter
Govindarajulu, Zakkula
Griffith, William Schuler
O'Connor, Carol Alf
Smith, Timothy Andre
Vaziri, Menoucher

LOUISIANA
Baldwin, Virgil Clark, Jr
Bennett, Lonnie Truman
Boullion, Thomas L
Bradley, Marshall Rice
Burford, Roger Lewis
Crump, Kenny Sherman
Dickinson, Peter Charles
Guzman Foresti, Miguel Angel
Kinchen, David G
Koonce, Kenneth Lowell
Kordoski, Edward William
Kuo, Hui-Hsiung
Montalvo, Joseph G, Jr
Schilling, Prentiss Edwin
Timon, William Edward, Jr
Watkins, Terry Anderson
Young, John Coleman

MAINE
McMillan, Brockway

MARYLAND
Abbey, Robert Fred, Jr
Alling, David Wheelock
Austin, Homer Wellington
Babcock, Anita Kathleen
Bartko, John Jaroslav
Bayless, David Lee
Behforooz, Ali
Blanche, Ernest Evred
Brant, Larry James
Bryant, Edward Clark
Cameron, Joseph Marion
Carroll, Raymond James
Celmins, Aivars Karlis Richards
Chacko, George Kuttickal
Chase, Gary Andrew
Chiacchierini, Richard Philip
DeArmon, Ira Alexander, Jr
Diamond, Earl Louis
Douglas, Lloyd Evans
Dubey, Satya D(eva)
Duncan, Acheson Johnston
Eberhardt, Keith Randall
Eisenhart, Churchill
Engel, Joseph H(enry)
Fairweather, William Ross
Gart, John Jacob
Geller, Harvey
Gilford, Leon
Goktepe, Omer Faruk
Gordon, Chester Murray
Gordon, Tavia
Greenhouse, Samuel William
Grubbs, Frank Ephraim
Hansen, Morris Howard
Hanson, Robert Harold
Hogben, David
Kacker, Raghu N
Kaplan, David L
Karr, Alan Francis
Kattakuzhy, George Chacko
Kligman, Ronald Lee
Kotz, Samuel
Ku, Harry Hsien Hsiang
Lachin, John Marion, III
Land, Charles Even
Langenberg, Patricia Warrington
Lao, Chang Sheng
Lechner, James Albert
Lee, Young Jack
Leone, Fred Charles
Lepson, Benjamin
Levine, Eugene
Liggett, Walter Stewart, Jr
Linder, Forrest Edward
Locke, Ben Zion
Lundegard, Robert James
Maar, James Richard
Maisel, Herbert
Maloney, Clifford Joseph
Mandel, John
Marzetti, Lawrence Arthur
Merrill, Warner Jay, Jr
Meyer, Charles Franklin
Mikulski, Piotr W
Miller, Robert Gerard
Milton, Roy Charles
Moshman, Jack
Murphy, Thomas James
Navarro, Joseph Anthony
Odle, John William
Paule, Robert Charles
Pickle, Linda Williams
Ponnapalli, Ramachandramurty
Raff, Morton Spencer
Rastogi, Suresh Chandra
Robinson, Richard Carleton, Jr
Robock, Alan
Royall, Richard Miles
Rubinstein, Lawrence Victor
Rukhin, Andrew Leo
Serfling, Robert Joseph
Shapiro, Sam
Sirken, Monroe Gilbert
Smith, Paul John
Steinberg, Joseph
Steinberg, Seth Michael
Tarone, Robert Ernest
Tepping, Benjamin Joseph
Thiebaux, Helen Jean
Tissue, Eric Bruce
Tripathi, Satish K
Waksberg, Joseph
Weir, Edward Earl, II
Weiss, Michael David
Wells, William T
Wierman, John C
Williford, William Olin
Yang, Grace L
Zoellner, John Arthur

MASSACHUSETTS
Abbott, Robert Classie
Arnold, Kenneth James
Ash, Arlene Sandra
Ben-Akiva, Moshe E
Bert, Mark Henry
Bussgang, J(ulian) J(akob)
Campbell, Cathy
Cardello, Armand Vincent
Chernoff, Herman
Chisholm, Donald Alexander
D'Agostino, Ralph B
Diaconis, Persi

Doerfler, Thomas Eugene
Dudley, Richard Mansfield
Esch, Robin E
Foulis, David James
Freeman, Harold Adolph
Giordano, Arthur Anthony
Gutoff, Edgar B(enjamin)
Hoaglin, David Caster
Hsieh, Hui-Kuang
Hyatt, Raymond R, Jr
Kitchin, John Francis
Kolakowski, Donald Louis
Lagakos, Stephen William
Lavin, Philip Todd
Lemeshow, Stanley Alan
Levin, Leonid A
Lo, Andrew W
Marino, Richard Matthew
Mehta, Mahendra
Mosteller, C Frederick
Ross, Edward William, Jr
Rubin, Donald Bruce
Satterthwaite, Franklin Eves
Siegel, Armand
Skibinsky, Morris
Stanley, H(arry) Eugene
Stanley, Kenneth Earl
Tsiatis, Anastasios A
Ware, James H
Waud, Douglas Russell
Wei, Lee-Jen
Welsch, Roy Elmer
Westwick, Carmen R
Wolock, Fred Walter
Zelen, Marvin

MICHIGAN
Arnold, Harvey James
Chakravarthy, Srinivasaraghavan
Crary, Selden Bronson
Cress, Charles Edwin
Dijkers, Marcellinus P
Dressel, Paul Leroy
Dubes, Richard C
Ericson, William Arnold
Fox, Martin
Gilliland, Dennis Crippen
Hannan, James Francis
Hill, Bruce M
Jebe, Emil H
Johnson, Whitney Larsen
Jones, Donald Akers
Kincaid, Wilfred Macdonald
Knappenberger, Herbert Allan
Koo, Delia Wei
Koul, Hira Lal
Kowalski, Charles Joseph
Kshirsagar, Anant Madhav
Lamberson, Leonard Roy
Leabo, Dick Albert
Lenker, Susan Stamm
LePage, Raoul
Lin, Paul Kuang-Hsien
McKean, Joseph Walter, Jr
Marchand, Margaret O
Muirhead, Robb John
Pandit, Sudhakar Madhavrao
Powell, James Henry
Root, William L(ucas)
Shantaram, Rajagopal
Sievers, Gerald Lester
Spivey, Walter Allen
Stapleton, James H
Starr, Norman
Stoline, Michael Ross
Tanis, Elliot Alan
Tetreault, Florence G
Tsao, Chia Kuei
Whalen, Thomas J(ohn)
Zemach, Rita

MINNESOTA
Alders, C Dean
Ayeni, Babatunde J
Balcziak, Louis William
Bingham, Christopher
Britz, Galen C
Buehler, Robert Joseph
Busch, Robert Henry
Eaton, Morris Leroy
Franta, William Roy
Geisser, Seymour
Giese, David Lyle
Hansen, Leslie Bennett
Hicks, Dale R
Hurwicz, Leonid
Larntz, Kinley
Le, Chap Than
McHenry, Kelly David
Martin, Frank Burke
Perry, Robert Leonard
Rust, Lawrence Wayne, Jr
Sentz, James Curtis
Smith, John Henry
Smith, Robert Elijah
Sudderth, William David
Weisberg, Sanford
White, John S(pencer)
Wolf, Frank Louis

MISSISSIPPI
Butler, Charles Morgan
Sharpe, Thomas R

MISSOURI
Bain, Lee J
Baltz, Howard Burl
Barnett, William Arnold
Basu, Asit Prakas
Carlson, Roger
Davis, James Wendell
Flora, Jairus Dale, Jr
Franck, Wallace Edmundt
Hewett, John Earl
Katti, Shriniwas Keshav
King, Terry Lee
Moore, James Thomas
Nemeth, Margaret Ann
Ott, R(ay) Lyman, Jr
Oviatt, Charles Dixon
Spitznagel, Edward Lawrence, Jr
Stalling, David Laurence
Tsutakawa, Robert K
Wright, Farroll Tim

MONTANA
Loftsgaarden, Don Owen
Reinhardt, Howard Earl
Taylor, William Robert

NEBRASKA
Ahlschwede, William T
Gessaman, Margaret Palmer
Schutz, Wilfred M

NEVADA
Carr, James Russell
Holdbrook, Ronald Gail
Kinnison, Robert Ray
Lowenthal, Douglas H
Miller, Forest Leonard, Jr

NEW HAMPSHIRE
Albert, Arthur Edward
Lamperti, John Williams
Tourgee, Ronald Alan

NEW JERSEY
Ahsanullah, Mohammad
Beaton, Albert E
Becker, Richard Alan
Brandinger, Jay J
Chambers, John McKinley
Chassan, Jacob Bernard
Cheung, Shiu Ming
Chien, Victor
Church, Eugene Lent
Cinlar, Erhan
Cleveland, William Swain
Clifford, Paul Clement
Coale, Ansley J
Dalal, Siddhartha Ramanlal
DeMarco, Ralph Richard
Denby, Lorraine
Eckler, Albert Ross
Fask, Alan S
Frey, William Carl
Gabbe, John Daniel
Gale, William Arthur
Gethner, Jon Steven
Gilmer, George Hudson
Gnanadesikan, Ramanathan
Gold, Daniel Howard
Green, Edwin James
Holland, Paul William
Hsu, Chin-Fei
Hunter, John Stuart
Hwang, Frank K
Jain, Aridaman Kumar
Johnston, Christian William
Jurkat, Martin Peter
Kleiner, Beat
Kruskal, Joseph Bernard
Labianca, Frank Michael
LaMastro, Robert Anthony
Lambert, Diane
Landwehr, James M
Layzer, Arthur James
Leon, Ramon V
Loader, Clive Roland
Lucantoni, David Michael
Magliola-Zoch, Doris
Mallows, Colin Lingwood
Mihram, George Arthur
Miller, Irwin
Murphy, Ray Bradford
Murphy, Thomas Daniel
Naus, Joseph Irwin
Ordille, Carol Maria
Pasteelnick, Louis Andre
Phadke, Madhav Shridhar
Poor, Harold Vincent
Radner, Roy
Rainal, Attilio Joseph
Ramaswami, Vaidyanathan
Riggs, Richard
Robbins, Herbert Ellis
Robbins, Naomi Bograd
Sak, Joseph
Sarakwash, Michael
Schneider, Alfred Marcel
Siegel, Andrew Francis
Smouse, Peter Edgar
Steiger, William Lee
Sykes, Donald Joseph
Teicher, Henry
Thomas, Ronald Emerson
Tukey, John Wilder

Turner, Edwin Lewis
Vardi, Yehuda
Verdu, Sergio
Vernon, Russel
Watson, Geoffrey Stuart
Wood, Eric F
Wu, Lancelot T L

NEW MEXICO
Bruckner, Lawrence Adam
Campbell, Katherine Smith
Cardenas, Manuel
Cobb, Loren
Cogburn, Robert Francis
Faulkenberry, Gerald David
Finkner, Morris Dale
George, Timothy Gordon
Goldman, Aaron Sampson
Gutjahr, Allan L
Johnson, Mark Edward
Julian, William H
Koopmans, Lambert Herman
Kraichnan, Robert Harry
Lohrding, Ronald Keith
McKay, Michael Darrell
Martz, Harry Franklin, Jr
Ortiz, Melchor, Jr
Qualls, Clifford Ray
Rogers, Gerald Stanley
Waller, Ray Albert
Zeigler, Royal Keith

NEW YORK
Altland, Henry Wolf
Anderson, Allan George
Aroian, Leo Avedis
Arvesen, James Norman
Banerjee, Kali Shankar
Bechhofer, Robert E(ric)
Beil, Gary Milton
Bialas, Wayne Francis
Blessing, John A
Bross, Irwin Dudley Jackson
Capobianco, Michael F
Cargo, Gerald Thomas
Chatterjee, Samprit
Clatworthy, Willard Hubert
Collins, James Joseph
Crooke, James Stratton
Cunia, Tiberius
Danziger, Lawrence
Derman, Cyrus
Desu, Manavala Mahamunulu
Ehrenfeld, Sylvain
Emrich, Lawrence James
Emsley, James Alan Burns
Farrell, Roger Hamlin
Ferguson, David Lawrence
Finch, Stephen Joseph
Fine, Terrence Leon
Freund, Richard A
Geller, Nancy L
Goel, Amrit Lal
Goldfarb, Nathan
Gordon, Florence S
Gordon, Sheldon P
Hachigian, Jack
Hahn, Gerald John
Hall, William Jackson
Hanson, David Lee
Harris, Stewart
Herbach, Leon Howard
Hunte, Beryl Eleanor
Jackson, James Edward
Kana, Alfred Jan
Kao, Samuel Chung-Siung
Karatzas, Ioannis
Kaufman, Sol
Keilson, Julian
King, Alan Jonathan
Kolesar, Peter John
Kurnow, Ernest
Kuzmak, Joseph Milton
Laska, Eugene
Lawton, William Harvey
Lerman, Steven I
Levene, Howard
Levin, Bruce
Li, Chou H(siung)
Lieberman, Burton Barnet
Lininger, Lloyd Lesley
Link, Richard Forest
Locke, Charles Stephen
Lorenzen, Jerry Alan
McCarthy, Philip John
Mahtab, M Ashraf
Marshall, Clifford Wallace
Mehrotra, Kishan Gopal
Melnick, Edward Lawrence
Mike, Valerie
Morrison, Nathan
Mudholkar, Govind S
Mundel, August B(aer)
Nagel, Harry Leslie
Nelson, Wayne Bryce
Norton, Larry
Oakes, David
Owen, Joel
Paulson, Edward
Prabhu, Narahari Umanath
Rao, Poduri S R S
Richter, Donald
Riffenburgh, Robert Harry
Rosberg, Zvi

Roshwalb, Irving
Rothenberg, Ronald Isaac
Ruppert, David
Schilling, Edward George
Schmee, Josef
Schoefer, Ernest A(lexander)
Schroeder, Anita Gayle
Searle, Shayle Robert
Sedransk, Joseph Henry
Shah, Bhupendra K
Shane, Harold D
Sherman, Malcolm J
Sievers, Sally Riedel
Simon, Gary Albert
Singh, Chanchal
Slezak, Jane Ann
Sommer, Charles John
Star, Martin Leon
Stedinger, Jery Russell
Stein, Arthur
Stephany, Edward O
Stiteler, William Merle, III
Straight, H Joseph
Tanner, Martin Abba
Terzuoli, Andrew Joseph
Turnbull, Bruce William
Weiss, Lionel Ira
Welch, Peter D
Whitby, Owen
Wilkinson, John Wesley
Wong, Edward Chor-Cheung
Yablon, Marvin
Zacks, Shelemyahu

NORTH CAROLINA
Benrud, Charles Harris
Bergsten, Jane Williams
Bhattacharyya, Bibhuti Bhushan
Boos, Dennis Dale
Burdick, Donald Smiley
Carpenter, Benjamin H(arrison)
Dinse, Gregg Ernest
Dotson, Marilyn Knight
Eisen, Eugene J
Farthing, Barton Roby
Gerig, Thomas Michael
Glenn, William Alexander
Gordy, Thomas D(aniel)
Hader, Robert John
Hafley, William LeRoy
Haynie, Fred Hollis
Herr, David Guy
Hoeffding, Wassily
Holbert, Donald
Hooke, Robert
Horner, Theodore Wright
Horvitz, Daniel Goodman
Johnson, Norman Lloyd
Johnson, Thomas
Koehler, Truman L, Jr
Koop, John C
Kozelka, Robert M
Kuebler, Roy Raymond, Jr
Leduc, Sharon Kay
Lessler, Judith Thomasson
Linnerud, Ardell Chester
McVay, Francis Edward
Manson, Allison Ray
Margolin, Barry Herbert
Mason, David Dickenson
Mason, Robert Edward
Monroe, Robert James
Nair, Sreekantan S
Nelson, A Carl, Jr
Norgaard, Nicholas J
Papadopoulos, Alex Spero
Peterson, David West
Poole, William Kenneth
Potthoff, Richard Frederick
Powers, William Allen, III
Quade, Dana Edward Anthony
Quesenberry, Charles P
Reinfurt, Donald William
Rhyne, A Leonard
Sen, Pranab Kumar
Shah, Babubhai Vadilal
Singh, Kanhaya Lal
Smith, Walter Laws
Steel, Robert George Douglas
Van Der Vaart, Hubertus Robert
Wesler, Oscar
Whipkey, Kenneth Lee
Whitmore, Roy Walter

NORTH DAKOTA
Gill, Dhanwant Singh
Hertsgaard, Doris M Fisher
Landry, Richard Georges
Williams, John Delane
Winger, Milton Eugene

OHIO
Adorno, David Samuel
Barr, David Ross
Berens, Alan Paul
Beyer, William Hyman
Bhargava, Triloki Nath
Blumenthal, Saul
Caldito, Gloria C
Chern, Wen Shyong
Choung, Hun Ryang
Chrissis, James W
Cooper, Tommye
Daschbach, James McCloskey

Mathematical Statistics (cont)

Davis, Wilford Lavern
Deddens, James Albert
Disko, Mildred Anne
Goel, Prem Kumar
Guo, Shumei
Gupta, Arjun Kumar
Gustafson, Steven Carl
Hull, David Lee
Khamis, Harry Joseph
Klein, John Peter
Low, Leone Yarborough
McCloskey, John W
McLemore, Benjamin Henry, Jr
McNichols, Roger J(effrey)
Moeschberger, Melvin Lee
Muller, Mervin Edgar
Notz, William Irwin
Orban, John Edward
Park, Won Joon
Patton, Jon Michael
Powers, Jean D
Ray-Chaudhuri, Dwijendra Kumar
Reilly, Charles Bernard
Reynolds, David Stephen
Rohatgi, Vijay
Ross, Louis
Santner, Joseph Frank
Santner, Thomas Joseph
Schutta, James Thomas
Seabaugh, Pyrtle W
Silverman, Robert
Sterrett, Andrew
Sweet, Leonard
Whitney, Donald Ransom
Willke, Thomas Aloys
Yiamouyiannis, John Andrew

OKLAHOMA

Anderson, Paul Dean
Carr, Meg Brady
Chambers, Richard Lee
Folk, Earl Donald
Folks, John Leroy
Koller, Glenn R
Lee, Lloyd Lieh-Shen
Parker, Don Earl
Pope, Paul Terrell
Testerman, Jack Duane
Vasicek, Daniel J
Weeks, David Lee

OREGON

Ahuja, Jagdish C
Andrews, Fred Charles
Birkes, David Spencer
Brunk, Hugh Daniel
Calvin, Lyle D
Enneking, Eugene A
Magwire, Craig A
Murphy, Allan Hunt
Nelsen, Roger Bain
Peterson, Reider Sverre
Ramsey, Fred Lawrence
Rowe, Kenneth Eugene
Seely, Justus Frandsen
Tate, Robert Flemming
Truax, Donald R
Urquhart, N Scott

PENNSYLVANIA

Adams, John William
Antle, Charles Edward
Bai, Zhi Dong
Beggs, William John
Blaisdell, Ernest Atwell, Jr
Block, Henry William
Callen, Herbert Bernard
Chelemer, Harold
Clelland, Richard Cook
De Cani, John Stapley
DeGroot, Morris H
Demskey, Sidney
Duncan, George Thomas
Eddy, William Frost
Eisenberg, Bennett
Emlemdi, Hasan Bashir
Esmen, Nurtan A
Feigelson, Eric Dennis
Fienberg, Stephen Elliott
Fryling, Robert Howard
Galambos, Janos
Ghosh, Bhaskar Kumar
Ginsburg, Herbert
Gleser, Leon Jay
Grandage, Arnold Herbert Edward
Hargrove, George Lynn
Harkness, William Leonard
Helmbold, Robert Lawson
Hensler, Gary Lee
Hettmansperger, Thomas Philip
Hildebrand, David Kent
Hultquist, Robert Allan
Iglewicz, Boris
Isett, Robert David
Iversen, Gudmund R
Jeruchim, Michel Claude
Kadane, Joseph Born
Kassam, Saleem Abdulali
Ko, Frank K
Koller, Robert Dene
Krishnaiah, Paruchuri Rama
Kushner, Harvey

Kussmaul, Keith
Larsen, Russell D
Lautenberger, William J
Lehoczky, John Paul
Lindsay, Bruce George
Lustbader, Edward David
McCune, Duncan Chalmers
McLaughlin, Francis X(avier)
McLennan, James Alan, Jr
Mehta, Jatinder S
Merrill, Samuel, III
Meyer, John Sigmund
Michalik, Edmund Richard
Morrison, Donald Franklin
Murray, Donald Shipley
Nair, K Aiyappan
Parnes, Milton N
Patil, Ganapati P
Pickands, James, III
Pierson, Ellery Merwin
Pinski, Gabriel
Richards, Winston Ashton
Rigby, Paul Herbert
Ryan, Thomas Arthur, Jr
Schervish, Mark John
Seltzer, James Edward
Shaman, Paul
Singh, Jagbir
Sokolowski, Henry Alfred
Sorensen, Frederick Allen
Stewart, William Charles
Stoyle, Judith
Sullivan, George Allen
Treadwell, Kenneth Myron
Vaux, James Edward, Jr
Ventura, Jose Antonio
Wagner, Clifford Henry
Wei, William Wu-Shyong
Westlake, Wilfred James

RHODE ISLAND

Carney, Edward J
Grenander, Ulf
Hanumara, Ramachandra Choudary
Hemmerle, William J
Jarrett, Jeffrey E
Lawing, William Dennis
McClure, Donald Ernest

SOUTH CAROLINA

Byrd, Wilbert Preston
Dobbins, James Gregory Hall
Drane, John Wanzer
Felling, William E(dward)
Glocker, Edwin Merriam
Gross, Alan John
Holmes, Paul Thayer
Hubbell, Douglas Osborne
Hunt, Hurshell Harvey
Ling, Robert Francis
Padgett, William Jowayne
Russell, Charles Bradley
Surak, John Godfrey
Turner, Albert Joseph, Jr
Visvanathan, T R
Walker, James Wilson
Yu, Kai Fun

TENNESSEE

Bayne, Charles Kenneth
Beauchamp, John J
Bowman, Kimiko Osada
Cross, James Thomas
Gardiner, Donald Andrew
Ginnings, Gerald Keith
Gosslee, David Gilbert
Lasater, Herbert Alan
Lowe, James Urban, II
McHenry, Hugh Lansden
McNutt, Charles Harrison
Redman, Charles Edwin
Samejima, Fumiko
Sherman, Gordon R
Somes, Grant William
Tan, Wai-Yuan
Upadhyaya, Belle Raghavendra
Uppuluri, V R Rao
Walker, Grayson Howard
Williams, Robin O('Dare)
Younger, Mary Sue

TEXAS

Adams, Jasper Emmett, Jr
Atkinson, Edward Neely
Bedgood, Dale Ray
Bhat, Uggappakodi Narayan
Bland, Richard P
Bratcher, Thomas Lester
Brockett, Patrick Lee
Broemeling, Lyle David
Brooks, Dorothy Lynn
Brown, Bradford S
Carroll, Raymond James
Cecil, David Rolf
Chen, Peter
Conover, William Jay
Cooke, William Peyton, Jr
Cruse, Carl Max
Duran, Benjamin S
Eubank, Randall Lester
Freund, Rudolf Jakob
Gehan, Edmund A
Gentle, James Eddie
Hallum, Cecil Ralph

Han, Chien-Pai
Harrist, Ronald Baxter
Hatch, John Phillip
Hinkley, David Victor
Hocking, Ronald Raymond
John, Peter William Meredith
Johnson, Richard Leon
Johnston, Walter Edward
Kimeldorf, George S
Kirk, Roger E
Lee, Eun Sul
Matis, James Henry
Maute, Robert Edgar
Maxwell, Donald A
Michalek, Joel Edmund
Moore, Edwin Forrest
Moore, Kris
Newton, Howard Joseph
Nickel, James Alvin
Odell, Patrick L
Owen, Donald Bruce
Parzen, Emanuel
Prabhu, Vasant K
Ringer, Larry Joel
Rosenblatt, Judah Isser
Sager, Thomas William
Schucany, William Roger
Schuster, Eugene F
Sielken, Robert Lewis, Jr
Smith, James Douglas
Smith, Patricia Lee
Smith, William Boyce
Sobol, Marion Gross
Stein, William Edward
Styblinski, Maciej A
Thompson, James Robert
Turner, Danny William
Turner, John Charles
Vance, Joseph Francis
Van Ness, John Winslow
Walling, Derald Dee
Webster, John Thomas
Wehrly, Thomas Edward
Wichern, Dean William
Wiorkowski, John James
Wright, Henry Albert
Wyllys, Ronald Eugene

UTAH

Bryce, Gale Rex
Faulkner, James Earl
Hilton, Horace Gill
Hurst, Rex LeRoy
Nielson, Howard Curtis
Oyler, Dee Edward
Paegle, Julia Nogues
Reading, James Cardon
Rencher, Alvin C
Richards, Dale Owen
Turner, David L(ee)
Walker, LeRoy Harold

VERMONT

Costanza, Michael Charles
Emerson, John David
Haugh, Larry Douglas
Hill, David Byrne
Wooding, William Minor

VIRGINIA

Arnold, Jesse Charles
Bailar, Barbara Ann
Bauer, David Francis
Berger, Beverly Jane
Bolstein, Arnold Richard
Bush, Norman
Corwin, Thomas Lewis
Davidson, John Richard
Dick, Ronald Stewart
Drew, Lawrence James
Evans, David Arthur
Fink, Lester Harold
Fisher, Gail Feimster
Gardenier, John Stark
Good, Irving John
Greenblatt, Seth Alan
Harshbarger, Boyd
Healy, John Joseph
Heller, Agnes S
Henry, Neil Wylie
Herd, G Ronald
Hinkelmann, Klaus Heinrich
Irick, Paul Eugene
Jensen, Donald Ray
Kramer, Clyde Young
Krutchkoff, Richard Gerald
Kullback, Joseph Henry
Kupperman, Morton
Lawrence, Philip Signor
Leczynski, Barbara Ann
Loatman, Robert Bruce
McLaughlin, Gerald Wayne
Mikhail, Nabih N
Minton, Paul Dixon
Moyer, Leroy
Myers, Raymond Harold
Narula, Subhash Chander
O'Neill, Brian
Orcutt, Bruce Call
Perry, Charles Rufus, Jr
Pirie, Walter Ronald
Ramirez, Donald Edward
Reynolds, Marion Rudolph, Jr
Richardson, Henry Russell

Rowe, William David
Schneider, Philip Allen David
Solomon, Vasanth Balan
Stroup, Cynthia Roxane
Tang, Victor Kuang-Tao
Walpole, Ronald Edgar
Wegman, Edward Joseph
Wine, Russell Lowell

WASHINGTON

Bennett, Carl Allen
Bogyo, Thomas P
Bolme, Mark W
Chapman, Roger Charles
Chiu, John Shih-Yao
Conquest, Loveday Loyce
Farnum, Peter
Ghurye, Sudhish G
Hansen, Rodney Thor
Kappenman, Russell Francis
Nicholson, Wesley Lathrop
Perlman, Michael David
Pina, Eduardo Isidorio
Prentice, Ross L
Pyke, Ronald
Riedel, Eberhard Karl
Roberts, Norman Hailstone
Russell, Thomas Solon
Saunders, Sam Cundiff
Van Belle, Gerald

WEST VIRGINIA

Brady, Dennis Patrick
Wearden, Stanley

WISCONSIN

Bates, Douglas Martin
Berthouex, Paul Mac
Bhattacharyya, Gouri Kanta
Bisgaard, Søren
Box, George Edward Pelham
Chang, Tsong-How
Draper, Norman Richard
Farrow, John H
Girard, Dennis Michael
Gopal, Raj
Gurland, John
Harris, Bernard
Hickman, James Charles
Hill, William Joseph
Hunter, William G(ordon)
Johnson, Charles Henry
Jowett, David
Kaye, Norman Joseph
Lawrence, Willard Earl
Lochner, Robert Herman
Miller, Paul Dean
Miller, Robert Burnham
Ogden, Robert Verl
Reinsel, Gregory Charles
Rimm, Alfred A
Silverman, Franklin Harold
Stoneman, David McNeel
Wahba, Grace
Walter, Gilbert G

WYOMING

Borgman, Leon Emry
George, John Harold
Guenther, William Charles

PUERTO RICO

Bangdiwala, Ishver Surchand

ALBERTA

Blais, J A Rodrigue
Consul, Prem Chandra
Enns, Ernest Gerhard
Kelker, Douglas
Scheinberg, Eliyahu
Wani, Jagannath K

BRITISH COLUMBIA

Booth, Andrew Donald
Eaves, David Magill
Hall, John Wilfred
Kaller, Cecil Louis
Lam, Gabriel Kit Ying
Odeh, Robert Eugene
Peterson, Raymond Glen
Promnitz, Lawrence Charles
Routledge, Richard Donovan
Srivastava, Rekha
Stephens, Michael A
Villegas, Cesareo
Zidek, James Victor

MANITOBA

Chan, Lai Kow
Kocherlakota, Kathleen
Kocherlakota, Subrahmaniam
Paul, Gilbert Ivan
Sinha, Snehesh Kumar
Zeiler, Frederick

NEW BRUNSWICK

Basilevsky, Alexander
Mureika, Roman A
Pham-Gia, Thu

NEWFOUNDLAND

Hiscott, Richard Nicholas

NOVA SCOTIA
Grant, Douglass Lloyd
Kabe, Dattatraya G

ONTARIO
Andrews, David F
Bacon, David W
Balakrishnan, Narayanaswany
Batra, Tilak Raj
Behara, Minketan
Blyth, Colin Ross
Brainerd, Barron
Dale, Douglas Keith
Dawson, Donald Andrew
DeLury, Daniel Bertrand
Denzel, George Eugene
Fraser, Donald Alexander Stuart
Gentleman, Jane Forer
Glew, David Neville
Godson, Warren Lehman
Graham, John Elwood
Haq, M Safiul
Hipel, Keith William
Jamieson, Derek Maitland
Jardine, John McNair
Kalbfleisch, James G
MacNeill, Ian B
Mohanty, Sri Gopal
Mullen, Kenneth
Nash, John Christopher
Paull, Allan E
Raktoe, B Leo
Rao, Jonnagadda Nalini Kanth
Reilly, Park McKnight
Rust, Velma Irene
Schaufele, Ronald A
Shklov, Nathan
Sprott, David Arthur
Srivastava, Muni Shanker
Stroud, Thomas William Felix
Tan, Peter Ching-Yao
Thompson, Mary Elinore
Tracy, Derrick Shannon
Wasan, Madanlal T
Watts, Donald George
West, Eric Neil
Wolynetz, Mark Stanley
Wong, Chi Song
Wonnacott, Thomas Herbert

QUEBEC
Allen, Harold Don
Châtillon, Guy
Giri, Narayan C
Jumarie, Guy Michael
Lefebvre, Mario
Lehman, Eugene H
Maag, Urs Richard
Mathai, Arak M
Morgera, Salvatore Domenic
O'Meara, Desmond
Plotkin, Eugene Isaak
Rahn, Armin
Sankoff, David
Seshadri, Vanamamalai
Srivastava, Trilokv N
St-Pierre, Jacques
Yalovsky, Morty A

SASKATCHEWAN
O'Shaughnessy, Charles Dennis
Singh, Rajinder

OTHER COUNTRIES
Allen, Robin Leslie
Camiz, Sergio
Cheng, Kuang-Fu
Chesson, Peter Leith
Das Gupta, Somesh
Deely, John Joseph
Durling, Frederick Charles
Foglio, Susana
Francis, Ivor Stuart
Garcia-Santesmases, Jose Miguel
Hochberg, Kenneth J
Konijn, Hendrik Salomon
McCue, Edmund Bradley
Malik, Mazhar Ali Khan
Mara, Michael Kelly
Mengali, Umberto
Philippou, Andreas Nicolaou
Samuel-Cahn, Ester
Siddiqui, Mohammed Moinuddin
Snyder, Mitchell
Stephens, Kenneth S
Sunahara, Yoshifumi
Van Eeden, Constance

Mathematics, General

ALABAMA
Ainsworth, Oscar Richard
Ball, Richard William
Benington, Frederick
Blake, John Archibald
Bond, Albert F
Burton, Leonard Pattillo
Cappas, C
Chow, Bryant
Cook, Frederick Lee
Crittenden, Richard James
Easterday, Kenneth E
Fiesler, Emile

Fitzpatrick, Ben, Jr
Gibson, Peter Murray
Hill, Paul Daniel
Hobby, Charles R
Howell, James Levert
Hutchison, Jeanne S
Kiser, Lola Frances
Locker, John L
Lovingood, Judson Allison
McGill, Suzanne
Mishra, Satya Narayan
Moore, Robert Laurens
Neggers, Joseph
Padulo, Louis
Peeples, William Dewey, Jr
Rice, Barbara Slyder
Rodgers, Aubrey
Roush, Fred William
Segal, Arthur Cherny
Shipman, Jerry
Stone, Max Wendell
Vaughan, Loy Ottis, Jr
Vinson, Richard G
Wilson, Walter Lucien, Jr

ALASKA
Van Veldhuizen, Philip Androcles

ARIZONA
Bedient, Jack DeWitt
Buell, Carleton Eugene
Cheema, Mohindar Singh
Clay, James Ray
Dougherty, John Joseph
Franz, Otto Gustav
Goldstein, Myron
Gray, Allan, Jr
Grove, Larry Charles
Hodges, Carl Norris
Johnson, Lee Murphy
Kelly, John Beckwith
Larney, Violet Hachmeister
Leonard, John Lander
Little, Charles Edward
Lovelock, David
McDonald, John N
McMahon, Douglas Charles
Meserve, Bruce Elwyn
Micklich, John R
Moore, Charles Godat
Moore, John Douglas
Myers, Donald Earl
Nering, Evar Dare
Pollin, Jack Murph
Rieke, Carol Anger
Rund, Hanno
Sansone, Fred J
Savage, Nevin William
Sherman, Thomas Lawrence
Smarandache, Florentin
Smith, Hal Leslie
Smith, Harvey Alvin
Smith, Lehi Tingen
Stewart, Donald George
Strebe, David Diedrich
Thompson, Richard Bruce
Walter, Everett L
Wood, Bruce

ARKANSAS
Fuller, Roy Joseph
Hudson, Frank M
Keesee, John William
Kimura, Naoki
Linnstaedter, Jerry Leroy
Madison, Bernard L
Meek, James Latham
Moon, Wilchor David
Moore, Marvin G
Oldham, Bill Wayne
Orton, William R, Jr
Scroggs, James Edward
Smith, Robert Paul
Starling, Albert Gregory
Walls, Walter Clarence
Wild, Wayne Grant

CALIFORNIA
Abraham, Ralph Herman
Adams, Ralph Melvin
Aissen, Michael Israel
Albert, Eugene
Allen, Arnold Oral
Alperin, Morton
Amemiya, Frances (Louise) Campbell
Ames, Dennis Burley
Anderson, A Keith
Anderson, Donald Werner
Apostol, Tom M
Arens, Richard Friederich
Armacost, William L
Arnold, Hubert Andrew
Arnold, Robert Fairbanks
Arveson, William Barnes
Astrachan, Max
Astromoff, Andrew
Austin, Charles Ward
Bade, William George
Bars, Itzhak
Basinger, Richard Craig
Becker, Gerald Anthony
Beechler, Barbara Jean
Bell, Charles Bernard, Jr
Bender, Edward Anton

Benson, Donald Charles
Berlekamp, Elwyn R(alph)
Biles, Charles Morgan
Biriuk, George
Birnbaum, Sidney
Birnir, Björn
Bishop, Errett A
Blackwell, David (Harold)
Block, Richard Earl
Borges, Carlos Rego
Bourne, Samuel G
Bressler, David Wilson
Brown, George William
Calvert, Ralph Lowell
Cantor, David Geoffrey
Ceder, Jack G
Chang, I-Dee
Chatland, Harold
Chellappa, Ramalingam
Chernick, Michael Ross
Chicks, Charles Hampton
Chinn, Phyllis Zweig
Chorin, Alexandre J
Chow, Tseng Yeh
Christianson, Deann, Jr
Christopher, John
Chung, Kai Lai
Church, Alonzo
Clark, Charles Lester
Clarke, Wilton E L
Coddington, Earl Alexander
Cohen, Paul Joseph
Cottle, Richard W
Cullen, Theodore John
Culley, Benjamin Hays
Cummings, Frederick W
Cutler, Leonard Samuel
David, Florence N
Davison, Walter Francis
Dean, Richard Albert
Deaton, Edmund Ike
DeLand, Edward Charles
De Phillis, John
Dimsdale, Bernard
Douglas, Robert James
Drobnies, Saul Isaac
Dubins, Lester Eli
Dugundji, James
Duncan, Donald Gordon
Dye, Henry Abel
Edelson, Allan L
Edelson, Sidney
Effros, Edward George
Ellis, Wade
Emerson, Thomas James
Ernest, John Arthur
Fan, Ky
Farrell, Edward Joseph
Fateman, Richard J
Feldman, Jerome A
Feldman, Leonard
Fendel, Daniel
Fisher, Newman
Fitting-Gifford, Marjorie Ann Premo
Fitzgerald, Carl
Ford, Lester Randolph, Jr
Foster, Lorraine L
Frankel, Theodore Thomas
Freund, Roland Wilhelm
Friedman, Henry David
Gale, David
Garsia, Adriano Mario
Gaver, Donald Paul
Gavin, Lloyd Alvin
Getoor, Ronald Kay
Gibson, James (Benjamin)
Gillman, David
Gittleman, Arthur P
Golomb, Solomon Wolf
Gordon, Samuel Robert
Goss, Robert Nichols
Gould, William E
Gragg, William Bryant, Jr
Graves, Glenn William
Gray, A(ugustine) H(eard), Jr
Green, John Willie
Greenberg, Marvin Jay
Greenspan, Bernard
Grossman, Nathaniel
Gruenberger, Fred J(oseph)
Halberg, Charles John August, Jr
Hales, Alfred Washington
Halkin, Hubert
Haltiner, George Joseph
Hardy, John W, Jr
Hart, Garry Dewaine
Hartshorne, Robert (Robin) Cope
Harvey, Albert Raymond
Hawley, Newton Seymour, Jr
Hayes, Charles Amos, Jr
Hayes, Robert Mayo
Healy, Paul William
Hedstrom, Gerald Walter
Henriksen, Melvin
Heynick, Louis Norman
Hirsch, Peter M
Hobbs, Billy Frankel
Hochschild, Gerhard P
Holl, Manfred Matthias
Holmes, Calvin Virgil
Holsztynski, Wlodzimierz
Hopponen, Jerry Dale
Householder, Alston Scott
Householder, James Earl

Hsu, Yu-Sheng
Hu, Sze-Tsen
Hudson, George Elbert
Hughart, Stanley Parlett
Hughes, Thomas Rogers, Jr
Hung, John Wenchung
Hurd, Cuthbert Corwin
Huskey, Harry D(ouglas)
Hyers, Donald Holmes
Jaffa, Robert E
James, Ralph L
Janos, Ludvik
Javaher, James N
Jayachandran, Toke
Johnson, Mark Scott
Juncosa, Mario Leon
Kahan, William M
Kalisch, Gerhard Karl
Kanarik, Rosella
Kaplansky, Irving
Karush, William
Kato, Tosio
Kelley, Allen Frederick, Jr
Kelley, John Le Roy
Kelly, Paul J
Kevorkian, Aram K
Kimme, Ernest Godfrey
King, R Maurice, Jr
Kipps, Thomas Charles
Knuth, Donald Ervin
Kobayashi, Shoshichi
Konheim, Alan G
Koval, Daniel
Krakowski, Fred
Kreith, Kurt
Krom, Melven R
Lai, Tze Leung
Landesman, Edward Milton
Lane, Richard Neil
Lange, Lester Henry
Lanski, Charles Philip
Larmore, Lawrence Louis
Lashof, Richard Kenneth
Latta, Gordon
Lawson, Charles L
Lax, Melvin David
Lehman, R Sherman
Lehmer, Derrick Henry
Leipnik, Roy Bergh
Lender, Adam
Lengyel, Bela Adalbert
Lenstra, Hendrik W
Lin, Tung-Po
Liu, Chi-Sheng
Lockhart, Brooks Javins
Lovaglia, Anthony Richard
Luttmann, Frederick William
Lutz, Donald Alexander
Lynch, William C
McCready, Thomas Arthur
Mach, George Robert
McKenzie, Ralph Nelson
McLeod, Edward Blake
Maksoudian, Y Leon
Manjarrez, Victor M
Marble, Frank E(arl)
Marcus, Bryan Harry
Mardellis, Anthony
Margulies, William George
Marley, Gerald C
Marsden, Jerrold Eldon
Matlow, Sheldon Leo
Mendis, Devamitta Asoka
Michael, William Alexander
Miech, Ronald Joseph
Milgram, Richard James
Mitchem, John Alan
Monier, Louis Marcel
Moore, Calvin C
Morgan, John Clifford, II
Morgan, Kathryn A
Morris, George William
Moser, Louise Elizabeth
Mumme, Judith E
Murphy, James Lee
Myers, William Howard
Najar, Rudolph Michael
Neustadter, Siegfried Friedrich
Newcomb, Robert Lewis
Newman, Morris
Norton, Donald Alan
Nothnagel, Eugene Alfred
Nower, Leon
Olshen, Richard Allen
Orchard, Henry John
Orland, George H
Ornstein, Donald Samuel
Osher, Stanley Joel
Pacholka, Kenneth
Pease, Marshall Carleton, III
Perry, William James
Pfeffer, Washek F
Pflugfelder, Hala
Phillips, Ralph Saul
Pignataro, Augustus
Port, Sidney C
Potts, Donald Harry
Preston, Gerald Cowles
Price, Charles Morton
Proskurowski, Wlodzimierz
Protter, Murray Harold
Purcell, Everett Wayne
Purvis, Colbert Thaxton
Radnitz, Alan

Mathematics, General (cont)

Ralston, James Vickroy, Jr
Ramsey, O C, Jr
Rebman, Kenneth Ralph
Redheffer, Raymond Moos
Reed, Irving Stoy
Rhodes, John Lewis
Riley, James Daniel
Rinehart, Robert Fross
Rissanen, Jorma Johannes
Ritchie, Robert Wells
Robertson, James Byron
Robertson, Thomas N
Robinson, Raphael Mitchel
Rodin, Burton
Rohrl, Helmut
Rose, Gene Fuerst
Rosenfeld, Melvin
Rosenlicht, Maxwell
Rothschild, Bruce Lee
Rozsnyai, Balazs
Ruggles, Ivan Dale
Rush, David Eugene
Sabharwal, Ranjit Singh
Saltz, Daniel
Samelson, Hans
Samuelson, Donald James
Sanderson, Judson
Sarason, Donald Erik
Savitch, Walter John
Scheinok, Perry Aaron
Schindler, Guenter Martin
Schlesinger, Stewart Irwin
Schneider, Harold O
Schoen, Richard M(elvin)
Schoenfeld, Alan Henry
Seidenberg, Abraham
Seppi, Edward Joseph
Shapley, Lloyd Stowell
Sharpe, Michael John
Shen, Nien-Tsu
Shen, Y(ung) C(hung)
Shepherd, William Lloyd
Sherry, Edwin J
Shiffman, Max
Shiflett, Ray Calvin
Short, Donald Ray, Jr
Shreve, David Carr
Shubert, Bruno Otto
Simon, Arthur Bernard
Sloss, James M
Smale, Stephen
Smith, Donald Ray
Smith, James Thomas
Smith, Larry
Smith, Marianne (Ruth) Freundlich
Smith, Marion Bush, Jr
Smith, William R
Solovay, Robert M
Sowder, Larry K
Specht, Robert Dickerson
Sprecher, David A
Stanek, Peter
Stannard, William A
Stavroudis, Orestes Nicholas
Stein, Charles M
Stein, James D, Jr
Stein, Sherman Kopald
Steinberg, Robert
Stiel, Edsel Ford
Stillman, Richard Ernest
Stone, Alexander Glattstein
Stone, Charles Joel
Stralka, Albert R
Strauch, Ralph Eugene
Sturgis, Howard Ewing
Summers, Audrey Lorraine
Swann, Howard Story Gray
Swenson, Donald Adolph
Takesaki, Masamichi
Tang, Hwa-Tsang
Tatum, Freeman A(rthur)
Taylor, Angus Ellis
Terry, Raymond Douglas
Thompson, Robert Charles
Thorp, Edward O
Tolsted, Elmer Beaumont
Topp, William Robert
Trahan, Donald Herbert
Tromba, Anthony Joseph
Tropp, Henry S
Tucker, Roy Wilbur
Tully, Edward Joseph, Jr
Ung, Man T
Van de Wetering, Richard Lee
Vaught, Robert L
Veigele, William John
Venit, Stewart Mark
Verdina, Joseph
Wagoner, Ronald Lewis
Waiter, Serge-Albert
Wales, David Bertram
Wallen, Clarence Joseph
Warten, Ralph Martin
Waterman, Michael S
Watkins, William
Weidlich, John Edward, Jr
Weihe, Joseph William
Weinbaum, Carl Martin
Welch, Lloyd Richard
Welmers, Everett Thomas
Wernick, Robert J
Wets, Roger J B

White, Alvin Murray
White, Paul A
Whiteman, Albert Leon
Whitmore, William Francis
Wilde, Carroll Orville
Winzenread, Marvin Russell
Wong, K(wee) C
Woo, Norman Tzu Teh
Wooldridge, Kent Ernest
Wrede, Robert C, Jr
Wright, Elisabeth Muriel Jane
Wulff, John Leland
Yale, Paul B
Yeh, James Jui-Tin
Zeitlin, Joel Loeb

COLORADO

Allan, David Wayne
Allgower, Eugene L
Baggett, Lawrence W
Bardwell, George
Bennett, W Scott
Boes, Ardel J
Brown, Austin Robert, Jr
Brown, Gordon Elliott
Butler, Walter Cassius
Buzbee, Billy Lewis
Cavanagh, Timothy D
Cicerone, Carol Mitsuko
Cohen, Joel Seymour
Colvis, John Paris
Crow, Edwin Louis
Darst, Richard B
Deal, Ervin R
DeMeyer, Frank R
Dorn, William S
Easton, Robert Walter
Eberhard, Wynn Lowell
Ellis, Homer Godsey
Elsey, Margaret Grace
Eyman, Earl Duane
Farnell, Albert Bennett
Fisch, Forest Norland
Fischer, Irwin
Fox, Bennett L
Frisinger, H Howard, II
Fulks, Watson
Gaines, Robert Earl
Goodrich, Robert Kent
Gudder, Stanley Phillip
Gustafson, Karl Edwin
Hansman, Margaret Mary
Hodges, John Herbert
Hoffman, Ruth I
Holley, Frieda Koster
Holley, Richard Andrew
Hoots, Felix R
Hufford, George (Allen)
Hundhausen, Joan Rohrer
Kafadar, Karen
Kelman, Robert Bernard
Klopfenstein, Kenneth F
Levinger, Bernard Werner
Litwhiler, Daniel W
Lundell, Albert Thomas
MacRae, Robert E
Magnus, Arne
Meyer, Burnett Chandler
Mielke, Paul W, Jr
Milligan, Merle Wallace
Monk, James Donald
Montague, Patricia Tucker
Mycielski, Jan
Niemann, Ralph Henry
Painter, Richard J
Payne, Stanley E
Phillips, David Lowell
Popejoy, William Dean
Raab, Joseph A
Rechard, Ottis William
Roeder, David William
Roth, Richard Lewis
Sather, Duane Paul
Schmidt, Wolfgang M
Simmons, George Finlay
Spencer, Donald Clayton
Srivastava, Jagdish Narain
Stein, Frederick Max
Taylor, Walter Fuller
Thomas, James William
Vilms, Jaak
Warren, Peter
Windholz, Walter M
Wouk, Arthur
Wyss, Walter
Zirakzadeh, Aboulghassem

CONNECTICUT

Aaboe, Asger (Hartvig)
Abikoff, William
Bannister, Peter Robert
Brunell, Gloria Florette
Calehuff, Girard Lester
Comfort, William Wistar
De Gray, Ronald Willoughby
Eby, Edward Stuart, Sr
Eby, Edward Stuart, Sr
Eisenberg, Sheldon Merven
Feit, Walter
Garland, Howard
Hedlund, Gustav Arnold
Hirsch, Warren Maurice
Ho, Grace Ping-Poo
Jacobson, Florence Dorfman

Jacobson, Nathan
Kagan, Joel (David)
Kakutani, Shizuo
Klimczak, Walter John
Lang, Serge
Linton, Fred E J
Locke, Stanley
MacDonnell, Joseph Francis
Markham, Elizabeth Mary
Mostow, George Daniel
Neuwirth, Jerome H
Nigam, Lakshmi Narayan
Perlis, Alan Jay
Poliferno, Mario Joseph
Raney, George Neal
Rickart, Charles Earl
Rosenbaum, Robert Abraham
Roulier, John Arthur
Schmerl, James H
Schrader, Dorothy Virginia
Seligman, George Benham
Sells, Jean Thurber
Shaffer, Dorothy Browne
Singleton, Robert Richmond
Spencer, Domina Eberle (Mrs Parry Moon)
Sternberg, Robert Langley
Stewart, Robert Clarence
Tollefson, Jeffrey L
Vinsonhaler, Charles I
Walde, Ralph Eldon
Welna, Cecilia
Whitley, William Thurmon
Whittlesey, Emmet Finlay
Wolk, Elliot Samuel
Woods, Jimmie Dale

DELAWARE

Baxter, Willard Ellis
Colton, David L
Doerner, William A(llen)
Gilbert, Robert Pertsch
Marquardt, Donald Wesley
Mishoe, Luna I
Nashed, Mohammed Zuhair Zaki
Pellicciaro, Edward Joseph
Roselle, David Paul
Thickstun, William Russell, Jr
Wenger, Ronald Harold

DISTRICT OF COLUMBIA

Aufenkamp, Darrel Don
Barrett, Eamon Boyd
Blankfield, Alan
Bobo, Edwin Ray
Bogdan, Victor Michael
Bram, Joseph
Dame, Richard Edward
Donaldson, James A
Egan, Howard L
Gordon, William Bernard
Gray, Mary Wheat
Hayek, Lee-Ann Collins
Hensel, Gustav
Hoffman, Kenneth Myron
Katz, Irving
Kenyon, Hewitt
Krause, Ralph M
Levy, Mark B
Lukacs, Eugene
McDonald, Bernard Robert
Matson, Michael
Moller, Raymond William
Osgood, Charles Freeman
Rose, Raymond Edward
Rosier, Ronald Crosby
Sanchez, David A
Siry, Joseph William
Stein, Marjorie Leiter
Stokes, Arnold Paul
Tucker, John Richard
Voytuk, James A
Williams, Leland Hendry

FLORIDA

Agins, Barnett Robert
Ahmad, Shair
Andregg, Charles Harold
Bagley, Robert Waller
Bednarek, Alexander R
Blake, Robert George
Brooks, James Keith
Bryant, John Logan
Bullis, George LeRoy
Butson, Alton Thomas
Byrkit, Donald Raymond
Chae, Soo Bong
Chen, Wayne H(wa-Wei)
Clark, William Edwin
Cleaver, Frank L
Cowling, Vincent Frederick
Debnath, Lokenath
DeSua, Frank Crispin
Eachus, Joseph Jackson
Ehrlich, Paul Ewing
Fagin, Karin
Fass, Arnold Lionel
Fine, Rana Arnold
Fitzgerald, Lawrence Terrell
Fulton, John David
Garman, Brian Lee
Garrett, James Richard
Gleyzal, Andre
Goldberg, Estelle Maxine

Goldberg, Lee Dresden
Goodman, Adolph Winkler
Grams, Anne P
Heerema, Nickolas
Heil, Wolfgang Heinrich
Hertzig, David
Jones, Roy Carl, Jr
Kalman, Rudolf Emil
Keedy, Mervin Laverne
Kelley, John Ernest
Kerr, Donald R
Kreimer, Herbert Frederick, Jr
Lakshmikantham, Vangipuram
Leja, Stanislaw
Levitz, Hilbert
Liang, Joseph Jen-Yin
Lighterman, Mark S
Livingood, John N B
Lofquist, George W
McArthur, Charles Wilson
McClure, Clair Wylie
McDougle, Paul E
Maddox, Billy Hoyte
Manougian, Manoug N
Martin, Kenneth Edward
Meacham, Robert Colegrove
Medlin, Gene Woodard
Miles, Ernest Percy, Jr
Moore, Theral Orvis
Nichols, Eugene Douglas
Nielsen, Kaj Leo
Norman, Edward
Novosad, Robert S
Palmer, Kenneth
Parr, William Chris
Pirkle, Fredric Lee
Pop-Stojanovic, Zoran Rista
Rose, Donald Clayton
Ryan, Peter Michael
Selfridge, Ralph Gordon
Sigmon, Kermit Neal
Sinclair, Annette
Smith, William K
Smythe, Robert C
Snider, Arthur David
Snover, James Edward
Squier, Donald Platte
Stiles, Wilbur F
Taylor, Michael Dee
Turner, Ralph Waldo
Ungvichian, Vichate
Van Zee, Richard Jerry
Varma, Arun Kumar
Villemure, M Paul James
Wahab, James Hatton
Ward, James Audley
Winton, Charles Newton
Zame, Alan

GEORGIA

Aust, Catherine Cowan
Barnes, Earl Russell
Bompart, Billy Earl
Brahana, Thomas Roy
Buccino, Alphonse
Clemmons, John B
Cliett, Otis Jay
Devries, David J
Duncan, Donald Lee
Duquette, Alfred L
Edwards, Charles Henry, Jr
Evans, Trevor
Evans, William Buell
Ford, David A
George, Dick Leon
Goslin, Roy Nelson
Hahn, Hwa Suk
Hale, Jack Kenneth
Hamrick, Anna Katherine Barr
Hardy, Flournoy Lane
Hollingsworth, John Gressett
Holtum, Alfred G(erard)
Horne, James Grady, Jr
Hudson, Anne Lester
Jackson, Prince A, Jr
Johnson, Roger D, Jr
Kammerer, William John
Karam, Ratib A(braham)
Kasriel, Robert H
Kilpatrick, Jeremy
Kunze, Ray A
Lavender, DeWitt Earl
Line, John Paul
McKinney, Max Terral
Mahavier, William S
Neff, John David
Oliker, Vladimir
Pittman, Chatty Roger
Pomerance, Carl
Robinson, Robert W
Rutledge, Dorothy Stallworth
Schleusner, John William
Shonkwiler, Ronald Wesley
Skypek, Dora Helen
Sledd, Marvin Banks
Smith, William Allen
Sparks, Arthur Godwin
Taylor, Howard Edward
Taylor, Robert Lee
Tong, Yung Liang
Tonne, Philip Charles
Wall, Donald Dines
Walton, Harriett J
Wilson, James William

Winkler, Peter Mann
Worth, Roy Eugene
Wyse, Frank Oliver
Zander, Vernon Emil

HAWAII
Allday, Christopher J
Barker, William Hamblin, II
Bear, Herbert S, Jr
Freese, Ralph
Fuller, Francis Brock
Heacox, William Dale
Luther, Norman Y
Oliphant, Malcolm William
Pager, David
Pitcher, Tom Stephen

IDAHO
Benson, Dean Clifton
Ferguson, David John
Henry, Boyd (Herbert)
Higdem, Roger Leon
Juola, Robert C
Mech, William Paul
Richardson, Lee S(pencer)
Wang, Ya-Yen Lee
Ward, Frederick Roger

ILLINOIS
Agnew, Robert A
Allard, Nona Mary
Allen, Frank B
Alperin, Jonathan L
Amick, Charles James
Angotti, Rodney
Ash, J Marshall
Ash, Robert B
Atkins, Ferrel
Austin, Donald Guy
Baily, Walter Lewis, Jr
Bank, Steven Barry
Barratt, Michael George
Bateman, Felice Davidson
Bateman, Paul Trevier
Becker, Jerry Page
Berg, Ira David
Berk, Kenneth N
Berman, Joel David
Biesterfeldt, Herman John, Jr
Billingsley, Patrick Paul
Bohrer, Robert Edward
Braunfeld, Peter George
Brigham, Nelson Allen
Brown, John Wesley
Brown, Joseph Ross
Burkholder, Donald Lyman
Burton, Theodore Allen
Butler, Margaret K
Calderon, Alberto Pedro
Catrambone, Joseph Anthony, Sr
Chen, Kuo-Tsai
Clough, Robert Ragan
Coon, Lewis Hulbert
Cormier, Romae Joseph
Cramton, Thomas James
Curtis, Herbert John
D'Angelo, John Philip
Darsow, William Frank
Davis, Stephen H(oward)
Deliyannis, Platon Constantine
Devinatz, Allen
Doob, Joseph Leo
Dornhoff, Larry Lee
Dwinger, Philip
Dwyer, Thomas Aloysius Walsh, III
Ecker, Edwin D
Evens, Leonard
Ferguson, William Allen
Fisher, Stephen D
Foland, Neal Eugene
Foulser, David A
Francis, George Konrad
Frank, Maurice Jerome
Franz, Edgar Arthur
Gates, Leslie Dean, Jr
Georgakis, Constantine
Goebel, Maristella
Goldberg, Samuel I
Graves, Robert Lawrence
Gray, Brayton
Gray, John Walker
Greenstein, David Snellenburg
Griffith, Phillip A
Grimmer, Ronald Calvin
Gundersen, Roy Melvin
Haken, Wolfgang
Hamstrom, Mary-Elizabeth
Hasdal, John Allan
Hattemer, Jimmie Ray
Hedayat, A Samad
Helms, Lester LaVerne
Hsu, Nai-Chao
Insel, Arnold J
Johnson, Robert Leroy
Jones, R E Douglas
Jordan, Steven Lee
Kahn, Daniel Stephen
Koch, Charles Frederick
Kuller, Robert G
Kwong, Man Kam
Langebartel, Ray Gartner
Langenhop, Carl Eric
Larson, Richard Gustavus
Lay, Steven R

Leonard, Henry Siggins, Jr
Lin, Sue Chin
Loeb, Peter Albert
McAlpin, John Harris
MacLane, Saunders
McLinden, Lynn
Marcus, Philip Selmar
Matlis, Eben
Maxwell, Charles Neville
Miles, Joseph Belsley
Mock, Gordon Duane
Monrad, Ditlev
Moore, Robert Alonzo
Morris, Herbert Allen
Moschandreas, Demetrios J
Mount, Kenneth R
Nafoosi, A Aziz
Nanda, Jagdish L
Nelson, Harry Ernest
O'Malley, Mary Therese
Osborn, Howard Ashley
Oursler, Clellie Curtis
Parker, E T
Parr, Phyllis Graham
Pearson, Hans Lennart
Peck, Emily Mann
Pedersen, Franklin D
Pinsky, Mark A
Pless, Vera Stepen
Porta, Horacio A
Potter, Meredith Woods
Pranger, Walter
Priddy, Stewart Beauregard
Putnam, Alfred Lunt
Radford, David Eugene
Raghavan, Thirukkannamangai E S
Reiner, Irma Moses
Reingold, Haim
Reisel, Robert Benedict
Rickert, Neil William
Ringenberg, Lawrence Albert
Robinson, Derek John Scott
Rodine, Robert Henry
Rotman, Joseph Jonah
Rubel, Lee Albert
Rutledge, Robert B
Sawinski, Vincent John
Schumaker, John Abraham
Schupp, Paul Eugene
Scott, Edward Joseph
Shelton, Ronald M
Shively, Ralph Leland
Shryock, A Jerry
Soare, Robert I
Sons, Linda Ruth
Srinivasan, Bhama
Stein, Michael Roger
Stipanowich, Joseph Jean
Stolarsky, Kenneth B
Sturley, Eric Avern
Sullivan, Michael Joseph, Jr
Suzuki, Michio
Swan, Richard Gordon
Szeto, George
Ting, Tsuan Wu
Tondeur, Philippe
Troy, Daniel Joseph
Troyer, Robert James
Tuzar, Jaroslav
Uhl, John Jerry, Jr
Vaughan, Herbert Edward
Wagreich, Philip Donald
Wantland, Evelyn Kendrick
Weichsel, Paul M
Weinberg, Elliot Carl
Weiner, Louis Max
Weinzweig, Avrum Israel
Welland, Robert Roy
Weller, Glenn Peter
Wendt, Arnold
Williams, Eddie Robert
Wirszup, Izaak
Wong, Yuen-Fat
Yntema, Mary Katherine
Zaring, Wilson Miles
Zygmund, Antoni

INDIANA
Abhyankar, Shreeram
Anderson, John R
Anderson, Virgil Lee
Bailey, Herbert R
Balka, Don Stephen
Bechtel, Robert D
Beineke, Lowell Wayne
Berkovitz, Leonard David
Bittinger, Marvin Lowell
Branges, Louis de
Brown, Arlen
Brown, Lawrence G
Buesking, Clarence W
Carlson, Kermit Howard
Chess, Karin V T
Cooley, Robert Lee
Cooney, Miriam Patrick
Cunningham, Ellen M
Dalphin, John Francis
DeMillo, Richard A
Diekhans, Herbert Henry
Drazin, Michael Peter
Drufenbrock, Diane
Dunham, William Wade
Finco, Arthur A
Fishback, William Thompson

Forbes, Jack Edwin
Fry, Cleota Gage
Fuller, William Richard
Gambill, Robert Arnold
Gass, Clinton Burke
Goetz, Abraham
Golomb, Michael
Goodman, Victor Wayne
Gottlieb, Daniel Henry
Gupta, Shanti Swarup
Haas, Felix
Hahn, Alexander J
Hamblen, John Wesley
Hanes, Harold
Hill, Robert Joe
Hood, Rodney Taber
Humberd, Jesse David
Hutton, Lucreda Ann
Jerison, Meyer
Kallman, Ralph Arthur
Kane, Robert B
Kaplan, Samuel
Keller, Marion Wiles
Kinney, John James
Kinsey, David Webster
Klinger, William Russell
Lipman, Joseph
Lowengrub, Morton
Ludwig, Hubert Joseph
McCormick, Roy L
Machtinger, Lawrence Arnold
Mansfield, Maynard Joseph
Mastrototaro, John Joseph
Mielke, Paul Theodore
Miller, John Grier
Minassian, Donald Paul
Morrel, Bernard Baldwin
Morrill, John Elliott
Morrison, Clarence C
Neugebauer, Christoph Johannes
Neuhouser, David Lee
Novodvorsky, Mark Benjamin
O'Meara, O Timothy
Otter, Richard Robert
Peterson, Harold LeRoy
Putnam, Calvin Richard
Reed, Ellen Elizabeth
Regan, Francis
Rhoades, Billy Eugene
Robold, Alice Ilene
Rubin, Herman
Rueve, Charles Richard
Schober, Glenn E
Schultz, Reinhard Edward
Schurle, Arlo Willard
Seifert, Ralph Louis
Sherman, Gary Joseph
Silverman, Edward
Smith, Peter David
Smith, Shelby Dean
Springer, George
Stanton, Nancy Kahn
Stephens, Stanley LaVerne
Svoboda, Rudy George
Swift, William Clement
Synowiec, John A
Townsend, Douglas Wayne
Tucker, Robert C, Jr
Vuckovic, Vladeta
Waldman, Alan S
Webster, Merritt Samuel
Weitsman, Allen William
Williams, Lynn Roy
Wilson, David E
Yackel, James W
Ziemer, William P
Zimmerman, Lester J
Zwick, Earl J

IOWA
Abian, Alexander
Athreya, Krishna Balasundaram
Atkinson, Kendall E
Barnes, Wilfred E
Camillo, Victor Peter
Canfield, Earle Lloyd
Dahiya, Rajbir Singh
Fink, Arlington M
Friedell, John C
Fuller, Kent Ralph
Geraghty, Michael A
Gillam, Basil Early
Graber, Leland D
Hart, Lawrence Alan
Hentzel, Irvin Roy
Herman, Eugene Alexander
Hill, Edward T
Hinrichsen, John James Luett
Homer, Roger Harry
Jenkins, Terry Lloyd
Khurana, Surjit Singh
Kirk, William Arthur
Kleiner, Alexander F, Jr
Kleinfeld, Margaret Humm
Kosier, Frank J
Lambert, Robert J
Lediaev, John P
Levine, Howard Allen
Lin, Bor-Luh
Lindsay, Charles McCown
Lott, Fred Wilbur, Jr
Mathews, Jerold Chase
Oehmke, Robert H
Oxley, Theron D, Jr

Peglar, George W
Pilgrim, Donald
Randell, Richard
Rothlisberger, Hazel Marie
Rudolph, William Brown
Russo, Ralph P
Schurrer, Augusta
Sprague, Richard Howard
Steiner, Eugene Francis
Tondra, Richard John
Trytten, George Norman
Waltmann, William Lee
Wilkinson, Jack Dale
Wright, Fred Marion
Yohe, James Michael
Zettel, Larry Joseph

KANSAS
Adams, Robert D
Arteaga, Lucio
Beougher, Elton Earl
Brady, Stephen W
Brown, Robert Dillon
Buser, Mary Paul
Chaney, George L
Chopra, Dharam-Vir
Conrad, Paul
Creese, Thomas Morton
Dixon, Lyle Junior
Eberhart, Paul
Flusser, Peter R
Foreman, Calvin
Fuller, Leonard Eugene
Galvin, Fred
Greechie, Richard Joseph
Haggard, J D
Hanna, Martin Slafter
Hearne, Horace Clark, Jr
Hight, Donald Wayne
Johnson, Dallas Eugene
Johnston, Andrea
Kriegsman, Helen
Laws, Leonard Stewart
McClendon, James Fred
Marr, John Maurice
Parker, Sidney Thomas
Parker, Willard Albert
Paulie, M Catherine Therese
Perel, William Morris
Price, Griffith Baley
Schweppe, Earl Justin
Stahl, Saul
Stamey, William Lee
Strecker, George Edison
Stromberg, Karl Robert
Szeptycki, Pawel
Thomas, Harold Lee
Upmeier, Harald
Van Vleck, Fred Scott
Votaw, Charles Isac
Wedel, Arnold Marion
Young, Paul McClure

KENTUCKY
Alter, Ronald
Barksdale, James Bryan, Jr
Boyce, Stephen Scott
Cantrell, Grady Leon
Cheatham, Thomas J
Coleman, Donald Brooks
Cox, Raymond H
Davitt, Richard Michael
Elder, Harvey Lynn
Enochs, Edgar Earle
Fugate, Joseph B
Gariepy, Ronald F
Geeslin, Roger Harold
Govindarajulu, Zakkula
Harris, Henson
Hayden, Thomas Lee
Howard, Aughtum Smith
King, Amy P
Lane, Bennie Ray
Lowman, Bertha Pauline
McFadden, Robert B
McGlasson, Alvin Garnett
Mostert, Paul Stallings
Newbery, A Chris
Oppelt, John Andrew
Rishel, Raymond Warren
Royster, Wimberly Calvin
Sabel, Clara Ann
Scorsone, Francesco G
Simpson, James Edward
Sitaraman, Yegnaseshan
Smith, Joe K
Stokes, Joseph Franklin
Wallace, Kyle David
Watson, Martha F
Wells, James Howard

LOUISIANA
Agarwal, Arun Kumar
Anderson, Richard Davis
Andrew, David Robert
Andrews, Bethlehem Kottes
Attebery, Billy Joe
Authement, Ray Paul
Bennett, Lonnie Truman
Birtel, Frank T
Boullion, Thomas L
Butts, Hubert S
Chadick, Stan R
Collins, Heron Sherwood

Mathematics, General (cont)

Conner, Pierre Euclide, Jr
Corley, Glyn Jackson
Crump, Kenny Sherman
Dearth, James Dean
Dupree, Daniel Edward
Ellis, James Watson
Ford, Patrick Lang
Fritsche, Richard T
Goldstein, Jerome Arthur
Griffin, Ernest Lyle
Keisler, James Edwin
Kelly, Edgar Preston, Jr
Koch, Robert Jacob
Lawson, Jimmie Don
Lefton, Lew Edward
Ligh, Steve
Looney, Stephen Warwick
Maxfield, John Edward
Mislove, Michael William
Mulcrone, Thomas Francis
Newman, Rogers J
Ohm, Jack Elton
Ohme, Paul Adolph
Ohmer, Merlin Maurice
Quigley, Frank Douglas
Rees, Charles Sparks
Rees, Paul Klein
Retherford, James Ronald
Rogers, James Ted, Jr
Rosencrans, Steven I
Ryan, Donald Edwin
Scholz, Dan Robert
Simmons, David Rae
Smith, Charles R
Spaht, Carlos G, II
Stoltzfus, Neal W

MAINE

Carter, William Caswell
Chittim, Richard Leigh
Haines, David Clark
Hsu, Yu Kao
Johnson, Robert Wells
Leonard, William Wilson
Lick, Dale W
Locke, Philip M
Mairhuber, John Carl
Munroe, Marshall Evans
Myers, Basil R
Northam, Edward Stafford
Norton, Karl Kenneth
Page, Robert Leroy
Raisbeck, Gordon
Small, Donald Bridgham
Ward, James Edward, III

MARYLAND

Atchison, William Franklin
Auslander, Joseph
Austin, Homer Wellington
Berger, William J
Bernstein, Allen Richard
Bernstein, Barbara Elaine
Betz, Ebon Elbert
Bhatia, Nam Parshad
Blanche, Ernest Evred
Blum, Joseph
Boisvert, Ronald Fernand
Cacciamani, Eugene Richard, Jr
Carasso, Alfred Samuel
Chang, Elizabeth B
Chiacchierini, Richard Philip
Chu, Sherwood Cheng-Wu
Cohen, Edgar A, Jr
Colvin, Burton Houston
Crampton, Theodore Henry Miller
Cuthill, Elizabeth
Dauber, Edwin George
Di Rienzi, Joseph
Dixon, John Kent
Dowling, Marie Augustine
Dubey, Satya D(eva)
Eastham, James Norman
Ehrlich, Gertrude
Elder, Alexander Stowell
Embree, Earl Owen
Engel, Joseph H(enry)
Faust, William R
Fitzpatrick, Patrick Michael
Franks, David A
Fusaro, Bernard A
Gabriele, Thomas L
Giese, John H
Gleissner, Gene Heiden
Gonzalez-Fernandez, Jose Maria
Gordon, Chester Murray
Gorman, John Richard
Gramick, Jeannine
Greenberg, James M
Gross, Fred
Gulick, Sidney L, III
Hanson, James Edward
Hanson, Robert Harold
Henkelman, James Henry
Henney, Dagmar Renate
Herrmann, Robert Arthur
Horelick, Brindell
Horn, Roger Alan
Hummel, James Alexander
Igusa, Jun-Ichi
Jablonski, Felix Joseph
Jones, John Paul

Kalme, John S
Kascic, Michael Joseph, Jr
Komm, Horace
Koppelman, Elaine
Kowalski, Richard
Lawson, Mildred Wiker
Lepson, Benjamin
Li, Ming Chiang
Lightner, James Edward
Lomonaco, Samuel James, Jr
Lopez-Escobar, Edgar George Kenneth
Marotta, Charles Rocco
May, Everette Lee, Jr
Meyer, Charles Franklin
Miller, A Eugene
Miller, Robert Gerard
Moulis, Edward Jean, Jr
Nanzetta, Philip Newcomb
O'Brien, Robert
Odle, John William
Parker, Rodger D
Pearl, Martin Herbert
Penn, Howard Lewis
Pierce, Harry Frederick
Pittenger, Arthur O
Ponnapalli, Ramachandramurty
Prewitt, Judith Martha Shimansky
Reed, Coke S
Rheinstein, Peter Howard
Rich, Robert Peter
Rubinstein, Lawrence Victor
Rudolph, Ray Ronald
Rukhin, Andrew Leo
Sagan, Leon Francis
Sampson, Joseph Harold
Saworotnow, Parfeny Pavolich
Schafer, James A
Schleiter, Thomas Gerard
Schneck, Paul Bennett
Siegel, Martha J
Slepian, Paul
Smith, Paul John
Sokoloski, Martin Michael
Spohn, William Gideon, Jr
Strauser, Wilbur Alexander
Strohl, George Ralph, Jr
Stuelpnagel, John Clay
Syski, Ryszard
Tepper, Morris
Theilheimer, Feodor
Toller, Gary Neil
Truesdell, Clifford Ambrose, III
Van Tuyl, Andrew Heuer
Varnhorn, Mary Catherine
Von Bun, Friedrich Otto
Wagner, Daniel Hobson
Wardlaw, William Patterson
Washington, Lawrence C
Watters, Edward C(harles), Jr
Weissberg, Alfred
Wierman, John C
Willcox, Alfred Burton
Wilson, John Phillips
Wolfe, Carvel Stewart
Wunderlich, Marvin C
Young, James Howard
Zelkowitz, Marvin Victor
Zucker, Steven Mark

MASSACHUSETTS

Ahlfors, Lars Valerian
Allen, Stephen Ives
Armacost, David Lee
Artin, Michael
Auslander, Bernice Liberman
Azpeitia, Alfonso Gil
Baglivo, Jenny Antoinette
Bandes, Dean
Bennett, Mary Katherine
Berger, Melvyn Stuart
Birkhoff, Garrett
Blackett, Donald Watson
Blackett, Shirley Allart
Borrego, Joseph Thomas
Bott, Raoul H
Buchsbaum, David Alvin
Burke, Leonarda
Catlin, Donald E
Chaiken, Jan Michael
Chen, Yu Why
Cohen, David Warren
Cohen, Haskell
Cole, Nancy
Crabtree, Douglas Everett
Cullen, Helen Frances
Darling, Eugene Merrill, Jr
Dickinson, David (James)
Doyle, Jon
Durfee, William Hetherington
Dyer, James Arthur
Eckstein, Jonathan
Evans, Thomas George
Fishman, Robert Sumner
Foulis, David James
Freedman, Marvin I
Freier, Jerome Bernard
Freyre, Raoul Manuel
Galmarino, Alberto Raul
Gleason, Andrew Mattei
Guillemin, Victor W
Hajian, Arshag B
Hardell, William John
Hawkins, Thomas William, Jr
Hayes, Dallas T

Helgason, Sigurdur
Hildebrand, Francis Begnaud
Hoffer, Alan R
Hogan, Guy T
Holst, Jewel Magee
Hurlburt, Douglas Herendeen
Hutzenlaub, John F
Isles, David Frederick
Jacob, Henry George, Jr
Jekeli, Christopher
Jensen, Barbara Lynne
Kamowitz, Herbert M
Kaplan, James
Kazhdan, David
Kearns, Donald Allen
Kennison, John Frederick
Killam, Eleanor
Kleitman, Daniel J
Kostant, Bertram
Levine, Harold
Lukas, Joan Donaldson
Luther, Holger Martin
Lutts, John A
McBrien, Vincent Owen
McCabe, Robert Lyden
MacDonnell, John Joseph
McGuigan, Robert Alister, Jr
Mackey, George W
Magee, John Francis
Magnanti, Thomas L
Manes, Ernest Gene
Martindale, Wallace S
Mauldon, James Grenfell
Mazur, Barry
Mendelson, Bert
Menzin, Margaret Schoenberg
Miller, William Brunner
Minsky, Marvin Lee
Mumford, David Bryant
Nitecki, Zbigniew
Orr, Richard Clayton
Ozimkoski, Raymond Edward
Palais, Richard Sheldon
Papert, Seymour A
Parrott, Stephen Kinsley
Perkins, Peter
Peters, Stefan
Peterson, Franklin Paul
Porter, Richard D
Ramras, Mark Bernard
Randall, Charles Hamilton
Resnikoff, Howard L
Rogers, Hartley, Jr
Rota, Gian-Carlo
Schafer, Richard Donald
Scheid, Francis
Schlesinger, James William
Schmid, Wilfried
Schoen, Kenneth
Schweizer, Berthold
Segal, Irving Ezra
Semon, Mark David
Senechal, Lester John
Sethares, George C
Shahin, Jamal Khalil
Shanahan, Patrick
Singer, Isadore Manual
Smith, John Howard
Smith, Luther W
Spencer, Guilford Lawson, II
Staknis, Victor Richard
Starr, Norton
Sternberg, Vita Shlomo
Stockton, Doris S
Stone, Arthur Harold
Stone, Dorothy Maharam
Stone, Marshall Harvey
Stone, Samuel Arthur
Strang, W(illiam) Gilbert
Strimling, Walter Eugene
Strother, Wayman L
Stubbe, John Sunapee
Sullivan, Joseph Arthur
Sulski, Leonard C
Sutcliffe, Samuel
Tanimoto, Taffee Tadashi
Tomlinson, Michael Bangs
Vogan, David A, Jr
Wagner, Robert Wanner
Wattenberg, Franklin Arvey
Whitehead, George William
Whitman, Philip Martin
Wilcox, Howard Joseph
Williamson, Susan
Wilson, F Wesley, Jr
Winston, Arthur William
Yau, Shing-Tung
Yonda, Alfred William

MICHIGAN

Adney, Joseph Elliott, Jr
Albaugh, A Henry
Arlinghaus, Sandra Judith Lach
Bartels, Robert Christian Frank
Bouwsma, Ward D
Bragg, Louis Richard
Brown, Morton
Buckeye, Donald Andrew
Buckley, Joseph Thaddeus
Caldwell, William V
Chartrand, Gary
Clarke, Allen Bruce
Eisenstadt, Bertram Joseph
Feeman, George Franklin

Feldman, Chester
Flanders, Harley
Folkert, Jay Ernest
Frame, James Sutherland
Froemke, Jon
Fryxell, Ronald C
Fuerst, Carlton Dwight
Funkenbusch, Walter W
Fyhrie, David Paul
Galler, Bernard Aaron
Gehring, Frederick William
Gerlach, Eberhard
Gilliland, Dennis Crippen
Gilpin, Michael James
Goldsmith, Donald Leon
Hay, George Edward
Heikkinen, Donald D
Heins, Albert Edward
Heuvers, Konrad John
Higman, Donald Gordon
Hocking, John Gilbert
Holter, Marvin Rosenkrantz
Hsieh, Po-Fang (Philip)
Irwin, John McCormick
Jones, Phillip Sanford
Kaplan, Wilfred
Kapoor, S F
Kelly, Leroy Milton
Kincaid, Wilfred Macdonald
Klaasen, Gene Allen
Koo, Delia Wei
Kosanovich, Robert Joseph
Krause, Eugene Franklin
Kugler, Lawrence Dean
Kuipers, Jack
Kwun, Kyung Whan
Le Beau, Stephen Edward
LePage, Raoul
Lewis, Donald John
Li, Tien-Yien
Lindsay, Robert Kendall
Longyear, Judith Querida
Ludden, Gerald D
Lyjak, Robert Fred
Lyndon, Roger Conant
McCoy, Thomas LaRue
McCully, Joseph C
McKay, James Harold
McNeill, Robert Bradley
Maffett, Andrew L
Marchand, Margaret O
Montgomery, Hugh Lowell
Moore, Warren Keith
Morash, Ronald
Murphy, Brian Boru
Nelson, James Donald
Nemeth, Abraham
Nemitz, William Charles
Nielson, George Marius
Nyman, Melvin Andrew
Nyquist, Gerald Warren
Palmer, Edgar M
Papp, F(rancis) J(oseph)
Pixley, Emily Chandler
Plotkin, Jacob Manuel
Rajnak, Stanley L
Ramanujan, Melapalayam Srinivasan
Schochet, Claude Lewis
Schreiner, Erik Andrew
Schuur, Jerry D
Shields, Allen Lowell
Simon, Carl Paul
Sinha, Indranand
Sinke, Carl
Skaff, Michael Samuel
Sledd, William T
Smith, Thomas Jefferson
Smoller, Joel A
Sonneborn, Lee Meyers
Stevenson, Richard Marshall
Stewart, Bonnie Madison
Storer, Thomas
Stortz, Clarence B
Swank, Rolland Laverne
Swart, William Lee
Tanis, Elliot Alan
Tihansky, Diane Rice
Titus, Charles Joseph
Ullman, Joseph Leonard
Van Zwalenberg, George
Verhey, Roger Frank
Vichich, Thomas E
Waggoner, Wilbur J
Wasserman, Arthur Gabriel
Winter, David John
Wong, Pui Kei
Zwier, Paul J

MINNESOTA

Aagard, Roger L
Aeppli, Alfred
Anderson, Sabra Sullivan
Aronson, Donald Gary
Barany, George
Braden, Charles McMurray
Brauer, George Ulrich
Bridgman, George Henry
Burgstahler, Sylvan
Cameron, Robert Horton
Carlson, Philip R
Coonce, Harry B
Eagon, John Alonzo
Ellefson, Ralph Donald
Fleming, Walter

Fosdick, Roger L(ee)
Fox, David William
Friedman, Avner
Fristedt, Bert
George, Melvin Douglas
Gershenson, Hillel Halkin
Goblirsch, Richard Paul
Green, Leon William
Harper, Laurence Raymond, Jr
Heuer, Charles Vernon
Heuer, Gerald Arthur
Jarvinen, Richard Dalvin
Jenkins, Howard Bryner
Kahn, Donald W
Kinderlehrer, David (Samuel)
Konhauser, Joseph Daniel Edward
Krueger, Eugene Rex
Lindgren, Bernard William
Lindgren, Gordon Edward
Loud, Warren Simms
McGehee, Richard Paul
McLean, Jeffery Thomas
Markus, Lawrence
Meyers, Norman George
Miracle, Chester Lee
Mordue, Dale Lewis
Nau, Richard William
Ness, Linda Ann
Nitsche, Johannes Carl Christian
Orey, Steven
Pedoe, Daniel
Perisho, Clarence R
Prikry, Karel Libor
Reich, Edgar
Richards, Jonathan Ian
Richter, Wayne H
Roberts, A Wayne
Roberts, Joel Laurence
Roosenraad, Cris Thomas
Rust, Lawrence Wayne, Jr
Sattinger, David H
Schuster, Seymour
Seebach, J Arthur, Jr
Sell, George Roger
Serrin, James B
Sethna, Patarasp R(ustomji)
Sibuya, Yasutaka
Slagle, James R
Steen, Lynn Arthur
Stein, Marvin L
Storvick, David A
Sudderth, William David
Thomsen, Warren Jessen
Turner, Veras D
Varberg, Dale Elthon
Walczak, Hubert R

MISSISSIPPI
Barrett, Lida Kittrell
Colonias, John S
Garner, Jackie Bass
Heller, Robert, Jr
King, Robert William
McCall, Jerry C
Morris, Halcyon Ellen McNeil
Ottinger, Carol Blanche
Riggs, Schultz
Sheffield, Roy Dexter
Shive, Robert Allen, Jr
Smith, Gaston
Solomon, Jimmy Lloyd
Stokes, Russell Aubrey
Tilley, John Leonard
Truax, Robert Lloyd
Webster, Porter Grigsby

MISSOURI
Ahlbrandt, Calvin Dale
Balbes, Raymond
Blackwell, Paul K, II
Bodine, Richard Shearon
Boothby, William Munger
Burcham, Paul Baker
Campbell, Larry N
Chan, Albert Sun Chi
Erkiletian, Dickran Hagop, Jr
Flowers, Joe D
Foran, James Michael
Francis, Richard L
Goldberg, Merrill B
Grimm, Louis John
Hatfield, Charles, Jr
Hicks, Troy L
Hirschman, Isidore Isaac, Jr
Jacobs, Marc Quillen
Jenkins, James Allister
Johnson, Charles Andrew
Johnson, Larry K
Kenner, Morton Roy
Kezlan, Thomas Phillip
King, Terry Lee
Kubicek, John D
Lange, Leo Jerome
Levine, Jeffrey
Liebnitz, Paul W
Luke, Yudell Leo
McDowell, Robert Hull
McIntosh, William David
Martin, William L, Jr
Nicholson, Eugene Haines
Padberg, Harriet A
Pearson, Bennie Jake
Rant, William Howard
Reeder, John Hamilton

Schrader, Keith William
Schwartz, Alan Lee
Sentilles, F Dennis, Jr
Shiflett, Lilburn Thomas
Stanojevic, Caslav V
Stilwell, Kenneth James
Stumpff, Howard Keith
Thro, Mary Patricia
Weiss, Guido Leopold
Welland, Grant Vincent
White, Warren D
Wilke, Frederick Walter
Wright, David Lee
Wright, Farroll Tim

MONTANA
Barrett, Louis Carl
Bryan, Charles A
Daniel, Victor Wayne
Goebel, Jack Bruce
Jamison, William H
McAllister, Byron Leon
McKelvey, Robert William
McRae, Daniel George
Manis, Merle E
Swartz, William John
Taylor, Donald Curtis
Whitesitt, John Eldon
Wright, Alden Halbert

NEBRASKA
Chivukula, Ramamohana Rao
Cox, Henry Miot
Downing, John Scott
Gross, Mildred Lucile
Heisler, Joseph Patrick
Jackson, Lloyd K
Jaswal, Sitaram Singh
Keller, Roy Fred
Kramer, Earl Sidney
Lenz, Charles Eldon
Leonhardt, Earl A
Lewis, William James
McDougal, Robert Nelson
Masters, Christopher Fanstone
Meakin, John C
Meisters, Gary Hosler
Mientka, Walter Eugene
Miller, Donald Wright
Mordeson, John N
Nelson, Theodora S
Pao, Y(en)-C(hing)
Pickens, Charles Glenn
Scheerer, Anne Elizabeth
Sharp, Edward A
Shores, Thomas Stephen
Thornton, Melvin Chandler
Wallen, Stanley Eugene
Wampler, Joe Forrest
Zechmann, Albert W

NEVADA
Aizley, Paul
Beesley, Edward Maurice
Blackadar, Bruce Evan
Brady, Allen H
Davis, Robert Dabney
Graham, Malcolm
McMinn, Trevor James
Nielsen, John Palmer
Pfaff, Donald Chesley
Werth, John St Clair, Jr

NEW HAMPSHIRE
Batho, Edward Hubert
Brown, Edward Martin
Copeland, Arthur Herbert, Jr
Jacoby, Alexander Robb
Jeffery, Lawrence R
Meeker, Loren David
Moore, Berrien, III
Nelson, Lloyd Steadman
Norman, Robert Zane
Prosser, Reese Trego
Ross, Shepley Littlefield
Ryan, James Michael
Shore, Samuel David
Slesnick, William Ellis
Snell, James Laurie
Stancl, Donald Lee
Williamson, Richard Edmund
Wixson, Eldwin A, Jr

NEW JERSEY
Almgren, Frederick Justin, Jr
Anderson, Stanley William
Anton, Howard
Bean, Ralph J
Beckman, Frank Samuel
Bein, Donald
Benes, Vaclav Edvard
Bidwell, Leonard Nathan
Borel, Armand
Bredon, Glen E
Browder, William
Burrill, Claude Wesley
Bush, Charles Edward
Cai, Jin-yi
Canavan, Robert I
Charlap, Leonard Stanton
Clark, Frank Eugene
Crabtree, James Bruce
Davis, Morton David
Davis, Robert Benjamin

Dekker, Jacob Christoph Edmond
Denby, Lorraine
Elliott, Joanne
Epstein, Marvin Phelps
Faith, Carl Clifton
Flatto, Leopold
Fox, Phyllis
Geshner, Robert Andrew
Gilbert, Edgar Nelson
Gonshor, Harry
Gorenstein, Daniel
Graham, Ronald Lewis
Griffiths, Phillip A
Guilfoyle, Richard Howard
Gurk, Herbert Morton
Haenisch, Siegfried
Hausdoerffer, William H
Hazard, Katharine Elizabeth
Henrich, Christopher John
Houle, Joseph E
Hoyt, William Lind
Jagerman, David Lewis
Johnson, Donald Glen
Jurkat, Martin Peter
Kalmanson, Kenneth
Karel, Martin Lewis
Karol, Mark J
Kauffman, Ellwood
Keen, Linda
Keenan, Edward Milton
Kees, Kenneth Lewis
Kemperman, Johannes Henricus
    Bernardus
Kessler, Irving Jack
Kochen, Simon Bernard
Kohn, Joseph John
Kotin, Leon
Kovach, Ladis Daniel
Kruskal, Joseph Bernard
Kruskal, Martin David
Kuhn, Harold William
Kuntz, Richard A
Lagarias, Jeffrey Clark
Langlands, Robert P
Lapidus, Arnold
Leader, Solomon
Lechleider, J W
Lewis, Charles Joseph
McCabe, John Patrick
Magliola-Zoch, Doris
Makar, Boshra Halim
Maletsky, Evan M
Milnor, Tilla Savanuck Klotz
Nelson, Joseph Edward
Petryshyn, Walter Volodymyr
Poiani, Eileen Louise
Pollak, Henry Otto
Polonsky, Ivan Paul
Puri, Narindra Nath
Ramanan, V R V
Riggs, Richard
Rinaldi, Leonard Daniel
Rosenberg, Herman
Roth, Rodney J
Sams, Burnett Henry, III
Segers, Richard George
Selberg, Atle
Senter, Harvey
Shepp, Lawrence Alan
Shimura, Goro
Sloane, Neil James Alexander
Sloyan, Mary Stephanie
Spencer, Thomas
Stein, Elias M
Stevenson, Robert Louis
Stewart, Ruth Carol
Stoddard, James H
Sweeney, William John
Thurston, William P
Tong, Mary Powderly
Treves, Jean Francois
Trotter, Hale Freeman
Vasconcelos, Wolner V
Wagner, Richard Carl
Waldinger, Hermann V
Weil, Andre
Weinberger, Peter Jay
White, Myron Edward
Whitney, Hassler
Williams, Gareth Pierce
Williams, Vernon
Wolfson, Kenneth Graham
Zierler, Neal
Zimmerberg, Hyman Joseph

NEW MEXICO
Amos, Donald E
Bailey, Paul Bernard
Ball, Ralph Wayne
Bell, Stoughton
Beyer, William A
Brenton, June Grimm
Christensen, N(ephi) A(lbert)
Davis, Jeffrey Robert
DeMarr, Ralph Elgin
DePree, John Deryck
Giever, John Bertram
Ginsparg, Paul H
Griego, Richard Jerome
Guinn, Theodore
Hahn, Liang-Shin
Harris, Reece Thomas
Hillman, Abraham P
Johnson, Ralph T, Jr

Jones, Mark Wallon
Kist, Joseph Edmund
Kruse, Arthur Herman
Loustaunau, Joaquin
Mark, J Carson
Metzler, Richard Clyde
Mitchell, I(da) Merle
Morrison, Donald Ross
Nicolaenko, Basil
Novlan, David John
O'Rourke, Peter John
Peterson, Donald Palmer
Pimbley, George Herbert, Jr
Rice, Robert Bruce
Scharn, Herman Otto Friedrich
Schlosser, Jon A
Searcy, Charles Jackson
Swartz, Charles W
Thompson, Robert James
Trowbridge, George Cecil
Wackerle, Jerry Donald
Walker, Elbert Abner
Wells, Mark Brimhall
Wendroff, Burton
Willbanks, Emily West
Wilson, Carroll Klepper
Winkler, Max Albert
Zeigler, Royal Keith

NEW YORK
Abbott, John S
Adhout, Shahla Marvizi
Adler, Alfred
Adler, Roy Lee
Akst, Geoffrey R
Albicki, Alexander
Alex, Leo James
Alling, Norman Larrabee
Alt, Franz L
Anastasio, Salvatore
Anshel, Michael
Archibald, Julius A, Jr
Archibald, Ralph George
Arkin, Joseph
Asprey, Winifred Alice
Auslander, Louis
Bachman, George
Bailey, William T
Barback, Joseph
Barcus, William Dickson, Jr
Bass, Hyman
Baumslag, Gilbert
Bazer, Jack
Bell, Jonathan George
Beltrami, Edward J
Berkowitz, Jerome
Bernkopf, Michael
Bernstein, Irwin S
Bers, Lipman
Bick, Theodore A
Bing, Kurt
Birman, Joan Sylvia
Blackman, Jerome
Blake, Louis Harvey
Bloch, Alan
Brown, David T
Buckley, Fred Thomas
Bumcrot, Robert J
Burling, James P
Butler, Lewis Clark
Capobianco, Michael F
Carpenter, James Edward
Case, John William
Chako, Nicholas
Chen, Yuh-Ching
Chilton, Bruce L
Chow, Yuan Shih
Chuckrow, Vicki G
Cohen, Daniel Isaac Aryeh
Cohn, Harvey
Craft, George Arthur
Danese, Arthur E
Dauben, Joseph W
Davis, Edward Dewey
De Carlo, Charles R
Deming, Robert W
DePalma, James John
Desrochers, Alan Alfred
Di Paola, Robert Arnold
Donsker, Monroe David
Doss, Raouf
Douglas, Ronald George
Dowds, Richard E
Dristy, Forrest E
Durkovic, Russell George
Dutka, Jacques
Ebin, David G
Egan, Francis P
Eilenberg, Samuel
Eisele, Carolyn
Fadell, Albert George
Fairchild, William Warren
Farrell, F Thomas
Feingold, Alex Jay
Feroe, John Albert
Foisy, Hector B
Forman, William
Forray, Marvin Jules
Fox, William Cassidy
Freilich, Gerald
Freimer, Marshall Leonard
Fremont, Herbert Irwin
Fu, Lorraine Shao-Yen
Fuchs, Wolfgang Heinrich

Mathematics, General (cont)

Gallagher, Patrick Ximenes
Gans, David
Garabedian, Paul Roesel
Garcia, Mariano
Garnet, Hyman R
Gelbart, Abe
Gelbaum, Bernard Russell
Gilbert, Paul Wilner
Gilman, S(ister) John Frances
Godino, Charles F
Goldberg, Kenneth Philip
Goldberg, Samuel
Goldman, Malcolm
Goldstein, Max
Goldstein, Richard
Golumbic, Martin Charles
Grad, Arthur
Griesmer, James Hugo
Grosof, Miriam Schapiro
Gross, Leonard
Grossman, George
Grossman, Norman
Gucker, Gail
Guggenheimer, Heinrich Walter
Gunderson, Norman Gustav
Gutkin, Eugene
Hachigian, Jack
Hailpern, Raoul
Hall, Dick Wick
Hanson, David Lee
Harrison, Malcolm Charles
Hartmanis, Juris
Head, Thomas James
Heller, Alex
Hemmingsen, Erik
Herbach, Leon Howard
Hershenov, Joseph
Herwitz, Paul Stanley
Hilbert, Stephen Russell
Hilton, Peter John
Hirshon, Ronald
Hochberg, Murray
Hoffman, Alan Jerome
Hollingsworth, Jack W
Horton, Thomas Roscoe
Hunte, Beryl Eleanor
Hurwitz, Solomon
Isaac, Richard Eugene
Isaacson, Eugene
Isbell, John Rolfe
Iwaniec, Tadeusz
Jenks, Richard D
John, Fritz
Johnson, Sandra Lee
Kahn, Peter Jack
Kao, Samuel Chung-Siung
Kazarinoff, Nicholas D
Kelisky, Richard Paul
Kesten, Harry
Kibbey, Donald Eugene
Kim, Woo Jong
Klimko, Eugene M
Knee, David Isaac
Kolchin, Ellis Robert
Koranyi, Adam
Kra, Irwin
Kronk, Hudson V
Kurss, Herbert
Laderman, Julian David
Larsen, Charles McLoud
Laska, Eugene
Lawler, John Patrick
Lawson, Herbert Blaine, Jr
Lax, Anneli
Lax, Peter David
Leela, Srinivasa (G)
Lefkowitz, Ruth Samson
Lefton, Phyllis
Lemke, Carlton Edward
Levenson, Morris E
Levin, Gerson
Levine, Leo Meyer
Lieberman, Burton Barnet
Lisman, Henry
Lissner, David
Liu, Chamond
Lorch, Edgar Raymond
Lubell, David
Luchins, Edith Hirsch
Lutwak, Erwin
McAuley, Louis Floyd
McGloin, Paul Arthur
McKean, Henry P
McNaughton, Robert
Magill, Kenneth Derwood, Jr
Malkevitch, Joseph
Mansfield, Larry Everett
Marcus, Michael Barry
Maskit, Bernard
Melnick, Edward Lawrence
Merriell, David McCray
Meyer, Paul Richard
Meyer, Walter Joseph
Miller, Frederick
Miller, Kenneth Sielke
Milnor, John Willard
Miranker, Willard Lee
Moise, Edwin Evariste
Montague, Harriet Frances
Montgomery, Mabel D
Moore, John Coleman
Moreno, Carlos Julio

Mott, Thomas
Muench, Donald Leo
Mylroie, Victor L
Nachbin, Leopoldo
Neisendorfer, Joseph
Nevison, Christopher H
Newberger, Edward
Nirenberg, Louis
O'Brien, Redmond R
O'Brien, Ronald J
Ogawa, Hajimu
Olson, Frank R
Padalino, Stephen John
Pardee, Otway O'Meara
Peluso, Ada
Phillips, Esther Rodlitz
Piech, Margaret Ann
Polimeni, Albert D
Pollack, Richard
Pomeranz, Janet Bellcourt
Posamentier, Alfred S
Postman, Robert Derek
Pownall, Malcolm Wilmor
Rabson, Gustave
Ramaley, James Francis
Randol, Burton
Randolph, John Fitz
Rao, Poduri S R S
Ravenel, Douglas Conner
Rees, Mina S
Resnick, Sidney I
Rhee, Haewun
Rivlin, Theodore J
Robinson, Louis
Robusto, C Carl
Rockhill, Theron D
Rockoff, Maxine Lieberman
Rose, Israel Harold
Rosenbloom, Paul Charles
Rosenthal, John William
Roth, John Paul
Rothaus, Oscar Seymour
Rothenberg, Ronald Isaac
Sacksteder, Richard Carl
Sah, Chih-Han
Scalora, Frank Salvatore
Schaefer, Paul Theodore
Schanuel, Stephen Hoel
Schaumberger, Norman
Schindler, Susan
Schlissel, Arthur
Schwartz, Abraham
Seckler, Bernard David
Segal, Sanford Leonard
Seiken, Arnold
Sellers, Peter Hoadley
Shabanowitz, Harry
Shapiro, Jack Sol
Sheingorn, Mark Elliot
Shilepsky, Arnold Charles
Shimamoto, Yoshio
Shub, Michael I
Shulman, Harold
Silberger, Donald Morison
Sloan, Robert W
Small, William Andrew
Snow, Wolfe
Sohmer, Bernard
Spencer, Armond E
Spitzer, Frank L
Spoelhof, Charles Peter
Standish, Charles Junior
Steinberg, David H
Stemple, Joel G
Stephany, Edward O
Stephens, Clarence Francis
Sterling, Nicholas J
Stern, Samuel T
Sternstein, Martin
Stevenson, John Crabtree
Stoker, James Johnston
Stolfo, Salvatore Joseph
Stoner, George Green
Storm, Edward Francis
Straight, H Joseph
Strait, Peggy
Strand, Richard Carl
Strasser, Elvira Rapaport
Strichartz, Robert Stephen
Sullivan, Dennis P
Swick, Kenneth Eugene
Szusz, Peter
Tamari, Dov
Taylor, Alan D
Taylor, Francis B
Texter, John
Thomas, Edward Sandusky, Jr
Thorpe, John Alden
Tucciarone, John Peter
Tucker, Alan Curtiss
Tuller, Annita
Turner, Edward C
Ullman, Arthur William James
Uschold, Richard L
Vasquez, Alphonse Thomas
Victor, Jonathan David
Vogeli, Bruce R
Wagner, Eric G
Wallach, Sylvan
Waltcher, Azelle Brown
Waterman, Daniel
Watts, Charles Edward
Weintraub, Sol
Wernick, William

White, Albert George, Jr
Wilken, Donald Rayl
Williams, Scott Warner
Willoughby, Ralph Arthur
Wilson, Paul Robert
Wohlgelernter, Devora Kasachkoff
Wolf, Ira Kenneth
Wolsson, Kenneth
Wormser, Gary Paul
Yang, Chao-Hui
Yu, Kai-Bor
Ziebur, Allen Douglas
Zlot, William Leonard
Zuckerberg, Hyam L
Zuckerman, Israel

NORTH CAROLINA

Allard, William Kenneth
Bernard, Richard Ryerson
Blackburn, Thomas Henry
Bright, George Walter
Burch, Benjamin Clay
Cameron, Edward Alexander
Campbell, Stephen La Vern
Carlitz, Leonard
Cato, Benjamin Ralph, Jr
Chandler, Richard Edward
Chandra, Jagdish
Charlton, Harvey Johnson
Church, Charles Alexander, Jr
Cleland, John Gregory
Cooke, Henry Charles
Davis, Kenneth Joseph
Davis, Myrtis
Davis, Robert Lloyd
Dressel, Francis George
Eakin, Richard R
Epstein, George
Fletcher, William Thomas
Franke, John Erwin
Frazier, Robert Carl
Gentry, Ivey Clenton
Gentry, Karl Ray
Gilman, Albert F, III
Goodman, Sue Ellen
Gordh, George Rudolph, Jr
Gordy, Thomas D(aniel)
Graham, Ray Logan
Grau, Albert A
Gurganus, Kenneth Rufus
Halton, John Henry
Hansen, Donald Joseph
Harrington, Walter Joel
Herbst, Robert Taylor
Herr, David Guy
Hildebrandt, Theodore Ware
Hooke, Robert
Houston, Johnny Lee
Hoyle, Hughes Bayne, Jr
Jackson, Robert Bruce, Jr
Johnson, Phillip Eugene
Johnson, Wendell Gilbert
Kallianpur, Gopinath
Karlof, John Knox
Kerr, Sandria Neidus
King, Lunsford Richardson
Kneale, Samuel George
Koh, Kwangil
Levine, Jack
Macey, Wade Thomas
Manduley, Ilma Morell
Martin, LeRoy Brown, Jr
Meyer, Carl Dean, Jr
Mobley, Jean Bellingrath
Murray, Francis Joseph
Neidinger, Richard Dean
Nelson, A Carl, Jr
Niccolai, Nilo Anthony
Norris, Fletcher R
Page, Nelson Franklin
Paur, Sandra Orley
Peterson, Elmor Lee
Pfaltzgraff, John Andrew
Plemmons, Robert James
Posey, Eldon Eugene
Roberts, John Henderson
Rodney, Earnest Abram
Rose, Nicholas John
Sagan, Hans
Saibel, Edward
Sanders, Oliver Paul
Saunders, Frank Wendell
Schaeffer, David George
Scoville, Richard Arthur
Selgrade, James Francis
Shell, Donald Lewis
Shelly, James H
Shoenfield, Joseph Robert
Silber, Robert
Simms, Nathan Frank, Jr
Singer, Michael
Smith, David Alexander
Smith, Douglas D
Smith, Fred R, Jr
Sonner, Johann
Sowell, Katye Marie Oliver
Speece, Herbert E
Stone, Erika Mares
Struble, Raimond Aldrich
Sutton, Louise Nixon
Taylor, Jerry Duncan
Thomas, John Pelham
Tolle, Jon Wright
Ullrich, David Frederick

Van Alstyne, John Pruyn
Van Der Vaart, Hubertus Robert
Vaughan, Jerry Eugene
Wahl, Jonathan Michael
Wesler, Oscar
Whipkey, Kenneth Lee
Willett, Richard Michael
Wright, Fred Boyer

NORTH DAKOTA

Bzoch, Ronald Charles
Gill, Dhanwant Singh
McBride, Woodrow H
Mathsen, Ronald M
Robinson, Thomas John
Rue, James Sandvik
Uherka, David Jerome
Winger, Milton Eugene

OHIO

Adorno, David Samuel
Allen, Harry Prince
Andrews, George Harold
Archer, Lawrence H(arry)
Arnoff, E Leonard
Atalla, Robert E
Baum, John Daniel
Beane, Donald Gene
Bell, Harold
Beyer, William Hyman
Blau, Julian Herman
Bohn, Sherman Elwood
Bolger, Edward M
Bonar, Daniel Donald
Brockman, Harold W
Brown, Dean Raymond
Brown, Richard Kettel
Brown, Robert Bruce
Buchthal, David C
Buck, Charles (Carpenter)
Burden, Richard L
Capel, Charles Edward
Carson, Robert Cleland
Cassim, Joseph Yusuf Khan
Chalkley, Roger
Chidambaraswamy, Jayanthi
Clemens, Stanley Ray
Cummins, Kenneth Burdette
Deever, David Livingstone
Denbow, Carl (Herbert)
Diestel, Joseph
Disko, Mildred Anne
Drobot, Stefan
Duval, Leonard A
Egar, Joseph Michael
Eldridge, Klaus Emil
Ericksen, Wilhelm Skjetstad
Ferguson, Harry
Flaspohier, David
Foreman, Matthew D
Foulk, Clinton Ross
Fricke, Gerd
Fridy, John Albert
Garrow, Robert Joseph
Gass, Frederick Stuart
Gilligan, Lawrence G
Graue, Louis Charles
Gruber, H Thomas
Guderley, Karl Gottfried
Haber, Robert Morton
Hahn, Samuel Wilfred
Hales, Raleigh Stanton, Jr
Hampton, Charles Robert
Hern, Thomas Albert
Horner, James M
Huneke, John Philip
Hungerford, Thomas W
Isenecker, Lawrence Elmer
James, Thomas Ray
Jasper, Samuel Jacob
Jehn, Lawrence A
Jones, John, Jr
Kaufmann, Alvern Walter
Knight, Lyman Coleman
Koehler, Donald Otto
Kolopajlo, Lawrence Hugh
Kullman, David Elmer
Laha, Radha Govinda
Lair, Alan Van
Leach, Ernest Bronson
Leake, Lowell, Jr
Leetch, James Frederick
Leitzel, James Robert C
Leitzel, Joan Phillips
Levine, Maita Faye
Lipsich, H David
Little, Richard Allen
Long, Clifford A
Low, Marc E
MacDowell, Robert W
Markley, William A, Jr
Marty, Roger Henry
Maxwell, Glenn
Mehr, Cyrus B
Meyers, Leroy Frederick
Miller, James Roland
Mityagin, Boris Samuel
Moscovici, Henri
Nelson, Raymond John
Neuzil, John Paul
Nikolai, Paul John
Notz, William Irwin
Oestreicher, Hans Laurenz
Ogden, William Frederick

Osterburg, James
Park, Chull
Park, Won Joon
Peck, Lyman Colt
Pepper, Paul Milton
Pinzka, Charles Frederick
Piper, Ervin L
Powell, Robert Ellis
Ralley, Thomas G
Rattan, Kuldip Singh
Ray-Chaudhuri, Dwijendra Kumar
Reeves, Roy Franklin
Riedl, John Orth, Jr
Riggle, Timothy A
Riner, John William
Robinson, Stewart Marshall
Rolwing, Raymond H
Ross, Louis
Rutter, Edgar A, Jr
Sachs, David
Saltzer, Charles
Schaefer, Donald John
Schlegel, Donald Louis
Schraut, Kenneth Charles
Schultz, James Edward
Schuurmann, Frederick James
Semon, Warren Lloyd
Shields, Paul Calvin
Shoemaker, Richard W
Sholander, Marlow
Silberger, Allan Joseph
Smith, Frank A
Smith, James L
Smith, Mark Andrew
Smith, Robert Sefton
Spielberg, Stephen E
Spring, Ray Frederick
Staley, David H
Stanton, Robert Joseph
Steele, William F
Stein, Ivie, Jr
Steinlage, Ralph Cletus
Sterrett, Andrew
Stickney, Alan Craig
Tikson, Michael
Townsend, Ralph N
Tung, John Shih-Hsiung
Ungar, Gerald S
Vance, Elbridge Putnam
Vancko, Robert Michael
Varga, Richard S
Waits, Bert Kerr
Walum, Herbert
Wen, Shih-Liang
Wente, Henry Christian
Wilson, Eric Leroy
Wilson, Richard Michael
Worrell, John Mays, Jr
Yantis, Richard P
Yozwiak, Bernard James
Zilber, Joseph Abraham

OKLAHOMA
Agnew, Jeanne Le Caine
Albright, James Curtice
Andree, Richard Vernon
Bednar, Jonnie Bee
Bertholf, Dennis E
Boyce, Donald Joe
Briggs, Phillip D
Brixey, John Clark
Burchard, Hermann Georg
Cairns, Thomas W
Davies-Jones, Robert Peter
Ewing, George McNaught
Goff, Gerald K
Gray, Samuel Hutchison
Greer, Earl Vincent
Hibbs, Leon
Huneke, Harold Vernon
Kotlarski, Ignacy Icchak
Krattiger, John Trubert
Loman, M Laverne
McKellips, Terral Lane
Magid, Andy Roy
Naymik, Daniel Allan
Page, Rector Lee
Phelps, Jack
Pope, Paul Terrell
Reimer, Dennis D
Springer, Charles Eugene
Switzer, Laura Mae
Uehara, Hiroshi
Veatch, Ralph Wilson
Wagner, Harry Mahlon

OREGON
Ahuja, Jagdish C
Anderson, Frank Wylie
Anderson, Tera Lougenia
Arnold, Bradford Henry
Ballantine, C S
Balogh, Charles B
Barrar, Richard Blaine
Bhattacharyya, Ramendra Kumar
Bowman, Eugene W
Brunk, Hugh Daniel
Burdg, Donald Eugene
Butler, John Ben, Jr
Carter, David Southard
Cater, Frank Sydney
Chrestenson, Hubert Edwin
Civin, Paul
Curtis, Charles Whittlesey

Enneking, Marjorie
Firey, William James
Freeman, Robert
Ghent, Kenneth Smith
Goheen, Harry Earl
Graham, Beardsley
Harrison, David Kent
Haskell, Charles Thomson
Hunter, Larry Clifton
Iltis, Donald Richard
Jensen, Bruce A
Kaplan, Edward Lynn
Khalil, M Aslam Khan
Kunkle, Donald Edward
Maier, Eugene Alfred
Montgomery, Richard Glee
Nelsen, Roger Bain
Newberger, Stuart Marshall
Olum, Paul
Palmer, Theodore W
Rempfer, Robert Weir
Rio, Sheldon T
Roberts, Joseph Buffington
Ross, Kenneth Allen
Schmidt, Harvey John, Jr
Seely, Justus Frandsen
Seshu, Lilly Hannah
Simons, William Haddock
Smith, John Wolfgang
Smith, Kennan Taylor
Stanley, Robert Lauren
Swanson, Leonard George
Wilson, Howard Le Roy

PENNSYLVANIA
Adams, William Sanders
Anderle, Richard
Andrews, George Eyre
Archer, David Horace
Argabright, Loren N
Arnold, Leslie K
Ayoub, Christine Williams
Ayoub, Raymond G Dimitri
Baric, Lee Wilmer
Barnes, Wallace Edward
Bartley, Edward Francis
Bartlow, Thomas L
Bartoo, James Breese
Bates, Grace Elizabeth
Bedient, Phillip E
Benedicty, Mario
Bickel, Robert John
Bissinger, Barnard Hinkle
Block, I(saac) Edward
Brady, Wray Grayson
Buriok, Gerald Michael
Busby, Robert Clark
Calabi, Eugenio
Castellano, Salvatore, Mario
Chao, Chong-Yun
Chow, Shin-Kien
Coffman, Charles Vernon
Crawley, James Winston, Jr
Cullen, Charles G
Cunningham, Frederic, Jr
Dank, Milton
Deutsch, Frank Richard
Dice, Stanley Frost
Duffin, Richard James
Earl, Boyd L
Ehrenpreis, Leon
England, James Walton
Evans, Edward William
Fell, James Michael Gardner
Filano, Albert E
Freyd, Peter John
Glasner, Moses
Goldman, Oscar
Gould, William Allen
Gowdy, Spenser O
Greene, Curtis
Hall, James Emerson
Harrell, Ronald Earl
Hartmann, Frederick W
Hastings, Stuart
Hearsey, Bryan Vandiver
Heath, Robert Winship
Heckscher, Stevens
Helmbold, Robert Lawson
Herpel, Coleman
Hickman, Warren David
Hinton, Raymond Price
Holder, Leonard Irvin
Holzinger, Joseph Rose
Homann, Frederick Anthony
Hostetler, Robert Paul
Hsiung, Chuan Chih
Hunter, Robert P
Idowu, Elayne Arrington
Jacobson, Bernard
Kadison, Richard Vincent
Kamel, Hyman
Kamen, Edward Walter
Kelemen, Charles F
Kerr, Carl E
Khabbaz, Samir Anton
Khalil, Mohamed Thanaa
Kirk, David Blackburn
Kolman, Bernard
Kolodner, Ignace Izaak
Korfhage, Robert R
Krall, Allan M
Krall, Harry Levern
Kresh, J Yasha

Krishnan, Viakalathur Sankrithi
Kung, Hsiang-Tsung
Lancaster, Otis Ewing
Larson, Roland Edwin
Laush, George
Lee, Hua-Tsun
Levitan, Michael Leonard
Light, John Henry
Lindgren, William Frederick
McAllister, Gregory Thomas, Jr
McAllister, Marialuisa N
McCammon, Mary
MacCamy, Richard C
Machusko, Andrew Joseph, Jr
Maclay, Charles Wylie
Masani, Pesi Rustom
Maserick, Peter H
Mayer, Joerg Werner Peter
Mehta, Jatinder S
Metzger, Thomas Andrew
Meyer, John Sigmund
Miehle, William
Miller, G(erson) H(arry)
Misra, Dhirendra N
Mode, Charles J
Moore, Richard Allan
Nee, M Coleman
Nichols, John C
Noll, Walter
Ohl, Donald Gordon
Oler, Norman
Olsen, Glenn W
Ord, John Allyn
Ossesia, Michel Germain
Oxtoby, John Corning
Paine, Dwight Milton
Pederson, Roger Noel
Phelps, Dean G
Pickands, James, III
Pinter, Charles Claude
Pitcher, Arthur Everett
Porter, Gerald Joseph
Porter, John Robert
Portmann, Walter Oddo
Rabenstein, Albert Louis
Ray, David Scott
Raymon, Louis
Reed, Joel
Reich, Daniel
Riggle, John H
Rim, Dock Sang
Rosen, David
Rosenstein, George Morris, Jr
Rowlands, R(ichard) O(wen)
Saaty, Thomas L
Scandura, Joseph M
Schechter, Murray
Schiller, John Joseph
Schultz, Blanche Beatrice
Sebesta, Charles Frederick
Shaffer, Douglas Howerth
Shale, David
Shatz, Stephen S
Shuck, John Winfield
Sigler, Laurence Edward
Silver, Edward A
Skeath, J Edward
Slaughter, Frank Gill, Jr
Snyder, Andrew Kagey
Sutherland, Ivan Edward
Taylor, Floyd Heckman
Thomas, George Brinton, Jr
Thomas, Joseph Charles
Tillman, Stephen Joel
Tozier, John E
Warren, Richard Hawks
Wei, William Wu-Shyong
Weiss, Sol
Western, Donald Ward
Westlake, Wilfred James
Wilansky, Albert
Williams, William Orville
Wolf, Charles Trostle
Wolfe, Dorothy Wexler
Wong, Roman Woon-Ching
Wyler, Oswald
Yang, Chung-Tao
Yukich, Joseph E
Zaphyr, Peter Anthony
Zindler, Richard Eugene
Zitarelli, David Earl

RHODE ISLAND
Anderson, George Albert
Baum, Paul Frank
Bernstein, Dorothy Lewis
Bordelon, Derrill Joseph
Davis, Philip J
Fleming, Wendell Helms
Gabriel, Richard Francis
Geman, Staurt Alan
Howland, Richard A
King, Charles W(illis)
LeVeque, William Judson
McKillop, Lucille Mary
Rosen, Michael Ira
Santos, Eugene (Sy)
Sine, Robert C
Steward, Robert F
Stewart, Frank Moore
Vitter, Jeffrey Scott
Woolf, William Blauvelt

SOUTH CAROLINA
Abernathy, Robert O
Allen, Roger W, Jr
Ashy, Peter Jawad
Bell, Curtis Porter
Bose, Anil Kumar
Clanton, Donald Henry
Clark, Hugh Kidder
Cleaver, Charles E
Cohn, Leslie
Croft, George Thomas
Dobbins, James Gregory Hall
Dodson, Norman Elmer
Fray, Robert Dutton
Hammett, Michael E
Hare, William Ray, Jr
Harley, Peter W, III
Hedberg, Marguerite Zeigel
Hind, Alfred Thomas, Jr
Hodges, Billy Gene
Huff, Charles William
Kenelly, John Willis, Jr
Kirkwood, Charles Edward, Jr
Laskar, Renu Chakravarti
LaTorre, Donald Rutledge
Lytle, Raymond Alfred
Nicol, Charles Albert
Padgett, William Jowayne
Phillips, Robert Gibson
Poole, John Terry
Ringeisen, Richard Delose
Russell, Charles Bradley
Saxena, Subhash C
Stephenson, Robert Moffatt, Jr
Sutton, Charles Samuel
Turner, Albert Joseph, Jr
Ulmer, Millard B
Womble, Eugene Wilson
Yu, Kai Fun

SOUTH DAKOTA
Carlson, John W
Cook, Cleland V
Fors, Elton W
Grimm, Carl Albert
Haigh, William E
Kranzler, Albert William
Miller, Bruce Linn
Rognlie, Dale Murray

TENNESSEE
Alsmiller, Rufard G, Jr
Alvarez, Laurence Richards
Arenstorf, Richard F
Austin, Billy Ray
Baumeyer, Joel Bernard
Beauchamp, John J
Bowling, Floyd E
Bradley, John Spurgeon
Brooks, Sam Raymond
Bryant, Billy Finney
Carruth, James Harvey
Celauro, Francis L
Childress, Denver Ray
Croom, Frederick Hailey
Dauer, Jerald Paul
Dent, William Hunter, Jr
Dixon, Edmond Dale
Dobyns, Roy A
Evans, Joe Smith
Forrest, Thomas Douglas
Frandsen, Henry
Ginnings, Gerald Keith
Goldberg, Richard Robinson
Haddock, John R
Jamison, King W, Jr
Jordan, George Samuel
Kalin, Robert
Kaltenborn, Howard Scholl
Kerce, Robert H
King, Calvin Elijah
Kreiling, Daryl
Lu, Mary Kwang-Ruey Chao
McConnel, Robert Merriman
McCowan, Otis Blakely
McHenry, Hugh Lansden
Mathews, Harry T
Megibben, Charles Kimbrough
Nymann, DeWayne Stanley
Oliver, Carl Edward
Pleasant, James Carroll
Plummer, Michael David
Poole, George Douglas
Potter, Thomas Franklin
Ratner, Lawrence Theodore
Ross, Clay Campbell, Jr
Rousseau, Cecil Clyde
Sakhare, Vishwa M
Shobe, L(ouis) Raymon
Smith, James H
Spraker, Harold Stephen
Stallmann, Friedemann Wilhelm
Stephens, Harold W
Stevenson, Everett E
Textor, Robin Edward
Tirman, Alvin
Turner, Lincoln Hulley
Vanaman, Sherman Benton
Wade, William Raymond, II
Wagner, Carl George
Walker, David Tutherly
Ware, James Gareth
Webster, Curtis Cleveland
Wesson, James Robert

Poss, Richard Leon
Prielipp, Robert Walter
Robbin, Joel W
Rosser, John Barkley
Rudin, Mary Ellen
Rudin, Walter
Russell, David L
Shea, Daniel Francis
Solomon, Louis
Spitzbart, Abraham
Turner, Robert E L
Van Ryzin, Martina
Voichick, Michael
Wahlstrom, Lawrence F
Wainger, Stephen
Wasow, Wolfgang Richard
Weibel, Armella
Wilson, Robert Lee, Jr

WYOMING
Collins, James R
Husain, Syed Alamdar
Rowland, John H
Shader, Leslie Elwin
Varineau, Verne John

PUERTO RICO
Beck, Jonathon Mock
Francis, Eugene A
Laplaza, Miguel Luis
Montes, Maria Eugenia

ALBERTA
Consul, Prem Chandra
Fields, Jerry L
Garg, Krishna Murari
Hoo, Cheong Seng
Jones, J P
Klamkin, Murray S
Lowig, Henry Francis Joseph
Macki, Jack W
McKinney, Richard Leroy
Milner, Eric Charles
Mollin, Richard Anthony
Muldowney, James
Mungan, Necmettin
Narayana, Tadepalli Venkata
Rhemtulla, Akbar Hussein
Schaer, Jonathan
Stone, Michael Gates
Timourian, James Gregory
Varadarajan, Kalathoor
Wong, James Chin-Sze
Wyman, Max
Zvengrowski, Peter Daniel

BRITISH COLUMBIA
Belluce, Lawrence P
Bullen, Peter Southcott
Bures, Donald John (Charles)
Chacon, Rafael Van Severen
Chang, Bomshik
Christian, Robert Roland
Daykin, Philip Norman
Divinsky, Nathan Joseph
Fournier, John J F
Freedman, Allen Roy
Hewgill, Denton Elwood
Hinrichs, Lowell A
Hurd, Albert Emerson
MacDonald, John Lauchlin
Marshall, Albert Waldron
Murdoch, David Carruthers
Restrepo, Rodrigo Alvaro
Rieckhoff, Klaus E
Robertson, Malcolm Slingsby
Srivastava, Hari Mohan
Srivastava, Rekha
Thyer, Norman Harold
Whittaker, James Victor

MANITOBA
Armstrong, Kenneth William
Gratzer, George
Gupta, Narain Datt
Mendelsohn, Nathan Saul
Sichler, Jiri Jan
Williams, Hugh Cowie

NEW BRUNSWICK
Basilevsky, Alexander
Crawford, William Stanley Hayes
Dekster, Boris Veniamin
Sullivan, Donald
Thompson, Jon H

NEWFOUNDLAND
Burry, John Henry William
Fekete, Antal E
Lal, Mohan
Shawyer, Bruce L R

NOVA SCOTIA
Asadulla, Syed
Oliver, Leslie Howard
Pearson, Terrance Laverne
Snow, Douglas Oscar
Swaminathan, Srinivasa
Tingley, Arnold Jackson

ONTARIO
Abrham, Jaromir Vaclav
Allan, George B
Auer, Jan Willem

Barbeau, Edward Joseph, Jr
Behara, Minaketan
Berman, Gerald
Bower, Oliver Kenneth
Brainerd, Barron
Choi, Man-Duen
Cole, Randal Hudie
Coleman, Albert John
Cross, George Elliot
Csorgo, Miklos
Davis, Harry Floyd
Davison, Thomas Matthew Kerr
Dawson, Donald Andrew
DeLury, Daniel Bertrand
Dlab, Vlastimil
Drury, Malcolm John
Dunham, Charles Burton
Ellers, Erich Werner
Gadamer, Ernst Oscar
Gentleman, William Morven
Graham, James W
Graham, John Elwood
Griffith, John Sidney
Haruki, Hiroshi
Headley, Velmer Bentley
Heinig, Hans Paul
Hilldrup, David J
Jamieson, Derek Maitland
Kannappan, Palaniappan
Kaufman, Hyman
Kent, Clement F
Laframboise, Marc Alexander
LeBel, Jean Eugene
Linders, James Gus
Linis, Viktors
Lorch, Lee (Alexander)
McCallion, William James
McKiernan, Michel Amedee
MacNeill, Ian B
Macphail, Moray St John
Mueller, Bruno J W
Mullen, Kenneth
Muller, Eric Rene
Mullin, Ronald Cleveland
Murasugi, Kunio
Nesheim, Michael Ernest
Norman, Robert Daniel
Pandey, Jagdish Narayan
Pelletier, Joan Wick
Pope, Noel Kynaston
Pundsack, Arnold L
Puttaswamaiah, Bannikuppe M
Read, Ronald Cedric
Rice, Norman Molesworth
Roberts, Leslie Gordon
Robinson, Gilbert de Beauregard
Rooney, Paul George
Rosenthal, Peter (Michael)
Schubert, Cedric F
Shenitzer, Abe
Thierrin, Gabriel
Tutte, William Thomas
Ursell, John Henry
Vaillancourt, Remi Etienne
Vanstone, J R
Verner, James Hamilton
Woods, Edward James
Woodside, William
Younger, Daniel H

QUEBEC
Allen, Harold Don
Barr, Michael
Chalk, John
Châtillon, Guy
Driscoll, John G
Dubuc, Serge
Duncan, Richard Dale
Goodspeed, Frederick Maynard
(Cogswell)
Herschorn, Michael
Joffe, Anatole
Jonsson, Wilbur Jacob
Klemola, Tapio
Kotzig, Anton
Labute, John Paul
Lacroix, Norbert Hector Joseph
Lambek, Joachim
Lederman, Frank L
Leroux, Pierre
Lin, Paul C S
Rattray, Basil Andrew
Rosenberg, Ivo George
Sayeki, Hidemitsu
Schlomiuk, Norbert
Srivastava, Trilokv N
Turgeon, Jean

SASKATCHEWAN
Amundrud, Donald Lorne
Blum, Richard
Conlan, James
Rao, Veldanda Venugopal
Saini, Girdhari Lal

OTHER COUNTRIES
Adams, Helen Elizabeth
Andrea, Stephen Alfred
Bush, George Clark
Camiz, Sergio
Cheng, Kuang-Fu
Cherenack, Paul Francis
D'Ambrosio, Ubiratan
Durling, Frederick Charles

Eells, James
Foglio, Susana
Furstenberg, Hillel
Hofmann, Karl Heinrich
Iwasawa, Kenkichi
Keogh, Frank Richard
Lambert, William M, Jr
Lee, Chung N
McCue, Edmund Bradley
Maltese, George J
Miller, Charles Frederick, III
Mond, Bertram
Moser, Jurgen (Kurt)
Olive, Gloria
Pekeris, Chaim Leib
Quillen, Daniel G
Rabinowitz, Philip
Rich, Michael
Roseman, Joseph Jacob
Rosen, William G
Ryan, Robert Dean
Sakai, Shoichiro
Settles, Ronald Dean
Stocks, Douglas Roscoe, Jr
Stoneham, Richard George
Thompson, J G
Wallace, Andrew Hugh
Warne, Ronson Joseph
Wendland, Wolfgang Leopold
Yang, Chung-Chun

Number Theory

ALABAMA
Mattics, Leon Eugene

ARIZONA
Smarandache, Florentin
Velez, William Yslas

CALIFORNIA
Alder, Henry Ludwig
Alexanderson, Gerald Lee
Aschbacher, Michael
Drobot, Vladimir
Estes, Dennis Ray
Garrison, Betty Bernhardt
Gerstein, Larry J
Gruenberger, Fred J(oseph)
Guralnick, Robert Michael
Harris, Vincent Crockett
Hilliker, David Lee
Lenstra, Hendrik W
Levit, Robert Jules
Lu, Kau U
Sun, Hugo Sui-Hwan
Sweet, Melvin Millard
Taussky, Olga
Terras, Audrey Anne
Thomas, Paul Emery

COLORADO
Hightower, Collin James
Marsh, Donald Charles Burr
Mathys, Peter
Rearick, David F
Roeder, David William

CONNECTICUT
Eigel, Edwin George, Jr
Reid, James Dolan
Stein, Alan H

DISTRICT OF COLUMBIA
Thaler, Alvin Isaac
Vegh, Emanuel

FLORIDA
DeLeon, Morris Jack
Goodman, Adolph Winkler
Zame, Alan

GEORGIA
Bompart, Billy Earl
Pomerance, Carl

IDAHO
Girse, Robert Donald

ILLINOIS
Ballew, David Wayne
Berndt, Bruce Carl
Cormier, Romae Joseph
Diamond, Harold George
Earnest, Andrew George
Eggan, Lawrence Carl
Halberstam, Heini
Hildebrand, Adolf J
Janusz, Gerald Joseph
Kasube, Herbert Emil
Lamont, Patrick
Lazerson, Earl Edwin
Philipp, Walter V
Redmond, Donald Michael
Reznick, Bruce Arie
Ullom, Stephen Virgil
Vanden Eynden, Charles Lawrence

INDIANA
Dudley, Underwood
Hahn, Alexander J
Murphy, Catherine Mary
Shahidi, Freydoon

Wagstaff, Samuel Standfield, Jr
Yates, Richard Lee

IOWA
Kutzko, Philip C

KANSAS
Dressler, Robert Eugene
McCarthy, Paul Joseph

KENTUCKY
LeVan, Marijo O'Connor
Lowman, Bertha Pauline

LOUISIANA
Cook, Virginia A
Maxfield, Margaret Waugh

MAINE
Norton, Karl Kenneth
Pogorzelski, Henry Andrew

MARYLAND
Adams, William Wells
Buell, Duncan Alan
Federighi, Enrico Thomas
Ferguson, Helaman Rolfe Pratt
Goldstein, Larry Joel
Katz, Victor Joseph
Lakein, Richard Bruce
Massell, Paul Barry
Sadowsky, John
Spohn, William Gideon, Jr
Sunley, Judith S
Washington, Lawrence C
Wunderlich, Marvin C

MASSACHUSETTS
Ankeny, Nesmith Cornett
Gessel, Ira Martin
Hayes, David R
Kenney, Margaret June
Mazur, Barry
Ostrovsky, Rafail M
Senechal, Marjorie Lee
Stark, Harold Mead

MICHIGAN
Calloway, Jean Mitchener
Gioia, Anthony Alfred
Herzog, Fritz
Malm, Donald E G
Papp, F(rancis) J(oseph)

MINNESOTA
Gaal, Ilse Lisl Novak
Hejhal, Dennis Arnold

NEW JERSEY
Bumby, Richard Thomas
Endo, Youichi
Faltings, Gerd
Lagarias, Jeffrey Clark
Mills, William Harold

NEW YORK
Alex, Leo James
Arpaia, Pasquale J
Barshay, Jacob
Brzenk, Ronald Michael
Childs, Lindsay Nathan
Chudnovsky, Gregory V
Cohen, Herman Jacob
Cohn, Harvey
Cusick, Thomas William
Edwards, Harold M
Goldfeld, Dorian
Grabois, Neil
Grossman, George
Jacquet, Herve
Kaltofen, Erich L
Kleiman, Howard
Kwong, Yui-Hoi Harris
Lefton, Phyllis
Moritz, Roger Homer
Nathanson, Melvyn Bernard
Pechacek, Terry Frank
Segal, Sanford Leonard
Spiro, Claudia Alison

NORTH CAROLINA
Blanchet-Sadri, Francine
Haggard, Paul Wintzel
Hodel, Margaret Jones
Howard, Fredric Timothy
Vaughan, Theresa Phillips

OHIO
Flicker, Yuval Zvi
Gold, Robert
Hsia, John S
Leitzel, James Robert C
Parson, Louise Alayne
Paul, Jerome L
Ponomarev, Paul
Rubin, Karl C
Steiner, Ray Phillip
Theusch, Colleen Joan
Wang, Paul Shyh-Horng
Zassenhaus, Hans J

OREGON
Ecklund, Earl Frank, Jr
Flahibe, Mary Elizabeth

NEVADA
Fain, William Wharton
Flueck, John A

NEW HAMPSHIRE
Bogart, Kenneth Paul
Bossard, David Charles

NEW JERSEY
Blickstein, Stuart I
Boesch, Francis Theodore
Boylan, Edward S
Chasalow, Ivan G
Chien, Victor
Cullen, Daniel Edward
DeBaun, Robert Matthew
Eisner, Mark Joseph
Garber, H(irsh) Newton
Green, Edwin James
Leon, Ramon V
Lucantoni, David Michael
McCallum, Charles John, Jr
Magliola-Zoch, Doris
Maxim, Leslie Daniel
Mihram, George Arthur
Naus, Joseph Irwin
Prekopa, Andras
Roberts, Fred Stephen
Robertson, Jerry L(ewis)
Rothkopf, Michael H
Sarakwash, Michael
Segal, Moshe
Sengupta, Bhaskar
Tucker, Albert William
Vardi, Yehuda
Weibel, Charles Alexander
Whitt, Ward
Wright, Margaret Hagen

NEW MEXICO
Baty, Richard Samuel
Bell, Stoughton
Boland, W Robert
Farmer, William Michael
Gutjahr, Allan L
Hulme, Bernie Lee
Ross, Timothy Jack
Rypka, Eugene Weston
Summers, Donald Lee

NEW YORK
Balachandran, Kashi Ramamurthi
Ballou, Donald Pollard
Bencsath, Katalin A
Bienstock, Daniel
Billera, Louis Joseph
Bland, Robert Gary
Burke, Paul J
Chatterjee, Samprit
Conway, Richard Walter
Ecker, Joseph George
Goldfarb, Donald
Gomory, Ralph E
Green, Alwin Clark
Heath, David Clay
Helly, Walter S
Hunt, Everett Clair
Johnson, Ellis Lane
Kabak, Irwin William
Keilson, Julian
Kershenbaum, Aaron
King, Alan Jonathan
Kleiman, Howard
Klein, Morton
Kolesar, Peter John
Lyons, Joseph Paul
Nagel, Harry Leslie
Nevison, Christopher H
Pan, Victor
Richter, Donald
Riffenburgh, Robert Harry
Rosberg, Zvi
Rothenberg, Ronald Isaac
Schweitzer, Paul Jerome
Singh, Chanchal
Smith, Bernard
Stanfel, Larry Eugene
Stein, David Morris
Todd, Michael Jeremy
Van Slyke, Richard M
White, William Wallace
Wolfe, Philip
Yao, David D

NORTH CAROLINA
Aydlett, Louise Clate
Elmaghraby, Salah Eldin
Fishman, George Samuel
Hodgson, Thom Joel
Honeycutt, Thomas Lynn
Jambor, Paul Emil
McAllister, David Franklin
Morris, Peter Alan
Nair, Sreekantan S
Nelson, A Carl, Jr
Peterson, David West
Potthoff, Richard Frederick
Roth, Walter John
Stidham, Shaler, Jr
Wallingford, John Stuart

OHIO
Arnoff, E Leonard
Chen, Kuei-Lin

Chrissis, James W
Clark, Gordon Meredith
Duckstein, Lucien
Emmons, Hamilton
Flowers, Allan Dale
Gephart, Landis Stephen
Hales, Raleigh Stanton, Jr
Hurt, James Joseph
Jones, Chester George
Kantor, Paul B
Martino, Joseph Paul
Mills, Wendell Holmes, Jr
Nielsen, Philip Edward
Patton, Jon Michael
Piotrowski, Zbigniew
Reisman, Arnold
Schultz, Edwin Robert
Solow, Daniel
Staker, Robert D
Ventresca, Carol
Ward, Douglas Eric
Waren, Allan D(avid)
Wicke, Howard Henry
Yantis, Richard P

OKLAHOMA
Schuermann, Allen Clark, Jr

OREGON
Beck, John Robert
Linstone, Harold A
Murphy, Allan Hunt
Tam, Kwok-Wai
Tinney, William Frank

PENNSYLVANIA
Abramson, Stanley L
Antle, Charles Edward
Arnold, Leslie K
Balas, Egon
Beck, Robert Edward
Belkin, Barry
Biegler, Lorenz Theodor
Block, Henry William
Bolmarcich, Joseph John
Farquhar, Peter Henry
Lehoczky, John Paul
Mazumdar, Mainak
Mitchell, Theodore
Mutmansky, Jan Martin
Pierskalla, William P
Ranck, John Philip
Rosenshine, Matthew
Scranton, Bruce Edward
Stewart, William Charles
Szyld, Daniel Benjamin
Thompson, Gerald Luther
Vaserstein, Leonid, IV
Weindling, Joachim I(gnace)
Wolfe, Harvey
Wolfe, Reuben Edward

RHODE ISLAND
Dafermos, Stella
Jarrett, Jeffrey E
Leifman, Lev Jacob
Olson, David Gardels

SOUTH CAROLINA
Balintfy, Joseph L
Dobbins, James Gregory Hall
Kostreva, Michael Martin
Ringeisen, Richard Delose
Ruckle, William Henry

TENNESSEE
Demetriou, Charles Arthur
Umholtz, Clyde Allan

TEXAS
Bhat, Uggappakodi Narayan
Blank, Leland T
Brockett, Patrick Lee
Feldman, Richard Martin
Gillette, Kevin Keith
Greaney, William A
Harbaugh, Allan Wilson
Ignizio, James Paul
Jensen, Paul Allen
Johnston, Walter Edward
Kimeldorf, George S
Klingman, Darwin Dee
Lea, James Dighton
Lesso, William George
Matula, David William
Miikkulainen, Risto Pekka
Nickel, James Alvin
Noonan, James Waring
Patrick, Wesley Clare
Randolph, Paul Herbert
Self, Glendon Danna
Smith, Philip Wesley
Sobol, Marion Gross
Stein, William Edward
Weiser, Dan
Wu, Hsin-i

UTAH
Haslem, William Joshua
Richards, Dale Owen
Stearman, Roebert L(yle)
Walker, LeRoy Harold
Windham, Michael Parks

VERMONT
Brown, Robert Goodell
Olinick, Michael

VIRGINIA
Abbott, Stephen Dudley
Arnberg, Robert Lewis
Barmby, John G(lennon)
Berghoefer, Fred G
Bolstein, Arnold Richard
Bothwell, Frank Edgar
Cooper, Henry Franklyn, Jr
Coram, Donald Sidney
Cummings, Peter Thomas
Deal, George Edgar
Desiderio, Anthony Michael
Drew, John H
Emlet, Harry Elsworth, Jr
Englund, John Arthur
Golub, Abraham
Greenberg, Irwin
Greenblatt, Seth Alan
Gross, Donald
Haering, George
Hall, Otis F
Ho, S(hui)
Hudgins, Aubrey C, Jr
Hurley, William Jordan
Hwang, William Gaong
Kapos, Ervin
Karns, Charles W(esley)
Kydes, Andy Steve
LaBudde, Robert Arthur
Lawwill, Stanley Joseph
Lowry, Philip Holt
McNichols, Gerald Robert
Malcolm, Janet May
May, Donald Curtis, Jr
Michalowicz, Joseph Victor
Mikhail, Nabih N
Overholt, John Lough
Pokrant, Marvin Arthur
Rehm, Allan Stanley
Reierson, James (Dutton)
Reynolds, Marion Rudolph, Jr
Richards, Paul Bland
Richardson, Henry Russell
Schwartz, Benjamin L
Sessler, John Charles
Shrier, Stefan
Sorrell, Gary Lee
Stone, Lawrence David
Walker, Robert Paul
Wilkinson, William Lyle
Woisard, Edwin Lewis

WASHINGTON
Bare, Barry Bruce
Heilenday, Frank W (Tod)
Liu, Chen-Ching
Millham, Charles Blanchard
Moore, Richard Lee
Rockafellar, Ralph Tyrrell
Rustagi, Krishna Prasad

WISCONSIN
Robinson, Stephen Micháel
Schultz, Hilbert Kenneth
Stahl, Neil

ALBERTA
Enns, Ernest Gerhard
Silver, Edward Allan

BRITISH COLUMBIA
Lambe, Thomas Anthony
Lancaster, George Maurice

MANITOBA
Zeiler, Frederick

NEWFOUNDLAND
May, Sherry Jan

ONTARIO
Caron, Richard John
Colbourn, Charles Joseph
Davis, Ronald Stuart
Hopkins, Nigel John
Hudson, John Leslie
Lindsey, George Roy
Magazine, Michael Jay
Mayberry, John Patterson
Mohanty, Sri Gopal
Muller, Eric Rene
Nash, John Christopher
North, Henry E(rick) T(uisku)
Paull, Allan E
Posner, Morton Jacob
Pressman, Irwin Samuel
Stauffer, Allan Daniel
Steiner, George

QUEBEC
Clarke, Francis
Galibois, Andre
Loulou, Richard Jacques
Sankoff, David

OTHER COUNTRIES
Davis, Edward Alex
Garcia-Santesmases, Jose Miguel
Gritzmann, Peter
Hamacher, Horst W

Lindquist, Anders Gunnar
Philippou, Andreas Nicolaou

Physical Mathematics

ALABAMA
Chang, Mou-Hsiung
Cook, Frederick Lee
Fowler, Bruce Wayne
Lee, Ching Tsung

ARIZONA
Faris, William Guignard
Hoehn, A(lfred) J(oseph)
Iverson, A Evan
Lomont, John S
Rund, Hanno
Triffet, Terry
Twomey, Sean Andrew

ARKANSAS
Lieber, Michael

CALIFORNIA
Antolak, Arlyn Joe
Bernard, Douglas Alan
Birnir, Björn
Bleick, Willard Evan
Burgoyne, Peter Nicholas
Burke, James Edward
Cohen, E Richard
Dedrick, Kent Gentry
Derenzo, Stephen Edward
Finn, Robert
Freis, Robert P
Garrison, John Carson
Gillespie, Daniel Thomas
Hershey, Allen Vincent
Hovingh, Jack
Hudson, George Elbert
Hutchin, Richard Ariel
Klapman, Solomon Joel
Klein, Abel
Krause, Lloyd O(scar)
Landshoff, Rolf
Langdon, Allan Bruce
Lapidus, Michel Laurent
Lu, Kau U
Mayer, Meinhard Edwin
Meecham, William Coryell
Moore, Mortimer Norman
Morris, George William
Nachamkin, Jack
Neustadter, Siegfried Friedrich
O'Malley, Thomas Francis
Park, Chong Jin
Saffren, Melvin Michael
Scholl, James Francis
Shepanski, John Francis
Taub, Abraham Haskel
Toupin, Richard A
Unal, Aynur
Villeneuve, A(lfred) T(homas)
Wagner, Richard John
Wilson, Robert Norton
Yeh, Cavour W

COLORADO
Augusteijn, Marijke Francina
Kassoy, David R
Laks, David Bejnesh
Marchand, Jean-Paul
Poore, Aubrey Bonner
Walter, Martin Edward
Whitten, Barbara L

CONNECTICUT
Kohlmayr, Gerhard Franz
McCartin, Brian James
Sachdeva, Baldev Krishan
Wong, Maurice King Fan

DELAWARE
Angell, Thomas Strong
Hirsch, Albert Edgar
Mishoe, Luna I
Morgan, John Davis, III

DISTRICT OF COLUMBIA
Berger, Martin Jacob
Grossman, John Mark
Herron, Isom H
Langworthy, James Brian
Meijer, Paul Herman Ernst
Nachman, Arje
Yen, Nai-Chyuan

FLORIDA
Albright, John Rupp
Curtright, Thomas Lynn
Debnath, Lokenath
Dubbelday, Pieter Steven
Emch, Gerard G
Kelley, Robert Lee
Neuringer, Joseph Louis
Nunn, Walter M(elrose), Jr

GEORGIA
Mickens, Ronald Elbert
Warsi, Nazir Ahmed

ILLINOIS
Carhart, Richard Alan

Physical Mathematics (cont)

Carroll, Robert Wayne
Ceperley, David Matthew
Cook, Joseph Marion
Dykla, John J
Hafele, Joseph Carl
Ho, Chung-Wu
Jerome, Joseph Walter
Kaper, Hans G
Kath, William Lawrence
Matkowsky, Bernard J
Oono, Yoshitsugu
Prais, Michael Gene
Schonfeld, Jonathan Furth
Siegert, Arnold John Frederick
Smarr, Larry Lee
Tao, Rongjia
Twersky, Victor

INDIANA
Challifour, John Lee
Lenard, Andrew
Madan, Ved P
Penna, Michael Anthony

IOWA
Alexander, Roger Keith
Carlson, Bille Chandler
Hammer, Charles Lawrence
Jorgensen, Palle E T
Struck-Marcell, Curtis John
Ton-That, Tuong

KANSAS
Acker, Andrew French, III
Lerner, David Evan
Miller, Forest R
Ramm, Alexander G
Unz, Hillel

KENTUCKY
Van Winter, Clasine

LOUISIANA
Brans, Carl Henry
Goldstein, Jerome Arthur
Heatherly, Henry Edward
Herzberger, Maximilian Jacob
Knill, Ronald John
Kuo, Hui-Hsiung
Kyame, Joseph John
Runnels, Lynn Kelli

MARYLAND
Blue, James Lawrence
Criss, John W
Dehn, James Theodore
Fischel, David
Fisher, Michael Ellis
Gates, Sylvester J, Jr
Granger, Robert A, II
Herman, Richard Howard
Hornstein, John Stanley
Hornstein, John Stanley
Hu, Bei-Lok Bernard
Kuttler, James Robert
Massell, Paul Barry
Rehm, Ronald George
Rostamian, Rouben
Wallace, Alton Smith

MASSACHUSETTS
Bernstein, Joseph N
Cronin-Golomb, Mark
Holst, Jewel Magee
Humi, Mayer
Imbrie, John Z
Jaffe, Arthur Michael
Kon, Mark A
Melrose, Richard B
Parker, Lee Ward
Ruskai, Mary Beth
Schay, Geza

MICHIGAN
Khargonekar, Pramod P
Lomax, Ronald J(ames)
Marriott, Richard
Miklavcic, Milan
Murphy, Brian Boru
Rauch, Jeffrey Baron
Williams, David Noel

MINNESOTA
Bonne, Ulrich
Ericksen, Jerald LaVerne
Follingstad, Henry George
Luskin, Mitchell B
Miller, Willard, Jr
Olver, Peter John
Pour-El, Marian Boykan

MISSISSIPPI
Keulegan, Garbis
Rayborn, Grayson Hanks

MISSOURI
Beem, John Kelly
Bender, Carl Martin
Katz, Israel Norman
Ornstein, Wilhelm
Penico, Anthony Joseph
Pettey, Dix Hayes

Plummer, Otho Raymond
Retzloff, David George

NEW HAMPSHIRE
MacGillivray, Jeffrey Charles
Shepard, Harvey Kenneth

NEW JERSEY
Bargmann, Valentine
Benaroya, Haym
Budny, Robert Vierling
Carter, Ashley Hale
Chien, Victor
Dyson, Freeman John
Goldin, Gerald Alan
Koss, Valery Alexander
Lepowsky, James Ivan
Lieb, Elliott Hershel
McKenna, James
Rulf, Benjamin
Schrenk, George L
Seidl, Frederick Gabriel Paul
Sverdlove, Ronald
Taylor, Jean Ellen
White, Benjamin Steven
Wightman, Arthur Strong

NEW MEXICO
Baker, George Allen, Jr
Becker, Wilhelm
Biggs, Frank
Bolie, Victor Wayne
Campbell, John Raymond
Critchfield, Charles Louis
Devaney, Joseph James
Dietz, David
Griffin, Patrick J
Hurd, Jon Rickey
Hyman, James Macklin
Lee, Clarence Edgar
Louck, James Donald
Mark, J Carson
Muir, Patrick Fred
Norton, John Leslie
O'Dell, Ralph Douglas
Ross, Timothy Jack
Sandford, Maxwell Tenbrook, II
Wing, Janet E (Sweedyk) Bendt

NEW YORK
Bowick, Mark John
Boyce, William Edward
Chudnovsky, Gregory V
Dasgupta, Gautam
Deady, Matthew William
Deift, Percy Alec
Dicker, Daniel
Dickman, Steven Richard
Doering, Charles Rogers
Fagot, Wilfred Clark
Feigenbaum, Mitchell Jay
Fisher, Edward
Fleishman, Bernard Abraham
Geer, James Francis
Glimm, James Gilbert
Habetler, George Joseph
Harrington, Steven Jay
Isaacson, David
Mane, Sateesh Ramchandra
Marcus, Paul Malcolm
Marcuvitz, Nathan
Newman, Charles Michael
Nirenberg, Louis
Phong, Duong Hong
Roskes, Gerald J
Roytburd, Victor
Rubenfeld, Lester A
Schlitt, Dan Webb
Siegmann, William Lewis
Stell, George Roger
Tulczyjew, Wlodzimierz Marek
Wang, Xingwu
Weng, Wu Tsung
Wolf, Emil
Yarmush, David Leon

NORTH CAROLINA
Brown, J(ohn) David
Cotanch, Stephen Robert
Dankel, Thaddeus George, Jr
Dillard, Margaret Bleick
Dolan, Louise Ann
Fulp, Ronald Owen
Reed, Michael Charles
Teague, David Boyce
York, James Wesley, Jr

OHIO
Burggraf, Odus R
Butz, Donald Josef
Dumas, H Scott
Goad, Clyde Clarenton
Kaplan, Bernard
Kummer, Martin
Long, Clifford A
Plybon, Benjamin Francis
Seiger, Harvey N
Tamburino, Louis A
Weinstock, Robert
Wen, Shih-Liang
Wicke, Howard Henry

OKLAHOMA
Fowler, Richard Gildart

Lal, Manohar
Robinson, Enders Anthony

OREGON
Guenther, Ronald Bernard
Kocher, Carl A
Murphy, Lea Frances
Warner, David Charles
Wheeler, Nicholas Allan

PENNSYLVANIA
Ambs, William Joseph
Becker, Kurt H
Blank, Albert Abraham
Bona, Jerry Lloyd
Bradley, James Henry Stobart
Farouk, Bakhtier
Gorman, Arthur Daniel
Kadison, Richard Vincent
Porter, John Robert
Roman, Paul
Rorres, Chris
Sarram, Mehdi
Wayne, Clarence Eugene
Weiner, Brian Lewis

RHODE ISLAND
Childs, Donald Ray

SOUTH CAROLINA
Stevens, James Levon

TENNESSEE
Dory, Robert Allan

TEXAS
Auchmuty, Giles
Aucoin, Paschal Joseph, Jr
Bronaugh, Edwin Lee
Fairchild, Edward Elwood, Jr
Fix, George Joseph
Haberman, Richard
Heelis, Roderick Antony
Lewis, Ira Wayne
McCoy, Jimmy Jewell
Narcowich, Francis Joseph
Nordlander, Peter Jan Arne
Olenick, Richard Peter
Ozsvath, Istvan
Prabhu, Vasant K
Radin, Charles Lewis
Roberts, Thomas M
Voigt, Gerd-Hannes
Walton, Jay R
Wang, Chao-Cheng
Wu, Hsin-i

UTAH
Bates, Peter William
Wilcox, Calvin Hayden
Wu, Yong-Shi

VIRGINIA
Adam, John Anthony
Bowden, Robert Lee, Jr
Bragg, Lincoln Ellsworth
Dorning, John Joseph
Farrukh, Usamah Omar
Greenberg, William
Hagedorn, George Allan
Horgan, Cornelius Oliver
Howland, James Secord
Hurley, William Jordan
Lasiecka, Irena
Parvulescu, Antares
Raychowdhury, Pratip Nath
Renardy, Yuriko
Richards, Paul Bland
Simmonds, James G
Thomas, Lawrence E
Wheeler, Robert Lee
Wohl, Philip R
Zweifel, Paul Frederick

WASHINGTON
Burkhart, Richard Henry
Cochran, James Alan
Nievergelt, Yves

WISCONSIN
Cook, David Marsden
Durand, Loyal
Stewart, W(arren) E(arl)
Strikwerda, John Charles
Uhlenbrock, Dietrich A

PUERTO RICO
Khan, Winston

ALBERTA
Blais, J A Rodrigue
Capri, Anton Zizi
Epstein, Marcelo
Kunzle, Hans Peter
Ludwig, Garry (Gerhard Adolf)
Weaver, Ralph Sherman

BRITISH COLUMBIA
Shinbrot, Marvin
Srivastava, Hari Mohan

MANITOBA
Cohen, Harley
Wilk, Stanislas Francois Jean

ONTARIO
Duff, George Francis Denton
Elliott, George Arthur
Graham, John
Huschilt, John
Jackson, David Phillip
Langford, William Finlay
Lee-Whiting, Graham Edward
Lipshitz, Stanley Paul
Moltyaner, Grigory
Nerenberg, Morton Abraham
Paldus, Josef
Prugovecki, Eduard
Sen, Dipak Kumar
Tross, Ralph G

QUEBEC
Boivin, Alberic
Clarke, Francis
Fabrikant, Valery Isaak
Grmela, Miroslav
Kröger, Helmut Karl
Wallace, Philip Russell

SASKATCHEWAN
Law, Alan Greenwell

OTHER COUNTRIES
Hjertager, Bjorn Helge
Lanford, Oscar E, III
Liu, Ta-Jo
Nussenzveig, Herch Moyses
Rivier, Nicolas Yves
Roseman, Joseph Jacob
Rosenbaum, Marcos
Schutz, Bernard Frederick
Suger-Cofino, Jose-Eduardo
Williams, Michael Maurice Rudolph

Probability

ALABAMA
Chang, Mou-Hsiung
Hudson, William Nathaniel
Schroer, Bernard J

ARIZONA
Barckett, Joseph Anthony
Daniel, Sam Mordochai
Faris, William Guignard
Frieden, Bernard Roy
Maier, Robert S
Neuts, Marcel Fernand

CALIFORNIA
Anderson, Paul Maurice
Blischke, Wallace Robert
Book, Stephen Alan
Calabrese, Philip G
Chernick, Michael Ross
Feldman, Jacob
Gray, Robert Molten
Haight, Frank Avery
Harmen, Raymond A
Harris, Theodore Edward
Helstrom, Carl Wilhelm
Henning, Harley Barry
Hsu, Yu-Sheng
Iglehart, Donald Lee
Jacobs, Patricia Anne
Jewell, Nicholas Patrick
Krener, Arthur James
Krieger, Henry Alan
Kuhlmann, Karl Frederick
Langdon, Glen George, Jr
Lange, Kenneth L
Le Cam, Lucien Marie
Liggett, Thomas Milton
Milch, Paul R
Mo, Charles Tse Chin
Molnar, Imre
Ott, Wayne R
Patterson, Tim J
Rao, Jammalamadaka S
Rao, Malempati Madhusudana
Rosenblatt, Murray
Sharpe, Michael John
Tucker, Howard Gregory
Weaver, William Bruce
Weber, Charles L
Weiner, Howard Jacob
Wendel, James G
Wiggins, Alvin Dennie
Williams, Jack Rudolph

COLORADO
Brockwell, Peter John
Crow, Edwin Louis
Darst, Richard B
Etter, Delores Maria
Heine, George Winfield, III
Kafadar, Karen
Marchand, Jean-Paul
Mathys, Peter
Nesenbergs, Martin

CONNECTICUT
Blei, Ron Charles
Cordery, Robert Arthur
Giné-Masdéu, Evarist
Kleinfeld, Ira H
Mukhopadhyay, Nitis
Phillips, Peter Charles B

Rao, Valluru Bhavanarayana
Ruckebusch, Guy Bernard
Stuck, Barton W
Vitale, Richard Albert

DISTRICT OF COLUMBIA
Cohen, Michael Paul
Filliben, James John
Hammer, Carl
Nayak, Tapan Kumar
Robinson, E Arthur, Jr
Rosenblatt, David
Thall, Peter Francis
Wagner, Andrew James

FLORIDA
Ballerini, Rocco
Haase, Richard Henry
Keesling, James Edgar
Leysieffer, Frederick Walter
Lin, Y(u) K(weng)
Mukherjea, Arunava
Muth, Eginhard Joerg
Sethuraman, Jayaram
Zame, Alan

GEORGIA
Kutner, Michael Henry
Lavender, DeWitt Earl

IDAHO
Bickel, John Henry

ILLINOIS
Barron, Emmanuel Nicholas
Bellow, Alexandra
Burkholder, Donald Lyman
Calderon, Calixto Pedro
Clover, William John, Jr
Cox, Dennis Dean
Heller, Barbara Ruth
Ionescu Tulcea, Cassius
Jones, Roger L
Knight, Frank B
Laud, Purushottam Waman
Philipp, Walter V
Pursley, Michael Bader
Tang, Wilson H
Tier, Charles
Wacker, William Dennis
Wolter, Kirk Marcus

INDIANA
Beekman, John Alfred
Bhattacharya, Rabindra Nath
Bradley, Richard Crane, Jr
Carlson, Lee Arnold
Kinney, John James
Lyons, Russell David
Pimblott, Simon Martin
Protter, Philip Elliott
Rego, Vernon J
Rhoades, Billy Eugene

IOWA
Andersland, Mark Steven
Athreya, Krishna Balasundaram
Cressie, Noel A C
Georgakakos, Konstantine P
Reckase, Mark Daniel
Russo, Ralph P

KANSAS
Burckel, Robert Bruce
Ramm, Alexander G
Stewart, William Henry
Stockli, Martin P

KENTUCKY
Griffith, William Schuler

LOUISIANA
Abbott, James H
Kuo, Hui-Hsiung
Sundar, P

MAINE
McMillan, Brockway

MARYLAND
Abbey, Robert Fred, Jr
Austin, Homer Wellington
Cacciamani, Eugene Richard, Jr
Cherniavsky, Ellen Abelson
Cohen, Edgar A, Jr
Ephremides, Anthony
Gillis, James Thompson
Huddleston, Charles Martin
Huneycutt, James Ernest, Jr
Karr, Alan Francis
Kattakuzhy, George Chacko
Lao, Chang Sheng
Lechner, James Albert
Lundegard, Robert James
Robinson, Richard Carleton, Jr
Rubinstein, Lawrence Victor
Rukhin, Andrew Leo
Saworotnow, Parfeny Pavolich
Serfling, Robert Joseph
Siegel, Martha J
Smith, Paul John
Sponsler, George C
Weiss, Michael David
Wierman, John C

Zabronsky, Herman

MASSACHUSETTS
Bras, Rafael Luis
Dudley, Richard Mansfield
Elias, Peter
Ellis, Richard Steven
Entekhabi, Dara
Fougere, Paul Francis
Gacs, Peter
Giordano, Arthur Anthony
Hahn, Marjorie G
Horowitz, Larry Lowell
Imbrie, John Z
Kitchin, John Francis
Kon, Mark A
Lo, Andrew W
Schonhoff, Thomas Arthur
Servi, Leslie D
Skibinsky, Morris

MICHIGAN
Albaugh, A Henry
Chakravarthy, Srinivasaraghavan
Chow, Pao Liu
Clarke, Allen Bruce
Hill, Bruce M
LePage, Raoul
Tanis, Elliot Alan

MINNESOTA
Ayeni, Babatunde J
Gray, Lawrence Firman
Pruitt, William Edwin
Sudderth, William David

MISSOURI
King, Terry Lee
Sawyer, Stanley Arthur

NEBRASKA
Conway, James Joseph
Heaney, Robert Proulx

NEW HAMPSHIRE
Jolly, Stuart Martin
Lamperti, John Williams

NEW JERSEY
Albin, Susan Lee
Badalamenti, Anthony Francis
Benaroya, Haym
Boylan, Edward S
Cimini, Leonard Joseph, Jr
Kadota, T Theodore
Landwehr, James M
Leon, Ramon V
Loader, Clive Roland
Lucantoni, David Michael
McKenna, James
Naus, Joseph Irwin
Ott, Teunis Jan
Prekopa, Andras
Riggs, Richard
Saltzberg, Burton R
Steiger, William Lee
Vardi, Yehuda
Verdu, Sergio
Weiss, Alan
White, Benjamin Steven
Whitt, Ward

NEW MEXICO
Beyer, William A
Cobb, Loren
Cogburn, Robert Francis
Davis, James Avery
Gutjahr, Allan L
Silver, Richard N

NEW YORK
Anderson, Allan George
Balachandran, Kashi Ramamurthi
Berman, Simeon Moses
Berresford, Geoffrey Case
Blessing, John A
Boyce, William Edward
Chan, Jack-Kang
Cohen, Joel Ephraim
Debany, Warren Harding, Jr
Dynkin, Eugene B
Falk, Harold
Fine, Terrence Leon
Frank, Kozi
Haag, Fred George
Hanson, David Lee
Heath, David Clay
Heidelberger, Philip
Hirtzel, Cynthia S
Karatzas, Ioannis
Keepler, Manuel
Kyburg, Henry
Lawton, William Harvey
Lax, Melvin
Moritz, Roger Homer
Newman, Charles Michael
Oakes, David
Papanicolaou, George Constantine
Prabhu, Narahari Umanath
Rosberg, Zvi
Ruppert, David
Singh, Chanchal
Todaro, Michael P
Tucker, Thomas William

Yablon, Marvin

NORTH CAROLINA
Bergsten, Jane Williams
Bernhard, Richard Harold
Bishir, John William
Dankel, Thaddeus George, Jr
Guess, Harry Adelbert
Jordan, Edward Daniel
Lawler, Gregory Francis
Menius, Arthur Clayton, Jr
Nelson, A Carl, Jr
Petersen, Karl Endel
Stidham, Shaler, Jr

OHIO
Bhargava, Triloki Nath
Breitenberger, Ernst
Caldito, Gloria C
Cavaretta, Alfred S
De Acosta, Alejandro
Fan, Liang-Shih
Goel, Prem Kumar
Hern, Thomas Albert
Neaderhouser Purdy, Carla Cecilia
Rayner, John Norman
Stackelberg, Olaf Patrick
Sterrett, Andrew
Ward, Douglas Eric

OKLAHOMA
Chambers, Richard Lee
Hochhaus, Larry
Richardson, J Mark

OREGON
Chartier, Vernon L
Demarest, Harold Hunt, Jr
Engelbrecht, Rudolf S
Nelsen, Roger Bain
Tunturi, Archie Robert

PENNSYLVANIA
Arnold, Leslie K
Beggs, William John
Belkin, Barry
Blaisdell, Ernest Atwell, Jr
Block, Henry William
Bona, Jerry Lloyd
Eisenberg, Bennett
Fryling, Robert Howard
Galambos, Janos
Gleser, Leon Jay
Grinberg, Eric L
Helmbold, Robert Lawson
Jech, Thomas J
Jeruchim, Michel Claude
Lehoczky, John Paul
Mehta, Jatinder S
Mitra, Shashanka S
Schervish, Mark John
Vaserstein, Leonid, IV
Wagner, Clifford Henry
Wilson, Myron Allen
Yukich, Joseph E

RHODE ISLAND
Bartram, James F(ranklin)
Kushner, Harold J(oseph)
Leifman, Lev Jacob

SOUTH CAROLINA
Rust, Philip Frederick

TENNESSEE
Dobosy, Ronald Joseph
Groer, Peter Gerold
Perey, Francis George
Tan, Wai-Yuan

TEXAS
Bhat, Uggappakodi Narayan
Brockett, Patrick Lee
Cecil, David Rolf
Eubank, Randall Lester
Hensley, Douglas Austin
Johnson, Johnny Albert
Kimeldorf, George S
Mauldin, Richard Daniel
Pfeiffer, Paul Edwin
Randolph, Paul Herbert
Schuster, Eugene F
Stein, William Edward
Turner, Danny William
Vaaler, Jeffrey David
Ward-McLemore, Ethel
Wheeler, Joe Darr
Whitmore, Ralph M
Wilson, Thomas Leon
Wright, James Foley
Yao, James T-P

UTAH
Walker, LeRoy Harold

VERMONT
Emerson, John David

VIRGINIA
Bolstein, Arnold Richard
Foy, C Allan
Gray, F(estus) Gail
Greenblatt, Seth Alan
Hendricks, Walter James

Henry, Neil Wylie
Kratzke, Thomas Martin
Martin, Nathaniel Frizzel Grafton
Michalowicz, Joseph Victor
Minton, Paul Dixon
Mittal, Yashaswini Deval
Richardson, Henry Russell
Singh, Mahendra Pal
Stone, Lawrence David

WASHINGTON
Anderson, Richard Gregory
Band, William
Burkhart, Richard Henry
Conquest, Loveday Loyce
Perlman, Michael David
Pyke, Ronald
Saunders, Sam Cundiff
Schaerf, Henry Maximilian

WEST VIRGINIA
Brady, Dennis Patrick

WISCONSIN
Gurland, John
Harris, Bernard
Kaye, Norman Joseph
Keisler, Howard Jerome
Kurtz, Thomas Gordon
Reinsel, Gregory Charles

WYOMING
Tung, Yeou-Koung

PUERTO RICO
Hussain, Malek Gholoum Malek

ALBERTA
Elliott, Robert James
Johnson, Dudley Paul

BRITISH COLUMBIA
Anderson, R F V
Haussmann, Ulrich Gunther
Reed, William J

NEW BRUNSWICK
Basilevsky, Alexander
Mureika, Roman A
Pham-Gia, Thu

NEWFOUNDLAND
Burry, John Henry William

ONTARIO
Brainerd, Barron
Campbell, Louis Lorne
Denzel, George Eugene
Mark, Jon Wei
Mayberry, John Patterson
Schaufele, Ronald A

QUEBEC
Choksi, Jal R
Herz, Carl Samuel
Lefebvre, Mario
Rahn, Armin
Stathopoulos, Theodore
Todorovic, Petar N

SASKATCHEWAN
Tomkins, Robert James

OTHER COUNTRIES
Brown, Richard Harland
Chesson, Peter Leith
Hochberg, Kenneth J
Mengali, Umberto
Nestvold, Elwood Olaf
Philippou, Andreas Nicolaou
Shapiro, Jesse Marshall
Sunahara, Yoshifumi

Topology

ALABAMA
Plunkett, Robert Lee
Rogers, Jack Wyndall, Jr

ARIZONA
Grace, Edward Everett
McMahon, Douglas Charles
Thompson, Richard Bruce

ARKANSAS
Cochran, Allan Chester
Feldman, William A
Schein, Boris M

CALIFORNIA
Agoston, Max Karl
Brown, Robert Freeman
Clarke, Wilton E L
Davis, Alan Lynn
Day, Jane Maxwell
De Sapio, Rodolfo Vittorio
Elderkin, Richard Howard
Feldman, Louis A
Flegal, Robert Melvin
Freedman, Michael Hartley
Greever, John
Hagopian, Charles Lemuel
Hirsch, Morris William

QUEBEC
Alagar, Vangalur S

SASKATCHEWAN
Kaul, S K
Mezey, Paul G
Tymchatyn, Edward Dmytro

OTHER COUNTRIES
Kinoshita, Shin'ichi
Stocks, Douglas Roscoe, Jr
Thomason, Robert Wayne
Zagier, Don Bernard

Other Mathematics

ALABAMA
Goodall, Marcus Campbell
Shipman, Jerry

ARIZONA
Smarandache, Florentin

ARKANSAS
Thornton, Kent W

CALIFORNIA
Benesch, Samuel Eli
Cates, Marshall L
Cherlin, George Yale
Chu, Kai-Ching
Dalrymple, Stephen Harris
Desoer, Charles A(uguste)
Green, Sherry Merrill
Lang, Martin T
Levine, Howard Bernard
Liu, David Shiao-Kung
McCarthy, Mary Anne
Marder, Stanley
Milstein, Jaime
Painter, Jeffrey Farrar
Patel, Arvindkumar Motibhai
Rosenberg, Alex
Sansom, Richard E
Shestakov, Aleksei Ilyich
Simon, Horst D
Urquidi-MacDonald, Mirna
Woodriff, Roger L

COLORADO
Ablowitz, Mark Jay
Glover, Fred William
Holley, Frieda Koster
Lee, Kotik Kai
Walter, Martin Edward

CONNECTICUT
Fischer, Michael
Frankel, Martin Richard
Sharlow, John Francis

DELAWARE
Kittlitz, Rudolf Gottlieb, Jr

FLORIDA
Allen, Jon Charles
Clutterham, David Robert
Halpern, Leopold (Ernst)
Martinez, Jorge
Ritter, Gerhard X
Shershin, Anthony Connors
Taylor, Michael Dee

GEORGIA
Bevis, Jean Harwell
Smith, William Allen

HAWAII
Anderson, Gerald M

IDAHO
Hoagland, Gordon Wood
Maloof, Giles Wilson

ILLINOIS
Hannon, Bruce Michael
Heller, Barbara Ruth
Reiter, Stanley
Smith, Stephen D
Usiskin, Zalman P
Wood, John William

INDIANA
Hood, Rodney Taber
Madan, Ved P
Taylor, Gladys Gillman

IOWA
Maxey, E James
Ton-That, Tuong

KANSAS
Curtis, Wendell D
Strecker, George Edison

LOUISIANA
Davis, Lawrence H
Goldstein, Jerome Arthur
Kythe, Prem Kishore
Looney, Stephen Warwick
Sundar, P

MARYLAND
Austin, Homer Wellington
Boisvert, Ronald Fernand
Calinger, Ronald Steve
Draper, Richard Noel
Engel, Joseph H(enry)
Grebogi, Celso
Kowalski, Richard
Krishnaprasad, Perinkulam S
Lopez-Escobar, Edgar George Kenneth
Lozier, Daniel William
Preisman, Albert
Shiffman, Bernard

MASSACHUSETTS
Ankeny, Nesmith Cornett
Berkovits, Shimshon
Bezuszka, Stanley John
Cole, Nancy
Gacs, Peter
Grossberg, Stephen
Kenney, Margaret June
Lam, Kui Chuen
McOwen, Robert C
Myers, John Martin
Shum, Annie Waiching

MICHIGAN
Bickart, Theodore Albert
Field, David Anthony
Johnson, Harold Hunt
McLaughlin, Renate
Murphy, Brian Boru
Papp, F(rancis) J(oseph)
Parks, Terry Everett
Verhey, Roger Frank

MISSISSIPPI
Doblin, Stephen Alan

MISSOURI
Elliott, David LeRoy
Kubicek, John D
Seydel, Robert E
Silverstein, Martin L
Tikoo, Mohan L

NEBRASKA
Maloney, John P
Meakin, John C

NEW HAMPSHIRE
Meeker, Loren David

NEW JERSEY
Ahsanullah, Mohammad
Beals, R Michael
Belanger, David Gerald
Brand, William Wayne
Endo, Youichi
Gilman, Jane P
Goldstein, Sheldon
Hunter, John Stuart
Jacquin, Arnaud Eric
Leef, Audrey V
Mather, John Norman
Poiani, Eileen Louise
Rosenstein, Joseph Geoffrey
Sen, Tapas K

NEW MEXICO
Ahmed, Nasir
Argyros, Ioannis Konstantinos
Booth, Thomas Edward

NEW YORK
Ascher, Marcia
Chen, Chi-Tsong
Cox, John Theodore
Daly, John T
Edwards, Harold M
Feuer, Richard Dennis
Guckenheimer, John
Gustavson, Fred Gehrung
Haddock, Jorge
Hastings, Harold Morris
Kline, Morris
LeVine, Micheal Joseph
Moskowitz, Martin A
Pan, Victor
Pattee, Howard Hunt, Jr
Shevack, Hilda N
Stearns, Richard Edwin

NORTH CAROLINA
Burbeck, Christina Anderson
Katzin, Gerald Howard
Lambert, Alan L
Lucas, Carol N
Tyndall, John Raymond
Wu, Julian Juh-Ren

OHIO
Barger, Samuel Floyd
Burden, Richard L
Chiu, Victor
Davis, Sheldon W
Kertz, George J
St John, Ralph C
Schneider, Harold William
Steinlage, Ralph Cletus

OKLAHOMA
Briggs, Phillip D

Carr, Meg Brady
Weeks, David Lee

OREGON
Ballantine, C S
Seitz, Gary M

PENNSYLVANIA
Baker, Ronald G
Chase, Gene Barry
Harris, Zellig S
Klotz, Eugene Arthur
McWhirter, James Herman
Mehta, Jatinder S
Pommersheim, James Martin
Scott, Dana S
Szyld, Daniel Benjamin
Tozier, John E
Williams, Albert J, Jr

RHODE ISLAND
Laurence, Geoffrey Cameron
Rudolph, Lee

SOUTH CAROLINA
Henry, Harold Robert
Saxena, Subhash C

TENNESSEE
Adams, Dennis Ray
Beauchamp, John J
Davidson, Robert C
Hill, Hubert Mack

TEXAS
Carey, Graham Francis
Gladwell, Ian
Hernandez, Norma
Ladde, Gangaram Shivlingappa
Linn, John Charles
Renka, Robert Joseph
Semmes, Stephen William
Trench, William Frederick

UTAH
Alfeld, Peter
Cohen, Elaine

VIRGINIA
Asterita, Mary Frances
Burns, William, Jr
Crowley, James M
Kratzke, Thomas Martin
Lasiecka, Irena
Smith, Emma Breedlove
Southworth, Raymond W(illiam)
Thomas, Lawrence E

WASHINGTON
Lind, Douglas A

WISCONSIN
Meitler, Carolyn Louise
Parter, Seymour Victor
Schneider, Hans
Strikwerda, John Charles

ALBERTA
Andersen, Kenneth F
Blais, J A Rodrigue
Muldowney, James
Tait, Robert James

BRITISH COLUMBIA
Das, Anadijiban
Miura, Robert Mitsuru

ONTARIO
Carlisle, David Brez
Choi, Man-Duen
Davison, E(dward) J(oseph)
Denzel, George Eugene
Duever, Thomas Albert

QUEBEC
Alagar, Vangalur S
Hall, Richard L
Lin, Paul C S

OTHER COUNTRIES
Dougalis, Vassilios
Lambert, William M, Jr

MEDICAL & HEALTH SCIENCES

Audiology & Speech Pathology

ALABAMA
Sheeley, Eugene C
Wester, Derin C

ARIZONA
Nation, James Edward

COLORADO
Blager, Florence Berman
Scherer, Ronald Callaway

FLORIDA
Brown, William Samuel, Jr
Pichal, Henri Thomas, II
Rothman, Howard Barry

HAWAII
Yates, James T

ILLINOIS
Hoshiko, Michael S
Schultz, Martin C

INDIANA
Hoops, Richard Allen
Ritter, E Gene

IOWA
Moll, Kenneth Leon
Schwartz, Ralph Jerome
Small, Arnold McCollum, Jr

LOUISIANA
Berlin, Charles I

MARYLAND
Brownell, William Edward
Causey, G(eorge) Donald
Mościcki, Eve Karin

NEBRASKA
Cochran, John Rodney

NEW JERSEY
Prasad, Marehalli Gopalan

NEW MEXICO
Stewart, Joseph Letie

NEW YORK
Dickson, Stanley
Katz, Jack

NORTH CAROLINA
Royster, Julia Doswell

OHIO
Nixon, Charles William

TENNESSEE
Mendel, Maurice I
Nabelek, Anna K
Stewart, James Monroe

WASHINGTON
Lipscomb, David M

WEST VIRGINIA
Lass, Norman J(ay)

ONTARIO
Kunov, Hans

OTHER COUNTRIES
Bergman, Moe

Dentistry

ALABAMA
Baker, John Rowland
Feagin, Frederick F
Graves, Carl N
Hammons, Paul Edward
Jeffcoat, Marjorie K
Keller, Stanley E
Koulourides, Theodore I
Manson-Hing, Lincoln Roy
Menaker, Lewis
Ranney, Richard Raymond
Retief, Daniel Hugo
Ringsdorf, Warren Marshall, Jr
Schuessler, Carlos Francis
Speed, Edwin Maurice
Volker, Joseph Francis
Weatherford, Thomas Waller, III

ARIZONA
Sarid, Dror

ARKANSAS
Vesely, David Lynn

CALIFORNIA
Abrams, Albert Maurice
Allerton, Samuel E
Ast, David Bernard
Barber, Thomas King
Bernard, George W
Bertolami, Charles Nicholas
Chierici, George J
Dixon, Andrew Derart
Dummett, Clifton Orrin
Duperon, Donald Francis
Feller, Ralph Paul
Goldberg, Louis J
Greenspan, John Simon
Grossman, Richard C
Hansen, Louis Stephen
Ingle, John Ide
Jarvis, William Tyler
Jendresen, Malcolm Dan
Jones, Edward George
Kaplan, Herman
Kapur, Krishan Kishore
Krol, Arthur J
Leake, Donald L
Lugassy, Armand Amram
McCarthy, Frank Martin
Marshall, Grayson William, Jr

Dentistry (cont)

Miller, Arthur Joseph
Newbrun, Ernest
Reeves, Robert Lloyd
Robinson, Hamilton Burrows Greaves
Ryge, Gunnar
Schoen, Max H
Schonfeld, Steven Emanuel
Simmons, Norman Stanley
Slavkin, Harold Charles
Stark, Marvin Michael
Steinman, Ralph R
Walters, Roland Dick
Way, Jon Leong
Weinstock, Alfred
Wescott, William B
Westmoreland, Winfred William
White, Stuart Cossitt
Wirthlin, Milton Robert, Jr
Woolley, LeGrand H
Wycoff, Samuel John
Yagiela, John Allen
Zernik, Joseph

COLORADO
Eames, Wilmer B
Greene, David Lee
Krikos, George Alexander
Meskin, Lawrence Henry

CONNECTICUT
Burstone, Charles Justin
Celesk, Roger A
Coykendall, Alan Littlefield
Goldberg, A Jon
Hand, Arthur Ralph
Katz, Ralph Verne
Krutchkoff, David James
Macko, Douglas John
Mumford, George
Nalbandian, John
Nanda, Ravindra
Norton, Louis Arthur
Nuki, Klaus
Poole, Andrew E
Prasad, Arun
Rossomando, Edward Frederick
Tinanoff, Norman
Venham, Larry Lee
Weinstein, Sam
Zubal, I George

DELAWARE
Jefferies, Steven
Schweiger, James W

DISTRICT OF COLUMBIA
Calhoun, Noah Robert
Ferrigno, Peter D
Gallardo-Carpentier, Adriana
Gupta, Om Prakash
Littleton, Preston A, Jr
Niswander, Jerry David
Redman, Robert Shelton
Sinkford, Jeanne C
Tesk, John A
Tuckson, Coleman Reed, Jr
Woolridge, Edward Daniel

FLORIDA
Anusavice, Kenneth John
Bennett, Carroll G
Clark, William Burton, IV
Corn, Herman
Duany, Luis F, Jr
Garrington, George Everett
Gibbs, Charles Howard
Hassell, Thomas Michael
Hill, Clement Joseph
Kerr, I Lawrence
Lobene, Ralph Rufino
Mackenzie, Richard Stanley
Medina, Jose Enrique
Robinson, John E, Jr
Simring, Marvin
Wisotzky, Joel
Zamikoff, Irving Ira

GEORGIA
Benoit, Peter Wells
Burnett, George Wesley
Bustos-Valdes, Sergio Enrique
Ciarlone, Alfred Edward
Clark, James William
Dirksen, Thomas Reed
Fritz, Michael E
Gangarosa, Louis Paul, Sr
Hawkins, Isaac Kinney
Marble, Howard Bennett, Jr
Pashley, David Henry
Sharawy, Mohamed
Urbanek, Vincent Edward
Weatherred, Jackie G
Whitford, Gary M
Zwemer, Thomas J

ILLINOIS
Aduss, Howard
Bahn, Arthur Nathaniel
Cleall, John Frederick
Cohen, Gloria
Corruccini, Robert Spencer
Coy, Richard Eugene

Dahlberg, Albert A
Douglas, Bruce L
Engel, Milton Baer
Flores, Samson Sol
Gaik, Geraldine Catherine
Goepp, Robert August
Gowgiel, Joseph Michael
Graber, T(ouro) M(or)
Grandel, Eugene Robert
Jacobson, Marvin
Martinez-Lopez, Norman Petronio
Murphy, Richard Allan
Norman, Richard Daviess
Rao, Gopal Subba
Rosenberg, Henry Mark
Rosenstein, Sheldon William
Scapino, Robert Peter
Sharma, Brahma Dutta
Steffek, Anthony J
Teuscher, George William
Turner, Donald W
Waterhouse, John P
Yale, Seymour Hershel

INDIANA
Bixler, David
Bogan, Robert L
Garner, LaForrest D
Henderson, Hala Zawawi
Hennon, David Kent
Hine, Maynard Kiplinger
Lund, Melvin Robert
McDonald, Ralph Earl
Mitchell, David Farrar
Patterson, Samuel S
Potter, Rosario H Yap
Standish, Samuel Miles
Starkey, Paul Edward
Stookey, George K
Swenson, Henry Maurice
Tomich, Charles Edward
Wagner, Morris

IOWA
Bishara, Samir Edward
Bjorndal, Arne Magne
Chan, Kai Chiu
Grigsby, William Redman
Hammond, Harold Logan
Jacobs, Richard M
Johnson, Wallace W
Kremenak, Charles Robert
Laffoon, John
McLeran, James Herbert
Nowak, Arthur John
Olin, William (Harold), Sr
Sabiston, Charles Barker, Jr
Staley, Robert Newton
Thayer, Keith Evans

KENTUCKY
Boyer, Harold Edwin
Crim, Gary Allen
Elwood, William K
Farman, Allan George
Mann, Wallace Vernon, Jr
Nash, David Allen
Parkins, Frederick Milton
Podshadley, Arlon George
Saxe, Stanley Richard
Smith, Timothy Andre
Spedding, Robert H
Staat, Robert Henry
Stewart, Arthur Van
Wesley, Robert Cook
Wittwer, John William

LOUISIANA
Alam, Syed Qamar
Nakamoto, Tetsuo
Petri, William Henry, III

MARYLAND
Baum, Bruce J
Bowen, Rafael Lee
Brauer, Gerhard Max
Chow, Laurence Chung-Lung
Church, Lloyd Eugene
Cohen, Lois K
Cotton, William Robert
Davidson, William Martin
Dionne, Raymond A
Driscoll, William
Gartner, Leslie Paul
Gobel, Stephen
Grewe, John Mitchell
Joly, Olga G
Kendrick, Francis Joseph
Kinnard, Matthew Anderson
Kirkpatrick, Diana (Rorabaugh) M
Larson, Rachel Harris (Mrs John Watson Henry)
Loe, Harald
Loebenstein, William Vaille
McCauley, H(enry) Berton
Means, Craig Ray
Minah, Glenn Ernest
Myslinski, Norbert Raymond
Nylen, Marie Ussing
Ogden, Ingram Wesley
Patel, Prafull Raojibhai
Rizzo, Anthony Augustine
Rovelstad, Gordon Henry
Seibel, Werner

Takagi, Shozo
Valega, Thomas Michael
Whitehurst, Virgil Edwards

MASSACHUSETTS
Auskaps, Aina Marija
Baratz, Robert Sears
Brudevold, Finn
Caslavska, Vera Barbara
Dunning, James Morse
Gianelly, Anthony Alfred
Goldhaber, Paul
Goldman, Henry M(aurice)
Grodberg, Marcus Gordon
Gron, Poul
Harris, Melvyn H
Hein, John William
Henry, Joseph L
Howell, Thomas Howard
Johansen, Erling
Keith, David Alexander
Kramer, Gerald M
Kronman, Joseph Henry
Massler, Maury
Moorrees, Coenraad Frans August
Niederman, Richard
Nizel, Abraham Edward
Ruben, Morris P
Slavin, Ovid
Smith, Daniel James
Smulow, Jerome B
Sonis, Stephen Thomas
Susi, Frank Robert
Tenca, Joseph Ignatius
Turesky, Samuel Saul
Wells, Herbert
Williams, David Lloyd
Williams, Ray Clayton
Yen, Peter Kai Jen

MICHIGAN
Ash, Major McKinley, Jr
Avery, James Knuckey
Brooks, Sharon Lynn
Burdi, Alphonse R
Burt, Brian Aubrey
Carlson, David Sten
Charbeneau, Gerald T
Christiansen, Richard Louis
Craig, Robert George
Harris, James Edward
Hartsook, Joseph Thurman
Holmstedt, Jan Olle Valter
Lillie, John Howard
Loesche, Walter J
McNamara, James Alyn, Jr
Millard, Herbert Dean
Morawa, Arnold Peter
Ramfjord, Sigurd
Richards, Albert Gustav
Rowe, Nathaniel H
Steiman, Henry Robert
Strachan, Donald Stewart
Striffler, David Frank

MINNESOTA
Cervenka, Jaroslav
Douglas, William Hugh
Elzay, Richard Paul
Gibilisco, Joseph
Goodkind, Richard Jerry
Gorlin, Robert James
Martens, Leslie Vernon
Messer, Louise Brearley
Meyer, Maurice Wesley
Oliver, Richard Charles
Schachtele, Charles Francis
Schaffer, Erwin Michael
Speidel, T(homas) Michael
Till, Michael John
Witkop, Carl Jacob, Jr

MISSISSIPPI
Alexander, William Nebel

MISSOURI
Bensinger, David August
Berndt, Alan Fredric
Cobb, Charles Madison
Nikolai, Robert Joseph
Porter, Chastain Kendall
Stewart, Jack Lauren
Zoeller, Gilbert Norbert

NEBRASKA
Bradley, Richard E
James, Garth A
Kramer, William S
Kutler, Benton
Rinne, Vernon Wilmer
Sullivan, Robert Emmett

NEW HAMPSHIRE
Hill, Percy Holmes
Zumbrunnen, Charles Edward

NEW JERSEY
Alfano, Michael Charles
Baron, Hazen Jay
Bolden, Theodore Edward
Chasens, Abram I
Cinotti, William Ralph
Curro, Frederick A
Di Paolo, Rocco John

Evans, Ralph H, Jr
Gershon, Sol D
Mellberg, James Richard
Najjar, Talib A
Parker, LeRoy A, Jr
Pianotti, Roland Salvatore
Reussner, George Henry
Rivetti, Henry Conrad
Ross, Norton Morris
Schuback, Philip
Shovlin, Francis Edward
Yamane, George M

NEW MEXICO
Bleiweiss, Max Phillip
Yudkowsky, Elias B

NEW YORK
Adisman, I Kenneth
Archard, Howell Osborne
Baer, Paul Nathan
Blechman, Harry
Boucher, Louis Jack
Brewer, Allen A
Chance, Kenneth Bernard
Ciancio, Sebastian Gene
Di Salvo, Nicholas Armand
Durr, David P
Featherstone, John Douglas Bernard
Fischman, Stuart L
Gilmour, Marion Nyholm H
Golub, Lorne Malcolm
Gottsegen, Robert
Green, Larry J
Hanley, Kevin Joseph
Hoffman, Robert
Horowitz, Sidney Lester
Jones, Melvin D
Kahn, Norman
Kaslick, Ralph Sidney
Kaufman, Edward Godfrey
Kleinberg, Israel
Lamster, Ira Barry
Lenchner, Nathaniel Herbert
Leverett, Dennis Hugh
Linkow, Leonard I
Lucca, John J
McHugh, William Dennis
Mandel, Irwin D
Morris, Melvin Lewis
Moss-Salentijn, Letty
Mundorff, Sheila Ann
Oaks, J Howard
Ortman, Harold R
Osborne, John William
Powell, Richard Anthony
Quartararo, Ignatius Nicholas
Rifkin, Barry Richard
Roistacher, Seymour Lester
Schaaf, Norman George
Scopp, Irwin Walter
Singh, Inder Jit
Sirianni, Joyce E
Sreebny, Leo Morris
Stahl, S Sigmund
Staple, Peter Hugh
Stevens, Roy Harris
Zegarelli, Edward Victor
Zero, Domenick Thomas

NORTH CAROLINA
Barker, Ben D
Bawden, James Wyatt
Howell, Robert Mac Arthur
Hylander, William Leroy
Johnston, Malcolm Campbell
Lindahl, Roy Lawrence
Lundblad, Roger Lauren
Nelson, Robert Mellinger
Oldenburg, Theodore Richard
Proffit, William R
Quinn, Galen Warren
Shankle, Robert Jack
Via, William Fredrick, Jr
Warren, Donald W
Webster, William Phillip

NORTH DAKOTA
Vogel, Gerald Lee

OHIO
Alley, Keith Edward
Amjad, Zahid
Blozis, George G
Conroy, Charles William
Cooley, William Edward
Dew, William Calland
Ferencz, Nicholas
Goorey, Nancy Reynolds
Horton, John Edward
Israel, Harry, III
Jordan, Freddie L
Lydy, David Lee
Rossi, Edward P
Sciulli, Paul William
Temple, Robert Dwight
Williams, Benjamin Hayden
Woelfel, Julian Bradford

OKLAHOMA
Brown, William Ernest
Glass, Richard Thomas
Howes, Robert Ingersoll, Jr
Miranda, Frank Joseph

Shapiro, Stewart

OREGON
Bruckner, Robert Joseph
Buck, Douglas L
Hedtke, James Lee
Howard, William Weaver
Jastak, J Theodore
Kinersly, Thorn
Riviere, George Robert
Saffir, Arthur Joel
Savara, Bhim Sen
Sorenson, Fred M
Tracy, William E
Van Hassel, Henry John
Zingeser, Maurice Roy

PENNSYLVANIA
Ackerman, James L
Bassiouny, Mohamed Ali
Brendlinger, Darwin
Brightman, Vernon
Buchin, Irving D
Cohen, David Walter
Cooper, Stephen Allen
Cornell, John Alston
Farber, Paul Alan
Jurecic, Anton
Lindemeyer, Rochelle G
Listgarten, Max
Liu, Andrew T C
Mazaheri, Mohammad
Milligan, Wilbert Harvey, III
Oka, Seishi William
Oliet, Seymour
Rapp, Robert
Segal, Alan H
Seltzer, Samuel
Smudski, James W
Stiff, Robert H
Sweeney, Edward Arthur
Taichman, Norton Stanley
Vergona, Kathleen Anne Dobrosielski
Winkler, Sheldon
Yankell, Samuel L

SOUTH CAROLINA
Corcoran, John W
Cordova-Salinas, Maria Asuncion
Edwards, James Burrows
Fingar, Walter Wiggs
Lawson, Benjamin F
Ludwig, Theodore Frederick
Morris, Alvin Leonard
Nuckles, Douglas Boyd
Waldrep, Alfred Carson, Jr
White, Edward

TENNESSEE
Andrews, James Tucker
Behrents, Rolf Gordon
Gibbs, Samuel Julian
Gotcher, Jack Everett
Jurand, Jerry George
McKnight, James Pope
Nunez, Loys Joseph
Richardson, Elisha Roscoe
Sintes, Jorge Luis
Siskin, Milton
Smith, Roy Martin
Sticht, Frank Davis
Weber, Faustin N

TEXAS
Allen, Don Lee
Allen, Edward Patrick
Biggerstaff, Robert Huggins
Binnie, William Hugh
Blanton, Patricia Louise
Burch, William Paul
Byrd, David Lamar
Cottone, James Anthony
Del Rio, Carlos Eduardo
DePaola, Dominick Philip
De Rijk, Waldemar G
Englander, Harold Robert
Ennever, John Joseph
Fultz, R Paul
Gage, Tommy Wilton
Goaz, Paul William
Guentherman, Robert Henry
Harris, Norman Oliver
Hatch, John Phillip
Henry, Clay Allen
Hoskins, Sam Whitworth, Jr
Hurt, William Clarence
Kalkwarf, Kenneth Lee
Lambert, Joseph Parker
Langland, Olaf Elmer
Lynch, Denis Patrick
McConnell, Duncan
Madden, Richard M
Nelson, John Franklin
Ohlenbusch, Robert Eugene
Olson, John Victor
Rawls, Henry Ralph
Seliger, William George
Shillitoe, Edward John
Storey, Arthur Thomas
Suddick, Richard Phillips
Taylor, Paul Peak
Throckmorton, Gaylord Scott
Williams, Fred Eugene

VIRGINIA
Abbott, David Michael
Butler, James Hansel
Cochran, David Lee
Isaacson, Robert John
Sweeney, Thomas Patrick
Wiebusch, F B
Wittemann, Joseph Klaus

WASHINGTON
Bolender, Charles L
Canfield, Robert Charles
Hall, Stanton Harris
Hodson, Jean Turnbaugh
Johnson, Robert H
Moffett, Benjamin Charles, Jr
Moore, Alton Wallace
Page, Roy Christopher
Riedel, Richard Anthony
Stern, Irving B
Stibbs, Gerald Denike
Swoope, Charles C
Wehner, Alfred Peter

WEST VIRGINIA
Biddington, William Robert
Bouquot, Jerry Elmer
Calhoon, Donald Alan
Overberger, James Edwin

WISCONSIN
Goggins, John Francis
Lyon, Harvey William

PUERTO RICO
Ortiz, Araceli

ALBERTA
Eggert, Frank Michael
Haryett, Rowland D
Holland, Graham Rex
MacRae, Patrick Daniel
Osborn, Jeffrey Whitaker
Sperber, Geoffrey Hilliard
Thomas, Norman Randall
Wood, Norman Kenyon

BRITISH COLUMBIA
Beagrie, George Simpson
Khanna, Shadi Lall
Kraintz, Leon
Lear, Clement S C
Leung, So Wah
Neilson, John Warrington
Ogilvie, Alfred Livingston

NOVA SCOTIA
Bennett, Ian Cecil
McLean, James Douglas
Sykora, Oskar P

ONTARIO
Brooke, Ralph Ian
Galil, Khadry Ahmed
Hunter, William Stuart
Mamandras, Antonios H
Mayhall, John Tarkington
Melcher, Antony Henry
Nikiforuk, Gordon
Popovich, Frank
Sandham, Herbert James
Saunders, Shelley Rae
Sessle, Barry John
Speck, John Edward
Ten Cate, Arnold Richard
Woodside, Donald G
Wu, Alan SeeMing

QUEBEC
Chan, Eddie Chin Sun

SASKATCHEWAN
Ambrose, Ernest R

OTHER COUNTRIES
Stallard, Richard E
Wei, Stephen Hon Yin

Environmental Health

ALABAMA
Bailey, William C
Dalvi, Ramesh R
Hood, Ronald David
Knox, Francis Stratton, III
Lavelle, George Cartwright
Mundy, Roy Lee
Rajanna, Bettaiya
Singh, Jarnail

ALASKA
Mills, William J, Jr
Tilsworth, Timothy

ARIZONA
Barnes, Howard Clarence
Beck, John R
Colburn, Wayne Alan
Lebowitz, Michael David
Wu, Hofu

ARKANSAS
Brewster, Marjorie Ann

Chang, Louis Wai-Wah
Chowdhury, Parimal
Howell, Robert T
Kadlubar, Fred F
Smith, Carroll Ward
Snow, Donald L(oesch)
Townsend, James Willis
Weeks, Robert Joe

CALIFORNIA
Amoore, John Ernest
Anspaugh, Lynn Richard
Barnes, Paul Richard
Beard, Rodney Rau
Bernstein, Leslie
Bhatnagar, Rajendra Sahai
Bils, Robert F
Blejer, Hector P
Brantingham, Charles Ross
Brown, Harold Victor
Campbell, Kirby I
Cass, Glen R
Chang, Daniel P Y
Chatigny, Mark A
Collins, James Francis
Conway, John Bell
Cooper, Robert Chauncey
Coughlin, James Robert
Daisey, Joan M
Danse, Ilene H Raisfeld
DePass, Linval R
Detels, Roger
Dobson, R Lowry
Duhl, Leonard J
Embree, James Willard, Jr
Fabrikant, Irene Berger
Fallon, Joseph Greenleaf
Garland, Cedric Frank
Goldman, Marvin
Hackett, Nora Reed
Hallesy, Duane Wesley
Hanlon, John Joseph
Hubbard, Eric R
Hubert, Helen Betty
Jensen, Ronald Harry
Judd, Stanley H
Kleinman, Michael Thomas
Kohn, Henry Irving
Koschier, Francis Joseph
Larkin, Edward Charles
Last, Jerold Alan
Leo, Albert Joseph
Linnell, Robert Hartley
Low, Hans
McGaughey, Charles Gilbert
McKone, Thomas Edward
Mayron, Lewis Walter
Newell, Gordon Wilfred
Niklowitz, Werner Johannes
Pierce, James Otto, II
Porcella, Donald Burke
Rice, Robert Hafling
Richters, Arnis
Sagan, Leonard A
Salmon, Eli J
Schienle, Jan Hoops
Sheneman, Jack Marshall
Shvartz, Esar
Slote, Lawrence
Sparling, Rebecca Hall
Sterner, James Hervi
Swan, Shanna Helen
Ury, Hans Konrad
Vanderlaan, Martin
Victery, Winona Whitwell
Volz, Michael George
Wagner, Jeames Arthur
Whipple, Christopher George
Whittenberger, James Laverre
Wiley, Lynn M
Wood, Corinne Shear
Wyzga, Ronald Edward
Yager, Janice L Winter
Zinkl, Joseph Grandjean

COLORADO
Aldrich, Franklin Dalton
Andrews, Ken J
Audesirk, Gerald Joseph
Barnes, Allan Marion
Batchelder, Arthur Roland
Best, Jay Boyd
Boatman, Joe Francis
Brooks, Bradford O
Bunn, William Bernice, III
Irons, Richard Davis
Ivory, Thomas Martin, III
Wilson, Vincent L

CONNECTICUT
Borgia, Julian Frank
Cohen, Steven Donald
Freudenthal, Ralph Ira
Hinkle, Lawrence Earl, Jr
Matheson, Dale Whitney
Milzoff, Joel Robert
Nadel, Ethan Richard
Pelletier, Charles A
Rho, Jinnque
Spiers, Donald Ellis
Stitt, John Thomas
Stolwijk, Jan Adrianus Jozef

DELAWARE
Frawley, John Paul
Garland, Charles E
Krone, Lawrence James
Mikell, William Gaillard
Patterson, Gordon Derby, Jr
Pell, Sidney
Potrafke, Earl Mark
Staples, Robert Edward

DISTRICT OF COLUMBIA
Breen, Joseph John
Campbell, Carlos Boyd Godfrey
Chiazze, Leonard, Jr
Cotruvo, Joseph Alfred
Florig, Henry Keith
Goeringer, Gerald Conrad
Gould, Jack Richard
Hill, Richard Norman
Kimmel, Carole Anne
Lee, I P
Lunchick, Curt
McEwen, Gerald Noah, Jr
Nadolney, Carlton H
Prival, Michael Joseph
Rao, Mamidanna S
Rhoden, Richard Allan
Richardson, Allan Charles Barbour
Rogers, Senta S(tephanie)
Schwartz, Sorell Lee
Sexton, Ken
Smith, David Allen
Smith, David Allen
Tompkins, George Jonathan
Weinberg, Myron Simon
Wolff, Arthur Harold
Zook, Bernard Charles

FLORIDA
Abraham, William Michael
Billings, Charles Edgar
Clark, Mary Elizabeth
Dougherty, Ralph C
Echelberger, Wayne F, Jr
Gay, Don Douglas
Gittelman, Donald Henry
Grabowski, Casimer Thaddeus
Harrold, Gordon Coleson
Lowe, Ronald Edsel
Lu, Frank Chao
Myers, Richard Lee
Richards, Norman Lee
Teaf, Christopher Morris
Thoma, Richard William
Tiffany, William James, III
Wolsky, Sumner Paul

GEORGIA
Batra, Gopal Krishan
Berg, George G
Bernstein, Robert Steven
Bryan, Frank Leon
Curley, Winifred H
Falk, Henry
Fineman, Manuel Nathan
French, Jean Gilvey
Galeano, Sergio F(rancis)
Hargrove, Robert John
Hedden, Kenneth Forsythe
Inge, Walter Herndon, Jr
Johnson, Barry Lee
Kilbourne, Edwin Michael
McGrath, William Robert
Oakley, Godfrey Porter, Jr
Peake, Thaddeus Andrew, III
Pratt, Harry Davis
Reinhardt, Donald Joseph
Schliessmann, D(onald) J(oseph)
Singh, Harpal P
Stevenson, Dennis A
Stone, Connie J
Thacker, Stephen Brady
Urso, Paul
Wolven-Garrett, Anne M

HAWAII
Crane, Sheldon Cyr

IDAHO
Gesell, Thomas Frederick
Hillyard, Ira William

ILLINOIS
Benkendorf, Carol Ann
Boulos, Badi Mansour
Brenniman, Gary Russell
Carnow, Bertram Warren
Conibear, Shirley Ann
Cossairt, Jack Donald
Doege, Theodore Charles
Ehrlich, Richard
Fenters, James Dean
Foxx, Richard Michael
Francis, Bettina Magnus
Garvin, Paul Joseph, Jr
Golchert, Norbert William
Hubbard, Lincoln Beals
Huberman, Eliezer
Jacobson, Marvin
Konopinski, Virgil J
Lasley, Stephen Michael
Leikin, Jerrold Blair
McFee, Donald Ray
Olecko, William Anton

Hopps, Howard Carl
Klaunig, James E
Loper, John C
Lovett, Joseph
Manson, Jeanne Marie
Mukhtar, Hasan
Mumtaz, Mohammad Moizuddin
Murthy, Raman Chittaram
Nebert, Daniel Walter
Niemeier, Richard William
Nixon, Charles William
Oleinick, Nancy Landy
Pfeifer, Gerard David
Sabourin, Thomas Donald
Saltzman, Bernard Edwin
Schanbacher, Floyd Leon
Skelly, Michael Francis
Slonim, Arnold Robert
Smith, Carl Clinton
Stoner, Gary David
Suskind, Raymond Robert
Tabor, Marvin Wilson
Von Gierke, Henning Edgar
Weiss, Harold Samuel
Willeke, Klaus
Wong, Laurence Cheng Kong
Yeary, Roger A
Yoon, Jong Sik

OKLAHOMA
Branch, John Curtis
Coleman, Nancy Pees
Coleman, Ronald Leon
Melton, Carlton E, Jr
Miranda, Frank Joseph

OREGON
Elia, Victor John
Matarazzo, Joseph Dominic
Parrott, Marshall Ward
Spencer, Peter Simner

PENNSYLVANIA
Anderson, David Martin
Atkins, Patrick Riley
Baker, Paul Thornell
Chan, Ping-Kwong
Danchik, Richard S
Deckert, Fred W
DeHaven-Hudkins, Diane Louise
Dinman, Bertram David
Duryea, William R
Esmen, Nurtan A
Fletcher, Ronald D
Fogleman, Ralph William
Gottlieb, Karen Ann
Hager, Nathaniel Ellmaker, Jr
Kamon, Eliezer
Karol, Meryl Helene
Kleban, Morton H
Levine, Geoffrey
Mason, Thomas Joseph
Mattison, Donald Roger
Metry, Amir Alfi
Mount, Lloyd Gordon
Neufeld, Ronald David
Nobel, Joel J
Poss, Stanley M
Rothman, Alan Michael
Sagik, Bernard Phillip
Shank, Kenneth Eugene
Shapiro, Maurice A
Shoub, Earle Phelps
Sooky, Attila A(rpad)
Urbach, Frederick
Wald, Niel
Weiss, William
Zelac, Ronald Edward

RHODE ISLAND
Hutzler, Leroy, III
Shaikh, Zahir Ahmad
Worthen, Leonard Robert

SOUTH CAROLINA
Darnell, W(illiam) H(eaden)
Muska, Carl Frank
Oswald, Edward Odell
Patterson, C(leo) Maurice

SOUTH DAKOTA
Schlenker, Evelyn Heymann

TENNESSEE
Barton, Charles Julian, Sr
Close, David Matzen
Darden, Edgar Bascomb
Dudney, Charles Sherman
Dumont, James Nicholas
Gehrs, Carl William
Guerin, Michael Richard
Iglar, Albert Francis, Jr
Marchok, Ann Catherine
Meeks, Robert G
Morgan, Monroe Talton, Sr
Oakes, Thomas Wyatt
Schultz, Terry Wayne
Uziel, Mayo
Vo-Dinh, Tuan
Watson, Annetta Paule

TEXAS
Bailey, Everett Murl, Jr
Barnes, Charles M

Biles, Robert Wayne
Campbell, Dan Norvell
Davis, Ernst Michael
Dodson, Ronald Franklin
Domask, W(illiam) G(erhard)
Ellis, Kenneth Joseph
Fox, Donald A
Garcia, Hector D
Gobuty, Allan Harvey
Gothelf, Bernard
Gruber, George J
Guentzel, M Neal
Harbordt, C(harles) Michael
Heffley, James D
Hill, Reba Michels
Holguin, Alfonso Hudson
Jauchem, James Robert
Journeay, Glen Eugene
Kusnetz, Howard L
Kutzman, Raymond Stanley
Ledbetter, Joe O(verton)
Lengel, Robert Charles
Lowry, William Thomas
Nachtwey, David Stuart
Nash, Donald Robert
Nechay, Bohdan Roman
Pier, Stanley Morton
Powell, Michael Robert
Randerath, Kurt
Smolensky, Michael Hale
Thompson, Richard John
Ulrich, Roger Steffen
Ward, Jonathan Bishop, Jr
Williams, Calvit Herndon, Jr
Wolf, Harold William

UTAH
Archer, Victor Eugene
Bennett, Jesse Harland
Gortatowski, Melvin Jerome
Melton, Arthur Richard
Moser, Royce, Jr
Parker, Robert Davis Rickard
Schiager, Keith Jerome
Wrenn, McDonald Edward

VERMONT
Falk, Leslie Alan
Hussak, Robert Edward
Keenan, Robert Gregory
McIntosh, Alan William

VIRGINIA
Blanke, Robert Vernon
Boardman, Gregory Dale
Brown, Elise Ann Brandenburger
Brown, Stephen L(awrence)
Byrd, Daniel Madison, III
Cavender, Finis Lynn
Donohue, Joyce Morrissey
Furr, Aaron Keith
Gould, David Huntington
Gregory, Arthur Robert
Grunder, Fred Irwin
Hill, Jim T
Jennings, Allen Lee
Lee, James A
McCormick-Ray, M Geraldine
McNerney, James Murtha
Mahboubi, Ezzat
Martin, James Henry, III
Melton, Thomas Mason
Meyer, Alvin F, Jr
Rodricks, Joseph Victor
Rosenbaum, David Mark
Sahli, Brenda Payne
Stroup, Cynthia Roxane
Tardiff, Robert G
Ulvedal, Frode
Walker, Charles R
Wands, Ralph Clinton
Wasti, Khizar
Woo, Yin-tak

WASHINGTON
Ammann, Harriet Maria
Benditt, Earl Philip
Bhagat, Surinder Kumar
Boatman, Edwin S
Johnstone, Donald Lee
Koenig, Jane Quinn
Lovely, Richard Herbert
Luchtel, Daniel Lee
Murphy, Sheldon Douglas
Reid, Donald House
Sanders, Charles Leonard, Jr
Scribner, John David
Smith, Victor Herbert
Valanis, Barbara Mayleas
Wedlick, Harold Lee
Young, Francis Allan

WEST VIRGINIA
Bouquot, Jerry Elmer
McCall, Robert G
Olenchock, Stephen Anthony
Ong, Tong-man
Sheikh, Kazim

WISCONSIN
Arthur, Robert M
Gasiorkiewicz, Eugene Constantine
Hake, Carl (Louis)
Kitzke, Eugene David

Marx, James John, Jr
Raff, Hershel
Saryan, Leon Aram
Siegesmund, Kenneth A
Smith, Michael James

ALBERTA
Man, Shu-Fan Paul

BRITISH COLUMBIA
Banister, Eric Wilton
Bates, David Vincent
Davison, Allan John
Law, Francis C P

MANITOBA
Weeks, John Leonard

NOVA SCOTIA
Chatt, Amares

ONTARIO
Basu, Prasanta Kumar
Bend, John Richard
Bertell, Rosalie
Bioleau, Luc J R
Cummins, Joseph E
Goyer, Robert Andrew
Hickman, John Roy
Hopton, Frederick James
Lees, Ronald Edward
Liu, Dickson Lee Shen
Morris, Larry Arthur
Robertson, James McDonald
Roy, Marie Lessard
Sanderson, Henry Preston
Somers, Emmanuel
Uchida, Irene Ayako
Warren, Douglas Robson
Williams, Norman S W

PRINCE EDWARD ISLAND
Singh, Amreek

QUEBEC
Becklake, Margaret Rigsby
Eddy, Nelson Wallace
Feng, Rong
Fournier, Michel
Maruvada, P Sarma
Page, Michel
Siemiatycki, Jack
Thériault, Gilles P

SASKATCHEWAN
Blakley, Barry Raymond
Schiefer, H Bruno

OTHER COUNTRIES
Back, Kenneth Charles
Bernstein, David Maier
Brydon, James Emerson
Emmett, Edward Anthony
Goldsmith, John Rothchild
Greene, Velvl William
Haschke, Ferdinand
Hilbert, Morton S
Hoadley, Alfred Warner
Jokl, Miloslav Vladimir
Kenney, Gary Dale
Radford, Edward Parish
Shiraki, Keizo
Shuval, Hillel Isaiah
Stam, Jos
Strehlow, Clifford David
Thiessen, Jacob Willem
Valencia, Mauro Eduardo
Van Loveren, Henk
Winnett, George

Hospital Administration

ALABAMA
Clark, Charles Edward
Roe, James Maurice, Jr

ARIZONA
Burkholder, Peter M
Schneller, Eugene S

ARKANSAS
Lloyd, Harris Horton
Towbin, Eugene Jonas

CALIFORNIA
Affeldt, John E
Bodfish, Ralph E
Cretin, Shan
Hansen, Howard Edward
Haughton, James Gray
Hudgins, Arthur Judson
Makowka, Leonard
Merchant, Roland Samuel, Sr
Merchant, Roland Samuel, Sr
Mills, Gary Kenith
Nolan, Janiece Simmons
Sams, Bruce Jones, Jr
Scitovsky, Anne A
Shine, Kenneth I
Urbach, Karl Frederic
Weinstein, Irwin M

COLORADO
Bonn, Ethel May
Jafek, Bruce William

CONNECTICUT
Bondy, Philip K
Foster, James Henry
Klimek, Joseph John
Terenzio, Joseph V
Wetstone, Howard J

DISTRICT OF COLUMBIA
Stark, Nathan Julius

FLORIDA
Kessler, Richard Howard
Rosenkrantz, Jacob Alvin

HAWAII
Bixler, Otto C

ILLINOIS
Best, William Robert
Hand, Roger
Korn, Roy Joseph
Paynter, Camen Russell
Pine, Michael B
Prabhudesai, Mukund M
Schyve, Paul Milton
Singer, Irwin
Stromberg, LaWayne Roland
Tourlentes, Thomas Theodore

INDIANA
Evans, Michael Allen

KANSAS
Coles, Anna Bailey
Simpson, William Stewart

KENTUCKY
Ram, Madhira Dasaradhi

LOUISIANA
Grace, Marcellus
Threefoot, Sam Abraham

MARYLAND
Abramson, Fredric David
Grimes, Carolyn E
Hausch, H George
Klarman, Herbert E

MASSACHUSETTS
Buchanan, J Robert
Czeisler, Charles Andrew
Gentry, John Tilmon
Grossman, Jerome H
Lederman, David Mordecai
Lipinski, Boguslaw
Litsky, Bertha Yanis
Winkelman, James W

MICHIGAN
Dauphinais, Raymond Joseph
Gershon, Samuel
Griffith, John Randall
Kachhal, Swatantra Kumar
Rosenberg, Jerry C
Sahney, Vinod K
Sorscher, Alan J

MISSOURI
Hoover, Loretta White
Whitten, Elmer Hammond

NEW HAMPSHIRE
Kaiser, C William

NEW JERSEY
Brubaker, Paul Eugene, II
Dunikoski, Leonard Karol, Jr
McDonnell, William Vincent
Royce, Paul C

NEW MEXICO
Omer, George Elbert, Jr

NEW YORK
English, Joseph T
Famighetti, Louise Orto
Frosch, William Arthur
Gaintner, John Richard
Gerstman, Hubert Louis
Gibofsky, Allan
Goldman, Stephen Shepard
Joshi, Sharad Gopal
Krynicki, Victor Edward
Lichtig, Leo Kenneth
Meyer, George Wilbur
Reichgott, Michael Joel
Riehle, Robert Arthur, Jr
Rosner, Fred
Ruskin, Asa Paul
Sidel, Victor William
Spaulding, Stephen Waasa
Spear, Paul William
Wolin, Harold Leonard

NORTH CAROLINA
McMahon, John Alexander
Rosenfeld, Leonard Sidney
Weil, Thomas P
Wilson, Ruby Leila

Hospital Administration (cont)

OHIO
Altose, Murray D
Anderson, Ruth Mapes
Bates, Stephen Roger
Crespo, Jorge H
Morrow, Grant, III
Neuhauser, Duncan B
Rogers, Richard C
Schubert, William K
Vogel, Thomas Timothy

OKLAHOMA
Donahue, Hayden Hackney
Lynn, Thomas Neil, Jr
Thomas, Ellidee Dotson
Vanhoutte, Jean Jacques
Whitcomb, Walter Henry

OREGON
Clark, William Melvin, Jr

PENNSYLVANIA
Arce, A Anthony
Bondi, Amedeo
Deitrick, John E
Gray, Frieda Gersh
Jungkind, Donald Lee
Sovie, Margaret D

SOUTH CAROLINA
Johnson, James Allen, Jr

TENNESSEE
Christman, Luther P
Clark, Winston Craig
Ross, Joseph C
Skoutakis, Vasilios A

TEXAS
Aune, Janet
Brown, J(ack) H(arold) U(pton)
Bruhn, John G
Burdine, John Alton
Hall, Esther Jane Wood
Mendel, Julius Louis
Mitra, Sunanda
Myers, Paul Walter

UTAH
Lee, Jeffrey Stephen

VIRGINIA
Chaudhuri, Tapan K
Levine, Jules Ivan
Maturi, Vincent Francis
Sheldon, Donald Russell

WASHINGTON
Smith, Victor Herbert

WEST VIRGINIA
Victor, Leonard Baker

WISCONSIN
Treffert, Darold Allen

BRITISH COLUMBIA
Garg, Arun K

ONTARIO
Finlay, Joseph Bryan
Hansebout, Robert Roger
Rosenberg, Gilbert Mortimer

QUEBEC
Contandriopoulos, Andre-Pierre
Joly, Louis Philippe
Palayew, Max Jacob

OTHER COUNTRIES
Ichikawa, Shuichi

Medical Sciences, General

ALABAMA
Anderson, Peter Glennie
Borton, Thomas Ernest
Elson, Charles O, III
Fuller, Gerald M
Gay, Renate Erika
Gay, Steffen
Jeffcoat, Marjorie K
Lloyd, L Keith
McCombs, Candace Cragen
Montgomery, John R
Roozen, Kenneth James
Schafer, James Arthur
Turco, Jenifer
Wilborn, Walter Harrison

ALASKA
DeLapp, Tina Davis

ARIZONA
Birkby, Walter H
Blouin, Leonard Thomas
Bowen, Theodore
Dereniak, Eustace Leonard
Hendrix, Mary J C
Kazal, Louis Anthony
Koldovsky, Otakar

Lebowitz, Michael David
Peric-Knowlton, Wlatka

ARKANSAS
Chang, Louis Wai-Wah
Epstein, Joshua
Pasley, James Neville
Schwarz, Anton

CALIFORNIA
Adair, Suzanne Frank
Adams, Thomas Edwards
Aly, Raza
Anderson, Milo Vernette
Antoine, John Eugene
Arcadi, John Albert
Ascher, Michael S
Bailey, David Nelson
Bainton, Cedric R
Barr, Allan Ralph
Berk, Arnold J
Bertolami, Charles Nicholas
Bissell, Dwight Montgomery
Bloomfield, Harold H
Bolaffi, Janice Lerner
Braemer, Allen C
Brown, Warren Shelburne, Jr
Buehring, Gertrude Case
Carlson, Don Marvin
Chen, Benjamin P P
Choudary, Prabhakara Velagapudi
Chuong, Cheng-Ming
Clewe, Thomas Hailey
Cohen, Morris
Cooper, Howard K
Cosman, Bard Clifford
Courtney, Kenneth Oliver
Dasgupta, Asim
David, Gary Samuel
Demetrescu, M
Dobson, R Lowry
Donnelly, Timothy C
Dorfman, Leslie Joseph
Downing, Michael Richard
Epstein, John Howard
Erickson, Jeanne Marie
Ewing, Dean Edgar
Fackler, Martin L
Fallon, James Harry
Feigal, Ellen G
Felch, William C
Fischell, Tim Alexander
Flacke, Werner Ernst
Francke, Uta
Fuhrman, Frederick Alexander
Fukuda, Minoru
Garratty, George
Gee, Adrian Philip
Geiselman, Paula J
Geokas, Michael C
Giorgi, Janis V(incent)
Gong, William C
Gottlieb, Michael Stuart
Grody, Wayne William
Gupta, Rishab Kumar
Harrison, Michael R
Hayes, Thomas L
Hayflick, Leonard
Haynes, Alfred
Hecht, Elizabeth Anne
Heuser, Gunnar
Huang, Sung-Cheng
Huber, Wolfgang
Ingram, Marylou
Insel, Paul Anthony
Kang, Chang-Yuil
Kaslow, Arthur L
Kempter, Charles Prentiss
Kiprov, Dobri D
Knapp, Theodore Martin
Krauss, Ronald
Laiken, Nora Dawn
Largman, Corey
Lee, Paul L
Leffert, Hyam Lerner
Levine, Jon David
Lipsick, Joseph Steven
Lu, John Kuew-Hsiung
McCabe, Donald Lee
MacFarlane, Malcolm David
Mack, Thomas McCulloch
Mann, Lewis Theodore, Jr
Mason, Dean Towle
Miyada, Don Shuso
Moxley, John H, III
Najafi, Ahmad
Nelson, Gary Joe
Neville, James Ryan
Nicholson, Arnold Eugene
Novack, Gary Dean
Patton, John Stuart
Pearl, Ronald G
Peroutka, Stephen Joseph
Philpott, Delbert E
Pierschbacher, Michael Dean
Pitts, Robert Gary
Povzhitkov, Moysey Michael
Powanda, Michael Christopher
Powell, Frank Ludwig, Jr
Prusiner, Stanley Ben
Quismorio, Francisco P, Jr
Ram, Michael Jay
Reeve, Peter
Remington, Jack Samuel

Rho, Joon H
Rinderknecht, Heinrich
Rosenfeld, George
Rubenstein, Howard S
Ruggeri, Zaverio Marcello
Sams, Bruce Jones, Jr
Schienle, Jan Hoops
Schiller, Neal Leander
Schmid, Peter
Scott, W Richard
Seegmiller, Jarvis Edwin
Seppi, Edward Joseph
Shortliffe, Edward Hance
Shvartz, Esar
Slater, Grant Gay
Spiller, Gene Alan
Strautz, Robert Lee
Sweet, Benjamin Hersh
Taylor, Anna Newman
Tewari, Sujata
Thompson, John Frederick
Vernikos, Joan
Victery, Winona Whitwell
Wagner, Jeames Arthur
Wakefield, Caroline Leone
Wangler, Roger Dean
Weinreb, Robert Neal
Weinstein, Irwin M
Westfahl, Pamela Kay
Winet, Howard
Withers, Hubert Rodney
Wood, Corinne Shear
Yamini, Sohrab
Yildiz, Alaettin
Yu, David Tak Yan
Zuckekandl, Emile

COLORADO
Allen, Larry Milton
Dobersen, Michael J
Hager, Jean Carol
Lee, Lela A
Nelson, Bernard W
Powell, Robert Lee
Urban, Richard William

CONNECTICUT
Agulian, Samuel Kevork
Bondy, Philip K
Borgia, Julian Frank
Boron, Walter Frank
Cohen, Donald J
Conrad, Eugene Anthony
Dooley, Joseph Francis
Heyd, Allen
Hlavacek, Robert Allen
Kidd, Kenneth Kay
Letts, Lindsay Gordon
Loose, Leland David
McIlhenny, Hugh M
Manner, Richard John
Matovcik, Lisa M
Milne, George McLean, Jr
Nadel, Ethan Richard
O'Looney, Patricia Anne
O'Neill, James F
Pinson, Ellis Rex, Jr
Riker, Donald Kay
Rockwell, Sara Campbell
Ryan, Richard Patrick
Schaaf, Thomas Ken
Schafer, David Edward
Schmeer, Arline Catherine
Stadnicki, Stanley Walter, Jr
Stuart, James Davies
Taylor, Kenneth J W
Waxman, Stephen George
Zaret, Barry L

DELAWARE
Dautlick, Joseph X
Eleuterio, Marianne Kingsbury
Ginnard, Charles Raymond

DISTRICT OF COLUMBIA
Allen, Robert Erwin
Brown, W Virgil
Daza, Carlos Hernan
Harmon, John W
Ison-Franklin, Eleanor Lutia
Jose, Pedro A
Kettel, Louis John
Martensen, Todd Martin
Miller, Linda Jean
Minners, Howard Alyn
Rao, Mamidanna S
Rapisardi, Salvatore C
Vogel, Carl-Wilhelm E

FLORIDA
Abitbol, Carolyn Larkins
Baker, Carleton Harold
Ben, Max
Berlin, Nathaniel Isaac
Bliznakov, Emile George
Brown, Hugh Keith
Cameron, Don Frank
Carter, James Harrison, II
DeRousseau, C(arol) Jean
Drummond, Willa H
Emerson, Geraldine Mariellen
Farah, Alfred Emil
Fisher, Waldo Reynolds
Goihman-Yahr, Mauricio

Hampp, Edward Gottlieb
Hurst, Josephine M
Joftes, David Lion
Kline, Jacob
Lyman, Gary Herbert
McMillan, Donald Ernest
Murchison, Thomas Edgar
Nonoyama, Meihan
Perez, Guido O
Phillips, Michael Ian
Prahl, Helmut Ferdinand
Ringel, Samuel Morris
Simring, Marvin
Soliman, Karam Farag Attia
Spielberger, Charles Donald
Stevens, Bruce Russell
Storrs, Eleanor Emerett
Stroud, Robert Malone
Sypert, George Walter
Thornthwaite, Jerry T
Tsibris, John-Constantine Michael
Wilcox, Christopher Stuart
Williams, Joseph Francis

GEORGIA
Binnicker, Pamela Caroline
Colborn, Gene Louis
Crowe, Dennis Timothy, Jr
Dailey, Harry A
Herman, Chester Joseph
Hicks, Heraline Elaine
Innes, David Lyn
Kaiser, Robert L
Klein, Luella
Kolbeck, Ralph Carl
Krauss, Jonathan Seth
Thacker, Stephen Brady
Warren, McWilson
Warren, Stephen Theodore
Woodford, James

HAWAII
Marchette, Nyven John
Yang, Hong-Yi

IDAHO
Hillyard, Ira William
Preszler, Alan Melvin

ILLINOIS
Albrecht, Ronald Frank
Anderson, Kenning M
Aramini, Jan Marie
Bone, Roger C
Brody, Jacob A
Cabana, Veneracion Garganta
Chan, Yun Lai
Christensen, Mary Lucas
Cohen, Gloria
Doege, Theodore Charles
Ecanow, Bernard
Ernest, J Terry
Fang, Victor Shengkuen
Ferguson, James L
Friedman, Yochanan
Fuchs, Elaine V
Gaik, Geraldine Catherine
Glisson, Silas Nease
Goebel, Maristella
Gross, Thomas Lester
Hast, Malcolm Howard
Hjelle, Joseph Thomas
Hua, Ping
Johari, Om
Jones, Peter Hadley
Kang, David Soosang
Kass, Leon Richard
Kim, Yoon Berm
Kim, Yung Dai
Kirschner, Robert H
Leikin, Jerrold Blair
Lin, James Chih-I
Loeb, Jerod M
Lopez-Majano, Vincent
Moticka, Edward James
Needleman, Saul Ben
Nevalainen, David Eric
Ostrow, David Henry
Potempa, Lawrence Albert
Rao, Mrinalini Chatta
Reft, Chester Stanley
Rosen, Arthur Leonard
Shepherd, Herndon Guinn
Smith, Douglas Calvin
Snyder, Ann Knabb
Stevens, Joseph Alfred
Swartz, Harold M
Tobin, Martin John
Tomita, Joseph Tsuneki
Weyhenmeyer, James Alan
Yelich, Michael Ralph
Yoon, Ji-Won

INDIANA
Biggs, Homer Gates
Broxmeyer, Hal Edward
Goh, Edward Hua Seng
Gray, Peter Norman
Haak, Richard Arlen
Macchia, Donald Dean
McNeill, Kenny Earl
Paladino, Frank Vincent
Svoboda, Gordon H
Tacker, Willis Arnold, Jr

Vasko, Michael Richard
Voelz, Michael H
Yoder, John Menly

IOWA
Biller, Jose
Conn, P Michael
Fellows, Robert Ellis, Jr
Koontz, Frank P
Maruyama, George Masao
Montgomery, Rex
Schrott, Helmut Gunther
Siebes, Maria
Stratton, Donald Brendan
Thompson, Robert Gary

KANSAS
Davis, John Stewart
Gattone, Vincent H, II
Norris, Patricia Ann
Oehme, Frederick Wolfgang
Singhal, Ram P
Stevens, David Robert
Tomita, Tatsuo

KENTUCKY
Cohn, David Valor
Harris, Patrick Donald
Kasarskis, Edward Joseph
Porter, William Hudson
Rees, Earl Douglas
Tseng, Michael Tsung
Urano, Muneyasu
Williams, Walter Michael

LOUISIANA
Bagby, Gregory John
Bazan, Nicolas Guillermo
Chen, Hoffman Hor-Fu
Datta, Ratna
Dessauer, Herbert Clay
Frohlich, Edward David
Lang, Charles H
Messerli, Franz Hannes
Newsome, David Anthony
Peacock, Milton O
Saha, Subrata
Samuels, Arthur Seymour
Spitzer, Judy A

MAINE
Norton, James Michael
Weiss, John Jay

MARYLAND
Abramson, I Jerome
Adelstein, Robert Simon
Balinsky, Doris
Balow, James E
Baquet, Claudia
Baxter, James Hubert
Benevento, Louis Anthony
Bergey, Gregory Kent
Berman, Howard Mitchell
Berson, Alan
Biswas, Robin Michael
Blackman, Marc Roy
Blaustein, Mordecai P
Boyd, Virginia Ann Lewis
Bricker, Jerome Gough
Bünger, Rolf
Calabi, Ornella
Carlson, Drew E
Carski, Theodore Robert
Catt, Kevin John
Chang, Henry
Chen, Joseph Ke-Chou
Cushman, Samuel Wright
Darby, Eleanor Muriel Kapp
Dewys, William Dale
Diggs, John W
Fakunding, John Leonard
Feinstone, Stephen Mark
Fertziger, Allen Philip
Fex, Jorgen
Fischell, Robert E
Fitch, Kenneth Leonard
Flaherty, John Thomas
Foster, Willis Roy
Freeman, Julia Berg
Froehlich, Luz
Gardner, Jerry David
Garruto, Ralph Michael
Goldstein, David Stanley
Goldwater, William Henry
Gonda, Matthew Allen
Gorden, Phillip
Gordon, Stephen L
Green, Jerome George
Gregerman, Robert Isaac
Handwerger, Barry S
Haseltine, Florence Pat
Hearn, Henry James, Jr
Heyse, Stephen P
Holmes, Kathryn Voelker
Horan, Michael
Hoth, Daniel F
Irwin, David
Johnston, Gerald Samuel
Kakefuda, Tsuyoshi
Kendrick, Francis Joseph
Kety, Seymour S
Kinnard, Matthew Anderson
Klombers, Norman

Koch, Thomas Richard
Kramer, Barnett Sheldon
Lawrence, Dale Nolan
Lee-Ham, Doo Young
Lenfant, Claude J M
LeRoith, Derek
Levine, Alan Stewart
Linzer, Melvin
Lipicky, Raymond John
Litterst, Charles Lawrence
Longo, Dan L
Lymangrover, John R
MacVittie, Thomas Joseph
Magrath, Ian
Manganiello, Vincent Charles
Marban, Eduardo
Marsden, Halsey M
Maslow, David E
Massof, Robert W
Maumenee, Irene Hussels
Meryman, Harold Thayer
Miller, Charles A
Mishler, John Milton, IV
Moshell, Alan N
Murphy, Donald G
Myslinski, Norbert Raymond
Nelson, James Harold
Porter, Roger J
Price, Thomas R
Rawlings, Samuel Craig
Read, Merrill Stafford
Rhim, Johng Sik
Robbins, Jacob
Robbins, Jay Howard
Roddy, Martin Thomas
Rovelstad, Gordon Henry
Sack, George H(enry), Jr
Salcman, Michael
Sarin, Prem S
Savage, Peter
Schechter, Alan Neil
Shevach, Ethan Menahem
Shih, Thomas Y
Simpson, Ian Alexander
Slife, Charles W
Sommer, Alfred
Spooner, Peter Michael
Spradling, Allan C
Sztein, Marcelo Benjamin
Talbot, Bernard
Talbot, John Mayo
Tankersley, Donald
Taube, Sheila Efron
Thorp, James Wilson
Traub, Richard Kimberley
Valega, Thomas Michael
Wallace, Craig Kesting
Wasserheit, Judith Nina
Watson, James Dewey
Wolff, David A
Wortman, Bernard
Yanagihara, Richard

MASSACHUSETTS
Ackerman, Jerome Leonard
Anderson, Rosalind Coogan
Aroesty, Julian Max
Axelrod, Lloyd
Castronovo, Frank Paul, Jr
Chiu, Tak-Ming
Dacheux, Ramon F, II
Davis, Michael Allan
Dayal, Yogeshwar
DeSanctis, Roman William
Eastwood, Gregory Lindsay
El-Bermani, Al-Walid I
Federman, Daniel D
Feld, Michael S
Fox, Irving Harvey
Fox, Maurice Sanford
Fox, Thomas Oren
Furie, Bruce
Gange, Richard William
Ganley, Oswald Harold
Goldman, Harvey
Goldstein, Richard Neal
Grossman, Jerome H
Haimes, Howard B
Hallock, Gilbert Vinton
Hawiger, Jack Jacek
Hilgar, Arthur Gilbert
Holick, Michael Francis
Hurd, Richard Nelson
Jain, Rakesh Kumar
Jung, Lawrence Kwok Leung
Kaminskas, Edvardas
Khaw, Ban-An
Klempner, Mark Steven
Koul, Omanand
Lees, Robert S
Libby, Peter
Lloyd, Weldon S
Loscalzo, Joseph
Modest, Edward Julian
Moskowitz, Michael Arthur
Murray, Joseph E
Palubinskas, Felix Stanley
Pandian, Natesa G
Paskins-Hurlburt, Andrea Jeanne
Pfeffer, Janice Marie
Pinkus, Jack Leon
Pino, Richard M
Rigney, David Roth
Rosen, Fred Saul

Schaub, Robert George
Schorr, Lisbeth Bamberger
Spragg, Jocelyn
Szlyk, Patricia Carol
Takeuchi, Kiyoshi Hiro
Trinkaus-Randall, Vickery E
Vanderburg, Charles R
Villee, Dorothy Balzer
Waud, Douglas Russell
Weinberg, Crispin Bernard
Weinstein, Robert
Wong, Shan Shekyuk
Wurtman, Richard Jay
Zetter, Bruce Robert

MICHIGAN
Alcala, Jose Ramon
Beck, Ronald Richard
Berent, Stanley
Bouma, Hessel, III
Briggs, Josephine P
Cerny, Joseph Charles
Cohen, Bennett J
Dore-Duffy, Paula
Eisenstein, Barry I
Ellis, C N
Glover, Roy Andrew
Gray, Robert Howard
Hashimoto, Ken
Heinrikson, Robert L
Jackson, Craig Merton
Kagan, Fred
McCann, Daisy S
Mahajan, Sudesh K
Marcelo, Cynthia Luz
Medlin, Julie Anne Jones
Meites, Joseph
Mizukami, Hiroshi
Mohrland, J Scott
Moos, Walter Hamilton
Noe, Frances Elsie
Parfitt, A(lwyn) M(ichael)
Poulik, Miroslav Dave
Sorscher, Alan J
Swovick, Melvin Joseph
Viano, David Charles
Voorhees, John James
Wei, Wei-Zen
Zak, Bennie

MINNESOTA
Adams, Ernest Clarence
Bach, Marilyn Lee
Brown, David Mitchell
Johnson, Susan Bissette
Lambert, Edward Howard
Leon, Arthur Sol
Leung, Benjamin Shuet-Kin
Mikhail, Adel Ayad
O'Connor, Michael Kieran
Robb, Richard A
Sawchuk, Ronald John
Schwartz, Samuel
Severson, Arlen Raynold
Tamsky, Morgan Jerome
Thompson, Edward William

MISSISSIPPI
Adams, Junius Greene, III
Rockhold, Robin William
Shirley, Aaron
Steinberg, Martin H

MISSOURI
Baumann, Thiema Wolf
Blim, Richard Don
Counts, David Francis
Covey, Douglas Floyd
Coyer, Philip Exton
Emami, Bahman
Goetz, Kenneth Lee
Green, Eric Douglas
Hansen, Hobart Raymond
Karl, Michael M
Lin, Hsiu-san
Lyle, Leon Richards
Morley, John E
Munns, Theodore Willard
Quay, Wilbur Brooks
Sadler, J(asper) Evan
Schuck, James Michael
Strauss, Arnold Wilbur
Taylor, John Joseph
Thomas, Lewis Jones, Jr
Welply, Joseph Kevin

NEBRASKA
Dalley, Arthur Frederick, II
Johnson, John Arnold

NEVADA
Newmark, Stephen

NEW HAMPSHIRE
Brous, Don W
Edwards, Brian Ronald
Musiek, Frank Edward
Stibitz, George Robert

NEW JERSEY
Arcese, Paul Salvatore
Baker, Herman
Bonner, R Alan
Boyle, Joseph, III

Buchweitz, Ellen
Chao, Yu-Sheng
Chimes, Daniel
Clayton, John Mark
Controulis, John
Curro, Frederick A
Desjardins, Raoul
Duran, Walter Nunez
Emele, Jane Frances
Firschein, Hilliard E
Flam, Eric
Ford, Daniel Morgan
Gately, Maurice Kent
Gorodetzky, Charles W
Green, Erika Ana
Gullo, Vincent Philip
Jabbar, Gina Marie
Ji, Sungchul
Kamiyama, Mikio
Kapp, Robert Wesley, Jr
Kronenthal, Richard Leonard
Kumbaraci-Jones, Nuran Melek
Laemle, Lois K
Leeds, Morton W
Leung, Albert Yuk-Sing
Leventhal, Howard
Levy, Alan
Levy, Martin J Linden
Li, John Kong-Jiann
Li, Shu-Tung
McDowell, Wilbur Benedict
McGregor, Walter
McKinstry, Doris Naomi
Margolskee, Robert F
Massa, Tobias
Matthijssen, Charles
Meyer, Harry Martin, Jr
Moreland, Suzanne
Nichols, Joseph
Orlando, Carl
OSullivan, Joseph
Parsonnet, Victor
Peets, Edwin Arnold
Perhach, James Lawrence
Pestka, Sidney
Pobiner, Bonnie Fay
Ponzio, Nicholas Michael
Reuben, Roberta C
Rhoten, William Blocher
Ritchie, David Malcolm
Saiger, George Lewis
Shin, Seung-il
Sterbenz, Francis Joseph
Swislocki, Norbert Ira
Todd, Margaret Edna
Wasserman, Martin Allan
Weir, James Henry, III
Wilson, Alan C
Yard, Allan Stanley

NEW MEXICO
Cole, Edward Issac, Jr
Crawford, Michael Howard
Elias, Laurence
Reyes, Philip
Welford, Norman Traviss
Wild, Gaynor (Clarke)
Williams, Mary Carol
Woodfin, Beulah Marie

NEW YORK
Althuis, Thomas Henry
Antzelevitch, Charles
April, Ernest W
Aronson, Ronald Stephen
Ascensao, Joao L
Beal, Myron Clarence
Berger, Eugene Y
Berger, Frank Milan
Bhattacharya, Jahar
Blau, Lawrence Martin
Blumenthal, Rosalyn D
Borer, Jeffrey Stephen
Borg, Donald Cecil
Bottomley, Paul Arthur
Bystryn, Jean-Claude
Cassell, Eric J
Chadha, Kailash Chandra
Chanana, Arjun Dev
Chess, Leonard
Chou, Ting-Chao
Coderre, Jeffrey Albert
Conard, Robert Allen
Cort, Susannah
David, Oliver Joseph
Delmonte, Lilian
Devereux, Richard Blyton
DiMauro, Salvatore
Dole, Vincent Paul
Donner, David Bruce
Doty, Stephen Bruce
Eder, Howard Abram
Edwards, Adrian L
Egan, Edmund Alfred
Eisinger, Josef
Elbaum, Danek
Elsbach, Peter
Erickson, Wayne Francis
Factor, Stephen M
Famighetti, Louise Orto
Field, Michael
Finlay, Thomas Hiram
Finster, Mieczyslaw
Gambert, Steven Ross

Medical Sciences, General (cont)

Gardner, Daniel
Gelbard, Alan Stewart
Gerolimatos, Barbara
Gintautas, Jonas
Glassman, Armand Barry
Goldman, James Eliot
Golub, Lorne Malcolm
Gonzalez, Eulogio Raphael
Goodman, Jerome
Gordon, Harry William
Gotschlich, Emil C
Grimwood, Brian Gene
Ham, Richard John
Hatcher, Victor Bernard
Heller, Milton David
Hershey, Linda Ann
Isenberg, Henry David
Kaplan, John Ervin
Kappas, Attallah
Karr, James Presby
Kashin, Philip
Kaufman, Albert Irving
Kegeles, Lawrence Steven
Kent, Barbara
Khan, Abdul Jamil
Kohler, Constance Anne
Kream, Jacob
Lamster, Ira Barry
Lee, Ching-Tse
Levy, David Edward
Lieberman, Abraham N
Lutton, John D
Matthews, Dwight Earl
Moor-Jankowski, J
Morin, John Edward
Morris, John Emory
Natta, Clayton Lyle
Oakes, David
O'Brien, Anne T
Ornt, Daniel Burrows
Ottenbacher, Kenneth John
Papsidero, Lawrence D
Pentney, Roberta Pierson
Pesce, Michael A
Pittman, Kenneth Arthur
Pogo, A Oscar
Pollack, Robert Elliot
Rattazzi, Mario Cristiano
Reichberg, Samuel Bringeissen
Reichert, Leo E, Jr
Reit, Barry
Rich, Marvin R
Rivlin, Richard S
Roboz, John
Rothman, Francoise
Ruddell, Alanna
Ruggiero, David A
Schenck, John Frederic
Schwartz, Ernest
Sehgal, Pravinkumar B
Seon, Ben Kuk
Siemann, Dietmar W
Simmons, Daryl Michael
Sitarz, Anneliese Lotte
Spangler, Robert Alan
Spaulding, Stephen Waasa
Stanley, Evan Richard
Stark, John Howard
Stolfi, Robert Louis
Stryker, Martin H
Szabo, Piroska Ludwig
Thornborough, John Randle
Toth, Eugene J
Trust, Ronald I
Tyagi, Suresh C
Vacirca, Salvatore John
Vladutiu, Adrian O
Vladutiu, Georgirene Dietrich
Waksman, Byron Halstead
Weber, Julius
Weiner, Matei
Weitkamp, Lowell R
Wellner, Daniel
Wineburg, Elliot N
Wolfson, Leonard Louis
Wolin, Harold Leonard
Wollman, Leo
Yao, Alice C
Zumoff, Barnett

NORTH CAROLINA
Anderegg, Robert James
Beaty, Orren, III
Bentzel, Carl Johan
Bode, Arthur Palfrey
Butz, Robert Frederick
Caterson, Bruce
Chen, Michael Yu-Men
Dunnick, June K
Frisell, Wilhelm Richard
Garris, David Roy
Gruber, Kenneth Allen
Hershfield, Michael Steven
Hicks, T Philip
Kucera, Louis Stephen
Laupus, William E
Lawson, Edward Earle
Lee, Martin Jerome
Levine, Richard Joseph
Louie, Dexter Stephen

McManus, Thomas (Joseph)
Misek, Bernard
Murdock, Harold Russell
Mustafa, Syed Jamal
Phelps, Allen Warner
Putnam, Charles E
Reinhard, John Frederick, Jr
Sadler, Thomas William
Sassaman, Anne Phillips
Schiffman, Susan S
Spilker, Bert
Swift, Michael
Thubrikar, Mano J
Toole, James Francis
Walker, Cheryl Lyn

OHIO
Adragna, Norma C
Atkins, Charles Gilmore
Barnes, H Verdain
Bar-Yishay, Ephraim
Blanton, Ronald Edward
Boxenbaum, Harold George
Centner, Rosemary Louise
Clanton, Thomas L
Clifford, Donald H
Crafts, Roger Conant
Curry, John Joseph, III
Davis, Pamela Bowes
Dell'Osso, Louis Frank
Dirckx, John H
Dobelbower, Ralph Riddall, Jr
Ely, Daniel Lee
Fenoglio, Cecilia M
Francis, Marion David
Geha, Alexander Salim
Green, Ralph
Haas, Erwin
Harris, Richard Lee
Hinkle, George Henry
Holtkamp, Dorsey Emil
Homer, George Mohn
Jamasbi, Roudabeh J
Klein, LeRoy
Latimer, Bruce Millikin
Lepp, Cyrus Andrew
Lindower, John Oliver
MacIntyre, William James
Mannino, Joseph Robert
Nahata, Milap Chand
Ramsey, James Marvin
Reddy, Narender Pabbathi
Resnick, Martin I
Rodman, Harvey Meyer
Rosenberg, Howard C
Stuhlman, Robert August
Sutherland, James McKenzie

OKLAHOMA
Calvert, Jon Channing
Collins, William Edward
Corder, Clinton Nicholas
Cunningham, Madeleine White
Dubowski, Kurt M(ax)
Epstein, Robert B
Faulkner, Kenneth Keith
Fine, Douglas P
King, Mary Margaret
McKee, Patrick Allen
Melton, Carlton E, Jr
Snow, Clyde Collins
Watson, Gary Hunter

OREGON
Bussman, John W
Fields, R(ance) Wayne
Mela-Riker, Leena Marja
Mooney, Larry Albert
Prescott, Gerald H
Robinson, Arthur B
Severson, Herbert H
Swank, Roy Laver
Zimmerman, Earl Abram

PENNSYLVANIA
Agarwal, Jai Bhagwah
Albertine, Kurt H
Alhadeff, Jack Abraham
Alteveer, Robert Jan George
Baechler, Charles Albert
Bhaduri, Saumya
Biebuyck, Julien Francois
Bly, Chauncey Goodrich
Botelho, Stella Yates
Cavanagh, Peter R
Chapman, John Donald
Cheng, Hung-Yuan
Cooper, Stephen Allen
Davis, Richard A
DeRubertis, Frederick R
Diamond, Leila
Duryea, William R
Ellis, Demetrius
Fabrizio, Angelina Maria
Flynn, John Thomas
Freeman, Joseph Theodore
Glick, John H
Godfrey, John Carl
Goldberg, Martin
Henderson, George Richard
Hobby, Gladys Lounsbury
Hoffman, Eric Alfred
Johnson, Ernest Walter
Jones, Robert Leroy

Klein, Herbert A
Klurfeld, David Michael
Koelle, Winifred Angenent
Kowlessar, O Dhodanand
Ladda, Roger Louis
Laufer, Igor
Lentini, Eugene Anthony
Levine, Geoffrey
Liberti, Paul A
Liu, Andrew T C
McKenna, William Gillies
Magargal, Larry Elliot
Meek, Edward Stanley
Miller, Barry
Mulvihill, John Joseph
Murer, Erik Homann
Okunewick, James Philip
Oldham, James Warren
Orenstein, David M
Perchonock, Carl David
Pirone, Louis Anthony
Poupard, James Arthur
Pruss, Thaddeus P
Raikow, Radmila Boruvka
Revesz, George
Rosenblum, Irwin Yale
Rottenberg, Hagai
Ryan, James Patrick
Schneider, Henry Peter
Shiau, Yih-Fu
Six, Howard R
Smith, John Bryan
Smith, Roger Bruce
Strauss, Jerome Frank, III
Strom, Brian Leslie
Tosch, William Conrad
Vickers, Florence Foster
Whinnery, James Elliott
Wissow, Lennard Jay
Zagon, Ian Stuart

RHODE ISLAND
Brautigan, David L
Carpenter, Charles C J
Christenson, Lisa
Jackson, Ivor Michael David
Kresina, Thomas Francis
O'Leary, Gerard Paul, Jr
Shank, Peter R
Vezeridis, Michael Panagiotis

SOUTH CAROLINA
Ainsworth, Sterling K
Goust, Jean Michel
Horger, Edgar Olin, III
Kaiser, Charles Frederick
Pittman, Fred Estes
Rathbun, Ted Allan

TENNESSEE
Bhattacharya, Syamal Kanti
Chandler, Robert Walter
Costanzi, John J
Crawford, Lloyd V
Crofford, Oscar Bledsoe
Gillespie, Elizabeth
Goans, Ronald Earl
Harris, Thomas R(aymond)
Huggins, James Anthony
Hunt, James Calvin
Morrison, Martin
Nag, Subir
Nanney, Lillian Bradley
Pui, Ching-Hon
Richmond, Chester Robert
Robertson, David
Satcher, David
Skoutakis, Vasilios A
Stone, William Lawrence
Sundell, Hakan W

TEXAS
Andia-Waltenbaugh, Ana Maria
Andresen, Michael Christian
Bessman, Joel David
Burton, Karen Poliner
Butler, Bruce David
Cottone, James Anthony
Crass, Maurice Frederick, III
DeBakey, Lois
DeBakey, Selma
Deloach, John Rooker
Dorn, Gordon Lee
Entman, Mark Lawrence
Franklin, Thomas Doyal, Jr
Goldberg, Leonard H
Hecht, Gerald
Hudspeth, Albert James
Hudspeth, Emmett Leroy
Hussian, Richard A
Johnson, Raleigh Francis, Jr
Kit, Saul
Klein, Peter Douglas
LaBree, Theodore Robert
Lees, George Edward
Lehmkuhl, L Don
Levin, Harvey Steven
Liepa, George Uldis
Low, Morton David
Lynch, Denis Patrick
McCredie, Kenneth Blair
McPhail, Jasper Lewis
Mark, Lloyd K
Marshall, Gailen D, Jr

Mastromarino, Anthony John
Mundy, Gregory Robert
Murthy, Krishna K
Quarles, John Monroe
Reid, Michael Baron
Riser, Mary Elizabeth
Rolfe, Rial Dewitt
Rosenfeld, Charles Richard
Sampson, Herschel Wayne
Sant'Ambrogio, Giuseppe
Sato, Clifford Shinichi
Shannon, Wilburn Allen, Jr
Sheng, Hwai-Ping
Shotwell, Thomas Knight
Skelley, Dean Sutherland
Taegtmeyer, Heinrich
Tilbury, Roy Sidney
Trentin, John Joseph
Waldbillig, Ronald Charles
Wilcox, Roberta Arlene
Wilson, L Britt
Younes, Muazaz A
Zwaan, Johan Thomas

UTAH
Border, Wayne Allen
Galinsky, Raymond Ethan
Olson, Randall J
Westenfelder, Christof

VERMONT
Beers, Roland Frank, Jr
Ellis, David M

VIRGINIA
Asterita, Mary Frances
Barrett, Richard John
Bempong, Maxwell Alexander
Chaudhuri, Tapan K
Cooper, John Allen Dicks
DeLorenzo, Robert John
DeSesso, John Michael
Diwan, Bhalchandra Apparao
Elford, Howard Lee
Gargus, James L
Grider, John Raymond
Lepper, Mark H
Lobel, Steven A
Loria, Roger Moshe
Malik, Vedpal Singh
Seipel, John Howard
Thorner, Michael Oliver
Turner, Terry Tomo
Wei, Enoch Ping
Weiss, Daniel Leigh
Williams, Gerald Albert
Williams, Patricia Bell

WASHINGTON
Ammann, Harriet Maria
Artru, Alan Arthur
Bhansali, Praful V
Canfield, Robert Charles
Couser, William Griffith
Haviland, James West
Hildebrandt, Jacob
Ho, Lydia Su-yong
Huntsman, Lee L
Kenney, Gerald
Lakshminarayan, S
Maguire, Yu Ping
Mannik, Mart
Mendelson, Martin
Mitchell, Michael Ernst
Rosenblatt, Roger Alan
Salk, Darrell John
Schiffrin, Milton Julius
Spackman, Darrel H
Wehner, Alfred Peter
Whatmore, George Bernard

WEST VIRGINIA
Ballantyne, Bryan
Chang, William Wei-Lien
Konat, Gregory W
Pore, Robert Scott

WISCONSIN
Allred, Lawrence Ervin (Joe)
Bortin, Mortimer M
Buchanan-Davidson, Dorothy Jean
Edelheit, Lewis S
Engerman, Ronald Lester
Goldberg, Burton David
Hinkes, Thomas Michael
Kostreva, David Robert
Ladinsky, Judith L
Manning, James Harvey
Mazess, Richard B
Murray, Mary Patricia
Perry, Billy Wayne
Raff, Hershel
Roberts, Ronald C
Sieber, Fritz
Steinhart, Carol Elder
Stowe, David F
Sufit, Robert Louis
Terasawa, EI
Uno, Hideo
Walker, Duard Lee
Warltier, David Charles

PUERTO RICO
Banerjee, Dipak Kumar

Bangdiwala, Ishver Surchand
De La Sierra, Angell O
Ramirez-Ronda, Carlos Hector

ALBERTA
Hollenberg, Morley Donald
Hutchison, Kenneth James
Jones, Richard Lee
Koopmans, Henry Sjoerd
Lederis, Karolis (Karl)
Leung, Alexander Kwok-Chu
Lorscheider, Fritz Louis
Roth, Sheldon H
Stell, William Kenyon

BRITISH COLUMBIA
Applegarth, Derek A
Bates, David Vincent
Frohlich, Jiri J
Kraintz, Leon
Milsom, William Kenneth
Perry, Thomas Lockwood
Skala, Josef Petr
Wong, Norman L M

MANITOBA
Bertalanffy, Felix D
Choy, Patrick C
Gerrard, Jonathan M
Glavin, Gary Bertrun
Polimeni, Philip Iniziato
Rosenmann, Edward A
Schacter, Brent Allan

NEWFOUNDLAND
Michalak, Thomas Ireneusz

NOVA SCOTIA
Armour, John Andrew
Carruthers, Samuel George
Dickson, Douglas Howard
Fraser, Albert Donald
Murray, T J
Tan, Meng Hee

ONTARIO
Baker, Michael Allen
Barr, Ronald Duncan
Basu, Prasanta Kumar
Biro, George P
Campbell, James Fulton
Capaldi, Dante James
Cheng, Hazel Pei-Ling
Doane, Frances Whitman
Evans, John R
Forstner, Janet Ferguson
Froese, Alison Barbara
Garfield, Robert Edward
Ghista, Dhanjoo Noshir
Girvin, John Patterson
Goldenberg, Gerald J
Harwood-Nash, Derek Clive
Inman, Robert Davies
Jeejeebhoy, Khursheed Nowrojee
Kako, Kyohei Joe
Kerbel, Robert Stephen
Kinlough-Rathbone, Lorne R
Kooh, Sang Whay
MacRae, Andrew Richard
Mazurkiewicz-Kwilecki, Irena Maria
Merskey, Harold
Mishra, Ram K
Montemurro, Donald Gilbert
Osmond, Daniel Harcourt
Papp, Kim Alexander
Phillips, Robert Allan
Pross, Hugh Frederick
Renaud, Leo P
Rosenthal, Kenneth Lee
Schimmer, Bernard Paul
Somers, Emmanuel
Steiner, George
Szewczuk, Myron Ross
Tepperman, Barry Lorne
Tobin, Sidney Morris
Uchida, Irene Ayako
Vadas, Peter

QUEBEC
Aranda, Jacob Velasco
Bélanger, Luc
Braun, Peter Eric
Chang, Thomas Ming Swi
Davis, Martin Arnold
Freeman, Carolyn Ruth
Frojmovic, Maurice Mony
Gagnon, Claude
Goltzman, David
Heisler, Seymour
Kearney, Robert Edward
Lussier, Andre (Joseph Alfred)
Nair, Vasavan N P
Outerbridge, John Stuart
Palayew, Max Jacob
Ponka, Premysl
Regoli, Domenico
Roy, Claude Charles
Sandor, Thomas
Skamene, Emil
Stephens, Heather R
Thomson, David M P
Zidulka, Arnold

SASKATCHEWAN
Kalra, Jawahar

OTHER COUNTRIES
Agarwal, Shyam S
Binder, Bernd R
Bodmer, Walter Fred
Bojalil, Luis Felipe
Fitzgerald, John Desmond
Gomez, Arturo
Gorski, Andrzej
Ha, Tai-You
Hashimoto, Paulo Hitonari
Hirokawa, Katsuiku
Hounsfield, G Newbold
Huang, Kun-Yen
Ikehara, Yukio
Izui, Shozo
Johnsen, Dennis O
Jonzon, Anders
Kawamura, Hiroshi
Kreider, Eunice S
Lu, Guo-Wei
Merlini, Giampaolo
Miyasaka, Kyoko
Modabber, Farrokh Z
Monos, Emil
Prywes, Moshe
Querinjean, Pierre Joseph
Salas, Pedro Jose I
Sasaki, Hidetada
Spiess, Joachim
Strous, Ger J
Takai, Yasuyuki
Tanphaichitr, Vichai
Tung, Che-Se
Vane, John Robert
Van Furth, Ralph
Wrede, Don Edward
Wu, Albert M
Yamada, Eichi

Medicine

ALABAMA
Alford, Charles Aaron, Jr
Allison, Ronald C
Bailey, William C
Ball, Gene V
Bargeron, Lionel Malcolm, Jr
Batson, Oscar Randolph
Baxley, William Allison
Benton, John William, Jr
Bishop, Sanford Parsons
Blackburn, Will R
Boshell, Buris Raye
Bradley, Laurence A
Brascho, Donn Joseph
Brogdon, Byron Gilliam
Bryan, Thornton Embry
Butterworth, Charles E, Jr
Carlo, Waldemar Alberto
Cassady, George
Cassell, Gail Houston
Caulfield, James Benjamin
Ceballos, Ricardo
Cheraskin, Emanuel
Conrad, Marcel E
Cooper, Max Dale
Daniel, William A, Jr
Diasio, Robert Bart
Dillon, Hugh C, Jr
Durant, John Ridgway
Dyken, Paul Richard
Eddleman, Elvia Etheridge, Jr
Evanochko, William Thomas
Finklea, John F
Flowers, Charles E, Jr
Fraser, Robert Gordon
Furner, Raymond Lynn
Gardner, William Albert, Jr
Garrett, J Marshall
Gay, Renate Erika
Gay, Steffen
Geer, Jack Charles
Gleysteen, John Jacob
Griffin, Frank M, Jr
Hageman, Gilbert Robert
Halsey, James H, Jr
Hefner, Lloyd Lee
Hill, Samuel Richardson, Jr
Hirschowitz, Basil Isaac
Holley, Howard Lamar
Hughes, Edwin R
Lecklitner, Myron Lynn
Lee, Daisy Si
Levene, Ralph Zalman
Lobuglio, Albert Francis
Logic, Joseph Richard
Lorincz, Andrew Endre
Lupton, Charles Hamilton, Jr
Lutwak, Leo
McCann, William Peter
McMahon, John Martin
Meredith, Ruby Frances
Montgomery, John R
Moreno, Hernan
Myers, Ira Lee
Oberman, Albert
Oh, Shin Joong
Oparil, Suzanne
Osment, Lamar Sutton
Palmisano, Paul Anthony

Pass, Robert Floyd
Peterson, Raymond Dale August
Pittman, James Allen, Jr
Reque, Paul Gerhard
Richter, Joel Edward
Roth, Robert Earl
Russell, Richard Olney, Jr
Sams, Wiley Mitchell, Jr
Schmidt, Leon Herbert
Schnaper, Harold Warren
Segrest, Jere Palmer
Sheehy, Thomas W
Sheffield, L Thomas
Shin, Myung Soo
Shingleton, Hugh Maurice
Siegal, Gene Philip
Simpson-Herren, Linda
Sowell, John Gregory
Stagno, Sergio Bruno
Stover, Samuel Landis
Tiller, Ralph Earl
Winternitz, William Welch

ALASKA
Crook, James Richard
Mills, William J, Jr

ARIZONA
Arkins, John A
Bernhard, Victor Montwid
Bowden, George Timothy
Boyse, Edward A
Brandenburg, Robert O
Brosin, Henry Walter
Burrows, Benjamin
Butler, Byron C
Capp, Michael Paul
Cassidy, Suzanne Bletterman
Cherny, Walter B
Christian, Charles Donald
Corrigan, James John, Jr
Criswell, Bennie Sue
Daly, David DeRouen
Denny, William F
Evans, Tommy Nicholas
Fagan, Timothy Charles
Fischer, Harry William
Fisher, William Gary
Gordon, Richard Seymour
Grobe, James L
Heine, Melvin Wayne
Hersh, Evan Manuel
Holloway, G Allen, Jr
Huestis, Douglas William
Huffman, John William
Iserson, Kenneth Victor
Johnson, Sture Archie Mansfield
Joy, Vincent Anthony
Katz, Murray Alan
Kay, Marguerite M B
Kelly, John V
Klotz, Arthur Paul
Koelsche, Giles Alexander
Lewis, Alan Ervin
Marcus, Frank I
Mooradian, Arshag Dertad
Nugent, Charles Arter, Jr
Ogden, David Anderson
Ordway, Nelson Kneeland
Orient, Jane Michel
Parsons, William Belle, Jr
Patton, Dennis David
Pettit, George Robert
Pinnas, Jacob Louis
Quan, Stuart F
Quinlivan, William Leslie G
Rifkind, David
Sandberg, Avery Aba
Scott, David Bytovetzski
Sibley, William Austin
Smith, Josef Riley
Snyder, Richard Gerald
Stephens, Charles Arthur Lloyd, Jr
Tatelman, Maurice
Trifan, Deonisie
Weil, Andrew Thomas
Werner, Sidney Charles
Yakaitis, Ronald William

ARKANSAS
Abernathy, Robert Shields
Bard, David S
Barnhard, Howard Jerome
Bates, Joseph H
Beard, Owen Wayne
Berry, Daisilee H
Borden, Craig W
Diner, Wilma Canada
Doherty, James Edward, III
Dungan, William Thompson
Ebert, Richard Vincent
Elders, Minnie Joycelyn
Flanigan, Stevenson
Flanigan, William J
Harrison, Robert Walker, III
Haut, Arthur
Hawks, Byron Lovejoy
Hiller, Frederick Charles
Lester, Roger
Sanders, Louis Lee
Stead, William White
Texter, E Clinton, Jr
Towbin, Eugene Jonas
Vesely, David Lynn

CALIFORNIA
Abildgaard, Charles Frederick
Adams, Forrest Hood
Agras, William Stewart
Agre, Karl
Agress, Clarence M
Aikawa, Jerry Kazuo
Ainsworth, Cameron
Allen, Joseph Garrott
Alpert, Louis Katz
Amador, Elias
Ament, Marvin Earl
Amstutz, Harlan Cabot
Anger, Hal Oscar
Apple, Martin Allen
Archambeau, John Orin
Archibald, Kenneth C
Arieff, Allen I
Arnold, Harry L(oren), Jr
Ash, Lawrence Robert
Ashburn, William Lee
Austin, Donald Franklin
Babior, Bernard M
Bacchus, Habeeb
Bagshaw, Malcolm A
Bainton, Dorothy Ford
Balchum, Oscar Joseph
Barker, Wiley Franklin
Barnard, R James
Barnes, Asa, Jr
Barnett, Eugene Victor
Barrett, Peter Van Doren
Barron, Charles Irwin
Basham, Teresa
Baum, David
Baxter, John Darling
Beall, Gildon Noel
Beard, Rodney Rau
Beck, John Christian
Behrens, Herbert Charles
Behrman, Richard Elliot
Beigelman, Paul Maurice
Beljan, John Richard
Bennett, Cleaves M
Bennett, Leslie R
Berek, Jonathan S
Berk, Jack Edward
Berk, Robert Norton
Bernstein, Gerald Sanford
Bernstein, Leslie
Bernstein, Sol
Berryman, George Hugh
Bessman, Alice Neuman
Bessman, Samuel Paul
Bettman, Jerome Wolf
Beutler, Ernest
Bick, Rodger Lee
Bierman, Howard Richard
Biglieri, Edward George
Bikle, Daniel David
Billingham, John
Billings, Charles Edgar, Jr
Bing, Richard John
Bissell, Dwight Montgomery
Black, Robert L
Blahd, William Henry
Blankenhorn, David Henry
Blantz, Roland C
Blejer, Hector P
Block, Jerome Bernard
Blois, Marsden Scott, Jr
Bluestein, Harry Gilbert
Boak, Ruth Alice
Bodfish, Ralph E
Bolt, Robert James
Bookstein, Joseph Jacob
Bordin, Gerald M
Bozdech, Marek Jiri
Bragonier, John Robert
Brasel, Jo Anne
Braude, Abraham Isaac
Brecher, George
Breslow, Lester
Bricker, Neal S
Bristow, Lonnie Robert
Brook, Robert H
Brown, David Edward
Brown, Ellen
Brown, J Martin
Brown, James K
Brown, Stuart Irwin
Brunell, Philip Alfred
Buchsbaum, Monte Stuart
Burnham, Thomas K
Burrow, Gerard N
Byfield, John Eric
Cady, Lee (De), Jr
Canham, John Edward
Carbone, John Vito
Carlsson, Erik
Carmel, Ralph
Carregal, Enrique Jose Alvarez
Carroll, Walter William
Carson, Dennis A
Carter, Kenneth
Carter, Paul Richard
Cassan, Stanley Morris
Castles, James Joseph, Jr
Catlin, Don Hardt
Catz, Boris
Cavalieri, Ralph R
Chase, Robert A
Cheitlin, Melvin Donald
Cherry, James Donald

Medicine (cont)

Reilly, Emmett B
Reisner, Ronald M
Reitz, Richard Elmer
Remington, Jack Samuel
Resch, Joseph Anthony
Reynolds, Telfer Barkley
Rider, Joseph Alfred
Riemenschneider, Paul Arthur
Robbins, Robert
Robinson, William Sidney
Roemer, Milton Irwin
Rosenberg, Saul Allen
Rosenman, Ray Harold
Ross, Gordon
Ross, John, Jr
Rubenstein, Edward
Rubenstein, Howard S
Ruggeri, Zaverio Marcello
Rytand, David A
Sagan, Leonard A
Saha, Pamela S
Salk, Jonas Edward
Salkin, David
Samloff, I Michael
Sanazaro, Paul Joseph
Sandberg, Eugene Carl
Sandler, Harold
Sapico, Francisco L
Sartoris, David John
Sazama, Kathleen
Schacher, John Fredrick
Schaefer, George
Scheie, Harold Glendon
Schlegel, Robert John
Schmid, Rudi
Schneider, Edward Lewis
Schneiderman, Lawrence J
Schoenfield, Leslie Jack
Schrier, Stanley Leonard
Schroeder, John Speer
Schulkind, Martin Lewis
Schulman, Irving
Schulz, Jeanette
Schwabe, Arthur David
Schwartz, Herbert C
Scott, W Richard
Seaman, William E
Sebastian, Anthony
Seeger, Robert Charles
Selvester, Ronald H
Selzer, Arthur
Sencer, David Judson
Severinghaus, John Wendell
Shames, David Marshall
Sharma, Arjun D
Shearn, Martin Alvin
Sheline, Glenn Elmer
Shimkin, Michael Boris
Shinefield, Henry R
Shipp, Joseph Calvin
Shnider, Sol M
Shore, Nomie Abraham
Shortliffe, Edward Hance
Shulman, Ira Andrew
Siegel, Lawrence Sheldon
Siegel, Michael Elliot
Siemsen, Jan Karl
Simmons, Daniel Harold
Simon, Allan Lester
Simon, Harold J
Siperstein, Marvin David
Skoog, William Arthur
Slater, James Munro
Sleisenger, Marvin Herbert
Smith, Lloyd Hollingworth, Jr
Smith, Norman Ty
Smithwick, Elizabeth Mary
Snape, William J, Jr
Soeldner, John Stuart
Sokolow, Maurice
Soll, Andrew H
Solomon, David Harris
Sooy, Francis Adrian
Sossong, Norman D
Sparkes, Robert Stanley
Spector, Samuel
Spector, Sheldon Laurence
Spencer, E Martin
Spitler, Lynn E
Spivey, Bruce Eldon
Spivey, Gary H
Stamey, Thomas Alexander
Staple, Tom Weinberg
Steen, Stephen N
Stein, Myron
Steinfeld, Jesse Leonard
Stern, Robert
Sterner, James Hervi
Stiehm, E Richard
Stockdale, Frank Edward
Stone, Daniel Boxall
Stoughton, Richard Baker
Straatsma, Bradley Ralph
Strauss, Bernard
Stringfellow, Dale Alan
Strober, Samuel
Swan, Harold James Charles
Sweetman, Lawrence
Swerdloff, Ronald S
Swezey, Robert Leonard
Tager, Ira Bruce
Takasugi, Mitsuo
Talley, Robert Boyd
Tanaka, Kouichi Robert

Tarver, Harold
Tatter, Dorothy
Thaler, M Michael
Thomas, Heriberto Victor
Thompson, William Benbow, Jr
Thrupp, Lauri David
Tierney, Donald Frank
Tobis, Jerome Sanford
Tooley, William Henry
Topping, Norman Hawkins
Torrance, Daniel J
Tourtellotte, Wallace William
Towers, Bernard
Tranquada, Robert Ernest
Trelford, John D
Triche, Timothy J
Tripp, R(ussell) Maurice
Troy, Frederic Arthur
Tubis, Manuel
Tune, Bruce Malcolm
Tupper, Charles John
Turner, John Dean
Tyan, Marvin L
Udall, John Alfred
Valentine, William Newton
Van Den Noort, Stanley
Van Der Meulen, Joseph Pierre
Varki, Ajit Pothan
Varon, Myron Izak
Vaughan, John Heath
Vaughan, Victor Clarence, III
Vermund, Halvor
Vohs, James A
Von Leden, Hans Victor
Vredevoe, Lawrence A
Wallerstein, Ralph Oliver, Jr
Walter, Richard D
Waltz, Arthur G
Ward, Paul H
Warner, Willis L
Wasserman, Karlman
Wasserman, Stephen I
Wat, Bo Ying
Watts, Malcolm S M
Waxman, Alan David
Way, Walter
Webber, Milo M
Wehrle, Paul F
Weichsel, Morton E, Jr
Weil, Marvin Lee
Weiner, Herbert
Weinreb, Robert Neal
Weinstein, Irwin M
Weisman, Harold
Werbach, Melvyn Roy
Werner, Sanford Benson
West, John B
Weston, Raymond E
Wheeler, Henry Orson
Whittenberger, James Laverre
Whorton, M Donald
Wiederholt, Wigbert C
Wilbur, David Wesley
Wilbur, Dwight Locke
Williams, James Hutchison
Williams, Lawrence Ernest
Williams, Roger Lea
Wilson, Charles B
Wilson, Miriam Geisendorfer
Wilson, William Jewell
Winget, Charles M
Winkelstein, Warren, Jr
Wishner, Kathleen L
Witschi, Hanspeter R
Wolfe, Allan Marvin
Wong-Staal, Flossie
Woodruff, John H, Jr
Wright, Harry Tucker, Jr
Wu, Sing-Yung
Yamini, Sohrab
Young, Robert, Jr
Yow, Martha Dukes
Zatz, Leslie M
Zboralske, F Frank
Zeldis, Jerome B
Zeleznick, Lowell D
Zerez, Charles Raymond
Ziegler, Michael Gregory
Zimmerman, Emery Gilroy
Zvaifler, Nathan J

COLORADO
Aldrich, Franklin Dalton
Allen, Robert H
Anderson, Paul Nathaniel
Angleton, George M
Arend, William Phelps
Austin, James Henry
Ballinger, Carter M
Battaglia, Frederick Camillo
Berthrong, Morgan
Betz, George
Bock, S Allan
Bowling, Franklin Lee
Brettell, Herbert R
Bromage, Philip R
Bunn, Paul A, Jr
Chai, Hyman
Chan, Laurence Kwong-fai
Cherniack, Reuben Mitchell
Claman, Henry Neumann
Daves, Marvin Lewis
Eickhoff, Theodore C
Eisele, Charles Wesley

Eliot, Robert S
Ellis, Philip Paul
Falliers, Constantine J
Gelfand, Erwin William
Gerschenson, Lazaro E
Gersten, Jerome William
Gilden, Donald Harvey
Githens, John Horace, Jr
Glode, Leonard Michael
Hall, Alan H
Hammermeister, Karl E
Hine, Gerald John
Hofeldt, Fred Dan
Hollister, Alan Scudder
Horwitz, Kathryn Bloch
Horwitz, Lawrence D
Hudson, Bruce William
Hutt, Martin P
Jacobson, Eugene Donald
Kappler, John W
Katz, Fred H
Kern, Fred, Jr
Kinzie, Jeannie Jones
Kirkpatrick, Charles Harvey
Lee, Lela A
Liechty, Richard Dale
McCord, Joe M
McDowell, Marion Edward
Makowski, Edgar Leonard
Marine, William Murphy
Moore, George Eugene
Mueller, John Frederick
Naughton, Michael A
Nelson, Albert Wendell
Nelson, Harold Stanley
Nies, Alan Sheffer
Nora, Audrey Hart
Nora, James Jackson
O'Brien, Donough
Peters, Bruce Harry
Petty, Thomas Lee
Platt, Kenneth Allan
Poland, Jack Dean
Prasad, Kedar N
Reeves, John T
Repine, John E
Robinson, Arthur
Rosenwasser, Lanny Jeffery
Rudnick, Michael Dennis
Schooley, Robert T
Schrier, Robert William
Shepard, Richard Hance
Silver, Henry K
Sokol, Ronald Jay
Solomons, Clive (Charles)
Spaulding, Harry Samuel, Jr
Stanek, Karen Ann
Stephens, Gwen Jones
Sussman, Karl Edgar
Talmage, David Wilson
Taylor, Edward Stewart
Tofe, Andrew John
Webber, Mukta Mala (Maini)
Weil, John Victor
Werkman, Sidney Lee
Williams, Ronald Lee

CONNECTICUT
Allaudeen, Hameedsulthan Sheik
Andriole, Vincent T
Ariyan, Stephan
Aronson, Peter S
Bailie, Michael David
Barash, Paul G
Baumgarten, Alexander
Berliner, Robert William
Binder, Henry J
Blechner, Jack Norman
Boas, Norman Francis
Bollet, Alfred Jay
Bondy, Philip K
Boron, Walter Frank
Boulpaep, Emile L
Bove, Joseph Richard
Boyer, James Lorenzen
Boylan, John W
Brandt, J Leonard
Braverman, Irwin Merton
Breg, William Roy
Bunney, Benjamin Stephenson
Burchenal, Joseph Holland
Cadman, Edwin Clarence
Carter, David Martin
Cheney, Charles Brooker
Cohen, Lawrence Sorel
Conn, Harold O
Curnen, Edward Charles, Jr
DeGraff, Arthur C, Jr
Dillard, Morris, Jr
Donaldson, Robert M, Jr
Donta, Sam Theodore
Dorsky, David Isaac
Duff, Raymond Stanley
Ebbert, Arthur, Jr
Eisenberg, Sidney Edwin
Eisenfeld, Arnold Joel
Ezekowitz, Michael David
Feinstein, Alvan Richard
Fischer, James Joseph
Fleeson, William
Fong, Jack Sun-Chik
Freston, James W
Freymann, John Gordon
Gee, J Bernard L

Gefter, William Irvin
Genel, Myron
Glaser, Gilbert Herbert
Goodyer, Allan Victor
Gottschalk, Alexander
Greene, Nicholas Misplee
Greenspan, Richard H
Gross, Ian
Groszmann, Roberto Jose
Heller, John Herbert
Hendler, Ernesto Danilo
Hinkle, Lawrence Earl, Jr
Hinz, Carl Frederick, Jr
Hobbins, John Clark
Hoffer, Paul B
Horwitz, Ralph Irving
Hosain, Fazle
Howard, Rufus Oliver
Janeway, Charles Alderson, Jr
Jensen, Keith Edwin
Katz, Arnold Martin
Kirchner, John Albert
Klatskin, Gerald
Klaus, Sidney N
Klimek, Joseph John
Korn, Joseph Howard
Krause, Peter James
Lange, Robert Carl
Leonard, Martha Frances
Lerner, Aaron Bunsen
Levine, Robert John
Liao, Sung Jui
Lowman, Robert Morris
Lurie, Alan Gordon
Lytton, Bernard
McGuire, Joseph Smith, Jr
Macko, Douglas John
Malawista, Stephen E
Markowitz, Milton
Marks, Paul A
Martin, R Russell
Martin, Rufus Russell
Massey, Robert Unruh
Miller, Herbert Chauncey
Miller, I George, Jr
Morris, John McLean
Morse, Edward Everett
Murray, Francis Joseph
Naftolin, Frederick
Nanda, Ravindra
Niederman, James Corson
O'Rourke, James
Papac, Rose
Patterson, John Ward
Pearson, Howard Allen
Perillie, Pasquale E
Raisz, Lawrence Gideon
Randt, Clark Thorp
Rasmussen, Howard
Reiskin, Allan B
Richards, Frank Frederick
Rickles, Frederick R
Robinson, Donald Stetson
Rosenfield, Arthur Ted
Rothfield, Naomi Fox
Sachs, Frederick Lee
Sasaki, Clarence Takashi
Schmidt, James L
Sears, Marvin Lloyd
Senn, Milton J E
Shehadi, William Henry
Sikand, Rajinder S
Silver, George Albert
Smith, Brian Richard
Sobol, Bruce J
Solnit, Albert J
Spencer, Richard Paul
Spiro, Howard Marget
Sulavik, Stephen B
Sulzer-Azaroff, Beth
Talner, Norman Stanley
Tamborlane, William Valentine
Taub, Arthur
Taylor, Kenneth J W
Thompson, Hartwell Greene, Jr
Thornton, George Fred
Vinocur, Myron
Walker, James Elliot Cabot
Weiss, Robert Martin
Welch, Annemarie S
Wetstone, Howard J
Whittemore, Ruth
Zaret, Barry L

DELAWARE
Dow, Lois Weyman
Graham, David Tredway
Harley, Robison Dooling
Linna, Timo Juhani
Mullane, John F
Reinhardt, Charles Francis
Sarrif, Awni M
Zajac, Barbara Ann

DISTRICT OF COLUMBIA
Abramson, David C
Alving, Carl Richard
Archer, Juanita Almetta
Avery, Gordon B
Barnet, Ann B
Becker, Kenneth Louis
Bellanti, Joseph A
Bourne, Peter
Bowles, Lawrence Thompson

Medicine (cont)

Bulger, Roger James
Burger, Edward James, Jr
Burman, Kenneth Dale
Burris, James F
Caceres, Cesar A
Calcagno, Philip Louis
Callaway, Clifford Wayne
Canary, John J(oseph)
Carpentier, Robert George
Cheng, Tsung O
Chopra, Joginder Gurbux
Chung, Ed Baik
Coakley, Charles Seymour
Deutsch, Stanley
Diggs, Carter Lee
Dillard, Martin Gregory
Dunbar, Burdett Sheridan
Egeberg, Roger O
Ein, Daniel
Elliott, Larry Paul
Fabro, Sergio
Felts, William Robert
Filipescu, Nicolae
Fine, Ben Sion
Finkelstein, James David
Finnerty, Frank Ambrose, Jr
Fishbein, William Nichols
Foley, H Thomas
Fox, Samuel Mickle, III
Freis, Edward David
Fromm, Hans
Gelmann, Edward P
Geyer, Stanley J
Giannini, Margaret Joan
Gootenberg, Joseph Eric
Gordon, James Samuel
Grigsby, Margaret Elizabeth
Gronvall, John Arnold
Harvey, John Collins
Hellman, Louis M
Henry, Walter Lester, Jr
Himmelsbach, Clifton Keck
Horner, William Harry
Hugh, Rudolph
Jackson, Dudley Pennington
Jackson, Marvin Alexander
Kaminetzky, Harold Alexander
Katz, Paul K
Katz, Sol
Keliher, Thomas Francis
Kennedy, Charles
Kimmel, Carole Anne
Kimmel, Gary Lewis
Kobrine, Arthur
Kulczycki, Lucas Luke
Law, David H
Lessin, Lawrence Stephen
Lilienfield, Lawrence Spencer
Loo, Ti Li
Lorber, Mortimer
Luessenhop, Alfred John
McAfee, John Gilmour
McLean, Ian William
Macnamara, Thomas E
Manley, Audrey Forbes
Martin, Malcolm Mencer
Massaro, Donald John
Meyers, Wayne Marvin
Miller, Russell Loyd, Jr
Novello, Antonia Coello
O'Doherty, Desmond Sylvester
Parker, Richard H
Parrish, Alvin Edward
Patel, Dali Jehangir
Pearse, Warren Harland
Pellegrino, Edmund Daniel
Perry, Seymour Monroe
Petersdorf, Robert George
Peterson, Malcolm Lee
Pincus, Jonathan Henry
Pipberger, Hubert V
Pollack, Herbert
Potter, John F
Queenan, John T
Ramsey, Elizabeth Mapelsden
Rath, Charles E
Reba, Richard Charney
Recant, Lillian
Rennert, Owen M
Robertson, James Sydnor
Rockoff, Seymour David
Sabin, Albert B(ruce)
Sampson, Calvin Coolidge
Schechter, Geraldine Poppa
Scheele, Leonard Andrew
Schreiner, George E
Scott, Roland Boyd
Sever, John Louis
Silva, Omega Logan
Simopoulos, Artemis Panageotis
Sjogren, Robert W, Jr
Sly, Ridge Michael
Smith, Bruce H
Stapleton, John F
Stark, Nathan Julius
Stemmler, Edward J
Sullivan, Louis Wade
Swanson, August George
Tidball, Charles Stanley
Twigg, Homer Lee
Utz, John Philip
Vahouny, George V

Vick, James
Wagner, Henry George
Wartofsky, Leonard
Watkin, Donald M
Weingold, Allan Byrne
Weintraub, Herbert D
Werner, Mario
Whidden, Stanley John
Williams, James Thomas
Winchester, James Frank
Wray, H Linton
Wright, Daniel Godwin

FLORIDA

Abell, Murray Richardson
Aldrete, Jorge Antonio
Alexander, Ralph William
Allinson, Morris Jonathan Carl
Andersen, Thorkild Waino
Anderson, Douglas Richard
Ansbacher, Stefan
Awad, William Michel, Jr
Ayoub, Elia Moussa
Baiardi, John Charles
Balducci, Lodovico
Balkany, Thomas Jay
Befeler, Benjamin
Behnke, Roy Herbert
Ben, Max
Berman, Irwin
Bernstein, Stanley H
Berry, Maxwell (Rufus)
Biersdorf, William Richard
Bisno, Alan Lester
Blank, Harvey
Block, Edward R
Blum, Leon Leib
Bourgoignie, Jacques J
Boyce, Henry Worth, Jr
Bramante, Pietro Ottavio
Brightwell, Dennis Richard
Brooks, Stuart Merrill
Bucklin, Robert van Zandt
Bukantz, Samuel Charles
Cade, James Robert
Cavanagh, Denis
Challoner, David Reynolds
Chandler, J Ryan
Cintron, Guillermo B
Cluff, Leighton Eggertsen
Cody, D Thane
Conn, Jerome W
Cooke, Robert E
Craig, Stanley Harold
Craythorne, N W Brian
Culver, James F
Curran, John S
Cutler, Paul
Davis, Joseph Harrison
Del Regato, Juan A
Diaz, Pedro Miguel
Eitzman, Donald V
Elfenbein, Gerald Jay
Ellis, Elliot F
Enneking, William Fisher
Epstein, Murray
Evans, Arthur T
Fagin, Karin
Fayos, Juan Vallvey
Feingold, Alfred
Flipse, Martin Eugene
Fogel, Bernard J
Freund, Gerhard
Friedland, Fritz
Friedman, Ben I
Galindo, Anibal H
Gelband, Henry
Gessner, Ira Harold
Gibson, Joyce C
Ginsberg, Myron David
Gold, Mark Stephen
Gold, Martin I
Goldberg, Lee Dresden
Goldberg, Stephen
Goldman, Allan Larry
Goldsmith, Carl
Good, Robert Alan
Graven, Stanley N
Gravenstein, Joachim Stefan
Graybiel, Ashton
Greenfield, George B
Griffitts, James John
Groel, John Trueman
Groover, Marshall Eugene, Jr
Grunberg, Emanuel
Gubner, Richard S
Hadden, John Winthrop
Halprin, Kenneth M
Hartmann, Robert Carl
Healey, John Edward, Jr
Hebert, Clarence Louis
Hecht, Frederick
Henderson, John Woodworth
Hickman, Jack Walter
Hinkley, Robert Edwin, Jr
Hodes, Philip J
Hornick, Richard B
Howard, Guy Allen
Howell, David Sanders
Howell, Ralph Rodney
Hulet, William Henry
Inana, George
Itkin, Irving Herbert
Jensen, Wallace Norup

Kalser, Martin
Kessler, Richard Howard
Kinney, Michael J
Kiplinger, Glenn Francis
Klemperer, Martin R
Lasseter, Kenneth Carlyle
Lawson, Robert Barrett
Lawton, Alfred Henry
Lenes, Bruce Allan
Lessner, Howard E
Levy, Norman Stuart
Lian, Eric Chun-Yet
Little, William Asa
Lockey, Richard Funk
Loeser, Eugene William
Lombardi, Max H
Luhan, Joseph Anton
MacDonald, Richard Annis
McGuigan, James E
McIntosh, Henry Deane
McKenzie, John Maxwell
McMillan, Donald Ernest
Mallams, John Thomas
Malone, John Irvin
Marks, Meyer Benjamin
Marriott, Henry Joseph Llewellyn
Marston, Robert Quarles
Martinez, Luis Osvaldo
Michael, Max, Jr
Miller, Billie Lynn
Million, Rodney Reiff
Mindlin, Rowland L
Modell, Jerome Herbert
Moreland, Alvin Franklin
Muench, Karl Hugo
Murphy, William Parry, Jr
Murtagh, Frederick Reed
Nagel, Eugene L
Neims, Allen Howard
Nelson, Erland
Nelson, Russell Andrew
Nevis, Arnold Hastings
Noble, Nancy Lee
Norton, Edward W D
Noyes, Ward David
Okulski, Thomas Alexander
Olson, Kenneth B
Orkin, Lazarus Allerton
Palmer, Roger
Panush, Richard Sheldon
Papper, Emanuel Martin
Pepine, Carl John
Persky, Lester
Prather, Elbert Charlton
Prineas, Ronald James
Prockop, Leon D
Reich, George Arthur
Reiss, Eric
Rhodin, Johannes A G
Roberts, Norbert Joseph
Root, Allen William
Rosenbloom, Arlan Lee
Rosenkrantz, Jacob Alvin
Rosomoff, Hubert Lawrence
Rothschild, Marcus Adolphus
Rubin, Melvin Lynne
Ryan, James Walter
Ryan, Una Scully
Samet, Philip
Scheinberg, Peritz
Schiebler, Gerold Ludwig
Schiff, Leon
Schmidt, Paul Joseph
Sharp, John Turner
Shipley, Thorne
Shively, John Adrian
Singleton, George Terrell
Sinkovics, Joseph G
Smith, Donn Leroy
Smith, Richard Thomas
So, Antero Go
Sommer, Leonard Samuel
Spellacy, William Nelson
Storaasli, John Phillip
Streiff, Richard Reinhart
Streitfeld, Murray Mark
Szentivanyi, Andor
Taylor, William Jape
Thomas, William Clark, Jr
Turner, Robert Alexander
Vail, Edwin George
Valdes-Dapena, Marie A
Van Mierop, Lodewyk H S
Vasey, Frank Barnett
Victorica, Benjamin (Eduardo)
Waife, Sholom Omi
Wallach, Stanley
Wanner, Adam
Warren, Joel
Waterhouse, Keith R
Weidler, Donald John
Weiss, Charles Frederick
Westlake, Robert Elmer, Sr
Wilkerson, James Edward
Williams, Ralph C, Jr
Witte, John Jacob
Yeh, Billy Kuo-Jiun
Young, Lawrence Eugene
Yunis, Adel A
Zubrod, Charles Gordon

GEORGIA

Alonso, Kenneth B
Baisden, Charles Robert

Ballard, Marguerite Candler
Bernstein, Robert Steven
Blumberg, Richard Winston
Bonkovsky, Herbert Lloyd
Broome, Claire V
Brown, Herbert Eugene
Brown, Mark
Brown, Walter John
Brubaker, Leonard Hathaway
Bruner, Barbara Stephenson
Bryans, Charles Iverson, Jr
Butcher, Brian T
Byrd, J Rogers
Chew, William Hubert, Jr
Collipp, Platon Jack
Connell-Tatum, Elizabeth Bishop
Cooper, Gerald Rice
Copeland, Robert B
Corley, Charles Calhoun, Jr
Cox, Frederick Eugene
Curran, James W
Davey, Winthrop Newbury
Davidson, John Keay, III
de Jong, Rudolph H
Desai, Rajendra C
Dowda, F William
Edelhauser, Henry F
Egan, Robert L
Ellison, Lois Taylor
Engler, Harold S
Evans, Edwin Curtis
Evatt, Bruce Lee
Faguet, Guy B
Falk, Henry
Feldman, Daniel S
Feldman, Elaine Bossak
Feringa, Earl Robert
Flowers, Nancy Carolyn
Foege, William Herbert
Franch, Robert H
Frank, Martin J
Frederickson, Evan Lloyd
Galambos, John Thomas
Gallagher, Brian Boru
Garrettson, Lorne Keith
Glass, Roger I M
Golbey, Robert (Bruce)
Goldman, John Abner
Grayzel, Arthur I
Groth, Donald Paul
Guinan, Mary Elizabeth
Gunter, Bobby J
Haskins, Arthur L, Jr
Herman, Chester Joseph
Herrmann, Kenneth L
Hoffman, William Hubert
Hollowell, Joseph Gurney, Jr
Holzman, Gerald Bruce
Hopkins, Donald R
Hudson, James Bloomer
Huff, Thomas Allen
Hug, Carl Casimir, Jr
Hughes, James Mitchell
Huguley, Charles Mason, Jr
Hunt, Andrew Dickson
Jarvis, Jack Reynolds
Jeffery, Geoffrey Marron
Jones, Jimmy Barthel
Joyner, Ronald Wayne
Karp, Herbert Rubin
Keene, Willis Riggs
Kilbourne, Edwin Michael
Klein, Luella
Knobloch, Hilda
Little, Robert Colby
Lohuis, Delmont John
Long, Wilmer Newton, Jr
McDuffie, Frederic Clement
McPherson, James C, Jr
Martinez-Maldonado, Manuel
Middleton, Henry Moore, III
Millar, John Donald
Mitch, William Evans
Moores, Russell R
Nahmias, Andre Joseph
Nelson, George Humphry
Oliver, John Eoff, Jr
Orenstein, Walter Albert
Parks, John S
Per-Lee, John H
Pool, Winford H, Jr
Raiford, Morgan B
Ramos, Harold Smith
Rimland, David
Robertson, Alex F
Rodey, Glenn Eugene
Rogers, James Virgil, Jr
Roper, Maryann
Roper, William L
Savage, John Edward
Schlant, Robert C
Schroder, Jack Spalding
Schultz, Everett Hoyle, Jr
Schwartz, James F
Sellers, Thomas F, Jr
Sharp, John T(homas)
Shulman, Jones A(lvin)
Smith, Jesse Graham, Jr
Snider, Dixie Edward, Jr
Spitznagel, John Keith
Steinhaus, John Edward
Steinschneider, Alfred
Stevens, John Joseph
Talledo, Oscar Eduardo

Tedesco, Francis J
Tuttle, Elbert P, Jr
Walton, Kenneth Nelson
Ward, Richard S
Waring, George O, III
Warren, McWilson
Webster, Paul Daniel, III
Wege, William Richard
Wenger, Nanette Kass
Whitsett, Carolyn F
Wigh, Russell
Willis, Isaac
Wilson, Colon Hayes, Jr
Witham, Abner Calhoun
Yancey, Asa G

HAWAII
Bass, James W
Batkin, Stanley
Blaisdell, Richard Kekuni
Char, Donald F B
Furusawa, Eiichi
Gilbert, Fred Ivan, Jr
Hammar, Sherrel L
Lance, Eugene Mitchell
Linman, James William
Massey, Douglas Gordon
Mills, George Hiilani
Person, Donald Ames
Quisenberry, Walter Brown
Schatz, Irwin Jacob
Sharma, Santosh Devraj
Siddiqui, Wasim A
Smith, Richard A
Ueland, Kent
Yang, Hong-Yi

IDAHO
Schwartz, Theodore Benoni
Vestal, Robert Elden

ILLINOIS
Abraira, Carlos
Abramson, David Irvin
Adams, Evelyn Eleanor
Adams, John R
Adelson, Bernard Henry
Adler, Solomon Stanley
Al-Bazzaz, Faiq J
Ambre, John Joseph
Andersen, Burton
Anderson, Kenning M
Archer, John Dale
Arieff, Alex J
Arnason, Barry Gilbert Wyatt
Balagot, Reuben Castillo
Barclay, William R
Barghusen, Herbert Richard
Barton, Evan Mansfield
Baumeister, Carl Frederick
Beattie, Craig W
Beck, Robert Nason
Becker, Michael Allen
Beem, Marc O
Beer, Alan E
Benson, Donald Warren
Bergen, Donna Catherine
Berger, Sheldon
Berkson, David M
Bernstein, Joel Edward
Bernstein, Lionel M
Berry, Leonidas Harris
Best, William Robert
Betts, Henry Brognard
Bharati, Saroja
Birnholz, Jason Cordell
Bloomfield, Daniel Kermit
Bluefarb, Samuel M
Boehm, John Joseph
Bond, Howard Edward
Borkon, Eli Leroy
Boulos, Badi Mansour
Brandfonbrener, Martin
Brasitus, Thomas
Brues, Austin
Bruetman, Martin Edgardo
Brunner, Edward A
Bulkley, George
Burd, Laurence Ira
Carney, Robert Gibson
Celesia, Gastone G
Cera, Lee M
Childers, Roderick W
Christian, Joseph Ralph
Chudwin, David S
Cibils, Luis Angel
Clark, John Whitcomb
Coe, Fredric Lawrence
Cohen, Edward P
Cohen, Lawrence
Cohen, Sidney
Colbert, Marvin J
Cole, Madison Brooks, Jr
Collins, Vincent J
Conibear, Shirley Ann
Cotsonas, Nicholas John, Jr
Couch, James Russell, Jr
Cugell, David Wolf
Danforth, David Newton
Dau, Peter Caine
Davis, Floyd Asher
Davis, Joseph Richard
Day, Emerson
DeCosta, Edwin J

De Forest, Ralph Edwin
DeGroot, Leslie Jacob
del Greco, Francesco
DePeyster, Frederick A
Diamond, Seymour
Dmowski, W Paul
Dray, Sheldon
Duda, Eugene Edward
Earle, David Prince, Jr
Economou, Steven George
Egan, Richard L
Elble, Rodger Jacob
Elliott, William John
Erickson, James C, III
Farber, Marilyn Diane
Fellner, Susan K
Fennessy, John James
Finkel, Asher Joseph
Firlit, Casimir Francis
Fisher, Ben
Fisher, Leonard V
Fozzard, Harry A
Freedman, Philip
Freeman, Maynard Lloyd
Freese, Uwe Ernst
Freinkel, Norbert
Freinkel, Ruth Kimmelstiel
Fried, Walter
Friederici, Hartmann H R
Friedlander, Ira Ray
Frischer, Henri
Gartner, Lawrence Mitchel
Gerbie, Albert B
Gewertz, Bruce Labe
Giovacchini, Peter L
Glick, Gerald
Glisson, Silas Nease
Gold, Jay Joseph
Goldberg, Leon Isadore
Goldberg, Morton Falk
Golomb, Harvey Morris
Goodheart, Clyde Raymond
Gordon, Paul
Grayhack, John Thomas
Green, David
Green, Orville
Griem, Melvin Luther
Griem, Sylvia F
Grossman, Burton Jay
Gunnar, Rolf McMillan
Gurney, Clifford W
Hammond, James B
Hanauer, Stephen B
Hand, Roger
Haring, Olga M
Harris, Jules Eli
Hasegawa, Junji
Hastreiter, Alois Rudolf
Heck, Robert Skinrood
Heller, Paul
Hellman, Samuel
Hendrickson, Frank R
Herting, Robert Leslie
Hicks, James Thomas
Hirsch, Lawrence Leonard
Holmes, Albert William, Jr
Horwitz, David Larry
Huggins, Charles Brenton
Hyatt, Robert Eliot
Iber, Frank Lynn
Jackson, George G
Jacobson, Leon Orris
Joseph, Donald J
Juniper, Kerrison, Jr
Kaizer, Herbert
Kalyan-Raman, Krishna
Kang, David Soosang
Kantor, Harvey Sherwin
Kaplan, Ervin
Karinattu, Joseph J
Katz, Adrian I
Kernis, Marten Murray
Khalili, Ali A
Kim, Yoon Berm
Kimball, Chase Patterson
Kirsner, Joseph Barnett
Knospe, William H
Koenig, Harold
Kohrman, Arthur Fisher
Kraft, Sumner Charles
Kwaan, Hau Cheong
Lashner, Bret Auerbach
Lathrop, Katherine Austin
Lax, Louis Carl
Leff, Alan R
Leikin, Jerrold Blair
Lesinski, John Silvester
Levin, Murray Laurence
Levitan, Ruven
Levitsky, Lynne Lipton
Levy, Leo
Lichter, Edward A
Liebner, Edwin J
Limarzi, Louis Robert
Lindheimer, Marshall D
Littman, Armand
Lorincz, Allan Levente
Louis, John
Lounsbury, Franklin
Love, Leon
Luisada, Aldo Augusto
Lurain, John Robert, III
McCormick, Kenneth James
McGuire, Christine H

Mafee, Mahmood Forootan
Malkinson, Frederick David
Marr, James Joseph
Mayor, Gilbert Harold
Meadows, W Robert
Mendel, Gerald Alan
Meredith, Stephen Charles
Metzger, Boyd Ernest
Mewissen, Dieudonne Jean
Meyer, Richard Charles
Millichap, J(oseph) Gordon
Mirkin, Bernard Leo
Moawad, Atef H
Molitch, Mark F.
Morishima, Akira
Morrell, Frank
Morris, Charles Elliot
Moser, Robert Harlan
Moy, Richard Henry
Nadler, Charles Fenger
Nadler, Henry Louis
Nahrwold, David Lange
Nayyar, Rajinder
Neill, William Alexander
Newell, Frank William
Newman, Louis Benjamin
O'Conor, Vincent John, Jr
Olson, Stanley William
Ostrow, Jay Donald
Pachman, Daniel James
Pachman, Lauren M
Page, Malcolm I
Paterson, Philip Y
Patterson, Roy
Paul, Milton Holiday
Paynter, Camen Russell
Peraino, Carl
Perlman, Robert
Petersen, Edward S
Phair, John P
Pilz, Clifford G
Plotke, Frederick
Potter, Elizabeth Vaughan
Poznanski, Andrew K
Pullman, Theodore Neil
Rabinovich, Sergio Rospigliosi
Radwanska, Ewa
Ratz, John Louis
Reisberg, Boris Elliott
Resnekov, Leon
Rhodes, Mitchell Lee
Roddick, John William, Jr
Roguska-Kyts, Jadwiga
Romack, Frank Eldon
Roos, Raymond Philip
Rosenberg, Henry Mark
Rosenblum, Arthur H
Rosenfield, Robert Lee
Rosenthal, Ira Maurice
Rossi, Ennio Claudio
Rowley, Janet D
Rubenstein, Arthur Harold
Sabesin, Seymour Marshall
Said, Sami I
Sammons, James Harris
Samter, Max
Saper, Clifford B
Scanu, Angelo M
Schlager, Seymour I
Schmid, Frank Richard
Schmidt, Alexander Mackay
Schmitz, Robert L
Schneider, Arthur Sanford
Schoenberger, James A
Schulman, Sidney
Schumacher, Gebhard Friederich B
Schwartz, Steven Otto
Schweppe, John S
Sciarra, John J
Shambaugh, George E, III
Shoch, David Eugene
Shulman, Morton
Shulman, Stanford Taylor
Silverstein, Edward Allen
Simon, Norman M
Singer, Donald H
Singer, Ira
Singer, Irwin
Sisson, George Allen
Skosey, John Lyle
Smith, Theodore Craig
Soffer, Alfred
Solomon, Lawrence Marvin
Sorensen, Leif Boge
Spaeth, Ralph
Stamler, Jeremiah
Stanley, Malcolm McClain
Stefani, Stefano
Steigmann, Frederick
Stepto, Robert Charles
Straumfjord, Jon Vidalin, Jr
Suker, Jacob Robert
Swerdlow, Martin A
Taylor, D Dax
Taylor, Samuel G, III
Thilenius, Otto G
Thompson, Phebe Kirsten
Thorp, Frank Kedzie
Toback, F(rederick) Gary
Tobin, John Robert, Jr
Tobin, Martin John
Torok, Nicholas
Traisman, Howard Sevin
Tripathi, Brenda Jennifer

Tripathi, Ramesh Chandra
Trobaugh, Frank Edwin, Jr
Truong, Xuan Thoai
Tso, Mark On-Man
Ultmann, John Ernest
Valvassori, Galdino E
Van Fossan, Donald Duane
Van Pernis, Paul Anton
Vesselinovitch, Stan Dushan
Visotsky, Harold M
Vye, Malcolm Vincent
Waldstein, Sheldon Saul
Warnick, Edward George
Waterhouse, John P
Weil, Max Harry
Wessel, Hans U
Wied, George Ludwig
Wiener, Stanley L
Wilbur, Richard Sloan
Wilensky, Jacob T
Winter, Robert John
Wolf, James Stuart
Wolter, Janet
Wong, Paul Wing-Kon
Wu, Anna Fang
Yachnin, Stanley
Yang, Sen-Lian
Yunus, Muhammad Basharat
Zatuchni, Gerald Irving
Zeiss, Chester Raymond
Zsigmond, Elemer K

INDIANA
Aldo-Benson, Marlene Ann
Babbs, Charles Frederick
Bang, Nils Ulrik
Baxter, Neal Edward
Beering, Steven Claus
Besch, Henry Roland, Jr
Bhatti, Waqar Hamid
Blix, Susanne
Bond, William H
Brandt, Ira Kive
Carter, James Edward
Clark, Charles Malcolm, Jr
Daly, Walter J
Dansby, Doris
De Myer, William Erl
Dexter, Richard Newman
Dobben, Glen D
Duckworth, William Clifford
Dyke, Richard Warren
Dyken, Mark Lewis
Eckenhoff, James Edward
Eyring, Edward J
Feigenbaum, Harvey
Fisch, Charles
Fischer, A(lbert)Alan
FitzGerald, Joseph Arthur
Gailani, Salman
Garrett, Robert Austin
Goldstein, David Joel
Gordee, Robert Stouffer
Grayson, Merrill
Henry, David P, II
Hicks, Edward James
Hodes, Marion Edward
Irwin, Glenn Ward, Jr
Johnson, Charles F
Johnson, Irving Stanley
Johnston, Cyrus Conrad, Jr
Judson, Walter Emery
Kim, Kil Chol
Klatte, Eugene
Knoebel, Suzanne Buckner
Leser, Ralph Ulrich
Li, Ting Kai
Lumeng, Lawrence
McKinney, Thurman Dwight
Manders, Karl Lee
Martz, Bill L
Merritt, Doris Honig
Mocharla, Raman
Mulcahy, John Joseph
Munsick, Robert Alliot
Nelson, Robert Leon
Norins, Arthur Leonard
Oppenheim, Bernard Edward
Owen, George Murdock
Patrick, Edward Alfred
Powell, Richard Cinclair
Rhoads, Paul Spottswood
Ridolfo, Anthony Sylvester
Rohn, Robert Jones
Siddiqui, Aslam Rasheed
Stonehill, Robert Berrell
Surawicz, Borys
Test, Charles Edward
Thompson, Wilmer Leigh
Wakim, Khalil Georges
Wappner, Rebecca Sue
Weaver, David Dawson
Weber, George
Weinberger, Myron Hilmar
Wellman, Henry Nelson
White, Arthur C
Wright, Joseph William, Jr
Yune, Heun Yung
Zipes, Douglas Peter

IOWA
Abboud, Francois Mitry
Armstrong, Mark l
Bean, William Bennett

Medicine (cont)

Bedell, George Noble
Bell, William E
Beller, Fritz K
Biermann, Janet Sybil
Biller, Jose
Blodi, Frederick Christopher
Caplan, Richard Melvin
Christensen, James
Clark, Robert A
Clifton, James Albert
Crump, Malcolm Hart
DeGowin, Richard Louis
Eckhardt, Richard Dale
Eckstein, John William
Filer, Lloyd Jackson, Jr
Fomon, Samuel Joseph
Funk, David Crozier
Goplerud, Clifford P
Gurll, Nelson
Hardin, Robert Calvin
Heistad, Donald Dean
Hom, Ben Lin
Howard, Robert Palmer
Hubel, Kenneth Andrew
January, Lewis Edward
Kaelber, William Walbridge
Kasik, John Edward
Kerber, Richard E
Kieso, Robert Alfred
Kolder, Hansjoerg E
Kos, Clair Michael
Kunesh, Jerry Paul
Latourette, Howard Bennett
Levy, Leonard Alvin
Levy, Leonard Alvin
McCabe, Brian Francis
Marcus, Melvin L
Montgomery, Max Malcolm
Morgan, Donald Pryse
Morriss, Frank Howard, Jr
Moyers, Jack
Nauseef, William Michael
Niebyl, Jennifer Robinson
Peterson, Richard Elsworth
Ponseti, Ignacio Vives
Read, Charles H
Richerson, Hal Bates
Schedl, Harold Paul
Schrott, Helmut Gunther
Seebohm, Paul Minor
Sheets, Raymond Franklin
Smith, Ian Maclean
Solomons, Gerald
Soper, Robert Tunnicliff
Spector, Arthur Abraham
Strauss, John S
Strauss, Ronald George
Summers, Robert Wendell
Theilen, Ernest Otto
Thompson, Herbert Stanley
Thompson, Robert Gary
Tranel, Daniel T
Van Allen, Maurice Wright
Van Leeuwen, Gerard
Weinberger, Miles
Welsh, Michael James
Zellweger, Hans Ulrich
Ziegler, Ekhard E

KANSAS

Abdou, Nabih I
Allen, Max Scott
Bolinger, Robert E
Brown, Robert Wayne
Burket, George Edward
Calkins, William Graham
Chang, Chae Han Joseph
Chin, Hong Woo
Cho, Cheng T
Consigli, Richard Albert
Cook, James Dennis
Diederich, Dennis A
Dujovne, Carlos A
Dunn, Marvin I
Foster, David Bernard
Frenkel, Jacob Karl
Grantham, Jared James
Greenberger, Norton Jerald
Holder, Thomas M
Kennedy, James A
Kimler, Bruce Franklin
Kimmel, Joe Robert
Krantz, Kermit Edward
Lavia, Lynn Alan
Lemoine, Albert N, Jr
Liu, Chien
Mainster, Martin Aron
Manning, Robert Thomas
Meek, Joseph Chester, Jr
Pingleton, Susan Kasper
Price, James Gordon
Pugh, David Milton
Redford, John W B
Rhodes, James B
Rinsley, Donald Brendan
Roy, William R
Shaad, Dorothy Jean
Strafuss, Albert Charles
Templeton, Arch W
Tosh, Fred Eugene
Waxman, David
Wilson, Sloan Jacob

Ziegler, Dewey Kiper

KENTUCKY

Andersen, Roger Allen
Andrews, Billy Franklin
Aronoff, George R
Baumann, Robert Jay
Bosomworth, Peter Palliser
Burki, Nausherwan Khan
Carrasquer, Gaspar
Christopherson, William Martin
Clark, David Barrett
Cummings, Norman Allen
Digenis, George A
Espinosa, Enrique
Friedell, Gilbert H
Glenn, James Francis
Greene, John W, Jr
Haynes, Douglas Martin
Holland, Nancy H
James, Mary Frances
Keeney, Arthur Hail
Kraman, Steve Seth
Litvak, Austin S
McLeish, Kenneth R
McQuillen, Michael Paul
McRoberts, J William
Mandelstam, Paul
Marciniak, Ewa J
Markesbery, William Ray
Maruyama, Yosh
Mazzoleni, Alberto
Miller, Ralph English
Mullins, Richard James
Munson, Edwin Sterling
Noonan, Jacqueline Anne
Raff, Martin Jay
Redinger, Richard Norman
Rees, Earl Douglas
Rigor, Benjamin Morales, Sr
Rollo, Frank David
Rosenbaum, Harold Dennis
Schwab, John J
Shih, Wei Jen
Simmons, Guy Held, Jr
Steele, James Patrick
Steiner, Marion Rothberg
Straus, Robert
Tamburro, Carlo Horace
Thind, Gurdarshan S
Thompson, John S
Urbscheit, Nancy Lee
Weisskopf, Bernard
Whayne, Tom French
Williams, Walter Michael
Wolfe, Walter McIlhaney
Yaes, Robert Joel
Yam, Lung Tsiong
Zimmerman, Thom J

LOUISIANA

Andes, W(illard) Abe
Arrowsmith, William Rankin
Bellina, Joseph Henry
Berenson, Gerald Sanders
Bray, George A
Buechner, Howard Albert
Burch, Robert Emmett
Burch, Robert Ray
Clemetson, Charles Alan Blake
D'Alessandro-Bacigalupo, Antonio
Daniels, Robert (Sanford)
Davis, William Duncan, Jr
Dickey, Richard Palmer
Duncan, Margaret Caroline
Eichner, Edward Randolph
Eisenberg, Ronald Lee
Epstein, Arthur William
Ferriss, Gregory Stark
Frohlich, Edward David
Fruthaler, George James, Jr
Fulginiti, Vincent Anthony
Gagliano, Nicholas Charles
Gallant, Donald
Ganley, James Powell
George, Ronald Baylis
Giles, Thomas Davis
Glancy, David L
Gonzalez, Francisco Manuel
Gordon, Douglas Littleton
Gottlieb, Marise Suss
Gould, Harry J, III
Gum, Oren Berkley
Hackett, Earl R
Haik, George Michel
Hamrick, Joseph Thomas
Hastings, Robert Clyde
Heath, Robert Galbraith
Hejtmancik, Milton R
Hollis, Walter Jesse
Hyman, Edward Sidney
Hyslop, Newton Everett, Jr
Ivens, Mary Sue
Jung, Rodney Clifton
Kadan, Savitri Singh
Kastl, Peter Robert
Kaufman, Herbert Edward
Krahenbuhl, James Lee
Krupp, Iris M
Lang, Erich Karl
Lehrer, Samuel Bruce
Lindsey, Edward Stormont
Locke, William
Loewenstein, Joseph Edward

Lopez-Santolino, Alfredo
Lyons, George D
McHardy, George Gordon
McLaurin, James Walter
McMahon, Francis Gilbert
Marks, Charles
Meneely, George Rodney
Meyers, Philip Henry
Michalski, Joseph Potter
Mickal, Abe
Millikan, Larry Edward
Mizell, Merle
Mogabgab, William Joseph
Morgan, Lee Roy, Jr
Myers, Melvil Bertrand, Jr
Nice, Charles Monroe, Jr
Ochsner, Seymour Fiske
Osofsky, Howard J
Paddison, Richard Milton
Pankey, George Atkinson
Peacock, Milton O
Pearson, James Eldon
Phillips, John Hunter, Jr
Pou, Jack Wendell
Puyau, Francis A
Ray, Clarence Thorpe
Re, Richard N
Reed, Adrian Faragher
Reed, Richard Jay
Rigby, Perry G
Roberts, James Allen
Roheim, Paul Samuel
Rosenberg, Stuart A
Rosenthal, J William
Rothschild, Henry
Rutledge, Lewis James
Salvaggio, John Edmond
Sander, Gary Edward
Sappenfield, Robert W
Schimek, Robert Alfred
Schlegel, Jorgen Ulrik
Seltzer, Benjamin
Seto, Jane Mei-Chun
Shushan, Morris
Sodeman, William A
Stewart, William Huffman
Stuckey, Walter Jackson, Jr
Thornton, William Edgar
Threefoot, Sam Abraham
Trufant, Samuel Adams
Vanselow, Neal A
Vial, Lester Joseph, Jr
Wallin, John David
Walsh, John Joseph
Waring, William Winburn
Webster, Douglas B
Weed, John Conant
Weill, Hans
Weiss, Thomas E
White, Charles A, Jr
Wickstrom, Jack
Wilson, John T
Wolf, Robert E

MAINE

Barker, Jane Ellen
Collins, H Douglas
Cronkhite, Leonard Woolsey, Jr
Cross, Harold Dick
Doby, Tibor
Fellers, Francis Xavier
Gezon, Horace Martin
Hayes, Guy Scull
Korst, Donald Richardson
LaMarche, Paul H
McEwen, Currier
Phelps, Paulding
Rabe, Edward Frederick
Rand, Peter W
Welsch, Federico

MARYLAND

Abbott, Betty Jane
Abbrecht, Peter H
Adelstein, Robert Simon
Adkinson, Newton Franklin, Jr
Alexander, Duane Frederick
Allen, Willard M
Alling, David Wheelock
Alter, Harvey James
Amato, R Stephen S
Ambudkar, Indu S
Anderson, James Howard
Anderson, Lucy Macdonald
Andres, Reubin
Anhalt, Grant James
Asher, David Michael
Ashman, Michael Nathan
Aszalos, Adorjan
Aurbach, Gerald Donald
Baer, Harold
Baker, Timothy D
Barker, Lewellys Franklin
Barranger, John Arthur
Barry, Kevin Gerard
Beisel, William R
Bell, William Robert, Jr
Bennett, John E
Berendes, Heinz Werner
Bereston, Eugene Sydney
Bering, Edgar Andrew, Jr
Berkower, Ira
Berne, Bernard H
Berzofsky, Jay Arthur

Birx, Deborah L
Black, Emilie A
Blackman, Marc Roy
Blaese, R Michael
Blake, David Andrew
Blattner, William Albert
Bleecker, Eugene R
Blough, Herbert Allen
Bono, Vincent Horace, Jr
Borsos, Tibor
Bosma, James F
Brewer, H Bryan, Jr
Brown, Paul Wheeler
Brusilow, Saul W
Burg, Maurice B
Burnett, Joseph W
Burton, Benjamin Theodore
Bush, David E
Canfield, Craig Jennings
Carski, Theodore Robert
Catz, Charlotte Schifra
Cereghino, James Joseph
Chang, Kenneth Shueh-Shen
Chanock, Robert Merritt
Charache, Patricia
Charache, Samuel
Chase, Thomas Newell
Chaudhari, Anshumali
Chernoff, Amoz Immanuel
Cheson, Bruce David
Childs, Barton
Chisolm, James Julian, Jr
Choppin, Purnell Whittington
Clemmens, Raymond Leopold
Cogan, David Glendenning
Cohen, Sheldon Gilbert
Cohn, Robert
Conley, Carroll Lockard
Connor, Thomas Byrne
Corfman, Philip Albert
Cornblath, Marvin
Cotter, Edward F
Cowley, Benjamin D, Jr
Crout, John Richard
Crystal, Ronald George
Cummings, Martin Marc
Cummings, Nancy Boucot
Cutler, Gordon Butler, Jr
Cutler, Jeffrey Alan
Daniels, Worth B, Jr
Davidson, Jack Dougan
Davis, Leroy Thomas
De Carlo, John, Jr
Decker, John Laws
De Monasterio, Francisco M
Dennis, John Murray
Deslattes, Richard D, Jr
Dewys, William Dale
Dhindsa, Dharam Singh
Didisheim, Paul
Dipple, Anthony
Di Sant'Agnese, Paul Emilio Artom
Donner, Martin W
Dornette, William Henry Lueders
Dorst, John Phillips
Drachman, Daniel Bruce
Drucker, William Richard
Dubois, Andre T
Ducker, Thomas Barbee
Duncan, Katherine
Duncan, Leroy Edward, Jr
Durkan, James P
Edelman, Robert
Elin, Ronald John
Elisberg, Bennett La Dolce
Epstein, Stephen Edward
Evans, Charles Hawes
Evans, Virginia John
Faich, Gerald Alan
Fast, Patricia E
Fauci, Anthony S
Felsenthal, Gerald
Ferencz, Charlotte
Ferguson, James Joseph, Jr
Ferrans, Victor Joaquin
Finley, Joanne Elizabeth
Fischinger, Peter John
Fishman, Elliot Keith
Fleisher, Thomas A
Forbes, Allan Louis
Foster, Willis Roy
Fratantoni, Joseph Charles
Fraumeni, Joseph F, Jr
Fredrickson, Donald Sharp
Freeman, John Mark
Friedman, Lawrence
Frommer, Peter Leslie
Frost, John Kingsbury
Fuller, Richard Kenneth
Gajdusek, Daniel Carleton
Gallo-Torres, Hugo E
Genecin, Abraham
Gerber, Naomi Lynn Hurwitz
Gibson, Sam Thompson
Gill, John Russell, Jr
Giovacchini, Rubert Peter
Glick, Samuel Shipley
Gold, Philip William
Golden, Archie Sidney
Goldfine, Lewis John
Goldschmidt, Peter Graham
Goldstein, David Stanley
Goldstein, Robert Arnold
Golumbic, Norma

Goodman, Louis E
Gordis, Leon
Graff, Thomas D
Graham, George G
Granger, Donald Lee
Green, Jerome George
Greenwald, Peter
Gregerman, Robert Isaac
Greisman, Sheldon Edward
Grimley, Philip M
Groves, Eric Stedman
Guthrie, Eugene Harding
Haddy, Francis John
Hahn, Richard David
Hallett, Mark
Hamburger, Anne W
Hamill, Peter Van Vechten
Handler, Joseph S
Hardegree, Mary Carolyn
Harlan, William R, Jr
Harrell, George T
Harrison, Harold Edward
Harter, Donald Harry
Harvey, Abner McGehee
Heald, Felix Pierpont, Jr
Healy, Bernadine P
Heetderks, William John
Hegyeli, Ruth I E J
Heller, John Roderick
Helmsen, Ralph John
Helrich, Martin
Heman-Ackah, Samuel Monie
Henkin, Robert I
Heptinstall, Robert Hodgson
Heyssel, Robert M
Higgins, Millicent Williams Payne
Hill, Charles Earl
Hinman, Edward John
Hirshman, Carol A
Hoeg, Jeffrey Michael
Hoffman, Paul Ned
Holmes, Randall Kent
Holtzman, Neil Anton
Howie, Donald Lavern
Hoyer, Leon William
Hume, John Chandler
Hunter, Oscar Benwood, Jr
Hutchins, Grover MacGregor
Hyde, Henry Van Zile
Ihde, Daniel Carlyle
Iype, Pullolickal Thomas
Jacobs, Edwin M
Jacox, Ada K
Jaffe, Elaine Sarkin
Jewett, Hugh Judge
Johns, Richard James
Johnson, Richard T
Johnston, Gerald Samuel
Jordan, William Stone, Jr
Joy, Robert John Thomas
Kaltreider, D Frank
Kaplan, Richard Stephen
Kaslow, Richard Alan
Keiser, Harry R
Kennedy, Thomas James, Jr
Kessler, Irving Isar
Kickler, Thomas Steven
King, Theodore Matthew
Klee, Claude Blenc
Klee, Gerald D'Arcy
Kline, Ira
Klinefelter, Harry Fitch
Knazek, Richard Allan
Knox, David Lalonde
Knox, Gaylord Shearer
Kopin, Irwin J
Kowalewski, Edward Joseph
Kowarski, A Avinoam
Kramer, Barnett Sheldon
Kramer, Norman Clifford
Krause, Richard Michael
Kuff, Edward Louis
Kupfer, Carl
Kwiterovich, Peter O, Jr
Lake, Charles Raymond
Landon, John Campbell
Latham, Patricia Suzanne
Law, Lloyd William
Lawrence, Dale Nolan
Lederer, William Jonathan
Legters, Llewellyn J
Leonard, Edward Joseph
Leonard, James Joseph
LeRoith, Derek
Leventhal, Brigid Gray
Leventhal, Carl M
Levine, Arthur Samuel
Levine, David Morris
Levine, Paul Howard
Liang, Isabella Y S
Lichtenstein, Lawrence M
Lipicky, Raymond John
Littlefield, John Walley
Lockshin, Michael Dan
Longo, Dan L
Loriaux, D Lynn
Lowe, Charles Upton
McCune, Susan K
McHugh, Paul Rodney
McIndoe, Darrell W
McIntyre, Patricia Ann
McKenna, Thomas Michael
McKhann, Guy Mead
McKusick, Victor Almon

Magrath, Ian
Maher, John Francis
Malech, Harry Lewis
Marban, Eduardo
Mardiney, Michael Ralph, Jr
Margolis, Simeon
Martin, Edgar J
Marzella, Louis
Matanoski, Genevieve M
Mayer, Richard F
Menkes, Harold A
Meredith, Orsell Montgomery
Merril, Carl R
Metcalfe, Dean Darrel
Mickel, Hubert Sheldon
Migeon, Claude Jean
Miller, Laurence Herbert
Miller, Louis Howard
Miller, Neil Richard
Miller, Robert Warwick
Mills, James Louis
Mitchell, Thomas George
Mohler, William C
Monath, Thomas P
Mond, James Jacob
Monroe, Russell Ronald
Moss, Joel
Murphy, Patrick Aidan
Murphy, Robert Patrick
Mushinski, J Frederic
Myers, Charles
Nager, George Theodore
Naunton, Ralph Frederick
Nelson, Karin Becker
Neumann, Ronald D
Neva, Franklin Allen
Newcombe, David S
Niedermeyer, Ernst F
Nisula, Bruce Carl
Oldfield, Edward Hudson
O'Malley, Joseph Paul
Order, Stanley Elias
Orloff, Jack
Oski, Frank
Ottesen, Eric Albert
Owens, Albert Henry, Jr
Packard, Barbara B K
Page, Lot Bates
Paige, David M
Palestine, Alan G
Pamnani, Motilal Bhagwandas
Park, Lee Crandall
Parker, Margaret Maier
Parker, Robert Tarbert
Parkman, Paul Douglas
Patz, Arnall
Pearson, John William
Peck, Carl Curtis
Permutt, Solbert
Pierce, Nathaniel Field
Pilkerton, A Raymond
Pinter, Gabriel George
Pitcairn, Donald M
Platt, William Rady
Plotnick, Gary David
Plotz, Paul Hunter
Poirier, Miriam Christine Mohrhoff
Polinsky, Ronald John
Pollard, Thomas Dean
Poskanzer, David Charles
Post, Robert M
Proctor, Donald Frederick
Prout, Thaddeus Edmund
Quinnan, Gerald Vincent, Jr
Rapoport, Stanley I
Raskin, Joan
Reichelderfer, Thomas Elmer
Reid, Clarice D
Reinig, James William
Reuber, Melvin D
Rheinstein, Peter Howard
Richards, Richard Davison
Richardson, Edward Henderson, Jr
Robbins, Jay Howard
Rodbard, David
Rodrigues, Merlyn M
Rogawski, Michael Andrew
Rogers, Mark Charles
Romansky, Monroe James
Rose, John Charles
Rosen, Saul W
Ross, Richard Starr
Roth, Harold Philmore
Russell, Philip King
Ruzicka, Francis Frederick, Jr
Saba, George Peter, II
Sabol, Steven Layne
Sack, George H(enry), Jr
Sack, Richard Bradley
Salans, Lester Barry
Salik, Julian Oswald
Samelson, Lawrence Elliot
Santos, George Wesley
Sapir, Daniel Gustave
Saudek, Christopher D
Scheibel, Leonard William
Schoenberg, Bruce Stuart
Schoolman, Harold M
Schulman, Joseph Daniel
Schwartz, John T
Seidel, Henry Murray
Seltser, Raymond
Sharma, Gopal Chandra
Sheridan, Philip Henry

Sherman, Kenneth Eliot
Shimizu, Hiroshi
Shnider, Bruce I
Shulman, Lawrence Edward
Sidbury, James Buren, Jr
Sidell, Frederick R
Siegel, Jay Philip
Silverman, Robert Eliot
Simpson, David Gordon
Singewald, Martin Louis
Sinks, Lucius Frederick
Smith, Mark Stephen
Smith, Phillip Doyle
Solez, Kim
Solomon, Howard Fred
Sommer, Alfred
Song, Byoung-Joon
Soulen, Renate Leroi
Spivak, Jerry Lepow
Sporn, Michael Benjamin
Spragins, Melchijah
Stanton, Mearl Fredrick
Starfield, Barbara Helen
Steinberg, Alfred David
Sternberger, Ludwig Amadeus
Stetten, DeWitt, Jr
Stevens, Harold
Stolley, Paul David
Strickland, George Thomas
Stromberg, Kurt
Summers, Richard James
Tabor, Edward
Talbot, John Mayo
Taube, Sheila Efron
Taylor, Carl Ernest
Taylor, David Neely
Thomas, George Howard
Triantaphyllopoulos, Demetrios
Triantaphyllopoulos, Eugenie
Tschudy, Donald P
Tumulty, Philip A
Turkeltaub, Paul Charles
Vaitukaitis, Judith L
Valentine, Martin Douglas
Van Echo, David Andrew
Van Metre, Thomas Earle, Jr
Veech, Richard L
Vermund, Sten Halvor
Vorosmarti, James, Jr
Vydelingum, Nadarajen Ameerdanaden
Wagley, Philip Franklin
Wagner, Henry N, Jr
Waldmann, Thomas A
Waldrop, Francis N
Walker, Wilbur Gordon
Wallace, Craig Kesting
Wallach, Edward E
Walser, Mackenzie
Watkins, Paul Allan
Watson, James Dewey
Webster, Henry deForest
Weiger, Robert W
Weight, Forrest F
Weintraub, Bruce Dale
Weisburger, Elizabeth Kreiser
Weisfeldt, Myron Lee
Westphal, Heiner J
Whelton, Andrew
White, John David
Whitehorn, William Victor
Whitley, Joseph Efird
Whitley, Nancy O'Neil
Whitney, Robert Arthur, Jr
Williamson, Charles Elvin
Wolff, Frederick William
Woodruff, James Donald
Woodward, Theodore Englar
Work, Henry Harcus
Yaffe, Sumner J
Yarchoan, Robert
Yerg, Raymond A
Yin, Frank Chi-Pong
Young, Eric D
Yuspa, Stuart Howard
Zee, David Samuel
Zeligman, Israel
Zierler, Kenneth
Zimmerman, Hyman Joseph
Zinkham, William Howard
Zukel, William John

MASSACHUSETTS
Abelmann, Walter H
Agnello, Vincent
Ahmed, A Razzaque
Aisenberg, Alan Clifford
Alper, Chester Allan
Alper, Milton H
Altschule, Mark David
Anast, Constantine Spiro
Anderson, Thomas Page
Andres, Giuseppe A
Arias, Irwin Monroe
Arnaout, M Amin
Aroesty, Julian Max
Arvan, Peter
Atkins, Elisha
Ausiello, Dennis Arthur
Austen, K(arl) Frank
Avery, Mary Ellen
Avruch, Joseph
Axelrod, Lloyd
Baden, Howard Philip
Baratz, Robert Sears

Barger, A(braham) Clifford
Barlow, Charles F
Barnett, Guy Octo
Barry, Joan
Baum, Jules Leonard
Baumgartner, Leona
Beck, William Samson
Beitins, Inese Zinta
Bennison, Bertrand Earl
Benson, Herbert
Berenberg, William
Bernfield, Merton Ronald
Bernhard, Jeffrey David
Bistrian, Bruce Ryan
Black, Paul H
Blau, Monte
Bloomberg, Wilfred
Boyer, Markley Holmes
Braunwald, Eugene
Brenner, Barry Morton
Brill, A Bertrand
Broder, Martin Ivan
Brody, Jerome Saul
Brown, Robert Stephen
Bucher, Nancy L R
Bunn, Howard Franklin
Burwell, E Langdon
Canellos, George P
Caplan, Louis Robert
Carey, Martin Conrad
Carvalho, Angelina C A
Cassidy, Carl Eugene
Castle, William Bosworth
Caviness, Verne Strudwick, Jr
Chalmers, Thomas Clark
Chang, Te Wen
Charney, Evan
Chasin, Werner David
Chin, William W
Chivian, Eric Seth
Chobanian, Aram V
Chodosh, Sanford
Christlieb, Albert Richard
Clark, Sam Lillard, Jr
Coffman, Jay D
Cohen, Alan Seymour
Cone, Thomas E, Jr
Costanza, Mary E
Cotran, Ramzi S
Craven, Donald Edward
Crigler, John F, Jr
Crocker, Allen Carrol
Crowley, William Francis, Jr
Dammin, Gustave John
David, John R
Davidson, Charles Sprecher
Dawber, Thomas Royle
Dealy, James Bond, Jr
Delabarre, Everett Merrill, Jr
Dershwitz, Mark
Desforges, Jane Fay
Dickson, James Francis, III
Dohlman, Claes Henrik
Douglas, Pamela Susan
Drachman, David A
Eastwood, Gregory Lindsay
Ebert, Robert H
Ehlers, Mario Ralf Werner
Elkinton, J(oseph) Russell
Epstein, David L(ee)
Epstein, Franklin Harold
Federman, Daniel D
Field, James Bernard
Fine, Samuel
Fischbach, Gerald David
Fitzpatrick, Thomas Bernard
Fiumara, Nicholas J
Fox, Irving Harvey
Fox, James Gahan
Frazier, Howard Stanley
Freed, Murray Monroe
Freedberg, Abraham Stone
Frei, Emil, III
Friedman, Emanuel A
Friedman, Ephraim
Fulmer, Hugh Scott
Galaburda, Albert Mark
Garrard, Sterling Davis
Geha, Raif S
Gellis, Sydney Saul
Gerald, Park S
Gerety, Robert John
Gissen, Aaron J
Godine, John Elliott
Goldberg, Irvin H(yman)
Gorbach, Sherwood Leslie
Grace, Norman David
Grand, Richard Joseph
Grant, Walter Morton
Greenblatt, David J
Griffith, Robert W
Gross, Jerome
Grossman, William
Guinane, James Edward
Guralnick, Walter
Haber, Edgar
Hales, Charles A
Hallock, Gilbert Vinton
Halpern, Daniel
Handin, Robert I
Hanshaw, James Barry
Haudenschild, Christian C
Hawiger, Jack Jacek
Hedley-Whyte, John

Medicine (cont)

Heinrich, Gerhard
Herrmann, John Bellows
Heyda, Donald William
Hiatt, Howard Haym
Hiregowdara, Dananagoud
Hnatowich, Donald John
Hollander, William
Holman, B Leonard
Homburger, Freddy
Ingram, Roland Harrison
Irwin, Richard Stephen
Ishikawa, Sadamu
Isselbacher, Kurt Julius
Jameson, James Larry
Jandl, James Harriman
Johnson, Elsie Ernest
Kaminskas, Edvardas
Kannel, William B
Kaplan, Melvin Hyman
Kass, Edward Harold
Kassirer, Jerome Paul
Kazemi, Homayoun
Kelly, Dorothy Helen
Kelton, Diane Elizabeth
Kessner, David Morton
Keusch, Gerald Tilden
Kieff, Elliott Dan
Kim, Agnes Kyung-Hee
Kinsbourne, Marcel
Kitz, Richard J
Klempner, Mark Steven
Kliman, Allan
Knapp, Peter Hobart
Koch-Weser, Dieter
Koff, Raymond Steven
Kolodny, Gerald Mordecai
Kosasky, Harold Jack
Kramer, Philip
Krane, Stephen Martin
Kritzman, Julius
Kupchik, Herbert Z
Kupferman, Allan
Kwan, Paul Wing-Ling
Kyriakis, John M
Lamb, George Alexander
Lamont, John Thomas
Landowne, Milton
Laster, Leonard
Laurenzi, Gustave
Lawrence, Robert Swan
Leach, Robert Ellis
Leaf, Alexander
Lebowitz, Elliot
Leder, Philip
Lees, Robert S
Lele, Padmakar Pratap
Lenke, Roger Rand
Lessell, Simmons
Levine, Herbert Jerome
Levinsky, Norman George
Levinson, Gilbert E
Leviton, Alan
Levy, Harvey Louis
Li, James C C
Lin, Chi-Wei
Lipsitt, Don Richard
Livingston, David M
Lo, Clifford W
Lodwick, Gwilym Savage
Loewenstein, Matthew Samuel
London, Irving Myer
Loscalzo, Joseph
Lou, Peter Louis
Lowenstein, Edward
Lown, Bernard
McCabe, William R
McCaughan, Donald
McNally, Elizabeth Mary
McNeil, Barbara Joyce
Madias, Nicolaos E
Madoc-Jones, Hywel
Madoff, Morton A
Maloof, Farahe
Mann, James
Marble, Alexander
Marchant, Douglas J
Marcus, Elliot M
Mark, Lester Charles
Marks, Leon Joseph
Mathews-Roth, Micheline Mary
Medearis, Donald N, Jr
Melby, James Christian
Mellins, Harry Zachary
Mendelson, Jack H
Merrill, John Putnam
Mihm, Martin C, Jr
Mitus, Wladyslaw J
Moloney, William Curry
Monette, Francis C
Montgomery, William Wayne
Moolten, Frederick London
Moore-Ede, Martin C
Mordes, John Peter
Morgan, James Philip
Nadas, Alexander Sandor
Nadol, Bronislaw Joseph, Jr
Najjar, Victor Assad
Nandy, Kalidas
Navab, Farhad
Neer, Robert M
Nichols, George, Jr
Page, Irvine Heinly

Pathak, Madhukar
Patterson, William Bradford
Pauker, Stephen Gary
Paul, Oglesby
Paul, Robert E, Jr
Pechet, Liberto
Pfister, Richard Charles
Pino, Richard M
Plaut, Andrew George
Pochi, Peter E
Pontoppidan, Henning
Potts, John Thomas, Jr
Prout, Curtis
Prout, George Russell, Jr
Raam, Shanthi
Rabkin, Mitchell T
Rankin, Joel Sender
Ransil, Bernard J(erome)
Raviola d'Elia, Giuseppina E(nrica)
Raymond, Samuel
Reichlin, Seymour
Relman, Arnold Seymour
Rencricca, Nicholas John
Reynolds, Robert N
Richardson, George S
Richman, Justin Lewis
Richmond, Julius Benjamin
Riley, Richard Lord
Rock, Paul Bernard
Rosenberg, Irwin Harold
Rosenberg, Isadore Nathan
Rosenberg, Robert D
Rothmeier, Jeffrey
Roubenoff, Ronenn
Russell, Robert M
Ryan, John F
Sabin, Thomas Daniel
Sacks, David B
Salhanick, Hilton Aaron
Sandson, John Ivan
Sawin, Clark Timothy
Schimmel, Elihu Myron
Schlossman, Stuart Franklin
Schmidt, Kurt F
Schuknecht, Harold Frederick
Schur, Peter Henry
Schwartz, Bernard
Schwartz, Edith Richmond
Schwartz, Robert Stewart
Schwartz, William Benjamin
Segal, Rosalind A
Segarra, Joseph M
Senior, Boris
Shapiro, Howard Maurice
Shear, Leroy
Sheffer, Albert L
Shih, Vivian Ean
Sifneos, Peter E
Silva, Patricio
Slack, Warner Vincent
Smith, Thomas W
Snider, Gordon Lloyd
Snyder, Louis Michael
Sobel, Edna H
Spodick, David Howard
Stanbury, John Bruton
Steel, R Knight
Stoeckle, John Duane
Stolbach, Leo Lucien
Stollerman, Gene Howard
Stone, Peter H
Strom, Terry Barton
Strong, Mervyn Stuart
Suit, Herman Day
Swartz, Morton N
Talbot, Nathan Bill
Tarlov, Alvin Richard
Tauber, Alfred Imre
Taveras, Juan M
Taylor, Isaac Montrose
Thier, Samuel Osiah
Thomas, Clayton Lay
Thorn, George W
Tishler, Peter Verveer
Toomey, James Michael
Treves, S T
Trier, Jerry Steven
Twitchell, Thomas Evans
Tyler, H Richard
Tyler, Kenneth Laurence
Udall, John Nicholas, Jr
Ulfelder, Howard
Vandam, Leroy David
VanPraagh, Richard
Volicer, Ladislav
Von Essen, Carl Francois
Wacker, Warren Ernest Clyde
Wang, Chiu-Chen
Watkins, Elton, Jr
Waud, Barbara E
Wegman, David Howe
Weinberger, Steven Elliott
Weinstein, Louis
Weinstein, Robert
Weintraub, Lewis Robert
Welch, Gary William
Weller, Thomas Huckle
Wilgram, George Friederich
Wingate, Martin Bernard
Witherell, Egilda DeAmicis
Wodinsky, Isidore
Wolf, Gerald Lee
Wolff, Peter Hartwig

Wolff, Sheldon Malcolm
Wyman, Stanley M
Young, Robert Rice
Yunis, Edmond J
Zamecnik, Paul Charles

MICHIGAN

Al-Sarraf, Muhyi
Anderson, David G
Anderson, Marvin David
Ansbacher, Rudi
Axelrod, Arnold Raymond
Bacon, George Edgar
Bagchi, Mihir
Barr, Mason, Jr
Baublis, Joseph V
Bauer, Jere Marklee
Behrman, Samuel J
Beierwaites, William Henry
Bender, Leonard Franklin
Bissell, Grosvenor Willse
Block, Duane Llewellyn
Bole, Giles G
Brennan, Michael James
Buchanan, Robert Alexander
Burdi, Alphonse R
Caldwell, John R
Castor, Cecil William
Cerny, Joseph Charles
Chou, Ching-Chung
Christenson, Paul John
Clapper, Muir
Cohen, Flossie
Corbett, Thomas Hughes
Cuatrecasas, Pedro
Curtis, George Clifton
Davenport, Fred M
Davenport, Horace Willard
DeJong, Russell Nelson
Dekornfeld, Thomas John
DeMuth, George Richard
Diaz, Fernando G
Dick, Macdonald, II
Dilts, Preston Vine, Jr
Diokno, Ananias Cornejo
Duff, Ivan Francis
Ellis, C N
Enzer, Norbert Beverley
Fajans, Stefan Stanislaus
Fekety, F Robert, Jr
Fernandez-Madrid, Felix
Fitzgerald, Robert Hannon, Jr
Floyd, John Claiborne, Jr
Frank, Donald Joseph
Frank, Robert Neil
Freimanis, Atis K
Gabrielsen, Trygve O
Gelehrter, Thomas David
Gilman, Sid
Gilroy, John
Goldstein, Sidney
Goodman, Jay Irwin
Gosling, John Roderick Gwynne
Gossain, Ved Vyas
Green, Robert A
Gross, George Alvin
Grossman, Herbert Jules
Hashimoto, Ken
Heifetz, Carl Louis
Henley, Keith Stuart
Hodges, Robert Manley
Hoffert, Jack Russell
Holt, John Floyd
Horwitz, Jerome Philip
Howatt, William Frederick
Hsu, Chen-Hsing
Hubbard, William Neill, Jr
Huseby, Robert Arthur
Jacobson, Arnold P
Jampel, Robert Steven
Johnson, Joseph Eggleston, III
Johnson, Tom Milroy
Joseph, Ramon R
Julius, Stevo
Kaplan, Joseph
Kauffman, Carol A
King, Charles Miller
Kithier, Karel
Knopf, Ralph Fred
Kuhl, David Edmund
Lapides, Jack
Lentz, Paul Jackson, Jr
Lerner, Albert Martin
Lisak, Robert Philip
Livingood, Clarence Swinehart
Lovell, Robert Gibson
Luk, Gordon David
McCormick, John
McGrath, Charles Morris
Mack, Robert Emmet
McLean, James Amos
Mader, Ivan John
Magen, Myron S
Mahajan, Sudesh K
Maher, Veronica Mary
Martel, William
Matovinovic, Josip
Mattson, Joan C
Michelakis, Andrew M
Mikkelsen, William Mitchell
Miller, Orlando Jack
Millikan, Clark Harold
Moghissi, Kamran S
Mohrland, J Scott

Morley, George W
Murray, Raymond Harold
Nyboer, Jan
Oberman, Harold A
Oliver, William J
Osborn, June Elaine
Ownby, Dennis Randall
Palmer, Kenneth Charles
Patterson, Maria Jevitz
Perrin, Eugene Victor
Perry, Harold
Pinkus, Hermann (Karl Benno)
Pitt, Bertram
Pomerleau, Ovide F
Potchen, E James
Poulik, Miroslav Dave
Powsner, Edward R
Prasad, Ananda S
Puro, Donald George
Pysh, Joseph John
Rapp, Robert
Rebuck, John Walter
Redding, Foster Kinyon
Reed, Melvin LeRoy
Rhodes, Robert Shaw
Riley, Michael Verity
Rogers, William Leslie
Rovner, David Richard
Ruddon, Raymond Walter, Jr
Sander, C Maureen
Schteingart, David E
Schwartz, Stanley Allen
Senagore, Anthony J
Shafer, A William
Shanberge, Jacob N
Sherman, Alfred Isaac
Siegel, George Jacob
Simon, Michael Richard
Smith, Edwin Mark
Sokol, Robert James
Sorscher, Alan J
Sparks, Robert D
Spink, Gordon Clayton
Spitz, Werner Uri
Stein, Paul David
Stern, Aaron Milton
Strang, Ruth Hancock
Suhrland, Leif George
Sullivan, Donita B
Swisher, Scott Neil
Tannen, Richard L
Teasdall, Robert Douglas
Thoene, Jess Gilbert
Thompson, George Richard
Thornbury, John R
Tishkoff, Garson Harold
Triebwasser, John
Vaitkevicius, Vainutis K
Vaughn, Clarence Benjamin
Vavra, James Joseph
Voorhees, John James
Waggoner, Raymond Walter
Ward, Robert C
Weber, Wendell W
Weg, John Gerard
Wegman, Myron Ezra
Weil, William B, Jr
Weiss, Joseph Jacob
Weller, John Martin
Welsch, Clifford William, Jr
Whitehouse, Walter MacIntire, Sr
Wiley, John W
Wilhelm, Rudolf Ernst
Willis, Park Weed, III
Wollschlaeger, Gertraud
Wolter, J Reimer
Yang, Gene Ching-Hua
Zarafonetis, Chris John Dimiter
Zweifler, Andrew J

MINNESOTA

Amplatz, Kurt
Andersen, Howard Arne
Ansari, Azam U
Awad, Essam A
Bacaner, Marvin Bernard
Baldus, William Phillip
Balfour, Henry H, Jr
Bartholomew, Lloyd Gibson
Bayrd, Edwin Dorrance
Berge, Kenneth G
Berman, Reuben
Bieber, Irving
Bisel, Harry Ferree
Bowie, Edward John Walter
Brunning, Richard Dale
Buckley, Joseph J
Burchell, Howard Bertram
Burke, Edmund C
Cain, James Clarence
Carryer, Haddon McCutchen
Carter, Robert Eldred
Cervenka, Jaroslav
Ciriacy, Edward W
Cohn, Jay Norman
Connelly, Donald Patrick
Corbin, Kendall Brooks
Crowley, Leonard Vincent
Dai, Xue Zheng (Charlie)
Dalmasso, Agustin Pascual
Daugherty, Guy Wilson
Decker, David Garrison
Dewald, Gordon Wayne
Doe, Richard P

Dousa, Thomas Patrick
Du Shane, James William
Duvall, Arndt John, III
Eaton, John Wallace
Ebner, Timothy John
Ellwood, Paul M, Jr
Ettinger, Milton G
Etzwiler, Donnell D
Fairbairn, John F, II
Fairbanks, Virgil
Ferris, Thomas Francis
Fisch, Robert O
Fraley, Elwin E
Francis, Gary Stuart
Fuller, Benjamin Franklin, Jr
Gastineau, Clifford Felix
Gault, N(eal) L, Jr
Gebhard, Roger Lee
Geraci, Joseph E
Gerding, Dale Nicholas
Gilbertsen, Victor Adolph
Gleich, Gerald J
Gobel, Frederick L
Goetz, Frederick Charles
Goldstein, Norman Philip
Gorman, Colum A
Gray, Joel Edward
Gross, John Burgess
Haase, Ashley Thomson
Hagedorn, Albert Berner
Hall, Wendell Howard
Hanson, Russell Floyd
Harris, Jean Louise
Hebbel, Robert P
Henderson, John Warren
Hermans, Paul E
Hill, John Roger
Hollenhorst, Robert William
Holtzman, Jordan L
Housmans, Philippe Robert H P
Jacob, Harry S
Jenkins, Robert Brian
Jepson, William W
Jessen, Carl Roger
Johnson, Joseph Richard
Juergens, John Louis
Kaplan, Manuel E
Kearns, Thomas P
Kennedy, Byrl James
Kennedy, William Robert
Kiang, David Teh-Ming
King, Richard Allen
Kottke, Bruce Allen
Kottke, Frederic James
Krivit, William
Kurland, Leonard T
Kyle, Robert Arthur
LeBien, Tucker W
Levitan, Alexander Allen
Levitt, Michael D
Levitt, Seymour H
Logan, George Bryan
Loken, Merle Kenneth
Lucas, Alexander Ralph
Lucas, Russell Vail, Jr
Luthra, Harvinder Singh
Lynch, Peter John
McCollister, Robert John
McConahey, William McConnell, Jr
McCullough, John Jeffrey
McGill, Douglas B
McMahon, M Molly
McPherson, Thomas C(oatsworth)
Malkasian, George D, Jr
Martin, Gordon Mather
Martin, Harold Roland
Michael, Alfred Frederick, Jr
Michels, Lester David
Millhouse, Oliver Eugene
Milliner, Eric Killmon
Moertel, Charles George
Morlock, Carl G
Mulder, Donald William
Mulhausen, Robert Oscar
Murray, M(urray) John
Nelson, John Daniel
Nichols, Donald Richardson
Nuttall, Frank Q
Obrien, Peter Charles
Olsen, Arthur Martin
Oppenheimer, Jack Hans
Page, Arthur R
Paller, Mark Stephen
Palumbo, Pasquale John
Perry, Harold Otto
Peterson, Harold Oscar
Phillips, Sidney Frederick
Pierre, Robert V
Polzin, David J
Popkin, Michael Kenneth
Prem, Konald Arthur
Quie, Paul Gerhardt
Rahman, Yueh Erh
Randall, Raymond Victor
Reed, Charles E
Rehder, Kai
Reitemeier, Richard Joseph
Riggs, Byron Lawrence
Rohlfing, Stephen Roy
Romero, Juan Carlos
Rosenberg, Murray David
Rovelstad, Randolph Andrew
Sabath, Leon David
Schultz, Alvin Leroy

Schuman, Leonard Michael
Shorter, Roy Gerrard
Siekert, Robert George
Simon, Geza
Smith, Lucian Anderson
Spink, Wesley William
Spittell, John A, Jr
Sprague, Randall George
Stickler, Gunnar B
Stillwell, George Keith
Strong, Cameron Gordon
Theologides, Athanasios
Tobian, Louis
Torres, Fernando
Tuna, Naip
Ulstrom, Robert
Van Bergen, Frederick Hall
Vennes, Jack A
Verby, John E
Vernier, Robert L
Wang, Yang
Warwick, Warren J
Webster, David Dyer
Weinshilboum, Richard Merle
Weir, Edward Kenneth
Weissler, Arnold M
Westmoreland, Barbara Fenn
Whisnant, Jack Page
Wild, John Julian
Winchell, C Paul
Winkelmann, Richard Knisely
Wirtschafter, Jonathan Dine
Wood, Michael Bruce
Zanjani, Esmail Dabaghchian
Zieve, Leslie

MISSISSIPPI

Batson, Blair Everett
Batson, Margaret Bailly
Bell, Warren Napier
Blake, Thomas Mathews
Brooks, Thomas Joseph, Jr
Bruce, David Lionel
Corbett, James John
Currier, Robert David
Evers, Carl Gustav
Grenfell, Raymond Frederic
Haerer, Armin Friedrich
Hellems, Harper Keith
Hogue, Raymond Ellsworth
Hutchison, William Forrest
Jackson, John Fenwick
Johnson, Ben Butler
Johnson, Samuel Britton
Kiley, John Edmund
Langford, Herbert Gaines
Lockwood, William Rutledge
Long, Billy Wayne
Montani, Jean-Pierre
Morrison, John Coulter
Steinberg, Martin H
Tavassoli, Mehdi
Uzodinma, John E
Watson, David Goulding
Wiser, Winfred Lavern

MISSOURI

Alpers, David Hershel
Anderson, Philip Carr
Atkinson, John Patterson
Avioli, Louis
Bajaj, S Paul
Barbero, Giulio J
Bauer, John Harry
Bergmann, Steven R
Bier, Dennis Martin
Black, Samuel P W
Boyarsky, Saul
Braciale, Thomas Joseph, Jr
Brodeur, Armand Edward
Brown, Elmer Burrell
Burke, William Joseph
Burns, Thomas Wade
Chaplin, Hugh, Jr
Chapman, Ramona Marie
Chu, Jen-Yih
Cole, Barbara Ruth
Colten, Harvey Radin
Colwill, Jack M
Crosby, William Holmes, Jr
Cryer, Philip Eugene
Daly, James William
Danforth, William H
Danis, Peter Godfrey
Daughaday, William Hamilton
Davis, Bernard Booth
Diehl, Antoni Mills
Dimond, Edmunds Grey
Dodge, Philip Rogers
Donati, Robert M
Dueker, David Kenneth
Eggers, George W Nordholtz, Jr
Eliasson, Sven Gustav
Ericson, Avis J
Fabian, Leonard William
Feeney-Burns, Mary Lynette
Fernandez-Pol, Jose Alberto
Fitch, Coy Dean
Fletcher, Anthony Phillips
Fletcher, James W
Forker, E Lee
Frawley, Thomas Francis
Fredrickson, John Murray

Freeman, Arnold I
Gallagher, Neil Ignatius
Gantner, George E, Jr
Garner, Harold E
Gilula, Louis Arnold
Givler, Robert L
Graham, Robert
Griffin, William Thomas
Griggs, Douglas M, Jr
Grunt, Jerome Alvin
Guze, Samuel Barry
Hall, David Goodsell, III
Hall, Lynn Raymond
Harford, Carl Gayler
Harvey, Joseph Eldon
Hellerstein, Stanley
Herzig, Geoffrey Peter
Hillman, Richard Ephraim
Hodges, Glenn R(oss)
Horenstein, Simon
Horwitz, Edwin M
Hyers, Thomas Morgan
Ide, Carl Heinz
Karl, Michael M
Keenan, William Jerome
Keown, Kenneth K
King, Morris Kenton
Kinsella, Ralph A, Jr
Kipnis, David Morris
Kissane, John M
Klahr, Saulo
Knight, William Allen, Jr
Kornfeld, Stuart Arthur
Kovacs, Sandor J, Jr
Krogstad, Donald John
Kulczycki, Anthony, Jr
Lake, Lorraine Frances
Landau, William M
Lennartz, Michelle R
Loeb, Virgil, Jr
Lonigro, Andrew Joseph
Majerus, Philip W
Martin, Richard Harvey
Massie, Edward
Masters, William Howell
Melick, William F
Middelkamp, John Neal
Mudd, J Gerard
Nahm, Moon H
Noback, Richardson K
Nuetzel, John Arlington
Ogura, Joseph H
Oliver, G Charles
Olson, Lloyd Clarence
Ortwerth, Beryl John
Parker, Brent M
Parker, Charles W
Parker, Mary Langston
Pearlman, Alan L
Perez, Carlos A
Perkoff, Gerald Thomas
Perry, Horace Mitchell, Jr
Pestronk, Alan
Pierce, John Albert
Plapp, Frederick Vaughn
Powers, William John
Purkerson, Mabel Louise
Raichle, Marcus Edward
Reinhard, Edward Humphrey
Ridings, Gus Ray
Ritter, Hubert August
Royal, Henry Duval
Santiago, Julio Victor
Schonfeld, Gustav
Schultz, Irwin
Schweiss, John Francis
Senturia, Ben Harlan
Shank, Robert Ely
Shuter, Eli Ronald
Siegel, Barry Alan
Sirridge, Marjorie Spurrier
Slatopolsky, Eduardo
Slavin, Raymond Granam
Soule, Samuel David
Stephen, Charles Ronald
Stoneman, William, III
Stroud, Malcolm Herbert
Thoma, George Edward
Twardowski, Zbylut Jozef
Vietti, Teresa Jane
Welch, Michael John
Weldon, Virginia V
Wenner, Herbert Allan
Whyte, Michael Peter
Wilkinson, Charles Brock
Woodruff, Calvin Watts
Woolsey, Robert M
Yarbro, John Williamson

MONTANA

Boggs, Dane Ruffner
Opitz, John Marius
Priest, Jean Lane Hirsch
Swanson, John L
Williams, Roger Stewart

NEBRASKA

Aita, John Andrew
Andrews, Charles Edward
Angle, William Dodge
Baker, Robert Norton
Booth, Richard W
Brody, Alfred Walter
Cassidy, James T

Clifford, George O
Connolly, John Francis
Cromwell, Norman Henry
Dalrymple, Glenn Vogt
Davis, Richard Bradley
Engel, Toby Ross
Foley, John F
Fuenning, Samuel Isaiah
Fusaro, Ramon Michael
Gifford, Harold
Grissom, Robert Leslie
Heaney, Robert Proulx
Hunt, Howard Beeman
Kass, Irving
Keim, Lon William
Kessinger, Margaret Anne
Klassen, Lynell W
Kobayashi, Roger Hideo
Lehnhoff, Henry John, Jr
Luby, Robert James
Lynch, Henry T
McIntire, Matilda S
McWhorter, Clarence Austin
Malashock, Edward Marvin
Mirvish, Sidney Solomon
Mohiuddin, Syed M
Mooring, Paul K
Nagel, Donald Lewis
Nair, Chandra Kunju
O'Brien, Richard Lee
Paustian, Frederick Franz
Pearson, Paul (Hammond)
Sanders, W Eugene, Jr
Sethi, V Sagar
Skultety, Francis Miles
Sorrell, Michael Floyd
Sullivan, James F
Thomas, John Martin
Tobin, Richard Bruce
Toth, Bela
Waggener, Ronald E
Ware, Frederick
Yonkers, Anthony J

NEVADA

Buxton, Iain Laurie Offord
Dehné, Edward James
Ferguson, Roger K
Manalo, Pacita
Pierson, William R
Pool, Peter Edward
Whipple, Gerald Howard

NEW HAMPSHIRE

Almy, Thomas Pattison
Burack, Walter Richard
Chapman, Carleton Burke
Clendenning, William Edmund
Kelley, Maurice Leslie, Jr
McIntyre, Oswald Ross
Mudge, Gilbert Horton
Patek, Arthur Jackson, Jr
Ritzman, Thomas A
Rolett, Ellis Lawrence
Rous, Stephen N
Valtin, Heinz
Zubkoff, Michael

NEW JERSEY

Ahmed, S Sultan
Alger, Elizabeth A
Amory, David William
Auerbach, Oscar
Ayvazian, L Fred
Behrle, Franklin C
Bergen, Stanley S, Jr
Bierenbaum, Marvin L
Block, A Jay
Boksay, Istvan Janos Endre
Bonner, Daniel Patrick
Boyle, Joseph, III
Breen, James Langhorne
Carter, Stephen Keith
Carver, David Harold
Chilton, Neal Warwick
Chinard, Francis Pierre
Cinotti, Alfonse A
Ciosek, Carl Peter, Jr
Conn, Hadley Lewis, Jr
Cook, Stuart D
Coutinho, Claude Bernard
Cross, Richard James
Crump, Jesse Franklin
Cryer, Dennis Robert
Daly, John F
Das, Kiron Moy
Davies, Richard O
Day-Salvatore, Debra-Lynn
De Salva, Salvatore Joseph
Dunn, Jonathan C
Duvoisin, Roger C
Edelman, Norman H
Eisinger, Robert Peter
Ellison, Rose Ruth
Eng, Robert H K
Ertel, Norman H
Evans, Hugh E
Feltman, Reuben
Finch, Stuart Cecil
Ford, Neville Finch
Franciosa, Joseph Anthony
Gaudino, Mario
Ginsberg, Barry Howard
Goger, Pauline Rohm

Medicine (cont)

Goldenberg, David Milton
Goldstein, Bernard David
Goldstein, Gideon
Goodhart, Robert Stanley
Goodkind, Morton Jay
Gorodetzky, Charles W
Greenberg, Stanley
Gross, Peter A
Haft, Jacob I
Hall, Thomas Christopher
Hilal, Sadek K
Hill, George James, II
Holler, Jacob William
Hutcheon, Duncan Elliot
Hutter, Robert V P
Imaeda, Tamotsu
Jacobus, David Penman
Jaffe, Ernst Richard
Katsampes, Chris Peter
Khachadurian, Avedis K
Kirschner, Marvin Abraham
Krey, Phoebe Regina
Lane, Alexander Z
Lanzoni, Vincent
Lasfargues, Etienne Yves
Lawrason, F Douglas
Layman, William Arthur
Leevy, Carroll M
Lehr, David
Lehrer, Harold Z
Leibowitz, Michael Jonathan
Leitz, Victoria Mary
LeSher, Dean Allen
Levine, Robert
Levy, Robert I
Lourenco, Ruy Valentim
Louria, Donald Bruce
Marshall, Carter Lee
Medway, William
Merriam, George Rennell, Jr
Mezey, Kalman C
Moolten, Sylvan E
Morse, Bernard S
Motz, Robin Owen
Niederland, William G
Oleske, James Matthew
Perr, Irwin Norman
Pinsky, Carl Muni
Poorvin, David Walter
Prineas, John William
Quinones, Mark A
Raetz, Christian Rudolf Hubert
Raskova, Jana D
Regan, Timothy Joseph
Royce, Paul C
Rubanyi, Gabor Michael
Schingale, Erich F
Schleifer, Steven Jay
Scolnick, Edward M
Seneca, Harry
Singer, Robert Mark
Sisson, Thomas Randolph Clinton
Slater, Eve Elizabeth
Spector, Reynold
Stambaugh, John Edgar, Jr
Stern, Elizabeth Kay
Vaun, William Stratin
Vukovich, Robert Anthony
Wachs, Gerald N
Wallach, Jacques Burton
Walters, Thomas Richard
Wedeen, Richard P
Weiss, Stanley H
Weisse, Allen B
Wildnauer, Richard Harry
Winters, Robert Wayne
Woske, Harry Max
Zuckerman, Leo

NEW MEXICO

Abrams, Jonathan
Allison, David Coulter
Appenzeller, Otto
Brown, Harold
Cobb, John Candler
Cochran, Paul Terry
Crawford, Michael Howard
Davis, Larry Ernest
Douglas, William Kennedy
Florman, Alfred Leonard
Gardner, Kenneth Drake, Jr
Grizzard, Michael B
Leach, John Kline
Lindeman, Robert D
Martinez, J Ricardo
Mason, William van Horn
Palmer, Darwin L
Pena, Hugo Gabriel
Raju, Mudundi Ramakrishna
Rhodes, Buck Austin
Schottstaedt, William Walter
Sloan, Robert Dye
Snipes, Morris Burton
Tempest, Bruce Dean
Tobey, Robert Allen
Voelz, George Leo
Vorherr, Helmuth Wilhelm
Walker, A Earl
Wax, Joan
Weston, James T
Zamora, Paul O
Zauder, Howard L

NEW YORK

Abolhassni, Mohsen
Abramson, Allan Lewis
Ackerman, Bruce David
Ackerman, Norman Bernard
Adler, Alexandra
Ahrens, Edward Hamblin, Jr
Aisen, Philip
Al-Askari, Salah
Aldrich, Thomas K
Alexander, Leslie Luther
Alpert, Seymour
Altman, Lawrence Kimball
Altner, Peter Christian
Ambrus, Clara Maria
Ambrus, Julian Lawrence
Anderson, Albert Douglas
Appelbaum, Emanuel
Archibald, Reginald MacGregor
Aronow, Wilbert Solomon
Aronson, Ronald Stephen
Artusio, Joseph F, Jr
Ascensao, Joao L
Ashutosh, Kumar
Askanazi, Jeffrey
Atkins, Harold Lewis
Atlas, Steven Alan
Atwater, Edward Congdon
Auchincloss, Joseph Howland, Jr
Auld, Peter A McF
Bachvaroff, Radoslav J
Baer, Rudolf L
Baker, David H
Baldini, James Thomas
Balint, John Alexander
Bank, Arthur
Bank, Norman
Bannon, Robert Edward
Barka, Tibor
Barnett, Henry Lewis
Barondess, Jeremiah A
Barron, Bruce Albrecht
Barron, Kevin D
Bartal, Arie H
Baruch, Sulamita B
Barzel, Uriel S
Bateman, John Laurens
Baum, John
Baum, Stephen Graham
Bauman, Norman
Baxter, Donald Henry
Beal, Myron Clarence
Beam, Thomas Roger
Bearn, Alexander Gordon
Becker, David Victor
Becker, Joshua A
Beckerman, Barry Lee
Beebe, Richard Townsend
Beinfield, William Harvey
Bell, Alfred Lee Loomis, Jr
Bell, James Milton
Bennett, James Anthony
Bennett, John M
Beranbaum, Samuel Louis
Berger, Eugene Y
Berger, Lawrence
Bergstrom, William H
Berkman, James Israel
Berlyne, Geoffrey Merton
Bernstein, David
Bertino, Joseph R
Bertles, John F
Bhattacharya, Jahar
Biedler, June Lee
Biempica, Luis
Bigger, J Thomas, Jr
Bito, Laszlo Z
Blaufox, Morton D
Bleicher, Sheldon Joseph
Bloomfield, Dennis Alexander
Blumenstock, David A
Bockman, Richard Steven
Bogdonoff, Morton David
Boley, Scott Jason
Bond, Victor Potter
Borer, Jeffrey Stephen
Borkowsky, William
Bosniak, Morton A
Bradford, James Carrow
Bradner, William Turnbull
Brand, Leonard
Brandaleone, Harold
Brandriss, Michael W
Braunstein, Joseph David
Brennan, Murray F
Brentjens, Jan R
Brick, Irving B
Briehl, Robin Walt
Briggs, Donald K
Brightman, I Jay
Briscoe, William Alexander
Brodoff, Bernard Noah
Brody, Bernard B
Brown, Audrey Kathleen
Brown, David Frederick
Brown, Lawrence S, Jr
Bruck, Erika
Buhac, Ivo
Burgener, Francis Andre
Buschke, Herman
Bushinsky, David Allen
Butler, John Joseph
Butler, Vincent Paul, Jr
Buxbaum, Joel N

Byles, Peter Henry
Cabrera, Edelberto Jose
Cahill, Kevin M
Canfield, Robert E
Canizares, Orlando
Carlson, Harold Ernest
Carr, Edward Albert, Jr
Carr, Ronald E
Carruthers, Christopher
Carter, Anne Cohen
Case, Robert B
Chandra, Pradeep
Chang, Chu Huai
Chase, Norman E
Chase, Randolph Montieth, Jr
Chasis, Herbert
Chess, Leonard
Chilcote, Max Eli
Chodos, Robert Bruno
Choi, Tai-Soon
Christy, Nicholas Pierson
Chu, Florence Chien-Hwa
Chusid, Joseph George
Chutkow, Jerry Grant
Clark, Duncan William
Clark, Julian Joseph
Clarkson, Bayard D
Cochran, George Van Brunt
Cockett, Abraham Timothy K
Cohen, Alfred M
Cohen, Bernard
Cohen, Burton D
Cohen, Irving Allan
Cohen, Jordan J
Cohen, Jules
Cohen, Noel Lee
Cohlan, Sidney Quex
Cohn, Peter Frank
Cohn, Zanvil A
Cole, Harold S
Cole, Patricia Ellen
Coleman, Morton
Coller, Barry Spencer
Collins, George H
Colman, Neville
Cook, Charles Davenport
Cooper, Louis Zucker
Copley, Alfred Lewin
Cote, Lucien Joseph
Cowger, Marilyn L
Craig, Albert Burchfield, Jr
Craig, John Philip
Cronkite, Eugene Pitcher
Cropp, Gerd J A
Cucci, Cesare Eleuterio
Cunninham-Rundles, Charlotte
Dack, Simon
Dancis, Joseph
Dao, Thomas Ling Yuan
Day, Stacey Biswas
Dean, David Campbell
Deane, Norman
DeCosse, Jerome J
DeFelice, Eugene Anthony
Deibel, Rudolf
Delagi, Edward F
DeLand, Frank H
Demeter, Steven
Deming, Quentin Burritt
Demis, Dermot Joseph
Deuschle, Kurt W
DeVita, Vincent T, Jr
DiMauro, Salvatore
DJang, Arthur H K
Dolgin, Martin
Dornfest, Burton S
Dorsey, Thomas Edward
Douglas, Gordon Watkins
Douglas, Robert Gordon, Jr
Dow, Bruce MacGregor
Downey, John A
Doyle, Eugenie F
Doyle, Joseph Theobald
Dreizen, Paul
Drusin, Lewis Martin
Dubroff, Lewis Michael
Duffy, Philip
Duncalf, Deryck
Durr, Friedrich (E)
Dutton, Cynthia Baldwin
Dutton, Robert Edward, Jr
Dworetzky, Murray
Eagle, Harry
Edelmann, Chester M, Jr
Edelson, Paul J
Egan, Edmund Alfred
Eich, Robert
Eichna, Ludwig Waldemar
Eisenberg, M Michael
Eisenstein, Albert Bernard
Elizan, Teresita S
Elkin, Milton
Engel, George Libman
Engle, Mary Allen English
Engle, Ralph Landis, Jr
Escher, Doris Jane Wolf
Eschner, Edward George
Fahn, Stanley
Faloon, William Wassell
Farber, Saul Joseph
Feldman, B Robert
Felig, Philip
Felman, Yehudi M
Ferguson, Earl Wilson

Ferris, Philip
Ferrone, Soldano
Field, Michael
Finberg, Laurence
Fink, Austin Ira
Finkbeiner, John A
Fischel, Edward Elliot
Fischman, Donald A
Fish, Irving
Fleischmajer, Raul
Fleischman, Alan R
Foldes, Francis Ferenc
Foley, Kathleen M
Foley, William Thomas
Fontana, Vincent J
Forbes, Gilbert Burnett
Fox, Arthur Charles
Fraad, Lewis M
Frame, Paul S
Francis, Charles K
Frank, Lawrence
Frantz, Andrew Gibson
Frazer, John P
Freedberg, Irwin Mark
Freedman, Aaron David
Freeman, Richard B
Freitag, Julia Louise
French, Joseph H
Friedewald, William Thomas
Friedman, Alan Herbert
Friedman, Eli A
Friedman, Irwin
Frishman, William Howard
Fritts, Harry Washington, Jr
Froelich, Ernest
Frost, E A M
Fuchs, Fritz Friedrich
Fulop, Milford
Galdston, Morton
Galin, Miles A
Gambert, Steven Ross
Gardner, Lytt Irvine
Gary, Nancy E
Garza, Cutberto
Geffen, Abraham
Geiger, H Jack
Gellhorn, Alfred
Gerber, Donald Albert
German, James Lafayette, III
Gershberg, Herbert
Gershengorn, Marvin Carl
Gershman, Lewis C
Gershon, Anne A
Gertler, Menard M
Ghosh, Nirmal Kumar
Gibbs, David Lee
Giblin, Denis Richard
Gibofsky, Allan
Gidari, Anthony Salvatore
Gilbert, Harriet S
Gilbert, Robert
Gillies, Alastair J
Ginsberg-Fellner, Fredda Vita
Glabman, Sheldon
Glaser, Warren
Glass, George B Jerzy
Glassman, Armand Barry
Glomski, Chester Anthony
Goldensohn, Eli Samuel
Goldring, Roberta M
Goldschmidt, Bernard Morton
Goldsmith, Lowell Alan
Goldsmith, Michael Allen
Goldstein, Robert
Gollub, Seymour
Goodgold, Joseph
Goodman, DeWitt Stetten
Goodman, Richard S
Gootman, Norman Lerner
Gordon, Harry Haskin
Gorlin, Richard
Gottlieb, Arlan J
Gould, Lawrence A
Gramiak, Raymond
Granger, Carl V
Greene, David Gorham
Greene, William Allan
Gregory, Daniel Hayes
Griffiths, Raymond Bert
Griner, Paul F
Grob, David
Grollman, Arthur Patrick
Gross, Ludwik
Grossman, Jacob
Grynbaum, Bruce B
Gulotta, Stephen Joseph
Gupta, Sanjeer
Gurney, Ramsdell
Gurwara, Sweet K
Gusberg, Saul Bernard
Haas, Albert B
Habicht, Jean-Pierre
Hadley, Susan Jane
Haggerty, Robert Johns
Haines, Thomas Henry
Halevy, Simon
Hall, Charles A
Ham, Richard John
Hambrick, George Walter, Jr
Hamerman, David Jay
Hamill, Robert W
Hamilton, Byron Bruce
Hamilton, Leonard Derwent
Han, Jaok

Neuman, Michael R
Newton, William Allen, Jr
Oestreich, Alan Emil
Ooi, Boon Seng
Pansky, Ben
Parker, Robert Frederic
Patrick, James R
Paul, Richard Jerome
Pearson, Olof Hjalmer
Perkins, Robert Louis
Pitner, Samuel Ellis
Pollak, Victor Eugene
Prior, John Alan
Rakita, Louis
Ratnoff, Oscar Davis
Reif, Thomas Henry
Reiner, Charles Brailove
Rejali, Abbas Mostafavi
Ritterhoff, Robert J
Robbins, Frederick Chapman
Robinow, Meinhard
Roche, Alexander F
Rodman, Harvey Meyer
Rolf, Clyde Norman
Rosenberg, Howard C
Saenger, Eugene L
Saunders, William H
Schafer, Irwin Arnold
Schiff, Gilbert Martin
Schreiner, Albert William
Schubert, William K
Schwartz, Howard Julius
Scott, Ralph Carmen
Scott, Ralph Mason
Scott, William James, Jr
Selim, Mostafa Ahmed
Shelley, Walter Brown
Shields, George Seamon
Shirkey, Harry Cameron
Shumrick, Donald A
Silberstein, Edward B
Sjoerdsma, Albert
Skillman, Thomas G
Smith, Robert
Somani, Pitambar
Sotos, Juan Fernandez
Stanberry, Lawrence Raymond
Staubus, Alfred Elsworth
Stevenson, Thomas Dickson
Straffon, Ralph Atwood
Strohl, Kingman P
Suskind, Raymond Robert
Sutherland, James McKenzie
Tavill, Anthony Sydney
Tennebaum, James I
Teteris, Nicholas John
Troup, Stanley Burton
Tsang, Reginald C
Tubbs, Raymond R
Tucker, Harvey Michael
Turner, Edward V
Tzagournis, Manuel
Vertes, Victor
Vester, John William
Vignos, Paul Joseph, Jr
Vilter, Richard William
Wall, Robert Leroy
Warkany, Josef
Warren, James Vaughn
Washburn, Lee Cross
Waterson, John R
Weiner, Murray
Weisman, Russell
Wentz, William Budd
Wenzel, Richard Louis
West, Clark Darwin
Wexler, Bernard Carl
White, Peter
Will, John Junior
Winter, Chester Caldwell
Wolinsky, Emanuel
Worrell, John Mays, Jr
Wright, Francis Stuart
Wyman, Milton
Yates, Allan James
Zuspan, Frederick Paul

OKLAHOMA
Bogardus, Carl Robert, Jr
Bottomley, Richard H
Bottomley, Sylvia Stakle
Bourdeau, James Edward
Brandt, Edward Newman, Jr
Caddell, Joan Louise
Caldwell, Glyn Gordon
Calvert, Jon Channing
Campbell, John Alexander
Chrysant, Steven George
Clark, Mervin Leslie
Comp, Philip Cinnamon
Cooper, Richard Grant
Corder, Clinton Nicholas
Coston, Tullos Oswell
Crosby, Warren Melville
Czerwinski, Anthony William
Dille, John Robert
Dormer, Kenneth John
Epstein, Robert B
Everett, Mark Allen
Gable, James Jackson, Jr
Ganesan, Devaki
Gleaton, Harriet Elizabeth
Gunn, Chesterfield Garvin, Jr
Hampton, James Wilburn

Harder, Harold Cecil
Harley, John Barker
Hope, Ronald Richmond
Ice, Rodney D
Kinasewitz, Gary Theodore
Kosbab, Frederic Paul Gustav
Lakin, James D
Lazzara, Ralph
Lewis, C S, Jr
Lhotka, John Francis
Lynn, Thomas Neil, Jr
McKee, Patrick Allen
Massion, Walter Herbert
Maton, Paul Nicholas
Muchmore, Harold Gordon
Nettles, John Barnwell
Olson, Robert Leroy
Parry, William Lockhart
Pikler, George Maurice
Reichlin, Morris
Rhoades, Everett Ronald
Rich, Clayton
Riley, Harris D, Jr
Seely, J Rodman
Smith, Carl Walter, Jr
Smith, William Ogg
Taylor, Fletcher Brandon, Jr
Thomas, Ellidee Dotson
Thurman, William Gentry
Torres-Pinedo, Ramon
Vanhoutte, Jean Jacques
Wenzl, James E
Whang, Robert
Whitcomb, Walter Henry
Whitsett, Thomas L
Williams, George Rainey
Wilson, Michael Friend

OREGON
Bagby, Grover Carlton
Bardana, Emil John, Jr
Bennett, Robert M
Bennett, William M
Benson, John Alexander, Jr
Bergman, Norman
Bigley, Robert Harry
Bissonnette, John Maurice
Black, Franklin Owen
Bristow, John David
Brummett, Robert E
Buchan, George Colin
Buist, Neil R M
Bussman, John W
Campbell, John Richard
Campbell, Robert A
Carter, Charles Conrad
Connor, William Elliott
De Maria, F John
DeWeese, David D
Ducsay, Charles Andrew
Dyson, Robert Duane
Edwards, Miles John
Engel, Rudolf
Fraunfelder, Frederick Theodor
Frey, Bruce Edward
Grimm, Robert John
Grover, M Roberts, Jr
Hemenway, William Garth
Herndon, Robert McCulloch
Hill, John Donald
Hu, Funan
Isom, John B
Jacob, Stanley W
Jones, Richard Theodore
Kendall, John Walker, Jr
Kohler, Peter
Lees, Martin H
Lucas, Oscar Nestor
Malinow, Manuel R
Matarazzo, Joseph Dominic
Meechan, Robert John
Menashe, Victor D
Metcalfe, James
Miller, Stephen Herschel
Morris, James F
Norman, Douglas James
Novy, Miles Joseph
Osterud, Harold T
Patterson, James Fulton
Pirofsky, Bernard
Porter, George A
Portman, Oscar William
Reynolds, John Weston
Ritzmann, Leonard W
Schaeffer, Morris
Smith, Catherine Agnes
Stevens, Janice R
Swan, Kenneth Carl
Swank, Roy Laver
Underwood, Rex J
Walsh, John Richard
Watzke, Robert Coit
Williams, Christopher P S
Wuepper, Kirk Dean

PENNSYLVANIA
Abruzzo, John L
Adibi, Siamak A
Agarwal, Jai Bhagwah
Agus, Zalman S
Alexander, Fred
Alter, Milton
Andros, George James
Asakura, Toshio

Atkins, Paul C
Austrian, Robert
Baird, Henry W, III
Balin, Arthur Kirsner
Ballentine, Rudolph Miller
Bannon, James Andrew
Barba, William P, II
Barker, Earl Stephens
Barry, William Eugene
Bartosik, Delphine
Bartuska, Doris G
Batshaw, Mark Levitt
Baum, Stanley
Beerman, Herman
Beizer, Lawrence H
Bell, Robert Lloyd
Bennett, Hugh deEvereaux
Berger, Harvey J
Berk, Lawrence B
Berlin, Cheston Milton
Berney, Steven
Bierly, Mahlon Zwingli, Jr
Bilaniuk, Larissa Tetiana
Black, Martin
Bluemle, Lewis W, Jr
Blumberg, Baruch Samuel
Boden, Guenther
Boger, William Pierce
Bonakdarpour, Akbar
Bongiovanni, Alfred Marius
Borges, Wayne Howard
Bovee, Kenneth C
Bowden, Charles Ronald
Bowers, Paul Applegate
Bradley, Matthews Ogden
Bradley, Stanley Edward
Brady, Luther Weldon, Jr
Bralow, S Philip
Brennan, James Thomas
Brest, Albert N
Bridger, Wagner H
Brightman, Vernon
Brody, Jerome Ira
Brooks, John J
Brucker, Paul Charles
Buck, Clayton Arthur
Burholt, Dennis Robert
Burkle, Joseph S
Burns, H Donald
Caliguiri, Lawrence Anthony
Cander, Leon
Carbone, Robert James
Carey, William Bacon
Carpenter, Gary Grant
Chait, Arnold
Chambers, Richard
Charkes, N David
Charron, Martin
Chervenick, Paul A
Clark, James Edward
Clark, James Michael
Clark, Wallace Henderson, Jr
Cohen, Margo Nita Panush
Cohn, Robert M
Colman, Robert W
Comis, Robert Leo
Conn, Rex Boland
Cooper, David Young
Cornfeld, David
Crane, August Reynolds
Creech, Richard Hearne
Croce, Carlo Maria
Custer, Richard Philip
Cutler, John Charles
Cutler, Winnifred Berg
D'Angio, Giulio J
Daniele, Ronald P
Davidoff, Frank F
Day, Harvey James
Deas, Thomas C
DeHoratius, Raphael Joseph
Demopoulos, James Thomas
DeRubertis, Frederick R
Dickerson, William H
Dickman, Albert
Di George, Angelo Mario
Dinman, Bertram David
DiPalma, Joseph Rupert
Djerassi, Isaac
Doghramji, Karl
Dorwart, Bonnie Brice
Douglas, Steven Daniel
Dreifus, Leonard S
Duane, Thomas David
Earley, Laurence E
Edwards, McIver Williamson, Jr
Eggleston, Forrest Cary
Ehrlich, George Edward
Eisenberg, John Meyer
Ellis, Demetrius
Engleman, Karl
Epps, Joyce E
Erdman, William James, II
Erecinska, Maria
Erslev, Allan Jacob
Evans, Audrey Elizabeth
Farrar, George Elbert, Jr
Finestone, Albert Justin
Finkelstein, David
Fireman, Philip
Fischer, Grace Mae
Foley, Thomas Preston, Jr
Franke, Frederick Rahde
Frankl, William S

Fraser, David William
Freeman, Joseph Theodore
French, Gordon Nichols
Friedman, Arnold Carl
Fromm, Gerhard Hermann
Futcher, Palmer Howard
Gabuzda, Thomas George
Gaffney, Paul Cotter
Gambescia, Joseph Marion
Garcia, Celso-Ramon
Gilbert, Robert Pettibone
Glick, John H
Goldberg, Harry
Goldberg, Martin
Goldberg, Michael Ellis
Goldberger, Michael Eric
Goldstein, Franz
Goldwein, Manfred Isaac
Gorson, Robert O
Gosfield, Edward, Jr
Gray, Frank Davis, Jr
Greenstein, Jeffrey Ian
Grenvik, Ake N A
Gross, Peter George
Guggenheimer, James
Gurwith, Marc Joseph
Guthrie, Marshall Beck
Haase, Gunter R
Hann, Hie-Won L
Haskin, Marvin Edward
Haskin, Myra Ruth Singer
Haurani, Farid I
Hayashi, Teruo Terry
Hedges, Thomas Reed, Jr
Heller, Melvin S
Helwig, John, Jr
Herberman, Ronald Bruce
Hildreth, Eugene Augustus
Ho, Monto
Hodges, John Hendricks
Hollander, Joseph Lee
Holling, Herbert Edward
Holroyde, Christopher Peter
Holsclaw, Douglas S, Jr
Horner, George John
Horwitz, Orville
Hurley, Harry James
Huth, Edward J
Isard, Harold Joseph
Israel, Harold L
Jacob, Leonard Steven
Jacoby, Jay
Jeffries, Graham Harry
Jegasothy, Brian V
Jimenez, Sergio
Johnston, Richard Boles, Jr
Joseph, Rosaline Resnick
Joyner, Claude Reuben
Kade, Charles Frederick, Jr
Kallen, Roland Gilbert
Kaplan, Sandra Solon
Karafin, Lester
Kashatus, William C
Kauffman, Leon A
Kauffman, Stuart Alan
Kaye, Donald
Kaye, Robert
Kefalides, Nicholas Alexander
Kelley, William Nimmons
Kendall, Norman
Kerstein, Morris D
Khoury, George
Kiesewetter, William Burns
Kimbel, Philip
Kirby, Edward Paul
Kivlighn, Salah Dean
Kligman, Albert Montgomery
Kline, Irwin Kaven
Klinghoffer, June F
Klionsky, Bernard Leon
Knobler, Robert Leonard
Knudson, Alfred George, Jr
Koelle, Winifred Angenent
Kressel, Herbert Yehude
Krisch, Robert Earle
Krishna, Gollapudi Gopal
Kundel, Harold Louis
Ladda, Roger Louis
Laibson, Peter R
Lame, Edwin Lever
Laties, Alan M
Laufer, Igor
Lazarus, Gerald Sylvan
Leberman, Paul R
Lee, Yien-Hwei
Leeper, Dennis Burton
Leighton, Charles Cutler
Levey, Gerald Saul
Levinsky, Walter John
Levison, Matthew Edmund
Levison, Sandra Peltz
Levit, Edithe J
Lewis, George Campbell, Jr
Lewis, Jessica Helen
Likoff, William
Linnemann, Roger E
Lippa, Erik Alexander
Lippa, Linda Susan Mottow
Lipton, Allan
London, William Thomas
Long, Joseph Pote
Longnecker, David Eugene
Luscombe, Herbert Alfred
McCloskey, Richard Vensel

Medicine (cont)

Madaio, Michael P
Madow, Leo
Magargal, Larry Elliot
Maguire, Henry C, Jr
Mancall, Elliott L
Mansmann, Herbert Charles, Jr
Marinchak, Roger Alan
Martin, Christopher Michael
Martin, Donald Beckwith
Martin, John Harvey
Martin, Samuel Preston, III
Mastrangelo, Michael Joseph
Mastroianni, Luigi, Jr
Mateer, Frank Marion
Maurer, Alan H
Mayock, Robert Lee
Mazaheri, Mohammad
Meadows, Anna T
Meisler, Arnold Irwin
Michelson, Eric L
Miller, John Michael
Mills, Lewis Craig, Jr
Minn, Fredrick Louis
Montgomery, Hugh
Moore, E(arl) Neil
Moore, Mary Elizabeth
Moran, John J
Morganroth, Joel
Morris, Harold Hollingsworth, Jr
Mortel, Rodrigue
Moster, Mark Leslie
Motoyama, Etsuro K
Moyer, John Henry
Murphy, John Joseph
Myers, David
Myers, Eugene Nicholas
Myers, Jack Duane
Myerson, Ralph M
Naide, Meyer
Nelson, Edward Blake
Nestico, Pasquale Francesco
Newmark, Jonathan
Nicholas, Leslie
Nobel, Joel J
Oaks, Wilbur W
O'Brien, Joan A
Onik, Gary M
Orenstein, David M
Owen, Oliver Elon
Palmer, Robert Howard
Park, Chan H
Parsons, John Andresen
Peltier, Hubert Conrad
Penneys, Raymond
Perryman, Charles Richard
Pinkel, Donald Paul
Pleasure, David
Plotkin, Stanley Alan
Polgar, George
Pontarelli, Domenic Joseph
Powell, Robin Dale
Poydock, Mary Eymard
Price, Henry Locher
Prockop, Darwin J
Proctor, Julian Warrilow
Prystowsky, Harry
Raffensperger, Edward Cowell
Rashkind, William Jacobson
Reavey-Cantwell, Nelson Henry
Reimann, Hobart Ansteth
Reinecke, Robert Dale
Reinmuth, Oscar McNaughton
Reivich, Martin
Rike, Paul Miller
Rockey, John Henry
Rockwell, Harriet Esther
Rodan, Gideon Alfred
Rodnan, Gerald Paul
Rogers, Fred Baker
Rogers, Kenneth D
Rogers, Robert M
Rohner, Thomas John
Ronis, Max Lee
Rosenblatt, Michael
Rosenkranz, Herbert S
Rosenow, Edward Carl, Jr
Roth, James Luther Aumont
Rovera, Giovanni
Rubin, Alan
Rubin, Donald Howard
Rubin, Walter
Rucinska, Ewa J
Rycheck, Russell Rule
Sachs, Marvin Leonard
Safar, Peter
Samitz, M H
Sanders, Martin E
Schaedler, Russell William
Schein, Philip Samuel
Scher, Charles D
Schick, Paul Kenneth
Schlezinger, Nathan Stanley
Schmaier, Alvin Harold
Schmickel, Roy David
Schmidt, Richard Ralph
Schnabel, Truman Gross, Jr
Schneider, Henry C, Sr
Schofield, Richard Alan
Schrogie, John Joseph
Schumacher, H Ralph, Jr
Schwartz, Arthur Gerald
Schwartz, Elias

Schwartz, Emanuel Elliot
Schwartz, Irving Robert
Schwartzman, Robert M
Schweighardt, Frank Kenneth
Segal, Bernard L
Segal, Stanton
Sellers, Alfred Mayer
Senior, John Robert
Shapiro, Alvin Philip
Shapiro, Bernard
Shapiro, Sandor Solomon
Sherry, Sol
Ship, Irwin I
Shuman, Charles Ross
Shumway, Clare Nelson, (Jr)
Siegler, Peter Emery
Sigler, Miles Harold
Silberberg, Donald H
Silverstein, Alexander
Slotnick, Victor Bernard
Smith, David S
Smith, Hugo Dunlap
Smith, Jackson Bruce
Smith, Jan D
Snow, James Byron, Jr
Soloff, Louis Alexander
Soma, Lawrence R
Southam, Chester Milton
Spaeth, George L
Spritzer, Albert A
Stein, George Nathan
Stoloff, Irwin Lester
Strahs, Gerald
Strauss, Jerome Frank, III
Strom, Brian Leslie
Sunderman, Frederick William
Sutnick, Alton Ivan
Szabo, Kalman Tibor
Tasman, William S
Tauxe, Welby Newlon
Taylor, Paul M
Teplick, Joseph George
Tourtellotte, Charles Dee
Tristan, Theodore A
Troen, Philip
Tse, Rose (Lou)
Tuddenham, William J
Tulenko, Thomas Norman
Tulsky, Emanuel Goodel
Tumen, Henry Joseph
Turchi, Joseph J
Uitto, Jouni Jorma
Urbach, Frederick
Vagnucci, Anthony Hillary
Van Scott, Eugene Joseph
Van Thiel, David H
Von Beckh, Harald Johannes
Wald, Arnold
Waldman, Joseph
Walsh, Peter Newton
Wang, Yen
Warren, George Harry
Waxman, Herbert Sumner
Webster, John H
Weinhouse, Sidney
Weiss, Arthur Jacobs
Weiss, William
Weng, (Frank) Tzong-Ruey
Wesson, Laurence Goddard, Jr
Whereat, Arthur Finch
Wilpizeski, Chester Robert
Wilson, Frederick Allen
Wilson, Frederick Sutphen
Wilson, Marjorie Price
Winkelstein, Alan
Winter, Peter Michael
Wolf, Stewart George, Jr
Wolfson, Robert Joseph
Wollman, Harry
Woloshin, Henry Jacob
Wood, Francis Clark
Wood, Margaret Gray
Wright, Francis Howell
Young, Irving
Young, Lionel Wesley
Zatuchni, Jacob
Zelis, Robert Felix
Ziegra, Sumner Root
Zimmerman, Robert Allan
Zinsser, Harry Frederick
Zurier, Robert B
Zweiman, Burton

RHODE ISLAND

Arnold, Mary B
Aronson, Stanley Maynard
Barnes, Frederick Walter, Jr
Calabresi, Paul
Carleton, Richard Allyn
Carpenter, Charles C J
Constantine, Herbert Patrick
Cowett, Richard Michael
Crowley, James Patrick
Davis, Robert Paul
Estrup, Faiza Fawaz
Farnes, Patricia
Glicksman, Arvin Sigmund
Greer, David Steven
Hamolsky, Milton William
Jackson, Ivor Michael David
Kaplan, Stephen Robert
Lichtman, Herbert Charles
Meroney, William Hyde, III
Oh, William

Pueschel, Siegfried M
Schwartz, Robert
Senft, Alfred Walter
Silver, Alene Freudenheim
Tefft, Melvin
Thayer, Walter Raymond, Jr
Turner, Michael D
Yankee, Ronald August
Zawadzki, Zbigniew Apolinary
Zinner, Stephen Harvey

SOUTH CAROLINA

Barrett, O'Neill, Jr
Bates, G William
Boger, Robert Shelton
Brown, Arnold
Buse, John Frederick
Buse, Maria F Gordon
Colwell, John Amory
Cooper, George, IV
Corley, John Bryson
Curry, Hiram Benjamin
Dobson, Richard Lawrence
Farrar, William Edmund
Fudenberg, H Hugh
Gillette, Paul Crawford
Goust, Jean Michel
Hester, Lawrence Lamar, Jr
Hogan, Edward L
Holper, Jacob Charles
Humphries, J O'Neal
Kane, John Joseph
Kirtley, William Raymond
Kolb, Leonard H
Lazarchick, John
Legerton, Clarence W, Jr
LeRoy, Edward Carwile
Lin, Tu
McConnell, Jack Baylor
McFarland, Kay Flowers
Macpherson, Roderick Ian
Malone, Michael Joseph
Maricq, Hildegard Rand
Meredith, Howard Voas
Mitchell, Hugh Bertron
Murdaugh, Herschel Victor, Jr
Newberry, William Marcus
Pickett, Jackson Brittain
Pittman, Fred Estes
Ploth, David W
Postic, Bosko
Preedy, John Robert Knowlton
Putney, Floyd Johnson
Redding, Joseph Stafford
Roberts, Rufus Winston
Roof, Betty Sams
Rubin, Mitchell Irving
Sandifer, Samuel Hope
Singh, Inderjit
Spann, James Fletcher, (Jr)
Still, Charles Neal
Stokes, David Kershaw, Jr
Swanson, Arnold Arthur
Vallotton, William Wise
Virella, Gabriel T
Westphal, Milton C, Jr
Wohltmann, Hulda Justine
Woodward, Kent Thomas
Young, Gilbert Flowers, Jr

SOUTH DAKOTA

Flora, George Claude
Gregg, John Bailey
Hamm, Joseph Nicholas
Johnson, Carl J
Mellinger, George T
Ranney, Brooks
Schramm, Mary Arthur
Wegner, Karl Heinrich
Zawada, Edward T, Jr

TENNESSEE

Adams, Robert Walker, Jr
Allen, Joseph Hunter
Allison, Fred, Jr
Altemeier, William Arthur, III
Anderson, Robert Spencer
Batalden, Paul B
Beauchene, Roy E
Bell, Persa Raymond
Berard, Costan William
Berman, M Lawrence
Biaggioni, Italo
Blaser, Martin Jack
Brigham, Kenneth Larry
Burk, Raymond Franklin, Jr
Burr, William Wesley, Jr
Byrd, Benjamin Franklin, Jr
Calhoun, Calvin L
Camacho, Alvro Manuel
Cantrell, William Fletcher
Castelnuovo-Tedesco, Pietro
Charyulu, Komanduri K N
Chesney, Russell Wallace
Christopher, Robert Paul
Churchill, John Alvord
Clark, Winston Craig
Cooke, Charles Robert
Cox, Clair Edward, II
Crofford, Oscar Bledsoe
Cummins, Alvin J
Davis, David A
Davis, Harry L
Derryberry, Oscar Merton

Des Prez, Roger Moister
Donald, William David
Dugdale, Marion
Elam, Lloyd C
Elliott, James H
Feman, Stephen Sosin
Fenichel, Gerald M
Fields, James Perry
Flexner, John M
Foster, Henry Wendell
Franks, John Julian
Freemon, Frank Reed
Friesinger, Gottlieb Christian
Fuhr, Joseph Ernest
Ganote, Charles Edgar
Ginn, H Earl
Gompertz, Michael L
Goodwin, Robert Archer, Jr
Goswitz, Francis Andrew
Goswitz, Helen Vodopick
Granner, Daryl Kitley
Grossman, Laurence Abraham
Gupta, Ramesh C
Hansen, Axel C
Hara, Saburo
Harwood, Thomas Riegel
Hayes, Raymond Leroy
Hayes, Wayland Jackson, Jr
Heimberg, Murray
Helderman, J Harold
Holaday, Duncan Asa
Hubner, Karl Franz
Hughes, James Gilliam
Hughes, Walter T
Jabbour, J T
Jacobs, Richard Lewis
Jacobson, Harry R
James, Alton Everette, Jr
Jordan, Robert, Jr
Jordan, Willis Pope, Jr
Kaplan, Robert Joel
Kaplan, Stanley Baruch
Karzon, David T
King, Lloyd Elijah, Jr
Kirshner, Howard Stephen
Kitabchi, Abbas E
Kitchen, Hyram
Kossmann, Charles Edward
Krantz, Sanford B
Kraus, Alfred Paul
Lacy, William White
Lange, Robert Dale
Lawton, Alexander R, III
Lefkowitz, Lewis Benjamin, Jr
Little, Joseph Alexander
Luton, Edgar Frank
Mauer, Alvin Marx
Montalvo, Jose Miguel
Muirhead, E Eric
Nag, Subir
Neely, Charles Lea, Jr
Netsky, Martin George
Newman, John Hughes
North, William Charles
Oates, John Alexander
Oldham, Robert Kenneth
Paine, Thomas Fite, Jr
Pate, James Wynford
Phillips, Jerry Clyde
Pincus, Theodore
Pratt, Charles Benton
Pribor, Hugo C
Price, Robert Allen
Pui, Ching-Hon
Ramsey, Lloyd Hamilton
Reddy, Churku Mohan
Rendtorff, Robert Carlisle
Reynolds, Leslie Boush, Jr
Reynolds, Vernon H
Rhamy, Robert Keith
Roberts, Lyman Jackson
Robinson, Roscoe Ross
Ross, Joseph C
Runyan, John William, Jr
Sell, Sarah H Wood
Shenefelt, Ray Eldon
Simone, Joseph Vincent
Siskin, Milton
Smith, Bradley Edgerton
Smith, Raphael Ford
Snapper, James Robert
Solomon, Alan
Solomon, Solomon Sidney
Soper, Richard Graves
Stahlman, Mildred
Stinson, Joseph McLester
Sullivan, Jay Michael
Summitt, Robert L
Tarleton, Gadson Jack, Jr
Terzaghi, Margaret
Teschan, Paul E
Townes, Alexander Sloan
Van Middlesworth, Lester
Wasserman, Jack F
Webster, Burnice Hoyle
Wells, Charles Edmon
Wilson, Robert John
Wood, Alastair James Johnston
Yang, Wen-Kuang
Yoo, Tai-June
Zee, Paulus

TEXAS

Abell, Creed Wills

Abrams, Steven Allen
Alexander, Charles Edward, Jr
Alexander, Drew W
Alexander, James Kermott
Alford, Bobby R
Alfrey, Clarence P, Jr
Allen, Herbert Clifton, Jr
Allen, Lois Brenda
Anderson, Karl E
Anderson, Robert E
Anuras, Sinn
Appel, Stanley Hersh
Armstrong, George Glaucus, Jr
Arnold, Watson Caufield
Bailey, Byron James
Baine, William Brennan
Balch, Charles Mitchell
Baron, Samuel
Barranco, Sam Christopher, III
Bashour, Fouad A
Batsakis, John G
Beaudet, Arthur L
Becker, Frederick F
Beller, Barry M
Benjamin, Robert Stephen
Bennett, Michael
Bergstresser, Paul Richard
Black, Homer Selton
Blaw, Michael Ervin
Blomqvist, Carl Gunnar
Bodey, Gerald Paul, Sr
Bradley, Thomas Bernard, Jr
Bridges, Charles Hubert
Briggs, Arthur Harold
Bryan, George Thomas
Burdine, John Alton
Burzynski, Stanislaw Rajmund
Butler, William T
Calverley, John Robert
Cantrell, Elroy Taylor
Capra, J Donald
Carpenter, Robert James, Jr
Carpenter, Robert Raymond
Carr, Bruce R
Carr, David Turner
Cate, Thomas Randolph
Catlin, Francis I
Catterson, Allen Duane
Chan, Lawrence Chin Bong
Chapman, John S
Chaudhuri, Tuhin
Chiuten, Delia Fungshe
Cohen, Allen Barry
Coltman, Charles Arthur, Jr
Combes, Burton
Cooper, Melvin Wayne
Couch, Robert Barnard
Crawford, Stanley Everett
Cruz, Anatolio Benedicto, Jr
Cunningham, Glenn R
Currarino, Guido
Daeschner, Charles William, Jr
Danhof, Ivan Edward
Daniels, Jerry Claude
Dantzker, David Roy
Davis, Jefferson C
Davis, Joyce S
Deiss, William Paul, Jr
Derrick, William Sheldon
Desmond, Murdina MacFarquhar
Diddle, Albert W
Dietlein, Lawrence Frederick
Dietschy, John Maurice
Di Ferrante, Nicola Mario
Dobson, Harold Lawrence
Dodd, Gerald Dewey, Jr
Dodge, Warren Francis
Dougherty, Joseph C
Dowben, Robert Morris
Dreizen, Samuel
Duff, Fratis L
Duflot, Leo Scott
DuPont, Herbert Lancashire
Dyck, Walter Peter
Edeiken, Jack
Edlin, John Charles
Eichenwald, Heinz Felix
Eknoyan, Garabed
Engelhardt, Hugo Tristram
Fashena, Gladys Jeannette
Feigin, Ralph David
Ferguson, Edward C, III
Fernbach, Donald Joseph
Fidler, Isaiah J
Fields, William Straus
Fink, Chester Walter
Fish, Stewart Allison
Forbis, Orie Lester, Jr
Forland, Marvin
Foster, Daniel W
Franklin, Robert Ray
Freedman, David Asa
Freireich, Emil J
Frenger, Paul F
Frenkel, Eugene Phillip
Frimpter, George W
Gant, Norman Ferrell, Jr
Garber, Alan J
Gaulden, Mary Esther
Gehan, Edmund A
German, Victor Frederick
Giesecke, Adolph H
Giovanella, Beppino C
Glezen, William Paul

Goldberg, Leonard H
Goldman, Armond Samuel
Grant, Arthur E
Grant, John Andrew, Jr
Green, Hubert Gordon
Green, Ronald W
Greenberg, Frank
Greenberg, Stephen B
Gregory, Raymond (Leslie)
Griffin, James Emmett
Gunn, Albert Edward
Gutierrez, Guillermo
Haggard, Mary Ellen
Hall, Robert Joseph
Harle, Thomas Stanley
Harris, Herbert H
Harrison, Gunyon M
Hartman, James T
Hawkins, Willard Royce
Haynie, Thomas Powell, III
Haywood, Theodore J
Heimbach, Richard Dean
Held, Berel
Herndon, James Henry, Jr
Hickey, Robert Cornelius
Hill, L(ouis) Leighton
Hill, Reba Michels
Hillis, William Daniel, Sr
Hinck, Vincent C
Holguin, Alfonso Hudson
Hollinger, F(rederick) Blaine
Holly, Frank Joseph
Holman, Gerald Hall
Holt, Charlene Poland
Hoyumpa, Anastacio Maningo
Hsu, Katharine Han Kuang
Hug, Verena
Hussey, Hugh Hudson
Ihler, Garret Martin
Jackson, Carmault B, Jr
James, Thomas Naum
Jasin, Hugo E
Jenicek, John Andrew
Johanson, Waldemar Gustave, Jr
Johnson, Larry
Johnson, William Cone
Jordon, Robert Earl
Journeay, Glen Eugene
Kallus, Frank Theodore
Kaplan, Norman M
Kaufman, Raymond H
Keele, Doman Kent
Keeler, Martin Harvey
Kirkendall, Walter Murray
Kitay, Julian I
Klein, Gordon Leslie
Knight, Vernon
Kniker, William Theodore
Kraemer, Duane Carl
Krakoff, Irwin Harold
Kraus, William Ludwig
Kraychy, Stephen
Kronenberg, Richard Samuel
Krueger, Gerhard R F
Krusen, Edward Montgomery
Kumar, Vinay
Kuo, Peter Te
Kurzrock, Razelle
Lamb, James Francis
Lancaster, Malcolm
Lane, Montague
Langland, Olaf Elmer
Langsjoen, Per Harald
LeBlanc, Adrian David
Ledley, Fred David
LeMaistre, Charles Aubrey
Leon, Robert Leonard
Leroy, Robert Frederick
LeVeau, Barney Francis
Levin, Victor Alan
Levin, William Cohn
Lifschitz, Meyer David
Lin, Tz-Hong
Lipshultz, Larry I
Lockhart, Lillian Hoffman
Lopez-Berestein, Gabriel
Low, Morton David
Lupski, James Richard
Lutherer, Lorenz O
Lynch, Edward Conover
Lynn, John R
Lynn, William Sanford
McCall, Charles B
McCredie, Kenneth Blair
McGanity, William James
McGovern, John Phillip
McGuire, William L
McKelvey, Eugene Mowry
McLaughlin, Peter
Maclean, Graeme Stanley
McNamara, Dan Goodrich
Marcus, Donald M
Marshall, Gailen D, Jr
Martinez-Lopez, Jorge Ignacio
Martt, Jack M
Medina, Daniel
Merin, Robert Gillespie
Meyer, John Stirling
Micks, Don Wilfred
Miller, Edward Godfrey, Jr
Miller, Jarrell E
Miller, William Franklin
Mintz, A Aaron
Misenhimer, Harold Robert

Mitchell, Jere Holloway
Mitchell, Jerry R
Mize, Charles Edward
Moldawer, Marc
Montague, Eleanor D
Moody, Eric Edward Marshall
Moreton, Robert Dulaney
Morrow, Dean Huston
Moyer, Mary Pat Sutter
Murphy, Paul Henry
Musher, Daniel Michael
Nau, Carl August
Nelson, John D
Nelson, Robert S
Nichaman, Milton Z
Nichols, Buford Lee, Jr
Norman, Floyd (Alvin)
North, Richard Ralph
Nunneley, Sarah A
Nusynowitz, Martin Lawrence
Oishi, Noboru
Park, Myung Kun
Patsch, Wolfgang
Pauerstein, Carl Joseph
Paull, Barry Richard
Peake, Robert Lee
Periman, Phillip
Peters, Lester John
Peters, Paul Conrad
Petty, Charles Sutherland
Pierce, Alan Kraft
Pierce, Alexander Webster, Jr
Piziak, Veronica Kelly
Poffenbarger, Phillip Lynn
Pollard, Richard Byrd
Powell, Don Watson
Powell, Leslie Charles
Powell, Norborne Berkeley
Pritchard, Jack Arthur
Raber, Martin Newman
Ram, C Venkata S
Raymond, Lawrence W
Reeves, T Joseph
Reuter, Stewart R
Rich, Robert Regier
Riggs, Stuart
Rodgers, Lawrence Rodney, Sr
Roffwarg, Howard Philip
Rolston, Kenneth Vijaykumar Issac
Rose, George G
Ross, Griff Terry, Sr
Rossen, Roger Downey
Rowen, Burt
Rudolph, Arnold Jack
Runge, Thomas Marschall
Rutledge, Felix N
Sakakini, Joseph, Jr
Sandstead, Harold Hilton
Sanford, Jay Philip
Schenker, Steven
Schneider, Rose G
Schneider, Sandra Lee
Schnur, Sidney
Schreiber, Melvyn Hirsh
Scurry, Murphy Townsend
Sears, David Alan
Seldin, Donald Wayne
Severn, Charles B
Shapiro, William
Shearer, William Thomas
Shulman, Robert Jay
Smiley, James Donald
Smith, Frank E
Smith, John Leslie, Jr
Smith, Keith Davis
Smith, Reginald Brian
Smythe, Cheves McCord
Spencer, William Albert
Spira, Melvin
Sprague, Charles Cameron
Starr, Jason Leonard
Stone, Marvin J
Stout, Landon Clarke, Jr
Strong, Louise Connally
Suki, Wadi Nagib
Sullivan, Margaret P
Swischuk, Leonard Edward
Szczesniak, Raymond Albin
Taegtmeyer, Heinrich
Taylor, Fred M
Thompson, Edward Ivins Bradbridge
Thompson, Howard K, Jr
Tompsett, Ralph Raymond
Towler, Martin Lee
Travis, Luther Brisendine
Tyner, George S
Uhr, Jonathan William
Unger, Roger Harold
Vallbona, Carlos
Vanatta, John Crothers, III
Varma, Surendra K
Villacorte, Guillermo Vilar
Von Noorden, Gunter Konstantin
Wallace, Sidney
Wallace, Tracy I
Wang, Yeu-Ming Alexander
Weiss, Gary Bruce
Weser, Elliot
Wigodsky, Herman S
Wildenthal, Kern
Williams, Bryan
Williams, Darryl Marlowe
Williams, John F, Jr
Williams, Robert L

Wilson, Jean Donald
Wilson, McClure
Wimer, Bruce Meade
Witherspoon, Samuel McBridge
Wolinsky, Jerry Saul
Wolma, Fred J
Worthen, Howard George
Wright, Stephen E
Wu, Kenneth Kun-Yu
Yamauchi, Toshio
Yatsu, Frank Michio
Yeung, Katherine Lu
Yielding, K Lemone
Young, Sue Ellen
Ziff, Morris

UTAH
Abildskov, J(unior) A
Ajax, Ernest Theodore
Archer, Victor Eugene
Arhelger, Roger Boyd
Athens, John William
Bailey, Richard Elmore
Baringer, J Richard
Bilbao, Marcia Kepler
Bland, Richard David
Bliss, Eugene Lawrence
Book, Linda Sue
Bragg, David Gordon
Chan, Gary Mannerstedt
Clark, Lincoln Dufton
Dethlefsen, Lyle A
Done, Alan Kimball
Englert, Edwin, Jr
Ensign, Paul Roselle
Glasgow, Lowell Alan
Herbst, John J
Hill, Harry Raymond
Koehler, P Ruben
Kolff, Willem Johan
Kuida, Hiroshi
Lahey, M Eugene
Lee, Glenn Richard
Matsen, John Martin
Mavor, Huntington
Millar, C Kay
Nelson, Alan R
Nelson, Don Harry
Noehren, Theodore Henry
Overall, James Carney, Jr
Parkin, James Lamar
Petajan, Jack Hougen
Petty, William Clayton
Pincus, Seth Henry
Prescott, Stephen M
Rawson, Rulon Wells
Renzetti, Attilio D, Jr
Ruttenberg, Herbert David
Tsagaris, Theofilos John
Tyler, Frank Hill
Urry, Ronald Lee
Ward, John Robert
Williams, Roger Richard
Wilson, Dana E
Wilson, John F

VERMONT
Albertini, Richard Joseph
Bland, John (Hardesty)
Bouchard, Richard Emile
Cooper, Sheldon Mark
Dustan, Harriet Pearson
Forsyth, Ben Ralph
Gennari, F John
Gibson, Thomas Chometon
Goran, Michael I
Gump, Dieter W
Hanson, John Sherwood
Horton, Edward S
Houston, Charles Snead
Howland, William Stapleton
Kelleher, Philip Conboy
Kelley, Jason
Kunin, Arthur Saul
Levy, Arthur Maurice
Lipson, Richard L
Lucey, Jerold Francis
McKay, Robert James, Jr
Maeck, John Van Sicklen
Martin, Herbert Lloyd
Myers, Warren Powers Laird
Page, Robert Griffith
Phillips, Carol Fenton
Reichsman, Franz Karl
Ring, B Albert
Schumacher, George Adam
Sheldon, Huntington
Sims, Ethan Allen Hitchcock
Strauss, Bella S
Tabakin, Burton Samuel
Tampas, John Peter
Waller, Julian Arnold
Weed, Lawrence Leonard
Welsh, George W, III
Wolf, George Anthony, Jr

VIRGINIA
Ackell, Edmund Ferris
Adler, Robert Alan
Astruc, Juan
Ayers, Carlos R
Ayres, Stephen McClintock
Beller, George Allan
Blackard, William Griffith

Medicine (cont)

Blaylock, W Kenneth
Blizzard, Robert M
Board, John Arnold
Bornmann, Robert Clare
Boyce, William Henry
Bressler, Bernard
Brown, Loretta Ann Port
Burke, James Otey
Calvert, Richard John
Cawley, Edward Philip
Chan, James C M
Chaudhuri, Tapan K
Chevalier, Robert Louis
Chung, Choong Wha
Cone, Clarence Donald, Jr
Connell, Alastair McCrae
Craig, James William
Crispell, Kenneth Raymond
Croft, Barbara Yoder
Culpepper, Roy M
Davis, Charles Stewart
Davis, John Staige, IV
Dreifuss, Fritz Emanuel
Dunn, John Thornton
Edlich, Richard French
Elford, Howard Lee
Elmore, Stanley Mcdowell
Epstein, Robert Marvin
Estep, Herschel Leonard
Fariss, Bruce Lindsay
Farrar, John Thruston
Ferry, Andrew P
Fisher, Lyman McArthur
Flickinger, Charles John
Fu, Shu Man
Gillenwater, Jay Young
Glauser, Frederick Louis
Goldstein, Jerome Charles
Greenberg, Jerome Herbert
Greenlee, John Edward
Gröschel, Dieter Hans Max
Guerrant, John Lippincott
Guzelian, Philip Samuel
Gwaltney, Jack Merrit, Jr
Halstead, Charles Lemuel
Hamlin, James T, III
Heaton, William Andrew Lambert
Hendley, Joseph Owen
Hess, Michael L
Hook, Edward Watson, Jr
Hotta, Shoichi Steven
Howards, Stuart S
Hunter, Thomas Harrison
Hurt, Waverly Glenn
Irby, William Robert
James, George Watson, III
Johns, Thomas Richards, II
Jones, John R
Jordan, Robert Lawrence
Kamenetz, Herman Leo
Kaul, Sanjiv
Keats, Theodore Eliot
Kelly, Thaddeus Elliott
Kendig, Edwin Lawrence, Jr
Klumpp, Theodore George
Kontos, Hermes A
Koontz, Warren Woodson, Jr
Kory, Ross Conklin
Lee, Hyung Mo
Lester, Richard Garrison
Linden, Joel Morris
McCue, Carolyn M
McCurdy, Paul Ranney
Macdonald, Roderick, Jr
Mackowiak, Robert Carl
Mahboubi, Ezzat
Maio, Domenic Anthony
Makhlouf, Gabriel Michel
Mauck, Henry Page, Jr
Mayer, William Dixon
Miller, James Q
Miller, William Weaver
Moore, Edward Weldon
Moran, Thomas James
Mullinax, Perry Franklin
Nance, Walter Elmore
Neal, Marcus Pinson, Jr
Ornato, Joseph Pasquale
Owen, John Atkinson, Jr
Ozer, Mark Norman
Pariser, Harry
Park, Herbert William, III
Pastore, Peter Nicholas
Patterson, John Legerwood, Jr
Paulsen, Elsa Proehl
Pfeiffer, Carl J
Porter, Frederick Stanley, Jr
Prindle, Richard Alan
Regelson, William
Richardson, David W
Rochester, Dudley Fortescue
Rogol, Alan David
Scheld, William Michael
Scott, Robert Bradley
Smith, Maurice John Vernon
Smith, Wade Kilgore
Spencer, Frederick J
Sprinkle, Philip Martin
Stevenson, Ian
Sturgill, Benjamin Caleb
Tegtmeyer, Charles John
Telfer, Nancy

Thorner, Michael Oliver
Thorup, Oscar Andreas, Jr
Wasserman, Albert J
Watlington, Charles Oscar
Watson, Robert Fletcher
Weary, Peyton Edwin
Westervelt, Frederic Ballard, Jr
Wheby, Munsey S
Wilds, Preston Lea
Wilson, Lester A, Jr
Wise, Robert Irby
Wolf, Barry
Wood, Maurice
Young, Reuben B
Zfass, Alvin Martin

WASHINGTON

Aagaard, George Nelson
Albert, Richard K
Aldrich, Robert Anderson
Altman, Leonard Charles
Anderson, Robert Authur
Avner, Ellis David
Bateman, Joseph R
Beeson, Paul Bruce
Bergman, Abraham
Bierman, Edwin Lawrence
Blackmon, John R
Bonica, John Joseph
Bornstein, Paul
Bremner, William John
Bruce, Robert Arthur
Burnell, James McIndoe
Butler, John
Chapman, Warren Howe
Chard, Ronald Leslie, Jr
Chase, John David
Cheney, Frederick Wyman
Couser, William Griffith
Cullen, Bruce F
Deisher, Robert William
Deyo, Richard Alden
Dodge, Harold T
Donaldson, James Adrian
Elgee, Neil J
Eschbach, Joseph W
Fialkow, Philip Jack
Figge, David C
Figley, Melvin Morgan
Finch, Clement Alfred
Fink, Bernard Raymond
Foy, Hjordis M
Fujimoto, Wilfred Y
Geyman, John Payne
Giblett, Eloise Rosalie
Gold, Eli
Goodner, Charles Joseph
Graham, C Benjamin
Guntheroth, Warren G
Hakomori, Sen-Itiroh
Harlan, John Marshall
Haviland, James West
Hellstrom, Karl Erik Lennart
Holcenberg, John Stanley
Holmes, King Kennard
Hornbein, Thomas F
Inui, Thomas S
Johnson, Murray Leathers
Jones, Robert F
Kalina, Robert E
Karp, Laurence Edward
Karr, Reynold Michael, Jr
Karstens, Andres Ingver
Kelley, Vincent Charles
Kenney, Gerald
Kirby, William M M
Klebanoff, Seymour J
Knopp, Robert H
Krieger, John Newton
Kuo, Cho-Chou
Lane, Wallace
Laramore, George Ernest
Lehmann, Justus Franz
Lemire, Ronald John
Loeb, Lawrence Arthur
Loeser, John David
Loop, John Wickwire
Mackler, Bruce
Maloff, Bruce L(arrie)
Mannik, Mart
Mason, Beryl Troxell
Merrill, James Allen
Metz, Robert John Samuel
Meyer, Roger J
Moore, Daniel Charles
Morris, Lucien Ellis
Motulsky, Arno Gunther
Nelp, Will B
Ochs, Hans D
Odland, George Fisher
Paulsen, Charles Alvin
Pavlin, Edward George
Pendergrass, Thomas Wayne
Phillips, Leon A
Pious, Donald A
Pope, Charles Edward, II
Porte, Daniel, Jr
Purnell, Dallas Michael
Rasey, Janet Sue
Rees, William James
Robertson, H Thomas, II
Robertson, William O
Rosen, Henry
Rosenblatt, Roger Alan

Rosse, Cornelius
Rubin, Cyrus E
Ruvalcaba, Rogelio H A
Sauvage, Lester Rosaire
Schwartz, Stephen Mark
Scribner, Belding Hibbard
Shepard, Thomas H
Shurtleff, David B
Simpson, David Patten
Singer, Jack W
Siscovick, David Stuart
Smith, Nathan James
Spadoni, Leon R
Sparkman, Donal Ross
Stamm, Stanley Jerome
Strandness, Donald Eugene, Jr
Swanson, Phillip D
Templeton, Frederic Eastland
Thomas, David Bartlett
Thomas, Edward Donnall
Thompson, Alvin Jerome
Todaro, George Joseph
Turck, Marvin
Van Arsdel, Paul Parr, Jr
Volwiler, Wade
Ward, Louis Emmerson
Ward, Richard John
Weaver, W Douglas
Wedgwood, Ralph Josiah Patrick
Weir, Michael Ross
Weisberg, Herbert
Whatmore, George Bernard
Willson, Richard Atwood
Wood, Francis C, Jr
Wyrick, Ronald Earl
Yarington, Charles Thomas, Jr

WEST VIRGINIA

Albrink, Margaret Joralemon
Ballantyne, Bryan
Battigelli, Mario C
Bowdler, Anthony John
Einzig, Stanley
Flink, Edmund Berney
Fugo, Nicholas William
Gabriele, Orlando Frederick
Gutmann, Ludwig
Hales, Milton Reynolds
Headings, Verle Emery
Jones, John Evan
Kartha, Mukund K
Klingberg, William Gene
Lapp, N LeRoy
Malin, Howard Gerald
Milam, Franklin D
Morgan, David Zackquill
Sheikh, Kazim
Trotter, Robert Russell
Tryfiates, George P
Welton, William Arch

WISCONSIN

Albright, Edwin C
Alexander, Samuel Craighead, Jr
Ambuel, John Philip
Azen, Edwin Allan
Bamforth, Betty Jane
Barness, Lewis Abraham
Benforado, Joseph Mark
Bennett, E Maxine
Berlow, Stanley
Blodgett, Frederic Maurice
Boake, William Charles
Borden, Ernest Carleton
Brandenburg, James H
Bryan, George Terrell
Buchsbaum, Herbert Joseph
Cadmus, Robert R
Carbone, Paul P
Chun, Raymond Wai Mun
Cohen, Bernard Allan
Cooper, Garrett
Cripps, Derek J
Crummy, Andrew B
Davis, Starkey D
Dickie, Helen Aird
Donegan, William L
Drucker, William D
Ehrlich, Edward Norman
Eichman, Peter L
Engbring, Norman H
Erwin, Chesley Para
Esterly, Nancy Burton
Farley, Eugene Shedden, Jr
Fink, Jordan Norman
Folts, John D
Gallen, William J
Goldberg, Alan Herbert
Goodfriend, Theodore L
Goodwin, James Simeon
Gould, Michael Nathan
Graziano, Frank M
Gustafson, David Harold
Hansen, Marc F
Hanson, Peter
Harb, Joseph Marshall
Harkness, Donald R
Harsch, Harold H
Hirsch, Samuel Roger
Hong, Richard
Howard, Robert Bruce
Jacobs, Lois Jean
Jefferson, James Walter
Kabler, J D

Kalkhoff, Ronald Kenneth
Keller, Robert H
Kim, Zaezeung
Klein, Ronald
Kohler, Elaine Eloise Humphreys
Kolesari, Gary Lee
Kubinski, Henry A
Lanphier, Edward Howell
Larson, Frank Clark
Lehman, Roger H
Lennon, Edward Joseph
Lobeck, Charles Champlin
Long, Sally Yates
McCarty, Daniel J, Jr
MacKinney, Archie Allen, Jr
Madsen, Paul O
Mattingly, Richard Francis
Melvin, John L
Moffet, Hugh L
Morrissey, John F
Mosesson, Michael W
Mosher, Deane Fremont, Jr
Olsen, Ward Alan
Perloff, William H(arry), Jr
Peters, Henry A
Pisciotta, Anthony Vito
Ramirez, Guillermo
Rankin, John
Rankin, John Horsley Grey
Rieselbach, Richard Edgar
Ritter, Mark Alfred
Rosenzweig, David Yates
Roth, Donald Alfred
Rowe, George G
Rudman, Daniel
Rusy, Ben F
Schatten, Heide
Schilling, Robert Frederick
Schlueter, Donald Paul
Schmidt, Donald Henry
Schultz, Richard Otto
Scudamore, Harold Hunter
Segar, William Elias
Shahidi, Nasrollah Thomas
Shrago, Earl
Silverman, Ellen-Marie
Sims, John LeRoy
Sobkowicz, Hanna Maria
Soergel, Konrad H
Sondel, Paul Mark
Spritz, Richard Andrew
Steeves, Richard Allison
Stewart, Richard Donald
Stowe, David F
Stueland, Dean T
Temin, Howard Martin
Terry, Leon Cass
Tomar, Russell H
Tormey, Douglass Cole
Treffert, Darold Allen
Turkington, Roger W
Wang, Richard I H
Warltier, David Charles
Wen, Sung-Feng
Wolberg, William Harvey
Yale, Charles E
Youker, James Edward
Zenz, Carl
Zimmerman, Stephen William

PUERTO RICO

Adamsons, Karlis, Jr
Bertran, Carlos Enrique
Brubaker, Merlin L
Garcia-Palmieri, Mario R
Haddock, Lillian
Moreno, Esteban
Pico, Guillermo
Ramirez-Ronda, Carlos Hector
Shapiro, Douglas York
Sifontes, Jose E
Sifre, Ramon Alberto

ALBERTA

Biederman, Brian Maurice
Bowen, Peter
Cochrane, William
Dawson, J W
Dewhurst, William George
Dossetor, John Beamish
Dunlop, D L
Fraser, Robert Stewart
Gilbert, James Alan Longmore
Greenhill, Stanley E
Halloran, Philip Francis
Jones, Richard Lee
Larke, R(obert) P(eter) Bryce
Leung, Alexander Kwok-Chu
McCoy, Ernest E
MacDonald, R Neil
Man, Shu-Fan Paul
Molnar, George D
Noujaim, Antoine Akl
Paul, Leendert Cornelis
Pearce, Keith Ian
Pearson, James Gordon
Percy, John Smith
Read, John Hamilton
Rorstad, Otto Peder
Rossall, Richard Edward
Rowlands, Stanley
Rudd, Noreen L
Smith, Eldon Raymond
Sproule, Brian J

Taylor, William Clyne
Thomson, Alan B R
Watanabe, Mamoru
Whitelaw, William Albert
Wyse, John Patrick Henry

BRITISH COLUMBIA
Auersperg, Nelly
Baird, Patricia A
Band, Pierre Robert
Banister, Eric Wilton
Burton, Albert Frederick
Calne, Donald Brian
Cleator, Iain Morrison
Dower, Gordon Ewbank
Drance, S M
Dunn, Henry George
Eaves, Allen Charles Edward
Elliot, Alfred Johnston
Ferguson, Alexander Cunningham
Fishman, Sherold
Ford, Denys Kensington
Friedman, Sydney Murray
Garg, Arun K
Goldie, James Hugh
Hardwick, David Francis
Jenkins, Leonard Cecil
Ledsome, John R
MacDiarmid, William Donald
Margetts, Edward Lambert
Matheson, David Stewart
Mauk, Arthur Grant
Osborne, John Alan
Osoba, David
Ovalle, William Keith, Jr
Price, John David Ewart
Quamme, Gary Arthur
Rotem, Chava Eve
Salinas, Fernando A
Sanders, Harvey David
Silver, Hulbert Keyes Belford
Slade, H Clyde
Wada, Juhn A
Young, Maurice Durward
Zaleski, Witold Andrew

MANITOBA
Adhikari, P K
Angel, Aubie
Beamish, Robert Earl
Birt, Arthur Robert
Bose, Deepak
Chernick, Victor
Fine, Adrian
Friesen, Rhinehart F
Greenberg, Arnold Harvey
Grewar, David
Harding, Godfrey Kynard Matthew
Haworth, James C
Hershfield, Earl S
Israels, Lyonel Garry
Kordova, Nonna
Lewis, Marion Jean
Lyons, Edward Arthur
MacEwan, Douglas W
Mathewson, Francis Alexander Lavens
Moorhouse, John A
Ronald, Allan Ross
Roulston, Thomas Mervyn
Schacter, Brent Allan
Schoemperlen, Clarence Benjamin
Thomson, Ashley Edwin
Weeks, John Leonard
Winter, Jeremy Stephen Drummond

NEWFOUNDLAND
Barrowman, James Adams
Bieger, Detlef
Hillman, Donald Arthur
Hillman, Elizabeth S
Ozere, Rudolph L
Pryse-Phillips, William Edward Maiben
Rees, Rolf Stephen

NOVA SCOTIA
Allen, Alexander Charles
Carr, Ronald Irving
Carruthers, Samuel George
Downie, John William
Fernandez, Louis Agnelo
Finley, John P
Goldbloom, Richard B
Grantmyre, Edward Bartlett
Hammerling, James Solomon
Hirsch, Solomon
Langley, G R
McFarlane, Ellen Sandra
Moffitt, Emerson Amos
Murray, T J
Parker, William Arthur
Purkis, Ian Edward
Stewart, Chester Bryant
Tan, Meng Hee
Tupper, W R Carl
Williams, Christopher Noel
Woodbury, John F L
Wort, Arthur John

ONTARIO
Alberti, Peter W R M
Allard, Johane
Armstrong, John Briggs
Armstrong, John Buchanan
Aye, Maung Tin

Bain, Barbara
Baker, Michael Allen
Barnett, H J M
Barr, Ronald Duncan
Beck, Ivan Thomas
Bergsagel, Daniel Egil
Bird, Charles Edward
Biro, George P
Bondy, Donald Clarence
Boone, James Edward
Boston, Robert Wesley
Brien, Francis Staples
Bryans, Alexander (McKelvey)
Buck, Carol Whitlow
Bush, Raymond Sydney
Carlen, Peter Louis
Chambers, Ann Franklin
Chiong, Miguel Angel
Chui, David H K
Clark, David A
Clarke, W T W
Collins-Williams, Cecil
Connell, Walter Ford
Cook, Michael Arnold
Couture, Roger
Crawford, John S
Davidson, Ronald G
De Veber, Leverett L
Diosy, Andrew
Dirks, John Herbert
Dolovich, Jerry
Dorian, William D
Egan, Thomas J
El-Sharkawy, Taher Youssef
Farrar, John Keith
Firstbrook, John Bradshaw
Freedman, Melvin Harris
Gammal, Elias Bichara
Garfinkel, Paul Earl
Glaser, Frederick Bernard
Goldberg, David Myer
Goldenberg, Gerald J
Goodacre, Robert Leslie
Gunton, Ramsay Willis
Haberman, Herbert Frederick
Hachinski, Vladimir C
Hansebout, Robert Roger
Harwood-Nash, Derek Clive
Hawke, W A
Haynes, Robert Brian
Heagy, Fred Clark
Hegele, Robert A
Hinton, George Greenough
Hoffstein, Victor
Hollenberg, Charles H
Howe, Geoffrey Richard
Hozumi, Nobumichi
Hryniuk, William
Hutton, Elaine Myrtle
Inman, Robert Davies
Jeejeebhoy, Khursheed Nowrojee
Jennings, Donald B
Kalant, Harold
Kang, C Yong
Khera, Kundan Singh
Kovacs, Bela A
Kovacs, Eve Maria
Laham, Souheil
Lala, Peeyush Kanti
Laski, Bernard
Le Riche, William Harding
Liao, Shuen-Kuei
Little, James Alexander
Logothetopoulos, J
Lott, James Stewart
Low, James Alexander
McAninch, Lloyd Nealson
McCredie, John A
McCulloch, Ernest Armstrong
McDonald, John William David
McNeill, Kenneth Gordon
McWhinney, Ian Renwick
Malkin, Aaron
Maykut, Madelaine Olga
Meakin, James William
Merskey, Harold
Miller, Anthony Bernard
Milliken, John Andrew
Morgan, William Keith C
Morrin, Peter Arthur Francis
Murphy, Edward G
Napke, Edward
Nisbet-Brown, Eric Robert
Orrego, Hector
Papp, Kim Alexander
Parker, John Orval
Phillipson, Eliot Asher
Pickering, Richard Joseph
Pinkerton, Peter Harvey
Posen, Gerald
Poyton, Herbert Guy
Pross, Hugh Frederick
Pruzanski, Waldemar
Rabin, Elijah Zephania
Rapoport, Abraham
Reid, Robert Leslie
Richardson, John Clifford
Roach, Margot Ruth
Rock, Gail Ann
Rosenberg, Gilbert Mortimer
Roy, Marie Lessard
Saleh, Wasfy Seleman
Sellers, Frank Jamieson
Shane, Samuel Jacob

Shephard, Roy Jesse
Sieniewicz, David James
Simon, Jerome Barnet
Sinclair, Nicholas Roderick
Skoryna, Stanley C
Slutsky, Arthur
Sole, Michael Joseph
Spaulding, William Bray
Spoerel, Wolfgang Eberhart G
Steele, Robert
Stewart, Thomas Henry McKenzie
Stiller, Calvin R
Swyer, Paul Robert
Tannock, Ian Frederick
Tobin, Sidney Morris
Tugwell, Peter
Tyson, John Edward Alfred
Valberg, Leslie S
Wallace, Alexander Cameron
Walters, Jack Henry
Warren, Douglas Robson
Watson, William Crawford
Wherrett, John Ross
White, Denis Naldrett
Whitfield, James F
Whitmore, Gordon Francis
Wigle, Ernest Douglas
Wilson, Donald Laurence
Wolfe, Bernard Martin
Woolf, C R
Zsoter, Thomas

QUEBEC
Ackman, Douglas Frederick
Aguayo, Albert J
Alpert, Elliot
Antel, Jack Perry
Aranda, Jacob Velasco
Baxter, Donald William
Becklake, Margaret Rigsby
Benard, Bernard
Benfey, Bruno Georg
Bergeron, Michel
Bolte, Edouard
Briere, Normand
Brouillette, Robert T
Burgess, John Herbert
Cameron, Donald Forbes
Cameron, Douglas George
Carrière, Serge
Chicoine, Luc
Claveau, Rosario
Collin, Pierre-Paul
Cooper, Bernard A
Davignon, Jean
Delespesse, Guy Joseph
De Margerie, Jean-Marie
DeRoth, Laszlo
Ducharme, Jacques R
Dupont, Claire Hammel
Elhilali, Mostafa M
Ferguson, Ronald James
Garcia, Raul
Genest, Jacques
Gjedde, Albert
Gold, Allen
Gold, Phil
Goresky, Carl A
Guttmann, Ronald David
Guyda, Harvey John
Hamet, Pavel
Heller, Irving Henry
Hugon, Jean S
Joncas, Jean Harry
Jones, Geoffrey Melvill
Karpati, George
Kuchel, Otto George
Lamoureux, Gilles
Lander, Philip Howard
Lanthier, Andre
Lawrence, Donald Gilbert
Lefebvre, Rene
Letarte, Jacques
Little, A Brian
Lussier, Andre (Joseph Alfred)
McGregor, Maurice
Macklem, Peter Tiffany
Marliss, Errol Basil
Mathieu, Roger Maurice
Maughan, George Burwell
Mignault, Jean De L
Miller, Sandra Carol
Mortola, Jacopo Prospero
Nadeau, Reginald Antoine
Nawar, Tewfik
Nickerson, Mark
Osmond, Dennis Gordon
Palayew, Max Jacob
Pavilanis, Vytautas
Poirier, Louis
Polosa, Canio
Rasmussen, Theodore Brown
Rola-Pleszczynski, Marek
Rose, Bram
Rosenblatt, David Sidney
Roy, Claude Charles
Ruiz-Petrich, Elena
Sainte-Marie, Guy
Schiffrin, Ernesto Luis
Schulz, Jan Ivan
Scriver, Charles Robert
Segall, Harold Nathan
Shapiro, Lorne
Sherwin, Allan Leonard

Shuster, Joseph
Tahan, Theodore Wahba
Thériault, Gilles P
Thomson, David M P
Wainberg, Mark Arnold
Watters, Gordon Valentine

SASKATCHEWAN
Gerrard, John Watson
Horlick, Louis
Houston, Clarence Stuart
Kalra, Jawahar
Kent, Henry Peter
McQueen, John Donald
Prasad, Kailash
Shokeir, Mohamed Hassan Kamel
Skinnider, Leo F
Smith, Colin McPherson
Wyant, Gordon Michael

OTHER COUNTRIES
Allessie, Maurits
Angelini, Corrado Italo
Bachmann, Fedor W
Balderrama, Francisco E
Baum, Gerald L
Benatar, Solomon Robert
Blanpain, Jan E
Boller, Francois
Borst-Eilers, Else
Brown, Nigel Andrew
Brunser, Oscar
Bunker, John Philip
Buzina, Ratko
Chang, H K
Cho, Young Won
Choi, Yong Chun
Criss, Wayne Eldon
Dyrberg, Thomas Peter
Engel, Eric
Fernández-Cruz, Eduardo P
Gemsa, Diethard
Gomez, Arturo
Gontier, Jean Roger
Gorski, Andrzej
Guerrero-Munoz, Frederico
Guimaraes, Armenio Costa
Hall, Peter
Hansson, Goran K
Haschke, Ferdinand
Helander, Herbert Dick Ferdinand
Herrmann, Walter L
Hillcoat, Brian Leslie
Hirano, Toshio
Hsu, Clement C S
Huang, Kun-Yen
Ichikawa, Shuichi
Izui, Shozo
Jonzon, Anders
Kessler, Alexander
Koch-Weser, Jan
Lambrecht, Richard Merle
Lammers, Wim
Lever, Walter Frederick
Levy, David Alfred
Lewis, David Harold
Luckey, Egbert Hugh
Marquardt, Hans Wilhelm Joe
Maruyama, Koshi
Ochoa, Severo
Partington, Michael W
Perez-Tamayo, Ruheri
Pronove, Pacita
Revillard, Jean-Pierre Rémy
Rossier, Alain B
Samuelson, Bengt Ingemar
Sasaki, Hidetada
Schena, Francesco Paolo
Shimaoka, Katsutaro
Shubik, Philippe
Solomons, Noel Willis
Solter, Davor
Spiess, Joachim
Stern, Kurt
Takai, Yasuyuki
Tanphaichitr, Vichai
Thiessen, Jacob Willem
Torún, Benjamín
Van Furth, Ralph
Wakayama, Yoshihiro
Wang, Soo Ray
Watanabe, Takeshi
Watanabe, Yoshio
Weissgerber, Rudolph E

Nursing

ALASKA
DeLapp, Tina Davis

ARIZONA
Peric-Knowlton, Wlatka

CALIFORNIA
Dracup, Kathleen A
Estes, Carroll L
Gortner, Susan Reichert
Leone, Lucile P
Styles, Margretta M

COLORADO
Fuller, Barbara Fink Brockway

Parham, Ellen Speiden
Poling, Clyde Edward
Proctor, Jerry Franklin
Rowe, William Bruce
Siedler, Arthur James
Sitrin, Michael David
Stiles, Lucille E
Subbaiah, Papasani Venkata
Tumbleson, M(yron) E(ugene)
Visek, Willard James

INDIANA
Abernathy, Richard Paul
Akrabawi, Salim S
Ali, Rida A
Biggs, Homer Gates
Burns, Robert Alexander
Coburn, Stephen Putnam
Coolidge, Ardath Anders
Cordano, Angel
Cravens, William Windsor
Doyle, Margaret Davis
Goh, Edward Hua Seng
Heinicke, Herbert Raymond
Henderson, Hala Zawawi
Ivaturi, Rao Venkata Krishna
Kirksey, Avanelle
Leary, Harvey Lee, Jr
Litov, Richard Emil
McDonald, James Lee, Jr
Melliere, Alvin L
Moore, Jerry Lamar
Morré, Dorothy Marie
Owen, George Murdock
Schneider, Donald Louis
Story, Jon Alan
Weinrich, Alan Jeffrey
Wostmann, Bernard Stephan

IOWA
Al-Jurf, Adel Saada
Chaplin, Michael H
Chung, Ronald Aloysius
Filer, Lloyd Jackson, Jr
Fomon, Samuel Joseph
Garcia, Pilar A
Kaplan, Murray Lee
Maruyama, George Masao
Rhead, William James
Roderuck, Charlotte Elizabeth
Runyan, Thora J
Runyan, William Scottie
Simpson, Robert John
Speer, Vaughn C
Swan, Patricia B
Thomas, Byron Henry
Trenkle, Allen H
Whitehead, Floy Eugenia
Ziegler, Ekhard E

KANSAS
Deyoe, Charles W
Frakes, Elizabeth (McCune)
Fryer, Elsie Beth
Grunewald, Katharine Klevesahl
Hinshaw, Charles Theron, Jr
Koo, Sung Il
Newell, Kathleen
Parrish, Donald Baker
Reeves, Robert Donald
Smith, Meredith Ford
Tomita, Tatsuo
Tsen, Cho Ching
Ward, George Merrill

KENTUCKY
Anderson, James Wingo
Cantor, Austin H
Chen, Linda Li-Yueh Huang
Chen, Theresa S
Lang, Calvin Allen
Marlatt, Abby Lindsey
Packett, Leonard Vasco
Thornton, Paul A

LOUISIANA
Bhattacharyya, Ashim Kumar
Burch, Robert Emmett
Cason, James Lee
Clemetson, Charles Alan Blake
Guzman Foresti, Miguel Angel
Hegsted, Maren
Herbert, Jack Durnin
Jain, Sushil Kumar
Kokatnur, Mohan Gundo
Lewis, Harvye Fleming
Liuzzo, Joseph Anthony
Nakamoto, Tetsuo
Ory, Robert Louis
Windhauser, Marlene M
York, David Anthony
Younathan, Margaret Tims

MARYLAND
Ahrens, Richard August
Allison, Richard Gall
Anderson, Richard Allen
Anderson, Sue Ann Debes
Beaton, John Rogerson
Beliveau, Gale Patrice
Bieri, John Genther
Bodwell, Clarence Eugene
Boehne, John William
Burton, Benjamin Theodore

Danford, Darla E
Dewys, William Dale
Dupont, Jacqueline (Louise)
Forman, Michele R
Foster, Willis Roy
Gallo-Torres, Hugo E
Graham, George G
Hallfrisch, Judith
Hansard, Samuel Leroy
Howe, Juliette Coupain
Hubbard, Van Saxton
Kahn, Samuel George
Kelsay, June Lavelle
Khan, Mushtaq Ahmad
Kim, Sooja K
Kleinman, Hynda Karen
Low, Mary Alice
Mertz, Walter
Milner, Max
Moment, Gairdner Bostwick
Morris, Eugene Ray
Morris, Rosemary Shull
Morrissey, Robert Leroy
Moser-Veillon, Phylis B
Murphy, Elizabeth Wilcox
Osman, Jack Douglas
Paige, David M
Pilch, Susan Marie
Polansky, Marilyn MacArthur
Prather, Mary Elizabeth Sturkie
Read, Merrill Stafford
Rider, Agatha Ann
Schoene, Norberta Wachter
Slyter, Leonard L
Soares, Joseph Henry, Jr
Szepesi, Bela
Taylor, Martha Loeb
Thorp, James Wilson
Underwood, Barbara Ann
Vaughan, David Arthur
Veillon, Claude
Wannemacher, Robert, Jr
Watkins, Paul Allan
Welsh, Susan
White, Edward Austin
Wiech, Norbert Leonard
Williams, Eleanor Ruth
Wilson, Christine Shearer
Wolf, Wayne Robert
Wong, Dennis Mun
Wykes, Arthur Albert

MASSACHUSETTS
Askew, Eldon Wayne
Beal, Virginia Asta
Bistrian, Bruce Ryan
Blackburn, George L
Blumberg, Jeffrey Bernard
Brown, Robert Glenn
Carter, Edward Albert
Cohen, Saul Israel
Cunningham, John James
Damassa, David Allen
De Nisco, Stanley Gabriel
Deutsch, Marshall Emanuel
Dietz, William H
Dwyer, Johanna T
Fillios, Louis Charles
Flatt, Jean-Pierre
Frisch, Rose Epstein
Fuqua, Mary Elizabeth
Galler, Janina Regina
Gershoff, Stanley Norton
Goldman, Peter
Gordon, Philip Ray
Hayes, Kenneth Cronise
Hegsted, David Mark
Hoffer, Alan R
Jordan, Constance (Louise) Brine
LeBaron, Francis Newton
Lees, Robert S
Lloyd, Weldon S
Lo, Clifford W
Loew, Franklin Martin
Maher, Timothy John
Mason, Joel B
Mason, Marion
Mayer, Jean
Munro, Hamish N
Newberne, Paul M
Nizel, Abraham Edward
Rosenberg, Irwin Harold
Roubenoff, Ronenn
Russell, Robert M
Scrimshaw, Nevin Stewart
Shaw, James Headon
Stare, Fredrick J
Thomas, Miriam Mason Higgins
Vitale, Joseph John
Wurtman, Judith Joy
Young, Vernon Robert

MICHIGAN
Bartholmey, Sandra Jean
Bennink, Maurice Ray
Bernard, Rudy Andrew
Bivins, Brack Allen
Block, Walter David
Bond, Jenny Taylor
Carlotti, Ronald John
Chenoweth, Wanda L
Clarke, Steven Donald
Cossack, Zafrallah Taha
Cotter, Richard

Craig, Winston John
Cullum, Malford Eugene
Dunbar, Joseph C
Erdman, Anne Marie
Garn, Stanley Marion
Grinker, Joel A
Hawthorne, Victor Morrison
Hegarty, Patrick Vincent
Larkin, Frances Ann
Luecke, Richard William
Mahajan, Sudesh K
Marsh, Alice Garrett
Martin, John Lee
Mehring, Jeffrey Scott
Mutch, Patricia Black
Purvis, George Allen
Riggs, Thomas Rowland
Romsos, Dale Richard
Saldanha, Leila Genevieve
Sardesai, Vishwanath M
Sato, Paul Hisashi
Schemmel, Rachel A
Stanley, John Pearson
Stowe, Howard Denison
Tsai, Alan Chung-Hong
Whitehair, Charles Kenneth
Yokoyama, Melvin T

MINNESOTA
Addis, Paul Bradley
Anderson, Carl Frederick
Anderson, Ingrid
Celender, Ivy M
Evans, Gary William
Gallaher, Daniel David
Ganguli, Mukul Chandra
Gannon, Mary Carol
Gordon, Joan
Gormican, Annette
Holman, Ralph Theodore
Keys, Ancel (Benjamin)
Liener, Irvin Ernest
Longley, Robert W(illiam)
Moxness, Karen
Otterby, Donald Eugene
Palumbo, Pasquale John
Polzin, David J
Prohaska, Joseph Robert
Stockland, Wayne Luvern
Thenen, Shirley Warnock

MISSISSIPPI
Angel, Charles
Futrell, Mary Feltner
Kilgore, Lois Taylor
Long, Billy Wayne
Ousterhout, Lawrence Elwyn
Verlangieri, Anthony Joseph

MISSOURI
Asplund, John Malcolm
Baile, Clifton Augustus, III
Chi, Myung Sun
Flynn, Margaret A
Gordon, Dennis T
Greene, Daryle E
Hill, Gretchen Myers
Hoover, Loretta White
Horwitt, Max Kenneth
House, William Burtner
Lau, Brad W C
MacDonald, Ruth S
Makdani, Dhiren D
Moore, Aimee N
O'Dell, Boyd Lee
Theuer, Richard C
Unklesbay, Nan F
Vandepopuliere, Joseph Marcel
Whitley, James R

MONTANA
Pagenkopf, Andrea L

NEBRASKA
Beach, Betty Laura
Clark, Ralph B
Dam, Richard
Driskell, Judy Anne
Fox, Hazel Metz
Grandjean, Carter Jules
Kies, Constance
Littledike, Ernest Travis
Schaefer, Arnold Edward

NEVADA
Read, Marsha H

NEW JERSEY
Alfano, Michael Charles
Baker, Herman
Bendich, Adrianne
Bhagavan, Hemmige
Bieber, Mark Allan
Bierenbaum, Marvin L
Bogden, John Dennis
Brogle, Richard Charles
Brush, Miriam Kelly
Cordts, Richard Henry, Jr
Daggs, Ray Gilbert
Dumm, Mary Elizabeth
Evans, Joseph Liston
Feigenbaum, Abraham Samuel
Fisher, Hans
Frank, Oscar

Garro, Anthony Joseph
Gaut, Zane Noel
Griminger, Paul
Hager, Mary Hastings
Harkins, Robert W
Johnson, Dewey, Jr
Keane, Kenneth William
Lachance, Paul Albert
Leevy, Carroll M
Leveille, Gilbert Antonio
Liebowitz, Stephen Marc
Lin, Grace Woan-Jung
Mackway-Girardi, Ann Marie
Miller, Gary A
Morgan, Karen J
Newmark, Harold Leon
Oien, Helen Grossbeck
Reussner, George Henry
Rusoff, Irving Isadore
Sherman, Adria Rothman
Singh, Vishwa Nath
Smith, Robert Ewing
Stevenson, Nancy Roberta
Stillings, Bruce Robert
Yacowitz, Harold

NEW MEXICO
Aberle, Sophie D
Corcoran, George Bartlett, III
Duncan, Irma W
Garry, Philip J

NEW YORK
Apgar, Barbara Jean
Austic, Richard Edward
Awad, Atif B
Axelson, Marta Lynne
Baldini, James Thomas
Barac-Nieto, Mario
Bensadoun, Andre
Berg, Benjamin Nathan
Berlyne, Geoffrey Merton
Bigler, Rodney Errol
Bowering, Jean
Boykin, Lorraine Stith
Brennan, Murray F
Browe, John Harold
Campbell, Thomas Colin
Carroll, Harvey Franklin
Chauhan, Ved P S
Chutkow, Jerry Grant
Coccodrilli, Gus D, Jr
Colman, Neville
Combs, Gerald Fuson, Jr
Contento, Isobel
Deschner, Eleanor Elizabeth
Devine, Marjorie M
Dibble, Marjorie Veit
Fabry, Thomas Lester
Famighetti, Louise Orto
Frankle, Reva Treelisky
Fusco, Carmen Louise
Gambert, Steven Ross
Garza, Cutberto
Geliebter, Allan
Greenberg, Danielle
Haas, Jere Douglas
Habicht, Jean-Pierre
Hackler, Lonnie Ross
Hashim, Sami A
Himes, John Harter
Hirsh, Kenneth Roy
Ip, Margot Morris
Johnson, Patricia R
Kallfelz, Francis A
Karmali, Rashida A
Kaunitz, Hans
Kotler, Donald P
Kraft, Patricia Lynn
Kroenberg, Berndt
Latham, Michael Charles
Lebenthal, Emanuel
Leleiko, Neal Simon
Levenson, Stanley Melvin
Lieber, Charles Saul
Lowry, Stephen Frederick
Magliulo, Anthony Rudolph
Mansour, Moustafa M
Martin, Kumiko Oizumi
Matthews, Dwight Earl
Miller, Dennis Dean
Montefusco, Cheryl Marie
Munves, Elizabeth Douglass
Nesheim, Malden Charles
Olson, Robert Eugene
Pearson, Thomas Arthur
Pi-Sunyer, F Xavier
Prutkin, Lawrence
Rambaut, Paul Christopher
Rao, Yalamanchili A K
Rasmussen, Kathleen Maher
Reddy, Bandaru Sivarama
Rivlin, Richard S
Roe, Daphne A
Rosensweig, Norton S
Schachter, Michael Ben
Schwartz, Ernest
Schwartz, Ruth
Scott, Milton Leonard
Seifter, Eli
Sheffy, Ben Edward
Short, Sarah Harvey
Short, Sarah Harvey
Stephenson, Lani Sue

Gregg, James R
Haines, Richard Foster
Hawryluk, Andrew Michael
Hopping, Richard Lee
Mandell, Robert Burton
Marg, Elwin
Margach, Charles Boyd
Morgan, Meredith Walter
Randle, Robert James
Tyler, Christopher William

CONNECTICUT
Telfair, William Boys

FLORIDA
Borish, Irvin Max
Weber, James Edward

ILLINOIS
Lit, Alfred
Rosenbloom, Alfred A, Jr

INDIANA
Allen, Merrill James
Everson, Ronald Ward
Heath, Gordon Glenn
Hofstetter, Henry W
Soni, Prem Sarita

MARYLAND
Ferris, Frederick L
Levinson, John Z
Massof, Robert W
Michaelis, Otho Ernest, IV
Yau, King-Wai

MASSACHUSETTS
Carter, John Haas
Sekuler, Robert W

MICHIGAN
Zacks, James Lee

NEW JERSEY
Roy, M S

NEW MEXICO
Wong, Siu Gum

NEW YORK
Bannon, Robert Edward
Goldrich, Stanley Gilbert
Haffner, Alden Norman
Hainline, Louise
Howland, Howard Chase
Lewis, Alan Laird
Rosenberg, Robert

OHIO
Hebbard, Frederick Worthman
Schoessler, John Paul

OKLAHOMA
Goss, David A

OREGON
Levine, Leonard
Ludlam, William Myrton

PENNSYLVANIA
Dobson, Margaret Velma

RHODE ISLAND
Lehmkuhle, Stephen W

TENNESSEE
Elberger, Andrea June
Levene, John Reuben

TEXAS
Baldwin, William Russell
Flom, Merton Clyde
Hecht, Gerald
Pease, Paul Lorin
Pheiffer, Chester Harry
Pitts, Donald Graves
Stein, Jerry Michael

VIRGINIA
Gabriel, Barbra L

WASHINGTON
Young, Francis Allan

PUERTO RICO
Afanador, Arthur Joseph

ALBERTA
Thomson, Alan B R

BRITISH COLUMBIA
Cynader, Max Sigmund

ONTARIO
Hallett, Peter Edward
Lyle, William Montgomery
Remole, Arnulf
Sivak, Jacob Gershon
Woo, George Chi Shing

OTHER COUNTRIES
Hughes, Abbie Angharad

Parasitology

ALABAMA
Bailey, Wilford Sherrill
Boyles, James McGregor
Buckner, Richard Lee
Current, William L
Dixon, Carl Franklin
Eckman, Michael Kent
Frandsen, John Christian
Kayes, Stephen Geoffrey
Klesius, Phillip Harry
Schmidt, Leon Herbert

ALASKA
Esslinger, Jack Houston
Fay, Francis Hollis
Swartz, Leslie Gerard

ARIZONA
Collins, Richard Cornelius
Dewhirst, Leonard Wesley
Van Riper, Charles, III
Wilkes, Stanley Northrup

ARKANSAS
Bridgman, John Francis
Daly, James Joseph
Hinck, Lawrence Wilson
Johnson, Arthur Albin

CALIFORNIA
Amrein, Yost Ursus Lucius
Anderson, John Richard
Ash, Lawrence Robert
Bair, Thomas De Pinna
Baker, Norman Fletcher
Basch, Paul Frederick
Beck, Albert J
Chao, Jowett
Chi, Lois Wong
Cornford, Eain M
Cosgrove, Gerald Edward
Davis, Charles Edward
Demaree, Richard Spottswood, Jr
Dowell, Armstrong Manly
Furman, Deane Philip
Garcia, Richard
Gilbreath, Michael Joye
Golder, Thomas Keith
Good, Anne Haines
Hackney, Robert Ward
Harmon, Wallace Morrow
Heyneman, Donald
Ho, Ju-Shey
Hoke, Carolyn Kay
Horvath, Kalman
Hunderfund, Richard C
Hunter, George William, III
Jones, Ira
Krassner, Stuart M
Kuris, Armand Michael
MacInnis, Austin J
Mankau, Sarojam Kurudamannil
Markell, Edward Kingsmill
Mitchell, John Alexander
Nahhas, Fuad Michael
Nichols, Barbara Ann
Nickel, Phillip Arnold
Noble, Elmer Ray
Noble, Glenn Arthur
Olson, Andrew Clarence, Jr
Schacher, John Fredrick
Sherman, Irwin William
Silverman, Paul Hyman
Uyeda, Charles Tsuneo
Vickrey, Herta Miller
Wagner, Edward D
Wang, Ching Chung
Weinmann, Clarence Jacob
Westervelt, Clinton Albert, Jr
Widmer, Elmer Andreas
Wong, Ming Ming
Woodruff, David Scott
Young, Robert, Jr

COLORADO
Berens, Randolph Lee
Grieve, Robert B
Gubler, Duane J
Hathaway, Ronald Philip
McLean, Robert George
Marquardt, William Charles
Schmidt, Gerald D
Vetterling, John Martin
Voth, David Richard

CONNECTICUT
Eustice, David Christopher
Howes, Harold, Jr
James, Hugo A
Krause, Peter James
Krinsky, William Lewis
Levine, Harvey Robert
Lynch, John Edward
Newborg, Michael Foxx
O'Neill, James F
Patton, Curtis LeVerne
Richards, Frank Frederick

DELAWARE
Cubberley, Virginia

DISTRICT OF COLUMBIA
Cressey, Roger F
Diggs, Carter Lee
Gravely, Sally Margaret
Jackson, George John
Mason, James O
Phillips, Terence Martyn
Schneider, Imogene Pauline
Taylor, Diane Wallace
Ward, Ronald Anthony

FLORIDA
Brenner, Richard Joseph
Burridge, Michael John
CmejLa, Howard Edward
Couch, John Alexander
Courtney, Charles Hill
Crandall, Richard B
Ferguson, Malcolm Stuart
Forrester, Donald Jason
Fournie, John William
Friedl, Frank Edward
Lawrence, Pauline Olive
Porter, James Armer, Jr
Rodrick, Gary Eugene
Rossan, Richard Norman
Sengbusch, Howard George
Short, Robert Brown
Splitter, Earl John
Telford, Sam Rountree, Jr
Young, Martin Dunaway
Zam, Stephen G, III

GEORGIA
Agosin, Moises
Ah, Hyong-Sun
Brooke, Marion Murphy
Damian, Raymond T
French, Frank Elwood, Jr
Hanson, William Lewis
Healy, George Richard
Husain, Ansar
Jones, Gilda Lynn
Kagan, Irving George
Kaiser, Robert L
Lammie, Patrick J
McDougald, Larry Robert
McDowell, John Willis
Maddison, Shirley Eunice
Maples, William Paul
Mathews, Henry Mabbett
Melvin, Dorothy Mae
Nguyen-Dinh, Phuc
Oliver, James Henry, Jr
Pratt, Harry Davis
Prestwood, Annie Katherine
Reid, Willard Malcolm
Roberson, Edward Lee
Saladin, Kenneth S
Sulzer, Alexander Jackson
Tsang, Victor Chiu Wana
Walls, Kenneth W
Warren, McWilson

HAWAII
Desowitz, Robert
Noda, Kaoru
Siddiqui, Wasim A

IDAHO
Hibbert, Larry Eugene
Schell, Stewart Claude
Wootton, Donald Merchant

ILLINOIS
Auerbach, Earl
Chandran, Satish Raman
Chang, Kwang-Poo
Dyer, William Gerald
French, Allen Lee
Garoian, George
Huizinga, Harry William
Jaskoski, Benedict Jacob
Kniskern, Verne Burton
Larson, Ingemar W
Leathem, William Dolars
Levine, Norman Dion
Murrell, Kenneth Darwin
Myer, Donal Gene
Nollen, Paul Marion
Ridgeway, Bill Tom
Seed, Thomas Michael
Todd, Kenneth S, Jr
von Zellen, Bruce Walfred
Yogore, Mariano G, Jr

INDIANA
Bowen, Richard Eli
Denner, Melvin Walter
Diffley, Peter
Dusanic, Donald G
Gaafar, Sayed Mohammed
Headlee, William Hugh
Jeffers, Thomas Kirk
McCowen, Max Creager
McCracken, Richard Owen
Ott, Karen Jacobs
Saz, Howard Jay
Summers, William Allen, Sr
Thorson, Ralph Edward
Vatne, Robert Dahlmeier
Weinstein, Paul P
Welker, George W

IOWA
Cain, George D
Collins, Richard Francis
Greve, John Henry
Hsu, Hsi Fan
Hsu, Shu Ying Li
Marty, Wayne George
Maruyama, George Masao
Nolf, Luther Owen
Petri, Leo Henry

KANSAS
Coil, William Herschell
Frenkel, Jacob Karl
Hansen, Merle Fredrick
Johnson, John Christopher, Jr

KENTUCKY
Drudge, J Harold
Gleason, Larry Neil
Justus, David Eldon
Lyons, Eugene T
Oetinger, David Frederick
Sonnenfeld, Gerald
Uglem, Gary Lee
Whittaker, Frederick Horace

LOUISIANA
Abadie, Stanley Herbert
Abram, James Baker, Jr
Boertje, Stanley
Brown, Kenneth Michael
Christian, Frederick Ade
Corkum, Kenneth C
D'Alessandro-Bacigalupo, Antonio
Durio, Walter O'Neal
Font, William Francis
Harris, Alva H
Jung, Rodney Clifton
Klei, Thomas Ray
Krupp, Iris M
Lanners, H Norbert
Little, Maurice Dale
Lumsden, Robert Dale
Malek, Emile Abdel
Miller, Joseph Henry
Murad, John Louis
Orihel, Thomas Charles
Page, Clayton R, III
Sizemore, Robert Carlen
Sodeman, William A
Specian, Robert David
Stewart, T Bonner
Stueben, Edmund Bruno
Swartzwelder, John Clyde
Trapido, Harold
Turner, Hugh Michael
Vidrine, Malcolm Francis
Warren, Lionel Gustave
Williams, James Carl
Yaeger, Robert George

MAINE
Chute, Robert Maurice
Gibbs, Harold Cuthbert
Meyer, Marvin Clinton
Najarian, Haig Hagop
Turner, James Henry

MARYLAND
Anderson, Norman Leigh
Ansher, Sherry Singer
Augustine, Patricia C
Bala, Shukal
Bram, Ralph A
Carney, William Patrick
Cheever, Allen
Colglazier, Merle Lee
Diamond, Louis Stanley
Djurickovic, Draginja Branko
Dubey, Jitender Prakash
Dwyer, Dennis Michael
Erickson, Duane Gordon
Fayer, Ronald
Fisher, Pearl Davidowitz
Gasbarre, Louis Charles
Gibson, Colvin Lee
Graham, Charles Lee
Gwadz, Robert Walter
Haley, Albert James
Hollingdale, Michael Richard
Jackson, Peter Richard
James, Stephanie Lynn
Lawrence, Dale Nolan
Lichtenfels, James Ralph
Lillehoj, Hyun Soon
Lunde, Milford Norman
Lunney, Joan K
Mercado, Teresa I
Miller, Louis Howard
Munson, Donald Albert
O'Grady, Richard Terence
Otto, Gilbert Fred
Palmer, Timothy Trow
Porter, Clarence A
Powers, Kendall Gardner
Rew, Robert Sherrard
Ruff, Michael David
Scheibel, Leonard William
Schiller, Everett L
Sheffield, Harley George
Slonka, Gerald Francis
Stirewalt, Margaret Amelia
Strickland, George Thomas

Parasitology (cont)

MASSACHUSETTS (cont, first col header)
Sztein, Marcelo Benjamin
Tarshis, Irvin Barry
Taylor, David Neely
Weisbroth, Steven H
Zierdt, Charles Henry

MASSACHUSETTS
Bruce, John Irvin
Caulfield, John P
Chernin, Eli
Chishti, Athar Husain
Coleman, Robert Marshall
Dammin, Gustave John
Duncan, Stewart
Honigberg, Bronislaw Mark
Humes, Arthur Grover
Hurley, Francis Joseph
Meszoely, Charles Aladar Maria
Nutting, William Brown
Pan, Steve Chia-Tung
Rivera, Ezequiel Ramirez
Spielman, Andrew
Telford, Sam Rountree, III
West, Arthur James, II
Wyler, David J

MICHIGAN
Blankespoor, Harvey Dale
Bratt, Albertus Dirk
Chobotar, Bill
Conder, George Anthony
De Giusti, Dominic Lawrence
Elslager, Edward Faith
Folz, Sylvester D
Hupe, Donald John
Jensen, James Burt
Maxey, Brian William
Peebles, Charles Robert
Rizzo, Donald Charles
Shaver, Robert John
Steen, Edwin Benzel
Szanto, Joseph
Twohy, Donald Wilfred
Waffle, Elizabeth Lenora
Werner, John Kirwin
Williams, Jeffrey F
Yates, Jon Arthur

MINNESOTA
Ballard, Neil Brian
Bartel, Monroe H
Bemrick, William Joseph
Eaton, John Wallace
Gilbertson, Donald Edmund
Lange, Robert Echlin, Jr
McCue, John Francis
Meyer, Fred Paul
Paulson, Carlton
Preiss, Frederick John
Robinson, Edwin James, Jr
Schlotthauer, John Carl
Stromberg, Bert Edwin, Jr
Thompson, John Harold, Jr
Wagenbach, Gary Edward
Wallace, Franklin Gerhard
Wyatt, Ellis Junior

MISSISSIPPI
Hutchison, William Forrest
Lawler, Adrian Russell
Norris, Donald Earl, Jr
Overstreet, Robin Miles
White, Jesse Steven

MISSOURI
Bethel, William Mack
Bose, Subir Kumar
Green, Theodore James
Hall, Robert Dickinson
Harrington, Glenn William
Miles, Donald Orval
Poeschel, Gordon Paul
Shelton, George Calvin
Starling, Jane Ann
Train, Carl T
Twining, Linda Carol
Wagner, Joseph Edward
Weppelman, Roger Michael

MONTANA
Mitchell, Lawrence Gustave
Speer, Clarence Arvon
Thomas, Leo Alvon
Worley, David Eugene

NEBRASKA
Ferguson, Donald Leon
Nickol, Brent Bonner
Phares, Cleveland Kirk
Pritchard, Mary (Louise) Hanson

NEVADA
Babero, Bert Bell
Mead, Robert Warren

NEW HAMPSHIRE
Bullock, Wilbur Lewis
Harrises, Antonio Efthemios
Pfefferkorn, Elmer Roy, Jr
Strout, Richard Goold

NEW JERSEY
Bacha, William Joseph, Jr
Berger, Harold
Burrows, Robert Beck
Crane, Mark
Dennis, Emmet Adolphus
Egerton, John Richard
Foster, William Burnham
Griffo, James Vincent, Jr
Guerrero, Jorge
Hayes, Terence James
Kantor, Sidney
Katz, Frank Fred
Leaning, William Henry Dickens
Leibowitz, Michael Jonathan
Levine, Donald Martin
Ostlind, Dan A
Schenkel, Robert H
Schulman, Marvin David
Tierno, Philip M, Jr
Virkar, Raghunath Atmaram
Wang, Guang Tsan
Wood, Irwin Boyden

NEW MEXICO
Davis, Larry Ernest
Duszynski, Donald Walter
Reyes, Philip
Tonn, Robert James

NEW YORK
Ablin, Richard J
Asch, Harold Lawrence
Bell, Robin Graham
Bettencourt, Joseph S, Jr
Clarkson, Allen Boykin, Jr
Coderre, Jeffrey Albert
Concannon, Joseph N
Cross, George Alan Martin
D'Alesandro, Philip Anthony
Dean, David A
Dennis, Emery Westervelt
Despommier, Dickson
Drobeck, Hans Peter
Gold, Allen Morton
Gordon, Morris Aaron
Grimwood, Brian Gene
Habicht, Gail Sorem
Hutner, Seymour Herbert
Imperato, Pascal James
Ingalls, James Warren, Jr
Isseroff, Hadar
Kaplan, Eugene Herbert
Keithly, Janet Sue
Kim, Charles Wesley
Kohls, Robert E
Levin, Norman Lewis
Lo Verde, Philip Thomas
McGraw, James Carmichael
Mackiewicz, John Stanley
McNamara, Thomas Francis
Muller, Miklós
Neenan, John Patrick
Nigrelli, Ross Franco
Nussenzweig, Victor
Pike, Eileen Halsey
Preisler, Harvey D
Reuter, Gerald Louis
Ritterson, Albert L
Santiago, Noemi
Shields, Robert James
Soave, Rosemary
Spencer, Frank
Stephenson, Lani Sue
Styles, Twitty Junius
Trager, William
Vanderberg, Jerome Philip
Vasey, Carey Edward
Weiner, Matei
Weires, Richard William, Jr
Whitlock, John Hendrick
Wittner, Murray
Wood, Raymond Arthur
Yarinsky, Allen
Yolles, Tamarath Knigin

NORTH CAROLINA
Batte, Edward G
Behlow, Robert Frank
Coffey, James Cecil, Jr
Esch, Gerald Wisler
Estevez, Enrique Gonzalo
Eure, Herman Edward
George, Charles Redgenal
Goulson, Hilton Thomas
Hampton, Carolyn Hutchins
Hendricks, James Richard
Henson, Richard Nelson
Knuckles, Joseph Lewis
LaFon, Stephen Woodrow
Lewert, Robert Murdoch
Martin, Virginia Lorelle
Noga, Edward Joseph
Nusser, Wilford Lee
Pattillo, Walter Hugh, Jr
Perkins, Kenneth Warren
Pung, Oscar J
Radovsky, Frank Jay
Roberts, John Fredrick
Seed, John Richard
Shepperson, Jacqueline Ruth
Spuller, Robert L
Weatherly, Norman F

NORTH DAKOTA
Andrews, Myron Floyd
Holloway, Harry Lee, Jr
Larson, Omer R

OHIO
Aikawa, Masamichi
Arlian, Larry George
Barriga, Omar Oscar
Barrow, James Howell, Jr
Blanton, Ronald Edward
Crites, John Lee
Etges, Frank Joseph
Grassmick, Robert Alan
Hallberg, Carl William
Heck, Oscar Benjamin
Hink, Walter Fredric
Jarroll, Edward Lee, Jr
Kapral, Frank Albert
Kochanowski, Barbara Ann
Kreier, Julius Peter
Mahmoud, Adel A F
Malemud, Charles J
Myers, Betty June
Needham, Glen Ray
Pappas, Peter William
Rabalais, Francis Cleo
Schaefer, Frank William, III
Van Zandt, Paul Doyle

OKLAHOMA
Branch, John Curtis
Burr, John Green
Ewing, Sidney Alton
Ivey, Michael Hamilton
John, David Thomas
Jordan, Helen Elaine
McKnight, Thomas John
Powders, Vernon Neil
Rogers, Steffen Harold

OREGON
Bayne, Christopher Jeffrey
Carter, Richard Thomas
Clark, David Thurmond
McConnaughey, Bayard Harlow
Macy, Ralph William
Olson, Robert Eldon
Rossignol, Philippe Albert

PENNSYLVANIA
Anderson, David Robert
Boutros, Susan Noblit
Dollahon, Norman Richard
Donnelly, John James, III
Foor, W Eugene
Fremount, Henry Neil
Fried, Bernard
Harms, Clarence Eugene
Hendrix, Sherman Samuel
Humphreys, Jan Gordon
Iralu, Vichazelhu
McManus, Edward Clayton
McNair, Dennis M
Mandel, John Herbert
Moss, William Wayne
Ogren, Robert Edward
Rogers, William Edwin
Ryan, James Patrick
Salerno, Ronald Anthony
Schad, Gerhard Adam
Sillman, Emmanuel I
Slifkin, Malcolm
Theodorides, Vassilios John
Vaidya, Akhil Babubhai
Walton, Bryce Calvin
Ward, William Francis
Williams, Russell Raymond
Yurchenco, John Alfonso
Zarkower, Arian

RHODE ISLAND
Hyland, Kerwin Ellsworth, Jr
Jones, Kenneth Wayne
Knopf, Paul M
Kresina, Thomas Francis
Senft, Alfred Walter

SOUTH CAROLINA
Cheng, Thomas Clement
Kowalczyk, Jeanne Stuart
Sargent, Roger Gary
Schmidt, Roger Paul
Vernberg, Winona B

SOUTH DAKOTA
Hugghins, Ernest Jay
Peanasky, Robert Joseph

TENNESSEE
Bogitsh, Burton Jerome
Bradley, Richard E
Colley, Daniel George
Cypess, Raymond Harold
Elliott, Alice
Harrison, Robert Edwin
Hill, George Carver
McGavock, Walter Donald
Patton, Sharon
Prudhon, Rolland A, Jr
Risby, Edward Louis
Wilhelm, Walter Eugene

TEXAS
Berry, Jewel Edward
Box, Edith Darrow
Bristol, John Richard
Dronen, Norman Obert, Jr
Ewert, Adam
Fisher, Frank M, Jr
Fisher, William Francis
Galvin, Thomas Joseph
Hejtmancik, Kelly Erwin
Huffman, David George
Kuntz, Robert Elroy
McGraw, John Leon, Jr
Meade, Thomas Gerald
Meola, Roger Walker
Moore, Donald Vincent
Morrison, Eston Odell
Niederkorn, Jerry Young
Race, George Justice
Richerson, Jim Vernon
Roberts, Larry Spurgeon
Rolston, Kenneth Vijaykumar Issac
Schlueter, Edgar Albert
Scholl, Philip Jon
Shadduck, John Allen
Slagle, Wayne Grey
Smith, Jerome H
Sogandares-Bernal, Franklin
Standifer, Lonnie Nathaniel
Stewart, George Louis
Tamsitt, James Ray
Taylor, D(orothy) Jane
Trevino, Gilberto Stephenson
Tuff, Donald Wray
Tulloch, George Sherlock
Turco, Charles Paul
Ubelaker, John E
Valdivieso, Dario

UTAH
Andersen, Ferron Lee
Fitzgerald, Paul Ray
Grundmann, Albert Wendell
Healey, Mark Calvin
Heckmann, Richard Anderson
Jensen, Emron Alfred
Tipton, Vernon John
Warnock, Robert G

VERMONT
Schall, Joseph Julian

VIRGINIA
AcKerman, Larry Joseph
Burreson, Eugene M
Cruthers, Larry Randall
Eckerlin, Ralph Peter
Fisher, Elwood
Holliman, Rhodes Burns
Lincicome, David Richard
Osborne, Paul James
Russell, Catherine Marie
Schwartzman, Joseph David

WASHINGTON
Allen, Julia Natalia
Catts, Elmer Paul
Clark, Glen W
Krieger, John Newton
Lang, Bruce Z
Rausch, Robert Lloyd
Rigby, Donald W
Senger, Clyde Merle
Sibley, Carol Hopkins
Stuart, Kenneth Daniel
Wescott, Richard Breslich

WEST VIRGINIA
Hall, John Edgar
Hoffman, Glenn Lyle

WISCONSIN
Burns, William Chandler
Coggins, James Ray
Guilford, Harry Garrett
McDonald, Malcolm Edwin
Nash, Reginald George
Oaks, John Adams
Rouse, Thomas C
Wittrock, Darwin Donald
Yoshino, Timothy Phillip

WYOMING
Bergstrom, Robert Charles
Kingston, Newton

PUERTO RICO
Hillyer, George Vanzandt
Kozek, Wieslaw Joseph
Oliver-Gonzales, Jose

ALBERTA
Arai, Hisao Philip
Baron, Robert Walter
Befus, A(lbert) Dean
Holmes, John Carl
Mahrt, Jerome L
Samuel, William Morris
Wallis, Peter Malcolm

BRITISH COLUMBIA
Adams, James Russell
Bower, Susan Mae
Ching, Hilda

Margolis, Leo
Webster, John Malcolm

MANITOBA
Novak, Marie Marta

NEW BRUNSWICK
Burt, Michael David Brunskill
Smith, Harry John

NEWFOUNDLAND
Threlfall, William

NOVA SCOTIA
Angelopoulos, Edith W
Wiles, Michael

ONTARIO
Anderson, Roy Clayton
Bourns, Thomas Kenneth Richard
Brooks, Daniel Rusk
Chadwick, June Stephens
Desser, Sherwin S
Fernando, Constantine Herbert
Freeman, Reino Samuel
Mellors, Alan
Mettrick, David Francis
Slocombe, Joseph Owen Douglas
Tryphonas, Helen
Wright, Kenneth A

QUEBEC
Meerovitch, Eugene
Rau, Manfred Ernst
Stevenson, Mary M
Tanner, Charles E

OTHER COUNTRIES
Bojalil, Luis Felipe
Crompton, David William Thomasson
Davidson, David Edward, Jr
Franklin, Richard Morris
Guimaraes, Romeu Cardoso
Lamoyi, Edmundo
Muñoz, Maria De Lourdes
Roche, Marcel
Santoro, Ferrucio Fontes
Van Furth, Ralph

Pathology

ALABAMA
Abrahamson, Dale Raymond
Anderson, Peter Glennie
Bishop, Sanford Parsons
Blackburn, Will R
Caulfield, James Benjamin
Ceballos, Ricardo
Cole, George William
Cook, William Joseph
Dowling, Edmund Augustine
Edgar, Samuel Allen
Farnell, Daniel Reese
Foft, John William
Fullmer, Harold Milton
Garcia, Julio H
Gardner, William Albert, Jr
Gay, Renate Erika
Gay, Steffen
Geer, Jack Charles
Giambrone, Joseph James
Gillespie, George Yancey
Goldman, Ronald
Gore, Ira
Hardy, Robert W
Heath, James Eugene
Ho, Kang-Jey
Hoerr, Frederic John
Kayes, Stephen Geoffrey
Lindsey, James Russell
Liu, Paul Ishen
Lupton, Charles Hamilton, Jr
Martinez, Mario Guillermo, Jr
Matukas, Victor John
Morgan, Horace C
Mowry, Robert Wilbur
Polt, Sarah Stephens
Powers, Robert D
Siegal, Gene Philip
Smith, George Thomas
Smith, Paul Clay
Wilborn, Walter Harrison
Wolfe, Lauren Gene

ALASKA
Crook, James Richard
Van Pelt, Rollo Winslow, Jr

ARIZONA
Bertke, Eldridge Melvin
Bicknell, Edward J
Bjotvedt, George
Burkholder, Peter M
Cawley, Leo Patrick
Chin, Lincoln
Chvapil, Milos
Davis, John Robert
Golden, Alfred
Huestis, Douglas William
Layton, Jack Malcolm
McKelvie, Douglas H
Markle, Ronald A
Mulder, John Bastian

Nagle, Ray Burdell
Reed, Raymond Edgar
Shively, James Nelson
Taylor, Richard G
Tollman, James Perry
Watrach, Adolf Michael
Weinberg, Bernd
Yohem, Karin Hummell

ARKANSAS
Beasley, Joseph Noble
Bucci, Thomas Joseph
Chang, Louis Wai-Wah
Fink, Louis Maier
Hsu, Su-Ming
Jones, Joe Maxey
Jones, Robin Richard
Poirier, Lionel Albert
Sun, Chao Nien
Townsend, James Willis

CALIFORNIA
Abrams, Albert Maurice
Aggot, J Desmond
Agnew, Leslie Robert Corbet
Allen, Raymond A
Allison, Anthony Clifford
Altera, Kenneth P
Amador, Elias
Amromin, George David
Anderson, Marilyn P
Argyris, Thomas Stephen
Arnold, Harry L(oren), Jr
Arquilla, Edward R
Bailey, David Nelson
Bainton, Dorothy Ford
Baluda, Marcel A
Bangle, Raymond, Jr
Barajas, Luciano
Barnes, Asa, Jr
Bearer, Elaine L
Beckett, Ralph Lawrence
Benirschke, Kurt
Bennington, James Lynne
Berliner, Judith A
Bernick, Sol
Bhaskar, Surindar Nath
Bidlack, Donald Eugene
Bissell, Michael G
Bloor, Colin Mercer
Bordin, Gerald M
Bostick, Warren Lithgow
Bozdech, Marek Jiri
Braunstein, Herbert
Brown, William Jann
Bull, Brian S
Cardiff, Robert Darrell
Cardinet, George Hugh, III
Carnes, William Henry
Chen, Stephen Shi-Hua
Chenoweth, Dennis Edwin
Cherrick, Henry M
Chisari, Francis Vincent
Choi, Byung Ho
Corbeil, Lynette B
Cordy, Donald R
Cosgrove, Gerald Edward
Coulson, Walter F
Crawford, William Howard, Jr
DaMassa, Al John
Davis, Richard LaVerne
Deitch, Arline D
Dixon, James Francis Peter
Dorfman, Ronald F
Douglass, Robert L
Dowell, Armstrong Manly
Dungworth, Donald L
Dutra, Frank Robert
Earnest, Sue W
Edgington, Thomas S
Edmondson, Hugh Allen
Eisenson, Jon
Eng, Lawrence F
Fairchild, David George
Farquhar, Marilyn Gist
Feigal, Ellen G
Feldman, Joseph David
Foos, Robert Young
Friedberg, Errol Clive
Friedman, Nathan Baruch
Friend, Daniel S
Furthmayr, Heinz
Gardner, Murray Briggs
Garratty, George
Geller, Stephen Arthur
Gill, James Edward
Gilmore, Stuart Irby
Glenner, George Geiger
Good, Anne Haines
Gottfried, Eugene Leslie
Gray, Constance Helen
Greenspan, John Simon
Griffin, John Henry
Grody, Wayne William
Gruber, Helen Elizabeth
Hadley, William Keith
Hamm, Thomas Edward, Jr
Hansen, Louis Stephen
Harding-Barlow, Ingeborg
Hartmann, William Herman
Hepler, Opal Elsie
Heuser, Eva T
Highman, Benjamin
Hill, Rolla B, Jr

Hinton, David Earl
Holaday, William J
Holmberg, Charles Arthur
Hopkins, Gordon Bruce
Horowitz, Richard E
Hougie, Cecil
Howell, Francis V
Huntington, Robert (Watkinson), Jr
Itabashi, Hideo Henry
Jain, Naresh C
Jasper, Donald Edward
Jensen, Hanne Margrete
Jordan, Stanley Clark
Kaneko, Jiro Jerry
Kang, Tae Wha
Kan-Mitchell, June
Kelln, Elmer
Kennedy, Peter Carleton
Kern, William H
Kikkawa, Yutaka
King, Eileen Brenneman
Kiprov, Dobri D
Koobs, Dick Herman
Korn, David
Kornblum, Ronald Norman
Korpman, Ralph Andrew
Kraft, Lisbeth Martha
Landing, Benjamin Harrison
Larkin, Edward Charles
Latta, Harrison
Leighton, Joseph
Lewis, Alvin Edward
Lipsick, Joseph Steven
Lombard, Louise Scherger
Lubran, Myer Michael
Lucas, David Owen
Ludwig, Frederic C
McKay, Donald George
MacNamee, James K
Madden, Sidney Clarence
Madin, Stewart Harvey
Makker, Sudesh Paul
Makowka, Leonard
Malamud, Nathan
Margaretten, William
Mason, William Burkett
Merchant, Bruce
Miller, Gary Arthur
Minckler, Jeff
Minckler, Tate Muldown
Miyai, Katsumi
Moyer, Dean La Roche
Moyer, Geoffrey H
Murad, Turhon Allen
Myhre, Byron Arnold
Nakamura, Robert Motoharu
Nathwani, Bharat N
Nichols, Barbara Ann
Nielsen, Surl L
O'Brien, John S
Oh, Jang Ok
Olander, Harvey Johan
Olson, Albert Lloyd
Osburn, Bennie Irve
Parker, John William
Parsons, Robert Jerome
Peng, Shi-Kaung
Perkins, William Hughes
Peterson-Falzone, Sally Jean
Philippart, Michel Paul
Pilgeram, Laurence Oscar
Pool, Roy Ransom
Porter, David Dixon
Price, Harold M
Quismorio, Francisco P, Jr
Rasheed, Suraiya
Rather, Lelland Joseph
Rearden, Carole Ann
Reilly, Emmett B
Rice, Robert Hafling
Richters, Arnis
Roberts, James C, Jr
Rosenau, Werner
Rosenthal, Sol Roy
Roy-Burman, Pradip
Ruebner, Boris Henry
Sazama, Kathleen
Semancik, Joseph Stephen
Sherwin, Russell P
Shott, Leonard D
Shulman, Ira Andrew
Siegler, Richard E
Simmons, Norman Stanley
Slager, Ursula Traugott
Smuckler, Edward Aaron
Sommer, Noel Frederick
Spear, Gerald Sanford
Steinberg, Bernhard
Stenberg, Paula E
Stern, Robert
Stevens, Jack Gerald
Stevens, Lewis Axtell
Stowell, Robert Eugene
Strautz, Robert Lee
Suffin, Stephen Chester
Sunshine, Irving
Sussman, Howard H
Swanson, Virginia Lee
Taylor, Clive Roy
Terry, Robert Davis
Terry, Roger
Thaler, M Michael
Thurman, Wayne Laverne
Toreson, Wilfred Earl

Triche, Timothy J
Van Lancker, Julien L
Vazquez, Jacinto Joseph
Verity, Maurice Anthony
Volk, Bruno W
Walford, Roy Lee, Jr
Wang, Nai-San
Warner, Nancy Elizabeth
Warnock, Martha L
Weissman, Irving L
Wellington, John Sessions
Wescott, William B
Willhite, Calvin Campbell
Withers, Hubert Rodney
Wolf, Paul Leon
Wood, David Alvra
Wood, Nancy Elizabeth
Woolley, LeGrand H
Wu, Sing-Yung
Yuen, Ted Gim Hing
Zamboni, Luciano
Zettner, Alfred
Zinkl, Joseph Grandjean

COLORADO
Alexander, Archibald Ferguson
Beckwith, John Bruce
Benjamin, Stephen Alfred
Bosley, Elizabeth Caswell
Charney, Michael
Cockerell, Gary Lee
Cox, Alvin Joseph, Jr
Davis, Robert Wilson
De Martini, James Charles
Dobersen, Michael J
Firminger, Harlan Irwin
Franklin, Wilbur A
Gardner, David Godfrey
Glock, Robert Dean
Gorthy, Willis Charles
Hoerlein, Alvin Bernard
Hollister, Alan Scudder
Irons, Richard Davis
Jensen, Rue
Kotin, Paul
Krikos, George Alexander
Norrdin, Robert W
Pierce, Gordon Barry
Repine, John E
Snyder, Stanley Paul
Solomon, Gordon Charles
Speers, Wendell Carl
Swartzendruber, Douglas Edward
Takeda, Yasuhiko
Thomas, Louis Barton
Whiteman, Charles E
Wilber, Charles Grady
Young, Stuart

CONNECTICUT
Barthold, Stephen William
Baumgarten, Alexander
Benz, Edward John
Booss, John
Cronin, Michael Thomas Ignatius
Cutler, Leslie Stuart
Dembitzer, Herbert
Downing, S Evans
Goldschneider, Irving
Haas, Mark
Hasson, Jack
Howard, Robert Eugene
Hukill, Peter Biggs
Jacoby, Robert Ottinger
Janeway, Charles Alderson, Jr
Kaplow, Leonard Samuel
Kashgarian, Michael
Krutchkoff, David James
Lee, Sin Hang
Lobdell, David Hill
Madri, Joseph Anthony
Manuelidis, Elias Emmanuel
Marchesi, Vincent T
Moretz, Roger C
Morrow, Jon S
Nalbandian, John
Nielsen, Svend Woge
Post, John E
Schmeer, Arline Catherine
Seligson, David
Smith, Brian Richard
Snellings, William Moran
Spitznagle, Larry Allen
Stenn, Kurt S
Sunderman, F William, Jr
Vidone, Romeo Albert
Vinocur, Myron
Yang, Tsu-Ju (Thomas)
Yesner, Raymond

DELAWARE
Cubberley, Virginia
Dohms, John Edward
Fogg, Donald Ernest
Fries, Cara Rosendale
Hubben, Klaus
Pegg, Philip John
Stula, Edwin Francis

DISTRICT OF COLUMBIA
Arcos, Joseph (Charles)
Bowling, Lloyd Spencer, Sr
Bulger, Ruth Ellen
Campbell, Carlos Boyd Godfrey

Bernier, Joseph Leroy
Biswas, Robin Michael
Bloom, Sherman
Bodammer, Joel Edward
Boone, Charles Walter
Burch, Ronald Martin
Bush, David E
Cheever, Allen
Chu, Elizabeth Wann
Connor, Daniel Henry
Cork, Linda K Collins
Dannenberg, Arthur Milton, Jr
DeClaris, N(icholas)
Deringer, Margaret K
Dubey, Jitender Prakash
Eckels, Kenneth Henry
Eddy, Hubert Allen
Eggleston, Joseph C
Elin, Ronald John
Evarts, Ritva Poukka
Ferrans, Victor Joaquin
Fornace, Albert J, Jr
Fraumeni, Joseph F, Jr
Fried, Jerrold
Friedman, Robert Morris
Frost, John Kingsbury
Gardner, Alvin Frederick
Gauhar, Aruna
Gazdar, Adi F
Gordon, Lance Kenneth
Green, Marie Roder
Griffith, Linda M(ae)
Grimley, Philip M
Gross, Mircea Adrian
Hackett, Joseph Leo
Hamilton, Stanley R
Haspel, Martin Victor
Henson, Donald E
Heptinstall, Robert Hodgson
Hess, Helen Hope
Hilberg, Albert William
Holmes, Kathryn Voelker
Howley, Peter Maxwell
Hutchins, Grover MacGregor
Iseri, Oscar Akio
Jacobs, Myron Samuel
Jaffe, Elaine Sarkin
Kendrick, Francis Joseph
Kickler, Thomas Steven
Kimball, Paul Clark
Kirschstein, Ruth Lillian
Kirsten, Werner H
Kroll, Martin Harris
Latham, Patricia Suzanne
Lerman, Michael Isaac
Liang, Isabella Y S
Liebelt, Annabel Glockler
Lindberg, Donald Allan Bror
Lindenberg, Richard
Liotta, Lance
Lipkin, Lewis Edward
Longo, Dan L
Lunin, Martin
McDowell, Elizabeth Mary
Malmgren, Richard Axel
Marzella, Louis
Mercado, Teresa I
Mickel, Hubert Sheldon
Micozzi, Marc S
Mishler, John Milton, IV
Morrissey, Robert Leroy
Muma, Nancy A
O'Beirne, Andrew Jon
Olson, Jean L
Petrali, John Patrick
Platt, William Rady
Prendergast, Robert Anthony
Rabson, Alan S
Reuber, Melvin D
Rice, Jerry Mercer
Rifkin, Erik
Robison, Wilbur Gerald, Jr
Rodrigues, Merlyn M
Rosenberg, Saul H
Saffiotti, Umberto
Shamsuddin, Abulkalam Mohammad
Shin, Moon L
Sobel, Mark E
Solez, Kim
Squire, Robert Alfred
Stewart, Harold L
Striker, G E
Stromberg, Kurt
Strum, Judy May
Thapar, Nirwan T
Tigertt, William David
Ting, Chou-Chik
Toker, Cyril
Torres-Anjel, Manuel Jose
Trump, Benjamin Franklin
Tseng, Hsiang Len
Vickers, James Hudson
Von Euler, Leo Hans
Wahl, Sharon Knudson
Ward, Jerrold Michael
Weihing, Robert Ralph
Weisbroth, Steven H
White, John David
Yardley, John Howard
Yodaiken, Ralph Emile
Zierdt, Charles Henry
Zopf, David Arnold

MASSACHUSETTS
Abromson-Leeman, Sara R
Ahmed, A Razzaque
Andres, Giuseppe A
Appel, Michael Clayton
Badonnel, Marie-Claude H M
Balogh, Karoly
Benacerraf, Baruj
Busch, George Jacob
Carvalho, Angelina C A
Caulfield, John P
Christensen, Thomas Gash
Christian, Howard J
Cobb, Carolus M
Collier, Robert John
Colvin, Robert B
Connolly, James L
Corson, Joseph Mackie
Cotran, Ramzi S
Dammin, Gustave John
D'Amore, Patricia Ann
Dayal, Yogeshwar
DeGirolami, Umberto
DeLellis, Ronald Albert
Diamandopolus, George Th
Dittmer, John Edward
Dvorak, Ann Marie-Tompkins
Dvorak, Harold Fisher
Fallon, John T
Federman, Micheline
Flax, Martin Howard
Fleischman, Robert Werder
Foster, Eugene A
Freiman, David Galland
George, Harvey
Gherardi, Gherardo Joseph
Gill, David Michael
Gilles, Floyd Harry
Gimbrone, Michael Anthony, Jr
Goldman, Harvey
Goldman, Henry M(aurice)
Gottlieb, Leonard Solomon
Haimes, Howard B
Handler, Alfred Harris
Haudenschild, Christian C
Hawiger, Jack Jacek
Hayes, John A
Hedley-Whyte, Elizabeth Tessa
Hertig, Arthur Tremain
Hirsch, Carl Alvin
Homburger, Freddy
Hunt, Ronald Duncan
Jacobs, Jerome Barry
Johansen, Erling
Jonas, Albert Moshe
Jones, Thomas Carlyle
Karnovsky, Morris John
Kim, Agnes Kyung-Hee
King, Norval William, Jr
Koch-Weser, Dieter
Kocon, Richard William
Kubin, Rosa
Kulka, Johannes Peter
Kwan, Paul Wing-Ling
Lage, Janice M
Laposata, Michael
Leav, Irwin
Legg, Merle Alan
Leith, John Douglas
Libby, Peter
Like, Arthur A
Locke, John Lauderdale
McCluskey, Robert Timmons
McCombs, H Louis
McDonagh, Jan M
MacMahon, Harold Edward
Meyer, Irving
Mihm, Martin C, Jr
Mitus, Wladyslaw J
Murphy, James Clair
Newberne, Paul M
Robbins, Stanley L
Rogers, Adrianne Ellefson
Rosen, Seymour
Russfield, Agnes Burt
Sacks, David B
Schaub, Robert George
Schneeberger, Eveline E
Schoen, Frederick J
Scully, Robert Edward
Shklar, Gerald
Sidman, Richard Leon
Smulow, Jerome B
Susi, Frank Robert
Szabo, Sandor
Taft, Edgar Breck
Townsend, John Ford
VanPraagh, Richard
Von Lichtenberg, Franz
Warhol, Michael J
White, Harold J
Winkelman, James W
Wright, Thomas Carr, Jr
Yunis, Edmond J

MICHIGAN
Abrams, Gerald David
Aggarwal, Surinder K
Arnold, James Schoonover
Bernstein, Jay
Bollinger, Robert Otto
Brock, Donald R
Chason, Jacob Leon
Chopra, Dharam Pal

Chua, Balvin H-L
Claflin, Robert Malden
Clark, John Jefferson
Coye, Robert Dudley
Dapson, Richard W
De La Iglesia, Felix Alberto
Diglio, Clement Anthony
Eisenbrandt, David Lee
Elfont, Edna A
Elliott, George Algimon
Feenstra, Ernest Star
Fitzgerald, James Edward
Floyd, Alton David
Garg, Bhagwan D
Giacomelli, Filiberto
Gikas, Paul William
Gilbertsen, Richard B
Giraldo, Alvaro A
Goyings, Lloyd Samuel
Grant, Rhoda
Hanks, Carl Thomas
Hashimoto, Ken
Headington, John Terrence
Hicks, Samuel Pendleton
Hinerman, Dorin Lee
Hollenberg, Paul Frederick
Hunt, Charles E
Jenkins, Thomas William
Jersey, George Carl
Johnson, Herbert Gardner
Johnson, Kent J
Jones, Margaret Zee
Keahey, Kenneth Karl
Kiechle, Frederick Leonard
Kithier, Karel
Kociba, Richard Joseph
Koestner, Adalbert
Landers, James Walter
Langham, Robert Fred
Leader, Robert Wardell
McClatchey, Kenneth D
McKeever, Paul Edward
Mattson, Joan C
Moe, James Burton
Moses, Gerald Robert
Oberman, Harold A
Padgett, George Arnold
Palmer, Kenneth Charles
Perrin, Eugene Victor
Phan, Sem Hin
Piper, Richard Carl
Poulik, Miroslav Dave
Powsner, Edward R
Puro, Donald George
Raven, Clara
Regezi, Joseph Alberts
Rowe, Nathaniel H
Rupp, Ralph Russell
Sabes, William Ruben
Sander, C Maureen
Schmidt, Robert W
Schnitzer, Bertram
Shanberge, Jacob N
Sharf, Donald Jack
Sleight, Stuart Duane
Smart, James Blair
Smith, Robert James
Spitz, Werner Uri
Stowe, Howard Denison
Stromsta, Courtney Paul
Tasker, John B
Trapp, Allan Laverne
Ward, Peter A
Waxler, Glenn Lee
Webster, Harris Duane
Whitehair, Charles Kenneth
Wiener, Joseph
Wolman, Sandra R
Wolter, J Reimer
Zak, Bennie

MINNESOTA
Anderson, W Robert
Azar, Miguel M
Baggenstoss, Archie Herbert
Bahn, Robert Carlton
Barnes, Donald McLeod
Benson, Ellis Starbranch
Brown, David Mitchell
Brunning, Richard Dale
Case, Marvin Theodore
Connelly, Donald Patrick
Crowley, Leonard Vincent
Dahlin, David Carl
Duke, Gary Earl
Edwards, Jesse Efrem
Edwards, William Dean
Elzay, Richard Paul
Engel, Andrew G
Estensen, Richard D
Foley, William Arthur
Fraley, Elwin B
Gibbons, Donald Frank
Goble, Frans Cleon
Gorlin, Robert James
Hedgecock, Le Roy Darien
Henrikson, Ernest Hilmer
Homburger, Henry A
Jacobson, Joan
Johnson, Kenneth Harvey
Kammermeier, Martin A
Karlson, Alfred Gustav
Kersey, John H
Knabe, George W, Jr

Kottke, Bruce Allen
Kurtz, Harold John
LeBien, Tucker W
Lober, Paul Hallam
Low, Walter Cheney
McCarthy, James Benjamin
McDermott, Richard P
Manning, Patrick James
Mariani, Toni Ninetta
Mitchell, Roger Harold
Moller, Karlind Theodore
Moore, Sean Breanndan
Nelson, Robert D
Okazaki, Haruo
Osborne, Carl Andrew
Perman, Victor
Ramakrishan, S
Rao, Gundu Hirisave Rama
Romero, Juan Carlos
Sautter, Jay Howard
Sparnins, Velta L
Sung, Joo Ho
Taswell, Howard Filmore
Tihon, Claude
Titus, Jack L
Vallera, Daniel A
Van Kampen, Kent Rigby
Vernier, Robert L
Wallace, Larry J
Wattenberg, Lee Wolff
Witkop, Carl Jacob, Jr
Yunis, Jorge J

MISSISSIPPI
Bradley, Doris P
Conroy, James D
Cruse, Julius Major, Jr
Evers, Carl Gustav
Gatling, Robert Riddick
Johnson, Warren W
Lewis, Robert Edwin, Jr
Lockwood, William Rutledge
O'Neal, Robert Munger
Overstreet, Robin Miles
Read, Virginia Hall
Spann, Charles Henry
Verlangieri, Anthony Joseph

MISSOURI
Bari, Wagih A
Berrier, Harry Hilbourn
Blumenthal, Herman T
Braciale, Thomas Joseph, Jr
Drees, David T
Eyestone, Willard Halsey
Gantner, George E, Jr
Givler, Robert L
Green, Eric Douglas
Hall, William Francis
Hamilton, Thomas Reid
Harding, Clifford Vincent, III
Johnston, Marilyn Frances Meyers
Kapp-Pierce, Judith A
Lacy, Paul Eston
Lagunoff, David
Lau, Brad W C
McDivitt, Robert William
Morehouse, Lawrence G
Nahm, Moon H
Olney, John William
Olson, LeRoy David
Plapp, Frederick Vaughn
Porter, Chastain Kendall
Ribelin, William Eugene
Roodman, Stanford Trent
Schmidt, Donald Arthur
Shah, Sheila
Smith, Carl Hugh
Stewart, Wellington Buel
Taylor, John Joseph
Torack, Richard M
Vogel, Arthur Mark
Wagner, Joseph Edward
Watanabe, Itaru S
Welch, Lin
Yu, Shiu Yeh

MONTANA
Hadlow, William John
Williams, Roger Stewart
Young, David Marshall

NEBRASKA
Berton, William Morris
Cohen, Samuel Monroe
Cox, Robert Sayre, Jr
Grace, Oliver Davies
Linder, James
Purtilo, David Theodore
Quigley, Herbert Joseph, Jr
Schenken, Jerald R
Schmitz, John Albert
Sheehan, John Francis
Todd, Gordon Livingston
Toth, Bela
Wilson, Richard Barr

NEVADA
Anderson, Bernard A
Barger, James Daniel
Tibbits, Donald Fay

NEW HAMPSHIRE
Christie, Robert William

Brandt, John T
Capen, Charles Chabert
Carter, John Robert
Chang, Jae Chan
Chisolm, Guy M
Cole, Clarence Russell
Collins, William Thomas
Conroy, Charles William
Copeland, Bradley Ellsworth
Cornhill, John Fredrick
Cras, Patrick
Cross, Robert Franklin
Dahlgren, Robert R
Deegan, Michael J
Denine, Elliot Paul
Deodhar, Sharad Dinkar
Donnelly, Kenneth Gerald
Fenoglio, Cecilia M
Gerrity, Ross Gordon
Gibson, John Phillips
Hamilton, Thomas Alan
Hanson, Daniel James
Hartroft, Phyllis Merritt
Higgins, Jerry Mitchell
Hoff, Henry Frederick
Hopps, Howard Carl
Hutterer, Ferenc
Hyman, Melvin
Igel, Howard Joseph
Ingalls, William Lisle
Jennings, Frank Lamont
Johnson, Anne Bradstreet
Kerr, Kirklyn M
Kier, Ann B
Klaunig, James E
Kleinerman, Jerome
Kwak, Yun Sik
Lamm, Michael Emanuel
Langley, Albert E
Liebelt, Robert Arthur
Liss, Leopold
Lohse, Carleton Leslie
Long, John Frederick
McAdams, Arthur James
Macpherson, Colin Robertson
Mahaffey, Kathryn Rose
Malemud, Charles J
Menefee, Max Gene
Metz, David A
Mirkin, L David
Moorhead, Philip Darwin
Muir, William W, III
Myers, Ronald Elwood
Nagode, Larry Allen
Newberne, James Wilson
Newton, William Allen, Jr
Patrick, James R
Perry, George
Platt, Marvin Stanley
Pretlow, Thomas Garrett, II
Rabin, Erwin R
Rapp, John P
Ratliff, Norman B, Jr
Reagan, James W
Reiner, Charles Brailove
Kikihisa, Yasuko
Rodin, Alvin E
Rossi, Edward P
Rynbrandt, Donald Jay
Saul, Frank Philip
Schut, Herman A
Senhauser, Donald Albert
Shannon, Barry Thomas
Sharma, Hari M
Smith, Roger Dean
Speicher, Carl Eugene
Stoner, Gary David
Tarr, Melinda Jean
Thomas, Donald Charles
Tubbs, Raymond R
Tykocinski, Mark L
Wagner, Alan R
Washington, John A, II
Yates, Allan James

OKLAHOMA
Bell, Paul Burton, Jr
Breazile, James E
Confer, Anthony Wayne
Fahmy, Aly
Flournoy, Dayl Jean
Gillum, Ronald Lee
Glass, Richard Thomas
Glenn, Bertis Lamon
Kirkham, William R
Lambird, Perry Albert
Lhotka, John Francis
McClellan, Betty Jane
Miller, Paul George
Min, Kyung-Whan
Monlux, Andrew W
Nordquist, Robert Ersel
Panciera, Roger J
Roszel, Jeffie Fisher
Wickham, M Gary

OREGON
Bartley, Murray Hill, Jr
Beck, John Robert
Beckstead, Jay H
Brooks, Robert E
Buchan, George Colin
Coleman, Ralph Orval, Jr
Hutchens, Tyra Thornton

Jastak, J Theodore
Koller, Loren D
McNulty, Wilbur Palmer
Maynard, Russell Milton
Moore, Richard Donald
Niles, Nelson Robinson
Palotay, James Lajos
Prescott, Gerald H
Rickles, Norman Harold
Spencer, Peter Simner
Stroud, Richard Kim
Thompson, Charles Calvin
Young, Norton Bruce

PENNSYLVANIA
Albertine, Kurt H
Allen, Henry L
Andrews, Edwin Joseph
Ashley, Charles Allen
Bagasra, Omar
Balin, Arthur Kirsner
Baserga, Renato
Berkheiser, Samuel William
Berry, Richard G
Bly, Chauncey Goodrich
Bokelman, Delwin Lee
Brooks, John J
Brooks, John S J
Burgi, Ernest, Jr
Cameron, Alexander Menzies
Carlson, Arthur Stephen
Chacko, Samuel K
Chan, Ping-Kwong
Chen, Sow-Yeh
Child, Proctor Louis
Clark, Wallace Henderson, Jr
Cohen, Stanley
Colman, Robert W
Conn, Rex Boland
Crane, August Reynolds
Damjanov, Ivan
Dekker, Andrew
Demers, Laurence Maurice
Dolphin, John Michael
Donnelly, John James, III
Duker, Nahum Johanan
Farber, John Lewis
Fenderson, Bruce Andrew
Fisher, Edwin Ralph
Flickinger, George Latimore, Jr
Flynn, John Thomas
Fox, Karl Richard
Furth, John J
Garman, Robert Harvey
Gasic, Gabriel J
Gill, Thomas James, III
Gupta, Prabodh Kumar
Hall, Judy Dale (Mrs Richard Modafferi)
Hartley, Harold V, Jr
Hartman, John David
Haskins, Mark
Hummeler, Klaus
Jarett, Leonard
Jensen, Richard Donald
Joseph, Jeymohan
Kashatus, William C
Kelly, Alan
Kline, Irwin Kaven
Klionsky, Bernard Leon
Klurfeld, David Michael
Koffler, David
Koprowska, Irena
Kreider, John Wesley
Krieg, Arthur F
Leifer, Calvin
Little, Brian Woods
Macartney, Lawson
McGary, Carl T
McGrath, John Thomas
Maenza, Ronald Morton
Mancall, Elliott L
Maniglia, Rosario
Martinez, A Julio
Martinez-Hernandez, Antonio
Mattison, Donald Roger
Meek, Edward Stanley
Meisler, Arnold Irwin
Mifflin, Theodore Edward
Miller, Arthur Simard
Ming, Si-Chun
Modafferi, Judy Hall
Moossy, John
Murer, Erik Homann
Naeye, Richard L
Nass, Margit M K
Nowell, Peter Carey
Oels, Helen C
Parker, Leslie
Pietra, Giuseppe G
Poste, George Henry
Rabin, Bruce S
Ramasastry, Sai Sudarshan
Rao, Kalipatnapu Narasimha
Riley, Gene Alden
Rorke, Lucy Balian
Rothenbacher, Hansjakob
Rovera, Giovanni
Rubin, Emanuel
Rubin, Herbert
Russo, Jose
Salazar, Hernando
Saunders, Leon Z
Schofield, Richard Alan
Schwartz, Heinz (Georg)

Shinozuka, Hisashi
Singh, Gurmukh
Snyder, Robert LeRoy
Steplewski, Zenon
Stewart, Gwendolyn Jane
Sunderman, Frederick William
Taichman, Norton Stanley
Tauxe, Welby Newlon
Van Zwieten, Matthew Jacobus
Weber, Wilfried T
Weiss, Michael Stephen
Wheeler, James English
Whiteside, Theresa L
Whitlock, Robert Henry
Winchester, Richard Albert
Witzleben, Camillus Leo
Young, Donald Stirling
Young, Irving
Zimmerman, Michael Raymond
Zmijewski, Chester Michael

RHODE ISLAND
Chang, Pei Wen
Crowley, James Patrick
Fausto, Nelson
Kane, Agnes Brezak
Kuhn, Charles, III
Lichtman, Herbert Charles
McCully, Kilmer Serjus
McMaster, Philip Robert Bache
Plotz, Richard Douglas
Singer, Don B
Strauss, Elliott William
Wolke, Richard Elwood
Zacks, Sumner Irwin

SOUTH CAROLINA
Allen, Robert Carter
Balentine, J Douglas
Bank, Harvey L
Cannon, Albert
Fowler, Stanley D
Garvin, Abbott Julian
Hennigar, Gordon Ross, Jr
La Via, Mariano Francis
Lill, Patsy Henry
Powell, Harold
Pratt-Thomas, Harold Rawling
Rathbun, Ted Allan
Spicer, Samuel Sherman, Jr
Virella, Gabriel T
Willoughby, William Franklin

SOUTH DAKOTA
Bergeland, Martin E
Johnson, Carl J
Wegner, Karl Heinrich

TENNESSEE
Asp, Carl W
Atkinson, James Byron
Belew-Noah, Patricia W
Berard, Costan William
Clapp, Neal K
Congdon, Charles C
Coogan, Philip Shields
Erickson, Cyrus Conrad
Fields, James Perry
Francisco, Jerry Thomas
Ganote, Charles Edgar
Green, Louis Douglas
Handorf, Charles Russell
Harwood, Thomas Riegel
LeQuire, Virgil Shields
Lucas, Fred Vance
Lushbaugh, Clarence Chancelum
McCormick, William F
McGavin, Matthew Donald
Machado, Emilio Alfredo
Marchok, Ann Catherine
Massengill, Raymond
Meyrick, Barbara O
Mitchell, William Marvin
Muirhead, E Eric
Page, David L
Peterson, Harold Arthur
Pitcock, James Allison
Pribor, Hugo C
Price, Robert Allen
Rettenmier, Carl Wayne
Rogers, Stanfield
Scott, Robert Eugene
Shenefelt, Ray Eldon
Smith, Roy Martin
Soper, Richard Graves
Stone, Robert Edward, Jr
Woodward, Stephen Cotter
Young, Joseph Marvin

TEXAS
Achilles, Robert F
Adams, Leslie Garry
Allen, Robert Charles
Bailey, Everett Murl, Jr
Batsakis, John G
Becker, Frederick F
Bennett, Michael
Binnie, William Hugh
Bost, Robert Orion
Bridges, Charles Hubert
Burton, Karen Poliner
Burton, Russell Rohan
Butler, James Johnson
Coalson, Jacqueline Jones

Cooper, Ronda Fern
Cottone, James Anthony
Cowan, Daniel Francis
Dahl, Elmer Vernon
Davis, Joyce S
Denko, John V
Dodson, Ronald Franklin
Dudley, Alden Woodbury, Jr
Falck, Frank James
Farhi, Diane C
Fidler, Isaiah J
Finegold, Milton J
Folse, Dean Sydney
Freeman, Robert Glen
Ghidoni, John Joseph
Gibson, William Andrew
Gleiser, Chester Alexander
Graham, David Lee
Gratzner, Howard G
Greenberg, Stanly Donald
Hensley, John Coleman, II
Hill, Joseph MacGlashan
Hurt, William Clarence
Ibanez, Michael Louis
Jones, Larry Philip
Keene, Harris J
Kim, Han-Seob
Kirkpatrick, Joel Brian
Kriesberg, Jeffrey Ira
Krueger, Gerhard R F
Kurtz, Stanley Morton
Leibowitz, Julian Lazar
Lieberman, Michael Williams
Lindner, Luther Edward
Lynch, Denis Patrick
McGavran, Malcolm Howard
McGill, Henry Coleman, Jr
McGregor, Douglas D
McKenna, Robert Wilson
McManus, Linda Marie
Maurer, Fred Dry
Mendel, Julius Louis
Meuten, Donald John
Migliore, Philip Joseph
Milam, John D
Montgomery, Philip O'Bryan, Jr
Moore, Donald Vincent
Murthy, Krishna K
Ordonez, Nelson Gonzalo
Pakes, Steven P
Pinckard, Robert Neal
Race, George Justice
Rajaraman, Srinivasan
Ranney, David Francis
Reynolds, Rolland C
Sato, Clifford Shinichi
Schmidt, Waldemar Adrian
Schwartz, Colin John
Schwartz, William Lewis
Sell, Stewart
Shadduck, John Allen
Shillitoe, Edward John
Sisson, Joseph A
Smith, Alice Lorraine
Smith, David English
Smith, Jerome H
Smith, John Leslie, Jr
Spellman, Craig William
Spjut, Harlan Jacobson
Stembridge, Vernie A(lbert)
Storts, Ralph Woodrow
Stout, Landon Clarke, Jr
Sybers, Harley D
Tessmer, Carl Frederick
Townsend, Frank Marion
Trevino, Gilberto Stephenson
Troxler, Raymond George
Turner, Robert Atwood
Ullrich, Robert Leo
Venkatachalam, Manjeri A
Villarreal, Jesse James
Walker, David Hughes
Weitzner, Stanley
Wigodsky, Herman S
Wiig, Elisabeth Hemmersam
Wordinger, Robert James
Yang, Ovid Y H
Zedler, Empress Young
Zimmermann, Eugene Robert
Zunker, Heinz Otto Hermann

UTAH
Arhelger, Roger Boyd
Ash, Kenneth Owen
Chapman, Arthur Owen
Chuang, Hanson Yii-Kuan
Eichwald, Ernest J
Fujinami, Robert S
Hill, Harry Raymond
Lloyd, Ray Dix
Mecham, Merlin J
Miles, Charles P
Miner, Merthyr Leilani
Mohammad, Syed Fazal
Moody, David Edward
Olsen, Don B
Warren, Reed Parley
Wilson, John F

VERMONT
Bolton, Wesson Dudley
Clemmons, Jackson Joshua Walter
Coon, Robert William
Craighead, John Edward

Pathology (cont)

Korson, Roy
Luginbuhl, William Hossfeld
Mossman, Brooke T
Sheldon, Huntington
Toolan, Helene Wallace

VIRGINIA
AcKerman, Larry Joseph
Bobbitt, Oliver Bierne
Bruns, David Eugene
Burr, Helen Gunderson
Cheatham, William Joseph
Cordes, Donald Ormond
Faulconer, Robert Jamieson
Ferry, Andrew P
Fisher, Lyman McArthur
Gander, George William
Gross, Walter Burnham
Halstead, Charles Lemuel
Hard, Richard C, Jr
Herman, Mary M
Hossaini, Ali A
Hsu, Hsiu-Sheng
Johnston, Charles Louis, Jr
Kay, Saul
Lobel, Steven A
Luke, James Lindsay
Martens, Vernon Edward
Moran, Thomas James
Normansell, David E
Peery, Thomas Martin
Rosenblum, William I
Rubinstein, Lucien Jules
Russell, Catherine Marie
Russi, Simon
Salley, John Jones
Savory, John
Schwartzman, Joseph David
Sirica, Alphonse Eugene
Somlyo, Andrew Paul
Still, William James Sangster
Sturgill, Benjamin Caleb
Vennart, George Piercy
Vinores, Stanley Anthony
Voelker, Richard William
Weiss, Daniel Leigh

WASHINGTON
Alvord, Ellsworth Chapman, Jr
Bankson, Daniel Duke
Bean, Michael Arthur
Benditt, Earl Philip
Boatman, Edwin S
Butts, William Cunningham
Callis, James Bertram
Couser, William Griffith
Giddens, William Ellis, Jr
Gribble, David Harold
Haggitt, Rodger C
Hall, Stanton Harris
Henson, James Bond
Kachmar, John Frederick
Killingsworth, Lawrence Madison
Kramer, John William
Landolt, Marsha LaMerle
Leid, R Wes
Martin, George Monroe
Merrill, James Allen
Mottet, Norman Karle
Narayanan, A Sampath
Nelson, Karen Ann
Page, Roy Christopher
Palmer, John M
Prehn, Richmond Talbot
Prieur, David John
Purnell, Dallas Michael
Rabinovitch, Peter S
Ragan, Harvey Albert
Robinovitch, Murray R
Ross, Russell
Sanders, Charles Leonard, Jr
Schwartz, Stephen Mark
Shaw, Cheng-Mei
Smith, Dean Harley
Strandjord, Paul Edphil
Van Hoosier, Gerald L, Jr
Wilson, Robert Burton
Wolf, Norman Sanford
Zamora, Cesario Siasoco
Zander, Donald Victor

WEST VIRGINIA
Albrink, Wilhelm Stockman
Blundell, George Phelan
Bouquot, Jerry Elmer
Chang, William Wei-Lien
Hales, Milton Reynolds
Hooper, Anne Caroline Dodge
Iammarino, Richard Michael
Morgan, Winfield Scott
Quittner, Howard
Rodman, Nathaniel Fulford, Jr
Ruscello, Dennis Michael
Schochet, Sydney Sigfried, Jr
Victor, Leonard Baker

WISCONSIN
Altshuler, Charles Haskell
Angevine, Daniel Murray
Beckfield, William John
Bloodworth, James Morgan Bartow, Jr
Brown, Arnold Lanehart, Jr

Collins, Richard Andrew
Eisenstein, Reuben
Erwin, Chesley Para
Fodden, John Henry
Folts, John D
Goldberg, Burton David
Goldfarb, Stanley
Greenspan, Daniel S
Hartmann, Henrik Anton
Hinze, Harry Clifford
Hoerl, Bryan G
Hussey, Clara Veronica
Inhorn, Stanley L
Jaeschke, Walter Henry
Kalfayan, Bernard
Kuzma, Joseph Francis
Lalich, Joseph John
Larson, Frank Clark
Locke, Louis Noah
Neubecker, Robert Duane
Norback, Diane Hagemen
Oberley, Terry De Wayne
Olson, Carl
Pitot, Henry C, III
Reneau, John
Rossetti, Louis Michael
Rude, Theodore Alfred
Siegesmund, Kenneth A
Silverman, Ellen-Marie
Silverman, Franklin Harold
Uno, Hideo
Wussow, George C

WYOMING
Isaak, Dale Darwin
Walton, Thomas Edward

PUERTO RICO
Moreno, Esteban
Ortiz, Araceli

ALBERTA
Bain, Gordon Orville
Bainborough, Arthur Raymond
Bell, Harold E
Dick, Henry Marvin
Dixon, John Michael Siddons
Jimbow, Kowichi
Paul, Leendert Cornelis
Rewcastle, Neill Barry
Russell, James Christopher
Shnitka, Theodor Khyam
Stinson, Robert Anthony
Stirrat, James Hill
Wilson, Frank B
Wood, Norman Kenyon

BRITISH COLUMBIA
Bower, Susan Mae
Dunn, William Lawrie
Eaves, Allen Charles Edward
Frohlich, Jiri J
Garg, Arun K
Godolphin, William
Hardwick, David Francis
Hogg, James Cameron
Kelly, Michael Thomas
Kim, Seung U
Pearce, Richard Hugh
Salinas, Fernando A
Spitzer, Ralph
Thurlbeck, William Michael

MANITOBA
Adamson, Ian Young Radcliffe
Bowden, Drummond Hyde
Carr, Ian
Crowson, Charles Neville
Henderson, James Stuart
Hoogstraten, Jan
Persaud, Trivedi Vidhya Nandan
Pettigrew, Norman M

NEW BRUNSWICK
Smith, Harry John

NEWFOUNDLAND
Michalak, Thomas Ireneusz

NOVA SCOTIA
Aterman, Kurt
Fraser, Albert Donald
Ghose, Tarunendu
Tonks, Robert Stanley

ONTARIO
Asa, Sylvia L
Barr, Ronald Duncan
Baumal, Reuben
Bencosme, Sergio Arturo
Campbell, James Stewart
Chander, Satish
Cutz, Ernest
De Bold, Adolfo J
Dedhar, Shoukat
De Harven, Etienne
Farber, Emmanuel
Fournier, Pierre William
Frederick, George Leonard
Frei, Jaroslav Vaclav
Goldberg, David Myer
Gotlieb, Avrum I
Goyer, Robert Andrew
Haust, M Daria

Heggtveit, Halvor Alexander
Hill, Donald P
Huang, Shao-nan
Hulland, Thomas John
Kaushik, Azad Kumar
Kisilevsky, Robert
Kovacs, Kalman T
Kuroski-De Bold, Mercedes Lina
Langille, Brian Lowell
Main, James Hamilton Prentice
Morris, Gerald Patrick
Movat, Henry Zoltan
Mustard, James Fraser
Neufeld, Abram Herman
Nielsen, N Ole
Nisbet-Brown, Eric Robert
Orr, Frederick William
Phillips, Melville James
Ranadive, Narendranath Santuram
Richter, Maxwell
Ritchie, Alexander Charles
Robertson, David Murray
Rosenthal, Kenneth Lee
Rowsell, Harry Cecil
Sarma, Dittakavi S R
Silver, Malcolm David
Stanisz, Andrzej Maciej
Sturgess, Jennifer Mary
Thibert, Roger Joseph
Vadas, Peter
Wallace, Alexander Cameron
Waugh, Douglas Oliver William

PRINCE EDWARD ISLAND
Singh, Amreek

QUEBEC
Bendayan, Moise
Brassard, Andre
Cantin, Marc
Cote, Roger Albert
Duguid, William Paris
Jasmin, Gaetan
Karpati, George
Lussier, Andre (Joseph Alfred)
Lussier, Gilles L
Meisels, Alexander
Messier, Bernard
Moore, Sean
Murthy, Mahadi Raghavandrarao Ven
Rona, George
Simard, Rene
Stejskal, Rudolf
Tremblay, Gilles
Ventura, Joaquin Calvo
Weigensberg, Bernard Irving
Yousef, Ibrahim Mohmoud

SASKATCHEWAN
Blair, Donald George Ralph
Cates, Geoffrey William
Emson, Harry Edmund
Kalra, Jawahar
Mills, James Herbert Lawrence
Rozdilsky, Bohdan
Saunders, James Robert
Schiefer, H Bruno
Skinnider, Leo F

OTHER COUNTRIES
Baldwin, Robert William
Bernstein, David Maier
Blanc, William Andre
Collins, Vincent Peter
Crompton, David William Thomasson
Czernobilsky, Bernard
Dutz, Werner
Fahimi, Hossein Dariush
Friede, Reinhard L
Fujimoto, Shigeyoshi
Gabbiani, Giulio
Gertz, Samuel David
Gonzalez-Angulo, Amador
Gorski, Andrzej
Gowans, James L
Gresser, Ion
Gullino, Pietro M
Hansson, Goran K
Helander, Herbert Dick Ferdinand
Hirokawa, Katsuiku
Ide, Hiroyuki
Iossifides, Ioulios A
Izui, Shozo
Jakowska, Sophie
Karasaki, Shuichi
Kerjaschki, Dontscho
Lee, Joseph Chuen Kwun
Lever, Walter Frederick
Lewin, Lawrence M
Lin, Chin-Tarng
Ling, Chung-Mei
Lund, John Edward
Maruyama, Koshi
Minowada, Jun
Pan, In-Chang
Renaud, Serge
Rettori, Ovidio
Shubik, Philippe
Silberberg, Ruth
Stehbens, William Ellis
Stern, Kurt
Tachibana, Takehiko
Thiery, Jean Paul
Van Loveren, Henk

Wan, Abraham Tai-Hsin
Warren, Bruce Albert
Yakura, Hidetaka
Yoshida, Takeshi
Yovich, John V

Pharmacology

ALABAMA
Ayling, June E
Barker, Samuel Booth
Beaton, John McCall
Bennett, Leonard Lee, Jr
Benos, Dale John
Berecek, Kathleen Helen
Berger, Robert S
Besse, John C
Campbell, William Howard
Chang, Chi Hsiung
Christian, Samuel Terry
Clark, Carl Heritage
Coker, Samuel Terry
Dalvi, Ramesh R
Diasio, Robert Bart
Downey, James Merritt
El Dareer, Salah
Frings, Christopher Stanton
Furner, Raymond Lynn
Guarino, Anthony Michael
Harding, Thomas Hague
Harrison, Steadman Darnell, Jr
Hill, Donald Lynch
Hocking, George Macdonald
Kaplan, Ronald S
Lecklitner, Myron Lynn
Lindsay, Raymond H
McCann, William Peter
McCarthy, Dennis Joseph
Macmillan, William Hooper
Meezan, Elias
Mundy, Roy Lee
Nair, Madhavan G
Page, John Gardner
Palmisano, Paul Anthony
Pillion, Dennis Joseph
Schmidt, Leon Herbert
Schneyer, Charlotte A
Sowell, John Gregory
Strada, Samuel Joseph
Struck, Robert Frederick
Susina, Stanley V
Tan, Boen Hie
Tate, Laurence Gray
Vacik, James P
Williams, Byron Bennett, Jr

ALASKA
DeLapp, Tina Davis

ARIZONA
Aposhian, Hurair Vasken
Blanchard, James
Bowman, Douglas Clyde
Brendel, Klaus
Bressler, Rubin
Burks, Thomas F
Chin, Lincoln
Clayton, John Wesley, Jr
Colburn, Wayne Alan
Consroe, Paul F
Eckardt, Robert E
Fagan, Timothy Charles
Gandolfi, A Jay
Halonen, Marilyn Jean
Halpert, James Robert
Huxtable, Ryan James
Jeter, Wayburn Stewart
Krahl, Maurice Edward
Laird, Hugh Edward, II
Lindell, Thomas Jay
Lukas, Ronald John
McQueen, Charlene A
Milch, Lawrence Jacques
North-Root, Helen May
Palmer, John Davis
Peric-Knowlton, Wlatka
Picchioni, Albert Louis
Rowe, Verald Keith
Russell, Findlay Ewing
Schmid, Jack Robert
Simonian, Vartkes Hovanes
Sipes, Ivan Glenn
Tong, Theodore G
Weil, Andrew Thomas

ARKANSAS
Cornett, Lawrence Eugene
Ginzel, Karl-Heinz
Greenman, David Lewis
Hanna, Calvin
Johnson, Bob Duell
Jordin, Marcus Wayne
Kadlubar, Fred F
Leakey, Julian Edwin Arundell
Light, Kim Edward
Littlefield, Neil Adair
McMillan, Donald Edgar
Paule, Merle Gale
Ruwe, William David
Schieferstein, George Jacob
Seifen, Ernst
Sorenson, John R J
Stone, Joseph

Valentine, Jimmie Lloyd
Walker, Charles A
Wenger, Galen Rosenberger
Wessinger, William David
Winters, Ronald Howard
Young, John Falkner

CALIFORNIA
Agabian, Nina
Aggot, J Desmond
Agre, Karl
Ainsworth, Earl John
Alousi, Adawia A
Amer, Mohamed Samir
Anderson, Hamilton Holland
Apple, Martin Allen
Ashe, John Herman
Azarnoff, Daniel Lester
Baldwin, Robert Charles
Barnes, Paul Richard
Batterman, Robert Coleman
Bauer, Robert Oliver
Benet, Leslie Z
Bennett, C Frank
Bergman, Hyman Chaim
Berman, David Albert
Berteau, Peter Edmund
Bidlack, Wayne Ross
Black, Kirby Samuel
Blankenship, James William
Blau, Helen Margaret
Bokoch, Gary M
Brazier, Mary A B
Brown, Joan Heller
Brunton, Laurence
Burkhalter, Alan
Byard, James Leonard
Carson, Virginia Rosalie Gottschall
Casida, John Edward
Castles, Thomas R
Catlin, Don Hardt
Chan, Kenneth Kin-Hing
Chang, Freddy Wilfred Lennox
Chang, Yi-Han
Chin, Jane Elizabeth Heng
Cho, Arthur Kenji
Cho, Tae Mook
Chuang, Ronald Yan-Li
Ciaramitaro, David A
Clark, Charles Richard
Clark, Dennis Richard
Clarke, David E
Cohen, Kenneth Samuel
Cohn, Major Lloyd
Corbascio, Aldo Nicola
Correia, Maria Almira
Courtney, Kenneth Randall
Cronheim, Georg Erich
Cronin, Michael John
Crooke, Stanley T
Dang, Peter Hung-Chen
Danse, Ilene H Raisfeld
Deedwania, Prakash Chandra
DeMet, Edward Michael
Dismukes, Robert Key
Dixon, Ross
Doroshow, James Halpern
Duckles, Sue Piper
Eakins, Kenneth E
Eltherington, Lorne
Embree, James Willard, Jr
Everett, Guy M
Fain, Gordon Lee
Fairchild, David George
Felton, James Steven
Ferguson, Noel Moore
Ferguson, Samuel A
Fields, Howard Lincoln
Flacke, Joan Wareham
Flacke, Werner Ernst
Fraser, Ian McLennan
Freeman, Leon David
Freeman, Walter Jackson, III
Fuhrman, Frederick Alexander
Furst, Arthur
Gans, Joseph Herbert
Gehrmann, John Edward
George, Robert
Geyer, Mark Allen
Giri, Shri N
Goldberg, Mark Arthur
Golder, Thomas Keith
Goldstein, Avram
Goldstein, Dora Benedict
Gong, William C
Goode, John Wolford
Gordon, Adrienne Sue
Gorelick, Kenneth J
Gourzis, James Theophile
Green, Donald Eugene
Green, Sidney
Gumbiner, Barry M
Guzman, Ruben Joseph
Haley, Thomas John
Hammock, Bruce Dupree
Hance, Anthony James
Hansen, Eder Lindsay
Hecht, Elizabeth Anne
Hefti, Franz F
Henderson, Gary Lee
Hess, Robert William
Hessinger, David Alwyn
Hidalgo, John
Hines, Leonard Russell

Hintz, Marie I
Ho, Begonia Y
Hoener, Betty-ann
Hollinger, Mannfred Alan
Hondeghem, Luc M
Hsieh, Dennis P H
Ignarro, Louis Joseph
Insel, Paul Anthony
Jaanus, Siret Desiree
Jacobs, Robert Saul
Jain, Naresh C
Jardetzky, Oleg
Jardine, Ian
Jarvik, Murray Elias
Jelliffe, Roger Woodham
Jenden, Donald James
Jensen, Richard Arthur
Jesmok, Gary J
Johnson, Howard (Laurence)
Johnson, Randolph Mellus
Jordan, Mary Ann
Joy, Robert McKernon
Kaplan, Stanley A
Katzung, Bertram George
Keasling, Hugh Hilary
Kelly, Raymond Crain
Kendig, Joan Johnston
Kilgore, Wendell Warren
Killam, Eva King
Killam, Keith Fenton, Jr
Kodama, Jiro Kenneth
Koschier, Francis Joseph
Koths, Kirston Edward
Kuczenski, Ronald Thomas
Laiken, Nora Dawn
Lee, Chi-Ho
Lee, Peter Van Arsdale
Lehmann, A(ldo) Spencer
Leung, Peter
Levine, Jon David
Levy, Joseph Victor
Levy, Louis
Lewis, Richard John
Lippmann, Wilbur
Liu, David H W
Loew, Gilda Harris
Lomax, Peter
Lowinger, Paul
Lytle, Loy Denham
McCaman, Richard Eugene
MacFarlane, Malcolm David
McGaughey, Charles Gilbert
McKenna, Edward J
Malone, Marvin Herbert
Mansour, Tag Eldin
Marangos, Paul Jerome
Maronde, Robert Francis
Marshall, John Foster
Mason, Dean Towle
Melmon, Kenneth Lloyd
Mensah, Patricia Lucas
Menzel, Daniel B
Meyers, Frederick H
Milby, Thomas Hutchinson
Miljanich, George Paul
Miller, Jon Philip
Mitoma, Chozo
Mizuno, Nobuko S(himotori)
Morey-Holton, Emily Rene
Mule, Salvatore Joseph
Nelson, Eric Loren
Nimni, Marcel Efraim
Novack, Gary Dean
O'Benar, John DeMarion
Okun, Ronald
O'Malley, Edward Paul
Ono, Joyce Kazuyo
Ortiz de Montellano, Paul Richard
Ottoboni, M(inna) Alice
Painter, Ruth Coburn Robbins
Patton, John Stuart
Pearl, Ronald G
Peoples, Stuart Anderson
Peroutka, Stephen Joseph
Peters, John Henry
Peters, Marvin Arthur
Peterson, Charles Marquis
Pharriss, Bruce Bailey
Pieper, Gustav Rene
Place, Virgil Alan
Povzhitkov, Moysey Michael
Printz, Morton Philip
Reifenrath, William Gerald
Renkin, Eugene Marshall
Ridley, Peter Tone
Riedesel, Carl Clement
Riemer, Robert Kirk
Ritzmann, Ronald Fred
Rivier, Catherine L
Rivier, Jean E F
Roberts, Carmel Montgomery
Rooks, Wendell Hofma, II
Roszkowski, Adolph Peter
Sadee, Wolfgang
Saifer, Mark Gary Pierce
Sandmeyer, Esther E
Sargent, Thornton William, III
Sassenrath, Ethelda Norberg
Sattin, Albert
Saute, Robert E
Sawyer, Wilbur Henderson
Schienle, Jan Hoops
Scholler, Jean
Schulman, Howard

Segal, David S
Selassie, Cynthia R
Setler, Paulette Elizabeth
Sharma, Arjun D
Shaw, Jane E
Shen, Wei-Chiang
Shih, Jean Chen
Siggins, George Robert
Smith, Charles G
Smith, Martyn Thomas
Smith, Norman Ty
Spiller, Gene Alan
Stanton, Hubert Coleman
Stanton, Toni Lynn
Stark, Larry Gene
Steffey, Eugene P
Stein, Larry
Stone, Deborah Bennett
Strosberg, Arthur Martin
Strother, Allen
Sullivan, John Lawrence
Sunshine, Irving
Taylor, Dermot Brownrigg
Taylor, Palmer William
Tewari, Sujata
Thomas, Lyell Jay, Jr
Thompson, John Frederick
Thueson, David Orel
Tiedcke, Carl Heinrich Wilhelm
Tilton, Bernard Ellsworth
Tramell, Paul Richard
Trevor, Anthony John
Tsien, Richard Winyu
Tune, Bruce Malcolm
Tutupalli, Lohit Venkateswara
Tye, Arthur
Urquhart, John, III
Uyeno, Edward Teiso
Van Dyke, Craig
Vernikos, Joan
Wallach, Marshall Ben
Wan, Suk Han
Wang, Howard Hao
Wangler, Roger Dean
Waterbury, Lowell David
Way, E Leong
Way, Jon Leong
Way, Walter
Wechter, William Julius
Wei, Edward T
Weinreb, Robert Neal
Weissberg, Robert Murray
Wendel, Otto Theodore, Jr
Wester, Ronald Clarence
Willhite, Calvin Campbell
Williams, Roger Lea
Wilson, Leslie
Winters, Wallace Dudley
Woolley, Dorothy Elizabeth Schumann
Yagiela, John Allen
Yaksh, Tony Lee
Yang, William C T
Ziegler, Michael Gregory

COLORADO
Allen, Larry Milton
Alpern, Herbert P
Appelt, Glenn David
Barrett, C Brent
Beam, Kurt George, Jr
Breault, George Omer
Collins, Allan Clifford
Coomes, Richard Merril
Deitrich, Richard Adam
Diamond, Louis
Erwin, Virgil Gene
Fennessey, Paul V
Froede, Harry Curt
Gerber, John George
Glass, Howard George
Herin, Reginald Augustus
Hesterberg, Thomas William
Hollister, Alan Scudder
Ishii, Douglas Nobuo
Malkinson, Alvin Maynard
Murphy, Robert Carl
Nies, Alan Sheffer
Palmer, Michael Rule
Petersen, Dennis Roger
Richards, Edmund A
Ruth, James Allan
Sharpless, Seth Kinman
Stewart, John Morrow
Thompson, John Alec
Vernadakis, Antonia
Weiner, Norman
Weliky, Irving
Williams, Ronald Lee
Wilson, Vincent L
Winstead, Jack Alan

CONNECTICUT
Aghajanian, George Kevork
Anderson, Rebecca Jane
Bacopoulos, Nicholas G
Bickerton, Robert Keith
Bunney, Benjamin Stephenson
Buyniski, Joseph P
Byck, Robert
Cadman, Edwin Clarence
Century, Bernard
Chatt, Allen Barrett
Chello, Paul Larson
Conrad, Eugene Anthony

Cooper, Jack Ross
Dannies, Priscilla Shaw
Davis, Michael
Dix, Douglas Edward
Doherty, Niall Stephen
Doshan, Harold David
Douglas, James Sievers
Douglas, William Wilton
Eisenfeld, Arnold Joel
Epstein, Mary A Farrell
Escobar, Javier I
Feinstein, Maurice B
Fleming, James Stuart
Forssen, Eric Anton
Freudenthal, Ralph Ira
Gay, Michael Howard
Gerritsen, Mary Ellen
Ghosh, Dipak K
Gillis, Charles Norman
Gomoll, Allen W
Gunther, Jay Kenneth
Gylys, Jonas Antanas
Handschumacher, Robert Edmund
Harding, Matthew William
Henderson, Edward George
Herbette, Leo G
Heyd, Allen
Hirsh, Eva Maria Hauptmann
Hitchcock, Margaret
Hobbs, Donald Clifford
Janis, Ronald Allen
King, Theodore Oscar
Klimek, Joseph John
Langner, Ronald O
Larson, Jerry King
Letts, Lindsay Gordon
Levine, Robert John
Loose, Leland David
McIlhenny, Hugh M
Makriyannis, Alexandros
Mayol, Robert Francis
Merker, Philip Charles
Milne, George McLean, Jr
Milzoff, Joel Robert
Moore, Peter Francis
Mycek, Mary J
Nair, Sreedhar
Niblack, John Franklin
Nieforth, Karl Allen
Nightingale, Charles Henry
Oates, Peter Joseph
Otterness, Ivan George
Pappano, Achilles John
Pazoles, Christopher James
Pinson, Ellis Rex, Jr
Possanza, Genus John
Prusoff, William Herman
Redmond, Donald Eugene, Jr
Riblet, Leslie Alfred
Riker, Donald Kay
Ritchie, Joseph Murdoch
Robinson, Donald Stetson
Rockwell, Sara Campbell
Rosenberg, Philip
Roth, Robert Henry, Jr
Rudnick, Gary
Sartorelli, Alan Clayton
Schach von Wittenau, Manfred
Schenkman, John Boris
Schmeer, Arline Catherine
Schmidt, James L
Schurig, John Eberhard
Schwartz, Pauline Mary
Scriabine, Alexander
Sheard, Michael Henry
Smilowitz, Henry Martin
Snellings, William Moran
Snider, Ray Michael
Soyka, Lester F
Stroebel, Charles Frederick, III
Tallman, John Francis
Taylor, Duncan Paul
Taylor, Russell James, Jr
Tyler, Tipton Ransom
Urquilla, Pedro Ramon
Wardell, William Michael
Weiss, Robert Martin
Weissman, Albert
Welch, Annemarie S
Wong, Stewart
Yocca, Frank D

DELAWARE
Aharony, David
Aungst, Bruce J
Brown, Barry Stephen
Buckner, Carl Kenneth
Christoph, Greg Robert
Clark, Robert
Cook, Leonard
DiPasquale, Gene
Galbraith, William
Giles, Ralph E
Goldberg, Morton Edward
Goldstein, Jeffrey Marc
Herblin, William Fitts
Hoffman, Howard Edgar
Kerr, Janet Spence
Kinoshita, Florence Keiko
Krell, Robert Donald
Krivanek, Neil Douglas
Lai, Chii-Ming
Lam, Gilbert Nim-Car
Loveless, Scott E

Pharmacology (cont)

McCurdy, David Harold
Malick, Jeffrey Bevan
Malley, Linda Angevine
Mullane, John F
Nielsen, Susan Thomson
Patel, Jitendra Balkrishna
Patel, Narayan Ganesh
Perry, Kenneth W
Price, William Alrich, Jr
Quisenberry, Richard Keith
Quon, Check Yuen
Rubin, Alan A
Salzman, Steven Kerry
Sarrif, Awni M
Schneider, Jurg Adolf
Shotzberger, Gregory Steven
Smith, Jerry Morgan
Steinberg, Marshall
Stump, John M
Taber, Robert Irving
Turlapaty, Prasad
U'Prichard, David C
Vernier, Vernon George
Wheeler, Allan Gordon
Whitney, Charles Candee, Jr
Zuckerman, Joan Ellen

DISTRICT OF COLUMBIA

Abernathy, Charles Owen
Abramson, Fred Paul
Alving, Carl Richard
Arcos, Joseph (Charles)
Argus, Mary Frances
Balazs, Tibor
Beaver, William Thomas
Beliles, Robert Pryor
Brands, Allen J
Brooker, Gary
Buday, Paul Vincent
Burris, James F
Cheney, Darwin L
Chung, Ho
Cohn, Victor Hugo
Crawford, Lester M
Deutsch, Stanley
Dickson, Robert Brent
Doctor, Bhupendra P
Dretchen, Kenneth Lewis
Ellis, Sydney
Fabro, Sergio
Fanning, George Richard
Farber, Theodore Myles
Gallardo-Carpentier, Adriana
Gillis, Richard A
Hamarneh, Sami Khalaf
Herman, Barbara Helen
Herman, Eugene H
Himmelsbach, Clifton Keck
Hollinshead, Ariel Cahill
Horakova, Zdenka
Karle, Jean Marianne
Kellar, Kenneth Jon
Kennedy, Katherine Ash
Klubes, Philip
Lee, Cheng-Chun
Lee, I P
Lerner, Pauline
Loo, Ti Li
Mandel, Harold George
Massari, V John
Mazel, Paul
Milbert, Alfred Nicholas
Miller, Russell Loyd, Jr
Moore, John Arthur
Murphy, James John
Noble, John F
Perry, David Carter
Raines, Arthur
Rall, David Platt
Ramwell, Peter William
Reilly, Joseph F
Rhoads, Allen R
Rhoden, Richard Allan
Schwartz, Sorell Lee
Sherman, John Foord
Sjogren, Robert W, Jr
Sobotka, Thomas Joseph
Soller, Roger William
Straw, James Ashley
Thomas, Richard Dean
Vick, James
Wagstaff, David Jesse
Wainer, Irving William
West, William Lionel
Woosley, Raymond Leon
Wyatt, Richard Jed
Yen-Koo, Helen C

FLORIDA

Abou-Khalil, Samir
Abraham, William Michael
Baker, Stephen Phillip
Bassett, Arthur L
Battista, Sam P
Bell, John Urwin
Ben, Max
Booth, Nicholas Henry
Boxill, Gale Clark
Boyd, Eleanor H
Chappell, Elizabeth
Childers, Steven Roger
Cintron, Guillermo B

Coffey, Ronald Gibson
Cohen, Glenn Milton
Corbett, Michael Dennis
Crews, Fulton T
Curry, Stephen H
Deichmann, William Bernard
Denber, Herman C B
Dickison, Harry Leo
Edelson, Jerome
Epstein, Murray
Farah, Alfred Emil
Finger, Kenneth F
Fitzgerald, Thomas James
Frank, H Lee
Freyburger, Walter Alfred
Garg, Lal Chand
Gelband, Henry
Goldstein, Burton Jack
Greenberg, Michael John
Hackman, John Clement
Hackney, John Franklin
Hahn, Elliot F
Harbison, Raymond D
Haynes, Duncan Harold
Herrmann, Roy G
Himes, James Albert
Howes, John Francis
Hudson, Reggie Lester
Jewett, Robert Elwin
Katovich, Michael J
Kerrick, Wallace Glenn Lee
Kiplinger, Glenn Francis
Korol, Bernard
Krishan, Awtar
Krzanowski, Joseph John, Jr
Kunisi, Venkatasubban S
Lamba, Surendar Singh
Lasseter, Kenneth Carlyle
Leibman, Kenneth Charles
Lewis, John Reed
Light, Amos Ellis
Maren, Thomas Hartley
Martin, Frank Gene
Mayer, Richard Thomas
Menzer, Robert Everett
Merritt, Alfred M, II
Michie, David Doss
Neims, Allen Howard
Oberst, Fred William
Palmer, Roger
Pfaffman, Madge Anna
Polson, James Bernard
Porter, Lee Albert
Potter, James D
Rader, William Austin
Radomski, Jack London
Rennick, Barbara Ruth
Richelson, Elliott
Ringel, Samuel Morris
Russell, Diane Haddock
Silverman, David Norman
Soliman, Karam Farag Attia
Soliman, Magdi R I
Szentivanyi, Andor
Teaf, Christopher Morris
Thomas, Vera
Tusing, Thomas William
Tuttle, Ronald Ralph
Van Breeman, Cornelis
Vogh, Betty Pohl
Walker, Don Wesley
Wecker, Lynn
Weidler, Donald John
Weiss, Charles Frederick
Wilcox, Christopher Stuart
Williams, Joseph Francis
Yeh, Billy Kuo-Jiun
Young, Michael David

GEORGIA

Adams, Robert Johnson
Bain, James Arthur
Black, Asa C, Jr
Boudinot, Frank Douglas
Bowen, John Metcalf
Brady, Ullman Eugene, Jr
Bruckner, James Victor
Buccafusco, Jerry Joseph
Byrd, Larry Donald
Catravas, John D
Ciarlone, Alfred Edward
Curley, Winifred H
Denson, Donald D
Diamond, Bruce I
Ebert, Andrew Gabriel
Edwards, Gaylen Lee
Garrettson, Lorne Keith
Gatipon, Glenn Blaise
Geber, William Frederick
Glass, David Bankes
Greenbaum, Lowell Marvin
Hatch, Roger Conant
Hofman, Wendell Fey
Huber, Thomas Lee
Hug, Carl Casimir, Jr
Israili, Zafar Hasan
Iturrian, William Ben
Iuvone, Paul Michael
Karow, Armand Monfort, Jr
Kirby, Margaret Loewy
Kolbeck, Ralph Carl
Kuo, Jyh-Fa
Lopez, Antonio Vincent
McGrath, William Robert

Madden, John Joseph
Miceli, Joseph N
Minneman, Kenneth Paul
Mitch, William Evans
Mokler, Corwin Morris
Moran, Neil Clymer
Nishie, Keica
Norred, William Preston
Potter, David Edward
Proctor, Charles Darnell
Pruett, Jack Kenneth
Pruitt, Albert Wesley
Schaefer, Gerald J
Schramm, Lee Clyde
Schweri, Margaret Mary
Shlevin, Harold H
Stone, Connie J
Sutherland, James Henry Richardson
Wade, Adelbert Elton
Whitford, Gary M
Wolven-Garrett, Anne M
Woodhouse, Bernard Lawrence

HAWAII

Bailey, Leslie Edgar
Lenney, James Francis
Lum, Bert Kwan Buck
Morton, Bruce Eldine
Norton, Ted Raymond
Okita, George Torao
Read, George Wesley
Shibata, Shoji

IDAHO

Cole, Franklin Ruggles
Dodson, Robin Albert
Duke, Victor Hal
Hillyard, Ira William
Vestal, Robert Elden

ILLINOIS

Albrecht, Ronald Frank
Aldred, J Phillip
Ambre, John Joseph
Anderson, David John
Anderson, Edmund George
Archer, John Dale
Atkinson, Arthur John, Jr
Baron, David Alan
Becker, Bernard Abraham
Beecher, Christopher W W
Bennett, Donald Raymond
Berman, Eleanor
Bernstein, Joel Edward
Berry, Charles Arthur
Bevill, Richard F, Jr
Bhargava, Hemendra Nath
Bianchi, Robert George
Bingel, Audrey Susanna
Bosmann, Harold Bruce
Boulos, Badi Mansour
Bourgault, Priscilla C
Bousquet, William F
Brodie, Mark S
Brown, Daniel Joseph
Browning, Ronald Anthony
Brunner, Edward A
Buck, William Boyd
Calandra, Joseph Carl
Caspary, Donald M
Chan, Yun Lai
Chiou, Win Loung
Chu, Sou Yie
Clay, George A
Cline, William H, Jr
Contreras, Patricia Cristina
Cook, Donald Latimer
Cranston, Joseph William, Jr
Crystal, George Jeffrey
Dailey, John William
Dajani, Esam Zafer
Davis, Harvey Virgil
Davis, Joseph Richard
Davis, Oscar F
Dawson, M Joan
Dean, Richard Raymond
Dodge, Patrick William
Dubocovich, Margarita L
Eagle, Edward
Ehrenpreis, Seymour
Elliott, William John
Erdös, Ervin George
Erve, Peter Raymond
Estep, Charles Blackburn
Faingold, Carl L
Farnsworth, Norman R
Feigen, Larry Philip
Feinberg, Harold
Ferguson, Hugh Carson
Fitzloff, John Frederick
Foreman, Ronald Louis
Friedman, Alexander Herbert
Garthwaite, Susan Marie
Giacobini, Ezio
Gillette, Martha Ulbrick
Glisson, Silas Nease
Goldberg, Leon Isadore
Green, Richard D
Haigler, Henry James, Sr
Hanin, Israel
Harris, Stanley Cyril
Haugen, David Allen
Henkin, Jack
Herting, Robert Leslie

Hiles, Richard Allen
Hirsch, Lawrence Leonard
Hjelle, Joseph Thomas
Ho, Andrew Kong-Sun
Hoffmann, Philip Craig
Hosey, M Marlene
Javaid, Javaid Iqbal
Jensen, Robert Alan
Jobe, Phillip Carl
Kang, David Soosang
Karczmar, Alexander George
Katz, Norman L
Kebabian, John Willis
Khan, Mohammed Abdul Quddus
Kimura, Eugene Tatsuru
Klabunde, Richard Edwin
Kochman, Ronald Lawrence
Kohli, Jai Dev
Koritz, Gary Duane
Kosman, Mary Ellen
Kulkarni, Anant Sadashiv
Kyncl, J Jaroslav
Laddu, Atul R
Lambert, Glenn Frederick
Lampe, Kenneth Francis
Lane, Benjamin Clay
Lasley, Stephen Michael
Le Breton, Guy C
Lee, Tony Jer-Fu
Leff, Alan R
Leikin, Jerrold Blair
Lesher, Gary Allen
Lin, Reng-Lang
Linde, Harry Wight
Link, Roger Paul
Loeb, Jerod M
Louis, John
Marczynski, Thaddeus John
Martin, Yvonne Connolly
Masry, Souheir El Defrawy
Maynert, Everett William
Mayor, Gilbert Harold
Miller, Donald Morton
Mimnaugh, Michael Neil
Mirkin, Bernard Leo
Mitrius, Joan C
Moawad, Atef H
Moon, Byong Hoon
Morgan, Juliet
Morris, Ralph William
Murad, Ferid
Musa, Mahmoud Nimir
Nair, Velayudhan
Nakajima, Shigehiro
Napier, T(avye) Celeste
Narahashi, Toshio
Neff-Davis, Carol Ann
Neidle, Enid Anne
Nelson, Thomas Eustis, Jr
Nutting, Ehard Forrest
Pai, Sadanand V
Papaioannou, Stamatios E
Parvez, Zaheer
Perkins, William Eldredge
Peterson, Rudolph Nicholas
Plotnikoff, Nicholas Peter
Prabhu, Venkatray G
Proudfit, Carol Marie
Proudfit, Herbert K(err)
Quock, Raymond Mark
Radzialowski, Frederick M
Rao, Gopal Subba
Rasenick, Mark M
Retz, Konrad Charles
Ringham, Gary Lewis
Rotenberg, Keith Saul
Sabelli, Hector C
Said, Sami I
Salafsky, Bernard P
Sanner, John Harper
Schreider, Bruce David
Schyve, Paul Milton
Seiden, Lewis S
Seidenfeld, Jerome
Sennello, Lawrence Thomas
Sherrod, Theodore Roosevelt
Siegel, Ivens Aaron
Smith, Steven Joel
Smith, Theodore Craig
Somani, Satu M
Spikes, John Jefferson
Spring, Bonnie Joan
Steffek, Anthony J
Steger, Richard Warren
Stein, Herman H
Stern, Paula Helene
Stickney, Janice Lee
Struthers, Barbara Joan Oft
Sturtevant, Frank Milton
Suarez, Kenneth Alfred
Sukowski, Ernest John
Takahashi, Joseph S
Taylor, Julius David
Ten Eick, Robert Edwin
Thomas, Elizabeth Wadsworth
Thompson, Emmanuel Bandele
Thomson, John Ferguson
Tourlentes, Thomas Theodore
Ts'o, Timothy On-To
Van de Kar, Louis David
Vazquez, Alfredo Jorge
Visek, Willard James
Walsh, Gerald Michael
Wiegand, Ronald Gay

Wilensky, Jacob T
Williams, Michael
Woolverton, William L
Wu, Chau Hsiung
Zaroslinski, John F

INDIANA
Abdallah, Abdulmuniem Husein
Anderson, Robert Clarke
Babbs, Charles Frederick
Berger, James Edward
Beach, Henry Roland, Jr
Bonderman, Dean P
Borowitz, Joseph Leo
Brater, D(onald) Craig
Bumpus, John Arthur
Burek, Joe Dale
Carlson, Gary P
Carmichael, Ralph Harry
Chio, E(ddie) Hang
Christ, Daryl Dean
Clark, William C
Cohen, Marlene Lois
Coppoc, Gordon Lloyd
Dantzig, Anne H
Diller, Erold Ray
DiMicco, Joseph Anthony
Doedens, David James
Dube, Gregory P
Dungan, Kendrick Webb
Echtenkamp, Stephen Frederick
Emmerson, John Lynn
Evans, Michael Allen
Fayle, Harlan Downing
Fleisch, Jerome Herbert
Forney, Robert Burns
Franklin, Ronald
Fuller, Ray W
Gehring, Perry James
Gibson, William Raymond
Goh, Edward Hua Seng
Goldstein, David Joel
Greenspan, Kalman
Griffing, William James
Hahn, Richard Allen
Harned, Ben King
Hayes, Donald Charles
Henry, David P, II
Ho, Peter Peck Koh
Hynes, Martin Dennis, III
Isom, Gary E
Kauffman, Raymond F
Kim, Kil Chol
Leander, John David
Leeling, Jerry L
Lemberger, Louis
Lin, Tsung-Min
Lindstrom, Terry Donald
Macchia, Donald Dean
McKinney, Gordon R
McLaughlin, Jerry Loren
McNay, John Leeper
Maickel, Roger Philip
Mais, Dale E
Mannix, Edward T
Marshall, Franklin Nick
Mason, Norman Ronald
Matsumoto, Charles
Nelson, David L
Nelson, David Lloyd
Nelson, Robert Leon
Nichols, David Earl
Nickander, Rodney Carl
Odya, Charles E
Parli, C(arol) John
Pfeifer, Richard Wallace
Pollock, G Donald
Rebec, George Vincent
Richter, Judith Anne
Ridolfo, Anthony Sylvester
Roach, Peter John
Robertson, David Wayne
Root, Mary Avery
Rubin, Alan
Rutledge, Charles O
Saz, Howard Jay
Seidehamel, Richard Joseph
Shea, Philip Joseph
Silbaugh, Steven A
Sisson, George Maynard
Smith, Gerald Floyd
Spolyar, Louis William
Spratto, George R
Stark, Paul
Steinberg, Mitchell I
Sullivan, Hugh R, Jr
Thompson, Wilmer Leigh
Thor, Karl Bruce
Tyler, Varro Eugene
Van Tyle, William Kent
Vasko, Michael Richard
Vodicnik, Mary Jo
Wagle, Shreepad R
Weber, George
Weikel, John Henry, Jr
Willis, Lynn Roger
Wolen, Robert Lawrence
Wright, Walter Eugene
Yim, George Kwock Wah
Zabik, Joseph

IOWA
Ahrens, Franklin Alfred
Ascoli, Mario

Baron, Jeffrey
Benton, Byrl E
Bhatnagar, Ranbir Krishna
Brody, Michael J
Conn, P Michael
Dutton, Gary Roger
Dyer, Donald Chester
Ghoneim, Mohamed Mansour
Granberg, Charles Boyd
Hsu, Walter Haw
Husted, Russell Forest
Johnson, Wallace W
Jones, Leslie F
Kasik, John Edward
Kunesh, Jerry Paul
Long, John Paul
Makar, Adeeb Bassili
Nair, Vasu
Niebyl, Jennifer Robinson
Orcutt, James A
Rosazza, John N
Russi, Gary Dean
Shires, Thomas Kay
Spratt, James Leo
Steele, William John
Tephly, Thomas R
Teppert, William Allan, Sr
Williamson, Harold E

KANSAS
Alper, Richard H
Berman, Nancy Elizabeth Johnson
Browne, Ronald K
Bunag, Ruben David
Carlson, Arthur, Jr
Cheng, Chia-Chung
Chin, Hong Woo
Davis, John Stewart
Dixon, Walter Reginald
Doull, John
Dujovne, Carlos A
Eisenbrandt, Leslie Lee
Faiman, Morris David
Gordon, Michael Andrew
Greene, Nathan Doyle
Kadoum, Ahmed Mohamed
Klaassen, Curtis Dean
Lanman, Robert Charles
Michaelis, Mary Louise
Nelson, Stanley Reid
Norton, Stata Elaine
Parkinson, Andrew
Pazdernik, Thomas Lowell
Poisner, Alan Mark
Ringle, David Allan
Slavik, Milan
Spohn, Herbert Emil
Takemoto, Dolores Jean
Tessel, Richard Earl
Upson, Dan W
Uyeki, Edwin M
Walaszek, Edward Joseph
Wenzel, Duane Greve

KENTUCKY
Aronoff, George R
Benz, Frederick W
Carr, Laurence A
Chen, Theresa S
Dagirmanjian, Rose
Diedrich, Donald Frank
Edmonds, Harvey Lee, Jr
Elkes, Joel
Flesher, James Wendell
Gordon, Helmut Albert
Gupta, Ramesh C
Heinicke, Ralph Martin
Huang, Kee-Chang
Hurst, Harrell Emerson
Jarboe, Charles Harry
Kary, Christina Dolores
Knoefel, Peter Klerner
Kulkarni, Prasad Shrikrishna
Lang, Calvin Allen
Lubawy, William Charles
Luckens, Mark Manfred
McGraw, Charles Patrick
McNicholas, Laura T
Martin, William Robert
Massik, Michael
Matheny, James Lafayette
Miller, Frederick N
Myers, Steven Richard
Nerland, Donald Eugene
Petering, Harold George
Ramp, Warren Kibby
Rees, Earl Douglas
Rodriguez, Lorraine Ditzler
Rowell, Peter Putnam
Scharff, Thomas G
Schurr, Avital
Tobin, Thomas
Volp, Robert Francis
Vore, Mary Edith
Waddell, William Joseph
Waite, Leonard Charles
Williams, Walter Michael
Yokel, Robert Allen
Zimmerman, Thom J

LOUISIANA
Agrawal, Krishna Chandra
Barker, Louis Allen
Beckman, Barbara Stuckey

Bobbin, Richard Peter
Boertje, Stanley
Bourn, William M
Brizzee, Kenneth Raymond
Brown, Richard Don
Carter, Mary Kathleen
Domer, Floyd Ray
Dunn, Adrian John
Eickholt, Theodore Henry
Ferguson, Gary Gene
Fisher, James W
Frohlich, Edward David
Garcia, Meredith Mason
Geiger, Paul Frank
George, William Jacob
Grace, Marcellus
Guth, Paul Spencer
Hastings, Robert Clyde
Hernandez, Thomas
Jenkins, William L
Komiskey, Harold Louis
McNamara, Dennis B
Manno, Barbara Reynolds
Manno, Joseph Eugene
Morgan, Lee Roy, Jr
Morrissette, Maurice Corlette
Olson, Richard David
Pearson, James Eldon
Redetzki, Helmut M
Robie, Norman William
Short, Charles Robert
Smith, James Edward
Stewart, John Joseph
Wilson, John T
Wood, Charles Donald

MAINE
Bird, Joseph G
Fassett, David Walter
Gabbert, Paul George
Major, Charles Walter

MARYLAND
Adams, John George
Adamson, Richard H
Adkinson, Newton Franklin, Jr
Adler, Michael
Ahluwalia, Gurpreet S
Albuquerque, Edson Xavier
Alderdice, Marc Taylor
Alleva, Frederic Remo
Alvares, Alvito Peter
Anderson, Arthur O
Ansher, Sherry Singer
Aronow, Lewis
Asghar, Khursheed
Atrakchi, Aisar Hasan
Axelrod, Julius
Bachur, Nicholas R, Sr
Baker, Houston Richard
Bareis, Donna Lynn
Barrack, Evelyn Ruth
Barrett, James E
Baskin, Steven Ivan
Batson, David Banks
Bayer, Barbara Moore
Beall, James Robert
Bean, Barbara Louise
Beaven, Michael Anthony
Blake, David Andrew
Blaszkowski, Thomas P
Blomster, Ralph N
Blumberg, Peter Mitchell
Blumenthal, Herbert
Bogdanski, Donald Frank
Braude, Monique Colsenet
Bronaugh, Robert Lewis
Brookes, Neville
Bruck, Stephen Desiderius
Bünger, Rolf
Burch, Ronald Martin
Burgison, Raymond Merritt
Burt, David Reed
Canfield, Craig Jennings
Caplan, Yale Howard
Carr, Charles Jelleff
Carrico, Christine Kathryn
Chaudhari, Anshumali
Chisolm, James Julian, Jr
Chiueh, Chuang Chin
Chuang, De-Maw
Coffey, Donald Straley
Cohen, Jack Sidney
Cohen, Robert Martin
Colombani, Paul Michael
Commarato, Michael A
Commissiong, John Wesley
Cone, Edward Jackson
Contrera, Joseph Fabian
Cook, Ellsworth Barrett
Coomes, Marguerite Wilton
Cosmides, George James
Cott, Jerry Mason
Cox, Brian Martyn
Cox, George Warren
Creveling, Cyrus Robbins
Cueto, Cipriano, Jr
Dacre, Jack Craven
Davidian, Nancy McConnell
De Luca, Luigi Maria
Dingell, James V
Dionne, Raymond A
Eckardt, Michael Jon
Egorin, Merril Jon

Ehrreich, Stewart Joel
Eldefrawi, Amira T
Eldefrawi, Mohyee E
Enna, Salvatore Joseph
Epstein, Stephen Edward
Fakunding, John Leonard
Fenselau, Catherine Clarke
Finch, Robert Allen
Fitzgerald, Glenna Gibbs (Cady)
Fowler, Bruce Andrew
Fricke, Robert F
Friess, Seymour Louis
Gabay, Sabit
Gelboin, Harry Victor
Geller, Irving
Gillette, James Robert
Glazer, Robert Irwin
Gold, Philip William
Goldberg, Alan Marvin
Goldberg, Steven R
Goldstein, David Stanley
Gram, Theodore Edward
Gray, Allan P
Greenstein, Edward Theodore
Grieshaber, Charles K
Griffiths, Roland Redmond
Grunberg, Neil Everett
Gueriguian, John Leo
Gusovsky, Fabian
Hahn, Fred Ernst
Hanbauer, Ingeborg
Hanson, Robert C
Hardy, Lester B
Harmon, Alan Dale
Harwood, Clare Theresa
Hassett, Charles Clifford
Hattan, David Gene
Heiffer, Melvin Harold
Heinrich, Max Alfred, Jr
Helke, Cinda Jane
Hickey, Robert Joseph
Hienz, Robert Douglas
Hirshman, Carol A
Hoffmann, Conrad Edmund
Hunt, Walter Andrew
Ichniowski, Casimir Thaddeus
Irving, George Washington, Jr
Irwin, Richard Leslie
Jacobowitz, David
Jacobson, Keith Hazen
Jagadeesh, Gowra G
Jasinski, Donald Robert
Jasper, Robert Lawrence
Johanson, Chris Ellyn
Johns, David Garrett
Joshi, Sewa Ram
Kafka, Marian Stern
Kaiser, Joseph Anthony
Kalser, Sarah Chinn
Kapetanovic, Izet Michael
Karel, Leonard
Kauffman, Frederick C
Kaufman, Joyce J
Kawalek, Joseph Casimir, Jr
Keefer, Larry Kay
Kelsey, Frances Oldham
Khan, Mushtaq Ahmad
Khazan, Naim
Kincaid, Randall L
Kinnard, William J, Jr
Kitzes, George
Kohn, Kurt William
Kohn, Leonard David
Kopin, Irwin J
Koslow, Stephen Hugh
Kramer, Barnett Sheldon
Kramer, Stanley Phillip
Kramer, William Geoffrey
Kraybill, Herman Fink
Kuhar, Michael Joseph
Kunos, George
Kupferberg, Harvey J
Lake, Charles Raymond
Lathers, Claire M
Lazar-Wesley, Eliane M
Lee, Lyndon Edmund, Jr
Lee-Ham, Doo Young
Levitan, Herbert
Levy, Alan C
Lietman, Paul Stanley
Lin, Shin
Lipicky, Raymond John
Lish, Paul Merrill
Litterst, Charles Lawrence
Liu, Ching-Tong
Liu, Leroy Fong
Liu, Yung-Pin
London, Edythe D
Love, Raymond Charles
Lowensohn, Howard Stanley
MacCanon, Donald Moore
McCurdy, John Dennis
MacDonald, William E, Jr
Macri, Frank John
Maengwyn-Davies, Gertrude Diane
Magnani, John Louis
Mainigi, Kumar D
Majchrowicz, Edward
Manen, Carol-Ann
Marcus, Richard
Markey, Sanford Philip
Martin, Edgar J
Martin, George Reilly
Marwah, Joe

Pharmacology (cont)

Marzulli, Francis Nicholas
Middlebrook, John Leslie
Milman, Harry Abraham
Moreton, Julian Edward
Morton, Joseph James Pandozzi
Mulinos, Michael George
Munn, John Irvin
Musselman, Nelson Page
Myers, Charles
Myslinski, Norbert Raymond
Nemec, Josef
Nuite-Belleville, Jo Ann
O'Leary, Anne K
O'Leary, John Francis
Oliverio, Vincent Thomas
Orahovats, Peter Dimiter
Osterberg, Robert Edward
Owens, Albert Henry, Jr
Pace, Judith G
Palazzolo, Matthew Joseph
Park, Lee Crandall
Parkhie, Mukund Raghunathrao
Pert, Candace B
Petrella, Vance John
Petrucelli, Lawrence Michael
Pickworth, Wallace Bruce
Pilch, Susan Marie
Polinsky, Ronald John
Porter, Roger J
Post, Robert M
Powers, Marcelina Venus
Prasanna, Hullahalli Rangaswamy
Reed, Warren Douglas
Rehak, Matthew Joseph
Resnick, Charles A
Rew, Robert Sherrard
Rice, Kenner Cralle
Robinson, Cecil Howard
Robinson-White, Audrey Jean
Rogawski, Michael Andrew
Rogawski, Michael Andrew
Rosen, Gerald M
Rosenstein, Laurence S
Rubin, Robert Jay
Rudo, Frieda Galindo
Russell, James T
Saavedra, Juan M
Safer, Brian
Salem, Harry
Sarvey, John Michael
Sass, Neil Leslie
Sastre, Antonio
Scheibel, Leonard William
Schleimer, Robert P
Schoenfeld, Ronald Irwin
Schuster, Charles Roberts, Jr
Schwartz, Joan Poyner
Schwartz, Paul
Scott, Charles Covert
Seamon, Kenneth Bruce
Seifried, Harold Edwin
Sharma, Gopal Chandra
Shellenberger, Thomas E
Shih, Tsung-Ming Anthony
Sidell, Frederick R
Sieber-Fabro, Susan M
Silbergeld, Ellen K
Skolnick, Phil
Smith, Douglas Lee
Smith, Ralph Grafton
Snyder, Solomon H
Sonawane, Babasaheb R
Song, Byoung-Joon
Steinberg, George Milton
Steranka, Larry Richard
Suomi, Stephen John
Szara, Stephen Istvan
Tabakoff, Boris
Talalay, Paul
Tamminga, Carol Ann
Teske, Richard H
Thut, Paul Douglas
Titus, Elwood Owen
Tocus, Edward C
Uhde, Thomas Whitley
Umberger, Ernest Joy
Valdes, James John
Valley, Sharon Louise
Van Arsdel, William Campbell, III
Vaupel, Donald Bruce
Velletri, Paul A
Viswanathan, C T
Vocci, Frank Joseph
Walker, Michael Dirck
Walters, Judith R
Warnick, Jordan Edward
Watkins, Clyde Andrew
Watzman, Nathan
Weeks, Maurice Harold
Weight, Forrest F
Weimer, John Thomas
Weiner, Myron
Weisburger, Elizabeth Kreiser
Welch, Arnold D(eMerritt)
Whitehurst, Virgil Edwards
Wolff, Frederick William
Wolters, Robert John
Woodman, Peter William
Wykes, Arthur Albert
Yaffe, Sumner J
Yang, Shen Kwei
Yeh, Shu-Yuan

Yergey, Alfred L, III
Yu, Mei-ying Wong
Zaharko, Daniel Samuel
Zatz, Martin
Zeeman, Maurice George
Zimmerman, Hyman Joseph

MASSACHUSETTS

Adams, Jack H(erbert)
Alper, Milton H
Amdur, Mary Ochsenhirt
Anderson, Julius Horne, Jr
Assaykeen, Tatiana Anna
Baldessarini, Ross John
Benjamin, David Marshall
Bird, Stephanie J
Blumberg, Harold
Boisse, Norman Robert
Brown, Neal Curtis
Cameron, John Stanley
Carlson, Kristin Rowe
Castellot, John J, Jr
Castronovo, Frank Paul, Jr
Chase, Arleen Ruth
Chodosh, Sanford
Cohen, Jonathan Brewer
Cornman, Ivor
D'Amato, Henry Edward
Dershwitz, Mark
Dews, Peter Booth
Ellingboe, James
Essigmann, John Martin
Fielding, Stuart
Gambill, John Douglas
Gamzu, Elkan R
Gaudette, Leo Eward
Goldberg, Irvin H(yman)
Goldmacher, Victor S
Goldman, Peter
Greenblatt, David J
Griffith, Robert W
Hadidian, Zareh
Hadjian, Richard Albert
Hallock, Gilbert Vinton
Hartmann, Ernest Louis
Hauser, George
Hershenson, Benjamin R
Hodge, Harold Carpenter
Jenkins, Howard Jones
Johnson, Elsie Ernest
Kaul, Pushkar Nath
Kelleher, Roger Thomson
Kornetsky, Conan
Kosersky, Donald Saadia
Kuemmerle, Nancy Benton Stevens
Kupfer, David
Kupferman, Allan
Lasagna, Louis (Cesare)
Latz, Arje
Levine, Ruth R
Levy, Deborah Louise
Lloyd, Weldon S
Lorenzo, Antonio V
Ludlum, David Blodgett
McKearney, James William
McNair, Douglas McIntosh
Madras, Bertha K
Maher, Timothy John
Maran, Janice Wengerd
Marquis, Judith Kathleen
Masek, Bruce James
Mautner, Henry George
Miczek, Klaus A
Miller, Keith Wyatt
Miller, Tracy Bertram
Modest, Edward Julian
Morgan, James Philip
Morgan, Kathleen Greive
Morse, William Herbert
Moskowitz, Michael Arthur
Muni, Indu A
Osgood, Patricia Fisher
Pelikan, Edward Warren
Pober, Zalmon
Reinhard, John Frederick
Resnick, Oscar
Riggi, Stephen Joseph
Rosenkrantz, Harris
Ruelius, Hans Winfried
Ryser, Hugues Jean-Paul
Schaub, Robert George
Schildkraut, Joseph Jacob
Schneider, Frederick Howard
Shader, Richard Irwin
Shargel, Leon David
Shuster, Louis
Smith, Emil Richard
Smith, Wendy Anne
Snyder, Benjamin Willard
Szabo, Sandor
Tashjian, Armen H, Jr
Taylor, James A
Volicer, Ladislav
Waud, Barbara E
Waud, Douglas Russell
Wells, Herbert
Wenger, Christian Bruce
Wurtman, Richard Jay
Yesair, David Wayne

MICHIGAN

Abel, Ernest Lawrence
Aggarwal, Surinder K
Aiken, James Wavell

Anderson, Gordon Frederick
Anderson, Thomas Edward
Aston, Roy
Bach, Michael Klaus
Bailey, Harold Edwards
Braselton, Webb Emmett, Jr
Braughler, John Mark
Braun, Werner Heinz
Briggs, Josephine P
Brody, Theodore Meyer
Bruns, Robert Frederick
Buchanan, Robert Alexander
Cheney, B Vernon
Chenoweth, Maynard Burton
Cochran, Kenneth William
Cohen, Sanford Ned
Collins, Robert James
Conover, James H
Cooper, Theodore
Counsell, Raymond Ernest
Dambach, George Ernest
Data, Joann L
De La Iglesia, Felix Alberto
Deuben, Roger R
Domino, Edward Felix
Drach, John Charles
DuCharme, Donald Walter
Dutta, Saradindu
Ellis, C N
Fischer, Lawrence J
Frade, Peter Daniel
Freedman, Robert Russell
Freeman, Arthur Scott
Gebber, Gerard L
Gershon, Samuel
Gilbertson, Michael
Gilman, Martin Robert
Glazko, Anthony Joachim
Goldenthal, Edwin Ira
Goldman, Harold
Goodman, Jay Irwin
Hajratwala, Bhupendra R
Hall, Edward Dallas
Heffner, Thomas G
Herzig, David Jacob
Hodges, Robert Manley
Hollenberg, Paul Frederick
Hollingworth, Robert Michael
Hudson, Roy Davage
Hult, Richard Lee
Johnson, Garland A
Johnson, Herbert Gardner
Johnston, Raymond F
Kabara, Jon Joseph
Kaplan, Harvey Robert
Kauffman, Ralph Ezra
Kramer, Sherman Francis
Krueger, Robert John
La Du, Bert Nichols, Jr
Lahti, Robert A
Leopold, Wilbur Richard, III
Lindblad, William John
Lucchesi, Benedict Robert
McCarty, Leslie Paul
McGovren, James Patrick
McKenna, Michael Joseph
Marcelo, Cynthia Luz
Marcoux, Frank W
Marks, Bernard Herman
Marrazzi, Mary Ann
Maxwell, Donald Robert
Medzihradsky, Fedor
Mehring, Jeffrey Scott
Mohrland, J Scott
Moore, Kenneth Edwin
Mostafapour, M Kazem
Neubig, Richard Robert
Njus, David Lars
Novak, Raymond Francis
Pals, Donald Theodore
Philipsen, Judith Lynne Johnson
Piercey, Montford F
Piper, Walter Nelson
Poschel, Bruno Paul Henry
Puro, Donald George
Rao, Suryanarayana K
Rapundalo, Stephen T
Rech, Richard Howard
Reeves, Andrew Louis
Reinke, David Albert
Reitz, Richard Henry
Roslinski, Lawrence Michael
Roth, Robert Andrew, Jr
Ruddon, Raymond Walter, Jr
Ruwart, Mary Jean
Schoener, Eugene Paul
Schrier, Denis J
Scicli, Alfonso Guillermo
Sethy, Vimala Hiralal
Shellenberger, Melvin Kent
Shepard, Robert Stanley
Shipman, Charles, Jr
Siddiqui, Waheed Hasan
Slywka, Gerald William Alexander
Smith, Charles Bruce
Smith, Robert James
Smith, Thomas Charles
Spilman, Charles Hadley
Straw, Robert Niccolls
Swain, Henry Huntington
Tang, Andrew H
Theiss, Jeffrey Charles
Thornburg, John Elmer
Uhler, Michael David

Van Dyke, Russell Austin
Vatsis, Kostas Petros
VonVoigtlander, Philip Friedrich
Vostal, Jaroslav Joseph
Vrbanac, John James
Wagner, John Garnet
Watanabe, Philip Glen
Webb, R Clinton
Weber, Wendell W
Weeks, James Robert
Wetherell, Herbert Ranson, Jr
Winbury, Martin M
Yamazaki, Russell Kazuo
Zabik, Matthew John
Zannoni, Vincent G
Zins, Gerald Raymond

MINNESOTA

Abraham, Robert Thomas
Abuzzahab, Faruk S, Sr
Ames, Matthew Martin
Beck, Lloyd
Beitz, Alvin James
Bowers, Larry Donald
Brimijoin, William Stephen
Broderius, Steven James
Brown, David Robert
Cloyd, James C, III
Conard, Gordon Joseph
Cutkomp, Laurence Kremer
Dai, Xue Zheng (Charlie)
Dysken, Maurice William
Eisenberg, Richard Martin
Elde, Robert Philip
Francis, Gary Stuart
From, Arthur Harvey Leigh
Gray, Grace Warner
Hapke, Bern
Holtzman, Jordan L
Housmans, Philippe Robert H P
Kerley, Troy Lamar
Knych, Edward Thomas
Kvam, Donald Clarence
Lee, Nancy M
Leon, Arthur Sol
Loh, Horace H
Mabry, Paul Davis
Mannering, Gilbert James
Messing, Rita Bailey
Miller, Jack W
Miller, Richard Lynn
Murray, John Randolph
Nelson, Robert A
Ober, Robert Elwood
O'Dea, Robert Francis
Powis, Garth
Quebbemann, Aloysius John
Rao, Gundu Hirisave Rama
Roerig, Sandra Charlene
Romero, Juan Carlos
Sawchuk, Ronald John
Seybold, Virginia Susan (Dick)
Shoeman, Don Walter
Sladek, Norman Elmer
Sparber, Sheldon B
Staba, Emil John
Stowe, Clarence M
Swingle, Karl F
Szurszewski, Joseph Henry
Takemori, Akira Eddie
Tedeschi, David Henry
Toscano, William Agostino, Jr
Weaver, Lawrence Clayton
Weinshilboum, Richard Merle
Wilcox, George Latimer
Zaske, Darwin Erhard
Zimmerman, Ben George
Zimmerman, Cheryl Lea

MISSISSIPPI

Alford, Geary Simmons
Chaires, Jonathan Bradford
Chambers, Howard Wayne
Davis, Wilbur Marvin
Deneau, Gerald Antoine
Dorough, H Wyman
ElSohly, Mahmoud Ahmed
Farley, Jerry Michael
Hoskins, Beth
Hufford, Charles David
Hume, Arthur Scott
Klein, Richard Lester
Kodavanti, Prasada Rao S
Mehendale, Harihara Mahadeva
Mercer, Henry Dwight
Pace, Henry Buford
Rockhold, Robin William
Thureson-Klein, Asa Kristina
Verlangieri, Anthony Joseph
Wahba, Albert J
Waters, Irving Wade
Wilson, Marvin Cracraft

MISSOURI

Adams, Henry Richard
Adams, Max David
Ahmed, Mahmoud S
Badr, Mostafa
Barnes, Byron Ashwood
Bettonville, Paul John
Blaine, Edward Homer
Blake, Robert L
Boime, Irving
Burton, Robert Main

Byington, Keith H
Cenedella, Richard J
Chapnick, Barry M
Chiappinelli, Vincent A
Cicero, Theodore James
Coret, Irving Allen
Counts, David Francis
Covey, Douglas Floyd
Csaky, Tihamer Zoltan
Ferrendelli, James Anthony
Forte, Leonard Ralph
Gold, Alvin Hirsh
Green, Theodore James
Green, Vernon Albert
Hix, Elliott Lee
Howlett, Allyn C
Hudgins, Patricia Montague
Hunter, Francis Edmund, Jr
Ireland, Gordon Alexander
Jensen, Clyde B
Johnson, Eugene Malcolm, Jr
Johnson, Richard Dean
Kim, Yee Sik
Levinskas, George Joseph
Lonigro, Andrew Joseph
Lowry, Oliver Howe
McDougal, David Blean, Jr
Miller, Lowell D
Naeger, Leonard L
Needleman, Philip
Nickols, G Allen
Nyquist-Battie, Cynthia
Pandya, Krishnakant Hariprasad
Piepho, Robert Walter
Pierce, John Thomas
Pike, Linda Joy
Podrebarac, Eugene George
Russell, Robert Lee
St Omer, Vincent Victor
Shukla, Shivendra Dutt
Slatopolsky, Eduardo
Smith, Ronald Gene
Tuttle, Warren Wilson
Van Deripe, Donald R
Westfall, Thomas Creed
Woodhouse, Edward John
Wosilait, Walter Daniel
Yarbro, John Williamson
Zee-Cheng, Robert Kwang-Yuen
Zenser, Terry Vernon
Zepp, Edwin Andrew

MONTANA

Bryan, Gordon Henry
Johnson, Howard Ernest
Parker, Keith Krom
Van Horne, Robert Loren

NEBRASKA

Abel, Peter William
Berndt, William O
Brody, Alfred Walter
Brown, David G
Bylund, David Bruce
Carson, Steven
Crampton, James Mylan
Deupree, Jean Durley
Dowd, Frank, Jr
Ebadi, Manuchair
Elder, John Thompson, Jr
Engel, Toby Ross
Gessert, Carl F
Grandjean, Carter Jules
Haskell, Albert Russell
Heath, Eugene Cartmill
Hexum, Terry Donald
Issenberg, Phillip
Johnson, John Raymond
Lockridge, Oksana Maslivec
McClurg, James Edward
Magee, Donal Francis
Murrin, Leonard Charles
Phillips, Hugh Jefferson
Prioreschi, Plinio
Roche, Edward Browining
Scholar, Eric M
Scholes, Norman W
Schulz, John C
Sethi, V Sagar
Shaw, David Harold

NEVADA

Bierkamper, George G
Buxton, Iain Laurie Offord
Ferguson, Roger K
Pardini, Ronald Shields
Pool, Peter Edward
Potter, Gilbert David
Van Remoortere, Emile C
Westfall, David Patrick

NEW HAMPSHIRE

Borison, Herbert Leon
Dayton, Peter Gustav
Gosselin, Robert Edmond
Kensler, Charles Joseph
McCarthy, Lawrence E
Maudsley, David V
Mudge, Gilbert Horton
Pape, Brian Eugene
Smith, Roger Powell

NEW JERSEY

Abou-Gharbia, Magid

Abrutyn, Donald
Ahn, Ho-Sam
Amory, David William
Antonaccio, Michael John
Asaad, Magdi Mikhaeil
Aviado, Domingo Mariano
Babcock, George, Jr
Babcock, Philip Arnold
Bagdon, Robert Edward
Baker, Thomas
Balwierczak, Joseph Leonard
Barnett, Allen
Bautz, Gordon T
Bex, Frederick James
Blaiklock, Robert George
Boksay, Istvan Janos Endre
Borella, Luis Enrique
Breckenridge, Bruce (McLain)
Brezenoff, Henry Evans
Brodie, David Alan
Brogle, Richard Charles
Brostrom, Charles Otto
Brostrom, Margaret Ann
Buchweitz, Ellen
Bullock, Francis Jeremiah
Buyske, Donald Albert
Byrne, Jeffrey Edward
Cafruny, Edward Joseph
Capetola, Robert Joseph
Carlson, Richard P
Carrano, Richard Alfred
Caruso, Frank San Carlo
Cayen, Mitchell Ness
Chaikin, Philip
Chand, Naresh
Chang, Joseph Yoon
Chau, Thuy Thanh
Cheng, Kang
Chiu, Peter Jiunn-Shyong
Condouris, George Anthony
Conn, Hadley Lewis, Jr
Conney, Allan Howard
Conway, Paul Gary
Cotty, Val Francis
Coutinho, Claude Bernard
Crawley, Lantz Stephen
Creese, Ian N
Cressman, William Arthur
Crouthamel, William Guy
Curro, Frederick A
Cushman, David Wayne
Dain, Jeremy George
Dashman, Theodore
Davidson, Arnold R
Davies, Richard O
Davis, Harry R, Jr
Day-Salvatore, Debra-Lynn
Demarest, Keith Thomas
Dennis, Susana R K de
Desjardins, Raoul
Devlin, Richard Gerald, Jr
Diamantis, William
Di Carlo, Frederick Joseph
DiFazio, Louis T
Dubinsky, Barry
Dunikoski, Leonard Karol, Jr
Easterday, Otho Dunreath
Eckhardt, Eileen Theresa
Ellis, Daniel B(enson)
Emele, Jane Frances
Eng, Robert H K
Fernandes, Prabhavathi Bhat
Fiedler-Nagy, Christa
Flynn, Edward Joseph
Foley, James E
Frantz, Beryl May
Free, Charles Alfred
Freund, Matthew J
Frost, David
Gaut, Zane Noel
Geczik, Ronald Joseph
Geller, Herbert M
Gertner, Sheldon Bernard
Ghai, Geetha R
Gilfillan, Alasdair Mitchell
Gilman, Steven Christopher
Glaser, Keith Brian
Gogerty, John Harry
Gold, Barry Ira
Goldemberg, Robert Lewis
Goldstein, Ann L
Goodman, Frank R
Gorodetzky, Charles W
Greenberg, Roland
Greenslade, Forrest C
Grimes, David
Grosso, John A
Grover, Gary James
Gussin, Robert Z
Haft, Jacob I
Hageman, William E
Handley, Dean A
Harris, Don Navarro
Harrison-Johnson, Yvonne E
Hartig, Paul Richard
Harun, Joseph Stanley
Hassert, G(eorge) Lee, Jr
Heilman, William Paul
Hey, John Anthony
Hoffman, Clark Samuel, Jr
Horovitz, Zola Phillip
Howland, Richard David
Hubbard, John W
Huger, Francis P

Hutcheon, Duncan Elliot
Hyndman, Arnold Gene
Jacobson, Martin Michael
Ji, Sungchul
Jirkovsky, Ivo
Johnson, Melvin Clark
Kamm, Jerome J
Kapoor, Inder Prakash
Katz, Laurence Barry
King, John A(lbert)
Kirsten, Edward Bruce
Kissel, John Walter
Knoppers, Antonie Theodoor
Kocsis, James Joseph
Konikoff, John Jacob
Kreutner, William
Kripalani, Kishin J
Kuntzman, Ronald Grover
Lan, Shih-Jung
Lanzoni, Vincent
Lassman, Howard B
Laurencot, Henry Jules
Leeds, Morton W
Lehr, David
Leinweber, Franz Josef
Leitz, Frederick Henry
LeSher, Dean Allen
Lester, David
Letizia, Gabriel Joseph
Leung, Albert Yuk-Sing
Levenstein, Irving
Levin, Wayne
Lewis, Alan James
Lewis, Steven Craig
Liebman, Jeffrey Mark
Lin, Chin-Chung
Lin, Yi-Jong
Liu, Alice Yee-Chang
Lowndes, Herbert Edward
Lukas, George
McGuire, John L
McIlreath, Fred J
Mackerer, Carl Robert
McKinstry, Doris Naomi
Mackway-Girardi, Ann Marie
Mandel, Lewis Richard
Martin, Lawrence Leo
Massa, Tobias
Mehta, Atul Mansukhbhai
Meyers, Kenneth Purcell
Mezey, Kalman C
Mezick, James Andrew
Mitala, Joseph Jerrold
Moreland, Suzanne
Morris, N Ronald
Moyer, John Allen
Murthy, Vadiraja Venkatesa
Muschek, Lawrence David
Nemecek, Georgina Marie
Nicolau, Gabriela
Novick, William Joseph, Jr
O'Neill, Patrick Joseph
Orzechowski, Raymond Frank
Osborne, Melville
Oser, Bernard Levussove
Oshiro, George
Pace, Daniel Gannon
Pachter, Jonathan Alan
Palmer, Keith Henry
Panagides, John
Paterniti, James R, Jr
Perhach, James Lawrence
Pfeiffer, Carl Curt
Piala, Joseph J
Pitts, Barry James Roger
Plummer, Albert J
Pobiner, Bonnie Fay
Pohorecky, Larissa Alexandra
Prince, Herbert N
Puri, Surendra Kumar
Quinn, Gertrude Patricia
Rabii, Jamshid
Reiser, H Joseph
Ress, Rudyard Joseph
Ritchie, David Malcolm
Robson, Ronald D
Rodda, Bruce Edward
Rodman, Morton Joseph
Rosenthale, Marvin E
Rothstein, Edwin C(arl)
Rubanyi, Gabor Michael
Rubin, Bernard
Rudzik, Allan D
Saini, Ravinder Kumar
Salvador, Richard Anthony
Sanders, Marilyn Magdanz
Sapru, Hreday N
Saunders, Robert Norman
Schinagle, Erich F
Schoepke, Hollis George
Schwartz, Morton Allen
Segelman, Alvin Burton
Sepinwall, Jerry
Sequeira, Joel August Louis
Shakarjian, Michael Peter
Shapiro, Stanley Seymour
Shellenberger, Carl H
Singer, Arnold J
Singhvi, Sampat Manakchand
Sisenwine, Samuel Fred
Snyder, Robert
Sofia, R Duane
Solomon, Thomas Allan
Spaulding, Theodore Charles

Spector, Reynold
Sperlein, Marie Teresa
Stambaugh, John Edgar, Jr
Stefko, Paul Lowell
Steiner, Kurt Edric
Stockton, John Richard
Sugerman, Abraham Arthur
Sullivan, John William
Symchowicz, Samson
Szewczak, Mark Russell
Tabachnick, Irving I A
Thomas, Kenneth Alfred, Jr
Thompson, George Rex
Tierney, William John
Tolman, Edward Laurie
Tozzi, Salvatore
Traina, Vincent Michael
Triscari, Joseph
Tse, Francis Lai-Sing
Vanden Heuvel, William John Adrian, III
Vander Wende, Christina
Vane, Floie Marie
Vogel, Wolfgang Hellmut
Von Hagen, D Stanley
Vukovich, Robert Anthony
Waggoner, William Charles
Wagner, William Edward, Jr
Wang, Shih Chun
Wasserman, Martin Allan
Waugh, Margaret H
Weichman, Barry Michael
Weinstein, Stephen Henry
Weiss, George B
Weiss, Harvey Richard
Welton, Ann Frances
Whiteley, Phyllis Ellen
Wiggins, Jay Ross
Wolff, Donald John
Wu, Wen-hsien
Wulf, Ronald James
Yard, Allan Stanley

NEW MEXICO

Backer, Ronald Charles
Buss, William Charles
Coffey, John Joseph
Corcoran, George Bartlett, III
Crawford, Michael Howard
Damon, Edward G(eorge)
Garst, John Eric
Groblewski, Gerald Eugene
Hadley, William Melvin
Hobbs, Charles Henry
Hurwitz, Leon
Jones, Vernon Douglas
Leach, John Kline
Martinez, J Ricardo
Petersen, Donald Francis
Priola, Donald Victor
Rhodes, Buck Austin
Uhlenhuth, Eberhard Henry
Vorherr, Helmuth Wilhelm
Wax, Joan

NEW YORK

Acara, Margaret A
Aiello, Edward Lawrence
Alexander, George Jay
Almon, Richard Reiling
Altszuler, Norman
Altura, Bella T
Altura, Burton Myron
Ambrus, Julian Lawrence
Anders, Marion Walter
Andersen, Kenneth J
Andrews, John Parray
Antzelevitch, Charles
Aronson, Ronald Stephen
Babish, John George
Back, Nathan
Bagchi, Sakti Prasad
Baird, Malcolm Barry
Baker, Robert George
Balis, Moses Earl
Bard, Enzo
Bardos, Thomas Joseph
Barkai, Amiram I
Barletta, Michael Anthony
Barnerjee, Shailesh Prasad
Bartus, Raymond T
Bastos, Milton Lessa
Beinfield, William Harvey
Belford, Julius
Bell, Duncan Hadley
Berger, Frank Milan
Bernacki, Ralph J
Bernheimer, Alan Weyl
Bertino, Joseph R
Bickerman, Hylan A
Bigger, J Thomas, Jr
Bito, Laszlo Z
Bloch, Alexander
Blosser, James Carlisle
Borch, Richard Frederic
Borgstedt, Harold Heinrich
Brooks, Robert R
Brown, Oliver Monroe
Butler, Vincent Paul, Jr
Cabrera, Edelberto Jose
Carr, Edward Albert, Jr
Castellion, Alan William
Cervoni, Peter
Chan, Arthur Wing Kay

Pharmacology (cont)

Chan, Peter Sinchun
Chan, W(ah) Y(ip)
Chisholm, David R
Chou, Dorthy T C T
Chou, Ting-Chao
Chovan, James Peter
Chryssanthou, Chryssanthos
Chung, Eunyong
Ciancio, Sebastian Gene
Clarkson, Allen Boykin, Jr
Clarkson, Thomas William
Claus, Thomas Harrison
Comer, William Timmey
Conway, Walter Donald
Coppola, John Anthony
Cory-Slechta, Deborah Ann
Costa, Max
Costello, Christopher Hollet
Cosulich, Donna Bernice
Coupet, Joseph
Crain, Stanley M
Crandall, David L
Creaven, Patrick Joseph
Crew, Malcolm Charles
Crouch, Billy G
Darzynkiewicz, Zbigniew Dzierzykraj
Davidow, Bernard
Davis, Kenneth Leon
Day, Ivana Podvalova
Demis, Dermot Joseph
DeVita, Vincent T, Jr
DiStefano, Victor
Dixon, Robert Louis
Dole, Vincent Paul
Donner, David Bruce
Drakontides, Anna Barbara
Drayer, Dennis Edward
Ellis, Keith Osborne
Evans, Hugh Lloyd
Fahrenbach, Marvin Jay
Ferrari, Richard Alan
Ferris, Steven Howard
Fiala, Emerich Silvio
Fink, Max
Finster, Mieczyslaw
Foley, Dennis Joseph
Foster, Kenneth William
Foster, Thomas Scott
Frishman, William Howard
Furchgott, Robert Francis
Gaver, Robert Calvin
Gessner, Peter K
Gessner, Teresa
Giering, John Edgar
Gillies, Alastair J
Gintautas, Jonas
Glassman, Jerome Martin
Glick, Stanley Dennis
Goldenberg, Marvin M
Goldfarb, Joseph
Govier, William Miller
Gray, William David
Greco, William Robert
Green, Jack Peter
Greenberg, Jacob
Greenberger, Lee M
Greenblatt, Eugene Newton
Greene, Lloyd A
Greengard, Olga
Greizerstein, Hebe Beatriz
Griffith, Owen Wendell
Gringauz, Alex
Grob, David
Grollman, Arthur Patrick
Guideri, Giancarlo
Guttenplan, Joseph B
Hakala, Maire Tellervo
Halevy, Simon
Halpryn, Bruce
Hamilton, Byron Bruce
Hanna, Samir A
Harris, Alex L
Harris-Warrick, Ronald Morgan
Hershey, Linda Ann
Hiller, Jacob Moses
Hirsh, Kenneth Roy
Holaday, John W
Holland, Mary Jean Carey
Horwitz, Susan Band
Houde, Raymond Wilfred
Hough, Lindsay B
Hrushesky, William John Michael
Hudecki, Michael Stephen
Hulme, Norman Arthur
Hutner, Seymour Herbert
Inchiosa, Mario Anthony, Jr
Ingalls, James Warren, Jr
Inturrisi, Charles E
Iodice, Arthur Alfonso
Ip, Margot Morris
Jacquet, Yasuko F
Jayme, David Woodward
Jean-Baptiste, Emile
Kahn, Norman
Kalman, Thomas Ivan
Kalsner, Stanley
Kao, Chien Yuan
Kappas, Attallah
Kashin, Philip
Kass, Robert S
Katocs, Andrew Stephen, Jr
Keefe, Deborah Lynn

Kennedy, Margaret Wiener
Kestenbaum, Richard Steven
Khan, Abdul Jamil
Kinsolving, C Richard
Klein, Donald Franklin
Klingman, Gerda Isolde
Knaak, James Bruce
Kohler, Constance Anne
Kraft, Patricia Lynn
Kraus, Shirley Ruth
Kreis, Willi
Kristal, Mark Bennett
Kuperman, Albert Sanford
Lai, Fong M
Lapidus, Herbert
Laties, Victor Gregory
Latimer, Clinton Narath
Lau-Cam, Cesar A
Laychock, Suzanne Gale
Leach, Leonard Joseph
Levi, Roberto
Levine, Robert Alan
Levine, Walter (Gerald)
Levitt, Barrie
Levy, Gerhard
Liang, Chang-seng
Loullis, Costas Christou
Lynch, Vincent De Paul
McCabe, R Tyler
McGiff, John C
McGuire, John Joseph
McIsaac, Robert James
McNamara, Thomas Francis
Madissoo, Harry
Makman, Maynard Harlan
Malbon, Craig Curtis
Malitz, Sidney
Mallov, Samuel
Margolis, Renee Kleimann
Margolis, Richard Urdangen
Marois, Robert Leo
Mary, Nouri Y
Mayhew, Eric George
Megirian, Robert
Meyers, Marian Bennett
Mihich, Enrico
Miller, Richard Kermit
Minatoya, Hiroaki
Misra, Anand Lal
Mittag, Thomas Waldemar
Modell, Walter
Moorehead, Thomas J
Morgan, John P
Morris, Marilyn Emily
Morrison, John Agnew
Morrow, Paul Edward
Munigle, Jo Anne
Nair, Ramachandran Mukundalayam Sivarama
Nakatsugawa, Tsutomu
Namba, Tatsuji
Nasjletti, Alberto
Neenan, John Patrick
Neu, Harold Conrad
Neubort, Shimon
Oberdörster, Günter
Okamoto, Michiko
Pantuck, Eugene Joel
Parham, Margaret Payne
Pasternak, Gavril William
Pearl, William
Pine, Martin J
Pollock, John Joseph
Pong, Schwe Fang
Pool, William Robert
Prell, George D
Reidenberg, Marcus Milton
Reilly, Margaret Anne
Reit, Ben
Reith, Maarten E A
Reynard, Alan Mark
Rifkind, Arleen B
Riker, Walter Franklyn, Jr
Rivera-Calimlim, Leonor
Rizack, Martin A
Robinson, Joseph Douglass
Roboz, John
Rocci, Mario Louis, Jr
Roth, Jerome Allan
Rothman, Francoise
Rubin, Ronald Philip
Salley, John Jones, Jr
Schachter, E Neil
Schayer, Richard William
Schiff, Joel D
Schneider, Allan Stanford
Schwark, Wayne Stanley
Shamoian, Charles Anthony
Sharp, Richard Lee
Siever, Larry Joseph
Silver, Paul J
Sinha, Asru Kumar
Sirotnak, Francis Michael
Skulan, Thomas William
Sloboda, Adolph Edward
Slocum, Harry Kim
Smith, Cedric Martin
Smith, Frank Ackroyd
Soderlund, David Matthew
Sohn, David
Spoor, Ryk Peter
Squires, Richard Felt
Staple, Peter Hugh
Stoewsand, Gilbert Saari

Stoll, William Russell
Strand, James Cameron
Sufrin, Janice Richman
Swamy, Vijay Chinnaswamy
Szeto, Hazel Hon
Terhaar, Clarence James
Titeler, Milt
Traitor, Charles Eugene
Triggle, David J
Tuttle, Richard Suneson
Van der Hoeven, Theo A
Van Poznak, Alan
Venter, J Craig
Ventura, William Paul
Verebey, Karl G
Violante, Michael Robert
Vulliemoz, Yvonne
Wajda, Isabel
Wakade, Arun Ramchandra
Wang, Hsueh-Hwa
Warner, William
Weiner, Irwin M
Weinstein, Harel
Weisburger, John Hans
West, Norman Reed
Whitaker-Azmitia, Patricia Mack
Wilk, Sherwin
Williams, Curtis Alvin, Jr
Winter, Jerrold Clyne, Sr
Wit, Andrew Lewis
Wolf, P S
Wolin, Michael Stuart
Wong, Patrick Yui-Kwong
Yoburn, Byron Crocker
York, James Lester
Zedeck, Morris Samuel
Zukin, Stephen R

NORTH CAROLINA

Abou-Donia, Mohamed Bahie
Aronson, A L
Barnes, Donald Wesley
Barnett, John William
Beaty, Orren, III
Bentley, Peter John
Bernheim, Frederick
Birnbaum, Linda Silber
Bishop, Jack Belmont
Bond, James Anthony
Booth, Raymond George
Bradley, William Robinson
Brauer, Ralph Werner
Burford, Hugh Jonathan
Bus, James Stanley
Butterworth, Byron Edwin
Dutz, Robert Frederick
Camp, Haney Bolon
Capizzi, Robert L
Carr, Fred K
Carroll, Bernard James
Carroll, Robert Graham
Chapman, Sharon K
Cheng, Yung-Chi
Chhabra, Rajendra S
Chignell, Colin Francis
Conners, Carmen Keith
Costa, Daniel Louis
Dar, Mohammad Saeed
Dauterman, Walter Carl
DaVanzo, John Paul
Davidson, Ivan William Frederick
Davis, James N(orman)
De Miranda, Paulo
Devereux, Theodora Reyling
DiAugustine, Richard Patrick
Dibner, Mark Douglas
Dingledine, Raymond J
Dren, Anthony Thomas
Dudley, Kenneth Harrison
Dunlap, Claud Evans, III
Eichelman, Burr S, Jr
Eldridge, John Charles
Eling, Thomas Edward
Elion, Gertrude Belle
Ellinwood, Everett Hews, Jr
Ellis, Fred Wilson
Engle, Thomas William
Faber, James Edward
Fouts, James Ralph
Fuchs, James Claiborne Allred
Fyfe, James Arthur
Garrett, Ruby Joyce Burriss
Gatzy, John T, Jr
Ghanayem, Burhan I
Gibson, James Edwin
Glenn, Thomas M
Goldstein, Joyce Allene
Goz, Barry
Gralla, Edward Joseph
Grantham, Preston Hubert
Gray, Timothy Kenney
Hak, Lawrence J
Harper, Curtis
Hart, Larry Glen
Hastings, Felton Leo
Hatch, Gary Ephraim
Hatfield, George Michael
Hayes, Andrew Wallace
Heck, Henry d'Arcy
Heck, Jonathan Daniel
Hicks, Robert Eugene
Hicks, T Philip
Hirsch, Philip Francis
Hodgson, Ernest

Hofmann, Lorenz M
Hong, Jau-Shyong
Hook, Gary Edward Raumati
Howard, James Lawrence
Hueter, Francis Gordon
Ingenito, Alphonse J
Janowsky, David Steffan
Jones(Stover), Betsy
Jurgelski, William, Jr
Kaufman, William
Kaufmann, John Simpson
Kirksey, Donny Frank
Kirshner, Norman
Knapp, William Arnold, Jr
Koster, Rudolf
Krasny, Harvey Charles
Lazarus, Lawrence H
Lefkowitz, Robert Joseph
Lewis, Mark Henry
Little, Patrick Joseph
Little, William C
McBay, Arthur John
McClellan, Roger Orville
McConnaughey, Mona M
McCutcheon, Rob Stewart
McLachlan, John Alan
Mailman, Richard Bernard
Matthews, Hazel Benton, Jr
Maxwell, Robert Arthur
Mennear, John Hartley
Mills, Elliott
Minard, Frederick Nelson
Mitchell, Clifford L
Miya, Tom Saburo
Modrak, John Bruce
Moorefield, Herbert Hughes
Morrissey, Richard Edward
Moser, Virginia Clayton
Mueller, Robert Arthur
Munson, Paul Lewis
Murdock, Harold Russell
Myers, Robert Durant
Nadler, J Victor
Namm, Donald H
Neal, Robert A
Nelson, Donald J
Nemeroff, Charles Barnet
Nichol, Charles Adam
O'Callaghan, James Patrick
Ottolenghi, Athos
Padilla, George M
Paluch, Edward Peter
Peng, Tai-Chan
Poole, Doris Theodore
Posner, Herbert S
Price, Kenneth Elbert
Pritchard, John B
Putney, James Wiley, Jr
Quinn, Richard Paul
Reel, Jerry Royce
Reinhard, John Frederick, Jr
Rickert, Douglas Edward
Riviere, Jim Edmond
Rolofson, George Lawrence
Schanberg, Saul M
Schroeder, David Henry
Schwetz, Bernard Anthony
Siemens, Albert John
Sinha, Birandra Kumar
Slotkin, Theodore Alan
Somjen, George G
Spector, Thomas
Stacy, David Lowell
Steelman, Sanford Lewis
Stern, Warren C
Stevens, James T
Stiles, Gary L
Stolk, Jon Martin
Strandhoy, Jack W
Stumpf, Walter Erich
Tadepalli, Anjaneyulu S
Thakker, Dhiren R
Thurman, Ronald Glenn
Tice, Raymond Richard
Tilson, Hugh Arval
Toverud, Svein Utheim
Tschanz, Christian
Van Stee, Ethard Wendel
Watkins, W David
Weiden, Mathias Herman Joseph
Welch, Richard Martin
White, Helen Lyng
Winter, Irwin Clinton
Witt, Peter Nikolaus
Wooles, Wallace Ralph
Zimmerman, Thomas Paul
Zung, William Wen-Kwai

NORTH DAKOTA

Ary, Thomas Edward
Auyong, Theodore Koon-Hook
Belknap, John Kenneth
DeBoer, Benjamin
Hein, David William
Husain, Syed
Khalil, Shoukry Khalil Wahba
Messiha, Fathy S
Parmar, Surendra S
Rao, Nutakki Gouri Sankara
Tanner, Noall Stevan
Travis, David M

OHIO

Agamanolis, Dimitris

Akbar, Huzoor
Anton, Aaron Harold
Askari, Amir
Awad, Albert T
Bachmann, Kenneth Allen
Banerjee, Sipra
Banning, Jon Willroth
Beal, Jack Lewis
Bhattacharya, Amar Nath
Bianchine, Joseph Raymond
Bloomfield, Saul
Blumer, Jeffrey L
Boggs, Robert Wayne
Dohrman, Jeffrey Stephen
Boxenbaum, Harold George
Breglia, Rudolph John
Bryant, Shirley Hills
Buck, Stephen Henderson
Buehler, Edwin Vernon
Bunde, Carl Albert
Burkman, Allan Maurice
Cawein, Madison Julius
Chen, Chi-Po
Cheng, Hsien
Ching, Melvin Chung Hing
Corson, Samuel Abraham
Couri, Daniel
Crutcher, Keith A
Dage, Richard Cyrus
D'Ambrosio, Steven M
Denine, Elliot Paul
Dinerstein, Robert Joseph
Dodd, Darol Ennis
Doskotch, Raymond Walter
Dunbar, Philip Gordon
Feller, Dannis R
Gardier, Robert Woodward
Geha, Alexander Salim
Geho, Walter Blair
Gerald, Michael Charles
Giannini, A James
Goldstein, Sidney
Gossel, Thomas Alvin
Griffin, Claude Lane
Grupp, Gunter
Grupp, Ingrid L
Gwinn, John Frederick
Hagerman, Larry M
Halaris, Angelos
Hall, Madeline Molnar
Hammond, Paul B
Hayton, William Leroy
Heesch, Cheryl Miller
Hille, Kenneth R
Hilmas, Duane Eugene
Hinkle, George Henry
Hollander, Philip B
Holtkamp, Dorsey Emil
Hoss, Wayne Paul
Hudak, William John
Hurwitz, David Allan
Imondi, Anthony Rocco
Johnson, Carl Lynn
Johnston, John O'Neal
Khairallah, Philip Asad
Kimura, Kazuo Kay
Klaunig, James E
Koerker, Robert Leland
Krimmer, Edward Charles
Krontiris-Litowitz, Johanna Kaye
Kuhn, William Lloyd
Kunin, Calvin Murry
Lang, James Frederick
Levin, Jerome Allen
Lima, John J
Lindower, John Oliver
Lutton, John Kazuo
MacKichan, Janis Jean
Mamrack, Mark Donovan
Manson, Jeanne Marie
Mellgren, Ronald Lee
Messer, William Sherwood, Jr
Michael, William R
Michaelson, I Arthur
Mieyal, John Joseph
Millard, Ronald Wesley
Muir, William W, III
Mukhtar, Hasan
Mumtaz, Mohammad Moizuddin
Nahata, Milap Chand
Nasrallah, Henry A
Nigrovic, Vladimir
Nolen, Granville Abraham
O'Dell, Thomas Beniah
Patil, Popat N
Pereira, Michael Alan
Powers, Thomas E
Rahwan, Ralf George
Ray, Richard Schell
Recknagel, Richard Otto
Reuning, Richard Henry
Ritschel, Wolfgang Adolf
Robinson, Michael K
Rolf, Clyde Norman
Rosenberg, Howard C
Saffran, Murray
Schechter, Martin David
Schlender, Keith K
Schradie, Joseph
Schroeder, Friedhelm
Schwartz, Arnold
Shemano, Irving
Sherman, Gerald Philip
Shertzer, Howard Grant

Sigell, Leonard
Sinnhuber, Russell Otto
Sjoerdsma, Albert
Skau, Kenneth Anthony
Skelly, Michael Francis
Smith, Dennison A
Somani, Pitambar
Sperelakis, Nick
Standaert, Frank George
Staubus, Alfred Elsworth
Stokinger, Herbert Ellsworth
Tedeschi, Ralph Earl
Tejwani, Gopi Assudomal
Tepperman, Katherine Gail
Tjioe, Sarah Archambault
Tomei, L David
Truitt, Edward Byrd, Jr
Tsai, Tsui Hsien
Ursillo, Richard Carmen
Van Maanen, Evert Florus
Vorhees, Charles V
Walenga, Ronald W
Wallin, Richard Franklin
Webster, Leslie T, Jr
Wilkerson, Robert Douglas
Wilkin, Jonathan Keith
Wolfe, Seth August, Jr
Wolgemuth, Richard Lee
Wong, Laurence Cheng Kong
Woodward, James Kenneth
Wright, George Joseph
Zigmond, Richard Eric
Zimmerman, Ernest Frederick

OKLAHOMA
Brenner, George Marvin
Burrows, George Edward
Carney, John Michael
Christensen, Howard Dix
Corder, Clinton Nicholas
Crane, Charles Russell
Czerwinski, Anthony William
Ganesan, Devaki
Gass, George Hiram
Harder, Harold Cecil
Hornbrook, K Roger
Huber, Wolfgang Karl
Jones, Sharon Lynn
Koss, Michael Campbell
Maton, Paul Nicholas
Meier, Charles Frederick, Jr
Melton, Carlton E, Jr
Moore, Joanne Iweita
Pento, Joseph Thomas
Prabhu, Vilas Anandrao
Reinke, Lester Allen
Revzin, Alvin Morton
Rikans, Lora Elizabeth
Ringer, David P
Robinson, Casey Perry
Rolf, Lester Leo, Jr
Whitsett, Thomas L

OREGON
Bartos, Frantisek
Bennett, William M
Brummett, Robert E
Buhler, Donald Raymond
Cowan, Frederick Fletcher, Jr
Dickinson, John Otis
Dost, Frank Norman
Fink, Gregory Burnell
Fox, Kaye Edward
Gabourel, John Dustan
Gardner, Sara A
Hedtke, James Lee
Julien, Robert M
Keenan, Edward James
Larson, Robert Elof
Levine, Leonard
McCawley, Elton Leeman
Murray, Thomas Francis
North, R Alan
Olsen, George Duane
Riker, William Kay
Roberts, Michael Foster
Russell, Nancy Jeanne
Schneider, Phillip William, Jr
Sciuchetti, Leo A
Smith, John Robert
Swan, Kenneth Carl
Tanz, Ralph
Weber, Lavern J

PENNSYLVANIA
Actor, Paul
Adler, Martin E
Agersborg, Helmer Pareli Kjerschow, Jr
Aronson, Carl Edward
Aston-Jones, Gary Stephen
Baer, John Elson
Bagdon, Walter Joseph
Bailey, Denis Mahlon
Baker, Alan Paul
Baker, Walter Wolf
Bannon, James Andrew
Barone, Frank Carmen
Barry, Herbert, III
Bayne, Gilbert M
Bergey, James L
Berkowitz, Barry Alan
Berlin, Cheston Milton
Beyer, Karl Henry, Jr
Bianchi, Carmine Paul

Block, Lawrence Howard
Blum, Lee M
Boshart, Charles Ralph
Brooks, David Patrick
Calesnick, Benjamin
Cerreta, Kenneth Vincent
Chakrin, Lawrence William
Chan, Ping-Kwong
Chang, Raymond S L
Chen, Shih-Fong
Chernick, Warren Sanford
Chiang, Soong Tao
Clineschmidt, Bradley Van
Comis, Robert Leo
Connamacher, Robert Henle
Connor, John D
Constantinides, Panayiotis Pericleous
Coon, Julius Mosher
Cooper, Stephen Allen
Cordova, Vincent Frank
Corey, Sharon Eva
Cowan, Alan
Creasey, William Alfred
Dalton, Colin
Daniels-Severs, Anne Elizabeth
Davignon, J Paul
Davis, Hugh Michael
De Groat, William C
DeGroat, William Chesney, Jr
DeHaven-Hudkins, Diane Louise
Di Gregorio, Guerino John
DiPalma, Joseph Rupert
Dixit, Balwant N
Donnelly, Thomas Edward, Jr
Duggan, Daniel Edward
Edwards, David J
Ellison, Theodore
Engleman, Karl
Ertel, Robert James
Fanelli, George Marion, Jr
Ferko, Andrew Paul
Flynn, John Thomas
Fogleman, Ralph William
Fong, Kei-Lai Lau
Frazer, Alan
Freeman, Alan R
Fromm, Gerhard Hermann
Furgiuele, Angelo Ralph
Gabriel, Karl Leonard
Gardocki, Joseph F
Gautieri, Ronald Francis
Gero, Alexander
Gibson, Raymond Edward
Glauser, Elinor Mikelberg
Glauser, Stanley Charles
Goldberg, Paula Bursztyn
Goldstein, Frederick J
Gottlieb, Karen Ann
Gould, Robert James
Greene, Frank Eugene
Greenstein, Jeffrey Ian
Grego, Nicholas John
Grindel, Joseph Michael
Gross, Dennis Michael
Harakal, Concetta
Harvey, John Adriance
Haubrich, Dean Robert
Haugaard, Niels
Hava, Milos
Hemwall, Edwin Lyman
Henderson, George Richard
Hess, Marilyn E
Hieble, J Paul
Hite, Mark
Hitner, Henry William
Holcslaw, Terry Lee
Hook, Jerry Bruce
Hopper, Sarah Priestly
Hucker, Howard B(enjamin)
Iuliucci, John Domenic
Jackson, Edwin Kerry
Jacob, Leonard Steven
Jacoby, Henry I
Janssen, Frank Walter
Jaweed, Mazher
Jenney, Elizabeth Holden
Jim, Kam Fook
Kade, Charles Frederick, Jr
Kaji, Hideko (Katayama)
Kallelis, Theodore S
Kelliher, Gerald James
Kinter, Lewis Boardman
Kivlighn, Salah Dean
Knowles, John Appleton, III
Koelle, George Brampton
Koelle, Winifred Angenent
Kraatz, Charles Parry
Kramer, Stanley Zachary
Krenzelok, Edward Paul
Kuhnert, Betty R
Lacouture, Peter George
Lage, Gary Lee
Lambertsen, Christian James
Lazo, John Stephen
Lee, David K H
Lerner, Leonard Joseph
Levin, Robert Martin
Levin, Sidney Seamore
Levin, Simon Eugene
Lewis, Michael Edward
Ling, Alfred Soy Chou
Lippa, Erik Alexander
Loeb, Alex Lewis
Lotlikar, Prabhakar Dattaram

Lotti, Victor J
Lynch, Joseph J, Jr
MacDonald, James Scott
McDonald, Robert H, Jr
McElligott, James George
MacFarland, Harold Noble
McKenna, William Gillies
Mackowiak, Elaine DeCusatis
McManus, Edward Clayton
Mann, David Edwin, Jr
Mansmann, Herbert Charles, Jr
Martin, Gregory Emmett
Matthews, Richard John, Jr
Meacham, Roger Hening, Jr
Michelson, Eric L
Milakofsky, Louis
Minsker, David Harry
Mir, Ghulam Nabi
Molinoff, Perry Brown
Mong, Seymour
Murray, Rodney Brent
Muth, Eric Anthony
Myers, Howard M
Nagle, Barbara Tomassone
Nambi, Ponnal
Nass, Margit M K
Neil, Gary Lawrence
Nelson, Edward Blake
Oldham, James Warren
Overton, Donald A
Packman, Albert M
Packman, Elias Wolfe
Panasevich, Robert Edward
Papacostas, Charles Arthur
Pelleg, Amir
Pendleton, Robert Grubb
Perel, James Maurice
Piperno, Elliot
Pitt, Bruce R
Poste, George Henry
Price, Henry Locher
Pruss, Thaddeus P
Puglia, Charles David
Raffa, Robert B
Reitz, Allen Bernard
Rickels, Karl
Rieders, Fredric
Riley, Gene Alden
Ritter, Carl A
Roberts, Jay
Robishaw, Janet D
Rosenberg, Franklin J
Rosenblum, Irwin Yale
Rossi, George Victor
Rucinska, Ewa J
Ruffolo, Robert Richard, Jr
Ruggieri, Michael Raymond
Saelens, David Arthur
Salganicoff, Leon
Sands, Howard
Sawutz, David G
Schein, Philip Samuel
Schrogie, John Joseph
Seay, Patrick H
Severs, Walter Bruce
Sevy, Roger Warren
Shank, Richard Paul
Short, John Albert
Shriver, David A
Siegman, Marion Joyce
Silver, Melvin Joel
Smith, John Bryan
Smith, Roger Bruce
Smudski, James W
Snipes, Charles Andrew
Snipes, Wallace Clayton
Sprince, Herbert
Stiller, Richard L
Stone, Clement A
Storella, Robert J, Jr
Strom, Brian Leslie
Summy-Long, Joan Yvette
Sweet, Charles Samuel
Tarver, James H, Jr
Tepper, Lloyd Barton
Tew, Kenneth David
Tobia, Alfonso Joseph
Tocco, Dominick Joseph
Torchiana, Mary Louise
Torosian, George
Triolo, Anthony J
Trippodo, Nick Charles
Van Arman, Clarence Gordon
Van Kammen, Daniel Paul
Van Rossum, George Donald Victor
Venkataramanan, Raman
Vesell, Elliot S
Vickers, Florence Foster
Vickers, Stanley
Volle, Robert Leon
Vollmer, Regis Robert
Walkenstein, Sidney S
Walz, Donald Thomas
Wardell, Joe Russell, Jr
Wasserman, Martin Allan
Waterhouse, Barry D
Weinryb, Ira
Weinstock, Joseph
Wendt, Robert Leo
Werner, Gerhard
Winek, Charles L
Wingard, Lemuel Bell, Jr
Winkler, James David
Yellin, Tobias O

Pharmacology (cont)

Yu, Ruey Jiin
Zemaitis, Michael Alan
Zigmond, Michael Jonathan

RHODE ISLAND

Agarwal, Kailash C
Calabresi, Paul
Cha, Sungman
Crabtree, Gerald Winston
DeFanti, David R
DeFeo, John Joseph
Doolittle, Charles Herbert, III
Fischer, Glenn Albert
Flanagan, Thomas Raymond
Kaplan, Stephen Robert
Miech, Ralph Patrick
Oxenkrug, Gregory Faiva
Parks, Robert Emmett, Jr
Patrick, Robert L
Shaikh, Zahir Ahmad
Shimizu, Yuzuru
Stern, Leo
Stoeckler, Johanna D
Youngken, Heber Wilkinson, Jr

SOUTH CAROLINA

Allen, Donald Orrie
Baggett, Billy
Bagwell, Ervin Eugene
Benmaman, Joseph David
Daniell, Herman Burch
Davis, Craig Wilson
Freeman, John J
Gaffney, Thomas Edward
Gale, Glen Roy
Gillette, Paul Crawford
Golod, William Hersh
Gross, Stephen Richard
Halushka, Perry Victor
Knapp, Daniel Roger
Kosh, Joseph William
Lin, Tu
Lindenmayer, George Earl
Margolius, Harry Stephen
Newman, Walter Hayes
Privitera, Philip Joseph
Richardson, James Albert
Wagle, Gilmour Lawrence
Walter, Wilbert George
Wilson, Steven Paul
Woodard, Geoffrey Dean Leroy

SOUTH DAKOTA

Dwivedi, Chandradhar
Gross, Guilford C
Hagen, Arthur Ainsworth
Hietbrink, Bernard Edward
Johnson, Carl J
Kauker, Michael Lajos
Schramm, Mary Arthur
Yutrzenka, Gerald J
Zawada, Edward T, Jr
Zeigler, David Wayne

TENNESSEE

Ahmed, Nahed K
Akin, Frank Jerrel
Akiskal, Hagop S
Anjaneyulu, P S R
Ban, Thomas Arthur
Bass, Allan Delmage
Berman, M Lawrence
Berney, Stuart Alan
Biaggioni, Italo
Branch, Robert Anthony
Brent, Thomas Peter
Byron, Joseph Winston
Caldwell, Robert William
Cardoso, Sergio Steiner
Chapman, John E
Cook, George A
Crowley, William Robert
Daigneault, Ernest Albert
Farrington, Joseph Kirby
Fields, James Perry
Fridland, Arnold
Friedman, Daniel Lester
Gehrs, Carl William
Gillespie, Elizabeth
Guengerich, Frederick Peter
Hancock, John Charles
Hardman, Joel G
Heimberg, Murray
Hoover, Donald Barry
Houghton, Janet Anne
Kostrzewa, Richard Michael
Lasslo, Andrew
Lawrence, William Homer
Lawson, James W
Leach, Eddie Dillon
Leffler, Charles William
Lothstein, Leonard
McColl, John Duncan
MacEwen, James Douglas
Manley, Emmett S
Mayer, Steven Edward
Meeks, Robert G
Miyamoto, Michael Dwight
Morrell, George
Nash, Clinton Brooks
North, William Charles
Oates, John Alexander

TEXAS

Ohi, Seigo
Olson, Erik Joseph
Patel, Tarun B
Petrie, William Marshall
Pittz, Eugene P
Ray, Oakley S
Reed, Peter William
Roberts, DeWayne
Roberts, Lyman Jackson
Robertson, David
Sanders-Bush, Elaine
Sastry, Bhamidipaty Venkata Rama
Schmidt, Dennis Earl
Schreiber, Eric Christian
Shockley, Dolores Cooper
Skoutakis, Vasilios A
Snapper, James Robert
Spector, Sydney
Sticht, Frank Davis
Straughn, Arthur Belknap
Sulser, Fridolin
Sweetman, Brian Jack
Van Eldik, Linda Jo
Weinstein, Ira
Wells, Jack Nulk
White, Richard Paul
Wicks, Wesley Doane
Wilcox, Henry G
Wiley, Ronald Gordon
Wilkinson, Grant Robert
Wojciechowski, Norbert Joseph
Wood, Alastair James Johnston
Woodbury, Robert A
Woods, Maribelle

Acosta, Daniel, Jr
Alkadhi, Karim A
Allen, Julius Cadden
Altshuler, Harold Leon
Anderson, Karl E
Andresen, Michael Christian
Barnes, George Edgar
Benjamin, Robert Stephen
Besch, Paige Keith
Bjorndahl, Jay Mark
Blair, Murray Reid, Jr
Blum, Kenneth
Bost, Robert Orion
Brazeau, Gayle A
Briggs, Arthur Harold
Brodwick, Malcolm Stephen
Bryant, Stephen G
Buckley, Joseph Paul
Burton, Karen Poliner
Busch, Harris
Campbell, William Bryson
Cantrell, Elroy Taylor
Carroll, Paul Treat
Chiou, George Chung-Yih
Clark, Wesley Gleason
Clay, Michael M
Combs, Alan B
Cooper, Cary Wayne
Crass, Maurice Frederick, III
Crawford, Isaac Lyle
Dalterio, Susan Linda
Davis, Virginia Eischen
Decker, Walter Johns
Deloach, John Rooker
Dilsaver, Steven Charles
Dorris, Roy Lee
Ducis, Ilze
Edds, George Tyson
Emmett-Oglesby, Michael Wayne
Entman, Mark Lawrence
Erickson, Carlton Kuehl
Eugere, Edward Joseph
Euler, Kenneth L
Fabre, Louis Fernand, Jr
Fischer, Susan Marie
Fisher, Seymour
Foster, Donald Myers
Fox, Donald A
Frye, Gerald Dalton
Gage, Tommy Wilton
Gallagher, Joel Peter
Garriott, James Clark
German, Dwight Charles
Geumei, Aida M
Gilman, Alfred G
Gobuty, Allan Harvey
Goth, Andres
Gothelf, Bernard
Hartsfield, Sandee Morris
Heavner, James E
Heidelbaugh, Norman Dale
Heilman, Richard Dean
Henderson, George I
Hester, Richard Kelly
Hill, Reba Michels
Hillman, Gilbert R
Hilton, James Gorton
Ho, Beng Thong
Ho, Dah-Hsi
Hollister, Leo E
Holly, Frank Joseph
Horning, Marjorie G
Housholder, Glenn Etta (Tholen)
Huffman, Ronald Dean
Hughes, Maysie J H
Jauchem, James Robert
Johnson, Alice Ruffin
Johnson, Kenneth Maurice, Jr

Kay, David Cyril
Keats, Arthur Stanley
Keeton, Kent T
Kehrer, James Paul
Kempen, Rene Richard
Kenny, Alexander Donovan
Keplinger, Moreno Lavon
Kimball, Aubrey Pierce
Knotts, Glenn Richard
Knutson, Victoria P
Krivoy, William Aaron
Kroeger, Donald Charles
Lakoski, Joan Marie
Lal, Harbans
Lane, Montague
Langston, Jimmy Byrd
Leaders, Floyd Edwin, Jr
Lee, Shih-Shun
Leslie, Steven Wayne
Liehr, Joachim G
Lombardini, John Barry
Lorenzetti, Ole J
Ludden, Thomas Marcellus
Malin, David Herbert
Margolin, Solomon B
Marks, Gerald A
Medina, Miguel Angel
Merin, Robert Gillespie
Mitchell, Helen C
Mitchell, Jerry R
Montgomery, Edward Harry
Moore, Erin Colleen
Mukherjee, Amal
Mundy, Gregory Robert
Nechay, Bohdan Roman
Nelson, James Arly
Newman, Robert Alwin
Park, Myung Kun
Perkins, John Phillip
Peterson, Steven Lloyd
Pirch, James Herman
Plunkett, William Kingsbury
Potter, David Edward
Poulsen, Lawrence Leroy
Ramanujam, V M Sadagopa
Randerath, Kurt
Ranney, David Francis
Riffee, William Harvey
Robison, George Alan
Ro-Choi, Tae Suk
Rose, Kathleen Mary
Ross, Elliott M
Ryan, Charles F
Sallee, Verney Lee
Saunders, Jack Palmer
Saunders, Priscilla Prince
Schonbrunn, Agnes
Schoolar, Joseph Clayton
Schultz, Fred Henry, Jr
Sherry, Clifford Joseph
Shinnick-Gallagher, Patricia L
Slaga, Thomas Joseph
Smalley, Harry Edwin
Smith, David Joseph
Sommer, Kathleen Ruth
Sorensen, Elsie Mae (Boecker)
Stavinoha, William Bernard
Stein, Jerry Michael
Stull, James Travis
Sullivan, Gerald
Taylor, Samuel Edwin
Tenner, Thomas Edward, Jr
Thomas, John A
Ticku, Maharaj K
Traber, Daniel Lee
Vanhoutte, Paul Michel
Vick, Robert Lore
Walton, Charles Anthony
Way, James Leong
Wayner, Matthew John
Weisbrodt, Norman William
Wells, Patrick Roland
Williams, Betty L
Wilson, Carol Maggart
Wilson, L Britt
Yeoman, Lynn Chalmers
Yeung, Katherine Lu
Yorio, Thomas

UTAH

Bennett, Jesse Harland
Done, Alan Kimball
Fingl, Edward (George)
Franz, Donald Norbert
Galinsky, Raymond Ethan
Gibb, James Woolley
Goodman, Louis Sanford
Harvey, Stewart Clyde
Karler, Ralph
Kemp, John Wilmer
Lee, Steve S
Mason, Robert C
Mays, Charles William
Musick, James R
Nichols, William Kenneth
Reed, Donal J
Sharma, Raghubir Prasad
Simmons, Daniel L
Street, Joseph Curtis
Swinyard, Ewart Ainslie
Turkanis, Stuart Allen
Walker, Edward Bell Mar
Withrow, Clarence Dean
Wolf, Harold Herbert

Woodbury, Dixon Miles

VERMONT

Bevan, John Acton
Brown, Theodore Gates, Jr
Eckhardt, Shohreh B
Friedman, Matthew Joel
Hacker, Miles Paul
Hughes, John Russell
Jaffe, Julian Joseph
McCormack, John Joseph, Jr
Reit, Ernest Marvin I
Rosenstein, Robert
Sinclair, Peter Robert
Tritton, Thomas Richard

VIRGINIA

Aceto, Mario Domenico
Allison, Trenton B
Alphin, Reevis Stancil
Balster, Robert L(ouis)
Barrett, Richard John
Batra, Karam Vir
Biber, Margaret Clare Boadle
Blanke, Robert Vernon
Bleiberg, Marvin Jay
Bowe, Robert Looby
Bowman, Edward Randolph
Brand, Eugene Dew
Brase, David Arthur
Brodie, Bernard Beryl
Brown, Elise Ann Brandenburger
Byrd, Daniel Madison, III
Carchman, Richard A
Castle, Manford C
Cavender, Finis Lynn
Chandler, Jerry LeRoy
Collins, Galen Franklin
Cone, Clarence Donald, Jr
Dannenburg, Warren Nathaniel
DeLorenzo, Robert John
Dementi, Brian Armstead
DeSesso, John Michael
Dewey, William Leo
Dvorchik, Barry Howard
Egle, John Lee, Jr
Elford, Howard Lee
Eyre, Peter
Fallon, Harold Joseph
Fenner-Crisp, Penelope Ann
Fitzhugh, Oscar Garth
Franko, Bernard Vincent
Freer, Richard John
Freund, Jack
Garrison, James C
Gewirtz, David A
Gold, Paul Ernest
Gourley, Desmond Robert Hugh
Gregory, Arthur Robert
Guzelian, Philip Samuel
Hackett, John Taylor
Hanig, Joseph Peter
Harris, Louis Selig
Hart, Jayne Thompson
Haynes, Robert Clark, Jr
Hendricks, Albert Cates
Jackson, Benjamin A
Johnson, David Norseen
Krop, Stephen
Lamanna, Carl
Larner, Joseph
Larson, Paul Stanley
Linden, Joel Morris
Lipnick, Robert Louis
Lloyd, John Willie, III
McConnell, William Ray
McCuen, Robert William
McNerney, James Murtha
Martin, Billy Ray
Mathur, Pershottam Prasad
Munson, Albert Enoch
Nelson, Neal Stanley
Olson, William Arthur
Owen, Fletcher Bailey, Jr
Patrick, Graham Abner
Peach, Michael Joe
Pfeiffer, Carl J
Proakis, Anthony George
Rall, Theodore William
Reno, Frederick Edmund
Robinson, Susan Estes
Rogol, Alan David
Rosecrans, John A
Rutter, Henry Alouis, Jr
Sancilio, Lawrence F
Sando, Julianne J
Schulze, Gene Edward
Silvette, Herbert
Snow, Anne E
Sperling, Frederick
Stone, John Patrick
Talbot, Richard Burritt
Tardiff, Robert G
Taylor, Jean Marie
Thorner, Michael Oliver
Tucker, Anne Nichols
Turner, Carlton Edgar
Uwaydah, Ibrahim Musa
Venditti, John M
Villar-Palasi, Carlos
Vos, Bert John
Ward, John Wesley
Wasserman, Albert J
Watts, Daniel Thomas

Weir, Robert James, Jr
West, Bob
West, Fred Ralph, Jr
Williams, Patricia Bell
Woo, Yin-tak
Woods, Lauren Albert
Zavecz, James Henry

WASHINGTON
Aagaard, George Nelson
Baillie, Thomas Allan
Baksi, Samarendra Nath
Ballou, John Edgerton
Beavo, Joseph A
Blinks, John Rogers
Borchard, Ronald Eugene
Camerman, Arthur
Churchill, Lynn
Crampton, George H
Dreisbach, Robert Hastings
Felton, Samuel Page
Gibaldi, Milo
Gibson, Melvin Roy
Hall, Nathan Albert
Halpern, Lawrence Mayer
Harding, Joseph Warren, Jr
Hawkins, Richard L
Holcenberg, John Stanley
Horita, Akira
Johnson, William Everett
Juchau, Mont Rawlings
Klavano, Paul Arthur
Krebs, Edwin Gerhard
Levy, Rene Hanania
Loomis, Ted Albert
Lybrand, Terry Paul
Maloff, Bruce L(arrie)
Martin, Arthur Wesley
Murphy, Sheldon Douglas
Schiffrin, Milton Julius
Scribner, John David
Share, Norman N
Su, Judy Ya-Hwa Lin
Sullivan, Maurice Francis
Taylor, Eugene M
Varanasi, Usha
Vincenzi, Frank Foster
West, Theodore Clinton
Whelton, Bartlett David
Woods, James Sterrett
Yarbrough, George Gibbs
Youmans, William Barton

WEST VIRGINIA
Boyd, James Edward
Chambers, John William
Cochrane, Robert Lowe
Colasanti, Brenda Karen
Cole, Larry King
Craig, Charles Robert
Fleming, William Wright
Malanga, Carl Joseph
Mawhinney, Michael G
O'Connell, Frank Dennis
Rankin, Gary O'Neal
Reasor, Mark Jae
Robinson, Robert Leo
Stitzel, Robert Eli
Van Dyke, Knox

WISCONSIN
Alexander, Samuel Craighead, Jr
Allred, Lawrence Ervin (Joe)
Barboriak, Joseph Jan
Bass, Paul
Bekersky, Ihor
Blackwell, Barry M
Bloom, Alan S
Brown, Charles Eric
Bryan, George Terrell
Burke, David H
Dodson, Vernon N
Drill, Victor Alexander
Evenson, Merle Armin
Fahien, Leonard A
Fahl, William Edwin
Fischer, Paul Hamilton
Folts, John D
Fujimoto, James Masao
Gallenberg, Loretta A
Goldberg, Alan Herbert
Gross, Garrett John
Haavik, Coryce Ozanne
Hake, Carl (Louis)
Hardman, Harold Francis
Heath, Timothy Douglas
Hetzler, Bruce Edward
Hillard, Cecilia Jane
Hokin-Neaverson, Mabel
Hopwood, Larry Eugene
Jefcoate, Colin R
Kochan, Robert George
Lalley, Peter Michael
Lech, John James
Murthy, Vishnubhakta Shrinivas
Osterberg, Arnold Curtis
Perloff, William H(arry), Jr
Petering, David Harold
Poland, Alan P
Roerig, David L
Rusy, Ben F
Stewart, Richard Donald
Terry, Leon Cass
Thibodeau, Gary A

Trautman, Jack Carl
Tseng, Leon L F
Uno, Hideo
Wang, Richard I H
Warltier, David Charles
Will, James Arthur

WYOMING
Catalfomo, Philip
Ksir, Charles Joseph
Nelson, Robert B
Seese, William Shober

PUERTO RICO
Fernández-Repollet, Emma D
Hupka, Arthur Lee
Kaye, Sidney
Orkand, Richard K
Ramirez-Ronda, Carlos Hector
Santos-Martinez, Jesus
Wolstenholme, Wayne W

ALBERTA
Baker, Glen Bryan
Basu, Tapan Kumar
Biggs, David Frederick
Cook, David Alastair
Cottle, Merva Kathryn Warren
Dewhurst, William George
Frank, George Barry
Henderson, Ruth McClintock
Hollenberg, Morley Donald
Huston, Mervyn James
Jaques, Louis Barker
Lederis, Karolis (Karl)
Locock, Robert A
MacCannell, Keith Leonard
Moore, Graham John
Paton, David Murray
Pittman, Quentin J
Reiffenstein, Rhoderic John
Rorstad, Otto Peder
Roth, Sheldon H
Scott, Gerald William
Severson, David Lester
Wolowyk, Michael Walter

BRITISH COLUMBIA
Bellward, Gail Dianne
Foulks, James Grigsby
Jenkins, Leonard Cecil
Law, Francis C P
McNeill, John Hugh
Perks, Anthony Manning
Salari, Hassan
Sanders, Harvey David
Semba, Kazue
Sutter, Morley Carman
Tweeddale, Martin George

MANITOBA
Bihler, Ivan
Bose, Deepak
Choy, Patrick C
Dhalla, Naranjan Singh
Glavin, Gary Bertrun
Greenway, Clive Victor
Innes, Ian Rome
Kim, Ryung-Soon (Song)
Klaverkamp, John Frederick
LaBella, Frank Sebastian
Lautt, Wilfred Wayne
Pinsky, Carl
Polimeni, Philip Iniziato
Rollo, Ian McIntosh
Sitar, Daniel Samuel
Thomson, Ashley Edwin
Vitti, Trieste Guido
Weisman, Harvey

NEW BRUNSWICK
Haya, Katsuji

NEWFOUNDLAND
Bieger, Detlef
Hillman, Elizabeth S
Hoekman, Theodore Bernard
Neuman, Richard Stephen
Snedden, Walter
Virgo, Bruce Barton

NOVA SCOTIA
Armour, John Andrew
Carruthers, Samuel George
Connolly, John Francis
Downie, John William
Ferrier, Gregory R
Fraser, Albert Donald
Howlett, Susan Ellen
McDonald, Terence Francis
McKenzie, Gerald Malcolm
Renton, Kenneth William
Reynolds, Albert Keith
Szerb, John Conrad
White, Thomas David

ONTARIO
Appel, Warren Curtis
Ashwin, James Guy
Batth, Surat Singh
Bend, John Richard
Beninger, Richard J
Brien, James Frederick
Burba, John Vytautas

Buttar, Harpal Singh
Camerman, Norman
Casselman, Warren Gottlieb Bruce
Ciriello, John
Cook, Michael Arnold
Daniel, Edwin Embrey
Davis, Alan
Dyer, Allan Edwin
El-Sharkawy, Taher Youssef
Endrenyi, Laszlo
Epand, Richard Mayer
Ferguson, Alastair Victor
Ferguson, James Kenneth Wallace
Feuer, George
Frederick, George Leonard
Garfield, Robert Edward
Garfinkel, Paul Earl
Goldenberg, Gerald J
Gowdey, Charles Willis
Graham, Richard Charles Burwell
Grant, Donald Lloyd
Gupta, Radhey Shyam
Hagen, Paul Beo
Hirst, Maurice
Hrdina, Pavel Dusan
Israel, Yedy
Jhamandas, Khem
Kacew, Sam
Kadar, Dezso
Kalant, Harold
Kalow, Werner
Kapur, Bhushan M
Karmazyn, Morris
Khanna, Jatinder Mohan
Khera, Kundan Singh
Kovacs, Bela A
Kuiper-Goodman, Tine
Laham, Souheil
Lee, Tyrone Yiu-Huen
Lucis, Ojars Janis
MacLeod, Stuart Maxwell
Mahon, William A
Marks, Gerald Samuel
Maykut, Madelaine Olga
Mazurkiewicz-Kwilecki, Irena Maria
Mellors, Alan
Mishra, Ram K
Nakatsu, Kanji
Ogilvie, Richard Ian
Orrego, Hector
Pace Asciak, Cecil
Pang, Kim-Ching Sandy
Philp, Richard Blain
Pope, Barbara L
Renaud, Leo P
Roschlau, Walter Hans Ernest
Schimmer, Bernard Paul
Seeman, Philip
Sellers, Edward Alexander
Sellers, Edward Moncrieff
Singhal, Radhey Lal
Spoerel, Wolfgang Eberhart G
Stephenson, Norman Robert
Stiller, Calvin R
Sunahara, Fred Akira
Sweeney, George Douglas
Talesnik, Jaime
Thomas, Barry Holland
Trifaró, José María
Ulpian, Carla
Weaver, Lynne C
Wiberg, George Stuart

PRINCE EDWARD ISLAND
Amend, James Frederick

QUEBEC
Aranda, Jacob Velasco
Belleau, Bernard Roland
Benfey, Bruno Georg
Blouin, Andre
Capek, Radan
Collier, Brian
Cooper, Paul David
Cummings, John Rhodes
De Champlain, Jacques
Du Souich, Patrick
Ecobichon, Donald John
Escher, Emanuel
Esplin, Barbara
Ford-Hutchinson, Anthony W
Garcia, Raul
Goyer, Robert G
Hamet, Pavel
Heisler, Seymour
Jaramillo, Jorge
Kallai-Sanfacon, Mary-Ann
Lussier, Andre (Joseph Alfred)
MacIntosh, Frank Campbell
Momparler, Richard Lewis
Nair, Vasavan N P
Nickerson, Mark
Palaic, Djuro
Panisset, Jean-Claude
Pihl, Robert O
Plaa, Gabriel Leon
Powell, William St John
Regoli, Domenico
Rola-Pleszczynski, Marek
Sasyniuk, Betty Irene
Schiffrin, Ernesto Luis
Schiller, Peter Wilhelm
Sharkawi, Mahmoud
Sirois, Pierre

Srivastava, Ashok Kumar
Varma, Daya Ram
Villeneuve, Andre
Waithe, William Irwin
Zablocka-Esplin, Barbara

SASKATCHEWAN
Gupta, Vidya Sagar
Hickie, Robert Allan
Johnson, Dennis Duane
Johnson, Gordon E
Juorio, Augusto Victor
Prasad, Kailash
Richardson, J(ohn) Steven
Sisodia, Chaturbhuj Singh
Thornhill, James Arthur

OTHER COUNTRIES
Aboul-Enein, Hassan Yousseff
Akera, Tai
Allessie, Maurits
Back, Kenneth Charles
Bader, Hermann
Baggot, J Desmond
Barouki, Robert
Bernal-Llanas, Enrique
Boullin, David John
Brown, Nigel Andrew
Cho, Young Won
Davidson, David Edward, Jr
De Haën, Christoph
Dunn, Christopher David Reginald
Fitzgerald, John Desmond
Fleckenstein, Albrecht
Ganten, Detlev
Gemsa, Diethard
Ghiara, Paolo
Greene, Lewis Joel
Greichus, Yvonne A
Guerrero-Munoz, Frederico
Haegele, Klaus D
Hall, Peter
Hebborn, Peter
Hedglin, Walter L
Henderson, David Andrew
Holman, Richard Bruce
Honda, Kazuo
Ide, Hiroyuki
Ishida, Yukisato
Judis, Joseph
Kang, Sungzong
Kawamura, Hiroshi
Koch-Weser, Jan
Kuriyama, Kinya
Ladinsky, Herbert
Lee, Kwang Soo
Lee, Men Hui
Lieb, William Robert
Ling, George M
Livett, Bruce G
Mahgoub, Ahmed
Maragoudakis, Michael E
Markert, Claus O
Marquardt, Hans Wilhelm Joe
Martin, James Richard
Nabeshima, Toshitaka
Piña, Enrique
Reuter, Harald
Sampson, Sanford Robert
Schaeppi, Ulrich Hans
Schönbaum, Eduard
Segal, Mark
Stebbing, Nowell
Sung, Chen-Yu
Suria, Amin
Tung, Che-Se
Vane, John Robert
Welty, Joseph D
Wettstein, Joseph G

Pharmacy

ALABAMA
Barker, Kenneth Neil
Hocking, George Macdonald
King, William David
Riley, Thomas N
Susina, Stanley V
Thomasson, Claude Larry
Tice, Thomas Robert

ARIZONA
Blanchard, James
Brodie, Donald Crum
Burton, Lloyd Edward
Redman, Kenneth
Tong, Theodore G

ARKANSAS
Lattin, Danny Lee
McCowan, James Robert
Mittelstaedt, Stanley George
Pynes, Gene Dale

CALIFORNIA
Adair, Dennis Wilton
Ballard, Kenneth J
Barker, Donald Young
Benet, Leslie Z
Blankenship, James William
Boger, Dale L
Chaubal, Madhukar Gajanan
Eiler, John Joseph

Pharmacy (cont)

Gong, William C
Havemeyer, Ruth Naomi
Hoener, Betty-ann
Hong, Keelung
Jones, Richard Elmore
Kang, Chang-Yuil
Kaplan, Stanley A
King, James C
Koda, Robert T
Lee, Chi-Ho
McCarron, Margaret Mary
Malone, Marvin Herbert
Masover, Gerald K
Matuszak, Alice Jean Boyer
Maurice, David Myer
Mlodozeniec, Arthur Roman
Nicholson, Arnold Eugene
Ong, John Tjoan Ho
Oppenheimer, Phillip R
Phares, Russell Eugene, Jr
Poulsen, Boyd Joseph
Raff, Allan Maurice
Rowland, Ivan W
Runkel, Richard A
Saute, Robert E
Schwarz, Thomas Werner
Smith, Terry Douglas
Sorby, Donald Lloyd
Stanton, Hubert Coleman
Stimmel, Glen Lewis
Theeuwes, Felix
Thompson, John Frederick
Tozer, Thomas Nelson
Yakatan, Gerald Joseph
Young, James George

COLORADO
Bone, Jack Norman
Daunora, Louis George
Hammerness, Francis Carl
Williams, Ronald Lee

CONNECTICUT
Cardinal, John Robert
Cohen, Edward Morton
Cohen, Steven Donald
Fenney, Nicholas William
Forssen, Eric Anton
Gans, Eugene Howard
Ghosh, Dipak K
Heyd, Allen
Horhota, Stephen Thomas
Kelleher, William Joseph
Kramer, Paul Alan
Lange, Winthrop Everett
Milzoff, Joel Robert
Munies, Robert
Rosenberg, Philip
Rother, Ana
Schurig, John Eberhard
Shah, Shirish A
Shapiro, Warren Barry
Simonelli, Anthony Peter
Skauen, Donald M
Tranner, Frank

DELAWARE
Aungst, Bruce J
Baum, Martin David
Lam, Gilbert Nim-Car
Rudolph, Jeffrey Stewart
Turlapaty, Prasad

DISTRICT OF COLUMBIA
Griffenhagen, George Bernard
Hamarneh, Sami Khalaf
Horakova, Zdenka
Whidden, Stanley John
Zalucky, Theodore B

FLORIDA
Araujo, Oscar Eduardo
Hepler, Charles Douglas
Hill, Richard A
Hom, Foo Song
Huber, Joseph William, III
Lamba, Surendar Singh
Lindstrom, Richard Edward
Schwartz, Michael Averill
Seugling, Earl William, Jr
Soliman, Karam Farag Attia
Soliman, Magdi R I
Young, Michael David

GEORGIA
Ansel, Howard Carl
Cadwallader, Donald Elton
Entrekin, Durward Neal
Littlejohn, Oliver Marsilius
McGrath, William Robert
Price, James Clarence
Schramm, Lee Clyde
Waters, Kenneth Lee
Whitworth, Clyde W

HAWAII
Patterson, Harry Robert

IDAHO
Hillyard, Ira William
Nelson, Arthur Alexander, Jr

ILLINOIS
Blake, Martin Irving
Chun, Alexander Hing Chinn
Cranston, Joseph William, Jr
Dodge, Patrick William
Farnsworth, Norman R
Fong, Harry H S
Foreman, Ronald Louis
Hutchinson, Richard Allen
Kaistha, Krishan K
Kimura, Eugene Tatsuru
Kinghorn, Alan Douglas
Kunka, Robert Leonard
Lu, Matthias Chi-Hwa
Martinek, Robert George
Mrtek, Robert George
Musa, Mahmoud Nimir
Sause, H William
Schleif, Robert H
Thomas, Elizabeth Wadsworth
Witczak, Zbigniew J
Yunker, Martin Henry

INDIANA
Abdallah, Abdulmuniem Husein
Akers, Michael James
Belcastro, Patrick Frank
Born, Gordon Stuart
Duvall, Ronald Nash
Evanson, Robert Verne
Gold, Gerald
Green, Mark Alan
Hamlow, Eugene Emanuel
Hem, Stanley L
Jackson, Richard Lee
Kaufman, Karl Lincoln
Kessler, Wayne Vincent
Kildsig, Dane Olin
Kuzel, Norbert R
McLaughlin, Jerry Loren
Parke, Russell Frank
Peck, Garnet E
Robbers, James Earl
Rowe, Edward John
Shaw, Margaret Ann
Shaw, Stanley Miner
Sperandio, Glen Joseph
Thakkar, Arvind Lavji
Williams, Edward James

IOWA
Baron, Jeffrey
Benton, Byrl E
Carew, David P
Granberg, Charles Boyd
Guillory, James Keith
Jones, Leslie F
Kerr, Wendle Louis
Lach, John Louis
Levine, Phillip J
Makar, Adeeb Bassili
Parrott, Eugene Lee
Wurster, Dale E
Wurster, Dale Eric, Jr

KANSAS
Rose, Wayne Burl
Rytting, Joseph Howard
Stella, Valentino John

KENTUCKY
Crooks, Peter Anthony
Dittert, Lewis William
Kostenbauder, Harry Barr
Lesshafft, Charles Thomas, Jr
Parker, Martin Dale
Turco, Salvatore J
Yokel, Robert Allen

LOUISIANA
Daigle, Josephine Siragusa
Danti, August Gabriel
Grace, Marcellus
Shrader, Kenneth Ray

MARYLAND
Asghar, Khursheed
Augsburger, Larry Louis
Benson, Walter Roderick
Bhagat, Hitesh Rameshchandra
Caplan, Yale Howard
Cone, Edward Jackson
Cosmides, George James
Gallelli, Joseph F
Heller, William Mohn
Hollenbeck, Robert Gary
Jagadeesh, Gowra G
Knapp, David Allan
Kramer, William Geoffrey
Lamy, Peter Paul
Levchuk, John W
Moreton, Julian Edward
Oddis, Joseph Anthony
Osterberg, Robert Edward
Oxman, Michael Allan
Palumbo, Francis Xavier Bernard
Paul, Steven M
Poochikian, Guiragos K
Price, Thomas R
Rabussay, Dietmar Paul
Resnick, Charles A
Rodowskas, Christopher A, Jr
Sewell, Winifred
Shangraw, Ralph F

Tousignaut, Dwight R
Valley, Sharon Louise
Viswanathan, C T
Watzman, Nathan
Weiner, Myron
Woodman, Peter William
Young, Gerald A

MASSACHUSETTS
Bhargava, Hridaya Nath
Curtis, Robert Arthur
Gozzo, James J
Hughes, Richard Lawrence
Hurd, Richard Nelson
Leary, John Dennis
Mendes, Robert W
Morgan, James Philip
Pecci, Joseph
Shargel, Leon David
Siegal, Bernard
Silverman, Harold I
Smith, Pierre Frank
Tuckerman, Murray Moses

MICHIGAN
Anderson, Gordon Frederick
Barr, Martin
Bartlett, Janeth Marie
Bivins, Brack Allen
Brown, Michael
Burkholder, David Frederick
Dauphinais, Raymond Joseph
Hajratwala, Bhupendra R
Hamlin, William Earl
Hiestand, Everett Nelson
Hult, Richard Lee
Jacoby, Ronald Lee
Jensen, Erik Hugo
Kinkel, Arlyn Walter
Knoechel, Edwin Lewis
Koshy, K Thomas
Kramer, Sherman Francis
Lamb, Donald Joseph
McGovren, James Patrick
Moore, Willis Eugene
Paul, Ara Garo
Rowe, Englebert L
Rowe, Thomas Dudley
Swartz, Harry Sip
Wagner, John Garnet
Wilkinson, Paul Kenneth
Wormser, Henry C

MINNESOTA
Agyilirah, George Augustus
Banker, Gilbert Stephen
Cloyd, James C, III
Conard, Gordon Joseph
Grant, David James William
Miller, Nicholas Carl
Rahman, Yueh Erh
Shier, Wayne Thomas
Suryanarayanan, Raj Gopalan
Zimmerman, Cheryl Lea

MISSISSIPPI
Dodge, Austin Anderson
Ferguson, Mary Hobson
Gafford, Lanelle Guyton
Guess, Wallace Louis
Hufford, Charles David
Sindelar, Robert D
Wilson, Marvin Cracraft

MISSOURI
Banakar, Umesh Virupaksh
Bruns, Lester George
Ericson, Avis J
Godt, Henry Charles, Jr
Johnson, Richard Dean
Slotsky, Myron Norton
Stenseth, Raymond Eugene
Tansey, Robert Paul, Sr
Zienty, Ferdinand B

MONTANA
Medora, Rustem Sohrab
Van Horne, Robert Loren
Wailes, John Leonard

NEBRASKA
Abel, Peter William
Birt, Diane Feickert
Borchert, Harold R
Brockemeyer, Eugene William
Gerraughty, Robert Joseph
Gloor, Walter Thomas, Jr
Greco, Salvatore Joseph
Sethi, V Sagar
Ueda, Clarence Tad

NEVADA
Buxton, Iain Laurie Offord

NEW HAMPSHIRE
Illian, Carl Richard

NEW JERSEY
Agalloco, James Paul
Bathala, Mohinder S
Bodin, Jerome Irwin
Bornstein, Michael
Bowers, Roy Anderson
Brogle, Richard Charles

Caroselli, Robert Anthony
Chau, Thuy Thanh
Chertkoff, Marvin Joseph
Colaizzi, John Louis
Controulis, John
Conway, Paul Gary
Crouthamel, William Guy
Gidwani, Ram N
Gilfillan, Alasdair Mitchell
Goldstein, Harris Sidney
Gyan, Nanik D
Henderson, Norman Leo
Hong, Wen-Hai
Huber, Raymond C
Hutchins, Hastings Harold, Sr
Infeld, Martin Howard
Jaffe, Jonah
Jain, Nemichand B
Janssen, Richard William
Katz, Irwin Alan
Kline, Berry James
Kornblum, Saul S
Leeson, Lewis Joseph
Lordi, Nicholas George
Lukas, George
Maulding, Hawkins Valliant, Jr
Mehta, Atul Mansukhbhai
Nashed, Wilson
Nessel, Robert J
Reimann, Hans
Ress, Rudyard Joseph
Scarpone, A John
Sequeira, Joel August Louis
Sethi, Jitender K
Sheinaus, Harold
Simonoff, Robert
Singhvi, Sampat Manakchand
Staum, Muni M
Tierney, William John
Tojo, Kakuji
Tse, Francis Lai-Sing
Turi, Paul George
Wadke, Deodatt Anant
Wangemann, Robert Theodore
Wasserman, Martin Allan
Weintraub, Howard Steven
Willis, Carl Raeburn, Jr
Zatz, Joel L

NEW MEXICO
Hadley, William Melvin
Kabat, Hugh F
Lehrman, George Philip

NEW YORK
Abdelnoor, Alexander Michael
Alks, Vitauts
Bartilucci, Andrew J
Belmonte, Albert Anthony
Borisenok, Walter A
Casler, David Robert
Castellion, Alan William
Chandran, V Ravi
Chou, Dorthy T C T
Cooper, Robert Michael
Ehrke, Mary Jane
Eisen, Henry
Fields, Thomas Lynn
Fox, Chester David
Gagnon, Leo Paul
Goldberg, Arthur H
Hirsh, Kenneth Roy
Hunke, William Allen
Jusko, William Joseph
Lachman, Leon
Lai, Fong M
Larson, Allan Bennett
Levy, Gerhard
Martell, Michael Joseph, Jr
Morris, Marilyn Emily
Overweg, Norbert I A
Patel, Nagin K
Redalieu, Elliot
Reuter, Gerald Louis
Rich, Arthur Gilbert
Santiago, Noemi
Schnell, Lorne Albert
Sciarra, John J
Spiegel, Allen J
Stempel, Edward
Strauss, Steven
Taub, Aaron M
Tkacheff, Joseph, Jr
Urdang, Arnold
Violante, Michael Robert
Whitaker-Azmitia, Patricia Mack
White, Albert M

NORTH CAROLINA
Breese, George Richard
Dar, Mohammad Saeed
Dren, Anthony Thomas
Eckel, Frederick Monroe
Ghanayem, Burhan I
Grantham, Preston Hubert
Hadzija, Bozena Wesley
Hak, Lawrence J
Hall, William Earl
Hartzema, Abraham Gijsbert
Holstius, Elvin Albert
Hong, Donald David
Misek, Bernard
Olsen, James Leroy
Ramagopal, Mudumbi Vijaya

Sawardeker, Jawahar Sazro
Strandhoy, Jack W
Swarbrick, James
Taylor, William West
Wier, Jack Knight
Wiser, Thomas Henry
Yeowell, David Arthur

NORTH DAKOTA
Auyong, Theodore Koon-Hook
Husain, Syed
Rosenberg, Harry
Schnell, Robert Craig
Vincent, Muriel C

OHIO
Billups, Norman Frederick
Black, Curtis Doersam
Bope, Frank Willis
Boxenbaum, Harold George
Breglia, Rudolph John
Brueggemeier, Robert Wayne
Chen, Chi-Po
Coran, Aubert Y
Dunbar, Philip Gordon
Feller, Dannis R
Frank, Sylvan Gerald
Gerald, Michael Charles
Graef, Walter L
Hagman, Donald Eric
Hayton, William Leroy
Hinkle, George Henry
Hurwitz, David Allan
Leyda, James Perkins
Lima, John J
MacKichan, Janis Jean
McVean, Duncan Edward
Manudhane, Krishna Shankar
Messer, William Sherwood, Jr
Nahata, Milap Chand
Oesterling, Thomas O
Sherman, Gerald Philip
Shirkey, Harry Cameron
Skau, Kenneth Anthony
Sokoloski, Theodore Daniel
Stadler, Louis Benjamin
Stansloski, Donald Wayne
Theodore, Joseph M, Jr
Visconti, James Andrew
Webb, Norval Ellsworth, Jr
Zoglio, Michael Anthony

OKLAHOMA
Carubelli, Raoul
Corder, Clinton Nicholas
Ice, Rodney D
Jones, Sharon Lynn
Keller, Bernard Gerard, Jr
Nithman, Charles Joseph
Prabhu, Vilas Anandrao
Reinke, Lester Allen
Shough, Herbert Richard
Sommers, Ella Blanche
Yanchick, Victor A

OREGON
Ayres, James Walter
Block, John Harvey
Constantine, George Harmon, Jr
Sisson, Harriet E

PENNSYLVANIA
Appino, James B
Bahal, Neeta
Bahal, Surendra Mohan
Bannon, James Andrew
Bhatt, Padmamabh P
Cowan, Alan
Davis, Neil Monas
Goldstein, Frederick J
Herzog, Karl A
Hussar, Daniel Alexander
Kelly, Ernest L
King, Robert Edward
Krenzelok, Edward Paul
Kwan, King Chiu
Levine, Geoffrey
McGhan, William Frederick
Mackowiak, Elaine DeCusatis
Marshall, Keith
Motsavage, Vincent Andrew
Nedich, Ronald Lee
ONeill, Joseph Lawrence
Osei-Gyimah, Peter
Peterson, Charles Fillmore
Ravin, Louis Joseph
Restaino, Frederick A
Riley, Gene Alden
Rodell, Michael Byron
Schiff, Paul L, Jr
Schott, Hans
Smith, Roger Bruce
Sugita, Edwin T
Tice, Linwood Franklin
Torosian, George
Venkataramanan, Raman
Walkling, Walter Douglas
Walsh, Arthur Campbell
Wu, Wu-Nan
Zemaitis, Michael Alan

RHODE ISLAND
Needham, Thomas E, Jr
Osborne, George Edwin

Worthen, Leonard Robert

SOUTH CAROLINA
Davis, Craig Wilson
Fincher, Julian H
Lawson, Benjamin F
Niebergall, Paul J
Orcutt, Donald Adelbert
Putney, Blake Fuqua
Sumner, Edward D

SOUTH DAKOTA
Hietbrink, Bernard Edward

TENNESSEE
Autian, John
Avis, Kenneth Edward
Berney, Stuart Alan
Caponetti, James Dante
Elowe, Louis N
Fields, James Perry
Gupta, Brij Mohan
Hamner, Martin E
Lawrence, William Homer
Meyer, Marvin Chris
Nunez, Loys Joseph
Sheth, Bhogilal
Shukla, Atul J
Skoutakis, Vasilios A
Sticht, Frank Davis
Straughn, Arthur Belknap
Wojciechowski, Norbert Joseph

TEXAS
Brazeau, Gayle A
Bryant, Stephen G
Driever, Carl William
Eugere, Edward Joseph
Feldman, Stuart
Gerding, Thomas G
Gupta, Vishnu Das
Hall, Esther Jane Wood
Hecht, Gerald
Hickman, Eugene, Sr
Iskander, Felib Youssef
Kasi, Leela Peshkar
Kehrer, James Paul
Leaders, Floyd Edwin, Jr
Martin, Alfred
Monk, Clayborne Morris
Morgan, H Keith
Newburger, Jerold
Ryan, Charles F
Schermerhorn, John W
Sheffield, William Johnson
Stavchansky, Salomon Ayzenman
Sullivan, Gerald
Ticku, Maharaj K
Walton, Charles Anthony
Woller, William Henry

UTAH
Aldous, Duane Leo
Balandrin, Manuel F
Galinsky, Raymond Ethan
Lee, Steve S
Segelman, Florence H
Stuart, David Marshall
Wolf, Harold Herbert

VIRGINIA
Aceto, Mario Domenico
Barrett, Richard John
Batra, Karam Vir
Collins, Galen Franklin
Dvorchik, Barry Howard
Feldmann, Edward George
Kapadia, Abhaysingh J
Lowenthal, Werner
Martin, John Walter, Jr
Penna, Richard Paul
Smith, Harold Linwood
Snow, Anne E
Stone, John Patrick
Teng, Lina Chen
Uwaydah, Ibrahim Musa
Weaver, Warren Eldred
Wood, John Henry

WASHINGTON
Galpin, Donald R
Hammarlund, Edwin Roy
Meadows, Gary Glenn
Plein, Elmer Michael
Plein, Joy Bickmore
Yost, Garold Steven

WEST VIRGINIA
Lim, James Khai-Jin
Malanga, Carl Joseph
Rosenbluth, Sidney Alan

WISCONSIN
Bardell, Eunice Bonow
Bass, Paul
Blockstein, William Leonard
Krahnke, Harold C
Lalley, Peter Michael
Lemberger, August Paul
Mukerjee, Pasupati

WYOMING
Brunett, Emery W
Julian, Edward A

ALBERTA
Baker, Glen Bryan
Biggs, David Frederick
Rogers, James Albert
Roth, Sheldon H
Shysh, Alec
Wiebe, Leonard Irving

BRITISH COLUMBIA
Goodeve, Allan McCoy
Riedel, Bernard Edward

NOVA SCOTIA
Dudar, John Douglas
Mezei, Catherine
Parker, William Arthur
Tonks, Robert Stanley

ONTARIO
Baxter, Ross M
Hughes, Francis Norman
Lucas, Douglas M
Miller, Robert Alon
Nairn, John Graham
Segal, Harold Jacob
Somers, Emmanuel

QUEBEC
Braun, Peter Eric
Joly, Louis Philippe
Lamonde, Andre M
Larose, Roger
Plourde, J Rosaire

OTHER COUNTRIES
Hedglin, Walter L
Ishida, Yukisato
Nabeshima, Toshitaka
Yeon-Sook, Yun

Psychiatry

ALABAMA
Bradley, Laurence A
Froelich, Robert Earl
Garver, David L
Linton, Patrick Hugo
Ryan, William George

ARIZONA
Beigel, Allan
Cutts, Robert Irving
Felix, Robert Hanna
Finch, Stuart McIntyre
Levenson, Alan Ira
Offenkrantz, William Charles
Starr, Phillip Henry
Tapia, Fernando

ARKANSAS
Henker, Fred Oswald, III
McMillan, Donald Edgar
Newton, Joseph Emory O'Neal
Reese, William George

CALIFORNIA
Agras, William Stewart
Auerback, Alfred
Barchas, Jack D
Barondes, Samuel Herbert
Beach, Frank Ambrose
Beisser, Arnold Ray
Bloomfield, Harold H
Brodsky, Carroll M
Brown, Joan Heller
Buchwald, Jennifer S
Bunney, William E
Call, Justin David
Callaway, Enoch, III
Cohn, Jay Binswanger
Courchesne, Eric
Dracup, Kathleen A
Duhl, Leonard J
Efron, Robert
Epstein, Leon J
Estrada, Norma Ruth
Evans, Harrison Silas
Feifel, Herman
Ferguson, James Mecham
Fish, Barbara
Freedman, Daniel X
Galin, David
Gericke, Otto Luke
Gislason, I(rving) Lee
Gorney, Roderic
Gottschalk, Louis August
Gray, Gregory Edward
Green, Richard
Greenblatt, Milton
Haig, Pierre Vahe
Hansen, Howard Edward
Harrison, Saul I
Hecht, Elizabeth Anne
Hewitt, Robert T
Horowitz, Mardi J
Jarvik, Lissy F
Jarvik, Murray Elias
Jensen, Gordon D
Jones, Reese Tasker
Judd, Lewis Lund
Kaplan, Herman
Kaufman, Janice Norton
Keilin, Bertram

Kempler, Walter
Kline, Frank Menefee
Koran, Lorrin Michael
Kripke, Daniel Frederick
Lamb, H Richard
Leiderman, P Herbert
Leighton, Dorothea Cross
Lesse, Henry
Liberman, Robert Paul
London, Ray William
Lord, Elizabeth Mary
Lowinger, Paul
McCabe, Donald Lee
McGinnis, James F
Mandell, Arnold J
Marangos, Paul Jerome
Marcus, Joseph
Marshall, John Foster
May, Philip Reginald Aldridge
Mendel, Werner Max
Miller, Milton H
Mills, Gary Kenith
Motto, Jerome (Arthur)
Nowlis, David Peter
O'Malley, Edward Paul
Ostwald, Peter Frederic
Palmer, Beverly B
Pellegrini, Robert J
Procci, Warren R
Rappaport, Maurice
Redlich, Fredrick Carl
Riskin, Jules
Ritvo, Edward R
Ritzmann, Ronald Fred
Rubin, Robert Terry
Ruesch, Jurgen
Sargent, Thornton William, III
Sattin, Albert
Saunders, Virginia Fox
Schneider, Lon S
Schwartz, Donald Alan
Selzer, Melvin Lawrence
Serafetinides, Eustace A
Shapiro, David
Simmons, James Quimby, III
Simon, Alexander
Sloane, Robert Bruce
Solomon, George Freeman
Starr, Arnold
Stimmel, Glen Lewis
Storms, Lowell H
Tarjan, George
Teicher, Joseph D
Tewari, Sujata
Thomas, Claudewell Sidney
Thomas, Robert Eugene
Tupin, Joe Paul
Turrell, Eugene Snow
Van Dyke, Craig
Wallerstein, Robert Solomon
Wasserman, Martin S
Wayne, George Jerome
Weiner, Herbert
Werbach, Melvyn Roy
West, Louis Jolyon
Woods, Sherwyn Martin
Yamamoto, Joe
Zaidel, Eran

COLORADO
Coppolillo, Henry P
Gaskill, Herbert Stockton
Kulkosky, Paul Joseph
Laudenslager, Mark LeRoy
Macdonald, John Marshall
Margolin, Sydney Gerald
Sander, Louis W
Shore, James H
Suinn, Richard M

CONNECTICUT
Amdur, Millard Jason
Astrachan, Boris Morton
Bunney, Benjamin Stephenson
Byck, Robert
Charney, Dennis S
Cohen, Donald J
Coleman, Jules Victor
Comer, James Pierpont
Davis, Michael
Donnelly, John
Escobar, Javier I
Fleck, Stephen
Fleeson, William
Glueck, Bernard Charles
Heninger, George Robert
Henisz, Jerzy Emil
Katz, Jay
Lidz, Theodore
Mirabile, Charles Samuel, Jr
Price, Lawrence Howard
Reiser, Morton Francis
Rodin, Judith
Rounsaville, Bruce J
Senn, Milton J E
Sheard, Michael Henry
Snyder, Daniel Raphael
Stroebel, Charles Frederick, III
Webb, William Logan

DELAWARE
Davis, V Terrell

Psychiatry (cont)

DISTRICT OF COLUMBIA
Bever, Christopher Theodore
Brands, Alvira Bernice
Chafetz, Morris Edward
Cowdry, Rex William
Gordon, James Samuel
Herman, Barbara Helen
Hoffman, Julius
Jacobsen, Frederick Marius
Kellar, Kenneth Jon
Kleber, Herbert David
Lebensohn, Zigmond Meyer
Lieberman, Edwin James
Noshpitz, Joseph Dove
Perlin, Seymour
Reynolds, Thomas De Witt
Rittelmeyer, Louis Frederick, Jr
Sabshin, Melvin
Webster, Thomas G

FLORIDA
Adams, John Evi
Apter, Nathaniel Stanley
Belmont, Herman S
Brightwell, Dennis Richard
Cohen, Sanford I
Denber, Herman C B
Denker, Martin William
Eisdorfer, Carl
Gold, Mark Stephen
Goldstein, Burton Jack
Kemph, John Patterson
Knopp, Walter
Merritt, Henry Neyron
Nagera, Humberto
Pfeiffer, Eric A
Reading, Anthony John
Richelson, Elliott
Sussex, James Neil
Zusman, Jack

GEORGIA
Adams, David B
Curtis, John Russell
Ermutlu, Ilhan M
Hartlage, Lawrence Clifton
Jackson, William James
Loomis, Earl Alfred, Jr
McCranie, Erasmus James
Michael, Richard Phillip
Schweri, Margaret Mary
Shaffer, David
Ward, Richard S

HAWAII
Amundson, Mary Jane
Bolman, William Merton
Char, Walter F
McDermott, John Francis
Morton, Bruce Eldine

ILLINOIS
Adams, John R
Aramini, Jan Marie
Barter, James T
Beiser, Helen R
Boshes, Benjamin
Boshes, Louis D
Brockman, David Dean
Brown, Meyer
Chessick, Richard D
Crawford, James Weldon
Davis, John Marcell
Davis, Oscar F
Dyrud, Jarl Edvard
Falk, Marshall Allen
Fang, Victor Shengkuen
Freedman, Lawrence Zelic
Gill, Merton
Giovacchini, Peter L
Goebel, Maristella
Goldsmith, Jewett
Grinker, Roy Richard, Sr
Hirsch, Jay G
Hughes, Patrick Henry
Juhasz, Stephen Eugene
Kaplan, Maurice
Kimball, Chase Patterson
Langsley, Donald Gene
Levitt, LeRoy P
Levy, Leo
McArdle, Eugene W
Metz, John Thomas
Miller, Derek Harry
Mota de Freitas, Duarte Emanuel
Musa, Mahmoud Nimir
Novick, Rudolph G
Offer, Daniel
Pollock, George Howard
Rasenick, Mark M
Rudy, Lester Howard
Sabelli, Hector C
Schulman, Jerome Lewis
Schyve, Paul Milton
Smith, Kathleen
Spring, Bonnie Joan
Taylor, Michael Alan
Thomas, Joseph Erumappettical
Tourlentes, Thomas Theodore
Trosman, Harry
Ts'o, Timothy On-To
Visotsky, Harold M

INDIANA
Altman, Joseph
Bennett, Ivan Frank
Blix, Susanne
Churchill, Don W
DeMyer, Marian Kendall
FitzGerald, Joseph Arthur
Fitzhugh-Bell, Kathleen
Greist, John Howard
Grosz, Hanus Jiri
Klinghammer, Erich
McBride, Angela Barron
Nurnberger, John Ignatius, Sr
Rebec, George Vincent
Simmons, James Edwin
Small, Iver Francis
Small, Joyce G
Stark, Paul

IOWA
Andreasen, Nancy Coover
Benton, Arthur Lester
Cadoret, Remi Jere
Clancy, John
Comly, Hunter Hall
Jenkins, Richard Leos
Nelson, Herbert Leroy
Norris, Albert Stanley
Noyes, Russell, Jr
Stewart, Mark Armstrong
Tranel, Daniel T
Wildberger, William Campbell
Winokur, George

KANSAS
Bradwhaw, Samuel Lockwood, Jr
Chin, Hong Woo
Laybourne, Paul C
Menninger, Karl Augustus
Menninger, William Walter
Modlin, Herbert Charles
Rinsley, Donald Brendan
Simpson, William Stewart
Voth, Harold Moser

KENTUCKY
Elkes, Joel
Humphries, Laurie Lee
Landis, Edward Everett
Levy, Robert Sigmund
Moore, Kenneth Boyd
Sandifer, Myron Guy, Jr
Schwab, John J
Storrow, Hugh Alan
Surwillo, Walter Wallace
Teller, David Norton
Weisskopf, Bernard
Zolman, James F

LOUISIANA
Brauchi, John Tony
Daniels, Robert (Sanford)
Easson, William McAlpine
Epstein, Arthur William
Gallant, Donald
Heath, Robert Galbraith
Knight, James Allen
Miles, Henry Harcourt Waters
Norman, Edward Cobb
Osofsky, Howard J
Palotta, Joseph Luke
Samuels, Arthur Seymour
Schober, Charles Coleman
Scrignar, Chester Bruno
Seltzer, Benjamin
Shushan, Morris
Straumanis, John Janis, Jr
Strauss, Arthur Joseph Louis

MARYLAND
Andersen, Arnold E
Anderson, David Everett
Ascher, Eduard
Baker, Frank
Barrett, James E
Berlin, Fred S
Bernstein, Barbara Elaine
Berrettini, Wade Hayhurst
Blum, Robert Allan
Boslow, Harold Meyer
Brauth, Steven Earle
Brody, Eugene B
Burnham, Donald Love
Cantor, David S
Cohen, Robert Martin
Costa, Paul T, Jr
Coyle, Joseph T
Dingman, Charles Wesley, II
DuPont, Robert L, Jr
Eckardt, Michael Jon
Gershon, Elliot Sheldon
Glaser, Kurt
Godenne, Ghislaine D
Gold, Philip William
Gray, David Bertsch
Greenspan, Stanley Ira
Gruenberg, Ernest Matsner
Hendler, Nelson Howard
Hoehn-Saric, Rudolf
Holloway, Harry Charles
Imboden, John Baskerville
Jones, Franklin Del
Klee, Gerald D'Arcy
Koslow, Stephen Hugh

KRANTZ / MICHIGAN
Krantz, David S
Kurland, Albert A
Lake, Charles Raymond
Lemkau, Paul V
Marwah, Joe
Miller, Nancy E
Mosher, Loren Richard
Paré, William Paul
Park, Lee Crandall
Pickar, David
Pollin, William
Post, Robert M
Rankin, Joseph Eugene
Rapoport, Judith Livant
Resnik, Harvey Lewis Paul
Rioch, David McKenzie
Ryback, Ralph Simon
Saavedra, Juan M
Salzman, Leon
Schuster, Charles Roberts, Jr
Silbergeld, Sam
Suomi, Stephen John
Tamminga, Carol Ann
Uhde, Thomas Whitley
Wehr, Thomas A
Work, Henry Harcus
Wurmser, Leon
Zatz, Martin

MASSACHUSETTS
Ayers, Joseph Leonard, Jr
Baldessarini, Ross John
Berman, Marlene Oscar
Bloomberg, Wilfred
Bojar, Samuel
Chivian, Eric Seth
Coles, Robert
Craig, John Merrill
Eisenberg, Carola
Eisenberg, Leon
Epstein, Nathan Bernic
Frankel, Fred Harold
Galler, Janina Regina
Gamzu, Elkan R
Gold, Richard Michael
Goldings, Herbert Jeremy
Greenblatt, David J
Grinspoon, Lester
Gutheil, Thomas Gordon
Hartmann, Ernest Louis
Havens, Leston Laycock
Hebben, Nancy
Hobson, John Allan
Holzman, Philip S
Kagan, Jerome
Kleinman, Arthur Michael
Knapp, Peter Hobart
Kulka, Johannes Peter
Levy, Deborah Louise
Lipsitt, Don Richard
Locke, Steven Elliot
McCarley, Robert William
Marotta, Charles Anthony
Mello, Nancy K
Mendelson, Jack H
Orren, Merle Morrison
Pierce, Chester Middlebrook
Pillard, Richard Colestock
Poussaint, Alvin Francis
Reich, James Harry
Rexford, Eveoleen Naomi
Riemer-Rubenstein, Delilah
Rothenberg, Albert
Russo, Dennis Charles
Saxe, Leonard
Schildkraut, Joseph Jacob
Shader, Richard Irwin
Shore, Miles Frederick
Stotsky, Bernard A
Swartz, Jacob
Torda, Clara
Tsuang, Ming Tso

MICHIGAN
Berent, Stanley
Bernardez, Teresa
Blom, Gaston Eugene
Bloom, Victor
Curtis, George Clifton
Enzer, Norbert Beverley
Freedman, Robert Russell
Gershon, Samuel
Ging, Rosalie J
Greden, John F
Hawthorne, Victor Morrison
Hendrickson, Willard James
Luby, Elliot Donald
Margolis, Philip Marcus
Newman, Sarah Winans
Paschke, Richard Eugene
Pomerleau, Ovide F
Rainey, John Marion, Jr
Rosenzweig, Norman
Sarwer-Foner, Gerald
Schorer, Calvin E
Silverman, Albert Jack
Waggoner, Raymond Walter
Watson, Andrew Samuel
Werner, Arnold

MINNESOTA
Abuzzahab, Faruk S, Sr
Bieber, Irving
Clayton, Paula Jean

DYSKEN / NEW YORK
Dysken, Maurice William
Hanson, Daniel Ralph
Hartman, Boyd Kent
Jepson, William W
Levitan, Alexander Allen
Lucas, Alexander Ralph
Lykken, David Thoreson
Meier, Manfred John
Milliner, Eric.Killmon
Mortimer, James Arthur
Overmier, J Bruce
Popkin, Michael Kenneth
Roberts, Warren Wilcox
Schofield, William
Swanson, David Wendell
Tyce, Francis Anthony
Yellin, Absalom Moses

MISSISSIPPI
Alford, Geary Simmons
Anderson, Kenneth Verle
Batson, Margaret Bailly
Suess, James Francis
Tourney, Garfield

MISSOURI
Chapel, James L
Cloninger, Claude Robert
Favazza, Armando Riccardo
Graham, Charles
Guze, Samuel Barry
Hudgens, Richard Watts
Justesen, Don Robert
Lamberti, Joseph W
Murphy, George Earl
Natani, Kirmach
Olney, John William
O'Neal, Patricia L
Othmer, Ekkehard
Robins, Eli
Solomon, Kenneth
Thale, Thomas Richard
Weiss, James Moses Aaron
Wilkinson, Charles Brock

MONTANA
Lynch, Wesley Clyde

NEBRASKA
Bartholow, George William
Eaton, Merrill Thomas, Jr
Wigton, Robert Spencer

NEVADA
Hausman, William
Rahe, Richard Henry
Smith, Aaron

NEW HAMPSHIRE
Hermann, Howard T
Payson, Henry Edwards
Ritzman, Thomas A
Russell, Donald Hayes
Vaillant, George Eman

NEW JERSEY
Buchweitz, Ellen
Carlton, Peter Lynn
Chassan, Jacob Bernard
Gaudino, Mario
Goldstein, Leonide
Gorodetzky, Charles W
Graham-Ellis, Avis
Hankoff, Leon Dudley
Kohn, Herbert Myron
Layman, William Arthur
McGough, William Edward
Manowitz, Paul
Millner, Elaine Stone
Moyer, John Allen
Murphree, Henry Bernard Scott
Niederland, William G
Perr, Irwin Norman
Pollack, Irwin W
Schleifer, Steven Jay
Schwartz, Arthur Harold
Sugerman, Abraham Arthur
Swigar, Mary Eva
Weiss, Robert Jerome

NEW MEXICO
Berlin, Irving Norman
Elliott, Charles Harold
Rhodes, John Marshell
Uhlenhuth, Eberhard Henry
Van Orden, Lucas Schuyler, III

NEW YORK
Adler, Kurt Alfred
Arkin, Arthur Malcolm
Arlow, Jacob
Atkins, Robert W
Babineau, G Raymond
Baekeland, Frederick
Bartlett, James Williams, Jr
Bartus, Raymond T
Bell, James Milton
Bernstein, Alvin Stanley
Berry, Gail W(ruble)
Blass, John P
Blume, Sheila Bierman
Bovbjerg, Dana H
Brahen, Leonard Samuel
Brodie, Jonathan D

Butler, Robert Neil
Cancro, Robert
Carlson, Eric Theodore
Carr, Edward Gary
Christ, Adolph Ervin
Clark, Julian Joseph
Cramer, Joseph Benjamin
Cushman, Paul, Jr
Daly, Robert Ward
David, Oliver Joseph
Davis, Kenneth Leon
Doering, Charles Henry
Engel, George Libman
Engelhardt, David M
Erlenmeyer-Kimling, L
Esser, Aristide Henri
Ferris, Steven Howard
Fieve, Ronald Robert
Fink, Max
Fisher, Saul Harrison
Fisher, Seymour
Flach, Frederic Francis
Forquer, Sandra Lynne
Freedman, Alfred Mordecai
Friedhoff, Arnold Jerome
Frosch, William Arthur
Gaylin, Willard
Gibbs, James Gendron, Jr
Glass, David Carter
Glusman, Murray
Goldfried, Marvin R
Golub, Sharon Bramson
Gorzynski, Janusz Gregory
Ham, Richard John
Hamburg, Beatrix A M
Hamburg, David Alan
Hawkins, Linda Louise
Henn, Fritz Albert
Hershey, Linda Ann
Herz, Marvin Ira
Itil, Turan M
Joseph, Edward David
Kaplan, Harold Irwin
Kaplan, Helen Singer
Kernberg, Otto F
Kestenbaum, Richard Steven
Klein, Donald Franklin
Klerman, Gerald L
Kolb, Lawrence Coleman
Kraft, Alan M
Krauss, Herbert Harris
Krynicki, Victor Edward
Kupfermann, Irving
Lang, Enid Asher
Langner, Thomas S
Levis, Donald J
Levy, Norman B
Loullis, Costas Christou
Malitz, Sidney
Marks, Neville
Mattson, Marlin Roy Albin
Mesnikoff, Alvin Murray
Michels, Robert
Millman, Robert Barnet
Milman, Doris H
Nathan, Ronald Gene
Neugebauer, Richard
Olczak, Paul Vincent
Pasamanick, Benjamin
Pelicci, Pier Guiseppe
Phillips, Richard Hart
Rabe, Ausma
Rainer, John David
Ravitz, Leonard J, Jr
Reifler, Clifford Bruce
Rieder, Ronald Olrich
Rifkin, Arthur
Ritter, Walter Paul
Robinson, David Bancroft
Romano, John
Sackeim, Harold A
Sadock, Benjamin
Sadock, Virginia A
Sager, Clifford J
Sankar, D V Siva
Schachter, Michael Ben
Schmale, Arthur H, Jr
Schuster, Daniel Bradley
Schwartz, Allan James
Shamoian, Charles Anthony
Sharpe, Lawrence
Siever, Larry Joseph
Small, S(aul) Mouchly
Smith, Charles James
Squires, Richard Felt
Stein, Marvin
Stein, Ruth E K
Stern, Marvin
Stewart, J W
Stokes, Peter E
Szasz, Thomas Stephen
Taketomo, Yasuhiko
Thaler, Otto Felix
Thomas, Alexander
Tryon, Warren W
Van praag, Herman M
Verebey, Karl G
Volavka, Jan
Wadden, Thomas Antony
Wineburg, Elliot N
Wollman, Leo
Wynne, Lyman Carroll
Yolles, Stanley Fausst
Yozawitz, Allan

Zitrin, Arthur
Zitrin, Charlotte Marker
Zukin, Stephen R

NORTH CAROLINA
Anderson, William B
Blazer, Dan German
Brodie, Harlow Keith Hammond
Busse, Ewald William
Carmen, Elaine (Hilberman)
Carroll, Bernard James
Conners, Carmen Keith
Curtis, Thomas Edwin
Eason, Robert Gaston
Eichelman, Burr S, Jr
Ellinwood, Everett Hews, Jr
Ewing, John Alexander
Fowler, John Alvis
Green, Robert Lee, Jr
Hall, William Charles
Halleck, Seymour Leon
Harris, Harold Joseph
Hicks, Robert Eugene
Hine, Frederick Roy
Krishnan, K Ranga Rama
Laupus, William E
Lewis, Mark Henry
Lipkin, Mack
Lipton, Morris Abraham
Llewellyn, Charles Elroy, Jr
Logan, Cheryl Ann
Mathis, James L
Melges, Frederick Towne
Nemeroff, Charles Barnet
Nenno, Robert Peter
Parker, Joseph B, Jr
Prange, Arthur Jergen, Jr
Reinhart, John Belvin
Rhoads, John McFarlane
Schiffman, Susan S
Smith, Charles E
Stolk, Jon Martin
Toole, James Francis
Verwoerdt, Adrian
Wang, Hsioh-Shan
Weiner, Richard D
Williams, Redford Brown, Jr
Wilson, William Preston
Zung, William Wen-Kwai

NORTH DAKOTA
Gladue, Brian Anthony

OHIO
Block, Stanley L
Brugger, Thomas C
Cacioppo, John T
Dizenhuz, Israel Michael
Ferguson, Shirley Martha
Giannini, A James
Gregory, Ian (Walter De Grave)
Hagenauer, Fedor
Halaris, Angelos
Heller, Abraham
Hiatt, Harold
Kaplan, Stanley Meisel
Kramer, Milton
Lenkoski, L Douglas
McConville, Brian John
Macklin, Martin
Meltzer, Herbert Yale
Nasrallah, Henry A
Ross, William Donald
Rowland, Vernon
Schubert, Daniel Sven Paul
Steele, Jack Ellwood
Titchener, James Lampton
Whitman, Roy Milton

OKLAHOMA
Deckert, Gordon Harmon
Donahue, Hayden Hackney
Hochhaus, Larry
Kosbab, Frederic Paul Gustav
Lovallo, William Robert
Miner, Gary David
Sengel, Randal Alan
Toussieng, Povl Winning

OREGON
Ary, Dennis
Casey, Daniel Edward
Crawshaw, Ralph Shelton
Denney, Donald Duane
Lewy, Alfred James
Morgenstern, Alan Lawrence
Severson, Herbert H
Taubman, Robert Edward

PENNSYLVANIA
Arce, A Anthony
Auerbach, Arthur Henry
Ballentine, Rudolph Miller
Barry, Herbert, III
Baum, O Eugene
Beck, Aaron Temkin
Brady, John Paul
Bridger, Wagner H
Clark, Robert Alfred
Cohen, Richard Lawrence
Cohen, Sheldon A
Detre, Thomas Paul
Dinges, David Francis
Doghramji, Karl

Epstein, Alan Neil
Fagin, Claire M
Field, Howard Lawrence
Gottheil, Edward
Heller, Melvin S
Hill, Shirley Yarde
Josiassen, Richard Carlton
Kendall, Philip C
Kupfer, David J
Leavitt, Marc Laurence
Leopold, Robert L
Lieberman, Daniel
Lief, Harold Isaiah
Madow, Leo
Mendelson, Myer
Michelson, Larry
Miller, Barry
Morris, Harold Hollingsworth, Jr
Myers, Jacob Martin
Needleman, Herbert L
Oken, Donald
Orne, Martin Theodore
Overton, Donald A
Palmer, Larry Alan
Pinski, Gabriel
Plutzer, Martin David
Ragins, Naomi
Ray, William J
Reis, Walter Joseph
Reynolds, Charles F
Rickels, Karl
Roemer, Richard Arthur
Rucinska, Ewa J
Ruff, George Elson
Saul, Leon Joseph
Schachter, Joseph
Schechter, Marshall David
Settle, Richard Gregg
Shagass, Charles
Sonis, Meyer
Stunkard, Albert J
Turk, Dennis Charles
Van Kammen, Daniel Paul
Werner, Gerhard
Whybrow, Peter Charles
Wolford, Jack Arlington
Wolpe, Joseph
Zwerling, Israel

RHODE ISLAND
Carter, George H
McMaster, Philip Robert Bache
Oxenkrug, Gregory Faiva
Sanberg, Paul Ronald

SOUTH CAROLINA
Ballenger, James Caudell
Kaiser, Charles Frederick
McCurdy, Layton
Maricq, Hildegard Rand
Mellette, Russell Ramsey, Jr
Riggs, Benjamin C
Wittson, Cecil L

TENNESSEE
Aivazian, Garabed Hagpop
Akiskal, Hagop S
Ban, Thomas Arthur
Berney, Stuart Alan
Billig, Otto
Castelnuovo-Tedesco, Pietro
Friedman, Daniel Lester
Haywood, H Carl
Hollender, Marc Hale
Petrie, William Marshall
Puzantian, Vahe Ropen
Strupp, Hans H
Thakar, Jay H
Wells, Charles Edmon

TEXAS
Adams, Paul Lieber
Blackburn, Archie Barnard
Bruce, E Ivan, Jr
Bryant, Stephen G
Cantrell, William Allen
Davis, David
Dilsaver, Steven Charles
Fabre, Louis Fernand, Jr
Faillace, Louis A
Fisher, Seymour
Forbis, Orie Lester, Jr
Gaitz, Charles M
Gardner, Russell, Jr
Giffen, Martin Brener
Gilliland, Robert McMurtry
Guynn, Robert William
Hansen, Douglas Brayshaw
Hatch, John Phillip
Henry, Billy Wendell
Hussian, Richard A
Johnson, Patricia Ann J
Kane, Francis Joseph, Jr
Kay, David Cyril
Kraft, Irvin Alan
Lakoski, Joan Marie
Leon, Robert Leonard
Maas, James Weldon
Martin, Jack
Meyer, George G
Pennebaker, James W
Pokorny, Alex Daniel
Rafferty, Frank Thomas
Roessler, Robert L

Roffwarg, Howard Philip
Rosenthal, Saul Haskell
Sauerland, Eberhardt Karl
Schoenfeld, Lawrence Steven
Schoolar, Joseph Clayton
Schottstaedt, Mary Gardner
Simmons, Charles Edward
Ticku, Maharaj K
White, Robert B
Williams, Robert L
Wilmer, Harry A

UTAH
Bliss, Eugene Lawrence
Grosser, Bernard Irving
Ruhmann Wennhold, Ann Gertrude
Schenkenberg, Thomas
Wender, Paul H

VERMONT
Barton, Walter E
Brooks, George Wilson
Friedman, Matthew Joel
Huessy, Hans Rosenstock
Hughes, John Russell
Swett, Chester Parker

VIRGINIA
Abse, David Wilfred
Aldrich, Clarence Knight
Balster, Robert L(ouis)
Brand, Eugene Dew
Dovenmuehle, Robert Henry
Garnett, Richard Wingfield, Jr
Gullotta, Frank Paul
Johnson, David Norseen
King, Lucy Jane
Narasimhachari, Nedathur
Pribram, Karl Harry
Sgro, J A
Spradlin, Wilford W
Stevenson, Ian
Volkan, Vamik

WASHINGTON
Anderson, Robert Authur
Bakker, Cornelis Bernardus
Bowden, Douglas Mchose
Cole, Nyla J
Dunner, David Louis
Garcia, John
Heston, Leonard L
McSweeney, Frances Kaye
Mason, Beryl Troxell
Reichler, Robert Jay
Scher, Maryonda E
Tucker, Gary Jay

WEST VIRGINIA
Bateman, Mildred Mitchell
Hein, Peter Leo, Jr
Kelley, John Fredric

WISCONSIN
Cleeland, Charles Samuel
Glover, Benjamin Howell
Harsch, Harold H
Headlee, Raymond
Jefferson, James Walter
Kepecs, Joseph Goodman
Prosen, Harry
Roberts, Leigh M
Spiro, Herzl Robert
Treffert, Darold Allen
Westman, Jack Conrad

PUERTO RICO
Perez-Cruet, Jorge

ALBERTA
Baker, Glen Bryan
Dewhurst, William George
Pearce, Keith Ian
Sainsbury, Robert Stephen
Weckowicz, Thaddeus Eugene

BRITISH COLUMBIA
Beiser, Morton
Crockett, David James
Freeman, Roger Dante
Laye, Ronald Curtis
Marcus, Anthony Martin
Margetts, Edward Lambert
Slade, H Clyde
Yonge, Keith A
Zaleski, Witold Andrew

MANITOBA
Adamson, John Douglas
Adaskin, Eleanor Jean
LeBow, Michael David
Sisler, George C
Varsamis, Ioannis

NOVA SCOTIA
Connolly, John Francis
Doane, Benjamin Knowles
Flynn, Patrick
Gold, Judith Hammerling
Hirsch, Solomon
Leighton, Alexander Hamilton
Richman, Alex

Psychiatry (cont)

ONTARIO
Anderson, James Edward
Atcheson, J D
Beninger, Richard J
Boag, Thomas Johnson
Bonkalo, Alexander
Caplan, Paula Joan
Cleghorn, Robert Allen
Donald, Merlin Wilfred
Garfinkel, Paul Earl
Glaser, Frederick Bernard
Hobbs, George Edgar
Hrdina, Pavel Dusan
Kutcher, Stanley Paul
Lapierre, Yvon Denis
Lee, Tyrone Yiu-Huen
Levine, Saul
Lipowski, Zbigniew J
Merskey, Harold
Offord, David Robert
Persad, Emmanuel
Powles, William Earnest
Rae-Grant, Quentin A
Rakoff, Vivian Morris
Sattar, Syed Abdus
Seeman, Mary Violette
Shamsie, Jalal
Silberfeld, Michel
Smith, Selwyn Michael
Stancer, Harvey C
Steinhauer, Paul David
Szyrynski, Victor
Witelson, Sandra Freedman

QUEBEC
Boulanger, Jean Baptiste
Cormier, Bruno M
Davis, John F
Dongier, Maurice Henri
Ervin, Frank (Raymond)
Kravitz, Henry
Lal, Samarthji
Lehmann, Heinz Edgar
Lundell, Frederick Waldemar
Minde, Karl Klaus
Nair, Vasavan N P
Palmour, Roberta Martin
Pihl, Robert O
Tan, Ah-Ti Chu
Tousignant, Michel
Villeneuve, Andre
Young, Simon N

SASKATCHEWAN
Coburn, Frank Emerson
McDonald, Ian MacLaren
Richardson, J(ohn) Steven
Smith, Colin McPherson

OTHER COUNTRIES
Belmaker, Robert Henry
Bogoch, Samuel
Hartocollis, Peter
Holman, Richard Bruce
Rutter, Michael L
Savage, Charles
Wettstein, Joseph G

Public Health & Epidemiology

ALABAMA
Bailey, William C
Bridgers, William Frank
Charles, Edgar Davidson, Jr
Habtemariam, Tsegaye
Heath, James Eugene
Hoff, Charles Jay
Holston, James L
Jamison, Homer Claude
King, William David
Myers, Ira Lee
Oberman, Albert
Palmisano, Paul Anthony
Pass, Robert Floyd
Peacock, Peter N B
Schnaper, Harold Warren
Schnurrenberger, Paul Robert
Windsor, Richard Anthony

ALASKA
Alter, Amos Joseph
Crook, James Richard
Segal, Bernard

ARIZONA
Aickin, Mikel G
Bennett, Peter Howard
Burton, Lloyd Edward
Dezelsky, Thomas Leroy
Fritz, Roy Fredolin
Goodwin, Melvin Harris, Jr
Hansen, Jo Ann Brown
Lebowitz, Michael David
Moon, Thomas Edward
Myers, Lloyd E(ldridge)
Nicolls, Ken E
Reinhard, Karl R
Rittenbaugh, Cheryl K
Rowe, Verald Keith
Schneller, Eugene S
Scott, David Bytovetzski

ARKANSAS
Brewster, Marjorie Ann
Weeks, Robert Joe

CALIFORNIA
Allen, LeRoy Richard
Aly, Raza
Anspaugh, Lynn Richard
Arthur, Susan Peterson
Ascher, Michael S
Austin, Donald Franklin
Barrett-Connor, Elizabeth L
Basch, Paul Frederick
Beck, Albert J
Belsky, Theodore
Benenson, Abram Salmon
Bernstein, Leslie
Black, Robert L
Blejer, Hector P
Blumberg, Mark Stuart
Brantingham, Charles Ross
Breslow, Lester
Brilliant, Lawrence Brent
Brook, Robert H
Buehring, Gertrude Case
Carlson, James Reynold
Chatigny, Mark A
Cohen, Kenneth Samuel
Constantine, Denny G
Conway, John Bell
Cooke, Kenneth Lloyd
Cosman, Bard Clifford
Cretin, Shan
Daniels, Jeffrey Irwin
Davenport, Calvin Armstrong
Deedwania, Prakash Chandra
Detels, Roger
Dewey, Kathryn G
Duffey, Paul Stephen
Duhl, Leonard J
Dunn, Frederick Lester
Emmons, Richard William
Enstrom, James Eugene
Fackler, Martin L
Fallon, Joseph Greenleaf
Forghani-Abkenari, Bagher
Franti, Charles Elmer
Friedman, Gary David
Garland, Cedric Frank
Gill, Ayesha Elenin
Gofman, John William
Gong, William C
Gray, Gregory Edward
Green, Lawrence Winter
Hall, Thomas Livingston
Hanlon, John Joseph
Hansen, Howard Edward
Harris, Mary Styles
Haughton, James Gray
Heyneman, Donald
Hill, Annlia Paganini
Hird, David William
Holder, Harold Douglas
Holman, Halsted Reid
Horwitz, Marcus Aaron
Hubbard, Eric R
Hubert, Helen Betty
Hunt, Isabelle F
James, Kenneth Eugene
Janda, John Michael
Jarvis, William Tyler
Jewell, Nicholas Patrick
King, Mary-Claire
Kissinger, David George
Kocol, Henry
Kraus, Jess F
Kritz-Silverstein, Donna
Lennette, Edwin Herman
Liskey, Nathan Eugene
Low, Hans
Mack, Thomas McCulloch
Maddy, Keith Thomas
Madison, Roberta Solomon
Magie, Allan Rupert
Merchant, Roland Samuel, Sr
Miller, Sol
Montgomery, Theodore Ashton
Moxley, John H, III
Nilsson, William A
Paffenbarger, Ralph Seal, Jr
Palmer, Alan
Plank, Stephen J
Portnoy, Bernard
Puntenney, Dee Gregory
Rappaport, Stephen Morris
Reeves, William Carlisle
Reisen, William Kenneth
Rice, Dorothy Pechman
Robinson, Henry William
Roemer, Milton Irwin
Room, Robin Gerald Walden
Salmon, Eli J
Sazama, Kathleen
Schienle, Jan Hoops
Schwabe, Calvin Walter
Scudder, Harvey Israel
Sheneman, Jack Marshall
Shonick, William
Spivey, Gary H
Swan, Shanna Helen
Tager, Ira Bruce
Thomas, Claudewell Sidney
Silverstein, Martin Elliot

Thompson, David A(lfred)
Volz, Michael George
Voulgaropoulos, Emmanuel
Wallace, Helen M
Wang, Virginia Li
Washino, Robert K
Welch, James Lee
Werdegar, David
Whittenberger, James Laverre
Whorton, M Donald
Wingard, Deborah Lee
Winkelstein, Warren, Jr
Wood, Corinne Shear
Work, Telford Hindley
Wright, Harry Tucker, Jr
Wycoff, Samuel John
Wyzga, Ronald Edward

COLORADO
Barnes, Ralph Craig
Bunn, William Bernice, III
Cada, Ronald Lee
Chase, Gerald Roy
Francy, David Bruce
Gubler, Duane J
Hine, Gerald John
Jacobs, Barbara B
Kemp, Graham Elmore
McLean, Robert George
Nora, Audrey Hart
Pearlman, Sholom
Poland, Jack Dean
Savage, Eldon P
Vaseen, V(esper) Albert
Vetterling, John Martin

CONNECTICUT
Aitken, Thomas Henry Gardiner
Bailit, Howard L
Bhatt, Pravin Nanabhai
Cipriano, Ramon John
Cohart, Edward Maurice
Conrad, Eugene Anthony
Curnen, Mary G McCrea
Darling, George Bapst
Evans, Alfred Spring
Fischer, Diana Bradbury
Freudenthal, Ralph Ira
Hinkle, Lawrence Earl, Jr
Honeyman, Merton Seymour
Horstmann, Dorothy Millicent
Jekel, James Franklin
Katz, Ralph Verne
Klerman, Lorraine Vogel
Klimek, Joseph John
Levine, Harvey Robert
Lieberman, James
Lloyd, Douglas Seward
Messing, Simon D
Murray, Francis Joseph
Niederman, James Corson
Novick, Alvin
Ostfeld, Adrian Michael
Roos, Henry
Silver, George Albert
Stitt, John Thomas
Stolwijk, Jan Adrianus Jozef
Sulzer-Azaroff, Beth
Tesh, Robert Bradfield
Viseltear, Arthur Jack

DELAWARE
Pell, Sidney
Pierce, Edward Ronald

DISTRICT OF COLUMBIA
Ahmed, Susan Wolofski
Banta, James E
Brandt, Carl David
Brignoli Gable, Carol
Briscoe, John
Bryant, Thomas Edward
Burris, James F
Chiazze, Leonard, Jr
Chopra, Joginder Gurbux
Daza, Carlos Hernan
Dublin, Thomas David
Dysinger, Paul William
Ellett, William H
Hamarneh, Sami Khalaf
Hanft, Ruth S
Howell, Embry Martin
Jacobsen, Frederick Marius
Jerome, Norge Winifred
Joly, Daniel Jose
Landau, Emanuel
Lieberman, Edwin James
MacArthur, Donald M
Miles, Corbin I
Millar, Jack William
Miller, William Robert
Minners, Howard Alyn
Novello, Antonia Coello
Olson, James Gordon
Pellegrino, Edmund Daniel
Pennington, Jean A T
Peterson, Malcolm Lee
Rao, Mamidanna S
Ross, Donald Morris
Scheele, Leonard Andrew
Sexton, Ken
Soldo, Beth Jean
Speidel, John Joseph
Thorne, Melvyn Charles

Vick, James

FLORIDA
Bevis, Herbert A(nderson)
Block, Irving H
Burridge, Michael John
Dougherty, Ralph C
Le Duc, James Wayne
Lyman, Gary Herbert
McGray, Robert James
Mercer, Thomas T
Mindlin, Rowland L
Orthoefer, John George
Plato, Phillip Alexander
Pletsch, Donald James
Porter, James Armer, Jr
Prather, Elbert Charlton
Reich, George Arthur
Rosenbloom, Arlan Lee
Sinkovics, Joseph G
Sorensen, Andrew Aaron
Telford, Sam Rountree, Jr
Young, Martin Dunaway
Zusman, Jack

GEORGIA
Amirtharajah, Appiah
Baer, George M
Baranowski, Tom
Barbaree, James Martin
Bernstein, Robert Steven
Blank, Carl Herbert
Brachman, Philip Sigmund
Broome, Claire V
Bryan, Frank Leon
Cavallaro, Joseph John
Curran, James W
Dean, Andrew Griswold
Dever, G E Alan
Dobrovolny, Charles George
Ellison, Lois Taylor
Falk, Henry
Foege, William Herbert
Francis, Donald Pinkston
French, Jean Gilvey
Gangarosa, Eugene J
Glass, Roger I M
Good, Robert Campbell
Gullen, Warren Hartley
Gunn, Walter Joseph
Harris, Elliott Stanley
Hatheway, Charles Louis
Hawkins, Theo M
Heath, Clark Wright, Jr
Herrmann, Kenneth L
Hogue, Carol Jane Rowland
Hollowell, Joseph Gurney, Jr
Hughes, James Mitchell
Johnson, Donald Ross
Jones, Gilda Lynn
Kagan, Irving George
Kaiser, Robert L
Kappus, Karl Daniel
Kaufmann, Arnold Francis
Kilbourne, Edwin Michael
Kutner, Michael Henry
LaMotte, Louis Cossitt, Jr
Lisella, Frank Scott
Margolis, Harold Stephen
Martin, Linda Spencer
Mathews, Henry Mabbett
Millar, John Donald
Morse, Stephen Allen
Nguyen-Dinh, Phuc
Oakley, Godfrey Porter, Jr
Orenstein, Walter Albert
Pratt, Harry Davis
Preblud, Stephen Robert
Reinhardt, Donald Joseph
Schantz, Peter Mullineaux
Schultz, Myron Gilbert
Sellers, Thomas F, Jr
Shotts, Emmett Booker, Jr
Sudia, William Daniel
Thacker, Stephen Brady
Trowbridge, Frederick Lindsley
Warren, McWilson

HAWAII
Char, Donald F B
Gilbert, Fred Ivan, Jr
Goodman, Madelene Joyce
Hankin, Jean H
Hartung, G(eorge) Harley
Hertlein, Fred, III
Lee, Richard K C
Marchette, Nyven John
Person, Donald Ames
Reed, Dwayne Milton
Rosen, Leon
Simeon, George John
Waslien, Carol Irene
Worth, Robert McAlpine

ILLINOIS
Brody, Jacob A
Carnow, Bertram Warren
Cho, Yongock
Conibear, Shirley Ann
Davis, Harvey Virgil
Doege, Theodore Charles
Druyan, Mary Ellen
Dunn, Dorothy Fay
Ellis, Effie O'Neal

Farber, Marilyn Diane
Flay, Brian Richard
Foreman, Ronald Louis
Gastwirth, Bart Wayne
Hall, Stephen Kenneth
Hallenbeck, William Hackett
Hirsch, Lawrence Leonard
Jacobson, Marvin
Levy, Leo
Lichter, Edward A
McFee, Donald Ray
Martin, Russell James
Martinek, Robert George
Masi, Alfonse Thomas
Moy, Richard Henry
Oleckno, William Anton
Rodin, Miriam Beth
Rosenblatt, Karin Ann
Rotkin, Isadore David
Slutsky, Herbert L
Stamler, Jeremiah
Stebbings, James Henry
Stoffer, Robert Llewellyn
Synovitz, Robert J
Winsberg, Gwynne Roeseler
Woods, George Theodore
Yogore, Mariano G, Jr

INDIANA
Bland, Hester Beth
Brown, Edwin Wilson, Jr
Glickman, Lawrence Theodore
Henderson, Hala Zawawi
Henzlik, Raymond Eugene
Hopper, Samuel Hersey
Hui, Siu Lui
Ivaturi, Rao Venkata Krishna
Jones, Russell K
Reed, Terry Eugene
Steinmetz, Charles Henry
Weinrich, Alan Jeffrey

IOWA
Beran, George Wesley
Berry, Clyde Marvin
Breuer, George Michael
Collins, Richard Francis
Colloton, John W
Cressie, Noel A C
Hausler, William John, Jr
Krafsur, Elliot Scoville
Remington, Richard Delleraine
Thiermann, Alejandro Bories

KANSAS
Chin, Tom Doon Yuen
Metzler, Dwight F
Neuberger, John Stephen
Smith, Meredith Ford
Thompson, Max Clyde
Tosh, Fred Eugene

KENTUCKY
Baumann, Robert Jay
Blackwell, Floyd Oris
Garrity, Thomas F
Hamburg, Joseph
Hochstrasser, Donald Lee
Hutchison, George B
Reed, Kenneth Paul
Stevens, Alan Douglas
Tamburro, Carlo Horace
Vandiviere, H Mac
Villarejos, Victor Moises
Weber, Frederick, Jr

LOUISIANA
Beaver, Paul Chester
Bradford, Henry Bernard, Jr
Diem, John Edwin
Ganley, James Powell
Gottlieb, Marise Suss
Hamrick, Joseph Thomas
Jung, Rodney Clifton
Kern, Clifford H, III
Oalmann, Margaret Claire
Scott, Harold George
Sodeman, William A
Soike, Kenneth Fieroe
Thompson, James Joseph
Trapido, Harold
Vaughn, John B
Webber, Larry Stanford

MAINE
Doran, Peter Cobb
Gezon, Horace Martin
Paigen, Beverly Joyce
Stoudt, Howard Webster

MARYLAND
Abramson, Fredric David
Alavanja, Michael Charles Robert
Alexander, Duane Frederick
Baker, Frank
Baker, Susan Pardee
Barker, Lewellys Franklin
Bawden, Monte Paul
Beebe, Gilbert Wheeler
Berendes, Heinz Werner
Blattner, William Albert
Boice, John D, Jr
Bricker, Jerome Gough
Buck, Alfred A

Burton, George Joseph
Calabi, Ornella
Cantor, Kenneth P
Carney, William Patrick
Chen, Tar Timothy
Chiacchierini, Richard Philip
Chisolm, James Julian, Jr
Chou, Nelson Shih-Toon
Cohen, Bernice Hirschhorn
Comstock, George Wills
Conley, Veronica Lucey
Cornely, Paul Bertau
Costa, Paul T, Jr
Cullen, Joseph Warren
Curlin, George Tams
Cutler, Jeffrey Alan
Depue, Robert Hemphill
Diamond, Earl Louis
Dodd, Roger Yates
Dubey, Jitender Prakash
Ducker, Thomas Barbee
Duggar, Benjamin Charles
Eaton, Barbra L
Edelman, Robert
Elkin, William Futter
Ensor, Phyllis Gail
Feldman, Jacob J
Ferencz, Charlotte
Fernie, Bruce Frank
Filner, Barbara
Finley, Joanne Elizabeth
Fisher, Pearl Davidowitz
Foster, Willis Roy
Fozard, James Leonard
Ganaway, James Rives
Garruto, Ralph Michael
Gelberg, Alan
Gibson, Sam Thompson
Gluckstein, Fritz Paul
Goldberg, Irving David
Goldschmidt, Peter Graham
Goldwater, William Henry
Gordis, Leon
Gori, Gio Batta
Greenhouse, Samuel William
Greenwald, Peter
Gruenberg, Ernest Matsner
Guthrie, Eugene Harding
Hairstone, Marcus A
Hallfrisch, Judith
Hamill, Peter Van Vechten
Harlan, William R, Jr
Harper, Paul Alva
Harris, Maureen Isabelle
Haynes, Suzanne G
Henderson, Donald Ainslie
Heyse, Stephen P
Higgins, Millicent Williams Payne
Hollingsworth, Charles Glenn
Horn, Susan Dadakis
Hoth, Daniel F
Hubbard, Donald
Hume, John Chandler
Jessup, Gordon L, Jr
Johnson, Mary Frances
Joly, Olga G
Kahn, Harold A
Kapikian, Albert Zaven
Kaslow, Richard Alan
Kawata, Kazuyoshi
Kessler, Irving Isar
Knapp, David Allan
Kolbye, Albert Christian, Jr
Krakauer, Henry
Kramer, Morton
Kroll, Bernard Hilton
Lao, Chang Sheng
Lawrence, Dale Nolan
Lee, Yuen San
Legters, Llewellyn J
Lemkau, Paul V
Levine, David Morris
Lewis, Irving James
Locke, Ben Zion
Lundin, Frank E, Jr
McCauley, H(enry) Berton
Malone, Winfred Francis
Matanoski, Genevieve M
Meredith, Orsell Montgomery
Michelson, Edward Harlan
Micozzi, Marc S
Miller, Robert Warwick
Mills, James Louis
Molenda, John R
Monath, Thomas P
Monjan, Andrew Arthur
Monk, Mary Alice
Moriyama, Iwao Milton
Morris, Joseph Anthony
Mościcki, Eve Karin
O'Beirne, Andrew Jon
Omran, Abdel Rahim
Ory, Marcia Gail
Osborne, J Scott, III
Osman, Jack Douglas
Page, Norbert Paul
Paige, David M
Parrish, Dale Wayne
Payne, William Walker
Phillips, Grace Briggs
Picciolo, Grace Lee
Pickle, Linda Williams
Plato, Chris C
Prabhakar, Bellur Subbanna

Price, Thomas R
Purcell, Robert Harry
Reichelderfer, Thomas Elmer
Reinke, William Andrew
Rhoads, George Grant
Roberts, Doris Emma
Rosenberg, Saul H
Rovelstad, Gordon Henry
Russell, Philip King
Sansone, Eric Brandfon
Saxinger, W(illiam) Carl
Schlueberg, Ann Elizabeth Snider
Schonfeld, Hyman Kolman
Schwartz, John T
Seagle, Edgar Franklin
Seltser, Raymond
Shapiro, Sam
Sharrett, A Richey
Shelokov, Alexis
Sherman, Kenneth Eliot
Shore, Moris Lawrence
Silverman, Charlotte
Sirken, Monroe Gilbert
Smith, Douglas Lee
Sommer, Alfred
Starfield, Barbara Helen
Stein, Harvey Philip
Steinwachs, Donald Michael
Stolley, Paul David
Strickland, George Thomas
Tarone, Robert Ernest
Tayback, Matthew
Taylor, Carl Ernest
Taylor, David Neely
Torres-Anjel, Manuel Jose
Vaught, Jimmie Barton
Vermund, Sten Halvor
Waalkes, T Phillip
Wachs, Melvin Walter
Wallace, Gordon Dean
Walter, Eugene LeRoy, Jr
Walter, William Arnold, Jr
Wong, Dennis Mun
Yanagihara, Richard
Yerby, Alonzo Smythe
Yodaiken, Ralph Emile
Yoder, Robert E
Young, Alvin L
Zelnik, Melvin
Zierdt, Charles Henry

MASSACHUSETTS
Barry, Joan
Berlandi, Francis Joseph
Bistrian, Bruce Ryan
Bures, Milan F
Castelli, William Peter
Cohen, Joel Ralph
Craven, Donald Edward
Czeisler, Charles Andrew
Darity, William Alexander
Dietz, William H
Dunning, James Morse
Eisenberg, Leon
Federman, Daniel D
Fineberg, Harvey V
First, Melvin William
Fiumara, Nicholas J
Frechette, Alfred Leo
Gentry, John Tilmon
George, Harvey
Gerety, Robert John
Goldberg, R J
Hammond, Sally Katharine
Hattis, Dale B
Hershberg, Philip I
Hiatt, Howard Haym
Ingalls, Theodore Hunt
Kosowsky, David I
Langmuir, Alexander D
Lavin, Philip Todd
Lemeshow, Stanley Alan
Leviton, Alan
Li, Frederick P
McCusker, Jane
MacMahon, Brian
Madoff, Morton A
Mata, Leonardo J
Newhouse, Joseph Paul
Ozonoff, David Michael
Pauker, Stephen Gary
Rand, William Medden
Ratney, Ronald Steven
Reich, James Harry
Roberts, Louis W
Robinton, Elizabeth Dorothy
Roubenoff, Ronenn
Rubenstein, Abraham Daniel
Sartwell, Philip Earl
Sawin, Clark Timothy
Scrimshaw, Nevin Stewart
Silverman, Gerald
Skrable, Kenneth William
Snyder, John Crayton
Speizer, Frank Erwin
Strong, Robert Michael
Telford, Sam Rountree, III
Thomas, Clayton Lay
Tsuang, Ming Tso
VanPraagh, Richard
Wegman, David Howe
Wyon, John Benjamin
Yankauer, Alfred

MICHIGAN
Adrounie, V Harry
Becker, Marshall Hilford
Cohen, Michael Alan
Craig, Winston John
Davenport, Fred M
Donabedian, Avedis
Guskey, Louis Ernest
Hawthorne, Victor Morrison
Higashi, Gene Isao
Higgins, Ian T
House, James Stephen
Humphrey, Harold Edward Burton, Jr
James, Sherman Athonia
Jensen, James Burt
Jordan, Brigitte
Kay, Bonnie Jean
Krebs, William H
Maassab, Hunein Fadlo
Macriss, Robert A
Mitchell, John Richard
Monto, Arnold Simon
Musch, David C(harles)
Mutch, Patricia Black
Paneth, Nigel Sefton
Powitz, Robert W
Rose, Gordon Wilson
Shea, Fredericka Palmer
Smith, Ralph G
Striffler, David Frank
Tilley, Barbara Claire
Viano, David Charles
Weg, John Gerard
Wegman, Myron Ezra
Wolman, Eric
Yates, Jon Arthur
Zemach, Rita
Zucker, Robert Alpert

MINNESOTA
Ackerman, Eugene
Barber, Donald E
Blackburn, Henry Webster, Jr
Diesch, Stanley L
Fan, David P
Fetcher, E(dwin) S(tanton)
Foreman, Harry
Gatewood, Lael Cranmer
Jacobs, David R, Jr
Kurland, Leonard T
Lawrenz, Frances Patricia
Levitan, Alexander Allen
Martens, Leslie Vernon
Melton, Lee Joseph, III
Mortimer, James Arthur
Roessler, Charles Ervin
Schuman, Leonard Michael
Thawley, David Gorden
Verby, John E

MISSISSIPPI
Bond, Marvin Thomas
Davis, Karen Padgett
Rundel, Robert Dean
Watson, Robert Lee

MISSOURI
Bagby, John R, Jr
Black, Wayne Edward
Blenden, Donald C
Chapman, Ramona Marie
Domke, Herbert Reuben
Donnell, Henry Denny, Jr
Kahrs, Robert F
Novak, Alfred
Perry, Horace Mitchell, Jr
Pierce, John Thomas
Spurrier, Elmer R
Weiss, James Moses Aaron
Wenner, Herbert Allan

MONTANA
Lackman, David Buell

NEBRASKA
Pearson, Paul (Hammond)
Thorson, James A

NEVADA
Dehné, Edward James

NEW HAMPSHIRE
Densen, Paul M
Hatch, Frederick Tasker
McCollum, Robert Wayne
Wennberg, John E
Zumbrunnen, Charles Edward

NEW JERSEY
Baron, Hazen Jay
Bendich, Adrianne
Bowns, Beverly Henry
Canzonier, Walter J
Cross, Richard James
Elinson, Jack
Hearn, Ruby Puryear
Iglewicz, Raja
Leventhal, Howard
Mallison, George Franklin
Millner, Elaine Stone
Quinones, Mark A
Saiger, George Lewis
Schreibeis, William J
Taylor, Bernard Franklin

Public Health & Epidemiology (cont)

Weiss, Robert Jerome
Weiss, Stanley H

NEW MEXICO
Bechtold, William Eric
Crawford, Michael Howard
Eller, Charles Howe
Ettinger, Harry Joseph
Key, Charles R
Schottstaedt, William Walter
Tonn, Robert James
Wong, Siu Gum
Wood, Gerry Odell

NEW YORK
Abraham, Jerrold L
Agnew-Marcelli, G(ladys) Marie
Alderman, Michael Harris
Allen, Ross Lorraine
Altman, Lawrence Kimball
Axelrod, David
Bellak, Leopold
Blane, Howard Thomas
Brightman, I Jay
Bromet, Evelyn J
Brophy, Mary O'Reilly
Bross, Irwin Dudley Jackson
Brown, Lawrence S, Jr
Bullough, Vern L
Carpenter, David O
Chance, Kenneth Bernard
Cherkasky, Martin
Cohen, Irving Allan
Cohen, Joel Ephraim
Colby, Frank Gerhardt
Collins, James Joseph
Cort, Susannah
Coutchie, Pamela Ann
Craig, John Philip
Cravitz, Leo
Davies, Jack Neville Phillips
Deibel, Rudolf
Dornbush, Rhea L
Ely, Thomas Sharpless
Engle, Ralph Landis, Jr
Felman, Yehudi M
Finkel, Madelon Lubin
Forquer, Sandra Lynne
Frankle, Reva Treelisky
Freitag, Julia Louise
Frishman, William Howard
Garfinkel, Lawrence
Geiger, H Jack
Gerber, Linda M
Gibofsky, Allan
Goldstein, Inge F
Greenhall, Arthur Merwin
Haas, Jere Douglas
Habicht, Jean-Pierre
Harley, Naomi Hallden
Harris, Raymond
Hauser, Willard Allen
Himes, John Harter
Hipp, Sally Sloan
Hook, Ernest
Imperato, Pascal James
Jahiel, Rene
Johnson, Kenneth Gerald
Keefe, Deborah Lynn
Kelsey, Jennifer Louise
Kilbourne, Edwin Dennis
Kim, Charles Wesley
Kline, Jennie Katherine
Kogel, Marcus David
Landrigan, Philip J
Langner, Thomas S
Latham, Michael Charles
Lawton, Richard Woodruff
Leverett, Dennis Hugh
Levin, Morton Loeb
Lichtig, Leo Kenneth
Lipson, Steven Mark
McCarthy, Eugene Gregory
McCarthy, James Francis
McCord, Colin Wallace
McHugh, William Dennis
Matuszek, John Michael, Jr
May, Paul S
Mike, Valerie
Millman, Robert Barnet
Mooney, Richard T
Moor-Jankowski, J
Moss, Leo D
Neugebauer, Richard
Oates, Richard Patrick
Ottman, Ruth
Pasamanick, Benjamin
Pasternack, Bernard Samuel
Pearson, Thomas Arthur
Pechacek, Terry Frank
Piore, Nora
Priore, Roger L
Rabino, Isaac
Rambaut, Paul Christopher
Rasmussen, Kathleen Maher
Reader, George Gordon
Reifler, Clifford Bruce
Richie, John Peter, Jr
Robinson, Harry
Rogatz, Peter

Roman, Stanford A, Jr
Schecter, Arnold Joel
Schupf, Nicole
Shapiro, Raymond E
Sheehe, Paul Robert
Shore, Roy E
Sidel, Victor William
Siegel, Morris
Slack, Nelson Hosking
Smith, Thomas Harry Francis
Soave, Rosemary
Socolar, Sidney Joseph
Solon, Leonard Raymond
Spencer, Frank
Spiegel, Allen David
Stein, Ruth E K
Stephenson, Lani Sue
Stevens, Roy White
Sturman, Lawrence Stuart
Susser, Mervyn W
Tanner, Martin Abba
Thompson, John C, Jr
Tichauer, Erwin Rudolph
Trowbridge, Richard Stuart
Vianna, Nicholas Joseph
Vivona, Stefano
Wassermann, Felix Emil
Weiner, Matei
Weinstein, Abbott Samson
Weisburger, John Hans
Weissman, Myrna Milgram
Whelan, Elizabeth M
Wing, Edward Joseph
Winikoff, Beverly
Wood, Paul Mix
Wray, Joe D
Wynder, Ernst Ludwig
Yates, Jerome William
Yolles, Tamarath Knigin
Zimmer, James Griffith

NORTH CAROLINA
Anderson, Richmond K
Bergsten, Jane Williams
Blazer, Dan German
Boatman, Ralph Henry, Jr
Bradley, William Robinson
Chapman, John Franklin, Jr
Cole, Jerome Foster
Costa, Daniel Louis
Coulter, Elizabeth Jackson
Decker, Clifford Earl, Jr
Dinse, Gregg Ernest
Fletcher, Robert Hillman
Fortney, Judith A
Fraser, David Allison
Freymann, Moye Wicks
Frothingham, Thomas Eliot
Gilfillan, Robert Frederick
Guess, Harry Adelbert
Hartzema, Abraham Gijsbert
Huang, Eng-Shang(Clark)
Hulka, Barbara Sorenson
Ibrahim, Michel A
LaFon, Stephen Woodrow
Levine, Richard Joseph
McClellan, Roger Orville
McPherson, Charles William
Marcus, Allan H
Miller, C Arden
Moggio, Mary Virginia
Radovsky, Frank Jay
Richardson, Stephen H
Rogan, Walter J
Rosenfeld, Leonard Sidney
Salber, Eva Juliet
Savitz, David Alan
Stern, A(rthur) C(ecil)
Suk, William Alfred
Udry, Joe Richard
Vakilzadeh, Javad
Weil, Thomas P
Willhoit, Donald Gillmor

NORTH DAKOTA
Klevay, Leslie Michael

OHIO
Alter, Joseph Dinsmore
Badger, George Franklin
Buncher, Charles Ralph
Burg, William Robert
Caldito, Gloria C
Caplan, Paul E
Clark, C Scott
Cook, Warren Ayer
Cooper, Dale A
Croft, Charles Clayton
Daniel, Thomas Mallon
Farrier, Noel John
Frazier, Todd Mearl
Friedland, Joan Martha
Geldreich, Edwin E(mery)
Gerson, Lowell Walter
Guo, Shumei
Hopps, Howard Carl
Houser, Harold Byron
Jalil, Mazhar
Johnson, Ronald Gene
Jones, Paul Kenneth
Keller, Martin David
Kowal, Norman Edward
McNamara, Michael Joseph
Moeschberger, Melvin Lee

Neff, Raymond Kenneth
Neuhauser, Duncan B
Neville, Janice Nelson
Powers, Jean D
Reiches, Nancy A
Ross, John Edward
Rushforth, Norman B
Siervogel, Roger M
Smith, Jerome Paul
Sprafka, Robert J

OKLAHOMA
Asal, Nabih Rafia
Caldwell, Glyn Gordon
Canter, Larry Wayne
Coleman, Ronald Leon
Folk, Earl Donald
Grubb, Alan S
Lynn, Thomas Neil, Jr
Shapiro, Stewart
Silberg, Stanley Louis
Smith, Vivian Sweibel
Walker, Bailus S
Wolgamott, Gary

OREGON
Anderson, Carl Leonard
Ary, Dennis
Buist, Aline Sonia
Elliker, Paul R
Greenlick, Merwyn Ronald
Horton, Aaron Wesley
Johnson, Larry Reidar
Morton, William Edwards
Osterud, Harold T
Rossignol, Philippe Albert

PENNSYLVANIA
Allen, Michael Thomas
Beilstein, Henry Richard
Benarde, Melvin Albert
Buskirk, Elsworth Robert
Cohen, Pinya
Cronk, Christine Elizabeth
Denenberg, Herbert S
Drew, Frances L
Ferrer, Jorge F
Fisher, Edward Allen
Fletcher, Suzanne W
Fraser, David William
Houts, Peter Stevens
Hsu, Chao Kuang
Iralu, Vichazelhu
Jungkind, Donald Lee
Kade, Charles Frederick, Jr
Katz, Solomon H
Kendall, Norman
Lacouture, Peter George
LaPorte, Ronald E
Levin, Simon Eugene
Levison, Matthew Edmund
Logue, James Nicholas
MacFarland, Harold Noble
McGhan, William Frederick
Martin, Samuel Preston, III
Mason, Thomas Joseph
Mattison, Donald Roger
Meadows, Anna T
Miller, G(erson) H(arry)
Morehouse, Chauncey Anderson
Morgan, James Frederick
Mulvihill, John Joseph
Osborne, William Wesley
Pinski, Gabriel
Prier, James Eddy
Rogers, Fred Baker
Rubin, Benjamin Arnold
Rycheck, Russell Rule
Salazar, Hernando
Shank, Kenneth Eugene
Shear, Charles L
Shoub, Earle Phelps
Singh, Balwant
Six, Howard R
Strom, Brian Leslie
Tepper, Lloyd Barton
Tokuhata, George K
Wald, Niel
Waldron, Ingrid Lore
Weiss, Kenneth Monrad
Weiss, William
Wells, Adoniram Judson

RHODE ISLAND
Jones, Kenneth Wayne

SOUTH CAROLINA
Gelfand, Henry Morris
Johnson, James Allen, Jr
Keil, Julian E
Lewis, Robert Frank
McCutcheon, Ernest P
Oswald, Edward Odell
Parker, Richard Langley
Richards, Gary Paul
Robson, John Robert Keith
Schuman, Stanley Harold
Thompson, Shirley Jean

SOUTH DAKOTA
Hodges. Neal Howard, II
Johnson, Carl J

TENNESSEE
Blaser, Martin Jack
Cullen, Marion Permilla
Cypess, Raymond Harold
Derryberry, Oscar Merton
Dupont, William Dudley
Federspiel, Charles Foster
Meeks, Robert G
New, John Coy, Jr
Scholtens, Robert George
Sloan, Frank A
Somes, Grant William
Zeighami, Elaine Ann

TEXAS
Alexander, Charles Edward, Jr
Baine, William Brennan
Benedict, Irvin J
Binnie, William Hugh
Bryant, Stephen G
Buffler, Patricia Ann Happ
Bushong, Stewart Carlyle
Chow, Rita K
Coelho, Anthony Mendes, Jr
Crandell, Robert Allen
Dahl, Elmer Vernon
Eppright, Margaret
Fultz, R Paul
Goldberg, Leonard H
Green, Hubert Gordon
Hanis, Craig L
Heidelbaugh, Norman Dale
Heimbach, Richard Dean
Hillis, Argye Briggs
Holguin, Alfonso Hudson
Hollinger, F(rederick) Blaine
Jenkins, Vernon Kelly
Kerr, George R
Knotts, Glenn Richard
Kusnetz, Howard L
Labarthe, Darwin Raymond
Lee, Eun Sul
LeMaistre, Charles Aubrey
McCloy, James Murl
Moore, Donald Vincent
Oseasohn, Robert
Park, Myung Kun
Parker, Roy Denver, Jr
Philips, Billy Ulyses
Pope, Robert Eugene
Shekelle, Richard Barten
Smith, Jerome H
Steele, James Harlan
Troxler, Raymond George
Valdivieso, Dario
Vallbona, Carlos
Vernon, Ralph Jackson
Ward, Jonathan Bishop, Jr
Wilcox, Roberta Arlene

UTAH
Archer, Victor Eugene
Ensign, Paul Roselle
Gortatowski, Melvin Jerome
Hunt, Steven Charles
Hunt, Steven Charles
Lee, Jeffrey Stephen
Melton, Arthur Richard
Moser, Royce, Jr
Nelson, Kenneth William

VERMONT
Babbott, Frank Lusk, Jr
Costanza, Michael Charles
Falk, Leslie Alan
Haugh, Larry Douglas
Terris, Milton
Waller, Julian Arnold

VIRGINIA
Attinger, Ernst Otto
Bempong, Maxwell Alexander
Brown, Robert Don
Burkhalter, Barton R
Ehrlich, S Paul, Jr
Emlet, Harry Elsworth, Jr
Fagan, Raymond
Fisher, Gail Feimster
Held, Joe R
Hilcken, John Allen
Jurinski, Neil B(ernard)
Kilpatrick, S James, Jr
Kurtzke, John F
Lawrence, Philip Signor
Lee, James A
McCormick-Ray, M Geraldine
Mahboubi, Ezzat
Maturi, Vincent Francis
Norman, James Everett, Jr
O'Brien, William M
Potts, Malcolm
Rodricks, Joseph Victor
Sheldon, Donald Russell
Sohler, Katherine Berridge
Spencer, Frederick J
Wands, Ralph Clinton
Wasti, Khizar
Weiss, William
White, Kerr Lachlan

WASHINGTON
Chrisman, Noel Judson
Day, Robert Winsor
Deyo, Richard Alden

Droppo, James Garnet, Jr
Emanuel, Irvin
Foy, Hjordis M
Gale, James Lyman
Grayston, J Thomas
Haviland, James West
Henderson, Maureen McGrath
Hurlich, Marshall Gerald
Johnstone, Donald Lee
Kathren, Ronald Laurence
Kenny, George Edward
Krieger, John Newton
Kuo, Cho-Chou
Lane, Wallace
Lee, John Alexander Hugh
Mason, Herman Charles
Omenn, Gilbert S
Perrin, Edward Burton
Peterson, Donald Richard
Ravenholt, Reimert Thorolf
Rosenblatt, Roger Alan
Siscovick, David Stuart
Thomas, David Bartlett
Valanis, Barbara Mayleas
Weiss, Noel Scott
Woods, James Sterrett
Yamanaka, William Kiyoshi

WEST VIRGINIA
Bouquot, Jerry Elmer
Sheikh, Kazim

WISCONSIN
Barboriak, Joseph Jan
Blockstein, William Leonard
Dick, Elliot C
Farley, Eugene Shedden, Jr
Golubjatnikov, Rjurik
Klein, Ronald
Ladinsky, Judith L
Laessig, Ronald Harold
Lundquist, Marjorie Ann
McIntosh, Elaine Nelson
Peterson, Jack Edwin
Southworth, Warren Hilbourne
Walker, Duard Lee

WYOMING
Eddy, David Maxon
Holbrook, Frederick R
Tabachnick, Walter J

PUERTO RICO
Brubaker, Merlin L

ALBERTA
Albritton, William Leonard
Dixon, John Michael Siddons
Larke, R(obert) P(eter) Bryce
Munan, Louis
Thompson, Gordon William

BRITISH COLUMBIA
Anderson, Donald Oliver
Baird, Patricia A
Bates, David Vincent
Davison, Allan John
Mackenzie, Cortlandt John Gordon
McLean, Donald Millis

MANITOBA
Choi, Nung Won
Maniar, Atish Chandra

NEWFOUNDLAND
Orr, Robin Denise Moore
Segovia, Jorge

NOVA SCOTIA
Rautaharju, Pentti M
Richman, Alex
Stewart, Chester Bryant

ONTARIO
Beaton, George Hector
Bertell, Rosalie
Chambers, Larry William
Currie, Violet Evadne
Gibson, Rosalind Susan
Glaser, Frederick Bernard
Hachinski, Vladimir C
Hewitt, David
Howe, Geoffrey Richard
Kraus, Arthur Samuel
Lang, Gerhard Herbert
Last, John Murray
Lees, Ronald Edward
Miller, Anthony Bernard
Rawls, William Edgar
Rigby, Charlotte Edith
Robertson, James McDonald
Roy, Marie Lessard
Silberfeld, Michel
Steele, Robert
Stemshorn, Barry William
Till, James Edgar
Todd, Ewen Cameron David
Tugwell, Peter
Wu, Alan SeeMing

QUEBEC
Aranda, Jacob Velasco
Bailar, John Christian, III
Becklake, Margaret Rigsby

Contandriopoulos, Andre-Pierre
Dugre, Robert
Frappier, Armand
McDonald, Alison Dunstan
Page, Michel
Siemiatycki, Jack
Spitzer, Walter O
Talbot, Pierre J
Thériault, Gilles P
Tousignant, Michel
Trepanier, Pierre
Vobecky, Josef

OTHER COUNTRIES
Aalund, Ole
Back, Kenneth Charles
Benatar, Solomon Robert
Bryant, John Harland
Buzina, Ratko
Cohen, Daniel
Goldsmith, John Rothchild
Greene, Velvl William
Haschke, Ferdinand
Herman, Bertram
Hoadley, Alfred Warner
Jacobson, Bertil
Kaplan, Martin Mark
Kessler, Alexander
Kourany, Miguel
Mahler, Halfdan
Mølhave, Lars
Shimaoka, Katsutaro
Shuval, Hillel Isaiah
Siu, Tsunpui Oswald
Tanphaichitr, Vichai
Thiessen, Jacob Willem
Torún, Benjamín
Webster, Isabella Margaret
Xintaras, Charles

Surgery

ALABAMA
Blakemore, William Stephen
Curreri, P William
El Dareer, Salah
French, John Douglas
Gleysteen, John Jacob
Hankes, Gerald H
Heath, James Eugene
Horne, Robert D
Kahn, Donald R
Kirklin, John W
Kurzweg, Frank Turner
Lamon, Eddie William
Lloyd, L Keith
McCallum, Charles Alexander, Jr
Rodning, Charles Bernard
Smith, Frederick Williams
White, Lowell Elmond, Jr

ALASKA
Mills, William J, Jr

ARIZONA
Du Val, Merlin Kearfott
Hale, Harry W, Jr
Holloway, G Allen, Jr
Kleitsch, William Philip
Malone, James Michael
Peltier, Leonard Francis
Peric-Knowlton, Wlatka
Schiller, William R
Silverstein, Martin Elliot
Stein, John Michael
Zukoski, Charles Frederick

ARKANSAS
Bard, David S
Boop, Warren Clark, Jr
Caldwell, Fred T, Jr
Campbell, Gilbert Sadler
McGilliard, A Dare
Read, Raymond Charles

CALIFORNIA
Adams, John Edwin
Akeson, Wayne Henry
Andrews, Neil Corbly
Bailey, Leonard Lee
Benfield, John R
Bernstein, Eugene F
Bernstein, Leslie
Black, Kirby Samuel
Boyne, Philip John
Bradford, David S
Branson, Bruce William
Brantingham, Charles Ross
Brody, Garry Sidney
Carter, Paul Richard
Carton, Charles Allan
Chase, Robert A
Chatterjee, Satya Narayan
Cherrick, Henry M
Connolly, John E
Convery, F Richard
Cooper, Howard K
Cosman, Bard Clifford
Crowley, Lawrence Grandjean
DeBas, Haile T
Dev, Parvati
Eilber, Frederick Richard
Fackler, Martin L

Foltz, Eldon Leroy
Fonkalsrud, Eric W
Frey, Charles Frederick
Furnas, David William
Gaspar, Max Raymond
Gittes, Ruben Foster
Gius, John Armes
Gortner, Susan Reichert
Griffin, David William
Grimes, Orville Frank
Hadley, Henry Lee
Halasz, Nicholas Alexis
Hallenbeck, George Aaron
Harrison, Michael R
Harvey, J Paul, Jr
Hays, Daniel Mauger
Heifetz, Milton David
Higgins, George A, Jr
Hinshaw, David B
Hubbard, Eric R
Huertas, Jorge
Hunt, Thomas Knight
Jamplis, Robert W
Jewett, Don L
Kaufman, Joseph J
Kaye, Michael Peter
Kirsch, Wolff M
Leake, Donald L
Lee, Sun
Lipscomb, Paul Rogers
Long, David Michael
Makowka, Leonard
Maloney, James Vincent, Jr
Moore, Wesley Sanford
Morton, Donald Lee
Murray, William R
Nelsen, Thomas Sloan
Newman, Melvin Micklin
Nickel, Vernon L
Oberhelman, Harry Alvin, Jr
Orloff, Marshall Jerome
Parker, Harold R
Perry, Jacquelin
Peters, Richard Morse
Porter, Robert Willis
Preston, Frederick Willard
Ray, Robert Durant
Richards, Victor
Roe, Benson Bertheau
Sadler, Charles Robinson, Jr
Sasaki, Gordon Hiroshi
Schwartz, Marshall Zane
Serkes, Kenneth Dean
Shoemaker, William C
Singh, Iqbal
Siposs, George G
Smith, Louis Livingston
State, David
Stemmer, Edward Alan
Stern, W Eugene
Street, Dana Morris
Sugar, Oscar
Thomas, Arthur Norman
Thompson, Ralph J, Jr
Tompkins, Ronald K
Trento, Alfredo
Trummer, Max Joseph
Turley, Kevin
Von Leden, Hans Victor
Wareham, Ellsworth Edwin
Way, Lawrence Wellesley
Weinreb, Robert Neal
Winet, Howard
Wise, Burton Louis
Wolfman, Earl Frank, Jr
Yamini, Sohrab
Youmans, Julian Ray
Zarem, Harvey A
Zawacki, Bruce Edwin

COLORADO
Boswick, John A
Eiseman, Ben
Jafek, Bruce William
Kindt, Glenn W
Lilly, John Russell
Moore, George Eugene
Paton, Bruce Calder
Pickrell, Kenneth LeRoy

CONNECTICUT
Baldwin, John Charles
Ballantyne, Garth H
Brown, Paul Woodrow
Cole, Jack Westley
Collins, William Francis, Jr
Dumont, Allan E
Foster, James Henry
Glenn, William Wallace Lumpkin
Goldberg, Morton Harold
Hayes, Mark Allan
Hlavacek, Robert Allen
Lewis, Jonathan Joseph
Lindskog, Gustaf Elmer
McKhann, Charles Fremont
Modlin, Irvin M
Owens, Guy
Pickett, Lawrence Kimball
Shichman, D(aniel)
Southwick, Wayne Orin
Taylor, Kenneth J W
Topazian, Richard G
Wright, Hastings Kemper

DISTRICT OF COLUMBIA
Adams, John Pletch
Balkissoon, Basdeo
Calhoun, Noah Robert
Cavanagh, Harrison Dwight
Depalma, Ralph G
Feller, William
Foegh, Marie Ladefoged
Geelhoed, Glenn William
Hardaway, Ernest, II
Harmon, John W
Hufnagel, Charles Anthony
Kobrine, Arthur
Koop, Charles Everett
Laws, Edward Raymond, Jr
Leffall, LaSalle Doheny, Jr
Luessenhop, Alfred John
Potter, John F
Randolph, Judson Graves
Sjogren, Robert W, Jr

FLORIDA
Barnett, William Oscar
Bland, Kirby Isaac
Bunch, Wilton Herbert
Chandler, J Ryan
Cohen, Marc Singman
Colahan, Patrick Timothy
Connar, Richard Grigsby
Daicoff, George Ronald
Dunbar, Howard Stanford
Evans, Arthur T
Finlayson, Birdwell
Finney, Roy Pelham
Garcia-Bengochea, Francisco
Habal, Mutaz
Hayward, James Rogers
Kaiser, Gerard Alan
Ketcham, Alfred Schutt
Klopp, Calvin Trexler
Legaspi, Adrian
Malinin, Theodore I
Orkin, Lazarus Allerton
Persky, Lester
Peters, Thomas G
Pfaff, William Wallace
ReMine, William Hervey
Rhoton, Albert Loren, Jr
Ryan, Robert F
Smyth, Nicholas Patrick Dillon
Steenburg, Richard Wesley
Stephenson, Samuel Edward, Jr
Sypert, George Walter
Talbert, James Lewis
Wangensteen, Stephen Lightner
Waterhouse, Keith R
Woodward, Edward Roy
Zeppa, Robert

GEORGIA
Allen, Marshall B, Jr
Bliven, Floyd E, Jr
Brown, Robert Lee
Child, Charles Gardner, III
Crowe, Dennis Timothy, Jr
Ellison, Robert G
Engler, Harold S
Humphries, Arthur Lee, Jr
Jurkiewicz, Maurice J
Ku, David Nelson
McGarity, William Cecil
Manganiello, Louis O J
Mansberger, Arlie Roland, Jr
Marble, Howard Bennett, Jr
Moretz, William Henry
Nguyen-Dinh, Phuc
Parrish, Robert A, Jr
Rawlings, Clarence Alvin
Sherman, Roger Talbot
Skandalakis, John Elias
Symbas, Panagiotis N
Tindall, George Taylor
Yancey, Asa G

HAWAII
Kokame, Glenn Megumi
Lance, Eugene Mitchell
McNamara, Joseph Judson
Mamiya, Richard T

ILLINOIS
Ausman, Robert K
Birtch, Alan Grant
Block, George E
Bombeck, Charles Thomas, III
Brizio-Molteni, Loredana
Choe, Byung-Kil
Choukas, Nicholas C
Ebert, Paul Allen
Ernest, J Terry
Ferguson, Donald John
Greenlee, Herbert Breckenridge
Griffith, B Herold
Hanlon, C Rollins
Harper, Paul Vincent
Hejna, William Frank
Hekmatpanah, Javad
Hines, James R
Keeley, John L
Kettelkamp, Donald B
Kittle, Charles Frederick
Krizek, Thomas Joseph
Langston, Hiram Thomas
Levitsky, Sidney

Surgery (cont)

Lopez-Majano, Vincent
Lurain, John Robert, III
Lyon, Edward Spafford
McKinney, Peter
Mackler, Saul Allen
Majarakis, James Demetrios
Manohar, Murli
Mason, George Robert
Merkel, Frederick Karl
Moss, Gerald S
Mrazek, Rudolph G
Mullan, John F
Nahrwold, David Lange
Nyhus, Lloyd Milton
Paloyan, Edward
Raimondi, Anthony John
Replogle, Robert Lee
Requarth, William H
Schumer, William
Seed, Randolph William
Shields, Thomas William
Spurrier, Wilma A
Stein, Irving F, Jr
Stromberg, LaWayne Roland
Tobin, Martin John
Watne, Alvin Lloyd
Wetzel, Nicholas
Whisler, Walter William
Wolf, James Stuart
Zintel, Harold Albert
Zsigmond, Elemer K

INDIANA

Borgens, Richard Ben
Campbell, Robert Louis
Fryer, Minot Packer
Fuson, Robert L
Grayson, Merrill
Kalsbeck, John Edward
King, Harold
Lempke, Robert Everett
Mandelbaum, Isidore
Manders, Karl Lee
Mealey, John, Jr
Shumacker, Harris B, Jr
Weirich, Walter Edward
Yune, Heun Yung

IOWA

Al-Jurf, Adel Saada
Bonfiglio, Michael
Cooper, Reginald Rudyard
Ehrenhaft, Johann L
Grier, Ronald Lee
Gurll, Nelson
Lawton, Richard L
McLeran, James Herbert
Mason, Edward Eaton
Merkley, David Frederick
Pearson, Phillip T
Perret, George (Edward)
Rossi, Nicholas Peter
Wass, Wallace M
Watkins, David Hyder

KANSAS

Clawson, David Kay
Friesen, Stanley Richard
Hardin, Creighton A
Holder, Thomas M
Jewell, William R
Masters, Frank Wynne
Nelson, Stanley Reid
Schloerb, Paul Richard
Valk, William Lowell

KENTUCKY

Brower, Thomas Dudley
Costich, Emmett Rand
Dillon, Marcus Lunsford, Jr
Griffen, Ward O, Jr
Lansing, Allan M
Moore, Condict
Ochsner, John Lockwood
Polk, Hiram Carey, Jr
Ram, Madhira Dasaradhi
Shih, Wei Jen
Spratt, John Stricklin
Tobin, Gordon Ross

LOUISIANA

Albert, Harold Marcus
Boyce, Frederick Fitzherbert
Clemetson, Charles Alan Blake
Cohn, Isidore, Jr
DeCamp, Paul Trumbull
Etheredge, Edward Ezekiel
Hewitt, Robert Lee
Kline, David G
Krementz, Edward Thomas
Litwin, Martin Stanley
McDonald, John C
McKay, Douglas William
McLaurin, James Walter
Moulder, Peter Vincent, Jr
Nance, Francis Carter
Parkins, Charles Warren
Petri, William Henry, III
Richardson, Donald Edward
Rosenberg, Dennis Melville Leo
Rosenthal, J William
Schramel, Robert Joseph

Walton, Thomas Peyton, III
Webb, Watts Rankin

MAINE

Bredenberg, Carl Eric
Fackelman, Gustave Edward
Hiatt, Robert Burritt
Swenson, Orvar

MARYLAND

Absolon, Karel B
Badder, Elliott Michael
Baker, Ralph Robinson
Barbul, Adrian
Bering, Edgar Andrew, Jr
Blair, Emil
Church, Lloyd Eugene
Colombani, Paul Michael
Djurickovic, Draginja Branko
Dubois, Andre T
Faich, Gerald Alan
Gann, Donald Stuart
Geisler, Fred Harden
Goodman, Louis E
Gott, Vincent Lynn
Handelsman, Jacob C
Hoopes, John Eugene
Hubbard, T Brannon, Jr
Javadpour, Nasser
Lee, Lyndon Edmund, Jr
Letterman, Gordon Sparks
Liang, Isabella Y S
Linehan, William Marston
Long, Donlin Martin
Matthews, Leslie Scott
Moran, Walter Harrison, Jr
Morrow, Andrew Glenn
Riley, Lee Hunter, Jr
Rob, Charles G
Robinson, David Weaver
Rosenberg, Steven A
Sachs, David Howard
Salcman, Michael
Schmeisser, Gerhard, Jr
Scovill, William Albert
Siegel, John H
Smith, Gardner Watkins
Sommer, Alfred
Udvarhelyi, George Bela
Walker, Michael Dirck
Woods, Alan Churchill, Jr
Young, Eric D
Zimmerman, Jack McKay

MASSACHUSETTS

Abbott, William M
Auchincloss, Hugh, Jr
Austen, W(illiam) Gerald
Bougas, James Andrew
Brooks, John Robinson
Brown, Henry
Burke, John Francis
Byrne, John Joseph
Cady, Blake
Castaneda, Aldo Ricardo
Cohen, Jonathan
Cohn, Lawrence H
Cope, Oliver
Deluca, Carlo J
Demling, Robert Hugh
Deterling, Ralph Alden, Jr
Doku, Hristo Chris
Duhamel, Raymond C
Egdahl, Richard H
Ellis, Franklin Henry, Jr
Fisher, John Herbert
Folkman, Moses Judah
Gaensler, Edward Arnold
Glimcher, Melvin Jacob
Glowacki, Julie
Goldsmith, Harry Sawyer
Hall, John Emmett
Harris, Melvyn H
Herrmann, John Bellows
Huggins, Charles Edward
Hume, Michael
Jackson, Benjamin T
Kosasky, Harold Jack
Lavine, Leroy S
Leach, Robert Ellis
Lele, Padmakar Pratap
Localio, S Arthur
McDermott, William Vincent, Jr
Malt, Ronald A
Meyer, Irving
Micheli, Lyle Joseph
Minsinger, William Elliot
Monaco, Anthony Peter
Moore, Francis Daniels
Murnane, Thomas William
Murray, Joseph E
O'Keefe, Dennis D
Patterson, William Bradford
Rheinlander, Harold F
Rousou, J A
Russell, Paul Snowden
Salzman, Edwin William
Schwartz, Anthony
Silen, William
Spellman, Mitchell Wright
Sweet, William Herbert
Tilney, Nicholas Lechmere
Turnquist, Carl Richard
Ulfelder, Howard

VanPraagh, Richard
Walter, Carl
Watkins, Elton, Jr
White, Augustus Aaron, III
Yablon, Isadore Gerald

MICHIGAN

Arbulu, Agustin
Banks, Henry H
Bivins, Brack Allen
Cerny, Joseph Charles
Coon, William Warner
Coran, Arnold Gerald
DeWeese, Marion Spencer
Diaz, Fernando G
Fitzgerald, Robert Hannon, Jr
Fromm, David
Fry, William James
Greenfield, Lazar John
Kantrowitz, Adrian
Keller, Waldo Frank
Large, Alfred McKee
Lindenauer, S Martin
Mahajan, Sudesh K
Moosman, Darvan Albert
Rosenberg, Jerry C
Senagore, Anthony J
Silbergleit, Allen
Silverman, Norman A
Sloan, Herbert
Smith, William S
Taren, James A
Thomas, Llywellyn Murray
Toledo-Pereyra, Luis Horacio
Turcotte, Jeremiah G
Walt, Alexander Jeffrey
Wilson, Robert Francis
Zuidema, George Dale

MINNESOTA

Buchwald, Henry
Chou, Shelley Nien-Chun
Coventry, Mark Bingham
Dai, Xue Zheng (Charlie)
Danielson, Gordon Kenneth
Delaney, John P
Doughman, Donald James
French, Lyle Albert
Gilbertsen, Victor Adolph
Godec, Ciril J
Grage, Theodor B
Hitchcock, Claude Raymond
Humphrey, Edward William
Hunter, Samuel W
Johnson, Einer Wesley, Jr
Kane, William J
Kelly, Patrick Joseph
Kelly, William Daniel
Leonard, Arnold S
Lofgren, Karl Adolph
Lovestedt, Stanley Almer
McCaffrey, Thomas Vincent
McQuarrie, Donald G
Najarian, John Sarkis
Nicoloff, Demetre M
Noer, Rudolf Juul
Perry, John Francis, Jr
Ray, Charles Dean
Rohlfing, Stephen Roy
Ryan, Allan James
Sako, Yoshio
Sullivan, W Albert, Jr
Wallace, Larry J

MISSISSIPPI

Andy, Orlando Joseph
Hardy, James D
Jones, Eric Wynn
McCaa, Connie Smith
Nelson, Norman Crooks
Wiser, Winfred Lavern

MISSOURI

Allen, William Corwin
Ballinger, Walter F, II
Barner, Hendrick Boyer
Butcher, Harvey Raymond, Jr
Callow, Allan Dana
Catalona, William John
Codd, John Edward
Goldring, Sidney
Hall, Lynn Raymond
Hershey, Falls Bacon
Kaiser, George C
Kaminski, Donald Leon
Moore, David Lee
Scherr, David DeLano
Shealy, Clyde Norman
Silver, Donald
Smith, Kenneth Rupert, Jr
Stephenson, Hugh Edward, Jr
Stoneman, William, III
Ternberg, Jessie L
Tritschler, Louis George
Weeks, Paul Martin
Wells, Samuel Alonzo, Jr
Willman, Vallee L

MONTANA

Butler, Hugh C

NEBRASKA

Feldhaus, Richard Joseph
Hamsa, William Rudolph

Hinder, Ronald Albert
Hodgson, Paul Edmund
Lynch, Benjamin Leo
McLaughlin, Charles William, Jr
Yonkers, Anthony J

NEVADA

Little, Alex G

NEW HAMPSHIRE

Crandell, Walter Bain
Kaiser, C William
Karl, Richard C
Naitove, Arthur
Ritzman, Thomas A
Rous, Stephen N

NEW JERSEY

Bregman, David
Brolin, Robert Edward
Fisher, Robert George
Gardner, Bernard
Hill, George James, II
Kaplan, Harry Arthur
Lazaro, Eric Joseph
Leyson, Jose Florante Justininane
Lord, Geoffrey Haverton
MacKenzie, James W
Martin, Boston Faust
Najjar, Talib A
Neville, William E
Parsonnet, Victor
Rush, Benjamin Franklin, Jr
Salerno, Alphonse
Schinagle, Erich F
Serlin, Oscar
Stefko, Paul Lowell
Swan, Kenneth G

NEW MEXICO

Barbo, Dorothy M
Davila, Julio C
Doberneck, Raymond C
Edwards, W Sterling
Omer, George Elbert, Jr

NEW YORK

Ackerman, Norman Bernard
Adler, Richard H
Bakay, Louis
Ballantyne, Donald Lindsay
Barie, Philip S
Barker, Harold Grant
Bassett, Charles Andrew Loockerman
Battista, Arthur Francis
Beattie, Edward J
Boley, Scott Jason
Borden, Edward B
Breed, Ernest Spencer
Brennan, Murray F
Burdick, Daniel
Chaikin, Lawrence
Clauss, Roy H
Cochran, George Van Brunt
Cochran, George Van Brunt
Cook, Albert William
Day, Stacey Biswas
Del Guercio, Louis Richard M
Dennis, Clarence
DeWeese, James A
Dreiling, David A
Dutta, Purnendu
Eckert, Charles
Eisenberg, M Michael
Enquist, Irving Fritiof
Everhard, Martin Edward
Fein, Jack M
Feinman, Max L
Fortner, Joseph Gerald
Frankel, Victor H
Frater, Robert William Mayo
Friedmann, Paul
Furman, Seymour
Gage, Andrew Arthur
Gerst, Paul Howard
Gliedman, Marvin L
Goldstein, Louis Arnold
Golomb, Frederick M
Goulian, Dicran
Greengard, Olga
Grossi, Carlo E
Gump, Frank E
Gumport, Stephen Lawrence
Harris, Matthew N
Harvin, James Shand
Herter, Frederic P
Hinshaw, J Raymond
Holman, Cranston William
Housepian, Edgar M
Hugo, Norman Eliot
Imparato, Anthony Michael
Jacobs, Richard L
King, Robert (Bainton)
King, Thomas Creighton
Kinney, John Martin
Kottmeier, Peter Klaus
Landor, John Henry
Laufman, Harold
Lawson, William
Levenson, Stanley Melvin
Lilien, Otto Michael
Linkow, Leonard I
Litwak, Robert Seymour
Lord, Jere Williams, Jr

Lore, John M, Jr
Lourie, Herbert
Lowe, John Edward
Lowry, Stephen Frederick
McSherry, Charles K
Madden, Robert E
Malis, Leonard I
Menguy, Rene
Metcalf, William
Mindell, Eugene R
Mittelman, Arnold
Montefusco, Cheryl Marie
Morton, John Henderson
Nealon, Thomas F, Jr
Nelson, Curtis Norman
Nemoto, Takuma
Nicholas, James A
Olsson, Carl Alfred
Pankovich, Arsen M
Patterson, Russel Hugo, Jr
Peirce, Edmund Converse, II
Rabinowitz, Ronald
Ragins, Herzl
Ransohoff, Joseph
Rapaport, Felix Theodosius
Redo, Saverio Frank
Reemtsma, Keith
Riehle, Robert Arthur, Jr
Rogers, Lloyd Sloan
Rohman, Michael
Rosenfeld, Isadore
Rovit, Richard Lee
Sako, Kumao
Santulli, Thomas V
Sawyer, Philip Nicholas
Schenk, Worthington G, Jr
Schimert, George
Schlesinger, Edward Bruce
Schwartz, Seymour I
Shaftan, Gerald Wittes
Shedd, Donald Pomroy
Siffert, Robert S
Skinner, David Bernt
Slattery, Louis R
Smith, Donald Frederick
Soren, Arnold
Soroff, Harry S
Spotnitz, Henry Michael
Stark, Richard B
Stein, Bennett M
Stein, Theodore Anthony
Stern, Martin
Stinchfield, Frank E
Tice, David Anthony
Torre, Douglas Paul
Van De Water, Joseph M
Veith, Frank James
Waltzer, Wayne C
Whitsell, John Crawford, II
Wilder, Joseph R
Young, Morris Nathan

NORTH CAROLINA
Anlyan, William George
Baker, Lenox Dial
Bell, William Harrison
Bevin, A Griswold
Bollinger, Ralph R
Buckwalter, Joseph Addison
Bucy, Paul Clancy
Caruolo, Edward Vitangelo
Clippinger, Frank Warren, Jr
Eifrig, David Eric
Fuchs, James Claiborne Allred
Georgiade, Nicholas George
Goldner, Joseph Leonard
Hamit, Harold F
Hayes, John Terrence
Hightower, Felda
Howard, Donald Robert
Hudson, William Rucker
Johnson, George, Jr
Klitzman, Bruce
Lowe, James Edward
Lyerly, Herbert Kim
McCollum, Donald E
Machemer, Robert
Meredith, Jesse Hedgepeth
Nashold, Blaine S
Odom, Guy Leary
Peacock, Erle Ewart
Peete, William P J
Pennell, Timothy Clinard
Pories, Walter J
Postlethwait, Raymond Woodrow
Raney, Richard Beverly
Sabiston, David Coston, Jr
Scott, Stewart Melvin
Seigler, Hilliard Foster
Stickel, Delford LeFew
Taylor, B Gray
Thomas, Colin Gordon, Jr
Thomas, Francis T
Watts, Charles D
White, Raymond Petrie, Jr
Wilcox, Benson Reid
Wilson, Frank Crane
Young, William Glenn, Jr

OHIO
Alexander, James Wesley
Berggren, Ronald B
Brodkey, Jerald Steven
Bryant, Lester Richard

Carey, Larry Campbell
Clifford, Donald H
DeWall, Richard A
Dobyns, Brown M
Elliott, Dan Whitacre
Ferguson, Ronald M
Gabel, Albert A
Geha, Alexander Salim
Gonzalez, Luis L
Hanto, Douglas W
Hasselgren, Per-Olof J
Helmsworth, James Alexander
Herndon, Charles Harbison
Holden, William Douglas
Howard, John Malone
Hubay, Charles Alfred
Hull, Bruce Lansing
Hummel, Robert P
Hunt, William Edward
Husni, Elias A
Izant, Robert James, Jr
Jackson, Robert Henry
Jelenko, Carl, III
Joffe, Stephen N
Kilman, James William
McCauley, John Corran, Jr
McLaurin, Robert L
McQuarrie, Irvine Gray
Martin, Lester W
Montague, Drogo K
Pace, William Greenville
Peterson, Larry James
Rayport, Mark
Redman, Donald Roger
Resnick, Martin I
Rogers, Richard C
Rudy, Richard L
Shons, Alan R
Stevenson, Jean Moorhead
Szczepanski, Marek Michal
Tucker, Harvey Michael
Vasko, John Stephen
Vogel, Thomas Timothy
White, Robert J
Wilson, George Porter, III
Yashon, David

OKLAHOMA
Elkins, Ronald C
Miner, Gary David
Pollay, Michael
Williams, George Rainey

OREGON
Beals, Rodney K
Campbell, John Richard
DeWeese, David D
Fletcher, William Sigourney
Fry, Louis Rummel
Gallo, Anthony Edward, Jr
Hill, John Donald
Jacob, Stanley W
Krippaehne, William W
Peterson, Clare Gray
Starr, Albert

PENNSYLVANIA
Barker, Clyde Frederick
Beck, William Carl
Bell, Robert Lloyd
Beller, Martin Leonard
Black, Perry
Brighton, Carl T
Conger, Kyril Bailey
Cooper, Donald Russell
Daly, John M
Davis, Richard A
Deitrick, John E
Donawick, William Joseph
Edmunds, Louis Henry, Jr
Eggleston, Forrest Cary
Ferguson, Albert Barnett
Fineberg, Charles
Fisher, Bernard
Futrell, J William
Gee, William
Grenvik, Ake N A
Grotzinger, Paul John
Hamilton, Ralph West
Harrison, Timothy Stone
Jannetta, Peter Joseph
Kao, Race Li-Chan
Kurze, Theodore
Laing, Patrick Gowans
Langfitt, Thomas William
Laufer, Igor
Lotze, Michael T
Magargal, Larry Elliot
Magovern, George J
Miller, Leonard David
Moody, Robert Adams
Mullen, James L
Murphy, John Joseph
Nemir, Paul, Jr
O'Neill, James A, Jr
Pierce, William Schuler
Pirone, Louis Anthony
Raker, Charles W
Ramasastry, Sai Sudarshan
Randall, Peter
Ravitch, Mark Mitchell
Reichle, Frederick Adolph
Rhoads, Jonathan Evans
Ritchie, Wallace Parks, Jr

Rombeau, John Lee
Rosemond, George P
Rothman, Richard Harrison
Salmon, James Henry
Shenkin, Henry A
Sigel, Bernard
Simmons, Richard Lawrence
Spatz, Sidney S
Starzl, Thomas E
Steel, Howard Haldeman
Stevens, Lloyd Weakley
Templeton, John Y, III
Tyson, Ralph Robert
Waldhausen, John Anton
Wallace, Herbert William
Wilberger, James Eldridge
Wolfson, Sidney Kenneth, Jr

RHODE ISLAND
Caldwell, Michael D
Greenblatt, Samuel Harold
Hopkins, Robert West
Karlson, Karl Eugene
Randall, Henry Thomas
Simeone, Fiorindo Anthony
Vezeridis, Michael Panagiotis

SOUTH CAROLINA
Anderson, Marion C
Bradham, Gilbert Bowman
Fitts, Charles Thomas
Greene, Frederick Leslie
Horger, Edgar Olin, III
Kempe, Ludwig George
Kolb, Leonard H
LeVeen, Harry Henry
Othersen, Henry Biemann, Jr
Perot, Phanor L, Jr
Putney, Floyd Johnson
Rittenbury, Max Sanford
Waldrep, Alfred Carson, Jr
Weidner, Michael George, Jr

TENNESSEE
Benson, Ralph C, Jr
Benz, Edmund Woodward
Byrd, Benjamin Franklin, Jr
Cox, Clair Edward, II
Dale, William Andrew
Dooley, Wallace T
Feman, Stephen Sosin
Foster, John Hoskins
Giaroli, John Nello
Hall, Hugh David
Hansen, Axel C
Hendrix, James Harvey, Jr
Ingram, Alvin John
Lynch, John Brown
McCoy, Sue
Meacham, William Feland
Merrill, Walter Hilson
Nag, Subir
Perry, Frank Anthony
Reynolds, Vernon H
Robertson, James Thomas
Rosensweig, Jacob
Sawyers, John Lazelle
Scott, Henry William, Jr
Wyler, Allen Raymer

TEXAS
Aust, J Bradley
Balagura, Saul
Balch, Charles Mitchell
Baxter, Charles Rufus
Beall, Arthur Charles, Jr
Bobechko, Walter Peter
Burdette, Walter James
Clark, William Kemp
Conti, Vincent R
Cooley, Denton Arthur
Corriere, Joseph N, Jr
Crawford, Duane Austin
Cruz, Anatolio Benedicto, Jr
DeBakey, Michael Ellis
Derrick, John Rafter
Dudrick, Stanley John
El-Domeiri, Ali A H
Ellett, Edwin Willard
Feola, Mario
Flatt, Adrian Ede
Franklin, Thomas Doyal, Jr
Gildenberg, Philip Leon
Goldberg, Leonard H
Grossman, Robert G
Haley, Harold Bernard
Hall, Charles William
Hardaway, Robert M, III
Hartman, Albert William
Hartman, James T
Hecht, Gerald
Herbert, Morley Allen
Herndon, David N
Hinds, Edward C
Jackson, Francis Charles
Jennings, Paul Bernard, Jr
Jordan, George Lyman, Jr
Jordan, Paul H, Jr
Kahan, Barry D
Laros, Gerald Snyder, II
Lewis, Stephen Robert
Lotzová, Eva
McClelland, Robert Nelson
McFee, Arthur Storer

May, Sterling Randolph
Moody, Frank Gordon
Morris, George Cooper, Jr
Mountain, Clifton Fletcher
Musgrave, F Story
Myers, Paul Walter
Paulson, Donald Lowell
Pestana, Carlos
Peters, Paul Conrad
Playter, Robert Franklin
Poth, Edgar J
Pruitt, Basil Arthur, Jr
Radwin, Howard Martin
Rogers, Waid
Romsdahl, Marvin Magnus
Root, Harlan D
Roth, Jack A
Sakakini, Joseph, Jr
Shires, George Thomas
Sparkman, Robert Satterfield
Spira, Melvin
Story, Jim Lewis
Thompson, James Charles
Thompson, Jesse Eldon
Tyner, George S
Waite, Daniel Elmer
Wolma, Fred J
Yollick, Bernard Lawrence

UTAH
Albo, Dominic, Jr
Anderson, Richard Lee
Castleton, Kenneth Bitner
Dixon, John Aldous
Kwan-Gett, Clifford Stanley
Maxwell, John Gary
Nelson, Russell Marion
Olsen, Don B
Olson, Randall J
Parkin, James Lamar
Snyder, Clifford Charles
Windham, Carol Thompson
Wolcott, Mark Walton

VERMONT
Coffin, Laurence Haines
Donaghy, Raymond Madiford Peardon
Foster, Roger Sherman, Jr

VIRGINIA
Adamson, Jerome Eugene
Cohen, I Kelman
Dammann, John Francis
Devine, Charles Joseph, Jr
Edgerton, Milton Thomas, Jr
Elmore, Stanley Mcdowell
Hayes, George J
Haynes, Boyd W, Jr
Horsley, John Shelton, III
Hurt, Waverly Glenn
Jane, John Anthony
Laskin, Daniel M
Lawrence, Walter, Jr
Lee, Hyung Mo
Lefrak, Edward Arthur
Lower, Richard Rowland
Lynn, Hugh Bailey
Muller, William Henry, Jr
Nolan, Stanton Peelle
Rodgers, Bradley Moreland
Rudolf, Leslie E

WASHINGTON
Ansell, Julian Samuel
Bornstein, Paul
Brockenbrough, Edwin C
Dennis, Melvin B, Jr
Diliard, David Hugh
Edwards, James Mark
Gehrig, John D
Heimbach, David M
Hutchinson, William Burke
Kalina, Robert E
Karagianes, Manuel Tom
Kelly, William Albert
Kenney, Gerald
Krieger, John Newton
Loeser, John David
Malette, William Graham
Marchioro, Thomas Louis
Mitchell, Michael Ernst
Roberts, Theodore S
Schilling, John Albert
Strandness, Donald Eugene, Jr
Ward, Arthur Allen, Jr
Westrum, Lesnick Edward
White, Thomas Taylor
Winterscheid, Loren Covart

WEST VIRGINIA
Boland, James P
Malin, Howard Gerald
Nugent, George Robert
Warden, Herbert Edgar
Zimmermann, Bernard

WISCONSIN
Belzer, Folkert O
Condon, Robert Edward
Donegan, William L
Gingrass, Ruedi Peter
Javid, Manucher J
Longley, B Jack
MacCarty, Collin Stewart

Surgery (cont)

Meyer, Glenn Arthur
Mohs, Frederic Edward
Schulte, William John, Jr
Southwick, Harry W
Whiffen, James Douglass
Wolberg, William Harvey
Wussow, George C
Yale, Charles E
Young, William Paul

PUERTO RICO
Korchin, Leo

ALBERTA
Halloran, Philip Francis
Lakey, William Hall
Salmon, Peter Alexander
Scott, Gerald William

BRITISH COLUMBIA
Cleator, Iain Morrison
Harrison, Robert Cameron

MANITOBA
Ferguson, Colin C
Klass, Alan Arnold

NOVA SCOTIA
Perey, Bernard Jean Francois

ONTARIO
Baird, Ronald James
Bayliss, Colin Edward
Carroll, Samuel Edwin
Cooke, T Derek V
Deitel, Mervyn
Downie, Harry G
Evans, Geoffrey
Farkas, Leslie Gabriel
Ferguson, Gary Gilbert
Ghent, William Robert
Gurd, Fraser Newman
Gutelius, John Robert
Harwood-Nash, Derek Clive
Heimbecker, Raymond Oliver
Keon, Wilbert Joseph
Lynn, Ralph Beverley
Macbeth, Robert Alexander
McCorriston, James Roland
McCredie, John A
Morley, Thomas Paterson
Pang, Cho Yat
Rappaport, Aron M
Saleh, Wasfy Seleman
Salter, Robert Bruce
Sirek, Anna
Sistek, Vladimir
Skoryna, Stanley C
Tasker, Ronald Reginald
Tator, Charles Haskell
Zingg, Walter

QUEBEC
Bentley, Kenneth Chessar
Chiu, Chu Jeng (Ray)
Collin, Pierre-Paul
Cruess, Richard Leigh
De Margerie, Jean-Marie
Entin, Martin A
Fazekas, Arpad Gyula
Feindel, William Howard
Gagnon, Claude
Guttman, Frank Myron
MacLean, Lloyd Douglas
Mortola, Jacopo Prospero
Olivier, Andre
Shizgal, Harry M
Stratford, Joseph
Yamamoto, Y Lucas

SASKATCHEWAN
Paine, Kenneth William
Shokeir, Mohamed Hassan Kamel

OTHER COUNTRIES
Fujimoto, Shigeyoshi
Häyry, Pekka J
Hong, Pill Whoon
Kennedy, John Hines
Moscovici, Mauricio
Poswillo, D E
Shiraki, Keizo
Yovich, John V

Veterinary Medicine

ALABAMA
Anderson, Peter Glennie
Bishop, Sanford Parsons
Clark, Carl Heritage
Cloyd, Grover David
Dalvi, Ramesh R
El Dareer, Salah
Farnell, Daniel Reese
Gray, Bruce William
Griswold, Daniel Pratt, Jr
Habtemariam, Tsegaye
Hankes, Gerald H
Heath, James Eugene
Horne, Robert D
Klesius, Phillip Harry

Knecht, Charles Daniel
Laster, William Russell, Jr
Lauerman, Lloyd Herman, Jr
Powers, Robert D
Schnurrenberger, Paul Robert
Sharma, Udhishtra Deva
Siddique, Irtaza H
Smith, Paul Clay
Vaughan, John Thomas
Walker, Donald F
Williams, Raymond Crawford
Williams, Theodore Shields

ARIZONA
Bjotvedt, George
Bone, Jesse Franklin
Chalquest, Richard Ross
Fisher, William Gary
Herrick, John Berne
Mare, Cornelius John
Mulder, John Bastian
Shively, James Nelson
Shull, James Jay

ARKANSAS
Skeeles, John Kirkpatrick

CALIFORNIA
Adams, Thomas Edwards
Altera, Kenneth P
Bailey, Cleta Sue
Baker, Norman Fletcher
Bankowski, Raymond Adam
Bernoco, Domenico
Bidlack, Donald Eugene
Bissell, Glenn Daniel
Braemer, Allen C
Campbell, Kirby I
Cappucci, Dario Ted, Jr
Constantine, Denny G
Fowler, Murray Elwood
Giri, Shri N
Griffin, David William
Hamm, Thomas Edward, Jr
Hird, David William
Holmberg, Charles Arthur
Hughes, John P
Hyde, Dallas Melvin
Jasper, Donald Edward
Jones, James Henry
Lewis, Nina Alissa
McGowan, Blaine, Jr
Maddy, Keith Thomas
Morgan, Joe Peter
Muller, George Heinz
Parker, Harold R
Plymale, Harry Hambleton
Pritchard, William Roy
Rankin, Alexander Donald
Rhode, Edward A, Jr
Ridgway, Sam H
Riemann, Hans
Schwabe, Calvin Walter
Siegel, Edward T
Siegrist, Jacob C
Sinha, Yagya Nand
Smithcors, James Frederick
Steffey, Eugene P
Theilen, Gordon H
Yamamoto, Richard
Young, Robert, Jr
Zinkl, Joseph Grandjean

COLORADO
Allen, Larry Milton
Barber, Thomas Lynwood
Benjamin, Stephen Alfred
Bennett, Dwight G, Jr
Cockerell, Gary Lee
De Martini, James Charles
Gillette, Edward LeRoy
Glock, Robert Dean
Heath, Robert Bruce
Hoerlein, Alvin Bernard
Keefe, Thomas J
Kemp, Graham Elmore
Kingman, Harry Ellis, Jr
Larson, Kenneth Allen
Lebel, Jack Lucien
Lee, Arthur Clair
Lumb, William Valjean
Park, Richard Dee
Ranu, Rajinder S
Smith, Ralph E
Tucker, Alan
Vetterling, John Martin

CONNECTICUT
Holzworth, Jean
Kersting, Edwin Joseph
Nielsen, Svend Woge
Post, John E
Schaaf, Thomas Ken
Yang, Tsu-Ju (Thomas)

DELAWARE
Dohms, John Edward
Reed, Theodore H(arold)

DISTRICT OF COLUMBIA
Carlson, William Dwight
Crawford, Lester M
Dua, Prem Nath
Gorham, John Richard

Miller, William Robert
Wolfe, Thomas Lee
Zook, Bernard Charles

FLORIDA
Buss, Daryl Dean
Campbell, Clarence L, Jr
Carter, James Harrison, II
Clemmons, Roger M
Colahan, Patrick Timothy
Cooperrider, Donald Elmer
Courtney, Charles Hill
Decker, Winston M
Donovan, Edward Francis
Drummond, Willa H
Grafton, Thurman Stanford
Merritt, Alfred M, II
Moreland, Alvin Franklin
Murchison, Thomas Edgar
Orthoefer, John George
Porter, James Armer, Jr
Povar, Morris Leon
Samuelson, Don Arthur

GEORGIA
Brackett, Benjamin Gaylord
Brugh, Max, Jr
Clark, James Derrell
Cornelius, Larry Max
Crowe, Dennis Timothy, Jr
Edwards, Gaylen Lee
Finco, Delmar R
Hatch, Roger Conant
Jain, Anant Vir
Jones, Jimmy Barthel
Kaplan, William
Kaufmann, Arnold Francis
McClure, Harold Monroe
McMurray, Birch Lee
Oliver, John Eoff, Jr
Patterson, William Creigh, Jr
Ragland, William Lauman, III
Rawlings, Clarence Alvin
Roberson, Edward Lee
Schantz, Peter Mullineaux
Shotts, Emmett Booker, Jr
Trim, Cynthia Mary
Williams, David John, III

HAWAII
DeTray, Donald Ervin

IDAHO
Frank, Floyd William

ILLINOIS
Bennett, Basil Taylor
Brodie, Bruce Orr
Cera, Lee M
Davis, Lloyd Edward
Ferguson, James L
Flynn, Robert James
Ford, Thomas Matthews
Freeman, Arthur
Fritz, Thomas Edward
Gordon, Donovan
Hannah, Harold Winford
Hanson, Lyle Eugene
Hillidge, Christopher James
Khan, Mohammed Nasrullah
Koritz, Gary Duane
Manohar, Murli
Nath, Nrapendra
Patterson, Douglas Reid
Port, Curtis DeWitt
Reynolds, Harry Aaron, Jr
Ristic, Miodrag
Schiller, Alfred George
Small, Erwin
Tekeli, Sait
Thurmon, John C
Tripathy, Deoki Nandan
Upham, Roy Walter
Wagner, William Charles
Whitmore, Howard Lloyd
Woods, George Theodore

INDIANA
Amstutz, Harold Emerson
Babcock, William Edward
Brush, F(ranklin) Robert
Gale, Charles
Jones, Russell K
Kirkham, Wayne Wolpert
Matsuoka, Tats
Morter, Raymond Lione
Neher, George Martin
Pollard, Morris
Stump, John Edward
Weirich, Walter Edward
Williams, Robert Dee

IOWA
Adams, Donald Robert
Baker, Durwood L
Beran, George Wesley
Boney, William Arthur, Jr
Christensen, George Curtis
Dougherty, Robert Watson
Grier, Ronald Lee
Hughes, David Edward
Jackson, Larry LaVern
Jones, Rex H
Kluge, John Paul

Kohn, Frank S
Leman, Allen Duane
Lundvall, Richard
McClurkin, Arlan Wilbur
Merkley, David Frederick
Miller, Janice Margaret Lilly
O'Berry, Phillip Aaron
Packer, R(aymond) Allen
Pearson, Phillip T
Roth, James A
Sweat, Robert Lee
Thiermann, Alejandro Bories
Van Houweling, Cornelius Donald
Wass, Wallace M
Wood, Richard Lee

KANSAS
Anderson, Neil Vincent
Anthony, Harry D
Burch, Clark Wayne
Erickson, Howard Hugh
Gray, Andrew P
Kelley, Donald Clifford
Kennedy, George Arlie
Minocha, Harish C
Mosier, Jacob Eugene
Oehme, Frederick Wolfgang
Stevens, David Robert
Upson, Dan W

KENTUCKY
Gochenour, William Sylva
Gupta, Ramesh C
Olds, Durward
Stevens, Alan Douglas
Tobin, Thomas

LOUISIANA
Beadle, Ralph Eugene
Huxsoll, David Leslie
Stewart, T Bonner

MAINE
Donahoe, John Philip
Fackelman, Gustave Edward
Haynes, N Bruce
Myers, David Daniel

MARYLAND
Bingham, Gene Austin
Chou, Nelson Shih-Toon
Das, Naba Kishore
Dubey, Jitender Prakash
Earl, Francis Lee
Evarts, Ritva Poukka
Ganaway, James Rives
Gluckstein, Fritz Paul
Golway, Peter L
Greenstein, Edward Theodore
Groopman, John Davis
Hildebrandt, Paul Knud
Holman, John Ervin, Jr
Joshi, Sewa Ram
Kawalek, Joseph Casimir, Jr
Ladson, Thomas Alvin
Liebelt, Annabel Glockler
Marquardt, Warren William
Mohanty, Sashi B
Moore, Roscoe Michael, Jr
Moreira, Jorge Eduardo
Norcross, Marvin Augustus
Page, Norbert Paul
Parkhie, Mukund Raghunathrao
Peterson, Irvin Leslie
Potkay, Stephen
Robens, Jane Florence
Russell, Robert John
Sarma, Padman S
Schricker, Robert Lee
Stoskopf, Michael Kerry
Teske, Richard H
Thapar, Nirwan T
Thompson, Clarence Henry, Jr
Torres-Anjel, Manuel Jose
Vadlamudi, Sri Krishna
Vickers, James Hudson
Ward, Fraser Prescott
Weisbroth, Steven H
Whitehair, Leo A
Whitney, Robert Arthur, Jr

MASSACHUSETTS
Cook, William Robert
Cotter, Susan M
Denniston, Joseph Charles
Fleischman, Robert Werder
Foster, Henry Louis
Fox, James Gahan
Hinrichs, Katrin
Jonas, Albert Moshe
King, Norval William, Jr
Loew, Franklin Martin
Papaioannou, Virginia Eileen
Pratt, George Woodman, Jr
Ross, James Neil, Jr
Schwartz, Anthony
Sevoian, Martin
Siegmund, Otto Hanns
Snoeyenbos, Glenn Howard
Tashjian, Robert John

MICHIGAN
Branstetter, Daniel G
Brown, Roger E

Chang, Timothy Scott
Cohen, Bennett J
Dutta, Saradindu
Eisenbrandt, David Lee
Fadly, Aly Mahmoud
Haas, Kenneth Brooks, Jr
Hoffert, Jack Russell
Keller, Waldo Frank
Mather, Edward Chantry
Page, Edwin Howard
Piper, Richard Carl
Ringler, Daniel Howard
Robinson, Norman Edward
Sawyer, Donald C
Stinson, Al Worth
Stocking, Gordon Gary
Subramanian, Marappa G
Tasker, John B
Trapp, Allan Laverne
Watts, Jeffrey Lynn
Webster, Harris Duane
Witter, Richard L
Yancey, Robert John, Jr

MINNESOTA
Case, Marvin Theodore
Diesch, Stanley L
Fletcher, Thomas Francis
Hansen, Mike
Johnson, Donald W
Karlson, Alfred Gustav
Kumar, Mahesh C
Lange, Robert Echlin, Jr
Larson, Vaughn Leroy
Manning, Patrick James
Muscoplat, Charles Craig
Nelson, Robert A
Osborne, Carl Andrew
Polzin, David J
Pomeroy, Benjamin Sherwood
Schlotthauer, John Carl
Sorensen, Dale Kenwood
Spurrell, Francis Arthur
Stoner, John Clark
Stowe, Clarence M
Wallace, Larry J

MISSISSIPPI
Jennings, David Phipps
Jones, Eric Wynn
Purchase, Harvey Graham

MISSOURI
Adldinger, Hans Karl
Blenden, Donald C
Hahn, Allen W
Hessler, Jack Ronald
Kahrs, Robert F
Kohlmeier, Ronald Harold
McCune, Emmett L
Niemeyer, Kenneth H
Olson, Gerald Allen
Rosenquist, Bruce David
St Omer, Vincent Victor
Schmidt, Donald Arthur
Tritschler, Louis George
Wagner, Joseph Edward

MONTANA
Speer, Clarence Arvon

NEBRASKA
Clemens, Edgar Thomas
Conrad, Robert Dean
Kolar, Joseph Robert, Jr
Littledike, Ernest Travis
Schmitz, John Albert
Schneider, Norman Richard
Swift, Brinton L
Torres-Medina, Alfonso
White, Raymond Gene

NEVADA
Simmonds, Richard Carroll

NEW HAMPSHIRE
Allen, Fred Ernest
Langweiler, Marc

NEW JERSEY
Abrutyn, Donald
Babu, Uma Mahesh
Benson, Charles Everett
Fabry, Andras
Fraser, Clarence Malcolm
Hayes, Terence James
Jochle, Wolfgang
Johnson, Allen Neill
McCoy, John Roger
Medway, William
Saini, Ravinder Kumar
Shor, Aaron Louis
Tudor, David Cyrus
Wang, Guang Tsan
Willson, John Ellis

NEW MEXICO
Cummins, Larry Bill
Hahn, Fletcher Frederick
Hobbs, Charles Henry
Muggenburg, Bruce Al
Saunders, George Cherdron

NEW YORK
Babish, John George
Callis, Jerry Jackson
Center, Sharon Anne
Fox, Francis Henry
Georgi, Jay R
Gillespie, James Howard
Habel, Robert Earl
Hall, Charles E
Hillman, Robert B
Kallfelz, Francis A
Kirk, Robert Warren
Lowe, John Edward
Maestrone, Gianpaolo
Moise, Nancy Sydney
O'Donoghue, John Lipomi
Perper, Robert J
Phemister, Robert David
Pogue, John Parker
Poppensiek, George Charles
Postle, Donald Sloan
Quimby, Fred William
Reimers, Thomas John
Sack, Wolfgang Otto
Schlafer, Donald H
Scott, Fredric Winthrop
Smith, Donald Frederick
Som, Prantika
Stark, Dennis Michael
Tennant, Bud C
Thompson, John C, Jr
Tonelli, George

NORTH CAROLINA
Aronson, A L
Boorman, Gary Alexis
Breitschwerdt, Edward B
Clarkson, Thomas Boston
Coffin, David L
Curtin, Terrence M
Gilfillan, Robert Frederick
Gonder, Eric Charles
Hamner, Charles Edward, Jr
Howard, Donald Robert
Johnson, George Andrew
Kuo, Eric Yung-Huei
Lay, John Charles
Liddle, Charles George
McClellan, Roger Orville
McConnell, Ernest Eugene
McPherson, Charles William
Noga, Edward Joseph
Olson, Leonard Carl
Peiffer, Rober Louis, Jr
Pick, James Raymond
Popp, James Alan
Riviere, Jim Edmond
Sar, Madhabananda
Smallwood, James Edgar
Tucker, Walter Eugene, Jr
Webb, Alfreda Johnson
Wright, James Francis

OHIO
Alden, Carl L
Bertram, Timothy Allyn
Boyle, Michael Dermot
Burt, James Kay
Clifford, Donald H
Diesem, Charles D
Dorn, Charles Richard
Gabel, Albert A
Gibson, John Phillips
Hilmas, Duane Eugene
Hull, Bruce Lansing
Jones, James Edward
Kier, Ann B
Kreier, Julius Peter
Mattingly, Steele F
Muir, William W, III
Murdick, Philip W
Myer, Carole Wendy
Nokes, Richard Francis
Ray, Richard Schell
Redman, Donald Roger
Rudy, Richard L
Saif, Linda Jean
Silverman, Jerald
Stuhlman, Robert August
Szczepanski, Marek Michal
Tarr, Melinda Jean
Tharp, Vernon Lance
Wallin, Richard Franklin
Wilson, George Porter, III
Wyman, Milton

OKLAHOMA
Alexander, Joseph Walker
Buckner, Ralph Gupton
Confer, Anthony Wayne
Ewing, Sidney Alton
Faulkner, Lloyd (Clarence)
Friend, Jonathon D
Graves, Donald C
Miller, Paul George
Rolf, Lester Leo, Jr
Still, Edwin Tanner
Tennille, Newton Bridgewater

OREGON
Dickinson, Ernest Milton
Hill, John Donald

PENNSYLVANIA
Andrews, Edwin Joseph
Bovee, Kenneth C
Della-Fera, Mary Anne
Deubler, Mary Josephine
Eberhart, Robert J
Fogleman, Ralph William
Fowler, Edward Herbert
Garman, Robert Harvey
Haskins, Mark
Herlyn, Dorothee Maria
Hsu, Chao Kuang
Jezyk, Peter Franklin
Lang, C Max
Liebenberg, Stanley Phillip
Live, Israel
McFeely, Richard Aubrey
Marshak, Robert Reuben
Melby, Edward C, Jr
Miselis, Richard Robert
Moore, Earl Neil
Morrow, David Austin
O'Brien, Joan A
Onik, Gary M
Patterson, Donald Floyd
Peardon, David Lee
Pirone, Louis Anthony
Prasad, Suresh
Raker, Charles W
Reid, Charles Feder
Rhodes, William Harker
Rozmiarek, Harry
Scheidy, Samuel F
Scott, George Clifford
Simmons, Donald Glick
Whitlock, Robert Henry
Zarkower, Arian

SOUTH CAROLINA
Yager, Robert H

SOUTH DAKOTA
Kirkbride, Clyde Arnold

TENNESSEE
Armistead, Willis William
Bradley, Richard E
Cypess, Raymond Harold
Eiler, Hugo
Harvey, Ralph Clayton
Kant, Kenneth James
New, John Coy, Jr
Oliver, Jack Wallace
Sachan, Dileep Singh
Walburg, H E

TEXAS
Bailey, Everett Murl, Jr
Banks, William Joseph, Jr
Barnes, Charles M
Beaver, Bonnie Veryle
Bridges, Charles Hubert
Burghardt, Robert Casey
Coleman, John Dee
Crandell, Robert Allen
Dubose, Robert Trafton
Edds, George Tyson
Eichberg, Jorg Wilhelm
Ellett, Edwin Willard
Fidler, Isaiah J
Galvin, Thomas Joseph
Green, Ronald W
Haensly, William Edward
Hartsfield, Sandee Morris
Heavner, James E
Hightower, Dan
Jennings, Paul Bernard, Jr
Kiel, Johnathan Lloyd
Lees, George Edward
Loan, Raymond Wallace
Messersmith, Robert E
Meuten, Donald John
Moore, Gary Thomas
Murnane, Thomas George
Murthy, Krishna K
Pakes, Steven P
Pierce, Kenneth Ray
Playter, Robert Franklin
Pope, Robert Eugene
Price, Alvin Audis
Ray, Allen Cobble
Roenigk, William J
Rosborough, John Paul
Smith, Malcolm Crawford, Jr
Steele, James Harlan
Szabuniewicz, Michael
Tizard, Ian Rodney
Trevino, Gilberto Stephenson
Whitford, Howard Wayne

UTAH
Blake, Joseph Thomas
Healey, Mark Calvin
Hoopes, Keith Hale
Larsen, Austin Ellis
McGill, Lawrence David
Shupe, James LeGrande
Thomas, Don Wylie

VIRGINIA
AcKerman, Larry Joseph
Blood, Benjamin Donald
Brown, Elise Ann Brandenburger
Colmano, Germille

Cordes, Donald Ormond
Cruthers, Larry Randall
Held, Joe R
Hooper, Billy Ernest
Lee, Robert Jerome
McVicar, John West
Misra, Hara Prasad
Todd, Frank Arnold
Webb, Willis Keith

WASHINGTON
Allen, Julia Natalia
Baksi, Samarendra Nath
Brobst, Duane Franklin
Bustad, Leo Kenneth
Cho, Byung-Ryul
Dennis, Melvin B, Jr
Evermann, James Frederick
Gillis, Murlin Fern
Graves, Scott Stoll
Hostetler, Roy Ivan
McDonald, John Stoner
Noel, Jan Christina
Ott, Richard L
Prieur, David John
Ragan, Harvey Albert
Ratzlaff, Marc Henry
Sande, Ronald Dean
Smith, Dean Harley
Van Hoosier, Gerald L, Jr
Zamora, Cesario Siasoco
Zander, Donald Victor

WEST VIRGINIA
Dozsa, Leslie
Olson, Norman O

WISCONSIN
Anand, Amarjit Singh
Berman, David Theodore
Cook, Mark Eric
Czuprynski, Charles Joseph
Dueland, Rudolf (Tass)
Easterday, Bernard Carlyle
Fahning, Melvyn Luverne
Hall, Robert Everett
Hanson, Robert Paul
Howard, Thomas Hyland
Larson, Lester Leroy
Nichols, Roy Elwyn
Rao, Ghanta Nageswara
Splitter, Gary Allen
Will, James Arthur

WYOMING
Pier, Allan Clark
Walton, Thomas Edward

ALBERTA
Secord, David Cartwright

BRITISH COLUMBIA
Booth, Amanda Jane
Salinas, Fernando A
Whiteford, Robert Daniel

NEW BRUNSWICK
Smith, Harry John

ONTARIO
Archibald, James
Barnum, Donald Alfred
Batra, Tilak Raj
Belbeck, L W
Bouffard, Marie Alice
Bouillant, Alain Marcel
Fisher, Kenneth Robert Stanley
Frederick, George Leonard
Hulland, Thomas John
Kaushik, Azad Kumar
Lang, Gerhard Herbert
Miniats, Olgerts Pauls
Morrison, Spencer Horton
Nielsen, Klaus H B
Nielsen, N Ole
Robertson, Frederick John
Robertson, James McDonald
Ruckerbauer, Gerda Margareta
Samagh, Bakhshish Singh
Savan, Milton
Shewen, Patricia Ellen
Slocombe, Joseph Owen Douglas
Stemshorn, Barry William
Wilkie, Bruce Nicholson
Willoughby, Russell A

PRINCE EDWARD ISLAND
Amend, James Frederick
Singh, Amreek

QUEBEC
Barrette, Daniel Claude
Flipo, Jean
Lussier, Gilles L
Robert, Suzanne

SASKATCHEWAN
Blakley, Barry Raymond
Leighton, Frederick Archibald
Schiefer, H Bruno

OTHER COUNTRIES
Aalund, Ole
Baggot, J Desmond

Veterinary Medicine (cont)

Davidson, David Edward, Jr
Ishida, Yukisato
Johnsen, Dennis O
Karstad, Lars
Lund, John Edward
Lutsky, Irving
Makita, Takashi
Mitruka, Brij Mohan
Mullenax, Charles Howard
Pan, In-Chang
Schmeling, Sheila Kay
Shore, Laurence Stuart
Usenik, Edward A
Yovich, John V

Other Medical & Health Sciences

ALABAMA
Borton, Thomas Ernest
Briles, David Elwood
Clark, Charles Edward
Crispens, Charles Gangloff, Jr
Duque, Ricardo Ernesto
Funkhouser, Jane D
Gay, Renate Erika
Gelman, Simon
Grubbs, Clinton Julian
Hardy, Robert W
Lloyd, L Keith
Mundy, Roy Lee
Pass, Robert Floyd
Reeves, R C
Wilson, G Dennis

ALASKA
Mills, William J, Jr
Segal, Bernard

ARIZONA
Appelgren, Walter Phon
Bijou, Sidney William
Birkby, Walter H
Butler, Lillian Catherine
Colburn, Wayne Alan
Goldman, Steven
Jeter, Wayburn Stewart
Kahn, Marvin William
Kazal, Louis Anthony
LaPointe, Leonard Lyell
Lohman, Timothy George
McQueen, Charlene A
Mossman, Kenneth Leslie
Scott, Ralph Asa, Jr
Stini, William Arthur
Tong, Theodore G

ARKANSAS
Andreoli, Thomas E
Berky, John James
Burleigh, Joseph Gaynor
Hardin, James W
Kadlubar, Fred F
Light, Kim Edward
Sedelow, Walter Alfred, Jr
Slikker, William, Jr
Wenger, Galen Rosenberger
Wolff, George Louis

CALIFORNIA
Abrams, Herbert L
Abramson, Edward E
Akesson, Norman B(erndt)
Akin, Gwynn Collins
Andre, Michael Paul
Assali, Nicholas S
Astrahan, Melvin Alan
Baker, Joffre B
Barnhart, James Lee
Basham, Teresa
Bigler, William Norman
Brantingham, Charles Ross
Brasel, Jo Anne
Brubacher, Elaine Schnitker
Burnett, Bryan Reeder
Butcher, Larry L
Carlisle, Harry J
Cassel, Russell N
Chen, Fu-Tai Albert
Chisari, Francis Vincent
Choy, Wai Nang
Chu, William Tongil
Clever, Linda Hawes
Cohn, Stanton Harry
Conner, Brenda Jean
Cooper, Allen D
Cooper, Howard K
Cornford, Eain M
Courchesne, Eric
Cromwell, Florence S
Davis, William Thompson
Denis, Kathleen A
Derenzo, Stephen Edward
Dillmann, Wolfgang H
Ebbe, Shirley Nadine
Eger, Edmond I, II
Elias, Peter M
Ellestad, Myrvin H
Epstein, John Howard
Epstein, Lois Barth

Fackler, Martin L
Faleschini, Richard John
Fallon, James Harry
Feifel, Herman
Feigal, David William, Jr
Fessenden, Peter
Finn, James Crampton, Jr
Finston, Roland A
Fisher, Delbert A
Flacke, Joan Wareham
Friedman, Alan E
Fuchs, Victor Robert
Furtado, Victor Cunha
Gee, Adrian Philip
Geiselman, Paula J
Georghiou, George Paul
Giannetti, Ronald A
Gibson, Thomas Alvin, Jr
Gill, James Edward
Gilmore, Stuart Irby
Glassock, Richard James
Gold, Richard Horace
Golde, David William
Goode, John Wolford
Gooding, Gretchen Ann Wagner
Gould, Robert George
Greene, John Clifford
Greenhouse, Nathaniel Anthony
Greenleaf, John Edward
Griffin, John Henry
Griffiss, John McLeod
Grumbach, Melvin Malcolm
Gumbiner, Barry M
Haas, Gustav Frederick
Hahn, Theodore John
Hainski, Martha Barrionuevo
Hammerschlag, Richard
Harris, John Wayne
Hayden, Jess, Jr
Haymond, Herman Ralph
Heslep, John McKay
Hislop, Helen Jean
Hoffman, Arlene Faun
Hollenbeck, Clarie Beall
Holmes, Donald Eugene
Holtzman, Samuel
Hsieh, Jen-Shu
Hubbard, Eric R
Jackson, Crawford Gardner, Jr
Jacobs, S Lawrence
Jones, Joie Pierce
Jordan, Stanley Clark
Kaplan, Stanley A
Keene, Clifford H
Keith, Robert Allen
Kempler, Walter
Kempter, Charles Prentiss
Kessler, Seymour
Klivington, Kenneth Albert
Kohn, Henry Irving
Kortright, James McDougall
Kosow, David Phillip
Kraft, Lisbeth Martha
Kripke, Daniel Frederick
Krippner, Stanley Curtis
Kritz-Silverstein, Donna
Kuwahara, Steven Sadao
Ladisch, Stephan
Largman, Corey
Lee, Roland Robert
Lindsey, Dortha Ruth
London, Ray William
Luft, Harold Stephen
McBride, William H
McDermott, Daniel J
McKinley, Michael P
Makowka, Leonard
Marmor, Michael F
Marshall, Sally J
Mason, Dean Towle
Mastropaolo, Joseph
Mills, Gary Kenith
Miquel, Jaime
Mondon, Carl Erwin
Montgomery, Leslie D
Morewitz, Harry Alan
Moxley, John H, III
Mulla, Mir Subhan
Murphree, A Linn
Nelson, Walter Ralph
Nicholson, Arnold Eugene
Nickerson, Bruce Greenwood
Niklowitz, Werner Johannes
Nowlis, David Peter
Nydegger, Corinne Nemetz
Pallak, Michael
Palmer, Beverly B
Pardridge, William M
Pearson, Robert Bernard
Peper, Erik
Peterlin, Boris Matija
Peterson, Kendall Robert
Peterson-Falzone, Sally Jean
Phelps, Michael Edward
Pigiet, Vincent P
Pitas, Robert E
Poland, Russell E
Ramachandran, Vilayanur Subramanian
Rapaport, Elliot
Reitz, Richard Elmer
Remsen, Joyce F
Richters, Arnis
Roberts, James M
Root, Richard Kay

Rosemark, Peter Jay
Rosenthal, Allan Lawrence
Salmon, Eli J
Samuel, Charles Edward
Schwartz, Marshall Zane
Schweisthal, Michael Robert
Seibert, J A
Selle, Wilbur A
Shanks, Susan Jane
Shapiro, David
Silagi, Selma
Smiles, Kenneth Albert
Smith, Michael R
Snead, O Carter, III
Sohn, Yung Jai
Soule, Roger Gilbert
Spanis, Curt William
Sperling, Jacob L
Stanczyk, Frank Zygmunt
Steckel, Richard J
Steffey, Eugene P
Strauch, Ralph Eugene
Sullivan, Stuart F
Sweet, Benjamin Hersh
Syme, S Leonard
Taeusch, H William
Teng, Evelyn Lee
Teresi, Joseph Dominic
Thomas, Ralph Harold
Thompson, John Frederick
Tin-Wa, Maung
Toy, Arthur John, Jr
Ulmer, Raymond Arthur
Urquhart, John, III
Wade, Michael James
Walsh, John H
Weliky, Norman
Whalen, Carol Kupers
Wieland, Bruce Wendell
Withers, Hubert Rodney
Wright, James Edward
Yager, Janice L Winter
Yagiela, John Allen
Yu, Jen

COLORADO
Balke, Bruno
Berge, Trygve O
Conger, John Janeway
Cosgriff, Thomas Michael
Ellis, Frank Russell
Hager, Jean Carol
Hall, Alan H
Heim, Werner George
Khandwala, Atul S
Lee, Arthur Clair
Leung, Donald Yap Man
Pearson, John Richard
Platt, Kenneth Allan
Riches, David William Henry
Schonbeck, Niels Daniel
Schrier, Robert William
Suinn, Richard M
Tucker, Alan
Verdeal, Kathey Marie
Weston, William Lee
Wilber, Charles Grady

CONNECTICUT
Alpert, Nelson Leigh
Bartoshuk, Linda May
Berdick, Murray
Bohannon, Richard Wallace
Booss, John
Cardinal, John Robert
Carroll, J Gregory
Crean, Geraldine L
Doshan, Harold David
Freudenthal, Ralph Ira
Grattan, James Alex
Hardin, John Avery
Kaplow, Leonard Samuel
Kitahata, Luke Masahiko
Luborsky, Judith Lee
Lynch, John Edward
Miller, Neal Elgar
Myers, Maureen
Pelletier, Charles A
Siegel, Norman Joseph
Sittig, Dean Forrest
Spitznagle, Larry Allen
Steinmetz, Philip R
Sternberg, Robert Jeffrey
Sulzer-Azaroff, Beth
Verdi, James L
Vinocur, Myron
Warshaw, Joseph B
Zaret, Barry L
Zigler, Edward

DELAWARE
Craig, Alan Daniel
Davis, Leonard George
Frey, William Adrian
Johnson, Donald Richard
Krahn, Robert Carl
Law, Amy Stauber
Read, Robert E
Reinhardt, Charles Francis
Steinberg, Marshall

DISTRICT OF COLUMBIA
Arcos, Joseph (Charles)
Armaly, Mansour F

Atkins, James Lawrence
Bick, Katherine Livingstone
Cavanagh, Harrison Dwight
Chung, Ho
Costa, Erminio
Drew, Robert Taylor
Eisner, Gilbert Martin
Gelmann, Edward P
Hamarneh, Sami Khalaf
Hoffman, Julius
Jones, Stanley B
Kornhauser, Andrija
Kromer, Lawrence Frederick
Larsen, Lynn Alvin
Ledley, Robert Steven
Leto, Salvatore
Lippman, Marc Estes
Melnick, Vijaya L
Milbert, Alfred Nicholas
Nightingale, Elena Ottolenghi
Perry, David Carter
Perry, Seymour Monroe
Pinkerson, Alan Lee
Pittman, Margaret
Rosenquist, Glenn Carl
Salu, Yehuda
Shukla, Kamal Kant
Silver, Sylvia
Slaughter, Lynnard J
Stark, Nathan Julius
Tearney, Russell James
Todhunter, John Anthony
Vick, James
Weinberg, Myron Simon
Yeager, Henry, Jr
Zawisza, Julie Anne A
Ziemer, Paul L

FLORIDA
Abitbol, Carolyn Larkins
Amaranath, L
Arenberg, David (Lee)
Brim, Orville G, Jr
Convertino, Victor Anthony
Cowan, Frederick Pierce
Dewanjee, Mrinal K
Eagle, Donald Frohlichstein
Fitzgerald, Lawrence Terrell
Frank, H Lee
Goldstein, Mark Kane
Gosselin, Arthur Joseph
Gottlieb, Charles F
Habal, Mutaz
Halpern, Ephriam Philip
Hawk, William Andrew
Haynes, Duncan Harold
Henson, Carl P
Johnson, James Harmon
Kunce, Henry Warren
Lo, Hilda K
Manhold, John Henry, Jr
Margo, Curtis Edward
Mercer, Thomas T
Miller, James Woodell
Pargman, David
Patterson, Richard Sheldon
Peters, Thomas G
Pollock, Michael L
Routh, Donald Kent
Schein, Jerome Daniel
Silverstein, Herbert
Simpkins, James W
Simring, Marvin
Sneade, Barbara Herbert
Thornthwaite, Jerry T
Whitman, Victor
Wingo, Charles S

GEORGIA
Adams, David B
Allen, Delmas James
Allison, Jerry David
Baranowski, Tom
Bess, John Clifford
Brooks, John Bill
Caplan, Daniel Bennett
Colborn, Gene Louis
Copeland, Robert B
Crowe, Dennis Timothy, Jr
Faraj, Bahjat Alfred
Gordon, David Stewart
Gupton, Guy Winfred, Jr
Hatch, Roger Conant
Hersh, Theodore
Inge, Walter Herndon, Jr
King, Frederick Alexander
Kokko, Juha Pekka
Krauss, Jonathan Seth
Lawley, Thomas J
Loring, David William
McDade, Joseph Edward
Martin, David Edward
Morgan, Karl Ziegler
Porterfield, Susan Payne
Potter, David Edward
Rigdon, Raymond Harrison
Smith, David Fletcher
Sridaran, Rajagopala
Stevenson, Dennis A
Stone, Connie J
Van Assendelft, Onno Willem
Wolf, Steven L
Zaia, John Anthony

HAWAII
Diamond, Milton
Hammer, Ronald Page, Jr
Howarth, Francis Gard
Lance, Eugene Mitchell
Manly, Philip James
Nelson, Marita Lee
Yates, James T

IDAHO
Vestal, Robert Elden

ILLINOIS
Alexander, Kenneth Ross
Andreoli, Kathleen Gainor
Bakkum, Barclay W
Barenberg, Sumner
Bevan, William
Boshes, Benjamin
Boulos, Badi Mansour
Brutten, Gene J
Campbell, Suzann Kay
Cember, Herman
Cera, Lee M
Cohen, Louis
Constantinou, Andreas I
Cook, L Scott
Debus, Allen George
Doege, Theodore Charles
Dunn, Dorothy Fay
Ernest, J Terry
Fang, Victor Shengkuen
Flay, Brian Richard
Foxx, Richard Michael
Gastwirth, Bart Wayne
Giometti, Carol Smith
Glisson, Silas Nease
Goebel, Maristella
Goldman, Allen S
Grdina, David John
Gross, Thomas Lester
Helseth, Donald Lawrence, Jr
Hoffman, William E
Hoo, Joe Jie
Ivankovich, Anthony D
James, Karen K(anke)
Kinders, Robert James
Kleinman, Kenneth Martin
Klocke, Francis J
Kochman, Ronald Lawrence
Kornel, Ludwig
Kulkarni, Bidy
Lauterbur, Paul Christian
Lee, Jang Y
Lehman, Dennis Dale
Levy, Jerre Marie
Lurain, John Robert, III
Lyon, Edward Spafford
McWhinnie, Dolores J
Mehta, Rajendra G
Metz, Charles Edgar
Moray, Neville Peter
Neugarten, Bernice L
Onal, Ergun
Oscai, Lawrence B
Ratajczak, Helen Vosskuhler
Richman, David Paul
Robertson, Abel Alfred Lazzarini, Jr
Rosen, Steven Terry
Schmidt, Steven Paul
Sharma, Ram Ratan
Sherman, Laurence A
Solt, Dennis Byron
Spencer, Herta
Spokas, John J
Stevenson, G W
Stinchcomb, Thomas Glenn
Strain, James E
Taber, David
Taylor, D Dax
Thomas, Joseph Erumappettical
Triano, John Joseph
Vyborny, Carl Joseph
Wagner, Vaughn Edwin
Weber, Karl T
Woo, Lecon
Wynveen, Robert Allen
Zsigmond, Elemer K

INDIANA
Baird, William McKenzie
Bergstein, Jerry Michael
Blatt, Joel Martin
Bridges, C David
Carlson, James C
Friedman, Julius Jay
George, Robert Eugene
Gibbs, Frederic Andrews
Guth, S(herman) Leon
Hayes, J Scott
Hinsman, Edward James
Hui, Siu Lui
LaFuze, Joan Esterline
McIntyre, John A
McKinney, Gordon R
Nahrwold, Michael L
Quist, Raymond Willard
Schaible, Robert Hilton
Shoup, Charles Samuel, Jr
Smith, Gerald Duane
Tyhach, Richard Joseph
Van Etten, Robert Lee
Waller, Bruce Frank
Wellman, Henry Nelson

Whitmer, Jeffrey Thomas

IOWA
Al-Jurf, Adel Saada
Anderson, Charles V
Biermann, Janet Sybil
Biller, Jose
Bonney, William Wallace
Di Bona, Gerald F
Gergis, Samir D
Goff, Jesse Paul
Hunninghake, Gary W
Levy, Leonard Alvin
Moorcroft, William Herbert
Orcutt, James A
Rhead, William James
Schrott, Helmut Gunther
Stewart, Mary E
Tranel, Daniel T
Younszai, M Kabir
Ziegler, Ekhard E

KANSAS
Chin, Hong Woo
Dawson, James Thomas
Dykstra, Mark Allan
Ferraro, John Anthony
Hinshaw, Charles Theron, Jr
Krishnan, Engil Kolaj
Maguire, Helen M
Norris, Patricia Ann

KENTUCKY
Budde, Mary Laurence
Christensen, Ralph C(hresten)
Ehmann, William Donald
Espinosa, Enrique
Essig, Carl Fohl
Farman, Allan George
Garrity, Thomas F
Giammara, Berverly L Turner Sites
Henrickson, Charles Henry
Kennedy, John Elmo, Jr
Porter, William Hudson
Randall, David Clark
Shih, Wei Jen
Syed, Ibrahim Bijli
Terango, Larry
Tietz, Norbert W
Urano, Muneyasu
Zimmerman, Thom J

LOUISIANA
Bazan, Nicolas Guillermo
Beuerman, Roger Wilmer
Carmines, Pamela Kay
Cherry, Flora Finch
Christian, Frederick Ade
Datta, John
Fermin, Cesar Danilo
Jain, Sushil Kumar
Levitzky, Michael Gordon
Lindberg, David Seaman, Sr
McDonald, Marguerite Bridget
Manno, Barbara Reynolds
Manno, Joseph Eugene
Palmgren, Muriel Signe
Raisen, Elliott
Riopelle, Arthur J
Roberts, James Allen
Thompson, James Joseph
Wolf, Thomas Mark
Yaeger, Robert George

MAINE
Mead, Jere
Norton, James Michael
Roderick, Thomas Huston
Swazey, Judith P

MARYLAND
Augustine, Robertson J
Baquet, Claudia
Barrett, James E
Bartley, William Call
Becker, Bruce Clare
Becker, Lewis Charles
Beebe, Gilbert Wheeler
Blattner, William Albert
Blum, Bruce I
Blum, Robert Allan
Bowen, Donnell
Bresler, Jack Barry
Broder, Samuel
Brown, Kenneth Stephen
Bryant, Patricia Sand
Buchanan, John Donald
Bunyan, Ellen Lackey Spotz
Burton, George Joseph
Carr, Charles Jelleff
Chabner, Bruce A
Chiueh, Chuang Chin
Colburn, Theodore R
Coligan, John E
Cone, Richard Allen
Conrad, Daniel Harper
Contrera, Joseph Fabian
Counts, George W
Coursey, Bert Marcel
Craig, Paul N
Curt, Gregory
Cutler, Gordon Butler, Jr
David, Henry P
Domanski, Thaddeus John

Doppman, John L
Duggar, Benjamin Charles
Dunning, Herbert Neal
Eckardt, Michael Jon
Edinger, Stanley Evan
Efron, Herman Yale
Engel, Bernard Theodore
Fauci, Anthony S
Feldman, Robert H L
Ferris, Frederick L
Fex, Jorgen
Fisher, Dale John
Fishman, Elliot Keith
Ford, Leslie
Fozard, James Leonard
Friedman, Michael A
Friess, Seymour Louis
Glatstein, Eli J
Gleich, Carol S
Goldberg, Andrew Paul
Goldschmidt, Peter Graham
Gorden, Phillip
Gordis, Enoch
Grdis, Enoch
Grunberg, Neil Everett
Gutierrez, Peter Luis
Hackett, Joseph Leo
Hadley, Evan C
Hager, Gordon L
Hall, Beverly Fenton
Hamill, Peter Van Vechten
Harkins, Rosemary Knighton
Harris, Curtis C
Hawkins, Michael John
Hazzard, DeWitt George
Heilman, Carol A
Hoak, John C
Hoofnagle, Jay H
Horan, Michael
Hubbell, John Howard
Johnson, Carl Boone
Johnston, Gerald Samuel
Kaper, James Bennett
Kaplan, Ann Esther
Katz, Stephen I
Kaufman, Joyce J
Kelty, Miriam C
Kessler, David
Killen, John Young, Jr
Kimes, Brian Williams
Klarman, Herbert E
Klippel, John Howard
Kraemer, Kenneth H
Krantz, David S
Lachin, John Marion, III
Lakatta, Edward G
Lathers, Claire M
Lee, Che-Hung R
Lichtenstein, Lawrence M
Liebelt, Annabel Glockler
Linehan, William Marston
Linnoila, Markku
Lowy, Douglas R
Lubet, Ronald A
Ludlow, Christy L
Luecke, Donald H
McCarthy, Charles R
McKenna, Thomas Michael
McLaughlin, Jack A
McPherson, Robert W
Magrath, Ian
Maumenee, Irene Hussels
Mirsky, Allan Franklin
Mockrin, Stephen Charles
Mohla, Suresh
Murphy, Dennis L
Myslobodsky, M S
Nelson, James Harold
Nino, Hipolito V
Nissley, S Peter
Nussenblatt, Robert Burton
Osborne, J Scott, III
Osterberg, Robert Edward
Owellen, Richard John
Pikus, Anita
Pizzo, Philip A
Plaut, Marshall
Provine, Robert Raymond
Quraishi, Mohammed Sayeed
Ranney, J Buckminster
Roberts, William C
Roche, Lidia Alicia
Sachs, Murray B
Schifreen, Richard Steven
Shleien, Bernard
Shoemaker, Dale
Shoukas, Artin Andrew
Shulman, N Raphael
Simpson, Ian Alexander
Sliney, David H
Smith, Philip Lees
Sowers, Arthur Edward
Spector, Novera Herbert
Sporn, Eugene Milton
Spring, Susan B
Stolz, Walter S
Streicher, Eugene
Taylor, Lauriston Sale
Thompson, Donald Leroy
Tower, Donald Bayley
Triantaphyllopoulos, Eugenie
Tucker, Robert Wilson
Ulsamer, Andrew George, Jr
Wasserheit, Judith Nina

Watson, James Dewey
Watters, Robert Lisle
Wehr, Thomas A
Wilder, Ronald Lynn
Willoughby, Anne D
Wingate, Catharine L
Wintercorn, Eleanor Stiegler
Woods, Wilna Ann
Yaniv, Shlomo Stefan
Yoder, Robert E
Young, Eric D
Young, Gerald A

MASSACHUSETTS
Alpert, Joel Jacobs
Aronow, Saul
Behrman, Richard H
Berman, Marlene Oscar
Bing, Oscar H L
Blank, Irvin H
Brenner, John Francis
Brown, Robert
Caldini, Paolo
Campbell, Charlotte Catherine
Castile, Robert G
Charney, Evan
Chernosky, Edwin Jasper
Cormack, Allan MacLeod
Deutsch, Thomas F
Dexter, Lewis
Fairbanks, Grant
Fein, Rashi
Fencl, Vladimir
Foster, Charles Stephen
Frelinger, Andrew L
Fridovich-Keil, Judith Lisa
Frisch, Rose Epstein
Gaudette, Leo Eward
Geisterfer-Lowrance, Anja A T
Glowacki, Julie
Goldman, Harvey
Gordon, Philip Ray
Goyal, Raj K
Greenberger, Joel S
Gutheil, Thomas Gordon
Hebben, Nancy
Hichar, Joseph Kenneth
Holland, Christie Anna
Holmes, Helen Bequaert
Hoop, Bernard, Jr
Kagan, Jerome
Kanarek, Robin Beth
Kasper, Dennis Lee
Landowne, Milton
Lemeshow, Stanley Alan
Liu, John
Lucey, Edgar C
Lux, Samuel E, IV
Madias, Nicolaos E
Martin, Thomas George, III
Masek, Bruce James
Mason, Joel B
Milbocker, Michael
Milder, Fredric Lloyd
Modest, Edward Julian
Morris, Robert
Muni, Indu A
Murray, Joseph E
Myers, Richard Hepworth
Naimi, Shapur
Nathan, David G
Olsen, Robert Thorvald
Palek, Jiri Frant
Parrish, John Albert
Paskins-Hurlburt, Andrea Jeanne
Pfeffer, Marc Alan
Philbin, Daniel Michael
Prout, Curtis
Rabin, Monroe Stephen Zane
Raviola d'Elia, Giuseppina E(nrica)
Russo, Dennis Charles
Scheuchenzuber, H Joseph
Schorr, Lisbeth Bamberger
Siegal, Bernard
Sisterson, Janet M
Skrinar, Gary Stephen
Snyder, Benjamin Willard
Vogel, James Alan
Volpe, Joseph J
Wegman, David Howe
Wineman, Robert Judson
Wolfe, Harry Bernard
Wood, William C
Wyler, David J
Zapol, Warren Myron

MICHIGAN
Allen, Cheryl
Al-Sarraf, Muhyi
Anderson, Thomas Edward
Bach, Shirley
Balazovich, Kenneth J
Berent, Stanley
Betz, A Lorris
Buchtel, Henry Augustus, IV
Callewaert, Denis Marc
Dijkers, Marcellinus P
Dumke, Paul Rudolph
Ferraro, Douglas P
Fisher, Leslie John
Floyd, John Claiborne, Jr
Gilbertsen, Richard B
Goodwin, Jesse Francis
Hagen, John William

Other Medical & Health Sciences (cont)

Hakim, Margaret Heath
Harris, Gale Ion
Hawthorne, Victor Morrison
Howard, Charles Frank, Jr
James, Sherman Athonia
Kurtz, Margot
Leopold, Wilbur Richard, III
Loor, Rueyming
Louis-Ferdinand, Robert T
McClary, Andrew
MacDonald, John Robert
McEwen, Charles Milton, Jr
Motiff, James P
Musch, David C(harles)
Pelzer, Charles Francis
Pieper, David Robert
Piercey, Montford F
Pike, Marilyn Cecile
Punch, Jerry L
Rupp, Ralph Russell
Ruwart, Mary Jean
Senagore, Anthony J
Shaw, M(elvin) P
Sheagren, John Newcomb
Shlafer, Marshal
Siddiqui, Waheed Hasan
Sobota, Walter Louis
Timmis, Gerald C
Toledo-Pereyra, Luis Horacio
Whitehouse, Frank, Jr
Zacks, James Lee
Zucker, Robert Alpert

MINNESOTA
Allen, David West
Anderson, W Robert
Athelstan, Gary Thomas
Barber, Donald E
Bowers, Larry Donald
Brown, David Mitchell
Butt, Hugh Roland
Chesney, Charles Frederic
Chevalier, Peter Andrew
Dai, Xue Zheng (Charlie)
Darley, Frederic Loudon
Dewald, Gordon Wayne
Douglas, William Hugh
Hancock, Peter Adrian
Hovde, Ruth Frances
Iverson, Laura Himes
Joseph, Stephen C
Kornblith, Carol Lee
McCullough, John Jeffrey
Manning, Patrick James
Mauer, S Michael
Michenfelder, John D
Moore, Sean Breanndan
Mortimer, James Arthur
Regal, Jean Frances
Roessler, Charles Ervin
Schofield, William
Slavin, Joanne Louise
Swanson, Anne Barrett
Swingle, Karl Frederick
Taswell, Howard Filmore
Timm, Gerald Wayne
Vallera, Daniel A
Vernier, Robert L
Vetter, Richard J
Wade, Michael George
White, James George
Yellin, Absalom Moses

MISSISSIPPI
Alford, Geary Simmons
Fowler, Stephen C
Jones, Eric Wynn
Kallman, William Michael
Lawler, Adrian Russell
Wiser, Winfred Lavern

MISSOURI
Adams, Max David
Chapman, Ramona Marie
Clare, Stewart
Cook, Mary Rozella
Dilley, William G
Ferrendelli, James Anthony
Grubb, Robert Lee, Jr
Hageman, Gregory Scott
Hedlund, James L
Hurst, Robert R(owe)
Kaplan, Henry J
Lubin, Bernard
Moe, Aaron Jay
Mosher, Donna Patricia
Murrell, Hugh Jerry
Natani, Kirmach
Podrebarac, Eugene George
Reynolds, Robert D
Robins, Eli
Simchowitz, Louis
Smith, Carl Hugh
Smith, Robert Francis
Stenson, William F
Strunk, Robert Charles
Taub, John Marcus
Zehr, Martin Dale
Zepp, Edwin Andrew

MONTANA
Combie, Joan D

NEBRASKA
Badeer, Henry Sarkis
Benschoter, Reba Ann
Bernthal, John E(rwin)
Cavalieri, Ercole Luigi
Heaney, Robert Proulx
Kimberling, William J
Kutler, Benton
Leuschen, M Patricia
Macaluso, Sister Mary Christelle
Prioreschi, Plinio
Records, Raymond Edwin
Saski, Witold
Schneider, Norman Richard
Schulz, John C
Sonderegger, Theo Brown

NEVADA
McCalden, Thomas A
May, Jerry Russell
Smith, Aaron

NEW HAMPSHIRE
Ball, Edward D
Brous, Don W

NEW JERSEY
Albu, Evelyn D
Alfano, Michael Charles
Alson, Eli
Apuzzio, J J
Baker, Thomas
Bennett, Debra A
Bernstein, Alan D
Boyle, Joseph, III
Brubaker, Paul Eugene, II
Dugan, Gary Edwin
Eisinger, Robert Peter
Falk, John L
Grebenau, Mark David
Griffo, James Vincent, Jr
Haft, Jacob I
Hakimi, John
Halpern, Myron Herbert
Harris, Jane E
Haughey, Francis James
Heveran, John Edward
Hey, John Anthony
Hockel, Gregory Martin
Houston, Vern Lynn
Ingoglia, Nicholas Andrew
Jacobs, Barry Leonard
Kapp, Robert Wesley, Jr
Khavkin, Theodor
Krauthamer, George Michael
Kruh, Daniel
Lamdin, Ezra
Lappe, Rodney Wilson
Lehrer, Paul Michael
Lewis, Steven Craig
Leyson, Jose Florante Justininane
Lowndes, Herbert Edward
Machac, Josef
Mackerer, Carl Robert
Mechanic, David
Miller, George Armitage
Miner, Karen Mills
Natelson, Benjamin Henry
Newton, Paul Edward
O'Keefe, Eugene H
Petillo, Phillip J
Reuben, Roberta C
Rich, Kenneth Eugene
Schroeder, Steven A
Sen, Tapas K
Sherr, Allan Ellis
Siegel, Jeffry A
Slomiany, Amalia
Solis-Gaffar, Maria Corazon
Squibb, Robert E
Stanton, Robert E
Stein, Donald Gerald
Stern, William
Stillman, John Edgar
Stults, Frederick Howard
Tio, Cesario O
Unowsky, Joel
Van Pelt, Wesley Richard
Vogel, Veronica Lee
Watson, Richard White, Jr
Weintraub, Howard Steven
Wiggins, Jay Ross
Wislocki, Peter G

NEW MEXICO
Boecker, Bruce Bernard
Guilmette, Raymond Alfred
Guziec, Lynn Erin
Hansen, Wayne Richard
Lawrence, James Neville Peed
Tillery, Marvin Ishmael
Vogel, Kathryn Giebler
Werkema, George Jan
Yudkowsky, Elias B

NEW YORK
Ader, Robert
Alexander, Justin
Anderson, Albert Edward
Archibald, Reginald MacGregor
Baines, Ruth Etta

Barac-Nieto, Mario
Barish, Robert John
Barletta, Michael Anthony
Barzel, Uriel S
Baum, John W
Bendixen, Henrik H
Bernstein, Alvin Stanley
Bigler, Rodney Errol
Bizios, Rena
Bloodstein, Oliver
Bofinger, Diane P
Boomsliter, Paul Colgan
Bovbjerg, Dana H
Bradley, Francis J
Butler, Katharine Gorrell
Callahan, Daniel
Carr, Edward Gary
Carsten, Arland L
Centola, Grace Marie
Christman, David R
Chryssanthou, Chryssanthos
Cohen, Bernard
Cohen, Elias
Cohen, Michael I
Conway, Walter Donald
Cort, Susannah
Cunha, Burke A
Dahl, John Robert
Daly, Robert Ward
Davey, Frederick Richard
Dornbush, Rhea L
Dumoulin, Charles Lucian
Edelstein, William Alan
Edwards, John Anthony
Ehrke, Mary Jane
Elbadawi, Ahmad
Ellis, Albert
Ellner, Paul Daniel
Fabricant, Catherine G
Faden, Howard
Fahey, Charles J
Famighetti, Louise Orto
Farah, Fuad Salim
Fawwaz, Rashid
Feldman, Alan Sidney
Ferris, Steven Howard
Finando, Steven J
Fish, Jefferson
Fliedner, Leonard John, Jr
Forquer, Sandra Lynne
Friedberg, Carl E
Furlanetto, Richard W
Gering, Robert Lee
Gerstman, Hubert Louis
Gintautas, Jonas
Glass, David Carter
Goldberger, Robert Frank
Goldfried, Marvin R
Gollon, Peter J
Gordon, Arnold J
Gordon, Wayne Alan
Gray, William David
Greenhall, Arthur Merwin
Grieco, Michael H
Haber, Sonja B
Hainline, Louise
Halevy, Simon
Ham, Richard John
Haywood, Anne Mowbray
Huang, Alice Shih-Hou
Huang, Cheng-Chun
Hull, Andrew P
Jan, Kung-Ming
Johnson, Marie-Louise T
Jung, Chan Yong
Killian, Carl Stanley
Klemperer, Friedrich W
Komorek, Michael Joesph, Jr
Kotler, Donald P
Kotval, Pesho Sohrab
Krauss, Herbert Harris
Kronzon, Itzhak
Kuftinec, Mladen M
Lamberg, Stanley Lawrence
Lambertsen, Eleanor C
Law, Paul Arthur
Lebenthal, Emanuel
Lempert, Neil
Lepow, Martha Lipson
Levenson, Stanley Melvin
Lubic, Ruth Watson
Lythcott, George I
Maillie, Hugh David
Martin, Constance R
Matsuzaki, Masaji
Mausner, Leonard Franklin
Meinhold, Charles Boyd
Michaelson, Solomon M
Midlarsky, Elizabeth
Mizma, Edward John
Moline, Sheldon Walter
Movshon, J Anthony
Nathan, Ronald Gene
Neu, Harold Conrad
Nevid, Jeffrey Steven
Norvitch, Mary Ellen
Okaya, Yoshi Haru
Olczak, Paul Vincent
O'Reilly, Richard
Orsini, Frank R
Ottenbacher, Kenneth John
Overweg, Norbert I A
Parker, Herbert Myers
Penn, Arthur

Pizzarello, Donald Joseph
Rafla, Sameer
Ranck, James Byrne, Jr
Ranu, Harcharan Singh
Reit, Barry
Richardson, Robert Lloyd
Ritch, Robert
Rivlin, Richard S
Roberts, Carlyle Jones
Roberts, Richard B
Roizin, Leon
Ross, John Brandon Alexander
Sackeim, Harold A
Safai, Bijan
Schenk, Eric A
Schoenfeld, Cy
Shamos, Morris Herbert
Shank, Brenda Mae Buckhold
Shanzer, Stefan
Shapiro, Raymond E
Sidel, Victor William
Slatkin, Daniel Nathan
Smith, Thomas Harry Francis
Soave, Rosemary
Stein, Ruth E K
Stull, G(eorge) A
Tatarczuk, Joseph Richard
Tichauer, Erwin Rudolph
Verma, Ram S
Victor, Jonathan David
Villanueva, German Baid
Vladeck, Bruce C
Wadell, Lyle H
Wang, Theodore Sheng-Tao
Weber, David Alexander
Weinstein, Abbott Samson
Weinstein, Arthur
Weissman, Michael Herbert
Welle, Stephen Leo
Williams, Christine
Wollman, Leo
Wright, Herbert N
Yip, Kwok Leung
York, James Lester
Yozawitz, Allan
Zaleski, Marek Bohdan
Zolla-Pazner, Susan Beth

NORTH CAROLINA
Case, Verna Miller
Caterson, Bruce
Ciancolo, George J
Cobb, Frederick Ross
Cronenberger, Jo Helen
Dingledine, Raymond J
Dykstra, Linda A
Floyd, Carey E, Jr
Friedman, Mitchell
Gallagher, Margie Lee
Hallock, James A
Heck, Jonathan Daniel
Howard, Donald Robert
Jaszczak, Ronald Jack
Jones, Lyle Vincent
Jorizzo, Joseph L
Kaplan, Berton Harris
Kirkbride, L(ouis) D(ale)
Klitzman, Bruce
Klotman, Paul
Knobeloch, F X Calvin
Krishnan, K Ranga Rama
LaFon, Stephen Woodrow
Levine, Richard Joseph
Long, Robert Allen
Lowe, James Edward
Meredith, Jesse Hedgepeth
Misek, Bernard
Murdock, Harold Russell
Murray, William J
O'Callaghan, James Patrick
O'Keefe, Edward John
Prazma, Jiri
Putnam, Charles E
Shackelford, Ernest Dabney
Strauss, Harold C
Thomas, Francis T
Thomas, William Grady
Webber, Richard Lyle
Wiley, Albert Lee, Jr
Wilson, Ruby Leila
Young, Larry Dale
Yount, William J

OHIO
Anderson, Douglas K
Antar, Mohamed Abdelchany
Ball, Edward James
Bannan, Elmer Alexander
Bar-Yishay, Ephraim
Berger, Melvin
Boulant, Jack A
Breglia, Rudolph John
Broge, Robert Walter
Brunengraber, Henri
Brunner, Robert Lee
Cacioppo, John T
Carter, James R
Carter, R Owen, Jr
Chang, Jae Chan
Colbert, Charles
Coots, Robert Herman
Cope, Frederick Oliver
Cruz, Julio C
Daroff, Robert Barry

Dauchot, Paul J
Decker, Clarence Ferdinand
DeNelsky, Garland
Doershuk, Carl Frederick
Dougherty, John A
Ewall, Ralph Xavier
Frank, Thomas Paul
Fucci, Donald James
Glaser, Roger Michael
Graham, Robert M
Greenstein, Julius S
Hinkle, George Henry
Hinman, Channing L
Hohnadel, David Charles
Hollander, Philip B
Holzbach, R Thomas
Hunt, Carl E
Jamasbi, Roudabeh J
Kaplan, Arthur Lewis
Lesner, Sharon A
Levin, Jerome Allen
Lima, John J
Lydy, David Lee
McFadden, Edward Regis
McKee, Michael Geoffrey
McNamara, John Regis
Mantil, Joseph Chacko
Miraldi, Floro D
Muir, William W, III
Mulick, James Anton
Myers, William Graydon
Neiman, Gary Scott
Nigrovic, Vladimir
Osipow, Samuel Herman
Ottolenghi, Abramo Cesare
Patton, Nancy Jane
Phillips, Chandler Allen
Pinchak, Alfred Cyril
Resnick, Martin I
Rheins, Lawrence A
Roche, Alexander F
Roessmann, Uros
Ross, John Edward
Skelly, Michael Francis
Smith, Carl Clinton
Somerson, Norman L
Starchman, Dale Edward
Stockman, Charles H(enry)
Strauss, Richard Harry
Suskind, Raymond Robert
Tallman, Richard Dale (Junior)
Thomas, Donald Charles
Vorhees, Charles V
Wallin, Richard Franklin
Warner, Ann Marie
Weiss, Harold Samuel
Wilkin, Jonathan Keith
Wodarski, John Stanley
Zoglio, Michael Anthony

OKLAHOMA
Folk, Earl Donald
Harrison, Aix B
Hochhaus, Larry
Lee, Kyung No
Letcher, John Henry, III
Lovallo, William Robert
Martin, Loren Gene
Melton, Carlton E, Jr
Min, Kyung-Whan
Revzin, Alvin Morton
Schaefer, Carl Francis
Schaefer, Frederick Vail
Schroeder, David J Dean
Schultz, Russell Thomas
Steffey, Oran Dean
Switzer, Laura Mae
Thomas, Ellidee Dotson

OREGON
Ary, Dennis
Buist, Neil R M
Burry, Kenneth A
Dooley, Douglas Charles
Ellinwood, William Edward
Fay, Warren Henry
Greenberg, Barry H
Hackman, Robert Mark
Jastak, J Theodore
Johnson, Larry Reidar
Kay, Michael Aaron
Maksud, Michael George
Matarazzo, Joseph Dominic
Parrott, Marshall Ward
Severson, Herbert H
Smith, Alvin Winfred

PENNSYLVANIA
Allen, Michael Thomas
Aston, Richard
Axel, Leon
Bagasra, Omar
Beavers, Ellington McHenry
Black, Martin
Blumberg, Baruch Samuel
Bochenek, Wieslaw Janusz
Bove, Alfred Anthony
Cander, Leon
Chan, Ping-Kwong
Chapman, John Donald
Cohn, Ellen Rassas
Cole, James Edward
Couch, Jack Gary
Daly, John

Dekker, Andrew
Delivoria-Papadopoulos, Maria
Diorio, Alfred Frank
Dobson, Margaret Velma
Doghramji, Karl
Edelstone, Daniel I
Etherton, Terry D
Fallon, Michael David
Fish, James E
Fisher, Robert Stephen
Fraser, Nigel William
Friedman, Eitan
Friend, Judith H
Fromm, Gerhard Hermann
Fuscaldo, Kathryn Elizabeth
Gerich, John Edward
Glessner, Alfred Joseph
Glick, Mary Catherine
Gould, Anne Bramlee
Grenvik, Ake N A
Haddad, John George
Hale, Creighton J
Harris, Dorothy Virginia
Harty, Michael
Hirschman, Lynette
Hite, Mark
Hoek, Joannes (Jan) Bernardus
Holsclaw, Douglas S, Jr
Homan, Elton Richard
Hook, Jerry Bruce
Houts, Peter Stevens
Isett, Robert David
Johnson, Candace Sue
Jungkind, Donald Lee
Kalia, Madhu P
Keith, Jennie
Kendall, Philip C
Kidawa, Anthony Stanley
Klein, Herbert A
Knutson, David W
Kresh, J Yasha
Kressel, Herbert Yehude
Kunz, Heinz W
Leddy, Susan
Leighton, Charles Cutler
Levine, Geoffrey
Liebenberg, Stanley Phillip
Lin, Jiunn H
Magargal, Larry Elliot
Mansmann, Herbert Charles, Jr
Miller, Barry
Miller, Kenneth L
Minsker, David Harry
Moster, Mark Leslie
Moyer, Robert (Findley)
Mulvihill, John Joseph
Nambi, Ponnal
Needleman, Herbert L
Neufeld, Gordon R
Orenstein, David M
Pack, Allan I
Piperno, Elliot
Porter, Sydney W, Jr
Prout, James Harold
Pruett, Esther D R
Rao, Vijay Madan
Reilly, Christopher F
Rescorla, Robert A
Rinaldo, Charles R
Robertson, Robert James
Rosenberg, Henry
Rozmiarek, Harry
Ruggieri, Michael Raymond
Ryan, James Patrick
Sashin, Donald
Schreiber, Alan D
Shiau, Yih-Fu
Shoub, Earle Phelps
Smith, M Susan
Somers, Anne R
Sperling, Mark Alexander
Steplewski, Zenon
Studer, Rebecca Kathryn
Suzuki, Jon Byron
Tabachnick, Joseph
Tarver, James H, Jr
Tice, Linwood Franklin
Turk, Dennis Charles
Van Ausdal, Ray Garrison
Villafana, Theodore
Walsh, Arthur Campbell
Waterman, Frank Melvin

RHODE ISLAND
Amols, Howard Ira
Barker, Barbara Elizabeth
Barnes, Frederick Walter, Jr
Gilman, John Richard, Jr
Hutzler, Leroy, III
Lipsitt, Lewis Paeff
McIlwain, James Terrell
Rotman, Boris

SOUTH CAROLINA
Johnson, James Allen, Jr
Kaiser, Charles Frederick
Patterson, C(leo) Maurice
Powell, Donald Ashmore
Rathbun, Ted Allan
Reinig, William Charles
Sanders, Samuel Marshall, Jr
Sartiano, George Philip

TENNESSEE
Burish, Thomas Gerard
Cloutier, Roger Joseph
Darden, Edgar Bascomb
Deivanayagan, Subramanian
Fleischer, Arthur C
Gammage, Richard Bertram
Gupta, Ramesh C
Hubbell, Harry Hopkins, Jr
Jackson, Carl Wayne
Kaye, Stephen Vincent
Lawler, James E
Lenhard, Joseph Andrew
Long, Charles Joseph
Lott, Sam Houston, Jr
Love, Russell Jacques
McConnell, Freeman Erton
Nag, Subir
Partain, Clarence Leon
Rivas, Marian Lucy
Roberts, Lee Knight
South, Mary Ann
Stone, William John
Strupp, Hans H
Studebaker, Gerald A
Sun, Deming
Tai, Douglas L
Tanner, Raymond Lewis
Thurman, Gary Boyd
Watson, Evelyn E
Wiley, Ronald Gordon
Wondergem, Robert
Wyler, Allen Raymer

TEXAS
Ali, Yusuf
Alperin, Jack Bernard
Armstrong, George Glaucus, Jr
Ashby, Jon Kenneth
Azizi, Sayed Ausim
Barnes, Jeffrey Lee
Bencomo, José A
Benison, Betty Bryant
Biles, Robert Wayne
Bonner, Hugh Warren
Bonte, Frederick James
Brown, J(ack) H(arold) U(pton)
Buckalew, Louis Walter
Cardus, David
Chilian, William M
Cooper, John C, Jr
Cottone, James Anthony
David, Yadin B
Ellis, Kenneth Joseph
Feldman, Stuart
Fernandes, Gabriel
Geoghegan, William David
German, Dwight Charles
German, Victor Frederick
Godfrey, John Joseph
Gould, K Lance
Greeley, George H, Jr
Griffin, Kathleen (Mary)
Gwirtz, Patricia Ann
Harkless, Lawrence Bernard
Hatch, John Phillip
Henrich, Williams Lloyd
Hong, Bor-Shyue
Horton, Jureta
Horton, William A
Hurd, Eric R
Hussian, Richard A
Ickes, William K
Jenkinson, Stephen G
Johnson, John E, Jr
Johnson, Patricia Ann J
Kadekaro, Massako
Kalkwarf, Kenneth Lee
Karacan, Ismet
Kasi, Leela Peshkar
Kehrer, James Paul
Krohmer, Jack Stewart
Kuo, Peter Te
Kutzman, Raymond Stanley
Lal, Harbans
Lane, John D
Lehmkuhl, L Don
Levin, Harvey Steven
MacDonald, Paul Cloren
Malloy, Michael H
Mavligit, Giora M
Mayer, Tom G
Meltz, Martin Lowell
Mendel, Julius Louis
Minerbo, Grace Moffat
Minna, John
Neff, Richard D
Patel, Anil S
Pennebaker, James W
Perry, Robert Riley
Posey, Daniel Earl
Poston, John Ware
Prentice, Norman Macdonald
Rehm, Lynn P
Reid, Michael Baron
Riser, Mary Elizabeth
Roffwarg, Howard Philip
Rolston, Kenneth Vijaykumar Issac
Rudenberg, Frank Hermann
Ryan, Charles F
Schoenfeld, Lawrence Steven
Shulman, Robert Jay
Siler-Khodr, Theresa M
Smith, Michael Lew

Sontheimer, Richard Dennis
Spence, Dale William
Stedman, James Murphey
Tate, Charlotte Anne
Tomasovic, Stephen Peter
Tsin, Andrew Tsang Cheung
Walsh, Scott Wesley
Willerson, James Thornton
Wolfenberger, Virginia Ann
Younes, Muazaz A
Yung, W K Alfred

UTAH
Ailion, David Charles
Archer, Victor Eugene
Berenson, Malcolm Mark
Black, Dean
Kohler, R Ramon
Lee, Steve S
Mays, Charles William
Molteni, Richard A(loysius)
Moser, Royce, Jr
Olsen, Don B
Peters, Jeffrey L
Petty, William Clayton
Schiager, Keith Jerome
Stanley, Theodore H
Striefel, Sebastian
Westenskow, Dwayne R

VERMONT
Evans, John N
Falk, Leslie Alan
Halpern, William
Pope, Malcolm H
Rosen, James Carl
Wooding, William Minor

VIRGINIA
Arora, Narinder
Asterita, Mary Frances
Auerbach, Stephen Michael
Bolton, W Kline
Bonvillian, John Doughty
Burger, Carol J
Chaudhuri, Tapan K
Furr, Aaron Keith
Grunder, Hermann August
Hanig, Joseph Peter
Harrell, Ruth Flinn
Hicks, John T
Hoegerman, Stanton Fred
Howard-Peebles, Patricia Nell
Kim, Yong Il
Kripke, Bernard Robert
Kurtzke, John F
Lenhardt, Martin Louis
McDonald, Gerald O
Nagarkatti, Mitzi
Nixon, John V
Payton, Otto D
Reppucci, Nicholas Dickon
Richards, Paul Bland
Ruhling, Robert Otto
Scheld, William Michael
Severns, Matthew Lincoln
Talbot, Richard Burritt
Thames, Marc
Thorner, Michael Oliver
Wands, Ralph Clinton
Wells, John Morgan, Jr
Wilson, John William
Wist, Abund Ottokar

WASHINGTON
Anderson, Robert Authur
Bair, William J
Bichsel, Hans
De Tornyay, Rheba
Evermann, James Frederick
Frank, Andrew Julian
Gollnick, Philip D
Grootes-Reuvecamp, Grada Alijda
Hutchison, Nancy Jean
Johnson, John Richard
Jonsen, Albert R
Kalina, Robert E
Kelly, Gregory M
Kenney, Gerald
Lakshminarayan, S
Larsen, Lawrence Harold
Ledbetter, Jeffrey A
Murphy, Sheldon Douglas
Nardella, Francis Anthony
Nelson, Iral Clair
Richardson, Michael Lewellyn
Roth, Gerald J
Salk, Darrell John
Shapley, James Louis
Shriver, M Kathleen
Smith, Victor Herbert
Soldat, Joseph Kenneth
Turner, Judith Ann
Wedlick, Harold Lee
Wehner, Alfred Peter
Wilson, Christopher

WEST VIRGINIA
Ballantyne, Bryan
Douglass, Kenneth Harmon
Edelstein, Barry Allen
Olenchock, Stephen Anthony
Reasor, Mark Jae

Other Medical & Health Sciences (cont)

WISCONSIN
Barschall, Henry Herman
Boerboom, Lawrence E
Brazy, Peter C
Cohen, Bernard Allan
Diaz, Luis A
Emmerling, David Alan
Friend, Milton
Goldstein, Robert
Hamsher, Kerry de Sandoz
Haughton, Victor Mellet
Hekmat, Hamid Moayed
Holden, James Edward
Kaufman, Paul Leon
Kent, Raymond D
Komai, Hirochika
Lee, Ping-Cheung
Nagle, Francis J
Nellis, Stephen H
Shah, Dharmishtha V
Terry, Leon Cass
Tormey, Douglass Cole
Valdivia, Enrique
Vanderheiden, Gregg
Warltier, David Charles

WYOMING
Holbrook, Frederick R

PUERTO RICO
Lewis, Brian Kreglow

ALBERTA
Albritton, William Leonard
Cottle, Merva Kathryn Warren
Famiglietti, Edward Virgil, Jr
Jimbow, Kowichi
Jones, Richard Lee
Larke, R(obert) P(eter) Bryce
Leung, Alexander Kwok-Chu
Mash, Eric Jay
Ruth, Royal Francis
Urtasun, Raul C

BRITISH COLUMBIA
Craig, Kenneth Denton
Cynader, Max Sigmund
Eaves, Allen Charles Edward
Laszlo, Charles Andrew
Laye, Ronald Curtis
Lee, Chi-Yu Gregory
Milsum, John H
Singh, Raj Kumari
Thurlbeck, William Michael

MANITOBA
Adaskin, Eleanor Jean
Cowdrey, Eric John
Gerrard, Jonathan M
Rigatto, Henrique
Sinha, Helen Luella

NOVA SCOTIA
Jones, John Verrier
Murray, T J
Tonks, Robert Stanley
Walker, Joan Marion

ONTARIO
Andrew, George McCoubrey
Basu, Prasanta Kumar
Birnboim, Hyman Chaim
Cafarelli, Enzo Donald
Cutz, Ernest
Czuba, Margaret
Drake, Charles George
Ferguson, Alastair Victor
Girvin, John Patterson
Goldenberg, Gerald J
Hare, William Currie Douglas
Harvey, John Wilcox
Harwood-Nash, Derek Clive
Holford, Richard Moore
Ing, Harry
Kennedy, James Cecil
Leeper, Herbert Andrew, Jr
Lemaire, Irma
Logan, James Edward
McLachlan, Richard Scott
Mills, David Edward
Mustard, James Fraser
Nielsen, Klaus H B
Osborne, Richard Vincent
Papp, Kim Alexander
Pfalzner, Paul Michael
Quinsey, Vernon Lewis
Rabinovitch, Marlene
Rapoport, Anatol
Rock, Gail Ann
Rogers, David William Oliver
Roland, Charles Gordon
Sibbald, William John
Silverman, Melvin
Sobell, Linda Carter
Sobell, Mark Barry
Till, James Edgar
Tolnai, Susan
Tyson, John Edward Alfred

QUEBEC
Burgess, John Herbert
Cantin, Marc
Chang, Thomas Ming Swi
Dykes, Robert William
Grassino, Alejandro E
Karpati, George
Kelly, Paul Alan
Landry, Fernand
Levy, Samuel Wolfe
Lussier, Gilles L
Malo, Jacques
Marisi, Dan(iel) Quirinus
Melzack, Ronald
Milner, Brenda (Atkinson)
Palayew, Max Jacob
Paris, David Leonard
Roskies, Ethel
Serbin, Lisa Alexandra
Tousignant, Michel

SASKATCHEWAN
Shokeir, Mohamed Hassan Kamel

OTHER COUNTRIES
Allessie, Maurits
Bader, Hermann
Bernstein, David Maier
Chung, Chien
Doll, Richard
Dunn, Christopher David Reginald
Fleckenstein, Albrecht
Garrard, Christopher S
Gomez, Arturo
Laura, Patricio Adolfo Antonio
Lekeux, Pierre Marie
Lorenz, Konrad Zacharias
Lynn, Robert K
McBeath, Elena R
Mitruka, Brij Mohan
O'Neill, Patricia Lynn
Quastel, Michael Reuben
Sterpetti, Antonio Vittorio
Stingl, Georg
Tsolas, Orestes
Weil, Marvin Lee

PHYSICS & ASTRONOMY

Acoustics

ALABAMA
Borton, Thomas Ernest
Jones, Jess Harold

ALASKA
Roederer, Juan Gualterio
Thomas, Gary Lee

ARIZONA
Bowen, Theodore
Cathey, Everett Henry
Fliegel, Frederick Martin
Gatley, William Stuart
Lamb, George Lawrence, Jr
Walker, Charles Thomas
Wallace, C(harles) E(dward)

ARKANSAS
Horton, Philip Bish
Smith, E(astman)
Wold, Donald C

CALIFORNIA
Ahumada, Albert Jil, Jr
Armstrong, Donald B
Atchley, Anthony A
Aurand, Henry Spiese, Jr
Ayers, Raymond Dean
Bachman, Richard Thomas
Barmatz, Martin Bruce
Behravesh, Mohamad Martin
Booth, Newell Ormond
Bourke, Robert Hathaway
Brown, Maurice Vertner
Bucker, Homer Park, Jr
Bushnell, James Judson
Carterette, Edward Calvin Hayes
Cho, Alfred Chih-Fang
Cho, Young-chung
Chodorow, Marvin
Coppens, Alan Berchard
Cummings, William Charles
De, Gopa Sarkar
Dempsey, Martin E(wald)
Eargle, John Morgan
Edmonds, Peter Derek
Fisher, Frederick Hendrick
Fitzgerald, Thomas Michael
Flatte, Stanley Martin
Gales, Robert Sydney
Galloway, William Joyce
Garrett, Steven Lurie
Grace, O Donn
Green, Charles E
Green, Philip S
Greenfield, Eugene W(illis)
Gulkis, Samuel
Halley, Robert
Henning, Harley Barry
Holliday, Dale Vance
Johnson, Robert A
Jones, Joie Pierce

Kagiwada, Reynold Shigeru
Keller, Joseph Bishop
Khoshnevisan, Mohsen Monte
Kinnison, Gerald Lee
Kino, Gordon Stanley
Kirchner, Ernst Karl
Knollman, Gilbert Carl
Lax, Edward
Lee, Keun Myung
Lovett, Jack R
Lubman, David
McIntyre, Marie Claire
Mackenzie, Kenneth Victor
Marcellino, George Raymond
Marcus, John Stanley
Martin, Gordon Eugene
Medwin, Herman
Merritt, Thomas Parker
Metherell, Alexander Franz
Meyer, Norman Joseph
Mikes, Peter
Mitzner, Kenneth Martin
Moe, Chesney Rudolph
Monkewitz, Peter Alexis
Morrow, Charles Tabor
Munson, Albert G
Nelson, Rex Roland
Neubert, Jerome Arthur
Nichols, Rudolph Henry
Piersol, Allan Gerald
Pierucci, Mauro
Potter, David Samuel
Powers, John Patrick
Reiss, Diana
Reynolds, Michael David
Rott, Nicholas
Rudnick, Isadore
Salmon, Vincent
Sanders, James Vincent
Schacher, Gordon Everett
Scharton, Terry Don
Schenck, Harry Allen
Segal, Alexander
Slaton, Jack H(amilton)
Sutherland, Louis Carr
Thomas, Graham Havens
Thompson, Paul O
Thorpe, Howard A(lan)
Vent, Robert Joseph
Warshaw, Stephen I
Watson, Earl Eugene
White, Ray Henry
White, Richard Wallace
Wiedow, Carl Paul
Williams, Jack Rudolph
Wilson, Oscar Bryan, Jr
Young, Robert William

COLORADO
Brown, Edmund Hosmer
Clifford, Steven Francis
DeSanto, John Anthony
Geers, Thomas L
Schultz, Theodore John
Sommerfeld, Richard Arthur
Verschoor, J(ack) D(ahlstrom)
White, James Edward

CONNECTICUT
Apfel, Robert Edmund
Arnoldi, Robert Alfred
Bell, Thaddeus Gibson
Burkhard, Mahlon Daniel
Cable, Peter George
Carter, G Clifford
Eby, Edward Stuart, Sr
Erf, Robert K
Fehr, Robert O
Gay, Thomas John
Ghen, David C
Green, Eugene L
Hanrahan, John J
Hasse, Raymond William, Jr
Hemond, Conrad J(oseph), Jr
Howkins, Stuart D
Kraft, David Werner
Maples, Louis Charles
Mellen, Robert Harrison
Orphanoudakis, Stelios Constantine
Paterson, Robert W
Peracchio, Aldo Anthony
Plona, Thomas Joseph
Reed, Robert Willard
Richards, Roger Thomas
Rubega, Robert A
Sen, Pabitra Narayan
Shirer, Donald Leroy
Tobias, Jerry Vernon
Von Winkle, William A
Williams, David G(erald)

DELAWARE
Gulick, Walter Lawrence
Holt, Rush Dew, Jr
Mehl, James Bernard
Sohl, Cary Hugh

DISTRICT OF COLUMBIA
Baer, Ralph Norman
Berkson, Jonathan Milton
Borgiotti, Giorgio V
Corsaro, Robert Dominic
Gaumond, Charles Frank
Hershey, Robert Lewis

Ingenito, Frank Leo
Knight, John C (Ian)
Mayer, Walter Georg
Menton, Robert Thomas
Rojas, Richard Raimond
Rosenblum, Lawrence Jay
Rosenthal, F(elix)
Strasberg, Murray
Wolf, Stephen Noll
Yaniv, Simone Liliane
Yen, Nai-Chyuan

FLORIDA
Beguin, Fred P
Blue, Joseph Edward
Clarke, Thomas Lowe
DeFerrari, Harry Austin
Dubbelday, Pieter Steven
Elfner, Lloyd F
Flax, Lawrence
Groves, Ivor Durham, Jr
Henriquez, Theodore Aurelio
Hollien, Harry
Kinzey, Bertram Y(ork), Jr
Mahig, Joseph
Malocha, Donald C
Mooers, Christopher Northrup Kennard
Olmstead, Paul Smith
Pichal, Henri Thomas, II
Piquette, Jean Conrad
Proni, John Roberto
Ramer, Luther Grimm
Rhodes, Richard Ayer, II
Rona, Donna C
Rothman, Howard Barry
Shih, Hansen S T
Siebein, Gary Walter
Silverman, Sol Richard
Tam, Christopher K W
Tanner, Robert H
Thomas, Dan Anderson
Ting, Robert Yen-Ying
Van Buren, Arnie Lee
Veinott, Cyril G
Vidmar, Paul Joseph
Wiebusch, Charles Fred
Williamson, Donald Elwin
Zalesak, Joseph Francis

GEORGIA
Ginsberg, Jerry Hal
Lau, Jark Chong
Nave, Carl R
Patronis, Eugene Thayer, Jr
Rogers, Peter H
Zinn, Ben T

HAWAII
Frazer, L Neil
Learned, John Gregory
Yates, James T

IDAHO
Johnson, John Alan

ILLINOIS
Achenbach, Jan Drewes
Boggs, George Johnson
Butler, Robert Allan
Elliott, Lois Lawrence
Frizzell, Leon Albert
Guttman, Newman
Haugen, Robert Kenneth
Kessler, Lawrence W
Mintzer, David
O'Brien, William Daniel, Jr
Rey, Charles Albert
Sathoff, H John
Schultz, Martin C
Solie, Leland Peter
Twersky, Victor
Wey, Albert Chin-Tang
Yost, William A

INDIANA
Brach, Raymond M
Cruikshank, Donald Burgoyne, Jr
Eggleton, Reginald Charles
Farringer, Leland Dwight
Hansen, Uwe Jens
Kelly-Fry, Elizabeth
Kent, Earle Lewis
Meeks, Wilkison (Winfield)
Rickey, Martin Eugene
Scheiber, Donald Joseph
Stanley, Gerald R
Tree, David R
Tubis, Arnold
Watson, Charles S

IOWA
Buck, Otto
Hanson, Roger James
Korpel, Adrianus
Papadakis, Emmanuel Philippos
Schwartz, Ralph Jerome
Small, Arnold McCollum, Jr
Thompson, Robert Bruce
Titze, Ingo Roland
Voots, Richard Joseph
Wilson, Lennox Norwood

KANSAS
Farokhi, Saeed

Ferraro, John Anthony

KENTUCKY
Brill, Joseph Warren
Edwards, Richard Glenn
Naake, Hans Joachim

LOUISIANA
Beeson, Edward Lee, Jr
Bradley, Marshall Rice
Chin-Bing, Stanley Arthur
Collins, Mary Jane
Erath, Louis W
Farwell, Robert William
Grimsal, Edward George
Head, Martha E Moore
Lafleur, Louis Dwynn
Slaughter, Milton Dean
Wenzel, Alan Richard

MAINE
LaCasce, Elroy Osborne, Jr
Shonle, John Irwin
Walkling, Robert Adolph

MARYLAND
Achor, William Thomas
Andrulis, Marilyn Ann
Bell, James F(rederick)
Bennett, Charles L
Berger, Harold
Blake, William King
Blatstein, Ira M
Brownell, William Edward
Causey, G(eorge) Donald
Chimenti, Dale Everett
Cook, Richard Kaufman
Cornett, Richard Orin
Cornyn, John Joseph
Crump, Stuart Faulkner
Devin, Charles, Jr
Dickey, Joseph W
Eitzen, Donald Gene
Etter, Paul Courtney
Everstine, Gordon Carl
Feit, David
Gammell, Paul M
Gammon, Robert Winston
Gaunaurd, Guillermo C
Geneaux, Nancy Lynne
Goepel, Charles Albert
Gottwald, Jimmy Thorne
Greenspon, Joshua Earl
Gupta, Vaikunth N
Hansen, Robert Jack
Harrison, George H
Hartmann, Bruce
Ho, Louis T
Hodge, David Charles
Johnson, William Bowie
Kamrass, Murray
Linzer, Melvin
McNicholas, John Vincent
Madigosky, Walter Myron
Martin, John J(oseph)
Moses, Edward Joel
Munson, J(ohn) C(hristian)
Palmer, Charles Harvey
Peppin, Richard J
Pierce, Harry Frederick
Potocki, Kenneth Anthony
Raff, Samuel J
Reader, Wayne Truman
Scharnhorst, Kurt Peter
Schleidt, Wolfgang Matthias
Schulkin, Morris
Sevik, Maurice
Skinner, Dale Dean
South, Hugh Miles
Urick, Robert Joseph
Vande Kieft, Laurence John
Weinstein, Marvin Stanley
Williams, Robert Glenn
Zankel, Kenneth L

MASSACHUSETTS
Barger, James Edwin
Baxter, Lincoln, II
Beranek, Leo Leroy
Bers, Abraham
Bogert, Bruce Plympton
Bolt, Richard Henry
Bose, Amar G(opal)
Breton, J Raymond
Burke, Shawn Edmund
Butler, John Louis
Callerame, Joseph
Chase, David Marion
Cole, John E(mery), III
Cudworth, Allen L
Dyer, Ira
Dym, Clive L
Eldred, Kenneth M
Figwer, J(ozef) Jacek
Free, John Ulric
Frisk, George Vladimir
Garrelick, Joel Marc
Griesinger, David Hadley
Harrison, John Michael
Hawkins, W(illiam) Bruce
Hills, Robert, Jr
Hurlburt, Douglas Herendeen
Ivey, Elizabeth Spencer
Junger, Miguel C

Keast, David N(orris)
Kerwin, Edward Michael, Jr
Labate, Samuel
McElroy, Paul Tucker
Malme, Charles I(rving)
Milder, Fredric Lloyd
Morse, Robert Warren
Nelson, Donald Frederick
Nelson, Keith A
Noiseux, Claude Francois
Pope, Joseph
Ryan, Robert Pat
Schart, Bertram
Seifer, Arnold David
Singer, Howard Joseph
Spindel, Robert Charles
Stephen, Ralph A
Stern, Ernest
Stevens, Kenneth N(oble)
Stowe, David William
Sulak, Lawrence Richard
Thompson, Charles
Vignos, James Henry
White, James Victor

MICHIGAN
Akay, Adnan
Birdsall, Theodore G
Blaser, Dwight A
Brown, Steven Michael
Gaerttner, Martin R
Gessert, Walter Louis
Goldstein, Albert
Hemdal, John Frederick
Hickling, Robert
Johnson, Glen Eric
Johnson, Wayne Douglas
Moody, David Burritt
Nefske, Donald Joseph
Porter, James Colegrove
Reiher, Harold Frederick
Sawatari, Takeo
Schubring, Norman W(illiam)
Stebbins, William Cooper
Tosi, Oscar I
Weinreich, Gabriel
Wolf, Joseph A(llen), Jr
Wright, Wayne Mitchell
Yoder, Levon Lee

MINNESOTA
Anderson, Ronald Keith
Lea, Wayne Adair
Mueller, Rolf Karl
Van Doeren, Richard Edgerly
Ward, Wallace Dixon
Wild, John Julian

MISSISSIPPI
Bolen, Lee Napier, Jr
Breazeale, Mack Alfred
Bruce, John Goodall, Jr
Crum, Lawrence Arthur
Fleischer, Peter
Love, Richard Harrison
Morris, Gerald Brooks
Morris, Halcyon Ellen McNeil
Posey, Joe Wesley
Ramsdale, Dan Jerry

MISSOURI
Agarwal, Ramesh K
Conradi, Mark Stephen
Ellison, John Vogelsanger
Kibens, Valdis
Miller, James Gegan
Perisho, Clarence H(oward)
Rigden, John Saxby
Weissenburger, Jason T

NEW HAMPSHIRE
Baumann, Hans D
Frost, Albert D(enver)
Von-Recklinghausen, Daniel R

NEW JERSEY
Ballato, Arthur
Berkley, David A
Carter, Ashley Hale
Francis, Samuel Hopkins
Fretwell, Lyman Jefferson, Jr
Gaer, Marvin Charles
Goodfriend, Lewis S(tone)
Guernsey, Richard Montgomery
Hall, Joseph L
Harvey, Floyd Kallum
Joyce, William B(axter)
Kam, Gar Lai
Kennedy, Anthony John
Kessler, Frederick Melvyn
Labianca, Frank Michael
Levinson, Stephen
Lunsford, Ralph D
Miller, George Earl
Mitchell, Olga Mary Mracek
Parkins, Bowen Edward
Polak, Joel Allan
Prasad, Marehalli Gopalan
Quinlan, Daniel A
Raichel, Daniel R(ichter)
Rothleder, Stephen David
Scales, William Webb
Stickler, David Collier
Teaney, Dale T

Temkin, Samuel
Thurston, Robert Norton
Vahaviolos, Sotirios J
Wilson, Charles Elmer
Yi-Yan, Alfredo
Zipfel, George G, Jr

NEW MEXICO
Alers, George A
Frost, Harold Maurice, III
Kelly, Robert Emmett
Martin, Richard Alan
Mercer-Smith, James A
Reed, Jack Wilson
Roberts, Peter Morse
Schwarz, Ricardo
Sinha, Dipen N
Swingle, Donald Morgan
Visscher, William M

NEW YORK
Angevine, Oliver Lawrence
Bienvenue, Gordon Raymond
Blazey, Richard N
Brody, Burton Alan
Carleton, Herbert Ruck
Cheney, Margaret
Cheung, Lim H
Frankel, Julius
Ganley, W Paul
Gavin, Donald Arthur
George, A(lbert) R(ichard)
Gerstman, Hubert Louis
Haines, Daniel Webster
Harris, Cyril Manton
Hiller, Lejaren Arthur
Hollenberg, Joel Warren
Jacobson, Melvin Joseph
Katocs, Andrew Stephen, Jr
Khanna, Shyam Mohan
Klepper, David Lloyd
Kornreich, Philipp G
Kritz, J(acob)
Lang, William Warner
Lean, Eric Gung-Hwa
McDonald, Frank Alan
Maling, George Croswell, Jr
Melcher, Robert Lee
Meyer, Robert Jay
Nelson, David Elmer
Oliner, Arthur A(aron)
Ozimek, Edward Joseph
Sachse, Wolfgang H
Sackman, George Lawrence
Savkar, Sudhir Dattatraya
Shiren, Norman S
Siegmann, William Lewis
Smith, Lowell Scott
Talham, Robert J
Taylor, Kenneth Doyle
Tehon, Stephen Whittier
White, Frederick Andrew
Zwislocki, Jozef John

NORTH CAROLINA
Bowie, James Dwight
Brown, Richard Dean
Dixon, Norman Rex
Hageseth, Gaylord Terrence
Hutchins, William R(eagh)
Keltie, Richard Francis
Kremkau, Frederick William
Lawson, Dewey Tull
Soderquist, David Richard
Wilson, Geoffrey Leonard

NORTH DAKOTA
Dewar, Graeme Alexander

OHIO
Adler, Laszlo
Chu, Mamerto Loarca
Eby, Ronald Kraft
Ensminger, Dale
Ghering, Walter L
Greene, David C
Hagelberg, M(yron) Paul
Lai, Jai-Lue
Lehiste, Ilse
Marshall, Kenneth D
Martin, Daniel William
Mawardi, Osman Kamel
Neiman, Gary Scott
Nichols, William Herbert
Oyer, Herbert Joseph
Pearson, Jerome
Proctor, David George
Richards, Walter Bruce
Rodman, Charles William
Schuele, Donald Edward
Stumpf, Folden Burt
Von Gierke, Henning Edgar
Yun, Seung Soo
Zaman, Khairul B M Q

OKLAHOMA
Boade, Rodney Russett
Brown, Graydon L
Coffman, Moody Lee
Major, John Keene

OREGON
Chartier, Vernon L
White, Donald Harvey

PENNSYLVANIA
Coltman, John Wesley
DeSanto, Daniel Frank
Goldman, Robert Barnett
Grindall, Emerson Leroy
Harbold, Mary Leah
Harrold, Ronald Thomas
Josephs, Jess J
Lauchle, Gerald Clyde
Maynard, Julian Decatur
Meyer, Paul A
Panuska, Joseph Allan
Pearman, G Timothy
Pierce, Allan Dale
Prout, James Harold
Reid, John Mitchell
Rhodes, Donald Frederick
Schumacher, Robert Thornton
Shung, K Kirk
Sibul, Leon Henry
Smith, Wesley R
Spalding, George Robert
Stern, Richard
Thompson, William, Jr
Tittmann, Bernhard R
Toulouse, Jean
Varadan, Vasundara Venkatraman
Whitman, Alan M
Young, In Min

RHODE ISLAND
Bartram, James F(ranklin)
Becken, Bradford Albert
Beyer, Robert Thomas
Browning, David Gunter
Dietz, Frank Tobias
Dinapoli, Frederick Richard
Ehrlich, Stanley L(eonard)
Ellison, William Theodore
Lall, Prithvi C
Mellberg, Leonard Evert
Ng, Kam Wing
Paster, Donald L(ee)
Petrou, Panayiotis Polydorou
Sherman, Charles Henry
Stepanishen, Peter Richard

SOUTH CAROLINA
Manson, Joseph Richard
Uldrick, John Paul

TENNESSEE
Bell, Zane W
Harris, Wesley Leroy
Lewis, James W L
Nabelek, Anna K
Stewart, James Monroe
Swim, William B(axter)

TEXAS
Anderson, Aubrey Lee
Angona, Frank Anthony
Ashby, Jon Kenneth
Baker, Dudley Duggan, III
Barnard, Garland Ray
Barthel, Romard
Blackstock, David Theobald
Boyer, Lester Leroy
Carroll, Michael M
Chung, Jing Yan
Connor, William Keith
DiFoggio, Rocco
Few, Arthur Allen
Finch, Robert D
Fountain, Lewis Spencer
Frazer, Marshall Everett
Griffy, Thomas Alan
Gruber, George J
Hampton, Loyd Donald
Hartley, Craig Jay
Hayre, Harbhajan Singh
Horton, Claude Wendell, Sr
Ickes, William K
Klein, William Richard
Kostoff, Morris R
Lengel, Robert Charles
Lopez, Jorge Alberto
McKinney, Chester Meek
Mitchell, Stephen Keith
Moyer, William C, Jr
Muir, Thomas Gustave, Jr
Nimitz, Walter W von
Packard, Robert Gay
Peters, Randall Douglas
Plemons, Terry D
Pope, Larry Debbs
Powell, Alan
Rader, Dennis
Rosenbaum, Joseph Hans
Shooter, Jack Allen
Simpson, S(tephen) H(arbert), Jr
Smith, R Lowell
Sobey, Arthur Edward, Jr
Sparks, Cecil Ray
Truchard, James Joseph
Zemanek, Joseph, Jr

UTAH
Chabries, Douglas M
Dudley, J Duane
Strong, William J

VERMONT
Gould, Robert K

Acoustics (cont)

Nyborg, Wesley LeMars
Sachs, Thomas Dudley
Spillman, William Bert, Jr

VIRGINIA

Briscoe, Melbourne George
Brooks, Thomas Furman
Burley, Carlton Edwin
Cantrell, John H(arris)
Chi, Tsung-Chin
Claus, Richard Otto
Davis, Charles Mitchell, Jr
Desiderio, Anthony Michael
Doles, John Henry, III
Dozier, Lewis Bryant
Duke, John Christian, Jr
Fuller, Christopher Robert
Gatski, Thomas Bernard
Gerlach, A(lbert) A(ugust)
Gilbert, Arthur Charles
Graver, William Robert
Hansen, Robert J
Hardin, Jay Charles
Hargrove, Logan Ezral
Heyman, Joseph Saul
Hubbard, Harvey Hart
Hurdle, Burton Garrison
Kenyon, Stephen C
Kooij, Theo
Lenhardt, Martin Louis
Lord, Norman W
Maestrello, Lucio
Morrow, Richard Joseph
Myers, Michael Kenneth
Neubauer, Werner George
Orr, Marshall H
Parvulescu, Antares
Poon, Ting-Chung
Puster, Richard Lee
Raney, William Perin
Sauder, William Conrad
Sebastian, Richard Lee
Seiner, John Milton
Smith, Carey Daniel
Spivack, Harvey Marvin
Stusnick, Eric
Sykes, Alan O'Neil
Tatro, Peter Richard
Trott, Winfield James
Ugincius, Peter
Watson, Robert Barden
Wyeth, Newell Convers
Yu, James Chun-Ying

WASHINGTON

Atneosen, Richard Allen
Chace, Alden Buffington, Jr
Feldman, Henry Robert
Forster, Fred Kurt
Hildebrand, Bernard Percy
Ingalls, Paul D
Ishimaru, Akira
Jacobs, Loyd Donald
Klepper, John Richard
Lipscomb, David M
Malone, Stephen D
Marston, Philip Leslie
Merchant, Howard Carl
Oncley, Paul Bennett
Peterson, Arnold (Per Gustaf)
Richardson, Richard Laurel
Sandstrom, Wayne Mark
Sengupta, Gautam
Thorne, Richard Eugene

WEST VIRGINIA

Lass, Norman J(ay)

WISCONSIN

Eriksson, Larry John
Flax, Stephen Wayne
Hind, Joseph Edward
Kent, Raymond D
Levy, Moises
Pillai, Thankappan A K

WYOMING

Roark, Terry P

ALBERTA

Ulagaraj, Munivandy Seydunganallur

BRITISH COLUMBIA

Greenwood, Donald Dean
Hughes, Blyth Alvin
Marko, John Robert
Seymour, Brian Richard
Thomson, David James

MANITOBA

Lyons, Edward Arthur
Svenne, Juris Peteris

NOVA SCOTIA

Roger, William Alexander

ONTARIO

Brammer, Anthony John
Chu, Wing Tin
Clarke, Robert Lee
Embleton, Tony Frederick Wallace
Hodgson, Richard John Wesley

Keeler, John S(cott)
Krishnappa, Govindappa
Kunov, Hans
Lew, Hin
Lipshitz, Stanley Paul
McMahon, Garfield Walter
Prasad, S E
Ribner, Herbert Spencer
Richarz, Werner Gunter
Schneider, Bruce Alton
Shaw, Edgar Albert George
Stewart, Thomas William Wallace
Stouffer, James L
Toole, Floyd Edward
Wong, George Shoung-Koon

QUEBEC

Leblanc, Roger M

OTHER COUNTRIES

Bies, David Alan
Eisner, Edward
Hansson, Inge Lief
Kagawa, Yukio
Martin, Jim Frank
Miyano, Kenjiro
Sessler, Gerhard Martin
Sugaya, Hiroshi
Tolstoy, Ivan
Voutsas, Alexander Matthew
  (Voutsadakis)
Wasa, Kiyotaka
Weil, Raoul Bloch

Astronomy

ALABAMA

Abbas, Mian Mohammad
Baksay, Laszlo Andras
Bartlett, James Holly
Boardman, William Jarvis
Curott, David Richard
Davis, John Moulton
Decher, Rudolf
Hermann, R(udolf)
Hinata, Satoshi
Krause, Helmut G L
Moses, Ray Napoleon, Jr
Sulentic, Jack William
Tandberg-Hanssen, Einar Andreas
Weisskopf, Martin Charles

ARIZONA

Ables, Harold Dwayne
Abt, Helmut Arthur
Barr, Lawrence Dale
Belton, Michael J S
Black, John Harry
Burrows, Adam Seth
Burstein, David
Cameron, Winifred Sawtell
Capen, Charles Franklin, Jr
Chapman, Clark Russell
Christy, James Walter
Cowley, Anne Pyne
Coyne, George Vincent
Crawford, David Livingstone
Dahn, Conard Curtis
Davis, Donald Ray
Dunkelman, Lawrence
Edlin, Frank E
Fitch, Walter Stewart
Foltz, Craig Billig
Fountain, John William
Franz, Otto Gustav
Gatley, Ian
Gehrels, Tom
Giclas, Henry Lee
Gordon, Mark A
Grandi, Steven Aldridge
Grauer, Albert D
Greeley, Ronald
Greenberg, Richard Joseph
Gregory, Brooke
Guetter, Harry Hendrik
Hagelbarger, David William
Hall, Richard Chandler
Harris, Hugh Courtney
Hartmann, William K
Hayes, Donald Scott
Hege, E K
Hilliard, Ronnie Lewis
Hoag, Arthur Allen
Hoffmann, William Frederick
Howard, Robert Franklin
Hubbard, William Bogel
Hunten, Donald Mount
Irwin, John (Henry) Barrows
Jacobson, Michael Ray
Joyce, Richard Ross
Kieffer, Hugh Hartman
Kinman, Thomas David
Kreidl, Tobias Joachim
Larson, Harold Phillip
Lebofsky, Larry Allen
Liebert, James William
Livingston, William Charles
Lockwood, George Wesley
Lucchitta, Baerbel Koesters
Lynds, Clarence Roger
Mahalingam, Lalgudi Muthuswamy
Massey, Philip Louis
Masursky, Harold

Mayall, Nicholas Ulrich
Melosh, Henry Jay, IV
Millis, Robert Lowell
Odell, Andrew Paul
O'Leary, Brian Todd
Osmer, Patrick Stewart
Pilachowski, Catherine Anderson
Randall, Lawrence Kessler, Jr
Rieke, Carol Anger
Rieke, George Henry
Rieke, Marcia Jean
Roemer, Elizabeth
Schaber, Gerald Gene
Sinton, William Merz
Soderblom, Laurence Albert
Strittmatter, Peter Albert
Strom, Robert Gregson
Swann, Gordon Alfred
Tifft, William Grant
Ulich, Bobby Lee
Vrba, Frederick John
Wasserman, Lawrence Harvey
Wehinger, Peter Augustus
Weidenschilling, Stuart John
Whitaker, Ewen A
White, Nathaniel Miller
White, Raymond E
White, Simon David Manton
Wildey, Robert Leroy
Wyckoff, Susan

ARKANSAS

Engle, Paul Randal
Rollefson, Aimar Andre
Sears, Derek William George

CALIFORNIA

Ahrens, Thomas J
Appleby, John Frederick
Arens, John Frederic
Armstrong, John William
Aumann, Hartmut Hans-Georg
Babcock, Horace W
Backer, Donald Charles
Barry, Don Cary
Benemann, John Rudiger
Bergstralh, Jay Thor
Berman, Louis
Blinn, James Frederick
Bonsack, Walter Karl
Borucki, William Joseph
Bowyer, C Stuart
Bracewell, Ronald Newbold
Brodie, Jean Pamela
Brooks, Walter Lyda
Bunch, Theodore Eugene
Buratti, Bonnie J
Burbidge, Eleanor Margaret
Burbidge, Geoffrey
Carpenter, Roland LeRoy
Carrigan, Charles Roger
Castle, Karen G
Catura, Richard Clarence
Chambers, Robert J
Chapman, Gary Allen
Chow, Brian G(ee-Yin)
Cohen, Marshall Harris
Collard, Harold Rieth
Cominsky, Lynn Ruth
Cordell, Bruce Monteith
Corwin, Harold G, Jr
Cruikshank, Dale Paul
Cudaback, David Dill
Danielson, George Edward, Jr
Davies, Merton Edward
Doolittle, Robert Frederick, II
Drake, Frank Donald
Dressler, Alan Michael
Duke, Douglas
Edelson, Sidney
Elliott, Denis Anthony
Elson, Lee Stephen
Epps, Harland Warren
Epstein, Eugene Ethan
Faber, Sandra Moore
Filippenko, Alexei Vladimir
Fisher, Philip Chapin
Fliegel, Henry Frederick
Fraknoi, Andrew Gabriel
Fymat, Alain L
Gaposchkin, Peter John Arthur
Gault, Donald E
Gibson, James (Benjamin)
Gilvarry, John James
Gregorich, David Tony
Hagar, Charles Frederick
Haisch, Bernhard Michael
Hamill, Patrick James
Hansen, Richard (Thomas)
Hanson, Robert Bruce
Harris, Alan William
Heiles, Carl
Helin, Eleanor Kay
Herzog, Emil Rudolph
Hughes, Thomas Rogers, Jr
Hunziker, Rodney William
Ingersoll, Andrew Perry
Jackson, Bernard Vernon
Janssen, Michael Allen
Johnson, Hugh Mitchell
Johnson, Torrence Vaino
Jones, Burton Fredrick
Kammerud, Ronald Claire
King, Ivan Robert

Kirk, John Gallatin
Klein, Michael John
Klemola, Arnold R
Kliore, Arvydas J(oseph)
Koch, David Gilbert
Koo, David Chih-yuen
Kraft, Robert Paul
Kristian, Jerome
Krupp, Edwin Charles
Kuiper, Thomas Bernardus Henricus
Lampton, Michael Logan
Landecker, Peter Bruce
Lark, Neil LaVern
Lee, Keun Myung
Linford, Gary Joe
McCormick, Philip Thomas
McFadden, Lucy-Ann Adams
Macy, William Wray, Jr
Malin, Michael Charles
Markowitz, Allan Henry
Martin, Terry Zachry
Matson, Dennis Ludwig
Mauche, Christopher W
Meinel, Aden Baker
Meinel, Marjorie Pettit
Merritt, Thomas Parker
Miner, Ellis Devere, Jr
Moffet, Alan Theodore
Morris, Mark Root
Mottmann, John
Muhleman, Duane Owen
Murray, Bruce C
Nelson, Burt
Nelson, Robert M
Neugebauer, Gerry
Newburn, Ray Leon, Jr
Oke, John Beverley
Oliver, Bernard M(ore)
Olsen, Edward Tait
Orton, Glenn Scott
Osterbrock, Donald E(dward)
Palluconi, Frank Don
Pennypacker, Carlton Reese
Persson, Sven Eric
Peters, Geraldine Joan
Pollack, James Barney
Popper, Daniel Magnes
Preston, George W, III
Preston, Robert Arthur
Reaves, Gibson
Resch, George Michael
Reynolds, Ray Thomas
Rhodes, Edward Joseph, Jr
Rickard, James Joseph
Robinson, Lloyd Burdette
Rosen, Harold A
Rubin, Robert Howard
Rudnyk, Marian E
Russell, John Albert
Salanave, Leon Edward
Sandage, Allan Rex
Sandmann, William Henry
Sargent, Anneila Isabel
Saunders, Ronald Stephen
Scargle, Jeffrey D
Scattergood, Thomas W
Schaefer, Albert Russell
Schmidt, Maarten
Schopp, John David
Schulte, Daniel Herman
Schuster, William John
Scott, Elizabeth Leonard
Sekanina, Zdenek
Shimabukuro, Fred Ichiro
Silk, Joseph Ivor
Simon, Lee Will
Simpson, Richard Allan
Slade, Martin Alphonse, III
Smith, Harding Eugene
Smith, Otto J(oseph) M(itchell)
Sovers, Ojars Juris
Spence, Harlan Ernest
Spinrad, Hyron
Standish, E Myles, Jr
Stauffer, John Richard
Stern, Albert Victor
Stevens, John Charles
Tarbell, Theodore Dean
Tenn, Joseph S
Terrile, Richard John
Toon, Owen Brian
Treffers, Richard Rowe
Trimble, Virginia Louise
Tucker, Wallace Hampton
Twieg, Donald Baker
Tyler, George Leonard
Underwood, James Henry
Unwin, Stephen Charles
Vasilevskis, Stanislaus
Vedder, James Forrest
Villere, Karen R
Vogt, Steven Scott
Walker, Arthur Bertram Cuthbert, Jr
Walker, Merle F
Walker, Russell Glenn
Wang, Charles P
Wannier, Peter Gregory
Weaver, Harold Francis
Weissman, Paul Robert
Welch, William John
Wells, Ronald Allen
West, Robert A
Westphal, James Adolph
Weymann, Ray J

Whitcomb, Stanley Ernest
Wilson, William John
Wood, Fergus James
Woosley, Stanford Earl
Wray, James David
Wright, Melvyn Charles Harman
Yanow, Gilbert
Yeomans, Donald Keith
Yoder, Charles Finney
Young, Andrew Tipton
Young, Arthur
Zirin, Harold
Zook, Alma Claire
Zuckerman, Benjamin Michael

COLORADO
Bruns, Donald Gene
Conti, Peter Selby
Culver, Roger Bruce
Dietz, Richard Darby
Dulk, George A
Esposito, Larry Wayne
Everhart, Edgar
Garmany, Catharine Doremus
Linsky, Jeffrey L
Lynds, Beverly T
McCray, Richard A
Mansfield, Roger Leo
Munro, Richard Harding
Snow, Theodore Peck
Stencel, Robert Edward
Wackernagel, Hans Beat
Warwick, James Walter
Weihaupt, John George

CONNECTICUT
Downing, Mary Brigetta
Ford, Clinton Banker
Ftaclas, Christ
Garfinkel, Boris
Herbst, William
Hoffleit, Ellen Dorrit
Houston, Walter Scott
Larson, Richard Bondo
Lu, Phillip Kehwa
Menke, David Hugh
Peterson, Cynthia Wyeth
Upgren, Arthur Reinhold, Jr
Van Altena, William F
Zinn, Robert James

DELAWARE
Evenson, Paul Arthur
Herr, Richard Baessler
Howard, Edward George, Jr

DISTRICT OF COLUMBIA
Alexander, Joseph Kunkle
Bautz, Laura Patricia
Berendzen, Richard
Bohlin, John David
Bowers, Phillip Frederick
Boyce, Peter Bradford
Claussen, Mark J
Cook, John William
Corbin, Thomas Elbert
Cruddace, Raymond Gibson
DeVorkin, David Hyam
Donivan, Frank Forbes, Jr
Fiala, Alan Dale
Ford, William Kent, Jr
Friedman, Herbert
Fritz, Gilbert Geiger
Gursky, Herbert
Harrington, Robert Sutton
Hart, Richard Cullen
Hawkins, Gerald Stanley
Hughes, James Arthur
Janiczek, Paul Michael
Johnston, Kenneth John
Kaluzienski, Louis Joseph
Kaplan, George Harry
Klepczynski, William J(ohn)
Kroeger, Richard Alan
McCarthy, Dennis Dean
Mariska, John Thomas
Matsakis, Demetrios Nicholas
Matson, Michael
Meekins, John Fred
Ohring, George
Parker, Robert Allan Ridley
Pascu, Dan
Pilcher, Carl Bernard
Rickard, Lee J
Rubin, Vera Cooper
Schweizer, Francois
Seidelmann, Paul Kenneth
Shivanandan, Kandiah
Shulman, Seth David
Smith, Clayton Albert, Jr
Strand, Kaj Aage
Summers, Michael Earl
Turner, Kenneth Clyde
Van Flandern, Thomas C(harles)
Wang, Yi-Ming
Weiler, Edward John
Weiler, Kurt Walter
Westerhout, Gart
Zimbelman, James Ray

FLORIDA
Buchler, Jean-Robert
Burns, Michael J
Carr, Thomas Deaderick

Chen, Kwan-Yu
Cohen, Howard Lionel
Dermott, Stanley Frederick
Eichhorn, Heinrich Karl
Franklin, Kenneth Linn
Gottesman, Stephen T
Gustafson, Bo Ake Sture
Hett, John Henry
Hunter, James Hardin, Jr
Klock, Benny LeRoy
Leacock, Robert Jay
Lebo, George Robert
Markowitz, William
Merrill, John Ellsworth
Misconi, Nebil Yousif
Mulholland, John Derral
Oliver, John Parker
Porter, Lee Albert
Rester, Alfred Carl, Jr
Roberts, Leonidas Howard
Smith, Haywood Clark, Jr
Stefanakos, Elias Kyriakos
Webb, James Raymond
Weinberg, Jerry L
Williams, Carol Ann
Wilson, Robert E
Wood, Frank Bradshaw

GEORGIA
Howard, Sethanne
McAlister, Harold Alister
Powell, Bobby Earl
Sadun, Alberto Carlo
Shaw, James Scott

HAWAII
Boesgaard, Ann Merchant
Cowie, Lennox Lauchlan
Glaspey, John Warren
Gradie, Jonathan Carey
Hall, Donald Norman Blake
Heacox, William Dale
Herbig, George Howard
Kormendy, John
Krisciunas, Kevin L
Kron, Gerald Edward
McCord, Thomas Bard
McLaren, Robert Alexander
Morrison, David Douglas
Owen, Tobias Chant
Smith, Bradford Adelbert
Tokunaga, Alan Takashi
Tully, Richard Brent
Zisk, Stanley Harris

IDAHO
Parker, Barry Richard

ILLINOIS
Bahng, John Deuck Ryong
Bartlett, J Frederick
Carlson, Eric Dungan
Chamberlain, Joseph Miles
Chandrasekhar, Subrahmanyan
Crutcher, Richard Metcalf
Dickel, Hélène Ramseyer
Harper, Doyal Alexander, Jr
Kaler, James Bailey
Langebartel, Ray Gartner
Leibhardt, Edward
Lo, Kwok-Yung
McCormack, Elizabeth Frances
Mihalas, Barbara R Weibel
Nelson, Harry Ernest
Nissim-Sabat, Charles
Palmer, Patrick Edward
Palmore, Julian Ivanhoe, III
Peterson, Melbert Eugene
Roberts, Arthur
Saari, Donald Gene
Spinka, Harold M
Sweitzer, James Stuart
Taam, Ronald Everett
Truran, James Wellington, Jr
Ulmer, Melville Paul
Vandervoort, Peter Oliver
Webb, Harold Donivan
Yoss, Kenneth M

INDIANA
Beery, Dwight Beecher
Durisen, Richard H
Edmondson, Frank Kelley
Hrivnak, Bruce John
Jacobs, William Wescott
Poirier, John Anthony
Read, Robert H
Sprague, Newton G
Stanley, Gerald R
Stephenson, Edward James
Wilson, Olin C(haddock)

IOWA
Basart, John Philip
Beavers, Willet I
Bowen, George Hamilton, Jr
Cadmus, Robert Randall, Jr
Frank, Louis Albert
Gurnett, Donald Alfred
Hartung, Jack Burdair, Sr
Intemann, Gerald William
Jacob, Richard L
Mutel, Robert Lucien
Neff, John S

Riggs, Philip Shaefer
Smith, Paul Aiken
Spangler, Steven Randall
Spencer, Donald Lee
Struck-Marcell, Curtis John
Willson, Lee Anne Mordy

KANSAS
Anthony-Twarog, Barbara Jean
Shawl, Stephen Jacobs
Twarog, Bruce Anthony
Walters, Charles Philip
Williams, Dudley

KENTUCKY
Gwinn, Joel Alderson
Kielkopf, John F
Krogdahl, Wasley Sven
Powell, Smith Thompson, III

LOUISIANA
Landolt, Arlo Udell
Lee, Paul D
Perry, Charles Lewis
Ryan, Donald Edwin
Tohline, Joel Edward

MARYLAND
Ake, Thomas Bellis, III
Beal, Robert Carl
Behannon, Kenneth Wayne
Bennett, Charles L
Berg, Richard Allen
Bond, Howard Emerson
Brown, Robert Alan
Bruhweiler, Frederick Carlton, Jr
Brunk, William Edward
Buhl, David
Carbary, James F
Chaisson, Eric Joseph
Chartrand, Mark Ray, III
Chen, Hsing-Hen
Clark, Thomas Arvid
Conrath, Barney Jay
Davidsen, Arthur Falnes
Deprit, Andre A(lbert) M(aurice)
Di Rienzi, Joseph
Doggett, LeRoy Elsworth
Dolan, Joseph Francis
Donn, Bertram (David)
Dunham, David Waring
Erickson, William Clarence
Evans, Larry Gerald
Fallon, Frederick Walter
Feibelman, Walter A
Feldman, Paul Donald
Felten, James Edgar
Ford, Holland Cole
French, Bevan Meredith
Gilliland, Ronald Lynn
Gull, Theodore Raymond
Hall, R(oyal) Glenn, Jr
Harvel, Christopher Alvin
Haupt, Ralph Freeman
Hauser, Michael George
Heckathorn, Harry Mervin, III
Heckman, Timothy Martin
Henry, Richard Conn
Hershey, John Landis
Hill, Jesse King
Hintzen, Paul Michael
Hobbs, Robert Wesley
Hollis, Jan Michael
Holman, Gordon Dean
Hubbard, Donald
Jaffe, Walter Joseph
Johnson, Donald Rex
Kaiser, Michael Leroy
Kerr, Frank John
Killen, Rosemary Margaret
Kissell, Kenneth Eugene
Kondo, Yoji
Kostiuk, Theodor
Kowal, Charles Thomas
Kupperian, James Edward, Jr
Lovas, Francis John
Lowman, Paul Daniel, Jr
Lufkin, Daniel Harlow
Malitson, Harriet Hutzler
Mead, Jaylee Montague
Milkey, Robert William
O'Keefe, John Aloysius
Pearl, John Christopher
Perry, Peter M
Robertson, Douglas Scott
Roman, Nancy Grace
Rountree, Janet
Schreier, Ethan Joshua
Shara, Michael M
Shore, Steven Neil
Smith, Joseph Collins
Stecher, Theodore P
Stecker, Floyd William
Stockman, Hervey S, Jr
Sturch, Conrad Ray
Toller, Gary Neil
Tschunko, Hubert F A
Turnrose, Barry Edmund
Verschuur, Gerrit L
Walborn, Nolan Revere
Warren, Wayne Hutchinson, Jr
Weber, Richard Rand
Webster, William John, Jr
Weinreb, Sander

Wells, Eddie N
Wende, Charles David
White, Richard Allan
Whitmore, Bradley Charles
Wilson, Andrew Stephen
Wood, Howard John, III
Zellner, Benjamin Holmes

MASSACHUSETTS
Albers, Henry
Ash, Michael Edward
Ball, John Allen
Barrett, Alan H
Bauer, Wendy Hagen
Belserene, Emilia Pisani
Birney, Dion Scott, Jr
Canizares, Claude Roger
Cappallo, Roger James
Counselman, Charles Claude, III
Davis, Robert James
Dennis, Tom Ross
Downs, George Samuel
Edwards, Suzan
Elliot, James Ludlow
Ewen, Harold Irving
Franklin, Fred Aldrich
Geary, John Charles
Geller, Margaret Joan
Gingerich, Owen (Jay)
Goldsmith, Paul Felix
Gordon, Courtney Parks
Gordon, Kurtiss Jay
Greenstein, George
Grindlay, Jonathan Ellis
Harris, Daniel Everett
Hazen, Martha L(ocke)
Hoff, Darrel Barton
Huchra, John P
Huguenin, George Richard
Irvine, William Michael
Jacchia, Luigi Giuseppe
Jaffe, Miriam Walther
Joss, Paul Christopher
Kellogg, Edwin M
Kent, Stephen Matthew
Kirshner, Robert Paul
Kleinmann, Douglas Erwin
Kwitter, Karen Beth
Lada, Charles Joseph
Latham, David Winslow
Layzer, David
Lilley, Arthur Edward
Little, Stephen James
McCrosky, Richard Eugene
Marsden, Brian Geoffrey
Mattei, Janet Akyüz
Mendillo, Michael
Moody, Elizabeth Anne
Moran, James Michael, Jr
Mumford, George Saltonstall
Murray, Stephen S
Myers, Philip Cherdak
Niell, Arthur Edwin
Pasachoff, Jay M(yron)
Pedersen, Peder Christian
Pettengill, Gordon Hemenway
Ratner, Michael Ira
Roberts, David Hall
Rogers, Alan Ernest Exel
Salah, Joseph E
Schild, Rudolph Ernest
Schwartz, Daniel Alan
Seward, Frederick Downing
Shapiro, Irwin Ira
Shapiro, Ralph
Shawcross, William Edgerton
Silk, John Kevin
Snell, Ronald Lee
Staelin, David Hudson
Stone, Peter Hunter
Strom, Stephen
Taff, Laurence Gordon
Tonry, John Landis
Toomre, Alar
VanSpeybroeck, Leon Paul
Wardle, John Francis Carleton
White, Richard Edward
Willner, Steven P
Wisdom, Jack Leach
Withbroe, George Lund
Wood, David Belden

MICHIGAN
Aller, Margo Friedel
Berger, Beverly Kobre
Brockmeier, Richard Taber
Burek, Anthony John
Christian, Larry Omar
Cowley, Charles Ramsay
Dunifer, Gerald Leroy
England, Anthony W
Haddock, Frederick Theodore, Jr
Houk, Nancy (Mia)
Kauppila, Walter Eric
Leonard, Charles Grant
MacAlpine, Gordon Madeira
Miller, Freeman Devold
Mohler, Orren (Cuthbert)
Osborn, Wayne Henry
Richstone, Douglas Orange
Simkin, Susan Marguerite
Teske, Richard Glenn
Van Till, Howard Jay
Williams, John Albert

Astronomy (cont)

MINNESOTA
Davidson, Kris
Eckroth, Charles Angelo
Hakkila, Jon Eric
Humphreys, Roberta Marie
La Bonte, Anton Edward
Luyten, Willem Jacob
Mathews, Robert Thomas
Rudnick, Lawrence
Whitehead, Andrew Bruce

MISSISSIPPI
Lestrade, John Patrick

MISSOURI
Ashworth, William Bruce, Jr
Cowsik, Ramanath
Edwards, Terry Winslow
Geilker, Charles Don
Johnson, Douglas William
McKinnon, William Beall
Nelson, George Driver
Peterson, Charles John
Schmitt, John Leigh
Schwartz, Richard Dean
Smith, William Hayden
Wolf, George William

MONTANA
Jeppesen, Randolph H

NEBRASKA
Billesbach, David P
Clements, Gregory Leland
Leung, Kam-Ching
Schmidt, Edward George
Taylor, Donald James
Underhill, Glenn

NEVADA
Weistrop, Donna Etta

NEW HAMPSHIRE
Baker, James Gilbert
Beasley, Wayne M(achon)
Cole, Charles N
Mook, Delo Emerson, II
Ryan, James Michael

NEW JERSEY
Bahcall, Neta Assaf
Blair, Grant Clark
Boeshaar, Patricia Chikotas
Damen, Theo C
Fels, Stephen Brook
Fenstermacher, Robert Lane
Gott, J Richard, III
Hulse, Russell Alan
Jarvis, John Frederick
Knapp, Gillian Revill
Levine, Jerry David
MacLennan, Carol G
Neugebauer, Otto E
Peebles, Phillip J
Penzias, Arno A(llan)
Pfeiffer, Raymond John
Pregger, Fred Titus
Schwarzschild, Martin
Spitzer, Lyman, Jr
Taylor, Joseph Hooton, Jr
Turner, Edwin Lewis
Tyson, J Anthony
Wilkinson, David Todd
Williams, Theodore Burton
Wilson, Robert Woodrow

NEW MEXICO
Argo, Harold Virgil
Bignell, Richard Carl
Bleiweiss, Max Phillip
Burns, Jack O'Neal
Clark, Barry Gillespie
Clark, Robert Edward Holmes
Cox, Arthur Nelson
Crane, Patrick Conrad
Dunn, Richard B
Gisler, Galen Ross
Hankins, Timothy Hamilton
Heiken, Grant Harvey
Linsley, John
MacCallum, Crawford John
Neidig, Donald Foster
Olson, Gordon Lee
Owen, Frazer Nelson
Ozernoy, Leonid M
Peterson, Alan W
Price, Richard Marcus
Reeves, Geoffrey D
Sanders, Walter L
Simon, George Warren
Sramek, Richard Anthony
Strong, Ian B
Tombaugh, Clyde W(illiam)
Wade, Campbell Marion
Zeilik, Michael

NEW YORK
Aveni, Anthony
Baez, Silvio
Baker, Norman Hodgson
Baum, Parker Bryant
Beckwith, Steven Van Walter

Branley, Franklyn M
Burns, Joseph A
Campbell, Donald Bruce
Cheung, Lim H
Craft, Harold Dumont, Jr
Cummins, John Frances
Davis, Raymond, Jr
Dubisch, Russell John
Elmegreen, Debra Meloy
Epstein, Isadore
Forman, Miriam Ausman
Gaffey, Michael James
Gelman, Donald
Gierasch, Peter Jay
Goebel, Ronald William
Gold, Thomas
Gross, Stanley H
Hagfors, Tor
Haynes, Martha Patricia
Helfand, David John
Huggins, Patrick John
Knacke, Roger Fritz
Lattimer, James Michael
Matsushima, Satoshi
Meisel, David Dering
Meltzer, Alan Sidney
Millikan, Allan G
Nicholson, Thomas Dominic
Noonan, Thomas Wyatt
O'Donoghue, Aileen Ann
Philip, A G Davis
Prather, Michael John
Prendergast, Kevin Henry
Reddy, Reginald James
Sagan, Carl
Schreurs, Jan W H
Sharpless, Stewart Lane
Simon, Michal
Stull, John Leete
Teiger, Martin
Terzian, Yervant
Thorsen, Richard Stanley
Veverka, Joseph F
Yahil, Amos

NORTH CAROLINA
Carney, Bruce William
Christiansen, Wayne Arthur
Davis, Morris Schuyler
Jenzano, Anthony Francis
Oberhofer, Edward Samuel
Rhynsburger, Robert Whitman
Shapiro, Lee Tobey

NORTH DAKOTA
Rathmann, Franz Heinrich

OHIO
Anderson, Lawrence Sven
Annear, Paul Richard
Bidelman, William Pendry
Bishop, Edwin Vandewater
Bopp, Bernard William
Capriotti, Eugene Raymond
Collins, George W, II
Ferguson, Dale Curtis
Ferland, Gary Joseph
Frogel, Jay Albert
Goedicke, Victor Alfred
James, Philip Benjamin
Keenan, Philip Childs
Keller, Geoffrey
Luck, Richard Earle
Lumsdaine, Edward
Macklin, Philip Alan
Malcuit, Robert Joseph
Mickelson, Michael Eugene
Mitchell, Walter Edmund, Jr
Morrison, Nancy Dunlap
Newsom, Gerald Higley
Patton, Jon Michael
Pesch, Peter
Peterson, Bradley Michael
Protheroe, William Mansel
Richardson, David Louis
Rodman, James Purcell
Sanduleak, Nicholas
Sellgren, Kristen
Stanger, Philip Charles
Stephenson, Charles Bruce
Strickler, Thomas David
Wagoner, Glen
Wing, Robert Farquhar
Young, Warren Melvin

OKLAHOMA
Cowan, John James
Gudehus, Donald Henry
Rutledge, Carl Thomas
Shull, Peter Otto, Jr

OREGON
McDaniels, David K
McKinney, William Mark

PENNSYLVANIA
Augensen, Harry John
Bauer, Carl August
Blitzstein, William
Boughn, Stephen Paul
Bowersox, Todd Wilson
Carpenter, Lynn Allen
Chambliss, Carlson Rollin
Córdova, France Anne-Dominic

Couch, Jack Gary
Feigelson, Eric Dennis
Gainer, Michael Kizinski
Garmire, Gordon Paul
Gatewood, George David
Guinan, Edward F
Heintz, Wulff Dieter
Holzinger, Joseph Rose
Jenkins, Edward Felix
Kiewiet de Jonge, Joost H A
Koch, Robert Harry
Levitt, Israel Monroe
Lippincott, Sarah Lee
McCook, George Patrick
Partridge, Robert Bruce
Poss, Howard Lionel
Ramsey, Lawrence William
Seeds, Michael August
Sion, Edward Michael
Sitterly, Charlotte Moore
Stauffer, George Franklin
Usher, Peter Denis
Weedman, Daniel Wilson
Winkler, Louis
Zabriskie, Franklin Robert

RHODE ISLAND
Pieters, Carle M

SOUTH CAROLINA
Adelman, Saul Joseph
Rust, Philip Frederick

TENNESSEE
Cordell, Francis Merritt
Daunt, Stephen Joseph
Eaton, Joel A
Hall, Douglas Scott
Hardie, Robert Howie
Heiser, Arnold M
Lorenz, Philip Jack, Jr
Siomos, Konstadinos
Tolk, Norman Henry

TEXAS
Barker, Edwin Stephen
Bash, Frank Ness
Benedict, George Frederick
Blanford, George Emmanuel, Jr
Chamberlain, Joseph Wyan
De Vaucouleurs, Gerard Henri
Douglas, James Nathaniel
Dufour, Reginald James
Duncombe, Raynor Lockwood
Evans, David Stanley
Evans, Neal John, II
Gilmore, William Steven
Gomer, Richard Hans
Gott, Preston Frazier
Harrison, Marjorie Hall
Hemenway, Mary-Kay Meacham
Henize, Karl Gordon
Heymann, Dieter
Hildebrandt, Alvin Frank
Hinterberger, Henry
Hoffman, Alexander A J
Hoffman, Jeffrey Alan
Jefferys, William H, III
Kattawar, George W
King, Elbert Aubrey, Jr
Lewis, Ira Wayne
Mangum, Jeffrey Gary
Middlehurst, Barbara Mary
Morgan, Thomas Harlow
Nacozy, Paul E
Nather, Roy Edward
Nicks, Oran Wesley
O'Dell, Charles Robert
Plass, Gilbert Norman
Potter, Andrew Elwin, Jr
Robinson, Edward Lewis
Shapiro, Paul Robert
Smith, Harlan J
Synek, Miroslav (Mike)
Szebehely, Victor
Talent, David Leroy
Trafton, Laurence Munro
Tull, Robert Gordon
Winningham, John David
Zook, Herbert Allen

UTAH
Christensen, Clark G
McNamara, Delbert Harold
Morris, Elliot Cobia

VERMONT
MacArthur, John Wood
Rankin, Joanna Marie
Wolfson, Richard L T

VIRGINIA
Beardall, John Smith
Bridle, Alan Henry
Broderick, John
Bunner, Alan Newton
Carpenter, Martha Stahr
Carr, Roger Byington
Chevalier, Roger Alan
Condon, James Justin
Dennison, Brian Kenneth
DuPuy, David Lorraine
Evans, John C
Finke, Reinald Guy

Fredrick, Laurence William
Goldstein, Samuel Joseph, (Jr)
Greisen, Eric Winslow
Hart, Michael H
Heeschen, David Sutphin
Hogg, David Edward
Hohl, Frank
Hooper, William John, Jr
Howard, William Eager, III
Kerr, Anthony Robert
Knappenberger, Paul Henry, Jr
Knowles, Stephen H
Kumar, Shiv Sharan
Langston, Glen Irvin
Lindenblad, Irving Werner
Liszt, Harvey Steven
Margrave, Thomas Ewing, Jr
Mayer, Cornell Henry
Motill, Ronald Allen
O'Connell, Robert West
Oesterwinter, Claus
Rea, Donald George
Roberts, Morton Spitz
Roberts, William Woodruff, Jr
Sarazin, Craig L
Schwiderski, Ernst Walter
Shaffer, David Bruce
Singh, Jag Jeet
Smith, Elske van Panhuys
Staib, Jon Albert
Thompson, Anthony Richard
Tolbert, Charles Ray
Tolson, Robert Heath
Turner, Barry Earl
Vanden Bout, Paul Adrian
Webber, John Clinton
Wootten, Henry Alwyn

WASHINGTON
Baum, William Alvin
Bookmyer, Beverly Brandon
Bracher, Katherine
Brownlee, Donald E
Ellis, Fred E
Gammon, Richard Harriss
Hodge, Paul William
Jenner, David Charles
Kells, Lyman Francis
Krienke, Ora Karl, Jr
Lutz, Thomas Edward
Mitchell, Robert Curtis
Quigley, Robert James
Rydgren, A Eric
Stokes, Gerald Madison
Sullivan, Woodruff Turner, III
Wallerstein, George
White, William Charles
Wilkening, Laurel Lynn
Windsor, Maurice William

WEST VIRGINIA
Ghigo, Frank Dunnington
Hewitt, Anthony Victor
Seielstad, George A

WISCONSIN
Anderson, Christopher Marlowe
Bless, Robert Charles
Campbell, Warren Adams
Churchwell, Edward Bruce
Code, Arthur Dodd
Cudworth, Kyle McCabe
Hobbs, Lewis Mankin
Kojoian, Gabriel
Morgan, William Wilson
Reynolds, Ronald J
Savage, Blair DeWillis
Schroeder, Daniel John
Shepherd, John Patrick George
Vilkki, Erkki Uuno
Wahl, Eberhard Wilhelm
Woods, Robert Claude

WYOMING
Canterna, Ronald William
Grasdalen, Gary
Howell, Robert Richard
Roark, Terry P
Thronson, Harley Andrew, Jr

PUERTO RICO
Davis, Michael Moore
Giovanelli, Riccardo
Lewis, Brian Murray

ALBERTA
Clark, Thomas Alan
Hube, Douglas Peter
Kwok, Sun
Milone, Eugene Frank

BRITISH COLUMBIA
Batten, Alan Henry
Burke, J Anthony
Fahlman, Gregory Gaylord
Hartwick, Frederick David Alfred
Higgs, Lloyd Albert
Hutchings, John Barrie
McCutcheon, William Henry
Palmer, Leigh Hunt
Purton, Christopher Roger
Richardson, Eric Harvey
Richer, Harvey Brian
Scarfe, Colin David

Shuter, William Leslie Hazlewood
Van den Bergh, Sidney

NEW BRUNSWICK
Faig, Wolfgang

NEWFOUNDLAND
Rochester, Michael Grant

NOVA SCOTIA
Bishop, Roy Lovitt
Turner, David Gerald
Welch, Gary Alan

ONTARIO
Andrew, Bryan Haydn
Bakos, Gustav Alfons
Barker, Paul Kenneth
Bertram, Robert William
Blackwell, Alan Trevor
Bolton, Charles Thomas
Broten, Norman W
Caldwell, John James
Chow, Yung Leonard
Clarke, James Newton
Clarke, Thomas Roy
Clement, Christine Mary (Coutts)
Dyer, Charles Chester
Evans, Nancy Remage
Feldman, Paul Arnold
Fernie, John Donald
Garrison, Robert Frederick
Griffith, John Sidney
Hagberg, Erik Gordon
Harris, William Edgar
Hogarth, Jacke Edwin
Hogg, Helen (Battles) Sawyer
Hughes, Victor A
Legg, Thomas Harry
Lester, John Bernard
McCallion, William James
MacLeod, John Munroe
MacRae, Donald Alexander
Marlborough, John Michael
Millman, Peter MacKenzie
Moffat, John William
Percy, John Rees
Pfalzner, Paul Michael
Seaquist, Ernest Raymond
Sida, Derek William
Vallée, Jacques P
Wehlau, Amelia W
Wehlau, William Henry

QUEBEC
Borra, Ermanno Franco
Demers, Serge
Moffat, Anthony Frederick John
Pineault, Serge Rene
Racine, Rene

SASKATCHEWAN
Kos, Joseph Frank
Naqvi, Saiyid Ishrat Husain

OTHER COUNTRIES
Arp, Halton Christian
Blanco, Betty M
Blanco, Victor Manuel
Burnell, S Jocelyn Bell
Burton, William Butler
Butcher, Harvey Raymond, III
Contopoulos, George
Hewish, Antony
Heyden, Francis Joseph
Kalnajs, Agris Janis
Kesteven, Michael
Kopal, Zdenek
Kunkel, William Eckart
Liller, William
McCarthy, Martin
Munch, G
Novotny, Eva
Reitmeyer, William L
Robinson, Richard David, Jr
Schnopper, Herbert William
Shurman, Michael Mendelsohn
Van der Hulst, Jan Mathijs
Vander Vorst, Andre
Von Hoerner, Sebastian
Wampler, E Joseph

Astrophysics

ALABAMA
Abbas, Mian Mohammad
Curott, David Richard
Davis, John Moulton
Decher, Rudolf
Dunn, Anne Roberts
Fishman, Gerald Jay
Hathaway, David Henry
Hinata, Satoshi
Hoover, Richard Brice
Krause, Helmut G L
Lundquist, Charles Arthur
Meegan, Charles Anthony
Moore, Ronald Lee
Parnell, Thomas Alfred
Sulentic, Jack William
Tan, Arjun
Weisskopf, Martin Charles

ALASKA
Deehr, Charles Sterling
Kan, Joseph Ruce
Lee, Lou-Chuang
Martins, Donald Henry
Roederer, Juan Gualterio
Stenbaek-Nielsen, Hans C
Swift, Daniel W

ARIZONA
Abt, Helmut Arthur
Adel, Arthur
Arnett, William David
Black, John Harry
Blitzer, Leon
Boynton, William Vandegrift
Brault, James William
Burrows, Adam Seth
Burstein, David
Christy, James Walter
De Gaston, Alexis Neal
De Young, David Spencer
Fan, Chang-Yun
Foltz, Craig Billig
Franz, Otto Gustav
Gatley, Ian
Giampapa, Mark Steven
Grauer, Albert D
Greenberg, Richard Joseph
Harris, Hugh Courtney
Hartmann, William K
Harvey, John Warren
Herbert, Floyd Leigh
Hill, Henry Allen
Hoffmann, William Frederick
Hsieh, Ke Chiang
Huffman, Donald Ray
Jefferies, John Trevor
Jokipii, Jack Randolph
Kreidl, Tobias Joachim
Kyrala, Ali
Leibacher, John W
Levy, Eugene Howard
Liebert, James William
Low, Frank James
Lunine, Jonathan Irving
Lutz, Barry Lafean
Massey, Philip Louis
Mayall, Nicholas Ulrich
Melia, Fulvio
Pacholczyk, Andrzej Grzegorz
Pilachowski, Catherine Anderson
Sandel, Bill Roy
Shemansky, Donald Eugene
Starrfield, Sumner Grosby
Swihart, Thomas Lee
Thompson, Rodger Irwin
Tifft, William Grant
Weekes, Trevor Cecil
Wehinger, Peter Augustus
White, Simon David Manton
Wildey, Robert Leroy
Wolff, Sidney Carne
Woolf, Neville John
Yelle, Roger V

ARKANSAS
Harter, William George
Rollefson, Aimar Andre
Wold, Donald C

CALIFORNIA
Acton, Loren Wilber
Aitken, Donald W, Jr
Allamandola, Louis John
Aller, Lawrence Hugh
Anderson, Kinsey Amor
Anderson, Milo Vernette
Appleby, John Frederick
Armstrong, John William
Arons, Jonathan
Ashburn, Edward V
Aumann, Hartmut Hans-Georg
Barnes, Aaron
Barry, James Dale
Barwick, Steven William
Baum, Peter Joseph
Becker, Robert Adolph
Bell, Graydon Dee
Benford, Gregory A
Bergstralh, Jay Thor
Blake, J Bernard
Blandford, Roger David
Bloxham, Laurence Hastings
Blumenthal, George Ray
Bodenheimer, Peter Herman
Bonsack, Walter Karl
Bratenahl, Alexander
Bridges, Frank G
Brown, Wilbur K
Brown, William Arnold
Bruner, Marilyn E
Buffington, Andrew
Burbidge, Eleanor Margaret
Burbidge, Geoffrey
Carpenter, Roland LeRoy
Cassen, Patrick Michael
Castor, John I
Chapline, George Frederick, Jr
Chapman, Gary Allen
Chase, Lloyd Fremont, Jr
Chow, Tai-Low
Christy, Robert Frederick
Cloutman, Lawrence Dean

Cohen, Judith Gamora
Collard, Harold Rieth
Cominsky, Lynn Ruth
Coroniti, Ferdinand Vincent
Corwin, Harold G, Jr
Crooker, Nancy Uss
Daub, Clarence Theodore, Jr
Davis, Leverett, Jr
Davis, Marc
Dewitt, Hugh Edgar
Dinger, Ann St Clair
Doolittle, Robert Frederick, II
Duke, Douglas
Dyal, Palmer
Erickson, Edwin Francis
Erickson, Stanley Arvid
Esposito, Pasquale Bernard
Everitt, C W Francis
Farmer, Crofton Bernard
Faulkner, John
Feynman, Joan
Filippenko, Alexei Vladimir
Fisher, Philip Chapin
Frazier, Edward Nelson
Frerking, Margaret Ann
Friedman, Richard M
Frye, William Emerson
Garrett, Henry Berry
Gerola, Humberto Cayetano
Goldreich, Peter
Goorvitch, David
Gould, Robert Joseph
Greenstein, Jesse Leonard
Gregorich, David Tony
Grossman, Allen S
Gulkis, Samuel
Gwinn, William Dulaney
Hackwell, John A
Haisch, Bernhard Michael
Hall, Donald Eugene
Harris, Alan William
Heiles, Carl
Helliwell, R(obert) A(rthur)
Holzer, Robert Edward
Horowitz, Gary T
Imhof, William Lowell
Ingraham, Richard Lee
Intriligator, Devrie Shapiro
Irvine, Cynthia Emberson
Jackson, Bernard Vernon
Jacobson, Allan Stanley (Bud)
Janssen, Michael Allen
Johnson, Hugh Mitchell
Jura, Michael Alan
Kagiwada, Harriet Hatsune
Karp, Alan H
Kennel, Charles Frederick
Klumpar, David Michael
Knobloch, Edgar
Knox, William Jordan
Koch, David Gilbert
Koga, Rokutaro
Koo, David Chih-yuen
Kurz, Richard J
Landman, Donald Alan
Larmore, Lewis
Lawrence, John Keeler
Lea, Susan Maureen
Lee, Edward Prentiss
Leich, Douglas Albert
Leighton, Robert Benjamin
Lillie, Charles Frederick
Lin, Peter Peichung
Lingenfelter, Richard Emery
Litvak, Marvin Mark
Lombardi, Gabriel Gustavo
Lugmair, Guenter Wilhelm
Luthey, Joe Lee
McCormac, Billy Murray
McDonough, Thomas Redmond
McKee, Christopher Fulton
Maloy, John Owen
Marcus, Philip Stephen
Mark, James Wai-Kee
Markowitz, Allan Henry
Marti, Kurt
Mauche, Christopher W
Max, Claire Ellen
Mayfield, Earle Byron
Mazurek, Thaddeus John
Mendis, Devamitta Asoka
Metzger, Albert E
Meyerott, Roland Edward
Mihalov, John Donald
Miner, Ellis Devere, Jr
Morris, Mark Root
Mozer, Forrest S
Muller, Richard A
Nash, Douglas B
Nelson, Jerry Earl
Neu, John Ternay
Neugebauer, Gerry
Noerdlinger, Peter David
Norman, Eric B
Oder, Frederic Carl Emil
O'Keefe, John Dugan
Olsen, Edward Tait
Orth, Charles Douglas
Pang, Kevin Dit Kwan
Paulikas, George A
Payne, David Glenn
Pazich, Philip Michael
Peale, Stanton Jerrold
Peccei, Roberto Daniele

Pehl, Richard Henry
Peters, Geraldine Joan
Peterson, Laurence E
Phillips, Thomas Gould
Plavec, Mirek Josef
Pneuman, Gerald W
Preskill, John P
Price, Paul Buford, Jr
Primack, Joel Robert
Rauch, Richard Travis
Richards, Paul Linford
Rosen, Leonard Craig
Ross, Ronald Rickard
Rothschild, Richard Eiseman
Rouse, Carl Albert
Rubin, Robert Howard
Sackmann, I Juliana
Sadoulet, Bernard
Sargent, Anneila Isabel
Sargent, Wallace Leslie William
Schindler, Stephen Michael
Scholl, James Francis
Schulz, Michael
Sekanina, Zdenek
Silk, Joseph Ivor
Siscoe, George L
Slade, Martin Alphonse, III
Smith, Harding Eugene
Smith, Sheldon Magill
Smoot, George Fitzgerald, III
Spreiter, John R(obert)
Stella, Paul M
Stevenson, David John
Straka, William Charles
Sturrock, Peter Andrew
Swanson, Paul N
Talbot, Raymond James, Jr
Tarbell, Theodore Dean
Tarter, Curtis Bruce
Thompson, Thomas William
Thompson, Timothy J
Thorne, Kip Stephen
Thorne, Richard Mansergh
Timothy, John Gethyn
Toor, Arthur
Townes, Charles Hard
Trimble, Virginia Louise
Tucker, Wallace Hampton
Unwin, Stephen Charles
Vampola, Alfred Ludvik
Van Hoven, Gerard
Villere, Karen R
Vitello, Peter A
Vogt, Rochus E
Wagner, Raymond Lee
Wagoner, Robert Vernon
Walker, Arthur Bertram Cuthbert, Jr
Walker, Raymond John
Walt, Martin
Wasson, John Taylor
Wax, Robert LeRoy
Weaver, William Bruce
Weissman, Paul Robert
Werner, Michael Wolock
Wertz, James Richard
West, Harry Irwin, Jr
Whitcomb, Stanley Ernest
White, Robert Stephen
Whitford, Albert Edward
Wiggins, John Shearon
Wilson, James R
Wilson, William John
Witteborn, Fred Carl
Woodward, James Franklin
Woosley, Stanford Earl
Yen, Chen-Wan Liu
Young, Louise Gray
Zumberge, James Frederick
Zuppero, Anthony Charles
Zych, Allen Dale

COLORADO
Baur, Thomas George
Bornmann, Patricia L(ee)
Chasson, Robert Lee
Conti, Peter Selby
Cox, John Paul
Culp, Robert D(udley)
Curtis, George William
Dryer, Murray
Dulk, George A
Eddy, John Allen
Everhart, Edgar
Faller, James E
Firor, John William
Fisher, Richard R
Garstang, Roy Henry
Gilman, Peter A
Hansen, Carl John
Hansen, Richard Olaf
Herring, Jackson Rea
Hildner, Ernest Gotthold, III
Holzer, Thomas Edward
House, Lewis Lundberg
Hummer, David Graybill
Hundhausen, Arthur James
Joselyn, Jo Ann Cram
Linsky, Jeffrey L
Liu, Joseph Jeng-Fu
Low, Boon-Chye
Malville, John McKim
Mankin, William Gray
Mizushima, Masataka
Newkirk, Gordon Allen, Jr

Astrophysics (cont)

Rottman, Gary James
Sawyer, Constance B
Sime, David Gilbert
Skumanich, Andrew P
Smith, Dean Francis
Speiser, Theodore Wesley
Stanger, Andrew L
Stebbins, Robin Tucker
Stokes, Robert Allan
Thomas, Richard Nelson
Vande Noord, Edwin Lee
Zweibel, Ellen Gould

CONNECTICUT
Baierlein, Ralph Frederick
Demarque, Pierre
Ftaclas, Christ
Jakacky, John M
Krauss, Lawrence Maxwell
Larson, Richard Bondo
Rich, John Charles
Similon, Philippe Louis
Sofia, Sabatino

DELAWARE
Crawford, Michael Karl
Evenson, Paul Arthur
Holt, Rush Dew, Jr
MacDonald, James
Mullan, Dermott Joseph
Ness, Norman Frederick
Pomerantz, Martin Arthur
Shipman, Harry Longfellow

DISTRICT OF COLUMBIA
Apruzese, John Patrick
Beall, James Howard
Berman, Barry L
Boss, Alan Paul
Brueckner, Guenter Erich
Caroff, Lawrence John
Cheng, Chung-Chieh
Claussen, Mark J
Cook, John William
Corliss, Charles Howard
Epstein, Gerald Lewis
Feldman, Uri
Fritz, Gilbert Geiger
Goldstein, Jeffrey Jay
Harwit, Martin Otto
Hubbard, Richard Forest
Jones, William Vernon
Kaluzienski, Louis Joseph
Kaplan, George Harry
Kinzer, Robert Lee
Kroeger, Richard Alan
Kurfess, James Daniel
Lynch, John Thomas
McNeal, Robert Joseph
Mange, Phillip Warren
Mariska, John Thomas
Matsakis, Demetrios Nicholas
Meekins, John Fred
Oertel, Goetz K
Pascu, Dan
Patel, Vithalbhai L
Reeves, Edmond Morden
Rickard, Lee J
Schmerling, Erwin Robert
Schweizer, Francois
Serene, Joseph William
Share, Gerald Harvey
Shivanandan, Kandiah
Silberberg, Rein
Summers, Michael Earl
Wang, Yi-Ming
Weiler, Kurt Walter

FLORIDA
Buchler, Jean-Robert
Burns, Jay, III
Cohen, Howard Lionel
Detweiler, Steven Lawrence
Fry, James N
Gottesman, Stephen T
Gustafson, Bo Ake Sture
Halpern, Leopold (Ernst)
Hammond, Gordon Leon
Ipser, James Reid
Mulholland, John Derral
Oelfke, William C
Plendl, Hans Siegfried
Ramond, Pierre Michel
Rester, Alfred Carl, Jr
Smith, Alexander Goudy
Smith, Haywood Clark, Jr
Wang, Ru-Tsang
Webb, James Raymond
Wilson, Robert E

GEORGIA
Dod, Bruce Douglas
Howard, Sethanne
Marks, Dennis William
Sadun, Alberto Carlo
Wiita, Paul Joseph

HAWAII
Boesgaard, Ann Merchant
Cowie, Lennox Lauchlan
Geballe, Thomas Ronald
Henry, Joseph Patrick

Herbig, George Howard
Learned, John Gregory
McLaren, Robert Alexander
Mickey, Donald Lee
Orrall, Frank Quimby
Owen, Tobias Chant
Tully, Richard Brent

IDAHO
Hoagland, Gordon Wood
Parker, Barry Richard
Preszler, Alan Melvin

ILLINOIS
Anders, Edward
Barnothy, Jeno Michael
Buscombe, William
Cahn, Julius Hofeller
Chandrasekhar, Subrahmanyan
Crutcher, Richard Metcalf
Davids, Cary Nathan
Dykla, John J
Eichten, Estia Joseph
Fortner, Brand I
Garcia-Munoz, Moises
Grossman, Lawrence
Harvey, Jeffrey Alan
Hildebrand, Roger Henry
Hill, Christopher T
Iben, Icko, Jr
Karim, Khondkar Rezaul
Kolb, Edward William
Lamb, Donald Quincy, Jr
Lebovitz, Norman Ronald
L'Heureux, Jacques (Jean)
Lo, Kwok-Yung
McKibben, Robert Bruce
Meyer, Peter
Mihalas, Barbara R Weibel
Mihalas, Dimitri
Miller, Richard Henry
Mouschovias, Telemachos Charalambous
Muller, Dietrich
Nezrick, Frank Albert
Oka, Takeshi
Olson, Edward Cooper
Palmer, Patrick Edward
Pandharipande, Vijay Raghunath
Parker, Eugene Newman
Pines, David
Pyle, K Roger
Roberts, Arthur
Rosner, Robert
Schramm, David N
Skadron, George
Smarr, Larry Lee
Smither, Robert Karl
Snyder, Lewis Emil
Spinka, Harold M
Taam, Ronald Everett
Truran, James Wellington, Jr
Turner, Michael Stanley
Watson, William Douglas
Webbink, Ronald Frederick
York, Donald Gilbert

INDIANA
Aprahamian, Ani
Burkhead, Martin Samuel
Durisen, Richard H
Gaidos, James A
Heinz, Richard Meade
Heydegger, Helmut Roland
Hrivnak, Bruce John
Jacobs, William Wescott
Johnson, Hollis Ralph
Lipschutz, Michael Elazar
Loeffler, Frank Joseph
LoSecco, John M
Mufson, Stuart Lee
Poirier, John Anthony
Willmann, Robert B

IOWA
Beavers, Willet I
Bowen, George Hamilton, Jr
Fix, John Dekle
Goertz, Christoph Klaus
Gurnett, Donald Alfred
Hagelin, John Samuel
Intemann, Gerald William
Lamb, Richard C
Neff, John S
Spangler, Steven Randall
Struck-Marcell, Curtis John

KANSAS
Armstrong, Thomas Peyton
Chu, Shih I
Twarog, Bruce Anthony

KENTUCKY
Elitzur, Moshe
Hawkins, Charles Edward
Kielkopf, John F
Pearson, Earl F

LOUISIANA
Chanmugam, Ganesar
Grenhich, Raymond Thomas
Ho, Yew Kam
Lee, Paul D
Rense, William A
Tipler, Frank Jennings, III

Tohline, Joel Edward
Wefel, John Paul

MAINE
Beard, David Breed
Comins, Neil Francis
Hughes, William Taylor
McClymer, James P

MARYLAND
Acuna, Mario Humberto
Allen, John Edward, Jr
Baker, Daniel Neal
Balasubrahmanyan, Vriddhachalam K
Bass, Arnold Marvin
Behannon, Kenneth Wayne
Bell, Roger Alistair
Bettinger, Richard Thomas
Birmingham, Thomas Joseph
Boggess, Albert
Boggess, Nancy Weber
Bohlin, Ralph Charles
Boldt, Elihu (Aaron)
Bond, Howard Emerson
Bostrom, Carl Otto
Brasunas, John Charles
Bruhweiler, Frederick Carlton, Jr
Calame, Gerald Paul
Carbary, James F
Chaisson, Eric Joseph
Chan, Kwing Lam
Chapman, Robert DeWitt
Cheng, Andrew Francis
Chi, L K
Chiu, Hong-Yee
Cline, Thomas L
Connerney, John E P
Crannell, Carol Jo Argus
Davidsen, Arthur Falnes
Dolan, Joseph Francis
Doschek, George A
Downing, Robert Gregory
Dubin, Maurice
Earl, James Arthur
Endal, Andrew Samson
Fainberg, Joseph
Feibelman, Walter A
Felten, James Edgar
Fichtel, Carl Edwin
Fischel, David
Fitz, Harold Carlton, Jr
Flasar, F Michael
Ford, Holland Cole
Fuhr, Jeffrey Robert
Gebbie, Katharine Blodgett
Giacconi, Riccardo
Gilliland, Ronald Lynn
Goldberg, Richard Aran
Goldstein, Melvyn L
Gull, Theodore Raymond
Gutsche, Graham Denton
Harrington, James Patrick
Hartle, Richard Eastham
Hartman, Robert Charles
Hashmall, Joseph Alan
Hauser, Michael George
Heckathorn, Harry Mervin, III
Heckman, Timothy Martin
Henry, Richard Conn
Hildebrand, Bernard
Hills, Howard Kent
Hoffman, Robert A
Holman, Gordon Dean
Holt, Stephen S
Hu, Bei-Lok Bernard
Hunter, Stanley Dean
Jones, Frank Culver
Jordan, Stuart Davis
Kastner, Sidney Oscar
Kissell, Kenneth Eugene
Kniffen, Donald Avery
Kondo, Yoji
Kostiuk, Theodor
Krimigis, Stamatios Mike
Krolik, Julian H
Krueger, Arlin James
Kumar, Cidambi Krishna
Kupperian, James Edward, Jr
Kyle, Herbert Lee
Lasker, Barry Michael
Leckrone, David Stanley
Liu, Chuan Sheng
Lucke, Robert Lancaster
McEntire, Richard Willian
McIlrath, Thomas James
Maier, Eugene Jacob Rudolph
Maran, Stephen Paul
Mather, John Cromwell
Meredith, Leslie Hugh
Misner, Charles William
Mitler, Henri Emmanuel
Moos, Henry Warren
Mumma, Michael Jon
Mushotzky, Richard Fred
Ogilvie, Keith W
Omidvar, Kazem
Ormes, Jonathan Fairfield
Orwig, Larry Eugene
Ostaff, William A(llen)
Oster, Ludwig Friedrich
Paddack, Stephen J(oseph)
Papadopoulos, Konstantinos Dennis
Pellerin, Charles James, Jr
Perry, Peter M

Poland, Arthur I
Powell, Edward Gordon
Ramaty, Reuven
Reames, Donald Vernon
Rodney, William Stanley
Rose, William K
Rosenberg, Theodore Jay
Rountree, Janet
Routly, Paul McRae
Rust, David Maurice
Sari, James William
Schreier, Ethan Joshua
Scudder, Jack David
Shara, Michael M
Shore, Steven Neil
Silbergeld, Mae Driscoll
Sittler, Edward Charles, Jr
Smith, David Edmund
Smith, Richard Lloyd
Stecker, Floyd William
Stief, Louis J
Strobel, Darrell Fred
Teegarden, Bonnard John
Thomas, Roger Jerry
Thompson, David John
Toller, Gary Neil
Toton, Edward Thomas
Trombka, Jacob Israel
von Rosenvinge, Tycho Tor
Wagner, William John
Warren, Wayne Hutchinson, Jr
Webster, William John, Jr
Weller, Charles Stagg, Jr
Wende, Charles David
Wentzel, Donat Gotthard
West, Donald K
White, Richard Allan
Wilkerson, Thomas Delaney
Williams, Donald J
Woodgate, Bruce Edward
Wright, James P
Yoon, Peter Haesung
Zalubas, Romuald

MASSACHUSETTS
Ahlen, Steven Paul
Arny, Thomas Travis
Avrett, Eugene Hinton
Ball, John Allen
Barnes, Arnold Appleton, Jr
Barrett, Alan H
Bechis, Kenneth Paul
Bell, Barbara
Bradt, Hale Van Dorn
Brecher, Aviva
Brecher, Kenneth
Bridge, Herbert Sage
Brown, Benjamin Lathrop
Brynjolfsson, Ari
Cameron, Alastair Graham Walter
Canizares, Claude Roger
Carleton, Nathaniel Phillips
Carovillano, Robert L
Chang, Edward Shi Tou
Chernosky, Edwin Jasper
Clark, George Whipple
Cook, Allan Fairchild, II
Corey, Brian E
Dalgarno, Alexander
Dickman, Robert Laurence
Dupree, Andrea K
Edwards, Suzan
Eoll, John Gordon
Farhi, Edward
Fazio, Giovanni Gene
Field, George Brooks
Ford, Lawrence Howard
Freese, Katherine
Geller, Margaret Joan
Goldsmith, Paul Felix
Gorenstein, Marc Victor
Gorenstein, Paul
Grindlay, Jonathan Ellis
Guth, Alan Harvey
Habbal, Shadia Rifai
Harrison, Edward Robert
Hodges, Hardy M
Huchra, John P
Huguenin, George Richard
Joss, Paul Christopher
Kalkofen, Wolfgang
Kalman, Gabor J
Kellogg, Edwin M
Kent, Stephen Matthew
Kinoshita, Kay
Kirby, Kate Page
Kistiakowsky, Vera
Kohl, John Leslie
Krieger, Allen Stephen
Kurucz, Robert Louis
Kwan, John Ying-Kuen
Kwitter, Karen Beth
Lecar, Myron
Leiby, Clare C, Jr
Levine, Randolph Herbert
Lewin, Walter H G
Lightman, Alan Paige
Little, Stephen James
Little-Marenin, Irene Renate
McNutt, Ralph Leroy, Jr
Marsden, Brian Geoffrey
Mendillo, Michael
Michael, Irving
Milbocker, Michael

Moody, Elizabeth Anne
Murray, Stephen S
Noyes, Robert Wilson
Papagiannis, Michael D
Parkinson, William Hambleton
Parsignault, Daniel Raymond
Pasachoff, Jay M(yron)
Pilot, Christopher H
Press, William Henry
Ratner, Michael Ira
Raymond, John Charles
Reasenberg, Robert David
Roberts, David Hall
Rothwell, Paul L
Sagalyn, Rita C
Schattenburg, Mark Lee
Schwartz, Daniel Alan
Seward, Frederick Downing
Shapiro, Irwin Ira
Sheldon, Eric
Shuman, Bertram Marvin
Smith, Peter Lloyd
Snell, Ronald Lee
Strom, Stephen
Sulak, Lawrence Richard
Taff, Laurence Gordon
Tananbaum, Harvey Dale
Tapia, Santiago
Thaddeus, Patrick
Tonry, John Landis
Traub, Wesley Arthur
Vessot, Robert F C
Victor, George A
Vinti, John Pascal
Weber, Edward Joseph
Whipple, Fred Lawrence
Whitney, Charles Allen
Willner, Steven P
Wisdom, Jack Leach
Withbroe, George Lund
Yates, George Kenneth

MICHIGAN
Akerlof, Carl W
Berger, Beverly Kobre
Donahue, Thomas Michael
Fontheim, Ernest Gunter
Hegyi, Dennis
Hiltner, William Albert
Kaiser, Christopher B
Kingman, Robert Earl
Krisch, Jean Peck
Kuhn, William R
Linnell, Albert Paul
Longo, Michael Joseph
MacAlpine, Gordon Madeira
Samir, Uri
Sears, Richard Langley
Stein, Robert Foster
Tarle, Gregory
Teske, Richard Glenn
Weinberger, Doreen Anne
Zimring, Lois Jacobs

MINNESOTA
Cahill, Laurence James, Jr
Davidson, Kris
Erickson, Kenneth Neil
Flower, Terrence Frederick
Gehrz, Robert Douglas
Gregory, Stephen Albert
Hakkila, Jon Eric
Jones, Thomas Walter
Kuhi, Leonard Vello
Lysak, Robert Louis
Marshak, Marvin Lloyd
Rudaz, Serge
Rudnick, Lawrence
Stein, Wayne Alfred
Waddington, Cecil Jacob

MISSISSIPPI
Lestrade, John Patrick

MISSOURI
Bieniek, Ronald James
Binns, Walter Robert
Bragg, Susan Lynn
Cowsik, Ramanath
Edwards, Terry Winslow
Friedlander, Michael Wulf
Israel, Martin Henry
Katz, Jonathan Isaac
Kovacs, Sandor J, Jr
Manuel, Oliver K
Mashhoon, Bahram
Schwartz, Richard Dean
Walker, Robert Mowbray
Will, Clifford Martin

MONTANA
Caughlan, Georgeanne Robertson
Hiscock, William Allen

NEBRASKA
Leung, Kam-Ching
Sartori, Leo
Schmidt, Edward George
Simon, Norman Robert

NEVADA
Weistrop, Donna Etta

NEW HAMPSHIRE
Arnoldy, Roger L
Boley, Forrest Irving
Chupp, Edward Lowell
Forbes, Terry Gene
Hollweg, Joseph Vincent
Hudson, Mary Katherine
Lee, Martin Alan
Lockwood, John Alexander
Morris, Daniel Joseph
Ryan, James Michael
Webber, William R

NEW JERSEY
Arthur, Wallace
Bahcall, John Norris
Bahcall, Neta Assaf
Bechis, Dennis John
Becken, Eugene D
Blair, Grant Clark
Chen, Liu
Daly, Ruth Agnes
Dicke, Robert Henry
Draine, Bruce T
Flannery, Brian Paul
Gautreau, Ronald
Goode, Philip Ranson
Gott, J Richard, III
Grek, Boris
Groth, Edward John, III
Gunn, James Edward
Hut, Piet
Jenkins, Edward Beynon
Langer, William David
Lanzerotti, Louis John
Leventhal, Marvin
Levine, Jerry David
Matilsky, Terry Allen
Mikkelsen, David Robert
Molnar, Michael Robert
Motz, Robin Owen
Nardi, Vittorio
Ostriker, Jeremiah P
Owens, David Kingston
Paczynski, Bohdan
Pearton, Stephen John
Pfeiffer, Raymond John
Raghavan, Pramila
Rogerson, John Bernard, Jr
Schwarzschild, Martin
Shipley, Edward Nicholas
Spergel, David Nathaniel
Spitzer, Lyman, Jr
Stark, Antony Albert
Taylor, Harold Evans
Taylor, Joseph Hooton, Jr
Tishby, Naftali Z
Turner, Edwin Lewis
Zimmermann, R Erik

NEW MEXICO
Ahluwalia, Harjit Singh
Altrock, Richard Charles
Arion, Douglas
Bame, Samuel Jarvis, Jr
Blake, Richard L
Brownlee, Robert Rex
Burman, Robert L
Cartwright, David Chapman
Clark, Robert Edward Holmes
Cleveland, Bruce Taylor
Conner, Jerry Power
Coon, James Huntington
Cox, Arthur Nelson
Crane, Patrick Conrad
Davey, William Robert
Davis, Cecil Gilbert
Deupree, Robert G
Dumas, Herbert M, Jr
Evans, John Wainwright, Jr
Feldman, William Charles
Forslund, David Wallace
Freyer, Gustav John
Fu, Jerry Hui Ming
Gosling, John Thomas
Greene, Arthur Edward
Hills, Jack Gilbert
Iwan, DeAnn Colleen
Jarmie, Nelson
Jones, Eric Manning
Keil, Stephen Lesley
Kemic, Stephen Bruce
King, David Solomon
Kopp, Roger Alan
Lilley, John Richard
Meier, Michael McDaniel
Merts, Athel Lavelle
Moore, Elliott Paul
Mutschlecner, Joseph Paul
Newman, Michael J(ohn)
Olson, Gordon Lee
Ozernoy, Leonid M
Parker, Winifred Ellis
Petschek, Albert George
Reedy, Robert Challenger
Reeves, Geoffrey D
Rokop, Donald J
Sanders, Walter L
Sandford, Maxwell Tenbrook, II
Schultz, Rodney Brian
Simon, George Warren
Stellingwerf, Robert Francis
Stone, Sidney Norman
Sweeney, Mary Ann

Terrell, N(elson) James, Jr
Van Riper, Kenneth Alan
Wallace, Richard Kent
Wayland, James Robert, Jr
Weaver, Robert Paul
Westpfahl, David John
White, Paul C
Wing, Janet E (Sweedyk) Bendt
Winske, Dan
Zeilik, Michael
Zirker, Jack Bernard

NEW YORK
Alpher, Ralph Asher
Baker, Norman Hodgson
Baltz, Anthony John
Beckwith, Steven Van Walter
Bocko, Mark Frederick
Boyd, Robert William
Brink, Gilbert Oscar
Brown, Stanley Gordon
Cheung, Lim H
Chung, Kuk Soo
Craft, Harold Dumont, Jr
DeNoyer, Linda Kay
Douglass, David Holmes
Dubisch, Russell John
Edelstein, William Alan
Elmegreen, Debra Meloy
Epstein, Isadore
Feit, Julius
Fishbone, Leslie Gary
Forman, Miriam Ausman
Gill, Ronald Lee
Glassgold, Alfred Emanuel
Goebel, Ronald William
Grannis, Paul Dutton
Greisen, Kenneth I
Gross, Stanley H
Hahn, Richard Leonard
Hardorp, Johannes Christfried
Helfand, David John
Helfer, Herman Lawrence
Houck, James Richard
Huggins, Patrick John
Kegeles, Lawrence Steven
Knacke, Roger Fritz
Lasher, Gordon (Jewett)
Lattimer, James Michael
Lazareth, Otto William, Jr
Lovelace, Richard Van Evera
Madey, Robert W
Mansfield, Victor Neil
Matsushima, Satoshi
Meisel, David Dering
Mohlke, Byron Henry
Moniot, Robert Keith
Motz, Lloyd
Novick, Robert
O'Donoghue, Aileen Ann
Peak, David
Ratcliff, Keith Frederick
Remo, John Lucien
Rosenzweig, Carl
Salpeter, Edwin Ernest
Sato, Makiko
Savedoff, Malcolm Paul
Schmalberger, Donald C
Schulz, Helmut Wilhelm
Seiden, Philip Edward
Shapiro, Stuart Louis
Sheppard, Ronald John
Smolin, Lee
Solomon, Philip M
Spergel, Martin Samuel
Spiegel, Edward A
Teukolsky, Saul Arno
Thomas, John Howard
Toner, John Joseph
Tryon, Edward Polk
Ullman, Jack Donald
Van Horn, Hugh Moody
Wolf, Henry
Yahil, Amos

NORTH CAROLINA
Brown, J(ohn) David
Christiansen, Wayne Arthur
Danby, John Michael Anthony
Herbst, Eric
Jenkins, Alvin Wilkins, Jr
Siewert, Charles Edward
Simon, Sheridan Alan
York, James Wesley, Jr

NORTH DAKOTA
Berkey, Gordon Bruce

OHIO
Bidelman, William Pendry
Boyd, Richard Nelson
Chai, An-Ti
Chiu, Victor
Czyzak, Stanley Joachim
Delsemme, Armand Hubert
Ferguson, Dale Curtis
Ferland, Gary Joseph
Frogel, Jay Albert
Jenkins, Thomas Llewellyn
Luck, Richard Earle
Macklin, Philip Alan
Peterson, Bradley Michael
Ptak, Roger Leon

Raby, Stuart
Rao, Kandarpa Narahari
Rodman, James Purcell
Sellgren, Kristen
Slettebak, Arne
Snider, Joseph Lyons
Steigman, Gary
Stoner, Ronald Edward
Winters, Ronald Ross
Witt, Adolf Nicolaus

OKLAHOMA
Branch, David Reed
Buck, Richard F
Chincarini, Guido Ludovico
Cowan, John James
Gudehus, Donald Henry
Hill, Stephen James
Schroeder, Leon William
Shull, Peter Otto, Jr

OREGON
Kemp, James Chalmers
Siemens, Philip John

PENNSYLVANIA
Axel, Leon
Beier, Eugene William
Boughn, Stephen Paul
Carpenter, Lynn Allen
Cashdollar, Kenneth Leroy
Chambliss, Carlson Rollin
Cohen, Jeffrey M
Córdova, France Anne-Dominic
Donoghue, Timothy R
Duggal, Shakti Prakash
Feigelson, Eric Dennis
Frenklach, Michael Y
Garmire, Gordon Paul
Gaustad, John Eldon
Green, Louis Craig
Gross, Peter George
Guinan, Edward F
Hapke, Bruce W
Kendall, Bruce Reginald Francis
Lande, Kenneth
McCluskey, George E, Jr
Mahaffy, John Harlan
Nagy, Theresa Ann
Quinn, Robert George
Ramsey, Lawrence William
Shen, Benjamin Shih-Ping
Sion, Edward Michael
Sitterly, Charlotte Moore
Soberman, Robert K
Steinhardt, Paul Joseph
Tobias, Russell Lawrence
Vila, Samuel Campderros
Youtcheff, John Sheldon

RHODE ISLAND
Brandenberger, Robert H
Gilman, John Richard, Jr
Lanou, Robert Eugene, Jr
Rossner, Lawrence Franklin
Timbie, Peter T

SOUTH CAROLINA
Adelman, Saul Joseph
Clayton, Donald Delbert
Flower, Phillip John
Safko, John Loren
Zechiel, Leon Norris

TENNESSEE
Barnes, Ronnie C
Bartelt, John Eric
Blass, William Errol
Eaton, Joel A
Fox, Kenneth
MacQueen, Robert Moffat
Phaneuf, Ronald Arthur
Thonnard, Norbert
Tolk, Norman Henry
Yau, Cheuk Chung

TEXAS
Allum, Frank Raymond
Auchmuty, Giles
Badhwar, Gautam D
Bering, Edgar Andrew, III
Black, David Charles
Cloutier, Paul Andrew
Criswell, David Russell
Deeming, Terence James
Dessler, Alexander Jack
Dicus, Duane A
Duke, Michael SN
Duller, Nelson M, Jr
Edmonds, Frank Norman, Jr
Evans, Neal John, II
Fenyves, Ervin J
Freeman, John Wright, Jr
Graham, William Richard Montgomery
Harrison, Marjorie Hall
Haymes, Robert C
Heikkila, Walter John
Henize, Karl Gordon
Hill, Thomas Westfall
Hodges, Ralph Richard, Jr
Huebner, Walter F
Johnson, Francis Severin
Konradi, Andrei
Koplyay, Janos Bernath

Astrophysics (cont)

Lewis, Ira Wayne
Liang, Edison Park-Tak
Lopez, Jorge Alberto
Mahajan, Swadesh Mitter
Mangum, Jeffrey Gary
Michel, F Curtis
Modisette, Jerry L
Morgan, Thomas Harlow
Nacozy, Paul E
Nather, Roy Edward
Nyquist, Laurence Elwood
O'Dell, Charles Robert
Opal, Chet Brian
Page, Thornton Leigh
Palmeira, Ricardo Antonio Ribeiro
Robbins, Ralph Robert
Roeder, Robert Charles
Shapiro, Paul Robert
Smoluchowski, Roman
Su, Shin-Yi
Tajima, Toshiki
Talent, David Leroy
Tinsley, Brian Alfred
Trafton, Laurence Munro
Weinstein, Roy
Weisheit, Jon Carleton
Wheeler, John Craig
Wilson, Thomas Leon
Zook, Herbert Allen

UTAH
Bergeson, Haven Eldred
Elbert, Jerome William
Jones, Douglas Emron
Lind, Vance Gordon
McDonald, Keith Leon
McNamara, Delbert Harold
Price, Richard Henry
Schunk, Robert Walter
Taylor, Benjamin Joseph

VERMONT
Rankin, Joanna Marie
Winkler, Paul Frank
Wolfson, Richard L T

VIRGINIA
Adam, John Anthony
Atalay, Bulent Ismail
Blondin, John Michael
Boozer, Allen Hayne
Bunner, Alan Newton
Callo, Anthony John
Chevalier, Roger Alan
Chubb, Talbot Albert
Dennison, Brian Kenneth
Evans, John C
Gibson, Luther Ralph
Gowdy, Robert Henry
Harvey, Gale Allen
Holt, Alan Craig
Howard, Russell Alfred
Jacobs, Kenneth Charles
Jaffe, Leonard
Kafatos, Minas
Kelch, Walter L
Kellermann, Kenneth Irwin
Koomen, Martin J
Lanzano, Paolo
Ludwig, George H
Marshak, Robert Eugene
Mayer, Cornell Henry
Moore, Garry Edgar
Morgan, Thomas Edward
Mullen, Joseph Matthew
O'Connell, Robert West
Oesterwinter, Claus
Opp, Albert Geelmuyden
Radoski, Henry Robert
Rood, Robert Thomas
Sarazin, Craig L
Shapiro, Maurice Mandel
Thuan, Trinh Xuan
Tidman, Derek Albert
Turner, Barry Earl
Walker, David N(orton)
Wallace, Lance Arthur
Wootten, Henry Alwyn

WASHINGTON
Anderson, Hugh Riddell
Bardeen, James Maxwell
Bodansky, David
Bookmyer, Beverly Brandon
Brodzinski, Ronald Lee
Campbell, Malcolm John
Despain, Lewis Gail
Ellis, Stephen Dean
Ely, John Thomas Anderson
Gil, Salvador
Greene, Thomas Frederick
Haxton, Wick Christopher
Hoch, Richmond Joel
Holzworth, Robert H, II
Jenner, David Charles
Lake, George Russell
Laul, Jagdish Chander
Lutz, Julie Haynes
Margon, Bruce Henry
Olsen, Kenneth Harold
Peters, Philip Carl

WEST VIRGINIA
Ghigo, Frank Dunnington
Littleton, John Edward
Weldon, Henry Arthur

WISCONSIN
Cassinelli, Joseph Patrick
Code, Arthur Dodd
Doherty, Lowell Ralph
Fry, William Frederick
Gallagher, John Sill
Hobbs, Lewis Mankin
Lawler, James Edward
McCammon, Dan
March, Robert Herbert
Mathis, John Samuel
Parker, Leonard Emanuel
Reynolds, Ronald J
Savage, Blair DeWillis
Scherb, Frank
Woods, Robert Claude

WYOMING
Howell, Robert Richard
Roark, Terry P
Thronson, Harley Andrew, Jr

PUERTO RICO
Lewis, Brian Murray

ALBERTA
Cook, Frederick Ahrens
Leahy, Denis Alan
Milone, Eugene Frank
Pinnington, Eric Henry
Rostoker, Gordon
Sreenivasan, Sreenivasa Ranga
Venkatesan, Doraswamy
Weale, Gareth Pryce
Wilson, William James Fitzpatrick

BRITISH COLUMBIA
Aikman, George Christopher Lawrence
Auman, Jason Reid
Batten, Alan Henry
Burke, J Anthony
Climenhaga, John Leroy
Fahlman, Gregory Gaylord
Galt, John (Alexander)
Hesser, James Edward
Horita, Robert Eiji
Menon, Thuppalay K
Palmer, Leigh Hunt
Singh, Manohar
Underhill, Anne Barbara
Walker, Gordon Arthur Hunter
Wright, Kenneth Osborne

NEW BRUNSWICK
Hawkes, Robert Lewis

NOVA SCOTIA
Welch, Gary Alan

ONTARIO
Barker, Paul Kenneth
Bernath, Peter Francis
Bolton, Charles Thomas
Clement, Maurice James Young
Dove, John Edward
Ewan, George T
Feldman, Paul Arnold
FitzGerald, Maurice Pim Valter
Garrison, Robert Frederick
Halliday, Ian
Henriksen, Richard Norman
Hruska, Antonin
Innanen, Kimmo A
Kronberg, Philipp Paul
Landstreet, John Darlington
Lester, John Bernard
Lowe, Robert Peter
McDiarmid, Ian Bertrand
McIntosh, Bruce Andrew
McNamara, Allen Garnet
Martin, Peter Gordon
Millman, Peter MacKenzie
Mitalas, Romas
Moffat, John William
Morton, Donald Charles
Sutherland, Peter Gordon
Tremaine, Scott Duncan
Vallée, Jacques P
Varshni, Yatendra Pal
Wesson, Paul Stephen

PRINCE EDWARD ISLAND
Lin, Wei-Ching

QUEBEC
Aktik, Cetin
Borra, Ermanno Franco
Chaubey, Mahendra
Davies, Roger
Demers, Serge
Fontaine, Gilles Joseph
Michaud, Georges Joseph
Moffat, Anthony Frederick John
Pearson, John Michael
Pineault, Serge Rene
Tassoul, Jean-Louis
Tassoul, Monique

SASKATCHEWAN
Papini, Giorgio Augusto

OTHER COUNTRIES
Beckers, Jacques Maurice
Bekenstein, Jacob David
Burnell, S Jocelyn Bell
Butcher, Harvey Raymond, III
Cernuschi, Felix
Chau, Wai-Yin
Contopoulos, George
Crane, Philippe
Feinstein, Alejandro
Greenberg, Jerome Mayo
Icke, Vincent
Munch, G
Novotny, Eva
Pottasch, Stuart Robert
Rees, Martin J
Schnopper, Herbert William
Schutz, Bernard Frederick
Sugiura, Masahisa
Vrebalovich, Thomas
Weil, Raoul Bloch
Wildman, Peter James Lacey

Atomic & Molecular Physics

ALABAMA
Bauman, Robert Poe
Edlin, George Robert
Helminger, Paul Andrew
Henry, Ronald James Whyte
Howgate, David W
Izatt, Jerald Ray
Jones, Robert William
Pindzola, Michael Stuart
Rosenberger, Albert Thomas
Smith, Lewis Taylor
Stephens, Timothy Lee
Stettler, John Dietrich
Tidwell, Eugene Delbert
Tipping, Richard H
Varghese, Sankoorikal Lonappan

ALASKA
Deehr, Charles Sterling
Degen, Vladimir
Sheridan, John Roger
Wentink, Tunis, Jr

ARIZONA
Bashkin, Stanley
Black, John Harry
Brault, James William
Engleman, Rolf, Jr
Fan, Chang-Yun
Farr, William Morris
Garcia, Jose Dolores, Jr
Gibbs, Hyatt McDonald
Hight, Ralph Dale
Jones, Roger C(lyde)
Lamb, Willis Eugene, Jr
Leavitt, John Adams
Lutz, Barry Lafean
Rafelski, Johann
Shemansky, Donald Eugene
Steimle, Timothy C
Stoner, John Oliver, Jr
West, Robert Elmer
Wing, William Hinshaw
Yelle, Roger V

ARKANSAS
Bronco, Charles John
Gupta, Rajendra
Hughes, Raymond Hargett
Lieber, Michael
Vyas, Reeta

CALIFORNIA
Adams, Arnold Lucian
Alonso, Jose Ramon
Alvi, Zahoor M
Anderson, Charles Hammond
Antolak, Arlyn Joe
Armstrong, Baxter Hardin
Auerbach, Daniel J
Bach, David Rudolph
Ball, William Paul
Bardsley, James Norman
Baur, James Francis
Behringer, Robert Ernest
Benton, Eugene Vladimir
Berkner, Klaus Hans
Bethune, David Stimson
Bloom, Arnold Lapin
Boehm, Felix H
Booker, Henry George
Brewer, Richard George
Brown, Linda Rose
Brown, William Arnold
Burnett, Lowell Jay
Burton, Donald Eugene
Caird, John Allyn
Camparo, James Charles
Carman, Robert Lincoln, Jr
Chackerian, Charles, Jr
Chapline, George Frederick, Jr
Cheng, Kwok-Tsang
Chodos, Steven Leslie
Chu, Keh-Cheng
Chu, Steven

Chutjian, Ara
Clark, Arnold Franklin
Cohen, Lawrence Mark
Conway, John George, Jr
Cook, Charles J
Crosley, David Risdon
Cutler, Leonard Samuel
Davis, James Ivey
Dowell, Jerry Tray
Dowling, Jerome M
Dows, David Alan
Doyle, Walter M
Dulgeroff, Carl Richard
Edwards, David Franklin
Ehlers, Kenneth Warren
Feiock, Frank Donald
Feng, Joseph Shao-Ying
Finston, Roland A
Fisk, George Ayrs
Franks, Larry Allen
Ganas, Perry S
George, Simon
Gillen, Keith Thomas
Gillespie, George H
Gilmore, Forrest Richard
Glass, Alexander Jacob
Gould, Harvey Allen
Gwinn, William Dulaney
Haas, Peter Herbert
Hackel, Lloyd Anthony
Hagstrom, Stig Bernt
Hanna, Stanley Sweet
Harrach, Robert James
Harris, David Owen
Harris, Dennis George
Heestand, Glenn Martin
Hemminger, John Charles
Herm, Ronald Richard
Herman, Frank
Hershey, Allen Vincent
Heusinkveld, Myron Ellis
Hill, Robert Matteson
Hrubesh, Lawrence Wayne
Hu, Chi-Yu
Huestis, David Lee
Jacobs, Ralph R
Jaduszliwer, Bernardo
Jaklevic, Joseph Michael
Janney, Gareth Maynard
Jusinski, Leonard Edward
Karo, Arnold Mitchell
Kasai, Paul Haruo
Keiser, George McCurrach
Kelly, Paul Sherwood
Kelly, Raymond Leroy
Kim, Jinchoon
Kirby, Jon Allan
Klein, August S
Klein, David Joseph
Knight, Walter David
Knipe, Richard Hubert
Knize, Randall James
Kocol, Henry
Kulander, Kenneth Charles
Kunkel, Wulf Bernard
Kwok, Munson Arthur
Lam, Leo Kongsui
Landman, Donald Alan
Lapp, M(arshall)
Lee, Long Chi
Lee, Paul L
Lee, Yim Tin
Lester, William Alexander, Jr
Levenson, Marc David
Lichter, James Joseph
Linford, Gary Joe
Ling, Rung Tai
Lipeles, Martin
Litvak, Marvin Mark
Loew, Gilda Harris
Lombardi, Gabriel Gustavo
Lorents, Donald C
Luther, Marvin L
McCall, Richard C
McFarlane, Ross Alexander
McGowan, James William
McKay, Dale Robert
McKenzie, Robert Lawrence
McKoy, Basil Vincent
McMaster, William H
Magnuson, Gustav Donald
Mahadevan, Parameswar
Maker, Paul Donne
Marino, Lawrence Louis
Marrs, Roscoe Earl
Marrus, Richard
Mathis, Ronald Floyd
Meyerhof, Walter Ernst
Meyerott, Roland Edward
Milanovich, Fred Paul
More, Richard Michael
Muntz, Eric Phillip
Myers, Benjamin Franklin, Jr
Nanes, Roger
Neugebauer, Gerry
Newman, David Edward
Neynaber, Roy H(arold)
Olness, Dolores Urquiza
O'Malley, Thomas Francis
Palatnick, Barton
Park, Chul
Perel, Julius
Peterson, James Ray
Phillips, Edward

Plock, Richard James
Prag, Arthur Barry
Prior, Michael Herbert
Rast, Howard Eugene, Jr
Reisler, Hanna
Rescigno, Thomas Nicola
Roeder, Stephen Bernhard Walter
Rosen, Mordecai David
Ross, Marvin Franklin
Rotenberg, Manuel
Salisbury, Stanley R
Satten, Robert A
Schaefer, Albert Russell
Schlachter, Alfred Simon
Schmieder, Robert W
Scott, Paul Brunson
Senitzky, Israel Ralph
Shore, Bruce Walter
Shrivastava, Prakash Narayan
Shugart, Howard Alan
Simon, Barry Martin
Sinton, Steven Williams
Skomal, Edward N
Smith, Felix Teisseire
Snow, William Rosebrook
Snyder, Robert Gene
Stearns, John Warren
Stephens, Jeffrey Alan
Stephenson, David Allen
Stuart, George Wallace
Suchannek, Rudolf Gerhard
Swanson, William Paul
Syage, Jack A
Tang, Stephen Shien-Pu
Taylor, Howard S
Terhune, Robert William
Thoe, Robert Steven
Trajmar, Sandor
Trujillo, Stephen Michael
Tu, Charles Wuching
Utterback, Nyle Gene
Varney, Robert Nathan
Vroom, David Archie
Walker, Keith Gerald
Wallis, Richard Fisher
Weber, Marvin John
Weissbluth, Mitchel
Wessel, John Emmit
Whaling, Ward
Whetten, Robert Lloyd
Wieder, Harold
Wieder, Irwin
Wilson, Walter Davis
Woerner, Robert Leo
Woldseth, Rolf
Wood, Calvin Dale
Woodruff, Truman Owen
Woodward, Ervin Chapman, Jr
Worden, Earl Freemont, Jr
Young, Louise Gray
Zewail, Ahmed H
Zimmerman, Ivan Harold
Zook, Alma Claire

COLORADO
Adams, Gail Dayton
Amme, Robert Clyde
Beckmann, Petr
Beers, Yardley
Bills, Daniel Granville
Chappell, Willard Ray
Cooper, John (Jinx)
Daw, Glen Harold
Dunn, Gordon Harold
Evenson, Kenneth Melvin
Fox, Michael Henry
Furcinitti, Paul Stephen
Gallagher, Alan C
Garstang, Roy Henry
Greene, Christopher Henry
House, Lewis Lundberg
Itano, Wayne Masao
Jespersen, James
Kiehl, Jeffrey Theodore
Leone, Stephen Robert
Mizushima, Masataka
Morgan, Wm Lowell
Morris, George Ronald
Mosburg, Earl R, Jr
Neumann, Herschel
Norcross, David Warren
O'Callaghan, Michael James
Oneil, Stephen Vincent
Parson, Robert Paul
Phelps, Arthur Van Rensselaer
Schowengerdt, Franklin Dean
Smith, Ernest Ketcham
Smith, Stephen Judson
Sullivan, Donald Barrett
Wessel, William Roy
Whitten, Barbara L

CONNECTICUT
Akkapeddi, Prasad Rao
Antar, Ali A
Bartram, Ralph Herbert
Cable, Peter George
Chupka, William Andrew
Gau, John N
Greenberg, Jack Sam
Greenwood, Ivan Anderson
Hayden, Howard Corwin
Hinchen, John J(oseph)
Jakacky, John M

Keith, H(arvey) Douglas
Kessel, Quentin Cattell
King, George, III
Lubell, Michael S
Michels, H(orace) Harvey
Monce, Michael Nolen
Moreland, Parker Elbert, Jr
Morgan, Thomas Joseph
Nath, Ravinder Katyal
Peterson, Cynthia Wyeth
Petersson, George A
Pollack, Edward
Smith, Winthrop Ware
Staker, William Paul
Ultee, Casper Jan
Wong, Shek-Fu

DELAWARE
Crawford, Michael Karl
Fou, Cheng-Ming
Morgan, John Davis, III
Sharnoff, Mark
Szalewicz, Krzysztof
Woo, Shien-Biau

DISTRICT OF COLUMBIA
Beard, Charles Irvin
Berman, Barry L
Borras, Caridad
Brown, Charles Moseley
Burkhalter, Philip Gary
Campillo, Anthony Joseph
Chiu, Lue-Yung Chow
Corliss, Charles Howard
Crandall, David Hugh
Crisp, Michael Dennis
Feldman, Uri
Haftel, Michael Ivan
Halpern, Joshua Baruch
Hessel, Merrill
Jacobs, Verne Louis
Lucatorto, Thomas B
McMahon, John Michael
Mahon, Rita
Markevich, Darlene Julia
Misra, Prabhakar
Phillips, Gary Wilson
Ramaker, David Ellis
Reeves, Edmond Morden
Reiss, Howard R
Scheel, Nivard
Schoen, Richard Isaac
Stone, Philip M
Taylor, Ronald D
White, John Arnold
Wodarczyk, Francis John

FLORIDA
Bailey, Thomas L, III
Bartlett, Rodney Joseph
Brucat, Philip John
Burns, Jay, III
Burns, Michael J
Callan, Edwin Joseph
Darling, Byron Thorwell
Edmonds, Dean Stockett, Jr
Fitzgerald, Lawrence Terrell
Foner, Samuel Newton
Gibson, Henry Clay, Jr
Hanson, Harold Palmer
Harney, Robert Charles
Hooper, Charles Frederick, Jr
Killinger, Dennis K
Linder, Ernest G
Micha, David Allan
Monkhorst, Hendrik J
Oelfke, William C
Rhodes, Richard Ayer, II
Ross, John Stoner
Sabin, John Rogers
Sheldon, John William
Shelton, Wilford Neil
Sherwood, Jesse Eugene
Skofronick, James Gust
Sullivan, Neil Samuel
Szczepaniak, Krystyna
Van Zee, Richard Jerry
Zerner, Michael Charles

GEORGIA
Cheng, Wu-Chieh
Cooper, Charles Dewey
Edwards, Alan Kent
Flannery, Martin Raymond
Gole, James Leslie
Manson, Steven Trent
Menendez, Manuel Gaspar
Nave, Carl R
Palmer, Richard Carl
Rao, Pemmaraju Venugopala
Roy, Rajarshi
Steuer, Malcolm F
Thomas, Edward Wilfrid
Uzer, Ahmet Turgay
Wood, Robert Manning
Zangwill, Andrew

HAWAII
Geballe, Thomas Ronald
Holmes, John Richard
McLaren, Robert Alexander

IDAHO
Davis, Lawrence William, Jr

Harper, Henry Amos, Jr
Knox, John MacMurray
Shirts, Randall Brent

ILLINOIS
Ali, Naushad
Atoji, Masao
Averill, Frank Wallace
Barrall, Raymond Charles
Bennett, Edgar F
Berkowitz, Joseph
Berry, Henry Gordon
Bertoncini, Peter Joseph
Buck, Warren Louis
Childs, William Jeffries
Cloud, William Max
Court, Anita
Day, Michael Hardy
Dehmer, Joseph Leonard
Dehmer, Patricia Moore
Dolecek, Elwyn Haydn
Donnally, Bailey Lewis
Druger, Stephen David
Dunford, Robert Walter
Eden, James Gary
Ellis, Donald Edwin
Fano, Ugo
Garcia-Munoz, Moises
Gislason, Eric Arni
Greene, John Philip
Hessler, Jan Paul
Holzberlein, Thomas M
Huebner, Russell Henry, Sr
Ingram, Forrest Duane
Inokuti, Mitio
Kang, Ik-Ju
Karim, Khondkar Rezaul
Kimura, Mineo
McCormack, Elizabeth Frances
Malik, Fazley Bary
Mallow, Jeffry Victor
Melendres, Carlos Arciaga
Nayfeh, Munir Hasan
Nieman, George Carroll
Oka, Takeshi
Palmer, Patrick Edward
Pratt, Stephen Turnham
Rhodes, Charles Kirkham
Sanders, Frank Clarence, Jr
Schatz, George Chappell
Secrest, Donald H
Sibener, Steven Jay
Smither, Robert Karl
Snyder, Lewis Emil
Spence, David
Spokas, John J
Tomkins, Frank Sargent
Treadwell, Elliott Allen
Verdeyen, Joseph T
Woodruff, William Lee
Yuster, Philip Harold

INDIANA
Bentley, John Joseph
Bunker, Bruce Alan
Clikeman, Franklyn Miles
Grant, Edward R
Langhoff, Peter Wolfgang
Mozumder, Asokendu
Sapirstein, Jonathan Robert
Shorer, Philip
Shupe, Robert Eugene
Tripathi, Govakh Nath Ram

IOWA
Houk, Robert Samuel
Klemm, Richard Andrew
Lutz, Robert William
Robinson, David
Schofield, Robert Edwin
Stwalley, William Calvin

KANSAS
Andrew, Kenneth L
Bhalla, Chander P
Broersma, Sybrand
Chu, Shih I
Curnutte, Basil, Jr
Dreschhoff, Gisela Auguste-Marie
Gray, Tom J
Greene, Frank T
Hagmann, Siegbert Johann
Harmony, Marlin D
Legg, James C
Lin, Chii-Dong
McGuire, James Horton
Richard, Patrick
Stockli, Martin P
Weller, Lawrence Allenby

KENTUCKY
Bradley, Eugene Bradford
Clouthier, Dennis James
Kielkopf, John F
MacAdam, Keith Bradford

LOUISIANA
Chin-Bing, Stanley Arthur
Eidson, William Whelan
Gibbs, Richard Lynn
Head, Charles Everett
Head, Martha E Moore
Ho, Yew Kam
Klasinc, Leo

Kumar, Devendra
Meckstroth, George R
Meriwether, John R
Perdew, John Paul
Rau, A Ravi Prakash
Scott, John Delmoth
Stockbauer, Roger Lewis
Zander, Arlen Ray

MAINE
Kingsbury, Robert Freeman

MARYLAND
Allen, John Edward, Jr
Allen, John Edward, Jr
Armstrong, Lloyd, Jr
Bates, Lloyd M
Bauer, Ernest
Bax, Ad
Benesch, William Milton
Benson, Richard C
Bhatia, Anand K
Birnbaum, George
Bryden, Wayne A
Bur, Anthony J
Buslik, Arthur J
Cavanagh, Richard Roy
Cave, William Thompson
Celotta, Robert James
Chaisson, Eric Joseph
Clark, Charles Winthrop
Cohen, Steven Charles
Coplan, Michael Alan
Criss, John W
Crosswhite, Henry Milton, Jr
Danos, Michael
Dick, Charles Edward
Doschek, George A
Elder, Robert Lee
Epstein, Gabriel Leo
Feldman, Paul Donald
Fuhr, Jeffrey Robert
Gammell, Paul M
Giacchetti, Athos
Ginter, Marshall L
Goodman, Leon Judias
Grant, David Graham
Hellwig, Helmut Wilhelm
Holt, Helen Keil
Hougen, Jon T
Hubbell, John Howard
Hudson, David Frank
Hudson, Robert Douglas
Hurst, Wilbur Scott
Jacox, Marilyn Esther
Jette, Archelle Norman
Judd, Brian Raymond
Julienne, Paul Sebastian
Kaplan, Alexander E
Kastner, Sidney Oscar
Kaufman, Victor
Kessler, Ernest George, Jr
Kessler, Karl Gunther
Kim, Yong-Ki
Klose, Jules Zeiser
Korobkin, Irving
Kostiuk, Theodor
Krisher, Lawrence Charles
Kumar, Cidambi Krishna
Kundu, Mukul Ranjan
Kuyatt, Chris E(rnie Earl)
Lafferty, Walter J
Land, David J(ohn)
LaVilla, Robert E
Lide, David Reynolds, Jr
Lovas, Francis John
McCoubrey, Arthur Orlando
McIlrath, Thomas James
McLaughlin, William Lowndes
Major, Fouad George
Mandelberg, Hirsch I
Martin, William Clyde
Miller, Frank L
Moore, John Hays
More, Kenneth Riddell
Mulligan, Joseph Francis
Mumma, Michael Jon
Oertel, Goetz Kuno Heinrich
Omidvar, Kazem
Ott, William Roger
Oza, Dipak H
Parr, Albert Clarence
Pearl, John Christopher
Phillips, William Daniel
Pierce, Elliot Stearns
Placious, Robert Charles
Plotkin, Henry H
Rasberry, Stanley Dexter
Redmon, Michael James
Rhee, Moon-Jhong
Roszman, Larry Joe
Ruffa, Anthony Richard
Saloman, Edward Barry
Sattler, Joseph Peter
Silver, David Martin
Silverstone, Harris Julian
Sinnott, George
Strombotne, Richard L(amar)
Sugar, Jack
Swenberg, Charles Edward
Taylor, Barry Norman
Taylor, Lauriston Sale
Temkin, Aaron
Thomsen, John Stearns

Atomic & Molecular Physics (cont)

Tilford, Shelby G
Van Brunt, Richard Joseph
Varma, Matesh Narayan
Venkatesan, Thirumalai
Wagner, William John
Way, Kermit R
Weber, Alfons
Weiner, John
Weiss, Andrew W
White, Kevin Joseph
Wiese, Wolfgang Lothar
Wingate, Catharine L
Yaniv, Shlomo Stefan

MASSACHUSETTS
Allis, Willam Phelps
Andersen, Roy Stuart
Benson, Bruce Buzzell
Berney, Charles V
Bjarngard, Bengt E
Bradford, John Norman
Bradley, Lee Carrington, III
Branscomb, Lewis McAdory
Brown, Benjamin Lathrop
Champion, Kenneth Stanley Warner
Chang, Edward Shi Tou
Chen, Sow-Hsin
Clough, Shepard Anthony
Cohen, Howard David
Crampton, Stuart J B
Douglas-Hamilton, Diarmaid H
Feld, Michael S
Franzen, Wolfgang
Garing, John Seymour
George, James Z
Handelman, Eileen T
Haugsjaa, Paul O
Hawkins, Bruce
Hilborn, Robert Clarence
Huffman, Robert Eugene
Hunter, Larry Russel
Hyman, Howard Allan
Jones, Kevin McDill
Katayama, Daniel Hideo
Kelley, Paul Leon
Kellogg, Edwin M
King, John Gordon
Kirby, Kate Page
Kivel, Bennett
Kleppner, Daniel
Kohin, Barbara Castle
Kohl, John Leslie
Krotkov, Robert Vladimir
Landman, Alfred
Lin, Alice Lee Lan
Mandl, Alexander Ernst
Martin, Frederick Wight
Meal, Janet Hawkins
Michael, Irving
Morris, Robert Alan
Moulton, Peter Franklin
Oberteuffer, John Amiard
Oettinger, Peter Ernest
Parkinson, William Hambleton
Person, James Carl
Pichanick, Francis Martin
Pipkin, Francis Marion
Posner, Martin
Pritchard, David Edward
Ramsey, Norman Foster, Jr
Robinson, Howard Addison
Rothman, Laurence Sidney
Ruskai, Mary Beth
Schneider, Robert Julius
Sellers, Francis Bachman
Smith, Peter Lloyd
Soltysik, Edward A
Sternheim, Morton Maynard
Stoner, William Weber
Sun, Yan
Terry, James Layton
Traub, Wesley Arthur
Tuchman, Avraham
Vessot, Robert F C
Victor, George A
Viggiano, Albert
Von Hippel, Arthur R
Wachman, Harold Yehuda
Wang, Chia Ping
Winick, Jeremy Ross
Witherell, Egilda DeAmicis
Wittkower, Andrew Benedict
Yoshino, Kouichi
Zajonc, Arthur Guy

MICHIGAN
Beck, Donald Richardson
Bernius, Mark Thomas
Bernstein, Eugene Merle
Chupp, Timothy E
Davis, Lloyd Craig
Derby, Stanley Kingdon
Donahue, Thomas Michael
Duchamp, David James
Fekety, F Robert, Jr
Ferguson, Stephen Mason
Hase, William Louis
Kauppila, Walter Eric
Kim, Yeong Wook
Krimm, Samuel

Kubis, Joseph J(ohn)
Larsen, Jon Thorsten
Leroi, George Edgar
Miller, Carl Elmer
Parker, Paul Michael
Peterson, Lauren Michael
Phelps, Frederick Martin, III
Rajnak, Katheryn Edmonds
Rand, Stephen Colby
Ressler, Neil William
Rich, Arthur
Rimai, Lajos
Rogers, Jerry Dale
Rol, Pieter Klaas
Rothe, Erhard William
Rothschild, Walter Gustav
Sanders, Barbara A
Sands, Richard Hamilton
Sharp, William Edward, III
Slocum, Robert Richard
Smith, George Wolfram
Stein, Talbert Sheldon
Vaishnava, Prem P
Van Baak, David Alan
Waber, James Thomas
Weidman, Robert Stuart
Weinberger, Doreen Anne
Williams, William Lee
Zimring, Lois Jacobs
Zitzewitz, Paul William
Zorn, Jens Christian

MINNESOTA
Cederberg, James W
Giese, Clayton
Greenlee, Thomas Russell
Hakkila, Jon Eric
Johnson, Walter Heinrick, Jr
Khan, Faiz Mohammad
Knutson, Charles Dwaine
McClure, Benjamin Thompson
Sautter, Chester A
Thomas, Bruce Robert
Truhlar, Donald Gene
Valley, Leonard Maurice

MISSISSIPPI
Monts, David Lee
Ramsdale, Dan Jerry
Rayborn, Grayson Hanks
Rundel, Robert Dean

MISSOURI
Bieniek, Ronald James
Freerks, Marshall Cornelius
Leventhal, Jacob J
McFarland, Robert Harold
Madison, Don Harvey
Menne, Thomas Joseph
Palmer, Kent Friedley
Park, John Thornton
Schearer, Laird D
Schupp, Guy
Taub, Haskell Joseph
Thomas, Timothy Farragut

MONTANA
Carlsten, John Lennart

NEBRASKA
Blotcky, Alan Jay
Burns, Donal Joseph
Burrow, Paul David
Cipolla, Sam J
Fabrikant, Ilya I
Fairchild, Robert Wayne
Gale, Douglas Shannon, II
Jaecks, Duane H
Jones, Ernest Olin
Leichner, Peter K
Rudd, Millard Eugene
Samson, James Alexander Ross
Starace, Anthony Francis

NEVADA
Altick, Philip Lewis
Farley, John William
Moore, Edwin Neal

NEW HAMPSHIRE
Bel Bruno, Joseph James
Brilliant, Howard Michael
Laaspere, Thomas
Lambert, Robert Henry
Phelps, James Parkhurst
Rieser, Leonard M
Strohbehn, John Walter
Wright, John Jay

NEW JERSEY
Ashkin, Arthur
Bjorkholm, John Ernst
Bonin, Keith Donald
Cecchi, Joseph Leonard
Collett, Edward
De Planque, Gail
Feldman, Leonard Cecil
Fisanick, Georgia Jeanne
Fishburne, Edward Stokes, III
Freeman, Richard Reiling
Freund, Robert Stanley
Gabriel, Oscar V
Gottscho, Richard Alan
Hagstrum, Homer Dupre

Hall, Gene Stephen
Happer, William, Jr
Hill, Kenneth Wayne
Hulse, Russell Alan
Kaldor, Andrew
Kelsey, Edward Joseph
Kugel, Henry W
Leventhal, Marvin
Lieb, Elliott Hershel
McAfee, Kenneth Bailey, Jr
Martin, John David
Miles, Richard Bryant
Millman, Sidney
Mills, Allen Paine, Jr
Moses, Herbert A
Murnick, Daniel E
Pfeiffer, Raymond John
Phillips, Julia M
Post, Douglass Edmund
Ramsey, Alan T
Reder, Friedrich H
Redi, Olav
Rothberg, Lewis Josiah
Rousseau, Denis Lawrence
Roy, M S
Salwen, Harold
Seidl, Milos
Silfvast, William Thomas
Suckewer, Szymon
Tomaselli, Vincent Paul
Torrey, Henry Cutler
Wittenberg, Albert M
Yurke, Bernard
Zipf, Elizabeth M(argaret)

NEW MEXICO
Barnes, John Fayette
Bartlett, Roger James
Beauchamp, Edwin Knight
Beckel, Charles Leroy
Becker, Wilhelm
Bellum, John Curtis
Bieniewski, Thomas M
Bingham, Felton Wells
Blake, Richard L
Bryant, Howard Carnes
Burr, Alexander Fuller
Cano, Gilbert Lucero
Cartwright, David Chapman
Chamberlin, Edwin Phillip
Chow, Weng Wah
Clark, Robert Edward Holmes
Coats, Richard Lee
Cohen, James Samuel
Cowan, Robert Duane
Cox, Arthur Nelson
Cross, Jon Byron
Cuderman, Jerry Ferdinand
Czuchlewski, Stephen John
Damerow, Richard Aasen
Doolen, Gary Dean
Funsten, Herbert Oliver, III
Gerardo, James Bernard
Greene, Arthur Edward
Hadley, George Ronald
Hill, Ronald Ames
Holland, Redus Foy
Jason, Andrew John
Jennison, Dwight Richard
Judd, O'Dean P
Kunz, Walter Ernest
Kyrala, George Amine
Ladish, Joseph Stanley
Lockwood, Grant John
Lyman, John L
MacArthur, Duncan W
MacCallum, Crawford John
McGuire, Eugene J
McKenzie, James Montgomery
Maier, William Bryan, II
Mann, Joseph Bird, (Jr)
Mazarakis, Michael Gerassimos
Merts, Athel Lavelle
Mjolsness, Raymond C
Nogar, Nicholas Stephen
Pack, Russell T
Palmer, Byron Allen
Patterson, Christopher Warren
Peek, James Mack
Price, Robert Harold
Redondo, Antonio
Rinker, George Albert, Jr
Sattler, Allan R
Schneider, Barry I
Schneider, Jacob David
Shaner, John Wesley
Sheffield, Richard Lee
Smith, H Vernon, Jr
Snider, Donald Edward
Steinhaus, David Walter
Sze, Robert Chia-Ting
Tonks, Davis Loel
Yates, Mary Anne

NEW YORK
Acrivos, Andreas
Arroe, Hack
Avouris, Phaedon
Becker, Kurt Heinrich
Bederson, Benjamin
Bendler, John Thomas
Bergeman, Thomas H
Beri, Avinash Chandra
Berman, Paul Ronald

Borowitz, Sidney
Boyd, Robert William
Boyer, Donald Wayne
Bramlet, Roland C
Brink, Gilbert Oscar
Brown, Howard Howland, Jr
Budick, Burton
Chapman, Sally
Chaturvedi, Ram Prakash
Chen, James Ralph
Christensen, Robert Lee
Cooper, Philip Harlan
Costa, Lorenzo F
Creasy, William Russel
Das, Tara Prasad
Derman, Samuel
Deutsch, John Ludwig
De Zafra, Robert Lee
Dodd, Jack Gordon, (Jr)
Donnelly, Denis Philip
Dowben, Peter Arnold
Engelke, Charles Edward
Ezra, Gregory Sion
Fairchild, Ralph Grandison
Fischer, C Rutherford
Fitchen, Douglas Beach
Franco, Victor
Garvey, James F
Gavin, Donald Arthur
Gavin, Gerard Brennan
Gentner, Robert F
Gersten, Joel Irwin
Grover, James Robb
Gunter, Karlene Klages
Hagen, Jon Boyd
Halpern, Alvin M
Hameed, Sultan
Hartmann, Francis Xavier
Helbig, Herbert Frederick
Hendrie, Joseph Mallam
Hirsh, Merle Norman
Holbrow, Charles H
Honig, Arnold
Hurst, Robert Philip
Innes, Kenneth Keith
Jain, Duli Chandra
Janak, James Francis
Johnson, Brant Montgomery
Johnson, Joseph Andrew, III
Johnson, Philip M
Jones, Keith Warlow
Kaplan, Martin Charles
Kass, Robert S
Kato, Walter Yoneo
Kelly, John Henry
Koch, Peter M
Komorek, Michael Joesph, Jr
Kostroun, Vaclav O
Krieger, Joseph Bernard
Kwok, Hoi S
Lerner, Rita Guggenheim
Levine, Alfred Martin
Levine, Melvin Mordecai
Lipari, Nunzio Ottavio
Li-Scholz, Angela
Lounsbury, John Baldwin
Lowder, Wayne Morris
Lulla, Kotusingh
Malsky, Stanley Joseph
Mandel, Leonard
Marino, Robert Anthony
Messmer, Richard Paul
Metcalf, Harold
Moe, George Wylbur
Morris, William Guy
Muenter, John Stuart
Mukhopadhyay, Nimai Chand
Osgood, Richard Magee, Jr
Pomilla, Frank R
Privman, Vladimir
Robinson, Edward J
Roellig, Leonard Oscar
Rohrig, Norman
Rosenberg, Leonard
Rustgi, Om Prakash
Sahni, Viraht
Scarl, Donald B
Scholz, Wilfried
Schreurs, Jan W H
Schroeder, John
Schweitzer, Donald Gerald
Solomon, Philip M
Solon, Leonard Raymond
Sternheimer, Rudolph Max
Stroke, Hinko Henry
Sumberg, David A
Thieberger, Peter
Titone, Luke Victor
Tourin, Richard Harold
Uzgiris, Egidijus E
Vacirca, Salvatore John
Varanasi, Prasad
Venugopalan, Srinivasa I
Vuskovic01, Leposava
Walmsley, Ian Alexander
Walter, William Trump
White, Frederick Andrew
Wu, Zhen
Yaakobi, Barukh
Yencha, Andrew Joseph
Zaider, Marco A
Zhang, John Zeng Hui
Zucker, Martin Samuel

NORTH CAROLINA
Christian, Wolfgang C
Clegg, Thomas Boykin
Dixon, Robert Leland
Dotson, Allen Clark
Hegstrom, Roger Allen
Herbst, Eric
Hirsch, Robert George
Hubbard, Paul Stancyl, Jr
Johnson, Charles Edward
Joyce, James Martin
Lapicki, Gregory
Miller, Roger Ervin
Mowat, J Richard
Parker, George W
Patty, Richard Roland
Rabinowitz, James Robert
Risley, John Stetler
Robinson, Hugh Gettys
Rogosa, George Leon
Sayetta, Thomas C
Shafroth, Stephen Morrison
Stephenson, Harold Patty
Vermillion, Robert Everett

NORTH DAKOTA
Moore, Vaughn Clayton
Rao, B Seshagiri

OHIO
Agard, Eugene Theodore
Bahr, Gustave Karl
Bhattacharya, Rabi Sankar
Chen, Charles Chin-Tse
Dakin, James Thomas
De Lucia, Frank Charles
Earhart, Richard Wilmot
Ferland, Gary Joseph
Fox, Thomas Allen
Gangemi, Francis A
Hagee, George Richard
Heer, Clifford V
Hunter, William Winslow, Jr
Kepes, Joseph John
Mickelson, Michael Eugene
Newsom, Gerald Higley
Ross, Charles Burton
Ruegsegger, Donald Ray, Jr
Russell, James Edward
Schectman, Richard Milton
Schlosser, Herbert
Schlosser, Philip A
Schreiber, Paul J
Singleton, Edgar Bryson
Snider, Joseph Lyons
Stoner, Ronald Edward
Strickler, Thomas David
Swofford, Robert Lewis
Williamson, William, Jr
Wood, David Roy
Woodard, Ralph Emerson
Yaney, Perry Pappas

OKLAHOMA
Buchanan, Ronnie Joe
Fowler, Richard Gildart
Huffaker, James Neal
Miller, Thomas Marshall
Ryan, Stewart Richard
St John, Robert Mahard
Schmelling, Stephen Gordon
Waldrop, Morgan A
Westhaus, Paul Anthony

OREGON
Carmichael, Howard John
Casperson, Lee Wendel
Ch'en, Shang-Yi
Crasemann, Bernd
Drake, Charles Whitney
Engelking, Paul Craig
Gilbert, David Erwin
Girardeau, Marvin Denham, Jr
Hardwick, John Lafayette
Kim, Dae Mann
Kocher, Carl A
Moseley, John Travis
Mossberg, Thomas William

PENNSYLVANIA
Bajaj, Ram
Becker, Kurt H
Berkowitz, Harry Leo
Bernheim, Robert A
Biondi, Manfred Anthony
Brackmann, Richard Theodore
Brewer, LeRoy Earl, Jr
Cohen, Leonard David
Dai, Hai-Lung
Davies, D K
Ernst, Wolfgang E
Galey, John Apt
Garbuny, Max
Gerjuoy, Edward
Gur, David
Haun, Robert Dee, Jr
Herman, Roger Myers
Huennekens, John Patrick
Intemann, Robert Louis
Janda, Kenneth Carl
Kellman, Simon
Kim, Yong Wook
Liberman, Irving
Lowry, Jerald Frank

MacLennan, Donald Allan
Norris, Wilfred Glen
Polo, Santiago Ramos
Pratt, Richard Houghton
Ranck, John Philip
Rhodes, Donald Frederick
Sampson, Douglas Howard
Sashin, Donald
Sheers, William Sadler
Siegel, Melvin Walter
Smith, Wesley R
Smith, Winfield Scott
Suntharalingam, Nagalingam
Taylor, Lyle Herman
Tredicce, Jorge Raul
Weiner, Brian Lewis
Withstandley, Victor DeWyckoff, III
Yarosewick, Stanley J
Yuan, Jian-Min
Zelac, Ronald Edward

RHODE ISLAND
Mason, Edward Allen
Sherman, Charles Henry

SOUTH CAROLINA
Cathey, LeConte
Chaplin, Robert Lee, Jr
Driggers, Frank Edgar
Graves, William Ewing
Haile, James Mitchell
Honeck, Henry Charles
Kendall, David Nelson
Malstrom, Robert Arthur
Singleton, Chloe Joi
Steiner, Pinckney Alston, III
Topp, Stephen V

SOUTH DAKOTA
Duffey, George Henry

TENNESSEE
Albridge, Royal
Allison, Stephen William
Alton, Gerald Dodd
Appleton, B R
Beene, James Robert
Bemis, Curtis Elliot, Jr
Blass, William Errol
Burns, John Francis
Carlson, Thomas Arthur
Carman, Howard Smith, Jr
Chen, Chung-Hsuan
Christophorou, Loucas Georgiou
Compton, Robert Norman
Crawford, Oakley H
Datz, Sheldon
Daunt, Stephen Joseph
Dittner, Peter Fred
Driskill, William David
Elston, Stuart B
Ewbank, Wesley Bruce
Ewig, Carl Stephen
Feigerle, Charles Stephen
Fischer, Charlotte Froese
Fox, Kenneth
Garrett, William Ray
Goans, Ronald Earl
Gregory, Donald Clifford
Haglund, Richard Forsberg, Jr
Hefferlin, Ray (Alden)
Hubbell, Harry Hopkins, Jr
Hurst, G Samuel
Hutcherson, Joseph William
Isler, Ralph Charles
Jones, Charles Miller, Jr
Keefer, Dennis Ralph
Klots, Cornelius E
Krause, Herbert Francis
Krause, Manfred Otto
Langley, Robert Archie
Lewis, James W L
Longmire, Martin Shelling
Macek, Joseph
McGregor, Wheeler Kesey, Jr
Mason, Arthur Allen
Meyer, Fred Wolfgang
Miller, John Cameron
Miller, Philip Dixon
Moak, Charles Dexter
Nalley, Samuel Joseph
Nestor, C William, Jr
Painter, Linda Robinson
Patterson, Malcolm Robert
Payne, Marvin Gay
Pegg, David John
Phaneuf, Ronald Arthur
Ricci, Enzo
Schmitt, Harold William
Sellin, Ivan Armand
Sheffield, John
Siomos, Konstadinos
Snell, Arthur Hawley
Staats, Percy Anderson
Tanner, Raymond Lewis
Tellinghuisen, Joel Barton
Thonnard, Norbert
Turner, James Edward
Vander Sluis, Kenneth Leroy
Willis, Robert D
Wilson, Robert John
Young, Jack Phillip

TEXAS
Aldridge, Jack Paxton, III
Baker, Samuel I
Bengtson, Roger D
Berry, Michael James
Borst, Walter Ludwig
Breig, Edward Louis
Brooks, Philip Russell
Browne, James Clayton
Chao, Jing
Chopra, Dev Raj
Church, David Arthur
Colcgrove, Forrest Donald
Coulson, Larry Vernon
Decker, John P
De Rijk, Waldemar G
Diana, Leonard M
Dunning, Frank Barrymore
Fetzer, Homer D
Fink, Manfred
Ford, Albert Lewis, Jr
Frommhold, Lothar Werner
Golden, David E
Graham, William Richard Montgomery
Hance, Robert Lee
Harvey, Kenneth C
Hatfield, Lynn LaMar
Herczeg, John W
Hubisz, John Lawrence, Jr
Huebner, Walter F
Hulet, Randall Gardner
Johnson, Raleigh Francis, Jr
Jones, Charles E
Kenefick, Robert Arthur
Kouri, Donald Jack
Krohmer, Jack Stewart
Lane, Neal F
LeBlanc, Adrian David
Loyd, David Heron
McDaniel, Floyd Delbert, Sr
Mangum, Jeffrey Gary
Menzel, Erhard Roland
Meshkov, Sydney
Moore, C Fred
Parker, Cleofus Varren, Jr
Paske, William Charles
Powers, Darden
Quade, Charles Richard
Quarles, Carroll Adair, Jr
Redding, Rogers Walker
Robbins, Ralph Robert
Robinson, George Wilse
Schuessler, Hans A
Sharma, Suresh C
Sherrill, William Manning
Slocum, Robert Earle
Stuart, Joe Don
Synek, Miroslav (Mike)
Walters, Geoffrey King
Watson, Rand Lewis
Wehring, Bernard William
Weinstein, Roy
Weisheit, Jon Carleton
Weisman, R(obert) Bruce
Wells, Michael Byron
Woessner, Donald Edward
Wright, Ann Elizabeth

UTAH
Berggren, Michael J
Breckenridge, William H
Knight, Larry V
Larson, Everett Gerald

VIRGINIA
Agarwal, Suresh Kumar
Anderson, Richard John
August, Leon Stanley
Bloomfield, Louis Aub
Brill, Arthur Sylvan
Brown, Ellen Ruth
Church, Charles Henry
Copeland, Gary Earl
Cummings, Peter Thomas
Dardis, John G
Delos, John Bernard
Dharamsi, Amin N
Doverspike, Lynn D
Eggleston, John Marshall
Eubank, Harold Porter
Exton, Reginald John
Fox, Russell Elwell
Gallagher, Thomas Francis
Hoppe, John Cameron
Howard, Russell Alfred
Jalufka, Nelson Wayne
Johnson, Robert Edward
Junker, Bobby Ray
Kabir, Prabahan Kemal
Kelley, Ralph Edward
Kramer, Steven David
Larson, Daniel John
Lee, Ja H
Liszt, Harvey Steven
Loda, Richard Thomas
Long, Edward Richardson, Jr
Mullen, Joseph Matthew
Pilloff, Herschel Sydney
Rainis, Albert Edward
Ries, Richard Ralph
Roy, Donald H
Sauder, William Conrad
Sepucha, Robert Charles
Wexler, Bernard Lester

Whitehead, Walter Dexter, Jr
Williams, Willie, Jr
Ziock, Klaus Otto H

WASHINGTON
Adelberger, Eric George
Baird, Quincey Lamar
Baughcum, Steven Lee
Bichsel, Hans
Bierman, Sidney Roy
Braby, Leslie Alan
Braunlich, Peter Fritz
Carter, Leland LaVelle
Clark, Kenneth Courtright
Currah, Walter E
Dehmelt, Hans Georg
Dubois, Robert Dean
Eggers, David Frank, Jr
Feldman, Henry Robert
Forsman, Earl N
Fortson, Edward Norval
Geballe, Ronald
Glass, William A
Gouterman, Martin (Paul)
Hernandez, Gonzalo J
Jefferts, Keith Bartlett
McClure, J Doyle
McDermott, Mark Nordman
McDowell, Robin Scott
McElroy, William Nordell
Maki, Arthur George, Jr
Miller, John Howard
Palmer, Harvey Earl
Radziemski, Leon Joseph
Reinhardt, William Parker
Soldat, Joseph Kenneth
Toburen, Larry Howard
Veit, Jiri Joseph
West, Martin Luther
Wilson, Walter Ervin

WISCONSIN
Anderson, Louis Wilmer
Brandenberger, John Russell
Dobson, David A
Doherty, Lowell Ralph
Egbert, Gary Trent
England, Walter Bernard
Fonck, Raymond John
Fystrom, Dell O
Greenebaum, Ben
Greenler, Robert George
Kobiske, Ronald Albert
Kurey, Thomas John
Lawler, James Edward
Lin, Chun Chia
Mallmann, Alexander James
Paliwal, Bhudatt R
Roesler, Frederick Lewis
Tonner, Brian P
Woods, Robert Claude

WYOMING
Denison, Arthur B

ALBERTA
Ali, Keramat
Ali, Shahida Parvin
Bland, Clifford J
Freeman, Gordon Russel
Gee, Norman
Kisman, Kenneth Edwin
Klobukowski, Mariusz Andrzej
Lepard, David William
McClung, Ronald Edwin Dawson
Newbound, Kenneth Bateman
Pinnington, Eric Henry

BRITISH COLUMBIA
Balfour, Walter Joseph
Coope, John Arthur Robert
Galt, John (Alexander)
Hesser, James Edward
Malli, Gulzari Lal
Measday, David Frederick
Ozier, Irving
Rieckhoff, Klaus E
Snider, Robert Folinsbee
Watton, Arthur

MANITOBA
Duckworth, Henry Edmison
Ens, E(rich) Werner
Kerr, Donald Philip
Mathur, Maya Swarup
Standing, Kenneth Graham
Sunder, Sham
Tabisz, George Conrad

NEW BRUNSWICK
Grein, Friedrich
Kaiser, Reinhold
Lees, Ronald Milne
Sichel, John Martin
Thakkar, Ajit Jamnadas

NEWFOUNDLAND
Cho, Chung Won
Foltz, Nevin D
Reddy, Satti Paddi

NOVA SCOTIA
Coxon, John Anthony
Kusalik, Peter Gerard

Atomic & Molecular Physics (cont)

Latta, Bryan Michael

ONTARIO
Atkinson, John Brian
Baylis, William Eric
Bernath, Peter Francis
Brumer, Paul William
Campeanu, Radu Ioan
Carrington, Tucker
Chapman, George David
Colpa, Johannes Pieter
Costain, Cecil Clifford
Davies, John Arthur
Drake, Gordon William Frederic
Erickson, Lynden Edwin
Fraser, Peter Arthur
Ganza, Kresimir Peter
Garside, Brian K
Geiger, James Stephen
Goodings, John Martin
Gordon, Robert Dixon
Hackam, Reuben
Hagberg, Erik Gordon
Hebert, Gerard Rosaire
Herzberg, Gerhard
Hitchcock, Adam Percival
Holt, Richard A(rnold)
Howard-Lock, Helen Elaine
Hunt, James L
Jones, R Norman
Krause, Lucjan
Lepock, James Ronald
LeRoy, Robert James
Lew, Hin
Lowe, Robert Peter
McConnell, John Charles
McCourt, Frederick Richard Wayne
McLay, David Boyd
Marmet, Paul
Meath, William John
Morris, Derek
Mungall, Allan George
Murphy, William Frederick
Ollerhead, Robin Wemp
Penner, Glenn H
Poll, Jacobus Daniel
Ramsay, Donald Allan
Rogers, David William Oliver
Rosner, Sheldon David
Sears, Varley Fullerton
Shepherd, Gordon Greeley
Sinha, Bidhu Bhushan Prasad
Stauffer, Allan Daniel
Stoicheff, Boris Peter
Vosko, Seymour H

PRINCE EDWARD ISLAND
Madan, Mahendra Pratap

QUEBEC
Beique, Rene Alexandre
Bose, Tapan Kumar
Knystautas, Emile J
Lee, Jonathan K P
Salahub, Dennis Russell

OTHER COUNTRIES
Band, Yehuda Benzion
Bondybey, Vladimir E
Bunge, Carlos Federico
Cardona, Manuel
Cue, Nelson
Heinz, Ulrich Walter
Huang, Keh-Ning
Krumbein, Aaron Davis
Kyle, Thomas Gail
Lee, Tong-Nyong
Niv, Yehuda
Nygaard, Kaare Johann
Pitchford, Leanne Carolyn
Shafi, Mohammad
Snyder, Harold Lee
Toennies, Jan Peter
Wrede, Don Edward
Yellin, Joseph

Electromagnetism

ALABAMA
Baird, James Kern
Ellenburg, Janus Yentsch
Hubbell, Wayne Charles
Jones, Robert William
McDonald, Jack Raymond
Mookherji, Tripty Kumar
Olson, Willard Paul
Passino, Nicholas Alfred
Poularikas, Alexander D

ARIZONA
Bickel, William Samuel
Chamberlin, Ralph Vary
Hopf, Frederic A
Levy, Eugene Howard
McDaniel, Terry Wayne
Macleod, Hugh Angus
Reagan, John Albert
Stoner, John Oliver, Jr
Ziolkowski, Richard Walter

ARKANSAS
Eichenberger, Rudolph John
Mackey, James E

CALIFORNIA
Alvarez, Raymond Angelo, Jr
Arnold, James S(loan)
Astrahan, Melvin Alan
Baggerly, Leo L
Bajorek, Christopher Henry
Bardin, Russell Keith
Bethune, Donald Stimson
Bevc, Vladislav
Bucker, Homer Park, Jr
Bush, Gary Graham
Buskirk, Fred Ramon
Caspers, Hubert Henri
Cautis, C Victor
Caves, Carlton Morris
Chase, Jay Benton
Chen, Tu
Cheng, Tsen-Chung
Cho, Young-chung
Clark, Arnold Franklin
Clark, Leigh Bruce
Clover, Richmond Bennett
Conway, John George, Jr
Deacon, David A G
Dowell, Jerry Tray
Driscoll, Charles F
Duneer, Arthur Gustav, Jr
Eikrem, Lynwood Olaf
Evans, Todd Edwin
Fa'arman, Alfred
Farone, William Anthony
Feldman, Nathaniel E
Fialer, Philip A
Foster, Leigh Curtis
Fraser-Smith, Antony Charles
Fried, Burton David
Fried, Walter Rudolf
Gamo, Hideya
George, Simon
Glass, Nathaniel E
Good, Roland Hamilton, Jr
Greifinger, Carl
Harker, Kenneth James
Hausman, Arthur Herbert
Hayes, Claude Q C
Heinz, Otto
Hempstead, Robert Douglas
Herrmannsfeldt, William Bernard
Hirschfeld, Tomas Beno
Hoffman, Hanna J
Honey, Richard Churchill
Huang, C Yuan
Jackson, Gary Leslie
Jackson, John David
Jory, Howard Roberts
Jungerman, John (Albert)
Kalensher, Bernard Earl
Kinnison, Gerald Lee
Kirchner, Ernst Karl
Klein, Melvin Phillip
Konrad, Gerhard T(hies)
Kuehl, Hans H(enry)
Kulke, Bernhard
Lacey, Richard Frederick
Lambertson, Glen Royal
Lemke, James Underwood
Linford, Gary Joe
Ling, Rung Tai
Lipson, Joseph Issac
Lockhart, James Marcus
Love, Allan Walter
Ludwig, Claus Berthold
McCurdy, Alan Hugh
MacGregor, Malcolm Herbert
Margolis, Jack Selig
Mitzner, Kenneth Martin
Mo, Charles Tse Chin
Morris, Richard Herbert
Mueller, James Lowell
Musal, Henry M(ichael), Jr
Nefkens, Bernard Marie
Nesbit, Richard Allison
Nevins, William McCay
Newcomb, William A
Newman, David Edward
Newman, John Joseph
Olney, Ross David
Papas, Charles Herach
Passenheim, Burr Charles
Pellegrini, Claudio
Perry, Richard Lee
Peskin, Michael Edward
Poggio, Andrew John
Poynter, Robert Louis
Rabinowitz, Mario
Rahmat-Samii, Yahya
Romagnoli, Robert Joseph
Saito, Theodore T
Sarwinski, Raymond Edmund
Schermer, Robert Ira
Sentman, Davis Daniel
Sharp, Richard Dana
Smith, Sheldon Magill
Snowden, Donald Philip
Spencer, James Eugene
Sperling, Jacob L
Stafsudd, Oscar M, Jr
Steinmetz, Wayne Edward
Stenzel, Reiner Ludwig
Stewart, Richard William

Stinson, Donald Cline
Tewari, Sujata
Tilles, Abe
Torgow, Eugene N
Tricoles, Gus P
Tyler, George Leonard
Vickers, Roger Spencer
Villeneuve, A(lfred) T(homas)
Wachowski, Hillard M(arion)
Wagner, Richard John
Waterman, Alan T(ower), Jr
Weinstein, Berthold Werner
Weiss, Jeffrey Martin
Wickersheim, Kenneth Alan
Wiedow, Carl Paul
Winick, Herman
Wolff, Milo Mitchell
Woodward, James Franklin
Wyatt, Philip Joseph
Yeh, Cavour W
Yeh, Edward H Y
Yeh, Paul Pao
Young, Frank

COLORADO
Bartlett, David Farnham
Beers, Yardley
Bennett, W Scott
Bernstein, Elliot R
Bittner, Burt James
Bussey, Howard Emerson
Estin, Arthur John
Fickett, Frederick Roland
Galeener, Frank Lee
Goldfarb, Ronald B
Haydon, George William
Hill, David Allan
Hjelme, Dag Roar
Hogg, David Clarence
Howe, David Allan
Hufford, George (Allen)
Kerns, David Marlow
Kiehl, Jeffrey Theodore
Lubell, Jerry Ira
Ma, Mark T
Morris, George Ronald
Randa, James P
Rauch, Gary Clark
Stencel, Robert Edward
Walden, Jack M

CONNECTICUT
Alpert, Nelson Leigh
Brinen, Jacob Solomon
Cheng, David H S
Close, Richard Thomas
Colthup, Norman Bertram
Glenn, William Henry, Jr
Lindsay, Robert
Liu, Qing-Huo
Sachdeva, Baldev Krishan
Shirer, Donald Leroy
Snyder, John William
Zandy, Hassan F

DELAWARE
Jarrett, Howard Starke, Jr

DISTRICT OF COLUMBIA
Beard, Charles Irvin
Butcher, Raymond John
Colombant, Denis Georges
English, William Joseph
Florig, Henry Keith
Forester, Donald Wayne
Grossman, John Mark
Harris, William Charles
Jackson, William David
Jordan, Arthur Kent
Kelly, Francis Joseph
Lin-Chung, Pay-June
Manka, Charles K
Reilly, Michael Hunt
Reiss, Howard R
Schriever, Richard L
Simmons, Joe Denton
Williams, Conrad Malcolm

FLORIDA
Baird, Alfred Michael
Bass, Michael
Burdick, Glenn Arthur
Darling, Byron Thorwell
De Lorge, John Oldham
Gustafson, Bo Ake Sture
Hagedorn, Fred Bassett
Johnston, Milton Dwynell, Jr
Kelso, John Morris
Killinger, Dennis K
Leitner, Alfred
Millar, Gordon Halstead
Neuringer, Joseph Louis
Nunn, Waiter M(elrose), Jr
Ross, John Stoner
Ungvichian, Vichate
Vala, Martin Thorvald, Jr
Wang, Ru-Tsang

GEORGIA
Agrawal, Pradeep Kumar
Anantha Narayanan, Venkataraman
Carreira, Lionel Andrade
Currie, Nicholas Charles
Gaylord, Thomas Keith

Joy, Edward Bennett
Lawrence, Kurt C
Long, Maurice W(ayne)
Rodrigue, George Pierre
Ryan, Charles Edward, Jr
Simmons, James Wood
Turner, Janice Butler
Wang, Johnson Jenn-Hwa

HAWAII
Andermann, George
Hall, Donald Norman Blake
Holmes, John Richard

IDAHO
Davis, Lawrence William, Jr

ILLINOIS
Basile, Louis Joseph
Brussel, Morton Kremen
Camras, Marvin
Cardman, Lawrence Santo
Chew, Weng Cho
Cole, Francis Talmage
Cooper, William Edward
Curry, Bill Perry
Dreska, Noel
Evans, Kenneth, Jr
Ferraro, John Ralph
Gallagher, David Alden
Henderson, Giles Lee
Henneberger, Walter Carl
Hoshiko, Michael S
Knop, Charles M(ilton)
Koster, David F
Kustom, Robert L
Lari, Robert Joseph
Laurin, Pushpamala
Mechtly, Eugene A
Naylor, David L
Peters, Robert Edward
Shenoy, Gopal K
Shepard, Kenneth Wayne
Smith, Richard Paul
Taflove, Allen
Tollestrup, Alvin V
Tomkins, Frank Sargent
Treadwell, Elliott Allen
Twersky, Victor

INDIANA
Farringer, Leland Dwight
Poirier, John Anthony

IOWA
Clem, John R
Gurnett, Donald Alfred

KANSAS
Carpenter, Kenneth Halsey
Long, Larry L
Sapp, Richard Cassell
Unz, Hillel
Weller, Lawrence Allenby
Williams, Dudley

KENTUCKY
Bakanowski, Stephen Michael
Wilt, Paxton Marshall

LOUISIANA
Chin-Bing, Stanley Arthur
Hilburn, John L
Imhoff, Donald Wilbur
Oberding, Dennis George
Weiss, Louis Charles

MAINE
Kingsbury, Robert Freeman

MARYLAND
Baker, Francis Edward, Jr
Bennett, Charles L
Bostrom, Carl Otto
Chang, Alfred Tieh-Chun
Chen, Wenpeng
Connerney, John E P
Driscoll, Raymond L
Dubin, Henry Charles
Eccleshall, Donald
Gaunaurd, Guillermo C
Gordon, Daniel Israel
Granatstein, Victor Lawrence
Gray, Ernest Paul
Heller, Stephen Richard
Hochheimer, Bernard Ford
Hyde, Geoffrey
Jablonski, Daniel Gary
Jennings, Donald Edward
Johnson, Donald Rex
Johnson, William Bowie
Joiner, R(eginald) Gracen
Kuttler, James Robert
Ledley, Brian G
McDonald, Jimmie Reed
Martin, William Clyde
Marx, Egon
Mather, John Cromwell
Merkel, George
Neupert, Werner Martin
Parsegian, Vozken Adrian
Phillips, William Daniel
Pirraglia, Joseph A
Powell, John David

Raines, Jeremy Keith
Rosado, John Allen
Sattler, Joseph Peter
Schmugge, Thomas Joseph
Schumacher, Clifford Rodney
Sharma, Jagadish
Soln, Josip Zvonimir
Strenzwilk, Denis Frank
Taylor, Leonard S
Taylor, Robert Joseph
Tompkins, Robert Charles
Wasylkiwskyj, Wasyl
Whicker, Lawrence R

MASSACHUSETTS
Baird, Albert Washington, III
Bekefi, George
Bers, Abraham
Bugnolo, Dimitri Spartaco
Burke, William J
Cohen, David
Connors, Robert Edward
Dhar, Sachidulal
Dionne, Gerald Francis
Dorschner, Terry Anthony
Drane, Charles Joseph, Jr
Gelman, Harry
Gianino, Peter Dominic
Giger, Adolf J
Graneau, Peter
Green, Jerome Joseph
Gross, Thomas Alfred Otto
Hooper, Robert John
Kincaid, Thomas Gardiner
Kolm, Henry Herbert
Lees, Wayne Lowry
Loewenstein, Ernest Victor
McBee, W(arren) D(ale)
Mack, Richard Bruce
Mailloux, Robert Joseph
Maloney, William Thomas
Massa, Frank
Maxwell, Emanuel
Neuringer, Leo J
Newburgh, Ronald Gerald
Oberteuffer, John Amiard
Osepchuk, John M
Rabin, Monroe Stephen Zane
Ram-Mohan, L Ramdas
Rana, Ram S
Reed, F(lood) Everett
Rheinstein, John
Richter, Stephen L(awrence)
Rork, Eugene Wallace
Rotman, Walter
Schattenburg, Mark Lee
Schneider, Robert Julius
Sethares, James C(ostas)
Shane, John Richard
Shen, Hao-Ming
Silevich, Michael B
Silevitch, Michael B
Simmons, Alan J(ay)
Temkin, Richard Joel
Tzeng, Wen-Shian Vincent
Vanasse, George Alfred
Vignos, James Henry
Vittoria, Carmine
Weggel, Robert John
Weiss, Jerald Aubrey
Whitney, Cynthia Kolb
Wu, Tai Tsun
Yaghjian, Arthur David
Yngvesson, K Sigfrid

MICHIGAN
Banks, Peter Morgan
Beaman, Donald Robert
Brailsford, Alan David
Clauer, C Robert, Jr
Derby, Stanley Kingdon
Donohue, Robert J
Gordon, Morton Maurice
Gustafson, Herold Richard
Jones, Frederick Goodwin
King, Stanley Shih-Tung
Kingman, Robert Earl
LaHaie, Ivan Joseph
Nyquist, Richard Allen
Pollack, Gerald Leslie
Rand, Stephen Colby
Roessler, David Martyn
Rolnick, William Barnett
Segall, Stephen Barrett
Senior, Thomas Bryan Alexander
Smith, Kenneth Edward
Van Baak, David Alan

MINNESOTA
Brom, Joseph March, Jr
Carlson, Frederick Paul
Crawford, Bryce (Low), Jr
Follingstad, Henry George
Hakkila, Jon Eric
Heltemes, Eugene Casmir
Lutes, Olin S
Roska, Fred James
Walters, John Philip

MISSISSIPPI
Cook, Robert Lee
Crow, Terry Tom
Miller, David Burke
Smith, Charles Edward

Taylor, Clayborne D
Winton, Raymond Sheridan

MISSOURI
Adawi, Ibrahim (Hasan)
Alexander, Ralph William, Jr
Bard, James Richard
Bell, Robert John
Brown, Harry Allen
Burgess, James Harland
DuBroff, Richard Edward
Huang, Justin C
Huddleston, Philip Lee
Indeck, Ronald S
Leader, John Carl
Leader, John Carl
Lind, Arthur Charles
Menne, Thomas Joseph
Muller, Marcel Wettstein
Smith, Robert Francis
Spielman, Barry
White, Warren D

MONTANA
Graham, Raymond

NEVADA
Babb, David Daniel
Rawat, Banmali Singh
Scott, William Taussig

NEW HAMPSHIRE
Buffler, Charles Rogers
Frost, Albert D(enver)
Humphrey, Floyd Bernard

NEW JERSEY
Alig, Roger Casanova
Amitay, Noach
Amundson, Karl Raymond
Belohoubek, Erwin F
Chand, Naresh
Cioffi, Paul Peter
Daniels, James Maurice
Dillon, Joseph Francis, Jr
Dutta, Mitra
Fleury, Paul A
Fork, Richard Lynn
Freeman, Richard Reiling
Gammel, George Michael
Grek, Boris
Hall, Herbert Joseph
Hatkin, Leonard
Hill, Kenneth Wayne
Ikeda, Tatsuya
Koch, Thomas L
Mueller, Dennis
Ogden, Joan Mary
Passner, Albert
Rastani, Kasra
Reder, Friedrich H
Rossol, Frederick Carl
Rulf, Benjamin
Schrenk, George L
Schwering, Felix
Smith, Carl Hofland
Spencer, Edward G
Stevens, James Everell
Taylor, Geoff W
Tenzer, Rudolf Kurt
Tsaliovich, Anatoly
Turner, Edward Harrison
Wang, Chao Chen
Whitman, Gerald Martin
Yen, You-Hsin Eugene

NEW MEXICO
Adler, Richard John
Anderson, Robert Alan
Argo, Paul Emmett
Baum, Carl E(dward)
Bradshaw, Martin Daniel
Brook, Marx
Brownell, John Howard
Butler, Harold S
Chamberlin, Edwin Phillip
Chylek, Petr
Clark, Wallace Thomas, III
Cooper, Richard Kent
Cox, Lawrence Edward
Dressel, Ralph William
Farmer, William Michael
Flicker, Herbert
Freeman, Bruce L, Jr
Frost, Harold Maurice, III
Godfrey, Brendan Berry
Gurbaxani, Shyam Hassomal
Hoeft, Lothar Otto
Hurd, James William
Ladish, Joseph Stanley
Lindman, Erick Leroy, Jr
Lyo, SungKwun Kenneth
Maier, William Bryan, II
Miller, Edmund K(enneth)
Parker, Jerald Vawer
Potter, James Martin
Rach, Randolph Carl
Reeves, Geoffrey D
Rinker, George Albert, Jr
Schriber, Stanley Owen
Schwarz, Ricardo
Searls, Craig Allen
Simon, George Warren
Taylor, Raymond Dean

Toepfer, Alan James
VanDevender, John Pace
Vittitoe, Charles Norman
Wiggins, Carl M
Wright, Thomas Payne

NEW YORK
Adams, Arlon Taylor
Barreto, Ernesto
Bechtel, Marley E(lden)
Bernstein, Burton
Bittner, John William
Blumberg, Leroy Norman
Borrego, Jose M
Bottomley, Paul Arthur
Boyd, Robert William
Bradshaw, John Alden
Chang, Hsu
Chen, Inan
Cheney, Margaret
Chew, Herman W
Chrenko, Richard Michael
Costa, Lorenzo F
Cottingham, James Garry
Craft, Harold Dumont, Jr
Davis, Abram
Demerdash, Nabeel A O
Derman, Samuel
Diament, Paul
Fulmor, William
Garretson, Craig Martin
Geer, James Francis
Goodwin, Arthur VanKleek
Haas, Werner E L
Hagfors, Tor
Hall, Dennis Gene
Hammer, David Andrew
Hartwig, Curtis P
Haus, Joseph Wendel
Hurst, Robert Philip
Kane, William Theodore
Kermisch, Dorian
Kitchen, Sumner Wendell
Kolb, Frederick J(ohn), Jr
Krakow, Burton
Kramer, Stephen Leonard
Lambeth, David N
Lauber, Thornton Stuart
McDonald, Robert Skillings
Mandel, Leonard
Marshall, Thomas C
Mihran, Theodore Gregory
Nelson, John Keith
Nisteruk, Chester Joseph
Oh, Byungdu
Oliner, Arthur A(aron)
Ozimek, Edward Joseph
Packard, Karle Sanborn, Jr
Paul, David I
Plane, Robert Allen
Rezanka, Ivan
Roalsvig, Jan Per
Rohrlich, Fritz
Sample, Steven Browning
Schlesinger, S Perry
Sen, Ujjal
Senus, Walter Joseph
Shapiro, Paul Jonathon
Shiren, Norman S
Simpson, Murray
Viertl, John Ruediger Mader
Vortuba, Jan
Wolf, Edward Lincoln

NORTH CAROLINA
Ji, Chueng Ryong
McRee, Donald Ikerd
Mink, James Walter
Prater, John Thomas
Straley, Joseph Ward
Tischer, Frederick Joseph

NORTH DAKOTA
Dewar, Graeme Alexander

OHIO
Bales, Howard E
Barranger, John P
Becker, Roger Jackson
Boughton, Robert Ivan, Jr
Compton, Ralph Theodore, Jr
Falkenbach, George J(oseph)
Hill, James Stewart
Holt, James Franklin
Hurley, William Joseph
McLennan, Donald Elmore
Manning, Robert M
Mathews, Collis Weldon
Mathews, John David
Mettee, Howard Dawson
Pathak, Prabhakar H
Redlich, Robert Walter
Restemeyer, William Edward
Spitzer, Jeffrey Chandler
Strnat, Karl J
Stroud, David Gordon
Wigen, Philip E
Williamson, William, Jr

OKLAHOMA
Armoudian, Garabed
Baldwin, Bernard Arthur
Beasley, William Harold
Bilger, Hans Rudolf

Cohn, Jack
Dixon, George Sumter, Jr
Taylor, William L

OREGON
Bhattacharya, Pallab Kumar
Casperson, Lee Wendel
Chartier, Vernon L
Csonka, Paul L
Gilbert, David Erwin
Griffiths, David Jeffery
Yamaguchi, Tadanori

PENNSYLVANIA
Berry, Richard Emerson
Carpenter, Lynn Allen
Charap, Stanley H
Coren, Richard L
Eberhardt, Nikolai
Falconer, Thomas Hugh
Grimes, Dale M(ills)
Heald, Mark Aiken
Hyatt, Robert Monroe
Jaggard, Dwight Lincoln
Judd, Jane Harter
Ku, Robert Tien-Hung
Long, Howard Charles
Long, Lyle Norman
Matocha, Charles K
Miskovsky, Nicholas Matthew
Molter, Lynne Ann
Morabito, Joseph Michael
Okress, Ernest Carl
Robl, Robert F(redrick), Jr
Schenk, H(arold) L(ouis), Jr
Seltzer, James Edward
Suhr, Norman Henry
Tahir-Kheli, Raza Ali
Thompson, Ramie Herbert
Thornburg, Donald Richard
Van Roggen, Arend
Whitley, James Heyward
Williams, James Earl, Jr
Young, Frederick J(ohn)

RHODE ISLAND
Grossi, Mario Dario
Polk, C(harles)

SOUTH CAROLINA
McNulty, Peter J
Pearson, Lonnie Wilson
Sheppard, Emory Lamar
Steiner, Pinckney Alston, III

TENNESSEE
Child, Harry Ray
Deeds, William Edward
Fuson, Nelson
Moak, Charles Dexter
Mosko, Sigmund W
Murakami, Masanori
Schwenterly, Stanley William, III
Shaw, Robert Wayne
Tarpley, Anderson Ray, Jr
Trivelpiece, Alvin William
Wikswo, John Peter, Jr

TEXAS
Andrychuk, Dmetro
Beissner, Robert Edward
Bronaugh, Edwin Lee
Brown, Glenn Lamar
Bussian, Alfred Erich
Chamberlain, Nugent Francis
Decker, John P
Fields, Reuben Elbert
Fung, Adrian K
Gavenda, John David
Heelis, Roderick Antony
Hutchinson, Bennett B
Johnson, Francis Severin
Johnson, Raleigh Francis, Jr
Ling, Hao
Long, Stuart A
Loos, Karl Rudolf
Neikirk, Dean P
Nevels, Robert Dudley
Olenick, Richard Peter
Ranganayaki, Rambabu Pothireddy
Reiff, Patricia Hofer
Rester, David Hampton
Roberts, Thomas M
Rollwitz, William Lloyd
Sablik, Martin J
Sams, Lewis Calhoun, Jr
Sharp, A C, Jr
Shepley, Lawrence Charles
Sherry, Clifford Joseph
Slocum, Robert Earle
Srnka, Leonard James
Stephenson, Danny Lon

UTAH
Chabries, Douglas M
Earley, Charles Willard
Edwards, W Farrell
Jeffery, Rondo Nelden
Miller, Akeley

VERMONT
Golden, Kenneth Ivan

IDAHO
Luke, Robert A
Parker, Barry Richard

ILLINOIS
Abrams, Robert Jay
Albright, Carl Howard
Anderson, Kelby John
Ankenbrandt, Charles Martin
Appel, Jeffrey Alan
Auvil, Paul R, Jr
Ayres, David Smith
Baker, Winslow Furber
Bardeen, William A
Bart, George Raymond
Bartlett, J Frederick
Berge, Jon Peter
Berger, Edmond Louis
Block, Martin M
Bollinger, Lowell Moyer
Brown, Bruce Claire
Brown, Charles Nelson
Burnstein, Ray A
Carey, David Crockett
Carrigan, Richard Alfred, Jr
Carroll, John Terrance
Chang, Shau-Jin
Cooper, John Wesley
Cooper, William Edward
Cossairt, Jack Donald
Curtis, Cyril Dean
Derrick, Malcolm
Eartly, David Paul
Eichten, Estia Joseph
Eisenstein, Robert Alan
Elwyn, Alexander Joseph
Errede, Steven Michael
Fields, Thomas Henry
Frauenfelder, Hans (Emil)
Freedman, Stuart Jay
Frisch, Henry Jonathan
Gelfand, Norman Mathew
Gladding, Gary Earle
Goldberg, Howard S
Gormley, Michael Francis
Griffin, James Edward
Groves, Thomas Hoopes
Hagstrom, Ray Theodore
Harvey, Jeffrey Alan
Hedin, David Robert
Hildebrand, Roger Henry
Hill, Christopher T
Hoff, Gloria Thelma (Albuerne)
Hojvat, Carlos F
Holloway, Leland Edgar
Hyman, Lloyd George
Izen, Joseph N
Johnson, David
Johnson, Marvin Elroy
Johnson, Porter W
Johnson, Rolland Paul
Jones, Lorella Margaret
Jostlein, Hans
Kaplan, Daniel Moshe
Kephart, Robert David
Kirk, Thomas Bernard Walter
Koester, Louis Julius, Jr
Kolb, Edward William
Kovacs, Eve Veronika
Lennox, Arlene Judith
MacLachlan, James Angell
Malamud, Ernest I(lya)
Mantsch, Paul Matthew
Margulies, Seymour
Marriner, John P
Miller, Robert Carl
Moore, Craig Damon
Muller, Dietrich
Murphy, Charles Thornton
Nash, Edward Thomas
Nezrick, Frank Albert
Nodulman, Lawrence Jay
Oehme, Reinhard
O'Halloran, Thomas A
Papanicolas, Costas Nicolas
Peoples, John, Jr
Pewitt, Edward Gale
Phillips, Thomas James
Pilcher, James Eric
Preston, Richard Swain
Price, Lawrence Edward
Pruss, Stanley McQuaide
Quigg, Chris
Raja, Rajendran
Rey, Charles Albert
Ringo, George Roy
Roberts, Arthur
Rubinstein, Roy
Sard, Robert Daniel
Schluter, Robert Arvel
Schonfeld, Jonathan Furth
Schreiner, Philip Allen
Schultz, Peter Frank
Shochet, Melvyn Jay
Singer, Richard Alan
Sivers, Dennis Wayne
Smart, Wesley Mitchell
Smith, Richard Paul
Solomon, Julius
Spinka, Harold M
Stanfield, Kenneth Charles
Stutte, Linda Gail
Sugano, Katsuhito
Sukhatme, Uday Pandurang

Swallow, Earl Connor
Swamy, Padmanabha Narayana
Teng, Lee Chang-Li
Thaler, Jon Jacob
Theriot, Edward Dennis, Jr
Thron, Jonathan Louis
Tollestrup, Alvin V
Toppel, Bert Jack
Toy, William W
Treadwell, Elliott Allen
Tung, Wu-Ki
Turkot, Frank
Uretsky, Jack Leon
Wagner, Robert G
Ward, Charles Eugene Willoughby
Wattenberg, Albert
Wehmann, Alan Ahlers
White, Herman Brenner, Jr
Wicklund, Arthur Barry
Willis, Suzanne Eileen
Winston, Roland
Yamada, Ryuji
Yoh, John K
Yokosawa, Akihiko
Young, Donald Edward
Yovanovitch, Drasko D
Zachos, Cosmas K

INDIANA
Barnes, Virgil Everett, II
Bishop, James Martin
Brabson, Bennet Bristol
Cason, Neal M
Fischbach, Ephraim
Gaidos, James A
Garfinkel, Arthur Frederick
Gottlieb, Steven Arthur
Hanson, Gail G
Heinz, Richard Meade
Hendry, Archibald Wagstaff
Koetke, Donald D
Kuo, Tzee-Ke
Loeffler, Frank Joseph
LoSecco, John M
McIlwain, Robert Leslie, Jr
Martin, Hugh Jack, Jr
Miller, David Harry
Ogren, Harold Olof
Poirier, John Anthony
Ruchti, Randal Charles
Sapirstein, Jonathan Robert
Shephard, William Danks
Shibata, Edward Isamu
Tubis, Arnold
Westgard, James Blake
Willmann, Robert B
Wills, John G
Yoder, Neil Richard

IOWA
Firestone, Alexander
Hagelin, John Samuel
Hammer, Charles Lawrence
Intemann, Gerald William
Lamb, Richard C
Lassila, Kenneth Eino
Leacock, Robert A
Pursey, Derek Lindsay
Rizzo, Thomas Gerard
Rosenberg, Eli Ira
Smith, Paul Aiken
Young, Bing-Lin

KANSAS
Ammar, Raymond George
Davis, Robin Eden Pierre
Kwak, Nowhan

KENTUCKY
Elitzur, Moshe
France, Peter William
Yaes, Robert Joel

LOUISIANA
Courtney, John Charles
Haymaker, Richard Webb
Head, Charles Everett
Head, Martha E Moore
Imlay, Richard Larry
Slaughter, Milton Dean

MARYLAND
Brenner, Alfred Ephraim
Chien, Chih-Yung
Comiso, Josefino Cacas
Domokos, Gabor
Dragt, Alexander James
Gates, Sylvester J, Jr
Greenberg, Oscar Wallace
Groves, Eric Stedman
Hauser, Michael George
Hildebrand, Bernard
Hu, Bei-Lok Bernard
Kacser, Claude
Kovesi-Domokos, Susan
Marx, Egon
Messing, Fredric
Mohapatra, Rabindra Nath
Panvini, Robert S
Peaslee, David Chase
Richardson, Clarence Robert
Schumacher, Clifford Rodney
Snow, George Abraham
Soln, Josip Zvonimir

Stecker, Floyd William
Steinberg, Phillip Henry
Taragin, Morton Frank
Zorn, Gus Tom

MASSACHUSETTS
Abbott, Laurence Frederick
Bar-Yam, Zvi H
Bensinger, James Robert
Brehm, John Joseph
Brenner, John Francis
Busza, Wit
Button-Shafer, Janice
Carter, Ashton Baldwin
Celmaster, William Noah
Cook, LeRoy Franklin, Jr
Cormack, Allan MacLeod
Decowski, Piotr
Deser, Stanley
Dhar, Sachidulal
Donoghue, John Francis
Dowd, John P
Everett, Allen Edward
Faissler, William L
Farhi, Edward
Fazio, Giovanni Gene
Feld, Bernard Taub
Feldman, Gary Jay
Ford, Lawrence Howard
Freese, Katherine
Friedman, Jerome Isaac
Friedman, Marvin Harold
George, James Z
Ginsberg, Edward S
Glashow, Sheldon Lee
Glaubman, Michael Juda
Godine, John Elliott
Goldberg, Hyman
Goldstone, Jeffrey
Guertin, Ralph Francis
Guth, Alan Harvey
Hafen, Elizabeth Susan Scott
Hodges, Hardy M
Horwitz, Paul
Hulsizer, Robert Inslee, Jr
Hunter, Larry Russel
Jacobs, Laurence Alan
Jaffe, Robert Loren
Jagannathan, Kannan
Jensen, Douglas Andrew
Kafka, Tomas
Kern, Wolfhard
Kinoshita, Kay
Kreisler, Michael Norman
Law, Margaret Elizabeth
Lipshutz, Nelson Richard
Mann, William Anthony
Marino, Richard Matthew
Matthews, June Lorraine
Mellen, Walter Roy
Milbocker, Michael
Milburn, Richard Henry
Miller, James Paul
Nath, Pran
Nicholson, Howard White, Jr
Oliver, William Parker
Osborne, Louis Shreve
Pilot, Christopher H
Pipkin, Francis Marion
Rabin, Monroe Stephen Zane
Ram-Mohan, L Ramdas
Ramsey, Norman Foster, Jr
Rebbi, Claudio
Redwine, Robert Page
Reucroft, Stephen
Roberts, Bradley Lee
Rosenson, Lawrence
Rothwell, Paul L
Schneps, Jack
Shambroom, Wiliam David
Strauch, Karl
Strelzoff, Alan G
Sulak, Lawrence Richard
Ting, Samuel C C
Tuchman, Avraham
Uritam, Rein Aarne
Vaughn, Michael Thayer
Von Goeler, Eberhard
Wang, Chia Ping
Wijangco, Antonio Robles
Wilk, Leonard Stephen
Wu, Tai Tsun
Yamamoto, Richard Kumeo
Yamartino, Robert J

MICHIGAN
Abbasabadi, Alireza
Abolins, Maris Arvids
Akerlof, Carl W
Chen, Hsuan
Francis, William Porter
Gupta, Suraj Narayan
Gustafson, Herold Richard
Jones, Lawrence William
Krisch, Alan David
Krisch, Jean Peck
Kubis, Joseph J(ohn)
Longo, Michael Joseph
Miller, Robert Joseph
Neal, Homer Alfred
Overseth, Oliver Enoch
Pope, Bernard G
Raban, Morton
Reitan, Daniel Kinseth

Repko, Wayne William
Rolnick, William Barnett
Savit, Robert Steven
Tarle, Gregory
Terwilliger, Kent Melville

MINNESOTA
Austin, Donald Murray
Courant, Hans Wolfgang Julius
Erickson, Kenneth Neil
Gasiorowicz, Stephen G
Heller, Kenneth Jeffrey
Jones, Roger Stanley
Marshak, Marvin Lloyd
Peterson, Earl Andrew
Rudaz, Serge

MISSISSIPPI
Lott, Fred Wilbur, III
Reidy, James Joseph

MISSOURI
DeFacio, W Brian
Huang, Justin C
Huddleston, Philip Lee

NEBRASKA
Finkler, Paul
Joseph, David Winram

NEVADA
Hudson, James Gary

NEW HAMPSHIRE
Shepard, Harvey Kenneth

NEW JERSEY
Bechis, Dennis John
Brucker, Edward Byerly
Budny, Robert Vierling
Dao, Fu Tak
Devlin, Thomas J
Edwards, Christopher Andrew
Gabbe, John Daniel
Goldin, Gerald Alan
Goodman, Mark William
Hughes, James Sinclair
Kalelkar, Mohan Satish
Klebanov, Igor Romanovich
Koller, Earl Leonard
Lemonick, Aaron
Li, Yuan
Loos, James Stavert
McDonald, Kirk Thomas
Michelson, Leslie Paul
Mueller, Dennis
Ng, Tai-Kai
Plano, Richard James
Poucher, John Scott
Purohit, Milind Vasant
Raghavan, Ramaswamy Srinivasa
Raychaudhuri, Kamal Kumar
Reynolds, George Thomas
Rockmore, Ronald Marshall
Sannes, Felix Rudolph
Schivell, John Francis
Shapiro, Joel Alan
Shoemaker, Frank Crawford
Smith, Arthur John Stewart
Socolow, Robert H(arry)
Spergel, David Nathaniel
Stickland, David Peter
Taylor, Snowden
Watts, Terence Leslie
Weinrich, Marcel

NEW MEXICO
Berardo, Peter Antonio
Biggs, Albert Wayne
Bowman, James David
Brolley, John Edward, Jr
Bryant, Howard Carnes
Burleson, George Robert
Burman, Robert L
Coon, Sidney Alan
Cooper, Frederick Michael
Cooper, Martin David
Gillen, Kenneth Todd
Ginsparg, Paul H
Glass, George
Goldman, Terrence Jack
Gray, Edward Ray
Hoffman, Cyrus Miller
Hutson, Richard Lee
Jarmie, Nelson
Jones, James Jordan
King, Nicholas S P
Leeper, Ramon Joe
MacArthur, Duncan W
Macek, Robert James
McNaughton, Michael Walford
Mischke, Richard E
Parker, Winifred Ellis
Peng, Jen-chieh
Robertson, Robert Graham Hamish
Sanders, Gary Hilton
Shafer, Robert E
Shively, Frank Thomas
Siciliano, Edward Ronald
Simmons, James E
Simmons, Leonard Micajah, Jr
Sunier, Jules Willy
Tanaka, Nobuyuki
Thiessen, Henry Archer

Elementary Particle Physics (cont)

Toepfer, Alan James
Turner, Leaf
White, David Hywel
Williams, Robert Allen
Wolfe, David M
Zemach, Charles

NEW YORK

Amato, Joseph Cono
Aronson, Samuel Harry
Auerbach, Elliot H
Baltz, Anthony John
Behrends, Ralph Eugene
Berkelman, Karl
Bermon, Stuart
Blumberg, Leroy Norman
Bodek, Arie
Bowick, Mark John
Bozoki, George Edward
Brown, Stanley Gordon
Bunce, Gerry Michael
Cassel, David Giske
Chang, Ngee Pong
Chen, James Ralph
Chew, Herman W
Chrien, Robert Edward
Chung, Kuk Soo
Chung, Suh Urk
Chung, Victor
Cole, James A
Dahl, Per Fridtjof
Das, Ashok Kumar
Dean, Nathan Wesley
Dell, George F, Jr
Devons, Samuel
DeWire, John William
Duek, Eduardo Enrique
Eckert, Richard Raymond
Etkin, Asher
Feigenbaum, Mitchell Jay
Feingold, Alex Jay
Feldman, Lawrence
Ferbel, Thomas
Fernow, Richard Clinton
Foelsche, Horst Wilhelm Julius
Foley, Kenneth John
Franco, Victor
Friedberg, Carl E
Genova, James John
Gibbard, Bruce
Gittelman, Bernard
Goldberg, Marvin
Gollon, Peter J
Goodman, Gerald Joseph
Gordon, Howard Allan
Goulianos, Konstantin
Grannis, Paul Dutton
Greene, Arthur Franklin
Haggerty, John S
Hahn, Richard Leonard
Hand, Louis Neff
Harrington, Steven Jay
Herbert, Marc L
Hough, Paul Van Campen
Jain, Piyare Lal
Johnson, Joseph Andrew, III
Kalogeropoulos, Theodore E
Koplik, Joel
Kramer, Martin A
Kramer, Stephen Leonard
Kycia, Thaddeus F
Lawson, Kent DeLance
Lee, Wonyong
Lee, Yong Yung
Leipuner, Lawrence Bernard
Lindenbaum, S(eymour) J(oseph)
Lindquist, William Brent
Lissauer, David Arie
Littauer, Raphael Max
Lobkowicz, Frederick
Longacre, Ronald Shelley
Love, William Alfred
Lowenstein, Derek Irving
McDaniel, Boyce Dawkins
Mandel, Leonard
Mane, Sateesh Ramchandra
Marciano, William Joseph
Marx, Michael David
Mathur, Vishnu Sahai
Melissinos, Adrian Constantin
Mincer, Allen I
Mistry, Nariman Burjor
Moneti, Giancarlo
Morse, William M
Mukhopadhyay, Nimai Chand
Nelson, Charles Arnold
Okubo, Susumu
Olsen, Stephen Lars
Orear, Jay
Ortel, William Charles Gormley
Paige, Frank Eaton, Jr
Parsa, Zohreh
Peierls, Ronald F
Pile, Philip H
Polychronakos, Venetios Alexander
Protopopescu, Serban Dan
Rahm, David Charles
Ratner, Lazarus Gershon
Rau, R Ronald
Rehak, Pavel

Ring, James Walter
Rohrlich, Fritz
Rosenzweig, Carl
Roth, Benjamin
Ruggiero, Alessandro Gabriele
Sachs, Alan Maxwell
Sadoff, Ahren J
Sakitt, Mark
Sanford, James R
Satz, Helmut T G
Schamberger, Robert Dean
Schechter, Joseph M
Schewe, Phillip Frank
Schlitt, Dan Webb
Schwarz, Cindy Beth
Sciulli, Frank J
Shpiz, Joseph M
Siemann, Robert Herman
Slattery, Paul Francis
Smolin, Lee
Souder, Paul A
Spergel, Martin Samuel
Stachel, Johanna
Stearns, Brenton Fisk
Stoler, David
Stoler, Paul
Stone, Sheldon Leslie
Story, Harold S
Strand, Richard Carl
Sumner, Richard Lawrence
Sun, Chih-Ree
Swartz, Clifford Edward
Tannenbaum, Michael J
Thern, Royal Edward
Thorndike, Alan Moulton
Thorndike, Edward Harmon
Tigner, Maury
Trueman, Thomas Laurence
Tryon, Edward Polk
Tye, Sze-Hoi Henry
Vortuba, Jan
Wanderer, Peter John, Jr
Weinberg, Erick James
Wellner, Marcel
Weng, Wu Tsung
Woody, Craig L
Yamin, Samuel Peter
Yan, Tung-Mow
Yeh, Noel Kuei-Eng
Yuan, Luke Chia Liu
Zahed, Ismail

NORTH CAROLINA

Cotanch, Stephen Robert
Dayton, Benjamin Bonney
Dolan, Louise Ann
Dotson, Allen Clark
Fortney, Lloyd Ray
Frampton, Paul Howard
Fulp, Ronald Owen
Good, Wilfred Manly
Goshaw, Alfred Thomas
Han, Moo-Young
Ji, Chueng Ryong
Meisner, Gerald Warren
Petersen, Karl Endel
Spielvogel, Bernard Franklin
Wolf, Albert Allen

OHIO

Aubrecht, Gordon James, II
Duval, Leonard A
Fickinger, William Joseph
Giamati, Charles C, Jr
Huwe, Darrell O
Isbrandt, Lester Reinhardt
Johnson, Randy Allan
Kosel, Peter Bohdan
Kowalski, Kenneth L
Ling, Ta-Yung
Nussbaum, Mirko
Raby, Stuart
Reibel, Kurt
Robinson, Donald Keith
Rodman, James Purcell
Romanowski, Thomas Andrew
Suranyi, Peter
Tanaka, Katsumi
Taylor, Beverley Ann Price

OKLAHOMA

Burwell, James Robert
Carlstone, Darry Scott
Howard, Robert Adrian
Kalbfleisch, George Randolph
Milton, Kimball Alan
Skubic, Patrick Louis

OREGON

Alston-Garnjost, Margaret
Brau, James Edward
Csonka, Paul L
Deshpande, Nilendra Ganesh
Griffiths, David Jeffery
Hwa, Rudolph Chia-Chao
Khalil, M Aslam Khan
Moravcsik, Michael Julius
Olness, Fredrick Iver
Richert, Anton Stuart
Siemens, Philip John
Soper, Davison Eugene
White, Donald Harvey

PENNSYLVANIA

Barnes, Peter David
Beier, Eugene William
Carlitz, Robert D
Cleland, Wilfred Earl
Collins, John Clements
Coon, Darryl Douglas
Cutkosky, Richard Edwin
Fetkovich, John Gabriel
Frankel, Sherman
Franklin, Jerrold
Frati, William
Goldschmidt, Yadin Yehuda
Highland, Virgil Lee
Hones, Michael J
Kanofsky, Alvin Sheldon
Kraemer, Robert Walter
Lande, Kenneth
Langacker, Paul George
Loman, James Mark
McFarlane, Kenneth Walter
Mann, Alfred Kenneth
Potter, Douglas Marion
Ridener, Fred Louis, Jr
Roskies, Ralph Zvi
Rovelli, Carlo
Russ, James Stewart
Selove, Walter
Shepard, Paul Fenton
Siegel, Robert Ted
Smith, Gerald A
Steinhardt, Paul Joseph
Sutton, Roger Beatty
Thompson, Julia Ann
White, George Rowland
Williams, Hugh Harrison

RHODE ISLAND

Cutts, David
Lanou, Robert Eugene, Jr
Widgoff, Mildred

SOUTH CAROLINA

Darden, Colgate W, III
Gurr, Henry S
Rosenfeld, Carl

TENNESSEE

Bartelt, John Eric
Beene, James Robert
Bugg, William Maurice
Condo, George T
Crater, Horace William
Hart, Edward Leon
Loebbaka, David S
Miller, Philip Dixon
Plasil, Franz
Roos, Charles Edwin
Sears, Robert F, Jr
Trubey, David Keith
Ward, Bennie Franklin Leon
Weiler, Thomas Joseph
Young, Glenn Reid

TEXAS

Baggett, Millicent (Penny)
Baggett, Neil Vance
Barton, Henry Ruwe, Jr
Bryan, Ronald Arthur
Chiu, Charles Bin
Clark, Robert Beck
Corcoran, Marjorie
Coulson, Larry Vernon
Cutler, Roger T
Danburg, Jerome Samuel
Dicus, Duane A
Edwards, Helen Thom
Espinosa, James Manuel
Fenyves, Ervin J
Gaedke, Rudolph Meggs
Gleeson, Austin M
Huang, Huey Wen
Huson, Frederick Russell
Ko, Che Ming
Lichti, Roger L
Lopez, Jorge Alberto
McDonough, James Edward
Mahajan, Swadesh Mitter
Meshkov, Sydney
Mutchler, Gordon Sinclair
Northcliffe, Lee Conrad
Olenick, Richard Peter
Quarles, Carroll Adair, Jr
Riley, Peter Julian
Sadler, Michael Ervin
Stevenson, Paul Michael
Stiening, Rae Frank
Teplitz, Vigdor Louis
Toohig, Timothy E
Waggoner, James Arthur
Wagoner, David Eugene
Watson, Jerry M
Webb, Robert Carroll
Weinstein, Roy
Wilson, Thomas Leon

UTAH

Ball, James Stutsman
Bergeson, Haven Eldred
Elbert, Jerome William
Jolley, David Kent
Loh, Eugene C
Wu, Yong-Shi

VIRGINIA

Aitken, Alfred H
Bardon, Marcel
Barrois, Bertrand C
Blecher, Marvin
Bridgewater, Albert Louis, Jr
Brown, Stephen L(awrence)
Buck, Warren W, III
Chang, Lay Nam
Cox, Bradley Burton
Cox, Richard Harvey
Diebold, Robert Ernest
Eckhause, Morton
Evans, David Arthur
Ficenec, John Robert
Ijaz, Mujaddid A
Isgur, Nathan
Jenkins, David A
Kabir, Prabahan Kemal
Marshak, Robert Eugene
Minehart, Ralph Conrad
Peacock, Richard Wesley
Perdrisat, Charles F
Sager, Earl Vincent
Schremp, Edward Jay
Shapiro, Maurice Mandel
Sinclair, Charles Kent
Small, Timothy Michael
Spivack, Harvey Marvin
Sundelin, Ronald M
Thacker, Harry B
Trower, W(illiam) Peter
Weber, Hans Jurgen
Welsh, Robert Edward
Winter, Rolf Gerhard
Ziock, Klaus Otto H

WASHINGTON

Albers, James Ray
Barrett, William Louis
Brodzinski, Ronald Lee
Burnett, Thompson Humphrey
Ellis, Stephen Dean
Ely, John Thomas Anderson
Fortson, Edward Norval
Harris, Roger Mason
Harrison, W Craig
Haxton, Wick Christopher
Lam, John Ling-Yee
Lowenthal, Dennis David
Lubatti, Henry Joseph
Miller, Gerald Alan
Mockett, Paul M
Robinson, Barry Wesley
Ruby, Michael Gordon
Wilkes, Richard Jeffrey
Williams, Robert Walter
Young, Kenneth Kong

WEST VIRGINIA

Weldon, Henry Arthur

WISCONSIN

Barger, Vernon Duane
Durand, Bernice Black
Durand, Loyal
Ebel, Marvin Emerson
Erwin, Albert R
Hewett, Joanne Lea
March, Robert Herbert
Morse, Robert Malcolm
Mortara, Lorne B
Pondrom, Lee Girard
Prepost, Richard
Reeder, Don David
Snider, Dale Reynolds
Wu, Sau Lan Yu

WYOMING

Kunselman, A(rthur), Raymond

ALBERTA

Kamal, Abdul Naim
Leahy, Denis Alan
Torgerson, Ronald Thomas
Umezawa, Hiroomi

BRITISH COLUMBIA

Auld, Edward George
Bryman, Douglas Andrew
Häusser, Otto Friedrich
Jennings, Byron Kent
Mason, Grenville R
Measday, David Frederick
Petch, Howard Earle
Picciotto, Charles Edward
Robertson, Lyle Purmal
Viswanathan, Kadayam Sankaran

MANITOBA

Jovanovich, Jovan Vojislav
Van Oero, Willem Theodorus Hendricus

NEW BRUNSWICK

Armstrong, Robin L
Weil, Francis Alphonse

NOVA SCOTIA

Moriarty, Kevin Joseph

ONTARIO

Barton, Richard Donald
Buchanan, Gerald Wallace
Davis, Ronald Stuart

Edwards, Kenneth Westbrook
Ewan, George T
Frisken, William Ross
Hargrove, Clifford Kingston
Huschilt, John
Key, Anthony W
Lee, Hoong-Chien
Lindsey, George Roy
McDonald, Arthur Bruce
Mann, Robert Bruce
Mes, Hans
Nogami, Yukihisa
Prugovecki, Eduard
Resnick, Lazer
Romo, William Joseph
Sherman, Norman K
Storey, Robert Samuel
Sundaresan, Mosur Kalyanaraman

QUEBEC
Depommier, Pierre Henri Maurice
Fuchs, Vladimir
Ho-Kim, Quang
Kalman, Calvin Shea
Kröger, Helmut Karl
Maciejko, Roman
Palfree, Roger Grenville Eric
Patel, Popat-Lal Mulji-bhai
Ralston, John Peter
Ryan, David George
Taras, Paul
Terreault, Bernard J E J

SASKATCHEWAN
Kim, Dong Yun

OTHER COUNTRIES
Afnan, Iraj Ruhi
Blumenfeld, Henry A
Booth, Norman E
De Bruyne, Peter
Dixit, Madhu Sudan
Dominguez, Cesareo Augusto
Eberhard, Philippe Henri
Heinz, Ulrich Walter
Huang, Keh-Ning
Kauffmann, Steven Kenneth
Lipkin, Harry Jeannot
Luckey, Paul David, Jr
Martin, Jim Frank
Nussenzveig, Herch Moyses
Ozaki, Satoshi
Paschos, Emmanuel Anthony
Powers, Richard James
Riska, Dan Olof
Sakmar, Ismail Aydin
Salam, Abdus
Scott, Mary Jean
Settles, Ronald Dean

Fluids

ALABAMA
Davis, Anthony Michael John
Fowlis, William Webster
Hathaway, David Henry
Hermann, R(udolf)
Martin, James Arthur
Moore, Ronald Lee
Rosenberger, Franz
Schutzenhofer, Luke A
Shih, Cornelius Chung-Sheng
Suess, Steven Tyler
Thompson, Byrd Thomas, Jr
Turner, Charlie Daniel, Jr
Visscher, Pieter Bernard
Yeh, Pu-Sen

ALASKA
Gosink, Joan P

ARIZONA
Boyer, Don Lamar
Fink, Jonathan Harry
Heinle, Preston Joseph
Kessler, John Otto
Levy, Eugene Howard
Pearlstein, Arne Jacob
Rouse, Hunter
Sutton, George W(alter)
Wygnanski, Israel Jerzy

ARKANSAS
Fletcher, Alan G(ordon)

CALIFORNIA
Adam, Randall Edward
Alder, Berni Julian
Andersen, Wilford Hoyt
Aref, Hassan
Banks, Robert B(lackburn)
Barmatz, Martin Bruce
Berger, Stanley A(llan)
Berryman, James Garland
Bingham, Harry H, Jr
Blackwelder, Ron F
Boercker, David Bryan
Bohachevsky, Ihor O
Bott, Jerry Frederick
Bowyer, J(ames) M(arston), Jr
Buckingham, Alfred Carmichael
Burden, Harvey Worth
Bushnell, James Judson

Busse, Friedrich Hermann
Caflisch, Russel Edward
Cannell, David Skipwith
Catton, Ivan
Chahine, Moustafa Toufic
Chang, Howard
Chapman, Gary Theodore
Chen, Yang-Jen
Cheng, H(sien) K(ei)
Cheng, Ralph T(a-Shun)
Chin, Jin H
Chu, Pe Cheng
Cloutman, Lawrence Dean
Coles, Donald (Earl)
Corcos, Gilles M
Cross, Ralph Herbert, III
Cunningham, Mary Elizabeth
Daily, James W(allace)
Denn, Morton M(ace)
Driscoll, Charles F
Duff, Russell Earl
Ellis, Albert Tromly
Evans, Todd Edwin
Farber, Joseph
Foster, Theodore Dean
Fu, Lee Lueng
Garrett, Steven Lurie
Garwood, Roland William, Jr
Gat, Nahum
Gibson, Carl H
Gilmore, Forrest Richard
Goldstein, Selma
Hackett, Colin Edwin
Hahn, Kyoung-Dong
Hershey, Allen Vincent
Hesselink, Lambertus
Higgins, Brian Gavin
Ho, Hung-Ta
Hoffmann, Jon Arnold
Hoover, William Graham
Horn, Kenneth Porter
Hornig, Howard Chester
Hsia, Henry Tao-Sze
Hsu, En Yun
Ikawa, Hideo
Ingersoll, Andrew Perry
Jakubowski, Gerald S
Johnston, James P(aul)
Kamegai, Minao
Kaplan, Richard E
Kelly, Robert Edward
Khurana, Krishan Kumar
Knobloch, Edgar
Koh, Robert Cy
Kraus, Samuel
Kroeker, Richard Mark
Kulkarny, Vijay Anand
Kwok, Munson Arthur
Lange, Rolf
Langlois, William Edwin
Lapp, M(arshall)
Lapple, Charles E
Larock, Bruce E
Laurmann, John Alfred
Leith, Cecil Eldon, Jr
Lind, Maurice David
Lindley, Charles A(lexander)
Ling, Rung Tai
Liu, David Shiao-Kung
Lomax, Harvard
Loos, Hendricus G
McKenzie, Robert Lawrence
McMaster, William H
Malmuth, Norman David
Marcus, Bruce David
Marcus, Philip Stephen
Margolis, Stephen Barry
Marino, Lawrence Louis
Markowski, Gregory Ray
Martin, Elmer Dale
Massier, Paul Ferdinand
Mazurek, Thaddeus John
Mihalov, John Donald
Miller, James Angus
Mitchner, Morton
Munson, Albert G
Muntz, Eric Phillip
Naghdi, P(aul) M(ansour)
Nevins, William McCay
Newcomb, William A
Ozawa, Kenneth Susumu
Pierucci, Mauro
Platzer, Maximilian Franz
Plotkin, Allen
Powell, Thomas Mabrey
Putterman, Seth Jay
Radbill, John R(ussell)
Rathmann, Carl Erich
Robben, Franklin Arthur
Roberts, Paul Harry
Rosen, Mordecai David
Rosenkilde, Carl Edward
Roshko, Anatol
Rossow, Vernon J
Rudnick, Joseph Alan
Sabersky, Rolf H(einrich)
Sachs, Donald Charles
Sadeh, Willy Zeev
Saffman, Philip Geoffrey
Sanders, James Vincent
Schubert, Gerald
Seiff, Alvin
Singh, Rameshwar
Skalak, Richard

Somerville, Richard Chapin James
Spencer, Donald Jay
Spera, Frank John
Spreiter, John R(obert)
Steinberg, Daniel J
Sternberg, Joseph
Street, Robert L(ynnwood)
Stuart, George Wallace
Stuhmiller, James Hamilton
Sturtevant, Bradford
Surko, Clifford Michael
Synolakis, Costas Emmanuel
Tellep, Daniel M
Trigger, Kenneth Roy
Tulin, Marshall P(eter)
Utterback, Nyle Gene
Van Atta, Charles W
Van Dyke, Milton D(enman)
Vanyo, James Patrick
Vinokur, Marcel
Wang, H E Frank
Warren, Walter R(aymond), Jr
Wazzan, A R Frank
Wehausen, John Vrooman
Weichman, Peter Bernard
Welge, Henry John
Williams, Forman A(rthur)
Woolard, Henry W(aldo)
Wright, Frederick Hamilton
Wu, Theodore Yao-Tsu
Yuan, Sidney Wei Kwun
Ziegenhagen, Allyn James
Zukoski, Edward Edom

COLORADO
Arp, Vincent D
Benton, Edward Rowell
Blumen, William
Bruno, Thomas J
Clifford, Steven Francis
Dryer, Murray
Faddick, Robert Raymond
Gessler, Johannes
Gilman, Peter A
Goldburg, Arnold
Hanley, Howard James Mason
Hector, David Lawrence
Herring, Jackson Rea
Kantha, Lakshmi
Kassoy, David R
Kerr, Robert McDougall
Krantz, William Bernard
Krueger, David Allen
Lenschow, Donald Henry
Loehrke, Richard Irwin
Meiss, James Donald
Olien, Neil Arnold
O'Sullivan, William John
Owen, Robert Barry
Snyder, Howard Arthur
Tuttle, Elizabeth R
Weinstock, Jerome

CONNECTICUT
Apfel, Robert Edmund
Barnett, Mark
Burridge, Robert
Calehuff, Girard Lester
Chriss, Terry Michael
Foley, William M
Gutierrez-Miravete, Ernesto
Lin, Jia Ding
Nigam, Lakshmi Narayan
Paskausky, David Frank
Peracchio, Aldo Anthony
Ramakrishnan, Terizhandur S
Similon, Philippe Louis
Smooke, Mitchell D
Sreenivasan, Katepalli Raju
Tokita, Noboru
Verdon, Joseph Michael
Von Winkle, William A
Wegener, Peter Paul

DELAWARE
Advani, Suresh Gopaldas
Greenberg, Michael D(avid)
Holt, Rush Dew, Jr
Lauzon, Rodrigue Vincent
Miller, Douglas Charles
Schwartz, Leonard William
Seidel, Barry S(tanley)
Wu, Jin

DISTRICT OF COLUMBIA
Chubb, Scott Robinson
Coffey, Timothy
Colombant, Denis Georges
Deshpande, Mohan Dhondorao
Fan, Dah-Nien
Huber, Peter William
Kalnay, Eugenia
Kao, Timothy Wu
Klebanoff, P(hilip) S(amuel)
Lee, David Allan
Mied, Richard Paul
Oran, Elaine Surick
Pao, Hsien Ping
Picone, J Michael
Rockett, John A
Selden, Robert Wentworth
Shen, Colin Yunkang
Sununu, John Henry
Whang, Yun Chow

FLORIDA
Anderson, Roland Carl
Bober, William
Bradfield, W(alter) S(amuel)
Buchler, Jean-Robert
Collier, Melvin Lowell
Debnath, Lokenath
Dowdell, Rodger B(irtwell)
Howard, Louis Norberg
Huber, Wayne Charles
Hunter, James Hardin, Jr
Kersten, Robert D(onavon)
Krishnamurti, Ruby Ebisuzaki
Kumar, Pradeep
Kurzweg, Ulrich H(ermann)
Le Mehaute, Bernard J
Millsaps, Knox
Mooers, Christopher Northrup Kennard
Nersasian, Arthur
Neuringer, Joseph Louis
Olson, Donald B
Sheldon, John William
Sheng, Yea-Yi Peter
Smith, Jerome Allan
Stern, Melvin Ernest
Tam, Christopher K W
Wang, Hsiang
Weatherly, Georges Lloyd

GEORGIA
Giddens, Don P(eyton)
Hackett, James E
Lau, Jark Chong
Marks, Dennis William
Martin, Charles Samuel
Neitzel, George Paul
Strahle, Warren C(harles)
Wallace, James Robert
Zinn, Ben T

HAWAII
Cheng, Ping
Chuan, Raymond Lu-Po
Grace, Robert Archibald
Loomis, Harold George
Williams, John A(rthur)

IDAHO
Moore, Kenneth Virgil
Ramshaw, John David

ILLINOIS
Adams, John Rodger
Barcilon, Victor
Beard, Kenneth Van Kirke
Bond, Charles Eugene
Buckmaster, John David
Chung, Paul M(yungha)
Crawford, Roy Kent
Davis, Stephen H(oward)
Fields, Gerald S
Frenzen, Paul
Fultz, Dave
Garcia, Marcelo Horacio
Hull, John R
Kasper, Gerhard
Kath, William Lawrence
Kistler, Alan L(ee)
Kuo, Hsiao-Lan
Lebovitz, Norman Ronald
Matalon, Moshe
Maxwell, William Hall Christie
Michelson, Irving
Mihalas, Barbara R Weibel
Mokadam, Raghunath G(anpatrao)
Morkovin, Mark V(ladimir)
Olmstead, William Edward
Ottino, Julio Mario
Pearlstein, Arne Jacob
Pericak-Spector, Kathleen Anne
Peters, James Empson
Rosner, Robert
Sami, Sedat
Simmons, Ralph Oliver
Strehlow, Roger Albert
Taam, Ronald Everett
Tankin, Richard S
Yuen, Man-Chuen

INDIANA
Agee, Ernest M(ason)
Ascarelli, Gianni
Gad-el-Hak, Mohamed
Goldschmidt, Victor W
Hoffman, Joe Douglas
Kentzer, Czeslaw P(awel)
McComas, Stuart T
Mueller, Thomas J
Nee, Victor W
Ng, Bartholomew Sung-Hong
Snow, John Thomas
Thompson, Howard Doyle
Tiederman, William Gregg, Jr
Tree, David R
Widener, Edward Ladd, Sr

IOWA
Chwang, Allen Tse-Yung
Hering, Robert Gustave
Kennedy, John Fisher
Nariboli, Gundo A
Patel, Virendra C
Struck-Marcell, Curtis John
Wilson, Lennox Norwood

Fluids (cont)

KANSAS
Braaten, David A
Connor, Sidney G
Farokhi, Saeed
Greywall, Mahesh S
Jackson, Roscoe George, II
Yu, Yun-Sheng

KENTUCKY
Stewart, Robert Earl
Wood, Don James

LOUISIANA
Baw, Philemon S H
Callens, Earl Eugene, Jr
Collins, Michael Albert
Courter, R(obert) W(ayne)
Foote, Joe Reeder
Grimsal, Edward George
Hussey, Robert Gregory
Latorre, Robert George
Scott, Richard Royce
Slaughter, Milton Dean
Tohline, Joel Edward
Wenzel, Alan Richard

MAINE
Ingard, Karl Uno
McClymer, James P
McKay, Susan Richards

MARYLAND
Apel, John Ralph
Baer, Ferdinand
Bauer, Peter
Baum, Howard Richard
Bennett, Frederick Dewey
Blake, William King
Bournia, Anthony
Bungay, Peter M
Burns, Timothy John
Chadwick, Richard Simeon
Chan, Kwing Lam
Cleveland, William Grover, Jr
Cronvich, Lester Louis
Crump, Stuart Faulkner
Devin, Charles, Jr
Domokos, Gabor
Eaton, Alvin Ralph
Ethridge, Noel Harold
Ferrell, Richard Allan
Gessow, Alfred
Gray, Ernest Paul
Haberman, William L(awrence)
Hansen, Robert Jack
Huang, Thomas Tsung-Tse
Jacobs, Sigmund James
Kaiser, Jack Allen Charles
Kitaigorodskii, Sergei Alexander
Lugt, Hans Josef
McCormick, Michael Edward
McFadden, Geoffrey Bey
Mollo-Christensen, Erik Leonard
Morgan, William Bruce
Moulton, James Frank, Jr
Nash, Jonathon Michael
Obremski, Henry J(ohn)
O'Brien, Vivian
Olson, Peter Lee
Pai, S(hih) I
Pirraglia, Joseph A
Potzick, James Edward
Powell, John David
Randall, John Douglas
Rehm, Ronald George
Rogers, David Freeman
Sandusky, Harold William
Schemm, Charles Edward
Schmid, Lawrence Alfred
Sedney, R(aymond)
Semerjian, Hratch G
Sengers, Jan V
Shore, Steven Neil
Tachmindji, Alexander John
Traugott, Stephen C(harles)
Waldo, George Van Pelt, Jr
Wallace, James M
Watt, Mamadov Hame
Yeh, Tsyh Tyan
Yoon, Peter Haesung
Zien, Tse-Fou

MASSACHUSETTS
Abbott, Douglas E(ugene)
Abernathy, Frederick Henry
Appleton, John P(atrick)
Brenner, Howard
Brocard, Dominique Nicolas
Chase, David Marion
Chen, Sow-Hsin
Cipolla, John William, Jr
Cole, John E(mery), III
Cordero, Julio
Dewan, Edmond M
Dewey, C(larence) Forbes, Jr
Durgin, William W
Emmons, Howard W(ilson)
Fohl, Timothy
Goela, Jitendra Singh
Hecker, George Ernst
Hoffman, Richard Laird
Hoge, Harold James

Kautz, Frederick Alton, II
Kemp, Nelson Harvey
Kingston, George C
Klein, Milton M
Lien, Hwachii
Little, Sarah Alden
Louis, Jean Francois
Madsen, Ole Secher
Martin, Paul Cecil
Mei, Chiang C(hung)
Murman, Earll Morton
Murphy, Peter John
Muthukumar, Murugappan
Neue, Uwe Dieter
Noiseux, Claude Francois
Ogilvie, T(homas) Francis
Pian, Carlson Chao-Ping
Reeves, Barry L(ucas)
Reilly, James Patrick
Shapiro, Ascher H(erman)
Sonin, Ain A(nts)
Stickler, David Bruce
Trefethen, Lloyd MacGregor
Vignos, James Henry
Wallace, James
Widnall, Sheila Evans
Wu, Lei
Wyler, John Stephen
Yamartino, Robert J
Yener, Yaman
Zalosh, Robert Geoffrey

MICHIGAN
Beck, Robert Frederick
Boyd, John Philip
Chock, David Poileng
Cutler, Warren Gale
Habib, Izzeddin Salim
Hartman, John L(ouis)
Jacobs, Stanley J
Karkheck, John Peter
Kordyban, Eugene S
Miklavcic, Milan
Prasuhn, Alan Lee
Sichel, Martin
Smith, George Wolfram
Stansel, John Charles
Yih, Chia-Shun

MINNESOTA
Arndt, Roger Edward Anthony
Beavers, Gordon Stanley
Berman, Abraham S
Lysak, Robert Louis
Olson, Reuben Magnus
Reilly, Richard J
Silberman, Edward
Stefan, Heinz G
Uban, Stephen A
Valls, Oriol Tomas
Yuen, David Alexander
Zimmermann, William, Jr

MISSISSIPPI
Bernard, Robert Scales
Coleman, Neil Lloyd
Folse, Raymond Francis, Jr
Green, Albert Wise
Horton, Thomas Edward, Jr
Pandey, Ras Bihari

MISSOURI
Bogar, Thomas John
Bower, William Walter
Chen, Ta-Shen
Elgin, Robert Lawrence
Gardner, Richard A
Hakkinen, Raimo Jaakko
Kibens, Valdis
Kotansky, D(onald) R(ichard)
Liu, Henry
Preckshot, G(eorge) W(illiam)
Sajben, Miklos
Spaid, Frank William
Venezian, Giulio

MONTANA
Hiscock, William Allen
Wood, William Wayne

NEW HAMPSHIRE
Baumann, Hans D
Hollweg, Joseph Vincent
Kantrowitz, Arthur (Robert)
Morduchow, Morris
Runstadler, Peter William, Jr
Sonnerup, Bengt Ulf osten

NEW JERSEY
Abraham, John
Bachman, Walter Crawford
Barnes, Derek A
Blumberg, Alan Fred
Brown, Garry Leslie
Caffarelli, Luis Angel
Carlson, Curtis Raymond
Chaikin, Paul Michael
Choi, Byung Chang
Daly, Ruth Agnes
Dixon, Paul King
Doby, Raymond
Edwards, Christopher Andrew
Fornberg, Bengt
Hohenberg, Pierre Claude

Hrycak, Peter
Hudson, Steven David
Johnston, Christian William
Kolodner, Paul R
Kovats, Andre
Lee, Wei-Kuo
Levy, Marilyn
Lewellen, William Stephen
Miles, Richard Bryant
Nguyen, Hien Vu
Orszag, Steven Alan
Pierrehumbert, Raymond T
Princen, Henricus Mattheus
Raichel, Daniel R(ichter)
Salwen, Harold
Simpkins, Peter G
Temkin, Samuel
Tsou, F(u) K(ang)
Vann, Joseph M
Williams, Gareth Pierce
Yurke, Bernard
Zabusky, Norman J
Zebib, Abdelfattah M G

NEW MEXICO
Baggett, Lester Marchant
Bankston, Charles A, Jr
Benjamin, Robert Fredric
Bertin, John Joseph
Brackbill, Jeremiah U
Brownell, John Howard
Butler, Thomas Daniel
Cline, Michael Castle
Cox, Arthur Nelson
Daly, Bartholomew Joseph
Deal, William E, Jr
Deupree, Robert G
Dick, Richard Dean
Erpenbeck, Jerome John
Farnsworth, Archie Verdell, Jr
Fickett, Wildon
Fukushima, Eiichi
Glasser, Alan Herbert
Goldman, S Robert
Harlow, Francis Harvey, Jr
Hoffman, Nelson Miles, III
Johnson, James Daniel
Jones, Eric Manning
Julien, Howard L
Kashiwa, Bryan Andrew
Kopp, Roger Alan
Lam, Kin Leung
Lindman, Erick Leroy, Jr
Martin, Richard Alan
Mead, William C
Mjolsness, Raymond C
Mutschlecner, Joseph Paul
Orgill, Montie M
O'Rourke, Peter John
Price, Robert Harold
Ragan, Charles Ellis, III
Rose, Harvey Arnold
Sicilian, James Michael
Sierk, Arnold John
Thieme, Melvin T
Travis, John Richard
Wallace, Richard Kent
Wright, Thomas Payne
Zemach, Charles

NEW YORK
Ackerberg, Robert C(yril)
Acrivos, Andreas
Alpher, Ralph Asher
Baker, Norman Hodgson
Brower, William B, Jr
Busnaina, Ahmed A
Chang, Ching Ming
Chevray, Rene
Childress, William Stephen
De Boer, P(ieter) C(ornelis) Tobias
Feigenbaum, Mitchell Jay
Fesjian, Sezar
Foreman, Kenneth M
Fox, William
Gans, Roger Frederick
Geer, James Francis
George, William Kenneth, Jr
Glimm, James Gilbert
Goodman, Theodore R(obert)
Gould, Robert Kinkade
Halitsky, James
Hall, J(ohn) Gordon
Hartley, Charles LeRoy
Herczynski, Andrzej
Hollenberg, Joel Warren
Holmes, Philip John
Hwang, Shy-Shung
Irvine, Thomas Francis
Johnson, Joseph Andrew, III
Johnson, Wallace E
Kaplan, Paul
Keller, D Steven
Kocher, Haribhajan S(ingh)
Koplik, Joel
Kroeger, Peter G
Lai, W(ei) Michael
Lehman, August F(erdinand)
Leibovich, Sidney
Liu, Ching Shi
Liu, Philip L-F
Lumley, John L(eask)
McCrory, Robert Lee, Jr
Marrone, Paul Vincent

Moore, Franklin K(ingston)
Moretti, G(ino)
Nagamatsu, Henry T
O'Brien, Edward E
Parpia, Jeevak Mahmud
Reynolds, John Terrence
Rezanka, Ivan
Rudinger, George
Ruschak, Kenneth John
Savkar, Sudhir Dattatraya
Scala, Sinclaire M(aximilian)
Schwarz, Klaus W
Seyler, Charles Eugene
Shapiro, Paul Jonathon
Shepherd, Joseph Emmett
Siegmann, William Lewis
Siggia, Eric Dean
Stahl, Laddie L
Staub, Fred W
Thomas, John Howard
Thompson, Philip A
Turcotte, Donald Lawson
Wang, Lin-Shu
Ward, Lawrence W(aterman)
Webb, Richard
Weinbaum, Sheldon
Zakkay, Victor

NORTH CAROLINA
Amein, Michael
Arya, Satya Pal Singh
Chandra, Suresh
Christiansen, Wayne Arthur
Doster, Joseph Michael
Dotson, Allen Clark
Edwards, John Auert
Georgiadis, John G
Griffith, Wayland C(oleman)
Janowitz, Gerald S(aul)
McRee, Donald Ikerd
Mosberg, Arnold T

OHIO
Bodonyi, Richard James
Brazee, Ross D
Bridge, John F(loyd)
Burggraf, Odus R
Chung, Benjamin T
Deissler, Robert G(eorge)
Edwards, David Olaf
Fenichel, Henry
Fitzmaurice, Nessan
Foster, Michael Ralph
Franke, Milton Eugene
Ghia, Kirti N
Goldstein, Marvin E
Hamed, Awatef A
Hantman, Robert Gary
Herbert, Thorwald
Jaikrishnan, Kadambi Rajgopal
Jeng, Duen-Ren
Keith, Theo Gordon, Jr
Kennedy, Lawrence A
Lawler, Martin Timothy
Leonard, Brian Phillip
Lin, Kuang-Ming
Ostrach, Simon
Pinchak, Alfred Cyril
Polak, Arnold
Prahl, Joseph Markel
T'ien, James Shaw-Tzuu
Von Ohain, Hans Joachim
Wen, Shih-Liang
Zaman, Khairul B M Q

OKLAHOMA
Ackerson, Bruce J
Blais, Roger Nathaniel
Cerro, Ramon Luis
Davies-Jones, Robert Peter
Fitch, Ernest Chester, Jr
Howard, Robert Adrian
Omer, Savas
Shell, Francis Joseph

OREGON
Donnelly, Russell James
Perry, Robert W(illiam)
Photinos, Panos John
Shirazi, Mostafa Ayat
Warren, William Willard, Jr

PENNSYLVANIA
Bau, Haim Heinrich
Blank, Albert Abraham
Blythe, Philip Anthony
Carelli, Mario Domenico
Carmi, Shlomo
Chow, Shin-Kien
Cohen, Ira M
Crawford, John David
Cutler, Paul H
Dash, Sanford Mark
Daywitt, James Edward
DeSanto, Daniel Frank
Emrich, Raymond Jay
Farn, Charles Luh-Sun
Farouk, Bakhtier
Gollub, Jerry Paul
Hagen, Oskar
Hogenboom, David L
Jhon, Myung S
Kazakia, Jacob Yakovos
Kim, Yong Wook

Lauchle, Gerald Clyde
Lazaridis, Anastas
Liang, Shoudan
Long, Lyle Norman
McDowell, Maurice James
Macpherson, Alistair Kenneth
Mahaffy, John Harlan
Maher, James Vincent
Maynard, Julian Decatur
Merz, Richard A
Miskovsky, Nicholas Matthew
Parkin, Blaine R(aphael)
Pierce, Allan Dale
Ramezan, Massood
Ramos, Juan Ignacio
Reed, Joseph Raymond
Rockwell, Donald O
Settles, Gary Stuart
Smith, Wesley R
Whirlow, Donald Kent
Wyngaard, John C
Zemel, Jay N(orman)

RHODE ISLAND
Clarke, Joseph H(enry)
Dobbins, Richard Andrew
Evans, David L
Karlsson, Sture Karl Fredrik
Kowalski, Tadeusz
Liu, J(oseph) T(su) C(hieh)
McEligot, Donald M(arinus)
Parmentier, Edgar M(arc)
Paster, Donald L(ee)
Sibulkin, Merwin
Stratt, Richard Mark
Su, Chau-Hsing

SOUTH CAROLINA
Castro, Walter Ernest
Haile, James Mitchell
Prziembel, Christian E G
Yang, Tah Teh

TENNESSEE
Allison, Stephen William
Collins, Frank Gibson
Hayter, John Bingley
Hosker, Rayford Peter, Jr
Lee, Donald William
Lewis, James W L
Reddy, Kapuluru Chandrasekhara
Remenyik, Carl John
Siomos, Konstadinos
Swim, William B(axter)
Thomas, David Glen
Witten, Alan Joel

TEXAS
Anderson, Edward Everett
Aucoin, Paschal Joseph, Jr
Ball, Kenneth Steven
Bergman, Theodore L
Chang, Ping
Chevalier, Howard L
Chung, Jing Yan
Darby, Ronald
Dodge, Franklin Tiffany
Eichhorn, Roger
Frenzen, Christopher Lee
Goldman, Joseph L
Haberman, Richard
Harpavat, Ganesh Lal
Hrkel, Edward James
Hurlburt, H(arvey) Zeh
Hussain, A K M Fazle
Hwang, Neddy H C
Johnson, Darell James
Koschmieder, Ernst Lothar
Krishnamurthy, Subramanian
Liang, Edison Park-Tak
Lu, Frank Kerping
McCauley, Joseph Lee
Maerker, John Malcolm
Miller, Gerald E
Mitchell, James Emmett
Modisette, Jerry L
Porter, John William
Prats, Michael
Rajagopalan, Raj
Sample, Thomas Earl, Jr
Shapiro, Paul Robert
Sharma, Suresh C
Sifferman, Thomas Raymond
Sparks, Cecil Ray
Swinney, Harry Leonard
Swope, Richard Dale
Truxillo, Stanton George
Tsahalis, Demosthenes Theodoros
Voigt, Gerd-Hannes
Weatherford, W(illiam) D(ewey), Jr
Young, Fred M(ichael)

UTAH
Clyde, Calvin G(eary)
Hanks, Richard W(ylie)
Schunk, Robert Walter

VIRGINIA
Carlson, Harry William
Chang, Chieh Chien
Crowley, Patrick Arthur
Csanady, Gabriel Tibor
Evans, John C
Evans, John Stanton

Fang, Ching Seng
Gale, Harold Walter
Gatski, Thomas Bernard
Grabowski, Walter John
Grossmann, William
Hansen, Robert J
Hardin, Jay Charles
Haussling, Henry Jacob
Hidalgo, Henry
Hoppe, John Cameron
Jacobson, Ira David
Jaisinghani, Rajan A
Johnson, David Harley
Kroll, John Ernest
Kuo, Albert Yi-Shuong
Lai, Chintu (Vincent C)
Lee, Ja H
Lewis, Clark Houston
Lilleleht, L(embit) U(no)
Monacella, Vincent Joseph
Morton, Jeffrey Bruce
Muraca, Ralph John
Myers, Michael Kenneth
Pereira, Nino Rodrigues
Pierce, Felix J(ohn)
Pruett, Charles David
Reischman, Michael Mack
Renardy, Michael
Renardy, Yuriko
Roberts, William Woodruff, Jr
Roth, Allan Charles
Schwiderski, Ernst Walter
Seiner, John Milton
Selwyn, Philip Alan
Sepucha, Robert Charles
Simpson, Roger Lyndon
Spence, Thomas Wayne
Tieleman, Henry William
Witting, James M
Wood, Houston Gilleylen, III
Yates, Charlie Lee
Yu, James Chun-Ying

WASHINGTON
Ahlstrom, Harlow G(arth)
Band, William
Bernard, Eddie Nolan
Booker, John Ratcliffe
Brown, Robert Alan
Christiansen, Walter Henry
Cokelet, Edward Davis
Connell, James Roger
Criminale, William Oliver, Jr
Delisi, Donald Paul
Dixon, Robert Jerome, Jr
Duvall, George Evered
Dvorak, Frank Arthur
Erdmann, Joachim Christian
Forster, Fred Kurt
Fowles, George Richard
Fremouw, Edward Joseph
Harrison, Don Edmunds
Marston, Philip Leslie
Moore, Richard Lee
Orsborn, John F
Rattray, Maurice, Jr
Reyhner, Theodore Alison
Roetman, Ernest Levane
Russell, David A
Sanford, Thomas Bayes
Schick, Michael
Stegen, Gilbert Rolland
Tung, Ka-Kit

WEST VIRGINIA
Ness, Nathan
Squire, William

WISCONSIN
Bird, R(obert) Byron
Doshi, Mahendra R
Hoopes, John A
Kurath, Sheldon Frank
Meyer, Heribert
Schultz, David Harold
Woods, Frank Robert

PUERTO RICO
Khan, Winston

ALBERTA
Collins, David Albert Charles
De Krasinski, Joseph S
Freeman, Gordon Russel
Fu, Cheng-Tze
Gee, Norman
Rajaratnam, N(allamuthu)
Williams, Michael C(harles)

BRITISH COLUMBIA
Crozier, Edgar Daryl
Dewey, John Marks
Gartshore, Ian Stanley
Hughes, Blyth Alvin
Krauel, David Paul
Parkinson, G(eoffrey) Vernon
Salcudean, Martha Eva
Seymour, Brian Richard
Singh, Manohar

NEWFOUNDLAND
Morris, Claude C
Rochester, Michael Grant

NOVA SCOTIA
Kusalik, Peter Gerard

ONTARIO
Ahmed, Nasir Uddin
Baird, Malcolm Henry Inglis
Barker, Paul Kenneth
Barron, Ronald Michael
Boadway, John Douglas
Bragg, Gordon McAlpine
Chan, Yat Yung
Cowan, John Arthur
Donelan, Mark Anthony
Dunn, D(onald) W(illiam)
Egelstaff, Peter A
Irwin, Peter Anthony
James, David F
Krishnappan, Bommanna Gounder
Langford, William Finlay
Moffatt, William Craig
Moltyaner, Grigory
North, Henry E(rick) T(uisku)
Orlik-Ruckemann, Kazimierz Jerzy
Perlman, Maier
Pope, Noel Kynaston
Ray, Ajit Kumar
Ribner, Herbert Spencer
St Maurice, Jean-Pierre
Svensson, Eric Carl
Vijay, Mohan Madan
Williams, Norman S W
Wilson, Kenneth Charles

QUEBEC
Chan, Sek Kwan
Dickinson, Edwin John
Grmela, Miroslav
Lin, Sui
Marchand, Richard
Newman, B(arry) G(eorge)
Paidoussis, Michael Pandeli
Savage, Stuart B
Tassoul, Jean-Louis

SASKATCHEWAN
Deckker, B(asil) E(ardley) L(eon)
Tsang, Gee

OTHER COUNTRIES
Archie, Charles Neill
Breslin, John P
Eichenberger, Hans P
Hsieh, Din-Yu
Icke, Vincent
Liu, Ta-Jo
Romero, Alejandro F
Safran, Samuel A
Segel, Lee Aaron
Vrebalovich, Thomas

Mechanics

ALABAMA
Bailey, J Earl
Bartlett, James Holly
Bowden, Charles Malcolm
Perkins, James Francis
Schutzenhofer, Luke A
Turner, Charlie Daniel, Jr

ARIZONA
Blitzer, Leon
Fraser, Harvey R(eed)
Lamb, Willis Eugene, Jr
Richard, Ralph Michael
Thompson, Lee P(rice)

ARKANSAS
Harter, William George

CALIFORNIA
Abey, Albert Edward
Alperin, Morton
Baldwin, Barrett S, Jr
Barnett, David M
Behrens, H Wilhelm
Birkimer, Donald Leo
Buckingham, Alfred Carmichael
Burton, Donald Eugene
Cannon, Robert H, Jr
Caughey, Thomas Kirk
Chang, I-Dee
Chapman, Gary Theodore
Chen, Albert Tsu-Fu
Cheng, H(sien) K(ei)
Christensen, Richard Monson
Coles, Donald (Earl)
Corngold, Noel Robert David
Crebbin, Kenneth Clive
Dailey, C(harles) L(ee)
Fourney, M(ichael) E(ugene)
Gerpheide, John H
Gogolewski, Raymond Paul
Hall, Charles Frederick
Herrmann, Leonard R(alph)
Heusinkveld, Myron Ellis
Hoover, William Graham
Jennings, Paul C(hristian)
Kane, Thomas R(eif)
Kaplan, Abner
Keefe, Denis
Keller, Joseph Edward, Jr
Kelly, Paul Sherwood

Knollman, Gilbert Carl
Kroeker, Richard Mark
Kubota, Toshi
Lambertson, Glen Royal
Lee, Martin Joe
Lee, Vin-Jang
Leitmann, G(eorge)
Lichtenberg, Allan J
Liepmann, H(ans) Wolfgang
Lin, Tung Hua
Lubliner, J(acob)
Lynch, David Dexter
McNiven, Hugh D(onald)
Mar, James Wah
Marble, Frank E(arl)
Meadows, Mark Allan
Miklowitz, Julius
Miller, James Avery
Mingori, Diamond Lewis
Moe, Mildred Minasian
Moore, Edgar Tilden, Jr
Mow, Maurice
Muki, Rokuro
Naghdi, P(aul) M(ansour)
Nielsen, Helmer L(ouis)
Passenheim, Burr Charles
Peale, Stanton Jerrold
Pearson, John
Pellegrini, Claudio
Pister, Karl S(tark)
Reinhardt, Walter Albert
Reissner, Eric
Roberson, Robert Errol
Roberts, Paul Harry
Rockwell, Robert Lawrence
Root, L(eonard) Eugene
Rosenberg, Reinhardt M
Roshko, Anatol
Sedgwick, Robert T
Shen, Y(ung) C(hung)
Sherman, Frederick S
Shi, Yun Yuan
Snowden, William Edward
Steele, Charles Richard
Sternberg, Eli
Stevens, Aldred Lyman
Tatro, Clement A(ustin)
Tsai, Stephen W
Vernon, Frank Lee, Jr
Wang, Paul Keng Chieh
Westmann, Russell A
Williams, James Gerard
Willis, D(onald) Roger
Yang, An Tzu
Yang, Charles Chin-Tze
Yoder, Charles Finney
Zimmerman, Ivan Harold

COLORADO
Bartlett, David Farnham
Busemann, Adolf
Essenburg, F(ranklin)
Hoots, Felix R
Liu, Joseph Jeng-Fu
Meiss, James Donald
Webster, Larry Dale

CONNECTICUT
Arnoldi, Robert Alfred
Howkins, Stuart D
Jenney, David S
Johnston, E(lwood) Russell, Jr
Krahula, Joseph L(ouis)
Loo, Francis T C
Ojalvo, Irving U
Onat, E(min) T(uran)
Stetson, Karl Andrew
Strawderman, Wayne Alan

DELAWARE
Elliott, John Habersham
Fish, F(loyd) H(amilton), Jr
Huntsberger, James Robert
Nowinski, Jerzy L

DISTRICT OF COLUMBIA
Arkilic, Galip Mehmet
Bainum, Peter Montgomery
Charyk, Joseph Vincent
Eggenberger, Andrew Jon
French, Francis William
Kearsley, Elliot Armstrong
Kelly, George Eugene
Klein, Vladislav
Lee, David Allan
Liebowitz, Harold
Rockett, John A
Rosenthal, F(elix)
Seidelmann, Paul Kenneth
Toridis, Theodore George

FLORIDA
Bose, Subir Kumar
Case, Robert Oliver
Coldwell, Robert Lynn
Edson, Charles Grant
Eisenberg, Martin A(llan)
Fearn, Richard L(ee)
Foster, Irving Gordon
Hemp, Gene Willard
Huerta, Manuel Andres
Lindgren, E(rik) Rune
Roberts, Sanford B(ernard)
Sidebottom, Omar M(arion)

Wheeler, Lewis Turner
Yeh, Hsiang-Yueh

UTAH
Fife, Paul Chase
Vyas, Ravindra Kantilal

VIRGINIA
Burns, Grover Preston
Claus, Richard Otto
DeLauer, R(ichard) D(aniel)
Finke, Reinald Guy
Henneke, Edmund George, II
Jacobson, Ira David
Jennings, Richard Louis
Lewis, David W(arren)
Milligan, John H
Monacella, Vincent Joseph
Muraca, Ralph John
Nayfeh, Ali Hasan
Pilkey, Walter David
Porzel, Francis Bernard
Raju, Ivatury Sanyasi
Reifsnider, Kenneth Leonard
Renardy, Yuriko
Rokni, Mohammad Ali
Schmidt, Louis Vincent
Sparrow, D(avid) A
Telionis, Demetri Pyrros
Wang, C(hiao) J(en)
Watson, Nathan Dee

WASHINGTON
Bollard, R(ichard) John H
Cole, Robert Stephen
Ehlers, Francis Edward
Evans, Roger James
Filler, Lewis
Fyfe, I(an) Millar
Gupta, Yogendra M(ohan)
Kennet, Haim
Leaf, Boris
Parmerter, R Reid
Pilet, Stanford Christian
Riedel, Eberhard Karl
Roetman, Ernest Levane
Taya, Minoru
Walton, Vincent Michael
Wilson, Thornton Arnold

WISCONSIN
Brackenridge, John Bruce
Johnson, Arthur Franklin
McQuistan, Richmond Beckett
Rang, Edward Roy
Symon, Keith Randolph
Van den Akker, Johannes Archibald
Washa, George William

WYOMING
Boresi, Arthur Peter

PUERTO RICO
Blum, Lesser

ALBERTA
Ellyin, Fernand
Glockner, Peter G
Hoffer, J(oaquin) A(ndres)
Kumar, Shrawan
Murray, David William

BRITISH COLUMBIA
Graham, George Alfred Cecil
Lardner, Robin Willmott
Parkinson, G(eoffrey) Vernon
Singh, Manohar

MANITOBA
Cohen, Harley

NOVA SCOTIA
Calkin, Melvin Gilbert
Sastry, Vankamamidi VRN

ONTARIO
Baronet, Clifford Nelson
Chapman, George David
Lennox, William C(raig)
Lind, Niels Christian
Mansinha, Lalatendu
Olson, Allan Theodore
Roorda, John
Todd, Leonard

QUEBEC
Axelrad, D(avid) R(obert)
Fazio, Paul (Palmerino)
Li, Zi-Cai
Massoud, Monir Fouad
Paidoussis, Michael Pandeli
Savage, Stuart B
Subramanian, Sesha

OTHER COUNTRIES
Chau, Wai-Yin
Contopoulos, George
Hashin, Zvi
Kydoniefs, Anastasios D
Markovitz, Hershel

Nuclear Structure

ALABAMA
Alford, William Lumpkin
Regner, John LaVerne
Robinson, Edward Lee
Thaxton, George Donald
Thompson, Lewis Chisholm

ARIZONA
Comfort, Joseph Robert
Donahue, Douglas James
Fan, Chang-Yun
Fischer, David Lloyd
Hill, Henry Allen
McCullen, John Douglas
McIntyre, Laurence Cook, Jr
Nigam, Bishan Perkash
Roy, Radha Raman

ARKANSAS
Rollefson, Aimar Andre
Vyas, Reeta

CALIFORNIA
Adams, Ralph Melvin
Adelson, Harold Ely
Albergotti, Jesse Clifton
Alonso, Carol Travis
Alonso, Jose Ramon
Alvarez, Raymond Angelo, Jr
Ames, Lawrence Lowell
Anderson, John D
Bach, David Rudolph
Baisden, Patricia Ann
Ball, William Paul
Bardin, Tsing Tchao
Barnes, Charles Andrew
Bauer, Rudolf Wilhelm
Becker, John Angus
Bennett, Charles Lougheed
Bertin, Michael C
Bloom, Stewart Dave
Bodfish, Ralph E
Brady, Franklin Paul
Britt, Harold Curran
Brown, James Roy
Brown, Virginia Ruth
Burginyon, Gary Alfred
Buskirk, Fred Ramon
Cahill, Thomas A
Camp, David Conrad
Cerny, Joseph, III
Chadwick, George Brierley
Chalmers, Robert Anton
Chase, Lloyd Fremont, Jr
Chen, Chia Hwa
Chu, Keh-Cheng
Cladis, John Baros
Clark, Arnold Franklin
Clark, William Gilbert
Collard, Harold Rieth
Conzett, Homer Eugene
Corman, Emmett Gary
Crandall, Walter Ellis
Creutz, Edward (Chester)
Creutz, Edward (Chester)
Crittenden, Eugene Casson, Jr
Dally, Edgar B
Diamond, Richard Martin
Dickinson, Wade
Dolan, Kenneth William
Draper, James Edward
Dudziak, Walter Francis
Dunlop, William Henry
Dyer, John Norvell
Eccles, Samuel Franklin
Edwards, David Franklin
Eisberg, Robert Martin
Erdmann, John Hugo
Ferguson, James Malcolm
Fisher, Thornton Roberts
Fricke, Martin Paul
Ganas, Perry S
Gearhart, Roger A
Gibson, Edward F
Gilbert, William Spencer
Glick, Harold Alan
Goldberg, Eugene
Goodwin, Lester Kepner
Goosman, David R
Goss, John Douglas
Gould, Harvey Allen
Greiner, Douglas Earl
Groom, Donald Eugene
Groseclose, Byron Clark
Haddock, Roy P
Hafemeister, David Walter
Handler, Harry Elias
Hanna, Stanley Sweet
Hansen, Luisa Fernandez
Heckman, Harry Hughes
Heikkinen, Dale William
Heindl, Clifford Joseph
Hoff, Richard William
Holzer, Alfred
Howerton, Robert James
Hu, Chi-Yu
Hudson, Alvin Maynard
Hulet, Ervin Kenneth
James, Ralph Boyd
John, Joseph
Jordan, Willard Clayton
Jungerman, John (Albert)

Kavanagh, Ralph William
King, Harriss Thornton
Knox, William Jordan
Kovar, Frederick Richard
Kull, Lorenz Anthony
Lanier, Robert George
Lark, Neil LaVern
Lee, Paul L
Lingenfelter, Richard Emery
Ludin, Roger Louis
Lugmair, Guenter Wilhelm
Lutz, Harry Frank
McCarthy, Mary Anne
MacGregor, Malcolm Herbert
Mann, Lloyd Godfrey
Margaziotis, Demetrius John
Marino, Lawrence Louis
Marrs, Roscoe Earl
Marrus, Richard
Martinelli, Ernest A
Martz, Harry Edward, Jr
Melkanoff, Michel Allan
Mendelson, Robert Alexander, Jr
Milne, Edmund Alexander
Moody, Kenton J
Morgan, James Frederick
Morton, John Robert, III
Nefkens, Bernard Marie
Nero, Anthony V, Jr
Newman, David Edward
Nitschke, J Michael
Norman, Eric B
O'Dell, Jean Marland
Oehlberg, Richard N
Okrent, David
Olness, Robert James
Orphan, Victor John
Parker, Vincent Eveland
Peek, Neal Frazier
Pehl, Richard Henry
Perry, Joseph Earl, Jr
Peterson, Jack Milton
Poggenburg, John Kenneth, Jr
Poppe, Carl Hugo
Poskanzer, Arthur M
Proctor, Ivan D
Pugh, Howel Griffith
Putnam, Thomas Milton, Jr
Randrup, Jorgen
Redlich, Martin George
Reines, Frederick
Reynolds, Glenn Myron
Reynolds, Harry Lincoln
Richardson, John Reginald
Rosenblum, Stephen Saul
Rosenfeld, Arthur H
Salisbury, Stanley R
Schweizer, Felix
Segré, Emilio Gino
Serduke, Franklin James David
Sextro, Richard George
Singletary, John Boon
Singletary, Lillian Darlington
Smoot, George Fitzgerald, III
Sobel, Henry Wayne
Spieler, Helmuth
Stein, William Earl
Stokstad, Robert G
Struble, Gordon Lee
Sugihara, Thomas Tamotsu
Symons, Timothy J
Thomas, James H
Tipler, Paul A
Tracy, James Frueh
Tucker, Allen Brink
Urone, Paul Peter
Van Bibber, Karl Albert
Verbinski, Victor V
Vogel, Peter
Walt, Martin
Warshaw, Stephen I
Weber, Hans Josef
West, Harry Irwin, Jr
Whitehead, Marian Nedra
Whitten, Charles A, Jr
Wilson, Robert Gray
Woldseth, Rolf
Wong, Chun Wa
Wood, Calvin Dale
Worth, Donald Calhoun
Wright, Byron Terry
Yap, Fung Yen
Yearian, Mason Russell
Young, John Chancellor
Yu, David U L
Zender, Michael J

COLORADO
Dooley, John Raymond, Jr
Haas, Francis Xavier, Jr
Herring, Jackson Rea
Iona, Mario
Kraushaar, Jack Jourdan
Lind, David Arthur
McAllister, Robert Wallace
Moore, Benjamin LaBree
Morris, George Ronald
Peterson, Roy Jerome
Poirier, Charles Philip
Rost, Ernest Stephan
Shelton, Frank Harvey
Smythe, William Rodman
Spenny, David Lorin
Zafiratos, Chris Dan

CONNECTICUT
Adair, Robert Kemp
Albats, Paul
Antkiw, Stephen
Bockelman, Charles Kincaid
Bromley, David Allan
Cobern, Martin E
Firk, Frank William Kenneth
Greenberg, Jack Sam
Hill, David Lawrence
Hosain, Fazle
Kasha, Henry
Kendziorski, Francis Richard
King, George, III
Krisst, Raymond John
Lange, Robert Carl
Levenstein, Harold
Lichtenberger, Harold V
Lubell, Michael S
Morrison, Richard Charles
Nath, Ravinder Katyal
Parker, Peter Donald MacDougall
Pisano, Daniel Joseph, Jr
Schweitzer, Jeffrey Stewart
Simpson, Wilburn Dwain
Slack, Lewis
Stephenson, Kenneth Edward
Tao, Shu-Jen
Tittman, Jay
Wolf, Elizabeth Anne

DELAWARE
Fou, Cheng-Ming
Pittel, Stuart
Sawers, James Richard, Jr
Swann, Charles Paul

DISTRICT OF COLUMBIA
Bassel, Robert Harold
Berman, Barry L
Bernthal, Frederick Michael
Brown, Louis
Cooper, Benjamin Stubbs
Erskine, John Robert
Fagg, Lawrence Wellburn
Gossett, Charles Robert
Hendrie, David Lowery
Kelly, Francis Joseph
Keyworth, George A, II
Lambert, James Morrison
Lehman, Donald Richard
Phillips, Gary Wilson
Reiss, Howard R
Richardson, Allan Charles Barbour
Ritter, James Carroll
Schima, Francis Joseph
Seglie, Ernest Augustus
Shon, Frederick John
Silberberg, Rein
Sober, Daniel Isaac
Stolovy, Alexander
Treado, Paul A
Whetstone, Stanley L, Jr

FLORIDA
Chapman, Kenneth Reginald
Clark, Mary Elizabeth
Davis, Robert Houser
Dubbelday, Pieter Steven
Dunnam, Francis Eugene, Jr
Fletcher, Neil Russel
Fox, John David
Frawley, Anthony Denis
Harmon, G Lamar
Howell, John Foss
Jin, Rong-Sheng
Kemper, Kirby Wayne
Leies, Gerard M
Llewellyn, Ralph A
Mintz, Stephen Larry
Peterson, Lennart Rudolph
Petrovich, Fred
Philpott, Richard John
Plendl, Hans Siegfried
Randall, Charles Addison, Jr
Rester, Alfred Carl, Jr
Reynolds, Bruce G
Riley, Mark Anthony
Robson, Donald
Sheline, Raymond Kay
Shelton, Wilford Neil
Stetson, Robert Franklin
Tabor, Samuel Lynn
Weber, Alfred Herman

GEORGIA
Baker, Francis Todd
Bass, William Thomas
Bowsher, Harry Fred
Boyd, James Emory
Braden, Charles Hosea
Dangle, Richard L
Ezell, Ronnie Lee
Fink, Richard Walter
Love, William Gary
Nelson, William Henry
Owens, William Richard
Palms, John Michael
Prior, Richard Marion
Rao, Pemmaraju Venugopala
Russell, John Lynn, Jr
Steuer, Malcolm F
Wyly, Lemuel David, Jr

Nuclear Structure (cont)

Cochran, Donald Roy Francis
Coon, Sidney Alan
Cooper, Martin David
Cramer, James D
Davey, William George
Dingus, Ronald Shane
Diven, Benjamin Clinton
Doolen, Gary Dean
Drake, Darrell Melvin
Dressel, Ralph William
Dyer, Peggy Lynn
Ensslin, Norbert
Famularo, Kendall Ferris
Farrell, John A
Foster, Duncan Graham, Jr
Fowler, Malcolm McFarland
Friar, James Lewis
Froman, Darol Kenneth
Garvey, Gerald Thomas
Gibson, Benjamin Franklin, V
Ginocchio, Joseph Natale
Glasgow, Dale William
Gordon, James Wylie
Graves, Glen Atkins
Grover, George Maurice
Haight, Robert Cameron
Halbleib, John A
Hardekopf, Robert Allen
Hemmendinger, Arthur
Hopkins, John Chapman
Hull, McAllister Hobart, Jr
Hunter, Raymond Eugene
Hurd, James William
Hurd, Jon Rickey
Jain, Mahavir
Jarmie, Nelson
Jett, James Hubert
Johnson, Mikkel Borlaug
Keaton, Paul W, Jr
Keepin, George Robert, Jr
King, Nicholas S P
Koehler, Dale Roland
Krick, Merlyn Stewart
Lee, David Mallin
Leonard, Ellen Marie
Liu, Lon-Chang
MacArthur, Duncan W
Macek, Robert James
McNally, James Henry
Maggiore, Carl Jerome
Malanify, John Joseph
Manley, John Henry
Marlow, Keith Winton
Mazarakis, Michael Gerassimos
Melgard, Rodney
Mischke, Richard E
Moore, Michael Stanley
Moss, Calvin E
Motz, Henry Thomas
Muir, Douglas William
Neher, Leland K
Nelson, Gerald Clifford
Nicholson, Nicholas
Nickols, Norris Allan
Nix, James Rayford
Obst, Andrew Wesley
Orth, Charles Joseph
Parker, Jack Lindsay
Parker, Winifred Ellis
Peng, Jen-chieh
Perkins, Roger Bruce
Peterson, Robert W
Peterson, Rolf Eugene
Plassmann, Elizabeth Hebb
Plassmann, Eugene Adolph
Poore, Emery Ray Vaughn
Ragan, Charles Ellis, III
Rahal, Leo James
Renken, James Howard
Richter, John Lewis
Robertson, Robert Graham Hamish
Salmi, Ernest William
Sandmeier, Henry Armin
Sattler, Allan R
Schery, Stephen Dale
Schriber, Stanley Owen
Scolman, Theodore Thomas
Scott, Hugh Logan, III
Seagrave, John Dorrington
Seamon, Robert Edward
Shera, E Brooks
Siciliano, Edward Ronald
Sierk, Arnold John
Silbar, Richard R(obert)
Smith, H Vernon, Jr
Smith, Hastings Alexander, Jr
Southward, Harold Dean
Steffen, Rolf Marcel
Stelts, Marion Lee
Summers, Donald Lee
Sunier, Jules Willy
Talbert, Willard Lindley, Jr
Tanaka, Nobuyuki
Tapphorn, Ralph M
Tesmer, Joseph Ransdell
Thiessen, Henry Archer
Veeser, Lynn Raymond
Vieira, David John
Wagner, Robert Thomas
Warner, Laurance Bliss
Watt, Bob Everett
Wildenthal, Bryan Hobson
Yates, Mary Anne
Young, Lloyd Martin

Young, Phillip Gaffney

NEW YORK
Alburger, David Elmer
Auerbach, Elliot H
Baltz, Anthony John
Bennett, Gerald William
Bernabei, Austin M
Bhat, Mulki Radhakrishna
Block, Robert Charles
Blumberg, Leroy Norman
Bond, Albert Haskell, Jr
Bond, Peter Danford
Bozoki, George Edward
Braun-Munzinger, Peter
Brown, Hugh Needham
Brown, Stanley Gordon
Burrows, Thomas Wesley
Carew, John Francis
Casten, Richard Francis
Chasman, Chellis
Chou, Tzi Shan
Choudhury, Deo C
Chrien, Robert Edward
Clark, David Delano
Cline, Douglas
Cokinos, Dimitrios
Corelli, John C
Corson, Dale Raymond
Cumming, James B
Davis, Harold Larue
Deady, Matthew William
der Mateosian, Edward
Devons, Samuel
Dicello, John Francis, Jr
Divadeenam, Mundrathi
Donnelly, Denis Philip
Dreiss, Gerard Julius
Duek, Eduardo Enrique
Eddy, Jerry Kenneth
Engelke, Charles Edward
Favale, Anthony John
Feingold, Arnold Moses
Feldman, Lawrence
Fielder, Douglas Stratton
Foley, Kenneth John
Foreman, Bruce Milburn, Jr
Franco, Victor
Fulbright, Harry Wilks
Ganley, W Paul
Garber, Donald I
Garg, Jagadish Behari
Gavin, Gerard Brennan
Gill, Ronald Lee
Gonzalez, Robert Anthony
Graves, Robert Gage
Hansen, Ole
Harkavy, Allan Abraham
Havens, William Westerfield, Jr
Holbrow, Charles H
Holden, Norman Edward
Horoshko, Roger N
Huizenga, John Robert
Hurwitz, Henry, Jr
Kane, Walter Reilly
Kistner, Ottmar Casper
Kitchen, Sumner Wendell
Koltun, Daniel S
Kostroun, Vaclav O
Kubik, Peter W
Kwok, Hoi S
Lancman, Henry
Lattimer, James Michael
Lee, Linwood Lawrence, Jr
Liebenauer, Paul (Henry)
Lipson, Edward David
Li-Scholz, Angela
Lubitz, Cecil Robert
McGrath, Robert L
McGuire, Stephen Craig
McLane, Victoria
Medicus, Heinrich Adolf
Mendell, Rosalind B
Millener, David John
Min, Kongki
Motz, Lloyd
Muether, Herbert Robert
Mughabghab, Said F
Mukhopadhyay, Nimai Chand
Neeson, John Francis
Olness, John William
Padalino, Stephen John
Parsa, Zohreh
Parsegian, V(ozcan) Lawrence
Paul, Peter
Pearlstein, Sol
Piel, William Frederick
Pile, Philip H
Raboy, Sol
Rafla, Sameer
Reber, Jerry D
Reynolds, John Terrence
Rezanka, Ivan
Roalsvig, Jan Per
Rustgi, Moti Lal
Sailor, Vance Lewis
Schneid, Edward Joseph
Scholz, Wilfried
Schulte, Robert Lawrence
Schwarzschild, Arthur Zeiger
Shakin, Carl M
Shore, Ferdinand John
Silverman, Albert
Souder, Paul A

Sprouse, Gene Denson
Stachel, Johanna
Stoler, Paul
Stroke, Hinko Henry
Sun, Chih-Ree
Thieberger, Peter
Titone, Luke Victor
Trail, Carroll C
Tuli, Jagdish Kumar
Vortuba, Jan
Warburton, Ernest Keeling
Ward, Thomas Edmund
Wegner, Harvey E
Wilson, Fred Lee
Yergin, Paul Flohr
Zahed, Ismail
Zaider, Marco A

NORTH CAROLINA
Adelberger, Rexford E
Biedenharn, Lawrence Christian, Jr
Clator, Irvin Garrett
Clegg, Thomas Boykin
Cleveland, Gregor George
Cobb, Grover Cleveland, Jr
Cotanch, Stephen Robert
Ely, Ralph Lawrence, Jr
Floyd, Carey E, Jr
Good, Wilfred Manly
Gould, Christopher Robert
Hubbard, Paul Stancyl, Jr
Jaszczak, Ronald Jack
Kalbach, Constance
Lewis, Harold Walter
Lindsay, James Gordon, Jr
Ludwig, Edward James
Mitchell, Gary Earl
Oberhofer, Edward Samuel
Park, Jae Young
Roberson, Nathan Russell
Rogosa, George Leon
Rolland, William Woody
Tilley, David Ronald
Waldman, Bernard
Walker, William Delany
Walter, Richard L
Waltner, Arthur
Weller, Henry Richard

OHIO
Agard, Eugene Theodore
Anderson, Bryon Don
Becker, Lawrence Charles
Besancon, Robert Martin
Brient, Charles E
Dillman, Lowell Thomas
Dollhopf, William Edward
Finlay, Roger W
Giamati, Charles C, Jr
Grimes, Steven Munroe
Hagee, George Richard
Hancock, George Whitmore, Jr
Hathaway, Charles Edward
Hausman, Hershel J
Hemsky, Joseph William
Jalbert, Jeffrey Scott
Jastram, Philip Sheldon
Jha, Shacheenatha
Julian, Glenn Marcenia
Kepes, Joseph John
Klingensmith, Raymond W
Lane, Raymond Oscar
Lemming, John Frederick
MacIntyre, William James
Madey, Richard
Madia, William J
Mantil, Joseph Chacko
Mayers, Richard Ralph
Ploughe, William D
Poth, James Edward
Priest, Joseph Roger
Rapaport, Jacobo
Ruegsegger, Donald Ray, Jr
Schneider, Ronald E
Segelken, Warren George
Seyler, Richard G
Sugarbaker, Evan Roy
Sund, Raymond Earl
Warner, Robert Edson
Winters, Ronald Ross
Wood, Galen Theodore
Wright, Louis E

OKLAHOMA
Bell, Robert Edward
Buchanan, Ronnie Joe
Cook, Charles Falk
Huffaker, James Neal
Major, John Keene
Petry, Robert Franklin
Polson, William Jerry
Souder, Wallace William

OREGON
Easterday, Harry Tyson
Kelley, Raymond H
Krane, Kenneth Saul
Lefevre, Harlan W
McDaniels, David K
Madsen, Victor Arviel
Mather, Keith Benson
Orloff, Jonathan H
Overley, Jack Castle
Richert, Anton Stuart

Schecter, Larry
Siemens, Philip John
Wack, Paul Edward
Wetzel, Karl Joseph
White, Donald Harvey

PENNSYLVANIA
Ajzenberg-Selove, Fay
Balamuth, David P
Barnes, Peter David
Bartko, John
Bilaniuk, Oleksa-Myron
Bock, Charles Walter
Cohen, Leonard David
Connors, Donald R
Couch, Jack Gary
Daehnick, Wilfried W
Donoghue, Timothy R
Doub, William Blake
Dressler, Edward Thomas
Emmerich, Werner Sigmund
Feldmeier, Joseph Robert
Folk, Robert Thomas
Fortune, H Terry
Freed, Norman
Georgopulos, Peter Demetrios
Graetzer, Reinhard
Green, Lawrence
Guttmann, Mark
Hardy, Judson, Jr
Haskins, Joseph Richard
Homsey, Robert John
Kahler, Albert Comstock, III
Keller, Eldon Lewis
Laws, Priscilla Watson
Lippincott, Ezra Parvin
Lochstet, William A
Luce, Robert James
Middleton, Roy
Mooney, Edward, Jr
Ostrander, Peter Erling
Pacer, John Charles
Pinkerton, John Edward
Poss, Howard Lionel
Pratt, William Winston
Rehfield, David Michael
Rhodes, Jacob Lester
Saladin, Jurg X
Sarram, Mehdi
Shwe, Hla
Siems, Norman Edward
Smith, Winfield Scott
Snedegar, William H
Sorensen, Raymond Andrew
Stallwood, Robert Antony
Thwaites, Thomas Turville
Van Patter, Douglas Macpherson
Walkiewicz, Thomas Adam
Wicker, Everett E
Zurmuhle, Robert W

RHODE ISLAND
Lall, Prithvi C
Levin, Frank S
Mecca, Stephen Joseph
Riesenfeld, Peter William

SOUTH CAROLINA
Aull, Luther Bachman, III
Avignone, Frank Titus, III
Baumann, Norman Paul
Benjamin, Richard Walter
Blanpied, Gary Stephen
Brantley, William Henry
Edge, Ronald (Dovaston)
Hahn, Walter I
Hendrick, Lynn Denson
O'Neill, George Francis
Parks, Paul Blair
Preedom, Barry Mason
Roggenkamp, Paul Leonard
Saunders, Edward A
Turner, James David

SOUTH DAKOTA
Hein, Warren Walter

TENNESSEE
Akovali, Yurdanur A
Albridge, Royal
Alexeff, Igor
Appleton, B R
Bair, Joe Keagy
Ball, James Bryan
Beene, James Robert
Bemis, Curtis Elliot, Jr
Bertrand, Fred Edmond
Bingham, Carrol R
Blankenship, James Lynn
Burton, John Williams
Carter, Hubert Kennon
Collins, Warren Eugene
Davies, Kenneth Thomas Reed
De Saussure, Gerard
Dickens, Justin Kirk
Ferguson, Robert Lynn
Garrett, Jerry Dale
Gove, Norwood Babcock
Greenbaum, Elias
Gross, Edward Emanuel
Halbert, Melvyn Leonard
Hamilton, Joseph H, Jr
Harvey, John Arthur
Horen, Daniel J

Green, Barry A
Harris, Hugh Courtney
Haynes, Munro K
Hill, Henry Allen
Hilliard, Ronnie Lewis
Hirleman, Edwin Daniel, Jr
Hopf, Frederic A
Jacobson, Michael Ray
Lund, Mark Wylie
McDaniel, Terry Wayne
McKenney, Dean Brinton
Macleod, Hugh Angus
Marathay, Arwind Shankar
Mayall, Nicholas Ulrich
Palmer, James McLean
Peyghambarian, Nasser
Rasmussen, William Otto
Sargent, Murray, III
Sarid, Dror
Scholl, Marija Strojnik
Schowengerdt, Robert Alan
Shack, Roland Vincent
Shannon, Robert Rennie
Shoemaker, Richard Lee
Slater, Philip Nicholas
Smith, David John
Stamm, Robert Franz
Sutton, George W(alter)
Turner, Arthur Francis
Walker, Charles Thomas
Wing, William Hinshaw
Wolfe, William Louis, Jr
Wyant, James Clair

ARKANSAS
Eichenberger, Rudolph John
Gupta, Rajendra
Horton, Philip Bish
Leming, Charles William
Richardson, Charles Bonner
Salamō, Gregory Joseph
Singh, Surendra Pal
Smith, E(astman)

CALIFORNIA
Adams, Arnold Lucian
Ahumada, Albert Jil, Jr
Ameer, George Albert
Ammann, E(ugene) O(tto)
Ashby, Val Jean
Asmus, John Fredrich
Austin, Roswell W(allace)
Avizonis, Petras V
Babrov, Harold J
Baez, Albert Vinicio
Bagby, John P(endleton)
Bailey, Ian L
Bajaj, Jagmohan
Bardsley, James Norman
Barry, James Dale
Bartling, Judd Quentin
Baur, James Francis
Becker, Randolph Armin
Beer, Reinhard
Behringer, Robert Ernest
Bennett, Glenn Taylor
Bennett, Harold Earl
Bennett, Jean McPherson
Bernard, Douglas Alan
Bethune, Donald Stimson
Billings, Bruce Hadley
Billman, Kenneth William
Bjorklund, Gary Carl
Bliss, Erlan S
Bloom, Arnold Lapin
Bloxham, Laurence Hastings
Botez, Dan
Boynton, Robert M
Bracewell, Ronald Newbold
Breckinridge, James Bernard
Bridges, William B
Burge, Dennis Knight
Caird, John Allyn
Cannell, David Skipwith
Carman, Robert Lincoln, Jr
Carver, John Guill
Caves, Carlton Morris
Ceglio, Natale Mauro
Cho, Young-chung
Chu, Steven
Cooper, Alfred William Madison
Cooper, Donald Edward
Coufal, Hans-Jürgén
Cover, Ralph A
Crittenden, Eugene Casson, Jr
Danielson, George Edward, Jr
Davies, Merton Edward
Davis, James Ivey
Davis, Jeffrey Arthur
Davis, Thomas Pearse
Deacon, David A G
Deckert, Curtis Kenneth
Dember, Alexis Berthold
Dessel, Norman F
Donovan, Terence M
Dowling, Jerome M
Doyle, Walter M
Duffield, Jack Jay
Eckstrom, Donald James
Edwards, David Franklin
Elliott, Stuart Bruce
Enoch, Jay Martin
Erdmann, John Hugo
Errett, Daryl Dale

Evtuhov, Viktor
Fahlen, Theodore Stauffer
Fairand, Barry Philip
Farmer, Crofton Bernard
Feichtner, John David
Feiock, Frank Donald
Feit, Michael Dennis
Fenner, Wayne Robert
Fisher, Philip Chapin
Fisher, Robert Alan
Fleck, Joseph Amadeus, Jr
Fouquet, Julie
Frank, Alan M
Fried, David L
Gamo, Hideya
Gara, Aaron Delano
Gelbwachs, Jerry A
Geller, Myer
Gilmartin, Thomas Joseph
Gimlett, James I
Glass, Alexander Jacob
Glass, Nathaniel E
Goell, James E(manuel)
Goldsborough, John Paul
Goodman, Joseph Wilfred
Goosman, David R
Graham, Robert (Klark)
Green, Philip S
Grossman, Jack Joseph
Gruber, John B
Gudmundsen, Richard Austin
Gundersen, Martin Adolph
Hadeishi, Tetsuo
Halden, Frank
Hall, James Timothy
Han, Ki Sup
Harkins, Carl Girvin
Harris, Dennis George
Haugen, Gilbert R
Hellwarth, Robert Willis
Helstrom, Carl Wilhelm
Henderson, David Michael
Henning, Harley Barry
Hesselink, Lambertus
Hilbert, Robert S
Hodges, Dean T, Jr
Hoffman, Hanna J
Holmes, Dale Arthur
Holzrichter, John Frederick
Honey, Richard Churchill
Hopkins, George William, II
Hsia, Yukun
Huang, C Yuan
Hudson, Richard Delano, Jr
Hunt, Arlon Jason
Hutchin, Richard Ariel
Jacobs, Michael Moises
Jacobs, Ralph R
Janney, Gareth Maynard
Jansen, Michael
Johnston, Alan Robert
Johnston, George Taylor
Judd, Floyd L
Jusinski, Leonard Edward
Kahn, Frederic Jay
Kapany, Narinder Singh
Karp, Arthur
Karunasiri, Gamani
Kelly, Raymond Leroy
Khoshnevisan, Mohsen Monte
Kino, Gordon Stanley
Kirk, John Gallatin
Knight, Gordon Raymond
Knize, Randall James
Koehler, Wilbert Frederick
Krishnan, Kamala Sivasubramaniam
Kulander, Kenneth Charles
Lakkaraju, H S
Lampert, Carl Matthew
Lapp, M(arshall)
Larmore, Lewis
Leavy, Paul Matthew
Lee, Peter H Y
Lee, Wai-Hon
Lehovec, Kurt
Levenson, Marc David
Li, Chia-Chuan
Lieber, Richard L
Linford, Gary Joe
Liu, Hua-Kuang
Liu, Jia-ming
Lombardi, Gabriel Gustavo
Lurie, Norman A(lan)
Lynch, David Dexter
McFarlane, Ross Alexander
McFee, Raymond Herbert
McKenzie, Robert Lawrence
McQuillen, Howard Raymond
Maker, Paul Donne
Manes, Kennneth Rene
Martin, Terry Zachry
Massey, Gail Austin
Massie, N A (Bert)
Mathis, Ronald Floyd
Maydan, Dan
Medved, David Bernard
Meier, Rudolf H
Meinel, Aden Baker
Meinel, Marjorie Pettit
Menzies, Robert Thomas
Merritt, Thomas Parker
Meyers, Robert Allen
Mohanty, Nirode C
Morris, James Russell

Morris, Richard Herbert
Moss, Steven C
Mueller, James Lowell
Muller, Richard A
Muller, Rolf Hugo
Mundie, Lloyd George
Murray, John Roberts
Nanes, Roger
Newman, Roger
Nicodemus, Fred(erick) E(dwin)
Noble, Robert Hamilton
O'Keefe, John Dugan
O'Loane, James Kenneth
Palatnick, Barton
Pan, Yu-Li
Park, Edward C(ahill), Jr
Partanen, Jouni Pekka
Passenheim, Burr Charles
Pavlopoulos, Theodore G
Perry, Richard Lee
Perry, Robert Nathaniel, III
Pertica, Alexander José
Pignataro, Augustus
Pines, Alexander
Porteus, James Oliver
Powers, John Patrick
Randall, Charles McWilliams
Randle, Robert James
Rast, Howard Eugene, Jr
Rawson, Eric Gordon
Reinheimer, Julian
Rennilson, Justin J
Reynolds, Michael David
Rice, Dennis Keith
Richman, Isaac
Rickard, James Joseph
Robben, Franklin Arthur
Rockower, Edward Brandt
Rockwell, David Alan
Rothrock, Larry R
Russell, Philip Boyd
Saito, Theodore T
Salanave, Leon Edward
Saltzman, Max
Sandefur, Kermit Lorain
Satten, Robert A
Schaefer, Albert Russell
Schawlow, Arthur Leonard
Schmars, William Thomas
Schulte, Daniel Herman
Scifres, Donald Ray
Sclar, Nathan
Senitzky, Israel Ralph
Seppala, Lynn G
Shank, Charles Vernon
Shelby, Robert McKinnon
Sheridon, Nicholas Keith
Sherman, George Charles
Shykind, David
Siegman, A(nthony) E(dward)
Silverstein, Elliot Morton
Smiley, Vern Newton
Smith, Sheldon Magill
Smith, Warren James
Soffer, Bernard Harold
Speck, David Ralph
Speen, Gerald Bruce
Spencer, Donald Jay
Spiro, Irving J
Spitzer, William George
Sprague, Robert Arthur
Stapelbroek, Maryn G
Starkweather, Gary Keith
Stavroudis, Orestes Nicholas
Stegelmann, Erich J
Stierwalt, Donald L
Stokowski, Stanley E
Strand, Timothy Carl
Streifer, William
Suchard, Steven Norman
Sziklai, George C(lifford)
Talley, Robert Morrell
Tam, Andrew Ching
Tarbell, Theodore Dean
Terhune, Robert William
Titterton, Paul James
Tooley, Richard Douglas
Treves, David
Tricoles, Gus P
Trigger, Kenneth Roy
Trolinger, James Davis
Tubbs, Eldred Frank
Unti, Theodore Wayne Joseph
Urbach, John C
Vance, Dennis William
Wada, James Yasuo
Wagner, Richard John
Wallerstein, Edward Perry
Wang, Jon Y
Warren, Walter R(aymond), Jr
Watkins, Robert Arnold
Waugh, John Blake-Steele
Webber, Donald Salyer
Weber, Marvin John
Webster, Emilia
Weinstein, Berthold Werner
Weissbluth, Mitchel
Wells, Willard H
Welsh, David Edward
Whitcomb, Stanley Ernest
Whitefield, Rodney Joe
Whitney, William Merrill
Wieder, Harold
Wiedow, Carl Paul

Winter, Donald Charles
Wittry, David Beryle
Woerner, Robert Leo
Wolff, Milo Mitchell
Wunderman, Irwin
Wyatt, Philip Joseph
Yamakawa, Kazuo Alan
Yap, Fung Yen
Yariv, A(mnon)
Yeh, Cavour W
Yeh, Edward H Y
Yeh, Pochi Albert
Yeh, Yea-Chuan Milton
Young, Andrew Tipton
Young, James Forrest
Zhou, Simon Zheng
Zook, Alma Claire

COLORADO
Baur, Thomas George
Bruns, Donald Gene
Cathey, Wade Thomas, Jr
Cerni, Todd Andrew
Chang, Bunwoo Bertram
Chen, Di
Chisholm, James Joseph
Churnside, James H
Clifford, Steven Francis
Dana, Robert Watson
Daw, Glen Harold
Dichtl, Rudolph John
Eberhard, Wynn Lowell
Emery, Keith Allen
Fairbank, William Martin, Jr
Galeener, Frank Lee
Griboval, Paul
Hadley, Lawrence Nathan
Hjelme, Dag Roar
Itano, Wayne Masao
Jewell, Jack Lee
Johnson, Eric G, Jr
Kalma, Arne Haerter
Lawton, Robert Arthur
Lazzarini, Albert John
Lightsey, Paul Alden
McMahon, Thomas Joseph
Mankin, William Gray
Moddel, Garret R
O'Callaghan, Michael James
O'Sullivan, William John
Owen, Robert Barry
Phelan, Robert J, Jr
Sanmann, Everett Eugene
Sauer, Jon Robert
Schwiesow, Ronald Lee
She, Chiao-Yao
Sime, David Gilbert
Smith, Archibald William
Smith, Richard Cecil
Smith, Stephen Judson
Young, Matt

CONNECTICUT
Akkapeddi, Prasad Rao
Antar, Ali A
Astheimer, Robert W
Bartram, Ralph Herbert
Brienza, Michael Joseph
Cable, Peter George
Cheo, Peter K
Crawford, John Okerson
DeMaria, Anthony John
Dolan, James F
Dreyfus, Marc George
Erf, Robert K
Glenn, William Henry, Jr
Green, Eugene L
Haacke, Gottfried
Harrison, Irene R
Hufnagel, Robert Ernest
Hyde, Walter Lewis
Javidi, Bahram
Keith, H(arvey) Douglas
Leonberger, Frederick John
Montgomery, Anthony John
Morrison, Richard Charles
Nath, Dilip K
O'Brien, Brian
Pinsley, Edward Allan
Poultney, Sherman King
Rich, John Charles
Roychoudhuri, Chandrasekhar
Siegmund, Walter Paul
Snyder, John William
Stetson, Karl Andrew
Walker, Marshall John
Yoder, Paul Rufus, Jr

DELAWARE
Bierlein, John David
Daniels, William Burton
Garland, Charles E
Glasser, Leo George
Gulick, Walter Lawrence
Holland, Russell Sedgwick
Huppe, Francis Frowin
Ih, Charles Chung-Sen
Jansson, Peter Allan
Klemas, Victor V
Pontrelli, Gene J
Ross, William D(aniel)
Sharnoff, Mark

LaHaie, Ivan Joseph
LaRocca, Anthony Joseph
Leith, Emmett Norman
Miller, Herman Lunden
Montgomery, George Paul, Jr
Nazri, Gholam-Abbas
Peterson, Lauren Michael
Peterson, Richard Carl
Phelps, Frederick Martin, III
Prostak, Arnold S
Rand, Stephen Colby
Ressler, Neil William
Rimai, Lajos
Roessler, David Martyn
Sawatari, Takeo
Schrenk, Walter John
Segall, Stephen Barrett
Sharp, William Edward, III
Trentelman, George Frederick
Upatnieks, Juris
Vaishnava, Prem P
Van Baak, David Alan
Weil, Herschel
Weinberger, Doreen Anne
Yang, Wen Jei
Zissis, George John

MINNESOTA
Beck, Warren R(andall)
Carlson, Frederick Paul
French, William George
Heinisch, Roger Paul
Hewitt, Frederick George
Hill, Brian Kellogg
Hocker, George Benjamin
Johnson, Edgar Gustav
Kruse, Paul Walters, Jr
Lee, Pui Kum
Lo, David S(hih-Fang)
McGlauchlin, Laurence D(onald)
Mikkelson, Raymond Charles
Nelson, Kyler Fischer
Ready, John Fetsch
Robinson, Glen Moore, III
Valasek, Joseph
Valley, Leonard Maurice
Zeyen, Richard John

MISSISSIPPI
Breazeale, Mack Alfred
Cibula, William Ganley
Ferguson, Joseph Luther, Jr

MISSOURI
Anderson, Richard Alan
Bryan, David A
Leader, John Carl
Leader, John Carl
Leopold, Daniel J
Linder, Solomon Leon
Liu, Yu
Look, Dwight Chester, Jr
Palmer, Kent Friedley
Rigler, A Kellam
Schmitt, John Leigh
Walker, James Harris

MONTANA
Carlsten, John Lennart
Tynes, Arthur Richard

NEBRASKA
Woods, Joseph
Zepf, Thomas Herman

NEVADA
Dunipace, Donald William
Farley, John William

NEW HAMPSHIRE
Adjemian, Haroutioon
Baker, James Gilbert
Kidder, John Newell
King, Allen Lewis
Meyer, James Wagner
Perovich, Donald Kole
Smith, F(rederick) Dow(swell)

NEW JERSEY
Abeles, Joseph Hy
Andrekson, Peter A(vo)
Anthony, Philip John
Barrett, Joseph John
Bartolini, Robert Alfred
Bechis, Dennis John
Bechtle, Daniel Wayne
Behrens, Herbert Ernest
Berreman, Dwight Winton
Biswas, Dipak R
Bjorkholm, John Ernst
Bonin, Keith Donald
Boyd, Gary Delane
Broer, Matthijs Meno
Brucker, Edward Byerly
Burrus, Charles Andrew, Jr
Capasso, Federico
Carlson, Curtis Raymond
Castor, William Stuart, Jr
Celler, George K
Chang, Tao-Yuan
Channin, Donald Jones
Chraplyvy, Andrew R
Chung, Yun C
Church, Eugene Lent

Cody, George Dewey
Collett, Edward
Collier, Robert Jacob
Deri, Robert Joseph
Duclos, Steven J
Dutta, Mitra
Edwards, Christopher Andrew
Field, Norman J
Fischer, Russell Jon
Fleury, Paul A
Giordmaine, Joseph Anthony
Glass, Alastair Malcolm
Goldstein, Robert Lawrence
Gottscho, Richard Alan
Graf, Hans Peter
Gray, Russell Houston
Gualtieri, Devlin Michael
Hammer, Jacob Meyer
Hartman, Richard Leon
Hernqvist, Karl Gerhard
Hubbard, William Marshall
Hutter, Edwin Christian
Hwang, Cherng-Jia
Johnson, Leo Francis
Joyce, William B(axter)
Kaminow, Ivan Paul
Kaprelian, Edward Karnig
Kash, Kathleen
Kluver, J(ohan) W(ilhelm)
Knausenberger, Wulf H
Knight, Douglas Maitland
Koch, Thomas L
Koester, Charles John
Kolodner, Paul R
Kornstein, Edward
Koss, Valery Alexander
Levine, Barry Franklin
Li, Tingye
Liao, Paul Foo-Hung
Linke, Richard Alan
McKenna, James
Medley, Sidney S
Meyerhofer, Dietrich
Miles, Richard Bryant
Miller, Arthur
Miller, David A B
Mollenauer, Linn F
Morrow, Scott
Moshey, Edward A
Murnick, Daniel E
Nahory, Robert Edward
O'Gorman, James
Orlando, Carl
Palladino, Richard Walter
Partovi, Afshin
Passner, Albert
Personick, Stewart David
Pinczuk, Aron
Ramsey, Alan T
Rastani, Kasra
Raybon, Gregory
Rothberg, Lewis Josiah
Rubinstein, Charles B(enjamin)
Ruderman, Irving Warren
Schuler, Mathias John
Schulte, Harry John, Jr
Schumer, Douglas B
Shand, Michael Lee
Siebert, Donald Robert
Silfvast, William Thomas
Smith, Peter William E
Staebler, David Lloyd
Stiles, Lynn F, Jr
Stone, Julian
Taylor, Gary
Thomson, Michael George Robert
Thurston, Robert Norton
Todd, Terry Ray
Tomaselli, Vincent Paul
Tomlinson, Walter John, III
Treu, Jesse Isaiah
Whitman, Gerald Martin
Wiesenfeld, Jay Martin
Wittenberg, Albert M
Wittke, James Pleister
Wolff, Peter A
Wood, Thomas H
Woodward, Ted K
Wullert, John R, II
Xie, Ya-Hong
Yablonovitch, Eli
Yi-Yan, Alfredo
Yurke, Bernard
Zoltan, Bart Joseph

NEW MEXICO
Ackerhalt, Jay Richard
Anderson, L(awrence) K(eith)
Balog, George
Bean, Brent Leroy
Becker, Wilhelm
Bellum, John Curtis
Bieniewski, Thomas M
Bolie, Victor Wayne
Brannon, Paul J
Brueck, Steven Roy Julien
Bryant, Howard Carnes
Cahill, Paul A
Chen, Tuan Wu
Chow, Weng Wah
Chylek, Petr
Clark, Wallace Thomas, III
Czuchlewski, Stephen John
Depatie, David A

Devaney, Joseph James
Dumas, Herbert M, Jr
Farmer, William Michael
Flicker, Herbert
Gerardo, James Bernard
Gerstl, Siegfried Adolf Wilhelm
Giles, Michael Kent
Gourley, Paul Lee
Greiner, Norman Roy
Hahn, Yu Hak
Hanson, Kenneth Merrill
Hargis, Philip Joseph, Jr
Hessel, Kenneth Ray
Hill, Ronald Ames
Hillsman, Matthew Jerome
Hopkins, Alan Keith
Jahoda, Franz C
Jamshidi, Mohammad Mo
Johnston, Roger Glenn
Jones, Eric Daniel
Judd, O'Dean P
Kopp, Roger Alan
Krabec, Charles Frank, Jr
La Delfe, Peter Carl
Ladish, Joseph Stanley
Land, Cecil E(lvin)
Leland, Wallace Thompson
Loree, Thomas Robert
Lyo, SungKwun Kenneth
McInerney, John Gerard
Mansfield, Charles Robert
Mauro, Jack Anthony
Mueller, Marvin Martin
Nereson, Norris (George)
Osinski, Marek Andrzej
Overhage, Carl F J
Parker, Joseph R(ichard)
Peercy, Paul S
Piltch, Martin Stanley
Quigley, Gerard Paul
Reichert, John Douglas
Rice, James Kinsey
Rigrod, William W
Robinson, C Paul
Schappert, Gottfried T
Seagrave, John Dorrington
Sheffield, Richard Lee
Shively, Frank Thomas
Small, James Graydon
Snider, Donald Edward
Solem, Johndale Christian
Sollid, Jon Erik
Sorem, Michael Scott
Stone, Sidney Norman
Stratton, Thomas Fairlamb
Sze, Robert Chia-Ting
Tanaka, Nobuyuki
Telle, John Martin
Tsao, Jeffrey Yeenien
Turner, Leaf
Veeser, Lynn Raymond
Watt, Bob Everett
Wenzel, Robert Gale
Wiggins, Carl M
Yarger, Frederick Lynn
York, George William
Zardecki, Andrew

NEW YORK
Agrawal, Govind P(rasad)
Alfano, Robert R
Axelrod, Norman Nathan
Ballantyne, Joseph M(errill)
Barrekette, Euval S
Becker, Kurt Heinrich
Berman, Paul Ronald
Bernstein, Burton
Billmeyer, Fred Wallace, Jr
Blaker, J Warren
Blazey, Richard N
Blumenthal, Ralph Herbert
Borrelli, Nicholas Francis
Boyd, Robert William
Breed, Henry Eltinge
Brehm, Lawrence Paul
Breneman, Edwin Jay
Brink, Gilbert Oscar
Brock, Robert H, Jr
Brody, Burton Alan
Burgmaier, George John
Carleton, Herbert Ruck
Chen, B(enjamin) T(eh-Kung)
Chen, Ying-Chih
Cheung, Lim H
Chew, Herman W
Chiang, Fu-Pen
Choy, Daniel S J
Cohen, Martin Gilbert
Coleman, John Howard
Costa, Lorenzo F
Craxton, Robert Stephen
Crossmon, Germain Charles
Cummins, Herman Z
Cunningham, Michael Paul
Cupery, Kenneth N
Cusano, Dominic A
Delano, Erwin
Denk, Ronald H
DePalma, James John
Diamond, Fred Irwin
Doyle-Feder, Donald Perry
Dreyfus, Russell Warren
Duggin, Michael J
Egan, Walter George

Engelke, Charles Edward
Eyer, James Arthur
Fitchen, Douglas Beach
Folan, Lorcan Michael
Forsyth, James M
Foster, Kenneth William
Friedman, Helen Lowenthal
Gajjar, Jagdish T(rikamji)
Ganley, W Paul
Gelman, Donald
George, Nicholas
Gilmour, Hugh Stewart Allen
Goble, Alfred Theodore
Good, William E
Greenebaum, Michael
Grischkowsky, Daniel Richard
Hall, Dennis Gene
Hamblen, David Philip
Hanau, Richard
Harris, Jack Kenyon
Haus, Joseph Wendel
Herman, Irving Philip
Ho, Ping-Pei
Hoffman, Robert
Howard, Richard John
Howe, Dennis George
Hui, Sek Wen
Inhaber, Herbert
Jacobsen, Chris J
Jacobsen, Edward Hastings
Johnson, Brant Montgomery
Jupnik, Helen
Keck, Donald Bruce
Kelly, John Henry
Kermisch, Dorian
Khanna, Shyam Mohan
Kiang, Ying Chao
King, Marvin
Kingslake, Rudolf
Kinzly, Robert Edward
Kirtley, John Robert
Kirz, Janos
Kriss, Michael Allen
Kuehler, Jack D
Kurtz, Clark N
Lamberts, Robert L
LaMuth, Henry Lewis
La Russa, Joseph Anthony
Lax, Melvin
Lean, Eric Gung-Hwa
Lee, Yung-Chang
LeMay, Charlotte Zihlman
Levine, Alfred Martin
Levinson, Steven R
Lewis, Alan Laird
Lipson, Edward David
Litynski, Daniel Mitchell
Liu, Yung Sheng
Loy, Michael Ming-Tak
Lynk, Edgar Thomas
MacAdam, David Lewis
McCamy, Calvin S
McKinstrie, Colin J(ohn)
McMahon, Donald Howland
Mahler, David S
Makous, Walter L
Maldonado, Juan Ramon
Manassah, Jamal Tewfek
Mandel, Leonard
Marchand, Erich Watkinson
Marcus, Michael Alan
Mattar, Farres Phillip
Matulic, Ljubomir Francisco
Melcher, Robert Lee
Mendez, Emilio Eugenio
Metcalf, Harold
Millikan, Allan G
Milne, Gordon Gladstone
Moe, George Wylbur
Moore, Duncan Thomas
Morey, Donald Roger
Morris, George Michael
Nicolosi, Joseph Anthony
Nyyssonen, Diana
Ockman, Nathan
Oliner, Arthur A(aron)
Owens, James Carl
Parks, Harold George
Pearson, George John
Penney, Carl Murray
Pernick, Benjamin J
Persans, Peter D
Peterson, Otis G
Piech, Kenneth Robert
Pike, John Nazarian
Prasad, Paras Nath
Reddy, Reginald James
Remo, John Lucien
Robinson, David Zav
Robinson, Edward J
Roetling, Paul G
Rosenberg, Paul
Rosenberg, Robert
Rustgi, Om Prakash
Saunders, Burt A
Sayre, David
Scarl, Donald B
Schroeder, John
Scidmore, Wright H(arwood)
Sen, Ujjal
Setchell, John Stanford, Jr
Shaw, Rodney
Shenker, Martin
Shepherd, Joseph Emmett

Howard-Lock, Helen Elaine
Jones, Alister Vallance
King, Gerald Wilfrid
Lit, John Wai-Yu
Mandelis, Andreas
Measures, Raymond Massey
Moskovits, Martin
Nilson, John Anthony
Racey, Thomas James
Remole, Arnulf
Robertson, Alan Robert
Shepherd, Gordon Greeley
Sherman, Norman K
Stoicheff, Boris Peter
Tam, Wing-Gay
Vanderkooy, John
Van Driel, Henry Martin
Yevick, David Owen

QUEBEC
Arsenault, Henri H
Boivin, Alberic
Borra, Ermanno Franco
Bose, Tapan Kumar
Chin, See Leang
Delisle, Claude
Gagne, Jean-Marie
Ghosh, Sanjib Kumar
Izquierdo, Ricardo
Jordan, Byron Dale
Lessard, Roger Alain
Maciejko, Roman
Meunier, Michel
Richard, Claude
Schwelb, Otto
Subramanian, Sesha
Waksberg, Armand L

OTHER COUNTRIES
Beran, Mark Jay
Bryngdahl, Olof
Cardona, Manuel
Eisner, Edward
Eng, Sverre T(horstein)
Friesem, Albert Asher
Gordon, Jeffrey Miles
Hansch, Theodor Wolfgang
Kafri, Oded
Kildal, Helge
Kyle, Thomas Gail
Lueder, Ernst H
Malacara, Daniel
Marom, Emanuel
Martin, William Eugene
Nussenzveig, Herch Moyses
Nygaard, Kaare Johann
Rozzi, Tullio
Sirohi, Rajpal Singh
Steffen, Juerg
Thiel, Frank L(ouis)
Unger, Hans-Georg
Ushioda, Sukekatsu
Weil, Raoul Bloch

Physics, General

ALABAMA
Alexander, Chester, Jr
Bartlett, James Holly
Boardman, William Jarvis
Budenstein, Paul Philip
Carr, Howard Earl
Cooper, John Raymond
Curott, David Richard
Dalins, Ilmars
Decher, Rudolf
French, John Donald
Fullerton, Larry Wayne
Govil, Narendra Kumar
Harvey, Stephen Craig
Jones, Robert William
Jones, Stanley Tanner
Kaylor, Hoyt McCoy
Korsch, Dietrich G
MacRae, Robert Alexander
Mishra, Satya Narayan
Morton, Perry Wilkes, Jr
Parnell, Thomas Alfred
Passino, Nicholas Alfred
Pearson, Colin Arthur
Reid, William James
Rodgers, Aubrey
Roe, James Maurice, Jr
Rosenblum, William M
Rowell, Neal Pope
Speer, Fridtjof Alfred
Stewart, Dorathy Anne
Stuhlinger, Ernst
Sulentic, Jack William
Tan, Arjun
Tandberg-Hanssen, Einar Andreas
Tohver, Hanno Tiit
Walker, William Waldrum
Wolin, Samuel
Young, John H

ALASKA
Bates, Howard Francis
Johnson, Ralph Sterling, Jr
Martins, Donald Henry
Morack, John Ludwig

ARIZONA
Babcock, Clarence Lloyd
Bedwell, Thomas Howard
Blitzer, Leon
Boettner, Edward Alvin
Broadfoot, Albert Lyle
Call, Reginald Lessey
Christy, James Walter
Cowley, John Maxwell
Curtis, David William
Dahl, Randy Lynn
Dunkelman, Lawrence
Evans, Robley D(unglison)
Foss, Martyn (Henry)
Franz, Otto Gustav
Gates, Halbert Frederick
Gregory, Brooke
Gullikson, Charles William
Hartig, Elmer Otto
Hendrickson, Lester Ellsworth
Hodges, Carl Norris
Hoffmann, William Frederick
Hood, Lonnie Lamar
Hurt, James Edward
Iijima, Sumio
Jacobs, Stephen Frank
Jacobson, Michael Ray
Kebler, Richard William
Kessler, John Otto
Kevane, Clement Joseph
Lamb, George Lawrence, Jr
Layton, Richard Gary
Macleod, Hugh Angus
Marzke, Robert Franklin
Meieran, Eugene Stuart
O'Leary, Brian Todd
Pokorny, Gerold E(rwin)
Rassweiler, Merrill (Paul)
Robson, John William
Sarid, Dror
Scholl, Marija Strojnik
Senitzky, Benjamin
Smith, Harvey Alvin
Stark, Royal William
Swindle, Timothy Dale
Tillery, Bill W
Treat, Jay Emery, Jr
Tsong, Ignatius Siu Tung
Turner, Arthur Francis
Wangsness, Roald Klinkenberg
Weaver, Albert Bruce
Willis, William Russell
Wilska, Alvar P
Yelle, Roger V
Young, Robert A

ARKANSAS
Bond, Robert Levi
Eichenberger, Rudolph John
Engle, Paul Randal
Horton, Philip Bish
Leming, Charles William
McCarty, Clark William
Mackey, James E
Mink, Lawrence Albright
Prince, Denver Lee
Richardson, Charles Bonner
Rollefson, Aimar Andre
Sharrah, Paul Chester
Wild, Wayne Grant
Zinke, Otto Henry

CALIFORNIA
Abers, Ernest S
Abraham, Farid Fadlow
Abrams, Richard Lee
Adler, Ronald John
Agnew, Harold Melvin
Aitken, Donald W, Jr
Alfven, Hannes Olof Gosta
Allen, Lew, Jr
Allen, Matthew Arnold
Alvarez, Raymond Angelo, Jr
Ames, Lawrence Lowell
Anderson, Carl David
Anderson, Charles Hammond
Anderson, John Thomas
Anderson, Milo Vernette
Anderson, Victor Charles
Anderson, Weston Arthur
Anspaugh, Bruce Edward
Archer, Douglas Harley
Armstrong, Baxter Hardin
Armstrong, Donald B
Arnold, James S(loan)
Ashby, Val Jean
Assmus, Alexi Josephine
Atchison, F(red) Stanley
Atchley, Anthony A
Attwood, David Thomas, Jr
Augenstein, Bruno (Wilhelm)
Ayres, Wesley P
Bacher, Robert Fox
Backus, John (Graham)
Badash, Lawrence
Baez, Albert Vinicio
Baggerly, Leo L
Bajorek, Christopher Henry
Bakun, William Henry
Ballam, Joseph
Bangerter, Roger Odell
Banner, David Lee
Bar-Cohen, Yoseph
Barker, William Alfred

Barr, Frank T(homas)
Barrett, Paul Henry
Bars, Itzhak
Bartling, Judd Quentin
Bate, George Lee
Baum, Dennis Willard
Baum, Peter Joseph
Baumeister, Philip Werner
Bechtel, James Harvey
Behravesh, Mohamad Martin
Behringer, Robert Ernest
Bell, Alan Edward
Bell, Graydon Dee
Bennett, Ralph Decker
Bernhardt, Anthony F
Bershader, Daniel
Bhaumik, Mani Lal
Billman, Kenneth William
Binder, Daniel
Bingham, Harry H, Jr
Birge, Ann Chamberlain
Bjorken, James D
Blachman, Nelson Merle
Blades, John Dieterle
Bleakney, Walker
Blink, James Allen
Bohmer, Heinrich Everhard
Bonsack, Walter Karl
Bork, Alfred Morton
Bott, Jerry Frederick
Braun, Robert Leore
Braunstein, Rubin
Bridges, William B
Britt, Edward Joseph
Brodie, Jean Pamela
Brooks, Walter Lyda
Brouillard, Robert Ernest
Brown, Douglas Richard
Brown, George Stephen
Brown, Gordon Edgar, Jr
Brown, Robert James Sidford
Brown, Robert Theodore
Brown, Sheldon (Jack)
Bruck, George
Burcham, Donald Preston
Burnett, Lowell Jay
Burns, Fred Paul
Bush, George Edward
Cabrera, Blas
Caren, Robert Poston
Carlson, Richard Frederick
Carr, Robert H
Carron, Neal Jay
Carter, David
Carver, John Guill
Case, Lloyd Allen
Cassen, Patrick Michael
Catura, Richard Clarence
Chakrabarti, Supriya
Chandler, Donald Ernest
Chang, David Bing Jue
Channel, Lawrence Edwin
Chase, Lloyd Fremont, Jr
Chase, Lloyd Lee
Chen, Chia Hwa
Chen, Min
Chiao, Raymond Yu
Chinowsky, William
Cho, Young-chung
Chow, Richard H
Christman, Arthur Castner, Jr
Chu, William Tongil
Churchill, Dewey Ross, Jr
Clark, Arnold Franklin
Clauser, John Francis
Cleland, Laurence Lynn
Clemens, Jon K(aufmann)
Clewell, Dayton Harris
Cochran, Stephen G
Codrington, Robert Smith
Cohen, Marvin Lou
Cole, Robert Kleiv
Coleman, Lamar William
Coleman, Philip Lynn
Commins, Eugene David
Cook, Thomas Bratton, Jr
Cooper, Eugene Perry
Coppens, Alan Berchard
Corey, Victor Brewer
Corruccini, Linton Reid
Costantino, Marc Shaw
Coufal, Hans-Jürgen
Cowan, Eugene Woodville
Cox, Aaron J
Craig, Paul Palmer
Cranston, Frederick Pitkin, Jr
Crawford, Frank Stevens, Jr
Crawford, John Clark
Creighton, John Rogers
Crews, Robert Wayne
Crowe, Kenneth Morse
Curran, Donald Robert
Cutler, Cassius Chapin
Cutler, Leonard Samuel
Dahl, Harvey A
Dahlberg, Richard Craig
Dairiki, Ned Tsuneo
Dart, Sidney Leonard
D'Attorre, Leonardo
Davis, James Ivey
Davis, Jay C
Davis, Leverett, Jr
Davis, Sumner P
Davis, Thomas Pearse
Daybell, Melvin Drew

Debs, Robert Joseph
de Hoffmann, Frederic
Deleray, Arthur Loyd
Dember, Alexis Berthold
Denney, Joseph M(yers)
De Pangher, John
Dessel, Norman F
Devoe, Ralph Godwin
Dickinson, Dale Flint
Dimeff, John
Dirks, Leslie C
Donaldson, John Riley
Doolittle, Robert Frederick, II
Dowell, Jerry Tray
Driscoll, Charles F
Duggan, Michael J
Dulock, Victor A, Jr
Duneer, Arthur Gustav, Jr
Dunning, John Ray, Jr
Dyal, Palmer
Early, James M
Ebert, Paul Joseph
Eby, Frank Shilling
Eckert, Hans Ulrich
Eckhardt, Gisela (Marion)
Eckhardt, Wilfried Otto
Edelson, Sidney
Edwards, Byron N
Eggen, Donald T(ripp)
Eimerl, David
Einarsson, Alfred W
Eisen, Fred Henry
Elings, Virgil Bruce
Elleman, Daniel Draudt
Elliott, Shelden Dougless, Jr
Elms, James Cornelius
Ely, Robert P, Jr
Enderby, Charles Eldred
Enns, John Hermann
Erlich, David C
Eschenfelder, Andrew Herbert
Esposito, Pasquale Bernard
Evans, John Ellis
Everitt, C W Francis
Evett, Arthur A
Fa'arman, Alfred
Fairbank, William Martin
Farmer, Donald Jackson
Faulkner, John Edward
Feher, Elsa
Feher, George
Feinstein, Joseph
Ferris, Horace Garfield
Filippenko, Alexei Vladimir
Finkelstein, Robert Jay
Fisher, George Phillip
Fisher, John Crocker
Fisher, Philip Chapin
Fitzpatrick, Gary Owen
Fleming, Lawrence Thomas
Flinn, Paul Anthony
Folsom, Theodore Robert
Fonck, Eugene J
Forster, Harriet Herta
Forster, Kurt
Forward, Robert L
Foster, John Stuart, Jr
Foster, Theodore Dean
Fowler, Charles Arman, Jr
Fowler, William Alfred
Frank, Wilson James
Fredkin, Donald Roy
Fredrickson, John E
Fretter, William Bache
Freund, Roland Wilhelm
Fritchle, Frank Paul
Fu, Lee Lueng
Fullmer, George Clinton
Futch, Archer Hamner
Gailar, Owen H
Gardner, Wilford Robert
Garmire, Elsa Meints
Garrick, B(ernell) John
Garrison, John Dresser
Garwin, Edward Lee
Getting, Ivan Alexander
Ghiorso, Albert
Ghose, Rabindra Nath
Giannini, Gabriel Maria
Gibson, Atholl Allen Vear
Gibson, James (Benjamin)
Gibson, Thomas Alvin, Jr
Giles, John Crutchlow
Giles, Peter Cobb
Gill, Stephen Paschall
Gillespie, George H
Gilmartin, Thomas Joseph
Ginzton, Edward Leonard
Glaser, Donald Arthur
Godfrey, Charles S
Goerz, David Jonathan, Jr
Gold, Richard Robert
Goodkind, John M
Goodstein, David Louis
Goss, Wilbur Hummon
Gossard, Arthur Charles
Goubau, Wolfgang M
Greenberg, Allan S
Greenfield, Eugene W(illis)
Grinberg, Jan
Grismore, Roger
Hadley, James Warren
Hageman, Donald Henry
Hahn, Erwin Louis

Physics, General (cont)

Turner, Eugene Bonner
Tuul, Johannes
Twieg, Donald Baker
Unwin, Stephen Charles
Vajk, J(oseph) Peter
Van Atta, Lester Clare
Van Lint, Victor Anton Jacobus
Van Thiel, Mathias
Veigele, William John
Vernon, C(arl) Wayne
Vernon, Frank Lee, Jr
Violet, Charles Earl
Vogt, Rochus E
Voreades, Demetrios
Vreeland, John Allen
Waddell, Charles Noel
Wainwright, Thomas Everett
Waiter, Serge-Albert
Wall, Leonard Wong
Wang, Shyh
Waniek, Ralph Walter
Warshaw, Stephen I
Washburn, Harold W(illiams)
Waters, Rodney Lewis
Waters, William Edward
Watson, Kenneth Marshall
Watson, Velvin Richard
Waugh, John Blake-Steele
Weiss, Max Tibor
Weissler, Gerhard Ludwig
Wenk, Hans-Rudolf
Westberg, Karl Rogers
Wheaton, Elmer Paul
White, Richard Manning
Whitefield, Rodney Joe
Whitmer, Robert Morehouse
Whitney, Robert C
Whitney, William Merrill
Whittemore, William Leslie
Wilcox, Howard Albert
Wilcox, Jaroslava Zitkova
Wilcox, Thomas Jefferson
Wilkening, Dean Arthur
Williams, E(dgar) P
Williamson, Hugh A
Winkelmann, Frederick Charles
Wintroub, Herbert Jack
Witteborn, Fred Carl
Woehler, Karlheinz Edgar
Woerner, Robert Leo
Wong, K(wee) C
Wood, Calvin Dale
Woodbury, Eric John
Woodruff, Truman Owen
Woodworth, John George
Wooldridge, Dean E
Worden, Paul Wellman, Jr
Worth, Donald Calhoun
Wright, Frederick Hamilton
Wuerker, Ralph Frederick
Yadavalli, Sriramamurti Venkata
Yanow, Gilbert
Yodh, Gaurang Bhaskar
York, Carl Monroe, Jr
York, Herbert Frank
Young, Richard D
Yu, Chyang John
Yu, Simon Shin-Lun
Zhou, Simon Zheng
Zimmerman, Elmer Leroy
Zubeck, Robert Bruce

COLORADO
Aamodt, Richard E
Allan, David Wayne
Baird, Ramon Condie
Barnes, James Allen
Barrett, Charles Sanborn
Barth, Charles Adolph
Bartlett, Albert Allen
Beam, Kurt George, Jr
Beaty, Earl Claude
Beehler, Roger Earl
Bender, Peter Leopold
Bills, Daniel Granville
Bordner, Charles Albert, Jr
Bradley, Richard Crane
Bruce, Charles Robert
Bruns, Donald Gene
Burnett, Clyde Ray
Burnett, Jerrold J
Collins, Royal Eugene
Colvis, John Paris
Davis, Milford Hall
Daw, Glen Harold
Derbyshire, William Davis
Duray, John R
Duvall, Wilbur Irving
Ellis, Homer Godsey
Engen, Glenn Forrest
Faller, James E
Faris, John Jay
Fickett, Frederick Roland
Fox, Michael Henry
Franklin, Allan David
Gille, John Charles
Gilman, Peter A
Granzow, Kenneth Donald
Gray, James Edward
Griffith, Gordon Lamar
Hall, John L
Hamerly, Robert Glenn
Hansen, Carl John
Harris, Richard Elgin

Heller, Marvin W
Hilt, Richard Leighton
Hjelme, Dag Roar
Hord, Charles W
Horton, Clifford E(dward)
Hust, Jerome Gerhardt
Iona, Mario
Johnson, Richard Harlan
Jordan, Albert Raymond
Kamper, Robert Andrew
Kashnow, Richard Allen
Kearney, Philip Daniel
Kemper, William Alexander
Kennedy, Patrick James
Koldewyn, William A
Koontz, Philip G
Lawrence, George Melvin
Lazzarini, Albert John
Leinbach, F Harold
Leisure, Robert Glenn
Levenson, Leonard L
McGavin, Raymond E
Martz, Dowell Edward
Muscari, Joseph A
Netzel, Richard G
Nickerson, John Charles, III
Olson, John Richard
Owen, Robert Barry
Reichardt, John William
Sauer, Herbert H
Schramm, Raymond Eugene
Sime, David Gilbert
Simon, Nancy Jane
Sites, James Russell
Slaughter, Maynard
Smith, Richard Cecil
Snyder, Wilbert Frank
Spence, William J
Spenny, David Lorin
Stebbins, Robin Tucker
Stein, Samuel Richard
Stern, Raul A(ristide)
Sullivan, Donald Barrett
Swenson, Hugo Nathanael
Tanttila, Walter H
Trefny, John Ulric
Turi, Raymond A
Uhlenbeck, George Eugene
Violett, Theodore Dean
Wahr, John Cannon
Ware, Walter Elisha
Watkins, Sallie Ann
Wells, Joseph S
White, Franklin Estabrook
Wooldridge, Gene Lysle
Wright, Wilbur Herbert
Wyss, Walter
Zaidins, Clyde

CONNECTICUT
Anderson, Wilmer Clayton
Azaroff, Leonid Vladimirovich
Bannister, Peter Robert
Barker, Richard Clark
Bennett, William Ralph, Jr
Beringer, Robert
Best, Philip Ernest
Budnick, Joseph Ignatius
Burkhard, Mahlon Daniel
Chih, Chung-Ying
Cipriano, Ramon John
Cole, Henderson
Connell, Richard Allen
Cummerow, Robert Leggett
Cunningham, Frederick William
Damon, Dwight Hills
Davenport, Lee Losee
DeMaria, Anthony John
Eastman, Daniel Robert Peden
Fraser, J(ulius) T(homas)
Frueh, Alfred Joseph, Jr
Gilliam, Otis Randolph
Glenn, William Henry, Jr
Goldman, Jacob E
Green, Eugene L
Green, Milton
Grubin, Harold Lewis
Halpern, Howard S
Hamilton, Douglas Stuart
Howkins, Stuart D
Hughes, Vernon Willard
Kattamis, Theodoulos Zenon
King, George, III
Klemens, Paul Gustav
Kraft, David Werner
Krauss, Lawrence Maxwell
Krisst, Raymond John
Lichten, William Lewis
Liebson, Sidney Harold
Lindquist, Richard Wallace
Lipschultz, Frederick Phillip
Locke, Stanley
Lu, Phillip Kehwa
Ma, Tso-Ping
MacDowell, Samuel Wallace
Matthews, Lee Drew
Melcher, Charles L
Moran, Thomas Irving
Moreland, Parker Elbert, Jr
Nash, Harold Earl
Nelligan, William Bryon
Nudelman, Sol
Paolini, Francis Rudolph
Patterson, Elizabeth Chambers

Plona, Thomas Joseph
Pollock, Herbert Chermside
Poultney, Sherman King
Prober, Daniel Ethan
Rau, Richard Raymond
Rich, Leonard G
Richards, Roger Thomas
Rockwell, Sara Campbell
Roychoudhuri, Chandrasekhar
Russek, Arnold
Rydz, John S
Sachdev, Subir
Sandweiss, Jack
Seely, Samuel
Sherman, Harold
Siegel, Lester Aaron
Sivinski, John A
Spencer, Domina Eberle (Mrs Parry Moon)
Strough, Robert I(rving)
Tobin, Marvin Charles
Tokita, Noboru
Trousdale, William Latimer
Tucker, Edmund Belford
Waine, Martin
Wegener, Peter Paul
Wingfield, Edward Christian
Wolf, Elizabeth Anne
Zandy, Hassan F

DELAWARE
Barton, Randolph, Jr
Burgess, John S(tanley)
Chromey, Fred Carl
Cooper, Charles Burleigh
Davies, Robert Dillwyn
Ewing, Richard Dwight
Griffiths, David
Groff, Ronald Parke
Halprin, Arthur
Hardy, Henry Benjamin, Jr
Kent, Donald Wetherald, Jr
Knop, Harry William, Jr
Lutz, Bruce Charles
Maher, John Philip
Marcus, Sanford M
Meakin, Paul
Mehl, James Bernard
Mrozinski, Peter Matthew
Robertson, Charles William, Jr
Shipman, Harry Longfellow
Weiher, James F
Wilson, Frank Charles

DISTRICT OF COLUMBIA
Appleman, Daniel Everett
Aufenkamp, Darrel Don
Beasley, Edward Evans
Berlincourt, Ted Gibbs
Bernard, William
Blankfield, Alan
Brown, Harold
Bryant, Barbara Everitt
Budgor, Aaron Bernard
Carroll, Kenneth Girard
Chappell, Samuel Estelle
Chubb, Scott Robinson
Chung, David Yih
Clarke, John F
Cochran, Thomas B
Corden, Pierce Stephen
Corliss, Edith Lou Rovner
Daehler, Mark
Davis, Jack
Eagleson, Halson Vashon
Edelsack, Edgar Allen
Fainberg, Anthony
Flippen-Anderson, Judith Lee
Friedman, Herbert
Goldberg, Stanley
Haas, George Arthur
Hendrie, David Lowery
Herschman, Arthur
Heydemann, Peter Ludwig Martin
Hollinger, James Pippert
Hubler, Graham Kelder, Jr
Jhirad, David John
Johnson, Thomas Hawkins
Kahn, Arnold Herbert
Kalmus, Henry P(aul)
Karle, Jerome
Keeny, Spurgeon Milton, Jr
Kepple, Paul C
Litovitz, Theodore Aaron
McKay, Jack Alexander
Madden, Robert Phyfe
Mandula, Jeffrey Ellis
Marcus, Gail Halpern
Michels, Donald Joseph
Montrose, Charles Joseph
Moon, Deug Woon
Murday, James Stanley
Myers, Peter Briggs
Needels, Theodore S
O'Fallon, John Robert
Ondik, Helen Margaret
Oran, Elaine Surick
Oswald, Robert B(ernard), Jr
Palmadesso, Peter Joseph
Patten, Raymond Alex
Perkins, Floyd
Peters, Gerald Joseph
Pikus, Irwin Mark
Prinz, Dianne Kasnic

Rauckhorst, William H
Riegel, Kurt Wetherhold
Riemer, Robert Lee
Ritter, Enloe Thomas
Rittner, Edmund Sidney
Rogers, Kenneth Cannicott
Scheel, Nivard
Schildcrout, Michael
Schrack, Roald Amundsen
Schweizer, Francois
Shih, Arnold Shang-Teh
Sinclair, Rolf Malcolm
Siry, Joseph William
Spohr, Daniel Arthur
Taylor, Ronald D
Teitler, Sidney
Tousey, Richard
Tsao, Chen-Hsiang
Vardiman, Ronald G
Wagner, Andrew James
Wakelin, James Henry, Jr
Waterhouse, Richard (Valentine)
Weiler, Kurt Walter
Willard, Daniel
Wood, Lawrence Arnell
Zuchelli, Artley Joseph

FLORIDA
Adhav, Ratnakar Shankar
Alonso, Marcelo
Andregg, Charles Harold
Aubel, Joseph Lee
Ballard, Stanley Sumner
Becker, Gordon Edward
Beguin , Fred P
Biver, Carl John, Jr
Bloch, Sylvan C
Bolemon, Jay S
Bolte, John R
Buckwalter, Gary Lee
Bueche, Frederick Joseph
Burns, Jay, III
Calhoun, Ralph Vernon, Jr
Carr, Thomas Deaderick
Clapp, Roger Williams, Jr
Coggeshall, Norman David
Cooper, John Niessink
Cooper, Raymond David
Cox, Joseph Robert
Dalehite, Thomas H
Dam, Cecil Frederick
Edwards, Palmer Lowell
Fausett, Laurene van Camp
Fenna, Roger Edward
Ference, Michael, Jr
Field, Richard D
Fleddermann, Richard G(rayson)
Flowers, John Wilson
Forman, Guy
Gager, William Ballantine
Gianola, Umberto Ferdinando
Gielisse, Peter Jacob Maria
Green, Alex Edward Samuel
Hales, Everett Burton
Hamtil, Charles Norbert
Harrington, John Vincent
Heaps, Melvin George
Herriott, Donald R
Herzog, Richard (Franz Karl)
Hirschberg, Joseph Gustav
Hock, Donald Charles
Holloway, Dennis Michael
Huber, Oren John
Humphries, Jack Thomas
Hunt, Robert Harry
Ihas, Gary Gene
Jin, Rong-Sheng
Kendall, Harry White
Keuper, Jerome Penn
Kiewit, David Arnold
Knowles, Harold Loraine
Krausche, Dolores Smoleny
Krc, John, Jr
Krishnamurti, Ruby Ebisuzaki
Kruschwitz, Walter Hillis
Lambe, Edward Dixon
Lebo, George Robert
Llewellyn, Ralph A
Ludeke, Carl Arthur
McLeroy, Edward Glenn
Miller, Hillard Craig
Milton, James E(dmund)
Mione, Anthony J
Mitchell, Richard Warren
Moulton, William G
Munk, Miner Nelson
Nelson, Edward Bryant
Oelfke, William C
Ohrn, Nils Yngve
Oleson, Norman Lee
Omer, Guy Clifton, Jr
Owens, James Samuel
Palenik, Gus J
Pardo, William Bermudez
Perlmutter, Arnold
Pfahnl, Arnold
Potter, James Gregor
Reynolds, George William, Jr
Rhodes, Richard Aver, II
Riley, Mark Anthony
Robertson, Harry S(troud)
Rosenzweig, Walter
Rudgers, Anthony Joseph
Schmidt, Klaus H

Physics, General (cont)

Scott, Thomas A
Shih, Hansen S T
Skramstad, Harold Kenneth
Smith, Malcolm (Kinmonth)
Stewart, Gregory Randall
Sundaram, Swaminatha
Thomas, Dan Anderson
Timme, Robert William
Tterlikkis, Lambros
Vollmer, James
Weber, Alfred Herman
Weller, Richard Irwin
Wells, Daniel R
Zaukelies, David Aaron
Zimmer, Martin F
Zinn, Walter Henry

GEORGIA

Ahrens, Rudolf Martin (Tino)
Anantha Narayanan, Venkataraman
Anderson, Robert Lester
Bomar, Lucien Clay
Boyd, James Emory
Burkhard, Donald George
Chandra, Kailash
Clement, Joseph D(ale)
Clemmons, John B
Cooper, Charles Dewey
Cramer, John Allen
Crawford, Vernon
Eichholz, Geoffrey G(unther)
Family, Fereydoon
Gallagher, James J
Gallaher, Lawrence Joseph
Garrison, Allen K
Gersch, Harold Arthur
Goslin, Roy Nelson
Hardin, Clyde D
Harrison, Gordon R
Hart, Raymond Kenneth
Hartman, Nile Franklin
Howard, Sethanne
Hunt, Gary W
Hurst, Vernon James
Jackson, Prince A, Jr
Jarzynski, Jacek
Kiang, Chia Szu
Kollig, Heinz Philipp
McDaniel, Earl Wadsworth
Martin, David Willis
Nave, Carl R
Palms, John Michael
Patronis, Eugene Thayer, Jr
Petitt, Gus A
Price, Edward Warren
Purcell, James Eugene
Rohrer, Robert Harry
Roy, Rajarshi
Sadun, Alberto Carlo
Simmons, James Wood
Stevenson, James Rufus
Williams, Joel Quitman
Winer, Ward Otis
Woodward, LeRoy Albert
Young, Robert Alan

HAWAII

Cence, Robert J
Dorrance, William Henry
Hayes, Charles Franklin
Holm-Kennedy, James William
McAllister, Howard Conlee
McFee, Richard
Pakvasa, Sandip
Seff, Karl
Sogo, Power Bunmei
Steiger, Walter Richard
Stenger, Victor John

IDAHO

Browne, Michael Edwin
Carpenter, Stuart Gordon
Davis, Lawrence William, Jr
East, Larry Verne
Ford, Gilbert Clayton
Froes, Francis Herbert (Sam)
Hall, J(ames) A(lexander)
Harding, Samuel William
Harker, Yale Deon
Harmon, J Frank
Knox, John MacMurray
Lott, Layman Austin
Luke, Robert A
Majumdar, Debaprasad (Debu)
Marks, Darrell L
Murphey, Byron Freeze
Olson, Willard Orvin
Otting, William Joseph
Poenitz, Wofgang P
Price, Joseph Earl
Reich, Charles William
Telschow, Kenneth Louis
Tracy, Joseph Charles
Wood, Richard Ellet

ILLINOIS

Abels, Larry L
Adler, Robert
Allen, Frank B
Altpeter, Lawrence L, Jr
Anderson, Scott
Anderson, William Raymond

Avery, Robert
Bailyn, Martin H
Bardeen, John
Baur, Werner Heinz
Behof, Anthony F, Jr
Bellamy, David
Bixby, William Ellis
Block, Martin M
Bodmer, Arnold R
Boedeker, Richard Roy
Born, Harold Joseph
Borso, Charles S
Bowe, Joseph Charles
Bowen, Samuel Philip
Boyd, John William
Braid, Thomas Hamilton
Braundmeier, Arthur John, Jr
Brussel, Morton Kremen
Bushnell, David L
Butler, William Albert
Carney, Rose Agnes
Cloud, William Max
Coester, Fritz
Cohn, Gerald Edward
Cole, Francis Talmage
Cooper, Duane H(erbert)
Cornell, David Allan
Coster, Joseph Constant
Crew, John Edwin
Crewe, Albert Victor
Crosbie, Edwin Alexander
Curry, Bill Perry
Curtis, Cyril Dean
Davis, A Douglas
Day, Paul Palmer
Debrunner, Peter Georg
Dixon, Roger L
Doerner, Robert Carl
Drickamer, Harry George
Dudley, Horace Chester
Dunford, Robert Walter
Ebrey, Thomas G
Erber, Thomas
Erck, Robert Alan
Evans, William Paul
Fink, Joanne Krupey
Fish, Ferol F, Jr
Galayda, John Nicolas
Giese, Robert Frederick
Ginsberg, Donald Maurice
Goldwasser, Edwin Leo
Granato, Andrew Vincent
Green, Donald Wayne
Griem, Melvin Luther
Gupta, Nand K
Hafele, Joseph Carl
Halperin, William Paul
Hannon, Bruce Michael
Harrington, Joseph Anthony
Hasdal, John Allan
Hauser, Isidore
Heaton, LeRoy
Hess, David Clarence
Huft, Michael John
Hull, Harvard Leslie
Hull, John R
Hurych, Zdenek
Ivory, John Edward
Jackson, Edwin Atlee
Johnson, Rolland Paul
Jones, Leonard Clive
Kaminsky, Manfred Stephan
Kaufmann, Elton Neil
Keren, Joseph
Ketterson, John Boyd
Kinnmark, Ingemar Per Erland
Klein, Miles Vincent
Knox, Jack Rowles
Koehler, James Stark
Krasner, Sol H
Krauss, Alan Robert
Krein, Philip Theodore
Kruse, Ulrich Ernst
Lach, Joseph T
Lanzl, Lawrence Herman
Leonard, Byron Peter
LeSage, Leo G
Levi-Setti, Riccardo
Licht, Arthur Lewis
Linde, Ronald K(eith)
Liu, Chao-Han
Lyman, Ernest McIntosh
McAneny, Laurence Raymond
McCart, Bruce Ronald
Marcus, Jules Alexander
Martin, Richard McFadden
Meeker, Ralph Dennis
Meyer, Axel
Meyer, Stuart Lloyd
Mills, Frederick Eugene
Moore, Paul Brian
Muller, Dietrich
Nickell, William Everett
Noble, Gordon Albert
Noble, John Dale
O'Gallagher, Joseph James
Orr, John R
Pagnamenta, Antonio
Persiani, Paul J
Pigott, Miles Thomas
Prohammer, Frederick George
Propst, Franklin Moore
Redman, William Charles
Rehn, Lynn Eduard

Renneke, David Richard
Rey, Charles Albert
Ring, James George
Roberts, Arthur
Roberts, John England
Roll, Peter Guy
Roothaan, Clemens Carel Johannes
Rossing, Thomas D(ean)
Rubin, Howard Arnold
Saboungi, Marie-Louise Jean
Saporoschenko, Mykola
Schillinger, Edwin Joseph
Schmidt, Charles William
Segel, Ralph E
Shaffer, John Clifford
Sharma, Ram Ratan
Shen, Sin-Yan
Shih, Hsio Chang
Siegel, Jonathan Howard
Siegel, Stanley
Simpson, John Alexander
Singwi, Kundan Singh
Skaggs, Lester S
Smith, P Scott
Snyder, James Newton
Soule, David Elliot
Spight, Carl
Staunton, John Joseph Jameson
Stenberg, Charles Gustave
Stevens, Fred Jay
Stinchcomb, Thomas Glenn
Strahm, Norman Dale
Stutz, Conley I
Sukhatme, Uday Pandurang
Sutton, David C(hase)
Tao, Rongjia
Throw, Francis Edward
Tomaschke, Harry E
Tomkins, Marion Louise
Turkot, Frank
Tuzzolino, Anthony J
Van Ginneken, Andreas J
Waddell, Robert Clinton
Watson, Richard Elvis
Weaver, Allen Dale
Weertman, Johannes
Weissman, Herman Benjamin
Whalin, Edwin Ansil, Jr
Wolfson, James
Wolsky, Alan Martin
Wright, Sydney Courtenay
Wyld, Henry William, Jr
Wylie, Douglas Wilson
Yamanouchi, Taiji
Yang, Shi-Tien
Yntema, Jan Lambertus
Yoh, John K
Youtsey, Karl John
Zaromb, Solomon

INDIANA

Alyea, Ethan Davidson, Jr
Beery, Dwight Beecher
Bent, Robert Demo
Biswas, Nripendra Nath
Blackstead, Howard Allan
Brill, Wilfred G
Cohen, Herbert Daniel
Conklin, Richard Louis
Coomes, Edward Arthur
Cowan, Raymond
Cox, Martha
Crable, George Francis
Crittenden, Ray Ryland
Cruikshank, Donald Burgoyne, Jr
Di Lavore, Philip, III
Durisen, Richard H
Farringer, Leland Dwight
Feuer, Paula Berger
Flick, Cathy
Gaidos, James A
Gerritsen, Alexander Nicolaas
Gutay, Laszlo J
Hake, Richard Robb
Hale, Robert E
Hamilton-Steinraut, Jean A
Hancock, John Ogden
Hartzler, Harrod Harold
Henninger, Ernest Herman
Henry, Hugh Fort
Henry, Robert Ledyard
Howes, Ruth Hege
Hrivnak, Bruce John
Hults, Malcom E
Johnson, C(harles) Bruce
Johnson, Walter Richard
Jumper, Eric J
Kenney, Vincent Paul
Kesmodel, Larry Lee
Khorana, Brij Mohan
Klontz, Everett Earl
Kolata, James John
Leiter, Howard Allen
Loeffler, Frank Joseph
Lurie, Fred Marcus
Mastrototaro, John Joseph
Moloney, Michael J
Nann, Hermann
Palfrey, Thomas Rossman, Jr
Poorman, Lawrence Eugene
Reilly, James Patrick
Robinson, John Murrell
Schindler, Albert Isadore
Schlueter, Donald Jerome

Shaffer, Lawrence Bruce
Snyder, Donald DuWayne
Sprague, Newton G
Stanley, Robert Weir
Stein, Frank S
Taylor, Raymond Ellory
Vawter, Spencer Max
Vigdor, Steven Elliot
Vinson, James S
Vondrak, Edward Andrew
Wolber, William George

IOWA

Azbell, William
Bowen, George Hamilton, Jr
Clotfelter, Beryl Edward
Cook, Barnett C
Denny, Wayne Belding
Frank, Louis Albert
Goertz, Christoph Klaus
Hodges, Laurent
Jennings, Michael Leon
Kernan, William J, Jr
Kirkham, Don
Kopp, Jay Patrick
Leung, Wai Yan
Macomber, Hilliard Kent
Moen, Allen LeRoy
Ostenson, Jerome E
Peterson, Francis Carl
Poppy, Willard Joseph
Pursey, Derek Lindsay
Savage, William Ralph
Stebbins, Dean Waldo
Van Allen, James Alfred

KANSAS

Bearse, Robert Carleton
Dale, Ernest Brock
Ellsworth, Louis Daniel
Feaster, Gene R(ichard)
Folland, Nathan Orlando
Law, Bruce Malcolm
Munczeu, Herman J
Pruitt, Roger Arthur
Senecal, Gerard
Stockli, Martin P
Thomas, James E
Twarog, Bruce Anthony
Unruh, Henry, Jr
Valenzeno, Dennis Paul
Wiseman, Gordon G
Witten, Maurice Haden
Wong, Kai-Wai

KENTUCKY

Carpenter, Dwight William
Cochran, Lewis Wellington
Davis, Thomas Haydn
Duncan, Don Darryl
Feola, Jose Maria
Gwinn, Joel Alderson
Huang, Wei-Feng
Kadaba, Prasad Krishna
McClain, John William
Merker, Stephen Louis
Moulder, Jerry Wright
Naake, Hans Joachim
Powell, Smith Thompson, III
Purdom, Ray Caldwell
Rollo, Frank David
Russell, Marvin W
Sinai, John Joseph
Watkins, Nancy Chapman

LOUISIANA

Bedell, Louis Robert
Bergeron, Clyde J, Jr
Bernard, William Hickman
Dempesy, Colby Wilson
Edmonds, James D, Jr
Fischer, David John
Hamilton, William Oliver
Hoy, Robert C
Johnson, Larry Don
Keiffer, David Goforth
King, Creston Alexander, Jr
Matlock, Rex Leon
Mitcham, Donald
Naidu, Seetala V
Neuman, Charles Herbert
Perdew, John Paul
Reynolds, Joseph Melvin
Riess, Karlem
Robert, Kearny Quinn, Jr
Smith, Ronald E
Stephenson, Paul Bernard
Ward, Truman L
Wefel, John Paul

MAINE

Bancroft, Dennison
Bennett, Clarence Edwin
Carr, Edward Frank
Chonacky, Norman J
Dudley, John Minot
Emery, Guy Trask
Gordon, Geoffrey Arthur
Hoyt, Rosalie Chase
Hughes, William Taylor
Mitchell, H Rees
Otto, Fred Bishop
Ruff, George Antony
Smith, Charles William, Jr

Snow, Joseph William
Tarr, Charles Edwin
Viette, Michael Anthony

MARYLAND
Achor, William Thomas
Acuna, Mario Humberto
Alers, Perry Baldwin
Alley, Reuben Edward, Jr
Anderson, J Robert
Arking, Albert
Arndt, Richard Allen
Aronson, Casper Jacob
Avery, William Hinckley
Babrauskas, Vytenis
Balasubrahmanyan, Vriddhachalam K
Barbe, David Franklin
Bargeron, Cecil Brent
Barrows, Austin Willard
Bass, Arnold Marvin
Bay, Zoltan Lajos
Beiler, Adam Clarke
Bennett, Charles L
Bennett, Lawrence Herman
Berning, Warren Walt
Bhagat, Satindar M
Bjerkaas, Allan Wayne
Blake, Lamont Vincent
Blevins, Gilbert Sanders
Block, Stanley
Bonavita, Nino Louis
Bostrom, Carl Otto
Bowers, Robert Clarence
Branson, Herman Russell
Brown, Samuel Heffner
Burlaga, Leonard F
Calame, Gerald Paul
Carlon, Hugh Robert
Carp, Gerald
Carter, Robert Emerson
Casella, Russell Carl
Cassidy, Esther Christmas
Caulton, M(artin)
Cavalieri, Donald Joseph
Chen, Hsing-Hen
Chertock, George
Chu, Tak-Kin
Clark, Trevor H
Coates, R(obert) J(ay)
Coffman, John W
Cohen, Steven Charles
Conklin, Glenn Ernest
Conrath, Barney Jay
Cook, Richard Kaufman
Cornett, Richard Orin
Cramer, William Smith
Crump, Stuart Faulkner
Culver, William Howard
Daniel, Charles Dwelle, Jr
Daniel, John Harrison
Dauber, Edwin George
Dayhoff, Edward Samuel
Degnan, John James, III
Dehn, James Theodore
De Rosset, William Steinle
Deslattes, Richard D, Jr
DiMarzio, Edmund Armand
Di Rienzi, Joseph
Dixon, Jack Richard
Dixon, John Kent
Dixon, Peggy A
Dodge, William R
Dubin, Henry Charles
Dubin, Maurice
Dunlap, Brett Irving
Earhart, J Ronald
Earl, James Arthur
Eckerle, Kenneth Lee
Egner, Donald Otto
Eichelberger, Robert John
Elder, Samuel Adams
Elsasser, Walter M
Emch, George Frederick
Ethridge, Noel Harold
Falk, Charles Eugene
Fansler, Kevin Spain
Farrell, Richard Alfred
Field, Herbert Cyre
Finkelstein, Robert
Fiorito, Ralph Bruno
Fitzgerald, Edwin Roger
Fitzpatrick, Hugh Michael
Flook, William Mowat, Jr
Foster, Margaret C
Fowler, Howland Auchincloss
Fristrom, Robert Maurice
Garroway, Allen N
Gasparovic, Richard Francis
Geckle, William Jude
Gelles, Isadore Leo
Gheorghiu, Paul
Gilliland, Ronald Lynn
Gion, Edmund
Glasser, Robert Gene
Gloeckler, George
Gluckstern, Robert Leonard
Goldberg, Benjamin
Goldfinger, Andrew David
Gonano, John Roland
Gordon, Gary Donald
Greenstone, Reynold
Griem, Hans Rudolf
Griffin, James J
Gutsche, Graham Denton

Hafstad, Lawrence Randolph
Hall, Forrest G
Halpin, Joseph John
Harman, George Gibson, Jr
Harris, Forest K
Hart, Lynn W
Hartmann, Gregory Kemenyi
Hayward, Evans Vaughan
Heckman, Timothy Martin
Heinberg, Milton
Heller, Douglas Max
Henry, Richard Conn
Hirschmann, Erwin
Hobbs, Robert Wesley
Holman, Gordon Dean
Horn, William Everett
Hummel, Harry Horner
Hybl, Albert
Hynes, John Edward
Irwin, George Rankin
Jablonski, Felix Joseph
Jacobs, Sigmund James
Jefferson, Donald Earl
Jen, Chih Kung
Jensen, Richard Eugene
Jette, Archelle Norman
Johnston, Robert Ward
Kagarise, Ronald Eugene
Kahn, Jack Henry
Kaiser, Jack Allen Charles
Kalil, Ford
Katcher, David Abraham
Katsanis, D(avid) J(ohn)
Kay, Mortimer Isaia
Kazi, Abdul Halim
Keating, Patrick Norman
Kelly, William Clark
Killen, Rosemary Margaret
Kim, Boris Fincannon
Kinsey, James Humphreys
Koblinsky, Chester John
Korobkin, Irving
Kotter, F(red) Ralph
Kravitz, Lawrence C
Krumbein, Simeon Joseph
Kulp, Bernard Andrew
Kuriyama, Masao
Lashof, Theodore William
Lauriente, Mike
Lee, Ronald Norman
Leffel, Claude Spencer, Jr
Li, Ming Chiang
Link, John Clarence
Little, John Llewellyn
Liu, Han-Shou
Luebke, Emmeth August
Lundholm, J(oseph) G(ideon), Jr
Lyman, Ona Rufus
McCally, Russell Lee
McCoubrey, Arthur Orlando
McCutchen, Charles Walter
McDonald, Frank Bethune
McElhinney, John
McLeod, Norman Barrie
Mahoney, Francis Joseph
Majkrzak, Charles Francis
Mangum, Billy Wilson
Mann, Wilfrid Basil
Marcinkowski, M(arion) J(ohn)
Maxwell, Louis R
Messina, Carla Gretchen
Michaelis, Michael
Mielenz, Klaus Dieter
Miller, Gerald R
Moldover, Michael Robert
Monaldo, Francis Michael
Montgomery, David Carey
Moon, Milton Lewis
More, Kenneth Riddell
Morgan, Bruce Harry
Motz, Joseph William
Mozer, Bernard
Munro, Ronald Gordon
Nakada, Minoru Paul
Ney, Wilbert Roger
Nossal, Ralph J
Ogilvie, Keith W
Oldfield, Edward Hudson
Oroshnik, Jesse
Ostrofsky, Bernard
Paik, Ho Jung
Park, Robert L
Parr, Albert Clarence
Passman, Sidney
Peiser, Herbert Steffen
Perry, Peter M
Pevsner, Aihud
Piermarini, Gasper J
Pinkston, Earl Roland
Potocki, Kenneth Anthony
Potyraj, Paul Anthony
Potzick, James Edward
Pouring, Andrew A
Powell, Cedric John
Prange, Richard E
Radcliffe, Alec
Rathbun, Edwin Roy, Jr
Ravitsky, Charles
Rawlins, Stephen Last
Reader, Joseph
Reames, Donald Vernon
Rector, Charles Willson
Reisse, Robert Alan
Reneker, Darrell Hyson

Richardson, John Marshall
Riedl, H Raymond
Roberts, James Richard
Robinson, Richard Carleton, Jr
Rosado, John Allen
Rosenbaum, Ira Joel
Roth, Michael William
Rubenstein, Albert Marvin
Ruff, Arthur William, Jr
Schmugge, Thomas Joseph
Schroeder, Frank, Jr
Schwee, Leonard Joseph
Seigel, Arnold E(lliott)
Shah, Shirish
Shapiro, Anatole Morris
Shotland, Martin
Sicotte, Raymond L
Sikora, Jerome Paul
Slawsky, Zaka I
Snow, George Abraham
Soln, Josip Zvonimir
Solow, Max
Sorrows, Howard Earle
Spangler, Glenn Edward
Spornick, Lynna
Stecker, Floyd William
Stevens, Howard Odell, (Jr)
Stone, Albert Mordecai
Strauser, Wilbur Alexander
Strittmater, Richard Carlton
Stroh, William Richard
Strombotne, Richard L(amar)
Sutter, David Franklin
Swartzendruber, Lydon James
Sztankay, Zoltan Geza
Taylor, Barry Norman
Toller, Gary Neil
Townsend, John William, Jr
Tropf, William Jacob
Turnrose, Barry Edmund
Varma, Matesh Narayan
Venkatesan, Thirumalai
Vette, James Ira
Von Bun, Friedrich Otto
Walker, Robert Lee
Walker, Ronald Elliot
Walter, John Fitler
Wang, Theodore Joseph
Wang, Ting-I
Ward, Alford L
Wasilik, John H(uber)
Weber, Joseph
Weller, Charles Stagg, Jr
Whicker, Lawrence R
Whitmore, Bradley Charles
Willett, Colin Sidney
Wimenitz, Francis Nathaniel
Winslow, George Harvey
Wood, Howard John, III
Wortman, Roger Matthew
Wyckoff, Harold Orville
Wyckoff, James M
Yin, Lo I
Young, Frank Coleman
Young, Russell Dawson
Zimmermann, Mark Edward
Zink, Sandra
Zirkind, Ralph
Zucker, Paul Alan

MASSACHUSETTS
Abbott, Norman John
Abkowitz, Martin A(aron)
Afsar, Mohammed Nurul
Aisenberg, Sol
Aldrich, Ralph Edward
Allemand, Charly D
Altshuler, Edward E
Baker, Adolph
Band, Hans Eduard
Barbour, William E, Jr
Barrington, A(lfred) E(ric)
Baxter, Lincoln, II
Bedo, Donald Elro
Benedek, George Bernard
Bennett, C Leonard
Bennett, Stewart
Berko, Stephan
Bertozzi, William
Bjorkholm, Paul J
Bliss, David Francis
Bolt, Richard Henry
Boness, Michael John
Bowness, Colin
Brandt, Bruce Losure
Brown, Fielding
Burke, Bernard Flood
Burton, Charles Jewell
Butler, James Preston
Butterworth, George A M
Button, Kenneth J
Caley, Wendell J, Jr
Callerame, Joseph
Cardon, Bartley Lowell
Carr, George Leroy
Carroll, Thomas Joseph
Caswell, Randall Smith
Cavallaro, Mary Caroline
Celmaster, William Noah
Chalmers, Bruce
Champion, Kenneth Stanley Warner
Champion, Paul Morris
Chang, Tien Sun (Tom)
Chase, Charles Elroy, Jr

Chretien, Max
Clancy, Edward Philbrook
Clark, Melville, Jr
Clifton, Brian John
Cook, LeRoy Franklin, Jr
Coor, Thomas
Coppi, Bruno
Costello, Ernest F, Jr
Cox, Richard T(hre!keld)
Cromer, Alan H
Dale, Brian
Davies, John A
Davis, Charles Freeman, Jr
Decowski, Piotr
Demos, Peter Theodore
Deutsch, Martin
Dion, Andre R
Donoghue, John Francis
Doshi, Anil G
Douglas-Hamilton, Diarmaid H
Dunlap, William Crawford
Eklund, Karl E
Emshwiller, Maclellan
Emslie, Alfred George
Ernst, Martin L
Estin, Robert William
Faissler, William L
Fante, Ronald Louis
Feshbach, Herman
Fireman, Edward Leonard
Fletcher, Robert Chipman
Foulis, David James
Frankel, Richard Barry
Freedman, George
French, Robert Dexter
Freyre, Raoul Manuel
Friedman, Jerome Isaac
Frisk, George Vladimir
Gamota, George
Garelick, David Arthur
Gelman, Harry
Gentile, Thomas Joseph
Gericke, Otto Reinhard
Gettner, Marvin
Glassbrenner, Charles J
Glauber, Roy Jay
Goldsmith, George Jason
Goldstein, Jack Stanley
Golowich, Eugene
Gordon, Joel Ethan
Greenstein, George
Greytak, Thomas John
Grisaru, Marcus Theodore
Guernsey, Janet Brown
Haber-Schaim, Uri
Hager, Bradford Hoadley
Hallgren, Richard E
Hallock, Robert B
Harling, Otto Karl
Harrison, Edward Robert
Hawkins, Bruce
Hayes, Dallas T
Hearn, David Russell
Heller, Ralph
Hellman, William S
Henry, Allan Francis
Herlin, Melvin Arnold
Heroux, Leon J
Heyda, Donald William
Hinteregger, Hans Erich
Hohenemser, Christoph
Holton, Gerald
Holway, Lowell Hoyt, Jr
Horwitz, Paul
Hovorka, John
Hull, Gordon Ferrie, Jr
Hull, Robert Joseph
Humphrey, Charles Harve
Hutzenlaub, John F
Hyman, Howard Allan
Ingham, Kenneth R
Itzkan, Irving
Jackson, Francis J
Jagannathan, Kannan
Janes, George Sargent
Jarrell, Joseph Andy
Jasperse, John R
Javan, Ali
Jekeli, Christopher
Jensen, Barbara Lynne
Johnson, John Clark
Jones, Phillips Russell
Jones, Robert Allan
Kanter, Irving
Karas, John Athan
Karo, Douglas Paul
Keck, James Collyer
Keene, Wayne Hartung
Kelland, David Ross
Kendall, Henry Way
King, Ronold (Wyeth Percival)
King, William Connor
Kingston, Robert Hildreth
Kirsch, Lawrence Edward
Kleiman, Herbert
Klein, Claude A
Kleinmann, Douglas Erwin
Kleppner, Daniel
Kofsky, Irving Louis
Kohin, Roger Patrick
Koster, George Fred
Kovaly, John J
Krieger, Allen Stephen
Kuo, Lawrence C

Physics, General (cont)

Lai, Shu Tim
Lanza, Giovanni
Lanza, Richard Charles
Lees, Wayne Lowry
Leiby, Clare C, Jr
Lemnios, William Zachary
Lempicki, Alexander
Lewis, Henry Rafalsky
Lewis, Margaret Nast
Liuima, Francis Aloysius
Lowder, J Elbert
Lowndes, Robert P
Lynton, Ernest Albert
Lyons, Donald Herbert
McCarthy, Kathryn Agnes
McCune, James E
McElroy, Michael Brendan
Mack, Charles Lawrence, Jr
Mack, Michael E
McNutt, Ralph Leroy, Jr
Mahon, Harold P
Mailloux, Robert Joseph
Malkus, Willem Van Rensselaer
Maloney, William Thomas
Marino, Richard Matthew
Markstein, George Henry
Marrero, Hector G
Marshall, Donald James
Masi, James Vincent
Massa, Frank
Masso, Jon Dickinson
Mathieson, Alfred Herman
Mavroides, John George
Maxwell, Emanuel
May, John Elliott, Jr
Mazur, Eric
Michelson, Louis
Milburn, Richard Henry
Mittler, Arthur
Montgomery, Donald Bruce
Moody, Elizabeth Anne
Morrison, Philip
Morse, Richard Stetson
Murray, Stephen V
Myers, John Martin
Nebolsine, Peter Eugene
Nelson, Donald Frederick
Nowak, Welville B(erenson)
Obermayer, Arthur S
O'Connor, John Joseph
O'Neill, Edward Leo
Orowan, Egon
Osepchuk, John M
Owyang, Gilbert Hsiaopin
Padovani, Francois Antoine
Pandorf, Robert Clay
Patrick, Richard Montgomery
Payne, Richard Steven
Peterson, Gerald Alvin
Petrasso, Richard David
Petschek, Harry E
Pettengill, Gordon Hemenway
Phillips, Richard Arlan
Pirri, Anthony Nicholas
Pless, Irwin Abraham
Post, Benjamin
Pound, Robert Vivian
Preston, William M
Purcell, Edward Mills
Pyle, Robert Wendell, Jr
Raemer, Harold R
Ramsey, Norman Foster, Jr
Randall, Charles Hamilton
Rao, K V N
Rediker, Robert Harmon
Redner, Sidney
Renner, Gerard W
Rheinstein, John
Richardson, Robert Esplin
Ring, Paul Joseph
Riseberg, Leslie Allen
Roberts, Louis W
Robinson, Howard Addison
Rogers, Howard Gardner
Romer, Robert Horton
Rosato, Frank Joseph
Rose, Peter Henry
Rossi, Bruno B
Rudenberg, H(ermann) Gunther
Safford, Richard Whiley
Sampson, John Laurence
Sandler, Sheldon Samuel
Sauro, Joseph Pio
Schenck, Hilbert Van Nydeck, Jr
Schild, Rudolph Ernest
Schleif, Robert Ferber
Schorr, Marvin Gerald
Schwartz, Jack
Seibel, Frederick Truman
Semon, Mark David
Shapira, Yaacov
Shapiro, Irwin Ira
Shaughnessy, Thomas Patrick
Shepherd, Freeman Daniel, Jr
Shiffman, Carl Abraham
Shirn, George Aaron
Skrable, Kenneth William
Smith, Alan Bradford
Smith, Luther W
Sodickson, Lester A
Soltzberg, Leonard Jay
Stanbrough, Jesse Hedrick, Jr

Stephen, Ralph A
Stone, Richard Spillane
Storer, James E(dward)
Stratton, Julius Adams
Strauss, Bruce Paul
Strimling, Walter Eugene
Strong, John (Donovan)
Stump, Robert
Sullivan, James Douglas
Swift, Arthur Reynders
Taff, Laurence Gordon
Tancrell, Roger Henry
Tangherlini, Frank R
Taylor, Edwin Floriman
Tessman, Jack Robert
Thaddeus, Patrick
Tonry, John Landis
Toscano, William Michael
Tsipis, Kosta M
Tuchman, Avraham
Turchinetz, William Ernest
Tzeng, Wen-Shian Vincent
Van Meter, David
Vaughn, Michael Thayer
Vignos, James Henry
Wachman, Harold Yehuda
Wadey, Walter Geoffrey
Walker, Christopher Bland
Wang, An
Waymouth, John Francis
Weeks, Dorothy Walcott
Weiner, Stephen Douglas
Weiss, Jerald Aubrey
Weiss, Rainer
Weiss, Richard Jerome
Weisskopf, Victor Frederick
Whitaker, John Scott
White, Lawrence S
Wiederhold, Pieter Rijk
Wild, John Frederick
Willis, Charles Richard
Wilson, Richard
Winer, Bette Marcia Tarmey
Winston, Arthur William
Wong, Chuen
Worrell, Francis Toussaint
Yannas, Ioannis Vassilios
Yukon, Stanford P
Zamenhof, Robert G A
Zeiger, Herbert J
Zemon, Stanley Alan
Zimmerman, George Ogurek
Zombeck, Martin Vincent

MICHIGAN

Axelrod, Daniel
Baldwin, Keith Malcolm
Ball, George A(ppleton)
Bardwick, John, III
Barnes, James Milton
Baxter, William John
Becchetti, Frederick Daniel
Bleil, Carl Edward
Bornemeier, Dwight D
Bretz, Michael
Brockmeier, Richard Taber
Chang, Jhy-Jiun
Chen, Juei-Teng
Christian, Larry Omar
Chupp, Timothy E
Coucouvanis, Dimitri N
Cowen, Jerry Arnold
Crane, Horace Richard
DeGraaf, Donald Earl
Edwards, Thomas Harvey
Felmlee, William John
Foiles, Carl Luther
Fry, David Lloyd George
Gessert, Walter Louis
Graves, Bruce Bannister
Hagenlocker, Edward Emerson
Haidler, William B(ernard)
Haynes, Sherwood Kimball
Hazen, Wayne Eskett
Hendel, Alfred Z
Holter, Marvin Rosenkrantz
Inghram, Mark Gordon
Jacko, Michael George
Jamerson, Frank Edward
Janecke, Joachim Wilhelm
Johnson, Fred Tulloch
Johnson, Wayne Jon
Kaiser, Christopher B
Karkheck, John Peter
Kauppila, Walter Eric
Keeling, Rolland Otis, Jr
Keyes, Paul Holt
Kikuchi, Chihiro
Kim, Yeong Wook
King, John Swinton
Koistinen, Donald Peter
Kruglak, Haym
LaRocca, Anthony Joseph
Laukonis, Joseph Vainys
Levich, Calman
Li, Chi-Tang
Lindsay, Robert Kendall
Lippmann, Seymour A
Loeber, Adolph Paul
Loh, Edwin Din
Malhiot, Robert
Marburger, Richard Eugene
Mayer, Frederick Joseph
Meitzler, Allen Henry

Menard, Albert Robert, III
Meyer, Donald Irwin
Montgomery, Donald Joseph
Muench, Nils Lilienberg
Musinski, Donald Louis
Nadasen, Aruna
Nichols, Nathan Lankford
Nims, John Buchanan
Nine, Harmon D
Nitz, David F
O'Neal, Russell D
Peters, James John
Peterson, Edward Charles
Pollack, Gerald Leslie
Polmanteer, Keith Earl
Potnis, Vasant Raghunath
Robbins, John Alan
Roe, Byron Paul
Roessler, David Martyn
Rohrer, Douglas C
Rolnick, William Barnett
Roth, Richard Francis
Rouze, Stanley Ruple
Rowland, Sattley Clark
Sanders, Barbara A
Sanders, Theodore Michael, Jr
Schroeder, Peter A
Segall, Stephen Barrett
Sell, Jeffrey Alan
Shure, Fred C(harles)
Silver, Robert
Spence, Robert Dean
Stearns, Martin
Stewart, Melbourne George
Summerfield, George Clark
Tibbetts, Gary George
Tiffany, Otho Lyle
Trentelman, George Frederick
Turley, Sheldon Gamage
Vander Velde, John Christian
Van Putten, James D, Jr
Vaz, Nuno A
Vincent, Dietrich H(ermann)
Ward, John Frank
Wasielewski, Paul Francis
Weinreich, Gabriel
Williamson, Robert Marshall
Wong, Victor Kenneth
Wright, Wayne Mitchell
Zietlow, James Philip

MINNESOTA

Aagard, Roger L
Anderson, Frances Jean
Anderson, Willard Eugene
Bailey, Carl Leonard
Blair, John Morris
Burkstrand, James Michael
Campbell, Charles Edwin
Casserberg, Bo R
Cederberg, James W
Cohen, Arnold A
Cohen, Philip Ira
Collins, Robert Joseph
Cornelissen Guillaume, Germaine G
Dahlberg, Duane Arlen
DiBona, Peter James
Dickson, Arthur Donald
Dierssen, Gunther Hans
Eckroth, Charles Angelo
Flower, Terrence Frederick
Freier, Phyllis S
Fritts, Robert Washburn
Gallo, Charles Francis
Gerding, Dale Nicholas
Glick, Forrest Irving
Goldman, Allen Marshall
Graves, Wayne H(aigh)
Grimsrud, David T
Hamermesh, Morton
Hanson, Howard Grant
Hatcher, John Burton
Heinisch, Roger Paul
Heltemes, Eugene Casmir
Huang, Cheng-Cher
Johnson, Robert Glenn
Jorgenson, Gordon Victor
Keffer, Charles Joseph
Kroening, John Leo
Lesikar, Arnold Vincent
Lindgren, Gordon Edward
McClure, Benjamin Thompson
McKinney, James T
Mahmoodi, Parviz
Marquit, Erwin
Marshak, Marvin Lloyd
Mikkelson, Raymond Charles
Mitchell, Michael A
Mordue, Dale Lewis
Mueller, Rolf Karl
Nathan, Marshall I
Ney, Edward Purdy
Nier, Alfred Otto Carl
Norberg, Arthur Lawrence
Pepin, Robert Osborne
Pontinen, Richard Ernest
Pou, Wendell Morse
Reilly, Richard J
Reitz, Robert Alan
Rubens, Sidney Michel
Ruddick, Keith
Schmit, Joseph Lawrence
Scriven, L E(dward), (II)
Sipson, Roger Fredrick

Tang, Yau-Chien
Theisen, Wilfred Robert
Thornton, Arnold William
Van Der Ziel, Aldert
Waddington, Cecil Jacob
Warner, Raymond M, Jr
Watkins, Ivan Warren
Wehner, Gottfried Karl
Weyhmann, Walter Victor
Whitehead, Andrew Bruce
Winckler, John Randolph
Yasko, Richard N
Youngner, Philip Genevus
Zimmermann, William, Jr
Zoltai, Tibor

MISSISSIPPI

Britt, James Robert
Bruce, John Goodall, Jr
Butler, Dwain Kent
Colonias, John S
Cress, Daniel Hugg
Croft, Walter Lawrence
Cullen, Abbey Boyd, Jr
Green, Albert Wise
Grissinger, Earl H
Howell, Everette Irl
Jones, Gordon Ervin
Lestrade, John Patrick
Lewis, Arthur B
Miller, Donald Piguet
Reidy, James Joseph
Rundel, Robert Dean
Shields, Fletcher Douglas

MISSOURI

Anderson, Harold D
Ard, William Bryant, Jr
Banaszak, Leonard Jerome
Bogar, Thomas John
Carstens, John C
Chan, Albert Sun Chi
Corey, Eugene R
Dimmock, John O
Fedders, Peter Alan
Friedlander, Michael Wulf
Fuller, Harold Q
Gard, Janice Koles
Geilker, Charles Don
Gingrich, Newell Shiffer
Grayson, Michael A
Hale, Barbara Nelson
Henson, Bob Londes
Hill, John Joseph
Hilton, Wallace Atwood
Hohenberg, Charles Morris
Holroyd, Louis Vincent
Hultsch, Roland Arthur
Kenyon, Allen Stewart
Klarmann, Joseph
Leonhard, Frederick Wilhelm
Lowe, Forrest Gilbert
McFarland, Charles Elwood
McFarland, Robert Harold
McIntosh, Harold Leroy
Major, Schwab Samuel, Jr
Manson, Donald Joseph
Mashhoon, Bahram
Moss, Frank Edward
Nichols, Robert Ted
Nicholson, Eugene Haines
Norberg, Richard Edwin
Nothdurft, Robert Ray
Peery, Larry Joe
Phillips, James M
Phillips, Russell Allan
Philpot, John Lee
Querry, Marvin Richard
Rawlings, Gary Don
Riley, James A
Russell, George A
Scandrett, John Harvey
Schmidt, Bruno (Francis)
Schmidt, Paul Woodward
Shull, Franklin Buckley
Snoble, Joseph Jerry
Stacey, Larry Milton
Takano, Masaharu
Thro, Mary Patricia
Thurman, Robert Ellis, II
Tompson, Clifford Ware
Townsend, Jonathan
Tsoulfanidis, Nicholas
Waring, Richard C
White, Henry W

MONTANA

Caughlan, Georgeanne Robertson
Hayden, Richard John
Holter, Norman Jefferis
Kirkpatrick, Larry Dale
Lapeyre, Gerald J
Robinson, Clark Shove, Jr
Rugheimer, Norman MacGregor
Smith, Richard James

NEBRASKA

Byerly, Paul Robertson, Jr
Davies, Kennard Michael
French, Walter Russell, Jr
Guenther, Raymond A
Hansen, Bernard Lyle
Hardy, Robert J
Jaswal, Sitaram Singh

Katz, Robert
Kirby, Roger D
Leichner, Peter K
Sachtleben, Clyde Clinton
Surkan, Alvin John
Swartzendruber, Dale
Woollam, John Arthur
Zepf, Thomas Herman

NEVADA
Barnes, George
Case, Clinton Meredith
Dunipace, Donald William
Frazier, Thomas Vernon
Hudson, James Gary
Jobst, Joel Edward
Kliwer, James Karl
Levine, Paul Hersh
Marsh, David Paul
Messenger, George Clement
Monroe, Eugene Alan
Odencrantz, Frederick Kirk
Prater, C(harles) D(wight)

NEW HAMPSHIRE
Beasley, Wayne M(achon)
Comerford, Matthias F(rancis)
Davis, William Potter, Jr
Dawson, John Frederick
Doyle, William Thomas
Houston, Robert Edgar, Jr
Jolly, Stuart Martin
Kasper, Joseph F, Jr
Kaufmann, Richard L
Kidder, John Newell
King, Allen Lewis
Lockwood, John Alexander
MacGillivray, Jeffrey Charles
McMickle, Robert Hawley
Mellor, Malcolm
Meyer, James Wagner
Mower, Lyman
Mulhern, John E, Jr
Pipes, Paul Bruce
Taffe, William John
Temple, Peter Lawrence
Webber, William R

NEW JERSEY
Allen, Jonathan
Arunasalam, Vickramasingam
Baltzer, Philip Keene
Behrens, Herbert Ernest
Belt, Roger Francis
Bernstein, Alan D
Berreman, Dwight Winton
Berthold, Joseph Ernest
Bjorkholm, John Ernst
Blumberg, William Emil
Bobeck, Andrew H
Bomke, Hans Alexander
Boorse, Henry Abraham
Bornstein, Lawrence A
Brawer, Steven Arnold
Brennan, John Joseph
Bridges, Thomas James
Briggs, George Roland
Brown, Joe Ned, Jr
Buchsbaum, Solomon Jan
Butler, Herbert I
Carr, Herman Yaggi
Channin, Donald Jones
Cherry, William Henry
Cladis, Patricia Elizabeth Ruth
Cohen, Samuel Alan
Courtney-Pratt, Jeofry Stuart
Curran, Robert Kyran
Curtis, Thomas Hasbrook
DeMarco, Ralph Richard
Dimock, Dirck L
Easley, James W
Eisner, Philip Nathan
Elkholy, Hussein A
Ellis, Robert Anderson, Jr
Epstein, Seymour
Fajans, Jack
Farber, Elliot
Feinblum, David Alan
Feldman, Martin
Fischer, Traugott Erwin
Fitch, Val Logsdon
Fredericks, Robert Joseph
Gautreau, Ronald
Geschwind, Stanley
Geshner, Robert Andrew
Girit, Ibrahim Cem
Glaberson, William I
Goldstein, Philip
Goldstein, Sheldon
Gordon, Eugene Irving
Gordon, James Power
Gottlieb, Melvin Burt
Grauman, Joseph Uri
Greenberg, Arthur
Greene, Laura H
Grimes, Charles Cullen
Grove, Donald Jones
Gruner, Sol Michael
Gunther-Mohr, Gerard Robert
Halemane, Thirumala Raya
Hall, Herbert Joseph
Hamann, Donald Robert
Harrington, David Rogers
Hartung, John

Hemmendinger, Henry
Hillier, James
Hinnov, Einar
Horn, David Nicholas
Hu, Pung Nien
Huang, John S
Hutter, Edwin Christian
Irgon, Joseph
Irvine, Merle M
Jayaraman, Aiyasami
Jobes, Forrest Crossett, Jr
Johnson, Larry Claud
Kahng, Dawon
Kaminow, Ivan Paul
Kaugerts, Juris E
Kingery, Bernard Troy
Kogelnik, H W
Kojima, Haruo
Kraus, Hubert Adolph
Kronenberg, Stanley
Kudman, Irwin
Kurshan, Jerome
La, Sung Yun
Lang, David (Vern)
Langer, William David
Lau, Ngar-Cheung
Layzer, Arthur James
Lebowitz, Joel Louis
Leupold, Herbert August
Levi, Barbara Goss
Levine, Jerry David
Lin, Dong Liang
Ling, Hung Chi
Maa, Jer-Shen
McCumber, Dean Everett
MacLennan, Carol G
McPherson, Ross
McRae, Eion Grant
Manickam, Janardhan
Marcuse, Dietrich
Matey, James Regis
Matula, Richard Allen
Mayhew, Thomas R
Miller, David A B
Miller, Gabriel Lorimer
Miller, Robert Alan
Mims, William B
Misra, Raj Pratap
Mitchell, John Peter
Nahory, Robert Edward
Nardi, Vittorio
Nash, Franklin Richard
Natapoff, Marshall
Neidhardt, Walter Jim
Nelson, Terence John
Newsome, Ross Whitted
Nicholls, Gerald P
O'Gorman, James
O'Keeffe, David John
Ollom, John Frederick
O'Neill, Gerard Kitchen
Ostermayer, Frederick William, Jr
Owens, David Kingston
Papantonopoulou, Aigli Helen
Pargellis, Andrew Nason
Patel, Chandra Kumar Naranbhai
Patterson, Omar Leroy
Penzias, Arno A(llan)
Peterson, George Earl
Pinczuk, Aron
Piroue, Pierre Adrien
Poate, John Milo
Polak, Joel Allan
Polk, Conrad Joseph
Pollock, Franklin
Polye, William Ronald
Pregger, Fred Titus
Presby, Herman M
Ramanan, V R V
Rastani, Kasra
Reder, Friedrich H
Reisman, Otto
Riddle, George Herbert Needham
Robbins, Allen Bishop
Rogers, Eric Malcolm
Rose, Albert
Rose, Carl Martin, Jr
Roukes, Michael L
Salzarulo, Leonard Michael
Sauer, John A
Schilling, Gerd
Schlink, F(rederick) J(ohn)
Schmidt, George
Schneider, Martin V
Schneider, Sol
Schubert, Rudolf
Shah, Atul A
Shastry, B Sriram
Shelton, James Churchill
Shyamsunder, Erramilli
Siegel, Andrew Francis
Silbernagel, Bernard George
Silfvast, William Thomas
Simon, Ralph Emanuel
Slusher, Richart Elliott
Slusky, Susan E G
Smith, Stephen Roger
Snitzer, Elias
Southgate, Peter David
Spruch, Grace Marmor
Stamer, Peter Eric
Steben, John D
Stephen, Michael John
Strauss, Walter

Subramanyam, Dilip Kumar
Summerfield, Martin
Tell, Benjamin
Tenzer, Rudolf Kurt
Thornton, William Andrus, Jr
Thouret, Wolfgang E(mery)
Tomlinson, Walter John, III
Towner, Harry H
Trambarulo, Ralph
Treiman, Sam Bard
Trimmer, William S(tuart)
Tzanakou, M Evangelia
Ullrich, Felix Thomas
Van Der Veen, James Morris
Vossen, John Louis
Wachtell, George Peter
Walsh, Peter
Watson, Hugh Alexander
Webster, William Merle
Weimer, Paul Kessler
Weinrich, Marcel
Weissmann, Sigmund
Wieder, Sol
Wilkinson, David Todd
Wong, King-Lap
Wong, Yiu-Huen
Wood, Darwin Lewis
Woodward, J Guy
Worley, Robert Dunkle
Yevick, George Johannus

NEW MEXICO
Ackerhalt, Jay Richard
Adler, Richard John
Agnew, Lewis Edgar, Jr
Allred, John C(aldwell)
Alpert, Seymour Samuel
Anderson, Herbert L
Anderson, Robert Alan
Baldwin, George C
Balestrini, Silvio J
Banister, John Robert
Barfield, Walter David
Barker, Lynn Marshall
Barnaby, Bruce E
Beckner, Everet Hess
Begay, Frederick
Best, George Harold
Bleyl, Robert Lingren
Booth, Lawrence A(shby)
Bostick, Winston Harper
Bradbury, Norris Edwin
Brannon, Paul J
Brenton, June Grimm
Brook, Marx
Brower, Keith Lamar
Brownell, John Howard
Broyles, Carter D
Bruce, Charles Wikoff
Burr, Alexander Fuller
Caldwell, Stephen E
Campbell, Katherine Smith
Chabai, Albert John
Chan, Kwok-chi Dominic
Clements, William Earl
Cobb, Donald D
Coburn, Horace Hunter
Colgate, Stirling Auchincloss
Conner, Jerry Power
Cowan, Maynard, Jr
Crane, Patrick Conrad
Cuthrell, Robert Eugene
Damerow, Richard Aasen
Davey, William George
Davis, L(loyd) Wayne
Davis, William Chester
Daw, Harold Albert
Deal, William E, Jr
Dick, Jerry Joel
Dick, Richard Dean
Dingus, Ronald Shane
Dressel, Ralph William
Duncan, Richard H(enry)
Elliott, Paul M
Evans, Winifred Doyle
Ewing, Ronald Ira
Feldman, Donald William
Flynn, Edward Robert
Fowler, Clarence Maxwell
Freeman, Bruce L, Jr
George, Michael James
Gibbs, Terry Ralph
Gillespie, Claude Milton
Ginsparg, Paul H
Godfrey, Thomas Nigel King
Godwin, Robert Paul
Goldstein, Louis
Goldstone, Philip David
Good, William Breneman
Graeber, Edward John
Graham, Robert Albert
Green, John Root
Griffin, Patrick J
Guenther, Arthur Henry
Hagerman, Donald Charles
Hall, Jerome William
Harrison, Charles Wagner, Jr
Hartley, Danny L
Hatcher, Charles Richard
Heckman, Richard Cooper
Heller, John Philip
Henins, Ivars
Hillsman, Matthew Jerome
Hoeft, Lothar Otto

Hopson, John Wilbur, Jr
Houston, Jack E
Hoyt, Harry Charles
Huestis, Stephen Porter
Hurd, Jon Rickey
Iwan, DeAnn Colleen
Jacobson, Abram Robert
Janney, Donald Herbert
Jason, Andrew John
Johannes, Robert
Johnson, Ralph T, Jr
Jones, Mark Wallon
Jones, Merrill C(alvin)
Kelly, Robert Emmett
Kennedy, Jerry Dean
King, James Claude
King, L D Percival
Knapp, Edward Alan
Koehler, Dale Roland
Kraus, Alfred Andrew, Jr
Krohn, Burton Jay
Kunz, Kaiser Schoen
La Delfe, Peter Carl
Langner, Gerald Conrad
Lanter, Robert Jackson
Lawrence, James Neville Peed
Lawrence, Raymond Jeffery
Leavitt, Christopher Pratt
Lee, David Mallin
Lewis, H(arold) Ralph
Lincoln, Richard Criddle
Lindstrom, Ivar E, Jr
Linsley, John
Liu, Lon-Chang
Luehr, Charles Poling
Lysne, Peter C
McClure, Gordon Wallace
McKibben, Joseph L
McLeod, John
Maley, Martin Paul
Mansfield, Charles Robert
Matzen, Maurice Keith
Mautz, Charles William
Mead, William C
Menlove, Howard Olsen
Metzger, Daniel Schaffer
Miller, Glenn Houston
Mogford, James A
Mueller, Marvin Martin
Nagle, Darragh (Edmund)
Neeper, Donald Andrew
Newman, Michael J(ohn)
Nickel, George H(erman)
Nielson, Clair W
Norris, Carroll Boyd, Jr
Olson, Gordon Lee
Overton, William Calvin, Jr
Paciotti, Michael Anthony
Panitz, Janda Kirk Griffith
Parker, Jack Lindsay
Parker, Jerald Vawer
Paul, Robert Hugh
Payton, Daniel N, III
Peng, Jen-chieh
Peterson, Alan W
Petschek, Albert George
Poukey, James W
Price, Robert Harold
Rabie, Ronald Lee
Ragan, Charles Ellis, III
Ram, Budh
Raymond, David James
Regener, Victor H
Ribe, Fred Linden
Rinehart, John Sargent
Robertson, Robert Graham Hamish
Roof, Raymond Bradley, Jr
Rosen, Louis
Sanders, Walter L
Schappert, Gottfried T
Scharn, Herman Otto Friedrich
Schelberg, Arthur Daniel
Schery, Stephen Dale
Schiferl, David
Schillaci, Mario Edward
Schlosser, Jon A
Schriber, Stanley Owen
Schultz, Rodney Brian
Schwarz, Ricardo
Schwoebel, Richard Lynn
Scully, Marlan Orvil
Seay, Glenn Emmett
Simmons, James E
Smith, Carl Walter
Smith, William Conrad
Spalding, Dan Wesley
Spillman, George Raymond
Stellingwerf, Robert Francis
Stout, Virgil L
Stratton, William R
Stromberg, Thorsten Frederick
Strong, Ian B
Summers, Donald Lee
Swingle, Donald Morgan
Swinson, Derek Bertram
Sydoriak, Stephen George
Sze, Robert Chia-Ting
Talley, Thurman Lamar
Taschek, Richard Ferdinand
Taylor, Thomas Newton
Terrell, N(elson) James, Jr
Theimer, Otto
Thieme, Melvin T
Thompson, Joe David

Physics, General (cont)

Thorn, Robert Nicol
Trainor, Robert James
Travis, James Roland
Trela, Walter Joseph
Turner, Terry Earle
Van Domelen, Bruce Harold
Van Heuvelen, Alan
Van Riper, Kenneth Alan
Varnum, William Sloan
Venable, Douglas
Vook, Frederick Ludwig
Vortman, L(uke) J(erome)
Wackerle, Jerry Donald
Wallace, Jon Marques
Wallace, Richard Kent
Walsh, John M
Wangler, Thomas P
Warnes, Richard Harry
Weaver, Robert Paul
Weiss, Paul B
Wenzel, Robert Gale
White, Paul C
Wilkening, Marvin H
Wilkins, Ronald Wayne
Winkler, Max Albert
Wolfe, Peter Nord
Wright, Bradford Lawrence
Wright, Thomas Payne
Wu, Adam Yu
Wylie, Kyral Francis
Yarger, Frederick Lynn

NEW YORK

Adams, Peter D
Alpher, Ralph Asher
Altemose, Vincent O
Altman, Joseph Henry
Ames, Irving
Arons, Michael Eugene
Bak, Per
Bakis, Raimo
Barton, Mark Q
Batterman, Boris William
Bazer, Jack
Bean, Charles Palmer
Berry, Herbert Weaver
Bhargava, Rameshwar Nath
Bigelow, John Edward
Birnbaum, David
Blachman, Arthur Gilbert
Blessing, Robert Harry
Blewett, John P
Bloch, Alan
Blume, Martin
Bolon, Roger B
Borst, Lyle Benjamin
Bouyoucos, John Vinton
Box, Harold C
Boyer, Donald Wayne
Braslau, Norman
Breed, Henry Eltinge
Brehm, Lawrence Paul
Brinkman, William F
Brody, Burton Alan
Broers, Alec N
Brosious, Paul Romain
Brown, David Chester
Brown, Gerald E
Brown, Howard Howland, Jr
Burnham, Dwight Comber
Burns, Gerald
Buttiker, Markus
Cambey, Leslie Alan
Carrier, Gerald Burton
Carroll, Clark Edward
Casabella, Philip A
Case, John William
Cassel, David Giske
Chang, Leroy L
Chaudhari, Praveen
Chessin, Henry
Chi, Benjamin E
Chiang, Joseph Fei
Chow, Che Chung
Chung, Victor
Churchill, Melvyn Rowen
Clark, David Lee
Cleare, Henry Murray
Cohen, Ezechiel Godert David
Cohen, Martin O
Constable, James Harris
Cool, Terrill A
Coppens, Philip
Corfield, Peter William Reginald
Cotter, Maurice Joseph
Csermely, Thomas J(ohn)
Cusano, Dominic A
Damouth, David Earl
Danby, Gordon Thompson
Danielson, Lee Robert
Das, Ashok Kumar
Davey, Paul Oliver
Davitian, Harry Edward
Debiak, Ted Walter
Dempsey, Daniel Francis
DeNoyer, Linda Kay
Desilets, Brian H
DeTitta, George Thomas
Diamond, Joshua Benamy
Doering, Charles Rogers
Donaldson, Robert Rymal
Donnelly, Denis Philip

Donovan, John Leo
Douglass, David Holmes
Dove, Derek Brian
Drauglis, Edmund
Duax, William Leo
Dube, Roger Raymond
Durkovic, Russell George
Edelstein, William Alan
Eisenstadt, Maurice
Elder, Fred Kingsley, Jr
Engelke, Charles Edward
Erlbach, Erich
Esaki, Leo
Eshbach, John Robert
Espersen, George A
Evans, Wayne Russell
Fabricand, Burton Paul
Falk, Theodore J(ohn)
Feldman, William
Ferrari, Lawrence A
Finch, Thomas Lassfolk
Fleischer, Robert Louis
Fleisher, Harold
Foley, Henry Michael
Forgacs, Gabor
Fossan, David B
Fowler, Alan B
Fox, Herbert Leon
Fullwood, Ralph Roy
Furst, Milton
Galginaitis, Simeon Vitis
Garber, Meyer
Garretson, Craig Martin
Garrett, Charles Geoffrey Blythe
Gasparini, Francis Marino
Gavin, Donald Arthur
Gelernter, Herbert Leo
Gelman, Donald
Ghosh, Arup Kumar
Giaever, Ivar
Giese, Rossman Frederick, Jr
Glascock, Homer Hopson, II
Glaser, Herman
Glickman, Walter A
Goertzel, Gerald
Goldberg, Conrad Stewart
Goldberg, Joshua Norman
Golden, Robert K
Goldhaber, Gertrude Scharff
Goldhaber, Maurice
Goldman, Vladimir J
Goldstein, Herbert
Good, Myron Lindsay
Good, William E
Goodman, Gerald Joseph
Gould, Robert Henderson
Gove, Harry Edmund
Graf, Erlend Haakon
Greisen, Kenneth I
Grischkowsky, Daniel Richard
Grumet, Alex
Gundlach, Robert William
Gunn, John Battiscombe
Hafner, Theodore
Hagen, Carl Richard
Ham, Frank Slagle
Hanau, Richard
Harbison, Gerard Stanislaus
Harke, Douglas J
Harker, David
Harper, Richard Allan
Harris, Donald R, Jr
Hart, Hiram
Hart, Robert John
Hartill, Donald L
Hartmann, George Cole
Hartmann, Sven Richard
Harvey, Alexander Louis
Hauptman, Herbert Aaron
Headrick, Randall L
Heberle, Juergen
Hecht, Eugene
Helfand, David John
Hendrie, Joseph Mallam
Henkel, Elmer Thomas
Henshaw, Clement Long
Hinson, David Chandler
Ho, Ping-Pei
Hoard, James Lynn
Hodgson, Rodney
Hogan, William
Honig, Arnold
Hopkins, Robert Earl
Horwitz, Nahmin
Howard, Richard John
Howe, David Allen
Hsieh, Shih-Yung
Huang, Jack Shih Ta
Hunter, Lloyd Philip
Hurd, Jeffery L
Jack, Hulan E, Jr
Jacobsen, Edward Hastings
Jaffe, Bernard Mordecai
Jaffe, Morry
Janak, James Francis
Jarvis, James Gordon
Jekeli, Walter
Johnson, Peter Dexter
Johnson, Woodrow Eldred
Jona, Franco Paul
Jones, Keith Warlow
Jones, William Barclay
Joynson, Reuben Edwin, Jr
Kalikstein, Kalman

Kao, Yi-Han
Kaplon, Morton Fischel
Keck, Donald Bruce
Kelly, Martin Joseph
Keyes, Robert William
Khare, Bishun Narain
Kinsella, John J
Kinsey, Kenneth F
Kirchgessner, Joseph L
Kiszenick, Walter
Klahr, Carl Nathan
Klaiber, George Stanley
Kleinman, Chemia Jacob
Koch, Herman William
Koehler, Richard Frederick, Jr
Kohler, Robert Henry
Kouts, Herbert John Cecil
Kramer, Paul Robert
Krewer, Semyon E
Kriss, Michael Allen
Krongelb, Sol
Kuper, J B Horner
Ladd, John Herbert
Laibowitz, Robert (Benjamin)
Lam, Kai Shue
Lang, Harry George
La Tourrette, James Thomas
Lawson, James Llewellyn
Lea, Robert Martin
Lee, Charles Alexander
Lee, David Morris
Leigh, Richard Woodward
Lessor, Arthur Eugene, Jr
Levin, Eugene (Manuel)
Levinger, Joseph S
Levinstein, Henry
Levitas, Alfred Dave
Li, Kelvin K
Litke, John David
Litynski, Daniel Mitchell
Liu, Yung Sheng
Lloyd, James Newell
Love, William Alfred
Lovejoy, Derek R
Lowder, Wayne Morris
Lubkin, Gloria Becker
McGuire, Stephen Craig
McKinley, William Albert
Malone, Dennis P(hilip)
Marple, Dudley Tyng Fisher
Marrone, Paul Vincent
Mason, Max Garrett
Maurer, Robert Distler
May, John Walter
Mayer, James W(alter)
Mehran, Farrokh
Melkonian, Edward
Mendelsohn, Lawrence Barry
Mendez, Emilio Eugenio
Menes, Meir
Mermin, N David
Metcalf, Harold
Michael, Paul Andrew
Middleton, David
Miller, Allen H
Miller, Melvin J
Mishkin, Eli Absalom
Mistry, Nariman Burjor
Mitchell, Joan LaVerne
Mittleman, Marvin Harold
Mogro-Campero, Antonio
Moore, Kenneth Howard
Morris, Thomas Wendell
Mueller, Alfred H
Mutter, Walter Edward
Nankivell, John (Elbert)
Nation, John
Nedderman, Howard Charles
Nelson, Clarence Norman
Nethercot, Arthur Hobart, Jr
Nicholson, William Jamieson
Nisteruk, Chester Joseph
Noonan, Thomas Wyatt
Nordin, Paul
O'Connor, Cecilian Leonard
O'Donoghue, Aileen Ann
Ogden, Philip Myron
Oja, Tonis
Okaya, Yoshi Haru
Orenstein, Albert
Ortel, William Charles Gormley
Pai, Damodar Mangalore
Palmer, Robert B
Parratt, Lyman George
Parthasarathy, Rengachary
Paskin, Arthur
Passoja, Dann E
Pe, Maung Hla
Pearson, George John
Peckham, Donald Charles
Penfield, Robert Harrison
Pennebaker, William B, Jr
Phillips, Melba (Newell)
Pierucci, Olga
Pilcher, Valter Ennis
Pimbley, Walter Thornton
Piore, Emanuel Ruben
Platner, Edward D
Pomerantz, Melvin
Pomian, Ronald J
Pribil, Stephen
Prince, Jack
Pritchard, Robert Leslie
Prodell, Albert Gerald

Pullman, Ira
Rahm, David Charles
Raka, Eugene Cd
Range, R Michael
Rau, R Ronald
Rautenberg, Theodore Herman
Read, Albert James
Reilly, Edwin David, Jr
Reisinger, Joseph G
Reppy, John David
Resnick, Robert
Rezanka, Ivan
Richards, James Austin, Jr
Richardson, Robert Coleman
Rini, Frank John
Robinson, William Kirley
Robusto, C Carl
Roetling, Paul G
Rohr, Robert Charles
Romer, Alfred
Rosar, Madeleine E
Rosen, Paul
Rosenberg, Paul
Rowe, Irving
Rowe, Raymond Grant
Rubin, Kenneth
Ruddick, James John
Russell, Allan Melvin
Rutz, Richard Frederick
Ryan, Donald F
Sachs, Allan Maxwell
Sadoff, Ahren J
Salinger, Gerhard Ludwig
Samios, Nicholas Peter
Sampson, William B
Sarjeant, Walter James
Sato, Makiko
Scheib, Richard, Jr
Schick, Jerome David
Schubert, Walter L
Schulman, Lawrence S
Schultheis, James J
Schwall, Robert E
Schwartz, Ernest
Schwartz, Melvin
Schwartz, Melvin J
Seitz, Frederick
Seka, Wolf
Sells, Robert Lee
Senus, Walter Joseph
Serber, Robert
Sharbaugh, Amandus Harry
Shelupsky, David I
Shen, Samuel Yi-Wen
Siemann, Dietmar W
Silsbee, Henry Briggs
Silsbee, Robert Herman
Simon, William
Simpson, James Henry
Skiff, Peter Duane
Skorinko, George
Smart, James Samuel
Smilowitz, Bernard
Smith, Bernard
Smith, Gail Preston
Smith, George David
Smith, Lyle W
Smoyer, Claude B
Sobell, Henry Martinique
Sperber, Daniel
Spoelhof, Charles Peter
Springett, Brian E
Stachel, Johanna
Stannard, Carl R, Jr
Stearns, Robert L
Stenzel, Wolfram G
Stephenson, Thomas E(dgar)
Sternglass, Ernest Joachim
Stier, Paul Max
Stith, James Herman
Stolov, Harold L
Stoner, George Green
Stover, Raymond Webster
Strassenburg, Arnold Adolph
Strnisa, Fred V
Strome, Forrest C, Jr
Strong, Herbert Maxwell
Strozier, John Allen, Jr
Struzynski, Raymond Edward
Stull, John Leete
Sumberg, David A
Swartz, Charles Dana
Swartz, Clifford Edward
Szydlik, Paul Peter
Talman, Richard Michael
Tan, Yen T
Tang, Chung Liang
Taylor, Jack Eldon
Taylor, Theodore Brewster
Teegarden, Kenneth James
Teichmann, Theodor
Teiger, Martin
Tesser, Herbert
Thomas, Richard Alan
Thorndike, Edward Moulton
Tiersten, Martin Stuart
Titone, Luke Victor
Treanor, Charles Edward
Trexler, Frederick David
Triebwasser, Sol
Trischka, John Wilson
Vance, Miles Elliott
Van Steenbergen, Arie
Venugopalan, Srinivasa I

Vizy, Kalman Nicholas
Von Keszycki, Carl Heinrich
Vosburgh, Kirby Gannett
Vought, Robert Howard
Wagner, Peter Ewing
Wajda, Edward Stanley
Waldrop, Ann Lyneve
Wang, Charles T P
Wang, Wen I
Webster, Harold Frank
Wessel, Gunter Kurt
Whetten, Nathan Rey
White, Frederick Andrew
Wiesner, Leo
Williams, Graheme John Bramald
Williams, Ross Edward
Williamson, Robert Samuel
Wilson, Jack Martin
Wilson, Robert Rathbun
Winhold, Edward John
Wolf, Edward D
Wolfe, Robert Norton
Wolga, George Jacob
Worthington, Thomas Kimber
Wu, Chien-Shiung
Wu, Tsu Ming
Yalow, A(braham) Aaron
Yang, Chen Ning
Young, James Roger
Young, Ralph Howard
Yuan, Luke Chia Liu
Zadoff, Leon Nathan
Zajac, Alfred
Zeitz, Louis
Ziegler, Robert C(harles)
Zwick, Daan Marsh

NORTH CAROLINA
Baker, David Kenneth
Bilpuch, Edward George
Bowers, Wayne Alexander
Briscoe, Charles Victor
Byrd, James William
Christian, Wolfgang C
Clark, Chester William
Clement, John Reid, Jr
Collins, Clifford B
Connolly, Walter Curtis
Daggerhart, James Alvin, Jr
Davis, William Robert
D'Eustachio, Dominic
Dotson, Marilyn Knight
Evans, Lawrence Eugene
Evans, Ralph Aiken
Fairbank, Henry Alan
Freedman, Stuart Jay
Gould, Christopher Robert
Gray, Walter C(larke)
Griffith, Wayland C(oleman)
Haase, David Glen
Hageseth, Gaylord Terrence
Hall, Peter M
Hocken, Robert John
Horie, Yasuyuki
Hovis, Louis Samuel
Howard, Clarence Edward
Ingram, Peter
Jacobson, Kenneth Allan
Jenzano, Anthony Francis
Jones, Creighton Clinton
Kratz, Howard Russel
Lado, Fred
Lawless, Philip Austin
Lawson, Dewey Tull
Lontz, Robert Jan
McEnally, Terence Ernest, Jr
Manring, Edward Raymond
Mayes, Terrill W
Memory, Jasper Durham
Menius, Arthur Clayton, Jr
Meyer, Horst
Mitchell, Earl Nelson
Murray, Raymond L(eRoy)
Oates, Jimmie C
O'Foghludha, Fearghus Tadhg
Palmatier, Everett Dyson
Pollak, Victor Louis
Prater, John Thomas
Read, Floyd M
Reardon, Anna Joyce
Robinson, Hugh Gettys
Russell, Lewis Keith
Saby, John Sanford
Sayetta, Thomas C
Schetzina, Jan Frederick
Seagondollar, Lewis Worth
Sharkoff, Eugene Gibb
Smith, Fred R, Jr
Snodgrass, Rex Jackson
Stephenson, Harold Patty
Stroscio, Michael Anthony
Sumney, Larry W
Suttle, Jimmie Ray
Thomas, Henry Coffman
Thomas, Llewellyn Hilleth
Thompson, Nancy Lynn
Varlashkin, Paul
Vermillion, Robert Everett
Wallingford, John Stuart
Walters, Mark David
Way, Katharine
Werntz, James Herbert, Jr
Wills, James E, Jr
Wilson, Geoffrey Leonard

Wolfe, Hugh Campbell

NORTH DAKOTA
Bale, Harold David
Berkey, Gordon Bruce
Kraus, Olen
Lykken, Glenn Irven
McCarthy, Gregory Joseph
Muraskin, Murray
Sinha, Mahendra Kumar

OHIO
Agard, Eugene Theodore
Anderson, David Leonard
Andrus, Paul Grier
Anno, James Nelson
Becker, Lawrence Charles
Beer, Albert Carl
Bhattacharya, Ashok Kumar
Bishop, Edwin Vandewater
Bohn, Randy G
Boughton, Robert Ivan, Jr
Breitenberger, Ernst
Brown, L Carlton
Brown, Robert William
Burnside, Phillips Brooks
Byers, Walter Hayden
Campbell, Bernerd Eugene
Capriotti, Eugene Raymond
Carome, Edward F
Carrabine, John Anthony
Cassim, Joseph Yusuf Khan
Cobb, Thomas Berry
Cochran, Thomas Howard
Cochran, William Ronald
Coffin, Frances Dunkle
Conrad, Malcolm Alvin
Cook, James Robert
Cooke, Charles C
Crandall, Arthur Jared
Crandell, Merrell Edward
Crouch, Marshall Fox
Dahm, Arnold J
Damian, Carol G
Dickey, Frederick Pius
Diller, Violet Marion
Dillman, Lowell Thomas
Duncan, George Comer
Earhart, Richard Wilmot
Eck, Thomas G
Edwards, David Olaf
Elwell, David Leslie
Erickson, Richard Ames
Essig, Gustave Alfred
Farrell, David E
Fawcett, Sherwood Luther
Fernelius, Nils Conard
Fineberg, Herbert
Forman, Ralph
Foster, Edward Staniford, Jr
Frye, Glenn McKinley, Jr
Gaines, Gordon Bradford
Ganguly, Bishwa Nath
Garrow, Robert Joseph
Globe, Samuel
Greenberg, William Michael
Greene, David C
Greenslade, Thomas Boardman, Jr
Griffin, Charles Frank
Grimes, Steven Munroe
Gruber, H Thomas
Guth, Sylvester Karl
Hancock, George Whitmore, Jr
Hanzely, Stephen
Haque, Azeez C
Harpster, Joseph
Hart, John Birdsall
Hathaway, Charles Edward
Haybron, Ronald M
Henderson, H(oman) Thurman
Hsu, Hsiung
Hunt, Earle Raymond
Hunter, George Truman
Hunter, Joseph Lawrence
Hunter, William Winslow, Jr
Jones, Dale Robert
Jossem, Edmund Leonard
Kelly, Kevin Anthony
Kereiakes, James Gus
Kerr, Robert Lowell
Kime, Joseph Martin
Klein, Robert Herbert
Knecht, Walter Ludwig
Koch, Ronald Joseph
Kollen, Wendell James
Kolopajlo, Lawrence Hugh
Kornylak, Andrew T
Larson, Lee Edward
Laskowski, Edward L
Lutz, Arthur Leroy
McConville, George T
McCown, Robert Bruce
McGraw, Delford Armstrong
Machlup, Stefan
Mallozzi, Philip James
Manning, Robert M
Marshall, Kenneth D
Merchant, Mylon Eugene
Milford, Frederick John
Miller, Don Wilson
Miller, Franklin, Jr
Miller, Harry Galen
Miller, James Roland
Miller, Raymond Edwin

Mochel, Virgil Dale
Nam, Sang Boo
Nelson, Richard Carl
Nelson, William Frank
Nielsen, Carl Eby
Nikkel, Henry
Nolan, James Francis
Notz, William Irwin
Palffy-Muhoray, Peter
Palmer, William Franklin
Palmieri, Joseph Nicholas
Peterson, Leroy Eric
Piper, Ervin L
Proctor, David George
Pronko, Peter Paul
Rao, Kandarpa Narahari
Reynolds, Donald C
Richards, Walter Bruce
Riley, William Robert
Rollins, Roger William
Rothstein, Jerome
Rudy, Yoram
Rutledge, Wyman Coe
Sanderson, Richard Blodgett
Saupe, Alfred (Otto)
Schlegel, Donald Louis
Schneider, James Roy
Seldin, Emanuel Judah
Shaffer, Wave H
Shaw, John H
Shetterly, Donivan Max
Silvidi, Anthony Alfred
Snider, John William
Soules, Jack Arbuthnott
Spielberg, Nathan
Stephenson, Francis Creighton
Stewart, Charles Neil
Sticksel, Philip Rice
Strickler, Thomas David
Stumpf, Folden Burt
Sugarbaker, Evan Roy
Sundaralingam, Muttaiya
Tabakoff, Widen
Tanaka, Katsumi
Taylor, Beverley Ann Price
Thourson, Thomas Lawrence
Toman, Karel
Tough, James Thomas
Uhrich, David Lee
Uralil, Francis Stephen
Vanderburg, Vance Dilks
Vanderlind, Merwyn Ray
Van Horn, David Downing
Versic, Ronald James
Von Meerwall, Ernst Dieter
Wada, Walter W
Wagoner, Glen
Wallis, Robert L
Weimer, David
Weinberg, Irving
Wheeler, Samuel Crane, Jr
White, John Joseph, III
Wickersham, Charles Edward, Jr
Williamson, William, Jr
Wilson, Charles Woodson, III
Wilson, Kenneth Geddes
Wittebort, Jules I
Witten, Louis
Wolfe, Paul Jay
Wood, Van Earl
Yu, Greta
Yun, Seung Soo
Zdanis, Richard Albert

OKLAHOMA
Albright, James Curtice
Bilger, Hans Rudolf
Boade, Rodney Russett
Breedlove, James Robby, Jr
Crowe, Christopher
Davies-Jones, Robert Peter
Evens, F Monte
Fischbeck, Helmut J
Hartsuck, Jean Ann
Harwood, William H
Hill, Benny Joe
Hoover, Gary McClellan
Kantowski, Ronald
Kuenhold, Kenneth Alan
Lange, James Neil, Jr
Leivo, William John
McKeever, Stephen William Spencer
Marks, L Whit
Mintz, Esther Uress
Naymik, Daniel Allan
Renner, John Wilson
Rutledge, Carl Thomas
Thomsen, Leon
Troelstra, Arne
Weems, Malcolm Lee Bruce
Williams, Thomas Henry Lee
Yoesting, Clarence C

OREGON
Anderson, Tera Lougenia
Arthur, John Read, Jr
Bates, David James
Beall, Robert Allan
Belinfante, Frederik J
Bell, Anthony E
Biehl, Arthur Trew
Brady, James Joseph
Brodie, Laird Charles
Burch, David Stewart

Charbonnier, Francis Marcel
Chezem, Curtis Gordon
Decker, Fred William
Evett, Jay Fredrick
Graham, Beardsley
Griffiths, David John
Gurevitch, Mark
Hawkinson, Stuart Winfield
Henke, Burton Lehman
Hinrichs, Clarence H
Holmes, James Frederick
Jones, Robert Edward
Katen, Paul C
Kolar, Oscar Clinton
Kunkle, Donald Edward
Martin, Robert Leonard
Matthews, Brian Wesley
Montague, Daniel Grover
Nicodemus, David Bowman
Orrok, George Timothy
Overley, Jack Castle
Powell, John Leonard
Purbrick, Robert Lamburn
Rayfield, George W
Rempfer, Gertrude Fleming
Reynolds, Robert Eugene
Rosenthal, Jenny Eugenie
Sarles, Lynn Redmon
Smith, Robert Lloyd
Stephas, Paul
Stringer, Gene Arthur
Swenson, Leonard Wayne
Taylor, Carson William
Towe, George Coffin
Zimmerman, Robert Lyman

PENNSYLVANIA
Abramson, Stanley L
Acheson, Willard Phillips
Alexander, Frank Creighton, Jr
Allen, Edward Franklin
Alstadt, Don Martin
Anderle, Richard
Anderson, Carl Einar
Anderson, Owen Thomas
Angello, Stephen James
Auerbach, Leonard B
Augensen, Harry John
Ayyangar, Komanduri M
Banks, Grace Ann
Barmby, David Stanley
Bell, Raymond Martin
Bennett, Lee Cotton, Jr
Benson, Brent W
Berger, Martin
Berman, Martin
Blasie, J Kent
Bortz, Alfred Benjamin
Bose, Shyamalendu M
Boughn, Stephen Paul
Boyer, Robert Allen
Brody, Howard
Brown, Richard Leland
Bruns, Charles Alan
Burnett, Roger Macdonald
Butkiewicz, Edward Thomas
Buzzelli, Edward S
Camarda, Harry Salvatore
Carbone, Robert James
Carfagno, Salvatore P
Carr, Walter James, Jr
Carriker, Roy C
Cashdollar, Kenneth Leroy
Chan, Moses Hung-Wai
Cheng, Kuang Liu
Cheston, Warren Bruce
Christensen, S(abinus) H(oegsbro)
  (Chris)
Cohen, Anna (Foner)
Cole, Milton Walter
Cook, Donald Bowker
Cowan, David J
Crawford, John David
Creagan, Robert Joseph
Cross, Leslie Eric
Custer, Hubert Minter
Dahlke, Walter Emil
Davies, D K
Dick, George W
Diehl, Renee Denise
Dowling, John
Duffin, Richard James
Edelstein, Richard Malvin
Eisner, Robert Linden
Engheta, Nader
Engler, Arnold
Engstrom, Ralph Warren
Fabish, Thomas John
Fahey, Paul Farrell
Feuchtwang, Thomas Emanuel
Fite, Wade Lanford
Foderaro, Anthony Harolde
Foland, William Douglas
Friedberg, Simeon Adlow
Fuget, Charles Robert
Fugger, Joseph
Garfunkel, Myron Paul
Geller, Kenneth N
Georgopulos, Peter Demetrios
Gilstein, Jacob Burrill
Giltinan, David Anthony
Glickstein, Stanley S
Glusker, Jenny Pickworth
Goldberg, Norman

Sharber, James Randall
Shepherd, Jimmie George
Shieh, Paulinus Shee-Shan
Siegfried, Robert Wayne, II
Smith, Daniel Montague
Smith, R Lowell
Sobey, Arthur Edward, Jr
Spitznogle, Frank Raymond
Stark, J(ohn) P(aul), Jr
Stebbings, Ronald Frederick
Stefanski, Raymond Joseph
Stewart, Frank Edwin
Stubbs, Norris
Stucker, Harry T
Swink, Laurence N
Swinney, Harry Leonard
Sybert, James Ray
Synek, Miroslav (Mike)
Taboada, John
Taylor, Herbert Lyndon
Theobald, J Karl
Thomas, Estes Centennial, III
Thomas, Richard Garland
Thompson, James Chilton
Tittel, Frank K(laus)
Tittle, Charles William
Tompkins, Donald Roy, Jr
Trafton, Laurence Munro
Truxillo, Stanton George
Unterberger, Robert Ruppe
VantHull, LORIN L
Vant-Hull, Lorin Lee
Walker, Robert Hugh
Wallick, George Castor
Webb, Theodore Stratton, Jr
White, Gifford
Wiggins, James Wendell
Wilheit, Thomas Turner
Wilkerson, Robert C
Williams, Robert Leroy
Wisseman, William Rowland
Young, Bernard Theodore
Young, David Thad
Youngblood, Dave Harper
Zimmerman, Carol Jean

UTAH
Baker, Kay Dayne
Bryner, John C
Dibble, William E
Dick, Bertram Gale
Dudley, J Duane
Eastmond, Elbert John
Gibbs, Peter (Godbe)
Grow, Richard W
Hall, Howard Tracy
Harris, Franklin Stewart, Jr
Hill, Armin John
Jamison, Ronald D
Jones, Merrell Robert
Kadesch, Robert R
Lee, James Norman
Lind, Vance Gordon
Loh, Eugene C
Megill, Lawrence Rexford
Merrill, John Jay
Rasband, S Neil
Symko, Orest George
Ward, Roger Wilson
Wood, John Karl

VERMONT
Bertelsen, Bruce I(rving)
Blatt, Frank Joachim
Campbell, Donald Edward
Clark, Donald Lyndon
Detenbeck, Robert Warren
Golden, Kenneth Ivan
Lane, George H
MacArthur, John Wood
McIntire, Sumner Harmon
Scarfone, Leonard Michael
Walters, Virginia F

VIRGINIA
Adams, Clifford Lowell
Aitken, Alfred H
Anthony, Lee Saunders
Atkins, Marvin C(leveland)
Berman, Alan
Biberman, Lucien Morton
Bierly, James N, Jr
Bloomfield, Louis Aub
Bomse, Frederick
Boring, John Wayne
Bowker, David Edwin
Braddock, Joseph V
Brandt, Richard Gustave
Brown, Arlin James
Brown, Ellen Ruth
Bruno, Ronald C
Bunner, Alan Newton
Burner, Alpheus Wilson, Jr
Burton, Larry C(lark)
Calle, Carlos Ignacio
Cameron, Louis McDuffy
Carpenter, Delma Rae, Jr
Carr, Roger Byington
Carter, William Walton
Caton, Randall Hubert
Clay, Forrest Pierce, Jr
Coleman, Robert Vincent
Collins, George Briggs
Connolly, John Irving, Jr

Crawford, George William
Crowley, Patrick Arthur
Davey, John Edmund
Davidson, Robert Bellamy
Dennison, Brian Kenneth
Desiderio, Anthony Michael
Dockery, John T
Dolezalek, Hans
Donaghy, James Joseph
Duykers, Ludwig Richard Benjamin
Dylla, Henry Frederick
Edlund, Milton Carl
Ehrlich, Robert
Epstein, Samuel David
Featherston, Frank Hunter
Findlay, John Wilson
Flory, Thomas Reherd
Frost, Robert T
Gamble, Thomas Dean
Garrick, Isadore Edward
Gibson, Luther Ralph
Gillies, George Thomas
Goncz, John Henry
Goodwin, Francis E
Grandy, Charles Creed
Grant, Frederick Cyril
Graybeal, Walter Thomas
Grosch, Chester Enright
Gugelot, Piet C
Ham, William Taylor, Jr
Harries, Wynford Lewis
Harrison, Mark
Hartline, Beverly Karplus
Harvalik, Zaboj Vincent
Hereford, Frank Loucks, Jr
Herwig, Lloyd Otto
Hilger, James Eugene
Hodge, Donald Ray
Hohl, Frank
Hooper, William John, Jr
Hoppe, John Cameron
Hove, John Edward
Hoy, Gilbert Richard
Hurdle, Burton Garrison
Ifft, Edward M
Ivanetich, Richard John
Jacobs, Ira
Johnson, Charles Minor
Johnson, Charles Nelson, Jr
Jones, Terry Lee
Kalab, Bruno Marie
Keefe, William Edward
Kelly, Hugh P
Kelly, Peter Michael
Kernell, Robert Lee
Kiess, Edward Marion
Koomen, Martin J
Kossler, William John
Krafft, Joseph Martin
Krueger, Peter George
Kuhlmann-Wilsdorf, Doris
Landes, Hugh S(tevenson)
Lapsley, Alwyn Cowles
Leemann, Christoph Willy
Leiss, James Elroy
Lewis, Clark Houston
Lilly, Arnys Clifton, Jr
Lowance, Franklin Elta
Lowitz, David Aaron
Lowry, Ralph A(ddison)
McLucas, John L(uther)
McNutt, Douglas P
Major, Robert Wayne
Margrave, Thomas Ewing, Jr
Marsh, Howard Stephen
Marshall, Samuel Wilson
Melfi, Leonard Theodore, Jr
Mikesell, Jon L
Minkin, Jean Albert
Minnix, Richard Bryant
Misner, Robert David
Mitchell, John Wesley
Mitchell, Richard Scott
Mo, Luke Wei
Molloy, Charles Thomas
Morgan, Joseph
Moritz, Barry Kyler
Moyer, Leroy
Mulder, Robert Udo
Neher, Dean Royce
Neubauer, Werner George
Neuendorffer, Joseph Alfred
Newbolt, William Barlow
Ni, Chen-Chou
Nikolic, Nikola M
Outlaw, Ronald Allen
Padgett, Doran William
Peacock, Richard Wesley
Pettus, William Gower
Porzel, Francis Bernard
Raney, William Perin
Reisler, Donald Laurence
Reiss, Keith Westcott
Remler, Edward A
Reynolds, Peter James
Rice, Roy Warren
Ritter, Rogers C
Ritz, Victor Henry
Robertson, Randal McGavock
Rogers, Thomas F
Rosi, Fred
Rothenberg, Herbert Carl
Ruffine, Richard S
Ruvalds, John

Satterthwaite, Cameron B
Sauder, William Conrad
Saunders, Peter Reginald
Schone, Harlan Eugene
Seiler, Steven Wing
Sepucha, Robert Charles
Sessler, John Charles
Shapiro, Maurice Mandel
Shore, David
Simpson, John Arol
Singh, Jag Jeet
Sloope, Billy Warren
Sobottka, Stanley Earl
Spivack, Harvey Marvin
Staib, Jon Albert
Stauss, George Henry
Stewart, John Westcott
Stronach, Carey E
Taggart, Keith Anthony
Taylor, Gerald Reed, Jr
Taylor, Jackson Johnson
Thaler, William John
Thompson, Anthony Richard
Thorp, Benjamin A
Tidman, Derek Albert
Tiller, Calvin Omah
Tipsword, Ray Fenton
Ulrich, Dale V
Veazey, Sidney Edwin
Via, Giorgio G
Wagner, Richard Lorraine, Jr
Wallace, Lance Arthur
Warren, Holland Douglas
Watson, Robert Barden
Weik, Martin H, Jr
Welch, Jasper Arthur, Jr
Welsh, Robert Edward
Wessel, Paul Roger
Williams, Willie, Jr
Wilsdorf, Heinz G(erhard) F(riedrich)
Wolfhard, Hans Georg
Woods, Roy Alexander
Yeager, Paul Ray

WASHINGTON
Anderson, Roger Harris
Astley, Eugene Roy
Auth, David C
Barnes, Ross Owen
Bennett, Robert Bowen
Bichsel, Hans
Bodansky, David
Brown, Bert Elwood
Brown, Robert Reginald
Bunch, Wilbur Lyle
Buske, Norman L
Cannon, William Charles
Carter, John Lemuel, Jr
Center, Robert E
Clayton, Eugene Duane
Cook, Victor
Dash, Jay Gregory
DeRemer, Russell Jay
De Shazer, Larry Grant
Donaldson, Edward Enslow
Dunning, Kenneth Laverne
Dye, David L
Ellis, Fred E
Ewart, Terry E
Farwell, George Wells
Ghose, Subrata
Gil, Salvador
Gillis, Murlin Fern
Grootes, Pieter Meiert
Halpern, Isaac
Harrison, Melvin Arnold
Haslund, R L
Hayward, Thomas Doyle
Hildebrandt, Jacob
Hinman, George Wheeler
Hoverson, Sigmund John
Hunting, Alfred Curtis
Ingalls, Robert L
Kells, Lyman Francis
Kissinger, Homer Everett
Knapp, Robert Hazard, Jr
Krienke, Ora Karl, Jr
Lagergren, Carl Robert
Lathrop, Arthur LaVern
Leonard, Bowen Raydo, Jr
Lessor, Delbert Leroy
Livdahl, Philip V
Lloyd, Raymond Clare
Long, Daniel R
Lord, Jere Johns
Lorenz, Richard Arnold
Lowell, Sherman Cabot
Lundy, Richard Alan
McDermott, Lillian Christie
Mitchell, Robert Curtis
Mobley, Curtis Dale
Murphy, Stanley Reed
Nalos, Ervin Joseph
Nelson, Martin Emanuel
Nicholson, Richard Benjamin
Nornes, Sherman Berdeen
Park, James Lemuel
Parks, George Kung
Paul, Ronald Stanley
Peck, Edson Ruther
Penning, John Russell, Jr
Peters, Philip Carl
Reudink, Douglas O
Richmond, James Kenneth

Riehl, Jerry A
Rothberg, Joseph Eli
Sandstrom, Donald Richard
Schmid, Loren Clark
Schmidt, Fred Henry
Smith, David Clement, IV
Snodgrass, Herschel Roy
Snover, Kurt Albert
Soreide, David Christien
Spillman, Richard Jay
Spinrad, Bernard Israel
Stenkamp, Ronald Eugene
Strelb, John Fredrick
Tang, Kwong-Tin
Thomas, Montcalm Tom
Van Heerden, Pieter Jacobus
Vilches, Oscar Edgardo
Walske, Max Carl
White, William Charles
Young, James Arthur, Jr

WEST VIRGINIA
Anderson, Gary Don
Brimhall, James Elmore
Franz, Frank Andrew
Ghigo, Frank Dunnington
Henry, Donald Lee
Jefimenko, Oleg D
Leech, H(arry) William
Manakkil, Thomas Joseph
Meloy, Thomas Phillips
Oberly, Ralph Edwin
Schramm, Robert William
Sheppard, David W
Vehse, William E

WISCONSIN
Anway, Allen R
Auchter, Harry A
Bailey, Sturges Williams
Barclay, John Arthur
Barmore, Frank E
Beck, Donald Edward
Behroozi, Feredoon
Besch, Gordon Otto Carl
Brown, Bruce Elliot
Caston, Ralph Henry
Chander, Jagdish
Cybriwsky, Alex
Dittman, Richard Henry
Ekern, Ronald James
Fossum, Steve P
Friedman, William Albert
Gade, Sandra Ann
Garrett, Robert Ogden
Greene, Jack Bruce
Herb, Raymond George
Hershkowitz, Noah
Hyzer, William Gordon
Johnson, Arthur Franklin
Johnson, DeWayne Carl
Kobiske, Ronald Albert
Kraushaar, William Lester
Lichtman, David
Lodge, Arthur Scott
McQuistan, Richmond Beckett
Mallmann, Alexander James
Margaritondo, Giorgio
Meyer, Robert Paul
Myers, Vernon W
Nutter, Gene Douglas
Papastamatiou, Nicolas
Pillai, Thankappan A K
Reynolds, Ronald J
Rollefson, Ragnar
Roys, Paul Allen
Schmieg, Glenn Melwood
Schneiderwent, Myron Otto
Schultz, Frederick Herman Carl
Seshadri, Sengadu Rangaswamy
Shinners, Carl W
Sikdar, Dhirendra N
Smith, Roy E
Spangler, John David
Stekel, Frank D
Walters, William Le Roy
Webb, Maurice Barnett
Wilson, Volney Colvin
Woods, Frank Robert
Zimm, Carl B

WYOMING
Bessey, Robert John
Edwards, William Charles
Muller, Burton Harlow
Pepin, Theodore John
Rosen, James Martin

PUERTO RICO
Lewis, Brian Murray
Muir, James Alexander

ALBERTA
Bazett-Jones, David Paul
Challice, Cyril Eugene
Coombes, Charles Allan
Huzinaga, Sigeru
Krouse, Howard Roy
Paranjape, Bhalachandra Vishwanath
Pinnington, Eric Henry
Rankin, David
Rood, Joseph Lloyd

Physics, General (cont)

Lee, Peter H Y
Lee, Ping
Leonard, Stanley Lee
Leung, Ka-Ngo
Lichtenberg, Allan J
Litvak, Marvin Mark
Lohr, John Michael
McCurdy, Alan Hugh
McPherron, Robert Lloyd
McWilliams, Roger Dean
Maglich, Bogdan
Malmberg, John Holmes
Manes, Kennneth Rene
Marino, Lawrence Louis
Mark, James Wai-Kee
Marrs, Roscoe Earl
Matossian, Jesse N
Matsuda, Kyoko
Matsuda, Yoshiyuki
Matthews, Stephen M
Max, Claire Ellen
Maxon, Marshall Stephen
Mazurek, Thaddeus John
Mendis, Devamitta Asoka
Mihalov, John Donald
Mitchner, Morton
Mo, Charles Tse Chin
Moir, Ralph Wayne
Molvik, Arthur Warren
Morales, George John
More, Richard Michael
Nevins, William McCay
Newcomb, William A
Noerdlinger, Peter David
Olness, Robert James
Olney, Ross David
O'Neil, Thomas Michael
Orth, Charles Douglas
Osher, John Edward
Overskei, David Orvin
Papas, Charles Herach
Payne, David Glenn
Pellinen, Donald Gary
Pontau, Arthur E
Porter, Gary Dean
Post, Richard Freeman
Pritchett, Philip Lentner
Rathmann, Carl Erich
Reinheimer, Julian
Rensink, Marvin Edward
Rhodes, Edward Joseph, Jr
Rognlien, Thomas Dale
Rosen, Mordecai David
Rugge, Henry F
Russell, Christopher Thomas
Rynn, Nathan
Schulz, Michael
Scott, Franklin Robert
Sentman, Davis Daniel
Shaffner, Richard Owen
Shearer, James Welles
Shelley, Edward George
Sheridon, Nicholas Keith
Shestakov, Aleksei Ilyich
Simonen, Thomas Charles
Smith, Edward John
Smith, Gary Richard
Speck, David Ralph
Spence, Harlan Ernest
Sperling, Jacob L
Stallings, Charles Henry
Stearns, John Warren
Stenzel, Reiner Ludwig
Stevens, John Charles
Stewart, Gordon Ervin
Stewart, John Joseph, Jr
Stoeckly, Robert E
Stuart, George Wallace
Surko, Clifford Michael
Tamano, Teruo
Thomassen, Keith I
Thompson, W P(aul)
Thompson, William Bell
Thorne, Richard Mansergh
Toor, Arthur
Trigger, Kenneth Roy
Tsurutani, Bruce Tadashi
Turchan, Otto Charles
Unti, Theodore Wayne Joseph
Urquidi-MacDonald, Mirna
Van Hoven, Gerard
Vaslow, Dale Franklin
Vernazza, Jorge Enrique
Wada, James Yasuo
Wagner, Carl E
Walker, Raymond John
Walt, Martin
Waltz, Ronald Edward
Wang, Charles P
Wang, Paul Keng Chieh
Weinstein, Berthold Werner
Williams, Edward Aston
Wilson, Andrew Robert
Wong, Alfred Yiu-Fai
Yeh, Yea-Chuan Milton

COLORADO
Aamodt, Richard E
Catto, Peter James
Chappell, Willard Ray
Cooper, John (Jinx)
Kaufman, Harold Richard
Law, William Brough
Lee, Yung-Cheng

Meiss, James Donald
Morrison, Charles Freeman, Jr
Robertson, Scott Harrison
Stern, Raul A(ristide)
Weinstock, Jerome
Zhang, Bing-Rong

CONNECTICUT
Bernstein, Ira Borah
Cheng, David H S
Haught, Alan F
Hirshfield, Jay Leonard
Meyerand, Russell Gilbert, Jr
Scott, Joseph Hurlong
Similon, Philippe Louis
Stein, Richard Jay

DELAWARE
Evenson, Paul Arthur

DISTRICT OF COLUMBIA
Alexander, Joseph Kunkle
Apruzese, John Patrick
Armstrong, Carter Michael
Baker, Dennis John
Book, David Lincoln
Coffey, Timothy
Colombant, Denis Georges
Corliss, Charles Howard
Crandall, David Hugh
Dove, William Francis
Ellis, William Rufus
Feldman, Uri
Gold, Steven Harvey
Grossman, John Mark
Haas, Gregory Mendel
Hart, Richard Cullen
Hubbard, Richard Forest
Joyce, Glenn Russell
Kapetanakos, Christos Anastasios
Killion, Lawrence Eugene
Lampe, Martin
Lynch, John Thomas
Mahon, Rita
Manheimer, Wallace Milton
Manka, Charles K
Nelson, David Brian
Palmadesso, Peter Joseph
Patel, Vithalbhai L
Picone, J Michael
Reilly, Michael Hunt
Ripin, Barrett Howard
Robson, Anthony Emerson
Sethian, John Dasho
Stephanakis, Stavros John
Stone, Philip M
Taylor, Ronald D
Wang, Yi-Ming
Wieting, Terence James

FLORIDA
Baird, Alfred Michael
Dufty, James W
Flynn, Robert W
Hooper, Charles Frederick, Jr
Howard, Bernard Eufinger
Jones, William Denver
Kribel, Robert Edward
Lamborn, Bjorn N A
Linder, Ernest G
Loper, David Eric
Miller, Hillard Craig
Robertson, Harry S(troud)
Stetson, Robert Franklin
Vollmer, James
Weber, Alfred Herman

GEORGIA
Chen, Robert Long Wen
Little, Robert
Stacey, Weston Monroe, Jr

IDAHO
Shaw, Charles Bergman, Jr
Woodall, David Monroe

ILLINOIS
Evans, Kenneth, Jr
Fink, Charles Lloyd
Hatch, Albert Jerold
Henneberger, Walter Carl
Massel, Gary Alan
Mouschovias, Telemachos Charalambous
Pyle, K Roger
Raether, Manfred
Scheeline, Alexander
Skadron, George

INDIANA
Helrich, Carl Sanfrid, Jr
Parish, Jeffrey Lee

IOWA
Carpenter, Raymon T
Case, William Bleicher
D'Angelo, Nicola
Gurnett, Donald Alfred
Houk, Robert Samuel
Knorr, George E
Lonngren, Karl E(rik)
McClelland, John Frederick
Nicholson, Dwight Roy

KANSAS
Carpenter, Kenneth Halsey
Enoch, Jacob
Unz, Hillel

KENTUCKY
Hawkins, Charles Edward
McGinness, William George, III
Ulrich, Aaron Jack

LOUISIANA
Gibbs, Richard Lynn
Jung, Hilda Ziifle
Kumar, Devendra

MAINE
Beard, David Breed

MARYLAND
Behannon, Kenneth Wayne
Benson, Robert Frederick
Bingham, Robert Lodewijk
Birmingham, Thomas Joseph
Bodner, Stephen E
Boyd, Derek Ashley
Brace, Larry Harold
Carbary, James F
Dean, Stephen Odell
Decker, James Frederick
DeSilva, Alan W
Dragt, Alexander James
Embury, Janon Frederick, Jr
Evans, Larry Gerald
Faust, William R
Finn, John McMaster
Fuhr, Jeffrey Robert
Goldenbaum, George Charles
Goldstein, Melvyn L
Granatstein, Victor Lawrence
Gray, Ernest Paul
Grebogi, Celso
Greenblatt, Marshal
Greig, J Robert
Hartle, Richard Eastham
Heppner, James P
Hitchcock, Daniel Augustus
Holman, Gordon Dean
Huddleston, Charles Martin
King, Joseph Herbert
Liu, Chuan Sheng
Lupton, William Hamilton
McFadden, Geoffrey Bey
McLean, Edgar Alexander
Martin, John J(oseph)
Matthews, David LeSueur
Matthews, David LeSueur
Merkel, George
Northrop, Theodore George
Oktay, Erol
Ott, Edward
Oza, Dipak H
Papadopoulos, Konstantinos Dennis
Potemra, Thomas Andrew
Powell, John David
Rhee, Moon-Jhong
Rosado, John Allen
Rosenberg, Theodore Jay
Roszman, Larry Joe
Sari, James William
Scudder, Jack David
Singh, Amarjit
Sittler, Edward Charles, Jr
Skiff, Frederick Norman
Stamper, John Andrew
Stern, David P
Stone, Albert Mordecai
Turner, James Marshall
Turner, Robert
Wiese, Wolfgang Lothar
Wilkerson, Thomas Delaney
Wu, Ching-Sheng
Yoon, Peter Haesung
Young, Frank Coleman

MASSACHUSETTS
Allis, Willam Phelps
Bekefi, George
Berman, Robert Hiram
Bers, Abraham
Bugnolo, Dimitri Spartaco
Burke, William J
Cohn, Daniel Ross
Cooke, Richard A
Cordero, Julio
Denig, William Francis
Douglas-Hamilton, Diarmaid H
Fougere, Paul Francis
Gelman, Harry
Graneau, Peter
Guss, William C
Habbal, Shadia Rifai
Jacob, Jonah Hye
Johnston, George Lawrence
Jose, Jorge V
Kalman, Gabor J
Kautz, Frederick Alton, II
Knecht, David Jordan
Kreisler, Michael Norman
Krieger, Allen Stephen
Lai, Shu Tim
Lax, Benjamin
Leiby, Clare C, Jr
Litzenberger, Leonard Nelson
McNutt, Ralph Leroy, Jr

Marmar, Earl Sheldon
Marshall, Theodore
Oettinger, Peter Ernest
Petrasso, Richard David
Pian, Carlson Chao-Ping
Pilot, Christopher H
Plumb, John Laverne
Porkolab, Miklos
Post, Richard S
Ratner, Michael Ira
Reilly, James Patrick
Rogoff, Gerald Lee
Sigmar, Dieter Joseph
Silevich, Michael B
Silk, John Kevin
Singer, Howard Joseph
Spiegel, Stanley Lawrence
Temkin, Richard Joel
Terry, James Layton
Whitehouse, David R(empfer)
Wolfe, Stephen Mitchell
Woskov, Paul Peter

MICHIGAN
Ancker-Johnson, Betsy
Banks, Peter Morgan
Beglau, David Alan
Berger, Beverly Kobre
Berger, Richard Leo
Bleil, Carl Edward
Clauer, C Robert, Jr
Fontheim, Ernest Gunter
Getty, Ward Douglas
Harrison, Michael Jay
Johnson, Roy Ragnar
Larsen, Jon Thorsten
Lee, Anthony
Mayer, Frederick Joseph
Samir, Uri

MINNESOTA
Chanin, Lorne Maxwell
Heberlein, Joachim Viktor Rudolf
Kellogg, Paul Jesse
Lysak, Robert Louis

MISSISSIPPI
Ferguson, Joseph Luther, Jr
Johnston, Russell Shayne

MISSOURI
Leader, John Carl
Leader, John Carl
McFarland, Robert Harold
Menne, Thomas Joseph
Painter, James Howard
Prelas, Mark Antonio
Rode, Daniel Leon
Shrauner, Barbara Abraham
Willett, Joseph Erwin

MONTANA
Rosa, Richard John

NEW HAMPSHIRE
Boley, Forrest Irving
Forbes, Terry Gene
Hollweg, Joseph Vincent
Hudson, Mary Katherine
Lee, Martin Alan
Montgomery, David Campbell
Ryan, James Michael
Sonnerup, Bengt Ulf osten
Walsh, John Edmond
White, Willard Worster, III

NEW JERSEY
Anandan, Munisamy
Barnes, Derek A
Bernabei, Stefano
Bird, Harvey Harold
Bol, Kees
Bond, Robert Harold
Budny, Robert Vierling
Bush, Charles Edward
Cecchi, Joseph Leonard
Chang, T(ao)-Y(uan)
Channon, Stephen R
Chen, Liu
Cohen, Leonard George
Cohen, Samuel Alan
Davidson, Ronald Crosby
Ellis, Robert Anderson, Jr
Furth, Harold Paul
Gammel, George Michael
Green, Michael Philip
Grek, Boris
Grisham, Larry Richard
Hawryluk, Richard Janusz
Hendel, Hans William
Hosea, Joel Carlton
Hsuan, Hulbert C S
Hulse, Russell Alan
Jassby, Daniel Lewis
Jensen, Betty Klainminc
Johnson, John Lowell
Kaita, Robert
Karney, Charles Fielding Finch
Kugel, Henry W
Kulsrud, Russell Marion
LaMarche, Paul Henry, Jr
Langer, William David
Lanzerotti, Louis John
Machalek, Milton David

SASKATCHEWAN
Hirose, Akira

OTHER COUNTRIES
Beasley, Cloyd O, Jr
Boxman, Raymond Leon
Choi, Duk-In
Chu, Kwo Ray
Crawford, Frederick William
Daly, Patrick William
Dewar, Robert Leith
Dixit, Madhu Sudan
Kimblin, Clive William
Krumbein, Aaron Davis
Lee, Tong-Nyong
Pitchford, Leanne Carolyn
Romero, Alejandro F
Wasa, Kiyotaka
Yanabu, Satoru

Solid State Physics

ALABAMA
Anderson, Elmer Ebert
Chen, An-Ban
Christensen, Charles Richard
Davis, Jack H
Doyle, William David
Fromhold, Albert Thomas, Jr
Harrell, James W, Jr
Kinzer, Earl T, Jr
Kroes, Roger L
Kwon, Tai Hyung
Lal, Ravindra Behari
McDonald, Jack Raymond
Mookherji, Tripty Kumar
Mosley, Ronald Bruce
Naumann, Robert Jordan
Pontius, Duane Henry
Reynolds, Robert Ware
Thaxton, George Donald
Worley, S D

ARIZONA
Bao, Qingcheng
Benin, David B
Beyen, Werner J
Carpenter, Ray Warren
Chamberlin, Ralph Vary
De La Moneda, Francisco Homero
Delinger, William Galen
Dereniak, Eustace Leonard
Dow, John Davis
Emrick, Roy M
Falco, Charles Maurice
Farnsworth, Harrison E
Fliegel, Frederick Martin
Gibbs, Hyatt McDonald
Green, Barry A
Gregory, Brooke
Handy, Robert M(axwell)
Hanson, Roland Clements
Harper, Francis Edward
Hickernell, Fred Slocum
Hohl, Jakob Hans
Hooper, Henry Olcott
Huffman, Donald Ray
Iijima, Sumio
Johnson, Earnest J
Joyce, Richard Ross
Leavitt, John Adams
McDaniel, Terry Wayne
Mahalingam, Lalgudi Muthuswamy
Marzke, Robert Franklin
O'Keefe, Michael Adrian
Page, John Boyd, Jr
Parmenter, Robert Haley
Peyghambarian, Nasser
Ren, Shang Yuan
Rutledge, James Luther
Sarid, Dror
Saul, Robert H
Scholl, Marija Strojnik
Schroder, Dieter K
Scott, Alwyn C
Seraphin, Bernhard Otto
Smith, David John
Stark, Royal William
Stearns, Mary Beth Gorman
Stuart, Derald Archie
Tomizuka, Carl Tatsuo
Venables, John Anthony
Vuillemin, Joseph J
Walker, Charles Thomas
Wang, Edward Yeong

ARKANSAS
Horton, Philip Bish
McCloud, Hal Emerson, Jr
Turner, Thomas Jenkins
Wrobel, Joseph Stephen

CALIFORNIA
Adams, Arnold Lucian
Ahlers, Guenter
Akella, Jagannadham
Albert, Paul A(ndre)
Allen, Frederick Graham
Almason, Carmen Cristina
Alves, Ronald V
Amelio, Gilbert Frank
Ames, Lawrence Lowell
Archer, Robert James

Arias, Jose Manuel
Aston, Duane Ralph
Astrue, Robert William
Atwater, Harry Albert
Aukerman, Lee William
Avrin, William F
Ayers, Raymond Dean
Baboolal, Lal B
Bachrach, Robert Zelman
Bagus, Paul Saul
Bajaj, Jagmohan
Bak, Chan Soo
Banks, Thomas
Bardin, Russell Keith
Barmatz, Martin Bruce
Barnes, Gene A
Barnoski, Michael K
Baron, Robert
Barrera, Joseph S
Bartelink, Dirk Jan
Baskes, Michael I
Bate, Geoffrey
Bates, Clayton Wilson, Jr
Batra, Inder Paul
Bauer, Robert Steven
Bauer, Walter
Beardsley, Irene Adelaide
Beasley, Malcolm Roy
Becker, Milton
Bennett, Alan Jerome
Bennett, Glenn Taylor
Bennett, Harold Earl
Berdahl, Paul Hilland
Bernard, Douglas Alan
Berryman, James Garland
Bethune, Donald Stimson
Biegelsen, David K
Bienenstock, Arthur Irwin
Birecki, Henryk
Bliss, Erlan S
Blois, Marsden Scott, Jr
Blum, Fred A
Boster, Thomas Arthur
Botez, Dan
Bower, Frank H(ugo)
Bowman, Robert Clark, Jr
Bragg, Robert H(enry)
Bratt, Peter Raymond
Bridges, Frank G
Bron, Walter Ernest
Brown, Ronald Franklin
Bryden, John Heilner
Bube, Richard Howard
Bullis, William Murray
Burland, Donald Maxwell
Burmeister, Robert Alfred
Butler, Jack F
Calhoun, Bertram Allen
Cape, John Anthony
Carlan, Alan J
Caspers, Hubert Henri
Chase, Jay Benton
Chau, Cheuk-Kin
Chester, Marvin
Cheung, Jeffrey Tai-Kin
Childs, William Henry
Chin, Maw-Rong
Chodos, Steven Leslie
Chow, Paul C
Claassen, Richard Strong
Clark, William Gilbert
Clarke, John
Clemens, Bruce Montgomery
Cogan, Adrian Ilie
Cohen, Marvin Lou
Cohen, Richard Lewis
Coleman, Charles Clyde
Coleman, Lawrence Bruce
Colwell, Joseph F
Cooper, Donald Edward
Coufal, Hans-Jürgen
Crittenden, Eugene Casson, Jr
Curtis, Orlie Lindsey, Jr
Dagotto, Elbio Ruben
Dalven, Richard
Davies, John Tudor
Davis, Jeffrey Arthur
De, Gopa Sarkar
Dember, Alexis Berthold
Deshpande, Shivajirao M
Detwiler, Daniel Paul
DeWames, Roger
Dobrov, Wadim (Ivan)
Domb, Ellen Ruth (Colmer)
Donovan, Terence M
Duffy, William Thomas, Jr
Duneer, Arthur Gustav, Jr
Dynes, Robert Carr
Eagleton, Robert Don
Eck, Robert Edwin
Eden, Richard Carl
Edwall, Dennis Dean
Edwards, David Franklin
Emmett, John L
Eng, Genghmun
Engel, Jan Marcin
Everett, Glen Exner
Evtuhov, Viktor
Falicov, Leopoldo Maximo
Feng, Joseph Shao-Ying
Fenner, Wayne Robert
Ferber, Robert R
Fetter, Alexander Lees
Fetterman, Harold Ralph

Feuerstein, Seymour
Fink, Herman Joseph
Fisher, Robert Amos, Jr
Fletcher, Paul Chipman
Flinn, Paul Anthony
Foiles, Stephen Martin
Fong, Ching-Yao
Fouquet, Julie
Franks, Larry Allen
Frescura, Bert Louis
Fultz, Brent T
Gabriel, Cedric John
Gaines, Edward Everett
Gamble, Fred Ridley, Jr
Gathers, George Roger
Geballe, Theodore Henry
Gehman, Bruce Lawrence
Gershenzon, M(urray)
Gibson, Atholl Allen Vear
Gill, William D(elahaye)
Gilman, John Joseph
Gilvarry, John James
Glass, Nathaniel E
Goddard, William Andrew, III
Goldberg, Ira Barry
Goldstein, Selma
Good, Roland Hamilton, Jr
Goubau, Wolfgang M
Gould, Christopher M
Grant, Paul Michael
Grant, Ronald W(arren)
Greene, Alan Campbell
Gruber, John B
Gudmundsen, Richard Austin
Gundersen, Martin Adolph
Hackwood, Susan
Haegel, Nancy M
Hafemeister, David Walter
Hagenlocher, Arno Kurt
Hagstrom, Stig Bernt
Haitz, Roland Hermann
Hall, James Timothy
Haller, Eugene Ernest
Hammond, Robert Bruce
Hanna, Stanley Sweet
Harris, Charles Bonner
Harris, James Stewart, Jr
Harrison, Walter Ashley
Hartwick, Thomas Stanley
Heeger, Alan J
Hellman, Frances
Hemminger, John Charles
Hempstead, Robert Douglas
Herb, John A
Herman, Ray Michael
Herring, William Conyers
Hill, Robert Matteson
Hiskes, Ronald
Hoffman, Hanna J
Holzrichter, John Frederick
Hone, Daniel W
Honnold, Vincent Richard
Hopfield, John Joseph
Housley, Robert Melvin
Hsieh, Yu-Nian
Huang, C Yuan
Hubert, Jay Marvin
Hunt, Arlon Jason
Hurrell, John Patrick
Hygh, Earl Hampton
Ikezi, Hiroyuki
Irvine, Stuart James Curzon
Jaccarino, Vincent
Jacobson, Albert H(erman), Jr
Jansen, Michael
Jeanloz, Raymond
Jeffries, Carson Dunning
Johanson, William Richard
Johnson, David Leroy
Johnson, Oliver
Johnson, William Lewis
Junga, Frank Arthur
Kagiwada, Reynold Shigeru
Kahn, Frederic Jay
Kaplan, Daniel Eliot
Karunasiri, Gamani
Kay, Eric
Kedzie, Robert Walter
Kendrick, Hugh
Khan, Mahbub R
Khoshnevisan, Mohsen Monte
Kirchner, Ernst Karl
Kittel, Charles
Kittel, Peter
Klein, August S
Kleitman, David
Knight, Walter David
Knollman, Gilbert Carl
Koliwad, Krishna M
Kosai, Kenneth
Krikorian, Esther
Kroeker, Richard Mark
Kroemer, Herbert
Krupke, William F
Kumar, S
Kupfer, John Carlton
Kyser, David Sheldon
Lamb, Walter Robert
Lao, Binneg Yanbing
Larsen, Ted LeRoy
Lau, S S
Law, Hsiang-Yi David
Lax, Edward
Learn, Arthur Jay

Lee, Kenneth
Lee, Roland Robert
Lee, Ronald S
Lehovec, Kurt
Leifer, Herbert Norman
Lerner, Lawrence S
Leung, Charles Cheung-Wan
Leventhal, Edwin Alfred
Leverton, Walter Frederick
Lewis, Robert Taber
Lind, Maurice David
Lindau, Evert Ingolf
Litvak, Marvin Mark
Liu, Jia-ming
Lockhart, James Marcus
Logan, James Columbus
Longo, Joseph Thomas
Lonky, Martin Leonard
Louie, Steven Gwon Sheng
Luehrmann, Arthur Willett, Jr
McCaldin, J(ames) O(eland)
McColl, Malcolm
Macfarlane, Roger Morton
MacLaughlin, Douglas Earl
McNutt, Michael John
Magnuson, Gustav Donald
Maker, Paul Donne
Maki, Kazumi
Maloney, Timothy James
Marrello, Vincent
Maserjian, Joseph
Masters, Burton Joseph
Mataré, Herbert Franz
Maxfield, Bruce Wright
Mee, Jack Everett
Meike, Annemarie
Merz, James L
Meyer, Stephen Frederick
Miller, Allan Stephen
Miller, David Lee
Mills, Douglas Leon
Minkiewicz, Vincent Joseph
Moerner, William Esco
More, Richard Michael
Moriarty, John Alan
Morin, Francis Joseph
Moriwaki, Melvin M
Moss, Steven C
Muir, Arthur H, Jr
Naqvi, Iqbal Mehdi
Nedoluha, Alfred K
Newman, Roger
Newsam, John M
Novotny, Vlad Joseph
Olness, Dolores Urquiza
Olson, Roy E
Orbach, Raymond Lee
Page, D(errick) J(ohn)
Painter, Ronald Dean
Palmer, John Parker
Paoli, Thomas Lee
Paquette, David George
Parker, William Henry
Parrish, William
Partanen, Jouni Pekka
Pearson, James Joseph
Pehl, Richard Henry
Perkins, Kenneth L(ee)
Persky, George
Peselnick, Louis
Petroff, Pierre Marc
Phillips, Thomas Gould
Phillips, William Baars
Pianetta, Piero Antonio
Picus, Gerald Sherman
Pines, Alexander
Pollak, Michael
Pontau, Arthur E
Portis, Alan Mark
Powell, Ronald Allan
Rabolt, John Francis
Radosevich, Lee George
Rasor, Ned S(haurer)
Redfield, David
Rehn, Victor Leonard
Reinheimer, Julian
Reiss, Howard
Renda, Francis Joseph
Rice, Dennis Keith
Richards, Paul Linford
Richman, Isaac
Rockstad, Howard Kent
Rogers, Robert N
Rohy, David Alan
Rosenblum, Stephen Saul
Ross, Bernd
Ross, Marvin Franklin
Rothrock, Larry R
Rowell, John Martin
Rudge, William Edwin
Rudnick, Joseph Alan
Saffren, Melvin Michael
Sager, Ronald E
Sashital, Sanat Ramanath
Satten, Robert A
Scalapino, Douglas J
Schaefer, Albert Russell
Schechter, Daniel
Schein, Lawrence Brian
Schulte, Alfons F
Schultz, Sheldon
Sclar, Nathan
Scott, John Campbell
Seidel, Thomas Edward

Solid State Physics (cont)

Sham, Lu Jeu
Shank, Charles Vernon
Shaw, Charles Alden
Shea, Michael Joseph
Shelton, Robert Neal
Shen, Yuen-Ron
Shepanski, John Francis
Sher, Arden
Shin, Ernest Eun-Ho
Shin, Soo H
Shing, Yuh-Han
Shore, Herbert Barry
Siegel, Irving
Siegel, Sidney
Silver, Arnold Herbert
Sinclair, Robert
Slivinsky, Sandra Harriet
Smit, Jan
Smith, Darryl Lyle
Smith, Gordon Stuart
Smith, Joe Nelson, Jr
Snow, Edward Hunter
Snowden, Donald Philip
Somoano, Robert Bonner
Spicer, William Edward
Spieler, Helmuth
Spitzer, William George
Sridhar, Champa Guha
Srour, Joseph Ralph
Stahl, Frieda A
Stapelbroek, Maryn G
Stella, Paul M
Stevenson, David John
Stierwalt, Donald L
Stirn, Richard J
Stöhr, Joachim
Stokowski, Stanley E
Stolte, Charles
Stupian, Gary Wendell
Suits, James Carr
Surko, Clifford Michael
Tabak, Mark David
Talley, Robert Morrell
Taylor, William H, II
Tennant, William Emerson
Terhune, Robert William
Thompson, Richard Scott
Thompson, Thomas Eaton
Thorsen, Arthur C
Torrance, Jerry Badgley, Jr
Tsai, Chuang Chuang
Tsai, Ming-Jong
Tseng, Samuel Chin-Chong
Tu, Charles Wuching
Vagelatos, Nicholas
Van Thiel, Mathias
Vorhaus, James Louis
Wagner, Richard John
Walden, Robert Henry
Walker, William Charles
Wallis, Richard Fisher
Wang, Kang-Lung
Warburton, William Kurtz
Weber, Eicke Richard
Weber, Hans Josef
Weber, Marvin John
Webster, Emilia
Weichman, Peter Bernard
Weinberg, Elliot Hillel
Weissman, Robert Henry
Whaley, Katharine Birgitta
White, Ray Henry
Whitney, William Merrill
Wickersheim, Kenneth Alan
Wieder, Harold
Wilcox, Jaroslava Zitkova
Wild, Robert Lee
Wilkenfeld, Jason Michael
Will, Theodore A
Williams, Richard Stanley
Wilson, Barbara Ann
Wilson, William Dennis
Wittry, David Beryle
Woerner, Robert Leo
Wolf, Robert Peter
Wong, Joe
Wood, Charles
Woodruff, Truman Owen
Wooten, Frederick (Oliver)
Yamakawa, Kazuo Alan
Yang, Chung Ching
Yeh, Nai-Chang
Yessik, Michael John
Yu, Karl Ka-Chung
Yu, Peter Yound
Zeidler, James Robert
Zernow, Louis
Zucca, Ricardo
Zucker, Joseph
Zuleeg, Rainer
Zwerdling, Solomon

COLORADO

Adams, Richard Owen
Ahrenkiel, Richard K
Baldwin, Thomas O
Beale, Paul Drew
Betterton, Jesse O, Jr
Blakeslee, A Eugene
Brock, George William
Bruns, Donald Gene
Carpenter, Steve Haycock

Ciszek, Ted F
DeGrand, Thomas Alan
Emery, Keith Allen
Fickett, Frederick Roland
Froyen, Sverre
Furtak, Thomas Elton
Galeener, Frank Lee
Gardner, Edward Eugene
Geller, Seymour
Goldfarb, Ronald B
Harris, Richard Elgin
Hermann, Allen Max
Iyer, Ravi
Joenk, Rudolph John, Jr
Kalma, Arne Haerter
Kanda, Motohisa
Kazmerski, Lawrence L
Kremer, Russell Eugene
Laks, David Bejnesh
Lebiedzik, Jozef
Lefever, Robert Allen
Love, William F
McMahon, Thomas Joseph
Moddel, Garret R
Nahman, Norris S(tanley)
Noufi, Rommel
Nozik, Arthur Jack
O'Callaghan, Michael James
Olhoeft, Gary Roy
Pankove, Jacques I
Parson, Robert Paul
Patton, Carl E
Pitts, John Roland
Powell, Robert Lee
Price, John C
Raich, John Carl
Read, David Thomas
Scott, James Floyd
Siegwarth, James David
Sites, James Russell
Sliker, Todd Richard
Smith, David Reeder
Sullivan, Donald Barrett
Westfall, Richard Merrill
Winder, Dale Richard
Zhang, Bing-Rong

CONNECTICUT

Akkapeddi, Prasad Rao
Anderson, R(obert) M(orris), Jr
Balasinski, Artur
Bartram, Ralph Herbert
Best, Philip Ernest
Bogardus, Egbert Hal
Bursuker, Isia
Chang, Richard Kounai
Clapp, Philip Charles
Cordery, Robert Arthur
Dolan, James F
Gaertner, Wolfgang Wilhelm
Gaidis, Michael Christopher
Galligan, James M
Gardner, Fred Marvin
Haacke, Gottfried
Haag, Robert Edwin
Henrich, Victor E
Kappers, Lawrence Allen
Keith, H(arvey) Douglas
Kessel, Quentin Cattell
Kierstead, Henry Andrew
Klemens, Paul Gustav
Klokholm, Erik
Kraft, David Werner
Leonberger, Frederick John
Ma, Tso-Ping
Markowitz, David
Marvin, Philip Roger
Melcher, Charles L
Otter, Fred August
Peterson, Cynthia Wyeth
Peterson, Gerald A
Philo, John Sterner
Prober, Daniel Ethan
Rau, Richard Raymond
Reed, Robert Willard
Reinberg, Alan R
Rollefson, Robert John
Sachdev, Subir
Schor, Robert
Scott, Joseph Hurlong
Shalvoy, Richard Barry
Shuskus, Alexander J
Siegmund, Walter Paul
Solomon, Peter R
Stein, Richard Jay
Wolf, Werner Paul

DELAWARE

Barteau, Mark Alan
Bierlein, John David
Bindloss, William
Böer, Karl Wolfgang
Chase, David Bruce
Christie, Phillip
Cole, G(eorge) Rolland
Cooper, Charles Burleigh
Crawford, Michael Karl
Daniels, William Burton
Dasgupta, Sunil Priya
Flattery, David Kevin
Flippen, Richard Bernard
Hegedus, Steven Scott
Hunsperger, Robert G(eorge)
Huppe, Francis Frowin

Ih, Charles Chung-Sen
Jarrett, Howard Starke, Jr
Johnson, Robert Chandler
McNeely, James Braden
Moore, Arnold Robert
Murray, Richard Bennett
Nielsen, Paul Herron
Onn, David Goodwin
Pontrelli, Gene J
Rogers, Donald B
Sawers, James Richard, Jr
Sharnoff, Mark
Stiles, A(lvin) B(arber)
Swann, Charles Paul
Zumsteg, Fredrick C, Jr

DISTRICT OF COLUMBIA

Abraham, George
Anderson, Gordon Wood
Batra, Narendra K
Berlincourt, Ted Gibbs
Bermudez, Victor Manuel
Bishop, Stephen Gray
Borsuk, Gerald M
Campbell, Arthur B
Carlos, William Edward
Carter, Gesina C
Chubb, Scott Robinson
Chung, David Yih
Crouch, Roger Keith
Davis, John Litchfield
De Graaf, Adriaan M
Della Torre, Edward
Dragoo, Alan Lewis
Duesbery, Michael Serge
Edelstein, Alan Shane
Ehrlich, Alexander Charles
Feldman, Charles
Feldman, Joseph Louis
Ferguson, George Alonzo
Finger, Larry W
Forester, Donald Wayne
Francavilla, Thomas L(ee)
Friebele, Edward Joseph
Fuller, Wendy Webb
Gaumond, Charles Frank
Goff, James Franklin
Goldin, Edwin
Gottschall, Robert James
Griscom, David Lawrence
Gubser, Donald Urban
Haas, George Arthur
Haworth, William Lancelot
Hobbs, Herman Hedberg
Kabler, Milton Norris
Kaplan, Raphael
Kertesz, Miklos
Kitchens, Thomas Adren
Klick, Clifford C
Krebs, James John
Kroeger, Richard Alan
Lasser, Marvin Elliott
Leibowitz, Jack Richard
Lessoff, Howard
Lin-Chung, Pay-June
McAlister, Archie Joseph
McBride, Duncan Eldridge
McKay, Jack Alexander
Manning, Irwin
Mao, Ho-Kwang
Marquardt, Charles L(awrence)
Marrone, Michael Joseph
Meijer, Paul Herman Ernst
Meyer, Ralph O
Michel, David John
Milton, Albert Fenner
Moon, Deug Woon
Morris, Robert Carter
Ngai, Kia Ling
Nisenoff, Martin
OOsterhuis, William Tenley
Pande, Chandra Shekhar
Papaconstantopoulos, Dimitrios A
Parker, Robert Louis
Prince, Morton Bronenberg
Prinz, Gary A
Ramaker, David Ellis
Ratchford, Joseph Thomas
Reinecke, Thomas Leonard
Rittner, Edmund Sidney
Roitman, Peter
Schneider, Irwin
Senftle, Frank Edward
Serene, Joseph William
Shapero, Donald Campbell
Sibley, William Arthur
Skelton, Earl Franklin
Summers, Geoffrey P
Taggart, G(eorge) Bruce
Thompson, Phillip Eugene
Tsang, Tung
Vold, Carl Leroy
Webb, Alan Wendell
Weinberg, Donald Lewis
White, Robert Marshall
Wickman, Herbert Hollis
Wieting, Terence James
Williams, Conrad Malcolm
Willis, James Stewart, Jr
Wilsey, Neal David
Wolf, Stuart Alan

FLORIDA

Adams, Earnest Dwight

Aubel, Joseph Lee
Bass, Michael
Brooker, Hampton Ralph
Browder, James Steve
Burdick, Glenn Arthur
Burns, Jay, III
Burns, Michael J
Chu, Ting Li
Dacey, George Clement
Edwards, Palmer Lowell
Einspruch, Norman G(erald)
Faulkner, John Samuel
Foner, Samuel Newton
Fossum, Jerry G
Halder, Narayan Chandra
Halpern, Leopold (Ernst)
Huebner, Jay Stanley
Hummel, Rolf Erich
Kumar, Pradeep
Lade, Robert Walter
Leverenz, Humboldt Walter
Linder, Ernest G
Lindholm, Fredrik Arthur
Linz, Arthur
Lowdin, Per-Olov
McAlpine, Kenneth Donald
Malocha, Donald C
Manousakis, Efstratios
Melich, Michael Edward
Monkhorst, Hendrik J
Moulton, Grace Charbonnet
Moulton, William G
Mukherjee, Pritish
Murphy, Joseph
Patterson, James Deane
Pepinsky, Raymond
Reynolds, George William, Jr
Rhodes, Richard Ayer, II
Rikvold, Per Arne
Riley, Mark Anthony
Shih, Hansen S T
Simmons, Joseph Habib
Skofronick, James Gust
Stach, Joseph
Stegeman, George I
Sullivan, Neil Samuel
Testardi, Louis Richard
Trickey, Samuel Baldwin
Ulug, Esin M
Wade, Thomas Edward
Wang, Yung-Li
Weber, Alfred Herman
Wille, Luc Theo
Wu, Jin Zhong
Yon, E(ugene) T
Zalesak, Joseph Francis
Ziernicki, Robert S

GEORGIA

Anderson, Robert Lester
Bacon, Roger
Bishop, Thomas Parker
Blue, Marts Donald
Copeland, John Alexander
De Mayo, Benjamin
DuVarney, Raymond Charles
Family, Fereydoon
Gaylord, Thomas Keith
Griffiths, James Edward
Hartman, Nile Franklin
Hathcox, Kyle Lee
Hetzler, Morris Clifford, Jr
Hornbeck, John A
Hsu, Frank Hsiao-Hua
Jokerst, Nan Marie
Kenan, Richard P
Korda, Edward J(ohn)
Landau, David Paul
Long, Thomas Ross
McMillan, Robert Walker
Meltzer, Richard S
Pandey, Surendra Nath
Perkowitz, Sidney
Pollard, William Blake
Powell, Bobby Earl
Puri, Om Parkash
Ringel, Steven Adam
Rives, John Edgar
Rodrigue, George Pierre
Shand, Julian Bonham, Jr
Stanford, Augustus Lamar, Jr
Stombler, Milton Philip
Vail, Charles R(owe)
Yen, William Maoshung
Young, Robert Alan
Zangwill, Andrew

HAWAII

Gaines, James R
Pong, William

IDAHO

Buescher, Brent J
Kearney, Robert James
Sieckmann, Everett Frederick
Spencer, Paul Roger
Tracy, Joseph Charles

ILLINOIS

Albrecht, Edward Daniel
Aldred, Anthony T
Ali, Naushad
Anderson, Ansel Cochran
Askill, John

Atoji, Masao
Auvil, Paul R, Jr
Averill, Frank Wallace
Bader, Samuel David
Barnett, Scott A
Breig, Marvin L
Brodsky, Merwyn Berkley
Brown, Bruce Stilwell
Buck, Warren Louis
Bukrey, Richard Robert
Burnham, Robert Danner
Carpenter, John Marland
Chan, Sai-Kit
Chang, Yia-Chung
Chiang, Tai-Chang
Chung, Yip-Wah
Connor, Donald W
Crabtree, George William
Day, Michael Hardy
Dlott, Dana D
Druger, Stephen David
Dunlap, Bobby David
Dutta, Pulak
Eades, John Alwyn
Eden, James Gary
Ehrlich, Gert
Ellis, Donald Edwin
English, Floyd L
Ewald, Arno Wilfred
Felcher, Gian Piero
Flynn, Colin Peter
Fradin, Frank Yale
Frank, Robert Carl
Freeman, Arthur Jay
Fritzsche, Hellmut
Gallagher, David Alden
Garland, James W
Gilbert, Robert L
Gleckman, Philip Landon
Goldberg, Colman
Goodman, Gordon Louis
Guttman, Lester
Hess, Karl
Hessler, Jan Paul
Hill, Christopher T
Hines, Roderick Ludlow
Hinks, David George
Holonyak, N(ick), Jr
Hurych, Zdenek
Jerome, Joseph Walter
Jesse, Kenneth Edward
Jorgensen, James D
Kannewurf, Carl Raeside
Karim, Khondkar Rezaul
Kimball, Clyde William
Koelling, Dale Dean
Kouvel, James Spyros
Krauss, Alan Robert
Kuchnir, Moyses
Lam, Nghi Quoc
Lange, Yvonne
Lazarus, David
Levin, Kathryn J
Levi-Setti, Riccardo
Liu, Liu
Liu, Yung Yuan
Macrander, Albert Tiemen
Mapother, Dillon Edward
Marcus, Jules Alexander
Marks, Laurence D
Merkelo, Henri
Miller, Robert Carl
Mochel, Jack McKinney
Moncton, David Eugene
Mozurkewich, George
Mueller, Melvin H(enry)
Nagel, Sidney Robert
Naylor, David L
Noble, Gordon Albert
Olvera-dela Cruz, Monica
Preston, Richard Swain
Price, David Cecil Long
Primak, William L
Reft, Chester Stanley
Ren, Shang-Fen
Ring, James George
Rothman, Steven J
Rowland, Theodore Justin
Salamon, Myron B
Schreiber, David Seyfarth
Schroeer, Juergen Max
Schultz, Arthur Jay
Sharma, Ram Ratan
Shenoy, Gopal K
Shepard, Kenneth Wayne
Sibener, Steven Jay
Siegel, Richard W(hite)
Sill, Larry R
Simmons, Ralph Oliver
Sinha, Shome Nath
Skov, Charles E
Slichter, Charles Pence
Smither, Robert Karl
Snow, Joel A
Spector, Harold Norman
Stapleton, Harvey James
Stillman, Gregory Eugene
Stout, John Willard
Sturm, William James
Tuzzolino, Anthony J
Udler, Dmitry
Unlu, M Selim
Van Ostenburg, Donald Ora
Veal, Boyd William, Jr

Weertman, Julia Randall
Weissman, Michael Benjamin
Welsh, Lawrence B
Wey, Albert Chin-Tang
Wiedersich, H(artmut)
Wilson, Robert Steven
Windhorn, Thomas H
Wolfe, James Phillip
Wolfram, Thomas
Zitter, Robert Nathan

INDIANA
Ascarelli, Gianni
Bessey, William Higgins
Bray, Ralph
Bunch, Robert Maxwell
Bunker, Bruce Alan
Buschert, Robert Cecil
Colella, Roberto
Compton, Walter Dale
Croat, John Joseph
De Young, Donald Bouwman
Fan, Hsu Yun
Furdyna, Jacek K
Giordano, Nicholas J
Girvin, Steven M
Harland, Glen Eugene, Jr
Helrich, Carl Sanfrid, Jr
Kaplan, Jerome I
Kesmodel, Larry Lee
Kissinger, Paul Bertram
Kliewer, Kenneth L
Lundstrom, Mark Steven
MacDonald, Allan Hugh
MacKay, John Warwick
Melloch, Michael Raymond
Mullen, James G
Pearlman, Norman
Pierret, Robert Francis
Prohofsky, Earl William
Ramdas, Anant Krishna
Robinson, John Murrell
Rodriguez, Sergio
Skadron, Peter
Sladek, Ronald John
Tomasch, Walter J
Tripathi, Govakh Nath Ram
Van Zandt, Lonnie L
Western, Arthur Boyd
White, Samuel Grandford, Jr
Zimmerman, Walter Bruce

IOWA
Barnaal, Dennis E
Barnes, G Richard
Bevolo, Albert Joseph
Clem, John R
Dalal, Vikram
Finnemore, Douglas K
Fuchs, Ronald
Goertz, Christoph Klaus
Gschneidner, Karl A(lbert), Jr
Harmon, Bruce Norman
Hsu, David Kuei-Yu
Klemm, Richard Andrew
Luban, Marshall
Lunde, Barbara Kegerreis, (BK)
Lynch, David William
McClelland, John Frederick
Papadakis, Emmanuel Philippos
Schaefer, Joseph Albert
Schweitzer, John William
Spitzig, William Andrew
Swenson, Clayton Albert
Thompson, Donald Oscar
Wechsler, Monroe S(tanley)

KANSAS
Brothers, Alfred Douglas
Culvahouse, Jack Wayne
Dragsdorf, Russell Dean
Friauf, Robert J
Grosskreutz, Joseph Charles
Ho, James Chien Ming
Long, Larry L
Pfluger, Clarence Eugene
Rahman, Talat Shahnaz
Sapp, Richard Cassell
Thomas, James E

KENTUCKY
Bakanowski, Stephen Michael
Bordoloi, Kiron
Brill, Joseph Warren
Buckman, William Gordon
Connolly, John William Domville
Davis, Thomas Haydn
DeLong, Lance Eric
Fox, Mary Eleanor
France, Peter William
Reucroft, Philip J
Straley, Joseph Paul
Subbaswamy, Kumble R(amarao)

LOUISIANA
Azzam, Rasheed M A
Bogle, Tommy Earl
Brumage, William Harry
Dahlquist, Wilbur Lynn
Deck, Ronald Joseph
Goodrich, Roy Gordon
Grenier, Claude Georges
Hoy, Robert C
Kurtz, Richard Leigh

Lafleur, Louis Dwynn
Marshak, Alan Howard
Naidu, Seetala V
O'Connell, Robert F
Perdew, John Paul
Piller, Herbert
Stockbauer, Roger Lewis
Vashishta, Priya Darshan
Veith, Daniel A
Watkins, Steven F
Zebouni, Nadim H

MAINE
Camp, Paul R
McKay, Susan Richards
Pribram, John Karl
Schmiedeshoff, George M

MARYLAND
Abbundi, Raymond Joseph
Ahearn, John Stephen
Allgaier, Robert Stephen
Alperin, Harvey Albert
Andreadis, Tim Dimitri
Bailey, Glenn Charles
Bardasis, Angelo
Bargeron, Cecil Brent
Berg, Norman J
Bis, Richard F
Blessing, Gerald Vincent
Blue, James Lawrence
Blum, Norman Allen
Bohandy, Joseph
Boone, Bradley Gilbert
Broadhurst, Martin Gilbert
Bryden, Wayne A
Buchner, Stephen Peter
Byer, Norman Ellis
Cacciamani, Eugene Richard, Jr
Callen, Earl Robert
Catchings, Robert Merritt, III
Cavanagh, Richard Roy
Celotta, Robert James
Charles, Harry Krewson, Jr
Chen, Wenpeng
Chesser, Nancy Jean
Chiang, Chwan K
Chien, Chia-Ling
Chimenti, Dale Everett
Clark, Alan Fred
Clark, Arthur Edward
Clark, Bill Pat
Cohen, Julius
Cohen, Marvin Morris
Colwell, Jack Harold
Conrad, Edward Ezra
Corak, William Sydney
Crane, Langdon Teachout, Jr
Cullen, James Robert
Davis, Guy Donald
Davisson, Charlotte Meaker
DeWit, Roland
DiPirro, Michael James
Drew, Howard Dennis
Egelhoff, William Frederick, Jr
Embury, Janon Frederick, Jr
Feldman, Albert
Fontanella, John Joseph
Forbes, Jerry Wayne
Frazer, Benjamin Chalmers
Frederick, William George DeMott
Frederikse, Hans Pieter Roetert
Frey, Jeffrey
Gammon, Robert Winston
Garstens, Martin Aaron
Garver, Robert Vernon
Ghoshtagore, Rathindra Nath
Gleason, Thomas James
Glick, Arnold J
Glover, Rolfe Eldridge, III
Green, Robert E(dward), Jr
Greene, Richard L
Greene, Richard Lorentz
Herman, David S
Houston, Bland Bryan, Jr
Huang, Jacob Wen-Kuang
Jablonski, Daniel Gary
Jette, Archelle Norman
Johnson, William Bowie
Joseph, Richard Isaac
Kaiser, Quentin C
Kirkendall, Thomas Dodge
Kirkpatrick, Theodore Ross
Kopanski, Joseph J
Kumar, K Sharvan
Kuriyama, Masao
Langenberg, Donald Newton
Larrabee, R(obert) D(ean)
Lindmayer, Joseph
Lowney, Jeremiah Ralph
Lynn, Jeffrey Whidden
McCally, Russell Lee
MacDonald, Rosemary A
McLean, Flynn Barry
Magno, Richard
Majkrzak, Charles Francis
Malmberg, Philip Ray
Mangum, Billy Wilson
Manning, John Randolph
Maycock, John Norman
Metze, George W
Minkowski, Jan Michael
Montalvo, Ramiro A
Moorjani, Kishin

Munasinghe, Mohan P
Murphy, John Cornelius
Oettinger, Frank Frederic
Palik, Edward Daniel
Pande, Krishna P
Pierce, Daniel Thornton
Poehler, Theodore O
Potyraj, Paul Anthony
Powell, Cedric John
Powell, John David
Prask, Henry Joseph
Prince, Edward
Rabin, Herbert
Randolph, Lynwood Parker
Rasera, Robert Louis
Rector, Charles Willson
Reno, Robert Charles
Restorff, James Brian
Revesz, Akos George
Rosenstock, Herbert Bernhard
Rowe, John Michael
Ruffa, Anthony Richard
Rush, John Joseph
Scharnhorst, Kurt Peter
Schneider, Carl Stanley
Schulman, James Herbert
Schumacher, Clifford Rodney
Sharma, Jagadish
Shaw, Robert William, Jr
Silberglitt, Richard Stephen
Silverton, James Vincent
Simonis, George Jerome
Sokoloski, Martin Michael
Steiner, Bruce
Strenzwilk, Denis Frank
Sweeting, Linda Marie
Swenberg, Charles Edward
Swerdlow, Max
Taylor, Barry Norman
Thomson, Robb M(ilton)
Thurber, Willis Robert
Trevino, Samuel Francisco
Van Antwerp, Walter Robert
Vande Kieft, Laurence John
Van Vechten, Deborah
Venables, John Duxbury
Waldo, George Van Pelt, Jr
Walker, J Calvin
Wang, Frederick E
Warner, Brent A
Warren, Wayne Hutchinson, Jr
Wasilik, John H(uber)
Williams, Ellen D
Wittels, Mark C
Yedinak, Peter Demerton
Yockey, Hubert Palmer

MASSACHUSETTS
Addiss, Richard Robert, Jr
Adlerstein, Michael Gene
Aggarwal, Roshan Lal
Aldrich, Ralph Edward
Alexander, Michael Norman
Antal, John Joseph
Antonoff, Marvin M
Baird, Donald Heston
Bansil, Arun
Bell, Richard Oman
Berera, Geetha Poonacha
Berker, Ahmet Nihat
Birgeneau, Robert Joseph
Bloom, Jerome H(ershel)
Blum, John Bennett
Buss, Dennis Darcy
Button-Shafer, Janice
Carr, Paul Henry
Caslavsky, Jaroslav Ladislav
Chase, Charles Elroy, Jr
Chen, Joseph H
Clarke, Edward Nielsen
Clifton, Brian John
Collins, Aliki Karipidou
Cronin-Golomb, Mark
Damon, Richard Winslow
Davis, Charles Freeman, Jr
Davis, Luther, Jr
Di Bartolo, Baldassare
Dionne, Gerald Francis
Dorschner, Terry Anthony
Dresselhaus, Gene Frederick
Dresselhaus, Mildred S
Eby, John Edson
Ekman, Carl Frederick W
Fan, John C C
Feist, Wolfgang Martin
Fleming, Phyllis Jane
Foner, Simon
Free, John Ulric
Friedman, Lionel Robert
Gangulee, Amitava
Garth, John Campbell
Gianino, Peter Dominic
Gold, Albert
Gordon, Roy Gerald
Gould, Harvey A
Gregory, Bob Lee
Greytak, Thomas John
Gross, David John
Groves, Steven H
Guentert, Otto Johann
Guertin, Robert Powell
Gustafson, John C
Halperin, Bertrand Israel
Handelman, Eileen T

Solid State Physics (cont)

Dixon, Richard Wayne
Dodabalapur, Ananth
Dolny, Gary Mark
Downs, David S
Dresner, Joseph
Duclos, Steven J
Dutta, Mitra
Eisenberger, Peter Michael
Ensign, Thomas Charles
Enstrom, Ronald Edward
Feldman, Leonard Cecil
Fenstermacher, Robert Lane
Feuer, Mark David
Finke, Guenther Bruno
Fischer, Russell Jon
Fish, Gordon E
Fleming, Robert McLemore
Fleury, Paul A
Fonger, William Hamilton
Fork, Richard Lynn
Freeman, Richard Reiling
Fu, Hui-Hsing
Ghosh, Amal Kumar
Gibson, J Murray
Giordmaine, Joseph Anthony
Gittleman, Jonathan I
Glass, Alastair Malcolm
Golding, Brage, Jr
Goldstein, Bernard
Golin, Stuart
Gossmann, Hans Joachim
Gourary, Barry Sholom
Green, Michael Philip
Greenblatt, Martha
Greene, Laura H
Grest, Gary Stephen
Greywall, Dennis Stanley
Grimes, Charles Cullen
Grupen, William Brightman
Gualtieri, Devlin Michael
Guenzer, Charles S P
Gummel, Hermann K
Gurvitch, Michael
Gustafsson, Torgny Daniel
Gyorgy, Ernst Michael
Haase, Oswald
Hafner, Erich
Hagstrum, Homer Dupre
Hakim, Edward Bernard
Halpern, Teodoro
Harbison, James Prescott
Hartman, Richard Leon
Hasegawa, Ryusuke
Hebard, Arthur Foster
Henry, Charles H
Hensel, John Charles
Herber, Rolfe H
Hershenov, B(ernard)
Hirt, Andrew Michael
Hohenberg, Pierre Claude
Husa, Donald L
Hutson, Andrew Rhodes
Hwang, Cherng-Jia
Jackel, Lawrence David
Jansen, Frank
Jeck, Richard Kahr
Johnson, Leo Francis
Johnston, Wilbur Dexter, Jr
Josenhans, James Gross
Joyce, William B(axter)
Kahng, Dawon
Kaminow, Ivan Paul
Kampas, Frank James
Kane, Evan O
Kash, Kathleen
Keramidas, Vassilis George
Koch, Thomas L
Kohn, Jack Arnold
Kokotailo, George T
Koller, Noemie
Kramer, Bernard
Kunzler, John Eugene
Lamelas, Francisco Javier
Lang, David (Vern)
Langreth, David Chapman
Leath, Paul Larry
Leheny, Robert Francis
Levi, Anthony Frederic John
Levin, Edwin Roy
Li, Yuan
Liang, Keng-San
Libowitz, George Gotthart
Lindenfeld, Peter
Lines, Malcolm Ellis
Logan, Ralph Andre
Lunsford, Ralph D
Luryi, Serge
Lyons, Kenneth Brent
McLane, George Francis
MacRae, Alfred Urquhart
Madey, Theodore Eugene
Marezio, Massimo
Martinelli, Ramon U
Matey, James Regis
Mattheiss, Leonard Francis
Matula, Richard Allen
Meiboom, Saul
Mette, Herbert L
Meyerhofer, Dietrich
Miller, David A B
Misra, Raj Pratap
Mollenauer, Linn F
Morris, Robert Craig
Moustakas, Theodore D

Nahory, Robert Edward
Nassau, Kurt
Ng, Tai-Kai
O'Gorman, James
Olsen, Gregory Hammond
Ong, Nai-Phuan
Owens, Frank James
Panish, Morton B
Partovi, Afshin
Passner, Albert
Pearsall, Thomas Perine
Pearton, Stephen John
Pei, Shin-Shem
Pfeiffer, Loren Neil
Phillips, James Charles
Phillips, Julia M
Phillips, William
Pinczuk, Aron
Platzman, Philip M
Raghavan, Pramila
Ramanan, V R V
Ramirez, Arthur P
Ravindra, Nuggehalli Muthanna
Raynor, Susanne
Reboul, Theo Todd, III
Reed, William Alfred
Riddle, George Herbert Needham
Ropp, Richard C
Rosen, Carol Zwick
Rothberg, Gerald Morris
Rothberg, Lewis Josiah
Roukes, Michael L
Rowe, John Edward
Royce, Barrie Saunders Hart
Rudman, Peter S
Sabisky, Edward Stephen
Salwen, Harold
Schlam, Elliott
Schlueter, Michael Andreas
Schluter, Michael
Schneemeyer, Lynn F
Schubert, Rudolf
Seidl, Milos
Shacklette, Lawrence Wayne
Shah, Jagdeep C
Shand, Michael Lee
Shanefield, Daniel J
Shay, Joseph Leo
Shen, Tek-Ming
Sheng, Ping
Shumate, Paul William, Jr
Sinclair, William Robert
Sinha, Sunil K
Skalski, Stanislaus
Slusky, Susan E G
Smith, Carl Hofland
Smith, George Elwood
Smith, Neville Vincent
Smith, Robert Owens
Solin, Stuart Allan
Solla, Sara A
Spencer, Edward G
Spruch, Grace Marmor
Staebler, David Lloyd
Stiles, Lynn F, Jr
Stolen, Rogers Hall
Stormer, Horst Ludwig
Sulewski, Paul Eric
Swartz, George Allan
Tauber, Arthur
Taylor, Geoff W
Taylor, George William
Teaney, Dale T
Tecotzky, Melvin
Thaler, Barry Jay
Thomas, Gordon Albert
Thornton, C G
Tischler, Oscar
Tomaselli, Vincent Paul
Torrey, Henry Cutler
Tsui, Daniel Chee
Tully, John Charles
Tumelty, Paul Francis
Upton, Thomas Hallworth
Vanderbilt, David Hamilton
Van Der Ziel, Jan Peter
Van De Vaart, Herman
Van Roosbroeck, Willy Werner
Vella-Coleiro, George
Wachtman, John Bryan, Jr
Wagner, Sigurd
Walsh, Walter Michael, Jr
Walstedt, Russell E
Wang, Chen-Show
Wang, Tsuey Tang
Wenzel, John Thompson
Wertheim, Gunther Klaus
White, Alice Elizabeth
Wiegand, Donald Arthur
Wiesenfeld, Jay Martin
Willens, Ronald Howard
Williams, Brown F
Wilson, Barbara Ann
Wittenberg, Albert M
Wolfe, Raymond
Wolff, Peter A
Wong, King-Lap
Wood, Thomas H
Woodward, Ted K
Worlock, John M
Wu, Chung Pao
Xie, Ya-Hong
Yablonovitch, Eli
Yafet, Yako

Zaininger, Karl Heinz
Zucker, Melvin Joseph

NEW MEXICO
Albers, Robert Charles
Alldredge, Gerald Palmer
Anderson, Richard Ernest
Anderson, Robert Alan
Andrew, James F
Arko, Aloysius John
Arnold, George W
Barsis, Edwin Howard
Bartlett, Roger James
Beattie, Alan Gilbert
Beauchamp, Edwin Knight
Beckel, Charles Leroy
Benjamin, Robert Fredric
Borders, James Alan
Brice, David Kenneth
Brueck, Steven Roy Julien
Burr, Alexander Fuller
Butler, Michael Alfred
Campbell, David Kelly
Castle, John Granville, Jr
Claytor, Thomas Nelson
Clogston, Albert McCavour
Cole, Edward Issac, Jr
Cooke, David Wayne
Dowell, Flonnie
Dumas, Herbert M, Jr
Eckert, Juergen
Edwards, Leon Roger
Emin, David
Erickson, Dennis John
Ewing, Rodney Charles
Farnum, Eugene Howard
Follstaedt, David Martin
Fritz, Joseph N
Frost, Harold Maurice, III
Fukushima, Eiichi
Funsten, Herbert Oliver, III
Galt, John Kirtland
Gauster, Wilhelm Belrupt
Giovanielli, Damon V
Gobeli, Garth William
Gorham, Elaine Deborah
Gourley, Paul Lee
Grannemann, W(ayne) W(illis)
Green, Thomas Allen
Gubernatis, James Edward
Gurbaxani, Shyam Hassomal
Jennison, Dwight Richard
Johnson, Ralph T, Jr
Jones, Eric Daniel
Jones, Orval Elmer
Keller, William Edward
Kenkre, Vasudev Mangesh
Kepler, Raymond Glen
King, James Claude
Kreidl, Norbert J(oachim)
Kurtz, Steven Ross
Land, Cecil E(lvin)
Lawson, Andrew Cowper, II
Linford, Rulon Kesler
Llamas, Vicente Jose
Loree, Thomas Robert
Lyo, SungKwun Kenneth
McInerney, John Gerard
Maggiore, Carl Jerome
Maurer, Robert Joseph
Miranda, Gilbert A
Mitchell, Terence Edward
Morosin, Bruno
Morris, Charles Edward
Moss, Marvin
Mueller, Fred Michael
Mueller, Karl Hugo, Jr
Myers, David Richard
Myers, Samuel Maxwell, Jr
Narath, Albert
Narayanamurti, Venkatesh
O'Rourke, John Alvin
Osbourn, Gordon Cecil
Osinski, Marek Andrzej
Panitz, Janda Kirk Griffith
Parkin, Don Merrill
Peercy, Paul S
Pettit, Richard Bolton
Picraux, Samuel Thomas
Pike, Gordon E
Pynn, Roger
Rauber, Lauren A
Redondo, Antonio
Richards, Peter Michael
Romig, Alton Dale, Jr
Salzbrenner, Richard John
Samara, George Albert
Schiferl, David
Schirber, James E
Schwalbe, Larry Allen
Schwarz, Ricardo
Shaner, John Wesley
Silver, Richard N
Sinha, Dipen N
Smith, James Lawrence
Smith, William Conrad
Southward, Harold Dean
Switendick, Alfred Carl
Taylor, Raymond Dean
Thacher, Philip Duryea
Thompson, Joe David
Tonks, Davis Loel
Trela, Walter Joseph
Tsao, Jeffrey Yeenien

Visscher, William M
Von Dreele, Robert Bruce
Vook, Frederick Ludwig
Weaver, Harry Talmadge
Westpfahl, David John
Willis, Jeffrey Owen
Yarnell, John Leonard

NEW YORK
Abbas, Daniel Cornelius
Abkowitz, Martin Arnold
Adams, Peter D
Adler, Michael Stuart
Agrawal, Govind P(rasad)
Aitken, John Malcolm
Alben, Richard Samuel
Alfano, Robert R
Allen, Philip B
Alley, Phillip Wayne
Amato, Joseph Cono
Ambegaokar, Vinay
Amer, Nabil Mahmoud
Anantha, Narasipur Gundappa
Anderson, Wayne Arthur
Anthony, Thomas Richard
Arajs, Sigurds
Arnold, Emil
Aschner, Joseph Felix
Ashcroft, Neil William
Aven, Manuel
Avouris, Phaedon
Bak, Per
Ballantyne, Joseph M(errill)
Bari, Robert Allan
Basavaiah, Suryadevara
Batterman, Boris William
Bebb, Herbert Barrington
Bednowitz, Allan Lloyd
Benenson, Raymond Elliott
Benumof, Reuben
Beri, Avinash Chandra
Bermon, Stuart
Berry, B(rian) S(hepherd)
Beshers, Daniel N(ewson)
Bhargava, Rameshwar Nath
Bickford, Lawrence Richardson
Bilderback, Donald Heywood
Blaker, J Warren
Bloch, Aaron N
Bocko, Mark Frederick
Borrego, Jose M
Borrelli, Nicholas Francis
Bray, James William
Bright, Arthur Aaron
Brillson, Leonard Jack
Brinkman, William F
Brodsky, Marc Herbert
Brouillette, Walter
Brown, Dale Marius
Brown, Rodney Duvall, III
Brown, Ronald Alan
Budzinski, Walter Valerian
Buhrman, Robert Alan
Burkey, Bruce Curtiss
Burnham, Dwight Comber
Butler, Walter John
Buttiker, Markus
Cargill, G Slade, III
Carleton, Herbert Ruck
Carlson, Richard Oscar
Carroll, Clark Edward
Castner, Theodore Grant, Jr
Chang, Ifay F
Chang, L(eroy) L(i-Gong)
Chen, Inan
Chen, Ying-Chih
Chow, Tat-Sing Paul
Chrenko, Richard Michael
Cluxton, David H
Coburn, Theodore James
Collins, Franklyn
Comly, James B
Connelly, John Joseph, Jr
Conwell, Esther Marly
Corbett, James William
Corelli, John C
Corelli, John Charles
Costa, Lorenzo F
Cotts, Robert Milo
Cox, David Ernest
Cronemeyer, Donald Charles
Crowe, George Joseph
Cummins, Herman Z
Cusano, Dominic A
Daman, Harlan Richard
Damask, Arthur Constantine
Das, Pankaj K
Das, Tara Prasad
Davenport, James Whitman
DeBlois, Ralph Walter
Dell, George F, Jr
Dermit, George
Diamond, Joshua Benamy
Dienes, George Julian
Di Maria, Donelli Joseph
DiSalvo, Francis Joseph, Jr
DiStefano, Thomas Herman
Domingos, Henry
Dowben, Peter Arnold
Dreyfus, Russell Warren
Dropkin, John Joseph
Dudley, Michael
Eberhard, Jeffrey Wayne
Eckert, Richard Raymond

Pillai, Padmanabha S
Pinnick, Harry Thomas
Poynter, James William
Puglielli, Vincent George
Rai, Amarendra Kumar
Reynolds, Donald C
Rigney, David Arthur
Rollins, Roger William
Rosenblatt, Charles Steven
Sanford, Edward Richard
Schopler, Kenneth Lee
Schlosser, Herbert
Scholten, Paul David
Schreiber, Paul J
Schuele, Donald Edward
Segall, Benjamin
Seltzer, Martin S
Silvidi, Anthony Alfred
Simon, Henry John
Smith, Robert Emery
Spielberg, Nathan
Spry, Robert James
Steckl, Andrew Jules
Stevenson, Frank Robert
Stoner, Ronald Edward
Stroud, David Gordon
Styer, Daniel F
Suranyi, Peter
Swinehart, Philip Ross
Taylor, Barney Edsel
Taylor, Charles Emery
Taylor, Philip Liddon
Tenhover, Michael Alan
Thomas, Joseph Francis, Jr
Trivisonno, Joseph, Jr
Uralil, Francis Stephen
Vassamillet, Lawrence Francois
Wagoner, Glen
Weeks, Stephen P
White, John Joseph, III
Wiff, Donald Ray
Wigen, Philip E
Wilkes, William Roy
Williams, Wendell Sterling
Wolff, Gunther Arthur
Wong, Anthony Sai-Hung
Wood, Van Earl
Yaney, Perry Pappas
Yeo, Yung Kee
Yun, Seung Soo

OKLAHOMA
Batchman, Theodore E
Daniels, Raymond D(eWitt)
Dixon, George Sumter, Jr
Doezema, Ryan Edward
Dreiling, Mark Jerome
Halliburton, Larry Eugene
Hartman, Roger Duane
Hendrickson, John Robert
Huffaker, James Neal
King, John Paul
Lafon, Earl Edward
Lauffer, Donald Eugene
McKeever, Stephen William Spencer
Major, John Keene
Martin, Joel Jerome
Powell, Richard Conger
Scales, John Alan
Waldrop, Morgan A
Whitmore, Stephen Carr
Wilson, Timothy M

OREGON
Abrahams, Sidney Cyril
Bhattacharya, Pallab Kumar
Boedtker, Ölaf A
Cohen, John David
Cutler, Melvin
Demarest, Harold Hunt, Jr
Engelbrecht, Rudolf S
Ford, Wayne Keith
Gardner, John Arvy, Jr
Griffith, William Thomas
Howard, Donald Grant
McClure, Joel William, Jr
Nussbaum, Rudi Hans
Owen, Sydney John Thomas
Photinos, Panos John
Semura, Jack Sadatoshi
Van Vechten, James Alden
Warren, William Willard, Jr
Wasserman, Allen Lowell
Wolfe, Gordon A
Yamaguchi, Tadanori
Young, Richard Accipiter

PENNSYLVANIA
Adda, Lionel Paul
Angello, Stephen James
Arora, Vijay Kumar
Artman, Joseph Oscar
Banavar, Jayanth Ramarao
Baratta, Anthony J
Barsch, Gerhard Richard
Baxter, Ronald Dale
Becker, Kurt H
Berger, Luc
Bhalla, Amar S
Biondi, Manfred Anthony
Bitler, William Reynolds
Bloomfield, Philip Earl
Böer, Karl Wolfgang
Boltax, Alvin

Booth, Bruce L
Bose, Shyamalendu M
Bowlden, Henry James
Burstein, Elias
Callen, Herbert Bernard
Caspari, Max Edward
Chaplin, Norman John
Charap, Stanley H
Cherry, Leonard Victor
Choyke, Wolfgang Justus
Chung, Tze-Chiang
Cohen, Alvin Jerome
Cole, Milton Walter
Collings, Peter John
Coon, Darryl Douglas
Crow, Jack Emerson
Cutler, Paul H
Dai, Hai-Lung
Daniel, Michael Roger
Dank, Milton
Das, Mukunda B
Davis, R(obert) E(lliot)
Deis, Daniel Wayne
Diehl, Renee Denise
Donahoe, Frank J
Drum, Charles Monroe
Dubeck, Leroy W
Dunn, Charles Nord
Egami, Takeshi
Emtage, Peter Roesch
Fabish, Thomas John
Feigl, Frank Joseph
Feingold, Earl
Ferrone, Frank Anthony
Feuchtwang, Thomas Emanuel
Finegold, Leonard X
Fischer, John Edward
Fonash, Stephan J(oseph)
Fowler, Wyman Beall, Jr
Foxhall, George Frederic
Frankl, Daniel Richard
Freud, Paul J
Friedberg, Simeon Adlow
Fuchs, Walter
Gainer, Michael Kizinski
Garito, Anthony Frank
Girifalco, Louis A(nthony)
Gittler, Franz Ludwig
Goldburg, Walter Isaac
Goldey, James Mearns
Goldschmidt, Yadin Yehuda
Gollub, Jerry Paul
Goodwin, Charles Arthur
Griffiths, Robert Budington
Gürer, Emir
Harmer, Martin Paul
Harris, Arthur Brooks
Henisch, Heinz Kurt
Hensler, Donald H
Hess, Dennis William
Houlihan, John Frank
Hulm, John Kenneth
Ivey, Henry Franklin
Jain, Himanshu
Johnson, William Lewis
Jones, Clifford Kenneth
Kahn, David
Kakar, Anand Swaroop
Kang, Joohee
Kasowski, Robert V
Keffer, Frederic
Klein, Michael Lawrence
Kraitchman, Jerome
Krishnaswamy, S V
Kryder, Mark Howard
Lang, Lawrence George
Langer, Dietrich Wilhelm
Lannin, Jeffrey S
Larson, Donald Clayton
Lee, Richard J
Liang, Shoudan
Licini, Jerome Carl
Loman, James Mark
Lowe, Irving J
Lu, Chih Yuan
McCammon, Robert Desmond
McGowan, William Courtney, Jr
McKelvey, John Philip
McKinstry, Herbert Alden
Macpherson, Alistair Kenneth
Mahajan, Subhash
Maher, James Vincent
Malmberg, Paul Rovelstad
Matolyak, John
Maynard, Julian Decatur
Merrill, John Raymond
Messier, Russell
Mihalisin, Ted Warren
Miller, Paul
Miskovsky, Nicholas Matthew
Molter, Lynne Ann
Morris, Marlene Cook
Mulay, Laxman Nilakantha
Muldawer, Leonard
Nathanson, H(arvey) C(harles)
Nigh, Harold Eugene
Nilan, Thomas George
Noreika, Alexander Joseph
Oder, Robin Roy
Orehotsky, John Lewis
Pfrogner, Ray Long
Pierce, Russell Dale
Plummer, E Ward
Polo, Santiago Ramos

Post, Irving Gilbert
Przybysz, John Xavier
Racette, George William
Rao, V Udaya S
Ray, Siba Prasad
Repper, Charles John
Reynolds, Claude Lewis, Jr
Richardson, Ralph J
Roeder, Edward A
Sabnis, Anant Govind
Schumacher, Robert Thornton
Shibib, M Ayman
Silzars, Aris
Smits, Friedolf M
Spitznagel, John A
Steele, Martin Carl
Stein, Barry Fred
Steinhardt, Paul Joseph
Stewart, Glenn Alexander
Steyert, William Albert
Stroud, Jackson Swavely
Sucov, E(ugene) W(illiam)
Szepesi, Zoltan Paul John
Talvacchio, John
Thomas, H Ronald
Thomas, R Noel
Thompson, Eric Douglas
Tittmann, Bernhard R
Toulouse, Jean
Tsong, Tien Tzou
Varnerin, Lawrence J(ohn)
Vehse, Robert Chase
Wagner, George Richard
Wagner, Timothy Knight
Walkiewicz, Thomas Adam
Waring, Robert Kerr, Jr
Watkins, George Daniels
Weidner, Richard Tilghman
Weiner, Brian Lewis
Weiner, Robert Allen
Weiss, Paul Storch
Wemple, Stuart H(arry)
White, Marvin Hart
Whiteman, John David

RHODE ISLAND
Bray, Philip James
Estrup, Peder Jan Z
Glicksman, Maurice
Heller, Gerald S
Loferski, Joseph J
Mardix, Shmuel
Maris, Humphrey John
Nunes, Anthony Charles
Pelcovits, Robert Alan
Pickart, Stanley Joseph
Seidel, George Merle
Sherman, Charles Henry
Stiles, Phillip John
Tauc, Jan
Timbie, Peter T
Ying, See Chen

SOUTH CAROLINA
Bagchi, Amitabha
Bostock, Judith Louise
Chaplin, Robert Lee, Jr
Darnell, Frederick Jerome
Datta, Timir
Edge, Ronald (Dovaston)
Faust, John William, Jr
Hamilton, John Frederick
Hurren, Weiler R
Jones, Edwin Rudolph, Jr
Keller, Frederick Jacob
Lathrop, Jay Wallace
Manson, Joseph Richard
Mosley, Wilbur Clanton, Jr
Moyle, David Douglas
Payne, James Edward
Payne, Linda Lawson
Radford, Loren E
Saunders, Edward A
Sherrill, Max Douglas
Skove, Malcolm John
Steiner, Pinckney Alston, III
Stevens, James Levon
Stewart, William Hogue, Jr
Stillwell, Ephraim Posey, Jr
Vogel, Henry Elliott
Wood, James C, Jr

SOUTH DAKOTA
Ashworth, T
Redin, Robert Daniel
Ross, Keith Alan
Tunheim, Jerald Arden
Vander Lugt, Karel L

TENNESSEE
Abraham, Marvin Meyer
Achar, B N Narahari
Appleton, B R
Arakawa, Edward Takashi
Arlinghaus, Heinrich Franz
Ball, Raiford Mill
Blankenship, James Lynn
Boatner, Lynn Allen
Borie, Bernard Simon
Budai, John David
Butler, William H
Cable, Joe Wood
Callcott, Thomas Anderson
Cardwell, Alvin Boyd

Child, Harry Ray
Christen, David Kent
Cleland, John W
Close, David Matzen
Collins, Thomas C
Coltman, Ralph Read, Jr
Dudney, Nancy Johnston
Feigerle, Charles Stephen
Fernandez-Baca, Jaime Alberto
Franceschetti, Donald Ralph
Garland, Michael McKee
Gruzalski, Greg Robert
Haglund, Richard Forsberg, Jr
Haynes, Tony Eugene
Huebschman, Eugene Carl
Huray, Paul Gordon
Jellison, Gerald Earle, Jr
Jenkins, Leslie Hugh
Kennedy, Eldredge Johnson
Kerchner, Harold Richard
Langley, Robert Archie
Larson, Bennett Charles
Lenhert, P Galen
Lowndes, Douglas H, Jr
Lubell, Martin S
Lundy, Ted Sadler
McKee, Rodney Allen
Miller, James Robert
Moak, Charles Dexter
Mook, Herbert Arthur, Jr
Moon, Ralph Marks, Jr
Mostoller, Mark Ellsworth
Nicklow, Robert Merle
Noonan, John Robert
Oen, Ordean Silas
O'Neal, Thomas Norman
Painter, Gayle Stanford
Powell, Harry Douglas
Raudorf, Thomas Walter
Ritchie, Rufus Haynes
Roberto, James Blair
Robinson, Mark Tabor
Sales, Brian Craig
Schwenterly, Stanley William, III
Silberman, Enrique
Smith, Harold Glenn
Sonder, Edward
Springer, John Mervin
Srygley, Fletcher Douglas
Stephenson, Charles V
Stiegler, James O
Stocks, George Malcolm
Thompson, James R
Thomson, John Oliver
Tolk, Norman Henry
Trammell, Rex Costo
Wachs, Alan Leonard
Wang, Jia-Chao
Weeks, Robert A
Wendelken, John Franklin
Wilkinson, Michael Kennerly
Young, Frederick Walter, Jr
Zuhr, Raymond Arthur

TEXAS
Adair, Thomas Weymon, III
Anderson, Robert E
Aton, Thomas John
Banerjee, Sanjay Kumar
Bartels, Richard Alfred
Barton, James Brockman
Bate, Robert Thomas
Bencomo, José A
Birchak, James Robert
Black, Truman D
Bottom, Virgil Eldon
Brient, Samuel John, Jr
Bucy, J Fred
Caflisch, Robert Galen
Carbajal, Bernard Gonzales, III
Caswell, Gregory K
Chapman, Richard Alexander
Chivian, Jay Simon
Chopra, Dev Raj
Chu, Ching-Wu
Chu, Wei-Kan
Claiborne, Lewis T, Jr
Collins, Dean Robert
Collins, Francis Allen
Crossley, Peter Anthony
Deaton, Bobby Charles
De Wette, Frederik Willem
Dodds, Stanley A
Duncan, Walter Marvin, Jr
Dunning, Frank Barrymore
Dupuis, Russell Dean
Erskine, James Lorenzo
Esquivel, Agerico Liwag
Estle, Thomas Leo
Fair, Harry David, Jr
Frensley, William Robert
Gavenda, John David
Gleim, Paul Stanley
Glosser, Robert
Goodenough, John Bannister
Goodman, David Wayne
Graham, William Richard Montgomery
Gryting, Harold Julian
Hamill, Dennis W
Hasty, Turner Elilah
Heilmeier, George Harry
Heinze, William Daniel
Heller, Adam
Herczeg, John W

Solid State Physics (cont)

Holverson, Edwin LeRoy
Hu, Chia-Ren
Hudson, Hugh T
Ignatiev, Alex
Isham, Elmer Rex
Keister, Jamieson Charles
Kirk, Wiley Price
Kleinman, Leonard
Kroger, Harry
Lacy, Lewis L
Lawson, Juan (Otto)
Leamy, Harry John
LeMaster, Edwin William
Leuchtag, H Richard
Levy, Robert Aaron
Lichti, Roger L
McDavid, James Michael
McDonald, Perry Frank
Mackey, Henry James
Mahendroo, Prem P
Mao, Shing
Matzkanin, George Andrew
Mays, Robert, Jr
Medlin, William Louis
Moiz, Syed Abdul
Moss, Simon Charles
Mueller, Dennis W
Naugle, Donald
Neikirk, Dean P
Nordlander, Peter Jan Arne
Overmyer, Robert Franklin
Pandey, Raghvendra Kumar
Parker, Sidney G
Paterson, James Lenander
Penz, P Andrew
Peters, Randall Douglas
Portnoy, William M
Potter, Robert Joseph
Reed, Mark Arthur
Rollwitz, William Lloyd
Rorschach, Harold Emil, Jr
Rubins, Roy Selwyn
Sablik, Martin J
Saslow, Wayne Mark
Schroen, Walter
Sears, Raymond Eric John
Seiler, David George
Seitchik, Jerold Alan
Sharma, Suresh C
Singh, Vijay Pal
Smith, Daniel Montague
Smith, James Lynn
Smoluchowski, Roman
Spencer, William J
Stratton, Robert
Sullivan, Jerry Stephen
Sybert, James Ray
Tasch, Al Felix, Jr
Thompson, Arthur Howard
Thompson, Bonnie Cecil
Trehan, Rajender
Walters, Geoffrey King
Wang, Paul Weily
Wood, Lowell Thomas
Zrudsky, Donald Richard

UTAH

Ailion, David Charles
Barnett, John Dean
Brandt, Richard Charles
Brophy, James John
Cohen, Richard M
Decker, Daniel Lorenzo
DeFord, John W
Evenson, William Edwin
Galli, John Ronald
Hansen, Wilford Nels
Haymet, Anthony Douglas-John
Huber, Robert John
Jeffery, Rondo Nelden
Johnson, Owen W
Jones, Merrell Robert
Larson, Everett Gerald
Luty, Fritz
Mattis, Daniel Charles
Nelson, Homer Mark
Ohlsen, William David
Smith, Kent Farrell
Stokes, Harold T
Stringfellow, Gerald B
Taylor, Philip Craig
Vanfleet, Howard Bay
Ward, Roger Wilson
Williams, George Abiah
Wu, Yong-Shi

VERMONT

Adler, Eric
Anderson, R(ichard) L(ouis)
Ferris-Prabhu, Albert Victor Michael
Foley, Edward Leo
Fox, Bradley Alan
Furukawa, Toshiharu
Lambert, Lloyd Milton, Jr
Muir, Wilson Burnett
Pires, Renato Guedes
Quinn, Robert M(ichael)
Smith, David Young
Warfield, George
Williams, Ronald Wendell

VIRGINIA

Amith, Avraham
Amstutz, Larry Ihrig
Bahl, Inder Jit
Bensel, John Phillip
Bishop, Marilyn Frances
Bloomfield, Louis Aub
Bottka, Nicholas
Brandt, Richard Gustave
Brockman, Philip
Butler, Charles Thomas
Byvik, Charles Edward
Cantrell, John H(arris)
Catlin, Avery
Caton, Randall Hubert
Celli, Vittorio
Chang, Lay Nam
Chi, Tsung-Chin
Chun, Myung K(i)
Cook, Desmond C
Cooper, Larry Russell
Crane, Sara W
Crump, John C, III
Deaver, Bascom Sine, Jr
DeFotis, Gary Constantine
Dharamsi, Amin N
Dickinson, Stanley Key, Jr
Duke, John Christian, Jr
Eck, John Stark
Faraday, Bruce (John)
Forman, Richard Allan
Gilmer, Thomas Edward, Jr
Goodman, A(lvin) M(alcolm)
Hass, Georg
Hass, Marvin
Hendricks, Robert Wayne
Hess, George Burns
Hilger, James Eugene
Hoffman, Allan Richard
Holmes, David Kelley
Holt, William Henry
Hoy, Gilbert Richard
Hudgins, Aubrey C, Jr
Jena, Purusottam
Jesser, William Augustus
Johnson, Charles Minor
Johnson, Robert Alan
Johnson, Robert Edward
Johnston, William Cargill
Kapron, Felix Paul
Kellett, Claud Marvin
Kennedy, Andrew John
Kirkendall, Ernest Oliver
Kossler, William John
Kramer, Steven David
Lankford, William Fleet
Leinhardt, Theodore Edward
Liebenberg, Donald Henry
Loda, Richard Thomas
Long, Jerome R
Major, Robert Wayne
Marsh, Howard Stephen
Maycock, Paul Dean
Mayo, Thomas Tabb, IV
Mielczarek, Eugenie V
Mitchell, Dean Lewis
Nicoll, Jeffrey Fancher
Park, Yoon Soo
Peters, Philip Boardman
Phillips, Donald Herman
Pollard, John Henry
Pritchard, Wenton Maurice
Rado, George Tibor
Reynolds, Richard Alan
Rosenthal, Michael David
Rothenberg, Herbert Carl
Rothwarf, Frederick
Sauder, William Conrad
Schnatterly, Stephen Eugene
Schriempf, John Thomas
Serway, Raymond A
Shur, Michael
Sigler, Julius Alfred, Jr
Silver, Meyer
Sloope, Billy Warren
Streever, Ralph L
Stronach, Carey E
Williams, Clayton Drews
Wyeth, Newell Convers
Zallen, Richard

WASHINGTON

Baer, Donald Ray
Barchet, William Richard
Bender, Paul A
Blakemore, John Sydney
Braunlich, Peter Fritz
Brimhall, J(ohn) L
Brown, Frederick Calvin
Claudson, T(homas) T(ucker)
Collins, Gary Scott
Craig, Richard Anderson
Dahl, Roy Edward
Darling, Robert Bruce
Doran, Donald George
Dresser, Miles Joel
Dunham, Glen Curtis
Einziger, Robert E
Erdmann, Joachim Christian
Exarhos, Gregory James
Fain, Samuel Clark, Jr
Fischbach, David Bibb
Gillingham, Robert J

Glass, James Clifford
Gunsul, Craig J W
Gupta, Yogendra M(ohan)
Hinman, Chester Arthur
Howland, Louis Philip
Huang, Fan-Hsiung Frank
Johnson, Gordon Oliver
Johnson, Harlan Paul
Klepper, John Richard
Knotek, Michael Louis
Koizumi, Carl Jan
Lam, John Ling-Yee
Lambe, John Joseph
Laramore, George Ernest
Lytle, Farrel Wayne
Macksey, Harry Michael
Matlock, John Hudson
Miles, Maurice Howard
Miller, John Howard
Olmstead, Marjorie A
Olsen, Larry Carrol
Quigley, Robert James
Rehr, John Jacob
Rimbey, Peter Raymond
Schick, Michael
Smith, Willis Dean
Spanel, Leslie Edward
Stern, Edward Abraham
Stuve, Eric Michael
Thouless, David James
Tsang, Leung
Tutihasi, Simpei
Vilches, Oscar Edgardo
Warren, John Lucius
Weber, William J
Willardson, Robert Kent
Yoshikawa, Herbert Hiroshi

WEST VIRGINIA

Bleil, David F
Cooper, Bernard Richard
De Barbadillo, John Joseph
Franklin, Alan Douglas
Franz, Judy R
Pavlovic, Arthur Stephen
Seehra, Mohindar Singh
Shanholtzer, Wesley Lee
Tewksbury, Stuart K

WISCONSIN

Aita, Carolyn Rubin
Anderson, James Gerard
Barclay, John Arthur
Baum, Gary Allen
Behroozi, Feredoon
Bondeson, Stephen Ray
Burch, Thaddeus Joseph
Cerrina, Francesco
Deshotels, Warren Julius
Dexter, Richard Norman
Draeger, Norman Arthur
Ebel, Marvin Emerson
Edelheit, Lewis S
England, Walter Bernard
Gelatt, Charles Daniel, Jr
Gueths, James E
Huber, David Lawrence
Hyde, James Stewart
Johnson, Arthur Franklin
Lagally, Max Gunter
Levy, Moises
Lin, Chun Chia
Lindeberg, George Kline
Margaritondo, Giorgio
Mendelson, Kenneth Samuel
Moran, Paul Richard
Myers, Vernon W
Olson, Clifford Gerald
Rhyner, Charles R
Smith, Thomas Stevenson
Sorbello, Richard Salvatore
Stanwick, Glenn
Tonner, Brian P
Truszkowska, Krystyna
Van Sciver, Steven W
Wiley, John Duncan
Zimm, Carl B

PUERTO RICO

Bailey, Carroll Edward
Gomez-Rodriguez, Manuel

ALBERTA

Adler, John G
Apps, Michael John
Buckmaster, Harvey Allen
Chatterjee, Ramananda
Deegan, Ross Alfred
Egerton, Raymond Frank
Haslett, James William
Kubynski, Jadwiga
Kuntz, Garland Parke Paul
Rogers, James Stewart
Tuszynski, Jack A
Umezawa, Hiroomi
Weichman, Frank Ludwig
Woods, Stuart B

BRITISH COLUMBIA

Ballentine, Leslie Edward
Barker, Alfred Stanley, Jr
Bichard, J W
Boal, David Harold
Booth, Ian Jeremy

Clayman, Bruce Philip
Cochran, John Francis
Colbow, Konrad
Crozier, Edgar Daryl
Dahn, Jeffery Raymond
Deen, Mohamed Jamal
Frindt, Robert Frederick
Haering, Rudolph Roland
Hardy, Walter Newbold
Irwin, John Charles
Kirczenow, George
Marko, John Robert
Martin, Peter Wilson
Opechowski, Wladyslaw
Rieckhoff, Klaus E
Rogers, Douglas Herbert
Schwerdtfeger, Charles Frederick
Seth, Rajinder Singh
Tiedje, J Thomas
Viswanathan, Kadayam Sankaran
Watton, Arthur
Wortis, Michael

MANITOBA

Gaunt, Paul
Kao, Kwan Chi
Loly, Peter Douglas
Southern, Byron Wayne
Vail, John Moncrieff
Woo, Chung-Ho

NEW BRUNSWICK

Banerjee, R L
Girouard, Fernand E
Leblanc, Leonard Joseph
Vo-Van, Truong
Wells, David Ernest

NOVA SCOTIA

Betts, Donald Drysdale
El-Masry, Ezz Ismail
Latta, Bryan Michael
Moriarty, Kevin Joseph
Roger, William Alexander
Simpson, Antony Michael
Steinitz, Michael Otto

ONTARIO

Alfred, Louis Charles Roland
Anderson, Anthony
Bahadur, Birendra
Baird, D C
Berezin, Alexander A
Berry, Robert John
Bertram, Robert William
Brandon, James Kenneth
Brockhouse, Bertram Neville
Brodie, Dr Don E
Brown, Ian David
Buyers, William James Leslie
Celinski, Olgierd J(erzy) Z(dzislaw)
Chamberlain, S(avvas) G(eorgiou)
Cocivera, Michael
Collins, Malcolm Frank
Colpa, Johannes Pieter
Datars, William Ross
Davies, John Arthur
Davison, Sydney George
Dixon, Arthur Edward
Dyment, John Cameron
Eastman, Philip Clifford
Egelstaff, Peter A
Erickson, Lynden Edwin
Fawcett, Eric
Fenton, Edward Warren
Ferrier, Jack Moreland
Fujimoto, Minoru
Goodings, David Ambery
Griffin, Allan
Grindlay, John
Hawton, Margaret H
Hedgecock, Nigel Edward
Hitchcock, Adam Percival
Hoffstein, Victor
Holden, Thomas More
Holuj, Frank
Hunt, James L
Hurd, Colin Michael
Jackman, Thomas Edward
Jeffrey, Kenneth Robert
Joos, Béla
Jorch, Harald Heinrich
Kim, Soo Myung
Koffyberg, Francois Pierre
Lachaine, Andre Raymond Joseph
Lamarche, J L Gilles
Lee, Martin J G
Lee-Whiting, Graham Edward
Leslie, James D
Lipsett, Frederick Roy
Liu, Wing-Ki
Lockwood, David John
McAlister, Sean Patrick
Mandelis, Andreas
Martin, Douglas Leonard
Marton, John Peter
Mendis, Eustace Francis
Nagi, Anterdhyan Singh
Nerenberg, Morton Abraham
Ogata, Hisashi
Ollerhead, Robin Wemp
Ottensmeyer, Frank Peter
Pascual, Roberto
Perz, John Mark

Piercy, George Robert
Pintar, M(ilan) Mik
Popovic, Zoran
Powell, Brian M
Purbo, Onno Wídodo
Reddoch, Allan Harvey
Rolfe, John
Roulston, David J
Salvadori, Antonio
Sawicka, Barbara Danuta
Sayer, Michael
Schlesinger, Mordechay
Sears, Varley Fullerton
Segel, Stanley Lewis
Selvakumar, Chettypalayam Ramanathan
Shewchun, John
Smith, Kenneth Carless
Song, Kong-Sop Augustin
Sonnenberg, Hardy
Stager, Carl Vinton
Storey, Robert Samuel
Straus, Jozef
Sundaresan, Mosur Kalyanaraman
Svensson, Eric Carl
Szabo, Alexander
Tam, Wing-Gay
Taylor, David Ward
Taylor, Roger
Templeton, Ian M
Thomas, Raye Edward
Thompson, David Allan
Tong, Bok Yin
Vanderkooy, John
Van Driel, Henry Martin
Vincett, Paul Stamford
Vosko, Seymour H
Walton, Derek
Wang, Shao-Fu
Wei, L(ing) Y(un)
Woods, Alfred David Braine
Yevick, David Owen
Zukotynski, Stefan

PRINCE EDWARD ISLAND
Madan, Mahendra Pratap

QUEBEC
Aktik, Cetin
Banville, Marcel
Bartnikas, Ray
Caille, Alain Emeril
Champness, Clifford Harry
Chaubey, Mahendra
Cochrane, Robert W
Crine, Jean-Pierre C
Fuchs, Vladimir
Grant, Martin
Harris, Richard
Izquierdo, Ricardo
Jordan, Byron Dale
Kahrizi, Mojtaba
Kipling, Arlin Lloyd
Leblanc, Roger M
McIntyre, Robert John
Marchand, Richard
Masut, Remo Antonio
Meunier, Michel
Misra, Sushil
Morris, Stanley P
Salahub, Dennis Russell
Savard, Jean Yves
Sharma, Ramesh C
Strom-Olsen, John Olaf
Subramanian, Sesha
Terreault, Bernard J E J
Tremblay, André -Marie
Van Vliet, Carolyne Marina
Vijh, Ashok Kumar
Wallace, Philip Russell
Wertheimer, Michael Robert
Yelon, Arthur Michael

SASKATCHEWAN
Kos, Joseph Frank
Papini, Giorgio Augusto

OTHER COUNTRIES
Archie, Charles Neill
Azbel, Mark
Baratoff, Alexis
Bauer, Ernst Georg
Braginski, Aleksander Ignace
Cardona, Manuel
Chandrasekhar, B S
Collver, Michael Moore
Cue, Nelson
Economou, Eleftherios Nickolas
Edmonds, James W
Eisner, Edward
Eng, Sverre T(horstein)
Fischer, Albert G
Foglio, Mario Eusebio
Greenfield, Arthur Judah
Heiblum, Mordehai
Howard, Iris Anne
Hudson, Ralph P
Kar, Nikhiles
Katz, Gerald
Keller, Jaime
Koch, J Frederick
Koehler, Helmut A
Ling, Samuel Chen-Ying
Low, William
Maas, Peter

Merz, Walter John
Miyano, Kenjiro
Moser, Frank
Neel, Louis Eugene Felix
Nyburg, Stanley Cecil
Rivier, Nicolas Yves
Safran, Samuel A
Salaneck, William R
Sessler, Gerhard Martin
Tantraporn, Wirojana
Ushioda, Sukekatsu
Van Overstraeten, Roger Joseph
Von Klitzing, Klaus
Wasa, Kiyotaka
Weil, Raoul Bloch
Weinstock, Harold
Wiser, Nathan
Yanabu, Satoru
Zabel, Hartmut

Spectroscopy & Spectrometry

ALABAMA
Leonard, Kathleen Mary
McKannan, Eugene Charles
Miller, George Paul

ARIZONA
Bao, Qingcheng
Breed, William Joseph
Giampapa, Mark Steven
Marzke, Robert Franklin
Peyghambarian, Nasser
Venables, John Anthony
Yelle, Roger V

ARKANSAS
Cardwell, David Michael

CALIFORNIA
Almason, Carmen Cristina
Bajaj, Jagmohan
Bowman, Robert Clark, Jr
Bridges, Frank G
Caird, John Allyn
Camp, David Conrad
Castle, Peter Myer
Chambers, Robert J
Curtis, Earl Clifton, Jr
Donoho, David Leigh
Fouquet, Julie
Fowler, Charles Arman, Jr
Hovanec, B(ernard) Michael
Jacobs, Ralph R
Jeffries, Jay B
Jusinski, Leonard Edward
Kim, Jinchoon
Lakkaraju, H S
Leite, Richard Joseph
Lorents, Donald C
Moerner, William Esco
Monard, Joyce Anne
Palatnick, Barton
Passell, Thomas Oliver
Perel, Julius
Perkins, Willis Drummond
Pertica, Alexander José
Pianetta, Piero Antonio
Satten, Robert A
Seki, Hajime
Short, Michael Arthur
Shykind, David
Stöhr, Joachim
Strauss, Herbert L
Weber, Marvin John
Wilson, Barbara Ann
Wu, Robert Chung-Yung

COLORADO
Daw, Glen Harold
Eaton, Sandra Shaw
Goetz, Alexander Franklin Hermann
Kinsinger, James A
Murphy, Robert Carl
O'Callaghan, Michael James
Zhang, Bing-Rong

CONNECTICUT
Chang, Ted T
Dolan, James F
Melcher, Charles L

DELAWARE
Daniels, William Burton
Dwivedi, Anil Mohan
Szalewicz, Krzysztof

DISTRICT OF COLUMBIA
Brown, Charles Moseley
Campillo, Anthony Joseph
Cook, John William
Elton, Raymond Carter
Goldstein, Jeffrey Jay
Meekins, John Fred
Misra, Prabhakar
Oertel, Goetz K
OOsterhuis, William Tenley

FLORIDA
Block, Ronald Edward
Mukherjee, Pritish
Schlottmann, Pedro U J

GEORGIA
Anantha Narayanan, Venkataraman
De Sa, Richard John
Hart, Raymond Kenneth
McMillan, Robert Walker
Meltzer, Richard S
Ringel, Steven Adam
Yen, William Maoshung

IDAHO
Wheeler, Gilbert Vernon

ILLINOIS
Ali, Naushad
Curry, Bill Perry
Cutnell, John Daniel
Dunford, Robert Walter
Levi-Setti, Riccardo
McCormack, Elizabeth Frances
Melendres, Carlos Arciaga
Neuhalfen, Andrew J
Scheeline, Alexander
Schulz, Charles Emil

INDIANA
Ascarelli, Gianni

KANSAS
Andrew, Kenneth L
Long, Larry L
Macke, Gerald Fred

LOUISIANA
Kurtz, Richard Leigh
Meriwether, John R
Scott, John Delmoth

MARYLAND
Benson, Richard C
Berger, Robert Lewis
Boyd, Derek Ashley
Brasunas, John Charles
Criss, John W
Dick, Kenneth Anderson
Duignan, Michael Thomas
Ginter, Marshall L
Kastner, Sidney Oscar
Kostkowski, Henry John
Plotkin, Henry H
Saloman, Edward Barry
Sattler, Joseph Peter
Seliger, Howard Harold
Skiff, Frederick Norman
Suenram, Richard Dee
Warren, Wayne Hutchinson, Jr
Wiese, Wolfgang Lothar

MASSACHUSETTS
Bansil, Rama
Carleton, Nathaniel Phillips
Clough, Shepard Anthony
Collins, Aliki Karipidou
Handelman, Eileen T
Miniscalco, William J
Oettinger, Peter Ernest
Picard, Richard Henry
Pritchard, David Edward
Rao, Devulapalli V G L N
Rupich, Martin Walter
Schattenburg, Mark Lee
Schempp, Ellory
Tiernan, Robert Joseph
Wu, Lei

MICHIGAN
Kim, Yeong Wook
Vaishnava, Prem P

MINNESOTA
Steck, Daniel John

MISSISSIPPI
Monts, David Lee

MISSOURI
Bragg, Susan Lynn
Bryan, David A
Mori, Erik Jun

NEBRASKA
Billesbach, David P
Fabrikant, Ilya I

NEW HAMPSHIRE
Ge, Weikun

NEW JERSEY
Barrett, Joseph John
Bonin, Keith Donald
Broer, Matthijs Meno
Chiu, Tien-Heng
Chung, Yun C
Damen, Theo C
Duclos, Steven J
Harris, Alexander L
Harris, Leonce Everett
Hirt, Andrew Michael
Kash, Kathleen
Ong, Nai-Phuan
Platzman, Philip M
Sulewski, Paul Eric
Todd, Terry Ray
Wadlow, David
Woodward, Ted K

NEW MEXICO
Apel, Charles Turner
Bieniewski, Thomas M
Blake, Richard L
Close, Donald Alan
Kelley, Gregory M
Loree, Thomas Robert
Palmer, Byron Allen
Parker, Jack Lindsay
Rokop, Donald J
Taylor, Raymond Dean
Thompson, Jill Charlotte

NEW YORK
Becker, Kurt Heinrich
Clark, David Delano
Dreyfus, Russell Warren
Elder, Fred A
Folan, Lorcan Michael
Garvey, James F
Greenbaum, Steven Garry
Li, Hong
Mercer, Kermit R
Moy, Dan
Persans, Peter D
Philips, Laura Alma
Pliskin, William Aaron
Reddy, Reginald James
Varanasi, Prasad
Walmsley, Ian Alexander

OHIO
Berthold, John William, III
Biedenbender, Michael David
Cochran, William Ronald
Johnson, Ray O
Miyoshi, Kazuhisa
Yu, Thomas Huei-Chung

OREGON
Park, Kwangjai

PENNSYLVANIA
Allara, David Lawrence
Ernst, Wolfgang E
Gürer, Emir
Liberman, Irving
Marcus, Robert Troy
Ryan, Frederick Merk
Schwan, Herman Paul
Steinbruegge, Kenneth Brian
Tredicce, Jorge Raul

SOUTH CAROLINA
Berlinghieri, Joel Carl

TENNESSEE
Arlinghaus, Heinrich Franz
Hoffman, Kenneth Wayne
Kiech, Earl Lockett
Muly, Emil Christopher, Jr
Rice, Walter Wilburn
Young, Jack Phillip

TEXAS
Busch, Kenneth Walter
Church, David Arthur
DiFoggio, Rocco
Glosser, Robert
Golden, David E
Hulet, Randall Gardner
McDaniel, Floyd Delbert, Sr
Mangum, Jeffrey Gary
Maute, Robert Edgar
Pitts, David Eugene
Potter, Andrew Elwin, Jr
Quade, Charles Richard
Rabalais, John Wayne
Smith, R Lowell
Waggoner, James Arthur
Wang, Paul Weily
Wood, Lowell Thomas

UTAH
Grey, Gothard C
Powers, Linda Sue

VIRGINIA
Cook, Desmond C
Dharamsi, Amin N
Drew, Russell Cooper
Kramer, Steven David

WASHINGTON
Baer, Donald Ray
Brodzinski, Ronald Lee
McDowell, Robin Scott
Olsen, Kenneth Harold

WEST VIRGINIA
Wallace, William Edward, Jr

WISCONSIN
Herget, William F
Liang, Shoudeng
Roesler, Frederick Lewis
Wright, John Curtis

ALBERTA
Egerton, Raymond Frank
Weale, Gareth Pryce

BRITISH COLUMBIA
Irwin, John Charles

Spectroscopy & Spectrometry (cont)

Ozier, Irving

MANITOBA
Attas, Ely Michael
Ens, E(rich) Werner

ONTARIO
Anderson, Anthony
Boggs, Steven A
Carpenter, Graham John Charles
Dagg, Ian Ralph
French, J(ohn) Barry
Ganza, Kresimir Peter
Lockwood, David John
Ryan, Dave
Williams, Norman S W

QUEBEC
Kahrizi, Mojtaba
Waksberg, Armand L

SASKATCHEWAN
Durden, David Alan

OTHER COUNTRIES
Band, Yehuda Benzion
Butcher, Harvey Raymond, III
Kafri, Oded

Theoretical Physics

ALABAMA
Baird, James Kern
Coulter, Philip W
Fowler, Bruce Wayne
Hinata, Satoshi
Howgate, David W
Kinzer, Earl T, Jr
Krause, Helmut G L
Perez, Joseph Dominique
Pindzola, Michael Stuart
Raphael, Robert B
Rush, John Edwin, Jr
Smalley, Larry L
Smith, Charles Ray
Stettler, John Dietrich
Sung, Chi Ching
Visscher, Pieter Bernard
Wu, Xizeng

ARIZONA
Barrett, Bruce Richard
Benin, David B
Burrows, Adam Seth
Carruthers, Peter A
Green, Barry A
Hestenes, David
Hetrick, David LeRoy
Iverson, A Evan
Jacob, Richard John
Jokipii, Jack Randolph
Kyrala, Ali
Lovelock, David
Mahmoud, Hormoz Massoud
Maier, Robert S
Melia, Fulvio
Patrascioiu, Adrian Nicolae
Rafelski, Johann
Ren, Shang Yuan
Scadron, Michael David
Thews, Robert L(eroy)
West, Robert Elmer

ARKANSAS
Harter, William George
Hobson, Arthur Stanley
Lieber, Michael
Webb, Jerry Glen

CALIFORNIA
Abarbanel, Henry D I
Acheson, Louis Kruzan, Jr
Adler, Ronald John
Alonso, Carol Travis
Amster, Harvey Jerome
Antolak, Arlyn Joe
Aref, Hassan
Armstead, Robert Louis
Arnush, Donald
Aron, Walter Arthur
Asendorf, Robert Harry
Augenstein, Bruno (Wilhelm)
Bander, Myron
Banks, Thomas
Barnett, R(alph) Michael
Bars, Itzhak
Beardsley, Irene Adelaide
Beni, Gerardo
Bennett, Alan Jerome
Berdahl, Paul Hilland
Berman, Sam Morris
Bing, George Franklin
Birnir, Björn
Boley, Charles Daniel
Brodsky, Stanley Jerome
Brueckner, Keith Allan
Burke, William L
Burton, Donald Eugene
Buskirk, Fred Ramon

Byers, Nina
Carman, Robert Lincoln, Jr
Caves, Carlton Morris
Chandler, David
Chang, Howard How Chung
Chapline, George Frederick, Jr
Chen, Joseph Cheng Yih
Chew, Geoffrey Foucar
Chow, Brian G(ee-Yin)
Chow, Paul C
Christy, Robert Frederick
Cladis, John Baros
Cook, Robert Crossland
Cooper, Ralph Sherman
Corman, Emmett Gary
Cornwall, John Michael
Cummings, Frederick W
Dagotto, Elbio Ruben
Dalton, Francis Norbert
Dashen, Roger Frederick
Davies, John Tudor
De, Gopa Sarkar
Dedrick, Kent Gentry
Depp, Joseph George
Doniach, Sebastian
Dorn, David W
Drell, Sidney David
Dresden, Max
Drummond, William Eckel
Einhorn, Martin B
Erickson, Glen Walter
Estabrook, Frank Behle
Falicov, Leopoldo Maximo
Faulkner, John
Feiock, Frank Donald
Feit, Michael Dennis
Felber, Franklin Stanton
Fernbach, Sidney
Fetter, Alexander Lees
Finkelstein, Jerome
Fletcher, John George
Foiles, Stephen Martin
Fowler, Thomas Kenneth
Frahm, Charles Peter
Frazer, William Robert
Fried, Burton David
Fulco, Jose Roque
Gaillard, Mary Katharine
Galas, David John
Ganas, Perry S
Gaposchkin, Peter John Arthur
Garrison, John Carson
Garrod, Claude
Gaskell, Robert Weyand
Gell-Mann, Murray
Gillespie, George H
Glass, Nathaniel E
Glendenning, Norman Keith
Goldberger, Marvin Leonard
Goldman, Stanford
Goodjohn, Albert J
Goodman, Julius
Gould, Robert Joseph
Green, Joseph Matthew
Gregorich, David Tony
Greider, Kenneth Randolph
Gross, Mark Warren
Guest, Gareth E
Gunion, John Francis
Gyulassy, Miklos
Haber, Howard Eli
Haener, Juan
Halpern, Francis Robert
Halpern, Martin B
Hamill, Patrick James
Harker, Kenneth James
Harris, Charles Bonner
Hartle, James Burkett
Helliwell, Thomas McCaffree
Herman, Frank
Herman, Zelek Seymour
Herring, William Conyers
Hirschfelder, Joseph Oakland
Hiskes, John Robert
Hone, Daniel W
Horowitz, Gary T
Huberman, Bernardo Abel
Huddlestone, Richard H
Hundley, Richard O'Neil
Ingraham, Richard Lee
Jackson, John David
Johnson, Montgomery Hunt
Judd, David Lockhart
Jusinski, Leonard Edward
Kalensher, Bernard Earl
Karunasiri, Gamani
Kasameyer, Paul William
Kaus, Peter Edward
Kershaw, David Stanley
Kikuchi, Ryoichi
Kiskis, Joseph Edward, Jr
Koga, Toyoki
Koonin, Steven Elliot
Korringa, Jan
Krieger, Stephan Jacques
Kroll, Norman Myles
Kulander, Kenneth Charles
Lambropoulos, Peter Poulos
Langer, James Stephen
Latter, Albert L
Latter, Richard
Le Levier, Robert Ernest
Lepore, Joseph Vernon
Lewis, Harold Warren

Loos, Hendricus G
Louie, Steven Gwon Sheng
Lurie, Joan B
Lynch, David Dexter
McCormick, Philip Thomas
MacDonald, Gordon James Fraser
Maki, Kazumi
Mandelstam, Stanley
Maradudin, Alexei
Matlow, Sheldon Leo
Mayer, Harris Louis
Mayer, Meinhard Edwin
Mazurek, Thaddeus John
Meads, Philip Francis, Jr
Miller, Jack Culbertson
Moe, Mildred Minasian
Morawitz, Hans
More, Richard Michael
Moriarty, John Alan
Morrison, Harry Lee
Moszkowski, Steven Alexander
Murad, Emil Moise
Nedoluha, Alfred K
Nefkens, Bernard Marie
Nelson, Eldred (Carlyle)
Nesbet, Robert Kenyon
Nevins, William McCay
Newcomb, William A
Norton, Richard E
Nosanow, Lewis H
Noyes, H Pierre
Olness, Robert James
O'Malley, Thomas Francis
Painter, Ronald Dean
Pardee, William Joseph
Payne, David Glenn
Pearlstein, Leon Donald
Peccei, Roberto Daniele
Perkins, Kenneth L(ee)
Perkins, Sterrett Theodore
Peskin, Michael Edward
Pettit, John Tanner
Pincus, Philip A
Preskill, John P
Primack, Joel Robert
Pritchett, Philip Lentner
Rauch, Richard Travis
Redlich, Martin George
Ree, Francis H
Reid, Roderick Vincent, Jr
Rensink, Marvin Edward
Richardson, John Mead
Riddell, Robert James, Jr
Roberts, Charles A, Jr
Rockower, Edward Brandt
Rosenbluth, Marshall N
Rosenkilde, Carl Edward
Ross, Marvin Franklin
Rouse, Carl Albert
Rudnick, Joseph Alan
Runge, Richard John
Sachs, Rainer Kurt
Sadoulet, Bernard
Sarfatti, Jack
Sawyer, Raymond Francis
Scalettar, Richard
Schechter, Martin
Schlafly, Roger
Schwartz, Charles Leon
Schwarz, John Henry
Schweizer, Felix
Scofield, James Howard
Scott, Bruce L
Seki, Ryoichi
Senitzky, Israel Ralph
Sessler, Andrew M
Sham, Lu Jeu
Shaw, Gordon Lionel
Shenker, Scott Joseph
Shin, Ernest Eun-Ho
Shore, Herbert Barry
Silk, Joseph Ivor
Simon, Barry Martin
Smith, Gary Richard
Steinert, Leon ALbert
Stephens, Jeffrey Alan
Stevenson, David John
Stuart, George Wallace
Sugar, Robert Louis
Suzuki, Mahiko
Tamor, Stephen
Tarter, Curtis Bruce
Taylor, Howard S
Tenn, Joseph S
Thorne, Kip Stephen
Tracy, Craig Arnold
True, William Wadsworth
Tsai, Yung Su
Vega, Robeto
Waggoner, Jack Holmes, Jr
Wagoner, Robert Vernon
Waltz, Ronald Edward
Warnock, Robert Lee
Weichman, Peter Bernard
Wells, Willard H
Wheelon, Albert Dewell
Wichmann, Eyvind Hugo
Wilcox, Charles Hamilton
Williams, James Gerard
Winter, Nicholas Wilhelm
Wong, Chun Wa
Woodruff, Truman Owen
Woods, Roger David
Wright, Jon Alan

Wulfman, Carl E
Yano, Fleur Belle
Yura, Harold Thomas
Zachariasen, Fredrik
Zee, Anthony
Zeleny, William Bardwell
Zhou, Simon Zheng
Zimmerman, Ivan Harold
Zumino, Bruno

COLORADO
Ashby, Neil
Augusteijn, Marijke Francina
Barut, Asim Orhan
Beale, Paul Drew
Beeman, David Edmund, Jr
Bierman, Arthur
Brittin, Wesley E
Brown, Edmund Hosmer
Brown, James T, Jr
Chang, Bunwoo Bertram
Chappell, Willard Ray
Cole, Lee Arthur
DeGrand, Thomas Alan
DeSanto, John Anthony
Downs, Bertram Wilson, Jr
Froyen, Sverre
Geltman, Sydney
Hansen, Richard Olaf
Herring, Jackson Rea
Hundhausen, Arthur James
Iddings, Carl Kenneth
Krueger, David Allen
McCray, Richard A
Meiss, James Donald
Miller, Stanley Custer, Jr
Peterson, Robert Lee
Rainwater, James Carlton
Randa, James P
Sakakura, Arthur Yoshikazu
Smith, Earl W
Swanson, Lawrence Ray
Taylor, John Robert
Tuttle, Elizabeth R
Yeatts, Frank Richard

CONNECTICUT
Appelquist, Thomas
Bartram, Ralph Herbert
Cable, Peter George
Chun, Kee Won
Cordery, Robert Arthur
Fader, Walter John
Fenton, David George
Fowler, Michael
Ftaclas, Christ
Goldstein, Rubin
Gürsey, Feza
Hadjimichael, Evangelos
Hahn, Yukap
Haller, Kurt
Herzenberg, Arvid
Hill, David Lawrence
Islam, Muhammad Munirul
Johnson, David Linton
Klemens, Paul Gustav
MacDowell, Samuel Wallace
McIntosh, John Stanton
Michels, H(orace) Harvey
Newton, Victor Joseph
Peterson, Gerald A
Picker, Harvey Shalom
Porter, William Samuel
Rawitscher, George Heinrich
Russu, Irina Maria
Sachdev, Subir
Sen, Pabitra Narayan
Shankar, Ramamurti
Sommerfield, Charles Michael

DELAWARE
Chui, Siu-Tat
Glyde, Henry Russell
Hill, Robert Nyden
Kerner, Edward Haskell
Leung, Chung Ngoc
Mohapatra, Pramoda Kumar
Pittel, Stuart

DISTRICT OF COLUMBIA
Bergmann, Otto
Boss, Alan Paul
Brennan, James Gerard
Casper, Barry Michael
Chubb, Scott Robinson
Clinton, William L
Coffey, Timothy
Ehrlich, Alexander Charles
Eisenstein, Julian (Calvert)
Haftel, Michael Ivan
Jordan, Arthur Kent
Klein, Lewis S
Lampe, Martin
Langworthy, James Brian
Lehman, Donald Richard
Lin-Chung, Pay-June
McClure, Joseph Andrew, Jr
Manning, Irwin
Massey, Walter Eugene
Meijer, Paul Herman Ernst
Mendlowitz, Harold
Palmadesso, Peter Joseph
Picone, J Michael
Prats, Francisco

Rajagopal, Attipat Krishnaswamy
Reilly, Michael Hunt
Reinecke, Thomas Leonard
Reiss, Howard R
Rubin, Robert Joshua
Sáenz, Albert William
Serene, Joseph William
Shapero, Donald Campbell
Toll, John Sampson
Trubatch, Sheldon L
Uberall, Herbert Michael
Valenzuela, Gaspar Rodolfo
Werntz, Carl W
White, John Arnold

FLORIDA
Albright, John Rupp
Baird, Alfred Michael
Bose, Subir Kumar
Broyles, Arthur Augustus
Buchler, Jean-Robert
Curtright, Thomas Lynn
Darling, Byron Thorwell
Desloge, Edward Augustine
Detweiler, Steven Lawrence
Dufty, James W
Duke, Dennis Wayne
Edwards, Steve
Faulkner, John Samuel
Field, Richard D
Fry, James N
Halpern, Leopold (Ernst)
Hart, Robert Warren
Hogan, William Alfred
Hooper, Charles Frederick, Jr
Kazaks, Peter Alexander
Kimel, Jacob Daniel, Jr
Klauder, John Rider
Kumar, Pradeep
Kursunoglu, Behram N
Lindgren, E(rik) Rune
Lowdin, Per-Olov
Manousakis, Efstratios
Mintz, Stephen Larry
Monkhorst, Hendrik J
Muttalib, Khandker Abdul
Sabin, John Rogers
Schlottmann, Pedro U J
Spector, Richard M
Speisman, Gerald
Stegeman, George I
Sullivan, Neil Samuel
Thomas, Billy Seay
Thorn, Charles Behan, III
Trickey, Samuel Baldwin
Wille, Luc Theo
Williams, Gareth
Woodard, Richard P

GEORGIA
Abbe, Winfield Jonathan
Adomian, George
Batten, George L, Jr
Biritz, Helmut
Bowling, Arthur Lee, Jr
Duncan, Marion M, Jr
Family, Fereydoon
Finkelstein, David
Flannery, Martin Raymond
Fong, Peter
Ford, Joseph
Freiser, Marvin Joseph
Gallaher, Lawrence Joseph
Gatland, Ian Robert
Henkel, John Harmon
Jenkins, Hughes Brantley, Jr
Kenan, Richard P
Lee, M(onhe) Howard
Marks, Dennis William
Pandres, Dave, Jr
Pollard, William Blake
Sinha, Om Prakash
Strobel, George L
Tanner, James Mervil
Uzer, Ahmet Turgay
Uzes, Charles Alphonse
Valk, Henry Snowden
Wiita, Paul Joseph
Zangwill, Andrew

HAWAII
Tata, Xerxes Ramyar
Tuan, San Fu
Watanabe, Michael Satosi
Yamauchi, Hiroshi

IDAHO
Ivey, Jerry Lee
Majumdar, Debaprasad (Debu)
Morley, Gayle L
Parker, Barry Richard
Patsakos, George
Shirts, Randall Brent
Sieckmann, Everett Frederick

ILLINOIS
Avery, Robert
Axford, Roy Arthur
Bart, George Raymond
Baym, Gordon A
Benedek, Roy
Bodmer, Arnold R
Bowen, Samuel Philip
Brown, Laurie Mark

Carhart, Richard Alan
Ceperley, David Matthew
Chan, Sai-Kit
Chang, Shau-Jin
Cohen, Stanley
Davis, A Douglas
Day, Benjamin Downing
Day, Michael Hardy
Dutta, Pulak
Dykla, John J
Eichten, Estia Joseph
Eubanks, Robert Alonzo
Evans, Kenneth, Jr
Fano, Ugo
Freund, Peter George Oliver
Gabriel, John R
Gaylord, Richard J
Geroch, Robert Paul
Gilbert, Thomas Lewis
Golestaneh, Ahmad Ali
Grube, Geraldine Joyce Terenzoni
Hart, Harold Bird
Heaston, Robert Joseph
Henderson, George Asa
Hill, Christopher T
Inokuti, Mitio
Johnson, Porter W
Kadanoff, Leo P
Kimura, Mineo
Kogut, John Benjamin
Kurath, Dieter
Lamb, Donald Quincy, Jr
Lamb, Frederick Keithley
Lange, Yvonne
Lawson, Robert Davis
Lellouche, Gerald S
Maclay, G Jordan
Monahan, James Emmett
Myron, Harold William
Nambu, Yoichiro
Nedelsky, Leo
Oakes, Robert James
Oehme, Reinhard
Oono, Yoshitsugu
Pandharipande, Vijay Raghunath
Pappademos, John Nicholas
Peshkin, Murray
Pieper, Steven Charles
Pines, David
Quigg, Chris
Ravenhall, David Geoffrey
Rosner, Jonathan Lincoln
Sachs, Robert Green
Schonfeld, Jonathan Furth
Schult, Roy Louis
Sharma, Ram Ratan
Siegert, Arnold John Frederick
Smarr, Larry Lee
Snow, Joel A
Stack, John D
Sukhatme, Uday Pandurang
Swamy, Padmanabha Narayana
Teng, Lee Chang-Li
Thomas, Gerald H
Truran, James Wellington, Jr
Tung, Wu-Ki
Uretsky, Jack Leon
Uslenghi, Piergiorgio L
Wald, Robert Manuel
Witten, Thomas Adams, Jr
Wolf, Dieter
Wolynes, Peter Guy
Zachos, Cosmas K

INDIANA
Baker, Howard Crittendon
Balazs, Louis A P
Bose, Samir K
Capps, Richard H
Dixon, Henry Marshall
Fuchs, Norman H
Gartenhaus, Solomon
Girvin, Steven M
Gottlieb, Steven Arthur
Hanson, Andrew Jorgen
Harris, Samuel M(elvin)
Kim, Yeong Ell
Kliewer, Kenneth L
Langhoff, Peter Wolfgang
Lichtenberg, Don Bernett
Londergan, John Timothy
LoSecco, John M
MacDonald, Allan Hugh
Macfarlane, Malcolm Harris
McGlinn, William David
Mast, Cecil B
Meiere, Forrest T
Newton, Roger Gerhard
Overhauser, Albert Warner
Parish, Jeffrey Lee
Robinson, John Murrell
Salter, Lewis Spencer
Schaich, William Lee
Stephenson, Edward James
Swihart, James Calvin
Tubis, Arnold
Vasavada, Kashyap V
Walker, George Edward
Weston, Vaughan Hatherley
Wills, John G

IOWA
Byers, Ronald Elner
Carlson, Bille Chandler

Clem, John R
Jacob, Richard L
Klemm, Richard Andrew
Klink, William H
Lassila, Kenneth Eino
Leacock, Robert A
Luban, Marshall
Marker, David
Payne, Gerald Lew
Pursey, Derek Lindsay
Rizzo, Thomas Gerard
Ross, Dennis Kent
Ruedenberg, Klaus
Schweitzer, John William
Williams, Stanley A

KANSAS
Ling, Daniel Seth, Jr
Mainster, Martin Aron
Strecker, Joseph Lawrence
Weaver, Oliver Laurence
Welling, Daniel J

KENTUCKY
Chalmers, Joseph Stephen
Connolly, John William Domville
Elitzur, Moshe
Lehman, Guy Walter
MacKellar, Alan Douglas
Subbaswamy, Kumble R(amarao)
Van Winter, Clasine
Yaes, Robert Joel
Yost, Francis Lorraine

LOUISIANA
Brans, Carl Henry
Callaway, Joseph
Carter, James Clarence
Draayer, Jerry Paul
Head, Charles Everett
Ho, Yew Kam
Matese, John J
O'Connell, Robert F
Perdew, John Paul
Purrington, Robert Daniel
Rau, A Ravi Prakash
Rosensteel, George T
Slaughter, Milton Dean
Tipler, Frank Jennings, III
Vashishta, Priya Darshan
Witriol, Norman Martin

MAINE
Beard, David Breed
Csavinszky, Peter John
McKay, Susan Richards
Metz, Roger N
Morrow, Richard Alexander

MARYLAND
Atanasoff, John Vincent
Aviles, Joseph B
Bird, Joseph Francis
Bonavita, Nino Louis
Boye, Robert James
Brill, Dieter Rudolf
Campolattaro, Alfonso
Cawley, Robert
Chan, Kwing Lam
Cook, Richard Kaufman
Criss, Thomas Benjamin
Danos, Michael
Davis, Frederic I
Dickey, Joseph W
Dorfman, Jay Robert
Drachman, Richard Jonas
Dragt, Alexander James
Einstein, Theodore Lee
Engle, Irene May
Feldman, Gordon
Ferrell, Richard Allan
Fivel, Daniel I
Frankel, Michael Jay
Fulton, Thomas
Gadzuk, John William
Gates, Sylvester J, Jr
Gaunaurd, Guillermo C
Gilinsky, Victor
Glick, Arnold J
Gray, Ernest Neal
Grayson, William Curtis, Jr
Greenberg, Oscar Wallace
Griffin, James J
Guier, William Howard
Haig, Frank Rawle
Harris, Isadore
Hayward, Raymond (W)ebster
Hu, Bei-Lok Bernard
Jones, George R
Judd, Brian Raymond
Kacser, Claude
Kaplan, Alexander E
Keenan, Thomas Aquinas
Kim, Chung W
Kirkpatrick, Theodore Ross
Korenman, Victor
Krause, Thomas Otto
Land, David J(ohn)
Libelo, Louis Francis
MacDonald, Rosemary A
MacDonald, William
Markley, Francis Landis
Marx, Egon
Meckler, Alvin

Miller, James Taggert, Jr
Misner, Charles William
Mountain, Raymond Dale
O'Connell, James S
Oneda, Sadao
Pastine, D John
Pati, Jogesh Chandra
Redish, Edward Frederick
Robertson, Baldwin
Rosenstock, Herbert Bernhard
Roszman, Larry Joe
Rubin, Morton Harold
Sachs, Lester Marvin
Schmid, Lawrence Alfred
Schumacher, Clifford Rodney
Shore, Steven Neil
Sidhu, Deepinder Pal
Simmons, John Arthur
Soln, Josip Zvonimir
Steven, Alasdair C
Thirumalai, Devarajan
Van Vechten, Deborah
Wallace, Stephen Joseph
Wang, Ching-Ping Shih
Yedinak, Peter Demerton
Yoon, Peter Haesung
Zerilli, Frank J
Zucker, Paul Alan

MASSACHUSETTS
Aaron, Ronald
Altman, Albert
Antonoff, Marvin M
Argyres, Petros
Baker, Adolph
Bakshi, Pradip M
Baranger, Michel
Berker, Ahmet Nihat
Bernstein, Herbert J
Blankschtein, Daniel
Burke, Edward Aloysius
Calusdian, Richard Frank
Carpenter, Jack William
Carter, Ashton Baldwin
Charpie, Robert Alan
Chase, David Marion
Clapp, Roger Edge
Coleman, James Andrew
Coleman, Sidney Richard
Collins, Aliki Karipidou
Cook, LeRoy Franklin, Jr
Corinaldesi, Ernesto
Crowell, Julian
Dalgarno, Alexander
Deser, Stanley
Dewan, Edmond M
Donnelly, Thomas William
Donoghue, John Francis
Doyle, Jon
Drane, Charles Joseph, Jr
Durso, John William
Ehrenreich, Henry
Eyges, Leonard James
Farhi, Edward
Ford, Lawrence Howard
Freedman, Daniel Z
Freese, Katherine
Friedman, Marvin Harold
Georgi, Howard
Godine, John Elliott
Goldberg, Hyman
Goldstone, Jeffrey
Graneau, Peter
Guertin, Ralph Francis
Gunther, Leon
Guth, Alan Harvey
Halperin, Bertrand Israel
Harrison, Ralph Joseph
Hilsinger, Harold W
Hodges, Hardy M
Holstein, Barry Ralph
Horrigan, Frank Anthony
Huang, Kerson
Hyman, Howard Allan
Imbrie, John Z
Jackiw, Roman Wladimir
Jacobs, Laurence Alan
Jaffe, Robert Loren
Jagannathan, Kannan
Joannopoulos, John Dimitris
Johnson, Kenneth Alan
Johnston, George Lawrence
Jones, Robert Clark
Jose, Jorge V
Joss, Paul Christopher
Kalman, Gabor J
Kannenberg, Lloyd C
Kaplan, Irving
Karakashian, Aram Simon
Kelley, Paul Leon
Kerman, Arthur Kent
Kincaid, Thomas Gardiner
Kivel, Bennett
Klein, William
Krass, Allan S(hale)
Krumhansl, James Arthur
Kuhn, Thomas S
Lai, Shu Tim
Larsen, David M
Leiby, Clare C, Jr
Lightman, Alan Paige
Lomon, Earle Leonard
Low, Francis Eugene
Machacek, Marie Esther

Theoretical Physics (cont)

Machta, Jonathan Lee
Malenka, Bertram Julian
Mano, Koichi
Martin, Arthur Wesley, III
Martin, Paul Cecil
Mellen, Walter Roy
Melrose, Richard B
Mollow, Benjamin R
Moniz, Ernest Jeffrey
Moomaw, William Renken
Moses, Harry Elecks
Mullin, William Jesse
Negele, John William
Nelson, David Robert
Park, David Allen
Pendleton, Hugh Nelson, III
Pilot, Christopher H
Pratt, George Woodman, Jr
Press, William Henry
Rebbi, Claudio
Roberts, David Hall
Robertson, David C
Salzman, George
Santilli, Ruggero Maria
Savickas, David Francis
Saxon, David Stephen
Schloemann, Ernst
Schnitzer, Howard J
Schwebel, Solomon Lawrence
Schweber, Silvan Samuel
Segal, Irving Ezra
Semon, Mark David
Sheldon, Eric
Shimony, Abner
Sneddon, Leigh
Stachel, John Jay
Sternheim, Morton Maynard
Tambasco, Daniel Joseph
Tisza, Laszlo
Tomljanovich, Nicholas Matthew
Towne, Dudley Herbert
Vassell, Milton O
Vaughn, Michael Thayer
Victor, George A
Vilenkin, Alexander
Villars, Felix Marc Hermann
Walker, James Frederick, Jr
Weaver, David Leo
Whitney, Cynthia Kolb
Wisdom, Jack Leach
Wong, Po Kee
Wootters, William Kent
Wu, Fa Yueh
Wu, Tai Tsun
Yos, Jerrold Moore
Young, James Edward

MICHIGAN

Abbasabadi, Alireza
Agin, Gary Paul
Berger, Beverly Kobre
Blass, Gerhard Alois
Brailsford, Alan David
Carley, David Don
Chang, Jhy-Jiun
Chen, Min-Shih
Davis, Lloyd Craig
Denman, Harry Harroun
Devlin, John F
Favro, Lawrence Dale
Federbush, Paul Gerard
Fishman, Frank J, Jr
Ford, George Willard
Fradkin, David Milton
Francis, William Porter
Gay, Jackson Gilbert
Gordon, Morton Maurice
Gupta, Suraj Narayan
Harrison, Michael Jay
Karkheck, John Peter
Kaskas, James
Kemeny, Gabor
Kovacs, Julius Stephen
Kromminga, Albion Jerome
Kuo, Pao-Kuang
Lee, Sung Mook
Lewis, Robert Richards, Jr
McKinley, John McKeen
McManus, Hugh
Mahanti, Subhendra Deb
Morgan-Pond, Caroline G
Reitz, John Richard
Richter, Roy
Rolnick, William Barnett
Ross, Marc Hansen
Sander, Leonard Michael
Saperstein, Alvin Martin
Savit, Robert Steven
Schnitker, Jurgen H
Signell, Peter Stuart
Smith, John Robert
Tomozawa, Yukio
Wu, Alfred Chi-Tai
Yao, York-Peng Edward
Yun, Suk Koo

MINNESOTA

Austin, Donald Murray
Bayman, Benjamin
Campbell, Charles Edwin
Cassola, Robert Louis

Halley, James Woods, (Jr)
Jordan, Thomas Fredrick
Kim, Sung Kyu
Lysak, Robert Louis
Meyer, Frank Henry
Rudaz, Serge
Schuldt, Spencer Burt
Suura, Hiroshi
Tait, William Charles
Titus, William James
Valls, Oriol Tomas
Wesley, Walter Glen

MISSISSIPPI

Corben, Herbert Charles
Lestrade, John Patrick

MISSOURI

Adawi, Ibrahim (Hasan)
Cheng, Ta-Pei
Ching, Wai-Yim
Clark, John Walter
Cochran, Andrew Aaron
DeFacio, W Brian
Delaney, Robert Michael
Fisch, Ronald
Gammel, John Ledel
Hagen, Donald E
Huang, Justin C
Huddleston, Philip Lee
Jaynes, Edwin Thompson
Kovacs, Sandor J, Jr
Mashhoon, Bahram
Parks, William Frank
Parry, Myron Gene
Retzloff, David George
Shrauner, James Ely
Taub, Haskell Joseph

MONTANA

Hiscock, William Allen
Kinnersley, William Morris
Nordtvedt, Kenneth L
Swenson, Robert J

NEBRASKA

Chakkalakal, Dennis Abraham
Davies, Marcia A
Fabrikant, Ilya I
Flocken, John W
Goss, David
Kennedy, Robert E
Longley, William Warren, Jr
Underhill, Glenn
Zimmerman, Edward John

NEVADA

Levine, Paul Hersh
Scott, William Taussig
Winterberg, Friedwardt

NEW HAMPSHIRE

Huggins, Elisha R
Lawrence, Walter Edward
Shepard, Harvey Kenneth

NEW JERSEY

Adler, Stephen L
Alig, Roger Casanova
Anderson, James Leroy
Anderson, Philip Warren
Anderson, Terry Lee
Becken, Eugene D
Bernstein, Jeremy
Blount, Eugene Irving
Bronzan, John Brayton
Callan, Curtis
Chan, Chun Kin
Cohen, Morrel Herman
Derkits, Gustav
Dohnanyi, Julius S
Gale, William Arthur
Gautreau, Ronald
Goldin, Gerald Alan
Goode, Philip Ranson
Gora, Thaddeus F, Jr
Gott, J Richard, III
Grant, James J, Jr
Gross, David (Jonathan)
Helfand, Eugene
Hohenberg, Pierre Claude
Jackson, Shirley Ann
Joyce, William B(axter)
Katz, Sheldon Lane
Kelsey, Edward Joseph
Kijewski, Louis Joseph
Klebanov, Igor Romanovich
Langreth, David Chapman
Lieb, Elliott Hershel
Lines, Malcolm Ellis
Lohse, David John
Lovelace, Claud William Venton
Luryi, Serge
McAfee, Walter Samuel
Mattheiss, Leonard Francis
Nardi, Vittorio
Ng, Tai-Kai
Phillips, James Charles
Platzman, Philip M
Rockmore, Ronald Marshall
Salwen, Harold
Schiller, Ralph
Schluter, Michael
Shapiro, Joel Alan

Solla, Sara A
Spergel, David Nathaniel
Tishby, Naftali Z
Torrey, Henry Cutler
Vanderbilt, David Hamilton
Von Hippel, Frank
Weinstein, Marvin
Werthamer, N Richard
Wheeler, John Archibald
Wigner, Eugene Paul
Wilczek, Frank Anthony
Wojtowicz, Peter Joseph
Wolff, Peter A
Wong, Tang-Fong Frank
Zapolsky, Harold Saul
Zatzkis, Henry

NEW MEXICO

Albers, Robert Charles
Alldredge, Gerald Palmer
Baker, George Allen, Jr
Bartel, Lewis Clark
Beckel, Charles Leroy
Becker, Wilhelm
Bergeron, Kenneth Donald
Bolsterli, Mark
Brandow, Baird H
Campbell, David Kelly
Campbell, Laurence Joseph
Chandler, Colston
Chen, Tuan Wu
Chow, Weng Wah
Cobb, Donald D
Cohen, James Samuel
Coon, Sidney Alan
Cooper, Richard Kent
Coulter, Claude Alton
Davis, Cecil Gilbert
Devaney, Joseph James
Dietz, David
Doolen, Gary Dean
Du Bois, Donald Frank
England, Talmadge Ray
Erpenbeck, Jerome John
Evans, Foster
Feibelman, Peter Julian
Finley, James Daniel, III
Frank, Robert Morris
Friar, James Lewis
Gerstl, Siegfried Adolf Wilhelm
Gibbs, William Royal
Gibson, Benjamin Franklin, V
Ginocchio, Joseph Natale
Ginsparg, Paul H
Glasser, Alan Herbert
Goldman, Terrence Jack
Goldstein, John Cecil
Goldstein, Louis
Gordon, James Wylie
Gubernatis, James Edward
Gursky, Martin Lewis
Hadley, George Ronald
Heller, Leon
Hirt, Cyril William, Jr
Hull, McAllister Hobart, Jr
Jennison, Dwight Richard
Johnson, James Daniel
Johnson, Mikkel Borlaug
Jones, Eric Daniel
Kenkre, Vasudev Mangesh
Laney, Billie Eugene
Lazarus, Roger Ben
Leon, Melvin
Lilley, John Richard
Liu, Lon-Chang
Lomanitz, Ross
Longley, H(erbert) Jerry
Lyo, SungKwun Kenneth
Malone, Robert Charles
Mascheroni, P Leonardo
Metropolis, Nicholas Constantine
Newby, Neal Dow, Jr
Newman, Michael J(ohn)
Nieto, Michael Martin
Nix, James Rayford
Osinski, Marek Andrzej
Ozernoy, Leonid M
Peaslee, Alfred Tredway, Jr
Rach, Randolph Carl
Rauber, Lauren A
Reichert, John Douglas
Richards, Peter Michael
Rinker, George Albert, Jr
Schappert, Gottfried T
Schriber, Stanley Owen
Schultz, Rodney Brian
Sharp, David Howland
Siciliano, Edward Ronald
Sierk, Arnold John
Silver, Richard N
Simmons, Leonard Micajah, Jr
Slansky, Richard Cyril
Solem, Johndale Christian
Stephenson, Gerard J, Jr
Strottman, Daniel
Switendick, Alfred Carl
Thaler, Raphael Morton
Thompson, Samuel Lee
Turner, Leaf
Visscher, William M
Walker, James Joseph
Wallace, Richard Kent
Wienke, Bruce Ray
Winske, Dan

NEW YORK

Abrahamson, Adolf Avraham
Agrawal, Govind P(rasad)
Allen, Philip B
Alper, Ralph Asher
Ambegaokar, Vinay
Arin, Kemal
Aronson, Raphael (Friedman)
Ashcroft, Neil William
Ashtekar, Abhay Vasant
Auer, Peter Louis
Bak, Per
Balazs, Nandor Laszlo
Baltz, Anthony John
Beg, Mirza Abdul Baqi
Berger, Jay Manton
Bergmann, Peter Gabriel
Beth, Eric Walter
Bethe, Hans Albrecht
Birman, Joseph Leon
Boardman, John
Borowitz, Sidney
Bowick, Mark John
Boyer, Timothy Howard
Bray, James William
Brinkman, William F
Brown, Edmond
Brown, Ronald Alan
Burschka, Martin A
Butkov, Eugene
Buttiker, Markus
Carroll, Clark Edward
Chang, Ngee Pong
Chew, Herman W
Chung, Victor
Clark, Edward Aloysius
Courant, Ernest David
Craxton, Robert Stephen
Creutz, Michael John
Das, Ashok Kumar
Diamond, Joshua Benamy
Doering, Charles Rogers
Dover, Carl Bellman
Dreiss, Gerard Julius
Dubisch, Russell John
Duke, Charles Bryan
Eberly, Joseph Henry
Eger, F Martin
Eisenbud, Leonard
Emery, Victor John
Falk, Harold
Feigenbaum, Mitchell Jay
Feinberg, Gerald
Fishbone, Leslie Gary
Ford, Kenneth William
Forgacs, Gabor
Fox, David
Francis, Norman
Franco, Victor
Friedberg, Richard Michael
Fuda, Michael George
Fujita, Shigeji
Gelman, Donald
Geoghegan, Ross
Gersten, Joel Irwin
Gillespie, John
Glassgold, Alfred Emanuel
Goldhaber, Alfred Scharff
Gottfried, Kurt
Greenberg, Howard
Greenberg, Newton Isaac
Greenberger, Daniel Mordecai
Gutzwiller, Martin Charles
Harrington, Steven Jay
Hart, Edward Walter
Harvey, Alexander Louis
Haus, Joseph Wendel
Inomata, Akira
Jabbur, Ramzi Jibrail
Jackson, Andrew D, Jr
Jepsen, Donald William
Kahana, Sidney H
Kahn, Peter B
Kaku, Michio
Kalos, Malvin Howard
Kaplan, Harvey
Kaup, David James
Khuri, Nicola Najib
Kinoshita, Toichiro
Kirtley, John Robert
Kjeldaas, Terje, Jr
Klarfeld, Joseph
Knight, Bruce Winton, (Jr)
Koltun, Daniel S
Komar, Arthur Baraway
Koplik, Joel
Krieger, Joseph Bernard
Krinsky, Samuel
Kuo, Thomas Tzu Szu
Lang, Norton David
Lavine, James Philip
Lax, Melvin
Lee, Tsung Dao
Lee, Yung-Chang
Lemos, Anthony M
Liboff, Richard L
Lin, Duo-Liang
Lindquist, William Brent
Lowenstein, John Hood
Lubitz, Cecil Robert
Lucey, Carol Ann
Lustig, Harry
Luttinger, Joaquin Mazdak
McCoy, Barry

McDonald, Frank Alan
McIrvine, Edward Charles
Malin, Shimon
Manassah, Jamal Tewfek
Mane, Sateesh Ramchandra
Mansfield, Victor Neil
Marr, Robert B
Mathur, Vishnu Sahai
Meyer, Robert Jay
Millener, David John
Millet, Peter J
Mould, Richard A
Mueller, Alfred H
Mukhopadhyay, Nimai Chand
Nelkin, Mark
Nelson, Charles Arnold
Newman, Charles Michael
Newstein, Maurice
Noz, Marilyn E
O'Dwyer, John J
Ohanian, Hans C
Pais, Abraham
Pantelides, Sokrates Theodore
Parzen, George
Peak, David
Pearle, Philip Mark
Pease, Robert Louis
Peierls, Ronald F
Percus, Jerome K
Privman, Vladimir
Rafanelli, Kenneth R
Ram, Michael
Raman, Varadaraja Venkata
Ratcliff, Keith Frederick
Remo, John Lucien
Rice, Michael John
Richardson, Robert William
Rohrlich, Fritz
Rosenberg, Leonard
Rosenzweig, Carl
Roskes, Gerald J
Ruderman, Malvin Avram
Sachs, Mendel
Sakita, Bunji
Satz, Helmut T G
Schlitt, Dan Webb
Schultz, Theodore David
Schwartz, Brian B
Schwartz, Melvin J
Seiden, Philip Edward
Seyler, Charles Eugene
Shalloway, David Irwin
Shimamoto, Yoshio
Siggia, Eric Dean
Silverman, Benjamin David
Sirlin, Alberto
Smith, Gerrit Joseph
Smith, Jack Howard
Smith, John
Smolin, Lee
Sobel, Michael I
Spergel, Martin Samuel
Spruch, Larry
Srinivasan, G(urumakonda) R
Stern, Frank
Tavel, Morton
Toner, John Joseph
Trigg, George Lockwood
Troubetzkoy, Eugene Serge
Trueman, Thomas Laurence
Tryon, Edward Polk
Tye, Sze-Hoi Henry
Wali, Kameshwar C
Webb, Richard
Weingarten, Donald Henry
Weisberger, William I
Wellner, Marcel
Weneser, Joseph
Wilson, Fred Lee
Yennie, Donald Robert
Yoffa, Ellen June
Zwanziger, Daniel

NORTH CAROLINA
Adler, Carl George
Bernholc, Jerzy
Biedenharn, Lawrence Christian, Jr
Brehme, Robert W
Brown, J(ohn) David
Choudhury, Abdul Latif
Ciftan, Mikael
Cotanch, Stephen Robert
Coulter, Byron Leonard
Cusson, Ronald Yvon
D'Arruda, Jose Joaquim
Dolan, Louise Ann
Dotson, Allen Clark
Fulp, Ronald Owen
Griffing, George Warren
Hall, George Lincoln
Han, Moo-Young
Iafrate, Gerald Joseph
Ji, Chueng Ryong
Katzin, Gerald Howard
Kerr, William Clayton
Klenin, Marjorie A
Lado, Fred
Lapicki, Gregory
Merzbacher, Eugen
Park, Jae Young
Robl, Hermann R
Torquato, Salvatore
Whitlock, Richard T
Yang, Weitao

York, James Wesley, Jr

NORTH DAKOTA
Hassoun, Ghazi Qasim
Kurtze, Douglas Alan
Schwalm, Mizuho K
Schwalm, William A

OHIO
Allender, David William
Arnett, Jerry Butler
Breitenberger, Ernst
Dilley, James Paul
Ellis, David Greenhill
Esposito, F Paul
Foldy, Leslie Lawrance
Goodman, Bernard
Gunther, Marian W J
Hanson, Harvey Myron
Kelly, Donald C
Kowalski, Kenneth L
Lang, Joseph Edward
Lock, James Albert
Macklin, Philip Alan
McLennan, Donald Elmore
Mainland, Gordon Bruce
Mallory, Willam R
Manning, Robert M
Mills, Robert Laurence
Montgomery, Charles Gray
Moroi, David S
Mulligan, Bernard
Nam, Sang Boo
O'Hare, John Michael
Onley, David S
Palffy-Muhoray, Peter
Patton, Bruce Riley
Pytte, Agnar
Raby, Stuart
Schlosser, Herbert
Schneider, Ronald E
Schreiber, Paul J
Stroud, David Gordon
Styer, Daniel F
Suranyi, Peter
Taylor, Beverley Ann Price
Tenhover, Michael Alan
Tobocman, William
Tuan, Tai-Fu
Uralil, Francis Stephen
Whitesell, William James
Wiff, Donald Ray
Williamson, William, Jr
Wood, Van Earl
Wright, Louis E
Wright, William Robert

OKLAHOMA
Coffman, Moody Lee
Cohn, Jack
Cowan, John James
Huffaker, James Neal
Letcher, John Henry, III
Milton, Kimball Alan
Samuel, Mark Aaron
Scales, John Alan
Sigal, Richard Frederick
Thomsen, Leon

OREGON
Belinfante, Frederik J
Carmichael, Howard John
Csonka, Paul L
Elliott, Richard Amos
Fontana, Peter R
Girardeau, Marvin Denham, Jr
Goswami, Amit
Hwa, Rudolph Chia-Chao
Kerlick, George David
Khalil, M Aslam Khan
Liu-Ger, Tsu-Huei
McClure, Joel William, Jr
McCollum, Gin
Madsen, Victor Arviel
Olness, Fredrick Iver
Semura, Jack Sadatoshi
Siemens, Philip John
Skinner, Richard Emery
Tunturi, Archie Robert

PENNSYLVANIA
Albano, Alfonso M
Amado, Ralph
Anderson, James B
Armenti, Angelo, Jr
Austern, Norman
Banavar, Jayanth Ramarao
Baranger, Elizabeth Urey
Becker, Stephen Fraley
Bloomfield, Philip Earl
Bludman, Sidney Arnold
Böer, Karl Wolfgang
Borse, Garold Joseph
Carlitz, Robert D
Cohen, Jeffrey M
Cohen, Michael
Collins, John Clements
Coon, Darryl Douglas
Cutkosky, Richard Edwin
Davidon, William Cooper
Emtage, Peter Roesch
Engheta, Nader
Feng, DaHsuan
Fleming, Gordon N

Ford, William Frank
Franklin, Jerrold
Freed, Norman
Gerjuoy, Edward
Gibson, Gordon
Goldschmidt, Yadin Yehuda
Griffiths, Robert Budington
Grotch, Howard
Gunton, James D
Havas, Peter
Hoffman, Richard Bruce
Janis, Allen Ira
Jhon, Myung S
Kazes, Emil
Kisslinger, Leonard Sol
Klein, Abraham
Langacker, Paul George
Larsen, Sigurd Yves
Li, Ling-Fong
Liang, Shoudan
Lim, Teck-Kah
Lubensky, Tom C
Mezger, Fritz Walter William
Miller, Irvin Alexander
Neville, Donald Edward
Newman, Ezra
Novaco, Anthony Dominic
Pratt, Richard Houghton
Rabii, Sohrab
Roman, Paul
Rosen, Gerald Harris
Rothwarf, Allen
Rovelli, Carlo
Satz, Ronald Wayne
Segre, Gino C
Shaffer, Russell Allen
Sion, Edward Michael
Steinhardt, Paul Joseph
Tabakin, Frank
Taylor, Lyle Herman
Verhanovitz, Richard Frank
Weiner, Brian Lewis
Winicour, Jeffrey
Wolfenstein, Lincoln
Yuan, Jian-Min

RHODE ISLAND
Bonner, Jill Christine
Brandenberger, Robert H
Cooper, Leon N
Feldman, David
Fried, Herbert Martin
Gora, Edwin Karl
Kang, Kyungsik
Kaufman, Charles
Kirwan, Donald Frazier
Mellberg, Leonard Evert
Westervelt, Peter Jocelyn

SOUTH CAROLINA
Au, Chi-Kwan
Bostock, Judith Louise
Burt, Philip Barnes
Graben, Henry Willingham
Knight, James Milton
Lerner, Edward Clarence
Manson, Joseph Richard
Safko, John Loren

SOUTH DAKOTA
Duffey, George Henry
Hall, Harold Hershey
Jones, Robert William

TENNESSEE
Alsmiller, Rufard G, Jr
Barrett, John Harold
Becker, Richard Logan
Bloch, Ingram
Childers, Robert Wayne
Collins, Thomas C
Cook, James Marion
Davies, Kenneth Thomas Reed
Davis, Harold Lloyd
Deeds, William Edward
Groer, Peter Gerold
Halbert, Edith Conrad
Harris, Edward Grant
Lane, Eric Trent
Larson, Nancy Marie
Liu, Samuel Hsi-Peh
Mahan, Gerald Dennis
Mansur, Louis Kenneth
Nestor, C William, Jr
Oen, Ordean Silas
Painter, Gayle Stanford
Payne, Marvin Gay
Pinkston, William Thomas
Quinn, John Joseph
Ritchie, Rufus Haynes
Satchler, George Raymond
Turner, James Edward
Wang, Jia-Chao
Welton, Theodore Allen

TEXAS
Allen, Roland E
Antoniewicz, Peter R
Arnowitt, Richard Lewis
Bailey, James Stephen
Bassichis, William
Bohm, Arno
Brachman, Malcolm K
Breig, Edward Louis

Bryan, Ronald Arthur
Coker, William Rory
Cooke, James Horton
DeWitt, Bryce Seligman
DeWitt-Morette, Cecile
Dicus, Duane A
Doran, Robert Stuart
Englund, John Caldwell
Evans, Wayne Errol
Ford, Albert Lewis, Jr
Fulling, Stephen Albert
Gilman, Frederick Joseph
Griffy, Thomas Alan
Hannon, James Patrick
Hardcastle, Donald Lee
Herman, Robert
Hewett, Lionel Donnell
Hoffman, Alexander A J
Horton, Claude Wendell, Jr
Hu, Bambi
Hu, Chia-Ren
Huang, Huey Wen
Hubisz, John Lawrence, Jr
Ivash, Eugene V
Johnson, Darell James
Kim, Young Nok
Klein, Douglas J
Ko, Che Ming
Kobe, Donald Holm
Kouri, Donald Jack
Liang, Edison Park-Tak
Lipps, Frederick Wiessner
McCauley, Joseph Lee
McIntyre, Robert Gerald
Mahajan, Swadesh Mitter
Marlow, William Henry
Meshkov, Sydney
Miller, Bruce Neil
Miller, James Gilbert
Minerbo, Gerald N
Moorhead, William Dean
Myles, Charles Wesley
Nordlander, Peter Jan Arne
Pettitt, Bernard Montgomery
Pooch, Udo Walter
Rindler, Wolfgang
Robinson, Ivor
Roeder, Robert Charles
Ross, David Ward
Sablik, Martin J
Schieve, William
Shepley, Lawrence Charles
Skolnick, Malcolm Harris
Smith, Frederick Adair, Jr
Smith, Roger Alan
Stevenson, Paul Michael
Stratton, Robert
Sudarshan, Ennackel Chandy George
Swift, Jack Bernard
Trammell, George Thomas
Voigt, Gerd-Hannes
Weinberg, Steven
Wilson, Thomas Leon
Winbow, Graham Arthur
Zook, Herbert Allen

UTAH
Ball, James Stutsman
Chatelain, Jack Ellis
Evenson, William Edwin
Gardner, John Hale
Harrison, Bertrand Kent
Hatch, Dorian Maurice
McDonald, Keith Leon
Price, Richard Henry
Sutherland, Bill
Wu, Yong-Shi

VERMONT
Krizan, John Ernest
Oughstun, Kurt Edmund
Smith, David Young

VIRGINIA
Adam, John Anthony
Aitken, Alfred H
Asterita, Mary Frances
Atalay, Bulent Ismail
Barrois, Bertrand C
Bishop, Marilyn Frances
Brown, Ellen Ruth
Buck, Warren W, III
Burns, Grover Preston
Calle, Carlos Ignacio
Chambers, Charles MacKay
Chang, Lay Nam
Coopersmith, Michael Henry
Debney, George Charles, Jr
De Wolf, David Alter
Dworzecka, Maria
Gibson, Luther Ralph
Gowdy, Robert Henry
Gross, Franz Lucretius
Hobart, Robert H
Hurley, William Jordan
Jacobs, Kenneth Charles
Jena, Purusottam
Kabir, Prabahan Kemal
Kelley, Ralph Edward
Kiefer, Harold Milton
McKnight, John Lacy
Madan, Rabinder Nath
Mansfield, John E
Marshak, Robert Eugene

Robertson, Harry S(troud)
Sullivan, Neil Samuel
Veziroglu, T Nejat
Vollmer, James

GEORGIA
Anderson, Robert Lester
Batten, George L, Jr
Colwell, Gene Thomas
Family, Fereydoon
Fox, Ronald Forrest
Kezios, Stothe Peter
Vachon, Reginald Irenee

IDAHO
Ramshaw, John David
Richardson, Lee S(pencer)

ILLINOIS
Abraham, Bernard M
Brewster, Marcus Quinn
Chao, B(ei) T(se)
Cooper, William Edward
Farhadieh, Rouyentan
Klein, Max
Kuchnir, Moyses
Lee, Anthony L
Liu, Yung Yuan
Mintzer, David
Mokadam, Raghunath G(anpatrao)
Olvera-dela Cruz, Monica
Quay, Paul Michael
Schiffman, Robert A
Shepard, Kenneth Wayne
Simmons, Ralph Oliver
Snelson, Alan
Tang, James Juh-Ling
Yuen, Man-Chuen

INDIANA
DeWitt, David P
Giordano, Nicholas J
Ho, Cho-Yen
Keesom, Pieter Hendrik
Liley, Peter Edward
McComas, Stuart T
Viskanta, Raymond
Wark, Kenneth, Jr
Western, Arthur Boyd

IOWA
Hering, Robert Gustave
Intemann, Gerald William
Luban, Marshall
Swenson, Clayton Albert

KANSAS
Gosman, Albert Louis
Ho, James Chien Ming
Sapp, Richard Cassell
Snyder, Melvin H(enry), Jr
Sorensen, Christopher Michael

KENTUCKY
Bakanowski, Stephen Michael
DeLong, Lance Eric
Mills, Roger Edward

LOUISIANA
Baw, Philemon S H
Rotty, Ralph M(cGee)

MAINE
Schmiedeshoff, George M

MARYLAND
Baba, Anthony John
Battin, William James
Baum, Howard Richard
Bock, Arthur E(mil)
Caveny, Leonard Hugh
Cezairliyan, Ared
Chang, Ren-Fang
Colwell, Jack Harold
Dorfman, Jay Robert
Ferrell, Richard Allan
Furukawa, George Tadaharu
Gammon, Robert Winston
Griffin, James J
Haberman, William L(awrence)
Hsia, Jack Jinn-Goe
Kirkpatrick, Theodore Ross
MacDonald, Rosemary A
Marqusee, Jeffrey Alan
Mitler, Henri Emmanuel
Murphy, John Cornelius
Nash, Jonathon Michael
Parker, William James
Radermacher, Reinhard
Randall, John Douglas
Roszman, Larry Joe
Schmid, Lawrence Alfred
Schooley, James Frederick
Scudder, Jack David
Sengers, Johanna M H Levelt
Tong, Long Sun
Traugott, Stephen C(harles)
Waldo, George Van Pelt, Jr
Weckesser, Louis Benjamin
Williams, Ellen D
Wu, Chih
Zirkind, Ralph

MASSACHUSETTS
Appleton, John P(atrick)
Bolsaitis, Pedro
Bowman, H(arry) Frederick
Cipolla, John William, Jr
Day, Robert William
Delvaille, John Paul
Emmons, Howard W(ilson)
Hatsopoulos, George Nicholas
Hoercher, Henry E(rhardt)
Hoge, Harold James
Jacobs, Laurence Alan
Jirmanus, Munir N
Jose, Jorge V
Kaufman, Larry
Kemp, Nelson Harvey
Klein, William
Liau, Zong-Long
Machta, Jonathan Lee
Maxwell, Emanuel
Mikic, Bora
Phillies, George David Joseph
Picard, Richard Henry
Reeves, Barry L(ucas)
Silvers, J(ohn) P(hillip)
Sonin, Ain A(nts)
Tiernan, Robert Joseph
Wang, Chia Ping
Williamson, Richard Cardinal
Yener, Yaman

MICHIGAN
Boyer, Raymond Foster
Chock, David Poileng
Edgerton, Robert Howard
Hartman, John L(ouis)
Heremans, Joseph P
Koenig, Milton G
LaRocca, Anthony Joseph
Menard, Albert Robert, III
Piccirelli, Robert Anthony
Smith, George Wolfram
Weinreich, Gabriel
Westrum, Edgar Francis, Jr
Yang, Wen Jei

MINNESOTA
Bonne, Ulrich
Flower, Terrence Frederick
Goldman, Allen Marshall
Heberlein, Joachim Viktor Rudolf
Lutes, Olin S
Olson, Reuben Magnus
Valls, Oriol Tomas
Zimmermann, William, Jr

MISSISSIPPI
Eastland, David Meade
Smith, Allie Maitland

MISSOURI
Chen, Ta-Shen
Elgin, Robert Lawrence
Fisch, Ronald
Gardner, Richard A
Look, Dwight Chester, Jr
Van Camp, W(illiam) M(orris)
Whitney, Arthur Edwin, Jr

MONTANA
Wood, William Wayne

NEBRASKA
Kennedy, Robert E

NEW HAMPSHIRE
King, Allen Lewis

NEW JERSEY
Anthony, Philip John
Bachman, Walter Crawford
Bechis, Dennis John
Blair, David W(illiam)
Buteau, L(eon) J
Cody, George Dewey
Cotter, Martha Ann
Couchman, Peter Robert
Debenedetti, Pablo Gaston
Farkass, Imre
Geskin, Ernest S
Gittleman, Jonathan I
Glass, John Richard
Haindl, Martin Wilhelm
Holt, Vernon Emerson
Hrycak, Peter
Kolodner, Paul R
Lindenfeld, Peter
Lines, Malcolm Ellis
Makofske, William Joseph
Matey, James Regis
Raichel, Daniel R(ichter)
Rollino, John
Stillinger, Frank Henry
Tsonopoulos, Constantine
Tsou, F(u) K(ang)
Zebib, Abdelfattah M G

NEW MEXICO
Bankston, Charles A, Jr
Dingus, Ronald Shane
Farnsworth, Archie Verdell, Jr
Grilly, Edward Rogers
Houghton, Arthur Vincent, III
Johnson, James Daniel

Julien, Howard L
Laquer, Henry L
Martin, Richard Alan
Moss, Marvin
Richards, Peter Michael
Rose, Harvey Arnold
Shaner, John Wesley
Sinha, Dipen N
Trela, Walter Joseph

NEW YORK
Abbott, Michael McFall
Alben, Richard Samuel
Allen, Philip B
Amato, Joseph Cono
Barron, Saul
Begell, William
Budzinski, Walter Valerian
Buttiker, Markus
Daman, Harlan Richard
De Boer, P(ieter) C(ornelis) Tobias
Hall, J(ohn) Gordon
Hecht, Charles Edward
Henkel, Elmer Thomas
Holbrow, Charles H
Hurst, Robert Philip
Irvine, Thomas Francis
Kroeger, Peter G
LaPietra, Joseph Richard
Li, Chung-Hsiung
Loring, Roger Frederic
Pribil, Stephen
Privman, Vladimir
Rimai, Donald Saul
Ring, James Walter
Shepherd, Joseph Emmett
Socolar, Sidney Joseph
Staub, Fred W
Stell, George Roger
Stiel, Leonard Irwin
Thompson, Philip A
Torrance, Kenneth E(ric)
Tourin, Richard Harold
Webb, Richard
Zahed, Ismail

NORTH CAROLINA
Brown, J(ohn) David
Edwards, John Auert
Kratz, Howard Russel
Lado, Fred

NORTH DAKOTA
Kurtze, Douglas Alan

OHIO
Berg, James Irving
Boughton, Robert Ivan, Jr
Deissler, Robert G(eorge)
De Rocco, Andrew Gabriel
Edwards, David Olaf
Fenichel, Henry
Heer, Clifford V
Jacobs, Donald Thomas
Jones, Charles E(dward)
Kaufman, Miron
Keith, Theo Gordon, Jr
La Mers, Thomas Herbert
Lawler, Martin Timothy
Lemke, Ronald Dennis
Mayers, Richard Ralph
Moon, Tag Young
Petschek, Rolfe George
Pool, Monte J
Rockstroh, Todd Jay
Rosenblatt, Charles Steven
Sanford, Edward Richard
Schmitt, George Frederick, Jr
Scholten, Paul David
Schuele, Donald Edward
Styer, Daniel F
Thekdi, Arvind C
Walters, Craig Thompson
Yaney, Perry Pappas

OKLAHOMA
Foster, C(harles) Vernon
Harden, Darrel Grover
Love, Tom Jay, Jr
Rutledge, Delbert Leroy

OREGON
Brodie, Laird Charles
Girardeau, Marvin Denham, Jr
Gokcen, Nev A(ltan)
Griffith, William Thomas
Schaefer, Seth Clarence
Semura, Jack Sadatoshi

PENNSYLVANIA
Arnold, James Norman
Boughn, Stephen Paul
Brewer, LeRoy Earl, Jr
Callen, Herbert Bernard
Carmi, Shlomo
Chun, Sun Woong
Crow, Jack Emerson
Farouk, Bakhtier
Finegold, Leonard X
Friedberg, Simeon Adlow
Goldschmidt, Yadin Yehuda
Griffiths, Robert Budington
Hager, Nathaniel Ellmaker, Jr
Hertzberg, Martin

Hogenboom, David L
Lazaridis, Anastas
Maher, James Vincent
Nagle, John F
Ramezan, Massood
Steyert, William Albert
Sturdevant, Eugene J
Tang, Y(u) S(un)
Toulouse, Jean
Yao, Shi Chune
Zemel, Jay N(orman)

RHODE ISLAND
Fallieros, Stavros
McEligot, Donald M(arinus)
Timbie, Peter T
Wilson, Mason P, Jr

SOUTH CAROLINA
Chuang, Ming Chia
Gaines, Albert L(owery)

SOUTH DAKOTA
Ashworth, T

TENNESSEE
Boatner, Lynn Allen
Busey, Richard Hoover
Fontana, Mario H
Keefer, Dennis Ralph
Keshock, Edward G
LeVert, Francis E
Lewis, James W L
McGregor, Wheeler Kesey, Jr
Scott, Herbert Andrew

TEXAS
Arumi, Francisco Noe
Bergman, Theodore L
Brostow, Witold Konrad
Caflisch, Robert Galen
Chao, Jing
Eichhorn, Roger
Howell, John Reid
Hrkel, Edward James
Huang, Huey Wen
Kirk, Wiley Price
Kreglewski, Alexander
Leach, James Woodrow
McCauley, Joseph Lee
Mack, Russel Travis
Prats, Michael
Rajagopalan, Raj
Swope, Richard Dale
Young, Fred M(ichael)

UTAH
Batty, Joseph Clair
Brown, Billings
Henderson, Douglas J

VIRGINIA
Bartis, James Thomas
Dutt, Gautam Shankar
Goglia, Gennaro Louis
Johnson, David Harley
Liebenberg, Donald Henry
Milligan, John H
Reese, William
Reynolds, Peter James
Simpson, Roger Lyndon
Watson, Nathan Dee
Williams, Willie, Jr

WASHINGTON
Band, William
Erdmann, Joachim Christian
Johnson, B(enjamin) M(artineau)
Lam, John Ling-Yee
Leaf, Boris
Marston, Philip Leslie
Russell, David A

WISCONSIN
Behroozi, Feredoon
Boom, Roger Wright
Chang, Y Austin
Rooks, H Corbyn
Van den Akker, Johannes Archibald
Wilson, Volney Colvin
Zimm, Carl B

ALBERTA
Jessop, Alan Michael
Tuszynski, Jack A

BRITISH COLUMBIA
Salcudean, Martha Eva
Taylor, John Bryan

NOVA SCOTIA
Hunter, Douglas Lyle
Kreuzer, Han Jurgen

ONTARIO
Bedford, Ronald Ernest
Edwards, Martin Hassall
Hooper, F(rank) C(lements)
Kozicki, William
Kumaran, Mavinkal K
Lachaine, Andre Raymond Joseph
Mandelis, Andreas
Martin, Douglas Leonard
Olson, Allan Theodore

Thermal Physics (cont)

Pathria, Raj Kumar
Pei, David Chung-Tze
Tarasuk, John David
Tyler, R(onald) A(nthony)
Valleau, John Philip

PRINCE EDWARD ISLAND
Madan, Mahendra Pratap

QUEBEC
Caille, Alain Emeril
Chan, Sek Kwan
Cochrane, Robert W
Leigh, Charles Henry
Lin, Sui

SASKATCHEWAN
Deckker, B(asil) E(ardley) L(eon)
Kos, Joseph Frank

OTHER COUNTRIES
Archie, Charles Neill
Garcia-Colin, Leopoldo Scherer
Hudson, Ralph P
Jokl, Miloslav Vladimir
Romero, Alejandro F
Safran, Samuel A
Sarmiento, Gustavo Sanchez
Weinstock, Harold

Other Physics

ALABAMA
Fowlis, William Webster
Hathaway, David Henry
Tan, Arjun
Werkheiser, Arthur H, Jr
Wilhold, Gilbert A

ARIZONA
Bickel, William Samuel
Chu, William How-Jen
Galloway, Kenneth Franklin
O'Brien, Keran
Sonett, Charles Philip
Staley, Dean Oden
Twomey, Sean Andrew
Wheeler, Lawrence

CALIFORNIA
Adam, Randall Edward
Adams, Ralph Melvin
Adelson, Harold Ely
Alonso, Jose Ramon
Anderson, Leonard Mahlon
Anderson, Lloyd James
Andre, Michael Paul
Arnold, James Tracy
Ashford, Victor Aaron
Astrahan, Melvin Alan
Banner, David Lee
Barish, Barry C
Barnes, Charles Andrew
Basler, Roy Prentice
Berkner, Klaus Hans
Birge, Robert Walsh
Bogle, Robert Worthington
Bowers, John E
Bradner, Hugh
Brodsky, Stanley Jerome
Brown, Keith H
Bryan, James Bevan
Cantow, Manfred Josef Richard
Carlson, Gary Wayne
Case, Kenneth Myron
Chau, Cheuk-Kin
Cheng, David
Chester, Marvin
Chu, William Tongil
Clendenin, James Edwin
Clendenning, Lester M
Cohn, Stanton Harry
Cover, Ralph A
Deshpande, Shivajirao M
Dewitt, Hugh Edgar
Dickinson, William Clarence
Dobbins, John Potter
Donovan, P F
DuBridge, Lee Alvin
Elliott, David Duncan
Feldman, Jack L
Fillius, Walker
Fineman, Morton A
Frost, Robert Hartwig
Gaines, Edward Everett
Gajewski, Ryszard
Garwin, Charles A
Gillespie, Daniel Thomas
Good, Robert Howard
Goubau, Wolfgang M
Groce, David Eiben
Harris, James Stewart, Jr
Harrison, Don Edward, Jr
Haymond, Herman Ralph
Hebel, Louis Charles
Heflinger, Lee Opert
Hornung, Erwin William
Hubbard, Edward Leonard
James, Ralph Boyd
Johnson, Chris Alan
Jones, Joie Pierce

Judd, Floyd L
Judge, Roger John Richard
Kaelble, David Hardie
Kamegai, Minao
Kittel, Peter
Knowles, Harrold B
Kohler, Donald Alvin
Krivanek, Ondrej Ladislav
Lal, Devendra
Liu, Fook Fah
Long, H(ugh) M(ontgomery)
Lu, Adolph
Ma, Joseph T
Manalis, Melvyn S
Merker, Milton
Merriam, Marshal F(redric)
Miles, Ralph Fraley, Jr
Morewitz, Harry Alan
Muller, Richard A
Munch, Jesper
Murphy, Frederick Vernon
Myer, Jon Harold
Nelson, Raymond Adolph
Nelson, Walter Ralph
Newkirk, Lester Leroy
Oakley, David Charles
Ogier, Walter Thomas
Paige, David A
Parkins, William Edward
Parks, Lewis Arthur
Passenheim, Burr Charles
Phillips, Thomas Joseph
Phinney, Nanette
Piccioni, Oreste
Prag, Arthur Barry
Rabl, Veronika Ariana
Radosevich, Lee George
Randrup, Jorgen
Reisman, Elias
Richards, Lorenzo Adolph
Ridgway, Stuart L
Rockwood, Stephen Dell
Rodriguez, Andres F
Rothrock, Larry R
Sadoulet, Bernard
Sandmann, William Henry
Schermer, Robert Ira
Seibert, J A
Shacklett, Robert Lee
Shapiro, Mark Howard
Shepard, Roger N
Sheridon, Nicholas Keith
Slater, Donald Carlin
Speen, Gerald Bruce
Spencer, James Eugene
Stahl, Ralph Henry
Stevens, Aldred Lyman
Stevens, John Charles
Subramanya, Shiva
Thomas, James H
Thompson, John Robert
Toor, Arthur
Tsai, Chuang Chuang
Tuul, Johannes
Wahlig, Michael Alexander
Warren, Walter R(aymond), Jr
Weaver, Harry Edward, Jr
Woodward, James Franklin
Zumberge, James Frederick

COLORADO
Allen, Joe Haskell
Aly, Hadi H
Barrett, Charles Sanborn
Carpenter, Donald Gilbert
Cerni, Todd Andrew
Czanderna, Alvin Warren
Dahl, Adrian Hilman
Dunn, Gordon Harold
Jewell, Jack Lee
Noufi, Rommel
Nozik, Arthur Jack
Pitts, John Roland
Wait, David Francis
Zoller, Paul

CONNECTICUT
Hamblen, David Gordon
Heller, John Herbert
Keith, H(arvey) Douglas
Monce, Michael Nolen
Powers, Joseph
Sen, Pabitra Narayan
Stephenson, Kenneth Edward
Stevens, Joseph Charles
Tokita, Noboru

DELAWARE
Avakian, Peter
Cessna, Lawrence C, Jr
Coleman, Marcia Lepri
Cooper, Charles Burleigh
Gibson, Joseph W(hitton), Jr
Holland, Russell Sedgwick
Hsu, William Yang-Hsing
Longworth, Ruskin
Onn, David Goodwin
Yau, Wallace Wen-Chuan

DISTRICT OF COLUMBIA
Ambler, Ernest
Batra, Narendra K
Carlos, William Edward
Daehler, Mark

Fanconi, Bruno Mario
Faust, Walter Luck
Grossling, Bernardo Freudenburg
Gubser, Donald Urban
Haas, George Arthur
Holloway, John Thomas
Jordy, George Y
Knudson, Alvin Richard
Lombard, David Bishop
McKinney, John Edward
Rodgers, James Earl
Salu, Yehuda
Schima, Francis Joseph
Shapero, Donald Campbell
Soulen, Robert J, Jr
Wallenmeyer, William Anton

FLORIDA
Cain, Joseph Carter, III
Cohen, Martin Joseph
Fitzgerald, Lawrence Terrell
Hammer, Clarence Frederick, Jr
Kromhout, Robert Andrew
Mauderli, Walter
Telkes, Maria
Wolfson, Bernard T

GEORGIA
Kotliar, Abraham Morris
Morgan, Karl Ziegler
Quisenberry, Dan Ray
Stevenson, Dennis A

HAWAII
Weber, Louis Russell

IDAHO
Gasidlo, Joseph Michael

ILLINOIS
Abella, Isaac D
Abraham, Bernard M
Ascoli, Giulio
Bersted, Bruce Howard
Chang, Yia-Chung
De Volpi, Alexander
Eisenstein, Bob I
Elwyn, Alexander Joseph
Geil, Phillip H
Haugen, Robert Kenneth
Herzenberg, Caroline Stuart Littlejohn
Hojvat, Carlos F
Kaminsky, Manfred Stephan
Kephart, Robert David
Kubitschek, Herbert Ernest
Lennox, Arlene Judith
Lyman, Ernest McIntosh
McInturff, Alfred D
McLeod, Donald Wingrove
Nissim-Sabat, Charles
Perlow, Gilbert Jerome
Pigott, Miles Thomas
Reft, Chester Stanley
Schulz, Charles Emil
Shepard, Kenneth Wayne
Shroff, Ramesh N
Sill, Larry R
Smedskjaer, Lars Christian
Spokas, John J
Stanford, George Stailing
Stinchcomb, Thomas Glenn
Toohey, Richard Edward
Voyvodic, Louis
Vyborny, Carl Joseph
Winston, Roland

INDIANA
Bellina, Joseph James, Jr
Carlson, James C
Carmony, Donald Duane
George, Robert Eugene
Madden, Keith Patrick
Witt, Robert Michael

IOWA
Cruse, Richard M
Green, Robert Wood
Kasper, Joseph Emil
Macomber, Hilliard Kent

KANSAS
Grosskreutz, Joseph Charles

KENTUCKY
Almond, Peter R
Christensen, Ralph C(hresten)
Isihara, Akira
Syed, Ibrahim Bijli
Ulrich, Aaron Jack

LOUISIANA
Chen, Isaac I H
Erath, Louis W
Huggett, Richard William, Jr
Kyame, George John
Scott, L Max

MAINE
Snyder, Arnold Lee, Jr

MARYLAND
Barker, John L, Jr
Barnes, John David
Beal, Robert Carl

Bean, Vern Ellis
Benson, Robert Frederick
Brodsky, Allen
Brown, Robert Alan
Carbary, James F
Chen, Henry Lowe
Coble, Anna Jane
Congel, Frank Joseph
Crissman, John Matthews
Das Gupta, Aaron
Davisson, James W
Eckerman, Jerome
Faust, William R
Fulmer, Glenn Elton
Galloway, William Don
Gloersen, Per
Guilarte, Tomas R
Hartmann, Bruce
Holdeman, Louis Brian
Kacser, Claude
Maxwell, Louis R
Mazur, Jacob
Penner, Samuel
Peterlin, Anton
Richard, Jean-Paul
Schamp, Homer Ward, Jr
Seeger, Raymond John
Skrabek, Emanuel Andrew
Smith, Jack Carlton
Stern, David P
Sugai, Iwao
Temperley, Judith Kantack
Tholen, Albert David
Zorn, Gus Tom

MASSACHUSETTS
Abbott, Norman John
Bansil, Rama
Barish, Leo
Bar-Yam, Zvi H
Bechis, Kenneth Paul
Becker, Ulrich J
Behrman, Richard H
Cohen, Howard David
Corey, Brian E
Crew, Geoffrey B
Dasari, Ramachandra R
Garth, John Campbell
Ginsberg, Edward S
Goitein, Michael
Gottschalk, Bernard
Hardt, David Edgar
Hawkins, W(illiam) Bruce
Heroux, Leon J
Hersh, John Franklin
Herzfeld, Judith
Hillel, Daniel
Hsu, Shaw Ling
Huffman, Fred Norman
Janney, Clinton Dales
Jirmanus, Munir N
Kornfield, Jack I
Kupferberg, Lenn C
Laible, Roy C
Mazur, Eric
Mendelson, Robert Allen
Murayama, Takayuki
Post, Richard S
Quynn, Richard Grayson
Rosenson, Lawrence
Rubin, Lawrence G
Sisterson, Janet M
Strauss, Bruce Paul
Young, James Edward

MICHIGAN
Blosser, Henry Gabriel
Goodrich, Max
Harris, Gale Ion
Hunt, Thomas Kintzing
Johnson, Wayne Douglas
Kane, Gordon Leon
Larsen, Jon Thorsten
Lee, Robert W
Leffert, Charles Benjamin
Malaczynski, Gerard W
Meier, Dale Joseph
Savit, Robert Steven
Schreiber, Thomas Paul
Schumacher, Berthold Walter
Strickland, James Shive
Watson, John H L
Yeh, Gregory Soh-Yu
Zeller, Albert Fontenot

MINNESOTA
Gallo, Charles Francis
Gerlach, Robert Louis
Hobbie, Russell Klyver
King, Wendell L
Mitchell, William Cobbey
Premanand, Visvanatha
Szpilka, Anthony M

MISSOURI
Bender, Carl Martin
Burke, James Joseph
Ching, Wai-Yim
Hirsh, Ira Jean
Hurst, Robert R(owe)
Larson, James D
Leader, John Carl
Martin, Frank Elbert
Schmidt, Paul Woodward

Schmitt, John Leigh
Taub, Haskell Joseph

NEBRASKA
Borchert, Harold R

NEVADA
Jobst, Joel Edward
Potter, Gilbert David

NEW HAMPSHIRE
Meyer, James Wagner

NEW JERSEY
Ashkin, Arthur
Barr, William J
Bartnoff, Shepard
Burnett, Bruce Burton
Canter, Nathan H
Carlson, Curtis Raymond
Christman, Edward Arthur
Cox, Donald Clyde
De Planque, Gail
Forster, Eric Otto
Girit, Ibrahim Cem
Gray, Russell Houston
Hall, Herbert Joseph
Heimann, Peter Aaron
Jaffe, Michael
Johnson, Leo Francis
Keith, Harvey Douglas
Kosel, George Eugene
Lindberg, Craig Robert
Lohse, David John
Ma, Z(ee) Ming
Madey, Theodore Eugene
Meissner, Hans Walter
O'Malley, James Joseph
Ong, Chung-Jian Jerry
Padden, Frank Joseph, Jr
Patel, Jamshed R(uttonshaw)
Peebles, Phillip J
Polyakov, Alexander
Prevorsek, Dusan Ciril
Ransome, Ronald Dean
Shattes, Walter John
Shyamsunder, Erramilli
Sinclair, Brett Jason
Sprague, Basil Sheldon
Stanton, Robert E
Thelin, Lowell Charles
Tonelli, Alan Edward
Wadlow, David
Wilchinsky, Zigmond Walter
Witten, Edward

NEW MEXICO
Anderson, Robert Alan
Belian, Richard Duane
Bendt, Philip Joseph
Bradbury, James Norris
Curro, John Gillette
Fenstermacher, Charles Alvin
Figueira, Joseph Franklin
Gosling, John Thomas
Gray, Edward Ray
Heldman, Julius David
Holm, Dale M
Hurd, James William
Jackson, Jasper Andrew, Jr
Jameson, Robert A
Johnson, James Daniel
Johnston, Roger Glenn
Jones, James Jordan
Keller, Donald V
Kupferman, Stuart L
Ladish, Joseph Stanley
Lindman, Erick Leroy, Jr
MacArthur, Duncan W
Malik, John S
Mercer-Smith, James A
Miller, Guthrie
Miranda, Gilbert A
Oyer, Alden Tremaine
Pickrell, Mark M
Ramsay, John Barada
Ranken, William Allison
Reedy, Robert Challenger
Robertson, Merton M
Ryder, Richard Daniel
Schaefer, Dale Wesley
Scott, Bobby Randolph
Seitz, Wendell L
Shafer, Robert E
Sheffield, Richard Lee
Sherman, Robert Howard
Smith, H Vernon, Jr
Smith, Horace Vernon, Jr
Sprinkle, James Kent, Jr
Stokes, Richard Hivling
Taylor, Warren Egbert
Thiessen, Henry Archer
Thompson, Jill Charlotte
Young, Lloyd Martin

NEW YORK
Aronson, Raphael (Friedman)
Barish, Robert John
Baum, Paul M
Baumel, Philip
Bayer, Raymond George
Beck, Harold Lawrence
Bennett, Gerald William
Bond, Peter Danford

Bourdillon, Antony John
Bozoki-Gombosi, Eva S
Breneman, Edwin Jay
Bunch, Phillip Carter
Bundy, Francis P
Burschka, Martin A
Campbell, Larry Enoch
Chow, Tsu-Sen
Chung, Victor
Cluxton, David H
Cohen, Beverly Singer
Cooper, Douglas W
Dahl, Per Fridtjof
Deady, Matthew William
Dermit, George
Erhardt, Peter Franklin
Feinberg, Robert Jacob
Fields, Alfred E
Folan, Lorcan Michael
Forgacs, Gabor
Garwin, Richard Lawrence
Gavin, Donald Arthur
Gibson, Walter Maxwell
Goodwin, Paul Newcomb
Habicht, Ernst Rollemann, Jr
Halket, Thomas D
Hamblen, David Philip
Hanson, Albert L
Hartman, Paul Leon
Heinz, Tony F
Henry, Arnold William
Hickey, Roger
Hobbs, Stanley Young
Hodge, Ian Moir
Hudak, Michael J
Hughes, Harold K
Johnson, Karen Elise
Jona, Franco Paul
Kambour, Roger Peabody
Kaufman, Raymond
Knowles, Richard James Robert
Kramer, Martin A
Lee-Franzini, Juliet
Leipuner, Lawrence Bernard
LeVine, Micheal Joseph
Maldonado, Juan Ramon
Marcus, Michael Alan
Massa, Dennis Jon
Nicolosi, Joseph Anthony
O'Reilly, James Michael
Parks, Ronald Dee
Parsa, Zohreh
Pochan, John Michael
Prest, William Marchant, Jr
Ring, James Walter
Sarko, Anatole
Sciulli, Frank J
Smith, Kenneth Judson, Jr
Soures, John Michael
Stearns, Brenton Fisk
Sternstein, Sanford Samuel
Sullivan, Peter Kevin
Thieberger, Peter
Tuli, Jagdish Kumar
Verrillo, Ronald Thomas
Webb, Watt Wetmore
Wedding, Brent (M)
Wright, Herbert N
Zwislocki, Jozef John

NORTH CAROLINA
Boldridge, David William
Buchanan, David Royal
Crawford-Brown, Douglas John
Dayton, Benjamin Bonney
Fornes, Raymond Earl
Frey, William Francis
Good, Wilfred Manly
Heffelfinger, Carl John
Hocken, Robert John
Johnson, G Allan
Leder, Lewis Beebe
McPeters, Arnold Lawrence
Olf, Heinz Gunther

OHIO
Agard, Eugene Theodore
Andria, George D
Bohm, Georg G A
Bohn, Randy G
Carpenter, Robert Leland
Colbert, Charles
Conant, Floyd Sanford
Dembowski, Peter Vincent
Dickens, Elmer Douglas, Jr
Eley, Richard Robert
Forster, Michael Jay
Friedman, Emil Martin
Fujimura, Osamu
Gelerinter, Edward
Griffing, David Francis
Hantman, Robert Gary
Jaworowski, Andrzej Edward
Kanakkanatt, Sebastian Varghese
Keck, Max Johann
Lehr, Marvin Harold
Meadows, Brian T
Pillai, Padmanabha S
Ponter, Anthony Barrie
Ruegsegger, Donald Ray, Jr
Stanton, Noel Russell
Starchman, Dale Edward
Suranyi, Peter
Yun, Seung Soo

OKLAHOMA
Anderson, David Walter
Buchanan, Ronnie Joe
Burwell, James Robert
Crawford, Gerald James Browning
Ryan, Stewart Richard
Stacy, Carl J

OREGON
Orloff, Jonathan H

PENNSYLVANIA
Baldino, Frank, Jr
Banavar, Jayanth Ramarao
Behrens, Ernst Wilhelm
Bergmann, Ernest Eisenhardt
Bhalla, Amar S
DeBenedetti, Sergio
Fetkovich, John Gabriel
Froehlich, Fritz Edgar
Goldstein, E Bruce
Grannemann, Glenn Niel
Green, Barry George
Heybey, Otfried Willibald Georg
Kliman, Harvey Louis
Knight, Stephen
Kurland, Robert John
Laws, Kenneth Lee
Marcus, Robert Troy
Moscatelli, Frank A
Page, Lorne Albert
Parker, James Henry, Jr
Pfeiffer, Heinz Gerhard
Pfrogner, Ray Long
Phares, Alain Joseph
Raymund, Mahlon
Shwe, Hla
Slade, Paul Graham
Stefanou, Harry
Sterk, Andrew A
Van Ausdal, Ray Garrison
Villafana, Theodore
Zehr, Floyd Joseph

RHODE ISLAND
Amols, Howard Ira
Cutts, David
Loferski, Joseph J

SOUTH CAROLINA
Billica, Harry Robert
Goswami, Bhuvenesh C
Hay, Ian Leslie
Kimmel, Robert Michael
McCrosson, F Joseph
McNulty, Peter J
Manson, Joseph Richard
Parks, Paul Blair
Street, Jabez Curry

TENNESSEE
Alexeff, Igor
Bell, Zane W
Boye, Charles Andrew, Jr
Callihan, Dixon
Gant, Kathy Savage
Garber, Floyd Wayne
Gronemeyer, Suzanne Alsop
Johnson, Elizabeth Briggs
Painter, Linda Robinson
Shugart, Cecil G
Silver, Ernest Gerard
Watson, Evelyn E
Wendelken, John Franklin
Yalcintas, M Güven
Zuhr, Raymond Arthur

TEXAS
Badhwar, Gautam D
Bencomo, José A
Bering, Edgar Andrew, III
Bohm, Arno
Ellis, Kenneth Joseph
Henry, Arthur Charles
Hillery, Herbert Vincent
Hirsch, John Michele
Marlow, William Henry
Nyquist, Laurence Elwood
Patriarca, Peter
Perry, Robert Riley
Reed, Mark Arthur
Sanchez, Isaac Cornelius
Sharber, James Randall
Stehling, Ferdinand Christian

UTAH
Galli, John Ronald
Hansen, Wilford Nels
Loh, Eugene C
Mays, Charles William

VERMONT
Hayes, John William

VIRGINIA
Auton, David Lee
Barker, Robert Edward, Jr
Bruno, Ronald C
Chang, Chieh Chien
Davidson, Charles Nelson
Farmer, Barry Louis
Gabriel, Barbra L
Gibian, Gary Lee
Graham, Kenneth Judson

Kabir, Prabahan Kemal
Krafft, Geoffrey Arthur
Kranbuehl, David Edwin
Kubu, Edward Thomas
Leemann, Christoph Willy
Long, Dale Donald
Maune, David Francis
Sobottka, Stanley Earl
Sundelin, Ronald M
Wincklhofer, Robert Charles

WASHINGTON
Buske, Norman L
DeMoney, Fred William
Ely, John Thomas Anderson
Fisher, Darrell R
Johnson, John Richard
Kathren, Ronald Laurence
Kushmerick, Martin Joseph
Peterson, James Macon
Prunty, Lyle Delmar
Roesch, William Carl
Styris, David Lee
Warren, John Lucius
Zimmerer, Robert W

WEST VIRGINIA
Franklin, Alan Douglas
Knight, Alan Campbell

WISCONSIN
Aita, Carolyn Rubin
Attix, Frank Herbert
Fry, William Frederick
Huston, Norman Earl
Truszkowska, Krystyna
Zagzebski, James Anthony
Zei, Dino

WYOMING
Denison, Arthur B
Doerges, John E

ALBERTA
Anger, Clifford D
Apps, Michael John
Kounosu, Shigeru
Weaver, Ralph Sherman

BRITISH COLUMBIA
Crooks, Michael John Chamberlain
Fisher, Paul Douglas
Ward, Lawrence McCue

MANITOBA
Barnard, John Wesley

NEWFOUNDLAND
Irfan, Muhammad

ONTARIO
Beck, Alan Edward
Bigham, Clifford Bruce
Dastur, Ardeshir Rustom
Dunn, Andrew Fletcher
Greenstock, Clive Lewis
Harvey, John Wilcox
Hunt, John Wilfred
Iribarne, Julio Victor
Ivey, Donald Glenn
Jackman, Thomas Edward
Jones, Alun Richard
McDiarmid, Donald Ralph
Mann, Robert Bruce
Morris, Derek
Prasad, S E
Preston-Thomas, Hugh
Racey, Thomas James
Rogers, David William Oliver
Sawicka, Barbara Danuta
Sinha, Bidhu Bhushan Prasad
Sonnenberg, Hardy
Sundararajan, Pudupadi Ranganathan
Whippey, Patrick William
Wood, Gordon Harvey

QUEBEC
Bunge, Mario Augusto
Hitschfeld, Walter
Rioux, Claude
Utracki, Lechoslaw Adam

SASKATCHEWAN
Robertson, Beverly Ellis

OTHER COUNTRIES
Bednorz, Johannes Georg
Binnig, Gerd
Brown, Peter
Gardiner, Barry Alan
Lapostolle, Pierre Marcel
Mott, Nevill Francis
Mueller, Karl Alexander
Neher, Erwin
Rabl, Ari
Rohrer, Heinrich
Scheiter, B Joseph Paul

OTHER PROFESSIONAL FIELDS

Education Administration

ALABAMA
Bailey, William C
Chastain, Benjamin Burton
Goldman, Jay
Hill, Walter Andrew
Irwin, John David
Johnson, Brian John
Knecht, Charles Daniel
Lowry, James Lee
Meezan, Elias
Menaker, Lewis
Mullins, Dail W, Jr
Nair, Madhavan G
Ranney, Richard Raymond
Roozen, Kenneth James
Segner, Edmund Peter, Jr
Watson, Raymond Coke, Jr
Wilson, G Dennis

ALASKA
Behrend, Donald Fraser
Kessel, Brina
Tilsworth, Timothy

ARIZONA
Cassidy, Suzanne Bletterman
Childs, Orlo E
Cunningham, Richard G(reenlaw)
Goldner, Adreas M
Iserson, Kenneth Victor
Kaufman, C(harles) W(esley)
Metz, Donald C(harles)
Metzger, Darryl E
Mossman, Kenneth Leslie
Wagner, Kenneth
Weaver, Albert Bruce

ARKANSAS
Deaver, Franklin Kennedy
Reed, Hazell

CALIFORNIA
Allen, William Merle
Beljan, John Richard
Carlan, Audrey M
Carson, Virginia Rosalie Gottschall
Castles, James Joseph, Jr
Clague, William Donald
Clark, David Ellsworth
Cockrum, Richard Henry
Colson, Elizabeth F
Coyle, Bernard Andrew
Croft, Paul Douglas
Cromwell, Florence S
Crum, James Davidson
DeJongh, Don C
Dewey, Donald O
Dyer, John Norvell
Edamura, Fred Y
Estrin, Thelma A
Fischer, Robert Blanchard
Fisher, Leon Harold
Frank, Sidney Raymond
Greenspan, John Simon
Grobstein, Clifford
Hardy, Rolland L(ee)
Haughton, James Gray
Heldman, Morris J
Henkin, Leon (Albert)
Heyneman, Donald
Hisserich, John Charles
Hubbard, Richard W
Jackson, Crawford Gardner, Jr
Jakubowski, Gerald S
James, Philip Nickerson
Juster, Norman Joel
Kaplan, Richard E
Knapp, Theodore Martin
Krippner, Stanley Curtis
Lai, Kai Sun
Lanham, Richard Henry, Jr
L'Annunziata, Michael Frank
Laub, Alan John
Lave, Roy E(llis), Jr
Law, George Robert John
Leffler, Esther Barbara
Lein, Allen
Martin, Joseph B
Meriam, James Lathrop
Miller, Gary Arthur
Miller, William Frederick
Moye, Anthony Joseph
O'Neill, R(ussell) R(ichard)
Orloff, Neil
Phillips, William Baars
Picus, Gerald Sherman
Pierce, James Otto, II
Rebman, Kenneth Ralph
Robinson, Hamilton Burrows Greaves
Rogers, Robert N
Saunders, Robert M(allough)
Shields, Loran Donald
Smith, Wayne Earl
Spicher, Robert G
Spooner, Charles Edward, Jr
Stahl, Frieda A
Sugihara, Thomas Tamotsu
Triche, Timothy J

Vanderhoef, Larry Neil
Van Elswyk, Marinus, Jr
Wagner, William Gerard
Watson, John Alfred
Watters, Gary Z
Watts, Malcolm S M
Wendel, Otto Theodore, Jr
Winer, Arthur Melvyn
Woyski, Margaret Skillman
Yagiela, John Allen
Zumberge, James Herbert

COLORADO
Ansell, George S(tephen)
Beal, Richard Sidney, Jr
Beck, Steven R
Bond, Richard Randolph
Boudreau, Robert Donald
Briggs, William Egbert
Fraikor, Frederick John
Golden, John O(rville)
Hermann, Allen Max
Janes, Donald Wallace
Johnson, Arnold I(van)
Kano, Adeline Kyoko
Kassoy, David R
Netzel, Richard G
Niehaus, Merle Hinson
Ogg, James Elvis
Peters, Roger Paul
Regenbrecht, D(ouglas) E(dward)
Strom, Oren Grant
Thomas, William Robb
Williams, Denis R

CONNECTICUT
Alcantara, Victor Franco
Anderson, R(obert) M(orris), Jr
Barone, John A
Carroll, J Gregory
Clarke, George A
Darling, George Bapst
Eigel, Edwin George, Jr
Foster, James Henry
Genel, Myron
Haas, Thomas J
Hall, Newman A(rnold)
Hyde, Walter Lewis
Katz, Lewis
Kollar, Edward James
Krummel, William Marion
Parthasarathi, Manavasi Narasimhan
Strand, Richard Alvin
Wright, Hastings Kemper
Wynn, Charles Martin, Sr
Wystrach, Vernon Paul

DELAWARE
Murray, Richard Bennett
Terss, Robert H
Wetlaufer, Donald Burton

DISTRICT OF COLUMBIA
Baker, Louis Coombs Weller
Berendzen, Richard
Brownlee, Paula Pimlott
Bulger, Roger James
Coates, Anthony George
Fowler, Earle Cabell
Frair, Karen Lee
Gronvall, John Arnold
Hawkins, Gerald Stanley
Holt, James Allen
Ison-Franklin, Eleanor Lutia
Kettel, Louis John
Krauss, Robert Wallfar
Miller, Russell Loyd, Jr
O'Fallon, John Robert
Phillips, Don Irwin
Plowman, Ronald Dean
Raizen, Senta Amon
Schultze, Charles L
Smith, Thomas Elijah
Spurlock, Langley Augustine
Wallis, W Allen
Webster, Thomas G

FLORIDA
Anderson, Melvin W(illiam)
Baker, Carleton Harold
Baum, Werner A
Berger, Phyllis Belous
Bolte, John R
Brown, John Lott
Burdick, Glenn Arthur
Buss, Daryl Dean
Davison, Beaumont
Dickson, David Ross
Dubravcic, Milan Frane
Dwornik, Julian Jonathan
Fogel, Bernard J
Fry, Jack L
Harris, Lee Errol
Hartley, Craig Sheridan
Hinkley, Robert Edwin, Jr
Jacobsen, Neil Soren
Jamrich, John Xavier
Jewett, Robert Elwin
Johnsen, Russell Harold
Kline, Jacob
Korwek, Alexander Donald
Kribel, Robert Edward
Mason, D(onald) R(omagne)
Mecklenburg, Roy Albert

Nelson, Gordon Leigh
Nimmo, Bruce Glen
Noble, Nancy Lee
Noble, Robert Vernon
Norris, Dean Rayburn
Peters, Thomas G
Rabbit, Guy
Revay, Andrew W, Jr
Smith, Wayne H
Spector, Bertram
Thomas, Dan Anderson
Trubatch, Janett
Weller, Richard Irwin
Wiebush, Joseph Roy
Wiltbank, William Joseph
Yerger, Ralph William
Zachariah, Gerald L

GEORGIA
Adams, David B
Antolovich, Stephen D
Christenberry, George Andrew
Comeau, Roger William
Dragoin, William Bailey
Evans, Thomas P
Greenberg, Jerrold
Hawkins, Isaac Kinney
Hayes, John Thompson
Heric, Eugene LeRoy
Hooper, John William
Jen, Joseph Jwu-Shan
Little, Robert Colby
Moretz, William Henry
Papageorge, Evangeline Thomas
Tedesco, Francis J
Threadgill, Ernest Dale
Watts, Sherrill Glenn
Weatherred, Jackie G
Wepfer, William J

HAWAII
Carlson, John Gregory
Goodman, Madelene Joyce
Hunter, Cynthia L
Malmstadt, Howard Vincent
Watanabe, Daniel Seishi

IDAHO
Mech, William Paul
Price, Joseph Earl
Renfrew, Malcolm MacKenzie

ILLINOIS
Al-Khafaji, Amir Wadi Nasif
Alpert, Daniel
Andreoli, Kathleen Gainor
Auvil, Paul R, Jr
Bentley, Orville George
Bevan, William
Cleall, John Frederick
Dunn, Dorothy Fay
Egan, Richard L
Frankenberg, Julian Myron
Funk, Richard Cullen
Gilbert, Thomas Lewis
Giles, Eugene
Goldman, Jack Leslie
Haber, Meryl H
Heichel, Gary Harold
Hoshiko, Michael S
Jaffe, Philip Monlane
Jonas, Jiri
Jugenheimer, Robert William
Kanter, Manuel Allen
Kernis, Marten Murray
Kinsinger, Jack Burl
Kleinman, Kenneth Martin
Kraft, Sumner Charles
Laible, Jon Morse
Mapother, Dillon Edward
Martinec, Emil Louis
Munyer, Edward Arnold
Nair, Velayudhan
Pai, Anantha M
Pai, Mangalore Anantha
Sechrist, Chalmers Franklin, Jr
Sill, Larry R
Swoyer, Vincent Harry
Tang, Wilson H
Waterhouse, John P
Wilson, Gerald Gene
Wojcik, Anthony Stephen
Zehr, John E

INDIANA
Alspaugh, Dale W(illiam)
Beering, Steven Claus
Beery, Dwight Beecher
Eckert, Roger E(arl)
Fredrich, Augustine Joseph
Guthrie, George Drake
Haffley, Philip Gene
Hagen, Charles William, Jr
Hansen, Arthur G(ene)
Hawks, Keith Harold
Herron, James Dudley
Hickey, William August
Hicks, John W, III
Hullinger, Ronald Loral
Kessler, David Phillip
Lorentzen, Keith Eden
Nagle, Edward John
Pierson, Edward S
Rand, Leon

Schaap, Ward Beecher
Smith, Charles Edward, Jr
White, Samuel Grandford, Jr
Wilson, Raphael
Winicur, Daniel Henry

IOWA
Christensen, George Curtis
Collins, Richard Francis
Garfield, Alan J
Kelly, William H
Kluge, John Paul
Koppelman, Ray
Morriss, Frank Howard, Jr
Schafer, John William, Jr
Stebbins, Dean Waldo
Williams, Fred Devoe

KANSAS
Connor, Sidney G
Frost-Mason, Sally Kay
Jennings, Paul Harry
Larson, Vernon C
Lenzen, K(enneth) H(arvey)
Locke, Carl Edwin, Jr
Parizek, Eldon Joseph
Waxman, David

KENTUCKY
Boehms, Charles Nelson
Crumb, Glenn Howard
Dixon, Wallace Clark, Jr
Farman, Allan George
Fleischman, Marvin
Gerhard, Earl R(obert)
Hanley, Thomas Richard
Harris, Henson
Humphries, Asa Alan, Jr
Kraman, Steve Seth
Kupchella, Charles E
McClain, John William
Nash, David Allen
Podshadley, Arlon George
Saxe, Stanley Richard
Vittetoe, Marie Clare

LOUISIANA
Birtel, Frank T
Dommert, Arthur Roland
Frohlich, Edward David
Hankins, B(obby) E(ugene)
Lembeck, William Jacobs
Lindberg, David Seaman, Sr
Maxfield, John Edward
Morris, Everett Franklin
Smith, Grant Warren, II
Sterling, Arthur MacLean
Teate, James Lamar
Walker, John Robert
Wetzel, Albert John
Zander, Arlen Ray

MAINE
Bernstein, Seldon Edwin
Brown, Gregory Neil
Rosenfeld, Melvin Arthur

MARYLAND
Bateman, Barry Lynn
Bollum, Frederick James
Calinger, Ronald Steve
Caret, Robert Laurent
Chartrand, Mark Ray, III
Cotton, William Robert
Dupont, Jacqueline (Louise)
Gluckstern, Robert Leonard
Harlan, William R, Jr
Heller, Charles O(ta)
Henderson, Donald Ainslie
Hindle, Brooke
Howard, Joseph H
Johnston, Robert Ward
Jones, Franklin Del
Karlander, Edward P
Kelley, John Francis, Jr
Knapp, David Allan
Langenberg, Donald Newton
Levchuk, John W
Lymangrover, John R
Margolis, Simeon
Mathieu, Richard D(etwiler)
Mayor, John Roberts
Miller, Gerald Ray
Montgomery, David Carey
Moorjani, Kishin
More, Kenneth Riddell
Mumford, Willard R
Nanzetta, Philip Newcomb
Richardson, William C
Solomon, Howard Fred
Sporn, Eugene Milton
Suskind, Sigmund Richard
Wallace, James M
Yarmolinsky, Adam
Young, Harold Henry

MASSACHUSETTS
Adler, Richard B(rooks)
Baird, Ronald C
Barlas, Julie S
Bhargava, Hridaya Nath
Blatt, S Leslie
Bolz, R(ay) E(mil)
Brown, Cynthia Ann

Cavallaro, Mary Caroline
Coghlan, Anne Eveline
Cohen, Joel Ralph
Cohn, Lawrence H
Cynkin, Morris Abraham
Deck, Joseph Charles
Delaney, Patrick Francis, Jr
Ehrenfeld, John R(oos)
Fano, Robert M(ario)
Federman, Daniel D
Field, Hermann Haviland
Fuller, Richard H
Fuqua, Mary Elizabeth
Gold, Albert
Graham, Robert Montrose
Halle, Morris
Haworth, D(onald) R(obert)
Henry, Joseph L
Hershenson, Benjamin R
Hilferty, Frank Joseph
Hooper, James R(ipley)
Hulbert, Thomas Eugene
Hyatt, Raymond R, Jr
Kazimi, Mujid S
Kosersky, Donald Saadia
Loew, Franklin Martin
Lynton, Ernest Albert
Mark, Melvin
Maser, Morton D
Morse, Robert Warren
Murnane, Thomas William
Nelson, J(ohn) Byron
Papageorgiou, John Constantine
Reintjes, J Francis
Robinson, Howard Addison
Seamans, R(obert) C(hanning), Jr
Skolnikoff, Eugene B
Snyder, John Crayton
Stinchfield, Carleton Paul
Stratton, Julius Adams
Taylor, Edwin Floriman
Taylor, Isaac Montrose
Tenca, Joseph Ignatius
Volkmann, Frances Cooper
Wedlock, Bruce D(aniels)
Weller, Paul Franklin
West, Arthur James, II
Westwick, Carmen R
Wick, Emily Lippincott
Work, Clyde E(verette)

MICHIGAN
Adrounie, V Harry
Banks, Peter Morgan
Beaufait, Frederick W
Becher, William D(on)
Becker, Marshall Hilford
Bohm, Henry Victor
Bollinger, Robert Otto
Cerny, Joseph Charles
Conway, Lynn Ann
Courtney, Gladys (Atkins)
Cutler, Warren Gale
Dauphinais, Raymond Joseph
Deuben, Roger R
Eliezer, Isaac
Gilroy, John
Gray, Robert Howard
Johnson, John Harris
Karkheck, John Peter
Khasnabis, Snehamay
Kiechle, Frederick Leonard
King, Herman (Lee)
Kingman, Robert Earl
Lomax, Ronald J(ames)
Lorenz, John Douglas
Magen, Myron S
Mather, Edward Chantry
Miller, M(urray) H(enri)
Nash, Edmund Garrett
Nelson, Eldon Lane, Jr
Osborn, Wayne Henry
Reinke, David Albert
Rusch, Wilbert H, Sr
Schneider, Michael J
Smith, Paul Dennis
Sommers, Lawrence M
Soper, Jon Allen
Stocker, Donald V(ernon)
Strachan, Donald Stewart
Stynes, Stanley K
Tasker, John B
Weg, John Gerard
Whitten, Eric Harold Timothy
Williams, Fredrick David

MINNESOTA
Anderson, Sabra Sullivan
Barden, Roland Eugene
Cavert, Henry Mead
Conlin, Bernard Joseph
Giese, David Lyle
Lukasewycz, Omelan Alexander
Martens, Leslie Vernon
Miller, Gerald R
Miller, Jack W
Romero, Juan Carlos
Sherman, Patsy O'Connell
Wolfe, Barbara Blair

MISSISSIPPI
Doblin, Stephen Alan
Foil, Robert Rodney
McKee, J(ewel) Chester, Jr

Oswalt, Jesse Harrell
Roark, Bruce (Archibald)
Scott, Charley
Smith, Allie Maitland

MISSOURI
Axthelm, Deon D
Bader, Kenneth L
Banakar, Umesh Virupaksh
Baumann, Thiema Wolf
Brodmann, John Milton
Ericson, Avis J
Ferris, Deam Hunter
Festa, Roger Reginald
Grobman, Arnold Brams
Hines, Anthony Loring
Johnson, James W(inston)
Landiss, Daniel Jay
Malzahn, Ray Andrew
Meyer, William Ellis
Mitchell, Henry Andrew
Myers, Richard F
Nikolai, Robert Joseph
Oetting, Robert B(enfield)
Oviatt, Charles Dixon
Porter, Chastain Kendall
Richards, Graydon Edward
Rouse, Robert Arthur
Santiago, Julio Victor
Sauer, Harry John, Jr
Stoneman, William, III
Toom, Paul Marvin
Weiss, James Moses Aaron
Weldon, Virginia V
Wilson, Edward Nathan

MONTANA
Hess, Lindsay LaRoy
McRae, Robert James

NEBRASKA
Bahar, Ezekiel
Benschoter, Reba Ann
Clark, Ronald David
Gardner, Paul Jay
Roche, Edward Browining
Schmitz, John Albert
Thorson, James A
Wehrbein, William Mead

NEVADA
Bohmont, Dale W
Leedy, Clark D
May, Jerry Russell
Merdinger, Charles J(ohn)
Smith, Robert Bruce

NEW HAMPSHIRE
Bernstein, Abram Bernard
Davis, William Potter, Jr
Kaiser, C William
Melvin, Donald Walter
Price, Glenn Albert
Wixson, Eldwin A, Jr

NEW JERSEY
Albu, Evelyn D
Anderson, Robert E(dwin)
Arnold, Frederic G
Bartnoff, Shepard
Benedict, Joseph T
Burrill, Claude Wesley
Chizinsky, Walter
Clark, Mary Jane
Colicelli, Elena Jeanmarie
DeProspo, Nicholas Dominick
Derucher, Kenneth Noel
Edelman, Norman H
Fahrenholtz, Susan Roseno
Fenster, Saul K
Field, Norman J
Finch, Rogers B(urton)
Friedman, Edward Alan
Gewirtz, Allan
Graham-Ellis, Avis
Griffin, William Dallas
Griffo, James Vincent, Jr
Griminger, Paul
Hager, Mary Hastings
Houle, Joseph E
Leventhal, Howard
Lucas, Henry C, Jr
Lynde, Richard Arthur
Merritt, Richard Howard
Miller, Edward
Pignataro, Louis J(ames)
Poiani, Eileen Louise
Robin, Michael
Schwartz, Arthur Harold
Sutman, Frank X

NEW MEXICO
Barbo, Dorothy M
Bhada, Rohinton(Ron) K
Bradshaw, Martin Daniel
Cano, Gilbert Lucero
Cornell, Samuel Douglas
Maez, Albert R
Omer, George Elbert, Jr
Yudkowsky, Elias B

NEW YORK
Ahmad, Jameel
Alexander, Justin

Allentuch, Arnold
Aschner, Joseph Felix
Aull, Felice
Auston, David H
Baines, Ruth Etta
Bartilucci, Andrew J
Becker, Barbara
Berg, Daniel
Blackmon, Bobby Glenn
Bloch, Aaron N
Brenner, Egon
Brody, Harold
Buhks, Ephraim
Burge, Robert Ernest, Jr
Butler, Katharine Gorrell
Chabot, Brian F
Chance, Kenneth Bernard
Clark, Allan H
Cook, Robert Edward
Cooke, Lloyd Miller
Day, Stacey Biswas
De Carlo, Charles R
De France, Joseph J
DeLong, Stephen Edwin
Eppenstein, Walter
Feisel, Lyle Dean
Gabriel, Mordecai Lionel
Gary, Nancy E
George, A(lbert) R(ichard)
George, Nicholas
Gilmore, Arthur W
Glass, David Carter
Goldman, James Allan
Greaves, Bettina Bien
Greisen, Kenneth I
Grosewald, Peter
Gussin, Arnold E S
Hood, Donald C
Hunte, Beryl Eleanor
Huntington, David Hans
Irwin, Lynne Howard
Johnson, Charles Richard, Jr
Kampmeier, Jack A
Kennelley, James A
Kiser, Kenneth M(aynard)
Kletsky, Earl J(ustin)
Kogel, Marcus David
Kugel, Robert Benjamin
Kuperman, Albert Sanford
Lanford, William Armistead
Lawson, Kent DeLance
Litynski, Daniel Mitchell
Lustig, Harry
McDonald, Donald
McIrvine, Edward Charles
Matsushima, Satoshi
Mehaffey, Leathem, III
Mennitt, Philip Gary
Metz, Donald J
Monagle, John Joseph, Jr
Morris, Robert Gemmill
Mueller, Edward E(ugene)
Muench, Donald Leo
Murphy, James Joseph
Muschio, Henry M, Jr
Nathanson, Melvyn Bernard
Oates, Richard Patrick
Paldy, Lester George
Peters, James J
Petersen, Ingo Hans
Polowczyk, Carl John
Posamentier, Alfred S
Postman, Robert Derek
Prutkin, Lawrence
Ranu, Harcharan Singh
Reifler, Clifford Bruce
Richards, Paul Granston
Riehle, Robert Arthur, Jr
Robusto, C Carl
Roman, Stanford A, Jr
Rosenberg, Robert
Rosenfeld, Norman Samuel
Sample, Steven Browning
Sank, Diane
Savic, Michael I
Schneider, Allan Stanford
Schultze, Lothar Walter
Shapiro, Sidney
Simons, Gene R
Solomon, Seymour
Springer, Dwight Sylvan
Stephens, Lawrence James
Stith, James Herman
Streett, William Bernard
Thomas, Everett Dake
Thorpe, John Alden
Vidosic, J(oseph) P(aul)
Vinciguerra, Michael Joseph
Wiesenfeld, John Richard
Wilder, Joseph R
Wilkins, Bruce Tabor
Worden, David Gilbert
Yavorsky, John Michael

NORTH CAROLINA
Aronson, A L
Bailey, Kincheon Hubert, Jr
Bell, Norman R(obert)
Brotak, Edward Allen
Burnett, John Nicholas
Butts, Jeffrey A
Chin, Robert Allen
Daw, John Charles
Graham, Doyle Gene

Hallock, James A
Hayek, Dean Harrison
Henning, Emilie D
Herrell, Astor Y
Hersh, Solomon Philip
Howard, Donald Robert
Hugus, Z Zimmerman, Jr
King, Kendall Willard
Kranich, Wilmer Leroy
Land, Ming Huey
Lytle, Charles Franklin
Maier, Robert Hawthorne
Miller, Robert L
Mosier, Stephen R
Oblinger, Diana Gelene
Peirce, James Jeffrey
Richardson, Frances Marian
Robinson, Harold Frank
Smith, Norman Cutler
Turinsky, Paul Josef
Werntz, James Herbert, Jr
Wilson, Lauren R
Winston, Hubert
Yarbrough, Charles Gerald
Yarger, William E

NORTH DAKOTA
Bares, William Anthony
Brown, Guendoline
Kotch, Alex
Sugihara, James Masanobu
Vasey, Edfred H

OHIO
Anderson, Ruth Mapes
Baker, Saul Phillip
Barsky, Constance Kay
Collins, William John
Corbato, Charles Edward
Cox, Milton D
Curtis, Charles R
Daroff, Robert Barry
D'Azzo, John Joachim
Dobbelstein, Thomas Norman
Essman, Joseph Edward
Faiman, Robert N(eil)
Friedman, Morton Harold
Fucci, Donald James
Gamble, Francis Trevor
Gerald, Michael Charles
Glasser, Arthur Charles
Goel, Prem Kumar
Gorse, Joseph
Greenstein, Julius S
Hales, Raleigh Stanton, Jr
Hutchinson, Frederick Edward
Leonard, Billie Charles
Lumsdaine, Edward
Millard, Ronald Wesley
Nuenke, Richard Harold
Papadakis, Constantine N
Ray, John Robert
Redmond, Robert F(rancis)
Robe, Thurlow Richard
Rubin, Stanley G(erald)
Saul, Frank Philip
Stafford, Magdalen Marrow
Stansloski, Donald Wayne
Sweeney, Thomas L(eonard)

OKLAHOMA
Alexander, Joseph Walker
Batchman, Theodore E
Beasley, William Harold
Brandt, Edward Newman, Jr
Campbell, John R(oy)
Cassens, Patrick
Deckert, Gordon Harmon
Edmison, Marvin Tipton
Gunter, Deborah Ann
Hamm, Donald Ivan
Hibbs, Leon
Kimpel, James Froome
Klatt, Arthur Raymond
Knapp, Roy M
Kriegel, Monroe W(erner)
Langwig, John Edward
McCollom, Kenneth A(llen)
Middlebrooks, Eddie Joe
Miranda, Frank Joseph
Owens, G(lenda) Kay
Robinson, Jack Landy
Smith, Philip Edward
Swaim, Robert Lee
Turner, Wayne Connelly

OREGON
Allen, John Eliot
Arscott, George Henry
Brookhart, John Mills
Clark, William Melvin, Jr
Decker, Fred William
Griffith, William Thomas
Neiland, Bonita J
Owen, Sydney John Thomas
Scott, Peter Carlton
Van Vliet, Antone Cornelis
Weber, Lavern J
Zaerr, Joe Benjamin

PENNSYLVANIA
Andrews, Edwin Joseph
Ballentine, Rudolph Miller
Baum, Joseph Herman

Lerner, Lawrence S
Lofgren, Edward Joseph
Lombardi, Gabriel Gustavo
McCarthy, John Lockhart
McLaughlin, William Irving
Markowski, Gregory Ray
Matlow, Sheldon Leo
Maurer, E(dward) Robert
Merzbacher, Claude F
Miller, Donald Gabriel
Muller, Cornelius Herman
Mumme, Judith E
Pang, Kevin Dit Kwan
Pearl, Judea
Rabin, Jeffrey Mark
Rather, Lelland Joseph
Risse, Guenter Bernhard
Roth, Ariel A
Russell, Stuart Jonathan
Saunders, John Bertrand De Cusance
    Morant
Schelar, Virginia Mae
Schneiderman, Jill Stephanie
Shubert, Bruno Otto
Sime, Ruth Lewin
Sridharan, Natesa S
Stavroudis, Orestes Nicholas
Tenn, Joseph S
Thaw, Richard Franklin
Towers, Bernard
Tuul, Johannes
Vincenti, Walter G(uido)
Way, E Leong
Weeks, Dennis Alan
Wells, Patrick Harrington
Wells, Ronald Allen
Wise, Matthew Norton
Woodward, James Franklin

COLORADO
Franklin, Allan David
Olson, George Gilbert
Reinhardt, William Nelson
Watkins, Sallie Ann

CONNECTICUT
Downing, Mary Brigetta
Egler, Frank Edwin
Fenton, David George
Forbes, Thomas Rogers
Fraser, J(ulius) T(homas)
Fruton, Joseph Stewart
Hanson, Trevor Russell
Klein, Martin J(esse)
Kuslan, Louis Isaac
Levine, Robert John
Martinez, Robert Manuel
Menke, David Hugh
Patterson, Elizabeth Chambers
Salamon, Richard Joseph
Stevenson, Harlan Quinn
Summers, William Cofield
Suprynowicz, Vincent A
Viseltear, Arthur Jack

DELAWARE
Beer, John Joseph
Durbin, Paul Thomas
Ferguson, Raymond Craig
Hounshell, David A
Singleton, Rivers, Jr
Van Dyk, John William
Williams, Mary Bearden

DISTRICT OF COLUMBIA
Abler, Ronald Francis
Adomaitis, Vytautas Albin
Bellmer, Elizabeth Henry
Berendzen, Richard
Blanpied, William Antoine
Buzzelli, Donald Edward
Coates, Joseph Francis
Curtin, Frank Michael
DeVorkin, David Hyam
Earley, Joseph Emmet
Feldman, Martin Robert
Forman, Paul
Goldberg, Stanley
Harwit, Martin Otto
Hazen, Robert Miller
Jones, Daniel Patrick
Lieberman, Edwin James
McLean, John Dickson
O'Hern, Elizabeth Moot
Scheel, Nivard

FLORIDA
Albright, John Rupp
Cash, Rowley Vincent
Davidovits, Joseph
Hartman, John Paul
Leitner, Alfred
Linder, Ernest G
Mecklenburg, Roy Albert
Roberts, Leonidas Howard
Rubin, Richard Lee
Stern, William Louis
Vergenz, Robert Allan

GEORGIA
Dragoin, William Bailey
Loomis, Earl Alfred, Jr
Peacock, Lelon James
Saladin, Kenneth S

HAWAII
Krisciunas, Kevin L
Weems, Charles William

ILLINOIS
Amarose, Anthony Philip
Braunfeld, Peter George
Brown, Laurie Mark
Carozzi, Albert Victor
Debus, Allen George
Dinsmore, Charles Earle
Goldman, Jack Leslie
Gray, John Walker
Halberstam, Heini
Hirsch, Jerry
Mrtek, Robert George
Orland, Frank J
Pappademos, John Nicholas
Quay, Paul Michael
Rosen, Sidney
Schrage, Samuel
Spiroff, Boris E N
Spradley, Joseph Leonard
Stigler, Stephen Mack
Swanson, Don R
Sweitzer, James Stuart
Tuttle, Russell Howard
Wakefield, Ernest Henry
Yos, David Albert
Zusy, Dennis

INDIANA
Bellina, Joseph James, Jr
Cassidy, Harold Gomes
Dudley, Underwood
Dunham, William Wade
Helrich, Carl Sanfrid, Jr
Martin, Charles Wellington, Jr
Miller, Richard William
Zinsmeister, William John

IOWA
Howard, Robert Palmer
Patterson, John W(illiam)
Rider, Paul Edward, Sr
Schofield, Robert Edwin

KANSAS
Coates, Gary Joseph
Keller, Leland Edward
Underwood, James Ross, Jr

KENTUCKY
Davidson, John Edwin
Dixon, Wallace Clark, Jr
Engelberg, Joseph
Hamon, J Hill
Knoefel, Peter Klerner
McClain, John William

LOUISIANA
Andrus, Jan Frederick
Goodman, Alan Leonard
Kak, Subhash Chandra
Morris, Everett Franklin
Tipler, Frank Jennings, III

MAINE
Hall, Diana E
Hawthorne, Robert Montgomery, Jr
McEwen, Currier
Marcotte, Brian Michael
Whitten, Maurice Mason

MARYLAND
Anderson, Larry Douglas
Bell, James F(rederick)
Brieger, Gert Henry
Brush, Stephen George
Calinger, Ronald Steve
Darden, Lindley
Doggett, LeRoy Elsworth
Dowling, Marie Augustine
Eisenhart, Churchill
Embury, Janon Frederick, Jr
Felten, James Edgar
Fischer, Irene Kaminka
Gutsche, Graham Denton
Hanle, Paul Arthur
Heilprin, Laurence Bedford
Hindle, Brooke
Joy, Robert John Thomas
Katz, Victor Joseph
Kippenberger, Donald Justin
McCauley, H(enry) Berton
Macon, Nathaniel
Miles, Wyndham Davies
Mindel, Joseph
Mulligan, Joseph Francis
O'Grady, Richard Terence
Parascandola, John Louis
Parkinson, Claire Lucille
Reupke, William Albert
Stern, David P
Suppe, Frederick (Roy)
Temkin, Owsei
Tower, Donald Bayley

MASSACHUSETTS
Aldrich, Michele L
Andersen, Roy Stuart
Arganbright, Donald G
Breton, J Raymond
Brown, Gilbert J

Cavallaro, Mary Caroline
Cohen, I Bernard
Cohen, Robert Sonne
Cohn, Lawrence H
Costanza, Mary E
Frondel, Clifford
Futrelle, Robert Peel
Gingerich, Owen (Jay)
Graham, Loren R
Harrison, Edward Robert
Hawkins, Thomas William, Jr
Herda, Hans-Heinrich Wolfgang
Hiebert, Erwin Nick
Hill, Victor Ernst, IV
Holmes, Helen Bequaert
Holton, Gerald
Hovorka, John
Kaufman, Martin
Kuhn, Thomas S
Lockwood, Linda Gail
Maddin, Robert
Mayr, Ernst
OHara, Robert J
Ozonoff, David Michael
Platt, John Rader
Price, Robert
Ramakrishna, Kilaparti
Rosenkrantz, Barbara G
Sawin, Clark Timothy
Schweber, Silvan Samuel
Senechal, Marjorie Lee
Sheldon, Eric
Shimony, Abner
Silverman, Sam M
Smith, Cyril Stanley
Stachel, John Jay
Stonier, Tom Ted
Struik, Dirk Jan
Tyler, H Richard
Vankin, George Lawrence
Webster, Eleanor Rudd
Weiner, Charles
Weininger, Stephen Joel
Weinstein, Alexander
Winthrop, John T

MICHIGAN
Atkinson, James William
Davenport, Horace Willard
Hagen, John William
Jones, Phillip Sanford
Kaiser, Christopher B
Katz, Ernst
Kopperl, Sheldon Jerome
Lopushinsky, Theodore
McManus, Hugh
Ramsay, Ogden Bertrand
Rollins-Page, Earl Arthur
Root-Bernstein, Robert Scott
Ruffner, James Alan
Schmidt, Robert
Toledo-Pereyra, Luis Horacio
Trosko, James Edward
Warren, H(erbert) Dale
Weymouth, Patricia Perkins
Zimring, Lois Jacobs

MINNESOTA
Keys, Thomas Edward
Layton, Edwin Thomas, Jr
Marquit, Erwin
Meyer, Frank Henry
Norberg, Arthur Lawrence
O'Rourke, Richard Clair
Rudolf, Paul Otto
Sapakie, Sidney Freidin
Schwartz, A(lbert) Truman
Shapiro, Alan Elihu
Sperber, William H
Stuewer, Roger Harry
Theisen, Wilfred Robert
Wilson, Leonard Gilchrist

MISSISSIPPI
Seymour, Raymond B(enedict)

MISSOURI
Adawi, Ibrahim (Hasan)
Allen, Garland Edward, III
Ashworth, William Bruce, Jr
Bieniek, Ronald James
Drennan, Ollin Junior
Ewan, Joseph (Andorfer)
Klotz, John William
McCollum, Clifford Glenn
Miller, Jane Alsobrook
Peterson, Charles John
Twardowski, Zbylut Jozef
Watson, Richard Allan

MONTANA
McAllister, Byron Leon
McRae, Robert James

NEBRASKA
Friedlander, Walter Jay
Harris, Holly Ann
Kennedy, Robert E
Zimmerman, Edward John

NEVADA
Merdinger, Charles J(ohn)
Scott, William Taussig

NEW HAMPSHIRE
Beasley, Wayne M(achon)
Ermenc, Joseph John
King, Allen Lewis
Schneer, Cecil Jack

NEW JERSEY
Abeles, Francine
Berendsen, Peter Barney
Chen, Thomas Shih-Nien
Chinard, Francis Pierre
Church, John Armistead
Cromartie, William James, Jr
Gershenowitz, Harry
Grob, Gerald N
Haindl, Martin Wilhelm
Harvey, Floyd Kallum
Holzmann, Gerard J
House, Edward Holcombe
Mark, Robert
Mihram, George Arthur
Prasad, Marehalli Gopalan
Rowe, John Edward
Spruch, Grace Marmor
Tanner, Daniel
Widmer, Kemble
Woolf, Harry

NEW MEXICO
Bertin, John Joseph
Divett, Robert Thomas
Schufle, Joseph Albert
Zeilik, Michael

NEW YORK
Atwater, Edward Congdon
Boeck, William Louis
Brown, Patricia Stocking
Bullough, Vern L
Carlson, Elof Axel
Cummins, John Frances
Dahl, Per Fridtjof
Dauben, Joseph W
Devons, Samuel
Donnelly, Denis Philip
Donovan, Arthur L
Eisele, Carolyn
Eisinger, Josef
Erlichson, Herman
Fitzgerald, Patrick James
Gorenstein, Shirley Slotkin
Gortler, Leon Bernard
Greaves, Bettina Bien
Harris, Charles Leon
Henderson, David Wilson
Heppa, Douglas Van
Holden, Norman Edward
Jahiel, Rene
Kanzler, Walter Wilhelm
Kennedy, Kenneth Adrian Raine
Kingslake, Rudolf
Kyburg, Henry
Lawson, Kent DeLance
Lowrance, William Wilson, Jr
Lucey, Carol Ann
Lurkis, Alexander
Lutzker, Edythe
Macklin, Ruth
Mansfield, Victor Neil
Merton, Robert K
Morey, Donald Roger
Neugebauer, Richard
Newman, Stuart Alan
Olczak, Paul Vincent
Packard, Karle Sanborn, Jr
Pelicci, Pier Guiseppe
Pence, Harry Edmond
Phillips, Esther Rodlitz
Porter, Beverly Fearn
Pugh, Emerson William
Raman, Varadaraja Venkata
Rapaport, William Joseph
Ravitz, Leonard J, Jr
Rohrlich, Fritz
Romer, Alfred
Sauer, Charles William
Segal, Sanford Leonard
Sharkey, John Bernard
Sharpe, William D
Smith, Gerrit Joseph
Smolin, Lee
Sparberg, Esther Braun
Sturges, Stuart
Tesmer, Irving Howard
Trigg, George Lockwood
Voelcker, Herbert B(ernhard)
Waldbauer, Eugene Charles
Weart, Spencer Richard
Westbrook, J(ack) H(all)

NORTH CAROLINA
Adler, Carl George
Fishman, George Samuel
Garg, Devendra Prakash
Gifford, James Fergus
Miller, Robert L
Mitchell, Earl Nelson
Petroski, Henry J
Roland, Alex
Shapere, Dudley
Stern, A(rthur) C(ecil)
Troyer, James Richard

Larmore, Lewis
Lave, Roy E(llis), Jr
Lechtman, Max D
Lederer, C Michael
Legge, Norman Reginald
Levine, Mark David
Levinthal, Elliott Charles
Lindquist, Robert Henry
Linnell, Robert Hartley
Lisman, Perry Hall
Lissaman, Peter Barry Stuart
Little, John Clayton
Lofgren, Edward Joseph
London, Ray William
Lukasik, Stephen Joseph
McCall, Chester Hayden, Jr
McDermott, Daniel J
McDonough, Thomas Redmond
MacKay, Kenneth Donald
Marrone, Pamela Gail
Mattice, Jack Shafer
Meechan, Charles James
Merilo, Mati
Merten, Ulrich
Meyers, Robert Allen
Mikkelsen, Duane Soren
Miller, William Frederick
Morris, Fred(erick) W(illiam)
Mueller, James Lowell
Musser, John H
Mysels, Karol Joseph
Nesbit, Richard Allison
Newell, Gordon Wilfred
O'Benar, John DeMarion
O'Donnell, Joseph Allen, III
Orcutt, Harold George
Pake, George Edward
Parkinson, Bradford W
Payne, Holland I
Pearson, John
Pedersen, Leo Damborg
Peeters, Randall Louis
Perry, Robert Hood, Jr
Peterson, Laurence E
Pierce, James Otto, II
Piez, Karl Anton
Poppoff, Ilia George
Prim, Robert Clay
Pugh, Howel Griffith
Ram, Michael Jay
Ranftl, Robert M(atthew)
Rapp, Donald
Richter, George Neal
Riley, N Allen
Rossin, A David
Rowell, John Martin
Rowntree, Robert Fredric
Rudy, Thomas Philip
Rutan, Elbert L
Saffren, Melvin Michael
Sager, Ronald E
Sakagawa, Gary Toshio
Salsig, William Winter, Jr
Sanchez, Robert A
Saute, Robert E
Scheinok, Perry Aaron
Schonberg, Russell George
Schweiker, George Christian
Seibel, Erwin
Seppi, Edward Joseph
Sharp, Joseph Cecil
Shelton, Robert Neal
Silverman, Jacob
Smith, Charles G
Sommers, William P(aul)
Spinrad, Robert J(oseph)
Statler, Irving C(arl)
Stelzer, Lorin Roy
Stephanou, Stephen Emmanuel
Stirn, Richard J
Strena, Robert Victor
Sundberg, John Edwin
Sutton, Paul McCullough
Sweeney, Lawrence Earl, Jr
Syvertson, Clarence A
Szego, Peter A
Szymanski, Paul Stephen
Tarbell, Theodore Dean
Taylor, Robert William
Thatcher, Everett Whiting
Thueson, David Orel
Trevelyan, Benjamin John
Turner, George Cleveland
Van Schaik, Peter Hendrik
Vernikos, Joan
Walker, Jimmy Newton
Walsh, Don
Warner, Willis L
Waters, William E
Weir, William Carl
Welch, Robin Ivor
Werth, Glenn Conrad
Whitney, William Merrill
Wilbarger, Edward Stanley, Jr
Winick, Herman
Worker, George F, Jr
Young, Robert, Jr
Younoszai, Rafi
Zander, Andrew Thomas
Zarem, Abe Mordecai
Ziemer, Robert Ruhl

COLORADO
Barber, George Arthur

Boudreau, Robert Donald
Boyd, L(andis) L(ee)
Broome, Paul W(allace)
Bull, Stanley R(aymond)
Czanderna, Alvin Warren
Dusenberry, William Earl
Evans, Norman A(llen)
Fall, Michael William
Firor, John William
Fraikor, Frederick John
Gentry, Willard Max, Jr
Glock, Robert Dean
Griffith, Cecilia Girz
Harrison, Monty DeVerl
Heberlein, Douglas Garavel
Horak, Donald L
Hutto, Francis Baird, Jr
Johnson, Janice Kay
Kniebes, Duane Van
LaBounty, James Francis, Sr
Lameiro, Gerard Francis
McKown, Cora F
McLean, Francis Glen
Micheli, Roger Paul
Morrison, Charles Freeman, Jr
Nauenberg, Uriel
Olien, Neil Arnold
Purdom, James Francis Whitehurst
Scherer, Ronald Callaway
Schleicher, David Lawrence
Seibert, Michael
Shackelford, Scott Addison
Siuru, William D, Jr
Stencel, Robert Edward
Streib, W(illiam) C(harles)
Sunderman, Duane Neuman
West, Anita
Westfall, Richard Merrill
Wilson, David George

CONNECTICUT
Bentz, Alan P(aul)
Berdick, Murray
Bluhm, Leslie
Bormann, Barbara-Jean Anne
Bronsky, Albert J
Brooks, Douglas Lee
Buchanan, Ronald Leslie
Burke, Carroll N
Burwell, Wayne Gregory
Calehuff, Girard Lester
Carter, G Clifford
Clark, Hugh
Cobern, Martin E
Cole, Stephen H(ervey)
Cooke, Theodore Frederic
Darling, George Bapst
Davenport, Lee Losee
Doshan, Harold David
Duvivier, Jean Fernand
Eby, Edward Stuart, Sr
Emond, George T
Enell, John Warren
Epperly, W Robert
Erlandson, Paul M(cKillop)
Fleming, James Stuart
Gunther, Jay Kenneth
Haacke, Gottfried
Haas, Ward John
Hand, Arthur Ralph
Hayes, Richard J
Hermann, Robert J
Hobbs, Donald Clifford
Juby, Peter Frederick
Kemp, Gordon Arthur
Kotick, Michael Paul
Leonard, Robert F
Levin, S Benedict
Lichtenberger, Harold V
Lines, Ellwood LeRoy
Linke, William Finan
McKeon, James Edward
Marchand, Nathan
Megrue, George Henry
Meierhoefer, Alan W
Menkart, John
Milliken, Frank R
Moreland, Walter Thomas, Jr
Munies, Robert
Nelson, Roger Peter
O'Brien, Robert L
Ortner, Mary Joanne
Page, Edgar J(oseph)
Parthasarathi, Manavasi Narasimhan
Pierce, James Benjamin
Pinson, Ellis Rex, Jr
Pinson, Rex, Jr
Proctor, Alan Ray
Rauhut, Michael McKay
Ray, Verne A
Richards, Frederic Middlebrook
Rishell, William Arthur
Robinson, Donald W(allace), Jr
Ryan, Richard Patrick
Schrage, F Eugene
Schurig, John Eberhard
Scott, Joseph Hurlong
Shah, Shirish A
Shorr, Bernard
Smith, Douglas Stewart
Snellings, William Moran
Stott, Paul Edwin
Suen, T(zeng) J(iueq)
Tangel, O(scar) F(rank)

Wheeler, Edward Stubbs
Wiseman, Edward H
Wood, David Dudley

DELAWARE
Armbrecht, Frank Maurice, Jr
Baer, Donald Robert
Bancroft, Lewis Clinton
Cartwright, Aubrey Lee, Jr
Chiu, Jen
Franta, William Alfred
Ginnard, Charles Raymond
Greenblatt, Hellen Chaya
Hannell, John W(eldale), Jr
Hess, Richard William
Jackson, David Archer
Jefferson, Edward G
Johnson, Donald Richard
Knipmeyer, Hubert Elmer
Lin, Kuang-Farn
McClelland, Alan Lindsey
Mikell, William Gaillard
Montague, Barbara Ann
Naylor, Marcus A, Jr
Norling, Parry McWhinnie
Panar, Manuel
Phillips, Brian Ross
Repka, Benjamin C
Rogers, Donald B
Rosenberg, Richard Martin
Schiroky, Gerhard H
Schroeder, Herman Elbert
Schwartz, Jerome Lawrence
Shambelan, Charles
Shotzberger, Gregory Steven
Sloan, Martin Frank
Steinberg, Marshall
Williams, Ebenezer David, Jr
Woods, Thomas Stephen

DISTRICT OF COLUMBIA
Alexander, Joseph Kunkle
Ausubel, Jesse Huntley
Baker, D(onald) James
Barker, Winona Clinton
Beach, Harry Lee, Jr
Bender, James Arthur
Berley, David
Borsuk, Gerald M
Broadbent, Noel Daniel
Bryant, Barbara Everitt
Burka, Maria Karpati
Burris, James F
Bushman, John Branson
Charlton, Gordon Randolph
Chen, Davidson Tah-Chuen
Cherniavsky, John Charles
Chung, Ho
Cowdry, Rex William
Cox, Lawrence Henry
Diemer, F(erdinand) P(eter)
Fechter, Alan Edward
Fiske, Richard Sewell
Flora, Lewis Franklin
Fowler, Earle Cabell
Fricken, Raymond Lee
Gaffney, Paul G, II
Goldfield, Edwin David
Goldstein, Gerald
Gould, Phillip
Grafton, Robert Bruce
Green, Richard James
Greene, Michael P
Gress, Mary Edith
Haworth, William Lancelot
Hedlund, James Howard
Holt, James Allen
Howell, Embry Martin
Hussain, Syed Taseer
Ison-Franklin, Eleanor Lutia
Jackson, Curtis Rukes
Johnson, George Patrick
Johnson, Leonard Evans
Kelly, Henry Charles
Kimmel, Carole Anne
Kinney, Terry B, Jr
Koeppen, Robert Carl
Korkegi, Robert Hani
Lombard, David Bishop
Lyon, Robert Lyndon
Mange, Phillip Warren
Martensen, Todd Martin
Mellor, John Williams
Nadolney, Carlton H
Novello, Antonia Coello
O'Fallon, Nancy McCumber
Peake, Harold J(ackson)
Petricciani, John C
Plowman, Ronald Dean
Rabin, Robert
Raizen, Senta Amon
Ramsey, Jerry Warren
Raslear, Thomas G
Redish, Janice Copen
Reinecke, Thomas Leonard
Reisa, James Joseph, Jr
Reynik, Robert John
Riegel, Kurt Wetherhold
Rosen, Howard Neal
Rothman, Sam
Rubinoff, Roberta Wolff
Schmidt, Berlie Louis
Schriever, Richard L
Scott, Roland Boyd

Selden, Robert Wentworth
Smidt, Fred August, Jr
Smith, David Allen
Smith, David Allen
Smith, Leslie E
Smith, Thomas Elijah
Sobotka, Thomas Joseph
Stevens, Donald Keith
Tompkins, Daniel Reuben
Topper, Leonard
Wallenmeyer, William Anton
Way, George H
Weiss, Leonard
Whitmore, Frank Clifford, Jr
Wile, Howard P
Wilkniss, Peter Eberhard
Wilsey, Neal David
Wilson, George Donald
Wittmann, Horst Richard
Wolfe, Thomas Lee
Wright, Bill C
Zeizel, A(rthur) J(ohn)

FLORIDA
Abraham, William Michael
Amey, William G(reenville)
Aronson, M(oses)
Baker, Carleton Harold
Baum, Werner A
Berry, Robert Eddy
Biasell, Laverne R(obert)
Bottoms, Albert Maitland
Bruner, Harry Davis
Burridge, Michael John
Callen, Joseph Edward
Coulter, Wallace H
Dacey, George Clement
Dailey, Charles E(lmer), III
Davis, Duane M
De Lorge, John Oldham
Diaz, Nils Juan
Dubach, Leland L
Duke, Dennis Wayne
Einspruch, Norman G(erald)
Everett, Woodrow Wilson, Jr
Francis, Stanley Arthur
Friedlander, Herbert Norman
Gerber, John Francis
Gianola, Umberto Ferdinando
Groupe, Vincent
Hall, David Goodsell, IV
Hampp, Edward Gottlieb
Harper, Verne Lester
Heggie, Robert
Henkin, Hyman
Heying, Theodore Louis
Hirsch, Donald Earl
Joftes, David Lion
Jones, Albert Cleveland
Joyce, Edwin A, Jr
Kender, Walter John
Kiplinger, Glenn Francis
Lannutti, Joseph Edward
Long, Calvin H
McFadden, James Douglas
Malthaner, W(illiam) A(mond)
Medin, A(aron) Louis
Merritt, Alfred M, II
Moser, Robert E
Murchison, Thomas Edgar
Nunnally, Stephens Watson
Owens, James Samuel
Peacock, Hugh Anthony
Rabbit, Guy
Randazzo, Anthony Frank
Rathburn, Carlisle Baxter, Jr
Revay, Andrew W, Jr
Rich, Jimmy Ray
Riordan, Michael Davitt
Roesner, Larry A
Saunders, James Henry
Seugling, Earl William, Jr
Sharkey, William Henry
Sheppard, Albert Parker
Simeral, William Goodrich
Smith, Wayne H
Splitter, Earl John
Spurlock, Jack Marion
Stephens, John C(arnes)
Stewart, Harris Bates, Jr
Stroh, Robert Carl
Wade, Thomas Edward
Weinberg, Jerry L
Wiltbank, William Joseph
Wise, Raleigh Warren

GEORGIA
Alberts, James Joseph
Barbaree, James Martin
Barreras, Raymond Joseph
Bodnar, Donald George
Bonkovsky, Herbert Lloyd
Broerman, F S
Cassanova, Robert Anthony
Circeo, Louis Joseph, Jr
Dees, J W
Donahoo, Pat
Dragoin, William Bailey
Evans, Thomas P
Fink, Richard Walter
Frere, Maurice Herbert
Goldsmith, Edward
Gollob, Lawrence
Grace, Donald J

Research Administration (cont)

Gunn, Walter Joseph
Harris, Elliott Stanley
Herbst, John A
Herrmann, Kenneth L
Ike, Albert Francis
Innes, David Lyn
Keller, Oswald Lewin
Koerner, T J
Kraeling, Robert Russell
Loomis, Earl Alfred, Jr
McClure, Harold Monroe
Malcolm, Earl Walter
Martin, Linda Spencer
Oliver, James Henry, Jr
Peck, Harry Dowd, Jr
Pence, Ira Wilson, Jr
Pollard, William Blake
Prince, M(orris) David
Reiss, Errol
Schutt, Paul Frederick
Schwartz, Robert John
Shlevin, Harold H
Smith, David Fletcher
Spauschus, Hans O
Stalford, Harold Lenn
Stevens, Ann Rebecca
Swank, Robert Roy, Jr
Thompson, William Oxley, II
Threadgill, Ernest Dale
Walls, Nancy Williams
Wepfer, William J
Whitehead, Marvin Delbert

HAWAII
Boehlert, George Walter
Ching, Chauncey T K
Daniel, Thomas Henry
Duckworth, Walter Donald
Flanagan, Patrick William
Helfrich, Philip
Khan, Mohammad Asad
Malmstadt, Howard Vincent
Matsuda, Fujio
Moberly, Ralph M
Swindale, Leslie D
Williams, David Douglas F
Yates, James T

IDAHO
Griffith, W(illiam) A(lexander)
Kleinkopf, Gale Eugene
Laurence, Kenneth Allen
Majumdar, Debaprasad (Debu)
Sarett, Lewis Hastings
Slansky, Cyril M
Vestal, Robert Elden
Wood, Richard Ellet

ILLINOIS
Abrams, Israel Jacob
Abrams, Robert A
Afremow, Leonard Calvin
Akin, Cavit
Altpeter, Lawrence L, Jr
Asbury, Joseph G
Auvil, Paul R, Jr
Baker, Winslow Furber
Barenberg, Sumner
Bartimes, George F
Beasley, Thomas Miles
Bentley, Orville George
Bietz, Jerold Allen
Bollinger, Lowell Moyer
Brezinski, Darlene Rita
Brotherson, Donald E
Brutten, Gene J
Bullard, Clark W
Burhop, Kenneth Eugene
Cafasso, Fred A
Cannon, Howard S(uckling)
Cardman, Lawrence Santo
Chaszeyka, Michael A(ndrew)
Chu, Daniel Tim-Wo
Cole, Francis Talmage
Collier, Donald W(alter)
Conn, Arthur L(eonard)
Danzig, Morris Juda
Davis, Edward Nathan
Ehrlich, Richard
Fenters, James Dean
Fish, Ferol F, Jr
Flay, Brian Richard
Ford, Richard Earl
Frank, James Richard
French, Allen Lee
Gealy, William James
Gillette, Richard F
Glaser, Janet H
Hahn, Richard Ray
Halaby, George Anton
Harris, Ronald David
Herting, Robert Leslie
Heybach, John Peter
Howsmon, John Arthur
Huebner, Russell Henry, Sr
Jain, Ravinder Kumar
Jezl, James Louis
Johnson, Dale Howard
Jugenheimer, Robert William
Kabisch, William Thomas
Kasper, Gerhard
Kittel, J Howard

Krug, Samuel Edward
Lederman, Leon Max
Levin, Alfred A
Levy, Nelson Louis
Linde, Ronald K(eith)
McLaughlin, Robert Lawrence
Mapother, Dillon Edward
Marchaterre, John Frederick
Menke, Andrew G
Nair, Velayudhan
Nezrick, Frank Albert
Pai, Mangalore Anantha
Pardo, Richard Claude
Pariza, Richard James
Phelps, Creighton Halstead
Pietri, Charles Edward
Plautz, Donald Melvin
Prabhudesai, Mukund M
Prakasam, Tata B S
Rafelson, Max Emanuel, Jr
Reilly, Christopher Aloysius, Jr
Richmond, James M
Robinson, Charles J
Rothfus, John Arden
Rowe, William Bruce
Rudnick, Stanley J
Rue, Edward Evans
Rundo, John
Rusinko, Frank, Jr
Sachs, Robert Green
Schaefer, Wilbur Carls
Schriesheim, Alan
Shapiro, Stanley
Sheldon, Victor Lawrence
Sibbach, William Robert
Singiser, Robert Eugene
Smith, Leslie Garrett
Staats, William R
Stafford, Fred E
Theriot, Edward Dennis, Jr
Thomas, Joseph Erumappetical
Thompson, Charles William Nelson
Tourlentes, Thomas Theodore
Trimble, John Leonard
Unik, John Peter
Vander Velde, George
Wargel, Robert Joseph
Weinstein, Hyman Gabriel
Wolsky, Alan Martin
Yackel, Walter Carl
Yarger, James G
Zeffren, Emil
Zeiss, Chester Raymond

INDIANA
Amundson, Merle E
Amy, Jonathan Weekes
Andrews, Frederick Newcomb
Botero, J M
Boyd, Frederick Mervin
Brinkmeyer, Raymond Samuel
Castleberry, Ron M
Cerimele, Benito Joseph
Cook, David Allan
Cravens, William Windsor
DeSantis, John Louis
Fredrich, Augustine Joseph
Free, Helen M
Gunter, Claude Ray
Guthrie, George Drake
Ho, Cho-Yen
Jacobs, Martin Irwin
Janis, F Timothy
Jantz, O K
Johnson, Irving Stanley
Judd, William Robert
Kammeraad, Adrian
Katz, Frances R
Khorana, Brij Mohan
Leander, John David
Leap, Darrell Ivan
McIntyre, Thomas William
McKeehan, Charles Wayne
Mazac, Charles James
Merritt, Doris Honig
Millar, Wayne Norval
Odor, David Lee
Ortman, Eldon Emil
Price, Howard Charles
Probst, Gerald William
Pyle, James L
Sampson, Charles Berlin
Sannella, Joseph L
Stark, Paul
Strycker, Stanley Julian
Swanson, Lynn Allen
Tarwater, Oliver Reed
Thompson, Gerald Lee
Wright, William Leland

IOWA
Austin, Tom Al
Clark, Robert A
Gaertner, Richard F(rancis)
Haendel, Richard Stone
Kelly, William H
Kluge, John Paul
Markuszewski, Richard
Mather, Roger Frederick
O'Berry, Phillip Aaron
Reckase, Mark Daniel
Spitzig, William Andrew

KANSAS
Erickson, Howard Hugh
Gerhard, Lee C
Ham, George Eldon
Jacobs, Louis John
Lee, Joe
Leland, Stanley Edward, Jr
Lyles, Leon
Mayeux, Jerry Vincent
Schroeder, Robert Samuel
Stevens, David Robert
Sutherland, John B(ennett)
Von Rumker, Rosmarie

KENTUCKY
Broida, Theodore Ray
Cohn, David Valor
Deen, Robert C(urba)
Harris, Patrick Donald
Huffman, Gerald P
Merker, Stephen Louis
Mitchell, Maurice McClellan, Jr
Podshadley, Arlon George
Price, Martin Burton
Smith, Gilbert Edwin
Wells, William Lochridge

LOUISIANA
Arthur, Jett Clinton, Jr
Avent, Robert M
Barbee, R Wayne
Baril, Albert, Jr
Barras, Stanley J
Berni, Ralph John
Caffey, Horace Rouse
Conrad, Franklin
Davis, Bryan Terence
DeMonsabert, Winston Russel
Eaton, Harvill Carlton
Gautreaux, Marcelian Francis
Goerner, Joseph Kofahl
Harrison, Richard Miller
Hastings, Robert Clyde
Hensley, Sess D
Holloway, Frederic Ancrum Lord
Hough, Walter Andrew
Huggett, Richard William, Jr
Mills, Earl Ronald
Moulder, Peter Vincent, Jr
O'Barr, Richard Dale
Polack, Joseph A(lbert)
Ristroph, John Heard
Starrett, Richmond Mullins
Threefoot, Sam Abraham
Vail, Sidney Lee
Van Lopik, Jack Richard
Vashishta, Priya Darshan
Walker, John Robert
Wiewiorowski, Tadeusz Karol
Wollensak, John Charles
Zietz, Joseph R, Jr

MAINE
Hanks, Robert William
Hill, Marquita K
Kirkpatrick, Francis Hubbard
McKerns, Kenneth (Wilshire)
Marcotte, Brian Michael
Shipman, C(harles) William
Spencer, Claude Franklin

MARYLAND
Aamodt, Roger Louis
Abbey, Robert Fred, Jr
Al-Aish, Matti
Albrecht, G(eorge) H(enry)
Alderson, Norris Eugene
Andres, Scott Fitzgerald
Attaway, David Henry
Atwell, Constance Woodruff
Baker, Carl Gwin
Baker, Charles P
Bareis, Donna Lynn
Barnes, James Alford
Bastress, E(rnest) Karl
Batson, David Banks
Bawden, Monte Paul
Beattie, Donald A
Beaven, Vida Helms
Beisler, John Albert
Bellino, Francis Leonard
Benbrook, Charles M
Benson, Walter Roderick
Berman, Howard Mitchell
Bingham, Robert Lodewijk
Blaunstein, Robert P
Bollum, Frederick James
Boyd, Michael R
Brady, Robert Frederick, Jr
Brown, Paul Joseph
Bryan, Edward H
Buck, John Henry
Calabi, Ornella
Cave, William Thompson
Cereghino, James Joseph
Coates, Arthur Donwell
Coleman, James Stafford
Colvin, Burton Houston
Condell, William John, Jr
Condliffe, Peter George
Cornoni-Huntley, Joan Claire
Costa, Paul T, Jr
Crane, Langdon Teachout, Jr
Crout, John Richard

Crump, Stuart Faulkner
Daniels, William Fowler, Sr
Das, Naba Kishore
Dashiell, Thomas Ronald
Dean, Donna Joyce
Decker, James Federick
Diggs, John W
Dodd, Roger Yates
Domanski, Thaddeus John
Doschek, Wardella Wolford
Duggar, Benjamin Charles
Dupont, Jacqueline (Louise)
Eaton, Alvin Ralph
Eccleshall, Donald
Efron, Herman Yale
Elkins, Earleen Feldman
Ellenberg, Susan Smith
Ellingwood, Bruce Russell
Engle, Robert Rufus
Ephremides, Anthony
Erickson, Duane Gordon
Evans, John V
Finkelstein, Robert
Fischer, Eugene Charles
Fisher, Dale John
Fisher, E(arl) Eugene
Fowler, Howland Auchincloss
Fozard, James Leonard
Frazer, Benjamin Chalmers
Freden, Stanley Charles
Fried, Jerrold
Frommer, Peter Leslie
Ganz, Aaron
Garman, John Andrew
Geller, Ronald G
Gessow, Alfred
Golding, Leonard S
Goldstein, Robert Arnold
Goldwater, William Henry
Gonano, John Roland
Gonda, Matthew Allen
Gonzales, Ciriaco Q
Gordon, Stephen L
Graham, Bettie Jean
Granatstein, Victor Lawrence
Gravatt, Claude Carrington, Jr
Gray, David Bertsch
Gray, Paulette S
Green, Jerome George
Gupta, Ramesh K
Guss, Maurice Louis
Guttman, Helene Augusta Nathan
Haberman, William L(awrence)
Hacker, Peter Wolfgang
Hallett, Mark
Hanna, Edgar Ethelbert, Jr
Hartmann, Gregory Kemenyi
Haseltine, Florence Pat
Hausman, Steven J
Hayes, Joseph Edward, Jr
Hazzard, DeWitt George
Hellwig, Helmut Wilhelm
Herman, Samuel Sidney
Herrmann, Robert Arthur
Hiatt, Caspar Wistar, III
Hildebrand, Bernard
Hilmoe, Russell J(ulian)
Hindle, Brooke
Hinners, Noel W
Hodge, David Charles
Holt, Stephen S
Howie, Donald Lavern
Hubbard, Van Saxton
Huddleston, Charles Martin
Irwin, David
Jacobson, Martin
Johnston, Laurance S
Joiner, R(eginald) Gracen
Jones, Franklin Del
Kalt, Marvin Robert
Karlander, Edward P
Katsanis, D(avid) J(ohn)
Kearney, Philip C
Kehr, Clifton Leroy
Kendrick, Francis Joseph
Kennedy, Robert A
Ketley, Jeanne Nelson
Khachaturian, Zaven Setrak
Khan, Mushtaq Ahmad
Kirby, Ralph C(loudsberry)
Kolstad, George Andrew
Koslow, Stephen Hugh
Krakauer, Henry
Kroll, Bernard Hilton
Kushner, Lawrence Maurice
Lansdell, Herbert Charles
Ledney, George David
LeRoy, André François
Levine, Arthur Samuel
Lundegard, Robert James
Lymn, Richard Wesley
McElhinney, John
Mahle, Christoph E
Maier, Eugene Jacob Rudolph
Malotky, Lyle Oscar
Marsden, Halsey M
Maslow, David E
Masnyk, Ihor Jarema
Meredith, Leslie Hugh
Mermagen, William Henry
Metcalfe, Dean Darrel
Metzger, Henry
Mielenz, Klaus Dieter
Milkey, Robert William

Miller, Nancy E
Mishler, John Milton, IV
Mockrin, Stephen Charles
Mohla, Suresh
Mohler, Irvin C, Jr
Moloney, John Bromley
Moorjani, Kishin
Moran, David Dunstan
Mordfin, Leonard
Morgan, Arthur I, Jr
Morgan, John D(avis)
Morrissey, Robert Leroy
Muller, Robert E
Murphy, Donald G
Murray, William Sparrow
Myers, Lawrence Stanley, Jr
Nagle, Dennis Charles
Nelson, James Harold
Norvell, John Charles
Orlick, Charles Alex
Ory, Marcia Gail
Osborne, J Scott, III
Oster, Ludwig Friedrich
Oxman, Michael Allan
Paré, William Paul
Parkhie, Mukund Raghunathrao
Passman, Sidney
Peiser, Herbert Steffen
Phillips, Grace Briggs
Plotkin, Henry H
Polinsky, Ronald John
Postow, Elliot
Powers, Marcelina Venus
Quraishi, Mohammed Sayeed
Randall, John Douglas
Randolph, Lynwood Parker
Ranhand, Jon M
Read, Merrill Stafford
Remondini, David Joseph
Reuther, Theodore Carl, Jr
Richardson, Billy
Richardson, Clarence Robert
Richardson, John Marshall
Ringler, Robert L
Rochlin, Robert Sumner
Rodney, William Stanley
Rosenstein, Robert William
Rueger, Lauren J(ohn)
Saloman, Edward Barry
Sansone, Eric Brandfon
Saville, Thorndike, Jr
Schiaffino, Silvio Stephen
Schmidt, Jack Russell
Schneck, Paul Bennett
Schneider, Carl Stanley
Schoman, Charles M, Jr
Sexton, Thomas John
Shalowitz, Erwin Emmanuel
Shannon, Larry J(oseph)
Sharma, Dinesh C
Shaw, Warren Cleaton
Sheffield, Harley George
Sheridan, Philip Henry
Siegel, Elliot Robert
Singh, Amarjit
Smith, Joseph Collins
Snow, George Abraham
Sorrows, Howard Earle
Sporn, Eugene Milton
Stamper, Hugh Blair
Stein, Harvey Philip
Stoolmiller, Allen Charles
Stuebing, Edward Willis
Swann, Madeline Bruce
Szego, George C(harles)
Tabor, Edward
Tai, Tsze Cheng
Talbott, Edwin M
Taylor, Barry Norman
Theuer, Paul John
Thompson, Robert John, Jr
Thorp, James Wilson
Tilford, Shelby G
Tillman, Michael Francis
Tolman, Robert Alexander
Tower, Donald Bayley
Triantaphyllopoulos, Eugenie
Turbyfill, Charles Lewis
Ulbrecht, Jaromir Josef
Valdes, James John
Vanderryn, Jack
Vermund, Sten Halvor
Villet, Ruxton Herrer
Vocci, Frank Joseph
Wachs, Melvin Walter
Walker, Michael Dirck
Wallace, Craig Kesting
Walter, Donald K
Walter, Eugene LeRoy, Jr
Watkins, Clyde Andrew
Weisfeldt, Myron Lee
Wells, James Robert
Westwood, Albert Ronald Clifton
White, Harris Herman
Wolff, David A
Wolff, John B
Wolff, Roger Glen
Wortman, Roger Matthew
Wright, James Roscoe
Yarmolinsky, Adam
Young, George Anthony
Zornetzer, Steven F

MASSACHUSETTS
Ahearn, James Joseph, Jr
Alpert, Ronald L
Anderson, Edward Everett
Barr, James K
Bhargava, Hridaya Nath
Blatt, S Leslie
Busby, William Fisher, Jr
Camougis, George
Carleton, Nathaniel Phillips
Carton, Edwin Beck
Chase, Fred Leroy
Chivian, Eric Seth
Cohn, Lawrence H
Collins, Carolyn Jane
Cukor, Peter
Cushing, David H
Davis, Michael Allan
Denniston, Joseph Charles
Dorman, Craig Emery
Feakes, Frank
Flagg, John Ferard
Fohl, Timothy
Friedman, Orrie Max
Friedman, Raymond
Futrelle, Robert Peel
Ganley, Oswald Harold
Garing, John Seymour
Giarrusso, Frederick Frank
Girard, Francis Henry
Gregory, Eric
Griffith, Robert W
Grodberg, Marcus Gordon
Hallock, Gilbert Vinton
Harrill, Robert W
Heyerdahl, Eugene Gerhardt
Hodgins, George Raymond
Hunt, Ronald Duncan
Hurlburt, Douglas Herendeen
Ingle, George William
Jonas, Albert Moshe
Kelland, David Ross
Kerr, Donald M, Jr
King, Norval William, Jr
Koehler, Andreas Martin
Kornfield, Jack I
Kovatch, George
McBee, W(arren) D(ale)
McCrea, Peter Frederick
Mack, Richard Bruce
Martin, Frederick Wight
Maser, Morton D
Milburn, Richard Henry
Miller, James Albert, Jr
Morbey, Graham Kenneth
Nauss, Kathleen Minihan
Papageorgiou, John Constantine
Parker, Henry Seabury, III
Paskins-Hurlburt, Andrea Jeanne
Pharo, Richard Levers
Picard, Dennis J
Pulling, Nathaniel H(osler)
Riggi, Stephen Joseph
Roberts, Francis Donald
Rogers, William Irvine
Rohde, Richard Allen
Rosenkrantz, Harris
Roth, Harold
Rothenberg, Albert
Rowley, Durwood B
Schaffel, Gerson Samuel
Seamans, R(obert) C(hanning), Jr
Sethares, James C(ostas)
Shusman, T(evis)
Silver, Frank Morris
Smith, Raoul Normand
Smith, William Edward
Staelin, David Hudson
Taylor, Isaac Montrose
Vartanian, Leo
Wigley, Roland L
Williams, David Lloyd
Wilson, Linda S (Whatley)

MICHIGAN
Albers, Walter Anthony, Jr
Anderson, Amos Robert
Arrington, Jack Phillip
Ashcroft, Frederick H
Axen, Udo Friedrich
Becker, Marshall Hilford
Bennett, Carl Leroy
Bens, Frederick Peter
Bohm, Henry Victor
Bredeck, Henry E
Brown, William M(ilton)
Cantlon, John Edward
Caplan, John D(avid)
Conway, Lynn Ann
Cooper, Theodore
Cossack, Zafrallah Taha
Cotter, Richard
Decker, R(aymond) F(rank)
Dijkers, Marcellinus P
Elslager, Edward Faith
Eltinge, Lamont
Erbisch, Frederic H
Fawcett, Timothy Goss
Fonken, Gunther Siegfried
Forist, Arlington Ardeane
Galan, Louis
Graham, John
Grochoski, Gregory T
Gutzman, Philip Charles

Hagen, John William
Harris, Gale Ion
Harris, Kenneth
Harwood, Julius J
Hayes, Edward J(ames)
Herk, Leonard Frank
Herzig, David Jacob
Heyd, William Ernst
Hinman, Jack Wiley
Hoerger, Fred Donald
Howard, Charles Frank, Jr
Huebner, George J
Humphrey, Harold Edward Burton, Jr
Hunstad, Norman A(llen)
Isleib, Donald Richard
Janik, Borek
Jones, Frank Norton
Joshi, Mukund Shankar
Le Beau, Stephen Edward
Lustgarten, Ronald Krisses
Maasberg, Albert Thomas
McCann, Daisy S
Macriss, Robert A
Maxey, Brian William
Mohrland, J Scott
Moll, Russell Addison
Petrick, Ernest N(icholas)
Phillips, Charles W(illiam)
Punch, Jerry L
Rao, Suryanarayana K
Rausch, Doyle W
Reynolds, Herbert McGaughey
Reynolds, John Z
Saul, William Edward
Senagore, Anthony J
Shea, Fredericka Palmer
Shore, Joseph D
Sokol, Robert James
Solomon, David Eugene
Soper, Jon Allen
Spilman, Charles Hadley
Spreitzer, William Matthew
Stevenson, Robin
Stucki, Jacob Calvin
Thurber, William Samuels
Tuesday, Charles Sheffield
Van Rheenen, Verlan H
Walters, Stephen Milo
Weisbach, Jerry Arnold
Weissman, Eugene Y(ehuda)
Young, David Caldwell

MINNESOTA
Adams, Robert McLean
Atchison, Thomas Calvin, Jr
Beebe, George Warren
Belmont, Arthur David
Boelter, Don Howard
Case, Marvin Theodore
Chevalier, Peter Andrew
Davidson, Donald
Dickie, John Peter
Evans, Roger Lynwood
Holman, Ralph Theodore
Holmen, Reynold Emanuel
Janes, Donald Lucian
Jensen, Timothy B(erg)
Jones, Lester Tyler
Kohlstedt, Sally Gregory
Lawrenz, Frances Patricia
Leung, Benjamin Shuet-Kin
McDuff, Charles Robert
Norberg, Arthur Lawrence
Overmier, J Bruce
Paulson, David J
Pitcher, Eric John
Rambosek, G(eorge) M(orris)
Ray, Charles Dean
Reever, Richard Eugene
Schultz, Clifford W
Tautvydas, Kestutis Jonas
Thompson, Phillip Gerhard
Thompson, Roy Lloyd
Uban, Stephen A
Wolfe, Barbara Blair

MISSISSIPPI
Ballard, James Alan
Berry, Charles Dennis
Collins, Johnnie B
ElSohly, Mahmoud Ahmed
Ferer, Kenneth Michael
Foil, Robert Rodney
Helms, Thomas Joseph
Jones, Eric Wynn
Lloyd, Edwin Phillips
Morgan, Eddie Mack, Jr
Paulk, John Irvine
Ranney, Carleton David
Smith, Allie Maitland
Steinberg, Martin H
Thompson, Joe Floyd

MISSOURI
Alexander, Charles William
Baumann, Thiema Wolf
Brodsky, Philip Hyman
Burton, Robert Main
Caldwell, E Gerald
Cook, Mary Rozella
Forero, Enrique
Goldberg, Herbert Sam
Hakkinen, Raimo Jaakko
Hardwick, William Aubrey, Jr

Hedlund, James L
Hill, Jack Filson
Jester, Guy Earlscort
Keller, Robert Ellis
Kramer, Richard Melvyn
Lannert, Kent Philip
Leifield, Robert Francis
Lubin, Bernard
Lyle, Leon Richards
Moore, James Frederick, Jr
Munger, Paul R
Neubert, Ralph Lewis
Peters, Elroy John
Pfander, William Harvey
Richards, Graydon Edward
Robertson, David G C
Sauer, Harry John, Jr
Schott, Jeffrey Howard
Sharp, Dexter Brian
Sidoti, Daniel Robert
Teng, James
Thomas, Lewis Jones, Jr
Thompson, Clifton C
Thornton, Linda Wierman
Toom, Paul Marvin
Weiss, James Moses Aaron
Wochok, Zachary Stephen
Yanders, Armon Frederick

MONTANA
Jackson, Grant D
McRae, Robert James

NEBRASKA
Ellwein, Leon Burnell
Gardner, Paul Jay
Provencher, Gerald Martin
Vanderholm, Dale Henry

NEVADA
Brandstetter, Albin
Holdbrook, Ronald Gail
Klainer, Stanley M
Leipper, Dale F
Merdinger, Charles J(ohn)
Nauman, Charles Hartley
Poziomek, Edward John
Simmonds, Richard Carroll
Smith, Aaron
Snaper, Alvin Allyn

NEW HAMPSHIRE
Cotter, Robert James
Egan, John Frederick
Petrasek, Emil John
Wolff, Nikolaus Emanuel

NEW JERSEY
Addor, Roger Williams
Auerbach, Victor
Bahr, James Theodore
Barker, Richard Gordon
Baron, Hazen Jay
Barrett, Walter Edward
Bauer, Frederick William
Beaton, Albert E
Bendure, Raymond Lee
Besso, Michael M
Bullock, Francis Jeremiah
Burley, David Richard
Busk, Grant Curtis, Jr
Cardinale, George Joseph
Carney, Richard William James
Caroselli, Robert Anthony
Chang, Joseph Yoon
Ciccone, Patrick Edwin
Conway, Paul Gary
Cutler, Frank Allen, Jr
Daggs, Ray Gilbert
Davis, Lance A(lan)
Derucher, Kenneth Noel
De Sesa, Michael Anthony
DiSalvo, Walter A
Dugan, Gary Edwin
Eash, John T(rimble)
Eastman, David Willard
Eastwood, Abraham Bagot
Edelstein, Harold
English, Alan Taylor
Evers, William L
Fahey, John Leonard
Farrington, Thomas Allan
Feigenbaum, Abraham Samuel
Field, Norman J
Finch, Rogers B(urton)
Fishbein, William
Flam, Eric
Fost, Dennis L
Fowles, Patrick Ernest
Frankoski, Stanley P
Froyd, James Donald
Gabriel, Robert
Gavini, Muralidhara B
Giordmaine, Joseph Anthony
Goldenberg, David Milton
Good, Mary Lowe
Good, Ralph Edward
Gordziel, Steven A
Grauman, Joseph Uri
Gregory, Richard Alan, Jr
Grenda, Victor J
Harding, Maurice James Charles
Hartop, William Lionel, Jr
Hayworth, Curtis B

Research Administration (cont)

Henry, Sydney Mark
Herold, E(dward) W(illiam)
Heveran, John Edward
Hillier, James
Hirschberg, Erich
Hirt, Andrew Michael
Hlavacek, Robert John
Hockel, Gregory Martin
Hwang, Cherng-Jia
Iezzi, Robert Aldo
Iglesia, Enrique
Jaffe, Jonah
Johnston, John
Kachikian, Rouben
Kaufman, Arnold
Kaufman, Harold Alexander
Kaufmann, Thomas G(erald)
King, John A(lbert)
Kohn, Jack Arnold
Kohout, Frederick Charles, III
Kronenthal, Richard Leonard
Landau, Edward Frederick
Leaning, William Henry Dickens
Letizia, Gabriel Joseph
Leventhal, Howard
Levy, Martin J Linden
Lewis, Steven Craig
Lunn, Anthony Crowther
Luryi, Serge
McDermott, Kevin J
Malkin, Martin F
Mandel, Lewis Richard
Mark, David Fu-Chi
Markle, George Michael
Marlin, Robert Lewis
Massa, Tobias
Matthijssen, Charles
Meisel, Seymour Lionel
Meyer, Harry Martin, Jr
Michel, Gerd Wilhelm
Migdalof, Bruce Howard
Mihalisin, John Raymond
Miller, Albert Thomas
Miller, Irwin
Miller, Robert H(enry)
Millman, Sidney
Morck, Roland Anton
Munger, Stanley H(iram)
Neiswender, David Daniel
O'Neill, Patrick Joseph
Opie, William R(obert)
Paterniti, James R, Jr
Patrick, James Edward
Pawson, Beverly Ann
Peets, Edwin Arnold
Pheasant, Richard
Pickard, Porter Louis, Jr
Pignataro, Louis J(ames)
Pinsky, Carl Muni
Robertson, Nat Clifton
Rodda, Bruce Edward
Rohovsky, Michael William
Rolle, F Robert
Salvador, Richard Anthony
San Soucie, Robert Louis
Sawin, Steven P
Saxon, Robert
Schunn, Robert Allen
Slater, Eve Elizabeth
Smagorinsky, Joseph
Smith, Robert Ewing
Smith, Stephen Roger
Smith, Terry Edward
Smith, William Edgar
Solis-Gaffar, Maria Corazon
Squibb, Robert E
Steinberg, Eliot
Stockton, John Richard
Tanner, Daniel
Templeman, Gareth J
Toothill, Richard B
Torrey, Henry Cutler
Trice, William Henry
Trivedi, Nayan B
Tucci, Edmond Raymond
Vaughn, Robert Donald
Vincent, Gerald Glenn
Vukovich, Robert Anthony
Wanat, Stanley Frank
Wang, Chao Chen
Wang, Guang Tsan
Warters, William Dennis
Westoff, Charles F
Wiggins, Jay Ross
Wilbur, Robert Daniel
Wildnauer, Richard Harry
Williams, Myra Nicol
Willson, John Ellis
Winter, David Leon
Woodruff, Hugh Boyd
Woolf, Harry
Zager, Ronald

NEW MEXICO

Apt, Kenneth Ellis
Bahr, Thomas Gordon
Barr, Donald Westwood
Beckner, Everet Hess
Bhada, Rohinton(Ron) K
Birely, John H
Brabson, George Dana, Jr
Cano, Gilbert Lucero

Chambers, William Hyland
Chapin, Charles Edward
Clauser, Milton John
Colp, John Lewis
Conner, Jerry Power
Cropper, Walter V
Danen, Wayne C
Darnall, Dennis W
Davis, L(loyd) Wayne
Del Valle, Francisco Rafael
Dowdy, Edward Joseph
Dreicer, Harry
Gilbert, Alton Lee
Giovanielli, Damon V
Goldstone, Philip David
Hughes, Jay Melvin
Icerman, Larry
Jones, Orval Elmer
Lohrding, Ronald Keith
Macek, Robert James
Martinez, J Ricardo
Mauderly, Joe L
Maurer, Robert Joseph
Molecke, Martin A
Morse, Fred A
Neary, Michael Paul
Overhage, Carl F J
Price, Richard Marcus
Pynn, Roger
Seay, Glenn Emmett
Simmons, Gustavus James
Stokes, Richard Hivling
Taber, Joseph John
Tombes, Averett S
Venable, Douglas
Vook, Frederick Ludwig
Westpfahl, David John
White, Paul C
Wing, Janet E (Sweedyk) Bendt
Yonas, Gerold

NEW YORK

Abetti, Pier Antonio
Akers, Charles Kenton
Allentuch, Arnold
Amidon, Thomas Edward
Antzelevitch, Charles
Axelrod, David
Barrows, John Frederick
Bays, Jackson Darrell
Berg, Daniel
Blackmon, Bobby Glenn
Blane, Howard Thomas
Blank, Robert H
Bloch, Aaron N
Bogard, Terry L
Bolton, Joseph Aloysius
Borg, Donald Cecil
Brodsky, Marc Herbert
Brody, Harold
Butler, Katharine Gorrell
Carr, Edward Gary
Castellion, Alan William
Chanana, Arjun Dev
Cohen, Edwin
Comly, James B
Coppersmith, Frederick Martin
Cote, Wilfred Arthur, Jr
Cowell, James Leo
Dean, Nathan Wesley
DeVita, Vincent T, Jr
Dickerman, Herbert W
Dietz, Russell Noel
Dolak, Terence Martin
Drosdoff, Matthew
Dubnick, Bernard
Dumbaugh, William Henry, Jr
Eschbach, Charles Scott
Farley, Thomas Albert
Feldman, William
Fenrich, Richard Karl
Fiske, Milan Derbyshire
Florman, Monte
Gilmore, Arthur W
Goland, Allen Nathan
Goldsmith, Eli David
Gomory, Ralph E
Gordon, Arnold J
Gordon, Harry William
Gordon, Wayne Alan
Greenberg, Jacob
Greizerstein, Hebe Beatriz
Grobman, Warren David
Gutcho, Sidney J
Haase, Jan Raymond
Hall, William Myron, Jr
Hanna, Samir A
Harding, Homer Robert
Harrison, James Beckman
Hatzakis, Michael
Henderson, Thomas E
Hileman, Robert E
Hodgson, William Gordon
Holtzman, Seymour
Howell, Mary Gertrude
Huggins, Charles Marion
Hutchings, William Frank
Insalata, Nino F
Irwin, Lynne Howard
Jain, Anrudh Kumar
Joshi, Sharad Gopal
Kato, Walter Yoneo
Katz, Maurice Joseph
Kauder, Otto Samuel

Kelly, Thomas Michael
Kinzly, Robert Edward
Klotzbach, Robert J(ames)
Knutson, Lloyd Vernon
Kohrt, Carl Fredrick
Kolaian, Jack H
Koppelman, Lee Edward
Langworthy, Harold Frederick
Latimer, Clinton Narath
Lawson, Kent DeLance
Levine, Stephen Alan
Levy, Mortimer
Lhila, Ramesh Chand
Lommel, J(ames) M(yles)
Lyons, Joseph F
McCamy, Calvin S
McElligott, Peter Edward
McHugh, William Dennis
McIrvine, Edward Charles
Makous, Walter L
Matsushima, Satoshi
Maurer, Gernant E
Midlarsky, Elizabeth
Miller, Melvin J
Moe, Robert Anthony
Monagle, John Joseph, Jr
Morey, Donald Roger
Morris, Robert Gemmill
Morrison, John Agnew
Murphy, Kenneth Robert
Mustacchi, Henry
Nolan, John Thomas, Jr
Paldy, Lester George
Pall, David B
Pasco, Robert William
Pentney, Roberta Pierson
Perez-Alburene, Evelio A
Peterson, Otis G
Pittman, Kenneth Arthur
Porter, Beverly Fearn
Reich, Ismar M(eyer)
Rorer, David Cooke
Rubin, Isaac D
Salzberg, Bernard
Savic, Michael I
Schneider, Robert Fournier
Schroeder, Anita Gayle
Schwartzman, Leon
Siebert, Karl Joseph
Singer, Barry M
Smoyer, Claude B
Spiegel, Allen David
Steffen, Daniel G
Sterman, Samuel
Sternfeld, Leon
Sternlicht, B(eno)
Streett, William Bernard
Strnisa, Fred V
Suits, Chauncey Guy
Sutherland, Robert Melvin
Talley, Charles Peter
Taylor, James Earl
Tessieri, John Edward
Thomann, Gary C
Topp, William Carl
Trowbridge, Richard Stuart
Tuite, Robert Joseph
Turnblom, Ernest Wayne
Vogel, Philip Christian
Von Strandtmann, Maximillian
Walker, Theresa Anne
Weingarten, Norman C
Weinstein, Abbott Samson
Weiss, Martin Joseph
White, Donald Robertson
White, Irvin Linwood
Wolfe, Roger Thomas
Woodall, Jerry M
Woodmansee, Donald Ernest
Wulff, Wolfgang

NORTH CAROLINA

Ahearne, John Francis
Aronson, A L
Bartholow, Lester C
Blake, Thomas Lewis
Bogaty, Herman
Burnett, John Nicholas
Butz, Robert Frederick
Carpenter, Benjamin H(arrison)
Carr, Dodd S(tewart)
Cole, Jerome Foster
Cubberley, Adrian H
Cundiff, Robert Hall
Daniels, William Ward
Daumit, Gene Philip
Elder, Joe Allen
Glenn, Thomas M
Gross, Harry Douglass
Harrison, Stanley L
Haynes, John Lenneis
Heck, Walter Webb
Holcomb, George Ruhle
Horvitz, Daniel Goodman
Iafrate, Gerald Joseph
Johnson, William Hugh
King, Kendall Willard
Klarman, William L
Lewin, Anita Hana
Linton, Richard William
Mink, James Walter
Minnemeyer, Harry Joseph
Morse, Roy E
Mosier, Stephen R

Murphy, Robert T
Nader, John S(haheen)
Odell, Norman Raymond
Pagano, Joseph Stephen
Putnam, Charles E
Reeber, Robert Richard
Reel, Jerry Royce
Rees, John
Roblin, John M
Roth, Roy William
Schmiedeshoff, Frederick William
Schultz, Frederick John
Spilker, Bert
Stier, Howard Livingston
Sud, Ish
Swenberg, James Arthur
Tapp, William Jouette
Tucker, Walter Eugene, Jr
Udry, Joe Richard
Vandemark, Noland Leroy
Wiley, Albert Lee, Jr

NORTH DAKOTA

Anderson, Donald E
Freeman, Thomas Patrick
Kotch, Alex
Mathsen, Don Verden

OHIO

Adamczak, Robert L
Altenau, Alan Giles
Anderson, Ruth Mapes
Ault, George Mervin
Badertscher, Robert F(rederick)
Bahnfleth, Donald R
Baker, Saul Phillip
Bales, Howard E
Bertelson, Robert Calvin
Bidelman, William Pendry
Bisson, Edmond E(mile)
Blaser, Robert U
Bonner, David Calhoun
Broge, Robert Walter
Brooman, Eric William
Brunner, Gordon Francis
Carman, Charles Jerry
Christensen, James Henry
Ciricillo, Samuel F
Coleman, John Franklin
Connolly, Denis Joseph
Courchene, William Leon
Crable, John Vincent
Cramer, Michael Brown
Croxton, Frank Cutshaw
Curtis, Charles R
Daschbach, James McCloskey
Davis, Wilford Lavern
De Vries, Adriaan
Doershuk, Carl Frederick
Duga, Jules Joseph
Durand, Edward Allen
Evans, Thomas Edward
Fabris, Hubert
Faiman, Robert N(eil)
Fawcett, Sherwood Luther
Ferguson, John Allen
Finn, John Martin
Fordyce, James Stuart
Forestieri, Americo F
Frank, Thomas Paul
Geho, Walter Blair
Gerace, Michael Joseph
Germann, Richard P(aul)
Graham, Robert William
Grisaffe, Salvatore J
Hall, Franklin Robert
Heasley, James Henry
Hiles, Maurice
Hilmas, Duane Eugene
Hoekenga, Mark T
Hollis, William Frederick
Holtkamp, Dorsey Emil
Hurley, William Joseph
Huston, Keith Arthur
Hutchinson, Frederick Edward
Jackson, Robert Henry
Jaworowski, Andrzej Edward
Jayne, Theodore D
Kaufman, John Gilbert, Jr
Kawasaki, Edwin Pope
Kelley, Frank Nicholas
Kelly, Kevin Anthony
Keplinger, Orin Clawson
Kinsman, Donald Vincent
Landstrom, D(onald) Karl
Laskowski, Edward L
Lee, Kai-Fong
Leonard, Billie Charles
Leyda, James Perkins
Linkenheimer, Wayne Henry
Lydy, David Lee
McCracken, John David
McCune, Homer Wallace
McMillin, Carl Richard
McVean, Duncan Edward
Madia, William J
Martino, Joseph Paul
Massey, L(ester) G(eorge)
Meek, Violet Imhof
Millard, Ronald Wesley
Miraldi, Floro D
Mukherjee, Mukunda Dev
Myhre, David V
Nestor, Ontario Horia

Nixon, Charles William
Ockerman, Herbert W
Olesen, Douglas Eugene
Olson, Walter T
Redmond, Robert F(rancis)
Richley, E(dward) A(nthony)
Roha, Max Eugene
Salkind, Michael Jay
Schlegel, Donald Louis
Schoch, Daniel Anthony
Scozzie, James Anthony
Sedor, Edward Andrew
Semler, Charles Edward
Seufzer, Paul Richard
Silverman, Jerald
Smith, James Edward, Jr
Snyder, Donald Lee
Stephan, David George
Stevenson, James Francis
Stone, Kathleen Sexton
Sweeney, Thomas L(eonard)
Townley, Charles William
Ungar, Edward William
Versic, Ronald James
Versteegh, Larry Robert
Vlcek, Donald Henry
Weeks, Stephen P
Wiff, Donald Ray
Williams, Josephine Louise
Williams, Robert Mack
Williamson, Frederick Dale
Wodarski, John Stanley
Wolaver, Lynn E(llsworth)
Wolf, Warren Walter
Woodard, Ralph Emerson
Woodward, James Kenneth
Wright, George Joseph

OKLAHOMA
Breedlove, James Robby, Jr
Collins, William Edward
Coon, Julian Barham
Crafton, Paul A(rthur)
Davis, Robert Elliott
Edwards, Lewis Hiram
Faulkner, Lloyd (Clarence)
Gray, Samuel Hutchison
Green, Ray Charles
Hale, John Dewey
Henderson, Frederick Bradley, III
Hopkins, Thomas R (Tim)
Jayaraman, H
Johnson, Ronald Roy
Logan, R(ichard) S(utton)
McDevitt, Daniel Bernard
Middlebrooks, Eddie Joe
Millheim, Keith K
Miranda, Frank Joseph
Powell, Jerrel B
Robinson, Jack Landy
Rotenberg, Don Harris
Schroeder, David J Dean
Short, James N
Warren, Kenneth Wayne
Weakley, Martin LeRoy

OREGON
Arscott, George Henry
Barnett, Gordon Dean
Bullock, Richard Melvin
Cooper, Glenn Adair, Jr
Davis, John R(owland)
Ehrmantraut, Harry Charles
Ethington, Robert Loren
Frakes, Rodney Vance
Henderson, Robert Wesley
Jaeger, Charles Wayne
Krueger, William Clement
Landsberg, Johanna D (Joan)
Martin, Lloyd W
Murphy, Thomas A
Norris, Logan Allen
Sarles, Lynn Redmon
Schroeder, Warren Lee
Seil, Fredrick John
Severson, Herbert H
Smiley, Richard Wayne
Thielges, Bart A
Walstad, John Daniel
Wick, William Quentin
Zaerr, Joe Benjamin

PENNSYLVANIA
Alexander, Stuart David
Andrews, Edwin Joseph
Angeloni, Francis M
Angrist, Stanley W
Anouchi, Abraham Y
Antes, Harry W
Archer, David Horace
Bailey, Denis Mahlon
Bannon, James Andrew
Barpal, Isaac Ruben
Bartuska, Doris G
Beavers, Ellington McHenry
Beedle, Lynn Simpson
Beier, Eugene William
Berger, Harvey J
Berkoff, Charles Edward
Biebuyck, Julien Francois
Bockosh, George R
Bonewitz, Robert Allen
Bosshart, Robert Perry
Bramfitt, Bruce Livingston

Bronzini, Michael Stephen
Browning, Daniel Dwight
Bucher, John Henry
Burns, Allan Fielding
Burns, Denver P
Buskirk, Elsworth Robert
Carley, Harold Edwin
Champagne, Paul Ernest
Chen, Michael S K
Cherry, John Paul
Coleman, Robert E
Colman, Robert W
Corbett, Robert B(arnhart)
Daehnick, Wilfried W
Deckert, Fred W
Deitrick, John E
Derby, James Victor
De Tommaso, Gabriel Louis
Dow, Norris F
Dromgold, Luther D
Feero, William E
Fitzgerald, James Allen
Foley, Thomas Preston, Jr
Forbes, Martin
Foti, Margaret A
Frank, William Benson
Frazier, James Lewis
Fritz, Lawrence William
Fromm, Eli
Frumerman, Robert
Gilstein, Jacob Burrill
Glaser, Robert
Glessner, Alfred Joseph
Gollub, Jerry Paul
Golton, William Charles
Gottscho, Alfred M(orton)
Gould, Anne Bramlee
Graham, Roger Kenneth
Gundersen, Larry Edward
Haines, William Joseph
Hauptschein, Murray
Henderson, Robert E
Herman, Frederick Louis
Hollibaugh, William Calvert
Isett, Robert David
Jacobsen, Terry Dale
Jarrett, Noel
Johnson, Wayne Orrin
Kaufman, William Morris
Keairns, Dale Lee
Keller, Eldon Lewis
Kennedy, Flynt
Kidawa, Anthony Stanley
Klevans, Edward Harris
Kolmen, Samuel Norman
Kopchik, Richard Michael
Korchynsky, M(ichael)
Kornfield, A(lfred) T(heodore)
Kottcamp, Edward H, Jr
Kowalski, Conrad John
Krueger, Charles Robert
Kulakowski, Elliott C
Kunesh, Charles Joseph
Kwan, King Chiu
Leighton, Charles Cutler
Libsch, Joseph F(rancis)
Lidman, William G
Lopez, R C Gerald
Lynch, Thomas John
McCarthy, Raymond Lawrence
McCormick, Robert H(enry)
McEvoy, James Edward
Maclay, William Nevin
Mehrkam, Quentin D
Merrill, John Raymond
Michelson, Eric L
Miller, Barry
Morehouse, Chauncey Anderson
Mottur, George Preston
Nash, David Henry George
Neil, Gary Lawrence
Oder, Robin Roy
Osborn, Elburt Franklin
Owens, Frederick Hammann
Packman, Albert M
Pierson, Ellery Merwin
Podgers, Alexander Robert
Pool, James C T
Poos, George Ireland
Rackoff, Jerome S
Rall, Waldo
Recktenwald, Gerald William
Roman, Paul
Roseman, Arnold S(aul)
Ross, Sidney
Rushton, Brian Mandel
Saggiomo, Andrew Joseph
Santamaria, Vito William
Schipper, Arthur Louis, Jr
Schneider, Bruce E
Schweighardt, Frank Kenneth
Seiner, Jerome Allan
Settles, Gary Stuart
Smith, William Novis, Jr
Snow, Jean Anthony
Soboczenski, Edward John
Solt, Paul E
Steg, L(eo)
Suter, Stuart Ross
Talbot, Timothy Ralph, Jr
Tarver, James H, Jr
Taylor, Robert Morgan
Trippodo, Nick Charles
Van Raalte, John A

Voltz, Sterling Ernest
Wagner, J Robert
Walker, Eric A(rthur)
Warren, Alan
Wasylyk, John Stanley
Watson, William Martin, Jr
Weedman, Daniel Wilson
Weinryb, Ira
Whyte, Thaddeus E, Jr
Wilkins, Raymond Leslie
Wissow, Lennard Jay
Wohleber, David Alan
Wolff, Ivan A
Yohe, Thomas Lester

RHODE ISLAND
Gilman, John Richard, Jr
Hedlund, Ronald David
Jarrett, Jeffrey E
Josephson, Edward Samuel
Mayer, Garry Franklin
Pickart, Stanley Joseph
Sage, Nathaniel McLean, Jr
Shipp, William Stanley
Spero, Caesar A(nthony), Jr

SOUTH CAROLINA
Bauer, Richard M
Bennet, Archie Wayne
Brown, Arnold
Dickerson, Ottie J
Doherty, William Humphrey
Failla, Patricia McClement
Felling, William E(dward)
Franklin, Ralph E
Hopper, Michael James
Irwin, Lafayette K(ey)
Kittrell, Benjamin Upchurch
Krumrei, W(illiam) C(larence)
Miley, John Wulbern
Morris, J(ames) William
Nevitt, Michael Vogt
Smith, Theodore Isaac Jogues

SOUTH DAKOTA
Hamilton, Steven J
Smith, Paul Letton, Jr
Swiden, LaDell Ray
Sword, Christopher Patrick

TENNESSEE
Auerbach, Stanley Irving
Ayers, Jerry Bart
Benton, Charles Herbert
Bryan, Robert H(owell)
Callihan, Dixon
Coover, Harry Wesley, Jr
Cope, David Franklin
Costanzi, John J
Crompton, Charles Edward
Dabbs, John Wilson Thomas
Danko, Joseph Christopher
Dilworth, Robert Hamilton, III
Elowe, Louis N
Garber, David H
Gardiner, Donald Andrew
Gossett, Dorsey McPeake
Gosslee, David Gilbert
Hanley, Wayne Stewart
Harms, William Otto
Haywood, Frederick F
Horen, Daniel J
House, Robert W(illiam)
Hunt, James Calvin
Huray, Paul Gordon
Inman, Franklin Pope, Jr
Isley, James Don
Jones, Larry Warner
Jurand, Jerry George
Kozub, Raymond Lee
LeQuire, Virgil Shields
Lim, Alexander Te
Livingston, Robert Simpson
Lowndes, Douglas H, Jr
Lundy, Ted Sadler
Martini, Mario
Partain, Clarence Leon
Poutsma, Marvin Lloyd
Prairie, Michael L
Reed, Peter William
Richmond, Chester Robert
Robertson, David
Row, Thomas Henry
Schreiber, Eric Christian
Sittel, Chester Nachand
Southards, Carroll J
Thonnard, Norbert
Trivelpiece, Alvin William
Veigel, Jon Michael
Wagner, Conrad
Wilkinson, Michael Kennerly
Williamson, Handy, Jr
Witherspoon, John Pinkney, Jr
Wymer, Raymond George

TEXAS
Adamski, Robert J
Alexander, William Carter
Allison, Jean Batchelor
Anderson, George Boine
Anthony, Donald Barrett
Aspelin, Gary B(ertil)
Benedict, George Frederick
Berry, Michael James

Blankenship, Lytle Houston
Burton, Russell Rohan
Cannon, Dickson Y
Cargill, Robert Lee, Jr
Carlile, Robert Elliot
Carlton, Donald Morrill
Cassard, Daniel W
Chevalier, Howard L
Clark, D L
Coelho, Anthony Mendes, Jr
Cooper, Howard Gordon
Coulson, Larry Vernon
Cranberg, Lawrence
Curtin, Richard B
Curtin, Thomas J
Davis, Selby Brinker
Dear, Robert E A
Dietlein, Lawrence Frederick
Dvoretzky, Isaac
Earlougher, Robert Charles, Jr
Fisher, Gene Jordan
Foster, Terry Lynn
Fowler, Robert McSwain
Franklin, Thomas Doyal, Jr
Gatti, Anthony Roger
Gipson, Robert Malone
Goins, William C, Jr
Goodwin, John Thomas, Jr
Gray, John Malcolm
Grogan, Michael John
Gully, John Houston
Gum, Wilson Franklin, Jr
Haase, Donald J(ames)
Hadley, William Owen
Heavner, James E
Heggers, John Paul
Herbert, Stephen Aven
Hilde, Thomas Wayne Clark
Hirasaki, George J
Hoffman, Herbert I(rving)
Hollinger, F(rederick) Blaine
Iampietro, P(atsy) F
Johnson, Elwin L Pete
Jordan, Wayne Robert
Katz, Marvin L(averne)
Kennedy, Ken
Kirk, Ivan Wayne
Kochhar, Rajindar Kumar
Koppa, Rodger J
Krautz, Fred Gerhard
Kutzman, Raymond Stanley
Leaders, Floyd Edwin, Jr
Lehmkuhl, L Don
Lifschitz, Meyer David
Linder, John Scott
Loan, Raymond Wallace
Low, Morton David
McDonald, Lynn Dale
Mack, Mark Philip
Mahendroo, Prem P
Mastromarino, Anthony John
Mefford, Roy B, Jr
Mitchell, James Emmett
Morton, Robert Alex
Musa, Samuel A
Nelson, John Franklin
Nicks, Oran Wesley
Patel, Anil S
Patton, Alton DeWitt
Poddar, Syamal K
Poe, Richard D
Potter, Andrew Elwin, Jr
Purser, Paul Emil
Randolph, Philip L
Reinert, James A
Robinson, Alfred Green
Rollmann, Louis Deane
Russell, B Don
Rutford, Robert Hoxie
Schuh, Frank J
Secrest, Everett Leigh
Sheppard, Louis Clarke
Shotwell, Thomas Knight
Smith, Curtis William
Smith, Larry
Spencer, Alexander Burke
Stewart, Bobby Alton
Teller, Cecil Martin, II
Thompson, Granville Berry
Torp, Bruce Alan
Ulrich, Roger Steffen
Varsel, Charles John
Vitkovits, John A(ndrew)
Walton, Charles Michael
Weldon, William Forrest
Whorton, Elbert Benjamin
Williams, Darryl Marlowe
Wilson, Peggy Mayfield Dunlap
Wimpress, Gordon Duncan, Jr
Wolf, Harold William
Wurth, Thomas Joseph

UTAH
Anderson, Douglas I
Case, James B(oyce)
Cox, Benjamin Vincent
Cramer, Harrison Emery
Germane, Geoffrey James
Healey, Mark Calvin
Joklik, G Frank
Krieger, Carl Henry
Phillips, Lee Revell
Straight, Richard Coleman
Walker, Donald I

Research Administration (cont)

VERMONT

Forcier, Lawrence Kenneth
Gray, John Patrick
Hoagland, Mahlon Bush
Lawson, Robert Bernard
Lipson, Richard L
Spillman, William Bert, Jr

VIRGINIA

Albrecht, Herbert Richard
Ammerman, Charles R(oyden)
Auton, David Lee
Bardon, Marcel
Berg, John Richard
Blurton, Keith F
Borum, Olin H
Brandt, Richard Gustave
Brown, Stephen L(awrence)
Buchanan, Thomas Joseph
Casazza, John Andrew
Cetron, Marvin J
Chambers, Charles MacKay
Chang, George Chunyi
Chappelle, Thomas W
Charvonia, David Alan
Childress, Otis Steele, Jr
Chulick, Eugene Thomas
Corwin, Gilbert
Cross, Ernest James, Jr
Crowley, Patrick Arthur
Cruthers, Larry Randall
Dane, Charles Warren
Davis, Joel L
Deal, George Edgar
DePoy, Phil Eugene
Diness, Arthur M(ichael)
Dolezalek, Hans
Dorsey, Clark L(awler), Jr
Doyle, Frederick Joseph
Dvorchik, Barry Howard
Fabbi, Brent Peter
Fisher, Gail Feimster
Gargus, James L
Gaugler, Robert Walter
Gibbons, John Howard
Gibson, John E(gan)
Gillies, George Thomas
Gourley, Desmond Robert Hugh
Grunder, Hermann August
Gulrich, Leslie William, Jr
Hall, Harvey
Hansen, Robert J
Hart, Dabney Gardner
Hazlett, Robert Neil
Herwig, Lloyd Otto
Hill, Jim T
Holt, Alan Craig
Horsburgh, Robert Laurie
Hortick, Harvey J
Houghton, Kenneth Sinclair, Sr
Howard, William Eager, III
Huggett, Clayton (McKenna)
Ijaz, Lubna Razia
Kapron, Felix Paul
Kelley, Ralph Edward
Levy, Richard Allen
Lynch, Maurice Patrick
McAfee, Donald A
McClung, Andrew Colin
McLaughlin, Gerald Wayne
Madan, Rabinder Nath
Malcolm, John Lowrie
Mandelberg, Martin
Marcus, Stanley Raymond
Masterson, Kleber Sanlin, Jr
Maune, David Francis
Mayer, George
Milligan, John H
Mock, John E(dwin)
Morrison, David Lee
Murino, Vincent S
Murray, Jeanne Morris
Ossakow, Sidney Leonard
Padovani, Elaine Reeves
Patterson, Earl E(dgar)
Pulling, Ronald W
Raab, Harry Frederick, Jr
Robertson, Randal McGavock
Robinson, Bruce B
Romney, Carl Fredrick
Rosenthal, Michael David
Rossini, Frederick Anthony
Rozzell, Thomas Clifton
Sarkes, Louis A(nthony)
Schmidt, Louis Vincent
Shapiro, Maurice Mandel
Sink, David Scott
Sloope, Billy Warren
Small, Timothy Michael
Sobieszczanski-Sobieski, Jaroslaw
Sobol, Stanley Paul
Spence, Thomas Wayne
Stein, Bland Allen
Streeter, Robert Glen
Swiger, Louis Andre
Thorp, Benjamin A
Tipper, Ronald Charles
Tucker, Richard Frank
Van Reuth, Edward C
Van Tilborg, André Marcel
Ventre, Francis Thomas
Wallace, Lance Arthur

Wallace, Raymond Howard, Jr
Warfield, J(ohn) N(elson)
Webber, John Clinton
Weedon, Gene Clyde
Whitesides, John Lindsey, Jr
Williams, Patricia Bell
Wood, George Marshall
Wood, Leonard E(ugene)

WASHINGTON

Albaugh, Fred William
Bennett, Clifton Francis
Bourquin, Al Willis J
Brown, Donald Jerould
Cykler, John Freuler
De Goes, Louis
Dietz, Sherl M
Dixon, Kenneth Randall
Evans, Thomas Walter
Fetter, William Allan
Gaines, Edward M(cCulloch)
Gillis, Murlin Fern
Hannay, Norman Bruce
Hawkins, Neil Middleton
Hinman, George Wheeler
Hopkins, Horace H, Jr
Johnson, John Richard
Katz, Yale H
Krier, Carol Alnoth
Leng, Earl Reece
McMurtrey, Lawrence J
Mahlum, Daniel Dennis
Maloney, Thomas M
Monan, Gerald E
Nelson, Randall Bruce
Noel, Jan Christina
Norman, Joe G, Jr
Roos, John Francis
Schiffrin, Milton Julius
Schmid, Loren Clark
Smith, Orville Auverne
Smith, Samuel H
Tukey, Harold Bradford, Jr
Untersteiner, Norbert
Warren, John Lucius
Wolthuis, Roger A
Zuiches, James J

WEST VIRGINIA

Bowdler, Anthony John
Calzonetti, Frank J
Sherman, Paul Dwight, Jr
Smith, Joseph James

WISCONSIN

Bishop, Charles Joseph
Bock, Robert Manley
Davis, Thomas William
Dickas, Albert Binkley
Fleischer, Herbert Oswald
Freas, Alan D('Yarmett)
Friend, Milton
Goodman, Robert Merwin
Jache, Albert William
Knous, Ted R
Koval, Charles Francis
McClenahan, William St Clair
Manning, James Harvey
Moore, Leonard Oro
Moses, Gregory Allen
Oehmke, Richard Wallace
Owens, John Michael
Parnell, Donald Ray
Riegel, Ilse Leers
Schopler, Harry A
Sheppard, Erwin
Stackman, Robert W

WYOMING

Dorrence, Samuel Michael
Gloss, Steven Paul
Marchant, Leland Condo
Miller, Daniel Newton, Jr
Pier, Allan Clark
Smith, James Lee

PUERTO RICO

Bangdiwala, Ishver Surchand
Lluch, Jose Francisco
Sasscer, Donald S(tuart)

ALBERTA

Bird, Gordon Winslow
Bowman, C(lement) W(illis)
Butler, R(oger) M(oore)
Cormack, George Douglas
Krahn, Thomas Richard
McAndrew, David Wayne
McElgunn, James Douglas
Muir, Donald Ridley
Nasser, Tourai
Rennie, Robert John
Sonntag, Bernard H
Weaver, Ralph Sherman

BRITISH COLUMBIA

Copes, Parzival
Cox, Lionel Audley
Dewey, John Marks
Drew, T John
Duncan, Douglas Wallace
Franz, Norman Charles
Godolphin, William
Hatton, John Victor

Healey, Michael Charles
Hesser, James Edward
Struble, Dean L
Trussell, Paul Chandos
Wynne-Edwards, Hugh Robert

MANITOBA

Adaskin, Eleanor Jean
Angel, Aubie
Atkinson, Thomas Grisedale
Ayles, George Burton
Rosinger, Herbert Eugene
Singh, Ajit
Swierstra, Ernest Emke

NEW BRUNSWICK

Frantsi, Christopher
Kohler, Carl
Stuart, Ronald S

NEWFOUNDLAND

Muggeridge, Derek Brian

NOVA SCOTIA

Bowen, William Donald
Collins, William Beck
Hooper, Donald Lloyd
Jamieson, William David
Wilson, George Peter

ONTARIO

Andrew, Bryan Haydn
Anstey, Thomas Herbert
Asculai, Samuel Simon
Babcock, Elkanah Andrew
Barica, Jan M
Baronet, Clifford Nelson
Bennett, W Donald
Biro, George P
Blevis, Bertram Charles
Burger, Dionys
Cartier, Jean Jacques
Costain, Cecil Clifford
Dorrel, D Gordon
Dyne, Peter John
Elfving, Donald Carl
Evans, John R
Ghista, Dhanjoo Noshir
Gillham, Robert Winston
Glaser, Frederick Bernard
Gold, Lorne W
Gowe, Robb Shelton
Halstead, Ronald Lawrence
Hansson, Carolyn M
Heacock, Ronald A
Hepburn, John Duncan
Hickman, John Roy
Hopton, Frederick James
Ingratta, Frank Jerry
Kavanagh, Robert John
Keys, John David
Kundur, Prabha Shankar
Landolt, Jack Peter
Lawford, George Ross
Lindsey, George Roy
McAdie, Henry George
McGeer, James Peter
Macqueen, Roger Webb
Menzie, Elmer Lyle
Moo-Young, Murray
Morand, Peter
Morton, Donald Charles
Mustard, James Fraser
Mutton, Donald Barrett
Olson, Arthur Olaf
Pigden, Wallace James
Prasad, S E
Pullan, George Thomas
Reid, Lloyd Duff
Renfrew, Robert Morrison
Rubin, Leon Julius
Sherman, Norman K
Sobell, Mark Barry
Stasko, Aivars B
Tugwell, Peter
Tyson, John Edward Alfred
Vollenweider, Richard A
Wade, Robert Simson
Walton, Alan

QUEBEC

Bailar, John Christian, III
Bordeleau, Lucien Mario
Buckland, Roger Basil
Chang, Thomas Ming Swi
Cloutier, Gilles Georges
Contandriopoulos, Andre-Pierre
Croctogino, Reinhold Hermann
Davis, Martin Arnold
Fleming, Bruce Ingram
Giroux, Guy
Giroux, Yves M(arie)
Goldstein, Sandu M
Guyda, Harvey John
Hardy, Yvan J
Lavallee, Andre
McKyes, Edward
Maclachlan, Gordon Alistair
Marisi, Dan(iel) Quirinus
Maruvada, P Sarma
Mavriplis, F
Nash, Peter Howard
Phan, Cong Luan

Poirier, Louis
Rahn, Armin

SASKATCHEWAN

Kybett, Brian David
Strathdee, Graeme Gilroy

OTHER COUNTRIES

Baggot, J Desmond
Balk, Pieter
Baratoff, Alexis
Burkhardt, Walter H
Conover, Lloyd Hillyard
Daugherty, David M
Fitzgerald, John Desmond
Gelzer, Justus
Gowans, James L
Gregory, Peter
Hase, Donald Henry
Heiniger, Hans-Jorg
Humphrey, Albert S
Ince, A Nejat
Irgolic, Kurt Johann
Johnsen, Dennis O
Kessler, Alexander
Krishna, J Hari
Kuroyanagi, Noriyoshi
Lammers, Wim
Metzger, Gershon
Modabber, Farrokh Z
Onoe, Morio
Packer, Leo S
Ray, Prasanta K
Schallenberg, Elmer Edward
Stebbing, Nowell
Struzak, Ryszard G
Taylor, Howard Lawrence
Thissen, Wil A
Turner, Fred, Jr
Van Overstraeten, Roger Joseph
Ward, Gerald T(empleton)
Winn, Edward Barriere

Resource Management

ALABAMA

Guthrie, Richard Lafayette
McCarl, Henry Newton
Rosene, Walter, Jr
Warfield, Carol Larson

ALASKA

Thomas, Gary Lee

ARIZONA

Beck, John R
Dworkin, Judith Marcia
Harris, DeVerle Porter
Swalin, Richard Arthur
Wasson, James Walter

ARKANSAS

Mack, Leslie Eugene

CALIFORNIA

Austin, Carl Fulton
Borsting, Jack Raymond
Brooks, William Hamilton
Bush, George Edward
Chang, David Bing Jue
Dawson, Kerry J
Doede, John Henry
Greene, Elias Louis
Hamm, Thomas Edward, Jr
Hammond, Martin L
Hewston, John G
Hygh, Earl Hampton
Jansen, Henricus Cornelis
Kilgore, Bruce Moody
L'Annunziata, Michael Frank
Lee, William Wai-Lim
Little, John Clayton
Mathias, Mildred Esther
Mooz, William Ernst
Olshen, Abraham C
Orcutt, Harold George
Partain, Gerald Lavern
Rickard, James Joseph
Ripley, William Ellis
Sakagawa, Gary Toshio
Smedes, Harry Wynn
Squires, Dale Edward
Stewart, Brent Scott
Sweeney, James Lee
Szekely, Ivan J
Vilkitis, James Richard
Walker, Warren Elliott
Welch, Robin Ivor

COLORADO

Asherin, Duane Arthur
Barber, George Arthur
Fly, Claude Lee
Gentry, Donald William
LaBounty, James Francis, Sr
McKown, Cora F
Martin, Stephen George
Osborn, Ronald George
Parker, H Dennison
Steele, Timothy Doak
White, Gilbert Fowler

CONNECTICUT
Barone, John A
Blackwood, Andrew W
DeDecker, Hendrik Kamiel Johannes
Golden, Gerald Seymour
Gould, Ernest Morton, Jr
Walton, Alan George

DELAWARE
Andersen, Donald Edward
Busche, Robert M(arion)
Collette, John Wilfred
DeDominicis, Alex John
Habibi, Kamran
Kundt, John Fred
Suarez, Thomas H

DISTRICT OF COLUMBIA
Alter, Harvey
Beer, Charles
Blanchard, Bruce
Carey, William Daniel
Green, Rayna Diane
Nickum, John Gerald
Noonan, Norine Elizabeth
Radcliffe, S Victor
Rogers, John Patrick
Selin, Ivan
Smith, Richard S, Jr
Spruill, Nancy Lyon
Waters, Robert Charles

FLORIDA
Barile, Diane Dunmire
Berry, Leonard
Burrill, Robert Meredith
Detrick, Robert Sherman
Fluck, Richard Conard
Jones, Albert Cleveland
Joyce, Edwin A, Jr
Koblick, Ian
Kruczynski, William Leonard
Malina, Marshall Albert
Marion, Wayne Richard
Means, D Bruce
Parker, Howard Ashley, Jr
Prince, Eric D

GEORGIA
Bozeman, John Russell
Ike, Albert Francis
Krochmal, Jerome J(acob)
Schultz, Donald Paul
Ware, Kenneth Dale

HAWAII
Berg, Carl John, Jr
Duckworth, Walter Donald
Flachsbart, Peter George
Helfrich, Philip
Hufschmidt, Maynard Michael

IDAHO
MacFarland, Craig George
Ulliman, Joseph James

ILLINOIS
Casella, Alexander Joseph
Girard, G Tanner
Herendeen, Robert Albert
Herricks, Edwin E
Lloyd, Monte
Rabb, George Bernard
Sasman, Robert T
Southern, William Edward
Winsberg, Gwynne Roeseler
Wittman, James Smythe, III

INDIANA
Davis, Edgar Glenn
DeMillo, Richard A
Jantz, O K
McGowan, Michael James
Mercer, Walter Ronald

IOWA
Jones, Rex H
Karlen, Douglas Lawrence
Lane, Orris John, Jr

KANSAS
Choate, Jerry Ronald
Coates, Gary Joseph
Grant, Stanley Cameron

KENTUCKY
Kohnhorst, Earl Eugene

LOUISIANA
Bahr, Leonard M, Jr
Cordes, Carroll Lloyd

MAINE
Mobraaten, Larry Edward
Sherburne, James Auril

MARYLAND
Andres, Scott Fitzgerald
Bawden, Monte Paul
Berkson, Harold
Costanza, Robert
Goldenberg, Martin Irwin
Griswold, Bernard Lee
Hazzard, DeWitt George

Hodgdon, Harry Edward
Holland, Marjorie Miriam
Johnson, Frederick Carroll
Jones, LeeRoy G(eorge)
Karadbil, Leon Nathan
Kreysa, Frank Joseph
McCawley, Frank X(avier)
Morgan, John D(avis)
Mountford, Kent
Pattee, Oliver Henry
Ramsay, William Charles
Schmidt, Jack Russell
Shalowitz, Erwin Emmanuel
Shands, Henry Lee
Shaw, Robert William, Jr
Sholdt, Lester Lance
Soto, Gerardo H
Talbott, Edwin M
Wachs, Melvin Walter
Wortman, Roger Matthew
Yakowitz, Harvey

MASSACHUSETTS
Bond, Robert Sumner
Bowley, Donovan Robin
Broadus, James Matthew
Clark, William Cummin
Godfrey, Paul Jeffrey
Hovorka, John
Hulbert, Thomas Eugene
Lapkin, Milton
Leahy, Richard Gordon
Meal, Harlan C
Morris, Robert
Petrovic, Louis John
Smith, Tim Denis
White, Alan Whitcomb

MICHIGAN
Armstrong, John Morrison
Bulkley, Jonathan William
Chappelle, Daniel Eugene
Edgerton, Robert Howard
Haidler, William B(ernard)
Haynes, Dean L
Meteer, James William
Northup, Melvin Lee
Patterson, Richard L
Weisbach, Jerry Arnold

MINNESOTA
Adams, Robert McLean
Broderius, Steven James
Foose, Thomas John
Gregersen, Hans Miller
Grew, Priscilla Croswell Perkins
Johnson, Howard Arthur, Sr
Kuehn, Jerome H
Lange, Robert Echlin, Jr
Robertsen, John Alan
Silberman, Edward

MISSOURI
Axthelm, Deon D
Craddock, John Harvey
Johannsen, Frederick Richard

MONTANA
Deibert, Max Curtis
McClelland, Bernard Riley

NEVADA
Davis, Phillip Burton

NEW HAMPSHIRE
Dingman, Stanley Lawrence

NEW JERSEY
Ben-Israel, Adi
Curcio, Lawrence Nicholas
Graham, Margaret Helen
Hort, Eugene Victor
Jain, Sushil C
Kesselman, Warren Arthur
Liebig, William John
Maxim, Leslie Daniel
Migdalof, Bruce Howard
Miller, Robert H(enry)
Semenuk, Nick Sarden
Tolman, Richard Lee
Widmer, Kemble
Zipf, Elizabeth M(argaret)

NEW MEXICO
Huey, William S
Hughes, Jay Melvin
Jett, James Hubert
Seay, Glenn Emmett
Sitney, Lawrence Raymond
Thode, E(dward) F(rederick)
Wilson, Lee

NEW YORK
Brooks, Robert R
Cole, Roger M
Cusano, Dominic A
Dykstra, Thomas Karl
Fey, Curt F
Forquer, Sandra Lynne
Ginzberg, Eli
Hennigan, Robert Dwyer
Katz, Maurice Joseph
Keck, Donald Bruce
Kelisky, Richard Paul

Kiviat, Erik
Klotzbach, Robert J(ames)
Lommel, J(ames) M(yles)
Lougeay, Ray Leonard
Maynard, Charles Alvin
Rivkin, Maxcy
Rosen, Stephen
Speidel, David H
Streett, William Bernard
White, Irvin Linwood
Wilkins, Bruce Tabor

NORTH CAROLINA
Ahearne, John Francis
Andrews, Richard Nigel Lyon
Godschalk, David Robinson
Roblin, John M
Rulifson, Roger Allen
Smith, William Adams, Jr
Stephenson, Richard Allen
Sullivan, Arthur Lyon

OHIO
Adamczak, Robert L
Bisson, Edmond E(mile)
Blaser, Robert U
Chen, Kuei-Lin
Doershuk, Carl Frederick
Earhart, Richard Wilmot
Hauser, Edward J P
Hickman, Howard Minor
Jackson, Robert Henry
Kochanowski, Barbara Ann
Lewis, James Edward
Mathews, A L
Ray, John Robert
Roth, Robert Earl
Snyder, Donald Lee
Spanier, Edward J
Stambaugh, Edgel Pryce
Vassell, Gregory S
Vertrees, Robert Layman
Weeks, Stephen P

OKLAHOMA
Green, Ray Charles
Still, Edwin Tanner
Sturgeon, Edward Earl
Talent, Larry Gene
Wickham, M Gary

OREGON
Boyd, Dean Weldon
Demaree, Thomas L
Dews, Edmund
Pease, James Robert
Ripple, William John
Van Vliet, Antone Cornelis

PENNSYLVANIA
Franco, Nicholas Benjamin
Frick, Neil Huntington
Grindel, Joseph Michael
L'Esperance, Robert Louis
Lipsky, Stephen E
Mann, Stanley Joseph
Mezger, Fritz Walter William
Phelps, Lee Barry
Ramani, Raja Venkat
Sharer, Cyrus J
Waxman, Herbert Sumner
Wuest, Paul J

RHODE ISLAND
Burroughs, Richard

SOUTH CAROLINA
Felling, William E(dward)
Smith, Theodore Isaac Jogues
Wilson, John Neville

SOUTH DAKOTA
Draeger, William Charles

TENNESSEE
Fordham, James Lynn
Jordan, Andrew G
LeBlanc, Larry Joseph
McNutt, Charles Harrison
Nealy, David Lewis
Wagner, Aubrey Joseph
Wilson, James Larry

TEXAS
Arnold, J Barto, III
Crouse, Philip Charles
Darnell, Rezneat Milton
Eckelmann, Walter R
Howe, Richard Samuel
Jackson, Eugene Bernard
Lacewell, Ronald Dale
Lehmann, Elroy Paul
Murnane, Thomas George
Stanford, Geoffrey

UTAH
Downing, Kenton Benson
Krejci, Robert Henry

VERMONT
Smith, Thomas David

VIRGINIA
Awad, Elias M

Bayliss, John Temple
Berry, James G(ilbert)
Berryman, Jack Holmes
Casazza, John Andrew
Cetron, Marvin J
Cox, William Edward
Davis, William Spencer
Dillaway, Robert Beacham
Dobyns, Samuel Witten
Dragonetti, John Joseph
Gilbert, Arthur Charles
Hall, Otis F
Hamilton, Thomas Charles
Harowitz, Charles Lichtenberg
Hinckley, Alden Dexter
Hobeika, Antoine George
Jahn, Laurence R
Kiessling, Oscar Edward
Knap, James E(li)
Lynch, Maurice Patrick
McEwen, Robert B
Rhode, Alfred S
Ries, Richard Ralph
Sink, David Scott
Stone, Harris B(obby)
Streeter, Robert Glen

WASHINGTON
Brewer, William Augustus
Fluharty, David Lincoln
Gaines, Edward M(cCulloch)
Gilbert, Frederick Franklin
Hinman, George Wheeler
Miles, Edward Lancelot
Nash, Colin Edward
Roos, John Francis
Royce, William Francis
Strandjord, Paul Edphil

WEST VIRGINIA
Calzonetti, Frank J

WISCONSIN
Beatty, Marvin Theodore
Cain, John Manford
Sheppard, Erwin

WYOMING
Miller, Daniel Newton, Jr
Wheasler, Robert

PUERTO RICO
Wilson, Marcia Hammerquist

ALBERTA
Freeman, Milton Malcolm Roland
Lane, Robert Kenneth
Macpherson, Andrew Hall

BRITISH COLUMBIA
Copes, Parzival
Griffiths, George Motley
Healey, Michael Charles
Newroth, Peter Russell
Pomeroy, Richard James
Ross, William Michael
Weintraub, Marvin
Yorque, Ralf Richard

NEW BRUNSWICK
Kohler, Carl

NEWFOUNDLAND
Bajzak, Denes
Ni, I-Hsun

NOVA SCOTIA
Bowen, William Donald

ONTARIO
Banfield, Alexander William Francis
Burk, Cornelius Franklin, Jr
Dyment, John Cameron
Hoffman, Douglas Weir
Lavigne, David M
Marks, Charles Frank
Olson, Arthur Olaf
Siddall, Ernest
Stebelsky, Ihor
Taylor, Roger
Watson, Jeffrey

QUEBEC
Bourchier, Robert James
Hardy, Yvan J

OTHER COUNTRIES
Dahl, Arthur Lyon
Hase, Donald Henry
Khatib, Hisham M
Krishna, J Hari

Science Administration

ALABAMA
Doorenbos, Norman John
Jolley, Homer Richard
Naumann, Robert Jordan
Six, Norman Frank, Jr

ALASKA
Leon, Kenneth Allen
Proenza, Luis Mariano

Rodgers, Charles H
Roscoe, Henry George
Ross, Philip
Sansone, Eric Brandfon
Sass, Neil Leslie
Schweizer, Malvina
Shafer, W Sue
Shaw, Warren Cleaton
Shore, Moris Lawrence
Smith, Douglas Lee
Spooner, Peter Michael
Sprott, Richard Lawrence
Steinwachs, Donald Michael
Stevenson, Robert Edwin
Stroud, Robert Church
Tepper, Morris
Thompson, Warren Elwin
Tilford, Shelby G
Tyeryar, Franklin Joseph
Valega, Thomas Michael
Vermund, Sten Halvor
Vocci, Frank Joseph
Wagner, William John
Watzman, Nathan
Weissberg, Alfred
White, Harris Herman
White, Howard Julian, Jr
Williams, Thomas Franklin
Wingate, Catharine L
Wolff, David A
Wortman, Bernard
Yellin, Herbert
Yerby, Alonzo Smythe
Zimmerman, Eugene Munro

MASSACHUSETTS
Aldrich, Michele L
Allen, Thomas John
Askew, Eldon Wayne
Assaykeen, Tatiana Anna
Barlas, Julie S
Barnes, Arnold Appleton, Jr
Blaustein, Ernest Herman
Blendon, Robert Jay
Busby, William Fisher, Jr
Carlson, Herbert Christian, Jr
Ciappenelli, Donald John
Cynkin, Morris Abraham
Davidson, Gilbert
Ebert, Robert H
Garing, John Seymour
Gold, Albert
Hilferty, Frank Joseph
Hilgar, Arthur Gilbert
Kelley, William S
Kosersky, Donald Saadia
LeBaron, Francis Newton
MacDougall, Edward Bruce
Moniz, Ernest Jeffrey
Morse, Robert Warren
Poduska, John W, Sr
Reis, Arthur Henry, Jr
Riemer-Rubenstein, Delilah
Rosner, Anthony Leopold
Roth, Harold

MICHIGAN
Braughler, John Mark
Costley, Gary E
Eliezer, Isaac
Haff, Richard Francis
Hagen, John William
Heustis, Albert Edward
Mallinson, George Greisen
Nagler, Robert Carlton
Solomon, Allen M
Stowell, Ewell Addison
Strickland, James Shive
Volz, Paul Albert
Weiss, Mark Lawrence
West, Robert MacLellan

MINNESOTA
Haaland, John Edward
Miller, Willard, Jr
Mulhausen, Robert Oscar
Norris, William C
Swanson, Anne Barrett
Wertheimer, Albert I

MISSISSIPPI
Doblin, Stephen Alan
McClurkin, Iola Taylor
Regier, Lloyd Wesley
Sharpe, Thomas R

MISSOURI
Alexander, Charles William
Crosby, Marshall Robert
Hunn, Joseph Bruce
Lipkin, David
Rash, Jay Justen
Rosebery, Dean Arlo
Santiago, Julio Victor
Vaughan, William Mace
Whitten, Elmer Hammond

MONTANA
Fritts, Steven Hugh
Hosley, Robert James

NEBRASKA
Eisen, James David
McClurg, James Edward

NEW HAMPSHIRE
Meader, Ralph Gibson
Wixson, Eldwin A, Jr

NEW JERSEY
Chizinsky, Walter
Gale, George Osborne
Gund, Peter Herman Lourie
Hirsch, Robert L
Jaffe, Jonah
Kalm, Max John
Perhach, James Lawrence
Pinsky, Carl Muni
Shapiro, Stanley Seymour
Sutman, Frank X
Weiss, Marvin
White, Addison Hughson
Woolf, Harry

NEW MEXICO
Apt, Kenneth Ellis
Barbo, Dorothy M
Birely, John H
Cornell, Samuel Douglas
Downs, William Fredrick
Garcia, Carlos E(rnesto)
Howard, William Jack
Hughes, Jay Melvin
Pappas, Daniel Samuel
Shannon, Spencer Sweet, Jr
Simmons, Leonard Micajah, Jr
Sivinski, Jacek Stefan
Todsen, Thomas Kamp
Venable, Douglas
Vook, Frederick Ludwig
Whetten, John T

NEW YORK
Allen, William F, Jr
Asch, Harold Lawrence
Berg, Daniel
Blaker, J Warren
Blatt, Sylvia
Bugliarello, George
Carr, Edward Gary
Dolak, Terence Martin
Fickies, Robert H
Fox, Gerard F
Frechet, Jean M J
Glass, David Carter
Goland, Allen Nathan
Goldberger, Robert Frank
Goodman, Jerome
Gwilt, John Ruff
Herczynski, Andrzej
Holloman, John L S, Jr
Lichter, Robert (Louis)
Loev, Bernard
Lovejoy, Derek R
Lowder, Wayne Morris
Lubic, Ruth Watson
Lulla, Jack D
Lustig, Harry
McTernan, Edmund J
Manassah, Jamal Tewfek
Moloney, Thomas W
Monagle, John Joseph, Jr
Montefusco, Cheryl Marie
Naughton, John
Rau, R Ronald
Reifler, Clifford Bruce
Rosen, Harry Mark
Rothstein, Robert
Schearer, Sherwood Bruce
Schwartz, Brian B
Sharkey, John Bernard
Sheldon, Eleanor Bernert
Sternfeld, Leon
Szumski, Stephen Aloysius
Trasher, Donald Watson
Walker, Theresa Anne
Washton, Nathan Seymour
Yates, Jerome William

NORTH CAROLINA
Cochran, George Thomas
Collins, Jeffrey Jay
Dibner, Mark Douglas
Etzel, Howard Wesley
Fairchild, Homer Eaton
Flagg, Raymond Osbourn
Hadeen, Kenneth Doyle
Jameson, Charles William
Kelsey, John Edward
Lefkowitz, Issai
McPherson, Charles William
Maroni, Donna F
Morrissey, Richard Edward
Radovsky, Frank Jay
Rogosa, George Leon
Sharkoff, Eugene Gibb
Walters, Douglas Bruce

NORTH DAKOTA
Kotch, Alex

OHIO
Adamczak, Robert L
Beyer, William Hyman
Campbell, Colin
Dobbelstein, Thomas Norman
Kelley, Frank Nicholas
Sinner, Donald H
Wenzel, Richard Louis

White, Willis S, Jr

OKLAHOMA
Beasley, William Harold
Kimpel, James Froome
Lambird, Perry Albert
Neely, Stanley Carrell
Still, Edwin Tanner

OREGON
Beaulieu, John David
McQuate, Robert Samuel

PENNSYLVANIA
Arrington, Wendell S
Bartuska, Doris G
Beedle, Lynn Simpson
Braunstein, David Michael
Brendlinger, Darwin
Cross, Leslie Eric
Hess, Eugene Lyle
Hoberman, Alfred Elliott
Kerstein, Morris D
Kulakowski, Elliott C
Moss, Melvin Lane
Perchonock, Carl David
Rothman, Milton A
Sherman, John Walter
Skutches, Charles L
Smith, Gerald A
Smith, Stewart Edward
Spear, Jo-Walter
Sun, James Dean
Thomas, William J
Tobin, Thomas Vincent
Turoczi, Lester J
Tuthill, Harlan Lloyd
Waxman, Herbert Sumner
Weisfeld, Lewis Bernard
Wiggill, John Bentley
Wilson, Marjorie Price
Wissow, Lennard Jay

RHODE ISLAND
Gilman, John Richard, Jr
Kirwan, Donald Frazier

SOUTH CAROLINA
Aull, Luther Bachman, III
Blood, Elizabeth Reid
Jaco, Charles M, Jr
Nerbun, Robert Charles, Jr
Owen, John Harding
Thomas, Claude Earle
Willoughby, William Franklin

TENNESSEE
Campbell, William B(uford)
Horen, Daniel J
House, Robert W(illiam)
Lowe, James Urban, II
Martin, Richard Blazo
Ramsey, Lloyd Hamilton
Row, Thomas Henry
Wachtel, Stephen Shoel
Wu, Ying-Chu Lin (Susan)

TEXAS
Baggett, Neil Vance
Barnes, Charles M
Bogard, Donald Dale
Caillouet, Charles W, Jr
Crawford, Stanley Everett
Farhataziz, Mr
Huebner, Walter F
Merrill, Joseph Melton
Peterson, Lysle Henry
Scholl, Philip Jon
Spallholz, Julian Ernest

UTAH
Cramer, Harrison Emery
McKell, Cyrus Milo
Madsen, James Henry, Jr
Olson, Randall J
Taylor, Philip Craig

VERMONT
Lawson, Robert Bernard
Moyer, Walter Allen, Jr

VIRGINIA
Auerbach, Stephen Michael
Bardon, Marcel
Blood, Benjamin Donald
Byrd, Lloyd G
Cooper, Earl Dana
Corwin, Gilbert
Diebold, Robert Ernest
Dragonetti, John Joseph
Easter, Donald Philips
Fabbi, Brent Peter
Gardenier, John Stark
Grunder, Fred Irwin
Heebner, David Richard
Hill, Carl McClellan
Huang, H(sing) T(sung)
Knappenberger, Paul Henry, Jr
Lemp, John Frederick, Jr
Levine, Jules Ivan
McGee, Henry A(lexander), Jr
Marshall, Harold George
Masters, Charles Day
Maybury, Robert H

Maycock, Paul Dean
Miller, William Lawrence
Mock, John E(dwin)
Morgan, John Walter
Rankin, Douglas Whiting
Rosenbaum, David Mark
Rosenthal, Michael David
Schad, Theodore M(acNeeve)
Sheldon, Donald Russell
Simpson, John Arol
Spindel, William
Suiter, Marilyn J
Tiedemann, Albert William, Jr
Torio, Joyce Clarke
Troxell, Terry Charles
Weiss, Daniel Leigh
Wessel, Paul Roger
Willett, James Delos

WASHINGTON
Anderson, Donald Hervin
Curl, Herbert (Charles), Jr
Harlin, Vivian Krause
Holbrook, Karen Ann
Hutchinson, William Burke
Miles, Edward Lancelot
Rieke, William Oliver
Robertson, William O
Strandjord, Paul Edphil
Tenforde, Thomas SeBastian
Zuiches, James J

WEST VIRGINIA
Larson, Gary Eugene
Victor, Leonard Baker

WISCONSIN
Cadmus, Robert R
Cramer, Jane Harris
Dibben, Martyn James
Eckert, Alfred Carl, Jr
Huber, David Lawrence
Inman, Ross
Lowenstein, Michael Zimmer
Petersen, John Robert
Will, James Arthur

WYOMING
Meyer, Edmond Gerald

PUERTO RICO
Bonnet, Juan A, Jr
McDowell, Dawson Clayborn

ALBERTA
McLeod, Lionel Everett

BRITISH COLUMBIA
Lancaster, George Maurice

MANITOBA
Rosinger, Eva L J
Trick, Gordon Staples

NOVA SCOTIA
Cooke, Robert Clark
Simpson, Frederick James

ONTARIO
Bennett, W Donald
Blackwell, Alan Trevor
Bouffard, Marie Alice
Casselman, Warren Gottlieb Bruce
Effer, W R
Forsdyke, Donald Roy
Greenaway, Keith R(ogers)
Holtslander, William John
LeRoy, Donald James
Lister, Earl Edward
Lucas, Douglas M
Manchee, Eric Best
Michael, Thomas Hugh Glynn
Mustard, James Fraser
Solandt, Omond McKillop
Somers, Emmanuel
Stephenson, Norman Robert
Sturgess, Jennifer Mary
Watson, Michael Douglas

QUEBEC
Bourchier, Robert James
Burgess, John Herbert
Dykes, Robert William
Paulin, Gaston (Ludger)

OTHER COUNTRIES
Assousa, George Elias
Beckler, David (Zander)
Burnell, S Jocelyn Bell
Daugherty, David M
Davis, William Jackson
Johnsen, Dennis O
Kim, Wan H(ee)
Kreider, Eunice S
La Chance, Leo Emery
Minnick, Danny Richard
Vrebalovich, Thomas

Science Communications

ALABAMA
Small, Robert James

ALASKA
Gedney, Larry Daniel

ARIZONA
Giampapa, Mark Steven
Handy, Robert M(axwell)
Hutchinson, Charles S, Jr
Pacholczyk, Andrzej Grzegorz
Paylore, Patricia Paquita
Tillery, Bill W

CALIFORNIA
Arnold, Harry L(oren), Jr
Ash, William Wesley
Barnett, R(alph) Michael
Bengelsdorf, Irving Swem
Benson, Harriet
Berg, Henry Clay
Bibel, Debra Jan
Blakeslee, Dennis L(auren)
Bourne, Charles Percy
Callaham, Robert Zina
Callahan-Compton, Joan Rea
Cuadra, Carlos A(lbert)
Davis, Ward Benjamin
Hardy, Edgar Erwin
Hearn, Walter Russell
McDonough, Thomas Redmond
Meyer, Carl Beat
Meyer, Gregory Carl
Novack, Gary Dean
Poppoff, Ilia George
Pursglove, Laurence Albert
Rasmussen, Robert A
Rogers, Bruce Joseph
Rossbacher, Lisa Ann
Schneider, Meier
Spiller, Gene Alan
Stokley, James
Subramanya, Shiva

COLORADO
Allen, Joe Haskell
Barrett, Charles Sanborn
Craig, Roy Phillip
Davis, Milford Hall
Kreith, Frank
McKown, Cora F
Metzger, H Peter
Naeser, Nancy Dearien
Parker, H Dennison
Schneider, Stephen Henry
Thompson, Starley Lee

CONNECTICUT
Alpert, Nelson Leigh
Aylesworth, Thomas Gibbons
Corning, Mary Elizabeth
Downing, Mary Brigetta
Fraser, J(ulius) T(homas)
Gingold, Kurt
Grayson, Martin
Krauss, Lawrence Maxwell
Levine, Leon
Poincelot, Raymond Paul, Jr
Sullivan, Walter Seager
Weinstein, Curt David

DELAWARE
Bartkus, Edward Peter
Cubberley, Virginia
Frankenburg, Peter Edgar
Hayek, Mason
Knodel, Elinor Livingston
Phillips, Brian Ross
Slade, Arthur Laird

DISTRICT OF COLUMBIA
Allan, Frank Duane
Boots, Sharon G
Bowen, David Hywel Michael
Carey, William Daniel
Carter, Gesina C
Cocks, Gary Thomas
Fainberg, Anthony
Gilbert, Myron B
Green, Rayna Diane
Hart, Charles Willard, Jr
Herschman, Arthur
Hines, Pamela Jean
Kelly, Henry Charles
Lang, Roger H
Newman, Simon M(eier)
Redish, Janice Copen
Shaub, Walter M
Vetter, Betty M
Vietmeyer, Noel Duncan
Welt, Isaac Davidson
Zawisza, Julie Anne A

FLORIDA
Barile, Diane Dunmire
Copeland, Richard Franklin
Ewart, R Bradley
McKently, Alexandra H
Oehser, Paul Henry
Prebluda, Harry Jacob
Robitaille, Henry Arthur
Sanford, Malcolm Thomas

Stang, Louis George

GEORGIA
Fletcher, James Erving
Hawkins, Isaac Kinney
Nethercut, Philip Edwin
Perkowitz, Sidney
Price, Edward Warren
Riggsby, Ernest Duward

IDAHO
Gesell, Thomas Frederick

ILLINOIS
Barr, Susan Hartline
Berg, Dana B
Bernstein, Elaine Katz
Kinnmark, Ingemar Per Erland
Kittel, J Howard
Koehler, Henry Max
Mafee, Mahmood Forootan
Newton, Stephen Bruington
Ren, Shang-Fen
Schaefer, Wilbur Carls
Sinclair, Thomas Frederick
Swanson, Don R
Sweitzer, James Stuart
Throw, Francis Edward
Wilke, Robert Nielsen
Yos, David Albert

INDIANA
Cassidy, Harold Gomes
Conway, Hertsell S
Free, Helen M
Kory, Mitchell
Markee, Katherine Madigan
Nicholson-Guthrie, Catherine Shirley
Tacker, Martha McClelland

IOWA
Hansman, Robert H

KANSAS
Coates, Gary Joseph
Shortridge, Robert William

LOUISIANA
Courtney, John Charles
Davis, George Diament

MAINE
Comins, Neil Francis
Senders, John W
Spencer, Claude Franklin

MARYLAND
Altman, Philip Lawrence
Behar, Marjam Gojchlerner
Bishop, Walton B
Chartrand, Mark Ray, III
Chernoff, Amoz Immanuel
Cosmides, George James
Crampton, Janet Wert
Crosby, Edwin Andrew
Eisler, Ronald
Estrin, Norman Frederick
Gibson, Colvin Lee
Hader, Rodney N(eal)
Hammond, Allen Lee
Heilprin, Laurence Bedford
Heller, William Mohn
Josephs, Melvin Jay
Kissman, Henry Marcel
Lide, David Reynolds, Jr
Lockard, J David
Lufkin, Daniel Harlow
McCarn, Davis Barton
Masys, Daniel R
Morgan, Walter L(eroy)
Moulton, James Frank, Jr
Siegel, Elliot Robert
Spilhaus, Athelstan Frederick, Jr
Wyckoff, James M
Zwanzig, Frances Ryder

MASSACHUSETTS
Chernin, Eli
Chomsky, Noam
Claff, Chester Eliot, Jr
Cromer, Alan H
Erickson, Alan Eric
Goodell, Rae Simpson
Harris, Miles Fitzgerald
Illinger, Joyce Lefever
Karas, John Athan
Lightman, Alan Paige
Lipinski, Boguslaw
Mack, Charles Lawrence, Jr
Parker, Henry Seabury, III
Rickter, Donald Oscar
Shawcross, William Edgerton
Taylor, Edwin Floriman

MICHIGAN
Crowther, C Richard
Kamrin, Michael Arnold
Ruffner, James Alan
Rycheck, Mark Rule

MINNESOTA
Anderson, Frances Jean
Ansari, Azam U
Lea, Wayne Adair

MISSISSIPPI
Ferguson, Mary Hobson

MISSOURI
Carson, Bonnie L Bachert
Crawford, Susan Young
Festa, Roger Reginald
Ucko, David A

NEBRASKA
Benschoter, Reba Ann

NEW JERSEY
An, Linda Huang
Bendich, Adrianne
Colicelli, Elena Jeanmarie
Cromartie, William James, Jr
Daggs, Ray Gilbert
Frantz, Beryl May
Frost, David
Garone, John Edward
Hall, Luther Axtell Richard
Hay, Peter Marsland
Levi, Barbara Goss
Lorenz, Patricia Ann
Matula, Richard Allen
Schindler, Max J
Seifert, Laurence C
Semenuk, Nick Sarden
Thelin, Lowell Charles
Weil, Benjamin Henry
Weinstein, Stephen B
Zipf, Elizabeth M(argaret)

NEW MEXICO
Amacher, Peter
Cropper, Walter V
Forscher, Bernard Kronman
Hill, Mary Rae

NEW YORK
Asimov, Isaac
Boardman, John
Brown, Stanley Gordon
Cummins, John Frances
Dahl, Per Fridtjof
Daniels, John Maynard
Dauben, Joseph W
Day, Stacey Biswas
Feinberg, Robert Jacob
Fessenden-MacDonald, June Marion
Foreman, Bruce Milburn, Jr
Geiger, H Jack
Gerolimatos, Barbara
Goodman, Gerald Joseph
Goodman, Jerome
Hoover, Peter Redfield
Immergut, Edmund H(einz)
Ingalls, James Warren, Jr
Insalata, Nino F
Koenig, Michael Edward Davison
Kornberg, Fred
Leerburger, Benedict Alan
Lichtig, Leo Kenneth
Li-Scholz, Angela
Mackay, Lottie Elizabeth Bohm
Milton, Kirby Mitchell
Murphy, Eugene F(rancis)
Oliver, Gene Leech
Paldy, Lester George
Piel, Gerard
Rosen, Stephen
Rosenfeld, Jack Lee
Rothman, Francoise
Sage, Martin Lee
Schewe, Phillip Frank
Sharpe, William D
Spiegel, Robert
Stoner, George Green
Trasher, Donald Watson
Trigg, George Lockwood
Vroman, Leo
Washton, Nathan Seymour
White, Irvin Linwood

NORTH CAROLINA
Bailey, Kincheon Hubert, Jr
Boyers, Albert Sage
Carr, Dodd S(tewart)
Dixon, Norman Rex
Good, Wilfred Manly
Hartford, Winslow H
Hess, Daniel Nicholas
Smith, Bradley Richard
Stubblefield, Charles Bryan
Zahed, Hyder Ali

OHIO
Bates, Robert Latimer
Centner, Rosemary Louise
Dirckx, John H
Fortner, Rosanne White
Losekamp, Bernard Francis
Martin, Scott McClung
Olson, Walter T
Platau, Gerard Oscar
Raciszewski, Zbigniew
Smith, James Edward, Jr
Turkel, Rickey M(artin)
Zaye, David F

OKLAHOMA
Collins, William Edward
Gorin, George

OREGON
Allen, John Eliot

PENNSYLVANIA
Baechler, Charles Albert
Benfey, Otto Theodor
Casey, Adria Catala
Creasey, William Alfred
Fales, Steven Lewis
Fansler, Bradford S
Foti, Margaret A
Freeman, Joseph Theodore
Garfield, Eugene
Jim, Kam Fook
Judge, Joseph Malachi
Kravitz, Edward
Lacoste, Rene John
Ladman, Aaron J(ulius)
Russey, William Edward
Small, Henry Gilbert
Umen, Michael Jay
Vladutz, George E
Wolfe, Reuben Edward
Wuest, Paul J

RHODE ISLAND
Zuehlke, Richard William

SOUTH CAROLINA
Manley, Donald Gene
Rothrock, George Moore
Wilson, Jerry D(ick)

SOUTH DAKOTA
Hamilton, Steven J
Leslie, Jerome Russell

TENNESSEE
Braunstein, Helen Mentcher
Dickens, Justin Kirk
Grant, Peter Malcolm
Horton, Charles Abell
Keedy, Hugh F(orrest)
Lasslo, Andrew
Partridge, Lloyd Donald

TEXAS
Albach, Roger Fred
Ashby, Jon Kenneth
DeBakey, Lois
DeBakey, Selma
Jauchem, James Robert
Olenick, Richard Peter
Sakakini, Joseph, Jr

UTAH
Case, James B(oyce)
Downing, Kenton Benson

VIRGINIA
Baker, Jeffrey John Wheeler
Briscoe, Melbourne George
Chappelle, Thomas W
Chulick, Eugene Thomas
Feldmann, Edward George
Frye, Keith
Kelley, John Michael
Lynd, Langtry Emmett
Scattergood, Leslie Wayne
Streeter, Robert Glen
Von Baeyer, Hans Christian

WASHINGTON
Scheffer, Victor B

WISCONSIN
Brock, Katherine Middleton
Buchanan-Davidson, Dorothy Jean
Dunwoody, Sharon Lee
Moffet, Hugh L
Riegel, Ilse Leers
Steinhart, Carol Elder

MANITOBA
Grant, Cynthia Ann

ONTARIO
Banfield, Alexander William Francis
Campbell, James Fulton
Clarke, Thomas Roy
Cook, David Greenfield
Hill, Martha Adele
Mackay, Rosemary Joan
Sinha, Bidhu Bhushan Prasad
Smith, Lorraine Catherine
Wiebe, John Peter
Winter, Peter
Wood, Gordon Harvey

QUEBEC
Allen, Harold Don
Bordan, Jack
Buckland, Roger Basil
Châtillon, Guy
DuFresne, Albert Herman

SASKATCHEWAN
Irvine, Donald Grant
Larson, D Wayne

OTHER COUNTRIES
Biederman-Thorson, Marguerite Ann
De Bruyne, Peter
Greene, Lewis Joel

Hudson, Ralph P
Lloyd, Christopher Raymond
Siegel, Herbert

Science Education

ALABAMA
Baker, June Marshall
Baksay, Laszlo Andras
Dean, Susan Thorpe
Doorenbos, Norman John
Falls, William Randolph, Sr
Glover, Elsa Margaret
McCarl, Henry Newton
Mullins, Dail W, Jr
Norman, Billy Ray
Rosenberger, Albert Thomas
Summerlin, Lee R
Thomas, Joseph Calvin
Watson, Jack Ellsworth

ALASKA
Beebee, John Christopher
Fahl, Charles Byron
Gedney, Larry Daniel
Hawkins, Daniel Ballou
Morrison, John Albert
Ragle, Richard Harrison
Sheridan, John Roger
Smoker, William Williams

ARIZONA
Barnes, Charles Winfred
Bitter, Gary G
Cunningham, Richard G(reenlaw)
Gates, Halbert Frederick
Iserson, Kenneth Victor
Justice, Keith Evans
Lawson, Anton Eric
Layton, Richard Gary
Lisonbee, Lorenzo Kenneth
Munch, Theodore
Palmer, James McLean
Roemer, Elizabeth
Smith, David John
Steinbrenner, Arthur H
Swihart, Thomas Lee
Tillery, Bill W
Venables, John Anthony
Wagner, Kenneth
Willoughby, Stephen Schuyler

ARKANSAS
Anderson, Robbin Colyer
Barton, Harvey Eugene
Cockerham, Lorris G(ay)
Dodson, B C
Edson, James Edward, Jr
Eichenberger, Rudolph John
Fribourgh, James H
Leming, Charles William
Mackey, James E
Texter, E Clinton, Jr

CALIFORNIA
Appleman, M Michael
Atkin, J Myron
Atkinson, Russell H
Baez, Albert Vinicio
Baggerly, Leo L
Barnett, R(alph) Michael
Bate, George Lee
Beard, Jean
Bennett, John Francis
Bharadvaj, Bala Krishnan
Billig, Franklin A
Bird, Harold L(eslie), Jr
Bohm, Howard A
Bourne, Charles Percy
Buchsbaum, Ralph
Caldwell, David Orville
Chadwick, Nanette Elizabeth
Chivers, Hugh John
Cichowski, Robert Stanley
Clague, William Donald
Clements, Linda L
Cogan, Adrian Ilie
Cohen, Nathan Wolf
Collins, Robert C
Corwin, Harold G, Jr
Cox, John William
Creutz, Edward (Chester)
Crum, James Davidson
Davis, Ward Benjamin
Dean, Richard Albert
Demond, Joan
Dessel, Norman F
Donnelly, Timothy C
Dunning, John Ray, Jr
Eisberg, Robert Martin
Feher, Elsa
Feign, David
Fischer, Robert Blanchard
Fisher, Kathleen Mary Flynn
Fraknoi, Andrew Gabriel
Franklin, Gene F(arthing)
Fredenburg, Robert Love
Fuller, Milton E
Funk, Glenn Albert
Gardner, Marjorie Hyer
Garratty, George
Glick, Harold Alan
Gold, Marvin B

Gosselin, Edward Alberic
Greenstadt, Melvin
Gupta, Madhu Sudan
Haller, Eugene Ernest
Hand, Judith Latta
Hartsough, Walter Douglas
Hayes, Janan Mary
Helmholz, August Carl
Henkin, Leon (Albert)
Hiemenz, Paul C
Hoffmann, Jon Arnold
Hurd, Paul DeHart
Iltis, Wilfred Gregor
Irvine, Cynthia Emberson
Jarvis, William Tyler
Jendresen, Malcolm Dan
Johanson, William Richard
Johnson, Carl Emil, Jr
Johnson, Hugh Mitchell
Jungerman, John (Albert)
Juster, Norman Joel
Karplus, Robert
Keeports, David
Kenealy, Patrick Francis
Kim, Chung Sul (Sue)
Kocol, Henry
Korst, William Lawrence
Krippner, Stanley Curtis
L'Annunziata, Michael Frank
Lechtman, Max D
Leff, Harvey Sherwin
Lerner, Lawrence S
Liepman, H(ans) P(eter)
Lipson, Joseph Issac
Lorance, Elmer Donald
Luehrmann, Arthur Willett, Jr
McDonough, Thomas Redmond
McPherron, Robert Lloyd
Nadler, Gerald
Neidlinger, Hermann H
Norman, Eric B
Nunn, Robert Harry
Padian, Kevin
Pake, George Edward
Paselk, Richard Alan
Peterson, Laurence E
Petrucci, Ralph Herbert
Powell, Ronald Allan
Putz, Gerard Joseph
Quinn, Helen Rhoda
Ranftl, Robert M(atthew)
Rateaver, Bargyla
Remsen, Joyce F
Reynolds, Michael David
Rice, Robert Arnot
Romanowski, Christopher Andrew
Salanave, Leon Edward
Schelar, Virginia Mae
Schneiderman, Jill Stephanie
Schoenfeld, Alan Henry
Shull, Harrison
Siebert, Eleanor Dantzler
Sime, Ruth Lewin
Smith, Richard Avery
Spivey, Bruce Eldon
Sprain, Wilbur
Stokley, James
Stollberg, Robert
Strickmeier, Henry Bernard, Jr
Stringall, Robert William
Swartzlander, Earl Eugene, Jr
Tenn, Joseph S
Thomas, Barry
Torgow, Eugene N
Tribbey, Bert Allen
Tubbs, Eldred Frank
Turner, George Cleveland
Tuul, Johannes
Van Bibber, Karl Albert
Walker, Dan B
Walters, Richard Francis
Waters, William E
Webb, Leland Frederick
Wendel, Otto Theodore, Jr
White, Alvin Murray
White, Richard Manning
Whitney, Robert C
Wiley, Michael David
Will, Theodore A
Wulfman, Carl E
Yanow, Gilbert
Yu, Grace Wei-Chi Hu

COLORADO
Allen, Ernest E
Barber, Thomas Lynwood
Bard, Eugene Dwight
Benton, Edward Rowell
Boudreau, Robert Donald
Coates, Donald Allen
Dixon, Robert Clyde
Drueliner, Melvin L
Edwards, Kenneth Ward
Garstka, Walter U(rban)
Gottschall, W Carl
Heikkinen, Henry Wendell
Henkel, Richard Luther
Hess, Dexter Winfield
Iona, Mario
Johnson, Janice Kay
Koldewyn, William A
Krausz, Stephen
Lauer, B(yron) E(lmer)
Nauenberg, Uriel

Neumann, Herschel
Nisbet, Jerry J
Noble, Richard Daniel
Schonbeck, Niels Daniel
Shelton, Robert Wayne
Siuru, William D, Jr
Spenny, David Lorin
Trowbridge, Leslie Walter
West, Anita

CONNECTICUT
Braun, Phyllis C
Carroll, J Gregory
Connor, Lawrence John
Covey, Irene Mabel
Dix, Douglas Edward
Dolan, James F
Galston, Arthur William
Goodstein, Madeline P
Hanson, Earl Dorchester
Menke, David Hugh
Nelson, Roger Peter
Patterson, Elizabeth Chambers
Poincelot, Raymond Paul, Jr
Rachinsky, Michael Richard
Ross, Donald Joseph
Salamon, Richard Joseph
Schile, Richard Douglas
Seitelman, Leon Harold
Votaw, Robert Grimm
Wolf, Elizabeth Anne
Wystrach, Vernon Paul

DELAWARE
Cole, G(eorge) Rolland
Herron, Norman
Islam, Mir Nazrul
Singleton, Rivers, Jr
Skolnik, Herman
Tolman, Chadwick Alma

DISTRICT OF COLUMBIA
Callanan, Margaret Joan
Chamot, Dennis
Cohn, Victor Hugo
Dickens, Charles Henderson
Fechter, Alan Edward
Finn, Edward J
Frair, Karen Lee
Goldstein, Jeffrey Jay
Kaye, John
Malcom, Shirley Mahaley
Matyas, Marsha Lakes
Raizen, Senta Amon
Rose, Raymond Edward
Rosenberg, Edith E
Scheel, Nivard
Sukow, Wayne William
Watson, Robert Francis
Weintraub, Herbert D
Williams, Conrad Malcolm

FLORIDA
Ballard, Stanley Sumner
Bolte, John R
Bueche, Frederick Joseph
Burkman, Ernest
Busse, Robert Franklyn
Cintron, Guillermo B
Clark, John F
Cohen, Howard Lionel
Davis, Jefferson Clark, Jr
DeLap, James Harve
Dudley, Frank Mayo
Elliott, Paul Russell
Faber, Shepard Mazor
Fausett, Laurene van Camp
Frank, Stanley
Fuller, Harold Wayne
Gottesman, Stephen T
Griffith, Gail Susan Tucker
Harris, Lee Errol
Huebner, Jay Stanley
Jungbauer, Mary Ann
Kromhout, Robert Andrew
Leitner, Alfred
Lillien, Irving
Mellon, Edward Knox, Jr
Peters, Thomas G
Piquette, Jean Conrad
Posner, Gerald Seymour
Rowe, Mary Budd
Rumbach, William Ervin
Sandor, George N(ason)
Simring, Marvin
Spencer, John Lawrence
Stewart, Harris Bates, Jr
Stewart, Herbert
Vergenz, Robert Allan
Waife, Sholom Omi
Weatherly, Georges Lloyd
West, Felicia Emminger
Wiebush, Joseph Roy
Yost, Richard A
Young, Frank Glynn

GEORGIA
Aust, Catherine Cowan
Barreras, Raymond Joseph
Bass, William Thomas
Blanchet, Waldo W E
Brieske, Thomas John
Butts, David
Carter, Eloise B

Eiss, Albert Frank
Fay, Alice D Awtrey
Garmon, Lucille Burnett
Grace, Donald J
Greenberg, Jerrold
Hamrick, Anna Katherine Barr
Hayes, John Thompson
Hayes, Willis B
Hunt, Gary W
Kilpatrick, Jeremy
Loomis, Earl Alfred, Jr
Marks, Dennis William
Murray, Joan Baird
Nosek, Thomas Michael
Price, Edward Warren
Reese, Andy Clare
Riggsby, Ernest Duward
Shrum, John W
Shutze, John V
Skypek, Dora Helen
Stokes, Jimmy Cleveland
Wepfer, William J

HAWAII
Brantley, Lee Reed
Demanche, Edna Louise
Henry, Joseph Patrick
Malmstadt, Howard Vincent
Newhouse, W Jan

IDAHO
Dykstra, Dewey Irwin, Jr
Jobe, Lowell A(rthur)
Mech, William Paul
Ulliman, Joseph James

ILLINOIS
Alpert, Daniel
Ames, Edward R
Averill, Frank Wallace
Bond, Charles Eugene
Bowe, Joseph Charles
Braunfeld, Peter George
Brezinski, Darlene Rita
Burger, Ambrose William
Carlson, Eric Dungan
Casella, Alexander Joseph
Clemans, Kermit Grover
Curtis, Veronica Anne
Dietz, Mark Louis
Dornhoff, Larry Lee
Egan, Richard L
Fisher, Ben
Forman, G Lawrence
Frank, Forrest Jay
Garoian, George
Greenberg, Elliott
Greenberg, Stephen Robert
Greene, John Philip
Halberstam, Heini
Herskovits, Lily Eva
Herzenberg, Caroline Stuart Littlejohn
Hutchcroft, Alan Charles
Jaffe, Philip Monlane
Lederman, Leon Max
Lennox, Arlene Judith
Lietz, Gerard Paul
McGuire, Christine H
Malamud, Ernest I(lya)
Marcus, Jules Alexander
Mateja, John Frederick
Michael, Joel Allen
Munyer, Edward Arnold
O'Donnell, Terence J
Perkins, Alfred J
Peters, James Empson
Rakoff, Henry
Rappaport, David
Ren, Shang-Fen
Retzer, Kenneth Albert
Rosen, Sidney
Sawinski, Vincent John
Scheeline, Alexander
Schillinger, Edwin Joseph
Schmidt, Arthur Gerard
Schroeer, Juergen Max
Simmons, William Howard
Smith, Leslie Garrett
Solomon, Lawrence Marvin
Sweitzer, James Stuart
Treadway, William Jack, Jr
Uretsky, Jack Leon
Verderber, Nadine Lucille
Voelker, Alan Morris
Wagreich, Philip Donald
Wessler, Max Alden
Wirszup, Izaak
Wolff, Robert John
Wolter, Kirk Marcus

INDIANA
Abegg, Carl F(rank)
Alton, Elaine Vivian
Baker, David Thomas
Bauer, Dietrich Charles
Beckman, Jean Catherine
Beery, Dwight Beecher
Bodner, George Michael
Brooks, Austin Edward
Bumpus, John Arthur
Caskey, Jerry Allan
Cunningham, Ellen M
Foos, Kenneth Michael
Grabowski, Sandra Reynolds

Bartilucci, Andrew J
Beal, Myron Clarence
Belove, Charles
Blumenthal, Ralph Herbert
Bossen, Douglas C
Boynton, John E
Brady, James Edward
Braun, Ludwig
Brink, Frank, Jr
Bunce, Stanley Chalmers
Burge, Robert Ernest, Jr
Carroll, Harvey Franklin
Cluxton, David H
Cohn, Deirdre Arline
Cook, Robert Edward
Cote, Wilfred Arthur, Jr
Cummins, John Frances
Darrow, Frank William
Donaldson, Robert Rymal
Druger, Marvin
Eppenstein, Walter
Farb, Edith
Farmer, Walter Ashford
Farrell, Margaret Alice
Feldman, Lawrence
Ferguson, David Lawrence
Figueras, John
Fleishman, Bernard Abraham
Freedman, Aaron David
Geer, Ira W
Gering, Robert Lee
Gilmore, Arthur W
Greisen, Kenneth I
Greller, Andrew M
Gross, Jonathan Light
Grover, Paul L, Jr
Grunwald, Hubert Peter
Harke, Douglas J
Henderson, David Wilson
Henkel, Elmer Thomas
Henshaw, Clement Long
Hilborn, David Alan
Holbrow, Charles H
Horvat, Robert Emil
Hudson, John B(alch)
Irwin, Lynne Howard
Jacob, Klaus H
Jacobson, Willard James
Jaffe, Marvin Richard
Kaplan, Eugene Herbert
Katcoff, Seymour
Keepler, Manuel
Kelly, Richard Delmer
Kingslake, Rudolf
Klir, George Jiri
Knee, David Isaac
Knox, Robert Seiple
Law, David Martin
Lehrer, Gerard Michael
LeMay, Charlotte Zihlman
Lenchner, Nathaniel Herbert
Lewin, Seymour Z
Lindberg, Vern Wilton
Lindstrom, Gary J
Linkow, Leonard I
Lipsey, Sally Irene
Lustig, Harry
Mackay, Lottie Elizabeth Bohm
Madden, Robert E
Marino, Robert Anthony
Mayers, George Louis
Milmore, John Edward
Morrison, John Agnew
Murphy, James Joseph
Novak, Joseph Donald
Olczak, Paul Vincent
Orear, Jay
Osborn, H(arland) James
Paldy, Lester George
Parsegian, V(ozcan) Lawrence
Perry, J Warren
Petersen, Ingo Hans
Pierce, Carol S
Posamentier, Alfred S
Postman, Robert Derek
Prener, Robert
Raffensperger, Edgar M
Rapaport, William Joseph
Reed, George Farrell
Resnick, Robert
Reuss, Ronald Merl
Rich, Marvin R
Rockcastle, Verne Norton
Romer, Alfred
Romey, William Dowden
Rosenberg, Robert
Rosenberg, Warren L
Rousseau, Viateur
Russell, Charlotte Sananes
Sauer, Charles William
Scaife, Charles Walter John
Schwartz, Brian B
Segal, Sanford Leonard
Shannon, Jerry A, Jr
Sharkey, John Bernard
Shilepsky, Arnold Charles
Siebert, Karl Joseph
Siemankowski, Francis Theodore
Siew, Ernest L
Sipe, Harry Craig
Smith, Alden Ernest
Solon, Leonard Raymond
Spelke, Elizabeth Shilin
Spielman, Harold S

Springer, Dwight Sylvan
Stannard, Carl R, Jr
Stenzel, Wolfram G
Stephens, Lawrence James
Stith, James Herman
Strassenburg, Arnold Adolph
Talley, Charles Peter
Tannenbaum, Harold E
Tanzer, Charles
Thorpe, John Alden
Thygeson, Kenneth Helmer
Treat, Donald Fackler
Trimble, Robert Bogue
Unterman, Ronald David
Vacirca, Salvatore John
Wald, Francine Joy Weintraub
Waltzer, Wayne C
Washton, Nathan Seymour
Whang, Sung H
Whittingham, M(ichael) Stanley
Wieder, Grace Marilyn
Wilson, Jack Martin
Wolfson, Edward A
Wulff, Wolfgang
Zingaro, Joseph S

NORTH CAROLINA

Adelberger, Rexford E
Ayers, Caroline LeRoy
Bailey, Donald Etheridge
Beeler, Joe R, Jr
Boyers, Albert Sage
Bright, George Walter
Brotak, Edward Allen
Burford, Hugh Jonathan
Carroll, Felix Alvin, Jr
Clemens, Donald Faull
Dough, Robert Lyle, Sr
Fyfe, James Arthur
Gains, Lawrence Howard
Glenn, Thomas M
Groszos, Stephen Joseph
Gupta, Ajaya Kumar
Hadzija, Bozena Wesley
Hartford, Winslow H
Herrell, Astor Y
Hoelzel, Charles Bernard
Hugus, Z Zimmerman, Jr
Hulcher, Frank H
Iafrate, Gerald Joseph
Kuenzler, Edward Julian
Lamb, James C(hristian), III
Lytle, Charles Franklin
Mattheis, Floyd E
Mayer, Eugene Stephen
Meyer, John Richard
Oberhofer, Edward Samuel
Perfetti, Patricia F
Petersen, Karl Endel
Read, Floyd M
Rees, John
Risley, John Stetler
Robinson, Gertrude Edith
Robinson, Hugh Gettys
Robinson, Kent
Rowlett, Russell Johnston, III
Russell, Henry Franklin
Saeman, W(alter) C(arl)
Sams, Emmett Sprinkle
Sheppard, Moses Maurice
Simon, Sheridan Alan
Smith, Fred R, Jr
Tyndall, Jesse Parker
Werntz, James Herbert, Jr
Wilcox, Floyd Lewis
Willett, Hilda Pope
Wilson, Geoffrey Leonard

NORTH DAKOTA

Saari, Jack Theodore
Scoby, Donald Ray

OHIO

Barnes, H Verdain
Bernstein, Stanley Carl
Bidelman, William Pendry
Burrows, Kerilyn Christine
Carter, Carolyn Sue
Caughey, John Lyon, Jr
Cohen, Irwin
Damian, Carol G
Deal, Don Robert
DeVillez, Edward Joseph
Disinger, John Franklin
Dollhopf, William Edward
Fenner, Peter
Fortner, Rosanne White
Gamble, Francis Trevor
Gibbins, Betty Jane
Gonzalez, Paula
Gorse, Joseph
Gwinn, John Frederick
Hagelberg, M(yron) Paul
Hamed, Awatef A
Hauser, Edward J P
Heinke, Clarence Henry
Hern, Thomas Albert
Hill, James Stewart
Hoagstrom, Carl William
Hollis, William Frederick
Jacobs, Donald Thomas
Jaworowski, Andrzej Edward
John, George
Kagen, Herbert Paul

Kreidler, Eric Russell
Kritsky, Gene Ralph
Lamont, Gary Byron
Laughlin, Ethelreda R
Leonard, Billie Charles
Lichstein, Herman Carlton
Lutz, Arthur Leroy
McKenzie, Garry Donald
Mayer, Victor James
Meleca, C Benjamin
Miller, Franklin, Jr
Millis, John Schoff
Morgan, Alan Raymond
Murray, Thomas Henry
Pappas, Leonard Gust
Ploughe, William D
Pribor, Donald B
Proctor, David George
Richards, Walter Bruce
Ritchie, Austin E
Rogers, Charles Edwin
Saffran, Murray
Schaff, John Franklin
Schmidt, Mark Thomas
Sommer, John G
Stickney, Alan Craig
Taylor, Beverley Ann Price
Taylor, Charles Emery
Taylor, John Langdon, Jr
Utgard, Russell Oliver
Wallis, Robert L
Weed, Herman Roscoe
Westneat, David French
Williams, John Paul
Wismar, Beth Louise
Wolfe, Paul Jay
Wyman, Bostwick Frampton

OKLAHOMA

Choike, James Richard
Conway, Kenneth Edward
Crane, Charles Russell
Eddington, Carl Lee
Graves, Victoria
Kimpel, James Froome
Koh, Eunsook Tak
Marek, Edmund Anthony
Martin, Loren Gene
Menzie, Donald E
Owens, G(lenda) Kay
Renner, John Wilson
Richardson, Verlin Homer
Rutledge, Carl Thomas
Shmaefsky, Brian Robert
Stratton, Charles Abner
Weakley, Martin LeRoy
Yoesting, Clarence C

OREGON

Allen, John Eliot
Bandick, Neal Raymond
Clark, William Melvin, Jr
Cummins, Ernie Lee
Decker, Fred William
Griffith, William Thomas
Halko, Barbara Tomlonovic
Hancock, John Edward Herbert
Lendaris, George G(regory)
Long, James William
Matthes, Steven Allen
Moursund, David G
Nelson, Norman Neibuhr
Overley, Jack Castle
Pizzimenti, John Joseph
Towe, George Coffin
Wick, William Quentin

PENNSYLVANIA

Alteveer, Robert Jan George
Barus, Carl
Bates, Thomas Fulcher
Benfey, Otto Theodor
Benson, Brent W
Berty, Jozsef M
Bilaniuk, Oleksa-Myron
Botelho, Stella Yates
Brendlinger, Darwin
Brennan, Thomas Michael
Brungraber, Robert J
Burness, James Hubert
Byler, David Michael
Cahir, John Joseph
Carbonell, Jaime Guillermo
Carfagno, Salvatore P
Couch, Jack Gary
Diehl, Renee Denise
Felty, Wayne Lee
Foster, Norman Francis
Fowler, H(oratio) Seymour
Gainer, Michael Kizinski
Glaser, Robert
Godfrey, Susan Sturgis
Goff, Christopher Godfrey
Gold, Lewis Peter
Goldstein, Frederick J
Gol'ub, Jerry Paul
Greenberg, Herman Samuel
Grobstein, Paul
Helfferich, Friedrich G
Henderson, Robert E
Hershey, Nathan
Ho, Thomas Inn Min
Hostetler, Robert Paul
Johnston, Gordon Robert

Jones, Robert Leroy
Kang, Joohee
Kerstein, Morris D
Knaster, Tatyana
Knief, Ronald Allen
Kuserk, Frank Thomas
Larson, Russell Edward
Lerner, Leonard Joseph
Licini, Jerome Carl
Lippa, Linda Susan Mottow
Lipsky, Stephen E
Lokay, Joseph Donald
McFarlane, Kenneth Walter
Marshall, Keith
Milakofsky, Louis
Mitchell, John Murray, Jr
Moon, Thomas Charles
Moser, Gene Wendell
Mullins, Jeanette Somerville
Murphy, Clarence John
Owens, Frederick Hammann
Pitkin, Ruthanne B
Rehfield, David Michael
Rhodes, Jacob Lester
Riban, David Michael
Rothman, Milton A
Sagik, Bernard Phillip
Schattschneider, Doris Jean
Schearer, William Richard
Schmuckler, Joseph S
Sherwood, Bruce Arne
Shoemaker, Richard Nelson
Shrigley, Robert Leroy
Silver, Edward A
Snyder, Evan Samuel
Sommer, Holger Thomas
Stinson, Richard Floyd
Thompson, Julia Ann
Torop, William
Twiest, Gilbert Lee
Veening, Hans
Weedman, Daniel Wilson
Welliver, Paul Wesley
Williamson, Hugh A
Wismer, Robert Kingsley
Woodard, Robert Louis
Young, Hugh David

RHODE ISLAND

Barnes, Frederick Walter, Jr
Fasching, James Le Roy
Hazeltine, Barrett
Heywood, Peter
Kirwan, Donald Frazier
Pickart, Stanley Joseph
Sanberg, Paul Ronald

SOUTH CAROLINA

Allen, Joe Frank
Asleson, Gary Lee
Blevins, Maurice Everett
Carpenter, John Richard
Darnell, W(illiam) H(eaden)
Gurr, Henry S
Hahs, Sharon K
Hendrick, Lynn Denson
Longtin, Bruce
MacKnight, Franklin Collester
Nerbun, Robert Charles, Jr
Pike, LeRoy
Roldan, Luis Gonzalez
Scott, Jaunita Simons
Sweeney, John Robert

SOUTH DAKOTA

Draeger, William Charles
Einhellig, Frank Arnold
Hilderbrand, David Curtis
Landborg, Richard John
Peanasky, Robert Joseph
Riemenschneider, Albert Louis

TENNESSEE

Armistead, Willis William
Ayers, Jerry Bart
Ball, Raiford Mill
Baxter, John Edwards
Davidson, Robert C
Davis, Phillip Howard
Driskill, William David
Edwards, Thomas F
Elowe, Louis N
Gee, Charles William
Greenberg, Neil
Hanley, Wayne Stewart
Herring, Carey Reuben
Kirksey, Howard Graden, Jr
Kopp, Otto Charles
Kozub, Raymond Lee
Lane, Eric Trent
Loebbaka, David S
Lorenz, Philip Jack, Jr
Lowndes, Douglas H, Jr
Lyman, Beverly Ann
McGhee, Charles Robert
McNeely, Robert Lewis
Mathis, Philip Monroe
Mayfield, Melburn Ross
Moore, George Edward
Pafford, William N
Pegg, David John
Scott, Dan Dryden
Vo-Dinh, Tuan
Wilson, James Larry

Science Education (cont)

TEXAS
Archer, Cass L
Bachmann, John Henry, Jr
Baine, William Brennan
Baldwin, William Russell
Banerjee, Sanjay Kumar
Benedict, George Frederick
Bennett, Lloyd M
Blystone, Robert Vernon
Bryan, George Thomas
Buckalew, Louis Walter
Bullard, Truman Robert
Cargill, Robert Lee, Jr
Cody, John T
Craik, Eva Lee
Crawford, Gladys P
DeBakey, Lois
DeBakey, Selma
Duncombe, Raynor Lockwood
Elliott, Jerry Chris
Funkhouser, Edward Allen
Hademenos, James George
Hassell, Clinton Alton
Hemenway, Mary-Kay Meacham
Hernandez, Norma
Hewett, Lionel Donnell
Howe, Richard Samuel
Huebner, Walter F
Kasi, Leela Peshkar
Kesler, Oren Byrl
Koballa, Thomas Raymond, Jr
Krohmer, Jack Stewart
Lee, Addison Earl
Levy, Robert Aaron
Little, Robert Narvaez, Jr
Longacre, Susan Ann Burton
McBride, Raymond Andrew
Marcotte, Ronald Edward
Melton, Lynn Ayres
Mize, Charles Edward
Moy, Mamie Wong
Nelson, John Franklin
Paske, William Charles
Robbins, Ralph Robert
Roebuck, Isaac Field
Rudenberg, Frank Hermann
Schimelpfenig, Clarence William
Schroeder, Melvin Carroll
Shumate, Kenneth McClellan
Stranges, Anthony Nicholas
Stubbs, Donald William
Summers, Richard Lee
Sybert, James Ray
Tsin, Andrew Tsang Cheung
Urdy, Charles Eugene
Vest, Floyd Russell
Wendel, Carlton Tyrus
Westmeyer, Paul
Whittlesey, John R B
Wood, Lowell Thomas

UTAH
Bergeson, Haven Eldred
Earley, Charles Willard
Gates, Henry Stillman
Grey, Gothard C
Hess, Joseph W, Jr
Lombardi, Paul Schoenfeld
Wilson, Byron J

VERMONT
Moyer, Walter Allen, Jr

VIRGINIA
Auerbach, Stephen Michael
Balster, Robert L(ouis)
Bauer, Henry Hermann
Benton, Duane Marshall
Blair, Barbara Ann
Blatt, Jeremiah Lion
Clark, Mary Eleanor
Davis, William Arthur
Faulkner, Gavin John
Hartline, Beverly Karplus
Hill, Carl McClellan
Horsburgh, Robert Laurie
Huddle, Benjamin Paul, Jr
Hudgins, Aubrey C, Jr
Hufford, Terry Lee
Ijaz, Lubna Razia
Jacobs, Kenneth Charles
Jones, Franklin M
Jones, William F
Jordan, Wade H(ampton), Jr
Kernell, Robert Lee
Lilleleht, L(embit) U(no)
McCombs, Freda Siler
McEwen, Robert B
Mandell, Alan
Martin, Robert Bruce
Maybury, Robert H
Morrell, William Egbert
Moyer, Wayne A
Myers, William Howard
Shanklin, James Robert, Jr
Sigler, Julius Alfred, Jr
Slayden, Suzanne Weems
Smith, Elske van Panhuys
Smith, John Melvin
Spresser, Diane Mar
Starnes, William Herbert, Jr
Stronach, Carey E

Suiter, Marilyn J
Wilber, Joe Casley, Jr
Willis, Lloyd L, II

WASHINGTON
Armstrong, Frank Clarkson Felix
Arons, Arnold Boris
Artru, Alan Arthur
Axtell, Darrell Dean
Barnard, Kathryn E
Benson, Keith Rodney
Brown, Willard Andrew
Cellarius, Richard Andrew
Clement, Stephen LeRoy
Connell, James Roger
Daniels, Patricia D
Eustis, William Henry
Frasco, David Lee
Garland, John Kenneth
Gerhart, James Basil
Haight, Gilbert Pierce, Jr
Hall, Wayne Hawkins
Hardy, John Thomas
McDermott, Lillian Christie
McMichael, Kirk Dugald
Narayanan, A Sampath
Olstad, Roger Gale
Richardson, Richard Laurel
Schmid, Loren Clark
Shrader, John Stanley
Slesnick, Irwin Leonard
Smith, Sam Corry
Steckler, Bernard Michael
Tabbutt, Frederick Dean
Wade, Leroy Grover, Jr
Waltar, Alan Edward
Whitmer, John Charles
Wilson, Archie Spencer

WEST VIRGINIA
Becker, Stanley Leonard
Dils, Robert James
Fezer, Karl Dietrich
Knorr, Thomas George
Malin, Howard Gerald
Matthews, Virgil Edison
Richardson, Rayman Paul
Sperati, Carleton Angelo

WISCONSIN
Andrews, Oliver Augustus
Balsano, Joseph Silvio
Burgess, Ann Baker
Chander, Jagdish
Dibben, Martyn James
Fitch, Robert McLellan
Jeanmaire, Robert L
Johnson, J(ames) R(obert)
Jungck, John Richard
Koch, Rudy G
Lagally, Max Gunter
Lentz, Mark Steven
Melvin, John L
Moore, John Ward
O'Hearn, George Thomas
Pella, Milton Orville
Scheusner, Dale Lee
Schneiderwent, Myron Otto
Schroeder, Daniel John
Seeburger, George Harold
Smith, Roy E
Sommers, Raymond A
Southworth, Warren Hilbourne
Sprott, Julien Clinton
Stekel, Frank D
Walker, Duard Lee
Wright, Steven Martin

WYOMING
Thronson, Harley Andrew, Jr

PUERTO RICO
Lewis, Brian Kreglow
McDowell, Dawson Clayborn
Ramos, Lillian

ALBERTA
Hunt, Robert Nelson
Klassen, Rudolph Waldemar
Krueger, Peter J
McLeod, Lionel Everett
Martin, John Scott
Mash, Eric Jay
Peirce, John Wentworth
Plambeck, James Alan
Wieser, Helmut

BRITISH COLUMBIA
Godolphin, William
Paszner, Laszlo

MANITOBA
Carbno, William Clifford
Kao, Kwan Chi
LeBow, Michael David

NEW BRUNSWICK
Frantsi, Christopher
Whitla, William Alexander

NEWFOUNDLAND
Aldrich, Frederick Allen
Rayner-Canham, Geoffrey William
Rusted, Ian Edwin L H

NOVA SCOTIA
Dunn, Kenneth Arthur

ONTARIO
Atkinson, George Francis
Bevier, Mary Lou
Buchanan, George Dale
Burton, John Heslop
Clarke, Thomas Roy
Gray, David Robert
Jarzen, David MacArthur
Key, Anthony W
Levere, Trevor Harvey
Lower, Stephen K
Mackay, Rosemary Joan
Montemurro, Donald Gilbert
Nesheim, Michael Ernest
Nirdosh, Inderjit
Osmond, Daniel Harcourt
Parr, James Gordon
Percy, John Rees
Rothman, Arthur I
Sandler, Samuel
Shurvell, Herbert Francis
Smith, Donald Alan
Tolnai, Susan
Usselman, Melvyn Charles
VandenHazel, Bessel J
Vervoort, Gerardus
Waugh, Douglas Oliver William
Whippey, Patrick William
Wightman, Robert Harlan
Winter, Peter
Yan, Maxwell Menuhin

QUEBEC
Alward, Ron E
Chagnon, Jean Yves
Cochrane, Robert W
Cook, Robert Douglas
Croctogino, Reinhold Hermann
Davidson, Colin Henry
Escher, Emanuel
Gurudata, Neville
Johns, Kenneth Charles
Marisi, Dan(iel) Quirinus
Pekau, Oscar A
Tomas, Francisco

SASKATCHEWAN
Larson, D Wayne

OTHER COUNTRIES
Ahn, Tae In
Berenson, Lewis Jay
Cartwright, Hugh Manning
Craig, John Frank
D'Ambrosio, Ubiratan
Davis, William Jackson
Fernández-Cruz, Eduardo P
Fettweis, Alfred Leo Maria
Foecke, Harold Anthony
Gomez, Arturo
Lee, Joseph Chuen Kwun
Ling, Samuel Chen-Ying
Revillard, Jean-Pierre Rémy
Rinehart, Frank Palmer
Robinson, Berol (Lee)
Scheiter, B Joseph Paul
Schwartz, Peter Larry
Stephens, Kenneth S
Toshio, Makimoto
Unger, Hans-Georg

Science Policy

ALABAMA
Bridgers, William Frank

ARIZONA
Bartocha, Bodo
Bradley, Michael Douglas
Dworkin, Judith Marcia
Leibacher, John W

CALIFORNIA
Akin, Gwynn Collins
Atkin, J Myron
Black, Robert L
Capron, Alexander Morgan
Chartock, Michael Andrew
Chow, Brian G(ee-Yin)
Craig, Paul Palmer
Denning, Peter James
Dismukes, Robert Key
Djerassi, Carl
Gortner, Susan Reichert
Hardy, Edgar Erwin
Holdren, John Paul
Holtz, David
Howd, Robert A
Isherwood, Dana Joan
Kevles, Daniel Jerome
London, Ray William
Long, Franklin A
McKone, Thomas Edward
Moore, Andrew Brooks
Orloff, Neil
Primack, Joel Robert
Revelle, Roger (Randall Dougan)
Rossin, A David
Rowell, John Martin
Schipper, Lee (Leon Jay)

Shin, Ernest Eun-Ho
Shull, Harrison
Smith, Ora E
Springer, Bernard G
Suman, Daniel Oscar
Sutcliffe, William George
Swearengen, Jack Clayton
Sweedler, Alan R
Szego, Peter A
Vajk, J(oseph) Peter
Whipple, Christopher George
York, Herbert Frank

COLORADO
Burnett, Jerrold J
Chamberlain, A(drian) R(amond)
Cohen, Robert
Giller, E(dward) B(onfoy)
Johnson, Clark E, Jr
Lameiro, Gerard Francis
McCray, Richard A
Schneider, Stephen Henry
Thompson, Starley Lee

CONNECTICUT
Bloom, Barry Malcolm
Brooks, Douglas Lee
Corning, Mary Elizabeth
Fraser, J(ulius) T(homas)
Lubell, Michael S
Novick, Alvin
Reid, James Dolan

DELAWARE
Holt, Rush Dew, Jr
Quisenberry, Richard Keith

DISTRICT OF COLUMBIA
Alic, John A
Alter, Harvey
Argus, Mary Frances
Ausubel, Jesse Huntley
Baker, Louis Coombs Weller
Beall, James Howard
Berlincourt, Ted Gibbs
Bernthal, Frederick Michael
Blanpied, William Antoine
Blockstein, David Edward
Boyce, Peter Bradford
Budgor, Aaron Bernard
Burley, Gordon
Burris, John Edward
Butter, Stephen Allan
Carey, William Daniel
Chamot, Dennis
Chapin, Douglas Scott
Civiak, Robert L
Cox, Lawrence Henry
Dickens, Charles Henderson
Epstein, Gerald Lewis
Fainberg, Anthony
Fechter, Alan Edward
Florig, Henry Keith
Gaffney, Paul G, II
Gough, Michael
Green, Rayna Diane
Greenfield, Richard Sherman
Hart, Richard Cullen
Hawkins, Gerald Stanley
Hazelrigg, George Arthur, Jr
Hemily, Philip Wright
Hershey, Robert Lewis
Hoagland, K Elaine
Holt, James Allen
Houck, Frank Scanland
Hudis, Jerome
Jackson, Curtis Rukes
Jacobson, Michael F
Johnson, George Patrick
Keitt, George Wannamaker, Jr
Kelly, Henry Charles
Korkegi, Robert Hani
Kovach, Eugene George
Lai, David Ying-lun
Lee, Alfred M
Logsdon, John Mortimer, III
Marcus, Michael Jay
Matyas, Marsha Lakes
Mellor, John Williams
Menzel, Joerg H
Myers, Peter Briggs
Nadolney, Carlton H
Nightingale, Elena Ottolenghi
Noonan, Norine Elizabeth
Novello, Antonia Coello
Orlans(Morton), F Barbara
Parham, Walter Edward
Peake, Harold J(ackson)
Phillips, Don Irwin
Pikus, Irwin Mark
Policansky, David J
Ratchford, Joseph Thomas
Reisa, James Joseph, Jr
Riemer, Robert Lee
Robbins, Robert John
Rogers, Kenneth Cannicott
Setlow, Valerie Petit
Shaub, Walter M
Shykind, Edwin B
Taylor, Ronald D
Topper, Leonard
Tucker, John Richard
Uman, Myron F
Watson, Robert Francis

Weiss, Leonard
Werdel, Judith Ann
Whidden, Stanley John
Wolfle, Thomas Lee
Zawisza, Julie Anne A
Zwilsky, Klaus M(ax)
Zwolenik, James J

FLORIDA
Arnold, Henry Albert
Chartrand, Robert Lee
Harwell, Mark Alan
Lannutti, Joseph Edward

GEORGIA
Boone, Donald Joe
Clifton, David S, Jr
Greenbaum, Lowell Marvin
Porter, Alan Leslie
Schutt, Paul Frederick

HAWAII
Carpenter, Richard A
Craven, John P

IDAHO
Mills, James Ignatius

ILLINOIS
Alpert, Daniel
Amarose, Anthony Philip
Ban, Stephen Dennis
Bentley, Orville George
Bevan, William
Bullard, Clark W
Changnon, Stanley A, Jr
Collier, Donald W(alter)
Cranston, Joseph William, Jr
Egan, Richard L
Fossum, Robert Merle
Gross, David Lee
Hubbard, Lincoln Beals
Huebner, Russell Henry, Sr
Jain, Ravinder Kumar
Kanter, Manuel Allen
Lehman, Dennis Dale
Lushbough, Channing Harden
Pesyna, Gail Marlane
Phelps, Creighton Halstead
Rabinowitch, Victor
Shen, Sin-Yan
Snow, Joel A

INDIANA
Katz, Frances R
Leonard, Jack E
Ortman, Eldon Emil

KANSAS
Mingle, John O(rville)

MARYLAND
Allgaier, Robert Stephen
Bartley, William Call
Bawden, Monte Paul
Beaven, Vida Helms
Berger, Robert Elliott
Berkson, Harold
Bingham, Robert Lodewijk
Carroll, James Barr
Cassidy, Esther Christmas
Chen, Philip Stanley, Jr
Chernoff, Amoz Immanuel
Clark, Joseph E(dward)
Clark, Trevor H
Daly, John Anthony
Decker, James Federick
Dunning, Herbert Neal
Ewing, June Swift
Falk, Charles Eugene
Filner, Barbara
Friedman, Abraham S(olomon)
Fry, John Craig
Gary, Robert
Geller, Ronald G
Goldwater, William Henry
Gonzales, Ciriaco Q
Greer, William Louis
Heinrich, Kurt Francis Joseph
Herman, Samuel Sidney
Herrett, Richard Allison
Hildebrand, Bernard
Hilmoe, Russell J(ulian)
Holland, Marjorie Miriam
Holloway, Caroline T
Johnston, Laurance S
Kalt, Marvin Robert
Katcher, David Abraham
Kushner, Lawrence Maurice
Langenberg, Donald Newton
Leshner, Alan Irvin
McCarn, Davis Barton
McRae, Vincent Vernon
Michaelis, Michael
Mielenz, Klaus Dieter
Miller, Hugh Hunt
Miller, Nancy E
Mills, William Andy
Pahl, Herbert Bowen
Pitlick, Frances Ann
Promisel, Nathan E
Rheinstein, Peter Howard
Richardson, John Marshall
Rosen, Milton W(illiam)

Rosenblum, Annette Tannenholz
Schulman, James Herbert
Schultz, Warren Walter
Seifried, Adele Susan Corbin
Siegel, Elliot Robert
Silbergeld, Ellen K
Simon, Robert Michael
Singh, Amarjit
Smedley, William Michael
Smith, John Henry
Sorrows, Howard Earle
Spilhaus, Athelstan Frederick, Jr
Stanley, Ronald Alwin
Starfield, Barbara Helen
Stonehill, Elliott H
Tabor, Edward
Tropf, Cheryl Griffiths
Ulbrecht, Jaromir Josef
Vogl, Thomas Paul
Weiss, Charles, Jr
Wortman, Bernard
Yakowitz, Harvey
Yarmolinsky, Adam
Young, Alvin L
Zimbelman, Robert George
Zornetzer, Steven F

MASSACHUSETTS
Bernstein, Herbert J
Bertera, James H
Bird, Stephanie J
Branscomb, Lewis McAdory
Breuning, Siegfried M
Broadus, James Matthew
Brooks, Harvey
Carnesale, Albert
Carter, Ashton Baldwin
Clark, William Cummin
De Neufville, Richard Lawrence
Fitch, John Henry
Foster, Charles Henry Wheelwright
Ganley, Oswald Harold
Goodell, Rae Simpson
Goodman, Philip
Graham, Loren R
Holmes, Helen Bequaert
Hornig, Donald Frederick
Karo, Douglas Paul
Krass, Allan S(hale)
Laster, Leonard
Lee, Kai Nien
Loew, Franklin Martin
McKnight, Lee Warren
Moore-Ede, Martin C
Ramakrishna, Kilaparti
Skolnikoff, Eugene B
Stonier, Tom Ted
Tuler, Floyd Robert
Weiner, Charles
Wilson, Linda S (Whatley)

MICHIGAN
Ancker-Johnson, Betsy
Bohm, Henry Victor
Evans, David Arnold
Hoerger, Fred Donald
Humphrey, Harold Edward Burton, Jr
Kamrin, Michael Arnold
McLaughlin, Renate
Mayer, Frederick Joseph
Montgomery, Donald Joseph
Saperstein, Alvin Martin
Schwing, Richard C
Viano, David Charles
Weissman, Eugene Y(ehuda)

MINNESOTA
Adams, Robert McLean
Flower, Terrence Frederick
Perry, James Alfred
Steen, Lynn Arthur
Thomborson, Clark D

MISSISSIPPI
Alexander, William Nebel
ElSohly, Mahmoud Ahmed

MISSOURI
Johannsen, Frederick Richard
Lawless, Edward William
Morgan, Robert P
Wilson, Edward Nathan

NEW HAMPSHIRE
Bernstein, Abram Bernard
Kantrowitz, Arthur (Robert)

NEW JERSEY
Brown, William Stanley
Brubaker, Paul Eugene, II
Good, Mary Lowe
Hall, Homer James
Hay, Peter Marsland
Jensen, Betty Klainminc
Klapper, Jacob
Levi, Barbara Goss
Miller, Robert Alan
Pinsky, Carl Muni
Shrier, Adam Louis
Smagorinsky, Joseph
Socolow, Robert H(arry)
Steinberg, Eliot
Stiles, Lynn F, Jr
Tanner, Daniel

Temmer, Georges Maxime
Turner, Edwin Lewis

NEW MEXICO
Birely, John H
Freese, Kenneth Brooks
Icerman, Larry
Lohrding, Ronald Keith
Otway, Harry John
Shipley, James Parish, Jr

NEW YORK
Allentuch, Arnold
Althuis, Thomas Henry
Bedford, Barbara Lynn
Beyea, Jan Edward
Bulloff, Jack John
Colligan, John Joseph
Fox, William
Geiger, H Jack
Goldfarb, Theodore D
Hardy, Ralph Wilbur Frederick
Jasanoff, Sheila Sen
Katz, Maurice Joseph
Komorek, Michael Joesph, Jr
Lowrance, William Wilson, Jr
Lubic, Ruth Watson
McNeil, Richard Jerome
Mahoney, Margaret E
Manassah, Jamal Tewfek
Mike, Valerie
Morris, Robert Gemmill
Mowshowitz, Abbe
O'Connor, Joel Sturges
Panem, Sandra
Pechacek, Terry Frank
Porter, Beverly Fearn
Rambaut, Paul Christopher
Richards, Paul Granston
Robertson, William, IV
Robinson, David Zav
Shaw, David Elliot
Swanson, Robert Lawrence
Swartz, Harry
Thorpe, John Alden
Voelcker, Herbert B(ernhard)
Walker, Theresa Anne
White, Irvin Linwood
Wilkins, Bruce Tabor
Wilson, Fred Lee

NORTH CAROLINA
Ahearne, John Francis
Andrews, Richard Nigel Lyon
Bishop, Jack Belmont
Gemperline, Margaret Mary Cetera
Mosier, Stephen R
Newell, Nanette
Rees, John
Schroeer, Dietrich

OHIO
Chan, Yupo
Duga, Jules Joseph
Mulick, James Anton
Sheskin, Theodore Jerome

OKLAHOMA
Crafton, Paul A(rthur)
Lambird, Perry Albert

OREGON
Bullock, Richard Melvin
Dews, Edmund
Linstone, Harold A
Moravcsik, Michael Julius

PENNSYLVANIA
Burns, Denver P
Camp, Frederick William
Córdova, France Anne-Dominic
Daehnick, Wilfried W
Fritz, Lawrence William
Fromm, Eli
Gerjuoy, Edward
Hager, Nathaniel Ellmaker, Jr
Humphrey, Watts S
Lu, Chih Yuan
Morgan, M(illett) Granger
Pool, James C T
Ray, Eva K
Shen, Benjamin Shih-Ping
Small, Henry Gilbert
Steg, L(eo)
Tuthill, Harlan Lloyd
Van der Werff, Terry Jay

SOUTH DAKOTA
Sword, Christopher Patrick
Wegman, Steven M

TENNESSEE
Cope, David Franklin
Garber, David H
Huray, Paul Gordon
Miller, Paul Thomas
Phung, Doan Lien
Veigel, Jon Michael

TEXAS
Albach, Roger Fred
Cranberg, Lawrence
Dilsaver, Steven Charles
Elliott, Jerry Chris

Heavner, James E
Neureiter, Norman Paul
Shotwell, Thomas Knight
Thomas, Estes Centennial, III
Walker, Laurence Colton

VIRGINIA
Barker, Robert Henry
Bartis, James Thomas
Berger, Beverly Jane
Burstyn, Harold Lewis
Carpenter, James E(ugene)
De Simone, Daniel V
Dragonetti, John Joseph
Featherston, Frank Hunter
Feldmann, Edward George
Fisher, Gail Feimster
Gibbons, John Howard
Hart, Dabney Gardner
Ifft, Edward M
Jennings, Allen Lee
Kolsrud, Gretchen Schabtach
Kramish, Arnold
Krassner, Jerry
Ling, James Gi-Ming
Lynch, Maurice Patrick
Morrison, David Lee
Munday, John Clingman, Jr
Newton, Elisabeth G
Owczarski, William A(nthony)
Pages, Robert Alex
Potts, Malcolm
Powell, Justin Christopher
Pry, Robert Henry
Rehm, Allan Stanley
Reischman, Michael Mack
Ries, Richard Ralph
Rodricks, Joseph Victor
Rossini, Frederick Anthony
Sabadell, Alberto Jose
Schad, Theodore M(acNeeve)
Schremp, Edward Jay
Smith, Bertram Bryan, Jr
Smith, Marcia Sue
Spindel, William
Sudarshan, T S
Talbot, Lee Merriam
Teichler-Zallen, Doris
Tipper, Ronald Charles
Tolin, Sue Ann
Troxell, Terry Charles
Ventre, Francis Thomas
Wilkinson, Christopher Foster
Wollan, David Strand
Wright, Robert Raymond
Zimmerman, Peter David

WASHINGTON
Fickeisen, Duane H
Miles, Edward Lancelot
Wenk, Edward, Jr
Wolfle, Dael (Lee)

WISCONSIN
Bleicher, Michael Nathaniel
Greger, Janet L
Inman, Ross
Lang, Norma M
Steinhart, John Shannon

WYOMING
Meyer, Edmond Gerald
Thronson, Harley Andrew, Jr

BRITISH COLUMBIA
Dewey, John Marks
LeBlond, Paul Henri

MANITOBA
Hall, Donald Herbert
Rosinger, Eva L J
Trick, Gordon Staples

ONTARIO
Cartier, Jean Jacques
Costain, Cecil Clifford
Cupp, Calvin R
French, J(ohn) Barry
Godson, Warren Lehman
Heacock, Ronald A
Jackson, Ray Weldon
Kates, Josef
Keys, John David
Kruus, Jaan
LeRoy, Donald James
McTaggart-Cowan, Patrick Duncan
Munroe, Eugene Gordon
Pigden, Wallace James
Read, D E
Sly, Peter G
Solandt, Omond McKillop
Sturgess, Jennifer Mary
Uffen, Robert James
Walsh, John Heritage
Winder, Charles Gordon
Wright, Douglas Tyndall

QUEBEC
Bailar, John Christian, III
L'Ecuyer, Jacques
McQueen, Hugh J

SASKATCHEWAN
Katz, Leon

Science Policy (cont)

OTHER COUNTRIES
Assousa, George Elias
Campbell, Virginia Wiley
D'Ambrosio, Ubiratan
Gowans, James L
Huckaba, Charles Edwin
Luo, Peilin
Metzger, Gershon
Okamura, Sogo
Packer, Leo S
Richards, Adrian F
Saba, Shoichi
Schmell, Eli David
Schopper, Herwig Franz
Stebbing, Nowell
Thissen, Wil A
Van Andel, Tjeerd Hendrik

Technical Management

ALABAMA
Arnold, David Walker
Cloyd, Grover David
Cochran, John Euell, Jr
Costes, Nicholas Constantine
Cowsar, Donald Roy
Davis, John Moulton
Dean, David Lee
Fowler, Bruce Wayne
Haak, Frederik Albertus
Hart, David R
Howgate, David W
Kandhal, Prithvi Singh
Lovingood, Judson Allison
Mookherji, Tripty Kumar
Morgan, Bernard S(tanley), Jr
Piwonka, Thomas Squire
Raymond, Dale Rodney
Regner, John LaVerne
Rekoff, M(ichael) G(eorge), Jr
Vaughan, William Walton

ALASKA
Bennett, F Lawrence
Hulsey, J Leroy
Leon, Kenneth Allen

ARIZONA
Adamson, Albert S, Jr
Adickes, H Wayne
Armstrong, Dale Dean
Carr, Clide Isom
Dickinson, Robert Earl
Dybvig, Paul Henry
Gatley, William Stuart
Greenberg, Milton
Hoehn, A(lfred) J(oseph)
Jungermann, Eric
McGuirk, William Joseph
Milnes, Dale J
North-Root, Helen May
Pokorny, Gerold E(rwin)
Ramler, W(arren) J(oseph)
Randall, Lawrence Kessler, Jr
Sears, Donald Richard
Swalin, Richard Arthur
Ulich, Bobby Lee
Upchurch, Jonathan Everett
Wasson, James Walter
Willson, Donald Bruce

ARKANSAS
Berky, John James
Burchard, John Kenneth
Glendening, Norman Willard
Havener, Robert D
Townsend, James Willis

CALIFORNIA
Adelman, Barnet Reuben
Adelson, Harold Ely
Adicoff, Arnold
Albert, Richard David
Alper, Marshall Edward
Anderson, George William
Anjard, Ronald P, Sr
Aster, Robert Wesley
Atkinson, Russell H
Augenstein, Bruno (Wilhelm)
Baba, Paul David
Baker, Andrew Newton, Jr
Baugh, Ann Lawrence
Baumann, Frederick
Beebe, Robert Ray
Bhanu, Bir
Bharadvaj, Bala Krishnan
Billig, Franklin A
Bilow, Norman
Birecki, Henryk
Block, Richard B
Bode, Donald Edward
Bonsack, Walter Karl
Brewer, George R
Brown, Tony Ray
Camp, David Conrad
Campbell, Bonita Jean
Campbell, Thomas Cooper
Canning, T(homas) F
Caren, Robert Poston
Carlan, Alan J
Carman, Robert Lincoln, Jr

Carr, Robert Charles
Casey, John Edward, Jr
Caswell, John N(orman)
Chan, David S
Chan, Kwoklong Roland
Chang, Chih-Pei
Channel, Lawrence Edwin
Chase, Lloyd Fremont, Jr
Chen, Kao
Chern, Ming-Fen Myra
Chester, Arthur Noble
Chivers, Hugh John
Chodos, Steven Leslie
Cipriano, Leonard Francis
Clazie, Ronald N(orris)
Cleary, Michael E
Clegg, Frederick Wingfield
Cluff, Lloyd Sterling
Cockrum, Richard Henry
Cohen, Jack
Coile, Russell Cleven
Cole, Richard
Coleman, Lamar William
Compton, Leslie Ellwyn
Conant, Curtis Terry
Cook, Glenn Melvin
Copper, John A(lan)
Crandall, Ira Carlton
Crossley, Frank Alphonso
Culler, F(loyd) L(eRoy), Jr
Current, Jerry Hall
Dahlberg, Richard Craig
Dally, Edgar B
Danielson, George Edward, Jr
Davis, Louis E(lkin)
Davis, William Ellsmore, Jr
Davison, John Blake
Deckert, Curtis Kenneth
Depp, Joseph George
Detz, Clifford M
Digby, James F(oster)
Docks, Edward Leon
Dowdy, William Louis
Dulock, Victor A, Jr
Durbeck, Robert C(harles)
Eck, Robert Edwin
Edelson, Robert Ellis
Ehrlich, Richard
Elverum, Gerard William, Jr
Englert, Robert D
Erickson, Stanley Arvid
Ernst, Roberta Dorothea
Farber, Joseph
Farmer, Crofton Bernard
Fasang, Patrick Pad
Feather, A(lan) L(ee)
Feiock, Frank Donald
Ferber, Robert R
Feuerstein, Seymour
Finn, James Crampton, Jr
Fischer, George K
Fishman, Norman
Fitzpatrick, Gary Owen
Fleischer, Allan A
Flora, Edward B(enjamin)
Forrest, James Benjamin
Forsberg, Kevin
Frazier, Edward Nelson
Frederick, David Eugene
Frenzel, Lydia Ann Melcher
Fricke, Martin Paul
Funkhouser, John Tower
Gaynor, Joseph
Gearhart, Roger A
Gehman, Bruce Lawrence
Geminder, Robert
Gerwick, Ben Clifford, Jr
Gesner, Bruce D
Gilbreath, Michael Joye
Gilkeson, M(urray) Mack
Gill, James Edward
Gill, William D(elahaye)
Gillette, Dean
Glaser, Harold
Glover, Leon Conrad, Jr
Goddard, Terrence Patrick
Gogolewski, Raymond Paul
Goode, John Wolford
Goodjohn, Albert J
Graham, Dee McDonald
Greene, Elias Louis
Greenleaf, John Edward
Grier, Herbert E
Gritton, Eugene Charles
Groner, Gabriel F(rederick)
Guest, Gareth E
Gür, Turgut M
Gyorey, Geza Leslie
Hadley, James Warren
Hadley, Jeffery A
Hahn, Harold Thomas
Hainski, Martha Barrionuevo
Hall, James Edison
Ham, Lee Edward
Hamilton, Carole Lois
Hansen, Grant Lewis
Hansmann, Douglas M
Harrison, George Conrad, Jr
Hartsough, Walter Douglas
Hartwick, Thomas Stanley
Harvey, A(lexander) C
Hausmann, Werner Karl
Hayes, Thomas Jay, III
Heil, John F, Jr

Heitkamp, Norman Denis
Hilbert, Robert S
Hill, Robert
Hill, Russell John
Hinkley, Everett David
Hintz, Marie I
Hoagland, Jack Charles
Hoch, Paul Edwin
Hodges, Dean T, Jr
Hodson, William Myron
Hoff, Richard William
Hoffman, Howard Torrens
Hoffman, Marvin Morrison
Holl, Richard Jacob
Holliday, Dale Vance
Hong, Ki C(hoong)
Hopponen, Jerry Dale
Hornig, Howard Chester
Horvath, Kalman
Howerton, Robert James
Hubert, Jay Marvin
Hudson, Cecil Ivan, Jr
Hughes, Robert Alan
Hugunin, Alan Godfrey
Hwang, Li-San
Ide, Roger Henry
Jacobs, Ralph R
Jacobs, S Lawrence
Jacobson, Albert H(erman), Jr
Jacobson, Allan Stanley (Bud)
Jacobson, Marcus
Jaffe, Leonard David
James, Philip Nickerson
Janney, Gareth Maynard
Johannessen, George Andrew
Johnson, Conor Deane
Johnson, Mark Scott
Jones, E(dward) M(cClung) T(hompson)
Jones, Richard Lee
Jorgensen, Paul J
Judd, Stanley H
Kalvinskas, John J(oseph)
Kaplan, Richard E
Kassakhian, Garabet Haroutioun
Keaton, Michael John
Keller, Joseph Edward, Jr
Kennedy, James Vern
Kennel, John Maurice
Kimball, Ralph B
King, John Mathews
King, Ray J(ohn)
Kissel, Charles Louis
Klestadt, Bernard
Knight, Patricia Marie
Kramer, David
Kropp, John Leo
Kukkonen, Carl Allan
Kull, Lorenz Anthony
Lagarias, John S(amuel)
Lampert, Carl Matthew
Lampert, Seymour
Larson, Edward William, Jr
Lavernia, Enrique Jose
Lechtman, Max D
LeFave, Gene M(arion)
Legge, Norman Reginald
Leibowitz, Lewis Phillip
Lewis, Arthur Edward
Lewis, Nina Alissa
Lewis, Robert Allen
Lichter, James Joseph
Liepman, H(ans) P(eter)
Lillie, Charles Frederick
Lippitt, Louis
Lissaman, Peter Barry Stuart
Logan, James Columbus
Long, H(ugh) M(ontgomery)
Love, Sydney Francis
Lovelace, Alan Mathieson
Ludwig, Frank Arno
Lukasik, Stephen Joseph
Luongo, Cesar Augusto
Lurie, Norman A(lan)
McCall, Chester Hayden, Jr
McCarthy, John Lockhart
McCreight, Louis R(alph)
McDonald, David William
McDonald, William True
McEwen, C(assius) Richard
MacKay, Kenneth Donald
Mackin, Robert James, Jr
McLaughlin, William Irving
Malouf, George M
Manes, Kennneth Rene
Mantey, Patrick E(dward)
Marcus, Bruce David
Marcus, Rudolph Julius
Massier, Paul Ferdinand
Mathews, W(arren) E(dward)
May, Bill B
Meechan, Charles James
Meghreblian, Robert V(artan)
Meiners, Henry C(ito)
Merilo, Mati
Michaelson, Jerry Dean
Miller, Arnold
Miller, Jon Philip
Miller, William Frederick
Moore, Edgar Tilden, Jr
Morey, Booker Williams
Morris, Fred(erick) W(illiam)
Mow, C(hao) C(how)
Mulroy, Thomas Wilkinson
Nacht, Sergio

Nadler, Gerald
Neidlinger, Hermann H
Ning, Robert Y
Nishi, Yoshio
North, Harper Qua
O'Donnell, Ashton Jay
O'Donohue, Cynthia H
Oehlberg, Richard N
Offen, George Richard
O'Keefe, John Dugan
Opfell, John Burton
Orth, Charles Douglas
Page, D(errick) J(ohn)
Parks, Lewis Arthur
Parsons, Michael L
Paszyc, Aleksy Jerzy
Perkins, Willis Drummond
Perlman, T(heodore)
Perloff, David Steven
Perry, Joseph Earl, Jr
Persky, George
Peterson, Victor Lowell
Philipson, Lloyd Lewis
Phillips, Russell C(ole)
Pietrzak, Lawrence Michael
Pitts, Robert Gary
Poggenburg, John Kenneth, Jr
Potter, David Samuel
Pressman, Ada Irene
Pretzer, Donavon Donald
Putz, Gerard Joseph
Rabl, Veronika Ariana
Rahmat-Samii, Yahya
Ram, Michael Jay
Ramspott, Lawrence Dewey
Randolph, James E
Ranftl, Robert M(atthew)
Rapp, Donald
Rasor, Ned S(haurer)
Rast, Howard Eugene, Jr
Rauch, Richard Travis
Reardon, Edward Joseph, Jr
Remer, Donald Sherwood
Remley, Marlin Eugene
Resch, George Michael
Rice, Dennis Keith
Richardson, John L(loyd)
Ripley, William Ellis
Ritzman, Robert L
Roberts, Charles Sheldon
Robin, Allen Maurice
Robinson, Lewis Howe
Rohde, Richard Whitney
Rohm, C E Tapie, Jr
Romanowski, Christopher Andrew
Roney, Robert K(enneth)
Rosen, Leonard Craig
Rossen, Joel N(orman)
Rowe, Lawrence A
Rowell, John Martin
Rowntree, Robert Fredric
Rudavsky, Alexander Bohdan
Ruskin, Arnold M(ilton)
Sachdev, Suresh
Sager, Ronald E
Saito, Theodore T
Salsig, William Winter, Jr
Saltman, William Mose
Saute, Robert E
Schjelderup, Hassel Charles
Schmidt, Raymond LeRoy
Schooley, Caroline Naus
Schreiner, Robert Nicolas, Jr
Schurmeier, Harris McIntosh
Schwab, Michael
Schwartz, Daniel M(ax)
Schweiker, George Christian
Scott, Franklin Robert
Seling, Theodore Victor
Selinger, Patricia Griffiths
Serbia, George William
Sharpe, Roland Leonard
Shaw, Charles Alden
Shlanta, Alexis
Siegel, Brock Martin
Siegfried, William
Siegman, Fred Stephen
Silverman, Jacob
Singer, Solomon Elias
Siposs, George G
Skavdahl, R(ichard) E(arl)
Smiley, Vern Newton
Smith, Ora E
Snow, John Thomas
Snyder, Charles Thomas
Snyder, Nathan W(illiam)
Sovish, Richard Charles
Spitzer, Irwin Asher
Sporer, Alfred Herbert
Srour, Joseph Ralph
Staudhammer, Peter
Steinberg, Gunther
Stevenson, Robin
Stockel, Ivar H(oward)
Stonebraker, Peter Michael
Strahl, Erwin Otto
Suri, Ashok
Sutcliffe, William George
Swain, Robert James
Swanson, Paul N
Swearengen, Jack Clayton
Sweeney, Lawrence Earl, Jr
Syvertson, Clarence A
Taft, David Dakin

Taylor, Charles Joel
Taylor, Robert William
Thomas, Frank J(oseph)
Torgow, Eugene N
Trevelyan, Benjamin John
Trippe, Anthony Philip
Uhl, Arthur E(dward)
Vagelatos, Nicholas
Vane, Arthur B(ayard)
Van Klaveren, Nico
Varon, Myron Izak
Viswanathan, R
Vlay, George John
Vroom, David Archie
Walker, Kelsey, Jr
Wallerstein, Edward Perry
Walsh, Don
Warburton, William Kurtz
Weber, Hans Josef
Werth, Glenn Conrad
Wertheim, Robert Halley
Wertz, James Richard
Westerman, Edwin J(ames)
Wilcox, Charles Hamilton
Wildes, Peter Drury
Williams, Donald Spencer
Winick, Herman
Wolfe, Bertram
Woodbury, Eric John
Wright, Edward Kenneth
Zarem, Abe Mordecai
Zeren, Richard William
Zwerdling, Solomon

COLORADO
Adams, James Russell
Barber, Robert Edwin
Bauer, Charles Edward
Broome, Paul W(allace)
Burdge, David Newman
Cada, Ronald Lee
Chamberlain, A(drian) R(amond)
Chisholm, James Joseph
Cole, Lee Arthur
Crockett, Allen Bruce
Davidson, Darwin Ervin
Dichtl, Rudolph John
Dixon, Robert Clyde
Dunn, Richard Lee
Evans, David Lane
Fickett, Frederick Roland
Fraikor, Frederick John
Gentry, Donald William
Giller, E(dward) B(onfoy)
Harlan, Ronald A
Harrington, Robert D(ean)
Havlick, Spenser Woodworth
Henkel, Richard Luther
Hill, Walter Edward, Jr
Johnson, Arnold I(van)
Kazmerski, Lawrence L
Keller, Frederick Albert, Jr
Koldewyn, William A
Ladd, Conrad Mervyn
Lameiro, Gerard Francis
Legal, Casimer Claudius, Jr
McLean, Francis Glen
Mandics, Peter Alexander
Miner, Frend John
Munro, Richard Harding
Plows, William Herbert
Reichardt, John William
Rothfeld, Leonard B(enjamin)
Steele, Timothy Doak
Tischendorf, John Allen
Vande Noord, Edwin Lee
Wonsiewicz, Bud Caesar

CONNECTICUT
Agarwal, Vipin K
Anderson, R(obert) M(orris), Jr
Andrade, Manuel
Auerbach, Michael Howard
Barney, James Earl, II
Berry, Richard C(hisholm)
Bronsky, Albert J
Burwell, Wayne Gregory
Calehuff, Girard Lester
Carter, G Clifford
Collins, John Barrett
Corwin, H(arold) E(arl)
DeDecker, Hendrik Kamiel Johannes
Dobbs, Gregory Melville
Dooley, Joseph Francis
Duvivier, Jean Fernand
Emond, George T
Epperly, W Robert
Falkenstein, Gary Lee
Feinstein, Myron Elliot
Fletcher, Frederick Bennett
Gans, Eugene Howard
Ghen, David C
Glomb, Walter L
Grattan, James Alex
Grayson, Martin
Greenwood, Ivan Anderson
Grossman, Leonard N(athan)
Haas, Ward John
Hanson, Trevor Russell
Harding, R(onald) H(ugh)
Hayes, Richard J
Howard, Phillenore Drummond
Howard, Robert Eugene
Johnson, Loering M

Juran, Joseph M
Kawaters, Woody H
Landsman, Douglas Anderson
Lines, Ellwood LeRoy
Linke, William Finan
Meschino, Joseph Albert
Mickley, Harold S(omers)
Moreland, Parker Elbert, Jr
Nath, Dilip K
Paterson, Robert W
Peracchio, Aldo Anthony
Pinson, Rex, Jr
Rauhut, Michael McKay
Robinson, Donald W(allace), Jr
Sarada, Thyagaraja
Saunders, Donald Roy
Scherr, Allan L
Schwartz, Michael H
Shah, Shirish A
Shapiro, Warren Barry
Shuey, Merlin Arthur
Sinha, Vinod T(arkeshwar)
Staker, William Paul
Storrs, Charles Lysander
Stott, Paul Edwin
Suen, T(zeng) J(iueq)
Tangel, O(scar) F(rank)
Thomas, Alvin David, Jr
Tittman, Jay
Troutman, Ronald R
Wernau, William Charles
Wingfield, Edward Christian
Ziegler, William Arthur

DELAWARE
Abrahamson, Earl Arthur
Andersen, Donald Edward
Bartkus, Edward Peter
Baum, Martin David
Bingham, Richard Charles
Boettcher, F Peter
Borden, James B
Brown, Stewart Cliff
Busche, Robert M(arion)
Cantwell, Edward N(orton), Jr
Carr, John B
Chait, Edward Martin
Cherkofsky, Saul Carl
Chorvat, Robert John
Cone, Michael McKay
DeDominicis, Alex John
Delp, Charles Joseph
Dubas, Lawrence Francis
Eissner, Robert M
Fleming, Richard Allan
Ginnard, Charles Raymond
Glasgow, Louis Charles
Greenblatt, Hellen Chaya
Harvey, John, Jr
Holob, Gary M
Huppe, Francis Frowin
Jack, John James
Johnson, Alexander Lawrence
Johnson, David Russell
Johnson, David Russell
Kane, Edward R
Kaplan, Ralph Benjamin
Knipmeyer, Hubert Elmer
Kolber, Harry John
Kramer, Brian Dale
Krone, Lawrence James
Lauzon, Rodrigue Vincent
McKay, Sandra J
Marquardt, Donald Wesley
Maynes, Gordon George
Read, Robert E
Sammak, Emil George
Sayre, Clifford M(orrill), Jr
Schappell, Frederick George
Schiewetz, D(on) B(oyd)
Schiroky, Gerhard H
Schwartz, Jerome Lawrence
Shambelan, Charles
Shellenbarger, Robert Martin
Shotzberger, Gregory Steven
Stewart, Edward William
Tanner, David
Taylor, Robert Burns, Jr
Thayer, Chester Arthur
Ulery, Dana Lynn
Urquhart, Andrew Willard
Waritz, Richard Stefen
Webb, Charles Alan
Wermus, Gerald R
Zinser, Edward John

DISTRICT OF COLUMBIA
Adey, Walter Hamilton
Alexander, Benjamin H
April, Robert Wayne
Berlincourt, Ted Gibbs
Bernard, Dane Thomas
Bigelow, Sanford Walker
Bloch, Robert J
Borsuk, Gerald M
Breen, Joseph John
Chen, Davidson Tah-Chuen
Cotruvo, Joseph Alfred
Flora, Lewis Franklin
Forester, Donald Wayne
Gardner, Sherwin
Gilbreath, William Pollock
Goldfield, Edwin David
Green, Richard James

Gunn, John William, Jr
Haas, Gregory Mendel
Hammersmith, John L(eo)
Hartzler, Alfred James
Holloway, John Thomas
Jolly, Janice Laurene Willard
Justus, Philip Stanley
Lapporte, Seymour Jerome
Larsen, Lynn Alvin
Lombard, David Bishop
McBride, Gordon Williams
Miles, Corbin l
Oertel, Goetz K
Peake, Harold J(ackson)
Penhollow, John O
Peters, Gerald Joseph
Reeves, Edmond Morden
Reisa, James Joseph, Jr
Selden, Robert Wentworth
Shaub, Walter M
Shaw, Robert R
Spurlock, Langley Augustine
Sterrett, Kay Fife
Stroup, Cynthia Roxane
Tokar, Michael
Trubatch, Sheldon L
Urban, Peter Anthony
Wade, Clarence W R
Waters, Robert Charles
Weinberg, Myron Simon
Wilkniss, Peter Eberhard
Wilsey, Neal David
Wilson, George Donald
Zwilsky, Klaus M(ax)

FLORIDA
Amey, William G(reenville)
Aronson, M(oses)
Baumann, Arthur Nicholas
Berry, Robert Eddy
Bingham, Richard S(tephen), Jr
Bishop, Charles Anthony
Black, William Bruce
Callan, Edwin Joseph
Chao, Raul Edward
Chryssafopoulos, Nicholas
Clark, Allen Varden
Clark, Mary Elizabeth
Coppoc, William Joseph
Coulter, Wallace H
Dailey, Charles E(lmer), III
Daniel, Donald Clifton
Dannenberg, E(li) M
Davis, Duane M
DeHart, Arnold O'Dell
Doyon, Leonard Roger
Dreves, Robert G(eorge)
Everett, Woodrow Wilson, Jr
Frazier, Stephen Earl
Givens, Paul Edward
Gold, Edward
Griffith, John E(dward)
Haase, Richard Henry
Haeger, Beverly Jean
Hagenmaier, Robert Doller
Halpern, Ephriam Philip
Hammond, Charles Eugene
Hannan, Roy Barton, Jr
Haug, Arthur John
Heggie, Robert
Heying, Theodore Louis
Hirsch, Donald Earl
Jackel, Simon Samuel
Jones, Walter H(arrison)
Karp, Abraham E
Kiewit, David Arnold
Klock, Benny LeRoy
Kolhoff, M(arvin) J(oseph)
Long, Calvin V
Lowe, Ronald Edsel
McGray, Robert James
McKently, Alexandra H
McSwain, Richard Horace
Mendell, Jay Stanley
Migliaro, Marco William
Milton, Robert Mitchell
Moser, Robert E
Myers, Richard Lee
Nimmo, Bruce Glen
Normile, Hubert Clarence
Nunnally, Stephens Watson
Oline, Larry Ward
Owen, Gwilym Emyr, Jr
Parker, Howard Ashley, Jr
Peterson, James Robert
Poet, Raymond B
Rathjen, Warren Francis
Rosenthal, Henry Bernard
Seugling, Earl William, Jr
Sherman, Edward
Stern, Milton
Swart, William W
Tebbe, Dennis Lee
Van Ligten, Raoul Fredrick
Wiseman, Robert S
Wolsky, Sumner Paul
Yon, E(ugene) T
Young, Charles Gilbert
Ziernicki, Robert S

GEORGIA
Blanton, Jackson Orin
Camp, Ronnie Wayne
Coyne, Edward James, Jr

Derrick, Robert P
Donahoo, Pat
Evans, Thomas P
Fineman, Manuel Nathan
Gambrell, Carroll B(lake), Jr
Goldsmith, Edward
Grace, Donald J
Griffin, Clayton Houstoun
Jacobs, R(oy) K(enneth)
Johnson, Donald Ross
Krajca, Kenneth Edward
Krochmal, Jerome J(acob)
Leffingwell, John C
Love, Jimmy Dwane
Meyer, Leon Herbert
Myers, Dirck V
Nethercut, Philip Edwin
Orofino, Thomas Allan
Owens, William Richard
Porter, Alan Leslie
Russell, John Lynn, Jr
Schulz, David Arthur
Shakun, Wallace
Simmons, George Allen
Sommers, Jay Richard
Starr, Thomas Louis
Swanson, David Henry
Thompson, William Oxley, II
Vachon, Reginald Irenee
Vicory, William Anthony
Ware, Kenneth Dale
Wiltse, James Cornelius
Woodall, William Robert, Jr
Yeske, Ronald A

HAWAII
Bixler, Otto C
Cruickshank, Michael James
Curtis, George Darwin
Hallanger, Lawrence William
Khan, Mohammad Asad
Wylly, Alexander

IDAHO
Bickel, John Henry
Bills, Charles Wayne
Carpenter, Stuart Gordon
Christian, Jerry Dale
East, Larry Verne
Engen, Ivar A
Gesell, Thomas Frederick
Griffith, W(illiam) A(lexander)
Lambuth, Alan Letcher
Lawroski, Harry
Tracy, Joseph Charles
Wood, Richard Ellet

ILLINOIS
Agarwal, Ashok Kumar
Anderson, Arnold Lynn
Ankenbrandt, Charles Martin
Assanis, Dennis N
Ban, Stephen Dennis
Bartimes, George F
Bartlett, J Frederick
Bays, Karl D
Bell, Charles Eugene, Jr
Benzinger, William Donald
Booth, David Layton
Bowles, William Allen
Brezinski, Darlene Rita
Broach, Robert William
Carrier, Steven Theodore
Christianson, Clinton Curtis
Chu, Daniel Tim-Wo
Clark, John Peter, III
Clark, Stephen Darrough
Collier, Donald W(alter)
Conn, Arthur L(eonard)
Danzig, Morris Juda
Davis, Edward Nathan
Deitrich, L(awrence) Walter
Devgun, Jas S
Dolgoff, Abraham
Dybel, Michael Wayne
Ehrlich, Richard
Empen, Joseph A
Feldman, Fred
Flinn, James Edwin
Flowers, Russell Sherwood, Jr
Galvin, Robert W
Gemmer, Robert Valentine
Gibbons, Larry V
Gillette, Richard F
Glaser, Milton Arthur
Goldman, Arthur Joseph
Griffith, James H
Hamer, Martin
Harris, Ronald David
Hatfield, Efton Everett
Heaston, Robert Joseph
Henning, Lester Allan
Hersh, Herbert N
Herskovits, Lily Eva
Hoffman, Gerald M
Hojvat, Carlos F
Irvin, Howard H
Ivins, Richard O(rville)
James, Karen K(anke)
Jarke, Frank Henry
Johari, Om
Johnson, Calvin Keith
Johnson, Dale Howard
Kamm, Gilbert G(eorge)

Technical Management (cont)

Kennedy, Albert Joseph
King, S(anford) MacCallum
Klein, Morton Joseph
Krein, Philip Theodore
Kyle, Martin Lawrence
Lemper, Anthony Louis
Linde, Ronald K(eith)
Linder, Louis Jacob
Lustig, Stanley
Mantsch, Paul Matthew
Marchaterre, John Frederick
Marcus, Mark
Marmer, Gary James
Martinec, Emil Louis
Massel, Gary Alan
Mateles, Richard I
Maynard, Theodore Roberts
Menke, Andrew G
Meter, Donald M(ervyn)
Moore, John Fitzallen
Moser, Kenneth Bruce
Muhlenbruch, Carl W(illiam)
Perino, Janice Vinyard
Pesyna, Gail Marlane
Pewitt, Edward Gale
Pietri, Charles Edward
Plautz, Donald Melvin
Prais, Michael Gene
Princen, Lambertus Henricus
Richmond, James M
Richmond, Patricia Ann
Rusinko, Frank, Jr
Scheie, Carl Edward
Sheppard, John Richard
Sill, Larry R
Smith, Richard Paul
Smittle, Richard Baird
Spector, Leo Francis
Staats, William R
Steindler, Martin Joseph
Stetter, Joseph Robert
Studtmann, George H
Stull, Elisabeth Ann
Sutton, Lewis McMechan
Swoyer, Vincent Harry
Theriot, Edward Dennis, Jr
Thompson, Charles William Nelson
Thompson, David Jerome
Tomkins, Marion Louise
Travelli, Armando
Trevillyan, Alvin Earl
Unik, John Peter
Wargel, Robert Joseph
Weil, Thomas Andre
Wensch, Glen W(illiam)
Wetegrove, Robert Lloyd
Whaley, Thomas Patrick
Wiegand, Ronald Gay
Wingender, Ronald John
Wittman, James Smythe, III
Wolf, Richard Eugene
Wolsky, Alan Martin
Wynveen, Robert Allen
Yao, Neng-Ping
Yarger, James G
Zeffren, Eugene

INDIANA

Case, Vernon Wesley
Contario, John Joseph
DeSantis, John Louis
English, Robert E
Free, Helen M
Gantzer, Mary Lou
Gorman, Eugene Francis
Gresham, Robert Marion
Halpin, Daniel William
Huber, Don Morgan
Hunt, Michael O'Leary
Jackson, John Eric
Jacobs, Martin Irwin
Jefferies, Michael John
Kammann, Karl Philip, Jr
Kendall, Perry E(ugene)
Khorana, Brij Mohan
Klein, Howard Joseph
Koch, Kay Frances
Lazaridis, Nassos A(thanasius)
McAleece, Donald John
McIntyre, Thomas William
Markee, Katherine Madigan
Mazac, Charles James
Mercer, Walter Ronald
Morris, David Alexander Nathaniel
Morris, John F
Nirschl, Joseph Peter
Orr, William H(arold)
Osuch, Mary Ann V
Palmer, Robert Gerald
Pielet, Howard M
Putnam, Thomas Milton
Ranade, Madhukar G
Robertson, David Wayne
Sannella, Joseph L
Shoup, Charles Samuel, Jr
Sieloff, Ronald F
Somerville, Ronald Lamont
Stroube, William Bryan, Jr
Tennyson, Richard Harvey
Tyhach, Richard Joseph
White, Samuel Grandford, Jr

IOWA

Edelson, Martin Charles
Egger, Carl Thomas
Hauenstein, Jack David
Hekker, Roeland M T
Kiser, Donald Lee
Kniseley, Richard Newman
Kohn, Frank S
Papadakis, Emmanuel Philippos
Riley, William F(ranklin)

KANSAS

Cohen, Jules Bernard
Connor, Sidney G
Crowther, Robert Hamblett
Frazier, A Joel
Grant, Stanley Cameron
Jacobs, Louis John
McBean, Robert Parker
Sutherland, John B(ennett)

KENTUCKY

Crowe, Richard Godfrey, Jr
Datta, Vijay J
Field, Jay Ernest
Gold, Harold Eugene
Goldknopf, Ira Leonard
Kary, Christina Dolores
Kohnhorst, Earl Eugene
Lauterbach, John Harvey
Merker, Stephen Louis
Mihelich, John L
Squires, Robert Wright
Tran, Long Trieu
Wells, William Lochridge

LOUISIANA

Bauer, Dennis Paul
Callens, Earl Eugene, Jr
Conrad, Franklin
DeLeon, Ildefonso R
Fischer, David John
Gibson, David Michael
Holloway, Frederic Ancrum Lord
Keys, L Ken
Kingrea, C(harles) L(eo)
Kordoski, Edward William
Mills, Earl Ronald
Overton, Edward Beardslee
Parish, Richard Lee
Polack, Joseph A(lbert)
Rosene, Robert Bernard
Shubkin, Ronald Lee
Smith, Robert Leonard
Tewell, Joseph Robert
Wiewiorowski, Tadeusz Karol
Wrobel, William Eugene
Zaweski, Edward F
Zietz, Joseph R, Jr

MAINE

Founds, Henry William
Hill, Marquita K
Shipman, C(harles) William
Wilms, Hugo John, Jr

MARYLAND

Abramson, Fredric David
Adamantiades, Achilles G
Anderson, W(endell) L
Atia, Ali Ezz Eldin
Baker, Francis Edward, Jr
Beach, Eugene Huff
Bersch, Charles Frank
Blackwell, Lawrence A
Blevins, Gilbert Sanders
Bournia, Anthony
Brady, Robert Frederick, Jr
Brenner, Alfred Ephraim
Bryan, Edward H
Buck, John Henry
Bugenhagen, Thomas Gordon
Butler, Louis Peter
Carrico, Christine Kathryn
Carver, Gary Paul
Cave, William Thompson
Chimenti, Dale Everett
Chmurny, Alan Bruce
Clark, Joseph E(dward)
Conklin, James Byron, Jr
Corak, William Sydney
Cornyn, John Joseph
Cotton, Ira Walter
Daniels, William Fowler, Sr
Decker, James Federick
Degnan, John James, III
Deutsch, Robert W(illiam)
Eades, James B(everly), Jr
Edinger, Stanley Evan
Egner, Donald Otto
Eng, Leslie
Estrin, Norman Frederick
Evans, Larry Gerald
Fabic, Stanislav
Fairweather, William Ross
Feldman, Alfred Philip
Feldmesser, Julius
Fink, Daniel J
Fischel, David
Fleig, Albert J, Jr
Frank, Carolyn
Frederick, William George DeMott
Gann, Richard George
Gartrell, Charles Frederick

Garver, Robert Vernon
Gelberg, Alan
Gevantman, Lewis Herman
Gilmore, Thomas Meyer
Gleissner, Gene Heiden
Glick, J Leslie
Goldberg, Michael Ian
Goldschmidt, Peter Graham
Gonano, John Roland
Graebert, Eric W
Gupta, Ramesh K
Hader, Rodney N(eal)
Haffner, Richard William
Hall, Kimball Parker
Hamilton, Bruce King
Handwerker, Thomas Samuel
Hartmann, Gregory Kemenyi
Heath, George A(ugustine)
Heller, Charles O(ta)
Hellwig, Helmut Wilhelm
Hertz, Harry Steven
Holt, Stephen S
Hopenfield, Joram
Hopkins, Homer Thawley
Hsia, Jack Jinn-Goe
Humphrey, S(idney) Bruce
Huneycutt, James Ernest, Jr
Johnson, David Lee
Johnson, David Simonds
Jones, Theodore Charles
Kascic, Michael Joseph, Jr
Kayser, Richard Francis
Kazi, Abdul Halim
Kehr, Clifton Leroy
Keily, Hubert Joseph
Kiebler, John W(illiam)
Kirby, Ralph C(loudsberry)
Kirkpatrick, Diana (Rorabaugh) M
Knapp, Harold Anthony, Jr
Koch, William Frederick
Kutik, Leon
Lankford, J(ohn) L(lewellyn)
Leight, Walter Gilbert
Lettieri, Thomas Robert
Lundholm, J(oseph) G(ideon), Jr
McElhinney, John
McIlvain, Jess Hall
Mandelberg, Hirsch I
Marciani, Dante Juan
Marple, Stanley Lawrence, Jr
Michaelis, Michael
Miller, A Eugene
Miller, Raymond Earl
Mohler, William C
Morgan, Walter L(eroy)
Morgan, William Bruce
Morton, Joseph James Pandozzi
Moshman, Jack
Muller, Robert E
Mumford, Willard R
Murray, William Sparrow
Nagle, Dennis Charles
Nash, Jonathon Michael
Orlick, Charles Alex
Ostaff, William A(llen)
Peppin, Richard J
Pierce, Harry Frederick
Poltorak, Andrew Stephen
Pomerantz, Irwin Herman
Potocki, Kenneth Anthony
Pratt, Thomas Herring, Jr
Prentice, Geoffrey Allan
Quinn, Thomas Patrick
Randolph, Lynwood Parker
Rasberry, Stanley Dexter
Renne, David Smith
Reuther, Theodore Carl, Jr
Rochlin, Robert Sumner
Rommel, Frederick Allen
Rosado, John Allen
Rosen, Milton W(illiam)
Roth, Michael William
Rountree, Janet
Saltman, Roy G
Schemm, Charles Edward
Schleiter, Thomas Gerard
Schmidt, Jack Russell
Schroeder, Frank, Jr
Schulman, James Herbert
Schultz, Warren Walter
Schuman, William John, Jr
Schwoerer, F(rank)
Sherman, Anthony Michael
Smith, Joseph Collins
Statt, Terry G
Steiner, Henry M
Strombotne, Richard L(amar)
Stuebing, Edward Willis
Stuelpnagel, John Clay
Sztankay, Zoltan Geza
Talbott, Edwin M
Theuer, Paul John
Thompson, Robert John, Jr
Tropf, Cheryl Griffiths
Troutman, James Scott
Turnrose, Barry Edmund
Wacker, George Adolf
Wallace, Alton Smith
Walsh, James Paul
Walter, John Fitler
Wang, Shou-Ling
Wende, Charles David
Wing, James
Wortman, Roger Matthew

Yoho, Clayton W

MASSACHUSETTS

Adler, Ralph Peter Isaac
Ahern, David George
Allen, Thomas John
Avila, Charles Francis
Bahr, Karl Edward
Bailey, Fred Coolidge
Bass, Arthur
Bhargava, Hridaya Nath
Brackett, John Washburn
Brandler, Philip
Breslau, Barry Richard
Brown, Walter Redvers John
Buckler, Sheldon A
Buono, John Arthur
Burke, John Michael
Callerame, Joseph
Carton, Edwin Beck
Cathou, Renata Egone
Clopper, Herschel
Close, R(ichard) N(orcross)
Cohen, Fredric Sumner
Colby, George Vincent, Jr
Deckert, Cheryl A
Dorman, Craig Emery
Dow, Paul C(rowther), Jr
Dudgeon, Dan Ed
Duffy, Robert A
Eaves, Reuben Elco
Eby, Robert Newcomer
Eklund, Karl E
Fennelly, Paul Francis
Fix, Richard Conrad
Freedman, George
Furbish, Francis Scott
Gardner, Donald Murray
Gelb, Arthur
Gersh, Michael Elliot
Giger, Adolf J
Gordon, Bernard M
Gray, Douglas Carmon
Gregory, Bob Lee
Griffin, Paul Joseph
Griffith, Robert W
Guidice, Donald Anthony
Haber, Stephen B
Hadlock, Charles Robert
Hanselman, Raymond Bush
Hearn, David Russell
Hertweck, Gerald
Herz, Matthew Lawrence
Heyerdahl, Eugene Gerhardt
Hills, Robert, Jr
Hodgins, George Raymond
Holmes, Douglas Burnham
Jalan, Vinod Motilal
Kachinsky, Robert Joseph
Karo, Douglas Paul
Keast, David N(orris)
Kelley, William S
Kepper, Robert Edgar
Kitchin, John Francis
Kleinmann, Douglas Erwin
Kocon, Richard William
Kolodzy, Paul John
Kovatch, George
Kozma, Adam
Kramer, Charles Edwin
Krause, Irvin
Kumar, Kaplesh
Lavin, Philip Todd
Lederman, David Mordechai
Lee, D(on) William
Lee, Eric Kin-Lam
Levine, Randolph Herbert
Lien, Hwachii
Liggero, Samuel Henry
Luft, Ludwig
McClatchey, Robert Alan
McCrea, Peter Frederick
McGrath, W Patrick
Mack, Richard Bruce
Manley, Harold J(ames)
Manning, Monis George
Maser, Morton D
Massa, Frank
Matsuda, Seigo
Meyer, Vincent D
Miffitt, Donald Charles
Milligan, Terry Wilson
Muni, Indu A
Nelson, J(ohn) Byron
Neuringer, Leo J
Nickerson, Richard G
Oberteuffer, John Amiard
Olsen, Kenneth
Plummer, William Torsch
Poirier, Victor L
Riggi, Stephen Joseph
Roberts, Francis Donald
Rogers, William Irvine
Rosenthal, Richard Alan
Roth, Harold
Rubin, Lawrence G
Sagalyn, Paul Leon
Sahatjian, Ronald Alexander
Salant, Abner Samuel
Schaffel, Gerson Samuel
Schempp, Ellory
Scherer, Harold Nicholas, Jr
Seamans, R(obert) C(hanning), Jr
Servi, I(talo) S(olomon)

Shaffer, Harry Leonard
Shu, Larry Steven
Shusman, T(evis)
Silevich, Michael B
Silver, Frank Morris
Sioui, Richard Henry
Sonnichsen, Harold Marvin
Sosnowski, Thomas Patrick
Spergel, Philip
Stanbrough, Jesse Hedrick, Jr
Stevens, J(ames) I(rwin)
Straub, Wolf Deter
Strelzoff, Alan G
Thornton, Richard D(ouglas)
Topping, Richard Francis
Towns, Donald Lionel
Tsandoulas, Gerasimos Nicholas
Turnquist, Carl Richard
Van Meter, David
Wald, Fritz Veit
Wasserman, Jerry
Wentworth, Stanley Earl
Wheeler, Ned Brent
Whitehouse, David R(empfer)
Winett, Joel M
Winkelman, James W
Wood, David Belden
Wyler, John Stephen
Yannoni, Nicholas

MICHIGAN

Aaron, Charles Sidney
Anderson, Amos Robert
Atwell, William Henry
Axen, Udo Friedrich
Ayres, Robert Allen
Bailey, Donald Leroy
Becher, William D(on)
Belmont, Daniel Thomas
Bens, Frederick Peter
Blaser, Dwight A
Borcherts, Robert H
Braughler, John Mark
Brown, Steven Michael
Castenson, Roger R
Chandra, Grish
Chen, Kan
Chon, Choon Taik
Crummett, Warren B
Cunningham, Fay Lavere
Cutler, Warren Gale
Evaldson, Rune L
Everett, Robert Line
Ezzat, Hazem Ahmed
Felmlee, William John
Flynn, John M(athew)
Francis, Ray Llewellyn
Frank, Max
Galan, Louis
Garvin, Donald Frank
Gutzman, Philip Charles
Halberstadt, Marcel Leon
Harris, Kenneth
Hart, Robert G
Hartman, John L(ouis)
Harwood, Julius J
Jensen, David James
Jones, Frederick Goodwin
Joshi, Mukund Shankar
Kimball, Frances Adrienne
Klimisch, Richard L
Knoll, Alan Howard
Langvardt, Patrick William
Larson, John Grant
Lentz, Charles Wesley
Lorch, Steven Kalman
Macriss, Robert A
Malchick, Sherwin Paul
Maley, Wayne A
Medick, Matthew A
Moore, Eugene Roger
Murphy, William R
Petrick, Ernest N(icholas)
Phillips, Charles W(illiam)
Pincus, Jack Howard
Reynolds, John Z
Riley, Bernard Jerome
Rohde, Steve Mark
Rollinger, Charles N(icholas)
Rosinski, Michael A
Roth, Norman Gilbert
Salinger, Rudolf Michael
Sanders, Barbara A
Sard, Richard
Schumacher, Berthold Walter
Shahabuddin, Syed
Spurgeon, William Marion
Stark, Forrest Otto
Stempel, Robert C
Stevens, Violete L
Stowe, Robert Allen
Thompson, James Lowry
Thurber, William Samuels
Trachman, Edward G
Turley, June Williams
Van Rheenen, Verlan H
Walters, Stephen Milo
Weyenberg, Donald Richard
Whitfield, Richard George
Williams, Sam B
Wims, Andrew Montgomery
Wolman, Eric
Wymore, Charles Elmer

MINNESOTA

Atchison, Thomas Calvin, Jr
Belmont, Arthur David
Bohon, Robert Lynn
Bolles, Theodore Frederick
Britz, Galen C
Buchholz, Allan C
Case, Marvin Theodore
Davidson, Donald
Dinneen, Gerald Paul
Durnick, Thomas Jackson
Fleming, Peter B
Glewwe, Carl W(illiam)
Greene, Christopher Storm
Hagemark, Kjell Ingvar
Hapke, Bern
Haun, J(ames) W(illiam)
Heberlein, Joachim Viktor Rudolf
Hoffmann, Thomas Russell
Holmen, Reynold Emanuel
Jansen, Bernard Joseph
Johnson, Bryce Vincent
Johnson, Howard Arthur, Sr
Johnson, Lennart Ingemar
Joseph, Earl Clark, II
Katz, William
Keitel, Glenn H(oward)
Owens, Boone Bailey
Pahl, Walter Henry
Penney, William Harry
Plovnick, Ross Harris
Premanand, Visvanatha
Pryor, Gordon Roy
Rambosek, G(eorge) M(orris)
Reich, Charles
Reinhart, Richard D
Reynolds, James Harold
Rodgers, Nelson Earl
Sackett, W(illiam) T(ecumseh), Jr
Sadjadi, Firooz Ahmadi
Sherck, Charles Keith
Stearns, Eugene Marion, Jr
Subach, Daniel James
Talbott, Richard Lloyd
Tamsky, Morgan Jerome
Weiss, Douglas Eugene
Yapel, Anthony Francis, Jr
Zeyen, Richard John

MISSISSIPPI

Brunton, George Delbert
Cotton, Frank Ethridge, Jr
Ferer, Kenneth Michael
George, Clifford Eugene
Gupta, Shyam Kirti
Morris, Halcyon Ellen McNeil

MISSOURI

Anderson, David W, Jr
Babcock, Daniel Lawrence
Bachman, Gerald Lee
Baltz, Howard Burl
Beaver, Earl Richard
Bodine, Richard Shearon
Bosanquet, Louis Percival
Carter, Don E
Craver, Clara Diddle (Smith)
Craver, John Kenneth
Dahl, William Edward
Deam, James Richard
Gash, Virgil Walter
Glauz, William Donald
Godt, Henry Charles, Jr
Griffith, Virgil Vernon
Hoffman, Jerry C
Holloway, Thomas Thornton
Johnson, Richard Dean
Keller, Robert Ellis
Kozlowski, Don Robert
Kramer, Richard Melvyn
Krueger, Paul A
Kurz, James Eckhardt
Lannert, Kent Philip
Linder, Solomon Leon
Major, Schwab Samuel, Jr
Malik, Joseph Martin
Morgan, Wyman
Munger, Paul R
Porter, Clark Alfred
Poynton, Joseph Patrick
Privott, Wilbur Joseph, Jr
Proctor, Stanley Irving, Jr
Radke, Rodney Owen
Rash, Jay Justen
Rinehart, Walter Arley
Rueppel, Melvin Leslie
Ryan, Carl Ray
Sarchet, Bernard Reginald
Scallet, Barrett Lerner
Schott, Jeffrey Howard
Schwartz, Henry Gerard, Jr
Sidoti, Daniel Robert
Thornton, Linda Wierman
Volz, William K(urt)
Wehrmann, Ralph F(rederick)
Wochok, Zachary Stephen
Wohl, Martin H
Worley, Jimmy Weldon

NEBRASKA

James, Merlin Lehn
Kolar, Joseph Robert, Jr
Wittke, Dayton D

NEVADA

Holdbrook, Ronald Gail
Klainer, Stanley M
Manhart, Robert (Audley)
Wegst, Walter F, Jr

NEW HAMPSHIRE

Block, James A
Carreker, R(oland) P(olk), Jr
Creagh-Deyter, Linda T
Egan, John Frederick
Ge, Weikun
Petrasek, Emil John
Wolff, Nikolaus Emanuel

NEW JERSEY

Adams, James Mills
Aldrich, Haven Scott
Andose, Joseph D
Andrews, Ronald Allen
Auerbach, Victor
Bartnoff, Shepard
Battisti, Angelo James
Beck, John Louis
Belohoubek, Erwin F
Bendure, Raymond Lee
Bernstein, Lawrence
Bieber, Harold H
Bieber, Herman
Bodin, Jerome Irwin
Bonacci, John C
Bond, Robert Harold
Borkan, Harold
Boucher, Thomas Owen
Bowen, J Hartley, Jr
Bowers, Klaus Dieter
Boyle, Richard James
Brandman, Harold A
Brill, Yvonne Claeys
Buckley, R Russ
Burley, David Richard
Burrill, Claude Wesley
Buyske, Donald Albert
Cardarelli, Joseph S
Carlson, Curtis Raymond
Carnes, James Edward
Carter, Ashley Hale
Castenschiold, Rene
Chance, Ronald Richard
Chester, Arthur Warren
Chynoweth, Alan Gerald
Cipriani, Cipriano
Coombs, Robert Victor
Cornell, W(arren) A(lvan)
Cox, Donald Clyde
Crane, Laura Jane
Curcio, Lawrence Nicholas
Curry, Michael Joseph
Daly, Daniel Francis, Jr
Davis, Lance A(lan)
Diegnan, Glenn Alan
DiSalvo, Walter A
Domeshek, S(ol)
Donovan, Richard C
Dorros, Irwin
Edelstein, Harold
Eisenberger, Peter Michael
Eisner, Mark Joseph
Erckel, Ruediger Josef
Erdmann, Duane John
Falconer, Warren Edgar
Feldman, Leonard Cecil
Finch, Rogers B(urton)
Fischell, David R
Flam, Eric
Fowles, Patrick Ernest
Francis, Samuel Hopkins
Frankoski, Stanley P
Franks, Richard Lee
Freedman, Henry Hillel
Friedman, Edward Alan
Froyd, James Donald
Garber, H(irsh) Newton
Garfinkel, Harmon Mark
Gierer, Paul L
Gitlitz, Melvin H
Goffman, Martin
Goldemberg, Robert Lewis
Gourary, Barry Sholom
Grauman, Joseph Uri
Gregory, Richard Alan, Jr
Grenda, Victor J
Groeger, Theodore Oskar
Grubman, Wallace Karl
Grumer, Eugene Lawrence
Guenzer, Charles S P
Gund, Peter Herman Lourie
Gyan, Nanik D
Hagel, Robert B
Halik, Raymond R(ichard)
Harding, Maurice James Charles
Harris, James Ridout
Hartung, John
Hatch, Charles Eldridge, III
Hay, Peter Marsland
Heath, Carl E(rnest), Jr
Heck, James Virgil
Heirman, Donald N
Hendrickson, Tom A(llen)
Herdklotz, John Key
Hlavacek, Robert John
Hubbard, William Marshall
Hundert, Murray Bernard
Hymans, William E

Inglis, James
Jaffe, Jonah
Jayant, Nuggehally S
Johnston, John Eric
Kagan, Harvey Alexander
Kahn, Donald Jay
Kaufmann, Thomas G(erald)
Kesselman, Warren Arthur
King, John A(lbert)
Knapp, Malcolm Hammond
Kobrin, Robert Jay
Kronenthal, Richard Leonard
Kruh, Daniel
Ladner, David William
LaMastro, Robert Anthony
Landau, Edward Frederick
Langseth, Rollin Edward
Lasky, Jack Samuel
Lawson, James Robert
Leaning, William Henry Dickens
Lee, Peter Chung-Yi
Leitz, Victoria Mary
Letterman, Herbert
Levine, Aaron William
Lewis, Arnold D
Liebig, William John
London, Mark David
Lorenz, Patricia Ann
Lyding, Arthur R
McCallum, Charles John, Jr
McCumber, Dean Everett
McDermott, Kevin J
McGuire, David Kelty
Machalek, Milton David
Mackerer, Carl Robert
Mandel, Andrea Sue
Manz, August Frederick
Martin, John David
Melveger, Alvin Joseph
Michaelis, Paul Charles
Michel, Gerd Wilhelm
Mikolajczak, Alojzy Antoni
Montana, Anthony J
Montgomery, Richard Millar
Moros, Stephen Andrew
Moshey, Edward A
Munger, Stanley H(iram)
Murthy, Srinivasa K R
Nelson, Roger Edwin
Opie, William R(obert)
Pawson, Beverly Ann
Pebly, Harry E
Perry, Erik David
Peters, Dale Thompson
Philipp, Ronald E
Popper, Robert David
Procknow, Donald Eugene
Putter, Irving
Rankin, Sidney
Rau, Eric
Reboul, Theo Todd, III
Rein, Alan James
Ren, Peter
Rich, Kenneth Eugene
Robinson, Jerome David
Robrock, Richard Barker, II
Rorabaugh, Donald T
Roukes, Michael L
Rubinstein, Charles B(enjamin)
Rulf, Benjamin
Sachs, Harvey Maurice
Sagal, Matthew Warren
Sausville, Joseph Winston
Sawin, Steven P
Scharfstein, Lawrence Robert
Schmidle, Claude Joseph
Schnapf, Abraham
Schoenfeld, Theodore Mark
Schumer, Douglas B
Schutz, Donald Frank
Schwarz, Maurice Jacob
Schwenker, Robert Frederick, Jr
Sharp, Hugh T
Sherman, Ronald
Shrier, Adam Louis
Siebert, Donald Robert
Sipress, Jack M
Slagel, Robert Clayton
Smith, William Bridges
Snowman, Alfred
Sorenson, Wayne Richard
Spohn, Ralph Joseph
Stinson, Mary Krystyna
Strausberg, Sanford I
Sweed, Norman Harris
Tauber, Arthur
Thelin, Lowell Charles
Toothill, Richard B
Tucci, Edmond Raymond
Vaughn, Robert Donald
Venuto, Paul B
Vogt, Herwart Curt
Wangemann, Robert Theodore
Warters, William Dennis
Watson, Richard White, Jr
Weinzimmer, Fred
Werthamer, N Richard
Wesner, John William
West, John M(aurice)
Westerdahl, Raymond P
White, Alice Elizabeth
Wildman, George Thomas
Wildnauer, Richard Harry
Wittick, James John

Technical Management (cont)

Zager, Ronald

NEW MEXICO
Altrock, Richard Charles
Arion, Douglas
Asbridge, John Robert
Baker, Floyd B
Barr, Donald Westwood
Bartlit, John R(ahn)
Berrie, David William
Burick, Richard Joseph
Butler, Harold S
Caldwell, John Thomas
Catlett, Duane Stewart
Chambers, William Hyland
Clark, Robert Paul
Clough, Richard H(udson)
Colp, John Lewis
Conner, Jerry Power
Cramer, James D
Cunningham, Paul Thomas
Danen, Wayne C
Davison, Lee Walker
Diegle, Ronald Bruce
Dowdy, Edward Joseph
Dreicer, Harry
Dumas, Herbert M, Jr
Erikson, Mary Jane
Farnum, Eugene Howard
Finch, Thomas Wesley
Franzak, Edmund George
Freese, Kenneth Brooks
Gancarz, Alexander John
Garcia, Carlos E(rnesto)
Gibbs, Terry Ralph
Gibson, Benjamin Franklin, V
Gillespie, Claude Milton
Goldstone, Philip David
Gover, James E
Gwyn, Charles William
Hansen, Wayne Richard
Hartley, Danny L
Henderson, Dale Barlow
Hill, Ronald Ames
Holmes, John Thomas
Hopkins, Alan Keith
Howell, Gregory A
Icerman, Larry
Jameson, Robert A
Janney, Donald Herbert
Jeffries, Robert Alan
Johnson, Ralph T, Jr
Jones, Kirkland Lee
Jones, Merrill C(alvin)
Jones, Orval Elmer
Keenan, Thomas K
King, Nicholas S P
Kjeldgaard, Edwin Andreas
Krupka, Milton Clifford
Kuswa, Glenn Wesley
Lohrding, Ronald Keith
Lyons, Peter Bruce
McCulla, William Harvey
Marcy, Willard
Mason, Allen Smith
Merritt, Melvin Leroy
Molecke, Martin A
Neeper, Donald Andrew
Newman, Michael J(ohn)
Osbourn, Gordon Cecil
Otway, Harry John
Pappas, Daniel Samuel
Peercy, Paul S
Perkins, Roger Bruce
Phister, Montgomery, Jr
Pierce, Robert Wesley
Pynn, Roger
Renken, James Howard
Robinson, Ross Utley
Ryder, Richard Daniel
Sandstrom, Donald James
Savage, Charles Francis
Schneider, Jacob David
Seay, Glenn Emmett
Seiler, Fritz A
Shipley, James Parish, Jr
Sivinski, Jacek Stefan
Solem, Johndale Christian
Sullivan, J Al
Tapp, Charles Millard
Tesmer, Joseph Ransdell
Thode, E(dward) F(rederick)
Toepfer, Alan James
Van Domelen, Bruce Harold
Venable, Douglas
Wallace, Terry Charles
Welber, Irwin
White, Paul C
Williamson, Kenneth Donald, Jr

NEW YORK
Abetti, Pier Antonio
Andersen, Kenneth J
Armstrong, John A
Bachman, Henry L(ee)
Bahary, William S
Batter, John F(rederic), Jr
Bays, Jackson Darrell
Benzinger, James Robert
Birnbaum, David
Birnstiel, Charles
Blackmon, Bobby Glenn

Blaker, J Warren
Bolon, Roger B
Bottger, Gary Lee
Bray, James William
Breuer, Charles B
Britton, Marvin Gale
Buchel, Johannes A
Buhks, Ephraim
Burnham, Dwight Comber
Calabrese, Carmelo
Chi, Chao Shu
Chiang, Yuen-Sheng
Chiulli, Angelo Joseph
Christensen, Robert Lee
Christiansen, Donald David
Cohen, Edwin
Collins, James Joseph
Comer, William Timmey
Conn, Paul Kohler
Coutchie, Pamela Ann
Dahm, Douglas Barrett
Daniels, John Maynard
Dehm, Richard Lavern
Dieterich, David Allan
Drozdowicz, Zbigniew Marian
Duke, David Allen
Eldin, Hamed Kamal
Eschbach, Charles Scott
Farley, Thomas Albert
Feit, Irving N
Feldman, Larry Howard
Fey, Curt F
Fiedler, Harold Joseph
Figueras, Patricia Ann McVeigh
Fiske, Milan Derbyshire
Flannery, John B, Jr
Foley, Dennis Joseph
Formanek, R(obert) J(oseph)
Foucher, W(alter) D(avid)
Fuerst, Adolph
Gale, Stephen Bruce
Galginaitis, Simeon Vitis
Gambs, Gerard Charles
Gaudioso, Stephen Lawrence
Genin, Dennis Joseph
George, Philip Donald
Gerace, Paul Louis
Goldman, James Allan
Goldstein, Lawrence Howard
Golibersuch, David Clarence
Gomory, Ralph E
Goodwin, Arthur VanKleek
Gray, D Anthony
Green, Robert I
Grossman, Norman
Haas, Werner E L
Haase, Jan Raymond
Haid, D(avid) A(ugustus)
Halket, Thomas D
Hall, Richard Travis
Hall, William Myron, Jr
Hammer, Richard Benjamin
Harris, Bernard
Harrison, James Beckman
Hauth, Willard E(llsworth), III
Haynes, Martha Patricia
Hedman, Dale E
Henkel, Elmer Thomas
Hersh, Charles K(arrer)
Himpsel, Franz J
Hollenberg, Joel Warren
Horton, Thomas Roscoe
Hull, Andrew P
Hunt, Everett Clair
Indusi, Joseph Paul
Insalata, Nino F
James, Daniel Shaw
Janik, Gerald S
Jensen, Betty Klainminc
Kagetsu, T(adashi) J(ack)
Kahn, Elliott H
Kammer, Paul A
Kasprzak, Lucian A
Kato, Walter Yoneo
Katz, Maurice Joseph
Kauder, Otto Samuel
Kelisky, Richard Paul
Kelly, Thomas Joseph
Kelly, Thomas Michael
Kenett, Ron
Ketchen, Mark B
Khan, Paul
Kirchmayer, Leon K
Kitchen, Sumner Wendell
Klijanowicz, James Edward
Klotzbach, Robert J(ames)
Kolaian, Jack H
Kolb, Frederick J(ohn), Jr
Korwin-Pawlowski, Michael Lech
Krieger, Gary Lawrence
Kuchar, Norman Russell
Kun, Kenneth Allan
Landzberg, Abraham H(arold)
Langworthy, Harold Frederick
Laubach, Gerald D
Law, Paul Arthur
Leff, Judith
Lerman, Steven I
Light, Thomas Burwell
Litz, Lawrence Marvin
Logan, Joseph Skinner
Logue, J(oseph) C(arl)
Longobardo, Anna Kazanjian
Lorenzo, J George Albert

Lowder, Wayne Morris
Lurkis, Alexander
MacDowell, John Fraser
McElligott, Peter Edward
McGee, James Patrick
Madey, Robert W
Marr, David Henry
Martens, Alexander E(ugene)
Mason, John Hugh
Meenan, Peter Michael
Merten, Alan Gilbert
Milton, Kirby Mitchell
Minnear, William Paul
Mort, Joseph
Mortenson, Kenneth Ernest
Mullen, Patricia Ann
Nelson, Robert Andrew
Nicolosi, Joseph Anthony
Ning, Tak Hung
O'Mara, Michael Martin
Orlando, Charles M
Palmedo, Philip F
Parham, Margaret Payne
Patmore, Edwin Lee
Pearson, Glen Hamilton
Pedroza, Gregorio Cruz
Penwell, Richard Carlton
Randall, John D(el)
Rau, R Ronald
Reber, Raymond Andrew
Redington, Rowland Wells
Reich, Ismar M(eyer)
Richards, Jack Lester
Riffenburgh, Robert Harry
Riggle, J(ohn) W(ebster)
Rosen, Stephen
Rubin, Isaac D
Rubloff, Gary W
Rudin, Bernard D
Sanford, Karl John
Scala, Sinclaire M(aximilian)
Schachter, Rozalie
Schanker, Jacob Z
Schindler, Joe Paul
Schottmiller, John Charles
Schroeder, Anita Gayle
Schucker, Gerald D
Schulz, Helmut Wilhelm
Schwan, Thomas James
Schwartz, Herbert Mark
Setchell, John Stanford, Jr
Seward, Thomas Philip, III
Shaw, David Elliot
Sheeran, Stanley Robert
Sherman, Seymour
Shevack, Hilda N
Shuey, R(ichard) L(yman)
Sieg, Albert Louis
Simons, Gene R
Siuta, Gerald Joseph
Slusarek, Lidia
Smith, Alan Jerrard
Solomon, Jack
Springett, Brian E
Sterman, Samuel
Stover, Raymond Webster
Strasser, Alfred Anthony
Strnisa, Fred V
Stupp, Edward Henry
Sullivan, Michael Francis
Thieberger, Peter
Thompson, David Allen
Thompson, David Fred
Trucker, Donald Edward
Tuite, Robert Joseph
Tuli, Jagdish Kumar
Turnblom, Ernest Wayne
Uretsky, Myron
Vogel, Philip Christian
Von Bacho, Paul Stephan, Jr
Vosburgh, Kirby Gannett
Wachtell, Richard L(loyd)
Wallis, Thomas Gary
Wang, Shu Lung
Wertheimer, Alan Lee
Wien, Richard W, Jr
Wiese, Warren M(elvin)
Wolfe, Roger Thomas
Wood, Paul Mix
Wyman, Donald Paul
Yasar, Tugrul
York, Raymond A
Zdan, William

NORTH CAROLINA
Ahearne, John Francis
Bockstahler, Theodore Edwin
Bode, Arthur Palfrey
Bursey, Joan Tesarek
Carlson, William Theodore
Chin, Robert Allen
Cubberley, Adrian H
Dibner, Mark Douglas
Dickerson, James Perry
Elder, James Franklin, Jr
Etter, Robert Miller
Fair, Richard Barton
Fytelson, Milton
Golden, Carole Ann
Gonder, Eric Charles
Harrison, Stanley L
Huber, William Richard, III
Hurwitz, Melvin David
Iverson, F Kenneth

Jennings, Carl Anthony
Jones, Rufus Sidney
Jordan, Edward Daniel
Kanopoulos, Nick
Kirk, Wilber Wolfe
Lontz, Robert Jan
Min, Tony C(harles)
Moody, Max Dale
Mosier, Stephen R
O'Connor, Lila Hunt
Odell, Norman Raymond
Pruett, Roy L
Reisman, Arnold
Rideout, Janet Litster
Roblin, John M
Roth, Roy William
Sargeant, Peter Barry
Spilker, Bert
Stenger, William J(ames), Sr
Stubblefield, Charles Bryan
Sud, Ish
Tartt, Thomas Edward
Trenholm, Andrew Rutledge
Wasson, John R
Wehner, Philip
Yeowell, David Arthur
Young, William Anthony

NORTH DAKOTA
Bares, William Anthony
Kotch, Alex
Watt, David Milne, Jr

OHIO
Adamczak, Robert L
Alden, Carl L
Amjad, Zahid
Antler, Morton
Ault, George Mervin
Badertscher, Robert F(rederick)
Bahnfleth, Donald R
Baker, Frank Weir
Barsky, Constance Kay
Battershell, Robert Dean
Becker, Carter Miles
Bement, A(rden) L(ee), Jr
Bierlein, James A(llison)
Boggs, Robert Wayne
Brandhorst, Henry William, Jr
Broge, Robert Walter
Brunner, Gordon Francis
Burte, Harris M(erl)
Campbell, Robert Wayne
Carman, Charles Jerry
Chamis, Alice Yanosko
Chan, Yupo
Christian, John B
Colby, Edward Eugene
Copeland, James Clinton
Cordea, James Nicholas
Coulter, Paul David
Crable, John Vincent
Dewar, Norman Ellison
Easterday, Jack L(eroy)
Erbacher, John Kornel
Faiman, Robert N(eil)
Feldman, Julian
Ferguson, John Allen
Fordyce, James Stuart
Forestieri, Americo F
Forsyth, Thomas Henry
Fraker, John Richard
Frank, Thomas Paul
Freeberg, Fred E
Gage, Frederick Worthington
Gerace, Michael Joseph
Ginning, P(aul) R(oll)
Golden, John Terence
Gorland, Sol H
Graef, Walter L
Grasselli, Jeanette Gecsy
Hannah, Sidney Allison
Hantman, Robert Gary
Harrington, Roy Victor
Hartsough, Robert Ray
Healy, James C
Hein, Richard William
Henderson, Courtland M
Henry, William Mellinger
Holubec, Zenowie Michael
Hurley, William Joseph
Ingram, David Christopher
Innes, John Edwin
Jackson, Curtis A(aitland)
Jayne, Theodore D
Katzen, Raphael
Kelley, Frank Nicholas
Keplinger, Orin Clawson
Kinsman, Donald Vincent
Kircher, John Frederick
Kochanowski, Barbara Ann
Laughon, Robert Bush
Lenhart, Jack G
Letton, James Carey
Lewis, James Edward
Leyda, James Perkins
Lydy, David Lee
McCune, Homer Wallace
Macklin, Martin
McMillin, Carl Richard
McVean, Duncan Edward
Magee, Thomas Alexander
Marks, Alfred Finlay
Marshall, Kenneth D

Martino, Joseph Paul
Mathews, A L
Metcalfe, Joseph Edward, III
Michael, William R
Mielke, Robert L
Moncrief, Eugene Charles
Morgan, William R(ichard)
Mosser, John Snavely
Olesen, Douglas Eugene
Paynter, John, Jr
Pearson, Jerome
Piper, Ervin L
Platau, Gerard Oscar
Pugh, John W(illiam)
Purdon, James Ralph, Jr
Purvis, John Thomas
Razgaitis, Richard A
Replogle, Clyde R
Richley, E(dward) A(nthony)
Roblin, John M
Rodman, Charles William
Roehrig, Frederick Karl
Ross, John Edward
Schoch, Daniel Anthony
Schultz, Edwin Robert
Schumm, Brooke, Jr
Schutta, James Thomas
Scozzie, James Anthony
Sedor, Edward Andrew
Sharma, Shri C
Sinner, Donald H
Soska, Geary Victor
Stadler, Louis Benjamin
Stambaugh, Edgel Pryce
Standish, Norman Weston
Steichen, Richard John
Stephan, David George
Stephens, Marvin Wayne
Stotz, Robert William
Temple, Robert Dwight
Thomas, Alexander Edward, III
Tipton, C(lyde) R(aymond), Jr
Toeniskoetter, Richard Henry
Turchi, Peter John
Turner, Andrew
Ungar, Edward William
Updegrove, Louis B
Vassell, Gregory S
Ventresca, Carol
Versteegh, Larry Robert
Voigt, Charles Frederick
Wadelin, Coe William
Wakefield, Shirley Lorraine
Warshay, Marvin
Weeks, Stephen P
Weeks, Thomas Joseph, Jr
Westerman, Arthur B(aer)
Wharton, H(arry) Whitney
Wickham, William Terry, Jr
Wigington, Ronald L
Wilkes, William Roy
Williamson, Frederick Dale
Wohlfort, Sam Willis
Womack, Edgar Allen, Jr
Wong, Anthony Sai-Hung
Yellin, Wilbur

OKLAHOMA
Bruner, Ralph Clayburn
Ciriacks, Kenneth W
Cowley, Thomas Gladman
Crafton, Paul A(rthur)
Davis, Robert Elliott
Evens, F Monte
Ferguson, William Sidney
Green, Ray Charles
Hale, John Dewey
Hale, William Henry, Jr
Hopkins, Thomas R (Tim)
Johnston, Harlin Dee
Koerner, E(rnest) L(ee)
Lambird, Perry Albert
McCollom, Kenneth A(llen)
Monn, Donald Edgar
Paxson, John Ralph
Ripley, Dennis L(eon)
Shell, Francis Joseph
Still, Edwin Tanner
Terrell, Marvin Palmer
West, John B(ernard)
Whitfill, Donald Lee
Zuech, Ernest A

OREGON
Autrey, Robert Luis
Barnett, Gordon Dean
Brinkley, John Michael
Collins, John A(ddison)
Demaree, Thomas L
Fields, R(ance) Wayne
Hackleman, David E
Hansen, Hugh J
Hoffman, Michael G
Humphrey, J Richard
Jones, Donlan F(rancis)
Kocaoglu, Dundar F
Lendaris, George G(regory)
Orloff, Jonathan H
Preston, Eric Miles
Sarles, Lynn Redmon
Schroeder, Warren Lee
Scott, Peter Carlton
Shilling, Wilbur Leo
Wagner, David Henry

PENNSYLVANIA
Abate, Kenneth
Angeloni, Francis M
Bartish, Charles Michael Christopher
Bauman, Bernard D
Bauman, Dwight Maylon Billy
Baxter, Ronald Dale
Beedle, Lynn Simpson
Beilstein, Henry Richard
Beitchman, Burton David
Bell, Stanley C
Berkoff, Charles Edward
Berty, Jozsef M
Box, Larry
Bradley, James Henry Stobart
Braunstein, David Michael
Brendlinger, Darwin
Brewer, LeRoy Earl, Jr
Browning, Daniel Dwight
Bucher, John Henry
Burk, David Lawrence
Camp, Frederick William
Capaldi, Eugene Carmen
Chen, Michael S K
Christie, Michael Allen
Colella, Donald Francis
Connor, James Edward, Jr
Connors, Donald R
Cook, Charles S
Corbett, Robert B(arnhart)
Craig, James Clifford, Jr
D'Alisa, Rose M
Dalton, Augustine Ivanhoe, Jr
Davis, Brian Clifton
Davis, Burl Edward
Derby, James Victor
Detwiler, John Stephen
Dicciani, Nancy Kay
Druffel, Larry Edward
Dumbri, Austin C
Durkee, Jackson L
Dyott, Thomas Michael
Einhorn, Philip A
Ellis, Jeffrey Raymond
Falcone, James Salvatore, Jr
Feero, Nicholas F
Feinman, J(erome)
Fiore, Nicholas F
Foster, Norman Francis
Frumerman, Robert
Garfinkle, Barry David
Garrett, Thomas Boyd
Gertz, Steven Michael
Giannovario, Joseph Anthony
Giffen, Robert H(enry)
Gluntz, Martin L
Golton, William Charles
Graybill, Donald Lee
Griffith, Michael Grey
Hahn, Peter Mathias
Haines, William Joseph
Harkins, Thomas Regis
Haun, Robert Dee, Jr
Hauptschein, Murray
Herman, Frederick Louis
Herzog, Karl A
Herzog, Leonard Frederick, II
Hinton, Raymond Price
Homsey, Robert John
Howard, Wilmont Frederick, Jr
Howe, Richard Hildreth
Humphrey, Watts S
Jaffe, Donald
Jarrett, Noel
Johnson, Littleton Wales
Jones, Roger Franklin
Judge, Joseph Malachi
Katz, Lewis E
Kaufman, William Morris
Kazahaya, Masahiro Matt
Kingrea, James I
Kleinschuster, Jacob John
Knief, Ronald Allen
Koch, Ronald N
Korchak, Ernest I(an)
Kottcamp, Edward H, Jr
Krishnaswamy, S V
Kunesh, Charles Joseph
Lacoste, Rene John
Laemmle, Joseph Thomas
Lantos, P(eter) R(ichard)
Lavin, J Gerard
Leas, J(ohn) W(esley)
LeBlanc, Norman Francis
Lenox, Ronald Sheaffer
Levy, Bernard, Jr
Loprest, Frank James
MacDonald, Hubert C, Jr
McDonnell, Leo F(rancis)
McGee, Charles E
Manuel, Thomas Asbury
Marmer, William Nelson
Mayer, George Emil
Merner, Richard Raymond
Merrill, John Raymond
Meyer, Paul A
Mezger, Fritz Walter William
Michaelis, Arthur Frederick
Miller, Dale L
Molinari, Robert James
Moore, Walter Calvin
Mount, Lloyd Gordon
Moyer, Robert (Findley)
Murrin, Thomas J

Nedwick, John Joseph
Orphanides, Gus George
Owens, Frederick Hammann
Papariello, Gerald Joseph
Parish, Roger Cook
Penman, Paul D
Petrich, Robert Paul
Pierson, Ellery Merwin
Pittman, G(eorge) F(rank), Jr
Prasad, Suresh
Purdy, David Lawrence
Ramani, Raja Venkat
Redmount, Melvin B(ernard)
Richardson, Ralph J
Robinson, Charles Albert
Roseman, Arnold S(aul)
Rubin, Benjamin Arnold
Rush, James E
Rushton, Brian Mandel
Saggiomo, Andrew Joseph
Saluja, Jagdish Kumar
Santamaria, Vito William
Schaller, Edward James
Schiffman, Louis F
Schulman, Marvin
Schwartz, Albert B
Scigliano, J Michael
Shellenberger, Donald J(ames)
Sheridan, John Joseph, III
Shibib, M Ayman
Shoaf, Mary La Salle
Silzars, Aris
Smith, Warren LaVerne
Snider, Albert Monroe, Jr
Sokolowski, Henry Alfred
Soler, Alan I(srael)
Spielvogel, Lawrence George
Spitznagel, John A
Stelting, Kathleen Marie
Stingelin, Ronald Werner
Stone, Gregory Michael
Stone, Herman
Stryker, Lynden Joel
Sucov, E(ugene) W(illiam)
Sykes, James Aubrey, Jr
Thomas, William J
Thompson, Ansel Frederick, Jr
Thompson, Sheldon Lee
Tobias, Philip E
Tobias, Russell Lawrence
Trumpler, Paul R(obert)
Tuba, I Stephen
Tzodikov, Nathan Robert
Uher, Richard Anthony
Umen, Michael Jay
Verhanovitz, Richard Frank
Vidt, Edward James
Viest, Ivan M
Vorchheimer, Norman
Walker, Augustus Chapman
Wardell, Joe Russell, Jr
Watson, William Martin, Jr
Weber, Frank L
Weddell, George G(ray)
Weisfeld, Lewis Bernard
Weldes, Helmut H
White, George Rowland
Whiteman, John David
Wiehe, William Henry
Wiggill, John Bentley
Williams, James Earl, Jr
Wilcox, William Jenkins, Jr
Zeigler, A(lfred) G(eyer)
Zweben, Carl Henry

RHODE ISLAND
Donnelly, Grace Marie
Jarrett, Jeffrey E
Johnson, Gordon Carlton
Poon, Calvin P C
Spero, Caesar A(nthony), Jr
Von Riesen, Daniel Dean

SOUTH CAROLINA
Avegeropoulos, G
Benjamin, Richard Walter
Berger, Richard S
Billica, Harry Robert
Boyd, George Edward
Garzon, Ruben Dario
Godfrey, W Lynn
Goldstein, Herman Bernard
Graulty, Robert Thomas
Hopper, Michael James
Jaco, Charles M, Jr
Joseph, J Walter, Jr
Krumrei, W(illiam) C(larence)
McCrosson, F Joseph
McDonell, William Robert
Morris, J(ames) William
O'Neill, John H(enry), Jr
Owen, John Harding
Pike, LeRoy
Przirembel, Christian E G
Robinson, Robert Earl
Stone, John Austin
Terry, Stuart Lee
Tolbert, Thomas Warren
Townes, George Anderson
Wagner, William Sherwood

SOUTH DAKOTA
Cannon, Patrick Joseph
Koepsell, Paul L(oel)
Swiden, LaDell Ray

Watkins, Allen Harrison
Wegman, Steven M

TENNESSEE
Adams, Robert Edward
Appleton, B R
Armentrout, Daryl Ralph
Baillargeon, Victor Paul
Baker, Merl
Baxter, John Edwards
Besser, John Edwin
Betz, Norman Leo
Bontadelli, James Albert
Boston, Charles Ray
Boudreau, William F(rancis)
Bretz, Philip Eric
Brewington, Percy, Jr
Commerford, John D
Cope, David Franklin
Cordell, Francis Merritt
Cowser, Kenneth Emery
Davidson, Robert C
Eads, B G
Fordham, James Lynn
Gant, Kathy Savage
Garber, Floyd Wayne
Gardiner, Donald Andrew
Gat, Uri
Grant, Peter Malcolm
Gray, Allen G(ibbs)
Gray, Theodore Flint, Jr
Guerin, Michael Richard
Hanley, Wayne Stewart
Harmer, David Edward
Haywood, Frederick F
Horak, James Albert
House, Robert W(illiam)
Jones, Charles Miller, Jr
Jones, Glenn Clark
Jordan, Andrew G
Keller, Charles A(lbert)
Kidd, George Joseph, Jr
Kraft, Edward Michael
Larkin, William (Joseph)
Lim, Alexander Te
Lotts, Adolphus Lloyd
Mansur, Louis Kenneth
Morrow, Roy Wayne
Pandeya, Prakash N
Pfuderer, Helen A
Potter, John Leith
Prairie, Michael L
Quist, Arvin Sigvard
Rash, Fred Howard
Salk, Martha Scheer
Scherpereel, Donald E
Schreiber, Eric Christian
Scroggie, Lucy E
Sittel, Chester Nachand
Sizemore, Douglas Reece
Smith, Michael James
Sorrells, Frank Douglas
Stiegler, James O
Stone, Robert Sidney
Thiessen, William Ernest
Van Winkle, Webster, Jr
Vaughen, Victor C(ornelius) A(dolph)
Weir, James Robert, Jr
Wessenauer, Gabriel Otto
West, Colin Douglas
Wilcox, William Jenkins, Jr
Witherspoon, John Pinkney, Jr
Yarbro, Claude Lee, Jr
Zerby, Clayton Donald

TEXAS
Abernathy, Bobby F
Addy, Tralance Obuama
Anthony, Donald Barrett
Aspelin, Gary B(ertil)
Bachmann, John Henry, Jr
Badhwar, Gautam D
Bare, Charles L
Barrow, Thomas D
Basham, Jerald F
Bendapudi, Kasi Visweswararao
Bhuva, Rohit L
Bland, William M, Jr
Bronaugh, Edwin Lee
Brown, James Michael
Burton, Russell Rohan
Campbell, William M
Cannon, Dickson Y
Canon, Roy Frank
Carlile, Robert Elliot
Carlton, Donald Morrill
Cassard, Daniel W
Caswell, Gregory K
Chen, Peter
Cleland, Franklin Andrew
Collins, Dean Robert
Collipp, Bruce Garfield
Cory, William Eugene
Coulson, Larry Vernon
Cruse, Carl Max
Cupples, Barrett L(eMoyne)
Danburg, Jerome Samuel
Darwin, James T, Jr
David, Yadin B
Dear, Robert E A
Dillard, Robert Garing, Jr
Dimitroff, Edward
Eckles, Wesley Webster, Jr
Elliott, Jerry Chris

Technical Management (cont)

Engle, Damon Lawson
Ervin, Hollis Edward
Fisher, Gene Jordan
Frank, Steven Neil
French, Robert Leonard
Friberg, Emil Edwards
Gallagher, John Joseph, Jr
Gatti, Anthony Roger
Gipson, Robert Malone
Glaspie, Donald Lee
Gray, John Malcolm
Grogan, Michael John
Gum, Wilson Franklin, Jr
Haase, Donald J(ames)
Hebert, Joel J
Hirsch, Robert Louis
Ho, Clara Lin
Hoffman, Herbert I(rving)
Holmes, Larry A
Ivey, E(dwin) H(arry), Jr
Johnson, Elwin L Pete
Johnson, Malcolm Pratt
Justice, James Horace
Kingsley, Henry A(delbert)
Krautz, Fred Gerhard
Kusnetz, Howard L
Kutzman, Raymond Stanley
Lawrence, Joseph D, Jr
Leaders, Floyd Edwin, Jr
Lichtenstein, Harris Arnold
Linder, John Scott
Lord, Samuel Smith, Jr
Ludwig, Allen Clarence, Sr
Mack, Mark Philip
McKee, Herbert C(harles)
Mai, Klaus L(udwig)
Mansfield, Clifton Tyler
Mao, Shing
Matzkanin, George Andrew
Maute, Robert Lewis
Myers, Paul Walter
Naismith, James Pomeroy
Noonan, James Waring
O'Connor, Rod
Overmyer, Robert Franklin
Perez, Ricardo
Petersen, Donald H
Poddar, Syamal K
Posey, Daniel Earl
Poska, F(orrest) L(ynn)
Potts, Mark John
Powell, Alan
Purser, Paul Emil
Purvis, Merton Brown
Quisenberry, Karl Spangler, Jr
Rhodes, Allen Franklin
Ripperger, Eugene Arman
Robinson, Alfred Green
Roeger, Anton, III
Rosenquist, Edward P
Russell, B Don
Scinta, James
Shilstone, James Maxwell, Jr
Shoré, Fred L
Smith, R Lowell
Smith, Rolf C, Jr
Southern, Thomas Martin
Spencer, Alexander Burke
Strieter, Frederick John
Stubbeman, Robert Frank
Summers, William Allen, Jr
Swink, Laurence N
Tannahill, Mary Margaret
Teller, Cecil Martin, II
Torp, Bruce Alan
VantHull, LORIN L
Waid, Margaret Cowsar
Wakeland, William Richard
Ward, Donald Thomas
Watson, Jerry M
Widerquist, V(ernon) R(oberts)
Wilde, Garner Lee
Witterholt, Edward John
Wulfers, Thomas Frederick
Wurth, Thomas Joseph
Young, H(enry) Ben, Jr
Zaczepinski, Sioma

UTAH

Ball, James Stutsman
Bramlett, William
Burton, Frederick Glenn
Butterfield, Veloy Hansen, Jr
Germane, Geoffrey James
Gortatowski, Melvin Jerome
Haslem, William Joshua
Kelsey, Stephen Jorgensen
Krejci, Robert Henry
Lee, Jeffrey Stephen
Lewis, Lawrence Guy
Madsen, James Henry, Jr
Nelson, Mark Adams
Prater, John D
Rothenberg, Mortimer Abraham
Wood, O Lew

VERMONT

Arns, Robert George
Gray, John Patrick
Smith, Thomas David
Spillman, William Bert, Jr

VIRGINIA

Aitken, Alfred H
Alcaraz, Ernest Charles
Almeter, Frank M(urray)
Awad, Elias M
Barmby, John G(lennon)
Beale, Guy Otis
Beardall, John Smith
Blurton, Keith F
Bolender, Carroll H
Brewer, Dana Alice
Briscoe, Melbourne George
Brown, Ellen Ruth
Buchanan, Thomas Joseph
Burley, Carlton Edwin
Burns, William, Jr
Bush, Norman
Carstea, Dumitru
Casazza, John Andrew
Cetron, Marvin J
Charvonia, David Alan
Cheng, George Chiwo
Childress, Otis Steele, Jr
Church, Charles Henry
Cooper, Earl Dana
Cooper, Henry Franklyn, Jr
Cortright, Edgar Maurice
Cross, Ernest James, Jr
Curtin, Brian Thomas
Davidson, Charles Nelson
Davidson, Robert Bellamy
Davis, William Spencer
Dillaway, Robert Beacham
Dorsey, Clark L(awler), Jr
Entzminger, John N
Estes, Edward Richard
Evans, David Arthur
Everett, Warren S
Fabrycky, Wolter J
Flory, Thomas Reherd
Forman, Richard Allan
Foy, C Allan
Gee, Sherman
Gerlach, A(lbert) A(ugust)
Gibbons, John Howard
Gibson, John E(gan)
Gilbert, Arthur Charles
Godbole, Sadashiva Shankar
Goodwin, Francis E
Greenberg, Arthur Bernard
Greinke, Everett D
Hager, Chester Bradley
Hanneman, Rodney E
Hansen, Robert J
Harris, Paul Robert
Hart, Dabney Gardner
Hartline, Beverly Karplus
Hauxwell, Gerald Dean
Heebner, David Richard
Holt, Alan Craig
Hortick, Harvey J
Howard, William Eager, III
Hunt, V Daniel
Ijaz, Lubna Razia
Jacobs, Theodore Alan
James, W(ilbur) Gerald
Jimeson, Robert M(acKay), Jr
Kahn, Robert Elliot
Kelch, Walter L
Knap, James E(li)
Knudsen, Dennis Ralph
Krassner, Jerry
Krueger, Peter George
Lassiter, William Stone
Lerner, Norman Conrad
McCutchen, Samuel P(roctor)
McGean, Thomas J
McGregor, Dennis Nicholas
McHale, Edward Thomas
MacKinney, Arland Lee
Mansfield, John E
Marcus, Stanley Raymond
Marsh, Howard Stephen
Matthews, R(obert) B(ruce)
Mehta, Gurmukh D
Mengenhauser, James Vernon
Milici, Robert Calvin
Mock, John E(dwin)
Morrison, David Lee
Mulder, Robert Udo
Mullen, Joseph Matthew
Myers, Donald Albin
Myers, Thomas DeWitt
Neil, George Randall
Ossakow, Sidney Leonard
Owczarski, William A(nthony)
Pages, Robert Alex
Perry, Dennis Gordon
Pokrant, Marvin Arthur
Price, Byron Frederick
Pry, Robert Henry
Rader, Louis T(elemacus)
Rainer, Norman Barry
Reischman, Michael Mack
Reynolds, Richard Alan
Reynolds, Richard Alan
Rhode, Alfred S
Rosenberg, Murray David
Rossini, Frederick Anthony
Sabadell, Alberto Jose
Sedriks, Aristide John
Seward, William Davis
Sheldon, Donald Russell
Shelkin, Barry David

Siebentritt, Carl R, Jr
Sink, David Scott
Small, Timothy Michael
Smith, Carey Daniel
Sobel, Robert Edward
Spokes, G(ilbert) Neil
Stein, Bland Allen
Strong, Jerry Glenn
Stroup, Cynthia Roxane
Summers, George Donald
Thorp, Benjamin A
Tiedemann, Albert William, Jr
VanDerlaske, Dennis P
Ventre, Francis Thomas
Waesche, R(ichard) H(enley) Woodward
Wallace, Raymond Howard, Jr
Waxman, Ronald
Weedon, Gene Clyde
Welch, Jasper Arthur, Jr
Wessel, Paul Roger
West, Bob
Wilson, Herbert Alexander, Jr
Woisard, Edwin Lewis
Wood, Albert D(ouglas)
Wood, Leonard E(ugene)
Zakrzewski, Thomas Michael
Zilczer, Janet Ann

WASHINGTON

Anderson, Harlan John
Anderson, Richard Gregory
Brehm, William Frederick
Brown, Donald Jerould
Bunch, Wilbur Lyle
Buonamici, Rino
Burns, Robert Ward
Butts, William Cunningham
Claudson, T(homas) T(ucker)
Clement, Stephen LeRoy
Cooper, Martin Jacob
Craine, Lloyd Bernard
Dahl, Roy Edward
Einziger, Robert E
Fetter, William Allan
Frank, Andrew Julian
Gahler, Arnold Robert
Goheen, Steven Charles
Goodstein, Robert
Gravitz, Sidney I
Gray, Robert H
Greager, Oswald Herman
Gunderson, Leslie Charles
Hinthorne, James Roscoe
Hirschfelder, John Joseph
Hueter, Theodor Friedrich
Johnson, B(enjamin) M(artineau)
Katz, Yale H
Kent, Ronald Allan
Krier, Carol Alnoth
Kruger, Owen L
Lago, James
Liemohn, Harold Benjamin
McClure, J Doyle
McMurtrey, Lawrence J
MacVicar-Whelan, Patrick James
Marshall, Robert P(aul)
Martin, George C(oleman)
Matlock, John Hudson
Meeder, Jeanne Elizabeth
Merchant, Howard Carl
Morton, Randall Eugene
Nelson, Randall Bruce
Newman, Darrell Francis
Nickerson, Robert Fletcher
Padilla, Andrew, Jr
Pilet, Stanford Christian
Riley, Robert Gene
Sanford, Thomas Bayes
Schiffrin, Milton Julius
Schmid, Loren Clark
Smilen, Lowell I
Stokes, Gerald Madison
Sutherland, Earl C
Templeton, William Lees
Thorne, Charles M(orris)
Tichy, Robert J
Warner, Ray Allen
Welliver, Albertus
Widmaier, Robert George
Yoshikawa, Herbert Hiroshi
Zimmermann, Charles Edward

WEST VIRGINIA

Beeson, Justin Leo
Breneman, William C
Bryant, George Macon
Chiang, Han-Shing
De Barbadillo, John Joseph
Decker, Quintin William
Derderian, Edmond Joseph
Lantz, Thomas Lee
Smith, Joseph James
Tompkins, Curtis Johnston

WISCONSIN

Baetke, Edward A
Becker, Edward Brooks
Bevelacqua, Joseph John
Bisgaard, Søren
Bishop, Charles Joseph
Cheng, Shun
Dasgupta, Rathindra
Daub, Edward E
Dwyer, Sean G

Eckert, Alfred Carl, Jr
Ekern, Ronald James
Eriksson, Larry John
Gopal, Raj
Gopal, Raj
Gottschalk, Robert Neal
Hinkes, Thomas Michael
Horntvedt, Earl W
Hummert, George Thomas
Huss, Ronald John
Hynek, Robert James
Jared, Alva Harden
Johnson, J(ames) R(obert)
Kester, Dennis Earl
Kinsinger, Richard Estyn
Liebe, Donald Charles
Lowenstein, Michael Zimmer
McMurray, David Claude
Mahadeva, Madhu Narayan
Makela, Lloyd Edward
Naik, Tarun Ratilal
Oehmke, Richard Wallace
Price, Walter Van Vranken
Rechtin, Michael David
Savereide, Thomas J
Schultz, Jay Ward
Smith, Michael James
Spiegelberg, Harry Lester
Sprott, Julien Clinton
Stackman, Robert W
Steudel, Harold Jude
Trischan, Glenn M
Whitehead, Howard Allan
Whyte, Donald Edward

WYOMING

Marchant, Leland Condo
Odasz, F(rancis) B(ernard), Jr
Tigner, James Robert

PUERTO RICO

Bonnet, Juan A, Jr
Lluch, Jose Francisco

ALBERTA

Butler, R(oger) M(oore)
Fisher, Harold M
Neuwirth, Maria
Sears, Timothy Stephen
Silver, Edward Allan

BRITISH COLUMBIA

Becker, Edward Samuel
Cox, Lionel Audley
Mackenzie, George Henry
Wynne-Edwards, Hugh Robert

MANITOBA

Rosinger, Herbert Eugene
Saluja, Preet Pal Singh
Trick, Gordon Staples

ONTARIO

Agar, G(ordon) E(dward)
Bertram, Robert William
Biggs, Ronald C(larke)
Blevis, Bertram Charles
Cohn-Sfetcu, Sorin
Crocker, Iain Hay
Cupp, Calvin R
Dawes, David Haddon
Dunford, Raymond A
Effer, W R
Fewer, Darrell R(aymond)
Ford, Richard Westaway
Graham, Richard Charles Burwell
Hancox, William Thomas
Handa, V(irender) K(umar)
Hemmings, Raymond Thomas
Henry, Roger P
Jackson, David Phillip
Kruus, Jaan
Kundur, Prabha Shankar
Lawford, George Ross
McGeer, James Peter
Morton, Donald Charles
Pundsack, Arnold L
Rauch, Sol
Reid, Lloyd Duff
Renfrew, Robert Morrison
Robertson, J(ohn) A(rchibald) L(aw)
Rummery, Terrance Edward
Sonnenberg, Hardy
Tunnicliffe, Philip Robert
Van Driel, Henry Martin
Vincett, Paul Stamford
Wright, Joseph D
Yan, Maxwell Menuhin

QUEBEC

Mavriplis, F
Sood, Vijay Kumar

SASKATCHEWAN

Mikle, Janos J

OTHER COUNTRIES

Assousa, George Elias
Balk, Pieter
Burkhardt, Walter H
Butcher, Harvey Raymond, III
Eaton, Robert James
Edmonds, James W
Ginis, Asterios Michael

Hase, Donald Henry
Hedglin, Walter L
Huckaba, Charles Edwin
Kenney, Gary Dale
Kimblin, Clive William

Kropp, William A
Kurokawa, Kaneyuki
Lahiri, Syamal Kumar
Mou, Duen-Gang
Mullenax, Charles Howard

Packer, Leo S
Schmell, Eli David
Shafi, Mohammad
Steffen, Juerg
Stephens, Kenneth S

Struzak, Ryszard G
Voutsas, Alexander Matthew
  (Voutsadakis)
Ward, John Edward
Winn, Edward Barriere